2012 IRC®

CODE AND COMMENTARY

Volume 1

INTERNATIONAL CODE COUNCIL®

2012 International Residential Code®—Code and Commentary–Volume I

First Printing: November 2011

ISBN: 978-1-60983-064-9 (soft-cover edition)

39859-T017539

PRINTED IN THE U.S.A.

PREFACE

The principal purpose of the Commentary is to provide a basic volume of knowledge and facts relating to building construction as it pertains to the regulations set forth in the 2012 *International Residential Code*®. The person who is serious about effectively designing, constructing and regulating buildings and structures will find the Commentary to be a reliable data source and reference to almost all components of the built environment.

As a follow-up to the *International Residential Code*, we offer a companion document, the *International Residential Code–Code and Commentary—Volume I.* Volume I covers Chapters 1 through 11 of the 2012 *International Residential Code*. The basic appeal of the Commentary is that it provides, in a small package and at reasonable cost, thorough coverage of many issues likely to be dealt with when using the *International Residential Code* — and then supplements that coverage with historical and technical background. Reference lists, information sources and bibliographies are also included.

Throughout all of this, effort has been made to keep the vast quantity of material accessible and its method of presentation useful. With a comprehensive yet concise summary of each section, the Commentary provides a convenient reference for regulations applicable to the construction of buildings and structures. In the chapters that follow, discussions focus on the full meaning and implications of the code text. Guidelines suggest the most effective method of application, and the consequences of not adhering to the code text. Illustrations are provided to aid understanding; they do not necessarily illustrate the only methods of achieving code compliance.

The format of the Commentary includes the full text of each section, table and figure in the code, followed immediately by the commentary applicable to that text. At the time of printing, the Commentary reflects the most up-to-date text of the 2012 *International Residential Code*. Each section's narrative includes a statement of its objective and intent, and usually includes a discussion about why the requirement commands the conditions set forth. Code text and commentary text are easily distinguished from each other. All code text is shown as it appears in the *International Residential Code*, and all commentary is indented below the code text and begins with the symbol ❖. All code figures and tables are reproduced as they appear in the IRC. Commentary figures and tables are identified in the text by the word "Commentary" (as in "see Commentary Figure 704.3"), and each has a full border.

Readers should note that the Commentary is to be used in conjunction with the *International Residential Code* and not as a substitute for the code. **The Commentary is advisory only;** the code official alone possesses the authority and responsibility for interpreting the code.

Comments and recommendations are encouraged, for through your input, we can improve future editions. Please direct your comments to the Codes and Standards Development Department at the Chicago District Office.

TABLE OF CONTENTS

Part I—Administrative

Chapter 1: Scope And Administration

General Comments

Chapter 1 of the code is largely concerned with maintaining due process of law in enforcing the performance criteria contained in the body of the code. Only through careful observation of the administrative provisions can the building official reasonably hope to demonstrate that equal protection under the law has been provided. While it is generally assumed that the administrative and enforcement section of a code is addressed to the building official, this is not entirely true. The provisions also establish the rights and privileges of the design professional, the contractor and the building owner. The position of the building official is to review the proposed and completed work and to determine whether the residential structure conforms to the code requirements. The design professional, if one is used, is responsible for the design of the structure. The contractor is responsible for constructing the building in strict accordance with the code and any approved construction documents.

During the course of the construction of a building, the building official reviews the activity to make certain that the intent and letter of the law are being met and that the structure will provide adequate protection for the health, safety and welfare of the users. As a public servant, the building official enforces the code in an unbiased, professional and honest manner. Every individual is guaranteed equal enforcement of the code. Furthermore, design professionals, contractors and building owners have the right of due process for any requirement in the code.

Section R101 establishes the title, scope and purpose of the document. Section R102 establishes the applicability of the code. Section R103 establishes the Department of Building Safety. Section R104 establishes the duties and powers of the building official. Section R105 addresses the requirements for permits. Section R106 establishes the requirements for construction documents. Section R107 addresses the topic of temporary structures and uses. Section R108 establishes permit fees, payment of fees, building permit valuations, and related fees and refunds. Section R109 establishes the requirements for inspections. Section R110 establishes the requirements for occupancy, as well as the issuance and revocation of occupancy certificates. Section R111 regulates the connection and disconnection of utilities. Section R112 establishes the board of appeals and its authority. Section R113 addresses the topic of violations of the code. Section R114 establishes the authority for the building official to stop work.

Purpose

A construction code is intended to be adopted as a legally enforceable document that will safeguard health, safety, property and public welfare. A code cannot be effective without adequate provisions for its administration and enforcement. The building official charged with the administration and enforcement of construction regulations has a great responsibility, and with this responsibility goes authority. No matter how detailed the code may be, the building official must exercise judgement in determining code compliance. He or she is responsible for assuring that the homes in which the citizens of the community reside are designed and constructed to be reasonably free from hazards associated with the building's use. The code establishes a minimum acceptable level of safety.

PART 1—SCOPE AND APPLICATION

SECTION R101
GENERAL

R101.1 Title. These provisions shall be known as the *Residential Code for One- and Two-family Dwellings* of [NAME OF JURISDICTION], and shall be cited as such and will be referred to herein as "this code."

❖ The code is formally known as the *International Residential Code*® (IRC®) for *One- and Two-family Dwellings*, generally referred to as the *International Residential Code* or IRC for short. Upon adoption by the jurisdiction, it is known as the *Residential Code for One- and Two-family Dwellings* of the adopting jurisdiction, and in the document is often referred to as "the code." It is offered for adoption as a model document of prescriptive provisions to jurisdictions as a stand-alone residential code that establishes minimum regulations for one- and two-family dwellings and townhouses. The forum under which the code is developed encourages consistency of application of its provisions, and it is offered ready for adoption by all communities, large and small, internationally.

R101.2 Scope. The provisions of the *International Residential Code for One- and Two-family Dwellings* shall apply to the construction, *alteration*, movement, enlargement, replacement, repair, equipment, use and occupancy, location, removal and demolition of detached one- and two-family dwellings and townhouses not more than three stories above *grade plane* in height with a separate means of egress and their *accessory structures*.

Exceptions:

1. Live/work units complying with the requirements of Section 419 of the *International Building Code* shall be permitted to be built as one- and two-family *dwellings* or townhouses. Fire suppression required by Section 419.5 of the *International Building Code* when constructed under the *International Residential Code for One- and Two-family Dwellings* shall conform to Section P2904.
2. Owner-occupied lodging houses with five or fewer guestrooms shall be permitted to be constructed in accordance with the *International Residential Code for One- and Two-family Dwellings* when equipped with a fire sprinkler system in accordance with Section P2904.

❖ The provisions of the code apply to all aspects of construction for detached one- and two-family dwellings; multiple single-family dwellings, defined as townhouses; and all structures accessory to the dwellings and townhouses. This section sets a limitation in its scope of application to include only those townhouses and dwellings that are up to and including three stories above grade. Additionally, the provisions require each two-family dwelling or townhouse to have separate egress systems for each of the dwelling units. Where a dwelling or townhouse exceeds the allowed height in stories, does not provide individual egress for each dwelling unit or does not conform to the prescriptive provisions of the code, the structures are then beyond the scope of the code, and the provisions of the code cannot be applied. The building must then meet the provisions of the *International Building Code*® (IBC®) or other legally adopted building code of the jurisdiction. The actual limiting height of the building, measured in feet and as applied to the height of each story, is limited by the governing provisions for each specific material as found in Chapter 6 of the code. The user of the code will discover that, depending upon which material is selected for the wall construction, the result may be buildings of different permitted heights. For instance, where the wall system is of insulating concrete form construction as prescribed in Section R611, the building is limited to two stories above grade and each story is limited to 10 feet (3048 mm) in height. If wood stud wall framing is used pursuant to the requirements of Section R602, the allowable story height and overall building height will greatly exceed those permitted for the insulating concrete form wall construction method. The code does not limit the area of the building.

The provisions address all aspects of constructing, altering, repairing, maintaining, using, occupying, enlarging, locating, removing or demolishing any one-family dwelling, two-family dwelling, townhouse or accessory structure. The code regulates any and all activities that modify the buildings, as well as any structures that are of incidental use to the main buildings and that are also located on the same lot. The code regulates construction, plumbing, mechanical, electrical, equipment, fixture and gas piping installations that are done to the building and its operating systems, as well as to other structures incidental to the main building and on the same lot. Even work that is specifically exempted from permits must comply with the requirements of the code.

Exception 1 addresses live/work units, which are designed to comply with Section 419 of the IBC and are equipped with an automatic sprinkler system complying with Section P2904. As stated in Section 419 of the IBC, a live/work unit is a dwelling unit in which a significant portion includes a nonresidential use such as an office, a hair styling shop or barbershop or small store. Section 419 of the IBC states that if the nonresidential portion of the building is an office that comprises less than 10 percent of the building area, the unit does not need to be made to comply with the provisions of Section 419 of the IBC for live/work units. Section 419 of the IBC places limitations on live/work units, including limitations on the nonresidential occupancies that can be included in a live/work unit and addresses specific issues regarding means of egress, accessibility, ventilation, fire safety and structural requirements.

Exception 2 allows small bed and breakfasts to be constructed according to the IRC. A definition of "Lodging house" is included in Chapter 2 to generally encompass rental lodging within dwelling units, distinct from hotels and boarding houses which are "not occupied as a single-family unit."

R101.3 Intent. The purpose of this code is to establish minimum requirements to safeguard the public safety, health and general welfare through affordability, structural strength, means of egress facilities, stability, sanitation, light and ventilation, energy conservation and safety to life and property from fire and other hazards attributed to the built environment and to provide safety to fire fighters and emergency responders during emergency operations.

❖ With the adoption and establishment of a set of minimum construction standards, a community can impose reasonable standards for construction that will maintain the livability of the community while reducing factors that contribute to substandard and hazardous conditions that risk public health, safety, welfare or contribute to undue risk to fire fighters and emergency responders. Adoption of a modern construction code, such as this one, increases the level of safety and quality in the built environment, and is a necessary instrument used to reduce substandard conditions or construction by establishing minimum levels of acceptable construction practice. A reduction in blighted and slum conditions benefits the general public welfare and contributes toward maintenance of a consistent base for the property tax assessments that local governments typically use to fund their general budgets. By applying minimum structural, health, sanitation, fire safety and life safety criteria that must be met through the prescriptive or performance provisions of the code, a standard is set that ensures the public and individual building occupants they will not be exposed to construction that has gone unchecked or unregulated. The regulation and inspection of plumbing, electrical and mechanical installations also enhances safety for the public's health and welfare. The imposition of construction requirements that are in excess of the minimum standards would, in most cases, be considered unreasonable and would encounter a lack of support, which in turn could undermine the purpose of construction regulations.

SECTION R102 APPLICABILITY

R102.1 General. Where there is a conflict between a general requirement and a specific requirement, the specific requirement shall be applicable. Where, in any specific case, different sections of this code specify different materials, methods of construction or other requirements, the most restrictive shall govern.

❖ This section provides guidance to both building officials and other code users on the application of the code when different sections of the code specify different materials, methods of construction or other requirements. The importance of this section should not be understated. It resolves the question of how to handle conflicts between the general and specific provisions found in the code or those instances where different sections specify different requirements. This section provides a necessary hierarchy for the application of code provisions and clarifies code applications that would otherwise leave persistent questions and lead to debate. The code requires that where different sections of the code apply, but contain different requirements, the most restrictive provisions govern. The code also resolves conflicts between the general requirements of any particular issue with any specific requirements of the same issue by indicating that the specific requirements take precedence over the general requirements.

The following example illustrates the principle. Section R311.7 applies to all stairway types within the purview of the code. Section R311.7.5.1 limits the maximum height of risers to $7^3/_4$ inches (196 mm), thus providing a general requirement for stairway riser height. Section R311.7.10.1 limits risers within a spiral stairway to a maximum height of $9^1/_2$ inches (241 mm). This provision is specific to spiral stairways. At first it may appear that these two sections have requirements that are in conflict with one another. However, Sections R311.7.5.1 and R311.7.10.1 are subordinate requirements of Section R311.7. In this case, the specific requirements of Section R311.7.10.1 take precedence over the general requirements of Section R311.7.5.1 in those applications specific to spiral stairways.

Another example would be in relating the requirements for foam plastics to the requirements for wall and ceiling finishes. The code might be interpreted to state that foam plastic boards meeting the requirements of Section R316.3, with a maximum flame-spread rating of 75, could be used as the final surface finish for walls and ceilings because Section R302.9 allows a flame spread classification for wall and ceiling finishes with a rating of up to 200. This, however, would be a mistake. The provisions of Section R316.4 require the foam plastic to be covered by a finish material equivalent to a thermal barrier that limits the average temperature rise of the unexposed surface to no less than 250°F (139°C) after 15 minutes of fire exposure in accordance with ASTM E 119 standard time-temperature curve, or to be covered with minimum $^1/_2$-inch (12.7 mm) gypsum wallboard. In this case, the uncovered foam plastic must be covered to meet the requirements of Section R316.4 and have the thermal barrier installed. Additionally, the final surface finish material that is chosen must comply with Section R316.4, and it must also meet the required flame spread rating of 200 or less, as specified in Section R302.9.

To summarize, where several code sections apply to the use of a material or a method of construction, the most restrictive requirements apply.

R102.2 Other laws. The provisions of this code shall not be deemed to nullify any provisions of local, state or federal law.

❖ Compliance with the requirements of the code does not entail authorization, approval or permission to violate the regulations of other local, state or federal laws. Other laws, ordinances and regulations not administered or enforced by the building official could be in existence and enforced by another authority having jurisdiction over those provisions. Although the requirements may have similar provisions to those of the code, the work must be in conformance to the other regulations.

R102.3 Application of references. References to chapter or section numbers, or to provisions not specifically identified by number, shall be construed to refer to such chapter, section or provision of this code.

❖ There are many instances in the code where a reference is merely a chapter number, section number or, in some cases, a provision not specified by number. In all such situations, these references are to the code and not some other code or publication.

R102.4 Referenced codes and standards. The codes and standards referenced in this code shall be considered part of the requirements of this code to the prescribed extent of each such reference and as further regulated in Sections R102.4.1 and R102.4.2.

> **Exception:** Where enforcement of a code provision would violate the conditions of the *listing* of the *equipment* or *appliance*, the conditions of the *listing* and manufacturer's instructions shall apply.

❖ A referenced code, standard or portion thereof is an enforceable extension of the code as if the content of the standard were included in the body of the code. For example, Section R314.2 references NFPA 72 in its entirety for the installation of household fire alarms. In those cases when the code references only portions of a standard, the use and application of the referenced standard is limited to those portions that are specifically identified. If conflicts occur because of scope or purpose, the code text governs.

R102.4.1 Differences. Where differences occur between provisions of this code and referenced codes and standards, the provisions of this code shall apply.

❖ The use of referenced codes and standards to cover certain aspects of residential occupancy and operations rather than write parallel or competing requirements into the code is a long-standing code development principle. Often, however, questions and potential conflicts in the use of referenced codes and standards can arise which can lead to inconsistent enforcement of the code.

R102.4.2 Provisions in referenced codes and standards. Where the extent of the reference to a referenced code or standard includes subject matter that is within the scope of this code, the provisions of this code, as applicable, shall take precedence over the provisions in the referenced code or standard.

❖ Section R102.4.2 expands upon the provisions of Section R102.4.1 by making it clear that, even if a referenced standard contains requirements that parallel the IRC in the standard's own duly referenced section(s), the provisions of the IRC will always take precedence. This section does not intend to take the place of carefully scoped and referenced text for written standards for the IRC.

R102.5 Appendices. Provisions in the appendices shall not apply unless specifically referenced in the adopting ordinance.

❖ Provisions of the appendix do not apply unless the jurisdiction has adopted the appendix by statute or ordinance.

R102.6 Partial invalidity. In the event any part or provision of this code is held to be illegal or void, this shall not have the effect of making void or illegal any of the other parts or provisions.

❖ There may be a situation where one or more specific provisions of the code are found to be void or illegal.

This may be because a local, state or federal ordinance, statute or law has precedence over the adopted construction provisions. Under such conditions, only those specific provisions found to be void or illegal are affected; the rest of the code remains in force.

R102.7 Existing structures. The legal occupancy of any structure existing on the date of adoption of this code shall be permitted to continue without change, except as is specifically covered in this code, the *International Property Maintenance Code* or the *International Fire Code*, or as is deemed necessary by the *building official* for the general safety and welfare of the occupants and the public.

❖ Buildings that exist legally at the time the code is adopted are allowed to have their existing use and occupancy continued if the use or occupancy of the structure was also legally in existence. This means that as long as a structure or building remains in a safe and sanitary condition it need not be upgraded to meet the more current standards. However, any new construction, addition or remodeling will require such work to conform to the requirements of the new code. A change of occupancy of the building also will force the building to conform to the new standards.

The existence of a building prior to the adoption of a new edition of the code does not grant it the status of a legal existence. A building is thought of as being "grandfathered" under prior rules and not needing to be brought up to current requirements when there are records to show that it was constructed to meet the regulations of the jurisdiction in force at the time it was built. The most common way to demonstrate legal compliance with the construction codes of a community is through the public records. Copies of past building permits can be researched at the jurisdictional archives. Upon discovery that a building does not have a legal existence, corrective actions will be needed in order to bring the structure into compliance with the regulations of the jurisdiction at the time the building was built.

R102.7.1 Additions, alterations or repairs. *Additions*, *alterations* or repairs to any structure shall conform to the requirements for a new structure without requiring the existing structure to comply with all of the requirements of this code, unless otherwise stated. *Additions*, *alterations* or repairs shall not cause an existing structure to become unsafe or adversely affect the performance of the building.

❖ An addition, alteration or repair is required to meet the provisions of this code for new materials, but the remainder of the building is not required to comply with the requirements of this code. However, another measure of the viability of the addition, alteration or repair is that it does not cause the existing structure to be adversely affected or made unsafe. The application of this provision can often be confusing regarding what code requirements apply to the addition, alteration, or repair. For instance, removal and replacement of a fiberglass roof covering with the same type of roof covering would require that the new shingles, fasteners and underlayment would need to meet the standards and methods specified in the code. However, this does not mean that the roof structure and decking would need to be designed to resist the snow loads in the code.

Another example is window replacement. Often, in very old homes, the bedroom windows were not required to have the dimensions for emergency escape windows. If a window in a bedroom is being replaced by a new window, the new window would not need to meet the requirements for emergency escape. However, if the window installed was not as energy efficient, then the existing structure is adversely affected.

PART 2—ADMINISTRATION AND ENFORCEMENT

SECTION R103 DEPARTMENT OF BUILDING SAFETY

R103.1 Creation of enforcement agency. The department of building safety is hereby created and the official in charge thereof shall be known as the *building official*.

❖ This section establishes a building department that provides plan review and inspections for buildings regulated by the code. It also establishes the position of the building official, who will be the administrator for the enforcement of the jurisdictional codes. The employees of the department may be given varying degrees of authority by the building official. The Department of Building Safety is charged with the responsibility for enforcing the provisions of the code.

R103.2 Appointment. The *building official* shall be appointed by the chief appointing authority of the *jurisdiction*.

❖ The building official is an appointed officer of the jurisdiction, charged with the administrative responsibilities of the Department of Building Safety. The building official is appointed to the position by the chief appointing authority of the jurisdiction. Typically, the appointment is made by the mayor, council or commission and carried out through the city manager or other administrative authority.

R103.3 Deputies. In accordance with the prescribed procedures of this *jurisdiction* and with the concurrence of the appointing authority, the *building official* shall have the authority to appoint a deputy *building official*, the related technical officers, inspectors, plan examiners and other employees. Such employees shall have powers as delegated by the *building official*.

❖ The building official has the authority, acting in conjunction and in agreement with the appointing authority of the jurisdiction, to appoint officers and employees of the department. The building official can delegate certain powers of his or her authority to a deputy building official, as well as to all technical officers and employees of the Department of Building Safety. This group of deputies typically includes inspectors and plans examiners.

SECTION R104 DUTIES AND POWERS OF THE BUILDING OFFICIAL

R104.1 General. The *building official* is hereby authorized and directed to enforce the provisions of this code. The *building official* shall have the authority to render interpretations of this code and to adopt policies and procedures in order to clarify the application of its provisions. Such interpretations, policies and procedures shall be in conformance with the intent and purpose of this code. Such policies and procedures shall not have the effect of waiving requirements specifically provided for in this code.

❖ The building official is appointed by the legislative body of the jurisdiction to serve as the employee with the authority and responsibility for the proper administration of the code enforcement agency. The building official establishes policies and procedures that will clarify the applications of the code. The development of those policies and procedures should not be simply for the convenience of the jurisdiction's employees, but should be viewed as a way to effectively communicate to all interested parties involved in the construction process how the department will process applications, review construction documents, make inspections, approve projects, and determine and clarify the application of the code provisions. Properly developed, these policies and procedures can make the code enforcement department more predictable for those who are regulated and will also establish improved code compliance and public relations.

When interpretation of the code is needed, the building official is the one individual of the jurisdiction with the legal authority to interpret the code and determine how the provisions should be applied, in both general and specific cases. Some departments formalize the interpretation process and require the person with a question to submit their question in writing. Departments are encouraged to develop policies for both formal (written) and informal (verbal) requests for code interpretations. Any such interpretations must be in conformance with the intent and letter of the code and may not waive any requirements. It may be necessary in some cases for the building official to write these code interpretations into the permit.

R104.2 Applications and permits. The *building official* shall receive applications, review *construction documents* and issue permits for the erection and alteration of buildings and structures, inspect the premises for which such permits have been issued and enforce compliance with the provisions of this code.

❖ This section states that the building official must receive applications, review construction documents, issue permits, conduct inspections and enforce the provisions of the code. She or he is to provide the services required to carry the project from application for the permit to final approval. The building official is to accept all properly completed applications and not refuse the receipt of an application that meets the policy requirements. This same principle holds for the review of the construction documents, issuance of permits, inspections and for the enforcement of the code's provisions. The requirements of the code must be met, and approval will be granted only when compliance is verified.

R104.3 Notices and orders. The *building official* shall issue all necessary notices or orders to ensure compliance with this code.

❖ Building officials are to communicate, in writing, the disposition of their findings regarding code compliance. If an inspection shows that the work fails to comply with the code provisions, the building official or technical officer who conducted the inspection must issue a written report noting the corrections that are needed. A copy of the report is to be provided to the permit holders or their agent.

R104.4 Inspections. The *building official* is authorized to make all of the required inspections, or the *building official* shall have the authority to accept reports of inspection by *approved agencies* or individuals. Reports of such inspections shall be in writing and be certified by a responsible officer of such *approved* agency or by the responsible individual. The *building official* is authorized to engage such expert opinion as deemed necessary to report upon unusual technical issues that arise, subject to the approval of the appointing authority.

❖ The code gives the building official the authority to conduct all required inspections. The building official also has the authority to accept reports from other inspection agencies or private inspectors who have been granted prior approval by the building official to conduct inspections and provide reports. The reports submitted by the approved inspection agencies or individuals must be in writing and must be certified by only those individuals who have been approved by the building official as being qualified to submit the reports. When unusual technical issues arise during the course of construction, the building official has the authority to hire the services of an expert to report on the conditions and technical issues germane to the subject at hand. Prior to hiring the expert for consultation services, the building official must seek approval from the appointing authority.

R104.5 Identification. The *building official* shall carry proper identification when inspecting structures or premises in the performance of duties under this code.

❖ When the building official and other employees of the jurisdiction are performing their duties and inspecting structures or premises of construction, they are required to carry and display identification that will identify them as employees of the jurisdiction. Commentary Figure R104.5 is an example of the proper identification.

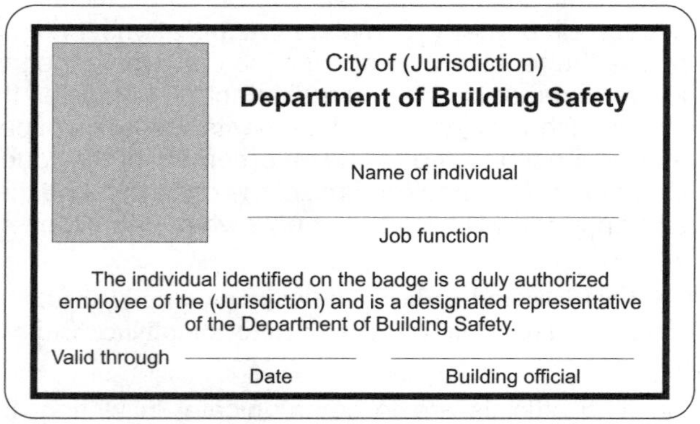

Figure R104.5
PERSONNEL IDENTIFICATION BADGE

R104.6 Right of entry. Where it is necessary to make an inspection to enforce the provisions of this code, or where the *building official* has reasonable cause to believe that there exists in a structure or upon a premises a condition which is contrary to or in violation of this code which makes the structure or premises unsafe, dangerous or hazardous, the *building official* or designee is authorized to enter the structure or premises at reasonable times to inspect or to perform the duties imposed by this code, provided that if such structure or premises be occupied that credentials be presented to the occupant and entry requested. If such structure or premises be unoccupied, the *building official* shall first make a reasonable effort to locate the owner or other person having charge or control of the structure or premises and request entry. If entry is refused, the *building official* shall have recourse to the remedies provided by law to secure entry.

❖ The building official and employees of the jurisdiction are authorized to conduct inspections in order to enforce the provisions of the jurisdiction's construction codes. Section R109 identifies specific progress points for inspections that must be conducted for both the intermediate phases of the work being done and the final inspection and approval. The officials of the jurisdiction also have the authority to conduct an inspection when the building official has reasonable cause to believe that conditions exist that constitute a violation of the code. The officials of the jurisdiction must take the necessary precautions so the constitutional rights of the owner or tenants are not violated.

The code authorizes the building official to enter and conduct an inspection at reasonable times.

Unless there is an immediate life-threatening condition or an immediate safety concern, the building official must secure an invitation or be requested to conduct a scheduled inspection. Additionally, the building official can conduct an inspection when permission has been granted by the person with immediate control of the premises.

When permission to conduct an inspection is denied, the building official has the authority to seek an inspection warrant.

R104.7 Department records. The *building official* shall keep official records of applications received, permits and certificates issued, fees collected, reports of inspections, and notices and orders issued. Such records shall be retained in the official records for the period required for the retention of public records.

❖ An important function of the Department of Building Safety is the record keeping associated with the department's functions. It is the responsibility of the building official to maintain adequate and accurate records that can be referenced for a variety of reasons. For example, staffing levels and other budgetary concerns can be better addressed through a historical review of department activity. Also, the department can provide a wealth of information to assist in community development and planning procedures, and a sound record-keeping process best supports legal actions.

Appropriate and complete department records must be maintained for an adequate period of time, based on the value of the information being retained. In all cases, those public records that must be retained by state or local law must be kept for the minimum time period required by the law. Any questions or concerns about the minimum time period for the retention of official records should be directed to the legal representative for the jurisdiction.

R104.8 Liability. The *building official*, member of the board of appeals or employee charged with the enforcement of this code, while acting for the *jurisdiction* in good faith and without malice in the discharge of the duties required by this code or other pertinent law or ordinance, shall not thereby be rendered liable personally and is hereby relieved from personal liability for any damage accruing to persons or property as a result of any act or by reason of an act or omission in the discharge of official duties. Any suit instituted against an officer or employee because of an act performed by that officer or employee in the lawful discharge of duties and under the provisions of this code shall be defended by legal representative of the *jurisdiction* until the final termination of the proceedings. The *building official* or any subordinate shall not be liable for cost in any action, suit or proceeding that is instituted in pursuance of the provisions of this code.

❖ The building official, members of the appeals board and other employees charged with the duty to enforce the provisions of the code are relieved from personal liability by the jurisdiction when they are acting in good faith their official duties. This provision does not grant absolute immunity from all tort liability for employees of the jurisdiction in all cases; an employee acting maliciously is not relieved from his or her personal liability and will most likely not be defended by the jurisdiction. Building officials should not fear lawsuits even if their state does not guarantee them absolute immunity for their actions. Rather, the building official should understand those elements that a plaintiff must show in a lawsuit in order to prevail.

Public officials should familiarize themselves with the laws of their state regarding their exposure to tort liability. Only a few states grant absolute immunity from liability for any actions taken by public officials when those actions are within the scope of their employment.

R104.9 Approved materials and equipment. Materials, *equipment* and devices *approved* by the *building official* shall be constructed and installed in accordance with such approval.

❖ The code is a compilation of criteria with which materials, equipment, devices and systems must comply to be suitable for a particular application. Where the building official grants approval for the use of specific materials, equipment or devices as a part of the construction process, it is important that the approved items be constructed or installed in a manner consistent with the approval. For example, the manufacturer's instructions and recommendations are to be followed if the approval of the material was based, even in part, on those instructions and recommendations.

The approval authority given the building official is a significant responsibility and is a key to code compliance. The approval process is first technical and then administrative. For example, if data to determine code compliance are required, the data should be in the form of test reports or engineering analysis and not simply taken from a sales brochure.

R104.9.1 Used materials and equipment. Used materials, *equipment* and devices shall not be reused unless *approved* by the *building official.*

❖ In keeping with the authority of the building official to evaluate construction materials based on their equivalency to those specified in the code, the use of used materials and equipment is limited to those approved by the building official. Testing and materials technology has permitted the development of new criteria that old materials may not satisfy. As a result, used materials are often evaluated by the building official in the same way as new materials. It is a common practice to require that used materials and equipment be equivalent to those required by the code if they are to be used in a new installation.

R104.10 Modifications. Wherever there are practical difficulties involved in carrying out the provisions of this code, the *building official* shall have the authority to grant modifications for individual cases, provided the *building official* shall first find that special individual reason makes the strict letter of this code impractical and the modification is in compliance with the intent and purpose of this code and that such modification does not lessen health, life and fire safety or structural requirements. The details of action granting modifications shall be recorded and entered in the files of the department of building safety.

❖ The building official has the authority to accept modifications of the code provisions in specific cases. For the building official to allow a modification, he or she must first determine that the strict application of the code is impractical for a specific reason. When the building official grants a modification, it is not a waiver from the requirements. It should be thought of as fulfilling the requirements to the greatest extent possible, but deviating from the requirements slightly to satisfy the intent of the provisions. The modification must not lessen the health, fire safety, life safety or structural requirements of the code. All modification actions must be recorded in the files of the building department.

R104.10.1 Flood hazard areas. The *building official* shall not grant modifications to any provision related to flood hazard areas as established by Table R301.2(1) without the granting of a variance to such provisions by the board of appeals.

❖ Section R322 contains provisions for determining flood hazard areas. A modification cannot be granted by the building official for structures located in areas that are prone to flooding without the board of appeals first granting a variance to the provisions. The regulations of the National Flood Insurance Program (NFIP) 44 CFR 60.3 require that proposals meet or exceed the minimum provisions of the program. Requests for modifications to any provision related to flood hazard areas are to be handled as formal variances. The criteria for issuance of variances are given in Section R112.2.2.

R104.11 Alternative materials, design and methods of construction and equipment. The provisions of this code are not intended to prevent the installation of any material or to prohibit any design or method of construction not specifically prescribed by this code, provided that any such alternative has been *approved.* An alternative material, design or method of construction shall be *approved* where the *building official* finds that the proposed design is satisfactory and complies with the intent of the provisions of this code, and that the material, method or work offered is, for the purpose intended, at least the equivalent of that prescribed in this code. Compliance with the specific performance-based provisions of the International Codes in lieu of specific requirements of this code shall also be permitted as an alternate.

❖ Although the code reflects current technologies, it is impossible to foresee all potential applications of new materials, construction techniques or design methods. The code encourages the use of new materials and technologies by allowing them to be presented to the building official for approval. The building official must approve a proposed alternative when it is found to be satisfactory and in compliance with the intent of the provisions of the code and is equivalent to that prescribed by the code. Approval may also be granted for the use of any alternative that is in compliance with the performance-based provisions of the *International Codes*®.

R104.11.1 Tests. Whenever there is insufficient evidence of compliance with the provisions of this code, or evidence that a material or method does not conform to the requirements of this code, or in order to substantiate claims for alternative materials or methods, the *building official* shall have the authority to require tests as evidence of compliance to be made at no expense to the *jurisdiction.* Test methods shall be as specified in this code or by other recognized test standards. In the absence of recognized and accepted test methods, the *building official* shall approve the testing procedures. Tests shall be performed by an *approved* agency. Reports of such tests shall be retained by the *building official* for the period required for retention of public records.

❖ The building official has the authority to require tests to substantiate the claim that an alternative is equivalent and meets the intent of the code. Any tests must be in compliance with those specified in the code or other recognized test standards approved by the building official. The cost of any tests will be borne by the proponent seeking the approval of the alternative.

SECTION R105
PERMITS

R105.1 Required. Any owner or authorized agent who intends to construct, enlarge, alter, repair, move, demolish or change the occupancy of a building or structure, or to erect, install, enlarge, alter, repair, remove, convert or replace any electrical, gas, mechanical or plumbing system, the installation of which is regulated by this code, or to cause any such work to be done, shall first make application to the *building official* and obtain the required *permit.*

❖ This section lists the types of work or installations of equipment or utilities that will require an owner or authorized agent to obtain permits, which are to be acquired before work begins. In general, a permit is required for all activities that are regulated by the code, and these activities cannot begin until the permit is issued.

R105.2 Work exempt from permit. *Permits* shall not be required for the following. Exemption from *permit* requirements of this code shall not be deemed to grant authorization for any work to be done in any manner in violation of the provisions of this code or any other laws or ordinances of this *jurisdiction.*

Building:

1. One-story detached *accessory structures* used as tool and storage sheds, playhouses and similar uses, provided the floor area does not exceed 200 square feet (18.58 m^2).
2. Fences not over 7 feet (2134 mm) high.
3. Retaining walls that are not over 4 feet (1219 mm) in height measured from the bottom of the footing to the top of the wall, unless supporting a surcharge.
4. Water tanks supported directly upon *grade* if the capacity does not exceed 5,000 gallons (18 927 L) and the ratio of height to diameter or width does not exceed 2 to 1.
5. Sidewalks and driveways.
6. Painting, papering, tiling, carpeting, cabinets, counter tops and similar finish work.
7. Prefabricated swimming pools that are less than 24 inches (610 mm) deep.
8. Swings and other playground equipment.
9. Window awnings supported by an exterior wall which do not project more than 54 inches (1372 mm) from the exterior wall and do not require additional support.
10. Decks not exceeding 200 square feet (18.58 m^2) in area, that are not more than 30 inches (762 mm) above *grade* at any point, are not attached to a *dwelling* and do not serve the exit door required by Section R311.4.

Electrical:

1. *Listed* cord-and-plug connected temporary decorative lighting.
2. Reinstallation of attachment plug receptacles but not the outlets therefor.
3. Replacement of branch circuit overcurrent devices of the required capacity in the same location.

4. Electrical wiring, devices, *appliances,* apparatus or *equipment* operating at less than 25 volts and not capable of supplying more than 50 watts of energy.
5. Minor repair work, including the replacement of lamps or the connection of *approved* portable electrical *equipment* to *approved* permanently installed receptacles.

Gas:

1. Portable heating, cooking or clothes drying *appliances.*
2. Replacement of any minor part that does not alter approval of *equipment* or make such *equipment* unsafe.
3. Portable-fuel-cell *appliances* that are not connected to a fixed piping system and are not interconnected to a power grid.

Mechanical:

1. Portable heating *appliances.*
2. Portable ventilation *appliances.*
3. Portable cooling units.
4. Steam, hot- or chilled-water piping within any heating or cooling *equipment* regulated by this code.
5. Replacement of any minor part that does not alter approval of *equipment* or make such *equipment* unsafe.
6. Portable evaporative coolers.
7. Self-contained refrigeration systems containing 10 pounds (4.54 kg) or less of refrigerant or that are actuated by motors of 1 horsepower (746 W) or less.
8. Portable-fuel-cell *appliances* that are not connected to a fixed piping system and are not interconnected to a power grid.

The stopping of leaks in drains, water, soil, waste or vent pipe; provided, however, that if any concealed trap, drainpipe, water, soil, waste or vent pipe becomes defective and it becomes necessary to remove and replace the same with new material, such work shall be considered as new work and a *permit* shall be obtained and inspection made as provided in this code.

The clearing of stoppages or the repairing of leaks in pipes, valves or fixtures, and the removal and reinstallation of water closets, provided such repairs do not involve or require the replacement or rearrangement of valves, pipes or fixtures.

❖ This section of the code lists the types of work in five categories—building, electrical, gas, mechanical and plumbing—that do not require permits. However, all work, even work that does not require a permit, must be done in a manner that will comply with the code requirements.

R105.2.1 Emergency repairs. Where *equipment* replacements and repairs must be performed in an emergency situation, the *permit* application shall be submitted within the next working business day to the *building official.*

❖ Occasionally repairs or replacement work must be done under emergency conditions. The code does not intend that such emergency work be held up until the necessary permits are secured. It is important, however, that the permit application be obtained as quickly as possible once the emergency has been controlled. Any required permit must be applied for within the next business day following the emergency repair or replacement.

R105.2.2 Repairs. Application or notice to the *building official* is not required for ordinary repairs to structures, replacement of lamps or the connection of *approved* portable electrical *equipment* to *approved* permanently installed receptacles. Such repairs shall not include the cutting away of any wall, partition or portion thereof, the removal or cutting of any structural beam or load-bearing support, or the removal or change of any required means of egress, or rearrangement of parts of a structure affecting the egress requirements; nor shall ordinary repairs include *addition* to, *alteration* of, replacement or relocation of any water supply, sewer, drainage, drain leader, gas, soil, waste, vent or similar piping, electric wiring or mechanical or other work affecting public health or general safety.

❖ There is a variety of ordinary repair, replacement or connection work that is exempt from the permit application process. This section identifies a number of general situations in which a permit is not required. The provisions then state the types of repairs for which a permit is required. Repair work done without a permit must still comply with the applicable provisions of the code.

R105.2.3 Public service agencies. A *permit* shall not be required for the installation, alteration or repair of generation, transmission, distribution, metering or other related *equipment* that is under the ownership and control of public service agencies by established right.

❖ When the ownership and control of equipment is held by a public service agency, such as a county water district, permits are not required for any work that might be done on that equipment. The scope of this provision includes not only repair activities, but also any installation or alteration work. It is clear from this section that public service agencies are self-regulating when it comes to work involving equipment used for generation, transmission, distribution and metering.

R105.3 Application for permit. To obtain a *permit*, the applicant shall first file an application therefor in writing on a form furnished by the department of building safety for that purpose. Such application shall:

1. Identify and describe the work to be covered by the *permit* for which application is made.
2. Describe the land on which the proposed work is to be done by legal description, street address or similar description that will readily identify and definitely locate the proposed building or work.
3. Indicate the use and occupancy for which the proposed work is intended.

4. Be accompanied by *construction documents* and other information as required in Section R106.1.
5. State the valuation of the proposed work.
6. Be signed by the applicant or the applicant's authorized agent.
7. Give such other data and information as required by the *building official*.

❖ The code lists the minimum information required in an application for a permit. The owner or agent is to fully describe the location of the site, the type and nature of the work to be done and all other pertinent information regarding the job. This provides the jurisdiction with a clear understanding of what will actually be done under the permit. It is a common belief by some owners that all that is required of them is to complete an application form, and then the permit will be issued over the counter. Although issuance of over-the-counter permits may be practical when a water heater is to be installed or for other minor work, this is not the case for more complex work. The general public is often surprised when they discover a complete application includes the requirements to provide construction drawings as required by Section R106, detailed information about the property, engineering data and other information as required by the building department. The department should develop informational fliers and packets that will assist the applicant with understanding the requirements for obtaining a permit. Recognizing that incomplete applications do not serve the applicant's best interest (and create unnecessary delays with processing applications), many building departments have personnel who specialize in providing assistance to customers to ease them through the process.

R105.3.1 Action on application. The *building official* shall examine or cause to be examined applications for permits and amendments thereto within a reasonable time after filing. If the application or the *construction documents* do not conform to the requirements of pertinent laws, the *building official* shall reject such application in writing stating the reasons therefor. If the *building official* is satisfied that the proposed work conforms to the requirements of this code and laws and ordinances applicable thereto, the *building official* shall issue a *permit* therefor as soon as practicable.

❖ When an application is determined to be incomplete, or when the construction documents do not show compliance with the code provisions or other requirements of the jurisdiction, the building official is to reject them. When the construction documents are rejected, the building official is required to list the reasons for the rejection for the applicant explaining why the plans have been rejected, so the applicant will know what action to take for subsequent approval.

When the application and construction documents are determined to be in compliance with the requirements of the code and other regulations of the jurisdiction, the building official must issue a permit. This section mandates the issuance of a permit when the plans and other data show compliance with all regulations of the jurisdiction, and it mandates that the permit be issued without delay.

R105.3.1.1 Determination of substantially improved or substantially damaged existing buildings in flood hazard areas. For applications for reconstruction, rehabilitation, *addition* or other improvement of existing buildings or structures located in a flood hazard area as established by Table R301.2(1), the *building official* shall examine or cause to be examined the *construction documents* and shall prepare a finding with regard to the value of the proposed work. For buildings that have sustained damage of any origin, the value of the proposed work shall include the cost to repair the building or structure to its predamaged condition. If the *building official* finds that the value of proposed work equals or exceeds 50 percent of the market value of the building or structure before the damage has occurred or the improvement is started, the finding shall be provided to the board of appeals for a determination of substantial improvement or substantial damage. Applications determined by the board of appeals to constitute substantial improvement or substantial damage shall require all existing portions of the entire building or structure to meet the requirements of Section R322.

❖ The definitions of "Substantial damage" and "Substantial improvement" are included in federal regulations (see 44 CFR 59.1, Definitions). The following is from the NFIP regulations:

> Substantial damage means damage of any origin sustained by a structure whereby the cost of restoring the structure to its before-damaged condition would equal or exceed 50 percent of the market value of the structure before the damage occurred.
>
> Substantial improvement means any reconstruction, rehabilitation, addition or other improvement of a structure, the cost of which equals or exceeds 50 percent of the market value of the structure before the "start of construction" of the improvement. This term includes structures [that] have incurred "substantial damage," regardless of the actual repair work performed. The term does not, however, include either:
>
> 1. Any project for improvement of a structure to correct existing violations of state or local health, sanitary or safety code specifications that have been identified by the local code enforcement official and that are the minimum necessary to ensure safe living conditions, or
> 2. Any alteration of a "historic structure," provided that the alteration will not preclude the structure's continued designation as an "historic structure."

These definitions are consistent with the definitions in Section 1612.2 of the IBC. "Substantial improvement" is also defined in Section R112.2.1 of the IBC.

Long-term reduction in exposure to flood hazards, including exposure of older buildings, is one of the purposes for regulating flood plain development. Existing buildings or structures located in flood hazard areas are to be brought into compliance with the flood-resistance provisions of Section R322 when the value of improvements or the repair of damage exceeds a certain value.

Section R105.3 requires the applicant to state the valuation of the proposed work as part of the information submitted to obtain a permit. If the proposed work will be performed on existing buildings or structures in flood hazard areas, including restoration of damage from any cause, the building official is to prepare a finding to determine the value of the proposed work, including the value of the property owner's labor, as well as the value of donated labor and materials. For damaged buildings, the value of the proposed work is the value of work necessary to restore the building to its predamage condition, even if the applicant is proposing less work.

To make a determination about whether a proposed repair, reconstruction, rehabilitation, addition or improvement of a building or structure will constitute a substantial improvement or correction of substantial damage, the cost of the proposed work is to be compared to the market value of the building or structure before the work is started. To determine market value, the building official may require the applicant to provide such information as allowed under Section R105.3. For additional guidance, refer to FEMA 213, *Answers to Questions about Substantially Damaged Buildings*, and FEMA 311, *Guidance on Estimating Substantial Damage Using the NFIP Residential Substantial Damage Estimator.*

The building official's finding is to be provided to the board of appeals if the finding is that the value of the proposed work equals or exceeds 50 percent of the market value of the building, less land value. The board of appeals is to determine whether the proposed work constitutes a substantial improvement of the existing building or whether a damaged building sustained substantial damage. If the board finds that the work is a substantial improvement, the existing building is to be brought into compliance with the flood-resistance provisions of Section R322.

R105.3.2 Time limitation of application. An application for a *permit* for any proposed work shall be deemed to have been abandoned 180 days after the date of filing unless such application has been pursued in good faith or a *permit* has been issued; except that the *building official* is authorized to grant one or more extensions of time for additional periods not exceeding 180 days each. The extension shall be requested in writing and justifiable cause demonstrated.

❖ Applications for permits are considered valid for 180 days. The permit application and review process must be done in a timely manner within that period. The applicant must be responsive to requests for additional information made by the building department. The 180-day limitation is not intended to penalize an applicant for the lack of action on the part of the jurisdiction. It is merely a measure that is used to void an application when it is no longer reasonable to keep it active because the applicant is delaying the process and is not responding to legitimate requests for information. The building official can extend the time limit of an application in increments of 180 days, provided the applicant can show a valid reason for an extension. The applicant must make this request in writing.

R105.4 Validity of permit. The issuance or granting of a *permit* shall not be construed to be a *permit* for, or an *approval* of, any violation of any of the provisions of this code or of any other ordinance of the *jurisdiction*. Permits presuming to give authority to violate or cancel the provisions of this code or other ordinances of the *jurisdiction* shall not be valid. The issuance of a *permit* based on *construction documents* and other data shall not prevent the *building official* from requiring the correction of errors in the *construction documents* and other data. The *building official* is also authorized to prevent occupancy or use of a structure where in violation of this code or of any other ordinances of this *jurisdiction*.

❖ A permit authorizes the permit holder to proceed with work that complies with the code requirements. A permit is not valid if it is issued for work that does not comply with the code requirements. If, after the permit has been issued, errors in the plans or construction are discovered, the building official has the authority to require the correction of the plans or the work to comply with the requirements of the code.

R105.5 Expiration. Every *permit* issued shall become invalid unless the work authorized by such *permit* is commenced within 180 days after its issuance, or if the work authorized by such *permit* is suspended or abandoned for a period of 180 days after the time the work is commenced. The *building official* is authorized to grant, in writing, one or more extensions of time, for periods not more than 180 days each. The extension shall be requested in writing and justifiable cause demonstrated.

❖ Once a permit has been issued, the permit holder has 180 days to begin the construction work; otherwise, the permit will become invalid. If at any time the work stops for a period of 180 days or more, the permit is invalid. Extensions can be granted by the building official, in writing, in increments of 180 days when the applicant makes a written request.

R105.6 Suspension or revocation. The *building official* is authorized to suspend or revoke a *permit* issued under the provisions of this code wherever the *permit* is issued in error or on the basis of incorrect, inaccurate or incomplete information, or in violation of any ordinance or regulation or any of the provisions of this code.

❖ Any permit that has been issued based on false, misleading or incorrect information can be revoked or suspended by the authority of the building official. The building official also has the authority to suspend or revoke any permit issued when a violation exists with regard to the code or any other ordinance, code, law or regulation that is legally in effect in the jurisdiction.

R105.7 Placement of permit. The building *permit* or copy thereof shall be kept on the site of the work until the completion of the project.

❖ The permit must be displayed at the work site until the certificate of occupancy has been issued. Because paperwork at a job site is sometimes lost, the code allows a copy of the permit to be kept at the site and the original to be retained in a more secure place. Keeping a record of permits at the project location satisfies the legal requirements set forth in the code, and any interested party can verify that a valid permit has been obtained.

R105.8 Responsibility. It shall be the duty of every person who performs work for the installation or repair of building, structure, electrical, gas, mechanical or plumbing systems, for which this code is applicable, to comply with this code.

❖ This is one of several code provisions that emphasize the required compliance for every aspect of the project. Although the permit holder is designated as having the primary responsibility for overall code compliance, it is the responsibility of each and every person working on the job to adhere to the requirements of the code.

R105.9 Preliminary inspection. Before issuing a *permit,* the *building official* is authorized to examine or cause to be examined buildings, structures and sites for which an application has been filed.

❖ This section provides the building official with a useful tool in the permit process, especially in cases of permits being issued for an existing building. While the construction documents may show the scope and nature of work to be done, there may be other existing conditions in the building that could affect the continued safety profile of the building and the approval of a permit, which could only be discovered by inspection.

SECTION R106 CONSTRUCTION DOCUMENTS

R106.1 Submittal documents. Submittal documents consisting of *construction documents*, and other data shall be submitted in two or more sets with each application for a *permit.* The *construction documents* shall be prepared by a registered *design professional* where required by the statutes of the *jurisdiction* in which the project is to be constructed. Where special conditions exist, the *building official* is authorized to require additional *construction documents* to be prepared by a registered *design professional.*

Exception: The *building official* is authorized to waive the submission of *construction documents* and other data not required to be prepared by a registered *design professional* if it is found that the nature of the work applied for is such that reviewing of *construction documents* is not necessary to obtain compliance with this code.

❖ This section provides the minimum requirements for construction documents that an applicant must provide along with the permit application form for the application package to be considered complete. Construction documents are not just a set of drawings. Construction documents are the entire set of all submitted forms and information necessary to accurately communicate the scope of the construction. The submittals may include written special inspection and structural observation programs, construction drawings, and details, reports, calculations, specifications, shop drawings, manufacturer's installation instructions, site plans and other graphic and written forms that will describe the proposed work in detail. The building official can waive the submission of construction documents for types of work where the review of documents is not necessary to show compliance with the requirements of the code.

The code is prescriptive and makes possible the design of a dwelling or townhouse without the requirement for a licensed design professional. However, the construction documents must be prepared by a licensed design professional when required by the statutes of the state or jurisdiction. Additionally, the building official has the authority to require that plans be prepared by licensed design professionals when not otherwise required by the statutes of the jurisdiction if, in the opinion of the building official, special or unique conditions exist or if the design of a building does not meet the prescriptive provisions of the code.

R106.1.1 Information on construction documents. *Construction documents* shall be drawn upon suitable material. Electronic media documents are permitted to be submitted when *approved* by the *building official. Construction documents* shall be of sufficient clarity to indicate the location, nature and extent of the work proposed and show in detail that it will conform to the provisions of this code and relevant laws, ordinances, rules and regulations, as determined by the *building official.* Where required by the *building official*, all braced wall lines, shall be identified on the *construction documents* and all pertinent information including, but not limited to, bracing methods, location and length of braced wall panels, foundation requirements of braced wall panels at top and bottom shall be provided.

❖ The emphasis of this section is on the clarity, completeness and accuracy of the construction documents. A wide variety of individuals will be using the construction documents to perform their specific tasks. Therefore, it is critical that there be no confusion about the intent of the designer based on the information in the plans and other documents.

Electronic submittal of construction documents is rapidly gaining popularity. Where the Department of Public Safety has the means to review plans electronically, it may request such submittals. Many departments continue to use hard-copy documents for the plan review process, with a request for electronic media copies of documents for archival purposes.

This section also requires that the relevant wall bracing information required in Section R602.10 be clearly marked on the construction drawings. Those parts of the wall system that are designated bracing panels have potentially different panel attachment schedules, foundation requirements and specific connection requirements to other parts of the building than do other exterior and interior walls. Not only are braced walls often different from other walls, they are always required in every code-conforming structure. An extra burden is placed on our building officials and plan checkers when they have to try to figure out which walls the designer has intended for use as bracing.

Requiring such details, as many professional home designers already provide, on all submittals will ensure that bracing is being considered during the design process. It will further ensure that the building's structural detailing is being done by the person who draws the plans and not by the plan checker. It will also make it easier for the builder to properly construct the required bracing on the job site when the details are clearly spelled out on the drawings.

R106.1.2 Manufacturer's installation instructions. Manufacturer's installation instructions, as required by this code, shall be available on the job site at the time of inspection.

❖ Throughout the code, it directs that materials or equipment be installed in accordance with the manufacturer's installation instructions. An example is the installation of modified bitumen roofing, as set forth in Section R905.11.3. The code recognizes that the manufacturer can best relate the specific installation requirements applicable to its specific product. Where the code mandates that the manufacturer's installation instructions be followed, those instructions must be available at the job site. In this way, both the installers and inspector are able to see that the directives of the manufacturer are being followed.

R106.1.3 Information for construction in flood hazard areas. For buildings and structures located in whole or in part in flood hazard areas as established by Table R301.2(1), *construction documents* shall include:

1. Delineation of flood hazard areas, floodway boundaries and flood zones and the design flood elevation, as appropriate;
2. The elevation of the proposed lowest floor, including *basement*; in areas of shallow flooding (AO Zones), the height of the proposed lowest floor, including *basement*, above the highest adjacent *grade*;
3. The elevation of the bottom of the lowest horizontal structural member in coastal high hazard areas (V Zone); and
4. If design flood elevations are not included on the community's Flood Insurance Rate Map (FIRM), the *building official* and the applicant shall obtain and reasonably utilize any design flood elevation and floodway data available from other sources.

❖ This section details the information to be included in an application for a permit to build within a flood hazard area. The site plan is to show sufficient detail and information about the designated flood hazard area, including floodway and flood zones, to allow for a complete review of the proposed activities. Flood Insurance Rate Maps (FIRMs) are flood hazard maps prepared by the Federal Emergency Management Agency (FEMA). FIRMs may show specific base flood elevations (BFE). If the community adopts a flood hazard map other than the FIRM, the design flood elevations (DFE) must be at least as high as the BFEs [see Table R301.2(1)].

Construction documents are to include the proposed elevation of the lowest floor and the elevation of the bottom of the lowest horizontal structural member, which is to be at or above the minimum as given in Section R322. In flood hazard areas except coastal high-hazard areas, applicants may propose placing fill with the intent of later constructing buildings with excavated basements. When excavated into fill, basements may be subject to damage, especially where waters remain high for more than a few hours. Fill materials can become saturated and provide inadequate support, or water pressure can collapse below-grade walls. Basements below residential buildings are not to be constructed below the DFE, even if excavated into fill that is placed above the DFE.

If elevated on individual fill pads in flood hazard areas, buildings will be surrounded by water during general conditions of flooding. Local emergency personnel responsible for evacuations should be consulted during the subdivision approval process. Many states and communities have provisions that require uninterrupted access to all buildings during a flood. These requirements may be administered by agencies responsible for permitting the construction of public rights-of-way or may be imposed as part of subdivision approval.

Many FIRMs show flood hazard areas without specifying the BFEs, indicating that FEMA has not prepared engineering analyses for those areas. These flood hazard areas are often referred to as unnumbered A Zones. An important step in regulating the development of these areas is the determination of the DFE. The building official and applicant are to search for, and use data from, other sources, which may include the U.S. Army Corps of Engineers, the Natural Resources Conservation Service, the state, a local flood control agency or district, or historical records. If flood elevation information is not available, the building official may require the applicant to develop the DFE in accordance with accepted engineering practices. Local officials unfamiliar with establishing DFEs are encouraged to contact the NFIP State Coordinator or the appropriate FEMA regional office. For additional guidance, refer to FEMA 265, *Managing Floodplain Development in Approximate Zone A Areas: A Guide for Obtaining and Developing Base (100-Year) Flood Elevations*.

R106.2 Site plan or plot plan. The *construction documents* submitted with the application for *permit* shall be accompanied by a site plan showing the size and location of new construction and existing structures on the site and distances from *lot lines*. In the case of demolition, the site plan shall show construction to be demolished and the location and size of existing structures and construction that are to remain on the site or plot. The *building official* is authorized to waive or modify the requirement for a site plan when the application for permit is for alteration or repair or when otherwise warranted.

❖ One valuable part of the construction documents is the site plan. As a part of reviewing the building's location on the site for conformance with the code, the building official must know the size and location of any other structures on the lot, as well as their physical relationship to the new structure. The proximity of the new structure to lot lines and any public ways must also be shown. The distance between a building and the lot lines may trigger a variety of code requirements, the most notable being exterior wall fire-resistance rating and opening prohibition (see Section R302).

The code also requires that a site plan be submitted when a building or structure is to be demolished. The plan must identify the location of the building to be demolished, as well as any surrounding structures that will remain in place.

R106.3 Examination of documents. The *building official* shall examine or cause to be examined *construction documents* for code compliance.

❖ The building official must review the construction documents, or have qualified employees or consultants examine the plans, to determine whether they comply with the requirements of the jurisdiction.

R106.3.1 Approval of construction documents. When the *building official* issues a *permit*, the *construction documents* shall be *approved* in writing or by a stamp which states "REVIEWED FOR CODE COMPLIANCE." One set of *construction documents* so reviewed shall be retained by the *building official*. The other set shall be returned to the applicant, shall be kept at the site of work and shall be open to inspection by the *building official* or his or her authorized representative.

❖ Approval of the construction documents is the first in a series of reviews and approvals throughout the design and construction process. The building official or authorized representative must indicate that the construction documents are approved for construction, in writing or by a stamp, which specifically states "APPROVED PLANS PER IRC R106.3.1" and including any additional information that is necessary for the project. A set of the documents must be retained by the building department as its record for the life of the project. It is not uncommon throughout the job for questions to arise that require referencing the approved set of plans in the office. An efficient filing system should be developed to make the retrieval of construction documents a simple process.

In addition to the set of construction documents retained by the Department of Building Safety, at least one set of approved plans is to be returned to the permit holder. These construction documents must be maintained at the job site for reference purposes throughout the project. To avoid confusion, they must duplicate the documents that were approved and stamped. They must be available for review by building department personnel during the numerous inspections that may take place. Additionally, the contractor cannot determine compliance with approved construction documents unless those approved documents are available. Another reason to have the documents available is that these plans will generally indicate any special items or issues identified by the plans examiner that may not be shown on any other construction documents. The approved plans usually must be available at the job site prior to an inspection.

R106.3.2 Previous approvals. This code shall not require changes in the *construction documents*, construction or designated occupancy of a structure for which a lawful *permit* has been heretofore issued or otherwise lawfully authorized, and the construction of which has been pursued in good faith within 180 days after the effective date of this code and has not been abandoned.

❖ The code does not require work to be brought up to the requirements of a code that was adopted since the edition of the code under which the existing valid permit was issued. If the work is abandoned for 180 days or more, and an extension of the permit is not granted, the original permit is no longer valid, and a new permit must be sought under the provisions of the latest edition of the code.

R106.3.3 Phased approval. The *building official* is authorized to issue a *permit* for the construction of foundations or any other part of a building or structure before the *construction documents* for the whole building or structure have been submitted, provided that adequate information and detailed statements have been filed complying with pertinent requirements of this code. The holder of such *permit* for the foundation or other parts of a building or structure shall proceed at the holder's own risk with the building operation and without assurance that a *permit* for the entire structure will be granted.

❖ Phased approval is needed for projects that use the "fast track" construction method, which allows construction to begin before completion of all of the plans and specifications. Although it is preferable to issue permits for projects in their entirety, the building official has the authority to issue, at his or her discretion, a permit for a portion of the construction. The building official must be satisfied that the information provided shows, in satisfactory detail, that the partial construction will conform to the requirements of the code. In such a case, the permit holder proceeds at his or her own risk with the construction and has no assurance of the entire permit ever being issued.

R106.4 Amended construction documents. Work shall be installed in accordance with the *approved construction documents*, and any changes made during construction that are not in compliance with the *approved construction documents* shall be resubmitted for approval as an amended set of *construction documents*.

❖ The code requires that all work to be done in accordance with the approved plans and other construction documents. Where the construction will not conform to the approved construction documents, the documents must be revised, and they must be resubmitted to the building official for review and approval. The building official must retain one set of the amended and approved plans. The other set is to be kept at the construction site, ready for use by the jurisdiction's inspection staff.

R106.5 Retention of construction documents. One set of *approved construction documents* shall be retained by the *building official* for a period of not less than 180 days from date of completion of the permitted work, or as required by state or local laws.

❖ Construction documents must be retained in case a dispute arises after the completion of the project. Unless modified because of state or local statutes, the retention period for the approved construction documents is a minimum of 180 days following the completion of the work, typically the date the certificate of occupancy is issued. Any further retention of plans by the jurisdiction as an archival record of construction activity in the community is not required by the code.

SECTION R107 TEMPORARY STRUCTURES AND USES

R107.1 General. The *building official* is authorized to issue a *permit* for temporary structures and temporary uses. Such permits shall be limited as to time of service, but shall not be permitted for more than 180 days. The *building official* is authorized to grant extensions for demonstrated cause.

❖ The building official can authorize temporary use permits when the applicant has met all regulations governing such use. Permits for temporary structures may also be granted subject to the provisions of this section. Although the permit is to be granted for a time period consistent with the temporary use, 180 days is the maximum period of time for which a temporary use or structure can be valid. This code section often applies to structures that are commonly used for a short duration of time at a specific location, then easily dismantled and removed, often to be reconstructed at a different site. It is common for this type of structure to be used at events like street fairs, carnivals, circuses, parades, sporting events, weddings, concerts or revivals, or they may also be used to house construction offices. These structures range in size from small tents or shade structures, which may house a vendor at a street fair, to large tents for circuses or religious revivals. Because of the large size, large occupant load and complexity that can result from smaller structures being grouped together, the fire safety and life safety concerns as covered in Section R107.2 must be duly considered before a permit is issued.

R107.2 Conformance. Temporary structures and uses shall conform to the structural strength, fire safety, means of egress, light, ventilation and sanitary requirements of this code as necessary to ensure the public health, safety and general welfare.

❖ This section gives the building official discretion in determining the specific criteria for conformance. The issues of structural strength, fire safety, egress, light, ventilation and sanitation are mentioned in this section as the key areas of concern for temporary structures or uses. The levels of required conformance must be determined by the building official to achieve the jurisdiction's required level of safety, health and welfare. Full compliance with the code is not required for a temporary structure or temporary use, but it is clear that the level of performance set forth by the code must be considered in the development of the requirements.

R107.3 Temporary power. The *building official* is authorized to give permission to temporarily supply and use power in part of an electric installation before such installation has been fully completed and the final certificate of completion has been issued. The part covered by the temporary certificate shall comply with the requirements specified for temporary lighting, heat or power in NFPA 70.

❖ The NFPA 70 is referenced regarding temporary lighting, heat and power. The code allows temporary electrical service to be provided prior to the completion of the entire electrical system if the building official approves such a connection.

R107.4 Termination of approval. The *building official* is authorized to terminate such *permit* for a temporary structure or use and to order the temporary structure or use to be discontinued.

❖ Where the use of a temporary structure is not consistent with that approved by the building official, or where an unsafe condition exists, the building official is authorized to terminate the temporary permit.

SECTION R108 FEES

R108.1 Payment of fees. A *permit* shall not be valid until the fees prescribed by law have been paid. Nor shall an amendment to a *permit* be released until the additional fee, if any, has been paid.

❖ This section addresses the costs necessary to operate a Department of Building Safety by creating a mechanism for fee collection. Such fees are typically set to provide enough funds to adequately pay for the costs of operating the various department functions, including administration, plans examination and inspection. All fees, including those for changes to the permit, are to be collected before a valid permit is issued.

R108.2 Schedule of permit fees. On buildings, structures, electrical, gas, mechanical and plumbing systems or *alterations* requiring a *permit*, a fee for each *permit* shall be paid as required, in accordance with the schedule as established by the applicable governing authority.

❖ The code states that fees for activities regulated by the jurisdiction and administered by the building official should be equitably assessed and adequate to fund the administration, inspection and plan review services required by the code. Exorbitant fees are unreasonable and can often be a basis for some citizens' distrust of government agencies, while unrealistically low fees serve little purpose. In either situation, the effectiveness of the department suffers from either distrust and poor relations with the public or undertrained and understaffed departments that are hopelessly unable to retain the best-qualified employees. Either fee schedule does not serve the best interests of the public, which depends on the jurisdiction to enforce the code provisions effectively to maintain a minimum level of safety in the built environment. An example of a building department fee schedule is shown in Appendix L.

R108.3 Building permit valuations. Building *permit* valuation shall include total value of the work for which a *permit* is being issued, such as electrical, gas, mechanical, plumbing equipment and other permanent systems, including materials and labor.

❖ Most jurisdictions develop a fee schedule based on the projected construction cost of the work to be done. Two methods are used to determine this cost valuation: (1) a "per-square-foot" factor based on the use or occupancy of the building and the type of construction involved or (2) the "bid cost" factor based on the total accepted bid price for doing the work. The valuation is determined by including the value of the construction process, including both materials and labor. It is important that a realistic valuation be determined for every project so that permit fees are applied fairly and accurately.

R108.4 Related fees. The payment of the fee for the construction, alteration, removal or demolition for work done in connection with or concurrently with the work authorized by a building *permit* shall not relieve the applicant or holder of the *permit* from the payment of other fees that are prescribed by law.

❖ A building permit and the fees attached to that permit do not necessarily cover all aspects of the work to be performed. All fees of the jurisdiction, including those for additional permits, reinspections, investigations or other departmental functions, must be paid along with those for a building permit.

R108.5 Refunds. The *building official* is authorized to establish a refund policy.

❖ The building official has the authority to develop policies for refunding permit fees. The refund policy should retain fees for which the jurisdiction has provided services; only that portion of a fee for which no service has been rendered should be refunded.

R108.6 Work commencing before permit issuance. Any person who commences work requiring a *permit* on a building, structure, electrical, gas, mechanical or plumbing system before obtaining the necessary permits shall be subject to a fee established by the applicable governing authority that shall be in addition to the required *permit* fees.

❖ The purpose of the permit process is to insure that the proposed building complies with the law before funds are expended toward actual construction. If construction is completed before a noncomplying element is found on the plans, costly repairs or some unsafe condition could be created. This section provides for a penalty fee to be imposed when construction is started before approval.

SECTION R109
INSPECTIONS

R109.1 Types of inspections. For onsite construction, from time to time the *building official*, upon notification from the *permit* holder or his agent, shall make or cause to be made any necessary inspections and shall either approve that portion of the construction as completed or shall notify the *permit* holder or his or her agent wherein the same fails to comply with this code.

❖ Inspections are necessary to verify that the construction conforms to the code requirements, and this section outlines the minimum required inspections. Besides the minimum required inspections that are specifically listed, the building official has the authority to require additional inspections so that compliance with the code can be determined. It is the duty of the permit holder or an authorized agent of the permit holder to notify the building department that some or all of the work covered by the permit is ready and available for inspection. At that point, the appropriate jurisdictional inspector performs the necessary on-site inspection. The inspector then must inform the permit holder or agent that the work has been inspected. This may be by a telephone call, an electronic message or, in many cases, a written record of the inspection posted at the job site.

R109.1.1 Foundation inspection. Inspection of the foundation shall be made after poles or piers are set or trenches or *basement* areas are excavated and any required forms erected and any required reinforcing steel is in place and supported prior to the placing of concrete. The foundation inspection shall include excavations for thickened slabs intended for the support of bearing walls, partitions, structural supports, or *equipment* and special requirements for wood foundations.

❖ The foundation inspection is typically the first inspection of the job site by a representative of the Building Safety Department. At that time, the inspector will verify that the foundation is located as shown on the approved plans and in accordance with the jurisdictional requirements for building setbacks and easements. Where a footing system is to be used, the footing trenches must be excavated and any required reinforcing steel must be in place. For foundation walls, the wall forms must be completely erected with the appropriate steel reinforcement placed within the forms.

Column pads, thickened slabs and other foundation work must also be ready for inspection prior to concrete placement.

R109.1.2 Plumbing, mechanical, gas and electrical systems inspection. Rough inspection of plumbing, mechanical, gas and electrical systems shall be made prior to covering or concealment, before fixtures or *appliances* are set or installed, and prior to framing inspection.

> **Exception:** Backfilling of ground-source heat pump loop systems tested in accordance with Section M2105.1 prior to inspection shall be permitted.

❖ The various trade rough-in inspections must be completed prior to the framing inspection. The electrical, gas, mechanical and plumbing systems that are to be concealed must be inspected and approved prior to their concealment. In addition, the effect of these systems on the structural integrity of the building must be reviewed after they are completed and ready to be covered. All of the trades inspections must be completed and approvals granted prior to the inspection of the framing system.

R109.1.3 Floodplain inspections. For construction in flood hazard areas as established by Table R301.2(1), upon placement of the lowest floor, including *basement*, and prior to further vertical construction, the *building official* shall require submission of documentation, prepared and sealed by a registered *design professional*, of the elevation of the lowest floor, including *basement*, required in Section R322.

❖ For any building constructed in a flood-prone area, the elevation of the lowest floor level must be established immediately after the placement of the floor. Documentation of the elevation of the lowest floor is evidence of compliance (see commentary, Section R322.3.6). If the documentation is submitted when the lowest floor level is established and before further vertical construction, errors in elevation may be corrected at the least cost. The elevation of the lowest floor is used by insurance agents to compute flood insurance premium costs. An error in the minimum required elevation may significantly increase the cost of flood insurance, which may be required by mortgage lenders. If the elevation of the lowest floor goes uncorrected, all future owners will incur the additional increase in cost.

R109.1.4 Frame and masonry inspection. Inspection of framing and masonry construction shall be made after the roof, masonry, all framing, firestopping, draftstopping and bracing are in place and after the plumbing, mechanical and electrical rough inspections are *approved*.

❖ The framing inspection is usually the final opportunity for the inspector to view all of the items that will be concealed within the structure. The inspection includes the structural frame work of the building, as well as any fireblocking or draftstopping that will be contained within concealed spaces. All of the electrical, mechanical, gas and plumbing inspections must be completed and approved prior to the framing inspection. This allows any framing members to be repaired while they are accessible.

R109.1.5 Other inspections. In addition to the called inspections above, the *building official* may make or require any other inspections to ascertain compliance with this code and other laws enforced by the *building official*.

❖ A variety of other inspections, such as those for gypsum board or insulation, may be mandated by the building department to verify compliance with the code or other city ordinances. These inspections are generally established as a result of local concern regarding a specific portion of the construction process. The procedure for these inspections should be consistent with the other provisions of this section.

R109.1.5.1 Fire-resistance-rated construction inspection. Where fire-resistance-rated construction is required between *dwelling units* or due to location on property, the *building official* shall require an inspection of such construction after all lathing and/or wallboard is in place, but before any plaster is applied, or before wallboard joints and fasteners are taped and finished.

❖ There are a limited number of situations in which fire-resistance-rated construction is required by the code. For example, where two or more dwelling units are located within the same structure, they are required by Section R302 to be completely separated from each other by a specified level of fire resistance. This is the case for both two-family dwellings and townhouses. A second example would be the proximity of the building to an adjoining property line. Section R302 requires exterior walls located less than 3 feet (914 mm) from a property line (unless abutting a public way) to be of minimum 1-hour fire-resistance-rated construction. If either of these conditions should occur, an inspection of the fire-resistance-rated construction is required.

The inspection for compliance with the fire-resistance requirements of the code should be made at a point of construction when the membrane materials are in place, but the fasteners are still exposed. This allows the inspector to verify the appropriate fastener type and location based on the specific fire-resistance listing of the portion of the building under consideration.

R109.1.6 Final inspection. Final inspection shall be made after the permitted work is complete and prior to occupancy.

❖ The final inspection should occur after all of the work addressed by the code is complete, but prior to occupancy of the building. The issues addressed in the final inspection cover all aspects of construction, including fire safety, life safety and structural safety, as well as electrical, plumbing, gas and mechanical items. The final inspection must be approved before a certificate of occupancy can be issued.

R109.1.6.1 Elevation documentation. If located in a flood hazard area, the documentation of elevations required in Section R322.1.10 shall be submitted to the *building official* prior to the final inspection.

❖ This section is intended to serve as a reminder that a documentation of elevations is required before final inspection for houses constructed in a flood hazard area. See Section R322.1.10 for information about this documentation.

R109.2 Inspection agencies. The *building official* is authorized to accept reports of *approved* agencies, provided such agencies satisfy the requirements as to qualifications and reliability.

❖ It is not uncommon for the building official to rely on other agencies for informational or inspection reports regarding various aspects covering methods of the construction process materials. This reliance should be based on the building official's approval of the qualifications and reliability possessed by the third-party inspection or testing service.

R109.3 Inspection requests. It shall be the duty of the *permit* holder or their agent to notify the *building official* that such work is ready for inspection. It shall be the duty of the person requesting any inspections required by this code to provide access to and means for inspection of such work.

❖ The individual doing the authorized work has the responsibility for notifying the building department when the work is ready for inspection. Each building department establishes its own procedures on how and when requests should be made. Once an inspection has been scheduled, access to the area ready for inspection must be provided. The individuals performing the work should make the inspection process run as smoothly as possible.

R109.4 Approval required. Work shall not be done beyond the point indicated in each successive inspection without first obtaining the approval of the *building official*. The *building official* upon notification, shall make the requested inspections and shall either indicate the portion of the construction that is satisfactory as completed, or shall notify the *permit* holder or an agent of the *permit* holder wherein the same fails to comply with this code. Any portions that do not comply shall be corrected and such portion shall not be covered or concealed until authorized by the *building official*.

❖ Work must not continue past the point of a required inspection until that inspection has been approved by the building department. It is possible that if the work progresses beyond this point and is not in total compliance with the code, some of the work may have to be removed. It is critical that each individual stage of the project be approved prior to continuance of construction.

As indicated in Section R109.1, inspections must be performed when requested, and the inspector must indicate whether the construction is satisfactory or is not compliant. If the work is not approved, it must be corrected, and a reinspection must be requested. No work may be concealed until the building department approves it.

SECTION R110 CERTIFICATE OF OCCUPANCY

R110.1 Use and occupancy. No building or structure shall be used or occupied, and no change in the existing occupancy classification of a building or structure or portion thereof shall be made until the *building official* has issued a certificate of occupancy therefor as provided herein. Issuance of a certificate of occupancy shall not be construed as an approval of a violation of the provisions of this code or of other ordinances of the *jurisdiction*. Certificates presuming to give authority to violate or cancel the provisions of this code or other ordinances of the *jurisdiction* shall not be valid.

Exceptions:

1. Certificates of occupancy are not required for work exempt from permits under Section R105.2.
2. Accessory buildings or structures.

❖ The tool the building official employs to control the uses and occupancies of the various buildings in a jurisdiction is the certificate of occupancy. This section establishes the conditions of a certificate of occupancy and identifies the information the certificate must contain. The building official must be satisfied that the structure meets the requirements of the code before a certificate of occupancy can be given, and the structure cannot be legally occupied until a certificate has been issued. If the occupancy classification of an existing building has changed, such as changing an old Victorian home to a small office building, a new certificate of occupancy must be issued. The certificate of occupancy is the legal notification from the Department of Building Safety that the building may be occupied for its intended purpose.

The granting of a certificate of occupancy does not necessarily indicate that no violations of the code or other jurisdictional laws exist. The building official should make every effort to determine compliance with all applicable code provisions and other ordinances of the jurisdiction. It is important that violations be corrected. It is possible that the certificate of occupancy will be revoked if it is found that the certificate was issued in error.

R110.2 Change in use. Changes in the character or use of an existing structure shall not be made except as specified in Sections 3408 and 3409 of the *International Building Code*.

❖ When an existing building's character or use is modified, the provisions of Sections 3408 and 3409 of the IBC dealing with change of occupancy and historic buildings must be met. Because the code is limited in scope to specific residential occupancies, any change of occupancy to a use beyond the scope of Section R101.2 will be governed by the IBC.

R110.3 Certificate issued. After the *building official* inspects the building or structure and finds no violations of the provisions of this code or other laws that are enforced by the department of building safety, the *building official* shall issue a certificate of occupancy which shall contain the following:

1. The building *permit* number.
2. The address of the structure.
3. The name and address of the owner.
4. A description of that portion of the structure for which the certificate is issued.
5. A statement that the described portion of the structure has been inspected for compliance with the requirements of this code.
6. The name of the *building official*.

7. The edition of the code under which the *permit* was issued.
8. If an automatic sprinkler system is provided and whether the sprinkler system is required.
9. Any special stipulations and conditions of the building *permit*.

❖ Prior to the use or occupancy of the building, the building official shall perform a final inspection as addressed in Section R109.1.6. If the official finds no violations of the code and other laws enforced by the Department of Building Safety, the building official is required to issue a certificate of occupancy. Commentary Figure R110.3 illustrates the information that must be provided on the certificate of occupancy.

R110.4 Temporary occupancy. The *building official* is authorized to issue a temporary certificate of occupancy before the completion of the entire work covered by the *permit*, provided that such portion or portions shall be occupied safely. The *building official* shall set a time period during which the temporary certificate of occupancy is valid.

❖ Where a portion of a building is intended to be occupied prior to the occupancy of the entire structure, the building official may issue a temporary certificate of occupancy. Prior to the issuance of a temporary certificate of occupancy, it is critical that the building official determine that the portions to be occupied provide the minimum levels of safety required by the code. In addition, the building official must establish a definitive length of time for the temporary certificate of occupancy to be valid.

R110.5 Revocation. The *building official* shall, in writing, suspend or revoke a certificate of occupancy issued under the provisions of this code wherever the certificate is issued in error, or on the basis of incorrect information supplied, or where it is determined that the building or structure or portion thereof is in violation of any ordinance or regulation or any of the provisions of this code.

❖ In essence, the certificate of occupancy certifies that the described building or portion of that building complies with the requirements of the code for the intended use. However, any certificate of occupancy may be suspended or revoked by the building official under one of three conditions: (1) when the certificate is issued in error, (2) when incorrect information is supplied to the building official, or (3) when it is determined that the building or a portion of the building is shown to be in violation of the code or any other ordinance or regulation of the jurisdiction.

SECTION R111
SERVICE UTILITIES

R111.1 Connection of service utilities. No person shall make connections from a utility, source of energy, fuel or power to any building or system that is regulated by this code for which a *permit* is required, until *approved* by the *building official*.

❖ This section addresses the connection and disconnection, either permanent or temporary, of any utilities that service a building or structure regulated by the code. The building official is authorized to control the connection for any service utility when the connection is to a

Figure R110.3
CERTIFICATE OF OCCUPANCY

building that is regulated by the code and requires a permit. Prior to the connection of a utility, source of energy, fuel or power, all conditions for the connection must be met and verified by required inspections.

R111.2 Temporary connection. The *building official* shall have the authority to authorize and approve the temporary connection of the building or system to the utility, source of energy, fuel or power.

❖ Temporary service utility connections, such as for temporary electrical service during the construction process, are permitted when approved by the building official after the necessary inspections have been performed and any additional conditions met.

R111.3 Authority to disconnect service utilities. The *building official* shall have the authority to authorize disconnection of utility service to the building, structure or system regulated by this code and the referenced codes and standards set forth in Section R102.4 in case of emergency where necessary to eliminate an immediate hazard to life or property or when such utility connection has been made without the approval required by Section R111.1 or R111.2. The *building official* shall notify the serving utility and whenever possible the owner and occupant of the building, structure or service system of the decision to disconnect prior to taking such action if not notified prior to disconnection. The owner or occupant of the building, structure or service system shall be notified in writing as soon as practical thereafter.

❖ When an immediate hazard to life or property exists, the building official has the authority to order disconnection of the utility services. This can also occur when the utility service has been connected without the necessary approvals required by the code. Whenever possible, the building owner and the building occupant or occupants should be notified prior to the disconnection of the services. Then at the first practical opportunity, the building official is to formally notify the building owner in writing of the disconnection activities. As with all administrative functions, all aspects of due process must be followed.

SECTION R112
BOARD OF APPEALS

R112.1 General. In order to hear and decide appeals of orders, decisions or determinations made by the *building official* relative to the application and interpretation of this code, there shall be and is hereby created a board of appeals. The *building official* shall be an ex officio member of said board but shall have no vote on any matter before the board. The board of appeals shall be appointed by the governing body and shall hold office at its pleasure. The board shall adopt rules of procedure for conducting its business, and shall render all decisions and findings in writing to the appellant with a duplicate copy to the *building official*.

❖ This section holds that any aggrieved party with a material interest in the decision of the building official may appeal such a decision before a board of appeals. This provides a forum other than the court of the jurisdiction in which the building official's action can be reviewed.

A board of appeals is to be created by the jurisdiction. The primary function of the board of appeals is hearing and acting on the appeal of orders or decisions the building official has made on the application or interpretation of the code. The building official is an exofficio member of the board of appeals, but he or she has no vote on any issue that comes before the board. The appellant is to be provided with a written copy of the board's findings and decisions, with a duplicate copy sent to the building official. Appendix B of the IBC supplies example rules of the procedure.

R112.2 Limitations on authority. An application for appeal shall be based on a claim that the true intent of this code or the rules legally adopted thereunder have been incorrectly interpreted, the provisions of this code do not fully apply, or an equally good or better form of construction is proposed. The board shall have no authority to waive requirements of this code.

❖ The code does not grant authority to the board to hear appeals regarding the administrative provisions of the code, nor does it grant the board the authority to waive any code requirements. An appeal must be based on the claim that the provisions of the code have been misinterpreted, that the provisions do not apply to the appellant's circumstances, or that where an alternative method or technique of construction is used it has been shown to be at least equal to the code requirements.

R112.2.1 Determination of substantial improvement in flood hazard areas. When the *building official* provides a finding required in Section R105.3.1.1, the board of appeals shall determine whether the value of the proposed work constitutes a substantial improvement. A substantial improvement means any repair, reconstruction, rehabilitation, *addition* or improvement of a building or structure, the cost of which equals or exceeds 50 percent of the market value of the building or structure before the improvement or repair is started. If the building or structure has sustained substantial damage, all repairs are considered substantial improvement regardless of the actual repair work performed. The term does not include:

1. Improvements of a building or structure required to correct existing health, sanitary or safety code violations identified by the *building official* and which are the minimum necessary to assure safe living conditions; or
2. Any alteration of an historic building or structure, provided that the alteration will not preclude the continued designation as an historic building or structure. For the purpose of this exclusion, an historic building is:
   - 2.1. *Listed* or preliminarily determined to be eligible for *listing* in the National Register of Historic Places; or
   - 2.2. Determined by the Secretary of the U.S.Department of Interior as contributing to the historical significance of a registered historic district or a district preliminarily determined to qualify as an historic district; or

2.3. Designated as historic under a state or local historic preservation program that is *approved* by the Department of Interior.

❖ "Substantial improvement," found in Sections R105.3.1.1 and R112.2.1, is defined in this section and is consistent with the definition used by the NFIP and Section 1612.2 of the IBC. One of the long-range objectives of the NFIP is to reduce the exposure of older buildings that were built in flood hazard areas before local jurisdictions adopted flood hazard area maps and regulations. Section R105.3 of the code directs the applicant to state the valuation of the proposed work as part of the application for a permit. To determine whether a proposed alteration, repair, addition or improvement of a building or structure constitutes a substantial improvement, the cost of the proposed work is to be compared to the market value of the building or structure before the work is started. To determine market value, the building official may require the applicant to provide the information listed in Section R105.3. For additional guidance, refer to FEMA 213, *Answers to Questions About Substantially Damaged Buildings*, and FEMA 311, *Guidance on Estimating Substantial Damage Using the NFIP Residential Substantial Damage Estimator*.

When the board of appeals makes its determination, certain items are not included in the valuation of proposed work. Specifically, if certain health-, sanitation- or safety-code violations have been cited previously, the cost of the minimum repairs required to correct those violations is not included in the determination.

Alteration of a historic structure is not considered a substantial improvement or repair of substantial damage, provided the proposed work does not alter the building to the extent that it would no longer qualify as a historic structure. The building official may require applicants to consult an appropriate historic preservation authority to determine whether a proposed alteration would jeopardize a structure's historic designation. The exception for historic structures does not apply to structures located within designated historic districts, unless those structures are individually listed as historic structures.

R112.2.2 Criteria for issuance of a variance for flood hazard areas. A variance shall be issued only upon:

1. A showing of good and sufficient cause that the unique characteristics of the size, configuration or topography of the site render the elevation standards in Section R322 inappropriate.
2. A determination that failure to grant the variance would result in exceptional hardship by rendering the *lot* undevelopable.
3. A determination that the granting of a variance will not result in increased flood heights, additional threats to public safety, extraordinary public expense, cause fraud on or victimization of the public, or conflict with existing local laws or ordinances.
4. A determination that the variance is the minimum necessary to afford relief, considering the flood hazard.
5. Submission to the applicant of written notice specifying the difference between the design flood elevation and the elevation to which the building is to be built, stating that the cost of flood insurance will be commensurate with the increased risk resulting from the reduced floor elevation, and stating that construction below the design flood elevation increases risks to life and property.

❖ All the criteria set forth in this section must be met in order for the board of appeals to: (1) consider granting a variance for construction in flood-prone areas, and (2) provide relief from selected provisions for flood-resistant construction. Granting of the variance must not cause additional public safety concerns beyond those already present.

The board of appeals is empowered to hear requests for variances from the flood hazard provisions of the code. Variances from these provisions may place people and property at significant risk. Therefore, communities are cautioned to carefully evaluate the impacts of issuing a variance, particularly variances to the requirements for elevating buildings to the BFE. The elements that are to be evaluated include impacts on the site, the applicant and other parties who may be affected; such as adjacent property owners and the community as a whole. Flood plain development that is not undertaken in accordance with the flood-resistance provisions of the code will be exposed to increased flood damages. As a consequence, flood insurance premium rates will be significantly higher. Variance decisions made by the board of appeals are to be based solely on technical justifications outlined in this section and not on the personal circumstances of an owner or applicant.

Applicants sometimes request variances to the minimum elevation requirements for the lowest floor of buildings in flood hazard areas to improve access for the disabled and elderly. Generally, variances of this nature are not to be granted because these are personal circumstances that will change as the property changes ownership. Not only would persons of limited mobility be at risk from flooding, but a building built below the BFE would continue to be exposed to flood damage long after the personal need for a variance ends. More appropriate alternatives are to be considered to serve the needs of disabled or elderly persons, such as varying setbacks to allow construction on less flood-prone portions of sites or the installation of personal elevators.

All variances are to be the minimum necessary to afford relief. The board of appeals will address each listed condition for a variance, especially the requirement that it determine whether the failure to grant the variance would result in exceptional hardship by rendering the lot undevelopable. By itself, this determination may be insufficient to result in an exceptional hardship if other conditions for issuing a variance cannot be met. The determination of hardship is to be based on the unique characteristics of the site and not on the personal circumstances of the applicant.

In guidance materials, FEMA cautions that economic hardship alone is not an exceptional hardship. Building officials and boards of appeals are cautioned that granting a variance does not affect how the building will be rated for the purposes of NFIP flood insurance. Even if circumstances justify granting a variance to build a lowest floor that is below the DFE, the rate used to calculate the cost of a flood insurance policy will be based on the risk to the building. Flood insurance, required by certain mortgage lenders, may be extremely expensive. Although the applicant may not be required to purchase flood insurance, the requirement may be imposed on subsequent owners. The building official is to provide the applicant a written notice to this effect, along with the other cautions listed in this section.

R112.3 Qualifications. The board of appeals shall consist of members who are qualified by experience and training to pass on matters pertaining to building construction and are not employees of the *jurisdiction*.

❖ The members of the board of appeals are to be selected by the appointing authority of the jurisdiction; the members must be qualified to pass judgment on appeals associated with building construction. The purpose of the board is to provide a review of the appeal independent from that of the building department, so no jurisdictional employee is permitted to be a member. Additional information on suggested qualifications of members of the board of appeals is contained in Appendix B of the IBC.

R112.4 Administration. The *building official* shall take immediate action in accordance with the decision of the board.

❖ Decisions made by the board of appeals must be enacted as quickly as possible. The building official is to take whatever action is necessary to see that the orders of the board are carried out.

SECTION R113 VIOLATIONS

R113.1 Unlawful acts. It shall be unlawful for any person, firm or corporation to erect, construct, alter, extend, repair, move, remove, demolish or occupy any building, structure or *equipment* regulated by this code, or cause same to be done, in conflict with or in violation of any of the provisions of this code.

❖ This section describes the citing, recording and subsequent actions to be taken when code violations are found.

R113.2 Notice of violation. The *building official* is authorized to serve a notice of violation or order on the person responsible for the erection, construction, alteration, extension, repair, moving, removal, demolition or occupancy of a building or structure in violation of the provisions of this code, or in violation of a detail statement or a plan *approved* thereunder, or in violation of a *permit* or certificate issued under the provisions of this code. Such order shall direct the discontinuance of the illegal action or condition and the abatement of the violation.

❖ The building official is required to notify the person responsible for the construction or use of a building found to be in violation of the code. The section of the code that is being violated must be cited so that the responsible party can respond to the notice.

R113.3 Prosecution of violation. If the notice of violation is not complied with in the time prescribed by such notice, the *building official* is authorized to request the legal counsel of the *jurisdiction* to institute the appropriate proceeding at law or in equity to restrain, correct or abate such violation, or to require the removal or termination of the unlawful occupancy of the building or structure in violation of the provisions of this code or of the order or direction made pursuant thereto.

❖ When the building owner, owner's agent or tenant does not correct the condition causing the violation as directed, the building official must pursue, through the use of legal counsel of the jurisdiction, legal means to correct the violation. This is not optional.

Any extensions that allow the violations to be corrected voluntarily must be for a reasonable, bona fide cause, or the building official may be subject to criticism for "arbitrary and capricious" actions. In general, it is better to have a standard time limitation for correction of violations. Departures from this standard must be for a clear and reasonable purpose, usually stated in writing by the violator.

R113.4 Violation penalties. Any person who violates a provision of this code or fails to comply with any of the requirements thereof or who erects, constructs, alters or repairs a building or structure in violation of the *approved construction documents* or directive of the *building official,* or of a *permit* or certificate issued under the provisions of this code, shall be subject to penalties as prescribed by law.

❖ The jurisdiction must establish penalties for a variety of violations that may occur. Violations specifically addressed by the code include failure to comply with: (1) the code, (2) approved plans or (3) directives of the building official.

SECTION R114 STOP WORK ORDER

R114.1 Notice to owner. Upon notice from the *building official* that work on any building or structure is being prosecuted contrary to the provisions of this code or in an unsafe and dangerous manner, such work shall be immediately stopped. The stop work order shall be in writing and shall be given to the owner of the property involved, or to the owner's agent or to the person doing the work and shall state the conditions under which work will be permitted to resume.

❖ The stop work order is a tool authorized by the code that enables the building official to demand that work on a building or structure be temporarily suspended.

Typically used under rare circumstances, this order may be issued where the work being performed is dangerous, unsafe or significantly contrary to the provisions of the code.

The stop work order is to be a written document indicating the reason or reasons for the suspension of work, identifying those conditions where compliance is necessary before work is allowed to resume. All work addressed by the order must cease immediately. The stop work order must be presented to either the owner of the subject property, the agent of the owner or the individual doing the work. Commentary Figure R114.1 is an example of a stop work order.

R114.2 Unlawful continuance. Any person who shall continue any work in or about the structure after having been served with a stop work order, except such work as that person is directed to perform to remove a violation or unsafe condition, shall be subject to penalties as prescribed by law.

❖ The only activity permitted in a building or a portion of a building subject to a stop work order is that work necessary to eliminate the violation or unsafe condition. Otherwise, penalties prescribed by laws of the jurisdiction must be imposed for illegal construction activity in defiance of a stop work order.

Bibliography

The following resource materials were used in the preparation of the commentary for this chapter of the code:

ASTM D 266-06, *Specification for Asphalt-saturated (Organic Felt) Used in Roofing and Waterproofing.* West Conshohocken, PA: ASTM International, 2006.

ASTM E 119-07, *Test Methods for Fire Tests of Building Construction and Materials.* West Conshohocken, PA: ASTM International, 2007.

FEMA 213, *Answers to Questions About Substantially Damaged Buildings.* Washington, DC: Federal Emergency Management Agency, 1991. (For ordering information, see the Bibliography for Chapter 3.)

FEMA 257, *Mitigation of Flood and Erosion Damage to Residential Buildings in Coastal Areas.* Washington, DC: Federal Emergency Management Agency, 1994.

FEMA 265, *Managing Floodplain Development in Approximate Zone A Areas: A Guide for Obtaining and Developing Base (100-year) Flood Elevations.* Washington, DC: Federal Emergency Management Agency, 1994.

FEMA 311, *Guidance on Estimating Substantial Damage Using the NFIP Residential Substantial Damage Estimator.* Washington, DC: Federal Emergency Management Agency, 1995.

FEMA 348, *Protecting Building Utilities from Flood Damage.* Washington, DC: Federal Emergency Management Agency, 1999.

IBC-12, *International Building Code.* Washington, DC: International Code Council, Inc., 2011.

IFC-12, *International Fire Code.* Washington, DC: International Code Council, Inc., 2011.

IPMC-12, *International Property Maintenance Code.* Washington, DC: International Code Council, Inc., 2011.

Legal Aspects of Code Administration. Country Club Hills, IL: Building Officials and Code Administrators International, Inc.; Whittier, CA: International Conference of Building Officials; Birmingham, AL: Southern Building Code Congress International, 1984.

NFPA 70-08, *National Electrical Code.* Quincy, MA: National Fire Protection Association, 2008.

LEGAL NOTICE

Date ________________

WHEREAS, VIOLATIONS OF Article ________, Section ________ of the Zoning Ordinance
Article ________, Section ________ of the Building Code
Article ________, Section ________ of the ____________ Code
have been found on these premises, IT IS HEREBY ORDERED in accordance with the above Code that all persons cease, desist from, and

STOP WORK

at once pertaining to construction, alterations or repairs on these premises known as ______________________________

All persons acting contrary to this order or removing or mutilating this notice are liable to arrest unless such action is authorized by the Department.

____
CODE OFFICIAL

Figure R114.1
STOP WORK ORDER

O'Bannon, Robert E. *Building Department Administration.* Whittier, CA: International Conference of Building Officials, 1973.

Readings in Code Administration, Volume 1: History/Philosophy/Law. Country Club Hills, IL: Building Officials and Code Administrators International, Inc., 1974.

Part II—Definitions

Chapter 2: Definitions

General Comments

The code user should be familiar with the terms in this chapter because: (1) the definitions are essential to the correct interpretation of the *International Residential Code*® (IRC®), and (2) the user might not be aware that a particular term encountered in the text has the special definition found herein.

Section R201.1 contains the scope of the chapter. Section R201.2 establishes the interchangeability of the terms in the code. Section R201.3 establishes the use of terms defined in other chapters. Section R201.4 establishes the use of undefined terms, and Section R202 lists terms and their definition according to this code.

Purpose

Codes are technical documents, so literally every word, term and punctuation mark can add to or change the meaning of the provision. Furthermore, the IRC, with its broad scope of applicability, includes terms inherent in a variety of construction disciplines. These terms can often have multiple meanings depending on the context or discipline being used at the time. For these reasons it is necessary to maintain a consensus on the specific meaning of terms contained in the IRC. Chapter 2 performs this function by stating clearly what specific terms mean for the purpose of this code.

SECTION R201 GENERAL

R201.1 Scope. Unless otherwise expressly stated, the following words and terms shall, for the purposes of this code, have the meanings indicated in this chapter.

❖ This section clarifies the terminology used in the IRC. The terms defined in the IRC often have very specific meanings, which can be different from their typical meanings. This section gives guidance to the use of the defined words relative to tense and gender and also provides the means to resolve those terms not defined.

R201.2 Interchangeability. Words used in the present tense include the future; words in the masculine gender include the feminine and neuter; the singular number includes the plural and the plural, the singular.

❖ Although the definitions contained in Chapter 2 are to be taken literally, gender and tense are considered to be interchangeable; thus, any gender and tense inconsistencies within the code text should not hinder the understanding or enforcement of the requirements.

R201.3 Terms defined in other codes. Where terms are not defined in this code such terms shall have meanings ascribed to them as in other code publications of the International Code Council.

❖ When a word or term appears in the code that is not defined in this chapter, other code publications of the International Code Council® (ICC®) may be used to find its definition. These code documents include the *International Building Code*® (IBC®), *International Mechanical Code*® (IMC®), *International Plumbing Code*® (IPC®), *International Fuel Gas Code*® (IFGC®), *International Fire Code*® (IFC®) and others. These codes contain additional definitions (some parallel and duplicated) which may be used in the enforcement of this code or in the enforcement of the other ICC codes by reference. When using a definition from another code, keep in mind the admonition from "Purpose": "These terms can often have multiple meanings depending on the context or discipline being used at the time."

R201.4 Terms not defined. Where terms are not defined through the methods authorized by this section, such terms shall have ordinarily accepted meanings such as the context implies.

❖ Another possible source for the definitions of words or terms not defined in this code or in other codes is their "ordinarily accepted meanings." Dictionary definitions may suffice, provided that the definitions are in context.

Sometimes construction terms used throughout the code may not be defined in Chapter 2, in another code, or in a dictionary. In such cases, one would first turn to the definitions contained in the referenced standards (see Chapter 43) and then to textbooks on the subject in question.

SECTION R202
DEFINITIONS

ACCESSIBLE. Signifies access that requires the removal of an access panel or similar removable obstruction.

❖ In general, where immediate access is not required because of the low level of hazard involved, an accessible method is all that is mandated, such as the removal of an access panel. The electrical definitions found in Chapter 34 contain two additional definitions for accessible that apply specifically to electrical wiring methods and electrical equipment. See Section 1102 of the IBC for the meaning of "accessible" as it applies to the requirements for "accessible" dwelling units in Section R320.

ACCESSIBLE, READILY. Signifies access without the necessity for removing a panel or similar obstruction.

❖ Where this term is designated as a requirement by other sections of the code, it is intended that access to the device, controls, shut-off valves or other element be extremely easy. "Readily accessible" means that the device must be reachable directly without a panel, door or equipment needing to be moved to gain access. Chapter 24, "Fuel Gas," has a similar definition noted as "Ready access (to)."

ACCESSORY STRUCTURE. A structure not greater than 3,000 square feet (279 m^2) in floor area, and not over two stories in height, the use of which is customarily accessory to and incidental to that of the dwelling(s) and which is located on the same *lot*.

❖ As it applies to the scope in Section R101.2, this term describes structures that are designed for accessory use to one- or two-family dwellings and multiple single-family townhouses. These structures are commonly used as garages, carports, cabanas, storage sheds, tool sheds, playhouses and garden structures.

The structures all house uses that are incidental to the primary use, which is the dwelling unit, and the activities that take place in accessory structures occur as a result of the primary building. Their use is secondary or minor in importance to the primary residence.

ADDITION. An extension or increase in floor area or height of a building or structure.

❖ Where a structure is increased in size vertically or horizontally, the increase is considered an addition. The addition will typically contribute additional floor area to the building, although an increase in roof height only is also an addition.

ADHERED STONE OR MASONRY VENEER. Stone or masonry veneer secured and supported through the adhesion of an *approved* bonding material applied to an *approved* backing.

❖ This is a common type of material used for a building façade. It is simply defined in order to allow use of the description to mean a specific type of material.

AIR ADMITTANCE VALVE. A one-way valve designed to allow air into the plumbing drainage system when a negative pressure develops in the piping. This device shall close by gravity and seal the terminal under conditions of zero differential pressure (no flow conditions) and under positive internal pressure.

❖ An air admittance valve is an alternative to traditional vent piping configurations. Such valves allow the venting of fixtures where conventional venting may be impractical. The valve is designed to open and admit air into the drainage system when negative pressures occur. The valve closes by gravity and seals the vent terminal when the internal drain pressure is equal to or exceeds atmospheric pressure. The seal prevents sewer gas from entering the building [see Commentary Figures R202(1) and R202(2)].

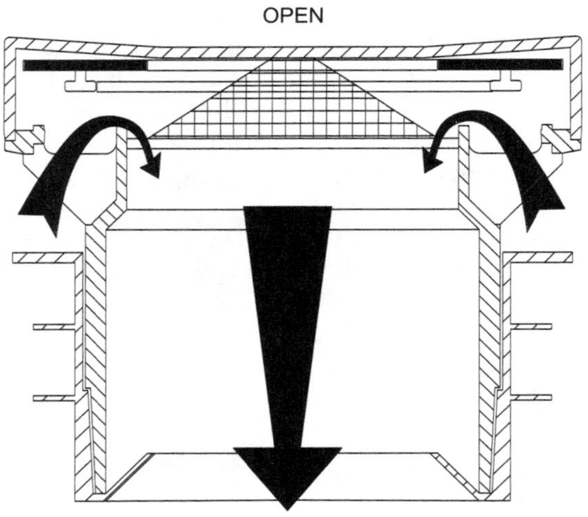

Figure R202(1)
AIR ADMITTANCE VALVE OPENS TO ADMIT AIR TO RELIEVE NEGATIVE PRESSURE

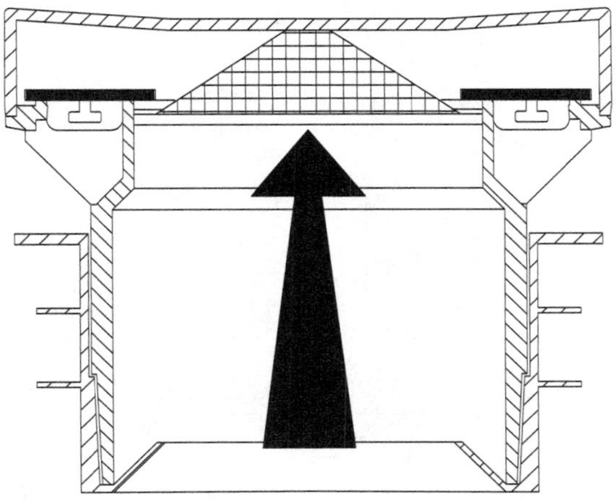

Figure R202(2)
AIR ADMITTANCE VALVE CLOSES AND SEALS VENT UNDER ZERO OR POSITIVE PRESSURE

AIR BARRIER. See Section N1101.9 for definition applicable in Chapter 11.

❖ Air barriers are an integral part of the thermal envelope of a building for energy conservation.

AIR BREAK (DRAINAGE SYSTEM). An arrangement in which a discharge pipe from a fixture, *appliance* or device drains indirectly into a receptor below the flood-level rim of the receptor, and above the trap seal.

❖ An air break is an indirect drainage method by which waste discharges to the drainage system through piping that terminates below the flood level rim of an approved receptor. An air break is commonly used to protect mechanical equipment from sewage backup in the event that stoppage occurs. It also protects the drainage system from adverse pressure conditions caused by pumped discharge [see Commentary Figure R202(3)].

AIR CIRCULATION, FORCED. A means of providing space conditioning utilizing movement of air through ducts or plenums by mechanical means.

❖ Forced-air systems use a central heating or cooling unit equipped with a fan to return unconditioned air through return ducts and to deliver conditioned air through supply ducts or plenums to occupied spaces.

AIR-CONDITIONING SYSTEM. A system that consists of heat exchangers, blowers, filters, supply, exhaust and return-air systems, and shall include any apparatus installed in connection therewith.

❖ This definition is limited to the components commonly used in a mechanical air-conditioning system. Additional parts of the air-conditioning system include thermostats, humidistats, dampers and any other controls needed for the system to operate properly.

AIR GAP, DRAINAGE SYSTEM. The unobstructed vertical distance through free atmosphere between the outlet of a waste pipe and the flood-level rim of the fixture or receptor into which it is discharging.

❖ In a drainage system air gap, waste is discharged to the drainage system through piping that terminates at a specified distance above the flood level rim of an approved receptor.

The air gap serves as an impossible barrier for sewage to overcome in the event that stoppage occurs in the receptor drain, because sewage backup would overflow in the receptor drain flood level before it comes in contact with the drain line above [see Commentary Figure R202(3)].

AIR GAP, WATER-DISTRIBUTION SYSTEM. The unobstructed vertical distance through free atmosphere between the lowest opening from a water supply discharge to the flood-level rim of a plumbing fixture.

❖ An air gap is the most reliable and effective means of backflow protection. It is simply the vertical air space between the potable water supply outlet and the possible source of contamination. This air gap prevents possible contamination of the potable water supply by preventing the supply outlet, such as a faucet, from backsiphoning waste water from a basin, for example. Many manufacturers have developed air gap fittings that provide a rigid connection to the drainage system while maintaining the minimum level of protection from contamination.

AIR-IMPERMEABLE INSULATION. An insulation having an air permanence equal to or less than 0.02 L/s-m^2 at 75 Pa pressure differential tested according to ASTM E 2178 or E 283.

❖ In all buildings, an air-impermeable layer is required to prevent air leakage into the building. In some cases, the insulation will serve that purpose, if it is insulation as defined here.

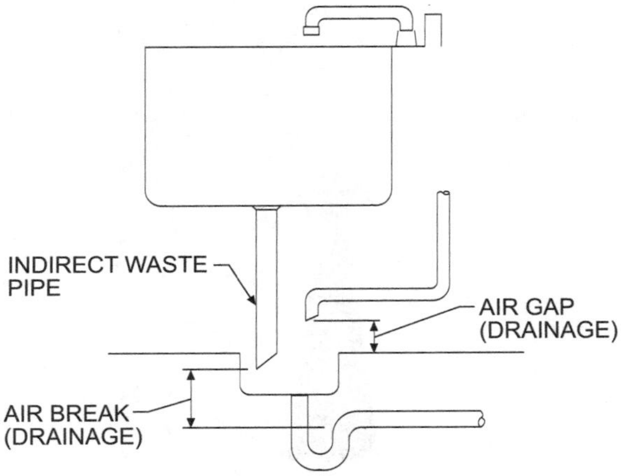

Figure R202(3)
AIR GAP, AIR BREAK

ALTERATION. Any construction or renovation to an existing structure other than repair or addition that requires a *permit.* Also, a change in a mechanical system that involves an extension, addition or change to the arrangement, type or purpose of the original installation that requires a *permit.*

❖ The modification of an existing structure without adding any floor area or height to the structure is an alteration. Section R105 of the code specifies that a permit for the alteration work is required before work begins. The term "alteration" also applies to mechanical work where the original installation is altered in a manner requiring a permit. The repairs described in Section R105.2.2 are not alterations because a permit is not required.

ANCHORED STONE OR MASONRY VENEER. Stone or masonry veneer secured with *approved* mechanical fasteners to an approved backing.

❖ This is a common type of material used for a building façade. It is simply defined in order to allow use of the description to mean a specific type of material.

ANCHORS. See "Supports."

❖ Anchors are the same types of devices as supports. Any device, such as a hanger used to support or secure piping and fixtures, is an anchor.

ANTISIPHON. A term applied to valves or mechanical devices that eliminate siphonage.

❖ The term refers to the function of certain valves and devices used to break or prevent the siphon effect that can be created in plumbing systems.

APPLIANCE. A device or apparatus that is manufactured and designed to utilize energy and for which this code provides specific requirements.

❖ An appliance is a manufactured component or assembly of components that converts one form of energy into a different form of energy to serve a specific purpose. The term "appliance" generally refers to residential and commercial equipment that is manufactured in standardized sizes or types. The term is generally not associated with industrial equipment. For the application of the code provisions, the terms "appliance" and "equipment" are mutually exclusive.

Examples of appliances include furnaces; boilers; water heaters; room heaters; refrigeration units; cooking equipment; clothes dryers; wood stoves; pool, spa and hot tub heaters; unit heaters ovens; and similar fuel-fired or electrically operated appliances. See the definition of "Equipment."

APPROVED. Acceptable to the *building official*.

❖ Throughout the code, the term "approved" is used to describe a specific material or method of construction, such as the approved drainage system mentioned in Section R408.5. Where "approved" is used, it means that the design, material or method of construction is acceptable to the building official. It is imperative that the building officials base their decision of approval on the result of investigations, tests or accepted principles or practices.

APPROVED AGENCY. An established and recognized agency regularly engaged in conducting tests or furnishing inspection services, when such agency has been *approved* by the building official.

❖ The building official will occasionally rely on the expertise of others to assist in plan review and inspection activities, or the owner will employ an individual or company to provide testing services for portions of the construction work. An agency qualified to perform such activities is an "approved agency." The building official will review an agency's qualifications, expertise and reliability to determine whether the agency should be granted approval for a specific activity.

ASPECT RATIO. The ratio of longest to shortest perpendicular dimensions, or for wall sections, the ratio of height to length.

❖ The shear wall or braced wall panel height divided by the width or length of the shear wall or braced wall panel is defined as the aspect ratio. Generally, the code places limits on shear wall aspect ratios to limit shear wall deformation.

ATTIC. The unfinished space between the ceiling assembly of the top *story* and the roof assembly.

❖ The unfinished space between the ceiling joists of the top story and the roof rafters.

Several provisions apply to the attic area of a building, such as those relating to ventilation of the attic space. The code identifies an attic as the unfinished space between the ceiling joists of the top story and the roof assembly. Such a space would be the top story, rather than the attic, if it is finished and occupiable.

ATTIC, HABITABLE. A finished or unfinished area, not considered a *story*, complying with all of the following requirements:

1. The occupiable floor area is at least 70 square feet (17 m^2), in accordance with Section R304,
2. The occupiable floor area has a ceiling height in accordance with Section R305, and
3. The occupiable space is enclosed by the roof assembly above, knee walls (if applicable) on the sides and the floor-ceiling assembly below.

❖ The code allows that an attic can contain habitable space. In such cases, the attic is not required to be considered as a story. Note that there is no limitation given for the area of the attic that contains habitable space.

BACKFLOW, DRAINAGE. A reversal of flow in the drainage system.

❖ Backflow is a condition found in plumbing systems where the contents of the piping flow in a direction opposite of the intended direction.

BACKFLOW PREVENTER. A device or means to prevent backflow.

❖ Backflow preventers are designed for different applications, depending on the pressures and the degree of hazard posed by the cross connection. There are six basic types of backflow preventers that can be used to correct cross connections: air gaps; barometric loops; vacuum breakers, both atmospheric and pressure type; double check valve with intermediate atmospheric vent; double check-valve assembly and reduced-pressure-principle devices.

BACKFLOW PREVENTER, REDUCED-PRESSURE-ZONE TYPE. A backflow-prevention device consisting of two independently acting check valves, internally force loaded to a normally closed position and separated by an intermediate chamber (or zone) in which there is an automatic relief means of venting to atmosphere internally loaded to a normally open position between two tightly closing shutoff valves and with means for testing for tightness of the checks and opening of relief means.

❖ This device is one of the most reliable mechanical devices (second only to an air gap) for the prevention of backflow. It uses two spring-loaded check valves with a relief valve in between that monitors the system pressure upstream and downstream of the device. The zone between the check valves is maintained at a pressure that is less than the water supply pressure.

BACKFLOW, WATER DISTRIBUTION. The flow of water or other liquids into the potable water-supply piping from any sources other than its intended source. Backsiphonage is one type of backflow.

❖ There are primarily three types of backflow conditions that could occur in a water distribution system: (1) backpressure; (2) backpressure, low-head; and (3) backsiphonage.

1. Backpressure. Liquid in a pipe flows from a high pressure condition to a low pressure condition, such as when the pressure within the water distribution system is greater than that in the water service. The water will reverse its intended direction of flow and return to the water supply unless a device to prevent backflow has been installed in the line. Elevated pressure can be created through mechanical means, such as a pumping system, or it can be the result of a loss of pressure within the main water supply. Backpressure often occurs in a system or a portion of a system that is closed (not open to the atmosphere).
2. Backpressure, low-head. This is the back-pressure created where the source of backpressure comes from a hose elevated to a level that produces 10 feet (3048 mm) or less of water column at the outlet. Low-head backpressure typically occurs at a hose connection, such as a wall hydrant or sillcock. Certain types of backflow assemblies and devices are designed specifically for either backpressure or low-head backpressure.
3. Backsiphonage. See the definition of Backsiphonage.

BACKPRESSURE. Pressure created by any means in the water distribution system, which by being in excess of the pressure in the water supply mains causes a potential backflow condition.

❖ Liquid in a pipe flows from a high-pressure condition to a low-pressure condition, such as when the pressure within the water distribution system is greater than that in the water service. The water will reverse its intended direction of flow and return to the water supply, unless a device to prevent backflow has been installed in the line. Elevated pressures can be created through mechanical means, such as a pumping system, or can be the result of a loss of pressure within the main water supply. Backpressure will often occur in a system or a portion of a system that is closed, i.e., it is not open to the atmosphere.

BACKPRESSURE, LOW HEAD. A pressure less than or equal to 4.33 psi (29.88 kPa) or the pressure exerted by a 10-foot (3048 mm) column of water.

❖ Low-head backpressure is created where the source of backpressure comes from a hose elevated to a level that produces 10 feet (3048 mm) or less of water column at the outlet. A hose connection, such as a wall hydrant or faucet, is where "low-head backpressure" is typically experienced. Also see the commentary to "Backpressure."

BACKSIPHONAGE. The flowing back of used or contaminated water from piping into a potable water-supply pipe due to a negative pressure in such pipe.

❖ Backsiphonage occurs when the pressure within a potable water distribution system drops below atmospheric pressure or negative gauge pressure. It, like backpressure, allows the normal direction of flow to reverse. A siphon can result in an unprotected cross connection, causing contamination or pollution of the potable water supply. The main difference between backsiphonage and backpressure is the pressure in the system. Water distribution systems having a pressure less than zero gauge have the potential to create a siphon, resulting in backsiphonage. Backpressure typically occurs in a closed system, and backsiphonage typically occurs in an open system (open to the atmosphere).

BACKWATER VALVE. A device installed in a drain or pipe to prevent backflow of sewage.

❖ A backwater valve is a type of check valve designed for installation in drainage piping. The valve has a lower invert downstream of the flapper than upstream to help prevent solids from interfering with valve closure.

BASEMENT. A *story* that is not a *story above grade plane* (see "*Story above grade plane*").

❖ Defined as "a story that is not a story above grade," a basement is further identified in the provisions in Section R202 for a "story above grade." Specific provisions, including the requirement for emergency escape and rescue openings in Section R310.1, are applicable for those floor levels meeting the criteria for basements. The presence of occupiable space below grade level causes various concerns that are addressed by the code.

BASEMENT WALL. The opaque portion of a wall that encloses one side of a *basement* and has an average below *grade* wall area that is 50 percent or more of the total opaque and non-opaque area of that enclosing side.

❖ A basement wall is defined in respect to the energy efficiency provisions of Chapter 11. A wall that does not qualify as an exterior wall is a basement wall.

BASIC WIND SPEED. Three-second gust speed at 33 feet (10 058 mm) above the ground in Exposure C (see Section R301.2.1) as given in Figure R301.2(4)A.

❖ Basic wind speed, the basis for a number of structural decisions in the IRC, is the 3-second gust speed measured at a point 33 feet (10058 mm) above the ground. The assumed exposure used in determining the basic wind speed is that for generally open terrain having scattered obstructions.

BATHROOM GROUP. A group of fixtures, including or excluding a bidet, consisting of a water closet, lavatory, and bathtub or shower. Such fixtures are located together on the same floor level.

❖ Special consideration is given to a bathroom group when sizing water distribution systems and drain, waste and vent piping because of fixture usage. "Bathroom group" has historically referred to a single water closet, a lavatory and a bathtub or shower, all located within a single room. Such an arrangement would normally allow only one occupant in the room; therefore, the likelihood of simultaneous fixture discharge is remote. For the fixture-usage theory of the bathroom group to be valid, the fixtures in the group must be located within the same bathroom. A bidet has been added to the group of fixtures constituting a bathroom group.

BEND. A drainage fitting, designed to provide a change in direction of a drain pipe of less than the angle specified by the amount necessary to establish the desired slope of the line (see "Elbow" and "Sweep").

❖ This term denotes a one-way fitting (often referred to as an elbow) that is typically used in a drainage system to make a change in direction within a single run of pipe. Although bends (or elbows) are used in other plumbing and mechanical piping systems, in this instance the term refers to a drainage fitting that will meet the required radius of turn.

BOILER. A self-contained *appliance* from which hot water is circulated for heating purposes and then returned to the boiler, and which operates at water pressures not exceeding 160 pounds per square inch gage (psig) (1102 kPa gauge) and at water temperatures not exceeding 250°F (121°C).

❖ Boilers are usually manufactured of steel or cast iron, and are used to transfer heat (from the combustion of a fuel or from an electric-resistance element) to water for supplying steam or hot pressurized water for heating or other process or power purposes.

Boilers are usually installed in closed systems where the heat transfer medium is recirculated and retained within the system. Hot water supply boilers are normally part of open systems where the heated water is supplied and used externally to the boiler. Large domestic (potable) water heating systems often employ hot water supply boilers.

Boilers must be labeled and installed in accordance with the manufacturer's installation instructions and the applicable sections of the code. Boilers are rated in accordance with standards published by the

American Society of Mechanical Engineers (ASME), the Hydronics Institute, the Steel Boiler Institute (SBI), the American Gas Association (AGA) and the American Boiler Manufacturers Association (ABMA). Boilers can be classified in accordance with the following: working temperature, working pressure, type of fuel used (or electric boilers), materials of construction and whether or not the heat transfer medium changes phase from a liquid to a vapor.

BOND BEAM. A horizontal grouted element within masonry in which reinforcement is embedded.

❖ Composed of masonry, grout and reinforcement, a bond beam is a horizontal structural member constructed to transfer loads while tying the wall together.

BRACED WALL LINE. A straight line through the building plan that represents the location of the lateral resistance provided by the wall bracing.

❖ Commonly referred to as shear wall, a braced wall line is used to resist the lateral racking forces created by seismic or wind loading. A number of braced wall panels work together to form a braced wall line. The code has specific requirements for the method of constructing a braced wall line. Where a building does not comply with the bracing requirements of Chapter 6, those portions of the building not meeting the requirements must be designed in accordance with accepted engineering practice.

BRACED WALL LINE, CONTINUOUSLY SHEATHED. A *braced wall line* with structural sheathing applied to all sheathable surfaces including the areas above and below openings.

❖ A series of braced wall panels in a single story constructed in accordance with Section R602.10 for wood framing or Section R603.7 or R301.1.1 for cold-formed steel framing to resist racking from seismic and wind forces.

Commonly referred to as a shear wall, a braced wall line is used to resist the lateral racking forces created by seismic or wind loading. A number of braced wall panels work together to form a braced wall line. The code has specific requirements for the method of constructing a braced wall line. Where a building does not comply with the bracing requirements of Chapter 6, those portions of the building not meeting the requirements must be designed in accordance with accepted engineering practice.

BRACED WALL PANEL. A full-height section of wall constructed to resist in-plane shear loads through interaction of framing members, sheathing material and anchors. The panel's length meets the requirements of its particular bracing method, and contributes toward the total amount of bracing required along its *braced wall line* in accordance with Section R602.10.1.

❖ A braced wall line is composed of a series of braced wall panels constructed in accordance with the provisions of Chapter 6. A braced wall panel must extend the full height of the wall.

BRANCH. Any part of the piping system other than a riser, main or stack.

❖ A branch refers to piping that connects to a riser, main or stack in a plumbing system. Branch piping is generalized as horizontal piping, can include vertical offsets and is smaller in size or capacity than the main artery or trunk from which it extends. It is analogous to branches of a tree.

BRANCH, FIXTURE. See "Fixture branch, drainage."

❖ See the commentary to "Fixture branch, drainage."

BRANCH, HORIZONTAL. See "Horizontal branch, drainage."

❖ See the commentary to "Horizontal branch, drainage."

BRANCH INTERVAL. A vertical measurement of distance, 8 feet (2438 mm) or more in *developed length*, between the connections of horizontal branches to a drainage stack. Measurements are taken down the stack from the highest horizontal branch connection.

❖ A branch interval is the vertical distance between a drainage stack's branch connections. In typical construction practice, the branch drain will connect to a stack near the floor level; therefore, the branch interval will correspond to the height of the story in which it is located. Depending on the design of the building, however, a branch interval may not correspond to the height of the story because stories are not limited in height. Therefore, the code requires a branch interval to be at least 8 feet (2438 mm) in developed length between the connections of the horizontal branches to the drainage stack.

Branch intervals are a design factor in drainage pipe sizing and venting design. Drain, waste and vent system design must consider the nature of waste and airflow in a stack and the effects that branch connections have on that flow.

BRANCH, MAIN. A water-distribution pipe that extends horizontally off a main or riser to convey water to branches or fixture groups.

❖ This applies to the primary water supply system piping that extends horizontally off a main or riser to supply water to branches or groups of fixtures.

BRANCH, VENT. A vent connecting two or more individual vents with a vent stack or stack vent.

❖ See Commentary Figure R202(4).

BTU/H. The *listed* maximum capacity of an *appliance*, absorption unit or burner expressed in British thermal units input per hour.

❖ This term stands for British thermal units per hour. Fuel-fired appliances and equipment are rated based on their Btu/h (W) input or output.

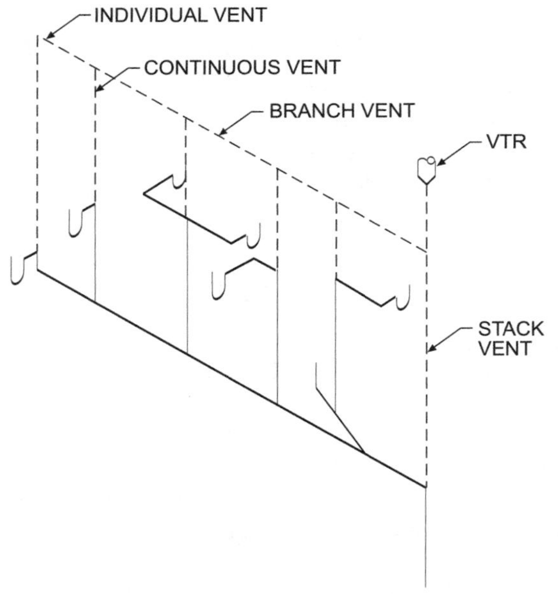

Figure R202(4)
BRANCH VENT

BUILDING. Building shall mean any one- and two-family dwelling or portion thereof, including *townhouses*, that is used, or designed or intended to be used for human habitation, for living, sleeping, cooking or eating purposes, or any combination thereof, and shall include accessory structures thereto.

❖ A building according to the IRC may be a single-family dwelling, a two-family dwelling, a townhouse or an accessory structure to such buildings. The use of a building, excluding an accessory structure, is human habitation, which specifically includes living, sleeping, cooking or eating.

BUILDING DRAIN. The lowest piping that collects the discharge from all other drainage piping inside the house and extends 30 inches (762 mm) in *developed length* of pipe, beyond the *exterior walls* and conveys the drainage to the *building sewer*.

❖ A building drain is usually the main drain of a system within a structure. Building drains are horizontal (including vertical offsets) and are the portion of the drainage system that is at the lowest elevation in the structure. All horizontal drains above the elevation of the building drain are horizontal branches. The building drain terminates at the point where it exits the building [see Commentary Figure R202(5)].

BUILDING, EXISTING. Existing building is a building erected prior to the adoption of this code, or one for which a legal building *permit* has been issued.

❖ There are two definitions for an existing building. The most general use of the term is for a building that has been constructed prior to the adoption of the 2000 IRC. In other words, any building that is in existence at the time the current code is adopted by the jurisdiction is "existing." A second definition relates to those buildings that have been issued a building permit. Once a building is under construction, it is an "existing building" and is regulated by the code under which the permit was issued.

BUILDING LINE. The line established by law, beyond which a building shall not extend, except as specifically provided by law.

❖ The building line is not established by the code but rather by the jurisdiction as the point beyond which the construction of a building is not permitted. Often known as a setback line or a building setback, the building line is established to maintain order within the community's building stock. The IRC does not regulate construction with regard to any jurisdictional building line.

BUILDING OFFICIAL. The officer or other designated authority charged with the administration and enforcement of this code.

❖ Regardless of title, the individual designated by the jurisdiction as the person who administers and enforces the IRC is the building official. In addition, the building official may appoint various other individuals to assist in the activities of the Department of Building Safety. In many jurisdictions, the authority of the building official is extended to plans examiners and inspectors to some degree. Section R104 sets forth the duties and responsibilities of the building official.

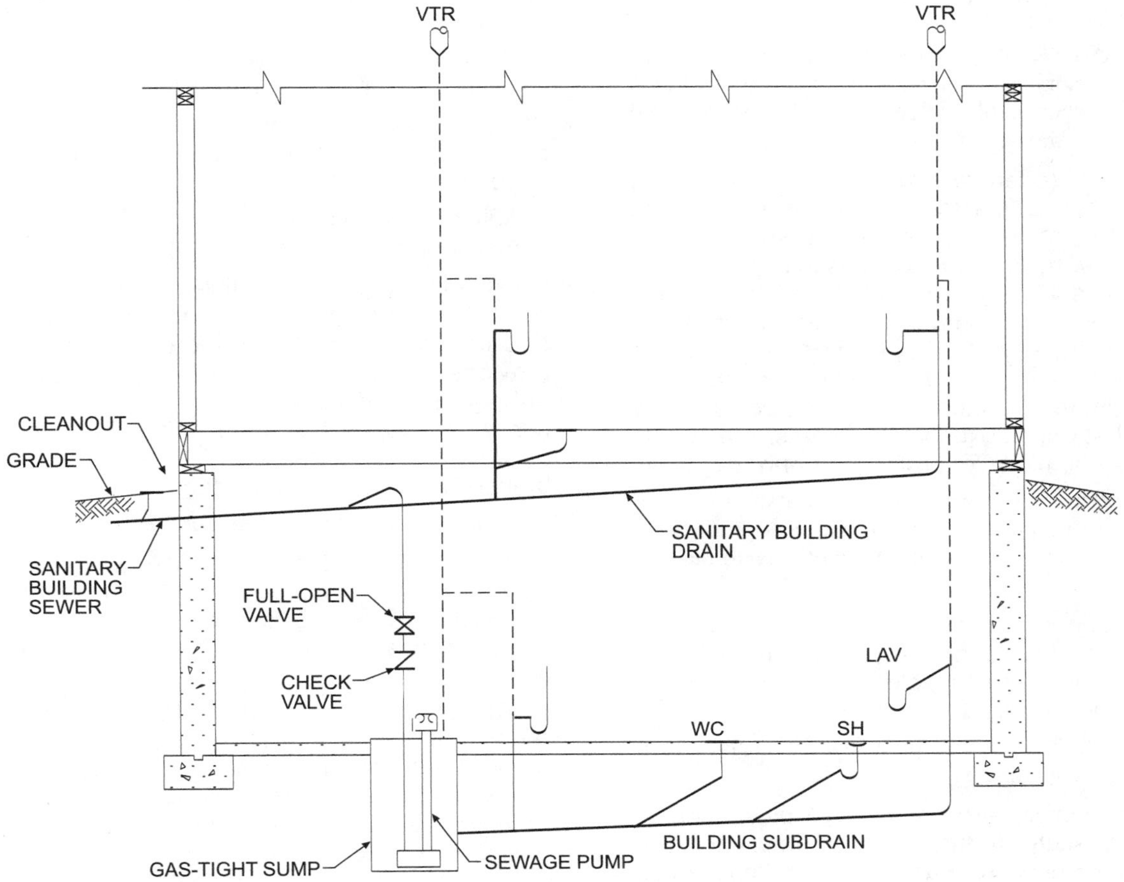

Figure R202(5)
DRAINAGE SYSTEM COMPONENTS

BUILDING SEWER. That part of the drainage system that extends from the end of the *building drain* and conveys its discharge to a public sewer, private sewer, individual sewage-disposal system or other point of disposal.

❖ The building sewer is the extension of the building drain and is located entirely outside the building exterior walls.

BUILDING THERMAL ENVELOPE. The *basement walls*, *exterior walls*, floor, roof and any other building element that enclose *conditioned spaces*.

❖ The building thermal envelope includes the roof/ceiling assembly, wall assemblies and floor assemblies that surround a conditioned area, which is the space that is being intentionally heated and/or cooled. The building envelope is the assembly that separates conditioned space from unconditioned space or the outdoors. For example, a wall between a conditioned dwelling and an unconditioned garage is part of the building thermal envelope. Other elements of the building envelope include attic kneewalls, the perimeter joist between two conditioned floors and skylight wells.

BUILT-UP ROOF COVERING. Two or more layers of felt cemented together and surfaced with a cap sheet, mineral aggregate, smooth coating or similar surfacing material.

❖ A common roof covering for buildings with relatively flat roofs, built-up roof covering uses two or more felt layers with a cap-sheet surfacing. A number of materials standards regulate the wide variety of built-up roofing materials that are available.

CAP PLATE. The top plate of the double top plates used in structural insulated panel (SIP) construction. The cap plate is cut to match the panel thickness such that it overlaps the wood structural panel facing on both sides.

❖ Structural insulated panels (SIPs) have been in use in building construction for some time, but are relatively new. This term is defined to facilitate requirements for SIPs.

CEILING HEIGHT. The clear vertical distance from the finished floor to the finished ceiling.

❖ Ceiling height as regulated by Section R305 is measured from the finished floor surface to the finished ceiling. Depending on the space under consideration, the minimum ceiling height required by the IRC can vary greatly.

CEMENT PLASTER. A mixture of portland or blended cement, portland cement or blended cement and hydrated lime, masonry cement or plastic cement and aggregate and other *approved* materials as specified in this code.

❖ Cement plaster (often referred to as "stucco") is a cementitious-based plaster material with excellent water-resistant properties. It is the only type of plaster that is permitted by the code to be used as an exterior wall covering and as a base coat and finish coat. Cement plaster can also be used as an interior finish wall covering, but is required by the code to be used in all interior wet areas, such as toilet rooms, showers, saunas, steam rooms, indoor swimming pools or any other area that will be exposed to excessive amounts of moisture or humidity for prolonged periods of time.

CHIMNEY. A primary vertical structure containing one or more flues, for the purpose of carrying gaseous products of combustion and air from a fuel-burning *appliance* to the outside atmosphere.

❖ Chimneys differ from vents in the materials they are made of and the type of appliance they are designed to serve. Chimneys are capable of venting flue gases of much higher temperatures than vents.

CHIMNEY CONNECTOR. A pipe that connects a fuel-burning *appliance* to a chimney.

❖ Chimney connectors are sections of pipe used to convey combustion products from an appliance flue outlet to a chimney inlet. Factory-built chimneys can connect directly to some appliances without the need for a connector; however, masonry chimneys cannot connect directly to an appliance because of the chimney's weight.

CHIMNEY TYPES.

Residential-type appliance. An *approved* chimney for removing the products of combustion from fuel-burning, residential-type *appliances* producing combustion gases not in excess of 1,000°F (538°C) under normal operating conditions, but capable of producing combustion gases of 1,400°F (760°C) during intermittent forces firing for periods up to 1 hour. All temperatures shall be measured at the *appliance* flue outlet. Residential-type *appliance* chimneys include masonry and factory-built types.

❖ The chimneys addressed in the IRC are generally for use with residential-type appliances. Such chimneys are limited in the temperature of the combustion gases they can exhaust. Both masonry and factory-built chimneys can be residential type.

CIRCUIT VENT. A vent that connects to a horizontal drainage branch and vents two traps to a maximum of eight traps or trapped fixtures connected into a battery.

❖ A circuit vent vents multiple fixtures using only one or two vents. The horizontal drainage branch actually serves as a wet vent for the fixtures located downstream of the circuit vent connection. See Section P3110.

CLADDING. The exterior materials that cover the surface of the building envelope that is directly loaded by the wind.

❖ Sheathing, siding or other materials used on the exterior envelope of a building are "cladding" for the purpose of wind design.

CLEANOUT. An accessible opening in the drainage system used for the removal of possible obstruction.

❖ Cleanouts are broadly defined as any access point into the drainage system that allows the removal of a clog or other obstruction in the drain line. A cleanout provides convenient access to the piping interior without significant disassembly of the plumbing installation. A "P" trap with a removable U-bend is typically acceptable as a cleanout for the fixture drain or branch to which the trap discharges.

CLOSET. A small room or chamber used for storage.

❖ A closet is simply a small storage room. Closets are generally found throughout dwelling units but are not considered habitable spaces. The typical hazard related to closets is the probable storage of combustible materials.

COMBINATION WASTE AND VENT SYSTEM. A specially designed system of waste piping embodying the horizontal wet venting of one or more sinks, lavatories or floor drains by means of a common waste and vent pipe adequately sized to provide free movement of air above the flow line of the drain.

❖ This special system is employed when no other conventional system of separate waste and vent is practical. The combination waste and vent system is used only for venting floor drains, standpipes, sinks and lavatories. The drainage piping is larger than required for draining purposes only, and the drainage branch and stack should be provided with vent piping. The system is commonly used where floor drains are installed in large open areas that cannot accommodate vertical vent risers from the floor drains. See Section P3111.

COMBUSTIBLE MATERIAL. Any material not defined as noncombustible.

❖ Any material that (ASTM E 136) does not qualify as a noncombustible material is considered combustible. The presence of sizable quantities of combustible material increases the potential fire hazard. Buildings containing or constructed of combustible materials are typically more highly regulated than those that are primarily noncombustible.

COMBUSTION AIR. The air provided to fuel-burning *equipment* including air for fuel combustion, draft hood dilution and ventilation of the *equipment* enclosure.

❖ The process of combustion requires a specific amount of oxygen to initiate and sustain the combustion reaction. Combustion air includes primary air, secondary air, draft hood dilution air and excess air.

Combustion air is the amount of atmospheric air required for complete combustion of a fuel and is related to the molecular composition of the fuel being burned, the design of the fuel-burning equipment and the percentage of oxygen in the combustion air. Too little combustion air will result in incomplete combustion of a fuel and the possible formation of carbon deposits (soot), carbon monoxide, toxic alcohols, ketones, aldehydes, nitrous oxides and other byproducts. The required amount of combustion air is usually stated in terms of cubic feet per minute (m^3/s) or pounds per hour (kg/h).

[CE] COMMERCIAL, BUILDING. See Section N1101.9.

COMMON VENT. A single pipe venting two trap arms within the same *branch interval*, either back-to-back or one above the other.

❖ Where two or more fixtures are connected to the same drainage pipe, a common vent could be installed. This common vent is sized and classified as an individual vent. Any two fixtures may be common vented to either a vertical or horizontal drainage pipe.

A typical form of common venting connects two fixtures at the same level. Two fixtures connecting at different levels but within the same story are also common vented. When one of the fixtures connected is at a different level, the vertical drain between the fixtures is oversized because it functions as a drain for the upper fixture and a vent for the lower fixture.

CONDENSATE. The liquid that separates from a gas due to a reduction in temperature, e.g., water that condenses from flue gases and water that condenses from air circulating through the cooling coil in air conditioning *equipment*.

❖ Condensate forms when the temperature of a vapor is lowered to its dew point temperature. Air conditioning systems produce condensate when an airstream contacts cooling coils. The moisture in the air condenses on the cold surface of the coils and the air is "dehumidified." Condensate also forms within improperly designed chimneys and vents when the products of combustion (which contain water vapor) contact the colder walls of the flue. If the temperature of the products of combustion is lowered to the dew point temperature of the water vapor, condensate will form on the inside walls of the flue. Condensed steam in hydronic systems is also referred to as "condensate."

CONDENSING APPLIANCE. An *appliance* that condenses water generated by the burning of fuels.

❖ Such units include compressors and condenser heat exchangers; they convert refrigerant vapor to liquid.

CONDITIONED AIR. Air treated to control its temperature, relative humidity or quality.

❖ Conditioned air is air that is heated, cooled, humidified, dehumidified or decontaminated.

CONDITIONED AREA. That area within a building provided with heating and/or cooling systems or *appliances* capable of maintaining, through design or heat loss/gain, 68°F (20°C) during the heating season and/or 80°F (27°C) during the cooling season, or has a fixed opening directly adjacent to a conditioned area.

❖ The conditioned area is the space within the building that receives conditioned air. The space can be directly conditioned or indirectly conditioned by an adjacent conditioned space.

CONDITIONED FLOOR AREA. The horizontal projection of the floors associated with the *conditioned space*.

❖ The conditioned floor area is the floor area in square feet (m^2) of a conditioned space.

CONDITIONED SPACE. For energy purposes, space within a building that is provided with heating and/or cooling *equipment* or systems capable of maintaining, through design or heat loss/gain, 50°F (10°C) during the heating season and 85°F (29°C) during the cooling season, or communicates directly with a *conditioned space*. For mechanical purposes, an area, room or space being heated or cooled by any *equipment* or *appliance*.

❖ If a space is heated, cooled or humidified/dehumidified, it is conditioned space. The building must have a system that is capable of maintaining the space at 50°F (10°C) or above for heating and 85°F (29°C) or below for cooling during normal operation. Also, a basement is "conditioned" if the floor/ceiling assembly between the basement and the conditioned first floor is uninsulated and there is enough heat transfer across the floor to maintain the space at 50°F (10°C) or above. In addition, uninsulated duct systems located in an unconditioned basement may raise the temperature of the space to 50°F (10°C) and above during normal operation of the system. The basement would then be conditioned space. If the space is conditioned, the building envelope that surrounds the space must meet the thermal requirements in Table N1102.1.

CONSTRUCTION DOCUMENTS. Written, graphic and pictorial documents prepared or assembled for describing the design, location and physical characteristics of the elements of a project necessary for obtaining a building *permit*. Construction drawings shall be drawn to an appropriate scale.

❖ Construction documents are the necessary drawings, specifications and support materials created for the design and construction of a building. Addressed in Section R106, construction documents are required at the time of permit application unless the work is of such a minor nature that the documents are not necessary to provide compliance with the code. The construction documents must be drawn to an appropriate scale so they may be easily interpreted.

CONTAMINATION. An impairment of the quality of the potable water that creates an actual hazard to the public health through poisoning or through the spread of disease by sewage, industrial fluids or waste.

❖ The Environmental Protection Agency (EPA) sets maximum levels for various chemicals and bacteria in drinking water. When an unacceptable level of one or more contaminants is present, the water is considered to be nonpotable. One of the primary purposes of the code is to protect potable water supplies from contamination. Contamination consists of either hazardous chemicals or raw sewage, and is considered by the code to represent a high hazard when present in the water supply. Refer to Table P2902.2 for the required potable protection requirements related to high-hazard situations.

The code and other related regulations, such as those from the EPA, are intended to reduce, if not totally eliminate, the risk to the general public of poisoning or health impairment through sewage, industrial fluids or waste entering the potable water supply. Note that these contaminants would have a harmful effect on the occupants of a building should the water be consumed. Contamination is quite different from pollution, which is a situation where the water supply is still considered to be nonpotable, but would not kill or harm if ingested. Refer to the definition of "Pollution" for additional information on that term.

CONTINUOUS WASTE. A drain from two or more similar adjacent fixtures connected to a single trap.

❖ Continuous waste is the piping described and regulated in Section P3201.6. The most common occurrences are where multiple compartments of a kitchen sink or double lavatories are connected to a single trap.

CONTROL, LIMIT. An automatic control responsive to changes in liquid flow or level, pressure, or temperature for limiting the operation of an *appliance*.

❖ Limit controls are safety devices used to protect equipment, appliances, property, and persons. Such controls act at their set point to limit a condition such as temperature or pressure and include high and low limits. This definition applies to controls for all types of fuel: gas, liquid and solid.

CONTROL, PRIMARY SAFETY. A safety control responsive directly to flame properties that senses the presence or absence of flame and, in event of ignition failure or unintentional flame extinguishment, automatically causes shutdown of mechanical equipment.

❖ This is a device designed to shut down mechanical equipment in response to changes in liquid or gas flow rates that could extinguish a flame.

CONVECTOR. A system-incorporating heating element in an enclosure in which air enters an opening below the heating element, is heated and leaves the enclosure through an opening located above the heating element.

❖ See Commentary Figure R202(6) for an illustration of an electric hydronic convector.

CORE. The light-weight middle section of the structural insulated panel composed of foam plastic insulation, which provides the link between the two facing shells.

❖ SIPs have been in use in building construction for some time, but are relatively new. This term is defined to facilitate requirements for SIPs.

CORROSION RESISTANCE. The ability of a material to withstand deterioration of its surface or its properties when exposed to its environment.

❖ Under those conditions where corrosion of materials is detrimental to the integrity of the building's construction and use, the code mandates corrosion-resistant materials. Corrosion resistant materials control or withstand the deterioration of their surfaces and retain their physical properties in their intended environment. Corrosion resistance can be provided by the material or by a process applied to the material.

COURT. A space, open and unobstructed to the sky, located at or above *grade* level on a *lot* and bounded on three or more sides by walls or a building.

❖ An exterior area is a court if it is enclosed on at least three sides by exterior walls of the building or other enclosing elements and is open and unobstructed to the sky above. By virtue of being substantially open to the exterior and to the sky, a court may be used to obtain natural light and ventilation for the building.

CRIPPLE WALL. A framed wall extending from the top of the foundation to the underside of the floor framing of the first *story above grade plane*.

❖ Cripple walls are built on the top of footings or foundation walls. They can typically be found along the top of stepped foundation walls where the grade adjoining the structure changes height. Cripple walls must be properly braced to resist lateral forces. They are often treated the same as a first-story wall. Provisions for the bracing of cripple walls are in Section R602.10.2.

CROSS CONNECTION. Any connection between two otherwise separate piping systems whereby there may be a flow from one system to the other.

❖ Cross connections are the links through which it is possible for contaminated materials to enter a potable water supply. The contaminant enters the potable water supply when the pressure of the polluted source exceeds the pressure of the potable source.

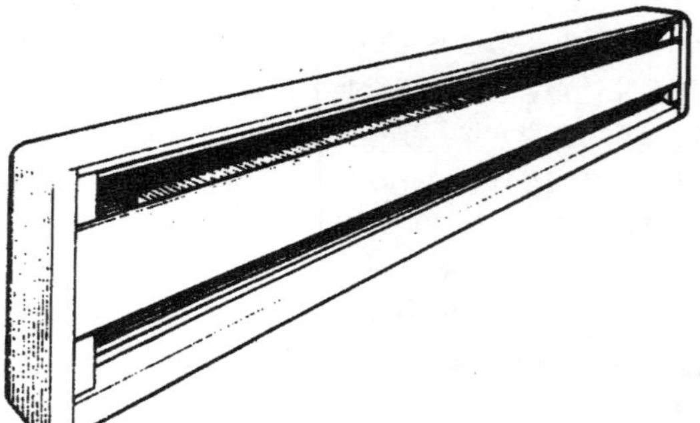

Figure R202(6)
ELECTRIC HYDRONIC CONVECTOR

The action may be called backsiphonage or backpressure.

Many serious outbreaks of illnesses and disease have been traced to cross connections. The intent of the code is to eliminate cross connections or prevent backflow where cross connections cannot be eliminated. Refer to the definitions of and the commentary to "Backflow" and "Backsiphonage."

CURTAIN WALL. See Section N1101.9 for definition applicable in Chapter 11.

DALLE GLASS. A decorative composite glazing material made of individual pieces of glass that are embedded in a cast matrix of concrete or epoxy.

❖ Various types of decorative glazing materials are exempt from the safety glazing provisions when installed in side-hinged doors, adjacent to doors, or in large glazed openings. Dalle glass is one such material. Made up of multiple individual pieces of glass, dalle glass is exempt because of its decorative function.

DAMPER, VOLUME. A device that will restrict, retard or direct the flow of air in any duct, or the products of combustion of heat-producing *equipment*, vent connector, vent or chimney.

❖ A damper is used in an air distribution system as a restrictor to regulate airflow through duct work. When used in flues venting combustion gases, a damper is used to regulate draft.

DEAD END. A branch leading from a DWV system terminating at a *developed length* of 2 feet (610 mm) or more. Dead ends shall be prohibited except as an *approved* part of a rough-in for future connection.

❖ "Dead end" refers to horizontal piping that does not conduct waste flow but is connected to piping that does conduct such flow. A dead end will collect solid waste as a result of the normal flow depth in the drain pipe to which it is connected. A dead end will also collect solids as a result of drainage system stoppages.

DEAD LOADS. The weight of all materials of construction incorporated into the building, including but not limited to walls, floors, roofs, ceilings, stairways, built-in partitions, finishes, cladding, and other similarly incorporated architectural and structural items, and fixed service *equipment*.

❖ Dead loads are considered in the structural design of a building and are used in the span tables for sizing floor joists, ceiling joists and rafters. The weight of the materials of construction joists, sheathing, studs and gypsum board is counted as the dead load, as are stairways, cladding and fixed service equipment. The dead loads make up the fixed load placed in the building and are independent from the live loads.

DECORATIVE GLASS. A carved, leaded or Dalle glass or glazing material whose purpose is decorative or artistic, not functional; whose coloring, texture or other design qualities or components cannot be removed without destroying the glazing material; and whose surface, or assembly into which it is incorporated, is divided into segments.

❖ Decorative glass is not designed for functional purposes such as a vision panel, but rather is used for artistic reasons. Some types of decorative glass include carved, leaded, and dalle glass. Decorative glass is typically made up of multiple pieces of glass arranged in some design or pattern. Fixed-in-place decorative glass is exempt from some of the requirements for glazing in hazardous locations.

DEMAND RECIRCULATION WATER SYSTEM. See Section N1101.9 for definition applicable in Chapter 11.

DESIGN PROFESSIONAL. See "*Registered design professional.*"

❖ "Design professional" is another term for registered design professional. See the definition of "Registered design professional."

DEVELOPED LENGTH. The length of a pipeline measured along the center line of the pipe and fittings.

❖ This term identifies a concept necessary for computing the actual length of piping in a plumbing system. The developed length is measured along the actual flow path of piping and includes the piping lengths in all offsets and changes in direction. Several code requirements are dependent on the actual length of plumbing piping.

DIAMETER. Unless specifically stated, the term "diameter" is the nominal diameter as designated by the *approved* material standard.

❖ "Diameter" is the length of a line passing though the center of a pipe from one side to the other. Depending on the pipe standard, this distance can be measured to the inside portions (inside diameter, or ID) or the outside portions (outside diameter, or OD) of the pipe.

DIAPHRAGM. A horizontal or nearly horizontal system acting to transmit lateral forces to the vertical resisting elements. When the term "*diaphragm*" is used, it includes horizontal bracing systems.

❖ In the load path for forces transmitted through a building or structure, the diaphragm is that horizontal portion of the path where lateral forces are transferred to the vertical resisting elements. Roof diaphragms and floor diaphragms carry lateral loads to the shear walls.

DILUTION AIR. Air that enters a draft hood or draft regulator and mixes with flue gases.

❖ Dilution air is associated with draft-hood-equipped Category I appliances. Fan-assisted Category I appliances do not use dilution air; all of the air brought in by the fan passes directly through the combustion chamber of the appliance. Dilution air causes lowering of the dew point and cooling of flue gases, but to a point that is still above the point when condensation will occur inside the vent.

DIRECT-VENT APPLIANCE. A fuel-burning *appliance* with a sealed combustion system that draws all air for combustion from the outside atmosphere and discharges all flue gases to the outside atmosphere.

❖ Such appliances are equipped with independent exhaust and intake pipes, or they have concentric pipes that vent through the inner pipe and convey combustion air in the annular space between pipe walls. See the commentary to Sections R304.1 and R804.1.

DRAFT. The pressure difference existing between the *appliance* or any component part and the atmosphere, that causes a continuous flow of air and products of combustion through the gas passages of the *appliance* to the atmosphere.

Induced draft. The pressure difference created by the action of a fan, blower or ejector, that is located between the *appliance* and the chimney or vent termination.

Natural draft. The pressure difference created by a vent or chimney because of its height, and the temperature difference between the flue gases and the atmosphere.

❖ Draft is the negative static pressure measured relative to atmospheric pressure that is developed in chimneys and vents and in the flue-ways of fuel-burning appliances. Draft can be produced by hot flue-gas buoyancy ("stack effect"), mechanically by fans and exhausters or by a combination of both natural and mechanical means.

DRAFT HOOD. A device built into an *appliance*, or a part of the vent connector from an *appliance*, which is designed to provide for the ready escape of the flue gases from the *appliance* in the event of no draft, backdraft or stoppage beyond the draft hood; prevent a backdraft from entering the *appliance*; and neutralize the effect of stack action of the chimney or gas vent on the operation of the *appliance*.

❖ Draft hoods are integral to or supplied with natural-draft, atmospheric-burner gas-fired appliances. Appliances equipped with draft hoods are classified by the manufacturer as Category I appliances. Because of new minimum-efficiency standards and the popularity of mechanical and other special proprietary venting systems, draft-hood-equipped appliances are becoming rare in the marketplace.

DRAFT REGULATOR. A device that functions to maintain a desired draft in the *appliance* by automatically reducing the draft to the desired value.

❖ Excessive draft reduces the efficiency of an appliance because of energy loss in the form of heat loss out the vent. Draft regulators automatically adjust the draft by allowing cooler air to enter the vent at a preset flow rate, reducing the draft.

DRAFT STOP. A material, device or construction installed to restrict the movement of air within open spaces of concealed areas of building components such as crawl spaces, floor-ceiling assemblies, roof-ceiling assemblies and *attics*.

❖ Draft stops divide a large concealed space into smaller compartments in wood-frame floor construction. They limit the movement of air within the cavity, reducing the potential for rapid fire spread. Draftstopping materials include gypsum board, wood structural panels, particleboard or other substantial materials that are adequately supported.

DRAIN. Any pipe that carries soil and water-borne wastes in a building drainage system.

❖ A drain is any pipe in a plumbing system that carries sanitary waste, clear-water waste or storm water. In the case of wet vents, common vents, waste stack vents, circuit vents and combination drain and vents, the drain may also be conducting airflow.

DRAINAGE FITTING. A pipe fitting designed to provide connections in the drainage system that have provisions for establishing the desired slope in the system. These fittings are made from a variety of both metals and plastics. The methods of coupling provide for required slope in the system (see "Durham fitting").

❖ These fittings will accommodate the necessary slope and sweep required for a drain line. See Section P3005.1.

DUCT SYSTEM. A continuous passageway for the transmission of air which, in addition to ducts, includes duct fittings, dampers, plenums, fans and accessory air-handling *equipment* and *appliances*. For definition applicable in Chapter 11, see Section N1101.9.

❖ Duct systems are part of an air distribution system and include supply, return and relief/exhaust air systems.

DURHAM FITTING. A special type of drainage fitting for use in the Durham systems installations in which the joints are made with recessed and tapered threaded fittings, as opposed to bell and spigot lead/oakum or solvent/cemented or soldered joints. The tapping is at an angle (not 90 degrees) to provide for proper slope in otherwise rigid connections.

❖ This is a specific type of fitting used in Durham piping systems, which employ specially designed cast-iron threaded fittings and threaded galvanized steel piping.

DURHAM SYSTEM. A term used to describe soil or waste systems where all piping is of threaded pipe, tube or other such rigid construction using recessed drainage fittings to correspond to the types of piping.

❖ See the definition of "Durham fittings."

DWELLING. Any building that contains one or two *dwelling units* used, intended, or designed to be built, used, rented, leased, let or hired out to be occupied, or that are occupied for living purposes.

❖ A dwelling is a building that contains either one or two dwelling units. The purpose of a dwelling is occupation for living purposes, regardless of the manner of ownership. Single-family houses and duplexes fall under the definition of dwelling. See also "Dwelling unit."

DWELLING UNIT. A single unit providing complete independent living facilities for one or more persons, including permanent provisions for living, sleeping, eating, cooking and sanitation.

❖ The specific purpose of a dwelling unit is to provide the essential amenities necessary for complete and independent facilities. Commonly, dwelling units are thought of as single-family houses or individual living units in duplexes or townhouses.

DWV. Abbreviated term for drain, waste and vent piping as used in common plumbing practice.

❖ This term is used extensively to identify systems and materials suitable for drain, waste and vent installations. It will often be labeled on specific materials approved for such use.

EFFECTIVE OPENING. The minimum cross-sectional area at the point of water-supply discharge, measured or expressed in terms of diameter of a circle and if the opening is not circular, the diameter of a circle of equivalent cross-sectional area. (This is applicable to air gap.)

❖ An effective opening is used to determine the minimum air gap required between a potable water supply outlet or opening and the flood level rim of a receptacle, fixture or other potential source of contamination.

ELBOW. A pressure pipe fitting designed to provide an exact change in direction of a pipe run. An elbow provides a sharp turn in the flow path (see "Bend" and "Sweep").

❖ See definition of "Bend."

EMERGENCY ESCAPE AND RESCUE OPENING. An operable exterior window, door or similar device that provides for a means of escape and access for rescue in the event of an emergency.

❖ In the case of an emergency, particularly a fire, immediate action must be taken. Quick evacuation of the dwelling unit is often required. If occupants are sleeping at the time of the incident, the time for evacuation is extended, often to the point where normal egress is not possible. In such situations, a door or window to the exterior can be used. It is also possible to use such an exterior opening for rescue. See Section R310 for code requirements.

EQUIPMENT. All piping, ducts, vents, control devices and other components of systems other than *appliances* that are permanently installed and integrated to provide control of environmental conditions for buildings. This definition shall also include other systems specifically regulated in this code.

❖ Throughout the code, the terms "equipment" and "appliance" have been used as necessary to match the terms with the intent and context of the code text. Appliances are not referred to as equipment and vice versa. Traditionally, the term "equipment" meant large machinery and specialized hardware not thought of as an "appliance." See the definition of "Appliance."

EQUIVALENT LENGTH. For determining friction losses in a piping system, the effect of a particular fitting equal to the friction loss through a straight piping length of the same nominal diameter.

❖ A straight run of piping will contain a friction loss [typically figured in friction loss per every 100 feet (30 480 mm)]. Where fittings are included, the result is additional friction loss. Although it is possible to find friction loss through calculations, the prescriptive method of water supply sizing in Chapter 29 uses a multiplier of 1.2 to provide an average friction-loss calculation. Therefore, the equivalent length of a 100-foot (30 480 mm) run of piping with fittings will be 120 feet (36 576 mm) to accommodate the friction loss expected. The equivalent length will be used in applying the sizing tables in Chapter 29. See Section P2903.7, Item 2.

ESCARPMENT. With respect to topographic wind effects, a cliff or steep slope generally separating two levels or gently sloping areas.

❖ This is a term to describe a topographical area where wind speed-up can occur. In cases where there is known to be structural damage due to this type of wind speed-up, the code requires consideration of topographical wind effects. See Section 301.2.1.5.

ESSENTIALLY NONTOXIC TRANSFER FLUIDS. Fluids having a Gosselin rating of 1, including propylene glycol; mineral oil; polydimethy oil oxane; hydrochlorofluorocarbon, chlorofluorocarbon and hydrofluorocarbon refrigerants; and FDA-*approved* boiler water additives for steam boilers.

❖ Transfer fluids are liquids or gases that transfer heat to or remove heat from another fluid, such as water. The exchange of heat energy takes place through a heat exchange material, which separates the transfer fluid from the fluid being heated or cooled.

ESSENTIALLY TOXIC TRANSFER FLUIDS. Soil, water or gray water and fluids having a Gosselin rating of 2 or more including ethylene glycol, hydrocarbon oils, ammonia refrigerants and hydrazine.

❖ See the commentary for the definition of "Essentially nontoxic transfer fluids."

EVAPORATIVE COOLER. A device used for reducing air temperature by the process of evaporating water into an airstream.

❖ Also known as "swamp coolers," such units are used in arid climates and use water as a refrigerant. Such units substantially increase the humidity of the air being conditioned.

EXCESS AIR. Air that passes through the combustion chamber and the *appliance* flue in excess of that which is theoretically required for complete combustion.

❖ The nature of natural-draft, fuel-fired appliances is such that the introduction and efficient intermixing of primary and secondary combustion air is not precise and is coupled with the induced air inflow caused by the internal draft. To achieve complete oxidation (combustion), there will always be more air (excess) introduced into the appliances than is theoretically necessary for the complete combustion of the fuel.

EXHAUST HOOD, FULL OPENING. An exhaust hood with an opening at least equal to the diameter of the connecting vent.

❖ See Commentary Figure R202(7) for an illustration of a full opening exhaust hood.

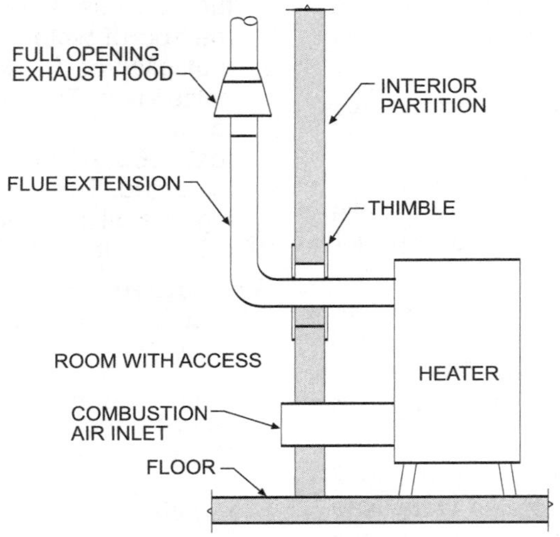

Figure R202(7)
FULL OPENING EXHAUST HOOD

EXISTING INSTALLATIONS. Any plumbing system regulated by this code that was legally installed prior to the effective date of this code, or for which a *permit* to install has been issued.

❖ "Existing installations" is a term that applies to all plumbing work that has been legally installed prior to the effective date of the code, as well as to all plumbing work for which a permit to install has been issued prior to the effective date of the code. Plumbing that has been illegally installed prior to the effective date of the code is not "existing" and is subject to all of the code requirements for new installations.

EXTERIOR INSULATION AND FINISH SYSTEMS (EIFS). EIFS are nonstructural, nonload-bearing *exterior wall* cladding systems that consist of an insulation board attached either adhesively or mechanically, or both, to the substrate; an integrally reinforced base coat; and a textured protective finish coat.

❖ A variety of popular exterior finish systems have been developed, with the typical construction consisting of five layers of material. The code has relatively few provisions regulating EIFS; most of the requirements are those of the manufacturer.

EXTERIOR INSULATION AND FINISH SYSTEMS (EIFS) WITH DRAINAGE. An EIFS that incorporates a means of drainage applied over a water-resistive barrier.

❖ This type of EIFS system typically contains a water impermeable barrier underneath to prevent accumulation of water in the exterior façade.

EXTERIOR WALL. An above-*grade* wall that defines the exterior boundaries of a building. Includes between-floor spandrels, peripheral edges of floors, roof and *basement* knee walls, dormer walls, gable end walls, walls enclosing a mansard roof and *basement walls* with an average below-*grade* wall area that is less than 50 percent of the total opaque and nonopaque area of that enclosing side.

❖ An exterior wall is any wall located above grade level enclosing conditioned space. Regulated by this definition for the purpose of energy efficiency, a basement wall is exterior where less than 50 percent of its enclosing surface area is below grade.

EXTERIOR WALL COVERING. A material or assembly of materials applied on the exterior side of exterior walls for the purpose of providing a weather-resistive barrier, insulation or for aesthetics, including but not limited to, veneers, siding, exterior insulation and finish systems, architectural trim and embellishments such as cornices, soffits, and fascias.

❖ The definition of "Exterior wall covering" from Chapter 14 of the IBC is introduced to the IRC for appropriate and consistent usage.

Section R602.3 applies this definition to clarify requirements for sheathing installation on exterior walls. Wall sheathing that is used for structural purposes (e.g., bracing) is addressed in Chapter 6, "Wall Construction," while wall sheathing that is used solely for exterior wall covering purposes is appropriately addressed in Chapter 7, "Wall Covering." The special reference to wood structural panels at the exclusion of listing specific requirements for other sheathing types is deleted because the requirements for applicable wall sheathing materials, including wood structural panels, are adequately addressed by reference to Tables R602.3(1) through R602.3(4).

FACING. The wood structural panel facings that form the two outmost rigid layers of the structural insulated panel.

❖ Structural insulated panels are relatively new to the code. Terminology describing components of the panel are necessary to facilitate provisions to regulate the material.

FACTORY-BUILT CHIMNEY. A *listed* and *labeled* chimney composed of factory-made components assembled in the field in accordance with the manufacturer's instructions and the conditions of the listing.

❖ A factory-built chimney is a manufactured, listed and labeled chimney that has been tested by an approved agency to determine its performance characteristics. Factory-built chimneys are manufactured in two basic designs: a double-wall insulated design or a triple-wall air-cooled design. Both designs use stainless steel inner liners to resist the corrosive effects of combustion products.

FENESTRATION. Skylights, roof windows, vertical windows (whether fixed or moveable); opaque doors; glazed doors; glass block; and combination opaque/glazed doors. For definition applicable in Chapter 11, see Section N1101.9.

❖ Exterior windows and doors are fenestrations in the exterior wall. Skylights and other sloped glazing are also included in the definition, as are glass doors, opaque doors and opaque/glazed doors. Chapter 11 regulates fenestrations for energy efficiency.

FIBER-CEMENT SIDING. A manufactured, fiber-reinforcing product made with an inorganic hydraulic or calcium silicate binder formed by chemical reaction and reinforced with discrete organic or inorganic nonasbestos fibers, or both. Additives which enhance manufacturing or product performance are permitted. Fiber-cement siding products have either smooth or textured faces and are intended for *exterior wall* and related applications.

❖ Fiber cement siding is produced from flat nonasbestos fiber cement sheets. It is intended for exterior application and consists of fiber-reinforced cement which is formed either with or without pressure and cured either under natural or accelerated conditions. The surface of the sheet to be exposed is either smooth, granular or textured. This code permits its use as either panel siding or horizontal lap siding.

FIREBLOCKING. Building materials or materials *approved* for use as fireblocking, installed to resist the free passage of flame to other areas of the building through concealed spaces.

❖ Fireblocking materials include lumber; wood structural panels, particleboard, gypsum board, cement-based millboard, batts or blankets of mineral wool or glass fiber, or any other approved material that will resist the passage of flame from one concealed area to another. Regulated by Section R602.8 for wood-frame construction, fireblocking is often used to isolate vertical cavities such as stud spaces from horizontal concealed areas such as attics or floor-ceiling assemblies.

FIREPLACE. An assembly consisting of a hearth and fire chamber of noncombustible material and provided with a chimney, for use with solid fuels.

❖ Fireplaces burn solid fuels such as wood and coal and are not referred to as appliances in the code. Both masonry and factory-built fireplaces are regulated by the IRC.

Factory-built fireplace. A *listed* and *labeled* fireplace and chimney system composed of factory-made components, and assembled in the field in accordance with manufacturer's instructions and the conditions of the listing.

❖ Factory-built fireplaces are solid-fuel burning units having a fire chamber that is either open to the room or, if equipped with doors, operated with the doors either open or closed. The term "fireplace" describes a complete assembly, which includes the hearth, fire chamber and chimney. A factory-built fireplace is composed of factory-built components representative of the prototypes tested, and it is to be installed in accordance with the manufacturer's installation instructions to form a completed fireplace.

Masonry chimney. A field-constructed chimney composed of solid masonry units, bricks, stones or concrete.

❖ Masonry chimneys can have one or more flues within them and are field constructed of brick, stone, concrete and fire-clay materials. A masonry chimney can stand alone or be part of a masonry fireplace.

Masonry fireplace. A field-constructed fireplace composed of solid masonry units, bricks, stones or concrete.

❖ Masonry fireplaces must be constructed in accordance with the requirements found in Section R1003. These specific requirements are based on tradition and field experience and describe the conventional fireplace that has proven to be reliable where properly constructed, used and maintained.

FIREPLACE STOVE. A free-standing, chimney-connected solid-fuel-burning heater designed to be operated with the fire chamber doors in either the open or closed position.

❖ Fireplace stoves are generally of the free-standing type and heat a space by direct radiation. There are various types of fireplace stoves, and their installation must be in compliance with the listing of the stove.

FIREPLACE THROAT. The opening between the top of the firebox and the smoke chamber.

❖ The fireplace throat is the point where the smoke and heat from the fireplace pass into the chimney.

FIRE-RETARDANT-TREATED WOOD. Pressure-treated lumber and plywood that exhibit reduced surface burning characteristics and resist propagation of fire.

Other means during manufacture. A process where the wood raw material is treated with a fire-retardant formulation while undergoing creation as a finished product.

Pressure process. A process for treating wood using an initial vacuum followed by the introduction of pressure above atmospheric.

❖ Fire-retardant-treated wood (FRTW) is a specific type of material that is made to be fire retardant by the infusion of chemicals into the wood using a pressure process. It is important to apply the code provisions that allow FRTW to products that are pressure treated in accordance with referenced standards. Other products will not necessarily perform in the same manner.

FIRE SEPARATION DISTANCE. The distance measured from the building face to one of the following:

1. To the closest interior *lot line*; or
2. To the centerline of a street, an alley or public way; or
3. To an imaginary line between two buildings on the *lot*.

The distance shall be measured at a right angle from the face of the wall.

❖ This is the distance between the exterior surface of a building and one of the following three locations: the nearest interior lot line; the centerline of a street, alley, or public way; or an imaginary line placed between two buildings on the same lot. The measurement is perpendicular to the lot line. See Commentary Figure R202(8). The fire separation distance is important in determining exterior wall protection and the potential prohibition of exterior openings based on the proximity to the lot lines.

FIXTURE. See "Plumbing fixture."

FIXTURE BRANCH, DRAINAGE. A drain serving two or more fixtures that discharges into another portion of the drainage system.

❖ See Commentary Figure R202(9).

FIXTURE BRANCH, WATER-SUPPLY. A water-supply pipe between the fixture supply and a main water-distribution pipe or fixture group main.

❖ See the definition of "Branch."

FIXTURE DRAIN. The drain from the trap of a fixture to the junction of that drain with any other drain pipe.

❖ Commonly referred to as a trap arm, a fixture drain is the horizontal section of pipe connecting the outlet weir of a trap to a stack, fixture branch or any other drain [see Commentary Figures R202(9) and (10)].

FIXTURE FITTING.

Supply fitting. A fitting that controls the volume and/or directional flow of water and is either attached to or accessible from a fixture or is used with an open or atmospheric discharge.

Waste fitting. A combination of components that conveys the sanitary waste from the outlet of a fixture to the connection of the sanitary drainage system.

❖ This includes the various portions of a faucet from its connecting supply lines to the termination point where water is delivered. When applied to waste components, it includes tubing, fittings and traps used for the connection to the drainage system.

FIXTURE GROUP, MAIN. The main water-distribution pipe (or secondary branch) serving a plumbing fixture grouping such as a bath, kitchen or laundry area to which two or more individual fixture branch pipes are connected.

❖ This is the main or branch water-distribution pipe that extends to a group of fixtures (i.e. bathroom, laundry or kitchen).

FIXTURE SUPPLY. The water-supply pipe connecting a fixture or fixture fitting to a fixture branch.

❖ A fixture supply is a water supply pipe serving a single fixture.

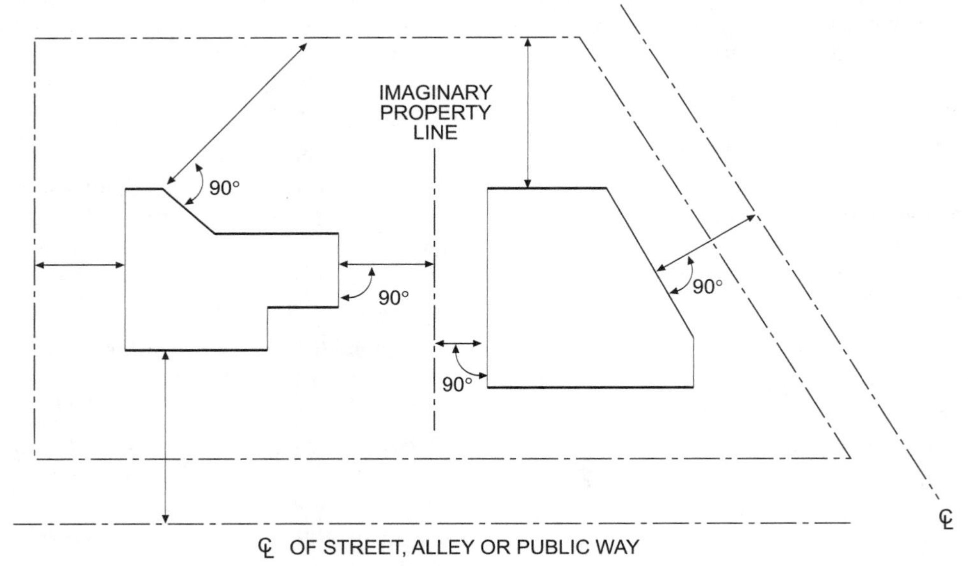

For SI: 90 degrees = 1.57 rad.

Figure R202(8)
FIRE-SEPARATION DISTANCE

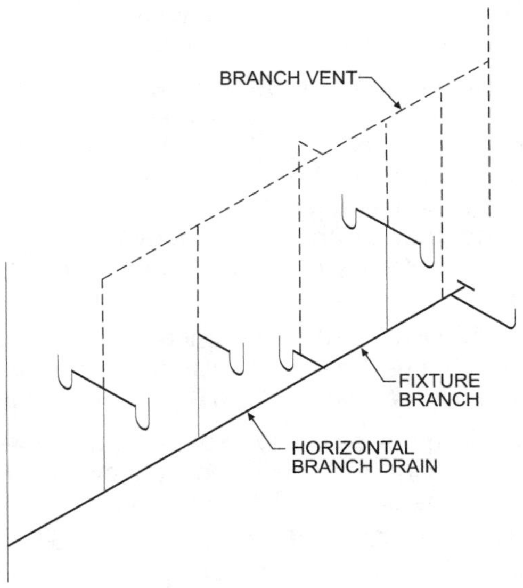

Figure R202(9)
FIXTURE AND HORIZONTAL BRANCH DRAIN

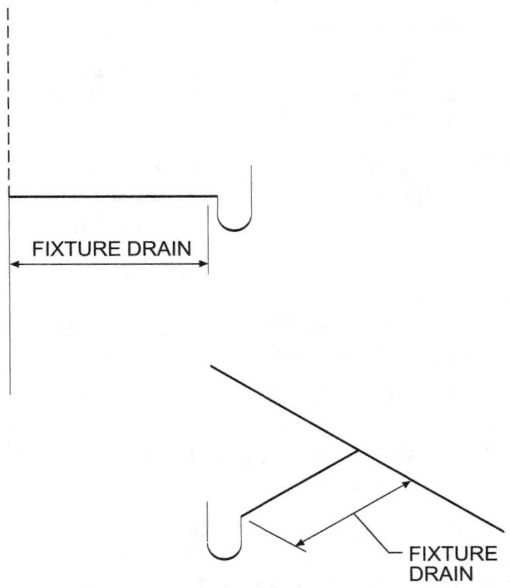

Figure R202(10)
FIXTURE DRAIN

FIXTURE UNIT, DRAINAGE (d.f.u.). A measure of probable discharge into the drainage system by various types of plumbing fixtures, used to size DWV piping systems. The drainage fixture-unit value for a particular fixture depends on its volume rate of drainage discharge, on the time duration of a single drainage operation and on the average time between successive operations.

❖ The conventional method of designing a sanitary drainage system is based on drainage fixture unit (d.f.u.) load values. The fixture-unit approach takes into consideration the probability of load on a drainage system. The d.f.u. is an arbitrary loading factor assigned to each fixture relative to its impact on the drainage system. D.f.u. values are determined based on the average rate of water discharge by a fixture, the duration of a single operation, and the frequency of use or interval between each operation.

Because d.f.u. values have a built-in probability factor, they cannot be directly translated into flow rates or discharge rates.

A d.f.u. is not the same as the water supply fixture unit (w.s.f.u.) described in Chapter 29 and below.

FIXTURE UNIT, WATER-SUPPLY (w.s.f.u.). A measure of the probable hydraulic demand on the water supply by various types of plumbing fixtures used to size water-piping systems. The water-supply fixture-unit value for a particular fixture depends on its volume rate of supply, on the time duration of a single supply operation and on the average time between successive operations.

❖ When estimating peak demand for water supply systems, sizing methods use water supply fixture units (w.s.f.u.). This is a numerical factor on an arbitrary scale assigned to intermittently used fixtures to calculate their load-producing effects on the water supply system.

The use of fixture units makes it possible to simplify the difficult task of calculating load-producing characteristics of any fixture to a common basis. The fixture units of different kinds of fixtures can be applied to a single basic probability curve found in various sizing methods.

FLAME SPREAD. The propagation of flame over a surface.

❖ During a fire, flames will travel across a building's interior surfaces in different forms and at different speeds. The movement of the flames along the surface material is the flame spread.

FLAME SPREAD INDEX. A comparative measure, expressed as a dimensionless number, derived from visual measurements of the spread of flame versus time for a material tested in accordance with ASTM E 84 or UL 723.

❖ To regulate the spread of fire over an interior surface material, limitations are imposed on the characteristics of the material. The standardized test procedure described in ASTM E 84 or UL 723 measures the flame spread under specified conditions, and a numerical value is assigned to the material being tested. This value is the flame spread index.

FLIGHT. A continuous run of rectangular treads or winders or combination thereof from one landing to another.

❖ Two points of clarification for stairways have been addressed by the definition of flight. One, a flight is made up of the treads and risers that occur between landings. Therefore, a stairway connecting two stories that includes an intermediate landing consists of two flights. Secondly, the inclusion of winders within a stairway does not create multiple flights. Winders are simply treads within a flight and are often combined with rectangular treads within the same flight.

FLOOD-LEVEL RIM. The edge of the receptor or fixture from which water overflows.

❖ The flood-level rim is the highest elevation that liquid can be contained in a receptacle without spilling over the side of the fixture or receptor.

FLOOR DRAIN. A plumbing fixture for recess in the floor having a floor-level strainer intended for the purpose of the collection and disposal of waste water used in cleaning the floor and for the collection and disposal of accidental spillage to the floor.

❖ Floor drains are typically installed as emergency fixtures, preventing the flooding of a room or space.

FLOOR FURNACE. A self-contained furnace suspended from the floor of the space being heated, taking air for combustion from outside such space, and with means for lighting the *appliance* from such space.

❖ Such units supply heat through a floor grille placed directly over the unit's heat exchanger. Typically floor furnaces are classified as gravity-type furnaces. The circulation of air from a gravity-type floor furnace is accomplished by convection. Floor furnaces may be equipped with factory fans to circulate the air.

FLOW PRESSURE. The static pressure reading in the water-supply pipe near the faucet or water outlet while the faucet or water outlet is open and flowing at capacity.

❖ Each fixture requiring water has a minimum flow pressure necessary for the fixture to operate properly and to help protect the potable water supply from contamination. "Flow pressure" is a factor used in sizing the water supply system. The flow pressure of a plumbing system is less than the static pressure because energy is lost in putting the fluid in motion.

FLUE. See "Vent."

FLUE, APPLIANCE. The passages within an *appliance* through which combustion products pass from the combustion chamber to the flue collar.

❖ This is the passage through an appliance used for the removal of combustion products.

FLUE COLLAR. The portion of a fuel-burning *appliance* designed for the attachment of a draft hood, vent connector or venting system.

❖ See the commentary on the definition of "Draft hood."

FLUE GASES. Products of combustion plus excess air in *appliance* flues or heat exchangers.

❖ The exact composition of flue gases depends on the materials being burned. The primary components of flue gases are nitrogen, carbon dioxide, water vapor, particulates, carbon monoxide and myriad compounds and trace elements that vary with the nature of the fuel.

FLUSH VALVE. A device located at the bottom of a flush tank that is operated to flush water closets.

❖ A flush valve consists of the flapper and the flush-valve seat in a flush-tank type water closet.

FLUSHOMETER TANK. A device integrated within an air accumulator vessel that is designed to discharge a predetermined quantity of water to fixtures for flushing purposes.

❖ This device is a hybrid between a flushometer valve and a gravity flush tank. A flushometer tank employs a compression tank that holds water under a pressure equivalent to the water-supply pressure. The flushing action created by the flushometer tank is similar to that of the flushometer valve, and the amount of water used is limited as it is with a gravity flush tank.

FLUSHOMETER VALVE. A flushometer valve is a device that discharges a predetermined quantity of water to fixtures for flushing purposes and is actuated by direct water pressure.

❖ This flushing device is a type of metering valve that is used primarily in public occupancy applications because of its powerful flushing action.

FOAM BACKER BOARD. Foam plastic used in siding applications where the foam plastic is a component of the siding.

❖ Products that combine an exterior siding material and foam plastic are available in the marketplace. The foam plastic component used in these products is referred to as a foam backer board. By using a foam layer in addition to the siding the thermal properties can be enhanced.

FOAM PLASTIC INSULATION. A plastic that is intentionally expanded by the use of a foaming agent to produce a reduced-density plastic containing voids consisting of open or closed cells distributed throughout the plastic for thermal insulating or acoustic purposes and that has a density less than 20 pounds per cubic foot (320 kg/m^3) unless it is used as interior trim.

❖ An expanded plastic produced through use of a foaming agent that has a reduced density. This plastic contains voids consisting of open or closed cells distributed throughout the plastic, which provide thermal insulation and/or acoustic control. The density of the material is to be less than 20 pounds per cubic foot (320 kg/m^3) unless it is used as interior trim. Foam plastic used in siding applications where the foam plastic is a component of the siding.

FOAM PLASTIC INTERIOR TRIM. Exposed foam plastic used as picture molds, chair rails, crown moldings, baseboards, handrails, ceiling beams, door trim and window trim and similar decorative or protective materials used in fixed applications.

❖ An expanded plastic used as a decorative trim which is produced through use of a foaming agent. This foam plastic contains voids consisting of open or closed cells distributed throughout the material with a minimum density of at least 20 pounds (320 kg/m^3) per cubic foot (pcf). The foam plastic trim is left exposed in applications such as picture moldings, chair rails, crown moldings, baseboards, handrails, ceiling beams, door trim, window trim and similar decorative or protective applications.

FUEL-PIPING SYSTEM. All piping, tubing, valves and fittings used to connect fuel utilization *equipment* to the point of fuel delivery.

❖ As used in this code, this term includes the tubing and pipe used to convey fuel gas from the point of delivery to the appliance.

FULLWAY VALVE. A valve that in the full open position has an opening cross-sectional area equal to a minimum of 85 percent of the cross-sectional area of the connecting pipe.

❖ Gate valves and ball valves are typical fullway (full-open) valves. Unlike a globe valve, a fullway valve will provide the most open and straight run for the contents to flow through because the closing portion is perpendicular to the run of pipe.

FURNACE. A vented heating *appliance* designed or arranged to discharge heated air into a *conditioned space* or through a duct or ducts.

❖ The single most distinguishing characteristic of furnaces is that they use air as the heat transfer medium. Furnaces can be fueled by gas, oil, solid fuel or electricity and can use fans, blowers and gravity (convection) to circulate the heated air to and from the unit. In the context of the code, the primary use of the term "furnace" refers to heating appliance units that combine a combustion chamber with related components, one or more heat exchangers and an air-handling system.

GLAZING AREA. The interior surface area of all glazed fenestration, including the area of sash, curbing or other framing elements, that enclose *conditioned space*. Includes the area of glazed fenestration assemblies in walls bounding conditioned *basements*.

❖ The glazing area includes not only the surface area of the exposed glazing but also the framing elements, including the sash and curbing. The amount of glazing area is regulated for natural light by Section R303; however, this definition applies to the energy efficiency provisions of Chapter 11.

GRADE. The finished ground level adjoining the building at all *exterior walls*.

❖ This is the point at which the finished exterior ground level intersects the exterior wall of the building. The grade around a building may remain relatively constant, such as on a flat site, or may change dramatically from one point to the next if the site is steeply sloping.

GRADE FLOOR OPENING. A window or other opening located such that the sill height of the opening is not more than 44 inches (1118 mm) above or below the finished ground level adjacent to the opening.

❖ In the requirements for emergency escape and rescue openings found in Section R310, the size of the openings may be reduced if they are grade floor openings. These are windows or other openings that are located within close proximity to the finished ground level. The sill of a grade floor opening may be located either above or below the adjacent ground level, provided it is located no more than 44 inches (1118 mm) vertically from the level of the ground.

GRADE, PIPING. See "Slope."

GRADE PLANE. A reference plane representing the average of the finished ground level adjoining the building at all *exterior walls*. Where the finished ground level slopes away from the *exterior walls*, the reference plane shall be established by the lowest points within the area between the building and the *lot line* or, where the *lot line* is more than 6 feet (1829 mm) from the building between the structure and a point 6 feet (1829 mm) from the building.

❖ This definition can be important in determining the number of stories within a building as well as its height in feet. In some cases, the finished surface of the ground may be artificially raised with imported fill to create a higher grade plane around a building to decrease the number of stories or height. The definition requires that the lowest elevation within 6 feet (1829 mm) of the exterior wall be used to determine the grade plane.

GRAY WATER. Waste discharged from lavatories, bathtubs, showers, clothes washers and laundry trays.

❖ This definition was added for the 2012 edition to support the new Section 3009, "Gray Water Recycling Systems." The collection of gray water must be kept separate from blackwater (waste water from water closets, urinals and kitchen sinks), therefore requiring separate drainage and vent systems. Although the code does not require marking of gray-water collection piping, it may be prudent to mark such systems to avoid future cross connection to blackwater collection piping.

GRIDDED WATER DISTRIBUTION SYSTEM. A water distribution system where every water distribution pipe is interconnected so as to provide two or more paths to each fixture supply pipe.

❖ These systems offer the advantage of a simplistic design, typically smaller sized distribution lines and aid water conservation. In a traditional water distribution system, the water contained in the larger diameter piping is wasted when the line is opened and the user has to wait until the water reaches the desired temperature.

Parallel or gridded water distribution systems differ from branch systems which have individual supply pipes that extend to each fixture or outlet from a central supply point [see Commentary Figure R202(1)]. The central supply point is a multiple-outlet manifold to which the distribution lines connect [see Commentary Figure R202(2)].

GROSS AREA OF EXTERIOR WALLS. The normal projection of all *exterior walls*, including the area of all windows and doors installed therein.

❖ The calculation for determining the gross area of exterior walls for energy efficiency purposes is based on the total area of the entire exterior surface, including openings such as windows and doors.

GROUND-SOURCE HEAT PUMP LOOP SYSTEM. Piping buried in horizontal or vertical excavations or placed in a body of water for the purpose of transporting heat transfer liquid to and from a heat pump. Included in this definition are closed loop systems in which the liquid is recirculated and open loop systems in which the liquid is drawn from a well or other source.

❖ This definition assists the user with a ready means of distinguishing ground-source heat pump loop systems from other hydronic systems.

GUARD. A building component or a system of building components located near the open sides of elevated walking surfaces that minimizes the possibility of a fall from the walking surface to the lower level.

❖ A guard is a component or system of components whose function is the prevention of falls from an elevated area. Placed adjacent to an elevation change, a guard must be of adequate height, strength and configuration to help prevent people, especially small children, from falling over or through the guard to the area below.

GUESTROOM. Any room or rooms used or intended to be used by one or more guests for living or sleeping purposes.

❖ Section 101.2 now provides an exception to allow lodging houses to be built in accordance with the code. A common example of a lodging house is a bed and breakfast. To support this new provision (Section 101.2, Exception 2), the code now includes a definition of "Lodging house," which includes the term "guestroom." Therefore, it is necessary to define guestroom as well.

HABITABLE SPACE. A space in a building for living, sleeping, eating or cooking. Bathrooms, toilet rooms, closets, halls, storage or utility spaces and similar areas are not considered *habitable spaces*.

❖ An area within a building used for living, sleeping, dining or cooking is a habitable space. Those areas not meeting this definition include bathrooms, closets, hallways and utility rooms. Habitable spaces are typically occupied, and as such they are more highly regulated than accessory use areas.

HANDRAIL. A horizontal or sloping rail intended for grasping by the hand for guidance or support.

❖ Typically used in conjunction with a ramp or stairway, a handrail provides support for the user along the travel path. Through its height, size, shape, continuity and structural stability, a handrail assists users at elevation changes inside and outside a building. A handrail may also be used as a guide to direct the user in a specified direction.

HANGERS. See "Supports."

HAZARDOUS LOCATION. Any location considered to be a fire hazard for flammable vapors, dust, combustible fibers or other highly combustible substances.

❖ The environment in which mechanical equipment and appliances operate plays a significant role in the safe performance of the equipment installation. Locations that may contain ignitable or explosive atmospheres are classified as hazardous locations with respect to the installation of mechanical equipment and appliances. For example, repair garages can be classified as hazardous locations because they can contain gasoline vapors from vehicles stored within them as well as other volatile chemicals. Public and private garages can be considered as hazardous locations because of the presence of motor vehicles and because of the owners' propensity for storing paint, varnish, thinners, lawn and home maintenance products and other chemicals within the space.

HEAT PUMP. An *appliance* having heating or heating/cooling capability and that uses refrigerants to extract heat from air, liquid or other sources.

❖ Heat pumps are referred to as reverse-cycle refrigeration systems. Special controls and valves allow the system to be used as a comfort heating system, comfort cooling system and potable water heating system. Auxiliary heat is often installed in heat pumps in the form of electric resistance heat or fuel-fired furnaces. In most cases, the design and installation of heat pumps is more critical than for other types of comfort air-conditioning systems because of the heat-transfer process and the amount of air that must be circulated. Heat-pump types include water source (hydronic), ground source (earth loop), water source (wells) and air source.

HEATING DEGREE DAYS (HDD). The sum, on an annual basis, of the difference between 65°F (18°C) and the mean temperature for each day as determined from "NOAA Annual Degree Days to Selected Bases Derived from the 1960-1990 Normals" or other weather data sources acceptable to the code official.

❖ Heating Degree Days are a measure of the heating requirements for buildings in a particular climate. The greater the number of HDD, the colder the climate. HDD is calculated by first determining the average temperature for each day of the year over a 24-hour period. The average temperature for each day is then subtracted from a base temperature of 65°F (18°C) to determine the HDD value for the day. For days on which the average temperature is over 65°F (18°C), the HDD is zero. The HDD for each day are then totaled to determine the HDD value for the particular area. The average temperatures are calculated over a number of years. These values are published in the ASHRAE *Handbook of Fundamentals*. They can also be found in the NOAA Annual Degree Days to Selected Bases Derived from the 1960-1990 Normals.

Climate Zone numbers listed in Table N1101.2 can also be used to approximate the HDD value for a particular county. Each climate zone is equal to 500 HDD. For example, Climate Zone 1 is equal to 0-499 HDD, and Climate Zone 2 is equal to 500-999 HDD.

HEIGHT, BUILDING. The vertical distance from *grade plane* to the average height of the highest roof surface.

❖ Once the elevation of the grade plane has been calculated, it is possible to determine a building's height, which is measured vertically from the grade plane to the average height of the highest roof surface. Commentary Figure R202(11) contains examples of this measurement.

If the building is stepped or terraced, it is logical that the height is the maximum height of any segment of the building. It may be appropriate under certain circumstances that the number of stories in a building be determined in the same manner. Each case should be judged individually based on the characteristics of the site and construction.

HEIGHT, STORY. The vertical distance from top to top of two successive tiers of beams or finished floor surfaces; and, for the topmost *story*, from the top of the floor finish to the top of the ceiling joists or, where there is not a ceiling, to the top of the roof rafters.

❖ Within the building itself, the height of an individual story is determined by measuring the vertical distance between two successive finished floor surfaces. In a single-story building or for the uppermost floor in a multistory structure, the measurement is taken from the finished floor surface to the top of the ceiling joists or to the underside of the roof deck (top of the roof rafters) where a ceiling is not provided [see Commentary Figure R202(12)].

HIGH-EFFICACY LAMPS. See Section N1101.9 for definition applicable in Chapter 11.

❖ The energy conservation requirements of this code now contain a requirement to use high-efficacy lamps in permanently installed lighting fixtures. These lamps are considerably more energy efficient than normal lamps.

HIGH-TEMPERATURE (H.T.) CHIMNEY. A high temperature chimney complying with the requirements of UL 103. A Type H.T. chimney is identifiable by the markings "Type H.T." on each chimney pipe section.

❖ This is a factory-built chimney used to vent solid-fuel-burning appliances when a masonry chimney is not available. See Commentary Figure R202(13) for an illustration of a Type H.T. chimney.

HILL. With respect to topographic wind effects, a land surface characterized by strong relief in any horizontal direction.

❖ This is a term to describe a topographical area where wind speed-up can occur. In cases where there is known to be structural damage due to this type of wind speed-up, the code requires consideration of topographical wind effects. See Section R301.2.1.5.

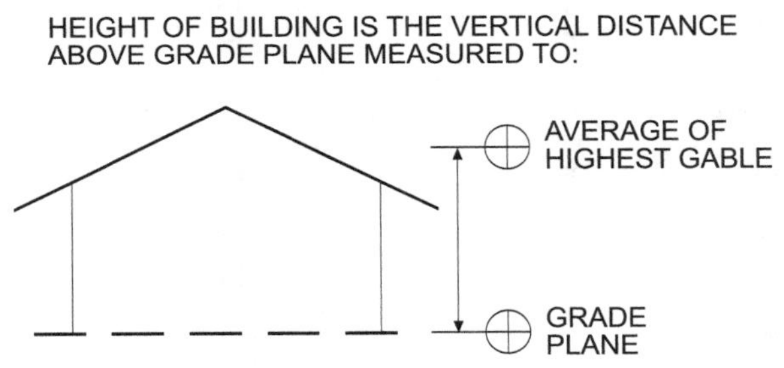

Figure R202(11)
HEIGHT OF BUILDING

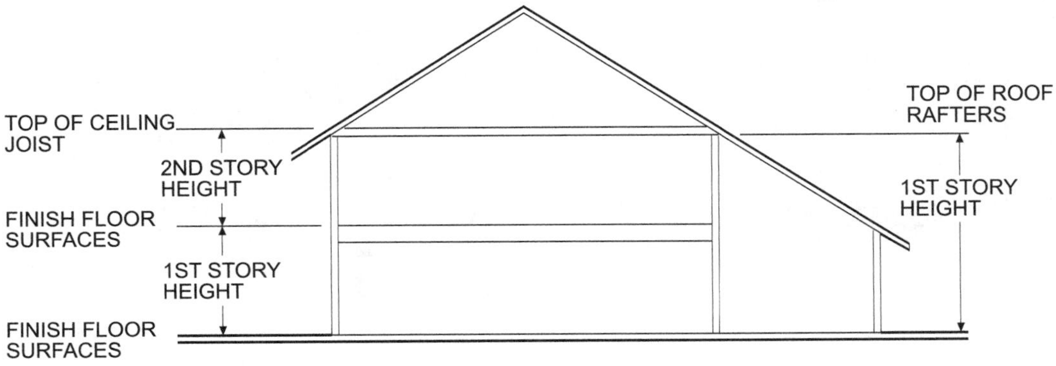

Figure R202.(12)
STORY HEIGHT

HORIZONTAL BRANCH, DRAINAGE. A drain pipe extending laterally from a soil or waste stack or *building drain*, that receives the discharge from one or more *fixture drains*.

❖ See Commentary Figure R202(9).

HORIZONTAL PIPE. Any pipe or fitting that makes an angle of less than 45 degrees (0.79 rad) with the horizontal.

❖ This definition is needed for the application of many code provisions that apply only to horizontal piping.

HOT WATER. Water at a temperature greater than or equal to 110°F (43°C).

❖ The code does not specify a maximum temperature for hot water; however, it should be limited to a temperature that will minimize: (1) the risk of a user's burn and (2) thermal stress to plumbing system components. The designer, installer or owner has the choice of determining the temperature. Most water heaters are now shipped with a factory-set temperature of 120°F (49°C). The minimum temperature is established to determine what is considered as hot water. Water temperature below the minimum temperature is either tempered or cold. Care must be used when installing a system that has temperature capability of exceeding 120°F (49°C) because of the possibility of a user being scalded.

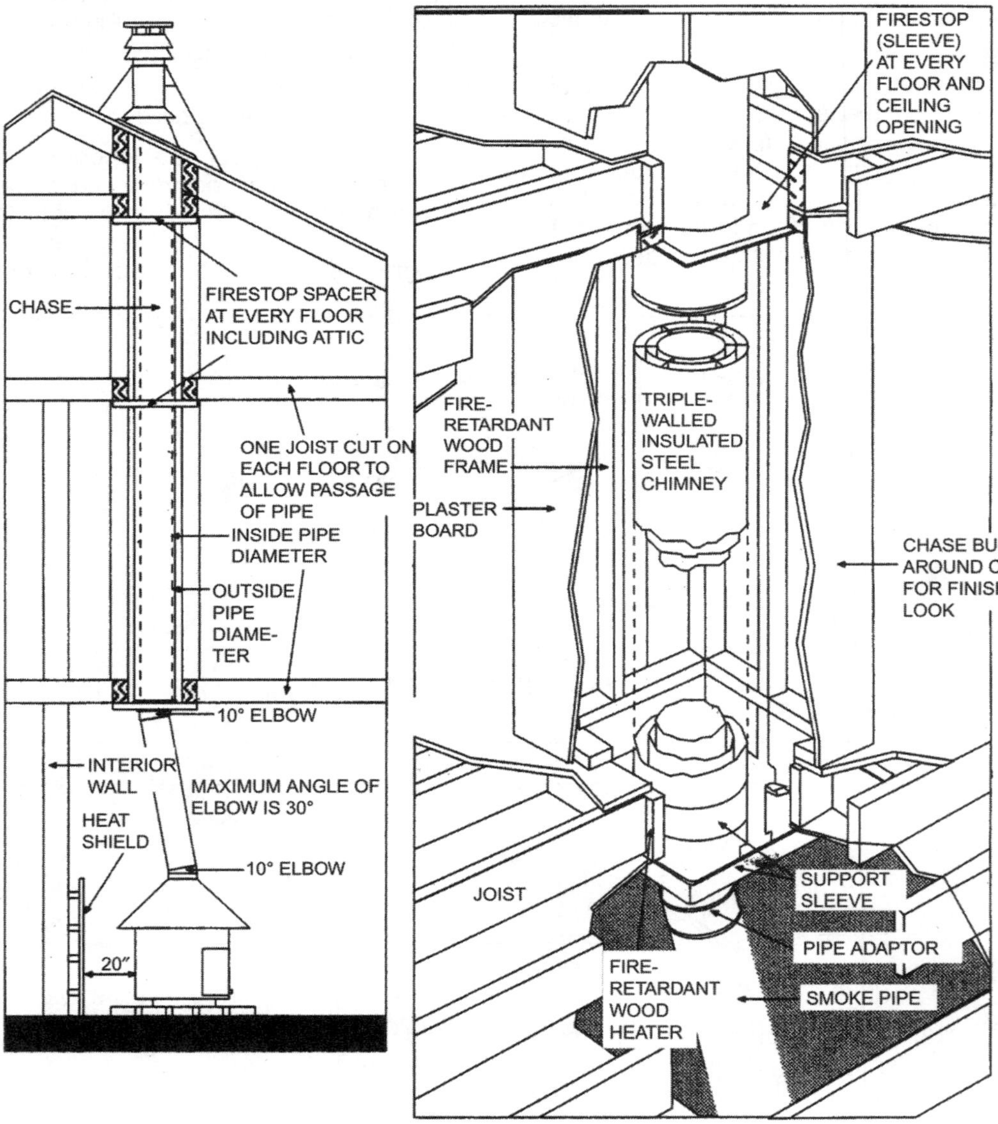

For SI: 1 inch = 25.4 mm, 1 degree = 0.01745 rad.

Figure R202(13)
HIGH-TEMPERATURE CHIMNEY

HURRICANE-PRONE REGIONS. Areas vulnerable to hurricanes, defined as the U.S. Atlantic Ocean and Gulf of Mexico coasts where the basic wind speed is greater than 90 miles per hour (40 m/s), and Hawaii, Puerto Rico, Guam, Virgin Islands, and America Samoa.

❖ Those geographical areas subject to hurricanes are "hurricane-prone regions." The specific areas include those having a minimum basic wind speed exceeding 110 miles per hour (177 km/h or 49 m/s) along the United States coastline of the Atlantic Ocean and the coasts bordering the Gulf of Mexico, as well as the indicated islands.

HYDROGEN GENERATING APPLIANCE. A self-contained package or factory-matched packages of integrated systems for generating gaseous hydrogen. Hydrogen generating *appliances* utilize electrolysis, reformation, chemical, or other processes to generate hydrogen.

❖ Hydrogen is being generated to fuel a new line of developmental vehicles and to generate electric power in fuel cell power systems. Several different processes are being used to generate hydrogen. One such process removes hydrogen from tap water (H_2O) in an electrically powered appliance.

Installation requirements for these appliances are found in Chapter 7 of the IFGC and ventilation requirements for spaces containing the appliances are found in Section M1307.4 of this code.

IGNITION SOURCE. A flame, spark or hot surface capable of igniting flammable vapors or fumes. Such sources include *appliance* burners, burner ignitions and electrical switching devices.

❖ Any energized portion of an electrical system that can generate a spark or that can produce significant heat (often referred to as "glow," as in a heating element) is an ignition source should it come in contact with flammable gasses or liquids. Understandably, any open flame (such as a pilot light) is also an ignition source.

INDIRECT WASTE PIPE. A waste pipe that discharges into the drainage system through an air gap into a trap, fixture or receptor.

❖ Often referred to as "open site" or "safe waste," indirect waste pipe installations protect fixtures, equipment and systems from the backflow or backsiphonage of waste.

INDIVIDUAL SEWAGE DISPOSAL SYSTEM. A system for disposal of sewage by means of a septic tank or mechanical treatment, designed for use apart from a public sewer to serve a single establishment or building.

❖ When installed in accordance with the *International Private Sewage Disposal Code*® (IPSDC®) and approved by the administrative authority, these systems are acceptable as a point of disposal for sewage when a public sewer system is not available.

INDIVIDUAL VENT. A pipe installed to vent a single-*fixture drain* that connects with the vent system above or terminates independently outside the building.

❖ An individual vent is the simplest and most common method of venting a single plumbing fixture.

INDIVIDUAL WATER SUPPLY. A supply other than an *approved* public water supply that serves one or more families.

❖ This type of water supply includes wells, springs, streams and cisterns. Surface bodies of water and land cisterns can be used when properly treated and approved.

INSULATING CONCRETE FORM (ICF). A concrete forming system using stay-in-place forms of rigid foam plastic insulation, a hybrid of cement and foam insulation, a hybrid of cement and wood chips, or other insulating material for constructing cast-in-place concrete walls.

❖ This is a relatively new construction technique for foundation/basement walls and above-grade walls that uses stay-in-place insulation forms and concrete. A wide variety of systems are available, with several types of insulating materials used for the formwork. Sections R404.4 (foundation walls) and R611 (above-grade walls) address three different systems consisting of insulating concrete forms: flat, waffle grid, and screen grid.

INSULATING SHEATHING. An insulating board having a minimum thermal resistance of R-2 of the core material. For definition applicable in Chapter 11, see Section N1101.9.

❖ A sheathing board having an insulating core material is insulating sheathing if its thermal resistance is at least R-2.

JURISDICTION. The governmental unit that has adopted this code under due legislative authority.

❖ The governmental entity that adopts and enforces the IRC is the "jurisdiction." A jurisdiction can take on a variety of forms such as a state, city, town, county, fire district, improvement district or federal department. The jurisdiction appoints a building official to administer and enforce the provisions of the code.

KITCHEN. Kitchen shall mean an area used, or designated to be used, for the preparation of food.

❖ This is an area designed to be used for the preparation of food. A kitchen is a habitable space and is subject to a number of requirements.

LABEL. An identification applied on a product by the manufacturer which contains the name of the manufacturer, the function and performance characteristics of the product or material, and the name and identification of an *approved agency* and that indicates that the representative sample of the product or material has been tested and evaluated by an *approved agency*. (See also "Manufacturer's designation" and "Mark.")

❖ A label identifies a product or material and provides other information that can be investigated if there is a question about the suitability of the product or material for a specific installation.

The applicable reference standard often states the minimum identifying information required on a label.

The information on a label as mandated by the code includes the name of the manufacturer, the product's function or performance characteristics, the name and identification of the approved labeling agency and the approval of the testing agency. The product name, serial number, installation specifications or applicable tests and standards might be additional information that is provided.

LABELED. *Equipment*, materials or products to which have been affixed a *label*, seal, symbol or other identifying *mark* of a nationally recognized testing laboratory, inspection agency or other organization concerned with product evaluation that maintains periodic inspection of the production of the above-*labeled* items and whose labeling indicates either that the *equipment*, material or product meets identified standards or has been tested and found suitable for a specified purpose.

❖ When a product is labeled, the label indicates that the material has been tested for conformance to an applicable standard and that the component is subject to a third-party inspection that verifies that the minimum level of quality required by the appropriate standard is maintained. Labeling provides a readily available source of information that is useful for field inspection of installed products.

The labeling agency performing the third-party inspection must be approved by the building official, and the basis for this approval may include, but is not necessarily limited to, the capacity and capability of the agency to perform the specific testing and inspection.

LIGHT-FRAME CONSTRUCTION. A type of construction whose vertical and horizontal structural elements are primarily formed by a system of repetitive wood or cold-formed steel framing members.

❖ Light-framed construction is by far the predominant construction technique used for dwelling units. Consisting primarily of repetitive framing members such as studs, floor joists, ceiling joists and rafters, light-framed construction can use wood or light-gage steel as framing members. The repetitive members work together as a system to provide the structural integrity of the building. The IRC devotes most of its structural provisions to structures of light-framed construction.

LISTED. *Equipment*, materials, products or services included in a list published by an organization acceptable to the code official and concerned with evaluation of products or services that maintains periodic inspection of production of *listed equipment* or materials or periodic evaluation of services and whose listing states either that the *equipment*, material, product or service meets identified standards or has been tested and found suitable for a specified purpose.

❖ As stated in the definition, the listing states that either the equipment or material meets nationally recognized standards or it has been found suitable for use in a specific manner. The listing becomes part of the documentation the building official can use to approve or disapprove the equipment or appliance.

LIVE LOADS. Those loads produced by the use and occupancy of the building or other structure and do not include construction or environmental loads such as wind load, snow load, rain load, earthquake load, flood load or dead load.

❖ A variety of loads are imposed on a structure that must be considered in the building design. One group is live loads, produced through the use of the building. Different types of uses or occupancies result in different live loads, due primarily to the number of people involved, the type of activity that takes place and the movable furnishings and equipment that are typically a part of the use.

Design loads that are not live loads include environmental loads such as wind loads, snow loads, rain loads, earthquake loads and flood loads, as well as construction loads. Fixed-in-place equipment is a dead load rather than a live load.

LIVING SPACE. Space within a *dwelling unit* utilized for living, sleeping, eating, cooking, bathing, washing and sanitation purposes.

❖ The living space within a dwelling extends beyond what is defined as habitable space. Living space includes not only those areas used for living, sleeping, eating and cooking but also includes bathing and washing areas such as bathrooms. Storage areas such as closets and garages are not a portion of the living space.

LOCAL EXHAUST. An exhaust system that uses one or more fans to exhaust air from a specific room or rooms within a dwelling.

❖ The most common example of local exhaust in a dwelling is the bathroom fan that exhausts air directly to the outside of the house. This definition was installed to clarify the intent of the exception to Section R303.3. The exception allows for a light and local exhaust to be used in lieu of the required amount of glazing to the exterior in a bathroom.

LODGING HOUSE. A one-family dwelling where one or more occupants are primarily permanent in nature, and rent is paid for guestrooms.

❖ Section 101.2 now provides an exception to allow lodging houses to be built in accordance with this code. A common example of a lodging house is a bed and breakfast. To support this new provision (Section 101.2, Exception 2), the code now includes this definition of "Lodging house." Note that this definition includes the term "guestroom," which is also a defined term in this code.

LOT. A portion or parcel of land considered as a unit.

❖ A lot is a piece of property regulated as a single unit. The code regulates the construction of one or more buildings based on their location on a single lot. Owners are expected to control what occurs on their own lots, with no control over any adjacent property. Therefore, several provisions are based upon the proximity of the building to the lot line.

LOT LINE. A line dividing one *lot* from another, or from a street or any public place.

❖ The lot line defines the exterior perimeter of the lot. It is used in several provisions of the code as the basis for requirements dealing with the proximity of an exterior wall or opening to the lot line.

MACERATING TOILET SYSTEMS. A system comprised of a sump with macerating pump and with connections for a water closet and other plumbing fixtures, that is designed to accept, grind and pump wastes to an *approved* point of discharge.

❖ See Section P2723.

MAIN. The principal pipe artery to which branches may be connected.

❖ This generic term applies to the primary or principal pipe artery in either water-supply systems or drainage systems to which branches may be connected.

MAIN SEWER. See "Public sewer."

MANIFOLD WATER DISTRIBUTION SYSTEMS. A fabricated piping arrangement in which a large supply main is fitted with multiple branches in close proximity in which water is distributed separately to fixtures from each branch.

❖ A manifold is a multiple-opening header to which one or more branch lines connect. See Section P2903.8.

MANUFACTURED HOME. *Manufactured home* means a structure, transportable in one or more sections, which in the traveling mode is 8 body feet (2438 body mm) or more in width or 40 body feet (12 192 body mm) or more in length, or, when erected on site, is 320 square feet (30 m^2) or more, and which is built on a permanent chassis and designed to be used as a *dwelling* with or without a permanent foundation when connected to the required utilities, and includes the plumbing, heating, air-conditioning and electrical systems contained therein; except that such term shall include any structure that meets all the requirements of this paragraph except the size requirements and with respect to which the manufacturer voluntarily files a certification required by the secretary (HUD) and complies with the standards established under this title. For mobile homes built prior to June 15, 1976, a *label* certifying compliance to the Standard for Mobile Homes, NFPA 501, in effect at the time of manufacture is required. For the purpose of these provisions, a mobile home shall be considered a *manufactured home*.

❖ Appendix E contains provisions for the use of manufactured housing as dwellings. Appendix E is applicable only if so specified in the adoption ordinance of the jurisdiction. The provisions of this appendix are limited to three areas: (1) the construction, alteration or repair of a foundation system used to support a manufactured home unit; (2) the construction, installation, addition, alteration, repair or maintenance of building services equipment necessary for connecting the manufactured home to water, fuel, power supplies or sewage systems; and (3) alterations, additions or repairs to existing manufactured homes.

The code describes a manufactured home as being transportable in one or more sections, sets forth minimum dimensions for the home and indicates that a manufactured home is constructed on a permanent chassis. Under the provisions of the IRC, a mobile home is a manufactured home.

MANUFACTURER'S DESIGNATION. An identification applied on a product by the manufacturer indicating that a product or material complies with a specified standard or set of rules. (See also "*Mark*" and "*Label*.")

❖ Identification applied on a product by the manufacturer indicating that a product or material complies with a specified standard or set of rules.

MANUFACTURER'S INSTALLATION INSTRUCTIONS. Printed instructions included with *equipment* as part of the conditions of listing and labeling.

❖ A set of instructions provided by the manufacturer of a product that guides the user or installer through the steps necessary to ensure a complying installation. These instructions help ensure that an installer does not violate the listing of a product.

MARK. An identification applied on a product by the manufacturer indicating the name of the manufacturer and the function of a product or material. (See also "Manufacturer's designation" and "*Label*.")

❖ Identification applied on a product by the manufacturer indicating the name of the manufacturer and the function of a product or material.

MASONRY CHIMNEY. A field-constructed chimney composed of solid masonry units, bricks, stones or concrete.

❖ Masonry chimneys can have one or more flues within them and are field-constructed of brick, stone, concrete, or fire-clay materials. Masonry chimneys can stand alone or be part of a masonry fireplace.

MASONRY HEATER. A masonry heater is a solid fuel burning heating *appliance* constructed predominantly of concrete or solid masonry having a mass of at least 1,100 pounds (500 kg), excluding the chimney and foundation. It is designed to absorb and store a substantial portion of heat from a fire built in the firebox by routing exhaust gases through internal heat exchange channels in which the flow path downstream of the firebox includes at least one 180-degree (3.14-rad) change in flow direction before entering the chimney and which deliver heat by radiation through the masonry surface of the heater.

❖ Masonry heaters are appliances designed to absorb and store heat from a relatively small fire and to radiate that heat into the building interior. They are thermally more efficient than traditional fireplaces because of their design. Interior passageways through the heater allow hot exhaust gases from the fire to transfer heat into the masonry, which then radiates into the building.

MASONRY, SOLID. Masonry consisting of solid masonry units laid contiguously with the joints between the units filled with mortar.

❖ In solid masonry construction, walls are made up of solid masonry units. The joints between the units are also solid because they must be filled with mortar. Solid masonry walls may be either single-wythe or multiple-wythe construction. Where multiple wythes are used, the space between the wythes is filled with mortar or grout.

MASONRY UNIT. Brick, tile, stone, glass block or concrete block conforming to the requirements specified in Section 2103 of the *International Building Code*.

❖ The IBC contains the material requirements for masonry units. Various materials are used to construct masonry units, including brick, tile, stone, glass and concrete.

Clay. A building unit larger in size than a brick, composed of burned clay, shale, fire clay or mixtures thereof.

❖ Clay masonry units are composed of burned clay, shale, fire clay or a combination of these materials. A clay masonry unit is somewhat larger than a brick.

Concrete. A building unit or block larger in size than 12 inches by 4 inches by 4 inches (305 mm by 102 mm by 102 mm) made of cement and suitable aggregates.

❖ Masonry units constructed of cement, aggregates, water and suitable admixtures are concrete masonry units. The minimum size for a concrete masonry unit is 4 inches by 4 inches by 12 inches.

Glass. Nonload-bearing masonry composed of glass units bonded by mortar.

❖ Glass masonry units are unique in that they are permitted for use only in nonload-bearing conditions. The units, either solid or hollow, are connected with mortar in the manner described in Section R610.

Hollow. A masonry unit whose net cross-sectional area in any plane parallel to the loadbearing surface is less than 75 percent of its gross cross-sectional area measured in the same plane.

❖ The net cross-sectional area of a hollow masonry unit is less than 75 percent of its gross cross-sectional area measured in the same plane and parallel to the load-bearing surface. A unit having a greater ratio of net area to gross area than a hollow masonry unit is considered a solid masonry unit.

Solid. A masonry unit whose net cross-sectional area in every plane parallel to the loadbearing surface is 75 percent or more of its cross-sectional area measured in the same plane.

❖ The net cross-sectional area of a solid masonry unit is at least 75 percent of its gross cross-sectional area measured in the same plane and parallel to the load-bearing surface. A unit having a ratio of net area to gross area below 75 percent is a hollow masonry unit.

MASS WALL. Masonry or concrete walls having a mass greater than or equal to 30 pounds per square foot (146 kg/m^2), solid wood walls having a mass greater than or equal to 20 pounds per square foot (98 kg/m^2), and any other walls having a heat capacity greater than or equal to 6 Btu/ft^2 · °F [266 J/(m^2 · K)].

❖ A mass wall must have a heat capacity of 6 Btu/ft^2 · °F. In general terms, the heat capacity is a measure of how well a material stores heat. The higher the heat capacity the greater amount of heat stored. For example, a 6-inch, heavyweight concrete wall has a heat capacity of 14 Btu/ft^2 · °F (286 kJ/m^2 · K) compared to a conventional 2-inch by 4-inch wood framed wall with a heat capacity of approximately 3 Btu/ft^2 · °F (63 kJ/m^2 · K). The code defines a mass wall as a masonry or concrete wall having a mass greater than or equal to 30 lb/ft^2, a solid wood wall (e.g., a log house) having a mass of greater than 20 lb/ft^2, or any other walls having a heat capacity of greater than or equal to 6 Btu/ft^2 · °F (123 kJ/m^2 · K).

MEAN ROOF HEIGHT. The average of the roof eave height and the height to the highest point on the roof surface, except that eave height shall be used for roof angle of less than or equal to 10 degrees (0.18 rad).

❖ For a relatively flat roof (up to 10 degrees slope), the mean roof height is the height of the eave. For steeper roofs, the mean roof height is measured to the average point between the eave and the highest point on the roof surface, typically the highest ridge.

MECHANICAL DRAFT SYSTEM. A venting system designed to remove flue or vent gases by mechanical means, that consists of an induced draft portion under nonpositive static pressure or a forced draft portion under positive static pressure.

❖ Mechanical draft systems do not depend on draft; they use fans or blowers.

Forced-draft venting system. A portion of a venting system using a fan or other mechanical means to cause the removal of flue or vent gases under positive static pressure.

❖ Power exhausters and some power-burner systems are examples of forced-draft systems. Vents and chimneys must be listed for positive pressure applications where used with forced-draft systems.

Induced draft venting system. A portion of a venting system using a fan or other mechanical means to cause the removal of flue or vent gases under nonpositive static vent pressure.

❖ Induced-draft venting is commonly accomplished with field-installed inducer fans designed to supplement natural-draft chimneys or vents.

Power venting system. A portion of a venting system using a fan or other mechanical means to cause the removal of flue or vent gases under positive static vent pressure.

❖ See the commentary to "Forced-draft venting system" above.

MECHANICAL EXHAUST SYSTEM. A system for removing air from a room or space by mechanical means.

❖ A mechanical exhaust system uses a fan or other air-handling equipment to exhaust air to the outdoors. Mechanical exhaust systems include those used for hazardous exhaust and commercial kitchen exhaust. Mechanical exhaust systems may or may not use ductwork as part of the system and may include air-cleaning or air-filtering equipment and fire-suppression equipment.

MECHANICAL SYSTEM. A system specifically addressed and regulated in this code and composed of components, devices, *appliances* and *equipment*.

❖ Mechanical systems include, among others, refrigeration systems, air-conditioning systems, exhaust systems, piping systems, duct systems, venting systems, hydronic systems and ventilation systems. A mechanical system may be any of the above systems or incorporate one or more of the systems. The system includes all of the equipment and appliances required to perform the function for which the system is designed.

METAL ROOF PANEL. An interlocking metal sheet having a minimum installed weather exposure of at least 3 square feet (0.28 m^2) per sheet.

❖ A metal roof panel is specifically defined here in order to identify its difference from a metal roof shingle. Metal roof panels are each at least 3 square feet in surface area and interlock to provide a suitable roof covering. The provisions of Section R905.10 address roof coverings consisting of metal roof panels.

METAL ROOF SHINGLE. An interlocking metal sheet having an installed weather exposure less than 3 square feet (0.28 m^2) per sheet.

❖ Section R905.4 regulates roof coverings consisting of metal roof shingles, which are smaller than 3 square feet (0.28 m^2) per sheet.

MEZZANINE, LOFT. An intermediate level or levels between the floor and ceiling of any *story* with an aggregate floor area of not more than one-third of the area of the room or space in which the level or levels are located.

❖ Floor levels consisting of relatively small floor areas opening into larger rooms are loft mezzanines. Located at an intermediate level within a high-ceiling space, a loft mezzanine is limited to one-third of the area of the room in which it is located. Because of its openness and relatively small size, a loft mezzanine is not an additional story.

MODIFIED BITUMEN ROOF COVERING. One or more layers of polymer modified asphalt sheets. The sheet materials shall be fully adhered or mechanically attached to the substrate or held in place with an *approved* ballast layer.

❖ Polymer-modified asphalt sheets, applied in one or more layers, may be used as a roof covering when installed in accordance with Section R905.11. There are several approved methods for attaching the asphalt sheets, including: (1) full adhesion to the substrate, (2) mechanical attachment to the substrate, and (3) use of a layer of ballast.

MULTIPLE STATION SMOKE ALARM. Two or more single station alarm devices that are capable of interconnection such that actuation of one causes all integral or separate audible alarms to operate.

❖ A multiple station smoke alarm consists of two or more interconnected single station smoke alarms. When one of the smoke alarms is actuated, all of the audible alarms sound. Most dwelling units will require several smoke alarms throughout the unit; Section R314.1 requires that they be interconnected.

NATURAL DRAFT SYSTEM. A venting system designed to remove flue or vent gases under nonpositive static vent pressure entirely by natural draft.

❖ Natural-draft chimneys and vents do not rely on any mechanical means to convey combustion products to the outdoors. Draft is produced by the temperature difference between the combustion gases (flue gases) and the ambient atmosphere. Hot gases are less dense and more buoyant; therefore, they rise or produce a draft.

NATURALLY DURABLE WOOD. The heartwood of the following species with the exception that an occasional piece with corner sapwood is permitted if 90 percent or more of the width of each side on which it occurs is heartwood.

❖ Because of their natural ability to resist deterioration, the harder portions of some species of wood are considered to be naturally durable. The code specifies that "occasional" sapwood is permitted if heartwood constitutes 90 percent of each side.

Decay resistant. Redwood, cedar, black locust and black walnut.

❖ Redwood, cedar, black locust and black walnut lumber are known to resist deterioration caused by the action of microbes that enter the wood fibers. The code defines these species of lumber as being decay resistant.

Termite resistant. Alaska yellow cedar, redwood, Eastern red cedar and Western red cedar including all sapwood of Western red cedar.

❖ Redwood and eastern red cedar are considered to be resistant to infestation by termites and are thus listed as naturally durable. The Formosan Termite, however, is capable of destroying all naturally durable species of wood.

NONCOMBUSTIBLE MATERIAL. Materials that pass the test procedure for defining noncombustibility of elementary materials set forth in ASTM E 136.

❖ A material that successfully passes the ASTM E 136 test is considered to be noncombustible. The test determines whether a building material will act to aid combustion or add appreciable heat to a fire. A material may have a limited combustible content but not contribute appreciably to a fire; thus, it may still qualify as noncombustible.

NONCONDITIONED SPACE. A space that is not a *conditioned space* by insulated walls, floors or ceilings.

❖ Defined for the purpose of regulating energy efficiency, this is an area of a building that does not qualify as conditioned. A nonconditioned space is not enclosed with the appropriate insulating materials of the kind that are considered in the definition of conditioned space.

NOSING. The leading edge of treads of stairs and of landings at the top of stairway flights.

❖ The code must limit the depth of stair tread nosing. Limiting the extent of the tread nosings results in a stair that is easy to use. If too large, they are a tripping hazard when walking up a stairway, and reduce the effective tread depth when walking down the stairway.

OCCUPIED SPACE. The total area of all buildings or structures on any *lot* or parcel of ground projected on a horizontal plane, excluding permitted projections as allowed by this code.

❖ The entire floor area of a building is generally the occupied space. It encompasses all floor levels, including basements. Any exterior space without exterior walls that is covered by a roof or floor above is occupied space. Only those areas beneath projections permitted by Section R302 do not contribute to the total occupied space.

OFFSET. A combination of fittings that makes two changes in direction bringing one section of the pipe out of line but into a line parallel with the other section.

❖ Offsets are necessary to route piping around or through structural elements such as beams, joists, trusses and columns. An offset always involves at least two changes in direction; this is necessary to keep the piping heading in the same direction as the pipe before the offset. Any other arrangement would be considered a bend.

OWNER. Any person, agent, firm or corporation having a legal or equitable interest in the property.

❖ The code places certain responsibilities on the owner in the construction and maintenance of a building. The owner is considered the individual, firm or corporation who, under applicable laws, has a legal interest in the property.

PAN FLASHING. Corrosion-resistant flashing at the base of an opening that is integrated into the building exterior wall to direct water to the exterior and is premanufactured, fabricated, formed or applied at the job site.

❖ Pan flashing is the code-prescribed method of flashing for fenestration. It is required if there are no manufacturer's instructions for any other type of flashing (see Section 703.8).

PANEL THICKNESS. Thickness of core plus two layers of structural wood panel facings.

❖ Structural insulated panels (SIPs) have been in use in building construction for some time, but are relatively new. This term is defined to facilitate requirements for SIPs.

PELLET FUEL-BURNING APPLIANCE. A closed combustion, vented *appliance* equipped with a fuel feed mechanism for burning processed pellets of solid fuel of a specified size and composition.

❖ This is an appliance that uses solid fuel in the form of pellets in a closed combustion system. The pellets consist of a blend of wood waste products (sawdust and chips) mixed with resin (binders) and compressed into small pellets. The pellets are fed into the appliance in a regulated manner to provide consistent burning.

PELLET VENT. A vent *listed* and *labeled* for use with a *listed* pellet fuel-burning *appliance*.

❖ See the commentary for the definition for "Vent."

PERFORMANCE CATEGORY. A designation of wood structural panels as related to the panel performance used in Chapters 4, 5, 6 and 8.

❖ This is a nomenclature that reflects the newest versions of National Standards DOC PS 1 and PS 2 for wood structural panels. Wood structural panels are required to be in conformance to DOC PS 1 and PS 2 in the code. The DOC PS 1 and PS 2 consensus standards now include the terminology of "performance category" to reference the thicknesses tolerance consistent with the nominal panel thicknesses in the IBC.

PERMIT. An official document or certificate issued by the authority having *jurisdiction* that authorizes performance of a specified activity.

❖ The department of building safety is authorized to require a permit according to Section R105. The permit is the official document issued by the jurisdiction authorizing construction-related activity to take place.

PERSON. An individual, heirs, executors, administrators or assigns, and also includes a firm, partnership or corporation, its or their successors or assigns, or the agent of any of the aforesaid.

❖ As used in the IRC, a person is not only an individual but also any heirs, executors, administrators, firm, partnership or corporation, including its or their successors or assigns. An agent for any person as described is also a "person."

PHOTOVOLTAIC MODULES/SHINGLES. A roof covering composed of flat-plate photovoltaic modules fabricated into shingles.

❖ New technologies for generation of electricity from solar power include photovoltaic modules or shingles (see Section 905.16).

PITCH. See "Slope."

❖ The pitch of a pipe run is the fall of the pipe in relationship to a horizontal plane. It is also known as the slope and is measured as a ratio between change in horizontal units and vertical units along a length of pipe.

PLATFORM CONSTRUCTION. A method of construction by which floor framing bears on load bearing walls that are not continuous through the *story* levels or floor framing.

❖ In light-frame construction, platform construction is the technique of stacking floor systems on load-bearing stud walls. As opposed to walls in balloon construction, the load bearing walls in platform construction do not continue beyond each individual story level. Platform construction is by far the most popular framing method in use today.

PLENUM. A chamber that forms part of an air-circulation system other than the *occupied space* being conditioned.

❖ A plenum can also be a room or space that supplies conditioned air to another room or space. Corridors cannot be used as plenums.

PLUMBING. For the purpose of this code, plumbing refers to those installations, repairs, maintenance and *alterations* regulated by Chapters 25 through 33.

❖ This term refers to the collective system of piping, fixtures, fittings, components, devices or appurtenances that transport potable water and liquid and solid wastes associated with cleaning, washing, bathing, food preparation, drinking and the elimination of bodily wastes. Plumbing also includes appliances, equipment and systems such as water conditioners, water heaters, storm drainage systems, water coolers, water filters and waste treatment systems.

The word "plumbing" comes from the Latin word for the element lead (plumbum), because lead was used extensively in the construction of piping systems. The practice of installing piping systems and the materials used in such systems became known as plumbing.

Piping systems such as fuel gas, fuel oil, heating, cooling, air conditioning, lawn irrigation and fire suppression systems are not plumbing, even though plumbers often perform the work necessary to install these systems. Chapters 25 to 32 regulate plumbing.

PLUMBING APPLIANCE. An energized household *appliance* with plumbing connections, such as a dishwasher, food-waste grinder, clothes washer or water heater.

❖ Examples of plumbing appliances include water heaters, hot water dispensers, garbage disposals, dishwashers, clothes washers, water purifiers and water softeners. Typically, an appliance is dependent on a source of energy for its operation.

PLUMBING APPURTENANCE. A device or assembly that is an adjunct to the basic plumbing system and demands no additional water supply nor adds any discharge load to the system. It is presumed that it performs some useful function in the operation, maintenance, servicing, economy or safety of the plumbing system. Examples include filters, relief valves and aerators.

❖ Examples of plumbing appurtenances include water closet seats, hand-held showers, manifolds, backflow preventers, water-hammer arrestors, strainers and filters. An appurtenance by default becomes part of the entire fixture.

PLUMBING FIXTURE. A receptacle or device that is connected to a water supply system or discharges to a drainage system or both. Such receptacles or devices require a supply of water; or discharge liquid waste or liquid-borne solid waste; or require a supply of water and discharge waste to a drainage system.

❖ This definition was changed for the 2012 edition in order to fully encompass the variety of plumbing fixtures that could be installed in a building. The most common type of plumbing fixture has a supply of water connected to it and is connected to the drain system. Examples are water closets, lavatories and bathtubs. Nonwater-supplied (waterless) urinals, floor sinks and floor drains are fixtures that only require connection to the drain system. An outdoor shower discharging to the ground is an example of a plumbing fixture that is connected only to a supply of water.

PLUMBING SYSTEM. Includes the water supply and distribution pipes, plumbing fixtures, supports and appurtenances; soil, waste and vent pipes; sanitary drains and *building sewers* to an *approved* point of disposal.

❖ The plumbing system includes all piping, fixtures and components that transport potable water and convey liquid and liquid-borne solid wastes (see "Plumbing"). Devices that treat the water prior to its use are also included.

POLLUTION. An impairment of the quality of the potable water to a degree that does not create a hazard to the public health but that does adversely and unreasonably affect the aesthetic qualities of such potable water for domestic use.

❖ Books have been written trying to define the term "pollution." For use in the code, pollution is anything that reduces the quality of the potable water supply so that it is undesirable for consumption or use. A pollutant in the water supply causes the water to look, smell or taste bad, but drinking the water would not be harmful. This does not change the fact that polluted water is still considered to be nonpotable. Pollutants are thus not considered potentially harmful, whereas a contaminant would be considered potentially harmful. Refer to the definition of the term "Contamination" to find additional information and the differences between pollution and contamination.

The methods for determining potable and nonpotable water supplies are very closely regulated by the EPA. For the purposes of protecting a water supply, the code considers a polluted source to represent a low hazard when selecting a backflow preventer. Refer to Table P2902.2 for requirements related to low-hazard situations in a potable water supply.

PORTABLE-FUEL-CELL APPLIANCE. A fuel cell generator of electricity, which is not fixed in place. A portable-fuel-cell *appliance* utilizes a cord and plug connection to a grid-isolated load and has an integral fuel supply.

❖ These appliances are a smaller version of the stationary fuel cell power plants in Section M1903 of this code. Fuel cells generate electricity from hydrocarbon fuels and, when pure hydrogen is used, release water and heat as the only by-products. Portable fuel cell appliances are generally fueled by gas tanks or some other local source and are connected to the house current or are used only to charge the batteries of electric cars. Connecting the appliance to the local electrical power grid or hard-piping fuel to the appliance would constitute a permanent installation and require permitting in accordance with Section 105.

POSITIVE ROOF DRAINAGE. The drainage condition in which consideration has been made for all loading deflections of the roof deck, and additional slope has been provided to ensure drainage of the roof within 48 hours of precipitation.

❖ Unless a roof is designed to support a specified amount of water, adequate drainage must be provided to approved locations. In considering the appropriate drainage pattern, all of the possible roof deck deflections should be included. The code mandates that the water be removed from the roof within a period of 48 hours from the time of deposit.

POTABLE WATER. Water free from impurities present in amounts sufficient to cause disease or harmful physiological effects and conforming in bacteriological and chemical quality to the requirements of the public health authority having *jurisdiction*.

❖ In the United States, the EPA Clean Drinking Water-Act defines the quality requirements for potable water. "Potable" means fit to drink.

PRECAST CONCRETE. A structural concrete element cast elsewhere than its final position in the structure.

❖ Precast concrete members may be cast on site and subsequently lifted into place (e.g., tilt-up wall construction) or cast off site, usually in a plant out of the outside weather and transported to the construction site when needed. Precast concrete which has been manufactured under plant conditions has closer controlled tolerances than cast-in-place concrete construction. This results in the need for lesser minimum concrete cover for reinforcement versus that needed for cast-in-place concrete (see ACI 318, Sections 7.7.1 and 7.7.2).

PRECAST CONCRETE FOUNDATION WALLS. Preengineered, precast concrete wall panels that are designed to withstand specified stresses and used to build below-*grade* foundations.

❖ Foundation walls are required to carry specific loads. Often, the foundation wall can be constructed of precast concrete. Precast concrete foundation walls therefore are addressed specifically in several areas of the code.

PRESSURE-RELIEF VALVE. A pressure-actuated valve held closed by a spring or other means and designed to automatically relieve pressure at the pressure at which it is set.

❖ This valve is designed to be actuated by excessive pressures for the purpose of preventing the rupture or explosion of tanks, vessels or piping. The degree of valve opening is directly proportional to the pressure acting on the valve disk.

PUBLIC SEWER. A common sewer directly controlled by public authority.

❖ A public sewer is the piping system that receives the discharge from building sewers and conveys it to treatment facilities.

PUBLIC WATER MAIN. A water-supply pipe for public use controlled by public authority.

❖ A water main is owned and operated by municipalities, rural water districts, privately owned water purveyors and other such entities.

PUBLIC WAY. Any street, alley or other parcel of land open to the outside air leading to a public street, which has been deeded, dedicated or otherwise permanently appropriated to the public for public use and that has a clear width and height of not less than 10 feet (3048 mm).

❖ A public way is essentially a street, alley or any parcel of land that is permanently appropriated to the public for public use. Therefore, the public's right to use such a parcel of land is guaranteed. A public way is open to the exterior and available for use as an egress path if necessary. To qualify as a public way, the space must be at least 10 feet (3048 mm) wide and must have a clear height of at least 10 feet (3048 mm).

PURGE. To clear of air, gas or other foreign substances.

❖ This term applies to a process used for clearing piping systems of impurities, making the system pure prior to the introduction of a desired substance. Although the practice of flushing a system to remove debris is sometimes referred to as purging the system, true purging involves a specific process of forcing a product into a system until impurities are completely removed. One such method would be the introduction of nitrous gas within a copper system serving a medical gas installation. Such a purge is required to clear the system of impurities that may have been left during installation.

QUICK-CLOSING VALVE. A valve or faucet that closes automatically when released manually or controlled by mechanical means for fast-action closing.

❖ A quick-closing valve is any type of solenoid-actuated valve; spring-loaded, self-closing faucets; or any other device capable of instantaneously reducing water flow from full flow to no flow. This rapid change from a full-flow to a no-flow condition can cause water-hammer-related damage. Although most faucets can be manually closed fast enough to produce water hammer, the code does not designate manually closed valves or faucets as "quick closing."

Typical quick-closing valves include electrically actuated valves such as those found in dishwashing machines, clothes washing machines and boiler makeup-water feeders. Some self-closing valves and faucets are not quick closing. For example, typical metering faucets and typical flushometer valves are slowly, automatically closed by a diaphragm/bleed orifice or pilot valve assembly. Also, a typical float-actuated valve is automatic yet slow in closing.

***R*-VALUE, THERMAL RESISTANCE.** The inverse of the time rate of heat flow through a *building thermal envelope* element from one of its bounding surfaces to the other for a unit temperature difference between the two surfaces, under steady state conditions, per unit area (h · ft^2 · °F/Btu).

❖ The *R*-value is a unit of thermal resistance used for comparing insulating values of different materials. The higher the *R*-value of a material, the greater its insulating properties and the slower heat flows through it. *R*-values are used to rate roof/ceiling, wall and floor insulation.

RAMP. A walking surface that has a running slope steeper than 1 unit vertical in 20 units horizontal (5-percent slope).

❖ A pedestrian travel surface that has a slope of 1:20 (5 percent) or less is a walking surface. A surface that is steeper than 1 unit vertical in 20 units horizontal is a ramp. Because steeply sloping walking surfaces present a considerable hazard to users, ramps are regulated for maximum slope, handrails and landings by Section R313.

RECEPTOR. A fixture or device that receives the discharge from indirect waste pipes.

❖ This term applies to a fixture used for collecting drainage from fixtures that are not directly connected to the drainage system. A floor sink is the most common type of receptor. However, a laundry sink that is used for collecting the drainage from a clothes washer could be defined as a receptor.

REFRIGERANT. A substance used to produce refrigeration by its expansion or evaporation.

❖ The refrigerant is the working fluid in refrigeration and air-conditioning systems. In vapor-refrigeration cycles, refrigerants absorb heat from the load side at the evaporator and reject heat at the condenser. The selection of a refrigerant must take into consideration suitable thermodynamic properties, chemical stability, flammability, toxicity and environmental compatibility. Refrigeration is a result of the physical laws of vaporization (evaporation) of liquids. Basically, evaporation of liquid refrigerant is an endothermic process, and condensing of vapors is an exothermic process.

REFRIGERANT COMPRESSOR. A specific machine, with or without accessories, for compressing a given refrigerant vapor.

❖ A compressor is the heart of mechanical refrigeration systems. It is used in a vapor-refrigeration cycle to raise the pressure and enthalpy of the refrigerant into the superheated vapor state, at which point the refrigerant vapor enters the condenser and transfers heat energy to a cooler medium.

REFRIGERATING SYSTEM. A combination of interconnected parts forming a closed circuit in which refrigerant is circulated for the purpose of extracting, then rejecting, heat. A direct refrigerating system is one in which the evaporator or condenser of the refrigerating system is in direct contact with the air or other substances to be cooled or heated. An indirect refrigerating system is one in which a secondary coolant cooled or heated by the refrigerating system is circulated to the air or other substance to be cooled or heated.

❖ Such systems include at minimum a pressure-imposing element or generator, an evaporator, a condenser and interconnecting piping. A single piece of equipment can contain multiple refrigeration systems (circuits).

REGISTERED DESIGN PROFESSIONAL. An individual who is registered or licensed to practice their respective design profession as defined by the statutory requirements of the professional registration laws of the state or *jurisdiction* in which the project is to be constructed.

❖ All states have imposed some degree of registration or licensing requirements on design professionals, such as architects and engineers, who practice in their state. An individual who meets the statutory requirements in the laws of the state, and possibly even an individual jurisdiction, is a registered design professional.

RELIEF VALVE, VACUUM. A device to prevent excessive buildup of vacuum in a pressure vessel.

❖ This device will automatically open to allow air into a vessel or system in the event of backsiphonage. Such a device is used where a water heater without an antisiphon drip tube is installed above the fixtures served. Where the water supply to the water heater is closed and the lower fixtures opened, the vacuum relief valve will open, thus breaking the siphoning effect.

REPAIR. The reconstruction or renewal of any part of an existing building for the purpose of its maintenance. For definition applicable in Chapter 11, see Section N1101.9

❖ When an existing building undergoes some form of remodel, reconstruction or renewal, the work is defined as a repair. A repair maintains a portion of the building in safe and sound working order, without reducing the intended level of protection or structural safety. Major repair work may be subject to a permit where required by Section R105.

REROOFING. The process of recovering or replacing an existing roof covering. See "Roof recover."

❖ The description of reroofing includes two different processes: (1) recovering an existing roof covering, and (2) replacing an existing roof covering. Recovering involves the installation of an additional roof covering over an existing roof covering without the removal of the existing roof covering. Replacement addresses those cases where the existing roof covering is completely removed prior to installation of the new roof covering.

RETURN AIR. Air removed from an *approved conditioned space* or location and recirculated or exhausted.

❖ Return air is air that is being returned to the air handler. Only air in excess of the required ventilation air can be recirculated.

RIDGE. With respect to topographic wind effects, an elongated crest of a hill characterized by strong relief in two directions.

❖ This is a term to describe a topographical area where wind speed-up can occur. In cases where there is known to be structural damage due to this type of wind speed-up, the code requires consideration of topographical wind effects. See Section 301.2.1.5.

RISER.

1. The vertical component of a *step* or *stair*.
2. A water pipe that extends vertically one full *story* or more to convey water to branches or to a group of fixtures.

❖ For Item 1, see the commentary to Section 311.7.5. For Item 2, see the commentary to "Water pipe, riser."

ROOF ASSEMBLY. A system designed to provide weather protection and resistance to design loads. The system consists of a roof covering and roof deck or a single component serving as both the roof covering and the roof deck. A roof assembly includes the roof deck, vapor retarder, substrate or thermal barrier, insulation, vapor retarder, and roof covering.

❖ A roof assembly may be either a single component serving as both the roof covering and the deck or, as typically occurs, a combination of individual roof deck and roof covering components used together to form a complete assembly. A roof assembly may include the roof deck, vapor retarder, substrate, thermal barrier, insulation and roof covering.

ROOF COVERING. The covering applied to the roof deck for weather resistance, fire classification or appearance.

❖ Roof covering provides a building with weather protection, fire retardancy or decoration.

ROOF COVERING SYSTEM. See "Roof assembly."

❖ "Roof covering system" is another term for "Roof assembly."

ROOF DECK. The flat or sloped surface not including its supporting members or vertical supports.

❖ Often composed of wood structural panels, a roof deck is the flat or sloped surface on which roof covering is typically attached. Roof decks assist in both the weather protection and structural integrity of the building. The rafters and associated structural roof members are not a part of the roof deck, nor is the insulation or roof covering that is placed on the deck.

ROOF RECOVER. The process of installing an additional roof covering over a prepared existing roof covering without removing the existing roof covering.

❖ Under the provisions of Section R907, it is acceptable under specific conditions to install new roof covering over an existing roof without first removing the old roof covering. In a roof recover, the existing roof covering must be prepared in an approved manner prior to installation of the new roof covering.

ROOF REPAIR. Reconstruction or renewal of any part of an existing roof for the purposes of its maintenance.

❖ A roof repair occurs when only minor reconstruction or renewal work to an existing roof is necessary. The extent of a roof repair is limited by Section R907.1.

ROOFTOP STRUCTURE. An enclosed structure on or above the roof of any part of a building.

❖ Often used to house mechanical equipment or installed for decorative purposes, a rooftop structure is any enclosed structure on or above the roof. A rooftop structure must be adequately supported and comply with the construction provisions applicable to the remainder of the building.

ROOM HEATER. A freestanding heating *appliance* installed in the space being heated and not connected to ducts.

❖ Room heaters are space-heating appliances that are not connected to ducts. They are also called "unvented room heaters."

ROUGH-IN. The installation of all parts of the plumbing system that must be completed prior to the installation of fixtures. This includes DWV, water supply and built-in fixture supports.

❖ Rough-in refers to that stage of construction prior to the installation of any building materials that would conceal plumbing installations in the building structure. This term generally refers to the installation of drain, waste, vent and water supply piping that will be concealed in floor, ceiling and wall cavities or concealed underground or under slab. At this stage of construction, the plumbing systems are tested for leaks and the entire plumbing "rough-in" is inspected for code compliance.

RUNNING BOND. The placement of masonry units such that head joints in successive courses are horizontally offset at least one-quarter the unit length.

❖ A common method of placing masonry units, running bond describes the type of masonry construction where the units are horizontally offset on each successive course. The length of the offset must be at least one-fourth the length of the masonry unit, otherwise it is considered as stack bond. Commentary Figure R202(14) shows an example of running bond.

SANITARY SEWER. A sewer that carries sewage and excludes storm, surface and groundwater.

❖ A sanitary sewer extends from the sanitary building drain to the public sewer system or to a private sewage disposal system. Most jurisdictions arbitrarily set a distance such as 3 or 5 feet (914 or 1524 mm) outside of the building foundation as the point at which the building drain becomes the building sewer.

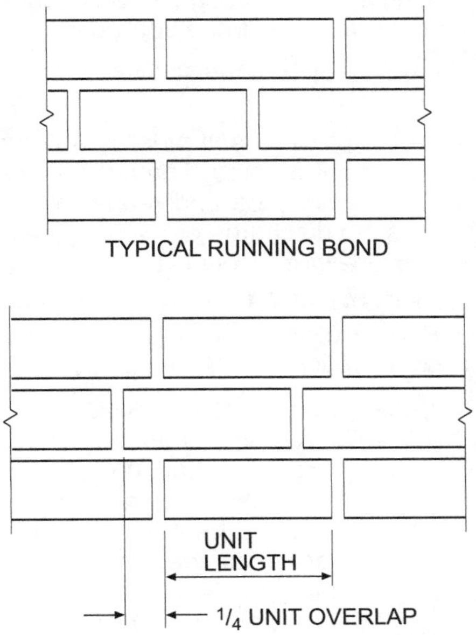

Figure R202(14)
RUNNING BOND MASONRY

SCUPPER. An opening in a wall or parapet that allows water to drain from a roof.

❖ Various design methods are used to drain water from a roof. A common approach is to slope the roof toward one or more exterior walls and provide drainage over the edge of the roof, often into a gutter and downspout system. Another method is the use of roof drains that capture water from numerous low points of the roof and use interior drain lines to remove the water. A third procedure is the use of scuppers, which are openings in the exterior wall surface that allow water to flow off the roof.

Where scuppers are used, the code specifies their size and height above the roof. Scuppers are often included as a portion of the overflow or emergency drain system, a system required by the code and designed to shed water when the primary roof drains are blocked.

SEISMIC DESIGN CATEGORY (SDC). A classification assigned to a structure based on its occupancy category and the severity of the design earthquake ground motion at the site.

❖ The structural provisions of the IRC vary based on several factors, one of which is the probability of an earthquake causing damage to the building. The term used by the code to assign a rating or level of seismic risk is "Seismic design category." Classification into one of the categories is based on two factors: (1) the Seismic Group, and (2) the severity of the ground motion at the site. The Seismic Group is the classification assigned based upon use of the structure. Residential construction is considered in Seismic Group I, with Groups II and III assigned to structures whose loss would create a substantial public hazard or those structures considered as essential facilities. The ground-motion criterion used in the IRC is based on Short Period Design Spectral Response Accelerations (0.2 second). Sections 1613 through 1623 of the IBC contain further explanation of the seismic considerations.

SEPTIC TANK. A water-tight receptor that receives the discharge of a building sanitary drainage system and is constructed so as to separate solids from the liquid, digest organic matter through a period of detention, and allow the liquids to discharge into the soil outside of the tank through a system of open joint or perforated piping or a seepage pit.

❖ This type of tank is typically used in private sewage disposal systems. See Appendix I.

SEWAGE. Any liquid waste containing animal matter, vegetable matter or other impurity in suspension or solution.

❖ This term refers to the discharge from all plumbing fixtures, which includes human bodily wastes and the wastes associated with cleaning, washing, bathing and food preparation.

SEWAGE PUMP. A permanently installed mechanical device for removing sewage or liquid waste from a sump.

❖ Sewage pumps are designed to transport liquid/solid mixtures without clogging and are typically equipped with large suction and discharging ports.

SHALL. The term, when used in the code, is construed as mandatory.

❖ The term "shall" means that the requirement is a mandatory provision. Where a provision is allowed but not required, the term "is permitted" generally appears in the code.

SHEAR WALL. A general term for walls that are designed and constructed to resist racking from seismic and wind by use of masonry, concrete, cold-formed steel or wood framing in accordance with Chapter 6 of this code and the associated limitations in Section R301.2 of this code.

❖ Shear walls are walls designed to carry lateral loads such as seismic or wind loads. Shear walls may be of a variety of materials; however, they must be designed in the specific manner required by the code. Shear walls are typically used in conjunction with diaphragms to provide a load path for transmitting lateral loads through the structure.

SIDE VENT. A vent connecting to the drain pipe through a fitting at an angle less than 45 degrees (0.79 rad) to the horizontal.

❖ A typical application of a side vent would be where the drain from a water closet flange extends downward vertically with the vent for the water closet, connecting to the vertical drain and extending horizontally, perpendicular to the drain.

SINGLE PLY MEMBRANE. A roofing membrane that is field applied using one layer of membrane material (either homogeneous or composite) rather than multiple layers.

❖ In contrast to a built-up roof, a single-ply membrane roof system consists of one layer of membrane material. Chapter 9 addresses both thermoset and thermoplastic single-ply roofing systems.

SINGLE STATION SMOKE ALARM. An assembly incorporating the detector, control *equipment* and alarm sounding device in one unit that is operated from a power supply either in the unit or obtained at the point of installation.

❖ A single-station smoke alarm contains all components for operation of the device, including the power supply in battery-operated units. For a hard-wired unit, the power is available directly at the alarm device. Where two or more single-station smoke alarms are interconnected, they are multiple-station alarms.

SKYLIGHT. See Section N1101.9 for definition applicable in Chapter 11.

SKYLIGHT AND SLOPED GLAZING. See Section R308.6.1.

❖ Any glass or other transparent or translucent glazing installed in the building construction at a slope greater than 15 degrees from the vertical is sloped glazing. Skylights are the most common type of sloped glazing, which may also include solariums, sun rooms, roofs and sloped walls.

SKYLIGHT, UNIT. See Section R308.6.1.

SLEEPING UNIT. See Section N1101.9 for definition applicable in Chapter 11.

SLIP JOINT. A mechanical-type joint used primarily on fixture traps. The joint tightness is obtained by compressing a friction-type washer such as rubber, nylon, neoprene, lead or special packing material against the pipe by the tightening of a (slip) nut.

❖ A slip joint is commonly found in fixture tailpieces and in “P” traps. See the commentary for Section P2704.

SLOPE. The fall (pitch) of a line of pipe in reference to a horizontal plane. In drainage, the slope is expressed as the fall in units vertical per units horizontal (percent) for a length of pipe.

❖ Commonly referred to as pitch, fall or grade, slope is the gradual change in elevation of horizontal piping. Slope is necessary to cause waste flow by gravity.

SMOKE-DEVELOPED INDEX. A comparative measure, expressed as a dimensionless number, derived from measurements of smoke obscuration versus time for a material tested in accordance with ASTM E 84 or UL 723.

❖ The ASTM E 84 (or UL 723) test standard for flame spread also addresses the potential for smoke development of specific materials. The numerical value assigned to the material being tested, based on the density of the smoke produced during the test, is the smoke-developed rating, which is typically regulated in conjunction with the flame spread index.

SOIL STACK OR PIPE. A pipe that conveys sewage containing fecal material.

❖ In the plumbing trade, “soil pipe” is a common name for cast-iron drainage pipe. The term, however, is not specific to cast-iron pipe and refers to any drainpipe that conveys the discharge of water closets, urinals or any other fixture that receives human waste.

SOLAR HEAT GAIN COEFFICIENT (SHGC). The solar heat gain through a fenestration or glazing assembly relative to the incident solar radiation (Btu/h · ft^2 · °F).

❖ The solar heat gain coefficient (SHGC) measures the ratio of the solar radiation that passes through the window to the total solar radiation that strikes the window. Sunlight, or solar radiation, striking a window contains visible light and heat. As the sunlight strikes the window, a portion of the visible light and heat pass through into the building. A portion of the heat is absorbed by the glass and window sash. Moreover, a portion of the visible light and heat is reflected to the outdoors. The lower the SHGC, the lower the solar radiation that is allowed to pass through the window. For example, a window with an SHGC value of 0.40 allows only 40 percent of the total solar radiation that strikes the window to pass through.

SOLID MASONRY. Load-bearing or nonload-bearing construction using masonry units where the net cross-sectional area of each unit in any plane parallel to the bearing surface is not less than 75 percent of its gross cross-sectional area. Solid masonry units shall conform to ASTM C 55, C 62, C 73, C 145 or C 216.

❖ Solid masonry can consist of either load-bearing or nonload-bearing construction made up of masonry units. Although not required to be totally of solid material, solid masonry must have a net cross-sectional area equal to at least 75 percent of its gross cross-sectional area. The percentage must be maintained in any plane parallel to the bearing surface.

SPLINE. A strip of wood structural panel cut from the same material used for the panel facings, used to connect two structural insulated panels. The strip (spline) fits into a groove cut into the vertical edges of the two structural insulated panels to be joined. Splines are used behind each facing of the structural insulated panels being connected as shown in Figure R613.8.

❖ Structural insulated panels are relatively new to the code. Terminology describing components of the panel are necessary to facilitate provisions to regulate the material.

STACK. Any main vertical DWV line, including offsets, that extends one or more stories as directly as possible to its vent terminal.

❖ Traditionally, the term “stack” has been used to describe vertical piping [90 degrees (1.57 rad) from the horizontal] that extends through one or more stories of a building and that constitutes a main to which branch piping connects. Vertical vent branch piping that extends to the open air or through two or more stories is a stack.

STACK BOND. The placement of masonry units in a bond pattern is such that head joints in successive courses are vertically aligned. For the purpose of this code, requirements for stack bond shall apply to all masonry laid in other than running bond.

❖ Where the head joints in successive courses of masonry construction are not offset horizontally by at least one-fourth of the unit length, the method of placement is a stack bond. Vertical joints directly aligned on each course are typical in stack bond masonry.

STACK VENT. The extension of soil or waste stack above the highest horizontal drain connected.

❖ A stack vent is the dry extension of a soil or waste stack. Generally, a stack vent extends to the open air and can serve as a main vent to which branch vents connect.

STACK VENTING. A method of venting a fixture or fixtures through the soil or waste stack without individual fixture vents.

❖ This is a method of wet venting where a uniformly oversized vertical drain serves as the vent for one or more fixtures. The fixtures being vented are connected to the stack independently. The use of this venting method is limited to waste fixtures. It is also referred to as waste stack venting; see Section P3109.

STAIR. A change in elevation, consisting of one or more risers.

❖ All steps, even a single step, are defined as a stair. This makes the stair requirements applicable to all steps unless specifically exempt in the code.

STAIRWAY. One or more flights of stairs, either interior or exterior, with the necessary landings and platforms connecting them to form a continuous and uninterrupted passage from one level to another within or attached to a building, porch or deck.

❖ It is important to note that this definition characterizes a stairway as connecting one level to another. The term "level" is not to be confused with "story." Steps that connect two levels, one of which is not considered a "story" of the structure, would be considered a stairway. For example, a set of steps between the basement level in an areaway and the outside ground level would be considered a stairway. A series of steps between the floor of a story and a mezzanine within that story would also be considered a stairway.

STANDARD TRUSS. Any construction that does not permit the roof/ceiling insulation to achieve the required *R*-value over the *exterior walls*.

❖ The term "standard truss" describes typical roof truss construction where the insulation cannot achieve the necessary *R*-value mandated for the roof-ceiling assembly.

STATIONARY FUEL CELL POWER PLANT. A self-contained package or factory-matched packages which constitute an automatically-operated assembly of integrated systems for generating useful electrical energy and recoverable thermal energy that is permanently connected and fixed in place.

❖ Fuel cells generate electricity from hydrocarbon fuels and, when pure hydrogen is used, release water and heat as the only by-products (see commentary to Section M1903 for a description of the conversion process). Stationary fuel cell appliances cannot exceed 1000 kW of power output, listed and tested in accordance with ANSI Z21.83 and installed in accordance with NFPA 853 and the manufacturer's instructions. These appliances may be independent of or connected to the local electrical power grid and may be fueled by fuel tanks or permanent piping systems.

STORM SEWER, DRAIN. A pipe used for conveying rainwater, surface water, subsurface water and similar liquid waste.

❖ This is the general term used to describe the piping that conducts rainwater or surface-water waste from structures to the point of disposal.

STORY. That portion of a building included between the upper surface of a floor and the upper surface of the floor or roof next above.

❖ A story is that portion of a building from a floor surface to the floor surface or roof above. In the case of the topmost story, the height of the story is measured from the floor surface to the top of the ceiling joists of an attic. Where a ceiling does not create an attic, such as a cathedral ceiling, the story height is measured to the top of the roof rafters. See "Story above grade."

STORY ABOVE GRADE PLANE. Any *story* having its finished floor surface entirely above *grade plane*, or in which the finished surface of the floor next above is:

1. More than 6 feet (1829 mm) *above grade plane; or*
2. More than 12 feet (3658 mm) above the finished ground level at any point.

❖ The code defines a "Story above grade" as any story having its finished floor surface entirely above grade. However, the critical part of the definition involves whether or not a basement is a story above grade. A level that is a story above grade may be either an inhabited story or unused under-floor space. Two criteria are important to the determination of whether a given floor level is either a story above grade or a basement:

1. If the finished floor above the level under consideration or above the under-floor space is more than 6 feet (1829 mm) above the grade plane as defined in Section R202, the level under consideration is a story above grade.

2. If the finished floor level above the level under consideration or above the under-floor space is more than 12 feet (3658 mm) above the finished ground level at any point, the floor level under consideration or the under-floor space is a story above grade.

Conversely, if the finished floor level above the level under consideration is 6 feet (1829 mm) or less above the grade plane, and is 6 feet (1829 mm) or less above the finished ground level for more than 50 percent of the perimeter and does not exceed 12 feet (3658 mm) at any point, the floor level under consideration is a basement. Or, described a bit differently, a basement is a floor level that does not qualify as a story above grade. Commentary Figure R202(15) illustrates the definition of "story above grade."

STRUCTURAL COMPOSITE LUMBER. Structural members manufactured using wood elements bonded together with exterior adhesives.

Examples of structural composite lumber are:

Laminated veneer lumber (LVL). A composite of wood veneer elements with wood fibers primarily oriented along the length of the member, where the veneer element thicknesses are 0.25 inches (6.4 mm) or less.

Parallel strand lumber (PSL). A composite of wood strand elements with wood fibers primarily oriented along the length of the member, where the least dimension of the wood strand elements is 0.25 inch (6.4 mm) or less and their average lengths are a minimum of 300 times the least dimension of the wood strand elements.

Laminated strand lumber (LSL). A composite of wood strand elements with wood fibers primarily oriented along the length of the member, where the least dimension of the wood strand elements is 0.10 inch (2.54 mm) or less and their average lengths are a minimum of 150 times the least dimension of the wood strand elements.

Oriented strand lumber (OSL). A composite of wood strand elements with wood fibers primarily oriented along the length of the member, where the least dimension of the wood strand elements is 0.10 inch (2.54 mm) or less and

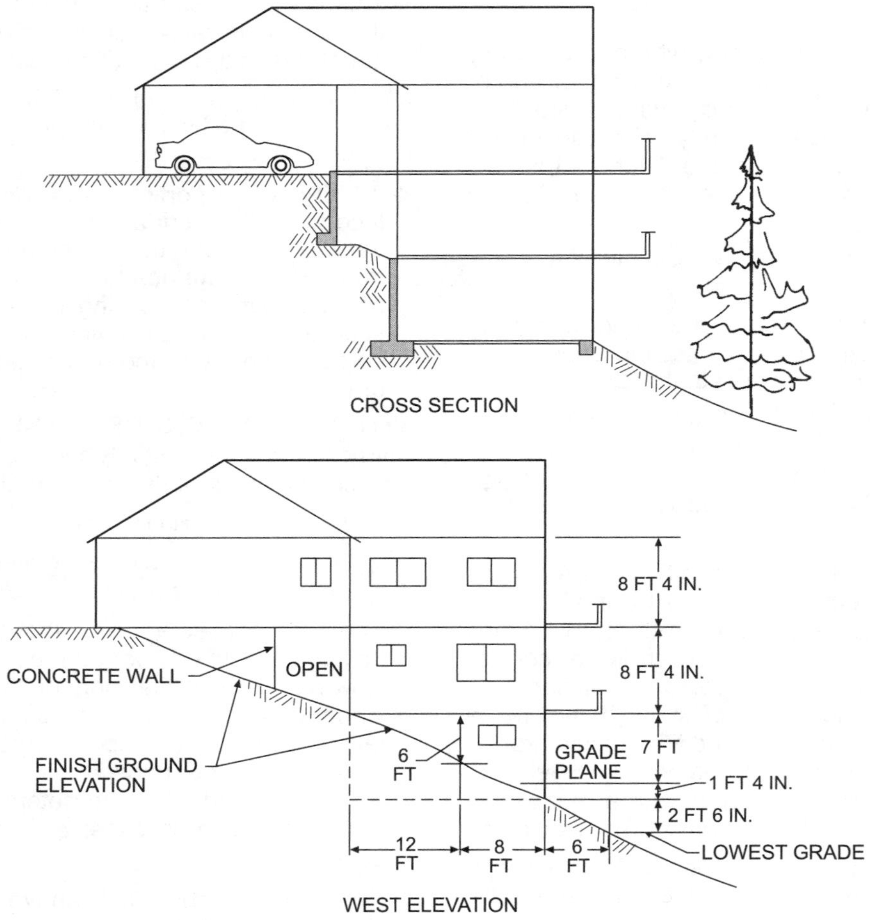

For SI: 1 inch = 25.4 mm, 1 foot = 304.8 mm.

Figure R202(15)
THREE-STORY BUILDING

their average lengths are a minimum of 75 times and less than 150 times the least dimension of the wood strand elements.

❖ ASTM D 5456-09 is the standard by which structural composite lumber is evaluated. This definition is consistent with the types of structural composite lumber defined in ASTM D 5456. These products are being used as beams, headers, long length studs, floor and roof framing and other applications where high strength, long length and/or dimensional stability make the use of structural composite lumber more desirable than sawn lumber.

STRUCTURAL INSULATED PANEL (SIP). A structural sandwich panel that consists of a light-weight foam plastic core securely laminated between two thin, rigid wood structural panel facings.

❖ Structural insulated panels are construction elements composed of solid-core insulation panels enclosed within structural wood-panel membranes. These panels are fabricated at the factory, then brought to the job site and installed.

STRUCTURE. That which is built or constructed.

❖ This definition is intentionally broad so as to include within its scope—and therefore the scope of the code (see Section R101.2)—everything that is built as an improvement to real property.

SUBSOIL DRAIN. A drain that collects subsurface water or seepage water and conveys such water to a place of disposal.

❖ These drains are generally installed adjacent to the foundation footing of a building. They are intended to alleviate problems caused by subsurface (ground) water. Additionally, any other piping that collects either subsurface water or seepage would be termed a subsoil drain as well.

SUMP. A tank or pit that receives sewage or waste, located below the normal *grade* of the gravity system and that must be emptied by mechanical means.

❖ This is the receiving tank or vessel used to collect and store waste from drainage systems that are incapable of draining by gravity. A sump generally houses an ejector or pump used to evacuate the contents. A sump can refer to a receiver for either waste water or storm water.

SUMP PUMP. A pump installed to empty a sump. These pumps are used for removing storm water only. The pump is selected for the specific head and volume of the load and is usually operated by level controllers.

❖ A typical sump pump [see Commentary Figure R202(16)] receives its primary power from the building primary wiring system; however, some are models equipped with a battery or natural-gas-activated backup system. This backup system is helpful during electrical storms or any condition where there is an unexpected loss of power.

SUNROOM. A one-story structure attached to a *dwelling* with a *glazing area* in excess of 40 percent of the gross area of the structure's *exterior walls* and roof. For definition applicable in Chapter 11, see Section N1101.9.

❖ Sunrooms are unique attached rooms that create a space which differs in character from that provided by conventional additions. Sunrooms and other highly glazed structures are sometimes called conservatories or solariums. This definition distinguishes a sunroom from other conventional structures because they are limited to one-story in height and the total glazing area needs to be at least 40 percent of the exterior wall and roof area. This definition is important because Section R303.2 permits a sunroom addition to be considered for light and ventilation. Sunroom additions are discussed in Section R303.8.1. See the commentary to "Thermal isolation."

SUPPLY AIR. Air delivered to a *conditioned space* through ducts or plenums from the heat exchanger of a heating, cooling or ventilating system.

❖ As opposed to return air, supply air is delivered to the conditioned space by an air handler and may or may not be returned to the air handler. Supply air can include ventilation air.

SUPPORTS. Devices for supporting, hanging and securing pipes, fixtures and *equipment*.

❖ Types of supports include hangers, anchors, braces, brackets, strapping and any other material or device used to support plumbing piping and components.

SWEEP. A drainage fitting designed to provide a change in direction of a drain pipe of less than the angle specified by the amount necessary to establish the desired slope of the line. Sweeps provide a longer turning radius than bends and a less turbulent flow pattern (see "Bend" and "Elbow").

❖ See the definition of "Bend."

TEMPERATURE- AND PRESSURE-RELIEF (T AND P) VALVE. A combination relief valve designed to function as both a temperature-relief and pressure-relief valve.

❖ These devices combine the components of pressure-relief and temperature-relief valves and are designed to discharge automatically at excessive pressures or temperatures or both.

TEMPERATURE-RELIEF VALVE. A temperature-actuated valve designed to discharge automatically at the temperature at which it is set.

❖ This valve is designed to discharge at excessive temperatures to prevent dangerously high water temperatures that can cause tank, vessel or pipe explosions.

TERMITE-RESISTANT MATERIAL. Pressure-preservative treated wood in accordance with the AWPA standards in Section R318.1, naturally durable termite-resistant wood, steel, concrete, masonry or other *approved* material.

❖ Section 318.1 allows for termite-resistant materials as protection against termites. Typical materials include steel, concrete and masonry.

THERMAL ISOLATION. Physical and space conditioning separation from *conditioned space(s)* consisting of existing or new walls, doors and/or windows. The *conditioned space(s)* shall be controlled as separate zones for heating and cooling or conditioned by separate *equipment*. For definition applicable in Chapter 11, see Section N1101.9.

❖ This term is conceptually somewhat similar to the separation provided by the "building envelope" but instead of being between conditioned space and the exterior or conditioned space and unconditioned space, this separation occurs between two conditioned spaces, the separation between the existing dwelling and the new sunroom addition. Unless a new wall is constructed, the existing "exterior wall" of the dwelling will generally provide the thermal isolation between the dwelling and the sunroom.

THERMAL RESISTANCE, *R*-VALUE. The inverse of the time rate of heat flow through a body from one of its bounding surfaces to the other for a unit temperature difference between the two surfaces, under steady state conditions, per unit area ($h \cdot ft^2 \cdot °F/Btu$) ($m^2 \cdot K$)/W.

❖ *R*-value is a unit of thermal resistance used for comparing insulating values of different materials. The greater the *R*-value of a material, the greater its insulating properties and the slower heat flows through it. *R*-values are used to rate roof/ceiling, wall and floor insulation.

THERMAL TRANSMITTANCE, *U*-FACTOR. The coefficient of heat transmission (air to air) through a building envelope component or assembly, equal to the time rate of heat flow per unit area and unit temperature difference between the warm side and cold side air films ($Btu/h \cdot ft^2 \cdot °F$) $W/(m^2 \cdot K)$.

❖ *U*-factor is the rate of heat transfer in Btu per hour through a square foot of a surface when the difference between the air temperatures on either side is one Fahrenheit degree. The *U*-factor is the reciprocal of the *R*-value. Vertical windows, skylights and doors are rated with *U*-factors. The lower the *U*-factor, the more thermally efficient the material or assembly.

THIRD-PARTY CERTIFICATION AGENCY. An approved agency operating a product or material certification system that incorporates initial product testing, assessment and surveillance of a manufacturer's quality control system.

❖ The code contains requirements for materials such as pipe, fittings, mechanical equipment and structural items that require either testing by a third-party agency or certification by a third-party agency. A "third party" is an agency independent of the product manufacturer.

THIRD PARTY CERTIFIED. Certification obtained by the manufacturer indicating that the function and performance characteristics of a product or material have been determined by testing and ongoing surveillance by an approved third-party certification agency. Assertion of certification is in the form of identification in accordance with the requirements of the third-party certification agency.

❖ The code contains requirements for materials such as pipe, fittings, mechanical equipment and structural items that require either testing by a third-party agency or certification by a third-party agency. A

Figure R202(16)
SUMP PUMP

"third party" is an agency independent of the product manufacturer.

THIRD-PARTY TESTED. Procedure by which an approved testing laboratory provides documentation that a product material or system conforms to specified requirements.

❖ The code contains requirements for materials such as pipe, fittings, mechanical equipment and structural items that require either testing by a third-party agency or certification by a third-party agency. A "third party" is an agency independent of the product manufacturer.

TOWNHOUSE. A single-family *dwelling unit* constructed in a group of three or more attached units in which each unit extends from foundation to roof and with a *yard* or public way on at least two sides.

❖ A configuration of three or more single-family dwellings attached together in a single structure constitutes a townhouse if both of the following conditions exist: (1) each unit extends vertically from the foundation to the roof (townhouses cannot be stacked), and (2) each unit is open to the exterior on at least two sides, providing some degree of independence from other units.

For townhouses to be designed and constructed under the code, they must also comply with the limitations of Section R101.2. Townhouses are limited to three stories, and each unit must have independent egress to the exterior.

TRAP. A fitting, either separate or built into a fixture, that provides a liquid seal to prevent the emission of sewer gases without materially affecting the flow of sewage or waste water through it.

❖ The sole purpose of a trap is to isolate the interior of occupiable spaces from the sanitary drainage and vent system. It traps liquid waste and retains it to form a seal or barrier through which gases and vapors in the drainage and vent system cannot pass under normal operating conditions. Traps are either separate devices or are integral with fixtures. Other than a grease trap or a trap that is integral with a fixture, the only type of trap permitted by the code is the "P" type. See Commentary Figure R202(17) and "Fixture drain."

TRAP ARM. That portion of a *fixture drain* between a trap weir and the vent fitting.

❖ This is the common name for a "fixture drain."

TRAP PRIMER. A device or system of piping to maintain a water seal in a trap, typically installed where infrequent use of the trap would result in evaporation of the trap seal, such as floor drains.

❖ This is a device or system of piping that maintains a water seal in a trap. See Section P3201.2.

TRAP SEAL. The trap seal is the maximum vertical depth of liquid that a trap will retain, measured between the crown weir and the top of the dip of the trap.

❖ A trap seal is the vertical depth of liquid held in the dip of a trap. Trap seals are required to have a minimum depth.

TRIM. Picture molds, chair rails, baseboards, handrails, door and window frames, and similar decorative or protective materials used in fixed applications.

❖ Where limited amounts of decorative or protective materials are installed as a part of the building construction, they are regulated as trim. Examples include chair rails, door and window frames, handrails, baseboards and cove molding.

TRUSS DESIGN DRAWING. The graphic depiction of an individual truss, which describes the design and physical characteristics of the truss.

❖ If a floor or roof truss system is to be used in a building's construction, truss design drawings must be prepared and submitted to the building official prior to truss installation. The truss design drawings should graphically illustrate each specific type of truss to be installed and include a variety of other information as required in the code.

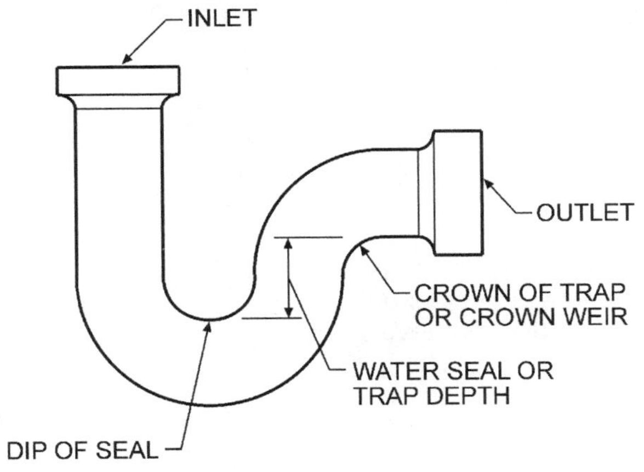

Figure R202(17)
DEPTH OF WATER SEAL

TYPE L VENT. A *listed* and *labeled* vent conforming to UL 641 for venting oil-burning *appliances listed* for use with Type L vents or with gas *appliances listed* for use with Type B vents.

❖ See the commentary to the definition of "Vent."

***U*-FACTOR, THERMAL TRANSMITTANCE.** See Section N1101.9 for definition applicable in Chapter 11.

❖ See "Thermal transmittance, *U*-factor."

UNDERLAYMENT. One or more layers of felt, sheathing paper, nonbituminous saturated felt, or other *approved* material over which a roof covering, with a slope of 2 to 12 (17-percent slope) or greater, is applied.

❖ Underlayment for roof covering is the initial layer of weather protection installed over a roof deck. The underlayment may consist of a single or multiple layers of material and is to be used on roofs having a minimum slope of 2 to 12. Examples of underlayment include felt, sheathing paper and nonbituminous saturated felt.

VACUUM BREAKERS. A device which prevents backsiphonage of water by admitting atmospheric pressure through ports to the discharge side of the device.

❖ A vacuum breaker is a device used as a backflow preventer. The device neutralizes (breaks) the negative pressures (vacuum) in closed systems by allowing atmospheric pressure to enter the system.

VAPOR PERMEABLE. The property of having a moisture vapor permeance rating of 5 perms (2.9×10^{-10} kg/Pa · s · m^2) or greater, when tested in accordance with the desiccant method using Procedure A of ASTM E 96. A vapor permeable material permits the passage of moisture vapor.

❖ A vapor permeable membrane can consist of breather paper or other materials that allow the passage of moisture vapor. An acceptable vapor permeable membrane can be any material that has a vapor permeance of no less than 5 perm ($2.9 \cdot 10^{-10}$ kg/Pa · s · m^2) when tested in accordance with the dessicant method using Procedure A of ASTM E 96.

VAPOR RETARDER CLASS. A measure of the ability of a material or assembly to limit the amount of moisture that passes through that material or assembly. Vapor retarder class shall be defined using the desiccant method with Procedure A of ASTM E 96 as follows:

Class I: 0.1 perm or less

Class II: $0.1 <$ perm ≤ 1.0 perm

Class III: $1.0 <$ perm ≤ 10 perm

❖ The code now contains detail requirements for moisture control in Section R702.7. This section calls out requirements for vapor retarders with different levels of vapor permeance. This defines each class.

VENT. A passageway for conveying flue gases from fuel-fired *appliances*, or their vent connectors, to the outside atmosphere.

❖ In code terminology, vents are distinguished from chimneys and usually are constructed of factory-made listed and labeled components intended to function as a system. Type B and BW vents are constructed of galvanized steel and aluminum sheet metal and are double-wall and air insulated. Such vents are designed to vent gas-fired appliances and equipment that are equipped with draft hoods or specifically listed (labeled) for use with Type B or BW vents. Type L vents are typically constructed of sheet steel and stainless steel. They are double-wall, air insulated and designed to vent gas and oil-fired appliances and equipment. Mid- to high-efficiency appliances are designed for use with corrosion-resistant vents such as those made of plastic pipe and special alloys of stainless steel. Pellet vents (PL) are specialized vents similar in design and construction to Type L vents. Pellet vents must not be used with any appliances other than pellet-burning appliances.

VENT COLLAR. See "Flue collar."

VENT CONNECTOR. That portion of a venting system which connects the flue collar or draft hood of an *appliance* to a vent.

❖ In most cases, appliances are not located directly in line with the vertically rising chimney or vent; therefore, a vent connector is necessary to connect the appliance flue outlet with the vent. Vent connectors can be single- or double-wall pipes and are usually made from galvanized steel, stainless steel or aluminum sheet metal. In many installations, the vent connectors must be constructed of the same material as the vent, as is typically done with Type B vent systems.

VENT DAMPER DEVICE, AUTOMATIC. A device intended for installation in the venting system, in the outlet of an individual, automatically operated fuel burning *appliance* and that is designed to open the venting system automatically when the *appliance* is in operation and to close off the venting system automatically when the *appliance* is in a standby or shutdown condition.

❖ The purpose of this device is the conservation of energy by preventing the cooling of the combustion chamber of the appliance. This is accomplished by preventing the flow of air (by means of connection) through the combustion chamber when the appliance is not in operation.

VENT GASES. Products of combustion from fuel-burning *appliances*, plus excess air and dilution air, in the venting system above the draft hood or draft regulator.

❖ These products of combustion are the result of burning fuel gas.

VENT STACK. A vertical vent pipe installed to provide circulation of air to and from the drainage system and which extends through one or more stories.

❖ This is a dry vent that extends from the base of a waste or soil stack and typically runs parallel and adjacent to the waste or soil stack it serves. The vent stack can connect to the horizontal drain downstream of the drain stack, or it can connect directly to the drainage stack.

VENT SYSTEM. Piping installed to equalize pneumatic pressure in a drainage system to prevent trap seal loss or blow-back due to siphonage or back pressure.

❖ This system is a piping arrangement designed to maintain pressure fluctuation at fixture traps to within plus or minus 2 inches of water column (249 Pa).

VENTILATION. The natural or mechanical process of supplying conditioned or unconditioned air to, or removing such air from, any space. For definition applicable in Chapter 11, see Section N1101.9.

❖ Ventilation can be used for comfort cooling, the control of air contaminants, equipment cooling and replenishing oxygen levels.

VENTING. Removal of combustion products to the outdoors.

❖ This is the desired result of the venting system.

VENTING SYSTEM. A continuous open passageway from the flue collar of an *appliance* to the outside atmosphere for the purpose of removing flue or vent gases. A venting system is usually composed of a vent or a chimney and vent connector, if used, assembled to form the open passageway.

❖ Venting systems operate by gravity (natural draft) or mechanical means. Venting systems include both vents and chimneys.

VERTICAL PIPE. Any pipe or fitting that makes an angle of 45 degrees (0.79 rad) or more with the horizontal.

❖ Vertical pipes are angled 45 degrees (0.79 rad) or more above the horizontal.

VINYL SIDING. A shaped material, made principally from rigid polyvinyl chloride (PVC), that is used to cover exterior walls of buildings.

❖ Vinyl siding is a manufactured plastic product whose primary component is polyvinyl chloride (PVC). Vinyl siding is either co-extruded in layers or thermoformed, depending on the style or profile of the product. Additives are used to improve weathering performance. Vinyl siding products are available in a wide range of sizes, shapes, colors and textures and are intended for exterior wall and related applications.

WALL, RETAINING. A wall not laterally supported at the top, that resists lateral soil load and other imposed loads.

❖ Retaining walls are discussed in Section R404.5 of this code. These walls retain soil that would otherwise slump or cave to a more natural slope. The lateral pressure of soil against a retaining wall is greatly influenced by soil moisture and it is this lateral pressure that constitutes the largest part of the load that the wall must withstand.

WALLS. Walls shall be defined as follows:

Load-bearing wall. A wall supporting any vertical load in addition to its own weight.

Nonbearing wall. A wall which does not support vertical loads other than its own weight.

❖ Walls are either load-bearing or nonbearing. A load-bearing wall supports not only its own weight but also an additional vertical load. An example is a wood stud wall supporting roof rafters and ceiling joists. A nonbearing wall does not support any vertical load other than its own weight.

WASTE. Liquid-borne waste that is free of fecal matter.

❖ Waste is any plumbing fixture discharge other than the discharge from a water closet or urinal. This material does not contain human fecal matter.

WASTE PIPE OR STACK. Piping that conveys only liquid sewage not containing fecal material.

❖ A waste pipe or stack is any waste-conducting pipe not falling under the definition of soil pipe. Waste pipes do not convey human fecal matter.

WATER DISTRIBUTION SYSTEM. Piping which conveys water from the service to the plumbing fixtures, *appliances*, appurtenances, *equipment*, devices or other systems served, including fittings and control valves.

❖ This is all the water piping within a building from the point of connection to the water service to the various connections of the fixture supplies or outlets.

WATER HEATER. Any heating *appliance* or *equipment* that heats potable water and supplies such water to the potable hot water distribution system.

❖ The types of water heaters include storage (tank), circulating and instantaneous. Point-of-use water heaters typically supply hot water to a single fixture or outlet and are located in close proximity to the fixture or outlet.

WATER MAIN. A water supply pipe for public use.

❖ Water mains are owned and operated by municipalities, rural water districts, privately owned water purveyors or other such entities.

WATER OUTLET. A valved discharge opening, including a hose bibb, through which water is removed from the potable water system supplying water to a plumbing fixture or plumbing *appliance* that requires either an air gap or backflow prevention device for protection of the supply system.

❖ A water outlet is a discharge opening through which water is supplied to a fixture, into the atmosphere (except into an open tank that is part of the water supply system), to a boiler or heating system or to any device or equipment requiring water to operate but which is not part of the plumbing system.

WATER-RESISTIVE BARRIER. A material behind an *exterior wall* covering that is intended to resist liquid water that has penetrated behind the exterior covering from further intruding into the *exterior wall* assembly.

❖ Protection of the building envelope is a primary concern. The ability of the water-resistive barrier to provide weather resistance and maintain the integrity of the building envelope is key to controlling water based issues like mold, decay and deterioration of a structure. Water-resistive barriers are discussed in Section R703.2 of the code and in Section 1404.2 of the IBC.

WATER SERVICE PIPE. The outside pipe from the water main or other source of potable water supply to the water distribution system inside the building, terminating at the service valve.

❖ Water-service pipe is any acceptable piping that conveys potable water.

WATER SUPPLY SYSTEM. The water service pipe, the water-distributing pipes and the necessary connecting pipes, fittings, control valves and all appurtenances in or adjacent to the building or premises.

❖ The water supply system includes all piping and components that convey potable water from the public main or other source to the points of water usage.

WET VENT. A vent that also receives the discharge of wastes from other fixtures.

❖ Although Section P3108 describes "wet venting" systems, the term "wet vent" applies to any pipe that is used concurrently as both a drain and a vent. This arrangement can be found in the following systems: Common Vent (Section P3107), Wet Venting (Section P3108), Waste Stack Vent (Section P3109), Circuit Venting (Section P3110), and Combination Waste and Vent System (Section P3111).

WHOLE-HOUSE MECHANICAL VENTILATION SYSTEM. An exhaust system, supply system, or combination thereof that is designed to mechanically exchange indoor air for outdoor air when operating continuously or through a programmed intermittent schedule to satisfy the whole-house ventilation rate. For definition applicable in Chapter 11, see Section N1101.9.

❖ See Section 1507.3.

WIND-BORNE DEBRIS REGION. Areas within *hurricane-prone regions* as designated in accordance with Figure R301.2(4)C.

❖ The code regulates the exterior openings of a building where there is a concern about the extensive damage that can be caused by debris carried through the air by extremely high winds. Wind-borne debris regions are those areas where the potential for such damage is high. The code identifies these areas in Figure R301.2(4)C.

WINDER. A tread with nonparallel edges.

❖ A winder is a tread in a winding stairway. Winder treads are used as components of stairs that change direction, just as straight treads are components in straight stairs. A winder serves as a tread but its shape allows the additional function of a gradual turning of the stairway direction. The tread depth of a winder at the walk line and the minimum tread depth at the narrow end can control the turn made by each winder.

WOOD/PLASTIC COMPOSITE. A composite material made primarily from wood or cellulose-based materials and plastic.

❖ The code now contains requirements for wood/plastic composite materials, which are primarily that they are manufactured to ASTM D 7032 (see Section 317.4). These are a material used in exterior applications that are weather and water resistant.

WOOD STRUCTURAL PANEL. A panel manufactured from veneers; or wood strands or wafers; bonded together with waterproof synthetic resins or other suitable bonding systems. Examples of wood structural panels are plywood, OSB or composite panels.

❖ Wood structural panels are sheet products commonly known as plywood, wafer board, oriented strand board (OSB) or chip board manufactured to meet the United States Department of Commerce (DOC) standards for Construction and Industrial Plywood, PS 1, and Performance Standards for Wood-Based Structural-Use Panels, PS 2.

YARD. An open space, other than a court, unobstructed from the ground to the sky, except where specifically provided by this code, on the *lot* on which a building is situated.

❖ Used throughout the code to describe an open space at the exterior of a building, a yard must be unobstructed from the ground to the sky and located on the same lot on which the building is situated. The openness of a yard allows it to be used to supply buildings with natural light and ventilation. A court, which is bounded on three or more sides by the exterior walls of the building, is not a yard.

Bibliography

The following resource materials were used in the preparation of the commentary for this chapter of the code:

ANSI Z21.83-1998, *Fuel Cell Power Plants.* New York, NY: American National Standards Institute, 1998.

ASTM C 55-03, *Standard Specification for Concrete Brick.* West Conshohocken, PA: ASTM International, 2003.

ASTM C 62-03, *Standard Specification for Building Brick (Solid Masonry Units Made from Clay or Shale).* West Conshohocken, PA: ASTM International, 2003.

ASTM C 73-99a, *Standard Specification for Calcium Silicate Face Brick (Sand Lime Brick).* West Conshohocken, PA: ASTM International, 1999.

ASTM C 90-03, Standard Specification for Load-bearing Concrete Masonry Units. West Conshohocken, PA: ASTM International, 2003.

ASTM C 216-04a, *Standard Specification for Facing Brick (Solid Masonry Units Made from Clay or Shale).* West Conshohocken, PA: ASTM International, 2004.

ASTM D 7032-08, *Standard Specification for Establishing Performance Ratings for Wood-plastic Composite Deck Boards and Guardrail Systems (Guards or Handrails).* West Conshohocken, PA: ASTM International, 2008.

ASTM E 84-04, *Standard Test Method for Surface Burning Characteristics of Building Materials.* West Conshohocken, PA: ASTM International, 2004.

ASTM E 96-00e04, *Standard Test Method for Water Vapor Transmission of Materials.* West Conshohocken, PA: ASTM International, 2004.

ASTM E 136-99e01, *Standard Test Method for Behavior of Materials in a Vertical Tube Furnace at 750°C.* West Conshohocken, PA: ASTM International, 2001.

DOC PS 1-95, *Construction and Industrial Plywood.* Gaithersburg, MD: U.S. Department of Commerce, 1995.

DOC PS 2-93, *Performance Standard for Wood-based Structural-use Panels.* Gaithersburg, MD: U.S. Department of Commerce, 1993.

IBC-12, *International Building Code.* Washington, DC: International Code Council, Inc., 2011.

IFC-12, *International Fire Code.* Washington, DC: International Code Council, Inc., 2011.

IFGC-12 *International Fuel Gas Code.* Washington, DC: International Code Council, Inc. 2011.

IMC-12, *International Mechanical Code.* Washington, DC: International Code Council, Inc., 2011.

IPC-12, *International Plumbing Code.* Washington, DC: International Code Council, Inc., 2011.

NFPA 501-03, *Standard on Manufactured Housing.* Quincy, MA: National Fire Protection Association, 2003.

NFPA 853-03, *Installation of Stationary Fuel Power Power Plants.* Quincy, MA: National Fire Protection Association, 2003.

UL 103-2001, *Chimneys, Factory-built, Residential Type and Building Heating Appliance—with Revision through March 1999.* Northbrook, IL: Underwriters Laboratories Inc., 2001.

UL 641-99, *Type L, Low-temperature Venting Systems—with Revisions through April 1999.* Northbrook, IL: Underwriters Laboratories Inc., 1999.

Part III—Building Planning and Construction

Chapter 3: Building Planning

General Comments

Chapter 3 is a compilation of the code requirements specific to the building planning sector of the design and construction process. The provisions address a wide variety of issues important to designing a building that is both safe and usable. The limitations placed on the materials and methods of construction contribute to the development of a structurally sound building. Snow, wind and seismic design and flood-resistant construction are regulated, as are the live and dead loads, in Chapter 3.

Fire-resistance-rated assemblies are necessary under two different conditions: (1) where a building is situated very close to a property line, the code addresses the concern for radiant heat exposure in a fire; (2) where two or more dwelling units are housed in a single structure, the code mandates a minimum level of fire separation between all units. Other concerns related to fires include the limitations on wall and ceiling finishes, the requirement for emergency escape and rescue openings, the required installation of smoke alarms throughout the dwelling unit and limitations on the use of foam plastics and other insulation materials. In addition, the specific construction requirements for the common wall between the house and garage and the ceiling assembly between the garage and habitable space are addressed in Chapter 3.

This chapter sets forth traditional code requirements dealing with light, ventilation, sanitation, room size, ceiling height and environmental comfort. Life-safety provisions include limitations on glazing used in hazardous areas, specifications on the use of guards at elevated surfaces, fall protection for open windows, and basic rules for the egress system. This chapter also contains most of the regulations found in the code that deal with the planning and design of dwelling units.

- Section R301 establishes the design criteria, including dead loads, live loads, roof loads, floor loads, snow loads, wind loads and seismic loads.
- Section R302 identifies the requirements for the fire-resistant construction for residential buildings.
- Section R303 establishes the light, ventilation and heating requirements for dwelling units.
- Section R304 establishes the minimum requirements for room areas in dwelling units.
- Section R305 establishes the ceiling height requirements for dwelling units.
- Section R306 contains requirements for sanitation.
- Section R307 contains requirements for toilets and bath and shower spaces.
- Section R308 contains requirements for glazing, hazardous locations of glazing, site-built windows and skylights.
- Section R309 contains provisions for garages and carports.
- Section R310 contains provisions for emergency escape and rescue openings.
- Section R311 establishes the means of egress requirements, including provisions for egress doors, hallways, stairways and ramps.
- Section R312 addresses guards and fall prevention for open windows.
- Section R313 provides requirements for an automatic sprinkler system with the option of complying with NFPA13D or Section P2904.
- Section R314 contains the requirements for smoke alarms.
- Section R315 provides criteria for the installation and location of carbon monoxide (CO) alarms.
- Section R316 addresses the use of foam plastic.
- Section R317 contains requirements for decay protection for wood and wood-based products.
- Section R318 contains requirements for termite protection.
- Section R319 provides the requirements for premise identification (site address).
- Section R320 provides a reference to the *International Building Code*® (IBC®) for accessibility requirements.
- Section R321 addresses elevators and platform lifts.
- Section R322 establishes flood-resistant construction provisions.
- Section R323 references ICC/NSSA 500 for when someone chooses to construct a storm shelter within their home or building.

Purpose

Chapter 3 provides guidelines for a minimum level of structural integrity, life safety, fire safety and livability for inhabitants of dwelling units regulated by the code. The chapter sets forth the requirements that affect the most basic planning and design aspects of dwelling construction. It identifies the various structural loads that are imposed on a building, and it establishes criteria that address each of the imposed loads. In the design of residential structures scoped by the code, there are many climatic and geographical issues that must be considered. This chapter provides guidance in the determination of all appropriate design criteria. In addition, it sets forth the limiting conditions under which a building may be designed and constructed using the code.

Fundamental issues of livability and sanitation are satisfied through the regulation of minimum room sizes and ceiling heights, as well as basic requirements for toilet rooms and kitchens. Life safety concerns are addressed in a number of areas, including provisions regulating emergency escape and rescue openings, glazing in areas subject to human impact and exiting. The chapter establishes minimum specifications for a number of different building components, including stairways, ramps, landings, handrails, guards and fall protection for open windows. It deals with fire-safety issues, such as automatic sprinkler systems, early fire detection by smoke alarms, exterior wall protection for proximity to property lines, separation of dwelling units in multiple-family buildings, and control of fire spread across wall and ceiling finishes. Other life safety concerns are dealt with by requirements for CO detectors and guidance for the design of storm shelters. Property protection is also a concern, with provisions established for protection against decay and termites.

SECTION R301 DESIGN CRITERIA

R301.1 Application. Buildings and structures, and all parts thereof, shall be constructed to safely support all loads, including dead loads, live loads, roof loads, flood loads, snow loads, wind loads and seismic loads as prescribed by this code. The construction of buildings and structures in accordance with the provisions of this code shall result in a system that provides a complete load path that meets all requirements for the transfer of all loads from their point of origin through the load-resisting elements to the foundation. Buildings and structures constructed as prescribed by this code are deemed to comply with the requirements of this section.

❖This section specifies the minimum design loads required for structures built in accordance with the provisions of the code. In structural design, loads are generally divided into two categories: gravity loads, which act vertically; and lateral loads, which act horizontally. Lateral loads typically result from either wind (see Section R301.2.1), earthquakes (see Section R301.2.2) or flood loads (see Section R301.2.4). Although wind, flood and earthquake design may concern themselves with lateral loads, there are also vertical force components that should be considered.

All structures must be designed to support these loads and provide a complete load path capable of transferring these loads from their point of origin through the appropriate load-resisting elements and foundation and, ultimately, to the supporting soil. The charging statement specifically states that any building or structure that has been built in strict compliance with the code provides a complete load path that meets all requirements for load transfer from the point of origin to the foundation. A load path that is either incomplete or inadequate will expose the structure to damage just as surely as an undersized structural member will. The concept of a complete load path is a fundamental principle in structural engineering, and the code makes it clear that a complete load path must be provided.

R301.1.1 Alternative provisions. As an alternative to the requirements in Section R301.1 the following standards are permitted subject to the limitations of this code and the limitations therein. Where engineered design is used in conjunction with these standards, the design shall comply with the *International Building Code*.

1. AF&PA *Wood Frame Construction Manual* (WFCM).
2. AISI *Standard for Cold-Formed Steel Framing—Prescriptive Method for One- and Two-Family Dwellings* (AISI S230).
3. ICC *Standard on the Design and Construction of Log Structures* (ICC 400).

❖This section permits the use of alternative prescriptive framing methods. Wood framing is permitted to comply with the provisions of the American Forest and Paper Association's (AF&PA)WFCM, *Wood Frame Construction Manual for One- and Two-family Dwellings.* Cold-formed steel framing is permitted to comply with American Iron and Steel Institute's (AISI) S230, *Standard for Cold-formed Steel Framing-prescriptive Method for One- and Two-family Dwellings.* Log homes can be constructed using ICC 400, *Standard on the Design and Construction of Log Structures.* Engineered design in accordance with the IBC is required when a building is beyond (or exceeds) the applicability limits of these standards.

R301.1.2 Construction systems. The requirements of this code are based on platform and balloon-frame construction for light-frame buildings. The requirements for concrete and masonry buildings are based on a balloon framing system. Other framing systems must have equivalent detailing to ensure force transfer, continuity and compatible deformations.

❖The requirements of the code are based on platform or balloon-frame construction for light-frame buildings (see the definitions of "Platform construction" and

"Light-frame construction" in Chapter 2) and on a balloon-framing system for concrete and masonry buildings.

R301.1.3 Engineered design. When a building of otherwise conventional construction contains structural elements exceeding the limits of Section R301 or otherwise not conforming to this code, these elements shall be designed in accordance with accepted engineering practice. The extent of such design need only demonstrate compliance of nonconventional elements with other applicable provisions and shall be compatible with the performance of the conventional framed system. Engineered design in accordance with the *International Building Code* is permitted for all buildings and structures, and parts thereof, included in the scope of this code.

❖Generally, proper application of the code requires a clear understanding of and adherence to its prescriptive limitations, which are based on conventional construction. However, a building may contain structural elements that are either unconventional or exceed the prescriptive limitations of the code. This is acceptable, if these elements are designed in accordance with accepted engineering practice by a design professional.

R301.2 Climatic and geographic design criteria. Buildings shall be constructed in accordance with the provisions of this code as limited by the provisions of this section. Additional criteria shall be established by the local *jurisdiction* and set forth in Table R301.2(1).

❖This section establishes the design criteria that vary based on location and/or climate. Some of the criteria reflect loading, such as earthquake, flood and wind; others reflect susceptibility to damage from hazards, such as weather exposure or termites. Additional criteria may be established by local jurisdictions as necessary. These would include, for example, whether a site is within a wind-borne debris region as described in Section R301.2.1.2, of this commentary. Table R301.2(1) lists the criteria that must be established within each jurisdiction for any project constructed under the code. The table must be filled in by the jurisdiction adopting the code for their particular area. Note that some of these criteria (e.g., wind exposure category or flood hazard) can vary within a given jurisdiction and may need to be established on a site-by-site (or project-by-project) basis. The table serves as a useful reminder for code enforcement personnel, builders, designers and owners. Verifying this information up front aids compliance with the code.

Table R301.2(1). See below.

❖Table R301.2(1) is designed so that jurisdictions recognize certain climatic and geographic design criteria

TABLE R301.2(1)
CLIMATIC AND GEOGRAPHIC DESIGN CRITERIA

GROUND SNOW LOAD	WIND DESIGN		SEISMIC DESIGN CATEGORY[f]	SUBJECT TO DAMAGE FROM			WINTER DESIGN TEMP[e]	ICE BARRIER UNDERLAYMENT REQUIRED[h]	FLOOD HAZARDS[g]	AIR FREEZING INDEX[i]	MEAN ANNUAL TEMP[j]
	Speed[d] (mph)	Topographic effects[k]		Weathering[a]	Frost line depth[b]	Termite[c]					

For SI: 1 pound per square foot = 0.0479 kPa, 1 mile per hour = 0.447 m/s.

a. Weathering may require a higher strength concrete or grade of masonry than necessary to satisfy the structural requirements of this code. The weathering column shall be filled in with the weathering index (i.e., "negligible," "moderate" or "severe") for concrete as determined from the Weathering Probability Map [Figure R301.2(3)]. The grade of masonry units shall be determined from ASTM C 34, C 55, C 62, C 73, C 90, C 129, C 145, C 216 or C 652.

b. The frost line depth may require deeper footings than indicated in Figure R403.1(1). The jurisdiction shall fill in the frost line depth column with the minimum depth of footing below finish grade.

c. The jurisdiction shall fill in this part of the table to indicate the need for protection depending on whether there has been a history of local subterranean termite damage.

d. The jurisdiction shall fill in this part of the table with the wind speed from the basic wind speed map [Figure R301.2(4)A]. Wind exposure category shall be determined on a site-specific basis in accordance with Section R301.2.1.4.

e. The outdoor design dry-bulb temperature shall be selected from the columns of 971/2-percent values for winter from Appendix D of the *International Plumbing Code*. Deviations from the Appendix D temperatures shall be permitted to reflect local climates or local weather experience as determined by the building official.

f. The jurisdiction shall fill in this part of the table with the seismic design category determined from Section R301.2.2.1.

g. The jurisdiction shall fill in this part of the table with (a) the date of the jurisdiction's entry into the National Flood Insurance Program (date of adoption of the first code or ordinance for management of flood hazard areas), (b) the date(s) of the Flood Insurance Study and (c) the panel numbers and dates of all currently effective FIRMs and FBFMs or other flood hazard map adopted by the authority having jurisdiction, as amended.

h. In accordance with Sections R905.2.7.1, R905.4.3.1, R905.5.3.1, R905.6.3.1, R905.7.3.1 and R905.8.3.1, where there has been a history of local damage from the effects of ice damming, the jurisdiction shall fill in this part of the table with "YES." Otherwise, the jurisdiction shall fill in this part of the table with "NO."

i. The jurisdiction shall fill in this part of the table with the 100-year return period air freezing index (BF-days) from Figure R403.3(2) or from the 100-year (99 percent) value on the National Climatic Data Center data table "Air Freezing Index-USA Method (Base 32°F)" at www.ncdc.noaa.gov/fpsf.html.

j. The jurisdiction shall fill in this part of the table with the mean annual temperature from the National Climatic Data Center data table "Air Freezing Index-USA Method (Base 32°F)" at www.ncdc.noaa.gov/fpsf.html.

k. In accordance with Section R301.2.1.5, where there is local historical data documenting structural damage to buildings due to topographic wind speed-up effects, the jurisdiction shall fill in this part of the table with "YES." Otherwise, the jurisdiction shall indicate "NO" in this part of the table.

that vary from location to location. Communities are directed to complete the table with a variety of factors. See the table footnotes for the sources of the information to be determined by the local jurisdiction to complete the table.

The required information includes the date of the jurisdictions entry into the National Flood Insurance Program (NFIP), which is the date of adoption of the first code or ordinance for management of flood hazard areas. With respect to the official map that shows flood hazard areas, the community inserts the date of the currently effective Flood Insurance Study and the panel numbers and dates of all currently effective Flood Insurance Rate Maps (FIRM) (and Flood Boundary and Floodway Maps, if applicable) or the date of other maps that are adopted. Another flood hazard map may be specified if it shows flood hazard areas that are larger than those shown on the FIRM, as may be the case if a community elects to define its flood plains based on higher standards, such as the "flood of record," "ultimate development" or "no-rise" rules to define the floodway.

From time to time, the Federal Emergency Management Agency's (FEMA) Flood Insurance Rate Maps (FIRM), Flood Insurance Studies (FIS) may be revised and republished. In recent years, revised FIRMs have been produced in a digital format for some communities (referred to as DFIRM). Communities that prefer to cite the digital data should obtain a legal opinion. DFIRMs are registered to the primary coordinate system of the state or community. FEMA advises that the horizontal location of flood hazard areas relative to specific sites should be determined using the coordinate grid, rather than planimetric base map features, such as streets.

When maps are revised and flood hazard areas are changed, FEMA involves the community and provides a formal opportunity to review the documents. Once the revisions are finalized, FEMA requires adoption of the new maps by the community. Communities may be able to minimize having to adopt each revision by referencing the date of the map and study, as amended. This is a method by which subsequent revisions to flood maps and studies may be adopted administratively without requiring legislative action on the part of the community. Communities will need to determine whether this adoption-by-reference approach is allowed under their state's enabling authority and due process requirements. If not allowed, communities are to follow their state's requirements, which normally require public notices, hearings and specific adoption of revised maps by the community's legislative body.

Table R301.2(1) requires the local municipality or code user to insert a frost line depth entry for the particular geographical area they are located in. Commentary Figure R301.2(8) (see page 3-21) provides some guidance to do this more accurately based on the US Weather Bureau information. Since elevation above sea level may increase frost depth, another local resource for frost depth is often the local graveyard owners.

In Table R301.2(1), under the title of Wind Design, Topographic effects, a jurisdiction will enter "yes" or "no." This is in consideration of historical information indicating unusually high wind speeds due to local topography (see commentary, Section R301.2.1.5).

R301.2.1 Wind design criteria. Buildings and portions thereof shall be constructed in accordance with the wind provisions of this code using the basic wind speed in Table R301.2(1) as determined from Figure R301.2(4)A. The structural provisions of this code for wind loads are not permitted where wind design is required as specified in Section R301.2.1.1. Where different construction methods and structural materials are used for various portions of a building, the applicable requirements of this section for each portion shall apply. Where not otherwise specified, the wind loads listed in Table R301.2(2) adjusted for height and exposure using Table R301.2(3) shall be used to determine design load performance requirements for wall coverings, curtain walls, roof coverings, exterior windows, skylights, garage doors and exterior doors. Asphalt shingles shall be designed for wind speeds in accordance with Section R905.2.4. A continuous load path shall be provided to transmit the applicable uplift forces in Section R802.11.1 from the roof assembly to the foundation.

❖Buildings must be capable of withstanding the wind loads based on the wind speed specified in Table R301.2(1). Wind speeds used for design are determined from Figure R301.2(4)A. The structural provisions in the code do not apply where the basic wind speed equals or exceeds 110 miles per hour as shown in Figure R301.2(4)A or where wind design is required in accordance with the code because the prescriptive provisions and tables are based on loads determined for basic wind speeds less than 110 miles per hour. Wall coverings, curtain walls, roof coverings, exterior windows, skylights, garage doors and exterior doors must be capable of withstanding the component and cladding wind pressures of Table R301.2(2) adjusted by the height and exposure coefficients given in Table R301.2(3).

Section R905.2.6 addresses the attachment details for asphalt shingles. Roofs with higher slopes or in areas subject to higher wind speeds may require special methods of attachment (see commentary, Section R905.2.6).

Wind loads are a major consideration in designing a structure's lateral-force-resisting system. See Commentary Figure R301.2.1 for a schematic representation of the lateral component of wind loading on a building. Wind loads affect more than the lateral load system as evidenced by provisions such as roof tie-downs in Section R802.11. A continuous load path must be provided to transmit the roof uplift forces to the foundation.

TABLE R301.2(2) COMPONENT AND CLADDING LOADS FOR A BUILDING WITH A MEAN ROOF HEIGHT OF 30 FEET LOCATED IN EXPOSURE B (psf)[a, b, c, d, e]

	ZONE	EFFECTIVE WIND AREA (feet²)	BASIC WIND SPEED (mph-3-second gust)																							
			85		90		100		105		110		120		125		130		140		145		150		170	
Roof > 0 to 10 degrees	1	10	10.0	-13.0	10.0	-14.6	10.0	-18.0	10.0	-19.8	10.0	-21.8	10.5	-25.9	11.4	-28.1	12.4	-30.4	14.3	-35.3	15.4	-37.8	16.5	-40.5	21.1	-52.0
	1	20	10.0	-12.7	10.0	-14.2	10.0	-17.5	10.0	-19.3	10.0	-21.2	10.0	-25.2	10.7	-27.4	11.6	-29.6	13.4	-34.4	14.4	-36.9	15.4	-39.4	19.8	-50.7
	1	50	10.0	-12.2	10.0	-13.7	10.0	-16.9	10.0	-18.7	10.0	-20.5	10.0	-24.4	10.0	-26.4	10.6	-28.6	12.3	-33.2	13.1	-35.6	14.1	-38.1	18.1	-48.9
	1	100	10.0	-11.9	10.0	-13.3	10.0	-18.5	10.0	-18.2	10.0	-19.9	10.0	-23.7	10.0	-25.7	10.0	-27.8	11.4	-32.3	12.2	-34.6	13.0	-37.0	16.7	-47.6
	2	10	10.0	-21.8	10.0	-24.4	10.0	-30.2	10.0	-33.3	10.0	-36.5	10.5	-43.5	11.4	-47.2	12.4	-51.0	14.3	-59.2	15.4	-63.5	16.5	-67.9	21.1	-87.2
	2	20	10.0	-19.5	10.0	-21.8	10.0	-27.0	10.0	-29.7	10.0	-32.6	10.0	-38.8	10.7	-42.1	11.6	-45.6	13.4	-52.9	14.4	-56.7	15.4	-60.7	19.8	-78.0
	2	50	10.0	-16.4	10.0	-18.4	10.0	-22.7	10.0	-25.1	10.0	-27.5	10.0	-32.7	10.0	-35.5	10.6	-38.4	12.3	-44.5	13.1	-47.8	14.1	-51.1	18.1	-65.7
	2	100	10.0	-14.1	10.0	-15.8	10.0	-19.5	10.0	-21.5	10.0	-23.6	10.0	-28.1	10.0	-30.5	10.0	-33.0	11.4	-38.2	12.2	-41.0	13.0	-43.9	16.7	-56.4
	3	10	10.0	-32.8	10.0	-36.8	10.0	-45.4	10.0	-50.1	10.0	-55.0	10.5	-65.4	11.4	-71.0	12.4	-76.8	14.3	-89.0	15.4	-95.5	16.5	102.2	21.1	131.3
	3	20	10.0	-27.2	10.0	-30.5	10.0	-37.6	10.0	-41.5	10.0	-45.5	10.0	-54.2	10.7	-58.8	11.6	-63.6	13.4	-73.8	14.4	-79.1	15.4	-84.7	19.8	108.7
	3	50	10.0	-19.7	10.0	-22.1	10.0	-27.3	10.0	-30.1	10.0	-33.1	10.0	-39.3	10.0	-42.7	10.6	-46.2	12.3	-53.5	13.1	-57.4	14.1	-61.5	18.1	-78.9
	3	100	10.0	-14.1	10.0	-15.8	10.0	-19.5	10.0	-21.5	10.0	-23.6	10.0	-28.1	10.0	-30.5	10.0	-33.0	11.4	-38.2	12.2	-41.0	13.0	-43.9	16.7	-56.4
Roof > 10 to 30 degrees	1	10	10.0	-11.9	10.0	-13.3	10.4	-16.5	11.4	-18.2	12.5	-19.9	14.9	-23.7	16.2	-25.7	17.5	-27.8	20.3	-32.3	21.8	-34.6	23.3	-37.0	30.0	-47.6
	1	20	10.0	-11.6	10.0	-13.0	10.0	-16.0	10.4	-17.6	11.4	-19.4	13.6	-23.0	14.8	-25.0	16.0	-27.0	18.5	-31.4	19.9	-33.7	21.3	-36.0	27.3	-46.3
	1	50	10.0	-11.1	10.0	-12.5	10.0	-15.4	10.0	-17.0	10.0	-18.6	11.9	-22.2	12.9	-24.1	13.9	-26.0	16.1	-30.2	17.3	-32.4	18.5	-34.6	23.8	-44.5
	1	100	10.0	-10.8	10.0	-12.1	10.0	-14.9	10.0	-16.5	10.0	-18.1	10.5	-21.5	11.4	-23.3	12.4	-25.2	14.3	-29.3	15.4	-31.4	16.5	-33.6	21.1	-43.2
	2	10	10.0	-25.1	10.0	-28.2	10.4	-34.8	11.4	-38.3	12.5	-42.1	14.9	-50.1	16.2	-54.3	17.5	-58.7	20.3	-68.1	21.8	-73.1	23.3	-78.2	30.0	100.5
	2	20	10.0	-22.8	10.0	-25.6	10.0	-31.5	10.4	-34.8	11.4	-38.2	13.6	-45.4	14.8	-49.3	16.0	-53.3	18.5	-61.8	19.9	-66.3	21.3	-71.0	27.3	-91.2
	2	50	10.0	-19.7	10.0	-22.1	10.0	-27.3	10.0	-30.1	10.0	-33.0	11.9	-39.3	12.9	-42.7	13.9	-46.1	16.1	-53.5	17.3	-57.4	18.5	-61.4	23.8	-78.9
	3	20	10.0	-22.8	10.0	-25.6	10.0	-31.5	10.4	-34.8	11.4	-38.2	13.6	-45.4	14.8	-49.3	16.0	-53.3	18.5	-61.8	19.9	-66.3	21.3	-71.0	27.3	-91.2
	3	50	10.0	-19.7	10.0	-22.1	10.0	-27.3	10.0	-30.1	10.0	-33.0	11.9	-39.3	12.9	-42.7	13.9	-46.1	16.1	-53.5	17.3	-57.4	18.5	-61.4	23.8	-78.9
	3	100	10.0	-17.4	10.0	-19.5	10.0	-24.1	10.0	-26.6	10.0	-29.1	10.5	-34.7	11.4	-37.6	12.4	-40.7	14.3	-47.2	15.4	-50.6	16.5	-54.2	21.1	-69.6
Roof > 30 to 45 degrees	1	10	11.9	-13.0	13.3	-14.6	16.5	-18.0	18.2	-19.8	19.9	-21.8	23.7	-25.9	25.7	-28.1	27.8	-30.4	32.3	-35.3	34.6	-37.8	37.0	-40.5	47.6	-52.0
	1	20	11.6	-12.3	13.0	-13.8	16.0	-17.1	17.6	-18.8	19.4	-20.7	23.0	-24.6	25.0	-26.7	27.0	-28.9	31.4	-33.5	33.7	-35.9	36.0	-38.4	46.3	-49.3
	1	50	11.1	-11.5	12.5	-12.8	15.4	-15.9	17.0	-17.5	18.6	-19.2	22.2	-22.8	24.1	-24.8	26.0	-25.8	30.2	-31.1	32.4	-33.3	34.6	-35.7	44.5	-45.8
	1	100	10.8	-10.8	12.1	-12.1	14.9	-14.9	16.5	-16.5	18.1	-18.1	21.5	-21.5	23.3	-23.3	25.2	-25.2	29.3	-29.3	31.4	-31.4	33.6	-33.6	43.2	-43.2
	2	10	11.9	-15.2	13.3	-17.0	16.5	-21.0	18.2	-23.2	19.9	-25.5	23.7	-30.3	25.7	-32.9	27.8	-35.6	32.3	-41.2	34.6	-44.2	37.0	-47.3	47.6	-60.8
	2	20	11.6	-14.5	13.0	-16.3	16.0	-20.1	17.6	-22.2	19.4	-24.3	23.0	-29.0	25.0	-31.4	27.0	-34.0	31.4	-39.4	33.7	-42.3	36.0	-45.3	46.3	-58.1
	2	50	11.1	-13.7	12.5	-15.3	15.4	-18.9	17.0	-20.8	18.6	-22.9	22.2	-27.2	24.1	-29.5	26.0	-32.0	30.2	-37.1	32.4	-39.8	34.6	-42.5	44.5	-54.6
	2	100	10.8	-13.0	12.1	-14.6	14.9	-18.0	16.5	-19.8	18.1	-21.8	21.5	-25.9	23.3	-28.1	25.2	-30.4	29.3	-35.3	31.4	-37.8	33.6	-40.5	43.2	-52.0
	3	10	11.9	-15.2	13.3	-17.0	16.5	-21.0	18.2	-23.2	19.9	-25.5	23.7	-30.3	25.7	-32.9	27.8	-35.6	32.3	-41.2	34.6	-44.2	37.0	-47.3	47.6	-60.8
	3	20	11.6	-14.5	13.0	-16.3	16.0	-20.1	17.6	-22.2	19.4	-24.3	23.0	-29.0	25.0	-31.4	27.0	-34.0	31.4	-39.4	33.7	-42.3	36.0	-45.3	46.3	-58.1
	3	50	11.1	-13.7	12.5	-15.3	15.4	-18.9	17.0	-20.8	18.6	-22.9	22.2	-27.2	24.1	-29.5	26.0	-32.0	30.2	-37.1	32.4	-39.8	34.6	-42.5	44.5	-54.5
	3	100	10.8	-13.0	12.1	-14.6	14.9	-18.0	16.5	-19.8	18.1	-21.8	21.5	-25.9	23.3	-28.1	25.2	-30.4	29.3	-35.3	31.4	-37.8	33.6	-40.5	43.2	-52.0
Wall	4	10	13.0	-14.1	14.6	-15.8	18.0	-19.5	19.8	-21.5	21.8	-23.6	25.9	-28.1	28.1	-30.5	30.4	-33.0	35.3	-38.2	37.8	-41.0	40.5	-43.9	52.0	-56.4
	4	20	12.4	-13.5	13.9	-15.1	17.2	-18.7	18.9	-20.6	20.8	-22.6	24.7	-26.9	26.8	-29.2	29.0	-31.6	33.7	-36.7	36.1	-39.3	38.7	-42.1	49.6	-54.1
	4	50	11.6	-12.7	13.0	-14.3	16.1	-17.6	17.8	-19.4	19.5	-21.3	23.2	-25.4	25.2	-27.5	27.2	-29.8	31.6	-34.6	33.9	-37.1	36.2	-39.7	46.6	-51.0
	4	100	11.1	-12.2	12.4	-13.6	15.3	-16.8	16.9	-18.5	18.5	-20.4	22.0	-24.2	23.9	-26.3	25.9	-28.4	30.0	-33.0	32.2	-35.4	34.4	-37.8	44.2	-48.6
	5	10	13.0	-17.4	14.6	-19.5	18.0	-24.1	19.8	-26.6	21.8	-29.1	25.9	-34.7	28.1	-37.6	30.4	-40.7	35.3	-47.2	37.8	-50.6	40.5	-54.2	52.0	-69.6
	5	20	12.4	-16.2	13.9	-18.2	17.2	-22.5	18.9	-24.8	20.8	-27.2	24.7	-32.4	26.8	-35.1	29.0	-38.0	33.7	-44.0	36.1	-47.2	38.7	-50.5	49.6	-64.9
	5	50	11.6	-14.7	13.0	-16.5	16.1	-20.3	17.8	-22.4	19.5	-24.6	23.2	-29.3	25.2	-31.8	27.2	-34.3	31.6	-39.8	33.9	-42.7	36.2	-45.7	46.6	-58.7
	5	100	11.1	-13.5	12.4	-15.1	15.3	-18.7	16.9	-20.6	18.5	-22.6	22.0	-26.9	23.9	-29.2	25.9	-31.6	30.0	-36.7	32.2	-39.3	34.4	-42.1	44.2	-54.1

For SI: 1 foot = 304.8 mm, 1 square foot = 0.0929 m², 1 mile per hour = 0.447 m/s, 1 pound per square foot = 0.0479 kPa.

a. The effective wind area shall be equal to the span length multiplied by an effective width. This width shall be permitted to be not be less than one-third the span length. For cladding fasteners, the effective wind area shall not be greater than the area that is tributary to an individual fastener.

b. For effective areas between those given above, the load may be interpolated; otherwise, use the load associated with the lower effective area.

c. Table values shall be adjusted for height and exposure by multiplying by the adjustment coefficient in Table R301.2(3).

d. See Figure R301.2(7) for location of zones.

e. Plus and minus signs signify pressures acting toward and away from the building surfaces.

Table 301.2(2). See page 3-5.

❖This table lists wind pressures for components or cladding building elements. The term "cladding" is defined in Section R202. Although the basic wind speed from Figure R301.2(4)A is applicable to the structure as a whole, this table applies to elements of the building that typically are not part of the wind-force-resisting system. Wind pressures on these elements vary depending on the tributary area, the location/orientation (i.e., a wall that is typically vertical versus a roof and its slope), the region or zone, as well as the basic wind speed.

"Zones" refer to various portions of a structure as illustrated in Figure R301.2(7). Some zones, such as at roof edges and the ends of a wall, represent discontinuities in the exterior building surfaces. Discontinuities interrupt the flow of wind and result in higher wind pressures at these locations. How to determine the appropriate "effective wind area" is defined in Note a. This approach is taken from ASCE 7 and modified editorially to fit the table.

For each combination of these various factors there are two entries in the table, one positive and one negative. The positive number denotes the pressure acting inward or toward the exterior surface of the building; the negative number denotes the pressure acting outward or away from the building surface. The latter is of interest in designing roof elements and their hold-downs that must resist wind uplift.

Unlike the IBC, the code does not elaborate on the distinctions between wind loads on the main wind-force-resisting system versus components and cladding wind loads. Instead, compliance with the components and cladding pressures is necessary only where code provisions explicitly refer to this table. This occurs in the establishment of design wind loads for skylights, exterior windows and doors (see Sections R301.2.1 and R612.2), and setting the threshold for requiring rafter ties (see Sections R611.9.3, R802.11 and R804.3.9).

TABLE R301.2(3)
HEIGHT AND EXPOSURE ADJUSTMENT COEFFICIENTS FOR TABLE R301.2(2

MEAN ROOF HEIGHT	EXPOSURE		
	B	C	D
15	1.00	1.21	1.47
20	1.00	1.29	1.55
25	1.00	1.35	1.61
30	1.00	1.40	1.66
35	1.05	1.45	1.70
40	1.09	1.49	1.74
45	1.12	1.53	1.78
50	1.16	1.56	1.81
55	1.19	1.59	1.84
60	1.22	1.62	1.87

❖This table provides adjustment factors for the components and cladding wind pressure of Table R301.2(2), which are based on a mean roof height of 30 feet (9144 mm) and wind Exposure Category B (see the commentary to Section R301.2.1.4 for a discussion of exposure). For mean roof heights and exposure categories that vary from those assumed in Table R301.2(2), the resulting pressure will change accordingly. Thus, adjustments to the pressures are required by the factors given in this table.

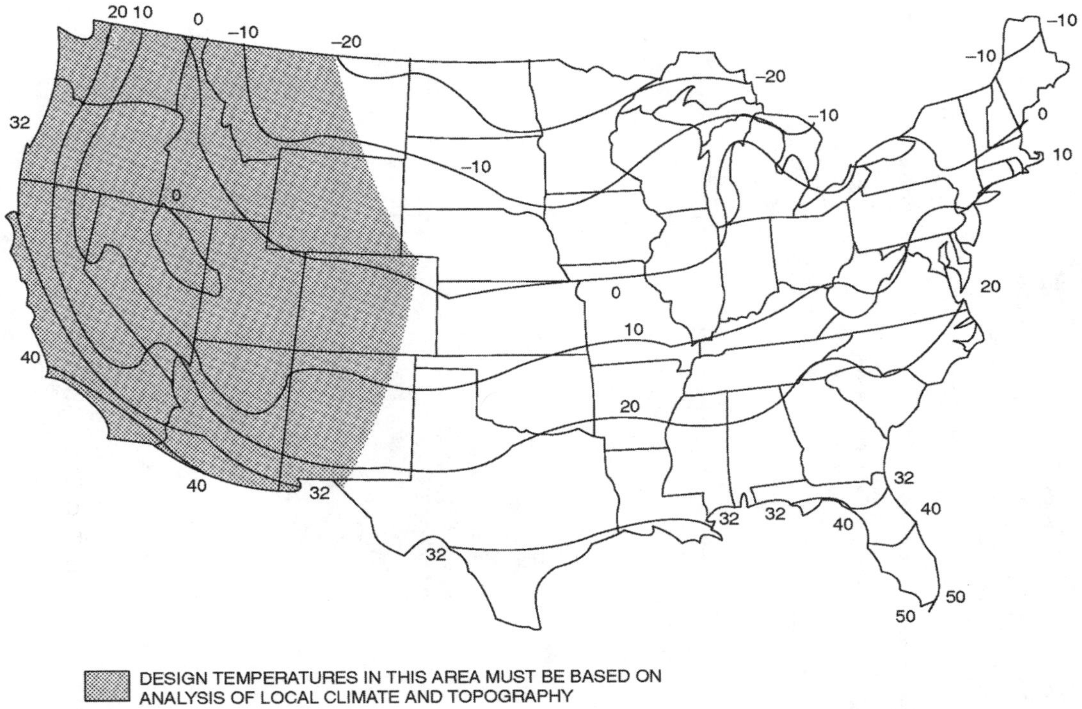

For SI: °C = [(°F) - 32]/1.8.

FIGURE R301.2(1)
ISOLINES OF THE 97$^{1}/_{2}$ PERCENT WINTER (DECEMBER, JANUARY AND FEBRUARY) DESIGN TEMPERATURES (°F)

❖This figure establishes the winter design temperature, which is a criteria for determining the need for dwelling unit heating (see Section R303.8), as well as determining the need for freeze protection of piping (see Sections M2301.2.5, P2603.6 and P3001.2).

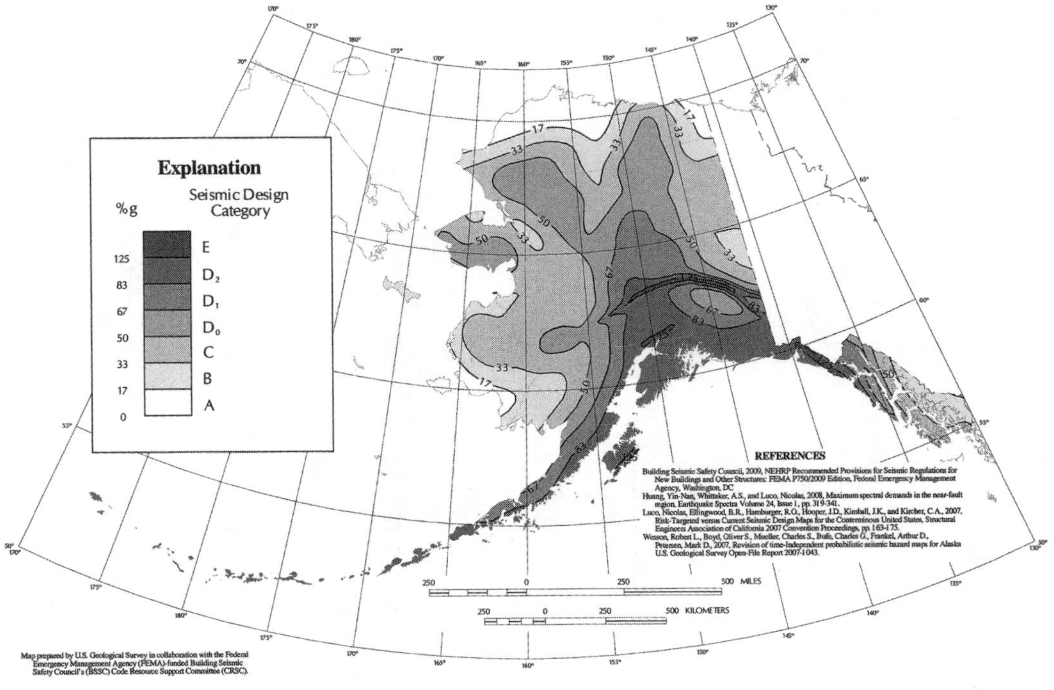

For SI: 1 mile = 1.61 km.

FIGURE R301.2(2)
SEISMIC DESIGN CATEGORIES—SITE CLASS D

(continued)

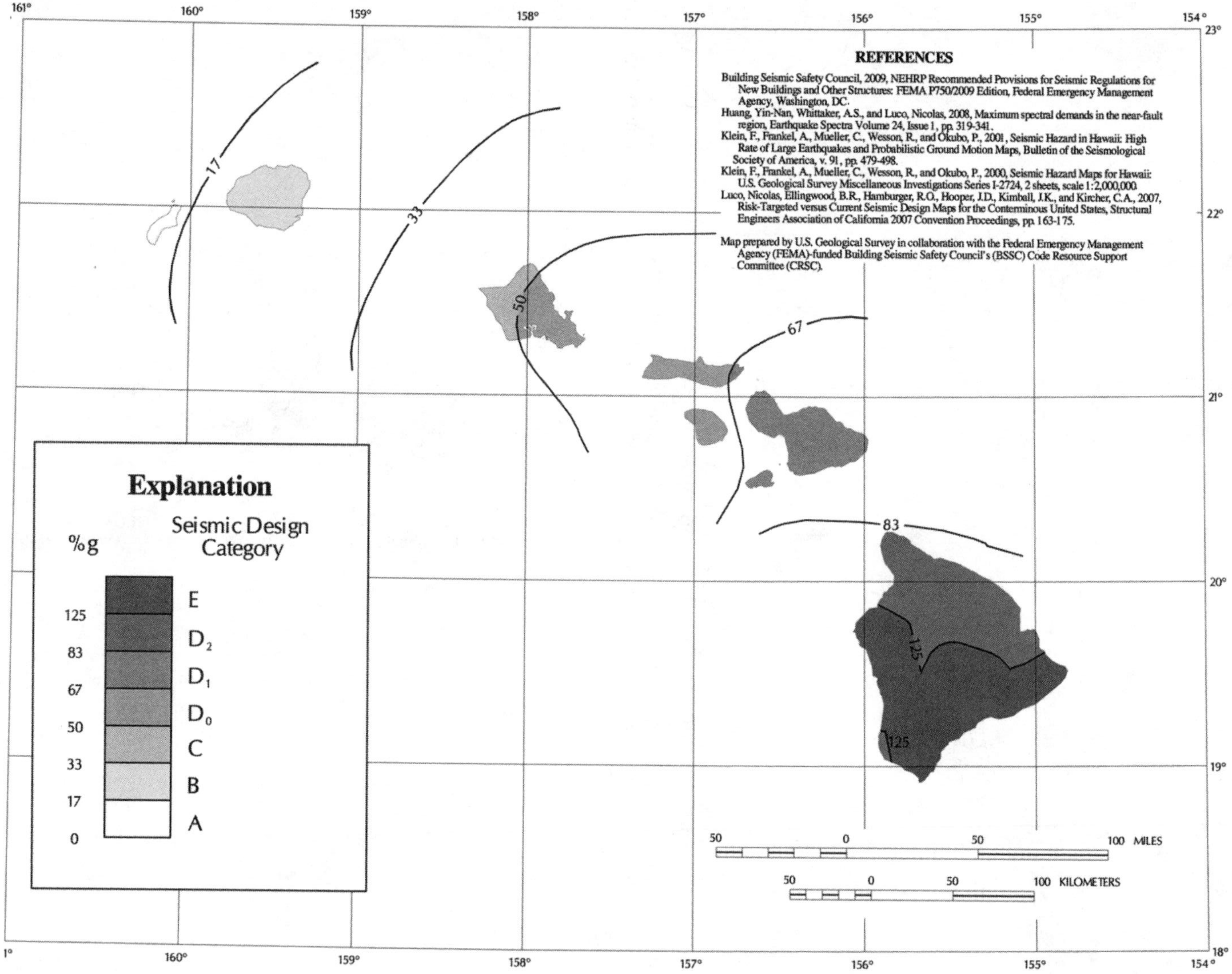

For SI: 1 mile = 1.61 km.

FIGURE R301.2(2)—continued
SEISMIC DESIGN CATEGORIES—SITE CLASS D

(continued)

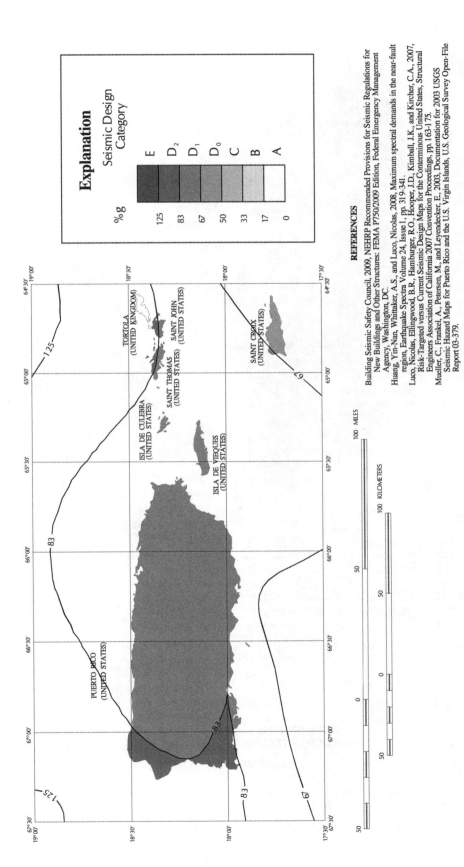

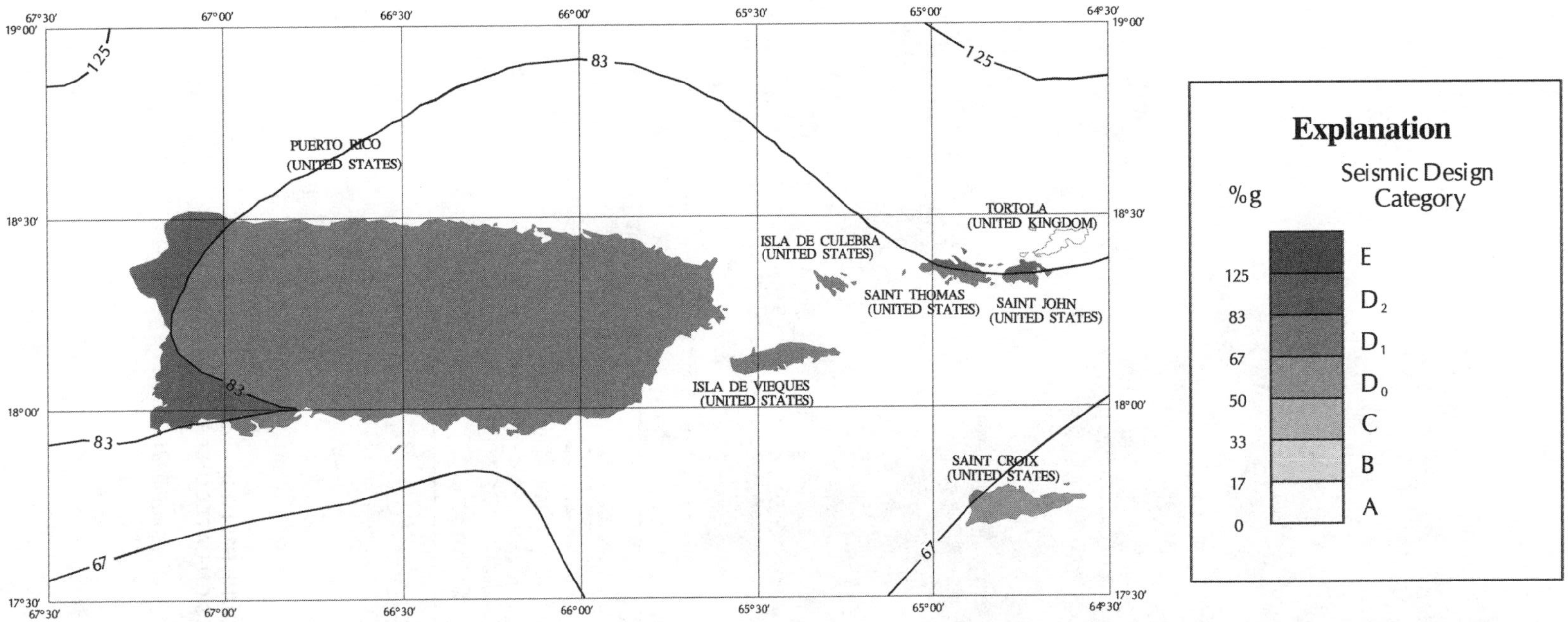

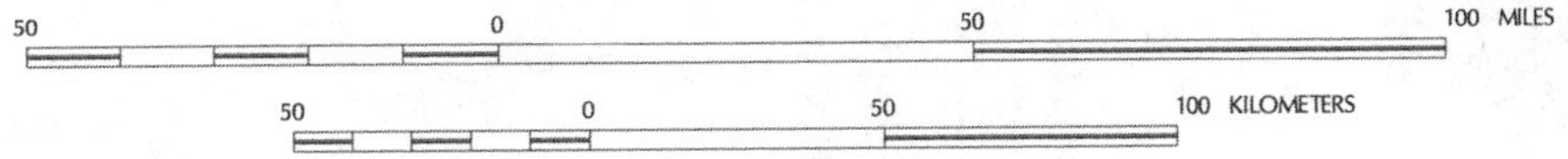

REFERENCES

Building Seismic Safety Council, 2009, NEHRP Recommended Provisions for Seismic Regulations for New Buildings and Other Structures: FEMA P750/2009 Edition, Federal Emergency Management Agency, Washington, DC.

Huang, Yin-Nan, Whittaker, A.S., and Luco, Nicolas, 2008, Maximum spectral demands in the near-fault region, Earthquake Spectra Volume 24, Issue 1, pp. 319-341.

Luco, Nicolas, Ellingwood, B.R., Hamburger, R.O., Hooper, J.D., Kimball, J.K., and Kircher, C.A., 2007, Risk-Targeted versus Current Seismic Design Maps for the Conterminous United States, Structural Engineers Association of California 2007 Convention Proceedings, pp. 163-175.

Mueller, C., Frankel, A., Petersen, M., and Leyendecker, E., 2003, Documentation for 2003 USGS Seismic Hazard Maps for Puerto Rico and the U.S. Virgin Islands, U.S. Geological Survey Open-File Report 03-379.

Map prepared by U.S. Geological Survey in collaboration with the Federal Emergency Management Agency (FEMA)-funded Building Seismic Safety Council's (BSSC) Code Resource Support Committee (CRSC).

FIGURE R301.2(2)—continued
SEISMIC DESIGN CATEGORIES—SITE CLASS D

(continued)

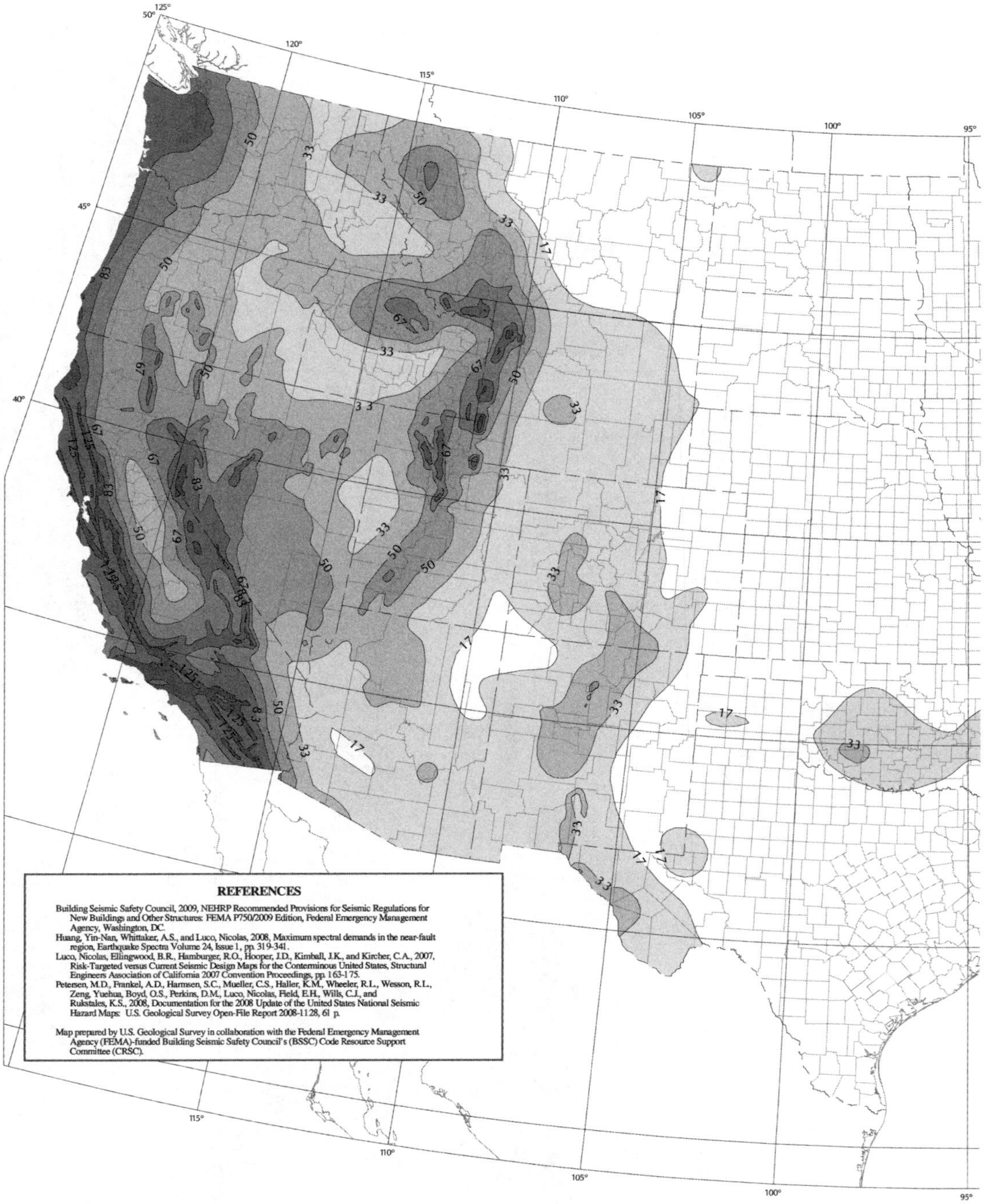

FIGURE R301.2(2)—continued
SEISMIC DESIGN CATEGORIES—SITE CLASS D

(continued)

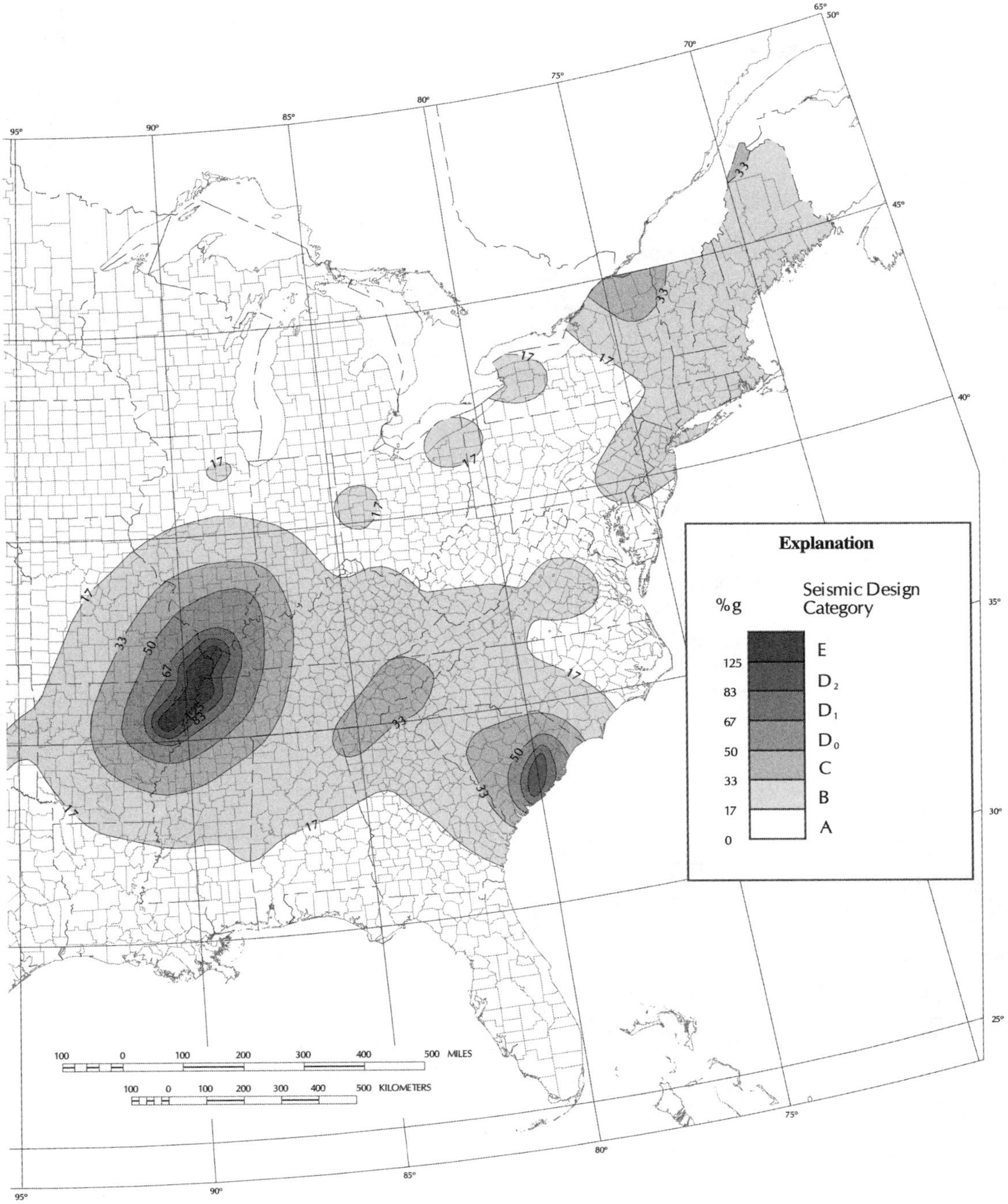

FIGURE R301.2(2)—continued
SEISMIC DESIGN CATEGORIES—SITE CLASS D

❖This figure establishes the seismic design category for a site as presented in Section R301.2.2.1. It provides the most direct determination of the seismic design category in the code. The earthquake-related provisions of the code state requirements as a function of the seismic design category.

This figure reflects new seismic hazard data developed by the U.S. Geological Survey (USGS) as part of its National Seismic Hazard Mapping Project and related technical changes developed by the Building Seismic Safety Council's (BSSC) Seismic Design Procedures Reassessment Group (SDPRG) as part of its work for the Federal Emergency Management Agency (FEMA).

The USGS and the FEMA-funded SDPRG worked together to update the seismic design maps and procedures for the 2009 edition of the NEHRP (National Earthquake Hazards Reduction Program) *Recom-*

mended Seismic Provisions for New Buildings and Other Structures. The new design maps are based on USGS updates to their seismic hazard data and ground motion attenuation formulas as well as the SDPRG's use of risk-targeted ground motions, maximum direction ground motions, and near-source 84th percentile ground motions.

These new IRC maps are different from earlier versions in that the division between Seismic Design Categories D_2 and E has been changed from 118 percent g to 125 percent g. The 125 percent g contour would have been used in earlier maps but the mapping technology then available for drawing the IRC maps did not permit this to be done. The result of this change and the improved seismic hazard data generated by the USGS over the past 10 years is that the geographic region affected by the Seismic Design Category E designation is smaller. This occurs primarily in the region around Charleston, South Carolina, but is also evident in Seismic Design Category E regions in other parts of the United States. Maps developed on the same basis have been incorporated into the IBC which will allow engineers to design components of the building that are outside of the scope of the IRC with compatible seismic loads.

R301.2.1.1 Wind limitations and wind design required. The wind provisions of this code shall not apply to the design of buildings where wind design is required in accordance with Figure R301.2(4)B or where the basic wind speed from Figure R301.2(4)A equals or exceeds 110 miles per hour (49 m/s).

Exceptions:

1. For concrete construction, the wind provisions of this code shall apply in accordance with the limitations of Sections R404 and R611.
2. For structural insulated panels, the wind provisions of this code shall apply in accordance with the limitations of Section R613.

In regions where wind design is required in accordance with Figure R301.2(4)B or where the basic wind speed shown on Figure R301.2(4)A equals or exceeds 110 miles per hour (49 m/s), the design of buildings for wind loads shall be in accordance with one or more of the following methods:

1. AF&PA *Wood Frame Construction Manual* (WFCM); or
2. ICC *Standard for Residential Construction in High-Wind Regions* (ICC 600); or
3. ASCE *Minimum Design Loads for Buildings and Other Structures* (ASCE 7); or
4. AISI *Standard for Cold-Formed Steel Framing—Prescriptive Method For One- and Two-Family Dwellings* (AISI S230); or
5. *International Building Code.*

The elements of design not addressed by the methods in Items 1 through 5 shall be in accordance with the provisions

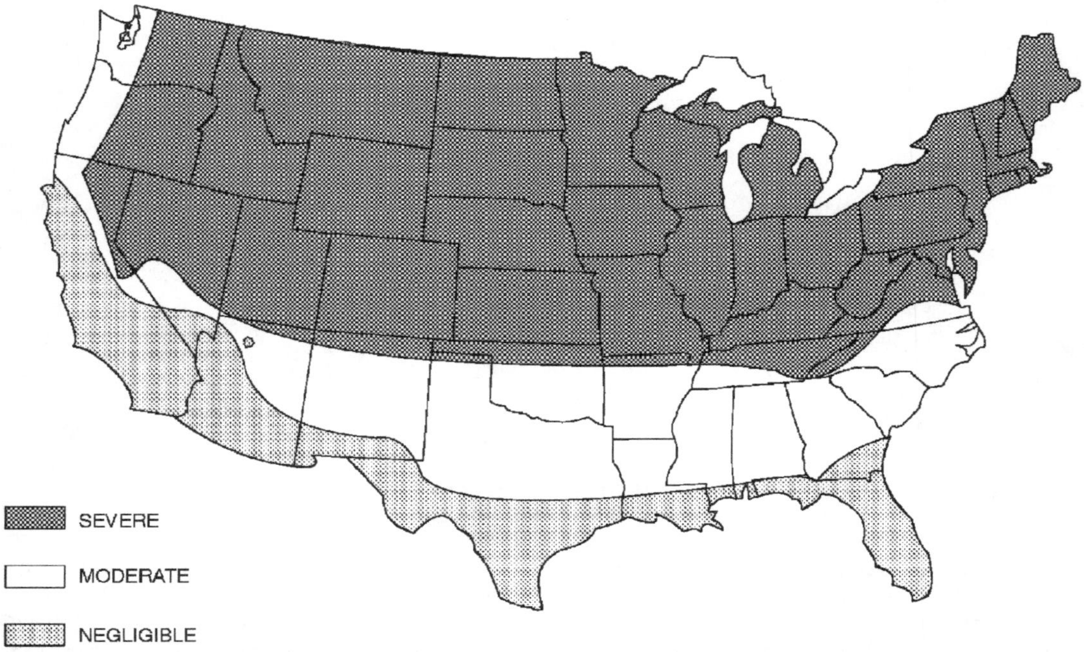

a. Alaska and Hawaii are classified as severe and negligible, respectively.

b. Lines defining areas are approximate only. Local conditions may be more or less severe than indicated by region classification. A severe classification is where weather conditions result in significant snowfall combined with extended periods during which there is little or no natural thawing causing deicing salts to be used extensively.

FIGURE R301.2(3)
WEATHERING PROBABILITY MAP FOR CONCRETE

of this code. When ASCE 7 or the *International Building Code* is used for the design of the building, the wind speed map and exposure category requirements as specified in ASCE 7 and the *International Building Code* shall be used.

❖With the exception of concrete construction and structural insulated panel (SIP) construction, the prescriptive structural provisions in the code do not apply where the basic wind speed equals or exceeds 110 miles per hour (49 m/s) as shown in Figure R301.2(4)A or where wind design is required in accordance with Figure R301.2(4)B. The prescriptive concrete provisions are applicable in regions with wind speeds up to 130 mph (58 m/s), where the site is classified as Exposure B, 110 mph (49 m/s) Exposure C and 100 mph Exposure D (see commentary, Section R611.2). The prescriptive SIP provisions are applicable in regions with wind speeds less than or equal to 120 mph (54 m/s), where the site is classified as Exposure A or B and 110 mph (49 m/s) where the site is classified as Exposure C (see commentary, Section R613.2).

Where the basic wind speed exceeds the limitations given in the code, structures must be designed for wind loads. For this purpose, the code expressly allows the use of AF&PA WFCM; ICC 600, *Standard for Residential Construction in High-wind Regions*; ASCE 7, *Minimum Design Loads for Buildings and Other Structures*; AISI S230, *Standard for Cold-formed Steel Framing—Prescriptive Method for One-and Two-family Dwellings for International Building Code®* (IBC®).

The building elements not addressed by Methods 1 through 5 must be in accordance with the provisions of the IRC. Only the structural design of the building to resist wind loads or seismic loads, and the selection of certain critical components such as window or roofing that is prone to wind damage, must be performed in accordance with Methods 1 through 5. The remaining provisions of the IRC including, but not limited to, architectural, mechanical, electrical and plumbing still apply to the building.

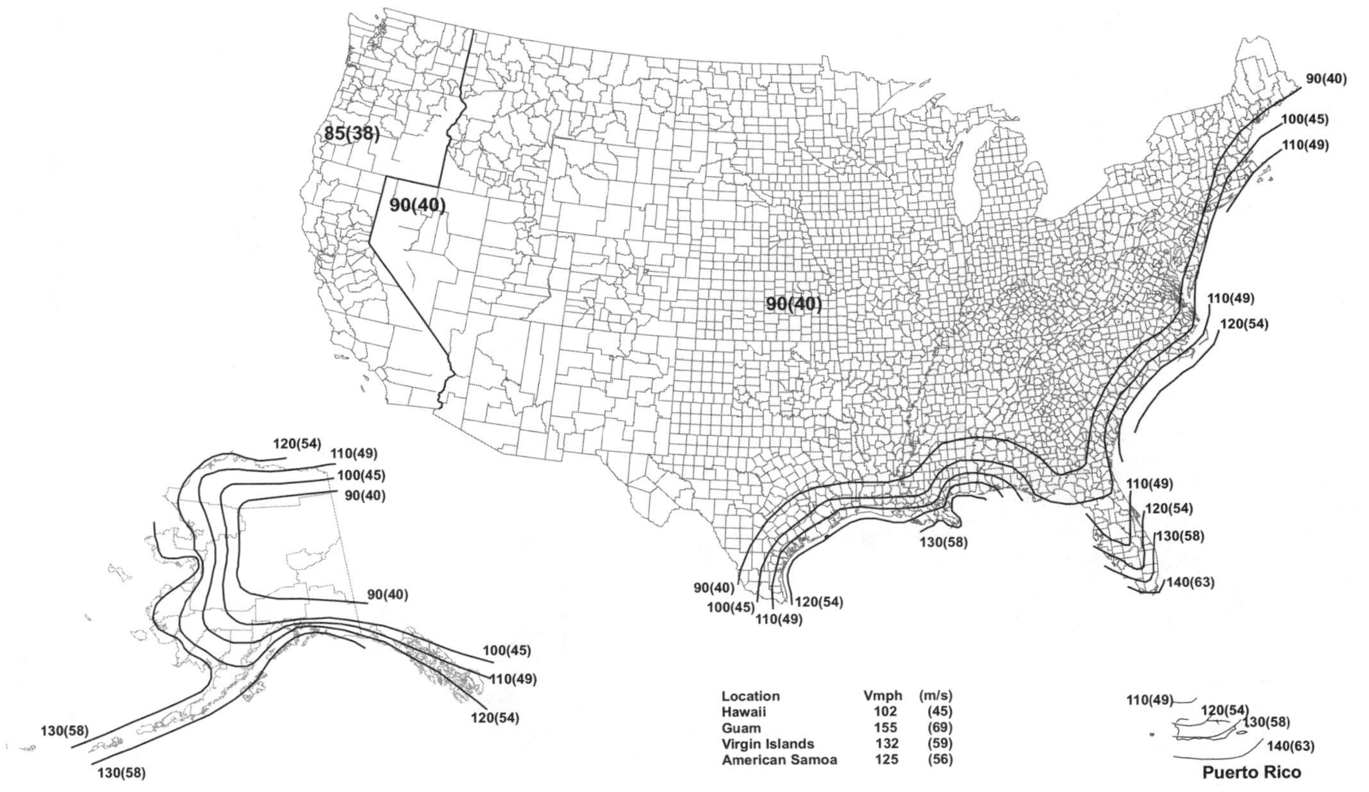

Notes:
1. Values are nominal design 3-second gust wind speeds in miles per hour (m/s) at 33 ft (10 m) above ground for Exposure C category.
2. Linear interpolation between contours is permitted.
3. Islands and coastal areas outside the last contour shall use the last wind speed contour of the coastal area.
4. Mountainous terrain, gorges, ocean promontories, and special wind regions shall be examined for unusual wind conditions.

FIGURE R301.2(4)A
BASIC WIND SPEEDS

❖See page 3-16 for commentary.

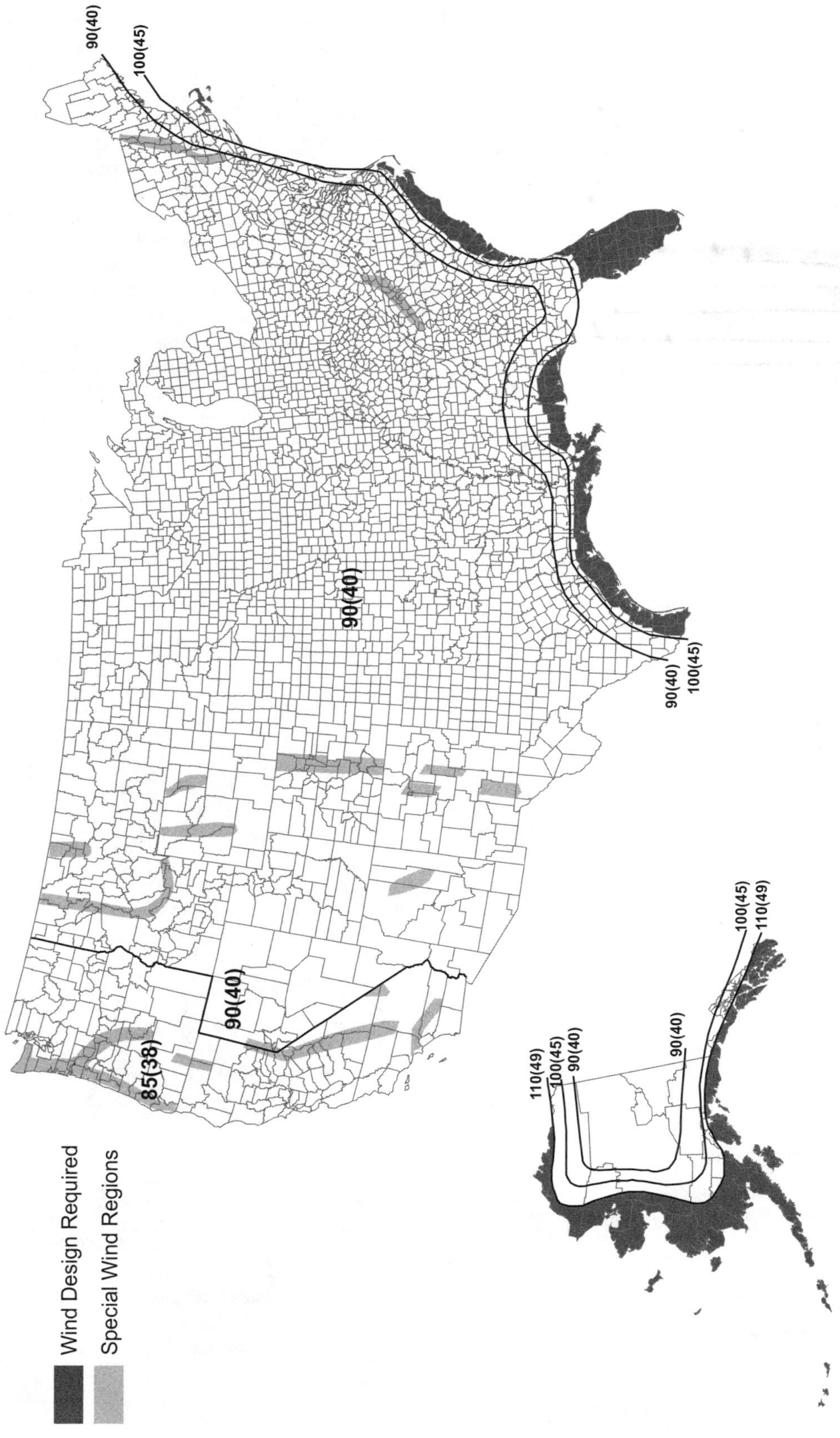

Other Locations where Wind Design Required
Puerto Rico
Guam
Virgin Islands
American Samoa
Hawaii - Special Wind Regions

**FIGURE R301.2(4)B
REGIONS WHERE DESIGN IS REQUIRED**

❖See page 3-16 for commentary.

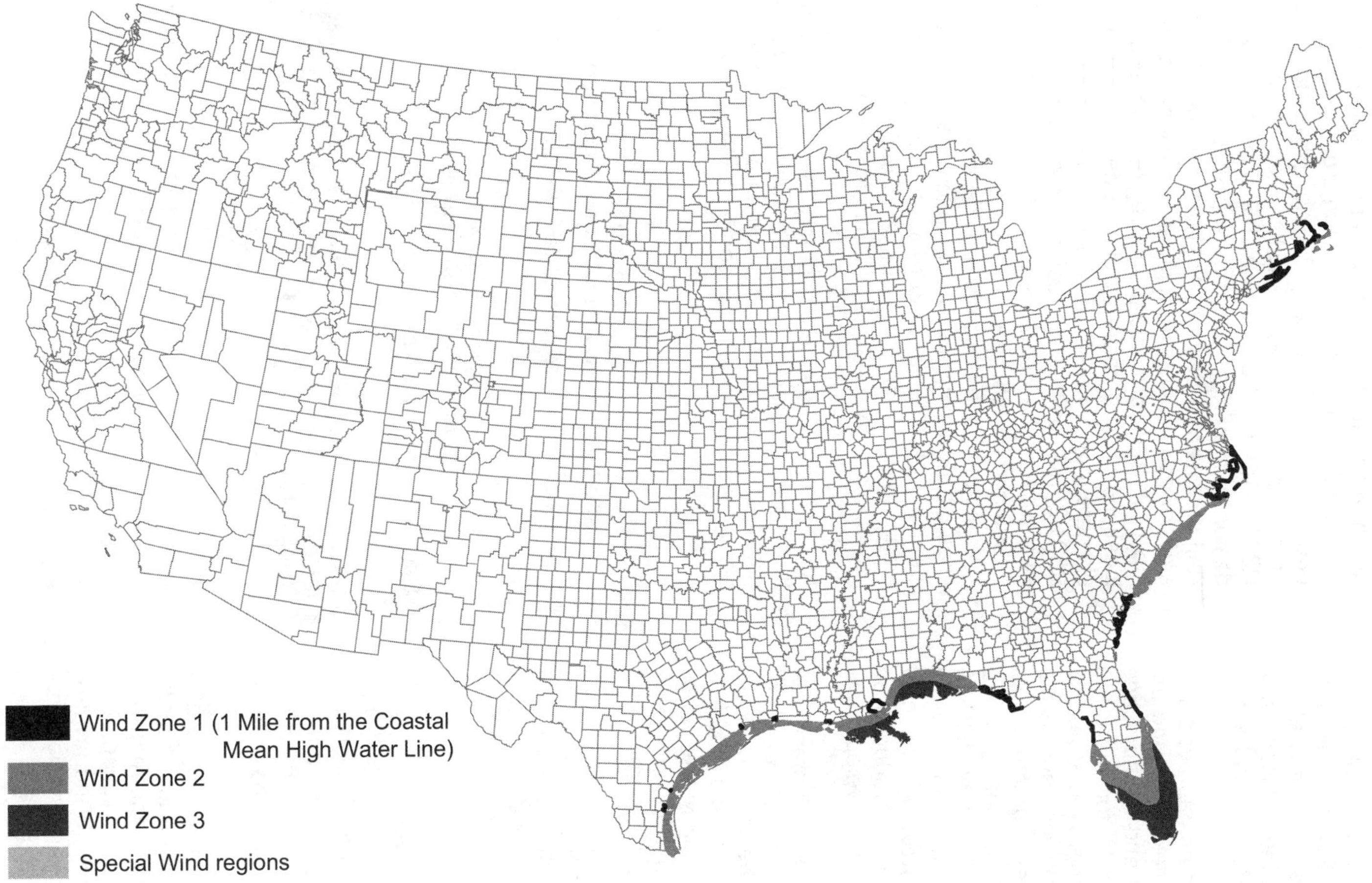

Note:
Wind Zone 3 applies for:
Guam
Virgin Islands
American Samoa
Puerto Rico

Note: Wind Zone 3 applies in Wind Zone 2 areas that are within a mile of the Costal Mean High Water Line

Note: Wind Zone 1 applies in Hawaii - Special Wind Regions

FIGURE R301.2(4)C
WIND-BORNE DEBRIS REGIONS

❖See page 3-16 for commentary.

FIGURE R301.2(4)A
BASIC WIND SPEEDS

❖Over the past 10 years, new data and research has been performed that indicated that the hurricane wind speeds provided in the maps of the previous editions of the IRC and ASCE-7 are too conservative and needed to be adjusted downward. Significantly more hurricane data have become available thereby allowing for substantial improvements in the hurricane simulation model that is used to create the wind speed maps.

These new data have resulted in an improved representation of the hurricane wind field, including the modeling of the sea-land transition and the hurricane boundary layer height; new models for hurricane weakening after landfall; and an improved statistical model for the Holland B parameter which controls the wind pressure relationship. The new hurricane hazard model yields hurricane wind speeds that are lower than those given in ASCE 7-05 and IRC-09 even though the overall rate of intense storms (as defined by central pressure) produced by the new model is increased compared to those produced by the hurricane simulation model used to develop previous maps.

In preparing the new maps, the ASCE 7 standards committee decided to use multiple ultimate event or strength design maps, based on the different occupancy categories in conjunction with a wind load factor of 1.0 for strength design—for allowable stress design, the factor was reduced from 1.0 to 0.6. Several factors that are important to an accurate wind load standard led to this decision:

- An ultimate event or strength design wind speed map makes the overall approach consistent with that used in seismic design in that they both map ultimate events and use a load factor of 1.0 for strength design.
- Utilizing different maps for the different occupancy categories eliminates the problems associated with using "importance factors" that vary with category. The difference in the importance factors in hurricane prone and nonhurricane-prone regions for Category I structures prompted many questions and has been removed from ASCE 7-10.
- The use of multiple maps eliminates the confusion associated with the recurrence interval associated with the existing map—the map was not a uniform 50 year return period map. This therefore created a situation where the level of safety provided for within the overall design was not consistent along the hurricane coast.

Because of the prescriptive nature of the IRC and the considerable number of embedded wind speed triggers throughout the code, integrating the new wind speed map into the IRC necessitated a different approach. For ease of the users of the IRC, it was decided to scale down the ultimate map or strength design map to a nominal or design level basic wind speed map. This new map, Figure R301.2(4)A, is the ultimate map from ASCE 7-10 with the wind speeds divided by the square root of the load factor ($V/\sqrt{1.6}$) with contours corresponding to whole numbers. The use of a scaled-down map was necessary due to the significant number of wind speed triggers embedded throughout the IRC that are based on the old nominal or design level map. Another new map, Figure R301.2(4)B, is introduced which indicates where wind design is required. This map replaces the 100 mph limit specified in Section R301.2.1.1 in the 2009 IRC and corresponds to 130 mph on the ultimate map for most of the hurricane-prone region. Because the locations of wind-borne debris regions are tied to the ultimate maps in ASCE 7-10, a new map Figure R301.2(4)C has been introduced to delineate the various wind-borne debris regions for use with ASTM E1996 and E1886.

While the code does not provide any specific guidance on the topic, one should take note of the areas identified as "special wind regions," as well as other areas described in the footnote where wind anomalies are observed. As the footnote explains, these areas must be examined for unusual wind conditions. ASCE 7 discusses the use of regional climatic data in these areas. Section R301.2.1.5 does provide greater detail for when topography causes a wind speed-up effect, which may overlap some of the special wind region considerations.

FIGURE R301.2(4)B
REGIONS WHERE WIND DESIGN IS REQUIRED

❖See the commentary for Figure R301.2(4)A.

FIGURE R301.2(4)C
WIND-BORNE DEBRIS REGIONS

❖See the commentary for Figure R301.2(4)A.

R301.2.1.2 Protection of openings. Exterior glazing in buildings located in windborne debris regions shall be protected from windborne debris. Glazed opening protection for windborne debris shall meet the requirements of the Large Missile Test of ASTM E 1996 and ASTM E 1886 referenced therein. The applicable wind zones for establishing misile types in ASTM E 1996 are shown on Figure R301.2(4)C. Garage door glazed opening protection for windborne debris shall meet the requirements of an *approved* impact-resisting standard or ANSI/DASMA 115.

> **Exception:** Wood structural panels with a minimum thickness of $^{7}/_{16}$ inch (11 mm) and a maximum span of 8 feet (2438 mm) shall be permitted for opening protection in one- and two-story buildings. Panels shall be precut and attached to the framing surrounding the opening containing the product with the glazed opening. Panels shall be predrilled as required for the anchorage method and shall be secured with the attachment hardware provided. Attach-

ments shall be designed to resist the component and cladding loads determined in accordance with either Table R301.2(2) or ASCE 7, with the permanent corrosion-resistant attachment hardware provided and anchors permanently installed on the building. Attachment in accordance with Table R301.2.1.2 is permitted for buildings with a mean roof height of 33 feet (10 058 mm) or less where located in Wind Zones 1 and 2 in accordance with Figure R301.2(4)C.

❖Exterior glazing in buildings located in wind-borne debris regions must have openings that provide protection from wind-borne debris. The opening protection, usually in the form of permanent shutters or laminated glass, must meet the requirements of the Large Missile Test of ASTM E 1996 and ASTM E 1886, listed in Chapter 44. Glazing in garage doors must meet the protection requirements for garage doors in ANSI/DSMA 115, Standard Method for Testing Sectional Garage Doors and Rolling Doors: Determination of Structural Performance Under Missile Impact and Cyclic Wind Pressure. The exception provides a prescriptive and more economical approach for one- and two-story buildings. This method of opening protection uses wood structural panels that are limited to a maximum span of 8 feet (2438 mm), with fastening as specified in Table R301.2.1.2 for buildings with a mean roof height of 33 feet (10 058 mm) or less where located in Wind Zones 1 and 2 in accordance with Figure R301.2(4)C.

The builder must precut these to fit each glazed opening and provide the necessary attachment hardware so that panels can be adequately anchored and installed quickly.

To determine where this requirement applies, refer to the definition in Chapter 2, which states that wind-borne debris regions are areas within hurricane-prone regions as designated in accordance with Figure R301.2(4)C.

TABLE R301.2.1.2
WINDBORNE DEBRIS PROTECTION FASTENING SCHEDULE FOR WOOD STRUCTURAL PANELS[a, b, c, d]

FASTENER TYPE	FASTENER SPACING (inches)[a, b]		
	Panel span ≤ 4 feet	4 feet < panel span ≤ 6 feet	6 feet < panel span ≤ 8 feet
No. 8 wood screw based anchor with 2-inch embedment length	16	10	8
No. 10 wood screw based anchor with 2-inch embedment length	16	12	9
$^{1}/_{4}$-inch lag screw based anchor with 2-inch embedment length	16	16	16

For SI: 1 inch = 25.4 mm, 1 foot = 304.8 mm, 1 pound = 4.448 N, 1 mile per hour = 0.447 m/s.

a. This table is based on 130 mph wind speeds and a 33-foot mean roof height.

b. Fasteners shall be installed at opposing ends of the wood structural panel. Fasteners shall be located a minimum of 1 inch from the edge of the panel.

c. Anchors shall penetrate through the exterior wall covering with an embedment length of 2 inches minimum into the building frame. Fasteners shall be located a minimum of $2^{1}/_{2}$ inches from the edge of concrete block or concrete.

d. Where panels are attached to masonry or masonry/stucco, they shall be attached using vibration-resistant anchors having a minimum ultimate withdrawal capacity of 1500 pounds.

❖The fastening schedule provided in this table allows for the selection of the appropriate fasteners size and spacing when the prescriptive alternative is applied to a tested assembly, as required by Section R301.2.1.2. Installing the specified fasteners at both ends of the wood structural panel provides the necessary panel support. Hardware must be permanently mounted when using wood structural panel shutters for window protection for new construction. Using wood structural panels as window protection is basically an emergency option for the protection of existing buildings where the homeowner does not have some permanent shutter system in place.

While the code requires the panels to be precut and the attachment hardware provided, there are potentially many logistical problems with homeowners actually installing the panels when attachment hardware is not permanent. It is not clear that the homeowners will be sufficiently instructed on (or remember at a later date) how to attach the panels, in particular using the prescribed minimum spacing. Additionally, it can be extremely cumbersome to attempt to nail a sheet of plywood over a window, particularly on the second story of a building. Additionally, there are concerns about the capacity of nailed connections where the nails are installed in the same hole repeatedly.

The minimum required capacity of masonry anchors is 1,500 pounds (6672 N). Evaluation reports (ICC, NES and SBCCI) for masonry anchors require a Factor of Safety (FS) of 4.0 if a special inspection is performed on the anchor installation. The required capacity of the masonry anchors to 1,500 pounds (6672 N) provides an FS more in line with the evaluation reports for masonry anchors.

FIGURE R301.2(5)
GROUND SNOW LOADS, P_g, FOR THE UNITED STATES (lb/ft^2)

(continued)

For SI: 1 foot = 304.8 mm, 1 pound per square foot = 0.0479 kPa.

FIGURE R301.2(5)-continued
GROUND SNOW LOADS, P_g, FOR THE UNITED STATES (lb/ft^2)

❖The ground snow-load map is taken from the ASCE 7 snow-load provisions and is based on statistical analysis of ground snow data. This map provides ground snow loads for direct use in the prescriptive provisions and tables of the code. Snow loads may increase due to lake effect or elevations. Check with local authorities for more refined snow-load maps.

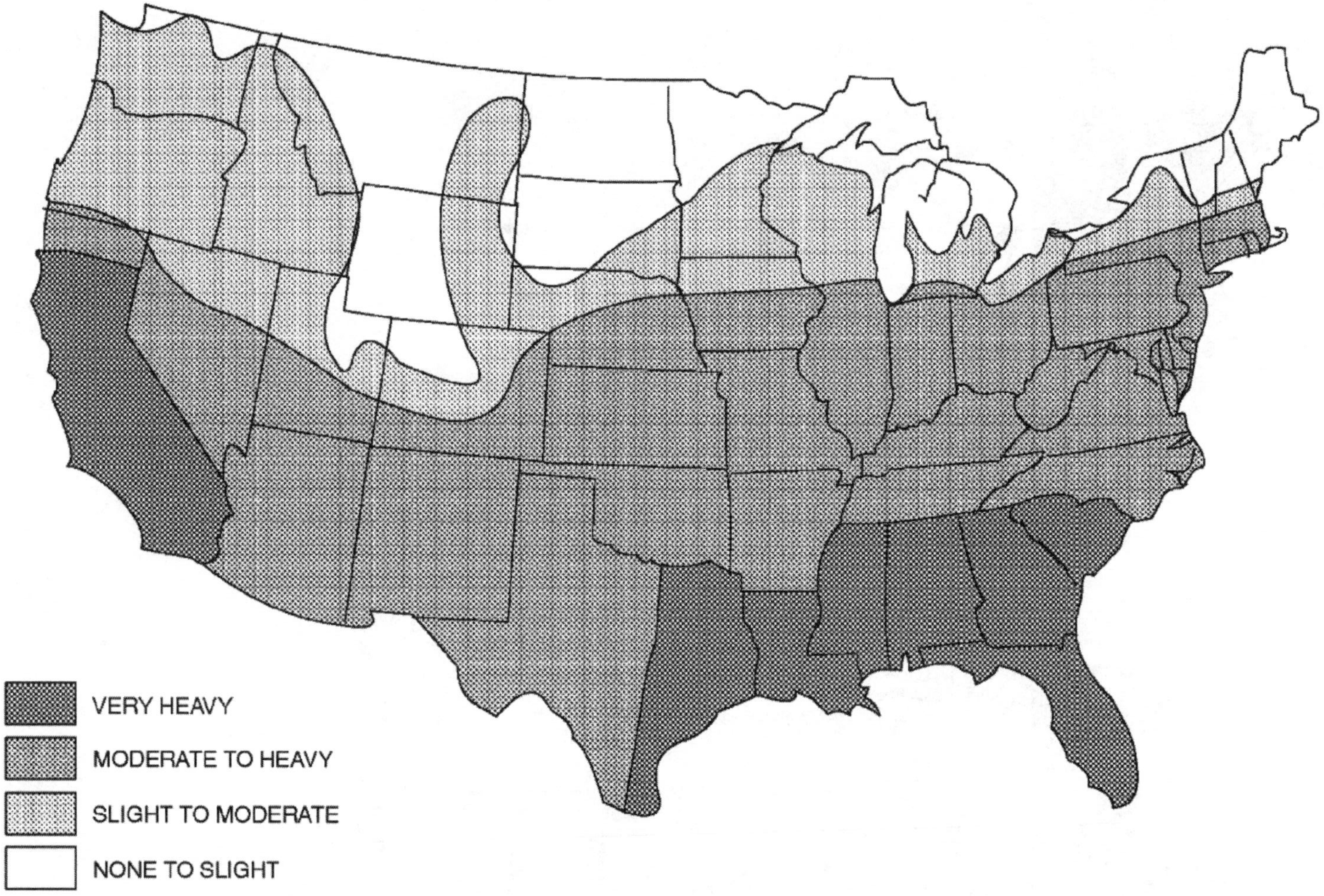

NOTE: Lines defining areas are approximate only. Local conditions may be more or less severe than indicated by the region classification.

FIGURE R301.2(6)
TERMITE INFESTATION PROBABILITY MAP

❖This figure establishes the potential for termite damage in areas of the continental United States. The boundaries for the various classifications are approximate, and local experience should be relied on either to affirm or to modify the classification determined from this map.

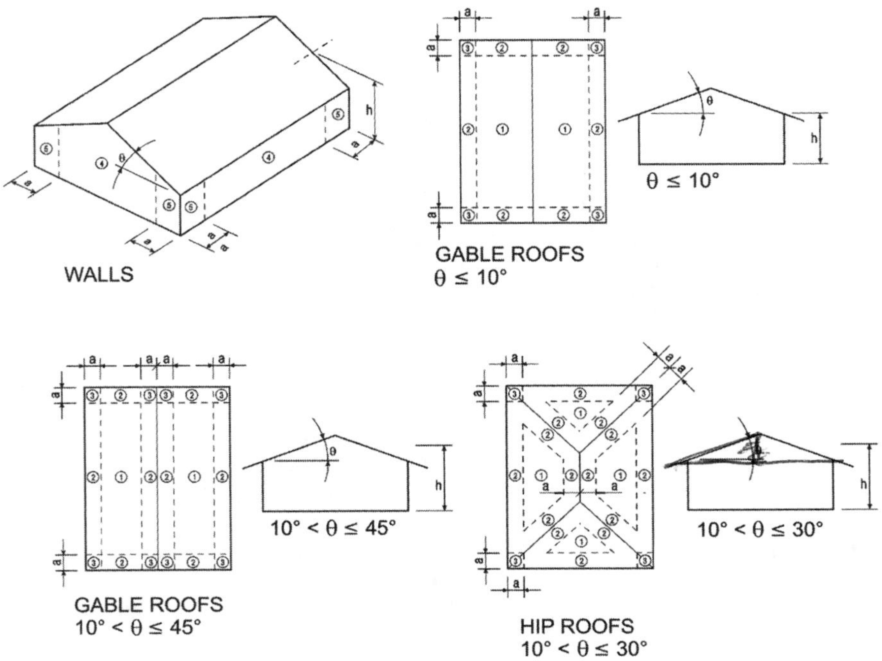

For SI: 1 foot = 304.8 mm, 1 degree = 0.0175 rad.
Note: a = 4 feet in all cases.

FIGURE R301.2(7)
COMPONENT AND CLADDING PRSSURE ZONES

❖This figure depicts the zones for the components and cladding wind pressures that are given in Table R301.2(2). See the commentary to this table for a discussion of components and cladding.

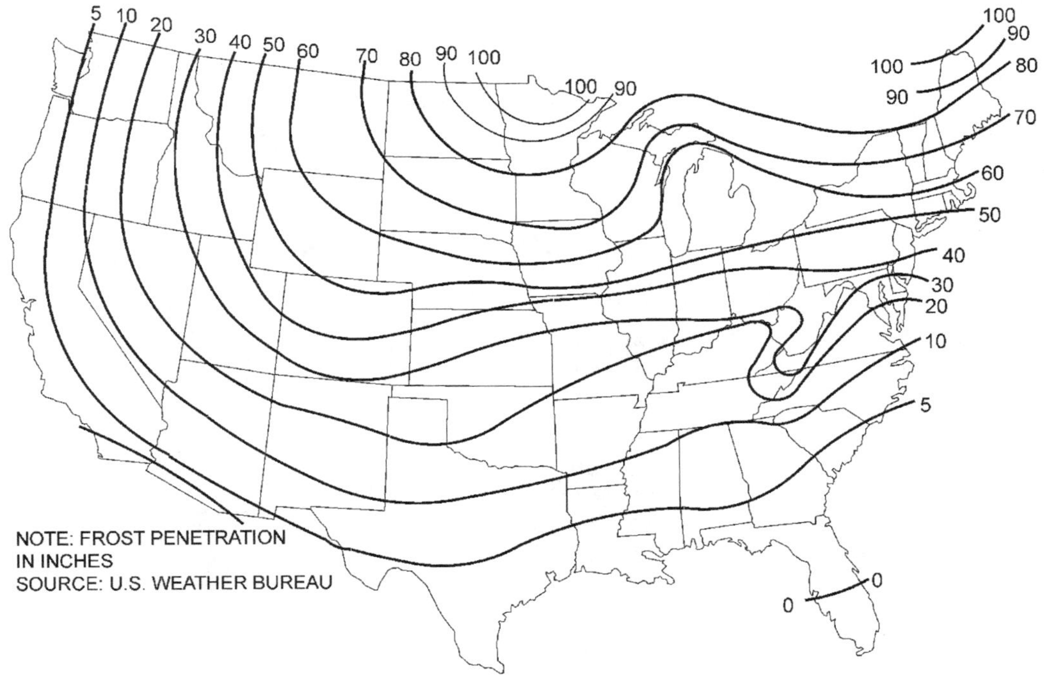

For SI:1 inch = 25.4 mm.

Figure R301.2(8)
FROST LINE DEPTH

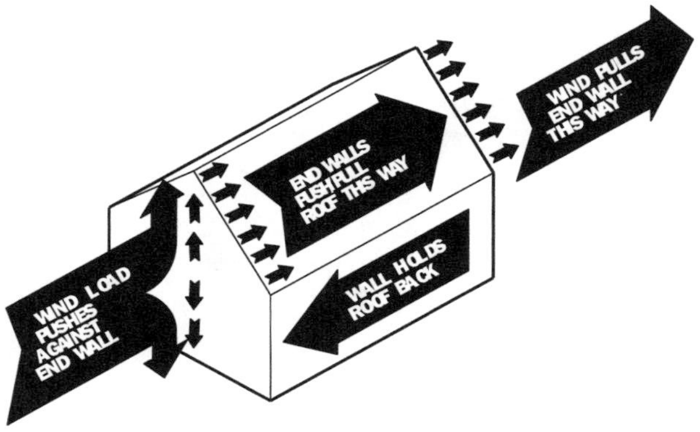

Figure R301.2.1
WIND FORCES ACTING ON AREA

R301.2.1.3 Wind speed conversion. When referenced documents are based on fastest mile wind speeds, the three-second gust basic wind speeds, V_{3s}, of Figure R301.2(4) shall be converted to fastest mile wind speeds, V_{fm}, using Table R301.2.1.3.

❖Prior to publication of the 2002 edition of ASCE 7, most wind-related code provisions were based on fastest-mile wind velocity, V_{fm}. Because many of the documents referenced in the code still use this criterion as a basis for wind design rather than the 3-second gust wind velocities, V_{3s}, of Figure R301.2(4), Table R301.2.1.3 provides the necessary conversion to fastest-mile wind velocities used in those documents.

R301.2.1.4 Exposure category. For each wind direction considered, an exposure category that adequately reflects the characteristics of ground surface irregularities shall be determined for the site at which the building or structure is to be constructed. For a site located in the transition zone between categories, the category resulting in the largest wind forces shall apply. Account shall be taken of variations in ground surface roughness that arise from natural topography and vegetation as well as from constructed features. For a site where multiple detached one- and two-family dwellings, *townhouses* or other structures are to be constructed as part of a subdivision, master-planned community, or otherwise designated as a developed area by the authority having jurisdiction, the exposure category for an individual structure shall be based upon the site conditions that will exist at the time when all adjacent structures on the site have been constructed, provided their construction is expected to begin within one year of the start of construction for the structure for which the exposure category is determined. For any given wind direction, the exposure in which a specific building or other structure is sited shall be assessed as being one of the following categories:

1. Exposure A. Large city centers with at least 50 percent of the buildings having a height in excess of 70 feet (21 336 mm). Use of this exposure category shall be limited to those areas for which terrain representative of Exposure A prevails in the upwind direction for a distance of at least 0.5 mile (0.8 km) or 10 times the height of the building or other structure, whichever is greater. Possible channeling effects or increased velocity pressures due to the building or structure being located in the wake of adjacent buildings shall be taken into account.
2. Exposure B. Urban and suburban areas, wooded areas, or other terrain with numerous closely spaced obstructions having the size of single-family dwellings or larger. Exposure B shall be assumed unless the site meets the definition of another type exposure.

TABLE R301.2.1.3
EQUIVALENT BASIC WIND SPEEDS[a]

3-second gust, V_{3s}	85	90	100	105	110	120	125	130	140	145	150	160	170
Fastest mile, V_{fm}	71	76	85	90	95	104	109	114	123	128	133	142	152

For SI: 1 mile per hour = 0.447 m/s.
a. Linear interpolation is permitted.

❖When referenced documents are based on fastest-mile wind velocities, the appropriate wind speed can be selected directly from this table using the basic wind speed determined from Figure R301.2(4). For example, if the basic wind speed for a site is determined to be 110 mph (49.2 m/s), and it is necessary to design using one of the referenced standards in Section R301.2.1.1, the corresponding fastest-mile wind velocity would be 90 mph (40.2 m/s).

3. Exposure C. Open terrain with scattered obstructions, including surface undulations or other irregularities, having heights generally less than 30 feet (9144 mm) extending more than 1,500 feet (457 m) from the building site in any quadrant. This exposure shall also apply to any building located within Exposure B type terrain where the building is directly adjacent to open areas of Exposure C type terrain in any quadrant for a distance of more than 600 feet (183 m). This category includes flat, open country and grasslands.

4. Exposure D. Flat, unobstructed areas exposed to wind flowing over open water for a distance of at least 1 mile (1.61 km). Shorelines in Exposure D include inland waterways, the Great Lakes, and coastal areas of California, Oregon, Washington and Alaska. This exposure shall apply only to those buildings and other structures exposed to the wind coming from over the water. Exposure D extends inland from the shoreline a distance of 1500 feet (457 m) or 10 times the height of the building or structure, whichever is greater.

❖Wind loading on structures is a function of a site exposure category, which reflects the characteristics of ground surface irregularities and accounts for variations in ground surface roughness that arise from natural topography and vegetation, as well as from constructed features. These categories range from Exposure A for the least severe exposure category to Exposure D for the most severe exposure category. Exposure B should be used unless the site meets the definition for another category of exposure. Exposure B includes urban and suburban areas, wooded areas and other terrain with numerous, closely spaced obstructions having the size of single-family dwellings or larger. The commentary of ASCE 7 contains a detailed discussion of wind exposure categories. The exposure category of homes constructed on a large residential development site should be based upon the exposure condition that will exist for the vast majority of the structure's life, not the temporary condition that may exist during build-out of the development. Residential construction frequently involves the construction of multiple detached homes and/or townhouses on a large residential development site. Such sites are often cleared creating a temporary situation where Exposure Category C exists until a significant number of the homes are constructed.

ASCE 7 defines Exposure Category B as "Urban and suburban areas, wooded areas, or other terrain with numerous closely spaced obstructions having the size of single-family dwellings or larger." The commentary to ASCE 7 indicates that in a recent study, the majority of buildings are in Exposure Category B-as many as 60 to 80 percent. Even a higher percentage is likely for residential construction as jurisdictions push for higher density development and homes are constructed on smaller lots to reduce the impact of rapidly escalating land costs.

R301.2.1.5 Topographic wind effects. In areas designated in Table R301.2(1) as having local historical data documenting structural damage to buildings caused by wind speed-up at isolated hills, ridges and escarpments that are abrupt changes from the general topography of the area, topographic wind effects shall be considered in the design of the building in accordance with Section R301.2.1.5.1 or in accordance with the provisions of ASCE 7. See Figure R301.2.1.5.1(1) for topographic features for wind speed-up effect.

In these designated areas, topographic wind effects shall apply only to buildings sited on the top half of an isolated hill, ridge or escarpment where all of the following conditions exist:

1. The average slope of the top half of the hill, ridge or escarpment is 10 percent or greater.
2. The hill, ridge or escarpment is 60 feet (18 288 mm) or greater in height for Exposure B, 30 feet (9144 mm) or greater in height for Exposure C, and 15 feet (4572 mm) or greater in height for Exposure D.

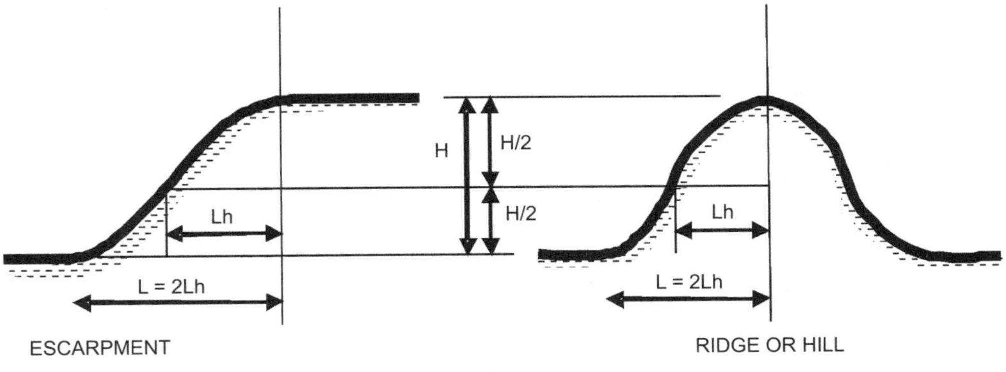

Note:H/2 determines the measurement point for Lh.L is twice Lh.

FIGURE R301.2.1.5.1(1)
TOPOGRAPHIC FEATURES FOR WIND SPEED-UP EFFECT

❖This figure provides guidance to aid builders and designers in designated areas requiring consideration of topographic wind effects on where wind speed-up will likely occur and needs to be considered (see commentary, Sections R301.2.1.5 and R301.2.1.5.1).

3. The hill, ridge or escarpment is isolated or unobstructed by other topographic features of similar height in the upwind direction for a distance measured from its high point of 100 times its height or 2 miles, whichever is less. See Figure R301.2.1.5.1(3) for upwind obstruction.
4. The hill, ridge or escarpment protrudes by a factor of two or more above the height of other upwind topographic features located in any quadrant within a radius of 2 miles measured from its high point.

❖These requirements are for consideration of topographic wind effects. Historically, consideration of wind speed-up due to topographic effects has been ignored in the United States for the design and construction of new homes. Provisions calling for consideration of topographic effects on wind were first included in the 1995 edition of ASCE 7.

The procedures for determining wind speed-up due to topographic effects require the services of a structural engineer with experience in wind analysis and design. Topographic configurations can be quite complex and any resulting speed-up can be difficult, at best, to determine. In fact, modeling the ground topography and testing in a wind tunnel is probably the only means of getting an accurate estimate of the wind speed-up. Using engineering calculations will likely give an overly conservative estimate of the resulting speed-up for an idealized topographic configuration.

A jurisdiction would require an investigation of topographic wind effects when historical data for the area documents structural damage to buildings due to wind speed-up at isolated hills, ridges and escarpments. The jurisdiction would indicate the need for this additional design requirement in Table R301.2(1).

Additionally, Figures R301.2.1.5.1(1), R301.2.1.5.1(2) and R301.2.1.5.1(3) provide guidance to aid builders and designers in designated areas requiring consideration of topographic wind effects on where wind speed-up will likely occur and needs to be considered. These provisions are based on the requirements of ASCE 7, with one exception. ASCE 7 requires consideration of wind speed-up for a building located on the top half of a hill, ridge or escarpment in Exposure Category C when its height is 15 feet (4572 mm) or greater.

This section sets the height at 30 feet (9144 mm) or greater for Exposure C, which is consistent with earlier editions of ASCE 7. If left at 15 feet (4572 mm) or greater, every isolated farmhouse located on a small hill surrounded by flat, cleared farmland would be subjected to wind speed-up due to topographic effects. The 15-foot (4572 mm) trigger was considered to be simply too low and does not reflect what is occurring in the built environment.

An alternative simplified method of designing for topographic wind speed-up effects is offered in Section R301.2.1.5.1, which can be used with the prescriptive methods, and thus does not require engineering design.

R301.2.1.5.1 Simplified topographic wind speed-up method. As an alternative to the ASCE 7 topographic wind provisions, the provisions of Section R301.2.1.5.1 shall be permitted to be used to design for wind speed-up effects, where required by Section R301.2.1.5.

Structures located on the top half of isolated hills, ridges or escarpments meeting the conditions of Section R301.2.1.5 shall be designed for an increased basic wind speed as determined by Table R301.2.1.5.1. On the high side of an escarpment, the increased basic wind speed shall extend horizontally downwind from the edge of the escarpment 1.5 times the horizontal length of the upwind slope (1.5L) or 6 times the height of the escarpment (6H), whichever is greater.

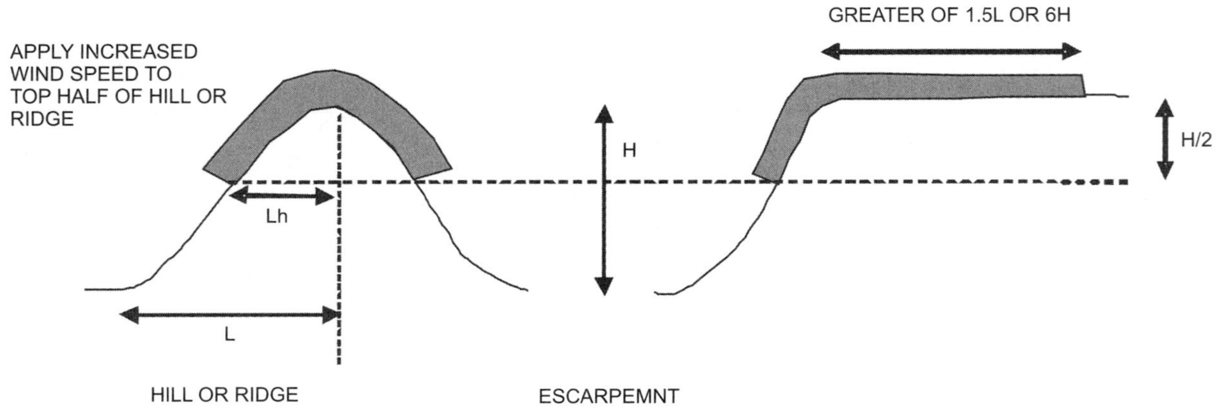

FIGURE R301.2.1.5.1(2)
ILLUSTRATION OF WHERE ON A TOPOGRAPHIC FEATURE, WIND SPEED INCREASE IS APPLIED

❖This figure provides guidance to aid builders and designers in designated areas requiring consideration of topographic wind effects on where wind speed-up will likely occur and needs to be considered (see commentary, Sections R301.2.1.5 and R301.2.1.5.1).

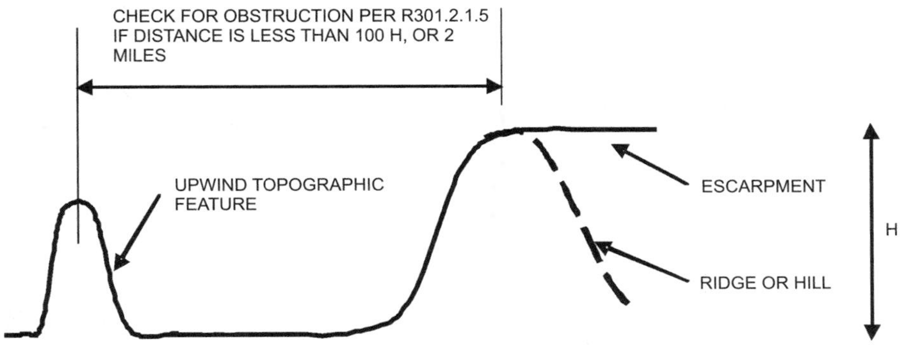

FIGURE R301.2.1.5.1(3)
ILLUSTRATION OF UPWIND OBSTRUCTION

❖This figure provides guidance to aid builders and designers in designated areas requiring consideration of topographic wind effects on where wind speed-up will likely occur and needs to be considered (see commentary, Sections R301.2.1.5 and R301.2.1.5.1).

TABLE R301.2.1.5.1
BASIC WIND MODIFICATION FOR TOPOGRAPHIC WIND EFFECT

BASIC WIND SPEED FROM FIGURE R301.2(4) (mph)	AVERAGE SLOPE OF THE TOP HALF OF HILL, RIDGE OR ESCARPMENT (percent)						
	0.10	0.125	0.15	0.175	0.20	0.23	0.25 or greater
	Required basic wind speed-up, modified for topographic wind speed up (mph)						
85	100	100	100	110	110	110	120
90	100	100	110	110	120	120	120
100	110	120	120	130	130	130	140
110	120	130	130	140	140	150	150
120	140	140	150	150	N/A	N/A	N/A
130	150	N/A	N/A	N/A	N/A	N/A	N/A

For SI: 1 mile per hour = 0.447 m/s.

❖See the commentary to Sections R301.2.1.5 and R301.2.1.5.1.

See Figure R301.2.1.5.1(2) for where wind speed increase is applied.

❖Section R301.2.1.5.1 is an alternative for compliance with ASCE 7 as specified in Section R301.2.1.5. The methodology proposed is an outgrowth of the wind simplification process developed by the Structural Engineers of Washington (SEAW) and the Applied Technology Council (ATC), and is an attempt at simplifying the wind design process in the code for areas with extreme topography.

R301.2.2 Seismic provisions. The seismic provisions of this code shall apply as follows:

1. Townhouses in Seismic Design Categories C, D_0, D_1 and D_2.
2. Detached one- and two-family dwellings in Seismic Design Categories, D_0, D_1 and D_2.

❖Earthquakes generate internal forces in a structure caused by inertia. This is depicted in Commentary Figure R301.2.2. These forces can cause a building to be distorted and severely damaged. The objective of earthquake-resistant construction is to resist these forces and the resulting distortions.

Providing for the adequate earthquake performance of any structure begins with assessing the degree of hazard at the proposed site. As in the IBC, this is accomplished in the code by assigning a seismic design category (see Section R301.2.2.1 and the definition in Chapter 2). These range from Seismic Design Category A, representing the lowest level of seismic hazard, to Seismic Design Category E, representing the highest hazard. Buildings in Seismic Design Category E must be designed in accordance with the IBC, except as allowed in Section R301.2.2.1. Under the code, the seismic design category depends on the building location according to Figure R301.2(2), unless one of the options permitting a more precise determination is exercised.

The code's seismic provisions apply to buildings classified as Seismic Design Category C, D_0, D_1 or D_2. Detached one- and two-family dwellings need to comply with only the seismic requirements of the code when classified as Seismic Design Category D_0, D_1 or D_2. In other words, only townhouses constructed under Section R101.2 need to comply with Seismic Design Category C requirements. This indicates that all buildings classified as Seismic Design Category A or B, as well as detached one- and two-family dwellings classified as Seismic Design Category C, will perform satisfactorily when constructed in accordance with the basic prescriptive requirements of the code.

Values for seismic loads on structures are based on a percentage of the dead load, and the code prescriptive provisions for structural elements that resist seismic forces are based on an assumed dead load. Generally, the building dead loads as described in Section R301.4 are reflected in the prescriptive design tables throughout the code.

FEMA 232, *Home Builder's Guide to Seismic-resistant Construction*, is an excellent source of additional information on seismic-resistant construction, particularly as it pertains to residential buildings.

R301.2.2.1 Determination of seismic design category. Buildings shall be assigned a seismic design category in accordance with Figure R301.2(2).

❖Unlike the IBC, the code does not require a calculation to determine the seismic design category of a structure. Instead, the seismic design category is taken directly from Figure R301.2(2). The seismic design categories shown in Figure R301.2(2) are based solely on the Short Period Design Spectral Response Accelerations, SDS, assuming soil Site Class D, as defined in Section 1613 of the IBC.

A convenient alternative to the maps of Figure R301.2(2) and the calculations described in Section R301.2.2.1.1 is the Seismic Design Parameters program prepared by USGS and available USGS National Seismic Hazard Mapping Project web site (http://earthquake.usgs.gov/hazards). Either of these allows a computer user to view the seismic hazard map of the code in more detail. The Seismic Design Parameters program also provides determination of a site's seismic design category using either the zip code or the longitude and latitude of the site. Because use of longitude and latitude is more precise, this approach is preferable.

The seismic design category classification provides a relative scale of earthquake risk to structures. The seismic design category considers not only the seismicity of the site in terms of the mapped spectral response accelerations, but also the site soil profile and the nature of the structure's occupancy category. Housing, in areas that have a low seismic risk and are classified as Seismic Design Categories A and B, need not go through the seismic design consideration. Structures in Seismic Design Categories C and D are considered moderate to high risk and should be designed in accordance with this section. In an area that is close to a major active fault, where S_1 is greater than or equal to 1.17g, there is deemed to be considerable seismic risk, and the seismic design category is classified as E. Dwellings in Seismic Category E must be designed in accordance with the IBC (see Section R301.2.2.4). The seismic design category classification is a key criterion in using and understanding the seismic requirements because the analysis method, general design, structural material detailing, and the structure's component and system design requirements are determined, at least in part, by the seismic design category.

R301.2.2.1.1 Alternate determination of seismic design category. The seismic design categories and corresponding short period design spectral response accelerations, S_{DS} shown in Figure R301.2(2) are based on soil Site Class D, as defined in Section 1613.5.2 of the *International Building Code*. If soil conditions are other than Site Class D, the short period design spectral response accelerations, S_{DS}, for a site can be determined according to Section 1613.5 of the *Inter-*

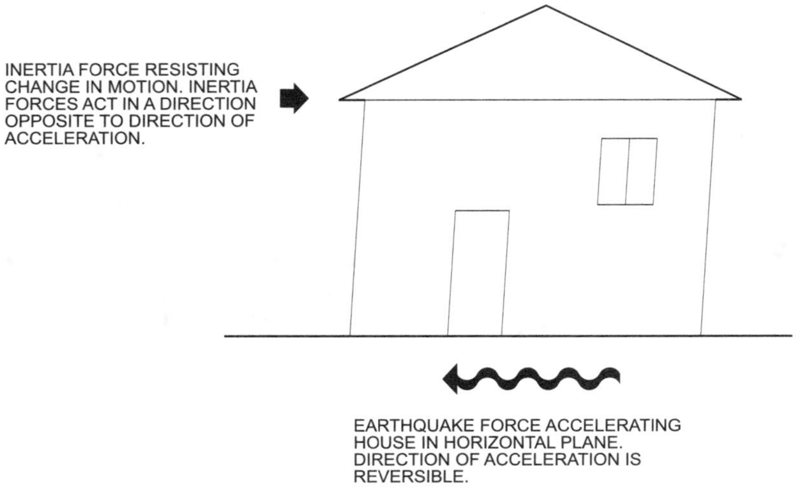

INERTIA FORCE RESISTING CHANGE IN MOTION. INERTIA FORCES ACT IN A DIRECTION OPPOSITE TO DIRECTION OF ACCELERATION.
EARTHQUAKE FORCE ACCELERATING HOUSE IN HORIZONTAL PLANE. DIRECTION OF ACCELERATION IS REVERSIBLE.

Figure R301.2.2
SEISMIC FORCE

national Building Code. The value of S_{DS} determined according to Section 1613.5 of the *International Building Code* is permitted to be used to set the seismic design category according to Table R301.2.2.1.1, and to interpolate between values in Tables R602.10.3(3), R603.9.2(1) and other seismic design requirements of this code.

❖In recognition that there may be an advantage to determining a building's seismic design category using the more detailed procedure of the IBC, the code allows this as an option. If soil conditions are other than Site Class D, for instance, the Short Period Design Spectral Response Acceleration (S_{DS}) for a site may be determined using Section 1613 of the IBC. This value of S_{DS} can then be used to establish the seismic design category according to Table R301.2.2.1.1. Determining S_{DS} in this manner, although requiring more work, will provide the benefit of reducing the amount of wall bracing that would otherwise be required by the provisions in Chapter 6, based solely on seismic design category. This is accomplished, in Table R602.10.3(3) for instance, by interpolating between the tabulated values using the values of S_{DS} determined as described above.

TABLE R301.2.2.1.1
SEISMIC DESIGN CATEGORY DETERMINATION

CALCULATED S_{DS}	SEISMIC DESIGN CATEGORY
$S_{DS} \leq 0.17g$	A
$0.17g < S_{DS} \leq 0.33g$	B
$0.33g < S_{DS} \leq 0.50g$	C
$0.50g < S_{DS} \leq 0.67g$	D_0
$0.67g < S_{DS} \leq 0.83g$	D_1
$0.83g < S_{DS} \leq 1.17g$	D_2
$1.17g < S_{DS}$	E

❖This table lists the seismic design category for ranges of Short Period Design Spectral Response Acceleration, S_{DS}. These numbers correspond exactly to the legend in the Seismic Design Category Map, Figure R301.2(2). Use of this table is necessary only if the alternative seismic design category determination is used.

R301.2.2.1.2 Alternative determination of Seismic Design Category E. Buildings located in Seismic Design Category E in accordance with Figure R301.2(2) are permitted to be reclassified as being in Seismic Design Category D_2 provided one of the following is done:

1. A more detailed evaluation of the seismic design category is made in accordance with the provisions and maps of the *International Building Code*. Buildings located in Seismic Design Category E per Table R301.2.2.1.1, but located in Seismic Design Category D per the *International Building Code*, may be designed using the Seismic Design Category D_2 requirements of this code.
2. Buildings located in Seismic Design Category E that conform to the following additional restrictions are permitted to be constructed in accordance with the provisions for Seismic Design Category D_2 of this code:
   - 2.1. All exterior shear wall lines or *braced wall panels* are in one plane vertically from the foundation to the uppermost story.
   - 2.2. Floors shall not cantilever past the exterior walls.
   - 2.3. The building is within all of the requirements of Section R301.2.2.2.5 for being considered as regular.

❖The code permits construction of buildings classified as Seismic Design Category E in Figure R301.2(2) in accordance with the provisions for Seismic Design Category D_2 if the building has no irregular features as described in Section R301.2.2.2.5. In addition, out-of-plane offsets in the exterior shear-wall lines or braced wall panels are not allowed, and floors are not permitted to cantilever beyond the exterior walls. In other words, this reclassification to a lower seismic design category is limited to very regular, "box-like" structures.

Also, a building classified as Seismic Design Category E using Figure R301.2(2) may possibly be reclassified to Seismic Design Category D_2 if a more detailed evaluation in accordance with the provisions of the IBC is undertaken. If application of the IBC provisions results in a classification of Seismic Design Category D, then the building may be constructed as Seismic Design Category D_2 under the code.

R301.2.2.2 Seismic Design Category C. Structures assigned to Seismic Design Category C shall conform to the requirements of this section.

❖This section establishes the limitations for weights, irregularity and structures in Seismic Design Category C. This would also be applicable for Seismic Design Categories D_0, D_1 and D_2 by the reference in Section R301.2.2.3.

R301.2.2.2.1 Weights of materials. Average dead loads shall not exceed 15 pounds per square foot (720 Pa) for the combined roof and ceiling assemblies (on a horizontal projection) or 10 pounds per square foot (480 Pa) for floor assemblies, except as further limited by Section R301.2.2. Dead loads for walls above *grade* shall not exceed:

1. Fifteen pounds per square foot (720 Pa) for exterior light-frame wood walls.
2. Fourteen pounds per square foot (670 Pa) for exterior light-frame cold-formed steel walls.
3. Ten pounds per square foot (480 Pa) for interior light-frame wood walls.
4. Five pounds per square foot (240 Pa) for interior light-frame cold-formed steel walls.
5. Eighty pounds per square foot (3830 Pa) for 8-inch-thick (203 mm) masonry walls.
6. Eighty-five pounds per square foot (4070 Pa) for 6-inch-thick (152 mm) concrete walls.
7. Ten pounds per square foot (480 Pa) for SIP walls.

Exceptions:

1. Roof and ceiling dead loads not exceeding 25 pounds per square foot (1190 Pa) shall be permitted provided the wall bracing amounts in Chapter 6 are increased in accordance with Table R301.2.2.2.1.
2. Light-frame walls with stone or masonry veneer shall be permitted in accordance with the provisions of Sections R702.1 and R703.
3. Fireplaces and chimneys shall be permitted in accordance with Chapter 10.

❖As stated in Section R301.2.2, the dead loads of structures in seismic design categories regulated by the code must be within the limits listed in this section. Exceeding these limits would necessitate a design of the lateral-force-resisting system as described in Section R301.1.2. In some cases, these dead load limits are more restrictive than the dead load allowances in the prescriptive tables of the code [e.g., Table R502.3.1(1) for floor joist and Table R802.5(1) for rafters].

Exception 1 permits the use of roof/ceiling assemblies that exceed the 15 pounds per square foot (psf) (720 Pa) dead load limit if additional wall bracing is installed as determined using the adjustment factors of Table R301.2.2.2.1. This would permit the use of heavier roof coverings, such as tile. The additional wall bracing compensates for the additional seismic force that would be attributed to the added dead load.

Exception 2 directly references Sections R702.1 and R703. Section R702.1 specifically addresses interior stone and masonry veneer and Section R703 contains the requirements for the support and anchorage of masonry veneer.

Exception 3 references Chapter 10 for the seismic limitations for chimneys and fireplaces. Seismic reinforcing requirements for masonry fireplaces are found in Section R1001.3 and seismic anchorage of masonry fireplaces is covered in Section R1001.4.

Seismic reinforcing of masonry chimneys is addressed in Section R1003.3.

TABLE R301.2.2.2.1
WALL BRACING ADJUSTMENT FACTORS BY ROOF COVERING DEAD LOAD[a]

WALL SUPPORTING	ROOF/CEILING DEAD LOAD	
	15 psf or less	25 psf
Roof only	1.0	1.2
Roof plus one or two stories	1.0	1.1

For SI: 1 pound per square foot = 0.0479 kPa.
a. Linear interpolation shall be permitted.

❖The adjustment factors in this table must be used to increase the amount of wall bracing if the exception is used from Section R301.2.2.2.1, which allows the installation of a roof/ceiling assembly exceeding the 15 psf (720 Pa) dead load limit. The resulting additional wall bracing accounts for the additional seismic forces to avoid overloading the lateral-force-resisting system.

R301.2.2.2.2 Stone and masonry veneer. Anchored stone and masonry veneer shall comply with the requirements of Sections R702.1 and R703.

❖Stone and masonry veneer are addressed in Chapter 7 of the code. Table R703.7(1) lists the stone and masonry veneer limitations based on the seismic design category.

R301.2.2.2.3 Masonry construction. Masonry construction shall comply with the requirements of Section R606.12.

❖This section provides a cross reference to the masonry sections in the code for seismic requirements.

R301.2.2.2.4 Concrete construction. Detached one- and two-family *dwellings* with exterior above-*grade* concrete walls shall comply with the requirements of Section R611, PCA 100 or shall be designed in accordance with ACI 318. *Townhouses* with above-*grade* exterior concrete walls shall comply with the requirements of PCA 100 or shall be designed in accordance with ACI 318.

❖This section provides a cross reference to the prescriptive provisions for concrete construction in Seismic Design Category C. Townhouses, by nature of being multiple-family housings, and exterior above-grade concrete walls, may not use the simplified approach in Section R611. As an alternative, concrete construction may be designed in accordance with accepted engineering practice.

R301.2.2.2.5 Irregular buildings. The seismic provisions of this code shall not be used for irregular structures located in Seismic Design Categories C, D_0, D_1 and D_2. Irregular portions of structures shall be designed in accordance with accepted engineering practice to the extent the irregular features affect the performance of the remaining structural system. When the forces associated with the irregularity are resisted by a structural system designed in accordance with accepted engineering practice, design of the remainder of the building shall be permitted using the provisions of this code. A building or portion of a building shall be considered to be irregular when one or more of the following conditions occur:

1. When exterior shear wall lines or *braced wall panels* are not in one plane vertically from the foundation to the uppermost *story* in which they are required.

   Exception: For wood light-frame construction, floors with cantilevers or setbacks not exceeding four times the nominal depth of the wood floor joists are permitted to support *braced wall panels* that are out of plane with *braced wall panels* below provided that:

   1. Floor joists are nominal 2 inches by 10 inches (51 mm by 254 mm) or larger and spaced not more than 16 inches (406 mm) on center.
   2. The ratio of the back span to the cantilever is at least 2 to 1.
   3. Floor joists at ends of *braced wall panels* are doubled.

4. For wood-frame construction, a continuous rim joist is connected to ends of all cantilever joists. When spliced, the rim joists shall be spliced using a galvanized metal tie not less than 0.058 inch (1.5 mm) (16 gage) and $1^1/_2$ inches (38 mm) wide fastened with six 16d nails on each side of the splice or a block of the same size as the rim joist of sufficient length to fit securely between the joist space at which the splice occurs fastened with eight 16d nails on each side of the splice; and

5. Gravity loads carried at the end of cantilevered joists are limited to uniform wall and roof loads and the reactions from headers having a span of 8 feet (2438 mm) or less.

2. When a section of floor or roof is not laterally supported by shear walls or *braced wall lines* on all edges.

Exception: Portions of floors that do not support shear walls or *braced wall panels* above, or roofs, shall be permitted to extend no more than 6 feet (1829 mm) beyond a shear wall or *braced wall line*.

3. When the end of a *braced wall panel* occurs over an opening in the wall below and ends at a horizontal distance greater than 1 foot (305 mm) from the edge of the opening. This provision is applicable to shear walls and *braced wall panels* offset in plane and to *braced wall panels* offset out of plane as permitted by the exception to Item 1 above.

Exception: For wood light-frame wall construction, one end of a *braced wall panel* shall be permitted to extend more than 1 foot (305 mm) over an opening not more than 8 feet (2438 mm) wide in the wall below provided that the opening includes a header in accordance with the following:

1. The building width, loading condition and framing member species limitations of Table R502.5(1) shall apply; and
2. Not less than one 2 × 12 or two 2 × 10 for an opening not more than 4 feet (1219 mm) wide; or
3. Not less than two 2 × 12 or three 2 × 10 for an opening not more than 6 feet (1829 mm) wide; or
4. Not less than three 2 × 12 or four 2 × 10 for an opening not more than 8 feet (2438 mm) wide; and
5. The entire length of the *braced wall panel* does not occur over an opening in the wall below.

4. When an opening in a floor or roof exceeds the lesser of 12 feet (3658 mm) or 50 percent of the least floor or roof dimension.

5. When portions of a floor level are vertically offset.

Exceptions:

1. Framing supported directly by continuous foundations at the perimeter of the building.
2. For wood light-frame construction, floors shall be permitted to be vertically offset when the floor framing is lapped or tied together as required by Section R502.6.1.

6. When shear walls and *braced wall lines* do not occur in two perpendicular directions.

7. When stories above *grade* plane partially or completely braced by wood wall framing in accordance with Section R602 or steel wall framing in accordance with Section R603 include masonry or concrete construction.

Exception: Fireplaces, chimneys and masonry veneer as permitted by this code. When this irregularity applies, the entire *story* shall be designed in accordance with accepted engineering practice.

❖Conventional light-frame construction typically allows cantilevers, offsets, etc. within certain limits in order to accommodate common design features and options. These features are not well suited to resisting earthquake forces. This becomes more of a concern in areas of higher seismic hazard. Where this is the case, it is preferable that the building, or more importantly the lateral-force-resisting system, be more "box-like" or regular.

Accordingly, this section identifies building features that are irregular and are not permitted under the concrete or conventional light-frame construction provisions in Seismic Design Categories C, D_0, D_1 and D_2. These irregular portions of structures must be designed in accordance with accepted engineering practice. A portion of a building is classified as "irregular" when one of the several conditions described in the code occurs.

Out-of-plane offsets in exterior braced wall lines have been recognized as a factor in buildings damaged by earthquakes. Where an offset occurs, the earthquake forces must be transferred through the floor framing. Thus, Item 1 stipulates that exterior shear wall lines or braced wall panels that are not in one plane vertically from the foundation to the uppermost story in which they are required constitute an irregularity as illustrated in Commentary Figure R301.2.2.2.5(1). The exception, illustrated in Commentary Figure R301.2.2.2.5(2), allows out-of-plane offsets not exceeding four times the nominal depth of the wood floor joists supporting the braced wall line above. There are five additional limitations that must be met to assure the adequacy of the supporting floor framing. Use of this exception is limited to wood light-frame construction.

Typically, braced wall lines are located around the perimeter of floors and walls where forces are greatest. As Item 2 indicates, an irregularity occurs where a portion of a floor or roof is not laterally supported by shear walls or braced wall lines on all edges. This is shown in Commentary Figure R301.2.2.2.5(3). The exception permits a floor or roof to extend beyond a braced wall line as illustrated in Commentary Figure R301.2.2.2.5(4).

Where the end of a braced wall panel occurs over an opening in the wall below, an irregularity will be present if the panel extends a horizontal distance greater than 1 foot (305 mm) from the edge of the opening. As Commentary Figure R301.2.2.2.5(5) illustrates, this applies to braced wall panels offset out of plane, as well as in plane. An exception for wood light-frame construction permits a braced wall panel to extend more than 1 foot (305 mm) over an opening in the wall below if the opening includes a header, in accordance with this exception, and the entire length of the braced wall panel does not occur over the opening.

Openings in floors are required in most buildings to provide for stairways, equipment chases, etc. If the size of the opening is large compared to the size of the floor diaphragm, earthquake forces may overstress portions of the diaphragm. Because this could result in premature failure of the diaphragm and, thus, constitute an incomplete load path, the code limits the size of such openings allowed under conventional light-frame construction. Commentary Figure R301.2.2.2.5(6) illustrates the irregularity created by an opening in a floor or roof that exceeds the lesser of 12 feet (3657 mm) or 50 percent of the least floor or roof dimension.

Commentary Figure R301.2.2.2.5(7) illustrates Item 5. An irregularity results where portions of a floor level are vertically offset. This limitation does not apply to framing that is supported directly by continuous foundations at the perimeter of the building. Also, in wood light-frame construction, floors may be vertically offset if the floor framing is lapped or tied together as required by Section R502.6.1.

More often than not, the shape of conventional light-frame buildings is regular with wall lines perpendicular to one another. Commentary Figure R301.2.2.2.5(8) depicts a nonorthogonal system. This type of irregularity occurs where shear walls or braced wall lines are not oriented perpendicular to each other.

In addition, an irregularity occurs if shear walls or braced wall lines are constructed of dissimilar bracing systems on any story level above grade plane. This requirement applies to the underlying framing system material, such as steel-framed shear walls or wood-framed braced wall panels combined with masonry or concrete construction in the same story.

Where this condition occurs, the entire story must be designed in accordance with accepted engineering practice.

Masonry or concrete fireplaces, chimneys and masonry veneer can be combined with steel or wood framing and not be considered irregular.

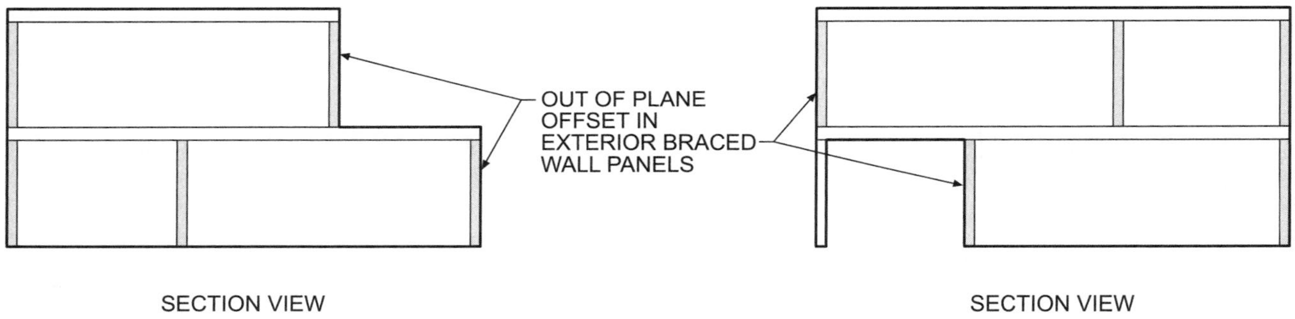

Figure R301.2.2.2.5(1)
BRACED WALL PANELS OUT OF PLANE

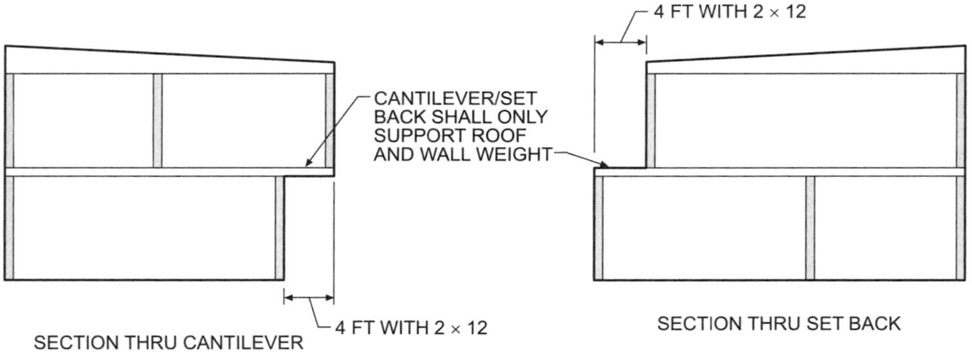

For SI:1 inch = 25.4 mm, 1 foot = 304.8 mm.

Figure R301.2.2.2.5(2)
BRACED WALL PANELS SUPPORTED BY CANTILEVER OR SET BACK

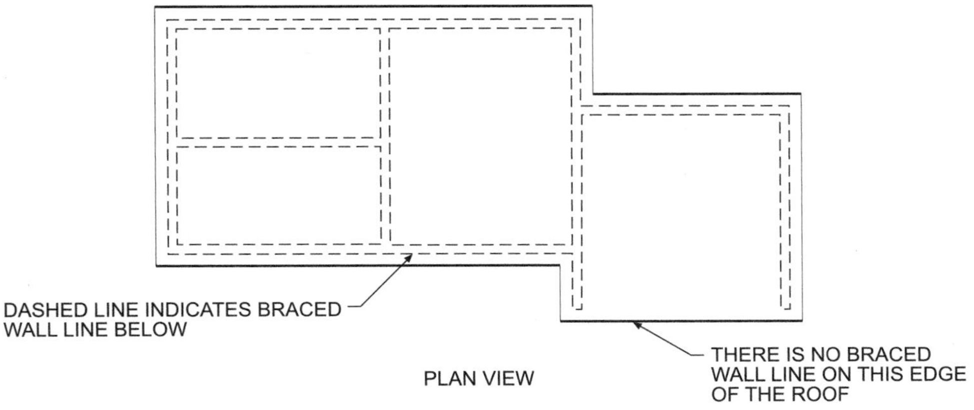
DASHED LINE INDICATES BRACED WALL LINE BELOW
THERE IS NO BRACED WALL LINE ON THIS EDGE OF THE ROOF
PLAN VIEW

Figure R301.2.2.2.5(3)
FLOOR OR ROOF NOT SUPPORTED AT ALL EDGES

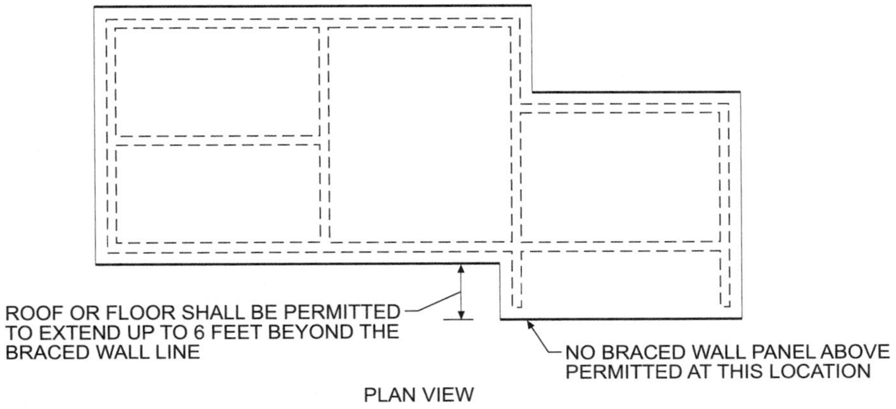
ROOF OR FLOOR SHALL BE PERMITTED TO EXTEND UP TO 6 FEET BEYOND THE BRACED WALL LINE
NO BRACED WALL PANEL ABOVE PERMITTED AT THIS LOCATION
PLAN VIEW

For SI:1 foot = 304.8 mm.

Figure R301.2.2.2.5(4)
ROOF OR FLOOR EXTENSION BEYOND BRACED WALL LINE

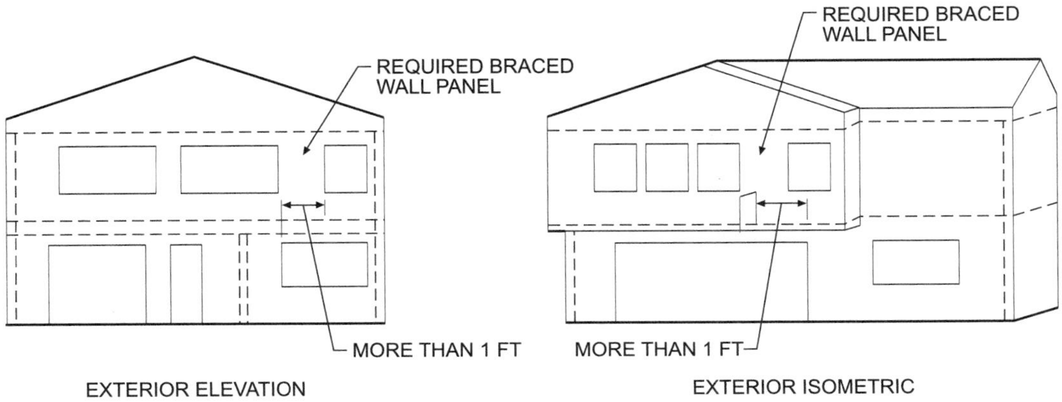
REQUIRED BRACED WALL PANEL
REQUIRED BRACED WALL PANEL
MORE THAN 1 FT
MORE THAN 1 FT
EXTERIOR ELEVATION
EXTERIOR ISOMETRIC

For SI:1 foot = 304.8 mm.

Figure R301.2.2.2.5(5)
BRACED WALL PANEL EXTENSION OVER OPENING

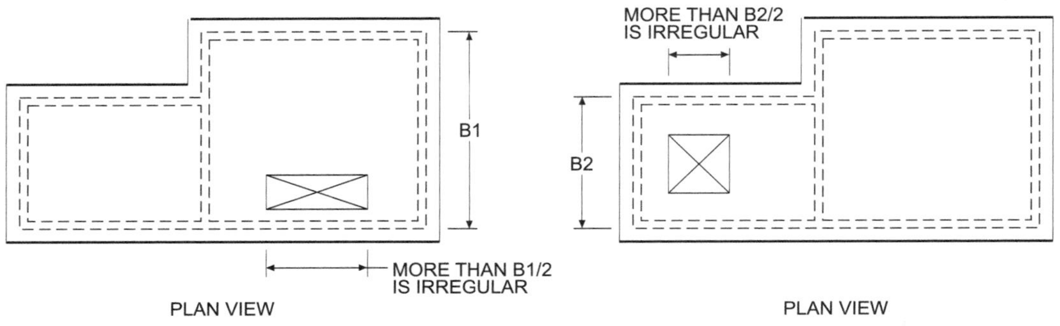

Figure R301.2.2.2.5(6)
OPENING LIMITATIONS FOR FLOOR AND ROOF DIAPHRAGMS

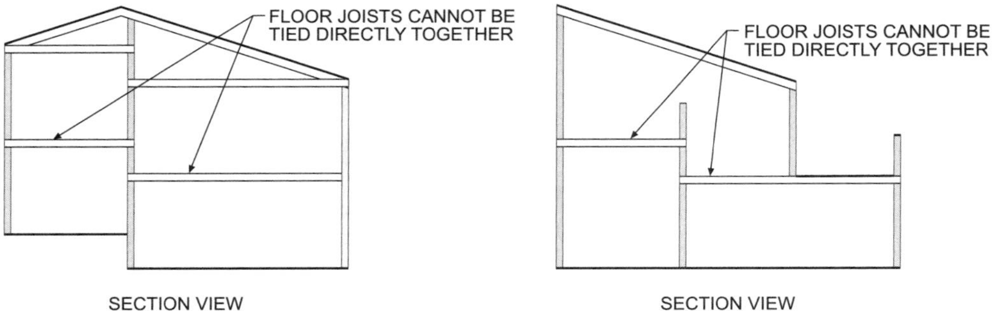

Figure 301.2.2.2.5(7)
PORTIONS OF FLOOR LEVEL OFFSET VERTICALLY

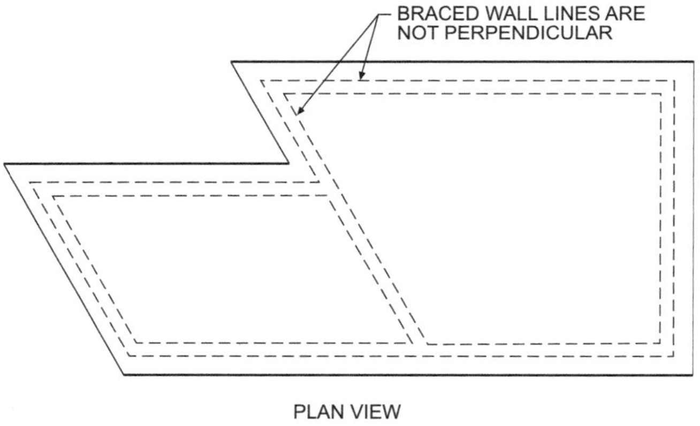

Figure R301.2.2.2.5(8)
BRACED WALL LINES NOT PERPENDICULAR

R301.2.2.3 Seismic Design Categories D_0, D_1 and D_2. Structures assigned to Seismic Design Categories D_0, D_1 and D_2 shall conform to the requirements for Seismic Design Category C and the additional requirements of this section.

❖Seismic requirements are cumulative as the seismic design category becomes higher. A structure that is assigned to Seismic Design Category D_0, D_1 or D_2 must comply with all the seismic requirements for Seismic Design Category C structures (see Section R301.2.2.2, in addition to those for Seismic Design Categories D_0, D_1 and D_2).

R301.2.2.3.1 Height limitations. Wood-framed buildings shall be limited to three stories above *grade* plane or the limits given in Table R602.10.3(3). Cold-formed, steel-framed buildings shall be limited to less than or equal to three stories above *grade* plane in accordance with AISI S230. Mezzanines as defined in Section R202 shall not be considered as stories. Structural insulated panel buildings shall be limited to two stories above *grade* plane.

❖This section limits the building height so the lateral-force-resisting elements will not be overloaded during an earthquake. It provides a cross reference to other sections that limit the height of structures. Wood-framed buildings are generally limited to three stories. Floor levels meeting the limitations for mezzanines (see the definition in Chapter 2) need not be counted as a story.

R301.2.2.3.2 Stone and masonry veneer. Anchored stone and masonry veneer shall comply with the requirements of Sections R702.1 and R703.

❖This section limits the height of stone and masonry veneer so the bracing capacity will not be exceeded during an earthquake. This section references Sections R702.1 and R703. Tables R703.7(1) and R703.7(2) provide specific information on the maximum allowed height of veneer and the maximum allowed thickness and weight of veneer.

R301.2.2.3.3 Masonry construction. Masonry construction in Seismic Design Categories D_0 and D_1 shall comply with the requirements of Section R606.12.1. Masonry construction in Seismic Design Category D_2 shall comply with the requirements of Section R606.12.4.

❖This is a cross reference to the section applicable to masonry seismic requirements.

R301.2.2.3.4 Concrete construction. Buildings with exterior above-*grade* concrete walls shall comply with PCA 100 or shall be designed in accordance with ACI 318.

❖In Seismic Design Categories D_0, D_1 and D_2, exterior above-grade concrete walls must be engineered in accordance with PCA 100, *Prescriptive Design for Exterior Concrete Walls for One- and Two-family Dwellings*, or ACI 318, *Building Code Requirements for Structural Concrete*.

R301.2.2.3.5 Cold-formed steel framing in Seismic Design Categories D_0, D_1 and D_2. In Seismic Design Categories D_0, D_1 and D_2 in addition to the requirements of this code, cold-formed steel framing shall comply with the requirements of AISI S230.

❖The prescriptive requirements for cold-formed steel framing in Seismic Design Category C are contained in the code. In addition to the seismic requirements found in the code, the cold-formed steel framing must comply with AISI S230.

R301.2.2.3.6 Masonry chimneys. Masonry chimneys shall be reinforced and anchored to the building in accordance with Sections R1003.3 and R1003.4.

❖This is a cross reference to the section applicable to masonry chimney requirements.

R301.2.2.3.7 Anchorage of water heaters. Water heaters shall be anchored against movement and overturning in accordance with Section M1307.2.

❖This is a cross reference to the section applicable to water heater anchorage.

R301.2.2.4 Seismic Design Category E. Buildings in Seismic Design Category E shall be designed to resist seismic loads in accordance with the *International Building Code*, except when the seismic design category is reclassified to a lower seismic design category in accordance with Section R301.2.2.1. Components of buildings not required to be designed to resist seismic loads shall be constructed in accordance with the provisions of this code.

❖In an area that is close to a major active fault, where S_1 is greater than or equal to 1.25g, there is deemed to be considerable seismic risk, and the seismic design category is classified as E. Dwellings in Seismic Design Category E must be designed in accordance with Section 1613 of the IBC. The seismic design category classification is a key criterion in using and understanding the seismic requirements because the analysis method, general design, structural material detailing, and the structure's component and system design requirements are determined, at least in part, by the seismic design category. See the commentary to Section R301.2.2.1.2 for the alternative design of buildings in Seismic Design Category E.

R301.2.3 Snow loads. Wood-framed construction, cold-formed, steel-framed construction and masonry and concrete construction, and structural insulated panel construction in regions with ground snow loads 70 pounds per square foot (3.35 kPa) or less, shall be in accordance with Chapters 5, 6 and 8. Buildings in regions with ground snow loads greater than 70 pounds per square foot (3.35 kPa) shall be designed in accordance with accepted engineering practice.

❖This section specifies the maximum ground snow load where the prescriptive provisions in Chapters 5, 6 and 8 apply.

The prescriptive provisions for floors, walls and roofs in Chapters 5, 6 and 8 apply in regions with ground snow loads of 70 psf (3.35 kPa) or less. Structures in regions with ground snow loads exceeding 70 psf (3.35 kPa) are beyond the limitations of these prescriptive provisions and, as stated in Section R301.1, a design would be required for all elements that carry snow loads. The ground snow load is to be determined using Figure R301.2(5) or by site-specific case studies as discussed in Section 7.0 of ASCE 7.

R301.2.4 Floodplain construction. Buildings and structures constructed in whole or in part in flood hazard areas (including A or V Zones) as established in Table R301.2(1) shall be designed and constructed in accordance with Section R322. Buildings and structures located in whole or in part in identified floodways shall be designed and constructed in accordance with ASCE 24.

❖Buildings in flood hazard areas must meet certain requirements. The requirements are intended to reduce flood damage and improve resistance to flood loads and other effects of flooding, such as saturation. Some riverine flood plains are mapped to show floodways. Because floodways typically are characterized by deeper and faster moving water, it is appropriate that the design of buildings in these areas explicitly account for flood loads as called for in ASCE 24. See commentary for Section R322.

R301.2.4.1 Alternative provisions. As an alternative to the requirements in Section R322.3 for buildings and structures located in whole or in part in coastal high-hazard areas (V Zones) and coastal A Zones, if delineated, ASCE 24 is permitted subject to the limitations of this code and the limitations therein.

❖This section provides an alternative for buildings and structures in certain parts of flood hazard areas (coastal high-hazard areas and coastal A zones) to be designed and constructed according to ASCE 24. See commentary for Section R322.3.

R301.3 Story height. The wind and seismic provisions of this code shall apply to buildings with story heights not exceeding the following:

1. For wood wall framing, the laterally unsupported bearing wall stud height permitted by Table R602.3(5) plus a height of floor framing not to exceed 16 inches (406 mm).

   Exception: For wood-framed wall buildings with bracing in accordance with Tables R602.10.3(1) and R602.10.3(3), the wall stud clear height used to determine the maximum permitted *story height* may be increased to 12 feet (3658 mm) without requiring an engineered design for the building wind and seismic force-resisting systems provided that the length of bracing required by Table R602.10.3(1) is increased by multiplying by a factor of 1.10 and the length of bracing required by Table R602.10.3(3) is increased by multiplying by a factor of 1.20. Wall studs are still subject to the requirements of this section.

2. For steel wall framing, a stud height of 10 feet (3048 mm), plus a height of floor framing not to exceed 16 inches (406 mm).

3. For masonry walls, a maximum bearing wall clear height of 12 feet (3658 mm) plus a height of floor framing not to exceed 16 inches (406 mm).

   Exception: An additional 8 feet (2438 mm) is permitted for gable end walls.

4. For insulating concrete form walls, the maximum bearing wall height per *story* as permitted by Section R611 tables plus a height of floor framing not to exceed 16 inches (406 mm).

5. For structural insulated panel (SIP) walls, the maximum bearing wall height per *story* as permitted by Section R613 tables shall not exceed 10 feet (3048 mm) plus a height of floor framing not to exceed 16 inches (406 mm).

Individual walls or walls studs shall be permitted to exceed these limits as permitted by Chapter 6 provisions, provided *story heights* are not exceeded. Floor framing height shall be permitted to exceed these limits provided the *story height* does not exceed 11 feet 7 inches (3531 mm). An engineered design shall be provided for the wall or wall framing members when they exceed the limits of Chapter 6. Where the *story height* limits of this section are exceeded, the design of the building, or the noncompliant portions thereof, to resist wind and seismic loads shall be in accordance with the *International Building Code*.

❖This section defines the story height and wall height limits. These story heights are the basis for the wind- and seismic-force-resisting system. The wall heights are based on the materials used for wall construction. The wall height limits are prescribed for wood, steel, masonry and concrete. The story height permitted is the wall height including floor framing [not to exceed 16 inches (406 mm)]. The story height may vary depending on the material used for wall construction. Individual walls or walls studs shall be permitted to exceed these limits as permitted by Chapter 6 provisions, provided story heights are not exceeded. An engineered design shall be provided for the wall or wall framing members when they exceed the limits of Chapter 6. However, if the story heights of this section are exceeded, an engineered design in accordance with the IBC must be provided for the building or non-compliant portion for the wind and seismic lateral forces.

R301.4 Dead load. The actual weights of materials and construction shall be used for determining dead load with consideration for the dead load of fixed service *equipment*.

❖The actual weights of materials of construction must be used to determine the dead loads. The code requires the weight of fixed service equipment (see the definition in Chapter 2) to be included, as well. It is important to verify that the actual dead loads of proposed buildings are within the limits of the dead load allowance in the prescriptive design tables of subsequent chapters. Otherwise, in accordance with Section R301.1, a design is required. Also, the dead load limits of Section R301.2.2.2 must be met where required by a structure's seismic design category as explained in the commentary to Section R301.2.2.

The dead load of a building or other structure is the weight of all permanent construction, such as floors, roofs, permanent partitions, stairways and walls. The actual weights of materials and construction should be considered in the building design along with the dead load of fixed service equipment. Commentary Table R301.4 lists the weight of typical residential building components. Sources of information for weights of fixed service equipment include manufacturer's literature and trade association publications.

Table R301.4
MINIMUM DESIGN DEAD LOADS FOR TYPICAL RESIDENTIAL COMPONENTS

COMPONENT	LOAD (psf)
CEILINGS	
Acoustical fiber tile	1
Gypsum board (per $^{1}/_{8}$-inch thickness)	0.55
Mechanical duct allowance	4
Plaster on tile or concrete	5
Plaster on wood lath	8
Suspended steel channel system	2
Suspended metal lath and cement plaster	15
Suspended metal lath and gypsum plaster	10
Wood-furring suspension system	2.5
COVERINGS, ROOF AND WALL	
Asbestos-cement shingles	4
Asphalt shingles	2
Cement tile	16
Clay tile (for mortar add 10 lb)	
Book tile, 2-inch	12
Book tile, 3-inch	20
Ludowici	10
Roman	12
Spanish	19
Composition:	
Three-ply ready roofing	1
Four-ply felt and gravel	5.5
Five-ply felt and gravel	6
Copper or tin	1
Corrugated asbestos-cement roofing	4
Deck, metal 20 gage	2.5
Deck, metal, 18 gage	3
Decking, 2-inch wood (Douglas fir)	5
Decking, 3-inch wood (Douglas fir)	8
Fiberboard, 2-inch	0.75
Gypsum sheathing, 2-inch	2
Insulation, roof boards (per inch thickness)	
Cellular glass	0.7
Fibrous glass	1.1
Fiberboard	1.5
Perlite	0.8
Polystyrene foam	0.2
Urethane foam with skin	0.5
Plywood (per $^{1}/_{8}$-inch thickness)	0.4
Rigid insulation, 2-inch	0.75
Skylight, metal frame, $^{3}/_{8}$-inch wired glass	8
Slate, $^{3}/_{16}$-inch	7
Slate, $^{1}/_{4}$-inch	10
Structural insulated panelsw (SIPs)	
$4^{1}/_{2}$-inch thick	3
$6^{1}/_{2}$-inch thick	3.5
$8^{1}/_{2}$-inch thick	3.5
$10^{1}/_{2}$-inch thick	4.0
$12^{1}/_{2}$-inch thick	4.5
Waterproofing members	
Bituminous, gravel-covered	5.5
Bituminous, smooth surface	1.5
Liquid applied	1.0
Single-ply, sheet	0.7
Wood sheathing (per 1-inch thickness	3
Wood shingles	3

COMPONENT	LOAD (psf)
FLOORS AND FLOOR FINISHES	
Asphalt block (2-inch), $^{1}/_{2}$-inch mortar	30
Cement finish (1-inch) on stone-concrete fill	32
Ceramic or quarry tile ($^{3}/_{4}$-inch) on 1/2-inch motar bed	16
Ceramic or quarry tile ($^{3}/_{4}$-inch) on 1-inch motar bed	23
Concrete fill finish (per thickness)	12
Hardwood flooring, $^{7}/_{8}$-inch	4
Linoleum or asphalt tile, $^{1}/_{4}$-inch	1
Marble and motar on stone-concrete fill	33
Slate (per inch thickness)	15
Solid flat tile on 1-inch motar base	23
Subflooring, $^{3}/_{4}$-inch	3
Terrazzo (12-inch) directly on slab	19
Terrazzo (1-inch) on stone-concrete fill	32
Terrazzo (1-inch) 2-inch stone concrete	32
Wood block (3-inch) on mastic, no fill	10
Wood block (3-inch) on 2-inch mortar base	16

FLOOR, WOOD-JOIST (NO PLASTER) DOUBLE WOOD FLOOR

Joint Sizes (inches	12-inch spacing (lb/ft^2)	16-inch spacing (1b/ft^2)	24-inch spacing (lb/ft^2)
2 × 6	6	5	5
2 × 8	6	6	5
2 × 10	7	6	6
2 × 12	8	7	6

COMPONENT	LOAD (psf)
FRAME PARTITIONS	
Moveable steel partitions	4
Wood or steel studs, 2-inch gypsum board each side	8
Wood studs, 2 × 4 unplastered	4
Wood studs, 2 × 4, plastered one side	12
Wood studs, 2 × 4, plastered two sides	20
FRAME WALLS	
Exterior stud walls:	
2 × 4 @ 16 inches, $^{5}/_{8}$-inch gypsum insulated, $^{3}/_{8}$-inch siding	11
2 × 6 @ 16 inches, $^{5}/_{8}$-inch gypsum insulated, $^{3}/_{8}$-inch siding	12
Exterior stud walls, with brick veneer	48
Windows, glass, frame and sash	8
MASONRY PARTITIONS AND WALLS	
Clay tile:	
4-inch	18
6-inch	24
8-inch	24
Concrete block, heavy aggregate:	
4-inch	30
6-inch	42
8-inch	55
12-inch	85
Concrete block, light aggregate:	
4-inch	20
6-inch	28
8-inch	38
12-inch	55

For SI: 1 pound per square foot = 0.0479 kPa, 1 inch = 25.4 mm, 1 pound = 0.4536 kg.

Note: Weights of masonry include mortar but nor plaster. For plaster add 5 lb/ft^2 for each inch of plaster. Values give represent averages. In some cases there is considerable range for the same construction.

R301.5 Live load. The minimum uniformly distributed live load shall be as provided in Table R301.5.

❖Table R301.5 lists the minimum uniformly distributed live loads (see the definition in Chapter 2) required for design of various portions of a residence. These loads are the basis for the prescriptive tables for floor systems in Chapter 5.

TABLE R301.5
MINIMUM UNIFORMLY DISTRIBUTED LIVE LOADS
(in pounds per square foot)

USE	LIVE LOAD
Uninhabitable attics without storage[b]	10
Uninhabitable attics with limited storage[b, g]	20
Habitable attics and attics served with fixed stairs	30
Balconies (exterior) and decks[e]	40
Fire escapes	40
Guardrails and handrails[d]	200[h]
Guardrail in-fill components[f]	50[h]
Passenger vehicle garages[a]	50[a]
Rooms other than sleeping room	40
Sleeping rooms	30
Stairs	40[c]

For SI: 1 pound per square foot = 0.0479 kPa, 1 square inch = 645 mm^2, 1 pound = 4.45 N.

a. Elevated garage floors shall be capable of supporting a 2,000-pound load applied over a 20-square-inch area.

b. Uninhabitable attics without storage are those where the maximum clear height between joists and rafters is less than 42 inches, or where there are not two or more adjacent trusses with web configurations capable of accommodating an assumed rectangle 42 inches high by 24 inches in width, or greater, within the plane of the trusses. This live load need not be assumed to act concurrently with any other live load requirements.

c. Individual stair treads shall be designed for the uniformly distributed live load or a 300-pound concentrated load acting over an area of 4 square inches, whichever produces the greater stresses.

d. A single concentrated load applied in any direction at any point along the top.

e. See Section R502.2.2 for decks attached to exterior walls.

f. Guard in-fill components (all those except the handrail), balusters and panel fillers shall be designed to withstand a horizontally applied normal load of 50 pounds on an area equal to 1 square foot. This load need not be assumed to act concurrently with any other live load requirement.

g. Uninhabitable attics with limited storage are those where the maximum clear height between joists and rafters is 42 inches or greater, or where there are two or more adjacent trusses with web configurations capable of accommodating an assumed rectangle 42 inches in height by 24 inches in width, or greater, within the plane of the trusses.

The live load need only be applied to those portions of the joists or truss bottom chords where all of the following conditions are met:

1. The attic area is accessible from an opening not less than 20 inches in width by 30 inches in length that is located where the clear height in the attic is a minimum of 30 inches.
2. The slopes of the joists or truss bottom chords are no greater than 2 inches vertical to 12 units horizontal.
3. Required insulation depth is less than the joist or truss bottom chord member depth.

The remaining portions of the joists or truss bottom chords shall be designed for a uniformly distributed concurrent live load of not less than 10 lb/ft^2.

h. Glazing used in handrail assemblies and guards shall be designed with a safety factor of 4. The safety factor shall be applied to each of the concentrated loads applied to the top of the rail, and to the load on the in-fill components. These loads shall be determined independent of one another, and loads are assumed not to occur with any other live load.

❖The uniform live loads of this table are consistent with Table 1607.1 of the IBC, which in turn is similar to ASCE 7. The table provides the minimum loads based on the use of a particular area or portion of a structure that must be considered for the design of corresponding structural elements of any residence constructed under the code. For instance, bedrooms (sleeping rooms) required use of a 30-psf (1.48 kN/m^2) live load, while all other rooms must be designed for 40-psf (1915 kN/m^2) uniform live load.

Attics with limited storage must be designed for 20-psf (957.6 N/m^2) live loads, while a 10-psf (0.48 kN/m^2) live load is required without storage. Note b provides criteria for when an attic is to be considered too small to provide adequate space for storage. Also, attics not required to have access in accordance with Section R807 may be considered as attics without storage. Habitable attics and attics accessed by fixed stairs (so they may be finished later) must be designed for a 30 psf (1.44 kN/m^2). Note g provides criteria for when the attic space is considered large enough to become potential storage or living space.

R301.6 Roof load. The roof shall be designed for the live load indicated in Table R301.6 or the snow load indicated in Table R301.2(1), whichever is greater.

❖The basic 20-psf (0.96 kN/m^2) roof live load is a severe enough loading condition on the roof to support live loads created by maintenance workers, including their equipment and materials. The code permits reduction of this basic live load based on the tributary area supported by any structural member of the roof. The rationale for this reduction of the roof live load is that it is highly improbable that structural members with large tributary areas would be loaded over the entire area with the full live load. In addition, the reduction of the basic roof live load for roofs is also a function of the slope of the roof because it becomes less probable that the loads on the roof members would be at maximum levels as the roof slope increases.

The code requires that the roof be designed to resist snow loading and does not permit a reduction of snow loads based on the tributary area, as it does for the roof live load. Be aware, however, that snow on the roof rarely accumulates evenly. The design should account for unbalanced snow loading. One case would be the loading of one slope of a gable roof with snow while the other slope is unloaded. Many roofs fail from accumulation of snow at valleys, parapets, roof structures and offsets in roofs. However, the code does not specify criteria for the determination of how these potential accumulations are to be handled. Snow loading provisions in the IBC may be consulted for these loading scenarios.

TABLE R301.6
MINIMUM ROOF LIVE LOADS IN POUNDS-FORCE PER SQUARE FOOT OF HORIZONTAL PROJECTION

ROOF SLOPE	TRIBUTARY LOADED AREA IN SQUARE FEET FOR ANY STRUCTURAL MEMBER		
	0 to 200	201 to 600	Over 600
Flat or rise less than 4 inches per foot (1:3)	20	16	12
Rise 4 inches per foot (1:3) to less than 12 inches per foot (1:1)	16	14	12
Rise 12 inches per foot (1:1) and greater	12	12	12

For SI: 1 square foot = 0.0929 [illegible] pound per square foot = 0.0479 kPa,
1 inch per foot = 83.3 [illegible].

❖See the commen[illegible] Section R301.6.

R301.7 Deflectio[illegible] [illegible] allowable deflection of any structural member u[illegible] the live load listed in Sections R301.5 and R301.6 or [illegible] loads determined by Section R301.2.1 shall not exceed the values in Table R301.7.

❖The allowable deflection of structural members from the design live load must not exceed the values in Table R301.7. These limits are expressed in terms of the span length. Brittle finishes, such as plaster ceilings and exterior stucco walls, are protected by limiting the deflection of those elements.

Commentary Figure R301.7 shows a simply supported beam with dead load deflection before the live load has been applied. It also shows the same beam after the live load has been applied. The vertical distance that the center of the beam has moved from the initial dead load deflection position is called the live load deflection. If a finish material is applied to this beam, the finish material is subject to distortion in proportion to this deflection.

TABLE R301.7
ALLOWABLE DEFLECTION OF STRUCTURAL MEMBERS[b, c]

STRUCTURAL MEMBER	ALLOWABLE DEFLECTION
Rafters having slopes greater than 3:12 with no finished ceiling attached to rafters	*L*/180
Interior walls and partitions	*H*/180
Floors/ceilings with plaster or stucco finish	*L*/360
All other structural members	*L*/240
Exterior walls—wind loads[a] with plaster or stucco finish	*H*/360
Exterior walls with other brittle finishes	*H*/240
Exterior walls with flexible finishes	*H*/120[d]
Lintels supporting masonry veneer walls[e]	*L*/600

Note: *L* = span length, *H* = span height.

a. The wind load shall be permitted to be taken as 0.7 times the Component and Cladding loads for the purpose of the determining deflection limits herein.

b For cantilever members, *L* shall be taken as twice the length of the cantilever.

c. For aluminum structural members or panels used in roofs or walls of sunroom additions or patio covers, not supporting edge of glass or sandwich panels, the total load deflection shall not exceed *L*/60. For continuous aluminum structural members supporting edge of glass, the total load deflection shall not exceed *L*/175 for each glass lite or *L*/60 for the entire length of the member, whichever is more stringent. For sandwich panels used in roofs or walls of sunroom additions or patio covers, the total load deflection shall not exceed *L*/120.

d. Deflection for exterior walls with interior gypsum board finish shall be limited to an allowable deflection of *H*/180.

e. Refer to Section R703.7.2.

❖Generally, it is not necessary to calculate the deflection indicated in Table R301.7, since the deflection limit is typically accounted for in the prescriptive design tables of the code. Since this is not the case for elements that required design in accordance with Section R301.1.2, the following example illustrates this requirement.

Example:

What is the maximum allowable deflection of a 20-foot (6096 mm) span with the following conditions?

a) Rafter with no ceiling load having a slope of 5 units vertical in 12 units horizontal (42 percent)

b) Rafters with no ceiling load having a slope of 2 units in 12 units horizontal (17 percent)

c) A floor joist supporting a finished floor Solution:

Length = Span = 20-feet (6096 mm)

a) D MAX = L/180 = 20 feet $\times$ 12 inches/180 = 1.33 inches or approx. $1^3/_8$ inches (35 mm)

b) D MAX = L/240 = 20 feet $\times$ 12 inches/240 = 1.00 inches or approx. 1 inch (25.4 mm)

c) D MAX = L/360 = 20 feet $\times$ 12 inches/360 = 0.67 inches or approx. $^5/_8$ inch (15.9 mm)

R301.8 Nominal sizes. For the purposes of this code, where dimensions of lumber are specified, they shall be deemed to be nominal dimensions unless specifically designated as actual dimensions.

❖Because solid sawn lumber sizes are normally referred to using the nominal lumber dimensions, this clarification explains that any code references to lumber dimensions are to be taken as nominal dimensions unless explicitly stated otherwise.

SECTION R302
FIRE-RESISTANT CONSTRUCTION

❖This section groups the fire-resistant construction requirements for between and within dwelling units. This section addresses exterior wall location; townhouse separation; two-family dwellings separation; rated penetrations; garage penetrations; garage separation; under-stair protection; flame spread and smoke development; insulation; fireblocking; draftstopping required and insulation clearance from heat-producing devices.

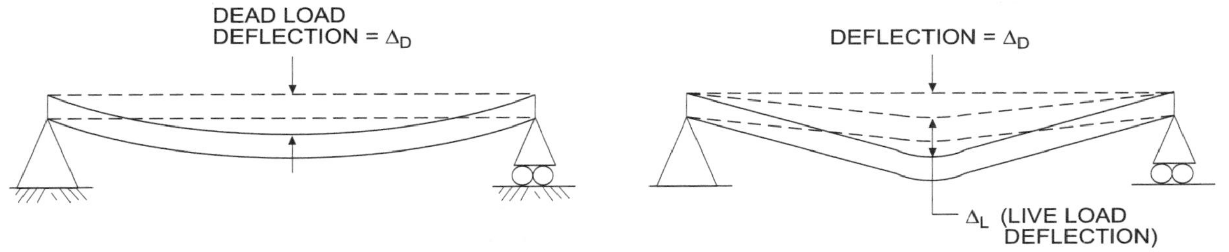

Figure R301.7
BEAM DEFLECTION

R302.1 Exterior walls. Construction, projections, openings and penetrations of *exterior walls* of *dwellings* and accessory buildings shall comply with Table R302.1(1); or *dwellings* equipped throughout with an *automatic sprinkler system* installed in accordance with Section P2904 shall comply with Table R302.1(2).

Exceptions:

1. Walls, projections, openings or penetrations in walls perpendicular to the line used to determine the *fire separation distance*.
2. Walls of *dwellings* and *accessory structures* located on the same *lot*.
3. Detached tool sheds and storage sheds, playhouses and similar structures exempted from permits are not required to provide wall protection based on location on the *lot*. Projections beyond the *exterior wall* shall not extend over the *lot line*.
4. Detached garages accessory to a *dwelling* located within 2 feet (610 mm) of a *lot line* are permitted to have roof eave projections not exceeding 4 inches (102 mm).
5. Foundation vents installed in compliance with this code are permitted.

❖This section provides details for issues related to building location on the property, including the fire rating of exterior walls, permitted openings and projections. Tables R302.1(1) and R302.1(2) provide a tabular overview of the requirements of this section.

Concerning exterior wall protection, the code assumes that an owner has no control over an adjoining property. Thus, the location of buildings on the owner's property relative to the property line requires regulation. In addition, Section R302.6, which lists the separation requirements for garages and carports, specifically requires garages located less than 3 feet (914 mm) from a dwelling unit on the same lot to have not less than $^1/_2$-inch (12.7 mm) gypsum board applied to the interior side of the walls. Opening protection for these walls is regulated by Section R302.5.

The property line concept is a convenient means of protecting one building from another as far as exposure is concerned. Exposure is the potential for heat to be transmitted from one building to another during a fire in the exposing building. Radiation is the primary means of heat transfer.

Table R302.1(1) specifies the exterior wall elements, fire separation distance and fire-resistance rating for dwellings without sprinkler systems. Walls less than 5 feet (1525 mm) from the property line must be of 1-hour fire-resistant construction. The fire-resistance rating also requires the rating exposure to be for both sides. The exterior rated walls should be tested in accordance with either ASTM E 119 or UL 263. This is not intended to limit fire-resistance-rated assemblies solely to the test criteria contained in these standards. Section R104.11 still allows the building official to approve alternative fire-resistance methodologies, such as those described in Section 703.3 of the IBC. This would still allow a builder to use acceptable engineering analysis, calculations in accordance with Section 721 of the IBC or prescriptive assemblies permitted by Section 720 of the IBC as alternatives to the standards contained within the code.

Projections must not extend more than 12 inches (305 mm) into the area where openings are prohibited. Therefore, projections cannot be closer than 2 feet (610 mm) from the lot line. Projections that are less than 5 feet (1525 mm) from the lot line are required to be protected on the underside with 1-hour fire-resistant construction [see Commentary Figure R302.1(1)].

Unlike the IBC, the code does not set a distance from the property line at which openings must be protected. Openings are not permitted in exterior walls where the exterior wall has a fire separation distance of less than 3 feet (914 mm) from the lot line. Openings in a wall located at a distance equal to or greater than 3 feet (914 mm), but less than 5 feet (1525 mm) from the lot line cannot exceed 25 percent of the maximum wall area [see Commentary Figures R302.1(2) and R302.1(3)]. The consensus as to the minimum distance necessary to provide a sufficient buffer against the spread of fire has changed somewhat over the years. For example, the 2000 and 2003 editions of the IRC required a 3-foot (914 mm) minimum fire separation distance for unrated exterior walls. In the 2006 edition, that distance was increased to 5 feet (1525 mm) to provide a higher level of safety and to correlate with the provisions for residential occupancies regulated by the IBC. The 2009 IRC introduced requirements for automatic fire sprinkler systems in all new one- and two-family dwellings and townhouses. Table R302.1(2) permits nonrated walls that have a 3-foot (914 mm)

TABLE R302.1(1)
EXTERIOR WALLS

EXTERIOR WALL ELEMENT		MINIMUM FIRE-RESISTANCE RATING	MINIMUM FIRE SEPARATION DISTANCE
Walls	Fire-resistance rated	1 hour—tested in accordance with ASTM E 119 or UL 263 with exposure from both sides	< 5 feet
	Not fire-resistance rated	0 hours	≥ 5 feet
Projections	Fire-resistance rated	1 hour on the underside	≥ 2 feet to < 5 feet
	Not fire-resistance rated	0 hours	≥ 5 feet
Openings in walls	Not allowed	N/A	< 3 feet
	25% maximum of wall area	0 hours	3 feet
	Unlimited	0 hours	5 feet
Penetrations	All	Comply with Section R302.4	< 5 feet
		None required	5 feet

For SI: 1 foot = 304.8 mm.
N/A = Not Applicable.

❖See the commentary to Section R302.1.

TABLE R302.1(2)
EXTERIOR WALLS—DWELLINGS WITH FIRE SPRINKLERS

EXTERIOR WALL ELEMENT		MINIMUM FIRE-RESISTANCE RATING	MINIMUM FIRE SEPARATION DISTANCE
Walls	Fire-resistance rated	1 hour—tested in accordance with ASTM E 119 or UL 263 with exposure from the outside	0 feet
	Not fire-resistance rated	0 hours	3 feet[a]
Projections	Fire-resistance rated	1 hour on the underside	2 feet[a]
	Not fire-resistance rated	0 hours	3 feet
Openings in walls	Not allowed	N/A	< 3 feet
	Unlimited	0 hours	3 feet[a]
Penetrations	All	Comply with Section R302.4	< 3 feet
		None required	3 feet[a]

For SI: 1 foot = 304.8 mm.
N/A = Not Applicable

a. For residential subdivisions where all dwellings are equipped throughout with an automatic sprinkler systems installed in accordance with Section P2904, the fire separation distance for nonrated exterior walls and rated projections shall be permitted to be reduced to 0 feet, and unlimited unprotected openings and penetrations shall be permitted, where the adjoining lot provides an open setback yard that is 6 feet or more in width on the opposite side of the property line.

❖See the commentary to Section R302.1.

minimum fire separation distance, a dimension previously prescribed in earlier editions of the code. The 3-foot (914 mm) dimension specified in Table R302.1(2) is the new threshold for exterior wall construction, projections, openings and penetrations for dwellings sprinklered in accordance with Section P2904 or NFPA 13D. For dwellings without sprinkler systems, the 5-foot (1525 mm) separation distance still applies.

The reduced clearances intend to provide design flexibility and reduce costs associated with fire-resistant construction, while maintaining a reasonable level of safety based on past performance of dwelling fire sprinkler systems. A dwelling automatic sprinkler system installed in accordance with Section P2904 or NFPA 13D aids in the detection and control of fires in residential occupancies regulated by the IRC. The design criteria of these sprinkler systems are for life safety to buy time for occupants to escape a fire; dwelling fire sprinklers are not designed for property protection. Sprinklers in accordance with Section P2904 or NFPA 13D are not required throughout the dwelling—they generally may be omitted in concealed spaces, closets, bathrooms, garages, and attics and crawl spaces without gas-fired appliances, for example. However, the automatic sprinkler system is expected to prevent total fire involvement (flashover) in the room of fire origin if the room is sprinklered. In addition to increasing the likelihood of occupants escaping or being evacuated, sprinklers often provide some measure of property protection as well.

Footnote a to Table R302.1(2) allows exterior walls of dwellings equipped with sprinkler systems to be placed on the lot line if the adjacent lot maintains a 6-foot (1829 mm) setback for buildings on the opposite side of the lot line. This provision allows flexibility in placing buildings on the lot for maximum effective use of the buildable area while still maintaining a minimum 6 feet (1829 mm) of clearance between buildings. Commentary Table R302.1(1) summarizes the new fire separation distance requirements for exterior walls that are not fire-resistance rated.

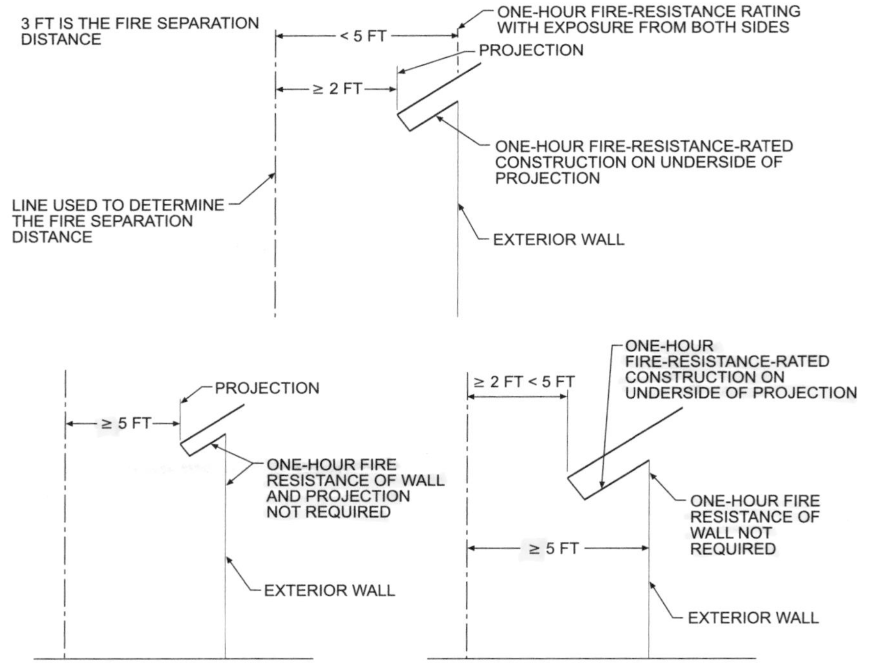

For SI: 1 foot = 304.8 mm.

Figure R302.1(1)
EXTERIOR WALLS AND PROJECTIONS FOR DWELLINGS WITHOUT AUTOMATIC SPRINKLER SYSTEM

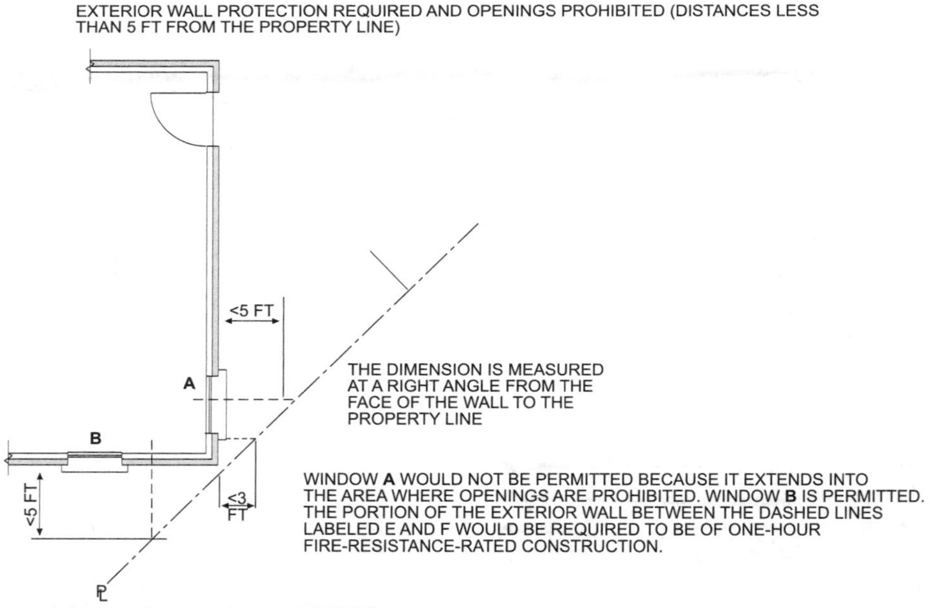

For SI: 1 foot = 304.8 mm.

Figure R302.1(2)
EXTERIOR WALL FIRE-RESISTANCE-RATED REQUIREMENTS AND PROHIBITED OPENINGS FOR DWELLINGS WITHOUT AUTOMATIC SPRINKLER SYSTEM

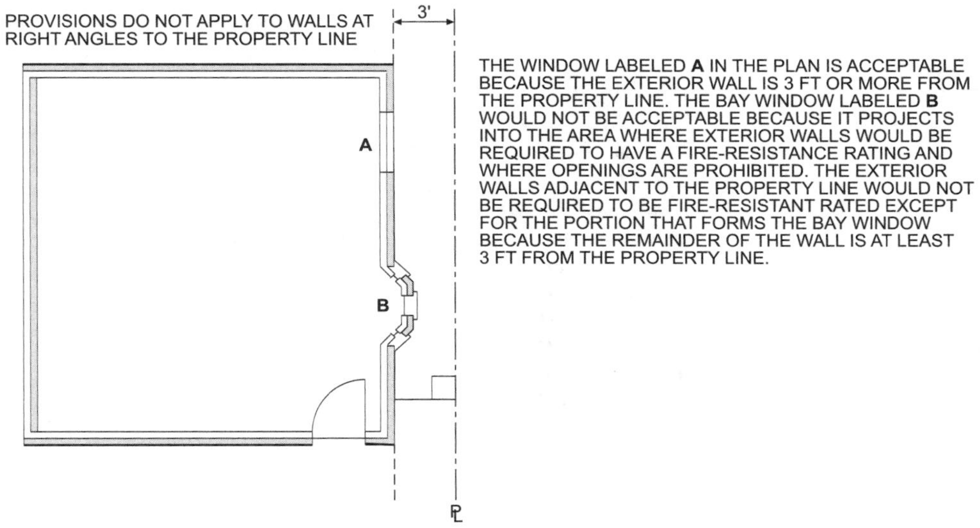

PROVISIONS DO NOT APPLY TO WALLS AT RIGHT ANGLES TO THE PROPERTY LINE
3'
A
B
THE WINDOW LABELED A IN THE PLAN IS ACCEPTABLE BECAUSE THE EXTERIOR WALL IS 3 FT OR MORE FROM THE PROPERTY LINE. THE BAY WINDOW LABELED B WOULD NOT BE ACCEPTABLE BECAUSE IT PROJECTS INTO THE AREA WHERE EXTERIOR WALLS WOULD BE REQUIRED TO HAVE A FIRE-RESISTANCE RATING AND WHERE OPENINGS ARE PROHIBITED. THE EXTERIOR WALLS ADJACENT TO THE PROPERTY LINE WOULD NOT BE REQUIRED TO BE FIRE-RESISTANT RATED EXCEPT FOR THE PORTION THAT FORMS THE BAY WINDOW BECAUSE THE REMAINDER OF THE WALL IS AT LEAST 3 FT FROM THE PROPERTY LINE.
PL

Figure R302.1(3)
EXTERIOR WALL FIRE-RESISTANCE-RATED REQUIREMENTS AND PROHIBITED OPENINGS FOR DWELLINGS WITHOUT AUTOMATIC SPRINKLER SYSTEM

Table R302.1
MINIMUM FIRE SEPARATION DISTANCE COMPARISON (NONRATED CONSTRUCTION)

EXTERIOR WALL ELEMENT (Not fire-resistance rated)	MINIMUM FIRE SEPARATION DISTANCE		
	Without Sprinkler System	With Sprinkler System	With Sprinkler System in all Dwellings of Subdivision and 6-foot Setback for Building on Adjoining Lot
Walls	5 feet	3 feet	0 feet
Projections	5 feet	3 feet	0 feet
Unlimited openings in walls	5 feet	3 feet	0 feet
Penetrations (no restrictions)	5 feet	3 feet	0 feet

For SI: 1 foot = 304.8 mm.

Exception 1 permits walls, openings, projections or penetrations that are 90 degrees (1.57 rad) (perpendicular) to the line used to determine the fire separation distance to be exempt from the requirements of Table R302.1. Section R302.4 describes through penetrations and membrane penetrations in detail (see the definition of "Fire separation distance" in Chapter 2) [see Commentary Figures R302.1(2) and R302.1(3)].

Exception 2 allows dwellings and accessory structures, on the same lot, to be considered one building and the requirements of Table R302.1 will not apply to the exterior walls facing each other. This exception eliminates the imaginary line between two buildings on the lot when measuring the fire separation distance. Table R302.1 will apply to the other exterior walls of the buildings.

Exception 3 applies to detached tool and storage sheds, playhouses and similar structures that are exempt from permits. Projections from these structures, however, are not permitted to extend over the property line.

Exception 4 will allow roof eave projection for detached garages to be closer than 2 feet (610 mm) from the lot line, but limits the roof eave projection to 4 inches (102 mm). This projection cannot extend over the property line.

Exception 5 allows foundation vents installed in compliance with the code in areas where openings are otherwise prohibited. ~2 separate 1hr walls

R302.2 Townhouses. Each *townhouse* shall be considered a separate building and shall be separated by fire-resistance-rated wall assemblies meeting the requirements of Section R302.1 for exterior walls.

Exception: A common 1-hour fire-resistance-rated wall assembly tested in accordance with ASTM E 119 or UL 263 is permitted for townhouses if such walls do not contain plumbing or mechanical equipment, ducts or vents in the cavity of the common wall. The wall shall be rated for fire exposure from both sides and shall extend to and be tight against exterior walls and the underside of the roof sheathing. Electrical installations shall be installed in accordance with Chapters 34 through 43. Penetrations of electrical outlet boxes shall be in accordance with Section R302.4.

❖The application of this section has its basis in the exterior wall requirements found in Section R302.1 that deal with the building's location on the lot. The definition of a townhouse in Section R202 should be reviewed, as well as the requirement for structural independence in Section R302.2.4. In general, because the "exterior wall" of the townhouse is essentially being constructed with no fire separation distance where one townhouse adjoins another, the code requires, by Section R302. 1, that the wall have not less than a 1-hour fire-resistance rating. The adjacent townhouse would have the same requirement. Therefore, the general requirement at this location (based on Sections R302.1 and R302.3) would be that each townhouse has its own "exterior wall." This would result in the construction of two separate 1-hour walls located side by side where one townhouse adjoins another.

Because of the difficulties involved in construction and the potential for unnecessary duplication, the exception offers an alternative to the two separate 1-hour walls by permitting the construction of a shared or "common" 2-hour-rated wall between the townhouses.

See Commentary Figure R302.2 for an illustration of the two separate 1-hour walls and the common 2-hour wall. This exception has its basis in the actions of many building officials who permit this type of common wall as an alternative method of construction using provisions similar to those found in Section R104.11. Because the common wall has the potential to create an interconnection between the adjacent dwelling units and reduce the clear separation that would exist if two separate walls were constructed, the code places limits on services being located within the wall. This exception does not permit the inclusion of any type of plumbing, mechanical equipment, ducts or vents within the cavity of the common wall. This prohibition is applicable even if the penetrations or openings are protected by the penetration provisions of Section R302.4 or if a damper is installed in the duct or vent. The prohibition on plumbing includes all types of plumbing materials and systems, as well as water supply and drainage piping of either combustible or noncombustible materials. However, the exception permits the cavity of the wall to be used for electrical installations if they comply with the electrical provisions of the code and the penetrations are properly protected.

R302.2.1 Continuity. The fire-resistance-rated wall or assembly separating *townhouses* shall be continuous from the foundation to the underside of the roof sheathing, deck or slab. The fire-resistance rating shall extend the full length of the wall or assembly, including wall extensions through and separating attached enclosed *accessory structures*.

❖This section addresses the continuity of the fire-resistance-rated wall or assembly separating townhouses using the exception in Section R302.2. The requirements are conceptually similar to the continuity issues that exist in Section R302.3. These provisions, by regulating the extensions and terminations of the wall, can make possible the separation of dwelling units from each other.

R302.2.2 Parapets. Parapets constructed in accordance with Section R302.2.3 shall be constructed for *townhouses* as an extension of exterior walls or common walls in accordance with the following:

1. Where roof surfaces adjacent to the wall or walls are at the same elevation, the parapet shall extend not less than 30 inches (762 mm) above the roof surfaces.

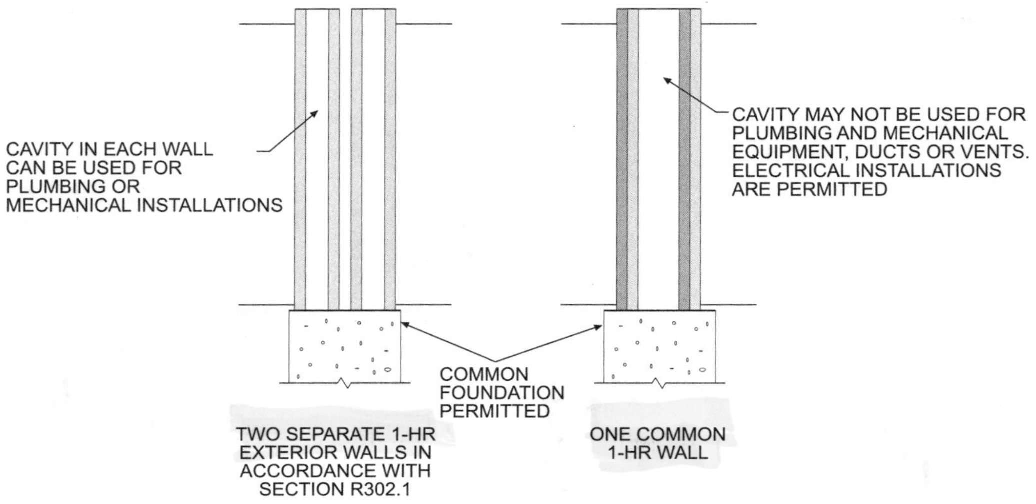
CAVITY IN EACH WALL CAN BE USED FOR PLUMBING OR MECHANICAL INSTALLATIONS
CAVITY MAY NOT BE USED FOR PLUMBING AND MECHANICAL EQUIPMENT, DUCTS OR VENTS. ELECTRICAL INSTALLATIONS ARE PERMITTED
COMMON FOUNDATION PERMITTED
TWO SEPARATE 1-HR EXTERIOR WALLS IN ACCORDANCE WITH SECTION R302.1
ONE COMMON 1-HR WALL

Figure R302.2
DWELLING UNIT SEPARATION FOR TOWNHOUSES

2. Where roof surfaces adjacent to the wall or walls are at different elevations and the higher roof is not more than 30 inches (762 mm) above the lower roof, the parapet shall extend not less than 30 inches (762 mm) above the lower roof surface.

   Exception: A parapet is not required in the two cases above when the roof is covered with a minimum class C roof covering, and the roof decking or sheathing is of noncombustible materials or *approved* fire-retardant-treated wood for a distance of 4 feet (1219 mm) on each side of the wall or walls, or one layer of $^{5}/_{8}$-inch (15.9 mm) Type X gypsum board is installed directly beneath the roof decking or sheathing, supported by a minimum of nominal 2-inch (51 mm) ledgers attached to the sides of the roof framing members, for a minimum distance of 4 feet (1219 mm) on each side of the wall or walls and there are no openings or penetrations in the roof within 4 feet (1219 mm) of the common walls.

3. A parapet is not required where roof surfaces adjacent to the wall or walls are at different elevations and the higher roof is more than 30 inches (762 mm) above the lower roof. The common wall construction from the lower roof to the underside of the higher roof deck shall have not less than a 1-hour fire-resistance rating. The wall shall be rated for exposure from both sides.

❖This section provides for the continuation of the dwelling-unit separation by requiring that a parapet be constructed above the common wall or the two "exterior walls" that occur between adjacent townhouses. This parapet requirement is applicable only to the wall or walls between townhouses and does not apply to the separation in a two-family dwelling, nor does it apply to other exterior walls on either townhouses or dwellings.

The code states three requirements that address the details of the parapet depending on the height of the adjacent roofs. The code also has an exception that can be used to eliminate the need for the parapet in the two conditions that require them. In general, parapets must extend at least 30 inches (762 mm) above the roof surfaces of the adjacent townhouses. This standard requirement is found in Item 1 and is applicable where the roof surfaces of the adjacent dwelling units are at the same level. The second item addresses the requirement for a parapet where the roofs are at different levels, but the difference is less than 30 inches (762 mm). Under this condition, the parapet height must still be 30 inches (762 mm), but it is measured only from the roof surface of the lower roof. The third item requires a parapet, but it may not be apparent to an observer because there is no requirement for anything to extend above the roof surfaces. In this case, the height difference between the roofs must be more than 30 inches (762 mm), and the code requires the common wall between the units to be rated to the height of the upper roof deck. The rating of this portion of the wall is a 1-hour fire-resistance rating and not the typical 2-hour-rated wall (or two separate 1-hour walls) that separates the actual dwelling units. See Commentary Figure R302.2.2(1) for an illustration of the three requirements.

The exception found in this section applies only to the first two items of the section. See Commentary Figure R302.2.2(2) for an illustration of the exception. Openings or penetrations through the roof is prohibited within 4 feet of the wall or walls. It does not apply where the height difference between the roofs is more than 30 inches (762 mm) and the provisions of Item 3 have been applied.

R302.2.3 Parapet construction. Parapets shall have the same fire-resistance rating as that required for the supporting wall or walls. On any side adjacent to a roof surface, the parapet shall have noncombustible faces for the uppermost 18 inches (457 mm), to include counterflashing and coping materials. Where the roof slopes toward a parapet at slopes greater than 2 units vertical in 12 units horizontal (16.7-percent slope), the parapet shall extend to the same height as any portion of the roof within a distance of 3 feet (914 mm), but in no case shall the height be less than 30 inches (762 mm).

❖In addition to having the same degree of fire resistance as required for the wall or walls below, the surface of the parapet that faces the roof must be of noncombustible materials for the upper 18 inches (457 mm). Thus, a fire that might travel along the roof and reach the parapet would not be able to continue upward along the face of the parapet and over the top, exposing the adjacent building. The requirement applies to the upper 18 inches (457 mm) of the parapet to allow for extending the roof covering up the base of the parapet so that it can be effectively flashed. The 18-inch (457 mm) figure is based on a parapet height of at least 30 inches (762 mm) and would not be applicable to the face of the parapet that was toward the higher roof deck as constructed using Section R302.2.2, Item 2, nor would it be applicable where the parapet is not required based on the exception in Section R302.2.2 or on Item 3 of that section.

The 30-inch (762 mm) requirement is measured "above the roof surface" as stated in Section R302.2.2. When a cricket or other element is installed adjacent to the parapet, the 30-inch (762 mm) dimension must be taken from the top of the roof surface on the cricket so the parapet truly extends 30 inches (762 mm) above the roof.

In those cases where the roof slopes upward away from the parapet and the slope exceeds 2 units vertical in 12 units horizontal (16.7-percent slope), the parapet must be extended to the same height as any portion of the roof that is within a distance of 3 feet (914 mm). However, in no case can the height of the parapet be less than 30 inches (762 mm). See Commentary Figure R302.2.3 for an illustration of this requirement.

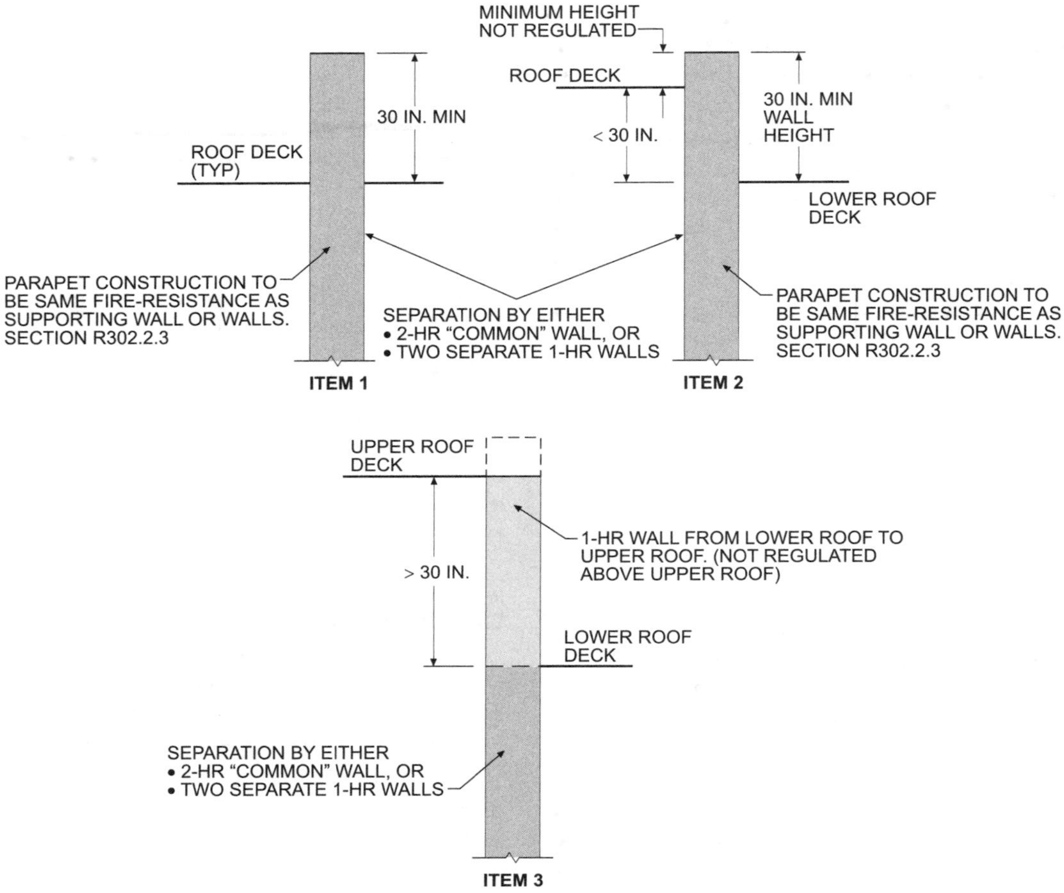

For SI: 1 inch = 25.4 mm.

Figure R302.2.2(1)
PARAPET REQUIREMENTS FOR WALL BETWEEN TOWNHOUSES

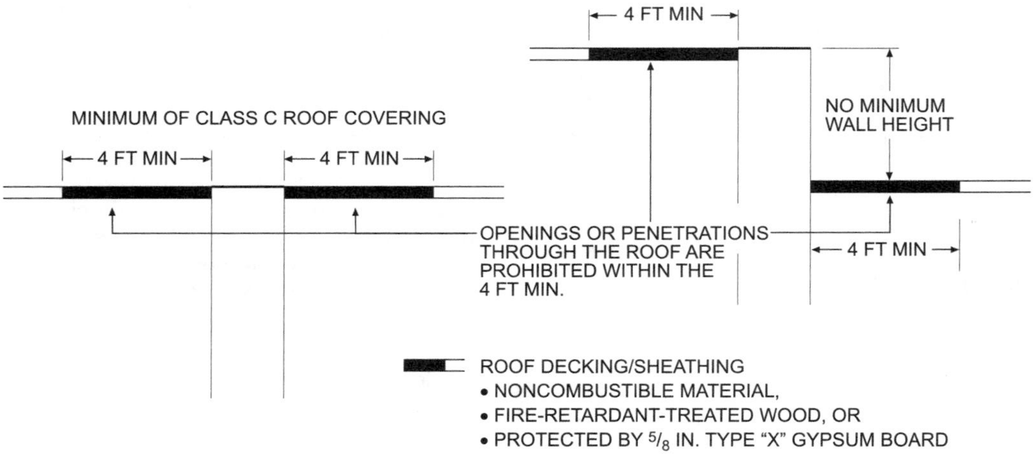

For SI: 1 inch = 25.4 mm, 1 foot = 304.8 mm.

Figure R302.2.2(2)
EXCEPTION TO ELIMINATE PARAPET

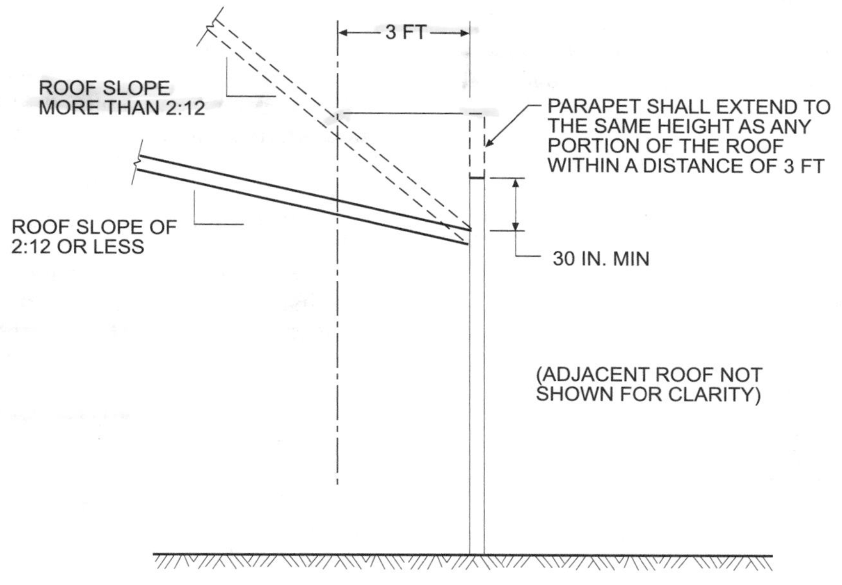

For SI: 1 inch = 25.4 mm, 304.8 mm.

Figure R302.2.3
PARPET REQUIREMENTS

R302.2.4 Structural independence. Each individual *townhouse* shall be structurally independent.

Exceptions:

1. Foundations supporting *exterior walls* or common walls.
2. Structural roof and wall sheathing from each unit may fasten to the common wall framing.
3. Nonstructural wall and roof coverings.
4. Flashing at termination of roof covering over common wall.
5. *Townhouses* separated by a common 1-hour fire-resistance-rated wall as provided in Section R302.2.

❖Each townhouse must be structurally independent and capable of being removed without affecting the adjacent dwelling unit. This provision is applicable only to townhouses, not two-family dwellings. This independence is useful not only in the event of a fire within one unit, but also during any remodeling or alteration. The objective of this structural independence is that a complete burnout could occur on one side of the wall without causing the collapse of the adjacent townhouse. This condition occurs rarely. The provision also helps if there is ever a fire or other problem by creating a clear separation between the units. With separate ownership and each owner having a different insurance company, the ability to gain access or get repairs made can be difficult and time consuming. By having clearly separated units, it is much easier to determine who is responsible and to make any needed repairs.

The code lists five exceptions that waive the structural independence. A quick review of the exceptions shows that they generally deal with items that will not structurally affect the townhouses should a problem develop in the adjacent dwelling unit. Exception 1 is based on the norm within the industry for foundation construction. In the code, Section R402 lists only wood and concrete within the foundation materials section, although Section R404 accepts masonry foundation walls. In general, concrete and masonry are the most common types of foundations; wood foundations are viewed as unique. Given the performance of both masonry and concrete, and the fact that these foundation systems must sustain loads from both the structure and the adjacent soils, it is reasonable to assume that the foundation will not be the item that fails in most situations. Permitting a common foundation also helps solve other problems that would arise if the structural independence issue were taken as an absolute. An example where requiring separate foundations would probably create more problems or difficulty is in the dampproofing or waterproofing of below-grade foundation walls.

If a wood foundation is used between adjacent units, what is the level of fire protection that may be needed? Because concrete and masonry foundations are the norm, it would be easy to forget or overlook protecting the foundation when it is constructed of wood. In these cases, it would seem appropriate to deal with the foundation as any other wall, and protect it on any exposed side. The level of fire resistance should be equal to that of the wall or walls that the foundation supports.

R302.3 Two-family dwellings. *Dwelling units* in two-family dwellings shall be separated from each other by wall and/or floor assemblies having not less than a 1-hour fire-resistance rating when tested in accordance with ASTM E 119 or UL 263. Fire-resistance-rated floor/ceiling and wall assemblies shall extend to and be tight against the *exterior wall*, and wall assemblies shall extend from the foundation to the underside of the roof sheathing.

Exceptions:

1. A fire-resistance rating of $^1/_2$ hour shall be permitted in buildings equipped throughout with an automatic sprinkler system installed in accordance with NFPA 13.
2. Wall assemblies need not extend through *attic* spaces when the ceiling is protected by not less than $^5/_8$-inch (15.9 mm) Type X gypsum board and an *attic* draft stop constructed as specified in Section R302.12.1 is provided above and along the wall assembly separating the *dwellings*. The structural framing supporting the ceiling shall also be protected by not less than $^1/_2$-inch (12.7 mm) gypsum board or equivalent.

❖Most of the nation's fires occur in residential buildings, particularly one- and two-family dwellings. These fires account for more than 80 percent of all deaths from fire in residential uses (including hotels, apartments, dormitories, etc.) and about two-thirds of all fire fatalities in any type of building. One- and two-family dwellings also account for more than 80 percent of residential property losses and more than one-half of all property losses from fire. Despite this poor fire record, there is wide-spread resistance to mandating much in the way of fire protection systems or methods because of our society's belief that people's homes are their castles. This viewpoint has limited the types of protection that are imposed on these private homes to the installation of smoke alarms and the more recent requirement of dwelling unit separation. Section R302.3 provides a separation for protection of the occupants of one dwelling unit from the actions of their neighbor. The requirements of this section pertain to any structure regulated by the code other than a single-family dwelling unit. To accomplish this protection, the code addresses separation between the units, structural support and any openings or penetrations of the separation.

Depending on the layout of the various dwelling units, Section R302.3 requires that the walls and/or floor assemblies that divide one dwelling unit from the adjacent unit is of at least 1-hour fire-resistant construction. See Commentary Figure R302.3 for examples of the separation. The separation rating is to be determined by either ASTM E 119 or UL 263, which is the normal test used for determining fire resistance. Many tested assemblies are available for use in these locations. The provisions of the section also address the continuity of the separation, so that one dwelling unit is completely divided from the other. The horizontal aspect of the separation, which requires that the assemblies extend to and be tight against the exterior wall, is not difficult to comply with. It is most likely the vertical aspect (continuing a wall assembly to the underside of the roof sheathing) that will require some detailed planning, careful construction and careful inspection for the units to be separated.

Exception 1 grants a reduction in the required separation for those cases in which the building is equipped with an automatic sprinkler system. In these cases, a rating of $^1/_2$ hour is permitted versus a 1-hour fire-resistance rating. The sprinkler system must be "installed in accordance with NFPA 13," and is to be installed "throughout" the building. The type of sprinkler system used must meet NFPA 13 and

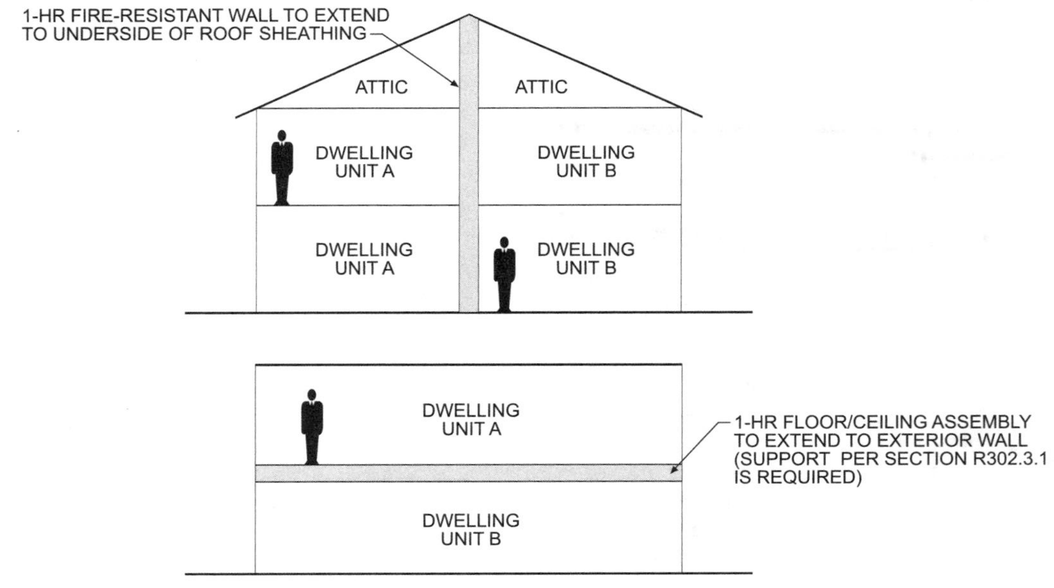

Figure R302.3
DWELLING UNIT SEPARATIONS

may not be installed to either the NFPA 13D or 13R, even though those two standards do address certain types of residential uses. The word "throughout" requires that the sprinkler system be installed in all portions of both dwelling units and any common spaces. The provisions of NFPA 13 that permit omitting sprinklers in certain areas, such as small concealed spaces, are applicable. Therefore, the provision requires a complying sprinkler system "throughout" the building (that is, in all areas of the building that must be protected according to the standard), and it does not accept any partial system, such as one installed in only one dwelling unit or only in the basement level of both units.

Exception 2 addresses separation in the area of the attic of two-family dwellings or duplexes. As long as the attic draft stop is present and meets the requirements in Section R302.12.1, there is a provision for the 1-hour fire separation to stop at a ceiling constructed of $^5/_8$-inch (15.9 mm) Type X gypsum board. Many times the type of truss or attic rafter and rafter tie/collar tie configuration will prohibit continuing construction of the 1-hour separation wall all the way up to the roof sheathing.

R302.3.1 Supporting construction. When floor assemblies are required to be fire-resistance rated by Section R302.3, the supporting construction of such assemblies shall have an equal or greater fire-resistance rating.

❖This provision applies to only the floor assemblies that form the separation between different dwelling units. When either all or portions of a dwelling unit separation are provided by a floor assembly, the code requires that the structural supports for the separation have a rating equal to or higher than the floor. This is conceptually similar to the garage separation of Section R302.6. Without the supporting construction being protected, a fire on the lower level could lead to an early failure of the dwelling unit separation (see Commentary Figure R302.3.1).

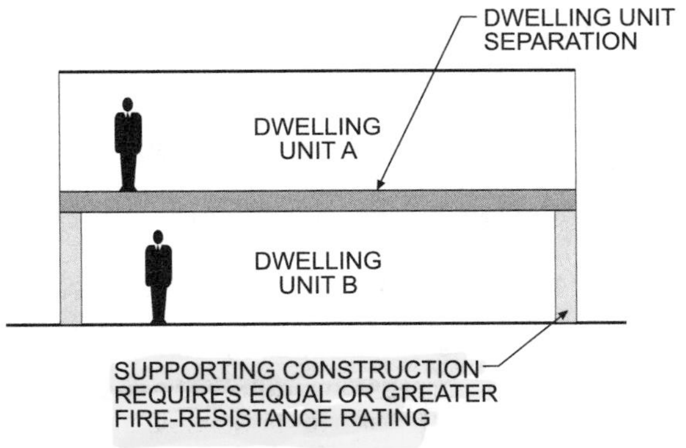

Figure R302.3.1
SUPPORT OF DWELLING UNIT SEPARATION

R302.4 Dwelling unit rated penetrations. Penetrations of wall or floor/ceiling assemblies required to be fire-resistance rated in accordance with Section R302.2 or R302.3 shall be protected in accordance with this section.

❖This section addresses the specific requirements for maintaining the integrity of fire-resistance-rated assemblies at penetrations. If the penetration of a rated assembly is not properly constructed, the assembly itself is jeopardized and may not perform as intended. The provisions of this section apply to penetrations of fire-resistance-rated walls and floor/ceiling assemblies that are a part of the dwelling unit separation in either two-family dwellings or townhouses. Penetrations of the rated assemblies range from combustible pipe and tubing to noncombustible wiring with combustible covering to noncombustible items, such as pipe, tube, conduit and ductwork.

Each type of penetration requires a specific method of protection, which is based on the type of fire-resistance-rated assembly penetrated and the size and type of the penetrating item. The first step in determining the type of penetration protection required is to identify whether a wall or floor/ceiling assembly is being penetrated. The next step is to determine the type of penetrating item and whether it is a membrane or through penetration. Once these factors are known, then the applicable section must be applied and the applicable method of protection must be decided upon.

R302.4.1 Through penetrations. Through penetrations of fire-resistance-rated wall or floor assemblies shall comply with Section R302.4.1.1 or R302.4.1.2.

Exception: Where the penetrating items are steel, ferrous or copper pipes, tubes or conduits, the annular space shall be protected as follows:

1. In concrete or masonry wall or floor assemblies, concrete, grout or mortar shall be permitted where installed to the full thickness of the wall or floor assembly or the thickness required to maintain the fire-resistance rating, provided:
   1.1. The nominal diameter of the penetrating item is a maximum of 6 inches (152 mm); and
   1.2. The area of the opening through the wall does not exceed 144 square inches (92 900 mm^2).
2. The material used to fill the annular space shall prevent the passage of flame and hot gases sufficient to ignite cotton waste where subjected to ASTM E 119 or UL 263 time temperature fire conditions under a minimum positive pressure differential of 0.01 inch of water (3 Pa) at the location of the penetration for the time period equivalent to the fire-resistance rating of the construction penetrated.

❖This section contains the general requirements for through penetrations, which are openings that pass through an entire assembly. A through penetration is in contrast to a membrane penetration, which creates an opening through only one side of an assembly. Membrane penetrations are addressed later in Section R302.4.2. See Commentary Figure R302.4.1 for an illustration of these two types of penetrations.

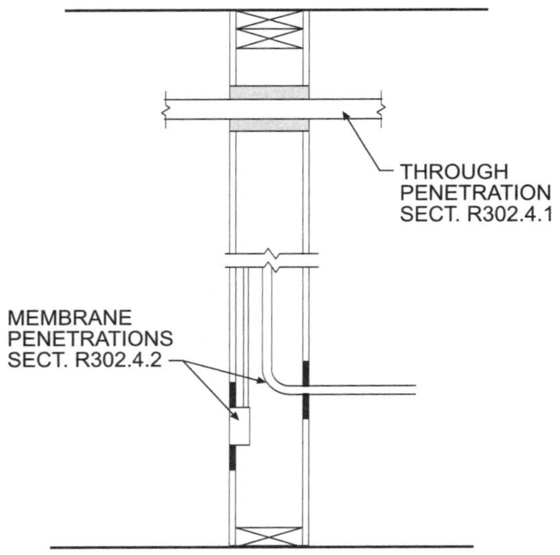

Figure R302.4.1
TYPES OF PENETRATIONS

Through penetrations must be protected to maintain the fire resistance of the penetrated assembly. The code states two methods, found in Sections R302.4.1.1 and R302.4.1.2, which can be used to assure the adequacy of the penetration protection. The difference between these two is the test methodology used, but they both provide essentially the same results. The commentary for those sections is additional discussion of the differences.

Based on the history of these provisions and on the wealth of fire test data that exists concerning items such as conduit, water piping and other similar penetrations, the code provides two exceptions that permit protection by methods other than those generally required. The first permits the use of concrete, grout or mortar to protect certain penetrations of concrete and masonry wall or floor assemblies. The concrete, grout or mortar must be applied for the full thickness of the assembly unless evidence can be produced demonstrating that the required fire-resistance rating can be achieved with a lesser depth. Concrete, grout and mortar have traditionally been used as protection for the annular space in penetrations of concrete and masonry assemblies. Experience has shown this form of protection to be viable. However, caution must be used any time something, such as a water pipe or conduit, is placed in concrete or masonry. Sections P2603.3 and P2603.5 contain examples of protection of plumbing systems.

Exception 2 addresses the space between the penetrating item and the original assembly construction. This gap is called the annular space, and this exception provides a method to simply evaluate the performance of the material used to fill that space. It is often mistakenly believed that this exception permits a variety of untested items, but as can be seen from the provision itself, the materials need to meet a specific performance level. This exception requires that the ability of the material to prevent the passage of flame and hot gases sufficient to ignite cotton when subjected to the time-temperature criteria of the ASTM E 119 test standard be prequalified. This requirement is similar to provisions found within both ASTM E 119 and ASTM E 814, the standards used to evaluate fire-resistant assemblies and penetration protection. Because it is very likely that the penetration in the actual fire will be exposed to a positive pressure, this section specifies that the test-fire exposure include a positive pressure of 0.01 inch (0.25 mm) of water column as a further means to verify the performance of this protection method. Thus the protection will not be blown out or moved from its place during a fire.

R302.4.1.1 Fire-resistance-rated assembly. Penetrations shall be installed as tested in the *approved* fire-resistance-rated assembly.

❖This section addresses situations in which the penetration is tested as a part of the regular full-scale test for the wall or floor/ceiling assembly. The penetration and proposed type of protection are evaluated as a part of the regular ASTM E 119 test, which evaluates the wall or floor/ceiling rating. This section and the option it provides are not used frequently because of the cost of conducting such full-scale tests and the limitations placed on the application of the tested assembly. Because of these issues, penetrations are most often protected in accordance with one of the exceptions in Section R302.4.1 or the provisions of Section R302.4.1.2.

R302.4.1.2 Penetration firestop system. Penetrations shall be protected by an *approved* penetration firestop system installed as tested in accordance with ASTM E 814 or UL 1479, with a minimum positive pressure differential of 0.01 inch of water (3 Pa) and shall have an F rating of not less than the required fire-resistance rating of the wall or floor/ceiling assembly penetrated.

❖Through-penetration firestop systems consist of specific materials or an assembly of materials that are designed to restrict the passage of fire and hot gases for a prescribed period of time through openings made in fire-resistance-rated assemblies. To determine the effectiveness of a through-penetration firestop system in restricting the passage of fire, and to determine that the penetration has not jeopardized the original fire-resistant assembly, firestop systems must be subjected to fire testing using the ASTM E 814 or UL 1479 test standard. This is a small-scale test method developed specifically for the evaluation of a firestop system's ability to resist the passage of flame and hot gases, withstand thermal stresses and restrict the transfer of heat through the penetrated assembly. There are hundreds if not thousands of tested through-penetration firestop systems available today. The actual type of system used will depend on the type and construction of the assembly being penetrated, the material makeup and size of the penetrating item, and the size of the annular space that exists between the penetrating item and the original assembly. Because there are a multitude of products available, and there is no "one size fits all" system available, it is helpful if the methods of protection are included on the construction documents as covered by Section R106.1.1.

The actual rating of the through-penetration firestop system is generated from the results of the testing and is reported as an "F" (flame) rating and a "T" (temperature) rating. The code requires only an F rating. The F rating indicates the period of time, in hours, that the through-penetration firestop system remained in place without allowing the passage of fire during the fire exposure test, or the passage of water during the hose stream portion of the test. The required F rating of a must be equal to the fire-resistance rating of the wall or floor/ceiling assembly that is being penetrated. This means either a 1- or 2-hour rating, depending on the dwelling unit separation.

Two of the most common materials used in through-penetration firestop systems are intumescent and endothermic materials. Intumescent materials expand approximately 8 to 10 times their original volume when exposed to temperatures exceeding 250°F (121°C). The expansion of the material fills the voids or openings within the penetration to resist the passage of flame, while the outer layer of the expanded intumescent material forms an insulating charred layer that assists in limiting the transfer of heat. The expansion properties of intumescent materials allow them to seal openings left by combustible penetrating items that burn away during a fire, but they do not retard heat as well as endothermic materials. Intumescent materials are typically used with combustible penetrating items or where a higher T rating is not required.

Endothermic materials provide protection through chemically bound water released in the form of steam when exposed to temperatures exceeding 600°F (316°C). This released water cools the penetration and retards heat transfer through the penetration. Endothermic materials tend to be superior in heat-transfer resistance and have higher T ratings, but they do not expand to fill voids left by combustible penetrating items that burn away during a fire. Therefore, endothermic materials are typically used with noncombustible penetrating items and where a higher T rating is required.

R302.4.2 Membrane penetrations. Membrane penetrations shall comply with Section R302.4.1. Where walls are required to have a fire-resistance rating, recessed fixtures shall be installed so that the required fire-resistance rating will not be reduced.

Exceptions:

1. Membrane penetrations of maximum 2-hour fire-resistance-rated walls and partitions by steel electrical boxes that do not exceed 16 square inches (0.0103 m^2) in area provided the aggregate area of the openings through the membrane does not exceed 100 square inches (0.0645 m^2) in any 100 square feet (9.29 m)^2 of wall area. The annular space between the wall membrane and the box shall not exceed $^1/_8$ inch (3.1 mm). Such boxes on opposite sides of the wall shall be separated by one of the following:
   - 1.1. By a horizontal distance of not less than 24 inches (610 mm) where the wall or partition is constructed with individual noncommunicating stud cavities;
   - 1.2. By a horizontal distance of not less than the depth of the wall cavity when the wall cavity is filled with cellulose loose-fill, rockwool or slag mineral wool insulation;
   - 1.3. By solid fire blocking in accordance with Section R302.11;
   - 1.4. By protecting both boxes with listed putty pads; or
   - 1.5. By other listed materials and methods.
2. Membrane penetrations by listed electrical boxes of any materials provided the boxes have been tested for use in fire-resistance-rated assemblies and are installed in accordance with the instructions included in the listing. The annular space between the wall membrane and the box shall not exceed $^1/_8$ inch (3.1 mm) unless listed otherwise. Such boxes on opposite sides of the wall shall be separated by one of the following:
   - 2.1. By the horizontal distance specified in the listing of the electrical boxes;
   - 2.2. By solid fireblocking in accordance with Section R302.11;
   - 2.3. By protecting both boxes with listed putty pads; or
   - 2.4. By other listed materials and methods.
3. The annular space created by the penetration of a fire sprinkler provided it is covered by a metal escutcheon plate.

❖This section deals with instances where only a single side of the fire-resistance-rated assembly is penetrated. This would be the situation for items such as electrical outlet boxes or plumbing fixtures located on one side of the wall only. Commentary Figure R302.4.1 shows this type of penetration. For the most part, a membrane penetration is to be protected by one of the previously described methods established for through penetrations. However, there are some penetrations that are allowed without a specific fire-stopping material in the annular space around them. These are addressed by the exceptions. This section also deals with the installation of recessed luminaires in fire-resistance-rated assemblies and states that their installation may not reduce the assembly's protection. Although these fixtures are common, they do represent a penetration of the assembly's protection and must be installed so that the assembly is not compromised.

Exception 1 allows penetrations of steel electrical outlet boxes under certain conditions. The criteria of this section limit the size of the box to 16 square inches (0.0103 m^2) or less in area and to an aggregate area not to exceed 100 square inches (64 500 mm^2) in 100 square feet (9.3 m^2). Commentary Figure R302.4.2(1) shows some of the requirements of this section. The area limitations are consistent with the criteria from fire tests, which have shown that within these limitations, these penetrations will not adversely affect the fire-resistance rating of the assembly. However, the boxes are assumed to be installed as they were during the fire tests. In general, the test requirements match the limitations shown by the code regarding their size and the need to be offset. An additional requirement, one that does not appear in the code, regulates the size of the annular space created around the outlet boxes. Both the Underwriters Laboratory's (UL) *Fire-Resistance Directory* and the *Gypsum Association's Fire-resistance Design Manual* specify a maximum over-cut of $^1/_8$ inch (3 mm) for the annular space around the outlet boxes. Additionally, Article 373 of the *National Electrical Code* (NEC) (also known as NFPA 70) includes the size limitation of the over-cut. Therefore, the exception applies only when the boxes are installed as they were during the original fire tests, including the limited annular space. Because outlet boxes on both sides of a wall create penetrations of both layers of a wall assembly's protection, the code provides five methods to address this problem. This gives code users several options and does not limit them to the usual 24-inch (610 mm) offset.

Exception 2 permits using outlet boxes of nonmetallic materials if they have been specifically tested. Because many different types of nonmetallic boxes are available, it is important to determine that the boxes being used in the rated dwelling unit separation have been tested. Although the exception applies to nonmetallic electrical outlet boxes, the same concept would apply to steel boxes that exceed the sizes specified in Exception 1.

Exception 3 provides an alternative to the annular space protection provisions for a fire sprinkler that penetrates a single membrane. This exception is available if the annular space around the sprinkler is completely covered by an escutcheon plate of noncombustible material. The nature of the hazard posed by single-membrane penetrations of the sprinkler is limited by the size of the opening, the potential number of openings present and the presence of a sprinkler system. The installation of a noncombustible escutcheon provides protection against the free passage of fire through the annular space and allows for the movement of the sprinkler piping without breaking during a seismic event [see Commentary Figure R302.4.2(2)].

R302.5 Dwelling/garage opening/penetration protection. Openings and penetrations through the walls or ceilings separating the *dwelling* from the garage shall be in accordance with Sections R302.5.1 through R302.5.3.

❖Openings to sleeping rooms from garages are not allowed because a person might not wake up in time if there was a hazard from CO fumes or smoke from the garage. The three subsections address doors, ducts and pipes. For wall and ceiling separation requirements, see Section R302.6 and Table R302.6.

R302.5.1 Opening protection. Openings from a private garage directly into a room used for sleeping purposes shall not be permitted. Other openings between the garage and residence shall be equipped with solid wood doors not less than 1$^3/_8$ inches (35 mm) in thickness, solid or honeycomb-core steel doors not less than 1$^3/_8$ inches (35 mm) thick, or 20-minute fire-rated doors, equipped with a self-closing device.

❖Openings from the garage are permitted only into rooms that are not used for sleeping. These openings must be protected by the installation of a door com-

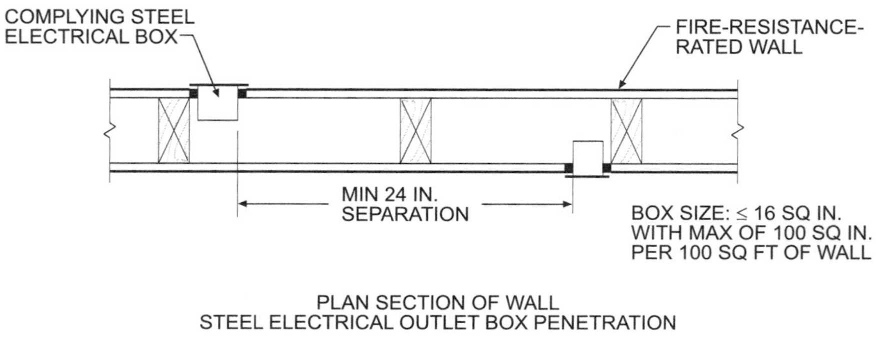

For SI: 1 inch = 25.4 mm, 1 square inch = 645 mm^2, 1 square foot = 0.0929 m^2.

Figure R302.4.2(1)
MEMBRANE PENETRATION BY OUTLET BOX

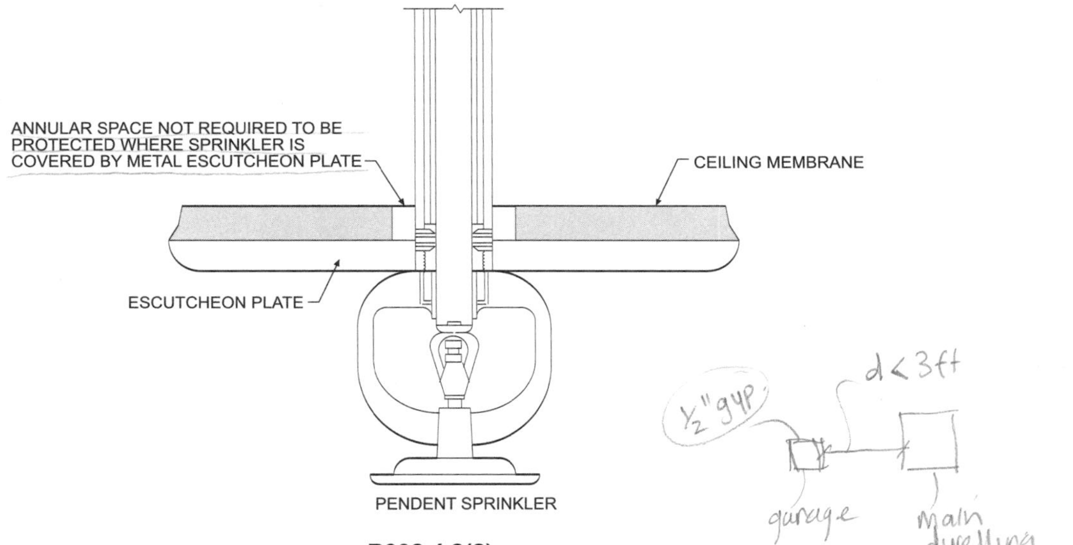

R302.4.2(2)
EXCEPTION TO ANNULAR SPACE PROTECTION

plying with the provisions of this section. The most common situation is the door between the garage and the inside of the home. Solid wood doors $1^3/_8$-inch (35 mm) thick, solid or honeycomb steel doors and 20-minute fire-rated doors are required for use in the opening between the garage and dwelling unit. A self-closing device must be installed on these doors as a safeguard to limit free flow of carbon monoxide or other products of combustion into the living area.

R302.5.2 Duct penetration. Ducts in the garage and ducts penetrating the walls or ceilings separating the *dwelling* from the garage shall be constructed of a minimum No. 26 gage (0.48 mm) sheet steel or other *approved* material and shall have no openings into the garage. -cannot open into garage.

❖Ducts are permitted to penetrate the required separation (see Section R302.6) between the garage and dwelling unit when the ducts within the garage and the portion of the duct penetrating the wall are of No. 26 gage (0.48 mm) sheet steel or other materials acceptable to the building official. Steel ducts are required to help prevent the passage of an undetected fire within the garage to the dwelling unit (see Commentary Figure R302.5.2). The opening limitation in the garage is to limit the path for smoke to enter the dwelling unit.

R302.5.3 Other penetrations. Penetrations through the separation required in Section R302.6 shall be protected as required by Section R302.11, Item 4. ?

❖This section addresses the annular space that results from a penetration of the common wall by pipes, conduits or ductwork. It is important that the building official verify that these spaces are properly filled and do not compromise the protection offered by the common wall between the residence and garage against the free passage of smoke, fire, noxious gases and odors.

R302.6 Dwelling/garage fire separation. The garage shall be separated as required by Table R302.6. Openings in garage walls shall comply with Section R302.5. This provision does not apply to garage walls that are perpendicular to the adjacent *dwelling unit* wall.

❖Numerous potential hazards exist within garages because occupants of dwelling units tend to store a variety of hazardous materials there. Along with this and the potential for CO build-up within the garage, the code requires that the garage be separated from the dwelling unit and attic as indicated in Table R302.6. Garage walls and ceilings that do not form a separation from the dwelling unit are not required to be rated unless they are an extension of a rated assembly.

TABLE R302.6. See page 3-53.

❖This table specifies when and how the garage must be separated from the dwelling unit and any attic space. Walls between the residence and the attached garage, or ceilings between the garage and an attic space, must have at least $^1/_2$-inch (12.7 mm) gypsum board on the garage side. If a habitable room is above the garage, the ceiling must be at least $^5/_8$-inch (15.9 mm) Type X gypsum board on the garage side. Additionally, the exterior walls of the garage are required to have $^1/_2$-inch (12.7 mm) gypsum board on the interior face where they support floors separating all or part of a dwelling unit above the garage.

Detached garages located less than 3 feet (305 mm) from an adjacent dwelling unit must be protected with at least $^1/_2$-inch (12.7 mm) gypsum board applied to the interior side of the garage. The close proximity

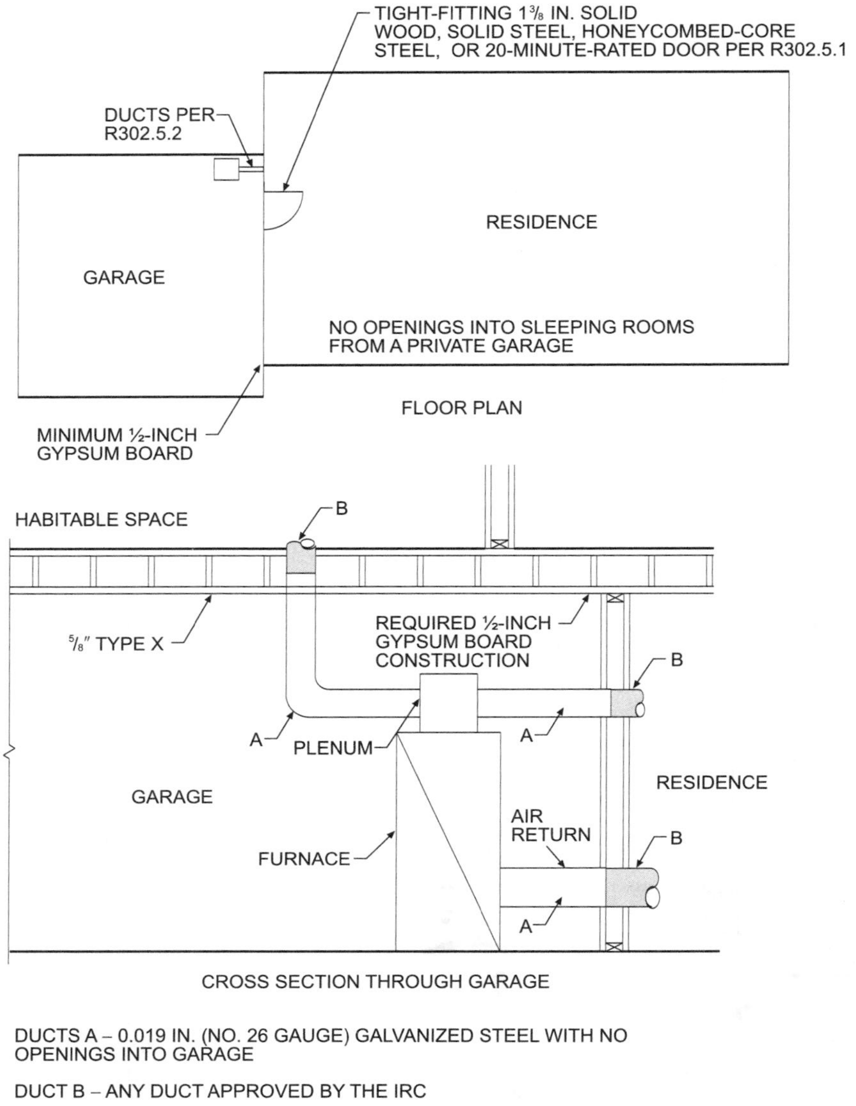

For SI: 1 inch = 25.4 mm.

Figure R302.5.2
GARAGE/DWELLING SEPARATION

to adjacent dwellings requires the additional protection. This $^1/_2$-inch (12.7 mm) gypsum board is required even when the exterior walls are exempted by Section R302.1.

The term "or equivalent" under each option allows for alternative means consistent with Section R104.11.

There are two primary reasons for the enhanced fire endurance of a garage ceiling located beneath a habitable room. First, a fire occurring in a garage may well go undetected for an extended period prior to activation of a detector or other visual alerting.

Second, the inherent fire load and hazardous household activities associated with a garage necessitate this additional level of protection if fire suppression forces are to have a reasonable opportunity to contain a garage fire to the area of origin.

The single layer of $^5/_8$-inch (15.9 mm) Type X gypsum board at the garage ceiling increases the fire endurance of the assembly considerably, from 15 minutes for a $^1/_2$-inch (12.7 mm) layer, to at least 40 minutes, or a 167-percent increase in endurance. When added to the rating for floor joists and certain subflooring combinations, the final endurance is close to 1 hour.

Commentary Figure R302.6 shows two locations of gypsum wallboard; each achieves the protection required by the code.

TABLE R302.6
DWELLING/GARAGE SEPARATION

SEPARATION	MATERIAL
From the residence and attics	Not less than $^1/_2$-inch gypsum board or equivalent applied to the garage side
From all habitable rooms above the garage	Not less than $^5/_8$-inch Type X gypsum board or equivalent
Structure(s) supporting floor/ceiling assemblies used for separation required by this section	Not less than $^1/_2$-inch gypsum board or equivalent
Garages located less than 3 feet from a dwelling unit on the same lot	Not less than $^1/_2$-inch gypsum board or equivalent applied to the interior side of exterior walls that are within this area

For SI: 1 inch = 25.4 mm, 1 foot = 304.8 mm.

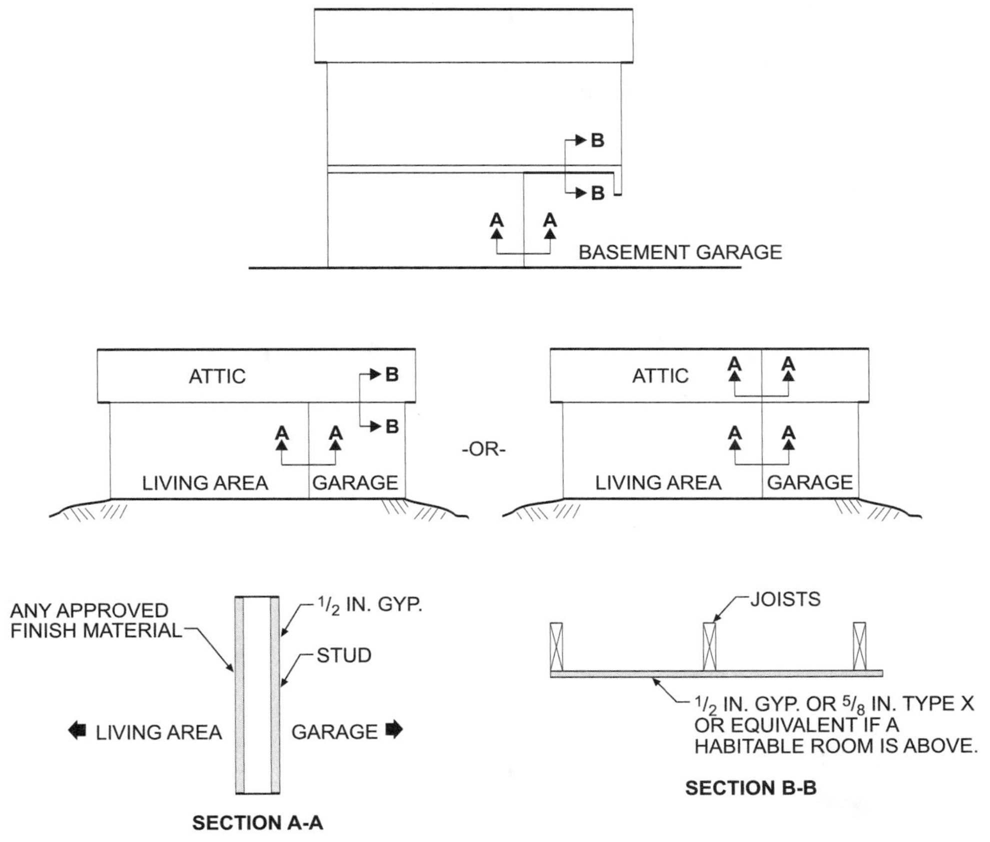

For SI: 1 inch = 25.4 mm.

Figure R302.6
SEPARATION BETWEEN ATTACHED GARAGES AND DWELLING UNITS

R302.7 Under-stair protection. Enclosed accessible space under stairs shall have walls, under-stair surface and any soffits protected on the enclosed side with $^1/_2$-inch (12.7 mm) gypsum board.

❖Often times the space under a stairway is used for storage because this space is often of little use for other purposes. The code permits the use of an open space beneath a stair without the need for any additional protection. Additionally, if the space is walled off and there is no access to the area, the code is also not concerned. If, however, the area beneath the stairway is enclosed and any type of access is provided into the space, the walls, soffits and ceilings of the enclosed space must be protected on the enclosed side with at least $^1/_2$-inch (12.7 mm) gypsum board.

R302.8 Foam plastics. For requirements for foam plastics see Section R316.

❖This section provides a cross reference to the foam plastic fire-resistant requirements for when this product is used within a home. Foam plastics are most commonly used in applications where there thermal qualities can be a cost savings for the home owner over time.

R302.9 Flame spread index and smoke-developed index for wall and ceiling finishes. Flame spread and smoke index for wall and ceiling finishes shall be in accordance with Sections R302.9.1 through R302.9.4.

❖Wall and ceiling finishes have requirements for flame spread and smoke developed indexes, and testing, which are addressed in the following subsections.

R302.9.1 Flame spread index. Wall and ceiling finishes shall have a flame spread index of not greater than 200.

Exception: Flame spread index requirements for finishes shall not apply to trim defined as picture molds, chair rails, baseboards and handrails; to doors and windows or their frames; or to materials that are less than $^{1}/_{28}$ inch (0.91 mm) in thickness cemented to the surface of walls or ceilings if these materials exhibit flame spread index values no greater than those of paper of this thickness cemented to a noncombustible backing.

❖The control of interior finishes is an important aspect of fire protection. Section R315.9 contains the requirements for controlling fire growth within buildings by restricting interior finish materials. The dangers of unregulated interior finish include both the rapid spread of fire and the contribution of additional fuel to the fire. The rapid spread of fire presents a threat to the occupants of a building by limiting or denying their use of exitways within and outside the building. This can be caused by the rapid spread of the fire itself or by the production of large quantities of dense, black smoke, which obscures the exit path or makes movement difficult. Unregulated finish materials also have the potential for adding fuel to the fire, thus increasing its intensity and shortening the time available for the occupants to exit safely. However, based on the test standard that is used, the code does not regulate the fuel contribution of interior finish materials.

The code regulates interior finish materials on both walls and ceilings. These provisions do not address the floor or any coverings applied to the floor. The level of performance the code establishes for finish materials is a flame spread index of 200 or less. The flame spread index for a material is based on reviewing its performance under the test standard specified in Section R302.9.3. The limitation of 200 for the flame spread index matches a "Class C" material in the IBC. Therefore, any material that has a flame spread index of less than 200 may be used as a finish material on both walls and ceilings. When foam plastics are used for interior wall and ceiling finishes, they must comply with Section R316.5.10.

The exception will permit the installation of materials that will not significantly contribute to a fire. This includes various types of trim, such as chair rails, door and window frames, and baseboards, which because of their quantities and locations, do not cause great concern. The actual size and quantity of such trim is not specified in the code because it has not been a concern. However, if some type of foam plastic is being used instead of other types of combustible trim, Section R316.5.9 does limit those situations. The exception also does not regulate thin materials, such as wallpaper, which are less than 1/28 inch (0.907 mm) thick when they are properly installed. These thin materials, when cemented to the surface of the wall or ceiling, behave essentially as the backing to which they are applied and, as a result, are not regulated.

R302.9.2 Smoke-developed index. Wall and ceiling finishes shall have a smoke-developed index of not greater than 450.

❖The development of smoke affects the occupants' safety. This section places a limitation of 450 on the density of smoke that is allowed from the wall and ceiling finishes. A product's smoke-developed index is established for the finish materials when they are tested under the standard specified in Section R302.9.3. The test measures only the obscurity caused by the smoke and does not consider the toxic content within the smoke.

R302.9.3 Testing. Tests shall be made in accordance with ASTM E 84 or UL 723.

❖This section establishes that the standard test for flame spread and smoke-development characteristics is either ASTM E 84 or UL 723, commonly known as the Steiner Tunnel Test. ASTM E 84 and UL 723 determine the relative burning behavior of materials on exposed surfaces, such as ceilings and walls, by visually observing the flame spread along the test specimen. Flame spread and smoke density are reported. The test method renders measurements of surface flame spread and smoke density in comparison with test results obtained by using select red oak and asbestos-cement board as control materials. Red oak is used for the furnace calibration because it is a fairly uniform grade of lumber that is readily available nationally, is uniform in thickness and moisture content, and generally gives consistent and reproducible results. Asbestos-cement board has a flame spread index of zero, while the red oak is assigned a flame spread index of 100. All other materials are then given an index based on a comparison with these two materials. Therefore, the flame spread index of 200 that is permitted by Section R302.9.1 essentially means that the maximum flame spread index for any finish material is twice that of the sample specimen of red oak.

R302.9.4 Alternative test method. As an alternative to having a flame spread index of not greater than 200 and a smoke-developed index of not greater than 450 when tested in accordance with ASTM E 84 or UL 723, wall and ceiling finishes shall be permitted to be tested in accordance with NFPA 286. Materials tested in accordance with NFPA 286 shall meet the following criteria:

flame + smoke development
→ E84 ASTM Steiner Tunnel test

The interior finish shall comply with the following:

1. During the 40 kW exposure, flames shall not spread to the ceiling.
2. The flame shall not spread to the outer extremity of the sample on any wall or ceiling.
3. Flashover, as defined in NFPA 286, shall not occur.
4. The peak heat release rate throughout the test shall not exceed 800 kW.
5. The total smoke released throughout the test shall not exceed 1,000 m^2.

❖This section allows the use of NFPA 286 instead of ASTM E 84 or UL 723 for testing of wall and ceiling finishes other than textiles. NFPA 286 is known as a "room corner" fire test. In this test, a fire source consisting of a wood crib is placed in the corner of a compartment. The materials tested are then placed on the walls of the compartment (see Commentary Figure R302.9.4). This generally provides a more realistic understanding of the hazards involved with the materials.

Two levels of exposures are used during an NFPA 286 fire test to better represent a growing fire. The first is a 40-kW fire size for 5 minutes and then a 160-kW exposure for 10 minutes. The 40-kW exposure represents the beginning of a fire where the initial spread is critical. Therefore, the stated criterion is that the fire cannot spread to the ceiling. The 160-kW exposure is obviously a more intense fire and the criterion relates to preventing flashover (as defined by NFPA 286) and the extent of flame spread throughout the entire test assembly. There is also a peak heat release limit of 800 kw and a total smoke production limit of 1,000 square feet (92.9 m^2) for both levels of exposure.

It should be noted that the flashover criteria for NFPA 286 are as follows:

- Heat release exceeds 1 MW,
- Heat flux at the floor exceeds 20 kW/m^2,
- Average upper layer temperature exceeds 600°C (1112°F),
- Flames exit the doorway, and
- Auto ignition of paper target on the floor occurs.

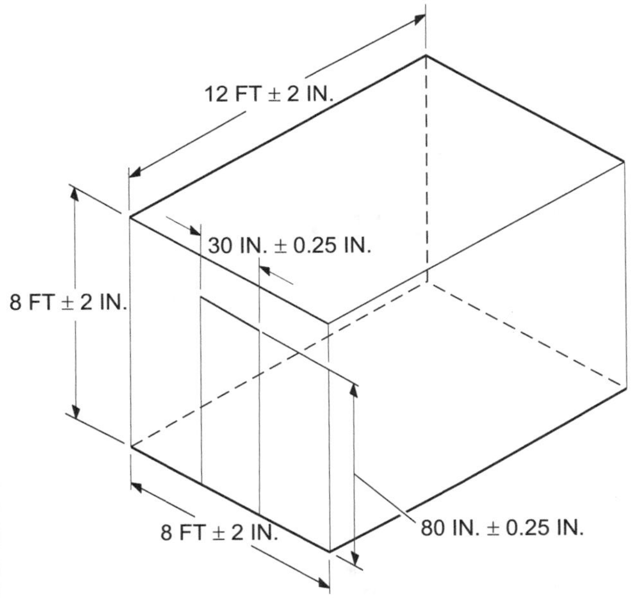

For SI: 1 inch = 25.4 mm, 1 foot = 304.8 mm.

Figure R302.9.4
INTERIOR FIRE TEST ROOM DIMENSIONS

R302.10 Flame spread index and smoke-developed index for insulation. Flame spread and smoke-developed index for insulation shall be in accordance with Sections R302.10.1 through R302.10.5.

❖Section R302.10 addresses insulating materials installed in building spaces. Insulating materials can affect fire development and fire spread and, therefore, are regulated. Insulation has requirements for flame spread, smoke development, critical radiant flux and testing, which are addressed in the following subsections. There are unique testing requirements for loose-fill insulations.

R302.10.1 Insulation. Insulation materials, including facings, such as vapor retarders and vapor-permeable membranes installed within floor/ceiling assemblies, roof/ceiling assemblies, wall assemblies, crawl spaces and *attics* shall have a flame spread index not to exceed 25 with an accompanying smoke-developed index not to exceed 450 when tested in accordance with ASTM E 84 or UL 723.

Exceptions:

1. When such materials are installed in concealed spaces, the flame spread index and smoke-developed index limitations do not apply to the facings, provided that the facing is installed in substantial contact with the unexposed surface of the ceiling, floor or wall finish.
2. Cellulose loose-fill insulation, which is not spray applied, complying with the requirements of Section R302.10.3, shall only be required to meet the smoke-developed index of not more than 450.
3. Foam plastic insulation shall comply with Section R316.

❖Section R302.10.1 addresses the various insulating materials that may be installed in building spaces, including insulating batts, blankets, fills (including vapor barriers and vapor-permeable membranes) and other coverings. Exposed insulating materials represent the same fire exposure hazard as any other exposed material, such as an interior finish. The provisions of Sections R302.10.2, R302.10.3 and R302.10.4, as well as the foam plastic provisions of Section R316, should also be reviewed based on the actual type of insulation and how it is installed. As a general requirement, insulation, including facings used as vapor retarders or as breather papers, must have a flame spread index not in excess of 25 and a smoke-developed index not in excess of 450. These values limit the contribution of the insulation to a fire. The flame spread requirement of 25 for the insulation will be more limiting than the 200, which is accepted for interior finishes by Section R302.9. The test method used to establish these limits is either ASTM E 84 or UL 723. See the commentary to Section R302.9.3 for additional information.

Exceptions 1 and 2 address situations where, because of the way the material is installed or because of other imposed regulations, the material does not make any significant contribution to a fire. The first exception eliminates the flame spread and smoke-developed indexes for the facing portion of the insulation if it is installed “in substantial contact” with the unexposed surface of the ceiling, floor or wall finish. For example, when paper-backed insulation is placed directly on top of a ceiling, the paper facing is not required to meet the smoke and flame spread limits. If the same material is applied to the underside of a roof deck and the paper facing is exposed to the attic space, the paper facing must then meet the general criteria. The potential for flame spread is greatly diminished when the facings are installed in direct contact with the finish material because of the lack of airspace to support a fire if the facing were to be exposed to a source of ignition. See Commentary Figure R302.10.1 for an example of the various facing provisions.

The second exception addresses the fact that cellulose loose-fill insulation is federally regulated by the Consumer Product Safety Commission (CPSC). Parts 1209 and 1404 of CPSC 16 CFR contain various requirements that regulate the product to avoid excessive flammability or significant fire hazards. The smoke-developed index for cellulose loose-fill insulation must be determined by either the ASTM E 84 or UL 723 test and must be 450 or less. This section requires that the smoke-developed index be measured using the either the ASTM E 84 or UL 723 test rather than the test procedures specified in Sections R302.10.2 and R302.10.3.

Exception 3 points the user to Section R316 for foam plastic insulation. Foam plastic materials pose some different problems that are addressed in Section R316.

R302.10.2 Loose-fill insulation. Loose-fill insulation materials that cannot be mounted in the ASTM E 84 or UL 723 apparatus without a screen or artificial supports shall comply with the flame spread and smoke-developed limits of Section R302.10.1 when tested in accordance with CAN/ULC S102.2.

> **Exception:** Cellulose loose-fill insulation shall not be required to be tested in accordance with CAN/ULC S102.2, provided such insulation complies with the requirements of Section R302.10.1 and Section R302.10.3.

❖The main provision establishes that CAN/U LC-S1 02.2 is used as the test standard to determine the flame spread and smoke-developed indexes for loose-fill insulation materials that cannot be tested using the normal ASTM E 84 or UL 723 test method. The exception makes a distinction between cellulose insulation that is spray applied using a water-mist spray applicator, and cellulose loose-fill insulation

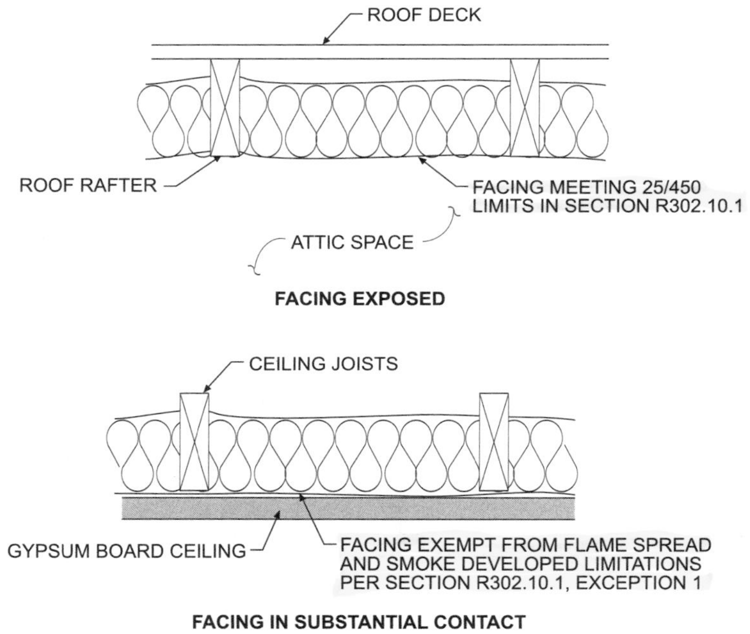

Figure R302.10.1
INSULATION FACING

that is poured or blown into place. Spray-applied cellulose insulation can be exposed on vertical and horizontal ceiling-type surfaces, so it is tested like any other insulating material. Cellulose loose-fill insulation that is poured or commentary to Section R302.10.1, Exception 2), and is exempt from the test procedure described in this section.

R302.10.3 Cellulose loose-fill insulation. Cellulose loose-fill insulation shall comply with CPSC 16 CFR, Parts 1209 and 1404. Each package of such insulating material shall be clearly *labeled* in accordance with CPSC 16 CFR, Parts 1209 and 1404.

❖Because cellulose loose-fill insulation is federally regulated, this section provides the reference to the various federal regulations that are to be used. Because the federal regulations have precedence, the code cannot impose other requirements on this product. Therefore, this section and the reference to it from Section R302.10.1, Exception 2, establish the requirements for this product and provide the exemption from the normal ASTM E 84 or UL 723 test standards.

R302.10.4 Exposed attic insulation. All exposed insulation materials installed on *attic* floors shall have a critical radiant flux not less than 0.12 watt per square centimeter.

❖This section provides the performance requirements for the test exposure that insulation must meet when it is exposed on the floor of an attic. It is tied to the testing provisions found in Section R302.10.5, which specifies that the ASTM E 970 test is to be used for determining the critical radiant flux. See the commentary to Section R302.10.5 regarding the application of this requirement to cellulose loose-fill insulation.

R302.10.5 Testing. Tests for critical radiant flux shall be made in accordance with ASTM E 970.

❖ASTM E 970 is a test method developed by the insulation industry to evaluate the fire hazard of exposed attic insulation and is referenced in the material standards for insulation. Cellulose loose-fill insulation must comply with CPSC 16 CFR, Part 1209 (see Section R302.10.3), which requires testing by this standard. Spray-applied cellulose insulation that is not subject to the CPSC standard (see Section R302.10.1 and Exception 2 in that section) is also subject to the ASTM E 970 testing through Section R302.10.4.

R302.11 Fireblocking. In combustible construction, fireblocking shall be provided to cut off all concealed draft openings (both vertical and horizontal) and to form an effective fire barrier between stories, and between a top *story* and the roof space.

Fireblocking shall be provided in wood-frame construction in the following locations:

1. In concealed spaces of stud walls and partitions, including furred spaces and parallel rows of studs or staggered studs, as follows:
   1.1. Vertically at the ceiling and floor levels.
   1.2. Horizontally at intervals not exceeding 10 feet (3048 mm).
2. At all interconnections between concealed vertical and horizontal spaces such as occur at soffits, drop ceilings and cove ceilings.
3. In concealed spaces between stair stringers at the top and bottom of the run. Enclosed spaces under stairs shall comply with Section R302.7.
4. At openings around vents, pipes, ducts, cables and wires at ceiling and floor level, with an *approved* material to resist the free passage of flame and products of combustion. The material filling this annular space shall not be required to meet the ASTM E 136 requirements.
5. For the fireblocking of chimneys and fireplaces, see Section R1003.19.
6. Fireblocking of cornices of a two-family *dwelling* is required at the line of *dwelling unit* separation.

❖To restrict the movement of flame and gasses to other areas of a building through concealed passages in building components, such as floors, walls and stairs, fireblocking of these concealed combustible spaces is required to form a barrier between stories, and between a top story and the roof space. For example, the following locations must be firestopped in wood frame construction:

- In concealed spaces of stud walls and partitions, including spaces at the ceiling and floor levels [see Commentary Figures R302.11(1) and R302.11(2)].
- At all interconnections between concealed vertical and horizontal spaces such as soffits [see Commentary Figure R302.11(3)], dropped ceilings [see Commentary Figure R302.11(4)] and cove ceilings [see Commentary Figure R302.11(5)]. Interconnections shown in Commentary Figure R302.11(6) for a bathtub installation must also be firestopped.
- In concealed spaces between stair stringers at the top and bottom of the run [see Commentary Figure R302.11(7)].
- At openings around vent pipes, ducts, chimneys and fireplaces at ceiling and floor levels with noncombustible materials [see Commentary Figures R302.11(8) and R302.11(9)]. By not requiring compliance with ASTM E 136, Item 4 indicates that the fireblocking material can be of combustible materials. Commentary Figure R302.11(9) illustrates fireblocking at chimneys and fireplaces. The fireblocking at the ductwork would be similar.

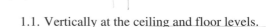

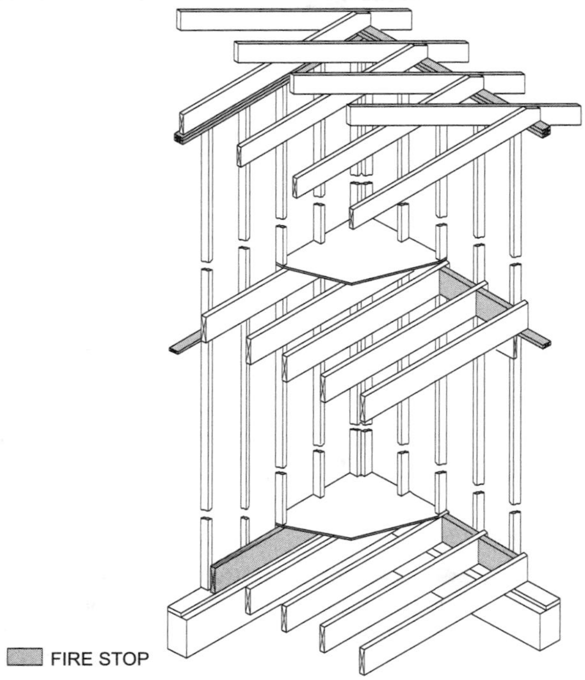

Figure R302.11(1)
FIRESTOPPING—BALLOON FRAMING

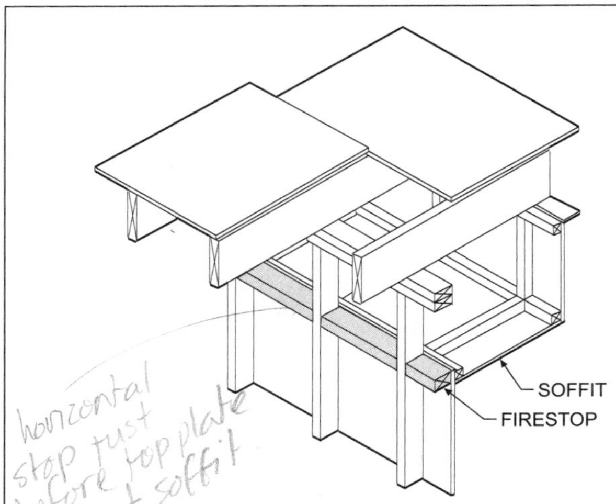

Figure R302.11(3)
FIRESTOPPING—FURRED SOFFIT

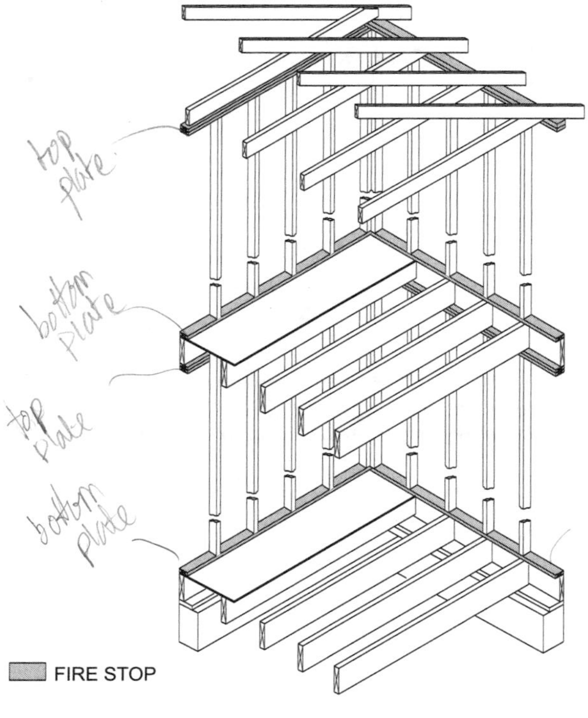

Figure R302.11(2)
FIRESTOPPING—PLATFORM FRAMING

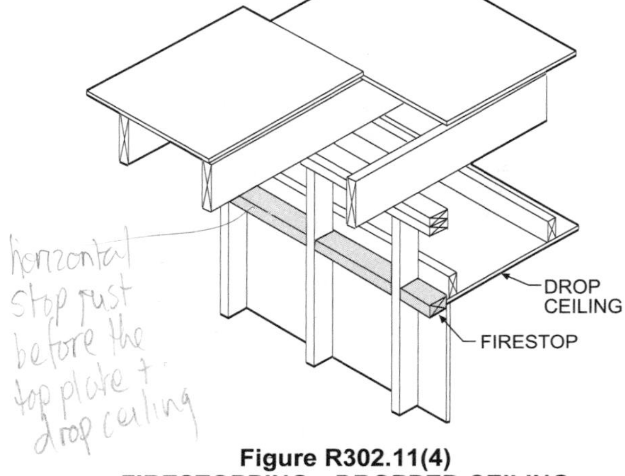

Figure R302.11(4)
FIRESTOPPING—DROPPED CEILING

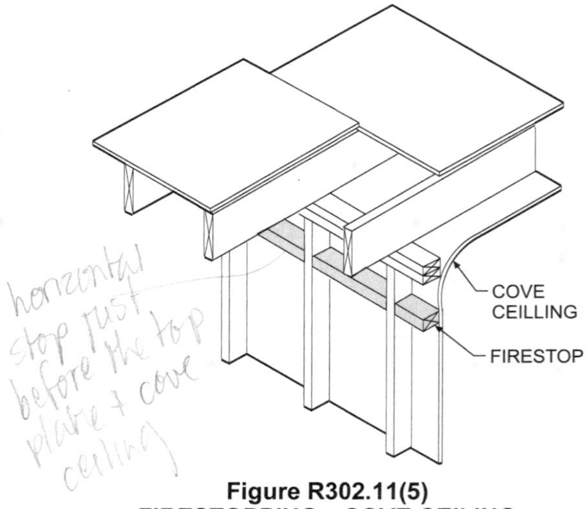

Figure R302.11(5)
FIRESTOPPING—COVE CEILING

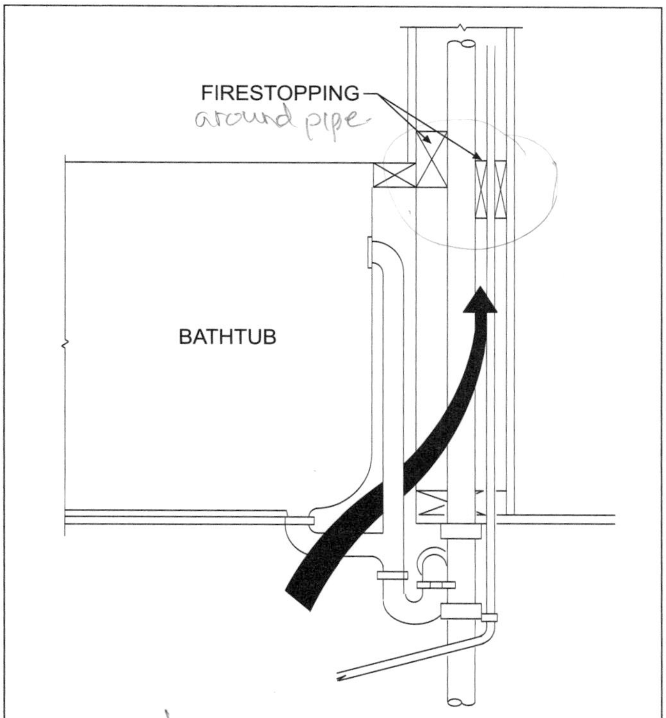

Figure R302.11(6)
FIRESTOPPING—AT TUB

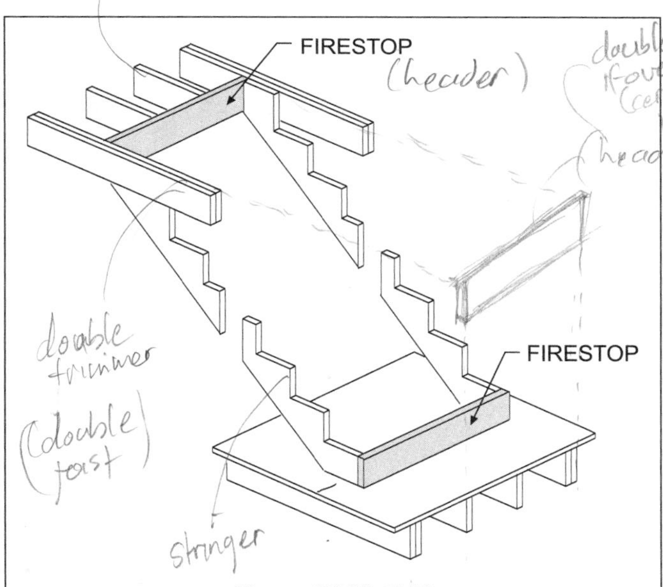

Figure R302.11(7)
FIRESTOPPING—AT STAIRWAYS

R302.11.1 Fireblocking materials. Except as provided in Section R302.11, Item 4, fireblocking shall consist of the following materials.

1. Two-inch (51 mm) nominal lumber.
2. Two thicknesses of 1-inch (25.4 mm) nominal lumber with broken lap joints.

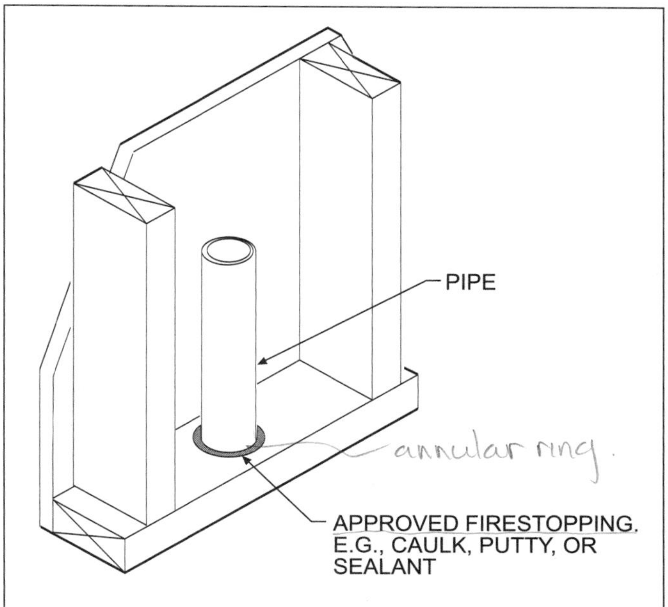

Figure R302.11(8)
FIRESTOPPING—AROUND PIPING

3. One thickness of $^{23}/_{32}$-inch (18.3 mm) wood structural panels with joints backed by $^{23}/_{32}$-inch (18.3 mm) wood structural panels.
4. One thickness of $^{3}/_{4}$-inch (19.1 mm) particleboard with joints backed by $^{3}/_{4}$-inch (19.1 mm) particleboard.
5. One-half-inch (12.7 mm) gypsum board.
6. One-quarter-inch (6.4 mm) cement-based millboard.
7. Batts or blankets of mineral wool or glass fiber or other *approved* materials installed in such a manner as to be securely retained in place.
8. Cellulose insulation installed as tested for the specific application.

❖This section specifies the material required for fireblocking. Fireblocking around chimneys and fireplaces must be noncombustible according to Section R1003.19 as specified in Item 5 of Section R302.11. The fireblocking at openings around vents, pipes, ducts and wires at the ceiling and floor level is required only to be adequate to resist the free passage of flames and smoke according to Item 4 of Section R302.11; however, this fireblocking can be constructed from combustible materials. So that it is not accidentally displaced, all fireblocking material must be securely fastened in place.

R302.11.1.1 Batts or blankets of mineral or glass fiber. Batts or blankets of mineral or glass fiber or other *approved* nonrigid materials shall be permitted for compliance with the 10-foot (3048 mm) horizontal fireblocking in walls constructed using parallel rows of studs or staggered studs.

❖Batts or blankets may serve as horizontal fireblocking in walls if installed in accordance with Sections R302.11.1.2 through R302.11.1.3.

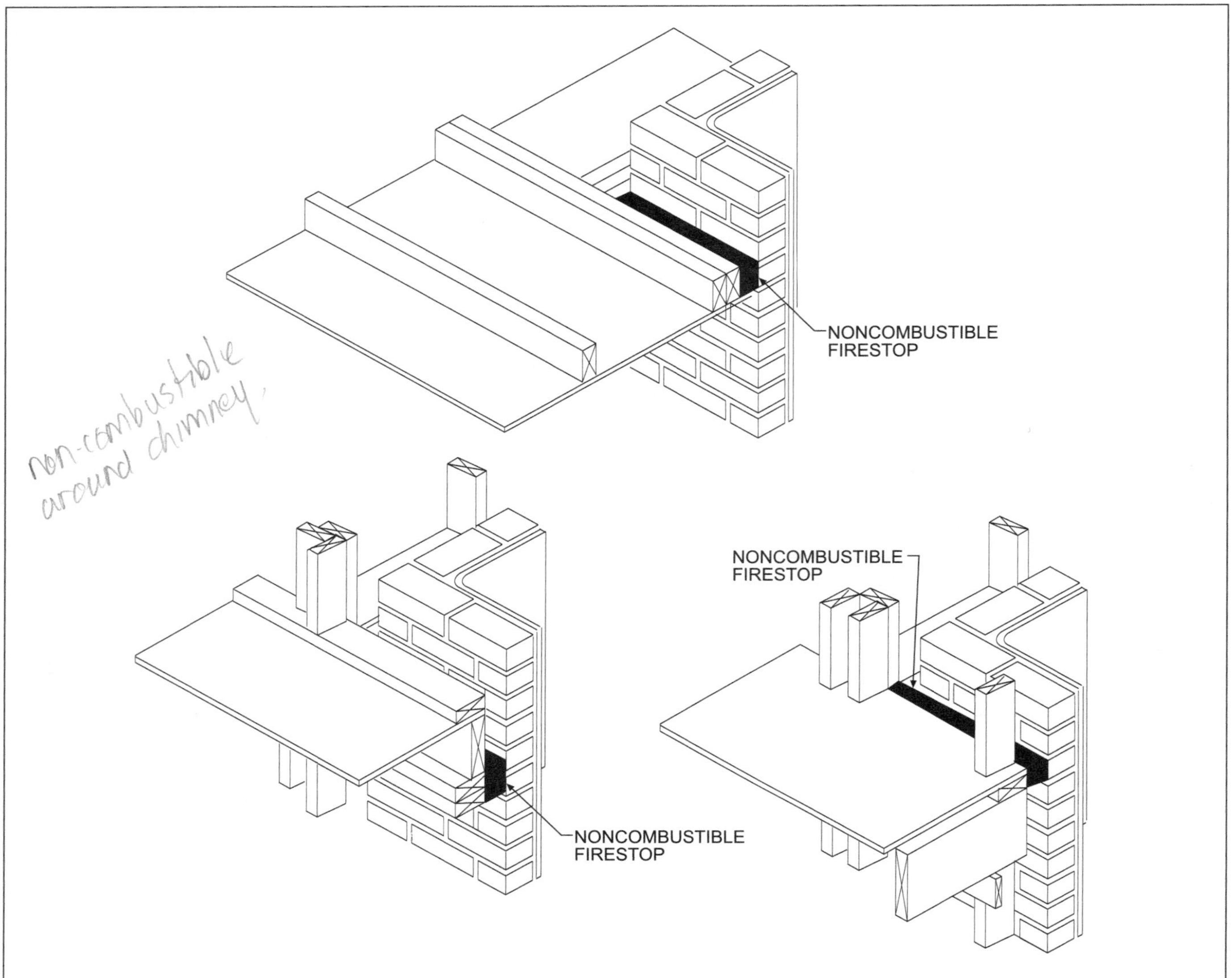

Figure R302.11(9)
FIRESTOPPING—AROUND CHIMNEYS AND FIREPLACES

R302.11.1.2 Unfaced fiberglass. Unfaced fiberglass batt insulation used as fireblocking shall fill the entire cross section of the wall cavity to a minimum height of 16 inches (406 mm) measured vertically. When piping, conduit or similar obstructions are encountered, the insulation shall be packed tightly around the obstruction.

❖When batts or blankets are used for fireblocking they must fill the wall cavity from stud to stud and be at least 16 inches (406 mm) measured from the bottom to the top of the fill.

R302.11.1.3 Loose-fill insulation material. Loose-fill insulation material shall not be used as a fireblock unless specifically tested in the form and manner intended for use to demonstrate its ability to remain in place and to retard the spread of fire and hot gases.

❖Not every type of batt or blanket insulation is acceptable as a fireblocking material. Testing must show that the type utilized will provide the intended protection.

R302.11.2 Fireblocking integrity. The integrity of all fireblocks shall be maintained.

❖Piping, ducts or other similar items that pass through firestops must be installed so that the integrity of the firestop is maintained. This may be accomplished by packing an oversized hole with an acceptable fireblocking material.

R302.12 Draftstopping. In combustible construction where there is usable space both above and below the concealed space of a floor/ceiling assembly, draftstops shall be installed so that the area of the concealed space does not exceed 1,000 square feet (92.9 m^2). Draftstopping shall divide the concealed space into approximately equal areas. Where the assembly is enclosed by a floor membrane above and a ceiling membrane below, draftstopping shall be provided in floor/ceiling assemblies under the following circumstances:

1. Ceiling is suspended under the floor framing.
2. Floor framing is constructed of truss-type open-web or perforated members.

❖Draftstopping is required to limit the spread of fire through combustible spaces in floor/ceiling assemblies when such spaces create a connected area beyond the normal joist cavity. For example, draftstopping is required when the ceiling is suspended under the floor framing, as illustrated in Commentary Figure R302.12(1), or when the floor framing is constructed of truss-type or open-web perforated members, as illustrated in Commentary Figure R302.12(2). When such space exceeds 1,000 square feet (92.9 m^2) in area, the code requires the space between the ceiling membrane and the floor to be divided into two approximately equal areas with no area larger than 1,000 square feet (92.9 m^2). A floor/ceiling assembly having a space exceeding 2,000 square feet (185.8 m^2), but not exceeding 3,000 square feet (278.7 m^2), would have to be divided into a minimum of three approximately equal areas.

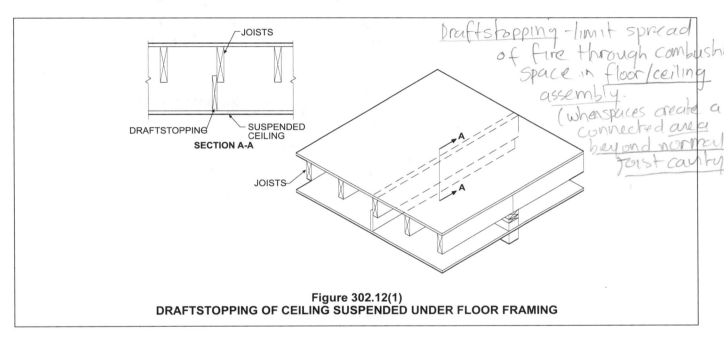

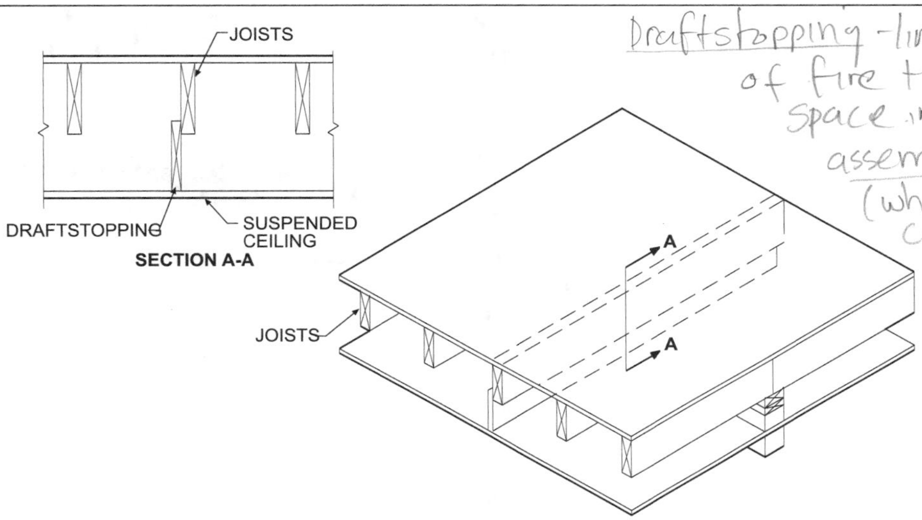

Draftstopping - limit spread of fire through combustible space in floor/ceiling assembly. (when spaces create a connected area beyond normal joist cavity)

Figure 302.12(1)
DRAFTSTOPPING OF CEILING SUSPENDED UNDER FLOOR FRAMING

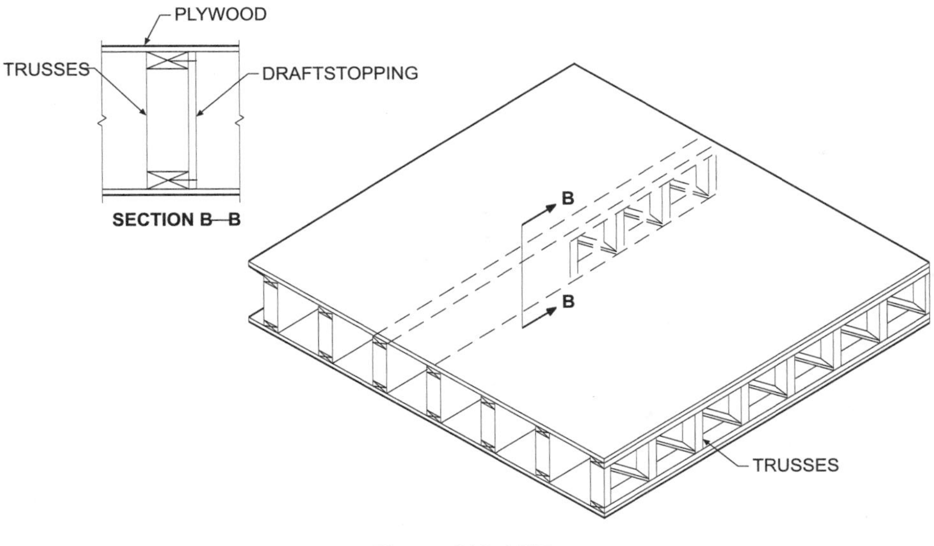

Figure 302.12(2)
DRAFTSTOPPING OF TRUSS-TYPE FLOOR FRAMING

R302.12.1 Materials. Draftstopping materials shall not be less than $^1/_2$-inch (12.7 mm) gypsum board, $^3/_8$-inch (9.5 mm) wood structural panels or other *approved* materials adequately supported. Draftopping shall be installed parallel to the floor framing members unless otherwise *approved* by the *building official*. The integrity of the draftstops shall be maintained.

❖Materials used to meet draftstopping requirements are typically $^1/_2$-inch (12.7 mm) gypsum board or $^3/_8$-inch (9.5 mm) plywood adequately attached to the supporting members. Other materials may be used for draftstopping, if those materials produced an equivalent barrier.

The code places an additional restriction on the specific location of the draftstopping, mandating that the draftstopping be placed parallel to the main floor framing members. Commentary Figures R302.12(1) and R302.12(2) illustrate two examples of draftstopping orientation placed as required.

When a penetration through the draftstopping is necessary, the integrity of the draftstop must be maintained in a manner similar to firestopping penetrations.

R302.13 Combustible insulation clearance. Combustible insulation shall be separated a minimum of 3 inches (76 mm) from recessed luminaires, fan motors and other heat-producing devices.

Exception: Where heat-producing devices are listed for lesser clearances, combustible insulation complying with the listing requirements shall be separated in accordance with the conditions stipulated in the listing.

Recessed luminaires installed in the *building thermal envelope* shall meet the requirements of Section N1102.4.4 of this code.

❖The minimum clearance required between combustible insulation and light fixtures, fan motors, etc., minimizes the chance of accidental ignition of the insulation. If a heat-producing device is listed for less than the minimum clearance specified in this section and all conditions of the listing are satisfied, the reduced clearance would apply.

SECTION R303
LIGHT, VENTILATION AND HEATING

R303.1 Habitable rooms. All habitable rooms shall have an aggregate glazing area of not less than 8 percent of the floor area of such rooms. Natural *ventilation* shall be through windows, doors, louvers or other *approved* openings to the outdoor air. Such openings shall be provided with ready access or shall otherwise be readily controllable by the building occupants. The minimum openable area to the outdoors shall be 4 percent of the floor area being ventilated.

Exceptions:

1. The glazed areas need not be openable where the opening is not required by Section R310 and a whole-house mechanical *ventilation* system is installed in accordance with Section M1507.
2. The glazed areas need not be installed in rooms where Exception 1 above is satisfied and artificial light is provided capable of producing an average illumination of 6 footcandles (65 lux) over the area of the room at a height of 30 inches (762 mm) above the floor level.
3. Use of sunroom and patio covers, as defined in Section R202, shall be permitted for natural *ventilation* if in excess of 40 percent of the exterior sunroom walls are open, or are enclosed only by insect screening.

❖All habitable rooms are to be constructed with an aggregate glazing area of not less than 8 percent of the floor area. The code does not require the glazed area to be openable; however, it must be placed as required by Section R303.7. This section also provides the standard of natural ventilation for all habitable rooms. Openings to the outdoor air, such as doors, windows and louvers, provide natural ventilation. The openable area must be not less than 4 percent of the floor area being ventilated. This section does not, however, state or intend that the doors, windows or openings actually be constantly open. The intent is that they be maintained in an operable condition so that they are available for use at the discretion of the occupant. In addition to the natural ventilation provisions of this section, a whole-house mechanical ventilation system may be required in accordance with the provisions of Section R303.4.

Although the glazed area may be operable, Exception 1 permits the glazed area for all openings, except the ones required for emergency escape and rescue, to be fixed or "not openable" when a whole-house mechanical ventilation system is installed in accordance with Section M1507.

Exception 2 to Section R303.1 allows glazing to be completely deleted from habitable rooms where artificial light is available that is capable of producing an average illumination of 6 footcandles (65 lux) over the room at a height of 30 inches (762 mm) above the floor level and where a whole-house mechanical ventilation system complying with Exception 1 to Section R303.1 is installed.

Exception 3 allows for the openings in a sunroom (those in which more than 40 percent of the exterior sunroom walls is open) to provide natural ventilation when the exterior walls of the sunroom are open. Insect screening for these seasonal rooms is also acceptable to make use of this exception.

Calculations should be submitted for review for each of the exceptions. Various handbooks are available from lighting manufacturers for calculating the required illumination.

R303.2 Adjoining rooms. For the purpose of determining light and *ventilation* requirements, any room shall be considered as a portion of an adjoining room when at least one-half of the area of the common wall is open and unobstructed and provides an opening of not less than one-tenth of the floor area of the interior room but not less than 25 square feet (2.3 m^2).

> **Exception:** Openings required for light and/or *ventilation* shall be permitted to open into a sunroom with thermal isolation or a patio cover, provided that there is an openable area between the adjoining room and the sunroom or patio cover of not less than one-tenth of the floor area of the interior room but not less than 20 square feet (2 m^2). The minimum openable area to the outdoors shall be based upon the total floor area being ventilated.

❖Where rooms do not have access to an exterior wall as part of an adjoining room if they are open to that room as shown in Commentary Figure R303.2.

The exception deals with a very common circumstance, especially in residential construction. As long as the sunroom addition is large enough and is thermally isolated, the building owner need not move openings for lighting and/or ventilation when installing an addition that meets the definition of "Sunroom addition."

R303.3 Bathrooms. Bathrooms, water closet compartments and other similar rooms shall be provided with aggregate glazing area in windows of not less than 3 square feet (0.3 m^2), one-half of which must be openable.

> **Exception:** The glazed areas shall not be required where artificial light and a local exhaust system are provided. The minimum local exhaust rates shall be determined in accordance with Section M1507. Exhaust air from the space shall be exhausted directly to the outdoors.

❖Like habitable rooms, bathrooms must be provided with natural light and ventilation or adequate artificial light and mechanical ventilation. For natural light, exterior glazing of at least 3 square feet (0.3 m^2) must be provided. To meet the natural ventilation requirement, at least 1.5 square feet (0.140 m^2) must be openable.

For local exhaust the exhaust rates must be determined in accordance with Section M1507.4. The exhaust must be directly to the outdoors.

R303.4 Mechanical ventilation. Where the air infiltration rate of a dwelling unit is less than 5 air changes per hour when tested with a blower door at a pressure of 0.2 inch w.c (50 Pa) in accordance with Section N1102.4.1.2, the dwelling unit shall be provided with whole-house mechanical ventilation in accordance with Section M1507.3.

❖Section R303.1 requires natural ventilation through windows, doors or other approved openings, with an option for mechanical ventilation. Multiple studies have shown that natural ventilation alone is not sufficient for dwellings that are tightly sealed such that their infiltration rate is below 5 air changes per hour (ACH). Section R303.4 sets the threshold for tightly sealed at 5 ACH at a 50 Pascal pressure differential between the indoors and outdoors (5 ACH50). For perspective, 50 Pa = 0.2 inch water column; 1 inch water column = 250 Pa. Traditionally, 0.35 ACH or 15 cfm per occupant has been the required mechanical ventilation rate allowed as an alternative to natural ventilation. An ACH of 0.35 at typical ambient pressure differentials is roughly equivalent to 7 to 10 ACH at a 50 Pa differential, thus the threshold of 5 ACH50 is comparable to the traditional ventilation rate. If the blower door test required by Section N1102.4.1.2 of the code indicates that the infiltration rate is less than 5 ACH50, the dwelling must be provided with whole-house mechanical ventilation. Note that Section N1102.4.1.2 intends for the infiltration rate to be 5

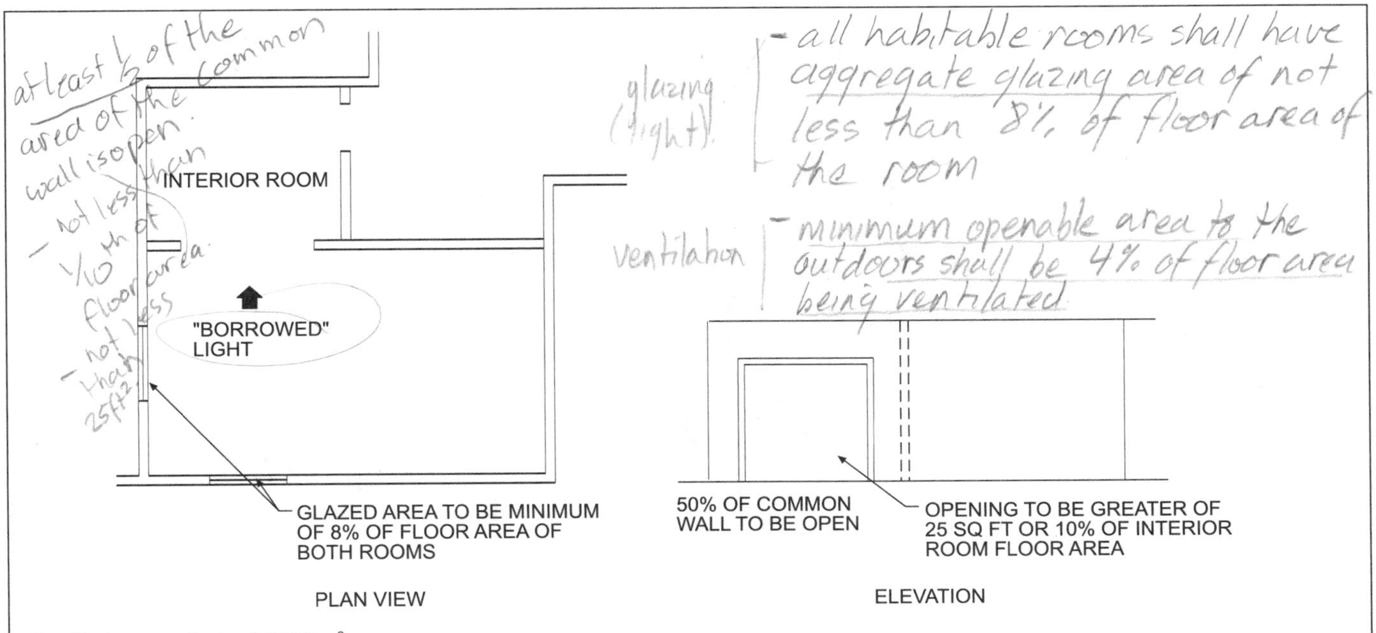

INTERIOR ROOM
"BORROWED" LIGHT
GLAZED AREA TO BE MINIMUM OF 8% OF FLOOR AREA OF BOTH ROOMS
PLAN VIEW
50% OF COMMON WALL TO BE OPEN
OPENING TO BE GREATER OF 25 SQ FT OR 10% OF INTERIOR ROOM FLOOR AREA
ELEVATION

For SI: 1 square foot = 0.0929 m^2.

Figure R303.2
NATURAL LIGHT

ACH50 or less, consistent with the trend for tighter building envelopes. Where a whole-house mechanical ventilation system is provided in accordance with Section R303.1, Exception 1, the intent of this section is satisfied and the ACH rate need not be known because Exception 1 short-cuts directly to Section M1507. As dwelling envelopes become more air tight, there is evidence that indoor contaminant levels are rising. Poor indoor air quality, the inability to predict ventilation rates from natural ventilation and the decreasing rates of infiltration have all led to this requirement for mechanical ventilation in dwellings. The model for the requirements of this section is ASHRAE 62.2, the standard for ventilation and indoor air quality in residential buildings, which is similar to that in Appendix B of ICC 700, *National Green Building Standard*. Also, several state codes now mandate mechanical ventilation in dwellings. This section applies whether or not the natural ventilation provisions of Section R303.1 are applied. In other words, the requirement of this section is in addition to the provisions of Section R303.1 (see commentary, Section M1507).

R303.5 Opening location. Outdoor intake and exhaust openings shall be located in accordance with Sections R303.5.1 and R303.5.2.

❖This section specifies the locations of outdoor intake and exhaust openings. These locations are intended to prevent the introduction of contaminants into the ventilation air of a building or to avoid the exhausting of contaminants onto areas that may be occupied by people or into other buildings.

R303.5.1 Intake openings. Mechanical and gravity outdoor air intake openings shall be located a minimum of 10 feet (3048 mm) from any hazardous or noxious contaminant, such as vents, chimneys, plumbing vents, streets, alleys, parking lots and loading docks, except as otherwise specified in this code. Where a source of contaminant is located within 10 feet (3048 mm) of an intake opening, such opening shall be located a minimum of 3 feet (914 mm) below the contaminant source.

For the purpose of this section, the exhaust from *dwelling* unit toilet rooms, bathrooms and kitchens shall not be considered as hazardous or noxious.

❖In the context of this section, intake openings include windows, doors, gravity air intakes, soffit vents, combustion air intake openings, outside air intakes for air handlers, makeup air intakes and similar openings that naturally or mechanically draw in air from the building exterior. This section identifies specific locations that are known to generate or emit noxious contaminants, and requires that both mechanical and gravity air intake openings be located a minimum of 10 feet (3048 mm) from such hazards to avoid introducing contaminants into the building. As an alternative, mechanical and gravity air intakes can be located within 10 feet (3048 mm) of such sources of contamination if the intakes are located at least 3 feet (914 mm) below the contaminant source. A 3-foot (914 mm) vertical separation distance will allow the noxious gases and contaminants to disperse into the atmosphere before they can be drawn into an air intake opening. Placing the source of contamination above an air intake takes advantage of the fact that normally encountered sources of contamination are lighter (less dense) than the surrounding air and, therefore, will rise above the vicinity of an air intake located below. Commentary Figure R303.4.1 shows an example of the relative locations for intake air

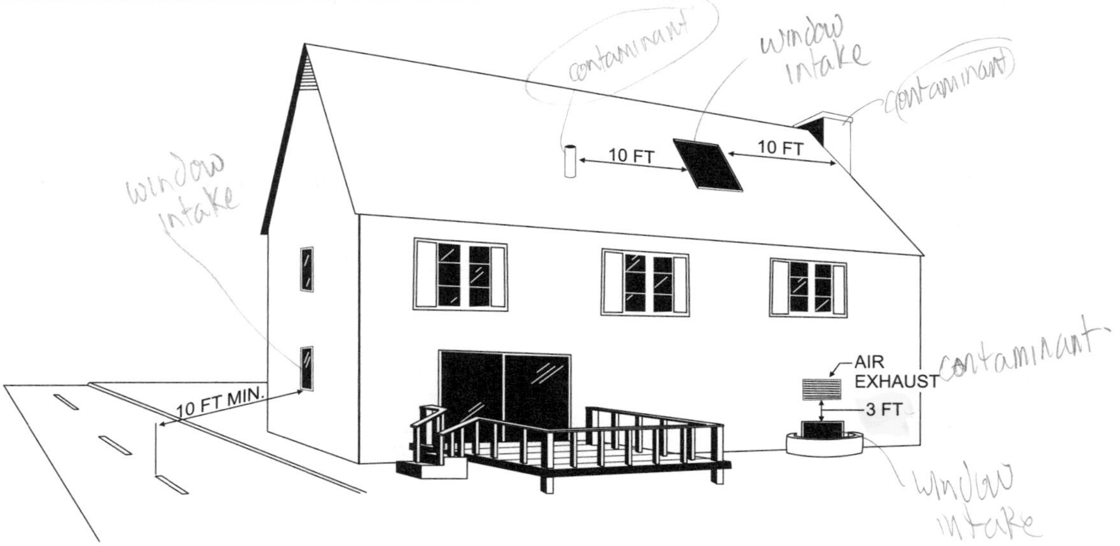

For SI: 1 foot = 304.8 mm.

Figure R303.4.1
AIR INTAKE OPENING LOCATIONS

openings for a building where sources of contaminants are present.

Particular types of exhausts may have more specific restrictions on their location that would supersede this section (see Section M1602.2, for example).

The air exhausts discharging from a dwelling unit (clothes dryer, kitchen and bathroom) are not considered to be significantly hazardous or noxious and are of low volume. In these situations, the building official must determine an appropriate distance or location for the relative placement of intake and exhaust openings. In evaluating each installation, consideration should be given to the orientation of the exhaust or intake louver and its spatial relationship to any source of contaminant or adjacent intake opening, as well as to the direction of the prevailing winds at the location.

R303.5.2 Exhaust openings. Exhaust air shall not be directed onto walkways.

❖For obvious health reasons, exhaust air cannot be directed onto walkways in a manner that subjects the users of the walkway to the exhaust airstream.

Ideally, mechanical exhaust air openings should not be located anywhere they may create a nuisance. A "Nuisance" is defined in Chapter 2 of the *International Plumbing Code*® (IPC®) as ". . . whatever is dangerous to human life or detrimental to health; whatever structure or premises is not sufficiently ventilated, sewered, drained, cleaned or lighted, with respect to its intended occupancy; and whatever renders the air, or human food, drink or water supply unwholesome." A nuisance is defined as much more than or much worse than simply "bothersome." Unfortunately, it is not an easy task to determine whether or not a nuisance will be present because the conditions under which an exhaust system performs vary considerably with the change of seasons, the ambient temperatures and the prevailing winds. As much information as possible should be gathered regarding the installation in order to evaluate the hypothetical worst-case scenario. This would include the characteristics and geometry of the installation, as well as the local ambient conditions, so that an educated guess may be made to determine the "nuisance effect" of the exhaust outlet.

R303.6 Outside opening protection. Air exhaust and intake openings that terminate outdoors shall be protected with corrosion-resistant screens, louvers or grilles having a minimum opening size of $^1/_4$ inch (6 mm) and a maximum opening size of $^1/_2$ inch (13 mm), in any dimension. Openings shall be protected against local weather conditions. Outdoor air exhaust and intake openings shall meet the provisions for *exterior wall* opening protectives in accordance with this code.

❖Outside air exhaust and intake openings must have corrosion-resistant screens, grilles or louvers to prevent foreign objects (such as insects or debris) from entering the system or building. Also, such openings must be protected against the entry of falling or wind-driven water, snow and ice. Exhaust systems sometimes incorporate rotating hoods over the opening to prevent high winds from restricting the flow of exhaust gases out of the system. The hoods align themselves with the direction of the wind to allow the unimpeded, and actually induced, discharge from the exhaust outlet. Rotating turbines are also used as both weather protection and a means of inducing airflow.

The opening sizes for louvers and grilles and the mesh sizes for screens must be within the specified range as indicated. The opening size must be large enough to inhibit blockage by debris, to prevent significant resistance to airflow, and still be small enough to keep out what must be kept out. A screen of such mesh size would restrict the passage of rodents and large insects and would be resistant to blockage by lint, debris and plant fibers.

R303.7 Stairway illumination. All interior and exterior stairways shall be provided with a means to illuminate the stairs, including the landings and treads. Interior stairways shall be provided with an artificial light source located in the immediate vicinity of each landing of the stairway. For interior stairs the artificial light sources shall be capable of illuminating treads and landings to levels not less than 1 footcandle (11 lux) measured at the center of treads and landings. Exterior stairways shall be provided with an artificial light source located in the immediate vicinity of the top landing of the stairway. Exterior stairways providing access to a *basement* from the outside *grade* level shall be provided with an artificial light source located in the immediate vicinity of the bottom landing of the stairway.

Exception: An artificial light source is not required at the top and bottom landing, provided an artificial light source is located directly over each stairway section.

❖Interior and exterior stairways may be illuminated in two ways. The first option is to install artificial lighting in the vicinity of each landing. This would include top, intermediate and bottom landings. For interior stairways, the artificial light must be capable of illuminating treads and landings to not less than 1 footcandle (11 lux). The measurement of 1 footcandle (11 lux) is to be taken at the center of landings and treads. Exterior stairways require illumination only at the top landing (see Commentary Figure R303.6).

Exterior stairways to a basement must have artificial illumination near the bottom landing.

The exception allows the light source to be installed over each individual stairway section, thus eliminating the lighting over the landings.

R303.7.1 Light activation. Where lighting outlets are installed in interior stairways, there shall be a wall switch at each floor level to control the lighting outlet where the stairway has six or more risers. The illumination of exterior stairways shall be controlled from inside the *dwelling* unit.

Exception: Lights that are continuously illuminated or automatically controlled.

❖The location of the light activation is based on the need to illuminate an area before it is used. Interior stairway lighting control is required at each floor level of the stairway. There is an exception for lights that are continuously illuminated or automatically activated. When manual switches are installed, they should be accessible without the switch operator stepping onto the stairway. Exterior stairway light control must be from the interior of the structure. This not only helps provide security for the dwelling, but enables the occupants to illuminate the stairway for guests without first having to traverse the stairway in darkness to do so.

R303.8 Required glazed openings. Required glazed openings shall open directly onto a street or public alley, or a *yard* or court located on the same *lot* as the building.

Exceptions:

1. Required glazed openings may face into a roofed porch where the porch abuts a street, *yard* or court and the longer side of the porch is at least 65 percent unobstructed and the ceiling height is not less than 7 feet (2134 mm).
2. Eave projections shall not be considered as obstructing the clear open space of a *yard* or court.
3. Required glazed openings may face into the area under a deck, balcony, bay or floor cantilever provided a clear vertical space at least 36 inches (914 mm) in height is provided.

❖Glazed openings must be placed so that natural light will be available even after future adjacent construction occurs. Where glazing is required by the code, it must face onto a street, public alley, yard or court of the lot on which the building is located. Glazed openings are not permitted in walls that must be of 1-hour fire-resistant construction in accordance with Section R302.1.

Exception 1 allows for an open porch where natural light will be available. Glazed openings required by the code are permitted to open onto a porch if it abuts a yard, street or court and the longer side is open at least 65 percent. The porch ceiling height must be at least 7 feet (2134 mm).

Yards and courts are defined as being open and unobstructed from the ground to the sky. Interpreted literally, this would prevent any projection above a required window. Yet eaves are commonly placed above windows and in apartment buildings it is common to have decks over required windows. Exception 2 removes that problem. Concern that an eave could project to a lot line and, thus, block all natural light is unfounded as eave projections are limited from extending to a lot line so there will always be some open space above them.

Exception 3 addresses windows that open under balconies, decks, bays, etc. Section R310.5 permits emergency escape openings to open "under decks and porches provided the location of the deck allows the emergency escape window to be fully opened and provides a path not less than 36 inches (914 mm) in height to a yard or court." This allowance will make the two types of windows consistent.

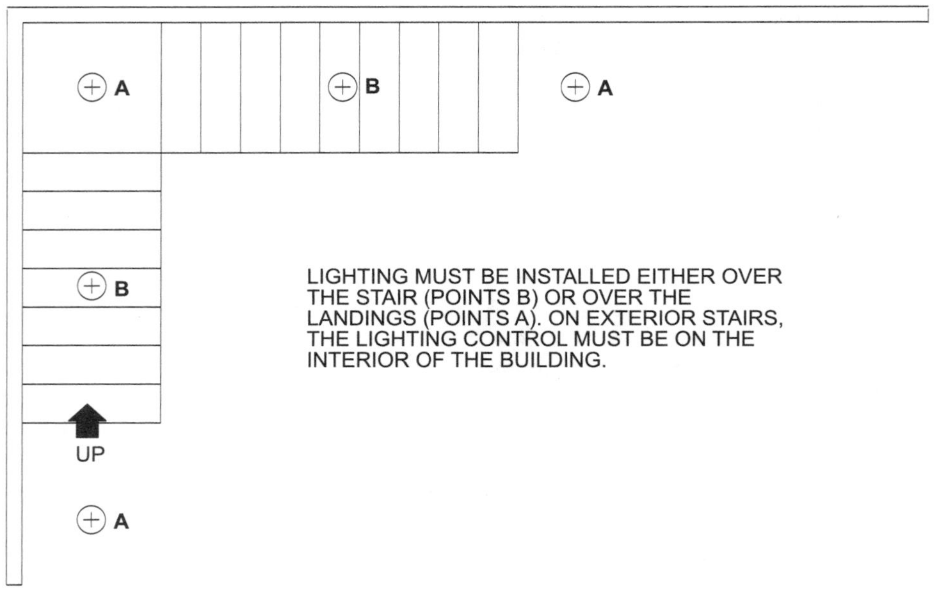
A
B
A
B
UP
A
LIGHTING MUST BE INSTALLED EITHER OVER THE STAIR (POINTS B) OR OVER THE LANDINGS (POINTS A). ON EXTERIOR STAIRS, THE LIGHTING CONTROL MUST BE ON THE INTERIOR OF THE BUILDING.

Figure R303.6
STAIRWAY ILLUMINATION

R303.8.1 Sunroom additions. Required glazed openings shall be permitted to open into sunroom *additions* or patio covers that abut a street, *yard* or court if in excess of 40 percent of the exterior sunroom walls are open, or are enclosed only by insect screening, and the ceiling height of the sunroom is not less than 7 feet (2134 mm).

❖This section permits openings for lighting to open into a sunroom that meets the criterion of being more than 40 percent open. Both enclosed porches and sunrooms must be a minimum of 7 feet (2134 mm) high and abut a court, yard or public space. Open areas near the sunroom enable the sunlight to effectively reach the structure. Sunrooms are those which are not conditioned and are either open or enclosed by screening or plastic.

R303.9 Required heating. When the winter design temperature in Table R301.2(1) is below 60°F (16°C), every *dwelling unit* shall be provided with heating facilities capable of maintaining a minimum room temperature of 68°F (20°C) at a point 3 feet (914 mm) above the floor and 2 feet (610 mm) from exterior walls in all habitable rooms at the design temperature. The installation of one or more portable space heaters shall not be used to achieve compliance with this section.

❖Minimum heating requirements are for health reasons. In areas where the design temperature is based on Table R301.2(1) and is below 60°F (16°C), dwelling units must be provided with heating facilities capable of maintaining a minimum room temperature of 68°F (20°C). The primary need here is that of human comfort. The minimum temperature is measured at 3 feet (914 mm) above the floor and 2 feet (610 mm) from the exterior walls of all habitable rooms (see Commentary Figure R303.9). Portable space heaters cannot be used to achieve compliance with this section.

SECTION R304
MINIMUM ROOM AREAS

R304.1 Minimum area. Every *dwelling* unit shall have at least one habitable room that shall have not less than 120 square feet (11 m^2) of gross floor area.

❖The interior living environment is affected by a number of issues. Among these are the size of the room, tightness of construction, ceiling height, number of occupants and ventilation. These all interact and impact the interior living conditions including odors, moisture and disease transmission. The code regulates room sizes to assist in maintaining a safe and comfortable interior environment (see Commentary Figure R304.1). At least one habitable room must be at least 120 square feet (11.2 m^2) of gross floor area. Because the definition of habitable space in Section R202 includes rooms used for living, sleeping, eating or cooking, any one of these rooms can be used to meet the requirement.

R304.2 Other rooms. Other habitable rooms shall have a floor area of not less than 70 square feet (6.5 m^2).

Exception: Kitchens.

❖One habitable room must comply with the provisions of Section R304.1. The remainder of the habitable rooms, except kitchens, are required to have a floor area of 70 square feet (6.5 m^2) (see Commentary Figure R304.1). Kitchens are exempt from the minimum floor area requirement.

R304.3 Minimum dimensions. Habitable rooms shall not be less than 7 feet (2134 mm) in any horizontal dimension.

Exception: Kitchens.

❖Except for kitchens, all habitable rooms must have a minimum horizontal dimension in any direction of at least 7 feet (2134 mm) (see Commentary Figure R304.1).

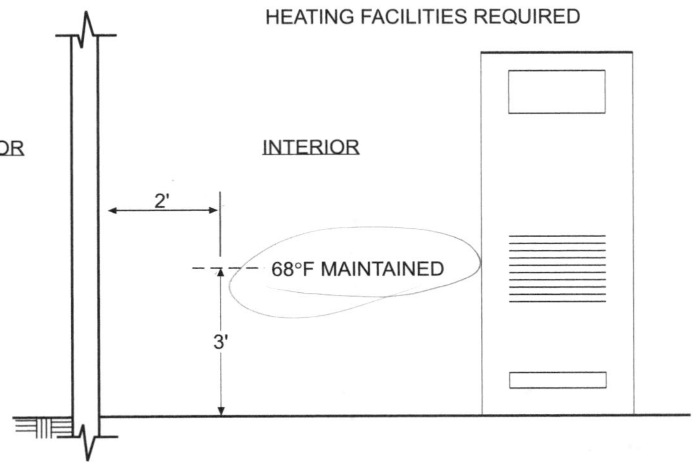

For SI: 1 foot = 304.8 mm, °C = [(°F)-32]/1.8.

Figure R303.9
REQUIRED HEATING

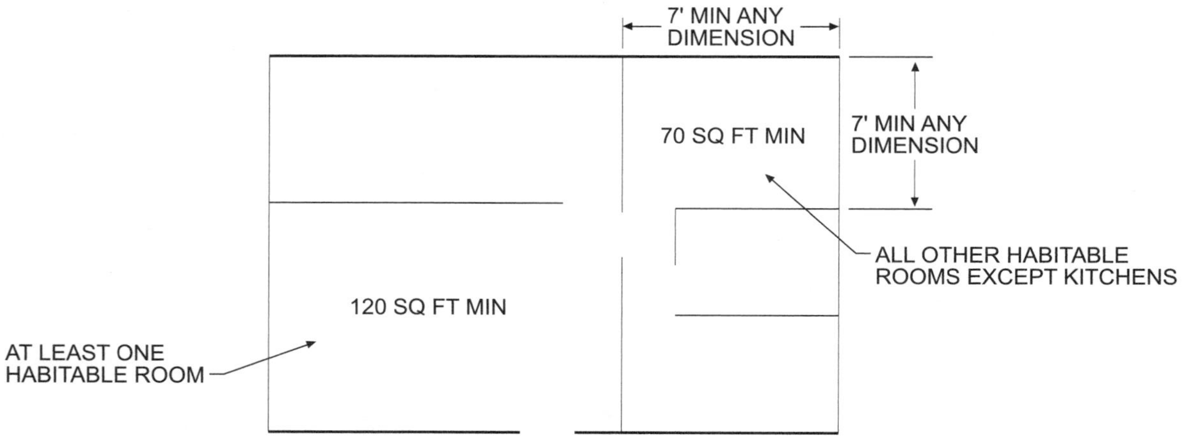

For SI: 1 foot = 304.8 mm, 1 square foot = 0.0929 m^2.

Figure R304.1
FLOOR AREA

R304.4 Height effect on room area. Portions of a room with a sloping ceiling measuring less than 5 feet (1524 mm) or a furred ceiling measuring less than 7 feet (2134 mm) from the finished floor to the finished ceiling shall not be considered as contributing to the minimum required habitable area for that room.

❖In a room with a sloping ceiling, any portion of the room with a vertical ceiling height less than 5 feet (1524 mm) from the finished floor does not provide the minimum required floor area for habitation. Likewise, the area under a furred ceiling with a vertical height less than 7 feet (2134 mm) from the finished floor is not part of the habitable area. An example of the first case would be an A-frame structure, which consists of a sloping roof and no or minimal exterior walls. This condition could also exist in any room that has a sloping ceiling. The low height makes those portions of the room generally unusable for adults (see Commentary Figure R304.4).

SECTION R305
CEILING HEIGHT

R305.1 Minimum height. *Habitable space*, hallways, bathrooms, toilet rooms, laundry rooms and portions of *basements* containing these spaces shall have a ceiling height of not less than 7 feet (2134 mm).

Exceptions:

1. For rooms with sloped ceilings, at least 50 percent of the required floor area of the room must have a ceiling height of at least 7 feet (2134 mm) and no portion of the required floor area may have a ceiling height of less than 5 feet (1524 mm).
2. Bathrooms shall have a minimum ceiling height of 6 feet 8 inches (2032 mm) at the center of the front clearance area for fixtures as shown in Figure R307.1. The ceiling height above fixtures shall be such that the fixture is capable of being used for its intended purpose. A shower or tub equipped with a showerhead shall have a minimum ceiling height of 6 feet 8 inches (2032 mm) above a minimum area 30 inches (762 mm) by 30 inches (762 mm) at the showerhead.

❖Minimum ceiling heights are required for habitable space, hallways, bathrooms, toilet rooms, laundry rooms, as well as portions of basements that contain the areas listed. The minimum required height of 7 feet (2134 mm) helps maintain a healthy interior environment. The dimension must be measured to the lowest projection from the ceiling.

For rooms with sloped ceilings, in accordance with Exception 1, the code requires only that the prescribed ceiling height be maintained in one-half the area of the room. However, no portion of the room that has a ceiling height of less than 5 feet (1524 mm) must be used in the computations for minimum floor area (see Section R304.4).

Exception 2 defines the required minimum ceiling height over toilet, bath and shower fixtures. This exception would allow a sloping ceiling over toilet, bath or shower fixtures if the minimum ceiling height of 6 feet, 8 inches (2036 mm) is maintained over the front clearance area (see Figure R307.1). If the fixture can still be used effectively, the ceiling height can be lower over the fixture itself. For example, the ceiling height over the tank and bowl of the toilet can be below 6 feet, 8 inches (2033 mm), provided that the clearance was high enough to allow someone to sit on the toilet.

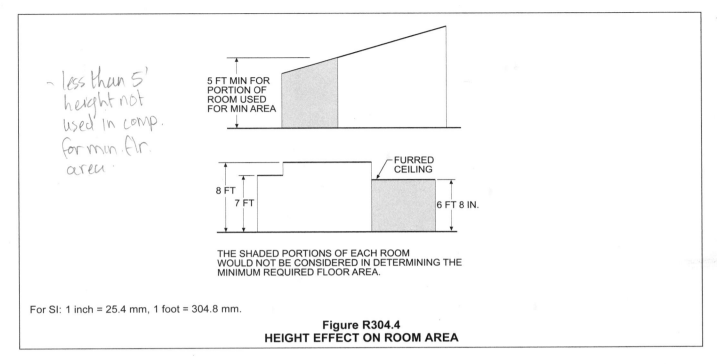

- less than 5' height not used in comp. for min. flr. area.

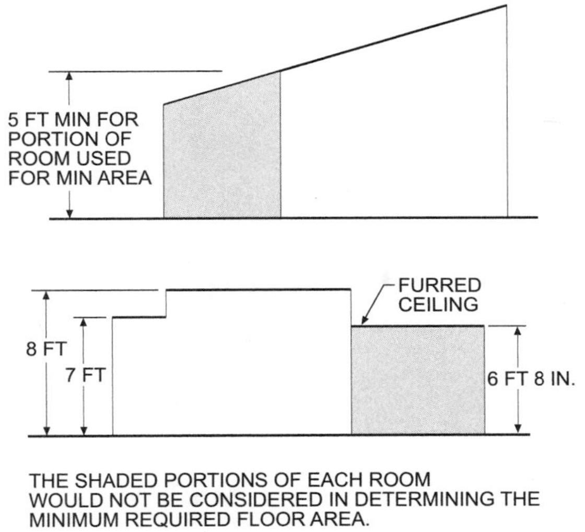

For SI: 1 inch = 25.4 mm, 1 foot = 304.8 mm.

Figure R304.4
HEIGHT EFFECT ON ROOM AREA

R305.1.1 Basements. Portions of *basements* that do not contain *habitable space*, hallways, bathrooms, toilet rooms and laundry rooms shall have a ceiling height of not less than 6 feet 8 inches (2032 mm).

Exception: Beams, girders, ducts or other obstructions may project to within 6 feet 4 inches (1931 mm) of the finished floor.

❖Portions of basements that are not addressed in Section R305.1 need to have a ceiling height of only 6 feet, 8 inches (2033 mm) or more, with at least 6 feet, 4 inches (1932 mm) of clear height under beams, girders, ducts and similar obstructions.

SECTION R306
SANITATION

R306.1 Toilet facilities. Every *dwelling* unit shall be provided with a water closet, lavatory, and a bathtub or shower.

❖Dwelling units have at least one of each of the fixtures indicated in the code, and the fixtures must be connected to an approved sanitary sewer or private sewage disposal system. A water closet, lavatory and either a bathtub or shower are the minimum fixtures needed to maintain the occupant's health and cleanliness. For additional information on plumbing requirements, see Chapters 25 through 33.

R306.2 Kitchen. Each *dwelling* unit shall be provided with a kitchen area and every kitchen area shall be provided with a sink.

❖Dwelling units must have a kitchen area with a sink for the basic preparation of food.

R306.3 Sewage disposal. All plumbing fixtures shall be connected to a sanitary sewer or to an *approved* private sewage disposal system.

❖To maintain sanitary conditions in the dwelling, plumbing fixtures must be attached to either a sanitary sewer or approved private sewage disposal system.

R306.4 Water supply to fixtures. All plumbing fixtures shall be connected to an *approved* water supply. Kitchen sinks, lavatories, bathtubs, showers, bidets, laundry tubs and washing machine outlets shall be provided with hot and cold water.

❖To provide proper sanitation for occupants of dwelling units, each plumbing fixture must be connected to an approved water supply. Additionally, specific fixtures must have both a hot and cold water supply.

SECTION R307
TOILET, BATH AND SHOWER SPACES

R307.1 Space required. Fixtures shall be spaced in accordance with Figure R307.1, and in accordance with the requirements of Section P2705.1.

❖Fixtures require certain clearances to be accessible and usable. Figure R307.1 shows the minimum fixture clearances. The diagram is consistent with the requirements in Section P2705.1, Item 5. In addition, Item 5 also specifies that adjacent fixtures must be at least 30 inches (762 mm) center-to-center. A common fixture configuration in a bathroom that could be affected by this would be two adjacent lavatories in one vanity or a lavatory adjacent to a water closet.

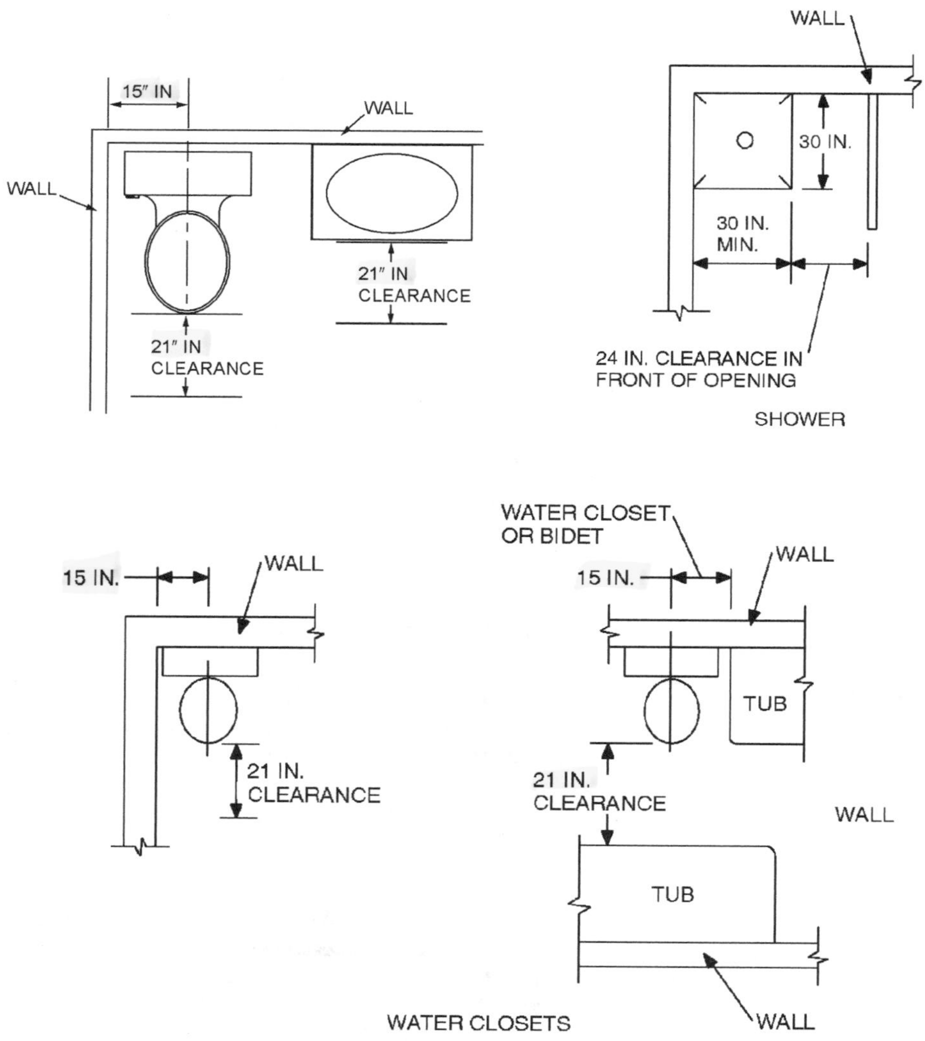

For SI: 1 inch = 25.4 mm.

FIGURE R307.1
MINIMUM FIXTURE CLEARANCES

R307.2 Bathtub and shower spaces. Bathtub and shower floors and walls above bathtubs with installed shower heads and in shower compartments shall be finished with a nonabsorbent surface. Such wall surfaces shall extend to a height of not less than 6 feet (1829 mm) above the floor.

❖Wall surfaces subject to water spray by showerheads must be protected with a nonabsorbent surface to a height of at least 6 feet (1829 mm) above the floor of a bathtub or shower.

SECTION R308
GLAZING

R308.1 Identification. Except as indicated in Section R308.1.1 each pane of glazing installed in hazardous locations as defined in Section R308.4 shall be provided with a manufacturer's designation specifying who applied the designation, designating the type of glass and the safety glazing standard with which it complies, which is visible in the final installation. The designation shall be acid etched, sandblasted, ceramic-fired, laser etched, embossed, or be of a type which once applied cannot be removed without being destroyed. A *label* shall be permitted in lieu of the manufacturer's designation.

Exceptions:

1. For other than tempered glass, manufacturer's designations are not required provided the *building official* approves the use of a certificate, affidavit or other evidence confirming compliance with this code.

2. Tempered spandrel glass is permitted to be identified by the manufacturer with a removable paper designation.

❖Once glass is installed in a window frame by a manufacturer, whether that glass is safety glazing or not is not easily determined. In theory, this can be established only by breaking the particular piece of glass, in which case the glass is no longer useable. Thus, the code requires that safety glazing be marked with a manufacturer's designation that is visible during the final building inspection. Except for tempered glass labels, labels may be omitted where approved by the building official and an affidavit, certificate or other evidence is submitted indicating compliance with the code. A manufacturer can identify safety glazing with a removable paper designation, provided it is destroyed during removal. This ensures that the designation will not be applied to a noncomplying piece of glass.

R308.1.1 Identification of multiple assemblies. Multipane assemblies having individual panes not exceeding 1 square foot (0.09 m^2) in exposed area shall have at least one pane in the assembly identified in accordance with Section R308.1. All other panes in the assembly shall be *labeled* "CPSC 16 CFR 1201" or "ANSI Z97.1" as appropriate.

❖Multipane assemblies of glass need identification for the same reasons noted in Section R308.1. This provision allows labeling of only one pane of glass in accordance with Section R308.1.1, when the exposed area of each pane is 1 square foot (0.09 m^2) or less. All other panes must be labeled either "16 CFR 1201" or "ANSI Z97.1."

R308.2 Louvered windows or jalousies. Regular, float, wired or patterned glass in jalousies and louvered windows shall be no thinner than nominal $^3/_{16}$ inch (5 mm) and no longer than 48 inches (1219 mm). Exposed glass edges shall be smooth.

❖The requirements for louvered windows exist because there is no edge support on the longitudinal edges of these panes. The code requires that the exposed edges be smooth for safety. The minimum thickness and maximum span are specified so that the glass has sufficient resistance to human impact loads.

R308.2.1 Wired glass prohibited. Wired glass with wire exposed on longitudinal edges shall not be used in jalousies or louvered windows.

❖Wired glass is not permitted if the wire is exposed on the longitudinal edge because it would be a hazard.

R308.3 Human impact loads. Individual glazed areas, including glass mirrors in hazardous locations such as those indicated as defined in Section R308.4, shall pass the test requirements of Section R308.3.1. must be safety glazed

Exceptions:

1. Louvered windows and jalousies shall comply with Section R308.2.
2. Mirrors and other glass panels mounted or hung on a surface that provides a continuous backing support.
3. Glass unit masonry complying with Section R610.

❖The code requires that glazing in hazardous locations subject to human impact pass the impact tests. Criteria is based on size and location [see Section R308.3.1, and Tables R308.3.1(1) and R308.3.1(2)]. The exceptions provide for three types of glazing that have alternative means of offering protection when used in hazardous locations: louvered windows and jalousies meeting the thickness and length limitations in Section R308.2; mirrors or glass hung on a wall or fitted with a backing; and glass block constructed in accordance with Section R610. Glass block is becoming more prevalent in the design of homes. One of the more common uses is enclosures for walk-in showers.

R308.3.1 Impact test. Where required by other sections of the code, glazing shall be tested in accordance with CPSC 16 CFR 1201. Glazing shall comply with the test criteria for Category II unless otherwise indicated in Table R308.3.1(1).

Exception: Glazing not in doors or enclosures for hot tubs, whirlpools, saunas, steam rooms, bathtubs and showers shall be permitted to be tested in accordance with ANSI Z97.1. Glazing shall comply with the test criteria for Class A unless indicated in Table R308.3.1 (2).

❖Section R308.4 lists seven different situations where there is a chance of someone falling or reaching into a piece of glass and possibly injuring themselves; therefore, testing is required to resist these human impact loads. Glazing in hazardous locations must pass the test requirements of CPSC 16 CFR, Part 1201 and meet the test criteria of Category Class II. Table R308.3.1(1) permits some glazing in hazardous locations to be less than Category Class II. The exception allows for another standard, ANSI Z97.1, as an alternative test for glazing that is located other than in a door or serving as part of a bathing enclosure. This limitation in the exception is due to the mandatory location requirements in CPSC 16 CFR, Part 1201. The exception requires the glazing to pass ANSI Z97.1 and meet the test criteria of Category A. Table R303.3.1(2) then permits some glazing in hazardous locations to be less than Category A.

Set forth below are the more significant differences between these two standards, both of which are applicable to safety glazing materials used in architectural applications. This statement makes no attempt to summarize all pertinent provisions of the two standards, only their significant differences.

The principal differences between CPSCs 16 CFR, Part 1201 and ANSI Z97.1 relate to their scope and function. The CPSC standard is not only a test method and a procedure for determining the safety performance of architectural glazing, but also a federal standard that mandates where and when safety glazing materials must be used in architectural applications and preempts any nonidentical state or local standard. In contrast, ANSI Z97.1 is only a voluntary safety performance specification and test method. It does not purport to indicate where and when safety glazing materials must be used, leaving those determinations up to the building codes and to glass and fenestration specifiers. In this instance, the IBC provides the requirements regarding the safety performance of architectural glazing beyond that which is covered by the federal standard.

The CPSC standard requires the installation of safety glazing materials meeting 16 CFR, Part 1201 only in storm doors, combination doors, entrance-exit doors, sliding patio doors, closet doors, and shower and tub doors and enclosures. See Section 308.4 of the IBC for additional locations where the code requires safety glazing.

Test Specimens: For impact testing, the CPSC standard requires only one specimen of each nominal thickness be submitted for testing and specifies it must be the largest size the manufacturer produces up to a maximum size of 34-inches by 76 inch (864 mm by 1830 mm). ANSI Z97.1 requires that four specimens of each nominal thickness and size must be impact tested. The manufacturer has the option of testing either 34-inch by 76-inch (864 mm by 1830 mm) specimens or the largest size it commercially produces less than 34 inches by 76 inches (864 mm by 1830 mm), but with a minimum size of 24 inches by 30 inches (610 mm by 762 mm).

Types of Glass: The CPSC standard has no performance tests for plastics or for bent glass. ANSI Z97.1 has specific tests for both.

The CPSC standard does not prohibit the use of ordinary annealed glass in hazardous locations as long as it passes the appropriate impact tests, consistent with the concept of a performance-based impact

TABLE R308.3.1(1)
MINIMUM CATEGORY CLASSIFICATION OF GLAZING USING CPSC 16 CFR 1201

EXPOSED SURFACE AREA OF ONE SIDE OF ONE LITE	GLAZING IN STORM OR COMBINATION DOORS (Category Class)	GLAZING IN DOORS (Category Class)	GLAZED PANELS REGULATED BY SECTION R308.4.3 (Category Class)	GLAZED PANELS REGULATED BY SECTION R308.4.2 (Category Class)	GLAZING IN DOORS AND ENCLOSURES REGULATED BY SECTION 308.4.5 (Category Class)	SLIDING GLASS DOORS PATIO TYPE (Category Class)
9 square feet or less	I	I	NR	I	II	II
More than 9 square feet	II	II	II	II	II	II

For SI: 1 square foot = 0.0929 m².
NR means "No Requirement."

❖In 1977, the CPSC adopted as a mandatory federal safety regulation *Safety Standard for Architectural Glazing Materials*, codified at 16 CFR, Part 1201. The CPSC amended the regulation on several occasions subsequent to its initial adoption, the last time on June 28, 1982. For additional information, see the commentary to Section R308.3.1.

Impact Categories or Levels: The CPSC standard has two distinct impact levels or categories, Category I and Category II, and specifies which defined hazardous location must contain Category II safety glazing materials and which may use Category I glazing materials. Glazing material successfully passing the impact test of a 48-inch (1219 mm) drop height a 400 foot-pound (940 J) impact, is classified as Category II glass. Glazing material passing the 18-inch (457 mm) drop height, at 150 foot-pounds (203.4 J) impact, is classified as Category I glass.

TABLE R308.3.1(2)
MINIMUM CATEGORY CLASSIFICATION OF GLAZING USING ANSI Z97.1

EXPOSED SURFACE AREA OF ONE SIDE OF ONE LITE	GLAZED PANELS REGULATED BY SECTION R308.4.3 (Category Class)	GLAZED PANELS REGULATED BY SECTION R308.4.2 (Category Class)	DOORS AND ENCLOSURES REGULATED BY SECTION R308.4.5[a] (Category Class)
9 square feet or less	No requirement	B	A
More than 9 square feet	A	A	A

For SI:1 square foot = 0.0929 m².
a. Use is permitted only by the exception to Section R308.3.1.

❖ANSI Z97.1 was revised in 2004. Utilization of this testing for glazing in hazardous locations is limited by the exception in Section R308.3.1. For additional information, see the commentary to Section R308.3.1.

Impact Categories or Levels: ANSI Z97.1 has adopted three separate impact categories or classes, based upon impact performance. ANSI Z97.1's Class A glazing materials are comparable to the CPSC's Category II glazing materials, passing a 48-inch (1219 mm) drop-height test, and its Class B glazing materials are comparable to the CPSC's Category I glazing materials, passing the 18-inch (457 mm) drop-height test. ANSI Z97.1 also has a product-specific Class C impact test, a 12-inch (305 mm) drop-height test, applicable only for fire-resistant glazing materials. However, Table R308.3.1(2) does not identify Class C as an acceptable product for use in hazardous locations.

test. [Thick, heavy annealed glass is likely to pass the CPSC 18-inch (457 mm) drop-height and 48-inch (1219 mm) drop-height impact tests for Category I and II locations.] ANSI Z97.1 contains an express limitation on annealed glass: "Monolithic annealed in any thickness is not considered safety glazing material under this standard."

Asymmetrical Glazing Material: The CPSC standard requires all asymmetrical glazing materials to be impacted on both sides of each specimen and then evaluated under the pass-fail criteria. There is no exception. ANSI Z97.1 requires that, with the exception of mirror glazing, all asymmetrical glass specimens must be impacted on both sides, two on one side and two on the other. With respect to mirror glazing products using reinforced or nonreinforced organic adhesive backing, all four specimens must be impacted only on the nonreinforced side "and with no other material applied."

Impact Categories or Levels: See the commentary to Tables R308.3.1(1) and R308.3.1(2).

Pass-fail Impact Criteria: The CPSC standard, like the ANSI standard, offers alternative criteria for evaluating whether a test specimen passes the impact test. The CPSC standard considers the specimen a pass if a 3-inch-diameter (76 mm) solid steel ball, weighing 4 pounds (18 N), will not pass through the opening when placed on the specimen for 1 second. ANSI uses the 3-inch (76 mm) sphere measure, but does not require the sphere to be a steel ball and does not specify its weight. It does require that the sphere not pass freely through the opening when a force of 4 pounds (18 N) is applied to the sphere. There is no time element associated with this alternative.

A second alternative pass-fail criterion under the CPSC standard involves weighing the 10 largest particles selected within 5 minutes after the impact test-they must weigh no more than the equivalent weight of 10 square inches (6452 m^2) of the original specimen. The ANSI standard has an almost identical criterion, except the 10 largest particles must be "crack free." It also includes additional product-specific qualifications applicable solely to selecting the 10 largest particles of tempered glass and offers a formula for determining the weight of 10 square inches (6452 m^2) of the original specimen.

The CPSC standard has no separate pass-fail impact criteria for the scenario in which the glass specimen separates from the frame after impact and breaks or produces a hole in the glass. The ANSI standard has a special criterion for that scenario—to pass, the glass is subjected to the same 3-inch (76 mm) sphere measure or to the weight criterion for the 10 largest crack-free particles.

The CPSC standard involves impact testing of only a single specimen of each nominal glass thickness. Accordingly, if that specimen passes, all glass of that type and thickness is deemed to pass. Under the ANSI standard, four specimens of each type, size and thickness must be impact tested, and if any one of the four specimens fails, there is a failure of that specific type, thickness and size.

Impact Testing Apparatus: Relatively minor technical differences exist between the test frames and impactors specified in the CPSC standard and those in ANSI Z97.1. The ANSI standard prescribes special test frame and subframe configurations for impact testing bent glass; the CPSC standard has no provisions for testing bent glass. The ANSI standard includes detailed specifications for the impactor suspension device and traction and release system, and for their operation; the CPSC standard does not.

Weathering Tests: The CPSC standard requires a weathering test only for organic-coated glass. ANSI requires a weathering test for laminated glass and plastics, as well as for organic-coated glass.

The CPSC accelerated weathering test (only for organic-coated glass) uses the xenon arc Weatherometer. The ANSI standard gives the manufacturer the choice of one of three weathering exposure alternatives: the xenon arc exposure, the enclosed twin carbon arc exposure or the 1-year outdoor exposure in South Florida. The ANSI prescribed xenon are apparatus and procedure are the more current versions of the pertinent ASTM standards, ASTM D 2565 and ASTM G 155, than the versions referenced in the CPSC standard. The CPSC's xenon arc procedure for interpreting results of the adhesion test requires an average adhesion value or pull force of no less than 90 percent of the average of the unexposed organic-coated glass specimens in order to "pass," whereas the ANSI standard requires no less than 75 percent of the average of the unexposed specimens.

Indoor Aging Tests: The CPSC standard does not prescribe any indoor aging test; the ANSI standard requires specified indoor aging tests for plastics and organic-coated glass intended for indoor-use only, followed by impact tests.

R308.4 Hazardous locations. The locations specified in Sections R308.4.1 through R308.4.7 shall be considered specific hazardous locations for the purposes of glazing.

❖Section R308.4 lists seven specific hazardous locations (Sections R308.4.1 through R308.4.7) where safety glazing is required. Listed under each location are the exceptions specific to that location. Some of these locations are shown in Commentary Figures R308.4(1) through R308.4(11). In addition to the hazardous locations shown in the 11 drawings, safety glazing is also required in a number of other locations, including fixed and sliding panels of sliding door assemblies, storm doors and glass railings.

Commentary Figure R308.4(1) illustrates several locations where safety glazing may or may not be required. To facilitate discussion, each glazed panel has been numbered. Panel 1 is not required to have safety glazing because a protective bar has been installed in compliance with Exception 2 to Section R308.4.3, the details of which are illustrated in Commentary Figure R308.4(2). Panels 4 and 7 require safety glazing because they are door sidelights. Exception 2 to Section R308.4.3 does not apply to panels adjacent to a door, so even though Panel 7 has a protective bar, safety glazing is still required (see Section R308.4.2).

Commentary Figures R308.4(3) and R308.4(4) illustrate where safety glazing is required for panels adjacent to a door (see Section R308.4.2). This requirement applies to both fixed and operable panels. Where there is an intervening wall or permanent barrier, as shown in Commentary Figure R308.4(5), safety glazing would not be required (see Section R308.4.2, Exception 2). Commentary Figure R308.4(6) illustrates Exception 3 to Section R308.4.2, which applies to glazing positioned perpendicular to the plane of the door when it is in the closed position and the perpendicular glazing is on the latch side. Only one side is considered to be the hazardous location, the side that the door swings toward. The other side need not have safety glazing. This wall has a much lower risk of problems. When a door swings

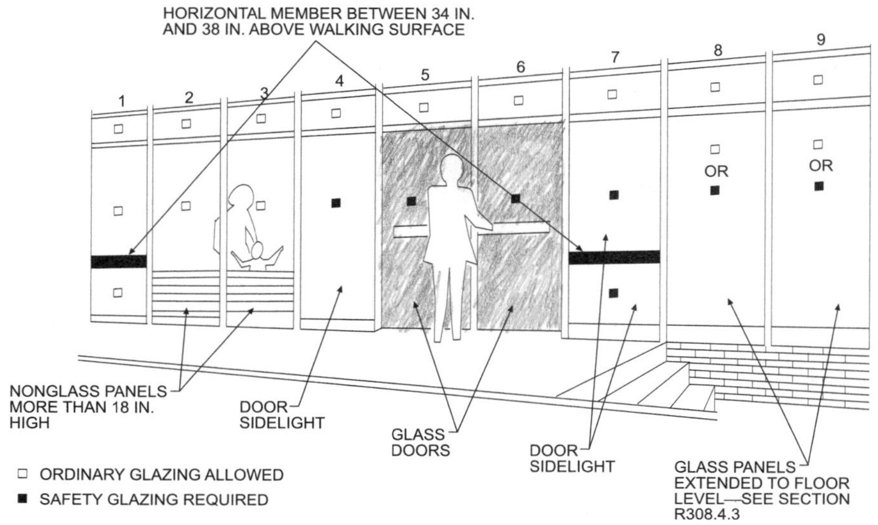

For SI: 1 inch = 25.4 mm.

Figure R308.4(1)
HAZARDOUS LOCATIONS

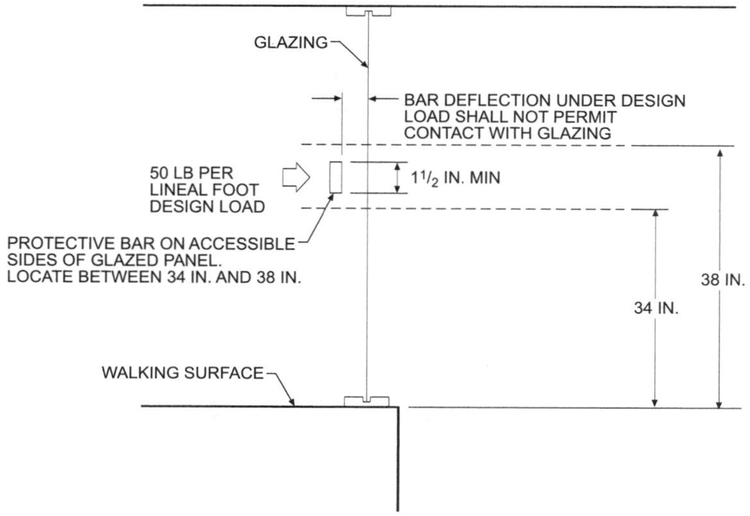

For SI: 1 inch = 25.4 mm, 1 pound per linear foot = 2.46 N/m.

Figure R308.4(2)
PROTECTIVE BAR ALTERNATIVE
(Section R308.4.3, Exception 2)

open to a perpendicular wall with glazing within 24 inches (610 mm), it is possible that if the door were caught by a strong wind it could slam into the wall and break the glass, or the doorknob could hit the glass and break it. There is also the possibility that someone could be caught behind the door when it is opened and they could be pushed into/through the glass. Thus, this would be an appropriate area to have the required safety glazing to protect the occupants.

Panels 8 and 9, as well as Panels 2 and 3, fall under Section R308.4.3. Under this section, all four stated conditions must occur before safety glazing is required. These conditions are as follows:

1. The area of an individual pane must be more than 9 square feet (0.84 m^2);
2. The bottom edge must be less than 18 inches (457 mm) above the floor;
3. The top edge must be more than 36 inches (914 mm) above the floor; and
4. One or more walking surfaces must be within 36 inches (914 mm), measured horizontally from the glazed panel.

However, Panels 2 and 3 do not require safety glazing because their bottom edges are not less than 18 inches (457 mm) from the floor.

If Panels 8 and 9 have a walking surface within 36 inches (914 mm) horizontally of the interior, safety glazing would be required. From the exterior side, as shown in Commentary Figure R308.4(1), the bottom of the panel appears to be more than 18 inches (457 mm) above the exterior walking surface, so the exterior condition would have no bearing on the determination. Panels 5 and 6 are glass doors, which require safety glazing based on the provisions of Section R308.4.1. Glazing in doors (except louvered or jalousies in accordance with Exception 1 to Section R308.3) requires safety glazing, but there are two exceptions. If openings in a door will not pass a 3-inch-diameter (76 mm) sphere, the glazing is exempt (see Section R308.4.1, Exception 1), as are assem-

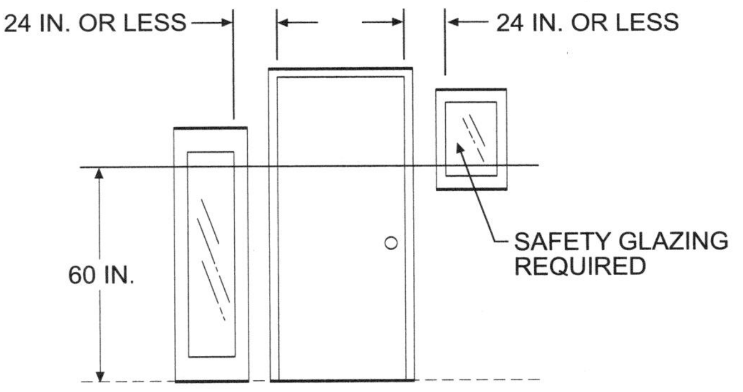

For SI: 1 inch = 25.4 mm.

Table R308.4(3)
GLAZING IN PANELS ADJACENT TO DOORS—ELEVATION
(Section R308.4.2)

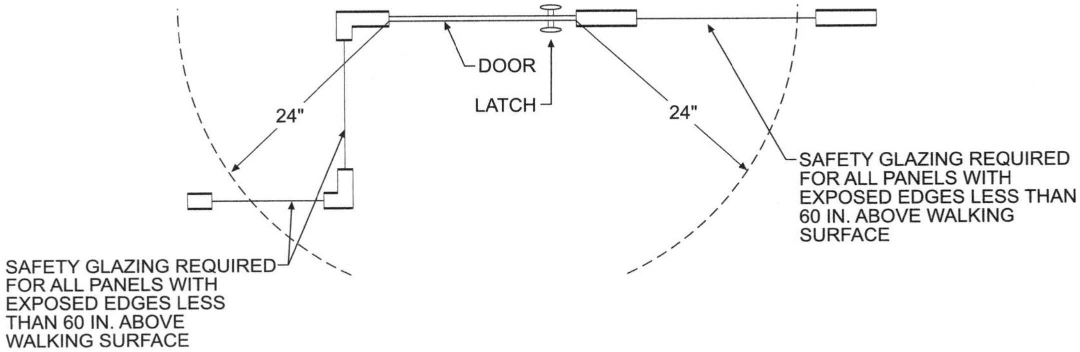

For SI: 1 inch = 25.4 mm.

Figure R308.4(4)
GLAZING IN PANELS ADJACENT TO DOORS—PLAN
(Section R308.4.2)

blies of decorative glass which is defined in Chapter 2 (see Section R308.4.1, Exception 2). The latter exception applies not only to doors but also to sidelights and other glazed panels (see Section R308.4.2, Exception 1).

Glazing in railings, balusters, panels and nonstructural in-fill panels, regardless of their height above a walking surface, requires safety glazing (see Section R308.4.4). Because of the high probability that people will strike guards, it is critical that an increased level of protection be provided.

Section R308.4.5 addresses glazing near water or wet areas. Safety glazing is required adjacent to hot tubs, spas, whirlpools, saunas, steam rooms, bathtubs, showers and swimming pools. Because of the presence of water, all of these locations represent slip hazards and need safety glazing to prevent injury in case of a fall. Glazing adjacent to these areas must be safety glazed if the glazing is less than 60 inches above any standing or walking surface. Commentary Figure R308.4(7) illustrates the condition where a window occurs within a shower enclosure. Commentary Figure R308.4(8) illustrates the requirements of Section R308.4.5. Glazing that is more than 60 inches (1524 mm) from the water's edge of a bathtub, hot tub, spa, whirlpool or swimming pool is not a hazardous location in accordance with the exception to Section R308.4.5. Commentary Figure R308.4(9) illustrates the exception to Section R308.4.5.

Sections R308.4.6 and R308.4.7 address the hazardous locations to be considered for stairways, landings and ramps. Stairways and ramps present users with a greater risk for injury caused by falling than a flat surface. Not only is the risk of falling greater when

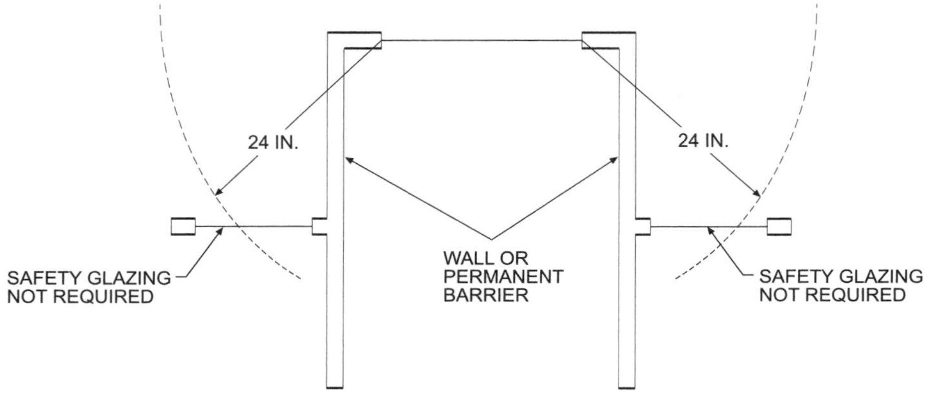

For SI: 1 inch = 25.4 mm.

Figure R308.4(5)
BARRIER BETWEEN GLAZING AND DOOR
(Section R308.4.2, Exception 2)

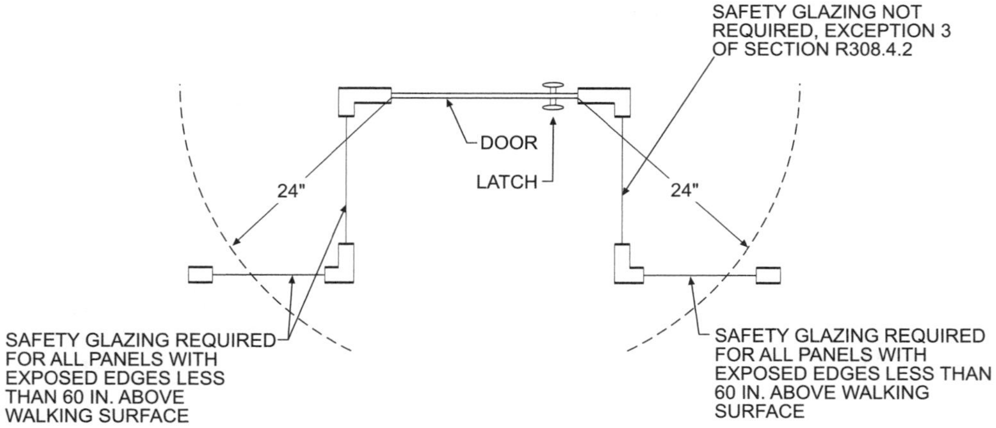

For SI: 1 degree = 0.01745 rad.

Figure R308.4(6)
GLAZING IN PANELS ADJACENT TO DOORS—PLAN
(Section R308.4.2, Exception 3)

using a stair, but the injuries are generally more severe. Unlike falling on a flat surface where the floor will, for the most part, break a person's fall, there is nothing to stop someone from continuing to fall until he or she reaches the bottom of the stair. The increased risks inherent in stairways, as well as attempting to be consistent with other chapters in the code that mandate more restrictive requirements when addressing safety issues involving stairways and ramps, account for the more restrictive requirements for glazing in and around stairways and ramps.

Section R308.4.6 includes any glazing when the exposed surface of that glazing is within 36 inches (914 mm) above the plane of the adjacent walking surface. The walking surface in question would be part of a stair or ramp itself, including intermediate landings [see Commentary Figure R308.4(10)]. Safety glazing is not required for Exception 1 of Sec-

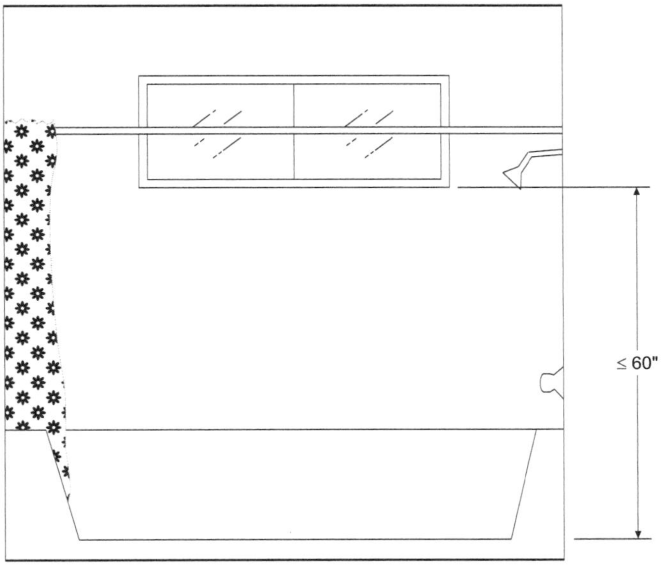

For SI: 1 inch = 25.4 mm.

Figure R308.4(7)
GLAZING WITHIN A SHOWER ENCLOSURE

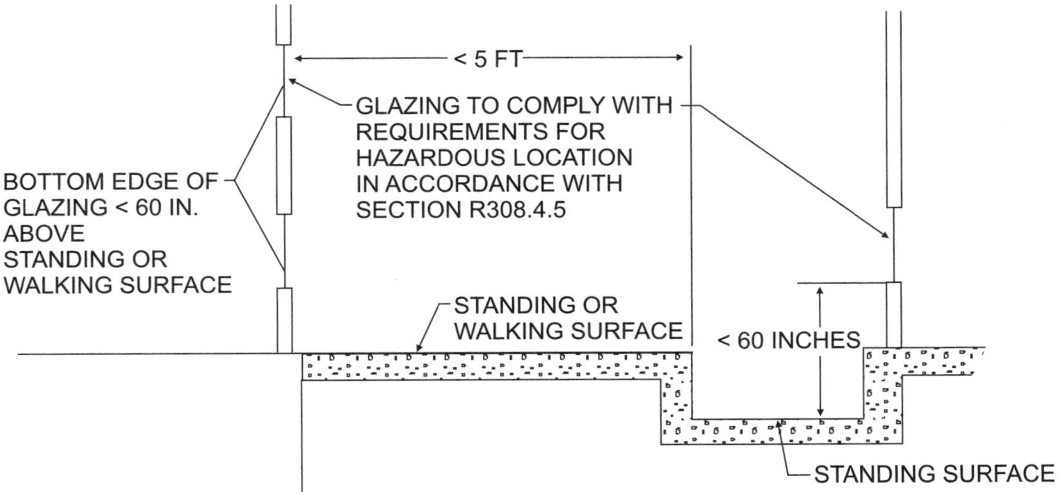

For SI: 1 inch = 25.4 mm, 1 foot = 304.8 mm.

Figure R308.4(8)
GLAZING IN WET AREAS ADJACENT TO HOT TUBS, SPAS, WHIRLPOOLS, SAUNAS, STEAM ROOMS, BATHTUBS, SHOWERS AND SWIMMING POOLS

tion R308.4.6 where the side of the stairway, landing or ramp has a rail, which can be part of the guard or handrail, which has a load resistance of 50 pounds per linear foot (730 N/m) (for additional loading criteria for handrails and guards, see Table R301.5).

In Section R308.4.7, the concern is glazing that may be located within 60 inches (1524 mm) horizontally of the bottom tread of the stairway and within 36 inches (914 mm) vertically above the bottom landing of a stairway. The 60-inch (1524 mm) dimension is from any point on the bottom tread, horizontally in any direction to any surface of any glazing within that range [see Commentary Figure R308.4(11)]. The exception to Section R308.4.7 will permit nonsafety glazing where a guard is installed and the actual plane of the glazing is located at least 18 inches (457 mm) from the guard.

R308.4.1 Glazing in doors. Glazing in all fixed and operable panels of swinging, sliding and bifold doors shall be considered a hazardous location.

Exceptions:

1. Glazed openings of a size through which a 3-inch-diameter (76 mm) sphere is unable to pass.
2. Decorative glazing.

❖See the commentary for Section R308.4.

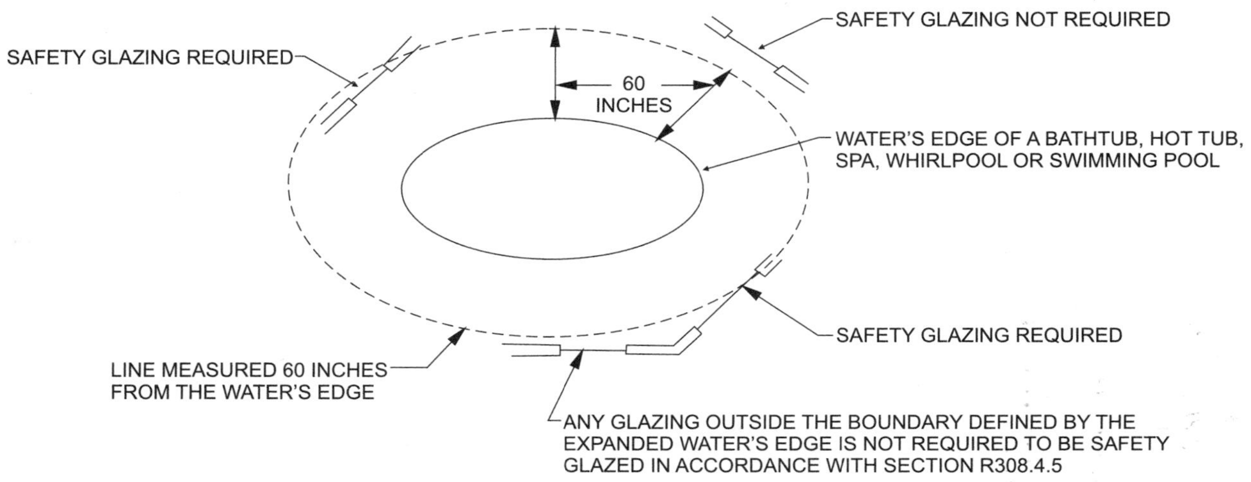

For SI: 1 inch = 25.4 mm.

Figure R308.4(9)
GLAZING AND WET SURFACES—PLAN VIEW
(Section R308.4.5, Exception)

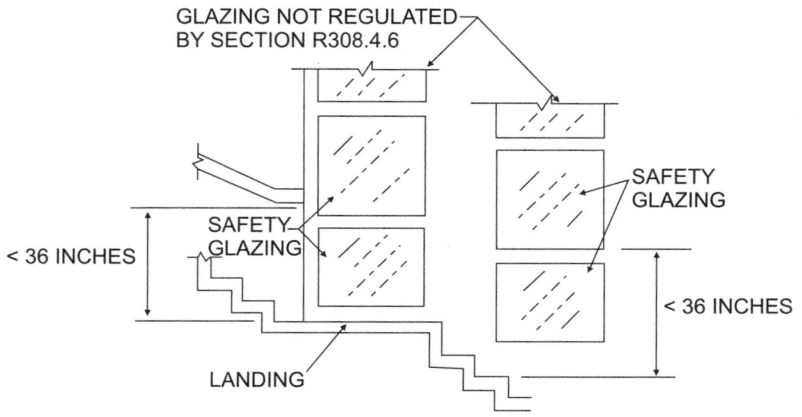

For SI: 1 inch = 25.4 mm, 1 foot = 304.8 mm.

Figure R308.4(10)
GLAZING ADJACENT TO STAIRWAYS
(Section R308.4.6)

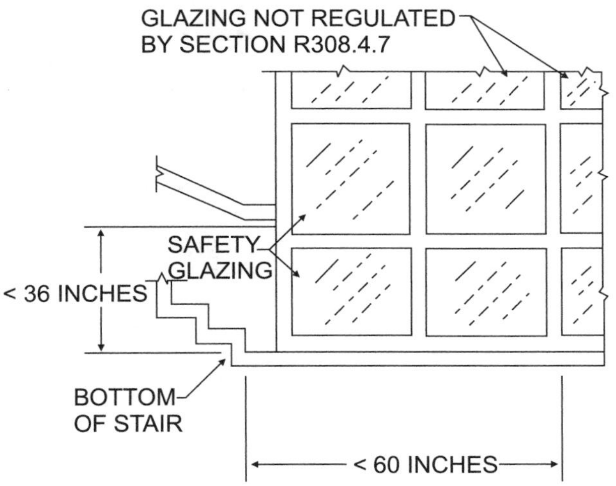

For SI: 1 inch = 25.4 mm.

Table R308.4(11)
GLAZING ADJACENT TO THE BOTTOM STAIR LANDING
(Section R308.4.7)

R308.4.2 Glazing adjacent doors. Glazing in an individual fixed or operable panel adjacent to a door where the nearest vertical edge of the glazing is within a 24-inch (610 mm) arc of either vertical edge of the door in a closed position and where the bottom exposed edge of the glazing is less than 60 inches (1524 mm) above the floor or walking surface shall be considered a hazardous location.

Exceptions:

1. Decorative glazing.
2. When there is an intervening wall or other permanent barrier between the door and the glazing.
3. Glazing in walls on the latch side of and perpendicular to the plane of the door in a closed position.
4. Where access through the door is to a closet or storage area 3 feet (914 mm) or less in depth. Glazing in this application shall comply with section R308.4.3.
5. Glazing that is adjacent to the fixed panel of patio doors.

❖See the commentary for Section R308.4.

R308.4.3 Glazing in windows. Glazing in an individual fixed or operable panel that meets all of the following conditions shall be considered a hazardous location:

1. The exposed area of an individual pane is larger than 9 square feet (0.836 m²);
2. The bottom edge of the glazing is less than 18 inches (457 mm) above the floor;
3. The top edge of the glazing is more than 36 inches (914 mm) above the floor; and
4. One or more walking surfaces are within 36 inches (914 mm), measured horizontally and in a straight line, of the glazing.

-24 inch arc -bottom of glazing is <60in above floor.

Exceptions:

1. Decorative glazing.
2. When a horizontal rail is installed on the accessible side(s) of the glazing 34 to 38 inches (864 to 965 mm) above the walking surface. The rail shall be capable of withstanding a horizontal load of 50 pounds per linear foot (730 N/m) without contacting the glass and be a minimum of $1^1/_2$ inches (38 mm) in cross sectional height.
3. Outboard panes in insulating glass units and other multiple glazed panels when the bottom edge of the glass is 25 feet (7620 mm) or more above *grade*, a roof, walking surfaces or other horizontal [within 45 degrees (0.79 rad) of horizontal] surface adjacent to the glass exterior.

❖See the commentary for Section R308.4.

R308.4.4 Glazing in guards and railings. Glazing in guards and railings, including structural baluster panels and nonstructural in-fill panels, regardless of area or height above a walking surface shall be considered a hazardous location.

❖See the commentary for Section R308.4.

R308.4.5 Glazing and wet surfaces. Glazing in walls, enclosures or fences containing or facing hot tubs, spas, whirlpools, saunas, steam rooms, bathtubs, showers and indoor or outdoor swimming pools where the bottom exposed edge of the glazing is less than 60 inches (1524 mm) measured vertically above any standing or walking surface shall be considered a hazardous location. This shall apply to single glazing and all panes in multiple glazing.

Exception: Glazing that is more than 60 inches (1524 mm), measured horizontally and in a straight line, from the water's edge of a bathtub, hot tub, spa, whirlpool, or swimming pool.

❖See the commentary for Section R308.4.

R308.4.6 Glazing adjacent stairs and ramps. Glazing where the bottom exposed edge of the glazing is less than 36 inches (914 mm) above the plane of the adjacent walking surface of stairways, landings between flights of stairs and ramps shall be considered a hazardous location.

Exceptions:

1. When a rail is installed on the accessible side(s) of the glazing 34 to 38 inches (864 to 965 mm) above the walking surface.The rail shall be capable of withstanding a horizontal load of 50 pounds per linear foot (730 N/m) without contacting the glass and be a minimum of $1^1/_2$ inches (38 mm) in cross sectional height.
2. Glazing 36 inches (914 mm) or more measured horizontally from the walking surface.

❖See the commentary for Section R308.4.

R308.4.7 Glazing adjacent to the bottom stair landing. Glazing adjacent to the landing at the bottom of a stairway where the glazing is less than 36 inches (914 mm) above the landing and within 60 inches (1524 mm) horizontally of the bottom tread shall be considered a hazardous location.

Exception: The glazing is protected by a guard complying with Section R312 and the plane of the glass is more than 18 inches (457 mm) from the guard.

❖See the commentary for Section R308.4.

R308.5 Site built windows. Site built windows shall comply with Section 2404 of the *International Building Code.*

❖Because site-built windows are not constructed in a manufacturing facility that follows industry standards, they must be constructed in accordance with Section 2404 of the IBC, which sets forth the wind, snow, seismic and dead loads on glass.

R308.6 Skylights and sloped glazing. Skylights and sloped glazing shall comply with the following sections.

❖Sloped glazing and skylights consist of glazing installed in roofs or walls that are on a slope 15 degrees (0.26 rad) or more from the vertical. The provisions of the code address loads normally attributed to roofs. The provisions also enhance the protection of the occupants of a building from the possibility of falling glazing materials.

R308.6.1 Definitions.

SKYLIGHT, UNIT. A factory assembled, glazed fenestration unit, containing one panel of glazing material, that allows for natural daylighting through an opening in the roof assembly while preserving the weather-resistant barrier of the roof.

❖Unit skylights are a specific type of sloped glazing assembly, which is factory assembled. The IBC and the code contain specific provisions that are appropriate for this type of building component. Factory-assembled units, as opposed to site-built skylights, can be designed, tested and rated as one component, which incorporates both glazing and framing, if applicable. The individual components of site-built glazing must be designed to resist the design loads of the codes individually, and are not usually rated as an assembly.

SKYLIGHTS AND SLOPED GLAZING. Glass or other transparent or translucent glazing material installed at a slope of 15 degrees (0.26 rad) or more from vertical. Glazing materials in skylights, including unit skylights, tubular daylighting devices, solariums, sunrooms, roofs and sloped walls are included in this definition.

❖The failure of skylights and sloped glazing could result in injury and building damage. This definition establishes the criteria to which the code requirements of Section R308.6 are to apply.

TUBULAR DAYLIGHTING DEVICE (TDD). A nonoperable fenestration unit primarily designed to transmit daylight from a roof surface to an interior ceiling via a tubular conduit. The basic unit consists of an exterior glazed weathering surface, a light-transmitting tube with a reflective interior surface, and an interior-sealing device such as a translucent ceiling panel. The unit may be factory assembled, or field assembled from a manufactured kit.

❖This definition provides the distinction from a unit skylight. Although tubular daylighting devices (TDDs) and unit skylights are similar and subjected to the same testing and labeling requirements, there are some differences. A TDD is typically field assembled from a manufactured kit, unlike a unit skylight which is typically shipped as a factory-assembled unit. If the unit skylight definition is applied to TDDs, it would imply that TDDs be entirely assembled in the factory. Also, the dome of a TDD is not necessarily constructed out of a single panel of glazing material. As such, a separate definition from that of a unit skylight is needed. The definition is adapted from the definition in AAMA/WDMA A440.

R308.6.2 Permitted materials. The following types of glazing may be used:

1. Laminated glass with a minimum 0.015-inch (0.38 mm) polyvinyl butyral interlayer for glass panes 16 square feet (1.5 m^2) or less in area located such that the highest point of the glass is not more than 12 feet (3658 mm) above a walking surface or other accessible area; for higher or larger sizes, the minimum interlayer thickness shall be 0.030 inch (0.76 mm).
2. Fully tempered glass.
3. Heat-strengthened glass.
4. Wired glass.
5. *Approved* rigid plastics.

❖The provisions of this section limit glazing materials in skylights and sloped glazing to those specified, and they outline glazing materials and protective measures for sloped glazing and skylights. The materials and their characteristics and limitations are as follows:

Laminated glass. Laminated glass is usually constructed with an inner layer of polyvinyl butyral, which has a minimum thickness of 0.030 inch (0.76 mm). Such glass is highly resistant to impact and, as a result, requires no further protection below. When used within dwelling units, laminated glass is permitted to have a 0.015-inch (0.38 mm) polyvinyl butyral inner layer if each pane of glass is 16 square feet (1.5 m^2) or less in area, and the highest point of the glass is no more than 12 feet (3658 mm) above a walking surface or other accessible area.

Fully tempered glass. Tempered glass has been specifically heat-treated or chemically treated to obtain high strength. When broken, the entire piece of glass immediately breaks into numerous small granular pieces. Because of its high strength and manner of breakage, tempered glass has been considered in the past to be a desirable glazing material for skylights that have no protective screens. However, as a result of studies by the industry that show that tempered glass is subject to spontaneous breakage that can result in large chunks of glass falling, the code requires screen protection below tempered glass.

Heat-strengthened glass. Heat-strengthened glass that has been reheated to just below its melting point and then cooled. This process forms a compression on the outer surface and increases the strength of the glass. However, heat-strengthened glass requires screen protection below the skylight to protect the occupants from falling shards.

Wired glass. Wired glass is resistant to impact and, when used as a single-layer glazing, requires no additional protection below.

Approved rigid plastics. Rigid plastics are fairly durable as a glazing material.

Annealed glass. Annealed glass is not allowed because it is subject to breakage by impact and has very low strength. Annealed glass is also unsatisfactory for use as a skylight because it breaks up under impact into large sharp shards, which, when they fall, are hazardous to occupants of a building.

R308.6.3 Screens, general. For fully tempered or heat-strengthened glass, a retaining screen meeting the requirements of Section R308.6.7 shall be installed below the glass, except for fully tempered glass that meets either condition listed in Section R308.6.5.

❖As a general rule, single-layer glazing of heat-strengthened glass and fully tempered glass must be fitted with screens below the glazing material.

R308.6.4 Screens with multiple glazing. When the inboard pane is fully tempered, heat-strengthened or wired glass, a retaining screen meeting the requirements of Section R308.6.7 shall be installed below the glass, except for either condition listed in Section R308.6.5. All other panes in the multiple glazing may be of any type listed in Section R308.6.2.

❖As does Section R308.6.3, this section states that screens are required for the inbound plane of glazing when it is fully tempered, heat-strengthened or wired glass. The screen must comply with Section R308.6.7. Screens are not required for either approved laminated glass or approved rigid plastics.

R308.6.5 Screens not required. Screens shall not be required when fully tempered glass is used as single glazing or the inboard pane in multiple glazing and either of the following conditions are met:

1. Glass area 16 square feet (1.49 m^2) or less. Highest point of glass not more than 12 feet (3658 mm) above a walking surface or other accessible area, nominal glass thickness not more than $^3/_{16}$ inch (4.8 mm), and (for multiple glazing only) the other pane or panes fully tempered, laminated or wired glass.
2. Glass area greater than 16 square feet (1.49 m^2). Glass sloped 30 degrees (0.52 rad) or less from vertical, and highest point of glass not more than 10 feet (3048 mm) above a walking surface or other accessible area.

❖Section R308.6.5 states two exceptions to the provisions of Sections R308.6.3 and R308.6.4. The first exception applies to a glazing area that is no larger than 16 square feet (1.49 m^2) and no more than 12 feet (3658 mm) above a walking surface or other accessible area. The second exception applies to sloped glazing with a maximum slope of 30 degrees (0.52 rad) from vertical that is not larger than 16 square feet (1.49 m^2) in area and is no greater than 10 feet (3048 mm) above a walking surface or other accessible area. Generally, installed skylights and sloped glazing will meet one of these exceptions, so screens are not required.

R308.6.6 Glass in greenhouses. Any glazing material is permitted to be installed without screening in the sloped areas of greenhouses, provided the greenhouse height at the ridge does not exceed 20 feet (6096 mm) above *grade*.

❖The glazing regulations for greenhouses are less stringent because greenhouses are seldom occupied during storms that might break the glass. These provisions also explain an exception to the screening provisions of Sections R308.6 and R308.6.4, specifically for the sloped glazing areas within greenhouses. Screens are not required for sloped areas of greenhouses if the ridge of the greenhouse is not more than 20 feet (6096 mm) above grade.

R308.6.7 Screen characteristics. The screen and its fastenings shall be capable of supporting twice the weight of the glazing, be firmly and substantially fastened to the framing members, and have a mesh opening of no more than 1 inch by 1 inch (25 mm by 25 mm).

❖It is critical that screens be installed in a manner that will adequately support the weight of the glass. In using a safety factor of 2, the screen and its fastenings must be capable of supporting twice the weight of the glazing. To accomplish this, the screen is to be fastened firmly to the framing members.

R308.6.8 Curbs for skylights. All unit skylights installed in a roof with a pitch flatter than three units vertical in 12 units horizontal (25-percent slope) shall be mounted on a curb extending at least 4 inches (102 mm) above the plane of the roof unless otherwise specified in the manufacturer's installation instructions.

❖Skylights installed on low-sloped roofs are more susceptible to leaking than those on higher pitched roofs because water does not drain as quickly on low-sloped roofs. For these skylights to be properly flashed to prevent leakage, they must be placed on a 4-inch-high (102 mm) curb unless the manufacturer's installation instructions indicate otherwise.

R308.6.9 Testing and labeling. Unit skylights and tubular daylighting devices shall be tested by an *approved* independent laboratory, and bear a *label* identifying manufacturer, performance *grade* rating and *approved* inspection agency to indicate compliance with the requirements of AAMA/WDMA/CSA 101/I.S.2/A440.

❖The referenced standard, ANSI/AAMA/WDMA 101/I.S.2/NAFS: *Voluntary Performance Specification for Windows, Skylights and Glass Doors*, includes a separate rating system for positive and negative pressure on skylights and TDDs which allows the manufacturer to design and fabricate products that are best suited for the climate in which they will be used. The standard establishes the performance requirements for skylights and TDDs based on the desired performance grade rating which includes minimum requirements for resistance to air leakage, water infiltration and the design load pressures. The resulting performance grade rating states the design load pressure used to rate the product, but it also includes consideration of these additional performance characteristics. For skylights and TDDs certified for only one performance grade, the rating is based on the minimum requirements met for both positive and negative design pressure. Skylights and TDDs certified for two performance grades are rated separately for positive and negative design pressure.

Skylights and TDDs must be capable of withstanding the component and cladding wind pressures of Table R301.2(2) adjusted by the height and exposure coefficients given in Table R301.2(3).

The most critical load on a skylight or TDD is determined by the climate in which it is installed. In a colder climate with heavier snow loads and moderate design wind speeds, the positive load on a skylight or TDD from the combined snow and dead load will be more critical than the negative load from wind uplift. The opposite will be the case in warmer, coastal climates with higher design wind speeds and little or no snow load.

SECTION R309
GARAGES AND CARPORTS

R309.1 Floor surface. Garage floor surfaces shall be of *approved* noncombustible material.

The area of floor used for parking of automobiles or other vehicles shall be sloped to facilitate the movement of liquids to a drain or toward the main vehicle entry doorway.

❖Garage floor surfaces must be of an approved noncombustible material such as concrete. Additionally, the floor surface must either slope toward the garage door opening or slope to an approved drain. This allows grease, flammable liquids or other hazardous materials that might drain from an automobile to drain from the garage.

R309.2 Carports. Carports shall be open on at least two sides. Carport floor surfaces shall be of *approved* noncombustible material. Carports not open on at least two sides shall be considered a garage and shall comply with the provisions of this section for garages.

Exception: Asphalt surfaces shall be permitted at ground level in carports.

The area of floor used for parking of automobiles or other vehicles shall be sloped to facilitate the movement of liquids to a drain or toward the main vehicle entry doorway.

❖The area of floor used for the parking of automobiles or other vehicles must be sloped to facilitate the movement of liquids associated with motor vehicles such as oil, gasoline or antifreeze. These are toxic or flammable materials that should not be allowed to collect in the building.

Carports must have at least two sides open to outside air. If two sides are not open, the structure is a garage and comes under the provisions of Sections R302.5, R302.6, R309.1 and R309.3.

R309.3 Flood hazard areas. For buildings located in flood hazard areas as established by Table R301.2(1), garage floors shall be:

1. Elevated to or above the design flood elevation as determined in Section R322; or
2. Located below the design flood elevation provided they are at or above *grade* on at least one side, are used solely for parking, building access or storage, meet the requirements of Section R322 and are otherwise constructed in accordance with this code.

❖Garage floors of buildings in flood hazard areas must meet one of two requirements. The first option is to construct the garage floor above the design flood elevation. The second option allows the floor of the garage to be below the design flood elevation if construction is compliant with the applicable provisions of Section R322 for enclosures, including: (1) the floor is at or above finished exterior grade on at least one side (not a basement); (2) it complies with provisions for enclosed areas below the design flood elevation (see Section R322.2.2 in A Zones and Section

R322.3.5 in V Zones); (3) materials and finishes below the design flood elevation are flood resistant (see Section R322.1.8); and (4) service equipment and systems comply with Section R322.1.6. Because of these limitations, garages must not be converted to accommodate other uses.

R309.4 Automatic garage door openers. Automatic garage door openers, if provided, shall be listed and labeled in accordance with UL 325.

❖The code does not require an automatic garage door opener. However, if one is installed, it must be listed and labeled in accordance with UL 325. Federal law requires automatic residential garage door openers to conform to the entrapment protection requirements of UL 325.

R309.5 Fire sprinklers. Private garages shall be protected by fire sprinklers where the garage wall has been designed based on Table R302.1(2), Footnote a. Sprinklers in garages shall be connected to an automatic sprinkler system that complies with Section P2904. Garage sprinklers shall be residential sprinklers or quick-response sprinklers, designed to provide a density of 0.05 gpm/ft^2. Garage doors shall not be considered obstructions with respect to sprinkler placement.

❖Section R309.5 provides a limitation on the application of Table R302.1(2) by only allowing use of sprinkler incentives in areas where sprinklers are provided. Normally, garages aren't required to have sprinklers; however, where a designer chooses to take advantage of reduced separation requirements for a garage wall, it is appropriate for the garage to be provided with sprinklers as a means of property protection. Proposed design criteria for sprinklers were derived from NFPA 13R Section 6.8.3.3, which addresses sprinkler protection for garages in buildings protected by NFPA 13R sprinkler systems. Often, garage protection is provided by dry pendant or dry sidewall sprinklers connected to a wet pipe sprinkler system.

SECTION R310
EMERGENCY ESCAPE AND RESCUE OPENINGS

R310.1 Emergency escape and rescue required. *Basements*, habitable attics and every sleeping room shall have at least one operable emergency escape and rescue opening. Where *basements* contain one or more sleeping rooms, emergency egress and rescue openings shall be required in each sleeping room. Where emergency escape and rescue openings are provided they shall have a sill height of not more than 44 inches (1118 mm) measured from the finished floor to the bottom of the clear opening. Where a door opening having a threshold below the adjacent ground elevation serves as an emergency escape and rescue opening and is provided with a bulkhead enclosure, the bulkhead enclosure shall comply with Section R310.3. The net clear opening dimensions required by this section shall be obtained by the normal operation of the emergency escape and rescue opening from the inside. Emergency escape and rescue openings with a finished sill height below the adjacent ground elevation shall be provided with a window well in accordance with Section R310.2. Emergency escape and rescue openings shall open directly into a public way, or to a *yard* or court that opens to a public way.

Exception: *Basements* used only to house *mechanical equipment* and not exceeding total floor area of 200 square feet (18.58 m^2).

❖Because so many fire deaths occur as a result of occupants being asleep in a residential building during a fire, the code requires that all basements, habitable attics and sleeping rooms have windows or doors that may be used for emergency escape or rescue. These emergency openings must open directly into a public street, public alley, yard or court. The requirement for emergency escape and rescue openings in sleeping rooms exists because a fire will usually have spread before the occupants are aware of the problem, and the normal exit channels may be blocked. The requirement for basements and habitable attics exists because they are so often used as sleeping rooms. For example, a fire in a mechanical room adjacent to a stairway could engulf the only means of egress for the basement without the egress window or door.

Openings required for emergency escape or rescue must be located on the exterior of the building so that rescue can be performed from the exterior. Alternatively, occupants may escape through that opening to the exterior of the building without having to travel through the building itself. Therefore, where openings are required, they should open directly into a public street, public alley, yard or court. After the occupants pass through the emergency escape and rescue opening, their continued egress is essential. Where a basement contains sleeping rooms and a habitable space, an emergency escape and rescue opening is required in each sleeping room, but is not required in adjoining areas of the basement. The same would hold true for a subdivided habitable attic, since there would be an emergency escape window on that level.

There is an exception for basements used only to house mechanical equipment with a total floor area not exceeding 200 square feet (18.58 m^2). Attics that housed only mechanical equipment would not be considered habitable attics.

The dimensions prescribed in the code and as illustrated in Commentary Figure R310.1, for exterior wall openings used for emergency egress and rescue, are based, in part, on extensive testing by the San Diego Building and Fire Departments to determine the proper relationships of the height and width of window openings to adequately serve for both rescue and escape. The minimum of 20 inches (508 mm) for the width is based on two criteria: the width necessary to place a ladder within the window opening and the width necessary to admit a fire fighter with full rescue equipment, including a breathing apparatus. The minimum 24-inch (610 mm) height is based on the minimum size necessary to admit a fire fighter with full rescue equipment. By requiring a minimum net clear opening size of the least 5.7 square feet (0.53 m^2), the code provides for an opening of adequate dimensions.

To be accessible from the interior of the sleeping room, attic or basement, the emergency escape and rescue opening cannot be located more than 44 inches (1118 mm) above the floor. The measurement is to be taken from the floor to the bottom of the clear opening.

The required opening dimensions must be achieved by the normal operation of the window, door or hatch from the inside without the use of keys, tools or special knowledge. The window industry is a highly competitive market. Manufacturers are constantly developing new products that are easier to clean and possess higher thermal protection properties. It is important to keep in mind that no special knowledge for operation of the egress window is a key operational constraint. It is impractical to assume that all occupants can operate a window that requires a special sequence of operations to achieve the required opening size. Although most occupants are familiar with the normal operation to open the window, children and guests are frequently unfamiliar with special procedures necessary to remove the sashes. The time spent comprehending special operations unnecessarily delays egress from the bedroom and could lead to panic and further confusion. Thus, windows that achieve the required opening dimensions only by performing a special sequence of operations, such as the removal of sashes or mullions, are not permitted. For example, if a specific area of the window has to be depressed or manipulated to allow the sash to be removed or released to achieve the open area requirement of 5.7 square feet (0.53 m^2), the window does not qualify as an egress window.

R310.1.1 Minimum opening area. All emergency escape and rescue openings shall have a minimum net clear opening of 5.7 square feet (0.530 m^2).

Exception: *Grade* floor openings shall have a minimum net clear opening of 5 square feet (0.465 m^2).

❖Where an emergency escape and rescue window is located at grade level, the opening size requirement is reduced to 5 square feet (0.46 m^2). This results from the increased ease of access from the exterior and the probability that a ladder will not be needed (see Commentary Figure R310.1).

R310.1.2 Minimum opening height. The minimum net clear opening height shall be 24 inches (610 mm).

❖The minimum opening height for emergency space and rescue is 24 inches (610 mm), based on the minimum dimension of a fire fighter with full rescue equipment (see Commentary Figure R310.1).

R310.1.3 Minimum opening width. The minimum net clear opening width shall be 20 inches (508 mm).

❖This section establishes a minimum width of 20 inches (508 mm) for emergency space and rescue openings, based on the minimum dimension of a fire fighter with full rescue equipment (see Commentary Figure R310.1).

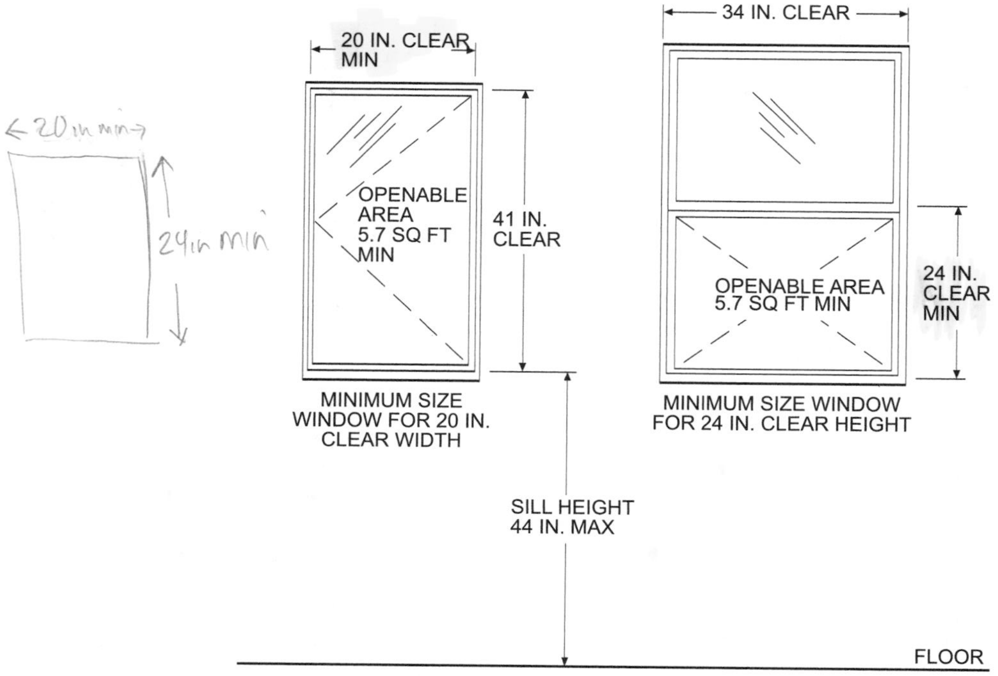

For SI: 1 inch = 25.4 mm, 1 square foot = 0.0929 m^2.

Figure R310.1
EMERGENCY ESCAPE AND RESCUE WINDOW

R310.1.4 Operational constraints. Emergency escape and rescue openings shall be operational from the inside of the room without the use of keys, tools or special knowledge.

❖Openings for emergency escape and rescue must be operational from the inside. Keys, tools or special knowledge must not be needed to operate these openings. If keys or tools were necessary, they might not be readily available in an emergency or panic situation, and an individual might not be able to use them, so the opening would be unusable. Section R310.1 also requires the opening size to be obtained by the normal operation of the window (see commentary, Section R310.1).

R310.2 Window wells. The minimum horizontal area of the window well shall be 9 square feet (0.9 m^2), with a minimum horizontal projection and width of 36 inches (914 mm). The area of the window well shall allow the emergency escape and rescue opening to be fully opened.

> **Exception:** The ladder or steps required by Section R310.2.1 shall be permitted to encroach a maximum of 6 inches (152 mm) into the required dimensions of the window well.

❖Window wells in front of emergency escape and rescue openings also have minimum size requirements. These provisions address those emergency escape windows that occur below grade. Just applying the standard emergency escape window criteria to these windows will result in an opening that occupants can get through, but the window well may actually trap the occupants against the building without providing for their escape from the window well or providing for a fire fighter to enter the residence.

The minimum size requirements in cross section are similar to the emergency escape and opening criteria; that is, they are sufficient to provide a nominal size to allow for the escape of occupants or the entry of fire fighters (see Commentary Figure R310.2). The ladder or steps requirement is the main difference.

R310.2.1 Ladder and steps. Window wells with a vertical depth greater than 44 inches (1118 mm) shall be equipped with a permanently affixed ladder or steps usable with the window in the fully open position. Ladders or steps required by this section shall not be required to comply with Sections R311.7 and R311.8. Ladders or rungs shall have an inside width of at least 12 inches (305 mm), shall project at least 3 inches (76 mm) from the wall and shall be spaced not more than 18 inches (457 mm) on center vertically for the full height of the window well.

❖When the depth of a window well exceeds 44 inches (1118 mm), a ladder or steps from the window is required. The details for construction of steps are not identified in the provisions; however, the design of the ladder is specifically addressed. Because ladders and steps in window wells are provided for emergency use only, they are not required to comply with the provisions for stairways found in Section R311.7.

R310.2.2 Drainage. Window wells shall be designed for proper drainage by connecting to the building's foundation drainage system required by Section R405.1 or by an approved alternative method.

> **Exception:** A drainage system for window wells is not required when the foundation is on well-drained soil or sand-gravel mixture soils according to the United Soil Classification System, Group I Soils, as detailed in Table R405.1.

❖This section requires the window well to be properly drained. Improper window well drainage could create a water accumulation level that could cause the window to become inoperable, or even break due to the pressure and blow out into the occupied room causing serious injury. Improper window well drainage may cause an emergency escape window to become

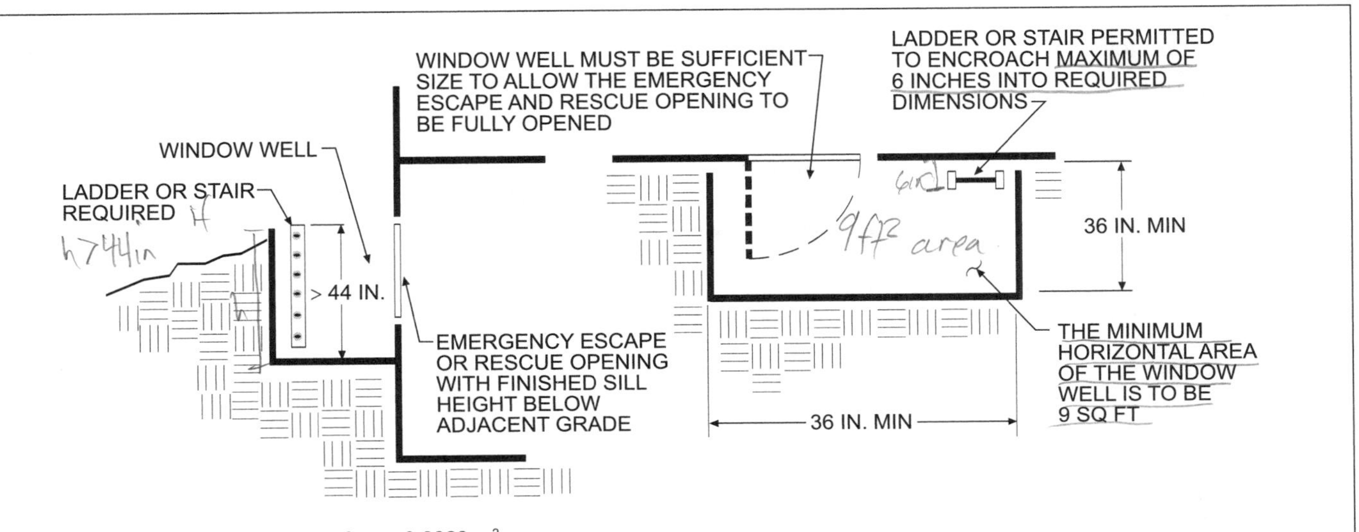

For SI: 1 inch = 25.4 mm, 1 square foot = 0.0929 m^2.

Figure R310.2
WINDOW WELLS

a hazard to the occupants. If the foundation is on well-drained soil, the exception permits deletion of the drainage system.

cellar.

R310.3 Bulkhead enclosures. Bulkhead enclosures shall provide direct access to the *basement*. The bulkhead enclosure with the door panels in the fully open position shall provide the minimum net clear opening required by Section R310.1.1. Bulkhead enclosures shall also comply with Section R311.7.10.2.

❖Bulkhead enclosures, when provided for access to below-grade openings used for emergency escape and rescue, must meet the net openable area provisions of Section R310.1.1. Also, bulkhead enclosures must comply with Section R311.7.10.2.

R310.4 Bars, grilles, covers and screens. Bars, grilles, covers, screens or similar devices are permitted to be placed over emergency escape and rescue openings, bulkhead enclosures, or window wells that serve such openings, provided the minimum net clear opening size complies with Sections R310.1.1 to R310.1.3, and such devices shall be releasable or removable from the inside without the use of a key, tool, special knowledge or force greater than that which is required for normal operation of the escape and rescue opening.

❖The ever-increasing concern for security, particularly in residential buildings, has created a fairly large demand for security devices, such as grilles, bars and steel shutters. Unless properly designed and constructed, the security devices over bedroom windows can completely defeat the purpose of the emergency escape and rescue opening. Therefore, the code makes provisions for security devices if the release mechanism has been approved and is operable from the inside without the use of a key, tool or force greater than that required for the normal operation of the escape and rescue opening.

The essence of the requirement for emergency escape openings is that a person must be able to effect escape or be rescued in a short period of time because the fire might have spread to a point where all other exit routes are blocked. Thus, time cannot be wasted in figuring out means of opening rescue windows or obtaining egress through them. Any impediment to escape or rescue caused by security devices, inadequate window size or difficult operating mechanisms is not permitted by the code.

R310.5 Emergency escape windows under decks and porches. Emergency escape windows are allowed to be installed under decks and porches provided the location of the deck allows the emergency escape window to be fully opened and provides a path not less than 36 inches (914 mm) in height to a *yard* or court.

❖The design of some homes makes the underside of decks the only location where an emergency escape and rescue window can be located. The 36-inch (914 mm) minimum height requirement allows a usable means of egress pathway. The 36-inch (914 mm) minimum height was based on the minimum window well size of 3 feet by 3 feet (914 mm by 914 mm).

SECTION R311
MEANS OF EGRESS

R311.1 Means of egress. All *dwellings* shall be provided with a means of egress as provided in this section. The means of egress shall provide a continuous and unobstructed path of vertical and horizontal egress travel from all portions of the *dwelling* to the exterior of the *dwelling* at the required egress door without requiring travel through a garage.

❖Sections R311.2 through R311.8 contain the requirements for the exit and the means of egress components. All dwelling units must have at least one egress door that complies with the provisions of Sections R311.2 and R311.3. This door must also access the exterior without the dwelling's occupants traveling through a garage where hazards could prevent a suitable means of egress. A ramp or stairway is required for access to habitable areas not having an exit on that level (see Section R311.4). The emergency escape and rescue openings in Section 310 are not considered an exit.

R311.2 Egress door. At least one egress door shall be provided for each *dwelling* unit. The egress door shall be side-hinged, and shall provide a minimum clear width of 32 inches (813 mm) when measured between the face of the door and the stop, with the door open 90 degrees (1.57 rad). The minimum clear height of the door opening shall not be less than 78 inches (1981 mm) in height measured from the top of the threshold to the bottom of the stop. Other doors shall not be required to comply with these minimum dimensions. Egress doors shall be readily openable from inside the *dwelling* without the use of a key or special knowledge or effort.

6'5"

❖All dwelling units must have at least one exit door that complies with the provisions of Sections R311.2 through R311.3.3. Other exterior doors need not comply with the provisions of this section.

The required egress door must be side hinged. A sliding patio door would not count as the required egress door.

Typically, a 36-inch-wide (914 mm) door slab would be required to achieve a minimum 32-inch-wide (813 mm) width opening. Door slabs are manufactured in width increments of 2 inches (51 mm) [32 inches (813 mm), 34 inches (864 mm), 36 inches (914 mm), etc.]. Once the thickness of the door slab [usually $1^3/_4$ inches (44 mm) for exterior doors], thickness of the door stop and allowance for hinges or other hardware are combined, the difference between the width of the door slab and the resultant opening size is greater than 2 inches (51 mm). Therefore, a 34-inch-wide (864 mm) door slab would not provide the 32-inch-wide (813 mm) door opening required, and a 36-inch-wide (914 mm) slab would need to be used.

In a similar fashion, the 80-inch (2032 mm) door height requirement is replaced with a 78-inch (1981 mm) height of opening requirement, with the height of the opening measured from the bottom of the door-stop to the top of the threshold. Since door slabs are also manufactured in height increments of 2 inches required egress door — must be side-hinged

(51 mm), it is not anticipated that this proposal would result in a reduction in the actual door size.

So that egress is always available, any locks provided on the egress door must be openable from the inside without the use of a key or special knowledge. While this would not limit the number of locks someone could install on the egress door, this will prohibit the type of dead bolt lock that is key operated from both the inside and outside. These lock provisions are not applicable to other exterior doors, however, just in case the way to the required egress door was blocked, following these provisions for other exterior doors may provide an extra level of safety.

R311.3 Floors and landings at exterior doors. There shall be a landing or floor on each side of each exterior door. The width of each landing shall not be less than the door served. Every landing shall have a minimum dimension of 36 inches (914 mm) measured in the direction of travel. Exterior landings shall be permitted to have a slope not to exceed $^1/_4$ unit vertical in 12 units horizontal (2-percent).

> **Exception:** Exterior balconies less than 60 square feet (5.6 m^2) and only accessible from a door are permitted to have a landing less than 36 inches (914 mm) measured in the direction of travel.

❖Landings are required for exterior doors and must be constructed on both the exterior and interior side of the door. Landings must be the same width as the door they serve and must be at least 36 inches (914 mm) in length. The length of a landing is measured in the direction of travel. The exterior landing must be reasonably level (a slope not exceeding 25:12) while still allowing enough of a slope for proper drainage.

The exception allows for smaller landings when an exterior door only leads out onto a small balcony. An example of this would be the french balconies commonly found in New Orleans' style architecture.

R311.3.1 Floor elevations at the required egress doors. Landings or finished floors at the required egress door shall not be more than $1^1/_2$ inches (38 mm) lower than the top of the threshold.

> **Exception:** The landing or floor on the exterior side shall not be more than $7^3/_4$ inches (196 mm) below the top of the threshold provided the door does not swing over the landing or floor.

Where exterior landings or floors serving the required egress door are not at *grade*, they shall be provided with access to *grade* by means of a ramp in accordance with Section R311.8 or a stairway in accordance with Section R311.7.

❖Thresholds should not be higher than $1^1/_2$ inches (38 mm) above interior floor level (see Commentary Figure R311.3.1). The exception permits the exterior landing of an exterior egress door, to be a maximum of $7^3/_4$ inches (196 mm) below the top of the threshold. The threshold height represents an important element in building construction. It has to be high enough to keep out snow accumulation and driving rain, yet low enough not to represent a tripping hazard or become a barrier to entry. The exception is limited to locations where the door swings in. Since this is the typical design for single-family homes and townhouses, the step down for the exterior landing will typically be permitted. A screen door that swung out in front of an egress door that swung in would not violate this exception (see Section R311.3.2).

If the door landing is not at grade level, any stairway or ramp leading from the door landing to grade must comply with the stairway and ramp provisions in Sections R311.7 and R311.8.

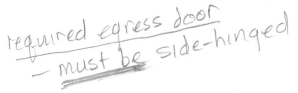

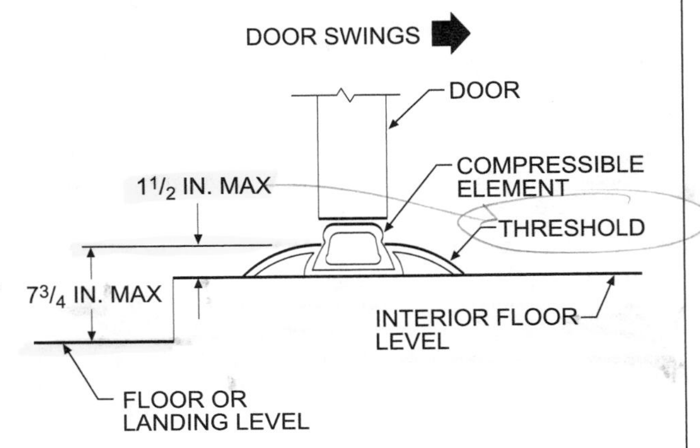

For SI: 1 inch = 25.4 mm.

Figure R311.3.1
THRESHOLD HEIGHTS

R311.3.2 Floor elevations for other exterior doors. Doors other than the required egress door shall be provided with landings or floors not more than $7^3/_4$ inches (196 mm) below the top of the threshold.

> **Exception:** A landing is not required where a stairway of two or fewer risers is located on the exterior side of the door, provided the door does not swing over the stairway.

❖For other egress doors, the top of the threshold can be up to $7^3/_4$ inches (196 mm) above both the inside and outside floor surface. Since this could be a tripping hazard, this is not a common practice and most exterior doors follow the limitations in Section R311.3.1. Thresholds are sometimes a little higher at patio sliding doors.

The exception would not require a landing when two or fewer stair risers are on the exterior side of a door (for the required egress exit door, see Section R311.3.1). However, the door is not allowed to swing over the steps to take advantage of this exception. A screen door that swung out in front of an exterior door that swung in would not violate this exception (see Section R311.3.2).

R311.3.3 Storm and screen doors. Storm and screen doors shall be permitted to swing over all exterior stairs and landings.

❖Sections R311.3.1 and R311.3.2 both have allowances for landing elevations when the door does not swing out over the landing. A common scenario is to have both a screen door and a wood/metal door at the front and/or rear entrance to the home. The screen door swinging out over the landing would not violate these exceptions.

R311.4 Vertical egress. Egress from habitable levels including habitable attics and *basements* not provided with an egress door in accordance with Section R311.2 shall be by a ramp in accordance with Section R311.8 or a stairway in accordance with Section R311.7.

❖All dwelling units must have at least one egress door that complies with the provisions of Sections R311.2 and R311.3. In a multiple-level home, a ramp or stairway is required for access from basements, upper floors, mezzanines, split levels or habitable attics that do not have an exit on that level. The emergency escape and rescue openings in Section 310 are not considered an exit.

R311.5 Construction.

R311.5.1 Attachment. Exterior landings, decks, balconies, stairs and similar facilities shall be positively anchored to the primary structure to resist both vertical and lateral forces or shall be designed to be self-supporting. Attachment shall not be accomplished by use of toenails or nails subject to withdrawal.

❖Exterior exit balconies, stairs and similar exit facilities must be properly attached to the primary structure so that their reaction to vertical and lateral forces will not cause separation from the structure. This is reiterating the requirement for a complete load path in Section R301.1. The reason for doing so is the need to maintain key elements of the egress system that are needed for emergency evacuations and the increased possibility of overlooking such connection. The requirement for positive anchor to the primary structure applies to all exterior means of egress components used as part of an egress system whether part of the required exit or not.

R311.6 Hallways. The minimum width of a hallway shall be not less than 3 feet (914 mm).

❖Hallways must be a minimum of 3 feet (914 mm) wide to accommodate moving furniture into rooms off the hallway and for safe egress from the structure. The code uses the term "hallway" instead of corridors to avoid confusion with the IBC. In the IBC, "corridor" is a defined term, and corridors may have a required fire-resistance rating.

R311.7 Stairways.

❖The requirements for stairways are contained in Sections R311.7.1 through R311.7.9. The provisions address a wide variety of issues important to designing a stairway that is both safe and usable.

R311.7.1 Width. Stairways shall not be less than 36 inches (914 mm) in clear width at all points above the permitted handrail height and below the required headroom height. Handrails shall not project more than 4.5 inches (114 mm) on either side of the stairway and the minimum clear width of the stairway at and below the handrail height, including treads and landings, shall not be less than $31^1/_2$ inches (787 mm) where a handrail is installed on one side and 27 inches (698 mm) where handrails are provided on both sides.

Exception: The width of spiral stairways shall be in accordance with Section R311.7.10.1.

❖Section R311.7.1 requires a minimum stairway width of 36 inches (914 mm). Generally, when the code specifies a required width of a component in the egress system, the width will be the clear, net, usable, unobstructed width. In this case, however, the width is specified as applying only to the area "above the permitted handrail height and below the required headroom height."

At and below the handrail height, the required width for the stairway, including treads and landings, is 27 inches (686 mm) if handrails are provided on each side, and $31^1/_2$ inches (800 mm) if there is a handrail installed on only one side. In essence, the code is not concerned about elements such as trim, stringers or other items that may be found below the level of the handrail, as long as they do not exceed the handrail's projection. This reduced width below the handrail is based on a body's movements as a person walks on a stair or other surface (see Commentary Figure R311.7.1). The exception and the provisions of Section R311.7.10.1 will permit a minimum width of 26 inches (660 mm) for spiral stairways.

It is important to note that three of the key elements in the means of egress, hallways, stairways and the egress door, have a required minimum width.

R311.7.2 Headroom. The minimum headroom in all parts of the stairway shall not be less than 6 feet 8 inches (2032 mm) measured vertically from the sloped line adjoining the tread nosing or from the floor surface of the landing or platform on that portion of the stairway.

Exception: Where the nosings of treads at the side of a flight extend under the edge of a floor opening through which the stair passes, the floor opening shall be allowed to project horizontally into the required headroom a maximum of $4^3/_4$ inches (121 mm).

❖A minimum headroom clearance of 6 feet, 8 inches (2032 mm) is required in connection with every stairway. This includes not only the above-the-tread portion, but also above any landings serving the stairway. The clearance is to be measured vertically above a plane that connects the stair nosings and also vertically above any landing or floor surface that is a part of the stairway [see Commentary Figure R311.7.2(1)]. This specific height requirement overrides the general ceiling height limitations of Section R305 and is modified for spiral stairways by Section R311.7.10.1.

The exception clarifies interpretation and practice by recognizing the common method of stairwell construction in which the open side of a stair is supported by the same structure as the side of the opening through which the stairway passes. The exception allows for common stairways that are slightly wider at the bottom and narrow in width as they ascend through a smaller width opening in the floor above. In this case, the plane of the nosings, from which headroom is determined, at the side of the stairs, extends under the ceiling or joist above at the edge of the floor opening. The exception allows this offset to be a maximum of $4^3/_4$ inches (121 mm) without being considered a projection into the required headroom. It is important to note that this exception only applies at the side of stairs [see Commentary Figure R311.7.2(2)].

R311.7.3 Vertical rise. A flight of stairs shall not have a vertical rise larger than 12 feet (3658 mm) between floor levels or landings.

❖ Between landings and platforms, the vertical rise is to be measured from one landing walking surface to another. The limited height provides a reasonable interval for users with physical limitations to rest on a level surface and also serves to alleviate potential negative psychological effects of long and uninterrupted stairway flights.

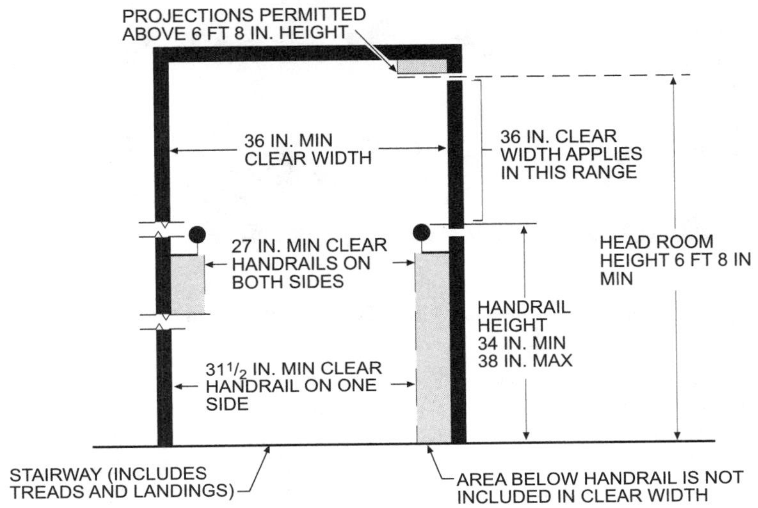

For SI: 1 inch = 25.4 mm, 1 foot = 304.8 mm.

Figure R311.7.1
STAIRWAY CLEARANCES

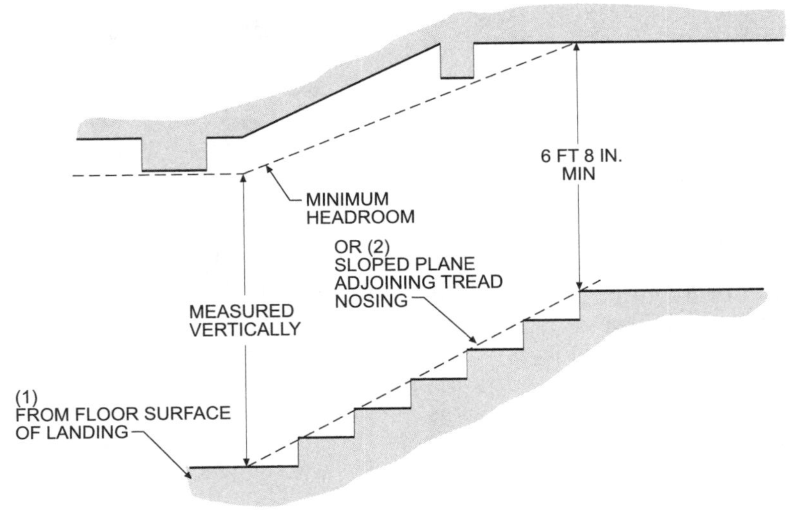

For SI: 1 inch = 25.4 mm, 1 foot = 304.8 mm.

Figure R311.7.2(1)
MINIMUM HEADROOM

Figure R311.7.2(2)
EXAMPLE OF EXCEPTION TO SECTION R311.7.2

R311.7.4 Walkline. The walkline across winder treads shall be concentric to the curved direction of travel through the turn and located 12 inches (305 mm) from the side where the winders are narrower. The 12-inch (305 mm) dimension shall be measured from the widest point of the clear stair width at the walking surface of the winder. If winders are adjacent within the flight, the point of the widest clear stair width of the adjacent winders shall be used.

❖This requirement is essential for smooth, consistent travel on stairs that turn with winder treads. It provides a standard location for the regulation of the uniform tread depth of winders. Due to the wide range of anthropometrics of stairway users, there is no one line that all persons will travel on stairs. However, the code recognizes that a standard location of a walkline is essential to design and enforcement. Each footfall of the user through the turn can be connected in an arc to describe the path traveled. As a user ascends or descends the flight, the turning at each step should be consistent through the turn. The walkline is established concentric, having the same center, or approximately parallel to the arc of travel of the user. The tread depth dimension at the walkline is one of two tread depths across the width of the stair at which winder tread depth is regulated. The second is the minimum tread depth. Regulation at these two points controls the angularity of the turn and the configuration of the flight. In order to establish consistently shaped winders, tread depths must always be measured concentric to the arc of travel. The walkline is unique as the only line or path of travel where winder tread depth is controlled by the same minimum tread depth as rectangular treads. However, Section R311.7.5.2 recognizes winder tread depth need not be compared to rectangular tread depths for dimensional uniformity in the same flight because the location of the walkline is chosen for the purpose of providing a standard and cannot be specific to the variety of actual paths followed by all users. This specific line location is determined by measuring along each nosing edge 12 inches (305 mm) from the extreme of the clear width of the stair at the surface of the winder tread or the limit of where the foot might be placed in use of the stair [see Commentary Figures R311.7.5.2.1(1) and R311.7.5.2.1(2)]. If adjacent winders are present the point of the widest clear stair width at the surface of the tread in the group of adjacent consecutive winders is used to provide the reference from which the 12-inch (305 mm) dimension will be measured along each nosing. The tread depth may be determined by measuring between adjacent nosings at these determined intersections of the nosings with the walkline. It is important to note that the clear stair width is only that portion of the stair width that is clear for passage. Portions of the stair beyond the clear width are not consequential to use of the stair, consistent travel or location of the walkline.

R311.7.5 Stair treads and risers. Stair treads and risers shall meet the requirements of this section. For the purposes of this section all dimensions and dimensioned surfaces shall be exclusive of carpets, rugs or runners.

❖The riser height, tread depth and profile requirements for stairways are specified in Sections R311.7.5.1 through R311.7.5.3. These provisions facilitate smooth and consistent travel. This section provides dimensional ranges and tolerances for the component elements to allow the flexibility required to design and construct a stair or a flight of stairs, which are elements of a stairway. The allowed proportion of maximum riser height and minimum tread depth provides for a maximum angle of ascent, but there is no maximum tread depth or minimum riser height that would define a minimum angle for a stairway. Nor is the proportion of riser height to tread depth compared with the limitations of the length of the users stride on stairways that is significantly foreshortened from the users stride on the level. For this reason, care should be taken when incorporating large tread depths or short risers to proportion the riser height and tread depth to avoid a step that is wide enough to require more than one step to cross or a short narrow step, which can be easily stepped over. With these same limitations for proportion in mind, by controlling the minimum depth of rectangular treads and the minimum depth and angularity of winder treads, these components can control the configuration of the plan of a flight of stairs to provide for smooth and consistent travel. Carpets, rugs and runners, like furniture, are frequently changed by the occupants and are not regulated by the code. For this reason, it is essential that the riser height and tread depth be regulated exclusive of these transitory surfaces to provide an enforceable standard. This practice minimizes the possible variation due to the removal of nonpermanent carpeting throughout the life of a structure and provides a standard enforcement methodology that will provide consistency across the built environment for all users. When owners or occupants add carpeting, rugs or runners they need to be able to add it to all tread and landing surfaces in the stairway. It is important that the tread and landing surfaces are consistent and comply with the code prior to the addition of carpet. This methodology of enforcement makes it unnecessary to reconstruct floor and stair elevations in the stairway when nonpermanent carpet surfaces are changed that do not require a building permit and eliminates the resulting variations in the built environment that will not comply with the tolerance in Sections R311.7.5.1 and R311.7.5.2.

R311.7.5.1 Risers. The maximum riser height shall be $7^3/_4$ inches (196 mm). The riser shall be measured vertically between leading edges of the adjacent treads. The greatest riser height within any flight of stairs shall not exceed the smallest by more than $^3/_8$ inch (9.5 mm). Risers shall be vertical or sloped from the underside of the nosing of the tread above at an angle not more than 30 degrees (0.51 rad) from the vertical. Open risers are permitted provided that the opening between treads does not permit the passage of a 4-inch-diameter (102 mm) sphere.

Exception: The opening between adjacent treads is not limited on stairs with a total rise of 30 inches (762 mm) or less.

❖The code establishes that the maximum riser height is $7^3/_4$ inches (197 mm). The provisions specify how the riser height is to be measured [see Commentary Figure R311.7.5.1(1)]. The uniformity of risers and treads is a safety factor in any flight of stairs. The section of a stairway leading from one landing to the next is defined as a flight of stairs. This is important because variations in excess of the $^3/_8$-inch (9.5 mm) tolerance could interfere with the rhythm of the stair user. It is true that adequate attention to the use of the stair can compensate for substantial variations in risers and treads; however, the stair user does not always give the necessary attention.

To obtain the best uniformity possible in a flight of stairs, the maximum variation between the highest and lowest risers is limited to $^3/_8$ inch (9.5 mm). This tolerance is not to be used as a design variation, but its inclusion is in recognition that normal construction practices give rise to variables that make it impossible to get exactly identical riser heights and tread dimensions in constructing a stairway. Therefore, the

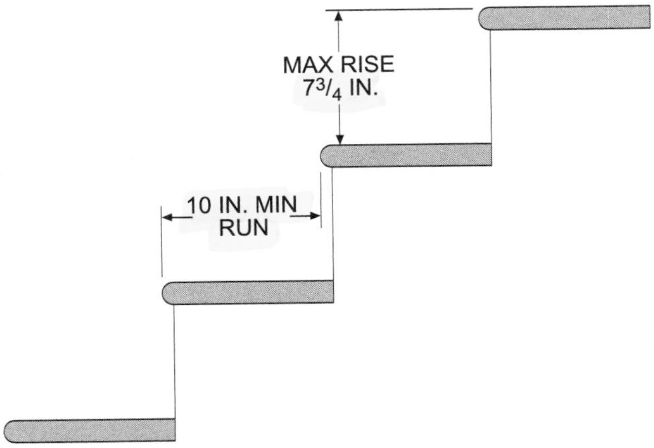

For SI: 1 inch = 25.4 mm.

Figure R311.7.5.1(1)
CONVENTIONAL STAIRWAY

code allows the variation indicated in Commentary Figure R311.7.5.1(2).

The risers must be vertical or slope back, effectively providing a wider overall tread depth. The code does not require solid risers, but where the height of the stairway exceeds 30 inches (762 mm), either solid risers or another method to limit the opening between adjacent treads is needed. This is consistent with the guard provisions of Section R312, where a 4-inch (102 mm) sphere is used to determine compliance.

R311.7.5.2 Treads. The minimum tread depth shall be 10 inches (254 mm). The tread depth shall be measured horizontally between the vertical planes of the foremost projection of adjacent treads and at a right angle to the tread's leading edge. The greatest tread depth within any flight of stairs shall not exceed the smallest by more than $^3/_8$ inch (9.5 mm).

❖The code establishes that the minimum tread depth is 10 inches (254 mm).The provisions specify how the tread depth is to be measured [see Commentary Figure R311.7.5.1(1)]. To obtain the best uniformity possible in a flight of stairs, the maximum variation between the greatest and smallest tread depth is limited to $^3/_8$ inch (9.5 mm). See the commentary to Section R311.7.5.1 for the discussion on uniformity.

R311.7.5.2.1 Winder treads. Winder treads shall have a minimum tread depth of 10 inches (254 mm) measured between the vertical planes of the foremost projection of adjacent treads at the intersections with the walkline. Winder treads shall have a minimum tread depth of 6 inches (152 mm) at any point within the clear width of the stair. Within any flight of stairs, the largest winder tread depth at the walkline shall not exceed the smallest winder tread by more than $^3/_8$ inch (9.5 mm). Consistently shaped winders at the walkline shall be allowed within the same flight of stairs as rectangular treads and do not have to be within $^3/_8$ inch (9.5 mm) of the rectangular tread depth.

❖The same criterion for rectangular treads applies to winder treads. However, the depth is to be measured as the horizontal distance between the points where the nosing of the adjacent treads intersects with the "walkline." The location of the "walkline" is defined in Section R311.7.4. Winder treads must have a minimum depth of 6 inches (152 mm) at any point. A stairway may have straight treads and winder treads within the same flight. If winders are used, they can either be used for an entire flight of a stairway, as a portion of a flight to provide a change of direction or to form a curved stairway. Because winder treads are used to change the direction of the stair it is important that winders comply with the specified dimensional criteria. See Commentary Figure R311.7.5.2.1(1) for examples of winders used as a portion of a stairway at a change of direction. See Commentary Figure R311.7.5.2.1(2) for an example of winders used to form a circular stairway. Rectangular treads can be used in combination with winder treads. The goal is to allow the foot placement along the walkline to be consistent along the length of the flight.

R311.7.5.3 Nosings. The radius of curvature at the nosing shall be no greater than $^9/_{16}$ inch (14 mm). A nosing not less than $^3/_4$ inch (19 mm) but not more than $1^1/_4$ inches (32 mm) shall be provided on stairways with solid risers. The greatest nosing projection shall not exceed the smallest nosing projection by more than $^3/_8$ inch (9.5 mm) between two stories, including the nosing at the level of floors and landings. Beveling of nosings shall not exceed $^1/_2$ inch (12.7 mm).

> **Exception:** A nosing is not required where the tread depth is a minimum of 11 inches (279 mm).

❖The sectional parameters of the components of a step or stair contribute to stairway safety. The radius or bevel of the nosing eases the otherwise square edge of the tread and prevents irregular chipping that can become a maintenance issue seriously affecting the

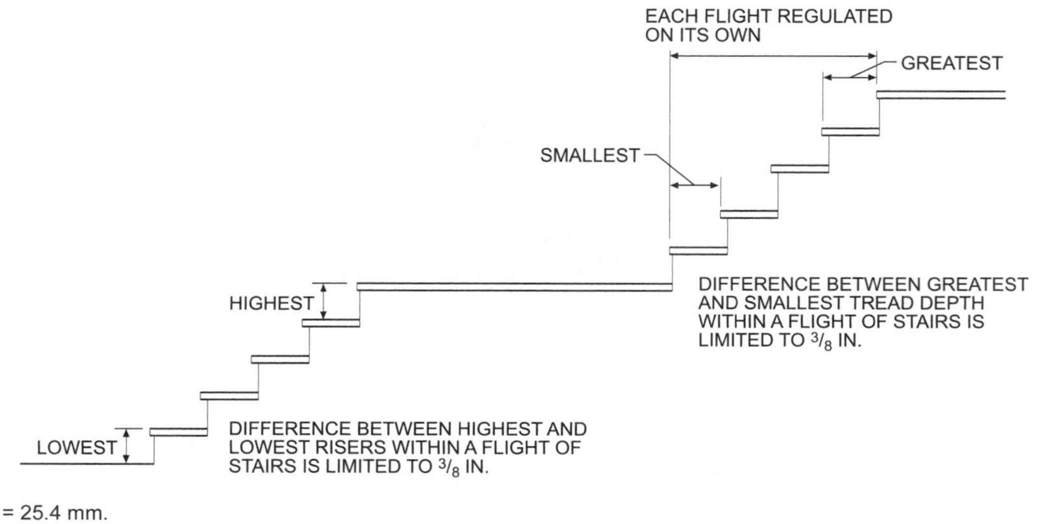
EACH FLIGHT REGULATED ON ITS OWN
GREATEST
SMALLEST
HIGHEST
DIFFERENCE BETWEEN GREATEST AND SMALLEST TREAD DEPTH WITHIN A FLIGHT OF STAIRS IS LIMITED TO 3/8 IN.
LOWEST
DIFFERENCE BETWEEN HIGHEST AND LOWEST RISERS WITHIN A FLIGHT OF STAIRS IS LIMITED TO 3/8 IN.

For SI: 1 inch = 25.4 mm.

Figure R311.7.5.1(2)
STAIR TOLERANCES

3/8" variation for horizontal, - vertical riser - nosing projection nosing = 3/4 to 1 1/4 in (for solid risers) nosing not required for tread depth > 11 in.

safe use of the stair; and eliminates a sharp square edge that will cause greater injury in falls. A radius or bevel allows light modeling, reflecting light at various angles, providing a certain contrast from the other surfaces of the stair allowing easier visual location of the start of the tread surface. The maximum radius of curvature at the leading edge of the tread is intended to allow descending foot placement on a surface that does not pitch the foot forward or allow the ball of the foot to slide off the treads and ascending foot placement to slide on to the tread without catching on a square edge. If a stairway design uses a beveled nosing configuration, the bevel is limited to a depth of $^1/_2$ inch (12.7 mm). A nosing projection allows the descending foot to be placed further forward on the tread and the heel to then clear the nosing of the tread above as it swings down in an arc landing further away from the riser on the tread that is effectively deeper than if no nosing projection is used. Nosing projections are so common in stair design that they are noticed by users when absent as affecting their gait and anticipated clearance for their heels from the riser in descent. A nosing projection may also be accommodated by slanting the riser under the tread above. The nosing projection is between $^3/_4$ inch minimum (19 mm) and $1^1/_4$ inches (32 mm) maximum (see Commentary Figure R311.7.5.3). It is critical that all tread and landing nosings associated with each step in the stairway be uniform to ensure that the user does not experience an effective change in tread depth outside the $^3/_8$ inch (9.5 mm) between stories. Critical to this understanding is that tread depth is regulated by measuring between the nosing edges of treads or a tread and a landing. The lack of a uniform nosing at the top landing of a flight is a serious safety problem that may not be apparent in initial rough inspections. Treads with a tread depth of at least 11 inches (279 mm) are allowed with or without a nosing projection.

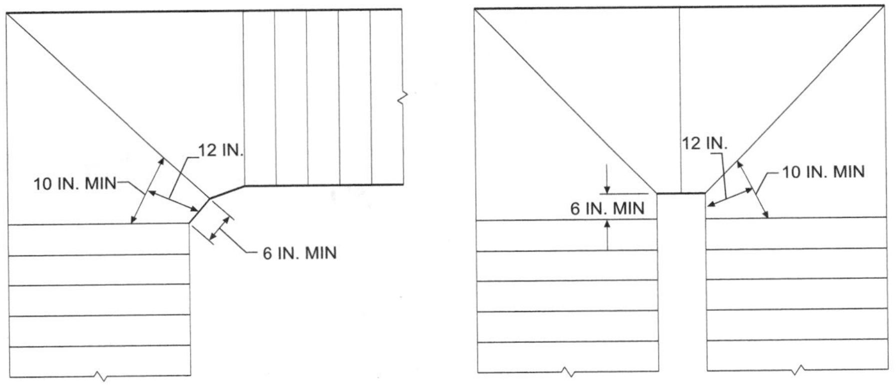

For SI: 1 inch = 25.4 mm.

Figure R311.7.5.2.1(1)
WINDERS

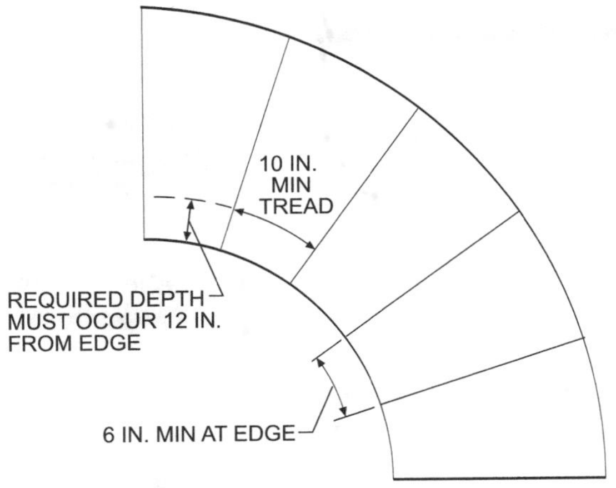

For SI: 1 inch = 25.4 mm.

Figure R311.7.5.2.1(2)
WINDERS USED FOR CIRCULAR STAIRWAY

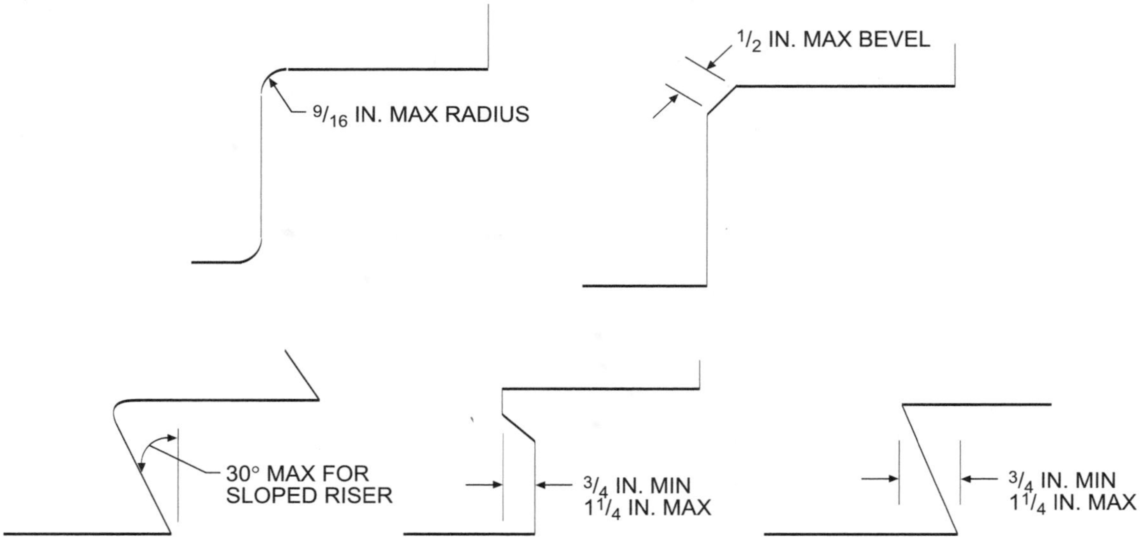

For SI: 1 inch = 25.4 mm, 1 degree = 0.01749 rad.

Figure R311.7.5.3
TREAD PROFILE

R311.7.5.4 Exterior wood/plastic composite stair treads. Wood/plastic composite stair treads shall comply with the provisions of Section R507.3.

❖Stair treads made of wood/plastic composite materials must meet the requirements for installation, labeling and compliance with ASTM D 7032 stated in Section R507.3, in addition to the requirements of Section R311.7.5.

R311.7.6 Landings for stairways. There shall be a floor or landing at the top and bottom of each stairway. The minimum width perpendicular to the direction of travel shall be no less than the width of the flight served. Landings of shapes other than square or rectangular shall be permitted provided the depth at the walk line and the total area is not less than that of a quarter circle with a radius equal to the required landing width. Where the stairway has a straight run, the minimum depth in the direction of travel shall be not less than 36 inches (914 mm).

Exception: A floor or landing is not required at the top of an interior flight of stairs, including stairs in an enclosed garage, provided a door does not swing over the stairs.

❖A landing is required at the top and bottom of each stairway; however, a landing is not required at the top of interior stairways, including an enclosed garage, if a door does not swing over the stairway [see Commentary Figure R311.7.6(1)]. Section R311.7.3 states that flights must be interrupted by a landing or floor such that they do not have a total rise of more than 12 feet (3658 mm).

The width of landings for stairways is measured perpendicular to the direction of travel. It is not the intent to require specifically shaped landings. Landings may have curved or segmented periphery edges provided the width perpendicular to the direction of travel is not less than the width of the stairway served. For a straight stairway run, the minimum dimension of 36 inches (914 mm) in the direction of travel is intended to provide a minimum depth at the landing that cannot be overstepped in descent of a straight run stairway. Landings that are square or rectangular serve to limit the minimum angle of turn of a landing to 90 degrees (1.57 rad) for at least 36 inches (914 mm). Landings for turns of 90 degrees (1.57 rad) or less or of any shape are permitted provided the depth of the tread at the walkline and the total area of the landing are the same as a quarter circle with a radius the same as the stairway width. See Commentary Figure R311.7.6(2) for an example.

R311.7.7 Stairway walking surface. The walking surface of treads and landings of stairways shall be sloped no steeper than one unit vertical in 48 inches horizontal (2-percent slope).

❖The slope of the walking surfaces must provide drainage to stairs and landings that may be subjected to accumulation of liquids, such as water, rain or melting snow. The use of such a slope, called a "wash," is a common technique used on all stairs to allow the nosing to be at a lower elevation than the remainder of the tread surface. This technique of building the flight to a slightly shorter total rise than the actual condition slopes the entire flight forward and better accommodates the placement of the user's foot as it slides onto the tread. It also serves to prevent long-term wear and tear at the nosing limiting problematic maintenance and safety issues. This section provides a limit of the slope to maintain a safe walking surface. This requirement applies to all stairs and landings, both exterior and interior.

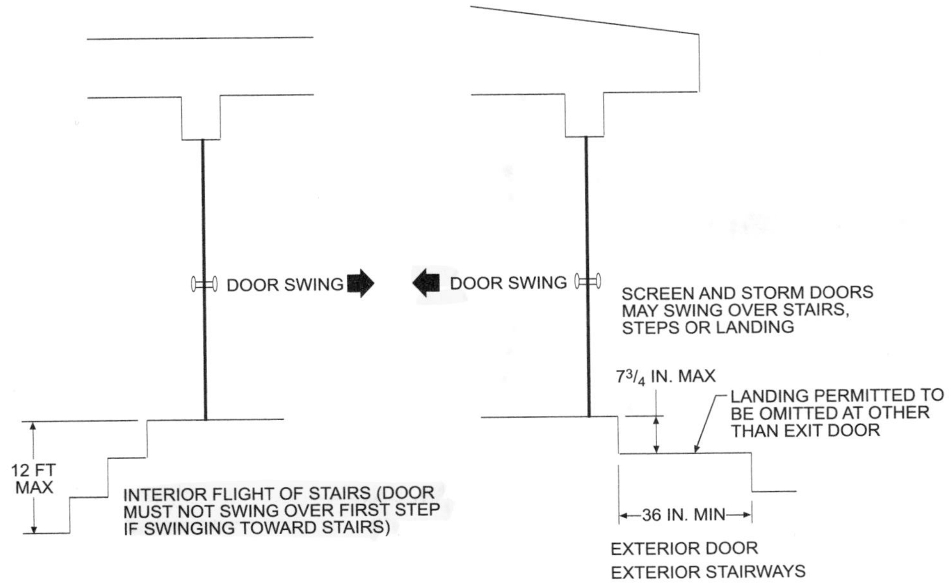

For SI: 1 inch = 25.4 mm, 1 foot = 304.8 mm.

Figure R311.7.6(1)
LANDINGS AT DOORS

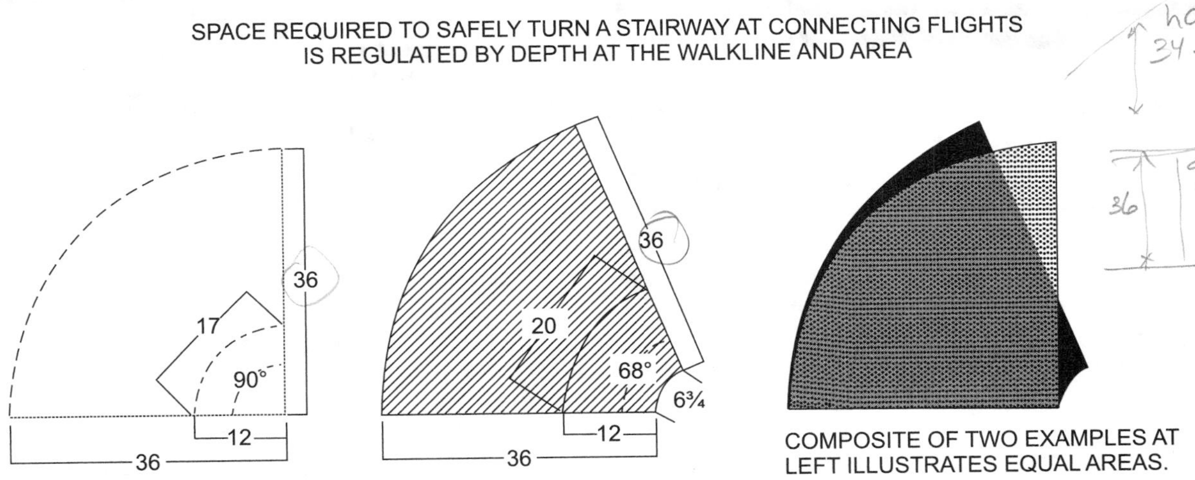

Figure R311.7.6(2)
EXAMPLE OF LANDING OF SHAPES OTHER THAN SQUARE OR RECTANGULAR

R311.7.8 Handrails. Handrails shall be provided on at least one side of each continuous run of treads or flight with four or more risers.

❖The provision of handrails increases the level of safety when used by the occupants while ascending and descending stairs. Handrails are used for guidance, stabilization, pulling and to assist in arresting a fall. This section states that a handrail must be provided on at least one side of flights of four or more risers. Handrails may be provided on both sides and this eliminates choosing the best side to securely attach the handrail. Otherwise, the generally preferred location is for use by the right hand in descent when feasible. Sections R311.7.8.1 through R311.7.8.3 contain provisions essential to the height, continuity and grip size of the handrail provided.

R311.7.8.1 Height. Handrail height, measured vertically from the sloped plane adjoining the tread nosing, or finish surface of ramp slope, shall be not less than 34 inches (864 mm) and not more than 38 inches (965 mm).

Exceptions:

1. The use of a volute, turnout or starting easing shall be allowed over the lowest tread.
2. When handrail fittings or bendings are used to provide continuous transition between flights, transitions at winder treads, the transition from handrail to guardrail, or used at the start of a flight, the handrail height at the fittings or bendings shall be permitted to exceed the maximum height.

❖Where handrails are required, they must be installed at a height within the limits of at least 34 inches (864 mm) and not more than 38 inches (965 mm). This height is to be measured vertically to the top of the handrail from the plane adjoining the tread nosings of the flight or the surface of the ramp slope. Exception 1 allows common starting fittings used as terminals over the lowest tread to fall outside the required height range. Exception 2 allows transition fittings to exceed the required height when used to provide a continuous rail at changes in the pitch of the rail within the stairway.

R311.7.8.2 Continuity. Handrails for stairways shall be continuous for the full length of the flight, from a point directly above the top riser of the flight to a point directly above the lowest riser of the flight. Handrail ends shall be returned or shall terminate in newel posts or safety terminals. Handrails adjacent to a wall shall have a space of not less than $1^1/_2$ inch (38 mm) between the wall and the handrails.

Exceptions:

1. Handrails shall be permitted to be interrupted by a newel post at the turn.
2. The use of a volute, turnout, starting easing or starting newel shall be allowed over the lowest tread.

❖This required handrail is to be continuous for the length of the flight. Where stairway flights are separated by landings or floor levels, handrails are not required (see Commentary Figure R311.7.8.2). The term "continuous" means not only that a single handrail must run from the top riser to the bottom riser, but it also indicates that users should be able to grasp the handrail and maintain their grasp without having to release the rail where it is supported. There is no requirement within the code for installation of a second handrail, but depending on the design and the placement of the required handrail, the requirement for a guard should be reviewed. The two exceptions to this section create situations where the graspable portion of the handrail may not be completely continuous from the top riser to the bottom riser. These traditional situations are well known to the occupants and have not been shown to represent a safety hazard requiring their restriction.

The ends of handrails are to be returned to the wall or floor, or to end in some type of terminal that will not catch clothing or limbs. A clear space of at least $1^1/_2$ inches (38 mm) is necessary between the handrail and any abutting wall. This distance will permit the fingers to slide past any adjacent rough surface that may cause injury, and it will provide an adequate distance so that the handrail may be quickly grabbed as an assist in the arrest of a fall.

R311.7.8.3 Grip-size. All required handrails shall be of one of the following types or provide equivalent graspability.

1. Type I. Handrails with a circular cross section shall have an outside diameter of at least $1^1/_4$ inches (32 mm) and not greater than 2 inches (51 mm). If the handrail is not circular, it shall have a perimeter dimension of at least 4 inches (102 mm) and not greater than $6^1/_4$ inches (160 mm) with a maximum cross section of dimension of $2^1/_4$ inches (57 mm). Edges shall have a minimum radius of 0.01 inch (0.25 mm).
2. Type II. Handrails with a perimeter greater than $6^1/_4$ inches (160 mm) shall have a graspable finger recess area on both sides of the profile. The finger recess shall begin within a distance of $^3/_4$ inch (19 mm) measured vertically from the tallest portion of the profile and achieve a depth of at least $^5/_{16}$ inch (8 mm) within $^7/_8$ inch (22 mm) below the widest portion of the profile.

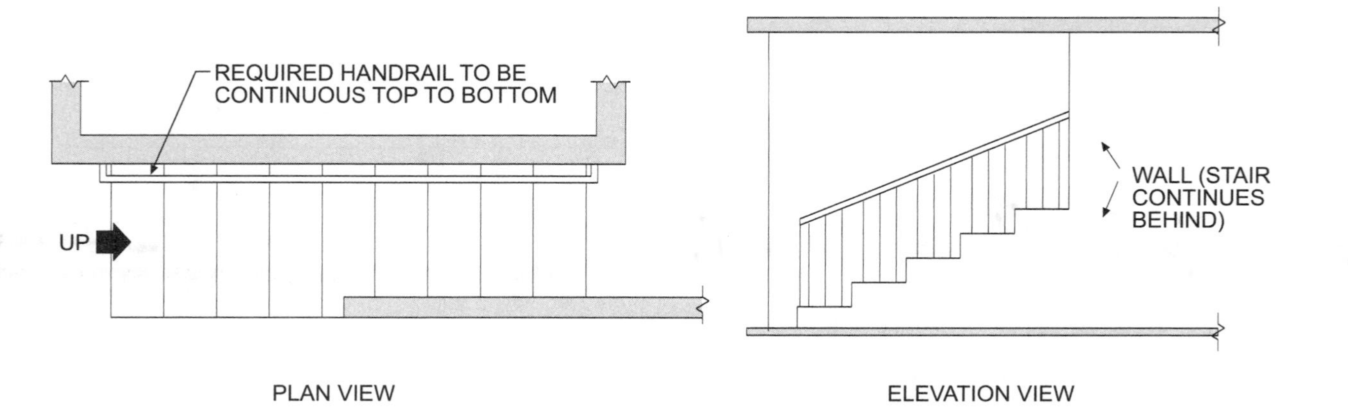

Figure R311.7.8.2
HANDRAILS

This required depth shall continue for at least 3/8 inch (10 mm) to a level that is not less than 1 3/4 inches (45 mm) below the tallest portion of the profile. The minimum width of the handrail above the recess shall be 1 1/4 inches (32 mm) to a maximum of 2 3/4 inches (70 mm). Edges shall have a minimum radius of 0.01 inch (0.25 mm).

❖To be effective, a handrail must be easily grasped by the vast majority of users. If it is too large, it is difficult for a user to get a strong enough grip to provide the needed support. If it is too small the fingers wrap and interfere with the thumb and palm and cannot close in a sufficient grip. For this reason Type I rails have minimum and maximum perimeters to restrict their use to the effective size range. Tests have proven it is beneficial to have graspable recesses for the fingers and opposing thumb such that wider and taller shapes can provide graspability comparable to rails within the Type I size range limitations. The Type II handrail code provides specifics to the location and depth of the recess as it relates to the variables of crown height and width to ensure the design is of a graspable shape. The mountings of smaller profiles can cause interference, as well. Care should be taken to minimize the interference caused by brackets and balusters supporting profiles that require the bottom mounting surface to be grasped.

The code specifies that the handrail be either Type I or Type II, or be equivalently graspable. A Type I can be either circular or noncircular in shape. See Commentary Figure R311.7.8.3(1) for examples of Type I handrails.

A Type II handrail has a perimeter larger than 6 1/4 inches (160 mm) with graspable finger recess area on both sides of the profile. See Commentary Figure R311.7.8.3(2) for the limitations of a Type II handrail.

R311.7.8.4 Exterior wood/plastic composite handrails. Wood/plastic composite handrails shall comply with the provisions of Section R507.3.

❖Handrails made of wood/plastic composite materials must meet the requirements for installation, labeling and compliance with ASTM D 7032 stated in Section R507.3, in addition to the general requirements for handrails in this section.

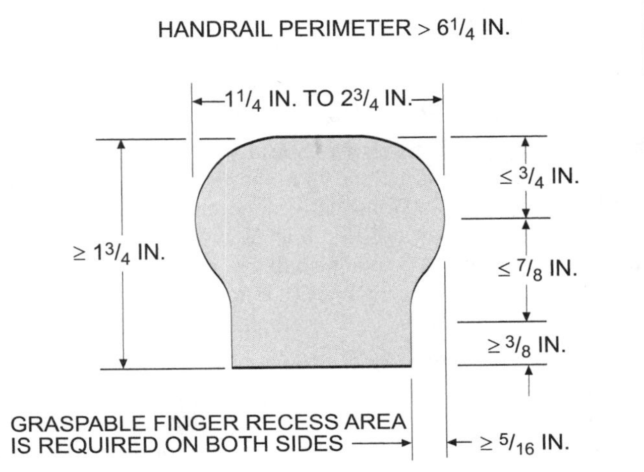

For SI: 1 inch = 25.4 mm.

Figure R311.7.8.3(2)
TYPE II HANDRAIL

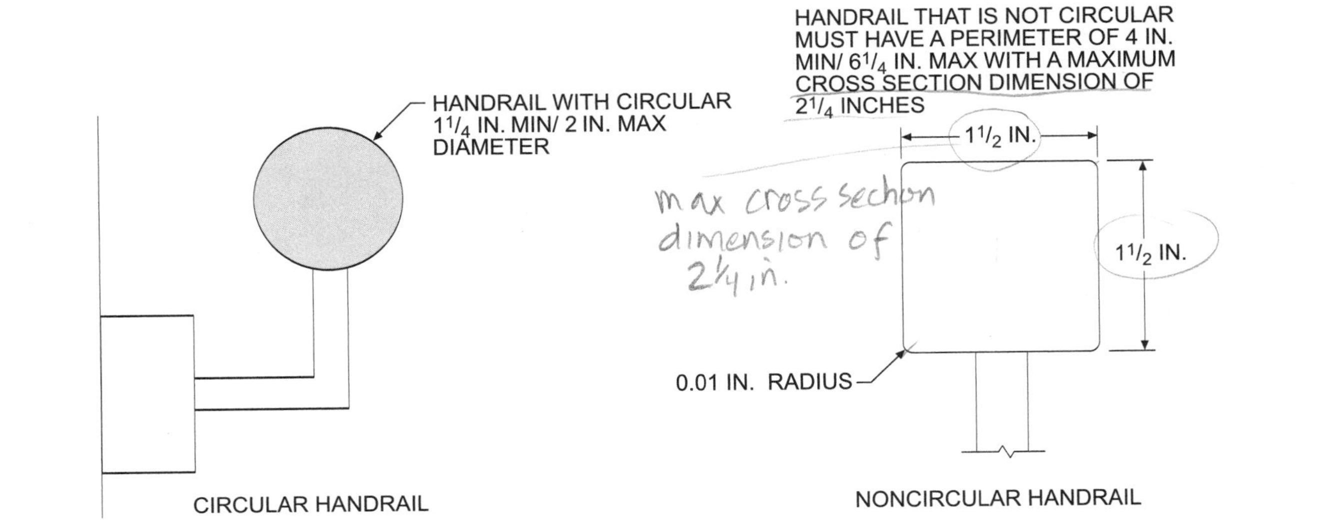

For SI: 1 inch = 25.4 mm.

Figure R311.7.8.3(1)
TYPE 1 HANDRAIL

R311.7.9 Illumination. All stairs shall be provided with illumination in accordance with Section R303.6.

❖This section contains a reference to the illumination provisions of Section R303.6. The proper illumination of stairways is an important part of stairway safety. This lighting can assist users by making sure the level changes do not occur in areas with shadows or in contrasting light, which would therefore make them difficult to see. See the commentary to Section R303.6 for additional information.

R311.7.10 Special stairways. Spiral stairways and bulkhead enclosure stairways shall comply with all requirements of Section R311.7 except as specified below.

❖Sections R311.7.10.1 and R311.7.10.2 are exceptions to the general requirements for stairways as prescribed in Section R311.7.

R311.7.10.1 Spiral stairways. Spiral stairways are permitted, provided the minimum clear width at and below the handrail shall be 26 inches (660 mm) with each tread having a $7^1/_2$-inch (190 mm) minimum tread depth at 12 inches (914 mm) from the narrower edge. All treads shall be identical, and the rise shall be no more than $9^1/_2$ inches (241 mm). A minimum headroom of 6 feet 6 inches (1982 mm) shall be provided.

❖A spiral stairway is one of two types of special stairs that the code permits. Although a spiral stair may be difficult to use to move furniture from one level to another, the code places no limitations on its use within the egress system if it meets the size requirements of this section. A spiral stairway that meets these requirements may provide the only means of egress from a level within an individual dwelling regardless of the occupant load or size of area served.

A spiral stairway is one in which the treads radiate from a central pole. Such a stair must provide a clear width of at least 26 inches (660 mm) at and below the handrail. Each tread must be identical and have a minimum dimension of $7^1/_2$ inches (191 mm) at a point 12 inches (305 mm) from its narrow end. The stair must have at least 6 feet, 6 inches (1981 mm) of headroom measured vertically from the leading edge of the tread. The rise between treads can be as much as, but not more than, $9^1/_2$ inches (241 mm). Commentary Figure R311.7.10.1 shows the required dimensions of a spiral stairway.

R311.7.10.2 Bulkhead enclosure stairways. Stairways serving bulkhead enclosures, not part of the required building egress, providing access from the outside *grade* level to the *basement* shall be exempt from the requirements of Sections R311.3 and R311.7 where the maximum height from the *basement* finished floor level to *grade* adjacent to the stairway does not exceed 8 feet (2438 mm) and the *grade* level opening to the stairway is covered by a bulkhead enclosure with hinged doors or other *approved* means.

❖This section exempts exterior "bulkhead enclosure stairways" from the landing stairway and handrail requirements found in Chapter 3, and it therefore permits a situation that has been fairly common in some areas.

See Commentary Figure R311.7.10.2 for an illustration of the requirements. Because these stairways are not a part of the building's egress system and serve only as a convenient way to access the basement from the exterior, the code exemption will not greatly affect the occupants' safety. Through this exemption, the size of the enclosure that is needed to provide weather protection for the stairway is greatly reduced.

R311.8 Ramps.

❖Section R311.8 states the code requirements for ramps when they are used to access, or within, a dwelling.

"Ramps" are defined in Section 202 as being a walking surface that has a running slope steeper than one unit vertical in 20 units horizontal (5-percent slope).

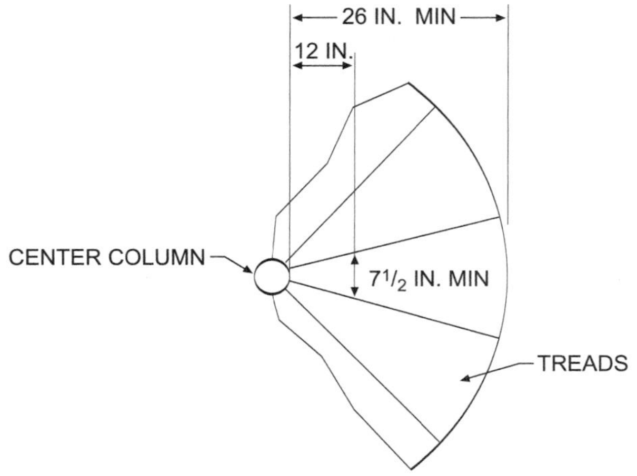

For SI: 1 inch = 25.4 mm.

Figure R311.7.10.1
SPIRAL STAIRS

ramps min - 1:20 ~ 5%
max - 1:12 ~ 8.3%
exception may be 1:8 ~ 12.5%
(if infeasible)

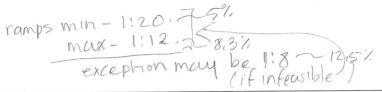

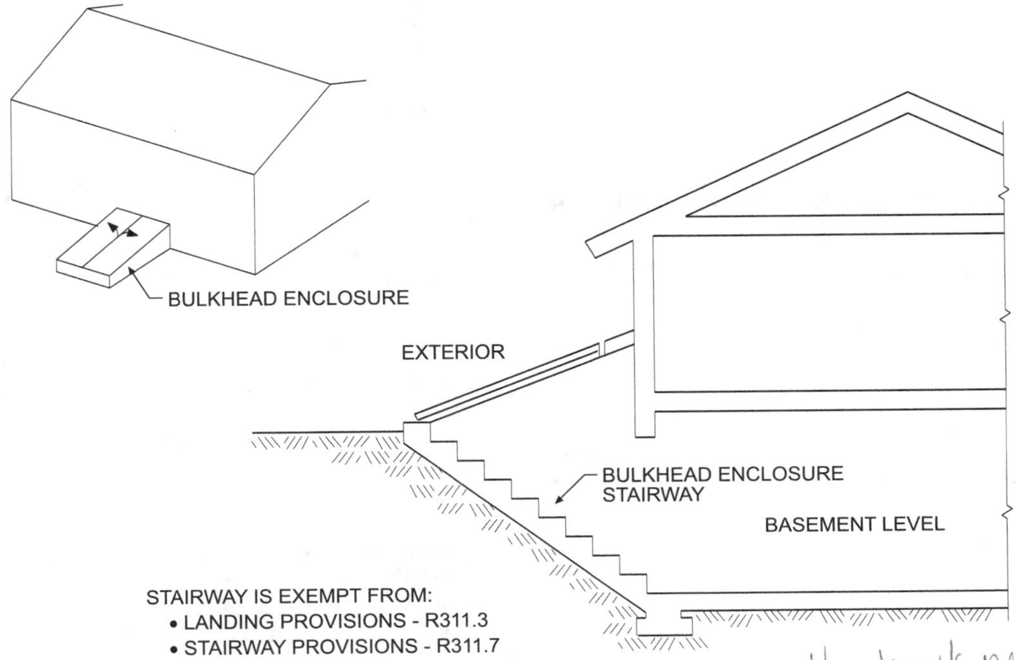

Figure R311.7.10.2
BULKHEAD ENCLOSURE STAIRWAY

Handrails needed for ramp if slope(m) >1:12

R311.8.1 Maximum slope. Ramps shall have a maximum slope of 1 unit vertical in 12 units horizontal (8.3-percent slope).

Exception: Where it is technically infeasible to comply because of site constraints, ramps may have a maximum slope of one unit vertical in eight horizontal (12.5-percent slope).

❖Section R311.8.1 places a maximum slope of one unit vertical in 12 units horizontal (8.3-percent slope) on ramps. This requirement applies to all ramps, including those on circulation routes and those leading to and from an exit. This maximum slope matches what is permitted by the IBC for ramps that are not a part of the means of egress. Egress ramps under the IBC also have a maximum slope limit of 1:12 so that the requirements are consistent with accessibility provisions.

R311.8.2 Landings required. A minimum 3-foot-by-3-foot (914 mm by 914 mm) landing shall be provided:

1. At the top and bottom of ramps.
2. Where doors open onto ramps.
3. Where ramps change direction.

❖The code requires a minimum 3-foot-by-3-foot (914 mm by 914 mm) landing at three specific locations on ramps. Landings should be provided at the top and bottom of each ramp run. When a ramp leaves or approaches a door, there needs to be a level landing to allow someone to open the door from a level surface. A change in direction could be any angle; however, these provisions are not intended to prohibit curved ramps. These dimensions are not tied to the actual width of the ramp. Item 2, dealing with doors that open onto ramps, calls for a larger size landing if it is also required by Section R311.3. The specified landing dimensions coordinate with the requirements for nonaccessible dwelling units, which are found in exceptions in the IBC. While not a requirement, if the ramp is intended to serve as part of an accessible route, the landing should be sized as indicated in ICC A117.1 in order to allow full wheelchair access.

R311.8.3 Handrails required. Handrails shall be provided on at least one side of all ramps exceeding a slope of one unit vertical in 12 units horizontal (8.33-percent slope).

❖Where a ramp exceeds a slope of one unit vertical in 12 units horizontal (8.3-percent slope) the code requires that a handrail be installed on at least one side to assist ramp users. Therefore, ramps would require handrails when the exception to Section R311.8.1 was utilized. This provision differs from that of the IBC, where a slope of one unit vertical in 20 units horizontal (5-percent slope) and a ramp rise of 6 inches (152 mm) establishes the limits. A designer might choose to provide handrails, edge protection and/or guards on a ramp as a safety concern, even if it is not literally a requirement. If the purpose of the ramp is for wheelchair access, ICC A117.1 would be a good resource for information.

Guard 36" handrail (& same as guard) 34-38"

R311.8.3.1 Height. Handrail height, measured above the finished surface of the ramp slope, shall be not less than 34 inches (864 mm) and not more than 38 inches (965 mm).

❖Where handrails are required, they must be installed at a height within a range of at least 34 inches (864 mm) and not more than 38 inches (965 mm), measured vertically from the finished surface of the ramp slope. This height should be measured to the top of the handrail.

R311.8.3.2 Grip size. Handrails on ramps shall comply with Section R311.7.8.3.

❖The grip size for handrails along ramps is the same as that required for stairways (see commentary, Section R311.7.8.3).

R311.8.3.3 Continuity. Handrails where required on ramps shall be continuous for the full length of the ramp. Handrail ends shall be returned or shall terminate in newel posts or safety terminals. Handrails adjacent to a wall shall have a space of not less than $1^1/_2$ inches (38 mm) between the wall and the handrails.

❖The continuity requirement for the ramp handrail is similar to the continuity requirement for the stair handrail (see commentary, Section R311.7.8.2).

SECTION R312
GUARDS AND WINDOW FALL PROTECTION

R312.1 Guards. Guards shall be provided in accordance with Sections R312.1.1 through R312.1.4.

❖The guard provisions of the code address the issue of protecting occupants from falling from any type of elevated walking surface. The provisions in Section R312 provide the scoping requirements, as well as the general construction requirements for the guards. Besides this section, code users should be aware that Section R301.5 contains the design load criteria for guards.

R312.1.1 Where required. *Guards* shall be located along open-sided walking surfaces, including stairs, ramps and landings, that are located more than 30 inches (762 mm) measured vertically to the floor or *grade* below at any point within 36 inches (914 mm) horizontally to the edge of the open side. Insect screening shall not be considered as a *guard*.

❖Section R312.1.1 establishes stairs, ramps and landings as examples of open-sided walking surfaces, but this is not an all-inclusive list of locations where guards are required. This section gives further specifics, to define the minimum elevation of the walking surface as greater than 30 inches (762 mm) that requires a guard. It also recognizes that a guard is needed to minimize falls if the elevation exceeds the 30-inch (762 mm) height at any point within 36 inches (914 mm) of the edge of the walking surface in consideration of such conditions as a sloping site or sudden drop. The scoping requirement for guards along open sides of stairs only applies to that portion of the stairway that is more than 30 inches (762 mm) above the determined point on the grade or floor below (see Commentary Figure R312.1.1).

Insect screening lacks sufficient strength to prevent someone from falling under a top rail. For this reason a guard is required for porches and decks enclosed with insect screening where the walking surface is located more than 30 inches (762 mm) above a floor or grade below.

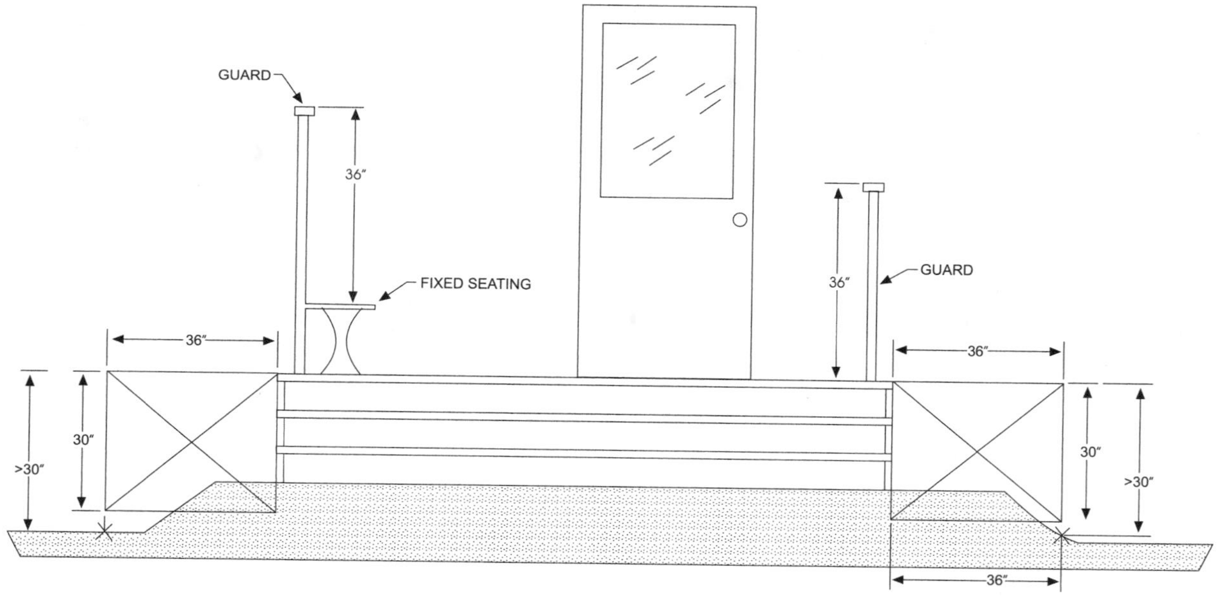

For SI: 1 inch = 25.4 mm.

Figure R312.1.1
DROP OFF AND GUARD HEIGHT REQUIREMENTS

R312.1.2 Height. Required *guards* at open-sided walking surfaces, including stairs, porches, balconies or landings, shall be not less than 36 inches (914 mm) high measured vertically above the adjacent walking surface, adjacent fixed seating or the line connecting the leading edges of the treads.

Exceptions:

1. *Guards* on the open sides of stairs shall have a height not less than 34 inches (864 mm) measured vertically from a line connecting the leading edges of the treads.
2. Where the top of the *guard* also serves as a handrail on the open sides of stairs, the top of the *guard* shall not be less than 34 inches (864 mm) and not more than 38 inches (965 mm) measured vertically from a line connecting the leading edges of the treads.

❖Where guards are required by Section R312.1.1, Section 312.1.2 specified a minimum height for those guards. The code provides for guards at open sides along walking surfaces and gives examples, but this list is not to be considered all inclusive. Required guards must be of an adequate height to minimize someone from falling off the edge of the walking surface. Therefore, the code establishes 36 inches (914 mm) as the minimum acceptable height for most walking surfaces. However, Exceptions 1 and 2 recognize that the minimum height for handrails along stairways is 34 inches (864 mm), therefore, there is a special allowance at the top of the guard along stairways that is consistent with the height of handrails.

Guard heights are determined by measuring vertically from the walking surface or the line connecting the nosings of the treads on stairways; however, when fixed seating is adjacent to a guard the height of the guard is to be measured from the seat where children might be inclined to stand or walk. See Commentary Figures R312.1.1, R312.1.2(1) and R312.1.2(2) for examples of how this provision is applied.

R312.1.3 Opening limitations. Required *guards* shall not have openings from the walking surface to the required *guard* height which allow passage of a sphere 4 inches (102 mm) in diameter.

Exceptions:

1. The triangular openings at the open side of stair, formed by the riser, tread and bottom rail of a guard, shall not allow passage of a sphere 6 inches (153 mm) in diameter.
2. *Guards* on the open side of stairs shall not have openings which allow passage of a sphere $4^3/_8$ inches (111 mm) in diameter.

❖Guards must be constructed so they prohibit smaller occupants, such as children, from falling through them. To prohibit people from slipping through a guard, any required guard would need to have supports, spindles, intermediate rails or some type of ornamental pattern so that a 4-inch (102 mm) sphere cannot pass through it. This spacing was chosen based on the head size and the chest depth of a child who had not yet developed an ability to crawl. The code does allow two exceptions for this spacing requirement. A $4^3/_8$-inch (111 mm) sphere rule is used for the guard on the open side of stair treads. This minor difference of just $^3/_8$ inch (9.5 mm) allows the use of just two balusters at each tread greatly reducing costs with no limitation of safety. A 6-inch (152 mm) sphere rule is used for the triangular area formed by the riser, tread and bottom rail of a guard along the open side of a stair because the triangular shape is more restrictive (see Commentary Figure R312.1.3).

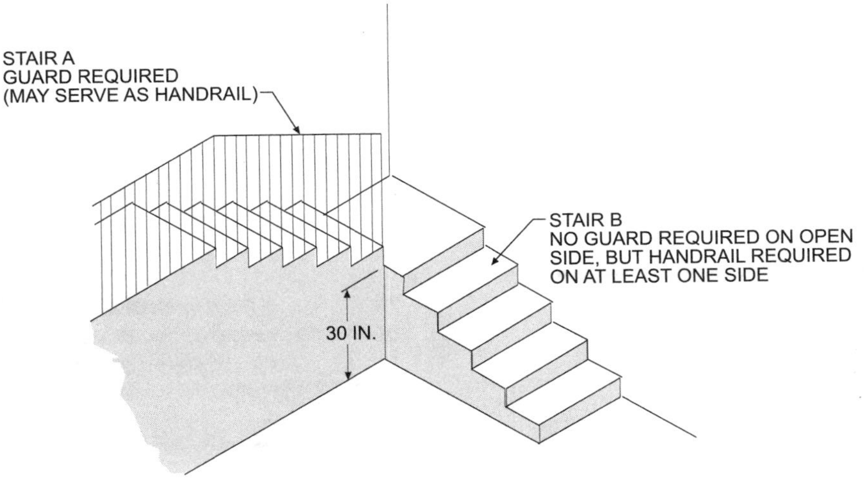

For SI: 1 inch = 25.4 mm.

Figure R312.1.2(1)
STAIRWAY GUARD

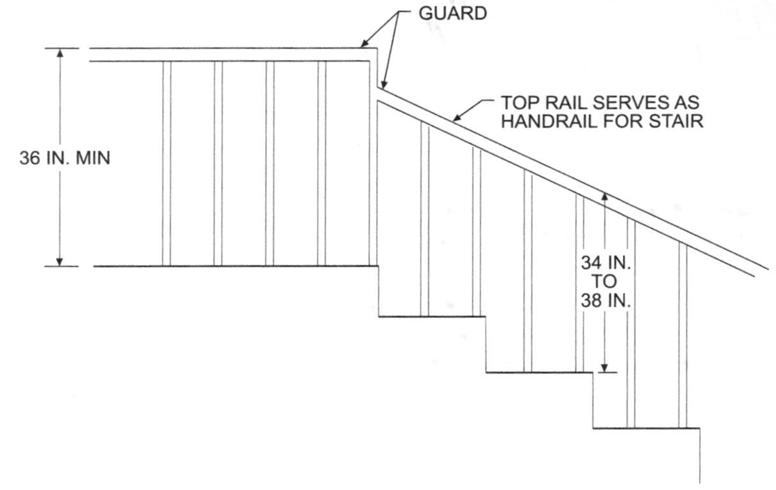

For SI: 1 inch = 25.4 mm.

Figure R312.1.2(2)
STAIRWAY GUARD

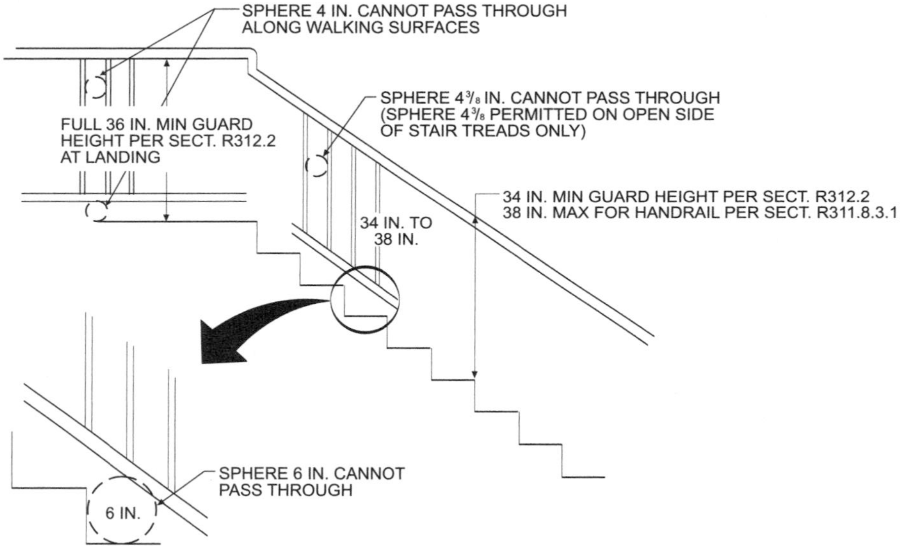

For SI: 1 inch = 25.4 mm.

Figure R312.1.3
GUARD REQUIREMENTS

R312.1.4 Exterior woodplastic composite guards. Wood-plastic composite *guards* shall comply with the provisions of Section R317.4.

❖Guards made of wood/plastic composite materials must meet the requirements for installation, labeling and compliance with ASTM D 7032 stated in Section R317.4, in addition to the general requirements for guards in this section.

R312.2 Window fall protection. Window fall protection shall be provided in accordance with Sections R312.2.1 and R312.2.2.

❖This section is not applicable to fixed or stationary windows. If any part of the clear opening area of an operable window is located more than 72 inches (1829 mm) above the finished grade, Section R312.2.1 requires that the lowest part of the clear opening be at least 24 inches (610 mm) above the floor surface of the room in which it is located. Windows may be located less than 24 inches (610 mm) above the interior floor surface only if they meet any one of the following criteria: 1) are fixed, 2) are located 72 inches (1829 mm) or less above grade, 3) have openings which will not allow passage of a 4-inch-diameter (102 mm) sphere, 4) are equipped with a window fall prevention device in accordance with ASTM F 2090, or 5) are equipped with opening control devices in accordance with Section R312.2.2 (see Commentary Figure R312.2.1).

The intent of these provisions is to prevent small children from falling out of open windows. The exceptions to Section R312.2.1 provide alternatives for fall prevention when the sill is lower than 24 inches (610 mm) above the floor—installing a barrier or limiting the dimensions of the window opening. The first exception permits installation of a window that is manufactured such that, when opened, it does not allow a 4-inch-diameter (102 mm) sphere to pass through. This option is not permitted in a location requiring an emergency escape and rescue opening. Window opening control devices, in accordance with Section R312.2.2 on the other hand, are now specifically approved for emergency escape and rescue windows when they meet ASTM F 2090. An opening control device installed on any window must have an emergency release device that is clearly identified and that operates without the need for a key, tool or special knowledge. These operation criteria match the language in the provisions for emergency escape and rescue openings. The other option for windows with sills lower than 24 inches (610 mm) above the floor is to provide a barrier at the window opening that does not permit passage of a 4-inch-diameter (102 mm) sphere.

The code references ASTM F 2090, *Window Fall Prevention Devices with Emergency Escape (Egress) Release Mechanisms* for the device requirements. The standard requires window fall prevention devices to be constructed such that a 4-inch-diameter (102 mm) sphere cannot pass through. Window fall prevention devices installed on any window must conform to ASTM F 2090, thereby complying with the operation provisions for emergency escape and rescue openings in Section R310.

R312.2.1 Window sills. In dwelling units, where the opening of an operable window is located more than 72 inches (1829 mm) above the finished grade or surface below, the lowest part of the clear opening of the window shall be a minimum of 24 inches (610 mm) above the fininshed floor of the room in which the window is located. Operable sections of windows shall not permit openings that allow passage of a 4-inch-diameter (102 mm) sphere where such openings are located within 24 inches (610 mm) of the finished floor.

Exceptions:

1. Windows whose openings will not allow a 4-inch-diameter (102 mm) sphere to pass through the opening when the opening is in its largest opened position.
2. Openings that are provided with window fall prevention devices that comply with ASTM F 2090.
3. Windows that are provided with window opening control devices that comply with Section R312.2.2.

❖See the commentary for Section R312.2.

R312.2.2 Window opening control devices. Window opening control devices shall comply with ASTM F 2090. The window opening control device, after operation to release the control device allowing the window to fully open, shall not reduce the minimum net clear opening area of the window unit to less than the area required by Section R310.1.1.

❖See the commentary for Section R312.2.

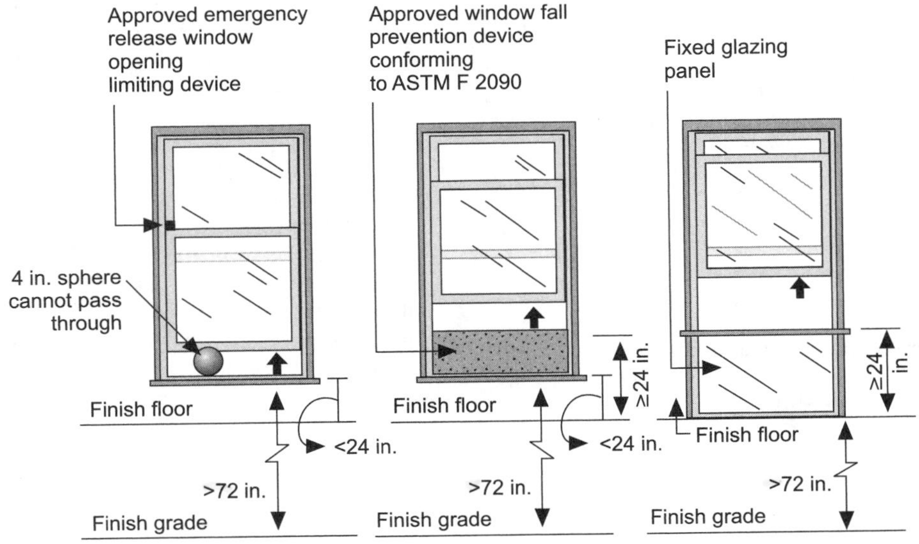

For SI: 1 inch = 25.4 mm.

Figure R312.2.1
WINDOW SILL HEIGHT

SECTION R313
AUTOMATIC FIRE SPRINKLER SYSTEMS

R313.1 Townhouse automatic fire sprinkler systems. An automatic residential fire sprinkler system shall be installed in *townhouses*.

> **Exception:** An automatic residential fire sprinkler system shall not be required when *additions* or *alterations* are made to existing *townhouses* that do not have an automatic residential fire sprinkler system installed.

❖Residential occupancies are the number one group of occupancies that suffer loss of life during fire events. A published study by the National Institute of Standards and Technology (NIST) entitled *Benefit-cost Analysis of Residential Fire Sprinkler Systems*, reports that, out of almost 2,000 fire incidents in homes equipped with fire sprinklers during the 4-year period 2002 to 2005, there were no fire-related fatalities. This statistic demonstrates the potential for sprinklers to save lives that would otherwise be lost in residential fires.

Since installation of a sprinkler system could be extensive in an existing dwelling unit, sprinkler systems are not required when an existing dwelling unit is being altered or has an addition. If the dwelling unit already has a sprinkler system, that system must be altered or added to as appropriate.

R313.1.1 Design and installation. Automatic residential fire sprinkler systems for *townhouses* shall be designed and installed in accordance with Section P2904.

❖While not stated in this section, Section P2904.1 allows for a designer to use either Section P2904 or NFPA 13D requirements for the design of a sprinkler system in a townhouse. Section P2904 includes requirements considered to provide an equivalent level of protection as an NFPA 13D sprinkler system. Section P2904 provides criteria for sprinklers, sprinkler piping, water supply, pipe sizing, instructions and inspection.

R313.2 One- and two-family dwellings automatic fire systems. An automatic residential fire sprinkler system shall be installed in one- and two-family *dwellings*.

> **Exception:** An automatic residential fire sprinkler system shall not be required for *additions* or *alterations* to existing buildings that are not already provided with an automatic residential sprinkler system.

❖See the commentary for Section R313.1.

R313.2.1 Design and installation. Automatic residential fire sprinkler systems shall be designed and installed in accordance with Section P2904 or NFPA 13D.

❖This section and Section P2904.1 allows for a designer to use either Section P2904 or NFPA 13D requirements for the design of a sprinkler system in a single-family home or duplex. Section P2904 includes requirements considered to provide an equivalent level of protection as an NFPA 13D sprinkler system. Section P2904 provides criteria for sprinklers, sprinkler piping, water supply, pipe sizing, instructions and inspection.

SECTION R314
SMOKE ALARMS

R314.1 Smoke detection and notification. All smoke alarms shall be listed and labeled in accordance with UL 217 and installed in accordance with the provisions of this code and the household fire warning *equipment* provisions of NFPA 72.

❖Section R314 provides the details of smoke detection and notification to alert occupants of potential problems. When asleep, the occupants of residential buildings will usually be unaware of a fire, and the fire will have an opportunity to spread before being detected. A majority of fire deaths occurring in residential buildings have occurred because of this delay in detection. It is for this reason that the code requires smoke alarms.

R314.2 Smoke detection systems. Household fire alarm systems installed in accordance with NFPA 72 that include smoke alarms, or a combination of smoke detector and audible notification device installed as required by this section for smoke alarms, shall be permitted. The household fire alarm system shall provide the same level of smoke detection and alarm as required by this section for smoke alarms. Where a household fire warning system is installed using a combination of smoke detector and audible notification device(s), it shall become a permanent fixture of the occupancy and owned by the homeowner. The system shall be monitored by an *approved* supervising station and be maintained in accordance with NFPA 72.

> **Exception:** Where smoke alarms are provided meeting the requirements of Section R314.4.

❖This detection and notification system provides early warning to occupants of the building in the event of a fire, thereby providing a greater opportunity for everyone in the building to evacuate or relocate to a safe area. Of all of the provisions for safety features that have been placed within the code over the past few decades, the provisions for these detection and alarm devices have probably offered the greatest benefit in increasing safety and reducing the loss of life when compared to their minor expense.

Requiring the system to become a permanent fixture of the occupancy and not be leased will prevent the system from being removed due to nonpayment. "Owned by the homeowner" is a good beginning and adds additional language that will ensure system reliability by requiring the owner to have the system electronically monitored and maintained in accordance with the referenced standard.

For larger homes, the only possible way to provide detection is through the use of a household fire warning system. NFPA 72, *National Fire Alarm Code*, has limits as to the number of smoke alarms that may be interconnected. Section 11.8.2.2 of the 2006 edition of the NFPA 72 allows only 12 smoke alarms to be interconnected if the interconnecting means is not supervised. Up to 42 smoke alarms may be interconnected if they are supervised. A number of homeowners prefer that their household fire warning systems be monitored by a supervising station. The listing of UL 217 smoke alarms prohibits them from being monitored.

R314.3 Location. Smoke alarms shall be installed in the following locations:

1. In each sleeping room.
2. Outside each separate sleeping area in the immediate vicinity of the bedrooms.
3. On each additional *story* of the *dwelling*, including *basements* and habitable attics but not including crawl spaces and uninhabitable *attics*. In *dwellings* or *dwelling units* with split levels and without an intervening door between the adjacent levels, a smoke alarm installed on the upper level shall suffice for the adjacent lower level provided that the lower level is less than one full *story* below the upper level.

❖So that all areas have at least some level of protection, and so that sleeping areas are adequately protected, Section R314.3 specifies where the devices are to be installed. The code requires that alarms be located within each sleeping room, as well as outside each separate sleeping area in the immediate vicinity of the bedrooms. The device within the bedroom will provide protection should the fire begin within that sleeping room, while the device outside of the room will provide early notification and protection should a problem develop in the area that generally will serve as the egress path for the bedroom. In addition, Item 3 will require installation of at least one smoke alarm on each story of the dwelling, including basements and habitable attics. The code does not require the installation of alarms within crawl spaces or within attics that are not habitable. This provides detection and notification within the areas of general occupancy, but ignores spaces that are not occupied. See Commentary Figure R314.3(1) for an illustration of the required alarm locations.

Where split levels occur in a dwelling and the adjacent levels openly communicate with each other, the alarm may be placed on the upper portion of the split level if it is not more than one full story different in elevation. Commentary Figure R314.3(2) is an example of this provision. This requirement is based on the fact that any fire initiating on the lower portion of the level will send products of combustion up to the upper portion and that a detector there will provide a quick response and early warning.

To assure that the audible alarm notification is loud enough to alert the occupants of any problem within the unit, the code requires that two or more smoke alarms be interconnected so that if one device is activated, all alarms within the dwelling unit will be activated. This interconnection is required so that no matter where the smoke first develops or is detected, occupants throughout the unit will be made aware of the situation. One of the main concerns of the code is occupants who may be asleep and unaware of any developing fire. The code requires that the alarm signal be "clearly audible" in the bedroom area. If smoke alarms are being installed in an existing building, see the commentary to Section R314.3.1, which contains an exception to the requirement for interconnection.

R314.3.1 Alterations, repairs and additions. When *alterations*, repairs or *additions* requiring a *permit* occur, or when one or more sleeping rooms are added or created in existing *dwellings*, the individual *dwelling unit* shall be equipped with smoke alarms located as required for new *dwellings*.

Exceptions:

1. Work involving the exterior surfaces of *dwellings*, such as the replacement of roofing or siding, or the *addition* or replacement of windows or doors, or the *addition* of a porch or deck, are exempt from the requirements of this section.

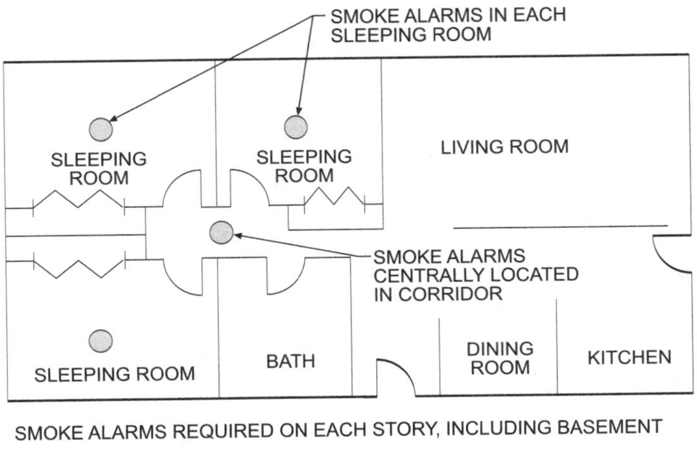

Figure R314.3(1)
LOCATION OF SMOKE ALARMS

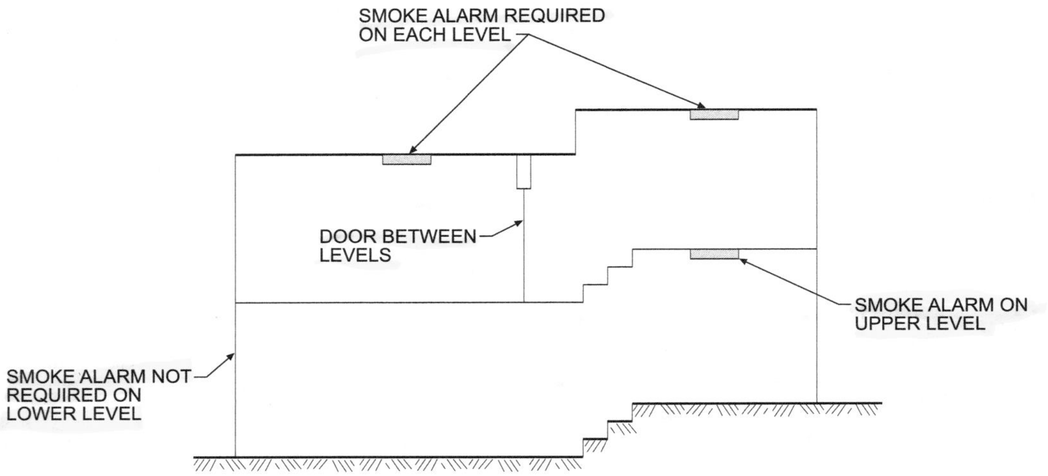

Figure R314.3(2)
SMOKE ALARM LOCATION

2. Installation, *alteration* or repairs of plumbing or mechanical systems are exempt from the requirements of this section.

❖This section contains a unique provision in the code, applying the smoke alarm provisions to existing buildings when an addition, alteration or repair is made that will require a permit, or if any sleeping rooms are added or created. See the commentary to Section R105.2 regarding what types of repairs or alterations require a permit. The smoke alarms in these existing buildings are to be installed in the same manner as required for new dwellings. This would not only require their installation in the same locations within the dwelling, but also that they be interconnected and receive their power from the building wiring. The commentary to Section R314.4 contains more discussion of the power source.

Two exceptions provide relief from the normal smoke alarm requirements in existing buildings that undergo some types of alteration, repair or addition.

The first exception exempts "exterior surface" repairs from initiating the requirement for smoke alarms being placed in an existing dwelling. This exception exempts work that is done on the exterior only. The final determination of what type of work is included is left to the building official, but this would generally be viewed as covering reroofing, siding repairs or siding replacement and could possibly include some window replacements.

The second exception exempts alterations that involve the replacement or repair of plumbing or mechanical equipment, fixtures or systems. This exception would allow replacement of items such as a furnace without expanding the project to include smoke alarms.

R314.4 Power source. Smoke alarms shall receive their primary power from the building wiring when such wiring is served from a commercial source, and when primary power is interrupted, shall receive power from a battery. Wiring shall be permanent and without a disconnecting switch other than those required for overcurrent protection.

Exceptions:

1. Smoke alarms shall be permitted to be battery operated when installed in buildings without commercial power.
2. Hard wiring of smoke alarms in existing areas shall not be required where the *alterations* or repairs do not result in the removal of interior wall or ceiling finishes exposing the structure, unless there is an *attic*, crawl space or *basement* available which could provide access for hard wiring without the removal of interior finishes.

❖Smoke alarms must use AC power as their primary source and battery power as a secondary source to enhance their reliability. For example, during a power outage, the probability of fire is increased because of the use of candles or lanterns for temporary light. Required backup battery power provides for continued performance of the smoke alarms. Smoke alarms are commonly designed to emit a recurring signal when batteries are low and need to be replaced. It is also for the reliability issue that the code does not permit the alarms to be on any type of circuit that could be disconnected or turned off, such as a lighting circuit with a switch. The only way to disconnect power to the smoke alarms should be through the electrical panel box by either flipping a circuit breaker or removing the circuit's fuse.

The exceptions acknowledge that the code does not require that smoke alarms in all existing buildings be served from a commercial power source. Battery-operated smoke alarms may be the only power source when a commercial power source is not available or when extensive alterations or repairs are not being made. Where permanent building wiring can be installed without the removal of interior finishes, this section recognizes the increased reliability that the "hardwired" commercial power source with battery back-up can provide. Therefore, where feasible, permanent wiring is to be installed.

R314.5 Interconnection. Where more than one smoke alarm is required to be installed within an individual dwelling unit in accordance with Section R314.3, the alarm devices shall be interconnected in such a manner that the actuation of one alarm will activate all of the alarms in the individual unit. Physical interconnection of smoke alarms shall not be required where listed wireless alarms are installed and all alarms sound upon activation of one alarm.

Exception: Interconnection of smoke alarms in existing areas shall not be required where alterations or repairs do not result in removal of interior wall or ceiling finishes exposing the structure, unless there is an attic, crawl space or basement available which could provide access for interconnection without the removal of interior finishes.

❖Smoke detectors within a dwelling are required to be interconnected so that activation of any of the smoke detectors on any level will alarm occupants. Section R314.5 will allow listed wireless interconnected alarms to substitute for wired interconnection of the smoke alarms in both new and existing construction.

SECTION R315 CARBON MONOXIDE ALARMS

R315.1 Carbon monoxide alarms. For new construction, an approved carbon monoxide alarm shall be installed outside of each separate sleeping area in the immediate vicinity of the bedrooms in *dwelling units* within which fuel-fired *appliances* are installed and in dwelling units that have attached garages.

❖Carbon monoxide (CO) is an odorless, colorless and toxic gas. Because it is impossible to see, taste or smell the toxic fumes, CO can kill occupants before they are aware it is in their home. At lower levels of exposure, CO causes mild effects that are often mistaken for the flu. These symptoms include headaches, dizziness, disorientation, nausea and fatigue. The effects of CO exposure can vary greatly from person to person depending on age, overall health, and the concentration and length of exposure. According to the Journal of the American Medical Association (JAMA), CO is the leading cause of accidental poisoning deaths in America.

Sources of CO include unvented kerosene and gas space heaters; leaking chimneys and furnaces; back-drafting from furnaces, gas water heaters, wood stoves and fireplaces; gas stoves; generators and other gasoline-powered equipment; and automobile exhaust from attached garages. Incomplete oxidation during combustion in gas ranges and unvented gas or kerosene heaters may cause high concentrations of CO in indoor air. Worn or poorly adjusted and maintained combustion devices (e.g., boilers, furnaces) can be significant sources, or if the flue is improperly sized, blocked, disconnected or is leaking. Auto, truck or bus exhaust from attached garages, nearby roads or parking areas can also be a source.

Section R315.1 requires that CO alarms be located outside each separate sleeping area in the immediate vicinity of the bedrooms. The carbon monoxide alarm can be located near the smoke alarm required in the immediate area of the bedrooms by Section R314.3. Section R315.1 does not require CO detectors to be hard wired or interconnected like smoke alarms (see Commentary Figure R315.1).

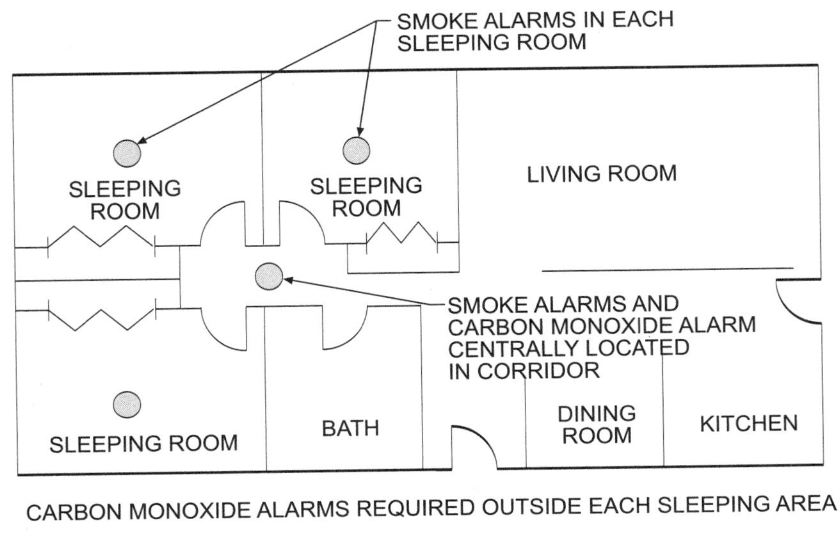

Figure R315.1
LOCATION OF CARBON MONOXIDE ALARMS

R315.2 Carbon monoxide detection systems. Carbon monoxide detection systems that include carbon monoxide detectors and audible notification appliances, installed and maintained in accordance with this section for carbon monoxide alarms and NFPA 720, shall be permitted. The carbon monoxide detectors shall be listed as complying with UL 2075. Where a household carbon monoxide detection system is installed, it shall become a permanent fixture of the occupancy, owned by the homeowner and shall be monitored by an approved supervising station.

Exception: Where carbon monoxide alarms are installed meeting the requirements of Section R315.1, compliance with Section 315.2 is not required.

❖This section permits carbon monoxide (CO) detection systems that include CO detectors and audible notification appliances to be installed. The performance and reliability of system-connected CO detectors have shown to be extremely high if they are listed and maintained to ANSI/UL 2075 and installed in accordance with NFPA 720. System-connected CO detectors designed to be part of a carbon monoxide detection system are required to be connected to an approved panel. The panel is required to be equipped with rechargeable batteries that keep the carbon monoxide detection system operating during a power outage and will communicate the power loss condition to the supervising station. When the primary power is restored, the control panel will fully recharge the standby batteries. An added feature of a carbon monoxide detection system is that the interconnecting wiring to system-connected CO detectors is supervised such that a wiring fault results in a trouble signal at the premises and the supervising station. Section 9.6.5 of NFPA 720 requires that when two or more carbon monoxide alarms are to be installed that they are interconnected. The rationale for this requirement is if a CO device is activated in the basement, the occupants on the second floor on the opposite end of the home are unable to hear the audible alarm if the devices are not interconnected. NFPA 720 requires CO devices to be installed on every level of a dwelling unit, including basements as well as outside each separate dwelling unit sleeping area in the immediate vicinity of the bedrooms.

The requirement for the household carbon monoxide detection system to be owned by the homeowner and to be monitored by an approved supervising station mirrors the household fire alarm system requirements in Section R314.2. The exception permits CO alarms to be provided in accordance with Section R315.1 in lieu of a CO detection system complying with Section R315.2.

R315.3 Where required in existing dwellings. Where work requiring a *permit* occurs in existing *dwellings* that have attached garages or in existing dwellings within which fuel-fired *appliances* exist, carbon monoxide alarms shall be provided in accordance with Section R315.1.

❖When any work requiring a permit is performed to existing townhouses, single-family homes or duplexes, CO alarms must be installed.

R315.4 Alarm requirements. Single-station carbon monoxide alarms shall be listed as complying with UL 2034 and shall be installed in accordance with this code and the manufacturer's installation instructions.

❖Average levels of CO in homes without gas stoves vary from 0.5 to 5 parts per million (ppm). Levels near properly adjusted gas stoves are often 5 to 15 ppm and those near poorly adjusted stoves may be 30 ppm or higher.

CO detectors must not be considered a replacement for the proper use and maintenance of fuel-burning appliances. The industry has addressed the issue of reliability by updating the requirements of UL 2034. All CO detectors available today meet the update requirements, which eliminated the false positive indications that occurred when CO detectors were first brought to market in the 1990s.

SECTION R316
FOAM PLASTIC

R316.1 General. The provisions of this section shall govern the materials, design, application, construction and installation of foam plastic materials.

❖Section R316 covers several topics related to the use and installation of various types of foam plastic materials used for insulation, trim and finishes. These requirements cover the acceptable uses of this combustible product and the associated protection needed to use it in building construction.

Section R316.1 lists the two basic issues that serve as the basis for foam plastic insulation requirements: the flame spread rating of the material and the separation of the foam plastic insulation from the interior of the building.

Commentary Figure R316.1 shows a simple flowchart to help the code user more easily comply with the foam plastics sections of the code.

R316.2 Labeling and identification. Packages and containers of foam plastic insulation and foam plastic insulation components delivered to the job site shall bear the *label* of an *approved agency* showing the manufacturer's name, the product listing, product identification and information sufficient to determine that the end use will comply with the requirements.

❖Foam plastics or packages of foam plastics delivered to the construction site must be labeled. Also, labels are required on containers [usually two components in 55-gallon (208 L) drums] of ingredients delivered for the production of foam plastic at the construction site. The label should include the name of the manufacturer or distributor, the type of foam plastic, the performance characteristics required to show code compliance and the name of the approved testing agency. The label may reference documents, such as

ICC-ES reports, approval agency certificates and other information, that can be used to determine code-required performance characteristics. [Note, although not required by the code, the Federal Trade Commission (FTC) also places specific labeling requirements regarding the insulation power or *R*-value on all insulations, including foam plastic insulation, used in residential applications.]

R316.3 Surface burning characteristics. Unless otherwise allowed in Section R316.5 or R316.6, all foam plastic or foam plastic cores used as a component in manufactured assemblies used in building construction shall have a flame spread index of not more than 75 and shall have a smoke-developed index of not more than 450 when tested in the maximum thickness intended for use in accordance with ASTM E 84 or UL 723. Loose-fill-type foam plastic insulation shall be tested as board stock for the flame spread index and smoke-developed index.

Exception: Foam plastic insulation more than 4 inches (102 mm) thick shall have a maximum flame spread index of 75 and a smoke-developed index of 450 where tested at a minimum thickness of 4 inches (102 mm), provided the end use is *approved* in accordance with Section R316.6 using the thickness and density intended for use.

❖Unless otherwise allowed in Section R316.5 or R316.6, foam plastic or foam plastic cores used as a component in manufactured assemblies used in building construction must have a flame spread index

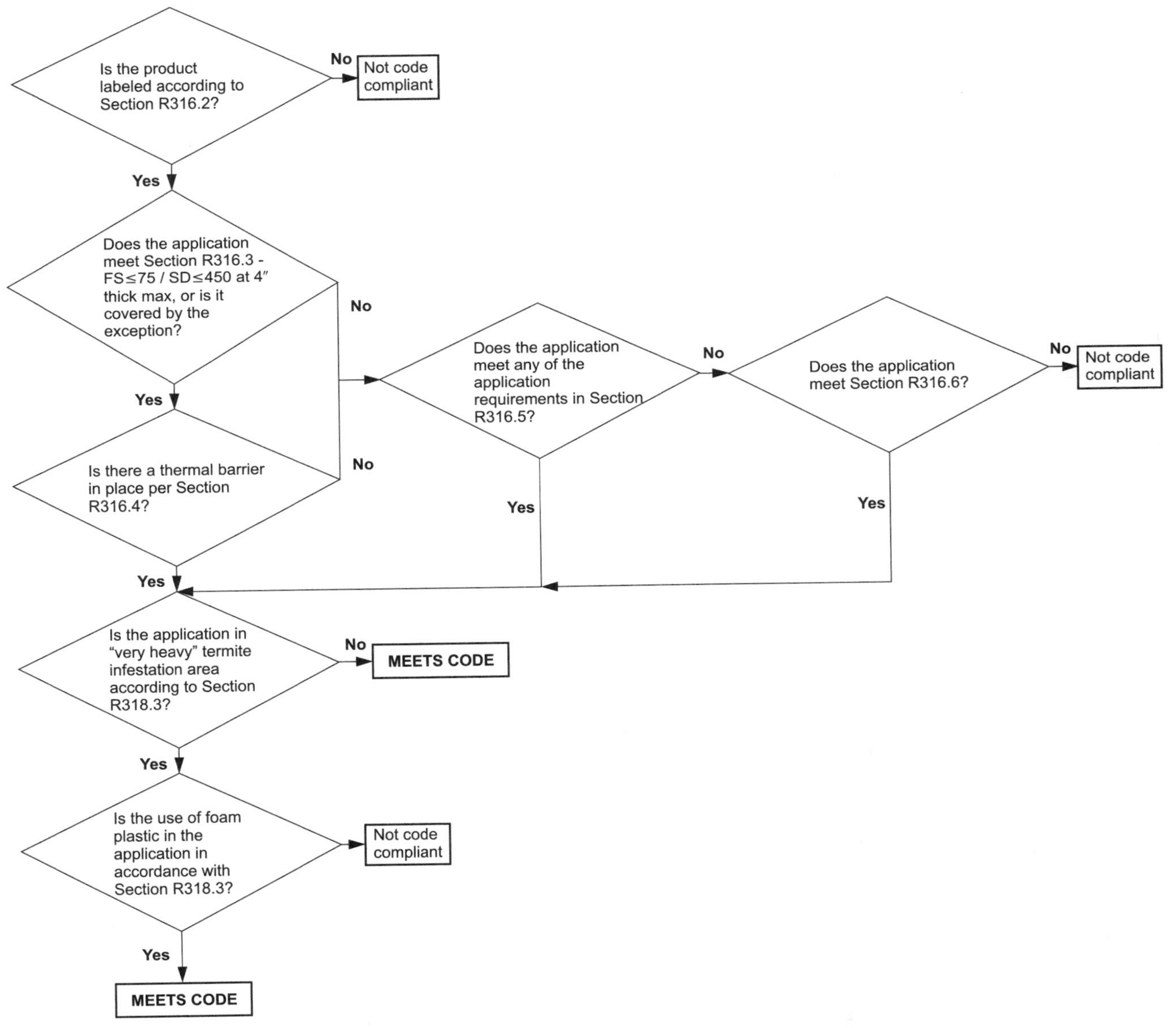

For SI: 1 inch = 25.4 mm.

Figure R316.1
USE AND INSTALLATION REQUIREMENTS OF FOAM PLASTIC IN BUILDING CONSTRUCTION

of not more than 75 and a smoke-developed index of not more than 450 when tested in the maximum thickness intended for use in accordance with ASTM E 84 or UL 723. Loose-fill-type foam plastic insulation must be tested as board stock for the flame spread index and smoke-developed index.

Foam plastic insulation or foam plastic cores used as a component in a manufactured assembly are combustible and must be assessed for flame spread index (FS) and smoke-developed indexes. Foam plastic materials must be tested in accordance with test method ASTM E 84 or UL 723, unless they are:

1. Specifically exempted in accordance with Section R316.3;
2. One of several applications listed in Section R316.5 (i.e., roofing, foam-filled exterior doors, foam, filled garage doors, interior trim, interior finish); or
3. Tested in an application that has been approved through Section R316.6 (testing under actual end-use configurations).

When testing in accordance with ASTM E 84 or UL 723 is required, the materials must be tested in the maximum thickness to be used [up to 4-inch (102 mm) thickness] with flame spread index results less than 75 and smoke-developed index results less than 450 SD, unless otherwise specified. If thicker foam is going to be used, it must still be tested to ASTM E 84 or UL 723 at 4-inch (102 mm) thickness with flame spread index results less than 75 and smoke-developed index results less than 450 results, as well as testing in accordance with Section R316.6 that is done at the actual foam thickness and density. If the material is loose-fill-type foam plastic, it must be tested as board stock for flame spread and smoke-development performances.

The maximum flame spread value of 75 was chosen on the basis that it was lower than untreated wood (which usually was 100 to 165). The maximum smoke-developed rating of 450 was selected because, at the time, the code permitted interior finish materials that gave off "smoke no more dense than that given off by untreated wood." In selecting the maximum flame spread and smoke developed values, it was believed that a conservative approach was being taken by requiring an insulation material to meet the same requirements as interior finish, even though the insulation was intended to be covered with an interior finish material. The requirements for surface-burning characteristics of foam plastic apply to foam plastics used as cores of manufactured assemblies. The intent is that, even though the finished assemblies might or might not require testing for surface-burning characteristics, the foam plastic core is not exempt from the general requirement; therefore, foam plastic is regulated in factory-manufactured assemblies the same as it is in field-fabricated applications.

R316.4 Thermal barrier. Unless otherwise allowed in Section R316.5 or Section R316.6, foam plastic shall be separated from the interior of a building by an *approved* thermal barrier of minimum $^{1}/_{2}$ inch (12.7 mm) gypsum wallboard or a material that is tested in accordance with and meets the acceptance criteria of both the Temperature Transmission Fire Test and the Integrity Fire Test of NFPA 275.

❖The use of an approved thermal barrier to separate foam plastics from the interior of a building is a basic requirement for the use of foam plastic as shown in this section of the code. The job of a thermal barrier is to isolate the foam plastic. An approved thermal barrier is defined as minimum $^{1}/_{2}$-inch (12.7 mm) gypsum wallboard or a material tested to NFPA 275. Before 1975, experience had shown that foam plastics covered with plaster or $^{1}/_{2}$-inch (12.7 mm) gypsum wallboard had performed satisfactorily in building fires. For this reason, $^{1}/_{2}$-inch (12.7 mm) gypsum wallboard was included in the code as a minimum requirement. It is recognized that specifying a single material is not desirable in a performance code; therefore, a material tested to NFPA 275 is permitted.

This section sets forth the test methods and performance criteria by which alternative thermal barriers are to be qualified. NFPA 275 *Standard Method of Fire Tests for the Evaluation of Thermal Barriers Used Over Foam Plastic Insulation* was developed to specifically address the testing of materials to qualify as a thermal barrier. The test method provides specific sample construction, fire exposures and acceptance criteria to qualify a material to be a 15-minute thermal barrier. The test methods address both the capability of the material to retard heat transfer via a fire-resistance test and to remain in place via a full-scale fire test.

Sections R316.5 and R316.6 describe circumstances where the requirement for a thermal barrier is modified or eliminated.

R316.5 Specific requirements. The following requirements shall apply to these uses of foam plastic unless specifically *approved* in accordance with Section R316.6 or by other sections of the code or the requirements of Sections R316.2 through R316.4 have been met.

❖This prescriptive section can be used as another path to code compliance for foam plastics. As the flowchart in the commentary to Section R316.1 points out, if an application, including any listed below, meets the requirements in Sections R316.2, R316.3, R316.4 and R316.7, or Sections R316.2, R316.6 and R316.7, that application is code compliant and the requirements spelled out below do not apply. It is only when the requirements of Section R316.3 or R316.4 are not met that the applications spelled out below can be used to show code compliance. Many of the applications below modify or remove the flame spread and smoke-developed requirements of Section R316.3 or modify or remove the need for the thermal barrier specified in Section R316.4.

Two applications that have caused confusion in the past are foam backer board and foam insulation used in residential applications. These two applications, by definition, are examples of foam plastic insulation used on the exterior of a wall assembly. If the foam plastic being used meets the requirements of Section R316.3 (requiring a flame spread index less that 75 and a smoke-developed index less than 450) and Section R316.4 [thermal barrier of $^1/_2$-inch (12.7 mm) gypsum board or an equivalent on the interior of the wall], the foam plastic insulation can be used up to the allowed thickness of 4 inches (102 mm), and Sections R316.5.7 and R316.5.8 do not apply.

R316.5.1 Masonry or concrete construction. The thermal barrier specified in Section R316.4 is not required in a masonry or concrete wall, floor or roof when the foam plastic insulation is separated from the interior of the building by a minimum 1-inch (25 mm) thickness of masonry or concrete.

❖No thermal barrier is required when 1 inch (25 mm) or more of masonry or concrete is placed between the foam plastic and the interior of the building. The intent is to accept 1-inch (25 mm) of masonry or concrete as adequate protection against ignition, even though the concrete does not necessarily meet the performance criteria for thermal barriers. This condition can arise when foam plastics are installed either within a wall or on one side of a wall. Some common examples are when foam plastics are installed:

- In the cavity of a hollow masonry wall,
- As the core of a concrete-faced panel,
- On the exterior face of a masonry wall and covered with an exterior finish, or
- Within the cores of hollow masonry units.

Encapsulated within a minimum of 1 inch (25 mm) concrete or masonry wall, floor or roof system, as in insulated tilt-up or pour-in-place concrete panels (see Commentary Figure R316.5.1).

R316.5.2 Roofing. The thermal barrier specified in Section R316.4 is not required when the foam plastic in a roof assembly or under a roof covering is installed in accordance with the code and the manufacturer's installation instructions and is separated from the interior of the building by tongue-and-groove wood planks or wood structural panel sheathing in accordance with Section R803, not less than $^{15}/_{32}$ inch (11.9 mm) thick bonded with exterior glue and identified as Exposure 1, with edges supported by blocking or tongue-and-groove joints or an equivalent material. The smoke-developed index for roof applications shall not be limited.

❖No thermal barrier is required when a foam plastic is incorporated into a roof assembly on the exterior side, over tongue-and-groove wood planks or wood structural panel sheathing, if the wood product meets all of the following:

- Used in accordance with Section R803,
- Identified as Exposure 1,
- Manufactured with exterior grade glue,
- Minimum $^{15}/_{32}$ inch thick (12 mm),
- Installed according to manufacturers instructions and
- Installed to provide adequate edge support (blocking when edges do not occur over framing members, tongue-and-groove joints or equivalent).

Also, the flame spread rating of the foam plastic used must comply with the requirements of Section R316.3, but the smoke-developed rating of the foam plastic is not limited.

R316.5.3 Attics. The thermal barrier specified in Section R316.4 is not required where all of the following apply:

1. *Attic* access is required by Section R807.1.
2. The space is entered only for purposes of repairs or maintenance.
3. The foam plastic insulation is protected against ignition using one of the following ignition barrier materials:
   3.1. $1^1/_2$-inch-thick (38 mm) mineral fiber insulation;

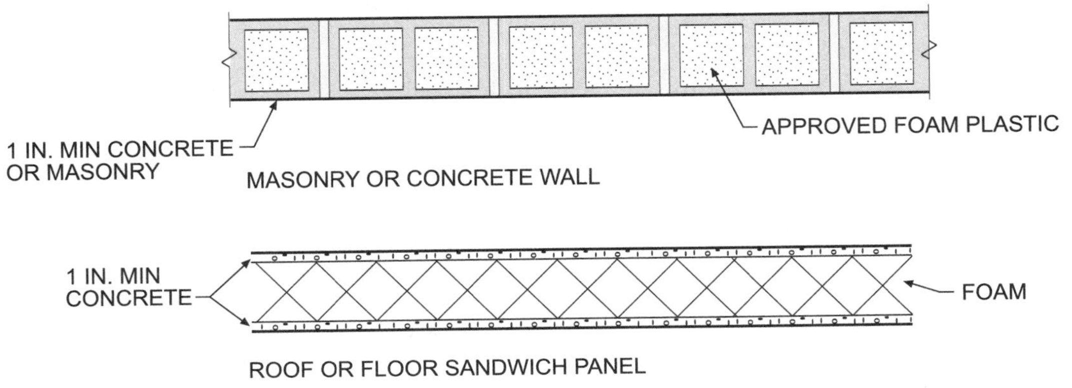

For SI: 1 inch =25.4 mm.

Figure R316.5.1
ENCAPSULATED FOAM PLASTIC

3.2. $^{1}/_{4}$-inch-thick (6.4 mm) wood structural panels;

3.3. $^{3}/_{8}$-inch (9.5 mm) particleboard;

3.4. $^{1}/_{4}$-inch (6.4 mm) hardboard;

3.5. $^{3}/_{8}$-inch (9.5 mm) gypsum board; or

3.6. Corrosion-resistant steel having a base metal thickness of 0.016 inch (0.406 mm);

3.7. $1^{1}/_{2}$-inch-thick (38 mm) cellulose insulation.

The above ignition barrier is not required where the foam plastic insulation has been tested in accordance with Section R316.6.

❖In an attic where access is required by Section R807.1 [where attic areas exceed 30 square feet (2.8 m^2) and have a vertical height of 30 inches (762 mm) or more], and entry is only for service of utilities, and when foam plastics are used, an ignition barrier may be used in place of a thermal barrier to cover the foam plastic. Multiple materials are listed that can be used as the ignition barrier (see Commentary Figure R316.5.3). The foam plastic material, covered with the ignition barrier can be on the floor, wall (often called a knee wall or gable end) or the ceiling of the attic. The phrase "purposes of repairs and maintenance" applies to attics that contain only mechanical equipment, electrical wiring, fans, plumbing, gas or electric hot water heaters, gas or electric furnaces, etc. The attic space cannot be used for storage. The reduced provision (from a thermal barrier to an ignition barrier) provides a barrier whose only purpose is to prevent the direct impingement of flame on the foam plastic insulation.

If the foam plastic insulation has passed testing, in the thickness and density intended for use, in accordance with Section R316.6, no thermal barrier or ignition barrier is required over the foam plastic insulation in an attic and this section of the code does not apply. It is important to note that the actual configuration must be tested. For example, a foam plastic insulation applied to the ceiling of the attic must be tested with the foam applied to the ceiling in a room corner test or in an assembly that reflects end use. The same restrictions would apply to those insulations applied to the walls, floors or combinations of surfaces.

R316.5.4 Crawl spaces. The thermal barrier specified in Section R316.4 is not required where all of the following apply:

1. Crawlspace access is required by Section R408.4
2. Entry is made only for purposes of repairs or maintenance.
3. The foam plastic insulation is protected against ignition using one of the following ignition barrier materials:

3.1. $1^{1}/_{2}$-inch-thick (38 mm) mineral fiber insulation;

3.2. $^{1}/_{4}$-inch-thick (6.4 mm) wood structural panels;

3.3. $^{3}/_{8}$-inch (9.5 mm) particleboard;

3.4. $^{1}/_{4}$-inch (6.4 mm) hardboard;

3.5. $^{3}/_{8}$-inch (9.5 mm) gypsum board; or

3.6. Corrosion-resistant steel having a base metal thickness of 0.016 inch (0.406 mm).

The above ignition barrier is not required where the foam plastic insulation has been tested in accordance with Section R316.6.

❖In a crawl space where access is required by Section R408.4 (access shall be provided to all under-floor spaces) and entry is only for service of utilities, and when foam plastics are used, an ignition barrier may be used in place of a thermal barrier to cover the foam plastic. Multiple materials are listed that can be

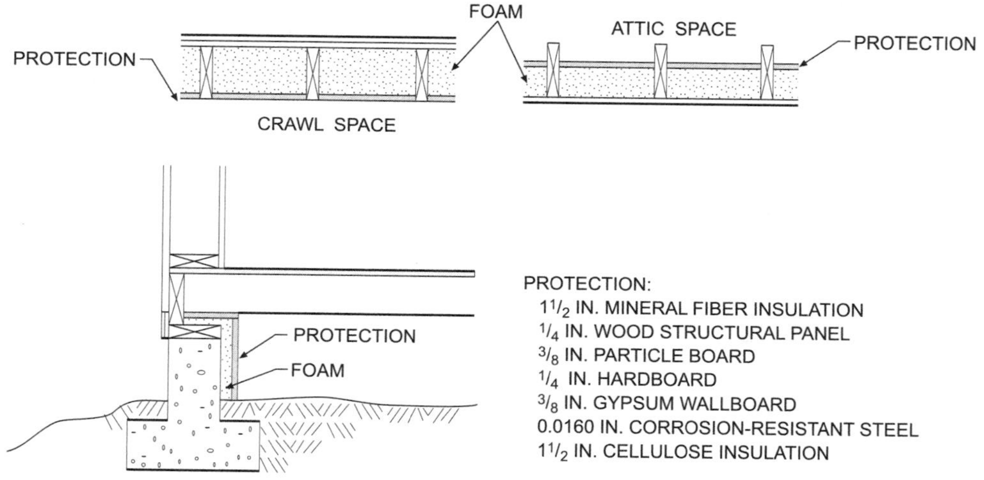

For SI: 1 inch = 25.4 mm.

Figure R316.5.3
FOAM PLASTIC, ATTIC AND CRAWL SPACES

used as the ignition barrier (see Commentary Figure R316.5.3). The foam plastic material, covered with the ignition barrier can be on the floor, wall or ceiling of the crawl space. The phrase "purposes of repairs or maintenance" applies to crawl spaces that contain only mechanical equipment, electrical wiring, fans, plumbing, gas, or electric hot water heaters, gas or electric furnaces, etc. The crawl space cannot be used for storage. The reduced requirement (from a thermal barrier to an ignition barrier) provides a barrier whose only purpose is to prevent the direct impingement of flame on the foam plastic.

If the foam plastic insulation has passed testing, in the thickness and density intended for use, in accordance with Section R316.6, no thermal barrier or ignition barrier is required over the foam plastic insulation in the crawl space and this section of the code does not apply. It is important to note that the actual configuration must be tested. For example, a foam plastic insulation applied to the ceiling of the crawl space must be tested with the foam applied to the ceiling in a room corner test or in an assembly that reflects end use. The same restrictions would apply to those insulations applied to the walls, floors or combinations of surfaces.

R316.5.5 Foam-filled exterior doors. Foam-filled exterior doors are exempt from the requirements of Sections R316.3 and R316.4.

❖No thermal barrier (see Section R316.4) or surface-burning characteristics testing (see Section R316.3) is required for foam-filled exterior doors.

R316.5.6 Foam-filled garage doors. Foam-filled garage doors in attached or detached garages are exempt from the requirements of Sections R316.3 and R316.4.

❖No thermal barrier (see Section R316.4) or surface-burning characteristics testing (see Section R316.3) is required for foam-filled garage doors in either an attached or detached garage.

R316.5.7 Foam backer board. The thermal barrier specified in Section R316.4 is not required where siding backer board foam plastic insulation has a maximum thickness of 0.5 inch (12.7 mm) and a potential heat of not more than 2000 Btu per square foot (22 720 kJ/m²) when tested in accordance with NFPA 259 provided that:

1. The foam plastic insulation is separated from the interior of the building by not less than 2 inches (51 mm) of mineral fiber insulation;
2. The foam plastic insulation is installed over existing *exterior wall* finish in conjunction with re-siding; or
3. The foam plastic insulation has been tested in accordance with Section R316.6.

❖The code contains a definition for "Foam backer board" (see Section R202). If these siding products are used on the exterior of a wall and the requirements of Sections R316.3 and R316.4 are met, this section of the code does not apply. If a thermal barrier is not used on the interior of the building, limitations are placed on the product and its use. In addition to the flame spread limitations of Section R316.3, other properties of the foam plastic portion of the product include maximum thickness of $^1/_2$ inch (12.7 mm) and potential heat of no more than 2000 British thermal units (Btu) per square foot (22 720 kJ/m²) when tested using NFPA 259. Limitation in siding/foam combination product use includes separation from the interior of the building by no less than 2 inches (51 mm) of mineral fiber insulation or installation over an existing wall finish as part of residing or the foam plastic insulation is tested in accordance with Section R316.6. The removal of the thermal barrier requirement in this section is reasonable considering the separation provided by the existing construction and the limitation of the potential heat of the foam plastic imposed by the code.

R316.5.8 Re-siding. The thermal barrier specified in Section R316.4 is not required where the foam plastic insulation is installed over existing *exterior wall* finish in conjunction with re-siding provided the foam plastic has a maximum thickness of 0.5 inch (12.7 mm) and a potential heat of not more than 2000 Btu per square foot (22 720 kJ/m²) when tested in accordance with NFPA 259.

❖Foam plastic is frequently used in residing applications to provide a leveling surface for new siding, while also bringing additional insulation value to the wall assembly. If these products are used in a wall assembly and the requirements of Sections R316.3 and R316.4 are met, this section of the code does not apply. If a thermal barrier is not used between the foam plastic and the interior of the building, the foam insulation must meet the flame spread requirements of Section R316.3, is limited to a maximum thickness of $^1/_2$ inch (12.7 mm) and potential heat of no less than 2,000 Btu per square foot (22 720 kJ/m²) when tested using NFPA 259. The removal of the thermal barrier requirement in this section is reasonable considering the separation provided by the existing construction and the limitation of the potential heat of the foam plastic imposed by the code.

R316.5.9 Interior trim. The thermal barrier specified in Section R316.4 is not required for exposed foam plastic interior trim, provided all of the following are met:

1. The minimum density is 20 pounds per cubic foot (320 kg/m³).
2. The maximum thickness of the trim is 0.5 inch (12.7 mm) and the maximum width is 8 inches (204 mm).
3. The interior trim shall not constitute more than 10 percent of the aggregate wall and ceiling area of any room or space.
4. The flame spread index does not exceed 75 when tested per ASTM E 84 or UL 723. The smoke-developed index is not limited.

❖Foam plastic interior trim is defined as exposed foam plastic used as picture molds, chair rails, crown moldings, baseboards, handrails, ceiling beams, door trim and window trim, and similar decorative or protective materials.

For a foam plastic to qualify as interior trim, each of the four criteria listed in this section must be met. Because foam plastic in this application is left exposed, these criteria limit its use.

1. The density of materials must be at least 20 pounds per cubic foot (pcf) (320 kg/m^3). The intent was to separate those materials used for trim from those intended for use as insulation. As a comparison, most foam plastic insulation is in the range of 1 to $2^1/_2$ pounds per cubic foot (16.02 to 40 kg/m^3) with very few materials over 5 pcf (81 kg/m^3).
2. Even though other nonfoam plastic trim materials are not limited in dimension, the maximum thickness and width of foam plastic trim is limited to $^1/_2$ inch (12.7 mm) and 4 inches (12.7 mm and 102 mm), respectively.
3. Foam plastic trim cannot constitute more than 10 percent of the aggregate area of the walls and ceiling of a room.
4. The flame spread index must not be higher than 75 when tested in accordance with ASTM E 84 or UL 723. The value of 75 was selected to be consistent with the requirement for foam plastic insulation, even though other materials used as trim are permitted to have flame spread indexes of up to 200 in many locations. The smoke-developed index is not regulated.

R316.5.10 Interior finish. Foam plastics shall be permitted as interior finish where *approved* in accordance with Section R316.6 Foam plastics that are used as interior finish shall also meet the flame spread index and smoke-developed index requirements of Sections R302.9.1 and R302.9.2.

❖Foam plastic used as interior finish must be approved through Section R316.6. This means that the foam plastic material has been tested to eliminate the thermal barrier in accordance with NFPA 286 and the acceptance criteria of Section R302.9.4, FM 4880, UL 723, UL 1040 or UL 1715, or fire tests related to actual end-use configurations (including the foam plastic thickness). The foam plastic must also meet the flame spread index requirements of Section R302.9 (flame spread index less than 200).

R316.5.11 Sill plates and headers. Foam plastic shall be permitted to be spray applied to a sill plate and header without the thermal barrier specified in Section R316.4 subject to all of the following:

1. The maximum thickness of the foam plastic shall be $3^1/_4$ inches (83 mm).
2. The density of the foam plastic shall be in the range of 0.5 to 2.0 pounds per cubic foot (8 to 32 kg/m^3).
3. The foam plastic shall have a flame spread index of 25 or less and an accompanying smoke-developed index of 450 or less when tested in accordance with ASTM E 84 or UL 723.

❖No thermal barrier is required when a foam plastic is spray applied to the sill plate and joist header when all of the conditions listed in Section R316.5.11 are met. Because foam plastic insulation in this application is left exposed, the three conditions listed [thickness less than $3^1/_4$ inches (82.6 mm), density 0.5 to 2.0 pcf (24 to 32 kg/m^3), flame spread index less than 25 and smoke-developed index less than 450] control the spray applied foam plastic used in this application (see Commentary Figure R316.5.11).

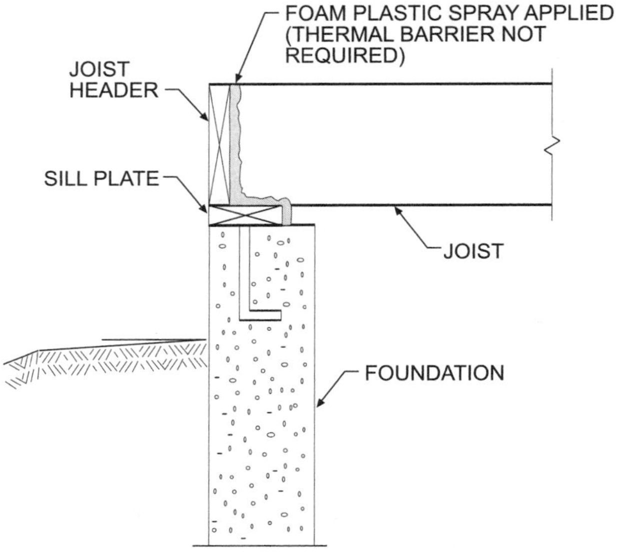

Figure R316.5.11
FOAM PLASTIC SPRAY APPLIED TO SILL PLATE AND HEADER

R316.5.12 Sheathing. Foam plastic insulation used as sheathing shall comply with Section R316.3 and Section R316.4. Where the foam plastic sheathing is exposed to the *attic* space at a gable or kneewall, the provisions of Section R316.5.3 shall apply.

❖Foam plastic used as sheathing is a very common application, adding insulation to the framing of the building. This section makes it clear that foam plastic insulation is code approved for use as a sheathing material. When used as a sheathing material, the foam plastic must meet the requirements of Sections R316.3 and R316.4. Often foam plastic sheathing is used on the outside of an exterior wall, continuously covering the wall. In this example, the living area of the building will have a thermal barrier, such as $^1/_2$-inch (12.7 mm) gypsum board, in place. That thermal barrier, combined with a foam plastic, which has met the surface burning requirements of less than 75 for a flame spread index and less than 450 for a smoke developed index gives a code compliant application.

The attic, in this example, will have foam plastic on the exterior side of the wall. When the foam plastic is exposed to the attic space at a gable or knee wall, the foam plastic insulation must meet the requirements of Section R316.5.3. For applications where foam plastic is used as sheathing, code approval follows the same paths outlined in Section R316.1.

If the foam plastic insulation has passed testing, in the thickness and density intended for use, in accordance with Section R316.6, no thermal barrier or ignition barrier is required over the foam plastic insulation and this section of the code does not apply. It is important to note that the actual configuration must be tested, including typical seams, joints and other details that will occur in the finished installation.

R316.5.13 Floors. The thermal barrier specified in Section R316.4 is not required to be installed on the walking surface of a structural floor system that contains foam plastic insulation when the foam plastic is covered by a minimum nominal $^1/_2$-inch-thick (12.7 mm) wood structural panel or equivalent. The thermal barrier specified in Section R316.4 is required on the underside of the structural floor system that contains foam plastic insulation when the underside of the structural floor system is exposed to the interior of the building.

❖In today's construction new types of products are being used which incorporate foam plastic insulation for energy reasons. One example is structural insulated panels where foam plastic is laminated between two structural wood facings. This type of panel can be used as a wall, floor or roof. Foam plastic is required to be protected by a thermal barrier which typically is $^1/_2$-inch (12.7 mm) gypsum wallboard. In the case of flooring, gypsum wallboard or other common thermal barrier materials cannot be used on the walking surfaces due to their friability, etc. Section R316.5.13 addresses this problem. The requirement for the $^1/_2$-inch-thick (12.7 mm) plywood or equivalent will provide sufficient protection to the foam plastic insulation. While $^1/_2$-inch (12.7 mm) plywood is not by itself a thermal barrier, in the case of a floor, it will provide sufficient protection since in the event of an interior fire, the floor is typically the last building element to be significantly exposed by the fire. If the foam plastic on the underside of the floor system is exposed to the interior of the building, then the foam plastic on the underside of the floor system must be covered by the required thermal barrier

R316.6 Specific approval. Foam plastic not meeting the requirements of Sections R316.3 through R316.5 shall be specifically *approved* on the basis of one of the following *approved* tests: NFPA 286 with the acceptance criteria of Section R302.9.4, FM4880, UL 1040, or UL 1715, or fire tests related to actual end-use configurations. Approval shall be based on the actual end use configuration and shall be performed on the finished foam plastic assembly in the maximum thickness intended for use. Assemblies tested shall include seams, joints and other typical details used in the installation of the assembly and shall be tested in the manner intended for use.

❖Foam plastic does not have to comply with the installation and use requirements of Sections R316.3 through R316.5 when specific approval is obtained in accordance with this section. This section lists examples of specific large-scale tests, such as FM 4880, NFPA 286, UL 1040 or UL 1715. Also, other large-scale fire tests related to actual end-use configuration can be used. The intent is to require testing based on the proposed end-use configuration of the foam plastic assembly with a fire exposure that is appropriate in size and location for the proposed application. These tests must be performed on full-scale assemblies. The tested assemblies must include typical seams, joints and other details that will occur in the finished installation. The foam plastic must be tested in the maximum thickness and density intended for use. Thorough testing provides an accurate depiction of the in-place fire performance of assemblies and systems using foam plastics.

There are two ways to show code compliance under Section R316.6. One method is to provide the actual test report that contains a description of the assembly and test results showing that the foam plastic, in the end use application, has passed the test. The second method is to obtain, from the ICC-ES, an evaluation report that covers the end-use application.

R316.7 Termite damage. The use of foam plastics in areas of "very heavy" termite infestation probability shall be in accordance with Section R318.4.

❖This section of the code refers to Section R318.4 that addresses the use of foam plastics in areas of "very heavy" termite infestation probability. When the structure is built in a area defined as "very heavy" termite infestation, Section R318.4 prohibits the use of foam plastics installed on the exterior face of below grade foundations walls or slab foundations, under exterior or interior foundation walls or slab foundations below-grade or where located within 6 inches (152 mm) of exposed earth. Section R318.4 states three exceptions where foam plastics are permitted:

1. Where the structural members of the building are either noncombustible or pressure-preservative-treated wood;
2. Where, in addition to the requirements of Section R318.1, the foam plastic is adequately protected from subterranean termite damage; or
3. On the interior side of basement walls.

SECTION R317
PROTECTION OF WOOD AND WOOD BASED PRODUCTS AGAINST DECAY

R317.1 Location required. Protection of wood and wood based products from decay shall be provided in the following locations by the use of naturally durable wood or wood that is preservative-treated in accordance with AWPA U1 for the species, product, preservative and end use. Preservatives shall be listed in Section 4 of AWPA U1.

1. Wood joists or the bottom of a wood structural floor when closer than 18 inches (457 mm) or wood girders when closer than 12 inches (305 mm) to the exposed ground in crawl spaces or unexcavated area located within the periphery of the building foundation.
2. All wood framing members that rest on concrete or masonry exterior foundation walls and are less than 8 inches (203 mm) from the exposed ground.
3. Sills and sleepers on a concrete or masonry slab that is in direct contact with the ground unless separated from such slab by an impervious moisture barrier.
4. The ends of wood girders entering exterior masonry or concrete walls having clearances of less than $^1/_2$ inch (12.7 mm) on tops, sides and ends.
5. Wood siding, sheathing and wall framing on the exterior of a building having a clearance of less than 6 inches (152 mm) from the ground or less than 2 inches (51 mm) measured vertically from concrete steps, porch slabs, patio slabs, and similar horizontal surfaces exposed to the weather.
6. Wood structural members supporting moisture-permeable floors or roofs that are exposed to the weather, such as concrete or masonry slabs, unless separated from such floors or roofs by an impervious moisture barrier.
7. Wood furring strips or other wood framing members attached directly to the interior of exterior masonry walls or concrete walls below *grade* except where an *approved* vapor retarder is applied between the wall and the furring strips or framing members.

❖This section addresses the need for minimum protection against decay damage for wood members located in certain locations.

For those portions of a wood-framed structure that are subject to damage by decay, the code mandates that the lumber be pressure-preservative treated or be naturally durable wood, or be of a species of wood having a natural resistance to decay. Naturally durable wood by definition is the heartwood of decay-resistant redwood, cedars, black locust and black walnut.

Item 1: Crawl spaces and unexcavated areas under a building usually contain moisture-laden air. These spaces must be ventilated in accordance with Section R408 to remove as much moisture as possible before it causes decay. Wood placed a minimum specified distance above grade in unexcavated under-floor areas or crawl spaces, as shown in Commentary Figure R317.1(1), need not be either preservative-treated wood or wood that is naturally decay-resistant durable wood. These clearances below floor joists and beams are deemed to be the minimum necessary to allow adequate circulation and removal of moisture from the air and from the wood framing members. Such clearances apply within the exterior wall line of the building foundation.

Item 2: Foundation walls will absorb moisture from the ground and by capillary action move it to framing members that are in contact with the foundation. Unless a minimum clearance of 8 inches (203 mm) is maintained from the finished grade to wood sills resting on concrete or masonry exterior foundation walls, decay-resistant or preservative-treated wood, as shown in Commentary Figure R317.1(2), must be used. The 8-inch (203 mm) clearance specified in this section has been determined to be large enough to prevent wetting of wood framing members under most circumstances.

Item 3: Concrete and masonry slabs that are in direct contact with the earth are very susceptible to

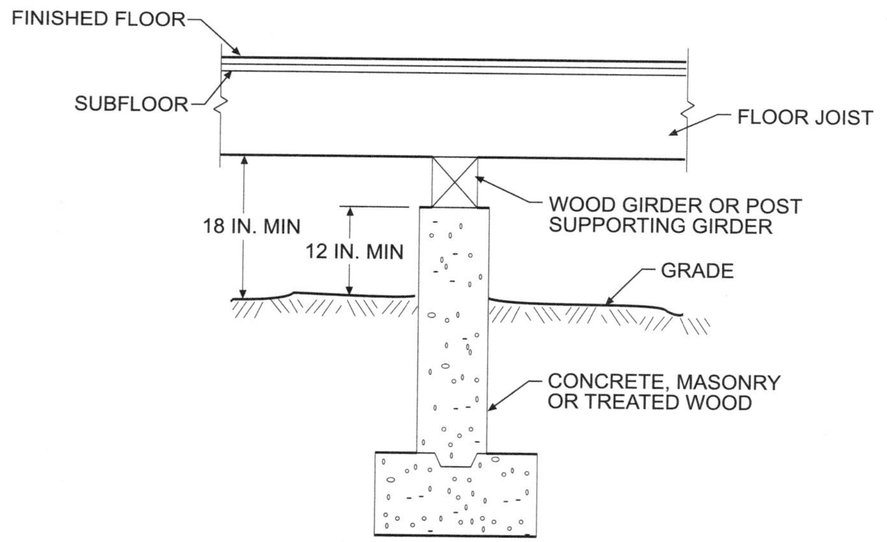

For SI: 1 inch = 25.4 mm.

Figure R317.1(1)
MINIMUM SEPARATION OF NONTREATED WOOD FROM GROUND

moisture because of absorption of ground water. This can occur on interior slabs, as well as at the perimeter. In the case of wood sills or sleepers placed on concrete or masonry slabs, decay-resistant wood or pressure-treated wood is required where the slabs are in direct contact with the ground, as illustrated in Commentary Figure R317.1(3). Concrete that is fully separated from the ground by a vapor barrier is not in direct contact with earth.

Item 4: A minimum $^1/_2$-inch (12.7 mm) clearance along the top, sides and ends of wood members projecting into exterior masonry or concrete walls must be maintained, as illustrated in Commentary Figure R317.1(4), unless the wood is treated or is of a species that is naturally decay resistant.

Item 5: Experience has shown that wood siding may extend below the sill plate to within 6 inches (152 mm) of the earth without decaying. Commentary Figure R317.1(5) shows the required minimum 6-inch (152 mm) clearance from the ground for wood siding, sheathing and wall framing on the exterior of a building. It should not be in direct contact with the foundation wall. If the sheathing is located over a concrete step or slab, the clearance may be reduced to 2 inches (51 mm) minimum.

Item 6: Concrete or masonry slabs that serve as roofs or floor systems that are exposed to the weather are very susceptible to moisture from rain or snow. If these slabs are supported by wood, the wood must be decay resistant or must be separated from the slabs by vapor barriers so there is not direct contact. This is similar to Item 3.

Item 7: When a basement area is finished, a common practice is to provide wood furring strips on top of the concrete or masonry walls for attachment of finishes. These furring strips must either be of decay-resistant material or must be separated from the wall by a vapor barrier. This will prevent the moisture from moving through the walls and rotting the furring strips, and even possibly transferring the moisture to the inside finish materials.

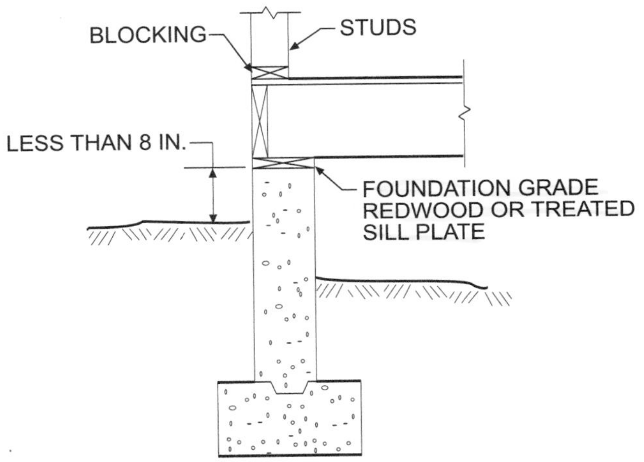

For SI: 1 inch = 25.4 mm.

Figure R317.1(2)
WOOD SILLS ON FOUNDATION WALLS

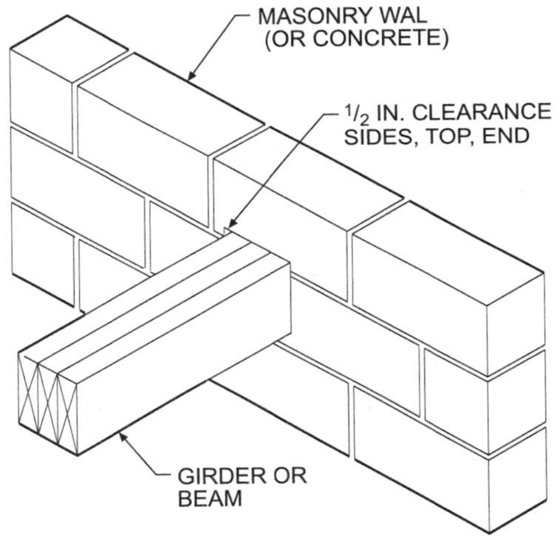

For SI: 1 inch = 25.4 mm.

Figure R317.1(4)
NONTREATED WOOD PROJECTING INTO BEAM POCKET

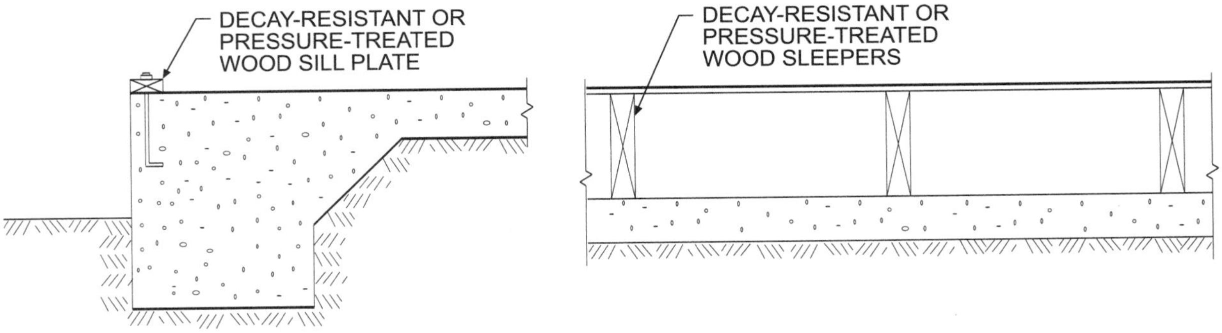

Figure R317.1(3)
WOOD SILLS AND SLEEPERS

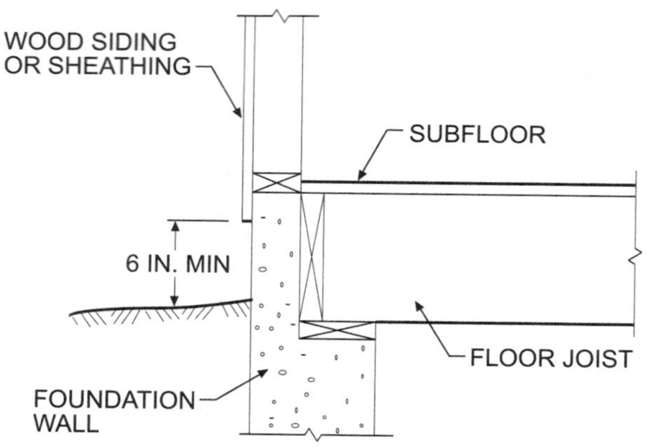

For SI: 1 inch = 25.4 mm.

Figure R317.1(5)
NONTREATED WOOD SIDING OR SHEATHING

R317.1.1 Field treatment. Field-cut ends, notches and drilled holes of preservative-treated wood shall be treated in the field in accordance with AWPA M4.

❖The requirement for field treatment of cuts and holes is duplicated from Section R318.1.2, which addresses the same type of situation in treated wood applications for termite protection. When pressure-preservative-treated wood is used for protection against decay, any cuts, notches or bored holes done in the field must be retreated. The retreatment methods and materials are to be in accordance with AWPA M4, which regulates the care of preservative-treated wood products.

R317.1.2 Ground contact. All wood in contact with the ground, embedded in concrete in direct contact with the ground or embedded in concrete exposed to the weather that supports permanent structures intended for human occupancy shall be *approved* pressure-preservative-treated wood suitable for ground contact use, except untreated wood may be used where entirely below groundwater level or continuously submerged in fresh water.

❖Wood members that are designed to be in contact with the ground and wood that is embedded in concrete in direct contact with the ground, or embedded in concrete exposed to the weather must be suitable for ground contact use. This provision applies to all wood members that support permanent structures designed for human occupancy. Untreated wood is permitted only where the wood members will be located below the ground water level or where the members are continuously submerged in fresh water because fungus that decays wood and termites cannot survive in a water-only environment with no oxygen (see Commentary Figure R317.1.2).

R317.1.3 Geographical areas. In geographical areas where experience has demonstrated a specific need, *approved* naturally durable or pressure-preservative-treated wood shall be used for those portions of wood members that form the structural supports of buildings, balconies, porches or similar permanent building appurtenances when those members are exposed to the weather without adequate protection from a roof, eave, overhang or other covering that would prevent moisture or water accumulation on the surface or at joints between members. Depending on local experience, such members may include:

1. Horizontal members such as girders, joists and decking.
2. Vertical members such as posts, poles and columns.
3. Both horizontal and vertical members.

❖This section gives the local jurisdiction the authority to require additional protection when the need can be documented by experience. Such experience may include high water tables, extended high-humidity conditions, etc.

Guidance is provided in this section for the protection of wood members other than those specifically identified elsewhere in this section that are exposed to the weather without benefit of protective elements. Included in the list of possible wood members that must be pressure-preservative treated or naturally

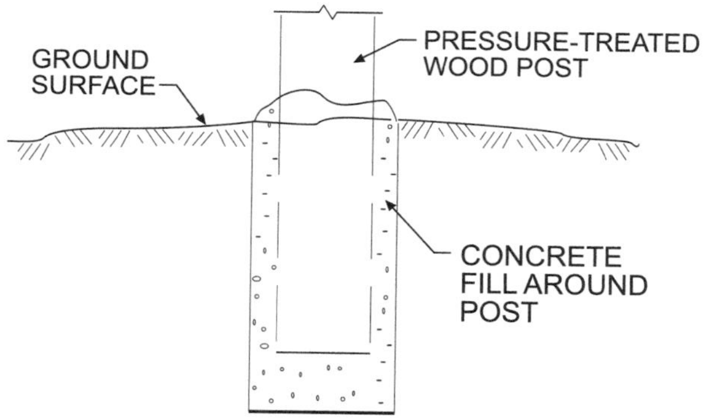

Figure R317.1.2
WOOD POST EMBEDDED IN CONCRETE

durable wood are girders, joists, decking, columns and similar structural members of porches, balconies and other exterior building elements.

R317.1.4 Wood columns. Wood columns shall be *approved* wood of natural decay resistance or *approved* pressure-preservative-treated wood.

Exceptions:

1. Columns exposed to the weather or in *basements* when supported by concrete piers or metal pedestals projecting 1 inch (25.4 mm) above a concrete floor or 6 inches (152 mm) above exposed earth and the earth is covered by an *approved* impervious moisture barrier.
2. Columns in enclosed crawl spaces or unexcavated areas located within the periphery of the building when supported by a concrete pier or metal pedestal at a height more than 8 inches (203 mm) from exposed earth and the earth is covered by an impervious moisture barrier.

❖Commentary Figure R317.1.4 shows where protection against decay is required for wood structural members on permeable surfaces in direct contact with the ground. Concrete or masonry that is in direct contact with the ground is susceptible to moisture because of the absorption of ground water. The exceptions permit the use of common framing lumber when the column is isolated from the earth or concrete. Generally, structural elements in basements or cellars are easier to inspect, which explains the lower clearances. Columns in crawl spaces must have a larger separation distance to safeguard against termites between inspections. In both exceptions, an impervious moisture barrier must be installed to ensure the column does not absorb water from the earth or concrete pier.

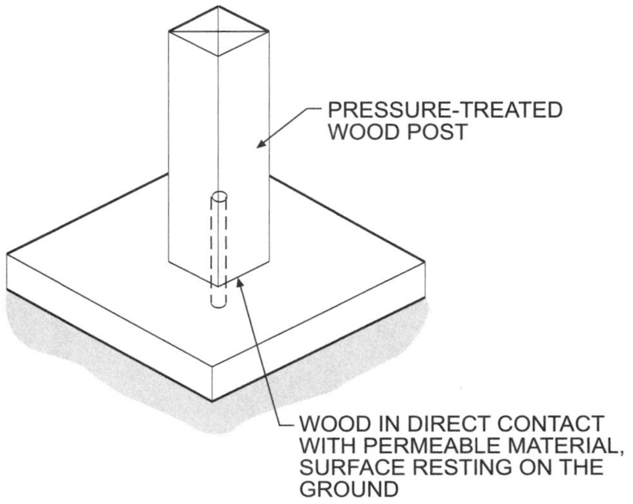

Figure R317.1.4
WOOD POST ON PERMEABLE PAD

R317.1.5 Exposed glued-laminated timbers. The portions of glued-laminated timbers that form the structural supports of a building or other structure and are exposed to weather and not properly protected by a roof, eave or similar covering shall be pressure treated with preservative, or be manufactured from naturally durable or preservative-treated wood.

❖It is common practice to design large glued-laminated arches that are connected to foundations near ground level. Exterior walls are built inside the span of these arches, leaving the initial few feet of the wood arches exposed to the weather. Experience has shown that covering the tops of these arches with metal or other water seals is not sufficient to prevent decay of the timber. Therefore, arches and other exposed wood members not protected by roofs or similar covers must be laminated of naturally durable or preservative-treated wood.

R317.2 Quality mark. Lumber and plywood required to be pressure-preservative-treated in accordance with Section R318.1 shall bear the quality *mark* of an *approved* inspection agency that maintains continuing supervision, testing and inspection over the quality of the product and that has been *approved* by an accreditation body that complies with the requirements of the American Lumber Standard Committee treated wood program.

❖The quality of lumber or plywood that has been pressure-preservative treated in accordance with the code is identifiable through a quality mark, such as shown in Commentary Figure R317.2.1. Without the proper identifying mark, it would be impossible to determine in the field that preservative-treated wood conforms to the applicable standards. The identifying mark must be from an approved inspection agency that has continuous follow-up services. Additionally, the inspection agency must be certified as being competent by an approved organization. The required identifying mark is in addition to the requirements for grade marks to identify the species and grade as specified elsewhere in the code.

R317.2.1 Required information. The required quality *mark* on each piece of pressure-preservative-treated lumber or plywood shall contain the following information:

1. Identification of the treating plant.
2. Type of preservative.
3. The minimum preservative retention.
4. End use for which the product was treated.
5. Standard to which the product was treated.
6. Identity of the *approved* inspection agency.
7. The designation "Dry," if applicable.

Exception: Quality *marks* on lumber less than 1 inch (25.4 mm) nominal thickness, or lumber less than nominal 1 inch by 5 inches (25.4 mm by 127 mm) or 2 inches by 4 inches (51 mm by 102 mm) or lumber 36 inches (914 mm) or less in length shall be applied by stamping the faces of exterior pieces or by end labeling not less than 25 percent of the pieces of a bundled unit.

❖This section provides the information that must be shown on a quality mark for pressure-preservative-treated lumber or plywood. Commentary Figure R317.2.1 illustrates how this information may be shown on a quality label.

R317.3 Fasteners and connectors in contact with preservative-treated and fire-retardant-treated wood. Fasteners, including nuts and washers, and connectors in contact with preservative-treated wood and fire-retardant-treated wood shall be in accordance with this section. The coating weights for zinc-coated fasteners shall be in accordance with ASTM A 153.

❖This section provides a charging statement for fasteners utilized in situations where the surroundings or the type of wood could corrode the fasteners over time. ASTM A 153 specifies criteria for hot-dipped zinc coating for iron and steel hardware.

R317.3.1 Fasteners for preservative-treated wood. Fasteners, including nuts and washers, for preservative-treated wood shall be of hot-dipped, zinc-coated galvanized steel, stainless steel, silicon bronze or copper. Coating types and weights for connectors in contact with preservative-treated wood shall be in accordance with the connector manufacturer's recommendations. In the absence of manufacturer's recommendations, a minimum of ASTM A 653 type G185 zinc-coated galvanized steel, or equivalent, shall be used.

Exceptions:

1. One-half-inch-diameter (12.7 mm) or greater steel bolts.
2. Fasteners other than nails and timber rivets shall be permitted to be of mechanically deposited zinc-coated steel with coating weights in accordance with ASTM B 695, Class 55 minimum.
3. Plain carbon steel fasteners in SBX/DOT and zinc borate preservative-treated wood in an interior, dry environment shall be permitted.

❖Except for large steel bolts, all fasteners used in conjunction with pressure-preservative-treated wood must be corrosion resistant. Some chemicals used to preservative-treat or fire-retardant-treated wood may have a corrosive effect on fasteners installed in locations that are high in moisture. Acceptable materials include stainless steel, silicon bronze, copper or hot-dipped zinc-coated galvanized steel. Such fasteners are also mandated for use with fire-retardant-treated wood members. Use of durable fasteners is required to decrease the likelihood that the load-carrying capacity of fasteners will be reduced by corrosion.

The second exception allows the use of a mechanical galvanization process in which the coating weight is mandatory as outlined in ASTM B 695.

The third exception permits the use of nonzinc-coated galvanized steel where used in a dry environment and in contact with SBX/DOT and zinc borate-treated wood.

R317.3.2 Fastenings for wood foundations. Fastenings, including nuts and washers, for wood foundations shall be as required in AF&PA PWF.

❖A permanent wood foundation (PWF) system is intended for light-frame construction including residential buildings. This AF&PA PWF — *Permanent Wood Foundation Design Specification*, primarily addresses the structural design requirements.

The PWF is a load-bearing wood frame wall and floor system designed for both above- and below-grade use as a foundation for light-frame construction. The PWF specifications are based on information developed cooperatively by the wood products industry and the U.S. Forest Service, with the advice and guidance of the Department of Housing and Urban Development's (HUD) Federal Housing Administration, and utilizing research findings of the National Association of Home Builders Research Center.

The PWF standard is also referenced in Sections R401.1 and R404.2.3.

	20_-20_[3]
	GROUND CONTACT[6]
ABC[1]	.40[5]
XXX[7]	AWPA—STDS[2]
	PRESERVATIVE[4]
	KDAT[8]
	X-XX[9]

1- THE IDENTIFYING SYMBOL, LOGO OR NAME OF THE ACCREDITED AGENCY.
2- THE APPLICABLE AMERICAN WOOD PRESERVER'S ASSOCIATION (AWPA) COMMODITY STANDARD.
3- THE YEAR OF TREATMENT IF REQUIRED BY AWPA STANDARD.
4- THE PRESERVATIVE USED, WHICH MAY BE ABBREVIATED.
5- THE PRESERVATIVE RETENTION.
6- THE EXPOSURE CATEGORY (E.G., ABOVE GROUND, GROUND CONTACT, ETC.).
7- THE PLANT NAME AND LOCATION; PLANT NAME AND NUMBER, OR PLANT NUMBER.
8- IF APPLICABLE, MOISTURE CONTENT AFTER TREATMENT.
9- IF APPLICABLE, LENGTH, AND/OR CLASS.

Figure R317.2.1
TYPICAL LABEL FOR PRESSURE-TREATED WOOD

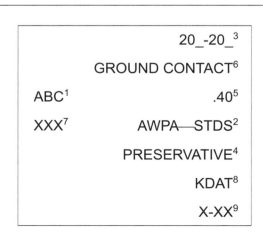

R317.3.3 Fasteners for fire-retardant-treated wood used in exterior applications or wet or damp locations. Fasteners, including nuts and washers, for fire-retardant-treated wood used in exterior applications or wet or damp locations shall be of hot-dipped, zinc-coated galvanized steel, stainless steel, silicon bronze or copper. Fasteners other than nails and timber rivets shall be permitted to be of mechanically deposited zinc-coated steel with coating weights in accordance with ASTM B 695, Class 55 minimum.

❖Sections R317.3.3 and R317.3.4 are intended to recognize the different exposures for fire-retardant-treated wood and the fastener requirements for that exposure. For wet conditions where corrosion may be a concern, this section allows the use of a mechanical galvanization process in which the coating weight is mandatory as outlined in ASTM B 695.

R317.3.4 Fasteners for fire-retardant-treated wood used in interior applications. Fasteners, including nuts and washers, for fire-retardant-treated wood used in interior locations shall be in accordance with the manufacturer's recommendations. In the absence of the manufacturer's recommendations, Section R317.3.3 shall apply.

❖Sections R317.3.3 and R317.3.4 are intended to recognize the different exposures for fire-retardant-treated wood and the fastener requirements for that exposure. The interior exposure for fire-retardant-treated wood is far less severe than the exposure for fire-retardant treated wood in wet, damp or exterior locations. Manufacturers of fire-retardant-treated wood make their recommendations for the appropriate fastener for fire-retardant-treated wood based on testing. If the manufacturer has no recommendations, the more restrictive provisions in Section R317.3.3 are applicable.

R317.4 Wood/plastic composites. Wood/plastic composites used in exterior deck boards, stair treads, handrails and guardrail systems shall bear a *label* indicating the required performance levels and demonstrating compliance with the provisions of ASTM D 7032.

❖Wood/plastic composite (WPC) materials, commonly used in exterior deck boards, guards and handrails, must be rated for appropriate performance criteria. These materials have a widespread acceptance for residential construction, and labeling requirements for WPCs will ensure the safe application of these materials in exterior deck systems. The referenced standard, ASTM D 7032, includes performance evaluations, such as flexural tests, ultraviolet resistance tests, freeze-thaw resistance tests, biodeterioration tests, fire performance tests, creep recovery tests, mechanical fastener holding tests and slip resistance tests. The standard also includes considerations of the effects of temperature and moisture, concentrated loads and fire propagation tests.

R317.4.1 Labeling. Deck boards and stair treads shall bear a label that indicates compliance to ASTM D 7032 and includes the allowable load and maximum allowable span. Handrails and guardrail systems or their packaging shall bear a label that indicates compliance to ASTM D 7032 and includes the maximum allowable span.

❖This section clarifies mandatory labeling requirements for wood/plastic composites in Section R314.4. Each deck board and stair tread, similar to pressure-preservative treated wood, is required to have a label. The required label would be applied on an end or on a face (side) of each board. Product labels will show verification of compliance with ASTM D 7032 and provide the appropriate performance information. For example, deck board labels would identify the allowable load and span [e.g., 40 psf load on a 16-inch (406 mm) span would be expressed as "16/40"]. Handrails and guardrail systems, which are more often supplied as "kits" in packages, require labels on the items or on the packaging. The maximum span (maximum vertical post spacing) is required to be on the label, as is verifying compliance to ASTM D 7032.

R317.4.2 Installation. Wood/plastic composites shall be installed in accordance with the manufacturer's instructions.

❖This section also includes a requirement for the installation of these WPC products in accordance with the manufacturer's instructions, which is the best way to ensure that WPCs perform to the required design loads, and installation instructions are an integral part of the manufacturers labeling program.

SECTION R318 PROTECTION AGAINST SUBTERRANEAN TERMITES

R318.1 Subterranean termite control methods. In areas subject to damage from termites as indicated by Table R301.2(1), methods of protection shall be one of the following methods or a combination of these methods:

1. Chemical termiticide treatment, as provided in Section R318.2.
2. Termite baiting system installed and maintained according to the *label*.
3. Pressure-preservative-treated wood in accordance with the provisions of Section R317.1.
4. Naturally durable termite-resistant wood.
5. Physical barriers as provided in Section R318.3 and used in locations as specified in Section R317.1.
6. Cold-formed steel framing in accordance with Sections R505.2.1 and R603.2.1.

❖This section establishes the rules for the protection of structures from damage caused by termites. The methods of protection address not only wood members, but also foam plastic and cold-formed steel materials.

Figure R301.2(6) illustrates those geographical areas where termite damage is probable. In those areas, the structure must be protected from termite damage in an appropriate manner. There are a number of methods permitted by the code to provide the necessary protection against termite damage. The most common method of termite control is soil poisoning. Alternatives include the use of pressure-preservative treated, naturally termite-resistant wood and barriers over perimeter walls. Often a combination of these methods is necessary to establish the required level of protection. The six acceptable methods of protection are broken out to make them easier to read.

R318.1.1 Quality mark. Lumber and plywood required to be pressure-preservative-treated in accordance with Section R318.1 shall bear the quality *mark* of an *approved* inspection agency which maintains continuing supervision, testing and inspection over the quality of the product and which has been *approved* by an accreditation body which complies with the requirements of the American Lumber Standard Committee treated wood program.

❖The required quality mark is similar to what is required for pressure-preservative-treated wood where decay is a concern (see commentary, Section R317.2 and Commentary Figure R317.2.1).

R318.1.2 Field treatment. Field-cut ends, notches, and drilled holes of pressure-preservative-treated wood shall be retreated in the field in accordance with AWPA M4.

❖When pressure-preservative-treated wood is used for protection against termites, any cuts, notches or bored holes done in the field must be retreated. The retreatment methods and materials are to be in accordance with AWPA M4, which regulates the care of preservative-treated wood products. This same requirement is found in Section R317.1.1 for pressure-preservative-treated wood used to prevent decay.

R318.2 Chemical termiticide treatment. Chemical termiticide treatment shall include soil treatment and/or field applied wood treatment. The concentration, rate of application and method of treatment of the chemical termiticide shall be in strict accordance with the termiticide *label*.

❖This language accurately reflects the current industry terminology in reference to termiticides and it emphasizes the necessity of using chemicals as required by their labels.

Where using a chemical poisoning of the soil for termite control, the chemical's instructions must be followed in detail. Treatment must be consistent with the recommendations of the chemical manufacturer.

R318.3 Barriers. *Approved* physical barriers, such as metal or plastic sheeting or collars specifically designed for termite prevention, shall be installed in a manner to prevent termites from entering the structure. Shields placed on top of an exterior foundation wall are permitted to be used only if in combination with another method of protection.

❖This section specifically lists several specific types of termite barriers. In addition, the proper use of a termite shield is addressed. Historically, termite shields were assumed to be effective in preventing termite penetration. However, the term "shield" is actually somewhat of a misnomer because they do not stop termites and were originally designed to make inspection and observation easier.

R318.4 Foam plastic protection. In areas where the probability of termite infestation is "very heavy" as indicated in Figure R301.2(6), extruded and expanded polystyrene, polyisocyanurate and other foam plastics shall not be installed on the exterior face or under interior or exterior foundation walls or slab foundations located below *grade*. The clearance between foam plastics installed above *grade* and exposed earth shall be at least 6 inches (152 mm).

Exceptions:

1. Buildings where the structural members of walls, floors, ceilings and roofs are entirely of noncombustible materials or pressure-preservative-treated wood.
2. When in *addition* to the requirements of Section R318.1, an *approved* method of protecting the foam plastic and structure from subterranean termite damage is used.
3. On the interior side of *basement walls*.

❖In those geographical areas where there is a very high risk of termite infestation, the code prohibits the use of foam plastic materials below or adjacent to ground level. This prohibition includes foam plastics installed on the exterior face of foundation walls or slab foundations, under exterior or interior foundation walls or slab foundations below grade or where located within 6 inches (152 mm) of exposed earth. Foam plastics are permitted where they are adequately protected or where the structural members are either noncombustible or of pressure-preservative-treated wood.

SECTION R319
SITE ADDRESS

R319.1 Address numbers. Buildings shall have *approved* address numbers, building numbers or *approved* building identification placed in a position that is plainly legible and visible from the street or road fronting the property. These numbers shall contrast with their background. Address numbers shall be Arabic numbers or alphabetical letters. Numbers shall be a minimum of 4 inches (102 mm) high with a minimum stroke width of $^1/_2$ inch (12.7 mm). Where access is by means of a private road and the building address cannot be viewed from the public way, a monument, pole or other sign or means shall be used to identify the structure.

❖The address requirements are consistent with the IBC, *International Fire Code*® (IFC®) and the code. The code requires buildings to have plainly visible and legible address numbers posted on the building or in such a place on the property that the building may be identified by emergency services, such as fire, medical and police. The primary concern is that emergency forces should be able to locate the building without going through a lengthy search procedure. In furthering the concept, the code states that the approved street numbers be placed in a location readily visible from the street or roadway fronting the property if a sign on the building would not be visible from the street. Address posting on a mailbox is not adequate for emergency responders when they are grouped or placed across the street from the dwelling.

SECTION R320
ACCESSIBILITY

R320.1 Scope. Where there are four or more *dwelling* units or sleeping units in a single structure, the provisions of Chapter 11 of the *International Building Code* for Group R-3 shall apply.

❖The IBC has been certified by HUD as a safe harbor document for complying with the accessibility provisions in the Fair Housing Act (FHA). Safe harbor means that the requirements in the IBC and ICC A117.1 meet or exceed the accessibility construction requirements in the Fair Housing Accessibility Guidelines (FHAG). Since FHA requirements start at when four units are constructed together, detached one-, two- and three-family homes are not required to go to the IBC for accessibility requirements however, townhouses and congregate residences are covered in Chapter 11 of the IBC. Such facilities should use the requirements in the IBC for Group R-3 occupancies to identify which units and facilities have accessibility requirements.

Section 1103.2.4 of the IBC reinforces the detached home exception by specifying that detached one- and two-family dwellings, their accessory structures and their associated sites and facilities need not be accessible.

Section 1107.5.4 of the IBC requires that, for Group R-3 with four or more dwelling or sleeping units in a single structure, every unit must meet Type B unit requirements. Notice the use of the term "structure" rather than "building." To determine the number of units in a single structure, the FHA considers all the units built under the same roof. The FHA does not recognize separating units with fire walls to create separate buildings. Neither are legal property lines considered to create separate buildings. All dwelling units built in a single structure are counted together. If a structure or individual unit has elevator service (passenger elevators, limited use/limited access (LULA) elevators or private residence elevators), the units served by the elevator must comply with the Type B accessibility requirements in Section 1004 of ICC A117.1.

Sections 1107.7 through 1107.7.5 of the IBC provide for a series of exceptions for Type B units. Section 1107.7.2 is an exception for multiple-story units without elevator service which would basically exempt most multiple-story townhouses. Since "multiple-story" is defined in Section 1102 as having living space on two or more levels, single-story townhouses with finished basements and two- and three-story townhouses would be exempted from complying with Type B unit requirements. In congregate residences without elevators, the grade level must contain at least one sleeping unit, and all shared spaces must meet Type B unit requirements. Section 1107.7.4 of the IBC exempts some structures in very hilly sites. Section 1107.7.5 of the IBC has allowances for homes constructed in areas with requirements for raised floor elevations due to flood concerns.

When there are units that must comply with Type B requirements, there are also provisions for the parking, building entrances and the site, including any recreational facilities. See the commentary to Chapter 11 and Section 1107 of the IBC for additional history and information on these requirements.

SECTION R321
ELEVATORS AND PLATFORM LIFTS

R321.1 Elevators. Where provided, passenger elevators, limited-use/limited-application elevators or private residence elevators shall comply with ASME A17.1.

❖ASME A17.1 is the safety code for elevators. All three types of elevators listed above are considered "passenger elevators." The vertical travel of the elevator car and which types can be used for public or private use is part of the ASME A17.1 safety requirements. Part V applies to elevators installed in or at a private residence. This part also applies to similar elevators installed in buildings as a means of access to private residences within such buildings if the elevators are installed so that they are not available to the general public or to other occupants of the building. All three types of elevators can serve as part of a required accessible route as they also comply with ICC A117.1 (see Section R320).

R321.2 Platform lifts. Where provided, platform lifts shall comply with ASME A18.1.

❖The previous technical standard (i.e., ASME A17.1) required key operation to platform lifts, which necessarily inhibits independent access by persons with physical disabilities. The requirements for platform lifts have been removed from the platform standard and now have their own standard, ASME 18.1, *Safety Standard for Platform Lifts and Stairway Chair Lifts*, which is referenced in this section.

A platform lift is an electrically operated mechanical device designed to transport a person who cannot use stairs over a short vertical distance. Platform lifts must be sized to accommodate a wheelchair user. Platform lifts can be used by wheelchair users and persons with limited mobility and are sometimes also equipped with folding seats. A fold-down seat that moves up the stairway is not a platform lift. While a stairway chair lift can be installed within a dwelling unit, it cannot serve as part of a required accessible route (see Section R320). A platform lift is most suitable for changes of elevation of one story or less where the installation of a ramp is not feasible.

There are two kinds of platform lifts: vertical lifts and inclined lifts. Vertical lifts are similar to elevators in that they travel only up and down in a fixed vertical space. Inclined platform lifts are usually installed in conjunction with a stairway and travel along the slope of the stairway. Inclined lifts are a design consideration for long flights of stairs where a vertical platform lift is not practical, where headroom is limited or where ceilings are low.

R321.3 Accessibility. Elevators or platform lifts that are part of an accessible route required by Chapter 11 of the *International Building Code*, shall comply with ICC A117.1.

❖Section 1107 of the IBC includes scoping requirements for three levels of accessibility: Accessible units, Type A units and Type B units. The types of facilities can include dwelling units or congregate residences constructed under the code. Accessible routes may also be required to public or common areas on a site, including recreational facilities, such as a pool or community building in a townhouse complex. Also see Section R320.1. This section requires that all elevators or platform lifts on an accessible route be useable by a person in a wheelchair without assistance. ICC A117.1 provides technical data on elevator and platform lift size, operation and controls. LULA elevators and private residence elevators are considered a type of passenger elevator that can serve as part of an accessible route.

SECTION R322
FLOOD-RESISTANT CONSTRUCTION

R322.1 General. Buildings and structures constructed in whole or in part in flood hazard areas (including A or V Zones) as established in Table R301.2(1) shall be designed and constructed in accordance with the provisions contained in this section. Buildings and structures located in whole or in part in identified floodways shall be designed and constructed in accordance with ASCE 24.

❖This section addresses additional requirements for all buildings and structures proposed to be located in whole or in part in areas designated as flood hazard areas. These areas are commonly referred to as "flood plains" and are shown on a community's FIRM prepared by FEMA or other adopted flood hazard map. Flood hazard areas are determined using the base flood, which is defined as having a 1-percent chance (one chance in 100) of occurring in any given year. The maps do not show the worst case flood, nor the "flood of record," which usually refers to the most severe flood in the history of the community. Although a 1-percent chance seems fairly remote, larger floods occur regularly throughout the United States. Application of the flood-resistant provisions of the code cannot prevent or eliminate all future flood damage. These provisions represent a reasonable balance of the knowledge and awareness of flood hazards, methods to guide development to less hazard-prone locations, methods of design and construction intended to resist flood damage, and each community's and landowner's reasonable expectations to use the land.

FEMA uses multiple designations for the flood hazard areas shown on each FIRM, including A, AO, AH, A1-30, AE, A99, AR, AR/A1-30, AR/AE, AR/AO, AR/AH, AR/A, VO, V1 -30, VE and V. Along many open coasts and lake shores, where wind-driven waves are predicted, the flood hazard area is commonly referred to as the "V Zone." Flood hazard areas that are inland of coastal high-hazard areas subject to high-velocity wave action, plus flood hazard areas along rivers and streams, are commonly referred to as "A Zones." Because of waves, the flood loads in areas subject to high-velocity wave action differ from those in other flood hazard areas. Some FIRMs in coastal communities show the "Limit of Moderate Wave Action," which is the inland extent of the $1^1/_2$-foot (457 mm) wave height. Post-flood field investigations and laboratory testing have indicated that significant structural damage can occur in areas shown as A/AE Zones that are inland of V Zones and inland of some shorelines without V Zones, and where breaking wave heights are less than 3 feet (914 mm), but more than $1^1/_2$ feet (457 mm) [areas with breaking waves 3 feet (914 mm) or higher are designated as V Zones]. See ASCE 24 for more guidance on these areas, which may be called "Coastal A Zones."

Some coastal communities have FIRMs that show areas that are designated as units of the Coastal Barrier Resource System (CBRS) established by the Coastal Barrier Resource Act (CoBRA) of 1982 and subsequent amendments. The NFIP is prohibited from offering flood insurance on new or substantially improved buildings in these areas. Jurisdictions are responsible for the application of the provisions of the code concerning flood resistance in all designated flood hazard areas, even if federal flood insurance is not available.

Communities that have AR Zones and A99 Zones shown on their FIRMs should consult with the appropriate state agency or FEMA regional office for guidance on requirements that must apply if the community participates in the NFIP. AR Zones are areas that result from the decertification of a previously accredited flood protection system (such as a levee) that is determined to be in the process of being restored to provide base flood protection. A99 Zones are areas subject to inundation by the 1-percent-annual-chance flood event, but which will ultimately be protected upon completion of an under-construction federal flood protection system.

Through the adoption of the code, communities meet some of the requirements necessary to participate in the NFIP. To learn more about how communities may meet the requirements by adoption of the code and the IBC (with Appendix G or a comparable ordinance), see *Reducing Flood Losses Through the International Code Series: Meeting the Requirements of the National Flood Insurance Program* (ICC and FEMA).

The NFIP was established to reduce flood losses, to better indemnify individuals from flood losses and to reduce federal expenditures for disaster assistance. A community that has flood hazard areas participates in the NFIP to protect health, safety and property, and so that its citizens can purchase federally-backed flood insurance. FEMA administers the NFIP and monitors community compliance with the flood plain management requirements of the NFIP.

Many states require that communities regulate flood plain development to a higher standard than the minimum standards established by the NFIP. Communities considering using the code and other *International Codes*® (I-Codes®) to meet the flood plain management requirements of the NFIP are advised to consult with their NFIP state coordinator or the appropriate FEMA regional office.

R322.1.1 Alternative provisions. As an alternative to the requirements in Section R322.3 for buildings and structures located in whole or in part in coastal high-hazard areas (V Zones) and Coastal A Zones, if delineated, ASCE 24 is permitted subject to the limitations of this code and the limitations therein.

❖Along many open coasts and lake shores where wind-driven waves are predicted to be 3 feet (914 mm) high and higher, flood hazard areas are called coastal high-hazard areas and are shown as "V Zones" on FEMA's Flood Insurance Rate Maps. Because of waves, flood loads in these areas differ from those in other flood hazard areas. This section allows building officials to accept designs based on ASCE 24 or to specify that buildings in these areas must be designed according to ASCE 24.

R322.1.2 Structural systems. All structural systems of all buildings and structures shall be designed, connected and anchored to resist flotation, collapse or permanent lateral movement due to structural loads and stresses from flooding equal to the design flood elevation.

❖New buildings, new structures and substantial improvements to existing buildings and structures in flood hazard areas must be designed and constructed to resist flood forces to minimize damage. Buildings must be connected to structural foundation systems that are capable of resisting flood forces, including flotation, lateral (hydrostatic) pressures, moving water (hydrodynamic) pressures, wave impact and debris impact.

See Section R322.3.3 for reference to FEMA 550, a publication that provides residential foundation designs for open foundations (pilings and piers) up to 15 feet (4572 mm) above ground level and for closed foundations (crawl spaces and stemwalls) up to 8 feet (2438 mm) above ground level, subject to certain limitations.

R322.1.3 Flood-resistant construction. All buildings and structures erected in areas prone to flooding shall be constructed by methods and practices that minimize flood damage.

❖In addition to the structural requirement of Section R322.1.2, this section establishes a broad requirement that techniques used for construction in flood hazard areas must contribute to the flood resistance by minimizing flood damage.

R322.1.4 Establishing the design flood elevation. The design flood elevation shall be used to define flood hazard areas. At a minimum, the design flood elevation is the higher of:

1. The base flood elevation at the depth of peak elevation of flooding (including wave height) which has a 1 percent (100-year flood) or greater chance of being equaled or exceeded in any given year; or
2. The elevation of the design flood associated with the area designated on a flood hazard map adopted by the community, or otherwise legally designated.

❖This section defines the "flood hazard area," which is the area predicted to be inundated during a flood that has a 1-in-100 (or 1-percent) chance of occurring in any given year. At a minimum, the flood plain is the Special Flood Hazard Area shown on the community's FIRM. This section also defines the term "design flood elevation," which is used throughout the code. The design flood elevation is the greater of the base flood elevation shown on the FIRM or other flood hazard area map identified in Table R301.2(1).

R322.1.4.1 Determination of design flood elevations. If design flood elevations are not specified, the *building official* is authorized to require the applicant to:

1. Obtain and reasonably use data available from a federal, state or other source; or
2. Determine the design flood elevation in accordance with accepted hydrologic and hydraulic engineering practices used to define special flood hazard areas. Determinations shall be undertaken by a registered *design professional* who shall document that the technical methods used reflect currently accepted engineering practice. Studies, analyses and computations shall be submitted in sufficient detail to allow thorough review and approval.

❖Many FIRMs show flood hazard areas without specifying the base flood elevations. These areas, often referred to as "approximate," are subject to the flood resistant construction requirements of the code even though the minimum elevation to which buildings and structures must be elevated has not been determined by FEMA. An important step in regulating these flood hazard areas is determination of the design flood elevation. As defined in Section R322.1.4, at a minimum the DFE is the base flood elevation shown on a community's FIRM identified in Table R301.2(1). In some instances, flood elevation information may have been developed by sources other than FEMA, including other federal or state agencies. The building official is to obtain the information if it is available or the building official can require the permit applicant to do so in accordance with accepted engineering practices.

Some communities develop flood hazard information and provide it to applicants so that all development in an area is based on the same level of risk. Local officials unfamiliar with establishing design flood elevations in unnumbered flood zones are encouraged to contact the state NFIP coordinator or the appropriate FEMA regional office. For additional guidance, refer to FEMA 265.

R322.1.4.2 Determination of impacts. In riverine flood hazard areas where design flood elevations are specified but floodways have not been designated, the applicant shall demonstrate that the effect of the proposed buildings and structures on design flood elevations, including fill, when combined with all other existing and anticipated flood hazard area encroachments, will not increase the design flood elevation more than 1 foot (305 mm) at any point within the jurisdiction.

❖Although FEMA has provided floodways along many rivers and streams shown on a community's FIRM, other riverine flood hazard areas have base flood elevation but do not have designated floodways. In these areas, the potential effects that flood plain activities may have on flood elevations have not been evaluated. If FEMA has not designated a regulatory floodway, the community is responsible for regulating development so as not to increase flood elevations by more than 1 foot (305 mm) at any point in the community. In effect, this means a community must either prepare a hydraulic analysis for proposed activities or require permit applicants to do so. These analyses should be prepared in accordance with accepted engineering practices by a qualified professional (see commentary, Section R322.1.4.1). Several states have more restrictive requirements, which can be determined by contacting the state NFIP coordinator.

R322.1.5 Lowest floor. The lowest floor shall be the floor of the lowest enclosed area, including *basement*, but excluding any unfinished flood-resistant enclosure that is useable solely for vehicle parking, building access or limited storage provided that such enclosure is not built so as to render the building or structure in violation of this section.

❖The lowest floor is the most important reference point when designing and constructing a building or structure in a flood hazard area. The term is specifically defined to include basements, which are any areas that are below grade on all sides. (This definition differs from the one in Chapter 2, which includes areas that are partly below grade.)

It is important to understand the lowest floor of a building or structure and how it relates to enclosures. The NFIP recognizes that elevated buildings may have certain enclosures below the elevated building that will be subject to flooding. Enclosures that meet certain provisions are deemed not to be the lowest floor, and federal flood insurance premium rates are based on the elevated portion of the building. For flood hazard areas known as "A Zones," enclosure provisions are set forth in Section R322.2.2. For coastal high-hazard areas including "V Zones," provisions for enclosures are set forth in Sections R322.3.4 and R322.3.5. In Chapter 2, the definition of "Habitable space" notes that certain spaces and areas are not considered habitable spaces (bathrooms, closets, hallway, storage or utility spaces, and similar areas). However, to comply with the requirements of Section 322, only three nonhabitable spaces are allowed as enclosures below elevated buildings: vehicle parking, building access and limited storage.

R322.1.6 Protection of mechanical and electrical systems. Electrical systems, *equipment* and components; heating, ventilating, air conditioning; plumbing *appliances* and plumbing fixtures; *duct systems*; and other service *equipment* shall be located at or above the elevation required in Section R322.2 (flood hazard areas including A Zones) or R322.3 (coastal high-hazard areas including V Zones). If replaced as part of a substantial improvement, electrical systems, *equipment* and components; heating, ventilating, air conditioning and plumbing *appliances* and plumbing fixtures; *duct systems*; and other service *equipment* shall meet the requirements of this section. Systems, fixtures, and *equipment* and components shall not be mounted on or penetrate through walls intended to break away under flood loads.

> **Exception:** Locating electrical systems, *equipment* and components; heating, ventilating, air conditioning; plumbing *appliances* and plumbing fixtures; *duct systems*; and other service *equipment* is permitted below the elevation required in Section R322.2 (flood hazard areas including A Zones) or R322.3 (coastal high-hazard areas including V Zones) provided that they are designed and installed to prevent water from entering or accumulating within the components and to resist hydrostatic and hydrodynamic loads and stresses, including the effects of buoyancy, during the occurrence of flooding to the design flood elevation in accordance with ASCE 24. Electrical wiring systems are permitted to be located below the required elevation provided they conform to the provisions of the electrical part of this code for wet locations.

❖This section sets a broad requirement that electrical and mechanical system elements, including electrical, heating, ventilating, air-conditioning, plumbing fixtures, duct systems and other service equipment are to be located above the elevations specified depending on whether the location is in a flood hazard area (A Zones) or coastal high-hazard areas (V Zones). The same requirement applies to the replacement of electrical and mechanical system elements that are included as part of substantial improvements to existing buildings. Attention should be paid to ductwork. Typical installation between floor joists, or joist spaces used as ducts, will not be in compliance unless the floor elevation is high enough to allow such installation above the required elevation. If a building is elevated to satisfy the minimum requirement, anything installed below that elevation is automatically not in compliance with this section unless it explicitly complies with the exception. For additional guidance, refer to FEMA 348.

The exception to this section provides criteria for placing specific equipment below the required elevation. It is important to be clear that to do so the equipment must be designed to prevent the entry or accumulation of water. For example, a typical single-family home has an exterior air-conditioning unit located on the ground adjacent to the structure. This unit is typically not designed to withstand the entry of water and would not meet the requirements of this exception.

Personal elevators may be allowed even though they extend below the design flood elevation provided the control equipment is above that elevation. In coastal high-hazard areas (V Zones), flood loads acting on the elevator components and any non-breakaway shaft walls must be accounted for in the building design. For guidance, see FEMA TB #4.

R322.1.7 Protection of water supply and sanitary sewage systems. New and replacement water supply systems shall be designed to minimize or eliminate infiltration of flood waters into the systems in accordance with the plumbing provisions of this code. New and replacement sanitary sewage systems shall be designed to minimize or eliminate infiltration of floodwaters into systems and discharges from systems into floodwaters in accordance with the plumbing provisions of this code and Chapter 3 of the *International Private Sewage Disposal Code*.

❖Health concerns arise when water supply systems are exposed to flood waters, and contamination from flooded sewage systems can pose health and environmental risks. To reduce these risks, water supply systems and sanitary sewage systems must be designed to minimize or eliminate infiltration and discharges under flood conditions. For on-site sewage disposal systems, the effects of flooding may be reduced by locating the systems on the highest available ground. State health regulations may apply to on-site sewage disposal systems proposed in flood hazard areas.

R322.1.8 Flood-resistant materials. Building materials used below the elevation required in Section R322.2 (flood hazard areas including A Zones) or R322.3 (coastal high-hazard areas including V Zones) shall comply with the following:

1. All wood, including floor sheathing, shall be pressure-preservative-treated in accordance with AWPA U1 for the species, product, preservative and end use or be the decay-resistant heartwood of redwood, black locust or cedars. Preservatives shall be listed in Section 4 of AWPA U1.
2. Materials and installation methods used for flooring and interior and *exterior walls* and wall coverings shall conform to the provisions of FEMA/FIA-TB-2.

❖To minimize flood damage, materials used below the elevation required in this section must resist damage caused by flood waters. Even though certain enclosures may be allowed under otherwise elevated buildings, the materials used in those enclosures must be resistant to water damage to minimize flood losses. For general flood-resistance requirements, see FEMA TB #2, and FEMA TB #8.

R322.1.9 Manufactured homes. New or replacement *manufactured homes* shall be elevated in accordance with Section R322.2 (flood hazard areas including A Zones) or Section R322.3 in coastal high-hazard areas (V Zones). The anchor and tie-down requirements of Sections AE604 and AE605 of Appendix E shall apply. The foundation and anchorage of *manufactured homes* to be located in identified floodways shall be designed and constructed in accordance with ASCE 24.

❖The placement of new manufactured homes is subject to the same elevation requirements as other new construction. Replacement of existing manufactured homes is considered "new construction" and must comply with the provisions of this section. Manufactured homes are to be elevated so that their lowest floors are at or above the design flood elevation. Because of anticipated flood forces, units must be anchored and tied down in accordance with Appendix E (see Sections AE 604 and AE 605). Within floodways, where water depths and velocities are anticipated to be greatest, foundation and anchorage must be designed in accordance with the ASCE 24. Some states and communities do not allow manufactured homes to be installed in coastal high-hazard areas (V Zones) or floodways. See FEMA P-85 for additional guidance and several pre-engineered foundation solutions. FEMA recommends that the bottom of the frame be at or above the required elevation to minimize damage to the floor system.

HUD's October 2007 final rule for Model Manufactured Home Installation Standard (24 CFR, Part 3285) includes some provisions related to flood hazard areas. Manufacturers must clearly specify that their installation instructions and foundation specifications have either (a) been designed for flood-resistant considerations [in which case the conditions are to be listed (velocities, depths, or wave action) and the design must be certified by a registered professional engineer or architect], or (b) not been designed to address flood loads, in which case the instructions must direct the installer to "obtain an alternate design prepared and certified by a registered professional engineer or registered architect for the support and anchorage." HUD places the burden on the installer to determine whether a home site is wholly or partly in a flood hazard area and to obtain additional designs, if needed.

R322.1.10 As-built elevation documentation. A registered *design professional* shall prepare and seal documentation of the elevations specified in Section R322.2 or R322.3.

❖Documentation of the "as-built" lowest floor elevation is to be submitted to verify compliance. Most communities require or request submission of FEMA's Elevation Certificate (FEMA Form 81-31), which is designed to obtain the information necessary to show compliance. Building owners will need elevation information to obtain federal flood insurance, and insurance agents use the information to compute the applicable flood insurance premium rates. The elevation certificate is available on-line at http://www.fema.gov/. For guidance on the elevation certificate, see FEMA P-467-1.

As called for in Section R109.1.3, as part of the flood plain inspection the elevation information is to be provided when the floor elevation is set and before further vertical construction occurs. If an error is discovered at the time, it will be significantly easier and less expensive to correct at this stage. Section R109.1.6.1 requires submission of "as-built" elevation documentation prior to the final inspection and before the certificate of occupancy is issued to determine that the building is in compliance with the flood-resistant provisions of the code. The building official must maintain copies of the elevation information on file and make elevation and related building information available for insurance rating and other purposes. Elevation documentation may be reviewed as part of periodic community assistance visits conducted by FEMA or states; FEMA considers buildings without elevation documentation to be violations.

R322.2 Flood hazard areas (including A Zones). All areas that have been determined to be prone to flooding but not subject to high-velocity wave action shall be designated as flood hazard areas. Flood hazard areas that have been delineated as subject to wave heights between $1^1/_2$ feet (457 mm) and 3 feet (914 mm) shall be designated as Coastal A Zones. All building and structures constructed in whole or in part in flood hazard areas shall be designed and constructed in accordance with Sections R322.2.1 through R322.2.3.

❖Flood hazard areas, often referred to as "A Zones," are found along nontidal waterways and lakes, and include the inland portions of coastal flood plains that are not subject to high velocity wave action [i.e., where waves are less than 3 feet (914 mm) in height]. The zone designation does not indicate anticipated velocities of river and stream flooding. Especially in areas where the slopes of streambeds are relatively steep, designers should check the Flood Insurance Study to see whether velocity information is available for use in considering the effects of hydrodynamic flood loads.

Some FIRMs in coastal communities show the "limit of moderate wave action," which is the inland extent of the $1^1/_2$-foot (457 mm) wave height. Post-flood field investigations and laboratory testing indicate that significant structural damage can occur in areas shown as A/AE Zones that are inland of V Zones and inland of some shorelines without V Zones, where wave heights are less than 3 feet (914 mm), but more than $1^1/_2$ feet (457 mm). These areas are commonly called "Coastal A Zones." Commentary Figure R322.2 shows how the limit of moderate wave action will be delineated on FIRMs.

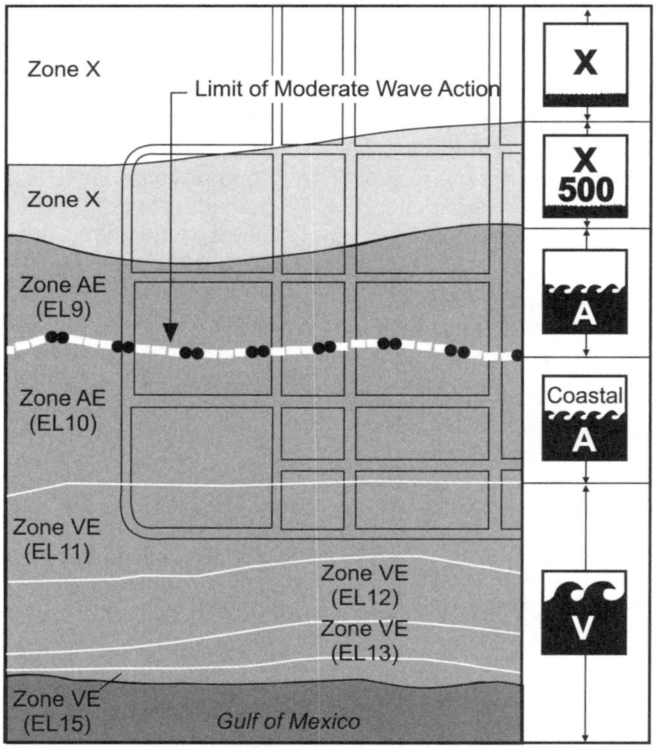

Figure R322.2
LIMIT OF MODERATE WAVE ACTION

R322.2.1 Elevation requirements.

1. Buildings and structures in flood hazard areas not designated as Coastal A Zones shall have the lowest floors elevated to or above the design flood elevation.
2. Buildings and structures in flood hazard areas designated as Coastal A Zones shall have the lowest floors elevated to or above the base flood elevation plus 1 foot (305 mm), or to the design flood elevation, whichever is higher.
3. In areas of shallow flooding (AO Zones), buildings and structures shall have the lowest floor (including *basement*) elevated at least as high above the highest adjacent *grade* as the depth number specified in feet on the FIRM, or at least 2 feet (610 mm) if a depth number is not specified.
4. Basement floors that are below *grade* on all sides shall be elevated to or above the design flood elevation.

Exception: Enclosed areas below the design flood elevation, including *basements* whose floors are not below *grade* on all sides, shall meet the requirements of Section R322.2.2.

❖The minimum requirement is that lowest floors must be elevated to or above the design flood elevation. If Coastal A Zones are designated, the lowest floor must be at least 1 foot (305 mm) higher than the base flood elevation to reduce the impact of waves on horizontal foundation elements.

AO Zones are areas of shallow flooding where FEMA has specified a "depth number" rather than a flood elevation that is referenced to a datum. Buildings in AO Zones must be elevated so the lowest floor, including the basement, is at least as high above grade as the depth number, or at least 2 feet (610 mm) if a depth number is not specified. The point where this depth is measured is at the highest natural grade adjacent to the proposed building. FEMA refers to some areas subject to shallow flooding as AH Zones. AH Zones have specified base flood elevations and they are treated the same as other A Zones with base flood elevations.

This section emphasizes that basements that are below grade on all sides must be elevated to or above the design flood elevation. The NFIP defines a basement not by its common use, but by whether it is below grade on all sides. Even if below grade on all sides by only 1 inch (25 mm), such an area is, technically, a basement.

In many flood hazard areas, it is common to place fill to elevate a building site above the design flood elevation. Fills that are exposed to flooding may become saturated and unstable. These fill slopes may be exposed to erosive velocities and waves during a flood. For additional guidance for development activities involving the placement of fill in a flood hazard area, refer to FEMA TB #10. Building officials may require applicants to submit documentation regarding the use of fill and whether structures on fill are reasonably safe from flooding.

The specification in Item 4 means that basements must not be excavated below the design flood elevation. The exception to this section allows enclosures that, if designed and constructed to meet the use limitations and the flood opening specifications of Section R322.2.2, are not the lowest floor. Allowable uses of such enclosures include parking of vehicles, building access, limited storage and crawl spaces.

R322.2.2 Enclosed area below design flood elevation. Enclosed areas, including crawl spaces, that are below the design flood elevation shall:

1. Be used solely for parking of vehicles, building access or storage.
2. Be provided with flood openings that meet the following criteria:
   - 2.1. There shall be a minimum of two openings on different sides of each enclosed area; if a building has more than one enclosed area below the design flood elevation, each area shall have openings on exterior walls.
   - 2.2. The total net area of all openings shall be at least 1 square inch (645 mm^2) for each square foot (0.093 m^2) of enclosed area, or the openings shall be designed and the *construction documents* shall include a statement by a registered *design professional* that the design of the openings will provide for equalization of hydrostatic flood forces on exterior walls by allowing for the automatic entry and exit of floodwaters as specified in Section 2.6.2.2 of ASCE 24.
   - 2.3. The bottom of each opening shall be 1 foot (305 mm) or less above the adjacent ground level.
   - 2.4. Openings shall be not less than 3 inches (76 mm) in any direction in the plane of the wall.
   - 2.5. Any louvers, screens or other opening covers shall allow the automatic flow of floodwaters into and out of the enclosed area.
   - 2.6. Openings installed in doors and windows, that meet requirements 2.1 through 2.5, are acceptable; however, doors and windows without installed openings do not meet the requirements of this section.

❖This section contains the use limitations and flood opening requirements that apply to enclosures below otherwise properly elevated buildings. If enclosures meet these requirements they are not considered the "lowest floor." The use limitation states that enclosures, including crawl spaces, are to be used only for parking of vehicles, building access or storage (see commentary to Section R322.1.5 regarding use limitations and the terms habitable and nonhabitable spaces that are used elsewhere in the code). Although the NFIP regulations do not define storage, most states interpret it to mean "limited" storage. Some states specify that hazardous materials must not be stored in enclosures. Experience has suggested that two design factors influence the use of enclosures and may discourage illegal conversions by owners: the size of the enclosed area and utility service. Some states and communities limit the size of enclosures. Plumbing service is not appropriate for the allowable uses, and only minimal electrical service should be provided (a light switch for a stairwell, for example, or electrical outlets dropped from the ceiling).

Openings intended to relieve differential hydrostatic pressure during the rise and fall of floodwaters must meet several specific requirements for location and size. Information about the number, size and location of flood openings is collected when FEMA's Elevation Certificate is completed; the information is used by insurance agents to determine the lowest floor and to compute the proper federal flood insurance rates (see FEMA P-467-1).

The requirements related to the location of flood openings are clear. Openings are to be provided on different sides of the enclosed area and they are to be no more than 12 inches (305 mm) above the higher of the interior grade or exterior grade under the opening. The size of total net open area of the flood openings is determined prescriptively (called "nonengineered openings") by computing the total area enclosed by the walls that extend below the design flood elevation and providing 1 square inch of net open area for every square foot (6944 mm^2 per m^2) of enclosed area. Alternatively, flood openings may be designed to automatically equalize hydrostatic flood forces (called "engineered openings"), in which case the construction documents must include a statement from a registered design professional that the openings are designed to meet this requirement. Engineered openings may be site-built and individually certified, or they may be products manufactured to meet this requirement, in which case they must have been issued an evaluation report by the ICC-ES. Devices intended for under-floor ventilation openings may be used as nonengineered openings. Care should be taken to determine the net open area provided by such devices, taking into account obstructions such as face plates, grilles, slats, louvers and other coverings. Airflow vent devices must be disabled in the "open" position to ensure automatic functioning in the event of a flood. For additional guidance on meeting the flood opening requirements of this section, see FEMA TB #1.

R322.2.3 Foundation design and construction. Foundation walls for all buildings and structures erected in flood hazard areas shall meet the requirements of Chapter 4.

Exception: Unless designed in accordance with Section R404:

1. The unsupported height of 6-inch (152 mm) plain masonry walls shall be no more than 3 feet (914 mm).
2. The unsupported height of 8-inch (203 mm) plain masonry walls shall be no more than 4 feet (1219 mm).
3. The unsupported height of 8-inch (203 mm) reinforced masonry walls shall be no more than 8 feet (2438 mm).

For the purpose of this exception, unsupported height is the distance from the finished *grade* of the under-floor space to the top of the wall.

❖Flood conditions impose additional loads on foundation walls, including hydrostatic and hydrodynamic loads. Under a cooperative agreement with FEMA, the American Society of Civil Engineers (ASCE) analyzed flood loads and flood-load combinations to examine the structural adequacy of the foundation walls prescribed in Chapter 4. A range of flood depths [4, 6, 8 and 9 feet (1219, 1829, 2438 and 2743 mm)] and velocities [3, 6 and 9 foot per second (fps) (0.91, 1.83 and 2.74 m/s)] was used to determine whether there were any height limitations for certain types of walls. The results suggested that unless the walls are designed in accordance with Section R404, it is appropriate to limit the unsupported height of certain masonry walls based on wall thickness and whether they are plain or reinforced.

R322.3 Coastal high-hazard areas (including V Zones). Areas that have been determined to be subject to wave heights in excess of 3 feet (914 mm) or subject to high-velocity wave action or wave-induced erosion shall be designated as coastal high-hazard areas. Buildings and structures constructed in whole or in part in coastal high-hazard areas shall be designed and constructed in accordance with Sections R322.3.1 through R322.3.6.

❖Buildings and structures located in flood hazard areas known as "V Zones," or coastal high-hazard areas, are exposed to flooding and wind-driven waves 3 feet (914 mm) or higher. Waves impose additional loads on foundations and can cause erosion. Note that Section R322.1.1 allows building officials to accept designs based on ASCE 24 or to specify that buildings in these areas must be designed according to ASCE 24. In addition, designers are referred to FEMA 55, and builders are referred to FEMA 499. Training on the design and construction of residential structures in coastal areas may be offered by the state NFIP coordinator or the appropriate FEMA regional office. Additional guidance about design and construction in coastal high-hazard areas can be found in FEMA TB #5, FEMA TB #8, and FEMA TB #9.

R322.3.1 Location and site preparation.

1. New buildings and buildings that are determined to be substantially improved pursuant to Section R105.3.1.1, shall be located landward of the reach of mean high tide.
2. For any alteration of sand dunes and mangrove stands the *building official* shall require submission of an

engineering analysis which demonstrates that the proposed *alteration* will not increase the potential for flood damage.

❖Buildings, including existing buildings for which a determination has been made that proposed work constitutes substantial improvement (or repair of substantial damage) must be sited such that they are not regularly affected by normal (mean) high tides. Many coastal locations benefit from natural storm protection because of the presence of sand dunes and mangrove stands. Those natural features must remain intact. However, alterations may be considered if specific analyses indicate that the alterations will not increase the potential for flood damage, including increasing the exposure of the site to greater flood depths and higher wave heights.

R322.3.2 Elevation requirements.

1. All buildings and structures erected within coastal high-hazard areas shall be elevated so that the lowest portion of all structural members supporting the lowest floor, with the exception of piling, pile caps, columns, grade beams and bracing, is:
   1.1. Located at or above the design flood elevation, if the lowest horizontal structural member is oriented parallel to the direction of wave approach, where parallel shall mean less than or equal to 20 degrees (0.35 rad) from the direction of approach, or
   1.2. Located at the base flood elevation plus 1 foot (305 mm), or the design flood elevation, whichever is higher, if the lowest horizontal structural member is oriented perpendicular to the direction of wave approach, where perpendicular shall mean greater than 20 degrees (0.35 rad) from the direction of approach.
2. Basement floors that are below *grade* on all sides are prohibited.
3. The use of fill for structural support is prohibited.
4. Minor grading, and the placement of minor quantities of fill, shall be permitted for landscaping and for drainage purposes under and around buildings and for support of parking slabs, pool decks, patios and walkways.

   Exception: Walls and partitions enclosing areas below the design flood elevation shall meet the requirements of Sections R322.3.4 and R322.3.5.

❖Buildings and structures to be located in coastal high-hazard areas (V Zones) must be elevated so that the bottom of the lowest horizontal structural members are not subject to flood loads during conditions of the base flood. Elevation is one of the most important requirements to provide resistance to flood damage. It is also the main factor used in determining federal flood insurance premium rates for buildings and structures.

For a specific building, the required elevation of the bottom of the lowest horizontal structural member depends on the proposed orientation of those members. Post-flood investigations have determined that more damage is sustained if those members are oriented perpendicular to the direction of wave approach because some waves exceed the predicted heights and impose wave loads on the members. Horizontal structural members that are parallel to the direction of wave approach are not exposed to the same wave loads. To determine orientation, members are parallel if they are within plus or minus 20 degrees (0.35 rad) of the primary direction. Therefore, if the orientation of the lowest horizontal structural member will be perpendicular to the primary direction from which waves are anticipated [more than 20 degrees (0.35 rad) from the primary direction], an additional foot of elevation above the base flood is required. Judgment is necessary to determine the likely direction of wave approach. In most cases, waves will approach from directly offshore (normal to the shoreline). In some locations, the likely direction of wave approach may not be easy to determine, in which case the higher elevation should be required.

In coastal high-hazard areas, the use of fill to provide structural support is prohibited because areas subject to wave action are more likely to experience flood-related erosion, which may lead to the failure of foundations on fill. Minor quantities of nonstructural fill may be allowed for the limited purposes of drainage and landscaping. For more guidance on acceptable uses of fill, see FEMA TB #5.

This section and the NFIP allow certain enclosures under buildings that otherwise meet the elevation requirements, and the exception notes that the requirements for such enclosures are found in Sections R322.3.4 and R322.3.5.

R322.3.3 Foundations. Buildings and structures erected in coastal high-hazard areas shall be supported on pilings or columns and shall be adequately anchored to such pilings or columns. The space below the elevated building shall be either free of obstruction or, if enclosed with walls, the walls shall meet the requirements of Section R322.3.4. Pilings shall have adequate soil penetrations to resist the combined wave and wind loads (lateral and uplift). Water-loading values used shall be those associated with the design flood. Wind-loading values shall be those required by this code. Pile embedment shall include consideration of decreased resistance capacity caused by scour of soil strata surrounding the piling. Pile systems design and installation shall be certified in accordance with Section R322.3.6. Spread footing, mat, raft or other foundations that support columns shall not be permitted where soil investigations that are required in accordance with Section R401.4 indicate that soil material under the spread footing, mat, raft or other foundation is subject to scour or erosion from wave-velocity flow conditions. If permitted, spread footing, mat, raft or other foundations that support columns shall be designed in accordance with ASCE 24. Slabs, pools, pool decks and walkways shall be located and constructed to be structurally independent of buildings and structures and their foundations to prevent transfer of flood loads to the buildings and structures during conditions of flooding, scour or erosion from wave-velocity flow conditions, unless the buildings and structures and their foundation are designed to resist the additional flood load.

❖Several provisions applicable to foundations for buildings and structures in coastal high-hazard areas (V Zones) are in this section. During coastal storms, both wind and water loads are significant, and buildings must be designed and constructed to withstand both loads. Buildings must be anchored to their foundations, and the foundations must withstand the anticipated forces. The requirement that the area under elevated buildings be free of obstructions is intended to minimize flood loads on foundations. Building elements that may be obstructions unless properly designed and constructed include stairs and ramps, decks and patios, elevators, equipment and cabinets mounted on foundation elements, foundation bracing (especially if perpendicular to the anticipate direction of wave approach), grade beams, shear walls, and slabs, including slabs used for parking (see FEMA TB #5).

Where soils are subject to scour and erosion from wave action, spread footing, mat, raft or other foundation elements cannot be used to support columns; if those foundation elements are used elsewhere, the requirements of ASCE 24 apply. It is common to have structures, such as pools, pool decks and slabs, built immediately adjacent to buildings in coastal communities. Such structures should be located and built so they do not obstruct waves and the flow of water or adversely impact buildings during a flood, including wave-induced scour and erosion. Additional guidance on these elements is found in FEMA TB #5.

After Hurricane Katrina, FEMA produced FEMA 550. The designs account for wind loads and flood loads and are applicable throughout the United States, with certain limitations. Selection of applicable designs depends on several factors, including flood depths, wind loads and soils. Designs are provided for open foundations (pilings and piers) up to 15 feet (4572 mm) above ground level.

R322.3.4 Walls below design flood elevation. Walls and partitions are permitted below the elevated floor, provided that such walls and partitions are not part of the structural support of the building or structure and:

1. Electrical, mechanical, and plumbing system components are not to be mounted on or penetrate through walls that are designed to break away under flood loads; and
2. Are constructed with insect screening or open lattice; or
3. Are designed to break away or collapse without causing collapse, displacement or other structural damage to the elevated portion of the building or supporting foundation system. Such walls, framing and connections shall have a design safe loading resistance of not less than 10 (479 Pa) and no more than 20 pounds per square foot (958 Pa); or
4. Where wind loading values of this code exceed 20 pounds per square foot (958 Pa), the *construction documents* shall include documentation prepared and sealed by a registered *design professional* that:
   4.1. The walls and partitions below the design flood elevation have been designed to collapse from a water load less than that which would occur during the design flood.
   4.2. The elevated portion of the building and supporting foundation system have been designed to withstand the effects of wind and flood loads acting simultaneously on all building components (structural and nonstructural). Water-loading values used shall be those associated with the design flood. Wind-loading values shall be those required by this code.

❖Buildings and structures in coastal high-hazard areas must be elevated to or above the design flood elevation. However, certain enclosures are allowed below otherwise elevated buildings if they meet certain specifications. These specifications allow the walls surrounding the enclosures to fail or break away under certain loads without causing damage to the foundation. Post-flood investigations indicate that breakaway walls do not perform as intended if wires, pipes and other utility components are mounted on or penetrate through the walls. Walls other than lattice, slats, shutters or insect screening must meet specific loading resistance; they are to be designed to fail under certain specified loads. If anticipated allowable loads exceed 20 pounds per square foot (960 Pa), the wall design is to be certified as meeting certain specifications. Wood or plastic lattice, slats or shutters installed between the foundation elements are not considered walls, provided at least 40 percent of the area is open to the passage of floodwaters and the material used is thin enough to break away, considered to be no thicker than $^1/_2$ inch (12.7 mm) for lattice and no thicker than 1 inch (25 mm) for slats and shutters. For additional guidance, refer to FEMA TB #5, FEMA 499 and FEMA 55. FEMA TB #9 is particularly useful because it includes three methods to design breakaway walls: a prescriptive design approach, a simplified design approach and a performance-based design approach.

R322.3.5 Enclosed areas below design flood elevation. Enclosed areas below the design flood elevation shall be used solely for parking of vehicles, building access or storage.

❖The limitations that are outlined in Section R322.2.2 on how enclosures below elevated buildings are used are repeated here. Enclosures are to be used only for parking, building access and storage. Although the NFIP regulations do not define storage, most states interpret it to mean "limited" storage. Some states specify that hazardous materials must not be stored in enclosures. Experience has suggested that two design factors influence the use of enclosures and may discourage illegal conversions by owners: the size of the enclosed area and utility service. Some communities limit the size of enclosures [federal flood insurance costs more if buildings in coastal high-hazard areas have enclosures, especially enclosures larger than 300 square feet (28 m^2)]. Another factor that influences enclosure use is the availability of utility services. Plumbing service is not associated with the allowable uses and in accordance with Section R322.1.6 must not be located below the design flood elevation. Only minimal electrical service should be provided (a light switch for a stairwell, for example, or electrical outlets that are dropped from the ceiling).

R322.3.6 Construction documents. The *construction documents* shall include documentation that is prepared and sealed by a registered *design professional* that the design and methods of construction to be used meet the applicable criteria of this section.

❖Documentation must be submitted to demonstrate that the design and proposed construction methods address the provisions of the code. This requirement recognizes that coastal high-hazard areas subject to waves and erosion are dynamic environments that impose significant loads on buildings. This documentation is different than the documentation of the elevation of the bottom of the lowest horizontal structural member that is required in Section 109.1.3 (flood plain inspections), Section 109.1.6.1 (prior to final inspection), and Section R322.1.10 (as-built).

SECTION R323 STORM SHELTERS

R323.1 General. This section applies to the construction of storm shelters when constructed as separate detached buildings or when constructed as safe rooms within buildings for the purpose of providing safe refuge from storms that produce high winds, such as tornados and hurricanes. In addition to other applicable requirements in this code, storm shelters shall be constructed in accordance with ICC/NSSA-500.

❖ICC 500 provides requirements for the design and construction of shelters to protect people from the violent winds of hurricanes and tornadoes. The standard includes special requirements for structural design, including wind loads that are considerably higher than the wind loads required by Section 301.2.1 for all structures.

Wind loads for storm shelters will be based upon wind speed contour maps developed specially for this standard. The wind load design requirements are relatively severe when compared to the wind speed maps in Chapter 3. Contour maps for wind speeds in hurricane-prone regions were determined based upon a 10,000-year mean return period. The map shows 200 mph (88 m/s) wind speeds on the coast of Florida and the Carolinas, and wind speeds higher than 200 mph (88 m/s) in some locations. These are wind speeds associated with a Category 5 hurricane. Shelter design wind speeds in the central part of the United States (a region called "tornado alley") are as high as 250 mph (110 m/s).

Such high wind speeds, of course, produce flying debris, turning construction materials into deadly missiles. The standard contains specific test methods and pass-fail criteria for window and door protection from flying debris.

Bibliography

The following resource materials were used in the preparation of the commentary for this chapter of the code:

AAMA/WDMA/CSA 101/I.S.2/A440-11, *North American Fenestration Standards/Specifications for Windows, Doors and Unit Skylights.* Schaumburg, IL: American Architectural Manufacturers Association, 2011.

ACI 318-11, *Building Code for Structural Concrete.* Farmington Hills, MI: American Concrete Institute, 2011.

AF&PA WFCM-2012, *Wood Frame Construction Manual for One- and Two-family Dwellings.* Washington, DC: American Forest and Paper Association, 2012.

AISI S230-07, *Standard for Cold-formed Steel Framing-prescriptive Method for One- and Two-family Dwellings with Supplement 2, dated 2008.* Washington, DC: American Iron and Steel Institute, 2007.

ANSI/AF&PA PWF-07, *Permanent Wood Foundation Design Specifications.* New York, NY: American National Standards Institute, 2007.

ANSI/DSMA 115-05, *Standard Method for Testing Sectional Garage Doors and Rolling Doors: Determination of Structural Performance Under Missile Impact and Cyclic Wind Pressure.* New York, NY: American National Standards Institute, 2005.

ANSI Z97. 1-09, *Safety Glazing Materials Used in Buildings—Safety Performance Specifications and Methods of Test.* New York, NY: American National Standards Institute, 2009.

ASCE 7-10, *Minimum Design Loads for Buildings and Other Structures.* Reston, VA: American Society of Civil Engineers, 2010.

ASCE 24-05, *Flood-resistant Design and Construction.* Reston, VA: American Society of Civil Engineers, 2005.

ASME A17.1/CSA B44-07, *Safety Code for Elevators and Escalators.* New York, NY: American Society of Mechanical Engineers, 2007.

ASME A18.1-08, *Safety Standard for Platforms and Stairway Chair Lifts.* New York, NY: American Society of Mechanical Engineers, 2008.

ASTM A 153/A 153M-05, *Standard Specification for Zinc Coating (Hot-Dip) on Iron and Steel Hardware.* West Conshohocken, PA: ASTM International, 2005.

ASTM B 695-04, *Standard Specification for Coatings of Zinc Mechanically Deposited on Iron and Steel.* West Conshohocken, PA: ASTM International, 2004.

ASTM D 2565-99, *Standard Practice for Xenon Arc Exposure of Plastics Intended for Outdoor Applications.* West Conshohocken, PA: ASTM International, 1999.

ASTM D 7032-08, *Standard Specification for Establishing Performance Ratings for Wood-plastic Composite Deck Boards and Guardrail Systems (Guards or Handrails).* West Conshohocken, PA: ASTM International, 2008.

ASTM E 84-09, *Test Method for Surface-burning Characteristics of Building Materials.* West Conshohocken, PA: ASTM International, 2009.

ASTM E 119-08a, *Test Method for Fire Tests of Building Construction and Materials.* West Conshohocken, PA: ASTM International, 2008a.

ASTM E 136-09, *Test Method for Behavior of Materials in a Vertical Tube Furnace at 750°C.* West Conshohocken, PA: ASTM International, 2009.

ASTM E 152-95, *Method of Fire Tests for Door Assemblies.* West Conshohocken, PA: ASTM International, 1995.

ASTM E 814-08b, *Test Method for Fire Tests of Through-penetration Firestops.* West Conshohocken, PA: ASTM International, 2008.

ASTM E 970-08a, *Test Method for Critical Radiant Flux of Exposed Attic Floor Insulation Using a Radiant Heat Energy Source.* West Conshohocken, PA: ASTM International, 2008.

ASTM E 1886-05, *Test Method for Performance of Exterior Windows, Curtain Walls, Doors and Storm Shutters Impacted by Missile(s) and Exposed to Cyclic Pressure Differentials.* West Conshohocken, PA: ASTM International, 2005.

ASTM E 1996-09, *Specification for Performance of Exterior Windows, Curtain Walls, Doors and Storm Shutters Impacted by Wind-borne Debris in Hurricanes.* West Conshohocken, PA: ASTM International, 2009.

ASTM F 2090-08, *Standard Specification for Window Fall Prevention Devices with Emergency Escape (Egress) Release Mechanisms.* West Conshohocken, PA: ASTM International, 2008

ASTM G 155-05a, *Standard Practice for Operating Xenon Arc light Apparatus for Exposure of Non-metallic Materials.* West Conshohocken, PA: ASTM International, 1999.

AWPA M4-08, *Standard for the Care of Preservative-treated Wood Products.* Granbury, TX: American Wood-Preservers' Association, 2008.

CAN/ULC-S102.2-1988, *Method of Test for Surface Burning Characteristics of Building Materials and Assemblies with 2000 Revisions.* Ontario, Canada: Underwriters Laboratories Canada, 2000.

CPSC 16 CFR, Part 1201-(2002), *Safety Standard for Architectural Glazing.* Bethesda, MD: Consumer Product Safety Commission, 2002.

CPSC 16 CFR, Part 1404-(2002), *Cellulose Insulation.* Bethesda, MD: Consumer Product Safety Commission, 2002.

CPSC 16 CFR, Part 1209-(2002), *Interim Safety Standard for Cellulose Insulation.* Bethesda, MD: Consumer Product Safety Commission, *2002.*

FEMA 44 CFR Parts 59-73: *National Flood Insurance Program (NFIP).*

FEMA 55CD (third edition): *Coastal Construction Manual.*

FEMA 232, *Home Builder's Guide to Seismic-resistant Construction.*

FEMA 265, *Managing Flood Plain Development in Approximate Zone Areas: A Guide for Obtaining and Developing Base (100-year) Flood Elevations.*

FEMA 348, *Protecting Building Utilities From Flood Damage: Principles and Practices for the Design and Construction of Flood-resistant Building Utility Systems.*

FEMA 499, *Home Builder's Guide to Coastal Construction: Technical Fact Sheet Series.*

FEMA 550, *Recommended Residential Construction for the Gulf Coast Building on Strong and Safe Foundations.*

FEMA TB #1, *Openings in Foundation Walls and Walls of Enclosures Below Elevated Buildings Located in Special Flood Hazard Areas.*

FEMA TB #2, *Flood Damage-resistant Materials Requirements for Buildings Located in Special Flood Hazard Areas in accordance with the National Flood Insurance Program.*

FEMA TB #4, *Elevator Installation for Buildings Located in Special Flood Hazard Areas.*

FEMA TB #5, *Free-of-obstruction Requirements for Buildings Located in Coastal High-Hazard Areas.*

FEMA TB #6, *Below-grade Parking Requirements for Buildings Located in Special Flood Hazard Areas.*

FEMA TB #8, *Corrosion Protection for Metal Connectors In Coastal Areas for Structures Located in Special Flood Hazard Areas.*

FEMA TB #9, *Design and Construction Guidance for Breakaway Walls Below Elevated Coastal Buildings in Coastal High-Hazard Areas.*

FEMA TB #10, *Ensuring that Structures Built on Fill In or Near Special Flood Hazard Areas are Reasonably Safe from Flooding.*

FEMA P-467-1, *Flood plain Management Bulletin on the Elevation Certificate.*

FEMA P-750; *National Earthquake Hazards Reduction Program Recommended Provisions for Seismic Regulations for New Buildings and Other Structures. 2009 Edition.* Washington, DC: Building Seismic Safety Council, 2009.

FEMA P-758, *Substantial Improvement/Substantial Damage Desk Reference.*

FEMA Form 81-31, *Elevation Certificate (2009).* Available at http://www.fema.gov/business/nfip/elvinst.shtm.

Fire Resistance Design Manual (GA-600-2009). Gypsum Association, Hyattsville, MD, 2009.

FM 4880-05, *American National Standard for Evaluating Insulated Wall or Wall and Roof/Ceiling Assemblies, Plastic Interior Finish Materials, Plastic Exterior Building Panels, Wall/Ceiling Coating Systems, Interior or Exterior Finish Systems.* Johnson, RI: Factory Mutual Global Research Standards Laboratories Department, 2005.

HUD 24 CFR Part 3285, *Model Manufactured Home Installation Standard.*

HUD Fair Housing Act.

HUD Fair Housing Accessibility Guidelines.

IBC-12, *International Building Code.* Washington, DC: International Code Council, 2011.

ICC 400-12, *Standard on The Design and Construction of Log Structures.* Washington, DC: International Code Council, 2011.

ICC 500-08, ICC/NSSA *Standard on the Design and Construction of Storm Shelters.* Washington, DC: International Code Council, 2008.

ICC 600-08, Standard for Residential Construction in High-wind Regions. Washington, DC: International Code Council, 2008.

ICC 700-08 *National Green Building Standard.* Washington, DC: International Code Council, 2008

ICC/ANSI A117.1-09, *Accessible and Usable Buildings and Facilities.* Washington, DC: International Code Council, 2009.

IFC-12, *International Fire Code.* Washington, DC: International Code Council, 2011.

I PC-12, *International Plumbing Code.* Washington, DC: International Code Council, 2011.

NFIP *Technical Bulletin Series.* Available at http://www.fema.gov/mit/techbul.htm

NFPA 13-10, *Installation of Sprinkler Systems.* Quincy, MA: National Fire Protection Association, 2010.

NFPA 13D-10, *Installation of Sprinkler Systems in One- and Two-family Dwellings and Manufactured Homes.* Quincy, MA: National Fire Protection Association, 2010.

NFPA 13R-10, *Installation of Sprinkler Systems in Residential Occupancies Up to and Including Four Stories in Height.* Quincy, MA: National Fire Protection Association, 2010.

NFPA 70-11, *National Electrical Code.* Quincy, MA: National Fire Protection Association, 2011.

NFPA 72-11, *National Fire Alarm Code.* Quincy, MA: National Fire Protection Association, 2011.

NFPA 259-08, *Standard Test Method for Potential Heat of Building Materials.* Quincy, MA: National Fire Protection Association, 2008.

NFPA 275-09, *Standard Method of Fire Tests for the Evaluation of Thermal Barriers Used Over Foam Plastic Insulation.* Quincy, MA: National Fire Protection Association, 2009

NFPA 286-11, *Standard Methods of Fire Tests for Evaluating Contribution of Wall and Ceiling Interior Finish to Room Fire Growth.* Quincy, MA: National Fire Protection Association, 2011.

NFPA 720-09 *Standard for The Installation of Carbon Monoxide (CO) Detection and Warning Equipment.* Quincy, MA: National Fire Protection Association, 2009

NIST 7451-07, *Benefit-cost Analysis of Residential Fire Sprinkler Systems.* Gaithersburg, MD: National Institute of Standards and Technology, 2007.

PCA 100-10, *Prescriptive Design for Exterior Concrete Walls for One- and Two-family Dwellings.* Skokie, IL: Portland Cement Organization, 2010.

Reducing Flood Losses Through the International Code Series: Meeting the Requirements of the National Flood Insurance Program. Washington, DC: International Code Council, 2008.

UL 217-06, *Single- and Multiple-station Smoke Alarms-with Revisions through August 2005.* Northbrook, IL: Underwriters Laboratories Inc., 2006.

UL 263-03, *Standard for Fire Test of Building Construction and Materials.* Northbrook, IL: Underwriters Laboratories Inc., 1996.

UL 325-02, *Door, Drapery, Gate, Louver and Window Operations and Systems—with Revisions through February 2006.* Northbrook, IL: Underwriters Laboratories Inc., 1996.

UL 723-03, *Standard for Test for Surface-burning Characteristics of Building Materials—with Revisions through May 2005.* Northbrook, IL: Underwriters Laboratories Inc., 1996.

UL 1040-96, *Fire Test of Insulated Wall Construction—with Revisions through April 2001.* Northbrook, IL: Underwriters Laboratories Inc., 1996.

UL 1479-03, *Fire Tests of Through-penetration Firestops.* Northbrook, IL: Underwriters Laboratories Inc., 2003.

UL 1715-97, *Fire Test of Interior Finish Material.* Northbrook, IL: Underwriters Laboratories Inc., 1997.

UL 2034-08, *Standard for Single- and Multiple-station Carbon Monoxide Alarms.* Northbrook, IL: Underwriters Laboratories Inc., 2008.

UL 2075-04, *Gas and Vapor Detectors and Sensors-with Revisions through September 28, 2007.* Northbrook, IL: Underwriters Laboratories Inc., 2004.

UL *Fire Resistance Directory.* Northbrook, IL: Underwriters Laboratories Inc., 2010.

USGS National Seismic Hazards Mapping Project, http://earthquake.usgs.gov/hazards. Reston, VA.

Chapter 4: Foundations

General Comments

Section R401 establishes the scope and applicability of the chapter. The criteria provide the details necessary for designing, building and inspecting most common foundation systems for homes. The selection of a particular type of foundation is based on many variables and it should result in the system that best suits the needs of a particular project. Also, the lot itself must be evaluated to determine the type of soil and the drainage pattern.

Section R402 specifies material requirements for wood, concrete and precast concrete foundations. Section R403 provides minimum requirements for concrete and masonry footings as well as footing for wood foundations and precast concrete foundations. Section R404 regulates foundation walls of concrete, masonry and wood as well as insulating concrete form foundation walls and precast concrete foundation walls. Section R405 has important criteria for maintaining the drainage of water away from foundations. The requirements for waterproofing and dampproofing in Section R406 protect below-grade habitable spaces from moisture. Section R407 provides requirements for underfloor areas and for protecting columns. Section R408 includes provisions for proper ventilation, adequate access, the removal of debris and flood resistance.

Purpose

Chapter 4 provides the requirements for the design and construction of foundation systems for buildings regulated by the code. Provisions for seismic load, flood load and frost protection are contained in this chapter. A foundation system consists of two interdependent components: the foundation structure itself and the supporting soil.

This chapter provides prescriptive requirements for constructing footings and walls for foundations of wood, masonry, concrete, and precast concrete. In addition to a foundation's ability to support the required design loads, this chapter addresses several other factors that can affect foundation performance. These include controlling surface water and subsurface drainage, requiring soil tests where conditions warrant, and evaluating proximity to slopes and minimum depth requirements. The chapter also provides requirements to minimize adverse effects of moisture, decay and pests in basements and crawl spaces.

SECTION R401 GENERAL

R401.1 Application. The provisions of this chapter shall control the design and construction of the foundation and foundation spaces for all buildings. In addition to the provisions of this chapter, the design and construction of foundations in flood hazard areas as established by Table R301.2(1) shall meet the provisions of Section R322. Wood foundations shall be designed and installed in accordance with AF&PA PWF.

Exception: The provisions of this chapter shall be permitted to be used for wood foundations only in the following situations:

1. In buildings that have no more than two floors and a roof.
2. When interior *basement* and foundation walls are constructed at intervals not exceeding 50 feet (15 240 mm).

Wood foundations in Seismic Design Category D_0, D_1 or D_2 shall be designed in accordance with accepted engineering practice.

❖This chapter contains all requirements relating to the design and construction of foundations and underfloor spaces. Also, an explicit link is provided to the flood-resistant construction provisions of Section R322. Thus, foundations for buildings to be located in flood hazard areas are to meet the provisions of Section R322 as well.

This section also specifically allows a wood foundation system designed and installed according to the American Forest & Paper Association (PWF) *Permanent Wood Foundation Design Specification.* When this system is installed under the prescriptive code requirements of this chapter, the code limits its use to structures where the loads exerted on the foundation are relatively light and to areas of low or moderate seismic hazard. This type of foundation construction is essentially a below-grade, load-bearing, wood-frame system capable of providing support for light frame structures. It generally would consist of the following components:

1. Walls consisting of plywood fastened to wood studs (Section R404.2).
2. A composite footing made up of a continuous wood plate set on a bed of granular materials, such as sand, gravel, or crushed stone, which in turn supports the foundation walls and transmits their loads to the bearing soil below (see Section R403.2).
3. Polyethylene film that serves as a vapor barrier and covers the exterior side of the plywood foundation walls from grade level down to the footing plate (see Section R406.3).
4. Caulking compounds used for sealing the joints in the plywood walls, bonding agents for attach-

ing the polyethylene film to the plywood, and for the film joints (see Section R406.3).

5. Metal fasteners made of silicon bronze, copper, or stainless steel or hot-dipped zinc-coated steel nails or staples (see Section R402.1).
6. Pressure-treated plywood and lumber to protect the foundation material against decay, termites, and other insects (see Section R402.1).

R401.2 Requirements. Foundation construction shall be capable of accommodating all loads according to Section R301 and of transmitting the resulting loads to the supporting soil. Fill soils that support footings and foundations shall be designed, installed and tested in accordance with accepted engineering practice. Gravel fill used as footings for wood and precast concrete foundations shall comply with Section R403.

❖In order to fulfill its role in the complete load path, the foundation must support the required design loads and transmit these to the soil. The phrase "accepted engineering practice" in this section means common practice that is acceptable to the code official of the jurisdiction.

R401.3 Drainage. Surface drainage shall be diverted to a storm sewer conveyance or other *approved* point of collection that does not create a hazard. *Lots* shall be graded to drain surface water away from foundation walls. The *grade* shall fall a minimum of 6 inches (152 mm) within the first 10 feet (3048 mm).

Exception: Where *lot lines*, walls, slopes or other physical barriers prohibit 6 inches (152 mm) of fall within 10 feet (3048 mm), drains or swales shall be constructed to ensure drainage away from the structure. Impervious surfaces within 10 feet (3048 mm) of the building foundation shall be sloped a minimum of 2 percent away from the building.

❖Along with the proper support for a structure through the foundation system, adequate preparation of the building site is necessary to keep water drainage away from the supporting foundations. Proper site drainage is an important element in preventing wet basements, damp crawl spaces, eroded banks, and possible failure of a foundation system.

One of the most important considerations is the arrangement of structures on a building site in a manner that retains natural drainage patterns and minimizes the alteration or disturbance to existing grades. If the designer keeps such factors in mind, the result will be a reduction of ground surface stabilization problems and opportunities for differential settlement through the reduction in the use of fills. A detailed treatment of drainage design is beyond the scope of this document; therefore, only rough guidelines can be provided for areas where a more comprehensive set of grading regulations does not exist.

As illustrated in Commentary Figures R401.3(1) and R401.3(2), drainage patterns should result in adequate slopes to approved drainage devices that are capable of carrying concentrated runoff. In some cases, control of concentrated roof runoff by gutters and downspouts may be needed, and if gutters and downspouts are used, provisions should be made to discharge runoff in order to prevent soil erosion. Refer also to Section R801.3.

Cross-lot drainage and drainage over graded slopes should generally be avoided. Slopes should be designed with as moderate a grade as possible to minimize instability and erosion. The minimum slope gradients which should be used are a function of the combined ground frost and moisture conditions, soil type, geological features, and geographic conditions. The slope away from the building is required to be 6 inches (152 mm) within the first 10 feet (3048 mm). The exception in this section allows for use of drains and swales if lot lines or physical barriers, such as a steeply sloping lot, do not allow for a 6-inch (152 mm) slope.

R401.4 Soil tests. Where quantifiable data created by accepted soil science methodologies indicate expansive, compressible, shifting or other questionable soil characteristics are likely to be present, the *building official* shall determine whether to require a soil test to determine the soil's character-

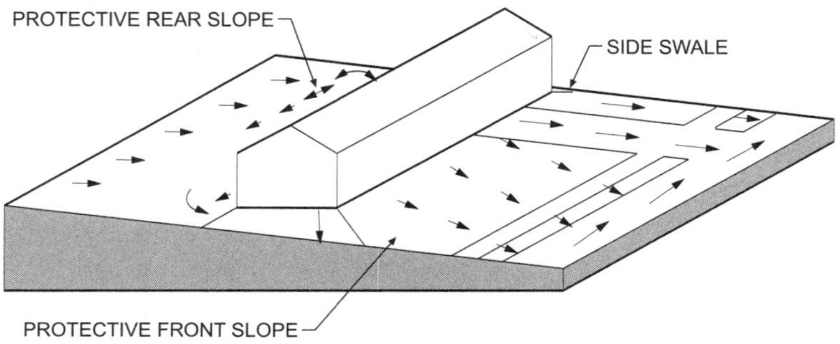

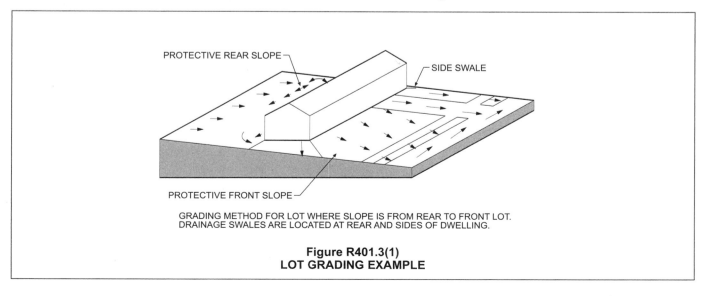

GRADING METHOD FOR LOT WHERE SLOPE IS FROM REAR TO FRONT LOT. DRAINAGE SWALES ARE LOCATED AT REAR AND SIDES OF DWELLING.

Figure R401.3(1)
LOT GRADING EXAMPLE

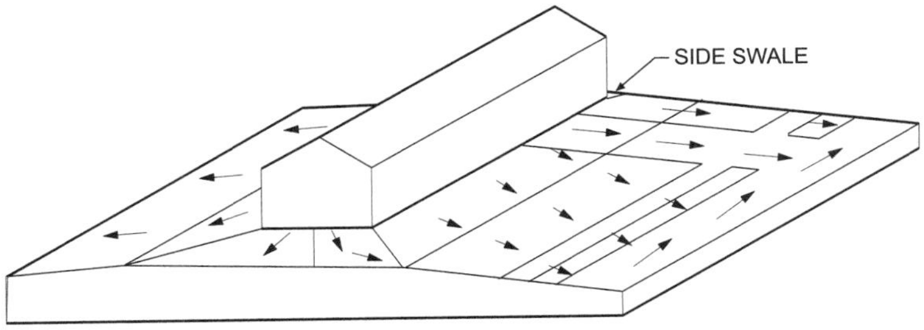
SIDE SWALE
GRADING METHOD FOR LOT WHERE DWELLING IS LOCATED ON A RIDGE.
DRAINAGE SWALES ARE LOCATED AT SIDE YARDS IN AREA OF POSITIVE SLOPES.

Figure R401.3(2)
LOT GRADING EXAMPLE

istics at a particular location. This test shall be done by an *approved agency* using an *approved* method.

❖Loading conditions beyond the normal live, dead, and wind or earthquake loading generally include the types of loading attributable to special regional conditions. Such conditions might include soil instability, forces generated on foundations by expansive soils, and increased lateral pressures due to a high water table or surcharge loads from adjacent structures. Accounting for such conditions in the structure is generally considered to be beyond the scope of the conventional methods of foundation construction specified within this chapter. In some cases, a soil test may be required to evaluate the existing conditions.

Where the bearing capacity of the soil has not been determined by geotechnical evaluation such as borings, field load tests, laboratory tests, and engineering analysis, it is a common practice to use presumptive bearing values for the design of the foundation system. Section R401.4.1 requires that the presumptive bearing values in Table R401.4.1 be used if a geotechnical evaluation has not been performed.

R401.4.1 Geotechnical evaluation. In lieu of a complete geotechnical evaluation, the load-bearing values in Table R401.4.1 shall be assumed.

❖See the commentary for Section R401.4.

R401.4.2 Compressible or shifting soil. Instead of a complete geotechnical evaluation, when top or subsoils are compressible or shifting, they shall be removed to a depth and width sufficient to assure stable moisture content in each active zone and shall not be used as fill or stabilized within each active zone by chemical, dewatering or presaturation.

❖A geotechnical engineer should be consulted by the owner or the owner's builder if there is a concern that a compressible or shifting soil condition exists. Where these conditions are present, one of two methods must be used. The first of these deals with removal and replacement of the soil, thus reducing the forces associated with soil movement. The second approach is soil stabilization by chemical means, dewatering or presaturation.

TABLE R401.4.1
PRESUMPTIVE LOAD-BEARING VALUES OF FOUNDATION MATERIALS[a]

CLASS OF MATERIAL	LOAD-BEARING PRESSURE (pounds per square foot)
Crystalline bedrock	12,000
Sedimentary and foliated rock	4,000
Sandy gravel and/or gravel (GW and GP)	3,000
Sand, silty sand, clayey sand, silty gravel and clayey gravel (SW, SP, SM, SC, GM and GC)	2,000
Clay, sandy clay, silty clay, clayey silt, silt and sandy silt (CL, ML, MH and CH)	1,500[b]

For SI: 1 pound per square foot = 0.0479 kPa.

a. When soil tests are required by Section R401.4, the allowable bearing capacities of the soil shall be part of the recommendations.

b. Where the building official determines that in-place soils with an allowable bearing capacity of less than 1,500 psf are likely to be present at the site, the allowable bearing capacity shall be determined by a soils investigation.

❖As explained in the commentary for Section R401.4, where a soils investigation is not required by the code, load-bearing values of Table R401.4.1 are to be used. The values vary by soil classifications, which are based on the Unified Soil Classification System. Table R405.1 also gives these soil descriptions in more detail. The allowable soil bearing values listed in the table are based on lengthy experience with the behavior of these materials in supporting loads from all types of structures.

SECTION R402
MATERIALS

R402.1 Wood foundations. Wood foundation systems shall be designed and installed in accordance with the provisions of this code.

❖The wood foundation system requirements are from the American Forest & Paper Association (PWF) *Per-*

manent Wood Foundation Design Specification. Figures R403.1(2) and R403.1(3) of the code illustrate some typical details of this system. Also refer to the commentary for Section R401.1.

R402.1.1 Fasteners. Fasteners used below *grade* to attach plywood to the exterior side of exterior *basement* or crawlspace wall studs, or fasteners used in knee wall construction, shall be of Type 304 or 316 stainless steel. Fasteners used above *grade* to attach plywood and all lumber-to-lumber fasteners except those used in knee wall construction shall be of Type 304 or 316 stainless steel, silicon bronze, copper, hot-dipped galvanized (zinc coated) steel nails, or hot-tumbled galvanized (zinc coated) steel nails. Electro-galvanized steel nails and galvanized (zinc coated) steel staples shall not be permitted.

❖Appropriate fasteners must be used in wood foundation construction because the presence of any moisture in combination with the preservative treatment can corrode incompatible fasteners.

R402.1.2 Wood treatment. All lumber and plywood shall be pressure-preservative treated and dried after treatment in accordance with AWPA U1 (Commodity Specification A, Use Category 4B and Section 5.2), and shall bear the *label* of an accredited agency. Where lumber and/or plywood is cut or drilled after treatment, the treated surface shall be field treated with copper naphthenate, the concentration of which shall contain a minimum of 2 percent copper metal, by repeated brushing, dipping or soaking until the wood absorbs no more preservative.

❖Performance of the wood foundation system is dependent on the use of properly treated materials; thus, the code provision emphasizes the use of properly treated lumber and plywood. Verification of the proper materials is provided by identification showing the approval of an accredited inspection agency. An example of such identification is shown in Commentary Figure R402.1.2. Where treated lumber or plywood is field cut (exposing wood that is untreated), the code specifies the method of treating the cut surface.

R402.2 Concrete. Concrete shall have a minimum specified compressive strength of f'_c, as shown in Table R402.2. Concrete subject to moderate or severe weathering as indicated in Table R301.2(1) shall be air entrained as specified in Table R402.2. The maximum weight of fly ash, other pozzolans, silica fume, slag or blended cements that is included in concrete mixtures for garage floor slabs and for exterior porches, carport slabs and steps that will be exposed to deicing chemicals shall not exceed the percentages of the total weight of cementitious materials specified in Section 4.2.3 of ACI 318. Materials used to produce concrete and testing thereof shall comply with the applicable standards listed in Chapter 3 of ACI 318 or ACI 332.

❖The code specifies minimum concrete compressive strengths ranging from 2,500 to 3,000 psi (17.2 to 24.1 Mpa). Table R402.2 specifies the required compressive strength of concrete based on a locale's weathering potential and a building element's exposure. Section R301 uses the Weathering Probability Map for Concrete to classify an area's weathering potential as negligible, moderate or severe. For concrete that will be subject to freezing and thawing in a moist condition (i.e., weathering) or subject to direct or indirect application of deicing chemicals, Table R402.2 requires the use of higher-strength, air-entrained concrete. Freezing and thawing cycles can be the most destructive weathering factors for concrete when it is wet. Water freezing in the cement matrix, in the aggregate, or both causes deterioration of concrete that is not air-entrained. Studies have documented that concrete with proper air entrainment is highly resistant to this deterioration.

All materials and testing must be in accordance with the American Concrete Institute's ACI 318, *Building Code Requirements for Structural Concrete*. Cementitious materials include portland cement (ASTM C

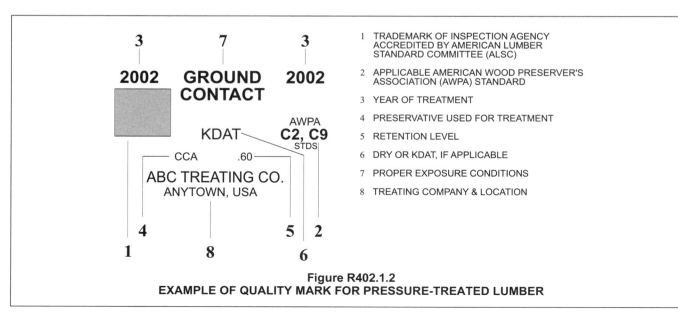

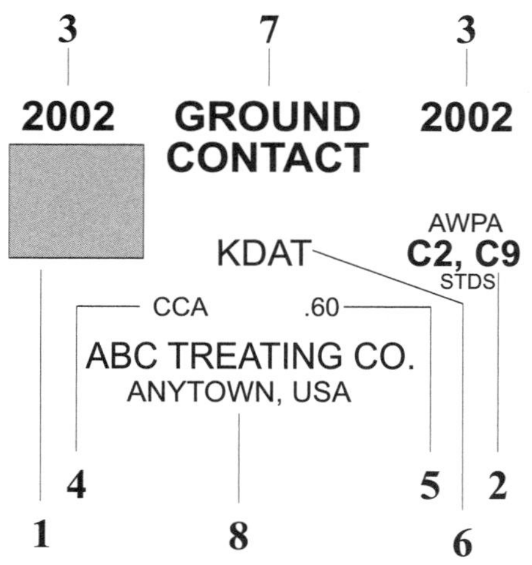

1 TRADEMARK OF INSPECTION AGENCY ACCREDITED BY AMERICAN LUMBER STANDARD COMMITTEE (ALSC)
2 APPLICABLE AMERICAN WOOD PRESERVER'S ASSOCIATION (AWPA) STANDARD
3 YEAR OF TREATMENT
4 PRESERVATIVE USED FOR TREATMENT
5 RETENTION LEVEL
6 DRY OR KDAT, IF APPLICABLE
7 PROPER EXPOSURE CONDITIONS
8 TREATING COMPANY & LOCATION

Figure R402.1.2
EXAMPLE OF QUALITY MARK FOR PRESSURE-TREATED LUMBER

150), blended hydraulic cements (ASTM C 595 and ASTM C 1157), and expansive cement (ASTM C 845). They can also include pozzolanic materials such as fly ash and other raw or calcined natural pozzolan (ASTM C 618), ground granulated blast-furnace slag (ASTM C 989), and/or silica fume (ASTM C 1240) when used in combination with the cements. Most concrete contains admixtures such as fly ash or other pozzolans that are added as a replacement for a portion of the cement. Although such substitutions can be for economic reasons, in some cases they can also be used to improve certain concrete properties.

TABLE R402.2. See below.

❖The code requires the use of higher-strength, air-entrained concrete for enclosed garage floor slabs in moderate and severe weathering regions. Even though these slabs may be enclosed and not subject to freezing and thawing conditions, they are likely to be subject to the deteriorating effects of deicing chemicals that drip from vehicles. The replacement of portland cement with other cementitious materials can be detrimental to the durability of the concrete if the concrete will be exposed to deicing chemicals. Thus, the code limits the total amount of these replacement materials to 50 percent of the total weight of all cementitious materials.

Note c to Table R402.2 requires air-entrained concrete for concrete elements not exposed to freezing and thawing conditions when the building is completed if it is likely that these elements will be subject to freezing and thawing during construction (also see commentary, Section R404.2).

Note f permits the air entrainment to be reduced to not less than 3 percent where the garage floor will have steel trowel finish. Field experience has shown that durable concrete may be obtained with a lesser amount of air-entrainment if the specified compressive strength of the concrete is increased to 4,000 psi (27.6 MPa). The higher concrete strength is accompanied by a denser cement-paste matrix, which results in the concrete being less permeable. This option has been successfully used on garage slabs with a steel trowel finish in areas of moderate and severe exposures.

R402.3 Precast concrete. Precast concrete foundations shall be designed in accordance with Section R404.5 and shall be installed in accordance with the provisions of this code and the manufacturer's installation instructions.

❖This section provides requirements for the design and installation of precast concrete foundations. See Section R404.5 for the design requirements for precast concrete walls. The materials used to produce precast concrete foundations must comply with Section R402.3.1. All requirements in the code for concrete foundations apply to precast concrete as well. Also, any manufacturer's instructions must be complied with.

R402.3.1 Precast concrete foundation materials. Materials used to produce precast concrete foundations shall meet the following requirements.

1. All concrete used in the manufacture of precast concrete foundations shall have a minimum compressive strength of 5,000 psi (34 470 kPa) at 28 days. Concrete exposed to a freezing and thawing environment shall be air entrained with a minimum total air content of 5 percent.
2. Structural reinforcing steel shall meet the requirements of ASTM A 615, A 706 or A 996. The minimum yield strength of reinforcing steel shall be 40,000 psi (Grade 40) (276 MPa). Steel reinforcement for precast concrete foundation walls shall have a minimum concrete cover of $^3/_4$ inch (19.1 mm).
3. Panel-to-panel connections shall be made with Grade II steel fasteners.

TABLE R402.2
MINIMUM SPECIFIED COMPRESSIVE STRENGTH OF CONCRETE

TYPE OR LOCATION OF CONCRETE CONSTRUCTION	MINIMUM SPECIFIED COMPRESSIVE STRENGTH[a] (f'_c)		
	Weathering Potential[b]		
	Negligible	Moderate	Severe
Basement walls, foundations and other concrete not exposed to the weather	2,500	2,500	2,500[c]
Basement slabs and interior slabs on grade, except garage floor slabs	2,500	2,500	2,500[c]
Basement walls, foundation walls, exterior walls and other vertical concrete work exposed to the weather	2,500	3,000[d]	3,000[d]
Porches, carport slabs and steps exposed to the weather, and garage floor slabs	2,500	3,000[d, e, f]	3,500[d, e, f]

For SI: 1 pound per square inch = 6.895 kPa.

a. Strength at 28 days psi.

b. See Table R301.2(1) for weathering potential.

c. Concrete in these locations that may be subject to freezing and thawing during construction shall be air-entrained concrete in accordance with Footnote d.

d. Concrete shall be air-entrained. Total air content (percent by volume of concrete) shall be not less than 5 percent or more than 7 percent.

e. See Section R402.2 for maximum cementitious materials content.

f. For garage floors with a steel-troweled finish, reduction of the total air content (percent by volume of concrete) to not less than 3 percent is permitted if the specified compressive strength of the concrete is increased to not less than 4,000 psi

4. The use of nonstructural fibers shall conform to ASTM C 1116.
5. Grout used for bedding precast foundations placed upon concrete footings shall meet ASTM C 1107.

❖Precast concrete foundations are preengineered systems. This section prescribes the minimum quality for materials to be used in the manufacture of precast concrete foundations.

SECTION R403
FOOTINGS

R403.1 General. All exterior walls shall be supported on continuous solid or fully grouted masonry or concrete footings, crushed stone footings, wood foundations, or other *approved* structural systems which shall be of sufficient design to accommodate all loads according to Section R301 and to transmit the resulting loads to the soil within the limitations as determined from the character of the soil. Footings shall be supported on undisturbed natural soils or engineered fill. Concrete footing shall be designed and constructed in accordance with the provisions of Section R403 or in accordance with ACI 332.

❖The provisions of this section include the general statement, also contained in Section R401.2, that a footing must be capable of supporting the required design loads. The section expands on this by referencing the character of the soil and the minimum extension below the frost line. The code mandates that footings be supported on undisturbed natural soil or engineered fill. The use of ACI 332 for concrete footing is permitted. ACI 332 provides additional technical details for the design and construction of concrete footings.

R403.1.1 Minimum size. Minimum sizes for concrete and masonry footings shall be as set forth in Table R403.1 and Figure R403.1(1). The footing width, W, shall be based on the load-bearing value of the soil in accordance with Table R401.4.1. Spread footings shall be at least 6 inches (152 mm) in thickness, T. Footing projections, P, shall be at least 2 inches (51 mm) and shall not exceed the thickness of the footing. The size of footings supporting piers and columns shall be based on the tributary load and allowable soil pressure in accordance with Table R401.4.1. Footings for wood foundations shall be in accordance with the details set forth in Section R403.2, and Figures R403.1(2) and R403.1(3).

❖Table R403.1 specifies minimum footing widths based on soil pressure and the number of stories supported. Figure R403.1(1) illustrates masonry and concrete footings. To avoid construction that might overstress the footing, this section also specifies the minimum footing thickness and places a limitation on the maximum projection of the footing beyond the face of the foundation wall as shown in Figure R403.1(1). The size of isolated footings for piers and columns should be determined from the tributary (design) load supported using the presumptive soil values of Table R401.4.1.

The limitation on the maximum projection of the footing beyond the face of the foundation wall is particularly important for 6-inch-thick (152 mm) footings of plain concrete subjected to soil-bearing pressures on the order of 2,000 pounds per square foot (95.8 kPa). Projections in excess of the maximum allowable could result in the footing being cracked in a plane coinciding with the face of the foundation wall. A crack could occur if the allowable flexural tension stress of the concrete is exceeded.

R403.1.2 Continuous footing in Seismic Design Categories D_0, D_1 and D_2. The *braced wall panels* at exterior walls of buildings located in Seismic Design Categories D_0, D_1 and D_2 shall be supported by continuous footings. All required interior *braced wall panels* in buildings with plan dimensions greater than 50 feet (15 240 mm) shall also be supported by continuous footings.

❖The code requires continuous footings for support of braced wall panels in exterior walls, a critical component of a structure's lateral-force-resisting system (see Chapter 6). Braced wall panels must be supported by a foundation that is continuous along the entire length of the braced wall line. For a structure classified as Seismic Design Category D_0, D_1 or D_2 the code requires continuous footings at all exterior walls as well as at interior braced wall lines in buildings with plan dimensions greater than 50 feet (15 240 mm). For buildings in Seismic Design Category D_2, additional continuous foundation requirements for interior braced wall lines must comply with Section R602.10.9 which requires continuous foundation for all interior braced wall lines for two-story buildings regardless of plan dimension (see commentary, Section R602.10.9).

TABLE R403.1
MINIMUM WIDTH OF CONCRETE, PRECAST OR MASONRY FOOTINGS (inches)[a]

	LOAD-BEARING VALUE OF SOIL (psf)			
	1,500	2,000	3,000	≥ 4,000
Conventional light-frame construction				
1-story	12	12	12	12
2-story	15	12	12	12
3-story	23	17	12	12
4-inch brick veneer over light frame or 8-inch hollow concrete masonry				
1-story	12	12	12	12
2-story	21	16	12	12
3-story	32	24	16	12
8-inch solid or fully grouted masonry				
1-story	16	12	12	12
2-story	29	21	14	12
3-story	42	32	21	16

For SI: 1 inch = 25.4 mm, 1 pound per square foot = 0.0479 kPa.
a. Where minimum footing width is 12 inches, use of a single wythe of solid or fully grouted 12-inch nominal concrete masonry units is permitted.

❖Table R403.1 specifies minimum footing widths where normal soil conditions are encountered. The minimum footing widths are based on the soil load-bearing values in Table R401.4.1. The minimum values were calculated based on 20 feet (6096 mm) of tributary roof area, 16 feet (4877 mm) of tributary floor areas, 10-foot-high (3048 mm) first-story walls, and 8-foot-high (2438 mm) second- and third-story walls. A 50-pounds-per-square foot (2.4 kPa) snow load was used.

A minimum footing width of 12 inches (305 mm) is required so that footings can span over weak locations in the soil and to allow for minor misalignment of foundations. The footnote permits solid or fully grouted 12-inch-nominal (305 mm) concrete masonry block to be used for this minimum width of footing. Section R403.1.1 specifies the minimum thickness of footings. These are intended to ensure that the structure transfers loads to the supporting soil without exceeding the capacity of the materials used to construct the footings. Table R403.1 prescribes the footing width required based on story height above grade, for buildings with "a story below grade," i.e., basement and stories above grade the additional load of the "story below grade" must be accounted for. The additional load will be the dead load of the basement wall plus any tributary load from the basement floor load. As an example, a one-story building with a basement could conservatively use the two-story requirement rather than provide an engineered design.

If any of the following conditions exist in the area of the foundation, conventional spread footings should

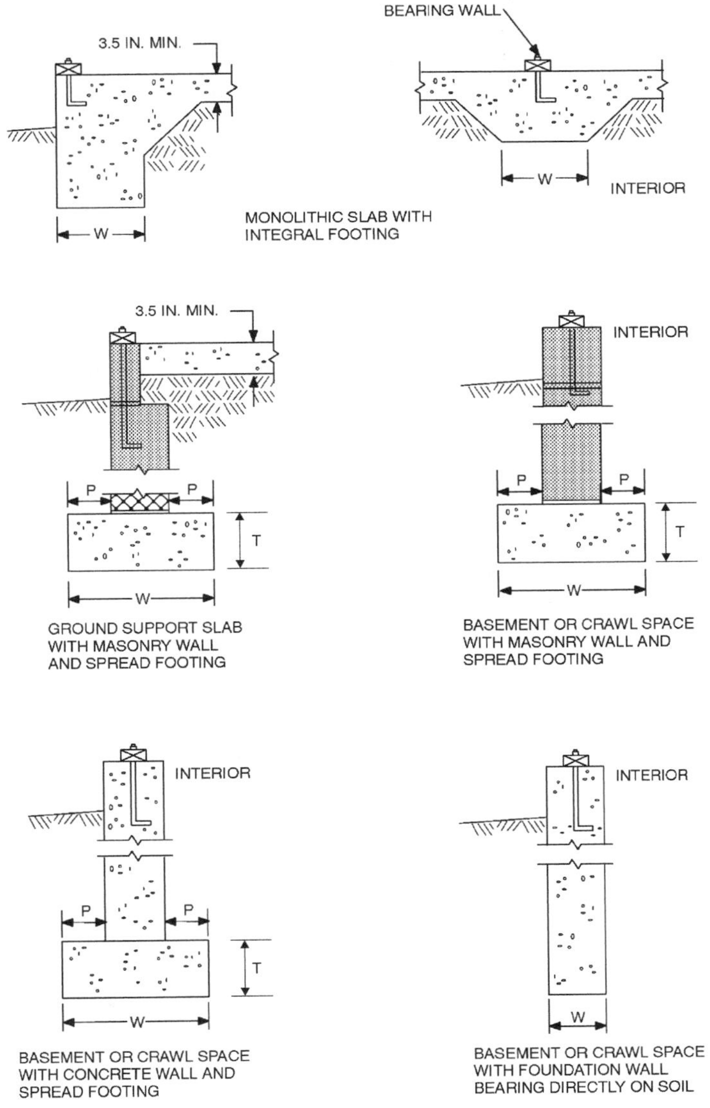

For SI: 1 inch = 25.4 mm.

FIGURE R403.1(1)
CONCRETE AND MASONRY FOUNDATION DETAILS

❖See the commentary to Section R403.1.

not be used, and a designed foundation may be necessary:

- Filled ground, except when properly compacted.
- Foundation soils subject to subsidence.
- Expansive soils such as those having a plasticity index of 15 or greater (see Section R403.1.8.1).
- Highly compressive clays.
- Unconfined sands and silts.

In cold weather, care must be taken that foundations are not placed on frozen soil, nor should such foundations be placed in freezing weather unless a method of ensuring that the underlying soil is free of frost is employed and the foundations are properly protected against the weather.

R403.1.3 Seismic reinforcing. Concrete footings located in Seismic Design Categories D_0, D_1 and D_2, as established in Table R301.2(1), shall have minimum reinforcement. Bottom reinforcement shall be located a minimum of 3 inches (76 mm) clear from the bottom of the footing.

In Seismic Design Categories D_0, D_1 and D_2 where a construction joint is created between a concrete footing and a stem wall, a minimum of one No. 4 bar shall be installed at not more than 4 feet (1219 mm) on center. The vertical bar shall extend to 3 inches (76 mm) clear of the bottom of the footing, have a standard hook and extend a minimum of 14 inches (357 mm) into the stem wall.

In Seismic Design Categories D_0, D_1 and D_2 where a grouted masonry stem wall is supported on a concrete footing and stem wall, a minimum of one No. 4 bar shall be installed at not more than 4 feet (1219 mm) on center. The vertical bar shall extend to 3 inches (76 mm) clear of the bottom of the footing and have a standard hook.

In Seismic Design Categories D_0, D_1 and D_2 masonry stem walls without solid grout and vertical reinforcing are not permitted.

Exception: In detached one- and two-family *dwellings* which are three stories or less in height and constructed with stud bearing walls, isolated plain concrete footings, supporting columns or pedestals are permitted.

❖This section specifies minimal reinforcement for buildings classified as Seismic Design Category D_0, D_1 or D_2. Interconnection of stem walls and supporting foundations is necessary to resist the tendency of the wall to slip in an earthquake. In grouted masonry stem walls supported on concrete footings and at construction joints between concrete stem walls and their supporting footing, a minimum of one vertical No. 4 bar at 48 inches (1219 mm) spacing must be provided. A monolithically cast foundation and stem

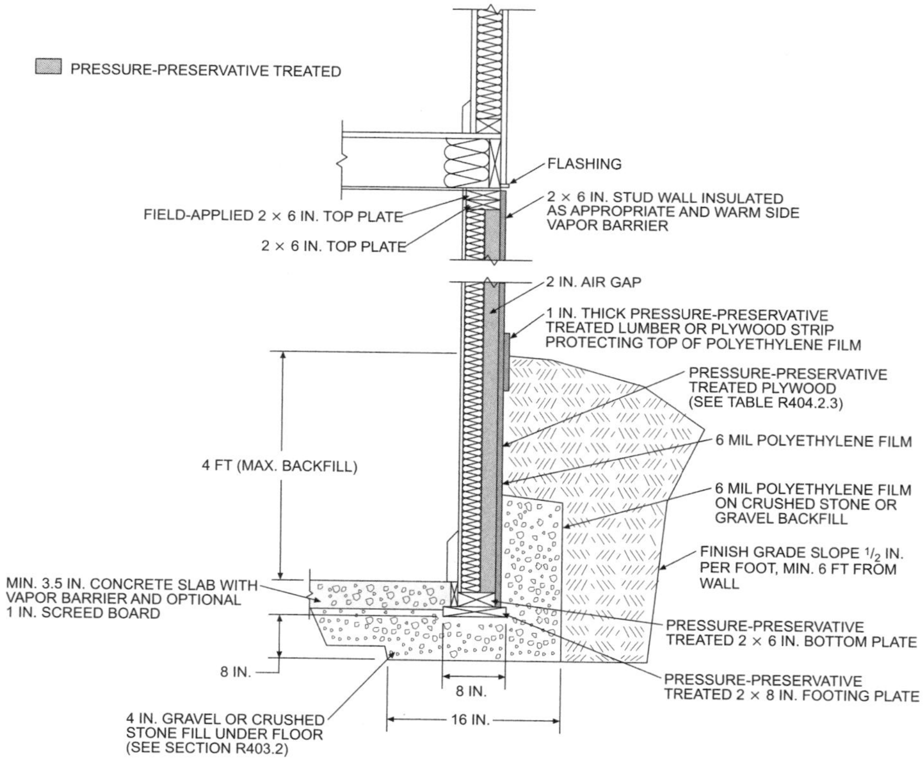

For SI: 1 inch = 25.4 mm, 1 foot = 304.8 mm, 1 mil = 0.0254.

FIGURE R403.1(2)
PERMANENT WOOD FOUNDATION BASEMENT WALL SECTION

❖See the commentary to Section R403.1.

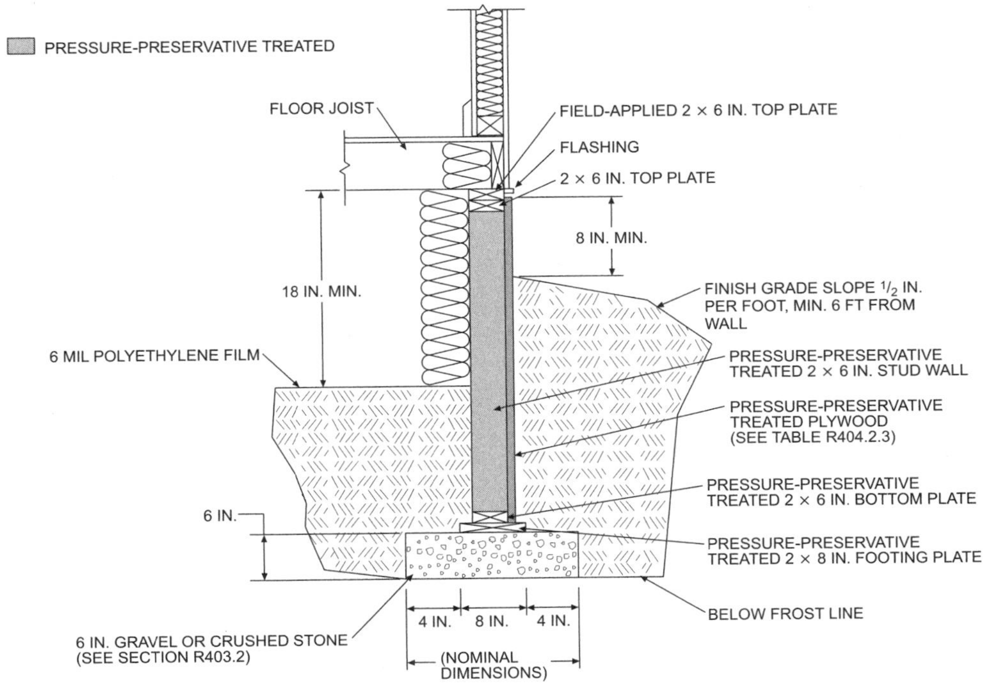

For SI: 1 inch = 25.4 mm, 1 foot = 304.8 mm, 1 mil = 0.0254 mm.

FIGURE R403.1(3)
PERMANENT WOOD FOUNDATION CRAWL SPACE SECTION

❖See the commentary to Section R403.1.

wall do not require this vertical reinforcement. Masonry stem walls must be grouted and provided with vertical reinforcement. The exception permits isolated plain concrete footings for columns or pedestals since these are not used to support or anchor braced walls unless engineered. If an alternate braced wall panel is used in accordance with Section R602.10.6, that section contains more specific foundation requirements that also must be met.

R403.1.3.1 Foundations with stemwalls. Foundations with stem walls shall have installed a minimum of one No. 4 bar within 12 inches (305 mm) of the top of the wall and one No. 4 bar located 3 inches (76 mm) to 4 inches (102 mm) from the bottom of the footing.

❖Where a stem wall exists, longitudinal reinforcement consisting of one No. 4 bar at the top of the wall and one No. 4 bar located 3 to 4 inches (76 to 102 mm) from the bottom of the footing must be provided for footings in Seismic Design Category D_0, D_1 or D_2.

R403.1.3.2 Slabs-on-ground with turned-down footings. Slabs on ground with turned down footings shall have a minimum of one No. 4 bar at the top and the bottom of the footing.

Exception: For slabs-on-ground cast monolithically with the footing, locating one No. 5 bar or two No. 4 bars in the middle third of the footing depth shall be permitted as an alternative to placement at the footing top and bottom.

Where the slab is not cast monolithically with the footing, No. 3 or larger vertical dowels with standard hooks on each end shall be provided in accordance with Figure R403.1.3.2. Standard hooks shall comply with Section R611.5.4.5.

❖For footings in Seismic Design Category D_0, D_1 or D_2, a slab-on-ground with a turned-down footing requires a No. 4 bar at the top and bottom of the footing. If cast monolithically with the slab, the exception permits either one No. 5 or two No. 4 bars to be placed in the middle third of the footing depth. Where a construction joint occurs, vertical dowels must be placed as shown in Figure R403.1.3.2. These dowels are required to prevent damage due to slippage at the construction joint during a seismic event.

R403.1.4 Minimum depth. All exterior footings shall be placed at least 12 inches (305 mm) below the undisturbed ground surface. Where applicable, the depth of footings shall also conform to Sections R403.1.4.1 through R403.1.4.2.

❖Footings are required to extend below the ground surface a minimum of 12 inches (305 mm). This is considered a minimum depth to protect the footing from movement of the soil caused by freezing and thawing in mild climate areas. See Section R403.1.4.1 for general frost protection requirements.

Footing depths may also be influenced by adjacent footings at different elevations; the footing at a higher elevation imposes surcharge pressures on lower adjacent footings. One method of alleviating the surcharge pressures on lower adjacent footings is the use of the guidelines illustrated in Commentary Figure R403.1.4. Other methods based on a soil and foundation analysis are also acceptable.

R403.1.4.1 Frost protection. Except where otherwise protected from frost, foundation walls, piers and other permanent supports of buildings and structures shall be protected from frost by one or more of the following methods:

1. Extended below the frost line specified in Table R301.2.(1);
2. Constructing in accordance with Section R403.3;
3. Constructing in accordance with ASCE 32; or
4. Erected on solid rock.

Exceptions:

1. Protection of freestanding *accessory structures* with an area of 600 square feet (56 m^2) or less, of light-frame construction, with an eave height of 10 feet (3048 mm) or less shall not be required.

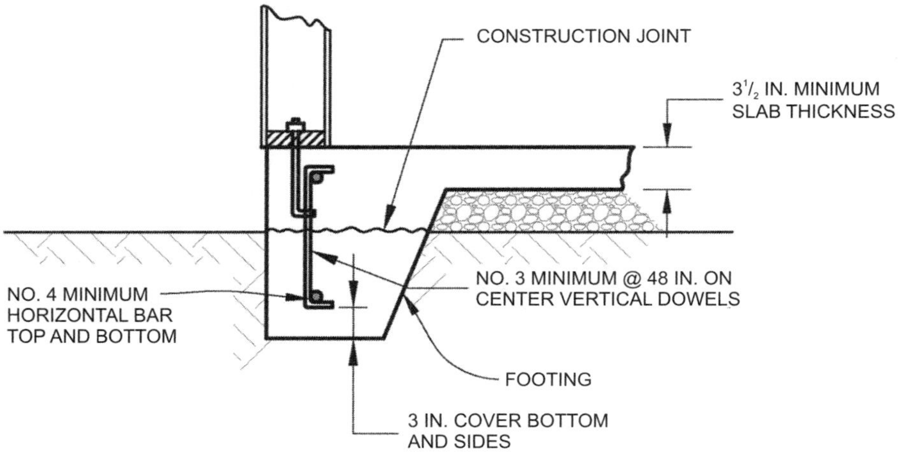

For SI: 1 inch = 25.4 mm.

FIGURE R403.1.3.2
DOWELS FOR SLABS-ON-GROUND WITH TURNED-DOWN FOOTINGS

❖See the commentary to Section R403.1.3.2.

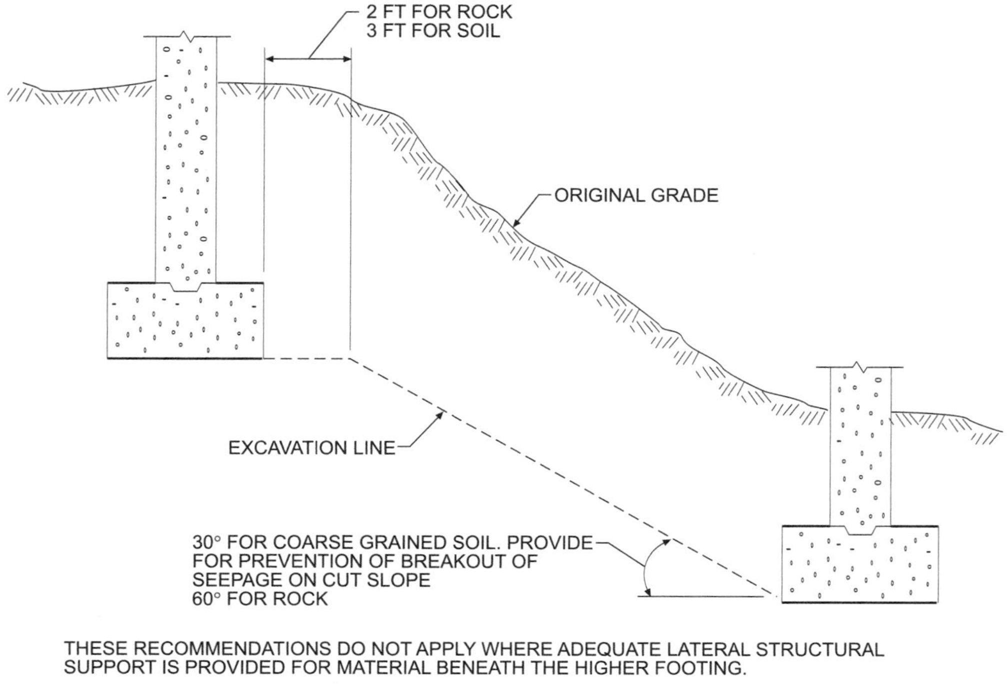

For SI: 1 foot = 304.8 mm, 1 degree = 0.01745 rad.

Figure R403.1.4
RECOMMENDATION FOR ADJACENT FOOTINGS AT DIFFERENT ELEVATIONS

2. Protection of freestanding *accessory structures* with an area of 400 square feet (37 m^2) or less, of other than light-frame construction, with an eave height of 10 feet (3048 mm) or less shall not be required.
3. Decks not supported by a dwelling need not be provided with footings that extend below the frost line.

Footings shall not bear on frozen soil unless the frozen condition is permanent.

❖Buildings must not be founded on or within frozen ground. This provision prevents damage to the exterior walls and other walls bearing on the frozen soil. The volume changes (frost heave) that take place during freezing and thawing produce excessive stresses in the foundations and, as a result, may cause extensive damage to the walls that are supported.

Section R403.1.4.1, Frost Protection, contains the three acceptable ways of protecting the footings and foundations from frost: (1) extend below the frost line; (2) insulate the foundation according to the code's prescriptive frost protected shallow foundation requirements or the ASCE 32, *Design and Construction of Frost-protected Shallow Foundation*, performance standard; and (3) erect the foundation on solid rock.

The frost-line depth is established by the authority having jurisdiction based on field experience and must be listed along with other design criteria in Table R301.2(1). Exterior foundations must extend below the frost line and a minimum of 12 inches (305 mm) below undisturbed soil. Foundations supported on rock or that are frost-protected in accordance with Section R403.3 are exempt from the extension below the frost line, but the 12-inch (305 mm) minimum still applies.

ASCE 32 contains requirements for:

1. A simplified method for heated buildings;
2. A more complicated performance method that permits trading off insulation for frost-susceptible fill or use of varying foundation depths;
3. Requirements for slabs-on-ground, crawl spaces (without outside vents), walk out basements and semi-heated buildings;
4. Requirements for unheated buildings that conserve heat from the ground with thick layers of insulation under the entire foundation and beyond it several feet; continuous walls and columns; and
5. Special design conditions.

All the ASCE standard requirements are consistent with theory, widespread practice in Scandinavia, and the successful construction of thousands of buildings in the U.S.

Foundations are not to be placed on frozen soil because when the ground thaws, uneven settlement of the structure is apt to occur, thereby causing structural damage. This section does, however, permit footings to be constructed on permanently frozen soil. In permafrost areas, special precautions are necessary to prevent heat from the structure from thawing the soil beneath the foundation.

R403.1.4.2 Seismic conditions. In Seismic Design Categories D_0, D_1 and D_2, interior footings supporting bearing or bracing walls and cast monolithically with a slab on *grade* shall extend to a depth of not less than 12 inches (305 mm) below the top of the slab.

❖In buildings classified as Seismic Design Category D_0, D_1 or D_2, the code establishes the minimum depth below the top of slab for interior footings that support bearing or bracing walls and are poured monolithically with the slab. This allows for the installation of anchor bolts, tie-downs, etc.

R403.1.5 Slope. The top surface of footings shall be level. The bottom surface of footings shall not have a slope exceeding one unit vertical in 10 units horizontal (10-percent slope). Footings shall be stepped where it is necessary to change the elevation of the top surface of the footings or where the slope of the bottom surface of the footings will exceed one unit vertical in ten units horizontal (10-percent slope).

❖The code requires that the top surface of footings for buildings be essentially level. However, a slope of 1 unit vertical in 10 units horizontal (10-percent slope) is permitted for the bottom bearing surface. If the slope is steeper, the foundation must be stepped. Commentary Figure R403.1.5 schematically shows a stepped foundation. Although the code places no restriction on a stepped foundation, the figure shows a recommended overlap of the top of the foundation wall beyond the step in the foundation. It is larger than the vertical step in the foundation wall at that point. This recommendation is based on possible crack propagation at an angle of 45 degrees (0.79 rad).

R403.1.6 Foundation anchorage. Sill plates and walls supported directly on continuous foundations shall be anchored to the foundation in accordance with this section.

Wood sole plates at all exterior walls on monolithic slabs, wood sole plates of *braced wall panels* at building interiors on monolithic slabs and all wood sill plates shall be anchored to the foundation with anchor bolts spaced a maximum of 6 feet (1829 mm) on center. Bolts shall be at least $^1/_2$ inch (12.7 mm) in diameter and shall extend a minimum of 7 inches (178 mm) into concrete or grouted cells of concrete masonry units. A nut and washer shall be tightened on each anchor bolt. There shall be a minimum of two bolts per plate section with one bolt located not more than 12 inches (305 mm) or less than seven bolt diameters from each end of the plate section. Interior bearing wall sole plates on monolithic slab foundation that are not part of a *braced wall panel* shall be positively anchored with *approved* fasteners. Sill plates and sole plates shall be protected against decay and termites where required by Sections R317 and R318. Cold-formed steel framing systems shall be fastened to wood sill plates or anchored directly to the foundation as required in Section R505.3.1 or R603.3.1.

Exceptions:

1. Foundation anchorage, spaced as required to provide equivalent anchorage to $^1/_2$-inch-diameter (12.7 mm) anchor bolts.
2. Walls 24 inches (610 mm) total length or shorter connecting offset *braced wall panels* shall be anchored to the foundation with a minimum of one anchor bolt located in the center third of the plate section and shall be attached to adjacent *braced wall panels* at corners as shown in item 8 of Table R602.3(1).
3. Connection of walls 12 inches (305 mm) total length or shorter connecting offset *braced wall panels* to the foundation without anchor bolts shall be permitted. The wall shall be attached to adjacent *braced wall panels* at corners as shown in item 8 of Table R602.3(1).

❖To prevent walls and floors from shifting under lateral loads, the code requires anchorage to the supporting foundation. Anchor bolts installed as specified in this section supply the minimum required capacity. Commentary Figure R403.1.6 illustrates anchorage of wood sill plates. The $^1/_2$-inch (12.7 mm) bolts must extend into concrete or grouted cells of concrete masonry units. This anchorage applies at exterior walls in any seismic design category. Braced wall panels at building interiors are designed to resist shear forces equal to braced wall panels at building exteriors, so equal anchorage is required. Use of approved fastening for anchorage is permitted where interior bearing-wall sole plates are not part of a braced wall panel. An exception explicitly allows other foundation anchorages if they are spaced to provide anchorage capacity equivalent to that of the $^1/_2$-inch (12.7 mm) diameter anchor bolts specified. In this case it would be necessary to use manufacturers' data, such as evaluation reports, to document the anchorage shear and tension capacities.

R403.1.6.1 Foundation anchorage in Seismic Design Categories C, D_0, D_1 and D_2. In addition to the requirements of Section R403.1.6, the following requirements shall apply to wood light-frame structures in Seismic Design Categories D_0, D_1 and D_2 and wood light-frame townhouses in Seismic Design Category C.

1. Plate washers conforming to Section R602.11.1 shall be provided for all anchor bolts over the full length of required *braced wall lines* except where *approved* anchor straps are used. Properly sized cut washers shall be permitted for anchor bolts in wall lines not containing *braced wall panels*.

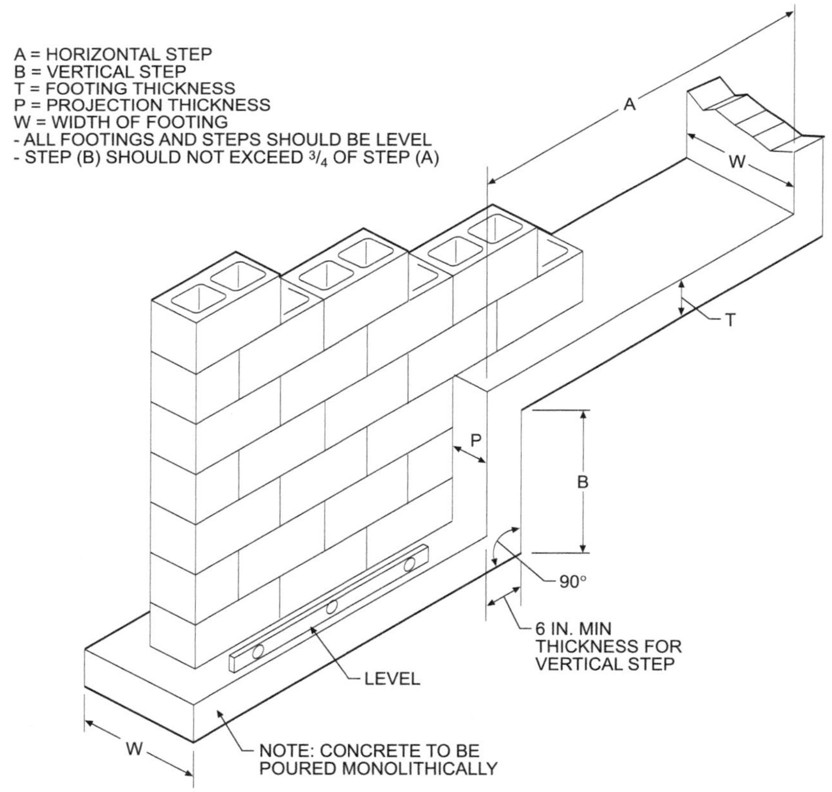

For SI: 1 inch = 25.4 mm, 1 degree = 0.01745 rad.

Figure R403.1.5
RECOMMENDATION FOR STEPPED WALL FOOTINGS

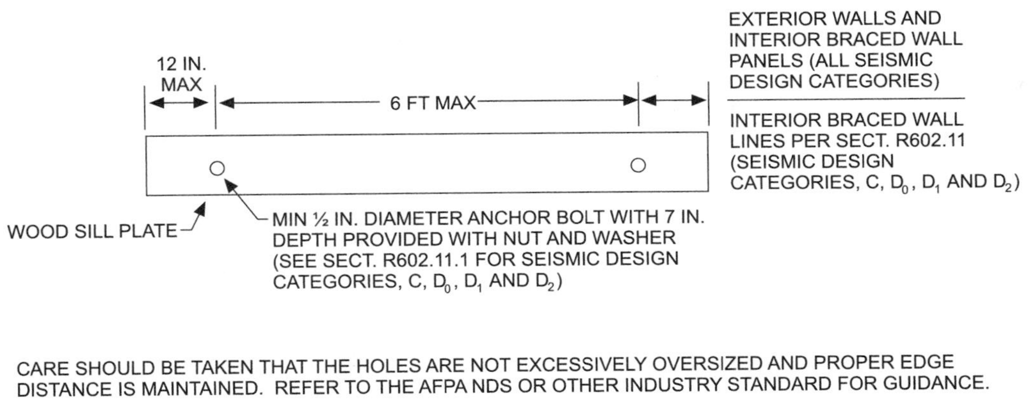

For SI: 1 inch = 25.4 mm, 1 foot = 304 mm.

Figure R403.1.6
WOOD SILL PLATE ANCHORAGE TO FOUNDATIONS

2. Interior braced wall plates shall have anchor bolts spaced at not more than 6 feet (1829 mm) on center and located within 12 inches (305 mm) of the ends of each plate section when supported on a continuous foundation.
3. Interior bearing wall sole plates shall have anchor bolts spaced at not more than 6 feet (1829 mm) on center and located within 12 inches (305 mm) of the ends of each plate section when supported on a continuous foundation.
4. The maximum anchor bolt spacing shall be 4 feet (1219 mm) for buildings over two stories in height.
5. Stepped cripple walls shall conform to Section R602.11.2.
6. Where continuous wood foundations in accordance with Section R404.2 are used, the force transfer shall have a capacity equal to or greater than the connections required by Section R602.11.1 or the *braced wall panel* shall be connected to the wood foundations in accordance with the *braced wall panel*-to-floor fastening requirements of Table R602.3(1).

❖The additional anchorage requirements of this section apply to light-frame wood structures classified as Seismic Design Category D_0, D_1 or D_2 and wood light-frame townhouses in Seismic Design Category C. These provisions were added following observations of past earthquakes, especially the 1994 Northridge earthquake in California. Considerable longitudinal splitting of sill plates occurred in that earthquake, resulting in substantial damage to the plates and subsequent loss of lateral load capacity. Plate washers are required on each bolt along the full length of the braced wall line (see commentary, Section R602.11.1). Approved anchor straps are permitted to be substituted for the anchor bolts with plate washers. The anchor straps must comply with Exception 1 of Section R403.1.6.

R403.1.7 Footings on or adjacent to slopes. The placement of buildings and structures on or adjacent to slopes steeper than one unit vertical in three units horizontal (33.3-percent slope) shall conform to Sections R403.1.7.1 through R403.1.7.4.

❖The provisions of the referenced sections apply to buildings placed on or adjacent to slopes steeper than 1 unit vertical in 3 units horizontal (33.3-percent slope) only.

R403.1.7.1 Building clearances from ascending slopes. In general, buildings below slopes shall be set a sufficient distance from the slope to provide protection from slope drainage, erosion and shallow failures. Except as provided in Section R403.1.7.4 and Figure R403.1.7.1, the following criteria will be assumed to provide this protection. Where the existing slope is steeper than one unit vertical in one unit horizontal (100-percent slope), the toe of the slope shall be assumed to be at the intersection of a horizontal plane drawn from the top of the foundation and a plane drawn tangent to the slope at an angle of 45 degrees (0.79 rad) to the horizontal. Where a retaining wall is constructed at the toe of the slope, the height of the slope shall be measured from the top of the wall to the top of the slope.

❖Figure R403.1.7.1 provides the criteria for the location of foundations adjacent to the toe of ascending slopes not exceeding one unit vertical in one unit horizontal. Commentary Figure R403.1.7.2(1) illustrates the criteria for the determination of the location of the toe of the slope where the slope exceeds one unit vertical in one unit horizontal.

R403.1.7.2 Footing setback from descending slope surfaces. Footings on or adjacent to slope surfaces shall be founded in material with an embedment and setback from the slope surface sufficient to provide vertical and lateral support for the footing without detrimental settlement. Except as provided for in Section R403.1.7.4 and Figure R403.1.7.1, the following setback is deemed adequate to meet the criteria. Where the slope is steeper than one unit vertical in one unit horizontal (100-percent slope), the required setback shall be measured from an imaginary plane 45 degrees (0.79 rad) to the horizontal, projected upward from the toe of the slope.

❖In this section, the code restricts the placement of footings adjacent to or on descending slopes so that both vertical and lateral support are provided. The criteria for this condition are as shown in Figure R403.1.7.1. It is possible to locate buildings closer to the slope than the indicated setback in Figure R403.1.7.1, and in fact it is possible to locate the footing of the structure on the slope itself. In these two cases, it will be necessary to provide an adequate depth of embedment of the footing so that the face of the footing at the bearing plane is set back from the edge of the slope at least the distance required by the code [H/3, but need not exceed 40 feet (12 192 mm)].

Commentary Figure R403.1.7.2(2) depicts the condition where the descending slope is steeper than one unit vertical in one unit horizontal and shows the proper location of the top of the slope as required by the code. The setback at the top of descending slopes will primarily provide lateral support for the foundations. The area so allocated also provides space for lot drainage away from the slope without creating too steep a drainage profile, which could create erosion problems. Furthermore, this space also provides for access around the building.

R403.1.7.3 Foundation elevation. On graded sites, the top of any exterior foundation shall extend above the elevation of the street gutter at point of discharge or the inlet of an *approved* drainage device a minimum of 12 inches (305 mm) plus 2 percent. Alternate elevations are permitted subject to the approval of the *building official*, provided it can be dem-

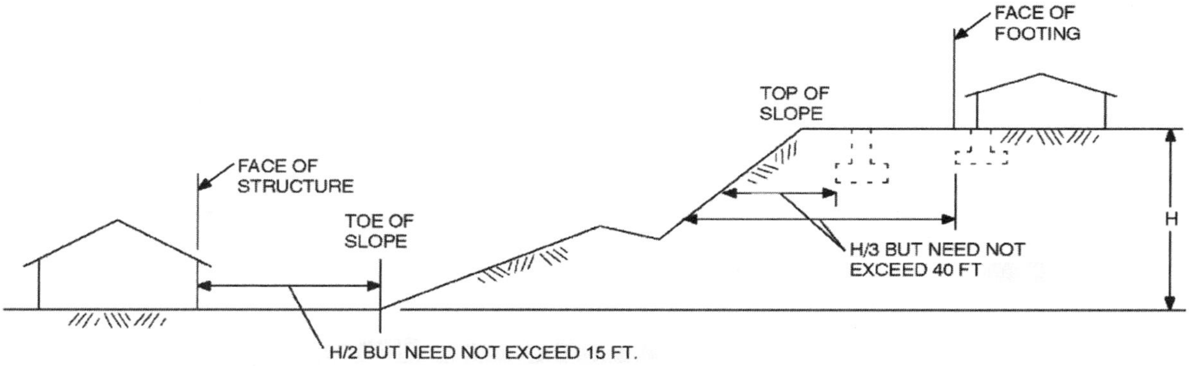

For SI: 1 foot = 304.8 mm.

FIGURE R403.1.7.1
FOUNDATION CLEARANCE FROM SLOPES

❖The setback required by the code provides protection to the structure from shallow failures (sometimes referred to as sloughing) and protection from erosion and slope drainage. Furthermore, the space provided by the setback provides access around the building and helps to create a light and open-air environment. The dimension from the toe of the slope to the face of the structure need not exceed 15 feet (4572 mm). Also refer to the commentary to Section R403.1.7.1.

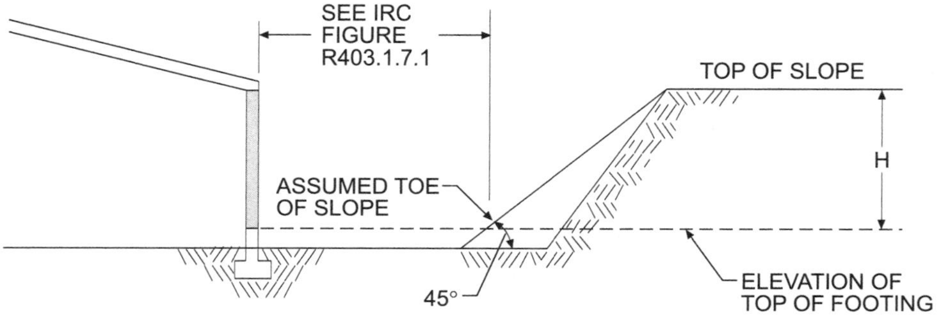

For SI: 1 degree = 0.01745 rad.

Figure R403.1.7.2(1)
BUILDINGS ADJACENT TO ASCENDING SLOPE EXCEEDING 1 TO 1

onstrated that required drainage to the point of discharge and away from the structure is provided at all locations on the site.

❖Commentary Figure R403.1.7.3 depicts the requirements of the code in this item for the elevation for exterior foundations with respect to the street, gutter or point of inlet of a drainage device. The elevation of the street or gutter shown is that point at which drainage from the site reaches the street or gutter.

This requirement protects the building from water encroachment in case of unusually heavy rains and may be modified on the approval of the building official if the building official finds that positive drainage slopes are provided to drain water away from the building and that the drainage pattern is not subject to temporary flooding due to landscaping or other impediments to drainage.

R403.1.7.4 Alternate setback and clearances. Alternate setbacks and clearances are permitted, subject to the approval of the *building official*. The *building official* is permitted to require an investigation and recommendation of a qualified engineer to demonstrate that the intent of this section has been satisfied. Such an investigation shall include consideration of material, height of slope, slope gradient, load intensity and erosion characteristics of slope material.

❖This item provides that the building official may approve alternate setbacks and clearances from slopes provided the building official is satisfied that the intent of this section has been met. The code gives the building official authority to require a foundation investigation by a qualified geotechnical engineer. This section also specifies which parameters must be considered by the geotechnical engineer in such an investigation.

R403.1.8 Foundations on expansive soils. Foundation and floor slabs for buildings located on expansive soils shall be designed in accordance with Section 1808.6 of the *International Building Code*.

Exception: Slab-on-ground and other foundation systems which have performed adequately in soil conditions simi-

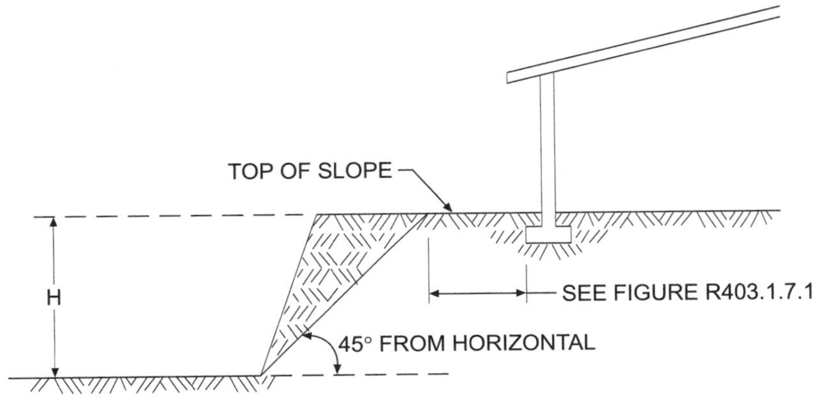

For SI: 1 degree = 0.01745 rad.

Figure R403.1.7.2(2)
BUILDINGS ADJACENT TO DESCENDING SLOPE EXCEEDING ONE TO ONE

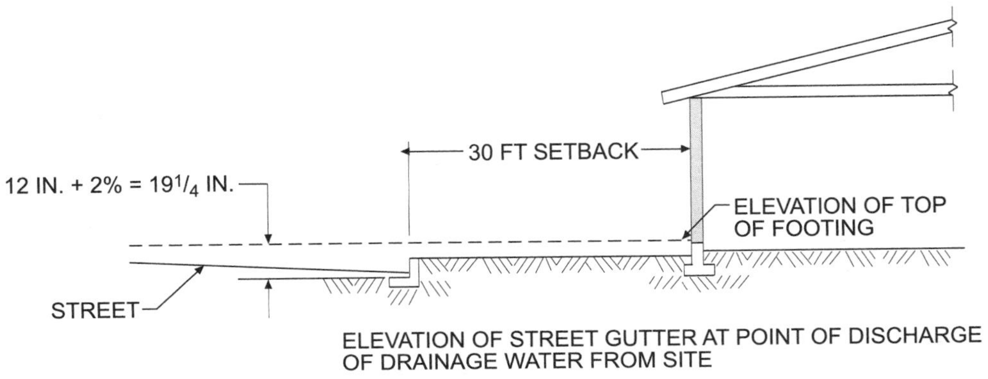

For SI: 1 degree = 0.07145 rad, 1 inch = 25.4 mm, 1 foot = 304.8 mm.

Figure R403.1.7.3
DETERMINATION OF FOUNDATION ELEVATION ON GRADED SITES

lar to those encountered at the building site are permitted subject to the approval of the *building official.*

❖Expansive soils are those that shrink and swell due to changes in moisture content. The resulting movements can be damaging to residential structures. The amount and depth of potential swelling that can occur in a clay material are to some extent functions of the cyclical moisture content in the soil. In dryer climates where the moisture content in the soil near the ground surface is low because of evaporation, there is a greater potential for extensive swelling than in the same soil in wetter climates where the variations of moisture content are not as severe.

When foundations or floor slabs are supported on soils determined to be expansive, the code defers to the *International Building Code*® (IBC®) provisions that address the design of foundations and slabs on expansive soil. The exception permits the use of systems that have demonstrated adequate performance with the approval of the building official.

R403.1.8.1 Expansive soils classifications. Soils meeting all four of the following provisions shall be considered expansive, except that tests to show compliance with Items 1, 2 and 3 shall not be required if the test prescribed in Item 4 is conducted:

1. Plasticity Index (PI) of 15 or greater, determined in accordance with ASTM D 4318.
2. More than 10 percent of the soil particles pass a No. 200 sieve (75 μm), determined in accordance with ASTM D 422.
3. More than 10 percent of the soil particles are less than 5 micrometers in size, determined in accordance with ASTM D 422.
4. Expansion Index greater than 20, determined in accordance with ASTM D 4829.

❖This section defines "expansive soil" as any plastic material with a PI of 15 or greater with more than 10 percent of the soil particles passing a No. 200 sieve and less than 5 micrometers in size. As an alternative, tests in accordance with ASTM D 4829 can be used to determine whether a soil is expansive. The expansion index is a measure of the swelling potential of the soil.

R403.2 Footings for wood foundations. Footings for wood foundations shall be in accordance with Figures R403.1(2) and R403.1(3). Gravel shall be washed and well graded. The maximum size stone shall not exceed $^3/_4$ inch (19.1 mm). Gravel shall be free from organic, clayey or silty soils. Sand shall be coarse, not smaller than $^1/_{16}$-inch (1.6 mm) grains and shall be free from organic, clayey or silty soils. Crushed stone shall have a maximum size of $^1/_2$ inch (12.7 mm).

❖See Figures R403.1(2) and R403.1(3) and refer to the commentary for Section R401.1.

R403.3 Frost-protected shallow foundations. For buildings where the monthly mean temperature of the building is maintained at a minimum of 64°F (18°C), footings are not required to extend below the frost line when protected from frost by insulation in accordance with Figure R403.3(1) and Table R403.3(1). Foundations protected from frost in accordance with Figure R403.3(1) and Table R403.3(1) shall not be used for unheated spaces such as porches, utility rooms, garages and carports, and shall not be attached to basements or crawl spaces that are not maintained at a minimum monthly mean temperature of 64°F (18°C).

Materials used below *grade* for the purpose of insulating footings against frost shall be *labeled* as complying with ASTM C 578.

❖This section provides an alternative method of protecting foundations against frost heave, thus allowing foundations to be constructed above the frost line. Frost-protected foundations use insulation to reduce the heat loss at the slab edge. By holding heat from the dwelling in the ground under the foundation, the insulation, in effect, raises the frost line around the foundation.

The provisions require that the building be heated and that insulation be installed in accordance with the criteria specified in Figure R403.3(1) and Table R403.3(1) based on the air-freezing index established by Figure R403.3(2) or Table R403.3(2). As Figure R403.3(1) illustrates, these provisions apply only to slab-on-ground floors. Frost-protected shallow foundation (FPSF) designs for heated buildings are not to be used for unheated spaces. The provisions also prohibit the attachment of additions with FPSF to basements or crawl spaces that are not heated because, although unlikely, in very cold climates frost could penetrate under FPSFs from unheated basements or ventilated crawl spaces. The insulation values specified in this section are only for the purposes of exercising this option, and the Chapter 11 energy conservation provisions could result in slab-edge insulation with a greater degree of thermal resistance. Conversely, if slab-edge insulation is used to comply with Chapter 11, a builder may take advantage of that insulation to reduce the foundation depth in accordance with this section.

R403.3.1 Foundations adjoining frost-protected shallow foundations. Foundations that adjoin frost-protected shallow foundations shall be protected from frost in accordance with Section R403.1.4.

❖This section requires that foundations that adjoin FPSFs be protected from frost according to the current requirements of Section R403.1.4, which requires that foundations extend below the frost line, be built on solid rock, or use FPSF designs. Each method prevents frost from heaving foundations. This language prevents the attachment of FPSFs to reinforced floating garage foundations, which may be permitted by some local codes to be placed at shallow depths without insulating for frost protection (i.e., the foundation moves up and down with frost).

R403.3.1.1 Attachment to unheated slab-on-ground structure. Vertical wall insulation and horizontal insulation of frost protected shallow foundations that adjoin a slab-on-ground foundation that does not have a monthly mean temperature maintained at a minimum of 64°F (18°C) shall be in accordance with Figure R403.3(3) and Table R403.3(1). Vertical wall insulation shall extend between the frost protected shallow foundation and the adjoining slab foundation. Required horizontal insulation shall be continuous under the adjoining slab foundation and through any foundation walls adjoining the frost protected shallow foundation. Where insulation passes through a foundation wall, it shall either be of a type complying with this section and having bearing capacity equal to or greater than the structural loads imposed by the building, or the building shall be designed and constructed using beams, lintels, cantilevers or other means of transferring building loads such that the structural loads of the building do not bear on the insulation.

❖This section requires that a building with a FPSF be insulated to current FPSF requirements even where an unheated garage or other slab-on-ground is attached to it. The vertical and horizontal insulation must be in accordance with Figure 403.3(3) and Table R403.3(1) and be continuous through the foundation of the unheated structure. The vertical wall insulation must extend between the FPSF and the adjoining slab foundation, and any required horizontal insulation must be continuous through any foundation walls adjoining the FPSF. No frost penetration or differential movement occurred with this construction in a HUD demonstration in Fargo, North Dakota. Deep foundations of unheated garages have been attached to hundreds of other buildings built on FPSFs in the U.S. In most cases the FPSF insulation has been placed continuously between the garage floor and the house FPSF; however, the insulation has not been continuous through the foundation walls of the unheated structures. In these buildings, no differential movement between the deep and shallow foundations has been reported.

Insulation passing through a foundation wall must comply with the requirements of Section R403.3, i.e., it must meet ASTM C 578 and the requirements of Table R403.3(1), and the insulation must be of a type that has sufficient compressive strength to carry the load of the building. Alternatively, the building must be designed to carry those loads over the insulation and not bear on it.

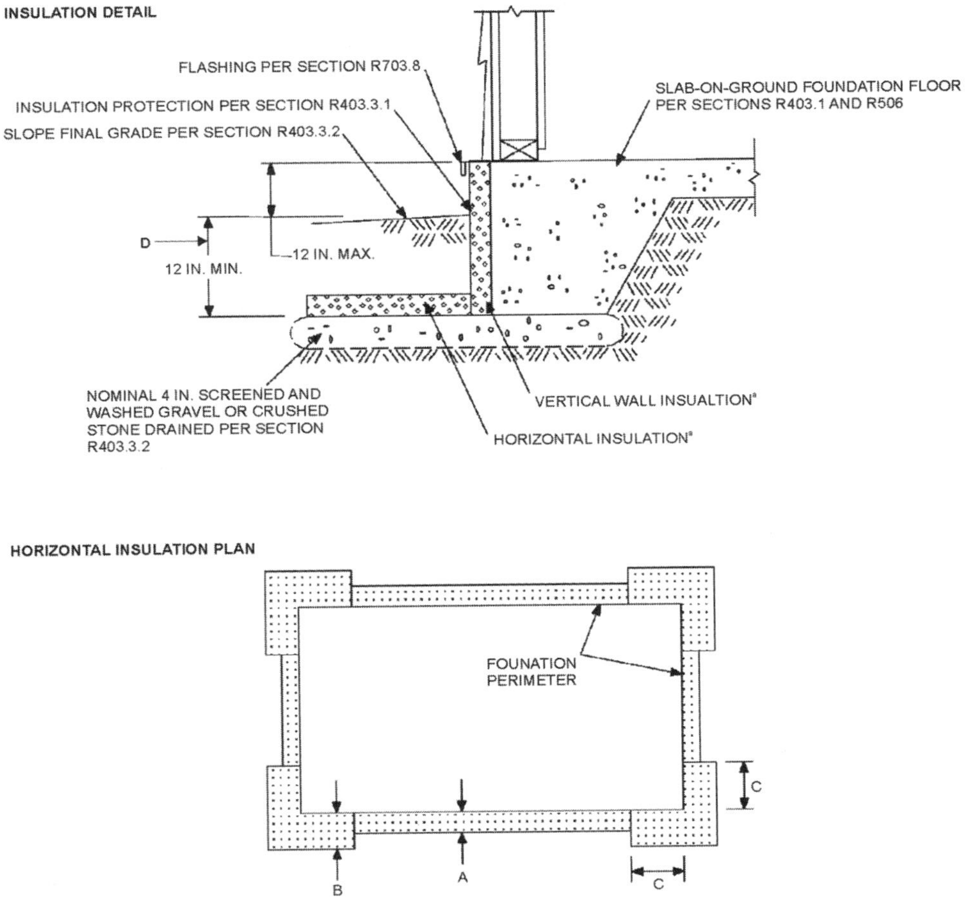

For SI: 1 inch = 25.4 mm.

a. See Table R403.3(1) for required dimensions and *R*-values for vertical and horizontal insulation and minimum footing depth.

FIGURE R403.3(1)
INSULATION PLACEMENT FOR FROST PROTECTED FOOTINGS IN HEATED BUILDINGS

❖See the commentary to Section R403.3.

TABLE R403.3(1)
MINIMUM FOOTING DEPTH AND INSULATION REQUIREMENTS FOR FROST-PROTECTED FOOTINGS IN HEATED BUILDINGS

AIR FREEZING INDEX (°F-days)[b]	MINIMUM FOOTING DEPTH, D (inches)	VERTICAL INSULATION R-VALUE[c, d]	HORIZONTAL INSULATION R-VALUE[c, e]		HORIZONTAL INSULATION DIMENSIONS PER FIGURE R403.3(1) (inches)		
			Along walls	At corners	A	B	C
1,500 or less	12	4.5	Not required	Not required	Not required	Not required	Not required
2,000	14	5.6	Not required	Not required	Not required	Not required	Not required
2,500	16	6.7	1.7	4.9	12	24	40
3,000	16	7.8	6.5	8.6	12	24	40
3,500	16	9.0	8.0	11.2	24	30	60
4,000	16	10.1	10.5	13.1	24	36	60

For SI: 1 inch = 25.4 mm, °C = [(°F) - 32]/1.8.

a. Insulation requirements are for protection against frost damage in heated buildings. Greater values may be required to meet energy conservation standards.

b. See Figure R403.3(2) or Table R403.3(2) for Air Freezing Index values.

c. Insulation materials shall provide the stated minimum *R*-values under long-term exposure to moist, below-ground conditions in freezing climates. The following *R*-values shall be used to determine insulation thicknesses required for this application: Type II expanded polystyrene-2.4*R* per inch; Type IV extruded polystyrene-4.5*R* per inch; Type VI extruded polystyrene-4.5*R* per inch; Type IX expanded polystyrene-3.2*R* per inch; Type X extruded polystyrene-4.5*R* per inch.

d. Vertical insulation shall be expanded polystyrene insulation or extruded polystyrene insulation.

e. Horizontal insulation shall be extruded polystyrene insulation.

❖This table provides the required *R*-value of horizontal and vertical insulation based on the air-freezing index. In addition, it gives the minimum footing depth and the plan dimensions required for the horizontal insulation in Figures R403.3(1) and R403.3(3) (see commentary, Sections R403.3 and R403.3.1.1).

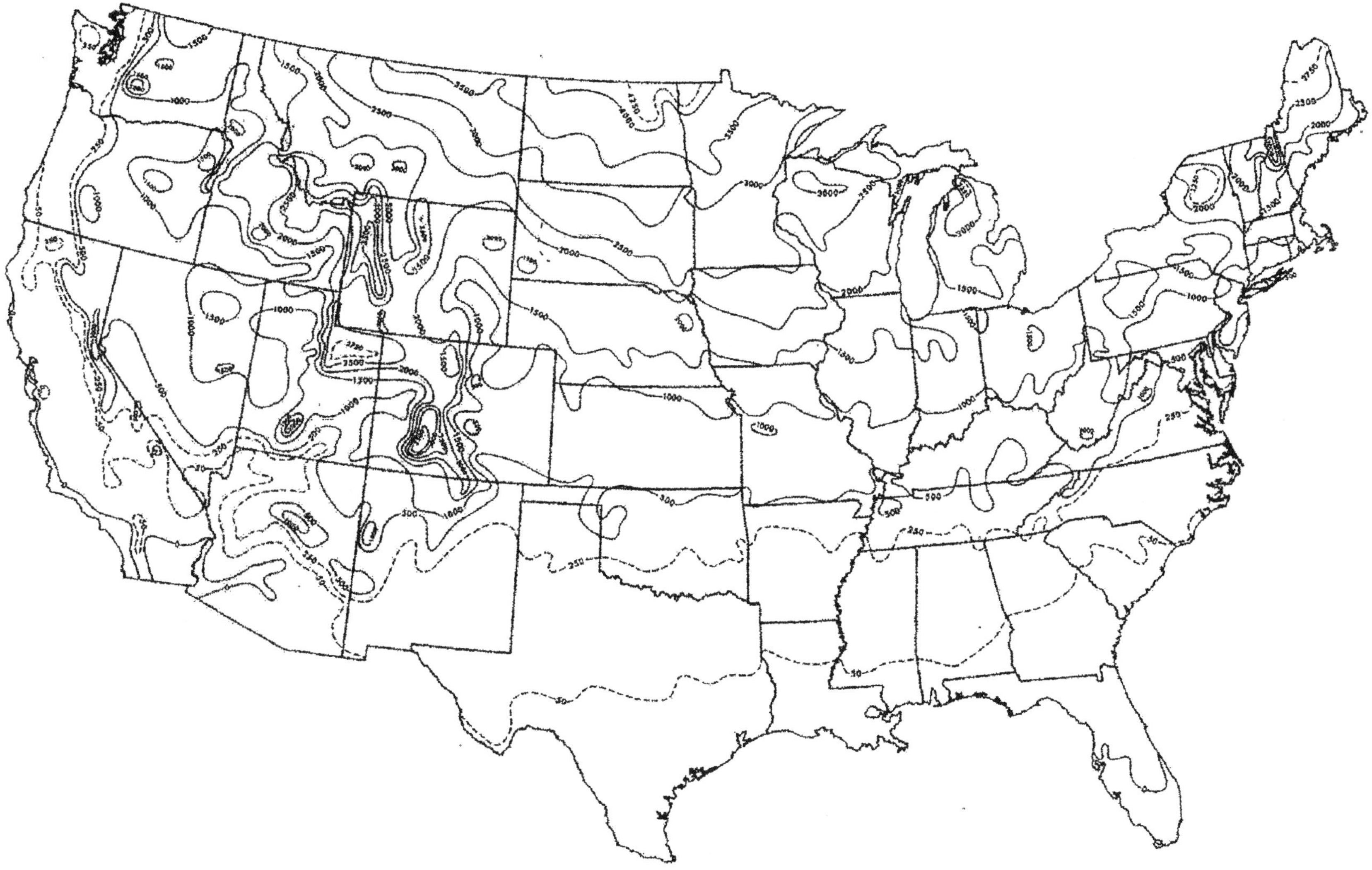

For SI: °C = [(°F) - 32]/1.8.

Note: The air-freezing index is defined as cumulative degree days below 32°F. It is used as a measure of the combined magnitude and duration of air temperature below freezing. The index was computed over a 12-month period (July-June) for each of the 3,044 stations used in the above analysis. Date from the 1951-80 period were fitted to a Weibull probability distribution to produce an estimate of the 100-year return period.

FIGURE R403.3(2)
AIR-FREEZING INDEX AN ESTMATE OF THE 100-YEAR RETURN PERIOD

❖This figure provides the criteria necessary to apply insulation requirements to Table R404.3. The note explains the term air-freezing index.

TABLE R403.3(2)
AIR-FREEZING INDEX FOR U.S. LOCATIONS BY COUNTY

STATE	AIR-FREEZING INDEX					
	1500 or less	2000	2500	3000	3500	4000
Alabama	All counties	—	—	—	—	—
Alaska	Ketchikan Gateway, Prince of Wales-Outer Ketchikan (CA), Sitka, Wrangell-Petersburg (CA)	—	Aleutians West (CA), Haines, Juneau, Skagway-Hoonah-Angoon (CA), Yakutat	—	—	All counties not listed
Arizona	All counties	—	—	—	—	—
Arkansas	All counties	—	—	—	—	—
California	All counties not listed	Nevada, Sierra	—	—	—	—
Colorado	All counties not listed	Archuleta, Custer, Fremont, Huerfano, Las Animas, Ouray, Pitkin, San Miguel	Clear Creek, Conejos, Costilla, Dolores, Eagle, La Plata, Park, Routt, San Juan, Summit	Alamosa, Grand, Jackson, Larimer, Moffat, Rio Blanco, Rio Grande	Chaffee, Gunnison, Lake, Saguache	Hinsdale, Mineral
Connecticut	All counties not listed	Hartford, Litchfield	—	—	—	—
Delaware	All counties	—	—	—	—	—
District of Columbia	All counties	—	—	—	—	—
Florida	All counties	—	—	—	—	—
Georgia	All counties	—	—	—	—	—
Hawaii	All counties	—	—	—	—	—
Idaho	All counties not listed	Adams, Bannock, Blaine, Clearwater, Idaho, Lincoln, Oneida, Power, Valley, Washington	Bingham, Bonneville, Camas, Caribou, Elmore, Franklin, Jefferson, Madison, Teton	Bear Lake, Butte, Custer, Fremont, Lemhi	Clark	—
Illinois	All counties not listed	Boone, Bureau, Cook, Dekalb, DuPage, Fulton, Grundy, Henderson, Henry, Iroquois, Jo Daviess, Kane, Kankakee, Kendall, Knox, La Salle, Lake, Lee, Livingston, Marshall, Mason, McHenry, McLean, Mercer, Peoria, Putnam, Rock Island, Stark, Tazewell, Warren, Whiteside, Will, Woodford	Carroll, Ogle, Stephenson, Winnebago	—	—	—
Indiana	All counties not listed	Allen, Benton, Cass, Fountain, Fulton, Howard, Jasper, Kosciusko, La Porte, Lake, Marshall, Miami, Newton, Porter, Pulaski, Starke, Steuben, Tippecanoe, Tipton, Wabash, Warren, White	—	—	—	—

(continued)

TABLE R403.3(2)—continued
AIR-FREEZING INDEX FOR U.S. LOCATIONS BY COUNTY

STATE	AIR-FREEZING INDEX					
	1500 or less	2000	2500	3000	3500	4000
Iowa	Appanoose, Davis, Fremont, Lee, Van Buren	All counties not listed	Allamakee, Black Hawk, Boone, Bremer, Buchanan, Buena Vista, Butler, Calhoun, Cerro Gordo, Cherokee, Chickasaw, Clay, Clayton, Delaware, Dubuque, Fayette, Floyd, Franklin, Grundy, Hamilton, Hancock, Hardin, Humboldt, Ida, Jackson, Jasper, Jones, Linn, Marshall, Palo Alto, Plymouth, Pocahontas, Poweshiek, Sac, Sioux, Story, Tama, Webster, Winnebago, Woodbury, Worth, Wright	Dickinson, Emmet, Howard, Kossuth, Lyon, Mitchell, O'Brien, Osceola, Winneshiek	—	—
Kansas	All counties	—	—	—	—	—
Kentucky	All counties	—	—	—	—	—
Louisiana	All counties	—	—	—	—	—
Maine	York	Knox, Lincoln, Sagadahoc	Androscoggin, Cumberland, Hancock, Kennebec, Waldo, Washington	Aroostook, Franklin, Oxford, Penobscot, Piscataquis, Somerset	—	—
Maryland	All counties	—	—	—	—	—
Massachusetts	All counties not listed	Berkshire, Frank-lin, Hampden, Worcester	—	—	—	—
Michigan	Berrien, Branch, Cass, Kalamazoo, Macomb, Ottawa, St. Clair, St. Joseph	All counties not listed	Alger, Charlevoix, Cheboygan, Chippewa, Crawford, Delta, Emmet, Iosco, Kalkaska, Lake, Luce, Mackinac, Menominee, Missaukee, Montmorency, Ogemaw, Osceola, Otsego, Roscommon, Schoolcraft, Wexford	Baraga, Dickinson, Iron, Keweenaw, Marquette	Gogebic, Houghton, Ontonagon	—
Minnesota	—	—	Houston, Winona	All counties not listed	Aitkin, Big Stone, Carlton, Crow Wing, Douglas, Itasca, Kanabec, Lake, Morrison, Pine, Pope, Stearns, Stevens, Swift, Todd, Wadena	Becker, Beltrami, Cass, Clay, Clearwater, Grant, Hubbard, Kittson, Koochiching, Lake of the Woods, Mahnomen, Marshall, Norman, Otter Tail, Pennington, Polk, Red Lake, Roseau, St. Louis, Traverse, Wilkin

(continued)

TABLE R403.3(2)—continued
AIR-FREEZING INDEX FOR U.S. LOCATIONS BY COUNTY

STATE	AIR-FREEZING INDEX					
	1500 or less	2000	2500	3000	3500	4000
Mississippi	All counties	—	—	—	—	—
Missouri	All counties not listed	Atchison, Mercer, Nodaway, Putnam	—	—	—	—
Montana	Mineral	Broadwater, Golden Valley, Granite, Lake, Lincoln, Missoula, Ravalli, Sanders, Sweet Grass	Big Horn, Carbon, Jefferson, Judith Basin, Lewis and Clark, Meagher, Musselshell, Powder River, Powell, Silver Bow, Stillwater, Westland	Carter, Cascade, Deer Lodge, Falcon, Fergus, Flathead, Gallanting, Glacier, Madison, Park, Petroleum, Ponder, Rosebud, Teton, Treasure, Yellowstone	Beaverhead, Blaine, Chouteau, Custer, Dawson, Garfield, Liberty, McCone, Prairie, Toole, Wibaux	Daniels, Hill, Phillips, Richland, Roosevelt, Sheridan, Valley
Nebraska	Adams, Banner, Chase, Cheyenne, Clay, Deuel, Dundy, Fillmore, Franklin, Frontier, Furnas, Gage, Garden, Gosper, Harlan, Hayes, Hitchcock, Jefferson, Kimball, Morrill, Nemaha, Nuckolls, Pawnee, Perkins, Phelps, Red Willow, Richardson, Saline, Scotts Bluff, Seward, Thayer, Webster	All counties not listed	Boyd, Burt, Cedar, Cuming, Dakota, Dixon, Dodge, Knox, Thurston	—	—	—
Nevada	All counties not listed	Elko, Eureka, Nye, Washoe, White Pine	—	—	—	—
New Hampshire	—	All counties not listed	—	—	—	Carroll, Coos, Grafton
New Jersey	All counties	—	—	—	—	—
New Mexico	All counties not listed	Rio Arriba	Colfax, Mora, Taos	—	—	—
New York	Albany, Bronx, Cayuga, Columbia, Cortland, Dutchess, Genessee, Kings, Livingston, Monroe, Nassau, New York, Niagara, Onondaga, Ontario, Orange, Orleans, Putnam, Queens, Richmond, Rockland, Seneca, Suffolk, Wayne, Westchester, Yates	All counties not listed	Clinton, Essex, Franklin, Hamilton, Herkimer, Jefferson, Lewis, St. Lawrence, Warren	—	—	—
North Carolina	All counties	—	—	—	—	—

(continued)

TABLE R403.3(2)—continued
AIR-FREEZING INDEX FOR U.S. LOCATIONS BY COUNTY

STATE	AIR-FREEZING INDEX					
	1500 or less	2000	2500	3000	3500	4000
North Dakota	—	—	—	Billings, Bowman	Adams, Dickey, Golden Valley, Hettinger, LaMoure, Oliver, Ransom, Sargent, Sioux, Slope, Stark	All counties not listed
Ohio	All counties not listed	Ashland, Crawford, Defiance, Holmes, Huron, Knox, Licking, Morrow, Paulding, Putnam, Richland, Seneca, Williams	—	—	—	—
Oklahoma	All counties	—	—	—	—	—
Oregon	All counties not listed	Baker, Crook, Grant, Harney	—	—	—	—
Pennsylvania	All counties not listed	Berks, Blair, Bradford, Cambria, Cameron, Centre, Clarion, Clearfield, Clinton, Crawford, Elk, Forest, Huntingdon, Indiana, Jefferson, Lackawanna, Lycoming, McKean, Pike, Potter, Susquehanna, Tioga, Venango, Warren, Wayne, Wyoming	—	—	—	—
Rhode Island	All counties	—	—	—	—	—
South Carolina	All counties	—	—	—	—	—
South Dakota	—	Bennett, Custer, Fall River, Lawrence, Mellette, Shannon, Todd, Tripp	Bon Homme, Charles Mix, Davison, Douglas, Gregory, Jackson, Jones, Lyman	All counties not listed	Beadle, Brookings, Brown, Campbell, Codington, Corson, Day, Deuel, Edmunds, Faulk, Grant, Hamlin, Kingsbury, Marshall, McPherson, Perkins, Roberts, Spink, Walworth	—
Tennessee	All counties	—	—	—	—	—
Texas	All counties	—	—	—	—	—
Utah	All counties not listed	Box Elder, Morgan, Weber	Garfield, Salt Lake, Summit	Carbon, Daggett, Duchesne, Rich, Sanpete, Uintah, Wasatch	—	—

❖This table is an alternative to Figure R403.3(2) for determining the air-freezing index. See the commentary to Section R403.3 and Table R403.3(1).

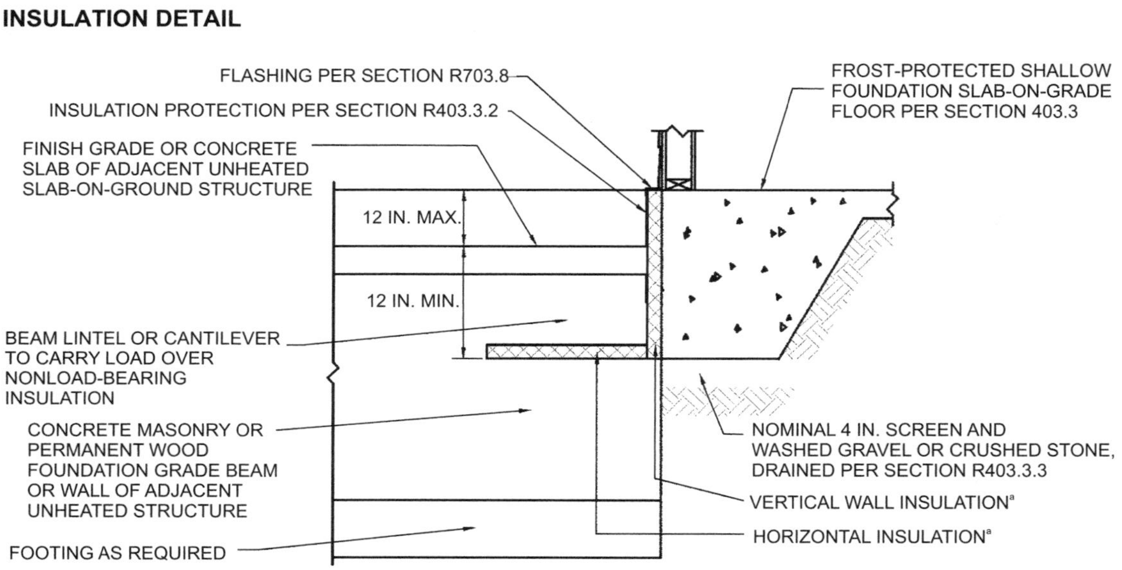

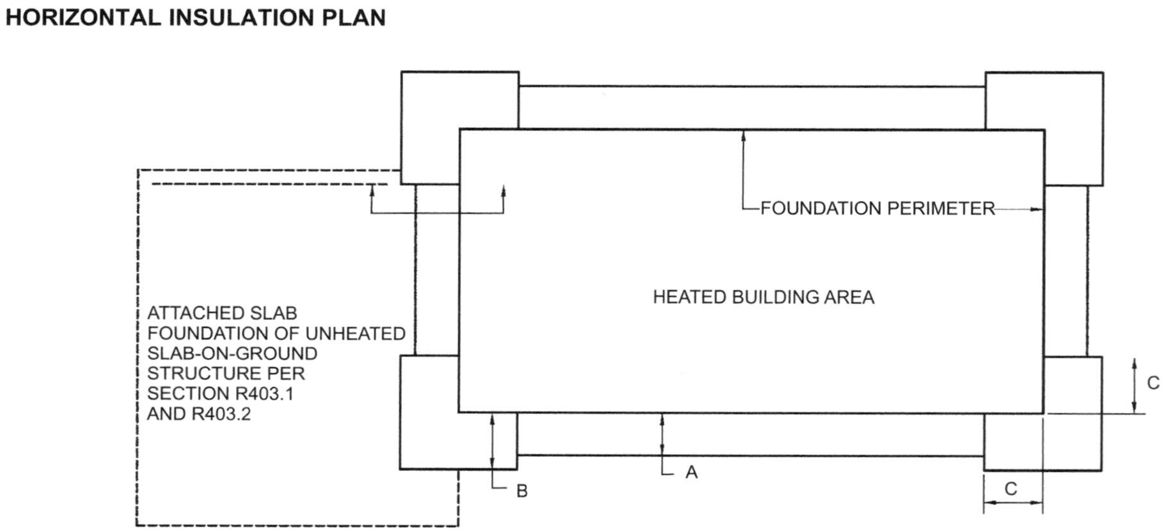

For SI: 1 inch = 25.4 mm.

a. See Table R403.3(1) for required dimensions and *R*-values for vertical and horizontal insulation.

FIGURE R403.3(3)
INSULATION PLACEMENT FOR FROST-PROTECTED FOOTINGS ADJACENT TO UNHEATED SLAB-ON-GROUND STRUCTURE

❖This figure of insulation detail shows (1) the maximum vertical distance between the top of the FPSF and the top of slab of adjacent unheated slab-on-ground structure and (2) the minimum vertical distance from the top of the slab and the bottom of the insulation. The horizontal insulation plan shows placement of a FPSF where it meets an unheated structure.

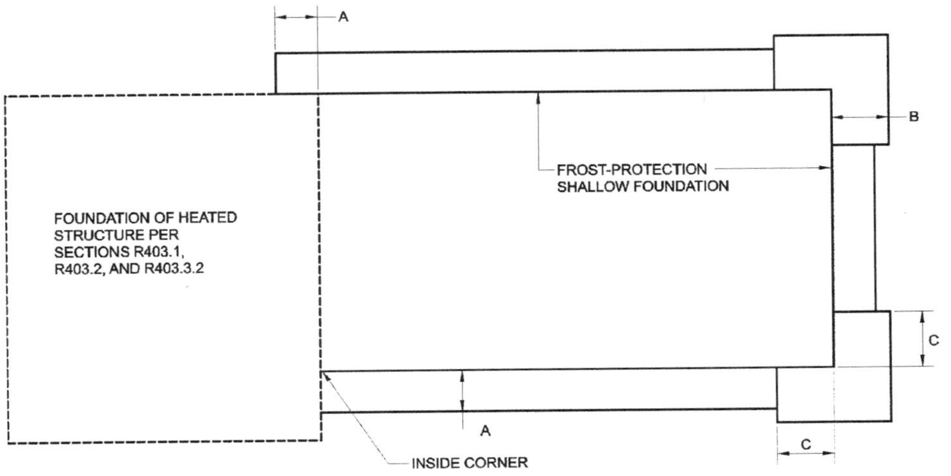

FIGURE R403.3(4)
INSULATION PLACEMENT FOR FROST-PROTECTED FOOTINGS ADJACENT TO HEATED STRUCTURE

❖This figure shows where extra insulation is required when a FPSF abuts a heated structure.

R403.3.1.2 Attachment to heated structure. Where a frost-protected shallow foundation abuts a structure that has a monthly mean temperature maintained at a minimum of 64°F (18°C), horizontal insulation and vertical wall insulation shall not be required between the frost-protected shallow foundation and the adjoining structure. Where the frost-protected shallow foundation abuts the heated structure, the horizontal insulation and vertical wall insulation shall extend along the adjoining foundation in accordance with Figure R403.3(4) a distance of not less than Dimension A in Table R403.3(1).

> **Exception:** Where the frost-protected shallow foundation abuts the heated structure to form an inside corner, vertical insulation extending along the adjoining foundation is not required.

❖This section stipulates that additions built on FPSFs need not be insulated where they join a heated structure because heat from both structures keeps the ground from freezing. However, where the FPSF abuts the heated building, the horizontal and vertical insulation must extend along the adjoining foundation to prevent cold from intruding under the slab through the foundation wall. The dimension of the insulation is related to the width of the horizontal insulation. This technique was shown to be effective by the U.S. Army Corps of Engineers Cold Regions Research and Engineering Laboratory (CRREL) on a FPSF addition to an airport control tower with a deep foundation in a severe Alaskan climate with a 13-foot (3962 mm) frost line.

An abutment at an inside corner is an exception. In this condition, both the building and the addition heat the ground, and extra insulation is not needed.

R403.3.2 Protection of horizontal insulation below ground. Horizontal insulation placed less than 12 inches (305 mm) below the ground surface or that portion of horizontal insulation extending outward more than 24 inches (610 mm) from the foundation edge shall be protected against damage by use of a concrete slab or asphalt paving on the ground surface directly above the insulation or by cementitious board, plywood rated for below-ground use, or other *approved* materials placed below ground, directly above the top surface of the insulation.

❖This protection for the insulation prevents damage due to excavating (e.g., for landscaping purposes).

R403.3.3 Drainage. Final *grade* shall be sloped in accordance with Section R401.3. In other than Group I Soils, as detailed in Table R405.1, gravel or crushed stone beneath horizontal insulation below ground shall drain to daylight or into an *approved* sewer system.

❖See Figure R403.3(1) and the commentary for Section R401.3.

R403.3.4 Termite damage. The use of foam plastic in areas of "very heavy" termite infestation probability shall be in accordance with Section R318.4.

❖Studies have shown that rigid board insulation installed below grade, particularly in areas where the hazard associated with termite infestation is very heavy, creates a pathway for termites that cannot be blocked with currently available termiticide treatments. Because it provides a place for termites to burrow in areas of very heavy termite infestation probability, the foam plastic insulation must be protected as stated in Section R320.5.

R403.4 Footings for precast concrete foundations. Footings for precast concrete foundations shall comply with Section R403.4.

❖Footings for precast concrete foundations may be crushed stone or concrete complying with Section R403.4.1 or R403.4.2.

R403.4.1 Crushed stone footings. Clean crushed stone shall be free from organic, clayey or silty soils. Crushed stone shall be angular in nature and meet ASTM C 33, with the maximum size stone not to exceed $^1/_2$ inch (12.7 mm) and the minimum stone size not to be smaller than $^1/_{16}$-inch (1.6 mm). Crushed stone footings for precast foundations shall be installed in accordance with Figure R403.4(1) and Table R403.4. Crushed stone footings shall be consolidated using a vibratory plate in a maximum of 8-inch lifts. Crushed stone footings shall be limited to Seismic Design Categories A, B and C.

❖This section prescribes the material and installation requirements for crushed stone footing to be used for precast concrete foundations, as shown in Figure R403.4(1). The soil or the interior should not be excavated below the elevation of the top of the footing. Crushed stone footings installed in accordance with this section are limited in use to areas of low or moderate seismic hazards.

R403.4.2 Concrete footings. Concrete footings shall be installed in accordance with Section R403.1 and Figure R403.4(2).

❖See the commentary for Section R403.1.

SECTION R404 FOUNDATION AND RETAINING WALLS

R404.1 Concrete and masonry foundation walls. Concrete foundation walls shall be selected and constructed in accordance with the provisions of Section R404.1.2. Masonry foundation walls shall be selected and constructed in accordance with the provisions of Section R404.1.1.

❖Masonry and concrete foundation walls must be designed and constructed in accordance with Sections R404.1.1 and R404.1.2.

R404.1.1 Design of masonry foundation walls. Masonry foundation walls shall be designed and constructed in accordance with the provisions of this section or in accordance with the provisions of TMS 402/ACI 530/ASCE 5 or NCMA TR68-A. When TMS 402/ACI 530/ASCE 5, NCMA TR68-A or the provisions of this section are used to design masonry foundation walls, project drawings, typical details and specifications are not required to bear the seal of the architect or engineer responsible for design, unless otherwise required by the state law of the *jurisdiction* having authority.

❖These provisions are for masonry foundation walls. Foundation walls should be in accordance with this section or designed under one of the following design documents:

- NCMA TR68-A: *Concrete Masonry Foundation Walls*
- TMS 402/ACI 530/ASCE 5: *Building Code Requirements for Masonry Structures*.

Foundation walls are usually designed and constructed to carry the vertical loads from the structure above, resist wind and any lateral forces transmitted to the foundations, and sustain earth pressures exerted against the walls, including any forces that may be imposed by frost action.

Most states do not require the seal of an architect or engineer for small residential buildings. Check with the state authorities in the location of the project for local regulations.

TABLE R403.4
MINIMUM DEPTH OF CRUSHED STONE FOOTINGS (D), (inches)

		LOAD-BEARING VALUE OF SOIL (psf)															
		1500				2000				3000				4000			
		MH, CH, CL, ML				SC, GC, SM, GM, SP, SW				GP, GW							
		Wall width (inches)				Wall width (inches)				Wall width (inches)				Wall width (inches)			
		6	8	10	12	6	8	10	12	6	8	10	12	6	8	10	12
Conventional light-frame construction																	
1-story	1100 plf	6	4	4	4	6	4	4	4	6	4	4	4	6	4	4	4
2-story	1800 plf	8	6	4	4	6	4	4	4	6	4	4	4	6	4	4	4
3-story	2900 plf	16	14	12	10	10	8	6	6	6	4	4	4	6	4	4	4
4-inch brick veneer over light-frame or 8-inch hollow concrete masonry																	
1-story	1500 plf	6	4	4	4	6	4	4	4	6	4	4	4	6	4	4	4
2-story	2700 plf	14	12	10	8	10	8	6	4	6	4	4	4	6	4	4	4
3-story	4000 plf	22	22	20	18	16	14	12	10	10	8	6	4	6	4	4	4
8-inch solid or fully grouted masonry																	
1-story	2000 plf	10	8	6	4	6	4	4	4	6	4	4	4	6	4	4	4
2-story	3600 plf	20	18	16	16	14	12	10	8	8	6	4	4	6	4	4	4
3-story	5300 plf	32	30	28	26	22	22	20	18	14	12	10	8	10	8	6	4

For SI:1 inch = 25.4 mm, 1 pound per square inch = 6.89 pounds per linear foot, 1 plf = 2.44 N/m, 1 pounds per square foot = 47.9 N/m^2.

❖This table and Figure R403.4(1) prescribe the crushed stone footing size based on story height, load-bearing value of the soil and the precast foundation wall thickness. Table R403.1 prescribes the footing width required based on story height above grade, for buildings with “a story below grade,” i.e., basement and stories above grade, the additional load of the “story below grade” must be accounted for. The additional load will be the dead load of the basement wall plus any tributary load from the basement floor load. As an example, a one-story building with a basement could conservatively use the two-story requirement rather than provide an engineered design.

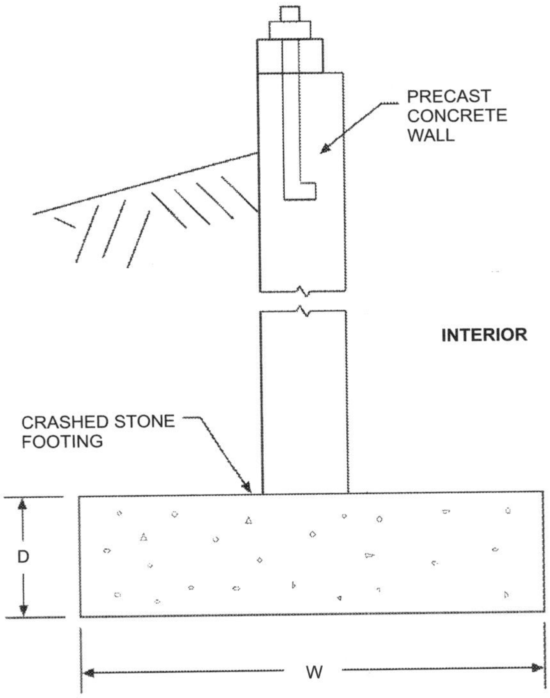

FIGURE R403.4(1)
BASEMENT OR CRAWL SPACE WITH PRECAST FOUNDATION WALL BEARING ON CRUSHED STONE

❖See the commentary to Section R403.4.1 and Table R403.4.

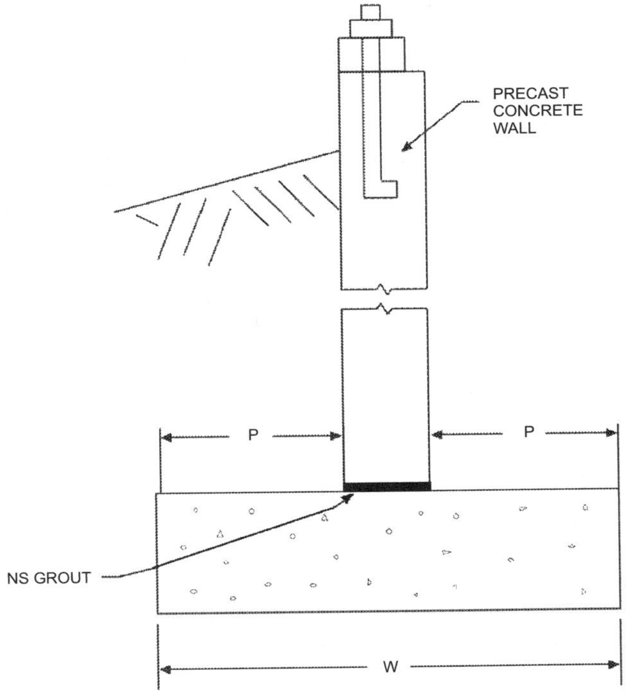

FIGURE R403.4(2)
BASEMENT OR CRAWL SPACE WITH PRECAST FOUNDATION WALL ON SPREAD FOOTING

❖See the commentary to Section R403.1.

R404.1.1.1 Masonry foundation walls. Concrete masonry and clay masonry foundation walls shall be constructed as set forth in Table R404.1.1(1), R404.1.1(2), R404.1.1(3) or R404.1.1(4) and shall also comply with applicable provisions of Sections R606, R607 and R608. In buildings assigned to Seismic Design Categories D_0, D_1 and D_2, concrete masonry and clay masonry foundation walls shall also comply with Section R404.1.4.1. Rubble stone masonry foundation walls shall be constructed in accordance with Sections R404.1.8 and R607.2.2. Rubble stone masonry walls shall not be used in Seismic Design Categories D_0, D_1 and D_2.

❖Minimum thickness of concrete masonry and clay masonry foundation walls listed in Table R404.1.1(1) are predicated on the type of soil (see Table R405.1), the maximum heights of wall and unbalanced backfill (see Commentary Figure R404.1.1), and the type of wall system that is to be supported. The tabulated values are based on the assumption that the lateral loads will be carried by the walls through bending in the vertical direction. The ASTM standards for masonry units regulate this minimum thickness of the units. The minimum thickness may be as much as $^1/_2$ inch (12.7 mm) less than the required nominal thickness specified in the table.

Example:

From Table R404.1.1(1): an 8-foot (2438 mm) high foundation wall constructed of hollow-unit masonry in a gravel-sand-clay mixture (SM) with a backfill height above the basement floor of 5 feet (1524 mm) would have to be minimum nominal thickness of 10 inches (254 mm). The actual wall thickness could be a minimum of $9^1/_2$ inches (241 mm).

The reinforcement specified in Tables R404.1.1(2) through R404.1.1(4) of the code is based on the nominal wall thickness, the type of soil, the height of the wall and the height of unbalanced fill. The height of unbalanced fill is measured as shown in Commentary Figure R404.1.1. The tabulated values are based on the assumption that lateral loads will be carried by the walls through bending, primarily in the vertical direction.

Additional (general) requirements for masonry construction found in Sections R606 through R608 must be followed also. See Section R404.1.4.1 for buildings in Seismic Design Category D_0, D_1 or D_2, and for rubble stone masonry, see the commentary to Section R404.1.8.

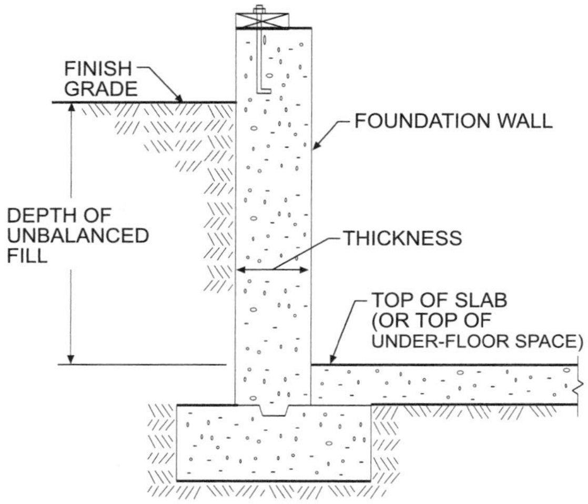

Figure R404.1.1
FOUNDATION WALL WITH UNBALANCED FILL

TABLE R404.1.1(1)
PLAIN MASONRY FOUNDATION WALLS

MAXIMUM WALL HEIGHT (feet)	MAXIMUM UNBALANCED BACKFILL HEIGHT[c] (feet)	PLAIN MASONRY[a] MINIMUM NOMINAL WALL THICKNESS (inches)		
		Soil classes[b]		
		GW, GP, SW and SP	GM, GC, SM, SM-SC and ML	SC, MH, ML-CL and inorganic CL
5	4	6 solid[d] or 8	6 solid[d] or 8	6 solid[d] or 8
	5	6 solid[d] or 8	8	10
6	4	6 solid[d] or 8	6 solid[d] or 8	6 solid[d] or 8
	5	6 solid[d] or 8	8	10
	6	8	10	12
7	4	6 solid[d] or 8	8	8
	5	6 solid[d] or 8	10	10
	6	10	12	10 solid[d]
	7	12	10 solid[d]	12 solid[d]
8	4	6 solid[d] or 8	6 solid[d] or 8	8
	5	6 solid[d] or 8	10	12
	6	10	12	12 solid[d]
	7	12	12 solid[d]	Footnote e
	8	10 solid[d]	12 solid[d]	Footnote e
9	4	6 solid[d] or 8	6 solid[d] or 8	8
	5	8	10	12
	6	10	12	12 solid[d]
	7	12	12 solid[d]	Footnote e
	8	12 solid[d]	Footnote e	Footnote e
	9	Footnote e	Footnote e	Footnote e

For SI: 1 inch = 25.4 mm, 1 foot = 304.8 mm, 1 pound per square inch = 6.895 Pa.

a. Mortar shall be Type M or S and masonry shall be laid in running bond. Ungrouted hollow masonry units are permitted except where otherwise indicated.

b. Soil classes are in accordance with the Unified Soil Classification System. Refer to Table R405.1.

c. Unbalanced backfill height is the difference in height between the exterior finish ground level and the lower of the top of the concrete footing that supports the foundation wall or the interior finish ground level. Where an interior concrete slab-on-grade is provided and is in contact with the interior surface of the foundation wall, measurement of the unbalanced backfill height from the exterior finish ground level to the top of the interior concrete slab is permitted.

d. Solid grouted hollow units or solid masonry units.

e. Wall construction shall be in accordance with either Table R404.1.1(2), Table R404.1.1(3), Table R404.1.1(4), or a design shall be provided.

❖See the commentary to Section R404.1.1.1.

TABLE R404.1.1(2)
8-INCH MASONRY FOUNDATION WALLS WITH REINFORCING WHERE d > 5 INCHES[a, c]

WALL HEIGHT	HEIGHT OF UNBALANCED BACKFILL[e]	MINIMUM VERTICAL REINFORCEMENT AND SPACING (INCHES)[b, c] Soil classes and lateral soil load[d] (psf per foot below grade)		
		GW, GP, SW and SP soils 30	GM, GC, SM, SM-SC and ML soils 45	SC, ML-CL and inorganic CL soils 60
6 feet 8 inches	4 feet (or less)	#4 at 48	#4 at 48	#4 at 48
	5 feet	#4 at 48	#4 at 48	#4 at 48
	6 feet 8 inches	#4 at 48	#5 at 48	#6 at 48
7 feet 4 inches	4 feet (or less)	#4 at 48	#4 at 48	#4 at 48
	5 feet	#4 at 48	#4 at 48	#4 at 48
	6 feet	#4 at 48	#5 at 48	#5 at 48
	7 feet 4 inches	#5 at 48	#6 at 48	#6 at 40
8 feet	4 feet (or less)	#4 at 48	#4 at 48	#4 at 48
	5 feet	#4 at 48	#4 at 48	#4 at 48
	6 feet	#4 at 48	#5 at 48	#5 at 48
	7 feet	#5 at 48	#6 at 48	#6 at 40
	8 feet	#5 at 48	#6 at 48	#6 at 32
8 feet 8 inches	4 feet (or less)	#4 at 48	#4 at 48	#4 at 48
	5 feet	#4 at 48	#4 at 48	#5 at 48
	6 feet	#4 at 48	#5 at 48	#6 at 48
	7 feet	#5 at 48	#6 at 48	#6 at 40
	8 feet 8 inches	#6 at 48	#6 at 32	#6 at 24
9 feet 4 inches	4 feet (or less)	#4 at 48	#4 at 48	#4 at 48
	5 feet	#4 at 48	#4 at 48	#5 at 48
	6 feet	#4 at 48	#5 at 48	#6 at 48
	7 feet	#5 at 48	#6 at 48	#6 at 40
	8 feet	#6 at 48	#6 at 40	#6 at 24
	9 feet 4 inches	#6 at 40	#6 at 24	#6 at 16
10 feet	4 feet (or less)	#4 at 48	#4 at 48	#4 at 48
	5 feet	#4 at 48	#4 at 48	#5 at 48
	6 feet	#4 at 48	#5 at 48	#6 at 48
	7 feet	#5 at 48	#6 at 48	#6 at 32
	8 feet	#6 at 48	#6 at 32	#6 at 24
	9 feet	#6 at 40	#6 at 24	#6 at 16
	10 feet	#6 at 32	#6 at 16	#6 at 16

For SI: 1 inch = 25.4 mm, 1 foot = 304.8 mm, 1 pound per square foot per foot = 0.157 kPa/mm.

a. Mortar shall be Type M or S and masonry shall be laid in running bond.

b. Alternative reinforcing bar sizes and spacings having an equivalent cross-sectional area of reinforcement per lineal foot of wall shall be permitted provided the spacing of the reinforcement does not exceed 72 inches.

c. Vertical reinforcement shall be Grade 60 minimum. The distance, *d*, from the face of the soil side of the wall to the center of vertical reinforcement shall be at least 5 inches.

d. Soil classes are in accordance with the Unified Soil Classification System and design lateral soil loads are for moist conditions without hydrostatic pressure. Refer to Table R405.1.

e. Unbalanced backfill height is the difference in height between the exterior finish ground level and the lower of the top of the concrete footing that supports the foundation wall or the interior finish ground level. Where an interior concrete slab-on-grade is provided and is in contact with the interior surface of the foundation wall, measurement of the unbalanced backfill height from the exterior finish ground level to the top of the interior concrete slab is permitted.

❖See the commentary to Section R404.1.1.1.

TABLE R404.1.1(3)
10-INCH MASONRY FOUNDATION WALLS WITH REINFORCING WHERE d > 6.75 INCHES[a, c]

WALL HEIGHT	HEIGHT OF UNBALANCED BACKFILL[e]	MINIMUM VERTICAL REINFORCEMENT AND SPACING (INCHES)[b, c]		
		Soil classes and later soil load[d] (psf per foot below grade)		
		GW, GP, SW and SP soils 30	GM, GC, SM, SM-SC and ML soils 45	SC, ML-CL and inorganic CL soils 60
6 feet 8 inches	4 feet (or less)	#4 at 56	#4 at 56	#4 at 56
	5 feet	#4 at 56	#4 at 56	#4 at 56
	6 feet 8 inches	#4 at 56	#5 at 56	#5 at 56
7 feet 4 inches	4 feet (or less)	#4 at 56	#4 at 56	#4 at 56
	5 feet	#4 at 56	#4 at 56	#4 at 56
	6 feet	#4 at 56	#4 at 56	#5 at 56
	7 feet 4 inches	#4 at 56	#5 at 56	#6 at 56
8 feet	4 feet (or less)	#4 at 56	#4 at 56	#4 at 56
	5 feet	#4 at 56	#4 at 56	#4 at 56
	6 feet	#4 at 56	#4 at 56	#5 at 56
	7 feet	#4 at 56	#5 at 56	#6 at 56
	8 feet	#5 at 56	#6 at 56	#6 at 48
8 feet 8 inches	4 feet (or less)	#4 at 56	#4 at 56	#4 at 56
	5 feet	#4 at 56	#4 at 56	#4 at 56
	6 feet	#4 at 56	#4 at 56	#5 at 56
	7 feet	#4 at 56	#5 at 56	#6 at 56
	8 feet 8 inches	#5 at 56	#6 at 48	#6 at 32
9 feet 4 inches	4 feet (or less)	#4 at 56	#4 at 56	#4 at 56
	5 feet	#4 at 56	#4 at 56	#4 at 56
	6 feet	#4 at 56	#5 at 56	#5 at 56
	7 feet	#4 at 56	#5 at 56	#6 at 56
	8 feet	#5 at 56	#6 at 56	#6 at 40
	9 feet 4 inches	#6 at 56	#6 at 40	#6 at 24
10 feet	4 feet (or less)	#4 at 56	#4 at 56	#4 at 56
	5 feet	#4 at 56	#4 at 56	#4 at 56
	6 feet	#4 at 56	#5 at 56	#5 at 56
	7 feet	#5 at 56	#6 at 56	#6 at 48
	8 feet	#5 at 56	#6 at 48	#6 at 40
	9 feet	#6 at 56	#6 at 40	#6 at 24
	10 feet	#6 at 48	#6 at 32	#6 at 24

For SI: 1 inch = 25.4 mm, 1 foot = 304.8 mm, 1 pound per square foot per foot = 0.157 kPa/mm.

a. Mortar shall be Type M or S and masonry shall be laid in running bond.

b. Alternative reinforcing bar sizes and spacings having an equivalent cross-sectional area of reinforcement per lineal foot of wall shall be permitted provided the spacing of the reinforcement does not exceed 72 inches.

c. Vertical reinforcement shall be Grade 60 minimum. The distance, *d*, from the face of the soil side of the wall to the center of vertical reinforcement shall be at least 6.75 inches.

d. Soil classes are in accordance with the Unified Soil Classification System and design lateral soil loads are for moist conditions without hydrostatic pressure. Refer to Table R405.1.

e. Unbalanced backfill height is the difference in height between the exterior finish ground level and the lower of the top of the concrete footing that supports the foundation wall or the interior finish ground level. Where an interior concrete slab-on-grade is provided and is in contact with the interior surface of the foundation wall, measurement of the unbalanced backfill height from the exterior finish ground level to the top of the interior concrete slab is permitted.

❖See the commentary to Section R404.1.1.1.

TABLE R404.1.1(4)
12-INCH MASONRY FOUNDATION WALLS WITH REINFORCING WHERE d > 8.75 INCHES[a, c]

WALL HEIGHT	HEIGHT OF UNBALANCED BACKFILL[e]	MINIMUM VERTICAL REINFORCEMENT AND SPACING (INCHES)[b, c]		
		Soil classes and lateral soil load[d] (psf per foot below grade)		
		GW, GP, SW and SP soils 30	GM, GC, SM, SM-SC and ML soils 45	SC, ML-CL and inorganic CL soils 60
6 feet 8 inches	4 feet (or less)	#4 at 72	#4 at 72	#4 at 72
	5 feet	#4 at 72	#4 at 72	#4 at 72
	6 feet 8 inches	#4 at 72	#4 at 72	#5 at 72
7 feet 4 inches	4 feet (or less)	#4 at 72	#4 at 72	#4 at 72
	5 feet	#4 at 72	#4 at 72	#4 at 72
	6 feet	#4 at 72	#4 at 72	#5 at 72
	7 feet 4 inches	#4 at 72	#5 at 72	#6 at 72
8 feet	4 feet (or less)	#4 at 72	#4 at 72	#4 at 72
	5 feet	#4 at 72	#4 at 72	#4 at 72
	6 feet	#4 at 72	#4 at 72	#5 at 72
	7 feet	#4 at 72	#5 at 72	#6 at 72
	8 feet	#5 at 72	#6 at 72	#6 at 64
8 feet 8 inches	4 feet (or less)	#4 at 72	#4 at 72	#4 at 72
	5 feet	#4 at 72	#4 at 72	#4 at 72
	6 feet	#4 at 72	#4 at 72	#5 at 72
	7 feet	#4 at 72	#5 at 72	#6 at 72
	8 feet 8 inches	#5 at 72	#7 at 72	#6 at 48
9 feet 4 inches	4 feet (or less)	#4 at 72	#4 at 72	#4 at 72
	5 feet	#4 at 72	#4 at 72	#4 at 72
	6 feet	#4 at 72	#5 at 72	#5 at 72
	7 feet	#4 at 72	#5 at 72	#6 at 72
	8 feet	#5 at 72	#6 at 72	#6 at 56
	9 feet 4 inches	#6 at 72	#6 at 48	#6 at 40
10 feet	4 feet (or less)	#4 at 72	#4 at 72	#4 at 72
	5 feet	#4 at 72	#4 at 72	#4 at 72
	6 feet	#4 at 72	#5 at 72	#5 at 72
	7 feet	#4 at 72	#6 at 72	#6 at 72
	8 feet	#5 at 72	#6 at 72	#6 at 48
	9 feet	#6 at 72	#6 at 56	#6 at 40
	10 feet	#6 at 64	#6 at 40	#6 at 32

For SI: 1 inch = 25.4 mm, 1 foot = 304.8 mm, 1 pound per square foot per foot = 0.157 kPa/mm.

a. Mortar shall be Type M or S and masonry shall be laid in running bond.

b. Alternative reinforcing bar sizes and spacings having an equivalent cross-sectional area of reinforcement per lineal foot of wall shall be permitted provided the spacing of the reinforcement does not exceed 72 inches.

c. Vertical reinforcement shall be Grade 60 minimum. The distance, *d*, from the face of the soil side of the wall to the center of vertical reinforcement shall be at least 8.75 inches.

d. Soil classes are in accordance with the Unified Soil Classification System and design lateral soil loads are for moist conditions without hydrostatic pressure. Refer to Table R405.1.

e. Unbalanced backfill height is the difference in height between the exterior finish ground level and the lower of the top of the concrete footing that supports the foundation wall or the interior finish ground levels. Where an interior concrete slab-on-grade is provided and in contact with the interior surface of the foundation wall, measurement of the unbalanced backfill height is permitted to be measured from the exterior finish ground level to the top of the interior concrete slab is permitted.

❖See the commentary to Section R404.1.1.1.

R404.1.2 Concrete foundation walls. Concrete foundation walls that support light-frame walls shall be designed and constructed in accordance with the provisions of this section, ACI 318, ACI 332 or PCA 100. Concrete foundation walls that support above-grade concrete walls that are within the applicability limits of Section R611.2 shall be designed and constructed in accordance with the provisions of this section, ACI 318, ACI 332 or PCA 100. Concrete foundation walls that support above-grade concrete walls that are not within the applicability limits of Section R611.2 shall be designed and constructed in accordance with the provisions of ACI 318, ACI 332 or PCA 100. When ACI 318, ACI 332, PCA 100 or the provisions of this section are used to design concrete foundation walls, project drawings, typical details and specifications are not required to bear the seal of the architect or engineer responsible for design, unless otherwise required by the state law of the *jurisdiction* having authority.

❖These provisions are for concrete foundation walls. Foundation walls should be in accordance with this section or designed in accordance with one of the following documents:

- ACI 318, *Building Code Requirements for Structural Concrete*
- ACI 332, *Requirements for Residential Concrete Construction*
- PCA100, *Prescriptive Design of Exterior Concrete Walls for One- and Two-family Dwellings*

Foundation walls are usually designed and constructed to carry the vertical loads from the structure above, resist wind and any lateral forces transmitted to the foundations, and sustain earth pressures exerted against the walls, including any forces that may be imposed by frost action.

Most states do not require the seal of an architect or engineer for small residential buildings. Check with the state authorities in the location of the project for local regulations.

R404.1.2.1 Concrete cross-section. Concrete walls constructed in accordance with this code shall comply with the shapes and minimum concrete cross-sectional dimensions required by Table R611.3. Other types of forming systems resulting in concrete walls not in compliance with this section and Table R611.3 shall be designed in accordance with ACI 318.

❖The minimum thickness and cross-section dimension of concrete walls must comply with Section R404.1.2 and Table R611.3. Other sizes must be designed in accordance with ACI 318.

Unlike masonry units, which must be manufactured in accordance with an ASTM standard, there is no standard to require the minimum thickness of nominal 6-, 8-, 10-, or 12-inch (127 mm, 203 mm, 254 mm, or 305 mm) flat concrete wall. The actual as-built minimum thickness is permitted to be no more than $^{1}/_{2}$ inch (12.7 mm) less than the nominal thickness. See Note d of Table R611.3.

The design strength of flat concrete walls are based on thicknesses of 5.5 (139.7 mm), 7.5 (180.5 mm), 9.5 (241.3 mm) and 11.5 inches (292.1 mm).

For waffle-grid and screen-grid minimum thickness, see Notes e, f and g of Table R611.3.

R404.1.2.2 Reinforcement for foundation walls. Concrete foundation walls shall be laterally supported at the top and bottom. Horizontal reinforcement shall be provided in accordance with Table R404.1.2(1). Vertical reinforcement shall be provided in accordance with Table R404.1.2(2), R404.1.2(3), R404.1.2(4), R404.1.2(5), R404.1.2(6), R404.1.2(7) or R404.1.2(8). Vertical reinforcement for flat *basement* walls retaining 4 feet (1219 mm) or more of unbalanced backfill is permitted to be determined in accordance with Table R404.1.2(9). For *basement* walls supporting above-grade concrete walls, vertical reinforcement shall be the greater of that required by Tables R404.1.2(2) through R404.1.2(8) or by Section R611.6 for the above-grade wall. In buildings assigned to Seismic Design Category D_0, D_1 or D_2, concrete foundation walls shall also comply with Section R404.1.4.2.

❖This section provides reinforcement requirements and tables for foundation walls constructed in accordance with Section R404.1.2. The tables provide the minimum required vertical and horizontal wall reinforcement for various lateral soil loads, wall heights, and unbalanced backfill heights. Vertical wall reinforcement tables are provided for foundation walls with unsupported wall heights up to 10 feet (3048 mm) (see Commentary Figure R404.1.1).

The wall must be laterally supported at the top and bottom. The tables were developed in accordance with ACI 318.

TABLE R404.1.2(1)
MINIMUM HORIZONTAL REINFORCEMENT FOR CONCRETE BASEMENT WALLS[a,]

MAXIMUM UNSUPPORTED HEIGHT OF BASEMENT WALL (feet)	LOCATION OF HORIZONTAL REINFORCEMENT
≤ 8	One No. 4 bar within 12 inches of the top of the wall story and one No. 4 bar near mid-height of the wall story.
> 8	One No. 4 bar within 12 inches of the top of the wall story and one No. 4 bar near third points in the wall story.

For SI: 1 inch = 25.4 mm, 1 foot = 304.8 mm, 1 pound per square inch = 6.895 kPa.

a. Horizontal reinforcement requirements are for reinforcing bars with a minimum yield strength of 40,000 psi and concrete with a minimum concrete compressive strength 2,500 psi.

b. See Section R404.1.2.2 for minimum reinforcement required for foundation walls supporting above-grade concrete walls.

❖See the commentary to Section R404.1.2.2.

The following design assumptions were used to analyze the walls:

1. Walls are simply supported at the top and bottom.
2. Walls contain no openings.
3. Lateral bracing is provided for the wall by the floors above and the floor slabs below.
4. Allowable deflection criterion is the laterally unsupported height of the wall, in inches, divided by 240.

Tables R404.1.2(2) through R404.1.2(7) require the vertical steel to be located at the centerline of the wall. For flat concrete walls, Table R404.1.2(8) requires the steel to be located with a maximum cover of $1^1/_4$ inches (31.75 mm) from the inside face of the wall (see Section R404.1.2.3.7.2). Tables R404.1.2(2) through R404.1.2(8) are based on reinforcing bars of Grade 60 (60,000 psi minimum yield strength). Table R404.1.2(9) provides for use of an alternate grade of steel and alternate bar size. See Section R404.1.4.2 for additional requirements for foundation walls in Seismic Design Categories D_0, D_1 and D_2.

TABLE R404.1.2(2)
MINIMUM VERTICAL REINFORCEMENT FOR 6-INCH NOMINAL FLAT CONCRETE BASEMENT WALLS[b, c, d, e, g, h, i, j]

MAXIMUM UNSUPPORTED WALL HEIGHT (feet)	MAXIMUM UNBALANCED BACKFILL HEIGHT[f] (feet)	MINIMUM VERTICAL REINFORCEMENT-BAR SIZE AND SPACING (inches)		
		Soil classes[a] and design lateral soil (psf per foot of depth)		
		GW, GP, SW, SP 30	GM, GC, SM, SM-SC and ML 45	SC, ML-CL and inorganic CL 60
8	4	NR	NR	NR
	5	NR	6 @ 39	6 @ 48
	6	5 @ 39	6 @ 48	6 @ 35
	7	6 @ 48	6 @ 34	6 @ 25
	8	6 @ 39	6 @ 25	6 @ 18
9	4	NR	NR	NR
	5	NR	5 @ 37	6 @ 48
	6	5 @ 36	6 @ 44	6 @ 32
	7	6 @ 47	6 @ 30	6 @ 22
	8	6 @ 34	6 @ 22	6 @ 16
	9	6 @ 27	6 @ 17	DR
10	4	NR	NR	NR
	5	NR	5 @ 35	6 @ 48
	6	6 @ 48	6 @ 41	6 @ 30
	7	6 @ 43	6 @ 28	6 @ 20
	8	6 @ 31	6 @ 20	DR
	9	6 @ 24	6 @ 15	DR
	10	6 @ 19	DR	DR

For SI: 1 foot = 304.8 mm; 1 inch = 25.4 mm; 1 pound per square foot per foot = 0.1571 kPa^2/m, 1 pound per square inch = 6.895 kPa.

NR = Not required.

a. Soil classes are in accordance with the Unified Soil Classification System. Refer to Table R405.1.

b. Table values are based on reinforcing bars with a minimum yield strength of 60,000 psi concrete with a minimum specified compressive strength of 2,500 psi and vertical reinforcement being located at the centerline of the wall. See Section R404.1.2.3.7.2.

c. Vertical reinforcement with a yield strength of less than 60,000 psi and/or bars of a different size than specified in the table are permitted in accordance with Section R404.1.2.3.7.6 and Table R404.1.2(9).

d. Deflection criterion is *L*/240, where *L* is the height of the basement wall in inches.

e. Interpolation is not permitted.

f. Where walls will retain 4 feet or more of unbalanced backfill, they shall be laterally supported at the top and bottom before backfilling.

g. NR indicates no vertical wall reinforcement is required, except for 6-inch-nominal walls formed with stay-in-place forming systems in which case vertical reinforcement shall be No. 4@48 inches on center.

h. See Section R404.1.2.2 for minimum reinforcement required for basement walls supporting above-grade concrete walls.

i. See Table R611.3 for tolerance from nominal thickness permitted for flat walls.

j. DR means design is required in accordance with the applicable building code, or where there is no code, in accordance with ACI 318.

❖See the commentary to Section R404.1.2.2.

TABLE R404.1.2(3)
MINIMUM VERTICAL REINFORCEMENT FOR 8-INCH (203 mm) NOMINAL FLAT CONCRETE BASEMENT WALLS[b, c, d, e, f, h, i]

MAXIMUM UNSUPPORTED WALL HEIGHT (feet)	MAXIMUM UNBALANCED BACKFILL HEIGHT[g] (feet)	MINIMUM VERTICAL REINFORCEMENT-BAR SIZE AND SPACING (inches)		
		Soil classes[a] and design lateral soil (psf per foot of depth)		
		GW, GP, SW, SP 30	GM, GC, SM, SM-SC and ML 45	SC, ML-CL and inorganic CL 60
8	4	NR	NR	NR
	5	NR	NR	NR
	6	NR	NR	6 @ 37
	7	NR	6 @ 36	6 @ 35
	8	6 @ 41	6 @ 35	6 @ 26
9	4	NR	NR	NR
	5	NR	NR	NR
	6	NR	NR	6 @ 35
	7	NR	6 @ 35	6 @ 32
	8	6 @ 36	6 @ 32	6 @ 23
	9	6 @ 35	6 @ 25	6 @ 18
10	4	NR	NR	NR
	5	NR	NR	NR
	6	NR	NR	6 @ 35
	7	NR	6 @ 35	6 @ 29
	8	6 @ 35	6 @ 29	6 @ 21
	9	6 @ 34	6 @ 22	6 @ 16
	10	6 @ 27	6 @ 17	6 @ 13

For SI: 1 foot = 304.8 mm; 1 inch = 25.4 mm; 1 pound per square foot per foot = 0.1571 kPa^2/m, 1 pound per square inch = 6.895 kPa.

NR = Not required.

a. Soil classes are in accordance with the Unified Soil Classification System. Refer to Table R405.1.

b. Table values are based on reinforcing bars with a minimum yield strength of 60,000 psi, concrete with a minimum specified compressive strength of 2,500 psi and vertical reinforcement being located at the centerline of the wall. See Section R404.1.2.3.7.2.

c. Vertical reinforcement with a yield strength of less than 60,000 psi and/or bars of a different size than specified in the table are permitted in accordance with Section R404.1.2.3.7.6 and Table R404.1.2(9).

d. NR indicates no vertical reinforcement is required.

e. Deflection criterion is *L*/240, where *L* is the height of the basement wall in inches.

f. Interpolation is not permitted.

g. Where walls will retain 4 feet or more of unbalanced backfill, they shall be laterally supported at the top and bottom before backfilling.

h. See Section R404.1.2.2 for minimum reinforcement required for basement walls supporting above-grade concrete walls.

i. See Table R611.3 for tolerance from nominal thickness permitted for flat walls.

❖See the commentary to Section R404.1.2.2.

TABLE R404.1.2(4)
MINIMUM VERTICAL REINFORCEMENT FOR 10-INCH NOMINAL FLAT CONCRETE BASEMENT WALLS[b, c, d, e, f, h, i]

MAXIMUM UNSUPPORTED WALL HEIGHT (feet)	MAXIMUM UNBALANCED BACKFILL HEIGHT[g] (feet)	MINIMUM VERTICAL REINFORCEMENT-BAR SIZE AND SPACING (inches)		
		Soil classes[a] and design lateral soil (psf per foot of depth)		
		GW, GP, SW, SP 30	GM, GC, SM, SM-SC and ML 45	SC, ML-CL and inorganic CL 60
8	4	NR	NR	NR
	5	NR	NR	NR
	6	NR	NR	NR
	7	NR	NR	NR
	8	6 @ 48	6 @ 35	6 @ 28
9	4	NR	NR	NR
	5	NR	NR	NR
	6	NR	NR	NR
	7	NR	NR	6 @ 31
	8	NR	6 @ 31	6 @ 28
	9	6 @ 37	6 @ 28	6 @ 24
10	4	NR	NR	NR
	5	NR	NR	NR
	6	NR	NR	NR
	7	NR	NR	6 @ 28
	8	NR	6 @ 28	6 @ 28
	9	6 @ 33	6 @ 28	6 @ 21
	10	6 @ 28	6 @ 23	6 @ 17

For SI: 1 foot = 304.8 mm; 1 inch = 25.4 mm; 1 pound per square foot per foot = 0.1571 kPa^2/m, 1 pound per square inch = 6.895 kPa.

NR = Not required.

a. Soil classes are in accordance with the Unified Soil Classification System. Refer to Table R405.1.

b. Table values are based on reinforcing bars with a minimum yield strength of 60,000 psi concrete with a minimum specified compressive strength of 2,500 psi and vertical reinforcement being located at the centerline of the wall. See Section R404.1.2.3.7.2.

c. Vertical reinforcement with a yield strength of less than 60,000 psi and/or bars of a different size than specified in the table are permitted in accordance with Section R404.1.2.3.7.6 and Table R404.1.2(9).

d. NR indicates no vertical reinforcement is required.

e. Deflection criterion is *L*/240, where *L* is the height of the basement wall in inches.

f. Interpolation is not permitted.

g. Where walls will retain 4 feet or more of unbalanced backfill, they shall be laterally supported at the top and bottom before backfilling.

h. See Section R404.1.2.2 for minimum reinforcement required for basement walls supporting above-grade concrete walls.

i. See Table R611.3 for tolerance from nominal thickness permitted for flat walls.

❖See the commentary to Section R404.1.2.2.

TABLE R404.1.2(5)
MINIMUM VERTICAL WALL REINFORCEMENT FOR 6-INCH WAFFLE-GRID BASEMENT WALLS[b, c, d, e, g, h, i]

MAXIMUM UNSUPPORTED WALL HEIGHT (feet)	MAXIMUM UNBALANCED BACKFILL HEIGHT[f] (feet)	MINIMUM VERTICAL REINFORCEMENT-BAR SIZE AND SPACING (inches)		
		Soil classes[a] and design lateral soil (psf per foot of depth)		
		GW, GP, SW, SP 30	GM, GC, SM, SM-SC and ML 45	SC, ML-CL and inorganic CL 60
8	4	4 @ 48	4 @ 46	6 @ 39
	5	4 @ 45	5 @ 46	6 @ 47
	6	5 @ 45	6 @ 40	DR
	7	6 @ 44	DR	DR
	8	6 @ 32	DR	DR
9	4	4 @ 48	4 @ 46	4 @ 37
	5	4 @ 42	5 @ 43	6 @ 44
	6	5 @ 41	6 @ 37	DR
	7	6 @ 39	DR	DR
	> 8	DR[i]	DR	DR
10	4	4 @ 48	4 @ 46	4 @ 35
	5	4 @ 40	5 @ 40	6 @ 41
	6	5 @ 38	6 @ 34	DR
	7	6 @ 36	DR	DR
	> 8	DR	DR	DR

For SI: 1 foot = 304.8 mm; 1 inch = 25.4 mm; 1 pound per square foot per foot = 0.1571 kPa^2/m, 1 pound per square inch = 6.895 kPa.

a. Soil classes are in accordance with the Unified Soil Classification System. Refer to Table R405.1.

b. Table values are based on reinforcing bars with a minimum yield strength of 60,000 psi concrete with a minimum specified compressive strength of 2,500 psi and vertical reinforcement being located at the centerline of the wall. See Section R404.1.2.3.7.2.

c. Maximum spacings shown are the values calculated for the specified bar size. Where the bar used is Grade 60 and the size specified in the table, the actual spacing in the wall shall not exceed a whole-number multiple of 12 inches (i.e., 12, 24, 36 and 48) that is less than or equal to the tabulated spacing. Vertical reinforcement with a yield strength of less than 60,000 psi and/or bars of a different size than specified in the table are permitted in accordance with Section R404.1.2.3.7.6 and Table R404.1.2(9).

d. Deflection criterion is *L*/240, where *L* is the height of the basement wall in inches.

e. Interpolation is not permitted.

f. Where walls will retain 4 feet or more of unbalanced backfill, they shall be laterally supported at the top and bottom before backfilling.

g. See Section R404.1.2.2 for minimum reinforcement required for basement walls supporting above-grade concrete walls.

h. See Table R611.3 for thicknesses and dimensions of waffle-grid walls.

i. DR means design is required in accordance with the applicable building code, or where there is no code, in accordance with ACI 318.

❖See the commentary to Section R404.1.2.2.

TABLE R404.1.2(6)
MINIMUM VERTICAL REINFORCEMENT FOR 8-INCH WAFFLE-GRID BASEMENT WALLS[b, c, d, e, f, h, i,]

MAXIMUM UNSUPPORTED WALL HEIGHT (feet)	MAXIMUM UNBALANCED BACKFILL HEIGHT[g] (feet)	MINIMUM VERTICAL REINFORCEMENT-BAR SIZE AND SPACING (inches)		
		Soil classes[a] and design lateral soil (psf per foot of depth)		
		GW, GP, SW, SP 30	GM, GC, SM, SM-SC and ML 45	SC, ML-CL and inorganic CL 60
8	4	NR	NR	NR
	5	NR	5 @ 48	5 @ 46
	6	5 @ 48	5 @ 43	6 @ 45
	7	5 @ 46	6 @ 43	6 @ 31
	8	6 @ 48	6 @ 32	6 @ 23
9	4	NR	NR	NR
	5	NR	5 @ 47	5 @ 46
	6	5 @ 46	5 @ 39	6 @ 41
	7	5 @ 42	6 @ 38	6 @ 28
	8	6 @ 44	6 @ 28	6 @ 20
	9	6 @ 34	6 @ 21	DR
10	4	NR	NR	NR
	5	NR	5 @ 46	5 @ 44
	6	5 @ 46	5 @ 37	6 @ 38
	7	5 @ 38	6 @ 35	6 @ 25
	8	6 @ 39	6 @ 25	DR
	9	6 @ 30	DR	DR
	10	6 @ 24	DR	DR

For SI: 1 foot = 304.8 mm; 1 inch = 25.4 mm; 1 pound per square foot per foot = 0.1571 kPa^2/m, 1 pound per square inch = 6.895 kPa.

NR = Not required.

a. Soil classes are in accordance with the Unified Soil Classification System. Refer to Table R405.1.

b. Table values are based on reinforcing bars with a minimum yield strength of 60,000 psi concrete with a minimum specified compressive strength of 2,500 psi and vertical reinforcement being located at the centerline of the wall. See Section R404.1.2.3.7.2.

c. Maximum spacings shown are the values calculated for the specified bar size. Where the bar used is Grade 60 (420 MPa) and the size specified in the table, the actual spacing in the wall shall not exceed a whole-number multiple of 12 inches (i.e., 12, 24, 36 and 48) that is less than or equal to the tabulated spacing. Vertical reinforcement with a yield strength of less than 60,000 psi and/or bars of a different size than specified in the table are permitted in accordance with Section R404.1.2.3.7.6 and Table R404.1.2(9).

d. NR indicates no vertical reinforcement is required.

e. Deflection criterion is *L*/240, where *L* is the height of the basement wall in inches.

f. Interpolation shall not be permitted.

g. Where walls will retain 4 feet or more of unbalanced backfill, they shall be laterally supported at the top and bottom before backfilling.

h. See Section R404.1.2.2 for minimum reinforcement required for basement walls supporting above-grade concrete walls.

i. See Table R611.3 for thicknesses and dimensions of waffle-grid walls.

j. DR means design is required in accordance with the applicable building code, or where there is no code, in accordance with ACI 318.

❖See the commentary to Section R404.1.2.2.

TABLE R404.1.2(7)
MINIMUM VERTICAL REINFORCEMENT FOR 6-INCH (152 mm) SCREEN-GRID BASEMENT WALLS[b, c, d, e, g, h, i]

MAXIMUM UNSUPPORTED WALL HEIGHT (feet)	MAXIMUM UNBALANCED BACKFILL HEIGHT[f] (feet)	MINIMUM VERTICAL REINFORCEMENT-BAR SIZE AND SPACING (inches)		
		Soil classes[a] and design lateral soil (psf per foot of depth)		
		GW, GP, SW, SP 30	GM, GC, SM, SM-SC and ML 45	SC, ML-CL and inorganic CL 60
8	4	4 @ 48	4 @ 48	5 @ 43
	5	4 @ 48	5 @ 48	5 @ 37
	6	5 @ 48	6 @ 45	6 @ 32
	7	6 @ 48	DR	DR
	8	6 @ 36	DR	DR
9	4	4 @ 48	4 @ 48	4 @ 41
	5	4 @ 48	5 @ 48	6 @ 48
	6	5 @ 45	6 @ 41	DR
	7	6 @ 43	DR	DR
	> 8	DR	DR	DR
10	4	4 @ 48	4 @ 48	4 @ 39
	5	4 @ 44	5 @ 44	6 @ 46
	6	5 @ 42	6 @ 38	DR
	7	6 @ 40	DR	DR
	> 8	DR	DR	DR

For SI: 1 foot = 304.8 mm; 1 inch = 25.4 mm; 1 pound per square foot per foot = 0.1571 kPa^2/m, 1 pound per square inch = 6.895 kPa.

a. Soil classes are in accordance with the Unified Soil Classification System. Refer to Table R405.1.

b. Table values are based on reinforcing bars with a minimum yield strength of 60,000 psi, concrete with a minimum specified compressive strength of 2,500 psi and vertical reinforcement being located at the centerline of the wall. See Section R404.1.2.3.7.2.

c. Maximum spacings shown are the values calculated for the specified bar size. Where the bar used is Grade 60 and the size specified in the table, the actual spacing in the wall shall not exceed a whole-number multiple of 12 inches (i.e., 12, 24, 36 and 48) that is less than or equal to the tabulated spacing. Vertical reinforcement with a yield strength of less than 60,000 psi and/or bars of a different size than specified in the table are permitted in accordance with Section R404.1.2.3.7.6 and Table R404.1.2(9).

d. Deflection criterion is $L/240$, where L is the height of the basement wall in inches.

e. Interpolation is not permitted.

f. Where walls will retain 4 feet or more of unbalanced backfill, they shall be laterally supported at the top and bottom before backfilling.

g. See Sections R404.1.2.2 for minimum reinforcement required for basement walls supporting above-grade concrete walls.

h. See Table R611.3 for thicknesses and dimensions of screen-grid walls.

i. DR means design is required in accordance with the applicable building code, or where there is no code, in accordance with ACI 318.

❖See the commentary to Section R404.1.2.2.

TABLE R404.1.2(8)
MINIMUM VERTICAL REINFORCEMENT FOR 6-, 8-, 10-INCH AND 12-INCH NOMINAL FLAT BASEMENT WALLS[b, c, d, e, f, h, i, k, n]

MAXIMUM WALL HEIGHT (feet)	MAXIMUM UNBALANCED BACKFILL HEIGHT[g] (feet)	MINIMUM VERTICAL REINFORCEMENT-BAR SIZE AND SPACING (inches)											
		Soil classes[a] and design lateral soil (psf per foot of depth)											
		GW, GP, SW, SP 30				GM, GC, SM, SM-SC and ML 45				SC, ML-CL and inorganic CL 60			
		Minimum nominal wall thickness (inches)											
		6	8	10	12	6	8	10	12	6	8	10	12
5	4	NR	NR	NR	NR	NR	NR	NR	NR	NR	NR	NR	NR
	5	NR	NR	NR	NR	NR	NR	NR	NR	NR	NR	NR	NR
6	4	NR	NR	NR	NR	NR	NR	NR	NR	NR	NR	NR	NR
	5	NR	NR	NR	NR	NR	NR[l]	NR	NR	4 @ 35	NR[l]	NR	NR
	6	NR	NR	NR	NR	5 @ 48	NR	NR	NR	5 @ 36	NR	NR	NR
7	4	NR	NR	NR	NR	NR	NR	NR	NR	NR	NR	NR	NR
	5	NR	NR	NR	NR	NR	NR	NR	NR	5 @ 47	NR	NR	NR
	6	NR	NR	NR	NR	5 @ 42	NR	NR	NR	6 @ 43	5 @ 48	NR[l]	NR
	7	5 @ 46	NR	NR	NR	6 @ 42	5 @ 46	NR[l]	NR	6 @ 34	6 @ 48	NR	NR
8	4	NR	NR	NR	NR	NR	NR	NR	NR	NR	NR	NR	NR
	5	NR	NR	NR	NR	4 @ 38	NR[l]	NR	NR	5 @ 43	NR	NR	NR
	6	4 @ 37	NR[l]	NR	NR	5 @ 37	NR	NR	NR	6 @ 37	5 @ 43	NR[l]	NR
	7	5 @ 40	NR	NR	NR	6 @ 37	5 @ 41	NR[l]	NR	6 @ 34	6 @ 43	NR	NR
	8	6 @ 43	5 @ 47	NR[l]	NR	6 @ 34	6 @ 43	NR	NR	6 @ 27	6 @ 32	6 @ 44	NR
9	4	NR	NR	NR	NR	NR	NR	NR	NR	NR	NR	NR	NR
	5	NR	NR	NR	NR	4 @ 35	NR[l]	NR	NR	5 @ 40	NR	NR	NR
	6	4 @ 34	NR[l]	NR	NR	6 @ 48	NR	NR	NR	6 @ 36	6 @ 39	NR[l]	NR
	7	5 @ 36	NR	NR	NR	6 @ 34	5 @ 37	NR	NR	6 @ 33	6 @ 38	5 @ 37	NR[l]
	8	6 @ 38	5 @ 41	NR[l]	NR	6 @ 33	6 @ 38	5 @ 37	NR[l]	6 @ 24	6 @ 29	6 @ 39	4 @ 48[m]
	9	6 @ 34	6 @ 46	NR	NR	6 @ 26	6 @ 30	6 @ 41	NR	6 @ 19	6 @ 23	6 @ 30	6 @ 39
10	4	NR	NR	NR	NR	NR	NR	NR	NR	NR	NR	NR	NR
	5	NR	NR	NR	NR	4 @ 33	NR[l]	NR	NR	5 @ 38	NR	NR	NR
	6	5 @ 48	NR[l]	NR	NR	6 @ 45	NR	NR	NR	6 @ 34	5 @ 37	NR	NR
	7	6 @ 47	NR	NR	NR	6 @ 34	6 @ 48	NR	NR	6 @ 30	6 @ 35	6 @ 48	NR[l]
	8	6 @ 34	5 @ 38	NR	NR	6 @ 30	6 @ 34	6 @ 47	NR[l]	6 @ 22	6 @ 26	6 @ 35	6 @ 45[m]
	9	6 @ 34	6 @ 41	4 @ 48	NR[l]	6 @ 23	6 @ 27	6 @ 35	4 @ 48[m]	DR	6 @ 22	6 @ 27	6 @ 34
	10	6 @ 28	6 @ 33	6 @ 45	NR	DR[j]	6 @ 23	6 @ 29	6 @ 38	DR	6 @ 22	6 @ 22	6 @ 28

For SI: 1 foot = 304.8 mm; 1 inch = 25.4 mm; 1 pound per square foot per foot = 0.1571 kPa^2/m, 1 pound per square inch = 6.895 kPa.

NR = Not required.

a. Soil classes are in accordance with the Unified Soil Classification System. Refer to Table R405.1.

b. Table values are based on reinforcing bars with a minimum yield strength of 60,000 psi.

c. Vertical reinforcement with a yield strength of less than 60,000 psi and/or bars of a different size than specified in the table are permitted in accordance with Section R404.1.2.3.7.6 and Table R404.1.2(9).

d. NR indicates no vertical wall reinforcement is required, except for 6-inch nominal walls formed with stay-in-place forming systems in which case vertical reinforcement shall be #4@48 inches on center.

e. Allowable deflection criterion is $L/240$, where L is the unsupported height of the basement wall in inches.

f. Interpolation is not permitted.

g. Where walls will retain 4 feet or more of unbalanced backfill, they shall be laterally supported at the top and bottom before backfilling.

h. Vertical reinforcement shall be located to provide a cover of 1.25 inches measured from the inside face of the wall. The center of the steel shall not vary from the specified location by more than the greater of 10 percent of the wall thickness or $^3/_8$-inch.

i. Concrete cover for reinforcement measured from the inside face of the wall shall not be less than $^3/_4$-inch. Concrete cover for reinforcement measured from the outside face of the wall shall not be less than $1^1/_2$ inches for No. 5 bars and smaller, and not less than 2 inches for larger bars.

j. DR means design is required in accordance with the applicable building code, or where there is no code in accordance with ACI 318.

k. Concrete shall have a specified compressive strength, f'_c, of not less than 2,500 psi at 28 days, unless a higher strength is required by footnote l or m.

l. The minimum thickness is permitted to be reduced 2 inches, provided the minimum specified compressive strength of concrete, f'_c, is 4,000 psi.

m. A plain concrete wall with a minimum nominal thickness of 12 inches is permitted, provided minimum specified compressive strength of concrete, f'_c, is 3,500 psi.

n. See Table R611.3 for tolerance from nominal thickness permitted for flat walls.

❖See the commentary to Section R404.1.2.2.

TABLE R404.1.2(9) MINIMUM SPACING FOR ALTERNATE BAR SIZE AND/OR ALTERNATE GRADE OF STEEL[a, b, c]

BAR SPACING FROM APPLICABLE TABLE IN SECTION R404.1.2.2 (inches)	BAR SIZE FROM APPLICABLE TABLE IN SECTION R404.1.2.2														
	#4					#5					#6				
	Alternate bar size and/or alternate grade of steel desired														
	Grade 60		Grade 40			Grade 60		Grade 40			Grade 60		Grade 40		
	#5	#6	#4	#5	#6	#4	#6	#4	#5	#6	#4	#5	#4	#5	#6
	Maximum spacing for alternate bar size and/or alternate grade of steel (inches)														
8	12	18	5	8	12	5	11	3	5	8	4	6	2	4	5
9	14	20	6	9	13	6	13	4	6	9	4	6	3	4	6
10	16	22	7	10	15	6	14	4	7	9	5	7	3	5	7
11	17	24	7	11	16	7	16	5	7	10	5	8	3	5	7
12	19	26	8	12	18	8	17	5	8	11	5	8	4	6	8
13	20	29	9	13	19	8	18	6	9	12	6	9	4	6	9
14	22	31	9	14	21	9	20	6	9	13	6	10	4	7	9
15	23	33	10	16	22	10	21	6	10	14	7	11	5	7	10
16	25	35	11	17	23	10	23	7	11	15	7	11	5	8	11
17	26	37	11	18	25	11	24	7	11	16	8	12	5	8	11
18	28	40	12	19	26	12	26	8	12	17	8	13	5	8	12
19	29	42	13	20	28	12	27	8	13	18	9	13	6	9	13
20	31	44	13	21	29	13	28	9	13	19	9	14	6	9	13
21	33	46	14	22	31	14	30	9	14	20	10	15	6	10	14
22	34	48	15	23	32	14	31	9	15	21	10	16	7	10	15
23	36	48	15	24	34	15	33	10	15	22	10	16	7	11	15
24	37	48	16	25	35	15	34	10	16	23	11	17	7	11	16
25	39	48	17	26	37	16	35	11	17	24	11	18	8	12	17
26	40	48	17	27	38	17	37	11	17	25	12	18	8	12	17
27	42	48	18	28	40	17	38	12	18	26	12	19	8	13	18
28	43	48	19	29	41	18	40	12	19	26	13	20	8	13	19
29	45	48	19	30	43	19	41	12	19	27	13	20	9	14	19
30	47	48	20	31	44	19	43	13	20	28	14	21	9	14	20
31	48	48	21	32	45	20	44	13	21	29	14	22	9	15	21
32	48	48	21	33	47	21	45	14	21	30	15	23	10	15	21
33	48	48	22	34	48	21	47	14	22	31	15	23	10	16	22
34	48	48	23	35	48	22	48	15	23	32	15	24	10	16	23
35	48	48	23	36	48	23	48	15	23	33	16	25	11	16	23
36	48	48	24	37	48	23	48	15	24	34	16	25	11	17	24
37	48	48	25	38	48	24	48	16	25	35	17	26	11	17	25
38	48	48	25	39	48	25	48	16	25	36	17	27	12	18	25
39	48	48	26	40	48	25	48	17	26	37	18	27	12	18	26
40	48	48	27	41	48	26	48	17	27	38	18	28	12	19	27
41	48	48	27	42	48	26	48	18	27	39	19	29	12	19	27
42	48	48	28	43	48	27	48	18	28	40	19	30	13	20	28
43	48	48	29	44	48	28	48	18	29	41	20	30	13	20	29
44	48	48	29	45	48	28	48	19	29	42	20	31	13	21	29
45	48	48	30	47	48	29	48	19	30	43	20	32	14	21	30
46	48	48	31	48	48	30	48	20	31	44	21	32	14	22	31
47	48	48	31	48	48	30	48	20	31	44	21	33	14	22	31
48	48	48	32	48	48	31	48	21	32	45	22	34	15	23	32

For SI: 1 inch = 25.4 mm, 1 pound per square inch = 6.895 kPa.

a. This table is for use with tables in Section R404.1.2.2 that specify the minimum bar size and maximum spacing of vertical wall reinforcement for foundation walls and above-grade walls. Reinforcement specified in tables in Sections R404.1.2.2 is based on Grade 60 steel reinforcement.

b. Bar spacing shall not exceed 48 inches on center and shall not be less than one-half the nominal wall thickness.

c. For Grade 50 steel bars (ASTM A 996, Type R), use spacing for Grade 40 bars or interpolate between Grades 40 and 60.

❖See the commentary to Section R404.1.2.2.

R404.1.2.2.1 Concrete foundation stem walls supporting above-grade concrete walls. Foundation stem walls that support above-grade concrete walls shall be designed and constructed in accordance with this section.

1. Stem walls not laterally supported at top.Concrete stem walls that are not monolithic with slabs-on-ground or are not otherwise laterally supported by slabs-on-ground shall comply with this section. Where unbalanced backfill retained by the stem wall is less than or equal to 18 inches (457 mm), the stem wall and above-grade wall it supports shall be provided with vertical reinforcement in accordance with Section R611.6 and Table R611.6(1), R611.6(2) or R611.6(3) for above-grade walls. Where unbalanced backfill retained by the stem wall is greater than 18 inches (457 mm), the stem wall and above-grade wall it supports shall be provided with vertical reinforcement in accordance with Section R611.6 and Table R611.6(4).
2. Stem walls laterally supported at top.Concrete stem walls that are monolithic with slabs-on-ground or are otherwise laterally supported by slabs-on-ground shall be vertically reinforced in accordance with Section R611.6 and Table R611.6(1), R611.6(2) or R611.6(3) for above-grade walls. Where the unbalanced backfill retained by the stem wall is greater than 18 inches (457 mm), the connection between the stem wall and the slab-on-ground, and the portion of the slab-on-ground providing lateral support for the wall shall be designed in accordance with PCA 100 or in accordance with accepted engineering practice. Where the unbalanced backfill retained by the stem wall is greater than 18 inches (457 mm), the minimum nominal thickness of the wall shall be 6 inches (152 mm).

❖A stem wall is a foundation wall that supports an above-grade wall and retains unbalanced backfill beneath the slab-on-ground of the first story above grade plane.

Stem walls supporting above-grade concrete walls must meet the following:

- Vertical reinforcement in accordance with Section R611.6 and Tables R611.6(1) through R611.6(3).
- Stem walls not laterally supported at the top and unbalanced backfill greater than 18 inches (457 mm), use Section R611.6 and Table R611.6(4).
- Stem walls laterally supported at the top and unbalanced backfill greater than 18 inches (457 mm), the connection to the slab on ground must be designed in accordance with PCA 100 or accepted engineering practice.

R404.1.2.2.2 Concrete foundation stem walls supporting light-frame above-grade walls. Concrete foundation stem walls that support light-frame above-grade walls shall be designed and constructed in accordance with this section.

1. Stem walls not laterally supported at top. Concrete stem walls that are not monolithic with slabs-on-ground or are not otherwise laterally supported by slabs-on-ground and retain 48 inches (1219 mm) or less of unbalanced fill, measured from the top of the wall, shall be constructed in accordance with Section R404.1.2. Foundation stem walls that retain more than 48 inches (1219 mm) of unbalanced fill, measured from the top of the wall, shall be designed in accordance with Sections R404.1.3 and R404.4.
2. Stem walls laterally supported at top. Concrete stem walls that are monolithic with slabs-on-ground or are otherwise laterally supported by slabs-on-ground shall be constructed in accordance with Section R404.1.2. Where the unbalanced backfill retained by the stem wall is greater than 48 inches (1219 mm), the connection between the stem wall and the slab-on-ground, and the portion of the slab-on-ground providing lateral support for the wall shall be designed in accordance with PCA 100 or in accordance with accepted engineering practice.

❖Stem walls supporting light-frame above-grade walls must meet the following:

- Constructed in accordance with Section R404.1.2.
- Not laterally supported at the top and unbalanced backfill greater than 48 inches (1219 mm), must be designed in accordance with Sections R404.1.3 and R404.4.
- Laterally supported at the top and unbalanced backfill greater than 48 inches (1219 mm), must be de-signed in accordance with PCA 100 or accepted engineering practice.

R404.1.2.3 Concrete, materials for concrete, and forms. Materials used in concrete, the concrete itself and forms shall conform to requirements of this section or ACI 318.

❖Concrete, materials for concrete and forms must comply with this section or ACI 318.

R404.1.2.3.1 Compressive strength. The minimum specified compressive strength of concrete, f'_c, shall comply with Section R402.2 and shall be not less than 2,500 psi (17.2 MPa) at 28 days in buildings assigned to Seismic Design Category A, B or C and 3000 psi (20.5 MPa) in buildings assigned to Seismic Design Category D_0, D_1 or D_2.

❖The minimum specified compressive strength of concrete of 2,500 psi (17.2 MPa) is based on the minimum current practice, which corresponds to minimum compressive strength permitted by ACI 318 and the code. Table R404.1.2(8) provides adjustment factors in the footnotes that recognize the benefits of using higher strength concrete. For all buildings assigned to Seismic Design Category D_0, D_1 or D_2, a minimum specified compressive strength of concrete of 3,000 psi (20.7 MPa) is required, which is consistent with Chapter 21 of ACI 318.

R404.1.2.3.2 Concrete mixing and delivery. Mixing and delivery of concrete shall comply with ASTM C 94 or ASTM C 685.

❖Concrete must be mixed and delivered in accordance with the requirements of *Specification for Ready-*

Mixed Concrete (ASTM C 94) or *Specification for Concrete Made by Volumetric Batching and Continuous Mixing* (ASTM C 685). Test methods for uniformity of mixing are given in ASTM C 94.

R404.1.2.3.3 Maximum aggregate size. The nominal maximum size of coarse aggregate shall not exceed one-fifth the narrowest distance between sides of forms, or three-fourths the clear spacing between reinforcing bars or between a bar and the side of the form.

Exception: When *approved*, these limitations shall not apply where removable forms are used and workability and methods of consolidation permit concrete to be placed without honeycombs or voids.

❖The maximum aggregate size is based on requirements in ACI 318. These limitations on aggregate size are intended to result in concrete that can be placed such that it properly encases reinforcement and minimizes honeycombing and voids. The exception, which is also based on ACI 318, is limited to use with removable forms because proper placement and encasement of reinforcement can be verified through visual inspection after removal of forms.

R404.1.2.3.4 Proportioning and slump of concrete. Proportions of materials for concrete shall be established to provide workability and consistency to permit concrete to be worked readily into forms and around reinforcement under conditions of placement to be employed, without segregation or excessive bleeding. Slump of concrete placed in removable forms shall not exceed 6 inches (152 mm).

Exception: When *approved*, the slump is permitted to exceed 6 inches (152 mm) for concrete mixtures that are resistant to segregation, and are in accordance with the form manufacturer's recommendations.

Slump of concrete placed in stay-in-place forms shall exceed 6 inches (152 mm). Slump of concrete shall be determined in accordance with ASTM C 143.

❖The maximum slump requirements are based on current practice. Considerations included in the prescribed maximums are ease of placement, ability to fill cavities thoroughly and limiting the pressures exerted on the form by fresh concrete.

Where removable wall forms are used, maximum slump of concrete is 6 inches (152 mm). The maximum slump is based on experience with ease of placement, ability to fill cavities thoroughly and limiting the pressures exerted on the form by fresh concrete. The 6-inch (152 mm) maximum slump refers to the characteristics of the specified mixture proportions based on water/cementitious materials ratio only. The exception envisions the use of mid-range or high-range water-reducing admixtures to increase the slump above 6 inches (152 mm), since their use does not adversely affect the tendency of the aggregate to segregate.

The minimum slump requirement for walls cast in stay-in-place forms must be greater than 6 inches (152 mm). It is important that the concrete mix design utilize proper materials in the proper proportions so that the specified strength is achieved at the specified slump [i.e., greater that 6 inches (152 mm)]. Ordering a concrete mix at a lower slump [e.g., 4 inches (102 mm)] and adding water at the job site to increase the slump to more than 6 inches (152 mm) will adversely affect the strength of the concrete, and possibly jeopardize the safety of the structure. Normally it would be expected that a concrete mixture with a specified slump in excess of 6 inches (152 mm) would contain mid-range or high-range water-reducing admixtures so as to mitigate the possibility of segregation of the aggregate.

R404.1.2.3.5 Consolidation of concrete. Concrete shall be consolidated by suitable means during placement and shall be worked around embedded items and reinforcement and into corners of forms. Where stay-in-place forms are used, concrete shall be consolidated by internal vibration.

Exception: When *approved* for concrete to be placed in stay-in-place forms, self-consolidating concrete mixtures with slumps equal to or greater than 8 inches (203 mm) that are specifically designed for placement without internal vibration need not be internally vibrated.

❖Where stay-in-place forms are used, concrete must be consolidated by internal vibration. A study by PCA determined that internal vibration is required for standard mix design with a slump between 4 inches (102 mm) and 8 inches (203 mm). Two of the mixtures were intended to be used without internal vibration and good results were obtained with both. One mixture utilized a high-range water-reducing admixture to achieve a slump between 8 inches (203 mm) and 10 inches (254 mm). The other contained a high-range water-reducing admixture and a viscosity-modifying admixture to achieve a self-consolidating concrete (SCC) with a slump in excess of 8 inches (203 mm). Slumps in excess of 8 inches (103 mm) should not be obtained by the use of water alone since segregation is likely.

R404.1.2.3.6 Form materials and form ties. Forms shall be made of wood, steel, aluminum, plastic, a composite of cement and foam insulation, a composite of cement and wood chips, or other *approved* material suitable for supporting and containing concrete. Forms shall provide sufficient strength to contain concrete during the concrete placement operation.

Form ties shall be steel, solid plastic, foam plastic, a composite of cement and wood chips, a composite of cement and foam plastic, or other suitable material capable of resisting the forces created by fluid pressure of fresh concrete.

❖The materials listed are based on currently available removable and stay-in-place forming systems. From a structural design standpoint, the material can be anything that has sufficient strength and durability to contain the concrete during pouring and curing.

R404.1.2.3.6.1 Stay-in-place forms. Stay-in-place concrete forms shall comply with this section.

1. Surface burning characteristics. The flame-spread index and smoke-developed index of forming material,

other than foam plastic, left exposed on the interior shall comply with Section R302. The surface burning characteristics of foam plastic used in insulating concrete forms shall comply with Section R316.3.

2. Interior covering. Stay-in-place forms constructed of rigid foam plastic shall be protected on the interior of the building as required by Section R316. Where gypsum board is used to protect the foam plastic, it shall be installed with a mechanical fastening system. Use of adhesives in addition to mechanical fasteners is permitted.
3. Exterior wall covering. Stay-in-place forms constructed of rigid foam plastics shall be protected from sunlight and physical damage by the application of an *approved* exterior wall covering complying with this code. Exterior surfaces of other stay-in-place forming systems shall be protected in accordance with this code.
4. Termite hazards. In areas where hazard of termite damage is very heavy in accordance with Figure R301.2(6), foam plastic insulation shall be permitted below *grade* on foundation walls in accordance with one of the following conditions:
   - 4.1. Where in addition to the requirements in Section R318.1, an *approved* method of protecting the foam plastic and structure from subterranean termite damage is provided.
   - 4.2. The structural members of walls, floors, ceilings and roofs are entirely of noncombustible materials or pressure-preservative-treated wood.
   - 4.3. On the interior side of *basement* walls.
5. Flat ICF wall system forms shall conform to ASTM E 2634.

❖Materials that stay in place after concrete cures must be able to withstand the rigors of the environment. From a thermal standpoint, most stay-in-place form systems and some removable form systems that incorporate insulation into the wall before the concrete is placed provide enough insulating value to meet code requirements. However, some forming systems will require the addition of insulation to the interior or exterior of the concrete wall to meet code requirements for insulating value. The form material, except foam plastic, can be anything that meets the flame spread and smoke-developed indices of Section R302. Foam plastic must comply with Section R316.3. Covering foam plastic that would otherwise be exposed to the interior of the building must comply with Section R316. Adhesively attaching the gypsum board to the foam plastic as the only means of fastening is not permitted out of concern that when the adhesive is exposed to higher temperatures expected in a fire, it may soften and allow the gypsum board to delaminate from the foam plastic.

It is generally accepted that a monolithic concrete wall is a solid wall through which water and air cannot readily flow; however, there is a possibility that the concrete wall may have drying shrinkage cracks through which water may enter. Small gaps between stay-in-place form blocks are inherent in current screen-grid stay-in-place form walls and may allow water to enter the structure. As a result, a moisture barrier on the exterior face of the stay-in-place form wall is generally required and should be considered minimum acceptable practice. See the requirements for concrete wall dampproofing and waterproofing in Sections R405 and R406.

Where the termite hazard is very heavy, foam plastic insulation may be installed below grade only if installed on the interior or if all walls, floors and ceilings are noncombustible material or treated wood or in accordance with Section R318.1 (see commentary, Section R403.3.4). Flat insulating concrete form (ICF) wall systems must comply with ASTM E 2634.

R404.1.2.3.7 Reinforcement.

❖This section prescribes the requirements for steel reinforcement for concrete foundation walls.

R404.1.2.3.7.1 Steel reinforcement. Steel reinforcement shall comply with the requirements of ASTM A 615, A 706, or A 996. ASTM A 996 bars produced from rail steel shall be Type R. In buildings assigned to Seismic Design Category A, B or C, the minimum yield strength of reinforcing steel shall be 40,000 psi (Grade 40) (276 MPa). In buildings assigned to Seismic Design Category D_0, D_1 or D_2, reinforcing steel shall comply with the requirements of ASTM A 706 for low-alloy steel with a minimum yield strength of 60,000 psi (Grade 60) (414 MPa).

❖The types of steel reinforcement permitted by this section are consistent with ACI 318. Type R is specified for ASTM A 996 (rail steel) because the ASTM standard permits two designations. The type designated R is specified because it is the only one of the two that complies with the more stringent bend requirements of ACI 318. ASTM A 615 Grade 60(420 MPa) reinforcing steel that complies with the additional criteria of Section 21.2.5 of ACI 318, or ASTM A 706 (low-alloy) Grade 60 (420 MPa) reinforcing steel is required in all buildings assigned to Seismic Design Category D_0, D_1 or D_2 for improved ductility. This too is consistent with ACI 318.

R404.1.2.3.7.2 Location of reinforcement in wall. The center of vertical reinforcement in *basement* walls determined from Tables R404.1.2(2) through R404.1.2(7) shall be located at the centerline of the wall. Vertical reinforcement in *basement* walls determined from Table R404.1.2(8) shall be located to provide a maximum cover of 1.25 inches (32 mm) measured from the inside face of the wall. Regardless of the table used to determine vertical wall reinforcement, the center of the steel shall not vary from the specified location by more than the greater of 10 percent of the wall thickness and $^3/_8$-inch (10 mm). Horizontal and vertical reinforcement shall be located in foundation walls to provide the minimum cover required by Section R404.1.2.3.7.4.

❖The amount of reinforcement required by Tables R404.1.2(2) through R404.1.2(7) for a given condition of wall type, height, backfill height and design lateral soil load was computed based on the centroid of the vertical reinforcement being at the center of the wall. In engineering terms, the effective depth, *d*, is one-half the wall thickness. Table R404.1.2(8) is based on the vertical reinforcement being placed with a cover of 1.25 inches (32 mm) from the inside face of the wall. This greatly increases the effective depth, *d*, for the thicker walls, thus significantly increasing the moment strength for a given amount of reinforcement. Use of Table R404.1.2(8) is restricted to situations where there are 4 feet (1.2 m) or more of unbalanced backfill to ensure that moment-reversal caused by outward acting wind pressure does not control the design. The vertical reinforcement must be placed within a tolerance of the larger of 10 percent of the wall thickness and $^3/_8$-inch (10 mm). The latter criterion is based on the tolerance on *d* in ACI 318 for members with effective depth, *d*, less than or equal to 8 inches (203 mm).

R404.1.2.3.7.3 Wall openings. Vertical wall reinforcement required by Section R404.1.2.2 that is interrupted by wall openings shall have additional vertical reinforcement of the same size placed within 12 inches (305 mm) of each side of the opening.

❖If a wall opening interrupts reinforcement, additional reinforcement of equivalent area must be provided. Although the code does not elaborate, a good practice is to distribute this equivalent reinforcing symmetrically on either side of the opening.

R404.1.2.3.7.4 Support and cover. Reinforcement shall be secured in the proper location in the forms with tie wire or other bar support system to prevent displacement during the concrete placement operation. Steel reinforcement in concrete cast against the earth shall have a minimum cover of 3 inches (75 mm). Minimum cover for reinforcement in concrete cast in removable forms that will be exposed to the earth or weather shall be $1^1/_2$ inches (38 mm) for No. 5 bars and smaller, and 2 inches (50 mm) for No. 6 bars and larger. For concrete cast in removable forms that will not be exposed to the earth or weather, and for concrete cast in stay-in-place forms, minimum cover shall be $^3/_4$ inch (19 mm). The minus tolerance for cover shall not exceed the smaller of one-third the required cover or $^3/_8$ inch (10 mm).

❖Reinforcement should be adequately supported in the forms to prevent displacement by concrete placement or workers. Concrete cover as protection of reinforcement against weather and other effects is measured from the concrete surface to the outermost surface of the steel to which the cover requirement applies. The condition "concrete surfaces exposed to earth or weather" refers to direct exposure to moisture changes and not just to temperature changes. Stay-in-place forms provide an alternative method of protection from earth or weather and the minimum cover for concrete not exposed to weather or earth is allowed. The tolerance for cover is based on ACI 318 requirements.

R404.1.2.3.7.5 Lap splices. Vertical and horizontal wall reinforcement shall be the longest lengths practical. Where splices are necessary in reinforcement, the length of lap splice shall be in accordance with Table R611.5.4.(1) and Figure R611.5.4(1). The maximum gap between noncontact parallel bars at a lap splice shall not exceed the smaller of one-fifth the required lap length and 6 inches (152 mm). See Figure R611.5.4(1).

❖Where continuous bars are not provided, lap splices are necessary to allow forces to be transferred between sections of rebar. The length of lap splices in Table R611.5.4(1) is based on ACI 318 requirements. Noncontact lap splices are permitted provided the spacing of lapped bars does not exceed the smaller of one-fifth the required lap length or 6 inches (152 mm). If individual bars in nonconcact lap splices are too widely spaced, an unreinforced section is created. Forcing a potential crack to follow a zigzag line (5:1 slope) is considered a minimum precaution. These requirements are based on ACI 318. Figure R611.5.4(1) illustrates the code text requirements.

R404.1.2.3.7.6 Alternate grade of reinforcement and spacing. Where tables in Section R404.1.2.2 specify vertical wall reinforcement based on minimum bar size and maximum spacing, which are based on Grade 60 (414 MPa) steel reinforcement, different size bars and/or bars made from a different grade of steel are permitted provided an equivalent area of steel per linear foot of wall is provided. Use of Table R404.1.2(9) is permitted to determine the maximum bar spacing for different bar sizes than specified in the tables and/or bars made from a different grade of steel. Bars shall not be spaced less than one-half the wall thickness, or more than 48 inches (1219 mm) on center.

❖See the commentary for Section R404.1.2.2.

R404.1.2.3.7.7 Standard hooks. Where reinforcement is required by this code to terminate with a standard hook, the hook shall comply with Section R611.5.4.5 and Figure R611.5.4(3).

❖The hook requirements shown in Figure R611.5.4(3) are consistent with the ACI 318 requirements.

R404.1.2.3.7.8 Construction joint reinforcement. Construction joints in foundation walls shall be made and located to not impair the strength of the wall. Construction joints in plain concrete walls, including walls required to have not less than No. 4 bars at 48 inches (1219 mm) on center by Sections R404.1.2.2 and R404.1.4.2, shall be located at points of lateral support, and a minimum of one No. 4 bar shall extend across the construction joint at a spacing not to exceed 24 inches (610 mm) on center. Construction joint reinforcement shall have a minimum of 12 inches (305 mm) embedment on both sides of the joint. Construction joints in reinforced concrete walls shall be located in the middle third of the span between lateral supports, or located and constructed as required for joints in plain concrete walls.

Exception: Use of vertical wall reinforcement required by this code is permitted in lieu of construction joint reinforcement provided the spacing does not exceed 24 inches (610 mm), or the combination of wall reinforcement and No.4 bars described above does not exceed 24 inches (610 mm).

❖Construction joints must be located where they will cause the least weakness in the structure. The reinforcement required by this section is required to provide for transfer of shear and other forces through the construction joint. These provisions are consistent with ACI 318.

R404.1.2.3.8 Exterior wall coverings. Requirements for installation of masonry veneer, stucco and other wall coverings on the exterior of concrete walls and other construction details not covered in this section shall comply with the requirements of this code.

❖The commentary for exterior wall coverings is contained in the commentary for Chapter 7 of the code.

R404.1.2.4 Requirements for Seismic Design Category C. Concrete foundation walls supporting above-grade concrete walls in townhouses assigned to Seismic Design Category C shall comply with ACI 318, ACI 332 or PCA 100 (see Section R404.1.2).

❖See the commentary for Section R404.1.2.

R404.1.3 Design required. Concrete or masonry foundation walls shall be designed in accordance with accepted engineering practice when either of the following conditions exists:

1. Walls are subject to hydrostatic pressure from groundwater.
2. Walls supporting more than 48 inches (1219 mm) of unbalanced backfill that do not have permanent lateral support at the top or bottom.

❖Walls subjected to hydrostatic pressure from groundwater and walls without permanent lateral support at the top or bottom are not specifically included in the code. This section specifies conditions under which foundation walls require design. Section R404.1 cites the standards that are commonly used for "accepted engineering practice."

R404.1.4 Seismic Design Category D_0, D_1 or D_2.

❖This section provides the requirement for masonry and concrete foundation walls in Seismic Design Category D_0, D_1 or D_2.

R404.1.4.1 Masonry foundation walls. In addition to the requirements of Table R404.1.1(1) plain masonry foundation walls in buildings assigned to Seismic Design Category D_0, D_1 or D_2, as established in Table R301.2(1), shall comply with the following.

1. Wall height shall not exceed 8 feet (2438 mm).
2. Unbalanced backfill height shall not exceed 4 feet (1219 mm).
3. Minimum nominal thickness for plain masonry foundation walls shall be 8 inches (203 mm).
4. Masonry stem walls shall have a minimum vertical reinforcement of one No. 3 (No. 10) bar located a maximum of 4 feet (1219 mm) on center in grouted cells. Vertical reinforcement shall be tied to the horizontal reinforcement in the footings.

Foundation walls in buildings assigned to Seismic Design Category D_0, D_1 or D_2, as established in Table R301.2(1), supporting more than 4 feet (1219 mm) of unbalanced backfill or exceeding 8 feet (2438 mm) in height shall be constructed in accordance with Table R404.1.1(2), R404.1.1(3) or R404.1.1(4). Masonry foundation walls shall have two No. 4 (No. 13) horizontal bars located in the upper 12 inches (305 mm) of the wall.

❖This section places four restrictions on the requirements in Table R404.1.1(1) for plain masonry foundation walls in Seismic Design Category D_0, D_1 or D_2.

The code requires minimum vertical reinforcement of masonry stem walls in order for them to resist lateral earthquake forces. Any wall over 8 feet (2338 mm) or supporting more than 4 feet (1219 mm) of unbalanced backfill must be constructed with minimum vertical reinforcement in accordance with Tables R404.1.1(2) through R404.1.1(4) and have the specified longitudinal reinforcement.

R404.1.4.2 Concrete foundation walls. In buildings assigned to Seismic Design Category D_0, D_1 or D_2, as established in Table R301.2(1), concrete foundation walls that support light-frame walls shall comply with this section, and concrete foundation walls that support above-grade concrete walls shall comply with ACI 318, ACI 332 or PCA 100 (see Section R404.1.2). In addition to the horizontal reinforcement required by Table R404.1.2(1), plain concrete walls supporting light-frame walls shall comply with the following.

1. Wall height shall not exceed 8 feet (2438 mm).
2. Unbalanced backfill height shall not exceed 4 feet (1219 mm).
3. Minimum thickness for plain concrete foundation walls shall be 7.5 inches (191 mm) except that 6 inches (152 mm) is permitted where the maximum wall height is 4 feet, 6 inches (1372 mm).

Foundation walls less than 7.5 inches (191 mm) in thickness, supporting more than 4 feet (1219 mm) of unbalanced backfill or exceeding 8 feet (2438 mm) in height shall be provided with horizontal reinforcement in accordance with Table R404.1.2(1), and vertical reinforcement in accordance with Table R404.1.2(2), R404.1.2(3), R404.1.2(4), R404.1.2(5), R404.1.2(6), R404.1.2(7) or R404.1.2(8). Where Tables R404.1.2(2) through R404.1.2(8) permit plain concrete walls, not less than No. 4 (No. 13) vertical bars at a spacing not exceeding 48 inches (1219 mm) shall be provided.

❖This section contains the provisions for concrete foundation walls supporting light-frame construction in Seismic Design Category D_0, D_1 or D_2. Concrete foundation walls supporting above-grade concrete walls must comply with ACI 318, ACI 332 or PCA 100.

In addition to the horizontal reinforcing, this section places three restrictions on the use of plain concrete foundation walls in Seismic Design Category D_0, D_1 or D_2.

Any wall over 8 feet (2338 mm) or supporting more than 4 feet (1219 mm) of unbalanced backfill must be constructed with minimum vertical reinforcement in accordance with Tables R404.1.2(2) through R404.1.2(8). Where no vertical reinforcement is required in Tables R404.1.2(2) through R404.1.2(8), No. 4 vertical bars at 48 inches on center are required.

R404.1.5 Foundation wall thickness based on walls supported. The thickness of masonry or concrete foundation walls shall not be less than that required by Section R404.1.5.1 or R404.1.5.2, respectively.

❖This section provides the requirement for minimum thickness of masonry and concrete foundation walls.

R404.1.5.1 Masonry wall thickness. Masonry foundation walls shall not be less than the thickness of the wall supported, except that masonry foundation walls of at least 8-inch (203 mm) nominal thickness shall be permitted under brick veneered frame walls and under 10-inch-wide (254 mm) cavity walls where the total height of the wall supported, including gables, is not more than 20 feet (6096 mm), provided the requirements of Section R404.1.1 are met.

❖Other than the two exceptions noted, masonry foundation walls should be equal to or greater than the thickness of the wall supported. The first exception permits brick veneered frame wall to be supported on 8-inch (203 mm) nominal thickness foundation walls. The second exception permits 10-inch-wide (254 mm) cavity walls with total height, including gables, of 20 feet (6096 mm) maximum to be supported on 8-inch (203 mm) nominal thickness foundation walls.

R404.1.5.2 Concrete wall thickness. The thickness of concrete foundation walls shall be equal to or greater than the thickness of the wall in the *story* above. Concrete foundation walls with corbels, brackets or other projections built into the wall for support of masonry veneer or other purposes are not within the scope of the tables in this section.

Where a concrete foundation wall is reduced in thickness to provide a shelf for the support of masonry veneer, the reduced thickness shall be equal to or greater than the thickness of the wall in the *story* above. Vertical reinforcement for the foundation wall shall be based on Table R404.1.2(8) and located in the wall as required by Section R404.1.2.3.7.2 where that table is used. Vertical reinforcement shall be based on the thickness of the thinner portion of the wall.

> **Exception:** Where the height of the reduced thickness portion measured to the underside of the floor assembly or sill plate above is less than or equal to 24 inches (610 mm) and the reduction in thickness does not exceed 4 inches (102 mm), the vertical reinforcement is permitted to be based on the thicker portion of the wall.

❖Other than the reduced thickness for a shelf for masonry veneer, concrete foundation walls must be equal to or greater than the thickness of the wall in the story above.

It is common practice to reduce the thickness of foundation walls approximately 4 inches (102 mm) to form a shelf for the support of masonry veneer. Generally the shelf is not located at the floor assembly that provides lateral support for the top of the foundation wall. The exception permits the vertical reinforcement for the foundation wall to be based on the thicker portion of the wall provided the shelf is located not more than 24 inches (610 mm) below the underside of the floor assembly and the reduction in thickness does not exceed 4 inches (104 mm). This assures that the location of the reduction in thickness is not near the point of maximum moment, which for a foundation wall is generally in the lower one-half of the wall height.

R404.1.5.3 Pier and curtain wall foundations. Use of pier and curtain wall foundations shall be permitted to support light-frame construction not more than two stories in height, provided the following requirements are met:

1. All load-bearing walls shall be placed on continuous concrete footings placed integrally with the exterior wall footings.
2. The minimum actual thickness of a load-bearing masonry wall shall be not less than 4 inches (102 mm) nominal or $3^3/_8$ inches (92 mm) actual thickness, and shall be bonded integrally with piers spaced in accordance with Section R606.9.
3. Piers shall be constructed in accordance with Section R606.6 and Section R606.6.1, and shall be bonded into the load-bearing masonry wall in accordance with Section R608.1.1 or R608.1.1.2.
4. The maximum height of a 4-inch (102 mm) load-bearing masonry foundation wall supporting wood-frame walls and floors shall not be more than 4 feet (1219 mm).
5. Anchorage shall be in accordance with Section R403.1.6, Figure R404.1.5(1), or as specified by engineered design accepted by the *building official*.
6. The unbalanced fill for 4-inch (102 mm) foundation walls shall not exceed 24 inches (610 mm) for solid masonry or 12 inches (305 mm) for hollow masonry.
7. In Seismic Design Categories D_0, D_1 and D_2, prescriptive reinforcement shall be provided in the horizontal and vertical direction. Provide minimum horizontal joint reinforcement of two No. 9 gage wires spaced not less than 6 inches (152 mm) or one $^1/_4$ inch (6.4 mm) diameter wire at 10 inches (254 mm) on center vertically. Provide minimum vertical reinforcement of one No. 4 bar at 48 inches (1220 mm) on center horizontally grouted in place.

❖Pier and curtain wall foundations may be used to support light-frame construction for buildings in Seismic Design Category A, B, C, D_0, D_1 or D_2, provided several conditions are met. A curtain wall typically refers to a nonload-bearing wall, but within the limits established in this section, it may be used to support up to two stories of light-frame construction. Four-inch (102 mm) nominal masonry walls are limited to 4 feet (1219 mm) in height so that a thin wall does not fail laterally due to the vertical loads on the wall. In addition, no more than 2 feet (610 mm) of unbalanced fill is permitted for solid masonry and 1 foot (305 mm) of unbalanced fill for hollow masonry to prevent failure due to the pressures created by unbalanced fill. Additional requirements are shown in Figure R404.1.5(1). This type of foundation wall must be reinforced when located in Seismic Design Category D_0, D_1 or D_2.

R404.1.6 Height above finished grade. Concrete and masonry foundation walls shall extend above the finished *grade* adjacent to the foundation at all points a minimum of 4 inches (102 mm) where masonry veneer is used and a minimum of 6 inches (152 mm) elsewhere.

❖The minimum distance above adjacent grade to which the foundation must be extended provides termite protection and minimizes the chance of decay resulting from moisture migrating to the wood framing. A reduced foundation extension is permitted when masonry veneer is used.

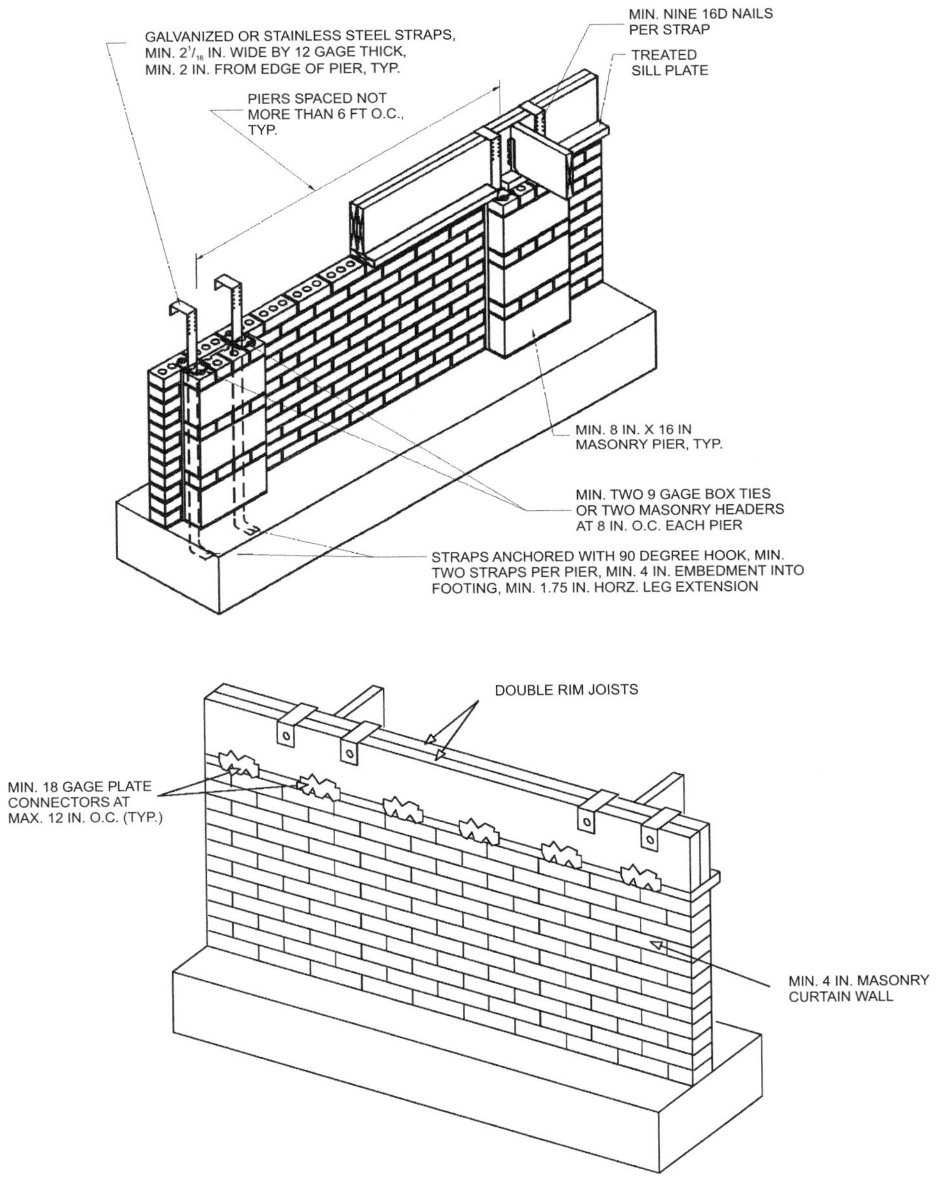

For SI: 1 inch = 25.4 mm, 1 foot = 304.8 mm, 1 degree = 0.0175 rad.

FIGURE R404.1.5(1)
FOUNDATION WALL CLAY MASONRY CURTAIN WALL WITH CONCRETE MASONRY PIERS

❖This figure illustrates several requirements for pier and curtain wall foundations described in Section R404.1.5.1. The pier spacing of not more than 6 feet, 0 inches (1829 mm) on center establishes minimum lateral (horizontal) wall support similar to that of Section R606.9 (also see commentary, Section R404.1.5.1).

R404.1.7 Backfill placement. Backfill shall not be placed against the wall until the wall has sufficient strength and has been anchored to the floor above, or has been sufficiently braced to prevent damage by the backfill.

Exception: Bracing is not required for walls supporting less than 4 feet (1219 mm) of unbalanced backfill.

❖Backfilling should not begin until a foundation wall is in suitable condition to resist lateral earth pressure. The wall must have sufficient strength, and the floor framing system should be in place and anchored to the foundation wall as shown in Commentary Figure R404.1.7 to minimize the chance of damage to the foundation wall resulting from placement of backfill against the foundation wall.

R404.1.8 Rubble stone masonry. Rubble stone masonry foundation walls shall have a minimum thickness of 16 inches (406 mm), shall not support an unbalanced backfill exceeding 8 feet (2438 mm) in height, shall not support a soil pressure greater than 30 pounds per square foot per foot (4.71 kPa/m), and shall not be constructed in Seismic Design Categories D_0, D_1, D_2 or townhouses in Seismic Design Category C, as established in Figure R301.2(2).

❖Rubble stone masonry is masonry walls constructed of roughly shaped stones. See the definitions in the IBC. Rubble stone masonry is almost never used for foundation walls. This type of wall construction is permitted only in Seismic Design Category A or B, and detached one- and two-family dwelling in Seismic Design Category C. Furthermore, it is restricted to a maximum unbalanced backfill height of 8 feet (2438 mm) and must not be subjected to lateral soil pressures more than 30 pounds per square foot (1436 Pa).

R404.1.9 Isolated masonry piers. Isolated masonry piers shall be constructed in accordance with this section and the general masonry construction requirements of Section R606. Hollow masonry piers shall have a minimum nominal thickness of 8 inches (203 mm), with a nominal height not exceeding four times the nominal thickness and a nominal length not exceeding three times the nominal thickness. Where hollow masonry units are solidly filled with concrete or grout, piers shall be permitted to have a nominal height not exceeding ten times the nominal thickness. Footings for isolated masonry piers shall be sized in accordance with Section R403.1.1.

❖Installing masonry pier foundations is a common construction method. This section provides prescriptive requirements for isolated masonry piers used as foundations for raised woodfloor systems.

The requirements of this section provide prescriptive guidance for isolated masonry piers constructed inside a basement or crawl space. The requirements of Section R404.1.9 are based on the empirical design limits contained in the TMS 402/ACI 530/ASCE 5. The language is adopted from the paragraph on foundation piers in the National Concrete Masonry Association's (NCMA) TEK Note 5-3A, *Concrete Masonry Foundation Wall Details.* Further limits are provided for piers supporting floor girders, braced wall panels and for piers in high-seismic or flood-hazard areas. Taller masonry piers supporting an elevated deck, sunroom or other substantially raised portion of a dwelling are relegated to engineered design.

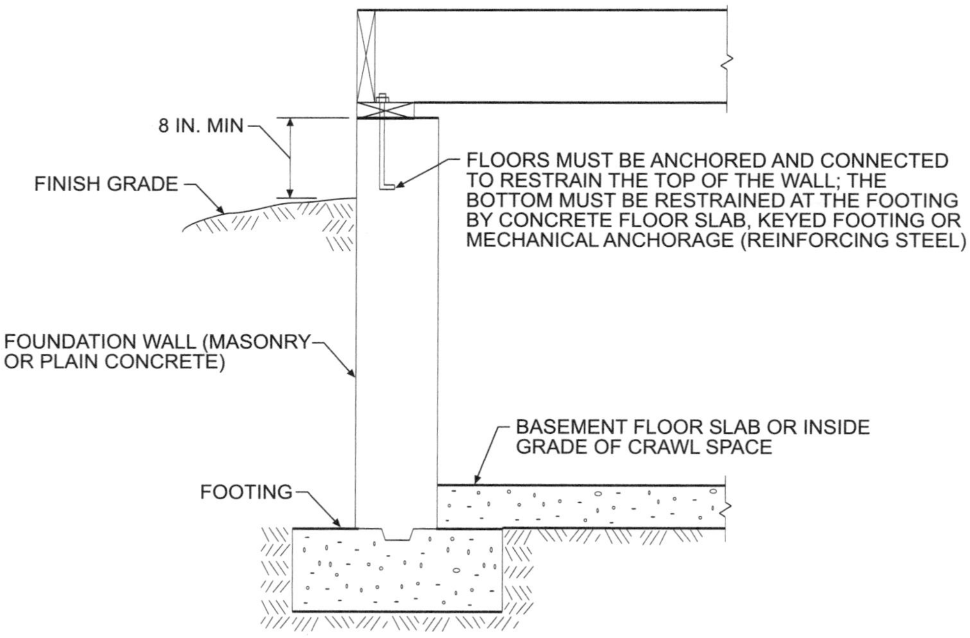
8 IN. MIN
FINISH GRADE
FLOORS MUST BE ANCHORED AND CONNECTED TO RESTRAIN THE TOP OF THE WALL; THE BOTTOM MUST BE RESTRAINED AT THE FOOTING BY CONCRETE FLOOR SLAB, KEYED FOOTING OR MECHANICAL ANCHORAGE (REINFORCING STEEL)
FOUNDATION WALL (MASONRY OR PLAIN CONCRETE)
BASEMENT FLOOR SLAB OR INSIDE GRADE OF CRAWL SPACE
FOOTING

Figure R404.1.7
BEARING FOUNDATION WALLS AGAINST LATERAL EARTH PRESSURE

R404.1.9.1 Pier cap. Hollow masonry piers shall be capped with 4 inches (102 mm) of solid masonry or concrete, a masonry cap block, or shall have cavities of the top course filled with concrete or grout. Where required, termite protection for the pier cap shall be provided in accordance with Section R318.

❖See the commentary to Section R404.1.9.

R404.1.9.2 Masonry piers supporting floor girders. Masonry piers supporting wood girders sized in accordance with Tables R502.5(1) and R502.5(2) shall be permitted in accordance with this section. Piers supporting girders for interior bearing walls shall have a minimum nominal dimension of 12 inches (305 mm) and a maximum height of 10 feet (3048 mm) from top of footing to bottom of sill plate or girder. Piers supporting girders for exterior bearing walls shall have a minimum nominal dimension of 12 inches (305 mm) and a maximum height of 4 feet (1220 mm) from top of footing to bottom of sill plate or girder. Girders and sill plates shall be anchored to the pier or footing in accordance with Section R403.1.6 or Figure R404.1.5(1). Floor girder bearing shall be in accordance with Section R502.6.

❖See the commentary to Section R404.1.9.

R404.1.9.3 Masonry piers supporting braced wall panels. Masonry piers supporting *braced wall panels* shall be designed in accordance with accepted engineering practice.

❖See the commentary to Section R404.1.9.

R404.1.9.4 Seismic design of masonry piers. Masonry piers in all *dwellings* located in Seismic Design Category D_0, D_1 or D_2, and townhouses in Seismic Design Category C, shall be designed in accordance with accepted engineering practice.

❖See the commentary to Section R404.1.9.

R404.1.9.5 Masonry piers in flood hazard areas. Masonry piers for *dwellings* in flood hazard areas shall be designed in accordance with Section R322.

❖See the commentary to Section R404.1.9.

R404.2 Wood foundation walls. Wood foundation walls shall be constructed in accordance with the provisions of Sections R404.2.1 through R404.2.6 and with the details shown in Figures R403.1(2) and R403.1(3).

❖This section covers the design and installation of a multicomponent wood foundation system. The construction is essentially a below-grade load-bearing wood frame system that serves not only as the enclosure for basements and crawl spaces but also as the structural foundation for the support of light-frame structures. See Figures R403.1(2) and R403.1(3) for typical installation details for wood foundation walls (also see commentary, Section R401.1).

Certain cold-weather precautions should be taken during the installation of the wood foundation system. The composite footing consisting of a wood plate supported on a bed of stone or sand fill should not be placed on frozen ground. While the bottom of the wood plate footing would normally be placed below the frost line (i.e., basement construction) under certain drainage conditions, AF&PA PWF permits the wood plate to be set above the frost level. The code official should verify that such an alternative design satisfies the intent of the code. Also important is the use of proper sealants during very cold weather. All manufacturers of sealants and bonding agents impose temperature restrictions on the use of their products. Only sealants and bonding agents specifically produced for cold weather conditions should be used.

R404.2.1 Identification. All load-bearing lumber shall be identified by the grade *mark* of a lumber grading or inspection agency which has been *approved* by an accreditation body that complies with DOC PS 20. In lieu of a grade *mark*, a certificate of inspection issued by a lumber grading or inspection agency meeting the requirements of this section shall be accepted. Wood structural panels shall conform to DOC PS 1 or DOC PS 2 and shall be identified by a grade *mark* or certificate of inspection issued by an *approved agency*.

❖See the commentary to Sections R502.1 and R503.2.1 as well as Figures R502.1 and R503.2.1.

R404.2.2 Stud size. The studs used in foundation walls shall be 2-inch by 6-inch (51 mm by 152 mm) members. When spaced 16 inches (406 mm) on center, a wood species with an F_b value of not less than 1,250 pounds per square inch (8619 kPa) as listed in AF&PA/NDS shall be used. When spaced 12 inches (305 mm) on center, an F_b of not less than 875 psi (6033 kPa) shall be required.

❖This section specifies the minimum bending stress, F_b, for studs based on their spacing in a foundation wall. The appropriate wood species that meet this requirement can be found in the supplement to the AF&PA/NDS *National Design Specification for Wood Construction*.

R404.2.3 Height of backfill. For wood foundations that are not designed and installed in accordance with AF&PA PWF, the height of backfill against a foundation wall shall not exceed 4 feet (1219 mm). When the height of fill is more than 12 inches (305 mm) above the interior *grade* of a crawl space or floor of a *basement*, the thickness of the plywood sheathing shall meet the requirements of Table R404.2.3.

❖Table R404.2.3 specifies requirements for plywood based on height of backfill and plywood span (stud spacing) where the height of retained earth is more than 12 inches (305 mm) [see Figure R403.1(2)]. The backfill limits of Table R404.2.3 do not apply to wood foundations designed and installed in accordance with AF&PA PWF.

R404.2.4 Backfilling. Wood foundation walls shall not be backfilled until the *basement* floor and first floor have been constructed or the walls have been braced. For crawl space construction, backfill or bracing shall be installed on the interior of the walls prior to placing backfill on the exterior.

❖These provisions require that lateral support be in place prior to backfilling on the exterior.

TABLE R404.2.3
PLYWOOD GRADE AND THICKNESS FOR WOOD FOUNDATION CONSTRUCTION (30 pcf equivalent-fluid weight soil pressure)

HEIGHT OF FILL (inches)	STUD SPACING (inches)	FACE GRAIN ACROSS STUDS			FACE GRAIN PARALLEL TO STUDS		
		Grade[a]	Minimum thickness (inches)	Span rating	Grade[a]	Minimum thickness (inches)[b, c]	Span rating
24	12	B	$^{15}/_{32}$	32/16	A	$^{15}/_{32}$	32/16
					B	$^{15}/_{32}$[c]	32/16
	16	B	$^{15}/_{32}$	32/16	A	$^{15}/_{32}$[c]	32/16
					B	$^{19}/_{32}$[c] (4, 5 ply)	40/20
36	12	B	$^{15}/_{32}$	32/16	A	$^{15}/_{32}$	32/16
					B	$^{15}/_{32}$[c] (4, 5 ply)	32/16
					B	$^{19}/_{32}$ (4, 5 ply)	40/20
	16	B	$^{15}/_{32}$[c]	32/16	A	$^{19}/_{32}$	40/20
					B	$^{23}/_{32}$	48/24
48	12	B	$^{15}/_{32}$	32/16	A	$^{15}/_{32}$[c]	32/16
					B	$^{19}/_{32}$[c](4, 5 ply)	40/20
	16	B	$^{19}/_{32}$	40/20	A	$^{19}/_{32}$[c]	40/20
					A	$^{23}/_{32}$	48/24

For SI: 1 inch = 25.4 mm, 1 foot = 304.8 mm, 1 pound per cubic foot = 0.1572 kN/m^3.

a. Plywood shall be of the following minimum grades in accordance with DOC PS 1 or DOC PS 2:

1. DOC PS 1 Plywood grades marked:
   - 1.1. Structural I C-D (Exposure 1).
   - 1.2. C-D (Exposure 1).
2. DOC PS 2 Plywood grades marked:
   - 2.1. Structural I Sheathing (Exposure 1).
   - 2.2. Sheathing (Exposure 1).
3. Where a major portion of the wall is exposed above ground and a better appearance is desired, the following plywood grades marked exterior are suitable:
   - 3.1. Structural I A-C, Structural I B-C or Structural I C-C (Plugged) in accordance with DOC PS 1.
   - 3.2. A-C Group 1, B-C Group 1, C-C (Plugged) Group 1 or MDO Group 1 in accordance with DOC PS 1.
   - 3.3. Single Floor in accordance with DOC PS 1 or DOC PS 2.

b. Minimum thickness $^{15}/_{32}$ inch, except crawl space sheathing may be $^{3}/_{8}$ inch for face grain across studs 16 inches on center and maximum 2-foot depth of unequal fill.

c. For this fill height, thickness and grade combination, panels that are continuous over less than three spans (across less than three stud spacings) require blocking 16 inches above the bottom plate. Offset adjacent blocks and fasten through studs with two 16d corrosion-resistant nails at each end.

❖This table specifies the minimum grade and thickness of plywood based on the height of retained earth (fill) and the plywood span.

R404.2.5 Drainage and dampproofing. Wood foundation basements shall be drained and dampproofed in accordance with Sections R405 and R406, respectively.

❖See Sections R405.2 and R406.3 of this commentary.

R404.2.6 Fastening. Wood structural panel foundation wall sheathing shall be attached to framing in accordance with Table R602.3(1) and Section R402.1.1.

❖The fastening schedule must be followed for wall sheathing attachment. Also see the commentary for Section R402.1.1.

R404.3 Wood sill plates. Wood sill plates shall be a minimum of 2-inch by 4-inch (51 mm by 102 mm) nominal lumber. Sill plate anchorage shall be in accordance with Sections R403.1.6 and R602.11.

❖Minimum 2-inch by 4-inch (51 mm by 102 mm) nominal sill plates are required, and should be anchored to the foundation as required by Section R403.1.6.

R404.4 Retaining walls. Retaining walls that are not laterally supported at the top and that retain in excess of 24 inches (610 mm) of unbalanced fill shall be designed to ensure stability against overturning, sliding, excessive foundation pressure and water uplift. Retaining walls shall be designed for a safety factor of 1.5 against lateral sliding and overturning.

❖This section provides design considerations for a retaining wall that retains in excess of 24 inches (610 mm) of unbalanced fill. The lateral pressure of the soil against the retaining wall is greatly influenced by soil moisture. Backfill is usually kept from being saturated for an extended length of time by placing drains near the base of the retaining wall to remove the water in the soil behind it.

R404.5 Precast concrete foundation walls.

❖This section prescribes the design, drawings and identification requirements for precast concrete foundation walls.

R404.5.1 Design. Precast concrete foundation walls shall be designed in accordance with accepted engineering practice. The design and manufacture of precast concrete foundation wall panels shall comply with the materials requirements of Section R402.3 or ACI 318. The panel design drawings shall be prepared by a registered design professional where required by the statutes of the *jurisdiction* in which the project is to be constructed in accordance with Section R106.1.

❖This section prescribes the minimum design and manufacture requirement for precast concrete foundation walls. The material required must comply with ACI 318 or Section R402.3 (see commentary, Section R402.3).

R404.5.2 Precast concrete foundation design drawings. Precast concrete foundation wall design drawings shall be submitted to the *building official* and *approved* prior to installation. Drawings shall include, at a minimum, the information specified below:

1. Design loading as applicable;
2. Footing design and material;
3. Concentrated loads and their points of application;
4. Soil bearing capacity;
5. Maximum allowable total uniform load;
6. Seismic design category; and
7. Basic wind speed.

❖This section prescribes the minimum design requirement for precast concrete foundations and must be shown on the design drawings that are submitted to the building official. Precast foundation systems are engineered products based on several design approaches including, but not limited to, stud and cavity, solid wall panel, composite panel and hollow core systems. This section provides minimum performance design criteria that all precast concrete foundation systems must meet.

R404.5.3 Identification. Precast concrete foundation wall panels shall be identified by a certificate of inspection *label* issued by an *approved* third party inspection agency.

❖This section requires identification by a certificate of inspection that should verify that precast foundation walls are designed to recognized engineering standards, manufactured in a plant under verified quality control, when installed will meet required performance criteria, and are neutral regarding the various design approaches and systems.

SECTION R405
FOUNDATION DRAINAGE

R405.1 Concrete or masonry foundations. Drains shall be provided around all concrete or masonry foundations that retain earth and enclose habitable or usable spaces located below *grade*. Drainage tiles, gravel or crushed stone drains, perforated pipe or other *approved* systems or materials shall be installed at or below the area to be protected and shall discharge by gravity or mechanical means into an *approved* drainage system. Gravel or crushed stone drains shall extend at least 1 foot (305 mm) beyond the outside edge of the footing and 6 inches (152 mm) above the top of the footing and be covered with an *approved* filter membrane material. The top of open joints of drain tiles shall be protected with strips of building paper. Perforated drains shall be surrounded with an *approved* filter membrane or the filter membrane shall cover the washed gravel or crushed rock covering the drain. Drainage tiles or perforated pipe shall be placed on a minimum of 2 inches (51 mm) of washed gravel or crushed rock at least one sieve size larger than the tile joint opening or perforation and covered with not less than 6 inches (152 mm) of the same material.

Exception: A drainage system is not required when the foundation is installed on well-drained ground or sand-gravel mixture soils according to the Unified Soil Classification System, Group I Soils, as detailed in Table R405.1.

❖To allow free groundwater that may be present adjacent to the foundation wall to be drained away, drains are usually placed around houses to remove water and prevent leakage into habitable or usable spaces below grade. The top of the gravel or crushed stone drain, with or without drainage tile or perforated pipe, must be covered with an approved filter membrane. Drainage tiles are extremely important in areas having moderate to heavy rainfall and in soils having a low percolation rate. Perforated pipe has perforations around the entire pipe. The perforations will be subject to clogging if not protected. For this reason the code requires an approved filter membrane to surround the perforated drain pipe. As an alternate the code permits the wash gravel or crushed stone covering the perforated pipe to be covered with an approved filter membrane. The filter membrane allows water to pass through the perimeter drain tiles or pipes without allowing, or at least greatly reducing, the possibility of fine soil materials entering the drainage system. A few examples of such systems are illustrated in Commentary Figures R405.1(1) and R405.1(2). The exception allows omission of the drainage system in Group I soils, which by definition exhibit good drainage characteristics as shown in Table R405.1.

R405.1.1 Precast concrete foundation. Precast concrete walls that retain earth and enclose habitable or useable space located below-*grade* that rest on crushed stone footings shall have a perforated drainage pipe installed below the base of the wall on either the interior or exterior side of the wall, at least one foot (305 mm) beyond the edge of the wall. If the exterior drainage pipe is used, an *approved* filter membrane material shall cover the pipe. The drainage system shall discharge into an *approved* sewer system or to daylight.

❖A perforated drainage pipe is required where crushed stone footing is used with precast concrete foundation walls that enclose habitable or usable space. It is important that the perforated drainage pipe be located at least 1 foot (305 mm) away from the edge of the wall. This is similar to the drainage requirement of Section R405.1 except it is limited to perforated drainage pipe and there is no exception for omission when installed in Group I soils.

TABLE R405.1
PROPERTIES OF SOILS CLASSIFIED ACCORDING TO THE UNIFIED SOIL CLASSIFICATION SYSTEM

SOIL GROUP	UNIFIED SOIL CLASSIFICATION SYSTEM SYMBOL	SOIL DESCRIPTION	DRAINAGE CHARACTERISTICS	FROST HEAVE POTENTIAL	VOLUME CHANGE POTENTIAL EXPANSION[b]
Group I	GW	Well-graded gravels, gravel sand mixtures, little or no fines	Good	Low	Low
	GP	Poorly graded gravels or gravel sand mixtures, little or no fines	Good	Low	Low
	SW	Well-graded sands, gravelly sands, little or no fines	Good	Low	Low
	SP	Poorly graded sands or gravelly sands, little or no fines	Good	Low	Low
	GM	Silty gravels, gravel-sand-silt mixtures	Good	Medium	Low
	SM	Silty sand, sand-silt mixtures	Good	Medium	Low
Group II	GC	Clayey gravels, gravel-sand-clay mixtures	Medium	Medium	Low
	SC	Clayey sands, sand-clay mixture	Medium	Medium	Low
	ML	Inorganic silts and very fine sands, rock flour, silty or clayey fine sands or clayey silts with slight plasticity	Medium	High	Low
	CL	Inorganic clays of low to medium plasticity, gravelly clays, sandy clays, silty clays, lean clays	Medium	Medium	Medium to Low
Group III	CH	Inorganic clays of high plasticity, fat clays	Poor	Medium	High
	MH	Inorganic silts, micaceous or diatomaceous fine sandy or silty soils, elastic silts	Poor	High	High
Group IV	OL	Organic silts and organic silty clays of low plasticity	Poor	Medium	Medium
	OH	Organic clays of medium to high plasticity, organic silts	Unsatisfactory	Medium	High
	Pt	Peat and other highly organic soils	Unsatisfactory	Medium	High

For SI: 1 inch = 25.4 mm.

a. The percolation rate for good drainage is over 4 inches per hour, medium drainage is 2 inches to 4 inches per hour, and poor is less than 2 inches per hour.

b. Soils with a low potential expansion typically have a plasticity index (PI) of 0 to 15, soils with a medium potential expansion have a PI of 10 to 35 and soils with a high potential expansion have a PI greater than 20.

❖This table provides soil properties based on the soil's classification in accordance with the United Soil Classification System. This method of identifying soil is determined in ASTM D 2487, *Standard Classification of Soils for Engineering Purposes*.

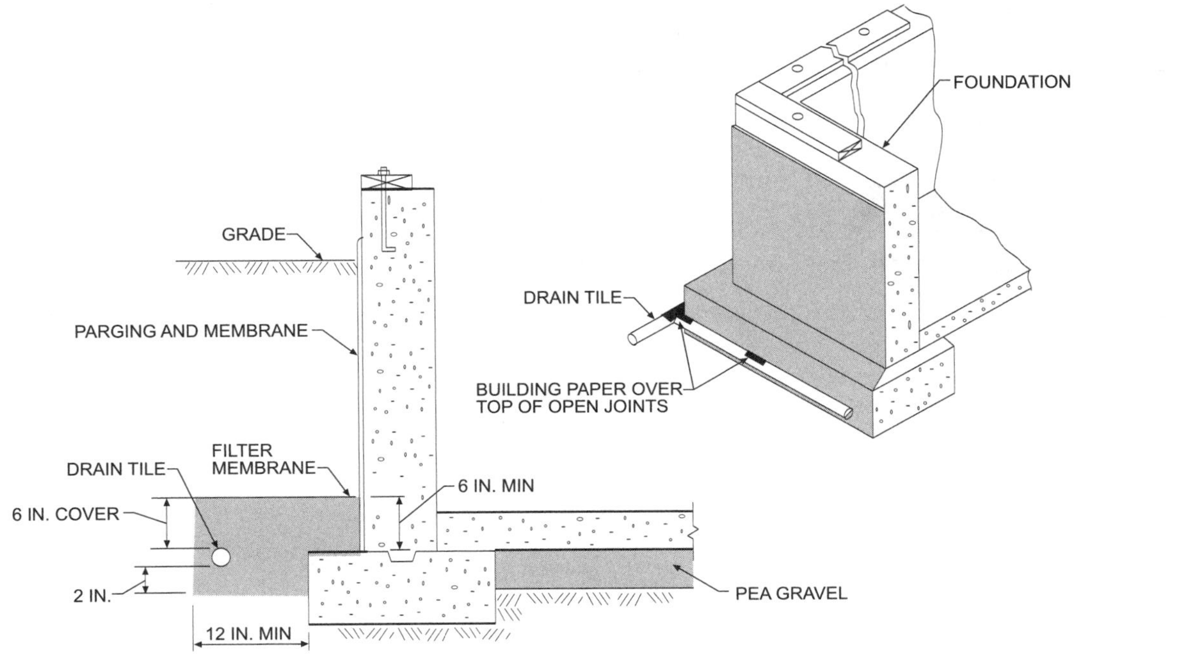

For SI: 1 inch = 25.4 mm.

Figure R405.1(1)
FOUNDATION DRAINAGE FOR HABITABLE SPACE BELOW GRADE

R405.2 Wood foundations. Wood foundations enclosing habitable or usable spaces located below *grade* shall be adequately drained in accordance with Sections R405.2.1 through R405.2.3.

❖As does Section R405.1, this section gives specific drainage requirements for wood foundations. Also refer to the commentary for Section R401.1.

R405.2.1 Base. A porous layer of gravel, crushed stone or coarse sand shall be placed to a minimum thickness of 4 inches (102 mm) under the *basement* floor. Provision shall be made for automatic draining of this layer and the gravel or crushed stone wall footings.

❖See Figure R403.1(2). This requirement for drainage under the basement floor is similar to those in Sections R504.2.1 (wood floor) and R506.2.2 (concrete floor).

R405.2.2 Vapor retarder. A 6-mil-thick (0.15 mm) polyethylene vapor retarder shall be applied over the porous layer with the *basement* floor constructed over the polyethylene.

❖See Figure R403.1(2). Also see the commentary to Section R506.2.3.

R405.2.3 Drainage system. In other than Group I soils, a sump shall be provided to drain the porous layer and footings. The sump shall be at least 24 inches (610 mm) in diameter or 20 inches square (0.0129 m^2), shall extend at least 24 inches (610 mm) below the bottom of the *basement* floor and shall be capable of positive gravity or mechanical drainage to remove any accumulated water. The drainage system shall discharge into an *approved* sewer system or to daylight.

❖Except for Group I soils, which exhibit good drainage (see Table R405.1), a sump is necessary to drain the footing (see Section R403.2) and porous layer (see Section R405.2.1). Also see Figure R403.1(2).

SECTION R406
FOUNDATION WATERPROOFING AND DAMPPROOFING

R406.1 Concrete and masonry foundation damp proofing. Except where required by Section R406.2 to be waterproofed, foundation walls that retain earth and enclose interior spaces and floors below *grade* shall be damp proofed from the top of the footing to the finished *grade*. Masonry walls shall have not less than $^3/_8$ inch (9.5 mm) portland cement parging applied to the exterior of the wall. The parging shall be damp proofed in accordance with one of the following:

1. Bituminous coating.
2. Three pounds per square yard (1.63 kg/m^2) of acrylic modified cement.
3. One-eighth inch (3.2 mm) coat of surface-bonding cement complying with ASTM C 887.
4. Any material permitted for waterproofing in Section R406.2.
5. Other *approved* methods or materials.

Exception: Parging of unit masonry walls is not required where a material is *approved* for direct application to the masonry.

Concrete walls shall be damp proofed by applying any one of the above listed damp proofing materials or any one of the waterproofing materials listed in Section R406.2 to the exterior of the wall.

❖To minimize moisture in the form of water vapor from entering below-ground spaces from the outside, damp proofing of the exterior foundation walls is necessary, unless waterproofing is required by Section R406.2. Although the terms "waterproofing" and "dampproofing" both relate to moisture protection, they are sometimes misapplied in the sense that "waterproofing" is used when "dampproofing" is really meant. Dampproofing does not give the same degree of moisture protection as does waterproofing. Methods of dampproofing for masonry and concrete foundation walls are shown in Commentary Figures R406.1(1) and R406.1(2).

Dampproofing installations generally consist of the application of one or more coatings of impervious compounds that are intended to prevent the passage of water vapor through walls or other building elements. Dampproofing may also restrict the flow of water under slight pressure.

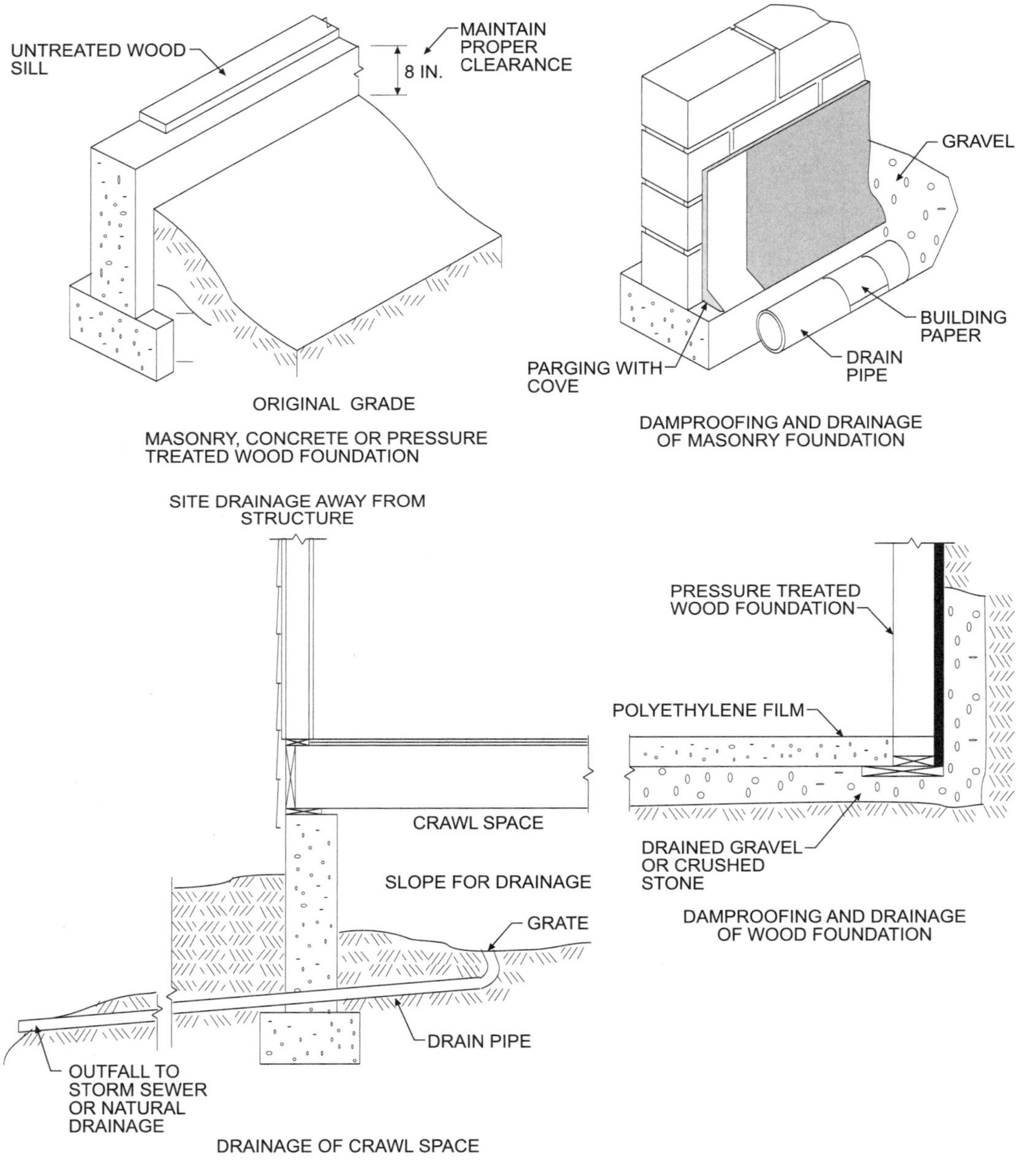

For SI: 1 inch = 25.4 mm.

Figure R405.1(2)
OTHER SUGGESTED METHODS OF FOUNDATION DRAINAGE

R406.2 Concrete and masonry foundation waterproofing. In areas where a high water table or other severe soil-water conditions are known to exist, exterior foundation walls that retain earth and enclose interior spaces and floors below *grade* shall be waterproofed from the top of the footing to the finished *grade*. Walls shall be waterproofed in accordance with one of the following:

1. Two-ply hot-mopped felts.
2. Fifty-five-pound (25 kg) roll roofing.
3. Six-mil (0.15 mm) polyvinyl chloride.
4. Six-mil (0.15 mm) polyethylene.
5. Forty-mil (1 mm) polymer-modified asphalt.
6. Sixty-mil (1.5 mm) flexible polymer cement.
7. One-eighth-inch (3 mm) cement-based, fiber-reinforced, waterproof coating.
8. Sixty-mil (0.22 mm) solvent-free liquid-applied synthetic rubber.

Exception: Organic-solvent-based products such as hydrocarbons, chlorinated hydrocarbons, ketones and esters shall not be used for ICF walls with expanded polystyrene form material. Use of plastic roofing cements, acrylic coatings, latex coatings, mortars and pargings to seal ICF walls is permitted. Cold-setting asphalt or hot asphalt shall conform to type C of ASTM D 449. Hot asphalt shall be applied at a temperature of less than 200°F (93°C).

All joints in membrane waterproofing shall be lapped and sealed with an adhesive compatible with the membrane.

❖Foundation walls that retain earth and enclose interior spaces and floors and extend below ground water level (seasonal or otherwise) require a positive

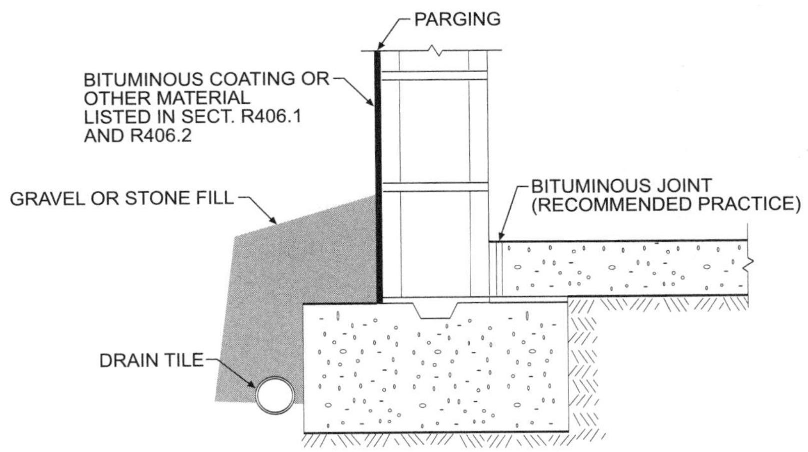

Figure R406.1(1)
DAMPPROOFING OF MASONRY FOUNDATION WALL
[see also Figure R405.1(1)]

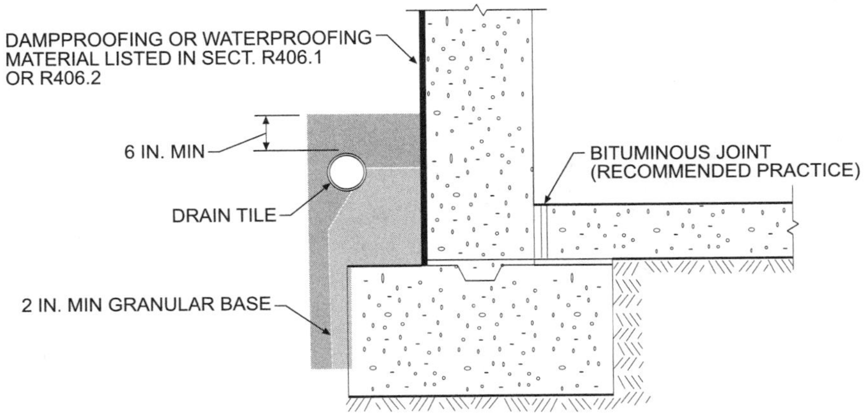

For SI: 1 inch = 25.4 mm.

Figure R406.1(2)
DAMPPROOFING OF CONCRETE FOUNDATION WALL
[see also Figure R405.1(1)]

means of preventing moisture migration. Waterproofing installations consist of the application of a combination of sealing materials and impervious coatings used on walls or other building elements to prevent the passage of moisture in either a vapor or liquid form under conditions of significant hydrostatic pressure. Methods for waterproofing concrete and masonry foundation walls are shown in Commentary Figure R406.2.

R406.3 Dampproofing for wood foundations. Wood foundations enclosing habitable or usable spaces located below *grade* shall be dampproofed in accordance with Sections R406.3.1 through R406.3.4.

❖The requirements for dampproofing wood foundation walls are shown in Figure R403.1(2). These requirements for sealing the plywood panel joints, installing a moisture barrier and installing porous fill can minimize the intrusion of water into a basement or usable space. Also refer to the commentary for Section R401.1.

R406.3.1 Panel joint sealed. Plywood panel joints in the foundation walls shall be sealed full length with a caulking compound capable of producing a moisture-proof seal under the conditions of temperature and moisture content at which it will be applied and used.

❖See the commentary to Section R406.3.

R406.3.2 Below-grade moisture barrier. A 6-mil-thick (0.15 mm) polyethylene film shall be applied over the below-*grade* portion of exterior foundation walls prior to backfilling. Joints in the polyethylene film shall be lapped 6 inches (152 mm) and sealed with adhesive. The top edge of the polyethylene film shall be bonded to the sheathing to form a seal. Film areas at *grade* level shall be protected from mechanical damage and exposure by a pressure preservatively treated lumber or plywood strip attached to the wall several inches above finish *grade* level and extending approximately 9 inches (229 mm) below *grade*. The joint between the strip and the wall shall be caulked full length prior to fastening the strip to the wall. Other coverings appropriate to the architectural treatment may also be used. The polyethylene film shall extend down to the bottom of the wood footing plate but shall not overlap or extend into the gravel or crushed stone footing.

❖See the commentary to Section R406.3.

R406.3.3 Porous fill. The space between the excavation and the foundation wall shall be backfilled with the same material used for footings, up to a height of 1 foot (305 mm) above the footing for well-drained sites, or one-half the total back-fill height for poorly drained sites. The porous fill shall be covered with strips of 30-pound (13.6 kg) asphalt paper or 6-mil (0.15 mm) polyethylene to permit water seepage while avoiding infiltration of fine soils.

❖See the commentary to Section R406.3.

R406.3.4 Backfill. The remainder of the excavated area shall be backfilled with the same type of soil as was removed during the excavation.

❖See the commentary to Sections R404.2.4 and R406.3.

R406.4 Precast concrete foundation system dampproofing. Except where required by Section R406.2 to be waterproofed, precast concrete foundation walls enclosing habitable or useable spaces located below *grade* shall be dampproofed in accordance with Section R406.1.

❖See the commentary to Sections R406.1 and R406.2.

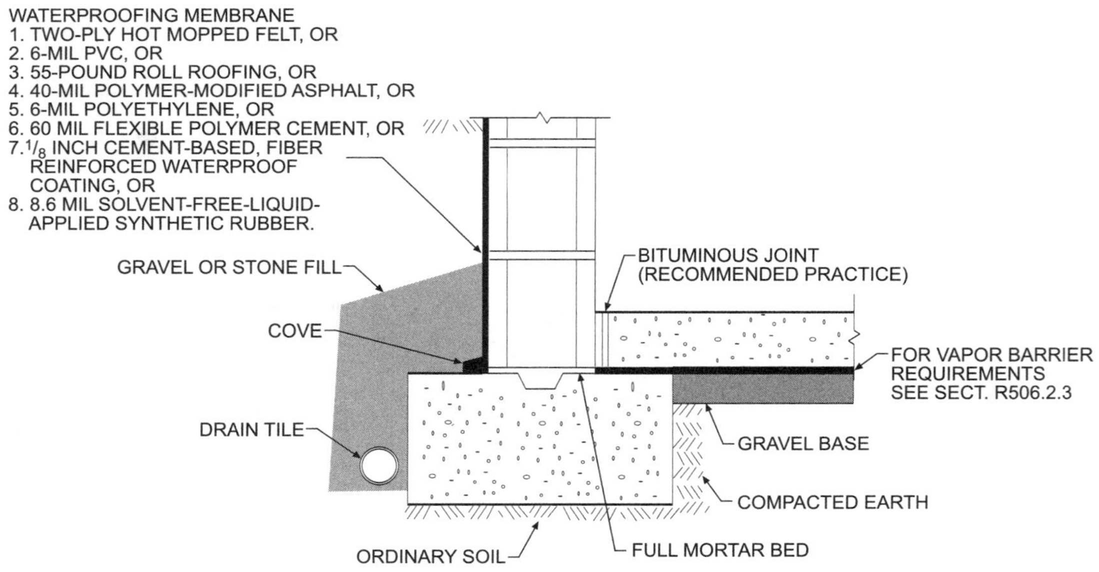

For SI: 1 inch = 25.4 mm.

Figure R406.2
METHODS OF WATERPROOFING BASEMENT WALLS
[see also Figures R405.1(1) and R405.1(2) for typical drainage details]

R406.4.1 Panel joints sealed. Precast concrete foundation panel joints shall be sealed full height with a sealant meeting ASTM C 920, Type S or M, *Grade* NS, Class 25, Use NT, M or A. Joint sealant shall be installed in accordance with the manufacturer's installation instructions.

❖All joints between precast concrete panels must be sealed in order to afford proper dampproofing or waterproofing. This section specifies the sealant material and the requirement to install the sealant in accordance with the manufacturer's instructions.

SECTION R407 COLUMNS

R407.1 Wood column protection. Wood columns shall be protected against decay as set forth in Section R317.

❖See the commentary for Section R319.

R407.2 Steel column protection. All surfaces (inside and outside) of steel columns shall be given a shop coat of rust-inhibitive paint, except for corrosion-resistant steel and steel treated with coatings to provide corrosion resistance.

❖The requirements of this section protect steel columns from the adverse effects of corrosion.

R407.3 Structural requirements. The columns shall be restrained to prevent lateral displacement at the bottom end. Wood columns shall not be less in nominal size than 4 inches by 4 inches (102 mm by 102 mm). Steel columns shall not be less than 3-inch-diameter (76 mm) Schedule 40 pipe manufactured in accordance with ASTM A 53 Grade B or *approved* equivalent.

Exception: In Seismic Design Categories A, B and C, columns no more than 48 inches (1219 mm) in height on a pier or footing are exempt from the bottom end lateral displacement requirement within under-floor areas enclosed by a continuous foundation.

❖Columns must be designed in accordance with accepted engineering practice. Minimum sizes for wood and steel columns are specified to reduce concerns about the structural capacity of slender columns. To minimize the chance of accidentally displacing columns supporting beams or girders, a means of mechanically anchoring a column is required [see Commentary Figures R407.3(1), R407.3(2) and R407.3(3)]. If prefabricated metal-wood connectors are used, they should be installed in accordance with the manufacturer's instructions.

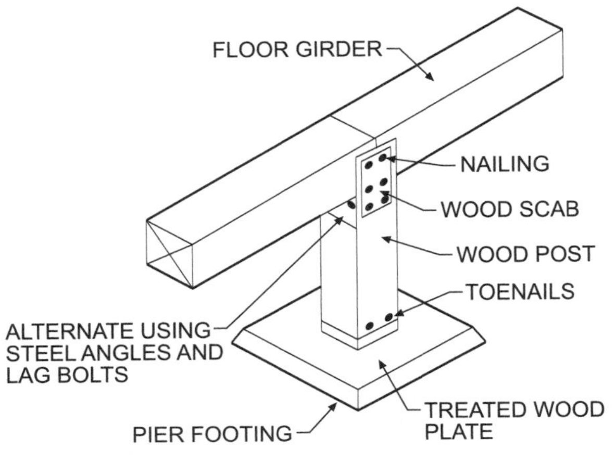

Figure R407.3(1)
COLUMN ANCHORAGE

SECTION R408 UNDER-FLOOR SPACE

R408.1 Ventilation. The under-floor space between the bottom of the floor joists and the earth under any building (except space occupied by a *basement*) shall have ventilation openings through foundation walls or exterior walls. The minimum net area of ventilation openings shall not be less than 1 square foot (0.0929 m^2) for each 150 square feet (14 m^2) of under-floor space area, unless the ground surface is

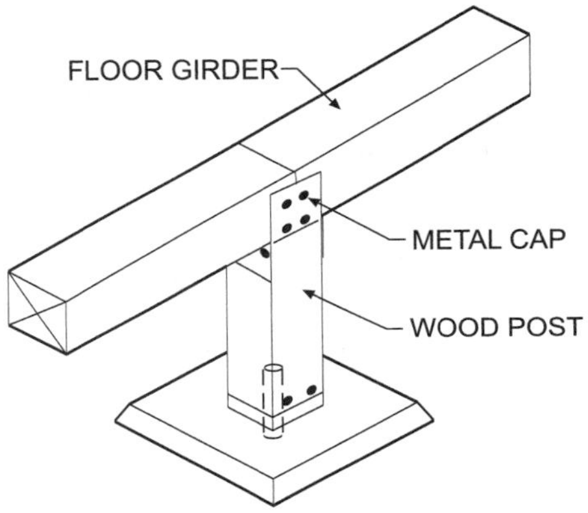

Figure R407.3(2)
COLUMN ANCHORAGE

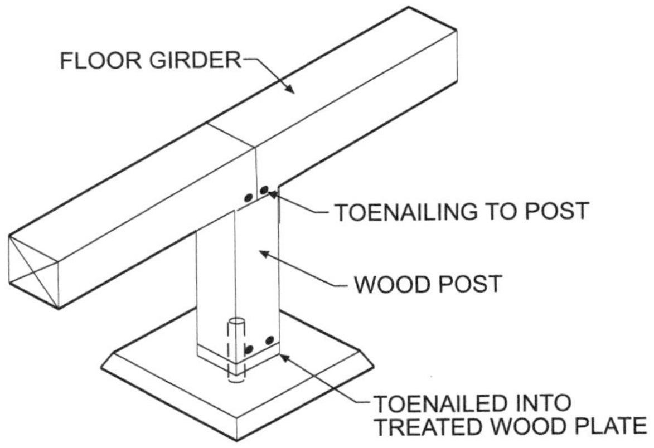

Figure R407.3(3)
COLUMN ANCHORAGE

covered by a Class 1 vapor retarder material. When a Class 1 vapor retarder material is used, the minimum net area of ventilation openings shall not be less than 1 square foot (0.0929 m^2) for each 1,500 square feet (140 m^2) of under-floor space area. One such ventilating opening shall be within 3 feet (914 mm) of each corner of the building.

❖Raised floor construction results in an under-floor space, commonly referred to as a "crawl space." To control condensation within crawl space areas and thus reduce the chance of dry rot, natural ventilation of such spaces by reasonably distributed openings through foundation walls or exterior walls is required. Condensation is a function of the geographical location and climatic conditions; therefore, the dependence on ventilating openings through the foundation wall or exterior wall may run counter to energy conservation measures. The use of a vapor retarder material on the ground surface inhibits the flow of moisture from the ground surface into the crawl space and thus reduces the need for ventilation. Commentary Figure R408.1 illustrates the use of openings through the foundation walls.

R408.2 Openings for under-floor ventilation. The minimum net area of ventilation openings shall not be less than 1 square foot (0.0929 m^2) for each 150 square feet (14 m^2) of under-floor area. One ventilation opening shall be within 3 feet (915 mm) of each corner of the building. Ventilation openings shall be covered for their height and width with any of the following materials provided that the least dimension of the covering shall not exceed $^1/_4$ inch (6.4 mm):

1. Perforated sheet metal plates not less than 0.070 inch (1.8 mm) thick.
2. Expanded sheet metal plates not less than 0.047 inch (1.2 mm) thick.
3. Cast-iron grill or grating.
4. Extruded load-bearing brick vents.
5. Hardware cloth of 0.035 inch (0.89 mm) wire or heavier.
6. Corrosion-resistant wire mesh, with the least dimension being $^1/_8$ inch (3.2 mm) thick.

Exception: The total area of ventilation openings shall be permitted to be reduced to $^1/_{1,500}$ of the under-floor area where the ground surface is covered with an *approved* Class I vapor retarder material and the required openings are placed to provide cross ventilation of the space. The installation of operable louvers shall not be prohibited.

❖Installing a covering material over the ventilation opening keeps animals such as rodents or vermin from entering the crawl space. Several options are provided and they all must have openings that have no dimension exceeding $^1/_4$ inch (6.4 mm). The area of the ventilation opening accounts for net free area of the covering material used to protect the ventilation opening. For some covering materials the net free area can be up to 50 percent less than the gross area of the ventilation opening.

The use of a vapor retarder material on the ground surface inhibits the flow of moisture from the ground surface into the crawl space and thus reduces, if not virtually eliminates, the need for ventilation. Therefore, the exception provides for a drastic reduction in the amount of ventilation openings required. While the vapor retarder may significantly reduce the moisture accumulation, ventilation openings are still required but may be equipped with manual dampers to permit them to be closed during the coldest weeks of the year in northern climates.

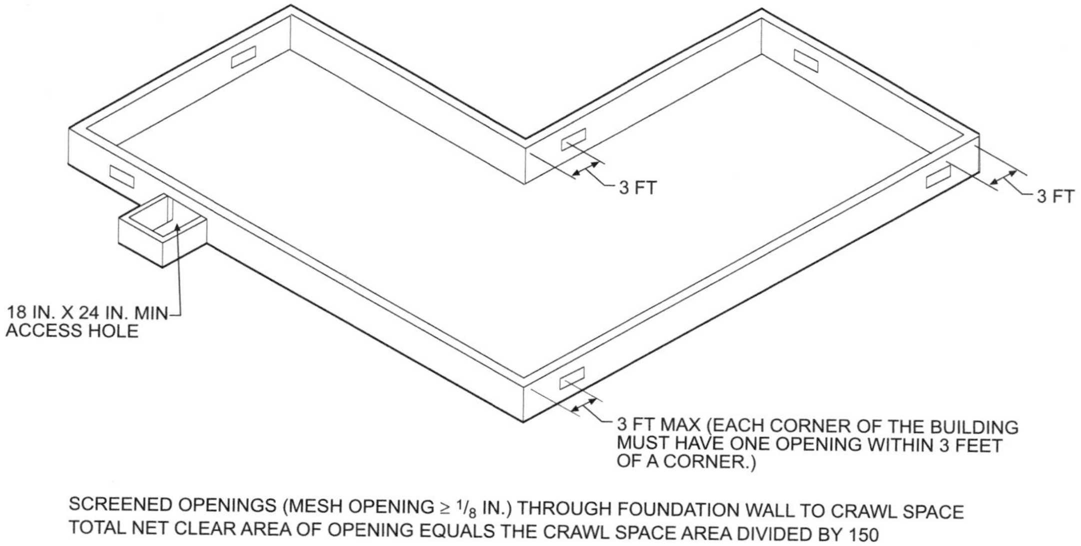

For SI: 1 inch = 25.4 mm, 1 foot = 304.8 mm.

Figure R408.1
CRAWL-SPACE VENTILATION

The following is an example of the area calculation:

Example 1:

A house has a crawl space area of 1,300 square feet (121 m^2). The amount of ventilation opening required is 1,300/150 = 8.7 square feet (0.805 m^2) × 144 square inches per square foot = 1,252 square inches (0.81 m^2). This is the total net clear area of opening required that must be distributed among all the openings. An 8-inch by 16-inch (203 mm by 406 mm) opening provides 128 square inches (0.83 m^2). Ten 8 by 16-inch (203 mm by 406 mm) openings will provide 1,280 square inches (0.83 m^2) which is greater than the 1,252 square inches (0.81 m^2) required.

The following is an example of using the exception:

Example 2:

In Example 1, the required amount opening area (1252 square inches) may be reduced by a factor of 10 to125.2 square inches (80 826 mm^2) if an approved Class I vapor retarder is used on the ground surface of the crawl space in accordance with the exception. Note that opening placement must provide cross ventilation.

R408.3 Unvented crawl space. Ventilation openings in under-floor spaces specified in Sections R408.1 and R408.2 shall not be required where:

1. Exposed earth is covered with a continuous Class I vapor retarder. Joints of the vapor retarder shall overlap by 6 inches (152 mm) and shall be sealed or taped. The edges of the vapor retarder shall extend at least 6 inches (152 mm) up the stem wall and shall be attached and sealed to the stem wall or insulation; and
2. One of the following is provided for the under-floor space:
   - 2.1. Continuously operated mechanical exhaust ventilation at a rate equal to 1 cubic foot per minute (0.47 L/s) for each 50 square feet (4.7m^2) of crawlspace floor area, including an air pathway to the common area (such as a duct or transfer grille), and perimeter walls insulated in accordance with Section N1103.2.1 of this code;
   - 2.2. *Conditioned air* supply sized to deliver at a rate equal to 1 cubic foot per minute (0.47 L/s) for each 50 square feet (4.7 m^2) of under-floor area, including a return air pathway to the common area (such as a duct or transfer grille), and perimeter walls insulated in accordance with Section N1102.2 of this code;
   - 2.3. Plenum in existing structures complying with Section M1601.5, if under-floor space is used as a plenum.

❖This section lists several conditions under which the ventilation openings through the foundation walls are not required. The exposed earth of the under-floor area must be covered with a continuous Class I vapor retarder. See the commentary to Section R202 for the definition of "Vapor retarder class." If the perimeter walls are insulated, either a continuously operating mechanical exhaust vent or conditioned air must be provided in order to eliminate the ventilation openings.

Elimination of the ventilation openings is permitted in existing structures when the under-floor space is used as a plenum. Under-floor plenums are prohibited in new construction (see commentary, Section M1601.5).

R408.4 Access. Access shall be provided to all under-floor spaces. Access openings through the floor shall be a minimum of 18 inches by 24 inches (457 mm by 610 mm). Openings through a perimeter wall shall be not less than 16 inches by 24 inches (407 mm by 610 mm). When any portion of the through-wall access is below *grade*, an areaway not less than 16 inches by 24 inches (407 mm by 610 mm) shall be provided. The bottom of the areaway shall be below the threshold of the access opening. Through wall access openings shall not be located under a door to the residence. See Section M1305.1.4 for access requirements where mechanical *equipment* is located under floors.

❖The provisions of this section require access to all under-floor spaces. Access is required for continuing maintenance of the building and for inspection and repair of such items as plumbing, mechanical systems or electrical system runs within the crawl space. This section addresses the different conditions encountered in accessing an under-floor space through the floor system or the foundation wall. Where the access opening is through the floor, an opening of not less than 18 inches by 24 inches (457 mm by 610 mm) is required. Where the access opening is through the foundation wall, an opening of not less than 16 inches by 24 inches (406 mm by 610 mm) is required. See Commentary Figure R408.1. The 16- and 24-inch (406 mm and 610 mm) dimensions are to work with standard CMU coursing. Where mechanical equipment is located under the floor, the minimum access opening must comply with Section M1305.1.4.

R408.5 Removal of debris. The under-floor *grade* shall be cleaned of all vegetation and organic material. All wood forms used for placing concrete shall be removed before a building is occupied or used for any purpose. All construction materials shall be removed before a building is occupied or used for any purpose.

❖Vegetation, stumps, roots and other matter left in an excavation around a building are major causes of termite infestation and moisture problems. As such material decays, the ground settles, negating the original drainage plan. Even before decay, the material provides pockets for water accumulation, which can have subsequent destructive impact on the structure. To eliminate a natural attraction to termites, insects or animals, all vegetation and organic material must be cleared.

R408.6 Finished grade. The finished *grade* of under-floor surface may be located at the bottom of the footings; however, where there is evidence that the groundwater table can rise to within 6 inches (152 mm) of the finished floor at the building perimeter or where there is evidence that the surface water does not readily drain from the building site, the *grade* in the under-floor space shall be as high as the outside finished *grade*, unless an *approved* drainage system is provided.

❖To circumvent moisture accumulation in the crawl space area, the grade in the under-floor space must be at the same elevation as grade outside of the building if the groundwater table can rise to within 6 inches (152 mm) of the finished floor elevation or the surface water does not readily drain from the site. An alternative would be an approved drainage system.

R408.7 Flood resistance. For buildings located in flood hazard areas as established in Table R301.2(1):

1. Walls enclosing the under-floor space shall be provided with flood openings in accordance with Section R322.2.2.
2. The finished ground level of the under-floor space shall be equal to or higher than the outside finished ground level on at least one side.

   Exception: Under-floor spaces that meet the requirements of FEMA/FIA TB 11-1.

❖To minimize hydrostatic loads by allowing the free inflow and outflow of floodwaters, buildings in flood hazard areas (A Zones) that have walls enclosing underfloor spaces are to have flood openings in those walls. This provision alerts the designer that flood openings are to be provided in addition to standard ventilation openings. This requirement applies only to buildings in flood hazard areas not subject to high-velocity wave action. Buildings in coastal high-hazard areas subject to high-velocity wave action may have enclosed areas below the design flood elevation, but only if those walls are intended to break away as specified in Section R324.3.4.

The provisions of this section require the elevation of the finished ground level of the under-floor space to be at or above the elevation of the outside finished ground level on at least one side. In flood hazard areas, if the floor of an enclosed area, including a crawl-space, is below the exterior grade on all sides, then the building is considered to have a basement. Note that an under-floor space may be partially below grade. As long as the interior finished ground level is at or above the exterior finished grade on at least one side, then a basement is not created. Section R408.6 requires the grade in under-floor spaces to be as high as the outside finished grade unless an approved drainage system is provided. Section R408.7, Item 2, clarifies that the arrangement is not allowed in flood hazard areas because a basement (below grade on all sides) would be created. The exception permits the use of FEMA TB-11-01, which permits the elevation of the under-floor space to be below the elevation of the outside finished ground level. Use of FEMA TB11-01 will result in flood insurance rates higher than for buildings that have the finished ground level of the under-floor space at or above the elevation of the outside finished ground level on at least one side.

Bibliography

The following resource materials were used in the preparation of the commentary for this chapter of the code:

ACI 318-11, *Building Code Requirements for Structural Concrete.* Farmington Hills, MI: American Concrete Institute, 2011.

ACI 332-10, Requirements for Residential Concrete Construction. Farmington Hills, MI: American Concrete Institute, 2010.

AFPA NDS-2012, *National Design Specification (NDS) for Wood Construction—with 2012 Supplement.* Washington, DC: American Forest and Paper Association, 2012.

AFPA PWF-2007, *Permanent Wood Foundation Design Specification.* Washington, DC: American Forest and Paper Association, 2007.

ASCE 32-01, *Design and Construction of Frost Protected Shallow Foundations.* Reston, VA: American Society of Civil Engineers, 2001.

ASTM A 615/A 615M-09, *Specification for Deformed-and Plain Billet Steel bars for Concrete Reinforcement.* West Conshohocken, PA: ASTM International, 2009.

ASTM A 706 M-09, *Specification for Low-alloy SteelDeformed Plain Bars for Concrete Reinforcement.* West Conshohocken, PA: ASTM International, 2009.

ASTM A 996/A 996M-09, *Specifications for Rail-steel and Axel-steel Deformed Bars for Concrete Reinforcement.* West Conshohocken, PA: ASTM International, 2009.

ASTM C 94/C 94M-09, *Specification for Ready-mixed Concrete.* West Conshohocken, PA: ASTM International, 2009.

ASTM C 150-07, *Specification for Portland Cement.* West Conshohocken, PA: ASTM International, 2007

ASTM C 595-08a, *Specification for Blended Hydraulic Cements.* West Conshohocken, PA: ASTM International, 2008.

ASTM C 618-08a, *Standard Specification for Coal Fly Ash and Raw or Calcined Natural Pozzolan for Concrete.* West Conshohocken, PA: ASTM International, 2008.

ASTM C 685-07, *Specification for Concrete Made by Volumetric Batching and Continuous Mixing.* West Conshohocken, PA: ASTM International, 2007.

ASTM C 845-04, *Standard Specification for Expansive Hydraulic Cement.* West Conshohocken, PA: ASTM International, 2004.

ASTM C 989-09, *Standard Specification for Slag Cement for Use in Concrete and Mortars.* West Conshohocken, PA: ASTM International, 2009.

ASTM C 1157-08a, *Standard Performance Specification for Hydraulic Cement.* West Conshohocken, PA: ASTM International, 2008.

ASTM C 1240-05, *Standard Specification for Silica Fume Used in Cementitious Mixtures.* West Conshohocken, PA: ASTM International, 2005.

ASTM D 2487-06e1, *Practice for Classification of Soils for Engineering Purposes (Unified Soil Classification System).* West Conshohocken, PA: ASTM International, 2006.

ASTM D 4829-08a, *Test Method for Expansion Index of Soils.* West Conshohocken, PA: ASTM International, 2008.

ASTM E 2634-08, *Standard Specification for Flat Wall Insulating Concrete Form (ICF) Systems.* West Conshohocken, PA: ASTM International, 2008.

FEMA TB-11-01, *Crawl space Construction for Buildings Located in Special Flood Hazard Area.* Washington, DC: Federal Emergency Management Agency, 2001.

IBC-12, *International Building Code.* Washington, DC: International Code Council, 2011.

PCA 100-07. *Prescriptive Design of Exterior Concrete Walls for One- and Two-family Dwellings (Pub. No. EB241).* Skokie, IL: Portland Cement Association, 2007.

TEK 5-3A, *Concrete Masonry Foundation Wall Details.* Herndon, VA: National Concrete Masonry Association, 2003.

TMS 402-11/ACI 530-11/ASCE 5-11, *Building Code Requirements for Masonry Structures.* Boulder, CO: The Masonry Society, 2011.

TR 68-A-75, *Design and Construction of Plain and Reinforced Concrete Masonry and Basement and Foundation Walls.* Herndon, VA: National Concrete Masonry Association, 1975.

Chapter 5: Floors

General Comments

Section R501 provides the scope of the chapter and states the general performance requirements for floor systems. The floor systems covered in the code consist of four different types: wood floor framing, wood floors on the ground, steel floor framing and concrete slabs on the ground. Section R502 addresses wood floor framing criteria and includes prescriptive tables from which the size of the joists can be determined based on the span and species of wood.

Section R503 specifies floor-sheathing requirements. Section R504 governs wood floors constructed on ground that would serve as part of a wood foundation system described in Chapter 4. This system relies on pressure-preservative-treated wood to resist the damage from exposure to moisture from contact with the ground. Section R505 contains steel floor framing provisions, which include prescriptive criteria that greatly simplify the use of steel framing in a floor system. Section R506 contains requirements for concrete slab-on-ground floors. Concrete slab-on-grade construction requires a minimum 3.5-inch-thick (89 mm) slab and a vapor retarder in most cases.

The primary consideration of this chapter is the structural integrity of the floor system. In a two-family dwelling, a floor that separates dwelling units must be a fire-resistance-rated assembly in accordance with Section R302.3. Such a floor must conform to the conditions of the floor assembly's listing in addition to the requirements of this chapter. Section R507 provides prescriptive requirements for attaching a deck to the main building. This section also addresses the use of wood/plastic composites for deck construction.

Purpose

Chapter 5 provides the requirements for the design and construction of floor systems that will be capable of supporting minimum required design loads. This chapter covers four different types: wood floor framing, wood floors on the ground, steel floor framing and concrete slabs on the ground. Allowable span tables are provided that greatly simplify the determination of joist, girder and sheathing sizes for raised floor systems of wood framing and cold-formed steel framing. This chapter also contains prescriptive requirements for attaching a deck to the main building.

The floor system must also serve as a diaphragm; therefore, it plays a key role in resisting lateral loads from earthquakes and/or wind. Although this chapter does not base floor system requirements on these loads, floor systems constructed as required in this chapter will perform this function. Chapter 6 of the code accounts for these loads by varying the spacing of the supporting braced wall lines or shear walls based on seismic design category and basic wind speed or, in selected cases, specifying additional diaphragm requirements.

SECTION R501 GENERAL

R501.1 Application. The provisions of this chapter shall control the design and construction of the floors for all buildings including the floors of *attic* spaces used to house mechanical or plumbing fixtures and *equipment.*

❖ Floors (including attic floors) that house mechanical equipment or plumbing fixtures (see the definition in Chapter 2) must comply with this chapter.

R501.2 Requirements. Floor construction shall be capable of accommodating all loads according to Section R301 and of transmitting the resulting loads to the supporting structural elements.

❖ This is a general performance statement that requires the floor system to support the design loads and provide the necessary load path to supporting elements.

R501.3 Fire protection of floors. Floor assemblies, not required elsewhere in this code to be fire-resistance rated, shall be provided with a $^1/_2$-inch (12.7 mm) gypsum wallboard membrane, $^5/_8$-inch (16 mm) wood structural panel membrane, or equivalent on the underside of the floor framing member.

Exceptions:

1. Floor assemblies located directly over a space protected by an automatic sprinkler system in accordance with Section P2904, NFPA13D, or other approved equivalent sprinkler system.
2. Floor assemblies located directly over a crawl space not intended for storage or fuel-fired appliances.
3. Portions of floor assemblies can be unprotected when complying with the following:
   - 3.1. The aggregate area of the unprotected portions shall not exceed 80 square feet per story
   - 3.2. Fire blocking in accordance with Section R302.11.1 shall be installed along the perimeter of the unprotected portion to separate the unprotected portion from the remainder of the floor assembly.
4. Wood floor assemblies using dimension lumber or structural composite lumber equal to or greater than 2-inch by 10-inch (50.8 mm by 254 mm) nominal dimension, or other approved floor assemblies demonstrating equivalent fire performance.

❖ During a fire unprotected floor assemblies of lightweight construction have a short time gap between untenable conditions and the structural collapse of the floor assembly. Section R501.3 requires protection to floors of lightweight construction that will provide the occupants additional time for self-evacuation and safety for fire fighters performing search and rescue. This section requires all floor assemblies, not fire rated, to be protected on the underside with $^{1}/_{2}$-inch (12.7 mm) gypsum board, $^{5}/_{8}$-inch wood structural panels or equivalent. This will apply mainly to floor assemblies located over nonsprinklered areas such as crawl spaces for storage or those containing fuel-fired appliances. There are four exceptions provided that specify where the protective membrane is not required.

Exception 1 permits floor assemblies above any space protected by an automatic sprinkler system to be exempt from the application of the protective membrane to the underside of the floor-framing members. Section R313 requires an automatic residential fire sprinkler system (ARFSS) for new construction and the floor assemblies at each story will have the space below protected; therefore, Section R501.3 is not required. For crawl spaces used for storage or containing fuel-fired appliances and alterations or additions to existing buildings that are not provided with an ARFSS, this exception will permit deletion of the protective membrane if the space is protected by an ARFSS in accordance with Section P2904, NFPA 13 D or other approved methods.

Exception 2 exempts floor assemblies over crawl spaces, unless the crawl space contains fuel-fired appliances or is intended for storage. However, Exception 1 may be used for crawl spaces intended for storage or containing fuel-fired appliances.

Exception 3 prescribes an allowable area where the protective membrane may be omitted.

Exception 4 permits floor framing of sawn lumber or structural composite lumber equal to or greater than 2 × 10 nominal dimension to be exempt from the application of the protective membrane to the underside of the floor-framing members. The basis for this exception is that tests conducted on floor framing constructed of 2 × 10 lumber and loaded to 50 percent of full design load showed that the assemblies provided adequate time for occupants to self-evacuate and safety for fire fighters performing search and rescue.

SECTION R502
WOOD FLOOR FRAMING

R502.1 Identification. Load-bearing dimension lumber for joists, beams and girders shall be identified by a grade *mark* of a lumber grading or inspection agency that has been *approved* by an accreditation body that complies with DOC PS 20. In lieu of a grade *mark*, a certificate of inspection issued by a lumber grading or inspection agency meeting the requirements of this section shall be accepted.

❖ Load-carrying wood members must be identified to verify their conformance to a minimum quality control standard. The required grade mark must reflect the following information:

1. The species.
2. The grade.
3. The moisture content (MC) at time of grading: S-Dry or S-Gm, MC15, KD15, or KD19. S-Grn indicates that the moisture content was in excess of 19 percent at the time of grading, and KD indicates kiln dried.
4. The grading agency.
5. The mill name or mill number.
6. In the case of independent agencies, the rules used for grading.

For examples of grade marks, see Commentary Figure R502.1.

R502.1.1 Preservative-treated lumber. Preservative treated dimension lumber shall also be identified as required by Section R317.2.

❖ Grade marks as described in Section R502.1 must also include the quality mark of an agency accredited

Figure R502.1
GRADE MAKE EXAMPLES

in the ALSC treated wood program. See Commentary Figure R502.1.1 for an example of a preservative-treated-lumber grade mark.

R502.1.2 Blocking and subflooring. Blocking shall be a minimum of utility grade lumber. Subflooring may be a minimum of utility grade lumber or No. 4 common grade boards.

❖ Allowing the use of lower lumber grades in blocking and subflooring in these instances results in possible economies.

R502.1.3 End-jointed lumber. *Approved* end-jointed lumber identified by a grade *mark* conforming to Section R502.1 may be used interchangeably with solid-sawn members of the same species and grade. End-jointed lumber used in an assembly required elsewhere in this code to have a fire-resistance rating shall have the designation "Heat Resistant Adhesive" or "HRA" included in its grade mark.

❖ End-jointed lumber is generally considered equivalent to solid sawn lumber, provided it is properly identified by a grade mark. The key is verifying proper end use. For instance, grade marks including the phrase "STUD USE ONLY" denote finger-jointed lumber that is limited to stud use and/or bending or tension stresses of short duration loads such as wind and seismic. Structural finger joints, on the other hand, allow horizontal as well as vertical applications. When end-jointed lumber is used in fire-resistance-rated assemblies, the adhesive used in the end-joint must be heat-resistant adhesive (HRA).

The American Lumber Standards Committee (ALSC) recently added elevated-temperature performance requirements for end-jointed lumber adhesives intended for use in fire-resistance-rated assemblies. End-jointed lumber manufactured with adhesives that meet the new requirements is being designated as "heat resistant adhesive" or "HRA" on the grade stamp. HRAs are required to be qualified in accordance with one of two new ASTM standards, ASTM D 7374-08, *Practice for Evaluating Elevated Temperature Performance of Adhesives Used in End-Jointed Lumber* and ASTM D 7470-08, *Practice for Evaluating Elevated Temperature Performance of End-Jointed Lumber Studs*. End-jointed lumber manufactured with HRAs under an auditing program of an ALSC-accredited grading agency is allowed to carry the HRA mark on the grade stamp. End-jointed lumber manufactured with an adhesive not qualified as an HRA will be designated as "non-heat resistant adhesive" or "non-HRA" on the grade stamp. Lumber carrying the HRA mark is permitted to be used interchangeably with solid-sawn members of the same species and grade in fire-rated applications.

R502.1.4 Prefabricated wood I-joists. Structural capacities and design provisions for prefabricated wood I-joists shall be established and monitored in accordance with ASTM D 5055.

❖ Prefabricated wood I-joists are some of the many engineered wood products currently available. These structural members consist of sawn lumber or structural composite lumber flanges bonded to wood structural panel webs, forming an I-shaped cross section. Because these products are widely used in residential floor construction ASTM D 5055, *Standard Specification for Establishing and Monitoring Structural Capacities of Prefabricated Wood I-joists*, is referenced as the applicable standard.

This standard does not specify the makeup of the component materials or how they must be assembled. Each manufacturer must develop product specifications and construction details for its products. These are subject to qualification tests, which are conducted or witnessed by qualified, independent agencies. The best results are used to establish the following design values:

- Allowable bending moment.
- Allowable shear.
- Member stiffness (EI).
- Sheer deflection factor (K).
- Allowable reactions.
- Bearing length requirements.
- Web stiffener requirements.
- A chart detailing the size, location, shape and spacing of any holes permitted in the web.

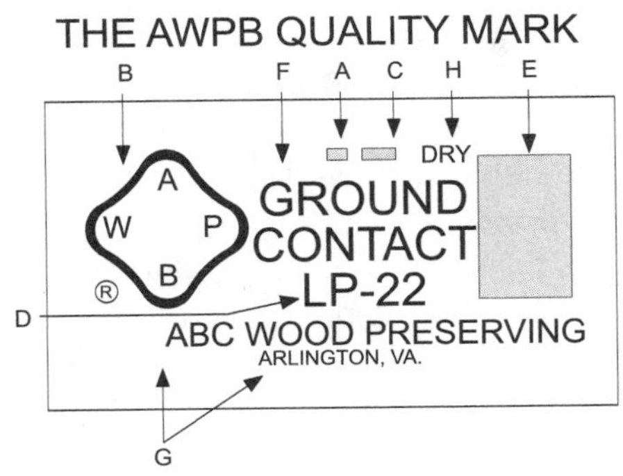

A. YEAR OF TREATMENT
B. AMERICAN WOOD PRESERVERS BUREAU TRADEMARK
C. THE PRESERVATIVE USED FOR TREATMENT
D. THE APPLICABLE AMERICAN WOOD PRESERVERS BUREAU QUALITY STANDARD
E. TRADEMARK OF THE AGENCY SUPERVISING THE TREATING PLANT
F. PROPER EXPOSURE CONDITIONS
G. TREATING COMPANY AND PLANT LOCATION
H. DRY OR KDAT IF APPLICABLE

Figure R502.1.1
TREATED LUMBER GRADE MARK

R502.1.5 Structural glued laminated timbers. Glued laminated timbers shall be manufactured and identified as required in ANSI/AITC A190.1 and ASTM D 3737.

❖ This section requires that glulam timbers be manufactured following ANSI/AITC A190.1 and ASTM D 3737. Knowing these standards make it easier to determine that the product found in the field will meet the design requirements.

R502.1.6 Structural log members. Stress grading of structural log members of nonrectangular shape, as typically used in log buildings, shall be in accordance with ASTM D 3957. Such structural log members shall be identified by the grade *mark* of an *approved* lumber grading or inspection agency. In lieu of a grade *mark* on the material, a certificate of inspection as to species and grade issued by a lumber-grading or inspection agency meeting the requirements of this section shall be permitted to be accepted.

❖ This section addresses grading requirements for logs used as structural members. This subsection specifies the reference for acceptable methods for establishing structural capacities of logs and specifies the requirement for a grading stamp or alternate certification on structural logs. The grading of structural log members must be in accordance with ASTM D 3957, *Standard Practices for Establishing Stress Grades for Structural Members Used in Log Buildings*.

R502.1.7 Structural composite lumber. Structural capacities for structural composite lumber shall be established and monitored in accordance with ASTM D 5456.

❖ ASTM D 5456 is the standard by which structural composite lumber is evaluated. Products manufactured to this standard are increasingly available in the market place and are being used in residential construction. These products are being used as beams, headers, long-length studs, floor and roof framing and other applications where high strength, long length and/or dimensional stability make sawn lumber unacceptable.

Recognition of the appropriate standard on the identification marks required by the code will provide the designer, builder, plans examiner and building inspector with the ensurance that structural composite lumber products are being manufactured with the appropriate quality control systems in place and that the design properties of the product are properly derived and maintained during production.

R502.2 Design and construction. Floors shall be designed and constructed in accordance with the provisions of this chapter, Figure R502.2 and Sections R317 and R318 or in accordance with AF&PA/NDS.

❖ Design and construction of wood floors must be in accordance with the prescriptive requirements of this section or where required in accordance with AF&PA/NDS, *National Design Specification for Wood Construction*. Figure R502.2 includes section references to provisions that apply to typical elements of floor systems.

R502.2.1 Framing at braced wall lines. A load path for lateral forces shall be provided between floor framing and *braced wall panels* located above or below a floor, as specified in Section R602.10.8.

❖ At braced wall panels, joists and blocking are an essential part of a complete lateral load path for either wind or seismic loads. These framing members are needed to accomplish the framing to framing connections specified in Table R602.3(1) and Section R602.10.8. Section R602.10.8 specifies the framing required for a complete load path for these conditions. Section R602.10.8 addresses the framing needed when braced wall panels occur above or below a floor and when floor framing is parallel to a braced wall panel.

R502.3 Allowable joist spans. Spans for floor joists shall be in accordance with Tables R502.3.1(1) and R502.3.1(2). For other grades and species and for other loading conditions, refer to the AF&PA Span Tables for Joists and Rafters.

❖ Tables R502.3.1(1) and R502.3.1(2) contain allowable floor joist spans for common lumber species and grades based on design loads and joist spacing. The tables provide spans for dead loads of 10 or 20 psf (479 Pa or 958 Pa). The weight of the floor joist is included in the 10 or 20 psf (479 Pa or 958 Pa) dead load.

The dead load of the floor assembly must not exceed 10 psf (479 Pa) for townhouses in Seismic Design Category C and all structures in Seismic Design Categories D_0, D_1 and D_2 (see commentary, Section R301.2.2.2.1).

The referenced standard may be used for grades and species of lumber not included in these tables.

Table R502.3.1(1) covers sleeping rooms where the design live load is 30 pounds per square foot (psf) (1.44 kPa) per Table R301.5. This includes attics with fixed stair access. It is important to identify the intended use because an attic space used as a bonus room, for example, could require use of a 40-pounds-per-square-foot (1.92 kPa) live load. Table 502.3.1(2) covers rooms other than sleeping rooms where the design load is 40 pounds per square foot (1.92 kPa) in accordance with Table R301.5.

Example:

A floor has a design live load of 40 psf (1.92 kPa) and a dead load of 10 psf (0.48 kPa). The joists span 17 feet, 6 inches (5334 mm). Find the required joist size using Douglas-Fir Larch #2 lumber and a joist spacing of 16 inches (406 mm).

Solution:

In Table R502.3.1(2), which is based on LL = 40 psf (1.92 kPa), find 16-inch (406 mm) joist spacing in the left hand column. Locate Douglas Fir-Larch #2 and under DL = 10 psf (0.48 kPa) look for an allowable span of 17 feet, 6 inches (5334 mm) or greater. Note that the allowable span for 2 inch by 12 inch (51 mm by 305 mm) is 17 feet, 10 inches (5436 mm). Therefore, use 2-inch by 12-inch (51 mm by 305 mm) joists at 16-inch (406 mm) spacing.

The span tables account for a uniform load condition. They will also permit isolated concentrated loads such as nonbearing partitions offset from a support by a distance less than or equal to the joist depth. They may not support large concentrated loads such as ones that result from an entire kitchen utility wall or bathtubs parallel to joists. In such instances, additional joists and other adequate supports must be installed.

R502.3.1 Sleeping areas and attic joists. Table R502.3.1(1) shall be used to determine the maximum allowable span of floor joists that support sleeping areas and *attics* that are accessed by means of a fixed stairway in accordance with Section R311.7 provided that the design live load does not exceed 30 pounds per square foot (1.44 kPa) and the design dead load does not exceed 20 pounds per square foot (0.96 kPa). The allowable span of ceiling joists that support *attics* used for limited storage or no storage shall be determined in accordance with Section R802.4.

❖ See the commentary to Section R502.3.

R502.3.2 Other floor joists. Table R502.3.1(2) shall be used to determine the maximum allowable span of floor joists that support all other areas of the building, other than sleeping rooms and *attics*, provided that the design live load does not exceed 40 pounds per square foot (1.92 kPa) and the design dead load does not exceed 20 pounds per square foot (0.96 kPa).

❖ See the commentary to Section R502.3.

R502.3.3 Floor cantilevers. Floor cantilever spans shall not exceed the nominal depth of the wood floor joist. Floor cantilevers constructed in accordance with Table R502.3.3(1) shall be permitted when supporting a light-frame bearing wall and roof only. Floor cantilevers supporting an exterior bal-

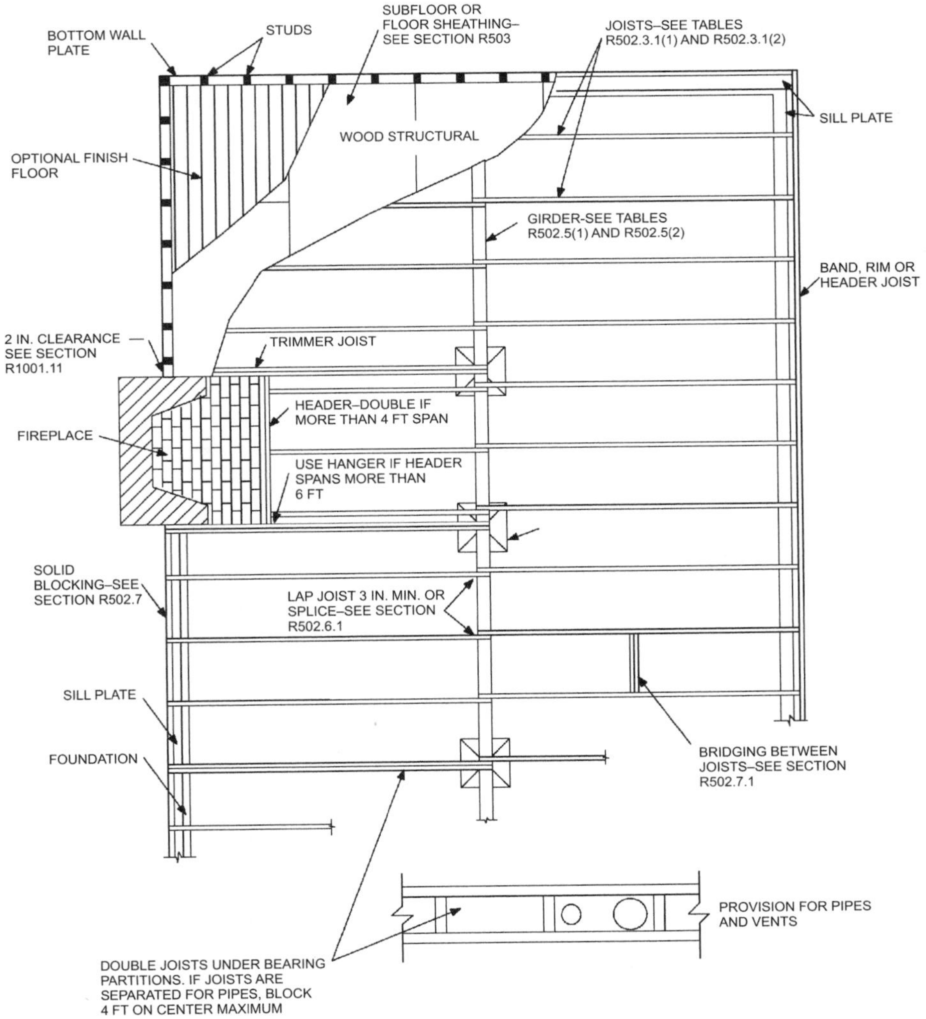

For SI: 1 inch = 25.4 mm, 1 foot = 304.8 mm.

FIGURE R502.2
FLOOR CONSTRUCTION

❖ This figure is a schematic plan of typical wood floor framing. It serves as a key, providing references to the applicable provisions of the code.

cony are permitted to be constructed in accordance with Table R502.3.3(2)

❖ The cantilever span is permitted to be equal to the nominal depth of the joist without additional limitations. This provides for load transfer to the support by direct bearing so that shear and bending of the joist is not a concern.

Larger cantilevers are permitted in accordance with the limitations of the appropriate table for floor joists supporting an exterior balcony or a light-frame bearing wall and roof.

TABLE R502.3.1(1)
FLOOR JOIST SPANS FOR COMMON LUMBER SPECIES
(Residential sleeping areas, live load = 30 psf, L/Δ = 360)[a]

JOIST SPACING (inches)	SPECIES AND GRADE		DEAD LOAD = 10 psf				DEAD LOAD = 20 psf			
			2 × 6	2 × 8	2 × 10	2 × 12	2 × 6	2 × 8	2 × 10	2 × 12
			Maximum floor joist spans							
			(ft - in.)	(ft - in.)	(ft - in.)	(ft - in.)	(ft - in.)	(ft - in.)	(ft - in.)	(ft - in.)
12	Douglas fir-larch	SS	12-6	16-6	21-0	25-7	12-6	16-6	21-0	25-7
	Douglas fir-larch	#1	12-0	15-10	20-3	24-8	12-0	15-7	19-0	22-0
	Douglas fir-larch	#2	11-10	15-7	19-10	23-0	11-6	14-7	17-9	20-7
	Douglas fir-larch	#3	9-8	12-4	15-0	17-5	8-8	11-0	13-5	15-7
	Hem-fir	SS	11-10	15-7	19-10	24-2	11-10	15-7	19-10	24-2
	Hem-fir	#1	11-7	15-3	19-5	23-7	11-7	15-2	18-6	21-6
	Hem-fir	#2	11-0	14-6	18-6	22-6	11-0	14-4	17-6	20-4
	Hem-fir	#3	9-8	12-4	15-0	17-5	8-8	11-0	13-5	15-7
	Southern pine	SS	12-3	16-2	20-8	25-1	12-3	16-2	20-8	25-1
	Southern pine	#1	12-0	15-10	20-3	24-8	12-0	15-10	20-3	24-8
	Southern pine	#2	11-10	15-7	19-10	24-2	11-10	15-7	18-7	21-9
	Southern pine	#3	10-5	13-3	15-8	18-8	9-4	11-11	14-0	16-8
	Spruce-pine-fir	SS	11-7	15-3	19-5	23-7	11-7	15-3	19-5	23-7
	Spruce-pine-fir	#1	11-3	14-11	19-0	23-0	11-3	14-7	17-9	20-7
	Spruce-pine-fir	#2	11-3	14-11	19-0	23-0	11-3	14-7	17-9	20-7
	Spruce-pine-fir	#3	9-8	12-4	15-0	17-5	8-8	11-0	13-5	15-7
16	Douglas fir-larch	SS	11-4	15-0	19-1	23-3	11-4	15-0	19-1	23-0
	Douglas fir-larch	#1	10-11	14-5	18-5	21-4	10-8	13-6	16-5	19-1
	Douglas fir-larch	#2	10-9	14-1	17-2	19-11	9-11	12-7	15-5	17-10
	Douglas fir-larch	#3	8-5	10-8	13-0	15-1	7-6	9-6	11-8	13-6
	Hem-fir	SS	10-9	14-2	18-0	21-11	10-9	14-2	18-0	21-11
	Hem-fir	#1	10-6	13-10	17-8	20-9	10-4	13-1	16-0	18-7
	Hem-fir	#2	10-0	13-2	16-10	19-8	9-10	12-5	15-2	17-7
	Hem-fir	#3	8-5	10-8	13-0	15-1	7-6	9-6	11-8	13-6
	Southern pine	SS	11-2	14-8	18-9	22-10	11-2	14-8	18-9	22-10
	Southern pine	#1	10-11	14-5	18-5	22-5	10-11	14-5	17-11	21-4
	Southern pine	#2	10-9	14-2	18-0	21-1	10-5	13-6	16-1	18-10
	Southern pine	#3	9-0	11-6	13-7	16-2	8-1	10-3	12-2	14-6
	Spruce-pine-fir	SS	10-6	13-10	17-8	21-6	10-6	13-10	17-8	21-4
	Spruce-pine-fir	#1	10-3	13-6	17-2	19-11	9-11	12-7	15-5	17-10
	Spruce-pine-fir	#2	10-3	13-6	17-2	19-11	9-11	12-7	15-5	17-10
	Spruce-pine-fir	#3	8-5	10-8	13-0	15-1	7-6	9-6	11-8	13-6

(continued)

TABLE R502.3.1(1)—continued
FLOOR JOIST SPANS FOR COMMON LUMBER SPECIES
(Residential sleeping areas, live load = 30 psf, L/Δ = 360)[a]

JOIST SPACING (inches)	SPECIES AND GRADE		DEAD LOAD = 10 psf				DEAD LOAD = 20 psf			
			2 × 6	2 × 8	2 × 10	2 × 12	2 × 6	2 × 8	2 × 10	2 × 12
			Maximum floor joist spans							
			(ft - in.)	(ft - in.)	(ft - in.)	(ft - in.)	(ft - in.)	(ft - in.)	(ft - in.)	(ft - in.)
19.2	Douglas fir-larch	SS	10-8	14-1	18-0	21-10	10-8	14-1	18-0	21-0
	Douglas fir-larch	#1	10-4	13-7	16-9	19-6	9-8	12-4	15-0	17-5
	Douglas fir-larch	#2	10-1	12-10	15-8	18-3	9-1	11-6	14-1	16-3
	Douglas fir-larch	#3	7-8	9-9	11-10	13-9	6-10	8-8	10-7	12-4
	Hem-fir	SS	10-1	13-4	17-0	20-8	10-1	13-4	17-0	20-7
	Hem-fir	#1	9-10	13-0	16-4	19-0	9-6	12-0	14-8	17-0
	Hem-fir	#2	9-5	12-5	15-6	17-1	8-11	11-4	13-10	16-1
	Hem-fir	#3	7-8	9-9	11-10	13-9	6-10	8-8	10-7	12-4
	Southern pine	SS	10-6	13-10	17-8	21-6	10-6	13-10	17-8	21-6
	Southern pine	#1	10-4	13-7	17-4	21-1	10-4	13-7	16-4	19-6
	Southern pine	#2	10-1	13-4	16-5	19-3	9-6	12-4	14-8	17-2
	Southern pine	#3	8-3	10-6	12-5	14-9	7-4	9-5	11-1	13-2
	Spruce-pine-fir	SS	9-10	13-0	16-7	20-2	9-10	13-0	16-7	19-6
	Spruce-pine-fir	#1	9-8	12-9	15-8	18-3	9-1	11-6	14-1	16-3
	Spruce-pine-fir	#2	9-8	12-9	15-8	18-3	9-1	11-6	14-1	16-3
	Spruce-pine-fir	#3	7-8	9-9	11-10	13-9	6-10	8-8	10-7	12-4
24	Douglas fir-larch	SS	9-11	13-1	16-8	20-3	9-11	13-1	16-2	18-9
	Douglas fir-larch	#1	9-7	12-4	15-0	17-5	8-8	11-0	13-5	15-7
	Douglas fir-larch	#2	9-1	11-6	14-1	16-3	8-1	10-3	12-7	14-7
	Douglas fir-larch	#3	6-10	8-8	10-7	12-4	6-2	7-9	9-6	11-0
	Hem-fir	SS	9-4	12-4	15-9	19-2	9-4	12-4	15-9	18-5
	Hem-fir	#1	9-2	12-0	14-8	17-0	8-6	10-9	13-1	15-2
	Hem-fir	#2	8-9	11-4	13-10	16-1	8-0	10-2	12-5	14-4
	Hem-fir	#3	6-10	8-8	10-7	12-4	6-2	7-9	9-6	11-0
	Southern pine	SS	9-9	12-10	16-5	19-11	9-9	12-10	16-5	19-11
	Southern pine	#1	9-7	12-7	16-1	19-6	9-7	12-4	14-7	17-5
	Southern pine	#2	9-4	12-4	14-8	17-2	8-6	11-0	13-1	15-5
	Southern pine	#3	7-4	9-5	11-1	13-2	6-7	8-5	9-11	11-10
	Spruce-pine-fir	SS	9-2	12-1	15-5	18-9	9-2	12-1	15-0	17-5
	Spruce-pine-fir	#1	8-11	11-6	14-1	16-3	8-1	10-3	12-7	14-7
	Spruce-pine-fir	#2	8-11	11-6	14-1	16-3	8-1	10-3	12-7	14-7
	Spruce-pine-fir	#3	6-10	8-8	10-7	12-4	6-2	7-9	9-6	11-0

For SI: 1 inch = 25.4 mm, 1 foot = 304.8 mm, 1 pound per square foot = 0.0479 kPa.

Note: Check sources for availability of lumber in lengths greater than 20 feet.

a. Dead load limits for townhouses in Seismic Design Category C and all structures in Seismic Design Categories D_0, D_1 and D_2 shall be determined in accordance with Section R301.2.2.2.1.

❖ This table provides allowable joist spans for common lumber species, grades and joist sizes. The spans are applicable to residential sleeping areas. Use of this table is similar to the use described in the example in Section R502.3 (see commentary, Section R502.3).

TABLE R502.3.1(2)
FLOOR JOIST SPANS FOR COMMON LUMBER SPECIES
(Residential living areas, live load = 40 psf, L/Δ = 360)

JOIST SPACING (inches)	SPECIES AND GRADE		DEAD LOAD = 10 psf				DEAD LOAD = 20 psf			
			2 × 6	2 × 8	2 × 10	2 × 12	2 × 6	2 × 8	2 × 10	2 × 12
			Maximum floor joist spans							
			(ft - in.)	(ft - in.)	(ft - in.)	(ft - in.)	(ft - in.)	(ft - in.)	(ft - in.)	(ft - in.)
12	Douglas fir-larch	SS	11-4	15-0	19-1	23-3	11-4	15-0	19-1	23-3
	Douglas fir-larch	#1	10-11	14-5	18-5	22-0	10-11	14-2	17-4	20-1
	Douglas fir-larch	#2	10-9	14-2	17-9	20-7	10-6	13-3	16-3	18-10
	Douglas fir-larch	#3	8-8	11-0	13-5	15-7	7-11	10-0	12-3	14-3
	Hem-fir	SS	10-9	14-2	18-0	21-11	10-9	14-2	18-0	21-11
	Hem-fir	#1	10-6	13-10	17-8	21-6	10-6	13-10	16-11	19-7
	Hem-fir	#2	10-0	13-2	16-10	20-4	10-0	13-1	16-0	18-6
	Hem-fir	#3	8-8	11-0	13-5	15-7	7-11	10-0	12-3	14-3
	Southern pine	SS	11-2	14-8	18-9	22-10	11-2	14-8	18-9	22-10
	Southern pine	#1	10-11	14-5	18-5	22-5	10-11	14-5	18-5	22-5
	Southern pine	#2	10-9	14-2	18-0	21-9	10-9	14-2	16-11	19-10
	Southern pine	#3	9-4	11-11	14-0	16-8	8-6	10-10	12-10	15-3
	Spruce-pine-fir	SS	10-6	13-10	17-8	21-6	10-6	13-10	17-8	21-6
	Spruce-pine-fir	#1	10-3	13-6	17-3	20-7	10-3	13-3	16-3	18-10
	Spruce-pine-fir	#2	10-3	13-6	17-3	20-7	10-3	13-3	16-3	18-10
	Spruce-pine-fir	#3	8-8	11-0	13-5	15-7	7-11	10-0	12-3	14-3
16	Douglas fir-larch	SS	10-4	13-7	17-4	21-1	10-4	13-7	17-4	21-0
	Douglas fir-larch	#1	9-11	13-1	16-5	19-1	9-8	12-4	15-0	17-5
	Douglas fir-larch	#2	9-9	12-7	15-5	17-10	9-1	11-6	14-1	16-3
	Douglas fir-larch	#3	7-6	9-6	11-8	13-6	6-10	8-8	10-7	12-4
16	Hem-fir	SS	9-9	12-10	16-5	19-11	9-9	12-10	16-5	19-11
	Hem-fir	#1	9-6	12-7	16-0	18-7	9-6	12-0	14-8	17-0
	Hem-fir	#2	9-1	12-0	15-2	17-7	8-11	11-4	13-10	16-1
	Hem-fir	#3	7-6	9-6	11-8	13-6	6-10	8-8	10-7	12-4
	Southern pine	SS	10-2	13-4	17-0	20-9	10-2	13-4	17-0	20-9
	Southern pine	#1	9-11	13-1	16-9	20-4	9-11	13-1	16-4	19-6
	Southern pine	#2	9-9	12-10	16-1	18-10	9-6	12-4	14-8	17-2
	Southern pine	#3	8-1	10-3	12-2	14-6	7-4	9-5	11-1	13-2
	Spruce-pine-fir	SS	9-6	12-7	16-0	19-6	9-6	12-7	16-0	19-6
	Spruce-pine-fir	#1	9-4	12-3	15-5	17-10	9-1	11-6	14-1	16-3
	Spruce-pine-fir	#2	9-4	12-3	15-5	17-10	9-1	11-6	14-1	16-3
	Spruce-pine-fir	#3	7-6	9-6	11-8	13-6	6-10	8-8	10-7	12-4

(continued)

TABLE R502.3.1(2)—continued
FLOOR JOIST SPANS FOR COMMON LUMBER SPECIES
(Residential living areas, live load = 40 psf, $L/\Delta = 360$)[b]

JOIST SPACING (inches)	SPECIES AND GRADE		DEAD LOAD = 10 psf				DEAD LOAD = 20 psf			
			2 × 6	2 × 8	2 × 10	2 × 12	2 × 6	2 × 8	2 × 10	2 × 12
			Maximum floor joist spans							
			(ft - in.)	(ft - in.)	(ft - in.)	(ft - in.)	(ft - in.)	(ft - in.)	(ft - in.)	(ft - in.)
19.2	Douglas fir-larch	SS	9-8	12-10	16-4	19-10	9-8	12-10	16-4	19-2
	Douglas fir-larch	#1	9-4	12-4	15-0	17-5	8-10	11-3	13-8	15-11
	Douglas fir-larch	#2	9-1	11-6	14-1	16-3	8-3	10-6	12-10	14-10
	Douglas fir-larch	#3	6-10	8-8	10-7	12-4	6-3	7-11	9-8	11-3
	Hem-fir	SS	9-2	12-1	15-5	18-9	9-2	12-1	15-5	18-9
	Hem-fir	#1	9-0	11-10	14-8	17-0	8-8	10-11	13-4	15-6
	Hem-fir	#2	8-7	11-3	13-10	16-1	8-2	10-4	12-8	14-8
	Hem-fir	#3	6-10	8-8	10-7	12-4	6-3	7-11	9-8	11-3
	Southern pine	SS	9-6	12-7	16-0	19-6	9-6	12-7	16-0	19-6
	Southern pine	#1	9-4	12-4	15-9	19-2	9-4	12-4	14-11	17-9
	Southern pine	#2	9-2	12-1	14-8	17-2	8-8	11-3	13-5	15-8
	Southern pine	#3	7-4	9-5	11-1	13-2	6-9	8-7	10-1	12-1
	Spruce-pine-fir	SS	9-0	11-10	15-1	18-4	9-0	11-10	15-1	17-9
	Spruce-pine-fir	#	8-9	11-6	14-1	16-3	8-3	10-6	12-10	14-10
	Spruce-pine-fir	#2	8-9	11-6	14-1	16-3	8-3	10-6	12-10	14-10
	Spruce-pine-fir	#3	6-10	8-8	10-7	12-4	6-3	7-11	9-8	11-3
24	Douglas fir-larch	SS	9-0	11-11	15-2	18-5	9-0	11-11	14-9	17-1
	Douglas fir-larch	#1	8-8	11-0	13-5	15-7	7-11	10-0	12-3	14-3
	Douglas fir-larch	#2	8-1	10-3	12-7	14-7	7-5	9-5	11-6	13-4
	Douglas fir-larch	#3	6-2	7-9	9-6	11-0	5-7	7-1	8-8	10-1
	Hem-fir	SS	8-6	11-3	14-4	17-5	8-6	11-3	14-4	16-10[a]
	Hem-fir	#1	8-4	10-9	13-1	15-2	7-9	9-9	11-11	13-10
	Hem-fir	#2	7-11	10-2	12-5	14-4	7-4	9-3	11-4	13-1
	Hem-fir	#3	6-2	7-9	9-6	11-0	5-7	7-1	8-8	10-1
	Southern pine	SS	8-10	11-8	14-11	18-1	8-10	11-8	14-11	18-1
	Southern pine	#1	8-8	11-5	14-7	17-5	8-8	11-3	13-4	15-11
	Southern pine	#2	8-6	11-0	13-1	15-5	7-9	10-0	12-0	14-0
	Southern pine	#3	6-7	8-5	9-11	11-10	6-0	7-8	9-1	10-9
	Spruce-pine-fir	SS	8-4	11-0	14-0	17-0	8-4	11-0	13-8	15-11
	Spruce-pine-fir	#1	8-1	10-3	12-7	14-7	7-5	9-5	11-6	13-4
	Spruce-pine-fir	#2	8-1	10-3	12-7	14-7	7-5	9-5	11-6	13-4
	Spruce-pine-fir	#3	6-2	7-9	9-6	11-0	5-7	7-1	8-8	10-1

For SI: 1 inch = 25.4 mm, 1 foot = 304.8 mm, 1 pound per square foot = 0.0479 kPa.

Note: Check sources for availability of lumber in lengths greater than 20 feet.

a. End bearing length shall be increased to 2 inches.

b. Dead load limits for townhouses in Seismic Design Category C and all structures in Seismic Design Categories D_0, D_1, and D_2 shall be determined in accordance with Section R301.2.2.2.1.

❖ See the commentary to Section R502.3.

TABLE R502.3.3(1)
CANTILEVER SPANS FOR FLOOR JOISTS SUPPORTING LIGHT-FRAME EXTERIOR BEARING WALL AND ROOF ONLY[a, b, c, f, g, h]
(Floor Live Load ≤ 40 psf, Roof Live Load ≤ 20 psf)

Member & Spacing	Maximum Cantilever Span (Uplift Force at Backspan Support in Lbs.)[d, e]											
	Ground Snow Load											
	≤ 20 psf			30 psf			50 psf			70 psf		
	Roof Width			Roof Width			Roof Width			Roof Width		
	24 ft	32 ft	40 ft	24 ft	32 ft	40 ft	24 ft	32 ft	40 ft	24 ft	32 ft	40 ft
2 × 8 @ 12″	20″ (177)	15″ (227)	—	18″ (209)	—	—	—	—	—	—	—	—
2 × 10 @ 16″	29″ (228)	21″ (297)	16″ (364)	26″ (271)	18″ (354)	—	20″ (375)	—	—	—	—	—
2 × 10 @ 12″	36″ (166)	26″ (219)	20″ (270)	34″ (198)	22″ (263)	16″ (324)	26″ (277)	—	—	19″ (356)	—	—
2 × 12 @ 16″	—	32″ (287)	25″ (356)	36″ (263)	29″ (345)	21″ (428)	29″ (367)	20″ (484)	—	23″ (471)	—	—
2 × 12 @ 12″	—	42″ (209)	31″ (263)	—	37″ (253)	27″ (317)	36″ (271)	27″ (358)	17″ (447)	31″ (348)	19″ (462)	—
2 × 12 @ 8″	—	48″ (136)	45″ (169)	—	48″ (164)	38″ (206)	—	40″ (233)	26″ (294)	36″ (230)	29″ (304)	18″ (379)

For SI: 1 inch = 25.4 mm, 1 foot = 304.8 mm, 1 pound per square foot = 0.0479 kPa.

a. Tabulated values are for clear-span roof supported solely by exterior bearing walls.

b. Spans are based on No. 2 Grade lumber of Douglas fir-larch, hem-fir, southern pine and spruce-pine-fir for repetitive (three or more) members.

c. Ratio of backspan to cantilever span shall be at least 3:1.

d. Connections capable of resisting the indicated uplift force shall be provided at the backspan support.

e. Uplift force is for a backspan to cantilever span ratio of 3:1. Tabulated uplift values are permitted to be reduced by multiplying by a factor equal to 3 divided by the actual backspan ratio provided (3/backspan ratio).

f. See Section R301.2.2.2.5, Item 1, for additional limitations on cantilevered floor joists for detached one- and two-family dwellings in Seismic Design Category D_0, D_1, or D_2 and townhouses in Seismic Design Category C, D_0, D_1 or D_2.

g. A full-depth rim joist shall be provided at the unsupported end of the cantilever joists. Solid blocking shall be provided at the supported end.

h. Linear interpolation shall be permitted for building widths and ground snow loads other than shown.

❖ This table provides the prescriptive span and uplift requirements for cantilever wood floor joists supporting a light-frame bearing wall and roof only. The permitted cantilever span ranges from 15 to 48 inches (331 to 1219 mm) based on the limitations prescribed in the table and in the notes.

TABLE R502.3.3(2)
CANTILEVER SPANS FOR FLOOR JOISTS SUPPORTING EXTERIOR BALCONY[a, b, e, f]

60psf=LL

Member Size	Spacing	Maximum Cantilever Span (Uplift Force at Backspan Support in lb)[c, d]		
		Ground Snow Load		
		≤ 30 psf	50 psf	70 psf
2 × 8	12″	42″ (139)	39″ (156)	34″ (165)
2 × 8	16″	36″ (151)	34″ (171)	29″ (180)
2 × 10	12″	61″ (164)	57″ (189)	49″ (201)
2 × 10	16″	53″ (180)	49″ (208)	42″ (220)
2 × 10	24″	43″ (212)	40″ (241)	34″ (255)
2 × 12	16″	72″ (228)	67″ (260)	57″ (268)
2 × 12	24″	58″ (279)	54″ (319)	47″ (330)

For SI: 1 inch = 25.4 mm, 1 pound per square foot = 0.0479 kPa.

a. Spans are based on No. 2 Grade lumber of Douglas fir-larch, hem-fir, southern pine and spruce-pine-fir for repetitive (three or more) members.

b. Ratio of backspan to cantilever span shall be at least 2:1.

c. Connections capable of resisting the indicated uplift force shall be provided at the backspan support.

d. Uplift force is for a backspan to cantilever span ratio of 2:1. Tabulated uplift values are permitted to be reduced by multiplying by a factor equal to 2 divided by the actual backspan ratio provided (2/backspan ratio).

e. A full-depth rim joist shall be provided at the unsupported end of the cantilever joists. Solid blocking shall be provided at the supported end.

f. Linear interpolation shall be permitted for ground snow loads other than shown.

❖ This table provides the prescriptive span and uplift requirements for cantilever wood floor joists supporting an exterior balcony. The table was developed based on an engineered design consistent with the requirements of the *International Building Code*® (IBC®). A live load of 60 psf (2.87 kPa) is used for the cantilevered portion of the floor joists and is combined with anticipated dead loads and snow loads. Snow loads vary up to 70 psf (3.35 kPa) and include consideration of drifted snow from an adjacent roof. Deflections are limited to the values specified in Table 1604.3 of the IBC. As indicated in Note b, the ratio of backspan to cantilever span is limited to a minimum of 2:1.

R502.4 Joists under bearing partitions. Joists under parallel bearing partitions shall be of adequate size to support the load. Double joists, sized to adequately support the load, that are separated to permit the installation of piping or vents shall be full depth solid blocked with lumber not less than 2 inches (51 mm) in nominal thickness spaced not more than 4 feet (1219 mm) on center. Bearing partitions perpendicular to joists shall not be offset from supporting girders, walls or partitions more than the joist depth unless such joists are of sufficient size to carry the additional load.

❖ Where floor joists support bearing partitions, the joists must be of adequate size [see Commentary Figure R502.4(1)] for an example. This provides added support for the additional load from the bearing partition above. Alternatively, a beam of adequate size may also be substituted for the double joist.

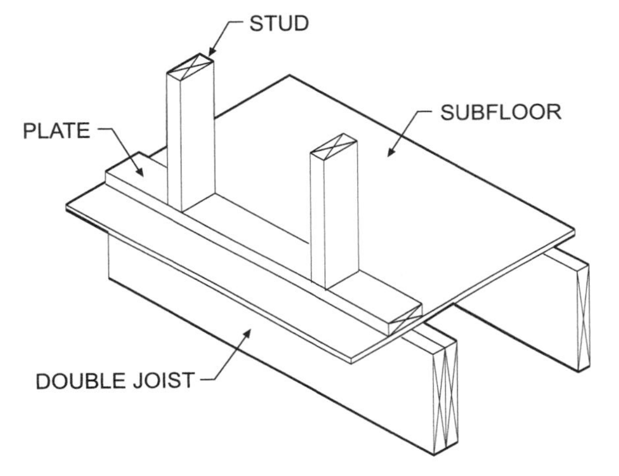

Figure R502.4(1)
DOUBLE JOIST UNDER BEARING PARTITION

Where piping or vents must penetrate the floor system, the double joists must be separated to allow the pipe or vent to pass through. In this case, full-depth solid blocking spaced a maximum of 4 feet (1219 mm) on center, as shown in Commentary Figure R502.4(2), is required so that the two joists will function as a combined member without twisting.

Bearing partitions oriented perpendicular to joists cannot be offset from supporting girders, walls or partitions more than the depth of the joists unless the joists are of sufficient size to carry the additional load.

R502.5 Allowable girder spans. The allowable spans of girders fabricated of dimension lumber shall not exceed the values set forth in Tables R502.5(1) and R502.5(2).

❖ Tables R502.5(1) and R502.5(2) list allowable spans for girders. These tables also apply to header spans and are referenced in Section R602.7 for that purpose. See Commentary Figure R502.5(1) and the following example for girder span-spacing relationships.

Example:

10 ft x 2 = 20 ft

Assuming two equal floor joist spans of 10 feet (3048 mm) as indicated in Figure R502.5(1) (only one joist span and the two supporting girders are shown for clarity), determine the maximum span for a double 2-inch by 10-inch (51 mm by 254 mm) interior girder supporting one floor only.

Solution:

Table R502.5(2) contains the information for an interior girder supporting one floor only. For a double 2-inch by 10-inch (51 mm by 254 mm) girder and a building width of 20 feet (6096 mm), the maximum allowable girder span is read directly from Table R502.5(2) as 7 feet (2134 mm).

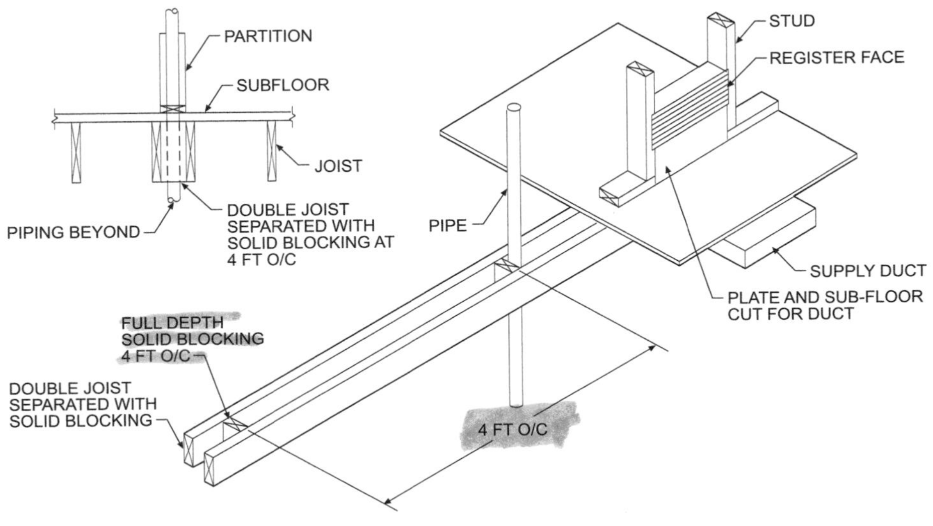

Figure R502.4(2)
SPACED JOISTS UNDER BEARING PARTITION WITH BLOCKING AT PENETRATION

TABLE R502.5(1)
GIRDER SPANS[a] AND HEADER SPANS[a] FOR EXTERIOR BEARING WALLS
(Maximum spans for Douglas fir-larch, hem-fir, southern pine and spruce-pine-fir[b] and required number of jack studs

GIRDERS AND HEADERS SUPPORTING	SIZE	GROUND SNOW LOAD (psf)[e]																	
		30						50						70					
		Building width[c] (feet)																	
		20		28		36		20		28		36		20		28		36	
		Span	NJ[d]	Span	NJ[d]	Span	NJ[d]	Span	NJ[d]	Span	NJ[d]	Span	NJ[d]	Span	NJ[d]	Span	NJ[d]	Span	NJ[d]
Roof and ceiling	2-2 × 4	3-6	1	3-2	1	2-10	1	3-2	1	2-9	1	2-6	1	2-10	1	2-6	1	2-3	1
	2-2 × 6	5-5	1	4-8	1	4-2	1	4-8	1	4-1	1	3-8	2	4-2	1	3-8	2	3-3	2
	2-2 × 8	6-10	1	5-11	2	5-4	2	5-11	2	5-2	2	4-7	2	5-4	2	4-7	2	4-1	2
	2-2 × 10	8-5	2	7-3	2	6-6	2	7-3	2	6-3	2	5-7	2	6-6	2	5-7	2	5-0	2
	2-2 × 12	9-9	2	8-5	2	7-6	2	8-5	2	7-3	2	6-6	2	7-6	2	6-6	2	5-10	3
	3-2 × 8	8-4	1	7-5	1	6-8	1	7-5	1	6-5	2	5-9	2	6-8	1	5-9	2	5-2	2
	3-2 × 10	10-6	1	9-1	2	8-2	2	9-1	2	7-10	2	7-0	2	8-2	2	7-0	2	6-4	2
	3-2 × 12	12-2	2	10-7	2	9-5	2	10-7	2	9-2	2	8-2	2	9-5	2	8-2	2	7-4	2
	4-2 × 8	9-2	1	8-4	1	7-8	1	8-4	1	7-5	1	6-8	1	7-8	1	6-8	1	5-11	2
	4-2 × 10	11-8	1	10-6	1	9-5	2	10-6	1	9-1	2	8-2	2	9-5	2	8-2	2	7-3	2
	4-2 × 12	14-1	1	12-2	2	10-11	2	12-2	2	10-7	2	9-5	2	10-11	2	9-5	2	8-5	2
Roof, ceiling and one center-bearing floor	2-2 × 4	3-1	1	2-9	1	2-5	1	2-9	1	2-5	1	2-2	1	2-7	1	2-3	1	2-0	1
	2-2 × 6	4-6	1	4-0	1	3-7	2	4-1	1	3-7	2	3-3	2	3-9	2	3-3	2	2-11	2
	2-2 × 8	5-9	2	5-0	2	4-6	2	5-2	2	4-6	2	4-1	2	4-9	2	4-2	2	3-9	2
	2-2 × 10	7-0	2	6-2	2	5-6	2	6-4	2	5-6	2	5-0	2	5-9	2	5-1	2	4-7	3
	2-2 × 12	8-1	2	7-1	2	6-5	2	7-4	2	6-5	2	5-9	3	6-8	2	5-10	3	5-3	3
	3-2 × 8	7-2	1	6-3	2	5-8	2	6-5	2	5-8	2	5-1	2	5-11	2	5-2	2	4-8	2
	3-2 × 10	8-9	2	7-8	2	6-11	2	7-11	2	6-11	2	6-3	2	7-3	2	6-4	2	5-8	2
	3-2 × 12	10-2	2	8-11	2	8-0	2	9-2	2	8-0	2	7-3	2	8-5	2	7-4	2	6-7	2
	4-2 × 8	8-1	1	7-3	1	6-7	1	7-5	1	6-6	1	5-11	2	6-10	1	6-0	2	5-5	2
	4-2 × 10	10-1	1	8-10	2	8-0	2	9-1	2	8-0	2	7-2	2	8-4	2	7-4	2	6-7	2
	4-2 × 12	11-9	2	10-3	2	9-3	2	10-7	2	9-3	2	8-4	2	9-8	2	8-6	2	7-7	2
Roof, ceiling and one clear span floor	2-2 × 4	2-8	1	2-4	1	2-1	1	2-7	1	2-3	1	2-0	1	2-5	1	2-1	1	1-10	1
	2-2 × 6	3-11	1	3-5	2	3-0	2	3-10	2	3-4	2	3-0	2	3-6	2	3-1	2	2-9	2
	2-2 × 8	5-0	2	4-4	2	3-10	2	4-10	2	4-2	2	3-9	2	4-6	2	3-11	2	3-6	2
	2-2 × 10	6-1	2	5-3	2	4-8	2	5-11	2	5-1	2	4-7	3	5-6	2	4-9	2	4-3	3
	2-2 × 12	7-1	2	6-1	3	5-5	3	6-10	2	5-11	3	5-4	3	6-4	2	5-6	3	5-0	3
	3-2 × 8	6-3	2	5-5	2	4-10	2	6-1	2	5-3	2	4-8	2	5-7	2	4-11	2	4-5	2
	3-2 × 10	7-7	2	6-7	2	5-11	2	7-5	2	6-5	2	5-9	2	6-10	2	6-0	2	5-4	2
	3-2 × 12	8-10	2	7-8	2	6-10	2	8-7	2	7-5	2	6-8	2	7-11	2	6-11	2	6-3	2
	4-2 × 8	7-2	1	6-3	2	5-7	2	7-0	1	6-1	2	5-5	2	6-6	1	5-8	2	5-1	2
	4-2 × 10	8-9	2	7-7	2	6-10	2	8-7	2	7-5	2	6-7	2	7-11	2	6-11	2	6-2	2
	4-2 × 12	10-2	2	8-10	2	7-11	2	9-11	2	8-7	2	7-8	2	9-2	2	8-0	2	7-2	2
Roof, ceiling and two center-bearing floors	2-2 × 4	2-7	1	2-3	1	2-0	1	2-6	1	2-2	1	1-11	1	2-4	1	2-0	1	1-9	1
	2-2 × 6	3-9	2	3-3	2	2-11	2	3-8	2	3-2	2	2-10	2	3-5	2	3-0	2	2-8	2
	2-2 × 8	4-9	2	4-2	2	3-9	2	4-7	2	4-0	2	3-8	2	4-4	2	3-9	2	3-5	2
	2-2 × 10	5-9	2	5-1	2	4-7	3	5-8	2	4-11	2	4-5	3	5-3	2	4-7	3	4-2	3
	2-2 × 12	6-8	2	5-10	3	5-3	3	6-6	2	5-9	3	5-2	3	6-1	3	5-4	3	4-10	3
	3-2 × 8	5-11	2	5-2	2	4-8	2	5-9	2	5-1	2	4-7	2	5-5	2	4-9	2	4-3	2
	3-2 × 10	7-3	2	6-4	2	5-8	2	7-1	2	6-2	2	5-7	2	6-7	2	5-9	2	5-3	2
	3-2 × 12	8-5	2	7-4	2	6-7	2	8-2	2	7-2	2	6-5	3	7-8	2	6-9	2	6-1	3
	4-2 × 8	6-10	1	6-0	2	5-5	2	6-8	1	5-10	2	5-3	2	6-3	2	5-6	2	4-11	2
	4-2 × 10	8-4	2	7-4	2	6-7	2	8-2	2	7-2	2	6-5	2	7-7	2	6-8	2	6-0	2
	4-2 × 12	9-8	2	8-6	2	7-8	2	9-5	2	8-3	2	7-5	2	8-10	2	7-9	2	7-0	2
Roof, ceiling, and two clear span floors	2-2 × 4	2-1	1	1-8	1	1-6	2	2-0	1	1-8	1	1-5	2	2-0	1	1-8	1	1-5	2
	2-2 × 6	3-1	2	2-8	2	2-4	2	3-0	2	2-7	2	2-3	2	2-11	2	2-7	2	2-3	2
	2-2 × 8	3-10	2	3-4	2	3-0	3	3-10	2	3-4	2	2-11	3	3-9	2	3-3	2	2-11	3

(continued)

TABLE R502.5(1)—continued
GIRDER SPANS[a] AND HEADER SPANS[a] FOR EXTERIOR BEARING WALLS
(Maximum spans for Douglas fir-larch, hem-fir, southern pine and spruce-pine-fir[b] and required number of jack studs

GIRDERS AND HEADERS SUPPORTING	SIZE	GROUND SNOW LOAD (psf)[e]																	
		30						50						70					
		Building width[c] (feet)																	
		20		28		36		20		28		36		20		28		36	
		Span	NJ[d]	Span	NJ[d]	Span	NJ[d]	Span	NJ[d]	Span	NJ[d]	Span	NJ[d]	Span	NJ[d]	Span	NJ[d]	Span	NJ[d]
Roof, ceiling, and two clear span floors	2-2 × 10	4-9	2	4-1	3	3-8	3	4-8	2	4-0	3	3-7	3	4-7	3	4-0	3	3-6	3
	2-2 × 12	5-6	3	4-9	3	4-3	3	5-5	3	4-8	3	4-2	3	5-4	3	4-7	3	4-1	4
	3-2 × 8	4-10	2	4-2	2	3-9	2	4-9	2	4-1	2	3-8	2	4-8	2	4-1	2	3-8	2
	3-2 × 10	5-11	2	5-1	2	4-7	3	5-10	2	5-0	2	4-6	3	5-9	2	4-11	2	4-5	3
	3-2 × 12	6-10	2	5-11	3	5-4	3	6-9	2	5-10	3	5-3	3	6-8	2	5-9	3	5-2	3
	4-2 × 8	5-7	2	4-10	2	4-4	2	5-6	2	4-9	2	4-3	2	5-5	2	4-8	2	4-2	2
	4-2 × 10	6-10	2	5-11	2	5-3	2	6-9	2	5-10	2	5-2	2	6-7	2	5-9	2	5-1	2
	4-2 × 12	7-11	2	6-10	2	6-2	3	7-9	2	6-9	2	6-0	3	7-8	2	6-8	2	5-11	3

For SI: 1 inch = 25.4 mm, 1 pound per square foot = 0.0479 kPa.

a. Spans are given in feet and inches.

b. Tabulated values assume #2 grade lumber.

c. Building width is measured perpendicular to the ridge. For widths between those shown, spans are permitted to be interpolated.

d. NJ - Number of jack studs required to support each end. Where the number of required jack studs equals one, the header is permitted to be supported by an approved framing anchor attached to the full-height wall stud and to the header.

e. Use 30 psf ground snow load for cases in which ground snow load is less than 30 psf and the roof live load is equal to or less than 20 psf.

❖ In addition to providing the allowable spans for girders and headers, this table lists the number of jack studs required to support the girder or header. The header studs (jack studs) on which the headers rest should be continuous from the header to the sill plate of the wall. Cutting the header stud to support a sill is not allowed. Headers should be adequately nailed together and to the wall studs. The table is broken down into five loading conditions, illustrated in Commentary Figure R502.5(2).

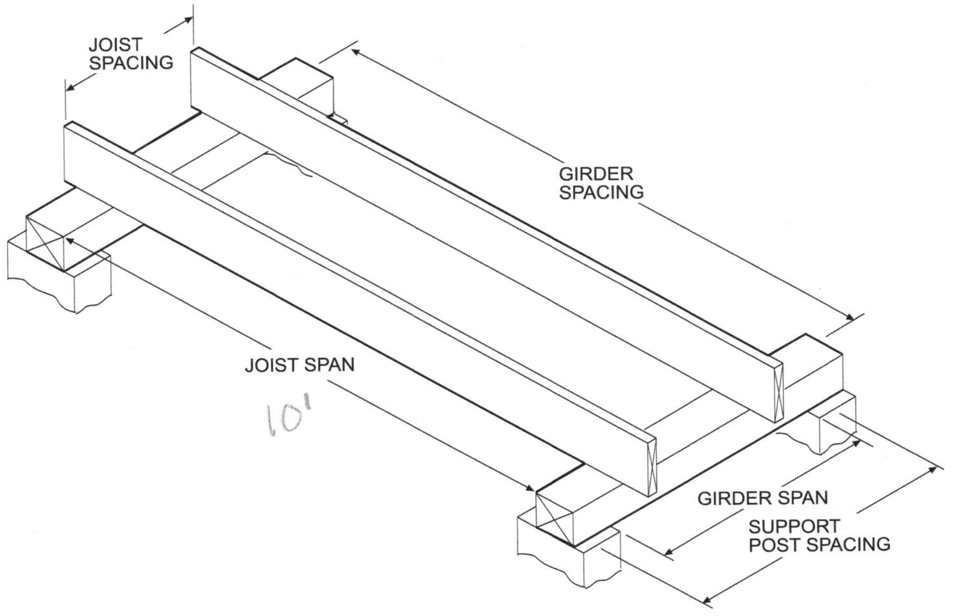

Figure R502.5(1)
GIRDER SPAN-SPACING RELATIONSHIP

TABLE R502.5(2)
GIRDER SPANS[a] AND HEADER SPANS[a] FOR INTERIOR BEARING WALLS
(Maximum spans for Douglas fir-larch, hem-fir, southern pine and spruce-pine-fir[b] and required number of jack studs)

HEADERS AND GIRDERS SUPPORTING	SIZE	BUILDING Width[c] (feet)					
		20		28		36	
		Span	NJ[d]	Span	NJ[d]	Span	NJ[d]
One floor only	2-2 × 4	3-1	1	2-8	1	2-5	1
	2-2 × 6	4-6	1	3-11	1	3-6	1
	2-2 × 8	5-9	1	5-0	2	4-5	2
	2-2 × 10	7-0	2	6-1	2	5-5	2
	2-2 × 12	8-1	2	7-0	2	6-3	2
	3-2 × 8	7-2	1	6-3	1	5-7	2
	3-2 × 10	8-9	1	7-7	2	6-9	2
	3-2 × 12	10-2	2	8-10	2	7-10	2
	4-2 × 8	9-0	1	7-8	1	6-9	1
	4-2 × 10	10-1	1	8-9	1	7-10	2
	4-2 × 12	11-9	1	10-2	2	9-1	2
Two floors	2-2 × 4	2-2	1	1-10	1	1-7	1
	2-2 × 6	3-2	2	2-9	2	2-5	2
	2-2 × 8	4-1	2	3-6	2	3-2	2
	2-2 × 10	4-11	2	4-3	2	3-10	3
	2-2 × 12	5-9	2	5-0	3	4-5	3
	3-2 × 8	5-1	2	4-5	2	3-11	2
	3-2 × 10	6-2	2	5-4	2	4-10	2
	3-2 × 12	7-2	2	6-3	2	5-7	3
	4-2 × 8	6-1	1	5-3	2	4-8	2
	4-2 × 10	7-2	2	6-2	2	5-6	2
	4-2 × 12	8-4	2	7-2	2	6-5	2

For SI: 1 inch = 25.4 mm, 1 foot = 304.8 mm.

a. Spans are given in feet and inches.

b. Tabulated values assume #2 grade lumber.

c. Building width is measured perpendicular to the ridge. For widths between those shown, spans are permitted to be interpolated.

d. NJ - Number of jack studs required to support each end. Where the number of required jack studs equals one, the header is permitted to be supported by an approved framing anchor attached to the full-height wall stud and to the header.

❖ In addition to providing the allowable spans for girders and headers, this table lists the number of jack studs required based on the bearing necessary for the girder or header. The table is broken down into two loading conditions, illustrated in Commentary Figure R502.5(3).

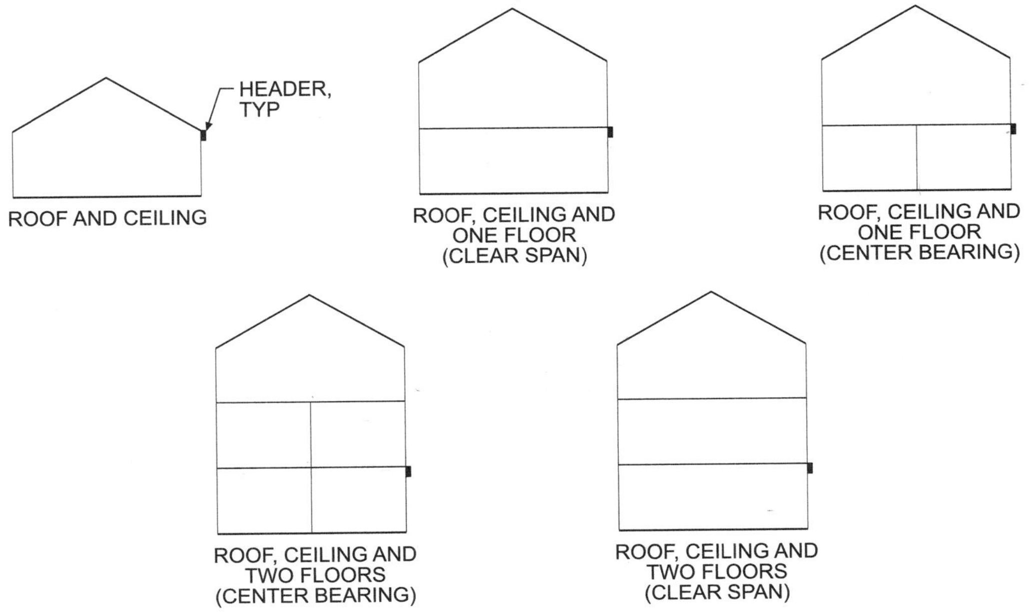

SOURCE: AF&PA WOOD FRAME CONSTRUCTION MANUAL FOR ONE- AND TWO-FAMILY DWELLINGS, 1995 SBC HIGH WIND EDITION

Figure R502.5(2)
HEADER SPANS FOR EXTERIOR LOAD BEARING WALLS

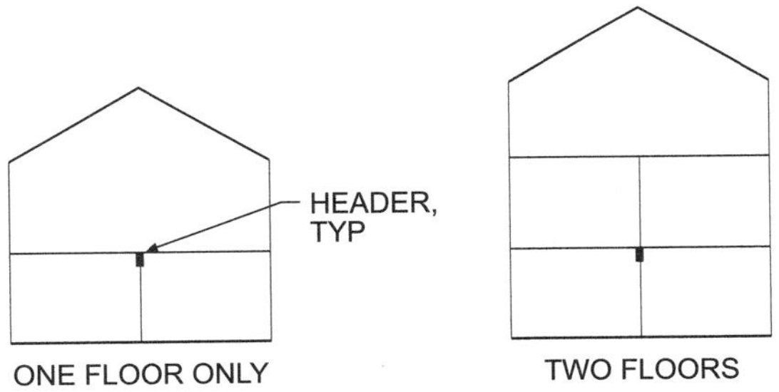

SOURCE: AF&PA WOOD FRAME CONSTRUCTION MANUAL FOR ONE- AND TWO-FAMILY DWELLINGS, 1995 SBC HIGH WIND EDITION

Figure R502.5(3)
HEADERS SPANS FOR INTERIOR LOAD BEARING WALLS

R502.6 Bearing. The ends of each joist, beam or girder shall have not less than 1.5 inches (38 mm) of bearing on wood or metal and not less than 3 inches (76 mm) on masonry or concrete except where supported on a 1-inch by 4-inch (25.4 mm by 102 mm) ribbon strip and nailed to the adjacent stud or by the use of approved joist hangers. The bearing on masonry or concrete shall be direct, or a sill plate of 2-inch-minimum (51 mm) nominal thickness shall be provided under the joist, beam or girder. The sill plate shall provide a minimum nominal bearing area of 48 square inches (30 865 mm^2).

❖ This section establishes minimum lengths of bearing for several alternative support systems to provide for the transfer of floor loads to supporting elements. These minimum requirements are based on anticipated loads and the allowable compressive stresses perpendicular to the grain for beam sizes and grades typical for wood-frame construction in addition to consideration of shear failure of the masonry. For joists, beams or girders bearing on wood or metal, the minimum bearing shown in Commentary Figure R502.6(1) must be provided.

As an alternative, the members may be supported by a 1-inch by 4-inch (25 mm by 102 mm) ribbon strip when joists are nailed to adjacent studs as shown in

Commentary Figure R502.6(3) or by the use of approved joist hangers as shown in Commentary Figure R502.6(4).

When joists, beams or girders bear on masonry, a minimum bearing of 3 inches (76 mm) is required, as shown in Commentary Figure R502.6(2).

Where joists, beams or girders span across masonry or concrete piers, full bearing of not less than 3 inches (76 mm) directly on the masonry or concrete pier is required. The code permits a sill plate of 2-inch (51 mm) minimum nominal thickness in lieu of bearing directly on the masonry or concrete. This will allow for the condition where the elevation of the top of the pier is below the elevation of the bottom of the joist, beam or girder and the joist, beam or girder is not bearing directly on the masonry or concrete. The minimum nominal bearing area of 48 square inches (0.03 m^2) is required to provide a bearing area on top of hollow masonry walls or piers.

R502.6.1 Floor systems. Joists framing from opposite sides over a bearing support shall lap a minimum of 3 inches (76 mm) and shall be nailed together with a minimum three 10d face nails. A wood or metal splice with strength equal to or greater than that provided by the nailed lap is permitted.

❖ To provide a concentric application of load from the joist to supporting beams or girders, joists framing from opposite sides of a beam or girder are required to lap at least 3 inches (76 mm) or the opposing joists are to be tied together with a wood or metal splice (see Commentary Figure R502.6.1).

R502.6.2 Joist framing. Joists framing into the side of a wood girder shall be supported by *approved* framing anchors or on ledger strips not less than nominal 2 inches by 2 inches (51 mm by 51 mm).

❖ For joists framed into the side of a wood beam or girder, the joists must be supported by approved framing anchors, as shown in Commentary Figure R502.6.2(1), or by ledger strips having a minimum nominal dimension of 2 inches (51 mm), as shown in Commentary Figure R502.6.2(2).

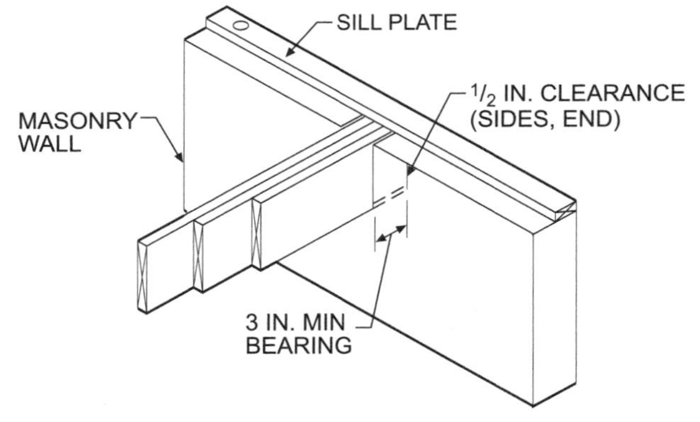

For SI: 1 inch = 25.4 mm.

Figure R502.6(2)
FLOOR JOIST BEARING ON MASONRY

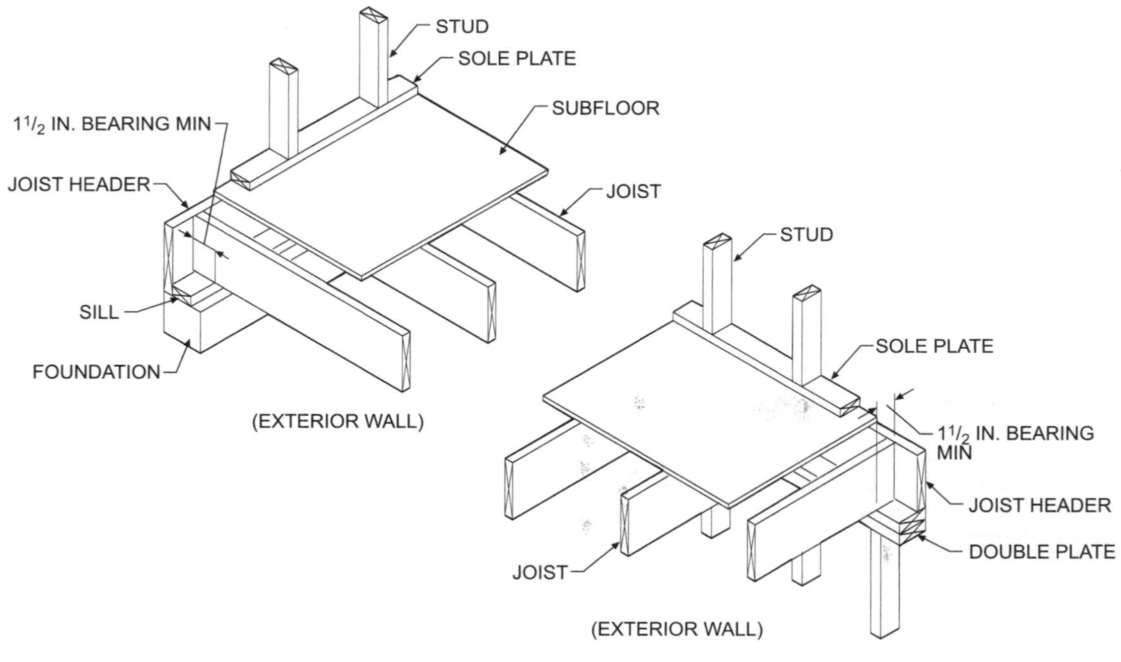

For SI: 1 inch = 25.4 mm.

Figure R502.6(1)
FLOOR JOIST BEARING ON WOOD

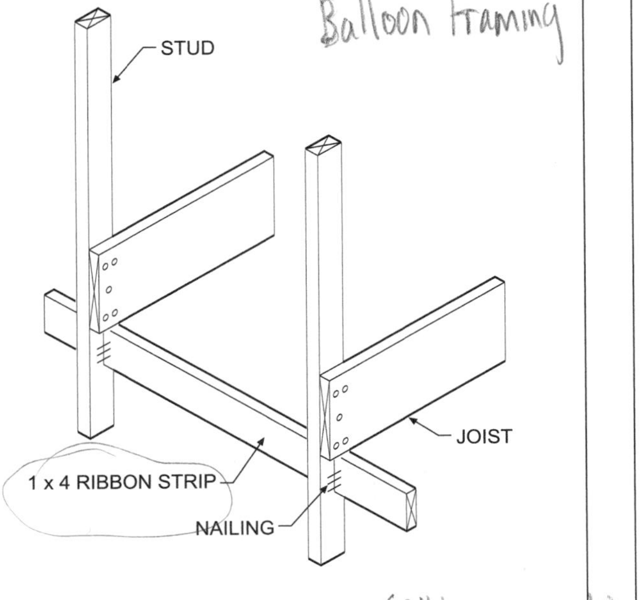

For SI: 1 inch = 25.4 mm.

Figure R502.6(3)
FLOOR JOIST BEARING ON RIBBON STRIP

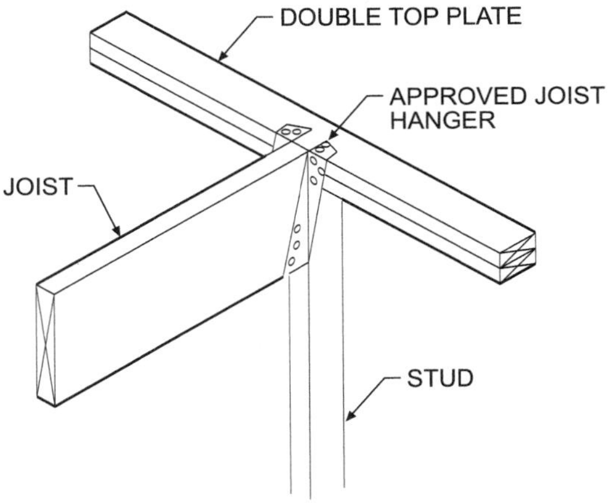

Figure R502.6(4)
JOIST HANGER AT DOUBLE TOP PLATE

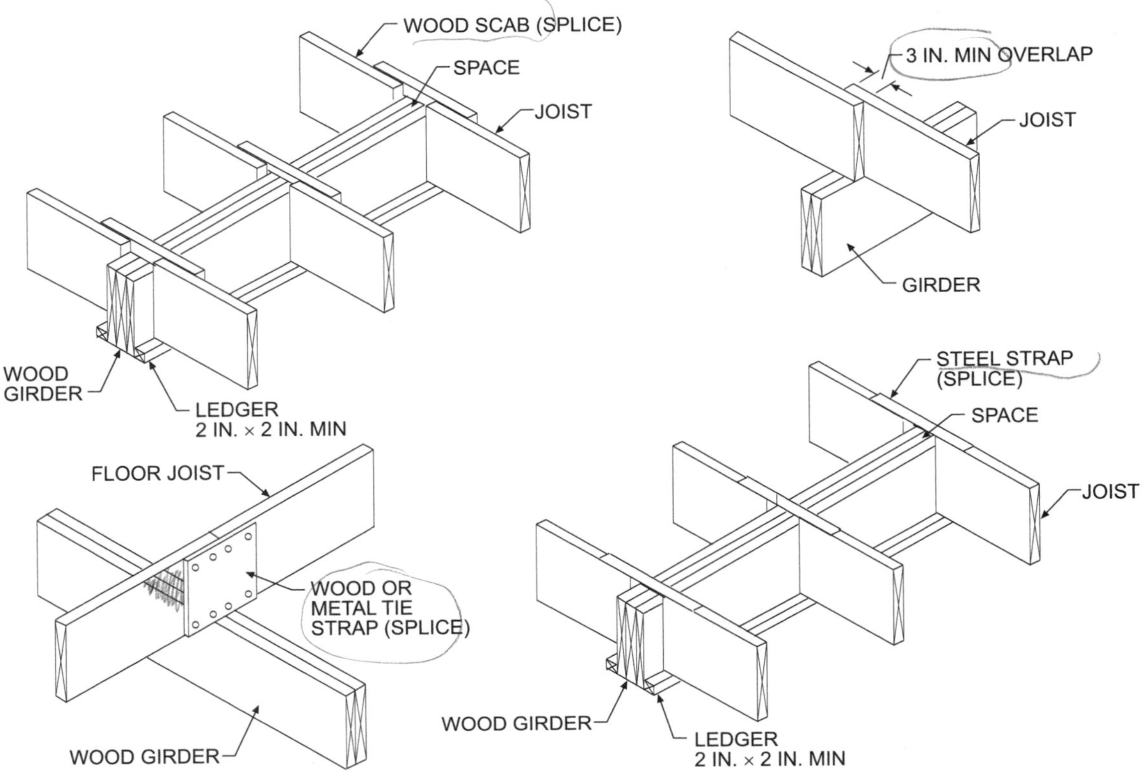

For SI: 1 inch = 25.4 mm.

Figure R502.6.1
JOIST AT GIRDER

Joists need.
1) blocking @ ends
2) for joists > 2x12" needs bridging

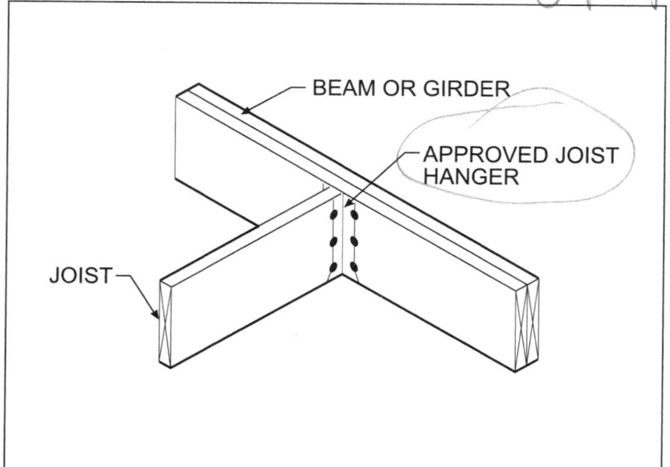

Figure R502.6.2(1)
JOIST6 HANGER AT GIRDER

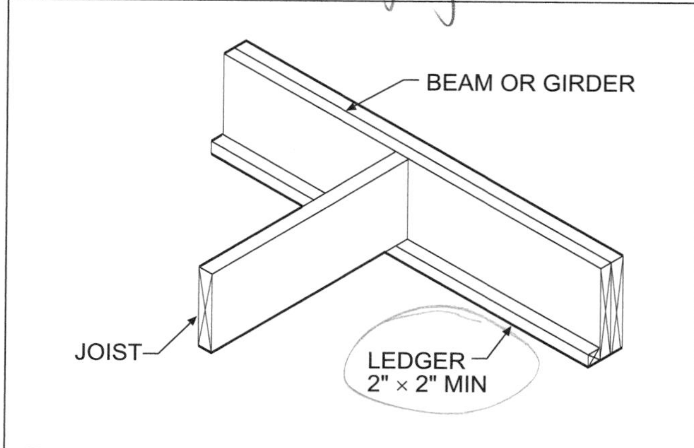

For SI: 1 inch = 25.4 mm.

Figure R502.6.2(2)
LEDGER STRIPS AT BEAM OR GIRDER

R502.7 Lateral restraint at supports. Joists shall be supported laterally at the ends by full-depth solid blocking not less than 2 inches (51 mm) nominal in thickness; or by attachment to a full-depth header, band or rim joist, or to an adjoining stud or shall be otherwise provided with lateral support to prevent rotation.

Exceptions:

1. Trusses, structural composite lumber, structural glued-laminated members and I-joists shall be supported laterally as required by the manufacturer's recommendations.

2. In Seismic Design Categories D_0, D_1 and D_2, lateral restraint shall also be provided at each intermediate support.

❖ Bridging, blocking or some other acceptable means of holding a joist in place is required so the floor joists do not twist out of the plane of the applied load. Lateral support at the ends of joists provide an additional function by transferring lateral loads to the supporting elements. Lateral support at ends may be provided by full-depth solid blocking not less than 2 inches (51 mm) in thickness, or the ends of joists may be nailed or bolted to a full-depth header, band or rim joist or to an adjoining stud as shown in Commentary Figure R502.7.

The requirements for lateral restraint of engineered wood products may require restraint other than that required for solid-sawn joists. Exception 1 requires restraint of the engineered wood products to be as required by the manufacturer's recommendations. Lateral restraint at each intermediate support is required for joists in Seismic Design Categories D_0, D_1 and D_2.

R502.7.1 Bridging. Joists exceeding a nominal 2 inches by 12 inches (51 mm by 305 mm) shall be supported laterally by solid blocking, diagonal bridging (wood or metal), or a continuous 1 inch by 3 inch (25.4 mm by 76 mm) strip nailed across the bottom of joists perpendicular to joists at intervals not exceeding 8 feet (2438 mm).

Exception: Trusses, structural composite lumber, structural glued-laminated members and I-joists shall be supported laterally as required by the manufacturer's recommendations.

❖ In addition to the lateral support at the ends, joists must have intermediate lateral support at intervals not exceeding 8 feet (2438 mm). Intermediate blocking is not required for joists 2 inches by 12 inches (51 mm by 305 mm) or smaller. The intermediate lateral support may be provided by solid blocking, diagonal bridging or wood bridging not less than 1 inch by 3 inches (25 mm by 76 mm) nominal, nailed to the bottom of the joist. Commentary Figure R502.7.1 illustrates the various alternatives for intermediate lateral support.

The requirements for bridging of engineered wood products may require bridging other than that required for solid-sawn joists. Exception 1 requires bridging of the engineered wood products to be as required by the manufacturer's recommendations.

R502.8 Cutting, drilling and notching. Structural floor members shall not be cut, bored or notched in excess of the limitations specified in this section. See Figure R502.8.

❖ Some designs and installation practices require that limited notching and cutting occur. Notching should be avoided when possible, and holes bored in beams and joists create the same problems as notches. When necessary, the holes should be located in areas with the least stress concentration, generally along the neutral axis of the joist. Limitations on the allowable cutting and notching of wood floor joists are meant to retain structural or functional integrity.

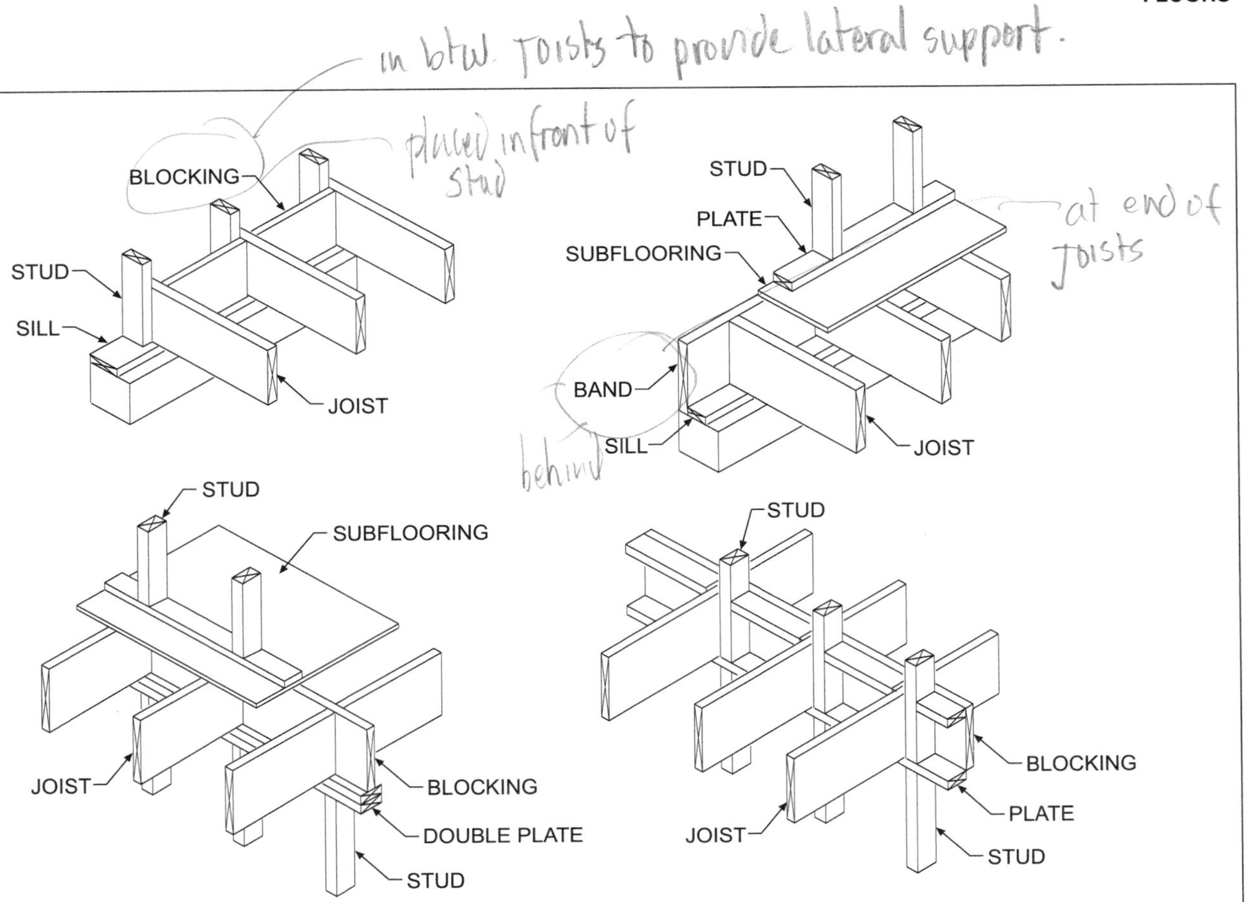

Figure R502.7
BLOCKING OF JOISTS

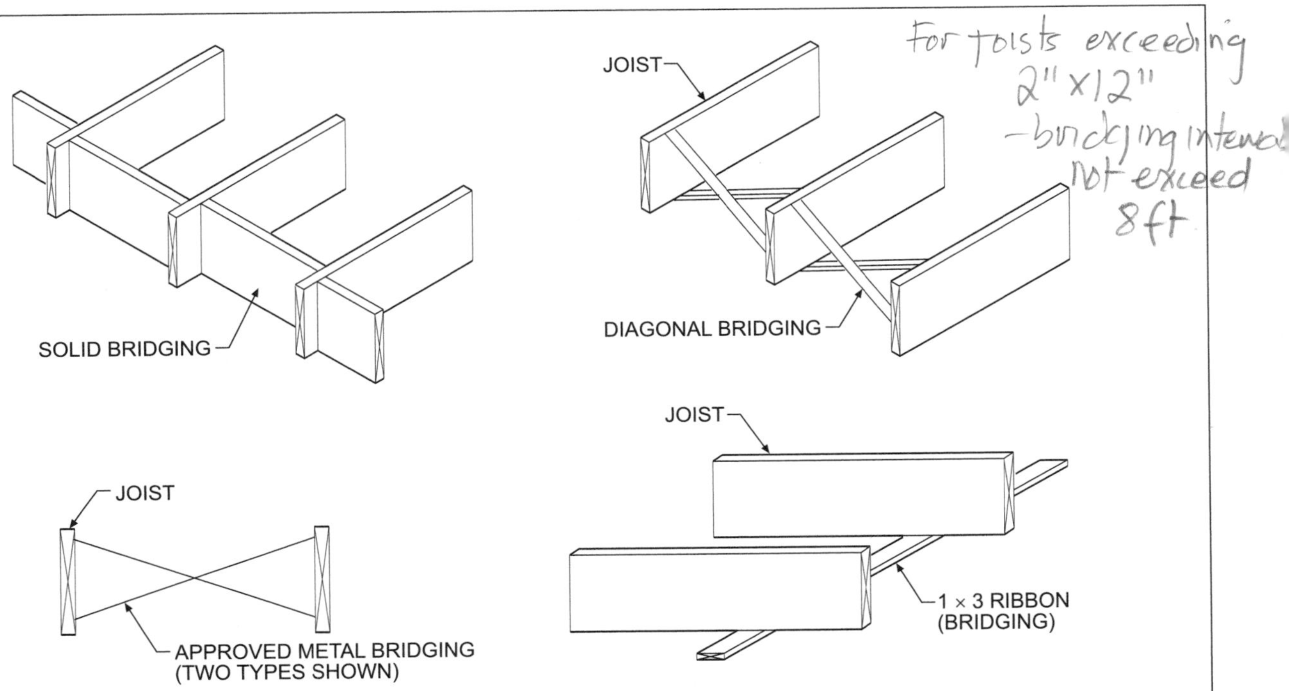

Figure R502.7.1
INTERMEDIATE LATERAL SUPPORT

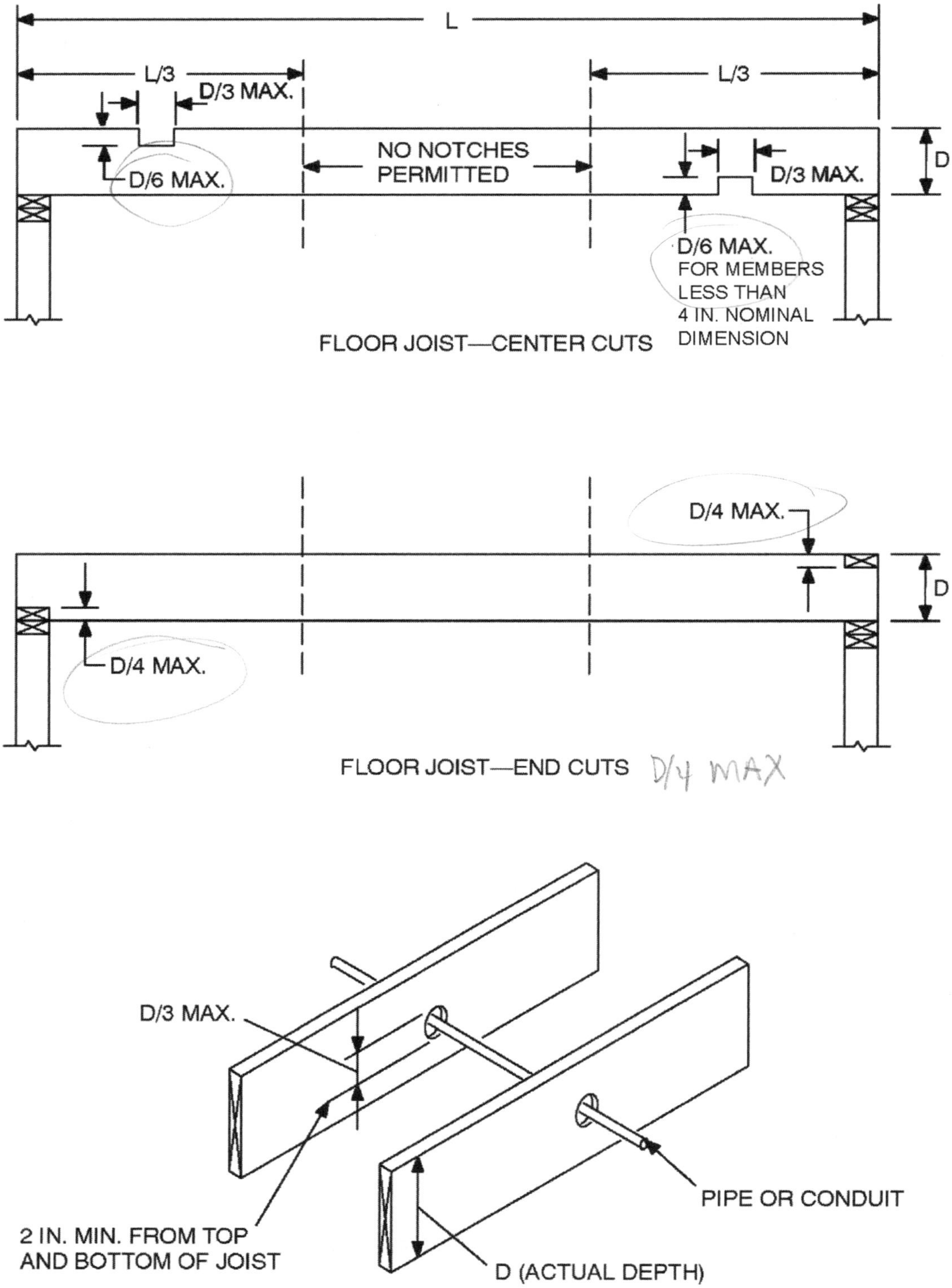

For SI: 1 inch = 25.4 mm.

FIGURE R502.8
CUTTING, NOTCHING AND DRILLING

R502.8.1 Sawn lumber. Notches in solid lumber joists, rafters and beams shall not exceed one-sixth of the depth of the member, shall not be longer than one-third of the depth of the member and shall not be located in the middle one-third of the span. Notches at the ends of the member shall not exceed one-fourth the depth of the member. The tension side of members 4 inches (102 mm) or greater in nominal thickness shall not be notched except at the ends of the members. The diameter of holes bored or cut into members shall not exceed one-third the depth of the member. Holes shall not be closer than 2 inches (51 mm) to the top or bottom of the member, or to any other hole located in the member. Where the member is also notched, the hole shall not be closer than 2 inches (51 mm) to the notch.

❖ Cutting and notching limitations are illustrated in Figure R502.8. Additionally, the tension side of members with a thickness of 4 inches (102 mm) or more can be notched at the ends only (see commentary, Section R502.8).

R502.8.2 Engineered wood products. Cuts, notches and holes bored in trusses, structural composite lumber, structural glue-laminated members or I-joists are prohibited except where permitted by the manufacturer's recommendations or where the effects of such alterations are specifically considered in the design of the member by a *registered design professional*.

❖ Cutting and notching limitations for sawn lumber do not apply to engineered wood products. Structural composite lumber is a generic term which encompasses a variety of engineered composite wood products including laminated veneer lumber (LVL). Also, included in the term are laminated strand lumber (LSL), parallel strand lumber (PSL), and oriented strand lumber (OSL). The prohibitions in this section apply to all of these products.

Engineered wood products must not be cut, notched or bored unless those alterations are considered in the design of the member. That consideration must come from either the manufacturer and be reflected in use recommendations (which is common in I-joists, permitting some limited alterations to webs) or from a registered design professional.

R502.9 Fastening. Floor framing shall be nailed in accordance with Table R602.3(1). Where posts and beam or girder construction is used to support floor framing, positive connections shall be provided to ensure against uplift and lateral displacement.

❖ Commentary Figure R502.9 shows various methods of accomplishing the mandatory positive connection between post and beam or girder construction.

R502.10 Framing of openings. Openings in floor framing shall be framed with a header and trimmer joists. When the header joist span does not exceed 4 feet (1219 mm), the header joist may be a single member the same size as the floor joist. Single trimmer joists may be used to carry a single header joist that is located within 3 feet (914 mm) of the trimmer joist bearing. When the header joist span exceeds 4 feet (1219 mm), the trimmer joists and the header joist shall be doubled and of sufficient cross section to support the floor joists framing into the header. *Approved* hangers shall be used for the header joist to trimmer joist connections when

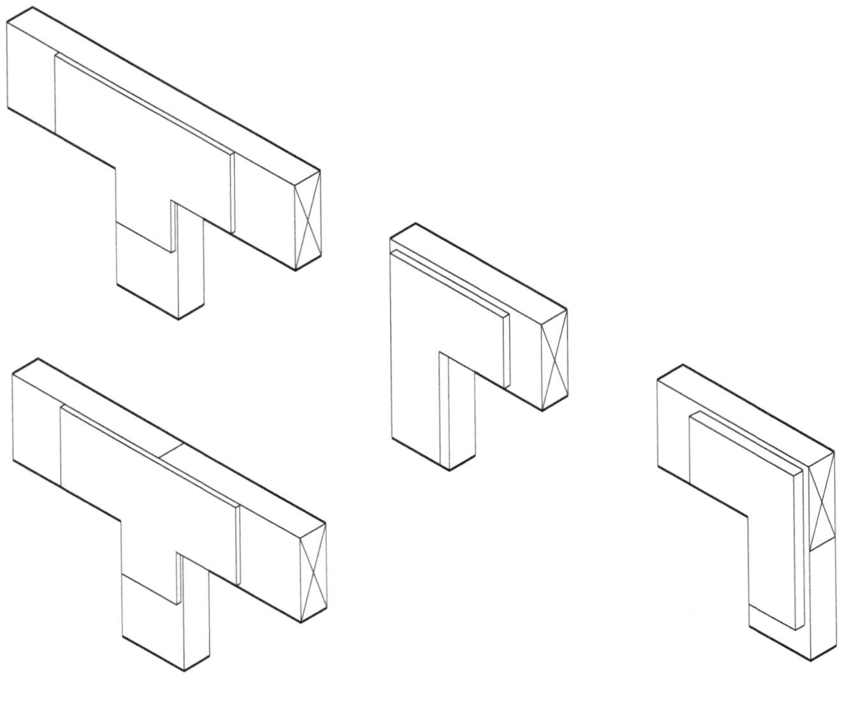

Figure R502.9
POSITIVE CONNECTION BETWEEN POSTS AND GIRDERS

the header joist span exceeds 6 feet (1829 mm). Tail joists over 12 feet (3658 mm) long shall be supported at the header by framing anchors or on ledger strips not less than 2 inches by 2 inches (51 mm by 51 mm).

❖ Where larger floor openings are necessary, adequate load-carrying transfer capability must be provided. Header joists the same size as floor joists may be used for spans not exceeding 4 feet (1219 mm), as shown in Commentary Figure R502.10(1). Trimmers must also be doubled to carry the additional load, unless the header joist is located within 3 feet (914 mm) of the trimmer joist bearing.

Where the header joist span exceeds 4 feet (1219 mm), a larger load-carrying capability is required. The header and trimmer joists must be doubled and be of sufficient size to support floor joist framing as illustrated in Commentary Figure R502.10(2).

In some cases, nailing would be insufficient to transfer vertical loads. Therefore, positive connections as shown in Commentary Figure R502.10(3) must be used. These hangers are required when the header joist span exceeds 6 feet (1829 mm).

R502.11 Wood trusses.

R502.11.1 Design. Wood trusses shall be designed in accordance with *approved* engineering practice. The design and manufacture of metal plate connected wood trusses shall comply with ANSI/TPI 1. The truss design drawings shall be prepared by a registered professional where required by the statutes of the *jurisdiction* in which the project is to be constructed in accordance with Section R106.1.

❖ The code contains no prescriptive provisions for the design and installation of wood trusses. A design that conforms to accepted engineering practice is required. Recognizing the extensive use of trusses in residential construction, the code references ANSI/TPI 1, *National Design Standard for Metal-plate-connected Wood Truss Construction.* The standard contains regulations for the design and installation of metal-plate-connected wood trusses, including the procedures for full-scale tests and testing methods for evaluating metal plate connectors. In addition to adequate design, it is important that the trusses be handled and erected properly so that the performance capability of the trusses is not compromised. A truss member should never be cut without approval from the design engineer.

Usually trusses are delivered in bundles, which reduces the potential for damage. Slings and spreader bars sized for the load should be used to reduce stresses caused by sway and bending.

Shop drawings showing the lumber schedule, design loads and panel point details (size, location and attachment of plates) should be filed with the building permit application and should be available at the time of inspection.

R502.11.2 Bracing. Trusses shall be braced to prevent rotation and provide lateral stability in accordance with the requirements specified in the *construction documents* for the building and on the individual truss design drawings. In the absence of specific bracing requirements, trusses shall be braced in accordance with accepted industry practices, such as, the SBCA *Building Component Safety Information (BCSI) Guide to Good Practice for Handling, Installing & Bracing of Metal Plate Connected Wood Trusses.*

❖ To prevent collapse during construction and until permanent bracing is installed, trusses should be adequately braced temporarily. When braced for use, trusses should be positioned as vertical as possible; tilted trusses will not perform as required.

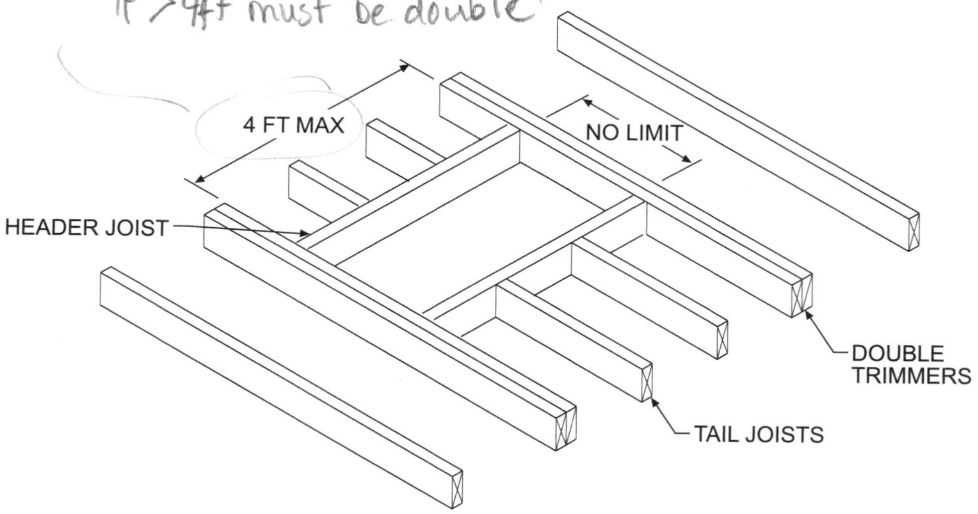

For SI: 1 foot = 304.8 mm.

Figure R502.10(1)
FLOOR FRAMING FOR MAXIMUM 4-FOOT OPENINGS

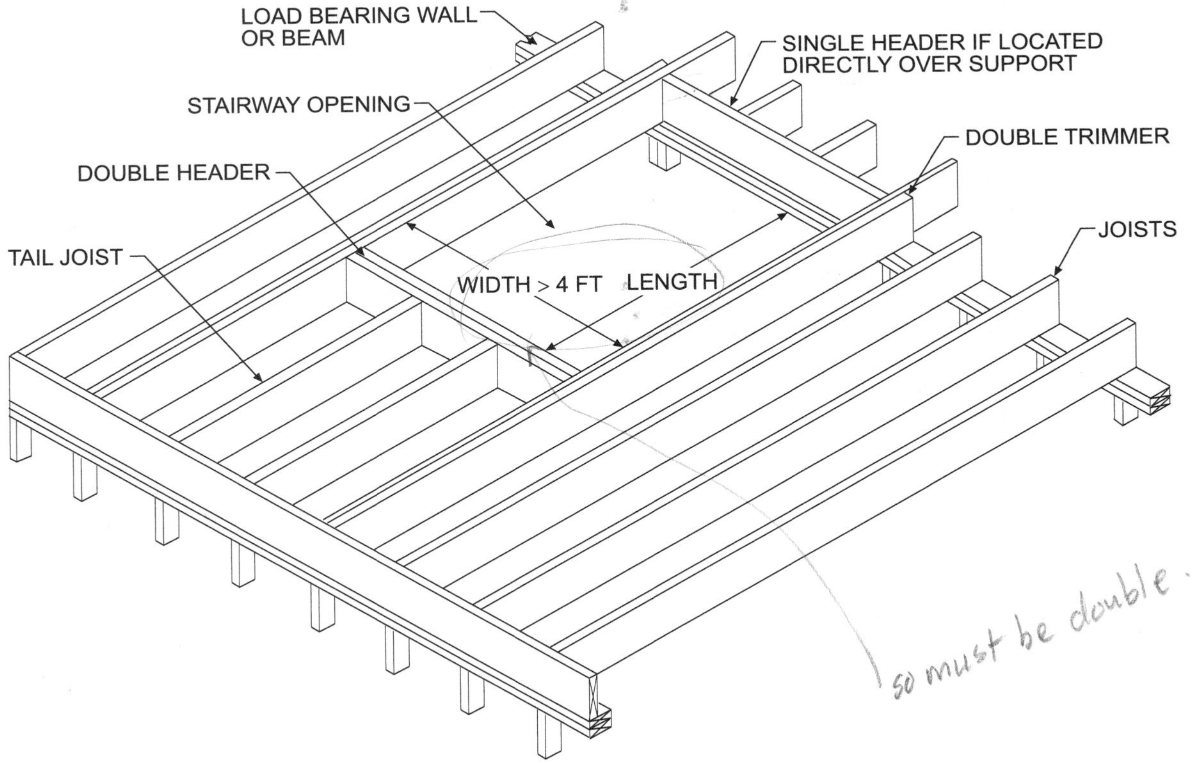

For SI: 1 foot = 304.8 mm.

Figure R502.10(2)
FLOOR FRAMING FOR GREATER THAN 4-FOOT OPENINGS

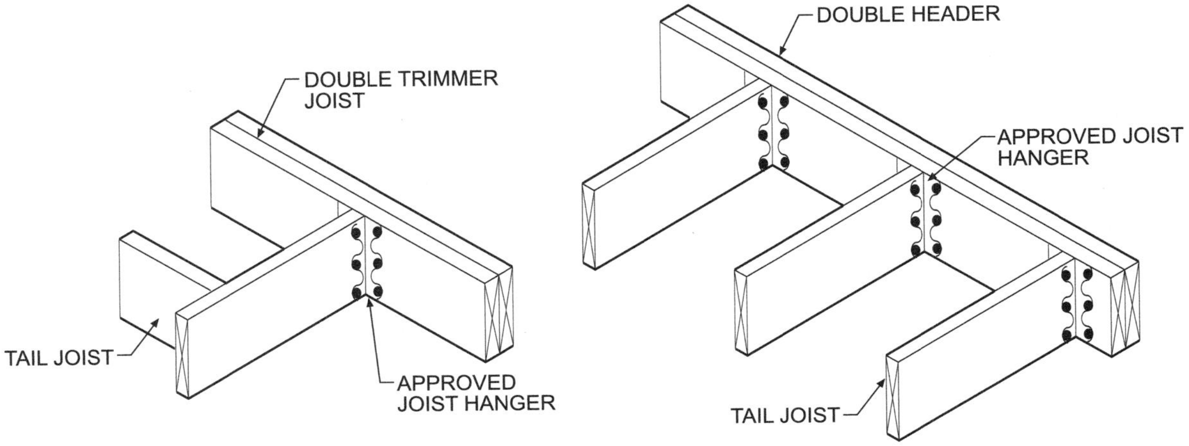

Figure R502.10(3)
HANGERS FOR JOIST-HEADER CONNECTION

R502.11.3 Alterations to trusses. Truss members and components shall not be cut, notched, spliced or otherwise altered in any way without the approval of a registered *design professional*. *Alterations* resulting in the addition of load (e.g., HVAC *equipment*, water heater, etc.), that exceed the design load for the truss, shall not be permitted without verification that the truss is capable of supporting the additional loading.

❖ Addition of loads in excess of the design load is allowed only if the additional capacity of the truss can be verified. Also see the commentary to Section R502.8.2.

R502.11.4 Truss design drawings. Truss design drawings, prepared in compliance with Section R502.11.1, shall be submitted to the *building official* and *approved* prior to installation. Truss design drawings shall be provided with the shipment of trusses delivered to the job site. Truss design drawings shall include, at a minimum, the information specified below:

1. Slope or depth, span and spacing.
2. Location of all joints.
3. Required bearing widths.
4. Design loads as applicable:
   - 4.1. Top chord live load;
   - 4.2. Top chord dead load;
   - 4.3. Bottom chord live load;
   - 4.4. Bottom chord dead load;
   - 4.5. Concentrated loads and their points of application; and
   - 4.6. Controlling wind and earthquake loads.
5. Adjustments to lumber and joint connector design values for conditions of use.
6. Each reaction force and direction.
7. Joint connector type and description, e.g., size, thickness or gauge, and the dimensioned location of each joint connector except where symmetrically located relative to the joint interface.
8. Lumber size, species and grade for each member.
9. Connection requirements for:
   - 9.1. Truss-to-girder-truss;
   - 9.2. Truss ply-to-ply; and
   - 9.3. Field splices.
10. Calculated deflection ratio and/or maximum description for live and total load.
11. Maximum axial compression forces in the truss members to enable the building designer to design the size, connections and anchorage of the permanent continuous lateral bracing. Forces shall be shown on the truss drawing or on supplemental documents.
12. Required permanent truss member bracing location.

❖ See the definition in Chapter 2 and the commentary to Section R502.11.1.

R502.12 Draftstopping required. Draftstopping shall be provided in accordance with Section R302.12.

❖ See the commentary to Section R302.12.

R502.13 Fireblocking required. Fireblocking shall be provided in accordance with Section R302.11.

❖ See the commentary to Section R302.11.

SECTION R503
FLOOR SHEATHING

R503.1 Lumber sheathing. Maximum allowable spans for lumber used as floor sheathing shall conform to Tables R503.1, R503.2.1.1(1) and R503.2.1.1(2).

❖ Table R503.1 sets forth the required thickness of lumber used as floor sheathing. The allowable spans are based on the floor sheathing thickness, joist spacing and the orientation of the sheathing with respect to the joist. Commentary Figures R503.1(1) and R503.1(2) show floor sheathing applications that are perpendicular and diagonal to the joist.

TABLE R503.1
MINIMUM THICKNESS OF LUMBER FLOOR SHEATHING

JOIST OR BEAM SPACING (inches)	MINIMUM NET THICKNESS	
	Perpendicular to joist	Diagonal to joist
24	$^{11}/_{16}$	$^{3}/_{4}$
16	$^{5}/_{8}$	$^{5}/_{8}$
48[a]	$1^{1}/_{2}$ T & G	N/A
54[b]		
60[c]		

For SI: 1 inch = 25.4 mm, 1 pound per square inch = 6.895 kPa.

N/A = Not applicable.

a. For this support spacing, lumber sheathing shall have a minimum F_b of 675 and minimum E of 1,100,000 (see AF&PA/NDS).

b. For this support spacing, lumber sheathing shall have a minimum F_b of 765 and minimum E of 1,400,000 (see AF&PA/NDS).

c. For this support spacing, lumber sheathing shall have a minimum F_b of 855 and minimum E of 1,700,000 (see AF&PA/NDS).

❖ See the commentary to Section R503.1.

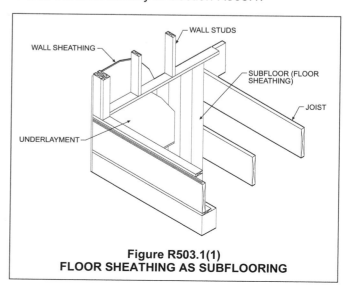

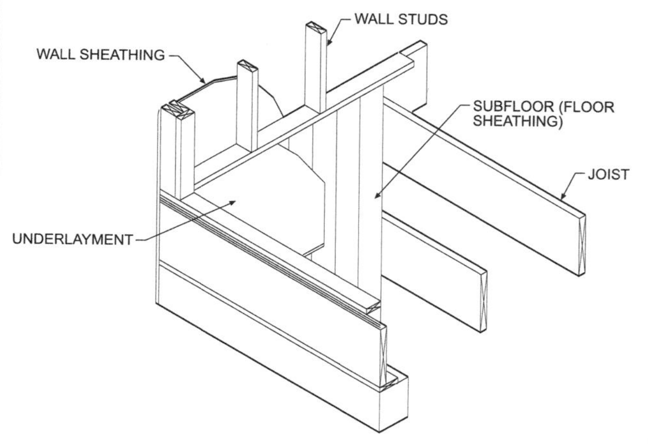

Figure R503.1(1)
FLOOR SHEATHING AS SUBFLOORING

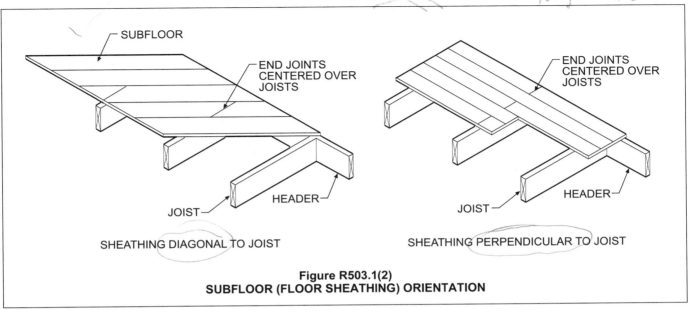

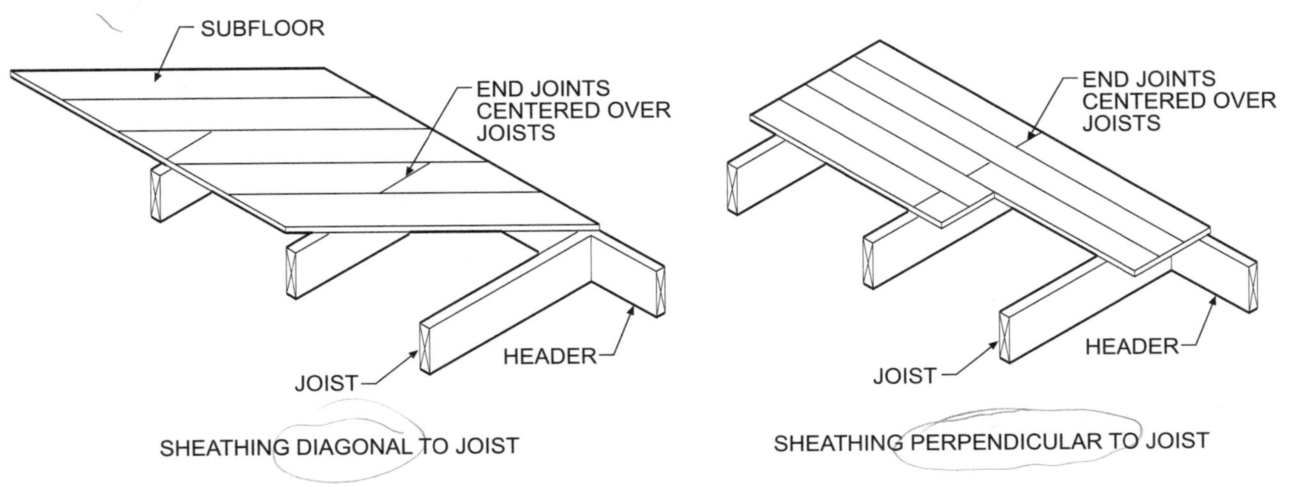

Figure R503.1(2)
SUBFLOOR (FLOOR SHEATHING) ORIENTATION

R503.1.1 End joints. End joints in lumber used as subflooring shall occur over supports unless end-matched lumber is used, in which case each piece shall bear on at least two joists. Subflooring may be omitted when joist spacing does not exceed 16 inches (406 mm) and a 1-inch (25.4 mm) nominal tongue-and-groove wood strip flooring is applied perpendicular to the joists.

❖ When lumber is to be used as subflooring, adequate load-carrying capability and continuity must be provided through either the proper placement of end joints over supports or the use of end-matched lumber as illustrated in Commentary Figure R503.1.1.

R503.2 Wood structural panel sheathing.

❖ Wood structural panels are manufactured with fully waterproof adhesive and include plywood, oriented strand board (OSB) and composite panels made up of a combination of wood veneers and reconstructed wood layers. Additional information on wood structural panels is available from the APA—The Engineered Wood Association (formerly the American Plywood Association), other trade groups and manufacturers.

Plywood is a wood structural panel manufactured by gluing three or more cross-laminated wood layers together. Plywood may be manufactured in accordance with PS 1 or PS 2. It is inspected, and if it is certified, the trademark of an approved testing and grading agency is stamped on it.

OSB panels are fabricated out of multiple layers of wood flakes or strands called "furnish." Like the layers in plywood, these layers of furnish are oriented 90 degrees (1.57 rad) to each other. This gives OSB panels properties that are very similar to those of plywood.

Composite panels are structural panels made up of a combination of wood veneers and wood-based materials. The wood veneer usually forms the two outer layers and may also be used in the core of the panel. These layers of veneer and furnish may be cross laminated.

Both OSB and composite panels are manufactured in accordance with PS 2. They are inspected, and if they are certified, the trademark of an approved testing and grading agency is stamped on them. The grade stamps of plywood, OSB and composite panels are almost identical in appearance, and in most cases, wood structural panels can be specified without regard to specific panel type.

Wood structural panels can be manufactured with Exterior, Exposure 1, Exposure 2 and Interior exposure durability classifications. Exterior and Exposure 1 are by far the most common classifications and both are manufactured with the same class of fully waterproof adhesives. Exterior panels are intended for applications subject to permanent exposure to the weather or moisture. Exposure 1 panels are intended for applications where long construction delays may be expected prior to protection being installed. Exposure 2 panels may be used for protected applications that are not continuously exposed to high humidity conditions. Interior panels may be used only for permanently protected interior applications. Exposure 2 and Interior panels are not readily available.

The intended end use of wood structural panels is designated in their grade-marking stamp.

R503.2.1 Identification and grade. Wood structural panel sheathing used for structural purposes shall conform to DOC PS 1, DOC PS 2 or, when manufactured in Canada, CSA O437 or CSA O325. All panels shall be identified for grade, bond classification, and Performance Category by a grade *mark* or certificate of inspection issued by an *approved agency*. The Performance Category value shall be used as the "nominal panel thickness" or "panel thickness" whenever referenced in this code.

❖ Plywood that performs a load-carrying function in floor construction must conform to a known quality control standard; therefore, a method of identifying the limitations under which the plywood may be used is needed. This is accomplished by a grade mark or certificate of inspection issued by an approved agency. Examples of plywood grade marks are shown in Commentary Figure R503.2.1.

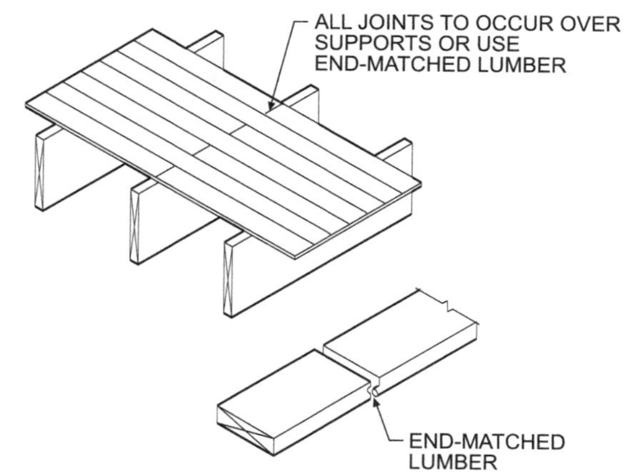

Figure R503.1.1
SUBFLOOR (FLOOR SHEATHING ORIENTATION)

R503.2.1.1 Subfloor and combined subfloor underlayment. Where used as subflooring or combination subfloor underlayment, wood structural panels shall be of one of the grades specified in Table R503.2.1.1(1). When sanded plywood is used as combination subfloor underlayment, the grade, bond classification, and Performance Category shall be as specified in Table R503.2.1.1(2).

❖ This section specifies grades of wood structural panels when they are used as subflooring or combination subflooring underlayment. It also addresses the use of sanded plywood as a combination subfloor underlayment.

A floor system could consist of three elements: (1) a subfloor, (2) an underlayment or a combination subfloor/underlayment system and (3) a finished floor surface material. A finished surfacing material may be wood-strip flooring; tongue-and-groove flooring or various types of resilient flooring coverings such as vinyl, tile or carpeting.

Wood structural panels are manufactured for use as either structural subfloor or combination subfloor underlayment. The allowable spans for structural subflooring and combination subfloor underlayment are based on the wood structural panels' face grain strength axis parallel to its supporting member, or they are based on the panels being continuous over two or more spans, with a face grain placed perpendicular to the supports. These qualifications are critical in determining the permissible spans. Most wood

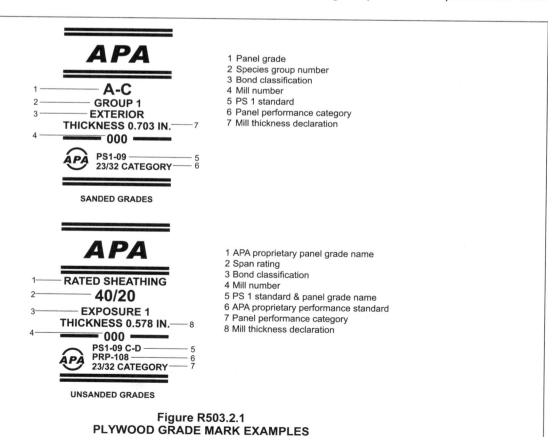

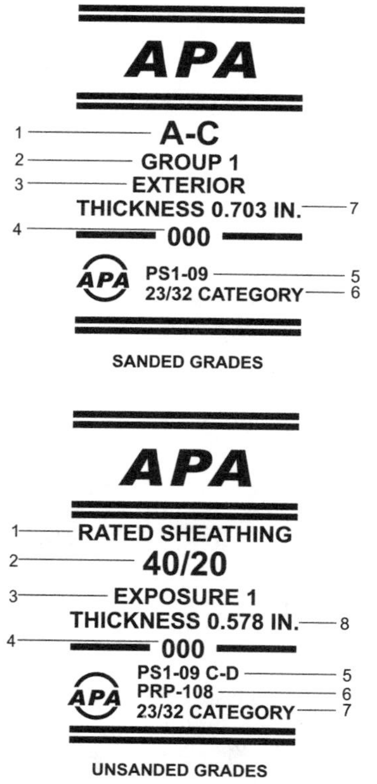

1 Panel grade
2 Species group number
3 Bond classification
4 Mill number
5 PS 1 standard
6 Panel performance category
7 Mill thickness declaration

1 APA proprietary panel grade name
2 Span rating
3 Bond classification
4 Mill number
5 PS 1 standard & panel grade name
6 APA proprietary performance standard
7 Panel performance category
8 Mill thickness declaration

Figure R503.2.1
PLYWOOD GRADE MARK EXAMPLES

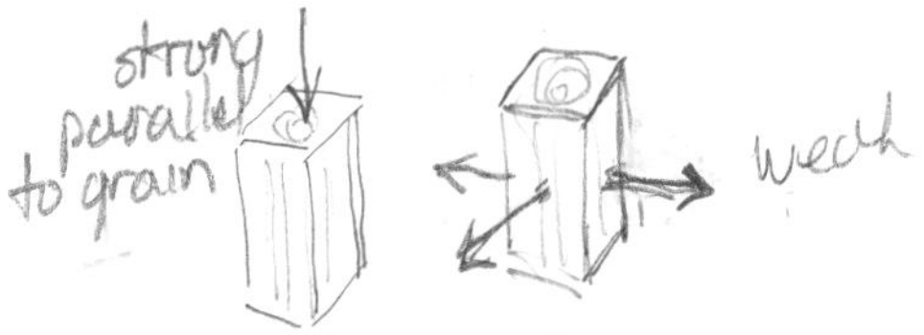

structural panels are considerably stronger when their face grain is parallel to the supports and continuous over two or more spans. Panels with multiple spans have greater capacity than those that are simply supported between the two joists.

TABLE R503.2.1.1(1)
ALLOWABLE SPANS AND LOADS FOR WOOD STRUCTURAL PANELS FOR ROOF AND SUBFLOOR SHEATHING AND COMBINATION SUBFLOOR UNDERLAYMENT[a, b, c]

SPAN RATING	MINIMUM NOMINAL PANEL THICKNESS (inch)	ALLOWABLE LIVE LOAD (psf)[h, l]		MAXIMUM SPAN (inches)		LOAD (pounds per square foot, at maximum span)		MAXIMUM SPAN (inches)
		SPAN @ 16″ o.c.	SPAN @ 24″ o.c.	With edge support[d]	Without edge support	Total load	Live load	
Sheathing[e]				Roof[f]				Subfloor[j]
16/0	3/8	30	—	16	16	40	30	0
20/0	3/8	50	—	20	20	40	30	0
24/0	3/8	100	30	24	20[g]	40	30	0
24/16	7/16	100	40	24	24	50	40	16
32/16	15/32, 1/2	180	70	32	28	40	30	16[h]
40/20	19/32, 5/8	305	130	40	32	40	30	20[h, i]
48/24	23/32, 3/4	—	175	48	36	45	35	24
60/32	7/8	—	305	60	48	45	35	32
Underlayment, C-C plugged, single floor[e]				Roof[f]				Combination subfloor underlayment[k]
16 o.c.	19/32, 5/8	100	40	24	24	50	40	16[i]
20 o.c.	19/32, 5/8	150	60	32	32	40	30	20[i, j]
24 o.c.	23/32, 3/4	240	100	48	36	35	25	24
32 o.c.	7/8	—	185	48	40	50	40	32
48 o.c.	1 3/32, 1 1/8	—	290	60	48	50	40	48

For SI: 1 inch = 25.4 mm, 1 pound per square foot = 0.0479 kPa.

a. The allowable total loads were determined using a dead load of 10 psf. If the dead load exceeds 10 psf, then the live load shall be reduced accordingly.

b. Panels continuous over two or more spans with long dimension (strength axis) perpendicular to supports. Spans shall be limited to values shown because of possible effect of concentrated loads.

c. Applies to panels 24 inches or wider.

d. Lumber blocking, panel edge clips (one midway between each support, except two equally spaced between supports when span is 48 inches), tongue-and-groove panel edges, or other approved type of edge support.

e. Includes Structural 1 panels in these grades.

f. Uniform load deflection limitation: $^{1}/_{180}$ of span under live load plus dead load, $^{1}/_{240}$ of span under live load only.

g. Maximum span 24 inches for $^{15}/_{32}$-and $^{1}/_{2}$-inch panels.

h. Maximum span 24 inches where $^{3}/_{4}$-inch wood finish flooring is installed at right angles to joists.

i. Maximum span 24 inches where 1.5 inches of lightweight concrete or approved cellular concrete is placed over the subfloor.

j. Unsupported edges shall have tongue-and-groove joints or shall be supported with blocking unless minimum nominal $^{1}/_{4}$-inch thick underlayment with end and edge joints offset at least 2 inches or 1.5 inches of lightweight concrete or approved cellular concrete is placed over the subfloor, or $^{3}/_{4}$-inch wood finish flooring is installed at right angles to the supports. Allowable uniform live load at maximum span, based on deflection of $^{1}/_{360}$ of span, is 100 psf.

k. Unsupported edges shall have tongue-and-groove joints or shall be supported by blocking unless nominal $^{1}/_{4}$-inch-thick underlayment with end and edge joints offset at least 2 inches or $^{3}/_{4}$-inch wood finish flooring is installed at right angles to the supports. Allowable uniform live load at maximum span, based on deflection of $^{1}/_{360}$ of span, is 100 psf, except panels with a span rating of 48 on center are limited to 65 psf total uniform load at maximum span.

l. Allowable live load values at spans of 16" o.c. and 24" o.c taken from reference standard APA E30, APA Engineered Wood Construction Guide. Refer to reference standard for allowable spans not listed in the table.

❖ The maximum spans for wood structural panel floor and roof sheathing are limited by the stresses and deflection imposed by the design live loads. For convenience, the trademarks of the inspection agencies include a span rating, which appears as two numbers separated by a slash (32/16 or 48/24, for example). The first number represents the maximum recommended span for roof sheathing when the panels are applied with the long dimension (strength axis) across three or more supports and the edges are blocked or when other support required by the table is provided. The second number indicates the maximum recommended span when the panel is used for structural floor sheathing with the panels applied with the long dimension (strength axis) across three or more supports. Single panels intended for single floor applications will be marked with a single number representing this span.

The edges of the wood structural panels between floor supports are prevented from moving relative to each other by tongue and groove panel edges, by the addition of wood blocking or by the use of an approved underlayment or structural finished floor system.

TABLE R503.2.1.1(2)
ALLOWABLE SPANS FOR SANDED PLYWOOD COMBINATION SUBFLOOR UNDERLAYMENT[a]

IDENTIFICATION	SPACING OF JOISTS (inches)		
	16	20	24
Species group[b]	—	—	—
1	$^1/_2$	$^5/_8$	$^3/_4$
2, 3	$^5/_8$	$^3/_4$	$^7/_8$
4	$^3/_4$	$^7/_8$	1

For SI: 1 inch = 25.4 mm, 1 pound per square foot = 0.0479 kPa.

a. Plywood continuous over two or more spans and face grain perpendicular to supports. Unsupported edges shall be tongue-and-groove or blocked except where nominal $^1/_4$-inch-thick underlayment or $^3/_4$-inch wood finish floor is used. Allowable uniform live load at maximum span based on deflection of $^1/_{360}$ of span is 100 psf.

b. Applicable to all grades of sanded exterior-type plywood.

❖ For exterior-type sanded plywood grades, the thickness required for a specific span is related to the spe cies grouping of the panel used. For example, species Group 1 may be $^1/_2$ inch (12.7 mm) for 16-inch (406 mm) joist spacing, but Group 4 will require a thickness of $^3/_4$ inch (19.1 mm) for the same span.

When the panels are used as a combination subfloor/underlayment, extra precautions should be observed in attaching the panels to the floor framing. Joints in adjacent panels should not be continuous and should not occur at locations where the orientation of the joist supports is different.

R503.2.2 Allowable spans. The maximum allowable span for wood structural panels used as subfloor or combination subfloor underlayment shall be as set forth in Table R503.2.1.1(1), or APA E30. The maximum span for sanded plywood combination subfloor underlayment shall be as set forth in Table R503.2.1.1(2).

❖ Table R503.2.1.1(1) indicates spans for plywood and wood structural panels used as subflooring. This table also covers roof sheathing and is referenced in Section R803.2.2 for that purpose. The span limitations are predicated on the grade of plywood used. In the case of rated sheathing used as structural subflooring, the maximum span is easily identified through the panel span rating (identification index). Where the panel span rating is stamped on the sheet of plywood, the denominator represents the allowable span of the plywood floor sheathing [see Commentary Figure R503.2.2(1)]. An illustration of the limitations imposed by Notes j and k to Table R503.2.1.1(1) on combination subfloor underlayment applications is illustrated in Commentary Figure R503.2.2(2). Allowable increases in the maximum span for certain span ratings are permitted as described in Notes h and i to Table R503.2.1.1(1).

In lieu of using Table R503.2.1.1(1), the code permits the use of APA E30, *Engineered Wood Construction Guide*.

When plywood combination subfloor underlayment is to be used, the allowable span as listed in Table R503.2.1.1(2) is applicable to underlayment grade C-C plugged and sanded exterior-type plywood of specific species.

R503.2.3 Installation. Wood structural panels used as subfloor or combination subfloor underlayment shall be attached to wood framing in accordance with Table R602.3(1) and shall be attached to cold-formed steel framing in accordance with Table R505.3.1(2).

❖ This section refers to the wood framing fastener schedule in Chapter 6 and the cold-formed steel floor-fastening schedule in Table R505.3.1(2).

R503.3 Particleboard.

❖ Particleboard is a generic term given to panels manufactured from cellulosic materials, usually wood in the form of discrete pieces and particles rather than fibers. The particles and pieces are combined with synthetic resins and other binders and bonded together under heat and pressure. Particleboard must conform to ANSI A208.1.

R503.3.1 Identification and grade. Particleboard shall conform to ANSI A208.1 and shall be so identified by a grade *mark* or certificate of inspection issued by an *approved agency*.

❖ Particleboard used in floor construction that performs a load-carrying function must conform to a known quality control standard. To this end, a grade mark or certificate of inspection issued by an approved agency identifies the limitations under which the particleboard may be used. Commentary Figure R503.3.1 shows an example of a particleboard grade mark.

The particleboard grades in ANSI A208.1 are identified by a letter followed by a hyphen and a digit or another letter. A letter, number or term following the grade designation identifies special performance characteristics. The first letter has the following meaning:

H: High density (above 50 lb/ft^3) (801 kg/m^3),

M: Medium density (40-50 lb/ft^3) (641-801 kg/m^3),

LD: Low density (below 40 lb/ft^3) (641 kg/m^3),

D: Manufactured home decking

PBU: Underlayment.

The digit or letter following the hyphen indicates the grade identification within a particular description. For example, 2 indicates medium-density particleboard, Grade 2.

R503.3.2 Floor underlayment. Particleboard floor underlayment shall conform to Type PBU and shall not be less than $^1/_4$ inch (6.4 mm) in thickness.

❖ Particleboard is often used over a structural subfloor to provide a smooth, even surface under textile or resilient-type finish floors. Particleboard must conform to Type P13U with a minimum thickness of $^1/_4$ inch (6.4 mm).

R503.3.3 Installation. Particleboard underlayment shall be installed in accordance with the recommendations of the manufacturer and attached to framing in accordance with Table R602.3(1).

❖ Particleboard used as underlayment only is not subject to span limitations because the plywood or lumber subflooring provides the load-carrying capability for the floor system; the particleboard has a nonstructural role. However, joints in particleboard should not occur over joints in the subfloor; the joints should be staggered.

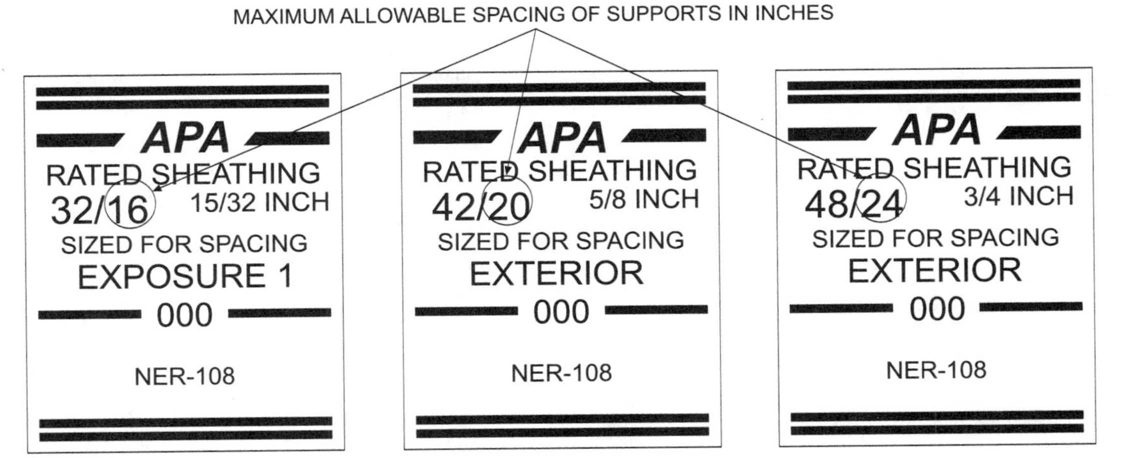

For SI: 1 inch = 25.4 mm.

Figure R503.2.2(1)
IDENTIFICATION OF PLYWOOD SUBFLOOR SPAN LIMITATIONS

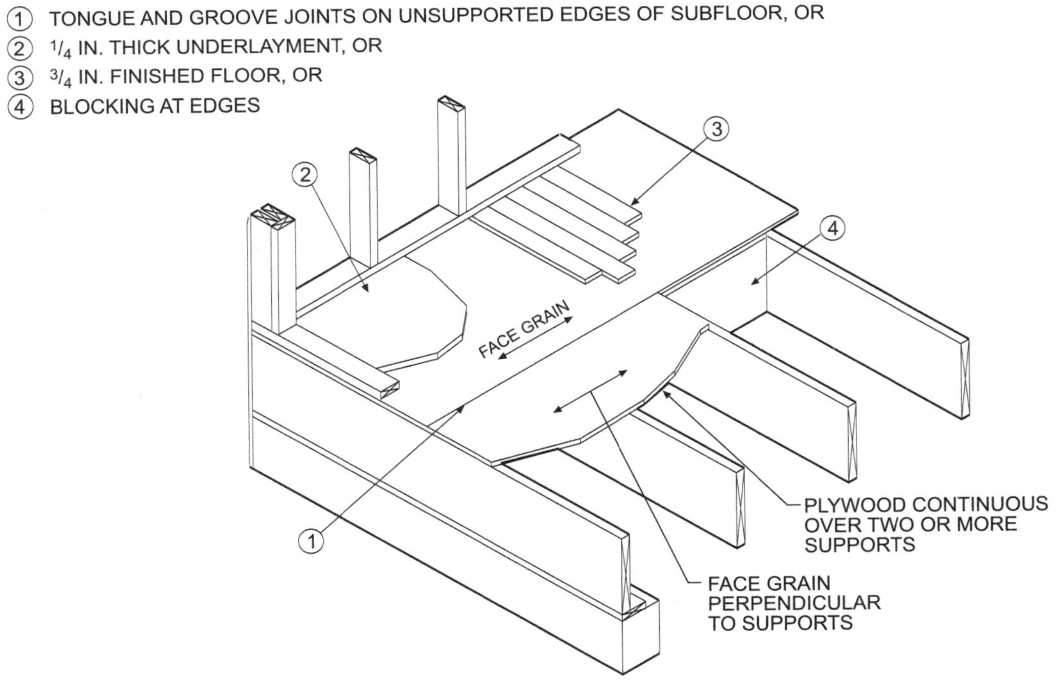

For SI: 1 inch = 25.4 mm.

Figure R503.2.2(2)
COMBINATION SUBFLOOR UNDERLAYMENT ALTERNATIVES

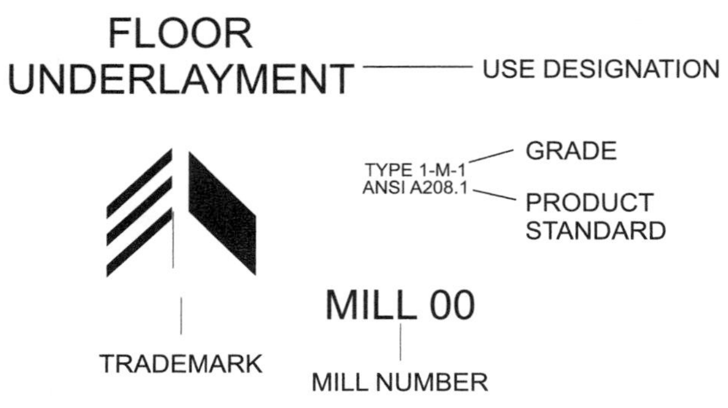

Figure R503.3.1
PARTICLEBOARD GRADE MARK EXAMPLE

SECTION R504
PRESSURE PRESERVATIVELY TREATED-WOOD FLOORS (ON GROUND)

R504.1 General. Pressure preservatively treated-wood *basement* floors and floors on ground shall be designed to withstand axial forces and bending moments resulting from lateral soil pressures at the base of the exterior walls and floor live and dead loads. Floor framing shall be designed to meet joist deflection requirements in accordance with Section R301.

❖ As indicated in the commentary for Section R404.1.1, lateral soil pressure loads are carried by the foundation walls through bending in the vertical direction. The resulting forces (reactions) are resisted by the basement floor and the first floor above. This section stipulates the requirements for basement wood-floor framing in order for it to resist the applied loads at the base of the foundation wall. It is implicit that this type of floor would be used in conjunction with wood foundations described in Sections R401.1.

R504.1.1 Unbalanced soil loads. Unless special provision is made to resist sliding caused by unbalanced lateral soil loads, wood *basement* floors shall be limited to applications where the differential depth of fill on opposite exterior foundation walls is 2 feet (610 mm) or less.

❖ The requirements of foundation walls permit up to 4 feet (1219 mm) of unbalanced fill (see Table R404.2.3). When exterior foundation walls on opposite sides of a building support an equal (or nearly equal) height of unbalanced fill, sliding is not a concern. When there is a difference in the amount of fill supported, however, there is a net horizontal force that must be resisted; otherwise, sliding may occur. To limit this unbalanced force, the height of differential fill is accordingly limited to 2 feet (610 mm) for this floor system.

R504.1.2 Construction. Joists in wood *basement* floors shall bear tightly against the narrow face of studs in the foundation wall or directly against a band joist that bears on the studs. Plywood subfloor shall be continuous over lapped joists or over butt joints between in-line joists. Sufficient blocking shall be provided between joists to transfer lateral forces at the base of the end walls into the floor system.

❖ These construction requirements provide a load path for the horizontal force described in the previous section.

R504.1.3 Uplift and buckling. Where required, resistance to uplift or restraint against buckling shall be provided by interior bearing walls or properly designed stub walls anchored in the supporting soil below.

❖ This is a general caution that the effects of uplift forces on the floor system, which could result from fluctuations in the water table, should be considered.

R504.2 Site preparation. The area within the foundation walls shall have all vegetation, topsoil and foreign material removed, and any fill material that is added shall be free of vegetation and foreign material. The fill shall be compacted to assure uniform support of the pressure preservatively treated-wood floor sleepers.

❖ This requirement is similar to Section R506.2. See the commentary to Section R506.2.2.

R504.2.1 Base. A minimum 4-inch-thick (102 mm) granular base of gravel having a maximum size of $^3/_4$ inch (19.1 mm) or crushed stone having a maximum size of $^1/_2$ inch (12.7 mm) shall be placed over the compacted earth.

❖ This requirement is similar to Section R506.2.2. See the commentary to Section R506.2.2.

R504.2.2 Moisture barrier. Polyethylene sheeting of minimum 6-mil (0.15 mm) thickness shall be placed over the granular base. Joints shall be lapped 6 inches (152 mm) and left unsealed. The polyethylene membrane shall be placed over the pressure preservatively treated-wood sleepers and shall not extend beneath the footing plates of the exterior walls.

❖ See the commentary to Section R506.2.2 concerning application of the moisture barrier.

R504.3 Materials. All framing materials, including sleepers, joists, blocking and plywood subflooring, shall be pressure-preservative treated and dried after treatment in accordance with AWPA U1 (Commodity Specification A, Use Category 4B and Section 5.2), and shall bear the *label* of an accredited agency.

❖ To resist potential decay, wood floor material is to be treated in accordance with AWPA U1, use category system: *User Specification for Treated Wood*. The wood materials must be treated as Use Category 413 (UC413), Commodity Specification A, in accordance with AWPA U1. Also the material must bear the label of an accredited agency to ensure that the treated wood is produced under an accredited quality auditing program.

SECTION R505
STEEL FLOOR FRAMING

R505.1 Cold-formed steel floor framing. Elements shall be straight and free of any defects that would significantly affect structural performance. Cold-formed steel floor framing members shall comply with the requirements of this section.

❖ The provisions of this section apply to the construction of floor systems using cold-formed steel framing, which is a type of construction made up in part or entirely of steel structural members cold formed to shape from sheet or strip steel. The general shape of the framing members is in the form of the letter "C" (C-shape) and is described in further detail in Section R505.2. The use of cold-formed steel-framed designs not consistent with these provisions is beyond the scope of this code, and an analysis by a design professional is recommended.

R505.1.1 Applicability limits. The provisions of this section shall control the construction of cold-formed steel floor framing for buildings not greater than 60 feet (18 288 mm) in length perpendicular to the joist span, not greater than 40 feet (12 192 mm) in width parallel to the joist span, and less than or equal to three stories above *grade* plane. Cold-formed steel floor framing constructed in accordance with the provisions of this section shall be limited to sites subjected to a maximum design wind speed of 110 miles per hour (49 m/s), Exposure B or C, and a maximum ground snow load of 70 pounds per square foot (3.35 kPa).

❖ Commentary Figure R505.1.1 depicts building limitations. These limits are repeated in Section R603.1.1 for walls and in Section R804.1.1 for roofs. Cold-formed steel floor framing constructed in accordance with the provisions of this section must be limited to sites with a maximum design wind speed of 110 miles per hour (49 m/s); Exposure B, or C and a maximum ground snow load of 70 pounds per square foot (3.35 kPa).

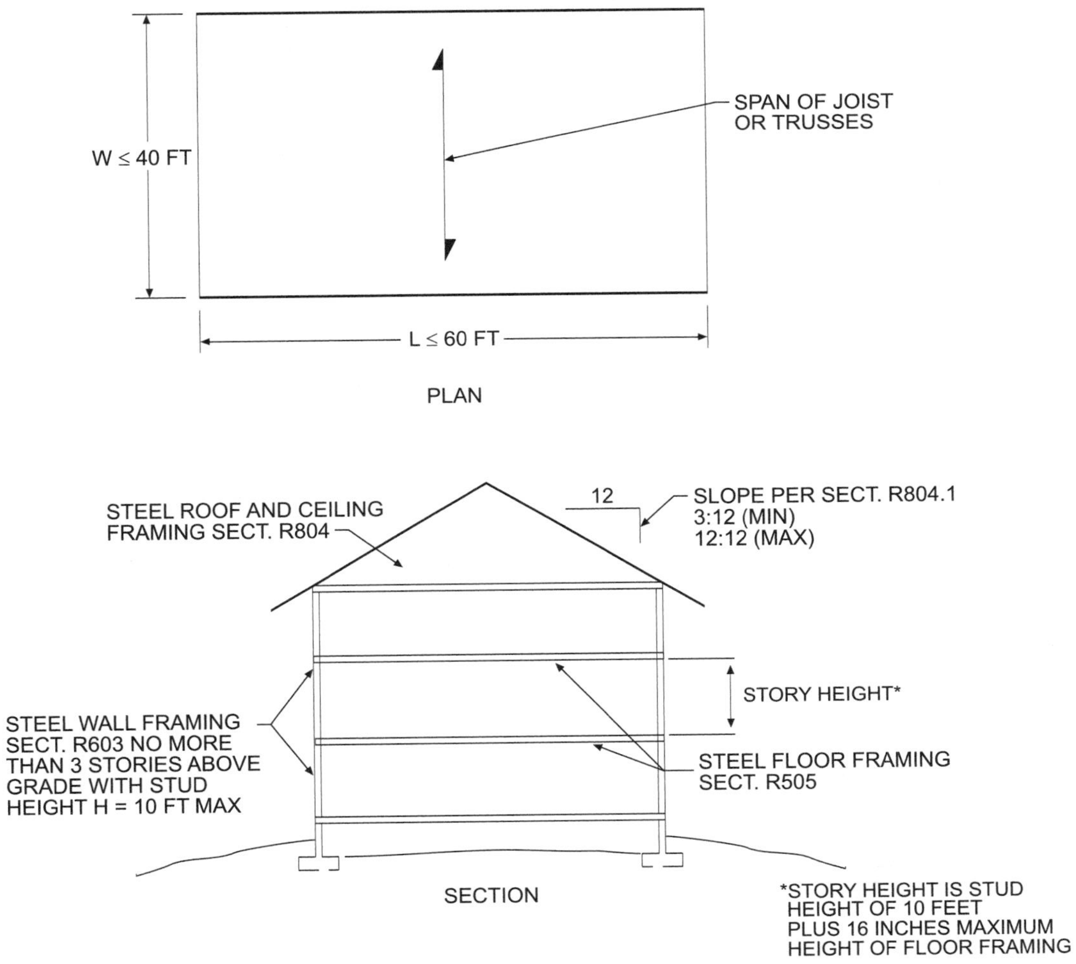

For SI: 1 foot = 304.8 mm.

Figure R505.1.1
STEEL FRAMING APPLICABILITY LIMITS

R505.1.2 In-line framing. When supported by cold-formed steel framed walls in accordance with Section R603, cold-formed steel floor framing shall be constructed with floor joists located in-line with load-bearing studs located below the joists in accordance with Figure R505.1.2 and the tolerances specified as follows:

1. The maximum tolerance shall be $^3/_4$ inch (19.1 mm) between the centerline of the horizontal framing member and the centerline of the vertical framing member.
2. Where the centerline of the horizontal framing member and bearing stiffener are located to one side of the centerline of the vertical framing member, the maximum tolerance shall be $^1/_8$ inch (3 mm) between the web of the horizontal framing member and the edge of the vertical framing member.

❖ In-line framing is the preferred framing method because it provides a direct load path for the transfer of forces from joists to studs. In-line framing maximizes framing alignment in order to minimize secondary moments on the framing members, taking into account that the track cannot function as a load-transfer member. The $^3/_4$-inch (19 mm) alignment value is accepted industry practice and addresses those conditions where two framing members (e.g., roof rafter and ceiling joist) are to be aligned over a wall stud. Where the bearing stiffener is located on the backside of the joist, the web of the joist cannot be more than $^1/_8$ inch (3 mm) from the edge of the wall stud. In the absence of in-line framing, a load distribution member, such as a structural track, may be required for this force transfer.

R505.1.3 Floor trusses. Cold-formed steel trusses shall be designed, braced and installed in accordance with AISI S100, Section D4. In the absence of specific bracing requirements, trusses shall be braced in accordance with accepted industry practices, such as the SBCA *Cold-Formed Steel Building Component Safety Information (CFSBCSI), Guide to Good Practice for Handling, Installing & Bracing of Cold-Formed Steel Trusses*. Truss members shall not be notched, cut or altered in any manner without an *approved* design.

❖ The code contains no prescriptive provisions for the design and installation of cold-formed steel trusses. A design is required that complies with accepted engineering practice. Recognizing the extensive use of trusses in residential construction, the code references AISI S100, Section D4. Section D4 of AISI S100 specifies the criteria for the design of cold-formed steel light-frame construction and requires trusses to be designed in accordance with AISI S214. The standard provides regulations for the design and installation of cold-formed steel trusses, including

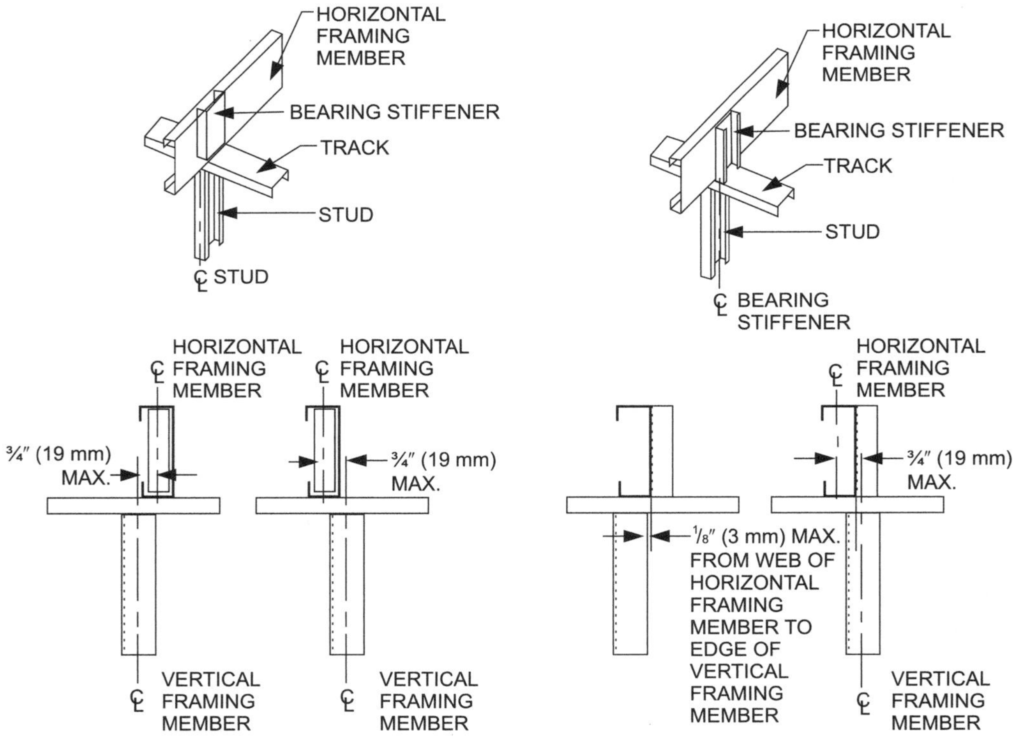

For SI: 1 inch = 25.4 mm.

FIGURE R505.1.2
IN-LINE FRAMING

❖ See the commentary to Section R505.1.2.

quality assurance and the procedures for full-scale tests. In addition to adequate design, it is important that the handling and erection of the trusses be performed properly so that the performance capability of the trusses is not compromised. A truss member must never be cut, notched or altered without approval from the design professional.

R505.2 Structural framing. Load-bearing cold-formed steel floor framing members shall comply with Figure R505.2(1) and with the dimensional and minimum thickness requirements specified in Tables R505.2(1) and R505.2(2). Tracks shall comply with Figure R505.2(2) and shall have a minimum flange width of $1^1/_4$ inches (32 mm).

❖ Joists must comply with the dimensional and minimum thickness requirements of Tables R505.2(1) and R505.2(2), respectively. Note a to Table R505.2(1) explains the meaning of the alphanumeric member designation used for all steel framing members in the code. This system replaces the varied designation approaches that were used by each individual manufacturer. In addition, the designation is used to identify not only a specific steel framing member, but also the section properties of that same member through the use of the product technical information document.

Steel thickness is expressed in mils, which pertains to the base metal thickness measured prior to painting or the application of metallic (corrosion-resistant) coatings. The base metal thickness is typically stamped or embossed on the member by the manufacturer. Table R505.2(2) also references the equivalent thickness in inches.

Steel track sections are to serve in nonload-bearing applications only and are shown in Figure R505.2(2). They must have a minimum flange width of $1^1/_4$ inches (32 mm).

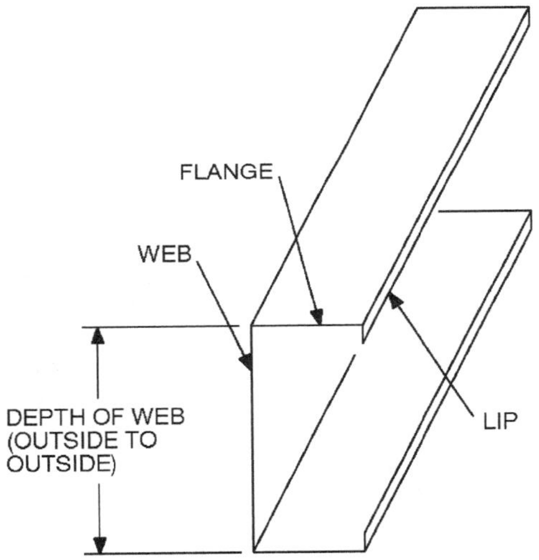

FIGURE R505.2(1)
C-SHAPED SECTION

❖ See the commentary to Section R505.2 for Figure R505.2(1).

TABLE R505.2(2)
MINIMUM THICKNESS OF COLD-FORMED STEEL MEMBERS

DESIGNATION THICKNESS (mils)	MINIMUM BASE STEEL THICKNESS (inches)
33	0.0329
43	0.0428
54	0.0538
68	0.0677
97	0.0966

For SI: inch = 25.4 mm, 1 mil = 0.0254 mm.

❖ See the commentary for Section R505.2.

R505.2.1 Material. Load-bearing cold-formed steel framing members shall be cold formed to shape from structural quality sheet steel complying with the requirements of one of the following:

1. ASTM A 653: Grades 33 and 50 (Class 1 and 3).
2. ASTM A 792: Grades 33 and 50A.
3. ASTM A 1003: Structural Grades 33 Type H and 50 Type H.

❖ Load-bearing steel framing members must have a legible label, stencil stamp or embossment. This identification allows for verification that materials installed are consistent with the design and meet the provisions of the code.

TABLE R505.2(1)
COLD-FORMED STEEL JOIST SIZES

MEMBER DESIGNATION[a]	WEB DEPTH (inches)	MINIMUM FLANGE WIDTH (inches)	MAXIMUM FLANGE WIDTH (inches)	MINIMUM LIP SIZE (inches)
550S162-t	5.5	1.625	2	0.5
800S162-t	8	1.625	2	0.5
1000S162-t	10	1.625	2	0.5
1200S162-t	12	1.625	2	0.5

For SI: 1 inch = 25.4 mm, 1 mil = 0.0254 mm.

a. The member designation is defined by the first number representing the member depth in 0.01 inch, the letter "S" representing a stud or joist member, the second number representing the flange width in 0.01 inch, and the letter "t" shall be a number representing the minimum base metal thickness in mils [See Table R505.2(2)].

❖ See the commentary to Section R505.2.

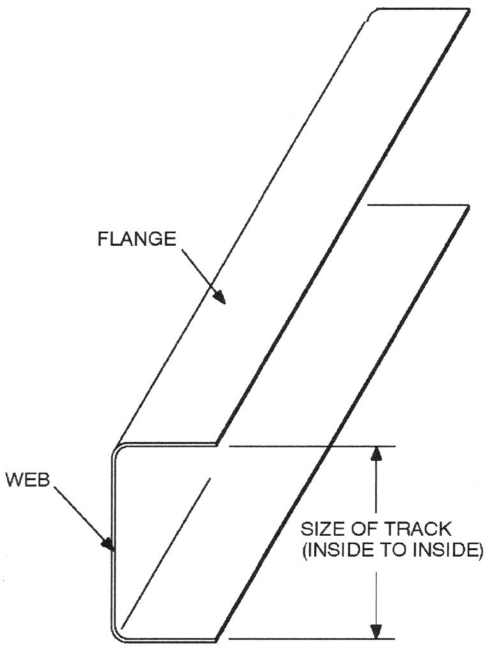

FIGURE R505.2(2)
TRACK SECTION

❖ See the commentary to Section R505.2 for Figure R505.2(2).

R505.2.2 Identification. Load-bearing cold-formed steel framing members shall have a legible *label*, stencil, stamp or embossment with the following information as a minimum:

1. Manufacturer's identification.
2. Minimum base steel thickness in inches (mm).
3. Minimum coating designation.
4. Minimum yield strength, in kips per square inch (ksi) (MPa).

❖ Load-bearing steel framing members must have a legible label, stencil, stamp or embossment. This allows for verification that the materials installed are consistent with the design and meet the code requirements.

R505.2.3 Corrosion protection. Load-bearing cold-formed steel framing shall have a metallic coating complying with ASTM A 1003 and one of the following:

1. A minimum of G 60 in accordance with ASTM A 653.
2. A minimum of AZ 50 in accordance with ASTM A 792.

❖ The specified metallic coatings correspond to requirements in the referenced material standards. The minimum coating designations assume normal exposure conditions that are best defined as having the framing members enclosed within a building envelope or wall assembly within a controlled environment. When more severe exposure conditions are probable, such as industrial or marine atmospheres, consideration should be given to specifying a heavier coating.

R505.2.4 Fastening requirements. Screws for steel-to-steel connections shall be installed with a minimum edge distance and center-to-center spacing of $^1/_2$ inch (12.7 mm), shall be self-drilling tapping, and shall conform to ASTM C 1513. Floor sheathing shall be attached to cold-formed steel joists with minimum No. 8 self-drilling tapping screws that conform to ASTM C 1513. Screws attaching floor-sheathing to cold-formed steel joists shall have a minimum head diameter of 0.292 inch (7.4 mm) with countersunk heads and shall be installed with a minimum edge distance of $^3/_8$ inch (9.5 mm). Gypsum board ceilings shall be attached to cold-formed steel joists with minimum No. 6 screws conforming to ASTM C 954 or ASTM C 1513 with a bugle head style and shall be installed in accordance with Section R702. For all connections, screws shall extend through the steel a minimum of three exposed threads. All fasteners shall have rust inhibitive coating suitable for the installation in which they are being used, or be manufactured from material not susceptible to corrosion.

Where No. 8 screws are specified in a steel-to-steel connection, the required number of screws in the connection is permitted to be reduced in accordance with the reduction factors in Table R505.2.4 when larger screws are used or when one of the sheets of steel being connected is thicker than 33 mils (0.84 mm). When applying the reduction factor, the resulting number of screws shall be rounded up.

❖ Fasteners meeting the referenced standards and installed in accordance with this section provide the necessary load capacity consistent with these prescriptive provisions. This section specifies self-drilling tapping screws conforming to ASTM C 1513, *Specification for Steel Tapping Screws for Cold-formed Steel Framing Connection,* for all steel-to-steel connections as well as for fastening floor sheathing to steel joists. All fasteners must have rust-inhibitive coating or be manufactured from material not susceptible to corrosion. Gypsum board must be attached to steel joists with No. 6 screws (minimum) conforming to ASTM C 954, *Standard Specification for Steel Drill Screws for the Application of Gypsum Panel Products or Metal Plaster Bases to Steel Studs from 0.033 in. (0.84 mm) to 0.112 in. (2.84 mm) in Thickness,* or ASTM C 1513. See Section R702 for installation. For all connections, screws must extend through the steel a minimum of three exposed threads.

TABLE R505.2.4
SCREW SUBSTITUTION FACTOR

SCREW SIZE	THINNEST CONNECTED STEEL SHEET (mils)	
	33	43
#8	1.0	0.67
#10	0.93	0.62
#12	0.86	0.56

For SI: 1 mil = 0.0254 mm.

❖ In providing prescriptive connection requirements, the code typically specifies the quantity of No. 8

screws necessary to resist the required design forces transferred through a given connection. If the quantity of screws required for a particular connection is large, it may be desirable to reduce the quantity of screws by substituting a larger screw size. Where No. 8 screws are specified for a steel-to-steel connection, the required quantity of screws may be reduced if a larger screw size is substituted in accordance with the reduction factors of Table R505.2.4. Likewise, if the thinnest of the connected sheets is at least 43 mils (1.1 mm), the required quantity of screws may be reduced.

R505.2.5 Web holes, web hole reinforcing and web hole patching. Web holes, web hole reinforcing, and web hole patching shall be in accordance with this section.

❖ Sections R505.2.5.1 through R505.2.5.1.3 specify the criteria for web holes, web hole reinforcing, and web hole patching in floor joists.

R505.2.5.1 Web holes. Web holes in floor joists shall comply with all of the following conditions:

1. Holes shall conform to Figure R505.2.5.1;
2. Holes shall be permitted only along the centerline of the web of the framing member;
3. Holes shall have a center-to-center spacing of not less than 24 inches (610 mm);
4. Holes shall have a web hole width not greater than 0.5 times the member depth, or $2^1/_2$ inches (64.5 mm);
5. Holes shall have a web hole length not exceeding $4^1/_2$ inches (114 mm); and
6. Holes shall have a minimum distance between the edge of the bearing surface and the edge of the web hole of not less than 10 inches (254 mm).

Framing members with web holes not conforming to the above requirements shall be reinforced in accordance with Section R505.2.5.2, patched in accordance with Section R505.2.5.3 or designed in accordance with accepted engineering practices.

❖ To allow for routing utilities through steel-framed floors, holes (also referred to as penetrations, utility holes and punchouts) are permitted in the webs of floor joists. To avoid adversely affecting the strength of floor joists, the web holes are limited to the locations along the centerline of the web of the framing member, as illustrated in Figure R505.2.5.1.

To avoid overstressing the joist web through high shear, these web holes must occur more than 10 inches (254 mm) from the edge of a load-bearing surface. Patching of web holes must extend at least 1 inch (25 mm) beyond all edges of the hole, and it must be fastened to the web with No. 8 screws having a minimum edge distance of $^1/_2$ inch (12.7 mm). Framing members with web holes exceeding these limits must be reinforced, patched or designed in accordance with accepted engineering practice (see commentary, Sections R505.2.5.2 and R505.2.5.3).

R505.2.5.2 Web hole reinforcing. Reinforcement of web holes in floor joists not conforming to the requirements of Section R505.2.5.1 shall be permitted if the hole is located fully within the center 40 percent of the span and the depth and length of the hole does not exceed 65 percent of the flat width of the web. The reinforcing shall be a steel plate or C-shape section with a hole that does not exceed the web hole size limitations of Section R505.2.5.1 for the member being reinforced. The steel reinforcing shall be the same thickness as the receiving member and shall extend at least 1 inch (25.4 mm) beyond all edges of the hole. The steel reinforcing shall be fastened to the web of the receiving member with No. 8 screws spaced no more than 1 inch (25.4 mm) center-to-center along the edges of the patch with minimum edge distance of $^1/_2$ inch (12.7 mm).

❖ Web holes that are not within the criteria of Section R505.2.5.1 are permitted if they are reinforced and comply with the criteria of this section.

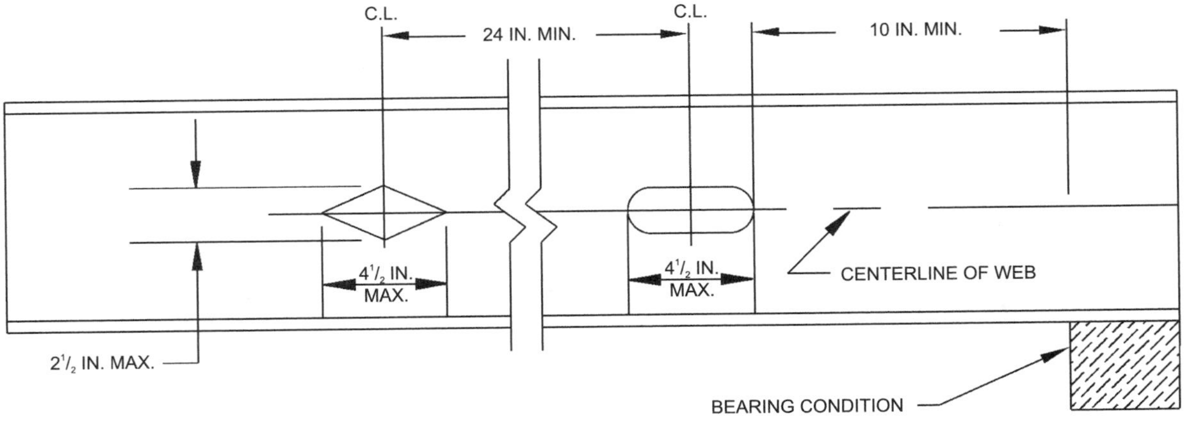

For SI: 1 inch = 25.4 mm.

FIGURE R505.2.5.1
FLOOR JOIST WEB HOLES

❖ See the commentary to Section R505.2.5.1.

The reinforcing provisions of this section were developed based on engineering judgement and accepted engineering practice and confirmed by testing.

R505.2.5.3 Hole patching. Patching of web holes in floor joists not conforming to the requirements in Section R505.2.5.1 shall be permitted in accordance with either of the following methods:

1. Framing members shall be replaced or designed in accordance with accepted engineering practices where web holes exceed the following size limits:
   1.1. The depth of the hole, measured across the web, exceeds 70 percent of the flat width of the web; or
   1.2. The length of the hole measured along the web, exceeds 10 inches (254 mm) or the depth of the web, whichever is greater.
2. Web holes not exceeding the dimensional requirements in Section R505.2.5.3, Item 1, shall be patched with a solid steel plate, stud section, or track section in accordance with Figure R505.2.5.3. The steel patch shall, as a minimum, be of the same thickness as the receiving member and shall extend at least 1 inch (25 mm) beyond all edges of the hole. The steel patch shall be fastened to the web of the receiving member with No. 8 screws spaced no more than 1 inch (25 mm) center-to-center along the edges of the patch with minimum edge distance of $^1/_2$ inch (13 mm).

❖ Framing members with web holes exceeding the limits, outlined in Item 1 of this section must be replaced or designed in accordance with accepted engineering practice. Framing members with web holes within the limits of Item 1, but exceeding the limits of Section R505.2.5.1, must be patched in accordance with Item 2 of this section (see commentary, Section R505.2.5.1).

R505.3 Floor construction. Cold-formed steel floors shall be constructed in accordance with this section.

❖ Load-bearing steel-framing members must be cold formed to shape from structural quality sheet steel complying with appropriate material standards. The steel grades specified have the ductility and strength to meet these provisions. For more detailed information refer to the AISI *Specification for Design of Cold-formed Steel Structural Members Commentary.*

R505.3.1 Floor to foundation or load-bearing wall connections. Cold-formed steel framed floors shall be anchored to foundations, wood sills or load-bearing walls in accordance with Table R505.3.1(1) and Figure R505.3.1(1), R505.3.1(2), R505.3.1(3), R505.3.1(4), R505.3.1(5) or R505.3.1(6). Anchor bolts shall be located not more than 12 inches (305 mm) from corners or the termination of bottom tracks. Continuous cold-formed steel joists supported by interior load-bearing walls shall be constructed in accordance with Figure R505.3.1(7). Lapped cold-formed steel joists shall be constructed in accordance with Figure R505.3.1(8). End floor joists constructed on foundation walls parallel to the joist span shall be doubled unless a C-shaped bearing stiffener, sized in accordance with Section R505.3.4, is installed web-to-web with the floor joist beneath each supported wall stud, as shown in Figure R505.3.1(9). Fastening of cold-formed steel joists to other framing members shall be in accordance with Section R505.2.4 and Table R505.3.1(2).

❖ This section provides minimum fastening requirements for attachment of steel floor framing to foundations, wood sills and both exterior and interior load-bearing walls. Figures R505.3.1(1) through R505.3.1(6) illustrate exterior wall conditions; thus, the corresponding connection requirements in Table R505.3.1(1) vary based on a site's basic wind speed and seismic design category. End joists must be doubled or have a C-shaped bearing stiffener beneath

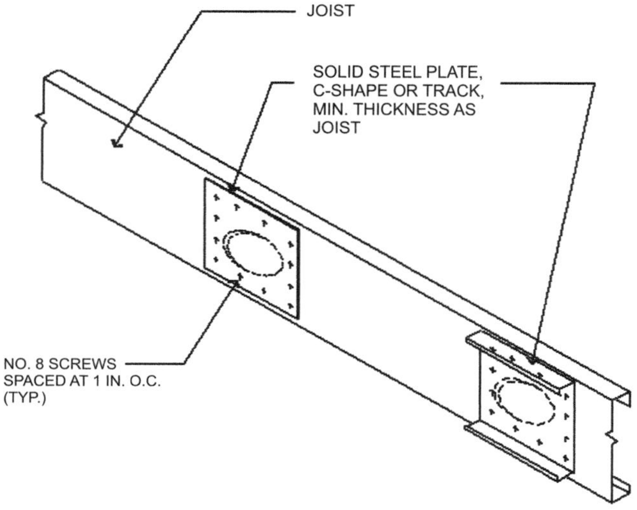

For SI: 1 inch = 25.4 mm.

FIGURE R505.2.5.3
WEB HOLE PATCH

❖ See the commentary to Section R505.2.5.3.

each wall stud as depicted in Figure R505.3.1(9). In the event these criteria give rise to requirements that are different from those in the table, the more restrictive would govern. Figures R505.3.1(7) and R505.3.1(8) illustrate the minimum connection requirements for floor joists supported by interior load-bearing walls in conjunction with the first line of Table R505.3.1(2).

TABLE R505.3.1(1)
FLOOR TO FOUNDATION OR BEARING WALL CONNECTION REQUIREMENTS[a, b]

FRAMING CONDITION	BASIC WIND SPEED (mph) AND EXPOSURE	
	85 mph Exposure C or less than 110 mph Exposure B	Less than 110 mph Exposure C
Floor joist to wall track of exterior wall per Figure R505.3.1(1)	2-No. 8 screws	3-No. 8 screws
Rim track or end joist to load-bearing wall top track per Figure R505.3.1(1)	1-No. 8 screw at 24 inches o.c.	1-No. 8 screw at 24 inches o.c.
Rim track or end joist to wood sill per Figure R505.3.1(2)	Steel plate spaced at 4 feet o.c. with 4-No. 8 screws and 4-10d or 6-8d common nails	Steel plate spaced at 2 feet o.c. with 4-No. 8 screws and 4-10d or 6-8d common nails
Rim track or end joist to foundation per Figure R505.3.1(3)	$^1/_2$ inch minimum diameter anchor bolt and clip angle spaced at 6 feet o.c. with 8-No. 8 screws	$^1/_2$ inch minimum diameter anchor bolt and clip angle spaced at 4 feet o.c. with 8-No. 8 screws
Cantilevered joist to foundation per Figure R505.3.1(4)	$^1/_2$ inch minimum diameter anchor bolt and clip angle spaced at 6 feet o.c. with 8-No. 8 screws	$^1/_2$ inch minimum diameter anchor bolt and clip angle spaced at 4 feet o.c. with 8-No. 8 screws
Cantilevered joist to wood sill per Figure R505.3.1(5)	Steel plate spaced at 4 feet o.c. with 4-No. 8 screws and 4-10d or 6-8d common nails	Steel plate spaced at 2 feet o.c. with 4-No. 8 screws and 4-10d or 6-8d common nails
Cantilevered joist to exterior load-bearing wall track per Figure R505.3.1(6)	2-No. 8 screws	3-No. 8 screws

For SI: 1 inch = 25.4 mm, 1 pound per square foot = 0.0479 kPa, 1 mile per hour = 0.447 m/s, 1 foot = 304.8 mm.

a. Anchor bolts are to be located not more than 12 inches from corners or the termination of bottom tracks (e.g., at door openings or corners). Bolts extend a minimum of 15 inches into masonry or 7 inches into concrete. Anchor bolts connecting cold-formed steel framing to the foundation structure are to be installed so that the distance from the center of the bolt hole to the edge of the connected member is not less than one and one-half bolt diameters.

b. All screw sizes shown are minimum.

❖ See the commentary to Section R505.3.1 as well as Figures R505.3.1(1) through R505.3.1(6).

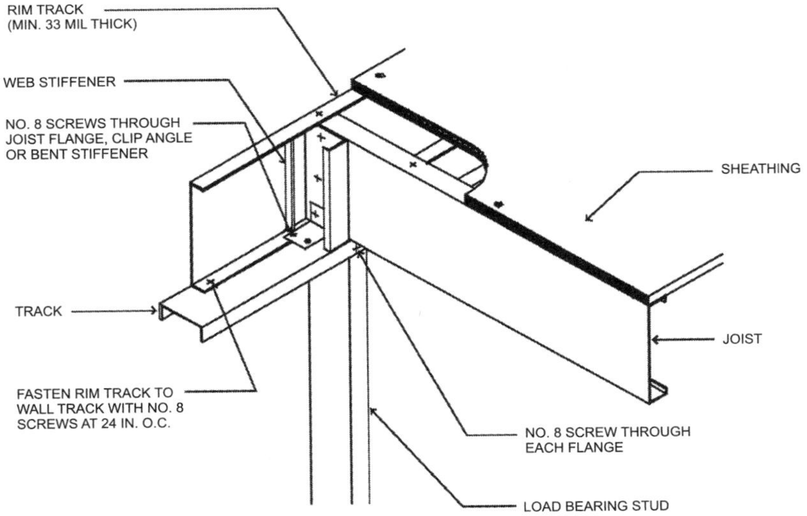

For SI: 1 mil = 0.0254 mm, 1 inch = 25.4 mm.

FIGURE 505.3.1(1)
FLOOR TO EXTERIOR LOAD-BEARING WALL STUD CONNECTION

❖ Figure R505.3.1(1) depicts connection requirements for steel floor joists at exterior steel bearing walls. The minimum number of screws through either the joist flange, clip angle or bent stiffener should be selected from the first row of Table R505.3.1(1) based on the seismic design category or wind speed and exposure. Similarly, if these joists cantilever beyond the wall as illustrated in Figure R505.3.1(7), the required number of screws should be selected from the seventh row of Table R505.3.1(1).

TABLE R505.3.1(2)
FLOOR FASTENING SCHEDULE[a]

DESCRIPTION OF BUILDING ELEMENTS	NUMBER AND SIZE OF FASTENERS	SPACING OF FASTENERS
Floor joist to track of an interior load-bearing wall per Figures R505.3.1(7) and R505.3.1(8)	2 No. 8 screws	Each joist
Floor joist to track at end of joist	2 No. 8 screws	One per flange or two per bearing stiffener
Subfloor to floor joists	No. 8 screws	6 in. o.c. on edges and 12 in. o.c. at intermediate supports

For SI: 1 inch = 25.4 mm.
a. All screw sizes shown are minimum.

❖ In addition to the other connections covered under Section R505.3.1, Table R505.3.1(2) provides the fastening requirements for tracks at the end of floor joists and for subfloor-to-floor joists.

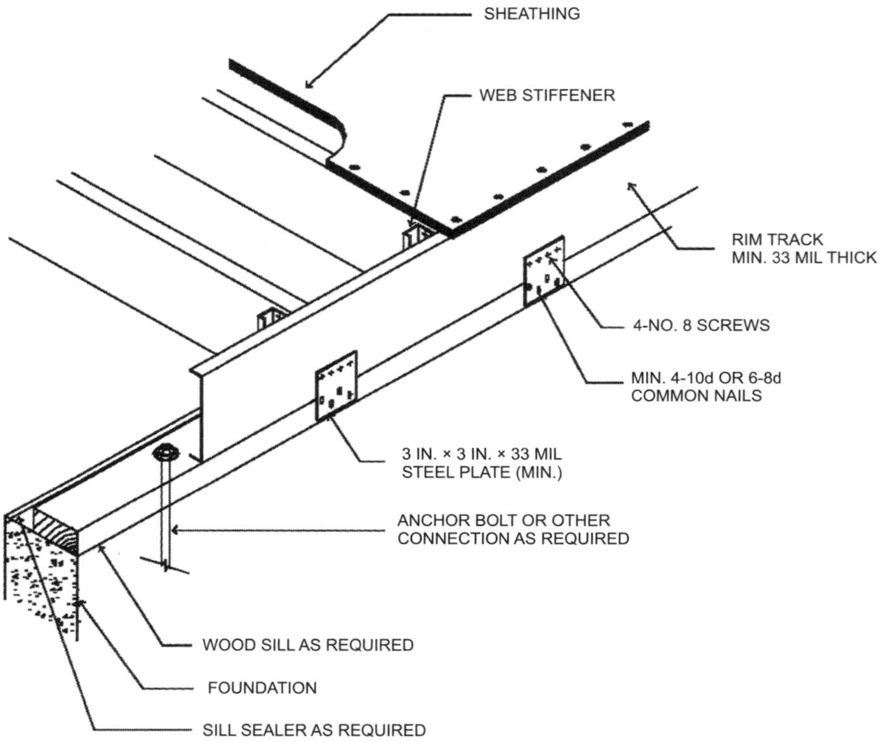

For SI: 1 mil = 0.0254 mm, 1 inch = 25.4 mm.

FIGURE R505.3.1(2)
FLOOR TO WOOD SILL CONNECTION

❖ Figures R505.3.1(2) and R505.3.1(5) depict the connection requirements of steel floor framing to a wood sill, which, in turn, is anchored to the foundation. Details such as the spacing of the clip angle and anchor bolt should be selected from the appropriate row of Table R505.3.1(1) [see Figure R505.3.1(1)].

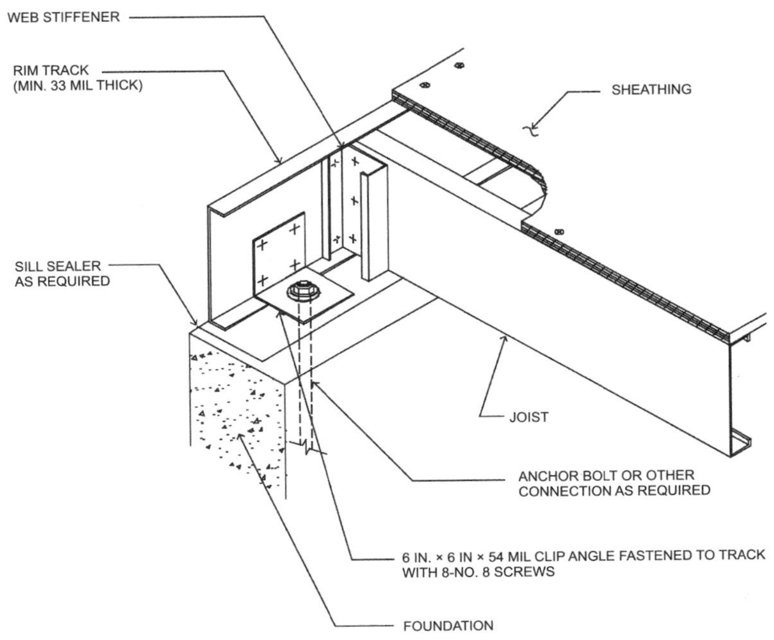

For SI: 1 mil = 0.0254 mm, 1 inch = 25.4 mm.

FIGURE R505.3.1(3)
FLOOR TO FOUNDATION CONNECTION

❖ Figures R505.3.1(3) and R505.3.1(4) depict requirements for anchoring steel floor framing directly to concrete or masonry foundations. Details such as the spacing of the clip angle and anchor bolt should be selected from the appropriate row of Table R505.3.1(1) [see Figure R505.3.1(1)].

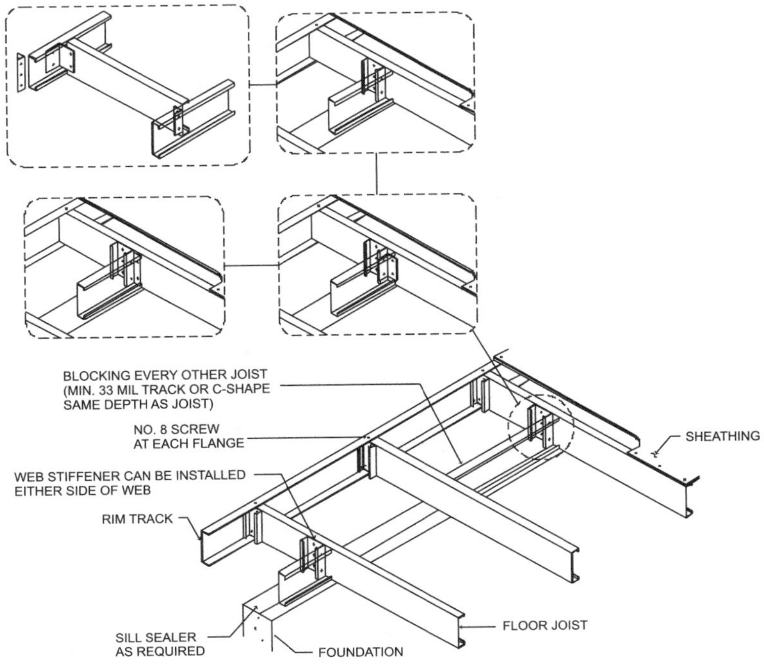

For SI: 1 mil = 0.0254 mm.

FIGURE R505.3.1(4)
CANTILEVERED FLOOR TO FOUNDATION CONNECTION

❖ See the commentary to Figure R505.3.1(1).

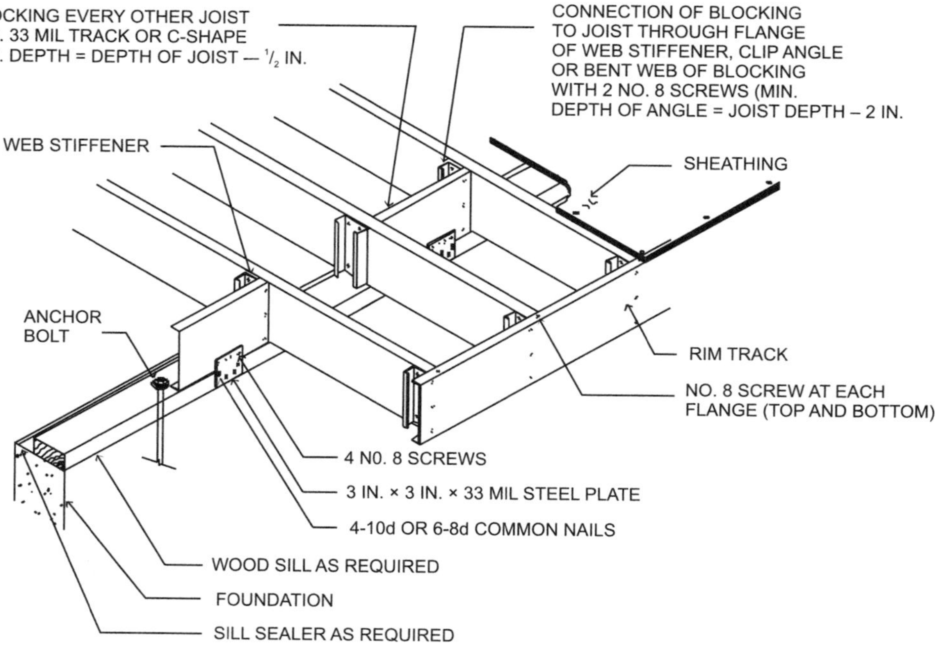

For SI: 1 mil = 0.0254 mm, 1 inch = 25.4 mm.

FIGURE R505.3.1(5)
CANTILEVERED FLOOR TO WOOD SILL CONNECTION

❖ See the commentary to Figures R505.3.1(1) and R505.3.1(2).

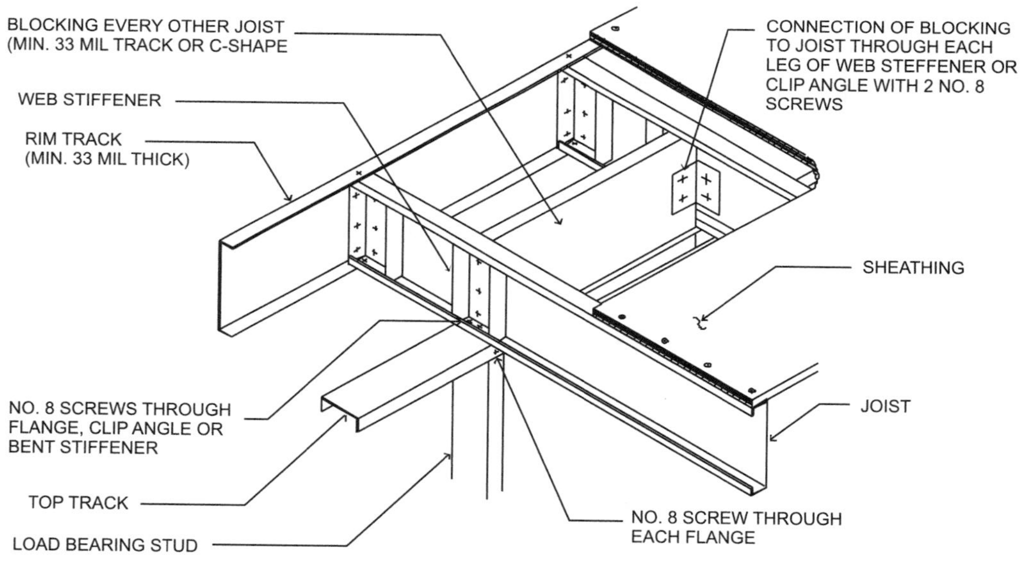

For SI: 1 mil = 0.0254 mm.

FIGURE R505.3.1(6)
CANTILEVERED FLOOR TO EXTERIOR LOAD-BEARING WALL CONNECTION

❖ See the commentary to Figures R505.3.1(1) and R505.3.1(3).

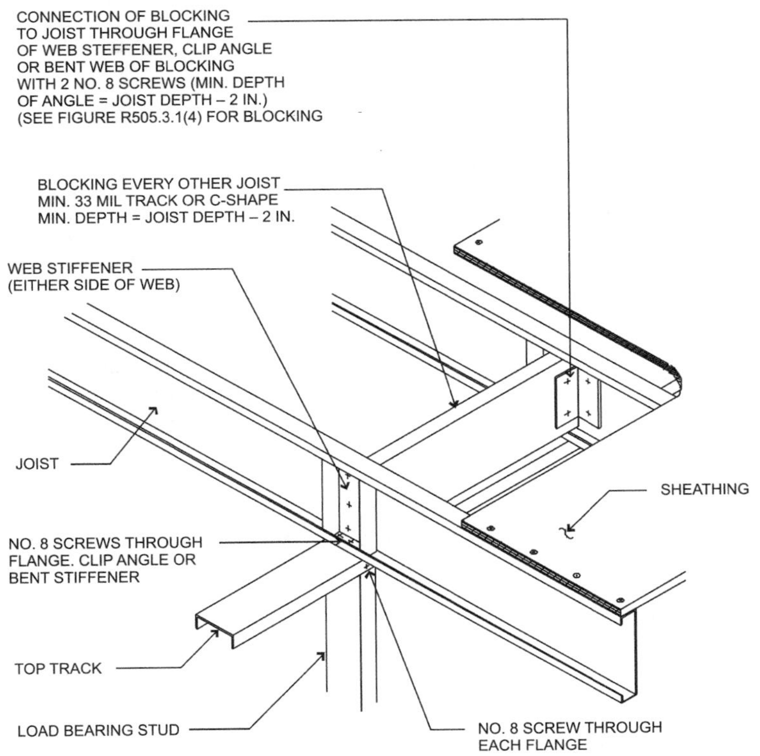

For SI: 1 mil = 0.0254 mm, 1 inch = 25.4 mm.

FIGURE R505.3.1(7)
CONTINUOUS SPAN JOIST SUPPORTED ON INTERIOR LOAD-BEARING WALL

❖ See the commentary to Section R505.3.1.

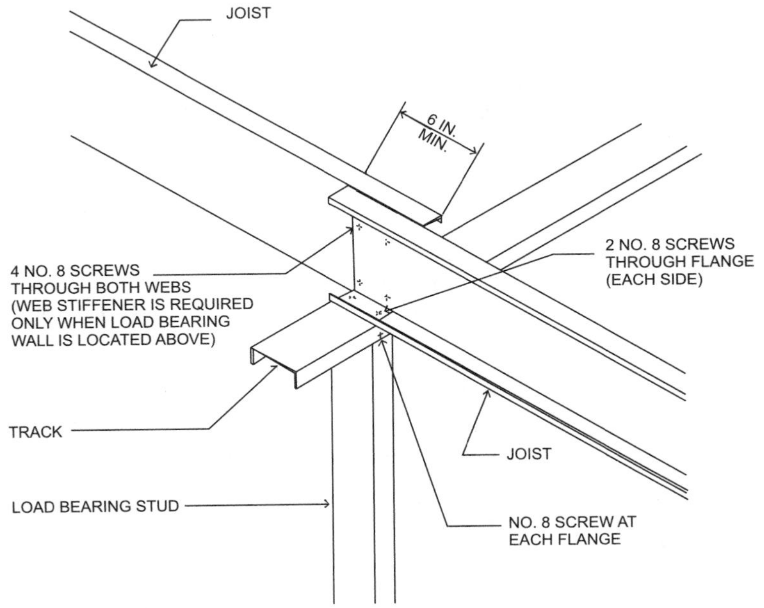

For SI: 1 inch = 25.4 mm.

FIGURE R505.3.1(8)
LAPPED JOISTS SUPPORTED ON INTERIOR LOAD-BEARING WALL

❖ See the commentary to Section R505.3.

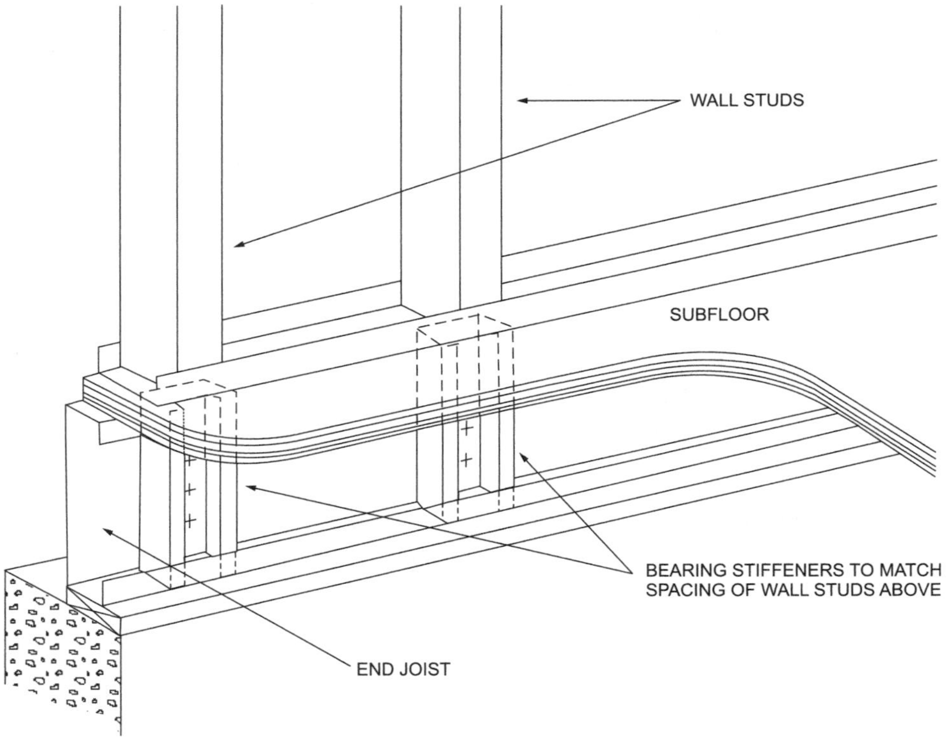

FIGURE R505.3.1(9)
BEARING STIFFENERS FOR END JOISTS

❖ See the commentary to Section R505.3.1.

R505.3.2 Minimum floor joist sizes. Floor joist size and thickness shall be determined in accordance with the limits set forth in Table R505.3.2(1) for single spans, and Tables R505.3.2(2) and R505.3.2(3) for multiple spans. When continuous joist members are used, the interior bearing supports shall be located within 2 feet (610 mm) of mid-span of the cold-formed steel joists, and the individual spans shall not exceed the spans in Table R505.3.2(2) or R505.3.2(3), as applicable. Floor joists shall have a bearing support length of not less than $1^1/_2$ inches (38 mm) for exterior wall supports and $3^1/_2$ inches (89 mm) for interior wall supports. Tracks shall be a minimum of 33 mils (0.84 mm) thick except when used as part of a floor header or trimmer in accordance with Section R505.3.8. Bearing stiffeners shall be installed in accordance with Section R505.3.4.

❖ Table R505.3.2(1) lists the allowable clear spans for single-span steel floor joists. Tables R505.3.2(2) and R505.3.2(3) list the allowable individual spans when the joists are continuous over an interior bearing support. The tabulated spans are based on dead loads of 10 psf (0.48 kPa) and design live loads of 30 psf (1.44 kPa) or 40 psf (1.92 kPa). The following example illustrates use of the tables.

Example:

Steel floor joists spaced at 16 inches (406 mm) on center support a floor live load of 40 psf (1.92 kPa), a dead load of 10 psf (0.48 kPa) and span 17 feet, 6 inches (5334 mm). Determine the required joist size.

Solution:

In Table R505.3.2(1) under 40 psf (1.92 kPa), find the live load column headed 16 inches (406 mm). Read down this column to find the first span greater than 17 feet, 6 inches (5334 mm), which is 18 feet, 2 inches (5537 mm). The required joist is 800S162-97.

When joists are continuous over interior bearing supports, the spans must be arranged as illustrated in Commentary Figure R505.3.2 and the spans must not exceed the limits in Tables R505.3.2(2) and R505.3.2(3).

R505.3.3 Joist bracing and blocking. Joist bracing and blocking shall be in accordance with this section.

❖ Sections R505.3.3.1 through R505.3.3.4 provide the requirements for joist top flange bracing, joist bottom flange bracing/blocking, blocking at interior bearing supports and blocking at cantilevers.

TABLE R505.3.2(1)
ALLOWABLE SPANS FOR COLD-FORMED STEEL JOISTS—SINGLE SPANS[a, b, c, d] 33 ksi STEEL

JOIST DESIGNATION	30 PSF LIVE LOAD				40 PSF LIVE LOAD			
	Spacing (inches)				Spacing (inches)			
	12	16	19.2	24	12	16	19.2	24
550S162-33	11′-7″	10′-7″	9′-6″	8′-6″	10′-7″	9′-3″	8′-6″	7′-6″
550S162-43	12′-8″	11′-6″	10′-10″	10′-2″	11′-6″	10′-5″	9′-10″	9′-1″
550S162-54	13′-7″	12′-4″	11′-7″	10′-9″	12′-4″	11′-2″	10′-6″	9′-9″
550S162-68	14′-7″	13′-3″	12′-6″	11′-7″	13′-3″	12′-0″	11′-4″	10′-6″
550S162-97	16′-2″	14′-9″	13′-10″	12′-10″	14′-9″	13′-4″	12′-7″	11′-8″
800S162-33	15′-8″	13′-11″	12′-9″	11′-5″	14′-3″	12′-5″	11′-3″	9′-0″
800S162-43	17′-1″	15′-6″	14′-7″	13′-7″	15′-6″	14′-1″	13′-3″	12′-4″
800S162-54	18′-4″	16′-8″	15′-8″	14′-7″	16′-8″	15′-2″	14′-3″	13′-3″
800S162-68	19′-9″	17′-11″	16′-10″	15′-8″	17′-11″	16′-3″	15′-4″	14′-2″
800S162-97	22′-0″	20′-0″	16′-10″	17′-5″	20′-0″	18′-2″	17′-1″	15′-10″
1000S162-43	20′-6″	18′-8″	17′-6″	15′-8″	18′-8″	16′-11″	15′-6″	13′-11″
1000S162-54	22′-1″	20′-0″	18′-10″	17′-6″	20′-0″	18′-2″	17′-2″	15′-11″
1000S162-68	23′-9″	21′-7″	20′-3″	18′-10″	21′-7″	19′-7″	18′-5″	17′-1″
1000S162-97	26′-6″	24′-1″	22′-8″	21′-0″	24′-1″	21′-10″	20′-7″	19′-1″
1200S162-43	23′-9″	20′-10″	19′-0″	16′-8″	21′-5″	18′-6″	16′-6″	13′-2″
1200S162-54	25′-9″	23′-4″	22′-0″	20′-1″	23′-4″	21′-3″	20′-0″	17′-10″
1200S162-68	27′-8″	25′-1″	23′-8″	21′-11″	25′-1″	22′-10″	21′-6″	21′-1″
1200S162-97	30′-11″	28′-1″	26′-5″	24′-6″	28′-1″	25′-6″	24′-0″	22′-3″

For SI: 1 inch = 25.4 mm, 1 foot = 304.8 mm, 1 pound per square foot = 0.0479 kPa.

a. Deflection criteria: *L*/480 for live loads, *L*/240 for total loads.

b. Floor dead load = 10 psf.

c. Table provides the maximum clear span in feet and inches.

d. Bearing stiffeners are to be installed at all support points and concentrated loads.

❖ See the commentary and example to Section R505.3.2.

TABLE R505.3.2(2)
ALLOWABLE SPANS FOR COLD-FORMED STEEL JOISTS—MULTIPLE SPANS[a, b, c, d, e, f] 33 ksi STEEL

JOIST DESIGNATION	30 PSF LIVE LOAD				40 PSF LIVE LOAD			
	Spacing (inches)				Spacing (inches)			
	12	16	19.2	24	12	16	19.2	24
550S162-33	12′-1″	10′-5″	9′-6″	8′-6″	10′-9″	9′-3″	8′-6″	7′-6″
550S162-43	14′-5″	12′-5″	11′-4″	10′-2″	12′-9″	11′-11″	10′-1″	9′-0″
550S162-54	16′-3″	14′-1″	12′-10″	11′-6″	14′-5″	12′-6″	11′-5″	10′-2″
550S162-68	19′-7″	17′-9″	16′-9″	15′-6″	17′-9″	16′-2″	15′-2″	14′-1″
550S162-97	21′-9″	19′-9″	18′-7″	17′-3″	19′-9″	17′-11″	16′-10″	15′-4″
800S162-33	14′-8″	11′-10″	10′-4″	8′-8″	12′-4″	9′-11″	8′-7″	7′-2″
800S162-43	20′-0″	17′-4″	15′-9″	14′-1″	17′-9″	15′-4″	14′-0″	12′-0″
800S162-54	23′-7″	20′-5″	18′-8″	16′-8″	21′-0″	18′-2″	16′-7″	14′-10″
800S162-68	26′-5″	23′-1″	21′-0″	18′-10″	23′-8″	20′-6″	18′-8″	16′-9″
800S162-97	29′-6″	26′-10″	25′-3″	22′-8″	26′-10″	24′-4″	22′-6″	20′-2″
1000S162-43	22′-2″	18′-3″	16′-0″	13′-7″	18′-11″	15′-5″	13′-6″	11′-5″
1000S162-54	26′-2″	22′-8″	20′-8″	18′-6″	23′-3″	20′-2″	18′-5″	16′-5″
1000S162-68	31′-5″	27′-2″	24′-10″	22′-2″	27′-11″	24′-2″	22′-1″	19′-9″
1000S162-97	35′-6″	32′-3″	29′-11″	26′-9″	32′-3″	29′-2″	26′-7″	23′-9″
1200S162-43	21′-8″	17′-6″	15′-3″	12′-10″	18′-3″	14′-8″	12′-8″	10′-6″
1200S162-54	28′-5″	24′-8″	22′-6″	19′-6″	25′-3″	21′-11″	19′-4″	16′-6″
1200S162-68	33′-7″	29′-1″	26′-6″	23′-9″	29′-10″	25′-10″	23′-7″	21′-1″
1200S162-97	41′-5″	37′-8″	34′-6″	30′-10″	37′-8″	33′-6″	30′-7″	27′-5″

For SI: 1 inch = 25.4 mm, 1 foot = 304.8 mm, 1 pound per square foot = 0.0479 kPa.

a. Deflection criteria: *L*/480 for live loads, *L*/240 for total loads.

b. Floor dead load = 10 psf.

c. Table provides the maximum clear span in feet and inches to either side of the interior support.

d. Interior bearing supports for multiple span joists consist of structural (bearing) walls or beams.

e. Bearing stiffeners are to be installed at all support points and concentrated loads.

f. Interior supports shall be located within 2 feet of mid-span provided that each of the resulting spans does not exceed the appropriate maximum span shown in the table above.

❖ See the commentary to Section R505.3.3.2.

TABLE R505.3.2(3)
ALLOWABLE SPANS FOR COLD-FORMED STEEL JOISTS—MULTIPLE SPANS[a, b, c, d, e, f] 50 ksi STEEL

JOIST DESIGNATION	30 PSF LIVE LOAD				40 PSF LIVE LOAD			
	Spacing (inches)				Spacing (inches)			
	12	16	19.2	24	12	16	19.2	24
550S162-33	13′-11″	12′-0″	11′-0″	9′-3″	12′-3″	10′-8″	9′-7″	8′-4″
550S162-43	16′-3″	14′-1″	12′-10″	11′-6″	14′-6″	12′-6″	11′-5″	10′-3″
550S162-54	18′-2″	16′-6″	15′-4″	13′-8″	16′-6″	14′-11″	13′-7″	12′-2″
550S162-68	19′-6″	17′-9″	16′-8″	15′-6″	17′-9″	16′-1″	15′-2″	14′-0″
550S162-97	21′-9″	19′-9″	18′-6″	17′-2″	19′-8″	17′-10″	16′-8″	15′-8″
800S162-33	15′-6″	12′-6″	10′-10″	9′-1″	13′-0″	10′-5″	8′-11″	6′-9″
800S162-43	22′-0″	19′-1″	17′-5″	15′-0″	19′-7″	16′-11″	14′-10″	12′-8″
800S162-54	24′-6″	22′-4″	20′-6″	17′-11″	22′-5″	19′-9″	17′-11″	15′-10″
800S162-68	26′-6″	24′-1″	22′-8″	21′-0″	24′-1″	21′-10″	20′-7″	19′-2″
800S162-97	29′-9″	26′-8″	25′-2″	23′-5″	26′-8″	24′-3″	22′-11″	21′-4″
1000S162-43	23′-6″	19′-2″	16′-9″	14′-2″	19′-11″	16′-2″	14′-0″	11′-9″
1000S162-54	28′-2″	23′-10″	21′-7″	18′-11″	24′-8″	20′-11″	18′-9″	18′-4″
1000S162-68	31′-10″	28′-11″	27′-2″	25′-3″	28′-11″	26′-3″	24′-9″	22′-9″
1000S162-97	35′-4″	32′-1″	30′-3″	28′-1″	32′-1″	29′-2″	27′-6″	25′-6″
1200S162-43	22′-11″	18′-5″	16′-0″	13′-4″	19′-2″	15′-4″	13′-2″	10′-6″
1200S162-54	32′-8″	28′-1″	24′-9″	21′-2″	29′-0″	23′-10″	20′-11″	17′-9″
1200S162-68	37′-1″	32′-5″	29′-4″	25′-10″	33′-4″	28′-6″	25′-9″	22′-7″
1200S162-97	41′-2″	37′-6″	35′-3″	32′-9″	37′-6″	34′-1″	32′-1″	29′-9″

For SI: 1 inch = 25.4 mm, 1 foot = 304.8 mm, 1 pound per square foot = 0.0479 kPa.

a. Deflection criteria: *L*/480 for live loads, *L*/240 for total loads.
b. Floor dead load = 10 psf.
c. Table provides the maximum clear span in feet and inches to either side of the interior support.
d. Interior bearing supports for multiple span joists consist of structural (bearing) walls or beams.
e. Bearing stiffeners are to be installed at all support points and concentrated loads.
f. Interior supports shall be located within 2 feet of mid-span provided that each of the resulting spans does not exceed the appropriate maximum span shown in the table above.

❖ See the commentary to Section R505.3.2.

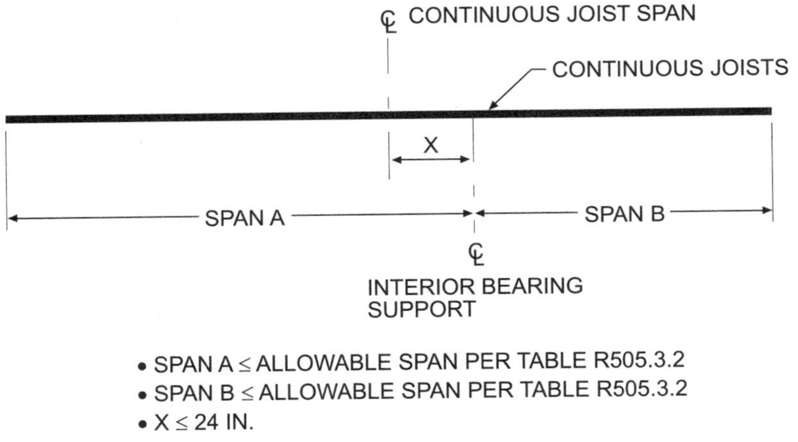

For SI: 1 inch = 25.4 mm.

Figure R505.3.2
SPAN ARRANGEMENT FOR FLOOR JOIST CONTINUOUS OVER SUPPORT

R505.3.3.1 Joist top flange bracing. The top flanges of cold-formed steel joists shall be laterally braced by the application of floor sheathing fastened to the joists in accordance with Section R505.2.4 and Table R505.3.1(2).

❖ The application of floor sheathing to steel floor joists in accordance with Table R505.3.1(2) and Section R505.2.4 of the code allows the transfer of loads to the joists and provides the necessary lateral bracing for the top flanges of those joists.

R505.3.3.2 Joist bottom flange bracing/blocking. Floor joists with spans that exceed 12 feet (3658 mm) shall have the bottom flanges laterally braced in accordance with one of the following:

1. Gypsum board installed with minimum No. 6 screws in accordance with Section R702.
2. Continuous steel straps installed in accordance with Figure R505.3.3.2(1). Steel straps shall be spaced at a maximum of 12 feet (3658 mm) on center and shall be at least $1^1/_2$ inches (38 mm) in width and 33 mils (0.84 mm) in thickness. Straps shall be fastened to the bottom flange of each joist with one No. 8 screw, fastened to blocking with two No. 8 screws, and fastened at each end (of strap) with two No. 8 screws. Blocking in accordance with Figure R505.3.3.2(1) or Figure R505.3.3.2(2) shall be installed between joists at each end of the continuous strapping and at a maximum spacing of 12 feet (3658 mm) measured along the continuous strapping (perpendicular to the joist run). Blocking shall also be located at the termination of all straps. As an alternative to blocking at the ends, anchoring the strap to a stable building component with two No. 8 screws shall be permitted.

❖ Steel floor joists with spans greater than 12 feet (3658 mm) are more prone to overturning and therefore must also have the bottom flanges laterally braced either with gypsum board or continuous steel strapping as illustrated in Figure R505.3.3.2(1) of the code. Strap bracing requires the installation of solid blocking or bridging (X-bracing) [see Figure R505.3.3.2(1) or R505.3.3.2(2)] in line with the straps at the termination of all straps and at a maximum spacing of 12 feet (3658 mm) measured perpendicular to the joist run.

R505.3.3.3 Blocking at interior bearing supports. Blocking is not required for continuous back-to-back floor joists at bearing supports. Blocking shall be installed between every other joist for single continuous floor joists across bearing supports in accordance with Figure R505.3.1(7). Blocking shall consist of C-shape or track section with a minimum thickness of 33 mils (0.84 mm). Blocking shall be fastened to each adjacent joist through a 33-mil (0.84 mm) clip angle, bent web of blocking or flanges of web stiffeners with two No. 8 screws on each side. The minimum depth of the blocking shall be equal to the depth of the joist minus 2 inches (51 mm). The minimum length of the angle shall be equal to the depth of the joist minus 2 inches (51 mm).

❖ Blocking must be installed between the joists, at the interior bearing support, for a single continuous floor joist. Blocking is not required at the bearing support for back-to-back floor joists [see Figure R505.3.1(7)].

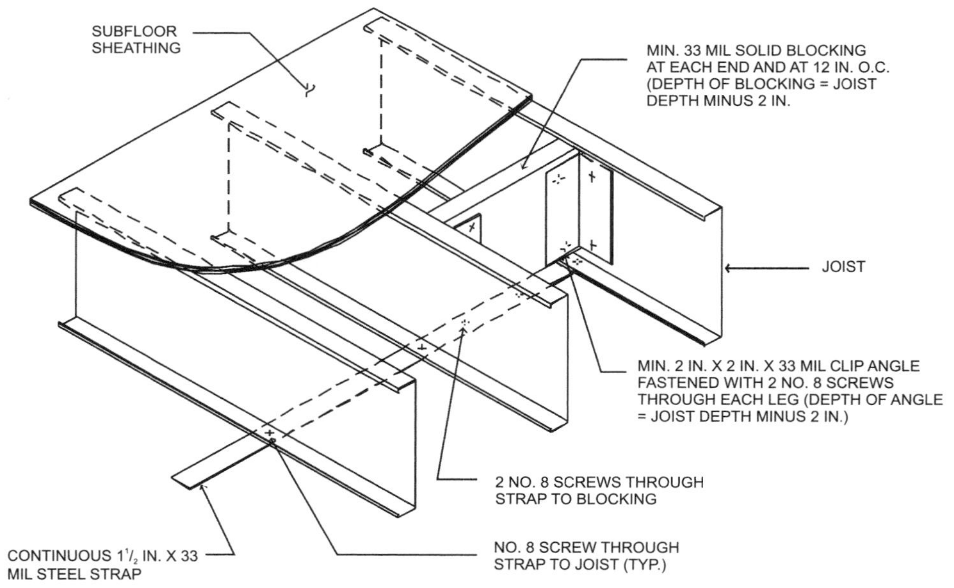

For SI: 1 mil = 0.0254, 1 inch = 25.4 mm.

FIGURE R505.3.3.2(1)
JOIST BLOCKING (SOLID)

❖ See the commentary to Section R505.3.3.2.

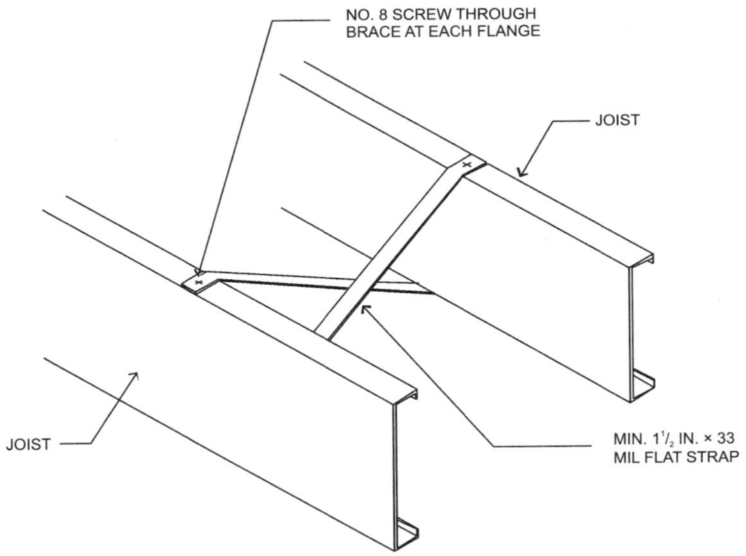

For SI: 1 mil = 0.0254, 1 inch = 25.4 mm.

FIGURE R505.3.3.2(2)
JOIST BLOCKING (STRAP)

❖ See the commentary to Section R505.3.3.2.

R505.3.3.4 Blocking at cantilevers. Blocking shall be installed between every other joist over cantilever bearing supports in accordance with Figure R505.3.1(4), R505.3.1(5) or R505.3.1(6). Blocking shall consist of C-shape or track section with minimum thickness of 33 mils (0.84 mm). Blocking shall be fastened to each adjacent joist through bent web of blocking, 33 mil clip angle or flange of web stiffener with two No. 8 screws at each end. The depth of the blocking shall be equal to the depth of the joist. The minimum length of the angle shall be equal to the depth of the joist minus 2 inches (51 mm). Blocking shall be fastened through the floor sheathing and to the support with three No. 8 screws (top and bottom).

❖ Blocking must be installed between every other joist at the cantilever support [see Figure R505.3.1(4), R505.3.1(5) or R505.3.1(6)].

R505.3.4 Bearing stiffeners. Bearing stiffeners shall be installed at each joist bearing location in accordance with this section, except for joists lapped over an interior support not carrying a load-bearing wall above. Floor joists supporting jamb studs with multiple members shall have two bearing stiffeners in accordance with Figure R505.3.4(1). Bearing stiffeners shall be fabricated from a C-shaped, track or clip angle member in accordance with the one of following:

1. C-shaped bearing stiffeners:
   - 1.1. Where the joist is not carrying a load-bearing wall above, the bearing stiffener shall be a minimum 33 mil (0.84 mm) thickness.
   - 1.2. Where the joist is carrying a load-bearing wall above, the bearing stiffener shall be at least the same designation thickness as the wall stud above.
2. Track bearing stiffeners:
   - 2.1. Where the joist is not carrying a load-bearing wall above, the bearing stiffener shall be a minimum 43 mil (1.09 mm) thickness.
   - 2.2. Where the joist is carrying a load-bearing wall above, the bearing stiffener shall be at least one designation thickness greater than the wall stud above.
3. Clip angle bearing stiffeners: Where the clip angle bearing stiffener is fastened to both the web of the member it is stiffening and an adjacent rim track using the fastener pattern shown in Figure R505.3.4(2), the bearing stiffener shall be a minimum 2 inch by 2 inch (51 mm by 51 mm) angle sized in accordance with Tables R505.3.4(1), R505.3.4(2), R505.3.4(3), and R505.3.4(4).

The minimum length of a bearing stiffener shall be the depth of member being stiffened minus $^3/_8$ inch (9.5 mm). Each bearing stiffener shall be fastened to the web of the member it is stiffening as shown in Figure R505.3.4(2). Each clip angle bearing stiffener shall also be fastened to the web of the adjacent rim track using the fastener pattern shown in Figure R505.3.4(2). No. 8 screws shall be used for C-shaped and track members of any thickness and for clip angle members with a designation thickness less than or equal to 54. No. 10 screws shall be used for clip angle members with a designation thickness greater than 54.

❖ Stiffeners (also referred to as transverse stiffeners or web stiffeners) at points of bearing provide resistance to web crippling in the joist. This section provides the option to use C-shaped track section or clip angle stiffening. These stiffeners are illustrated in Figure R505.3.4(2), but a stiffener can be installed on either side of the web of a joist as depicted in Figures R505.3.1(1) through R505.3.1(9). Two bearing stiffeners are required to be located where jamb studs are supported by the floor joist [see Figure R505.3.4(1)].

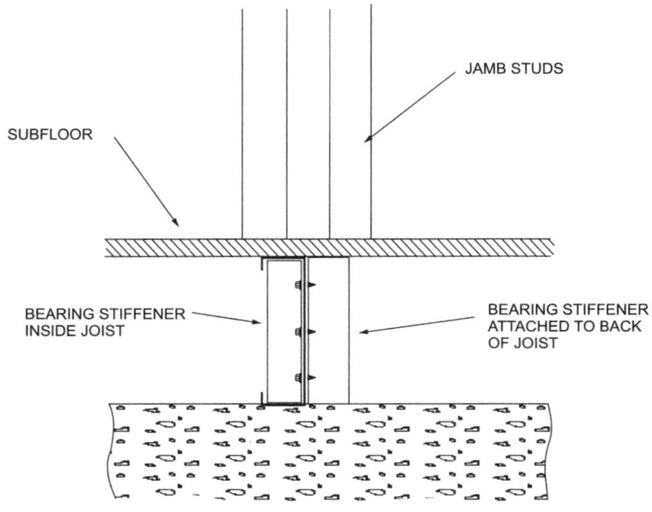

FIGURE R505.3.4(1)
BEARING STIFFENERS UNDER JAMB STUDS

❖ See the commentary to Section R503.3.4.

Tables R505.3.4(1) through R505.3.4(4) provide the design of the clip angles based on snow load of 20 psf to 70 psf and location of floor and joist spacing.

All bearing stiffeners are to be fastened in accordance with Figure R505.3.4(2); however, clip angles greater than 54 mils thick must be fastened with No. 10 screws.

R505.3.5 Cutting and notching. Flanges and lips of load-bearing cold-formed steel floor framing members shall not be cut or notched.

❖ Flanges and lips of load-bearing steel floor framing members must not be cut or notched because this would affect their structural integrity. The only holes permitted in steel floor joists are those that conform to Section R505.2.5.

R505.3.6 Floor cantilevers. Floor cantilevers for the top floor of a two- or three-story building or the first floor of a one-story building shall not exceed 24 inches (610 mm). Cantilevers, not exceeding 24 inches (610 mm) and supporting two stories and roof (i.e., first floor of a two-story building), shall also be permitted provided that all cantilevered joists are doubled (nested or back-to-back). The doubled cantilevered joists shall extend a minimum of 6 feet (1829 mm) toward the inside and shall be fastened with a minimum of two No. 8 screws spaced at 24 inches (610 mm) on center through the webs (for back-to-back) or flanges (for nested joists).

❖ The maximum permitted floor cantilever is 24 inches (610 mm). Floor cantilevers that support a roof only are permitted to be single joist conforming to the allowable span requirements of Section R505.3.2.

Where a cantilever floor supports two stories and a roof all cantilever joists must be doubled (nested or

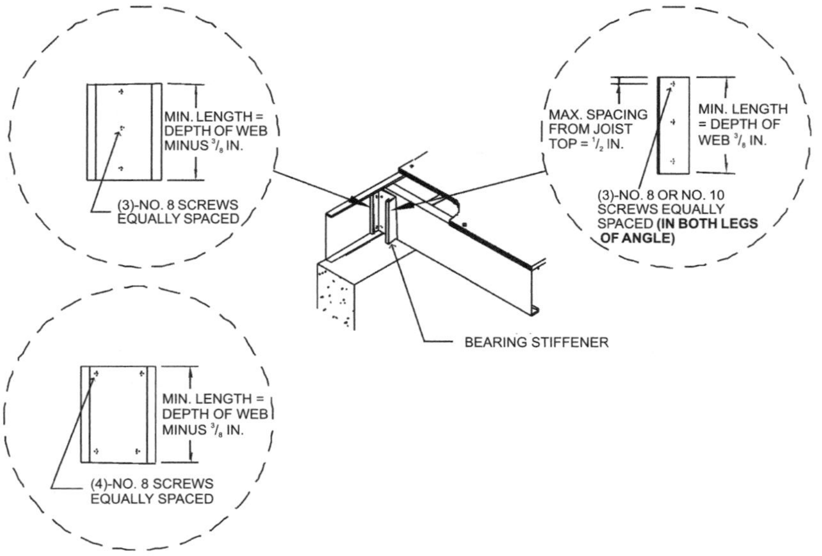

For SI: 1 inch = 25.4 mm.

FIGURE R505.3.4(2)
BEARING STIFFENER

❖ See the commentary to Section R505.3.4.

back to back). Double joist consists of members of the same size and material thickness as that for single joist in accordance with the allowable span requirements of Section R505.3.2. The double joist must extend a minimum of 6 feet (1829 mm) within the building.

TABLE R505.3.4(1)
CLIP ANGLE BEARING STIFFENERS (20 psf equivalent snow load)

JOIST DESIGNATION	MINIMUM THICKNESS (mils) OF 2 INCH × 2 INCH CLIP ANGLE											
	Top floor				Bottom floor in 2 story Middle floor in 3 story				Bottom floor in 3 story			
	Joist spacing (inches)				Joist spacing (inches)				Joist spacing (inches)			
	12	16	19.2	24	12	16	19.2	24	12	16	19.2	24
800S162-33	43	43	43	43	43	54	68	68	68	97	97	—
800S162-43	43	43	43	43	54	54	68	68	97	97	97	97
800S162-54	43	43	43	43	43	54	68	68	68	97	97	—
800S162-68	43	43	43	43	43	43	54	68	54	97	97	—
800S162-97	43	43	43	43	43	43	43	43	43	43	54	97
1000S162-43	43	43	43	43	54	68	97	97	97	—	—	—
1000S162-54	43	43	43	43	54	68	68	97	97	97	—	—
1000S162-68	43	43	43	43	54	68	97	97	97	—	—	—
1000S162-97	43	43	43	43	43	43	43	54	43	68	97	—
1200S162-43	43	54	54	54	97	97	97	97	—	—	—	—
1200S162-54	54	54	54	54	97	97	97	97	—	—	—	—
1200S162-68	43	43	54	54	68	97	97	97	—	—	—	—
1200S162-97	43	43	43	43	43	54	68	97	97	—	—	—

For SI:1 mil = 0.254 mm, 1 inch = 25.4 mm, 1 pound per square foot = 0.0479 kPa.

❖ See the commentary to Section R505.3.4.

TABLE R505.3.4(2)
CLIP ANGLE BEARING STIFFENERS (30 psf equivalent snow load)

JOIST DESIGNATION	MINIMUM THICKNESS (mils) OF 2 INCH × 2 INCH CLIP ANGLE											
	Top floor				Bottom floor in 2 story Middle floor in 3 story				Bottom floor in 3 story			
	Joist spacing (inches)				Joist spacing (inches)				Joist spacing (inches)			
	12	16	19.2	24	12	16	19.2	24	12	16	19.2	24
800S162-33	43	43	43	43	54	68	68	97	97	97	97	—
800S162-43	43	43	43	54	68	68	68	97	97	97	97	—
800S162-54	43	43	43	43	54	68	68	97	97	97	—	—
800S162-68	43	43	43	43	43	54	68	97	68	97	97	—
800S162-97	43	43	43	43	43	43	43	43	43	43	68	97
1000S162-43	54	54	54	54	68	97	97	97	97	—	—	—
1000S162-54	54	54	54	54	68	97	97	97	97	—	—	—
1000S162-68	43	43	54	68	68	97	97	—	97	—	—	—
1000S162-97	43	43	43	43	43	43	54	68	54	97	—	—
1200S162-43	54	68	68	68	97	97	97	—	—	—	—	—
1200S162-54	68	68	68	68	97	97	—	—	—	—	—	—
1200S162-68	68	68	68	68	97	97	97	—	—	—	—	—
1200S162-97	43	43	43	43	54	68	97	—	97	—	—	—

For SI:1 mil = 0.0254 mm, 1 inch = 25.4 mm, 1 pound per square foot = 0.0479 kPa.

❖ See the commentary to Section R505.3.4.

TABLE R505.3.4(3)
CLIP ANGLE BEARING STIFFENERS (50 psf equivalent snow load)

JOIST DESIGNATION	MINIMUM THICKNESS (mils) OF 2 INCH × 2 INCH CLIP ANGLE											
	Top floor				Bottom floor in 2 story Middle floor in 3 story				Bottom floor in 3 story			
	Joist spacing (inches)				Joist spacing (inches)				Joist spacing (inches)			
	12	16	19.2	24	12	16	19.2	24	12	16	19.2	24
800S162-33	54	54	54	54	68	97	97	97	97	—	—	—
800S162-43	68	68	68	68	97	97	97	97	—	—	—	—
800S162-54	54	68	68	68	97	97	97	97	—	—	—	—
800S162-68	43	43	54	54	68	97	97	97	97	—	—	—
800S162-97	43	43	43	43	43	43	43	54	54	68	97	—
1000S162-43	97	68	68	68	97	97	97	97	—	—	—	—
1000S162-54	97	97	68	68	97	97	97	—	—	—	—	—
1000S162-68	68	97	97	97	97	—	—	—	—	—	—	—
1000S162-97	43	43	43	43	54	68	97	97	—	—	—	—
1200S162-43	97	97	97	97	—	—	—	—	—	—	—	—
1200S162-54	—	97	97	97	—	—	—	—	—	—	—	—
1200S162-68	97	97	97	97	—	—	—	—	—	—	—	—
1200S162-97	54	68	68	97	97	—	—	—	—	—	—	—

For SI:1 mil = 0.0254 mm, 1 inch = 25.4 mm, 1 pound per square foot = 0.0479 kPa.

❖ See the commentary to Section R505.3.4.

TABLE R505.3.4(4)
CLIP ANGLE BEARING STIFFENERS (70 psf equivalent snow load)

JOIST DESIGNATION	MINIMUM THICKNESS (mils) OF 2 INCH × 2 INCH CLIP ANGLE											
	Top floor				Bottom floor in 2 story Middle floor in 3 story				Bottom floor in 3 story			
	Joist spacing (inches)				Joist spacing (inches)				Joist spacing (inches)			
	12	16	19.2	24	12	16	19.2	24	12	16	19.2	24
800S162-33	68	68	68	68	97	97	97	97	—	—	—	—
800S162-43	97	97	97	97	97	97	97	—	—	—	—	—
800S162-54	97	97	97	97	97	—	—	—	—	—	—	—
800S162-68	68	68	68	97	97	97	97	—	—	—	—	—
800S162-97	43	43	43	43	43	54	68	97	97	97	—	—
1000S162-43	97	97	97	97	—	—	—	—	—	—	—	—
1000S162-54	—	97	97	97	—	—	—	—	—	—	—	—
1000S162-68	97	97	—	—	—	—	—	—	—	—	—	—
1000S162-97	68	68	68	68	97	97	—	—	—	—	—	—
1200S162-43	97	97	97	97	—	—	—	—	—	—	—	—
1200S162-54	—	—	—	—	—	—	—	—	—	—	—	—
1200S162-68	—	—	—	—	—	—	—	—	—	—	—	—
1200S162-97	97	97	97	—	—	—	—	—	—	—	—	—

For SI:1 mil = 0.0254 mm, 1 inch = 25.4 mm, 1 pound per square foot = 0.0479 kPa.

❖ See the commentary to Section R505.3.4.

R505.3.7 Splicing. Joists and other structural members shall not be spliced. Splicing of tracks shall conform to Figure R505.3.7.

❖ Splicing of structural members is not permitted except when lapped joists occur at interior bearing points.

Splicing of tracks is allowed because they are not permitted to act as load-carrying members (see Figure R505.3.7). The exception to this would be under lateral loads from wind and earthquakes. In this case, the floor system must act as a diaphragm, with tracks serving as diaphragm chords. To provide continuity and carry the anticipated chord force, the splice must be as shown in Figure R505.3.7.

R505.3.8 Framing of floor openings. Openings in floors shall be framed with header and trimmer joists. Header joist spans shall not exceed 6 feet (1829 mm) or 8 feet (2438 mm) in length in accordance with Figure R505.3.8(1) or R505.3.8(2), respectively. Header and trimmer joists shall be fabricated from joist and track members, having a minimum size and thickness at least equivalent to the adjacent floor joists and shall be installed in accordance with Figures R505.3.8(1), R505.3.8(2), R505.3.8(3), and R505.3.8(4). Each header joist shall be connected to trimmer joists with four 2 inch by 2 inch (51 mm by 51 mm) clip angles. Each clip angle shall be fastened to both the header and trimmer joists with four No. 8 screws, evenly spaced, through each leg of the clip angle. The clip angles shall have a thickness not less than that of the floor joist. Each track section for a built-up header or trimmer joist shall extend the full length of the joist (continuous).

❖ Recognizing that floor openings are necessary for such things as stairways and utility chases, the code provides prescriptive criteria for framing these openings. This section provides prescriptive design for floor openings 6 feet (1829 mm) and 8 feet (2438 mm) in length. These requirements for openings in floors are illustrated in Figures R505.3.8(1) through R505.3.8(4).

SECTION R506 CONCRETE FLOORS (ON GROUND)

R506.1 General. Concrete slab-on-ground floors shall be designed and constructed in accordance with the provisions of this section or ACI 332. Floors shall be a minimum 3.5 inches (89 mm) thick (for expansive soils, see Section R403.1.8). The specified compressive strength of concrete shall be as set forth in Section R402.2.

❖ For weathering requirements, see the commentary to Sections R301.2 and R402.2. Commentary Figure R506.1 shows the requirements of this section.

R506.2 Site preparation. The area within the foundation walls shall have all vegetation, top soil and foreign material removed.

❖ Removal of construction debris and foreign materials, such as lumber formwork, stakes, tree stumps and other vegetation, limits the attraction of termites, insects and vermin. Top soil and soil vegetation should also be removed because such top soil is generally loosely compacted or so full of vegetation that soil settlement will occur when the vegetation decays. For concrete slabs placed on uncompacted fill or on large quantities of foreign materials, differential settlement may take place as a result of subsequent compaction of the soil, which can result in cracking of the floor slab and the interior wall/ceiling finishes.

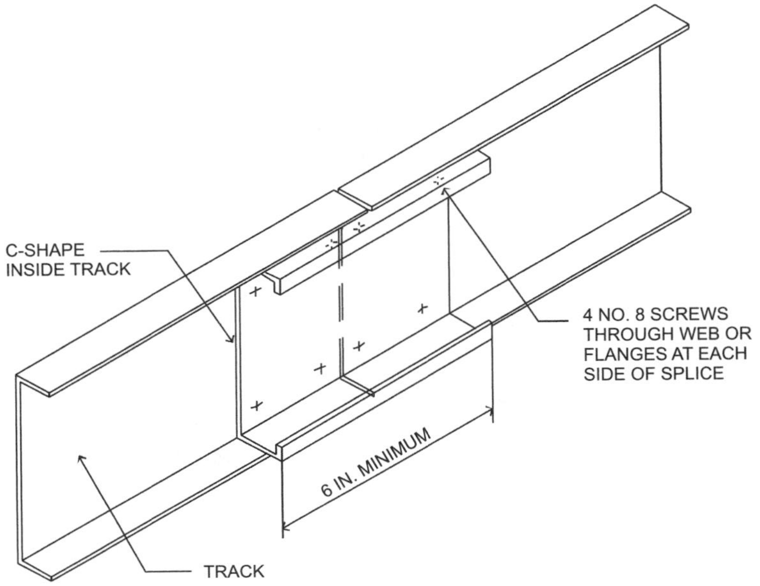

For SI: 1 inch = 25.4 mm.

FIGURE R505.3.7
TRACK SPLICE

❖ See the commentary to Section R505.3.7.

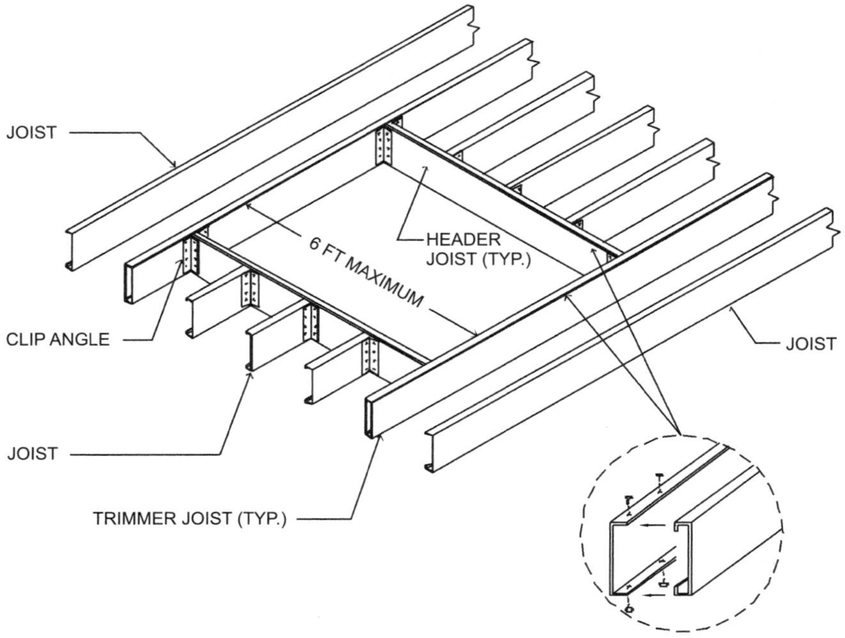

For SI: 1 foot = 304.8 mm.

FIGURE R505.3.8(1)
COLD-FORMED STEEL FLOOR CONSTRUCTION: 6-FOOT FLOOR OPENING

❖ See the commentary to Section R505.3.8.

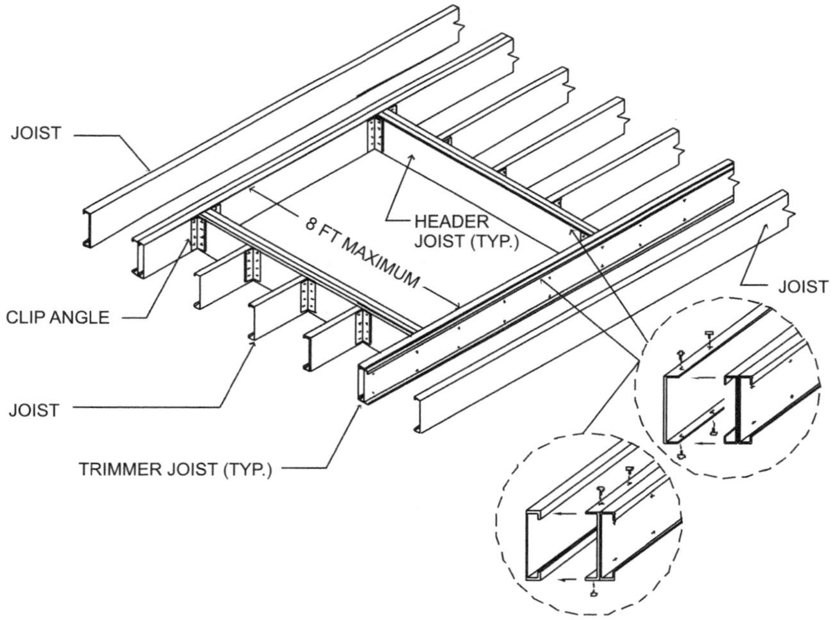

For SI: 1 foot = 304.8 mm.

FIGURE R505.3.8(2)
COLD-FORMED STEEL FLOOR CONSTRUCTION: 8-FOOT FLOOR OPENING

❖ See the commentary to Section R505.3.8.

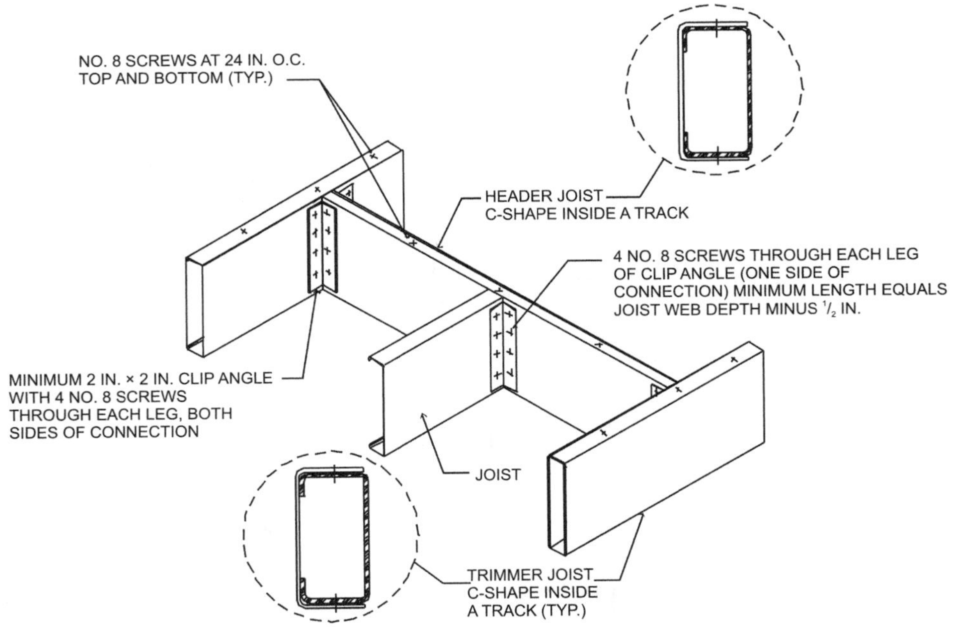

For SI: 1 inch = 25.4 mm, 1 foot = 304.8 mm.

FIGURE R505.3.8(3)
COLD-FORMED STEEL FLOOR CONSTRUCTION: FLOOR HEADER TO TRIMMER CONNECTION—6-FOOT OPENING

❖ See the commentary to Section R505.3.8.

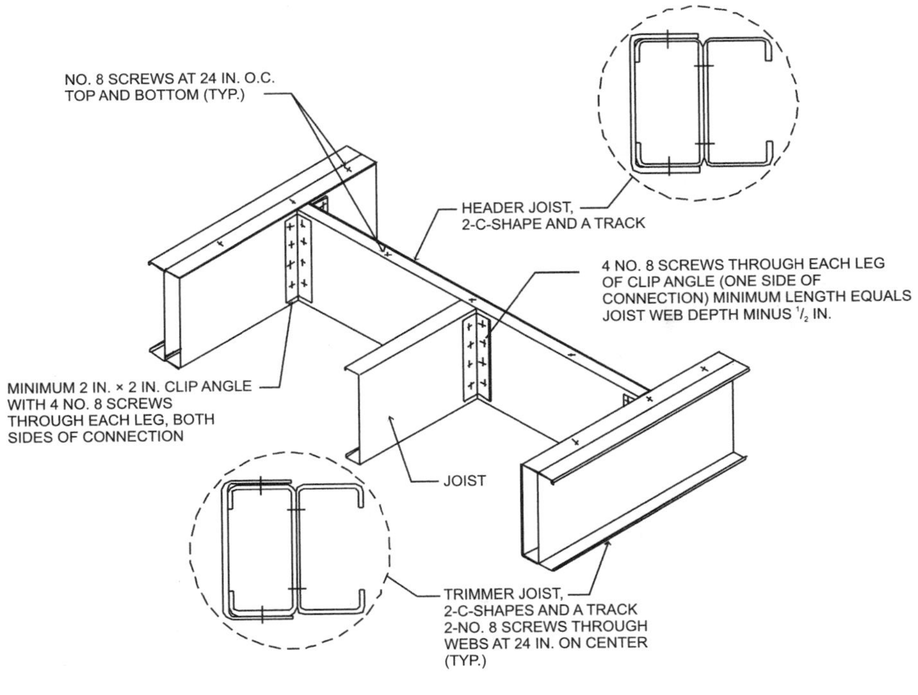

For SI: 1 inch = 25.4 mm, 1 foot = 304.8 mm.

FIGURE R505.3.8(4)
COLD-FORMED STEEL FLOOR CONSTRUCTION: FLOOR HEADER TO TRIMMER CONNECTION—8-FOOT OPENING

❖ See the commentary to Section R505.3.8.

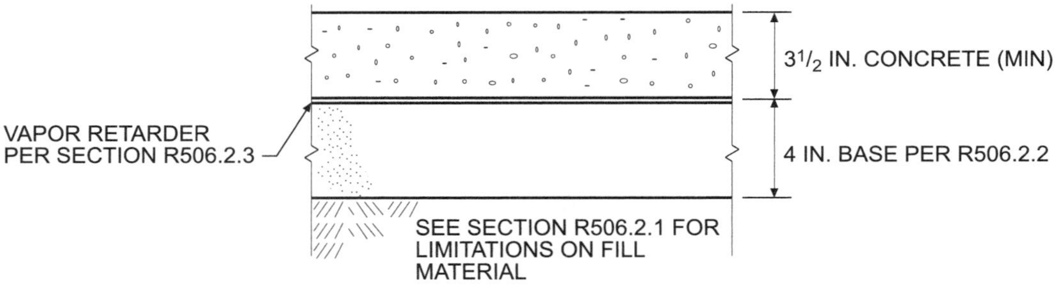

For SI: 1 inch = 25.4 mm.

Figure R506.1
CONCRETE SLAB-ON-GROUND REQUIREMENTS

R506.2.1 Fill. Fill material shall be free of vegetation and foreign material. The fill shall be compacted to assure uniform support of the slab, and except where *approved*, the fill depths shall not exceed 24 inches (610 mm) for clean sand or gravel and 8 inches (203 mm) for earth.

❖ To minimize differential settlement caused by consolidation of uncompacted fill and the problems associated with differential settlement, any fill beneath a concrete slab must be compacted. Properly compacted fill, besides minimizing settlement, increases the soil load-bearing characteristics and soil stability and reduces water penetration. The amount of compaction for fill is not specifically stipulated. Generally, fill soils should be compacted to 95-percent maximum density as determined by a Standard Proctor Test (ASTM D 698). Compaction requirements for expansive soils should be determined from an engineering analysis.

Soil may be compacted using equipment appropriate to the type of material being compacted, with lifts not exceeding 8 inches (203 mm). In general, thinner layers produce better compaction of fill, regardless of the type of soil being compacted.

R506.2.2 Base. A 4-inch-thick (102 mm) base course consisting of clean graded sand, gravel, crushed stone or crushed blast-furnace slag passing a 2-inch (51 mm) sieve shall be placed on the prepared subgrade when the slab is below *grade*.

Exception: A base course is not required when the concrete slab is installed on well-drained or sand-gravel mixture soils classified as Group I according to the United Soil Classification System in accordance with Table R405.1.

❖ Slabs need to be protected from the penetration of water and water vapor from below to prevent damage to interior finish material. Additionally, the effectiveness of thermal insulation at the slab may be adversely affected because of moisture. A 4-inch (102 mm) minimum granular base course is placed over the fill or undisturbed soil for slabs below grade to provide a capillary stop for water rising through the soil and into the slab. A base course is especially necessary when the site soil is other than gravel or clean sand. Both surface and groundwater problems must be addressed. Proper site preparation, selection of adequate fill and base course materials, and the installation of a vapor barrier minimize potential moisture problems associated with slab-on-grade construction.

R506.2.3 Vapor retarder. A 6-mil (0.006 inch; 152 mm) polyethylene or *approved* vapor retarder with joints lapped not less than 6 inches (152 mm) shall be placed between the concrete floor slab and the base course or the prepared subgrade where no base course exists.

Exception: The vapor retarder may be omitted:

1. From garages, utility buildings and other unheated *accessory structures*.
2. For unheated storage rooms having an area of less than 70 square feet (6.5 m^2) and carports.
3. From driveways, walks, patios and other flatwork not likely to be enclosed and heated at a later date.
4. Where *approved* by the *building official*, based on local site conditions.

❖ Although good quality, uncracked concrete is practically impermeable to the passage of water (unless the water is under a considerable pressure), concrete is not impervious to the passage of water vapors. If the surface of the slab is not sealed, water vapor will pass through the slab. If a floor finish such as linoleum, vinyl tile, wood flooring or any type of covering is placed on top of the slab, the moisture will be trapped in the slab. If the floor finish is adhered to the concrete, it may eventually loosen and buckle or blister.

Many of the moisture problems associated with enclosed slabs-on-ground can be minimized by installing a vapor retarder. When required, vapor retarders of either single- or multiple-layer membranes should be acceptable if the vapor retarder is properly installed with lapped joints and the barrier is not punctured during construction.

R506.2.4 Reinforcement support. Where provided in slabs on ground, reinforcement shall be supported to remain in place from the center to upper one third of the slab for the duration of the concrete placement.

❖ The code does not require reinforcement for the concrete slab-on-ground floors. When reinforcement is provided, common practice is to use welded wire fabric. The welded wire fabric frequently is left on the ground during the pour and not picked up. When lifted up by construction personnel, it is often not in a proper or consistent location, with some on the ground and some near the top of the slab. This section does not require installation of reinforcement; however, if it is installed, it must be properly located in the slab, or its benefit is lost.

SECTION R507 DECKS

R507.1 Decks. Where supported by attachment to an exterior wall, decks shall be positively anchored to the primary structure and designed for both vertical and lateral loads. Such attachment shall not be accomplished by the use of toenails or nails subject to withdrawal. Where positive connection to the primary building structure cannot be verified during inspection, decks shall be self-supporting. For decks with cantilevered framing members, connections to exterior walls or other framing members, shall be designed and constructed to resist uplift resulting from the full live load specified in Table R301.5 acting on the cantilevered portion of the deck.

❖ If an exterior wall is used to support a deck, the deck framing must be positively attached to the building structure. This connection design must include a consideration of both vertical and lateral loads, and the connection must be available for inspection. If it is not, this method of support is not permitted and the deck must be self-supporting.

If a deck has cantilevered framing, the framing must have a connection to its support that is designed to resist any uplift resulting from the full live load acting on the cantilevered span only. This load condition will produce maximum uplift at the support opposite the cantilevered end.

R507.2 Deck ledger connection to band joist. For decks supporting a total design load of 50 pounds per square foot (2394 Pa) [40 pounds per square foot (1915 Pa) live load plus 10 pounds per square foot (479 Pa) dead load], the connection between a deck ledger of pressure-preservative-treated Southern Pine, incised pressure-preservative-treated Hem-Fir or *approved* decay-resistant species, and a 2-inch (51 mm) nominal lumber band joist bearing on a sill plate or wall plate shall be constructed with $^1/_2$-inch (12.7 mm) lag screws or bolts with washers in accordance with Table R507.2. Lag screws, bolts and washers shall be hot-dipped galvanized or stainless steel.

❖ This section contains the prescriptive design for positive anchorage of a deck attached to the primary structure. The design is for vertical loads only [50 psf (2394 Pa) maximum] on the deck.

Researchers at Virginia Tech University and Washington State University have tested simulated deck-ledger to house-band-joist connections in their respective laboratories. A practical range of pressure-preservative-treated (PPT) deck ledger lumber (incised Hem-Fir and Southern Pine) was attached to a simulated Spruce-Pine-Fir band joist by $^1/_2$-inch (12.7 mm) lag screws or bolts with washers. The ledger connec-

TABLE R507.2
FASTENER SPACING FOR A SOUTHERN PINE OR HEM-FIR DECK LEDGER AND A 2-INCH-NOMINAL SOLID-SAWN SPRUCE-PINE-FIR BAND JOIST[c, f, g]
(Deck live load = 40 psf, deck dead load = 10 psf)

JOIST SPAN	6′ and less	6′1″ to 8′	8′1″ to 10′	10′1″ to 12′	12′1″ to 14′	14′1″ to 16′	16′1″ to 18′
Connection details	On-center spacing of fasteners[d, e]						
$^1/_2$ inch diameter lag screw with $^{15}/_{32}$ inch maximum sheathing[a]	30	23	18	15	13	11	10
$^1/_2$ inch diameter bolt with $^{15}/_{32}$ inch maximum sheathing	36	36	34	29	24	21	19
$^1/_2$ inch diameter bolt with $^{15}/_{32}$ inch maximum sheathing and $^1/_2$ inch stacked washers[b, h]	36	36	29	24	21	18	16

For SI: 1 inch = 25.4 mm, 1 foot = 304.8 mm. 1 pound per square foot = 0.0479 kPa.
a. The tip of the lag screw shall fully extend beyond the inside face of the band joist.
b. The maximum gap between the face of the ledger board and face of the wall sheathing shall be $^1/_2$ inch.
c. Ledgers shall be flashed to prevent water from contacting the house band joist.
d. Lag screws and bolts shall be staggered in accordance with Section R507.2.1.
e. Deck ledger shall be minimum 2 × 8 pressure-preservative-treated No. 2 grade lumber, or other approved materials as established by standard engineering practice.
f. When solid-sawn pressure-preservative-treated deck ledgers are attached to a minimum 1-inch-thick engineered wood product (structural composite lumber, laminated veneer lumber or wood structural panel band joist), the ledger attachment shall be designed in accordance with accepted engineering practice.
g. A minimum 1 × 9$^1/_2$ Douglas Fir laminated veneer lumber rimboard shall be permitted in lieu of the 2-inch nominal band joist.
h. Wood structural panel sheathing, gypsum board sheathing or foam sheathing not exceeding 1 inch in thickness shall be permitted. The maximum distance between the face of the ledger board and the face of the band joist shall be 1 inch

❖ See the commentary to Sections R507.2 and R507.2.1.

tion tests did not include carriage bolts; therefore, an engineered design of the ledger connection is required if carriage bolts are used. The deck ledger was separated from the house band joist by placing a piece of $^{15}/_{32}$-inch wall sheathing in the connection, and in another test case for bolts only, a $^{1}/_{2}$-inch (12.7 mm) stack of washers was inserted into the connection to produce a drainage plane. The specimens were tested to failure and the average test results ware divided by a factor of 3.0, intended to provide an adequate in-service safety factor, and further divided by 1.6 to convert from a "test duration" to a "normal duration" of 10 years recognized by the NDS and IBC as the proper duration for occupancy live load.

The test was made with two different band joists, a 2-inch (51 mm) nominal Spruce-Pine-Fir and a 1-inch by $9^{1}/_{2}$-inch Douglas Fir-Laminated (DFL) veneer. Note g of Table R507.2 permits the DFL band joist in lieu of the 2-inch (51 mm) nominal. Due to the limited investigation into the performance of composite-type band joists (only DFL was evaluated) and the possibility of band joists entering the market being a lower quality than what was tested at Washington State University, engineered wood band joists are not included in the scope of the fastener spacing table. Instead, Note f of Table R507.2 requires the attachment to be designed in accordance with accepted engineering practice when a minimum 1-inch (25 mm) engineered wood product is used as the band joist. The maximum distance between the face of the band joist and the deck ledger must not exceed 1-inch (25 mm). Note h of Table R507.2 will permit sheathing greater than $^{15}/_{32}$-inch if the 1-inch distance is not exceeded.

Penetration of moisture at the deck and the exterior house interface can be detrimental to the connectors. For this reason, the code requires the connectors to be hot-dipped galvanized or stainless steel.

R507.2.1 Placement of lag screws or bolts in deck ledgers and band joists. The lag screws or bolts in deck ledgers and band joists shall be placed in accordance with Table R507.2.1 and Figures R507.2.1(1) and R507.2.1(2).

❖ The on-center spacing of the fasteners as shown in Table R507.2.1 and Figures R507.2.1(1) and R507.2.1(2) is the closest spacing for the two cases of deck ledger lumber tested as referenced in the commentary for Section R502.7. In deck construction, a joist hanger or angle connector is commonly installed at the end of the ledger board to support the deck rim joist. Often these members are double 2x members (3 inch (76 mm) nominal width). The spacing of 2 inches to 5 inches (51 mm to 127 mm) from the ends will allow the installer flexibility to locate the lag screw or bolt so that it does not interfere with the installation of the joist hanger or structural connector. Five inches (127 mm) will accommodate an inverted flange double 2x joist hanger.

R507.2.2 Alternate deck ledger connections. Deck ledger connections not conforming to Table R507.2 shall be designed in accordance with accepted engineering practice.

TABLE R507.2.1
PLACEMENT OF LAG SCREWS AND BOLTS IN DECK LEDGERS AND BAND JOISTS

	MINIMUM END AND EDGE DISTANCES AND SPACING BETWEEN ROWS			
	TOP EDGE	BOTTOM EDGE	ENDS	ROW SPACING
Ledger[a]	2 inches[d]	$^{1}/_{4}$ inch	2 inches[b]	$1^{5}/_{8}$ inches[b]
Band Joist[c]	$^{3}/_{4}$ inch	2 inches	2 inches[b]	$1^{5}/_{8}$ inches[b]

For SI: 1 inch = 25.4 mm.

a. Lag screws or bolts shall be staggered from the top to the bottom along the horizontal run of the deck ledger in accordance with Figure R507.2.1(1).

b. Maximum 5 inches.

c. For engineered rim joists, the manufacturer's recommendations shall govern.

d. The minimum distance from bottom row of lag screws or bolts to the top edge of the ledger shall be in accordance with Figure R507.2.1(1).

❖ See the commentary to Section R507.2.1.

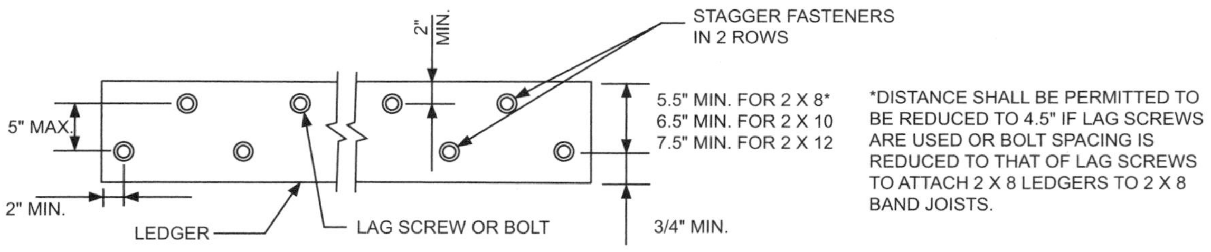

For SI: 1 inch = 25.4 mm.

FIGURE R507.2.1(1)
PLACEMENT OF LAG SCREWS AND BOLTS IN LEDGERS

❖ See the commentary to Section R507.2.1.

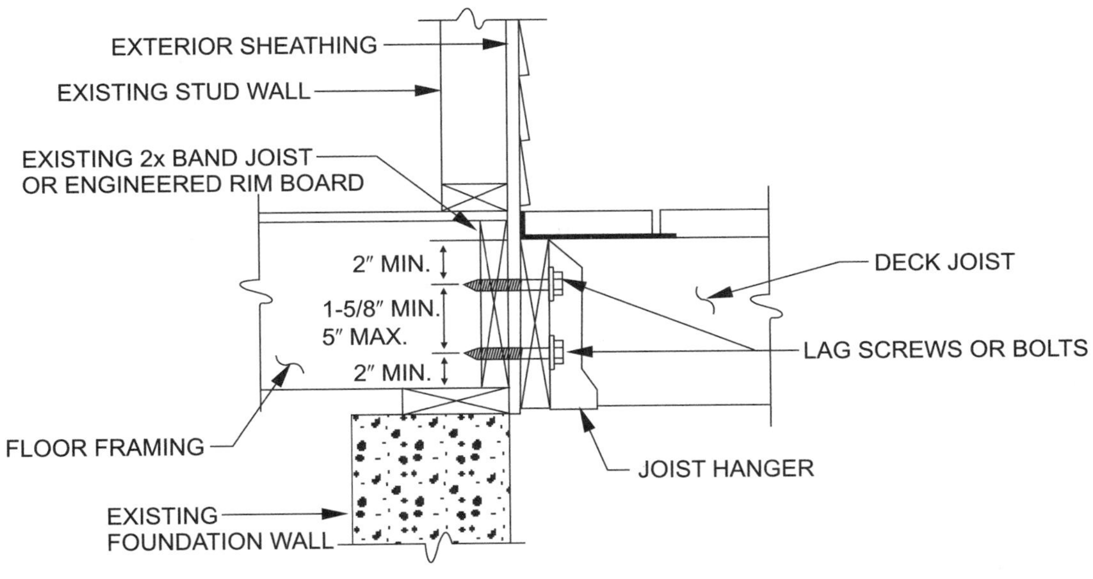

For SI: 1 inch = 25.4 mm.

FIGURE R507.2.1(2)
PLACEMENT OF LAG SCREWS AND BOLTS IN BAND JOISTS

❖ See the commentary to Section R507.2.1.

Girders supporting deck joists shall not be supported on deck ledgers or band joists. Deck ledgers shall not be supported on stone or masonry veneer.

❖ This section clarifies when an engineered design is required for the deck connections. Deck ledgers or band joists must not support the girders supporting deck joists. The deck ledger is prohibited from being supported by stone or masonry veneer.

R507.2.3 Deck lateral load connection. The lateral load connection required by Section R507.1 shall be permitted to be in accordance with Figure R507.2.3. Where the lateral load connection is provided in accordance with Figure 507.2.3, hold-down tension devices shall be installed in not less than two locations per deck, and each device shall have an allowable stress design capacity of not less than 1500 pounds (6672 N).

❖ The code requires decks that are supported by attachment to an exterior wall to be positively anchored to the primary structure, and be designed for both vertical and lateral loads. The vertical and lateral loads referred to are code-prescribed loads such as dead, live, wind and seismic loads. The magnitude of the lateral loads to be resisted is not specified and must be determined in accordance with accepted engineering practice.

Attachment only to the band joist may not be sufficient for the lateral loads. Positive anchorage of the deck joist to the floor framing addresses this potential failure. Figure R507.2.3 shows a typical hold-down device that provides a positive connection to the floor framing for the lateral loads. This figure is based on a similar figure from FEMA 232. The required number and actual design of the connection to resist the lateral loads must be determined in accordance with accepted engineering practice. Where Figure R507.2.3 is used, a minimum of two hold-down devices, with a design capacity of 1,500 pounds (6672 N) each, are required. Where Figure R507.2.3 is not used then the quantity and load capacity must comply with the engineered design.

R507.3 Wood/plastic composites. Wood/plastic composites used in exterior deck boards, stair treads, handrails and guardrail systems shall bear a label indicating the required performance levels and demonstrating compliance with the provisions of ASTM D 7032.

❖ Wood/plastic composite (WPC) materials, commonly used in exterior deck boards, guards and handrails, must be rated for appropriate performance criteria. These materials have a widespread acceptance for residential construction, and labeling requirements for WPCs will ensure the safe application of these materials in exterior deck systems. The referenced standard, ASTM D 7032, includes performance evaluations, such as flexural tests, ultraviolet-resistance tests, freeze-thaw-resistance tests, biodeterioration tests, fire-performance tests, creep-recovery tests, mechanical fastener holding tests and slip-resistance tests. The standard also includes considerations of the effects of temperature and moisture, concentrated loads and fire propagation tests.

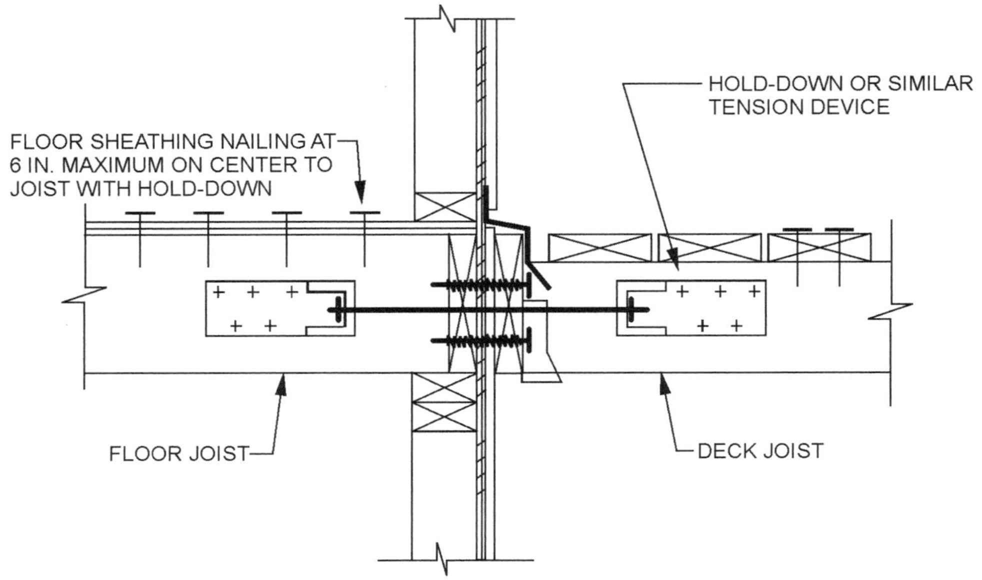

For SI: 1 inch = 25.4 mm.

FIGURE 507.2.3
DECK ATTACHMENT FOR LATERAL LOADS

❖ See the commentary to Section R507.2.3.

R507.3.1 Installation of wood/plastic composites. Wood/plastic composites shall be installed in accordance with the manufacturer's instructions.

❖ This section includes the requirement for the installation of these WPC products in accordance with the manufacturer's instructions, which is the best way to ensure that WPCs perform to the required design loads, and installation instructions are an integral part of the manufacturer's labeling program.

Bibliography

The following resource materials were used in the preparation of the commentary for this chapter of the code:

AISI S100-07/S1-10, *North American Specification for Design of Cold-formed Steel Structural Members with Supplement 1, dated 2010*. Washington, DC: American Iron and Steel Institute.

AISI S214-07/S2-08, *North American Standard for Cold-formed Steel Framing-Truss with Supplement 2 dated 2008.* Washington, DC: American Iron and Steel Institute, 2007.

AFPA/N DS-2012, *National Design Specification for Wood Construction with 2005 Supplement*. Washington, DC: American Forest and Paper Association, 2012.

ANSI/AITC A 190.1-07, *Structural Glued Laminated Timber*. Centennial, CO: American Institute of Timber Construction, 2007.

ANSI A208.1-2009, *Particleboard*. New York: American National Standards Institute, 2009.

APA E30-03, *Engineered Wood Construction Guide.* Tacoma, WA: APA-The Engineered Wood Association, 2003.

ASTM C 954-07, *Specification for Steel Drill Screws for the Application of Gypsum Panel Products or Metal Plaster Bases to Steel Studs from 0.033 in. (0.84 mm) to 0.112 in. (2.84 mm) in Thickness.* West Conshohocken, PA: ASTM International, 2007.

ASTM C 1513-04, *Standard Specification for Steel Tapping Screws for Cold-formed Steel Framing Connections.* West Conshohocken, PA: ASTM International, 2004.

ASTM D 698-07e1, *Standard Test Methods for Laboratory Compaction Characteristics of Soil Using Effort [12 400 ft-lbf/ft^3 (600 kN-m/m^3)].* West Conshohocken, PA: ASTM International, 2007.

ASTM D 3737-08 *Practice for Establishing Allowable Properties for Structural Glued Laminated Timber (Glulam).* West Conshohocken, PA: ASTM International, 2008.

ASTM D 3957-06, *Standard Practices for Establishing Stress Grades for Structural Members Used in Log Buildings.* West Conshohocken, PA: ASTM International, 2006.

ASTM D 5055-09, *Specification for Establishing and Monitoring Structural Capacities of Prefabricated Wood I-Joists.* West Conshohocken, PA: ASTM International, 2009.

ASTM D 5456-09, *Standard Specification for Evaluation of Structural Composite Lumber Products.* West Conshohocken, PA: ASTM International, 2009.

ASTM D 7032-08, *Standard Specification for Establishing Performance Ratings for Wood-Plastic Composite Deck Boards and Guardrail Systems (Guards or Handrails).* West Conshohocken, PA: ASTM International, 2008.

ASTM D 7374-08, *Standard Practice for Evaluating Temperature Performance of Adhesives Used In End-Jointed Lumber.* West Conshohocken, PA: ASTM International, 2008.

ASTM D 7470-08, *Standard Practice for Evaluating Elevated Temperature Performance of End-Jointed Lumber Studs.* West Conshohocken, PA: ASTM International, 2008.

AWPA U1-11, *Use Category System: User Specification for Treated Wood, except Section 6, Commodity Specification H.* Birmingham, AL: American Wood-Preservers Association, 2011.

FEMA 232-06, *Homebuilders' Guide to Earthquake Resistant Design and Construction.* Washington, DC: Federal Emergency Management Agency, 2006.

IBC-12, *International Building Code.* Washington, DC: International Code Council, 2011.

TPI 1-2007, *National Design Standard for Metal-plate-connected Wood Truss Construction.* Madison, WI: Truss Plate Institute, 2007.

Chapter 6: Wall Construction

General Comments

Chapter 6 contains provisions that regulate the design and construction of walls. The wall construction covered in Chapter 6 consists of five different types: wood framed, cold-formed steel framed, masonry, concrete and structural insulated panel (SIP). The primary concern of this chapter is the structural integrity of wall construction and transfer of all imposed loads to the supporting structure.

Section R601 contains the scope, as well as the performance expectations for walls. Section R602 addresses wood wall framing. Section R603 addresses cold-formed steel wall framing. Section R604 specifies requirements for wood-structural-panel wall sheathing. Section R605 provides requirements for particleboard used as wall sheathing. Section R606 states the general requirements pertaining to masonry wall construction. Section R607 addresses unit masonry walls. Section R608 covers walls of multiple-wythe masonry. Section R609 contains grouted masonry provisions. Section R610 specifies requirements for wall panels of glass unit masonry. Section R611 addresses prescriptive requirements for concrete wall construction with stay-in-place or removable forms. Section R612 contains the criteria for the performance of exterior windows and doors. Section R613 contains the provisions for SIP wall construction.

In certain instances a wall must have a fire-resistance rating. Where this is necessary, the wall system must be a tested assembly, and any conditions specific to the installation of the assembly must apply as well. Examples of these provisions include:

- Exterior walls based on location on property (see Section R302).
- Walls serving as dwelling unit separations in two-family dwellings (see Section R302.3).
- Common wall between townhouses (see Section R302.2, exception).

Purpose

This chapter provides the requirements for the design and construction of wall systems that are capable of supporting the minimum design vertical loads (dead, live and snow loads) and lateral loads (wind or seismic loads). This chapter contains the prescriptive requirements for wall bracing and/or shear walls to resist the imposed lateral loads due to wind and seismic. Chapter 6 also contains requirements for the use of vapor retarders for moisture control in walls.

Chapter 6 also regulates exterior windows and doors installed in walls. The chapter contains criteria for the performance of exterior windows and doors and includes provisions for window sill height, testing and labeling, vehicular access doors, wind-borne debris protection and anchorage details.

SECTION R601 GENERAL

R601.1 Application. The provisions of this chapter shall control the design and construction of all walls and partitions for all buildings.

❖ This section establishes the scope of the chapter.

R601.2 Requirements. Wall construction shall be capable of accommodating all loads imposed according to Section R301 and of transmitting the resulting loads to the supporting structural elements.

❖ This is a general performance statement, requiring that walls support the required design loads and provide an adequate load path to supporting elements.

R601.2.1 Compressible floor-covering materials. Compressible floor-covering materials that compress more than $^1/_{32}$ inch (0.8 mm) when subjected to 50 pounds (23 kg) applied over 1 inch square (645 mm) of material and are greater than $^1/_8$ inch (3 mm) in thickness in the uncompressed state shall not extend beneath walls, partitions or columns, which are fastened to the floor.

❖ Although it is preferable to fasten walls directly to the supporting structure, this provision allows them to be installed over finish floor materials that meet the specified criteria.

SECTION R602 WOOD WALL FRAMING

R602.1 Identification. Load-bearing dimension lumber for studs, plates and headers shall be identified by a grade mark of a lumber grading or inspection agency that has been *approved* by an accreditation body that complies with DOC PS 20. In lieu of a grade mark, a certification of inspection issued by a lumber grading or inspection agency meeting the requirements of this section shall be accepted.

❖ Wood materials performing a load-carrying function must conform to a minimum quality control standard. To verify this, load-bearing lumber must be properly identified to indicate that the grades and species meet the minimum requirements specified in this chapter. Commentary Figure R602.1 shows the information contained in a grade mark. For more examples, see Commentary Figure R502.1. For examples of marks for wood structural panels, see Commentary Figure R503.2.1.

R602.1.1 End-jointed lumber. Approved end-jointed lumber identified by a grade mark conforming to Section R602.1 may be used interchangeably with solid-sawn members of the same species and grade. End-jointed lumber used in an assembly required elsewhere in this code to have a fire-resistance rating shall have the designation "Heat Resistant Adhesive" or "HRA" included in its grade mark.

❖ End-jointed lumber that has been fabricated in accordance with nationally recognized standards can be used interchangeably with solid-sawn members, provided that the species and grade of the wood are comparable. Commentary Figure R602.1.1(1) shows end-jointed lumber, which is also referred to as "finger-jointed" (see commentary, Section R502.1.3).

End-jointed lumber carrying the HRA mark, indicating it has been joined using heat-resistant adhesive, is permitted to be used interchangeably with solid-sawn members of the same species and grade in fire-rated applications. End-jointed lumber manufactured with an adhesive not qualified as a heat-resistant adhesive will be designated as "non-heat resistant adhesive" or "non-HRA" on the grade stamp. Commentary Figure R602.1.1(2) shows the HRA mark included in a grade stamp.

R602.1.2 Structural glued laminated timbers. Glued laminated timbers shall be manufactured and identified as required in ANSI/AITC A190.1 and ASTM D 3737.

❖ Glued laminated (gluelam) timbers are engineered wood elements consisting of layers of plywood plies with varying grades of plywood through the thickness of the member. The strength of the members and the use of the members (beams, studs, etc.) are based on different combinations of plywood veneer grades, called layups. The design, testing and manufacture of different layups for different applications are standardized in the referenced standards. The standards also address the identification of these timbers, which is useful at the construction site during inspection.

R602.1.3 Structural log members. Stress grading of structural log members of nonrectangular shape, as typically used in log buildings, shall be in accordance with ASTM D 3957. Such structural log members shall be identified by the grade mark of an *approved* lumber grading or inspection agency. In lieu of a grade mark on the material, a certificate of inspection as to species and grade, issued by a lumber-grading or inspection agency meeting the requirements of this section, shall be permitted to be accepted.

❖ This section addresses grading requirements for logs used as structural members. This subsection speci-

INTERPRETING GRADE STAMPS

MOST GRADE STAMPS, EXCEPT THOSE FOR ROUGH LUMBER OR HEAVY TIMBERS, CONTAIN FIVE BASIC ELEMENTS:

A. THE TRADEMARK INDICATES AGENCY QUALITY SUPERVISION.

B. MILL IDENTIFICATION – FIRM NAME, BRAND OR ASSIGNED MILL NUMBER.

C. GRADE DESIGNATION – GRADE NAME, NUMBER OR ABBREVIATION.

D. SPECIAL IDENTIFICATION – INDICATES SPECIES INDIVIDUALLY OR IN COMBINATION.

E. CONDITION OF SEASONING AT TIME OF SURFACING:
- S-DRY – 19% MAX MOISTURE CONTENT
- MC 15 – 15% MAX MOISTURE CONTENT
- S-GRN – OVER 19% MOISTURE CONTENT (UNSEASONED)

Figure R602.1
GRADE STAMP EXAMPLE FOR DIMENSIONAL LUMBER

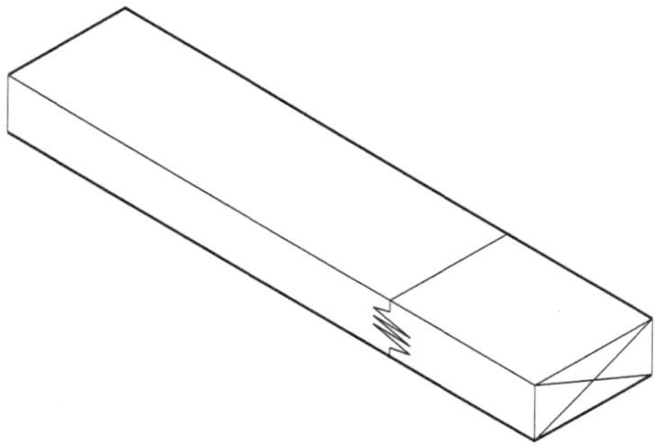

Figure R602.1.1(1)
END-JOINTED LUMBER—FINGER JOINT

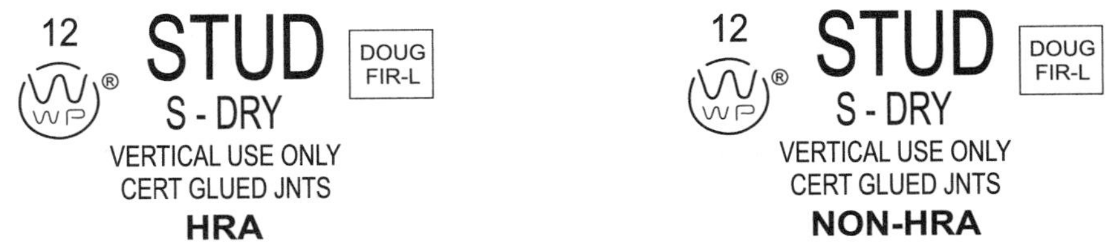

Figure R602.1.1(2)
GRADE STAMP INCLUDING HRA MARK

fies the reference for acceptable methods for establishing structural capacities of logs and specifies the requirement for a grading stamp or alternate certification on structural logs. Structural log members must be graded in accordance with ASTM D 3957, *Standard Practices for Establishing Stress Grades for Structural Members Used in Log Buildings*.

R602.1.4 Structural composite lumber. Structural capacities for structural composite lumber shall be established and monitored in accordance with ASTM D 5456.

❖ Structural composite lumber (SCL) is an engineered wood product using veneer sheets or strands of wood glued together to form structural beams, columns, studs, or sheathing. The strength and use of the members depends upon the size and thickness of the strands, the use of veneer sheets, and the orientation of the strand or veneer. Design, testing and manufacture of the engineered lumber are standardized in ASTM D 5456, the referenced standard for SCL.

R602.2 Grade. Studs shall be a minimum No. 3, standard or stud grade lumber.

Exception: Bearing studs not supporting floors and nonbearing studs may be utility grade lumber, provided the studs are spaced in accordance with Table R602.3(5).

❖ A minimum grade of No. 3, standard, or stud grade lumber is specified for studs in conventional wood-framed walls to maintain minimum load-carrying capabilities. The maximum size of standard grade lumber is limited to 2 to 4 inches (51 to 102 mm) nominal in thickness and 4 inches (102 mm) in width. Thus, a 4-inch by 6-inch (102 mm by 152 mm) piece of dimensional lumber would not be available in the standard grade.

To allow more economical use of lumber for bearing studs not supporting floors and nonbearing studs, the exception permits the use of utility grade lumber, provided that the studs are spaced according to the limits established in Table R602.3(5). Note "a" of the table places a more restrictive limit on utility grade stud spacing where permitted in a load-bearing application.

R602.3 Design and construction. Exterior walls of wood-frame construction shall be designed and constructed in accordance with the provisions of this chapter and Figures R602.3(1) and R602.3(2) or in accordance with AF&PA's NDS. Components of exterior walls shall be fastened in accordance with Tables R602.3(1) through R602.3(4). Wall sheathing shall be fastened directly to framing members and, when placed on the exterior side of an exterior wall, shall be capable of resisting the wind pressures listed in Table R301.2(2) adjusted for height and exposure using Table R301.2(3). Wood structural panel sheathing used for exterior walls shall conform to DOC PS 1, DOC PS 2 or, when manufactured in Canada, CSA O437 or CSA O325. All panels shall be identified for grade, bond classification, and Performance Category by a grade mark or certificate of inspection issued by an approved agency and shall conform to the requirements of Table R602.3(3). Wall sheathing used only for exterior wall covering purposes shall comply with Section R703.

Studs shall be continuous from support at the sole plate to a support at the top plate to resist loads perpendicular to the wall. The support shall be a foundation or floor, ceiling or roof diaphragm or shall be designed in accordance with accepted engineering practice.

Exception: Jack studs, trimmer studs and cripple studs at openings in walls that comply with Tables R502.5(1) and R502.5(2).

❖ Figures R602.3(1) and R602.3(2) in conjunction with the minimum nailing requirements of Tables R602.3(1) and R602.3(2) provide typical wall-framing details for construction of exterior walls. Walls must be in accordance with this section or American Forest & Paper Association's (AF&A) *National Design Specification* (NDS).

All exterior wall coverings must be capable of resisting the wind pressures required by Section R301.2.1.

Wall sheathing used only for exterior wall covering purposes shall comply with Section R703. Wall sheathing used for structural purposes, i.e. wall bracing, must comply with Section R602.10 or R602.12. All wood structural panels used for sheathing on exterior walls must be fastened in accordance with Table R602.3(3)

For wood structural panel sheathing used for interior walls, the fastening is in accordance with Table R602.3(1). The 6d common nail fastening for the wood structural panels in Table R602.3(1) is not adequate for the wind pressures required by Section R301.2.

The studs for exterior walls must be continuous from the support at the bottom to the support at the top, i.e., stacked framing that forms a "hinge" is not permitted. There is no prescriptive design in the code for stacked framing. Exterior wall framing that is stacked and forms a "hinge" must be engineered. The exception for continuity of studs is for jack studs, trimmer studs, and cripple studs at openings.

TABLE R602.3(1)
FASTENER SCHEDULE FOR STRUCTURAL MEMBERS

ITEM	DESCRIPTION OF BUILDING ELEMENTS	NUMBER AND TYPE OF FASTENER[a, b, c]	SPACING OF FASTENERS
		Roof	
1	Blocking between joists or rafters to top plate, toe nail	3-8d ($2^1/_2'' \times 0.113''$)	—
2	Ceiling joists to plate, toe nail	3-8d ($2^1/_2'' \times 0.113''$)	—
3	Ceiling joists not attached to parallel rafter, laps over partitions, face nail	3-10d	—
4	Collar tie to rafter, face nail or $1^1/_4'' \times 20$ gage ridge strap	3-10d ($3'' \times 0.128''$)	—
5	Rafter or roof truss to plate, toe nail	3-16d box nails ($3^1/_2'' \times 0.135''$) or 3-10d common nails ($3'' \times 0.148''$)	2 toe nails on one side and 1 toe nail on opposite side of each rafter or truss[j]
6	Roof rafters to ridge, valley or hip rafters: toe nail face nail	4-16d ($3^1/_2'' \times 0.135''$) 3-16d ($3^1/_2'' \times 0.135''$)	—
		Wall	
7	Built-up studs-face nail	10d ($3'' \times 0.128''$)	24″ o.c.
8	Abutting studs at intersecting wall corners, face nail	16d (3 ½" x 0.135")	12″ o.c.
9	Built-up header, two pieces with $^1/_2''$ spacer	16d ($3^1/_2'' \times 0.135''$)	16″ o.c. along each edge
10	Continued header, two pieces	16d ($3^1/_2'' \times 0.135''$)	16″ o.c. along each edge
11	Continuous header to stud, toe nail	4-8d ($2^1/_2'' \times 0.113''$)	—
12	Double studs, face nail	10d ($3'' \times 0.128''$)	24″ o.c.
13	Double top plates, face nail	10d ($3'' \times 0.128''$)	24″ o.c.
14	Double top plates, minimum 24-inch offset of end joints, face nail in lapped area	8-16d ($3^1/_2'' \times 0.135''$)	—
15	Sole plate to joist or blocking, face nail	16d ($3^1/_2'' \times 0.135''$)	16″ o.c.
16	Sole plate to joist or blocking at braced wall panels	3-16d ($3^1/_2'' \times 0.135''$)	16″ o.c.
17	Stud to sole plate, toe nail	3-8d ($2^1/_2'' \times 0.113''$) or 2-16d ($3^1/_2'' \times 0.135''$)	— —
18	Top or sole plate to stud, end nail	2-16d ($3^1/_2'' \times 0.135''$)	—
19	Top plates, laps at corners and intersections, face nail	2-10d ($3'' \times 0.128''$)	—
20	1″ brace to each stud and plate, face nail	2-8d ($2^1/_2'' \times 0.113''$) 2 staples $1^3/_4''$	— —
21	1″ × 6″ sheathing to each bearing, face nail	2-8d ($2^1/_2'' \times 0.113''$) 2 staples $1^3/_4''$	— —
22	1″ × 8″ sheathing to each bearing, face nail	2-8d ($2^1/_2'' \times 0.113''$) 3 staples $1^3/_4''$	— —
23	Wider than 1″ × 8″ sheathing to each bearing, face nail	3-8d ($2^1/_2'' \times 0.113''$) 4 staples $1^3/_4''$	— —
		Floor	
24	Joist to sill or girder, toe nail	3-8d ($2^1/_2'' \times 0.113''$)	—
25	Rim joist to top plate, toe nail (roof applications also)	8d ($2^1/_2'' \times 0.113''$)	6″ o.c.
26	Rim joist or blocking to sill plate, toe nail	8d ($2^1/_2'' \times 0.113''$)	6″ o.c.
27	1″ × 6″ subfloor or less to each joist, face nail	2-8d ($2^1/_2'' \times 0.113''$) 2 staples $1^3/_4''$	— —
28	2″ subfloor to joist or girder, blind and face nail	2-16d ($3^1/_2'' \times 0.135''$)	—
29	2″ planks (plank & beam - floor & roof)	2-16d ($3^1/_2'' \times 0.135''$)	at each bearing
30	Built-up girders and beams, 2-inch lumber layers	10d ($3'' \times 0.128''$)	Nail each layer as follows: 32″ o.c. at top and bottom and staggered. Two nails at ends and at each splice.
31	Ledger strip supporting joists or rafters	3-16d ($3^1/_2'' \times 0.135''$)	At each joist or rafter

(continued)

TABLE R602.3(1)—continued
FASTENER SCHEDULE FOR STRUCTURAL MEMBERS

ITEM	DESCRIPTION OF BUILDING MATERIALS	DESCRIPTION OF FASTENER[b, c, e]	SPACING OF FASTENERS	
			Edges (inches)[i]	Intermediate supports[c, e] (inches)
Wood structural panels, subfloor, roof and interior wall sheathing to framing and particleboard wall sheathing to framing				
32	$^3/_8$" - $^1/_2$"	6d common (2" × 0.113") nail (subfloor wall)[j] 8d common (2$^1/_2$" × 0.131") nail (roof)[f]	6	12[g]
33	$^{19}/_{32}$" - 1"	8d common nail (2$^1/_2$" × 0.131")	6	12[g]
34	1$^1/_8$" - 1$^1/_4$"	10d common (3" × 0.148") nail or 8d (2$^1/_2$" × 0.131") deformed nail	6	12
Other wall sheathing[h]				
35	$^1/_2$" structural cellulosic fiberboard sheathing	1$^1/_2$" galvanized roofing nail, $^7/_{16}$" crown or 1" crown staple 16 ga., 1$^1/_4$" long	3	6
36	$^{25}/_{32}$" structural cellulosic fiberboard sheathing	1$^3/_4$" galvanized roofing nail, $^7/_{16}$" crown or 1" crown staple 16 ga., 1$^1/_2$" long	3	6
37	$^1/_2$" gypsum sheathing[d]	1$^1/_2$" galvanized roofing nail; staple galvanized, 1$^1/_2$" long; 1$^1/_4$ screws, Type W or S	7	7
38	$^5/_8$" gypsum sheathing[d]	1$^3/_4$" galvanized roofing nail; staple galvanized, 1$^5/_8$" long; 1$^5/_8$" screws, Type W or S	7	7
	Wood structural panels, combination subfloor underlayment to framing			
39	$^3/_4$" and less	6d deformed (2" × 0.120") nail or 8d common (2$^1/_2$" × 0.131") nail	6	12
40	$^7/_8$" - 1"	8d common (2$^1/_2$" × 0.131") nail or 8d deformed (2$^1/_2$" × 0.120") nail	6	12
41	1$^1/_8$" - 1$^1/_4$"	10d common (3" × 0.148") nail or 8d deformed (2$^1/_2$" × 0.120") nail	6	12

For SI: 1 inch = 25.4 mm, 1 foot = 304.8 mm, 1 mile per hour = 0.447 m/s; 1 Ksi = 6.895 MPa.

a. All nails are smooth-common, box or deformed shanks except where otherwise stated. Nails used for framing and sheathing connections shall have minimum average bending yield strengths as shown: 80 ksi for shank diameter of 0.192 inch (20d common nail), 90 ksi for shank diameters larger than 0.142 inch but not larger than 0.177 inch, and 100 ksi for shank diameters of 0.142 inch or less.

b. Staples are 16 gage wire and have a minimum $^7/_{16}$-inch on diameter crown width.

c. Nails shall be spaced at not more than 6 inches on center at all supports where spans are 48 inches or greater.

d. Four-foot by 8-foot or 4-foot by 9-foot panels shall be applied vertically.

e. Spacing of fasteners not included in this table shall be based on Table R602.3(2).

f. For regions having basic wind speed of 110 mph or greater, 8d deformed (2$^1/_2$" × 0.120) nails shall be used for attaching plywood and wood structural panel roof sheathing to framing within minimum 48-inch distance from gable end walls, if mean roof height is more than 25 feet, up to 35 feet maximum.

g. For regions having basic wind speed of 100 mph or less, nails for attaching wood structural panel roof sheathing to gable end wall framing shall be spaced 6 inches on center. When basic wind speed is greater than 100 mph, nails for attaching panel roof sheathing to intermediate supports shall be spaced 6 inches on center for minimum 48-inch distance from ridges, eaves and gable end walls; and 4 inches on center to gable end wall framing.

h. Gypsum sheathing shall conform to ASTM C 1396 and shall be installed in accordance with GA 253. Fiberboard sheathing shall conform to ASTM C 208.

i. Spacing of fasteners on floor sheathing panel edges applies to panel edges supported by framing members and required blocking and at all floor perimeters only. Spacing of fasteners on roof sheathing panel edges applies to panel edges supported by framing members and required blocking. Blocking of roof or floor sheathing panel edges perpendicular to the framing members need not be provided except as required by other provisions of this code. Floor perimeter shall be supported by framing members or solid blocking.

j. Where a rafter is fastened to an adjacent parallel ceiling joist in accordance with this schedule, provide two toe nails on one side of the rafter and toe nails from the ceiling joist to top plate in accordance with this schedule. The toe nail on the opposite side of the rafter shall not be required.

❖ The fastener schedule provides minimum nailing requirements (i.e., size, spacing) for connecting building elements used in wood framed construction. For wood structural panels, both edge nailing and intermediate (field) nailing are specified. In addition to the nailing for wood structural panels, fasteners are specified for gypsum wall sheathing, cellulosic fiberboard wall sheathing and combination subfloor underlayment.

TABLE R602.3(2)
ALTERNATE ATTACHMENTS TO TABLE R602.3(1)

NOMINAL MATERIAL THICKNESS (inches)	DESCRIPTION[a, b] OF FASTENER AND LENGTH (inches)	SPACING[c] OF FASTENERS	
		Edges (inches)	Intermediate supports (inches)
Wood structural panels subfloor, roof[g] and wall sheathing to framing and particleboard wall sheathing to framing[f]			
Up to $^1/_2$	Staple 15 ga. $1^3/_4$	4	8
	0.097 - 0.099 Nail $2^1/_4$	3	6
	Staple 16 ga. $1^3/_4$	3	6
$^{19}/_{32}$ and $^5/_8$	0.113 Nail 2	3	6
	Staple 15 and 16 ga. 2	4	8
	0.097 - 0.099 Nail $2^1/_4$	4	8
$^{23}/_{32}$ and $^3/_4$	Staple 14 ga. 2	4	8
	Staple 15 ga. $1^3/_4$	3	6
	0.097 - 0.099 Nail $2^1/_4$	4	8
	Staple 16 ga. 2	4	8
1	Staple 14 ga. $2^1/_4$	4	8
	0.113 Nail $2^1/_4$	3	6
	Staple 15 ga. $2^1/_4$	4	8
	0.097 - 0.099 Nail $2^1/_2$	4	8
NOMINAL MATERIAL THICKNESS (inches)	**DESCRIPTION[a,b] OF FASTENER AND LENGTH (inches)**	**SPACING[c] OF FASTENERS**	
		Edges (inches)	**Body of panel[d] (inches)**
Floor underlayment; plywood-hardboard-particleboard[f]			
Plywood			
$^1/_4$ and $^5/_{16}$	$1^1/_4$ ring or screw shank nail-minimum $12^1/_2$ ga. (0.099″) shank diameter	3	6
	Staple 18 ga., $^7/_8$, $^3/_{16}$ crown width	2	5
$^{11}/_{32}$, $^3/_8$, $^{15}/_{32}$, and $^1/_2$	$1^1/_4$ ring or screw shank nail-minimum $12^1/_2$ ga. (0.099″) shank diameter	6	8[e]
$^{19}/_{32}$, $^5/_8$, $^{23}/_{32}$ and $^3/_4$	$1^1/_2$ ring or screw shank nail-minimum $12^1/_2$ ga. (0.099″) shank diameter	6	8
	Staple 16 ga. $1^1/_2$	6	8
Hardboard[f]			
0.200	$1^1/_2$ long ring-grooved underlayment nail	6	6
	4d cement-coated sinker nail	6	6
	Staple 18 ga., $^7/_8$ long (plastic coated)	3	6
Particleboard			
$^1/_4$	4d ring-grooved underlayment nail	3	6
	Staple 18 ga., $^7/_8$ long, $^3/_{16}$ crown	3	6
$^3/_8$	6d ring-grooved underlayment nail	6	10
	Staple 16 ga., $1^1/_8$ long, $^3/_8$ crown	3	6
$^1/_2$, $^5/_8$	6d ring-grooved underlayment nail	6	10
	Staple 16 ga., $1^5/_8$ long, $^3/_8$ crown	3	6

For SI: 1 inch = 25.4 mm.

a. Nail is a general description and may be T-head, modified round head or round head.
b. Staples shall have a minimum crown width of $^7/_{16}$-inch on diameter except as noted.
c. Nails or staples shall be spaced at not more than 6 inches on center at all supports where spans are 48 inches or greater. Nails or staples shall be spaced at not more than 12 inches on center at intermediate supports for floors.
d. Fasteners shall be placed in a grid pattern throughout the body of the panel.
e. For 5-ply panels, intermediate nails shall be spaced not more than 12 inches on center each way.
f. Hardboard underlayment shall conform to CPA/ANSI A135.4
g. Specified alternate attachments for roof sheathing shall be permitted for windspeeds less than 100 mph. Fasteners attaching wood structural panel roof sheathing to gable end wall framing shall be installed using the spacing listed for panel edges.

❖ This table offers alternatives to the nailing specified for wood structural panels in Table R602.3(1).

TABLE R602.3(3)
REQUIREMENTS FOR WOOD STRUCTURAL PANEL WALL SHEATHING USED TO RESIST WIND PRESSURES[a, b,]

MINIMUM NAIL		MINIMUM WOOD STRUCTURAL PANEL SPAN RATING	MINIMUM NOMINAL PANEL THICKNESS (inches)	MAXIMUM WALL STUD SPACING (inches)	PANEL NAIL SPACING		MAXIMUM WIND SPEED (mph)		
							Wind exposure category		
Size	Penetration (inches)				Edges (inches o.c.)	Field (inches o.c.)	B	C	D
6d Common (2.0″ × 0.113″)	1.5	24/0	$^3/_8$	16	6	12	110	90	85
8d Common (2.5″ × 0.131″)	1.75	24/16	$^7/_{16}$	16	6	12	130	110	105
				24	6	12	110	90	85

For SI: 1 inch = 25.4 mm, 1 mile per hour = 0.447 m/s.

a. Panel strength axis parallel or perpendicular to supports. Three-ply plywood sheathing with studs spaced more than 16 inches on center shall be applied with panel strength axis perpendicular to supports.

b. Table is based on wind pressures acting toward and away from building surfaces per Section R301.2. Lateral bracing requirements shall be in accordance with Section R602.10.

c. Wood structural panels with span ratings of Wall-16 or Wall-24 shall be permitted as an alternate to panels with a 24/0 span rating. Plywood siding rated 16 o.c. or 24 o.c. shall be permitted as an alternate to panels with a 24/16 span rating. Wall-16 and Plywood siding 16 o.c. shall be used with studs spaced a maximum of 16 inches on center.

❖ In 2009, this table updated the previous wood structural panel wall sheathing table to include requirements for the wind pressures specified in Section R301.2.1. The $^5/_{16}$-inch (8 mm) wood structural panels have been deleted since they are currently a very small fraction of the panels produced today. While they have been the minimum panel thickness specified for many applications over the years, the building industry has shifted away from them due to manufacturing efficiencies and marketplace demand. The de facto minimum has become $^3/_8$ inch (9.5 mm).

This table provides the minimum thickness, maximum wall stud spacing and fastening for wood structural panels used for sheathing on exterior walls. See Table R602.3(1) for wood structural panel sheathing for interior walls.

In 2009, this table updated the previous wood structural panel wall sheathing table to include requirements for the wind pressures specified in Section R301.2.1. The $^5/_{16}$-inch (8 mm) wood structural panels have been deleted since they are currently a very small fraction of the panels produced today. While they have been the minimum panel thickness specified for many applications over the years, the building industry has shifted away from them due to manufacturing efficiencies and marketplace demand. The de facto minimum has become $^3/_8$ inch (9.5 mm).

This table provides the minimum thickness, maximum wall stud spacing and fastening for wood structural panels used for sheathing on exterior walls. See Table R602.3(1) for wood structural panel sheathing for interior walls.

The previous Table R602.2(3) in the code gave recommended minimum panel thicknesses for wall panel sheathing. It was adequate most of the time but in higher wind regions (still within the range of the code) the panel thicknesses and orientations recommended in the table and footnotes may not provide the minimum protection to the home and inhabitants that is currently required in Section R301.2.1. Analysis conducted by the APA The Engineered Wood Association (APA) staff indicates that in the extreme wind regions covered by the code [less than 110 mph (49 m/s)] and with more severe exposures (C and D) the minimum thicknesses recommendations given in the previous Table R603.2(3), Wood Structural Panel Wall Sheathing, were insufficient in thickness and attachment. This new table provides the requirements to ensure that this important part of the structural system is correct. The analysis considered panel bending, stiffness, nail withdrawal and nail head pull through as well as the wind pressure requirements of Section R301.2.1. Note that the impact to most will be minimal because the most commonly used wood structural panel sheathing thickness in the US is $^7/_{16}$ inch (11 mm). As can be seen in the new table, this sheathing thickness is satisfactory for winds up to 110 mph (49 m/s) in all but Exposure D conditions. Most builders will only see the requirement for 8d nails as a change, and this is already the nail required for roof sheathing applications.

TABLE R602.3(4)
ALLOWABLE SPANS FOR PARTICLEBOARD WALL SHEATHING

THICKNESS (inch)	GRADE	STUD SPACING (inches)	
		When siding is nailed to studs	When siding is nailed to sheathing
$^{3}/_{8}$	M-1 Exterior glue	16	—
$^{1}/_{2}$	M-2 Exterior glue	16	16

For SI: 1 inch = 25.4 mm.

a. Wall sheathing not exposed to the weather. If the panels are applied horizontally, the end joints of the panel shall be offset so that four panels corners will not meet. All panel edges must be supported. Leave a $^{1}/_{16}$-inch gap between panels and nail no closer than $^{3}/_{8}$ inch from panel edges.

❖ The allowable spans for particleboard wall sheathing listed in Table R602.3(4) are for wall sheathing that is not exposed to weather. The panels may be applied with the long dimension parallel or perpendicular to the studs.

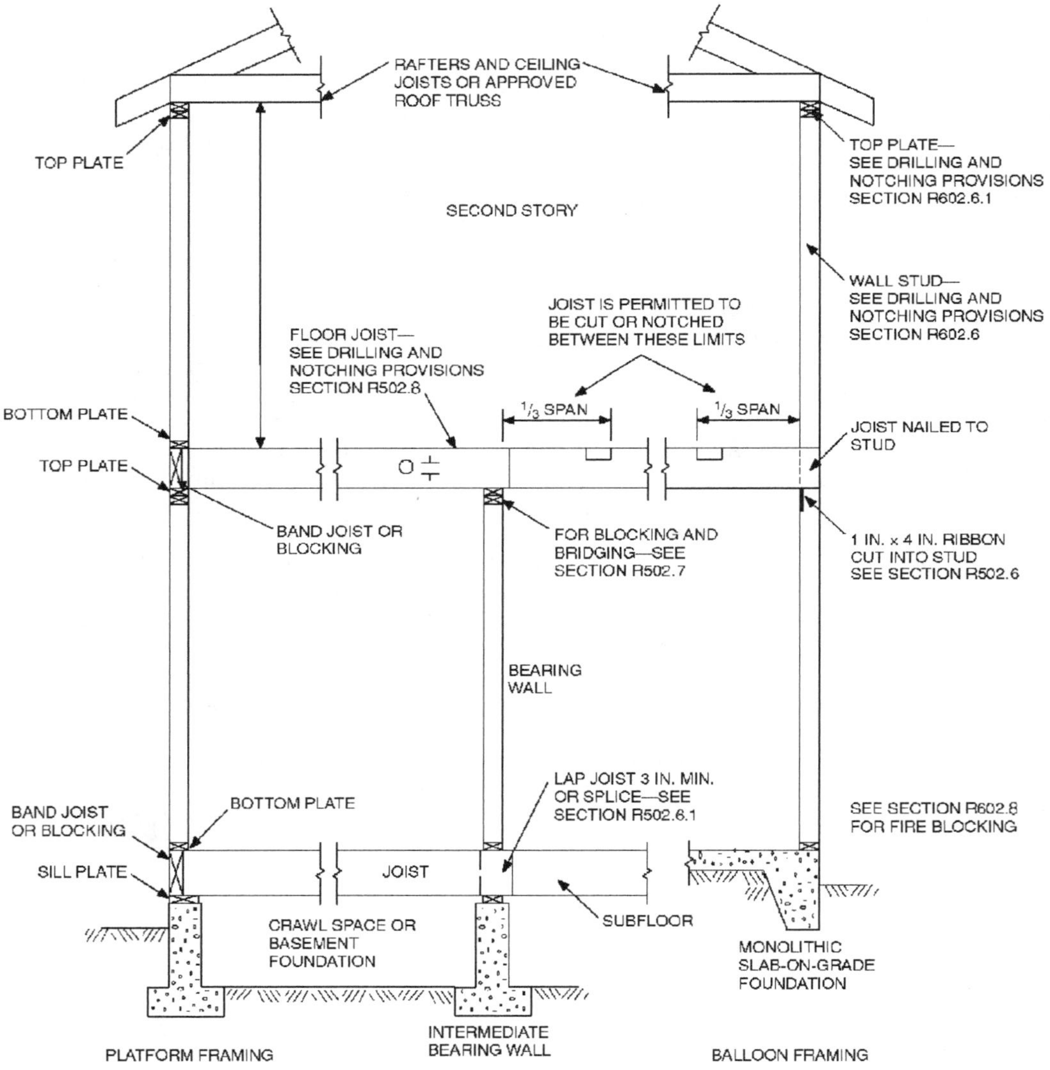

For SI: 1 inch = 25.4 mm.

FIGURE R602.3(1)
TYPICAL WALL, FLOOR AND ROOF FRAMING

❖ This is a schematic wood-framed building section that shows code provisions for wood floor systems as well as wall construction.

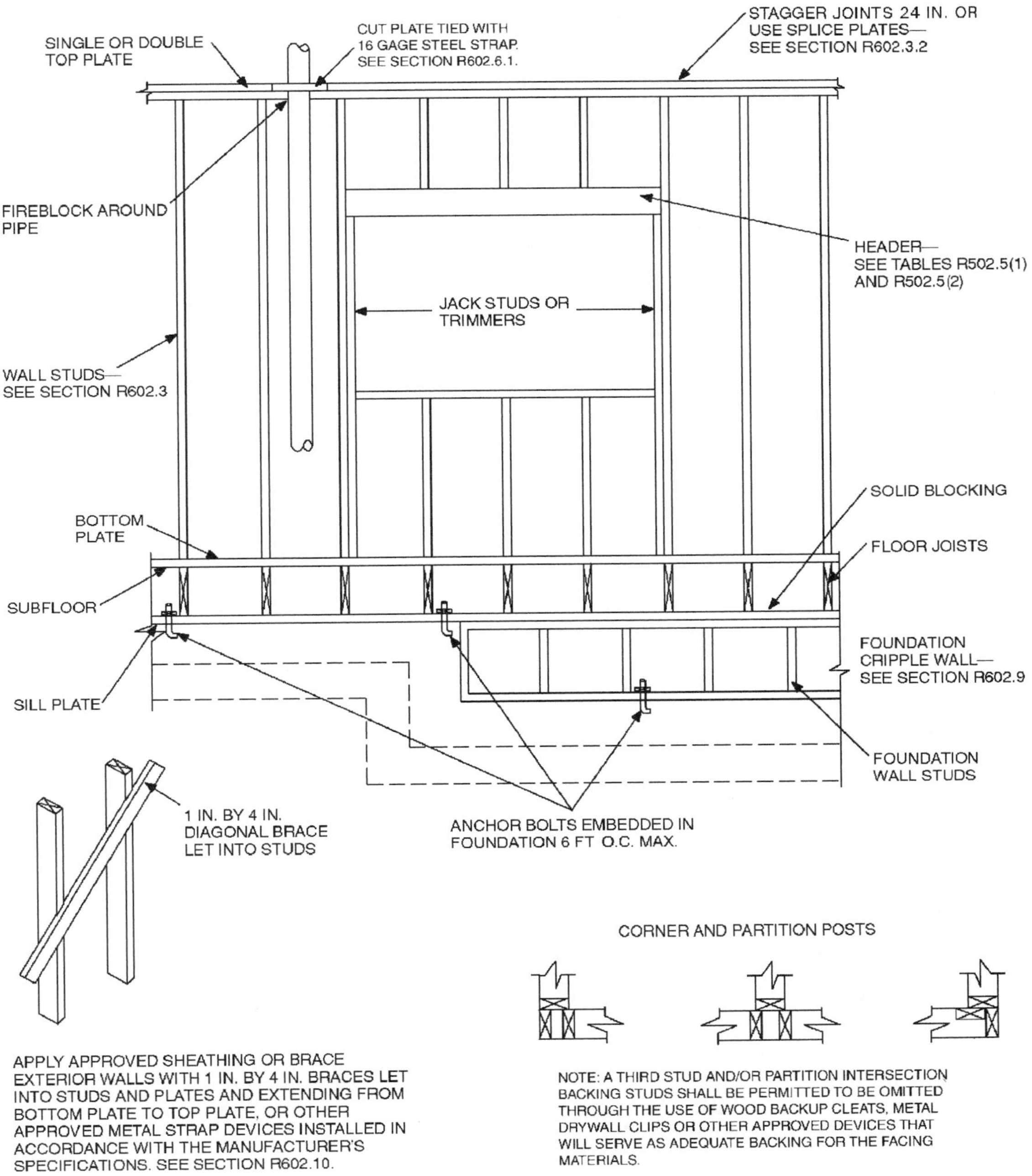

For SI: 1 inch = 25.4 mm, 1 foot = 304.8 mm.

FIGURE R602.3(2)
FRAMING DETAILS

❖ This figure is an elevation of typical wood-framed wall construction showing several code provisions as well as illustrating features such as cripple walls.

R602.3.1 Stud size, height and spacing. The size, height and spacing of studs shall be in accordance with Table R602.3.(5).

Exceptions:

1. Utility grade studs shall not be spaced more than 16 inches (406 mm) on center, shall not support more than a roof and ceiling, and shall not exceed 8 feet (2438 mm) in height for exterior walls and load-bearing walls or 10 feet (3048 mm) for interior non-load-bearing walls.
2. Studs more than 10 feet (3048 mm) in height which are in accordance with Table R602.3.1.

❖ Table R602.3(5) lists required stud spacing based on the stud size and loading condition.

Exception 1 addresses the fact that utility grade lumber is lower in strength and quality and therefore requires tighter spacing and shorter lengths in exterior walls. Note that the minimum grade lumber for application of Table R602.3(5) is, by implication, stud grade or standard grade.

As noted in Exception 2, if the bearing wall height exceeds 10 feet (3048 mm), stud size is determined using Table R602.3.1. A design is required if the limits given in the table's title are exceeded or if Note a applies.

TABLE R602.3(5)
SIZE, HEIGHT AND SPACING OF WOOD STUDS

STUD SIZE (inches)	BEARING WALLS					NONBEARING WALLS	
	Laterally unsupported stud height[a] (feet)	Maximum spacing when supporting a roof-ceiling assembly or a habitable attic assembly, only (inches)	Maximum spacing when supporting one floor, plus a roof-ceiling assembly or a habitable attic assembly (inches)	Maximum spacing when supporting two floors, plus a roof-ceiling assembly or a habitable attic assembly (inches)	Maximum spacing when supporting one floor height[a] (feet)	Laterally unsupported stud height[a] (feet)	Maximum spacing (inches)
		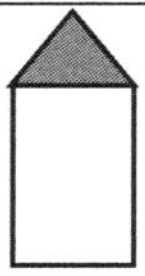			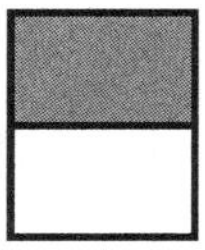		
2 × 3[b]	—	—	—	—	—	10	16
2 × 4	10	24[c]	16[c]	—	24	14	24
3 × 4	10	24	24	16	24	14	24
2 × 5	10	24	24	—	24	16	24
2 × 6	10	24	24	16	24	20	24

For SI: 1 inch = 25.4 mm, 1 foot = 304.8 mm, 1 square foot = 0.093 m^2.

a. Listed heights are distances between points of lateral support placed perpendicular to the plane of the wall. Increases in unsupported height are permitted where justified by analysis.

b. Shall not be used in exterior walls.

c. A habitable attic assembly supported by 2 × 4 studs is limited to a roof span of 32 feet. Where the roof span exceeds 32 feet, the wall studs shall be increased to 2 × 6 or the studs shall be designed in accordance with accepted engineering practice.

❖ This table provides the allowed stud spacing in bearing walls up to 10 feet (3048 mm) high as stated in Section R602.3.1. As described in Note a, increases in unsupported height are permitted where justified by an engineering analysis. The "unsupported height" of a stud is really the distance between points of lateral support, which generally occurs at each floor level.

This table was updated in 2009 to include a habitable attic assembly. See Section R202 for the definition of "Attic, habitable." This update permits the addition of a habitable attic without considering it as a story load on the studs. The basis of the limit of how much habitable attic load the wall can support is a 32-foot (9754 mm) span with 2 by 4 wall studs. Where the span exceeds 32 feet (9754 mm), the wall stud must be 2 by 6 or designed in accordance with accompanying engineering practice.

TABLE R602.3.1
MAXIMUM ALLOWABLE LENGTH OF WOOD WALL STUDS EXPOSED TO WIND SPEEDS OF 100 MPH OR LESS IN SEISMIC DESIGN CATEGORIES A, B, C, D_0, D_1 and D_2[b, c]

HEIGHT (feet)	ON-CENTER SPACING (inches)			
	24	16	12	8
Supporting a roof only				
> 10	2 × 4	2 × 4	2 × 4	2 × 4
12	2 × 6	2 × 4	2 × 4	2 × 4
14	2 × 6	2 × 6	2 × 6	2 × 4
16	2 × 6	2 × 6	2 × 6	2 × 4
18	NA[a]	2 × 6	2 × 6	2 × 6
20	NA[a]	NA[a]	2 × 6	2 × 6
24	NA[a]	NA[a]	NA[a]	2 × 6
Supporting one floor and a roof				
> 10	2 × 6	2 × 4	2 × 4	2 × 4
12	2 × 6	2 × 6	2 × 6	2 × 4
14	2 × 6	2 × 6	2 × 6	2 × 6
16	NA[a]	2 × 6	2 × 6	2 × 6
18	NA[a]	2 × 6	2 × 6	2 × 6
20	NA[a]	NA[a]	2 × 6	2 × 6
24	NA[a]	NA[a]	NA[a]	2 × 6
Supporting two floors and a roof				
> 10	2 × 6	2 × 6	2 × 4	2 × 4
12	2 × 6	2 × 6	2 × 6	2 × 6
14	2 × 6	2 × 6	2 × 6	2 × 6
16	NA[a]	NA[a]	2 × 6	2 × 6
18	NA[a]	NA[a]	2 × 6	2 × 6
20	NA[a]	NA[a]	NA[a]	2 × 6
22	NA[a]	NA[a]	NA[a]	NA[a]
24	NA[a]	NA[a]	NA[a]	NA[a]

For SI: 1 inch = 25.4 mm, 1 foot = 304.8 mm, 1 pound per square foot = 0.0479 kPa
1 pound per square inch = 6.895 kPa, 1 mile per hour = 0.447 m/s.

a. Design required.

b. Applicability of this table assumes the following: Snow load not exceeding 25 psf, f_b not less than 1310 psi determined by multiplying the AF&PA NDS tabular base design value by the repetitive use factor, and by the size factor for all species except southern pine, E not less than 1.6×10^6 psi, tributary dimensions for floors and roofs not exceeding 6 feet, maximum span for floors and roof not exceeding 12 feet, eaves not over 2 feet in dimension and exterior sheathing. Where the conditions are not within these parameters, design is required.

c. Utility, standard, stud and No. 3 grade lumber of any species are not permitted.

(continued)

TABLE R602.3.1—continued
MAXIMUM ALLOWABLE LENGTH OF WOOD WALL STUDS EXPOSED TO WIND SPEEDS OF 100 MPH OR LESS IN SEISMIC DESIGN CATEGORIES A, B, C, D_0, D_1 and D_2

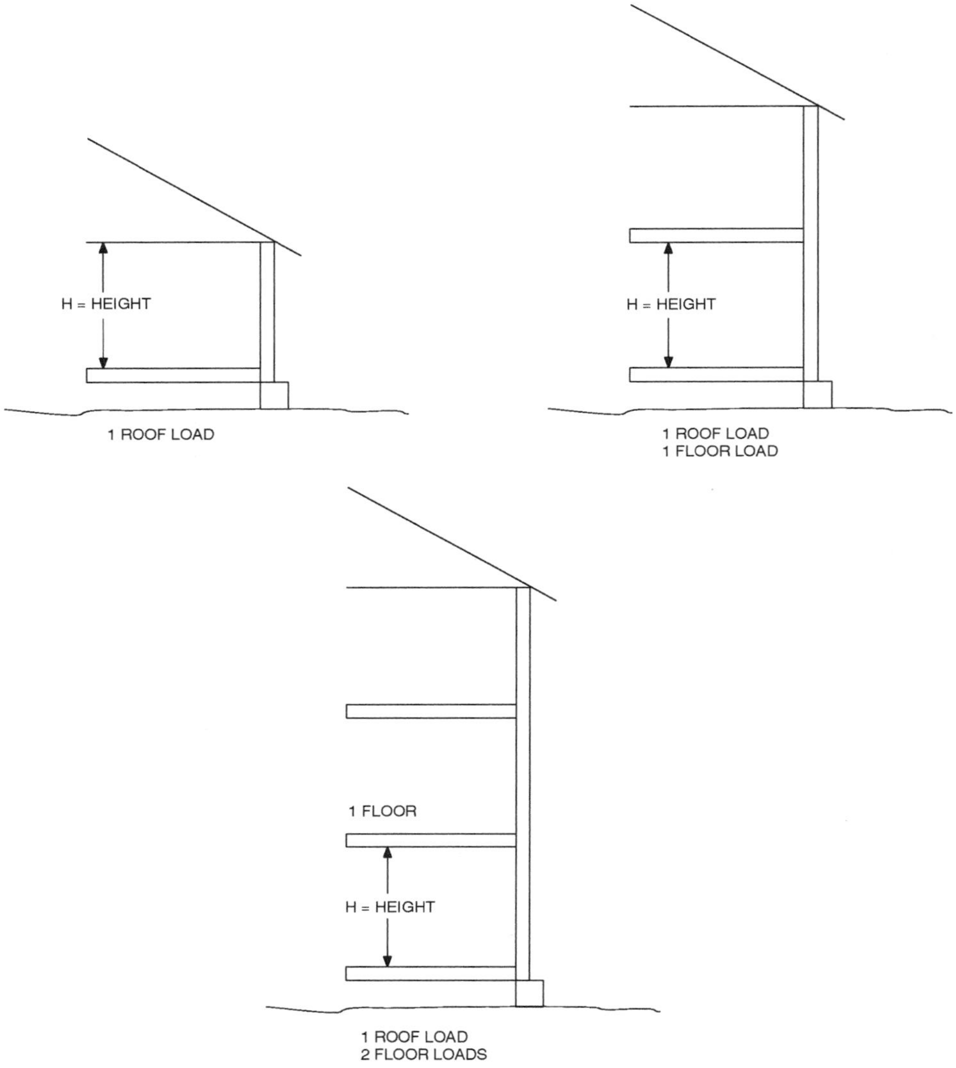

❖ This table provides the stud-spacing limits in bearing walls over 10 feet (3048 mm) high as stated in Section R602.3.1. Notes b and c make it clear that the lumber grades specified in Section R602.2 are not applicable but instead must be determined by using AF&PA NDS and applying the limitations of Note b. Conditions noted as "NA" in the table require a design per Note a.

R602.3.2 Top plate. Wood stud walls shall be capped with a double top plate installed to provide overlapping at corners and intersections with bearing partitions. End joints in top plates shall be offset at least 24 inches (610 mm). Joints in plates need not occur over studs. Plates shall be not less than 2-inches (51 mm) nominal thickness and have a width at least equal to the width of the studs.

Exception: A single top plate may be installed in stud walls, provided the plate is adequately tied at joints, corners and intersecting walls by a minimum 3-inch by 6-inch by a 0.036-inch-thick (76 mm by 152 mm by 0.914 mm) galvanized steel plate that is nailed to each wall or segment of wall by six 8d nails on each side, provided the rafters or joists are centered over the studs with a tolerance of no more than 1 inch (25 mm). The top plate may be omitted over lintels that are adequately tied to adjacent wall sections with steel plates or equivalent as previously described.

❖ Plates are the horizontal elements of walls capping the top of walls and framing the bottom of stud walls. They are called plates because they are laid flat, with the deeper dimension horizontal. In building construction applications, top plates must be a nominal 2-inch (51 mm) thickness.

Double top plates serve three major functions:

1. They overlap at corners and bearing wall intersections, providing a means of tying the building together. (Note that the overlap is not required at intersections with nonbearing interior walls.)
2. They serve as beams to support joists and rafters that are not located directly over the studs.
3. They serve as chords for floor and roof diaphragms.

Along with provisions for stud size and spacing limitations, this section requires the installation of double top plates to provide a continuous tie along the tops of the walls. With the advent of wider wall framing to accommodate increased thickness of insulation, a desire to save on material costs led to the allowance of a single top plate alternative. The exception permits the use of a single top plate in bearing and exterior walls as long as adequate top-plate ties are provided. In addition, joists or rafters framing into the wall must be placed more closely to the vertical stud below. This is necessary to limit the bending stress in the top plate. The single top plate exception is illustrated in Commentary Figure R602.3.2.

R602.3.3 Bearing studs. Where joists, trusses or rafters are spaced more than 16 inches (406 mm) on center and the bearing studs below are spaced 24 inches (610 mm) on center, such members shall bear within 5 inches (127 mm) of the studs beneath.

Exceptions:

1. The top plates are two 2-inch by 6-inch (38 mm by 140 mm) or two 3-inch by 4-inch (64 mm by 89 mm) members.
2. A third top plate is installed.
3. Solid blocking equal in size to the studs is installed to reinforce the double top plate.

❖ The bottom plate serves to anchor the wall to the floor, and studs are attached to it by end nailing or toe-nailing. The basic requirements of this section are shown in Commentary Figures R602.3.3(1) and R602.3.3(3). The code permits three options (exceptions) to account for loading that must be resisted by the double top plate when the stud spacing is 24 inches (610 mm) on center. The first option is an increase in the minimum top plate size. The second option is to provide a third plate. The third option is solid blocking used as reinforcement for a double plate, installed so that the load from the framing members above is transferred to the supporting studs. Commentary Figures R602.3.3(2) and R602.3.3(4) illustrate the first and third options.

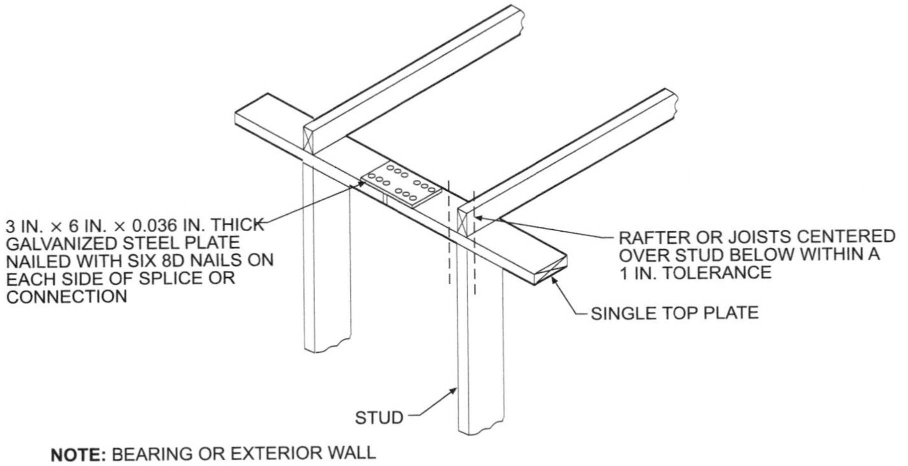

For SI: 1 inch = 25.4 mm.

Figure R602.3.2
SINGLE TOP-PLATE SPLICE

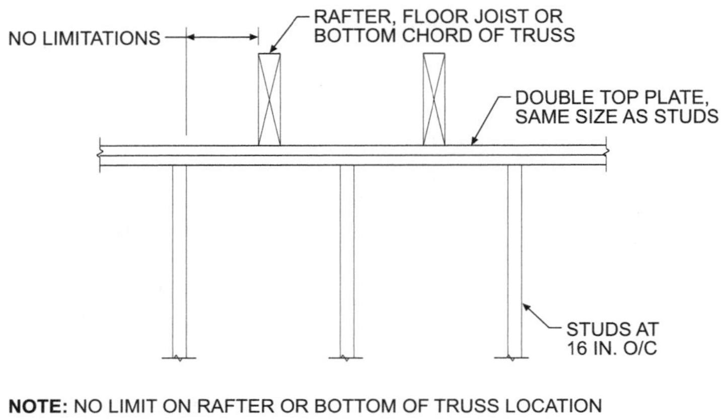

For SI: 1 inch = 25.4 mm.

Figure R602.3.3(1)
TOP PLATE WITH 16-INCH STUD SPACING

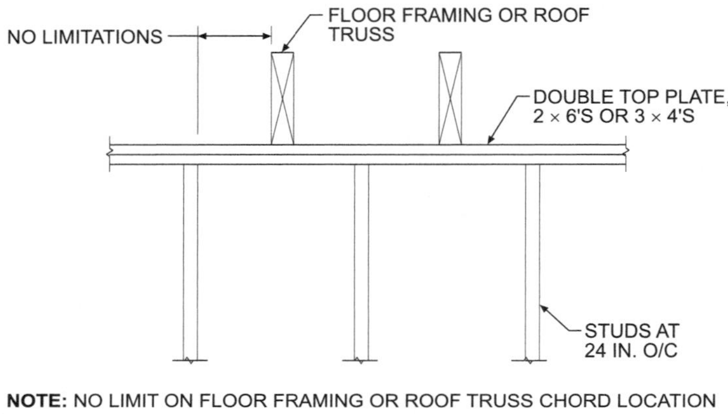

For SI: 1 inch = 25.4 mm.

Figure R602.3.3(2)
TOP PLATE WITH 24-INCH STUD SPACING

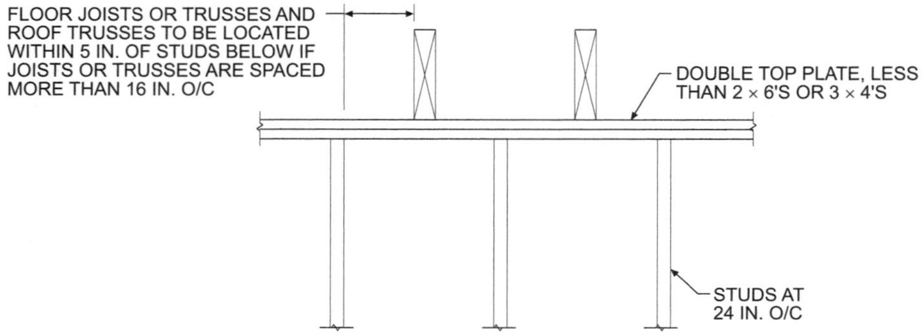

For SI: 1 inch = 25.4 mm.

Figure R602.3.3(3)
TOP PLATE WITH 24-INCH STUD SPACING AND BEARING POINT LIMITATIONS

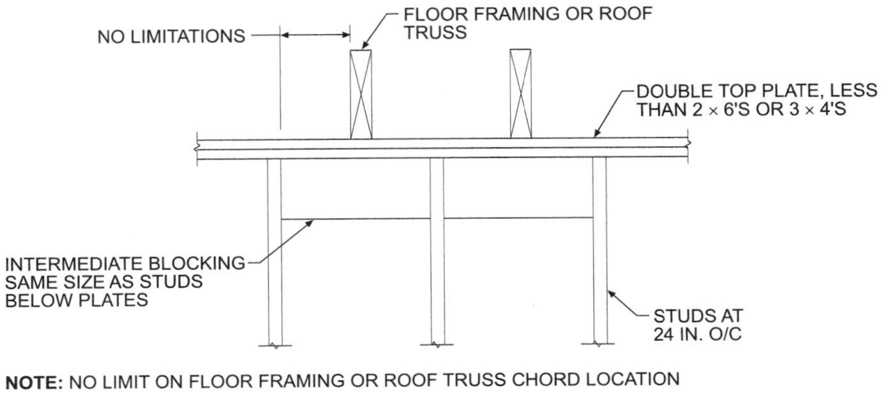

For SI: 1 inch = 25.4 mm.

Figure R602.3.3(4)
BLOCKED TOP PLATE WITH 24-INCH STUD SPACING

R602.3.4 Bottom (sole) plate. Studs shall have full bearing on a nominal 2-by (51 mm) or larger plate or sill having a width at least equal to the width of the studs.

❖ This provision requires that studs bear on an adequately sized bottom plate. This allows the bottom plate to serve as a nailing surface for the wall sheathing while controlling compression perpendicular to the grain on the sill. If too great, this stress could result in crushing.

R602.3.5 Braced wall panel uplift load path. *Braced wall panels* located at exterior walls that support roof rafters or trusses (including stories below top *story*) shall have the framing members connected in accordance with one of the following:

1. Fastening in accordance with Table R602.3(1) where:
   1.1. The basic wind speed does not exceed 90 mph (40 m/s), the wind exposure category is B, the roof pitch is 5:12 or greater, and the roof span is 32 feet (9754 mm) or less, or
   1.2. The net uplift value at the top of a wall does not exceed 100 plf. The net uplift value shall be determined in accordance with Section R802.11 and shall be permitted to be reduced by 60 plf (86 N/mm) for each full wall above.
2. Where the net uplift value at the top of a wall exceeds 100 plf (146 N/mm), installing *approved* uplift framing connectors to provide a continuous load path from the top of the wall to the foundation or to a point where the uplift force is 100 plf (146 N/mm) or less. The net uplift value shall be as determined in Item 1.2 above.
3. Wall sheathing and fasteners designed in accordance with accepted engineering practice to resist combined uplift and shear forces.

As Section R602.3.5 is currently written, braced wall panels on the exterior walls that support roof framing require an uplift load path. As braced wall panels on floors below support the upper braced wall panels, uplift requirements must also be met at the lower floors. The intention of this section is to provide an adequate load path for the lateral-force-resisting system of the structure (wall bracing) as a minimum. From an engineering perspective, however, this provision appears to be incomplete because uplift loads are possible at the end of every roof truss or rafter, whether it bears on a bracing panel or not. To provide for the appropriate attachment of roof framing to the supporting walls for the whole structure, the uplift requirements of Chapter 6 must be used in conjunction with the requirements of Section R802.11, as discussed below.

The prescriptive nailing provisions in Table R602.3(1), Item 5, specify three 16d toenails for the "rafter or roof truss to plate" connection. Based on historical performance, this prescriptive schedule is deemed to provide sufficient attachment for 90 mph (40 m/s) winds, Exposure B, 5:12 roof pitch or greater, and a roof span of 32 feet (9754 mm) or less. For this reason, Section R602.3.5, Item 1.1, exempts structures meeting these characteristics from the uplift load path requirements, provided the roof is attached in accordance with Table R602.3(1), Item 5.

From Section R602.3.5, Item 1.1, it can be inferred that the Table R602.3(1), Item 5, prescriptive nailing requirements provide 100 pounds (45 kg) of net uplift (plf) (146 N/mm) resistance to properly attached roofs. No additional attachment is required until the uplift loads specified in Table R802.11 exceed this amount. In addition, Section R602.3.5, Item 2, provides the effective weight of each full wall above the wall-to-floor connection in question as 60 plf (86 N/mm). This amount can be subtracted from the uplift amount, as the uplift requirement for each lower floor is calculated (see the example in this section). Given this information, the user is able to use the truss or rafter connection uplift force provided in Table

R802.11 and its notes to compute the net uplift requirement for a structure at a given location.

When the uplift resistance of the roof framing provided in Table R802.11 is reduced by the number of walls above the connection in question or the location in the structure, and the net result is less than 100 pounds per square foot (146 N/mm), the prescriptive nailing requirements of Table R602.3(1) are deemed to be sufficient to provide an adequate load path.

When the uplift resistance of the roof framing provided in Table R802.11 is reduced by the number of walls above the connection in question or the location in the structure, and the net result is greater than 100 plf (146 N/mm), the prescriptive nailing requirements of Table R602.3(1) are insufficient to provide an adequate load path and the following options are available:

1. Installation of an approved uplift framing anchor of sufficient capacity to resist the net uplift force.
2. Section R104.11 permits engineering design to be used to determine other nailing schedules or details that may provide sufficient uplift resistance.
3. Engineering analysis can be used to determine the wind uplift capacity in lieu of Table R802.11. Such analysis can take into account heavier roofing materials and other details not accounted for in the tables.
4. Other referenced documents, such as the American Wood Council's (AWC) *Wood Frame Construction Manual for One- and Two-family Dwellings*, can be used to generate uplift requirements and prescriptive hold-down requirements.

Example:

What is the net uplift capacity for each level of a two-story home located in a 110 mph Exposure B wind zone? The width of the house is 32 feet (9754 mm) the house has no roof overhangs, the roof pitch is 4:12 and the roof framing is at 24 inches (610 mm) on center.

Solution:

From Table R802.11, the wind uplift acting at the roof-to-second story wall is 504 pounds (229 kg) per connection.

- Capacity of the roof-to-second-story wall connection—There is no wall above this connection, so the uplift at this point is 504 pounds (229 kg) per connection. Table R602.3(1), Item 5 attachment schedule provides 100 pounds (45 kg) per connector of uplift resistance. If this attachment schedule is used, a connector with a minimum capacity of [504 pounds (229 kg] per connector—100 pounds [(45 kg) per connector =] 404 pounds (184 kg) per connector is required. If this attachment schedule is not used, a connector with the full 504-pound (229 kg) capacity is required.
- Capacity of the second-story to first-story connection-As each lineal foot of wall contributes 60 pounds of uplift (86 N/mm) resistance, and there are 2 feet (610 mm) of wall per connector (roof framing/connectors on 24-inch (610 mm) centers), the uplift requirement for the second-story to first-story connection will be [504 - (2 × 60) =] 384 pounds (175 kg) per connector.
- Capacity of the first-story to foundation connection—In this case, there are two stories above the connection, and each lineal foot of wall contributes 60 pounds of uplift (86 N/mm) resistance, and there are 2 feet (610 mm) of wall per connector (roof framing/connectors on 24-inch (610 mm) centers). The uplift requirement for the first-story to foundation connection will be [504 - (2 × 2 × 60) =] 264 pounds (120 kg) per connector.

While not specified in the code, it is reasonable to assume that Table R602.3(1) nailing schedules provide at least some of this uplift resistance at the second-to-first and first-to-foundation locations. The use of 100 pounds per foot (0.34 kg/m) for the capacity of the Table R602.3(1) connections at these two locations is probably as reasonable as the assumption that Table R602.3(1), Items 1-5 provide 100 plf (146 N/m) uplift resistance capacity. From a historical perspective, these second to first-story and first-story to foundation connections as specified have been as effective as the roof-to-top story connections. As noted elsewhere, deviations from the code provisions should be approved by the building official.

R602.4 Interior load-bearing walls. Interior load-bearing walls shall be constructed, framed and fireblocked as specified for exterior walls.

❖ Interior load-bearing walls must be designed and constructed using the criteria for exterior walls. Thus, Table R602.3(5) should be used to establish stud spacing of walls up to 10 feet (3048 mm) high, and Table R602.3.1 applies to walls over 10 feet (3048 mm).

R602.5 Interior nonbearing walls. Interior nonbearing walls shall be permitted to be constructed with 2 inch by 3 inch (51 mm by 76 mm) studs spaced 24 inches (610 mm) on center or, when not part of a *braced wall line*, 2 inch by 4 inch (51 mm by 102 mm) flat studs spaced at 16 inches (406 mm) on center. Interior nonbearing walls shall be capped with at least a single top plate. Interior nonbearing walls shall be fireblocked in accordance with Section R602.8.

❖ The code permits reduced stud sizes, increased stud spacing and capping with a single top plate in interior nonbearing partitions. These allowances are based on minimal superimposed vertical loads.

R602.6 Drilling and notching of studs. Drilling and notching of studs shall be in accordance with the following:

1. Notching. Any stud in an exterior wall or bearing partition may be cut or notched to a depth not exceeding 25 percent of its width. Studs in nonbearing partitions may be notched to a depth not to exceed 40 percent of a single stud width.
2. Drilling. Any stud may be bored or drilled, provided that the diameter of the resulting hole is no more than 60 percent of the stud width, the edge of the hole is no more than $^5/_8$ inch (16 mm) to the edge of the stud, and the hole is not located in the same section as a cut or notch. Studs located in exterior walls or bearing partitions drilled over 40 percent and up to 60 percent shall also be doubled with no more than two successive doubled studs bored. See Figures R602.6(1) and R602.6(2).

   Exception: Use of *approved* stud shoes is permitted when they are installed in accordance with the manufacturer's recommendations.

❖ This section addresses the allowable drilling and notching of studs used to frame partitions. See Figures R602.6(1) and R602.6(2) for examples of permitted drilling and notching of studs in exterior walls, bearing walls and nonbearing walls. These limitations retain the structural integrity of the studs. Where stud shoes are used, the exception allows drilling and notching to be in accordance with the approved manufacturer's instructions.

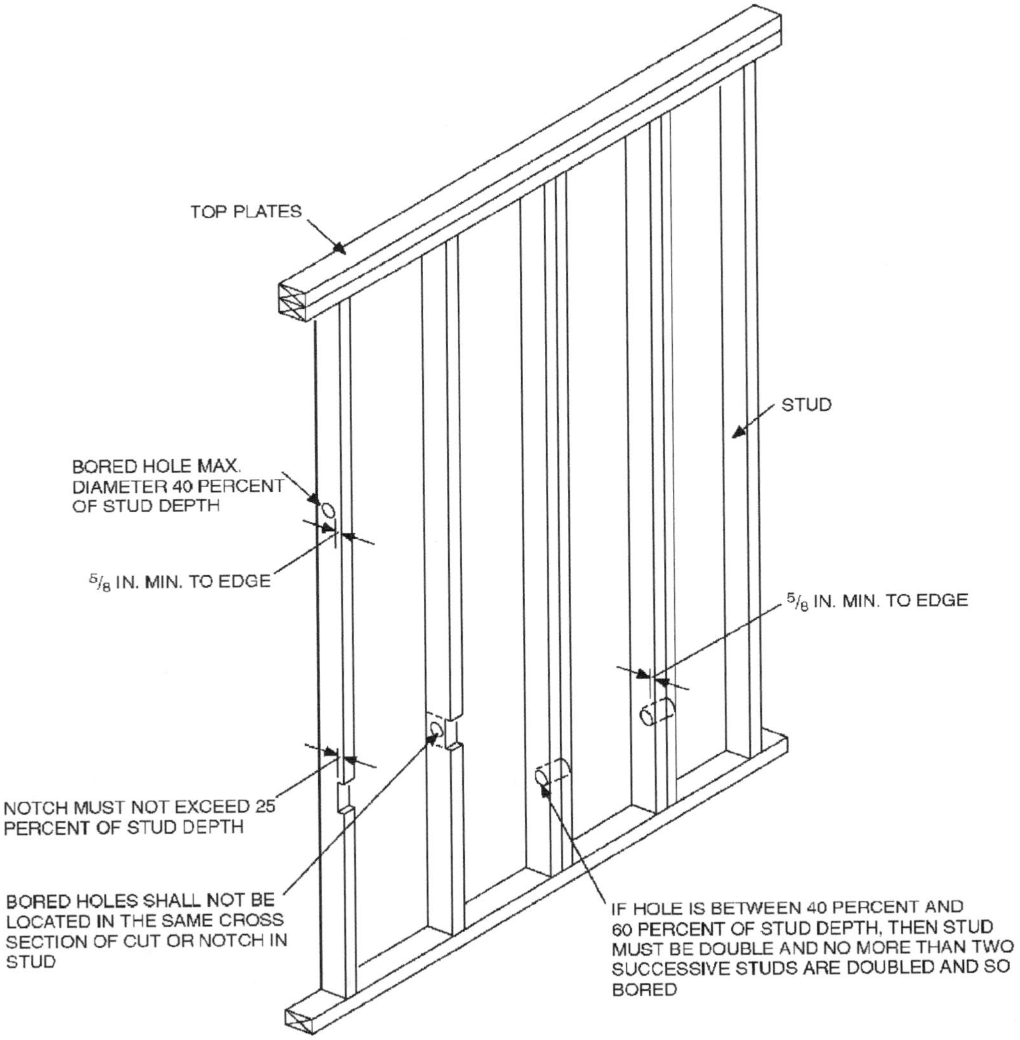
TOP PLATES
STUD
BORED HOLE MAX. DIAMETER 40 PERCENT OF STUD DEPTH
5/8 IN. MIN. TO EDGE
5/8 IN. MIN. TO EDGE
NOTCH MUST NOT EXCEED 25 PERCENT OF STUD DEPTH
BORED HOLES SHALL NOT BE LOCATED IN THE SAME CROSS SECTION OF CUT OR NOTCH IN STUD
IF HOLE IS BETWEEN 40 PERCENT AND 60 PERCENT OF STUD DEPTH, THEN STUD MUST BE DOUBLE AND NO MORE THAN TWO SUCCESSIVE STUDS ARE DOUBLED AND SO BORED

For SI: 1 inch = 25.4 mm.
Note: Condition for exterior and bearing wall

FIGURE R602.6(1)
NOTCHING AND BORED HOLE LIMITATIONS FOR EXTERIOR WALLS AND BEARING WALLS

❖ See the commentary for Section R602.6

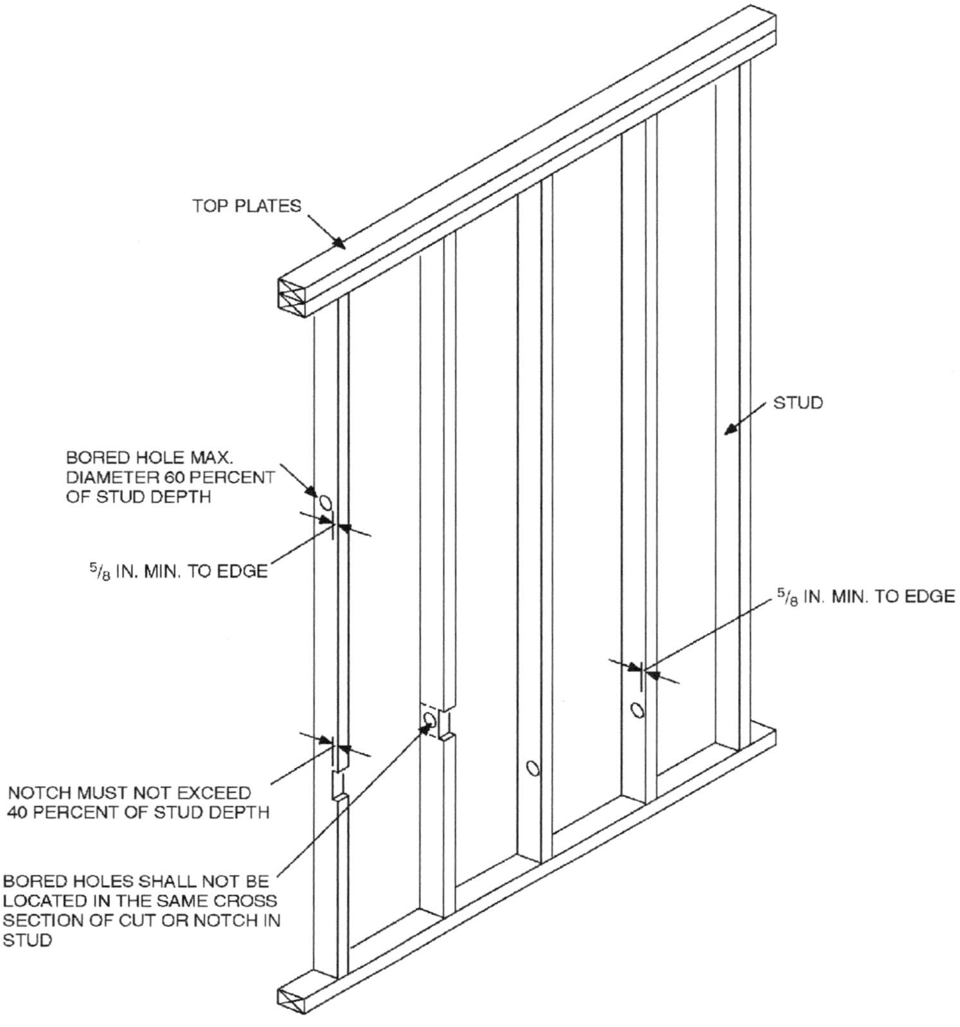

For SI: 1 inch = 25.4 mm.

FIGURE R602.6(2)
NOTCHING AND BORED HOLE LIMITATIONS FOR INTERIOR NONBEARING WALLS

❖ See the commentary for Section R602.6.

R602.6.1 Drilling and notching of top plate. When piping or ductwork is placed in or partly in an exterior wall or interior load-bearing wall, necessitating cutting, drilling or notching of the top plate by more than 50 percent of its width, a galvanized metal tie not less than 0.054 inch thick (1.37 mm) (16 ga) and $1^1/_2$ inches (38 mm) wide shall be fastened across and to the plate at each side of the opening with not less than eight 10d (0.148 inch diameter) having a minimum length of $1^1/_2$ inches (38 mm) at each side or equivalent. The metal tie must extend a minimum of 6 inches past the opening. See Figure R602.6.1.

Exception: When the entire side of the wall with the notch or cut is covered by wood structural panel sheathing.

❖ In many cases, drilling or notching of the top plate is necessary to allow plumbing, heating or other pipes to be placed within the exterior walls and load-bearing interior walls. When cutting the top plate by more than one-half its width is necessary, strapping across the plates as illustrated in Figure R602.6.1, is required to provide top plate continuity and to retain the structural integrity of the wall system as a whole. In the case of a double top plate, only the upper of the two plates needs to have a strap installed as described.

The top plate functions as a collector as part of the lateral load path. It collects the lateral loads, due to wind and seismic, from the roof or floor diaphragm and distributes them along the wall into the wall bracing.

In doing so, the top plate is subjected to tension or compression due to the lateral loads in the plane of the wall.

Removal of any top plate material due to cutting, drilling or notching will reduce the tension/compression capacity of the top plate.

This section provides a prescriptive design that allows removal of 50 percent or less of the top plate material without replacing the material.

If more than 50 percent is removed, a metal strap is required to replace the material removed as shown in Figure R602.6.1.

Common practice is to notch the top plate as shown in Figure R602.6.1.

This section does not specifically state the amount of top plate material that must remain. However, Section R602.6 requires $^5/_8$ inch (16 mm) minimum to edge for holes in studs. This can be used as a guideline since this is for the compression load and assuming the metal strap is for the tension load.

For notching as shown in Figure R602.6.1, at least $^5/_8$ inch (16 mm) of the top plate should remain.

For a bored hole, $^5/_8$ inch (16 mm) should remain on at least one edge. The edge with less than $^5/_8$ inch (16 mm) will require the tension strap.

Notching or boring holes outside those guidelines will require a design in accordance with accepted engineering practices.

R602.7 Headers. For header spans see Tables R502.5(1), R502.5(2) and R602.7.1.

❖ At a wall opening, headers transfer loads received from the wall and floors or roof to the foundation. Openings such as doors or windows within bearing walls must be framed with headers of sufficient size to span the opening and transfer loads to jamb studs (header studs). Commentary Figure R602.7 illustrates a double header over a door opening. Tables R502.5(1) and R502.5(2) of the code are used to determine allowable spans for headers as well as girders. These tables are applicable to openings in bearing walls but are not applicable to nonbearing walls. See Section R602.7.2 for headers in nonbearing walls. Table R502.5(1) is to be used for headers over openings in exterior bearing walls. Table R502.5(2) is to be used for headers over openings in interior bearing walls. Where wood girders are used in a basement of a one-story dwelling, Table R502.5(2) may be used to prescriptively size the basement girders (the space between supporting columns may be viewed as openings).

Example:

You wish to size a header in the exterior wall of a one-story house that is 30 feet (9144 mm) wide. The opening size is 8 feet (2438 mm). No. 2 grade southern pine is common for the area, and the ground-snow loading is 20 pounds per square foot (958 Pa).

Solution:

Table R502.5(1) lists, for a 30 psf (1436 Pa) snow load and a building width of 36 feet (10 973 mm), a triple 2 by 10 inch (51 mm by 254 mm) header; it will span 8 feet, 2 inches (2489 mm).

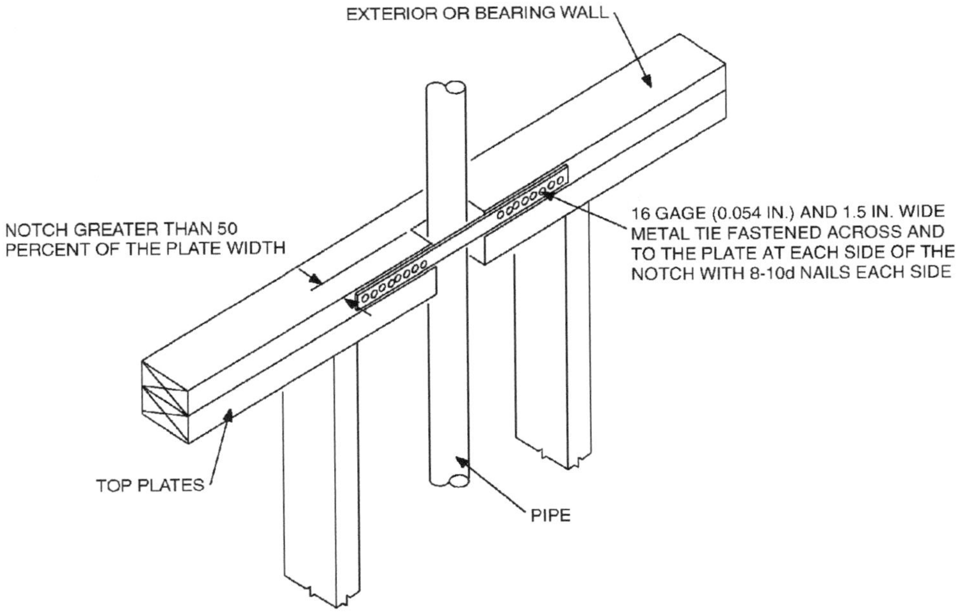

For SI: 1 inch = 25.4 mm.

FIGURE R602.6.1
TOP PLATE FRAMING TO ACCOMMODATE PIPING

❖ See the commentary for Section R602.6.1.

TABLE R602.7.1
SPANS FOR MINIMUM No.2 GRADE SINGLE HEADER[a, b, c, f]

SINGLE HEADERS SUPPORTING	SIZE	WOOD SPECIES	GROUND SNOW LOAD (psf)								
			≤ 20[d]			30			50		
			Building Width (feet)[e]								
			20	28	36	20	28	36	20	28	36
Roof and ceiling	2 × 8	Spruce-Pine-Fir	4-10	4-2	3-8	4-3	3-8	3-3	3-7	3-0	2-8
		Hem-Fir	5-1	4-4	3-10	4-6	3-10	3-5	3-9	3-2	2-10
		Douglas-Fir or Southern Pine	5-3	4-6	4-0	4-7	3-11	3-6	3-10	3-3	2-11
	2 × 10	Spruce-Pine-Fir	6-2	5-3	4-8	5-5	4-8	4-2	4-6	3-11	3-1
		Hem-Fir	6-6	5-6	4-11	5-8	4-11	4-4	4-9	4-1	3-7
		Douglas-Fir or Southern Pine	6-8	5-8	5-1	5-10	5-0	4-6	4-11	4-2	3-9
	2 × 12	Spruce-Pine-Fir	7-6	6-5	5-9	6-7	5-8	4-5	5-4	3-11	3-1
		Hem-Fir	7-10	6-9	6-0	6-11	5-11	5-3	5-9	4-8	3-8
		Douglas-Fir or Southern Pine	8-1	6-11	6-2	7-2	6-1	5-5	5-11	5-1	4-6
Roof, ceiling and one center-bearing floor	2 × 8	Spruce-Pine-Fir	3-10	3-3	2-11	3-9	3-3	2-11	3-5	2-11	2-7
		Hem-Fir	4-0	3-5	3-1	3-11	3-5	3-0	3-7	3-0	2-8
		Douglas-Fir or Southern Pine	4-1	3-7	3-2	4-1	3-6	3-1	3-8	3-2	2-9
	2 × 10	Spruce-Pine-Fir	4-11	4-2	3-8	4-10	4-1	3-6	4-4	3-7	2-10
		Hem-Fir	5-1	4-5	3-11	5-0	4-4	3-10	4-6	3-11	3-4
		Douglas-Fir or Southern Pine	5-3	4-6	4-1	5-2	4-5	4-0	4-8	4-0	3-7
	2 × 12	Spruce-Pine-Fir	5-8	4-2	3-4	5-5	4-0	3-6	4-9	3-6	2-10
		Hem-Fir	5-11	4-11	3-11	5-10	4-9	4-2	5-5	4-2	3-4
		Douglas-Fir or Southern Pine	6-1	5-3	4-8	6-0	5-2	4-10	5-7	4-10	4-3
Roof, ceiling and one clear span floor	2 × 8	Spruce-Pine-Fir	3-5	2-11	2-7	3-4	2-11	2-7	3-3	2-10	2-6
		Hem-Fir	3-7	3-1	2-9	3-6	3-0	2-8	3-5	2-11	2-7
		Douglas-Fir or Southern Pine	3-8	3-2	2-10	3-7	3-1	2-9	3-6	3-0	2-9
	2 × 10	Spruce-Pine-Fir	4-4	3-7	2-10	4-3	3-6	2-9	4-2	3-4	2-7
		Hem-Fir	4-7	3-11	3-5	4-6	3-10	3-3	4-4	3-9	3-1
		Douglas-Fir or Southern Pine	4-8	4-0	3-7	4-7	4-0	3-6	4-6	3-10	3-5
	2 × 12	Spruce-Pine-Fir	4-11	3-7	2-10	4-9	3-6	2-9	4-6	3-4	2-7
		Hem-Fir	5-6	4-3	3-5	5-6	4-2	3-3	5-4	3-11	3-1
		Douglas-Fir or Southern Pine	5-8	4-11	4-4	5-7	4-10	4-3	5-6	4-8	4-2

For SI: 1 inch=25.4 mm, 1 pound per square foot = 0.0479 kPa.

a. Spans are given in feet and inches.

b. Table is based on a maximum roof-ceiling dead load of 15 psf.

c. The header is permitted to be supported by an approved framing anchor attached to the full-height wall stud and to the header in lieu of the required jack stud.

d. The 20 psf ground snow load condition shall apply only when the roof pitch is 9:12 or greater. In conditions where the ground snow load is 30 psf or less and the roof pitch is less than 9:12, use the 30 psf ground snow load condition.

e. Building width is measured perpendicular to the ridge. For widths between those shown, spans are permitted to be interpolated.

f. The header shall bear on a minimum of one jack stud at each end.

❖ See the commentary for Section R602.7.1.

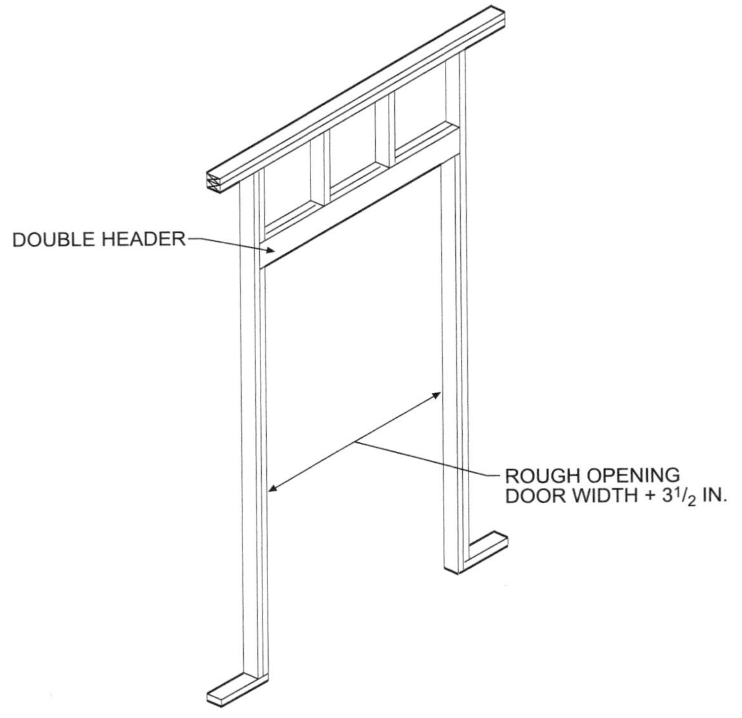

For SI: 1 inch = 25.4 mm.

Figure R602.7
HEADER DETAIL

R602.7.1 Single member headers. Single headers shall be framed with a single flat 2-inch-nominal (51 mm) member or wall plate not less in width than the wall studs on the top and bottom of the header in accordance with Figures R602.7.1(1) and R602.7.1(2).

❖ Headers prescribed in Tables R502.5(1) and R502.5(2) require a minimum of two members, each with a nominal 2-inch (51 mm) thickness. Single headers under limited loading conditions allow an increase in the energy efficiency of a dwelling. Installation of a single header results in a greater thickness of cavity insulation to reduce heat loss through the header in exterior walls.

The single header configuration prescribed in Table R602.7.1 may be used for small openings. Use is limited to areas with a maximum 50 psf (2394 N/mm^2) ground snow load and a maximum 15 psf (718 N/mm^2) roof-ceiling dead load. Table R602.7.1 gives maximum header lengths for 20 psf (958 N/mm^2), 30 psf (1436 N/mm^2), and 50 psf (2394 N/mm^2) snow loads. To use the 20 psf (958 N/mm^2) column, a roof pitch must be 9:12 or greater. For a roof pitch of 8:12 or less, the 30 psf (1436 N/mm^2) snow load column must be used for snow loads of 0 to 30 psf.

Two figures clarify the installation details for single headers with two different top plate conditions. In Figure R602.7.1(1), a cripple wall may be built above the header to finish out the space between the top plate and header. In Figure R602.7.1(2), the header fills the entire space between the top plate above the header and the plate below the header.

Table values are based on the National Design Specification for Wood Construction (NDS) and ASCE 7 *Minimum Design Loads for Buildings and Other Structures*.

R602.7.2 Wood structural panel box headers. Wood structural panel box headers shall be constructed in accordance with Figure R602.7.2 and Table R602.7.2.

❖ Wood structural panel box headers are used in some construction and are an efficient way of carrying load over an opening. The box header must be built in accordance with the criteria specified in Figure R602.7.2 and Table R602.7.2. Spans of box headers are established based on the construction of the header, the header depth, and the depth of the house. Note a to the table explains the depth criterion, which is based on a single story with a clear-span trussed roof or a two-story in which part of the load from the floors and roofs is supported by interior bearing walls. House depth is based on the span from the header to an interior bearing wall or the opposing exterior wall.

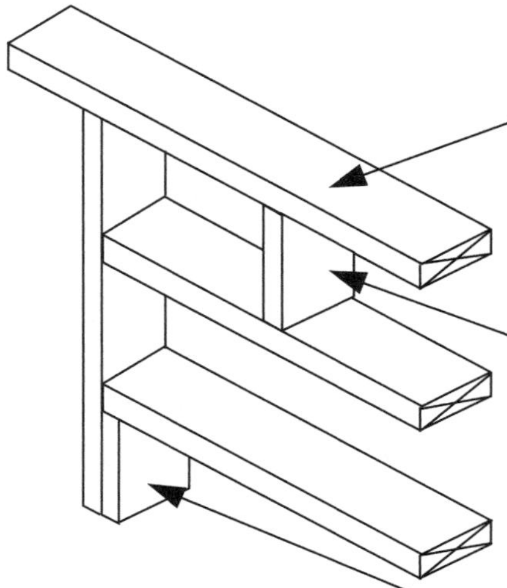

FIGURE 602.7.1(1)
SINGLE MEMBER HEADER IN EXTERIOR BEARING WALL

❖ See the commentary for Section R602.7.1.

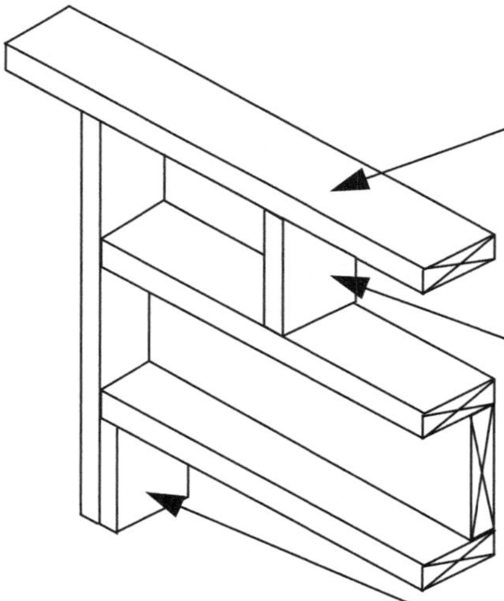

FIGURE R602.7.1(2)
ALTERNATIVE SINGLE MEMBER HEADER WITHOUT CRIPPLE

❖ See the commentary for Section R602.7.1.

TABLE R602.7.2
MAXIMUM SPANS FOR WOOD STRUCTURAL PANEL BOX HEADERS[a]

HEADER CONSTRUCTION[b]	HEADER DEPTH (inches)	HOUSE DEPTH (feet)				
		24	26	28	30	32
Wood structural panel–one side	9 15	4 5	4 5	3 4	3 3	— 3
Wood structural panel–both sides	9 15	7 8	5 8	5 7	4 7	3 6

For SI: 1 inch = 25.4 mm, 1 foot = 304.8 mm.

a. Spans are based on single story with clear-span trussed roof or two-story with floor and roof supported by interior-bearing walls.

b. See Figure R602.7.2 for construction details.

❖ See the commentary for Section R602.7.2.

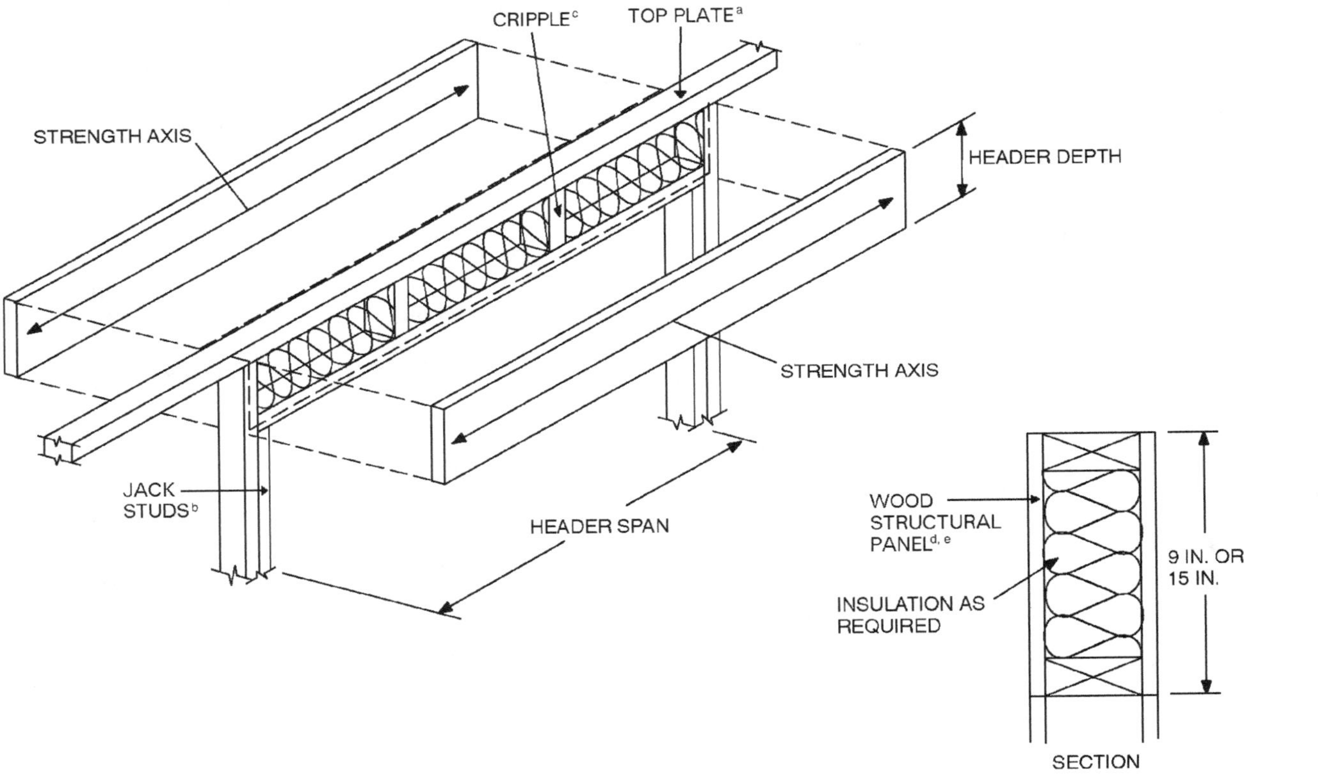

For SI: 1 inch = 25.4 mm, 1 foot = 304.8 mm.

NOTES:

a. The top plate shall be continuous over header.

b. Jack studs shall be used for spans over 4 feet.

c. Cripple spacing shall be the same as for studs.

d. Wood structural panel faces shall be single pieces of $^{15}/_{32}$-inch-thick Exposure 1 (exterior glue) or thicker, installed on the interior or exterior or both sides of the header.

e. Wood structural panel faces shall be nailed to framing and cripples with 8d common or galvanized box nails spaced 3 inches on center, staggering alternate nails $^{1}/_{2}$ inch. Galvanized nails shall be hot-dipped or tumbled.

FIGURE R602.7.2
TYPICAL WOOD STRUCTURAL PANEL BOX HEADER CONSTRUCTION

❖ See the commentary for Section R602.7.2.

R602.7.3 Nonbearing walls. Load-bearing headers are not required in interior or exterior nonbearing walls. A single flat 2-inch by 4-inch (51 mm by 102 mm) member may be used as a header in interior or exterior nonbearing walls for openings up to 8 feet (2438 mm) in width if the vertical distance to the parallel nailing surface above is not more than 24 inches (610 mm). For such nonbearing headers, no cripples or blocking are required above the header.

❖ In walls that are not supporting significant loads, the code provides a simple (more economical) option for constructing headers.

R602.8 Fireblocking required. Fireblocking shall be provided in accordance with Section R302.11.

❖ See the commentary for Section R302.11.

R602.9 Cripple walls. Foundation cripple walls shall be framed of studs not smaller than the studding above. When exceeding 4 feet (1219 mm) in height, such walls shall be framed of studs having the size required for an additional *story*.

Cripple walls with a stud height less than 14 inches (356 mm) shall be continuously sheathed on one side with wood structural panels fastened to both the top and bottom plates in accordance with Table R602.3(1), or the cripple walls shall be constructed of solid blocking.

All cripple walls shall be supported on continuous foundations.

❖ A cripple wall is a framed stud extending from the top of foundation to the underside of the floor framing above the foundation. Should these foundation studs exceed 4 feet (1219 mm) in height, the partial-height wall must, for purposes of stud sizing, be sized as if the partial-height wall were an additional story. Such a condition may occur in buildings of a split-level type with floor levels partially below grade or in buildings with under-floor crawl spaces larger than the minimum specified. See Figure R602.3(2). The rationale for the wall being considered an additional story is based on the load capacity of studs of a particular size changing as the unsupported length is increased.

The minimum length of 14 inches (356 mm) for cripple wall studs provides sufficient clear space for required nailing of the framing. If the wood-framed foundation wall does not permit the installation of foundation studs of such a length, wood structural panel sheathing must be applied, or a solid-blocking method of construction may be used.

R602.10 Wall bracing. Buildings shall be braced in accordance with this section or, when applicable, Section R602.12. Where a building, or portion thereof, does not comply with one or more of the bracing requirements in this section, those portions shall be designed and constructed in accordance with Section R301.1.

❖ All buildings must be braced to resist lateral loads due to wind and seismic forces. Wall construction must include bracing to resist imposed lateral loads resulting from wind or seismic loading and to provide stability to the structure. When subjected to wind loads, the upper portion of the structure moves horizontally while the lower portion is restrained at ground level. During an earthquake, the ground motion displaces the foundation while the top portion of the structure tries to remain stationary. In both of these cases the bracing resists the differential movement and thus prevents or limits damage to the building.

This section provides the prescriptive requirements for conventional construction of wall bracing to resist the lateral wind and seismic loads within the scope of the code. For buildings, or portions of buildings, that do not meet the bracing requirements of this section, an engineered design in accordance with the *International Building Code*® (IBC®) or other referenced standard is required in accordance with Section R301.1.

Wall bracing for lateral loads due to wind speed is required for all buildings. Wall bracing for lateral loads due to seismic loading is required for townhouses in Seismic Design Category (SDC) C and all buildings in SDC D_0, D_1, and D_2 as defined in Section R301.2.2. Buildings in Seismic Design Categories A, B or, for detached one- and two-family dwellings, SDC C, are only required to meet wind provisions. Dwellings and townhouses in these seismic design categories are exempt from all seismic provisions.

For walls to meet wall bracing requirements, a continuous lateral load path must be prescribed for the building. It is very important to understand the concept of a lateral load path because it helps make sense of the prescriptive requirements in the code. In short, the lateral load path is simply the path that the lateral or horizontal load takes as it passes through the structure, including components and connections, on its way to the ground.

The lateral load path for wind loads is simpler to visualize than the load path for seismic loads. Commentary Figure R602.10 provides a basic example of the lateral load path resulting from wind loading. The load is shown acting on a windward receiving wall, and its subsequent load path through the building. For simplicity, the suction pressure on the leeward receiving wall is not included in this illustration.

As shown in Commentary Figure R602.10, there are six critical parts in the load path for a simple rectangular structure. The critical parts of the lateral load path addressed in this section are Items 4, 5 and 6 of Commentary Figure R602.10.

From a lateral load perspective, the walls support the roof and floor diaphragms through the use of bracing panels. The type, amount and number of bracing panels are, of course, dependent on the magnitude of the lateral load. Stronger resistance (greater numbers of bracing panels) and reduced braced wall line spacing (interior braced wall lines) may be required in areas of high wind and/or seismic activity.

Failure of a braced wall line is evidenced by racking of the wall line. Racking occurs when a rectangular wall deforms to a parallelogram shape, in which the top and the bottom of the wall remain horizontal but the sides are no longer vertical. The purpose of wall bracing is to prevent such failures.

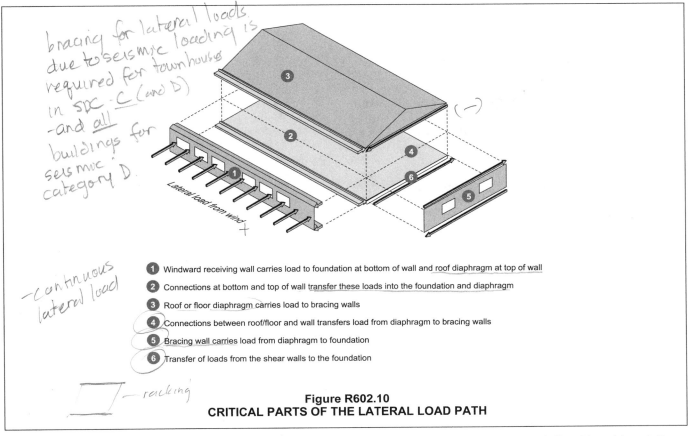

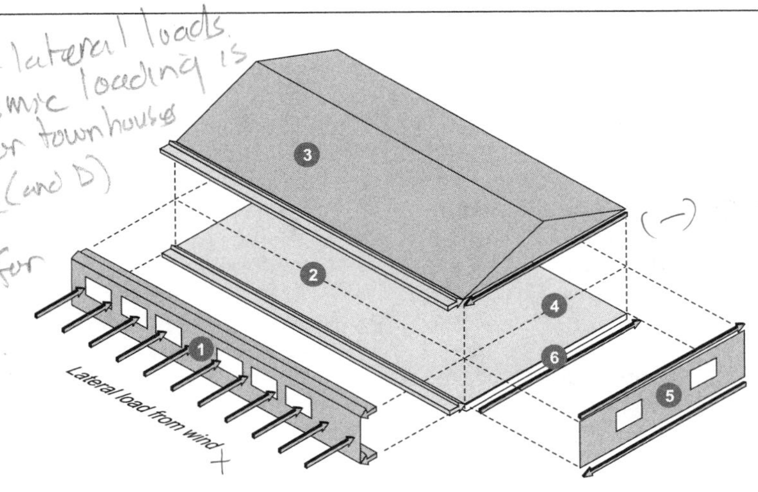

1. Windward receiving wall carries load to foundation at bottom of wall and roof diaphragm at top of wall
2. Connections at bottom and top of wall transfer these loads into the foundation and diaphragm
3. Roof or floor diaphragm carries load to bracing walls
4. Connections between roof/floor and wall transfers load from diaphragm to bracing walls
5. Bracing wall carries load from diaphragm to foundation
6. Transfer of loads from the shear walls to the foundation

Figure R602.10
CRITICAL PARTS OF THE LATERAL LOAD PATH

In the 2012 edition of the IRC, the wall bracing section is organized to address one bracing topic per subsection. Intermittent and continuous sheathing methods are combined into a single section. A new section, R602.12, offers a simplified prescriptive approach for wall bracing in low-wind, low-seismic regions.

Previously, in the 2009 code edition, the bracing provisions more than doubled in page count over the 2006 bracing provisions, due to additional bracing options and new figures that made the provisions easier to understand and apply.

In the 2009 edition of the code, many changes were made to the wood-framed bracing provisions:

- Most significant was the development of wind-bracing tables based on engineering principles.
- Separate tables for wind and seismic bracing were developed.
- The amount of bracing required changed from a "percentage" to an actual "length" of bracing.
- The code was reorganized to consolidate all of the bracing provisions for wood-framed construction into the Chapter 6 bracing section.
- New bracing methods were added to increase the choices available to the builder and to reflect ongoing product research.
- The bracing method options were clarified into two distinct classifications: "intermittent" and "continuous."
- Bracing methods were defined by abbreviations instead of method numbers. For example, wood structural panel bracing, formerly referred to as Method 3, became Method WSP. Gypsum board bracing became Method GB, let-in bracing became Method LIB, structural fiberboard sheathing became Method SFB, etc.
- The number of narrow wall bracing alternates grew.
- Method SFB (structural fiberboard sheathing) was recognized as a continuous sheathing method for use in areas of low wind and earthquake loads.

In the 2006 edition, the 32-inch-wide (813 mm) alternate (ABW) was changed to permit wall bracing elements as narrow as 28 inches (711 mm) for certain cases. A portal frame with hold-downs, a narrow vertical element that is attached at the plate level below and to the header above in a way that will allow it to act as a bracing unit, was also added, which permitted bracing elements as narrow as 16 inches (406 mm) in certain cases. In addition, the portal frame was permitted to be used without hold-downs for bracing next to garage doors (in SDCs A, B and C, for use with up to one story above) when used in conjunction with continuously sheathed wood structural panel walls (Table R602.10.5, Note c).

In the 2003 edition of the code, braced wall line spacing limits were added.

In the 2000 IRC, an option for continuous wood structural panel sheathing (based on the perforated shear wall method defined in the IBC) was available to allow narrower braced wall panels. This method required the braced wall lines to be fully sheathed with wood structural panels. A narrow bracing element to either side of the garage opening was allowed with continuous sheathing for support of a light frame roof with roof dead loads of 3 psf or less. The narrow braced wall panels were allowed to have a 4:1 aspect ratio.

A portion of the commentary on the wall bracing provisions is based upon information found in "*A Guide to the IRC Wood Wall Bracing Provisions.*" This guide, with versions written for the 2006, 2009, and 2012 IRC, is a joint publication by ICC and APA-The Engineered Wood Association and provides a more extensive explanation and application of the wall bracing provisions.

Commentary Table R602.10 lists topics in the 2012 IRC Wall Bracing section. It also shows the location of the subsections in earlier editions of the IRC. When an older code edition does not have an equivalent provision, the column is marked with NEP—No Equivalent Provision.

Table R602.10
CROSS REFERENCE OF 2012 IRC WALL BRACING TOPICS TO 2009/2006 IRC

TOPIC	2012	2009	2006
Corner Nailing	Table R602.3(1)	Figure R602.10.4.4(1)	Figure R602.10.5
BWP Uplift	R602.3.5	R602.10.1.2.1	NEP
Length of a BWL	R602.10.1.1	R602.10.1	R602.10.1
BWL Offsets	R602.10.1.2	R602.10.1.4	R602.10.1
Spacing of BWL	R602.10.1.3, Table R602.10.1.3	R602.10.1, R602.10.1.4, & R602.10.1.5, Tables R602.10.1.2(1) & (3), Figures 602.10.1.4(1), (3) & (4)	R602.10.1.1, R602.10.11.1
Angled Walls	R602.10.1.4, Figure R602.10.1.4	R602.10.1.3, Figure R602.10.1.3	NEP
BWP Uplift Load Path	R602.10.2.1 – details moved to R602.3.5	R602.10.1.2.1	NEP
Location BWPs	R602.10.2.2, Figure R602.10.2.2	R602.10.1.4, Figure R602.10.1.4(2)	R602.10.1
Location BWPs in SDC D_0, D_1 and D_2	R602.10.2.2.1	R602.10.1.4.1, Figure R602.10.1.4.1	R602.10.11.2
Min Number of BWPs	R602.10.2.3	NEP	NEP
Required Bracing Length, Wind & Seismic Application	R602.10.3	R602.10.1.2	R602.10.4, R602.10.6, & Table R602.10.1
Wind	Table R602.10.3(1)	Table R602.10.1.2(1)	Table R602.10.1
Wind Adjustment Factors	Table R602.10.3(2)	Table R602.10.1.2(1) footnotes	NEP
Seismic	Table R602.10.3(3)	Table R602.10.1.2(2)	Table R602.10.1
Seismic Adjustment Factors	Table R602.10.3(4)	Table R602.10.1.2(3)	Table R602.10.1 footnotes
Bracing Methods	Table R602.10.4	Tables R602.10.2, R602.10.4.1, & R602.10.5	R602.10.3, R602.10.5
Mixing Methods	R602.10.4.1	R602.10.1.1	NEP
Continuous Sheathing Methods	R602.10.4.2	R602.10.4, R602.10.5	R602.10.5
Interior Finish	R602.10.4.3	R602.10.2.1	NEP
Min Length BWP	Table R602.10.5	R602.10.1.2, R602.10.3, R602.10.4.2, R602.10.5.2, Tables R602.10.3, R602.10.3.1, R602.10.4.2 & R602.10.5.2	R602.10.4, R602.10.6.2, Tables R602.10.5 & R6012.10.6
ABW	Table R602.10.5	Table R602.10.3.2	Table R602.10.6
	Figure R602.10.6.1	Figure R602.10.3.2	R602.10.6 text
PFH	Table R602.10.5, Figure R602.10.6.2	Figure R602.10.3.3	R602.10.6.2 and Figure R602.10.6.2

(continued)

Table R602.10—continued
CROSS REFERENCE OF 2012 IRC WALL BRACING TOPICS TO 2009/2006 IRC

TOPIC	2012	2009	2006
PFG	Table R602.10.5, Figure R602.10.6.3	Figure R602.10.3.4	NEP
CS-PF	R602.10.6.4, Table R602.10.5 & Figure R602.10.6.4	R602.10.4.1.1, Figure R602.10.4.1.1	NEP
Tension Strap	Table R602.10.6.4	Table R602.10.4.1.1	NEP
Wall Bracing for Masonry Veneer in SDC D_0, D_1 and D_2	R602.10.6.5	R602.12	R703.7
End of Wall Line	R602.10.2.2, Figure R602.10.7	R602.10.1.4, R602.10.4.4	R602.10.1, R602.10.11.2
BWP to Roof Connections	R602.10.8, R602.10.8.2	R602.10.6, R602.10.6.2	R602.10.8
BWP Support	R602.10.8, R602.10.8.1, & R602.10.9	R602.10.6, R602.10.6.1 & R602.10.7	R602.10.8, R602.10.9
Panel Joints	R602.10.10	R602.10.8	R602.10.7
Cripple Walls	R602.9, R602.10.11, & R602.11.2	R602.9, R602.10. 9, & R602.11.2	R602.10.2, R602.10.11.4, & R602.11.3
Simplified Wall Bracing	R602.12	NEP	NEP

R602.10.1 Braced wall lines. For the purpose of determining the amount and location of bracing required in each story level of a building, *braced wall lines* shall be designated as straight lines in the building plan placed in accordance with this section.

❖ The definition of a "Braced wall line" from Section R202 is:

BRACED WALL LINE. A straight line through the building plan that represents the location of the lateral resistance provided by the wall bracing.

In the 2012 edition, information about braced wall lines—length, spacing, and orientation—is in Section R602.10.1. Commentary Figure R602.10.1(1) shows two braced wall lines (1 and 2) in the longitudinal direction and three braced wall lines (A, B and C) in the transverse direction. Each wall line is made up of a number of braced wall panels. Not every braced wall line has to be continuous, as shown in braced wall line B. Furthermore, the code permits offsets of up to 4 feet (1219 mm) from the braced wall line (see Section R602.10.1.2). If sections of a braced wall line are offset by 4 feet (1219 mm) or less, they are assumed to act in combination to resist lateral load. If the sections are farther than 4 feet (1219 mm) from the braced wall line, they must be considered as separate braced wall lines. Offsets in braced wall lines are further discussed in Section R602.10.1.2.

Braced wall lines may include interior walls, as braced wall line B in Commentary Figure R602.10.1(1) illustrates. An interior braced wall line may be required—depending on the size of the house, the wind speed, or seismic design category—to supplement the exterior braced wall lines. Interior braced wall lines have requirements similar to exterior wall lines in terms of bracing length, panel location, wall line offsets and attachments.

The effective (imaginary) braced wall line, a concept introduced in Commentary Figure R602.10.1(2), is based on the principle that a braced wall line does not have to coincide with a physical wall line. The designated wall line location may be within a grouping of wall line sections and assumed to provide a line of lateral resistance somewhere near the center of the wall sections. This concept is illustrated in Commentary Figure R602.10.1(2).

R602.10.1.1 Length of a braced wall line. The length of a *braced wall line* shall be the distance between its ends. The end of a *braced wall line* shall be the intersection with a perpendicular *braced wall line*, an angled *braced wall line* as permitted in Section R602.10.1.4 or an exterior wall as shown in Figure R602.10.1.1.

❖ In Section R602.10.1.1, the definition of a braced wall line (BWL) is now the distance between the most distant ends of the wall lines included in the braced wall line. If a wall line segment is angled and 8 feet (2438 mm) or less in length, the projected length of the angled wall segment may also count as part of the length of the braced wall line. There is no longer a requirement to draw projections of braced wall lines beyond the end of an intersecting wall line. The end of the building determines a braced wall line's end point.

In the 2006 edition of the code, a major challenge in working with braced wall lines was that if a building was not a simple rectangle, it was difficult to determine the beginning and end points of braced wall lines, which made measuring the line length problematic.

"The length of a braced wall line shall be measured as the distance between the ends of the wall line."

While this definition may seem simplistic, its significance becomes apparent as house plans deviate from simple rectangles. With permitted braced wall line offsets, enclosed porches, wall bump outs, etc., it can be difficult for all parties to agree on where the beginning and end points for a given braced wall line are. In the 2009 edition, this became increasingly confusing when incorporating the effective (imaginary) braced wall lines, a braced wall line that does not rest on a physical wall line in the plans.

Because of the effective braced wall line concept, the 2009 IRC defined the end of the braced wall line as the intersection with perpendicular exterior walls or projection thereof or the intersection with perpendicular braced wall lines. The end of the braced wall line was chosen such that the maximum length resulted.

The provision was meant to address the following condition which can arise from the addition of an effective braced wall line. When a perpendicular braced wall line is located beyond the end of a braced wall line, the BWL length was extended beyond the end of the physical wall line [see Commentary Figure R602.10.1.1(1)].

Issues arose with this definition. In some cases a short exterior wall line would have a braced wall line length more than twice its actual length because the longest of the length definitions was required. In the 2012 edition, the braced wall line length is the distance between the most distant ends of the wall lines included in the braced wall line. The braced wall line does not need to be extended to intersect other braced wall lines in the plans. Note BWL A in Commentary Figure R602.10.1.1(2). In the 2012 IRC provisions BWL A is not required to continue beyond the end of the physical wall line. Instead, the braced wall line and the physical wall line end at the perpendicular exterior wall on the left side of the building. Changes to this provision have allowed a decrease of nearly 50 percent for the BWL A length in Commentary Figure R602.10.1.1(2).

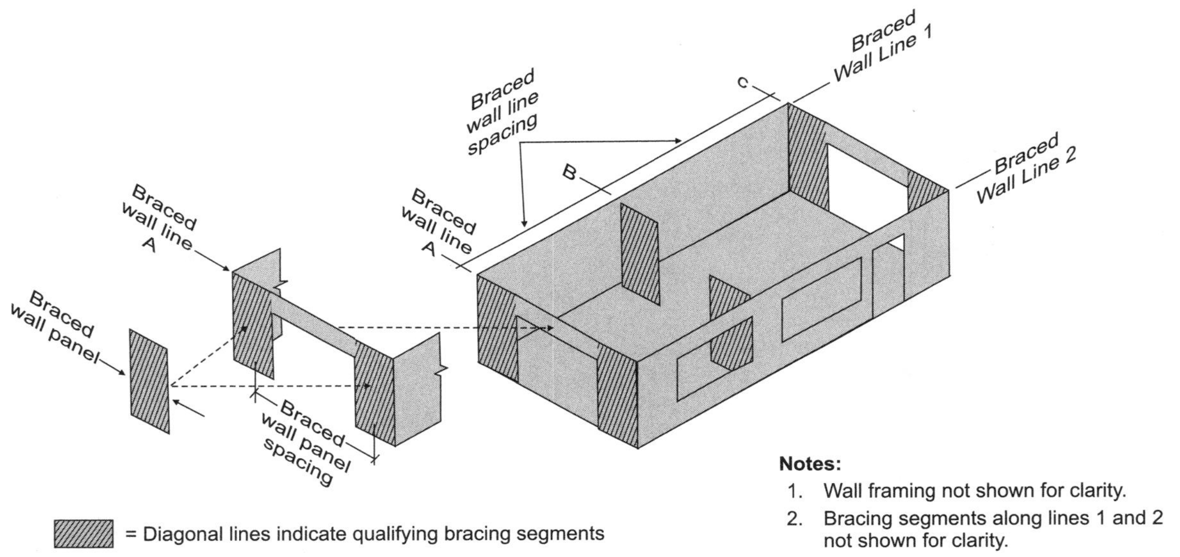

Figure R602.10.1(1)
WALL BRACING TERMINOLOGY

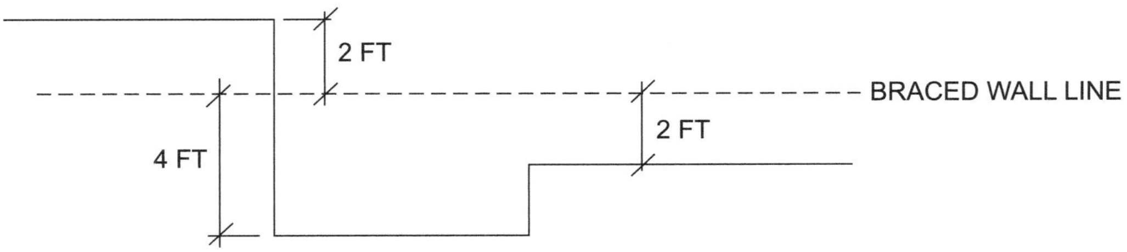

Figure R602.10.1(2)
EFFECTIVE (IMAGINARY) BRACED WALL LINES

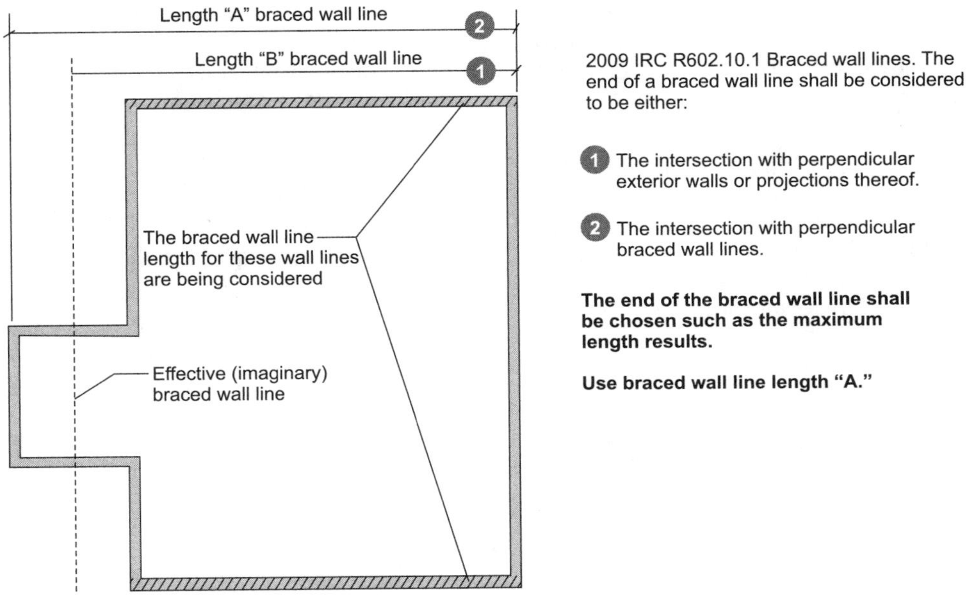

Figure R602.10.1.1(1)
2009 IRC EFFECTIVE (IMAGINARY) BRACED WALL LINES

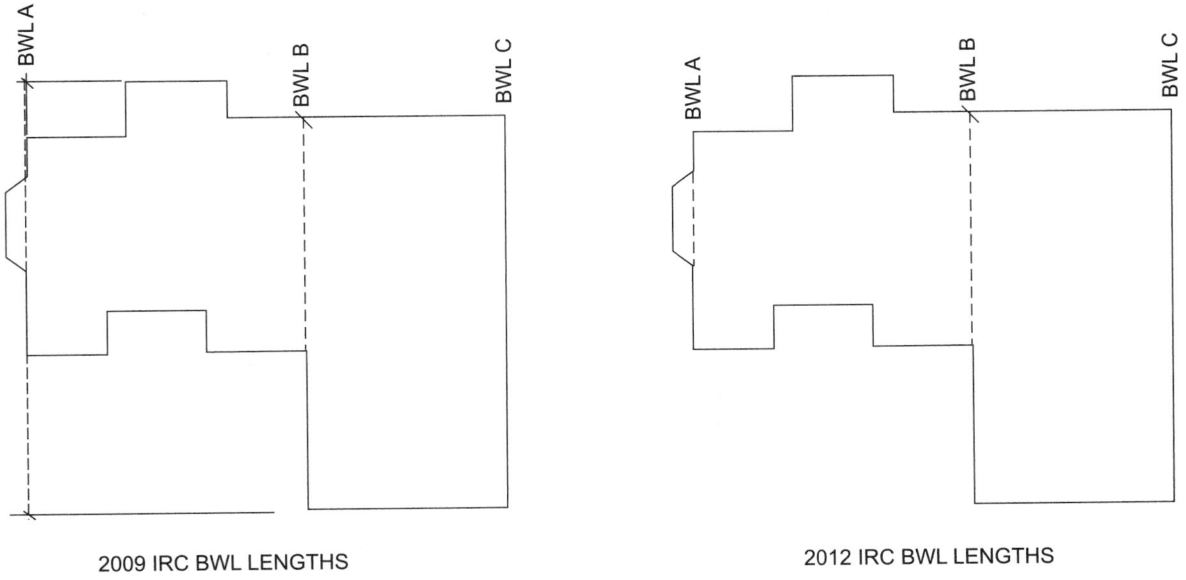

Figure R602.10.1.1(2)
COMPARISON OF 2009 AND 2012 IRC PROVISIONS FOR BRACED WALL LINE LENGTH

R602.10.1.2 Offsets along a braced wall line. All exterior walls parallel to a *braced wall line* shall be offset not more than 4 feet (1219 mm) from the designated *braced wall line* location as shown Figure R602.10.1.1. Interior walls used as bracing shall be offset not more than 4 feet (1219 mm) from a *braced wall line* through the interior of the building as shown in Figure R602.10.1.1.

❖ Many home designs feature offsets along the braced wall line length. Figure R602.10.1.1 is updated to show offsets, braced wall line spacing and placement. As in the 2009 IRC, a designated braced wall line is not required to be placed on a wall line in the plans. It may be placed between two or more wall lines. For wall line offsets, the former limit of an 8-foot (2438 mm) total offset has been deleted. The limit of a maximum of 4 feet (1219 mm) from the braced wall line to a physical wall line requiring a braced panel remains. The limit of 8 feet (2438 mm) was redundant as the maximum distance of a wall line segment to either side of the braced wall line is 4 feet (1219 mm).

R602.10.1.3 Spacing of braced wall lines. The spacing between parallel *braced wall lines* shall be in accordance with Table R602.10.1.3. Intermediate *braced wall lines* through the interior of the building shall be permitted.

❖ Minimum braced wall line (BWL) spacing in the 2009 edition was found in multiple sections and tables, including: Sections R602.10.1, R602.10.1.4, and R602.10.1.5, Tables R602.10.1.2(1) and (3), and Fig-

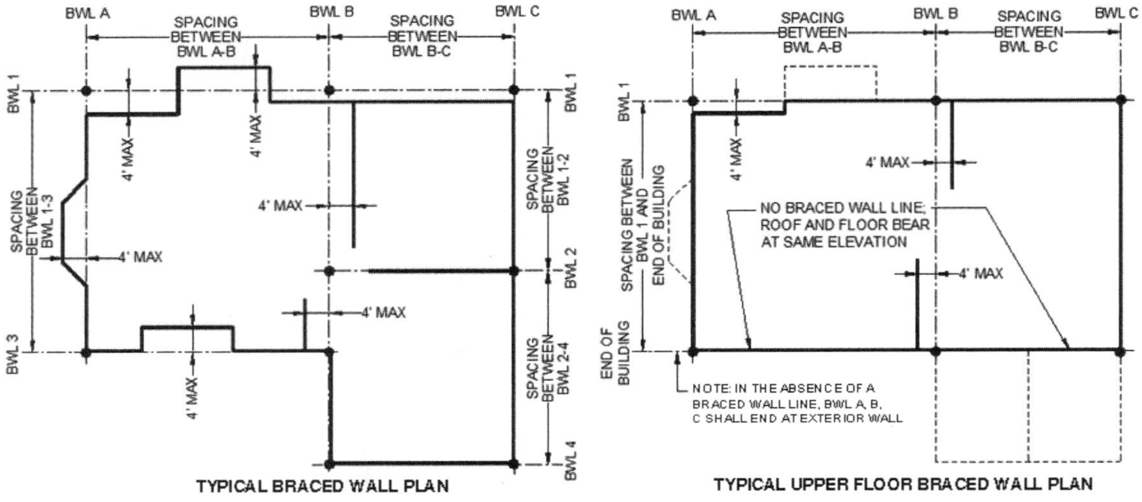

For SI: 1 foot = 304.8 mm.

FIGURE R602.10.1.1
BRACED WALL LINE SPACING

❖ See the commentary for Sections R602.10.1.1 and R602.10.1.2.

TABLE R602.10.1.3
BRACED WALL LINE SPACING

APPLICATION	CONDITION	BUILDING TYPE	BRACED WALL LINE SPACING CRITERIA	
			Maximum Spacing	Exception to Maximum Spacing
Wind bracing	85 mph to < 110 mph	Detached, townhouse	60 feet	None
Seismic bracing	SDC A – C	Detached	Use wind bracing	
	SDC A – B	Townhouse	Use wind bracing	
	SDC C	Townhouse	35 feet	Up to 50 feet when length of required bracing per Table R602.10.3(3) is adjusted in accordance with Table R602.10.3(4).
	SDC D_0, D_1, D_2	Detached, townhouses, one- and two-story only	25 feet	Up to 35 feet to allow for a single room not to exceed 900 square feet. Spacing of all other braced wall lines shall not exceed 25 feet.
	SDC D_0, D_1, D_2	Detached, townhouse	25 feet	Up to 35 feet when length of required bracing per Table R602.10.3(3) is adjusted in accordance with Table R602.10.3(4).

For SI: 1 foot = 304.8 mm, 1 square foot = 0.0929 m^2, 1 mile per hour = 0.447 m/s.

❖ See the commentary for Section R602.10.1.3.

ures 602.10.1.4(1), (3) and (4). All BWL spacing information is now located in Table R602.10.1.3. The limitation of 60 feet (18 287 mm) for wind requirements and 25 feet (7620 mm) for seismic requirements for BWL spacing is unchanged.

For townhouses in SDC C, braced wall lines must not be spaced more than 35 feet (10 668 mm) apart. However, an exception permits spacing up to 50 (15 239 mm) feet with certain limitations.

- The exception permits an increase up to 50 feet (15 239 mm) provided the length of the required wall bracing has been increased in accordance with Table R602.10.3(4) for braced wall lines spaced greater than 35 feet (10 668 mm).

For structures in SDC D_0, D_1 or D_2, braced wall lines must not be spaced more than 25 feet (7620 mm) apart. However, two exceptions permit a spacing of 35 feet (10 668 mm) with certain limitations.

- The first permits an increase to 35 feet (10 668 mm) for a single room up to 900 feet square (84 m^2) where the spacing between all other braced wall lines does not exceed 25 feet (7620 mm).
- The second permits an increase to 35 feet (10 668 mm) provided the length of the required wall bracing has been increased in accordance with Table R602.10.3(4) for the braced wall lines spaced greater than 25 feet (7620 mm), the aspect ratio of the floor above or roof diaphragm does not exceed 3:1, and the top plate, of the braced wall line is spliced as specified.

Table R602.10.3(4) provides bracing adjustments for braced wall line spacing up to 50 feet (15 239 mm) for SDC C townhouses and for BWL spacing of 30 and 35 feet (9144 and 10 668 mm) for structures in SDC D_0, D_1 or D_2.

R602.10.1.4 Angled walls. Any portion of a wall along a *braced wall line* shall be permitted to angle out of plane for a maximum diagonal length of 8 feet (2438 mm). Where the angled wall occurs at a corner, the length of the *braced wall line* shall be measured from the projected corner as shown in Figure R602.10.1.4. Where the diagonal length is greater than 8 feet (2438 mm), it shall be considered a separate *braced wall line* and shall be braced in accordance with Section R602.10.1.

❖ This provision allows the sheathing in a diagonal wall to be "counted" towards the total bracing length for a single braced wall line. Note that the angled portion of the wall line must be less than or equal to 8 feet (2438 mm) in length. If it is greater than 8 feet (2438 mm) in length, it must be considered its own braced wall line.

Note that the provision does not permit "double dipping": the bracing on an angled wall may be applied towards the total bracing length required for only one braced wall line connected to the angled wall line.

The section previously called "angled corners" has been renamed "angled walls" and modified to focus on wall length and the angle of the wall from the braced wall line being examined. As in the 2009 IRC, angled wall lines may be counted as part of a braced wall line when 8 feet (2438 mm) long or less. In the 2012 edition, the projected length of the braced wall line should be added to the total wall line length. Commentary Table R602.10.1.4 gives the projected braced wall line length to add to the braced wall line length for 4, 6, and 8-foot (1219, 1829, and 2438 mm) long angled wall lines. If an angled wall line is longer than 8 feet (2438 mm), it should be considered a separate braced wall line. See Commentary Figure R602.10.1.4(1) for the projected angled wall length.

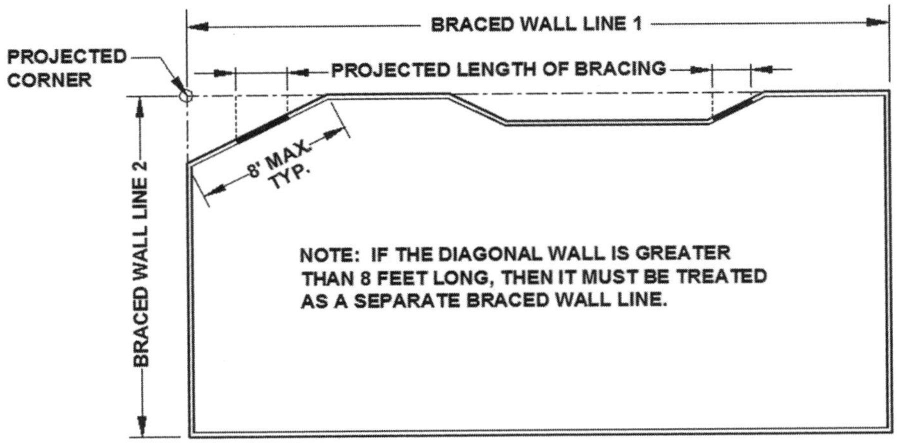

For SI: 1 foot = 304.8 mm.

FIGURE R602.10.1.4
ANGLED WALLS

❖ See the commentary for Section R602.10.1.4.

Table R602.10.1.4
PROJECTED BRACED WALL LINE LENGTH CONTRIBUTED BY THE ANGLED WALL

ANGLE FROM BWL (DEGREES)	PROJECTED BWL LENGTH FROM ANGLED WALL LINE (feet)		
	4	6	8
15	3.9	5.8	7.8
30	3.5	5.2	7.0
45	2.9	4.3	5.7

What if the angled wall contains the entryway of the house and there is not sufficient space for bracing panels? The length of the angled entryway still counts towards the length of the wall line, and the 10 foot (3048 mm) rule (see Section R602.10.2.2) also applies. When the length of the angled wall shown in Commentary Figure R602.10.1.4(1) is less than 10 feet (3048 mm) bracing panels are not required in the angled wall section, provided that the other requirements of Section R602.10.2.2 are met.

What happens if the angled wall segment is greater than 10 feet (3048 mm)? Whenever the length of the angled wall is greater than 8 feet (2438 mm), the angled wall provision of Section R602.10.1.4 no longer applies and the angled wall is considered its own wall line which must be braced accordingly, as shown in Commentary Figure R602.10.1.4(2). This can be a difficult proposition when using the 2012 code because of the need for "braced wall line spacing" in accordance with both the wind and seismic bracing tables. If no parallel angled line is available to measure between, then determining the correct distance to use for the braced wall line spacing becomes a matter of interpretation. A reasonable solution is to run a perpendicular line from the center of the angled wall until it intersects a braced wall line and use that distance as the braced wall line spacing.

If there are no bracing elements in the angled wall segment and the first braced wall panel is a distance greater than 10 feet (3048 mm) from the end of the wall line, an engineered solution will be required for the entire wall line.

R602.10.2 Braced wall panels. *Braced wall panels* shall be full-height sections of wall that shall have no vertical or horizontal offsets. *Braced wall panels* shall be constructed and placed along a *braced wall line* in accordance with this section and the bracing methods specified in Section R602.10.4.

❖ The definition of a braced wall panel (BWP), information on BWP location, and the minimum required number of BWPs in a wall line are now in one section. The BWP definition has been clarified to limit panels to areas of a wall line with no vertical or horizontal irregularities.

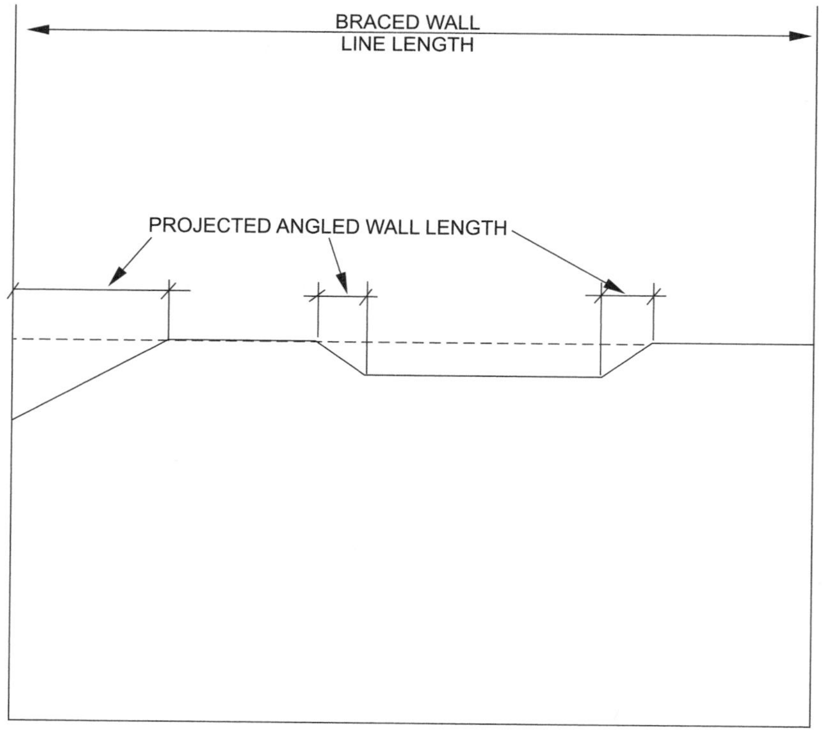

Figure R602.10.1.4(1)
PROJECTED ANGLE WALL LENGTH

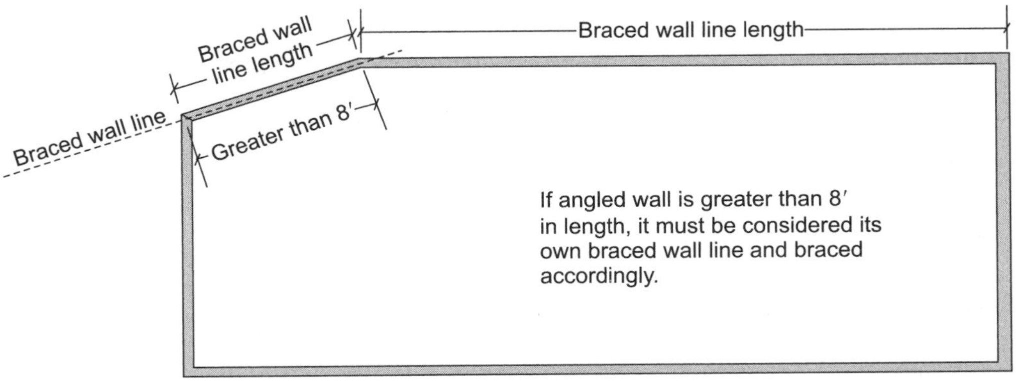

Figure R602.10.1.4(2)
ANGLED WALL GREATER THAN 8 FEET IN LENGTH

The definition of a braced wall panel, from Section R202, is:

BRACED WALL PANEL. A full-height section of wall constructed to resist in-plane shear loads through interaction of framing members, sheathing material and anchors. The panel's length meets the requirements of its particular bracing method and contributes toward the total amount of bracing required along its braced wall line in accordance with Section R602.10.2.

Put in simpler terms, braced wall panel describes a code-qualified bracing element. The name is probably derived from the fact that most of the recognized methods of bracing use panel-type products. Even let-in bracing is often referred to as a bracing panel. "Panel" is somewhat of a misnomer since it actually describes a wall "section," "segment" or "unit." As such, the terms braced wall section, braced wall segment and bracing unit are often used interchangeably.

Each braced wall panel must extend the full height of the wall—from the bottom plate to the top of the double top plates. A "panel" may be constructed from more than one piece of sheathing. For example, a 6-foot-long (1829 mm) braced wall panel may be constructed by joining a 4-foot-long (1219 mm) panel with a 2-foot-long (610 mm) panel.

Braced wall panels have a height and a length dimension. The permitted height of a braced wall panel ranges from 8 to 12 feet (2438 to 3658 mm). In some cases, the panel height is limited to 10 feet (3048 mm). The length dimension is measured parallel to the length of the wall. For example, the length of a 4 by 8 oriented strand board (OSB) bracing panel placed with the 8-foot (2438 mm) dimension in the up-and-down direction is 4 feet (1219 mm). Knowing the length of a braced wall panel is important because the various bracing methods have different required minimum lengths. Also, the combined length of individual braced wall panels must total or exceed a required minimum length for the braced wall line.

Although in previous editions of the code the terms "length" and "width" were often interchanged, the 2009 and 2012 editions of the code uses the term "length" more consistently throughout the bracing provisions.

R602.10.2.1 Braced wall panel uplift load path. The bracing lengths in Table R602.10.3(1) apply only when uplift loads are resisted in accordance with Section R602.3.5.

❖ The wall bracing section assumes that all wall lines with braced panels have been nailed to prescriptive requirements for the uplift load path. When uplift connection is not as specified in Section R602.3.5, the required wall bracing in Tables R602.10.3(1) and R602.10.3(3) does not apply and the wall lines will need engineering to determine bracing requirements. Refer to Section R602.3.5 for more information on the requirements of uplift load paths for braced wall panels.

R602.10.2.2 Locations of braced wall panels. A *braced wall panel* shall begin within 10 feet (3810 mm) from each end of a *braced wall line* as determined in Section R602.10.1.1. The distance between adjacent edges of *braced wall panels* along a *braced wall line* shall be no greater than 20 feet (6096 mm) as shown in Figure R602.10.2.2.

❖ Braced wall panel (BWP) location has been simplified. In the 2012 IRC, braced wall panels are to begin within 10 feet (3048 mm) of the end of a wall line in SDC A, B and C and all wind speeds for buildings are allowed to be designed in accordance with the IRC. The panels may be inset up to 10 feet (3048 mm) from both ends of a wall line for all intermittent and continuous sheathing bracing methods, as shown in Figure R602.10.2.2.

In the 2009 IRC, intermittent braced wall panels could begin up to $12^1/_2$ feet (3810 mm) from the end of a wall line in SDC A, B and C but the cumulative total distance was $12^1/_2$ feet (3810 mm) as well. This was a new provision for 2009 as previous provisions allowed insets of $12^1/_2$ feet (3810 mm) from both ends of a wall line. The 2012 code change removes the need to calculate total inset while reducing the allowable inset distance.

The 2009 requirement was added to prevent the 12.5 foot (3810 mm) end-distance provision from being used to eliminate a braced wall panel in a wall line. For example, if the 12.5-foot (3810 mm) rule is applied at each end of a 29-foot (8839 mm) long wall line, only a single panel is required in the wall line: one braced panel positioned 12.5 feet (3810 mm) from each end of the wall line [12.5 feet + 4 feet + 12.5 feet = 29 feet (8839 mm)]. The 2009 provision eliminated the possibility of such an interpretation. The original bracing provisions required bracing at each end and every 25 feet (7620 mm) on center. A single panel in a braced wall line violates the intent of the original provision to have a minimum of two braced panels, and may not provide sufficient stability to the roof or floor diaphragm above.

If bracing is located more than 10 feet (3048 mm) from either end of a braced wall line, the wall line requires an engineered design for lateral loads.

The maximum distance between two braced wall panels is now measured from edge to edge. Previous codes measured braced wall panel distances on center. The 25-foot (7620 mm) braced wall panel spacing requirement was straightforward when 4-foot (1219 mm) braced wall panels were used; however, the intent of the requirement needed to be considered when dealing with longer lengths of bracing or continuously sheathed bracing methods. The intent is that unbraced wall lengths longer than 21 feet (6401 mm) (measured between adjacent bracing panel edges) are not permitted. Given this interpretation, for example, the center of a 4-foot (1219 mm) section within an 8-foot (2438 mm) braced wall panel can be used as the measuring point for determining the 25-foot (7620 mm) maximum distance. In such a case, measuring 21 feet (6401 mm) between bracing panels may be the easier method of determining the maximum braced wall panel spacing.

Without this interpretation, the permissible distance between bracing segments would decrease as the length of the segments increased. This was not the intent of the code. As long as the distance between braced wall panels did not yield unbraced wall lengths of greater than 21 feet (6401 mm), the code's braced wall panel spacing requirements were satisfied.

Changing to a measurement of distance from panel edge to panel edge is intended to simplify calculations. The previous provision of braced wall panel spacing of 25 feet (7620 mm) on center is now a distance of 20 feet (6096 mm) edge to edge. This reduces maximum panel spacing by 1 foot (305 mm), but allows all measurements to be edge to edge distances.

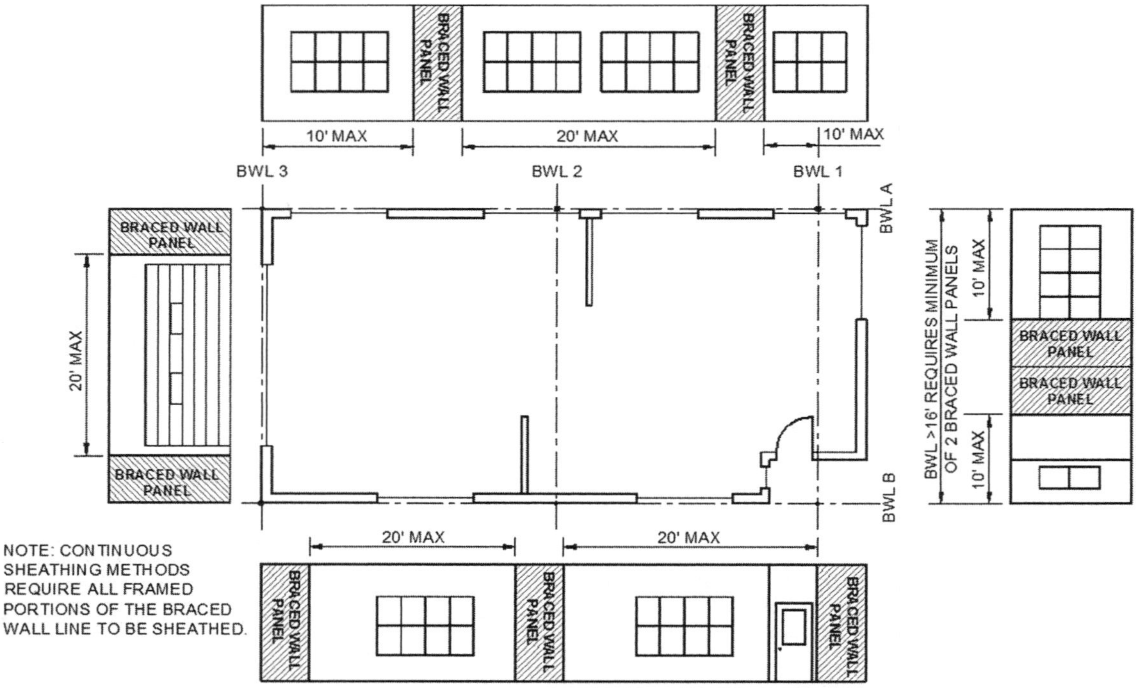

For SI: 1 foot = 304.8 mm.

FIGURE R602.10.2.2
LOCATION OF BRACED WALL PANELS

❖ See the commentary for Section R602.10.2.2.

R602.10.2.2.1 Location of braced wall panels in Seismic Design Categories D_0, D_1 and D_2. *Braced wall panels* shall be located at each end of a *braced wall line*.

Exception: *Braced wall panels* constructed of Methods WSP or BV-WSP and continuous sheathing methods as specified in Section R602.10.4 shall be permitted to begin no more than 10 feet (3048 mm) from each end of a *braced wall line* provided each end complies with one of the following.

1. A minimum 24-inch-wide (610 mm) panel for Methods WSP, BV-WSP, CS-WSP, CS-G, and CS-PF, and 32-inch-wide (813 mm) panel for Method CS-SFB is applied to each side of the building corner as shown in Condition 4 of Figure R602.10.7.
2. The end of each *braced wall panel* closest to the end of the *braced wall line* shall have an 1,800 lb (8 kN) hold-down device fastened to the stud at the edge of the *braced wall panel* closest to the corner and to the foundation or framing below as shown in Condition 5 of Figure R602.10.7.
3. For Method BV-WSP, hold-down devices shall be provided in accordance with Table R602.10.6.5 at the ends of each *braced wall panel.*

❖ To further simplify the provisions, in SDC D_0, D_1 and D_2, braced wall panels must be placed at the ends of the braced wall line. Using Method WSP, if a hold-down or end panel is used, an exception allows braced panels to begin up to 10 feet (3048 mm) from the end of the braced wall line. In previous codes, the allowable offset was 8 feet (2438 mm). To simplify the provisions, all braced wall panel inset distances are a maximum of 10 feet (3048 mm) from the end of the wall line. In all seismic design categories, the distance from wall end to braced wall panel is measured from wall end to panel edge.

Note that a 10-foot (3048 mm) maximum panel distance is permitted at each end if either a hold-down is placed at the panel edge or a 24-inch (610 mm) end panel is provided. If two braced wall lines meet at a corner and along one wall line the first panel is displaced from the corner, then both braced wall lines lose the structural effect of the return corner. In this case, both wall lines must be anchored with a 1,800 pound (8 kN) hold-down device.

For wind and seismic requirements, offset of braced wall panels may now be up to 10 feet (3048 mm) from each end of the braced wall line if all provisions are met.

R602.10.2.3 Minimum number of braced wall panels. *Braced wall lines* with a length of 16 feet (4877 mm) or less shall have a minimum of two *braced wall panels* of any length or one *braced wall panel* equal to 48 inches (1219 mm) or more. *Braced wall lines* greater than 16 feet (4877 mm) shall have a minimum of two *braced wall panels*.

❖ Minimum bracing length for short wall lines is now provided in the code. A braced wall line of 16 feet (4877 mm) or less in length may have one braced wall panel when it is at least 48 inches (1219 mm) long. A braced wall line of 16 feet (4877 mm) or less may alternatively have two qualifying braced wall panels from the tables in Section R602.10 that are less than 48 inches (1219 mm) in length. Braced wall panel spacing and maximum edge distances are illustrated in Figure R602.10.2.2. All distances are measured from a braced wall panel's edge.

R602.10.3 Required length of bracing. The required length of bracing along each *braced wall line* shall be determined as follows.

1. All buildings in Seismic Design Categories A and B shall use Table R602.10.3(1) and the applicable adjustment factors in Table R602.10.3(2).
2. Detached buildings in Seismic Design Category C shall use Table R602.10.3(1) and the applicable adjustment factors in Table R602.10.3(2).
3. Townhouses in Seismic Design Category C shall use the greater value determined from Table R602.10.3(1) or R602.10.3(3) and the applicable adjustment factors in Table R602.10.3(2) or R602.10.3(4) respectively.
4. All buildings in Seismic Design Categories D_0, D_1 and D_2 shall use the greater value determined from Table R602.10.3(1) or R602.10.3(3) and the applicable adjustment factors in Table R602.10.3(2) or R602.10.3(4) respectively.

Only *braced wall panels* parallel to the *braced wall line* shall contribute toward the required length of bracing of that *braced wall line*. *Braced wall panels* along an angled wall meeting the minimum length requirements of Tables R602.10.5 and R602.10.5.2 shall be permitted to contribute its projected length toward the minimum required length of bracing for the *braced wall line* as shown in Figure R602.10.1.4. Any *braced wall panel* on an angled wall at the end of a *braced wall line* shall contribute its projected length for only one of the *braced wall lines* at the projected corner.

Exception: The length of wall bracing for dwellings in Seismic Design Categories D_0, D_1 and D_2 with stone or masonry veneer installed per Section R703.7 and exceeding the first-story height shall be in accordance with Section R602.10.6.5.

❖ The concept of computing bracing from two separate tables was new to the 2009 code and is provided in this section. Minor changes have been made to the tables as well:

- Continuously Sheathed—Structural Fiberboard (CS-SFB) is added to the tables. The method is limited to regions with wind speeds of 100 mph (45 m/s) or less and SDC A-C.
- Seismic Design Category D_0 and D_1 are separated into two categories. Minimum bracing lengths for SDC D_0 are now listed in a separate section from SDC D_1.
- Method Gypsum Board (GB) has been redefined as a 4-foot (1219 mm) single-sided braced wall panel similar to the other bracing methods. The required length of bracing for a braced wall line with Method GB continues to assume a dou-

ble-sided application in Tables R602.10.3(1) and R602.10.3(3). Table R602.10.5 requires an adjustment factor of 0.5 times the physical braced wall panel length for single sided gypsum braced wall panels.

- The amount of bracing from a braced wall panel on an angled wall is defined.

The 2006 IRC and earlier code editions contained bracing tables which assumed a braced wall line spacing of 35 feet (10 668 mm) for wind speeds of 110 mph (49 m/s) or less and SDC A, B and C, and a braced wall line spacing of 25 feet (7620 mm) for SDC D_0, D_1 and D_2. Minimum required bracing length was determined by the length of the braced wall line.

Basing the bracing requirement on the width of the building (the braced wall line spacing) is appropriate for wind. Wind pushes against a building in the same manner that it pushes against a sail on a boat. When wind acts on the building width, the length of the building (dimension parallel to the wind) is irrelevant to determining the bracing required to resist that wind load. In other words, a short building that is 35 feet (10 668 mm) wide has the same "sail area"—and receives the same wind load—as a long building that is 35 feet (10 668 mm) wide. Therefore, if the short building requires 12 feet (3658 mm) of bracing, the long building also requires 12 feet (3658 mm) of bracing. This reasoning is contrary to previous code edition bracing tables that based the amount of wind bracing on a percentage of the braced wall line length, which required more bracing for the long building than the short building. These different approaches are illustrated in Commentary Figure R602.10.3(1). In the 2012 IRC wind-bracing table, Table R602.10.3(1), the user inputs braced wall line spacing to determine the required bracing length in feet, regardless of building length.

There are now two separate tables for computing bracing, because bracing for seismic loads is determined based on the length of the braced wall line. When determining the amount of bracing required to resist seismic forces, the length of the building parallel to the direction of loading is the most important consideration. This is because mass is generally evenly distributed along the length and width of a building. For a given building width, the long building has more mass—and thus receives greater earthquake forces—than the short building. As a result, the long building requires a greater amount of bracing. For this reason, in the 2012 seismic bracing table, [Table R602.10.3(3)], the user inputs the length of the braced wall line to determine the amount of bracing required. This is illustrated in Commentary Figure R602.10.3(2).

Previous joint wind and seismic bracing tables in the code were based on seismic loads, so the amount of required bracing increased as the braced wall line length increased. However, unlike previous editions of the code, beginning in the 2009 edition, the amount of required bracing was provided in feet rather than as a percentage of braced wall line length, eliminating the need for the user to compute the necessary feet of bracing.

Each process for determining required bracing length includes adjustment factors. These adjustment factors—modifications to the amount of bracing based on variations in the structural geometry—are in tables following the bracing requirements tables. Adjustment factors vary for the wind and seismic tables. It is important to note that neglecting an adjustment factor can

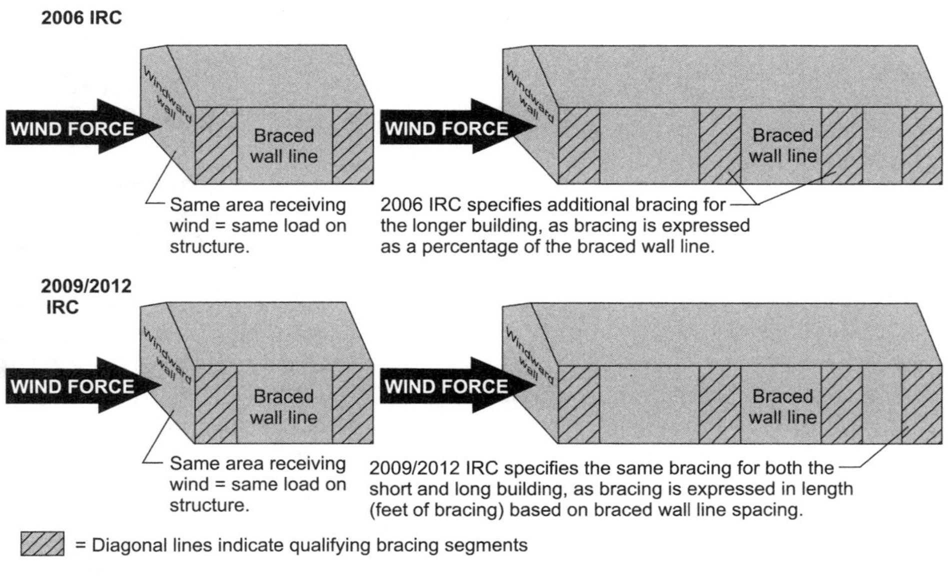

Figure R602.10.3(1)
WALL BRACING—WIND LOADS

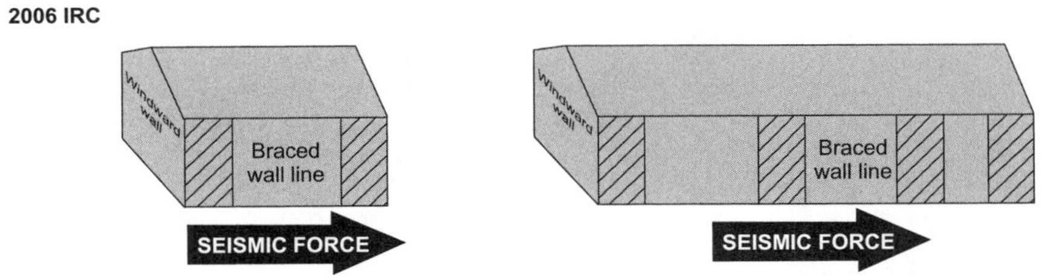

Longer building receives more seismic force when seismic forces act parallel to long side. 2006 IRC specifies additional bracing for the longer building, as bracing is expressed as a percentage of braced wall line length. Percentage bracing equal on both buildings above. 2006 IRC is correct in approach.

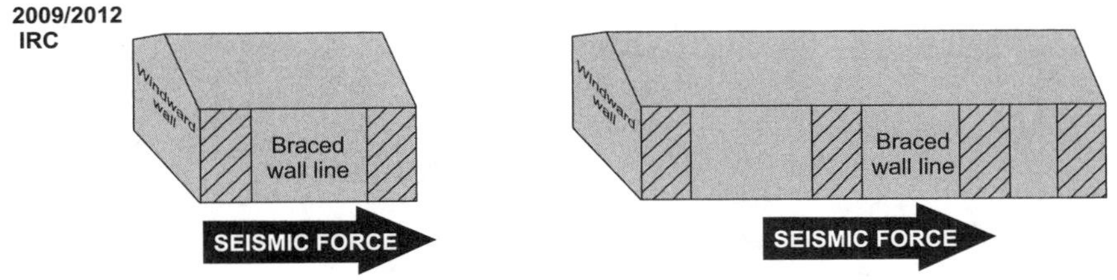

Longer buildings receive more seismic load when seismic forces act parallel to the long side. 2009/2012 IRC specifies additional bracing for the longer buildings, as bracing is expressed in length (feet of bracing) based on the braced wall line length. Both the 2006 IRC and 2009/2012 IRC yield the correct answer.

= Diagonal lines indicate qualifying bracing segments

Figure R602.10.3(2)
WALL BRACING—SEISMIC LOADS

result in insufficient bracing for a specific application. For example, the wind bracing tables are based on a roof eave-to-ridge height of 10 feet (3048 mm). If the roof height of a given single-story structure is 15 feet (4572 mm) and the adjustment factor is ignored, the wall bracing will be insufficient by 30 percent. For this reason, all adjustment factors in Tables R602.10.3(2) and R602.10.3(4) must be considered carefully.

As illustrated in Commentary Figure R602.10.3(3), the information in the wind bracing table is based on:

- Exposure Category B
- 30-foot (9144 mm) mean roof height
- 10-foot (3048 mm) eave-to-ridge height
- 10-foot (3048 mm) wall height per story
- Two braced wall lines per direction of wind

Footnotes to Tables R602.10.3(1), Bracing Requirements Based on Wind Speed, and R602.10.3(3), Bracing Requirements Based on Seismic Design Category, must be followed. After selection of a required bracing length from Table R602.10.3(1) or (3), check the footnotes at the bottom of the tables for additional requirements and then go to Table R602.10.3(2) or (4) for applicable adjustment factors.

In Table R602.10.3(1), Footnote b clarifies additional requirements for Method LIB. While all intermittent bracing methods shall have gypsum board on at least one side of the wall (see Section R602.10.4.3), Note b requires that Method LIB have gypsum board installed on at least one side in accordance with Table R602.3(1) for exterior sheathing or Table R702.3.5 for interior gypsum board at the Method LIB location on either side of the wall. Spacing of fasteners at panel edges may not exceed 8 inches (203 mm).

Footnote c reiterates the requirements for Method CS-SFB. This method is limited to regions where wind speeds are 100 mph (45 m/s) or less.

Table R602.10.3(2), Wind Adjustment Factors to the Required Length of Wall Bracing, contains adjustment factors to accommodate variations from the wind requirements table's assumptions for the wide range of residential structures covered by the code. A number of the adjustment factors are explained in the text below. The adjustments were in the 2009 edition as footnotes to Table R602.10.1.2(1), Bracing Requirements Based on Wind Speed. Many requirements had been taken from other sections and collected into a common location in the 2009 edition. With the 2012 edition, the adjustments are located in a separate table following the Bracing Requirements Based on Wind Speed table. The adjustment factors are described below:

- Exposure category — Note that the wind-bracing table is based on Exposure Category B. The exposure category for a given jurisdiction can be obtained from the local building department. [Refer to Table R301.2(1) of the locally adopted version of the IRC.]

As the number of stories increases (increasing the mean roof height); the adjustment factors increase. This is to accommodate the increased exposure and

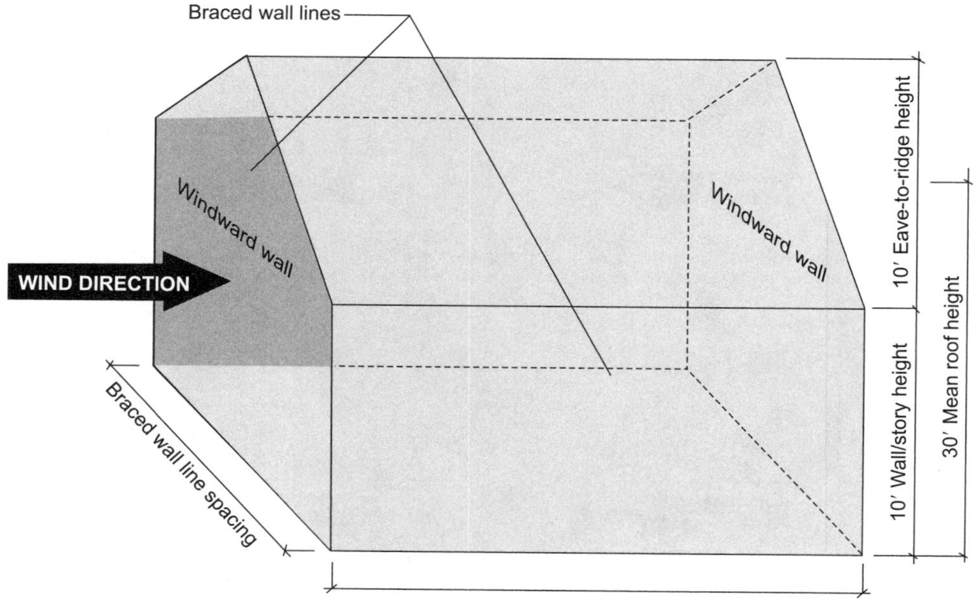

For SI: 1 foot = 304.8 mm.

Figure R602.10.3(3)
BASIS FOR WIND BRACING TABLE

larger sail area of the structure. The following two adjustment factors account for increased sail area.

- Eave-to-ridge height — Eave-to-ridge height is an important consideration because it increases the sail area of a structure, therefore increasing the wind load on the structure, as illustrated in Commentary Figure R602.10.3(4). Increasing eave-to-ridge height by as little as 5 feet (1524 mm), from 10 to 15 feet (3048 to 4572 mm) for example, can increase the required bracing panel length by up to 30 percent. Note that as the number of stories increases the contribution of the roof eave-to-ridge height to the sail area decreases.
- Wall height — Like eave-to-ridge height, increasing the sail area of the structure by making it taller increases the wind load on the structure, thus requiring more bracing. Because the bracing wind table assumes a 10-foot (3048 mm) wall height, taller walls require an adjustment to increase the amount of required bracing, while shorter walls are permitted an adjustment to decrease the amount of required bracing. This adjustment can be applied to each story individually. For example, in a house with a first story wall height of 9 feet (2743 mm) and a second-story wall height of 8 feet (2438 mm), the bracing amount for the first story can be reduced by 5 percent, and the bracing amount for the second story reduced by 10 percent.
- Number of braced wall lines (per plan direction) — At first glance this adjustment factor may give the impression that adding additional interior bracing walls is a disadvantage because the amount of bracing in the wall lines must increase. However, the values in Table R602.10.3(1) are based on the spacing of the braced wall lines. If, for example, the distance between two exterior walls in a home design requires a length of bracing that cannot be accommodated by the exterior wall lines alone, a braced wall line may be added through the interior of the structure. In this case, because the spacing of the braced wall lines decreases, the amount of bracing required for the braced wall lines also decreases. The adjustment factor is applied to the reduced amount of bracing for the braced wall lines. This method is conservative [see Commentary Figure R602.10.3(5)]. Only the interior wall lines receive loads from the floor or roof along both sides of the wall line. The exterior wall lines have tributary loads from only one side of the wall line. But at this time, all wall lines require the increase in bracing. Interior wall lines double the lateral load carried by adjacent exterior wall lines, but the required adjustment factors do not require a 200-percent increase in bracing. Rather, the dwelling is assumed to be tied together sufficiently by connections to allow loads to be transferred to both wall lines.

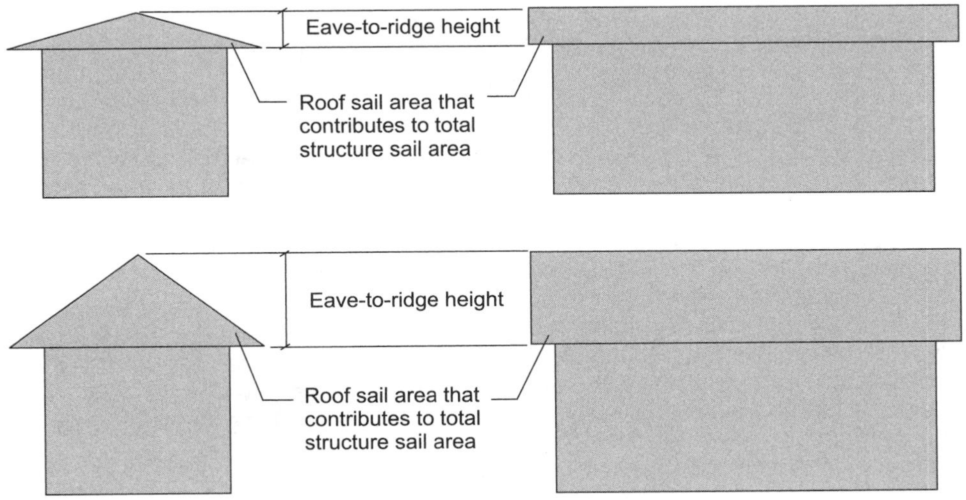

Figure R602.10.3(4)
EAVE-TO-RIDGE HEIGHT

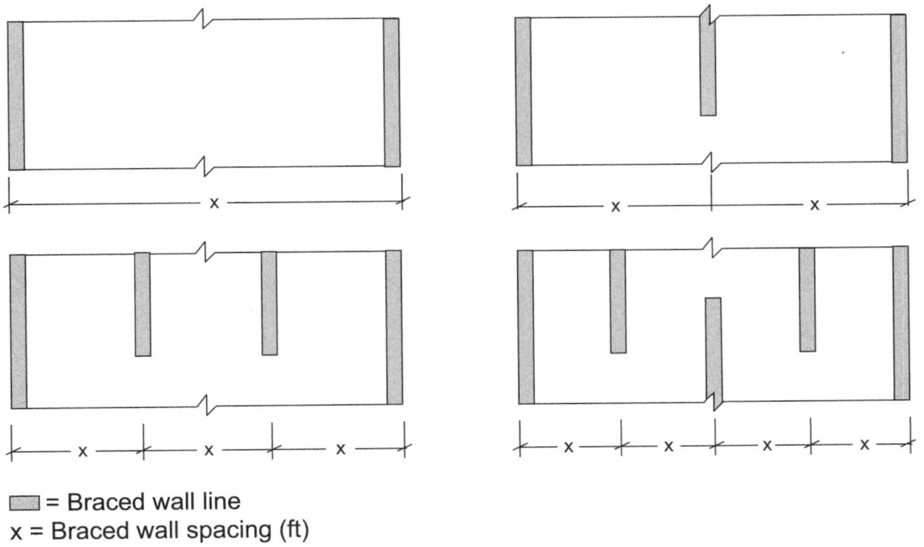

Figure R602.10.3(5)
BRACED WALL LINE SPACING

- Additional 800-pound (363 kg) hold-down device — In developing the new wind-bracing provisions, braced wall panels only supporting the roof above them (one-story buildings and the top story of multistory buildings) were recognized to be more effective in resisting wind forces when a hold-down was installed at the base of each panel. The use of an 800-pound (363 kg) hold-down device, installed in accordance with the manufacturer's recommendations, increases the capacity of bracing methods when supporting roof loads only. A 20-percent reduction in the length of bracing is permitted due to the increased capacity achieved by adding a hold-down device. Note that the hold-downs must be used at both ends of every bracing panel in the wall line and the hold-down anchorage must extend through all lower floors (if any) to provide a continuous load path until anchored into the foundation.

- Interior gypsum board finish — The wind bracing table was developed based on balancing the wind load acting on the structure against the strength of the bracing material providing resistance. The bracing lengths are based on the assumption that gypsum wall board is applied on the inside surface of braced wall panels. The addition of gypsum wall board, even though not attached with the same quantity of fasteners as Method GB bracing, does add strength and stiffness to the bracing; therefore, in cases in which gypsum board is not applied on the inside surface of braced wall panels, an adjustment must be made. In modern residential construction, the absence of gypsum board finish material is only likely to occur at gable end walls (above the plate) and at exterior garage walls. Typically, in either of these applications, it is not difficult to increase the amount of bracing because these walls are not likely to have many openings. Note this adjustment factor is used with Methods DWB, WSP, SFB, PBS, PCP, HPS, CS-WSP, CS-G and CS-SFB.
- Gypsum board fastening — Using the bracing lengths as stated from the Method GB column in Table R602.10.3(1) requires both sides of the braced wall line to be constructed to meet the requirements of Method GB bracing, including type and quantity of fasteners.

  When fastener spacing is reduced to 4 inches (102 mm) on center at all panel ends, edges and intermediate supports with horizontal joints blocked, the required length of Method GB bracing may be multiplied by a factor of 0.7 for wind applications only. The reduction may be applied to single-sided or double-sided Method GB.

To determine whether a dwelling requires seismic bracing, Section R301.2.2 should be reviewed. Section R301.2.2 exempts detached one- and two-family dwellings in SDC A, B and C and townhouses in SDC A and B from the seismic provisions of the code. In these cases, only the wind bracing requirements must be met.

As illustrated in Commentary Figure R602.10.3(6), the information in Table R602.10.3(3), the seismic bracing table, is based on:

- Soil site classification D
- 10-foot (3048 mm) wall height
- 10 psf (479 Pa) floor dead load
- 15 psf (718 Pa) roof and ceiling dead load
- Braced wall line spacing of 25 feet (7620 mm) or less

Table R602.10.3(3) is based on soil Site Class D, which is defined in Section 1613.5.2 of the IBC. When the soil properties are not known in sufficient detail to determine a specific soil site class, site Class D shall be used unless the building official or geotechnical data determines that Site Class E or F is present. If Site Class E or F is present, IBC Section 1613.5, provides a methodology for determining the site SDS and converting it into a SDC that can be used directly in Table R602.10.3(3).

Table R602.10.3(4) includes adjustment factors to accommodate variations from these assumptions for the wide range of residential structures covered by the code. While the adjustments are not new to the code, many were taken from other sections and collected into a common location in the 2009 edition of the IRC.

Each of the adjustments described in Table R602.10.3(4) is explained below:

- Story height (see Section R301.3) — The amount of bracing required for seismic loads is directly related to the mass/weight of the structure. As a wall gets taller, its mass increases, thus requiring more bracing to resist the resulting seismic loads.
- Braced wall line spacing, townhouses in SDC C — This adjustment factor applies to townhouses in SDC C. Townhouses in SDC C are permitted to have a 35- to 50-foot (10 668 mm to 15 240 mm) braced wall line spacing using this adjustment factor.
- Braced wall line spacing in SDC D_0, D_1 and D_2 — The 25-foot (7620 mm) braced wall line spacing applies to one- and two-family dwellings and townhouses in SDC D_0, D_1 and D_2. However, for one room only per dwelling unit, the 25-foot (7620 mm) braced wall line spacing for these structures can be increased to a maximum of 35 feet (10 668 mm).
- Wall dead load — Table R602.10.3(3) is based on a wall weight of 8 to 15 pounds per square feet (338 to 718 Pa). If a lighter weight wall is used, a reduction is permitted in the amount of bracing required. (Again, more mass requires more bracing and less mass requires less bracing.) Note that a standard wood-framed stud wall has a weight of 11 to 12 psf (527 to 575 Pa) (see Table C3-1 of ASCE 7).
- Roof/ceiling dead load for wall supporting — Table R602.10.3(3) is based on a roof/ceiling weight of 15 pounds per square feet (718 Pa) or less. As the roof weight/mass increases, so does the amount of bracing required. Note that a standard wood-framed roof ceiling with lightweight asphalt or wood shingles has a weight that varies between 10 and 15 psf (479 to 718 Pa) (Table C3-1 of ASCE 7).

  Roof/ceiling dead load adjustment factors vary depending upon how many floors exist below the roof. If the building is a single-story structure, the effect of the additional weight on the roof is greater than when the building is three stories tall. The adjustment factors are dependent upon the number of stories in the structure. From Table R602.10.3(4), for a single-story building the adjustment factor for roof loads greater than 15

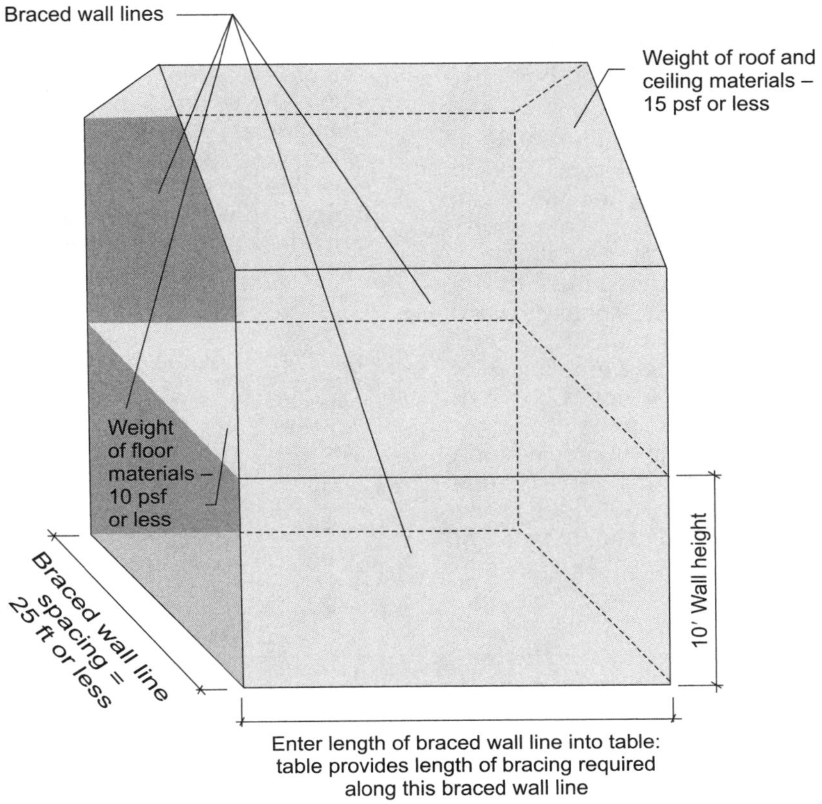

For SI: 1 foot = 304.8 mm,
1 pound per square foot = 47.89 Pa.

Figure R602.10.3(6)
BASIS FOR SEISMIC BRACING TABLE

psf (718 Pa) is 1.2. For a two-story or three-story building, the adjustment factor is 1.1.

- Walls with stone or masonry veneer, townhouses in SDC C — This is a case of more mass requiring more bracing. This adjustment factor applies only to townhouses in SDC C. The adjustment factor varies based on story height. Any intermittent or continuous sheathing method may be used.

  The adjustment factor for wall bracing length for the bottom of a two-story and bottom two stories of a three-story townhouse is 1.5. This increase applies only when veneer extends more than one story in height. The top story of a multistory house does not require an increase in bracing for masonry and stone veneer regardless of veneer height.

- Walls with stone or masonry veneer in SDC D_0, D_1 and D_2 — This is another case of more mass requiring more bracing. Section R602.10.6.5 describes the limits and requirements for wall bracing in SDC D_0-D_2 with walls with stone and masonry veneer in one- and two-family dwellings. Townhouses in SDC D_0, D_1 and D_2 with stone or masonry veneer exceeding the first story height must be designed in accordance with accepted engineering practice.

- Interior gypsum board finish — The seismic bracing table was developed based on balancing the seismic forces acting on the structure against the strength of the bracing material providing resistance. The bracing lengths are based on the assumption that gypsum wall board is applied on the inside surface of braced wall panels. The addition of gypsum wall board, even though not attached with the same quantity of fasteners as Method GB bracing, does add strength and stiffness to the bracing; therefore, in cases in which gypsum board is not applied on the inside surface of braced wall panels, an adjustment must be made. In modern residential construction, the absence of gypsum board finish material is only likely to occur at gable end walls (above the plate) and at exterior garage walls. Typically, in either of these applications, it is not difficult to increase the amount of bracing because these walls are not likely to have many openings. This factor is applied when the following methods are used without interior gypsum board finish - DWB, WSP, SFB, PBS, PCP, HPS, CS-WSP, CS-G, or CS-SFB [SDC C only and winds less than 100 mph (45 m/s)].

Other changes to the bracing requirements tables in the 2012 edition include the division of Seismic Design Category D_0 and D_1 into two categories. Minimum bracing lengths for SDC D_0 are now listed in a separate section from SDC D_1. Continuously Sheathed — Structural Fiberboard (CS-SFB) was added to the tables. Method CS-SFB is limited to regions with wind speeds of 100 mph (45 m/s) or less and SDC C.

Changes to bracing lengths include a redefinition of Method Gypsum Board (GB) as a 4-foot (1219 mm) single-sided braced wall panel similar to the other bracing methods. The required length of bracing for a braced wall line with Method GB continues to assume a double-sided application in Tables R602.10.3(1) and (3). Table R602.10.5 requires an adjustment factor of 0.5 times the physical braced wall panel length for single-sided gypsum braced wall panels.

Braced wall panels in angled walls may be counted as braced wall panels along a braced wall line if the requirements of Section R602.10.1.4 are followed. The projected length of the BWP rather than the actual length should be added to the total wall bracing length. Commentary Figure R602.10.3(7) shows an example of the projected length. Commentary Table R602.10.3 gives the projected braced wall panel length to add to the braced wall line for 4 foot to 8 foot (1219 to 2438 mm) long BWPs on angled wall lines.

Table R602.10.3
ADJUSTED ANGLED WALL BRACED WALL PANEL LENGTH

ANGLE FROM BWL (degrees)	BWP LENGTH ON ANGLED WALL LINE (feet)				
	4	5	6	7	8
15	3.8	4.8	5.7	6.7	7.7
30	3.4	4.3	5.1	6.0	6.9
45	2.8	3.5	4.2	4.9	5.6

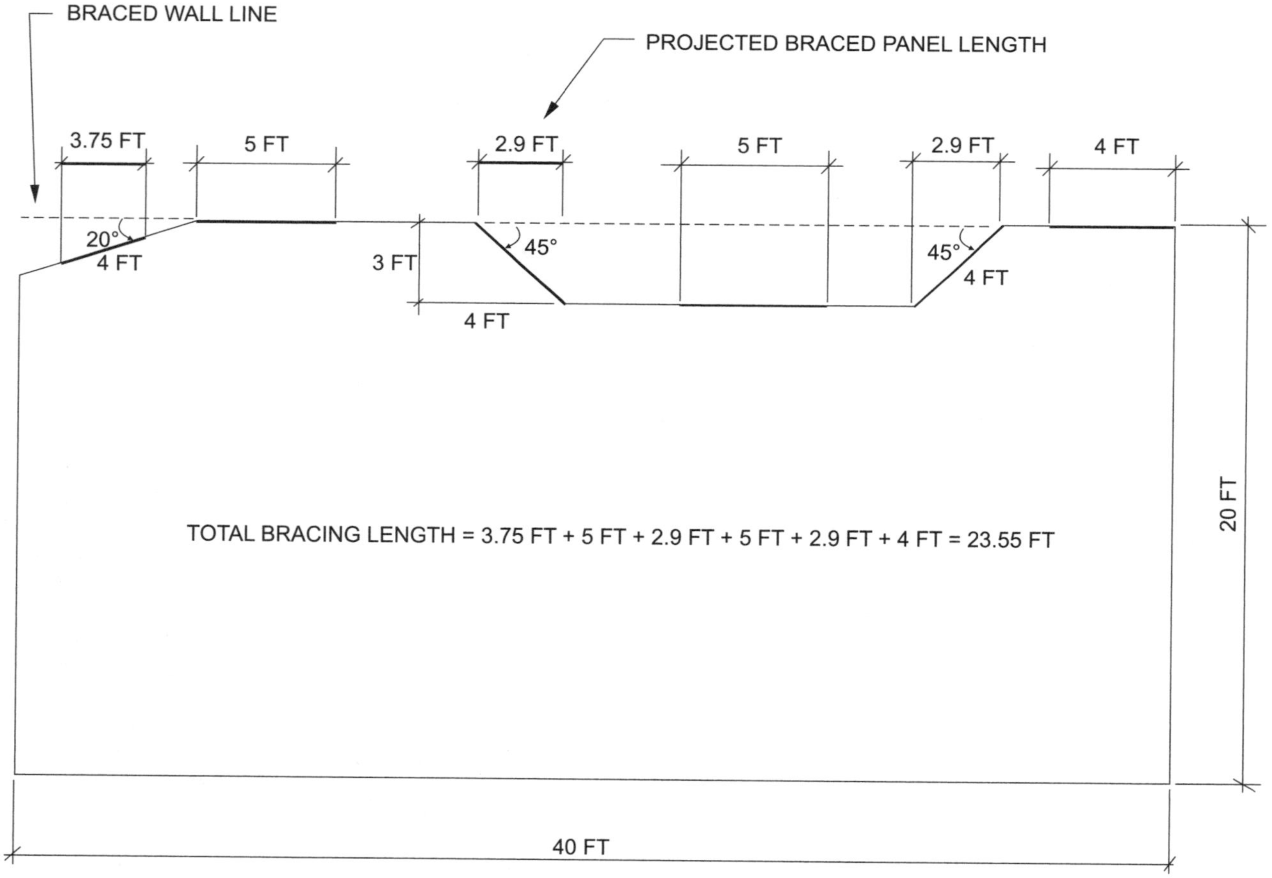

Figure R602.10.3(7)
PROJECTED LENGTH OF A BRACED WALL PANEL ON AN ANGLED WALL

TABLE R602.10.3(1)
BRACING REQUIREMENTS BASED ON WIND SPEED

• EXPOSURE CATEGORY B • 30 FOOT MEAN ROOF HEIGHT • 10 FOOT EAVE-TO-RIDGE HEIGHT • 10 FOOT WALL HEIGHT • 2 BRACED WALL LINES			MINIMUM TOTAL LENGTH (FEET) OF BRACED WALL PANELS REQUIRED ALONG EACH BRACED WALL LINE[a]			
Basic Wind Speed (mph)	Story Location	**Braced Wall Line Spacing** (feet)	Method LIB[b]	Method GB	**Methods** DWB, WSP, SFB, PBS, PCP, HPS, CS-SFB[c]	**Methods** CS-WSP, CS-G, CS-PF
≤ 85	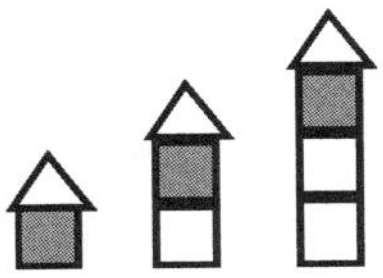	10	3.5	3.5	2.0	1.5
		20	6.0	6.0	3.5	3.0
		30	8.5	8.5	5.0	4.5
		40	11.5	11.5	6.5	5.5
		50	14.0	14.0	8.0	7.0
		60	16.5	16.5	9.5	8.0
	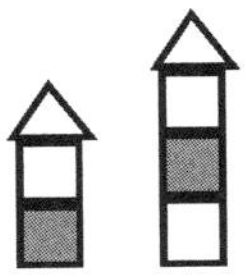	10	6.5	6.5	3.5	3.0
		20	11.5	11.5	6.5	5.5
		30	16.5	16.5	9.5	8.0
		40	21.5	21.5	12.5	10.5
		50	26.5	26.5	15.0	13.0
		60	31.5	31.5	18.0	15.5
		10	NP	9.0	5.5	4.5
		20	NP	17.0	10.0	8.5
		30	NP	24.5	14.0	12.0
		40	NP	32.0	18.0	15.5
		50	NP	39.0	22.5	19.0
		60	NP	46.5	26.5	22.5
≤ 90	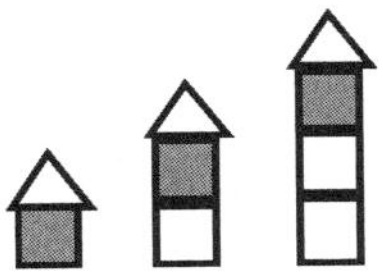	10	3.5	3.5	2.0	2.0
		20	7.0	7.0	4.0	3.5
		30	9.5	9.5	5.5	5.0
		40	12.5	12.5	7.5	6.0
		50	15.5	15.5	9.0	7.5
		60	18.5	18.5	10.5	9.0
	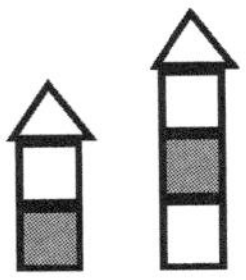	10	7.0	7.0	4.0	3.5
		20	13.0	13.0	7.5	6.5
		30	18.5	18.5	10.5	9.0
		40	24.0	24.0	14.0	12.0
		50	29.5	29.5	17.0	14.5
		60	35.0	35.0	20.0	17.0
		10	NP	10.5	6.0	5.0
		20	NP	19.0	11.0	9.5
		30	NP	27.5	15.5	13.5
		40	NP	35.5	20.5	17.5
		50	NP	44.0	25.0	21.5
		60	NP	52.0	30.0	25.5

(continued)

TABLE R602.10.3(1)—continued
BRACING REQUIREMENTS BASED ON WIND SPEED

• EXPOSURE CATEGORY B • 30 FOOT MEAN ROOF HEIGHT • 10 FOOT EAVE-TO-RIDGE HEIGHT • 10 FOOT WALL HEIGHT • 2 BRACED WALL LINES			MINIMUM TOTAL LENGTH (FEET) OF BRACED WALL PANELS REQUIRED ALONG EACH BRACED WALL LINE[a]			
Basic Wind Speed (mph)	**Story Location**	**Braced Wall Line Spacing** (feet)	**Method LIB[b]**	**Method GB**	**Methods DWB, WSP, SFB, PBS, PCP, HPS, CS-SFB[c]**	**Methods CS-WSP, CS-G, CS-PF**
≤ 100	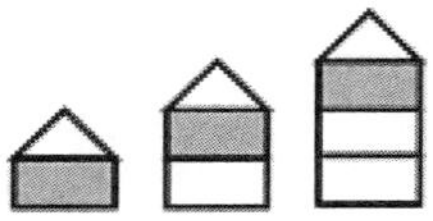	10	4.5	4.5	2.5	2.5
		20	8.5	8.5	5.0	4.0
		30	12.0	12.0	7.0	6.0
		40	15.5	15.5	9.0	7.5
		50	19.0	19.0	11.0	9.5
		60	22.5	22.5	13.0	11.0
	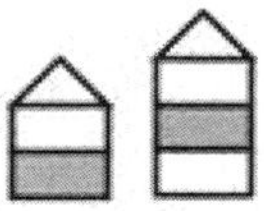	10	8.5	8.5	5.0	4.5
		20	16.0	16.0	9.0	8.0
		30	23.0	23.0	13.0	11.0
		40	29.5	29.5	17.0	14.5
		50	36.5	36.5	21.0	18.0
		60	43.5	43.5	25.0	21.0
		10	NP	12.5	7.5	6.0
		20	NP	23.5	13.5	11.5
		30	NP	34.0	19.5	16.5
		40	NP	44.0	25.0	21.5
		50	NP	54.0	31.0	26.5
		60	NP	64.0	36.5	31.0
< 110[c]		10	5.5	5.5	3.0	3.0
		20	10.0	10.0	6.0	5.0
		30	14.5	14.5	8.5	7.0
		40	18.5	18.5	11.0	9.0
		50	23.0	23.0	13.0	11.5
		60	27.5	27.5	15.5	13.5
	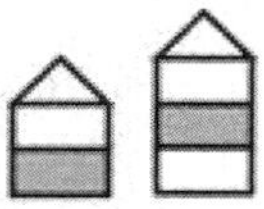	10	10.5	10.5	6.0	5.0
		20	19.0	19.0	11.0	9.5
		30	27.5	27.5	16.0	13.5
		40	36.0	36.0	20.5	17.5
		50	44.0	44.0	25.5	21.5
		60	52.5	52.5	30.0	25.5
		10	NP	15.5	9.0	7.5
		20	NP	28.5	16.5	14.0
		30	NP	41.0	23.5	20.0
		40	NP	53.0	30.5	26.0
		50	NP	65.5	37.5	32.0
		60	NP	77.5	44.5	37.5

For SI: 1 inch = 25.4 mm, 1 foot = 305 mm, 1 mile per hour = 0.447 m/s.

a. Linear interpolation shall be permitted.

b. Method LIB shall have gypsum board fastened to at least one side with nails or screws in accordance with Table R602.3(1) for exterior sheathing or Table R702.3.5 for interior gypsum board. Spacing of fasteners at panel edges shall not exceed 8 inches.

c. Method CS-SFB does not apply where the wind speed is greater than 100 mph.

❖ See the commentary for Section R602.10.3.

TABLE R602.10.3(2)
WIND ADJUSTMENT FACTORS TO THE REQUIRED LENGTH OF WALL BRACING

ADJUSTMENT BASED ON	STORY/ SUPPORTING	CONDITION	ADJUSTMENT FACTOR[a, b] [multiply length from Table R602.10.3(1) by this factor]	APPLICABLE METHODS
Exposure category	One-story structure	B	1.00	All methods
		C	1.20	
		D	1.50	
	Two-story structure	B	1.00	
		C	1.30	
		D	1.60	
	Three-story structure	B	1.00	
		C	1.40	
		D	1.70	
Roof eave-to-ridge height	Roof only	≤ 5 feet	0.70	
		10 feet	1.00	
		15 feet	1.30	
		20 feet	1.60	
	Roof + 1 floor	≤ 5 feet	0.85	
		10 feet	1.00	
		15 feet	1.15	
		20 feet	1.30	
	Roof + 2 floors	≤ 5 feet	0.90	
		10 feet	1.00	
		15 feet	1.10	
		20 feet	Not permitted	
Wall height adjustment	Any story	8 feet	0.90	
		9 feet	0.95	
		10 feet	1.00	
		11 feet	1.05	
		12 feet	1.10	
Number of braced wall lines (per plan direction)[c]	Any story	2	1.00	
		3	1.30	
		4	1.45	
		≤ 5	1.60	
Additional 800-pound hold-down device	Top story only	Fastened to the end studs of each braced wall panel and to the foundation or framing below	0.80	DWB, WSP, SFB, PBS, PCP, HPS
Interior gypsum board finish (or equivalent)	Any story	Omitted from inside face of braced wall panels	1.40	DWB, WSP, SFB,PBS, PCP, HPS, CS-WSP, CS-G, CS-SFB
Gypsum board fastening	Any story	4 inches o.c. at panel edges, including top and bottom plates, and all horizontal joints blocked	0.7	GB

For SI: 1 inch = 25.4 mm, 1 foot = 305 mm, 1 pound = 4.48 N.

a. Linear interpolation shall be permitted.

b. The total adjustment factor is the product of all applicable adjustment factors.

c. The adjustment factor is permitted to be 1.0 when determining bracing amounts for intermediate braced wall lines provided the bracing amounts on adjacent braced wall lines are based on a spacing and number that neglects the intermediate braced wall line.

❖ See the commentary for Section R602.10.3.

TABLE R602.10.3(3)
BRACING REQUIREMENTS BASED ON SEISMIC DESIGN CATEGORY

• SOIL CLASS D[b] • WALL HEIGHT = 10 FEET • 10 PSF FLOOR DEAD LOAD • 15 PSF ROOF/CEILING DEAD LOAD • BRACED WALL LINE SPACING ≤ 25 FEET			MINIMUM TOTAL LENGTH (FEET) OF BRACED WALL PANELS REQUIRED ALONG EACH BRACED WALL LINE[a]				
Seismic Design Category	**Story Location**	**Braced Wall Line Length (feet)**	**Method LIB[c]**	**Method GB**	**Methods DWB, SFB, PBS, PCP, HPS, CS-SFB[d]**	**Method WSP**	**Methods CS-WSP, CS-G**
C (townhouses only)		10	2.5	2.5	2.5	1.6	1.4
		20	5.0	5.0	5.0	3.2	2.7
		30	7.5	7.5	7.5	4.8	4.1
		40	10.0	10.0	10.0	6.4	5.4
		50	12.5	12.5	12.5	8.0	6.8
	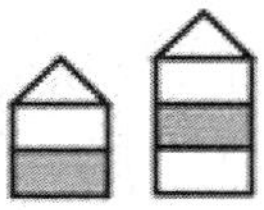	10	NP	4.5	4.5	3.0	2.6
		20	NP	9.0	9.0	6.0	5.1
		30	NP	13.5	13.5	9.0	7.7
		40	NP	18.0	18.0	12.0	10.2
		50	NP	22.5	22.5	15.0	12.8
		10	NP	6.0	6.0	4.5	3.8
		20	NP	12.0	12.0	9.0	7.7
		30	NP	18.0	18.0	13.5	11.5
		40	NP	24.0	24.0	18.0	15.3
		50	NP	30.0	30.0	22.5	19.1
D_0		10	NP	2.8	2.8	1.8	1.6
		20	NP	5.5	5.5	3.6	3.1
		30	NP	8.3	8.3	5.4	4.6
		40	NP	11.0	11.0	7.2	6.1
		50	NP	13.8	13.8	9.0	7.7
	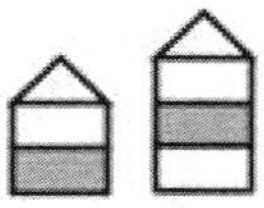	10	NP	5.3	5.3	3.8	3.2
		20	NP	10.5	10.5	7.5	6.4
		30	NP	15.8	15.8	11.3	9.6
		40	NP	21.0	21.0	15.0	12.8
		50	NP	26.3	26.3	18.8	16.0
	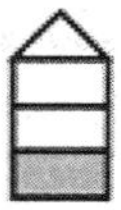	10	NP	7.3	7.3	5.3	4.5
		20	NP	14.5	14.5	10.5	9.0
		30	NP	21.8	21.8	15.8	13.4
		40	NP	29.0	29.0	21.0	17.9
		50	NP	36.3	36.3	26.3	22.3

(continued)

TABLE R602.10.3(3)—continued
BRACING REQUIREMENTS BASED ON SEISMIC DESIGN CATEGORY

• SOIL CLASS D[b] • WALL HEIGHT = 10 FEET • 10 PSF FLOOR DEAD LOAD • 15 PSF ROOF/CEILING DEAD LOAD • BRACED WALL LINE SPACING ≤ 25 FEET			MINIMUM TOTAL LENGTH (FEET) OF BRACED WALL PANELS REQUIRED ALONG EACH BRACED WALL LINE[a]				
Seismic Design Category	**Story Location**	**Braced Wall Line Length (feet)**	**Method LIB[c]**	**Method GB**	**Methods DWB, SFB, PBS, PCP, HPS, CS-SFB[d]**	**Method WSP**	**Methods CS-WSP, CS-G**
D_1		10	NP	3.0	3.0	2.0	1.7
		20	NP	6.0	6.0	4.0	3.4
		30	NP	9.0	9.0	6.0	5.1
		40	NP	12.0	12.0	8.0	6.8
		50	NP	15.0	15.0	10.0	8.5
	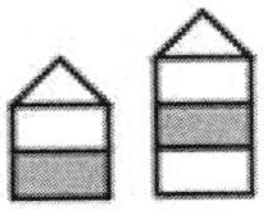	10	NP	6.0	6.0	4.5	3.8
		20	NP	12.0	12.0	9.0	7.7
		30	NP	18.0	18.0	13.5	11.5
		40	NP	24.0	24.0	18.0	15.3
		50	NP	30.0	30.0	22.5	19.1
		10	NP	8.5	8.5	6.0	5.1
		20	NP	17.0	17.0	12.0	10.2
		30	NP	25.5	25.5	18.0	15.3
		40	NP	34.0	34.0	24.0	20.4
		50	NP	42.5	42.5	30.0	25.5
D_2		10	NP	4.0	4.0	2.5	2.1
		20	NP	8.0	8.0	5.0	4.3
		30	NP	12.0	12.0	7.5	6.4
		40	NP	16.0	16.0	10.0	8.5
		50	NP	20.0	20.0	12.5	10.6
	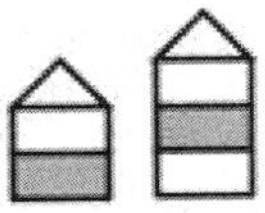	10	NP	7.5	7.5	5.5	4.7
		20	NP	15.0	15.0	11.0	9.4
		30	NP	22.5	22.5	16.5	14.0
		40	NP	30.0	30.0	22.0	18.7
		50	NP	37.5	37.5	27.5	23.4
	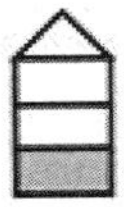	10	NP	NP	NP	NP	NP
		20	NP	NP	NP	NP	NP
		30	NP	NP	NP	NP	NP
		40	NP	NP	NP	NP	NP
		50	NP	NP	NP	NP	NP
	Cripple wall below one- or two-story dwelling	10	NP	NP	NP	7.5	6.4
		20	NP	NP	NP	15.0	12.8
		30	NP	NP	NP	22.5	19.1
		40	NP	NP	NP	30.0	25.5
		50	NP	NP	NP	37.5	31.9

For SI: 1 inch = 25.4 mm, 1 foot = 305 mm, 1 pound per square foot = 0.0479 kPa.

a. Linear interpolation shall be permitted.

b. Wall bracing lengths are based on a soil site class "D." Interpolation of bracing length between the S_{ds} values associated with the Seismic Design Categories shall be permitted when a site-specific S_{ds} value is determined in accordance with Section 1613.3 of the *International Building Code*.

c. Method LIB shall have gypsum board fastened to at least one side with nails or screws per Table R602.3(1) for exterior sheathing or Table R702.3.5 for interior gypsum board. Spacing of fasteners at panel edges shall not exceed 8 inches.

d. Method CS-SFB applies in SDC C only.

❖ See the commentary for Section R602.10.3.

TABLE R602.10.3(4)
SEISMIC ADJUSTMENT FACTORS TO THE REQUIRED LENGTH OF WALL BRACING

ADJUSTMENT BASED ON:	STORY/SUPPORTING	CONDITION	ADJUSTMENT FACTOR[a, b] [Multiply length from Table R602.10.3(1) by this factor]	APPLICABLE METHODS
Story height (Section 301.3)	Any story	≤ 10 feet	1.0	All methods
		> 10 feet and ≤ 12 feet	1.2	
Braced wall line spacing, townhouses in SDC C	Any story	≤ 35 feet	1.0	
		> 35 feet and ≤ 50 feet	1.43	
Braced wall line spacing, in SDC D_0, D_1, D_2[c]	Any story	> 25 feet and ≤ 30 feet	1.2	
		> 30 feet and ≤ 35 feet	1.4	
Wall dead load	Any story	> 8 psf and < 15 psf	1.0	
		< 8 psf	0.85	
Roof/ceiling dead load for wall supporting	Roof only or roof plus one or two stories	≤15 psf	1.0	
	Roof plus one or two stories	> 15 psf and ≤ 25 psf	1.1	
	Roof only	> 15 psf and ≤ 25 psf	1.2	
Walls with stone or masonry veneer, townhouses in SDC[d, e]	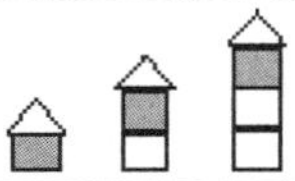	1.0		All intermittent and continuous methods
	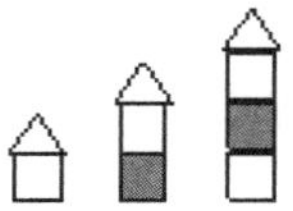	1.5		
	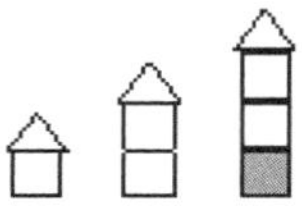	1.5		
Walls with stone or masonry veneer, detached one-and two-family dwellings in SDC $D_0 - D_2$[d]	Any story	See Table R602.10.6.5		BV-WSP
Interior gypsum board finish (or equivalent)	Any story	Omitted from inside face of braced wall panels	1.5	DWB, WSP, SFB, PBS, PCP, HPS, CS-WSP, CS-G, CS-SFB

For SI: 1 foot = 304.8 mm, 1 pound per square foot = 0.0479 kPa.

a. Linear interpolation shall be permitted.

b. The total length of bracing required for a given wall line is the product of all applicable adjustment factors.

c. The length-to-width ratio for the floor/roof *diaphragm* shall not exceed 3:1. The top plate lap splice nailing shall be a minimum of 12-16d nails on each side of the splice.

d. Applies to stone or masonry veneer exceeding the first story height. See Section R602.10.6.5 for requirements when stone or masonry veneer does not exceed the first story height.

e. The adjustment factor for stone or masonry veneer shall be applied to all exterior *braced wall lines* and all *braced wall lines* on the interior of the building, backing or perpendicular to and laterally supported veneered walls.

❖ See the commentary for Section R602.10.3.

R602.10.4 Construction methods for braced wall panels. Intermittent and continuously sheathed *braced wall panels* shall be constructed in accordance with this section and the methods listed in Table R602.10.4.

❖ Definitions of intermittent and continuously sheathed bracing methods have been combined into one table, Table R602.10.4. Alternate bracing methods have also been added to the table. Each method lists minimum thickness requirements and connection criteria separated into fastener type and spacing requirements.

Today there are 12 intermittent and four continuously sheathed bracing methods included in Section R602.10.4, each with its own minimum length and installation requirements. The first eight intermittent methods are Methods 1 to 8 from previous editions of the code; a new intermittent method has been added to braced wall lines when veneer is more than one story high, and the last three intermittent methods are alternate methods for bracing along a wall line or adjacent to a garage opening. The purpose of the alternate methods is to provide a narrow-length option for sections of wall line with less than 48 inches (1219 mm) of space requiring a braced panel.

The traditional method of referring to bracing methods by number was replaced in the 2009 code with abbreviations.

- Abbreviations for the intermittent methods come from the material used to make the braced wall panel.
- Alternate method abbreviations come from the braced panel names.
- Continuously sheathed methods each start with CS (e.g., Method CS-PF is the continuously sheathed portal frame bracing method).

The term "intermittent" is used to identify bracing methods that can be placed in discrete locations along a braced wall line, while the term "continuous" is used to identify bracing methods that require the whole wall line to be sheathed.

1. Let-in bracing (Method LIB) (formerly Method 1)
2. $^5/_8$-inch (16 mm) diagonal wood boards (Method DWB) (formerly Method 2)
3. Wood structural panel (plywood or OSB) (Method WSP) (formerly Method 3)
4. $^7/_{16}$-inch (11.1 mm) wood structural panel with stone or masonry veneer (Method BV-WSP) (new)
5. $^1/_2$-inch (12.7 mm) structural fiberboard sheathing (Method SFB) (formerly Method 4)
6. $^1/_2$-inch (12.7 mm) gypsum board (Method GB) (formerly Method 5)
7. Particleboard sheathing (Method PBS) (formerly Method 6)
8. Portland cement plaster (Method PCP) (formerly Method 7)
9. Hardboard panel siding (Method HPS) (formerly Method 8)

As the name implies, intermittent bracing methods are meant to be used intermittently along the wall line. For these methods, the minimum length required for a single braced wall panel segment ranges from 48 to 96 inches (1219 to 2438 mm).

Three alternative methods are also classified as intermittent bracing methods because, while structurally different from the eight traditional methods, these methods are designed to be used intermittently and can be substituted for any of the traditional bracing methods on a one-for-one basis (they may be intermixed with other intermittent bracing methods along the same wall line). The intermittent alternate braced methods are designated as:

1. Alternate braced wall (Method ABW) (formerly alternate braced wall panel)
2. Portal frame with hold-downs (Method PFH) (formerly alternate braced wall panel adjacent to a door or window opening)
3. Portal frame at garage (Method PFG) (new in the 2009 edition)

Methods ABW and PFH were both included in the 2006 code as alternative methods. The minimum lengths for these methods range from 16 to 40 inches (406 to 1016 mm). Either may be substituted for any intermittent bracing method permitted in Table R602.10.4. For the purposes of computing the required length of bracing, a single unit of either method is equivalent to 4 feet (1219 mm). Both utilize prefabricated metal hold-downs, in addition to anchor bolts, to connect the bracing panel to the foundation below.

The third alternative method, Method PFG, was new to the 2009 code. This bracing method is a type of portal frame that does not require hold-downs, has a minimum length of 24 inches (610 mm), may only be used in SDC A, B and C and may only be used adjacent to a garage door. For the purposes of computing the required length of bracing, a single unit of this method is equivalent to 1.5 times its actual length.

The four continuous bracing methods require continuous sheathing on all wall areas of the wall line. As walls with continuous sheathing are generally considered to be stiffer and stronger than similar walls with intermittent bracing, the continuous methods allow for bracing solutions that are narrower than the widths required for the intermittent methods.

Three of these bracing methods are for use with continuous wood structural panel sheathing.

Method CS-WSP, continuously sheathed wood structural panel, was originally introduced in the 2000 code and has not changed much for the 2012 code. The method requires all braced wall lines, including areas above and below openings and gable ends, to be fully sheathed with a minimum of $^{3}/_{8}$-inch (9.5 mm) wood structural panel sheathing.

Method CS-G, continuously sheathed wood structural panel adjacent to garage openings, is a narrow-length panel that can be used with Method CS-WSP. This alternate first appeared in the 2000 edition as a footnote to Table R602.10.5, Length Requirements for Braced Wall Panels in a Continuously Sheathed Wall. In the 2009 edition it was redefined as a bracing method and only permitted when using Method CS-WSP. This method permits the use of a panel length as short as 24 inches (610 mm) for an 8-foot (2438 mm) wall, but because the length of the wall is linked to an aspect ratio, the segment is required to get longer as the wall gets taller.

TABLE R602.10.4
BRACING METHODS

METHODS, MATERIAL		MINIMUM THICKNESS	FIGURE	CONNECTION CRITERIA[a]	
				Fasteners	**Spacing**
Intermittent Bracing Method	**LIB** Let-in-bracing	1 × 4 wood or approved metal straps at 45° to 60° angles for maximum 16″ stud spacing	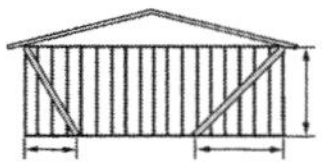	Wood: 2-8d common nails or 3-8d ($2^{1}/_{2}$″ long x 0.113″ dia.) nails	Wood: per stud and top and bottom plates
				Metal strap: per manufacturer	Metal: per manufacturer
	DWB Diagonal wood boards	$^{3}/_{4}$″(1″ nominal) for maximum 24″ stud spacing	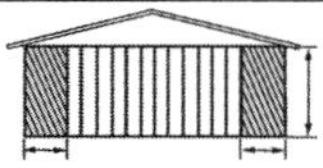	2-8d ($2^{1}/_{2}$″ long × 0.113″ dia.) nails or 2 - $1^{3}/_{4}$″ long staples	Per stud
	WSP Wood structural panel (See Section R604)	$^{3}/_{8}$″	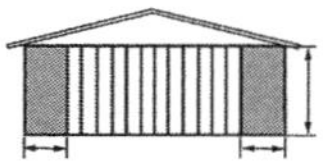	Exterior sheathing per Table R602.3(3)	6″ edges 12″ field
				Interior sheathing per Table R602.3(1) or R602.3(2)	Varies by fastener
	BV-WSP[e] Wood Structural Panels with Stone or Masonry Veneer (See Section R602.10.6.5)	$^{7}/_{16}$″	See Figure R602.10.6.5	8d common ($2^{1}/_{2}$″ × 0.131) nails	4″ at panel edges 12″ at intermediate supports 4″ at braced wall panel end posts
	SFB Structural fiberboard sheathing	$^{1}/_{2}$″ or $^{25}/_{32}$″ for maximum 16″ stud spacing	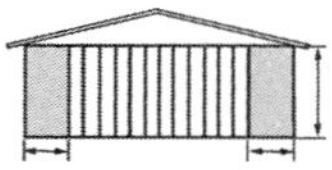	$1^{1}/_{2}$″ long × 0.12″ dia. (for $^{1}/_{2}$″ thick sheathing) $1^{3}/_{4}$″ long × 0.12″ dia. (for $^{25}/_{32}$″ thick sheathing) galvanized roofing nails or 8d common ($2^{1}/_{2}$″ long × 0.131″ dia.) nails	3″ edges 6″ field
	GB Gypsum board	$^{1}/_{2}$″	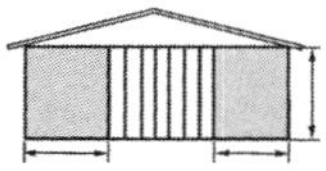	Nails or screws per Table R602.3(1) for exterior locations	For all braced wall panel locations: 7″ edges (including top and bottom plates) 7″ field
				Nails or screws per Table R702.3.5 for interior locations	
	PBS Particleboard sheathing (See Section R605)	$^{3}/_{8}$″ or $^{1}/_{2}$″ for maximum 16″ stud spacing	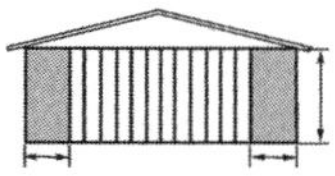	For $^{3}/_{8}$″, 6d common (2″ long × 0.113″ dia.) nails For $^{1}/_{2}$″, 8d common ($2^{1}/_{2}$″ long × 0.131″ dia.) nails	3″ edges 6″ field
	PCP Portland cement plaster	See Section R703.6 for maximum 16″ stud spacing	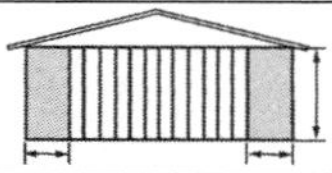	$1^{1}/_{2}$″ long, 11 gage, $^{7}/_{16}$″ dia. head nails or $^{7}/_{8}$″ long, 16 gage staples	6″ o.c. on all framing members
	HPS Hardboard panel siding	$^{7}/_{16}$″ for maximum 16″ stud spacing	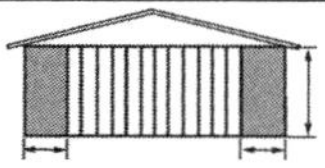	0.092″ dia., 0.225″ dia. head nails with length to accommodate $1^{1}/_{2}$″ penetration into studs	4″ edges 8″ field
	ABW Alternate braced wall	$^{3}/_{8}$″	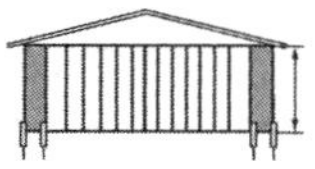	See Section R602.10.6.1	See Section R602.10.6.1

(continued)

TABLE R602.10.4—continued
BRACING METHODS

	METHODS, MATERIAL	MINIMUM THICKNESS	FIGURE	CONNECTION CRITERIA[a]	
				Fasteners	Spacing
Intermittent Bracing Methods	**PFH** Portal frame with hold-downs	3/8"	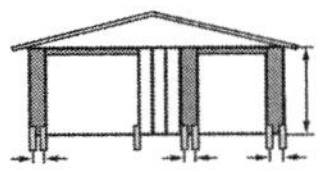	See Section R602.10.6.2	See Section R602.10.6.2
Intermittent Bracing Methods	**PFG** Portal frame at garage	7/16"	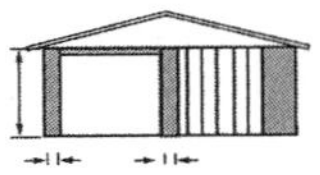	See Section R602.10.6.3	See Section R602.10.6.3
Continuous Sheathing Methods	**CS-WSP** Continuously sheathed wood structural panel	3/8"	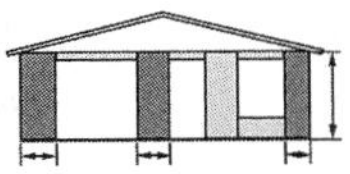	Exterior sheathing per Table R602.3(3)	6" edges 12" field
				Interior sheathing per Table R602.3(1) or R602.3(2)	Varies by fastener
Continuous Sheathing Methods	**CS-G**[b, c] Continuously sheathed wood structural panel adjacent to garage openings	3/8"		See Method CS-WSP	See Method CS-WSP
Continuous Sheathing Methods	**CS-PF** Continuously sheathed portal frame	7/16"	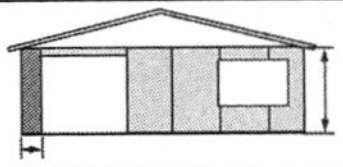	See Section R602.10.6.4	See Section R602.10.6.4
Continuous Sheathing Methods	**CS-SFB**[d] Continuously sheathed structural fiberboard	1/2" or 25/32" for maximum 16" stud spacing	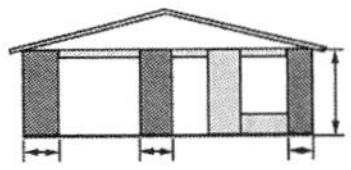	1 1/2" long × 0.12" dia. (for 1/2" thick sheathing) 1 3/4" long × 0.12" dia. (for 25/32" thick sheathing) galvanized roofing nails or 8d common (2 1/2" long × 0.131" dia.) nails	3" edges 6" field

For SI: 1 inch = 25.4 mm, 1 foot = 305 mm, 1 degree = 0.0175 rad, 1 pound per square foot = 47.8 N/m^2, 1 mile per hour = 0.447 m/s.

a. Adhesive attachment of wall sheathing, including Method GB, shall not be permitted in Seismic Design Categories C, D_0, D_1 and D_2.

b. Applies to panels next to garage door opening when supporting gable end wall or roof load only. May only be used on one wall of the garage. In Seismic Design Categories D_0, D_1 and D_2 roof covering dead load may not exceed 3 psf.

c. Garage openings adjacent to a Method CS-G panel shall be provided with a header in accordance with Table R502.5(1). A full height clear opening shall not be permitted adjacent to a Method CS-G panel.

d. Method CS-SFB does not apply in Seismic Design Categories D_0, D_1 and D_2 and in areas where the wind speed exceeds 100 mph.

e. Method applies to detached one- and two-family dwellings in Seismic Design Categories D_0 through D_2 only.

❖ See the commentary for Section R602.10.4

Pay particular attention to footnotes b and c in Table R602.10.4. Footnote b does not permit the use of 4:1 aspect ratio narrow wall panels on more than one wall of a garage. In other words, the braced panel may only be used around the garage door opening. If garage door openings occur in two walls of the garage, only one wall may use this braced panel. This should not be interpreted to mean that narrow wall segments cannot be used on both sides of a garage door within a single wall; rather, that is their designed use.

The footnote also restricts 4:1 aspect ratio narrow wall segments to supporting roofs only, and further restricts the dead load of the roof covering to 3 pounds per square foot (144 Pa) for D_0, D_1, and D_2. This minimizes the seismic weight (mass) of the roof in a seismic event by limiting the roof covering to relatively light roofing materials. Footnote c requires a header above a garage opening adjacent to Method CS-G panels. The opening is not permitted to be full height next to this panel. Use Table R502.5(1) to select a header for the opening.

For purposes of determining length of bracing in a wall line with Method CS-G, use the actual length of the narrow braced wall panels. Of course, the required length of bracing must be met for any wall in the structure: this provision simply allows the use of smaller segments to make up the required length of bracing.

Method CS-PF, continuously sheathed portal frame, has been in the code since the 2006 edition where it was listed as a footnote in Table R602.10.5, Length Requirements for Braced Wall Panels in a Continuously Sheathed Wall. In the 2009 edition it was redefined as a bracing method and only permitted when using Method CS-WSP.

The fourth continuously sheathed method, Method CS-SFB, was added to the 2009 edition of the code. It is very similar to Method CS-WSP except that there are restrictions as to where it may be used. Method CS-SFB is not permitted to be used in areas with wind speeds greater than 100 mph (45 m/s) or in SDC D_0, D_1 or D_2.

When using Method CS-SFB, the entire length of the braced wall line (including areas above and below openings and gable ends, if applicable) must be fully sheathed with a minimum of $^1/_2$-inch (13 mm) structural fiberboard sheathing.

The length of bracing for continuous sheathing is determined from the continuous sheathing column of Tables R602.10.3(1) and R602.10.3(3). Only full-height braced wall panels in accordance with Section R602.10.5 are to be counted toward the minimum required length of bracing.

Method CS-SFB bracing lengths are based on the adjacent clear opening height as shown in Figure R602.10.5. Note that Method CS-SFB may only be used in walls up to 10 feet high (3048 mm), and that the minimum panel length is aspect-ratio linked (the minimum bracing length increases as the wall height increases). This is done to ensure that as the wall gets taller, it remains sufficiently stiff.

Method CS-G and Method CS-PF, although continuously sheathed methods, are based on the use of wood structural panel sheathing and are not applicable for Method CS-SFB.

Lastly, Footnote a of Table R602.10.4 prohibits the use of adhesives in moderate-to-high seismic areas (SDC C, D_0, D_1 and D_2). The intent of this provision is that only sheathing used as a part of a braced wall segment, as described in Table R602.10.4, shall not be attached with adhesives. Note that detached one- and two-family dwellings are exempt from seismic rules in SDC C. Only townhouses in SDC C must meet this provision. This provision does not apply to Section R602.10.4.3 (braced wall panel interior finish material). While the interior finish is a required part of the bracing panel segment, its contribution to the overall bracing strength is relatively minor, and is actually predicated on the fastener spacing used in conjunction with the adhesive attachment of gypsum wall board.

R602.10.4.1 Mixing methods. Mixing of bracing methods shall be permitted as follows:

1. Mixing intermittent bracing and continuous sheathing methods from story to story shall be permitted.
2. Mixing intermittent bracing methods from *braced wall line* to *braced wall line* within a story shall be permitted. Within Seismic Design Categories A, B and C or in regions where the basic wind speed is less than or equal to 100 mph (45 m/s), mixing of intermittent bracing and continuous sheathing methods from *braced wall line* to *braced wall line* within a story shall be permitted.
3. Mixing intermittent bracing methods along a *braced wall line* shall be permitted in Seismic Design Categories A and B, and detached dwellings in Seismic Design Category C provided the length of required bracing in accordance with Table R602.10.3(1) or R602.10.3(3) is the highest value of all intermittent bracing methods used.
4. Mixing of continuous sheathing methods CS-WSP, CS-G and CS-PF along a *braced wall line* shall be permitted.
5. In Seismic Design Categories A and B, and for detached one- and two-family dwellings in Seismic Design Category C, mixing of intermittent bracing methods along the interior portion of a *braced wall line* with continuous sheathing methods CS-WSP, CS-G and CS-PF along the exterior portion of the same *braced wall line* shall be permitted. The length of required bracing shall be the highest value of all intermittent bracing methods used in accordance with Table R602.10.3(1) or R602.10.3(3) as adjusted by Tables R602.10.3(2) and R602.10.3(4), respectively. The requirements of Section R602.10.7 shall apply to each end of the continuously sheathed portion of the *braced wall line*.

❖ New to the 2012 edition, for detached one- and two-family dwellings and townhouses in SDC A and B, and detached dwellings in SDC C, braced wall lines which have sections of both exterior and interior wall line may now mix continuous sheathing made of wood structural panels on the exterior portion of the wall line with intermittent bracing methods along the interior of the braced wall line. This is commonly used for a continuously sheathed exterior of wood structural panel with interior wall lines sheathed in gypsum board intermittent braced wall panels.

Commentary Table R602.10.4.1 shows the mixing possibilities and their limitations.

The 2012 edition permits mixing of bracing methods as follows:

1. Story-to-story. Mixing any bracing methods from story-to-story is permitted. An example is using Method CS-WSP (continuously sheathed wood structural panel) bracing on the bottom two stories of a three-story building while using Method PBS (particleboard sheathing) or Method GB (gypsum board) bracing on the top story.
2. Braced wall line to braced wall line within one story. A builder is permitted to use different bracing methods on different walls within a story. For example, in a “window wall,” a designer may use Method WSP bracing along one wall line of a story because its lower bracing length requirements can help to accommodate additional windows. On the other walls having fewer windows, the designer may use Method GB or another intermittent bracing method.

Table R602.10.4.1
MIXING BRACING METHODS

MIXING LOCATIONS	MIXING LIMITATIONS	SDC A-B	SDC C DETACHED	SDC C TOWNHOUSES	SDC D_0-D_2
Story to Story	Mixing intermittent & continuously sheathed methods	X	X	X	X
BWL to BWL	Mixing intermittent methods	X	X	X	X
BWL to BWL	Mixing intermittent & continuously sheathed methods	X	X	X	NP
Within BWL	Mixing intermittent methods in a single wall line	X	X	NP	NP
Within BWL	Mixing continuously sheathed methods using wood structural panels only (Mixing CS-WSP, CS-G & CS-PF)	X	X	X	X
Within BWL	Mixing an intermittent method on an interior portion & CS-WSP, CS-PF, and CS-G on an exterior portion of a wall line	X	X	NP	NP

BWL = Braced Wall Line, NP = Not Permitted.

Mixing any bracing methods with continuously sheathed methods from one wall line to another is permitted in SDC A, B or C and where the basic wind speeds are 100 mph (45 m/s) or less.

In regions of SDC D_0-D_2 or wind speeds above 100 mph, intermittent bracing methods may change from wall line to wall line. When using continuously sheathed methods, all exterior wall lines must be continuously sheathed.

For the 2009 code, it was determined that in high seismic and high wind zones, all exterior walls of a given level must be continuously sheathed if a single wall on that level is continuously sheathed. However, walls inside the building or walls on other levels may use whatever bracing meets the requirements for that wall or level.

3. Mixing in one braced wall line. For one- and two-family dwellings in SDC A, B and C, and for townhouses in SDC A and B, mixing intermittent bracing methods within a braced wall line is permitted. The length of the required bracing for the braced wall line with mixed sheathing types must have the greatest required bracing length, with all applicable adjustment factors applied, of all types of bracing used in that wall line.
4. Mixing continuous sheathing methods in one wall line. Mixing of continuous sheathing methods CS-WSP, CS-G and CS-PF along a braced wall line is permitted as these methods assume use in a continuously sheathed wood structural panel wall line. For example, the braced wall line at the front of a house might have CS-WSP along the majority of the wall line with CS-PF at the garage opening.
5. Mixing intermittent and continuous sheathing methods in one wall line. In SDC A-C for detached one- and two-family dwellings and SDC A-B for townhouses, mixing of intermittent bracing methods along the interior portion of a braced wall line with continuous sheathing methods CS-WSP, CS-G and CS-PF along the exterior portion of the same braced wall line is permitted. The length of required bracing is the highest value of all intermittent bracing methods used with all applicable adjustment factors applied. The requirements of Section R602.10.7 apply to each end of the continuously sheathed portion of the braced wall line.

R602.10.4.2 Continuous sheathing methods. Continuous sheathing methods require structural panel sheathing to be used on all sheathable surfaces on one side of a *braced wall line* including areas above and below openings and gable end walls and shall meet the requirements of Section R602.10.7.

❖ The definition of a continuously sheathed braced wall line from Section R202 is:

BRACED WALL LINE, CONTINUOUSLY SHEATHED. A braced wall line with structural sheathing applied to all sheathable surfaces including the areas above and below openings.

In the 2006 code edition, the provisions required all walls on all stories to be continuously sheathed.

In the 2012 and 2009 editions, Section R602.10.4 requires all exterior walls on the same story to be continuously sheathed. For structures within SDC A, B and C, or in regions where the basic wind speed is less than or equal to 100 mph (45 m/s), there is no requirement for continuously sheathed bracing methods to be used on all exterior walls of a story using continuous sheathing. Continuous sheathing bracing methods are permitted to be used on a wall-by-wall, story-by-story basis, with other walls or stories of the structure utilizing any approved bracing method in the code.

In SDC D_0, D_1 and D_2, all exterior braced wall lines on the same story must be continuously sheathed before Method CS-WSP can be used. Note that Method CS-SFB is not permitted in SDC D_0, D_1 or D_2.

R602.10.4.3 Braced wall panel interior finish material. *Braced wall panels* shall have gypsum wall board installed on the side of the wall opposite the bracing material. Gypsum wall board shall be not less than $^1/_2$ inch (12.7 mm) in thickness and be fastened with nails or screws in accordance with Table R602.3(1) for exterior sheathing or Table R702.3.5 for interior gypsum wall board. Spacing of fasteners at panel edges for gypsum wall board opposite Method LIB bracing shall not exceed 8 inches (203 mm). Interior finish material shall not be glued in Seismic Design Categories D_0, D_1 and D_2.

Exceptions:

1. Interior finish material is not required opposite wall panels that are braced in accordance with Methods GB, BV-WSP, ABW, PFH, PFG and CS-PF, unless otherwise required by Section R302.6.
2. An approved interior finish material with an in-plane shear resistance equivalent to gypsum board shall be permitted to be substituted, unless otherwise required by Section R302.6.
3. Except for Method LIB, gypsum wall board is permitted to be omitted provided the required length of bracing in Tables R602.10.3(1) and R602.10.3(3) is multiplied by the appropriate adjustment factor in Tables R602.10.3(2) and R602.10.3(4) respectively, unless otherwise required by Section R302.6.

❖ In the 2009 code a requirement was added that, for intermittent braced wall panels, regular gypsum wall board must be installed as an interior finish material on the side of the wall opposite the bracing material. Note that this requirement is for the use of standard gypsum wall board, not Method GB bracing. Interior finish gypsum wall board can be attached in accordance with Section R702.3.5 (7 inches (178 mm) on center attachment required for Method GB does not apply). Narrow wall intermittent bracing methods (AWB, PFG, PFH and CS-PF) are exempt from this requirement because they were developed without gypsum wall board interior finish and they are typically used adjacent to garage door openings, where gypsum wall board is not commonly used. Wood structural panels with stone or masonry veneer (BV-WSP) do not require interior finish material. Additionally, there are exceptions that eliminate this requirement for gypsum wall board as an interior finish material if the wall bracing amount is increased.

Although the other bracing methods permit the gypsum board to be installed as a wall covering rather than a Method GB bracing material; when gypsum board is installed as Method GB bracing, the fasteners must be spaced 7 inches (178 mm) on center at all panel ends, edges and intermediate supports. The fastener schedule for gypsum board installed as a wall covering (see Section R702.3.5) is less stringent than when it is installed as a bracing material.

R602.10.5 Minimum length of a braced wall panel. The minimum length of a *braced wall panel* shall comply with Table R602.10.5. For Methods CS-WSP and CS-SFB, the minimum panel length shall be based on the adjacent clear opening height in accordance with Table R602.10.5 and Figure R602.10.5. When a panel has an opening on either side of differing heights, the taller opening height shall be used to determine the panel length.

❖ Braced wall panel minimum lengths are placed into one table for clarity. 2009 IRC Tables R602.10.3.1, R602.10.3.2, R602.10.4.2 and R602.10.5.2 are combined into 2012 IRC Table R602.10.5. A new column, contributing length, gives the equivalent intermittent braced panel length for alternate intermittent braced wall panels and the continuously sheathed braced panel length for continuous sheathing methods.

This section requires that the minimum intermittent braced panel length be at least 48 inches (1219 mm), except when employing alternate braced panels. Minimum continuously sheathed braced panel length is dependent on the height of the wall and adjacent opening; therefore, the braced panels vary in required minimum length.

Minimum braced panel length varies depending upon bracing method and in some cases wall height, seismic design category or adjacent opening height.

- Minimum bracing length is 48 inches (1219 mm) for Methods DWB, WSP, SFB, PBS, GB, PCP, HPS, and BV-WSP for walls 10 feet (3048 mm) tall or less. Walls taller than 10 feet (3048 mm) require longer minimum braced panel lengths.
- Method LIB requires a minimum length of 55 inches (1397 mm) for an 8-foot-tall (2438 mm) wall. The minimum length varies depending upon the wall height.
- Minimum length of a Method ABW panel varies depending upon wall height and seismic design category.
- Methods CS-WSP and CS-SFB minimum length depend upon wall height and the size of the opening next to the braced wall panel.
- Methods PFH (supporting one story and a roof), PFG and CS-G have a minimum braced panel length of 24 inches (610 mm).
- Method PFH (supporting a roof) and Method CS-PF have a minimum panel length of 16 inches (406 mm).

Methods ABW, PFG, CS-G and CS-PF have a bracing capacity related to the ratio of their panel height to length. Note that this aspect ratio adjustment is separate from the required bracing adjustment in Table R602.10.3(4). However, any increase in length as a result of the aspect ratio adjustment in Table R602.10.5 is permitted to be counted towards the amount of bracing required. In many cases, when a number of minimum-length panels are used for bracing, the aspect ratio adjustment will completely account for the length increase from Table R602.10.3(4).

Remember that, for all bracing methods, wall length is measured along the wall line in the horizontal direction.

TABLE R602.10.5
MINIMUM LENGTH OF BRACED WALL PANELS

METHOD (See Table R602.10.4)		MINIMUM LENGTH[a] (inches)					CONTRIBUTING LENGTH (inches)
		Wall Height					
		8 feet	9 feet	10 feet	11 feet	12 feet	
DWB, WSP, SFB, PBS, PCP, HPS, BV-WSP		48	48	48	53	58	Actual[b]
GB		48	48	48	53	58	Double sided = Actual Single sided = 0.5 × Actual
LIB		55	62	69	NP	NP	Actual[b]
ABW	SDC A, B and C, wind speed < 110 mph	28	32	34	38	42	48
	SDC D_o, D_1 and D_2, wind speed < 110 mph	32	32	34	NP	NP	
PFH	Supporting roof only	16	16	16	18[c]	20[c]	48
	Supporting one story and roof	24	24	24	27[c]	29[c]	48
PFG		24	27	30	33[d]	36[d]	1.5 × Actual[b]
CS-G		24	27	30	33	36	Actual[b]
CS-PF		16	18	20	22[e]	24[e]	Actual[b]
CS-WSP, CS-SFB	Adjacent clear opening height (inches)						
	≤ 64	24	27	30	33	36	Actual[b]
	68	26	27	30	33	36	
	72	27	27	30	33	36	
	76	30	29	30	33	36	
	80	32	30	30	33	36	
	84	35	32	32	33	36	
	88	38	35	33	33	36	
	92	43	37	35	35	36	
	96	48	41	38	36	36	
	100	—	44	40	38	38	
	104	—	49	43	40	39	
	108	—	54	46	43	41	
	112	—	—	50	45	43	
	116	—	—	55	48	45	
	120	—	—	60	52	48	
	124	—	—	—	56	51	
	128	—	—	—	61	54	
	132	—	—	—	66	58	
	136	—	—	—	—	62	
	140	—	—	—	—	66	
	144	—	—	—	—	72	

For SI: 1 inch = 25.4 mm, 1 foot = 304.8 mm, 1 mile per hour = 0.447 m/s.

NP = Not Permitted.

a. Linear interpolation shall be permitted.

b. Use the actual length when it is greater than or equal to the minimum length.

c. Maximum header height for PFH is 10 feet in accordance with Figure R602.10.6.2, but wall height may be increased to 12 feet with pony wall.

d. Maximum opening height for PFG is 10 feet in accordance with Figure R602.10.6.3, but wall height may be increased to 12 feet with pony wall.

e. Maximum opening height for CS-PF is 10 feet in accordance with Figure R602.10.6.4, but wall height may be increased to 12 feet with pony wall.

❖ See the commentary for Sections R602.10.5 and R602.10.5.1.

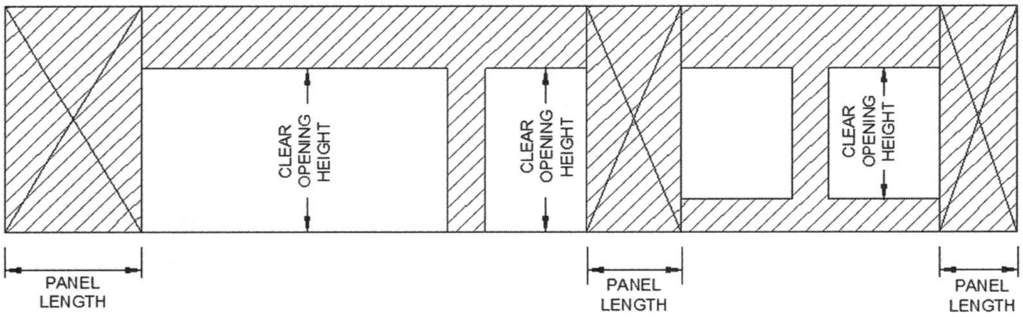

FIGURE R602.10.5
BRACED WALL PANELS WITH CONTINUOUS SHEATING

❖ See the commentary for Sections R602.10.5 and R602.10.5.1.

R602.10.5.1 Contributing length. For purposes of computing the required length of bracing in Tables R602.10.3(1) and R602.10.3(3), the contributing length of each *braced wall panel* shall be as specified in Table R602.10.5.

❖ This section clarifies what length should be counted for a specific bracing method when the actual length of the panel is not used.

The following bracing methods contribute the measured length of the braced panel to the total bracing length:

- LIB, DWB, WSP, SFB, PBS, PCP, HPS, and BV-WSP for intermittent bracing
- Double-sided GB
- CS-WSP, CS-G, CS-PF, and CS-SFB for continuous sheathing

Other methods have a fixed amount of bracing per braced panel or an adjustment factor that is applied to the measured length of braced panel.

- Method ABW panel and each Method PFH panel in a portal contribute 4 feet (1219 mm) of bracing.
- Method GB is worth half its length for single-sided applications, i.e., 20 feet (6096 mm) of single-sided GB equals 10 feet (3048 mm) of wall bracing.
- Method PFG panel is worth $1^1/_2$ times its physical length for wall bracing length, i.e., a 24-inch (610 mm) panel is worth 36 inches (914 mm) of wall bracing.

For continuous sheathing, the bracing length adjustment for Method CS-WSP was changed in 2009. Prior to that edition, the bracing length adjustment was calculated based on an aspect ratio of the size of the wall openings (door or window) to the height of the stud wall in the braced wall line. The ICC Ad Hoc Wall Bracing Committee (AHWBC) determined that the differences in wall opening sizes were of little enough impact that a single factor could be applied, regardless of the opening size. The continuous sheathing column in Tables R602.10.3(1) and R602.10.3(3) eliminates the need for an additional calculation that factored in the aspect ratio of opening size to stud wall height. Only full-height braced wall panels are to be counted toward the minimum required length of bracing. Table R602.10.5 provides minimum length requirements for Methods CS-WSP or CS-SFB for various wall heights.

The braced panel lengths are based on the adjacent clear opening height as shown in Figure R602.10.5. Note that minimum panel lengths are aspect-ratio linked: as the wall height increases, so does the minimum bracing panel length. This is necessary to ensure that, as the wall gets taller, it remains sufficiently stiff.

R602.10.5.2 Partial credit. For Methods DWB, WSP, SFB, PBS, PCP and HPS in Seismic Design Categories A, B and C, panels between 36 inches and 48 inches (914 mm and 1219 mm)) in length shall be considered a *braced wall panel* and shall be permitted to partially contribute toward the required length of bracing in Tables R602.10.3(1) and R602.10.3(3), and the contributing length shall be determined from Table R602.10.5.2.

❖ Section R602.10.5.2 in the 2012 IRC permits reductions in the minimum panel length for specific panel-type bracing methods (DWB, WSP, SFB, PBS, PCP and HPS) in SDC A, B and C as long as a length penalty is applied to the actual length of the wall panel in accordance with Table R602.10.5.2. For example, a 36-inch (914 mm) panel can be used in an 8-foot (2438 mm) wall instead of a full 48-inch (1219 mm) panel, but it will only count as 27 inches (686 mm) towards the required length of bracing.

TABLE R602.10.5.2
PARTIAL CREDIT FOR BRACED WALL PANELS LESS THAN 48 INCHES IN ACTUAL LENGTH

ACTUAL LENGTH OF BRACED WALL PANEL (inches)	CONTRIBUTING LENGTH OF BRACED WALL PANEL (inches)[a]	
	8-foot Wall Height	9-foot Wall Height
48	48	48
42	36	36
36	27	N/A

For SI: 1 inch = 25.4 mm, 1 foot = 304.8 mm.

N/A = Not Applicable.

a. Linear interpolation shall be permitted.

❖ See the commentary for Section R602.10.5.2.

R602.10.6 Construction of Methods ABW, PFH, PFG, CS-PF and BV-WSP. Methods ABW, PFH, PFG, CS-PF and BV-WSP shall be constructed as specified in Sections R602.10.6.1 through R602.10.6.5.

❖ This section places former intermittent and continuously sheathed alternate braced wall panel methods together. The text descriptions of the methods have been deleted with a figure replacing the text description. The figures contain the majority of the construction requirements for Methods ABW, PFH, PFG, and CS-PF. The section also includes a new method, BV-WSP, created from the requirements of the 2009 IRC Section R602.12 for masonry and stone veneer used in high seismic regions.

Allowed wall heights have been increased to 12 feet (3658 mm) with the portal header height limited to 10 feet (3048 mm) and a cripple wall built above. When a cripple wall is above the header, tension straps should be nailed to the cripple wall stud, header and the jack stud below creating a continuous load path to the story or foundation below.

In 2009 provisions were added for the use of a cripple wall (pony wall) directly over the continuously sheathed portal frame. These provisions have been included in all the portal frame alternate methods in the 2012 edition. The pony wall provisions are used often as a means of elevating the second story of the structure over the garage in a home with a split-level entry. Cripple walls are meant to address a problem in which a structural hinge is created over a door or window header that can result in the header bulging in or out due to wind loads blowing directly against the wall, or even differential moisture conditions. In severe cases, it can lead to structural problems. Often in garages, these areas are braced back to the ceiling with framing to prevent such an occurrence.

When "braced back" detailing is not desired, the information provided in Table R602.10.6.4 and Figure R602.10.6.2, R602.10.6.3, or R602.10.6.4 can be used to detail the portal frame to handle various design wind loads and exposures. While not specifically permitted in the code, from an engineering perspective, interpolation of the variables in the table is appropriate.

Note that the cripple wall information in Table R602.10.6.4 may be used for pony walls over any header within the scope of the table. For conventionally framed headers, the table is slightly conservative, but it is the only prescriptive guidance available for cripple walls over headers. The use of this table will ensure that the header/pony wall assembly has sufficient stiffness to prevent a hinge from forming at the header-to-pony wall joint when subjected to wind loads acting against the wall.

R602.10.6.1 Method ABW: Alternate braced wall panels. Method ABW *braced wall panel*s shall be constructed in accordance with Figure R602.10.6.1. The hold-down force shall be in accordance with Table R602.10.6.1.

❖ Method ABW (see Section R602.10.6.1) has been in the IRC in one form or another since 2000 and was often referred to as the "32-inch (813 mm) alternate with hold-downs." Method ABW panels are typically used when bracing is needed in a wall area that is not long enough to accommodate a 4-foot (1219 mm) braced wall panel.

For the 2012 code, Method ABW is grouped with the intermittent wall bracing methods because it may be substituted on a one-for-one basis for a braced wall panel from any other bracing method, and, as such, it is used intermittently. Figure R602.10.6.1 clarifies the method. In 2009 the text description of the construction method was first deleted leaving a figure detailing the method. The double-sided requirement for the first-of-two-story applications was also removed. Instead of double-sided wood structural panels—which make hold-down and anchor bolt placement hard to inspect—the 2009 code required nail spacing of 4 inches (102 mm) on center at the panel perimeter.

For purposes of computing the required length of bracing, a single Method ABW panel is considered to be equivalent to 4 feet (1219 mm) of bracing, regardless of its actual length.

The maximum height and minimum length are given in Table R602.10.5 while the required hold-down force is given in Table R602.10.6.1.

TABLE R602.10.6.1
MINIMUM HOLD-DOWN FORCES FOR METHOD ABW BRACED WALL PANELS

SEISMIC DESIGN CATEGORY AND WIND SPEED	SUPPORTING/STORY	HOLD DOWN FORCE (pounds)				
		Height of Braced Wall Panel				
		8 feet	9 feet	10 feet	11 feet	12 feet
SDC A, B and C Wind speed < 110 mph	One story	1,800	1,800	1,800	2,000	2,200
	First of two stories	3,000	3,000	3,000	3,300	3,600
SDC D_o, D_1 and D_2 Wind speed < 110 mph	One story	1,800	1,800	1,800	NP	NP
	First of two stories	3,000	3,000	3,000	NP	NP

For SI: 1 inch = 25.4 mm, 1 foot = 304.8 mm, 1 pound = 4.45 N, 1 mile per hour = 0.447 m/s.
NP = Not Permitted.

❖ See the Commentary for Section R602.10.6.1.

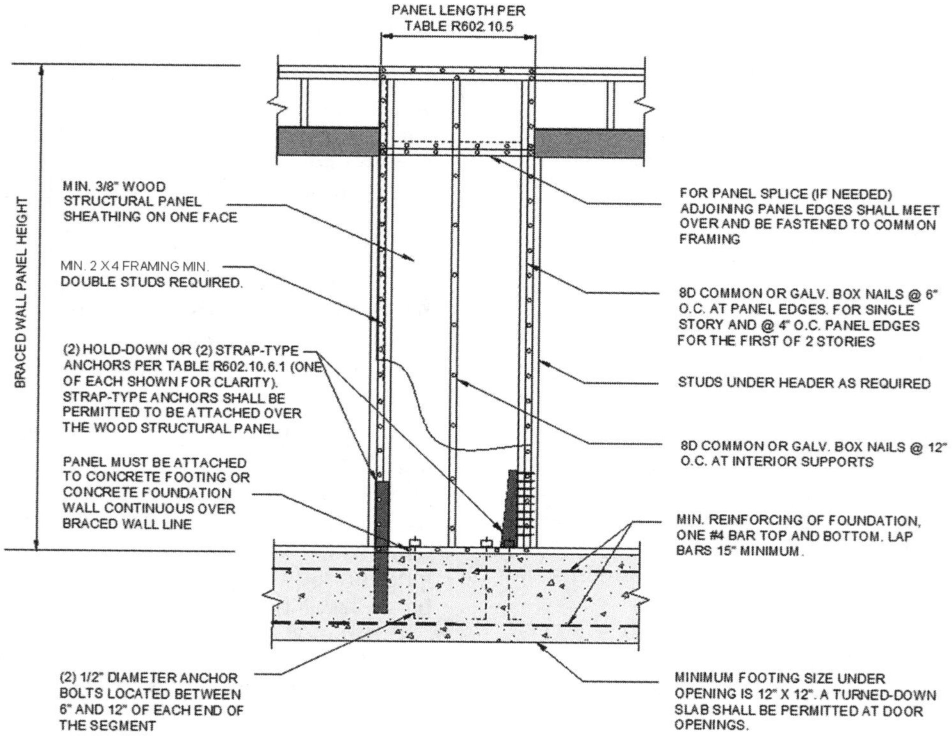

For SI: 1 inch = 25.4 mm.

FIGURE R602.10.6.1
METHOD ABW—ALTERNATE BRACED WALL PANEL

❖ See the commentary for Section R602.10.6.1.

R602.10.6.2 Method PFH: Portal frame with hold-downs. Method PFH *braced wall panels* shall be constructed in accordance with Figure R602.10.6.2.

❖ Method PFH has been in the code since 2006 (where it was called "alternate braced wall panel adjacent to a door or window opening." This method was developed primarily to maintain the traditional look of narrow wall segments on either side of garage door openings. Like Method ABW, Method PFH was included in the 2009 code as an intermittent bracing method because it may be substituted on a one-for-one basis for a braced wall panel from any other bracing method and, as such, it is used intermittently.

In the 2012 edition, the construction method of the intermittent portal frame with hold-downs has been updated.

- Steel headers are not permitted.
- A cripple wall or pony wall may be placed above the header to finish out the opening. The cripple wall may be up to 4 feet (1219 mm) tall as long as the total opening height does not exceed 12 feet (3658 mm). Top of header height is limited to 10 feet (3048 mm).

- Header to jack stud straps (formerly a 1000 pound (454 kg) strap) are sized according to Table R602.10.6.4.
- For a single portal, the minimum number of jack studs required for the post is determined using Tables R502.5(1) and (2).
- Minimum foundation reinforcement is one #4 bar at top and bottom of footing with bars lapped 15 inches (381 mm) minimum.
- The minimum footing size is 12 inches (305 mm) by 12 inches (305 mm) under the opening. A turned down slab may be used under door openings.
- The anchor bolt and plate washer assembly are clarified with a $^3/_{16}$" × 2" × 2" (5 mm × 51 mm × 51 mm) plate washer required.

For purposes of computing the required length of bracing, each leg of the portal frame is considered to be equivalent to 4 feet (1219 mm) of bracing. The portal frame can be built with one or two sides as required. This method can be used for wall heights up to 10 feet (3048 mm), or if a cripple wall (pony wall) is placed above the header, a total opening height may be 12 feet (3658 mm) tall. The top of the header is limited to a maximum height of 10 feet (3048 mm). Note that Method PFH is not aspect ratio based: all single-story applications are a minimum of 16 inches (406 mm) in length and all first of two-story applications are a minimum of 24 inches (610 mm) in length. If the opening height is greater than 10 feet (3048 mm) tall, then the required braced panel length increases (see Table R602.10.5).

R602.10.6.3 Method PFG: Portal frame at garage door openings in Seismic Design Categories A, B and C. Where supporting a roof or one story and a roof, a Method PFG *braced wall panel* constructed in accordance with Figure R602.10.6.3 shall be permitted on either side of garage door openings.

❖ Method PFG was new for the 2009 code edition and is a "light" variation of Method PFH that is limited to areas of low and moderate seismicity (SDC A, B and C). Like Method PFH, it is a portal frame bracing method, but it does not include hold-downs and is restricted to single-story and first of two-story applications. Method PFG is a 4:1 aspect ratio-based system with a 24-inch (610 mm) minimum length requirement [e.g., for an 8-foot (2438 mm) wall, the minimum length is 96 inches (2438 mm)/4 = 24 inches (610 mm); for a 10-foot (3048 mm) wall, the minimum length is 120 inches (3048 mm)/4 = 30 inches (762 mm)]. For the purposes of computing, its bracing length is equal to 1.5 times the length (horizontal dimension) of the vertical leg. Note the nailing at the

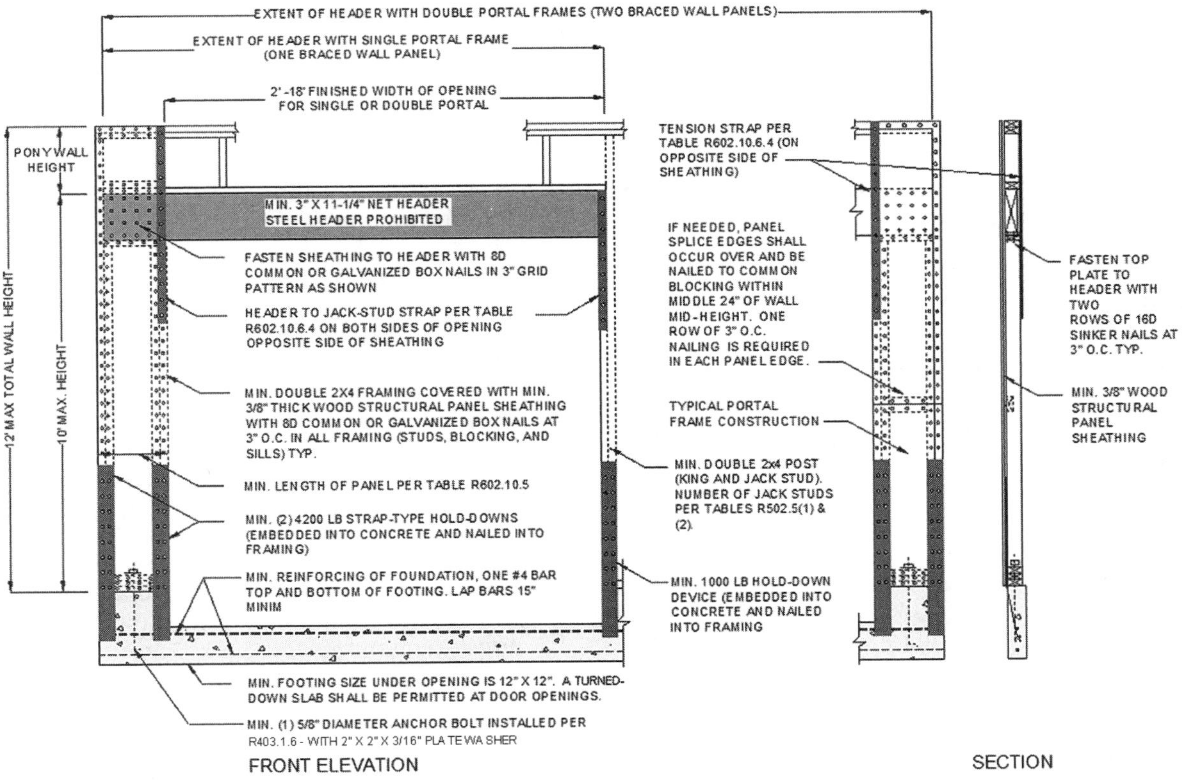

For SI: 1 inch = 25.4 mm, 1 foot = 304.8 mm.

FIGURE R602.10.6.2
METHOD PFH—PORTAL FRAME WITH HOLD-DOWNS

❖ See the commentary for Section R602.10.6.2.

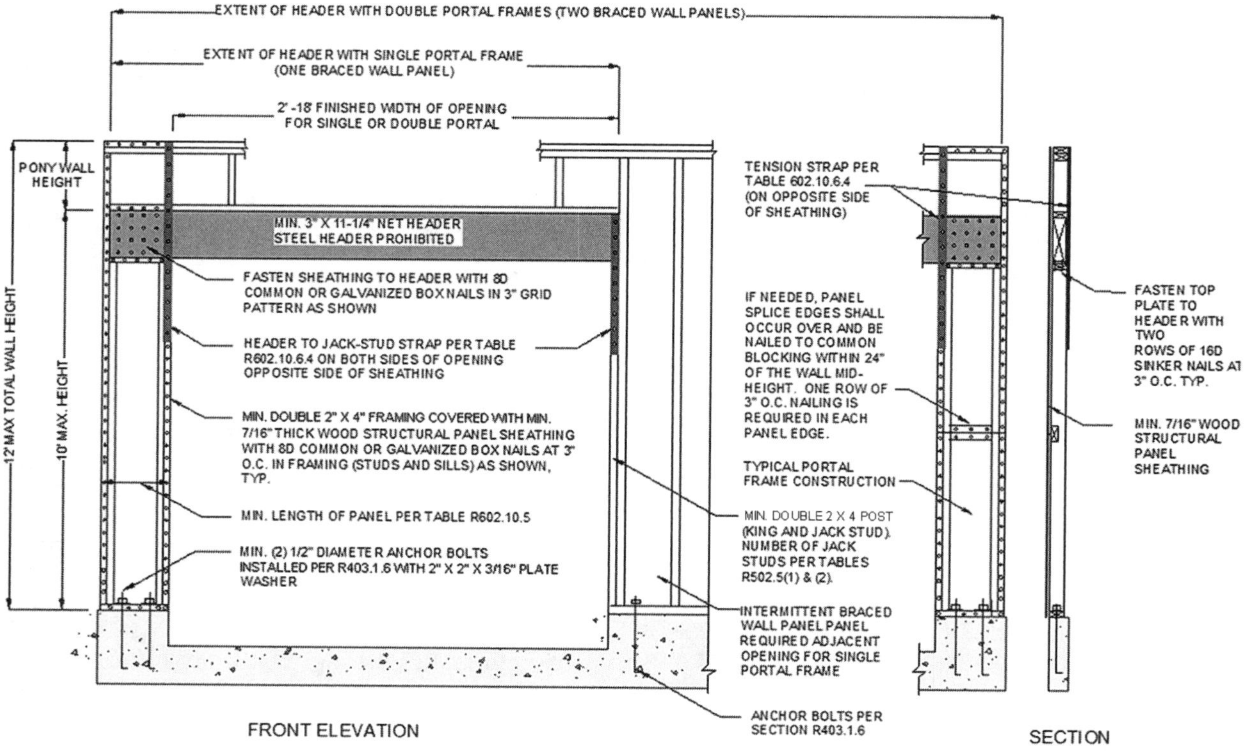

For SI: 1 inch = 25.4 mm, 1 foot = 304.8 mm.

FIGURE R602.10.6.3
METHOD PFG—PORTAL FRAME AT GARAGE DOOR OPENINGS IN SEISMIC DESIGN CATEGORIES A, B AND C

❖ See the commentary for Section R602.10.6.3.

sheathing-to-header overlap is the same as Method PFH but without nailing the interior studs of the doubled 2 by 4s at either end of the vertical legs.

In the 2012 edition, the construction method of intermittent portal frames at garage door openings has been updated.

- Steel headers are not permitted.
- A cripple wall or pony wall may be placed above the header to finish out the opening. The cripple wall may be up to 4 feet (1219 mm) tall as long as the total opening height does not exceed 12 feet (3658 mm). Top of header height is limited to 10 feet (3048 mm).
- The required plate washer is now a $^3/_{16}$ inch × 2 inch × 2 inch (5 mm × 51 mm × 51 mm) washer. This matches the requirements for Method PFH.

R602.10.6.4 Method CS-PF: Continuously sheathed portal frame. Continuously sheathed portal frame *braced wall panels* shall be constructed in accordance with Figure R602.10.6.4 and Table R602.10.6.4. The number of continuously sheathed portal frame panels in a single *braced wall line* shall not exceed four.

❖ Minimum thickness of wood structural panel sheathing for CS-PF is $^7/_{16}$ inch (11 mm) thickness. This is an increase from the 2009 IRC minimum of $^3/_8$ inch (10 mm) thickness.

In the 2006 code, this method was specified in Note c of Table R602.10.5 and it was known in the industry as "portal frame without hold-downs." In the 2009 code edition the scope of this method, now referred to as Method CS-PF, was expanded for use on any floor and in any wind zone or SDC covered by the code. The expansion to any floor comes as the result of the development of attachment details for raised wood floors in addition to foundation details.

For purposes of determining the length of bracing, the length of the vertical leg of the portal frame is used as the bracing length for the element. The minimum length is in Table R602.10.5. It is based on a 6:1 height-to-length ratio: for example: 16 inches (406 mm) minimum for 8 foot (2438 mm) height.

Method CS-PF can only be used with Method CS-WSP.

In the 2012 edition, the construction method of the continuously sheathed portal frame has been updated.

- Steel headers are not permitted.
- The minimum sheathing thickness is $^7/_{16}$ (11 mm) inches.

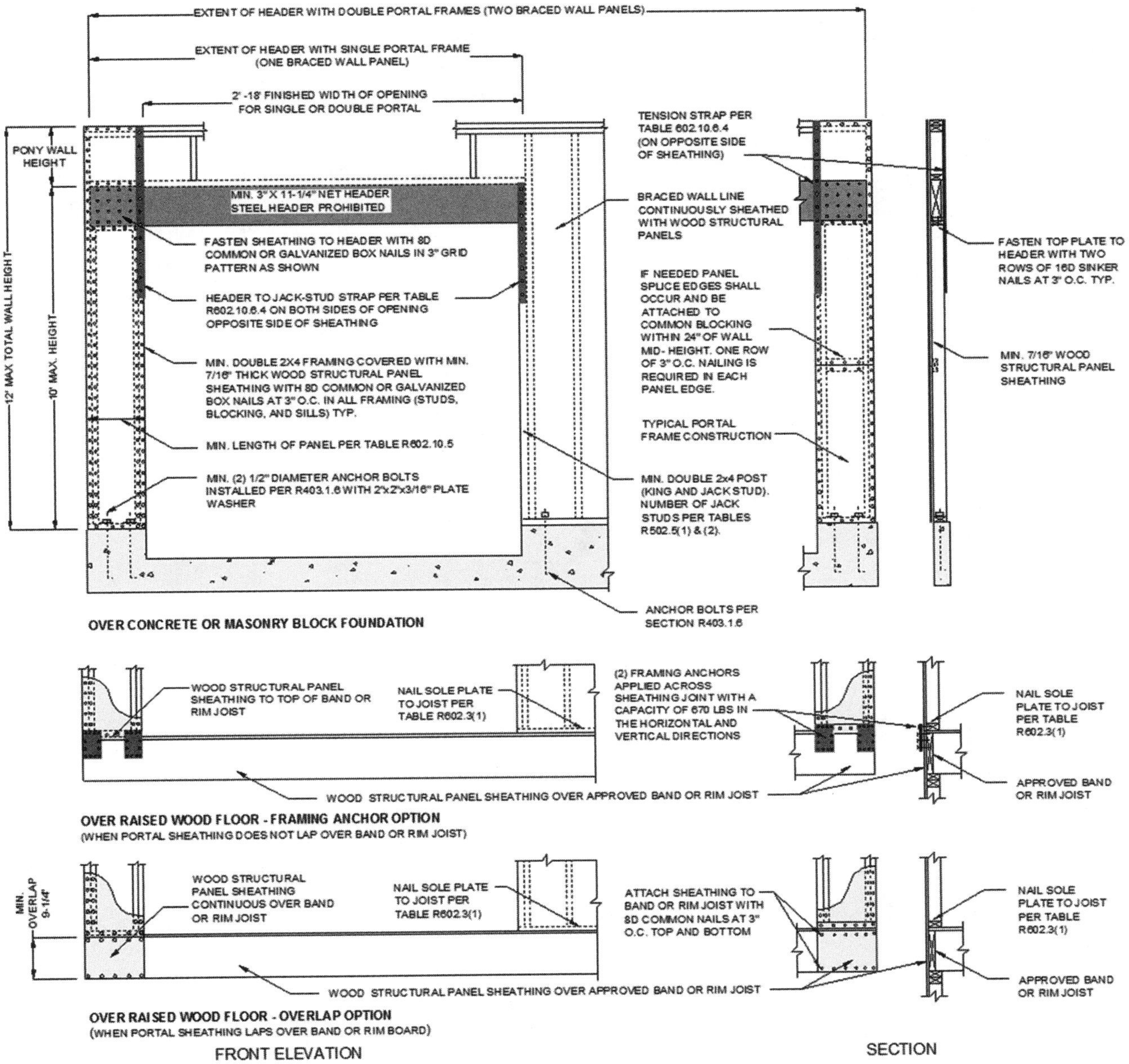

For SI: 1 inch = 25.4 mm, 1 foot = 304.8 mm.

FIGURE R602.10.6.4
METHOD CS-PF-CONTINUOUSLY SHEATHED PORTAL FRAME PANEL CONSTRUCTION

❖ See the commentary for Section R602.10.6.4.

R602.10.6.5 Wall bracing for dwellings with stone and masonry veneer in Seismic Design Categories D_0, D_1 and D_2. Where stone and masonry veneer are installed in accordance with Section R703.7, wall bracing on exterior *braced wall lines* and *braced wall lines* on the interior of the building, backing or perpendicular to and laterally supporting veneered walls shall comply with this section.

Where dwellings in Seismic Design Categories D_0, D_1 and D_2 have stone or masonry veneer installed in accordance with Section R703.7, and the veneer does not exceed the first-story height, wall bracing shall be in accordance with Section R602.10.3.

Where detached one- or two-family dwellings in Seismic Design Categories D_0, D_1 and D_2 have stone or masonry veneer installed in accordance with Section R703.7, and the veneer exceeds the first-*story height*, wall bracing at exterior *braced wall lines* and *braced wall lines* on the interior of the building shall be constructed using Method BV-WSP in accordance with this section and Figure R602.10.6.5. Cripple walls shall not be permitted, and required interior *braced wall lines* shall be supported on continuous foundations.

TABLE R602.10.6.4
TENSION STRAP CAPACITY REQUIRED FOR RESISTING WIND PRESSURES PERPENDICULAR TO METHOD PFH, PFG AND CS-PF BRACED WALL PANELS

MINIMUM WALL STUD FRAMING NOMINAL SIZE AND GRADE	MAXIMUM PONY WALL HEIGHT (feet)	MAXIMUM TOTAL WALL HEIGHT (feet)	MAXIMUM OPENING WIDTH (feet)	TENSION STRAP CAPACITY REQUIRED (pounds)[a, b]					
				Basic Wind Speed (mph)					
				85	90	100	85	90	100
				Exposure B			Exposure C		
2 × 4 No. 2 Grade	0	10	18	1,000	1,000	1,000	1,000	1,000	1,000
	1	10	9	1,000	1,000	1,000	1,000	1,000	1,275
			16	1,000	1,000	1,750	1,800	2,325	3,500
			18	1,000	1,200	2,100	2,175	2,725	DR
	2	10	9	1,000	1,000	1,025	1,075	1,550	2,500
			16	1,525	2,025	3,125	3,200	3,900	DR
			18	1,875	2,400	3,575	3,700	DR	DR
	2	12	9	1,000	1,200	2,075	2,125	2,750	4,000
			16	2,600	3,200	DR	DR	DR	DR
			18	3,175	3,850	DR	DR	DR	DR
	4	12	9	1,775	2,350	3,500	3,550	DR	DR
			16	4,175	DR	DR	DR	DR	DR
2 × 6 Stud Grade	2	12	9	1,000	1,000	1,325	1,375	1,750	2,550
			16	1,650	2,050	2,925	3,000	3,550	DR
			18	2,025	2,450	3,425	3,500	4,100	DR
	4	12	9	1,125	1,500	2,225	2,275	2,775	3,800
			16	2,650	3,150	DR	DR	DR	DR
			18	3,125	3,675	DR	DR	DR	DR

For SI: 1 inch = 25.4 mm, 1 foot = 304.8 mm, 1 pound = 4.45 N.

a. DR = design required.

b. Strap shall be installed in accordance with manufacturer's recommendations.

❖ See the commentary for Section R602.10.6.

Townhouses in Seismic Design Categories D_0, D_1 and D_2 with stone or masonry veneer exceeding the first-story height shall be designed in accordance with accepted engineering practice.

❖ This section includes provisions for wall bracing when stone or masonry veneer is installed as a wall covering. These provisions apply only to stone and masonry veneer, and not other wall coverings, because stone and masonry are heavy compared to other cladding materials. When ground motion from an earthquake puts the cladding in motion, the seismic weight (mass) of a cladding material produces load on the building's lateral-load-resisting structural elements. In the code, bracing panels are the structural elements that resist the motion of the heavy cladding. The seismic lateral load on a building increases with cladding weight and ground motion intensity; therefore, the requirements for this provision are more stringent than for other bracing methods.

This section was in Chapter 7 of the 2006 edition of the code (the wall covering chapter). For 2009, the stone and masonry veneer bracing provisions were relocated to the same chapter as the other bracing provisions where they were placed in Section R602.12. For the 2012 edition, the section has been moved into Section R602.10 and placed into two separate subsections. The bracing element used to brace wall lines has been listed as a separate bracing method.

The intent of Section R602.10 is to permit full-height or less stone or masonry veneer on first-story walls using the bracing requirements of Table R602.10.3(3) and R602.10.3(4). Townhouses in SDC C have additional bracing for veneer required in Table R602.10.3(4). Detached one- and two-family dwellings in SDC D_0, D_1 and D_2 with veneer one story or less in height also have additional bracing provided by Table R602.10.3(4).

One- and two-family dwellings in SDC A, B and C and townhouses in SDC A and B are exempt from seismic requirements (per Section R301.2.2). No additional bracing is required due to the use of stone or masonry veneer in these seismic design categories.

Wall veneers of one story or less are thought to resist seismic forces independently from the braced walls to which they are attached. Once these wall veneers exceed one story, however, they are subject to additional bracing requirements.

TABLE R602.10.6.5
METHOD BV-WSP WALL BRACING REQUIREMENTS

SEISMIC DESIGN CATEGORY	STORY	BRACED WALL LINE LENGTH (FEET)					SINGLE-STORY HOLD-DOWN FORCE (pounds)[a]	CUMULATIVE HOLD-DOWN FORCE (pounds)[b]
		10	20	30	40	50		
		MINIMUM TOTAL LENGTH (FEET) OF BRACED WALL PANELS REQUIRED ALONG EACH BRACED WALL LINE						
D_0		4.0	7.0	10.5	14.0	17.5	N/A	—
		4.0	7.0	10.5	14.0	17.5	1900	—
	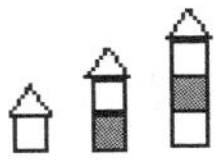	4.5	9.0	13.5	18.0	22.5	3500	5400
	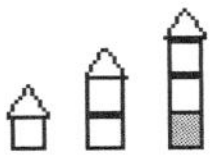	6.0	12.0	18.0	24.0	30.0	3500	8900
D_1		4.5	9.0	13.5	18.0	22.5	2100	—
	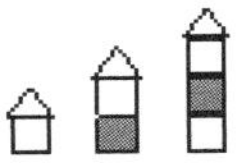	4.5	9.0	13.5	18.0	22.5	3700	5800
	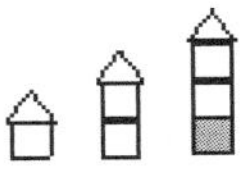	6.0	12.0	18.0	24.0	30.0	3700	9500
D_2		5.5	11.0	16.5	22.0	27.5	2300	—
	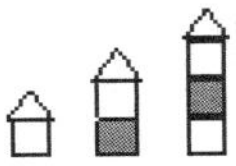	5.5	11.0	16.5	22.0	27.5	3900	6200
	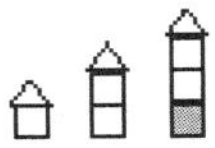	NP	NP	NP	NP	NP	N/A	N/A

For SI: 1 inch = 25.4 mm, 1 foot = 304.8 mm, 1 pound per square foot = 0.479 kPa, 1 pound-force = 4.448 N.

NP = Not Permitted.

N/A = Not Applicable.

a. Hold-down force is minimum allowable stress design load for connector providing uplift tie from wall framing at end of braced wall panel at the noted story to wall framing at end of braced wall panel at the story below, or to foundation or foundation wall. Use single-story hold-down force where edges of braced wall panels do not align; a continuous load path to the foundation shall be maintained.

b. Where hold-down connectors from stories above align with stories below, use cumulative hold-down force to size middle- and bottom-story hold-down connectors.

❖ See the commentary for Sections R602.10.6.5 and R602.10.6.5.1.

For detached one- and two-family dwellings in SDC D_0, D_1 and D_2 with stone and masonry veneer taller than the first story, Section R602.10.6.5 must be used to determine required bracing material, length and hold-down size. This is consistent with Section R301.2.2.2.5, Item 7, Exception, which permits the use of masonry veneer as permitted elsewhere in the code. This exception specifies "masonry" veneer only; however, the intent of the code is to permit the use of both stone and masonry veneer, as the two are grouped together elsewhere in the code. For example, the title of Section R703.7 is "Stone and Masonry Veneer."

Since Section R602.10.6.5 does not extend "permission" for the installation of stone and masonry veneer to townhomes in SDC D_0, D_1 and D_2, townhouses would be categorized as "irregular" under Section R301.2.2.2.5, Item 7. Townhouses in SDC D_0, D_1 and D_2 must be engineered.

The amount of bracing required for one- and two-family dwellings in SDC D_0, D_1 and D_2 that have stone and masonry veneer above the top of the first story is provided in Table R602.10.6.5. Note that Table R602.10.6.5 has bracing requirements for single-story buildings but when the veneer is one story less in height, use Tables R602.10.3(3) and R602.10.3(4).

The bracing requirements for stone and masonry veneer apply to both the interior and exterior braced wall lines. These requirements define specific bracing materials (Method BV-WSP) and mechanical hold-down devices required at the ends of each braced wall panel.

Method BV-WSP is required to be attached with 8d nails spaced 4 inches (102 mm) on center at panel ends and edges, and hold-downs are required at ends of each continuous length of wood structural panel bracing in accordance with Table R602.10.4. Requiring BV-WSP effectively prohibits the use of continuous sheathing. For walls with stone or masonry veneer above the first story in high seismic regions, BV-WSP is the only bracing method allowed.

Continuous sheathing is prohibited because the capacities of Methods CS-WSP, CS-G and CS-PF are considerably lower than fully restrained, 48-inch-long (1219 mm) panels constructed with $^7/_{16}$-inch-thick (11 mm) wood structural panel sheathing, attached with 8d common nails at 4 inches (1029 mm) on center and hold-downs at the end of each braced wall panel.

The hold-downs are typically required to be nailed into two full-height studs at each end of the braced wall panel element. Designers used to working with the IBC will recognize that these bracing elements are similar to shear walls.

Hold-down location and capacity requirements are provided in Table R602.10.6.5 and Figure

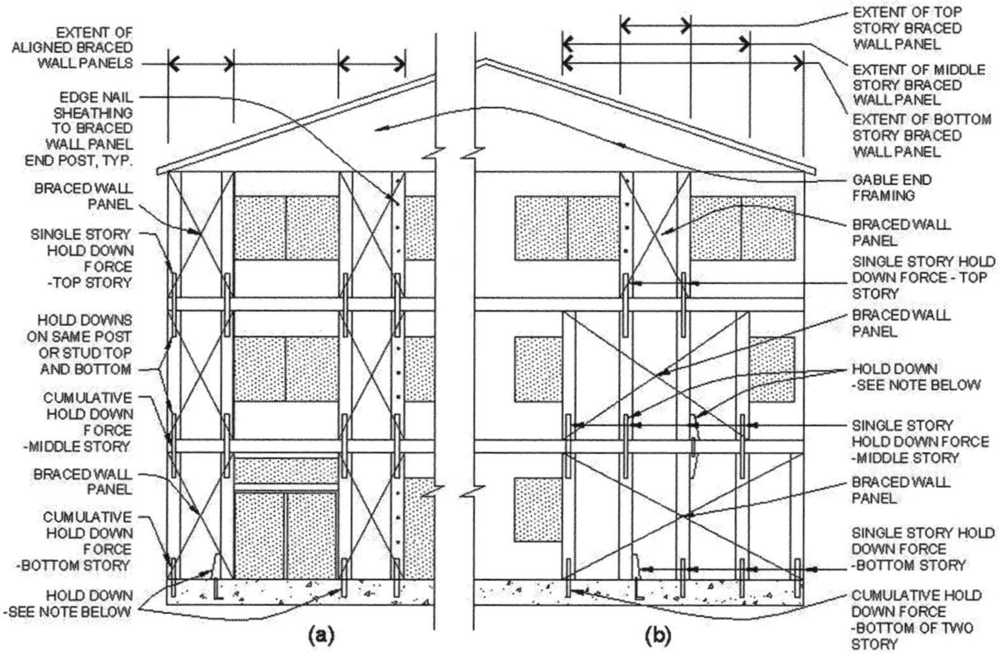

(a) Braced wall panels stacked (aligned story to story) Use cumulative hold down force.

(b) Braced wall panels mixed stacked and not stacked. Use hold down force as noted

Note: Hold downs should be strap ties, tension ties, or other approved hold down devices and shall be installed in accordance with the manufacturer's instructions.

FIGURE R602.10.6.5
METHOD BV-WSP—WALL BRACING FOR DWELLINGS WITH STONE AND MASONRY VENEER IN SEISMIC DESIGN CATEGORIES D_0, D_1 AND D_2.

❖ See the commentary for Section R602.10.6.5.

R602.10.6.5, respectively. As illustrated in Figure R602.10.6.5, the capacity of the hold-down is determined by the relationship of the ends of braced wall panels between stories. When BV-WSP panel ends are stacked above one another, the cumulative hold-down force from Table R602.10.6.5 is required for the lower braced wall panels.

When the ends of braced wall lines are not vertically aligned, i.e., one panel ends at a different location along the wall line than the panel below; the single-story hold-down capacity given in Table R602.10.6.5 is used.

When a BV-WSP occurs above the first story, the hold-down anchorage must be continued down to the foundation. For example, if a hold-down is installed on the second or third story, each story below must have a hold-down of the same capacity that is vertically aligned with the hold down on the story above. If a braced wall panel at the story below ends at the same location along the wall line, the hold-down capacity should be greater than or equal to the cumulative hold-down force.

Note that while wood structural panel sheathing may be used over the whole wall, above and below openings, hold-downs only need to be applied to the ends of those elements that are required for intermittent bracing. This is an important consideration when bracing lengths greater than 48 inches (1219 mm) are used. For example, if a wall line has a 6-foot (1829 mm) length of BV-WSP wall bracing, consisting of one 4-foot (1219 mm) and one 2-foot (610 mm) piece attached to a common stud, the hold-downs are only required at the start and end of the 6-foot (1829 mm) length and not at 4-foot (1219 mm) intervals.

The use of stone and masonry veneer in SDC D_0, D_1 and D_2 requires that interior braced wall lines be supported on continuous foundations regardless of plan dimension length. In addition, cripple walls are not permitted to be used.

R602.10.6.5.1 Length of bracing. The length of bracing along each *braced wall line* shall be the greater of that required by the design wind speed and *braced wall line* spacing in accordance with Table R602.10.3(1) as adjusted by the factors in the Table R602.10.3(2) or the Seismic Design Category and *braced wall line* length in accordance with Table R602.10.6.5. Angled walls shall be permitted to be counted in accordance with Section R602.10.1.4, and *braced wall panel* location shall be in accordance with Section R602.10.2.2. The seismic adjustment factors in Table R602.10.3(4) shall not be applied to the length of bracing determined using Table R602.10.6.5. In no case shall the minimum total length of bracing in a *braced wall line*, after all adjustments have been taken, be less than 48 inches (1219 mm) total.

❖ The minimum bracing length is determined by Table R602.10.6.5. This length is required to resist the increased loads produced by seismic forces resulting from the stone and masonry veneer. Table R602.10.6.5 supersedes the use of Tables R602.10.3(3) and R602.10.3(4) only when stone or masonry veneer exceeds the first story height. Note that the wind bracing requirements of Tables R602.10.3(1) and R602.10.3(2) must also be considered, although it is unlikely that the wind will control in higher seismic areas.

Braced wall panels may begin up to 10 feet (3048 mm) from the end of the braced wall line. The required minimum length of a braced wall panel used in SDC D_0, D_1 and D_2 is 48 inches (1219 mm) with the hold-down capacity given in Table R602.10.6.5.

The intent of the new bracing method is to prevent alternate braced wall panels or any method which reduces bracing length from being used for the purpose of reducing the required panel length or hold-down capacity.

R602.10.7 Ends of braced wall lines with continuous sheathing. Each end of a *braced wall line* with continuous sheathing shall have one of the conditions shown in Figure R602.10.7.

❖ The information in the 2009 IRC Figures R602.10.4.4(2), R602.10.4.4(3), R602.10.4.4(4), and R602.10.4.4(5), continuously sheathed braced wall panel locations, has been combined into one figure. A new end condition has also been added to the 2012 IRC Figure R602.10.7. Condition 3 does not require return panels or hold-downs if a 4-foot (1219 mm) braced wall panel is located at the end of a braced wall line.

For continuously sheathed bracing methods a corner as shown in Figure R602.10.7 or some other method of anchoring the wall is required at each end of the braced wall line.

For End Condition 1, a 24-inch (610 mm) corner return is required for continuously sheathed wall lines with wood structural panel in accordance with Figure R602.10.**7**. For wall lines with CS-SFB, the minimum corner return length is 32 inches (813 mm).

The purpose of the corner attachment for the continuously sheathed bracing methods is to connect the intersecting walls together to create a stronger, box-like structure that will perform better during high wind or seismic events. In accordance with Table R602.3(1), the corner studs require a minimum of a single row of 16d nails at 12 inches (305 mm) on center. A double row of 16d nails at 24 inches (610 mm) on center, framing member orientation permitting, is considered equivalent. It is important to note that the intent of the nailing requirement is to provide the specified amount of nailing between the two studs, each on adjacent walls, to which the adjacent wall sheathing is attached.

An 800-pound (3560 N) hold-down device required to be attached between the stud at the edge of the braced wall panel closest to the corner and the foundation below-can be used in lieu of the return corner (see End Condition 2, Figure R602.10.7). If two continuously sheathed braced wall lines meet at a corner, and the first braced panel is spaced away from the corner on one of the wall lines, then both braced wall lines lose the structural effect of the corner return; therefore, both braced wall lines must be anchored with an 800-pound (3560 N) hold-down device at the edge adjacent to the corner.

When a minimum of 4 feet (1219 mm) of braced panel is placed at a corner in a braced wall line, a corner return is not required on the perpendicular wall line (see End Condition 3, Figure R602.10.7).

For continuously sheathed wall panel and corner construction, the first full-height braced wall segment of the continuously sheathed braced wall line is allowed to be spaced away from the end of the braced wall line, in accordance with the exception to Section R602.10.2.2.1 and End Conditions 4 and 5, Figure R602.10.7.

Note that if two continuously sheathed wall lines meet at a corner, and along one of the two wall lines, the first bracing panel is displaced from this common corner, then both braced wall lines are denied the structural effect of the corner return. In this case, both wall lines must be anchored with an 800-pound (3560 N) hold-down device or have return panels.

- In End Condition 1 there are return panels/braced wall panels on each side of the corner.
- In End Condition 2 there is a braced wall panel and a hold-down replacing the return panel.
- In End Condition 3, no return panel is required.
- In End Condition 4, there is a return panel and a 24 or 32 inch panel at the corner.
- In End Condition 5, there is a hold-down on the braced wall line, there is also a hold-down at the first braced wall panel on the return wall to brace that end of the wall line.

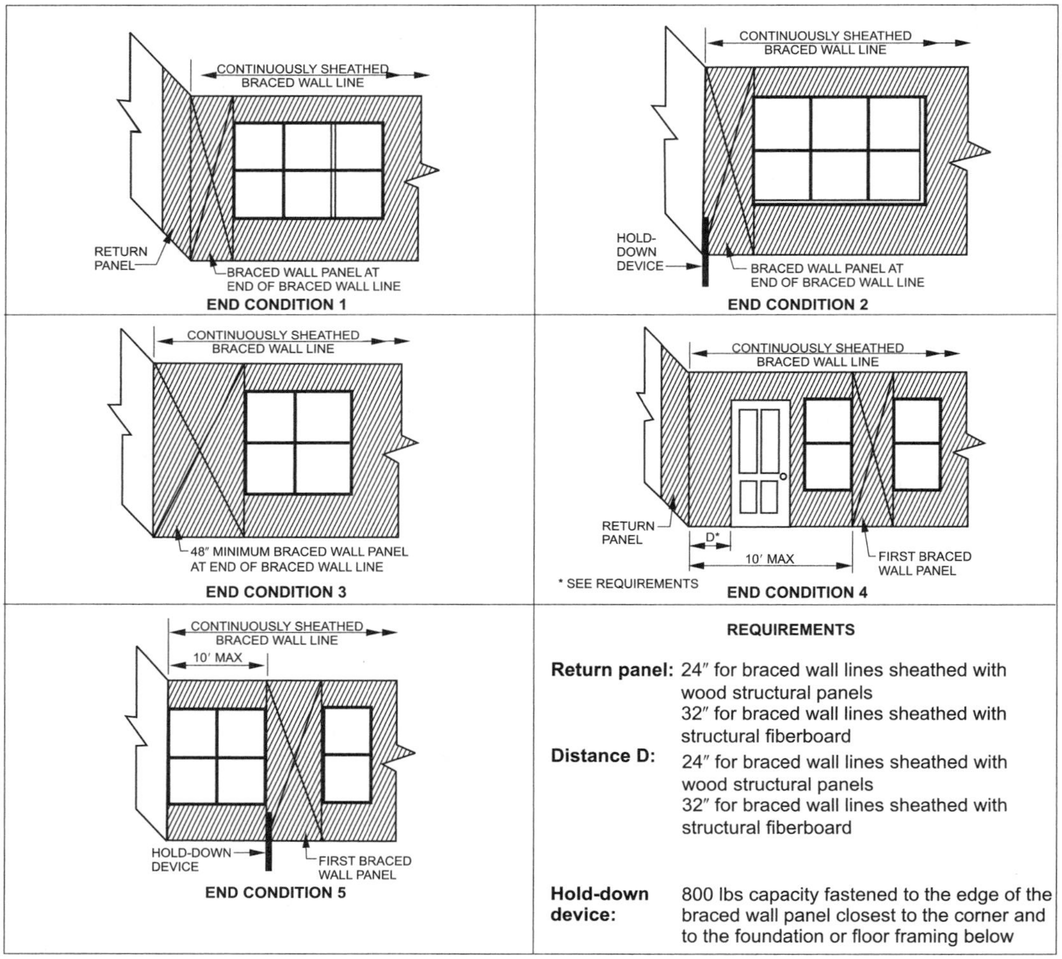

For SI: 1 inch = 25.4 mm, 1 foot = 304.8 mm, 1 pound = 4.45 N.

FIGURE R602.10.7
END CONDITIONS FOR BRACED WALL LINES WITH CONTINUOUS SHEATHING

❖ See the commentary for Section R602.10.7.

R602.10.8 Braced wall panel connections. *Braced wall panels* shall be connected to floor framing or foundations as follows:

1. Where joists are perpendicular to a *braced wall panel* above or below, a rim joist, band joist or blocking shall be provided along the entire length of the *braced wall panel* in accordance with Figure R602.10.8(1). Fastening of top and bottom wall plates to framing, rim joist, band joist and/or blocking shall be in accordance with Table R602.3(1).
2. Where joists are parallel to a *braced wall panel* above or below, a rim joist, end joist or other parallel framing member shall be provided directly above and below the *braced wall panel* in accordance with Figure R602.10.8(2). Where a parallel framing member cannot be located directly above and below the panel, full-depth blocking at 16-inch (406 mm) spacing shall be provided between the parallel framing members to each side of the *braced wall panel* in accordance with Figure R602.10.8(2). Fastening of blocking and wall plates shall be in accordance with Table R602.3(1) and Figure R602.10.8(2).
3. Connections of *braced wall panels* to concrete or masonry shall be in accordance with Section R403.1.6.

❖ This section describes the code connection requirements between braced wall panels and the portions of the structure both above and below. Without the connections required to complete the lateral load path, the purpose of wall bracing will not be realized and the structure will not be able to resist lateral loads.

Table R602.3(1) provides the basic nailing schedule for attaching braced wall panels to the floor below and the ceiling or floor above. As an example, the sole plate (or bottom plate) of the braced wall panels is required to be nailed with three 16d nails every 16 inches (406 mm) on center into the joist or blocking below the braced wall panels [see Item 16 of Table R602.3(1)].

Note that the nails do not have to be clustered every 16 inches (406 mm) on center: there just has to be three nails in every 16 inch-length (406 mm) of braced wall panel. In fact, clustering the nails may not be a good idea if, for example, they are going into a rim board. In other cases, clustering may be needed, such as an interior braced wall parallel to and situated between floor framing: clustering is necessary to attach the braced wall to the blocking.

For a braced wall panel to provide lateral load resistance to the structure, it is essential that lateral loads transfer into the panel at its top and out to the framing members below at its bottom. Correct connections are necessary to complete this lateral load path and are as shown in Figures R602.10.8(1) and R602.10.8(2).

Braced wall panel connections to foundation or floor slab must be in accordance with Section R403.1.6.

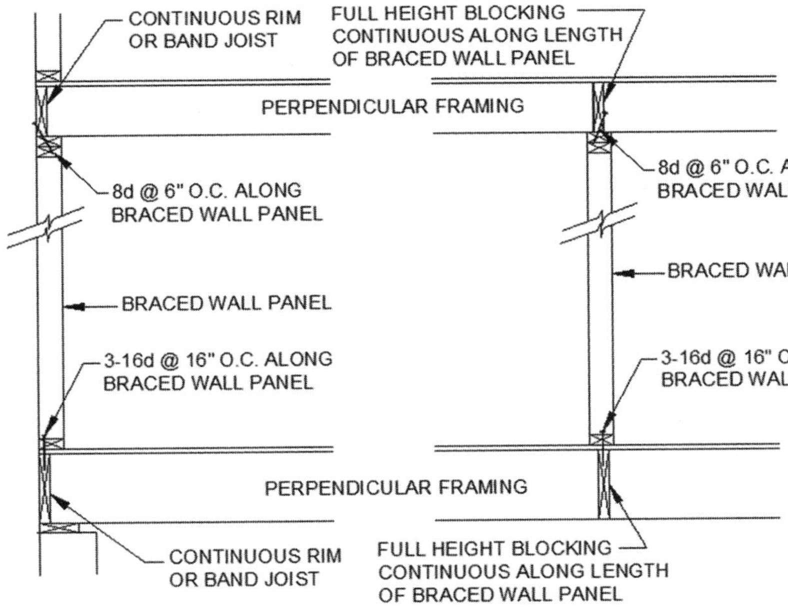

For SI: 1 inch = 25.4 mm.

FIGURE R602.10.8(1)
BRACED WALL PANEL CONNECTION WHEN PERPENDICULAR

❖ See the commentary for Section R602.10.8.

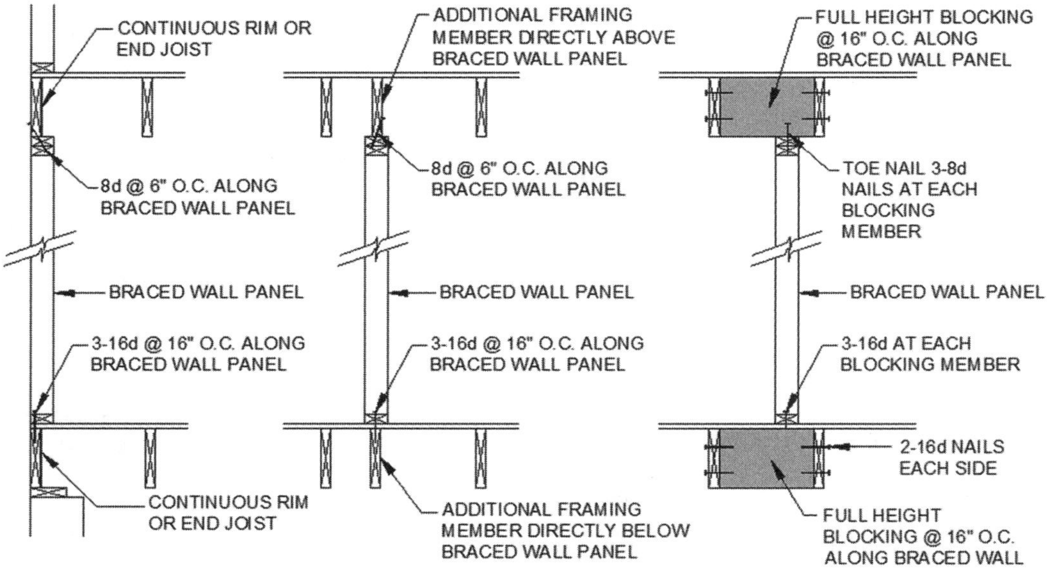

For SI: 1 inch = 25.4 mm.

FIGURE R602.10.8(2)
BRACED WALL PANEL CONNECTION WHEN PARALLEL TO FLOOR/CEILING FRAMING

❖ See the commentary for Section R602.10.8.

R602.10.8.1 Braced wall panel connections for Seismic Design Categories D_0, D_1 and D_2. *Braced wall panel*s shall be fastened to required foundations in accordance with Section R602.11.1, and top plate lap splices shall be face-nailed with at least eight 16d nails on each side of the splice.

❖ Structures in higher seismic zones need to have the capacity to withstand greater loads; therefore, additional provisions are required to ensure that these loads are transferred through the braced wall panels to the foundation. The requirements of Section R602.10.8.1 ensure the top plate lap splices and foundation connections have the needed higher load capacity.

The requirement for eight 16d nails on each side of the lap splice in SDC D_0, D_1 and D_2 is reiterated in Section R602.10.8.1. Note that Table R602.3(1), Item 14, requires a 24-inch (610 mm) overlap to provide sufficient length to put in the eight 16d nails.

R602.10.8.2 Connections to roof framing. Top plates of exterior *braced wall panels* shall be attached to rafters or roof trusses above in accordance with Table R602.3(1) and this section. Where required by this section, blocking between rafters or roof trusses shall be attached to top plates of *braced wall panel*s and to rafters and roof trusses in accordance with Table R602.3(1). A continuous band, rim, or header joist or roof truss parallel to the *braced wall panels* shall be permitted to replace the blocking required by this section. Blocking shall not be required over openings in continuously-sheathed *braced wall lines*. In addition to the requirements of this section, lateral support shall be provided for rafters and ceiling joists in accordance with Section R802.8 and for trusses in accordance with Section R802.10.3. Roof ventilation shall be provided in accordance with Section R806.1.

1. For Seismic Design Categories A, B and C and wind speeds less than 100 mph (45 m/s) where the distance from the top of the *braced wall panel* to the top of the rafters or roof trusses above is $9^1/_4$ inches (235 mm) or less, blocking between rafters or roof trusses shall not be required. Where the distance from the top of the *braced wall panel* to the top of the rafters or roof trusses above is between $9^1/_4$ inches (235 mm) and $15^1/_4$ inches (387 mm), blocking between rafters or roof trusses shall be provided above the *braced wall panel* in accordance with Figure R602.10.8.2(1).
2. For Seismic Design Categories D_0, D_1 and D_2 or wind speeds of 100 mph (45 m/s) or greater, where the distance from the top of the *braced wall panel* to the top of the rafters or roof trusses is $15^1/_4$ inches (387 mm) or less, blocking between rafters or roof trusses shall be provided above the *braced wall panel* in accordance with Figure R602.10.8.2(1).
3. Where the distance from the top of the *braced wall panel* to the top of rafters or roof trusses exceeds $15^1/_4$ inches (387 mm), the top plates of the *braced wall panel* shall be connected to perpendicular rafters or roof trusses above in accordance with one or more of the following methods:
   - 3.1. Soffit blocking panels constructed in accordance with Figure R602.10.8.2(2);
   - 3.2. Vertical blocking panels constructed in accordance with Figure R602.10.8.2(3);

3.3. Full-height engineered blocking panels designed in accordance with the AF&PA WFCM; or

3.4. Blocking, blocking panels, or other methods of lateral load transfer designed in accordance with accepted engineering practice.

❖ In the 2009 edition new provisions were added for the connection between braced wall panels and roof framing. Although previous versions of the code did not address this connection, complex roof shapes used in modern design have necessitated prescriptive connection details to ensure an effective load path exists.

These connection details represent a simple principle: braced wall lines extend from diaphragm to diaphragm and must be connected at both top and bottom. The roof and floor sheathing are the structural diaphragms of the building. Figures R602.10.8.2(1) through (3) illustrate the required connections and blocking between braced wall panels and roof framing, which are summarized in Commentary Table R602.10.8.2.

Section R602.10.8.2, Item 1, provides what is essentially an exemption from required blocking for SDC A, B and C and wind speeds less than 100 mph

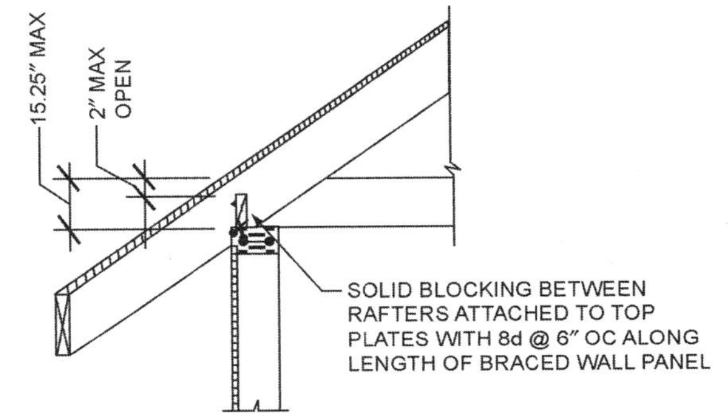

For SI: 1 inch = 25.4 mm.

FIGURE R602.10.8.2(1)
BRACED WALL PANEL CONNECTION TO PERPENDICULAR RAFTERS

❖ See the commentary for Section R602.10.8.2

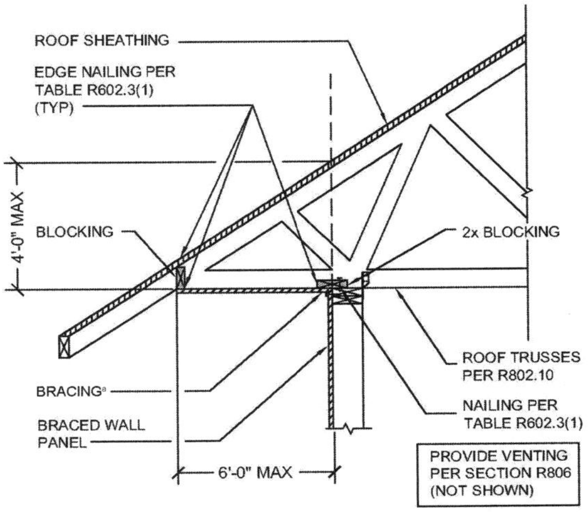

For SI: 1 inch = 25.4 mm, 1 foot = 304.8 mm.
a. Methods of bracing shall be as described in Section R602.10.4.

FIGURE R602.10.8.2(2)
BRACED WALL PANEL CONNECTION OPTION TO PERPENDICULAR RAFTERS OR ROOF TRUSSES

❖ See the commentary for Section R602.10.8.2

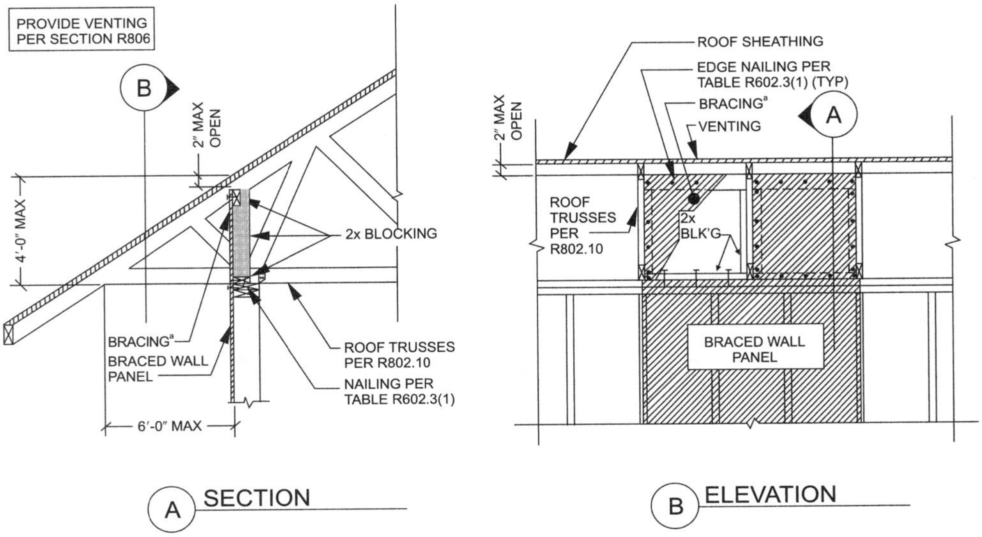

For SI: 1 inch = 25.4 mm, 1 foot = 304.8 mm.
a. Methods of bracing shall be as described in Section R602.10.4.

FIGURE R602.10.8.2(3)
BRACED WALL PANEL CONNECTION OPTION TO PERPENDICULAR RAFTERS OR ROOF TRUSSES

❖ See the commentary for Section R602.10.8.2.

(45 m/s), where the distance from the top of the rafters or roof trusses to the perpendicular top plates is 9.25 inches (235 mm) or less, provided the rafters or roof trusses are connected to the top plates of braced wall lines in accordance with Table R602.3(1).

Figure R602.10.8.2(1) illustrates the required detailing and connection for relatively low-heel trusses and rafters (less than or equal to 15.25 inches (387 mm) between the bottom of roof diaphragm sheathing and the top of double top plate). For such applications, the addition of solid lumber blocking is sufficient to prevent the trusses or conventional framing members from rolling over when subjected to wind and/or seismic loads. Attachment in accordance with Table R602.3(1), Items 1, 2 and 5-nailing at the bottom of the roof framing and blocking-is required to transfer the lateral forces (that push the roof framing along the top plate) from the roof sheathing (or diaphragm) into the braced wall line. The 2-inch (51 mm) gap at the top of the blocking permits required roof ventilation.

Figure R602.10.8.2(2) illustrates an alternate connection between a braced wall panel and roof sheathing (or diaphragm) that, while less direct, will provide adequate transfer of lateral forces from the roof diaphragm to the braced wall line. This detail does not provide lateral stability of the trusses at the bearing point (see truss installation diagrams) or the required roof venting. While the code permits various bracing methods for this application [see Figure R602.10.8(2), Note a], the user must be careful to specify only those methods that have sufficient weather resistance or provide a covering with an approved exterior finish. Note that the edge nailing requirement for the bracing method used at the soffit is provided in Table R602.10.4 and is not necessarily

Table R602.10.8.2
CONNECTION AND BLOCKING REQUIREMENTS BETWEEN BRACED WALL PANELS AND ROOF FRAMING

SEISMIC DESIGN CATEGORY AND WIND SPEED	DISTANCE (bottom of roof sheathing to top of top plate) (See Figure R602.10.8.2)	BLOCKING[a]
SDC A, B, C and wind speed less than 100 mph	9.25" or less	Not required per Section R60210.8.2, Item 1. Roof framing attached per Section R602.3(1)
	Greater than 9.25" to 15.25"	Required per Section R602.10.8.2, Item 1 and Figure R602.10.8.2(1)
SDC D_0, D_1, D_2 or wind speed 100 mph or greater	15.25" or less	Required per Section R602.10.8.2, Item 2 and Figure R602.10.8.2(1)
All SDCs and wind speeds	15.25" to 48"	Required per Section R602.10.8.(2) or Item 3 or Figures R602.10.8.2(2) or R602.10.8.2(3)

For SI: 1 inch = 25.4 mm, 1 mile per hour = 0.45 m/s.

a. Rafter or truss connection to top plate per Table R602.3(1).

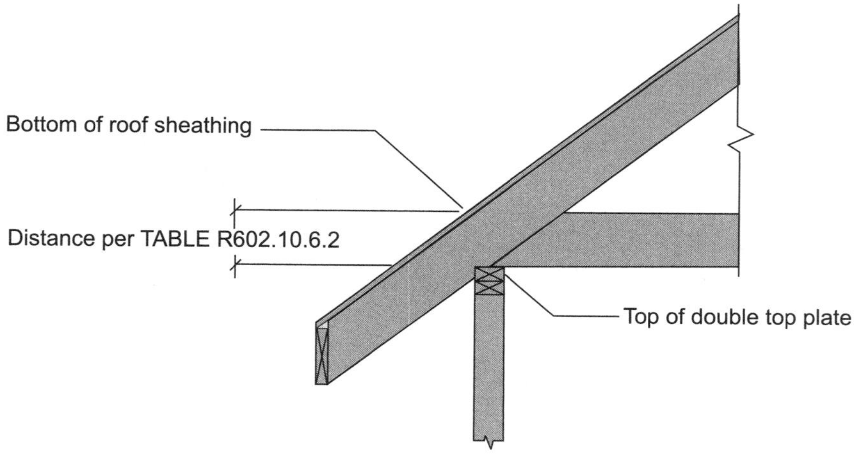

Figure R602.10.8.2
DISTANCE FROM TOP PLATE TO BOTTOM OF ROOF SHEATHING FOR PROVIDING CONNECTION REQUIREMENTS AND BLOCKING PER TABLE R602.10.8.2

the same as the roof sheathing edge nailing, unless the same material is used for both applications.

Figure R602.10.8.2(3) illustrates a connection applicable for raised-heel trusses of up to 4 feet (1219 mm) in height. In addition to transferring the lateral load from the roof sheathing (or diaphragm) to the braced wall line, this method does address lateral stability of the trusses at the bearing point. Ventilation is provided by a gap of up to 2 inches (51 mm) at the top of braced panels between trusses. While the code permits various bracing methods for this application [see Figure R602.10.8(3), Note a], the user must be careful to specify only those methods that have sufficient weather resistance or provide a covering with an approved exterior finish. Closed soffit construction may also be used to protect bracing materials at this location. Note that the edge nailing requirement for the specific bracing method used to construct this option is provided in Table R602.10.4 and is not necessarily the same as the roof sheathing edge nailing, unless the same material is used for both applications.

R602.10.9 Braced wall panel support. *Braced wall panel* support shall be provided as follows:

1. Cantilevered floor joists complying with Section R502.3.3 shall be permitted to support *braced wall panels*.
2. Elevated post or pier foundations supporting *braced wall panels* shall be designed in accordance with accepted engineering practice.
3. Masonry stem walls with a length of 48 inches (1219 mm) or less supporting *braced wall panels* shall be reinforced in accordance with Figure R602.10.9. Masonry stem walls with a length greater than 48 inches (1219 mm) supporting *braced wall panels* shall be constructed in accordance with Section R403.1 Methods ABW and PFH shall not be permitted to attach to masonry stem walls.
4. Concrete stem walls with a length of 48 inches (1219 mm) or less, greater than 12 inches (305 mm) tall and less than 6 inches (152 mm) thick shall have reinforcement sized and located in accordance with Figure R602.10.9.

❖ Section R602.10.9 contains requirements for several braced wall line special circumstances not addressed in editions of the code prior to 2009. Included are provisions for cantilevered floor joists, elevated post and pier foundations, masonry stem walls and narrow concrete stem walls (added to the 2012 edition).

1. Floor cantilevers. Section R502.3.3 provides the basic cantilever requirements, and Section R602.10.9 further clarifies those requirements for use with braced walls.

   Section R502.3.3 restricts floor cantilevers for general use to the depth of the floor joist. Thus a 2 by 10 floor joist can cantilever out $9^1/_4$ inches (235 mm) for all cases covered by the code without other considerations. Table R502.3.3(1) permits greater length cantilevers if the structure above the cantilever is limited to a single light-frame wall and roof. The length of these cantilevers can be as long as 48 inches (1219 mm) under certain circumstances. Note g of Table R502.3.3(1) requires both the blocking at the support and the continuous full-height rim board at the end of the cantilever, regardless of the basic wind speed or SDC.

2. Elevated post and pier foundations. Elevated post and pier foundations generally provide little lateral support to the structure above. Lateral support instead typically comes from diagonal bracing or other elements of the foundation; however, there are ways of cantilevering the posts of the foundation by burying the ends deeply into the ground, so that they develop lateral-load-resisting characteristics.

   It is difficult to accurately and fully present all of the post and pier lateral support methods in one set of prescriptive provisions. For this reason, post and pier foundations supporting braced wall panels must be engineered in accordance with the IBC or referenced documents, as stated in the Section R602.10.9, Item 2.

3. Masonry stem walls. Past field problems have made it clear that free-standing unreinforced masonry foundations adjacent to garage doors (or similar openings) may not perform well with narrow bracing options. So for the 2009 code, the ICC Ad Hoc Committee on Wall Bracing (AHCWB) and the National Concrete Masonry Association developed Figure R602.10.9, along with provisions for the reinforcement of masonry stem walls that support braced wall panels. The provisions are intended for masonry stem walls adjacent to garage doors or similar openings.

   The reinforcement details in Figure R602.10.9 are appropriate for masonry stem walls that are up to 4 feet (1219 mm) in length and not more than 4 feet (1219 mm) in height.

   - If the masonry stem walls are taller than 4 feet (1219 mm), an engineered design of the reinforcement is required.
   - If the masonry stem walls are longer than 4 feet (1219 mm), this specific reinforcement is not necessary (standard construction in accordance with Section R403.1 is sufficient).
   - Masonry stem walls are not permitted to support Method ABW and PFH bracing unless specifically engineered for such applications.

4. Concrete stem walls. In the 2012 code, the ICC Ad Hoc Committee on Wall Bracing (AHCWB) added narrow concrete stem walls less than 6

inches (152 mm) in thickness to this section. Narrow concrete stem walls less than 48 inches (1219 mm) in length and more than 12 inches (305 mm) tall must be reinforced using Figure R602.10.9 when supporting braced wall panels. The provisions are intended for concrete stem walls adjacent to garage doors or similar openings.

The additional reinforcement is not necessary (standard construction in accordance with Section R403.1 is sufficient) if the concrete stem walls are:

- 6 inches (152 mm) or greater in thickness.
- Less than 12 inches (305 mm) in height.
- Longer than 4 feet (1219 mm).

R602.10.9.1 Braced wall panel support for Seismic Design Category D_2. In one-story buildings located in Seismic Design Category D_2, *braced wall panels* shall be supported on continuous foundations at intervals not exceeding 50 feet (15 240 mm). In two-story buildings located in Seismic Design Category D_2, all *braced wall panels* shall be supported on continuous foundations.

Exception: Two-story buildings shall be permitted to have interior *braced wall panels* supported on continuous foundations at intervals not exceeding 50 feet (15 240 mm) provided that:

1. The height of cripple walls does not exceed 4 feet (1219 mm).
2. First-floor *braced wall panels* are supported on doubled floor joists, continuous blocking or floor beams.
3. The distance between bracing lines does not exceed twice the building width measured parallel to the *braced wall line*.

❖ When a braced wall line is not supported by a continuous foundation, the supporting floor (diaphragm) must distribute the lateral force to a resisting element (foundation). If a floor diaphragm is overloaded, it could lead to a failure, which in turn results in an

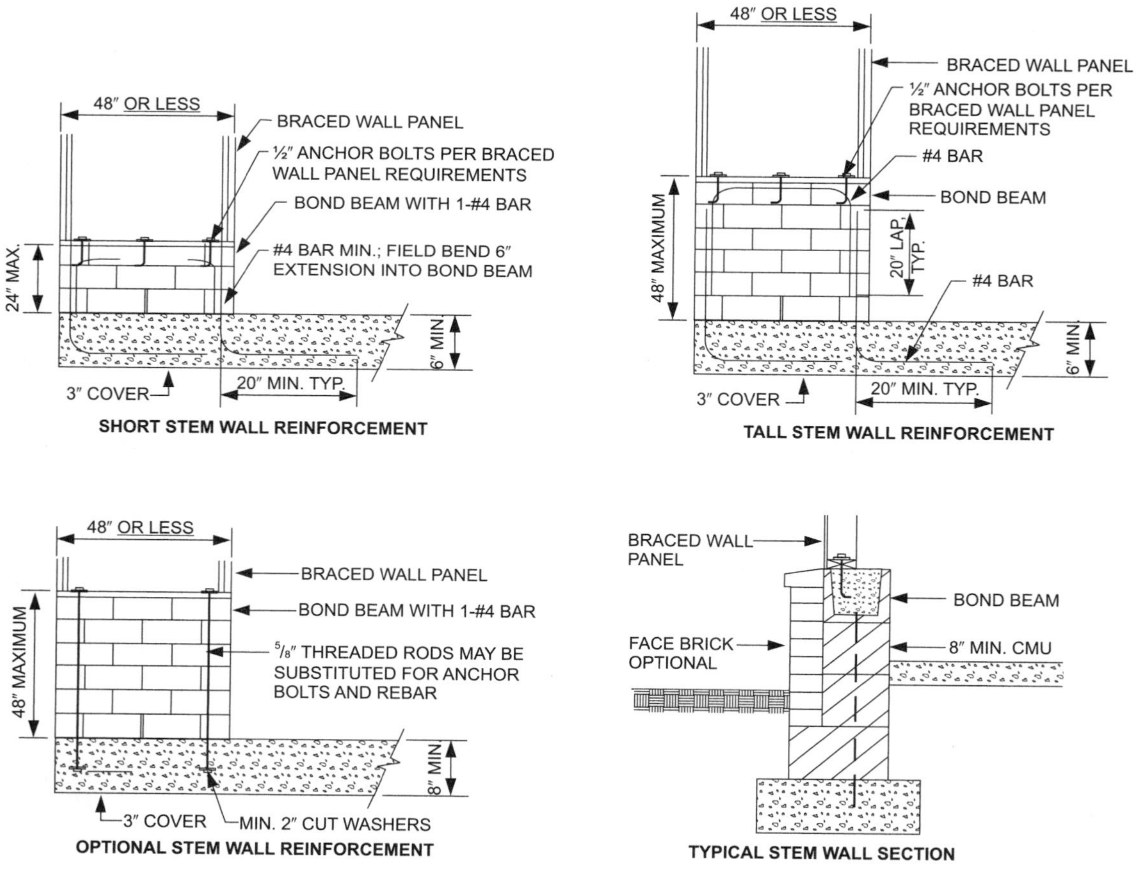

NOTE: GROUT BOND BEAMS AND ALL CELLS WHICH CONTAIN REBAR, THREADED RODS AND ANCHOR BOLTS.

For SI: 1 inch = 25.4 mm.

FIGURE R602.10.9
MASONRY STEM WALLS SUPPORTING BRACED WALL PANELS

❖ See the commentary for Section R602.10.9.

incomplete load path. This section limits this condition for buildings classified as SDC D_2. However, where any of the overall plan dimensions of the building are greater than 50 feet (15 240 mm), Section R403.1.2 supersedes this section (it is more restrictive) and requires that all interior braced wall panels be supported on continuous footings in any building located in SDC D_0, D_1 or D_2. Braced wall panels (as required by Section R602.10.9.1) in structures located in SDC D_2 must be supported on continuous footings except:

- One-story buildings where none of the overall plan dimensions exceed 50 feet (15 240 mm), or
- Two-story buildings where none of the overall plan dimensions exceed 50 feet (15 240 mm) and all three items of the exception to Section R602.10.9.1 are satisfied.

Commentary Figure R602.10.9.1(1) shows the limitations for one- and two-story buildings. Commentary Figure R602.10.9.1(2) shows the exception to the requirement for two-story buildings.

R602.10.10 Panel joints. All vertical joints of panel sheathing shall occur over, and be fastened to, common studs. Horizontal joints in *braced wall panels* shall occur over, and be fastened to, common blocking of a minimum $1^1/_2$ inch (38 mm) thickness.

Exceptions:

1. Vertical joints of panel sheathing shall be permitted to occur over double studs, where adjoining panel edges are attached to separate studs with the required panel edge fastening schedule, and the adjacent studs are attached together with two rows of 10d box nails [3 inches by 0.128 inch (76.2 mm by 3.25 mm)] at 10 inches o.c. (254 mm).
2. Blocking at horizontal joints shall not be required in wall segments that are not counted as *braced wall panels*.
3. Where the bracing length provided is at least twice the minimum length required by Table R602.10.3(1) and Table R602.10.3(3) blocking at horizontal joints shall not be required in *braced wall panels* constructed using Methods WSP, SFB, GB, PBS or HPS.
4. When Method GB panels are installed horizontally, blocking of horizontal joints is not required.

❖ Section R602.10.10 requires that all vertical and horizontal joints in panel sheathing used for bracing occur over and be attached to common framing. This is to ensure that the bracing performs as intended when subjected to lateral loads.

The requirement for vertical panel edges to be fastened to a common stud has an exception:

- The first exception in the code provision allows vertical abutting panel sheathing placed over double studs to have each panel attached to separate studs when the studs are attached with two rows of 10d box nails spaced a maximum of 10 inches (254 mm) on center.

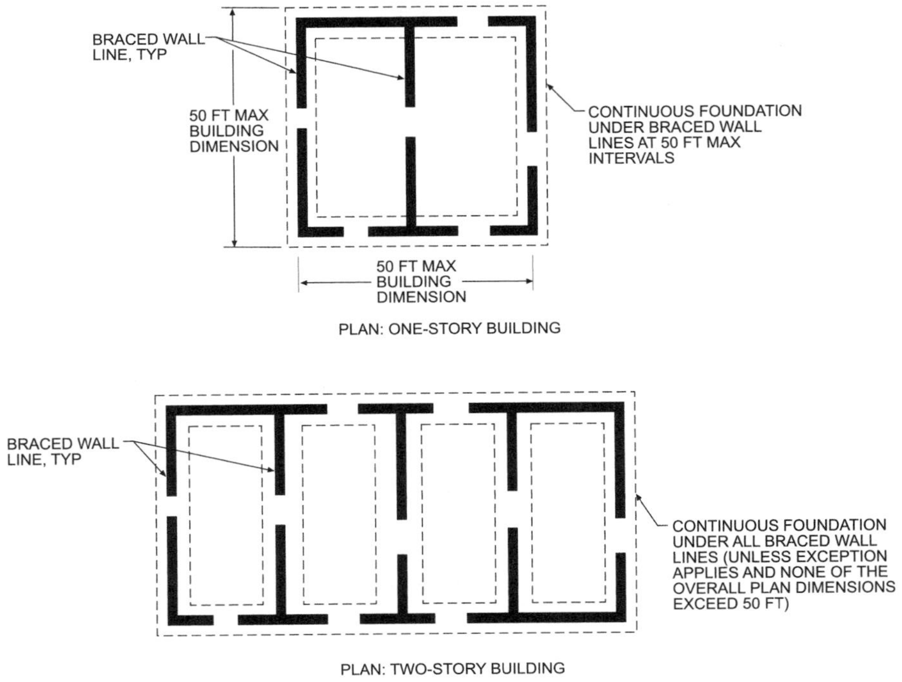

For SI: 1 foot = 304.8 mm.

Figure R602.10.9.1(1)
SEISMIC DESIGN CATEGORY D_2 INTERIOR BRACED WALL SUPPORT

The requirement for horizontal joints to be blocked when they occur between the top and bottom plates has some exceptions:

- The second exception states that the horizontal joints of sheathing panels not used as braced panels do not have to be blocked. However, panel manufacturers or panel associations may recommend blocking the horizontal joints of sheathing panels regardless of whether they are used as braced panels. For example, brittle finishes (such as stucco) over panel wall sheathing may require blocked horizontal panel edges to prevent cracking, even when the panels are not used for wall bracing.
- When horizontal joints of braced wall panels are not blocked, the effectiveness of the bracing is reduced and additional bracing is required. Consequently, for Methods WSP, SFB, GB, PBS and HPS, horizontal blocking of the braced wall panels may be eliminated, provided that twice the minimum required bracing length is used in the wall line. The purpose of this exception is to provide a construction alternative to installing horizontal blocking when circumstances permit.
- The required bracing length for Method GB is based on its vertical application with horizontal joints occurring over framing or blocking. Method GB bracing is considerably stronger when installed horizontally. The fourth exception permits Method GB bracing to be unblocked when it is installed horizontally.

R602.10.11 Cripple wall bracing. Cripple walls shall be constructed in accordance with Section R602.9 and braced in accordance with this section. Cripple walls shall be braced with the length and method of bracing used for the wall above in accordance with Tables R602.10.3(1) and R602.10.3(3), and the applicable adjustment factors in Table R602.10.3(2) or R602.10.3(4), respectivley, except that the length of cripple wall bracing shall be multiplied by a factor of 1.15. The distance between adjacent edges of *braced wall panels* shall be reduced from 20 feet (6096 mm) to 14 feet (4267 mm).

❖ A cripple wall is a less-than-full-height wall that is used to raise the elevation of a floor above the foundation.

Cripple walls are also often used in conjunction with a stepped foundation to maintain a common plate elevation when the foundation drops away from the plate line, accommodating a sloped building site, as shown in Figure R602.11.2. In either case, a cripple wall has the same limitations as any other stud wall in that it has no lateral load capacity without bracing.

Even though they are less than full height, cripple walls are still a part of the load path and are subject to the same vertical and horizontal loads as full-height walls. In fact, because gravity and lateral loads get

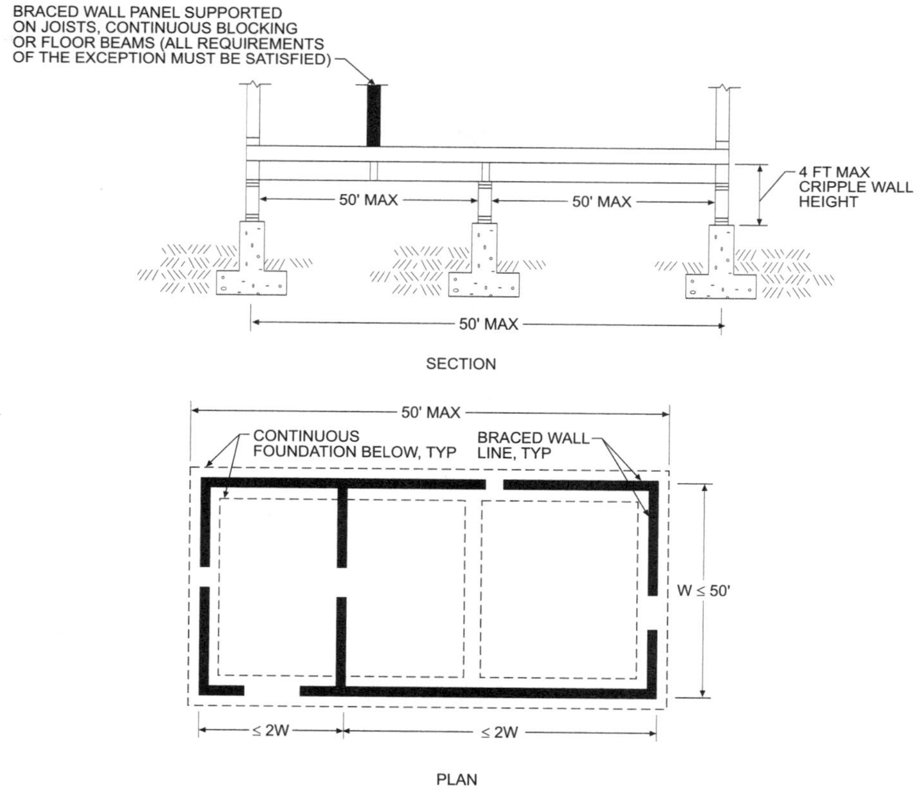

For SI: 1 foot = 304.8 mm.

Figure R602.10.9.1(2)
SEISMIC DESIGN CATEGORY D_2 INTERIOR BRACED WALL SUPPORT

larger as building height increases, cripple walls are subject to even greater loads than the walls in stories above them. Because of these greater loads, the code permits cripple walls to be braced using any method in Tables R602.10.3(1) and (3) in any SDC except D_2, as long as the length of bracing used for the cripple wall is 15 percent greater (multiplied by a factor of 1.15) and the maximum braced wall panel spacing in the cripple wall line is reduced from 20 to 14 feet (6096 to 4267 mm).

R602.10.11.1 Cripple wall bracing for Seismic Design Categories D_0 and D_1 and townhouses in Seismic Design Category C. In addition to the requirements in Section R602.10.11, the distance between adjacent edges of *braced wall panels* for cripple walls along a *braced wall line* shall be 14 feet (4267 mm) maximum.

Where *braced wall lines* at interior walls are not supported on a continuous foundation below, the adjacent parallel cripple walls, where provided, shall be braced with Method WSP or Method CS-WSP in accordance with Section R602.10.4. The length of bracing required in accordance with Table R602.10.3(3) for the cripple walls shall be multiplied by 1.5. Where the cripple walls do not have sufficient length to provide the required bracing, the spacing of panel edge fasteners shall be reduced to 4 inches (102 mm) on center and the required bracing length adjusted by 0.7. If the required length can still not be provided, the cripple wall shall be designed in accordance with accepted engineering practice.

❖ In SDC D_0, D_1 and townhouses in SDC C, Section R602.10.11.1 requires increased exterior cripple wall bracing when interior braced walls are not supported by a continuous foundation. (Note that this requirement is in addition to the requirements of Section R602.10.11, discussed above.) Without direct interior foundation support, these interior braced wall lines transfer the lateral forces into the floor diaphragm instead of the foundation. The floor diaphragm, in turn, transfers these lateral forces into the exterior cripple walls. Therefore, the wall bracing at the exterior cripple walls [specified in Table R602.10.3(3)] must be increased to accommodate the increased load. This is done by multiplying the length of the required exterior cripple wall bracing parallel to the unsupported interior braced wall by a factor of 1.5 and using Method WSP or CS-WSP.

If the length of wall bracing cannot be increased by a factor of 1.5 because the wall line is not long enough to accommodate the increased bracing length, the nail spacing along the perimeter wood structural panel sheathing may be reduced from 6 inches (152 mm) to 4 inches (102 mm) on center, in accordance with Section R602.10.11.1. Note that in this case it is not necessary to also increase the length of bracing by the 1.5 factor; the provision recognizes that the shear capacity of wood structural panels (Method WSP or CS-WSP) is increased sufficiently by increasing the perimeter nailing.

If there is insufficient wall line length for the required cripple wall bracing, an engineered design or other alternative bracing approved by the building official must be incorporated.

R602.10.11.2 Cripple wall bracing for Seismic Design Category D_2. In Seismic Design Category D_2, cripple walls shall be braced in accordance with Tables R602.10.3(3) and R602.10.3(4).

❖ Cripple wall bracing for SDC D_2 is listed in Table R602.10.3(3). Appropriate adjustment factors from Table R602.10.3(4) must also be applied. In the 2009 edition, cripple wall bracing in SDC D_2 was deleted from the seismic bracing requirements table. It has been added back into the table in the 2012 edition. Story height is limited to two stories with a cripple wall below.

R602.10.11.3 Redesignation of cripple walls. Where all cripple wall segments along a *braced wall line* do not exceed 48 inches (1219 mm) in height, the cripple walls shall be permitted to be redesignated as a first-*story* wall for purposes of determining wall bracing requirements. Where any cripple wall segment in a *braced wall line* exceeds 48 inches (1219 mm) in height, the entire cripple wall shall be counted as an additional *story*. If the cripple walls are redesignated, the stories above the redesignated *story* shall be counted as the second and third stories, respectively.

❖ In all wind or seismic applications, the building designer is permitted to redesignate or redefine the cripple wall as a story, when all segments are a maximum of 48 inches (1219 mm) tall, and then use the bracing requirements of Table R602.10.3(1) or (3) without the cripple wall adjustments required by Sections R602.10.11 and R602.10.11.1. When cripple wall segments are greater than 48 inches (1219 mm) tall, the entire cripple wall must be counted as an additional story.

When redesignation is used, the cripple walls are considered the first story, while the first full-height story becomes the second story, etc. Redesignation is optional ("shall be permitted" in code language) and the builder in any SDC may choose to exercise this option.

Note that in SDC D_2, Table R602.10.3(3) only permits buildings of up to two stories, thus limiting the redesignation option to one-story buildings. With redesignation, a one story becomes a two-story, which is the maximum number of stories permitted in SDC D_2. If a two-story structure is located above a redesignated cripple wall foundation, the second story would become the third story. A three-story structure in SDC D_2 is required to be designed in accordance with the IBC.

R602.11 Wall anchorage. *Braced wall line* sills shall be anchored to concrete or masonry foundations in accordance with Sections R403.1.6 and R602.11.1.

❖ When braced wall panels are supported directly by foundations (i.e., turned-down slab edge, thickened slab, masonry or concrete foundation wall), the wood sill plate must be anchored to the foundation with anchor bolts spaced a maximum of 6 feet (1829 mm) on center. There must be a minimum of two bolts per sill plate section, with a bolt located not more than 12 inches (305 mm) and not less than 7 bolt diameters (3.5 inches) (89 mm) from each end of the plate section (see Section R403.1.6). Bolts should be at least $^1/_2$ inch (13 mm) in diameter and should extend a minimum of 7 inches (178 mm) into the concrete or masonry foundation. A nut and washer are required on each bolt to hold the plate to the foundation.

Section R602.11 specifies how braced wall line sill plates are to be anchored to the foundation. Section R403.1.6 specifies the minimum number, size and spacing of anchor bolts.

R602.11.1 Wall anchorage for all buildings in Seismic Design Categories D_0, D_1 and D_2 and townhouses in Seismic Design Category C. Plate washers, a minimum of 0.229 inch by 3 inches by 3 inches (5.8 mm by 76 mm by 76 mm) in size, shall be provided between the foundation sill plate and the nut except where *approved* anchor straps are used. The hole in the plate washer is permitted to be diagonally slotted with a width of up to $^3/_{16}$ inch (5 mm) larger than the bolt diameter and a slot length not to exceed $1^3/_4$ inches (44 mm), provided a standard cut washer is placed between the plate washer and the nut.

❖ For one- and two-family dwellings in SDC A, B and C and townhouses in SDC A and B, braced wall lines are to be connected to the foundation with $^1/_2$-inch (12.7 mm) bolts using a nut and washer. For higher seismic requirements (townhouses in SDC C and all structures in SDC D_0, D_1 and D_2), braced wall lines are to be connected to the foundation with $^1/_2$-inch (12.7 mm) bolts using a 3-inch (76 mm) square plate washer and nut, in accordance with Section R602.11.1. The plate washer is permitted to have a slotted hole, which helps to locate the washer over the sill plate. If the plate washer has a slotted hole, a standard cut washer must be used between the plate washer and the nut. The slotted hole is permitted to be $^3/_{16}$ inch (5 mm) wider than the bolt diameter, and the slot length is limited to a maximum of $1^3/_4$ inches (44 mm).

Section R602.11.1 also permits the use of "approved anchor straps" in lieu of anchor bolts, washers and nuts.

The following provisions are required in Section R602.11.1, in addition to the requirements for SDC A, B and C as previously described:

- For all buildings in SDC D_0, D_1 and D_2 and townhouses in SDC C, the plate washers are required to be 0.229 inch by 3 inches by 3 inches (5.8 mm by 76 mm by 76 mm) between the nut and the sill plate except where approved anchor straps are used.
- The use of a diagonally slotted plate washer is permitted. The slot width is required to be equal to or less than $^3/_{16}$ inch (5 mm) larger than the bolt diameter and not more than $1^3/_4$ inches (44 mm) long. When a slotted plate washer is used, a standard cut washer must be used between the nut and plate washer.

R602.11.2 Stepped foundations in Seismic Design Categories D_0, D_1 and D_2. In all buildings located in Seismic Design Categories D_0, D_1 or D_2, where the height of a required *braced wall line* that extends from foundation to floor above varies more than 4 feet (1219 mm), the *braced wall line* shall be constructed in accordance with the following:

1. Where the lowest floor framing rests directly on a sill bolted to a foundation not less than 8 feet (2440 mm) in length along a line of bracing, the line shall be considered as braced. The double plate of the cripple stud wall beyond the segment of footing that extends to the lowest framed floor shall be spliced by extending the upper top plate a minimum of 4 feet (1219 mm) along the foundation. Anchor bolts shall be located a maximum of 1 foot and 3 feet (305 and 914 mm) from the step in the foundation. See Figure R602.11.2.
2. Where cripple walls occur between the top of the foundation and the lowest floor framing, the bracing requirements of Sections R602.10.11, R602.10.11.1 and R602.10.11.2 shall apply.
3. Where only the bottom of the foundation is stepped and the lowest floor framing rests directly on a sill bolted to the foundations, the requirements of Sections R403.1.6 and R602.11.1 shall apply.

❖ When a cripple wall in SDC, D_0, D_1 and D_2 is used in a stepped foundation with a vertical step exceeding 4 feet (1219 mm), the cripple wall must be constructed as follows:

Section R602.11.2, Item 1, contains multiple requirements:

- If 8 feet (2438 mm) or more of the lowest floor is anchored directly to the foundation (the cripple wall does not run under this 8-foot (2438 mm) portion of the wall line) as shown in Commentary Figure R602.11.2(1), the foundation provides complete bracing for the cripple wall, as long as the wall above is not longer than 25 feet (7620 mm). In order to gain this benefit, the cripple wall top plate must extend over the foundation at least 4 feet (1219 mm) and be anchored with at least two bolts, as shown in Commentary Figure R602.11.2(1). Note that the first story wall must be braced in accordance with the requirements of Tables R602.10.3(1) and (3). The intent of the code provision is that, if the first story wall is anchored directly to the foundation for at least 8 feet (2438 mm) and the cripple wall top plate is extended and spliced as directed, then the first story braced wall line is sufficiently attached to the foundation to complete the load path and no additional bracing is required.

  As shown in Figure R602.11.2, a metal tie can be used as an alternative to the cripple wall top plate's 4-foot (1219 mm) overlap of the foundation, if distance "A" is greater than 8 feet (2438 mm). No additional wall bracing is required in the cripple wall if the splice is properly made.

When the distance "A" is less than 8 feet (2438 mm) in a 25-foot-long (7620 mm) wall, there is insufficient attachment directly into the foundation to complete the lateral load path for the first-story braced wall line. As such, the total required length of bracing must be applied to the cripple wall end of the wall line for it to be considered braced. In this event, neither the metal splice plate nor the overlapping upper top plate is required, although either would be considered good practice.

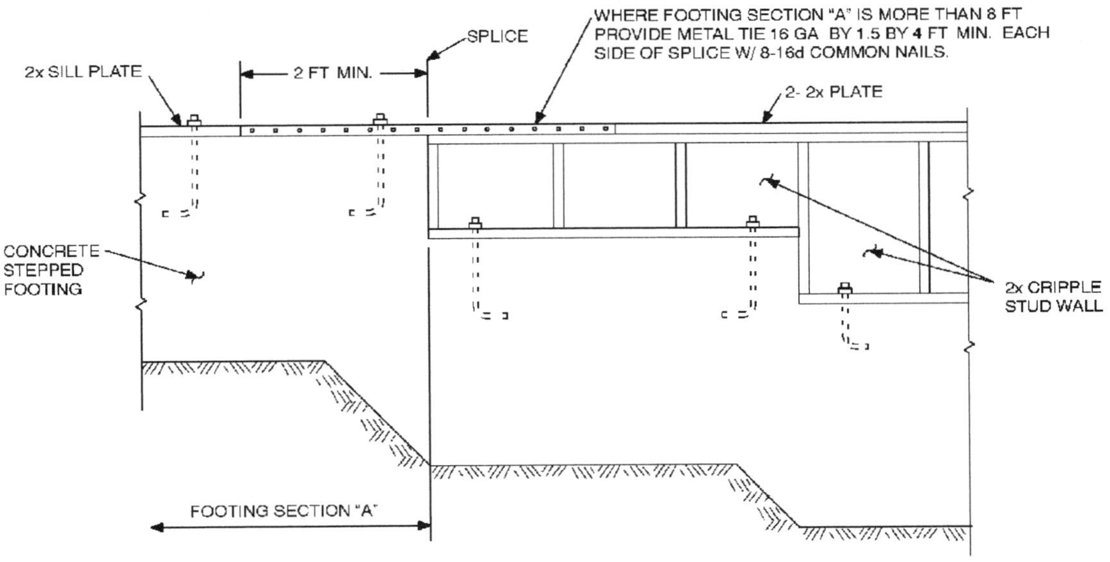

For SI: 1 inch = 25.4 mm, 1 foot = 304.8 mm.

Note: Where footing Section "A" is less than 8 feet long in a 25-foot-long wall, install bracing at cripple stud wall.

FIGURE R602.11.2
STEPPED FOUNDATION CONSTRUCTION

❖ See the commentary for Section R602.11.2.

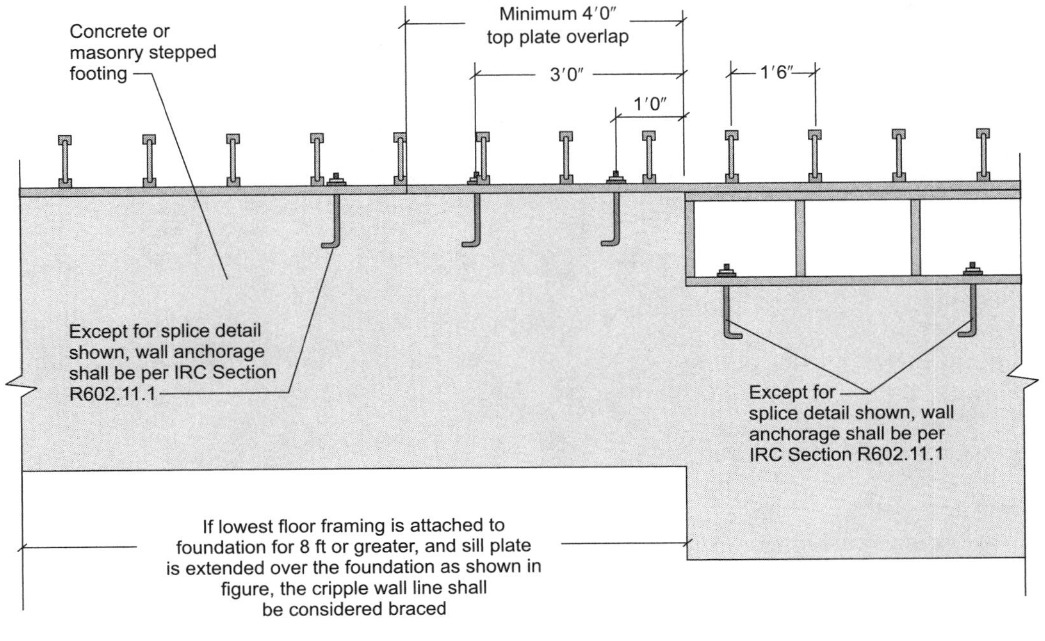

For SI: 1 inch = 25.4 mm, 1 foot = 304.8 mm.

Figure R602.11.2(1)
CRIPPLE WALL TOP PLATE EXTENDED 4 FEET OVER THE FOUNDATION

R602.12 Simplified wall bracing. Buildings meeting all of the conditions listed in items 1-8 shall be permitted to be braced in accordance with this section as an alternative to the requirements of Section R602.10. The entire building shall be braced in accordance with this section; the use of other bracing provisions of R602.10, except as specified herein, shall not be permitted.

1. There shall be no more than two stories above the top of a concrete or masonry foundation or basement wall. Permanent wood foundations shall not be permitted.
2. Floors shall not cantilever more than 24 inches (607 mm) beyond the foundation or bearing wall below.
3. Wall height shall not be greater than 10 feet (2743 mm).
4. The building shall have a roof eave-to-ridge height of 15 feet (4572 mm) or less.
5. All exterior walls shall have gypsum board with a minimum thickness of $^1/_2$ inch (12.7 mm) installed on the interior side fastened in accordance with Table R702.3.5.
6. The structure shall be located where the basic wind speed is less than or equal to 90 mph (40 m/s), and the Exposure Category is A or B.
7. The structure shall be located in Seismic Design Category A, B or C for detached one- and two-family dwellings or Seismic Design Category A or B for townhouses.
8. Cripple walls shall not be permitted in two-story buildings.

❖ Section R602.12 provides a prescriptive procedure for bracing wall lines in detached one- or two-family dwellings and townhouses with the following limits:

- Seismic Design Category A, B, or C for detached one- or two-family dwellings, SDC A or B for townhouses
- Basic wind speeds of 90 mph (40 m/s) or less with Wind Exposure Category A or B
- One- or two-story structure
- Wood structural panel or structural fiberboard sheathing is used on exterior walls with gypsum board fastened to the interior side.

When the requirements above are met, using the simplified method removes the need to check provisions for high seismic areas, to determine braced wall lines or braced wall line spacing.

A braced wall panel and its minimum length is a "bracing unit." The minimum number of bracing units is determined by drawing a rectangle around the building and then using the rectangle's dimensions to select the total bracing from Table R602.12.4.

Bracing units are required to be placed per the distribution requirements in Section R602.12.4 rather than along braced wall lines. When homes do not qualify or if the builder prefers, the traditional approach in Section R602.10 must be used.

See Section R602.10 for an explanation of lateral loads and the purpose of wall bracing. Failure of a braced wall line is evidenced by racking of the wall line. Racking occurs when a rectangular wall deforms to a parallelogram shape, in which the top and the bottom of the wall remain horizontal but the sides are no longer vertical. The purpose of wall bracing is to prevent such failures.

R602.12.1 Circumscribed rectangle. The bracing required for each building shall be determined by circumscribing a rectangle around the entire building on each floor as shown in Figure R602.12.1. The rectangle shall surround all enclosed offsets and projections such as sunrooms and attached

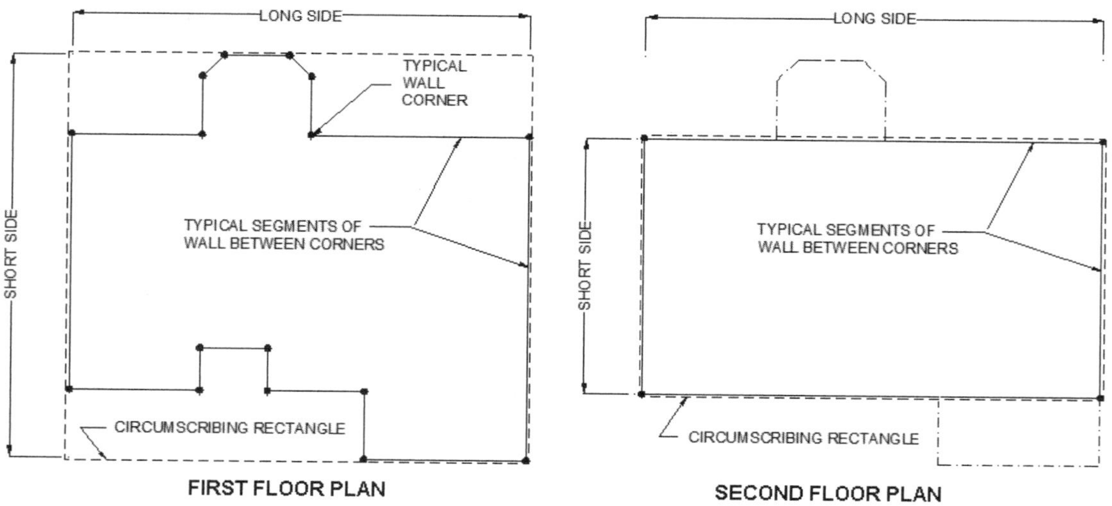

FIGURE R602.12.1
RECTANGLE CIRCUMSCRIBING AN ENCLOSED BUILDING

❖ See the commentary for Section R602.12.1.

garages. Open structures, such as carports and decks, shall be permitted to be excluded. The rectangle shall have no side greater than 60 feet (18 288 mm), and the ratio between the long side and short side shall be a maximum of 3:1.

❖ Required bracing length in the simplified wall bracing method is determined by drawing a rectangle around the building. Once the rectangle is drawn, all measurements for bracing are done down the length of the sides of the rectangle, not the length of the walls of the building.

Placement of braced wall lines is not necessary for this procedure. Rather, the drawing of the rectangle determines an equivalent "braced wall line." Note, this procedure is conservative as it assumes a rectangle side length equal to or greater than the building wall line length.

Once the rectangle is drawn, the longer side will be referred to in the provisions as the "long side." The shorter side will be referred to as the "short side." If the rectangle's sides are equal in length, assign one side the label "short side" and the perpendicular walls the label "long side."

R602.12.2 Sheathing materials. The following sheathing materials installed on the exterior side of exterior walls shall be used to construct a bracing unit as defined in Section R602.12.3. Mixing materials is prohibited.

1. Wood structural panels with a minimum thickness of $^3/_8$ inch (9.5 mm) fastened in accordance with Table R602.3(3).
2. Structural fiberboard sheathing with a minimum thickness of $^1/_2$ inch (12.7 mm) fastened in accordance with Table R602.3(1).

❖ For the simplified wall bracing procedure, only wood structural panels and structural fiberboard sheathing may be used to brace the exterior walls. All exterior walls must have minimum $^1/_2$-inch (13 mm) gypsum board on the interior side of the walls fastened in accordance with Table R702.3.5.

When using wood structural panels, narrow bracing panels may be substituted for bracing units as permitted by Section R602.12.6.

R602.12.3 Bracing unit. A bracing unit shall be a full-height sheathed segment of the exterior wall with no openings or vertical or horizontal offsets and a minimum length as specified herein. Interior walls shall not contribute toward the amount of required bracing. Mixing of Items 1 and 2 is prohibited on the same story.

1. Where all framed portions of all exterior walls are sheathed in accordance with Section R602.12.2, including wall areas between bracing units, above and below openings and on gable end walls, the minimum length of a bracing unit shall be 3 feet (914 mm).
2. Where the exterior walls are braced with sheathing panels in accordance with Section R602.12.2 and areas between bracing units are covered with other materials, the minimum length of a bracing unit shall be 4 feet (1219 mm).

❖ Bracing units are full-height sheathed wall segments similar to braced wall panels from Section R602.10. When buildings are continuously sheathed—sheathed over all wall areas—the minimum length of a braced unit is 3 feet (914 mm). When the bracing units are intermittent—placed along wall sections with other materials sheathing in between the bracing units—the minimum length of a braced unit is 4 feet (1219 mm).

Continuous and intermittent bracing units may be mixed from one story to another. For example, there may be intermittent sheathing on the top story and continuous sheathing on the bottom story of a building. Bracing unit methods may not be mixed within one story.

Mixing of wood structural panel and structural fiberboard sheathing in one building is not permitted in accordance with Section R602.12.2.

Commentary Table R602.12.3 contains the two materials available and the requirements for each material and bracing unit length.

R602.12.3.1 Multiple bracing units. Segments of wall compliant with Section R602.12.3 and longer than the minimum bracing unit length shall be considered as multiple bracing units. The number of bracing units shall be determined by dividing the wall segment length by the minimum bracing unit length. Full-height sheathed segments of wall narrower than the minimum bracing unit length shall not contribute toward a bracing unit except as specified in Section R602.12.6.

❖ When bracing units are longer than the minimum required length, the total length of bracing along one side of a rectangle may be added and divided by the minimum required bracing length to calculate the number of braced units on one side of the rectangle.

Table R602.12.3
AVAILABLE BRACING METHODS FOR SIMPLIFIED WALL BRACING

MATERIAL	STUD SPACING AND FASTENER CRITERIA	FASTENER SPACING CRITERIA	BRACING UNIT METHOD	MINIMUM BRACING UNIT LENGTH (feet)
Wood structural panel	Table R602.3(3)	6" edge, 12" field	Continuous	3
			Intermittent	4
Structural fiberboard sheathing	Maximum 16" spacing Table R602.3(1)	3" edge, 6" field	Continuous	3
			Intermittent	4

For example, a building has been intermittently braced with wood structural panel. There are three sections of bracing, 8, 9, and 6 feet (2438, 3048, and 1829 mm) long. Divide each wall segment length by 4 feet (1219 mm) (the minimum bracing unit length) and add the total number of bracing units together.

8 ft/4 ft + 9 ft/4 ft + 6 ft/4 ft =
2 + 2.25 + 1.5 = 5.75 full bracing units

The wall line has an equivalent 5.75 bracing units. Use Section R602.12.4 to determine if that is a sufficient number of bracing units for the side of the rectangle.

R602.12.4 Number of bracing units. Each side of the circumscribed rectangle, as shown in Figure R602.12.1, shall have, at a minimum, the number of bracing units in accordance with Table R602.12.4 placed on the parallel exterior walls facing the side of the rectangle. Bracing units shall then be placed using the distribution requirements specified in Section R602.12.5.

❖ Section R602.12.4 defines the required number of bracing units for each side of the rectangle drawn around the building. Note the minimum number of bracing units for one side of the rectangle is based on the length of the perpendicular side of the rectangle. For example, to determine the number of bracing units required on the long sides of the rectangle, use the length of the short side of the rectangle.

Interpolation between rectangle side lengths is not permitted. Take the length of one side of the rectangle and round it up to the nearest 10 feet (3048 mm) of length. For example, if the side of a rectangle is 23 feet (7010 mm), use the 30-foot (9144 mm) column in Table R602.12.4 to find the required number of bracing units.

Cripple walls are assumed to be the first story when using Table R602.12.4. Two-story buildings above a cripple wall are not permitted. If the building has two stories and a cripple wall, use Section R602.10 to determine wall bracing requirements.

R602.12.5 Distribution of bracing units. The placement of bracing units on exterior walls shall meet all of the following requirements as shown in Figure R602.12.5.

1. A bracing unit shall begin no more than 12 feet (3658 mm) from any wall corner.
2. The distance between adjacent edges of bracing units shall be no greater than 20 feet (6096 mm).
3. Segments of wall greater than 8 feet (2438 mm) in length shall have a minimum of one bracing unit.

❖ Bracing units shall be placed so that the first unit's edge begins within 12 feet (3658 mm) of each end of a wall line. The bracing unit edges shall be a maximum of 20 feet (6096 mm) apart. All segments of wall longer than 8 feet (2438 mm) shall have one or more bracing units.

R602.12.6 Narrow panels. The bracing methods referenced in Section R602.10 and specified in Sections R602.12.6.1 through R602.12.6.3 shall be permitted when using simplified wall bracing.

❖ See Section R602.10.6 for commentary on narrow braced wall panels.

R602.12.6.1 Method CS-G. *Braced wall panels* constructed as Method CS-G in accordance with Tables R602.10.4 and R602.10.5 shall be permitted for one-story garages when all framed portions of all exterior walls are sheathed with wood structural panels. Each CS-G panel shall be equivalent to 0.5 of a bracing unit. Segments of wall which include a Method CS-G panel shall meet the requirements of Section R602.10.4.2.

❖ See Section R602.10.4 for commentary on Method CS-G.

R602.12.6.2 Method CS-PF. *Braced wall panels* constructed as Method CS-PF in accordance with Section R602.10.6.4 shall be permitted when all framed portions of all exterior walls are sheathed with wood structural panels. Each CS-PF panel shall equal 0.5 bracing units. A maximum of four CS-PF panels shall be permitted on all segments of walls parallel to each side of the circumscribed rectangle. Segments of wall which include a Method CS-PF panel shall meet the requirements of Section R602.10.4.2.

❖ See Sections R602.10.4 and R602.10.6.4 for commentary on Method CS-PF.

R602.12.6.3 Methods PFH and PFG. *Braced wall panels* constructed as Method PFH and PFG shall be permitted when bracing units are constructed using wood structural panels. Each PFH panel shall equal one bracing unit and each PFG panel shall be equal to 0.75 bracing units.

❖ See Sections R602.10.4 and R602.10.6.2 for commentary on Method PFH. See Sections R602.10.4 and R602.10.6.3 for commentary on Method PFG.

R602.12.7 Lateral support. For bracing units located along the eaves, the vertical distance from the outside edge of the top wall plate to the roof sheathing above shall not exceed 9.25 inches (235 mm) at the location of a bracing unit unless lateral support is provided in accordance with Section R602.10.8.2.

❖ See Section R602.10.8.2 for commentary on required lateral support above braced wall panels and bracing units at the roof.

R602.12.8 Stem walls. Masonry stem walls with a height and length of 48 inches (1219 mm) or less supporting a bracing unit or a Method CS-G, CS-PF or PFG *braced wall panel* shall be constructed in accordance with Figure R602.10.9. Concrete stem walls with a length of 48 inches (1219 mm) or less, greater than 12 inches (305 mm) tall and less than 6 inches (152 mm) thick shall be reinforced sized and located in accordance with Figure R602.10.9.

❖ See Section R602.10.9, Items 3 and 4 for commentary on required reinforcement of stem walls less than 48 inches (1219 mm) in length. Method PFH is not permitted on masonry stem walls.

TABLE R602.12.4
MINIMUM NUMBER OF BRACING UNITS ON EACH SIDE OF THE CIRCUMSCRIBED RECTANGLE

STORY LEVEL	EAVE-TO-RIDGE HEIGHT (feet)	MINIMUM NUMBER OF BRACING UNITS ON EACH LONG SIDE[a, b] Length of short side (feet)[c]						MINIMUM NUMBER OF BRACING UNITS ON EACH SHORT SIDE[a, b] Length of long side (feet)[c]					
		10	20	30	40	50	60	10	20	30	40	50	60
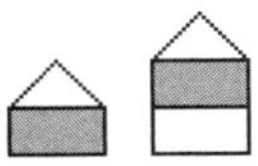	10	1	2	2	2	3	3	1	2	2	2	3	3
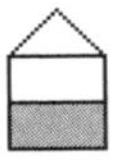	10	2	3	3	4	5	6	2	3	3	4	5	6
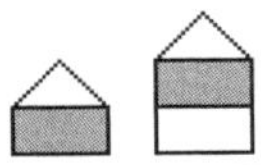	15	1	2	3	3	4	4	1	2	3	3	4	4
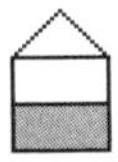	15	2	3	4	5	6	7	2	3	4	5	6	7

For SI: 1 inch = 25.4 mm, 1 foot = 304.8 mm.
a. Interpolation shall not be permitted.
b. Cripple walls or wood-framed basement walls in a walk-out condition of a one-story structure shall be designed as the first floor of a two-story house.
c. Actual lengths of the sides of the circumscribed rectangle shall be rounded to the next highest unit of 10 when using this table.

❖ See the commentary for Section R602.12.4.

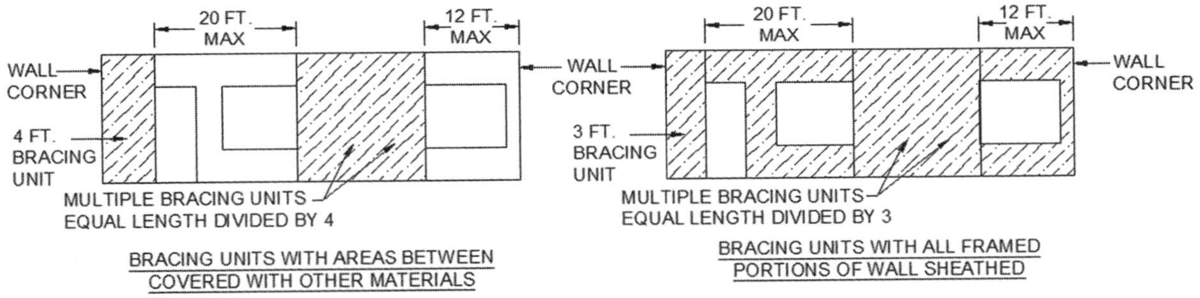

For SI: 1 foot = 304.8 mm.

FIGURE R602.12.5
BRACING UNIT DISTRIBUTION

❖ See the commentary for Section R602.12.5.

SECTION R603
STEEL WALL FRAMING

R603.1 General. Elements shall be straight and free of any defects that would significantly affect structural performance. Cold-formed steel wall framing members shall comply with the requirements of this section.

❖ The provisions of this section apply to the construction of load-bearing wall systems using cold-formed steel framing, which is that type of construction made up in part or entirely of steel structural members cold-formed to shape from sheet or strip steel. The framing members are generally in the form of the letter "C" (C-shape), which is described in further detail in Section R603.2. The use of cold-formed steel-framed designs not consistent with these provisions is beyond the scope of the code. In such cases, an analysis and design by a design professional is required.

R603.1.1 Applicability limits. The provisions of this section shall control the construction of exterior cold-formed steel wall framing and interior load-bearing cold-formed steel wall framing for buildings not more than 60 feet (18 288 mm) long perpendicular to the joist or truss span, not more than 40 feet (12 192 mm) wide parallel to the joist or truss span, and less than or equal to three stories above *grade plane*. All exterior walls installed in accordance with the provisions of this section shall be considered as load-bearing walls. Cold-formed steel walls constructed in accordance with the provisions of this section shall be limited to sites subjected to a maximum design wind speed of 110 miles per hour (49 m/s) Exposure B or C and a maximum ground snow load of 70 pounds per square foot (3.35 kPa).

❖ See the commentary for Section R505.1.1.

R603.1.2 In-line framing. Load-bearing cold-formed steel studs constructed in accordance with Section R603 shall be located in-line with joists, trusses and rafters in accordance with Figure R603.1.2 and the tolerances specified as follows:

1. The maximum tolerance shall be $^3/_4$ inch (19 mm) between the centerline of the horizontal framing member and the centerline of the vertical framing member.
2. Where the centerline of the horizontal framing member and bearing stiffener are located to one side of the centerline of the vertical framing member, the maximum tolerance shall be $^1/_8$ inch (3 mm) between the web of the horizontal framing member and the edge of the vertical framing member.

❖ In-line framing is the preferred framing method. The advantage is that it provides a direct load path for the transfer of forces from joists to studs. The $^3/_4$-inch (19 mm) tolerance in Item 1 is accepted industry practice, and addresses those conditions where two framing members (roof rafter and ceiling joist, for example) are to be aligned over a wall stud. The case of a bearing stiffener located on the back side of the horizontal member where the centerlines of both are to one side of the stud centerline must meet the additional limit in Item 2, as Figure R603.1.2 illustrates.

In-line framing maximizes framing alignment to minimize secondary moments on the framing members, taking into account that the prescriptive track requirements do not consider it to be a load transfer member. In the absence of in-line framing within these tolerances, a load distribution member, such as a structural track, may be required for this force transfer.

R603.2 Structural framing. Load-bearing cold-formed steel wall framing members shall comply with Figure R603.2(1) and with the dimensional and minimum thickness requirements specified in Tables R603.2(1) and R603.2(2). Tracks shall comply with Figure R603.2(2) and shall have a minimum flange width of $1^1/_4$ inches (32 mm).

❖ Stud framing must comply with the dimensional and minimum thickness requirements specified in Tables R603.2(1) and R603.2(2), respectively. Note a to Table R603.2(1) explains the meaning of the alphanumeric member designation used for all steel framing members in the code. This system replaces the varied designation approaches that had been produced by each individual manufacturer. In addition, the designation is used to identify not only a specific steel framing member but also the section properties of that same member through the use of the product technical information document.

Steel thickness is expressed in mils, which pertains to the base metal thickness measured prior to painting or the application of metallic (corrosion resistant) coatings. The base metal thickness is typically stamped or embossed on the member by the manufacturer. Table R603.2(2) also references the equivalent thickness in inches. The use of gage numbers to define the thickness (a specific mean decimal thickness for which rolling tolerances were established) of steel sheet is a practice that has been discontinued by the steel industry.

Steel track sections are to serve in nonload-bearing applications only and are shown in Figure R603.2(2). They must have a minimum flange width of $1^1/_4$ inches (32 mm).

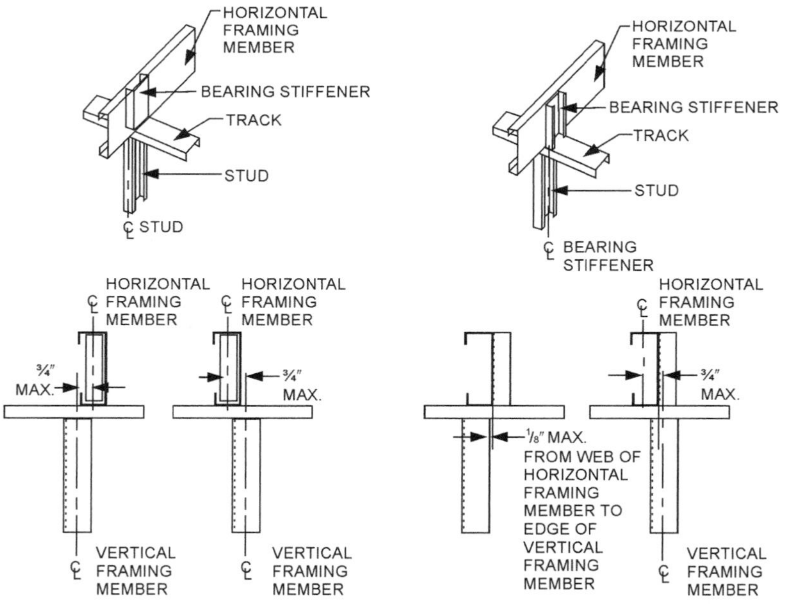

For SI: 1 inch = 25.4 mm.

FIGURE R603.1.2
IN-LINE FRAMING

❖ See the commentary for Section R603.1.2.

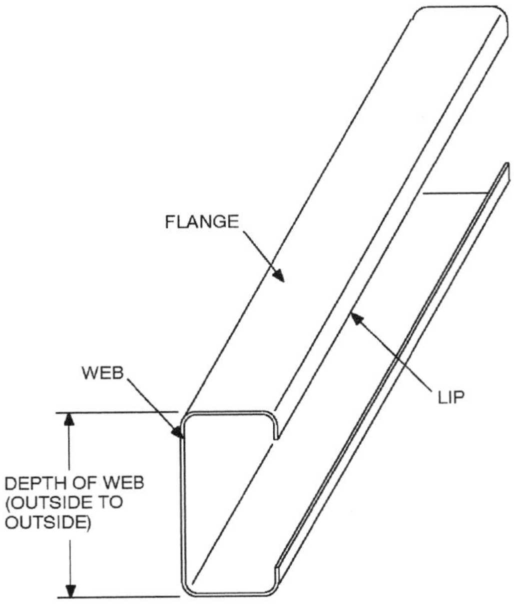

FIGURE R603.2(1)
C-SHAPED SECTION

❖ See the commentary for Section R603.2.

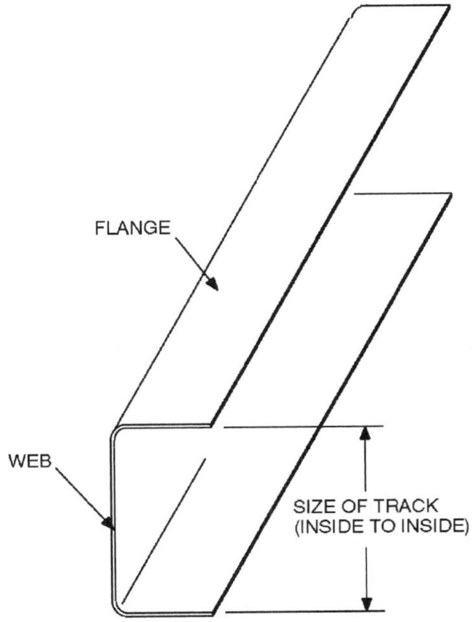

FIGURE R603.2(2)
TRACK SECTION

❖ See the commentary for Section R603.2.

TABLE R603.2(1)
LOAD-BEARING COLD-FORMED STEEL STUD SIZES

MEMBER DESIGNATION[a]	WEB DEPTH (inches)	MINIMUM FLANGE WIDTH (inches)	MAXIMUM FLANGE WIDTH (inches)	MINIMUM LIP SIZE (inch)
350S162-t	3.5	1.625	2	0.5
550S162-t	5.5	1.625	2	0.5

For SI: 1 inch = 25.4 mm; 1 mil = 0.0254 mm.

a. The member designation is defined by the first number representing the member depth in hundredths of an inch "S" representing a stud or joist member, the second number representing the flange width in hundredths of an inch, and the letter "t" shall be a number representing the minimum base metal thickness in mils [See Table R603.2(2)].

❖ See the commentary for Section R603.2.

TABLE R603.2(2)
MINIMUM THICKNESS OF COLD-FORMED STEEL MEMBERS

DESIGNATION THICKNESS (mils)	MINIMUM BASE STEEL THICKNESS (inch)
33	0.0329
43	0.0428
54	0.0538
68	0.0677
97	0.0966

For SI: 1 mil = 0.0254 mm, 1 inch = 25.4 mm.

❖ See the commentary for Section R603.2.

R603.2.1 Material. Load-bearing cold-formed steel framing members shall be cold-formed to shape from structural quality sheet steel complying with the requirements of one of the following:

1. ASTM A 653: Grades 33 and 50 (Class 1 and 3).
2. ASTM A 792: Grades 33 and 50A.
3. ASTM A 1003: Structural Grades 33 Type H, and 50 Type H.

❖ Load-bearing steel framing members must be cold-formed to shape from structural-quality sheet applicable material standards. The steel grades specified have the ductility and strength to meet the intent of these provisions.

R603.2.2 Identification. Load-bearing cold-formed steel framing members shall have a legible *label*, stencil, stamp or embossment with the following information as a minimum:

1. Manufacturer's identification.
2. Minimum base steel thickness in inches (mm).
3. Minimum coating designation.
4. Minimum yield strength, in kips per square inch (ksi) (MPa).

❖ Load-bearing steel framing identification provides verification that materials installed are consistent with the design and meet the intent of the code.

R603.2.3 Corrosion protection. Load-bearing cold-formed steel framing shall have a metallic coating complying with ASTM A 1003 and one of the following:

1. A minimum of G 60 in accordance with ASTM A 653.
2. A minimum of AZ 50 in accordance with ASTM A 792.

❖ The metallic coatings specified correspond to requirements in the referenced material standards. The minimum coating designations assume normal exposure conditions that are best defined as having the framing members enclosed within a building envelope or wall assembly within a controlled environment. While not common under the code when more severe exposure conditions are encountered, consideration should be given to specifying a heavier coating.

R603.2.4 Fastening requirements. Screws for steel-to-steel connections shall be installed with a minimum edge distance and center-to-center spacing of $^1/_2$ inch (12.7 mm), shall be self-drilling tapping and shall conform to ASTM C 1513. Structural sheathing shall be attached to cold-formed steel studs with minimum No. 8 self-drilling tapping screws that conform to ASTM C 1513. Screws for attaching structural sheathing to cold-formed steel wall framing shall have a minimum head diameter of 0.292 inch (7.4 mm) with countersunk heads and shall be installed with a minimum edge distance of $^3/_8$ inch (9.5 mm). Gypsum board shall be attached to cold-formed steel wall framing with minimum No. 6 screws conforming to ASTM C 954 or ASTM C 1513 with a bugle head style and shall be installed in accordance with Section R702. For all connections, screws shall extend through the steel a minimum of three exposed threads. All fasteners shall have rust inhibitive coating suitable for the installation in which they are being used, or be manufactured from material not susceptible to corrosion.

Where No. 8 screws are specified in a steel-to-steel connection, the required number of screws in the connection is permitted to be reduced in accordance with the reduction factors in Table R603.2.4, when larger screws are used or when one of the sheets of steel being connected is thicker than 33 mils (0.84 mm). When applying the reduction factor, the resulting number of screws shall be rounded up.

❖ Fasteners meeting the referenced standards and installed in accordance with this section will provide a load capacity consistent with these prescriptive provisions.

TABLE R603.2.4
SCREW SUBSTITUTION FACTOR

SCREW SIZE	THINNEST CONNECTED STEEL SHEET (mils)	
	33	43
#8	1.0	0.67
#10	0.93	0.62
#12	0.86	0.56

For SI: 1 mil = 0.0254 mm.

❖ In providing prescriptive connection requirements, the code typically specifies the quantity of No. 8 screws necessary to resist the required design forces transferred through a given connection. If the quantity of screws required for a particular connection is large, it may be desirable to reduce the quantity of screws by substituting a larger screw size. Where No. 8 screws are specified for a steel-to-steel connection, the required quantity of screws may be reduced if a larger screw size is substituted in accordance with the reduction factors of Table R603.2.4. Likewise, if the thinnest of the connected steel sheets is at least 43 mils (1.1 mm), the required quantity of screws may be reduced.

R603.2.5 Web holes, web hole reinforcing and web hole patching. Web holes, web hole reinforcing and web hole patching shall be in accordance with this section.

❖ To allow for routing utilities, web holes (also referred to as "penetrations," "utility holes" and "punchouts") are permitted in the webs of studs. To avoid adversely affecting the strength of wall framing, web holes must meet the limits of Section R605.2.1.

R603.2.5.1 Web holes. Web holes in wall studs and other structural members shall comply with all of the following conditions:

1. Holes shall conform to Figure R603.2.5.1;
2. Holes shall be permitted only along the centerline of the web of the framing member;
3. Holes shall have a center-to-center spacing of not less than 24 inches (610 mm);
4. Holes shall have a web hole width not greater than 0.5 times the member depth, or $1^1/_2$ inches (38 mm);

5. Holes shall have a web hole length not exceeding $4^1/_2$ inches (114 mm); and
6. Holes shall have a minimum distance between the edge of the bearing surface and the edge of the web hole of not less than 10 inches (254 mm).

Framing members with web holes not conforming to the above requirements shall be reinforced in accordance with Section R603.2.5.2, patched in accordance with Section R603.2.5.3 or designed in accordance with accepted engineering practice.

❖ Web holes are limited in size as well as to the locations along the centerline of the web of the framing member as illustrated in Figure R603.2.5.1. Otherwise, reinforcement or patching of the web hole is required.

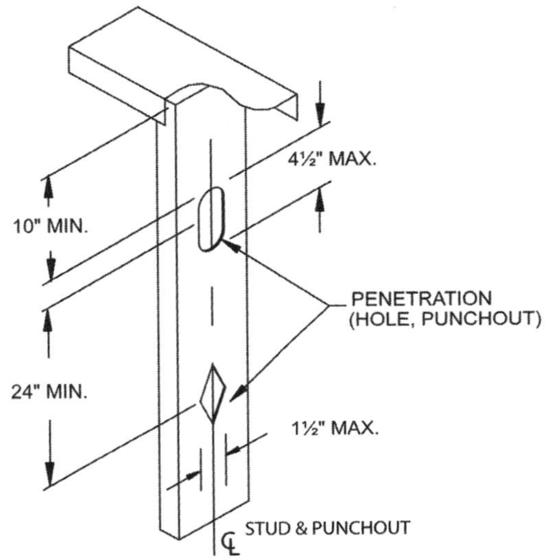

For SI: 1 inch = 25.4 mm.

FIGURE R603.2.5.1
WEB HOLES

❖ Holes are permitted in the web of steel wall studs as shown to allow for the installation of plumbing wiring, etc., in a wall. The limits on size, location and spacing of these holes accommodate these utilities without adversely affecting the stud's load-carrying capacity.

R603.2.5.2 Web hole reinforcing. Web holes in gable endwall studs not conforming to the requirements of Section R603.2.5.1 shall be permitted to be reinforced if the hole is located fully within the center 40 percent of the span and the depth and length of the hole does not exceed 65 percent of the flat width of the web. The reinforcing shall be a steel plate or C-shape section with a hole that does not exceed the web hole size limitations of Section R603.2.5.1 for the member being reinforced. The steel reinforcing shall be the same thickness as the receiving member and shall extend at least 1 inch (25.4 mm) beyond all edges of the hole. The steel reinforcing shall be fastened to the web of the receiving member with No.8 screws spaced no more than 1 inch (25.4 mm) center-to-center along the edges of the patch with minimum edge distance of $^1/_2$ inch (12.7 mm).

❖ Web holes in gable endwall studs that do not meet the limitations stated in Section R603.2.5.1 may be reinforced in accordance with this section. It permits utility installations as long as the reinforcement is installed in accordance with this section and the finished opening meets the limits of Section R603.2.5.1.

R603.2.5.3 Hole patching. Web holes in wall studs and other structural members not conforming to the requirements in Section R603.2.5.1 shall be permitted to be patched in accordance with either of the following methods:

1. Framing members shall be replaced or designed in accordance with accepted engineering practice when web holes exceed the following size limits:
   1.1. The depth of the hole, measured across the web, exceeds 70 percent of the flat width of the web; or
   1.2. The length of the hole measured along the web exceeds 10 inches (254 mm) or the depth of the web, whichever is greater.
2. Web holes not exceeding the dimensional requirements in Section R603.2.5.3, Item 1 shall be patched with a solid steel plate, stud section or track section in accordance with Figure R603.2.5.3. The steel patch shall, as a minimum, be the same thickness as the receiving member and shall extend at least 1 inch (25.4 mm) beyond all edges of the hole. The steel patch shall be fastened to the web of the receiving member with No. 8 screws spaced no more than 1 inch (25.4 mm) center-to-center along the edges of the patch with a minimum edge distance of $^1/_2$ inch (12.7 mm).

❖ Where web holes exceed the limits of Section R603.2.5.1 but are not greater than the size limits of Item 1 in this section, they may be patched as illustrated in Figure R603.2.5.3. Otherwise the patch must be designed or the framing member should be replaced.

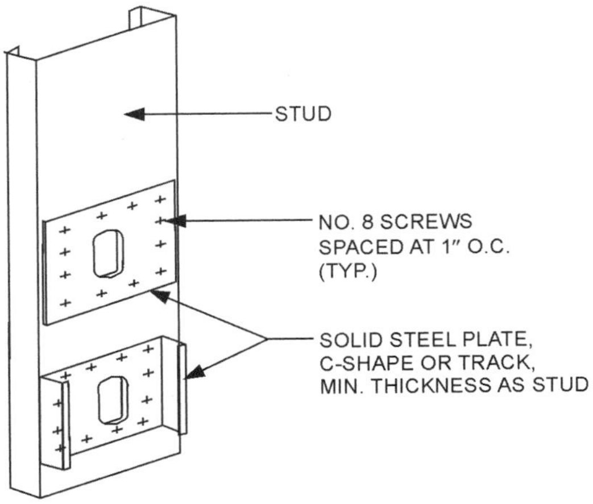

For SI: 1 inch = 25.4 mm.

FIGURE R603.2.5.3
STUD WEB HOLE PATCH

❖ See the commentary for Section R603.2.5.3.

R603.3 Wall construction. All exterior cold-formed steel framed walls and interior load-bearing cold-formed steel framed walls shall be constructed in accordance with the provisions of this section.

❖ The balance of this section consists of the prescriptive design provisions for steel-framed walls.

R603.3.1 Wall to foundation or floor connection. Cold-formed steel framed walls shall be anchored to foundations or floors in accordance with Table R603.3.1 and Figure R603.3.1(1), R603.3.1(2) or R603.3.1(3). Anchor bolts shall be located not more than 12 inches (305 mm) from corners or the termination of bottom tracks. Anchor bolts shall extend a minimum of 15 inches (381 mm) into masonry or 7 inches (178 mm) into concrete. Foundation anchor straps shall be permitted, in lieu of anchor bolts, if spaced as required to provide equivalent anchorage to the required anchor bolts and installed in accordance with manufacturer's requirements.

❖ The bottom tracks of cold-formed steel walls should be attached to the floor as illustrated in Figure R603.3.1(1) and in accordance with Table R603.3.1. Steel-framed walls anchored directly to foundations must be in accordance with Figure R603.3.1(2) with $^1/_2$-inch diameter (12.7 mm) anchor bolts spaced as required by Table R603.3.1 based on the wind speed and the exposure category. Wall tracks connected to wood sills must be in accordance with Figure R603.3.1(3) and Table R603.3.1.

R603.3.1.1 Gable endwalls. Gable endwalls with heights greater than 10 feet (3048 mm) shall be anchored to foundations or floors in accordance with Tables R603.3.1.1(1) or R603.3.1.1(2).

❖ The connection requirements for gable endwalls that are more than 10 feet (3048 mm) in height are tabulated in Tables R603.3.1.1(1) and R603.3.1.1(2) for floors and foundations, respectively. Gable endwalls up to 10 feet (3048 mm) in height are permitted to be connected in accordance with Section R603.3.1.

R603.3.2 Minimum stud sizes. Cold-formed steel walls shall be constructed in accordance with Figure R603.3.1(1), R603.3.1(2) or R603.3.1(3), as applicable. Exterior wall stud size and thickness shall be determined in accordance with the limits set forth in Tables R603.3.2(2) through R603.3.2(31). Interior load-bearing wall stud size and thickness shall be determined in accordance with the limits set forth in Tables R603.3.2(2) through R603.3.2(31) based upon an 85 miles per hour (38 m/s) Exposure A/B wind value and the building width, stud spacing and snow load, as appropriate. Fastening requirements shall be in accordance with Section R603.2.4 and Table R603.3.2(1). Top and bottom tracks shall have the same minimum thickness as the wall studs.

TABLE R603.3.1
WALL TO FOUNDATION OR FLOOR CONNECTION REQUIREMENTS[a, b]

FRAMING CONDITION	WIND SPEED (MPH) AND EXPOSURE					
	85 B	90 B	100 B 85 C	110 B 90 C	100 C	< 110 C
Wall bottom track to floor per Figure R603.3.1(1)	1-No. 8 screw at 12″ o.c.	1-No. 8 screw at 12″ o.c.	1-No. 8 screw at 12″ o.c.	1-No. 8 screw at 12″ o.c.	2-No. 8 screws at 12″ o.c.	2 No. 8 screws at 12″ o.c.
Wall bottom track to foundation per Figure R603.3.1(2)[d]	$^1/_2$″ minimum diameter anchor bolt at 6″ o.c.	$^1/_2$″ minimum diameter anchor bolt at 6″ o.c.	$^1/_2$″ minimum diameter anchor bolt at 4″ o.c.	$^1/_2$″ minimum diameter anchor bolt at 4″ o.c.	$^1/_2$″ minimum diameter anchor bolt at 4″ o.c.	$^1/_2$″ minimum diameter anchor bolt at 4″ o.c.
Wall bottom track to wood sill per Figure R603.3.1(3)	Steel plate spaced at 4″ o.c., with 4-No. 8 screws and 4-10d or 6-8d common nails	Steel plate spaced at 4″ o.c., with 4-No. 8 screws and 4-10d or 6-8d common nails	Steel plate spaced at 3″ o.c., with 4-No. 8 screws and 4-10d or 6-8d common nails	Steel plate spaced at 3″ o.c., with 4-No. 8 screws and 4-10d or 6-8d common nails	Steel plate spaced at 2″ o.c., with 4-No. 8 screws and 4-10d or 6-8d common nails	Steel plate spaced at 2″ o.c., with 4-No. 8 screws and 4-10d or 6-8d common nails
Wind uplift connector strength to 16″ stud spacing[c]	NR	NR	NR	NR	NR	65 lb per foot of wall length
Wind uplift connector strength for 24″ stud spacing[c]	NR	NR	NR	NR	NR	100 lb per foot of wall length

For SI: 1 inch = 25.4 mm, 1 mile per hour = 0.447 m/s, 1 foot = 304.8 mm, 1 pound = 4.45 N.

a. Anchor bolts are to be located not more than 12 inches from corners or the termination of bottom tracks (e.g., at door openings or corners). Bolts are to extend a minimum of 15 inches into masonry or 7 inches into concrete.

b. All screw sizes shown are minimum.

c. NR = uplift connector not required.

d. Foundation anchor straps are permitted in place of anchor bolts, if spaced as required to provide equivalent anchorage to the required anchor bolts and installed in accordance with manufacturer's requirements.

❖ This table specifies connections for wall tracks, which will provide appropriate load capacity based on the basic wind speed and exposure category.

Exterior wall studs shall be permitted to be reduced to the next thinner size, as shown in Tables R603.3.2(2) through R603.3.2(31), but not less than 33 mils (0.84 mm), where both of the following conditions exist:

1. Minimum of $^1/_2$ inch (12.7 mm) gypsum board is installed and fastened in accordance with Section R702 on the interior surface.
2. Wood structural sheathing panels of minimum $^7/_{16}$-inch-thick (11 mm) oriented strand board or $^{15}/_{32}$-inch-thick (12 mm) plywood is installed and fastened in accordance with Section R603.9.1 and Table R603.3.2(1) on the outside surface.

Interior load-bearing walls shall be permitted to be reduced to the next thinner size, as shown in Tables R603.3.2(2) through R603.3.2(31), but not less than 33 mils (0.84 mm), where a minimum of $^1/_2$-inch (12.7 mm) gypsum board is installed and fastened in accordance with Section R702 on both sides of the wall. The tabulated stud thickness for load-bearing walls shall be used when the *attic* load is 10 pounds per square feet (480 Pa) or less. A limited *attic* storage load of 20 pounds per square feet (960 Pa) shall be permitted provided that the next higher snow load column is used to select the stud size from Tables R603.3.2(2) through R603.3.2(31).

For two-story buildings, the tabulated stud thickness for walls supporting one floor, roof and ceiling shall be used when second floor live load is 30 pounds per square feet (1440 Pa). Second floor live loads of 40 psf (1920 pounds per square feet) shall be permitted provided that the next higher snow load column is used to select the stud size from Tables R603.3.2(2) through R603.3.2(21).

For three-story buildings, the tabulated stud thickness for walls supporting one or two floors, roof and ceiling shall be used when the third floor live load is 30 pounds per square feet (1440 Pa). Third floor live loads of 40 pounds per square feet (1920 Pa) shall be permitted provided that the next higher snow load column is used to select the stud size from Tables R603.3.2(22) through R603.3.2(31).

❖ Steel studs must comply with Tables R603.3.2(2) through R603.3.2(31). These tables specify the minimum steel stud thickness based on floor/ceilings supported, wall height, wind speed/exposure, ground snow load, building width and stud depth and spacing. Although the tabular values are calculated based upon attic live loads of 10 psf (479 Pa) or less, values for attic live loads of up to 20 psf (958 Pa) may be determined by simply moving to the next higher snow load column. Similarly, although the tabular values are based upon floor live loads of 30 psf (1436 Pa) or less, values for floor live loads of up to 40 psf (1915 Pa) may be determined by moving to the next higher snow load column.

In recognition of the stiffening effects of sheathing, the code allows the required stud thickness to be reduced to the next thinner size for exterior walls with minimum $^1/_2$-inch (12.7 mm) gypsum board installed on the interior surface and wood structural panels [minimum $^7/_{16}$-inch (11.1 mm) thick oriented strand board or $^{15}/_{32}$-inch (11.9 mm) thick plywood] installed on the exterior. Stud thickness must not be less than 33 mils (0.84 mm). Similarly, interior load-bearing walls with minimum $^1/_2$-inch (12.7 mm) gypsum board installed on both sides of the wall may use the next thinner stud, but not less than 33 mils (0.84 mm).

Fastening of studs to tracks and structural sheathing to studs must be as specified in Section R603.2.4 and Table R603.3.2(1). Track thickness must be no less than that of the wall studs.

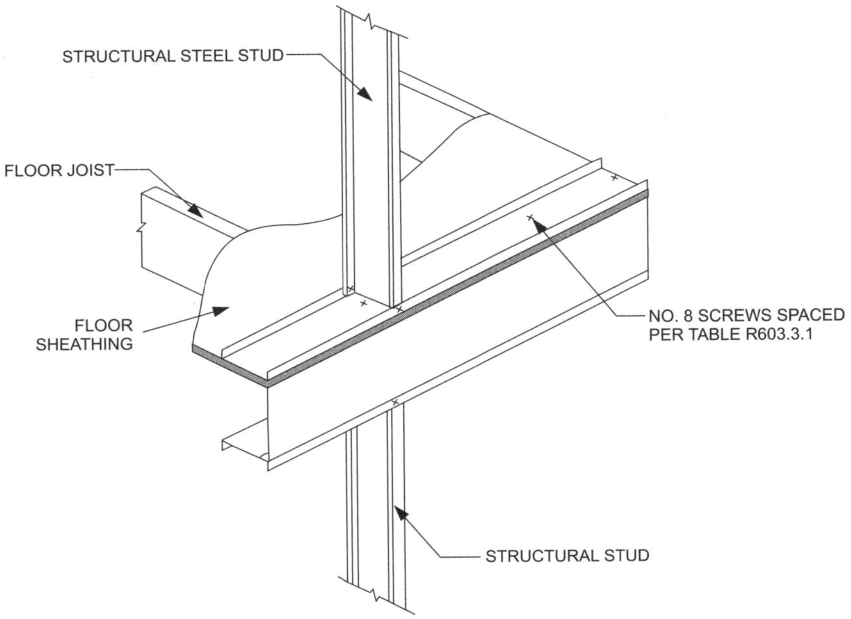

FIGURE R603.3.1(1)
WALL TO FLOOR CONNECTION

❖ This connection detail depicts the item on the first line of Table R603.3.1.

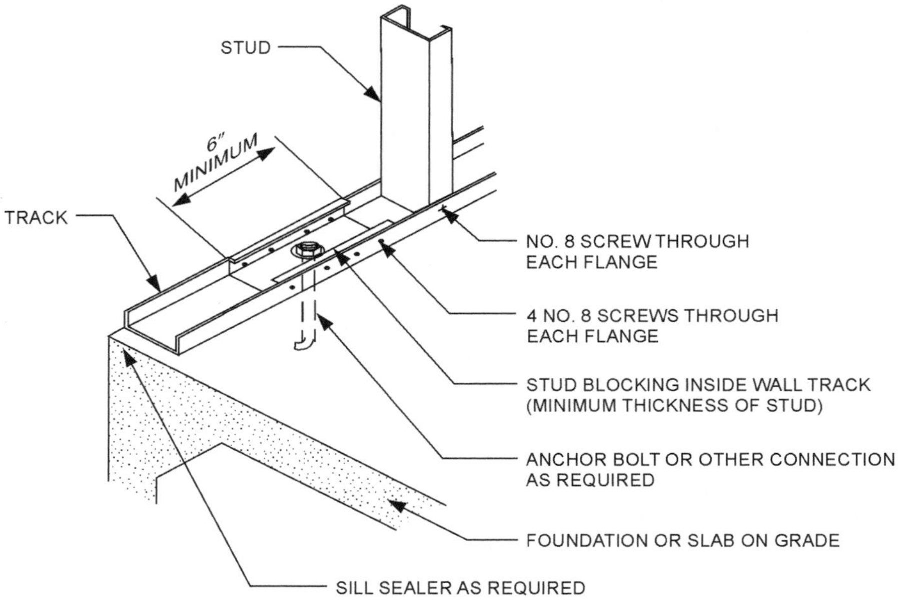

For SI: 1 inch = 25.4 mm.

FIGURE R603.3.1(2)
WALL TO FOUNDATION CONNECTION

❖ This connection detail depicts the item on the second line of Table R603.3.1.

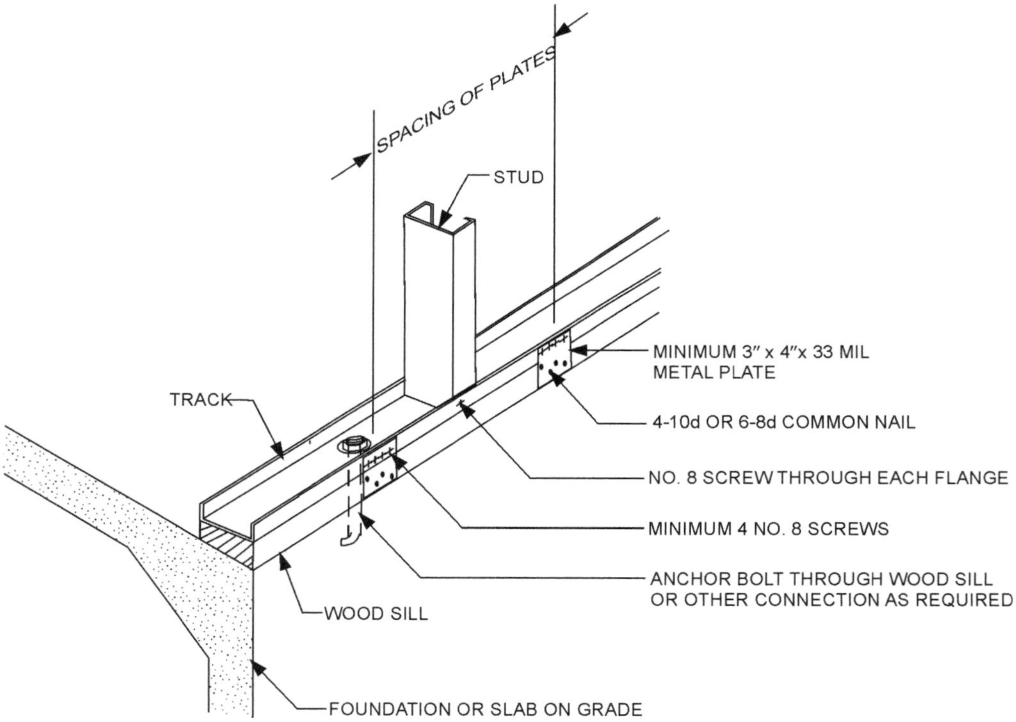

For SI: 1 mil = 0.0254 mm, 1 inch = 25.4.

FIGURE R603.3.1(3)
WALL TO WOOD SILL CONNECTION

❖ This connection detail illustrates the item on the third line of Table R603.3.1.

TABLE R603.3.1.1(1)
GABLE ENDWALL TO FLOOR CONNECTION REQUIREMENTS[a, b, c]

BASIC WIND SPEED (mph)		WALL BOTTOM TRACK TO FLOOR JOIST OR TRACK CONNECTION		
Exposure		Stud height, *h* (feet)		
B	C	$10 < h \leq 14$	$14 < h \leq 18$	$18 < h \leq 22$
85	—	1-No. 8 screw @ 12″ o.c.	1-No. 8 screw @ 12″ o.c.	1-No. 8 screw @ 12″ o.c.
90	—	1-No. 8 screw @ 12″ o.c.	1-No. 8 screw @ 12″ o.c.	1-No. 8 screw @ 12″ o.c.
100	85	1-No. 8 screw @ 12″ o.c.	1-No. 8 screw @ 12″ o.c.	1-No. 8 screw @ 12″ o.c.
110	90	1-No. 8 screw @ 12″ o.c.	1-No. 8 screw @ 12″ o.c.	2-No. 8 screws @ 12″ o.c.
—	100	1-No. 8 screw @ 12″ o.c.	2-No. 8 screws @ 12″ o.c.	1-No. 8 screw @ 8″ o.c.
—	110	2-No. 8 screws @ 12″ o.c.	1-No. 8 screw @ 8″ o.c.	2-No. 8 screws @ 8″ o.c.

For SI: 1 inch = 25.4 mm, 1 mile per hour = 0.447 m/s, 1 foot = 304.8 mm.

a. Refer to Table R603.3.1.1(2) for gable endwall bottom track to foundation connections.

b. Where attachment is not given, special design is required.

c. Stud height, *h*, is measured from wall bottom track to wall top track or brace connection height.

❖ This table specifies requirements for connecting wall tracks of gable endwalls to floors in order to provide appropriate load capacity based on the stud height as well as the basic wind speed and exposure category. Design is required for basic wind speeds exceeding 110 mph (49 m/s) in accordance with the general applicability limits of Section R603.1.1.

TABLE R603.3.1.1(2)
GABLE ENDWALL BOTTOM TRACK TO FOUNDATION CONNECTION REQUIREMENTS[a, b, c]

BASIC WIND SPEED (mph)		MINIMUM SPACING FOR $^1/_2$-INCH-DIAMETER ANCHOR BOLTS[d]		
Exposure		Stud height, *h* (feet)		
B	C	$10 < h \leq 14$	$14 < h \leq 18$	$18 < h \leq 22$
85	—	6′- 0″ o.c.	6′- 0″ o.c.	6′- 0″ o.c.
90	—	6′- 0″ o.c.	5′- 7″ o.c.	6′- 0″ o.c.
100	85	5′- 10″ o.c.	6′- 0″ o.c.	6′- 0″ o.c.
110	90	4′- 10″ o.c.	5′- 6″ o.c.	6′- 0″ o.c.
—	100	4′- 1″ o.c.	6′- 0″ o.c.	6′- 0″ o.c.
—	110	5′- 1″ o.c.	6′- 0″ o.c.	5′- 2″ o.c.

For SI: 1 inch = 25.4 mm, 1 mile per hour = 0.447 m/s, 1 foot = 304.8 mm.

a. Refer to Table R603.3.1.1(1) for gable endwall bottom track to floor joist or track connection connections.

b. Where attachment is not given, special design is required.

c. Stud height, *h*, is measured from wall bottom track to wall top track or brace connection height.

d. Foundation anchor straps are permitted in place of anchor bolts if spaced as required to provide equivalent anchorage to the required anchor bolts and installed in accordance with manufacturer's requirements.

❖ This table specifies requirements for connecting wall tracks of gable endwalls to foundations in order to provide appropriate load capacity based on the stud height as well as the basic wind speed and exposure category. Design is required for basic wind speeds exceeding 110 mph (49 m/s) in accordance with the general applicability limits of Section R603.1.1.

TABLE R603.3.2(1)
WALL FASTENING SCHEDULE

DESCRIPTION OF BUILDING ELEMENT	NUMBER AND SIZE OF FASTENERS[a]	SPACING OF FASTENERS
Floor joist to track of load-bearing wall	2-No. 8 screws	Each joist
Wall stud to top or bottom track	2-No. 8 screws	Each end of stud, one per flange
Structural sheathing to wall studs	No. 8 screws[b]	6″ o.c. on edges and 12″ o.c. at intermediate supports
Roof framing to wall	Approved design or tie down in accordance with Section R802.11.	

For SI: 1 inch = 25.4 mm.

a. All screw sizes shown are minimum.

b. Screws for attachment of structural sheathing panels are to be bugle-head, flat-head, or similar head styles with a minimum head diameter of 0.29 inch.

❖ This table provides fastener size and spacing for various portions of steel stud walls. See Figure R603.9 for an illustration of sheathing fastener locations indicated as "edge" and "field."

TABLE R603.3.2(2)
24-FOOT-WIDE BUILDING SUPPORTING ROOF AND CEILING ONLY[a, b, c]
33 KSI STEEL

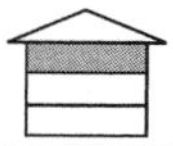

WIND SPEED		MEMBER SIZE	STUD SPACING (inches)	MINIMUM STUD THICKNESS (mils)											
				8-foot Studs				9-foot Studs				10-foot Studs			
				Ground Snow Load (psf)											
Exp. B	Exp. C			20	30	50	70	20	30	50	70	20	30	50	70
85 mph	—	350S162	16	33	33	33	33	33	33	33	33	33	33	33	33
			24	33	33	33	43	33	33	33	43	33	33	43	43
		550S162	16	33	33	33	33	33	33	33	33	33	33	33	33
			24	33	33	33	33	33	33	33	33	33	33	33	33
90 mph	—	350S162	16	33	33	33	33	33	33	33	33	33	33	33	33
			24	33	33	33	43	33	33	33	43	33	33	43	43
		550S162	16	33	33	33	33	33	33	33	33	33	33	33	33
			24	33	33	33	33	33	33	33	33	33	33	33	33
100 mph	85 mph	350S162	16	33	33	33	33	33	33	33	33	33	33	33	33
			24	33	33	33	43	33	33	33	43	43	43	43	43
		550S162	16	33	33	33	33	33	33	33	33	33	33	33	33
			24	33	33	33	43	33	33	33	33	33	33	33	43
110 mph	90 mph	350S162	16	33	33	33	33	33	33	33	33	33	33	33	33
			24	33	33	33	43	43	43	43	43	43	43	43	54
		550S162	16	33	33	33	33	33	33	33	33	33	33	33	33
			24	33	33	33	43	33	33	33	33	43	43	43	43
—	100 mph	350S162	16	33	33	33	33	33	33	33	33	43	43	43	43
			24	43	43	43	43	43	43	43	43	54	54	54	54
		550S162	16	33	33	33	33	33	33	33	33	33	33	33	33
			24	33	33	33	43	43	43	43	43	43	43	43	43
—	110 mph	350S162	16	33	33	33	33	43	43	43	43	43	43	43	43
			24	43	43	43	43	54	54	54	54	68	68	68	68
		550S162	16	33	33	33	33	33	33	33	33	33	33	33	33
			24	33	43	43	43	43	43	43	43	43	43	43	43

For SI: 1 inch = 25.4 mm, 1 foot = 304.8 mm, 1 mil = 0.0254 mm, 1 mile per hour = 0.447 m/s, 1 pound per square foot = 0.0479 kPa, 1 Ksi = 1,000 psi = 6.895 MPa.

a. Deflection criterion: *L*/240.

b. Design load assumptions:
Second floor dead load is 10 psf.
Second floor live load is 30 psf.
Roof/ceiling dead load is 12 psf.
Attic live load is 10 psf.

c. Building width is in the direction of horizontal framing members supported by the wall studs.

❖ See the commentary for Section R603.3.2.

TABLE R603.3.2(3)
24-FOOT-WIDE BUILDING SUPPORTING ROOF AND CEILING ONLY[a, b, c]
50 KSI STEEL

WIND SPEED		MEMBER SIZE	STUD SPACING (inches)	MINIMUM STUD THICKNESS (mils)											
				8-foot Studs				9-foot Studs				10-foot Studs			
				Ground Snow Load (psf)											
Exp. B	Exp. C			20	30	50	70	20	30	50	70	20	30	50	70
85 mph	—	350S162	16	33	33	33	33	33	33	33	33	33	33	33	33
			24	33	33	33	43	33	33	33	33	33	33	33	43
		550S162	16	33	33	33	33	33	33	33	33	33	33	33	33
			24	33	33	33	33	33	33	33	33	33	33	33	33
90 mph	—	350S162	16	33	33	33	33	33	33	33	33	33	33	33	33
			24	33	33	33	43	33	33	33	33	33	33	33	43
		550S162	16	33	33	33	33	33	33	33	33	33	33	33	33
			24	33	33	33	33	33	33	33	33	33	33	33	33
100 mph	85 mph	350S162	16	33	33	33	33	33	33	33	33	33	33	33	33
			24	33	33	33	43	33	33	33	33	33	33	33	43
		550S162	16	33	33	33	33	33	33	33	33	33	33	33	33
			24	33	33	33	33	33	33	33	33	33	33	33	33
110 mph	90 mph	350S162	16	33	33	33	33	33	33	33	33	33	33	33	33
			24	33	33	33	43	33	33	33	43	43	43	43	43
		550S162	16	33	33	33	33	33	33	33	33	33	33	33	33
			24	33	33	33	33	33	33	33	33	33	33	33	33
—	100 mph	350S162	16	33	33	33	33	33	33	33	33	33	33	33	33
			24	33	33	33	43	43	43	43	43	43	43	43	43
		550S162	16	33	33	33	33	33	33	33	33	33	33	33	33
			24	33	33	33	33	33	33	33	33	33	33	33	33
—	110 mph	350S162	16	33	33	33	33	33	33	33	33	33	33	33	33
			24	33	33	33	43	43	43	43	43	54	54	54	54
		550S162	16	33	33	33	33	33	33	33	33	33	33	33	33
			24	33	33	33	33	33	33	33	33	33	33	33	33

For SI: 1 inch = 25.4 mm, 1 foot = 304.8 mm, 1 mil = 0.0254 mm, 1 mile per hour = 0.447 m/s, 1 pound per square foot = 0.0479 kPa,
1 Ksi = 1,000 psi = 6.895 MPa.

a. Deflection criterion: *L*/240.

b. Design load assumptions:
Second floor dead load is 10 psf.
Second floor live load is 30 psf.
Roof/ceiling dead load is 12 psf.
Attic live load is 10 psf.

c. Building width is in the direction of horizontal framing members supported by the wall studs.

❖ See the commentary for Section R603.3.2.

TABLE R603.3.2(4)
28-FOOT-WIDE BUILDING SUPPORTING ROOF AND CEILING ONLY[a, b, c]
33 KSI STEEL

WIND SPEED		MEMBER SIZE	STUD SPACING (inches)	MINIMUM STUD THICKNESS (mils)											
				8-foot Studs				9-foot Studs				10-foot Studs			
				Ground Snow Load (psf)											
Exp. B	Exp. C			20	30	50	70	20	30	50	70	20	30	50	70
85 mph	—	350S162	16	33	33	33	33	33	33	33	33	33	33	33	33
			24	33	33	43	43	33	33	43	43	33	33	43	54
		550S162	16	33	33	33	33	33	33	33	33	33	33	33	33
			24	33	33	33	43	33	33	33	43	33	33	33	43
90 mph	—	350S162	16	33	33	33	33	33	33	33	33	33	33	33	33
			24	33	33	43	43	33	33	43	43	33	33	43	54
		550S162	16	33	33	33	33	33	33	33	33	33	33	33	33
			24	33	33	33	43	33	33	33	43	33	33	33	43
100 mph	85 mph	350S162	16	33	33	33	33	33	33	33	33	33	33	33	33
			24	33	33	43	43	33	33	43	43	43	43	43	54
		550S162	16	33	33	33	33	33	33	33	33	33	33	33	33
			24	33	33	33	43	33	33	33	43	33	33	33	43
110 mph	90 mph	350S162	16	33	33	33	33	33	33	33	33	33	33	33	43
			24	33	33	43	43	43	43	43	43	43	43	43	54
		550S162	16	33	33	33	33	33	33	33	33	33	33	33	33
			24	33	33	33	43	33	33	33	43	33	33	33	43
—	100 mph	350S162	16	33	33	33	33	33	33	33	33	43	43	43	43
			24	43	43	43	54	43	43	43	54	54	54	54	54
		550S162	16	33	33	33	33	33	33	33	33	33	33	33	33
			24	33	33	33	43	33	33	33	43	33	33	33	43
—	110 mph	350S162	16	33	33	33	33	43	43	43	43	43	43	43	43
			24	43	43	43	54	54	54	54	54	68	68	68	68
		550S162	16	33	33	33	33	33	33	33	33	33	33	33	33
			24	33	33	33	43	33	33	33	43	43	43	43	43

For SI: 1 inch = 25.4 mm, 1 foot = 304.8 mm, 1 mil = 0.0254 mm, 1 mile per hour = 0.447 m/s, 1 pound per square foot = 0.0479 kPa, 1 Ksi = 1,000 psi = 6.895 MPa.

a. Deflection criterion: *L*/240.

b. Design load assumptions:
Second floor dead load is 10 psf.
Second floor live load is 30 psf.
Roof/ceiling dead load is 12 psf.
Attic live load is 10 psf.

c. Building width is in the direction of horizontal framing members supported by the wall studs.

❖ See the commentary for Section R603.3.2.

TABLE R603.3.2(5)
28-FOOT-WIDE BUILDING SUPPORTING ROOF AND CEILING ONLY[a, b, c]
50 KSI STEEL

WIND SPEED		MEMBER SIZE	STUD SPACING (inches)	MINIMUM STUD THICKNESS (mils)											
				8-foot Studs				9-foot Studs				10-foot Studs			
				Ground Snow Load (psf)											
Exp. B	Exp. C			20	30	50	70	20	30	50	70	20	30	50	70
85 mph	—	350S162	16	33	33	33	33	33	33	33	33	33	33	33	33
			24	33	33	33	43	33	33	33	43	33	33	33	43
		550S162	16	33	33	33	33	33	33	33	33	33	33	33	33
			24	33	33	33	33	33	33	33	33	33	33	33	33
90 mph	—	350S162	16	33	33	33	33	33	33	33	33	33	33	33	33
			24	33	33	33	43	33	33	33	43	33	33	33	43
		550S162	16	33	33	33	33	33	33	33	33	33	33	33	33
			24	33	33	33	33	33	33	33	33	33	33	33	33
100 mph	85 mph	350S162	16	33	33	33	33	33	33	33	33	33	33	33	33
			24	33	33	33	43	33	33	33	43	33	33	43	43
		550S162	16	33	33	33	33	33	33	33	33	33	33	33	33
			24	33	33	33	33	33	33	33	33	33	33	33	33
110 mph	90 mph	350S162	16	33	33	33	33	33	33	33	33	33	33	33	33
			24	33	33	33	43	33	33	33	43	43	43	43	43
		550S162	16	33	33	33	33	33	33	33	33	33	33	33	33
			24	33	33	33	33	33	33	33	33	33	33	33	33
—	100 mph	350S162	16	33	33	33	33	33	33	33	33	33	33	33	33
			24	33	33	33	43	43	43	43	43	43	43	43	43
		550S162	16	33	33	33	33	33	33	33	33	33	33	33	33
			24	33	33	33	43	33	33	33	33	33	33	33	33
—	110 mph	350S162	16	33	33	33	33	33	33	33	33	33	33	33	33
			24	33	33	43	43	43	43	43	43	54	54	54	54
		550S162	16	33	33	33	33	33	33	33	33	33	33	33	33
			24	33	33	33	33	33	33	33	33	33	33	33	43

For SI: 1 inch = 25.4 mm, 1 foot = 304.8 mm, 1 mil = 0.0254 mm, 1 mile per hour = 0.447 m/s, 1 pound per square foot = 0.0479 kPa, 1 Ksi = 1,000 psi = 6.895 MPa.

a. Deflection criterion: *L*/240.

b. Design load assumptions:
  Second floor dead load is 10 psf.
  Second floor live load is 30 psf.
  Roof/ceiling dead load is 12 psf.
  Attic live load is 10 psf.

c. Building width is in the direction of horizontal framing members supported by the wall studs.

❖ See the commentary for Section R603.3.2.

TABLE R603.3.2(6)
32-FOOT-WIDE BUILDING SUPPORTING ROOF AND CEILING ONLY[a, b, c]
33 KSI STEE

WIND SPEED		MEMBER SIZE	STUD SPACING (inches)	MINIMUM STUD THICKNESS (mils)											
				8-foot Studs				9-foot Studs				10-foot Studs			
				Ground Snow Load (psf)											
Exp. B	Exp. C			20	30	50	70	20	30	50	70	20	30	50	70
85 mph	—	350S162	16	33	33	33	33	33	33	33	33	33	33	33	43
			24	33	33	43	54	33	33	43	43	33	33	43	54
		550S162	16	33	33	33	33	33	33	33	33	33	33	33	33
			24	33	33	33	43	33	33	33	43	33	33	33	43
90 mph	—	350S162	16	33	33	33	33	33	33	33	33	33	33	33	43
			24	33	33	43	54	33	33	43	43	33	33	43	54
		550S162	16	33	33	33	33	33	33	33	33	33	33	33	33
			24	33	33	33	43	33	33	33	43	33	33	33	43
100 mph	85 mph	350S162	16	33	33	33	33	33	33	33	33	33	33	33	43
			24	33	33	43	54	33	33	43	54	43	43	43	54
		550S162	16	33	33	33	33	33	33	33	33	33	33	33	33
			24	33	33	33	43	33	33	33	43	33	33	33	43
110 mph	90 mph	350S162	16	33	33	33	43	33	33	33	33	33	33	33	43
			24	33	33	43	54	43	43	43	54	43	43	43	54
		550S162	16	33	33	33	33	33	33	33	33	33	33	33	33
			24	33	33	33	43	33	33	33	43	33	33	43	43
—	100 mph	350S162	16	33	33	33	43	33	33	33	43	43	43	43	43
			24	43	43	43	54	43	43	43	54	54	54	54	54
		550S162	16	33	33	33	33	33	33	33	33	33	33	33	33
			24	33	33	43	43	33	33	33	43	33	33	43	43
—	110 mph	350S162	16	33	33	33	43	43	43	43	43	43	43	43	43
			24	43	43	43	54	54	54	54	54	68	68	68	68
		550S162	16	33	33	33	33	33	33	33	33	33	33	33	33
			24	33	33	43	43	33	33	43	43	43	43	43	43

For SI: 1 inch = 25.4 mm, 1 foot = 304.8 mm, 1 mil = 0.0254 mm, 1 mile per hour = 0.447 m/s, 1 pound per square foot = 0.0479 kPa, 1 Ksi = 1,000 psi = 6.895 MPa.

a. Deflection criterion: *L*/240.

b. Design load assumptions:
 Second floor dead load is 10 psf.
 Second floor live load is 30 psf.
 Roof/ceiling dead load is 12 psf.
 Attic live load is 10 psf.

c. Building width is in the direction of horizontal framing members supported by the wall studs.

❖ See the commentary for Section R603.3.2.

TABLE R603.3.2(7)
32-FOOT-WIDE BUILDING SUPPORTING ROOF AND CEILING ONLY[a, b, c]
50 KSI STEEL

WIND SPEED		MEMBER SIZE	STUD SPACING (inches)	MINIMUM STUD THICKNESS (mils)											
				8-foot Studs				9-foot Studs				10-foot Studs			
				Ground Snow Load (psf)											
Exp. B	Exp. C			20	30	50	70	20	30	50	70	20	30	50	70
85 mph	—	350S162	16	33	33	33	33	33	33	33	33	33	33	33	33
			24	33	33	33	43	33	33	33	43	33	33	43	43
		550S162	16	33	33	33	33	33	33	33	33	33	33	33	33
			24	33	33	33	43	33	33	33	33	33	33	33	43
90 mph	—	350S162	16	33	33	33	33	33	33	33	33	33	33	33	33
			24	33	33	33	43	33	33	33	43	33	33	43	43
		550S162	16	33	33	33	33	33	33	33	33	33	33	33	33
			24	33	33	33	43	33	33	33	33	33	33	33	43
100 mph	85 mph	350S162	16	33	33	33	33	33	33	33	33	33	33	33	33
			24	33	33	43	43	33	33	33	43	33	33	43	43
		550S162	16	33	33	33	33	33	33	33	33	33	33	33	33
			24	33	33	33	43	33	33	33	33	33	33	33	43
110 mph	90 mph	350S162	16	33	33	33	33	33	33	33	33	33	33	33	33
			24	33	33	43	43	33	33	33	43	43	43	43	54
		550S162	16	33	33	33	33	33	33	33	33	33	33	33	33
			24	33	33	33	43	33	33	33	33	33	33	33	43
—	100 mph	350S162	16	33	33	33	33	33	33	33	33	33	33	33	33
			24	33	33	43	43	43	43	43	43	43	43	43	54
		550S162	16	33	33	33	33	33	33	33	33	33	33	33	33
			24	33	33	33	43	33	33	33	43	33	33	33	43
—	110 mph	350S162	16	33	33	33	33	33	33	33	33	33	33	33	43
			24	33	33	43	43	43	43	43	43	54	54	54	54
		550S162	16	33	33	33	33	33	33	33	33	33	33	33	33
			24	33	33	33	43	33	33	33	43	33	33	33	43

For SI: 1 inch = 25.4 mm, 1 foot = 304.8 mm, 1 mil = 0.0254 mm, 1 mile per hour = 0.447 m/s, 1 pound per square foot = 0.0479 kPa, 1 Ksi = 1,000 psi = 6.895 MPa.

a. Deflection criterion: *L*/240.

b. Design load assumptions:
   Second floor dead load is 10 psf.
   Second floor live load is 30 psf.
   Roof/ceiling dead load is 12 psf.
   Attic live load is 10 psf.

c. Building width is in the direction of horizontal framing members supported by the wall studs.

❖ See the commentary for Section R603.3.2.

TABLE R603.3.2(8)
36-FOOT-WIDE BUILDING SUPPORTING ROOF AND CEILING ONLY[a, b, c]
33 KSI STEEL

WIND SPEED		MEMBER SIZE	STUD SPACING (inches)	MINIMUM STUD THICKNESS (mils)											
				8-foot Studs				9-foot Studs				10-foot Studs			
Exp. B	Exp. C			Ground Snow Load (psf)											
				20	30	50	70	20	30	50	70	20	30	50	70
85 mph	—	350S162	16	33	33	33	43	33	33	33	43	33	33	33	43
			24	33	33	43	54	33	33	43	54	33	43	43	54
		550S162	16	33	33	33	33	33	33	33	33	33	33	33	33
			24	33	33	43	43	33	33	43	43	33	33	43	43
90 mph	—	350S162	16	33	33	33	43	33	33	33	43	33	33	33	43
			24	33	33	43	54	33	33	43	54	33	43	43	54
		550S162	16	33	33	33	33	33	33	33	33	33	33	33	33
			24	33	33	43	43	33	33	43	43	33	33	43	43
100 mph	85 mph	350S162	16	33	33	33	43	33	33	33	43	33	33	33	43
			24	33	33	43	54	33	33	43	54	43	43	54	54
		550S162	16	33	33	33	33	33	33	33	33	33	33	33	33
			24	33	33	43	43	33	33	43	43	33	33	43	43
110 mph	90 mph	350S162	16	33	33	33	43	33	33	33	33	33	33	33	43
			24	33	33	43	54	43	43	43	43	43	43	54	68
		550S162	16	33	33	33	33	33	33	33	33	33	33	33	33
			24	33	33	43	43	33	33	43	43	33	33	43	43
—	100 mph	350S162	16	33	33	33	43	33	33	33	43	43	43	43	43
			24	43	43	43	54	43	43	43	54	54	54	54	68
		550S162	16	33	33	33	33	33	33	33	33	33	33	33	33
			24	33	33	43	43	33	33	43	43	33	33	43	43
—	110 mph	350S162	16	33	33	33	43	43	43	43	43	43	43	43	43
			24	43	43	54	54	54	54	54	54	68	68	68	68
		550S162	16	33	33	33	33	33	33	33	33	33	33	33	33
			24	33	33	43	54	33	33	43	43	43	43	43	54

For SI: 1 inch = 25.4 mm, 1 foot = 304.8 mm, 1 mil = 0.0254 mm, 1 mile per hour = 0.447 m/s, 1 pound per square foot = 0.0479 kPa, 1 Ksi = 1,000 psi = 6.895 MPa.

a. Deflection criterion: *L*/240.

b. Design load assumptions:
Second floor dead load is 10 psf.
Second floor live load is 30 psf.
Roof/ceiling dead load is 12 psf.
Attic live load is 10 psf.

c. Building width is in the direction of horizontal framing members supported by the wall studs.

❖ See the commentary for Section R603.3.2.

TABLE R603.3.2(9)
36-FOOT-WIDE BUILDING SUPPORTING ROOF AND CEILING ONLY[a, b, c]
50 KSI STEEL

WIND SPEED		MEMBER SIZE	STUD SPACING (inches)	MINIMUM STUD THICKNESS (mils)											
				8-foot Studs				9-foot Studs				10-foot Studs			
				Ground Snow Load (psf)											
Exp. B	Exp. C			20	30	50	70	20	30	50	70	20	30	50	70
85 mph	—	350S162	16	33	33	33	33	33	33	33	33	33	33	33	33
			24	33	33	43	43	33	33	43	43	33	33	43	54
		550S162	16	33	33	33	33	33	33	33	33	33	33	33	33
			24	33	33	33	43	33	33	33	43	33	33	33	43
90 mph	—	350S162	16	33	33	33	33	33	33	33	33	33	33	33	33
			24	33	33	43	43	33	33	43	43	33	33	43	54
		550S162	16	33	33	33	33	33	33	33	33	33	33	33	33
			24	33	33	33	43	33	33	33	43	33	33	33	43
100 mph	85 mph	350S162	16	33	33	33	33	33	33	33	33	33	33	33	33
			24	33	33	43	43	33	33	43	43	33	33	43	54
		550S162	16	33	33	33	33	33	33	33	33	33	33	33	33
			24	33	33	33	43	33	33	33	43	33	33	33	43
110 mph	90 mph	350S162	16	33	33	33	33	33	33	33	33	33	33	33	43
			24	33	33	43	54	33	33	33	43	43	43	43	54
		550S162	16	33	33	33	33	33	33	33	33	33	33	33	33
			24	33	33	33	43	33	33	33	43	33	33	33	43
—	100 mph	350S162	16	33	33	33	33	33	33	33	33	33	33	33	43
			24	33	33	33	54	43	43	43	43	43	43	43	54
		550S162	16	33	33	33	33	33	33	33	33	33	33	33	33
			24	33	33	33	43	33	33	33	43	33	33	33	43
—	110 mph	350S162	16	33	33	33	43	33	33	33	33	33	33	33	43
			24	33	33	43	54	43	43	43	54	54	54	54	54
		550S162	16	33	33	33	33	33	33	33	33	33	33	33	33
			24	33	33	33	43	33	33	33	43	33	33	33	43

For SI: 1 inch = 25.4 mm, 1 foot = 304.8 mm, 1 mil = 0.0254 mm, 1 mile per hour = 0.447 m/s, 1 pound per square foot = 0.0479 kPa, 1 Ksi = 1,000 psi = 6.895 MPa.

a. Deflection criterion: *L*/240.

b. Design load assumptions:
   Second floor dead load is 10 psf.
   Second floor live load is 30 psf.
   Roof/ceiling dead load is 12 psf.
   Attic live load is 10 psf.

c. Building width is in the direction of horizontal framing members supported by the wall studs.

❖ See the commentary for Section R603.3.2.

TABLE R603.3.2(10)
40-FOOT-WIDE BUILDING SUPPORTING ROOF AND CEILING ONLY[a, b, c]
33 KSI STEEL

WIND SPEED		MEMBER SIZE	STUD SPACING (inches)	MINIMUM STUD THICKNESS (mils)											
				8-foot Studs				9-foot Studs				10-foot Studs			
				Ground Snow Load (psf)											
Exp. B	Exp. C			20	30	50	70	20	30	50	70	20	30	50	70
85 mph	—	350S162	16	33	33	33	43	33	33	33	43	33	33	33	43
			24	33	33	43	54	33	33	43	54	43	43	54	68
		550S162	16	33	33	33	33	33	33	33	33	33	33	33	33
			24	33	33	43	54	33	33	43	43	33	33	43	54
90 mph	—	350S162	16	33	33	33	43	33	33	33	43	33	33	33	43
			24	33	33	43	54	33	33	43	54	43	43	54	68
		550S162	16	33	33	33	33	33	33	33	33	33	33	33	33
			24	33	33	43	54	33	33	43	43	33	33	43	54
100 mph	85 mph	350S162	16	33	33	33	43	33	33	33	43	33	33	33	43
			24	33	43	43	54	33	43	43	54	43	43	54	68
		550S162	16	33	33	33	43	33	33	33	33	33	33	33	33
			24	33	33	43	54	33	33	43	43	33	33	43	54
110 mph	90 mph	350S162	16	33	33	33	43	33	33	33	43	33	33	43	43
			24	33	43	43	54	43	43	43	54	43	43	54	68
		550S162	16	33	33	33	43	33	33	33	33	33	33	33	43
			24	33	33	43	54	33	33	43	43	33	33	43	54
—	100 mph	350S162	16	33	33	33	43	33	33	33	43	43	43	43	43
			24	43	43	54	68	43	43	54	54	54	54	54	68
		550S162	16	33	33	33	43	33	33	33	33	33	33	33	43
			24	33	33	43	54	33	33	43	54	33	33	43	54
—	110 mph	350S162	16	33	33	43	43	43	43	43	43	43	43	43	54
			24	43	43	54	68	54	54	54	68	68	68	68	68
		550S162	16	33	33	33	43	33	33	33	43	33	33	33	43
			24	33	33	43	54	33	33	43	54	43	43	43	54

For SI: 1 inch = 25.4 mm, 1 foot = 304.8 mm, 1 mil = 0.0254 mm, 1 mile per hour = 0.447 m/s, 1 pound per square foot = 0.0479 kPa, 1 Ksi = 1,000 psi = 6.895 MPa.

a. Deflection criterion: *L*/240.

b. Design load assumptions:
  Second floor dead load is 10 psf.
  Second floor live load is 30 psf.
  Roof/ceiling dead load is 12 psf.
  Attic live load is 10 psf.

c. Building width is in the direction of horizontal framing members supported by the wall studs.

❖ See the commentary for Section R603.3.2.

TABLE R603.3.2(11)
40-FOOT-WIDE BUILDING SUPPORTING ROOF AND CEILING ONLY[a, b, c]
50 KSI STEEL

WIND SPEED		MEMBER SIZE	STUD SPACING (inches)	MINIMUM STUD THICKNESS (mils)											
				8-foot Studs				9-foot Studs				10-foot Studs			
				Ground Snow Load (psf)											
Exp. B	Exp. C			20	30	50	70	20	30	50	70	20	30	50	70
85 mph	—	350S162	16	33	33	33	33	33	33	33	33	33	33	33	43
			24	33	33	43	54	33	33	43	43	33	33	43	54
		550S162	16	33	33	33	33	33	33	33	33	33	33	33	33
			24	33	33	33	43	33	33	33	43	33	33	33	43
90 mph	—	350S162	16	33	33	33	33	33	33	33	33	33	33	33	43
			24	33	33	43	54	33	33	43	43	33	33	43	54
		550S162	16	33	33	33	33	33	33	33	33	33	33	33	33
			24	33	33	33	43	33	33	33	43	33	33	33	43
100 mph	85 mph	350S162	16	33	33	33	43	33	33	33	33	33	33	33	43
			24	33	33	43	54	33	33	43	54	33	33	43	54
		550S162	16	33	33	33	33	33	33	33	33	33	33	33	33
			24	33	33	33	43	33	33	33	43	33	33	33	43
110 mph	90 mph	350S162	16	33	33	33	43	33	33	33	33	33	33	33	43
			24	33	33	43	54	33	33	43	54	43	43	43	54
		550S162	16	33	33	33	33	33	33	33	33	33	33	33	33
			24	33	33	33	43	33	33	33	43	33	33	33	43
—	100 mph	350S162	16	33	33	33	43	33	33	33	43	33	33	33	43
			24	33	33	43	54	43	43	43	54	43	43	54	54
		550S162	16	33	33	33	33	33	33	33	33	33	33	33	33
			24	33	33	43	43	33	33	33	43	33	33	43	43
—	110 mph	350S162	16	33	33	33	43	33	33	33	43	33	33	33	43
			24	33	33	43	54	43	43	43	54	54	54	54	68
		550S162	16	33	33	33	33	33	33	33	33	33	33	33	33
			24	33	33	43	43	33	33	33	43	33	33	43	43

For SI: 1 inch = 25.4 mm, 1 foot = 304.8 mm, 1 mil = 0.0254 mm, 1 mile per hour = 0.447 m/s, 1 pound per square foot = 0.0479 kPa, 1 Ksi = 1,000 psi = 6.895 MPa.

a. Deflection criterion: *L*/240.

b. Design load assumptions:
   Second floor dead load is 10 psf.
   Second floor live load is 30 psf.
   Roof/ceiling dead load is 12 psf.
   Attic live load is 10 psf.

c. Building width is in the direction of horizontal framing members supported by the wall studs.

❖ See the commentary for Section R603.3.2.

TABLE R603.3.2(12)
24-FOOT-WIDE BUILDING SUPPORTING ONE FLOOR, ROOF AND CEILING[a, b, c]
33 KSI STEEL

WIND SPEED		MEMBER SIZE	STUD SPACING (inches)	MINIMUM STUD THICKNESS (mils)											
				8-foot Studs				9-foot Studs				10-foot Studs			
				Ground Snow Load (psf)											
Exp. B	Exp. C			20	30	50	70	20	30	50	70	20	30	50	70
85 mph	—	350S162	16	33	33	33	33	33	33	33	33	33	33	33	43
			24	33	33	43	43	33	43	43	43	43	43	43	54
		550S162	16	33	33	33	33	33	33	33	33	33	33	33	33
			24	33	33	33	43	33	33	33	43	33	33	33	43
90 mph	—	350S162	16	33	33	33	33	33	33	33	33	33	33	33	43
			24	33	33	43	43	33	43	43	43	43	43	43	54
		550S162	16	33	33	33	33	33	33	33	33	33	33	33	33
			24	33	33	33	43	33	33	33	43	33	33	33	43
100 mph	85 mph	350S162	16	33	33	33	33	33	33	33	33	33	33	33	43
			24	33	43	43	43	43	43	43	43	43	43	43	54
		550S162	16	33	33	33	33	33	33	33	33	33	33	33	33
			24	33	33	33	43	33	33	33	43	33	33	33	43
110 mph	90 mph	350S162	16	33	33	33	43	33	33	33	33	33	33	43	43
			24	43	43	43	43	43	43	43	43	54	54	54	54
		550S162	16	33	33	33	33	33	33	33	33	33	33	33	33
			24	33	33	33	43	33	33	33	43	43	43	43	43
—	100 mph	350S162	16	33	33	33	43	33	33	33	43	43	43	43	43
			24	43	43	43	54	43	43	54	54	54	54	54	54
		550S162	16	33	33	33	33	33	33	33	33	33	33	33	33
			24	33	33	33	43	43	43	43	43	43	43	43	43
—	110 mph	350S162	16	33	33	33	43	43	43	43	43	43	43	43	43
			24	43	43	43	54	54	54	54	54	68	68	68	68
		550S162	16	33	33	33	33	33	33	33	33	33	33	33	33
			24	43	43	43	43	43	43	43	43	43	43	43	43

For SI: 1 inch = 25.4 mm, 1 foot = 304.8 mm, 1 mil = 0.0254 mm, 1 mile per hour = 0.447 m/s, 1 pound per square foot = 0.0479 kPa, 1 Ksi = 1,000 psi = 6.895 MPa.

a. Deflection criterion: *L*/240.

b. Design load assumptions:
   Second floor dead load is 10 psf.
   Second floor live load is 30 psf.
   Roof/ceiling dead load is 12 psf.
   Attic live load is 10 psf.

c. Building width is in the direction of horizontal framing members supported by the wall studs.

❖ See the commentary for Section R603.3.2.

TABLE R603.3.2(13)
24-FOOT-WIDE BUILDING SUPPORTING ONE FLOOR, ROOF AND CEILING[a, b, c]
50 KSI STEEL

WIND SPEED		MEMBER SIZE	STUD SPACING (inches)	MINIMUM STUD THICKNESS (mils)											
				8-foot Studs				9-foot Studs				10-foot Studs			
				Ground Snow Load (psf)											
Exp. B	Exp. C			20	30	50	70	20	30	50	70	20	30	50	70
85 mph	—	350S162	16	33	33	33	33	33	33	33	33	33	33	33	33
			24	33	33	33	43	33	33	33	43	33	33	43	43
		550S162	16	33	33	33	33	33	33	33	33	33	33	33	33
			24	33	33	33	33	33	33	33	33	33	33	33	33
90 mph	—	350S162	16	33	33	33	33	33	33	33	33	33	33	33	33
			24	33	33	33	43	33	33	33	43	33	33	43	43
		550S162	16	33	33	33	33	33	33	33	33	33	33	33	33
			24	33	33	33	33	33	33	33	33	33	33	33	33
100 mph	85 mph	350S162	16	33	33	33	33	33	33	33	33	33	33	33	33
			24	33	33	33	43	33	33	33	43	43	43	43	43
		550S162	16	33	33	33	33	33	33	33	33	33	33	33	33
			24	33	33	33	33	33	33	33	33	33	33	33	33
110 mph	90 mph	350S162	16	33	33	33	33	33	33	33	33	33	33	33	33
			24	33	33	43	43	33	33	43	43	43	43	43	43
		550S162	16	33	33	33	33	33	33	33	33	33	33	33	33
			24	33	33	33	33	33	33	33	33	33	33	33	33
—	100 mph	350S162	16	33	33	33	33	33	33	33	33	33	33	33	33
			24	33	33	43	43	43	43	43	43	43	43	43	54
		550S162	16	33	33	33	33	33	33	33	33	33	33	33	33
			24	33	33	33	43	33	33	33	33	33	33	33	43
—	110 mph	350S162	16	33	33	33	33	33	33	33	33	33	33	43	43
			24	43	43	43	43	43	43	43	43	54	54	54	54
		550S162	16	33	33	33	33	33	33	33	33	33	33	33	33
			24	33	33	33	43	33	33	33	33	33	33	33	43

For SI: 1 inch = 25.4 mm, 1 foot = 304.8 mm, 1 mil = 0.0254 mm, 1 mile per hour = 0.447 m/s, 1 pound per square foot = 0.0479 kPa, 1 Ksi = 1,000 psi = 6.895 MPa.

a. Deflection criterion: *L*/240.

b. Design load assumptions:
  Second floor dead load is 10 psf.
  Second floor live load is 30 psf.
  Roof/ceiling dead load is 12 psf.
  Attic live load is 10 psf.

c. Building width is in the direction of horizontal framing members supported by the wall studs.

❖ See the commentary for Section R603.3.2.

TABLE R603.3.2(14)
28-FOOT-WIDE BUILDING SUPPORTING ONE FLOOR, ROOF AND CEILING[a, b, c]
33 KSI STEEL

WIND SPEED		MEMBER SIZE	STUD SPACING (inches)	MINIMUM STUD THICKNESS (mils)											
				8-foot Studs				9-foot Studs				10-foot Studs			
				Ground Snow Load (psf)											
Exp. B	Exp. C			20	30	50	70	20	30	50	70	20	30	50	70
85 mph	—	350S162	16	33	33	33	43	33	33	33	43	33	33	33	43
			24	43	43	43	54	43	43	43	54	43	43	43	54
		550S162	16	33	33	33	33	33	33	33	33	33	33	33	33
			24	33	33	43	43	33	33	43	43	33	33	43	43
90 mph	—	350S162	16	33	33	33	43	33	33	33	43	33	33	33	43
			24	43	43	43	54	43	43	43	54	43	43	43	54
		550S162	16	33	33	33	33	33	33	33	33	33	33	33	33
			24	33	33	43	43	33	33	43	43	33	33	43	43
100 mph	85 mph	350S162	16	33	33	33	43	33	33	33	43	33	33	43	43
			24	43	43	43	54	43	43	43	54	43	43	54	54
		550S162	16	33	33	33	33	33	33	33	33	33	33	33	33
			24	33	33	43	43	33	33	43	43	33	33	43	43
110 mph	90 mph	350S162	16	33	33	33	43	33	33	33	43	43	43	43	43
			24	43	43	43	54	43	43	43	54	54	54	54	54
		550S162	16	33	33	33	33	33	33	33	33	33	33	33	33
			24	33	33	43	43	33	33	43	43	43	43	43	43
—	100 mph	350S162	16	33	33	33	43	33	33	43	43	43	43	43	43
			24	43	43	43	54	54	54	54	54	54	54	54	68
		550S162	16	33	33	33	33	33	33	33	33	33	33	33	33
			24	33	33	43	43	43	43	43	43	43	43	43	43
—	110 mph	350S162	16	33	33	43	43	43	43	43	43	43	43	43	54
			24	43	43	54	54	54	54	54	54	68	68	68	68
		550S162	16	33	33	33	33	33	33	33	33	33	33	33	33
			24	43	43	43	43	43	43	43	43	43	43	43	43

For SI: 1 inch = 25.4 mm, 1 foot = 304.8 mm, 1 mil = 0.0254 mm, 1 mile per hour = 0.447 m/s, 1 pound per square foot = 0.0479 kPa,
1 Ksi = 1,000 psi = 6.895 MPa.

a. Deflection criterion: *L*/240.

b. Design load assumptions:
Second floor dead load is 10 psf.
Second floor live load is 30 psf.
Roof/ceiling dead load is 12 psf.
Attic live load is 10 psf.

c. Building width is in the direction of horizontal framing members supported by the wall studs.

❖ See the commentary for Section R603.3.2.

TABLE R603.3.2(15)
28-FOOT-WIDE BUILDING SUPPORTING ONE FLOOR, ROOF AND CEILING[a, b, c]
50 KSI STEEL

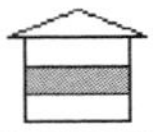

WIND SPEED		MEMBER SIZE	STUD SPACING (inches)	MINIMUM STUD THICKNESS (mils)											
				8-foot Studs				9-foot Studs				10-foot Studs			
				Ground Snow Load (psf)											
Exp. B	Exp. C			20	30	50	70	20	30	50	70	20	30	50	70
85 mph	—	350S162	16	33	33	33	33	33	33	33	33	33	33	33	33
			24	33	33	43	43	33	33	43	43	43	43	43	54
		550S162	16	33	33	33	33	33	33	33	33	33	33	33	33
			24	33	33	33	43	33	33	33	43	33	33	33	43
90 mph	—	350S162	16	33	33	33	33	33	33	33	33	33	33	33	33
			24	33	33	43	43	33	33	43	43	43	43	43	54
		550S162	16	33	33	33	33	33	33	33	33	33	33	33	33
			24	33	33	33	43	33	33	33	43	33	33	33	43
100 mph	85 mph	350S162	16	33	33	33	33	33	33	33	33	33	33	33	43
			24	33	33	43	43	33	33	43	43	43	43	43	54
		550S162	16	33	33	33	33	33	33	33	33	33	33	33	33
			24	33	33	33	43	33	33	33	43	33	33	33	43
110 mph	90 mph	350S162	16	33	33	33	33	33	33	33	33	33	33	33	43
			24	33	33	43	43	43	43	43	43	43	43	43	54
		550S162	16	33	33	33	33	33	33	33	33	33	33	33	33
			24	33	33	33	43	33	33	33	43	33	33	33	43
—	100 mph	350S162	16	33	33	33	33	33	33	33	33	33	33	33	43
			24	43	43	43	54	43	43	43	43	43	43	54	54
		550S162	16	33	33	33	33	33	33	33	33	33	33	33	33
			24	33	33	33	43	33	33	33	43	33	33	33	43
—	110 mph	350S162	16	33	33	33	43	33	33	33	33	43	43	43	43
			24	43	43	43	54	43	43	43	43	54	54	54	54
		550S162	16	33	33	33	33	33	33	33	33	33	33	33	33
			24	33	33	33	43	33	33	33	43	33	33	33	43

For SI: 1 inch = 25.4 mm, 1 foot = 304.8 mm, 1 mil = 0.0254 mm, 1 mile per hour = 0.447 m/s, 1 pound per square foot = 0.0479 kPa, 1 Ksi = 1,000 psi = 6.895 MPa.

a. Deflection criterion: *L*/240.

b. Design load assumptions:

Second floor dead load is 10 psf.

Second floor live load is 30 psf.

Roof/ceiling dead load is 12 psf.

Attic live load is 10 psf.

c. Building width is in the direction of horizontal framing members supported by the wall studs.

❖ See the commentary for Section R603.3.2.

TABLE R603.3.2(16)
32-FOOT-WIDE BUILDING SUPPORTING ONE FLOOR, ROOF AND CEILING[a, b, c]
33 KSI STEEL

WIND SPEED		MEMBER SIZE	STUD SPACING (inches)	MINIMUM STUD THICKNESS (mils)											
				8-foot Studs				9-foot Studs				10-foot Studs			
				Ground Snow Load (psf)											
Exp. B	Exp. C			20	30	50	70	20	30	50	70	20	30	50	70
85 mph	—	350S162	16	33	33	33	43	33	33	33	43	33	33	43	43
			24	43	43	43	54	43	43	43	54	43	43	54	54
		550S162	16	33	33	33	43	33	33	33	33	33	33	33	43
			24	33	43	43	54	33	33	43	43	33	33	43	43
90 mph	—	350S162	16	33	33	33	43	33	33	33	43	33	33	43	43
			24	43	43	43	54	43	43	43	54	43	43	54	54
		550S162	16	33	33	33	43	33	33	33	33	33	33	33	43
			24	33	43	43	54	33	33	43	43	33	33	43	43
100 mph	85 mph	350S162	16	33	33	33	43	33	33	33	43	33	43	43	43
			24	43	43	43	54	43	43	43	54	54	54	54	68
		550S162	16	33	33	33	43	33	33	33	33	33	33	33	43
			24	33	43	43	54	33	33	43	43	33	33	43	43
110 mph	90 mph	350S162	16	33	33	43	43	33	33	33	43	43	43	43	43
			24	43	43	54	54	43	43	54	54	54	54	54	68
		550S162	16	33	33	33	43	33	33	33	33	33	33	33	43
			24	33	43	43	54	33	33	43	43	43	43	43	54
—	100 mph	350S162	16	33	33	43	43	43	43	43	43	43	43	43	43
			24	43	43	54	54	54	54	54	54	54	54	54	54
		550S162	16	33	33	33	43	33	33	33	33	33	33	33	43
			24	33	43	43	54	43	43	43	43	43	43	43	54
—	110 mph	350S162	16	43	43	43	43	43	43	43	43	43	43	54	54
			24	54	54	54	68	54	54	54	68	68	68	68	68
		550S162	16	33	33	33	43	33	33	33	43	33	33	33	43
			24	43	43	43	54	43	43	43	43	43	43	43	54

For SI: 1 inch = 25.4 mm, 1 foot = 304.8 mm, 1 mil = 0.0254 mm, 1 mile per hour = 0.447 m/s, 1 pound per square foot = 0.0479 kPa, 1 Ksi = 1,000 psi = 6.895 MPa.

a. Deflection criterion: *L*/240.

b. Design load assumptions:

Second floor dead load is 10 psf.

Second floor live load is 30 psf.

Roof/ceiling dead load is 12 psf.

Attic live load is 10 psf.

c. Building width is in the direction of horizontal framing members supported by the wall studs.

❖ See the commentary for Section R603.3.2.

TABLE R603.3.2(17)
32-FOOT-WIDE BUILDING SUPPORTING ONE FLOOR, ROOF AND CEILING[a, b, c]
50 KSI STEEL

WIND SPEED Exp. B	WIND SPEED Exp. C	MEMBER SIZE	STUD SPACING (inches)	MINIMUM STUD THICKNESS (mils) 8-foot Studs, Ground Snow Load (psf) 20	30	50	70	9-foot Studs 20	30	50	70	10-foot Studs 20	30	50	70
85 mph	—	350S162	16	33	33	33	43	33	33	33	33	33	33	33	43
			24	33	33	43	54	33	33	43	43	43	43	43	54
		550S162	16	33	33	33	33	33	33	33	33	33	33	33	33
			24	33	33	43	43	33	33	33	43	33	33	33	43
90 mph	—	350S162	16	33	33	33	43	33	33	33	33	33	33	33	43
			24	33	33	43	54	33	33	43	43	43	43	43	54
		550S162	16	33	33	33	33	33	33	33	33	33	33	33	33
			24	33	33	43	43	33	33	33	43	33	33	33	43
100 mph	85 mph	350S162	16	33	33	33	43	33	33	33	33	33	33	33	43
			24	33	33	43	54	33	33	43	43	43	43	43	54
		550S162	16	33	33	33	33	33	33	33	33	33	33	33	33
			24	33	33	43	43	33	33	33	43	33	33	33	43
110 mph	90 mph	350S162	16	33	33	33	43	33	33	33	33	33	33	33	43
			24	43	43	43	54	43	43	43	54	43	43	54	54
		550S162	16	33	33	33	33	33	33	33	33	33	33	33	33
			24	33	33	43	43	33	33	33	43	33	33	33	43
—	100 mph	350S162	16	33	33	33	43	33	33	33	43	33	33	43	43
			24	43	43	43	54	43	43	43	54	54	54	54	54
		550S162	16	33	33	33	33	33	33	33	33	33	33	33	33
			24	33	33	43	43	33	33	33	43	33	33	43	43
—	110 mph	350S162	16	33	33	33	43	33	33	33	43	43	43	43	43
			24	43	43	43	54	43	43	43	54	54	54	54	54
		550S162	16	33	33	33	33	33	33	33	33	33	33	33	33
			24	33	33	43	43	33	33	33	43	33	33	43	43

For SI: 1 inch = 25.4 mm, 1 foot = 304.8 mm, 1 mil = 0.0254 mm, 1 mile per hour = 0.447 m/s, 1 pound per square foot = 0.0479 kPa, 1 Ksi = 1,000 psi = 6.895 MPa.

a. Deflection criterion: *L*/240.

b. Design load assumptions:
   Second floor dead load is 10 psf.
   Second floor live load is 30 psf.
   Roof/ceiling dead load is 12 psf.
   Attic live load is 10 psf.

c. Building width is in the direction of horizontal framing members supported by the wall studs.

❖ See the commentary for Section R603.3.2.

TABLE R603.3.2(18)
36-FOOT-WIDE BUILDING SUPPORTING ONE FLOOR, ROOF AND CEILING[a, b, c]
33 KSI STEEL

WIND SPEED		MEMBER SIZE	STUD SPACING (inches)	MINIMUM STUD THICKNESS (mils)											
				8-foot Studs				9-foot Studs				10-foot Studs			
				Ground Snow Load (psf)											
Exp. B	Exp. C			20	30	50	70	20	30	50	70	20	30	50	70
85 mph	—	350S162	16	33	33	43	43	33	33	43	43	33	33	43	43
			24	43	43	54	54	43	43	54	54	54	54	54	68
		550S162	16	33	33	33	43	33	33	33	43	33	33	33	43
			24	43	43	43	54	43	43	43	54	43	43	43	54
90 mph	—	350S162	16	33	33	43	43	33	33	43	43	33	33	43	43
			24	43	43	54	54	43	43	54	54	54	54	54	68
		550S162	16	33	33	33	43	33	33	33	43	33	33	33	43
			24	43	43	43	54	43	43	43	54	43	43	43	54
100 mph	85 mph	350S162	16	33	33	43	43	33	33	43	43	43	43	43	43
			24	43	43	54	68	43	43	54	54	54	54	54	68
		550S162	16	33	33	33	43	33	33	33	43	33	33	33	43
			24	43	43	43	54	43	43	43	54	43	43	43	54
110 mph	90 mph	350S162	16	33	33	43	43	33	33	43	43	43	43	43	54
			24	43	43	54	68	54	54	54	54	54	54	54	68
		550S162	16	33	33	33	43	33	33	33	43	33	33	33	43
			24	43	43	43	54	43	43	43	54	43	43	43	54
—	100 mph	350S162	16	33	33	43	43	43	43	43	43	43	43	43	54
			24	54	54	54	68	54	54	54	68	54	68	68	68
		550S162	16	33	33	33	43	33	33	33	43	33	33	33	43
			24	43	43	43	54	43	43	43	54	43	43	43	54
—	110 mph	350S162	16	43	43	43	43	43	43	43	43	43	54	54	54
			24	54	54	54	68	54	54	54	68	68	68	68	68
		550S162	16	33	33	33	43	33	33	33	43	33	33	33	43
			24	43	43	43	54	43	43	43	54	43	43	43	54

For SI: 1 inch = 25.4 mm, 1 foot = 304.8 mm, 1 mil = 0.0254 mm, 1 mile per hour = 0.447 m/s, 1 pound per square foot = 0.0479 kPa, 1 Ksi = 1,000 psi = 6.895 MPa.

a. Deflection criterion: *L*/240.

b. Design load assumptions:
  Second floor dead load is 10 psf.
  Second floor live load is 30 psf.
  Roof/ceiling dead load is 12 psf.
  Attic live load is 10 psf.

c. Building width is in the direction of horizontal framing members supported by the wall studs.

❖ See the commentary for Section R603.3.2.

TABLE R603.3.2(19)
36-FOOT-WIDE BUILDING SUPPORTING ONE FLOOR, ROOF AND CEILING[a, b, c]
50 KSI STEEL

WIND SPEED		MEMBER SIZE	STUD SPACING (inches)	MINIMUM STUD THICKNESS (mils)											
				8-foot Studs				9-foot Studs				10-foot Studs			
				Ground Snow Load (psf)											
Exp. B	Exp. C			20	30	50	70	20	30	50	70	20	30	50	70
85 mph	—	350S162	16	33	33	33	43	33	33	33	43	33	33	33	43
			24	43	43	43	54	33	33	43	54	43	43	43	54
		550S162	16	33	33	33	33	33	33	33	33	33	33	33	33
			24	33	33	43	43	33	33	43	43	33	33	43	43
90 mph	—	350S162	16	33	33	33	43	33	33	33	43	33	33	33	43
			24	43	43	43	54	33	33	43	54	43	43	43	54
		550S162	16	33	33	33	33	33	33	33	33	33	33	33	33
			24	33	33	43	43	33	33	43	43	33	33	43	43
100 mph	85 mph	350S162	16	33	33	33	43	33	33	33	43	33	33	33	43
			24	43	43	43	54	43	43	43	54	43	43	54	54
		550S162	16	33	33	33	33	33	33	33	33	33	33	33	33
			24	33	33	43	43	33	33	43	43	33	33	43	43
110 mph	90 mph	350S162	16	33	33	33	43	33	33	33	43	33	33	43	43
			24	43	43	43	54	43	43	43	54	43	43	54	54
		550S162	16	33	33	33	33	33	33	33	33	33	33	33	33
			24	33	33	43	43	33	33	43	43	33	33	43	43
—	100 mph	350S162	16	33	33	33	43	33	33	33	43	43	43	43	43
			24	43	43	43	54	43	43	43	54	54	54	54	68
		550S162	16	33	33	33	33	33	33	33	33	33	33	33	33
			24	33	33	43	43	33	33	43	43	33	33	43	43
—	110 mph	350S162	16	33	33	43	43	33	33	33	43	43	43	43	43
			24	43	43	54	54	43	43	54	54	54	54	54	68
		550S162	16	33	33	33	33	33	33	33	33	33	33	33	33
			24	33	33	43	43	33	33	43	43	43	43	43	43

For SI: 1 inch = 25.4 mm, 1 foot = 304.8 mm, 1 mil = 0.0254 mm, 1 mile per hour = 0.447 m/s, 1 pound per square foot = 0.0479 kPa, 1 Ksi = 1,000 psi = 6.895 MPa.

a. Deflection criterion: *L*/240.

b. Design load assumptions:
Second floor dead load is 10 psf.
Second floor live load is 30 psf.
Roof/ceiling dead load is 12 psf.
Attic live load is 10 psf.

c. Building width is in the direction of horizontal framing members supported by the wall studs.

❖ See the commentary for Section R603.3.2.

TABLE R603.3.2(20)
40-FOOT-WIDE BUILDING SUPPORTING ONE FLOOR, ROOF AND CEILING[a, b, c]
33 KSI STEE

WIND SPEED		MEMBER SIZE	STUD SPACING (inches)	MINIMUM STUD THICKNESS (mils)											
				8-foot Studs				9-foot Studs				10-foot Studs			
				Ground Snow Load (psf)											
Exp. B	Exp. C			20	30	50	70	20	30	50	70	20	30	50	70
85 mph	—	350S162	16	33	33	43	43	33	33	43	43	43	43	43	54
			24	43	43	54	68	43	43	54	68	54	54	54	68
		550S162	16	33	33	33	43	33	33	33	43	33	33	33	43
			24	43	43	54	54	43	43	43	54	43	43	43	54
90 mph	—	350S162	16	33	33	43	43	33	33	43	43	43	43	43	54
			24	43	43	54	68	43	43	54	68	54	54	54	68
		550S162	16	33	33	33	43	33	33	33	43	33	33	33	43
			24	43	43	54	54	43	43	43	54	43	43	43	54
100 mph	85 mph	350S162	16	33	33	43	43	33	33	43	43	43	43	43	54
			24	43	43	54	68	43	43	54	68	54	54	54	68
		550S162	16	33	33	33	43	33	33	33	43	33	33	33	43
			24	43	43	54	54	43	43	43	54	43	43	43	54
110 mph	90 mph	350S162	16	33	33	43	43	43	43	43	43	43	43	43	54
			24	43	43	54	68	54	54	54	68	54	54	68	68
		550S162	16	33	33	43	43	33	33	33	43	33	33	33	43
			24	43	43	54	54	43	43	43	54	43	43	43	54
—	100 mph	350S162	16	43	43	43	54	43	43	43	54	43	43	54	54
			24	54	54	54	68	54	54	54	68	68	68	68	97
		550S162	16	33	33	43	43	33	33	33	43	33	33	43	43
			24	43	43	54	54	43	43	43	54	43	43	54	54
—	110 mph	350S162	16	43	43	43	54	43	43	43	54	54	54	54	54
			24	54	54	54	68	54	54	68	68	68	68	68	97
		550S162	16	33	33	43	43	33	33	33	43	33	33	43	43
			24	43	43	54	54	43	43	43	54	43	43	54	54

For SI: 1 inch = 25.4 mm, 1 foot = 304.8 mm, 1 mil = 0.0254 mm, 1 mile per hour = 0.447 m/s, 1 pound per square foot = 0.0479 kPa,
1 Ksi = 1,000 psi = 6.895 MPa.

a. Deflection criterion: *L*/240.

b. Design load assumptions:
   Second floor dead load is 10 psf.
   Second floor live load is 30 psf.
   Roof/ceiling dead load is 12 psf.
   Attic live load is 10 psf.

c. Building width is in the direction of horizontal framing members supported by the wall studs.

❖ See the commentary for Section R603.3.2.

TABLE R603.3.2(21)
40-FOOT-WIDE BUILDING SUPPORTING ONE FLOOR, ROOF AND CEILING[a, b, c]
50 KSI STEEL

WIND SPEED		MEMBER SIZE	STUD SPACING (inches)	MINIMUM STUD THICKNESS (mils)											
				8-foot Studs				9-foot Studs				10-foot Studs			
				Ground Snow Load (psf)											
Exp. B	Exp. C			20	30	50	70	20	30	50	70	20	30	50	70
85 mph	—	350S162	16	33	33	33	43	33	33	33	43	33	33	43	43
			24	43	43	43	54	43	43	43	54	43	43	54	54
		550S162	16	33	33	33	43	33	33	33	33	33	33	33	33
			24	33	43	43	54	33	33	43	43	33	33	43	43
90 mph	—	350S162	16	33	33	33	43	33	33	33	43	33	33	43	43
			24	43	43	43	54	43	43	43	54	43	43	54	54
		550S162	16	33	33	33	43	33	33	33	33	33	33	33	33
			24	33	43	43	54	33	33	43	43	33	33	43	43
100 mph	85 mph	350S162	16	33	33	33	43	33	33	33	43	33	33	43	43
			24	43	43	54	54	43	43	43	54	43	43	54	68
		550S162	16	33	33	33	43	33	33	33	33	33	33	33	33
			24	33	43	43	54	33	33	43	43	33	33	43	43
110 mph	90 mph	350S162	16	33	33	43	43	33	33	33	43	33	33	43	43
			24	43	43	54	54	43	43	43	54	54	54	54	68
		550S162	16	33	33	33	43	33	33	33	33	33	33	33	43
			24	33	43	43	54	33	33	43	43	33	33	43	43
—	100 mph	350S162	16	33	33	43	43	33	33	33	43	43	43	43	43
			24	43	43	54	54	43	43	54	54	54	54	54	68
		550S162	16	33	33	33	43	33	33	33	33	33	33	33	43
			24	33	43	43	54	33	33	43	43	33	43	43	43
—	110 mph	350S162	16	33	33	43	43	33	33	43	43	43	43	43	54
			24	43	43	54	68	54	54	54	54	54	54	54	68
		550S162	16	33	33	33	43	33	33	33	33	33	33	33	43
			24	33	43	43	54	33	33	43	43	43	43	43	54

For SI: 1 inch = 25.4 mm, 1 foot = 304.8 mm, 1 mil = 0.0254 mm, 1 mile per hour = 0.447 m/s, 1 pound per square foot = 0.0479 kPa, 1 Ksi = 1,000 psi = 6.895 MPa.

a. Deflection criterion: *L*/240.

b. Design load assumptions:
Second floor dead load is 10 psf.
Second floor live load is 30 psf.
Roof/ceiling dead load is 12 psf.
Attic live load is 10 psf.

c. Building width is in the direction of horizontal framing members supported by the wall studs.

❖ See the commentary for Section R603.3.2.

TABLE R603.3.2(22)
24-FOOT-WIDE BUILDING SUPPORTING TWO FLOORS, ROOF AND CEILING[a, b, c]
33 KSI STEEL

WIND SPEED		MEMBER SIZE	STUD SPACING (inches)	MINIMUM STUD THICKNESS (mils)											
				8-foot Studs				9-foot Studs				10-foot Studs			
				Ground Snow Load (psf)											
Exp. B	Exp. C			20	30	50	70	20	30	50	70	20	30	50	70
85 mph	—	350S162	16	43	43	43	43	33	33	33	43	43	43	43	43
			24	54	54	54	54	43	43	54	54	54	54	54	54
		550S162	16	33	33	43	43	33	33	33	33	33	33	33	43
			24	43	43	54	54	43	43	43	43	43	43	43	54
90 mph	—	350S162	16	43	43	43	43	33	33	33	43	43	43	43	43
			24	54	54	54	54	43	43	54	54	54	54	54	54
		550S162	16	33	33	43	43	33	33	33	33	33	33	33	43
			24	43	43	54	54	43	43	43	43	43	43	43	54
100 mph	85 mph	350S162	16	43	43	43	43	33	33	33	43	43	43	43	43
			24	54	54	54	54	54	54	54	54	54	54	54	68
		550S162	16	33	33	43	43	33	33	33	33	33	33	33	43
			24	43	43	54	54	43	43	43	43	43	43	43	54
110 mph	90 mph	350S162	16	43	43	43	43	43	43	43	43	43	43	43	43
			24	54	54	54	54	54	54	54	54	54	54	68	68
		550S162	16	33	33	43	43	33	33	33	33	33	33	33	43
			24	43	43	54	54	43	43	43	43	43	43	43	54
—	100 mph	350S162	16	43	43	43	43	43	43	43	43	43	43	43	54
			24	54	54	54	54	54	54	54	54	68	68	68	68
		550S162	16	33	33	43	43	33	33	33	33	33	33	33	43
			24	43	43	54	54	43	43	43	43	43	43	43	54
—	110 mph	350S162	16	43	43	43	43	43	43	43	43	54	54	54	54
			24	54	54	54	68	54	54	68	68	68	68	68	97
		550S162	16	33	33	43	43	33	33	33	33	33	33	33	43
			24	43	43	54	54	43	43	43	43	43	43	43	54

For SI: 1 inch = 25.4 mm, 1 foot = 304.8 mm, 1 mil = 0.0254 mm, 1 mile per hour = 0.447 m/s, 1 pound per square foot = 0.0479 kPa, 1 Ksi = 1,000 psi = 6.895 MPa.

a. Deflection criterion: *L*/240.

b. Design load assumptions:
   Top and middle floor dead load is 10 psf.
   Top floor live load is 30 psf.
   Middle floor live load is 40 psf.
   Roof/ceiling dead load is 12 psf.
   Attic live load is 10 psf.

c. Building width is in the direction of horizontal framing members supported by the wall studs.

❖ See the commentary for Section R603.3.2.

TABLE R603.3.2(23)
24-FOOT-WIDE BUILDING SUPPORTING TWO FLOORS, ROOF AND CEILING[a, b, c]
33 KSI STEEL

WIND SPEED		MEMBER SIZE	STUD SPACING (inches)	MINIMUM STUD THICKNESS (mils)											
				8-foot Studs				9-foot Studs				10-foot Studs			
				Ground Snow Load (psf)											
Exp. B	Exp. C			20	30	50	70	20	30	50	70	20	30	50	70
85 mph	—	350S162	16	33	33	33	43	33	33	33	33	33	33	33	33
			24	43	43	54	54	43	43	43	43	43	43	43	54
		550S162	16	33	33	33	33	33	33	33	33	33	33	33	33
			24	43	43	43	43	43	43	43	43	43	43	43	43
90 mph	—	350S162	16	33	33	33	43	33	33	33	33	33	33	33	33
			24	43	43	54	54	43	43	43	43	43	43	43	54
		550S162	16	33	33	33	33	33	33	33	33	33	33	33	33
			24	43	43	43	43	43	43	43	43	43	43	43	43
100 mph	85 mph	350S162	16	33	33	33	43	33	33	33	33	33	33	33	33
			24	43	43	54	54	43	43	43	43	43	43	54	54
		550S162	16	33	33	33	33	33	33	33	33	33	33	33	33
			24	43	43	43	43	43	43	43	43	43	43	43	43
110 mph	90 mph	350S162	16	33	33	33	43	33	33	33	33	33	33	43	43
			24	43	43	54	54	43	43	43	43	54	54	54	54
		550S162	16	33	33	33	33	33	33	33	33	33	33	33	33
			24	43	43	43	43	43	43	43	43	43	43	43	43
—	100 mph	350S162	16	33	33	33	43	33	33	33	33	43	43	43	43
			24	43	43	54	54	43	43	54	54	54	54	54	54
		550S162	16	33	33	33	33	33	33	33	33	33	33	33	33
			24	43	43	43	43	43	43	43	43	43	43	43	43
—	110 mph	350S162	16	33	33	33	43	33	33	33	43	43	43	43	43
			24	54	54	54	54	54	54	54	54	54	54	54	68
		550S162	16	33	33	33	33	33	33	33	33	33	33	33	33
			24	43	43	43	43	43	43	43	43	43	43	43	43

For SI: 1 inch = 25.4 mm, 1 foot = 304.8 mm, 1 mil = 0.0254 mm, 1 mile per hour = 0.447 m/s, 1 pound per square foot = 0.0479 kPa, 1 Ksi = 1,000 psi = 6.895 MPa.

a. Deflection criterion: *L*/240.

b. Design load assumptions:
   - Top and middle floor dead load is 10 psf.
   - Top floor live load is 30 psf.
   - Middle floor live load is 40 psf.
   - Attic live load is 10 psf.

c. Building width is in the direction of horizontal framing members supported by the wall studs.

❖ See the commentary for Section R603.3.2.

TABLE R603.3.2(24) 28-FOOT-WIDE BUILDING SUPPORTING TWO FLOORS, ROOF AND CEILING[a, b, c] 33 KSI STEEL

WIND SPEED		MEMBER SIZE	STUD SPACING (inches)	MINIMUM STUD THICKNESS (mils)											
				8-foot Studs				9-foot Studs				10-foot Studs			
				Ground Snow Load (psf)											
Exp. B	Exp. C			20	30	50	70	20	30	50	70	20	30	50	70
85 mph	—	350S162	16	43	43	43	43	43	43	43	43	43	43	43	43
			24	54	54	54	68	54	54	54	54	54	54	54	68
		550S162	16	43	43	43	43	43	43	43	43	43	43	43	43
			24	54	54	54	54	54	54	54	54	54	54	54	54
90 mph	—	350S162	16	43	43	43	43	43	43	43	43	43	43	43	43
			24	54	54	54	68	54	54	54	54	54	54	54	68
		550S162	16	43	43	43	43	43	43	43	43	43	43	43	43
			24	54	54	54	54	54	54	54	54	54	54	54	54
100 mph	85 mph	350S162	16	43	43	43	43	43	43	43	43	43	43	43	43
			24	54	54	54	68	54	54	54	54	54	54	68	68
		550S162	16	43	43	43	43	43	43	43	43	43	43	43	43
			24	54	54	54	54	54	54	54	54	54	54	54	54
110 mph	90 mph	350S162	16	43	43	43	43	43	43	43	43	43	43	43	43
			24	54	54	54	68	54	54	54	54	68	68	68	68
		550S162	16	43	43	43	43	43	43	43	43	43	43	43	43
			24	54	54	54	54	54	54	54	54	54	54	54	54
—	100 mph	350S162	16	43	43	43	43	43	43	43	43	43	43	54	54
			24	54	54	54	68	54	54	68	68	68	68	68	97
		550S162	16	43	43	43	43	43	43	43	43	43	43	43	43
			24	54	54	54	54	54	54	54	54	54	54	54	54
—	110 mph	350S162	16	43	43	43	43	43	43	43	43	54	54	54	54
			24	54	68	68	68	68	68	68	68	68	68	97	97
		550S162	16	43	43	43	43	43	43	43	43	43	43	43	43
			24	54	54	54	54	54	54	54	54	54	54	54	54

For SI: 1 inch = 25.4 mm, 1 foot = 304.8 mm, 1 mil = 0.0254 mm, 1 mile per hour = 0.447 m/s, 1 pound per square foot = 0.0479 kPa, 1 Ksi = 1,000 psi = 6.895 MPa.

a. Deflection criterion: *L*/240.

b. Design load assumptions:

Top and middle floor dead load is 10 psf.

Top floor live load is 30 psf.

Middle floor live load is 40 psf.

Roof/ceiling dead load is 12 psf.

Attic live load is 10 psf.

c. Building width is in the direction of horizontal framing members supported by the wall studs.

❖ See the commentary for Section R603.3.2.

TABLE R603.3.2(25)
28-FOOT-WIDE BUILDING SUPPORTING TWO FLOORS, ROOF AND CEILING[a, b, c]
50 KSI STEEL

WIND SPEED		MEMBER SIZE	STUD SPACING (inches)	MINIMUM STUD THICKNESS (mils)											
				8-foot Studs				9-foot Studs				10-foot Studs			
				Ground Snow Load (psf)											
Exp. B	Exp. C			20	30	50	70	20	30	50	70	20	30	50	70
85 mph	—	350S162	16	43	43	43	43	33	33	33	43	43	43	43	43
			24	54	54	54	54	43	43	54	54	54	54	54	54
		550S162	16	33	33	33	43	33	33	33	33	33	33	33	33
			24	43	43	43	54	43	43	43	43	43	43	43	43
90 mph	—	350S162	16	43	43	43	43	33	33	33	43	43	43	43	43
			24	54	54	54	54	43	43	54	54	54	54	54	54
		550S162	16	33	33	33	43	33	33	33	33	33	33	33	33
			24	43	43	43	54	43	43	43	43	43	43	43	43
100 mph	85 mph	350S162	16	43	43	43	43	33	33	33	43	43	43	43	43
			24	54	54	54	54	43	43	54	54	54	54	54	54
		550S162	16	33	33	33	43	33	33	33	33	33	33	33	33
			24	43	43	43	54	43	43	43	43	43	43	43	43
110 mph	90 mph	350S162	16	43	43	43	43	33	33	33	43	43	43	43	43
			24	54	54	54	54	43	43	54	54	54	54	54	54
		550S162	16	33	33	33	43	33	33	33	33	33	33	33	33
			24	43	43	43	54	43	43	43	43	43	43	43	43
—	100 mph	350S162	16	43	43	43	43	33	33	33	43	43	43	43	43
			24	54	54	54	54	54	54	54	54	54	54	54	68
		550S162	16	33	33	33	43	33	33	33	33	33	33	33	33
			24	43	43	43	54	43	43	43	43	43	43	43	43
—	110 mph	350S162	16	43	43	43	43	43	43	43	43	43	43	43	43
			24	54	54	54	54	54	54	54	54	68	68	68	68
		550S162	16	33	33	33	43	33	33	33	33	33	33	33	33
			24	43	43	43	54	43	43	43	43	43	43	43	43

For SI: 1 inch = 25.4 mm, 1 foot = 304.8 mm, 1 mil = 0.0254 mm, 1 mile per hour = 0.447 m/s, 1 pound per square foot = 0.0479 kPa, 1 Ksi = 1,000 psi = 6.895 MPa.

a. Deflection criterion: *L*/240.

b. Design load assumptions:

Top and middle floor dead load is 10 psf.

Top floor live load is 30 psf.

Middle floor live load is 40 psf.

Roof/ceiling dead load is 12 psf.

Attic live load is 10 psf.

c. Building width is in the direction of horizontal framing members supported by the wall studs.

❖ See the commentary for Section R603.3.2.

TABLE R603.3.2(26)
32-FOOT-WIDE BUILDING SUPPORTING TWO FLOORS, ROOF AND CEILING[a, b, c]
33 KSI STEEL

WIND SPEED		MEMBER SIZE	STUD SPACING (inches)	MINIMUM STUD THICKNESS (mils)											
				8-foot Studs				9-foot Studs				10-foot Studs			
				Ground Snow Load (psf)											
Exp. B	Exp. C			20	30	50	70	20	30	50	70	20	30	50	70
85 mph	—	350S162	16	43	43	43	54	43	43	43	43	43	43	43	54
			24	68	68	68	68	54	54	68	68	68	68	68	68
		550S162	16	43	43	43	43	43	43	43	43	43	43	43	43
			24	54	54	54	68	54	54	54	54	54	54	54	54
90 mph	—	350S162	16	43	43	43	54	43	43	43	43	43	43	43	54
			24	68	68	68	68	54	54	68	68	68	68	68	68
		550S162	16	43	43	43	43	43	43	43	43	43	43	43	43
			24	54	54	54	68	54	54	54	54	54	54	54	54
100 mph	85 mph	350S162	16	43	43	43	54	43	43	43	43	43	43	43	54
			24	68	68	68	68	54	54	68	68	68	68	68	68
		550S162	16	43	43	43	43	43	43	43	43	43	43	43	43
			24	54	54	54	68	54	54	54	54	54	54	54	54
110 mph	90 mph	350S162	16	43	43	43	54	43	43	43	43	43	43	54	54
			24	68	68	68	68	54	54	68	68	68	68	68	68
		550S162	16	43	43	43	43	43	43	43	43	43	43	43	43
			24	54	54	54	68	54	54	54	54	54	54	54	54
—	100 mph	350S162	16	43	43	43	54	43	43	43	43	54	54	54	54
			24	68	68	68	68	68	68	68	68	68	68	97	97
		550S162	16	43	43	43	43	43	43	43	43	43	43	43	43
			24	54	54	54	68	54	54	54	54	54	54	54	54
—	110 mph	350S162	16	43	43	43	54	43	43	54	54	54	54	54	54
			24	68	68	68	68	68	68	68	68	97	97	97	97
		550S162	16	43	43	43	43	43	43	43	43	43	43	43	43
			24	54	54	54	68	54	54	54	54	54	54	54	54

For SI: 1 inch = 25.4 mm, 1 foot = 304.8 mm, 1 mil = 0.0254 mm, 1 mile per hour = 0.447 m/s, 1 pound per square foot = 0.0479 kPa, 1 Ksi = 1,000 psi = 6.895 MPa.

a. Deflection criterion: *L*/240.

b. Design load assumptions:
   Top and middle floor dead load is 10 psf.
   Top floor live load is 30 psf.
   Middle floor live load is 40 psf.
   Roof/ceiling dead load is 12 psf.
   Attic live load is 10 psf.

c. Building width is in the direction of horizontal framing members supported by the wall studs.

❖ See the commentary for Section R603.3.2.

TABLE R603.3.2(27)
32-FOOT-WIDE BUILDING SUPPORTING TWO FLOORS, ROOF AND CEILING[a, b, c]
50 KSI STEEL

WIND SPEED		MEMBER SIZE	STUD SPACING (inches)	MINIMUM STUD THICKNESS (mils)											
				8-foot Studs				9-foot Studs				10-foot Studs			
				Ground Snow Load (psf)											
Exp. B	Exp. C			20	30	50	70	20	30	50	70	20	30	50	70
85 mph	—	350S162	16	43	43	43	43	43	43	43	43	43	43	43	43
			24	54	54	54	68	54	54	54	54	54	54	54	68
		550S162	16	43	43	43	43	33	33	33	43	33	33	43	43
			24	54	54	54	54	43	43	43	54	43	43	54	54
90 mph	—	350S162	16	43	43	43	43	43	43	43	43	43	43	43	43
			24	54	54	54	68	54	54	54	54	54	54	54	68
		550S162	16	43	43	43	43	33	33	33	43	33	33	43	43
			24	54	54	54	54	43	43	43	54	43	43	54	54
100 mph	85 mph	350S162	16	43	43	43	43	43	43	43	43	43	43	43	43
			24	54	54	54	68	54	54	54	54	54	54	54	68
		550S162	16	43	43	43	43	33	33	33	43	33	33	43	43
			24	54	54	54	54	43	43	43	54	43	43	54	54
110 mph	90 mph	350S162	16	43	43	43	43	43	43	43	43	43	43	43	43
			24	54	54	54	68	54	54	54	54	54	54	54	68
		550S162	16	43	43	43	43	33	33	33	43	33	33	43	43
			24	54	54	54	54	43	43	43	54	43	43	54	54
—	100 mph	350S162	16	43	43	43	43	43	43	43	43	43	43	43	43
			24	54	54	54	68	54	54	54	54	68	68	68	68
		550S162	16	43	43	43	43	33	33	33	43	33	33	43	43
			24	54	54	54	54	43	43	43	54	43	43	54	54
—	110 mph	350S162	16	43	43	43	43	43	43	43	43	43	43	43	54
			24	54	54	54	68	54	54	54	54	68	68	68	68
		550S162	16	43	43	43	43	33	33	33	43	33	33	43	43
			24	54	54	54	54	43	43	43	54	43	43	54	54

For SI: 1 inch = 25.4 mm, 1 foot = 304.8 mm, 1 mil = 0.0254 mm, 1 mile per hour = 0.447 m/s, 1 pound per square foot = 0.0479 kPa, 1 Ksi = 1,000 psi = 6.895 MPa.

a. Deflection criterion: *L*/240.

b. Design load assumptions:
   Top and middle floor dead load is 10 psf.
   Top floor live load is 30 psf.
   Middle floor live load is 40 psf.
   Roof/ceiling dead load is 12 psf.
   Attic live load is 10 psf.

c. Building width is in the direction of horizontal framing members supported by the wall studs.

❖ See the commentary for Section R603.3.2.

TABLE R603.3.2(28)
36-FOOT-WIDE BUILDING SUPPORTING TWO FLOORS, ROOF AND CEILING[a, b, c]
33 KSI STEEL

WIND SPEED		MEMBER SIZE	STUD SPACING (inches)	MINIMUM STUD THICKNESS (mils)											
				8-foot Studs				9-foot Studs				10-foot Studs			
				Ground Snow Load (psf)											
Exp. B	Exp. C			20	30	50	70	20	30	50	70	20	30	50	70
85 mph	—	350S162	16	54	54	54	54	43	43	43	54	54	54	54	54
			24	68	68	68	97	68	68	68	68	68	68	68	97
		550S162	16	43	43	43	54	43	43	43	43	43	43	43	43
			24	68	68	68	68	54	54	54	68	54	54	68	68
90 mph	—	350S162	16	54	54	54	54	43	43	43	54	54	54	54	54
			24	68	68	68	97	68	68	68	68	68	68	68	97
		550S162	16	43	43	43	54	43	43	43	43	43	43	43	43
			24	68	68	68	68	54	54	54	68	54	54	68	68
100 mph	85 mph	350S162	16	54	54	54	54	43	43	43	54	54	54	54	54
			24	68	68	68	97	68	68	68	68	68	68	68	97
		550S162	16	43	43	43	54	43	43	43	43	43	43	43	43
			24	68	68	68	68	54	54	54	68	54	54	68	68
110 mph	90 mph	350S162	16	54	54	54	54	43	43	43	54	54	54	54	54
			24	68	68	68	97	68	68	68	68	68	68	97	97
		550S162	16	43	43	43	54	43	43	43	43	43	43	43	43
			24	68	68	68	68	54	54	54	68	54	54	68	68
—	100 mph	350S162	16	54	54	54	54	43	43	54	54	54	54	54	54
			24	68	68	68	97	68	68	68	68	97	97	97	97
		550S162	16	43	43	43	54	43	43	43	43	43	43	43	43
			24	68	68	68	68	54	54	54	68	54	54	68	68
—	110 mph	350S162	16	54	54	54	54	54	54	54	54	54	54	54	68
			24	68	68	68	97	68	68	68	97	97	97	97	97
		550S162	16	43	43	43	54	43	43	43	43	43	43	43	43
			24	68	68	68	68	54	54	54	68	54	54	68	68

For SI: 1 inch = 25.4 mm, 1 foot = 304.8 mm, 1 mil = 0.0254 mm, 1 mile per hour = 0.447 m/s, 1 pound per square foot = 0.0479 kPa, 1 Ksi = 1,000 psi = 6.895 MPa.

a. Deflection criterion: *L*/240.

b. Design load assumptions:
Top and middle floor dead load is 10 psf.
Top floor live load is 30 psf.
Middle floor live load is 40 psf.
Roof/ceiling dead load is 12 psf.
Attic live load is 10 psf.

c. Building width is in the direction of horizontal framing members supported by the wall studs.

❖ See the commentary for Section R603.3.2.

TABLE R603.3.2(29)
36-FOOT-WIDE BUILDING SUPPORTING TWO FLOORS, ROOF AND CEILING[a, b, c]
50 KSI STEEL

WIND SPEED		MEMBER SIZE	STUD SPACING (inches)	MINIMUM STUD THICKNESS (mils)											
				8-foot Studs				9-foot Studs				10-foot Studs			
				Ground Snow Load (psf)											
Exp. B	Exp. C			20	30	50	70	20	30	50	70	20	30	50	70
85 mph	—	350S162	16	43	43	43	54	43	43	43	43	43	43	43	43
			24	68	68	68	68	54	54	54	68	68	68	68	68
		550S162	16	43	43	43	43	43	43	43	43	43	43	43	43
			24	54	54	54	54	54	54	54	54	54	54	54	54
90 mph	—	350S162	16	43	43	43	54	43	43	43	43	43	43	43	43
			24	68	68	68	68	54	54	54	68	68	68	68	68
		550S162	16	43	43	43	43	43	43	43	43	43	43	43	43
			24	54	54	54	54	54	54	54	54	54	54	54	54
100 mph	85 mph	350S162	16	43	43	43	54	43	43	43	43	43	43	43	43
			24	68	68	68	68	54	54	54	68	68	68	68	68
		550S162	16	43	43	43	43	43	43	43	43	43	43	43	43
			24	54	54	54	54	54	54	54	54	54	54	54	54
110 mph	90 mph	350S162	16	43	43	43	54	43	43	43	43	43	43	43	43
			24	68	68	68	68	54	54	54	68	68	68	68	68
		550S162	16	43	43	43	43	43	43	43	43	43	43	43	43
			24	54	54	54	54	54	54	54	54	54	54	54	54
—	100 mph	350S162	16	43	43	43	54	43	43	43	43	43	43	43	54
			24	68	68	68	68	54	54	54	68	68	68	68	68
		550S162	16	43	43	43	43	43	43	43	43	43	43	43	43
			24	54	54	54	54	54	54	54	54	54	54	54	54
—	110 mph	350S162	16	43	43	43	54	43	43	43	43	43	54	54	54
			24	68	68	68	68	54	54	68	68	68	68	68	68
		550S162	16	43	43	43	43	43	43	43	43	43	43	43	43
			24	54	54	54	54	54	54	54	54	54	54	54	54

For SI: 1 inch = 25.4 mm, 1 foot = 304.8 mm, 1 mil = 0.0254 mm, 1 mile per hour = 0.447 m/s, 1 pound per square foot = 0.0479 kPa, 1 Ksi = 1,000 psi = 6.895 MPa.

a. Deflection criterion: *L*/240.

b. Design load assumptions:
   Top and middle floor dead load is 10 psf.
   Top floor live load is 30 psf.
   Middle floor live load is 40 psf.
   Roof/ceiling dead load is 12 psf.
   Attic live load is 10 psf.

c. Building width is in the direction of horizontal framing members supported by the wall studs.

❖ See the commentary for Section R603.3.2.

TABLE R603.3.2(30)
40-FOOT-WIDE BUILDING SUPPORTING TWO FLOORS, ROOF AND CEILING[a, b, c]
33 KSI STEEL

WIND SPEED		MEMBER SIZE	STUD SPACING (inches)	MINIMUM STUD THICKNESS (mils)											
				8-foot Studs				9-foot Studs				10-foot Studs			
				Ground Snow Load (psf)											
Exp. B	Exp. C			20	30	50	70	20	30	50	70	20	30	50	70
85 mph	—	350S162	16	54	54	54	54	54	54	54	54	54	54	54	54
			24	97	97	97	97	68	68	68	97	97	97	97	97
		550S162	16	54	54	54	54	43	43	54	54	43	43	54	54
			24	68	68	68	68	68	68	68	68	68	68	68	68
90 mph	—	350S162	16	54	54	54	54	54	54	54	54	54	54	54	54
			24	97	97	97	97	68	68	68	97	97	97	97	97
		550S162	16	54	54	54	54	43	43	54	54	43	43	54	54
			24	68	68	68	68	68	68	68	68	68	68	68	68
100 mph	85 mph	350S162	16	54	54	54	54	54	54	54	54	54	54	54	54
			24	97	97	97	97	68	68	68	97	97	97	97	97
		550S162	16	54	54	54	54	43	43	54	54	43	43	54	54
			24	68	68	68	68	68	68	68	68	68	68	68	68
110 mph	90 mph	350S162	16	54	54	54	54	54	54	54	54	54	54	54	54
			24	97	97	97	97	68	68	68	97	97	97	97	97
		550S162	16	54	54	54	54	43	43	54	54	43	43	54	54
			24	68	68	68	68	68	68	68	68	68	68	68	68
—	100 mph	350S162	16	54	54	54	54	54	54	54	54	54	54	54	54
			24	97	97	97	97	68	68	68	97	97	97	97	97
		550S162	16	54	54	54	54	43	43	54	54	43	43	54	54
			24	68	68	68	68	68	68	68	68	68	68	68	68
—	110 mph	350S162	16	54	54	54	54	54	54	54	54	54	54	68	68
			24	97	97	97	97	68	68	97	97	97	97	97	97
		550S162	16	54	54	54	54	43	43	54	54	43	43	54	54
			24	68	68	68	68	68	68	68	68	68	68	68	68

For SI: 1 inch = 25.4 mm, 1 foot = 304.8 mm, 1 mil = 0.0254 mm, 1 mile per hour = 0.447 m/s, 1 pound per square foot = 0.0479 kPa, 1 Ksi = 1,000 psi = 6.895 MPa.

a. Deflection criterion: *L*/240.

b. Design load assumptions:
   Top and middle floor dead load is 10 psf.
   Top floor live load is 30 psf.
   Middle floor live load is 40 psf.
   Roof/ceiling dead load is 12 psf.
   Attic live load is 10 psf.

c. Building width is in the direction of horizontal framing members supported by the wall studs.

❖ See the commentary for Section R603.3.2.

TABLE R603.3.2(31)
40-FOOT-WIDE BUILDING SUPPORTING TWO FLOORS, ROOF AND CEILING[a, b, c]
50 KSI STEEL

WIND SPEED		MEMBER SIZE	STUD SPACING (inches)	MINIMUM STUD THICKNESS (mils)											
				8-foot Studs				9-foot Studs				10-foot Studs			
				Ground Snow Load (psf)											
Exp. B	Exp. C			20	30	50	70	20	30	50	70	20	30	50	70
85 mph	—	350S162	16	54	54	54	54	43	43	43	43	43	54	54	54
			24	68	68	68	68	68	68	68	68	68	68	68	68
		550S162	16	43	43	43	43	43	43	43	43	43	43	43	43
			24	54	54	54	68	54	54	54	54	54	54	54	54
90 mph	—	350S162	16	54	54	54	54	43	43	43	43	43	54	54	54
			24	68	68	68	68	68	68	68	68	68	68	68	68
		550S162	16	43	43	43	43	43	43	43	43	43	43	43	43
			24	54	54	54	68	54	54	54	54	54	54	54	54
100 mph	85 mph	350S162	16	54	54	54	54	43	43	43	43	43	54	54	54
			24	68	68	68	68	68	68	68	68	68	68	68	68
		550S162	16	43	43	43	43	43	43	43	43	43	43	43	43
			24	54	54	54	68	54	54	54	54	54	54	54	54
110 mph	90 mph	350S162	16	54	54	54	54	43	43	43	43	43	54	54	54
			24	68	68	68	68	68	68	68	68	68	68	68	68
		550S162	16	43	43	43	43	43	43	43	43	43	43	43	43
			24	54	54	54	68	54	54	54	54	54	54	54	54
—	100 mph	350S162	16	54	54	54	54	43	43	43	43	43	54	54	54
			24	68	68	68	68	68	68	68	68	68	68	68	68
		550S162	16	43	43	43	43	43	43	43	43	43	43	43	43
			24	54	54	54	68	54	54	54	54	54	54	54	54
—	110 mph	350S162	16	54	54	54	54	43	43	43	43	54	54	54	54
			24	68	68	68	68	68	68	68	68	68	68	68	97
		550S162	16	43	43	43	43	43	43	43	43	43	43	43	43
			24	54	54	54	68	54	54	54	54	54	54	54	54

For SI: 1 inch = 25.4 mm, 1 foot = 304.8 mm, 1 mil = 0.0254 mm, 1 mile per hour = 0.447 m/s, 1 pound per square foot = 0.0479 kPa, 1 Ksi = 1,000 psi = 6.895 MPa.

a. Deflection criterion: *L*/240.

b. Design load assumptions:
   Top and middle floor dead load is 10 psf.
   Top floor live load is 30 psf.
   Middle floor live load is 40 psf.
   Roof/ceiling dead load is 12 psf.
   Attic live load is 10 psf.

c. Building width is in the direction of horizontal framing members supported by the wall studs.

❖ See the commentary for Section R603.3.2.

R603.3.2.1 Gable endwalls. The size and thickness of gable endwall studs with heights less than or equal to 10 feet (3048 mm) shall be permitted in accordance with the limits set forth in Table R603.3.2.1(1) or R603.3.2.1(2). The size and thickness of gable endwall studs with heights greater than 10 feet (3048 mm) shall be determined in accordance with the limits set forth in Table R603.3.2.1(3) or R603.3.2.1(4).

❖ The stud size for gable endwalls that are tabulated does not vary with the building width since the gable endwall is assumed to not provide support for the horizontal framing members. The tables provide requirements based on the stud spacing and height and they cover basic wind speeds up to 110 mph (49 m/s) for Exposure B or C. For stud heights up to 10 feet (3048 mm) the stud size is provided in Tables R603.3.2.1(1) and R603.3.2.1(2) for 33 ksi and 50 ksi steel, respectively. For stud heights greater than 10 feet (3048 mm) and up to 22 feet (6706 mm), the stud size is provided in Tables R603.3.2.1(3) and R603.3.2.1(4) for 33 ksi and 50 ksi steel, respectively.

TABLE R603.3.2.1(1)
ALL BUILDING WIDTHS GABLE ENDWALLS 8, 9 OR 10 FEET IN HEIGHT[a, b, c]
33 KSI STEEL

WIND SPEED		MEMBER SIZE	STUD SPACING (inches)	MINIMUM STUD THICKNESS (Mils)		
Exp. B	Exp. C			8-foot Studs	9-foot Studs	10-foot Studs
85 mph	—	350S162	16	33	33	33
			24	33	33	33
		550S162	16	33	33	33
			24	33	33	33
90 mph	—	350S162	16	33	33	33
			24	33	33	33
		550S162	16	33	33	33
			24	33	33	33
100 mph	85 mph	350S162	16	33	33	33
			24	33	33	43
		550S162	16	33	33	33
			24	33	33	33
110 mph	90 mph	350S162	16	33	33	33
			24	33	33	43
		550S162	16	33	33	33
			24	33	33	33
—	100 mph	350S162	16	33	33	43
			24	43	43	54
		550S162	16	33	33	33
			24	33	33	33
—	110 mph	350S162	16	33	43	43
			24	43	54	54
		550S162	16	33	33	33
			24	33	33	43

For SI: 1 inch = 25.4 mm, 1 foot = 304.8 mm, 1 mil = 0.0254 mm, 1 mile per hour = 0.447 m/s, 1 pound per square foot = 0.0479 kPa, 1 Ksi = 1,000 psi = 6.895 MPa.

a. Deflection criterion *L*/240.

b. Design load assumptions:
   Ground snow load is 70 psf.
   Roof/ceiling dead load is 12 psf.
   Floor dead load is 10 psf.
   Floor live load is 40 psf.
   Attic dead load is 10 psf.

c. Building width is in the direction of horizontal framing members supported by the wall studs.

❖ See the commentary for Section R603.3.2.1.

TABLE R603.3.2.1(2)
ALL BUILDING WIDTHS GABLE ENDWALLS 8, 9 OR 10 FEET IN HEIGHT[a, b, c]
50 KSI STEEL

WIND SPEED		MEMBER SIZE	STUD SPACING (inches)	MINIMUM STUD THICKNESS (Mils)		
Exp. B	Exp. C			8-foot Studs	9-foot Studs	10-foot Studs
85 mph	—	350S162	16	33	33	33
			24	33	33	33
		550S162	16	33	33	33
			24	33	33	33
90 mph	—	350S162	16	33	33	33
			24	33	33	33
		550S162	16	33	33	33
			24	33	33	33
100 mph	85 mph	350S162	16	33	33	33
			24	33	33	33
		550S162	16	33	33	33
			24	33	33	33
110 mph	90 mph	350S162	16	33	33	33
			24	33	33	43
		550S162	16	33	33	33
			24	33	33	33
—	100 mph	350S162	16	33	33	33
			24	33	33	43
		550S162	16	33	33	33
			24	33	33	33
—	110 mph	350S162	16	33	33	33
			24	33	43	54
		550S162	16	33	33	33
			24	33	33	33

For SI: 1 inch = 25.4 mm, 1 foot = 304.8 mm, 1 mil = 0.0254 mm, 1 mile per hour = 0.447 m/s, 1 pound per square foot = 0.0479 kPa, 1 Ksi = 1,000 psi = 6.895 MPa.

a. Deflection criterion *L*/240.

b. Design load assumptions:

Ground snow load is 70 psf.

Roof/ceiling dead load is 12 psf.

Floor dead load is 10 psf.

Floor live load is 40 psf.

Attic dead load is 10 psf.

c. Building width is in the direction of horizontal framing members supported by the wall studs.

❖ See the commentary for Section R603.3.2.1.

TABLE R603.3.2.1(3)
ALL BUILDING WIDTHS GABLE ENDWALLS OVER 10 FEET IN HEIGHT[a, b, c]
33 KSI STEEL

WIND SPEED		MEMBER SIZE	STUD SPACING (inches)	MINIMUM STUD THICKNESS (mils)					
				Stud Height, *h* (feet)					
Exp. B	Exp. C			10 < *h* ≤ 12	12 < *h* ≤ 14	14 < *h* ≤ 16	16 < *h* ≤ 18	18 < *h* ≤ 20	20 < *h* ≤ 22
85 mph	—	350S162	16	33	43	54	97	—	—
			24	43	54	97	—	—	—
		550S162	16	33	33	33	43	43	54
			24	33	33	43	54	68	97
90 mph	—	350S162	16	33	43	68	97	—	—
			24	43	68	97	—	—	—
		550S162	16	33	33	33	43	54	54
			24	33	33	43	54	68	97
100 mph	85 mph	350S162	16	43	54	97	—	—	—
			24	54	97	—	—	—	—
		550S162	16	33	33	43	54	54	68
			24	33	43	54	68	97	97
110 mph	90 mph	350S162	16	43	68	—	—	—	—
			24	68	—	—	—	—	—
		550S162	16	33	43	43	54	68	97
			24	43	54	68	97	97	—
—	100 mph	350S162	16	54	97	—	—	—	—
			24	97	—	—	—	—	—
		550S162	16	33	43	54	68	97	—
			24	43	68	97	97	—	—
—	110 mph	350S162	16	68	97	—	—	—	—
			24	97	—	—	—	—	—
		550S162	16	43	54	68	97	97	—
			24	54	68	97	—	—	—

For SI: 1 inch = 25.4 mm, 1 foot = 304.8 mm, 1 mil = 0.0254 mm, 1 mile per hour = 0.447 m/s, 1 pound per square foot = 0.0479 kPa, 1 Ksi = 1,000 psi = 6.895 MPa.

a. Deflection criterion *L*/240.

b. Design load assumptions:
Ground snow load is 70 psf.
Roof/ceiling dead load is 12 psf.
Floor dead load is 10 psf.
Floor live load is 40 psf.
Attic dead load is 10 psf.

c. Building width is in the direction of horizontal framing members supported by the wall studs.

❖ See the commentary for Section R603.3.2.1.

TABLE R603.3.2.1(4)
ALL BUILDING WIDTHS GABLE ENDWALLS OVER 10 FEET IN HEIGHT[a, b, c]
50 KSI STEEL

WIND SPEED		MEMBER SIZE	STUD SPACING (inches)	MINIMUM STUD THICKNESS (mils)					
				Stud Height, *h* (feet)					
Exp. B	Exp. C			10 < *h* ≤ 12	12 < *h* ≤ 14	14 < *h* ≤ 16	16 < *h* ≤ 18	18 < *h* ≤ 20	20 < *h* ≤ 22
85 mph	—	350S162	16	33	43	54	97	—	—
			24	33	54	97	—	—	—
		550S162	16	33	33	33	33	43	54
			24	33	33	33	43	54	97
90 mph	—	350S162	16	33	43	68	97	—	—
			24	43	68	97	—	—	—
		550S162	16	33	33	33	33	43	54
			24	33	33	43	43	68	97
100 mph	85 mph	350S162	16	33	54	97	—	—	—
			24	54	97	—	—	—	—
		550S162	16	33	33	33	43	54	68
			24	33	33	43	54	97	97
110 mph	90 mph	350S162	16	43	68	—	—	—	—
			24	68	—	—	—	—	—
		550S162	16	33	33	43	43	68	97
			24	33	43	54	68	97	—
—	100 mph	350S162	16	54	97	—	—	—	—
			24	97	—	—	—	—	—
		550S162	16	33	33	43	54	97	—
			24	43	54	54	97	—	—
—	110 mph	350S162	16	54	97	—	—	—	—
			24	97	—	—	—	—	—
		550S162	16	33	43	54	68	97	—
			24	43	54	68	97	—	—

For SI: 1 inch = 25.4 mm, 1 foot = 304.8 mm, 1 mil = 0.0254 mm, 1 mile per hour = 0.447 m/s, 1 pound per square foot = 0.0479 kPa, 1 Ksi = 1,000 psi = 6.895 MPa.

a. Deflection criterion *L*/240.

b. Design load assumptions:
Ground snow load is 70 psf.
Roof/ceiling dead load is 12 psf.
Floor dead load is 10 psf.
Floor live load is 40 psf.
Attic dead load is 10 psf.

c. Building width is in the direction of horizontal framing members supported by the wall studs.

❖ See the commentary for Section R603.3.2.1.

R603.3.3 Stud bracing. The flanges of cold-formed steel studs shall be laterally braced in accordance with one of the following:

1. Gypsum board on both sides, structural sheathing on both sides, or gypsum board on one side and structural sheathing on the other side of load-bearing walls with gypsum board installed with minimum No. 6 screws in accordance with Section R702 and structural sheathing installed in accordance with Section R603.9.1 and Table R603.3.2(1).
2. Horizontal steel straps fastened in accordance with Figure R603.3.3(1) on both sides at mid-height for 8-foot (2438 mm) walls, and at one-third points for 9-foot and 10-foot (2743 mm and 3048 mm) walls. Horizontal steel straps shall be at least 1.5 inches in width and 33 mils in thickness (38 mm by 0.84 mm). Straps shall be attached to the flanges of studs with one No. 8 screw. In-line blocking shall be installed between studs at the termination of all straps and at 12 foot (3658 mm) intervals along the strap. Straps shall be fastened to the blocking with two No. 8 screws.

3. Sheathing on one side and strapping on the other side fastened in accordance with Figure R603.3.3(2). Sheathing shall be installed in accordance with Item 1. Steel straps shall be installed in accordance with Item 2.

❖ To provide resistance against buckling, lateral bracing of steel-stud flanges must be provided by gypsum board installed according to Section R702, structural sheathing installed in accordance with Table R603.3.2(1), horizontal steel strapping as indicated in Figure R603.3.3(1) or sheathing in accordance with Item 1 on one side and steel straps in accordance with Item 2 on the other side [see Figure R603.3.3(2)]. Steel strap bracing must be installed at mid-height for 8-foot (2438 mm) walls and at one-third points for 9-foot (2743 mm) and 10-foot (3048 mm) walls (two rows of bracing). These straps must be attached to stud flanges with at least one No. 8 screw. In-line blocking must be installed between studs at the termination of all straps. Straps must be fastened to this blocking with at least two No. 8 screws.

R603.3.4 Cutting and notching. Flanges and lips of cold-formed steel studs and headers shall not be cut or notched.

❖ Flanges and lips of load-bearing wall framing, to retain standard integrity, must not be cut or notched. The only holes permitted in steel wall framing are those conforming to Section R603.2.5.

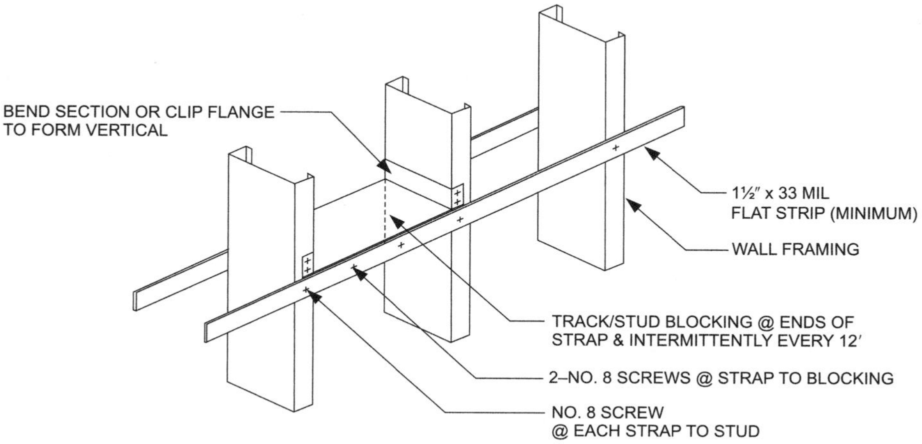

For SI: 1 mil = 0.0254 mm, 1 inch = 25.4 mm, 1 foot = 304.8 mm.

FIGURE R603.3.3(1)
STUD BRACING WITH STRAPPING ONLY

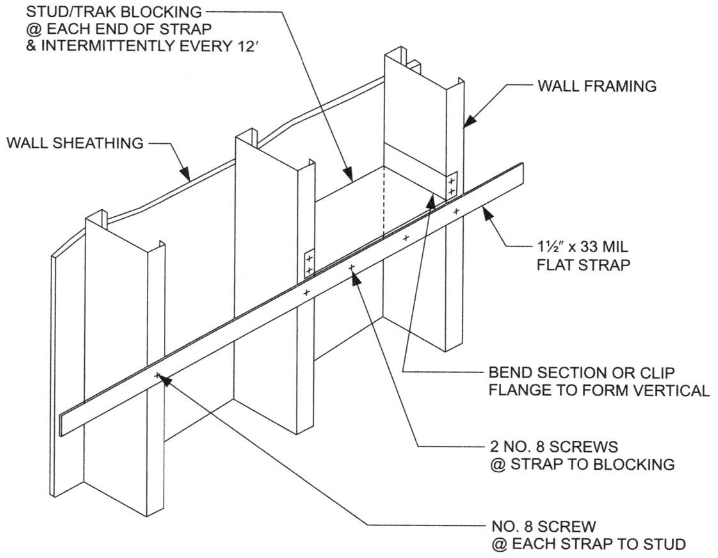

For SI: 1 mil = 0.0254 mm, 1 inch = 25.4 mm, 1 foot = 304.8 mm.

FIGURE R603.3.3(2)
STUD BRACING WITH STRAPPING AND SHEATHING MATERIAL

R603.3.5 Splicing. Steel studs and other structural members shall not be spliced. Tracks shall be spliced in accordance with Figure R603.3.5.

❖ Splicing of structural members, such as studs, is not permitted. Splicing of tracks is allowed since they do not serve as gravity load-carrying members. The splice shown in Figure R603.3.5 provides nominal load-carrying capability; however, it is necessary to resist the lateral loads produced by moderate wind and earthquake loads. In these cases, the floor system must act as a diaphragm with the tracks serving as the diaphragm chords.

R603.4 Corner framing. In exterior walls, corner studs and the top tracks shall be installed in accordance with Figure R603.4.

❖ Figure R603.4 illustrates the required configurations of corner studs and overlapping of tracks at corners of exterior walls.

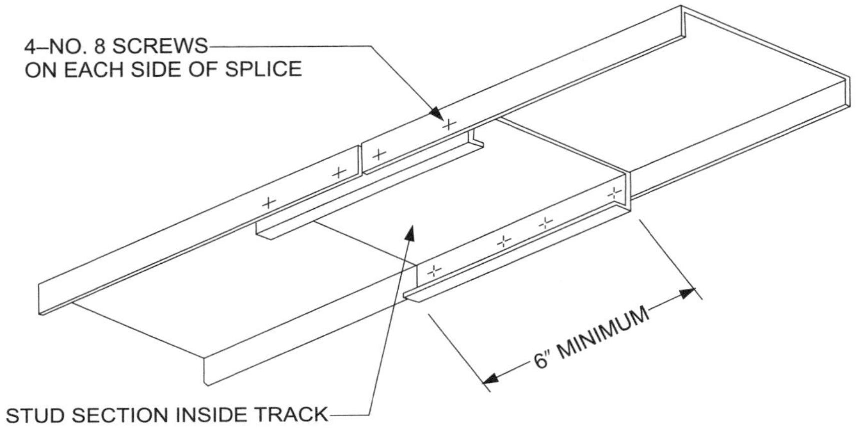

For SI: 1 inch = 25.4 mm.

FIGURE R603.3.5
TRACK SPLICE

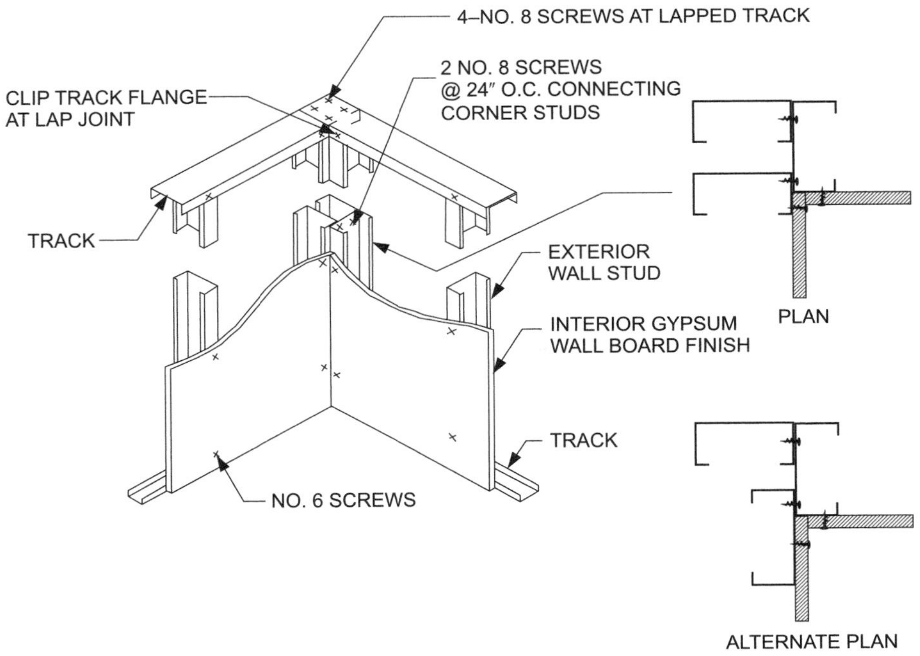

For SI: 1 inch = 25.4 mm.

FIGURE R603.4
CORNER FRAMING

R603.5 Exterior wall covering. The method of attachment of exterior wall covering materials to cold-formed steel stud wall framing shall conform to the manufacturer's installation instructions.

❖ Attachment of exterior wall coverings must be according to the manufacturer's instructions.

R603.6 Headers. Headers shall be installed above all wall openings in exterior walls and interior load-bearing walls. Box beam headers and back-to-back headers each shall be formed from two equal sized C-shaped members in accordance with Figures R603.6(1) and R603.6(2), respectively, and Tables R603.6(1) through R603.6(24). L-shaped headers shall be permitted to be constructed in accordance with AISI S230. Alternately, headers shall be permitted to be designed and constructed in accordance with AISI S100, Section D4.

❖ Headers must be located immediately below the ceiling or roof framing (at the top track, for example) above wall openings in all exterior walls and interior load-bearing walls in accordance with Figures R603.6(1) and R603.6(2), which illustrate the two types of prescriptive header options—either a box beam or back-to-back. Tables R603.6(1) through R603.6(24) specify allowable header spans for various combinations of loading and header sizes. Since these tables provide a somewhat limited range of design options, the code provides the option to design headers in accordance with AISI S100 or use AISI S230 for L-shaped headers.

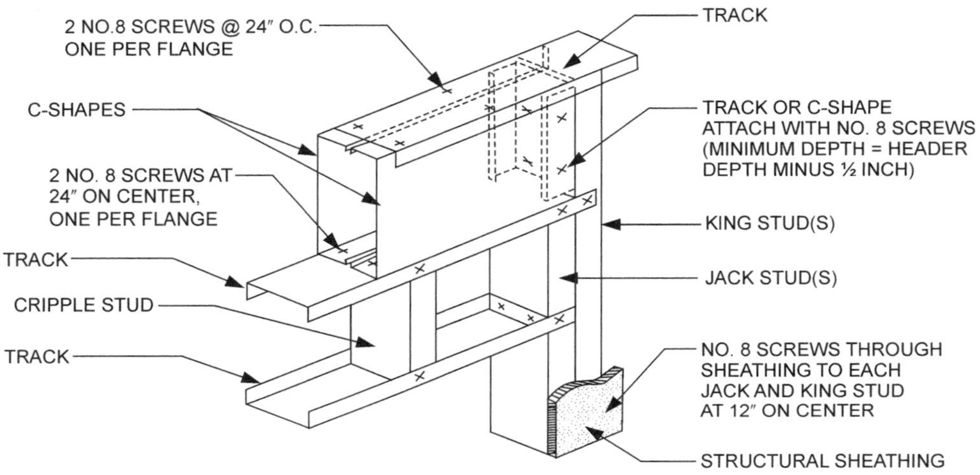

For SI: 1 inch = 25.4 mm.

FIGURE R603.6(1)
BOX BEAM HEADER

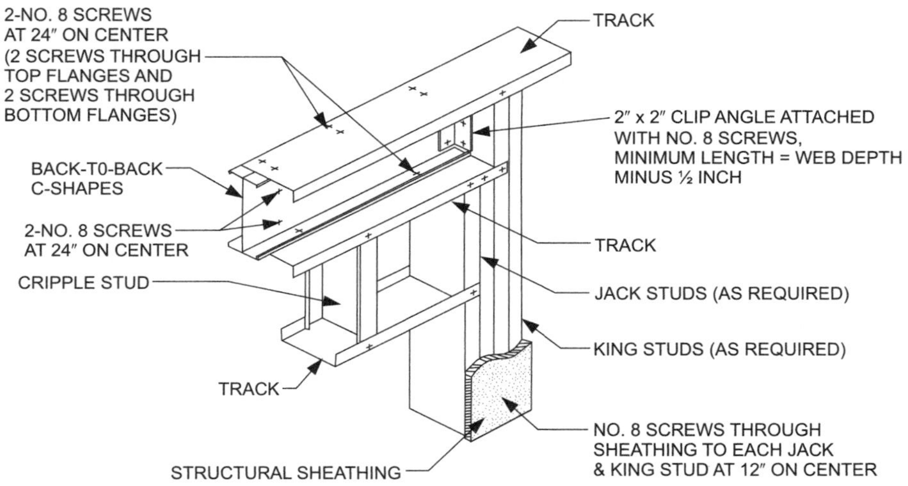

For SI: 1 inch = 25.4 mm.

FIGURE R603.6(2)
BACK-TO-BACK HEADER

TABLE R603.6(1)
BOX-BEAM HEADER SPANS
Headers Supporting Roof and Ceiling Only
(33 Ksi steel)[a, b]

MEMBER DESIGNATION	GROUND SNOW LOAD (20 psf)					GROUND SNOW LOAD (30 psf)				
	Building width[c] (feet)					Building width[c] (feet)				
	24	28	32	36	40	24	28	32	36	40
2-350S162-33	3′-3″	2′-8″	2′-2″	—	—	2′-8″	2′-2″	—	—	—
2-350S162-43	4′-2″	3′-9″	3′-4″	2′-11″	2′-7″	3′-9″	3′-4″	2′-11″	2′-7″	2′-2″
2-350S162-54	5′-0″	4′-6″	4′-1″	3′-8″	3′-4″	4′-6″	4′-1″	3′-8″	3′-3″	3′-0″
2-350S162-68	5′-7″	5′-1″	4′-7″	4′-3″	3′-10″	5′-1″	4′-7″	4′-2″	3′-10″	3′-5″
2-350S162-97	7′-1″	6′-6″	6′-1″	5′-8″	5′-3″	6′-7″	6′-1″	5′-7″	5′-3″	4′-11″
2-550S162-33	4′-8″	4′-0″	3′-6″	3′-0″	2′-6″	4′-1″	3′-6″	3′-0″	2′-6″	—
2-550S162-43	6′-0″	5′-4″	4′-10″	4′-4″	3′-11″	5′-5″	4′-10″	4′-4″	3′-10″	3′-5″
2-550S162-54	7′-0″	6′-4″	5′-9″	5′-4″	4′-10″	6′-5″	5′-9″	5′-3″	4′-10″	4′-5″
2-550S162-68	8′-0″	7′-4″	6′-9″	6′-3″	5′-10″	7′-5″	6′-9″	6′-3″	5′-9″	5′-4″
2-550S162-97	9′-11″	9′-2″	8′-6″	8′-0″	7′-6″	9′-3″	8′-6″	8′-0″	7′-5″	7′-0″
2-800S162-33	4′-5″	3′-11″	3′-5″	3′-1″	2′-10″	3′-11″	3′-6″	3′-1″	2′-9″	2′-3″
2-800S162-43	7′-3″	6′-7″	5′-11″	5′-4″	4′-10″	6′-7″	5′-11″	5′-4″	4′-9″	4′-3″
2-800S162-54	8′-10″	8′-0″	7′-4″	6′-9″	6′-2″	8′-1″	7′-4″	6′-8″	6′-1″	5′-7″
2-800S162-68	10′-5″	9′-7″	8′-10″	8′-2″	7′-7″	9′-8″	8′-10″	8′-1″	7′-6″	7′-0″
2-800S162-97	13′-1″	12′-1″	11′-3″	10′-7″	10′-0″	12′-2″	11′-4″	10′-6″	10′-0″	9′-4″
2-1000S162-43	7′-10″	6′-10″	6′-1″	5′-6″	5′-0″	6′-11″	6′-1″	5′-5″	4′-11″	4′-6″
2-1000S162-54	10′-0″	9′-1″	8′-3″	7′-7″	7′-0″	9′-2″	8′-4″	7′-7″	6′-11″	6′-4″
2-1000S162-68	11′-11″	10′-11″	10′-1″	9′-4″	8′-8″	11′-0″	10′-1″	9′-3″	8′-7″	8′-0″
2-1000S162-97	15′-3″	14′-3″	13′-5″	12′-6″	11′-10″	14′-4″	13′-5″	12′-6″	11′-9″	11′-0″
2-1200S162-54	11′-1″	10′-0″	9′-2″	8′-5″	7′-9″	10′-1″	9′-2″	8′-4″	7′-7″	7′-0″
2-1200S162-68	13′-3″	12′-1″	11′-2″	10′-4″	9′-7″	12′-3″	11′-2″	10′-3″	9′-6″	8′-10″
2-1200S162-97	16′-8″	15′-7″	14′-8″	13′-11″	13′-3″	15′-8″	14′-8″	13′-11″	13′-2″	12′-6″

For SI: 1 inch = 25.4 mm, 1 foot = 304.8 mm, 1 pound per square foot = 0.0479 kPa, 1 pound per square inch = 6.895 kPa, 1 Ksi = 1,000 psi = 6.895 MPa.

a. Deflection criterion: *L*/360 for live loads, *L*/240 for total loads.

b. Design load assumptions:
Roof/ceiling dead load is 12 psf.
Attic dead load is 10 psf.

c. Building width is in the direction of horizontal framing members supported by the header.

❖ See the commentary to Section R603.6.

TABLE R603.6(2)
BOX-BEAM HEADER SPANS
Headers Supporting Roof and Ceiling Only
(50 Ksi steel)[a, b]

MEMBER DESIGNATION	GROUND SNOW LOAD (20 psf)					GROUND SNOW LOAD (30 psf)				
	Building width[c] (feet)					Building width[c] (feet)				
	24	28	32	36	40	24	28	32	36	40
2-350S162-33	4′-4″	3′-11″	3′-6″	3′-2″	2′-10″	3′-11″	3′-6″	3′-1″	2′-9″	2′-5″
2-350S162-43	5′-6″	5′-0″	4′-7″	4′-2″	3′-10″	5′-0″	4′-7″	4′-2″	3′-10″	3′-6″
2-350S162-54	6′-2″	5′-10″	5′-8″	5′-3″	4′-10″	5′-11″	5′-8″	5′-2″	4′-10″	4′-6″
2-350S162-68	6′-7″	6′-3″	6′-0″	5′-10″	5′-8″	6′-4″	6′-1″	5′-10″	5′-8″	5′-6″
2-350S162-97	7′-3″	6′-11″	6′-8″	6′-5″	6′-3″	7′-0″	6′-8″	6′-5″	6′-3″	6′-0″
2-550S162-33	6′-2″	5′-6″	5′-0″	4′-7″	4′-2″	5′-7″	5′-0″	4′-6″	4′-1″	3′-8″
2-550S162-43	7′-9″	7′-2″	6′-7″	6′-1″	5′-8″	7′-3″	6′-7″	6′-1″	5′-7″	5′-2″
2-550S162-54	8′-9″	8′-5″	8′-1″	7′-9″	7′-3″	8′-6″	8′-1″	7′-8″	7′-2″	6′-8″
2-550S162-68	9′-5″	9′-0″	8′-8″	8′-4″	8′-1″	9′-1″	8′-8″	8′-4″	8′-1″	7′-10″
2-550S162-97	10′-5″	10′-0″	9′-7″	9′-3″	9′-0″	10′-0″	9′-7″	9′-3″	8′-11″	8′-8″
2-800S162-33	4′-5″	3′-11″	3′-5″	3′-1″	2′-10″	3′-11″	3′-6″	3′-1″	2′-9″	2′-6″
2-800S162-43	9′-1″	8′-5″	7′-8″	6′-11″	6′-3″	8′-6″	7′-8″	6′-10″	6′-2″	5′-8″
2-800S162-54	10′-10″	10′-2″	9′-7″	9′-0″	8′-5″	10′-2″	9′-7″	8′-11″	8′-4″	7′-9″
2-800S162-68	12′-8″	11′-10″	11′-2″	10′-7″	10′-1″	11′-11″	11′-2″	10′-7″	10′-0″	9′-6″
2-800S162-97	14′-2″	13′-6″	13′-0″	12′-7″	12′-2″	13′-8″	13′-1″	12′-7″	12′-2″	11′-9″
2-1000S162-43	7′-10″	6′-10″	6′-1″	5′-6″	5′-0″	6′-11″	6′-1″	5′-5″	4′-11″	4′-6″
2-1000S162-54	12′-3″	11′-5″	10′-9″	10′-2″	9′-6″	11′-6″	10′-9″	10′-1″	9′-5″	8′-9″
2-1000S162-68	14′-5″	13′-5″	12′-8″	12′-0″	11′-6″	13′-6″	12′-8″	12′-0″	11′-5″	10′-10″
2-1000S162-97	17′-1″	16′-4″	15′-8″	14′-11″	14′-3″	16′-5″	15′-9″	14′-10″	14′-1″	13′-6″
2-1200S162-54	12′-11″	11′-3″	10′-0″	9′-0″	8′-2″	11′-5″	10′-0″	9′-0″	8′-1″	7′-4″
2-1200S162-68	15′-11″	14′-10″	14′-0″	13′-4″	12′-8″	15′-0″	14′-0″	13′-3″	12′-7″	11′-11″
2-1200S162-97	19′-11″	18′-7″	17′-6″	16′-8″	15′-10″	18′-9″	17′-7″	16′-7″	15′-9″	15′-0″

For SI: 1 inch = 25.4 mm, 1 foot = 304.8 mm, 1 pound per square foot = 0.0479 kPa, 1 pound per square inch = 6.895 kPa, 1 Ksi = 1,000 psi = 6.895 MPa.

a. Deflection criterion: *L*/360 for live loads, *L*/240 for total loads.

b. Design load assumptions:
Roof/ceiling dead load is 12 psf.
Attic dead load is 10 psf.

c. Building width is in the direction of horizontal framing members supported by the header.

❖ See the commentary for Section R603.6.

TABLE R603.6(3)
BOX-BEAM HEADER SPANS
Headers Supporting Roof and Ceiling Only
(33 Ksi steel)[a, b]

MEMBER DESIGNATION	GROUND SNOW LOAD (50 psf)					GROUND SNOW LOAD (70 psf)				
	Building width[c] (feet)					Building width[c] (feet)				
	24	28	32	36	40	24	28	32	36	40
2-350S162-33	—	—	—	—	—	—	—	—	—	—
2-350S162-43	2′-4″	—	—	—	—	—	—	—	—	—
2-350S162-54	3′-1″	2′-8″	2′-3″	—	—	2′-1″	—	—	—	—
2-350S162-68	3′-7″	3′-2″	2′-8″	2′-3″	—	2′-6″	—	—	—	—
2-350S162-97	5′-1″	4′-7″	4′-3″	3′-11″	3′-7″	4′-1″	3′-8″	3′-4″	3′-0″	2′-8″
2-550S162-33	2′-2″	—	—	—	—	—	—	—	—	—
2-550S162-43	3′-8″	3′-1″	2′-6″	—	—	2′-3″	—	—	—	—
2-550S162-54	4′-7″	4′-0″	3′-6″	3′-0″	2′-6″	3′-3″	2′-8″	2′-1″	—	—
2-550S162-68	5′-6″	4′-11″	4′-5″	3′-11″	3′-6″	4′-3″	3′-8″	3′-1″	2′-7″	2′-1″
2-550S162-97	7′-3″	6′-7″	6′-1″	5′-8″	5′-3″	5′-11″	5′-4″	4′-11″	4′-6″	4′-1″
2-800S162-33	2′-7″	—	—	—	—	—	—	—	—	—
2-800S162-43	4′-6″	3′-9″	3′-1″	2′-5″	—	2′-10″	—	—	—	—
2-800S162-54	5′-10″	5′-1″	4′-6″	3′-11″	3′-4″	4′-3″	3′-6″	2′-9″	—	—
2-800S162-68	7′-2″	6′-6″	5′-10″	5′-3″	4′-8″	5′-7″	4′-10″	4′-2″	3′-7″	2′-11″
2-800S162-97	9′-7″	8′-9″	8′-2″	7′-7″	7′-0″	7′-11″	7′-2″	6′-7″	6′-0″	5′-7″
2-1000S162-43	4′-8″	4′-1″	3′-6″	2′-9″	—	3′-3″	2′-2″	—	—	—
2-1000S162-54	6′-7″	5′-10″	5′-1″	4′-5″	3′-9″	4′-10″	4′-0″	3′-2″	2′-3″	—
2-1000S162-68	8′-3″	7′-5″	6′-8″	6′-0″	5′-5″	6′-5″	5′-7″	4′-9″	4′-1″	3′-5″
2-1000S162-97	11′-4″	10′-5″	9′-8″	9′-0″	8′-5″	9′-5″	8′-6″	7′-10″	7′-2″	6′-7″
2-1200S162-54	7′-3″	6′-5″	5′-7″	4′-10″	4′-2″	5′-4″	4′-4″	3′-5″	2′-5″	—
2-1200S162-68	9′-2″	8′-2″	7′-5″	6′-8″	6′-0″	7′-1″	6′-2″	5′-4″	4′-6″	3′-9″
2-1200S162-97	12′-10″	11′-9″	10′-11″	10′-2″	9′-6″	10′-7″	9′-8″	8′-10″	8′-2″	7′-6″

For SI: 1 inch = 25.4 mm, 1 foot = 304.8 mm, 1 pound per square foot = 0.0479 kPa, 1 pound per square inch = 6.895 kPa, 1 Ksi = 1,000 psi = 6.895 MPa.

a. Deflection criterion: *L*/360 for live loads, *L*/240 for total loads.

b. Design load assumptions:
Roof/ceiling dead load is 12 psf.
Attic dead load is 10 psf.

c. Building width is in the direction of horizontal framing members supported by the header.

❖ See the commentary for Section R603.6.

TABLE R603.6(4)
BOX-BEAM HEADER SPANS
Headers Supporting Roof and Ceiling Only
(50 Ksi steel)[a, b]

MEMBER DESIGNATION	GROUND SNOW LOAD (50 psf)					GROUND SNOW LOAD (70 psf)				
	Building width[c] (feet)					Building width[c] (feet)				
	24	28	32	36	40	24	28	32	36	40
2-350S162-33	2′-7″	2′-2″	—	—	—	—	—	—	—	—
2-350S162-43	3′-8″	3′-3″	2′-10″	2′-6″	2′-1″	2′-8″	2′-3″	—	—	—
2-350S162-54	4′-8″	4′-2″	3′-9″	3′-5″	3′-1″	3′-7″	3′-2″	2′-9″	2′-5″	2′-0″
2-350S162-68	5′-7″	5′-2″	4′-9″	4′-4″	3′-11″	4′-7″	4′-1″	3′-7″	3′-2″	2′-10″
2-350S162-97	6′-2″	5′-11″	5′-8″	5′-6″	5′-4″	5′-8″	5′-5″	5′-3″	4′-11″	4′-7″
2-550S162-33	3′-11″	3′-4″	2′-10″	2′-4″	—	2′-7″	—	—	—	—
2-550S162-43	5′-4″	4′-10″	4′-4″	3′-10″	3′-5″	4′-2″	3′-7″	3′-1″	2′-7″	2′-1″
2-550S162-54	6′-11″	6′-3″	5′-9″	5′-3″	4′-9″	5′-6″	4′-11″	4′-5″	3′-11″	3′-5″
2-550S162-68	8′-0″	7′-6″	6′-11″	6′-5″	5′-11″	6′-9″	6′-1″	5′-6″	5′-0″	4′-7″
2-550S162-97	8′-11″	8′-6″	8′-2″	7′-11″	7′-8″	8′-1″	7′-9″	7′-6″	7′-1″	6′-7″
2-800S162-33	2′-8″	2′-4″	2′-1″	1′-11″	1′-9″	2′-0″	1′-9″	—	—	—
2-800S162-43	5′-10″	5′-2″	4′-7″	4′-2″	3′-10″	4′-5″	3′-11″	3′-6″	3′-0″	2′-6″
2-800S162-54	8′-0″	7′-3″	6′-8″	6′-1″	5′-7″	6′-5″	5′-9″	5′-1″	4′-7″	4′-0″
2-800S162-68	9′-9″	9′-0″	8′-3″	7′-8″	7′-1″	8′-0″	7′-3″	6′-7″	6′-0″	5′-6″
2-800S162-97	12′-1″	11′-7″	11′-2″	10′-8″	10′-2″	11′-0″	10′-4″	9′-9″	9′-2″	8′-7″
2-1000S162-43	4′-8″	4′-1″	3′-8″	3′-4″	3′-0″	3′-6″	3′-1″	2′-9″	2′-6″	2′-3″
2-1000S162-54	9′-1″	8′-2″	7′-3″	6′-7″	6′-0″	7′-0″	6′-2″	5′-6″	5′-0″	4′-6″
2-1000S162-68	11′-1″	10′-2″	9′-5″	8′-8″	8′-1″	9′-1″	8′-3″	7′-6″	6′-10″	6′-3″
2-1000S162-97	13′-9″	12′-11″	12′-2″	11′-7″	11′-1″	11′-11″	11′-3″	10′-7″	9′-11″	9′-4″
2-1200S162-54	7′-8″	6′-9″	6′-1″	5′-6″	5′-0″	5′-10″	5′-1″	4′-7″	4′-1″	3′-9″
2-1200S162-68	12′-3″	11′-3″	10′-4″	9′-7″	8′-11″	10′-1″	9′-1″	8′-3″	7′-6″	6′-10″
2-1200S162-97	15′-4″	14′-5″	13′-7″	12′-11″	12′-4″	13′-4″	12′-6″	11′-10″	11′-1″	10′-5″

For SI: 1 inch = 25.4 mm, 1 foot = 304.8 mm, 1 pound per square foot = 0.0479 kPa, 1 pound per square inch = 6.895 kPa,
1 Ksi = 1,000 psi = 6.895 MPa.

a. Deflection criterion: *L*/360 for live loads, *L*/240 for total loads.

b. Design load assumptions:
Roof/ceiling dead load is 12 psf.
Attic dead load is 10 psf.

c. Building width is in the direction of horizontal framing members supported by the header.

❖ See the commentary for Section R603.6.

TABLE R603.6(5)
BOX-BEAM HEADER SPANS
Headers Supporting One Floor, Roof and Ceiling
(33 Ksi steel)[a, b]

MEMBER DESIGNATION	GROUND SNOW LOAD (20 psf)					GROUND SNOW LOAD (30 psf)				
	Building width[c] (feet)					Building width[c] (feet)				
	24	28	32	36	40	24	28	32	36	40
2-350S162-33	—	—	—	—	—	—	—	—	—	—
2-350S162-43	2′-2″	—	—	—	—	2′-1″	—	—	—	—
2-350S162-54	2′-11″	2′-5″	—	—	—	2′-10″	2′-4″	—	—	—
2-350S162-68	3′-8″	3′-2″	2′-9″	2′-4″	-	3′-7″	3′-1″	2′-8″	2′-3″	—
2-350S162-97	4′-11″	4′-5″	4′-2″	3′-8″	3′-5″	4′-10″	4′-5″	4′-0″	3′-8″	3′-4″
2-550S162-33	—	—	—	—	—	—	—	—	—	—
2-550S162-43	3′-5″	2′-9″	2′-1″	—	—	3′-3″	2′-7″	—	—	—
2-550S162-54	4′-4″	3′-9″	3′-2″	2′-7″	2′-1″	4′-3″	3′-7″	3′-1″	2′-6″	—
2-550S162-68	5′-3″	4′-8″	4′-1″	3′-7″	3′-2″	5′-2″	4′-7″	4′-0″	3′-6″	3′-1″
2-550S162-97	7′-0″	6′-5″	5′-10″	5′-5″	5′-0″	6′-11″	6′-4″	5′-9″	5′-4″	4′-11″
2-800S162-33	2′-1″	—	—	—	—	—	—	—	—	—
2-800S162-43	4′-2″	3′-4″	2′-7″	—	—	4′-0″	3′-3″	2′-5″	—	—
2-800S162-54	5′-6″	4′-9″	4′-1″	3′-5″	2′-9″	5′-5″	4′-8″	3′-11″	3′-3″	2′-8″
2-800S162-68	6′-11″	6′-2″	5′-5″	4′-10″	4′-3″	6′-9″	6′-0″	5′-4″	4′-8″	4′-1″
2-800S162-97	9′-4″	8′-6″	7′-10″	7′-3″	6′-8″	9′-2″	8′-4″	7′-8″	7′-1″	6′-7″
2-1000S162-43	4′-4″	3′-9″	2′-11″	—	—	4′-3″	3′-8″	2′-9″	—	—
2-1000S162-54	6′-3″	5′-5″	4′-7″	3′-11″	3′-2″	6′-1″	5′-3″	4′-6″	3′-9″	3′-0″
2-1000S162-68	7′-11″	7′-0″	6′-3″	5′-6″	4′-10″	7′-9″	6′-10″	6′-1″	5′-4″	4′-9″
2-1000S162-97	11′-0″	10′-1″	9′-3″	8′-7″	8′-0″	10′-11″	9′-11″	9′-2″	8′-5″	7′-10″
2-1200S162-54	6′-11″	5′-11″	5′-1″	4′-3″	3′-5″	6′-9″	5′-9″	4′-11″	4′-1″	3′-3″
2-1200S162-68	8′-9″	7′-9″	6′-11″	6′-1″	5′-4″	8′-7″	7′-7″	6′-9″	5′-11″	5′-3″
2-1200S162-97	12′-4″	11′-5″	10′-6″	9′-8″	9′-0″	12′-3″	11′-3″	10′-4″	9′-6″	8′-10″

For SI: 1 inch = 25.4 mm, 1 foot = 304.8 mm, 1 pound per square foot = 0.0479 kPa, 1 pound per square inch = 6.895 kPa,
1 Ksi = 1,000 psi = 6.895 MPa.

a. Deflection criterion: *L*/360 for live loads, *L*/240 for total loads.

b. Design load assumptions:
  Second floor dead load is 10 psf.
  Roof/ceiling dead load is 12 psf.
  Second floor live load is 30 psf.
  Attic dead load is 10 psf.

c. Building width is in the direction of horizontal framing members supported by the header.

❖ See the commentary to Section R603.6.

TABLE R603.6(6)
BOX-BEAM HEADER SPANS
Headers Supporting One Floor, Roof and Ceiling
(50 Ksi steel)[a, b]

MEMBER DESIGNATION	GROUND SNOW LOAD (20 psf)					GROUND SNOW LOAD (30 psf)				
	Building width[c] (feet)					Building width[c] (feet)				
	24	28	32	36	40	24	28	32	36	40
2-350S162-33	2′-4″	—	—	—	—	2′-3″	—	—	—	—
2-350S162-43	3′-4″	2′-11″	2′-6″	2′-1″	—	3′-3″	2′-10″	2′-5″	2′-0″	—
2-350S162-54	4′-4″	3′-10″	3′-5″	3′-1″	2′-9″	4′-3″	2′-9″	3′-4″	3′-0″	2′-8″
2-350S162-68	5′-0″	4′-9″	4′-7″	4′-2″	3′-9″	4′-11″	4′-8″	4′-6″	4′-1″	3′-9″
2-350S162-97	5′-6″	5′-3″	5′-1″	4′-11″	2′-9″	5′-5″	5′-2″	5′-0″	4′-10″	4′-8″
2-550S162-33	3′-6″	2′-11″	2′-4″	—	—	3′-5″	2′-10″	2′-3″	—	—
2-550S162-43	5′-0″	4′-5″	3′-11″	3′-5″	3′-0″	4′-11″	4′-4″	3′-10″	3′-4″	2′-11″
2-550S162-54	6′-6″	5′-10″	5′-3″	4′-9″	4′-4″	6′-4″	5′-9″	5′-2″	4′-8″	4′-3″
2-550S162-68	7′-2″	6′-10″	6′-5″	5′-11″	5′-6″	7′-0″	6′-9″	6′-4″	5′-10″	5′-4″
2-550S162-97	7′-11″	7′-7″	7′-3″	7′-0″	6′-10″	7′-9″	7′-5″	7′-2″	6′-11″	6′-9″
2-800S162-33	2′-5″	2′-2″	1′-11″	1′-9″	—	2′-5″	2′-1″	1′-10″	1′-8″	—
2-800S162-43	5′-5″	4′-9″	4′-3″	3′-9″	3′-5″	5′-3″	4′-8″	4′-1″	3′-9″	3′-5″
2-800S162-54	7′-6″	6′-9″	6′-2″	5′-7″	5′-0″	7′-5″	6′-8″	6′-0″	5′-5″	4′-11″
2-800S162-68	9′-3″	8′-5″	7′-8″	7′-1″	6′-6″	9′-1″	8′-3″	7′-7″	7′-0″	6′-5″
2-800S162-97	10′-9″	10′-3″	9′-11″	9′-7″	9′-3″	10′-7″	10′-1″	9′-9″	9′-5″	9′-1″
2-1000S162-43	4′-4″	3′-9″	3′-4″	3′-0″	2′-9″	4′-3″	3′-8″	3′-3″	2′-11″	2′-8″
2-1000S162-54	8′-6″	7′-6″	6′-8″	6′-0″	5′-5″	8′-4″	7′-4″	6′-6″	5′-10″	5′-4″
2-1000S162-68	10′-6″	9′-7″	8′-9″	8′-0″	7′-5″	10′-4″	9′-5″	8′-7″	7′-11″	7′-3″
2-1000S162-97	12′-11″	12′-4″	11′-8″	11′-1″	10′-6″	12′-9″	12′-2″	11′-6″	10′-11″	10′-5″
2-1200S162-54	7′-1″	6′-2″	5′-6″	5′-0″	4′-6″	6′-11″	6′-1″	5′-5″	4′-10″	4′-5″
2-1200S162-68	11′-7″	10′-7″	9′-8″	8′-11″	8′-2″	11′-5″	10′-5″	9′-6″	8′-9″	8′-0″
2-1200S162-97	14′-9″	13′-9″	13′-0″	12′-4″	11′-9″	14′-7″	13′-8″	12′-10″	12′-3″	11′-8″

For SI: 1 inch = 25.4 mm, 1 foot = 304.8 mm, 1 pound per square foot = 0.0479 kPa, 1 pound per square inch = 6.895 kPa, 1 Ksi = 1,000 psi = 6.895 MPa.

a. Deflection criterion: *L*/360 for live loads, *L*/240 for total loads.

b. Design load assumptions:
   Second floor dead load is 10 psf.
   Roof/ceiling dead load is 12 psf.
   Second floor live load is 30 psf.
   Attic live load is 10 psf.

c. Building width is in the direction of horizontal framing members supported by the header.

❖ See the commentary for Section R603.6.

TABLE R603.6(7)
BOX-BEAM HEADER SPANS
Headers Supporting One Floor, Roof and Ceiling
(33 Ksi steel)[a, b]

MEMBER DESIGNATION	GROUND SNOW LOAD (50 psf)					GROUND SNOW LOAD (70 psf)				
	Building width[c] (feet)					Building width[c] (feet)				
	24	28	32	36	40	24	28	32	36	40
2-350S162-33	—	—	—	—	—	—	—	—	—	—
2-350S162-43	—	—	—	—	—	—	—	—	—	—
2-350S162-54	—	—	—	—	—	—	—	—	—	—
2-350S162-68	2′-8″	2′-3″	—	—	—	—	—	—	—	—
2-350S162-97	4′-0″	3′-7″	3′-3″	2′-11″	2′-7″	3′-4″	2′-11″	2′-6″	2′-2″	—
2-550S162-33	—	—	—	—	—	—	—	—	—	—
2-550S162-43	2′-0″	—	—	—	—	—	—	—	—	—
2-550S162-54	3′-1″	2′-6″	—	—	—	—	—	—	—	—
2-550S162-68	4′-1″	3′-6″	2′-11″	2′-5″	—	3′-1″	2′-5″	—	—	—
2-550S162-97	5′-10″	5′-3″	4′-10″	4′-5″	4′-0″	4′-11″	4′-5″	3′-11″	3′-6″	3′-2″
2-800S162-33	—	—	—	—	—	—	—	—	—	—
2-800S162-43	2′-6″	—	—	—	—	—	—	—	—	—
2-800S162-54	4′-0″	3′-3″	2′-6″	—	—	2′-8″	—	—	—	—
2-800S162-68	5′-5″	4′-8″	4′-0″	3′-4″	2′-8″	4′-2″	3′-4″	2′-6″	—	—
2-800S162-97	7′-9″	7′-1″	6′-6″	5′-11″	5′-5″	6′-7″	5′-11″	5′-4″	4′-10″	4′-4″
2-1000S162-43	2′-10″	—	—	—	—	—	—	—	—	—
2-1000S162-54	4′-7″	3′-8″	2′-9″	—	—	3′-0″	—	—	—	—
2-1000S162-68	6′-2″	5′-4″	4′-7″	3′-10″	3′-1″	4′-9″	3′-10″	2′-11″	—	—
2-1000S162-97	9′-3″	8′-5″	7′-8″	7′-1″	6′-6″	7′-10″	7′-1″	6′-5″	5′-9″	5′-2″
2-1200S162-54	5′-0″	4′-0″	3′-1″	—	—	3′-4″	—	—	—	—
2-1200S162-68	6′-10″	5′-11″	5′-0″	4′-3″	3′-5″	5′-3″	4′-3″	3′-2″	—	—
2-1200S162-97	10′-5″	9′-6″	8′-8″	8′-0″	7′-4″	8′-10″	8′-0″	7′-3″	6′-6″	5′-10″

For SI: 1 inch = 25.4 mm, 1 foot = 304.8 mm, 1 pound per square foot = 0.0479 kPa, 1 pound per square inch = 6.895 kPa, 1 Ksi = 1,000 psi = 6.895 MPa.

a. Deflection criterion: *L*/360 for live loads, *L*/240 for total loads.

b. Design load assumptions:
Second floor dead load is 10 psf.
Roof/ceiling dead load is 12 psf.
Second floor live load is 30 psf.
Attic live load is 10 psf.

c. Building width is in the direction of horizontal framing members supported by the header.

❖ See the commentary for Section R603.6.

TABLE R603.6(8)
BOX-BEAM HEADER SPANS
Headers Supporting One Floor, Roof and Ceiling
(50 Ksi steel)[a, b]

MEMBER DESIGNATION	GROUND SNOW LOAD (50 psf)					GROUND SNOW LOAD (70 psf)				
	Building width[c] (feet)					Building width[c] (feet)				
	24	28	32	36	40	24	28	32	36	40
2-350S162-33	—	—	—	—	—	—	—	—	—	—
2-350S162-43	2′-8″	—	—	—	—	—	—	—	—	—
2-350S162-54	3′-5″	3′-0″	2′-7″	2′-2″	-	2′-8″	2′-2″	—	—	—
2-350S162-68	4′-6″	4′-1″	3′-8″	3′-3″	2′-11″	3′-9″	3′-3″	2′-10″	2′-5″	2′-1″
2-350S162-97	5′-1″	4′-10″	4′-8″	4′-6″	4′-5″	4′-10″	4′-7″	4′-4″	4′-0″	3′-8″
2-550S162-33	2′-4″	—	—	—	—	—	—	—	—	—
2-550S162-43	3′-10″	3′-4″	2′-9″	2′-3″	—	2′-11″	2′-3″	—	—	—
2-550S162-54	5′-3″	3′-8″	4′-1″	3′-8″	3′-2″	4′-3″	3′-8″	3′-1″	2′-7″	2′-0″
2-550S162-68	6′-5″	5′-10″	5′-3″	4′-9″	4′-4″	5′-5″	4′-9″	4′-3″	3′-9″	3′-4″
2-550S162-97	7′-4″	7′-0″	6′-9″	6′-6″	6′-4″	6′-11″	6′-8″	6′-3″	5′-10″	5′-5″
2-800S162-33	1′-11″	1′-8″	—	—	—	—	—	—	—	—
2-800S162-43	4′-2″	3′-8″	3′-4″	2′-9″	2′-2″	3′-5″	2′-9″	—	—	—
2-800S162-54	6′-1″	5′-5″	4′-10″	4′-3″	3′-9″	4′-11″	4′-3″	3′-8″	3′-0″	2′-5″
2-800S162-68	7′-8″	6′-11″	6′-3″	5′-9″	5′-2″	6′-5″	5′-9″	5′-1″	4′-6″	4′-0″
2-800S162-97	9′-11″	9′-6″	9′-2″	8′-10″	8′-3″	9′-5″	8′-10″	8′-2″	7′-7″	7′-0″
2-1000S162-43	3′-4″	2′-11″	2′-7″	2′-5″	2′-2″	2′-8″	2′-5″	2′-2″	—	—
2-1000S162-54	6′-7″	5′-10″	5′-3″	4′-9″	4′-3″	5′-4″	4′-9″	4′-1″	3′-5″	2′-9″
2-1000S162-68	8′-8″	7′-10″	7′-2″	6′-6″	5′-11″	7′-4″	6′-6″	5′-9″	5′-1″	4′-6″
2-1000S162-97	11′-7″	10′-11″	10′-3″	9′-7″	9′-0″	10′-5″	9′-7″	8′-10″	8′-2″	7′-8″
2-1200S162-54	5′-6″	4′-10″	4′-4″	3′-11″	3′-7″	4′-5″	3′-11″	3′-6″	3′-2″	2′-11″
2-1200S162-68	9′-7″	8′-8″	7′-11″	7′-2″	6′-6″	8′-1″	7′-2″	6′-4″	5′-8″	5′-0″
2-1200S162-97	12′-11″	12′-2″	11′-6″	10′-8″	10′-0″	11′-8″	10′-9″	9′-11″	9′-2″	8′-6″

For SI: 1 inch = 25.4 mm, 1 foot = 304.8 mm, 1 pound per square foot = 0.0479 kPa, 1 pound per square inch = 6.895 kPa, 1 Ksi = 1,000 psi = 6.895 MPa.

a. Deflection criterion: *L*/360 for live loads, *L*/240 for total loads.

b. Design load assumptions:
Second floor dead load is 10 psf.
Roof/ceiling dead load is 12 psf.
Second floor live load is 30 psf.
Attic live load is 10 psf.

c. Building width is in the direction of horizontal framing members supported by the header.

❖ See the commentary for Section R603.6.

TABLE R603.6(9)
BOX-BEAM HEADER SPANS
Headers Supporting Two Floors, Roof and Ceiling
(33 Ksi steel)[a, b]

MEMBER DESIGNATION	GROUND SNOW LOAD (20 psf)					GROUND SNOW LOAD (30 psf)				
	Building width[c] (feet)					Building width[c] (feet)				
	24	28	32	36	40	24	28	32	36	40
2-350S162-33	—	—	—	—	—	—	—	—	—	—
2-350S162-43	—	—	—	—	—	—	—	—	—	—
2-350S162-54	—	—	—	—	—	—	—	—	—	—
2-350S162-68	—	—	—	—	—	—	—	—	—	—
2-350S162-97	3′-1″	2′-8″	2′-3″	—	—	3′-1″	2′-7″	2′-2″	—	—
2-550S162-33	—	—	—	—	—	—	—	—	—	—
2-550S162-43	—	—	—	—	—	—	—	—	—	—
2-550S162-54	—	—	—	—	—	—	—	—	—	—
2-550S162-68	2′-9″	—	—	—	—	2′-8″	—	—	—	—
2-550S162-97	4′-8″	4′-1″	3′-7″	3′-2″	2′-9″	4′-7″	4′-0″	3′-6″	3′-1″	2′-8″
2-800S162-33	—	—	—	—	—	—	—	—	—	—
2-800S162-43	—	—	—	—	—	—	—	—	—	—
2-800S162-54	2′-1″	—	—	—	—	—	—	—	—	—
2-800S162-68	3′-8″	2′-9″	—	—	—	3′-7″	2′-8″	—	—	—
2-800S162-97	6′-3″	5′-6″	4′-11″	4′-4″	3′-9″	6′-2″	5′-5″	4′-10″	4′-3″	3′-9″
2-1000S162-43	—	—	—	—	—	—	—	—	—	—
2-1000S162-54	2′-5″	—	—	—	—	2′-3″	—	—	—	—
2-1000S162-68	4′-3″	3′-2″	2′-0″	—	—	4′-2″	3′-1″	—	—	—
2-1000S162-97	7′-5″	6′-7″	5′-10″	5′-2″	4′-7″	7′-4″	6′-6″	5′-9″	5′-1″	4′-6″
2-1200S162-54	2′-7″	—	—	—	—	2′-6″	—	—	—	—
2-1200S162-68	4′-8″	3′-6″	2′-2″	—	—	4′-7″	3′-5″	2′-0″	—	—
2-1200S162-97	8′-5″	7′-5″	6′-7″	5′-10″	5′-2″	8′-3″	7′-4″	6′-6″	5′-9″	5′-1″

For SI: 1 inch = 25.4 mm, 1 foot = 304.8 mm, 1 pound per square foot = 0.0479 kPa, 1 pound per square inch = 6.895 kPa, 1 Ksi = 1,000 psi = 6.895 MPa.

a. Deflection criterion: *L*/360 for live loads, *L*/240 for total loads.

b. Design load assumptions:
   Second floor dead load is 10 psf.
   Roof/ceiling dead load is 12 psf.
   Second floor live load is 40 psf.
   Third floor live load is 30 psf.
   Attic live load is 10 psf.

c. Building width is in the direction of horizontal framing members supported by the header.

❖ See the commentary for Section R603.6.

TABLE R603.6(10)
BOX-BEAM HEADER SPANS
Headers Supporting Two Floors, Roof and Ceiling
(50 Ksi steel)[a, b]

MEMBER DESIGNATION	GROUND SNOW LOAD (20 psf)					GROUND SNOW LOAD (30 psf)				
	Building width[c] (feet)					Building width[c] (feet)				
	24	28	32	36	40	24	28	32	36	40
2-350S162-33	—	—	—	—	—	—	—	—	—	—
2-350S162-43	—	—	—	—	—	—	—	—	—	—
2-350S162-54	2′-5″	—	—	—	—	2′-4″	—	—	—	—
2-350S162-68	3′-6″	3′-0″	2′-6″	2′-1″	—	3′-5″	2′-11″	2′-6″	2′-0″	—
2-350S162-97	4′-9″	4′-6″	4′-1″	3′-8″	3′-4″	4′-8″	4′-5″	4′-0″	3′-8″	3′-4″
2-550S162-33	—	—	—	—	—	—	—	—	—	—
2-550S162-43	2′-7″	—	—	—	—	2′-6″	—	—	—	—
2-550S162-54	3′-11″	3′-3″	2′-8″	2′-0″	—	3′-10″	3′-3″	2′-7″	—	—
2-550S162-68	5′-1″	4′-5″	3′-10″	3′-3″	2′-9″	5′-0″	4′-4″	3′-9″	3′-3″	2′-9″
2-550S162-97	6′-10″	6′-5″	5′-10″	5′-5″	4′-11″	6′-9″	6′-4″	5′-10″	5′-4″	4′-11″
2-800S162-33	—	—	—	—	—	—	—	—	—	—
2-800S162-43	3′-1″	2′-3″	—	—	—	3′-0″	2′-2″	—	—	—
2-800S162-54	4′-7″	3′-10″	3′-1″	2′-5″	—	4′-6″	3′-9″	3′-0″	2′-4″	—
2-800S162-68	6′-0″	5′-3″	4′-7″	3′-11″	3′-4″	6′-0″	5′-2″	4′-6″	3′-11″	3′-3″
2-800S162-97	9′-2″	8′-4″	7′-8″	7′-0″	6′-6″	9′-1″	8′-3″	7′-7″	7′-0″	6′-5″
2-1000S162-43	2′-6″	2′-2″	—	—	—	2′-6″	2′-2″	—	—	—
2-1000S162-54	5′-0″	4′-4″	3′-6″	2′-9″	—	4′-11″	4′-3″	3′-5″	2′-7″	—
2-1000S162-68	6′-10″	6′-0″	5′-3″	4′-6″	3′-10″	6′-9″	5′-11″	5′-2″	4′-5″	3′-9″
2-1000S162-97	10′-0″	9′-1″	8′-3″	7′-8″	7′-0″	9′-10″	9′-0″	8′-3″	7′-7″	7′-0″
2-1200S162-54	4′-2″	3′-7″	3′-3″	2′-11″	—	4′-1″	3′-7″	3′-2″	2′-10″	—
2-1200S162-68	7′-7″	6′-7″	5′-9″	5′-0″	4′-2″	7′-6″	6′-6″	5′-8″	4′-10″	4′-1″
2-1200S162-97	11′-2″	10′-1″	9′-3″	8′-6″	7′-10″	11′-0″	10′-0″	9′-2″	9′-2″	7′-9″

For SI: 1 inch = 25.4 mm, 1 foot = 304.8 mm, 1 pound per square foot = 0.0479 kPa, 1 pound per square inch = 6.895 kPa, 1 Ksi = 1,000 psi = 6.895 MPa.

a. Deflection criterion: *L*/360 for live loads, *L*/240 for total loads.

b. Design load assumptions:
   Second floor dead load is 10 psf.
   Roof/ceiling dead load is 12 psf.
   Second floor live load is 40 psf.
   Third floor live load is 30 psf.
   Attic live load is 10 psf.

c. Building width is in the direction of horizontal framing members supported by the header.

❖ See the commentary for Section R603.6.

TABLE R603.6(11)
BOX-BEAM HEADER SPANS
Headers Supporting Two Floors, Roof and Ceiling
(33 Ksi steel)[a, b]

MEMBER DESIGNATION	GROUND SNOW LOAD (50 psf)					GROUND SNOW LOAD (70 psf)				
	Building width[c] (feet)					Building width[c] (feet)				
	24	28	32	36	40	24	28	32	36	40
2-350S162-33	—	—	—	—	—	—	—	—	—	—
2-350S162-43	—	—	—	—	—	—	—	—	—	—
2-350S162-54	—	—	—	—	—	—	—	—	—	—
2-350S162-68	—	—	—	—	—	—	—	—	—	—
2-350S162-97	2′-11″	2′-5″	2′-0″	—	—	2′-7″	2′-2″	—	—	—
2-550S162-33	—	—	—	—	—	—	—	—	—	—
2-550S162-43	—	—	—	—	—	—	—	—	—	—
2-550S162-54	—	—	—	—	—	—	—	—	—	—
2-550S162-68	2′-5″	—	—	—	—	—	—	—	—	—
2-550S162-97	4′-4″	3′-10″	3′-4″	2′-10″	2′-5″	4′-0″	3′-6″	3′-1″	2′-7″	2′-2″
2-800S162-33	—	—	—	—	—	—	—	—	—	—
2-800S162-43	—	—	—	—	—	—	—	—	—	—
2-800S162-54	—	—	—	—	—	—	—	—	—	—
2-800S162-68	3′-3″	2′-3″	—	—	—	2′-8″	—	—	—	—
2-800S162-97	5′-11″	5′-2″	4′-6″	4′-0″	3′-5″	5′-6″	4′-10″	4′-3″	3′-8″	3′-2″
2-1000S162-43	—	—	—	—	—	—	—	—	—	—
2-1000S162-54	—	—	—	—	—	—	—	—	—	—
2-1000S162-68	3′-9″	2′-7″	—	—	—	3′-1″	—	—	—	—
2-1000S162-97	7′-0″	6′-2″	5′-5″	4′-9″	4′-2″	6′-6″	5′-9″	5′-1″	4′-5″	3′-10″
2-1200S162-54	—	—	—	—	—	—	—	—	—	—
2-1200S162-68	4′-2″	2′-10″	—	—	—	3′-5″	2′-0″	—	—	—
2-1200S162-97	7′-11″	7′-0″	6′-2″	5′-5″	4′-8″	7′-4″	6′-6″	5′-9″	5′-0″	4′-4″

For SI: 1 inch = 25.4 mm, 1 foot = 304.8 mm, 1 pound per square foot = 0.0479 kPa, 1 pound per square inch = 6.895 kPa, 1 Ksi = 1,000 psi = 6.895 MPa.

a. Deflection criterion: *L*/360 for live loads, *L*/240 for total loads.

b. Design load assumptions:
Second floor dead load is 10 psf.
Roof/ceiling dead load is 12 psf.
Second floor live load is 40 psf.
Third floor live load is 30 psf.
Attic live load is 10 psf.

c. Building width is in the direction of horizontal framing members supported by the header.

❖ See the commentary for Section R603.6.

TABLE R603.6(12) BOX-BEAM HEADER SPANS Headers Supporting Two Floors, Roof and Ceiling (50 Ksi steel)[a,b]

MEMBER DESIGNATION	GROUND SNOW LOAD (50 psf)					GROUND SNOW LOAD (70 psf)				
	Building width[c] (feet)					Building width[c] (feet)				
	24	28	32	36	40	24	28	32	36	40
2-350S162-33	—	—	—	—	—	—	—	—	—	—
2-350S162-43	—	—	—	—	—	—	—	—	—	—
2-350S162-54	2′-2″	—	—	—	—	—	—	—	—	—
2-350S162-68	3′-3″	2′-9″	2′-3″	—	—	2′-11″	2′-5″	—	—	—
2-350S162-97	4′-6″	4′-3″	3′-10″	3′-6″	3′-2″	4′-3″	4′-0″	3′-7″	3′-3″	3′-0″
2-550S162-33	—	—	—	—	—	—	—	—	—	—
2-550S162-43	2′-3″	—	—	—	—	—	—	—	—	—
2-550S162-54	3′-7″	2′-11″	2′-3″	—	—	3′-3″	2′-7″	—	—	—
2-550S162-68	4′-9″	2′-1″	3′-6″	3′-0″	2′-5″	4′-4″	3′-9″	3′-2″	2′-8″	2′-1″
2-550S162-97	6′-5″	6′-1″	5′-7″	5′-1″	4′-8″	6′-3″	5′-10″	5′-4″	4′-10″	4′-5″
2-800S162-33	—	—	—	—	—	—	—	—	—	—
2-800S162-43	2′-8″	—	—	—	—	2′-2″	—	—	—	—
2-800S162-54	4′-3″	3′-5″	2′-8″	—	—	3′-9″	3′-0″	2′-3″	—	—
2-800S162-68	5′-8″	4′-11″	4′-2″	3′-7″	2′-11″	5′-3″	4′-6″	3′-10″	3′-3″	2′-7″
2-800S162-97	8′-9″	8′-0″	7′-3″	6′-8″	6′-2″	8′-4″	7′-7″	6′-11″	6′-4″	5′-10″
2-1000S162-43	2′-4″	2′-0″	—	—	—	2′-2″	—	—	—	—
2-1000S162-54	4′-8″	3′-11″	3′-1″	2′-2″	—	4′-3″	3′-5″	2′-7″	—	—
2-1000S162-68	6′-5″	5′-7″	4′-9″	4′-1″	3′-4″	5′-11″	5′-1″	4′-5″	3′-8″	2′-11″
2-1000S162-97	9′-6″	8′-8″	7′-11″	7′-3″	6′-8″	9′-0″	8′-3″	7′-6″	6′-11″	6′-4″
2-1200S162-54	3′-11″	3′-5″	3′-0″	2′-4″	—	3′-7″	3′-2″	2′-10″	—	—
2-1200S162-68	7′-1″	6′-2″	5′-3″	4′-6″	3′-8″	6′-6″	5′-8″	4′-10″	4′-0″	3′-3″
2-1200S162-97	10′-8″	9′-8″	8′-10″	8′-1″	7′-5″	10′-1″	9′-2″	8′-5″	7′-9″	7′-1″

For SI: 1 inch = 25.4 mm, 1 foot = 304.8 mm, 1 pound per square foot = 0.0479 kPa, 1 pound per square inch = 6.895 kPa, 1 Ksi = 1,000 psi = 6.895 MPa.

a. Deflection criterion: *L*/360 for live loads, *L*/240 for total loads.

b. Design load assumptions:

Second floor dead load is 10 psf.

Roof/ceiling dead load is 12 psf.

Second floor live load is 40 psf.

Third floor live load is 30 psf.

Attic live load is 10 psf.

c. Building width is in the direction of horizontal framing members supported by the header.

❖ See the commentary for Section R603.6.

TABLE R603.6(13)
BACK-TO-BACK HEADER SPANS
Headers Supporting Roof and Ceiling Only
(33 Ksi steel)[a,b]

MEMBER DESIGNATION	GROUND SNOW LOAD (20 psf)					GROUND SNOW LOAD (30 psf)				
	Building width[c] (feet)					Building width[c] (feet)				
	24	28	32	36	40	24	28	32	36	40
2-350S162-33	2′-11″	2′-4″	—	—	—	2′-5″	—	—	—	—
2-350S162-43	4′-8″	3′-10″	3′-5″	3′-1″	2′-9″	3′-11″	3′-5″	3′-0″	2′-8″	2′-4″
2-350S162-54	5′-3″	4′-9″	4′-4″	4′-1″	3′-8″	4′-10″	4′-4″	4′-0″	3′-8″	3′-4″
2-350S162-68	6′-1″	5′-7″	5′-2″	4′-10″	4′-6″	5′-8″	5′-3″	4′-10″	4′-6″	4′-2″
2-350S162-97	7′-3″	6′-10″	6′-5″	6′-0″	5′-8″	6′-11″	6′-5″	6′-0″	5′-8″	5′-4″
2-550S162-33	4′-5″	3′-9″	3′-1″	2′-6″	—	3′-9″	3′-2″	2′-6″	—	—
2-550S162-43	6′-2″	5′-7″	5′-0″	4′-7″	4′-2″	5′-7″	5′-0″	4′-6″	4′-1″	3′-8″
2-550S162-54	7′-5″	6′-9″	6′-3″	5′-9″	5′-4″	6′-10″	6′-3″	5′-9″	5′-4″	4′-11″
2-550S162-68	6′-7″	7′-11″	7′-4″	6′-10″	6′-5″	8′-0″	7′-4″	6′-10″	6′-5″	6′-0″
2-550S162-97	10′-5″	9′-8″	9′-0″	8′-6″	8′-0″	9′-9″	9′-0″	8′-6″	8′-0″	7′-7″
2-800S162-33	4′-5″	3′-11″	3′-5″	3′-1″	2′-4″	3′-11″	3′-6″	3′-0″	2′-3″	—
2-800S162-43	7′-7″	6′-10″	6′-2″	5′-8″	5′-2″	6′-11″	6′-2″	5′-7″	5′-1″	4′-7″
2-800S162-54	9′-3″	8′-7″	7′-11″	7′-4″	6′-10″	8′-8″	7′-11″	7′-4″	6′-9″	6′-3″
2-800S162-68	10′-7″	9′-10″	9′-4″	8′-10″	8′-5″	9′-11″	9′-4″	8′-10″	8′-4″	7′-11″
2-800S162-97	13′-9″	12′-9″	12′-0″	11′-3″	10′-8″	12′-10″	12′-0″	11′-3″	10′-7″	10′-0″
2-1000S162-43	7′-10″	6′-10″	6′-1″	5′-6″	5′-0″	6′-11″	6′-1″	5′-5″	4′-11″	4′-6″
2-1000S162-54	10′-5″	9′-9″	9′-0″	8′-4″	7′-9″	9′-10″	9′-0″	8′-4″	7′-9″	7′-2″
2-1000S162-68	12′-1″	11′-3″	10′-8″	10′-1″	9′-7″	11′-4″	10′-8″	10′-1″	9′-7″	9′-1″
2-1000S162-97	15′-3″	14′-3″	13′-5″	12′-9″	12′-2″	14′-4″	13′-5″	12′-8″	12′-1″	11′-6″
2-1200S162-54	11′-6″	10′-9″	10′-0″	9′-0″	8′-2″	10′-10″	10′-0″	9′-0″	8′-1″	7′-4″
2-1200S162-68	13′-4″	12′-6″	11′-9″	11′-2″	10′-8″	12′-7″	11′-10″	11′-2″	10′-7″	10′-1″
2-1200S162-97	16′-8″	15′-7″	14′-8″	13′-11″	13′-3″	15′-8″	14′-8″	13′-11″	13′-2″	12′-7″

For SI: 1 inch = 25.4 mm, 1 foot = 304.8 mm, 1 pound per square foot = 0.0479 kPa, 1 pound per square inch = 6.895 kPa, 1 Ksi = 1,000 psi = 6.895 MPa.

a. Deflection criterion: *L*/360 for live loads, *L*/240 for total loads.

b. Design load assumptions:
Second floor dead load is 12 psf.
Attic live load is 10 psf.

c. Building width is in the direction of horizontal framing members supported by header.

❖ See the commentary for Section R603.6.

TABLE R603.6(14)
BACK-TO-BACK HEADER SPANS
Headers Supporting Roof and Ceiling Only
(50 Ksi steel)[a,b]

MEMBER DESIGNATION	GROUND SNOW LOAD (20 psf)					GROUND SNOW LOAD (30 psf)				
	Building width[c] (feet)					Building width[c] (feet)				
	24	28	32	36	40	24	28	32	36	40
2-350S162-33	4′-2″	3′-8″	3′-3″	2′-10″	2′-6″	3′-8″	3′-3″	2′-10″	2′-5″	2′-1″
2-350S162-43	5′-5″	5′-0″	4′-6″	4′-2″	3′-10″	5′-0″	4′-7″	4′-2″	3′-10″	3′-6″
2-350S162-54	6′-2″	5′-10″	5′-8″	5′-4″	5′-0″	5′-11″	5′-8″	5′-4″	5′-0″	4′-8″
2-350S162-68	6′-7″	6′-3″	6′-0″	5′-10″	5′-8″	6′-4″	6′-1″	5′-10″	5′-8″	5′-6″
2-350S162-97	7′-3″	6′-11″	6′-8″	6′-5″	6′-3″	7′-0″	6′-8″	6′-5″	6′-3″	6′-0″
2-550S162-33	5′-10″	5′-3″	4′-8″	4′-3″	3′-9″	5′-3″	4′-9″	4′-2″	3′-9″	3′-3″
2-550S162-43	7′-9″	7′-2″	6′-7″	6′-1″	5′-8″	7′-3″	6′-7″	6′-1″	5′-8″	5′-3″
2-550S162-54	8′-9″	8′-5″	8′-1″	7′-9″	7′-5″	8′-6″	8′-1″	7′-9″	7′-5″	6′-11″
2-550S162-68	9′-5″	9′-0″	8′-8″	8′-4″	8′-1″	9′-1″	8′-8″	8′-4″	8′-1″	7′-10″
2-550S162-97	10′-5″	10′-0″	9′-7″	9′-3″	9′-0″	10′-0″	9′-7″	9′-3″	8′-11″	8′-8″
2-800S162-33	4′-5″	3′-11″	3′-5″	3′-1″	2′-10″	3′-11″	3′-6″	3′-1″	2′-9″	2′-6″
2-800S162-43	9′-1″	8′-5″	7′-8″	6′-11″	6′-3″	8′-6″	7′-8″	6′-10″	6′-2″	5′-8″
2-800S162-54	10′-10″	10′-2″	9′-7″	9′-1″	8′-8″	10′-2″	9′-7″	9′-0″	8′-7″	8′-1″
2-800S162-68	12′-8″	11′-10″	11′-2″	10′-7″	10′-1″	11′-11″	11′-2″	10′-7″	10′-0″	9′-7″
2-800S162-97	14′-2″	13′-6″	13′-0″	12′-7″	12′-2″	13′-8″	13′-1″	12′-7″	12′-2″	11′-9″
2-1000S162-43	7′-10″	6′-10″	6′-1″	5′-6″	5′-0″	6′-11″	6′-1″	5′-5″	4′-11″	4′-6″
2-1000S162-54	12′-3″	11′-5″	10′-9″	10′-3″	9′-9″	11′-6″	10′-9″	10′-2″	9′-8″	8′-11″
2-1000S162-68	14′-5″	13′-5″	12′-8″	12′-0″	11′-6″	13′-6″	12′-8″	12′-0″	11′-5″	10′-11″
2-1000S162-97	17′-1″	16′-4″	15′-8″	14′-11″	14′-3″	16′-5″	15′-9″	14′-10″	14′-1″	13′-6″
2-1200S162-54	12′-11″	11′-3″	10′-0″	9′-0″	8′-2″	11′-5″	10′-0″	9′-0″	8′-1″	7′-4″
2-1200S162-68	15′-11″	14′-10″	14′-0″	13′-4″	12′-8″	15′-0″	14′-0″	13′-3″	12′-7″	12′-0″
2-1200S162-97	19′-11″	18′-7″	17′-6″	16′-8″	15′-10″	18′-9″	17′-7″	16′-7″	15′-9″	15′-0″

For SI: 1 inch = 25.4 mm, 1 foot = 304.8 mm, 1 pound per square foot = 0.0479 kPa, 1 pound per square inch = 6.895 kPa, 1 Ksi = 1,000 psi = 6.895 MPa.

a. Deflection criterion: $L/360$ for live loads, $L/240$ for total loads.

b. Design load assumptions:
Roof/ceiling dead load is 12 psf.
Attic live load is 10 psf.

c. Building width is in the direction of horizontal framing members supported by the header.

❖ See the commentary for Section R603.6.

TABLE R603.6(15)
BACK-TO-BACK HEADER SPANS
Headers Supporting Roof and Ceiling Only
(33 Ksi steel)[a, b]

MEMBER DESIGNATION	GROUND SNOW LOAD (50 psf)					GROUND SNOW LOAD (70 psf)				
	Building width[c] (feet)					Building width[c] (feet)				
	24	28	32	36	40	24	28	32	36	40
2-350S162-33	—	—	—	—	—	—	—	—	—	—
2-350S162-43	2′-6″	—	—	—	—	—	—	—	—	—
2-350S162-54	3′-6″	3′-1″	2′-8″	2′-4″	2′-0″	2′-7″	2′-1″	—	—	—
2-350S162-68	4′-4″	3′-11″	3′-7″	3′-3″	2′-11″	3′-5″	3′-0″	2′-8″	2′-4″	2′-1″
2-350S162-97	5′-5″	5′-0″	4′-8″	4′-6″	4′-1″	4′-6″	4′-2″	3′-10″	3′-6″	3′-3″
2-550S162-33	—	—	—	—	—	—	—	—	—	—
2-550S162-43	3′-10″	3′-3″	2′-9″	2′-2″	—	2′-6″	—	—	—	—
2-550S162-54	5′-1″	4′-7″	4′-1″	3′-8″	3′-4″	3′-11″	3′-5″	2′-11″	2′-6″	2′-0″
2-550S162-68	6′-2″	5′-8″	5′-2″	4′-9″	4′-5″	5′-0″	4′-6″	4′-1″	3′-9″	3′-4″
2-550S162-97	7′-9″	7′-2″	6′-8″	6′-3″	5′-11″	6′-6″	6′-0″	5′-7″	5′-2″	4′-10″
2-800S162-33	—	—	—	—	—	—	—	—	—	—
2-800S162-43	4′-10″	4′-1″	3′-6″	2′-11″	2′-3″	3′-3″	2′-5″	—	—	—
2-800S162-54	6′-6″	5′-10″	5′-3″	4′-9″	4′-4″	5′-1″	4′-6″	3′-11″	3′-4″	2′-10″
2-800S162-68	8′-1″	7′-5″	6′-10″	6′-4″	5′-11″	6′-8″	6′-1″	5′-6″	5′-0″	4′-7″
2-800S162-97	10′-3″	9′-7″	8′-11″	8′-5″	7′-11″	8′-8″	8′-0″	7′-6″	7′-0″	6′-7″
2-1000S162-43	4′-8″	4′-1″	3′-8″	3′-4″	2′-8″	3′-6″	2′-10″	—	—	—
2-1000S162-54	7′-5″	6′-8″	6′-1″	5′-6″	5′-0″	5′-10″	5′-1″	4′-6″	3′-11″	3′-4″
2-1000S162-68	9′-4″	8′-7″	7′-11″	7′-4″	6′-10″	7′-8″	7′-0″	6′-4″	5′-10″	5′-4″
2-1000S162-97	11′-9″	11′-0″	10′-5″	9′-11″	9′-5″	10′-3″	9′-7″	8′-11″	8′-4″	7′-10″
2-1200S162-54	7′-8″	6′-9″	6′-1″	5′-6″	5′-0″	5′-10″	5′-1″	4′-7″	4′-1″	3′-9″
2-1200S162-68	10′-4″	9′-6″	8′-10″	8′-2″	7′-7″	8′-7″	7′-9″	7′-1″	6′-6″	6′-0″
2-1200S162-97	12′-10″	12′-1″	11′-5″	10′-10″	10′-4″	11′-2″	10′-6″	9′-11″	9′-5″	9′-0″

For SI: 1 inch = 25.4 mm, 1 foot = 304.8 mm, 1 pound per square foot = 0.0479 kPa, 1 pound per square inch = 6.895 kPa, 1 Ksi = 1,000 psi = 6.895 MPa.

a. Deflection criterion: *L*/360 for live loads, *L*/240 for total loads.

b. Design load assumptions:

Roof/ceiling dead load is 12 psf.

Attic live load is 10 psf.

c. Building width is in the direction of horizontal framing members supported by the header.

❖ See the commentary for Section R603.6.

TABLE R603.6(16)
BACK-TO-BACK HEADER SPANS
Headers Supporting Roof and Ceiling Only
(50 Ksi steel)[a, b]

MEMBER DESIGNATION	GROUND SNOW LOAD (50 psf)					GROUND SNOW LOAD (70 psf)				
	Building width[c] (feet)					Building width[c] (feet)				
	24	28	32	36	40	24	28	32	36	40
2-350S162-33	2′-3″	—	—	—	—	—	—	—	—	—
2-350S162-43	3′-8″	3′-3″	2′-10″	2′-6″	2′-2″	2′-8″	2′-3″	—	—	—
2-350S162-54	4′-9″	4′-4″	4′-0″	3′-8″	3′-8″	3′-10″	3′-5″	3′-1″	2′-9″	2′-5″
2-350S162-68	5′-7″	5′-4″	5′-2″	4′-11″	4′-7″	5′-1″	4′-8″	4′-3″	3′-11″	3′-8″
2-350S162-97	6′-2″	5′-11″	5′-8″	5′-6″	5′-4″	5′-8″	5′-5″	5′-3″	5′-0″	4′-11″
2-550S162-33	3′-6″	2′-10″	2′-3″	—	—	2′-0″	—	—	—	—
2-550S162-43	5′-5″	4′-10″	4′-4″	3′-11″	3′-6″	4′-2″	3′-8″	3′-2″	2′-8″	2′-3″
2-550S162-54	7′-2″	6′-6″	6′-0″	5′-7″	5′-2″	5′-10″	5′-3″	4′-10″	4′-5″	4′-0″
2-550S162-68	8′-0″	7′-8″	7′-3″	6′-11″	6′-6″	7′-2″	6′-7″	6′-1″	5′-8″	5′-4″
2-550S162-97	8′-11″	8′-6″	8′-2″	7′-11″	7′-8″	8′-1″	7′-9″	7′-6″	7′-2″	6′-11″
2-800S162-33	2′-8″	2′-4″	2′-1″	1′-11″	—	2′-0″	—	—	—	—
2-800S162-43	5′-10″	5′-2″	4′-7″	4′-2″	3′-10″	4′-5″	3′-11″	3′-6″	3′-2″	2′-9″
2-800S162-54	8′-4″	7′-8″	7′-1″	6′-7″	6′-1″	6′-10″	6′-3″	5′-8″	5′-2″	4′-9″
2-800S162-68	9′-9″	9′-2″	8′-8″	8′-3″	7′-10″	8′-6″	7′-11″	7′-4″	6′-10″	6′-5″
2-800S162-97	12′-1″	11′-7″	11′-2″	10′-8″	10′-2″	11′-0″	10′-4″	9′-9″	9′-3″	8′-10″
2-1000S162-43	4′-8″	4′-1″	2′-8″	3′-4″	3′-0″	3′-6″	10′-1″	2′-9″	2′-6″	2′-3″
2-1000S162-54	9′-3″	8′-2″	7′-3″	6′-7″	6′-0″	7′-0″	6′-2″	5′-6″	5′-0″	4′-6″
2-1000S162-68	11′-1″	10′-5″	9′-10″	9′-4″	8′-11″	9′-8″	9′-1″	8′-5″	7′-10″	7′-4″
2-1000S162-97	13′-9″	12′-11″	12′-2″	11′-7″	11′-1″	11′-11″	11′-3″	10′-7″	10′-1″	9′-7″
2-1200S162-54	7′-8″	6′-9″	6′-1″	5′-6″	5′-0″	5′-10″	5′-1″	4′-7″	4′-1″	3′-9″
2-1200S162-68	12′-3″	11′-6″	10′-11″	10′-4″	9′-11″	10′-8″	10′-0″	9′-2″	8′-4″	7′-7″
2-1200S162-97	15′-4″	14′-5″	13′-7″	12′-11″	12′-4″	13′-4″	12′-6″	11′-10″	11′-3″	10′-9″

For SI: 1 inch = 25.4 mm, 1 foot = 304.8 mm, 1 pound per square foot = 0.0479 kPa, 1 pound per square inch = 6.895 kPa, 1 Ksi = 1,000 psi = 6.895 MPa.

a. Deflection criterion: *L*/360 for live loads, *L*/240 for total loads.

b. Design load assumptions:

Roof/ceiling dead load is 12 psf.

Attic live load is 10 psf.

c. Building width is in the direction of horizontal framing members supported by the header.

❖ See the commentary for Section R603.6.

TABLE R603.6(17)
BACK-TO-BACK HEADER SPANS
Headers Supporting One Floor, Roof and Ceiling
(33 Ksi steel)[a, b]

MEMBER DESIGNATION	GROUND SNOW LOAD (20 psf)					GROUND SNOW LOAD (30 psf)				
	Building width[c] (feet)					Building width[c] (feet)				
	24	28	32	36	40	24	28	32	36	40
2-350S162-33	—	—	—	—	—	—	—	—	—	—
2-350S162-43	2′-2″	—	—	—	—	2′-1″	—	—	—	—
2-350S162-54	3′-3″	2′-9″	2′-5″	2′-0″	—	3′-2″	2′-9″	2′-4″	—	—
2-350S162-68	4′-4″	3′-8″	3′-3″	2′-11″	2′-8″	4′-0″	3′-7″	3′-2″	2′-11″	2′-7″
2-350S162-97	5′-2″	4′-9″	4′-4″	4′-1″	3′-9″	5′-1″	4′-8″	4′-4″	4′-0″	3′-9″
2-550S162-33	—	—	—	—	—	—	—	—	—	—
2-550S162-43	3′-6″	2′-10″	2′-3″	—	—	3′-5″	2′-9″	2′-2″	—	—
2-550S162-54	4′-9″	4′-2″	3′-9″	3′-3″	2′-10″	4′-8″	4′-1″	3′-8″	3′-2″	2′-9″
2-550S162-68	5′-10″	5′-3″	4′-10″	4′-5″	4′-1″	5′-9″	5′-3″	4′-9″	4′-4″	4′-0″
2-550S162-97	7′-4″	6′-9″	6′-4″	5′-11″	5′-6″	7′-3″	6′-9″	6′-3″	5′-10″	5′-5″
2-800S162-33	—	—	—	—	—	—	—	—	—	—
2-800S162-43	4′-4″	3′-8″	2′-11″	2′-3″	—	4′-3″	3′-6″	2′-10″	2′-1″	—
2-800S162-54	6′-1″	5′-5″	4′-10″	4′-4″	3′-10″	6′-0″	5′-4″	4′-9″	4′-3″	3′-9″
2-800S162-68	7′-8″	7′-0″	6′-5″	5′-11″	5′-5″	7′-7″	6′-11″	6′-4″	5′-10″	5′-4″
2-800S162-97	9′-10″	9′-1″	8′-5″	7′-11″	7′-5″	9′-8″	8′-11″	8′-4″	7′-10″	7′-4″
2-1000S162-43	4′-4″	3′-9″	3′-4″	2′-8″	—	4′-3″	3′-8″	3′-3″	2′-6″	—
2-1000S162-54	6′-11″	6′-2″	5′-6″	5′-0″	4′-5″	6′-10″	6′-1″	5′-5″	4′-10″	4′-4″
2-1000S162-68	8′-10″	8′-1″	7′-5″	6′-10″	6′-4″	8′-8″	7′-11″	7′-3″	6′-8″	6′-2″
2-1000S162-97	11′-3″	10′-7″	9′-11″	9′-5″	8′-10″	11′-2″	10′-5″	9′-10″	9′-3″	8′-9″
2-1200S162-54	7′-1″	6′-2″	5′-6″	5′-0″	4′-6″	6′-11″	6′-1″	5′-5″	4′-10″	4′-5″
2-1200S162-68	9′-10″	9′-0″	8′-3″	7′-7″	7′-0″	9′-8″	8′-10″	8′-1″	7′-6″	6′-11″
2-1200S162-97	12′-4″	11′-7″	10′-11″	10′-4″	9′-10″	12′-3″	11′-5″	10′-9″	10′-3″	9′-9″

For SI: 1 inch = 25.4 mm, 1 foot = 304.8 mm, 1 pound per square foot = 0.0479 kPa, 1 pound per square inch = 6.895 kPa,
1 Ksi = 1,000 psi = 6.895 MPa.

a. Deflection criterion: *L*/360 for live loads, *L*/240 for total loads.

b. Design load assumptions:
- Second floor dead load is 10 psf.
- Roof/ceiling dead load is 12 psf.
- Second floor live load is 30 psf.
- Attic live load is 10 psf.

c. Building width is in the direction of horizontal framing members supported by the header.

❖ See the commentary for Section R603.6.

TABLE R603.6(18)
BACK-TO-BACK HEADER SPANS
Headers Supporting One Floor, Roof and Ceiling
(50 Ksi steel)[a, b]

MEMBER DESIGNATION	GROUND SNOW LOAD (20 psf)					GROUND SNOW LOAD (30 psf)				
	Building width[c] (feet)					Building width[c] (feet)				
	24	28	32	36	40	24	28	32	36	40
2-350S162-33	—	—	—	—	—	—	—	—	—	—
2-350S162-43	3′-4″	2′-11″	2′-6″	2′-2″	—	3′-3″	2′-10″	2′-5″	2′-1″	—
2-350S162-54	4′-6″	4′-1″	3′-8″	3′-4″	3′-0″	4′-5″	4′-0″	3′-7″	3′-3″	2′-11″
2-350S162-68	5′-0″	4′-9″	4′-7″	4′-5″	4′-3″	4′-11″	4′-8″	4′-6″	4′-4″	4′-2″
2-350S162-97	5′-6″	5′-3″	5′-1″	4′-11″	4′-9″	5′-5″	5′-2″	5′-0″	4′-10″	4′-8″
2-550S162-33	3′-1″	2′-5″	—	—	—	3′-0″	2′-3″	—	—	—
2-550S162-43	5′-1″	4′-6″	4′-0″	3′-6″	3′-1″	4′-11″	4′-5″	3′-11″	3′-5″	3′-0″
2-550S162-54	6′-8″	6′-2″	5′-7″	5′-2″	4′-9″	6′-6″	6′-0″	5′-6″	5′-1″	4′-8″
2-550S162-68	7′-2″	6′-10″	6′-7″	6′-4″	6′-1″	7′-0″	6′-9″	6′-6″	6′-3″	6′-0″
2-550S162-97	7′-11″	7′-7″	7′-3″	7′-0″	6′-10″	7′-9″	7′-5″	7′-2″	6′-11″	6′-9″
2-800S162-33	2′-5″	2′-2″	1′-11″	—	—	2′-5″	2′-1″	1′-10″	—	—
2-800S162-43	5′-5″	4′-9″	4′-3″	3′-9″	3′-5″	5′-3″	4′-8″	4′-1″	3′-9″	3′-5″
2-800S162-54	7′-11″	7′-2″	6′-7″	6′-1″	5′-7″	7′-9″	7′-1″	6′-6″	6′-0″	5′-6″
2-800S162-68	9′-5″	8′-9″	8′-3″	7′-9″	7′-4″	9′-3″	8′-8″	8′-2″	7′-8″	7′-3″
2-800S162-97	10′-9″	10′-3″	9′-11″	9′-7″	9′-3″	10′-7″	10′-1″	9′-9″	9′-5″	9′-1″
2-1000S162-43	4′-4″	3′-9″	3′-4″	3′-0″	2′-9″	4′-3″	3′-8″	3′-3″	2′-11″	2′-8″
2-1000S162-54	8′-6″	7′-5″	6′-8″	6′-0″	5′-5″	8′-4″	7′-4″	6′-6″	5′-10″	5′-4″
2-1000S162-68	10′-8″	10′-0″	9′-5″	8′-11″	8′-4″	10′-7″	9′-10″	9′-4″	8′-9″	8′-3″
2-1000S162-97	12′-11″	12′-4″	11′-8″	11′-1″	10′-6″	12′-9″	12′-2″	11′-6″	10′-11″	10′-5″
2-1200S162-54	7′-1″	6′-2″	5′-6″	5′-0″	4′-6″	6′-11″	6′-1″	5′-5″	4′-10″	4′-5″
2-1200S162-68	11′-9″	11′-0″	10′-5″	9′-10″	9′-1″	11′-8″	10′-11″	10′-3″	9′-9″	8′-11″
2-1200S162-97	14′-9″	13′-9″	13′-0″	12′-4″	11′-9″	14′-7″	13′-8″	12′-10″	12′-3″	11′-8″

For SI: 1 inch = 25.4 mm, 1 foot = 304.8 mm, 1 pound per square foot = 0.0479 kPa, 1 pound per square inch = 6.895 kPa, 1 Ksi = 1,000 psi = 6.895 MPa.

a. Deflection criterion: *L*/360 for live loads, *L*/240 for total loads.

b. Design load assumptions:
 Second floor dead load is 10 psf.
 Roof/ceiling dead load is 12 psf.
 Second floor live load is 30 psf.
 Attic live load is 10 psf.

c. Building width is in the direction of horizontal framing members supported by the header.

❖ See the commentary for Section R603.6.

TABLE R603.6(19)
BACK-TO-BACK HEADER SPANS
Headers Supporting One Floor, Roof and Ceiling
(33 Ksi steel)[a, b]

MEMBER DESIGNATION	GROUND SNOW LOAD (50 psf)					GROUND SNOW LOAD (70 psf)				
	Building width[c] (feet)					Building width[c] (feet)				
	24	28	32	36	40	24	28	32	36	40
2-350S162-33	—	—	—	—	—	—	—	—	—	—
2-350S162-43	—	—	—	—	—	—	—	—	—	—
2-350S162-54	2′-4″	—	—	—	—	—	—	—	—	—
2-350S162-68	3′-3″	2′-10″	2′-6″	2′-2″	—	2′-7″	2′-2″	—	—	—
2-350S162-97	4′-4″	4′-0″	3′-8″	3′-4″	3′-1″	3′-9″	3′-4″	3′-1″	2′-9″	2′-6″
2-550S162-33	—	—	—	—	—	—	—	—	—	—
2-550S162-43	2′-2″	—	—	—	—	—	—	—	—	—
2-550S162-54	3′-8″	3′-2″	2′-8″	2′-3″	—	2′-10″	2′-3″	—	—	—
2-550S162-68	4′-9″	4′-4″	3′-11″	3′-6″	3′-2″	4′-0″	3′-6″	3′-1″	2′-9″	2′-4″
2-550S162-97	6′-3″	5′-9″	5′-4″	5′-0″	4′-8″	5′-6″	5′-0″	4′-7″	4′-3″	3′-11″
2-800S162-33	—	—	—	—	—	—	—	—	—	—
2-800S162-43	2′-11″	2′-0″	—	—	—	—	—	—	—	—
2-800S162-54	4′-9″	4′-2″	3′-7″	3′-1″	2′-7″	3′-9″	3′-1″	2′-5″	—	—
2-800S162-68	6′-4″	5′-9″	5′-3″	4′-9″	4′-4″	5′-4″	4′-9″	4′-3″	3′-10″	3′-4″
2-800S162-97	8′-5″	7′-9″	7′-3″	6′-9″	6′-4″	7′-4″	6′-9″	6′-3″	5′-10″	5′-5″
2-1000S162-43	3′-4″	2′-5″	—	—	—	—	—	—	—	—
2-1000S162-54	5′-6″	4′-10″	4′-2″	3′-7″	3′-0″	4′-4″	3′-7″	2′-11″	2′-2″	—
2-1000S162-68	7′-4″	6′-8″	6′-1″	5′-7″	5′-1″	6′-3″	5′-7″	5′-0″	4′-5″	4′-0″
2-1000S162-97	9′-11″	8′-3″	8′-7″	8′-1″	7′-7″	8′-9″	8′-1″	7′-6″	7′-0″	6′-6″
2-1200S162-54	5′-6″	4′-10″	4′-4″	3′-11″	3′-5″	4′-5″	3′-11″	3′-3″	2′-6″	—
2-1200S162-68	8′-2″	7′-5″	6′-9″	6′-3″	5′-8″	6′-11″	6′-3″	5′-7″	5′-0″	4′-6″
2-1200S162-97	10′-10″	10′-2″	9′-8″	9′-2″	8′-7″	9′-9″	9′-2″	8′-6″	7′-11″	7′-5″

For SI: 1 inch = 25.4 mm, 1 foot = 304.8 mm, 1 pound per square foot = 0.0479 kPa, 1 pound per square inch = 6.895 kPa, 1 Ksi = 1,000 psi = 6.895 MPa.

a. Deflection criterion: *L*/360 for live loads, *L*/240 for total loads.

b. Design load assumptions:
Second floor dead load is 10 psf.
Roof/ceiling dead load is 12 psf.
Second floor live load is 30 psf.
Attic live load is 10 psf.

c. Building width is in the direction of horizontal framing members supported by the header.

❖ See the commentary for Section R603.6.

TABLE R603.6(20)
BACK-TO-BACK HEADER SPANS
Headers Supporting One Floor, Roof and Ceiling
(50 Ksi steel)[a, b]

MEMBER DESIGNATION	GROUND SNOW LOAD (50 psf)					GROUND SNOW LOAD (70 psf)				
	Building width[c] (feet)					Building width[c] (feet)				
	24	28	32	36	40	24	28	32	36	40
2-350S162-33	—	—	—	—	—	—	—	—	—	—
2-350S162-43	2′-6″	2′-0″	—	—	—	—	—	—	—	—
2-350S162-54	3′-8″	3′-3″	2′-11″	2′-7″	2′-3″	3′-0″	2′-7″	2′-2″	—	—
2-350S162-68	4′-7″	4′-5″	4′-1″	3′-9″	3′-6″	4′-2″	3′-9″	3′-5″	3′-1″	2′-10″
2-350S162-97	5′-1″	4′-10″	4′-8″	4′-6″	4′-5″	4′-10″	4′-7″	4′-5″	4′-3″	4′-1″
2-550S162-33	—	—	—	—	—	—	—	—	—	—
2-550S162-43	3′-11″	3′-5″	2′-11″	2′-5″	—	3′-0″	2′-5″	—	—	—
2-550S162-54	5′-7″	5′-0″	4′-7″	4′-2″	3′-9″	4′-8″	4′-2″	3′-8″	3′-3″	2′-11″
2-550S162-68	6′-7″	6′-4″	5′-11″	5′-6″	5′-1″	6′-0″	5′-6″	5′-0″	4′-7″	4′-3″
2-550S162-97	7′-4″	7′-0″	6′-9″	6′-6″	6′-4″	6′-11″	6′-8″	6′-5″	6′-2″	6′-0″
2-800S162-33	1′-11″	—	—	—	—	—	—	—	—	—
2-800S162-43	4′-2″	3′-8″	3′-4″	3′-0″	2′-6″	3′-5″	3′-0″	2′-4″	—	—
2-800S162-54	6′-7″	5′-11″	5′-5″	4′-11″	4′-6″	5′-6″	4′-11″	4′-5″	3′-11″	3′-6″
2-800S162-68	8′-3″	7′-8″	7′-1″	6′-8″	6′-2″	7′-3″	6′-7″	6′-1″	5′-7″	5′-2″
2-800S162-97	9′-11″	9′-6″	9′-2″	8′-10″	8′-7″	9′-5″	9′-0″	8′-7″	8′-2″	7′-9″
2-1000S162-43	3′-4″	2′-11″	2′-7″	2′-5″	2′-2″	2′-8″	2′-5″	2′-2″	1′-11″	—
2-1000S162-54	6′-7″	5′-10″	5′-3″	4′-9″	4′-4″	5′-4″	4′-9″	4′-3″	3′-10″	3′-6″
2-1000S162-68	9′-4″	8′-9″	8′-1″	7′-7″	7′-1″	8′-3″	7′-7″	6′-11″	6′-5″	5′-11″
2-1000S162-97	11′-7″	10′-11″	10′-4″	9′-10″	9′-5″	10′-5″	9′-10″	9′-3″	8′-10″	8′-5″
2-1200S162-54	5′-6″	4′-10″	4′-4″	3′-11″	3′-7″	4′-5″	3′-11″	3′-6″	3′-2″	2′-11″
2-1200S162-68	10′-4″	9′-8″	8′-8″	7′-11″	7′-2″	8′-11″	7′-11″	7′-1″	6′-5″	5′-10″
2-1200S162-97	12′-11″	12′-2″	11′-6″	11′-0″	10′-6″	11′-8″	11′-0″	10′-5″	9′-10″	9′-5″

For SI: 1 inch = 25.4 mm, 1 foot = 304.8 mm, 1 pound per square foot = 0.0479 kPa, 1 pound per square inch = 6.895 kPa, 1 Ksi = 1,000 psi = 6.895 MPa.

a. Deflection criterion: *L*/360 for live loads, *L*/240 for total loads.

b. Design load assumptions:
   Second floor dead load is 10 psf.
   Roof/ceiling dead load is 12 psf.
   Second floor live load is 30 psf.
   Attic live load is 10 psf.

c. Building width is in the direction of horizontal framing members supported by the header.

❖ See the commentary for Section R603.6.

TABLE R603.6(21)
BACK-TO-BACK HEADER SPANS
Headers Supporting Two Floors, Roof and Ceiling
(33 Ksi steel)[a, b]

MEMBER DESIGNATION	GROUND SNOW LOAD (20 psf)					GROUND SNOW LOAD (30 psf)				
	Building width[c] (feet)					Building width[c] (feet)				
	24	28	32	36	40	24	28	32	36	40
2-350S162-33	—	—	—	—	—	—	—	—	—	—
2-350S162-43	—	—	—	—	—	—	—	—	—	—
2-350S162-54	—	—	—	—	—	—	—	—	—	—
2-350S162-68	2′-5″	—	—	—	—	2′-4″	—	—	—	—
2-350S162-97	3′-6″	3′-2″	2′-10″	2′-6″	2′-3″	3′-6″	3′-1″	2′-9″	2′-6″	2′-3″
2-550S162-33	—	—	—	—	—	—	—	—	—	—
2-550S162-43	—	—	—	—	—	—	—	—	—	—
2-550S162-54	2′-6″	—	—	—	—	2′-5″	—	—	—	—
2-550S162-68	3′-9″	3′-3″	2′-9″	2′-4″	—	3′-8″	3′-2″	2′-9″	2′-4″	—
2-550S162-97	5′-3″	4′-9″	4′-4″	3′-11″	3′-8″	5′-2″	4′-8″	4′-3″	3′-11″	3′-7″
2-800S162-33	—	—	—	—	—	—	—	—	—	—
2-800S162-43	—	—	—	—	—	—	—	—	—	—
2-800S162-54	3′-5″	2′-8″	—	—	—	3′-4″	2′-7″	—	—	—
2-800S162-68	5′-1″	4′-5″	3′-11″	3′-4″	2′-11″	5′-0″	4′-4″	3′-10″	3′-4″	2′-10″
2-800S162-97	7′-0″	6′-5″	5′-11″	5′-5″	5′-0″	7′-0″	6′-4″	5′-10″	5′-5″	5′-0″
2-1000S162-43	—	—	—	—	—	—	—	—	—	—
2-1000S162-54	3′-11″	3′-1″	2′-3″	—	—	3′-10″	3′-0″	2′-2″	—	—
2-1000S162-68	5′-10″	5′-2″	4′-6″	4′-0″	3′-5″	5′-9″	5′-1″	4′-6″	3′-11″	3′-4″
2-1000S162-97	8′-5″	7′-8″	7′-1″	6′-6″	6′-1″	8′-4″	7′-7″	7′-0″	6′-6″	6′-0″
2-1200S162-54	4′-2″	3′-6″	2′-7″	—	—	4′-1″	3′-5″	2′-6″	—	—
2-1200S162-68	6′-6″	5′-9″	5′-1″	4′-6″	3′-11″	6′-6″	5′-8″	5′-0″	4′-5″	3′-10″
2-1200S162-97	9′-5″	8′-8″	8′-0″	7′-5″	6′-11″	9′-5″	8′-7″	7′-11″	7′-4″	6′-10″

For SI: 1 inch = 25.4 mm, 1 foot = 304.8 mm, 1 pound per square foot = 0.0479 kPa, 1 pound per square inch = 6.895 kPa, 1 Ksi = 1,000 psi = 6.895 MPa.

a. Deflection criterion: *L*/360 for live loads, *L*/240 for total loads.

b. Design load assumptions:
   Second floor dead load is 10 psf.
   Roof/ceiling dead load is 12 psf.
   Second floor live load is 40 psf.
   Third floor live load is 30 psf.
   Attic live load is 10 psf.

c. Building width is in the direction of horizontal framing members supported by the header.

❖ See the commentary for Section R603.6.

TABLE R603.6(22)
BACK-TO-BACK HEADER SPANS
Headers Supporting Two Floors, Roof and Ceiling
(50 Ksi steel)[a, b]

MEMBER DESIGNATION	GROUND SNOW LOAD (20 psf)					GROUND SNOW LOAD (30 psf)				
	Building width[c] (feet)					Building width[c] (feet)				
	24	28	32	36	40	24	28	32	36	40
2-350S162-33	—	—	—	—	—	—	—	—	—	—
2-350S162-43	—	—	—	—	—	—	—	—	—	—
2-350S162-54	2′-9″	2′-3″	—	—	—	2′-8″	2′-3″	—	—	—
2-350S162-68	3′-11″	3′-6″	3′-2″	2′-10″	2′-6″	3′-11″	3′-6″	3′-1″	2′-9″	2′-6″
2-350S162-97	4′-9″	4′-6″	4′-4″	4′-1″	3′-10″	4′-8″	4′-6″	4′-4″	4′-1″	3′-9″
2-550S162-33	—	—	—	—	—	—	—	—	—	—
2-550S162-43	2′-9″	2′-0″	—	—	—	2′-8″	—	—	—	—
2-550S162-54	4′-5″	3′-10″	3′-4″	2′-11″	2′-5″	4′-4″	3′-9″	3′-3″	2′-10″	2′-5″
2-550S162-68	5′-8″	5′-2″	4′-8″	4′-3″	3′-11″	5′-8″	5′-1″	4′-8″	4′-3″	3′-10″
2-550S162-97	6′-10″	6′-6″	6′-3″	6′-0″	5′-7″	6′-9″	6′-5″	6′-3″	5′-11″	5′-6″
2-800S162-33	—	—	—	—	—	—	—	—	—	—
2-800S162-43	3′-2″	2′-7″	—	—	—	3′-1″	2′-6″	—	—	—
2-800S162-54	5′-2″	4′-7″	4′-0″	3′-6″	3′-0″	5′-2″	4′-6″	3′-11″	3′-5″	2′-11″
2-800S162-68	6′-11″	6′-3″	5′-8″	5′-2″	4′-9″	6′-10″	6′-2″	5′-7″	5′-2″	4′-8″
2-800S162-97	9′-3″	8′-8″	8′-3″	7′-9″	7′-4″	9′-2″	8′-8″	8′-2″	7′-9″	7′-4″
2-1000S162-43	2′-6″	2′-2″	2′-0″	—	—	2′-6″	2′-2″	1′-11″	—	—
2-1000S162-54	5′-0″	4′-4″	3′-11″	3′-6″	3′-2″	4′-11″	4′-4″	3′-10″	3′-6″	3′-2″
2-1000S162-68	7′-10″	7′-2″	6′-6″	5′-11″	5′-6″	7′-9″	7′-1″	6′-5″	5′-11″	5′-5″
2-1000S162-97	10′-1″	9′-5″	8′-11″	8′-6″	8′-0″	10′-0″	9′-5″	8′-10″	8′-5″	7′-11″
2-1200S162-54	—	—	—	—	—	—	—	—	—	—
2-1200S162-68	7′-4″	6′-8″	6′-1″	5′-6″	5′-1″	7′-3″	6′-7″	6′-0″	5′-6″	5′-0″
2-1200S162-97	9′-5″	8′-8″	8′-1″	7′-6″	7′-1″	9′-4″	8′-8″	8′-0″	7′-6″	7′-0″

For SI: 1 inch = 25.4 mm, 1 foot = 304.8 mm, 1 pound per square foot = 0.0479 kPa, 1 pound per square inch = 6.895 kPa, 1 Ksi = 1,000 psi = 6.895 MPa.

a. Deflection criterion: *L*/360 for live loads, *L*/240 for total loads.

b. Design load assumptions:
   Second floor dead load is 10 psf.
   Roof/ceiling dead load is 12 psf.
   Second floor live load is 40 psf.
   Third floor live load is 30 psf.
   Attic live load is 10 psf.

c. Building width is in the direction of horizontal framing members supported by the header.

❖ See the commentary for Section R603.6.

TABLE R603.6(23)
BACK-TO-BACK HEADER SPANS
Headers Supporting Two Floors, Roof and Ceiling
(33 Ksi steel)[a, b]

MEMBER DESIGNATION	GROUND SNOW LOAD (50 psf)					GROUND SNOW LOAD (70 psf)				
	Building width[c] (feet)					Building width[c] (feet)				
	24	28	32	36	40	24	28	32	36	40
2-350S162-33	—	—	—	—	—	—	—	—	—	—
2-350S162-43	—	—	—	—	—	—	—	—	—	—
2-350S162-54	—	—	—	—	—	—	—	—	—	—
2-350S162-68	2′-2″	—	—	—	—	—	—	—	—	—
2-350S162-97	3′-3″	3′-0″	2′-8″	2′-4″	2′-1″	3′-1″	2′-9″	2′-6″	2′-2″	—
2-550S162-33	—	—	—	—	—	—	—	—	—	—
2-550S162-43	—	—	—	—	—	—	—	—	—	—
2-550S162-54	2′-2″	—	—	—	—	—	—	—	—	—
2-550S162-68	3′-6″	3′-0″	2′-6″	2′-1″	—	3′-2″	2′-9″	2′-3″	—	—
2-550S162-97	5′-0″	4′-6″	4′-1″	3′-9″	3′-5″	4′-8″	4′-3″	3′-11″	3′-7″	3′-3″
2-800S162-33	—	—	—	—	—	—	—	—	—	—
2-800S162-43	—	—	—	—	—	—	—	—	—	—
2-800S162-54	3′-0″	2′-3″	—	—	—	2′-7″	—	—	—	—
2-800S162-68	4′-9″	4′-2″	3′-7″	3′-1″	2′-7″	4′-5″	3′-10″	3′-3″	2′-9″	2′-3″
2-800S162-97	6′-9″	6′-1″	5′-7″	5′-2″	4′-9″	6′-4″	5′-10″	5′-4″	4′-11″	4′-7″
2-1000S162-43	—	—	—	—	—	—	—	—	—	—
2-1000S162-54	3′-6″	2′-8″	—	—	—	3′-1″	2′-2″	—	—	—
2-1000S162-68	5′-6″	4′-10″	4′-2″	3′-7″	3′-1″	5′-1″	4′-6″	3′-10″	3′-4″	2′-9″
2-1000S162-97	8′-0″	7′-4″	6′-9″	6′-3″	5′-9″	7′-7″	7′-0″	6′-5″	5′-11″	5′-6″
2-1200S162-54	3′-11″	3′-0″	2′-0″	—	—	3′-5″	2′-6″	—	—	—
2-1200S162-68	6′-2″	5′-5″	4′-9″	4′-1″	3′-6″	5′-9″	5′-0″	4′-4″	3′-9″	3′-2″
2-1200S162-97	9′-1″	8′-4″	7′-8″	7′-1″	6′-7″	8′-8″	7′-11″	7′-4″	6′-9″	6′-3″

For SI: 1 inch = 25.4 mm, 1 foot = 304.8 mm, 1 pound per square foot = 0.0479 kPa, 1 pound per square inch = 6.895 kPa, 1 Ksi = 1,000 psi = 6.895 MPa.

a. Deflection criterion: *L*/360 for live loads, *L*/240 for total loads.

b. Design load assumptions:
 Second floor dead load is 10 psf.
 Roof/ceiling dead load is 12 psf.
 Second floor live load is 40 psf.
 Third floor live load is 30 psf.
 Attic live load is 10 psf.

c. Building width is in the direction of horizontal framing members supported by the header.

❖ See the commentary for Section R603.6.

TABLE R603.6(24)
BACK-TO-BACK HEADER SPANS
Headers Supporting Two Floors, Roof and Ceiling
(50 Ksi steel)[a, b]

MEMBER DESIGNATION	GROUND SNOW LOAD (50 psf)					GROUND SNOW LOAD (70 psf)				
	Building width[c] (feet)					Building width[c] (feet)				
	24	28	32	36	40	24	28	32	36	40
2-350S162-33	—	—	—	—	—	—	—	—	—	—
2-350S162-43	—	—	—	—	—	—	—	—	—	—
2-350S162-54	2′-6″	2′-1″	—	—	—	2′-3″	—	—	—	—
2-350S162-68	3′-9″	3′-4″	2′-11″	2′-7″	2′-4″	3′-6″	3′-1″	2′-9″	2′-5″	2′-2″
2-350S162-97	4′-6″	4′-4″	4′-2″	3′-11″	3′-8″	4′-4″	4′-2″	4′-0″	3′-9″	3′-6″
2-550S162-33	—	—	—	—	—	—	—	—	—	—
2-550S162-43	2′-5″	—	—	—	—	—	—	—	—	—
2-550S162-54	4′-1″	3′-7″	3′-1″	2′-7″	2′-2″	3′-10″	3′-3″	2′-10″	2′-4″	—
2-550S162-68	5′-5″	4′-11″	4′-5″	4′-0″	3′-8″	5′-1″	4′-7″	4′-2″	3′-10″	3′-5″
2-550S162-97	6′-5″	6′-2″	5′-11″	5′-9″	5′-4″	6′-3″	6′-0″	5′-9″	5′-6″	5′-2″
2-800S162-33	—	—	—	—	—	—	—	—	—	—
2-800S162-43	2′-11″	2′-2″	—	—	—	2′-6″	—	—	—	—
2-800S162-54	4′-11″	4′-3″	3′-8″	3′-2″	2′-8″	4′-6″	3′-11″	3′-5″	2′-11″	2′-4″
2-800S162-68	6′-7″	5′-11″	5′-4″	4′-11″	4′-6″	6′-2″	5′-7″	5′-1″	4′-8″	4′-3″
2-800S162-97	8′-9″	8′-5″	7′-11″	7′-6″	7′-0″	8′-5″	8′-1″	7′-9″	7′-3″	6′-10″
2-1000S162-43	2′-4″	2′-1″	—	—	—	2′-2″	1′-11″	—	—	—
2-1000S162-54	4′-8″	4′-1″	3′-8″	3′-3″	3′-0″	4′-4″	3′-10″	3′-5″	3′-1″	2′-9″
2-1000S162-68	7′-6″	6′-9″	6′-2″	5′-8″	5′-2″	7′-1″	6′-5″	5′-10″	5′-4″	4′-11″
2-1000S162-97	9′-9″	9′-2″	8′-7″	8′-2″	7′-8″	9′-5″	8′-10″	8′-5″	7′-11″	7′-5″
2-1200S162-54	—	—	—	—	—	—	—	—	—	—
2-1200S162-68	7′-0″	6′-4″	5′-9″	5′-3″	4′-9″	6′-7″	6′-0″	5′-5″	5′-0″	4′-6″
2-1200S162-97	9′-1″	8′-4″	7′-9″	7′-3″	6′-9″	8′-8″	8′-0″	7′-6″	7′-0″	6′-7″

For SI: 1 inch = 25.4 mm, 1 foot = 304.8 mm, 1 pound per square foot = 0.0479 kPa, 1 pound per square inch = 6.895 kPa, 1 Ksi = 1,000 psi = 6.895 MPa.

a. Deflection criterion: *L*/360 for live loads, *L*/240 for total loads.

b. Design load assumptions:
Second floor dead load is 10 psf.
Roof/ceiling dead load is 12 psf.
Second floor live load is 40 psf.
Third floor live load is 30 psf.
Attic live load is 10 psf.

c. Building width is in the direction of horizontal framing members supported by the header.

❖ See the commentary for Section R603.6.

R603.6.1 Headers in gable endwalls. Box beam and back-to-back headers in gable endwalls shall be permitted to be constructed in accordance with Section R603.6 or with the header directly above the opening in accordance with Figures R603.6.1(1) and R603.6.1(2) and the following provisions:

1. Two 362S162-33 for openings less than or equal to 4 feet (1219 mm).
2. Two 600S162-43 for openings greater than 4 feet (1219 mm) but less than or equal to 6 feet (1830 mm).
3. Two 800S162-54 for openings greater than 6 feet (1829 mm) but less than or equal to 9 feet (2743 mm).

❖ This section specifies the minimum size of box-beam and back-to-back headers in gable endwalls. Figures R603.6.1(1) and R603.6.1(2) illustrate these configurations and unlike Section R603.6 they may be installed directly above the opening. The required header size varies with the span and if the span exceeds 9 feet (2743 mm), the requirements of Section R603.6 should be applied.

R603.7 Jack and king studs. The number of jack and king studs installed on each side of a header shall comply with Table R603.7(1). King, jack and cripple studs shall be of the same dimension and thickness as the adjacent wall studs. Headers shall be connected to king studs in accordance with Table R603.7(2) and the following provisions:

1. For box beam headers, one-half of the total number of required screws shall be applied to the header and one half to the king stud by use of C-shaped or track member in accordance with Figure R603.6.1(1). The track or C-shape sections shall extend the depth of the header minus $^1/_2$ inch (12.7 mm) and shall have a minimum thickness not less than that of the wall studs.
2. For back-to-back headers, one-half the total number of screws shall be applied to the header and one-half to the

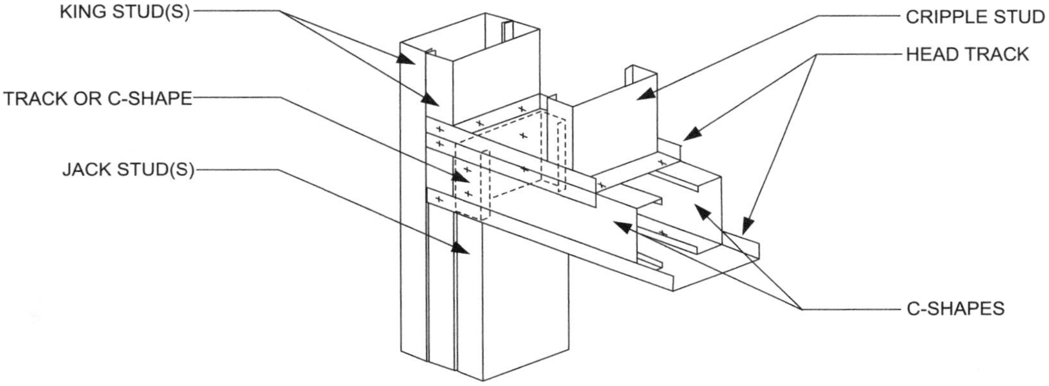

FIGURE R603.6.1(1)
BOX BEAM HEADER IN GABLE ENDWALL

❖ See the commentary for Section R603.6.1.

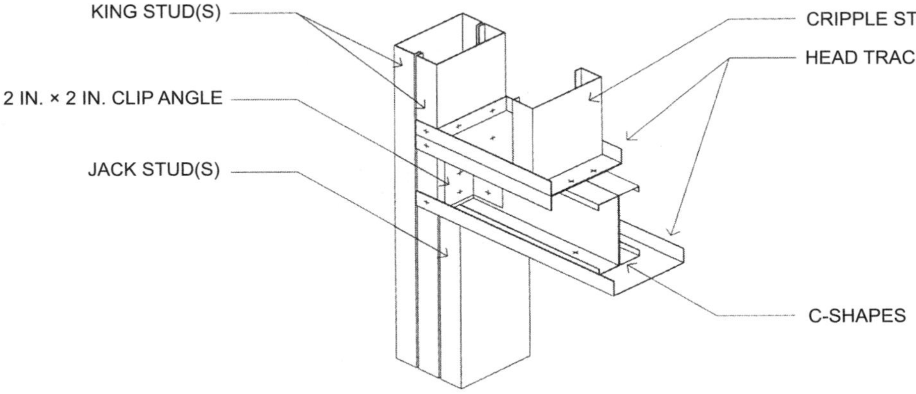

For SI: 1 inch = 25.4 mm.

FIGURE R603.6.1(2)
BACK-TO-BACK HEADER IN GABLE ENDWALL

❖ See the commentary for Section R603.6.1.

king stud by use of a minimum 2-inch by 2-inch (51 mm by 51 mm) clip angle in accordance with Figure R603.6.1(2). The clip angle shall extend the depth of the header minus $^1/_2$ inch (12.7 mm) and shall have a minimum thickness not less than that of the wall studs. Jack and king studs shall be interconnected with structural sheathing in accordance with Figures R603.6(1) and R603.6(2).

❖ Table R603.7(1) specifies the number of jack and king studs based on the header span. King and jack studs must be the same dimension as adjacent wall studs, with a steel thickness not less than that of the adjacent wall studs. The size and number of screws required for the header-to-king-stud connection are listed in Table R603.7(2). One-half of the screws are to be applied to the box-beam header and one-half to the king stud with a clip angle as shown in Figure R603.6.1(2). The clip angle must be a minimum 2-inch by 2-inch (51 mm by 51 mm) angle with a minimum steel thickness not less than the header members and the wall studs.

R603.8 Head and sill track. Head track spans above door and window openings and sill track spans beneath window openings shall comply with Table R603.8. For openings less than 4 feet (1219 mm) in height that have both a head track and a sill track, multiplying the spans by 1.75 shall be permitted in Table R603.8. For openings less than or equal to 6 feet (1829 mm) in height that have both a head track and a sill track, multiplying the spans in Table R603.8 by 1.50 shall be permitted.

❖ Head track spans must comply with Table R603.8 [see Figures R603.6.1(1) and R603.6.1(2) for illustrations]. Where both a top and a bottom track are provided, the tabular head track spans may be increased by the multipliers indicated.

TABLE R603.7(1)
TOTAL NUMBER OF JACK AND KING STUDS REQUIRED AT EACH END OF AN OPENING

SIZE OF OPENING (feet-inches)	24-INCH O.C. STUD SPACING		16-INCH O.C. STUD SPACING	
	No. of jack studs	No. of king studs	No. of jack studs	No. of king studs
Up to 3′-6″	1	1	1	1
> 3′-6″ to 5′-0″	1	2	1	2
> 5′-0″ to 5′-6″	1	2	2	2
> 5′-6″ to 8′-0″	1	2	2	2
> 8′-0″ to 10′-6″	2	2	2	3
> 10′-6″ to 12′-0″	2	2	3	3
> 12′-0″ to 13′-0″	2	3	3	3
> 13′-0″ to 14′-0″	2	3	3	4
> 14′-0″ to 16′-0″	2	3	3	4
> 16′-0″ to 18′-0″	3	3	4	4

For SI: 1 inch = 25.4 mm, 1 foot = 304.8 mm.

❖ See the commentary for Section R603.7.

TABLE R603.7(2)
HEADER TO KING STUD CONNECTION REQUIREMENTS[a, b, c, d]

HEADER SPAN (feet)	BASIC WIND SPEED (mph), EXPOSURE		
	85 B or Seismic Design Categories A, B, C, D_0, D_1 and D_2	85 C or less than 110 B	Less than 110 C
≤ 4′	4-No. 8 screws	4-No. 8 screws	6-No. 8 screws
> 4′ to 8′	4-No. 8 screws	4-No. 8 screws	8-No. 8 screws
> 8′ to 12′	4-No. 8 screws	6-No. 8 screws	10-No. 8 screws
> 12′ to 16′	4-No. 8 screws	8-No. 8 screws	12-No. 8 screws

For SI: 1 inch = 25.4 mm, 1 foot = 304.8 mm, 1 mile per hour = 0.447 m/s, 1 pound = 4.448 N.

a. All screw sizes shown are minimum.

b. For headers located on the first floor of a two-story building or the first or second floor of a three-story building, the total number of screws is permitted to be reduced by 2 screws, but the total number of screws shall be no less than 4.

c. For roof slopes of 6:12 or greater, the required number of screws may be reduced by half, but the total number of screws shall be no less than four.

d. Screws can be replaced by an uplift connector which has a capacity of the number of screws multiplied by 164 pounds (e.g., 12-No. 8 screws can be replaced by an uplift connector whose capacity exceeds 12 × 164 pounds = 1,968 pounds).

❖ See the commentary for Section R603.7.

TABLE R603.8
HEAD AND SILL TRACK SPAN F_y = 33 KSI

BASIC WIND SPEED (mph)		ALLOWABLE HEAD AND SILL TRACK SPAN[a, b, c] (feet-inches)					
EXPOSURE		TRACK DESIGNATION					
B	C	350T125-33	350T125-43	350T125-54	550T125-33	550T125-43	550T125-54
85	—	5′-0″	5′-7″	6′-2″	5′-10″	6′-8″	7′-0″
90	—	4′-10″	5′-5″	6′-0″	5′-8″	6′-3″	6′-10″
100	85	4′-6″	5′-1″	5′-8″	5′-4″	5′-11″	6′-5″
110	90	4′-2″	4′-9″	5′-4″	5′-1″	5′-7″	6′-1″
120	100	3′-11″	4′-6″	5′-0″	4′-10″	5′-4″	5′-10″
130	110	3′-8″	4′-2″	4′-9″	4′-1″	5′-1″	5′-7″
140	120	3′-7″	4′-1″	4′-7″	3′-6″	4′-11″	5′-5″
150	130	3′-5″	3′-10″	4′-4″	2′-11″	4′-7″	5′-2″
—	140	3′-1″	3′-6″	4′-1″	2′-3″	4′-0″	4′-10″
—	150	2′-9″	3′-4″	3′-10″	2′-0″	3′-7″	4′-7″

For SI: 1 inch = 25.4 mm, 1 foot = 304.8 mm, 1 mile per hour = 0.447 m/s.

a. Deflection limit: $L/240$.

b. Head and sill track spans are based on components and cladding wind speeds and 48-inch tributary span.

c. For openings less than 4 feet in height that have both a head track and sill track, the above spans are permitted to be multiplied by 1.75. For openings less than or equal to 6 feet in height that have both a head track and a sill track, the above spans are permitted to be multiplied by a factor of 1.5.

❖ See the commentary for Section R603.8.

R603.9 Structural sheathing. Structural sheathing shall be installed in accordance with Figure R603.9 and this section on all sheathable exterior wall surfaces, including areas above and below openings.

❖ Wood structural panel sheathing must be installed on all exterior wall surfaces as shown in Figure R603.9.

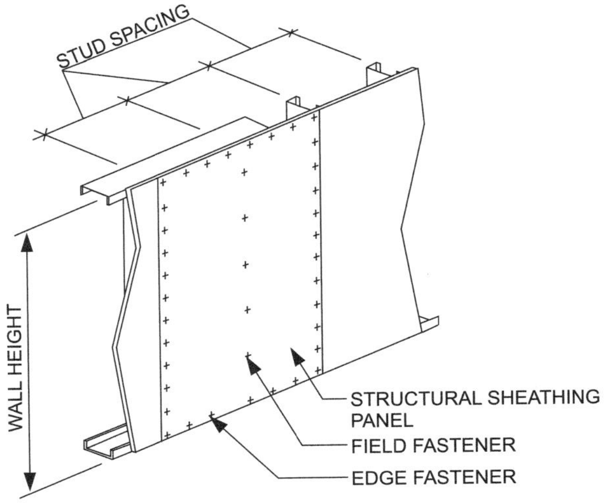

FIGURE R603.9
STRUCTURAL SHEATHING FASTENING PATTERN

❖ This illustrates the sheathing fastening and locations that are specified in Table R603.3.2(1) for structural sheathing. See the commentary for Section R603.9.3.

R603.9.1 Sheathing materials. Structural sheathing panels shall consist of minimum $^{7}/_{16}$-inch-thick (11 mm) oriented strand board or $^{15}/_{32}$-inch-thick (12 mm) plywood.

❖ Wood structural-panel sheathing must be a minimum $^{7}/_{16}$-inch-thick (11.1 mm) oriented strand board or $^{15}/_{32}$-inch-thick (11.9 mm) plywood fastened to studs and tracks in accordance with Table R603.3.2.

R603.9.2 Determination of minimum length of full height sheathing. The minimum length of full height sheathing on each *braced wall line* shall be determined by multiplying the length of the *braced wall line* by the percentage obtained from Table R603.9.2(1) and by the plan aspect-ratio adjustment factors obtained from Table R603.9.2(2). The minimum length of full height sheathing shall not be less than 20 percent of the *braced wall line* length.

To be considered full height sheathing, structural sheathing shall extend from the bottom to the top of the wall without interruption by openings. Only sheathed, full height wall sections, uninterrupted by openings, which are a minimum of 48 inches (1219 mm) wide, shall be counted toward meeting the minimum percentages in Table R603.9.2(1). In addition, structural sheathing shall comply with all of the following requirements:

1. Be installed with the long dimension parallel to the stud framing (i.e., vertical orientation) and shall cover the full vertical height of wall from the bottom of the bottom track to the top of the top track of each *story*. Installing the long dimension perpendicular to the stud framing or using shorter segments shall be permitted provided that the horizontal joint is blocked as described in Item 2.

2. Be blocked when the long dimension is installed perpendicular to the stud framing (i.e., horizontal orientation). Blocking shall be a minimum of 33 mil (0.84 mm) thickness. Each horizontal structural sheathing panel shall be fastened with No. 8 screws spaced at 6 inches (152 mm) on center to the blocking at the joint.
3. Be applied to each end (corners) of each of the exterior walls with a minimum 48-inch-wide (1219 mm) panel.

❖ The minimum length of full-height sheathing required on exterior walls is determined from Table R603.9.2(1). The minimum tabulated lengths are expressed as a percentage of the wall length. Because the table is based on an 8-foot (2438 mm) wall height, the lengths must be increased by the appropriate multiplier in Section R603.9.2.1 for 9-foot- (2743 mm) and 10-foot-high (3048 mm) walls as well as the building aspect ratio found in Table R603.9.2(2). Additional adjustments are permitted for buildings with hipped roofs and for hold-downs in accordance with Sections R603.9.2.2 and R603.9.2.3. After all adjustments are applied to these tabulated values, the resulting minimum percentage of full-height sheathing must not be less than 20 percent of the wall length.

Only sheathed wall sections that are a minimum of 48 inches (1219 mm) wide and uninterrupted by openings may be considered in meeting the minimum length requirement. This structural sheathing must be installed with the long dimension parallel to the stud framing, and it must cover the full vertical height of studs, from the bottom of the bottom track to the top of the top track of each story. Additionally, structural sheathing must be applied to each end (corners) of all exterior walls with a minimum 48-inch-wide (1219 mm) panel.

TABLE R603.9.2(1)
MINIMUM PERCENTAGE OF FULL HEIGHT STRUCTURAL SHEATHING ON EXTERIOR WALLS[a, b]

WALL SUPPORTING	ROOF SLOPE	BASIC WIND SPEED AND EXPOSURE (mph)					
		85 B	90 B	100 B / 85 C	< 110 B / 90 C	100 C	< 110 C
Roof and ceiling only (one story or top floor of two- or three-story building).	3:12	8	9	9	12	16	20
	6:12	12	13	15	20	26	35
	9:12	21	23	25	30	50	58
	12:12	30	33	35	40	66	75
One story, roof and ceiling (first floor of a two-story building or second floor of a three-story building).	3:12	24	27	30	35	50	66
	6:12	25	28	30	40	58	74
	9:12	35	38	40	55	74	91
	12:12	40	45	50	65	100	115
Two story, roof and ceiling (first floor of a three-story building).	3:12	40	45	51	58	84	112
	6:12	38	43	45	60	90	113
	9:12	49	53	55	80	98	124
	12:12	50	57	65	90	134	155

For SI: 1 mile per hour = 0.447 m/s.

a. Linear interpolation is permitted.

b. For hip-roofed homes the minimum percentage of full height sheathing, based upon wind, is permitted to be multiplied by a factor of 0.95 for roof slopes not exceeding 7:12 and a factor of 0.9 for roof slopes greater than 7:12.

❖ See the commentary for Section R603.9.2.

TABLE R603.9.2(2)
FULL HEIGHT SHEATHING LENGTH ADJUSTMENT FACTORS

PLAN ASPECT RATIO	LENGTH ADJUSTMENT FACTORS	
	Short wall	Long wall
1:1	1.0	1.0
1.5:1	1.5	0.67
2:1	2.0	0.50
3:1	3.0	0.33
4:1	4.0	0.25

❖ See the commentary for Section R603.9.2.

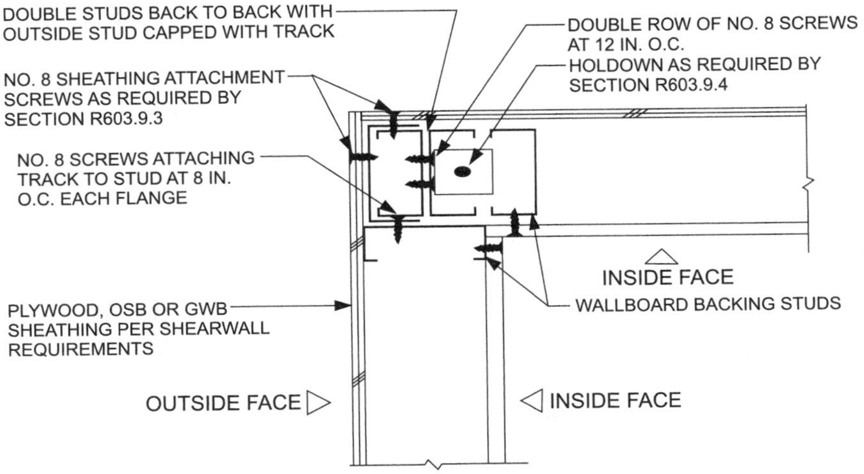

For SI: 1 inch = 25.4 mm.

FIGURE R603.9.2
CORNER STUD HOLD-DOWN DETAIL

❖ This figure illustrates a single hold-down that is permitted for corners by Section R603.9.4.2.

R603.9.2.1 Full height sheathing. The minimum percentage of full-height structural sheathing shall be multiplied by 1.10 for 9-foot-high (2743 mm) walls and multiplied by 1.20 for 10-foot-high (3048 mm) walls.

❖ See the commentary to Section R603.9.2.

R603.9.2.2 Full height sheathing in hip roof homes. For hip roofed homes, the minimum percentages of full height sheathing in Table R603.9.2(1), based upon wind, shall be permitted to be multiplied by a factor of 0.95 for roof slopes not exceeding 7:12 and a factor of 0.9 for roof slopes greater than 7:12.

❖ See the commentary to Section R603.9.2.

R603.9.2.3 Full height sheathing in lowest story. In the lowest *story* of a *dwelling*, multiplying the percentage of full height sheathing required in Table R603.9.2(1) by 0.6, shall be permitted provided hold down anchors are provided in accordance with Section R603.9.4.2.

❖ See the commentary to Section R603.9.2.

R603.9.3 Structural sheathing fastening. All edges and interior areas of structural sheathing panels shall be fastened to framing members and tracks in accordance with Figure R603.9 and Table R603.3.2(1). Screws for attachment of structural sheathing panels shall be bugle-head, flat-head, or similar head style with a minimum head diameter of 0.29 inch (8 mm).

For continuously-sheathed *braced wall lines* using wood structural panels installed with No. 8 screws spaced 4-inches (102 mm) on center at all panel edges and 12 inches (304.8 mm) on center on intermediate framing members, the following shall apply:

1. Multiplying the percentages of full height sheathing in Table R603.9.2(1) by 0.72 shall be permitted.
2. For bottom track attached to foundations or framing below, the bottom track anchor or screw connection spacing in Table R505.3.1(1) and Table R603.3.1 shall be multiplied by two-thirds

❖ The wall-fastening schedule [see Table R603.3.2(1)] specifies edge and field spacing requirements (see Figure R603.9). This section specifies the required type and size of screws for attaching structural sheathing to wall framing. A reduction in sheathing requirements is allowed where a reduced edge fastener spacing is utilized.

R603.9.4 Uplift connection requirements. Uplift connections shall be provided in accordance with this section.

❖ Uplift connections are required based on wind criteria in Section R603.9.4.1. When the optional sheathing reduction permitted by Section R603.9.2.3 is taken, a hold down must be provided in accordance with Section R603.9.4.2.

R603.9.4.1 Wind speeds greater than 100 mph. Where wind speeds are in excess of 100 miles per hour (45 m/s), Exposure C, walls shall be provided wind direct uplift connections in accordance with AISI S230, Section E13.3, and AISI S230, Section F7.2, as required for 110 miles per hour (49 m/s), Exposure C.

❖ Where the basic wind speed is greater than 100 mph (45 m/s) and the exposure is C, the user is referred to AISI S230 in order to determine the uplift connection requirements.

R603.9.4.2 Hold-down anchor. Where the percentage of full height sheathing is adjusted in accordance with Section R603.9.2.3, a hold-down anchor, with a strength of 4,300 pounds (19 kN), shall be provided at each end of each full-height sheathed wall section used to meet the minimum percent sheathing requirements of Section R603.9.2. Hold-down anchors shall be attached to back-to-back studs; structural sheathing panels shall have edge fastening to the studs, in accordance with Section R603.9.3 and AISI S230, Table E11-1.

A single hold-down anchor, installed in accordance with Figure R603.9.2, shall be permitted at the corners of buildings.

❖ Section R603.9.2.3 affords the option of using shorter braced-wall lengths and providing hold-down anchorage at the ends of the braced walls to resist uplift. The length of wall may be reduced by up to 40 percent where the hold-downs specified in this section are installed.

R603.9.5 Structural sheathing for stone and masonry veneer. In Seismic Design Category C, where stone and masonry veneer is installed in accordance with Section R703.7, the length of structural sheathing for walls supporting one *story*, roof and ceiling shall be the greater of the amount required by Section R603.9.2 or 36 percent, modified by Section R603.9.2 except Section R603.9.2.2 shall not be permitted.

❖ This section provides for a minimum amount of wall bracing of 36 percent in SDC C, which is considered a moderate level of earthquake risk. The general wall bracing requirements of Section R603.9.2 must also be met, but because that section bases the percentage of wall bracing on wind criteria only [see Table R603.9.2(1)], it is prudent to establish a minimum bracing amount where there is a higher risk due to earthquakes.

SECTION R604
WOOD STRUCTURAL PANELS

R604.1 Identification and grade. Wood structural panels shall conform to DOC PS 1, DOC PS 2 or ANSI/APA PRP 210 or, when manufactured in Canada, CSA O437 or CSA O325. All panels shall be identified by a grade mark or certificate of inspection issued by an *approved* agency.

❖ To verify the acceptability of wood structural panel products, this section requires labeling in accordance with the referenced standards. The purpose is the same as the purpose of Section R602.1 for dimension lumber. This label provides the inspector with the necessary information to determine the acceptability of the wood structural panel (see commentary, Sections R503.2 and R503.2.1).

R604.2 Allowable spans. The maximum allowable spans for wood structural panel wall sheathing shall not exceed the values set forth in Table R602.3(3).

❖ The allowable span for wood structural panel sheathing is a function of its thickness and grade. Table R602.3(3) in the code provides allowable spans based on the thickness and the panel index.

R604.3 Installation. Wood structural panel wall sheathing shall be attached to framing in accordance with Table R602.3(1) or R602.3(3). Wood structural panels marked Exposure 1 or Exterior are considered water-repellent sheathing under the code.

❖ Table R602.3(1) of the code contains the fastening requirements for proper wood structural panel installation for interior walls. Table R602.3(3) contain the fastening requirements for proper wood structural panel installation for exterior walls.

SECTION R605
PARTICLEBOARD

R605.1 Identification and grade. Particleboard shall conform to ANSI A208.1 and shall be so identified by a grade mark or certificate of inspection issued by an *approved* agency. Particleboard shall comply with the grades specified in Table R602.3(4).

❖ As with wood structural panels, particleboard must be labeled in accordance with the referenced standard. Table R602.3.(4) provides prescriptive requirements relative to allowable spans. As with wood structural panel spans, these spans are a function of the thickness and grade (see commentary, Section R503.3).

SECTION R606
GENERAL MASONRY CONSTRUCTION

R606.1 General. Masonry construction shall be designed and constructed in accordance with the provisions of this section, TMS 403 or in accordance with the provisions of TMS 402/ACI 530/ASCE 5.

❖ Masonry wall construction under the code must comply with this section or TMS 402/ACI 530/ASCE 5, *Building Code Requirements for Masonry Structures*.

R606.1.1 Professional registration not required. When the empirical design provisions of Chapter 5 of TMS 402/ACI 530/ASCE 5, the provisions of TMS 403, or the provisions of this section are used to design masonry, project drawings, typical details and specifications are not required to bear the seal of the architect or engineer responsible for design, unless otherwise required by the state law of the *jurisdiction* having authority.

❖ These empirical provisions are meant for use without the services of a professional engineer or architect. State law, however, may dictate otherwise and would take precedence.

R606.2 Thickness of masonry. The nominal thickness of masonry walls shall conform to the requirements of Sections R606.2.1 through R606.2.4.

❖ This section establishes minimum thicknesses for various masonry wall constructions.

R606.2.1 Minimum thickness. The minimum thickness of masonry bearing walls more than one *story* high shall be 8 inches (203 mm). *Solid masonry* walls of one-story *dwellings* and garages shall not be less than 6 inches (152 mm) in thickness when not greater than 9 feet (2743 mm) in height, provided that when gable construction is used, an additional 6 feet (1829 mm) is permitted to the peak of the gable. Masonry walls shall be laterally supported in either the horizontal or vertical direction at intervals as required by Section R606.9.

❖ The minimum thickness of bearing walls and exterior nonbearing walls of masonry construction more than one story high is limited to a nominal 8 inches (203 mm), which means a net thickness of not less than $7^1/_2$ inches (191 mm). A lesser thickness of 6-inch (152 mm) nominal or $5^1/_2$-inch (140 mm) net of solid masonry is permitted if the wall does not exceed 9 feet (2743 mm) in height other than at the peak of a gable, where such height may extend to 15 feet (4572 mm) as shown in Commentary Figure R606.2.1.

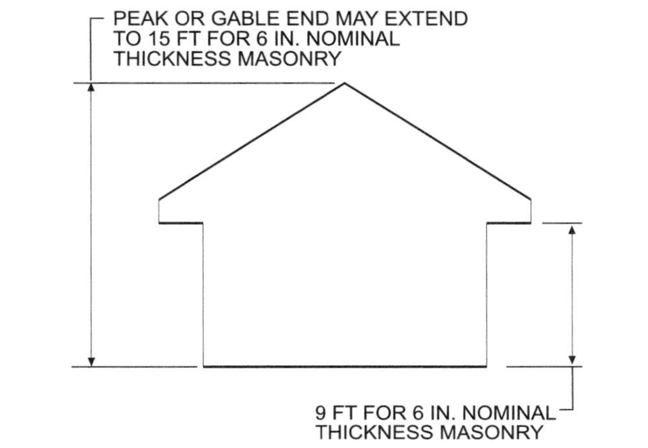

For SI: 1 inch = 25.4 mm, 1 foot = 304.8 mm.

Figure R606.2.1
MINIMUM THICKNESS—MASONRY BEARING AND EXTERIOR WALLS FOR A ONE-STORY DWELLING

R606.2.2 Rubble stone masonry wall. The minimum thickness of rough, random or coursed rubble stone masonry walls shall be 16 inches (406 mm).

❖ Rubble stone walls must be at least 16 inches (406 mm) thick (see commentary, Section R404.1.8).

R606.2.3 Change in thickness. Where walls of masonry of hollow units or masonry-bonded hollow walls are decreased in thickness, a course of *solid masonry* shall be constructed between the wall below and the thinner wall above, or special units or construction shall be used to transmit the loads from face shells or wythes above to those below.

❖ Where the thickness of hollow masonry wall construction is decreased, a solid masonry course provides bearing for the face shells of the thinner hollow unit above.

R606.2.4 Parapet walls. Unreinforced *solid masonry* parapet walls shall not be less than 8 inches (203 mm) thick and their height shall not exceed four times their thickness. Unreinforced hollow unit masonry parapet walls shall be not less than 8 inches (203 mm) thick, and their height shall not exceed three times their thickness. Masonry parapet walls in areas subject to wind loads of 30 pounds per square foot (1.44 kPa) located in Seismic Design Category D_0, D_1 or D_2, or on townhouses in Seismic Design Category C shall be reinforced in accordance with Section R606.12.

❖ A parapet is an extension of a wall that extends above the roof line. Generally, parapet walls are required to comply only with fire-resistance requirements of the code. Parapet walls are required to have a certain minimum thickness with height limitations based on the thickness and reinforcement in the wall.

R606.3 Corbeled masonry. Corbeled masonry shall be in accordance with Sections R606.3.1 through R606.3.3.

❖ This section prescribes the units, projection and conditions for supporting floor or roof framing for corbeled masonry.

R606.3.1 Units. *Solid masonry* units or masonry units filled with mortar or grout shall be used for corbeling.

❖ Units to be used for corbeling should include solid units or units filled with mortar or grout. Units filled solid with mortar or grout will enable the unit to act as a solid unit in supporting the corbel above. Solid units and units filled solid with mortar or grout will distribute the load adequately to the masonry wall or wythe below. Further, there are many instances where solid units are not available while units filled solid with mortar or grout can be readily made on the job site as they are needed.

R606.3.2 Corbel projection. The maximum projection of one unit shall not exceed one-half the height of the unit or one-third the thickness at right angles to the wall. The maximum corbeled projection beyond the face of the wall shall not exceed:

1. One-half of the wall thickness for multiwythe walls bonded by mortar or grout and wall ties or masonry headers, or
2. One-half the wythe thickness for single wythe walls, masonry-bonded hollow walls, multiwythe walls with open collar joints and veneer walls.

❖ Corbeling of masonry walls is permitted within certain limitations. The maximum projection for each unit is limited to one-third the unit bed depth *(D)* or one-half the unit height *(H)*, whichever is less, as shown in Commentary Figure R606.3(1). The total horizontal projection of the corbeled courses is limited to one-half the thickness of a solid wall or one-half the thickness of a wythe of a cavity wall as shown in Commentary Figure R606.3(2).

R606.3.3 Corbeled masonry supporting floor or roof-framing members. When corbeled masonry is used to support floor or roof-framing members, the top course of the corbel shall be a header course or the top course bed joint shall have ties to the vertical wall.

❖ The top course of the corbel must be a header course or the bed joints must have ties to the wall where a corbel is used to support floor or roof framing.

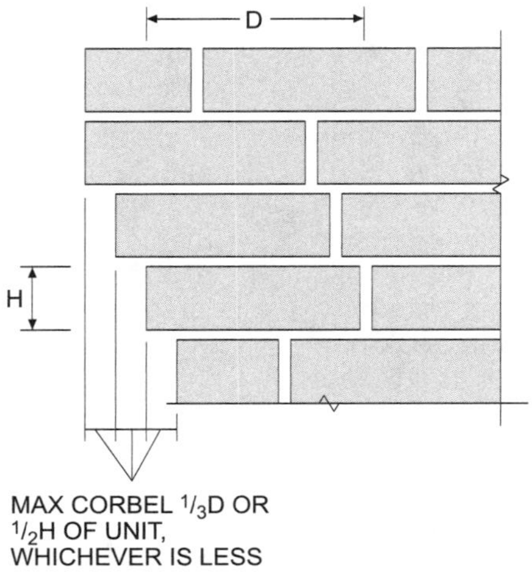

Figure R606.3(1)
MASONRY CORBELING LIMITATIONS PER COURSE

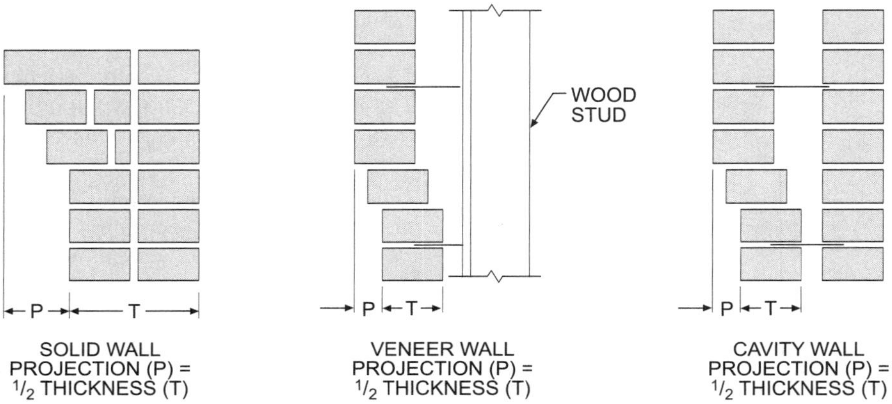

Figure R606.3(2)
MASONRY CORBELING LIMITATIONS PER COURSE

R606.4 Support conditions. Bearing and support conditions shall be in accordance with Sections R606.4.1 and R606.4.2.

❖ Masonry bearing and support conditions must be in accordance with the subsections of Section R606.4.

R606.4.1 Bearing on support. Each masonry wythe shall be supported by at least two-thirds of the wythe thickness.

❖ At least two-thirds the thickness of the bottom course of all masonry wythes must bear directly on the supporting construction (i.e., bear on the supporting foundation wall, lintel or header, etc.).

R606.4.2 Support at foundation. Cavity wall or masonry veneer construction may be supported on an 8-inch (203 mm) foundation wall, provided the 8-inch (203 mm) wall is corbeled to the width of the wall system above with masonry constructed of *solid masonry* units or masonry units filled with mortar or grout. The total horizontal projection of the corbel shall not exceed 2 inches (51 mm) with individual corbels projecting not more than one-third the thickness of the unit or one-half the height of the unit. The hollow space behind the corbeled masonry shall be filled with mortar or grout.

❖ Foundations supporting masonry cavity walls must be at least 8 inches (203 mm) thick, and where the wall supported is thicker, the foundation wall must be corbeled to attain a thickness at least equal to that of the wall supported. Corbeling limitations of this section are shown in Commentary Figure R606.4(2).

R606.5 Allowable stresses. Allowable compressive stresses in masonry shall not exceed the values prescribed in Table R606.5. In determining the stresses in masonry, the effects of all loads and conditions of loading and the influence of all forces affecting the design and strength of the several parts shall be taken into account.

❖ Masonry compressive stresses are determined from the type and compressive strength of the unit and the type of mortar used in construction. Table R606.5 lists the maximum permitted stresses. Stresses must be computed based on the actual dimensions of the masonry.

Example:

How does a Type S versus Type N mortar affect the allowable compressive stress in unit masonry for a wall composed of solid concrete brick having a unit compressive stress of 2,500 psi (17 237 kPa)?

Solution:

From Table R606.5, the allowable compressive strength is a function of the mortar and would be:

2,500 psi solid units (17 237 kPa): using Type S mortar = 160 psi (1103 kPa).

2,500 psi solid units (17 237 kPa): using Type N mortar = 140 psi (965 kPa).

R606.5.1 Combined units. In walls or other structural members composed of different kinds or grades of units, materials or mortars, the maximum stress shall not exceed the allowable stress for the weakest of the combination of units, materials and mortars of which the member is composed. The net thickness of any facing unit that is used to resist stress shall not be less than 1.5 inches (38 mm).

❖ Walls containing different grades of units, materials or mortars must be limited to the maximum compressive stresses for the weakest combination of units and mortar.

Example:

What is the allowable compressive stress in unit masonry for a wall constructed of 8,000 psi (55 152 kPa) solid brick units, 2,000 psi (13 788 kPa) hollow masonry units, and Type S mortar?

Solution:

From Table R606.5, the allowable compressiveness would be limited to the smaller of:

8,000 psi (55 158 kPa) solid brick units using Type S mortar = 350 psi (2413 kPa);

2,000 psi (13 789 kPa) hollow masonry units using Type S mortar = 140 psi (965 kPa).

The allowable unit masonry stress used in design would be the weaker of the two materials constructed with Type S mortar. Therefore, the allowable stress would be 140 psi (965 kPa).

R606.6 Piers. The unsupported height of masonry piers shall not exceed ten times their least dimension. When structural clay tile or hollow concrete masonry units are used for isolated piers to support beams and girders, the cellular spaces shall be filled solidly with concrete or Type M or S mortar, except that unfilled hollow piers may be used if their unsupported height is not more than four times their least dimension. Where hollow masonry units are solidly filled with concrete or Type M, S or N mortar, the allowable compressive stress shall be permitted to be increased as provided in Table R606.5.

❖ Isolated masonry piers conforming to this section may be used to support beams and girders. The minimum pier area should be based on the design load supported and the allowable stress of the construction per Table R606.5.

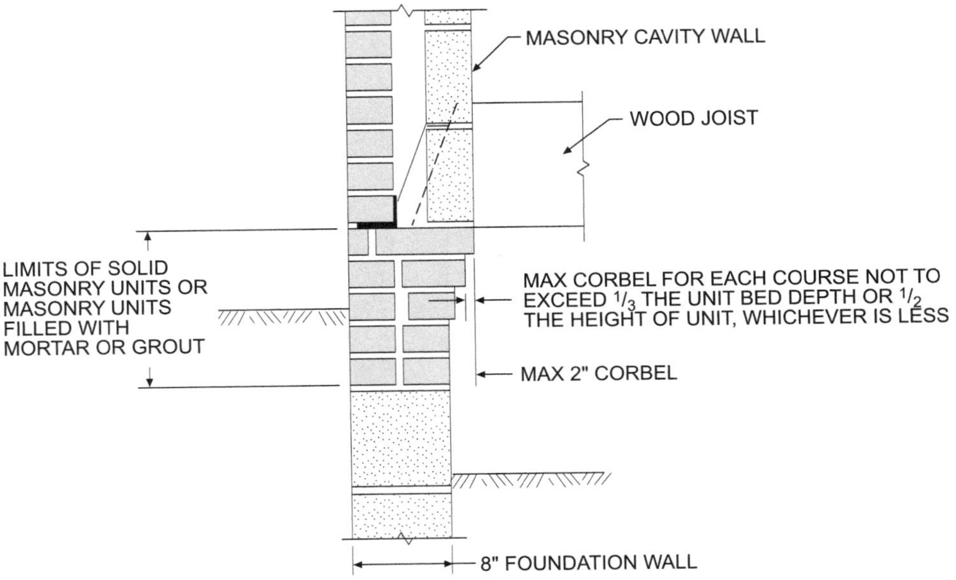

For SI: 1 inch = 25.4 mm.

Figure R606.4(2)
MASONRY CORBELING LIMITATIONS PER COURSE

TABLE R606.5
ALLOWABLE COMPRESSIVE STRESSES FOR EMPIRICAL DESIGN OF MASONRY

CONSTRUCTION; COMPRESSIVE STRENGTH OF UNIT, GROSS AREA	ALLOWABLE COMPRESSIVE STRESSES[a] GROSS CROSS-SECTIONAL AREA[b]	
	Type M or S mortar	Type N mortar
Solid masonry of brick and other solid units of clay or shale; sand-lime or concrete brick:		
8,000 + psi	350	300
4,500 psi	225	200
2,500 psi	160	140
1,500 psi	115	100
Grouted[c] masonry, of clay or shale; sand-lime or concrete:		
4,500 + psi	225	200
2,500 psi	160	140
1,500 psi	115	100
Solid masonry of solid concrete masonry units:		
3,000 + psi	225	200
2,000 psi	160	140
1,200 psi	115	100
Masonry of hollow load-bearing units:		
2,000 + psi	140	120
1,500 psi	115	100
1,000 psi	75	70
700 psi	60	55
Hollow walls (cavity or masonry bonded[d]) solid units:		
2,500 + psi	160	140
1,500 psi	115	100
Hollow units	75	70
Stone ashlar masonry:		
Granite	720	640
Limestone or marble	450	400
Sandstone or cast stone	360	320
Rubble stone masonry:		
Coarse, rough or random	120	100

For SI: 1 pound per square inch = 6.895 kPa.

a. Linear interpolation shall be used for determining allowable stresses for masonry units having compressive strengths that are intermediate between those given in the table.

b. Gross cross-sectional area shall be calculated on the actual rather than nominal dimensions.

c. See Section R608.

d. Where floor and roof loads are carried upon one wythe, the gross cross-sectional area is that of the wythe under load; if both wythes are loaded, the gross cross-sectional area is that of the wall minus the area of the cavity between the wythes. Walls bonded with metal ties shall be considered as cavity walls unless the collar joints are filled with mortar or grout.

❖ See the commentary for Section R606.5.

R606.6.1 Pier cap. Hollow piers shall be capped with 4 inches (102 mm) of *solid masonry* or concrete, a masonry cap block, or shall have cavities of the top course filled with concrete or grout.

❖ Beams and girders must be directly supported on a sill plate or the masonry piers to provide a complete load path to the foundation.

R606.7 Chases. Chases and recesses in masonry walls shall not be deeper than one-third the wall thickness, and the maximum length of a horizontal chase or horizontal projection shall not exceed 4 feet (1219 mm), and shall have at least 8 inches (203 mm) of masonry in back of the chases and recesses and between adjacent chases or recesses and the jambs of openings. Chases and recesses in masonry walls shall be designed and constructed so as not to reduce the required strength or required fire resistance of the wall and in no case shall a chase or recess be permitted within the required area of a pier. Masonry directly above chases or recesses wider than 12 inches (305 mm) shall be supported on noncombustible lintels.

❖ A chase is a continuous recess in a masonry wall that receives a pipe, conduit, etc. Generally, chases are vertical. It is best to construct chases as the masonry wall is built so the strength of the wall is not reduced.

R606.8 Stack bond. In unreinforced masonry where masonry units are laid in stack bond, longitudinal reinforcement consisting of not less than two continuous wires each with a minimum aggregate cross-sectional area of 0.017 square inch (11 mm^2) shall be provided in horizontal bed joints spaced not more than 16 inches (406 mm) on center vertically.

❖ To control cracking where masonry units are placed in a stack bond and where the wall is of unreinforced masonry construction, longitudinal reinforcement must be used in the horizontal bed joints as shown in Commentary Figure R606.8(1). Premanufactured ladder or truss-type reinforcement as shown in Commentary Figure R606.8(2) is generally used to meet the requirements for longitudinal steel.

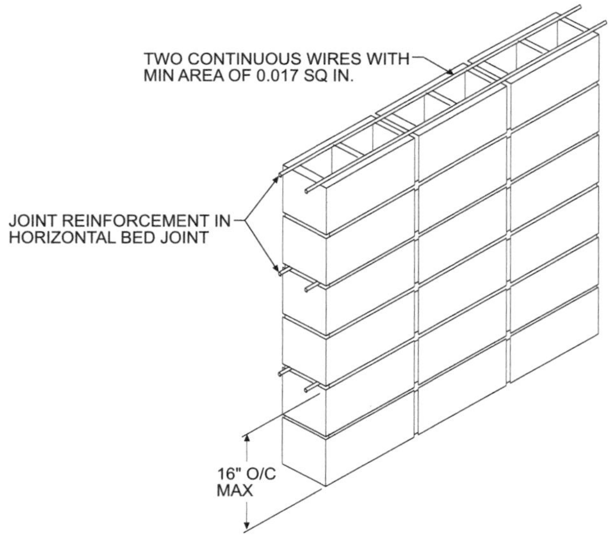

For SI: 1 square inch = 645.2 mm^2.

Figure R606.8(1)
MINIMUM HORIZONTAL JOINT REINFORCEMENT FOR STACK-BOND MASONRY

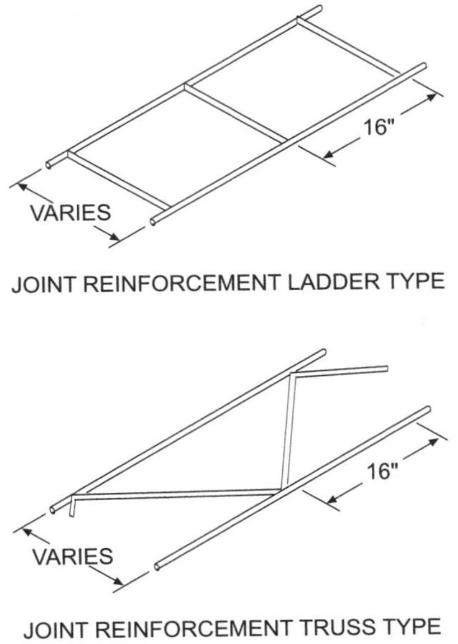

For SI: 1 inch = 25.4 mm.

Figure R606.8(2)
HORIZONTAL JOINT REINFORCEMENT-LADDER AND TRUSS TYPES

R606.9 Lateral support. Masonry walls shall be laterally supported in either the horizontal or the vertical direction. The maximum spacing between lateral supports shall not exceed the distances in Table R606.9. Lateral support shall be provided by cross walls, pilasters, buttresses or structural frame members when the limiting distance is taken horizontally, or by floors or roofs when the limiting distance is taken vertically.

❖ The limitations on the maximum unsupported height or length of masonry walls specified in Table R606.9 provides reasonable performance. For purposes of applying the unsupported height limitations, Figures R606.11(1), R606.11(2) and R606.11(3) provide details recognized as appropriate methods of anchorage. At the base of the wall, footings are a lateral support point. Thus, the unsupported height from the footing to the anchorage point at the floor or roof is the unsupported height, which must be limited to the values in Table R606.9.

Instead of unsupported height limitations being measured vertically from footing to supporting floor or roof, the span limitations in Table R606.9 of the code may be met with the use of pilasters, columns, piers, cross walls or similar elements whose relative stiffness is greater than that of the wall. These elements are anchored to the roof or floor structural elements in a manner that transmits imposed lateral forces.

Commentary Figure R606.9 illustrates the lateral support limitations specified in the table.

Example:

The height or length limitation between lateral supports (see Commentary Figure R606.9) for a solid-grouted masonry bearing wall of 8-inch (203 mm) nominal units would be 8 × 20 = 160 inches (4064 mm). For a nonbearing interior wall using two wythes of 4-inch (102 mm) brick, the allowable limitation would be (4 + 4) × 36 = 288 inches (7315 mm).

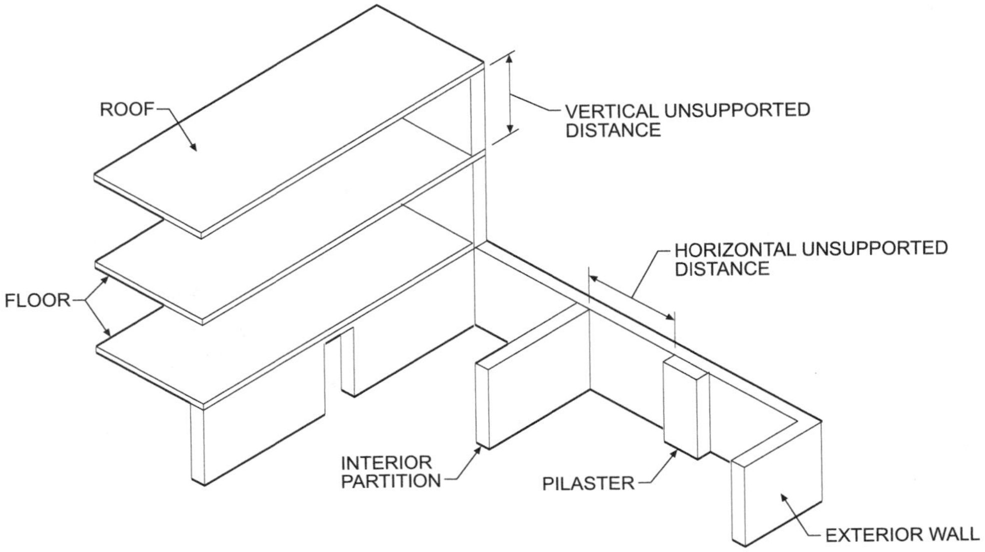

For SI: 1 inch = 25.4 mm.

Figure R606.9
BUILDING ELEMENTS PROVIDING LATERAL SUPPORT TO WALLS

TABLE R606.9
SPACING OF LATERAL SUPPORT FOR MASONRY WALLS

CONSTRUCTION	MAXIMUM WALL LENGTH TO THICKNESS OR WALL HEIGHT TO THICKNESS[a, b]
Bearing walls:	
Solid or solid grouted	20
All other	18
Nonbearing walls:	
Exterior	18
Interior	36

For SI: 1 foot = 304.8 mm.

a. Except for cavity walls and cantilevered walls, the thickness of a wall shall be its nominal thickness measured perpendicular to the face of the wall. For cavity walls, the thickness shall be determined as the sum of the nominal thicknesses of the individual wythes. For cantilever walls, except for parapets, the ratio of height to nominal thickness shall not exceed 6 for solid masonry, or 4 for hollow masonry. For parapets, see Section R606.2.4.

b. An additional unsupported height of 6 feet is permitted for gable end walls.

❖ Limitations on ratios of distance between lateral supports to wall thickness listed in this table are primarily traditional ratios based on successful performance. The thickness in this table is based on the nominal thickness of the wall for other than cavity wall construction and the sum of the nominal thickness of the wythes not including the cavity for cavity walls; see Note a. To allow for additional height at gable-end walls, Note b permits an additional 6 feet (1829 mm). This is similar to the allowance discussed in Commentary Figure R606.2.1.

R606.9.1 Horizontal lateral support. Lateral support in the horizontal direction provided by intersecting masonry walls shall be provided by one of the methods in Section R606.9.1.1 or Section R606.9.1.2.

❖ When masonry walls span horizontally between intersecting walls, anchorage is achieved using the method in either of the following sections.

R606.9.1.1 Bonding pattern. Fifty percent of the units at the intersection shall be laid in an overlapping masonry bonding pattern, with alternate units having a bearing of not less than 3 inches (76 mm) on the unit below.

❖ Using this method, anchorage is accomplished by off setting (overlapping) alternating courses of masonry.

R606.9.1.2 Metal reinforcement. Interior nonload- bearing walls shall be anchored at their intersections, at vertical intervals of not more than 16 inches (406 mm) with joint reinforcement of at least 9 gage [0.148 inch (4mm)], or $^{1}/_{4}$-inch (6 mm) galvanized mesh hardware cloth. Intersecting masonry walls, other than interior nonloadbearing walls, shall be anchored at vertical intervals of not more than 8 inches (203 mm) with joint reinforcement of at least 9 gage and shall extend at least 30 inches (762 mm) in each direction at the intersection. Other metal ties, joint reinforcement or anchors, if used, shall be spaced to provide equivalent area of anchorage to that required by this section.

❖ See Commentary Figure R606.9.1.2.

R606.9.2 Vertical lateral support. Vertical lateral support of masonry walls in Seismic Design Category A, B or C shall be provided in accordance with one of the methods in Section R606.9.2.1 or Section R606.9.2.2.

❖ Where walls span vertically in buildings classified as SDC A, B or C, masonry walls must be anchored to the floor and roof diaphragms to transmit the anticipated lateral forces. The unsupported height from the footing to the floor or roof must be within the limits of Table R606.9.

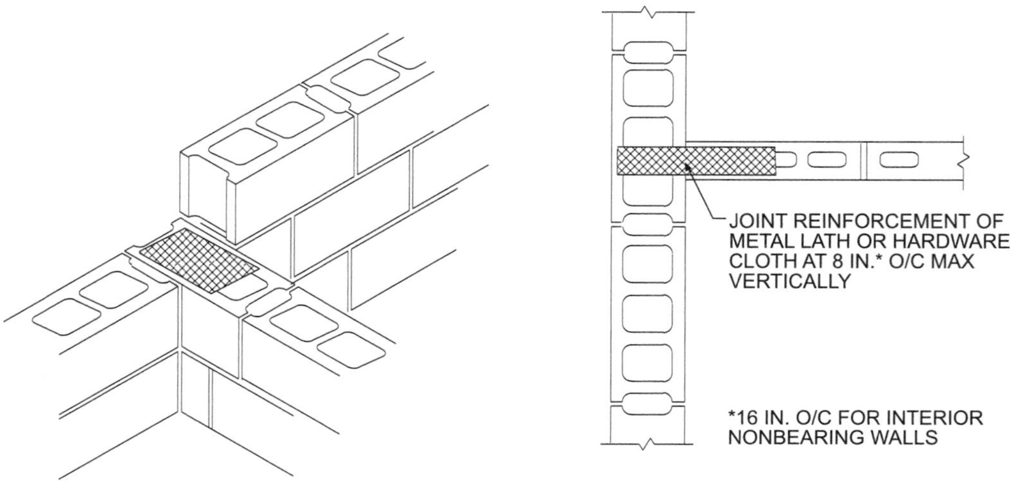

For SI: 1 inch = 25.4 mm.

Figure R606.9.1.2
INTERSECTING MASONRY WALLS

R606.9.2.1 Roof structures. Masonry walls shall be anchored to roof structures with metal strap anchors spaced in accordance with the manufacturer's instructions, $^1/_2$-inch (13 mm) bolts spaced not more than 6 feet (1829 mm) on center, or other *approved* anchors. Anchors shall be embedded at least 16 inches (406 mm) into the masonry, or be hooked or welded to bond beam reinforcement placed not less than 6 inches (152 mm) from the top of the wall.

❖ If $^1/_2$-inch (12.7 mm) diameter anchor bolts are used, they must be spaced no farther apart than 6 feet (1829 mm) on center. Where approved metal strap anchors are used, the spacing must be in accordance with the manufacturer's recommendations. Anchors are embedded in the masonry wall and connected to the wooden ledger as illustrated in Commentary Figures R606.9.2.1(1) and R606.9.2.1(2).

R606.9.2.2 Floor diaphragms. Masonry walls shall be anchored to floor *diaphragm* framing by metal strap anchors spaced in accordance with the manufacturer's instructions, $^1/_2$-inch-diameter (13 mm) bolts spaced at intervals not to exceed 6 feet (1829 mm) and installed as shown in Figure R606.11(1), or by other *approved* methods.

❖ Masonry walls must be anchored to the floor diaphragm with bolts spaced no farther apart than 6 feet (1829 mm) on center. Fasteners may be either metal strap anchors installed per the manufacturer's recommendations or bolts that are embedded in the masonry walls and connected to the wooden ledger. Ledge fasteners transfer shear forces when loading is in the plane of the wall; when forces are out of plane, ledge fasteners prevent the wall and flooring system from separating.

R606.10 Lintels. Masonry over openings shall be supported by steel lintels, reinforced concrete or masonry lintels or masonry arches, designed to support load imposed.

❖ Masonry wall openings require a structural member designed to support the masonry above. Note that wood is not permitted for a lintel supporting a masonry wall.

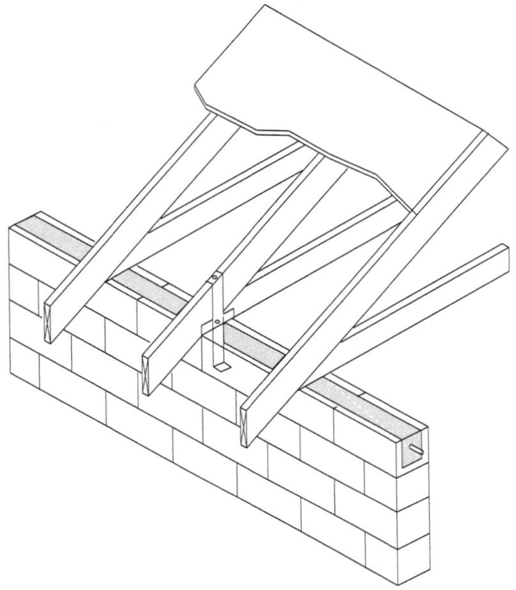

Figure R606.9.2.1(2)
ANCHORAGE OF ROOF STRUCTURES

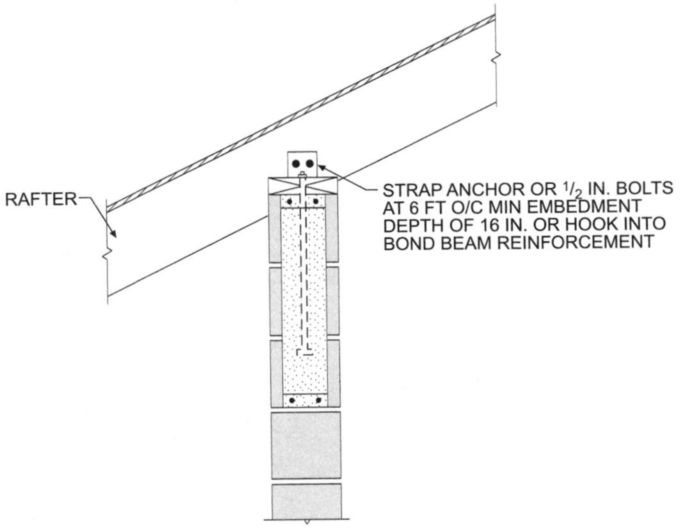

For SI: 1 inch = 25.4 mm, 1 foot = 304.8.

Figure R606.9.2.1(1)
ANCHORAGE OF ROOF STRUCTURES

R606.11 Anchorage. Masonry walls shall be anchored to floor and roof systems in accordance with the details shown in Figure R606.11(1), R606.11(2) or R606.11(3). Footings may be considered as points of lateral support.

❖ Masonry walls depend on floors and roofs for out-of-plane lateral support. Inadequate anchorage of masonry walls in areas of high, and even moderate, seismicity can be problematic. The referenced figures show anchorage requirements that vary based on seismic design category. They illustrate details that provide adequate load transfer under lateral loads.

R606.12 Seismic requirements. The seismic requirements of this section shall apply to the design of masonry and the construction of masonry building elements located in Seismic Design Category D_0, D_1 or D_2. Townhouses in Seismic Design Category C shall comply with the requirements of Section R606.12.2. These requirements shall not apply to glass unit masonry conforming to Section R610 or masonry veneer conforming to Section R703.7.

❖ Although the code generally permits unreinforced walls, this section requires reinforcement in walls of buildings classified as SDC C, D_0, D_1 or D_2.

R606.12.1 General. Masonry structures and masonry elements shall comply with the requirements of Sections R606.12.2 through R606.12.4 based on the seismic design category established in Table R301.2(1). Masonry structures and masonry elements shall comply with the requirements of Section R606.12 and Figures R606.11(1), R606.11(2) and R606.11(3) or shall be designed in accordance with TMS 402/ACI 530/ASCE 5 or TMS 403.

❖ Masonry walls must comply with this section unless a design is provided. The three cited figures show the reinforcing requirements as a function of the seismic design category and wind load.

R606.12.1.1 Floor and roof diaphragm construction. Floor and roof *diaphragms* shall be constructed of wood structural panels attached to wood framing in accordance with Table R602.3(1) or to cold-formed steel floor framing in accordance with Table R505.3.1(2) or to cold-formed steel roof framing in accordance with Table R804.3. Additionally, sheathing panel edges perpendicular to framing members shall be backed by blocking, and sheathing shall be connected to the blocking with fasteners at the edge spacing. For Seismic Design Categories C, D_0, D_1 and D_2, where the width-to-thickness dimension of the *diaphragm* exceeds 2-to-1, edge spacing of fasteners shall be 4 inches (102 mm) on center.

❖ This section reiterates the requirements of Chapters 5 and 8 for fastening structural wood sheathing to floor framing and roof framing. Additional blocking is specified for the structural wood sheathing edges perpendicular to the floor or roof framing, thus providing a blocked diaphragm. Edge fasteners as specified for panel edges must be used at this blocking. Furthermore, if a diaphragm's depth-to-span ratio is greater than 2 in a building classified as SDC C, D_0, D_1 or D_2, the edge nail spacing must be reduced to 4 inches (102 mm) on center. These requirements result in added strength and stiffness for diaphragms that support masonry walls.

R606.12.2 Seismic Design Category C. Townhouses located in Seismic Design Category C shall comply with the requirements of this section.

❖ Townhouses classified as SDC C must conform to the requirements of this section as well as Figures R606.11(1) and R606.11(2). The requirements are based on whether or not an element is part of the lateral-force-resisting system. They are aimed at providing ductility by specifying minimum reinforcement. This section is similar to Section 2106.4 of the IBC.

R606.12.2.1 Minimum length of wall without openings. Table R606.12.2.1 shall be used to determine the minimum required solid wall length without openings at each masonry exterior wall. The provided percentage of solid wall length shall include only those wall segments that are 3 feet (914 mm) or longer. The maximum clear distance between wall segments included in determining the solid wall length shall not exceed 18 feet (5486 mm). Shear wall segments required to meet the minimum wall length shall be in accordance with Section R606.12.2.2.3.

❖ The code prescribes length of bracing for wood, steel and concrete wall systems. Previous to the 2009 edition of the code, there was no regulation of the minimum length of bracing wall to be provided in masonry wall buildings. This section corrects this situation by basing the minimum wall length requirements on those developed for concrete walls, a comparable material both in terms of load requirements and capacity. These provisions are applicable to multistory construction without engineered design in SDC C and single story in SDC D_0, D_1 and D_2. The provisions are based on concrete requirements, providing interim guidance until more specific masonry requirements are developed.

R606.12.2.2 Design of elements not part of the lateral force-resisting system.

R606.12.2.2.1 Load-bearing frames or columns. Elements not part of the lateral force-resisting system shall be analyzed to determine their effect on the response of the system. The frames or columns shall be adequate for vertical load carrying capacity and induced moment caused by the design *story* drift.

❖ Load-bearing elements not part of a lateral-force-resisting system must be analyzed for their effect on the response of the system and must be capable of supporting all loads in combination with drift-induced bending stresses. These empirical masonry provisions do not cover the design of elements referred to in this section. They would require an engineering design using TMS 402/ACI 530/ASCE 5 (see the commentary for Section R606.12.2).

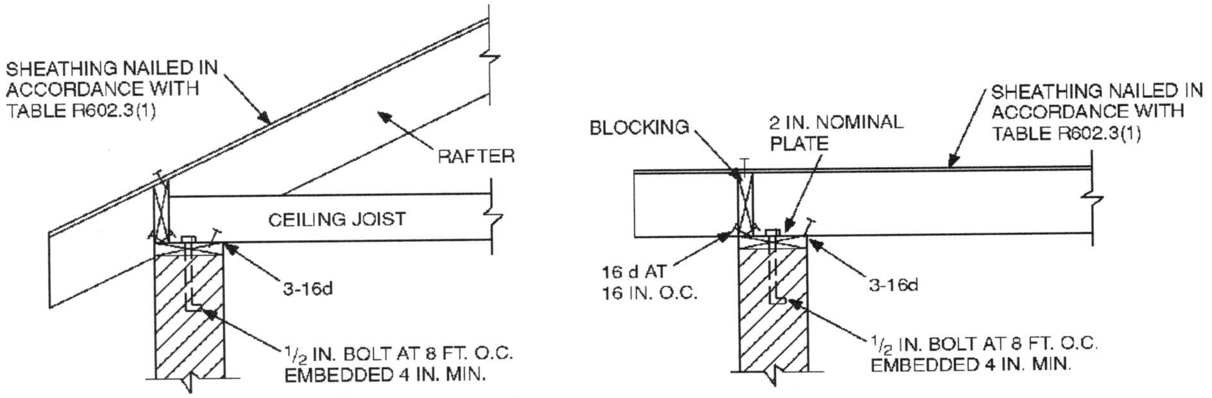

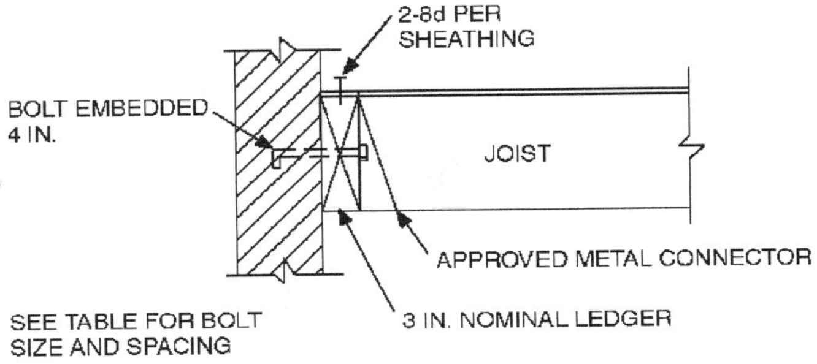

LEDGER BOLT SIZE AND SPACING

JOIST SPAN	BOLT SIZE AND SPACING	
	ROOF	FLOOR
10 FT.	1/2 AT 2 FT. 6 IN. 7/8 AT 3 FT. 6 IN.	1/2 AT 2 FT. 0 IN. 7/8 AT 2 FT. 9 IN.
10–15 FT.	1/2 AT 1 FT. 9 IN. 7/8 AT 2 FT. 6 IN.	1/2 AT 1 FT. 4 IN. 7/8 AT 2 FT. 0 IN.
15-20 FT.	1/2 AT 1 FT. 3 IN. 7/8 AT 2 FT. 0 IN.	1/2 AT 1 FT. 0 IN. 7/8 AT 1 FT. 6 IN.

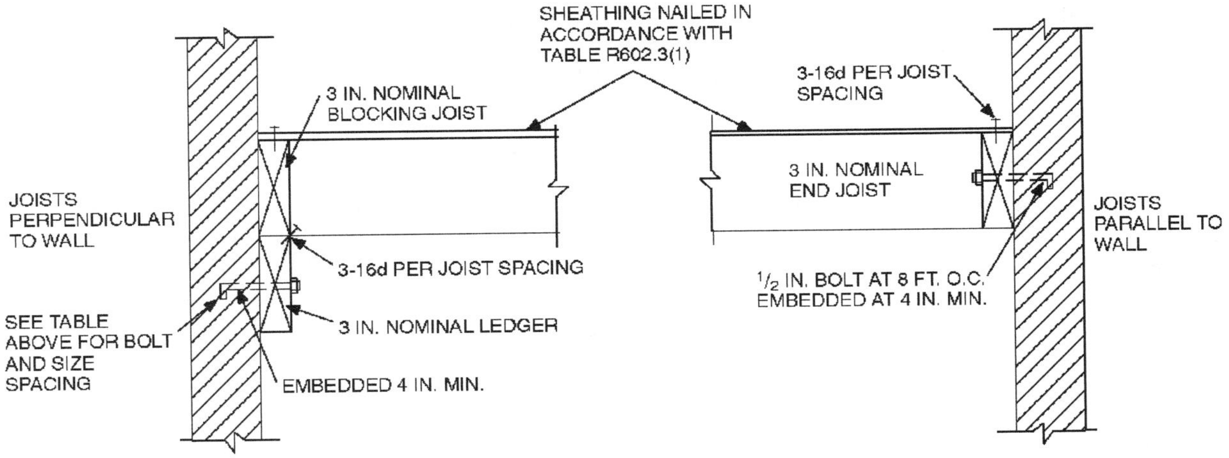

For SI: 1 inch = 25.4 mm, 1 foot = 304.8 mm, 1 pound per square foot = 0.0479 kPa.

Note: Where bolts are located in hollow masonry, the cells in the courses receiving the bolt shall be grouted solid.

FIGURE R606.11(1)
ANCHORAGE REQUIREMENTS FOR MASONRY WALLS LOCATED IN SEISMIC DESIGN CATEGORY A, B OR C AND WHERE WIND LOADS ARE LESS THAN 30 PSF

❖ See the commentary for Sections R606.11 and R606.11(1).

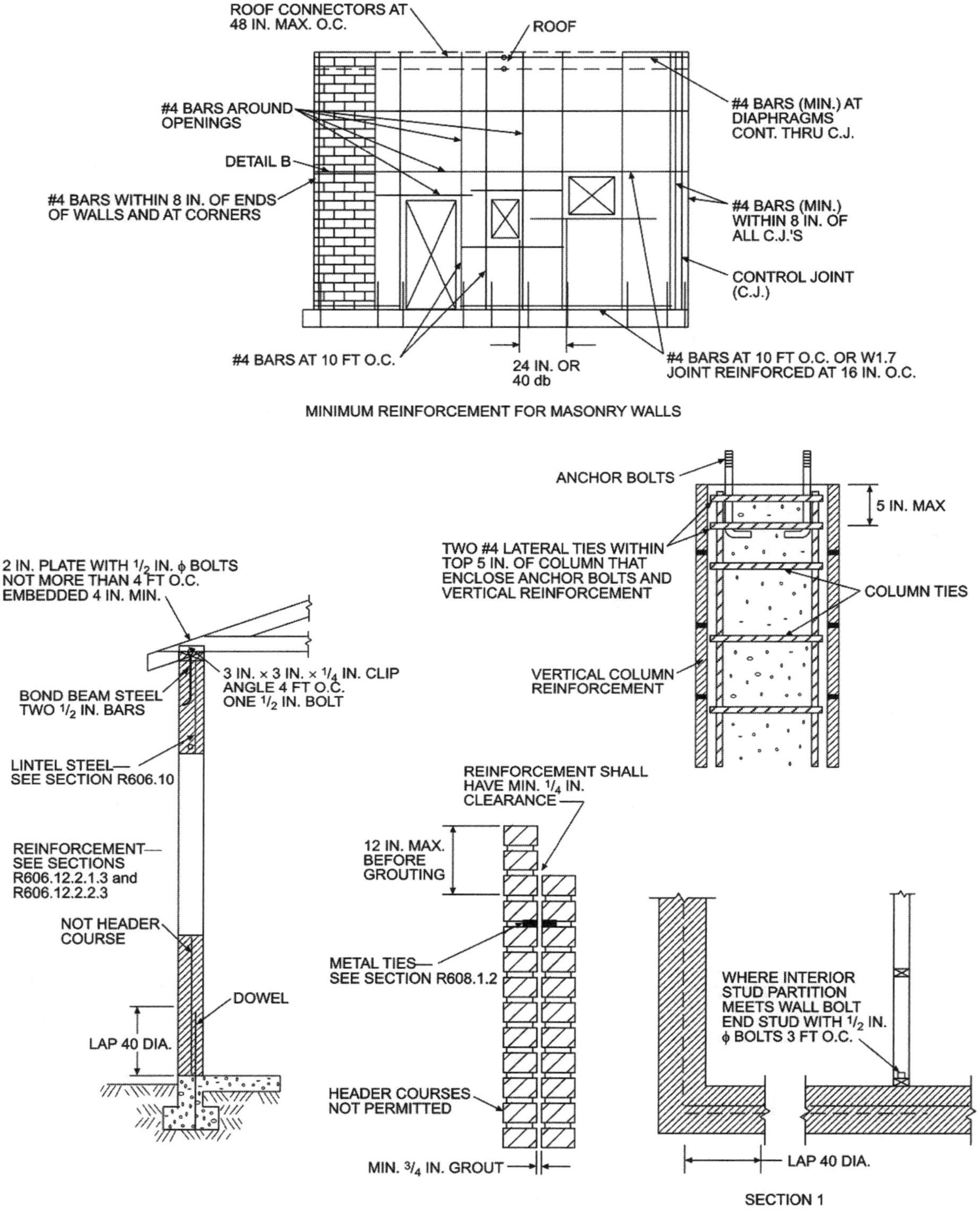

For SI: 1 inch = 25.4 mm, 1 foot = 304.8 mm.

FIGURE R606.11(2)
REQUIREMENTS FOR REINFORCED GROUTED MASONRY CONSTRUCTION IN SEISMIC DESIGN CATEGORY C

❖ See the commentary for Sections R606.11 and R606.12.1.

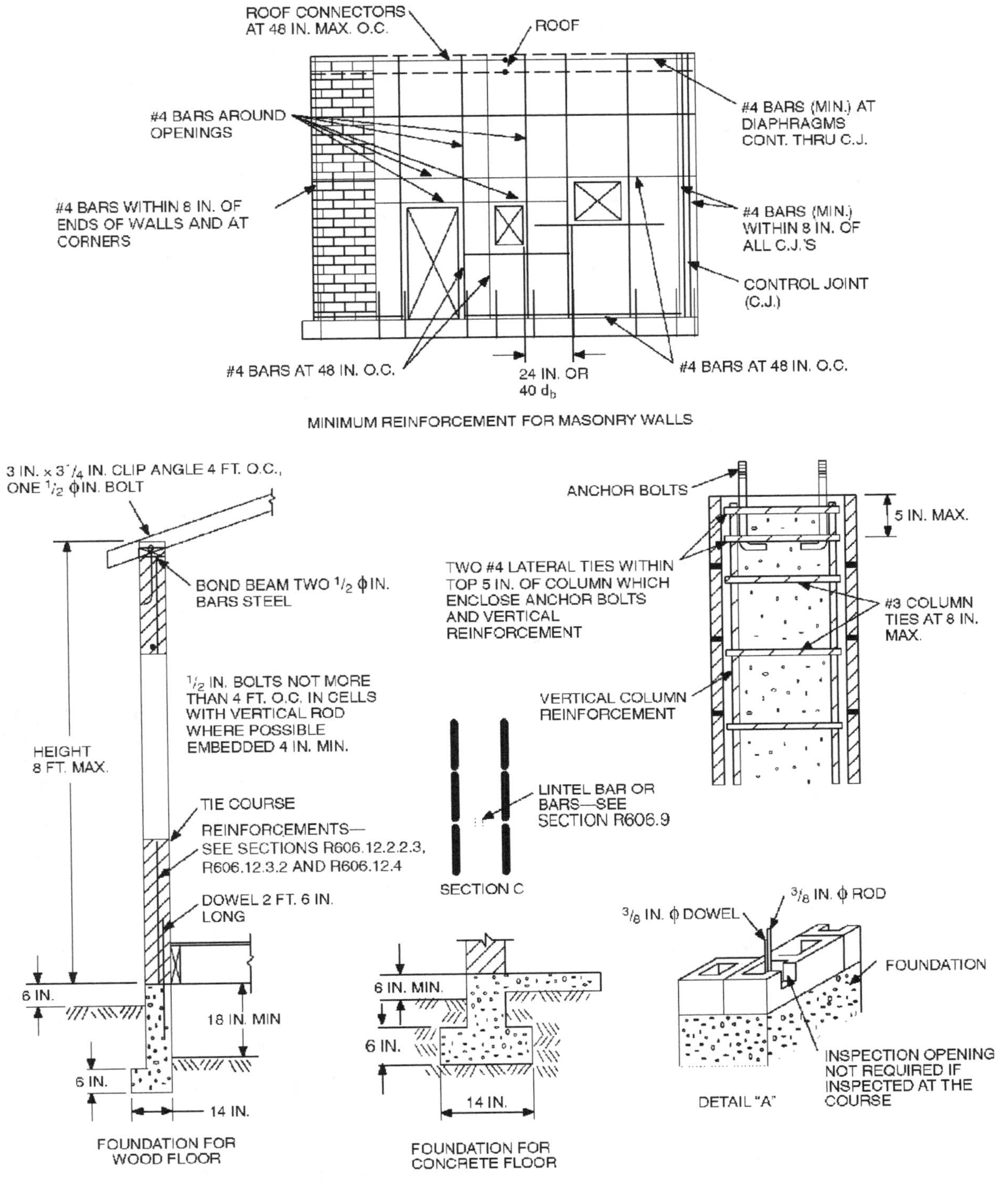

For SI: 1 inch = 25.4 mm, 1 foot = 304.8 mm.

Note: A full bed joint must be provided. All cells containing vertical bars are to be filled to the top of wall and provide inspection opening as shown on detail "A."

Horizontal bars are to be laid as shown on detail "B." Lintel bars are to be laid as shown on Section C.

FIGURE R606.11(3)
REQUIREMENTS FOR REINFORCED MASONRY CONSTRUCTION IN SEISMIC DESIGN CATEGORY D_0, D_1, OR D_2

❖ See the commentary for Sections R606.11 and R606.12.1.

TABLE R606.12.2.1
MINIMUM SOLID WALL LENGTH ALONG EXTERIOR WALL LINES

SESIMIC DESIGN CATEGORY	MINIMUM SOLID WALL LENGTH (percent)[a]		
	One story or top story of two story	Wall supporting light-framed second story and roof	Wall supporting masonry second story and roof
Townhouses in C	20	25	35
D_0 or D_1	25	NP	NP
D_2	30	NP	NP

NP = Not permitted, except with design in accordance with the *International Building Code*.

a. For all walls, the minimum required length of solid walls shall be based on the table percent multiplied by the dimension, parallel to the wall direction under consideration, of a rectangle inscribing the overall building plan.

❖ See the commentary for Section R606.12.2.1.

R606.12.2.2.2 Masonry partition walls. Masonry partition walls, masonry screen walls and other masonry elements that are not designed to resist vertical or lateral loads, other than those induced by their own weight, shall be isolated from the structure so that vertical and lateral forces are not imparted to these elements. Isolation joints and connectors between these elements and the structure shall be designed to accommodate the design *story* drift.

❖ Nonload-bearing elements that are not designed to resist vertical or lateral loads from earthquake effects on the building lateral force system must be isolated from the structure with isolation joints and connectors that can accommodate the design drift. The design story drift should be taken as 1 percent of the story height as allowed in Section 12.14.8.5 of ASCE 7 unless an analysis is performed.

R606.12.2.2.3 Reinforcement requirements for masonry elements. Masonry elements listed in Section R606.12.2.2.2 shall be reinforced in either the horizontal or vertical direction as shown in Figure R606.11(2) and in accordance with the following:

1. Horizontal reinforcement. Horizontal joint reinforcement shall consist of at least two longitudinal W1.7 wires spaced not more than 16 inches (406 mm) for walls greater than 4 inches (102 mm) in width and at least one longitudinal W1.7 wire spaced not more than 16 inches (406 mm) for walls not exceeding 4 inches (102 mm) in width; or at least one No. 4 bar spaced not more than 48 inches (1219 mm). Where two longitudinal wires of joint reinforcement are used, the space between these wires shall be the widest that the mortar joint will accommodate. Horizontal reinforcement shall be provided within 16 inches (406 mm) of the top and bottom of these masonry elements.
2. Vertical reinforcement. Vertical reinforcement shall consist of at least one No. 4 bar spaced not more than 48 inches (1219 mm). Vertical reinforcement shall be located within 16 inches (406 mm) of the ends of masonry walls.

❖ This section specifies reinforcement for masonry walls that are not part of the building lateral force system, as described in the previous section (see commentary, Section R606.12.2).

R606.12.2.3 Design of elements part of the lateral force-resisting system.

R606.12.2.3.1 Connections to masonry shear walls. Connectors shall be provided to transfer forces between masonry walls and horizontal elements in accordance with the requirements of Section 1.7.4 of TMS 402/ACI 530/ASCE 5. Connectors shall be designed to transfer horizontal design forces acting either perpendicular or parallel to the wall, but not less than 200 pounds per linear foot (2919 N/m) of wall. The maximum spacing between connectors shall be 4 feet (1219 mm). Such anchorage mechanisms shall not induce tension stresses perpendicular to grain in ledgers or nailers.

❖ The connection of horizontal elements such as diaphragms to masonry walls or columns must be designed in accordance with TMS 402/ACI 530/ASCE 5. These connections should also be in accordance with Figure R606.11(2) and meet the minimum requirements specified in this section. Because these connections have been a source of earthquake-related failures, the code does not allow diaphragm anchorage mechanisms that cause cross-grain bending. This addresses the problem of premature failure of wood ledgers (see commentary, Section R606.12.2).

R606.12.2.3.2 Connections to masonry columns. Connectors shall be provided to transfer forces between masonry columns and horizontal elements in accordance with the requirements of Section 1.7.4 of TMS 402/ACI 530/ASCE 5. Where anchor bolts are used to connect horizontal elements to the tops of columns, the bolts shall be placed within lateral ties. Lateral ties shall enclose both the vertical bars in the column and the anchor bolts. There shall be a minimum of two No. 4 lateral ties provided in the top 5 inches (127 mm) of the column.

❖ These empirical masonry provisions do not cover the design of masonry columns. A design in accordance with TMS 402/ACI 530/ASCE 5 must be provided. In that event, this section provides a reference to the appropriate section of the standard for the column-to-diaphragm connection design (see commentary, Section R606.12.2).

R606.12.2.3.3 Minimum reinforcement requirements for masonry shear walls. Vertical reinforcement of at least one

No. 4 bar shall be provided at corners, within 16 inches (406 mm) of each side of openings, within 8 inches (203 mm) of each side of movement joints, within 8 inches (203 mm) of the ends of walls, and at a maximum spacing of 10 feet (3048 mm).

Horizontal joint reinforcement shall consist of at least two wires of W1.7 spaced not more than 16 inches (406 mm); or bond beam reinforcement of at least one No. 4 bar spaced not more than 10 feet (3048 mm) shall be provided. Horizontal reinforcement shall also be provided at the bottom and top of wall openings and shall extend not less than 24 inches (610 mm) nor less than 40 bar diameters past the opening; continuously at structurally connected roof and floor levels; and within 16 inches (406 mm) of the top of walls.

❖ Masonry shear walls resist the lateral loads from wind or earthquakes. This section specifies reinforcing requirements, which are illustrated in Figure R606.11(2) (see commentary, Section R606.12.2).

R606.12.3 Seismic Design Category D_0 or D_1. Structures in Seismic Design Category D_0 or D_1 shall comply with the requirements of Seismic Design Category C and the additional requirements of this section.

❖ Buildings classified as SDC D_0 or D_1 must conform to the requirements of this section as well as the requirements for SDC C [see Section R606.12.2 and Figure R606.11(2)].

R606.12.3.1 Design requirements. Masonry elements other than those covered by Section R606.12.2.2.2 shall be designed in accordance with the requirements of Chapter 1 and Sections 2.1 and 2.3 of TMS 402, ACI 530/ASCE 5 and shall meet the minimum reinforcement requirements contained in Sections R606.12.3.2 and R606.12.3.2.1. Otherwise, masonry shall be designed in accordance with TMS 403.

> **Exception:** Masonry walls limited to one *story* in height and 9 feet (2743 mm) between lateral supports need not be designed provided they comply with the minimum reinforcement requirements of Sections R606.12.3.2 and R606.12.3.2.1.

❖ This section specifies the design criteria for masonry elements in SDC D_0 or D_1.

Except for nonload-bearing elements that are not part of the lateral-force-resisting system, masonry elements must be designed using TMS 402/ACI 530/ASCE 5. The exception allows certain one-story masonry walls to satisfy the specified prescriptive code requirements rather than being designed to other engineering standards.

R606.12.3.2 Minimum reinforcement requirements for masonry walls. Masonry walls other than those covered by Section R606.12.2.2.3 shall be reinforced in both the vertical and horizontal direction. The sum of the cross-sectional area of horizontal and vertical reinforcement shall be at least 0.002 times the gross cross-sectional area of the wall, and the minimum cross-sectional area in each direction shall be not less than 0.0007 times the gross cross-sectional area of the wall. Reinforcement shall be uniformly distributed. Table R606.12.3.2 shows the minimum reinforcing bar sizes required for varying thicknesses of masonry walls. The maximum spacing of reinforcement shall be 48 inches (1219 mm) provided that the walls are solid grouted and constructed of hollow open-end units, hollow units laid with full head joints or two wythes of solid units. The maximum spacing of reinforcement shall be 24 inches (610 mm) for all other masonry.

❖ The sum of horizontal and vertical reinforcement must be at least 0.2 percent of the cross-sectional area of the wall, and the reinforcing should be distributed as shown in Table R606.12.3.2 [see commentary, Figure R606.11(3) and Section R606.12.3].

R606.12.3.2.1 Shear wall reinforcement requirements. The maximum spacing of vertical and horizontal reinforcement shall be the smaller of one-third the length of the shear wall, one-third the height of the shear wall, or 48 inches (1219 mm). The minimum cross-sectional area of vertical reinforcement shall be one-third of the required shear reinforcement. Shear reinforcement shall be anchored around vertical reinforcing bars with a standard hook.

❖ See the commentary for Section R606.12.3.

TABLE R606.12.3.2
MINIMUM DISTRIBUTED WALL REINFORCEMENT FOR BUILDING ASSIGNED TO SEISMIC DESIGN CATEGORY D_0 or D_1

NOMINAL WALL THICKNESS (inches)	MINIMUM SUM OF THE VERTICAL AND HORIZONTAL REINFORCEMENT AREAS[a] (square inches per foot)	MINIMUM REINFORCEMENT AS DISTRIBUTED IN BOTH HORIZONTAL AND VERTICAL DIRECTIONS[b] (square inches per foot)	MINUMUM BAR SIZE FOR REINFORCEMENT SPACED AT 48 INCHES
6	0.135	0.047	#4
8	0.183	0.064	#5
10	0.231	0.081	#6
12	0.279	0.098	#6

For SI: 1 inch = 25.4 mm, 1 foot = 304.8 mm, 1 square inch per foot = 2064 mm^2/m.

a. Based on the minimum reinforcing ratio of 0.002 times the gross cross-sectional area of the wall.

b. Based on the minimum reinforcing ratio each direction of 0.0007 times the gross cross-sectional area of the wall.

❖ This table lists the minimum reinforcement areas specified in Section R606.12.3.2. The minimum bar size shown in the far right column satisfies the requirement for 0.0007 times the gross cross-sectional area of the wall. This must be provided as a minimum in both the vertical and horizontal directions. Additional reinforcement is necessary in either direction to meet the requirement for a total of 0.002 times the gross cross-sectional area.

R606.12.3.3 Minimum reinforcement for masonry columns. Lateral ties in masonry columns shall be spaced not more than 8 inches (203 mm) on center and shall be at least $^3/_8$-inch (9.5 mm) diameter. Lateral ties shall be embedded in grout.

❖ See Figure R606.11(3). Also see the commentary for Sections R606.12.2.2.2 and R606.12.3.

R606.12.3.4 Material restrictions. Type N mortar or masonry cement shall not be used as part of the lateral-force-resisting system.

❖ This restates the requirement of Section R607.1.3.

R606.12.3.5 Lateral tie anchorage. Standard hooks for lateral tie anchorage shall be either a 135-degree (2.4 rad) standard hook or a 180-degree (3.2 rad) standard hook.

❖ Lateral ties (required by Section R606.12.3.3) must be anchored as specified. Also see the commentary for Section R606.12.3.

R606.12.4 Seismic Design Category D_2. All structures in Seismic Design Category D_2 shall comply with the requirements of Seismic Design Category D_1 and to the additional requirements of this section.

❖ Buildings classified as SDC D_2 must conform to the requirements of this section as well as all the requirements for SDC D_0 and D_1 and Figure R606.11(3).

R606.12.4.1 Design of elements not part of the lateral force-resisting system. Stack bond masonry that is not part of the lateral force-resisting system shall have a horizontal cross-sectional area of reinforcement of at least 0.0015 times the gross cross-sectional area of masonry. Table R606.12.4.1 shows minimum reinforcing bar sizes for masonry walls. The maximum spacing of horizontal reinforcement shall be 24 inches (610 mm). These elements shall be solidly grouted and shall be constructed of hollow open-end units or two wythes of solid units.

❖ For elements that are not part of the lateral-force-resisting system, horizontal reinforcement must be at least 0.0015 times the cross-sectional area of the wall as shown in Table R606.12.4.1.

TABLE R606.12.4.1
MINIMUM REINFORCING FOR STACKED BONDED MASONRY WALLS IN SEISMIC DESIGN CATEGORY D_2

NOMINAL WALL THICKNESS (inches)	MINIMUM BAR SIZE SPACED AT 24 INCHES
6	#4
8	#5
10	#5
12	#6

For SI: 1 inch = 25.4 mm.

❖ See the commentary for Section R606.12.4.1.

R606.12.4.2 Design of elements part of the lateral force-resisting system. Stack bond masonry that is part of the lateral force-resisting system shall have a horizontal cross-sectional area of reinforcement of at least 0.0025 times the gross cross-sectional area of masonry. Table R606.12.4.2 shows minimum reinforcing bar sizes for masonry walls. The maximum spacing of horizontal reinforcement shall be 16 inches (406 mm). These elements shall be solidly grouted and shall be constructed of hollow open-end units or two wythes of solid units.

❖ Elements that are part of the lateral-force-resisting system must have horizontal reinforcement of at least 0.0025 times the cross-sectional area of the wall as shown in Table R606.12.4.2. These elements may be constructed of solid grouted hollow open-end units or of two wythes of solid units solidly grouted.

TABLE R606.12.4.2
MINIMUM REINFORCING FOR STACKED BONDED MASONRY WALLS IN SEISMIC DESIGN CATEGORY D_2

NOMINAL WALL THICKNESS (inches)	MINIMUM BAR SIZE SPACED AT 16 INCHES
6	#4
8	#5
10	#5
12	#6

For SI: 1 inch = 25.4 mm.

❖ See the commentary for Section R606.12.4.2.

R606.13 Protection for reinforcement. Bars shall be completely embedded in mortar or grout. Joint reinforcement embedded in horizontal mortar joints shall not have less than $^5/_8$-inch (15.9 mm) mortar coverage from the exposed face. All other reinforcement shall have a minimum coverage of one bar diameter over all bars, but not less than $^3/_4$ inch (19 mm), except where exposed to weather or soil, in which case the minimum coverage shall be 2 inches (51 mm).

❖ Commentary Figure R606.13 shows placement restrictions for reinforcement relative to exposed faces and coverage.

R606.14 Beam supports. Beams, girders or other concentrated loads supported by a wall or column shall have a bearing of at least 3 inches (76 mm) in length measured parallel to the beam upon *solid masonry* not less than 4 inches (102 mm) in thickness, or upon a metal bearing plate of adequate design and dimensions to distribute the load safely, or upon a continuous reinforced masonry member projecting not less than 4 inches (102 mm) from the face of the wall.

❖ To provide for the transfer of vertical loads from beams, girders or other elements to the masonry wall or column, a minimum bearing as shown in Commentary Figure R606.14 is required. An alternative to the minimum bearing is the use of metal bearing plates that are based on structural design.

R606.14.1 Joist bearing. Joists shall have a bearing of not less than $1^1/_2$ inches (38 mm), except as provided in Section R606.14, and shall be supported in accordance with Figure R606.11(1).

❖ Where joists bear on masonry wall or column elements, the minimum bearings as shown in Figure R606.11(1) and Commentary Figure R606.14.1 are considered to be adequate. The code does not provide guidance distinguishing whether a member is a joist or a beam; however, joists are generally considered to be members that are a nominal 2 inches (51 mm) maximum in thickness and placed not more than 24 inches (610 mm) on center.

R606.15 Metal accessories. Joint reinforcement, anchors, ties and wire fabric shall conform to the following: ASTM A 82 for wire anchors and ties; ASTM A 36 for plate, headed and bent-bar anchors; ASTM A 510 for corrugated sheet metal anchors and ties; ASTM A 951 for joint reinforcement; ASTM B 227 for copper-clad steel wire ties; or ASTM A 167 for stainless steel hardware.

❖ This section provides referenced standards for metal accessories used in masonry wall construction.

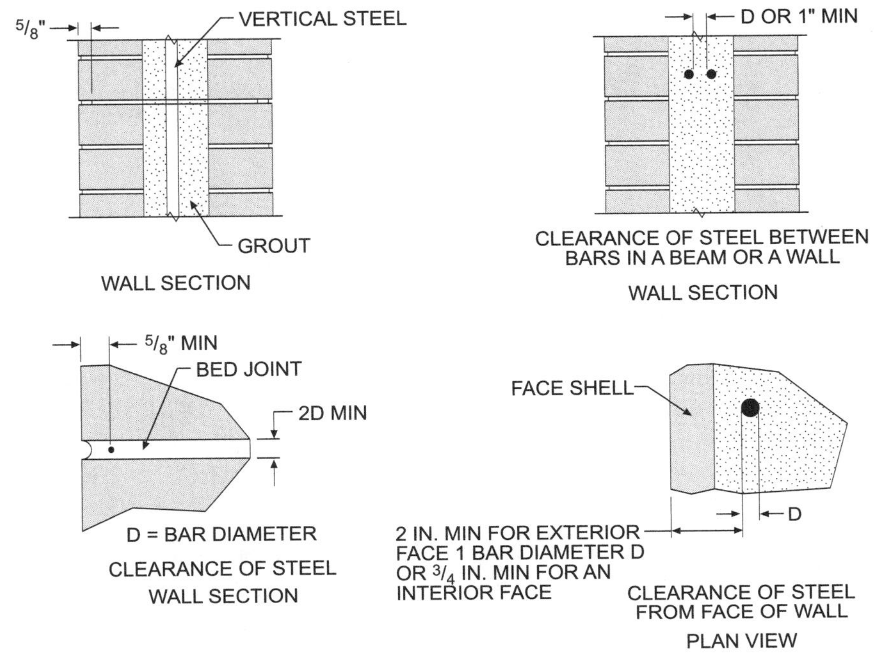

For SI: 1 inch = 25.4 mm.

Figure R606.13
REINFORCING STEEL CLEARANCES

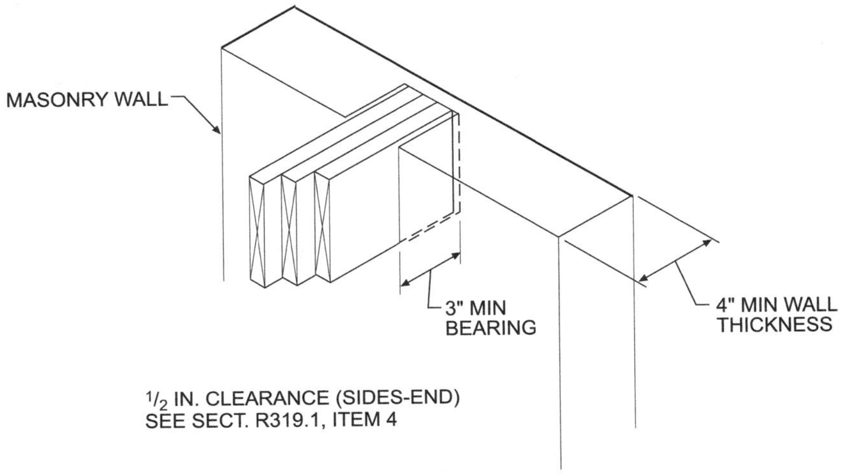

For SI:1 inch = 25.4 mm.

Figure R606.14
BEAM OR GIRDER BEARING ON MASONRY

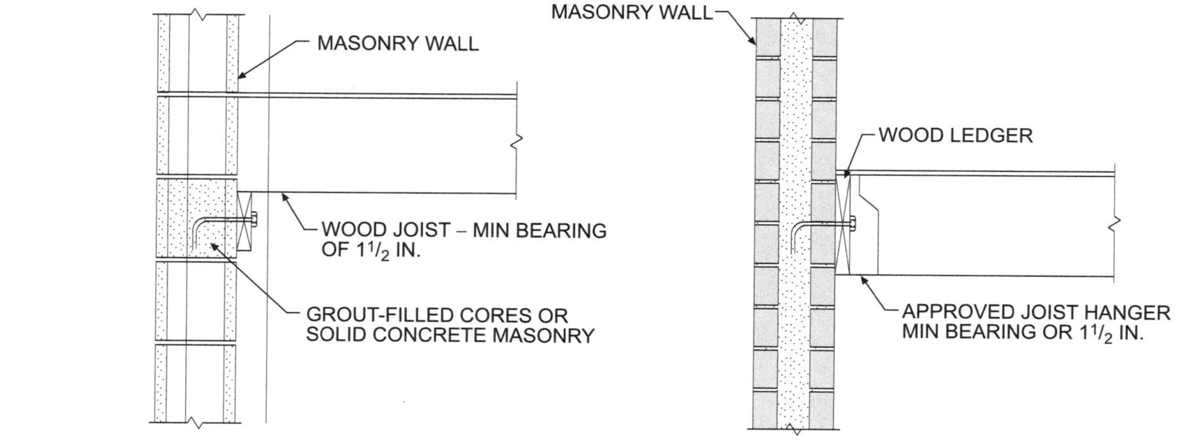

For SI: 1 inch = 25.4 mm.

Figure R606.14.1
JOIST-BEARING MINIMUMS

R606.15.1 Corrosion protection. Minimum corrosion protection of joint reinforcement, anchor ties and wire fabric for use in masonry wall construction shall conform to Table R606.15.1.

❖ Table R606.15.1 contains reference standards for protecting metal accessories used in masonry wall construction.

The corrosion protection requirements are dependent on type of steel and exposure. Because of its high resistance to corrosion, stainless steel need not be coated. Joint reinforcement, anchors, ties and accessories of other than stainless steel must be protected by zinc coatings (galvanizing). The protective value of the zinc coating increases with increasing coating thickness; therefore, the amount of galvanizing required increases with the potential severity of exposure.

TABLE R606.15.1
MINIMUM CORROSION PROTECTION

MASONRY METAL ACCESSORY	STANDARD
Joint reinforcement, interior walls	ASTM A 641, Class 1
Wire ties or anchors in exterior walls completely embedded in mortar or grout	ASTM A 641, Class 3
Wire ties or anchors in exterior walls not completely embedded in mortar or grout	ASTM A 153, Class B-2
Joint reinforcement in exterior walls or interior walls exposed to moist environment	ASTM A 153, Class B-2
Sheet metal ties or anchors exposed to weather	ASTM A 153, Class B-2
Sheet metal ties or anchors completely embedded in mortar or grout	ASTM A 653, Coating Designation G60
Stainless steel hardware for any exposure	ASTM A 167, Type 304

❖ See the commentary for Section R606.15.1.

SECTION R607
UNIT MASONRY

R607.1 Mortar. Mortar for use in masonry construction shall comply with ASTM C 270. The type of mortar shall be in accordance with Sections R607.1.1, R607.1.2 and R607.1.3 and shall meet the proportion specifications of Table R607.1 or the property specifications of ASTM C 270.

❖ This section specifies the requirements for mortar used in masonry construction. Mortar is the bonding agent that separates masonry units while bonding them together. It is an integral part of any masonry wall and must be strong, durable and capable of keeping the wall intact. In addition, it should help to create a moisture-resistant barrier.

Mortar makes up approximately 25 percent of a standard modular brick wall and about 10 percent of a block wall. It is therefore a significant part of the structure. ASTM C 270 discusses the materials and methods for mixing mortar. For purposes of economy, a good rule of thumb is to specify the lowest strength mortar that will satisfy the structural requirements of the project. Mortar consists of cementitious materials and well-graded sand with sufficient fines. Mortar is used for the following purposes:

- It is a bedding or seating material for the masonry unit. It allows the unit to be leveled and properly placed.
- It bonds the units together.
- It provides compressive strength.
- It provides shear strength, particularly parallel to the wall.
- It allows some movement and elasticity between units.
- It seals irregularities of the masonry unit and provides a weather-tight wall, preventing penetration of wind and water into and through the

wall. It can provide color to the wall when a mineral color additive is used.

- It can provide an architectural appearance by using various types of joints.

R607.1.1 Foundation walls. Masonry foundation walls constructed as set forth in Tables R404.1.1(1) through R404.1.1(4) and mortar shall be Type M or S.

❖ This section specifies the applicable tables in Chapter 4 that apply to masonry foundation walls.

R607.1.2 Masonry in Seismic Design Categories A, B and C. Mortar for masonry serving as the lateral-force-resisting system in Seismic Design Categories A, B and C shall be Type M, S or N mortar.

❖ Masonry construction in buildings classified as SDC A, B or C must use Type M, S or N mortar.

R607.1.3 Masonry in Seismic Design Categories D_0, D_1 and D_2. Mortar for masonry serving as the lateral-force-resisting system in Seismic Design Categories D_0, D_1 and D_2 shall be Type M or S portland cement-lime or mortar cement mortar.

❖ In buildings classified as Seismic Design Category D_0, D_1 or D_2, in walls which are intended to resist lateral forces, and in foundation walls in any building, only Type M or S mortar is acceptable.

TABLE R607.1
MORTAR PROPORTIONS[a, b]

		PROPORTIONS BY VOLUME (cementitious materials)								
MORTAR	TYPE	Portland cement or blended cement	Mortar cement			Masonry cement			Hydrated lime[c] or lime putty	Aggregate ratio (measured in damp, loose conditions)
			M	S	N	M	S	N		
Cement-lime	M	1	—	—	—	—	—	—	$^1/_4$	Not less than $2^1/_4$ and not more than 3 times the sum of separate volumes of lime, if used, and cement
	S	1	—	—	—	—	—	—	over $^1/_4$ to $^1/_2$	
	N	1	—	—	—	—	—	—	over $^1/_2$ to $1^1/_4$	
	O	1	—	—	—	—	—	—	over $1^1/_4$ to $2^1/_2$	
Mortar cement	M	1	—	—	1	—	—	—	—	
	M	—	1	—	—	—	—	—		
	S	$^1/_2$	—	—	1	—	—	—		
	S	—	—	1	—	—	—	—		
	N	—	—	—	1	—	—	—		
	O	—	—	—	1	—	—	—		
Masonry cement	M	1				—	—	1	—	
	M	—				1	—	—		
	S	$^1/_2$				—	—	1		
	S	—				—	1	—		
	N	—				—	—	1		
	O	—				—	—	1		

For SI: 1 cubic foot = 0.0283 m^3, 1 pound = 0.454 kg.

a. For the purpose of these specifications, the weight of 1 cubic foot of the respective materials shall be considered to be as follows:

Portland Cement	94 pounds	Masonry Cement	Weight printed on bag
Mortar Cement	Weight printed on bag	Hydrated Lime	40 pounds
Lime Putty (Quicklime)	80 pounds	Sand, damp and loose	80 pounds of dry sand

b. Two air-entraining materials shall not be combined in mortar.

c. Hydrated lime conforming to the requirements of ASTM C 207.

❖ The mortar required by Sections R607.1.1 through R607.1.3 must be proportioned as shown in this table or ASTM C 270. Type M mortar is suited for structures below or against grade such as retaining walls. It is also suited for masonry construction subject to high compressive loads, severe frost action or high lateral loads from earth pressures, hurricane winds or earthquakes. Type S mortar is appropriate for use in structures requiring high flexural bond strength that are subject to compressive and lateral loads. Type N mortar is meant for general use in above-grade masonry, residential basement construction, interior walls and partitions, masonry veneer and nonstructural masonry partitions.

R607.2 Placing mortar and masonry units.

R607.2.1 Bed and head joints. Unless otherwise required or indicated on the project drawings, head and bed joints shall be $^3/_8$ inch (10 mm) thick, except that the thickness of the bed joint of the starting course placed over foundations shall not be less than $^1/_4$ inch (7 mm) and not more than $^3/_4$ inch (19 mm).

❖ The provisions of Section R607 set forth specific requirements that are unique to construction using hollow-unit masonry. Hollow masonry units are to be laid and set in mortar. The properties of masonry walls, including strength and appearance, vary significantly depending on the thickness of the mortar joints. The initial bed joint, which is the bed joint between the first course of masonry and the foundation wall, is limited to a thickness of not less than $^1/_4$ inch (6.4 mm) and not greater than $^3/_4$ inch (19 mm). This larger variation is to allow for the inherent unevenness of concrete or concrete masonry foundation walls. The remainders of bed joints, which can be controlled by the mason, are limited to $^3/_8$ inch (9.5 mm), unless specified otherwise on the project drawings.

R607.2.1.1 Mortar joint thickness tolerance. Mortar joint thickness for load-bearing masonry shall be within the following tolerances from the specified dimensions:

1. Bed joint: + $^1/_8$ inch (3 mm).
2. Head joint: - $^1/_4$ inch (7 mm), + $^3/_8$ inch (10 mm).
3. Collar joints: - $^1/_4$ inch (7 mm), + $^3/_8$ inch (10 mm).

❖ Joint-thickness tolerances limit the eccentricity of applied loads and reflect typical industry practice. They also provide a more uniform appearance for exposed masonry. The bed joint is a horizontal joint; the head joint is a vertical joint.

R607.2.2 Masonry unit placement. The mortar shall be sufficiently plastic and units shall be placed with sufficient pressure to extrude mortar from the joint and produce a tight joint. Deep furrowing of bed joints that produces voids shall not be permitted. Any units disturbed to the extent that initial bond is broken after initial placement shall be removed and relaid in fresh mortar. Surfaces to be in contact with mortar shall be clean and free of deleterious materials.

❖ This section gives placement requirements meant to provide adequate bonding between mortar and masonry units. Mortar is the bonding agent that integrates masonry units into a wall; it is used to bind masonry units into a single element by developing a complete, strong, and durable bond. Mortar is usually placed between absorbent masonry units and loses water upon contact with the units. Mortars have a high water-cement ratio when mixed, but the ratio decreases when the mortar comes into contact with the absorbent unit.

Once the mortar has begun to set or harden, tapping or attempting to otherwise move masonry units can be detrimental to the bond. Movement at this time will break the bond between the masonry unit and mortar. The partially dried mortar will not have sufficient plasticity to re-adhere sufficiently to the masonry units. After the initial bond has been broken, the masonry units should be removed and replaced in fresh mortar.

R607.2.2.1 Solid masonry. *Solid masonry* units shall be laid with full head and bed joints and all interior vertical joints that are designed to receive mortar shall be filled.

❖ This section applies to solid masonry construction (see the definition in Section 202).

R607.2.2.2 Hollow masonry. For hollow masonry units, head and bed joints shall be filled solidly with mortar for a distance in from the face of the unit not less than the thickness of the face shell.

❖ See the definition for "Masonry unit, hollow" in Section R202. Normally, cross webs are not mortared. When individual cells are to be grouted, mortar should be placed on the webs on both sides of the cell to be grouted to prevent leakage of the grout.

R607.3 Installation of wall ties. The installation of wall ties shall be as follows:

1. The ends of wall ties shall be embedded in mortar joints. Wall ties shall have a minimum of $^5/_8$-inch (15.9 mm) mortar coverage from the exposed face.
2. Wall ties shall not be bent after being embedded in grout or mortar.
3. For solid masonry units, solid grouted hollow units, or hollow units in anchored masonry veneer, wall ties shall be embedded in mortar bed at least $1^1/_2$ inches (38 mm).
4. For hollow masonry units in other than anchored masonry veneer, wall ties shall engage outer face shells by at least $^1/_2$ inch (13 mm).

❖ Proper installation of masonry wall ties is necessary to achieve bonding of multiple wythes (also see Section R608.1.2.1). In addition, wall ties should not be bent after placement in the mortar joints. This can break the initial bond between the masonry unit and mortar and possibly lead to an increased amount of moisture infiltration into the wall assembly. Bending the wall ties after initial placement can also be detrimental to the strength of the wall system because the ties lose their ability to transfer lateral loads to the backing wall assembly.

Wall ties used with solid masonry units, solid grouted units and hollow units in anchored masonry veneer are embedded in the mortar bed at least $1^1/_2$ inches (38 mm) to ensure that there is adequate bond of the mortar to the veneer ties. Hollow units in anchored masonry veneer have been added to this provision. In the 2009 edition, hollow units in anchored masonry veneer were not separated from other hollow unit applications.

Wall ties for hollow units used in applications other than anchored masonry veneer continue to require embedment of the face shell by no less than $^1/_2$ inch (13 mm). This allows the cells of the unit to be later filled with grout.

SECTION R608
MULTIPLE-WYTHE MASONRY

R608.1 General. The facing and backing of multiple-wythe masonry walls shall be bonded in accordance with Section R608.1.1, R608.1.2 or R608.1.3. In cavity walls, neither the facing nor the backing shall be less than 3 inches (76 mm) nominal in thickness and the cavity shall not be more than 4 inches (102 mm) nominal in width. The backing shall be at least as thick as the facing.

Exception: Cavities shall be permitted to exceed the 4-inch (102 mm) nominal dimension provided tie size and tie spacing have been established by calculation.

❖ The provisions of Section R608 apply to masonry walls more than one masonry unit in thickness. In cases where a wall thickness is made up of multiple wythes of masonry units, bonding is required so that the wythes are tied together to act as a unit. Section R608 contains the specifics of what constitutes an acceptable bonding mechanism. Two basic types of bonding are permissible: the first involves various overlapping arrangements; the second is achieved by installation of corrosion-resistant metal ties as stipulated for cavity-wall masonry construction.

R608.1.1 Bonding with masonry headers. Bonding with solid or hollow masonry headers shall comply with Sections R608.1.1.1 and R608.1.1.2.

❖ There are two methods of bonding multiwythe walls through the use of headers. They can be solid or hollow unit, depending on the wall system in question.

R608.1.1.1 Solid units. Where the facing and backing (adjacent wythes) of *solid masonry* construction are bonded by means of masonry headers, no less than 4 percent of the wall surface of each face shall be composed of headers extending not less than 3 inches (76 mm) into the backing. The distance between adjacent full-length headers shall not exceed 24 inches (610 mm) either vertically or horizontally. In walls in which a single header does not extend through the wall, headers from the opposite sides shall overlap at least 3 inches (76 mm), or headers from opposite sides shall be covered with another header course overlapping the header below at least 3 inches (76 mm).

❖ See the commentary for Section R608.1.1 and Commentary Figure R608.1.1.1.

R608.1.1.2 Hollow units. Where two or more hollow units are used to make up the thickness of a wall, the stretcher courses shall be bonded at vertical intervals not exceeding 34 inches (864 mm) by lapping at least 3 inches (76 mm) over the unit below, or by lapping at vertical intervals not exceeding 17 inches (432 mm) with units that are at least 50 percent thicker than the units below.

❖ See the commentary for Section R608.1.1 and Commentary Figure R608.1.1.2.

R608.1.2 Bonding with wall ties or joint reinforcement. Bonding with wall ties or joint reinforcement shall comply with Sections R608.1.2.1 through R608.1.2.3.

❖ See the commentary for Section R608.1.

R608.1.2.1 Bonding with wall ties. Bonding with wall ties, except as required by Section R610, where the facing and backing (adjacent wythes) of masonry walls are bonded with $^3/_{16}$-inch-diameter (5 mm) wall ties embedded in the horizontal mortar joints, there shall be at least one metal tie for each 4.5 square feet (0.418 m^2) of wall area. Ties in alternate courses shall be staggered. The maximum vertical distance between ties shall not exceed 24 inches (610 mm), and the maximum horizontal distance shall not exceed 36 inches (914 mm). Rods or ties bent to rectangular shape shall be used with hollow masonry units laid with the cells vertical. In other walls, the ends of ties shall be bent to 90-degree (0.79 rad) angles to provide hooks no less than 2 inches (51 mm) long. Additional bonding ties shall be provided at all openings, spaced not more than 3 feet (914 mm) apart around the perimeter and within 12 inches (305 mm) of the opening.

❖ Multiwythe walls can be bonded with wire-type ties. Generally, this type of tie consists of metal Z ties or rectangular ties. The code requires certain maximum spacing for wire tires similar to that for masonry headers. This section provides minimum wire diameters and requires spacing based on a square footage area of the

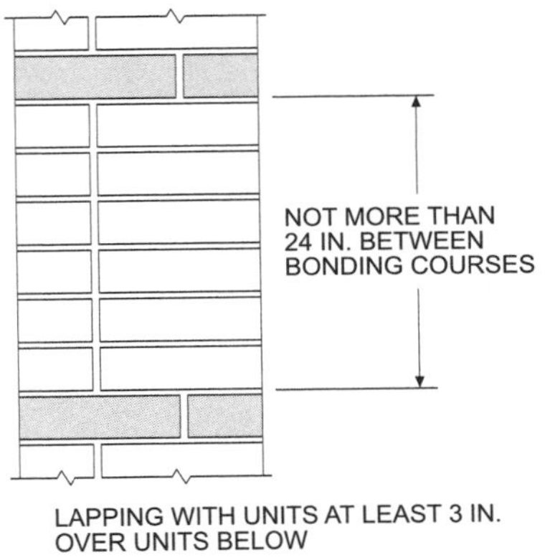

For SI: 1 inch = 25.4 mm.

Figure R608.1.1.1
BONDING WITH SOLID UNIT MASONRY

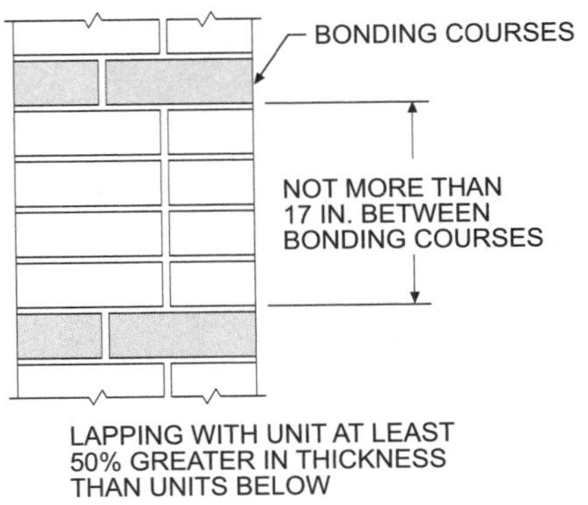

For SI: 1 inch = 25.4 mm.

Figure R608.1.1.2
BONDING WITH HOLLOW UNIT MASONRY

wall in question. Additional bonding ties are required around wall openings, such as windows or doors. These wire ties are used to anchor wythes of masonry together to resist loads or to transfer loads across air spaces to the backing wall material [see Commentary Figures R608.1.2.1(1) and R608.1.2.1(2)].

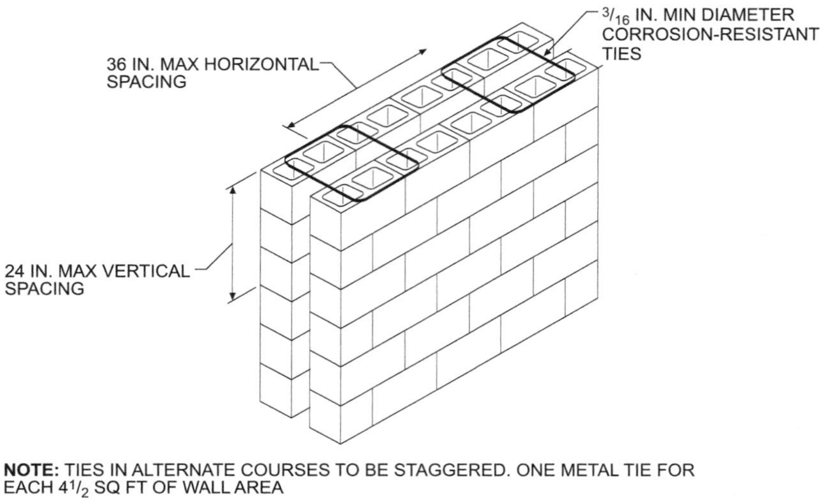

For SI: 1 inch = 25.4 mm, 1 square foot = 0.0929 m^2.

Figure R608.1.2.1(1)
MASONRY BONDING—METAL TIES

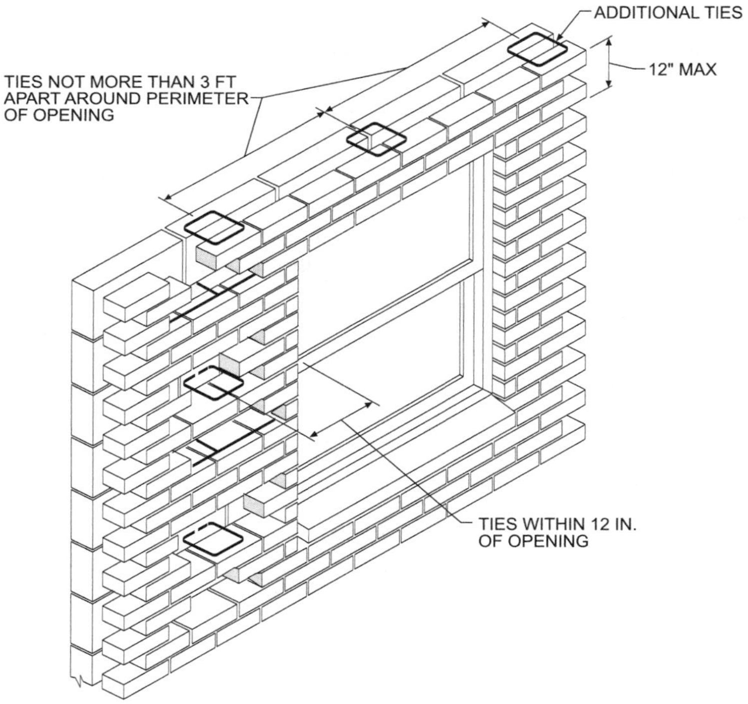

For SI: 1 inch = 25.4 mm, 1 foot = 304.8 mm.

Figure R608.1.2.1(2)
BONDING AROUND OPENING IN MASONRY WALLS

R608.1.2.2 Bonding with adjustable wall ties. Where the facing and backing (adjacent wythes) of masonry are bonded with adjustable wall ties, there shall be at least one tie for each 2.67 square feet (0.248 m^2) of wall area. Neither the vertical nor the horizontal spacing of the adjustable wall ties shall exceed 24 inches (610 mm). The maximum vertical offset of bed joints from one wythe to the other shall be 1.25 inches (32 mm). The maximum clearance between connecting parts of the ties shall be $^1/_{16}$ inch (2 mm). When pintle legs are used, ties shall have at least two $^3/_{16}$-inch-diameter (5 mm) legs.

❖ This section pertains to the use of adjustable ties for anchoring the facing to the backing. Reduced tie spacing is required when these types of ties are specified. Maximum vertical offsets are prescribed so that the adjustable tie will not disengage through use. Many forms of adjustable wall ties are available for masonry construction (see Commentary Figure R608.1.2.2).

R608.1.2.3 Bonding with prefabricated joint reinforcement. Where the facing and backing (adjacent wythes) of masonry are bonded with prefabricated joint reinforcement, there shall be at least one cross wire serving as a tie for each 2.67 square feet (0.248 m^2) of wall area. The vertical spacing of the joint reinforcement shall not exceed 16 inches (406 mm). Cross wires on prefabricated joint reinforcement shall not be smaller than No. 9 gage. The longitudinal wires shall be embedded in the mortar.

❖ Adjacent wythes of masonry could also be bonded with prefabricated joint reinforcements. They can be used in these wall types when air spaces exist between two wythes of masonry, or they can be used in double-wythe grouted-wall construction. Cross wires must be of minimum size to provide proper load transfer. This reinforcement is placed in the bed joints. Commentary Figure R608.1.2.3 illustrates two of the more common types available.

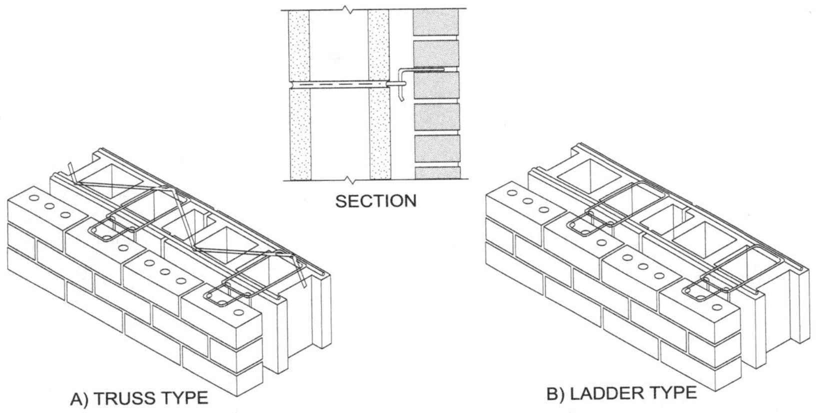

Figure R608.1.2.2
BONDING WITH ADJUSTABLE WALL TIES

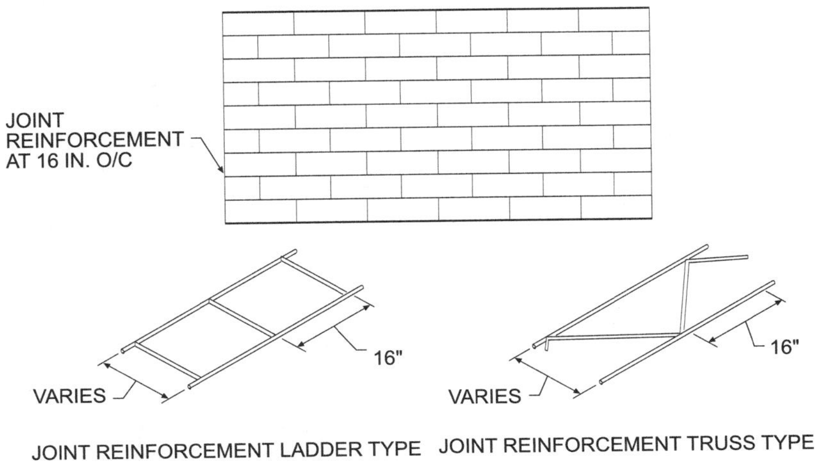

For SI: 1 inch = 25.4 mm.

Figure R608.1.2.3
BONDING WITH PREFABRICATED JOINT REINFORCEMENT

R608.1.3 Bonding with natural or cast stone. Bonding with natural and cast stone shall conform to Sections R608.1.3.1 and R608.1.3.2.

❖ Ashlar masonry and rubble stone masonry require bonding through masonry headers. Although quite different from the usual masonry headers for clay or concrete masonry walls, bonder stones are required based on the square footage of wall area. Bonder stones must be embedded in mortar and have the proper extension into the wall surfaces on both sides. Each masonry type has separate bonding stone spacing requirements based on the construction. The bonding requirements in this section are primarily drawn from TMS 402/ACI 530/ASCE 5.

R608.1.3.1 Ashlar masonry. In ashlar masonry, bonder units, uniformly distributed, shall be provided to the extent of not less than 10 percent of the wall area. Such bonder units shall extend not less than 4 inches (102 mm) into the backing wall.

❖ When one or both of the wythes in a multiwythe masonry wall system are laid in an ashlar pattern, the wythes must be tied together with masonry bonder units spaced as prescribed in this section.

R608.1.3.2 Rubble stone masonry. Rubble stone masonry 24 inches (610 mm) or less in thickness shall have bonder units with a maximum spacing of 3 feet (914 mm) vertically and 3 feet (914 mm) horizontally, and if the masonry is of greater thickness than 24 inches (610 mm), shall have one bonder unit for each 6 square feet (0.557 m^2) of wall surface on both sides.

❖ See the commentary for Section R608.1.3.

R608.2 Masonry bonding pattern. Masonry laid in running and stack bond shall conform to Sections R608.2.1 and R608.2.2.

❖ Masonry must be laid in running bond (see the definition in Chapter 2) or be reinforced horizontally as specified in this section.

R608.2.1 Masonry laid in running bond. In each wythe of masonry laid in running bond, head joints in successive courses shall be offset by not less than one-fourth the unit length, or the masonry walls shall be reinforced longitudinally as required in Section R608.2.2.

❖ When masonry block is placed with an offset, it is called running bond [see Commentary Figure R608.1.2.1(1)]. Running bond is the most commonly used bonding pattern for masonry wall construction.

R608.2.2 Masonry laid in stack bond. Where unit masonry is laid with less head joint offset than in Section R608.2.1, the minimum area of horizontal reinforcement placed in mortar bed joints or in bond beams spaced not more than 48 inches (1219 mm) apart, shall be 0.0007 times the vertical cross-sectional area of the wall.

❖ Blocks stacked vertically are in the stack bond pattern. Refer to the definition in Section R202 of the code and Commentary Figure R606.8(1). The amount of steel in this section is an arbitrary amount to provide continuity across the head joints.

SECTION R609
GROUTED MASONRY

R609.1 General. Grouted multiple-wythe masonry is a form of construction in which the space between the wythes is solidly filled with grout. It is not necessary for the cores of masonry units to be filled with grout. Grouted hollow unit masonry is a form of construction in which certain cells of hollow units are continuously filled with grout.

❖ The provisions of Section R609 pertain to grouted masonry. The two types of grouted masonry, grouted multiple-wythe and grouted hollow unit masonry, are defined in the wording of this section and illustrated in Commentary Figure R609.1. To provide for an adequate bond between the masonry units and the grout, the units must be free of excessive dust or dirt.

R609.1.1 Grout. Grout shall consist of cementitious material and aggregate in accordance with ASTM C 476 and the proportion specifications of Table R609.1.1. Type M or Type S mortar to which sufficient water has been added to produce pouring consistency can be used as grout.

❖ ASTM C 476 contains the requirements for grout. The code identifies two types of grout for masonry construction: fine grout and coarse grout. They differ primarily in the maximum allowable size of aggregates. The proportions in Table R609.1.1 provide a minimum compressive strength of 2,000 psi (13 789 kPa). Type M or Type S mortar (see Section R607.1) may also be used as grout if sufficient water is added to produce a pouring consistency.

R609.1.2 Grouting requirements. Maximum pour heights and the minimum dimensions of spaces provided for grout placement shall conform to Table R609.1.2. If the work is stopped for one hour or longer, the horizontal construction joints shall be formed by stopping all tiers at the same elevation and with the grout 1 inch (25 mm) below the top.

❖ The selection of fine or coarse grout is made based on the size of grout space and the height of grout pour. Table R609.1.2 covers requirements for selection of grout type. Fine grout is required where grout space is small, narrow or too congested with reinforcing steel. Fine grout also can be used where coarse grout is permitted. It is preferable, however, to use coarse grout, because it reduces shrinkage and is less expensive. See Commentary Figure R609.1.2 for an illustration of grout placement.

R609.1.3 Grout space (cleaning). Provision shall be made for cleaning grout space. Mortar projections that project more than $^1/_2$ inch (13 mm) into grout space and any other foreign matter shall be removed from grout space prior to inspection and grouting.

❖ See the commentary for Section R609.1.2.

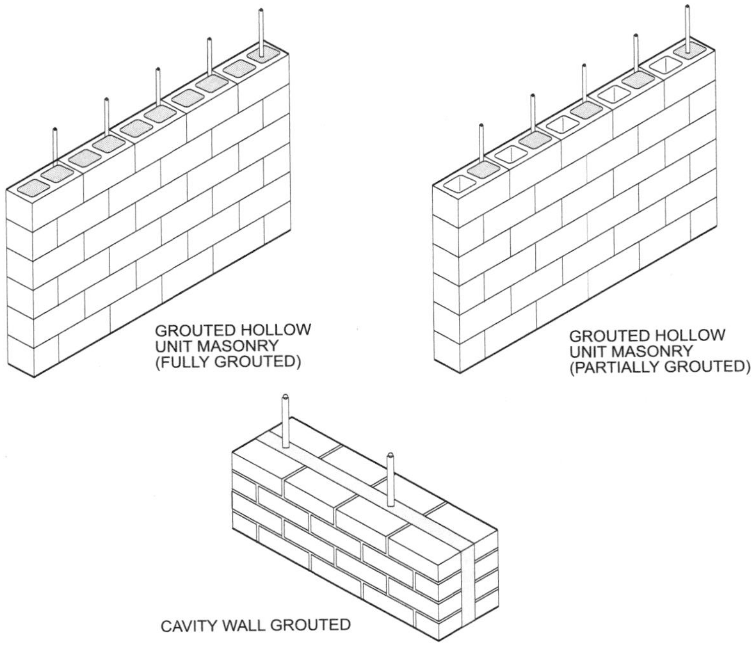

Figure R609.1
GROUTED MULTIPLE-WYTHE MASONRY

TABLE R609.1.1
GROUT PROPORTIONS BY VOLUME FOR MASONRY CONSTRUCTION

TYPE	PORTLAND CEMENT OR BLENDED CEMENT SLAG CEMENT	HYDRATED LIME OR LIME PUTTY	AGGREGATE MEASURED IN A DAMP, LOOSE CONDITION	
			Fine	Coarse
Fine	1	0 to 1/10	$2^1/_4$ to 3 times the sum of the volume of the cementitious materials	—
Coarse	1	0 to 1/10	$2^1/_4$ to 3 times the sum of the volume of the cementitious materials	1 to 2 times the sum of the volumes of the cementitious materials

❖ See the commentary for Section R609.1.1.

TABLE R609.1.2
GROUT SPACE DIMENSIONS AND POUR HEIGHTS

GROUT TYPE	GROUT POUR MAXIMUM HEIGHT (feet)	MINIMUM WIDTH OF GROUT SPACES[a, b] (inches)	MINIMUM GROUT[b, c] SPACE DIMENSIONS FOR GROUTING CELLS OF HOLLOW UNITS (inches × inches)
Fine	1	0.75	1.5 × 2
	5	2	2 × 3
	12	2.5	2.5 × 3
	24	3	3 × 3
Coarse	1	1.5	1.5 × 3
	5	2	2.5 × 3
	12	2.5	3 × 3
	24	3	3 × 4

For SI: 1 inch = 25.4 mm, 1 foot = 304.8 mm.

a. For grouting between masonry wythes.

b. Grout space dimension is the clear dimension between any masonry protrusion and shall be increased by the horizontal projection of the diameters of the horizontal bars within the cross section of the grout space.

c. Area of vertical reinforcement shall not exceed 6 percent of the area of the grout space.

❖ See the commentary for Section R609.1.2.

R609.1.4 Grout placement. Grout shall be a plastic mix suitable for pumping without segregation of the constituents and shall be mixed thoroughly. Grout shall be placed by pumping or by an *approved* alternate method and shall be placed before any initial set occurs and in no case more than $1^1/_2$ hours after water has been added. Grouting shall be done in a continuous pour, in lifts not exceeding 5 feet (1524 mm). It shall be consolidated by puddling or mechanical vibrating during placing and reconsolidated after excess moisture has been absorbed but before plasticity is lost.

❖ See Commentary Figure R609.1.2.

R609.1.4.1 Grout pumped through aluminum pipes. Grout shall not be pumped through aluminum pipes.

❖ Aluminum can adversely affect grout and unprotected reinforcing steel through galvanic action. This restriction protects against possible contamination during grout placement.

R609.1.5 Cleanouts. Where required by the *building official*, cleanouts shall be provided as specified in this section. The cleanouts shall be sealed before grouting and after inspection.

❖ Cleanouts should be furnished as described in this section. These aid the grout-spacing cleaning required by Section R609.1.3.

R609.1.5.1 Grouted multiple-wythe masonry. Cleanouts shall be provided at the bottom course of the exterior wythe at each pour of grout where such pour exceeds 5 feet (1524 mm) in height.

❖ See the commentary for Section R609.1.5.

R609.1.5.2 Grouted hollow unit masonry. Cleanouts shall be provided at the bottom course of each cell to be grouted at each pour of grout, where such pour exceeds 4 feet (1219 mm) in height.

❖ See the commentary for Section R609.1.5.

R609.2 Grouted multiple-wythe masonry. Grouted multiple-wythe masonry shall conform to all the requirements specified in Section R609.1 and the requirements of this section.

❖ Grouted multiple-wythe masonry walls must comply with this section.

R609.2.1 Bonding of backup wythe. Where all interior vertical spaces are filled with grout in multiple-wythe construction, masonry headers shall not be permitted. Metal wall ties shall be used in accordance with Section R608.1.2 to prevent spreading of the wythes and to maintain the vertical alignment of the wall. Wall ties shall be installed in accordance with Section R608.1.2 when the backup wythe in multiple-wythe construction is fully grouted.

❖ Masonry headers are not permitted in double-wythe grouted masonry construction because they span across the wythes and would interfere with the placement of the grout. Metal ties in accordance with Section R608.1.2 of the code prevent the spreading of wythes and maintain alignment during the grouting process.

R609.2.2 Grout spaces. Fine grout shall be used when interior vertical space to receive grout does not exceed 2 inches (51 mm) in thickness. Interior vertical spaces exceeding 2 inches (51 mm) in thickness shall use coarse or fine grout.

❖ Fine grout must be used for grout spaces not more than 2 inches (51 mm) wide. This clearance should

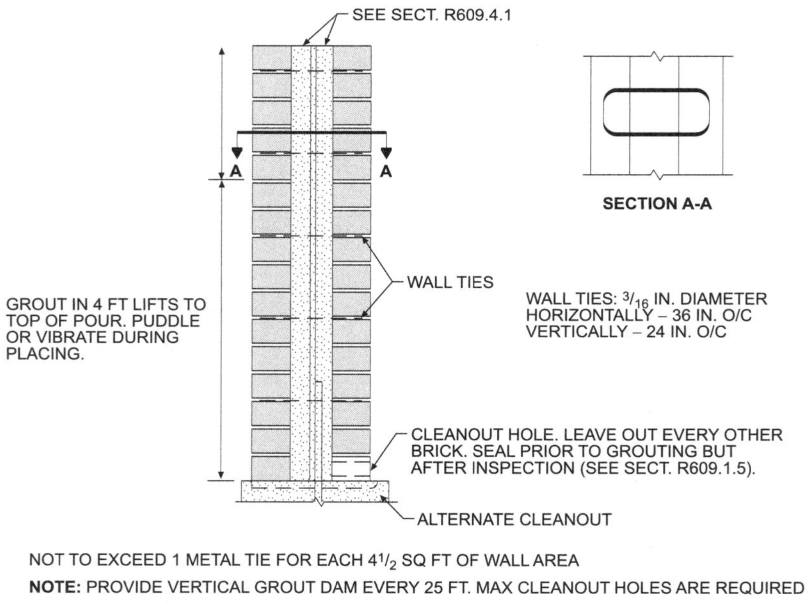

For SI: 1 inch = 25.4 mm, 1 foot = 304.8 mm.
1 square foot = 0.0929 m^2.

Figure R609.1.2
GROUTED MULTIPLE-WYTHE MASONRY

account for any horizontal wall reinforcing. ASTM C 476 differentiates between fine and coarse grout.

R609.2.3 Grout barriers. Vertical grout barriers or dams shall be built of *solid masonry* across the grout space the entire height of the wall to control the flow of the grout horizontally. Grout barriers shall not be more than 25 feet (7620 mm) apart. The grouting of any section of a wall between control barriers shall be completed in one day with no interruptions greater than one hour.

❖ Grout barriers allow large walls to be properly grouted in sections.

R609.3 Reinforced grouted multiple-wythe masonry. Reinforced grouted multiple-wythe masonry shall conform to all the requirements specified in Sections R609.1 and R609.2 and the requirements of this section.

❖ Reinforced grouted multiple-wythe masonry walls must comply with this section in addition to Section R609.2.

R609.3.1 Construction. The thickness of grout or mortar between masonry units and reinforcement shall not be less than $^1/_4$ inch (7 mm), except that $^1/_4$-inch (7 mm) bars may be laid in horizontal mortar joints at least $^1/_2$ inch (13 mm) thick, and steel wire reinforcement may be laid in horizontal mortar joints at least twice the thickness of the wire diameter.

❖ When reinforcing is used in grouted masonry, the dimensional constraints given in this section provide clearances necessary for the adequate placement of the grout within the cavity as well as the proper bond between the reinforcement and the grout. These minimums are illustrated in Commentary Figure R609.3.1.

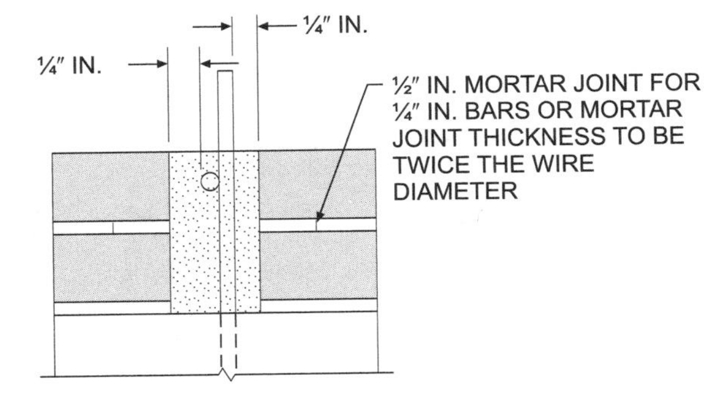

For SI: 1 inch = 25.4 mm.

Figure R609.3.1
REINFORCEMENT PLACEMENT LIMITATIONS

R609.4 Reinforced hollow unit masonry. Reinforced hollow unit masonry shall conform to all the requirements of Section R609.1 and the requirements of this section.

❖ Reinforced hollow-unit masonry uses hollow masonry units with reinforcement embedded in mortar or grout within certain cells throughout the wall length.

R609.4.1 Construction. Requirements for construction shall be as follows:

1. Reinforced hollow-unit masonry shall be built to preserve the unobstructed vertical continuity of the cells to be filled. Walls and cross webs forming cells to be filled shall be full-bedded in mortar to prevent leakage of grout. Head and end joints shall be solidly filled with mortar for a distance in from the face of the wall or unit not less than the thickness of the longitudinal face shells. Bond shall be provided by lapping units in successive vertical courses.
2. Cells to be filled shall have vertical alignment sufficient to maintain a clear, unobstructed continuous vertical cell of dimensions prescribed in Table R609.1.2.
3. Vertical reinforcement shall be held in position at top and bottom and at intervals not exceeding 200 diameters of the reinforcement.
4. Cells containing reinforcement shall be filled solidly with grout. Grout shall be poured in lifts of 8-foot (2438 mm) maximum height. When a total grout pour exceeds 8 feet (2438 mm) in height, the grout shall be placed in lifts not exceeding 5 feet (1524 mm) and special inspection during grouting shall be required.
5. Horizontal steel shall be fully embedded by grout in an uninterrupted pour.

❖ The general purpose of the five requirements contained in the section is to provide for the proper placement of reinforcing steel and the grouting of cells so that the construction will act structurally as a unit.

Item 1 provides for proper laying of hollow masonry units with appropriate mortar beds, which contain the grout (prevents leakage). The item also contains requirements for the grout space in which reinforcement may be placed [see Commentary Figure R609.4.1(1)].

Item 2 specifies minimum vertical cell-dimension limitations to provide for sufficient clear space in which to place the grout as illustrated in Commentary Figure R609.4.1(2).

Following the provision in Item 3 is necessary in order to maintain the vertical alignment of reinforcing steel during the grout pour. Also, the reinforcement will be maintained in a proper position relative to the block thickness as required by the design if the vertical reinforcement is secured at the top, bottom, and intervening points so that the distance between such anchorage does not exceed 200 bar diameters. Commentary Figure R609.4.1(3) illustrates methods of securing the reinforcement.

Items 4 and 5 provide limitations on the method of grouting the cells containing reinforcement so that the cells to be grouted will be filled solidly and the grout will be reasonably consolidated around the reinforcing steel. Refer to Commentary Figures R609.4.1(4) and R609.4.1(5) for illustrations of the provisions.

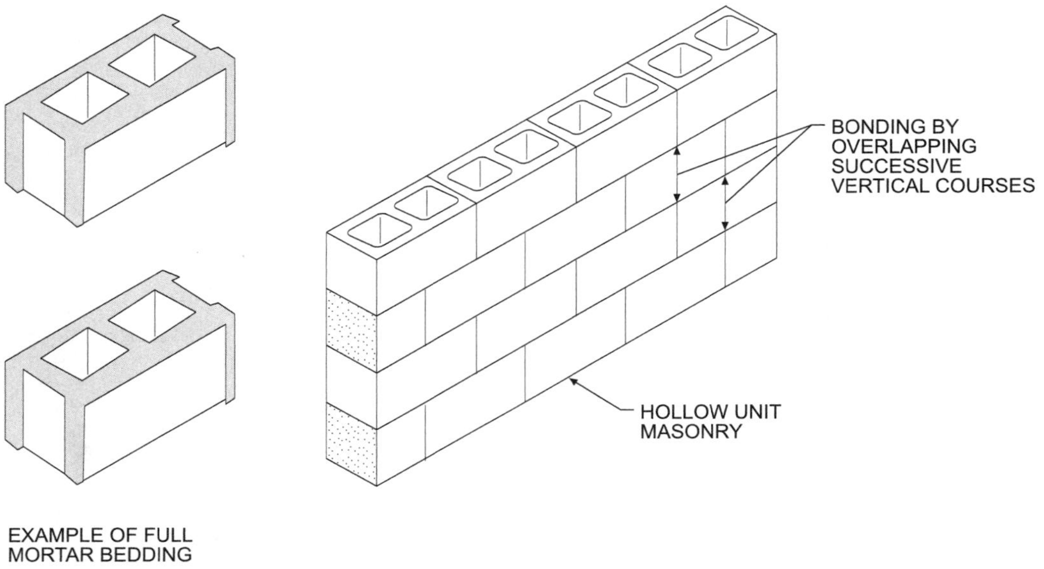

Figure R609.4.1(1)
REINFORCEMENT PLACEMENT LIMITATIONS

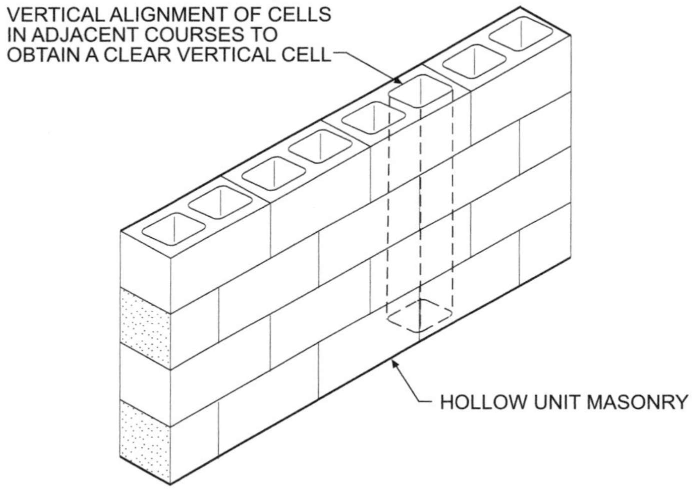

Figure R609.4.1(2)
REINFORCED HOLLOW UNIT MASONRY VERTICAL CELL ALIGNMENT

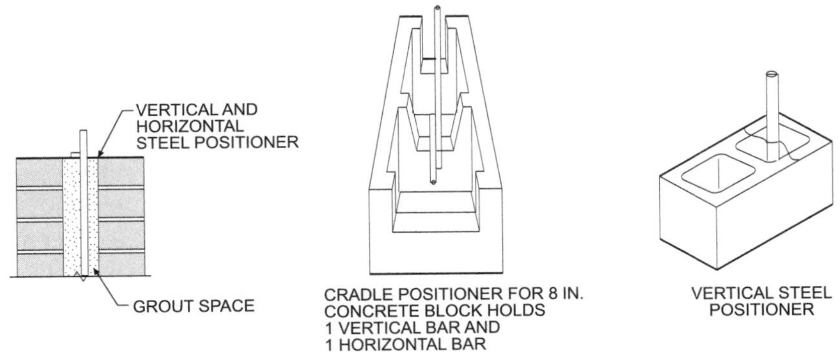

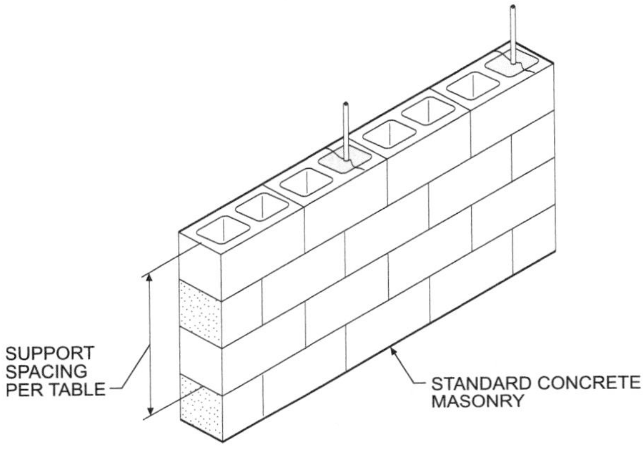

DISTANCE FOR SUPPORT OF VERTICAL STEEL

BAR NO.	BAR SIZE	200 BAR DIAM
3	$^3/_8$ IN.	75 IN.
4	$^1/_2$ IN.	100 IN.
5	$^5/_8$ IN.	125 IN.
6	$^3/_4$ IN.	150 IN.
7	$^7/_8$ IN.	175 IN.

For SI: 1 inch = 25.4 mm.

Figure R609.4.1(3)
MASONRY REINFORCING STEEL ALIGNMENT

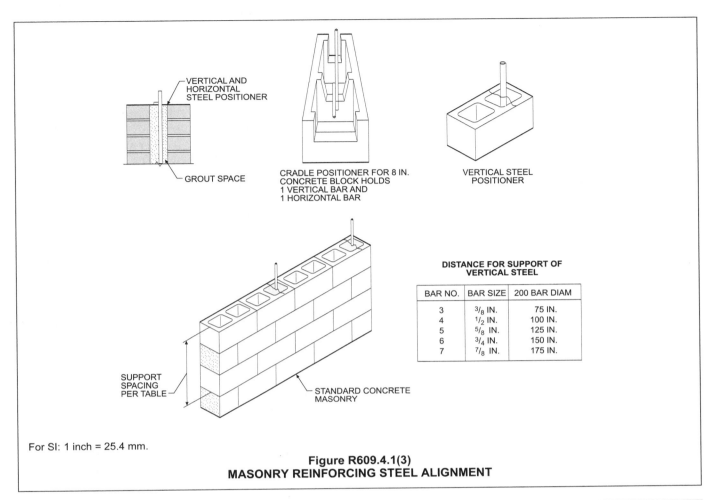

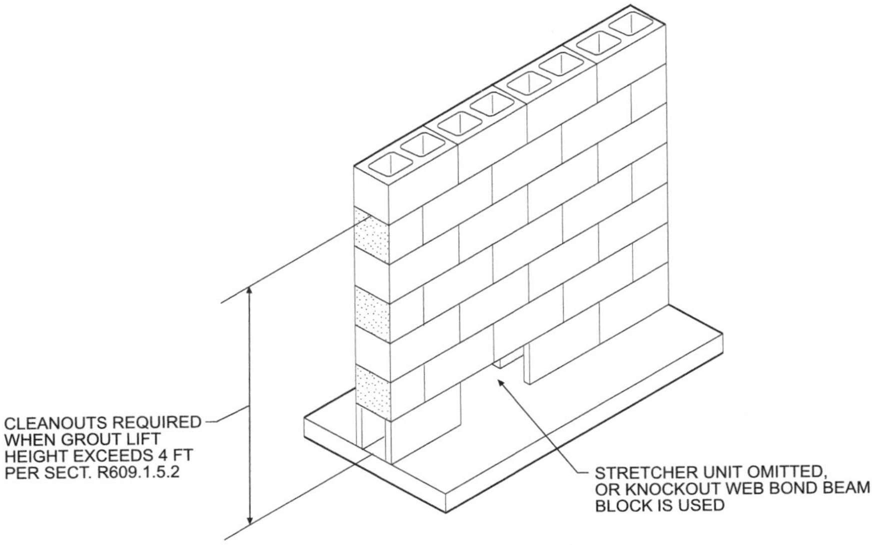

For SI: 1 inch = 25.4 mm, 1 foot = 304.8 mm.

Figure R609.4.1(4)
REINFORCED HOLLOW UNIT MASONRY CLEANOUTS

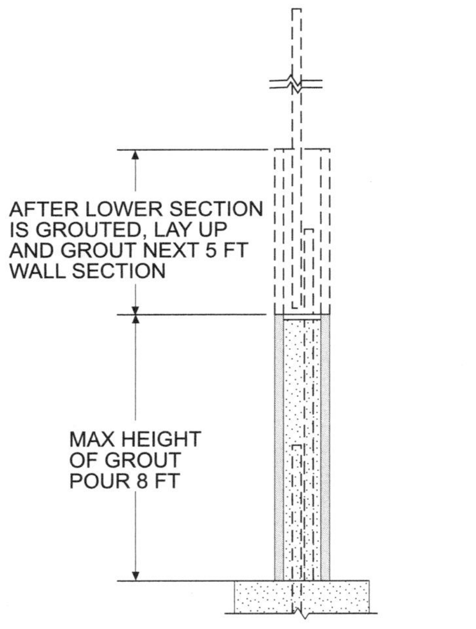
AFTER LOWER SECTION IS GROUTED, LAY UP AND GROUT NEXT 5 FT WALL SECTION
MAX HEIGHT OF GROUT POUR 8 FT
NOTE: PUDDLE ALL GROUT AFTER POURING

For SI: 1 foot = 304.8 mm.

Figure R609.4.1(5)
MASONRY REINFORCING STEEL ALIGNMENT

SECTION R610
GLASS UNIT MASONRY

R610.1 General. Panels of glass unit masonry located in load-bearing and nonload-bearing exterior and interior walls shall be constructed in accordance with this section.

❖ Code provisions for glass unit masonry are empirical. By definition (see the definition for "Masonry unit, glass" in Section R202) these units are nonload-bearing. Therefore, when they are used in a wall opening, a properly designed header or lintel must be installed. Furthermore, panels of glass unit masonry must be isolated from the structure so that in-plane loads are not transferred to the glass.

R610.2 Materials. Hollow glass units shall be partially evacuated and have a minimum average glass face thickness of $^3/_{16}$ inch (5 mm). The surface of units in contact with mortar shall be treated with a polyvinyl butyral coating or latex-based paint. The use of reclaimed units is prohibited.

❖ Glass masonry units may be hollow or solid. Hollow units must have the required minimum face thickness. The materials specified for treating edges in contact with mortar improve the bonding between glass blocks and mortar.

R610.3 Units. Hollow or solid glass block units shall be standard or thin units.

❖ Hollow or solid-glass units are classified as either standard units or thin units.

R610.3.1 Standard units. The specified thickness of standard units shall be at least $3^7/_8$ inches (98 mm).

❖ See the commentary for Section R610.3.

R610.3.2 Thin units. The specified thickness of thin units shall be at least $3^1/_8$ inches (79 mm) for hollow units and at least 3 inches (76 mm) for solid units.

❖ See the commentary for Section R610.3.

R610.4 Isolated panels. Isolated panels of glass unit masonry shall conform to the requirements of this section.

❖ The code limits hollow glass unit masonry panel sizes based on structural and performance considerations. Height limits are more restrictive than length limits, primarily as a result of historical requirements rather than actual field experience or engineering principles. Exterior panels constructed of standard units may be up to 144 square feet (13.4 m^2). Exterior panels constructed of thin units are limited to 85 square feet (7.9 m^2). Interior panels are limited to 250 square feet (23.2 m^2) for standard units and 150 square feet (13.9 m^2) for thin units. These limitations are derived from TMS 402/ACI 530/ASCE 5.

R610.4.1 Exterior standard-unit panels. The maximum area of each individual standard-unit panel shall be 144 square feet (13.4 m^2) when the design wind pressure is 20 psf (958 Pa). The maximum area of such panels subjected to design wind pressures other than 20 psf (958 Pa) shall be in accordance with Figure R610.4.1. The maximum panel dimension between structural supports shall be 25 feet (7620 mm) in width or 20 feet (6096 mm) in height.

❖ The area limit specified in this section corresponds to a wind pressure of 20 psf (958 Pa). The area should be adjusted using Figure R610.4.1 for other values of wind pressure (see the commentary, Section R610.4).

R610.4.2 Exterior thin-unit panels. The maximum area of each individual thin-unit panel shall be 85 square feet (7.9 m^2). The maximum dimension between structural supports shall be 15 feet (4572 mm) in width or 10 feet (3048 mm) in height. Thin units shall not be used in applications where the design wind pressure as stated in Table R301.2(1) exceeds 20 psf (958 Pa).

❖ See the commentary for Section R610.4.

R610.4.3 Interior panels. The maximum area of each individual standard-unit panel shall be 250 square feet (23.2 m^2). The maximum area of each thin-unit panel shall be 150 square feet (13.9 m^2). The maximum dimension between structural supports shall be 25 feet (7620 mm) in width or 20 feet (6096 mm) in height.

❖ See the commentary for Section R610.4.

R610.4.4 Curved panels. The width of curved panels shall conform to the requirements of Sections R610.4.1, R610.4.2 and R610.4.3, except additional structural supports shall be provided at locations where a curved section joins a straight section, and at inflection points in multicurved walls.

❖ Support is required for standard and thin units, exterior and interior panels, at locations where a curved panel joins a straight section, and at inflection points (see the commentary for Section R610.4).

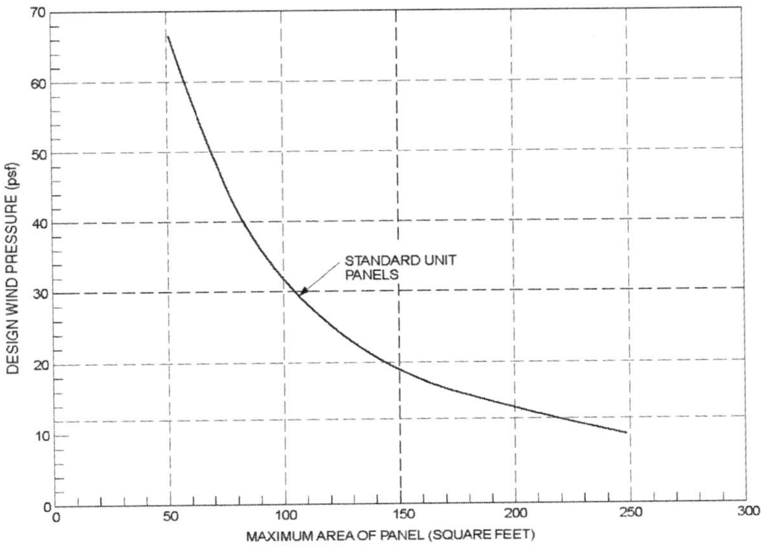

For SI: 1 square foot = 0.0929 m², 1 pound per square foot = 0.0479 kPa.

FIGURE R610.4.1
GLASS UNIT MASONRY DESIGN WIND LOAD RESISTANCE

❖ See the commentary for Section R610.4.1.

R610.5 Panel support. Glass unit masonry panels shall conform to the support requirements of this section.

❖ Structural members supporting glass-unit masonry panels must meet the deflection limit of this section to minimize the potential for cracking in the glass panels. Lateral support is required for the top and sides of panels and may be accomplished by anchors or channel-type restraints.

R610.5.1 Deflection. The maximum total deflection of structural members that support glass unit masonry shall not exceed $^1/_{600}$.

❖ See the commentary for Section R610.5.

R610.5.2 Lateral support. Glass unit masonry panels shall be laterally supported along the top and sides of the panel. Lateral supports for glass unit masonry panels shall be designed to resist a minimum of 200 pounds per lineal feet (2918 N/m) of panel, or the actual applied loads, whichever is greater. Except for single unit panels, lateral support shall be provided by panel anchors along the top and sides spaced a maximum of 16 inches (406 mm) on center or by channel-type restraints. Single unit panels shall be supported by channel-type restraints.

Exceptions:

1. Lateral support is not required at the top of panels that are one unit wide.
2. Lateral support is not required at the sides of panels that are one unit high.

❖ See the commentary for Section R610.5.

R610.5.2.1 Panel anchor restraints. Panel anchors shall be spaced a maximum of 16 inches (406 mm) on center in both jambs and across the head. Panel anchors shall be embedded a minimum of 12 inches (305 mm) and shall be provided with two fasteners so as to resist the loads specified in Section R610.5.2.

❖ See Commentary Figure R610.5.2.1.

R610.5.2.2 Channel-type restraints. Glass unit masonry panels shall be recessed at least 1 inch (25 mm) within channels and chases. Channel-type restraints shall be oversized to accommodate expansion material in the opening, packing and sealant between the framing restraints, and the glass unit masonry perimeter units.

❖ This type of lateral support is an option to the anchors described in the previous section. No specific material is specified for the channel, which means that any channel-type restraint providing the lateral support specified by Section R610.5.2 is acceptable.

R610.6 Sills. Before bedding of glass units, the sill area shall be covered with a water base asphaltic emulsion coating. The coating shall be a minimum of $^1/_8$ inch (3 mm) thick.

❖ Application of asphaltic emulsion is recommended by glass block manufacturers. See Commentary Figure R610.5.2.1.

R610.7 Expansion joints. Glass unit masonry panels shall be provided with expansion joints along the top and sides at all structural supports. Expansion joints shall be a minimum of $^3/_8$ inch (10 mm) in thickness and shall have sufficient thickness to accommodate displacements of the supporting structure. Expansion joints shall be entirely free of mortar and other debris and shall be filled with resilient material.

❖ See Commentary Figure R610.5.2.1.

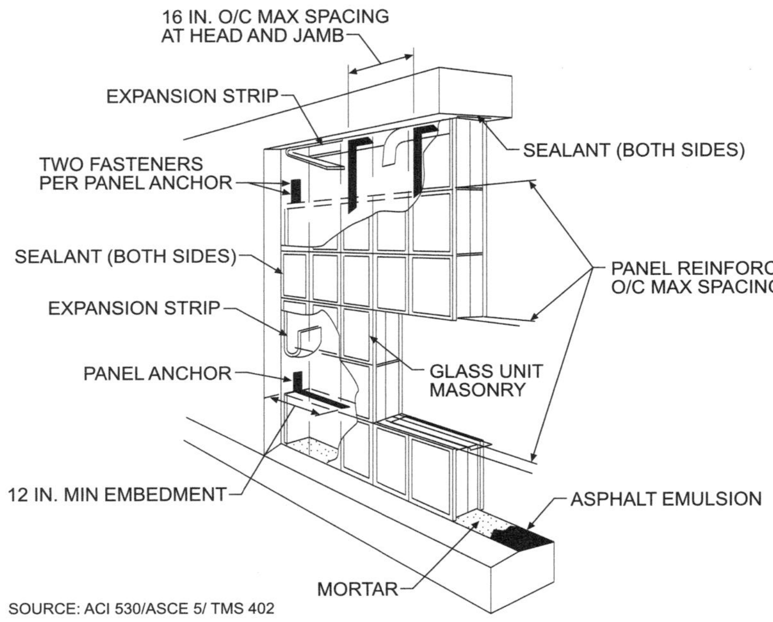

For SI: 1 inch = 25.4 mm.

Figure R610.5.2.1
GLASS UNIT MASONRY

R610.8 Mortar. Glass unit masonry shall be laid with Type S or N mortar. Mortar shall not be retempered after initial set. Mortar unused within $1^1/_2$ hours after initial mixing shall be discarded.

❖ See the commentary for Section R607.1.

R610.9 Reinforcement. Glass unit masonry panels shall have horizontal joint reinforcement spaced a maximum of 16 inches (406 mm) on center located in the mortar bed joint. Horizontal joint reinforcement shall extend the entire length of the panel but shall not extend across expansion joints. Longitudinal wires shall be lapped a minimum of 6 inches (152 mm) at splices. Joint reinforcement shall be placed in the bed joint immediately below and above openings in the panel. The reinforcement shall have not less than two parallel longitudinal wires of size W1.7 or greater, and have welded cross wires of size W1.7 or greater.

❖ See Commentary Figure R610.5.2.1.

R610.10 Placement. Glass units shall be placed so head and bed joints are filled solidly. Mortar shall not be furrowed. Head and bed joints of glass unit masonry shall be $^1/_4$ inch (6.4 mm) thick, except that vertical joint thickness of radial panels shall not be less than $^1/_8$ inch (3 mm) or greater than $^5/_8$ inch (16 mm). The bed joint thickness tolerance shall be minus $^1/_{16}$ inch (1.6 mm) and plus $^1/_8$ inch (3 mm). The head joint thickness tolerance shall be plus or minus $^1/_8$ inch (3 mm).

❖ This section specifies the joint thicknesses and tolerances consistent with TMS 402/ACI 530/ASCE 5.

SECTION R611
EXTERIOR CONCRETE WALL CONSTRUCTION

R611.1 General. Exterior concrete walls shall be designed and constructed in accordance with the provisions of this section or in accordance with the provisions of PCA 100 or ACI 318. When PCA 100, ACI 318 or the provisions of this section are used to design concrete walls, project drawings, typical details and specifications are not required to bear the seal of the architect or engineer responsible for design, unless otherwise required by the state law of the jurisdiction having authority.

❖ These provisions are for exterior concrete walls which should be designed in accordance with this section or under one of the following design documents:

ACI 318: *Building Code Requirements for Structural Concrete*

PCA100: *Prescriptive Design of Exterior Concrete Walls for One- and Two-family Dwellings*

Exterior concrete walls are usually designed and constructed to carry the vertical loads from the structure above and to resist wind and seismic lateral forces. Walls must be cast-in-place using either removable or stay-in-place forming systems. This section provides requirements, primarily in tables and figures, that will permit the exterior wall of typical homes to be constructed of concrete without the added expense of hiring a registered design professional.

Most states do not require the seal of an architect or engineer for small residential buildings. Check with the state authorities in the location of the project for local regulations.

R611.1.1 Interior construction. These provisions are based on the assumption that interior walls and partitions, both load-bearing and nonload-bearing, floors and roof/ceiling assemblies are constructed of *light-framed construction* complying with the limitations of this code and the additional limitations of Section R611.2. Design and construction of light-framed assemblies shall be in accordance with the applicable provisions of this code. Where second-story exterior walls are of *light-framed construction*, they shall be designed and constructed as required by this code.

Aspects of concrete construction not specifically addressed by this code, including interior concrete walls, shall comply with ACI 318.

❖ The provisions of Section R611 are based on the assumption that the exterior concrete walls would be used in combination with floors, roofs and interior walls of light-framed construction complying with the prescriptive requirements of the code. Light-framed construction includes members of wood or cold-formed steel.

While Section R611 does not address the use of interior walls of concrete construction, in some cases such walls can be used without adversely affecting the performance of buildings designed in accordance with this section. Interior concrete walls must be designed in accordance with the code or ACI 318.

R611.1.2 Other concrete walls. Exterior concrete walls constructed in accordance with this code shall comply with the shapes and minimum concrete cross-sectional dimensions of Table R611.3. Other types of forming systems resulting in concrete walls not in compliance with this section shall be designed in accordance with ACI 318.

❖ It is intended that portions of the building and aspects of concrete construction not specifically addressed by Section R611 be designed and constructed in accordance with ACI 318.

R611.2 Applicability limits. The provisions of this section shall apply to the construction of exterior concrete walls for buildings not greater than 60 feet (18 288 mm) in plan dimensions, floors with clear spans not greater than 32 feet (9754 mm) and roofs with clear spans not greater than 40 feet (12 192 mm). Buildings shall not exceed 35 feet (10 668 mm) in mean roof height or two stories in height above-grade. Floor/ceiling dead loads shall not exceed 10 pounds per square foot (479 Pa), roof/ceiling dead loads shall not exceed 15 pounds per square foot (718 Pa) and *attic* live loads shall not exceed 20 pounds per square foot (958 Pa). Roof overhangs shall not exceed 2 feet (610 mm) of horizontal projection beyond the exterior wall and the dead load of the overhangs shall not exceed 8 pounds per square foot (383 Pa).

Walls constructed in accordance with the provisions of this section shall be limited to buildings subjected to a maximum design wind speed of 130 miles per hour (58 m/s) Exposure B, 110 miles per hour (49 m/s) Exposure C and 100 miles per hour (45 m/s) Exposure D. Walls constructed in accordance with the provisions of this section shall be limited to detached one- and two-family *dwellings* and townhouses assigned to Seismic Design Category A or B, and detached one- and two-family *dwellings* assigned to Seismic Design Category C.

Buildings that are not within the scope of this section shall be designed in accordance with PCA 100 or ACI 318.

❖ The requirements set forth in Section R611 apply only to the construction of houses that meet the limits set forth in Section R611.2 (see Commentary Figure R611.2). The applicability limits are necessary for defining reasonable boundaries to the conditions that must be considered in developing prescriptive construction requirements. This section, however, does not limit the application of alternative methods or materials through engineering design by a registered design professional.

The applicability limits are based on industry convention and experience. Detailed applicability limits were documented in the process of developing prescriptive design requirements for various elements of the structure. In some cases, engineering sensitivity analyses were performed to help define appropriate limits.

The applicability limits strike a reasonable balance among engineering theory, available test data, and proven field practices for typical residential construction applications. They are intended to prevent misapplication while at the same time addressing a reasonably large percentage of new housing conditions.

The limits of Section R611.2 apply to detached one- or two-family dwellings, and townhouses, with maximum building plan dimensions of 60 feet (18 288 mm), that are not more than two stories in height above grade, excluding a basement that is not considered a story above grade plane.

The provisions for above-grade exterior concrete walls are limited to buildings where the maximum design wind speed is 130 miles per hour (58 m/s) Exposure B, 110 miles per hour (49 m/s) Exposure C and 100 miles per hour (45 m/s) Exposure D. Also, the provisions are limited to detached one- and two-family dwellings and townhouses in SDC A or B, and detached one- and two-family dwellings in SDC C. One- and two-family dwellings located in higher design wind speed regions and requiring seismic design will be required to be designed in accordance with PCA 100 or ACI 318. Townhouses in SDC C and all buildings in SDC D must meet the requirements in PCA Standard 100 or be designed in accordance with ACI 318.

R611.3 Concrete wall systems. Concrete walls constructed in accordance with these provisions shall comply with the shapes and minimum concrete cross-sectional dimensions of Table R611.3.

❖ The minimum thickness and cross-section dimension of concrete walls must comply with Table R611.3. This section prescribes the dimensional requirements of exterior concrete walls. The requirements for flat, waffle-grid and screen-gird wall systems are prescribed.

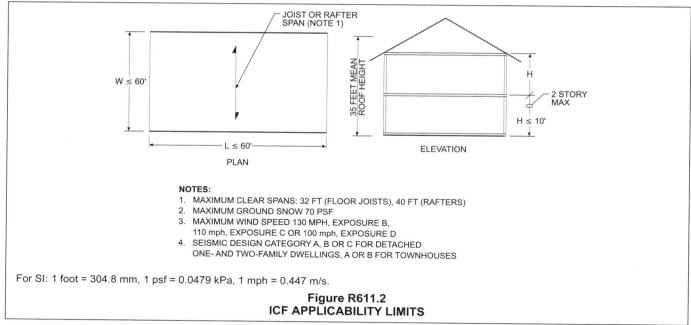

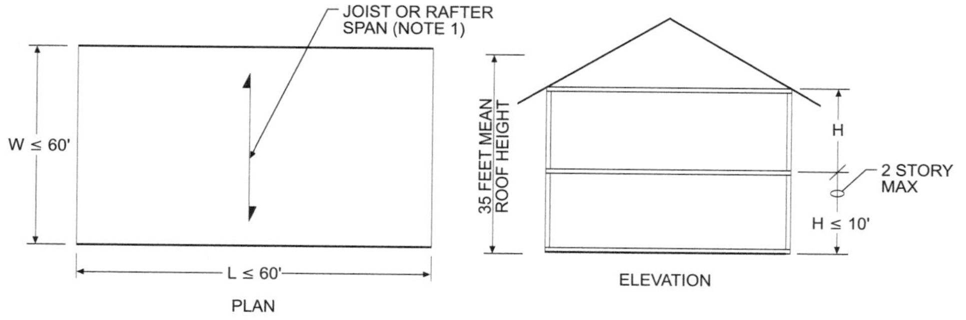

NOTES:
1. MAXIMUM CLEAR SPANS: 32 FT (FLOOR JOISTS), 40 FT (RAFTERS)
2. MAXIMUM GROUND SNOW 70 PSF
3. MAXIMUM WIND SPEED 130 MPH, EXPOSURE B, 110 mph, EXPOSURE C OR 100 mph, EXPOSURE D
4. SEISMIC DESIGN CATEGORY A, B OR C FOR DETACHED ONE- AND TWO-FAMILY DWELLINGS, A OR B FOR TOWNHOUSES

For SI: 1 foot = 304.8 mm, 1 psf = 0.0479 kPa, 1 mph = 0.447 m/s.

Figure R611.2
ICF APPLICABILITY LIMITS

For flat wall systems refer to Figure R611.3(1). Unlike masonry units, which must be manufactured in accordance with an ASTM standard, there is no standard to require the minimum thickness of nominal 6-, 8-, 10-, or 12-inch (152, 203, 254 or 305 mm) flat concrete walls. The actual as-built minimum thickness is permitted to be no more than $^1/_2$ inch (12.7 mm) less than the nominal thickness (see Note d of Table R611.3).

The design strength of flat concrete walls is based on thicknesses of 5.5, 7.5, 9.5 and 11.5 inches (140, 191, 241 and 292 mm).

For waffle-grid and screen-grid systems, refer to Figures R611.3(2) and R611.3(3). For waffle-grid and screen-grid minimum thickness, see Notes e, f and g of Table R611.3.

R611.3.1 Flat wall systems. Flat concrete wall systems shall comply with Table R611.3 and Figure R611.3(1) and have a minimum nominal thickness of 4 inches (102 mm).

❖ See the commentary for Section R611.3.

R611.3.2 Waffle-grid wall systems. Waffle-grid wall systems shall comply with Table R611.3 and Figure R611.3(2). and shall have a minimum nominal thickness of 6 inches (152 mm) for the horizontal and vertical concrete members (cores). The core and web dimensions shall comply with Table R611.3. The maximum weight of waffle-grid walls shall comply with Table R611.3.

❖ See the commentary for Section R611.3.

R611.3.3 Screen-grid wall systems. Screen-grid wall systems shall comply with Table R611.3 and Figure R611.3(3) and shall have a minimum nominal thickness of 6 inches (152 mm) for the horizontal and vertical concrete members (cores). The core dimensions shall comply with Table R611.3. The maximum weight of screen-grid walls shall comply with Table R611.3.

❖ See the commentary for Section R611.3.

R611.4 Stay-in-place forms. Stay-in-place concrete forms shall comply with this section.

❖ Materials that stay in place after concrete cures must be able to withstand the rigors of the environment. From a thermal standpoint, most stay-in-place form systems and some removable form systems that incorporate insulation into the wall before the concrete is placed provide enough insulating value to meet code requirements. However, some forming systems will require the addition of insulation to the interior or exterior of the concrete wall to meet code requirements for insulating value. The form material, except foam plastic, can be anything that meets the flame spread and smoke-developed indexes of Section R302.9. Foam plastic must comply with Section R316.3. Covering foam plastic that would otherwise be exposed to the interior of the building must comply with Sections R316.4 and R702.3.4. Adhesively attaching the gypsum board to the foam plastic as the only means of fastening is not permitted out of concern that when the adhesive is exposed to higher temperature expected in a fire, it may soften and allow the gypsum board to delaminate from the foam plastic.

It is generally accepted that a monolithic concrete wall is a solid wall through which water and air cannot readily flow; however, there is a possibility that the concrete wall may have drying shrinkage cracks through which water may enter. Small gaps between stay-in-place form blocks are inherent in current screen-grid stay-in-place form walls and may allow water to enter the structure. As a result, a moisture barrier on the exterior face of the stay-in-place form wall is generally required and should be considered minimum acceptable practice.

Flat wall insulating concrete form (ICF) systems must comply with ASTM E 2634.

TABLE R611.3
DIMENSIONAL REQUIREMENTS FOR WALLS[a, b]

WALL TYPE AND NOMINAL THICKNESS	MAXIMUM WALL WEIGHT[c] (psf)	MINIMUM WIDTH, W, OF VERTICAL CORES (inches)	MINIMUM THICKNESS, T, OF VERTICAL CORES (inches)	MAXIMUM SPACING OF VERTICAL CORES (inches)	MAXIMUM SPACING OF HORIZONTAL CORES (inches)	MINIMUM WEB THICKNESS (inches)
4″ Flat[d]	50	N/A	N/A	N/A	N/A	N/A
6″ Flat[d]	75	N/A	N/A	N/A	N/A	N/A
8″ Flat[d]	100	N/A	N/A	N/A	N/A	N/A
10″ Flat[d]	125	N/A	N/A	N/A	N/A	N/A
6″ Waffle-grid	56	8[e]	5.5[e]	12	16	2
8″ Waffle-grid	76	8[f]	8[f]	12	16	2
6″ Screen-grid	53	6.25[g]	6.25[g]	12	12	N/A

For SI: 1 inch = 25.4 mm; 1 pound per square foot = 0.0479 kPa, 1 pound per cubic foot = 2402.77 kg/m^3, 1 square inch = 645.16 mm^2, 1 inch^4 = 42 cm^4.

a. Width "W," thickness "T," spacing and web thickness, refer to Figures R611.3(2) and R611.3(3).

b. N/A indicates not applicable.

c. Wall weight is based on a unit weight of concrete of 150 pcf. For flat walls the weight is based on the nominal thickness. The tabulated values do not include any allowance for interior and exterior finishes.

d. Nominal wall thickness. The actual as-built thickness of a flat wall shall not be more than $^1/_2$-inch less or more than $^1/_4$-inch more than the nominal dimension indicated.

e. Vertical core is assumed to be elliptical-shaped. Another shape core is permitted provided the minimum thickness is 5 inches, the moment of inertia, *I*, about the centerline of the wall (ignoring the web) is not less than 65 inch^4, and the area, *A*, is not less than 31.25 in^2. The width used to calculate *A* and *I* shall not exceed 8 inches.

f. Vertical core is assumed to be circular. Another shape core is permitted provided the minimum thickness is 7 inches, the moment of inertia, *I*, about the centerline of the wall (ignoring the web) is not less than 200 in^4, and the area, *A*, is not less than 49 square inch. The width used to calculate *A* and *I* shall not exceed 8 inches.

g. Vertical core is assumed to be circular. Another shape core is permitted provided the minimum thickness is 5.5 inches, the moment of inertia, *I*, about the centerline of the wall is not less than 76 inch^4, and the area, *A*, is not less than 30.25 square inch. The width used to calculate *A* and *I* shall not exceed 6.25 inches.

❖ See the commentary for Section R611.3.

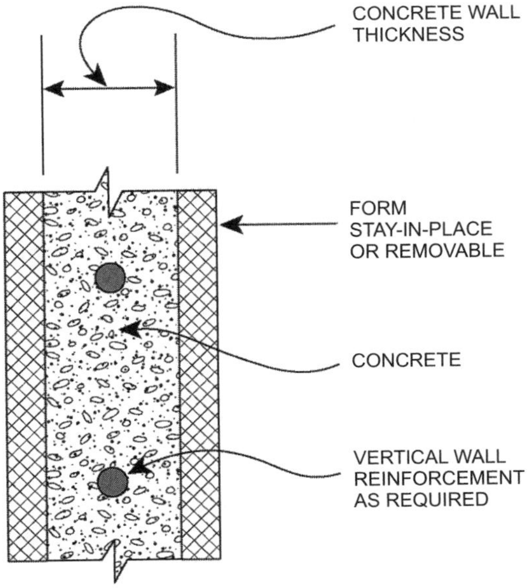

PLAN VIEW

SEE TABLE R611.3 FOR MINIMUM DIMENSIONS

FIGURE R611.3(1)
FLAT WALL SYSTEM

❖ See the commentary for Section R611.3.

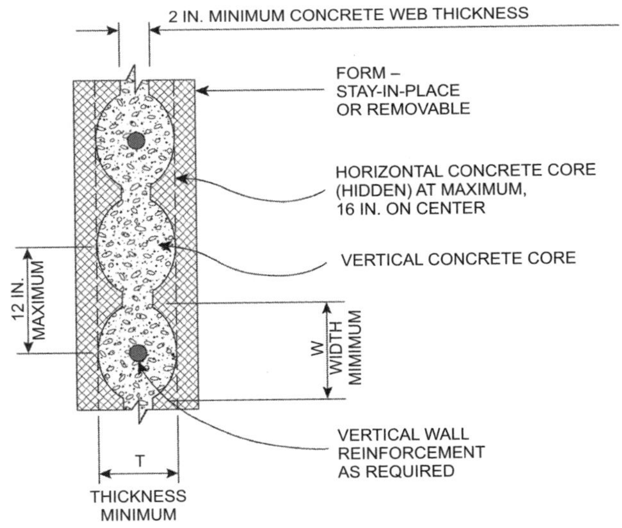

PLAN VIEW

SEE TABLE R611.3 FOR MINIMUM DIMENSIONS

For SI: 1 inch = 25.4 mm.

FIGURE R611.3(2)
WAFFLE-GRID WALL SYSTEM

❖ See the commentary for Section R611.3.

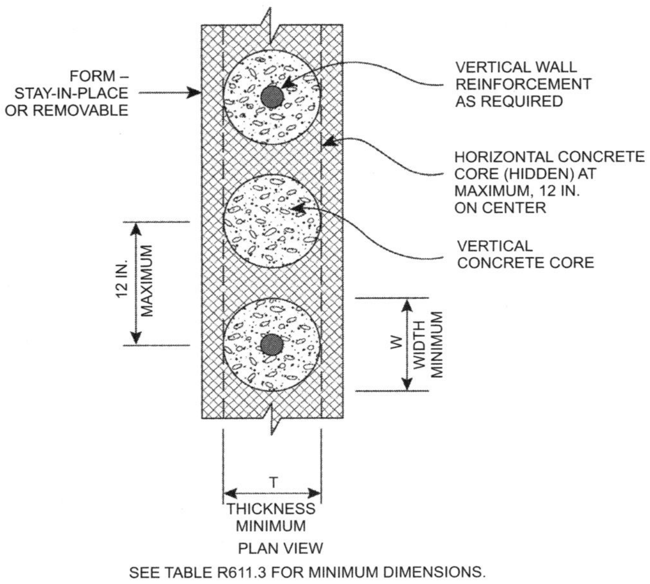

For SI: 1 inch = 25.4 mm.

FIGURE R611.3(3)
SCREEN-GRID SYSTEM

❖ See the commentary for Section R611.3.

R611.4.1 Surface burning characteristics. The flame spread index and smoke-developed index of forming material, other than foam plastic, left exposed on the interior shall comply with Section R302.9. The surface burning characteristics of foam plastic used in insulating concrete forms shall comply with Section R316.3.

❖ See the commentary for Section R611.4.

R611.4.2 Interior covering. Stay-in-place forms constructed of rigid foam plastic shall be protected on the interior of the building as required by Sections R316.4 and R702.3.4. Where gypsum board is used to protect the foam plastic, it shall be installed with a mechanical fastening system. Use of adhesives is permitted in addition to mechanical fasteners.

❖ See the commentary for Section R611.4.

R611.4.3 Exterior wall covering. Stay-in-place forms constructed of rigid foam plastics shall be protected from sunlight and physical damage by the application of an *approved* exterior wall covering complying with this code. Exterior surfaces of other stay-in-place forming systems shall be protected in accordance with this code.

Requirements for installation of masonry veneer, stucco and other finishes on the exterior of concrete walls and other construction details not covered in this section shall comply with the requirements of this code.

❖ See the commentary for Section R611.4.

R611.4.4 Flat ICF wall systems. Flat ICF wall system forms shall conform to ASTM E 2634.

❖ See the commentary for Section R611.4.

R611.5 Materials. Materials used in the construction of concrete walls shall comply with this section.

❖ Concrete walls are composed of three basic materials: concrete, reinforcing steel for concrete and form materials. These materials must comply with this section or ACI 318.

R611.5.1 Concrete and materials for concrete. Materials used in concrete, and the concrete itself, shall conform to requirements of this section, or ACI 318.

❖ Concrete and materials for concrete must comply with this section or ACI 318.

R611.5.1.1 Concrete mixing and delivery. Mixing and delivery of concrete shall comply with ASTM C 94 or ASTM C 685.

❖ Concrete must be mixed and delivered in accordance with requirements of *Specification for Ready-mixed Concrete*, ASTM C 94 or ASTM C 685, *Specification for Concrete Made by Volumetric Batching and Continuous Mixing*. Test methods for uniformity of mixing are given in ASTM C 94.

R611.5.1.2 Maximum aggregate size. The nominal maximum size of coarse aggregate shall not exceed one-fifth the narrowest distance between sides of forms, or three-fourths the clear spacing between reinforcing bars or between a bar and the side of the form.

Exception: When *approved*, these limitations shall not apply where removable forms are used and workability and methods of consolidation permit concrete to be placed without honeycombs or voids.

❖ The maximum aggregate size is based on requirements in ACI 318. These limitations on aggregate size are intended to result in concrete that can be placed such that it properly encases reinforcement and minimizes honeycombing and voids. The exception, which is also based on ACI 318, is limited to use with removable forms because proper placement and encasement of reinforcement can be verified through visual inspection after removal of forms.

R611.5.1.3 Proportioning and slump of concrete. Proportions of materials for concrete shall be established to provide workability and consistency to permit concrete to be worked readily into forms and around reinforcement under conditions of placement to be employed, without segregation or excessive bleeding. Slump of concrete placed in removable forms shall not exceed 6 inches (152 mm).

Exception: When *approved*, the slump is permitted to exceed 6 inches (152 mm) for concrete mixtures that are resistant to segregation, and are in accordance with the form manufacturer's recommendations.

Slump of concrete placed in stay-in-place forms shall exceed 6 inches (152 mm). Slump of concrete shall be determined in accordance with ASTM C 143.

❖ The maximum slump requirements are based on current practice. Considerations included in the prescribed maximums are ease of placement, ability to fill cavities thoroughly and limiting the pressures exerted on the form by fresh concrete.

Where removable wall forms are used, maximum slump of concrete is 6 inches (152 mm). The maximum slump is based on experience with ease of placement, ability to fill cavities thoroughly, and limiting the pressures exerted on the form by fresh concrete. The 6-inch (152 mm) maximum slump refers to the characteristics of the specified mixture proportions based on water/cementitious materials ratio only. The exception envisions the use of mid-range or high-range water-reducing admixtures to increase the slump above 6 inches (152 mm), since their use does not adversely affect the tendency of the aggregate to segregate.

The minimum slump requirement for walls cast in stay-in-place forms must be greater than 6 inches (152 mm). It is important that the concrete mix design utilize proper materials in the proper proportions so that the specified strength is achieved at the specified slump [i.e., greater that 6 inches (152 mm)]. Ordering a concrete mix at a lower slump (e.g., 4 inches [102 mm]) and adding water at the job site to increase the slump to more than 6 inches (152 mm) will adversely affect the strength of the concrete, and possibly jeopardize the safety of the structure. Normally it would be expected that a concrete mixture with a specified slump in excess of 6 inches (152 mm) would contain mid-range or high-range water-reducing admixtures so as to mitigate the possibility of segregation of the aggregate.

R611.5.1.4 Compressive strength. The minimum specified compressive strength of concrete, f'_c, shall comply with Section R402.2 and shall be not less than 2,500 pounds per square inch (17.2 MPa) at 28 days.

❖ The minimum specified compressive strength of concrete of 2,500 psi (17.2 MPa) is based on the minimum current practice, which corresponds to minimum compressive strength permitted by ACI 318 and Section R402.2 of the code (see commentary, Section R402.2).

R611.5.1.5 Consolidation of concrete. Concrete shall be consolidated by suitable means during placement and shall be worked around embedded items and reinforcement and into corners of forms. Where stay-in-place forms are used, concrete shall be consolidated by internal vibration.

Exception: When *approved*, self-consolidating concrete mixtures with slumps equal to or greater than 8 inches (203 mm) that are specifically designed for placement without internal vibration need not be internally vibrated.

❖ Where stay-in-place forms are used, concrete must be consolidated by internal vibration. A study by Portland Cement Association (PCA) determined that internal vibration is required. The study showed that with standard mix design with a slump between 4 inches (102 mm) and 8 inches (203 mm) that internal vibration is required. Two of the mixtures were intended to be used without internal vibration and good results were obtained with both. One mixture utilized a high-range water-reducing admixture to achieve a slump between 8 inches (203 mm) and 10 inches (254 mm). The other contained a high-range water-reducing admixture and a viscosity-modifying admixture to achieve a self-consolidating concrete (SCC) with a slump in excess of 8 inches (203 mm). Slumps in excess of 8 inches (203 mm) should not be obtained by the use of water alone since segregation is likely.

R611.5.2 Steel reinforcement and anchor bolts.

❖ This section prescribes the requirements for the steel reinforcement for concrete walls. Also, this section contains the requirements for anchor bolts and tension straps for connection details.

R611.5.2.1 Steel reinforcement. Steel reinforcement shall comply with ASTM A 615, ASTM A 706, or ASTM A 996. ASTM A 996 bars produced from rail steel shall be Type R.

❖ The types of steel reinforcement permitted by this section are consistent with ACI 318. The wall reinforcement tables, in Section R611.6.2, are based on Grade 60 [minimum yield strength of 60,000 psi (414 mPa)]. However, Section R611.5.4.7 has provisions for use of alternate grades. Type R is specified for ASTM A 996 (rail steel) because the ASTM standard permits two designations. The type designated R is specified because it is the only one of the two that complies with the more stringent bend requirements of ACI 318.

R611.5.2.2 Anchor bolts. Anchor bolts for use with connection details in accordance with Figures R611.9(1) through R611.9(12) shall be bolts with heads complying with ASTM A 307 or ASTM F 1554. ASTM A 307 bolts shall be Grade A (i.e., with heads). ASTM F 1554 bolts shall be Grade 36 minimum. Instead of bolts with heads, it is permissible to use rods with threads on both ends fabricated from steel complying with ASTM A 36. The threaded end of the rod to be embedded in the concrete shall be provided with a hex or square nut.

❖ All bolts in the prescriptive details of Figures R611.9(1) through R611.9(12) may be subject to tensile loading. Since the concrete breakout strength in tension is considerably greater for a headed bolt versus a hooked bolt of the same size and embedment depth, the prescriptive designs are based on the use of bolts with heads. In lieu of bolts with heads, the provisions permit the use of rods with threads on both ends with the end embedded in the concrete having a hex or square nut. ACI 318 requires that connections subjected to seismic forces be designed to fail in a ductile manner. Therefore, it is imperative that bolts with a higher yield strength or bolts larger than specified in the figures not be used in these connections since it is likely that in the event of a load being applied that is greater than that assumed in design, the concrete may fail prior to the bolt yielding.

R611.5.2.3 Sheet steel angles and tension tie straps. Angles and tension tie straps for use with connection details in accordance with Figures R611.9(1) through R611.9(12) shall be fabricated from sheet steel complying with ASTM A 653 SS,

ASTM A 792 SS, or ASTM A 875 SS. The steel shall be minimum Grade 33 unless a higher grade is required by the applicable figure.

❖ Grade 33 is the minimum grade of sheet steel specified for components of connections in Figures R611.9(1) through R611.9(12) because it is the lowest strength recognized under the standards for structural steel (Type SS). The requirement applies to angles and other components of connections that are field fabricated and is not intended to apply to proprietary clip angles and other proprietary connectors that are available.

R611.5.3 Form materials and form ties. Forms shall be made of wood, steel, aluminum, plastic, a composite of cement and foam insulation, a composite of cement and wood chips, or other *approved* material suitable for supporting and containing concrete. Forms shall provide sufficient strength to contain concrete during the concrete placement operation.

Form ties shall be steel, solid plastic, foam plastic, a composite of cement and wood chips, a composite of cement and foam plastic, or other suitable material capable of resisting the forces created by fluid pressure of fresh concrete.

❖ The materials listed are based on currently available removable and stay-in-place forming systems. From a structural design standpoint, the material can be anything that has sufficient strength and durability to contain the concrete during pouring and curing.

R611.5.4 Reinforcement installation details.

❖ This section contains the requirements for concrete wall reinforcement support and cover, location, lap splices, development lengths, hooks, and alternate grade.

R611.5.4.1 Support and cover. Reinforcement shall be secured in the proper location in the forms with tie wire or other bar support system such that displacement will not occur during the concrete placement operation. Steel reinforcement in concrete cast against the earth shall have a minimum cover of 3 inches (76 mm). Minimum cover for reinforcement in concrete cast in removable forms that will be exposed to the earth or weather shall be $1^1/_2$ inches (38 mm) for No. 5 bars and smaller, and 2 inches (50 mm) for No. 6 bars and larger. For concrete cast in removable forms that will not be exposed to the earth or weather, and for concrete cast in stay-in-place forms, minimum cover shall be $^3/_4$ inch (19 mm). The minus tolerance for cover shall not exceed the smaller of one-third the required cover and $^3/_8$ inch (10 mm). See Section R611.5.4.4 for cover requirements for hooks of bars developed in tension.

❖ Reinforcement should be adequately supported in the forms to prevent displacement by concrete placement or workers. Concrete cover as protection of reinforcement against weather and other effects is measured from the concrete surface to the outermost surface of the steel to which the cover requirement applies. The condition "concrete surfaces exposed to earth or weather" refers to direct exposure to moisture changes and not just to temperature changes. Stay-in-place forms provide an alternative method of protecting from earth or weather and the minimum cover for concrete not exposed to weather or earth is allowed. The tolerance for cover is based on ACI 318 requirements.

R611.5.4.2 Location of reinforcement in walls. For location of reinforcement in foundation walls and above-grade walls, see Sections R404.1.2.3.7.2 and R611.6.5, respectively.

❖ See the commentary for Section R611.6.5 for exterior concrete above-grade walls.

See the commentary for Section R404.1.2.3.7.2 for foundation walls.

R611.5.4.3 Lap splices. Vertical and horizontal wall reinforcement required by Sections R611.6 and R611.7 shall be the longest lengths practical. Where splices are necessary in reinforcement, the length of lap splices shall be in accordance with Table R611.5.4(1) and Figure R611.5.4 (1). The maximum gap between noncontact parallel bars at a lap splice shall not exceed the smaller of one-fifth the required lap length and 6 inches (152 mm). See Figure R611.5.4(1).

❖ Where continuous bars are not provided, lap splices are necessary to allow forces to be transferred between sections of rebar. The length of lap splices in Table R611.5.4(1) is based on ACI 318 requirements. Noncontact lap splices are permitted provided the spacing of lapped bars does not exceed the smaller of one fifth the required lap length or 6 inches (152 mm). If individual bars in noncontact lap splices are too widely spaced, an unreinforced section is created. Forcing a potential crack to follow a zigzag line (5:1 slope) is considered a minimum precaution. These requirements are based on ACI 318. Figure R611.5.4(1) illustrates the code text requirements.

R611.5.4.4 Development of bars in tension. Where bars are required to be developed in tension by other provisions of this code, development lengths and cover for hooks and bar extensions shall comply with Table R611.5.4(1) and Figure R611.5.4(2). The development lengths shown in Table R611.5.4(1) also apply to bundled bars in lintels installed in accordance with Section R611.8.2.2.

❖ Development lengths and cover for hooks and bar extensions must comply with Table R611.5.4(1) and Figure R611.5.4(2). These requirements are based on ACI 318.

R611.5.4.5 Standard hooks. Where reinforcement is required by this code to terminate with a standard hook, the hook shall comply with Figure R611.5.4(3).

❖ The hook requirements shown in Figure R611.5.4(3) are consistent with ACI 318 requirements.

R611.5.4.6 Webs of waffle-grid walls. Reinforcement, including stirrups, shall not be placed in webs of waffle-grid walls, including lintels. Webs are permitted to have form ties.

❖ Reinforcement must not be placed in the web of waffle-grid walls and lintels.

TABLE R611.5.4(1)
LAP SPLICE AND TENSION DEVELOPMENT LENGTHS

	BAR SIZE NO.	YIELD STRENGTH OF STEEL, f_y- psi (MPa)	
		40,000 (280)	60,000 (420)
		Splice length or tension development length (inches)	
Lap splice length-tension	4	20	30
	5	25	38
	6	30	45
Tension development length for straight bar	4	15	23
	5	19	28
	6	23	34
Tension development length for: a.90-degree and 180-degree standard hooks with not less than 2$^1/_2$ inches of side cover perpendicular to plane of hook, and b.90-degree standard hooks with not less than 2 inches of cover on the bar extension beyond the hook.	4	6	9
	5	7	11
	6	8	13
Tension development length for bar with 90-degree or 180-degree standard hook having less cover than required above.	4	8	12
	5	10	15
	6	12	18

For SI: 1 inch = 25.4 mm.

❖ See the commentary for Sections R611.5.4.3 and R611.5.4.4.

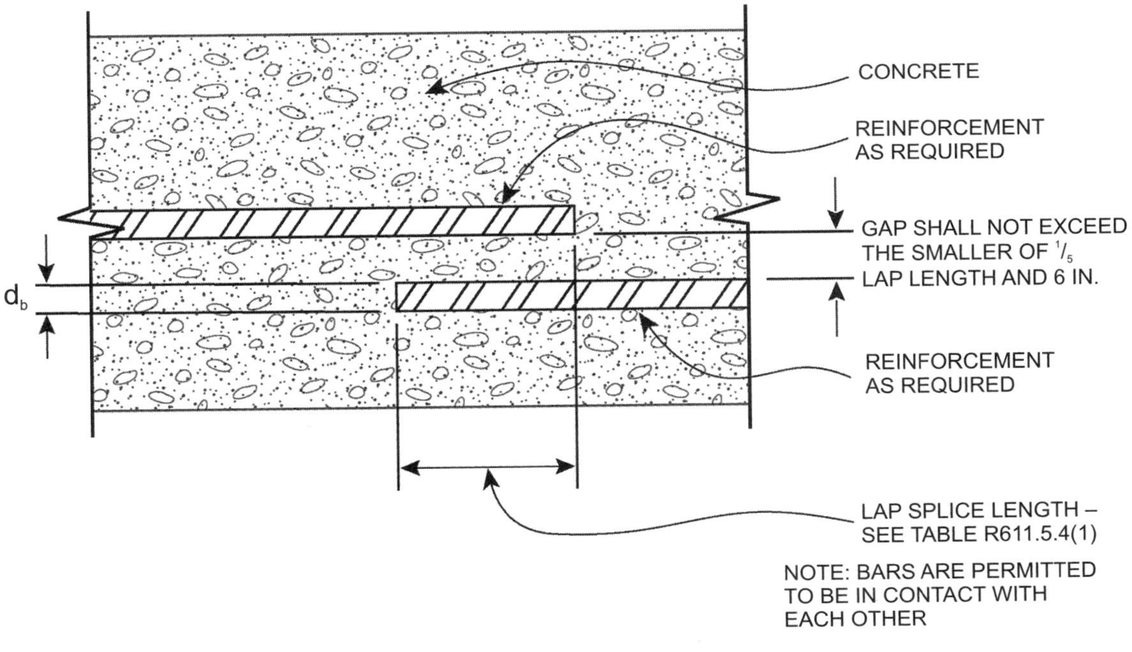

For SI: 1 inch = 25.4 mm.

FIGURE R611.5.4(1)
LAP SPLICES

❖ This figure illustrates the requirements of Section R611.5.4.3.

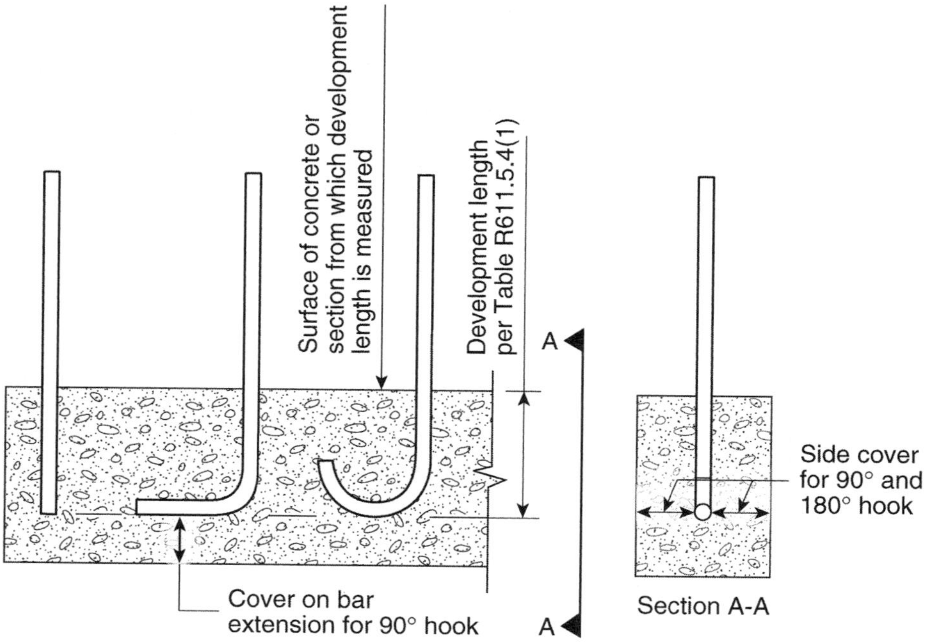

For SI: 1 degree = 0.0175 rad.

FIGURE R611.5.4(2)
DEVELOPMENT LENGTH AND COVER FOR HOOKS AND BAR EXTENSION

❖ See the commentary for Section R611.5.4.4.

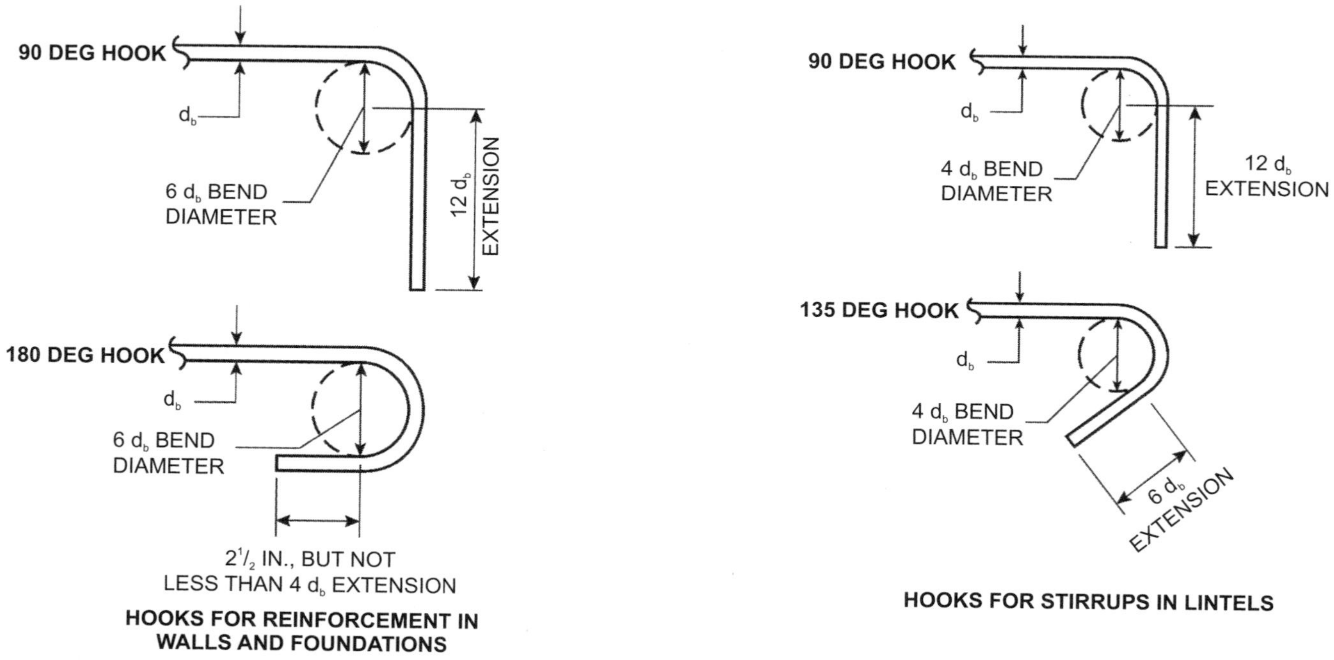

For SI: 1 inch = 25.4 mm, 1 degree = 0.0175 rad.

FIGURE R611.5.4(3)
STANDARD HOOKS

❖ See the commentary for Section R611.5.4.5.

TABLE R611.5.4(2) MAXIMUM SPACING FOR ALTERNATIVE BAR SIZE AND/OR ALTERNATIVE GRADE OF STEEL[a, b, c]

BAR SPACING FROM APPLICABLE TABLE IN SECTION R611.6 (inches)	BAR SIZE FROM APPLICABLE TABLE IN SECTION R611.6														
	#4					#5					#6				
	Alternate bar size and/or alternate grade of steel desired														
	Grade 60		Grade 40			Grade 60		Grade 40			Grade 60		Grade 40		
	#5	#6	#4	#5	#6	#4	#6	#4	#5	#6	#4	#5	#4	#5	#6
	Maximum spacing for alternate bar size and/or alternate grade of steel (inches)														
8	12	18	5	8	12	5	11	3	5	8	4	6	2	4	5
9	14	20	6	9	13	6	13	4	6	9	4	6	3	4	6
10	16	22	7	10	15	6	14	4	7	9	5	7	3	5	7
11	17	24	7	11	16	7	16	5	7	10	5	8	3	5	7
12	19	26	8	12	18	8	17	5	8	11	5	8	4	6	8
13	20	29	9	13	19	8	18	6	9	12	6	9	4	6	9
14	22	31	9	14	21	9	20	6	9	13	6	10	4	7	9
15	23	33	10	16	22	10	21	6	10	14	7	11	5	7	10
16	25	35	11	17	23	10	23	7	11	15	7	11	5	8	11
17	26	37	11	18	25	11	24	7	11	16	8	12	5	8	11
18	28	40	12	19	26	12	26	8	12	17	8	13	5	8	12
19	29	42	13	20	28	12	27	8	13	18	9	13	6	9	13
20	31	44	13	21	29	13	28	9	13	19	9	14	6	9	13
21	33	46	14	22	31	14	30	9	14	20	10	15	6	10	14
22	34	48	15	23	32	14	31	9	15	21	10	16	7	10	15
23	36	48	15	24	34	15	33	10	15	22	10	16	7	11	15
24	37	48	16	25	35	15	34	10	16	23	11	17	7	11	16
25	39	48	17	26	37	16	35	11	17	24	11	18	8	12	17
26	40	48	17	27	38	17	37	11	17	25	12	18	8	12	17
27	42	48	18	28	40	17	38	12	18	26	12	19	8	13	18
28	43	48	19	29	41	18	40	12	19	26	13	20	8	13	19
29	45	48	19	30	43	19	41	12	19	27	13	20	9	14	19
30	47	48	20	31	44	19	43	13	20	28	14	21	9	14	20
31	48	48	21	32	45	20	44	13	21	29	14	22	9	15	21
32	48	48	21	33	47	21	45	14	21	30	15	23	10	15	21
33	48	48	22	34	48	21	47	14	22	31	15	23	10	16	22
34	48	48	23	35	48	22	48	15	23	32	15	24	10	16	23
35	48	48	23	36	48	23	48	15	23	33	16	25	11	16	23
36	48	48	24	37	48	23	48	15	24	34	16	25	11	17	24
37	48	48	25	38	48	24	48	16	25	35	17	26	11	17	25
38	48	48	25	39	48	25	48	16	25	36	17	27	12	18	25
39	48	48	26	40	48	25	48	17	26	37	18	27	12	18	26
40	48	48	27	41	48	26	48	17	27	38	18	28	12	19	27
41	48	48	27	42	48	26	48	18	27	39	19	29	12	19	27
42	48	48	28	43	48	27	48	18	28	40	19	30	13	20	28
43	48	48	29	44	48	28	48	18	29	41	20	30	13	20	29
44	48	48	29	45	48	28	48	19	29	42	20	31	13	21	29
45	48	48	30	47	48	29	48	19	30	43	20	32	14	21	30
46	48	48	31	48	48	30	48	20	31	44	21	32	14	22	31
47	48	48	31	48	48	30	48	20	31	44	21	33	14	22	31
48	48	48	32	48	48	31	48	21	32	45	22	34	15	23	32

For SI: 1 inch = 25.4 mm.

a. This table is for use with tables in Section R611.6 that specify the minimum bar size and maximum spacing of vertical wall reinforcement for foundation walls and above-grade walls. Reinforcement specified in tables in Section R611.6 is based on Grade 60 (420 MPa) steel reinforcement.

b. Bar spacing shall not exceed 48 inches on center and shall not be less than one-half the nominal wall thickness.

c. For Grade 50 (350 MPa) steel bars (ASTM A 996, Type R), use spacing for Grade 40 (280 MPa) bars or interpolate between Grade 40 (280 MPa) and Grade 60 (420 MPa).

❖ See the commentary for Section R611.5.2.1.

R611.5.4.7 Alternate grade of reinforcement and spacing. Where tables in Sections R404.1.2 and R611.6 specify vertical wall reinforcement based on minimum bar size and maximum spacing, which are based on Grade 60 (420 MPa) steel reinforcement, different size bars and/or bars made from a different grade of steel are permitted provided an equivalent area of steel per linear foot of wall is provided. Use of Table R611.5.4(2) is permitted to determine the maximum bar spacing for different bar sizes than specified in the tables and/or bars made from a different grade of steel. Bars shall not be spaced less than one-half the wall thickness, or more than 48 inches (1219 mm) on center.

❖ See the commentary for Section R611.5.2.1.

R611.5.5 Construction joints in walls. Construction joints shall be made and located to not impair the strength of the wall. Construction joints in plain concrete walls, including walls required to have not less than No. 4 bars at 48 inches (1219 mm) on center by Section R611.6, shall be located at points of lateral support, and a minimum of one No. 4 bar shall extend across the construction joint at a spacing not to exceed 24 inches (610 mm) on center. Construction joint reinforcement shall have a minimum of 12 inches (305 mm) embedment on both sides of the joint. Construction joints in reinforced concrete walls shall be located in the middle third of the span between lateral supports, or located and constructed as required for joints in plain concrete walls.

> **Exception:** Vertical wall reinforcement required by this code is permitted to be used in lieu of construction joint reinforcement, provided the spacing does not exceed 24 inches (610 mm), or the combination of wall reinforcement and No. 4 bars described above does not exceed 24 inches (610 mm).

❖ Construction joints must be located where they will cause the least weakness in the structure. The reinforcement required by this section is required to provide for transfer of shear and other forces through the construction joint. These provisions are consistent with ACI 318.

R611.6 Above-grade wall requirements.

❖ This section provides the following requirements for above-grade concrete walls:

- General requirements
- Wall reinforcement
- Continuity of vertical reinforcement between stories
- Termination of vertical reinforcement in the topmost story
- Location of reinforcement in the wall

R611.6.1 General. The minimum thickness of load-bearing and nonload-bearing above-grade walls and reinforcement shall be as set forth in the appropriate table in this section based on the type of wall form to be used. Where the wall or building is not within the limitations of Section R611.2, design is required by the tables in this section, or the wall is not within the scope of the tables in this section, the wall shall be designed in accordance with ACI 318.

Above-grade concrete walls shall be constructed in accordance with this section and Figure R611.6(1), R611.6(2), R611.6(3) or R611.6(4). Above-grade concrete walls that are continuous with stem walls and not laterally supported by the slab-on-ground shall be designed and constructed in accordance with this section. Concrete walls shall be supported on continuous foundation walls or slabs-on-ground that are monolithic with the footing in accordance with Section R403. The minimum length of solid wall without openings shall be in accordance with Section R611.7. Reinforcement around openings, including lintels, shall be in accordance with Section R611.8. Lateral support for above-grade walls in the out-of-plane direction shall be provided by connections to the floor framing system, if applicable, and to ceiling and roof framing systems in accordance with Section R611.9. The wall thickness shall be equal to or greater than the thickness of the wall in the *story* above.

❖ Section R611.6 provides reinforcement tables for load-bearing and nonload-bearing above-grade walls constructed within the applicability limits of Section R611.2. Section R611.6.2 provides the minimum required vertical and horizontal wall reinforcement for different design wind pressures and three wall heights. Vertical wall reinforcement tables are limited to one- and two-story buildings for nonload-bearing and load-bearing walls, in accordance with Figure R611.6(1), R611.6(2), R611.6(3) or R611.6(4).

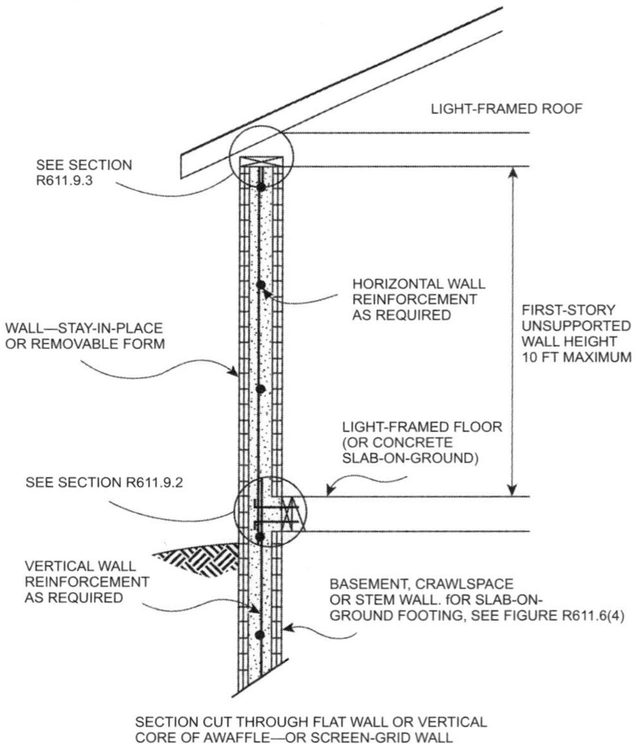

For SI: 1 foot = 304.8 mm.

FIGURE R611.6(1)
ABOVE-GRADE CONCRETE WALL CONSTRUCTION ONE STORY

❖ See the commentary for Section R611.6.1.

Section R611.6 completely replaces the prescriptive technical provisions for exterior concrete walls from the 2006 edition of the code. The new provisions reflect changes made to ACI 318 and ASCE 7 since the provisions in the previous code were developed. The new provisions cover horizontal and vertical reinforcement in Section R611.6, reinforcement and shear wall (solid walls) requirements for wind loads in Section R611.7 and revised reinforcement requirements around openings and lintels over openings in Section R611.8. Additionally, the provisions providing revised details for connecting wood frame assemblies (floors, ceilings and roofs) to exterior concrete walls and added details for connecting cold-formed steel framing assemblies to the exterior concrete walls are contained in Section R611.9.

R611.6.2 Wall reinforcement for wind. Vertical wall reinforcement for resistance to out-of-plane wind forces shall be determined from Table R611.6(1), R611.6(2), R611.6(3) or R611.6(4). Also, see Sections R611.7.2.2.2 and R611.7.2.2.3. There shall be a vertical bar at all corners of exterior walls. Unless more horizontal reinforcement is required by Section R611.7.2.2.1, the minimum horizontal reinforcement shall be four No. 4 bars [Grade 40 (280 MPa)] placed as follows: top bar within 12 inches (305 mm) of the top of the wall, bottom bar within 12 inches (305 mm) of the finish floor, and one bar each at approximately one-third and two-thirds of the wall height.

❖ The wall reinforcement must be in accordance with this section and Tables R611.6(1), R611.6(2), R611.6(3) and R611.6(4).

The following design assumptions were used in developing these tables.

1. Walls are simply supported at each floor and roof that provides lateral support, or by the exterior finish ground level in the case where walls are continuous with stem walls.
2. Lateral support is provided for the wall by the floors and roof, and by the slab-on-ground where the wall is monolithic with or anchored to the slab.
3. Allowable deflection criterion is the laterally unsupported height of the wall, in inches (mm), divided by 240.

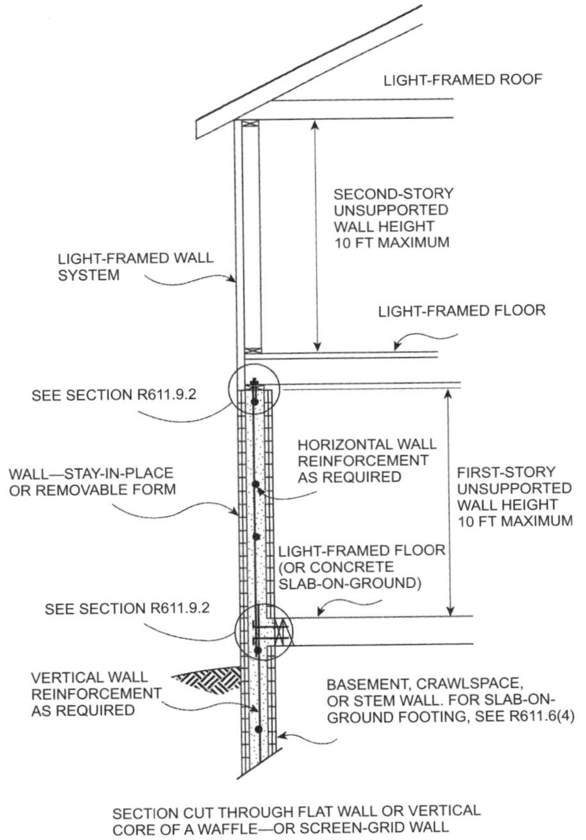

For SI: 1 foot = 304.8 mm.

FIGURE R611.6(2)
ABOVE-GRADE CONCRETE FIRST-STORY AND LIGHT-FRAMED SECOND-STORY

❖ See the commentary for Section R611.6.1.

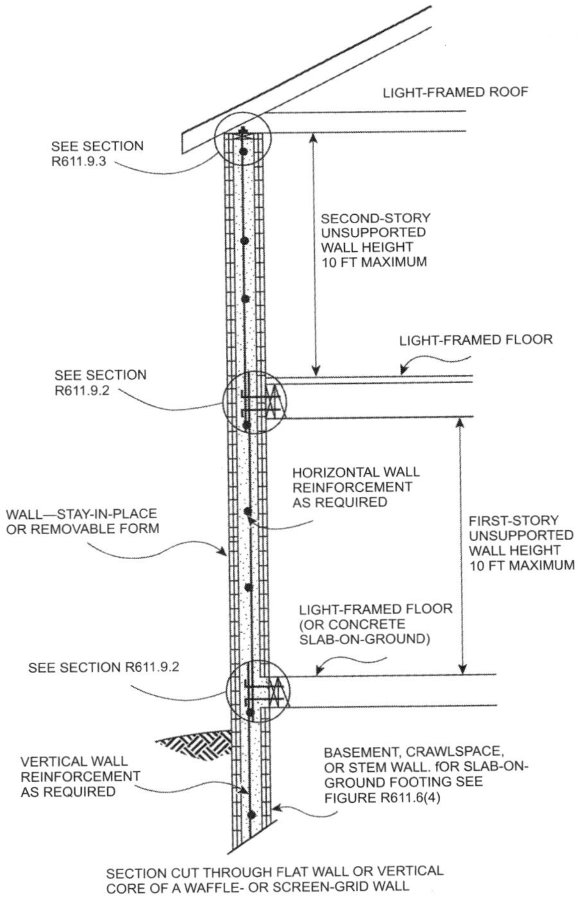

For SI: 1 foot = 304.8 mm.

FIGURE R611.6(3)
ABOVE-GRADE CONCRETE WALL CONSTRUCTION TWO-STORY

❖ See the commentary for Section R611.6.1.

4. The minimum possible axial load is considered for each case. For "top" bearing walls, no axial load was considered. For "side" bearing walls, the moment induced by the eccentricity of the dead and live load of the floor construction supported by the ledger board or cold-formed steel track bolted to the side of the wall was considered since it is additive with the moment induced by the outward-acting wind load. The axial load on the wall due to the floor construction supported by the ledger board or cold-formed steel track was not considered.

5. Wind loads were calculated in accordance with ASCE 7 for enclosed buildings using components and cladding pressure coefficients for the wall interior zone (designated number 4 in Figure 6-11A of ASCE 7), effective wind area of 10 square feet (0.929 m^2) and mean roof height of 35 feet (10.7 m).

R611.6.3 Continuity of wall reinforcement between stories. Vertical reinforcement required by this section shall be continuous between elements providing lateral support for the wall. Reinforcement in the wall of the *story* above shall be continuous with the reinforcement in the wall of the *story* below, or the foundation wall, if applicable. Lap splices, where required, shall comply with Section R611.5.4.3 and Figure R611.5.4(1). Where the above-grade wall is supported by a monolithic slab-on-ground and footing, dowel bars with a size and spacing to match the vertical above-grade concrete wall reinforcement shall be embedded in the monolithic slab-on-ground and footing the distance required to develop the dowel bar in tension in accordance with Section R611.5.4.4 and Figure R611.5.4(2) and lap-spliced with the above-grade wall reinforcement in accordance with Section R611.5.4.3 and Figure R611.5.4(1).

Exception: Where reinforcement in the wall above cannot be made continuous with the reinforcement in the wall below, the bottom of the reinforcement in the wall above shall be terminated in accordance with one of the following:

1. Extend below the top of the floor the distance required to develop the bar in tension in accordance with Section R611.5.4.4 and Figure R611.5.4(2).
2. Lap-spliced in accordance with Section R611.5.4.3 and Figure R611.5.4(1) with a dowel bar that extends into the wall below the distance required to develop the bar in tension in accordance with Section R611.5.4.4 and Figure R611.5.4(2).

Where a construction joint in the wall is located below the level of the floor and less than the distance required to develop the bar in tension, the distance required to develop the bar in tension shall be measured from the top of the concrete below the joint. See Section R611.5.5.

❖ The wall vertical reinforcement must be continuous through the basement, crawl space or stem wall and must be continuous through the stories above and terminate at the top of the top most story.

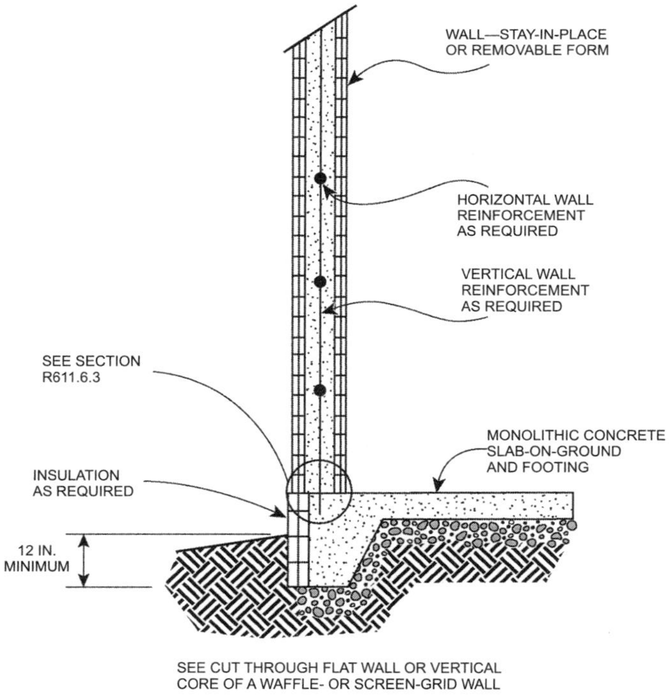

For SI: 1 inch = 25.4 mm.

FIGURE R611.6(4)
ABOVE-GRADE CONCRETE WALL SUPPORTED ON MONOLITHIC SLAB-ON-GROUND FOOTING

❖ See the commentary for Section R611.6.1.

Where there is no basement, crawl space or stem wall, the continuous vertical reinforcement must be doweled into a monolithic slab-on-ground footing in accordance with Section R611.5.4.4.

Where continuity cannot be made, the exception permits extending the wall reinforcing below the top of the floor to the required development length or lap-splice with a dowel bar in the wall below.

TABLE R611.6(1)
MINIMUM VERTICAL REINFORCEMENT FOR FLAT ABOVE-GRADE WALLS[a, b, c, d, e]

MAXIMUM WIND SPEED (mph)			MAXIMUM UNSUPPORTED WALL HEIGHT PER STORY (feet)	MINIMUM VERTICAL REINFORCEMENT-BAR SIZE AND SPACING (inches)[f, g]							
Exposure Category				Nominal[h] wall thickness (inches)							
				4		6		8		10	
B	C	D		Top[i]	Side[i]	Top[i]	Side[i]	Top[i]	Side[i]	Top[i]	Side[i]
85	—	—	8	4@48	4@48	4@48	4@48	4@48	4@48	4@48	4@48
			9	4@48	4@43	4@48	4@48	4@48	4@48	4@48	4@48
			10	4@47	4@36	4@48	4@48	4@48	4@48	4@48	4@48
90	—	—	8	4@48	4@47	4@48	4@48	4@48	4@48	4@48	4@48
			9	4@48	4@39	4@48	4@48	4@48	4@48	4@48	4@48
			10	4@42	4@34	4@48	4@48	4@48	4@48	4@48	4@48
100	85	—	8	4@48	4@40	4@48	4@48	4@48	4@48	4@48	4@48
			9	4@42	4@34	4@48	4@48	4@48	4@48	4@48	4@48
			10	4@34	4@34	4@48	4@48	4@48	4@48	4@48	4@48
110	90	85	8	4@44	4@34	4@48	4@48	4@48	4@48	4@48	4@48
			9	4@34	4@34	4@48	4@48	4@48	4@48	4@48	4@48
			10	4@34	4@31	4@48	4@37	4@48	4@48	4@48	4@48
120	100	90	8	4@36	4@34	4@48	4@48	4@48	4@48	4@48	4@48
			9	4@34	4@32	4@48	4@38	4@48	4@48	4@48	4@48
			10	4@30	4@27	4@48	5@48	4@48	4@48	4@48	4@48
130	110	100	8	4@34	4@34	4@48	4@48	4@48	4@48	4@48	4@48
			9	4@32	4@28	4@48	4@33	4@48	4@48	4@48	4@48
			10	4@26	4@23	4@48	5@43	4@48	4@48	4@48	4@48

For SI: 1 inch = 25.4 mm, 1 foot = 304.8 mm, 1 mile per hour = 0.447 m/s, 1 pound per square inch = 1.895 kPa, 1 square foot = 0.0929 m^2.

a. Table is based on ASCE 7 components and cladding wind pressures for an enclosed building using a mean roof height of 35 feet, interior wall area 4, an effective wind area of 10 square feet, and topographic factor, K_{zt}, and importance factor, I, equal to 1.0.

b. Table is based on concrete with a minimum specified compressive strength of 2,500 psi.

c. See Section R611.6.5 for location of reinforcement in wall.

d. Deflection criterion is $L/240$, where L is the unsupported height of the wall in inches.

e. Interpolation is not permitted.

f. Where No. 4 reinforcing bars at a spacing of 48 inches are specified in the table, use of bars with a minimum yield strength of 40,000 psi or 60,000 psi is permitted.

g. Other than for No. 4 bars spaced at 48 inches on center, table values are based on reinforcing bars with a minimum yield strength of 60,000 psi. Vertical reinforcement with a yield strength of less than 60,000 psi and/or bars of a different size than specified in the table are permitted in accordance with Section R611.5.4.7 and Table R611.5.4(2).

h. See Table R611.3 for tolerances on nominal thicknesses.

i. Top means gravity load from roof and/or floor construction bears on top of wall. Side means gravity load from floor construction is transferred to wall from a wood ledger or cold-formed steel track bolted to side of wall. Where floor framing members span parallel to the wall, use of the top bearing condition is permitted.

❖ See the commentary for Section R611.6.2. For flat walls, design strength (i.e., capacity) was based on thicknesses of 3.5, 5.5, 7.5 and 9.5 inches (89, 140, 196 and 241 mm). This approach will provide adequate strength regardless of whether the forms used are $^1/_2$-inch (12.7 mm) less than the stated nominal thickness of 4, 6, 8 or 10 inches (102, 152, 203 or 254 mm) or $^1/_4$-inch (6.4 mm) more than the stated nominal thickness.

TABLE R611.6(2)
MINIMUM VERTICAL REINFORCEMENT FOR WAFFLE-GRID ABOVE-GRADE WALLS[a, b, c, d, e]

MAXIMUM WIND SPEED (mph)			MAXIMUM UNSUPPORTED WALL HEIGHT PER STORY (feet)	MINIMUM VERTICAL REINFORCEMENT-BAR SIZE AND SPACING (inches)[f, g]			
Exposure Category				Nominal[h] wall thickness (inches)			
				6		8	
B	C	D		Top[i]	Side[i]	Top[i]	Side[i]
85	—	—	8	4@48	4@36, 5@48	4@48	4@48
			9	4@48	4@30, 5@47	4@48	4@45
			10	4@48	4@26, 5@40	4@48	4@39
90	—	—	8	4@48	4@33, 5@48	4@48	4@48
			9	4@48	4@28, 5@43	4@48	4@42
			10	4@31, 5@48	4@24, 5@37	4@48	4@36
100	85	—	8	4@48	4@28, 5@44	4@48	4@43
			9	4@31, 5@48	4@24, 5@37	4@48	4@36
			10	4@25, 5@39	4@24, 5@37	4@48	4@31, 5@48
110	90	85	8	4@33, 5@48	4@25, 5@38	4@48	4@38
			9	4@26, 5@40	4@24, 5@37	4@48	4@31, 5@48
			10	4@24, 5@37	4@23, 5@35	4@48	4@27, 5@41
120	100	90	8	4@27, 5@42	4@24, 5@37	4@48	4@33, 5@48
			9	4@24, 5@37	4@23, 5@36	4@48	4@27, 5@43
			10	4@23, 5@35	4@19, 5@30	4@48	4@23, 5@36
130	110	100	8	4@24, 5@37	4@24, 5@37	4@48	4@29, 5@45
			9	4@24, 5@37	4@20, 5@32	4@48	4@24, 5@37
			10	4@19, 5@30	4@17, 5@26	4@23, 5@36	4@20, 5@31

For SI: 1 inch = 25.4 mm, 1 foot = 304.8 mm, 1 mile per hour = 0.447 m/s, 1 pound per square inch = 6.895 kPa, 1 square foot = 0.0929 m^2.

a. Table is based on ASCE 7 components and cladding wind pressures for an enclosed building using a mean roof height of 35 feet, interior wall area 4, an effective wind area of 10 square feet, and topographic factor, K_{zt}, and importance factor, *I*, equal to 1.0.

b. Table is based on concrete with a minimum specified compressive strength of 2,500 psi.

c. See Section R611.6.5 for location of reinforcement in wall.

d. Deflection criterion is *L*/240, where *L* is the unsupported height of the wall in inches.

e. Interpolation is not permitted.

f. Where No. 4 reinforcing bars at a spacing of 48 inches are specified in the table, use of bars with a minimum yield strength of 40,000 psi or 60,000 psi is permitted.

g. Other than for No. 4 bars spaced at 48 inches on center, table values are based on reinforcing bars with a minimum yield strength of 60,000 psi. Maximum spacings shown are the values calculated for the specified bar size. Where the bar used is Grade 60 and the size specified in the table, the actual spacing in the wall shall not exceed a whole-number multiple of 12 inches (i.e., 12, 24, 36 and 48) that is less than or equal to the tabulated spacing. Vertical reinforcement with a yield strength of less than 60,000 psi and/or bars of a different size than specified in the table are permitted in accordance with Section R611.5.4.7 and Table R611.5.4(2).

h. See Table R611.3 for minimum core dimensions and maximum spacing of horizontal and vertical cores.

i. Top means gravity load from roof and/or floor construction bears on top of wall. Side means gravity load from floor construction is transferred to wall from a wood ledger or cold-formed steel track bolted to side of wall. Where floor framing members span parallel to the wall, the top bearing condition is permitted to be used.

❖ See the commentary for Section R611.6.2. For 6-inch (152 mm) and 8-inch (203 mm) waffle-grid walls, rectangular cross sections with through-the-wall thicknesses of 5 inches (127 mm) and 7 inch (178 mm) respectively, were used. The lengths of the rectangles for 6-inch (152 mm) and 8-inch (203 mm) walls were 6.25 and 7 inches (159 and 178 mm), respectively. The web was ignored for computing section proportions and resisting out-of-plane shear.

TABLE R611.6(3)
MINIMUM VERTICAL REINFORCEMENT FOR 6-INCH SCREEN-GRID ABOVE-GRADE WALLS[a, b, c, d, e]

MAXIMUM WIND SPEED (mph)			MAXIMUM UNSUPPORTED WALL HEIGHT PER STORY (feet)	MINIMUM VERTICAL REINFORCEMENT-BAR SIZE AND SPACING (inches)[f, g]	
Exposure Category				Nominal[h] wall thickness (inches)	
				6	
B	C	D		Top[i]	Side[i]
85	—	—	8	4@48	4@34, 5@48
			9	4@48	4@29, 5@45
			10	4@48	4@25, 5@39
90	—	—	8	4@48	4@31, 5@48
			9	4@48	4@27, 5@41
			10	4@30, 5@47	4@23, 5@35
100	85	—	8	4@48	4@27, 5@42
			9	4@30, 5@47	4@23, 5@35
			10	4@24, 5@38	4@22, 5@34
110	90	85	8	4@48	4@24, 5@37
			9	4@25, 5@38	4@22, 5@34
			10	4@22, 5@34	4@22, 5@34
120	100	90	8	4@26, 5@41	4@22, 5@34
			9	4@22, 5@34	4@22, 5@34
			10	4@22, 6@34	4@19, 5@26
130	110	100	8	4@22, 5@35	4@22, 5@34
			9	4@22, 5@34	4@20, 5@30
			10	4@19, 5@29	4@16, 5@25

For SI: 1 inch = 25.4 mm, 1 foot = 304.8 mm, 1 mile per hour = 0.447 m/s, 1 pound per square inch = 6.895 kPa, 1 square foot = 0.0929 m^2.

a. Table is based on ASCE 7 components and cladding wind pressures for an enclosed building using a mean roof height of 35 feet, interior wall area 4, an effective wind area of 10 square feet, and topographic factor, K_{zt}, and importance factor, I, equal to 1.0.

b. Table is based on concrete with a minimum specified compressive strength of 2,500 psi.

c. See Section R611.6.5 for location of reinforcement in wall.

d. Deflection criterion is $L/240$, where L is the unsupported height of the wall in inches.

e. Interpolation is not permitted.

f. Where No. 4 reinforcing bars at a spacing of 48 inches are specified in the table, use of bars with a minimum yield strength of 40,000 psi or 60,000 psi is permitted.

g. Other than for No. 4 bars spaced at 48 inches on center, table values are based on reinforcing bars with a minimum yield strength of 60,000 psi. Maximum spacings shown are the values calculated for the specified bar size. Where the bar used is Grade 60 and the size specified in the table, the actual spacing in the wall shall not exceed a whole-number multiple of 12 inches (i.e., 12, 24, 36 and 48) that is less than or equal to the tabulated spacing. Vertical reinforcement with a yield strength of less than 60,000 psi and/or bars of a different size than specified in the table are permitted in accordance with Section R611.5.4.7 and Table R611.5.4(2).

h. See Table R611.3 for minimum core dimensions and maximum spacing of horizontal and vertical cores.

i. Top means gravity load from roof and/or floor construction bears on top of wall. Side means gravity load from floor construction is transferred to wall from a wood ledger or cold-formed steel track bolted to side of wall. Where floor framing members span parallel to the wall, use of the top bearing condition is permitted.

❖ See the commentary for Section R611.6.2. For 6-inch (152 mm) screen-grid walls, a square cross section with 5.5-inch (140 mm) sides was used. The spacing of the rectangular resisting elements was 12 inches (305 mm) in all cases since this is the maximum spacing of vertical cores permitted by Table R611.3.

TABLE R611.6(4)
MINIMUM VERTICAL REINFORCEMENT FOR FLAT, WAFFLE- AND SCREEN-GRID ABOVE-GRADE WALLS DESIGNED CONTINUOUS WITH FOUNDATION STEM WALLS[a, b, c, d, e, k, l]

MAXIMUM WIND SPEED (mph)			HEIGHT OF STEM WALL[h, i] (feet)	MAXIMUM DESIGN LATERAL SOIL LOAD (psf/ft)	MAXIMUM UNSUPPORTED HEIGHT OF ABOVE-GRADE WALL (feet)	MINIMUM VERTICAL REINFORCEMENT-BAR SIZE AND SPACING (inches)[f, g]						
Exposure Category						Wall type and nominal thickness[j] (inches)						
						Flat				Waffle		Screen
B	C	D				4	6	8	10	6	8	6
85	—	—	3	30	8	4@33	4@39	4@48	4@48	4@24	4@28	4@22
					10	4@26	5@48	4@41	4@48	4@19	4@22	4@18
				60	10	4@21	5@40	5@48	4@44	4@16	4@19	4@15
			6	30	10	DR	5@22	6@35	6@43	DR	4@11	DR
				60	10	DR	DR	6@26	6@28	DR	DR	DR
90	—	—	3	30	8	4@30	4@36	4@48	4@48	4@22	4@26	4@21
					10	4@24	5@44	4@38	4@48	4@17	4@21	4@17
				60	10	4@20	5@37	4@48	4@41	4@15	4@18	4@14
			6	30	10	DR	5@21	6@35	6@41	DR	4@10	DR
				60	10	DR	DR	6@26	6@28	DR	DR	DR
100	85	—	3	30	8	4@26	5@48	4@42	4@48	4@19	4@23	4@18
					10	4@20	5@37	4@33	4@41	4@15	4@18	4@14
			6	60	10	4@17	5@34	5@44	4@36	4@13	4@17	4@12
				30	10	DR	5@20	6@35	6@38	DR	4@9	DR
				60	10	DR	DR	6@24	6@28	DR	DR	DR
110	90	85	3	30	8	4@22	5@42	4@37	4@46	4@16	4@20	4@16
					10	4@17	5@34	5@44	4@35	4@12	4@17	4@12
				60	10	4@15	5@34	5@39	5@48	4@11	4@17	4@11
			6	30	10	DR	5@18	6@35	6@35	DR	4@9	DR
				60	10	DR	DR	6@23	6@28	DR	DR	DR
120	100	90	3	30	8	4@19	5@37	5@48	4@40	4@14	4@17	4@14
					10	4@14	5@34	5@38	5@48	4@11	4@17	4@10
				60	10	4@13	5@33	6@48	5@43	4@10	4@16	4@9
			6	30	10	DR	5@16	6@33	6@32	DR	4@8	DR
				60	10	DR	DR	6@22	6@28	DR	DR	DR
130	110	100	3	30	8	4@17	5@34	5@44	4@36	4@12	4@17	4@10
					10	DR	5@32	6@47	5@42	4@9	4@15	DR
				60	10	DR	5@29	6@43	5@39	DR	4@14	DR
			6	30	10	DR	5@15	6@30	6@29	DR	4@7	DR
				60	10	DR	DR	6@21	6@27	DR	DR	DR

For SI: 1 inch = 25.4 mm, 1 foot = 304.8 mm, 1 mile per hour = 0.447 m/s, 1 pound per square inch = 6.895 kPa, 1 square foot = 0.0929 m^2.

a. Table is based on ASCE 7 components and cladding wind pressures for an enclosed building using a mean roof height of 35 feet, interior wall area 4, an effective wind area of 10 square feet, and topographic factor, K_{zt}, and importance factor, I, equal to 1.0.

b. Table is based on concrete with a minimum specified compressive strength of 2,500 psi.

c. See Section R611.6.5 for location of reinforcement in wall.

d. Deflection criterion is $L/240$, where L is the height of the wall in inches from the exterior finish ground level to the top of the above-grade wall.

e. Interpolation is not permitted. For intermediate values of basic wind speed, heights of stem wall and above-grade wall, and design lateral soil load, use next higher value.

f. Where No. 4 reinforcing bars at a spacing of 48 inches are specified in the table, use of bars with a minimum yield strength of 40,000 psi or 60,000 psi is permitted.

g. Other than for No. 4 bars spaced at 48 inches on center, table values are based on reinforcing bars with a minimum yield strength of 60,000 psi. Maximum spacings shown are the values calculated for the specified bar size. In waffle and screen-grid walls where the bar used is Grade 60 and the size specified in the table, the actual spacing in the wall shall not exceed a whole-number multiple of 12 inches (i.e., 12, 24, 36 and 48) that is less than or equal to the tabulated spacing. Vertical reinforcement with a yield strength of less than 60,000 psi and/or bars of a different size than specified in the table are permitted in accordance with Section R611.5.4.7 and Table R611.5.4(2).

(continued)

TABLE R611.6(4)—continued
MINIMUM VERTICAL REINFORCEMENT FOR FLAT, WAFFLE- AND SCREEN-GRID ABOVE-GRADE WALLS DESIGNED CONTINUOUS WITH FOUNDATION STEM WALLS[a, b, c, d, e, k, l]

h. Height of stem wall is the distance from the exterior finish ground level to the top of the slab-on-ground.
i. Where the distance from the exterior finish ground level to the top of the slab-on-ground is equal to or greater than 4 feet, the stem wall shall be laterally supported at the top and bottom before backfilling. Where the wall is designed and constructed to be continuous with the above-grade wall, temporary supports bracing the top of the stem wall shall remain in place until the above-grade wall is laterally supported at the top by floor or roof construction.
j. See Table R611.3 for tolerances on nominal thicknesses, and minimum core dimensions and maximum spacing of horizontal and vertical cores for waffle- and screen-grid walls.
k. Tabulated values are applicable to construction where gravity loads bear on top of wall, and conditions where gravity loads from floor construction are transferred to wall from a wood ledger or cold-formed steel track bolted to side of wall. See Tables R611.6(1), R611.6(2) and R611.6(3).
l. DR indicates design required.

❖ See the commentary for Section R611.6.2. In addition to the load combinations for dead, live and wind loads, also included is the lateral soil load [i.e., 30 and 60 psf/ft (4.7 and 9.4 kPa/m) of depth]. In addition, the floor live load, *L*, on the slab-on-ground [40 psf (1.9 kPa)] was included as a surcharge load. The surcharge load is the product of the floor live load and the coefficient of earth pressure, *K*, implied by the lateral soil load used in the design of the stem wall. Since lateral soil load is the product of *K* and the density of the soil, by assuming a soil density of 110 pcf (1762 kg/m^3), a value of *K* was computed for each soil class.

R611.6.4 Termination of reinforcement. Where indicated in Items 1 through 3, vertical wall reinforcement in the top-most *story* with concrete walls shall be terminated with a 90-degree (1.57 rad) standard hook complying with Section R611.5.4.5 and Figure R611.5.4(3).

1. Vertical bars adjacent to door and window openings required by Section R611.8.1.2.
2. Vertical bars at the ends of required solid wall segments. See Section R611.7.2.2.2.
3. Vertical bars (other than end bars, see Item 2) used as shear reinforcement in required solid wall segments where the reduction factor for design strength, R_3, used is based on the wall having horizontal and vertical shear reinforcement. See Section R611.7.2.2.3.

The bar extension of the hook shall be oriented parallel to the horizontal wall reinforcement and be within 4 inches (102 mm) of the top of the wall.

Horizontal reinforcement shall be continuous around the building corners by bending one of the bars and lap-splicing it with the bar in the other wall in accordance with Section R611.5.4.3 and Figure R611.5.4(1).

Exception: In lieu of bending horizontal reinforcement at corners, separate bent reinforcing bars shall be permitted provided that the bent bar is lap-spliced with the horizontal reinforcement in both walls in accordance with Section R611.5.4.3 and Figure R611.5.4(1).

In required solid wall segments where the reduction factor for design strength, R_3, is based on the wall having horizontal and vertical shear reinforcement in accordance with Section R611.7.2.2.1, horizontal wall reinforcement shall be terminated with a standard hook complying with Section R611.5.4.5 and Figure R611.5.4(3) or in a lap-splice, except at corners where the reinforcement shall be continuous as required above.

❖ This section prescribes the required termination of the vertical reinforcement at the top of the top-most story and the termination of the horizontal reinforcement at the building corners.

The vertical wall reinforcement in the top-most story must terminate with a 90-degree (1.57 rad) standard hook oriented parallel to the horizontal wall reinforcement and within 4 inches (102 mm) of the top of the wall.

This applies to the following vertical bars:

1. The required minimum vertical wall reinforcement.
2. Vertical bars adjacent to door and window openings.
3. Vertical bars at the ends of required solid wall segments.
4. Vertical shear reinforcing bars (except other than end bars for solid wall segments) in accordance with Section R611.7.2.2.3.

The requirement that vertical wall reinforcement be terminated with a standard hook is based on current standards for conventional masonry construction. The requirement has proven very effective in masonry construction conditions with wind speeds of 110 mph (49 m/s) or greater. In the 2006 code the requirement for hooks was triggered at a design wind pressure greater than 40 pounds per square foot (psf). The requirement for hooks in the 2009 code provides for a factored roof uplift force of 1000 plf. The hook allows the full tensile strength of the reinforcing bar to be developed and therefore provides additional tensile strength in the concrete wall to resist the large roof uplift forces in high wind areas.

The horizontal reinforcing must be continuous at the building corners. The continuity is achieved by bending the horizontal bar in one wall around the corner with sufficient length to lap splice with the horizontal bar in the other wall. The lap splice is to be in accordance with Section R611.5.4.3. The exception permits the use of a bent corner bar with sufficient length of each leg to lap splice with the horizontal bars in both walls.

R611.6.5 Location of reinforcement in wall. Except for vertical reinforcement at the ends of required solid wall segments, which shall be located as required by Section R611.7.2.2.2, the location of the vertical reinforcement shall not vary from the center of the wall by more than the greater of 10 percent of the wall thickness and $^3/_8$-inch (10 mm). Horizontal and vertical reinforcement shall be located to provide not less than the minimum cover required by Section R611.5.4.1.

❖ The amount of reinforcement required by the tables in Section R611.6 for a given condition of wall type, height, thickness and eccentricity of loads supported was computed based on the centroid of the vertical reinforcement being at the center of the wall. In engineering terms, the effective depth, d, was one-half the wall thickness. The 2006 code required the reinforcement to be located within the center one-third of the wall thickness. Depending upon how this was interpreted, this allowed the centroid of the vertical reinforcement to vary by up to one-sixth the wall thickness from the location assumed in the calculations. This meant that the effective depth for a 6-inch (152 mm) wall was reduced 1 inch (25 mm) if the reinforcement was positioned toward the flexural compression side of the wall. This amounted to a reduction in d of approximately 33 percent. Since design moment strength is approximately proportional to the effective depth, this misplacement of the reinforcement from where it was assumed in the design resulted in an approximately 33-percent reduction in design moment strength.

Section R611.6.5 requires that the vertical reinforcement be placed at the center of the wall with a tolerance of the larger of 10 percent of the wall thickness and $^3/_8$-inch (9.5 mm). The latter criterion is based on the tolerance on d in ACI 318 for members with effective depth, d, less than or equal to 8 inches. The ACI 318 tolerance of $^3/_8$-inch (9.5 mm) is approximately 10 percent of the wall thickness of a $3^1/_2$-inch (89 mm) wall. Based on this, it was decided that if 10 percent tolerance was permitted for a very thin wall, then it should also be acceptable for thicker walls.

The location for vertical reinforcement at the ends of solid wall segments must be in accordance with Section R611.7.2.2.2.

R611.7 Solid walls for resistance to lateral forces.

❖ The tables in Section R611.6 are based on concrete walls without door or window openings. This simplified approach rarely arises in residential construction since walls generally contain windows and doors to meet functional needs. The amount of openings affects the lateral (racking) strength of the building parallel to the wall, which resists wind and seismic forces. Section R611.7 addresses the minimum amount of solid wall required to resist in-plane shear loads from wind and seismic forces.

This section prescribes the required length of solid wall, the solid wall segments, minimum length of solid wall segments, maximum spacing of segments, required reinforcement in solid wall segments and required solid wall segments at the ends (corners) of all exterior walls.

R611.7.1 Length of solid wall. Each exterior wall line in each *story* shall have a total length of solid wall required by Section R611.7.1.1. A solid wall is a section of flat, waffle-grid or screen-grid wall, extending the full *story height* without openings or penetrations, except those permitted by Section R611.7.2. Solid wall segments that contribute to the total length of solid wall shall comply with Section R611.7.2.

❖ This section defines a solid wall. A solid wall is permitted to have only openings or penetrations as allowed by Section R611.7.2.

R611.7.1.1 Length of solid wall for wind. All buildings shall have solid walls in each exterior endwall line (the side of a building that is parallel to the span of the roof or floor framing) and sidewall line (the side of a building that is perpendicular to the span of the roof or floor framing) to resist lateral in-plane wind forces. The site-appropriate basic wind speed and exposure category shall be used in Tables R611.7(1A) through (1C) to determine the unreduced total length, *UR*, of solid wall required in each exterior endwall line and sidewall line. For buildings with a mean roof height of less than 35 feet (10 668 mm), the unreduced values determined from Tables R611.7(1A) though (1C) is permitted by multiplying by the applicable factor, R_1, from Table R611.7(2); however, reduced values shall not be less than the minimum values in Tables R611.7(1A) through (1C). Where the floor-to-ceiling height of a *story* is less than 10 feet (3048 mm), the unreduced values determined from Tables R611.7(1A) through (C), including minimum values, is permitted to be reduced by multiplying by the applicable factor, R_2, from Table R611.7(3). To account for different design strengths than assumed in determining the values in Tables R611.7(1A) through (1C), the unreduced lengths determined from Tables R611.7(1A) through (1C), including minimum values, are permitted to be reduced by multiplying by the applicable factor, R_3, from Table R611.7(4). The reductions permitted by Tables R611.7(2), R611.7(3) and R611.7(4) are cumulative.

The total length of solid wall segments, *TL*, in a wall line that comply with the minimum length requirements of Section R611.7.2.1 [see Figure R611.7(1)] shall be equal to or greater than the product of the unreduced length of solid wall from Tables R611.7(1A) through (1C), *UR* and the applicable reduction factors, if any, from Tables R611.7(2), R611.7(3) and R611.7(4) as indicated by Equation R6-1.

$$TL \geq R_1 \cdot R_2 \cdot R_3 \cdot UR \qquad \textbf{(Equation R6-1)}$$

where:

TL = Total length of solid wall segments in a wall line that comply with Section R611.7.2.1 [see Figure R611.7(1)];

R_1 = 1.0 or reduction factor for mean roof height from Table R611.7(2);

R_2 = 1.0 or reduction factor for floor-to-ceiling wall height from Table R611.7(3);

R_3 = 1.0 or reduction factor for design strength from Table R611.7(4), and

UR = Unreduced length of solid wall from Tables R611.7(1A) through (1C).

The total length of solid wall in a wall line, *TL*, shall not be less than that provided by two solid wall segments complying with the minimum length requirements of Section R611.7.2.1.

To facilitate determining the required wall thickness, wall type, number and *grade* of vertical bars at the each end of each solid wall segment, and whether shear reinforcement is required, use of Equation R6-2 is permitted.

$$R \leq \frac{TL}{R_1 \cdot R_2 \cdot UR} \qquad \textbf{(Equation R6-2)}$$

After determining the maximum permitted value of the reduction factor for design strength, R_3, in accordance with Equation R6-2, select a wall type from Table R611.7(4) with R_3 less than or equal to the value calculated.

❖ All buildings must have minimum solid wall length as described in Figure R611.7(1).

Each endwall and sidewall [as shown in Figure R611.7(1)] must have solid walls to resist the lateral in-plane wind forces.

The unreduced length, UR, of solid wall is determined from Tables R611.7(1A), R611.7(1B) and R611.7(1C) and then reduced by the appropriate reduction factors from Tables R611.7(2), R611.7(3) and R611.7(4). The minimum length cannot be less than that required by Section R611.7.2.1. The minimum length must have two solid wall segments complying with Section R611.7.2.1.

Table R611.7(4) provides reduction factors for increased design strength of solid wall segments. Previous prescriptive methods do not recognize that different lengths of solid wall segments and different amounts of vertical reinforcement near the end of a segment of the same length generally increase the design strength with regard to the in-plane force the segment is capable of resisting. In order to recognize these benefits, solid wall segments of different lengths, thicknesses, number and size of bars at the ends of the segments, and presence or absence of horizontal shear reinforcement were evaluated. Segments were evaluated as cantilevered flexural elements fixed at the base with a height of 8 feet (2438 mm). It was felt that 8 feet (2438 mm) was a reasonable height to use for this calculation, even where the floor-to-ceiling wall height may be 10 feet (3048 mm), because the tops of most wall openings are 8 feet (2438 mm) or less above the floor. Design moment strengths for solid wall segments in flat walls were computed based on the parameters listed below:

1. Solid wall segment length
2. Segment thickness (i.e., nominal thickness minus $^1/_2$ inch)
3. Number of reinforcing bars at end of segment and their location
4. Size of reinforcing bars at end of segment
5. Yield strength of reinforcement
6. Specified compressive strength of concrete

Using these parameters, design moment strengths were computed for each unique solid wall segment combination shown in Table R611.7(4).

For waffle and screen-grid walls, design moment strengths of solid wall segments were computed based on the assumption of a flanged section. The flange width in the through-the-wall dimension was assumed to be 5.5 inches (140 mm) for 6-inch (152 mm) waffle- and screen-grid walls, and 7.5 inches (191 mm) for 8-inch (203 mm) waffle-grid walls. The depth of the concrete compression zone, a, was computed to ensure that a sufficient depth of concrete (in the in-plane direction) is provided in the area of the segment under compression. Sufficient depth is ensured in the field by complying with note e of Table R611.7(4). In the 2006 code, the design shear strength (based on concrete alone) of solid wall segments was computed based on the area of the vertical core of concrete. In Table R611.7(4), the design shear strength (based on concrete alone) of solid wall segments was computed based on an assumed web thickness of 2.6 inches (66 mm) for 6-inch and 8-inch (152 mm and 203 mm) nominal waffle-grid walls and 2.2 inches (56 mm) for 6-inch (152 mm) nominal screen-grid walls. These values, which are based on recommendations contained in Testing and Design of Lintels Using Insulating Concrete Forms, give some credit to the flanges for their contribution in resisting shear stresses that is not accounted for in the provisions of ACI 318.

Tables R611.7(1A), R611.7(1B) and R611.7(1C) are based on the analytical procedure (Method 2) of the wind load provisions of ASCE 7 for low-rise buildings. ASCE 7 defines a low-rise building as an enclosed or partially enclosed building with a mean roof height of less than or equal to 60 feet (18 288 mm) in which the mean roof height does not exceed the least horizontal dimension. Commentary Table R611.7.1.1 shows the velocity pressures for the various basic wind speeds and exposure categories and how they were grouped for purposes of designs in the table. The velocities pressures are based on a building with a mean roof height of 35 feet (10 668 mm). The values in the shaded cells were used to compute the in-plane forces to be resisted by solid walls in each of the two exterior wall lines parallel to the direction of the wind.

To compute the forces, external pressure coefficients, GCpf, from Figure 6-10 of ASCE 7 were used to calculate the design wind pressures for various building surfaces and these pressures were applied to the building as shown in Figure 6-10. The following assumptions were used to calculate the forces.

1. Floor-to-ceiling wall heights (all stories)—10 feet (3048 mm)
2. Thickness of second-floor assembly, if applicable—16 inches (406 mm)

3. Thickness of attic floor/ceiling assembly above the top story ceiling—12 inches (305 mm)
4. Width of roof overhang (when computing wind forces perpendicular to ridge)—24 inches (610 mm)

ASCE 7 requires that the main wind-force-resisting system be designed for the larger of the forces computed as described above, and a force of 10 psf multiplied by the area of the building projected onto a vertical plane normal to the assumed wind direction. The minimum force generally governs for the design of buildings having lower sloped roofs [i.e., approximately 5 in 12 (23 degrees) or less] sited in areas with lower basic wind speeds, as illustrated by the shaded cells in Tables R611.7(1A), R611.7(1B) and R611.7(1C).

The unadjusted solid wall lengths are based on a design strength of 840 pounds per foot (132 kPa/m) of length of solid wall segment and the forces computed as described above were divided by 840 plf to obtain the unadjusted lengths of solid wall tabulated in Tables R611.7(1A), R611.7(1B) and R611.7(1C).

R611.7.2 Solid wall segments. Solid wall segments that contribute to the required length of solid wall shall comply with this section. Reinforcement shall be provided in accordance with Section R611.7.2.2 and Table R611.7(4). Solid wall segments shall extend the full story-height without openings, other than openings for the utilities and other building services passing through the wall. In flat walls and waffle-grid walls, such openings shall have an area of less than 30 square inches (19 355 mm^2) with no dimension exceeding $6^1/_4$ inches (159 mm), and shall not be located within 6 inches (152 mm) of the side edges of the solid wall segment. In screen-grid walls, such openings shall be located in the portion of the solid wall segment between horizontal and vertical cores of concrete and opening size and location are not restricted provided no concrete is removed.

❖ Provisions for limited openings in solid wall segments recognize the reality that mandating a segment with no openings is not practical. Prohibiting the opening within 6 inches (152 mm) of the edge of a solid wall segment is intended to prevent holes from reducing the area of concrete in the compression zone.

R611.7.2.1 Minimum length of solid wall segment and maximum spacing. Only solid wall segments equal to or greater than 24 inches (610 mm) in length shall be included in the total length of solid wall required by Section R611.7.1. In addition, no more than two solid wall segments equal to or greater than 24 inches (610 mm) in length and less than 48 inches (1219 mm) in length shall be included in the required total length of solid wall. The maximum clear opening width shall be 18 feet (5486 mm). See Figure R611.7(1).

❖ A minimum of two solid wall segments is required for each solid wall line, as shown in Figure R611.7(1). Only solid wall segments that are a minimum of 24 inches (610 mm) in length are to be counted in the total required length of solid wall. No more than two solid wall segments are permitted to be less than 48 inches (1219 mm). The maximum spacing (clear opening width) between segments is 18 feet (5486 mm).

R611.7.2.2 Reinforcement in solid wall segments.

❖ This section contains the provisions for horizontal and vertical shear reinforcement in solid wall segments where the option to provide the shear reinforcement from Table R611.7(4) is chosen.

Also, this section prescribes the location and termination of the required vertical bars at the ends of the solid wall segments.

This section also contains the provisions for solid wall segments at corners.

R611.7.2.2.1 Horizontal shear reinforcement. Where reduction factors for design strength, R_3, from Table R611.7(4) based on horizontal and vertical shear reinforcement being provided are used, solid wall segments shall have horizontal reinforcement consisting of minimum No. 4 bars. Horizontal shear reinforcement shall be the same grade of steel required for the vertical reinforcement at the ends of solid wall segments by Section R611.7.2.2.2.

The spacing of horizontal reinforcement shall not exceed the smaller of one-half the length of the solid wall segment, minus 2 inches (51 mm), and 18 inches (457 mm). Horizontal shear reinforcement shall terminate in accordance with Section R611.6.4.

❖ There are two options with respect to horizontal shear reinforcement. If the option to use no horizontal and vertical shear reinforcement is assumed in selecting the reduction factor (R_3) for design strength from Table R611.7(4), horizontal reinforcement must be provided in accordance with Section R611.6.2. On the other hand, if the option to use horizontal and vertical shear reinforcement is assumed in selecting the reduction factor (R_3) for design strength from Table R611.7.4, horizontal reinforcement must be installed in accordance with Section R611.7.2.2.1.

R611.7.2.2.2 Vertical reinforcement. Vertical reinforcement applicable to the reduction factor(s) for design strength, R_3, from Table R611.7(4) that is used, shall be located at each end of each solid wall segment in accordance with the applicable detail in Figure R611.7(2). The No. 4 vertical bar required on each side of an opening by Section R611.8.1.2 is permitted to be used as reinforcement at the ends of solid wall segments where installed in accordance with the applicable detail in Figure R611.7(2). There shall be not less than two No. 4 bars at each end of solid wall segments located as required by the applicable detail in Figure R611.7(2). One of the bars at each end of solid wall segments shall be deemed to meet the requirements for vertical wall reinforcement required by Section R611.6.

The vertical wall reinforcement at each end of each solid wall segment shall be developed below the bottom of the adjacent wall opening [see Figure R611.7(3)] by one of the following methods:

1. Where the wall height below the bottom of the adjacent opening is equal to or greater than 22 inches (559 mm) for No. 4 or 28 inches (711 mm) for No. 5 vertical wall reinforcement, reinforcement around openings in accordance with Section R611.8.1 shall be sufficient, or
2. Where the wall height below the bottom of the adjacent opening is less than required by Item 1 above, the vertical wall reinforcement adjacent to the opening shall extend into the footing far enough to develop the bar in tension in accordance with Section R611.5.4.4 and Figure R611.5.4(2), or shall be lap-spliced with a dowel that is embedded in the footing far enough to develop the dowel-bar in tension.

❖ Solid wall segments must have vertical rebars at each end of the segment. The location is to be in accordance with this section and Figure R611.7(2). The vertical rebars at each end of the segment must be developed below the adjacent opening in accordance with this section and Figure R611.7(3).

R611.7.2.2.3 Vertical shear reinforcement. Where reduction factors for design strength, R_3, from Table R611.7(4) based on horizontal and vertical shear reinforcement being provided are used, solid wall segments shall have vertical reinforcement consisting of minimum No. 4 bars. Vertical shear reinforcement shall be the same grade of steel required by Section R611.7.2.2.2 for the vertical reinforcement at the ends of solid wall segments. The spacing of vertical reinforcement throughout the length of the segment shall not exceed the smaller of one third the length of the segment, and 18 inches (457 mm). Vertical shear reinforcement shall be continuous between stories in accordance with Section R611.6.3, and shall terminate in accordance with Section R611.6.4. Vertical shear reinforcement required by this section is permitted to be used for vertical reinforcement required by Table R611.6(1), R611.6(2), R611.6(3) or R611.6(4), whichever is applicable.

❖ There are two options with respect to vertical shear reinforcement. If the option to use no horizontal and vertical shear reinforcement is assumed in selecting the reduction factor (R_3) for design strength from Table R611.7(4), vertical reinforcement must be provided in accordance with Sections R611.6.2 and R611.7.2.2.2. On the other hand, if the option to use horizontal and vertical shear reinforcement is assumed in selecting the reduction factor (R_3) for design strength from Table R611.7(4), vertical reinforcement must be installed in accordance with Sections R611.7.2.2.2 and R611.7.2.2.3.

R611.7.2.3 Solid wall segments at corners. At all interior and exterior corners of exterior walls, a solid wall segment shall extend the full height of each wall *story*. The segment shall have the length required to develop the horizontal reinforcement above and below the adjacent opening in tension in accordance with Section R611.5.4.4. For an exterior corner, the limiting dimension is measured on the outside of the wall, and for an interior corner the limiting dimension is measured on the inside of the wall. See Section R611.8.1. The length of a segment contributing to the required length of solid wall shall comply with Section R611.7.2.1.

The end of a solid wall segment complying with the minimum length requirements of Section R611.7.2.1 shall be located no more than 6 feet (1829 mm) from each corner.

❖ A full-height (story height) solid wall segment must be located at each corner (end) of an endwall line and each corner (end) of a sidewall line as shown in Figure R611.7(1).

The minimum length must be sufficient for the development length of the horizontal rebar above and below the adjacent opening and comply with Section R611.7.2.1.

The code permits the corner (end) segments, as shown in Figure R611.7(1), that comply with the minimum length requirements to be located no more than 6 feet (1829 mm) from each corner.

Table R611.7.1.1
VELOCITY PRESSURES

BASIC WIND SPEED (mph)			VELOCITY PRESSURE (psf)		
Exposure Category			Exposure Category		
B	C	D	B	C	D
85			11.51		
90			12.90		
100	85		15.93	15.95	
110	90	85	19.28	17.88	18.77
120	100	90	22.94	22.08	21.04
130	110	100	26.92	26.72	25.98

For SI: 1 mph = 0.4470 m/s, 1 psf = 0.0479 kN/m^2.

TABLE R611.7(1A)
UNREDUCED LENGTH, *UR*, OF SOLID WALL REQUIRED IN EACH EXTERIOR ENDWALL FOR WIND PERPENDICULAR TO RIDGE ONE STORY OR TOP STORY OF TWO STORY[a, c, d, e, f, g]

SIDEWALL LENGTH (feet)	ENDWALL LENGTH (feet)	ROOF SLOPE	UNREDUCED LENGTH, *UR*, OF SOLID WALL REQUIRED IN ENDWALLS FOR WIND PERPENDICULAR TO RIDGE (feet)						
			Basic Wind Speed (mph) Exposure						
			85B	90B	100B	110B	120B	130B	Minimum[b]
					85C	90C	100C	110C	
						85D	90D	100D	
15	15	< 1:12	0.90	1.01	1.25	1.51	1.80	2.11	0.98
		5:12	1.25	1.40	1.73	2.09	2.49	2.92	1.43
		7:12	1.75	1.96	2.43	2.93	3.49	4.10	1.64
		12:12	2.80	3.13	3.87	4.68	5.57	6.54	2.21
	30	< 1:12	0.90	1.01	1.25	1.51	1.80	2.11	1.09
		5:12	1.25	1.40	1.73	2.09	2.49	2.92	2.01
		7:12	2.43	2.73	3.37	4.08	4.85	5.69	2.42
		12:12	4.52	5.07	6.27	7.57	9.01	10.58	3.57
	45	< 1:12	0.90	1.01	1.25	1.51	1.80	2.11	1.21
		5:12	1.25	1.40	1.73	2.09	2.49	2.92	2.59
		7:12	3.12	3.49	4.32	5.22	6.21	7.29	3.21
		12:12	6.25	7.00	8.66	10.47	12.45	14.61	4.93
	60	< 1:12	0.90	1.01	1.25	1.51	1.80	2.11	1.33
		5:12	1.25	1.40	1.73	2.09	2.49	2.92	3.16
		7:12	3.80	4.26	5.26	6.36	7.57	8.89	3.99
		12:12	7.97	8.94	11.05	13.36	15.89	18.65	6.29
30	15	< 1:12	1.61	1.80	2.23	2.70	3.21	3.77	1.93
		5:12	2.24	2.51	3.10	3.74	4.45	5.23	2.75
		7:12	3.15	3.53	4.37	5.28	6.28	7.37	3.12
		12:12	4.90	5.49	6.79	8.21	9.77	11.46	4.14
	30	< 1:12	1.61	1.80	2.23	2.70	3.21	3.77	2.14
		5:12	2.24	2.51	3.10	3.74	4.45	5.23	3.78
		7:12	4.30	4.82	5.96	7.20	8.57	10.05	4.52
		12:12	7.79	8.74	10.80	13.06	15.53	18.23	6.57
	45	< 1:12	1.61	1.80	2.23	2.70	3.21	3.77	2.35
		5:12	2.24	2.51	3.10	3.74	4.45	5.23	4.81
		7:12	5.44	6.10	7.54	9.12	10.85	12.73	5.92
		12:12	10.69	11.98	14.81	17.90	21.30	25.00	9.00
	60	< 1:12	1.61	1.80	2.23	2.70	3.21	3.77	2.56
		5:12	2.24	2.51	3.10	3.74	4.45	5.23	5.84
		7:12	6.59	7.39	9.13	11.04	13.14	15.41	7.32
		12:12	13.58	15.22	18.82	22.75	27.07	31.77	11.43

(continued)

TABLE R611.7(1A)—continued
UNREDUCED LENGTH, UR, OF SOLID WALL REQUIRED IN EACH EXTERIOR ENDWALL FOR WIND PERPENDICULAR TO RIDGE ONE STORY OR TOP STORY OF TWO STORY[a, c, d, e, f, g]

SIDEWALL LENGTH (feet)	ENDWALL LENGTH (feet)	ROOF SLOPE	UNREDUCED LENGTH, UR, OF SOLID WALL REQUIRED IN ENDWALLS FOR WIND PERPENDICULAR TO RIDGE (feet)						
			Basic Wind Speed (mph) Exposure						
			85B	90B	100B	110B	120B	130B	Minimum[b]
					85C	90C	100C	110C	
						85D	90D	100D	
60	15	< 1:12	2.99	3.35	4.14	5.00	5.95	6.98	3.83
		5:12	4.15	4.65	5.75	6.95	8.27	9.70	5.37
		7:12	5.91	6.63	8.19	9.90	11.78	13.83	6.07
		12:12	9.05	10.14	12.54	15.16	18.03	21.16	8.00
	30	< 1:12	2.99	3.35	4.14	5.00	5.95	6.98	4.23
		5:12	4.15	4.65	5.75	6.95	8.27	9.70	7.31
		7:12	7.97	8.94	11.05	13.36	15.89	18.65	8.71
		12:12	14.25	15.97	19.74	23.86	28.40	33.32	12.57
	45	< 1:12	3.11	3.48	4.30	5.20	6.19	7.26	4.63
		5:12	4.31	4.84	5.98	7.23	8.60	10.09	9.25
		7:12	10.24	11.47	14.19	17.15	20.40	23.84	11.35
		12:12	19.84	22.24	27.49	33.23	39.54	46.40	17.14
	60	< 1:12	3.22	3.61	4.46	5.39	6.42	7.53	5.03
		5:12	4.47	5.01	6.19	7.49	8.91	10.46	11.19
		7:12	12.57	14.09	17.42	21.05	25.05	29.39	13.99
		12:12	25.61	28.70	35.49	42.90	51.04	59.90	21.71

For SI: 1 inch = 25.4 mm, 1 foot = 304.8 mm, 1 mile per hour = 0.447 m/s, 1 pound-force per linear foot = 0.146 kN/m, 1 pound per square foot = 47.88 Pa.

a. Tabulated lengths were derived by calculating design wind pressures in accordance with Figure 6-10 of ASCE 7 for a building with a mean roof height of 35 feet. For wind perpendicular to the ridge, the effects of a 2-foot overhang on each endwall are included. The design pressures were used to calculate forces to be resisted by solid wall segments in each endwall [Table R611.7(1A) or R611.7(1B) or sidewall (Table R611.7(1C)], as appropriate. The forces to be resisted by each wall line were then divided by the default design strength of 840 pounds per linear foot of length to determine the required solid wall length. The actual mean roof height of the building shall not exceed the least horizontal dimension of the building.

b. Tabulated lengths in the "minimum" column are based on the requirement of Section 6.1.4.1 of ASCE 7 that the main windforce-resisting system be designed for a minimum service level force of 10 psf multiplied by the area of the building projected onto a vertical plane normal to the assumed wind direction. Tabulated lengths in shaded cells are less than the "minimum" value. Where the minimum controls, it is permitted to be reduced in accordance with Notes c, d and e. See Section R611.7.1.1.

c. For buildings with a mean roof height of less than 35 feet, tabulated lengths are permitted to be reduced by multiplying by the appropriate factor, R_1, from Table R611.7(2). The reduced length shall not be less than the "minimum" value shown in the table.

d. Tabulated lengths for "one story or top story of two story" are based on a floor-to-ceiling height of 10 feet. Tabulated lengths for "first story of two story" are based on floor-to-ceiling heights of 10 feet each for the first and second story. For floor-to-ceiling heights less than assumed, use the lengths in Table R611.7(1A), (1B) or (1C), or multiply the value in the table by the reduction factor, R_2, from Table R611.7(3).

e. Tabulated lengths are based on the default design shear strength of 840 pounds per linear foot of solid wall segment. The tabulated lengths are permitted to be reduced by multiplying by the applicable reduction factor for design strength, R_3, from Table R611.7(4).

f. The reduction factors, R_1, R_2 and R_3, in Tables R611.7(2), R611.7(3), and R611.7(4), respectively, are permitted to be compounded, subject to the limitations of Note b. However, the minimum number and minimum length of solid walls segments in each wall line shall comply with Sections R611.7.1 and R611.7.2.1, respectively.

g. For intermediate values of sidewall length, endwall length, roof slope and basic wind speed, use the next higher value, or determine by interpolation.

❖ See the commentary for Section R611.7.1.1.

TABLE R611.7(1B)
UNREDUCED LENGTH, *UR*, OF SOLID WALL REQUIRED IN EACH EXTERIOR ENDWALL FOR WIND PERPENDICULAR TO RIDGE FIRST STORY OF TWO STORY[a, c, d, e, f, g]

SIDEWALL LENGTH (feet)	ENDWALL LENGTH (feet)	ROOF SLOPE	UNREDUCED LENGTH, *UR*, OF SOLID WALL REQUIRED IN ENDWALLS FOR WIND PERPENDICULAR TO RIDGE (feet)						
			Basic Wind Speed (mph) Exposure						
			85B	90B	100B	110B	120B	130B	Minimum[b]
					85C	90C	100C	110C	
						85D	90D	100D	
			Velocity pressure (psf)						
			11.51	12.90	15.95	19.28	22.94	26.92	
15	15	< 1:12	2.60	2.92	3.61	4.36	5.19	6.09	2.59
		5:12	3.61	4.05	5.00	6.05	7.20	8.45	3.05
		7:12	3.77	4.23	5.23	6.32	7.52	8.82	3.26
		12:12	4.81	5.40	6.67	8.06	9.60	11.26	3.83
	30	< 1:12	2.60	2.92	3.61	4.36	5.19	6.09	2.71
		5:12	3.61	4.05	5.00	6.05	7.20	8.45	3.63
		7:12	4.45	4.99	6.17	7.46	8.88	10.42	4.04
		12:12	6.54	7.33	9.06	10.96	13.04	15.30	5.19
	45	< 1:12	2.60	2.92	3.61	4.36	5.19	6.09	2.83
		5:12	3.61	4.05	5.00	6.05	7.20	8.45	4.20
		7:12	5.14	5.76	7.12	8.60	10.24	12.01	4.83
		12:12	8.27	9.27	11.46	13.85	16.48	19.34	6.55
	60	< 1:12	2.60	2.92	3.61	4.36	5.19	6.09	2.95
		5:12	3.61	4.05	5.00	6.05	7.20	8.45	4.78
		7:12	5.82	6.52	8.06	9.75	11.60	13.61	5.61
		12:12	9.99	11.20	13.85	16.74	19.92	23.37	7.90
30	15	< 1:12	4.65	5.21	6.45	7.79	9.27	10.88	5.16
		5:12	6.46	7.24	8.95	10.82	12.87	15.10	5.98
		7:12	6.94	7.78	9.62	11.62	13.83	16.23	6.35
		12:12	8.69	9.74	12.04	14.55	17.32	20.32	7.38
	30	< 1:12	4.65	5.21	6.45	7.79	9.27	10.88	5.38
		5:12	6.46	7.24	8.95	10.82	12.87	15.10	7.01
		7:12	8.09	9.06	11.21	13.54	16.12	18.91	7.76
		12:12	11.58	12.98	16.05	19.40	23.08	27.09	9.81
	45	< 1:12	4.65	5.21	6.45	7.79	9.27	10.88	5.59
		5:12	6.46	7.24	8.95	10.82	12.87	15.10	8.04
		7:12	9.23	10.35	12.79	15.46	18.40	21.59	9.16
		12:12	14.48	16.22	20.06	24.25	28.85	33.86	12.24
	60	< 1:12	4.65	5.21	6.45	7.79	9.27	10.88	5.80
		5:12	6.46	7.24	8.95	10.82	12.87	15.10	9.08
		7:12	10.38	11.63	14.38	17.38	20.69	24.27	10.56
		12:12	17.37	19.47	24.07	29.10	34.62	40.63	14.67

(continued)

TABLE R611.7(1B)—continued
UNREDUCED LENGTH, *UR*, OF SOLID WALL REQUIRED IN EACH EXTERIOR ENDWALL FOR WIND PERPENDICULAR TO RIDGE FIRST STORY OF TWO STORY[a, c, d, e, f, g]

SIDEWALL LENGTH (feet)	ENDWALL LENGTH (feet)	ROOF SLOPE	UNREDUCED LENGTH, *UR*, OF SOLID WALL REQUIRED IN ENDWALLS FOR WIND PERPENDICULAR TO RIDGE (feet)						
			Basic Wind Speed (mph) Exposure						Minimum[b]
			85B	90B	100B	110B	120B	130B	
					85C	90C	100C	110C	
						85D	90D	100D	
			Velocity pressure (psf)						
			11.51	12.90	15.95	19.28	22.94	26.92	
60	15	< 1:12	8.62	9.67	11.95	14.45	17.19	20.17	10.30
		5:12	11.98	13.43	16.61	20.07	23.88	28.03	11.85
		7:12	13.18	14.78	18.27	22.08	26.28	30.83	12.54
		12:12	16.32	18.29	22.62	27.34	32.53	38.17	14.48
	30	< 1:12	8.62	9.67	11.95	14.45	17.19	20.17	10.70
		5:12	11.98	13.43	16.61	20.07	23.88	28.03	13.79
		7:12	15.25	17.09	21.13	25.54	30.38	35.66	15.18
		12:12	21.52	24.12	29.82	36.05	42.89	50.33	19.05
	45	< 1:12	8.97	10.06	12.43	15.03	17.88	20.99	11.10
		5:12	12.46	13.97	17.27	20.88	24.84	29.15	15.73
		7:12	17.67	19.80	24.48	29.59	35.21	41.32	17.82
		12:12	27.27	30.56	37.79	45.68	54.35	63.78	23.62
	60	< 1:12	9.30	10.43	12.89	15.58	18.54	21.76	11.50
		5:12	12.91	14.47	17.90	21.63	25.74	30.20	17.67
		7:12	20.14	22.58	27.91	33.74	40.15	47.11	20.46
		12:12	33.19	37.19	45.99	55.59	66.14	77.62	28.19

For SI: 1 inch = 25.4 mm, 1 foot = 304.8 mm, 1 mile per hour = 0.447 m/s, 1 pound force per linear foot = 0.146 kN/m, 1 pound per square foot = 47.88 Pa.

a. Tabulated lengths were derived by calculating design wind pressures in accordance with Figure 6-10 of ASCE 7 for a building with a mean roof height of 35 feet. For wind perpendicular to the ridge, the effects of a 2-foot overhang on each endwall are included. The design pressures were used to calculate forces to be resisted by solid wall segments in each endwall [Table R611.7(1A) or R611.7(1B)] or sidewall [Table R611.7(1C)], as appropriate. The forces to be resisted by each wall line were then divided by the default design strength of 840 pounds per linear foot of length to determine the required solid wall length. The actual mean roof height of the building shall not exceed the least horizontal dimension of the building.

b. Tabulated lengths in the "minimum" column are based on the requirement of Section 6.1.4.1 of ASCE 7 that the main windforce-resisting system be designed for a minimum service level force of 10 psf multiplied by the area of the building projected onto a vertical plane normal to the assumed wind direction. Tabulated lengths in shaded cells are less than the "minimum" value. Where the minimum controls, it is permitted to be reduced in accordance with Notes c, d and e. See Section R611.7.1.1.

c. For buildings with a mean roof height of less than 35 feet, tabulated lengths are permitted to be reduced by multiplying by the appropriate factor, R_1, from Table R611.7(2). The reduced length shall not be less than the "minimum" value shown in the table.

d. Tabulated lengths for "one story or top story of two story" are based on a floor-to-ceiling height of 10 feet. Tabulated lengths for "first story of two story" are based on floor-to-ceiling heights of 10 feet each for the first and second story. For floor-to-ceiling heights less than assumed, use the lengths in Table R611.7(1A), (1B) or (1C), or multiply the value in the table by the reduction factor, R_2, from Table R611.7(3).

e. Tabulated lengths are based on the default design shear strength of 840 pounds per linear foot of solid wall segment. The tabulated lengths are permitted to be reduced by multiplying by the applicable reduction factor for design strength, R_3, from Table R611.7(4).

f. The reduction factors, R_1, R_2 and R_3, in Tables R611.7(2), R611.7(3), and R611.7(4), respectively, are permitted to be compounded, subject to the limitations of Note b. However, the minimum number and minimum length of solid walls segments in each wall line shall comply with Sections R611.7.1 and R611.7.2.1, respectively.

g. For intermediate values of sidewall length, endwall length, roof slope and basic wind speed, use the next higher value, or determine by interpolation.

❖ See the commentary for Section R611.7.1.1.

TABLE R611.7(1C)
UNREDUCED LENGTH, *UR*, OF SOLID WALL REQUIRED IN EACH EXTERIOR SIDEWALL FOR WIND PARALLEL TO RIDGE[a, c, d, e, f, g]

SIDEWALL LENGTH (feet)	ENDWALL LENGTH (feet)	ROOF SLOPE	UNREDUCED LENGTH, *UR*, OF SOLID WALL REQUIRED IN ENDWALLS FOR WIND PERPENDICULAR TO RIDGE (feet)						
			Basic Wind Speed (mph) Exposure						
			85B	90B	100B	110B	120B	130B	Minimum[b]
					85C	90C	100C	110C	
						85D	90D	100D	
			One story or top story of two story						
< 30	15	< 1:12	0.95	1.06	1.31	1.59	1.89	2.22	0.90
		5:12	1.13	1.26	1.56	1.88	2.24	2.63	1.08
		7:12	1.21	1.35	1.67	2.02	2.40	2.82	1.17
		12:12	1.43	1.60	1.98	2.39	2.85	3.34	1.39
	30	< 1:12	1.77	1.98	2.45	2.96	3.53	4.14	1.90
		5:12	2.38	2.67	3.30	3.99	4.75	5.57	2.62
		7:12	2.66	2.98	3.69	4.46	5.31	6.23	2.95
		12:12	3.43	3.85	4.76	5.75	6.84	8.03	3.86
	45	< 1:12	2.65	2.97	3.67	4.43	5.27	6.19	2.99
		5:12	3.98	4.46	5.51	6.66	7.93	9.31	4.62
		7:12	4.58	5.14	6.35	7.68	9.14	10.72	5.36
		12:12	6.25	7.01	8.67	10.48	12.47	14.63	7.39
	60	< 1:12	3.59	4.03	4.98	6.02	7.16	8.40	4.18
		5:12	5.93	6.65	8.22	9.93	11.82	13.87	7.07
		7:12	6.99	7.83	9.69	11.71	13.93	16.35	8.38
		12:12	9.92	11.12	13.75	16.62	19.77	23.21	12.00
60	45	< 1:12	2.77	3.11	3.84	4.65	5.53	6.49	2.99
		5:12	4.15	4.66	5.76	6.96	8.28	9.72	4.62
		7:12	4.78	5.36	6.63	8.01	9.53	11.18	5.36
		12:12	6.51	7.30	9.03	10.91	12.98	15.23	7.39
	60	< 1:12	3.86	4.32	5.35	6.46	7.69	9.02	4.18
		5:12	6.31	7.08	8.75	10.57	12.58	14.76	7.07
		7:12	7.43	8.32	10.29	12.44	14.80	17.37	8.38
		12:12	10.51	11.78	14.56	17.60	20.94	24.57	12.00

(continued)

TABLE R611.7(1C)—continued
UNREDUCED LENGTH, *UR*, OF SOLID WALL REQUIRED IN EACH EXTERIOR SIDEWALL FOR WIND PARALLEL TO RIDGE[a, c, d, e, f, g]

SIDEWALL LENGTH (feet)	ENDWALL LENGTH (feet)	ROOF SLOPE	UNREDUCED LENGTH, *UR*, OF SOLID WALL REQUIRED IN ENDWALLS FOR WIND PERPENDICULAR TO RIDGE (feet)						
			Basic Wind Speed (mph) Exposure						
			85B	90B	100B	110B	120B	130B	Minimum[b]
					85C	90C	100C	110C	
						85D	90D	100D	
			One story or top story of two story						
First story of two story									
< 30	15	< 1:12	2.65	2.97	3.67	4.44	5.28	6.20	2.52
		5:12	2.83	3.17	3.92	4.74	5.64	6.62	2.70
		7:12	2.91	3.26	4.03	4.87	5.80	6.80	2.79
		12:12	3.13	3.51	4.34	5.25	6.24	7.32	3.01
	30	< 1:12	4.81	5.39	6.67	8.06	9.59	11.25	5.14
		5:12	5.42	6.08	7.52	9.09	10.81	12.69	5.86
		7:12	5.70	6.39	7.90	9.55	11.37	13.34	6.19
		12:12	6.47	7.25	8.97	10.84	12.90	15.14	7.10
	45	< 1:12	6.99	7.83	9.69	11.71	13.93	16.35	7.85
		5:12	8.32	9.33	11.53	13.94	16.59	19.47	9.48
		7:12	8.93	10.01	12.37	14.95	17.79	20.88	10.21
		12:12	10.60	11.88	14.69	17.75	21.13	24.79	12.25
	60	< 1:12	9.23	10.35	12.79	15.46	18.40	21.59	10.65
		5:12	11.57	12.97	16.03	19.38	23.06	27.06	13.54
		7:12	12.63	14.15	17.50	21.15	25.17	29.54	14.85
		12:12	15.56	17.44	21.56	26.06	31.01	36.39	18.48
60	45	< 1:12	7.34	8.22	10.17	12.29	14.62	17.16	7.85
		5:12	8.72	9.77	12.08	14.60	17.37	20.39	9.48
		7:12	9.34	10.47	12.95	15.65	18.62	21.85	10.21
		12:12	11.08	12.41	15.35	18.55	22.07	25.90	12.25
	60	< 1:12	9.94	11.14	13.77	16.65	19.81	23.25	10.65
		5:12	12.40	13.89	17.18	20.76	24.70	28.99	13.54
		7:12	13.51	15.14	18.72	22.63	26.92	31.60	14.85
		12:12	16.59	18.59	22.99	27.79	33.06	38.80	18.48

For SI: 1 inch = 25.4 mm, 1 foot = 304.8 mm, 1 mile per hour = 0.447 m/s, 1 pound force per linear foot = 0.146 kN/m, 1 pound per square foot = 47.88 Pa.

a. Tabulated lengths were derived by calculating design wind pressures in accordance with Figure 6-10 of ASCE 7 for a building with a mean roof height of 35 feet. For wind perpendicular to the ridge, the effects of a 2-foot overhang on each endwall are included. The design pressures were used to calculate forces to be resisted by solid wall segments in each endwall [Table R611.7(1A) or R611.7(1B)] or sidewall [(Table R611.7(1C)], as appropriate. The forces to be resisted by each wall line were then divided by the default design strength of 840 pounds per linear foot of length to determine the required solid wall length. The actual mean roof height of the building shall not exceed the least horizontal dimension of the building.

b. Tabulated lengths in the "minimum" column are based on the requirement of Section 6.1.4.1 of ASCE 7 that the main windforce-resisting system be designed for a minimum service level force of 10 psf multiplied by the area of the building projected onto a vertical plane normal to the assumed wind direction. Tabulated lengths in shaded cells are less than the "minimum" value. Where the minimum controls, it is permitted to be reduced in accordance with Notes c, d and e. See Section R611.7.1.1.

c. For buildings with a mean roof height of less than 35 feet, tabulated lengths are permitted to be reduced by multiplying by the appropriate factor, R_1, from Table R611.7(2). The reduced length shall not be less than the "minimum" value shown in the table.

d. Tabulated lengths for "one story or top story of two story" are based on a floor-to-ceiling height of 10 feet. Tabulated lengths for "first story of two story" are based on floor-to-ceiling heights of 10 feet each for the first and second story. For floor-to-ceiling heights less than assumed, use the lengths in Table R611.7(1A), (1B) or (1C), or multiply the value in the table by the reduction factor, R_2, from Table R611.7(3).

e. Tabulated lengths are based on the default design shear strength of 840 pounds per linear foot of solid wall segment. The tabulated lengths are permitted to be reduced by multiplying by the applicable reduction factor for design strength, R_3, from Table R611.7(4).

f. The reduction factors, R_1, R_2 and R_3, in Tables R611.7(2), R611.7(3), and R611.7(4), respectively, are permitted to be compounded, subject to the limitations of Note b. However, the minimum number and minimum length of solid walls segments in each wall line shall comply with Sections R611.7.1 and R611.7.2.1, respectively.

g. For intermediate values of sidewall length, endwall length, roof slope and basic wind speed, use the next higher value, or determine by interpolation.

❖ See the commentary for Section R611.7.11.

TABLE R611.7(2)
REDUCTION FACTOR, R_1, FOR BUILDINGS WITH MEAN ROOF HEIGHT LESS THAN 35 FEET[a]

MEAN ROOF HEIGHT[b, c] (feet)	REDUCTION FACTOR R_1, FOR MEAN ROOF HEIGHT		
	Exposure category		
	B	C	D
< 15	0.96	0.84	0.87
20	0.96	0.89	0.91
25	0.96	0.93	0.94
30	0.96	0.97	0.98
35	1.00	1.00	1.00

For SI: 1 foot = 304.8 mm, 1 degree = 0.0175 rad.

a. See Section R611.7.1.1 and Note c to Table R611.7(1A) for application of reduction factors in this table. This reduction is not permitted for "minimum" values.

b. For intermediate values of mean roof height, use the factor for the next greater height, or determine by interpolation.

c. Mean roof height is the average of the roof eave height and height of the highest point on the roof surface, except that for roof slopes of less than or equal to $2^1/_8$:12 (10 degrees), the mean roof height is permitted to be taken as the roof eave height.

❖ See the commentary for Section R611.7.1.1.

TABLE R611.7(3)
REDUCTION FACTOR, R_2, FOR FLOOR-TO-CEILING WALL HEIGHTS LESS THAN 10 FEET[a, b]

STORY UNDER CONSIDERATION	FLOOR-TO-CEILING HEIGHT[c] (feet)	ENDWALL LENGTH (feet)	ROOF SLOPE	REDUCTION FACTOR, R_2
Endwalls—for wind perpendicular to ridge				
One story or top story of two story	8	15	< 5:12	0.83
			7:12	0.90
			12:12	0.94
		60	< 5:12	0.83
			7:12	0.95
			12:12	0.98
First story of two story	16 combined first and second story	15	< 5:12	0.83
			7:12	0.86
			12:12	0.89
		60	< 5:12	0.83
			7:12	0.91
			12:12	0.95
Sidewalls—for wind parallel to ridge				
One story or top story of two story	8	15	< 1:12	0.84
			5:12	0.87
			7:12	0.88
			12:12	0.89
		60	< 1:12	0.86
			5:12	0.92
			7:12	0.93
			12:12	0.95
First story of two story	16 combined first and second story	15	< 1:12	0.83
			5:12	0.84
			7:12	0.85
			12:12	0.86
		60	< 1:12	0.84
			5:12	0.87
			7:12	0.88
			12:12	0.90

For SI: 1 foot = 304.8 mm.

a. See Section R611.7.1.1 and Note d to Table R611.7(1A) for application of reduction factors in this table.

b. For intermediate values of endwall length, and/or roof slope, use the next higher value, or determine by interpolation.

c. Tabulated values in Table R611.7(1A) and (1C) for "one story or top story of two story" are based on a floor-to-ceiling height of 10 feet. Tabulated values in Table R611.7(1B) and (1C) for "first story of two story" are based on floor-to-ceiling heights of 10 feet each for the first and second story. For floor to ceiling heights between those shown in this table and those assumed in Table R611.7(1A), (1B) or (1C), use the solid wall lengths in Table R611.7(1A), (1B) or (1C), or determine the reduction factor by interpolating between 1.0 and the factor shown in this table.

❖ See the commentary for Section R611.7.1.1.

TABLE R611.7(4)
REDUCTION FACTOR FOR DESIGN STRENGTH, R_3, FOR FLAT, WAFFLE- AND SCREEN-GRID WALLS[a, c]

NOMINAL THICKNESS OF WALL (inches)	VERTICAL BARS AT EACH END OF SOLID WALL SEGMENT		VERTICAL REINFORCEMENT LAYOUT DETAIL [see Figure R611.7(2)]	REDUCTION FACTOR, R_3, FOR LENGTH OF SOLID WALL			
				Horizontal and vertical shear reinforcement provided			
				No		Yes[d]	
	Number of bars	Bar size		40,000[b]	60,000[b]	40,000[b]	60,000[b]
Flat walls							
4	2	4	1	0.74	0.61	0.74	0.50
	3	4	2	0.61	0.61	0.52	0.27
	2	5	1	0.61	0.61	0.48	0.25
	3	5	2	0.61	0.61	0.26	0.18
6	2	4	3	0.70	0.48	0.70	0.48
	3	4	4	0.49	0.38	0.49	0.33
	2	5	3	0.46	0.38	0.46	0.31
	3	5	4	0.38	0.38	0.32	0.16
8	2	4	3	0.70	0.47	0.70	0.47
	3	4	5	0.47	0.32	0.47	0.32
	2	5	3	0.45	0.31	0.45	0.31
	4	4	6	0.36	0.28	0.36	0.25
	3	5	5	0.31	0.28	0.31	0.16
	4	5	6	0.28	0.28	0.24	0.12
10	2	4	3	0.70	0.47	0.70	0.47
	2	5	3	0.45	0.30	0.45	0.30
	4	4	7	0.36	0.25	0.36	0.25
	6	4	8	0.25	0.22	0.25	0.13
	4	5	7	0.24	0.22	0.24	0.12
	6	5	8	0.22	0.22	0.12	0.08
Waffle-grid walls[e]							
6	2	4	3	0.78	0.78	0.70	0.48
	3	4	4	0.78	0.78	0.49	0.25
	2	5	3	0.78	0.78	0.46	0.23
	3	5	4	0.78	0.78	0.24	0.16
8	2	4	3	0.78	0.78	0.70	0.47
	3	4	5	0.78	0.78	0.47	0.24
	2	5	3	0.78	0.78	0.45	0.23
	4	4	6	0.78	0.78	0.36	0.18
	3	5	5	0.78	0.78	0.23	0.16
	4	5	6	0.78	0.78	0.18	0.13
Screen-grid walls[e]							
6	2	4	3	0.93	0.93	0.70	0.48
	3	4	4	0.93	0.93	0.49	0.25
	2	5	3	0.93	0.93	0.46	0.23
	3	5	4	0.93	0.93	0.24	0.16

For SI: 1 inch = 25.4 mm, 1,000 pounds per square inch = 6.895 MPa.

a. See Note e to Table R611.7(1A) for application of adjustment factors in this table.

b. Yield strength in pounds per square inch of vertical wall reinforcement at ends of solid wall segments.

c. Values are based on concrete with a specified compressive strength, f'_c, of 2,500 psi. Where concrete with f'_c of not less than 3,000 psi is used, values in shaded cells are permitted to be decreased by multiplying by 0.91.

d. Horizontal and vertical shear reinforcement shall be provided in accordance with Section R611.7.2.2.

e. Each end of each solid wall segment shall have rectangular flanges. In the through-the-wall dimension, the flange shall not be less than $5^1/_2$ inches for 6-inch-nominal waffle- and screen-grid walls, and not less than $7^1/_2$ inches for 8-inch-nominal waffle-grid walls. In the in-plane dimension, flanges shall be long enough to accommodate the vertical reinforcement required by the layout detail selected from Figure R611.7(2) and provide the cover required by Section R611.5.4.1. If necessary to achieve the required dimensions, form material shall be removed or use of flat wall forms is permitted.

❖ See the commentary for Sections R611.7.1.1, R611.7.2.2, R611.7.2.2.1 and R611.7.2.2.3.

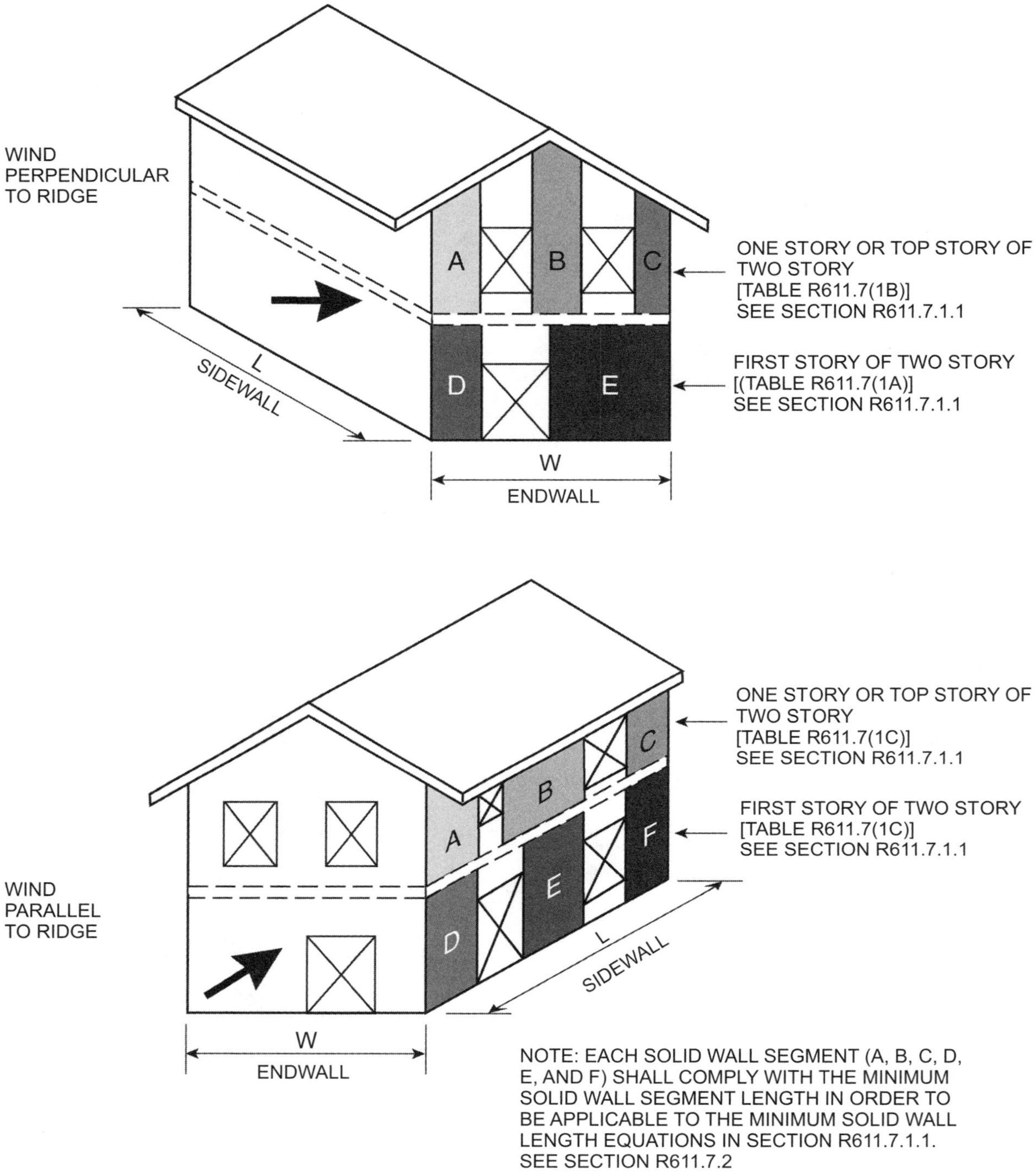

FIGURE R611.7(1)
MINIMUM SOLID WALL LENGTH

❖ See the commentary for Sections R611.7.1.1, R611.7.2.1 and R611.7.2.3.

Detail No.	Nom. wall thickness, in.	Reinforcement layout at ends of solid wall segments	Notes
1	4	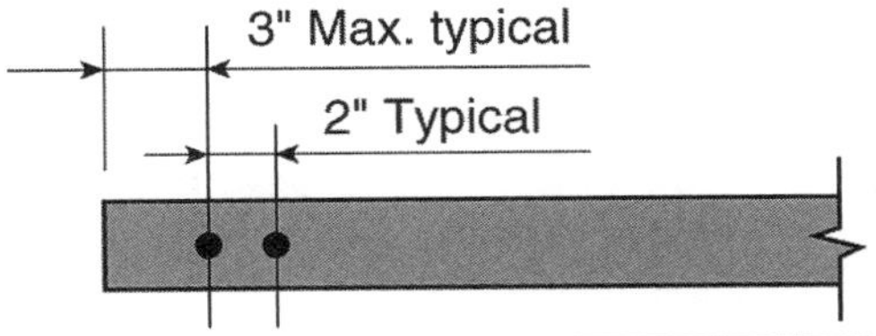	For SI: 1" = 25.4 mm 1. See Table R611.7(4) for use of details. 2. Minimum length of solid wall segment, and size and grade of reinforcement in each end of each solid wall segment shall be determined from Table R611.7(4). 3. For minimum cover requirements, see Section R611.5.4.1. 4. For details 3 – 8 where two or more bars are in the same row parallel to the end of the segment, place bars so that corner bars are as close to the sides of the wall segments as minimum cover requirements of Section R611.5.4.1 will permit. 5. For waffle- and screen-grid walls, each end of each solid wall segment shall have rectangular flanges. In the through-the-wall dimension, the flange shall not be less than 5.5 inches for 6-inch nominal waffle- and screen-grid forms, and not less than 7.5 inches for 8-inch nominal waffle-grid forms. In the in-plane dimension, flanges shall be long enough to accommodate the vertical reinforcement required by the layout detail selected and provide the cover required by Section R611.5.4.1. If necessary to achieve the required dimensions, form material shall be removed or flat wall forms are permitted to be used. See Table R611.7(4), Note e.
2	4		
3	6 8 10	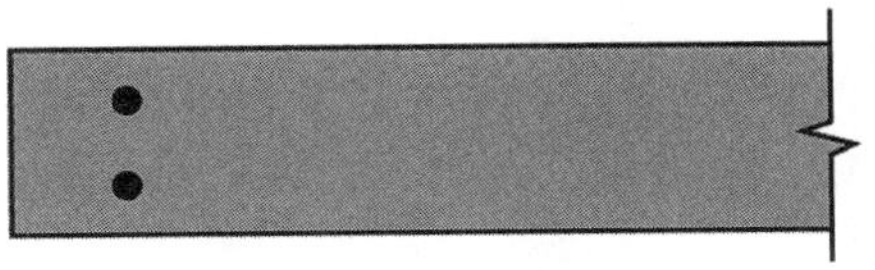	
4	6	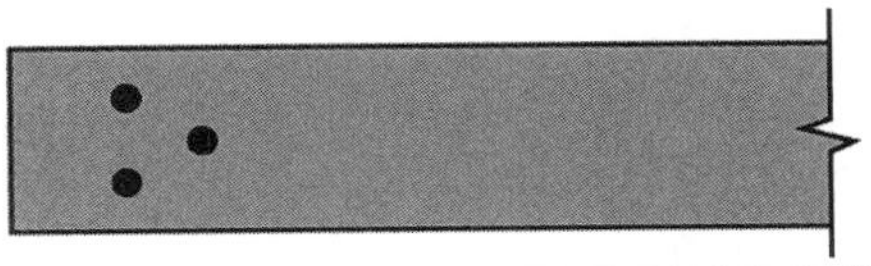	
5	8	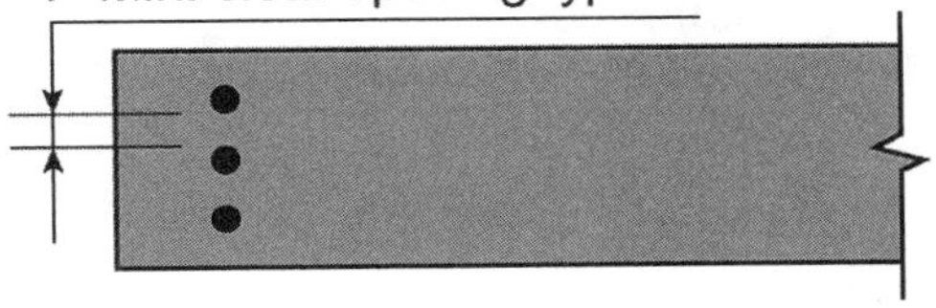	
6	8		
7	10		
8	10	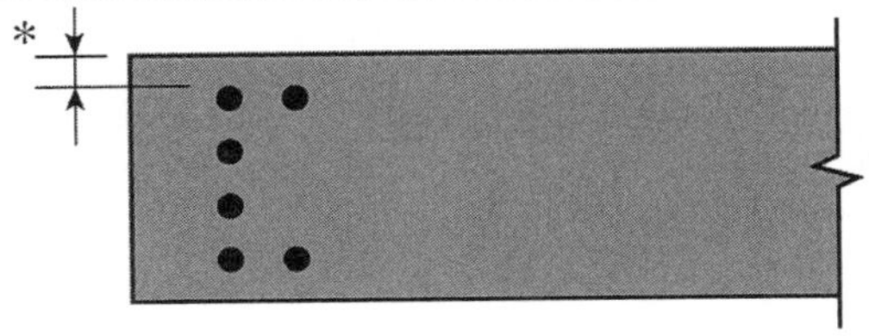 * For minimum cover see Section R611.5.4.1	

FIGURE R611.7(2)
VERTICAL REINFORCEMENT LAYOUT DETAIL

❖ See the commentary for Section R611.7.2.2.2.

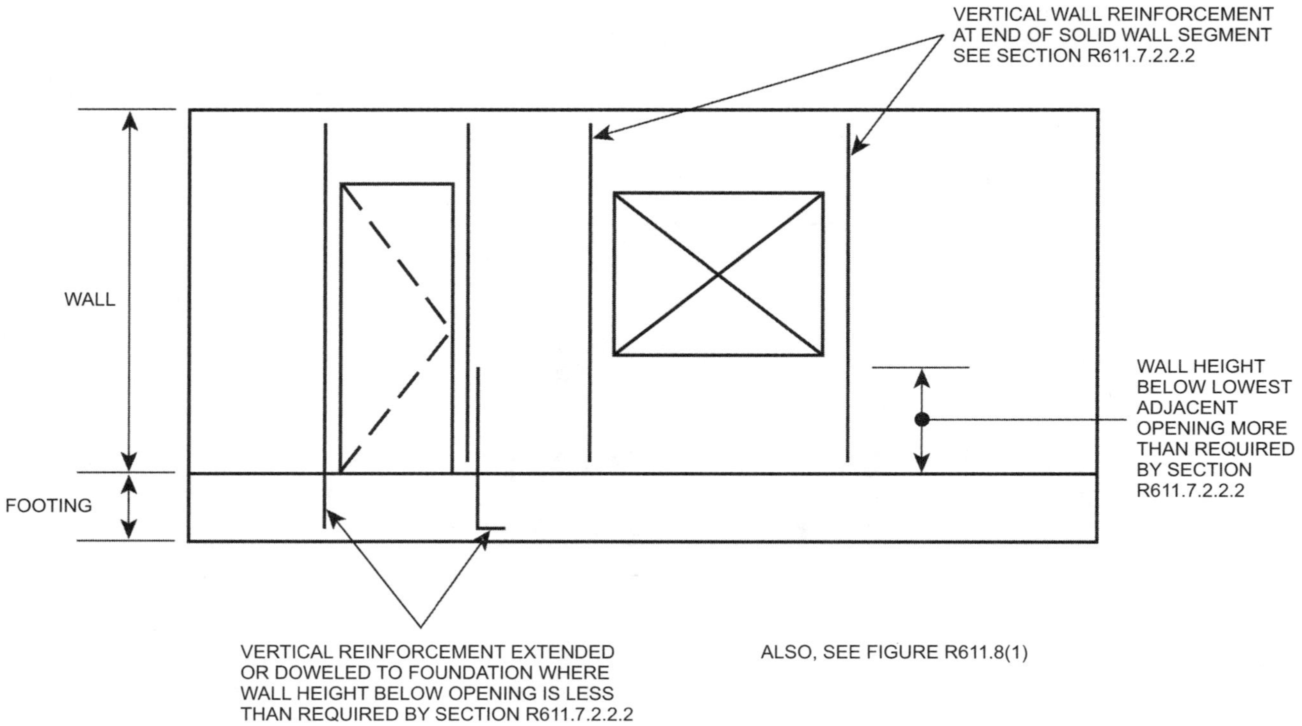

FIGURE R611.7(3)
VERTICAL WALL REINFORCEMENT ADJACENT TO WALL OPENINGS

❖ See the commentary for Section R611.7.2.2.2.

R611.8 Requirements for lintels and reinforcement around openings.

❖ This section contains the requirements for reinforcement around openings and lintels above openings as prescribed in Sections R611.8.1 and R611.8.2.

R611.8.1 Reinforcement around openings. Reinforcement shall be provided around openings in walls equal to or greater than 2 feet (610 mm) in width in accordance with this section and Figure R611.8(1), in addition to the minimum wall reinforcement required by Sections R404.1.2, R611.6 and R611.7. Vertical wall reinforcement required by this section is permitted to be used as reinforcement at the ends of solid wall segments required by Section R611.7.2.2.2 provided it is located in accordance with Section R611.8.1.2. Wall openings shall have a minimum depth of concrete over the width of the opening of 8 inches (203 mm) in flat walls and waffle-grid walls, and 12 inches (305 mm) in screen-grid walls. Wall openings in waffle-grid and screen-grid walls shall be located such that not less than one-half of a vertical core occurs along each side of the opening.

❖ Reinforcement of openings less than 2 feet (610 mm) in width is not required. Only openings equal to or greater than 2 feet (610 mm) in width require reinforcement. The required opening reinforcement is in addition to the minimum wall reinforcement required for concrete foundation walls, above-grade walls and solid wall segments. See Figure R611.8(1) for the minimum reinforcement required at wall openings.

R611.8.1.1 Horizontal reinforcement. Lintels complying with Section R611.8.2 shall be provided above wall openings equal to or greater than 2 feet (610 mm) in width.

Exception: Continuous horizontal wall reinforcement placed within 12 inches (305 mm) of the top of the wall *story* as required in Sections R404.1.2.2 and R611.6.2 is permitted in lieu of top or bottom lintel reinforcement required by Section R611.8.2 provided that the continuous horizontal wall reinforcement meets the location requirements specified in Figures R611.8(2), R611.8(3), and R611.8(4) and the size requirements specified in Tables R611.8(2) through R611.8(10).

Openings equal to or greater than 2 feet (610 mm) in width shall have a minimum of one No. 4 bar placed within 12 inches (305 mm) of the bottom of the opening. See Figure R611.8(1).

Horizontal reinforcement placed above and below an opening shall extend beyond the edges of the opening the dimension required to develop the bar in tension in accordance with Section R611.5.4.4.

❖ Horizontal reinforcement is required both above and below all openings equal to or greater than 2 feet (610 mm) in width.

The reinforcement above the opening must be a lintel complying with Section R611.8.2. Lintels are horizontal members used to transfer wall, floor, roof, and attic dead loads, snow and live loads above opening in walls to the sides of the opening. The exception permits the continuous horizontal wall reinforcement required for foundation walls and above-grade walls to serve as the top or bottom lintel reinforcement provided it meets the requirement of Section R611.8.2. The reinforcement below the opening must be a No. 4 bar placed within 12 inches (305 mm) of the bottom of the opening [see Figure R611.8(1)].

R611.8.1.2 Vertical reinforcement. Not less than one No. 4 bar [Grade 40 (280 MPa)] shall be provided on each side of openings equal to or greater than 2 feet (610 mm) in width. The vertical reinforcement required by this section shall extend the full height of the wall story and shall be located within 12 inches (305 mm) of each side of the opening. The vertical reinforcement required on each side of an opening by this section is permitted to serve as reinforcement at the ends of solid wall segments in accordance with Section R611.7.2.2.2, provided it is located as required by the applicable detail in Figure R611.7(2). Where the vertical reinforcement required by this section is used to satisfy the requirements of Section R611.7.2.2.2 in waffle- and screen-grid walls, a concrete flange shall be created at the ends of the solid wall segments in accordance with Table R611.7(4), note e. In the top-most story, the reinforcement shall terminate in accordance with Section R611.6.4.

❖ The minimum vertical reinforcement required on each side of the opening is to provide resistance to roof uplift loads. One No. 4 bar on each side of the opening is required and will resist roof uplift load based on the design wind speeds defined in Section R611.2. The vertical reinforcement must extend the full height of the wall story and be within 12 inches (305 mm) of each side of the opening as shown in Figure R611.8(1).

This minimum vertical reinforcement is permitted to serve as reinforcement at the ends of solid wall segments (see Section R611.7.2.2.2).

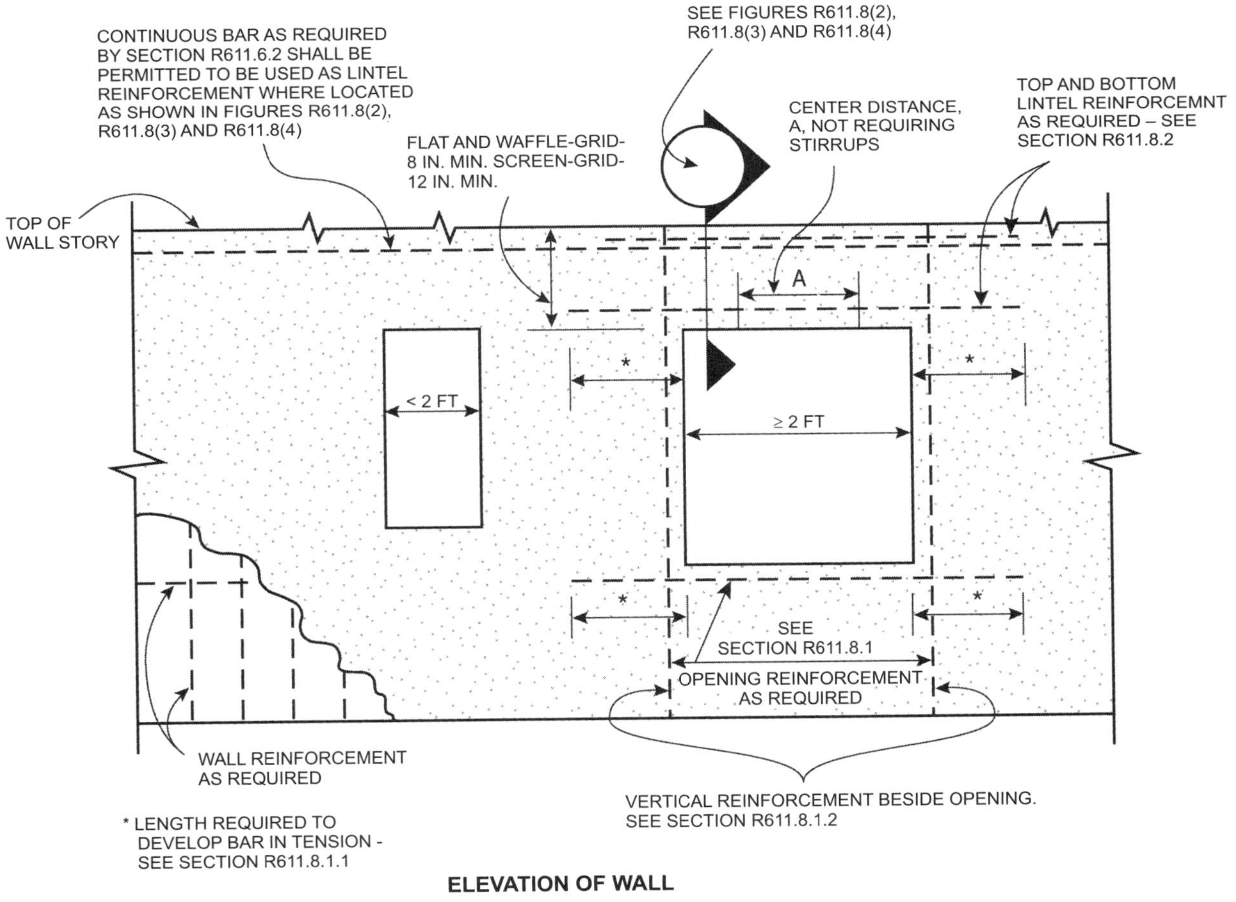

For SI: 1 inch = 25.4 mm, 1 foot = 304.8 mm.

FIGURE R611.8(1)
REINFORCEMENT OF OPENINGS

❖ See the commentary for Sections R611.8.1, R611.8.1.1 and R611.8.1.2.

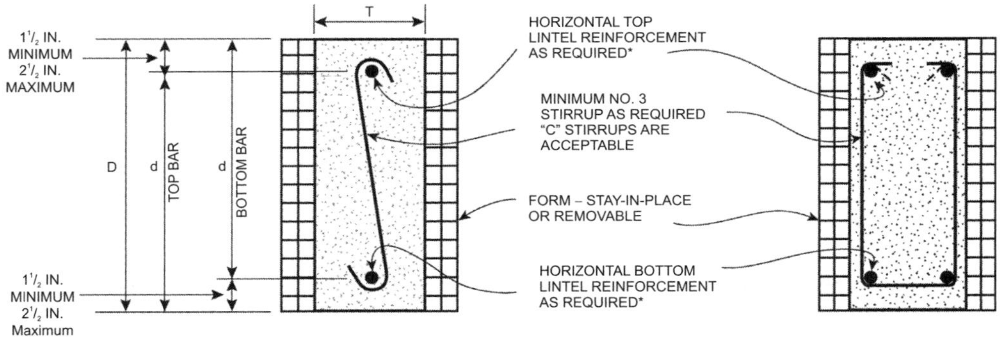

*FOR BUNDLED BARS, SEE SECTION R611.8.2.2.
SECTION CUT THROUGH FLAT WALL LINTEL

For SI: 1 inch = 25.4 mm.

FIGURE R611.8(2)
LINTEL FOR FLAT WALLS

❖ See the commentary for Sections R611.8.2 and R611.8.2.1.

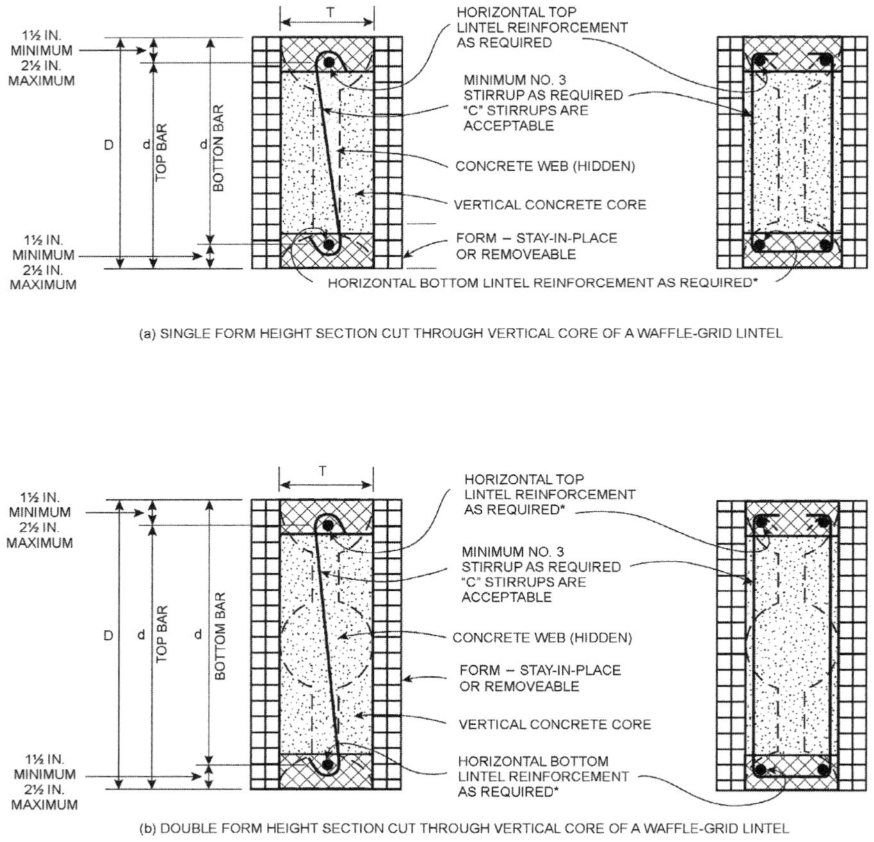

*FOR BUNDLED BARS, SEE SECTION R611.8.2.2.

NOTE: CROSS-HATCHING REPRESENTS THE AREA IN WHICH FORM MATERIAL SHALL BE REMOVED, IF NECESSARY, TO CREATE FLANGES CONTINUOUS THE LENGTH OF THE LINTEL. FLANGES SHALL HAVE A MINIMUM THICKNESS OF 3 IN., AND A MINIMUM WIDTH OF 5 IN. AND 7 IN. IN 6 IN. NOMINAL AND 8 IN. NOMINAL WAFFLE-GRID WALLS, RESPECTIVELY. SEE NOTE a TO TABLES R611.8(6) AND R611.8(10).

For SI:1 inch = 25.4 mm.

FIGURE R611.8(3)
LINTELS FOR WAFFLE-GRID WALLS

❖ See the commentary for Sections R611.8.2 and R611.8.2.1.

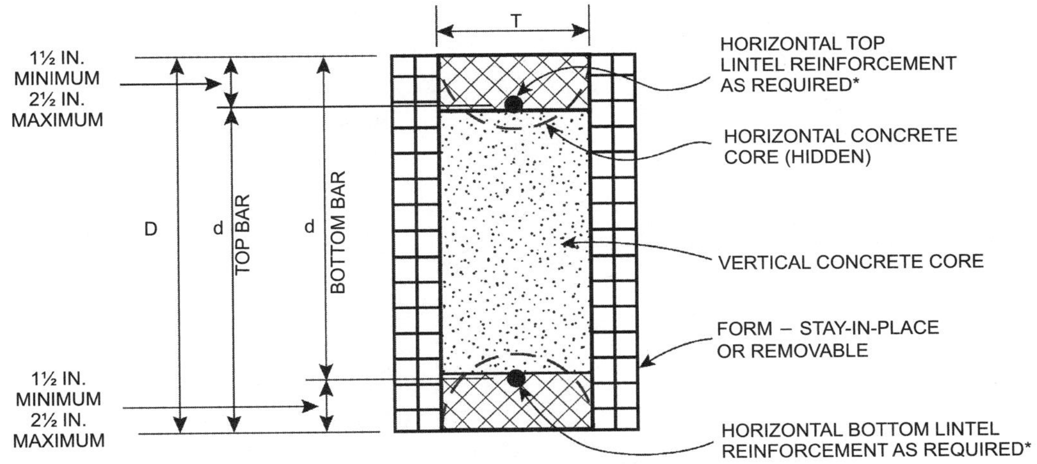

(a) SINGLE FORM HEIGHT SECTION CUT THROUGH VERTICAL CORE OF A SCREEN-GRID LINTEL

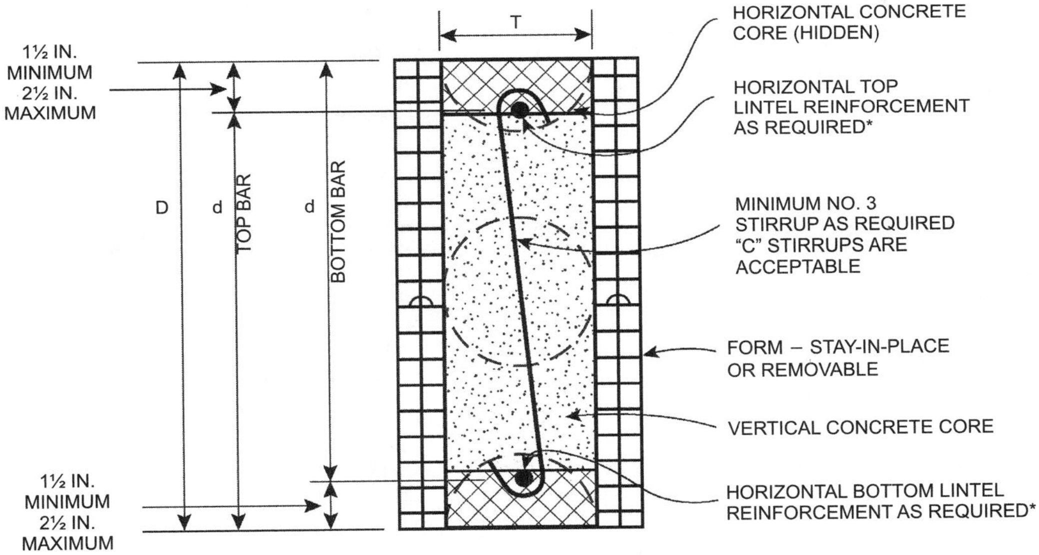

(b) DOUBLE FORM HEIGHT SECTION CUT THROUGH VERTICAL CORE OF A SCREEN-GRID LINTEL

*FOR BUNDLED BARS, SEE SECTION R611.8.2.2.

NOTE: CROSS-HATCHING REPRESENTS THE AREA IN WHICH FORM MATERIAL SHALL BE REMOVED, IF NECESSARY, TO CREATE FLANGES CONTINUOUS THE LENGTH OF THE LINTEL. FLANGES SHALL HAVE A MINIMUM THICKNESS OF 2.5 IN. AND A MINIMUM WIDTH OF 5 IN. SEE NOTE a TO TABLES R611.8(8) AND R611.8(10).

For SI: 1 inch = 25.4 mm.

FIGURE R611.8(4)
LINTELS FOR SCREEN-GRID WALLS

❖ See the commentary for Sections R611.8.2 and R611.8.2.1.

TABLE R611.8(1)
LINTEL DESIGN LOADING CONDITIONS[a, b, d]

DESCRIPTION OF LOADS AND OPENINGS ABOVE INFLUENCING DESIGN OF LINTEL			DESIGN LOAD CONDITION[c]
Opening in wall of top story of two-story building, or first story of one-story building			
Wall supporting loads from roof, including attic floor, if applicable, and	Top of lintel equal to or less than W/2 below top of wall		2
	Top of lintel greater than W/2 below top of wall		NLB
Wall not supporting loads from roof or attic floor			NLB
Opening in wall of first story of two-story building where wall immediately above is of concrete construction, or opening in basement wall of one-story building where wall immediately above is of concrete construction			
LB ledger board mounted to side of wall with bottom of ledger less than or equal to W/2 above top of lintel, and	Top of lintel greater than W/2 below bottom of opening in story above		1
	Top of lintel less than or equal to W/2 below bottom of opening in story above, and	Opening is entirely within the footprint of the opening in the story above	1
		Opening is partially within the footprint of the opening in the story above	4
LB ledger board mounted to side of wall with bottom of ledger more than W/2 above top of lintel			NLB
NLB ledger board mounted to side of wall with bottom of ledger less than or equal to W/2 above top of lintel, or no ledger board, and	Top of lintel greater than W/2 below bottom of opening in story above		NLB
	Top of lintel less than or equal to W/2 below bottom of opening in story above, and	Opening is entirely within the footprint of the opening in the story above	NLB
		Opening is partially within the footprint of the opening in the story above	1
Opening in basement wall of two-story building where walls of two stories above are of concrete construction			
LB ledger board mounted to side of wall with bottom of ledger less than or equal to W/2 above top of lintel, and	Top of lintel greater than W/2 below bottom of opening in story above		1
	Top of lintel less than or equal to W/2 below bottom of opening in story above, and	Opening is entirely within the footprint of the opening in the story above	1
		Opening is partially within the footprint of the opening in the story above	5
LB ledger board mounted to side of wall with bottom of ledger more than W/2 above top of lintel			NLB
NLB ledger board mounted to side of wall with bottom of ledger less than or equal to W/2 above top of lintel, or no ledger board, and	Top of lintel greater than W/2 below bottom of opening in story above		NLB
	Top of lintel less than or equal to W/2 below bottom of opening in story above, and	Opening is entirely within the footprint of the opening in the story above	NLB
		Opening is partially within the footprint of the opening in the story above	1
Opening in wall of first story of two-story building where wall immediately above is of light-framed construction, or opening in basement wall of one-story building, where wall immediately above is of light-framed construction			
Wall supporting loads from roof, second floor and top-story wall of light-framed construction, and	Top of lintel equal to or less than W/2 below top of wall		3
	Top of lintel greater than W/2 below top of wall		NLB
Wall not supporting loads from roof or second floor			NLB

a. LB means load bearing, NLB means nonload bearing, and W means width of opening.

b. Footprint is the area of the wall below an opening in the story above, bounded by the bottom of the opening and vertical lines extending downward from the edges of the opening.

c. For design loading condition "NLB" see Tables R611.8(9) and R611.8(10). For all other design loading conditions see Tables R611.8(2) through R611.8(8).

d. A NLB ledger board is a ledger attached to a wall that is parallel to the span of the floor, roof or ceiling framing that supports the edge of the floor, ceiling or roof.

❖ See the commentary for Section R611.8.2.1.

TABLE R611.8(2)
MAXIMUM ALLOWABLE CLEAR SPANS FOR 4-INCH-NOMINAL THICK FLAT LINTELS IN LOAD-BEARING WALLS[a, b, c, d, e, f, m]
ROOF CLEAR SPAN 40 FEET AND FLOOR CLEAR SPAN 32 FEET

LINTEL DEPTH, D[g] (inches)	NUMBER OF BARS AND BAR SIZE IN TOP AND BOTTOM OF LINTEL	STEEL YIELD STRENGTH[h], f_y (psi)	DESIGN LOADING CONDITION DETERMINED FROM TABLE R611.8(1)								
			1	2		3		4		5	
			Maximum ground snow load (psf)								
			—	30	70	30	70	30	70	30	70
			Maximum clear span of lintel (feet - inches)								
8	Span without stirrups[i, j]		3-2	3-4	2-4	2-6	2-2	2-1	2-0	2-0	2-0
	1-#4	40,000	5-2	5-5	4-1	4-3	3-10	3-7	3-4	2-9	2-9
		60,000	6-2	6-5	4-11	5-1	4-6	4-2	3-8	2-11	2-10
	1-#5	40,000	6-3	6-7	5-0	5-2	4-6	4-2	3-8	2-11	2-10
		60,000	DR	DR	DR	DR	DR	DR	DR	DR	DR
	Center distance A[k, l]		1-1	1-2	0-8	0-9	0-7	0-6	0-5	0-4	0-4
12	Span without stirrups[i, j]		3-4	3-7	2-9	2-11	2-8	2-6	2-5	2-2	2-2
	1-#4	40,000	6-7	7-0	5-4	5-7	5-0	4-9	4-4	3-8	3-7
		60,000	7-11	8-6	6-6	6-9	6-0	5-9	5-3	4-5	4-4
	1-#5	40,000	8-1	8-8	6-7	6-10	6-2	5-10	5-4	4-6	4-5
		60,000	9-8	10-4	7-11	8-2	7-4	6-11	6-2	4-10	4-8
	2-#4 1-#6	40,000	9-1	9-8	7-4	7-8	6-10	6-6	6-0	4-10	4-8
		60,000	DR	DR	DR	DR	DR	DR	DR	DR	DR
	Center distance A[k, l]		1-8	1-11	1-1	1-3	1-0	0-11	0-9	0-6	0-6
16	Span without stirrups[i, j]		4-7	5-0	3-11	4-0	3-8	3-7	3-4	3-1	3-0
	1-#4	40,000	6-8	7-3	5-6	5-9	5-2	4-11	4-6	3-10	3-8
		60,000	9-3	10-1	7-9	8-0	7-2	6-10	6-3	5-4	5-2
	1-#4	40,000	9-6	10-4	7-10	8-2	7-4	6-11	6-5	5-5	5-3
		60,000	11-5	12-5	9-6	9-10	8-10	8-4	7-9	6-6	6-4
	2-#4 1-#6	40,000	10-7	11-7	8-10	9-2	8-3	7-9	7-2	6-1	5-11
		60,000	12-9	13-10	10-7	11-0	9-10	9-4	8-7	6-9	6-6
	2-#5	40,000	13-0	14-1	10-9	11-2	9-11	9-2	8-2	6-6	6-3
		60,000	DR	DR	DR	DR	DR	DR	DR	DR	DR
	Center distance[k, l]		2-3	2-8	1-7	1-8	1-4	1-3	1-0	0-9	0-8
20	Span without stirrups[i, j]		5-9	6-5	5-0	5-2	4-9	4-7	4-4	3-11	3-11
	1-#4	40,000	7-5	8-2	6-3	6-6	5-10	5-7	5-1	4-4	4-2
		60,000	9-0	10-0	7-8	7-11	7-1	6-9	6-3	5-3	5-1
	1-#5	40,000	9-2	10-2	7-9	8-1	7-3	6-11	6-4	5-4	5-2
		60,000	12-9	14-2	10-10	11-3	10-1	9-7	8-10	7-5	7-3
	2-#4 1-#6	40,000	11-10	13-2	10-1	10-5	9-4	8-11	8-2	6-11	6-9
		60,000	14-4	15-10	12-1	12-7	11-3	10-9	9-11	8-4	8-1
	2-#5	40,000	14-7	16-2	12-4	12-9	11-4	10-6	9-5	7-7	7-3
		60,000	17-5	19-2	14-9	15-3	13-5	12-4	11-0	8-8	8-4
	2-#6	40,000	16-4	18-11	12-7	13-3	11-4	10-6	9-5	7-7	7-3
		60,000	DR	DR	DR	DR	DR	DR	DR	DR	DR
	Center distance A[k, l]		2-9	3-5	2-0	2-2	1-9	1-7	1-4	0-11	0-11

(continued)

TABLE R611.8(2)—continued
MAXIMUM ALLOWABLE CLEAR SPANS FOR 4-INCH-NOMINAL THICK FLAT LINTELS IN LOAD-BEARING WALLS[a, b, c, d, e, f, m]
ROOF CLEAR SPAN 40 FEET AND FLOOR CLEAR SPAN 32 FEET

LINTEL DEPTH, D[g] (inches)	NUMBER OF BARS AND BAR SIZE IN TOP AND BOTTOM OF LINTEL	STEEL YIELD STRENGTH[h], f_y (psi)	DESIGN LOADING CONDITION DETERMINED FROM TABLE R611.8(1)								
			1	2		3		4		5	
			Maximum ground snow load (psf)								
			—	30	70	30	70	30	70	30	70
			Maximum clear span of lintel (feet - inches)								
24	Span without stirrups[i, j]		6-11	7-9	6-1	6-3	5-9	5-7	5-3	4-9	4-8
	1-#4	40,000	8-0	9-0	6-11	7-2	6-5	6-2	5-8	4-9	4-8
		60,000	9-9	11-0	8-5	8-9	7-10	7-6	6-11	5-10	5-8
	1-#5	40,000	10-0	11-3	8-7	8-11	8-0	7-7	7-0	5-11	5-9
		60,000	13-11	15-8	12-0	12-5	11-2	10-7	9-10	8-3	8-0
	2-#4 1-#6	40,000	12-11	14-6	11-2	11-6	10-5	9-10	9-1	7-8	7-5
		60,000	15-7	17-7	13-6	13-11	12-7	11-11	11-0	9-3	9-0
	2-#5	40,000	15-11	17-11	13-7	14-3	12-8	11-9	10-8	8-7	8-4
		60,000	19-1	21-6	16-5	17-1	15-1	14-0	12-6	9-11	9-7
	2-#6	40,000	17-7	21-1	14-1	14-10	12-8	11-9	10-8	8-7	8-4
		60,000	DR	DR	DR	DR	DR	DR	DR	DR	DR
	Center distance A[k, l]		3-3	4-1	2-5	2-7	2-1	1-11	1-7	1-2	1-1

For SI: 1 inch = 25.4 mm, 1 foot = 304.8 mm, 1 pound per square foot = 0.0479 kPa, Grade 40 = 280 MPa, Grade 60 = 420 MPa.

a. See Table R611.3 for tolerances permitted from nominal thickness.

b. Table values are based on concrete with a minimum specified compressive strength of 2,500 psi. See Note j.

c. Table values are based on uniform loading. See Section R611.8.2 for lintels supporting concentrated loads.

d. Deflection criterion is $L/240$, where L is the clear span of the lintel in inches, or $^1/_2$-inch, whichever is less.

e. Linear interpolation is permitted between ground snow loads and between lintel depths.

f. DR indicates design required.

g. Lintel depth, D, is permitted to include the available height of wall located directly above the lintel, provided that the increased lintel depth spans the entire length of the lintel.

h. Stirrups shall be fabricated from reinforcing bars with the same yield strength as that used for the main longitudinal reinforcement.

i. Allowable clear span without stirrups applicable to all lintels of the same depth, D. Top and bottom reinforcement for lintels without stirrups shall not be less than the least amount of reinforcement required for a lintel of the same depth and loading condition with stirrups. All other spans require stirrups spaced at not more than $d/2$.

j. Where concrete with a minimum specified compressive strength of 3,000 psi is used, clear spans for lintels without stirrups shall be permitted to be multiplied by 1.05. If the increased span exceeds the allowable clear span for a lintel of the same depth and loading condition with stirrups, the top and bottom reinforcement shall be equal to or greater than that required for a lintel of the same depth and loading condition that has an allowable clear span that is equal to or greater than that of the lintel without stirrups that has been increased.

k. Center distance, A, is the center portion of the clear span where stirrups are not required. This is applicable to all longitudinal bar sizes and steel yield strengths.

l. Where concrete with a minimum specified compressive strength of 3,000 psi is used, center distance, A, shall be permitted to be multiplied by 1.10.

m. The maximum clear opening width between two solid wall segments shall be 18 feet. See Section R611.7.2.1. Lintel clear spans in the table greater than 18 feet are shown for interpolation and information only.

❖ See the commentary for Section R611.8.2.1.

TABLE R611.8(3)
MAXIMUM ALLOWABLE CLEAR SPANS FOR 6-INCH-NOMINAL THICK FLAT LINTELS IN LOAD-BEARING WALLS[a, b, c, d, e, f, m]
ROOF CLEAR SPAN 40 FEET AND FLOOR CLEAR SPAN 32 FEET

LINTEL DEPTH, D^g (inches)	NUMBER OF BARS AND BAR SIZE IN TOP AND BOTTOM OF LINTEL	STEEL YIELD STRENGTH[h], f_y (psi)	DESIGN LOADING CONDITION DETERMINED FROM TABLE R611.8(1)								
			1	2		3		4		5	
			Maximum ground snow load (psf)								
			—	30	70	30	70	30	70	30	70
			Maximum clear span of lintel (feet - inches)								
8	Span without stirrups[i, j]		4-2	4-8	3-1	3-3	2-10	2-6	2-3	2-0	2-0
8	1-#4	40,000	5-1	5-5	4-2	4-3	3-10	3-6	3-3	2-8	2-7
8	1-#4	60,000	6-2	6-7	5-0	5-2	4-8	4-2	3-11	3-3	3-2
8	1-#5	40,000	6-3	6-8	5-1	5-3	4-9	4-3	4-0	3-3	3-2
8	1-#5	60,000	7-6	8-0	6-1	6-4	5-8	5-1	4-9	3-8	3-6
8	2-#4 1-#6	40,000	7-0	7-6	5-8	5-11	5-3	4-9	4-5	3-8	3-6
8	2-#4 1-#6	60,000	DR	DR	DR	DR	DR	DR	DR	DR	DR
8	Center distance $A^{k, l}$		1-7	1-10	1-1	1-2	0-11	0-9	0-8	0-5	0-5
12	Span without stirrups[i, j]		4-2	4-8	3-5	3-6	3-2	2-11	2-9	2-5	2-4
12	1-#4	40,000	5-7	6-1	4-8	4-10	4-4	3-11	3-8	3-0	2-11
12	1-#4	60,000	7-9	8-6	6-6	6-9	6-1	5-6	5-1	4-3	4-1
12	1-#5	40,000	7-11	8-8	6-8	6-11	6-2	5-7	5-2	4-4	4-2
12	1-#5	60,000	9-7	10-6	8-0	8-4	7-6	6-9	6-3	5-2	5-1
12	2-#4 1-#6	40,000	8-11	9-9	7-6	7-9	6-11	6-3	5-10	4-10	4-8
12	2-#4 1-#6	60,000	10-8	11-9	8-12	9-4	8-4	7-6	7-0	5-10	5-8
12	2-#5	40,000	10-11	12-0	9-2	9-6	8-6	7-8	7-2	5-6	5-3
12	2-#5	60,000	12-11	14-3	10-10	11-3	10-1	9-0	8-1	6-1	5-10
12	2-#6	40,000	12-9	14-0	10-8	11-1	9-7	8-1	7-3	5-6	5-3
12	2-#6	60,000	DR	DR	DR	DR	DR	DR	DR	DR	DR
12	Center distance $A^{k, l}$		2-6	3-0	1-9	1-10	1-6	1-3	1-1	0-9	0-8
16	Span without stirrups[i, j]		5-7	6-5	4-9	4-11	4-5	4-0	3-10	3-4	3-4
16	1-#4	40,000	6-5	7-2	5-6	5-9	5-2	4-8	4-4	3-7	3-6
16	1-#4	60,000	7-10	8-9	6-9	7-0	6-3	5-8	5-3	4-4	4-3
16	1-#5	40,000	7-11	8-11	6-10	7-1	6-5	5-9	5-4	4-5	4-4
16	1-#5	60,000	11-1	12-6	9-7	9-11	8-11	8-0	7-6	6-2	6-0
16	2-#4 1-#6	40,000	10-3	11-7	8-10	9-2	8-3	7-6	6-11	5-9	5-7
16	2-#4 1-#6	60,000	12-5	14-0	10-9	11-1	10-0	9-0	8-5	7-0	6-9
16	2-#5	40,000	12-8	14-3	10-11	11-4	10-2	9-2	8-7	6-9	6-6
16	2-#5	60,000	15-2	17-1	13-1	13-7	12-3	11-0	10-3	7-11	7-7
16	2-#6	40,000	14-11	16-9	12-8	13-4	11-4	9-8	8-8	6-9	6-6
16	2-#6	60,000	DR	DR	DR	DR	DR	DR	DR	DR	DR
16	Center distance $A^{k, l}$		3-3	4-1	2-5	2-7	2-1	1-9	1-6	1-0	1-0

(continued)

TABLE R611.8(3)—continued
MAXIMUM ALLOWABLE CLEAR SPANS FOR 6-INCH-NOMINAL THICK FLAT LINTELS IN LOAD-BEARING WALLS[a, b, c, d, e, f, m]
ROOF CLEAR SPAN 40 FEET AND FLOOR CLEAR SPAN 32 FEET

LINTEL DEPTH, D[g] (inches)	NUMBER OF BARS AND BAR SIZE IN TOP AND BOTTOM OF LINTEL	STEEL YIELD STRENGTH[h], f_y (psi)	DESIGN LOADING CONDITION DETERMINED FROM TABLE R611.8(1)								
			1	2		3		4		5	
			Maximum ground snow load (psf)								
			—	30	70	30	70	30	70	30	70
			Maximum clear span of lintel (feet - inches)								
20	Span without stirrups[i, j]		6-11	8-2	6-1	6-3	5-8	5-2	4-11	4-4	4-3
	1-#5	40,000	8-9	10-1	7-9	8-0	7-3	6-6	6-1	5-1	4-11
		60,000	10-8	12-3	9-5	9-9	8-10	8-0	7-5	6-2	6-0
	2-#4 1-#6	40,000	9-11	11-4	8-9	9-1	8-2	7-4	6-10	5-8	5-7
		60,000	13-9	15-10	12-2	12-8	11-5	10-3	9-7	7-11	7-9
	2-#5	40,000	14-0	16-2	12-5	12-11	11-7	10-6	9-9	7-11	7-8
		60,000	16-11	19-6	15-0	15-6	14-0	12-7	11-9	9-1	8-9
	2-#6	40,000	16-7	19-1	14-7	15-3	13-1	11-3	10-2	7-11	7-8
		60,000	19-11	22-10	17-4	18-3	15-6	13-2	11-10	9-1	8-9
	Center distance A[k, l]		3-11	5-2	3-1	3-3	2-8	2-2	1-11	1-4	1-3
24	Span without stirrups[i, j]		8-2	9-10	7-4	7-8	6-11	6-4	5-11	5-3	5-2
	1-#5	40,000	9-5	11-1	8-7	8-10	8-0	7-3	6-9	5-7	5-5
		60,000	11-6	13-6	10-5	10-9	9-9	8-9	8-2	6-10	6-8
	2-#4 1-#6	40,000	10-8	12-6	9-8	10-0	9-0	8-2	7-7	6-4	6-2
		60,000	12-11	15-2	11-9	12-2	11-0	9-11	9-3	7-8	7-6
	2-#5	40,000	15-2	17-9	13-9	14-3	12-10	11-7	10-10	9-0	8-9
		60,000	18-4	21-6	16-7	17-3	15-6	14-0	13-1	10-4	10-0
	2-#6	40,000	18-0	21-1	16-4	16-11	14-10	12-9	11-8	9-2	8-11
		60,000	21-7	25-4	19-2	20-4	17-2	14-9	13-4	10-4	10-0
	Center distance A[k, l]		4-6	6-2	3-8	4-0	3-3	2-8	2-3	1-7	1-6

For SI: 1 inch = 25.4 mm, 1 foot = 304.8 mm, 1 pounds per square foot = 0.0479 kPa, Grade 40 = 280 MPa, Grade 60 = 420 MPa.

a. See Table R611.3 for tolerances permitted from nominal thickness.

b. Table values are based on concrete with a minimum specified compressive strength of 2,500 psi. See Note j.

c. Table values are based on uniform loading. See Section R611.8.2 for lintels supporting concentrated loads.

d. Deflection criterion is *L*/240, where *L* is the clear span of the lintel in inches, or $^1/_2$-inch, whichever is less.

e. Linear interpolation is permitted between ground snow loads and between lintel depths.

f. DR indicates design required.

g. Lintel depth, *D*, is permitted to include the available height of wall located directly above the lintel, provided that the increased lintel depth spans the entire length of the lintel.

h. Stirrups shall be fabricated from reinforcing bars with the same yield strength as that used for the main longitudinal reinforcement.

i. Allowable clear span without stirrups applicable to all lintels of the same depth, *D*. Top and bottom reinforcement for lintels without stirrups shall not be less than the least amount of reinforcement required for a lintel of the same depth and loading condition with stirrups. All other spans require stirrups spaced at not more than *d/2*.

j. Where concrete with a minimum specified compressive strength of 3,000 psi is used, clear spans for lintels without stirrups shall be permitted to be multiplied by 1.05. If the increased span exceeds the allowable clear span for a lintel of the same depth and loading condition with stirrups, the top and bottom reinforcement shall be equal to or greater than that required for a lintel of the same depth and loading condition that has an allowable clear span that is equal to or greater than that of the lintel without stirrups that has been increased.

k. Center distance, *A*, is the center portion of the clear span where stirrups are not required. This is applicable to all longitudinal bar sizes and steel yield strengths.

l. Where concrete with a minimum specified compressive strength of 3,000 psi is used, center distance, *A*, shall be permitted to be multiplied by 1.10.

m. The maximum clear opening width between two solid wall segments shall be 18 feet. See Section R611.7.2.1. Lintel clear spans in the table greater than 18 feet are shown for interpolation and information only.

❖ See the commentary for Section R611.8.2.1.

TABLE R611.8(4)
MAXIMUM ALLOWABLE CLEAR SPANS FOR 8-INCH-NOMINAL THICK FLAT LINTELS IN LOAD-BEARING WALLS[a, b, c, d, e, f, m]
ROOF CLEAR SPAN 40 FEET AND FLOOR CLEAR SPAN 32 FEET

LINTEL DEPTH, D[g] (inches)	NUMBER OF BARS AND BAR SIZE IN TOP AND BOTTOM OF LINTEL	STEEL YIELD STRENGTH[h], f_y (psi)	DESIGN LOADING CONDITION DETERMINED FROM TABLE R611.8(1)								
			1	2		3		4		5	
			Maximum ground snow load (psf)								
			—	30	70	30	70	30	70	30	70
			Maximum clear span of lintel (feet - inches)								
8	Span without stirrups[i, j]		4-4	4-9	3-7	3-9	3-4	2-10	2-7	2-1	2-0
	1-#4	40,000	4-4	4-9	3-7	3-9	3-4	2-11	2-9	2-3	2-2
		60,000	6-1	6-7	5-0	5-3	4-8	4-0	3-9	3-1	3-0
	1-#5	40,000	6-2	6-9	5-2	5-4	4-9	4-1	3-10	3-2	3-1
		60,000	7-5	8-1	6-2	6-5	5-9	4-11	4-7	3-9	3-8
	2-#4 1-#6	40,000	6-11	7-6	5-9	6-0	5-4	4-7	4-4	3-6	3-5
		60,000	8-3	9-0	6-11	7-2	6-5	5-6	5-2	4-2	4-1
	2-#5	40,000	8-5	9-2	7-0	7-3	6-6	5-7	5-3	4-2	4-0
		60,000	DR	DR	DR	DR	DR	DR	DR	DR	DR
	Center distance A[k, l]		2-1	2-6	1-5	1-6	1-3	0-11	0-10	0-6	0-6
12	Span without stirrups[i, j]		4-10	5-8	4-0	4-2	3-9	3-2	3-0	2-7	2-6
	1-#4	40,000	5-5	6-1	4-8	4-10	4-4	3-9	3-6	2-10	2-10
		60,000	6-7	7-5	5-8	5-11	5-4	4-7	4-3	3-6	3-5
	1-#5	40,000	6-9	7-7	5-9	6-0	5-5	4-8	4-4	3-7	3-6
		60,000	9-4	10-6	8-1	8-4	7-6	6-6	6-1	5-0	4-10
	2-#4 1-#6	40,000	8-8	9-9	7-6	7-9	7-0	6-0	5-8	4-7	4-6
		60,000	10-6	11-9	9-1	9-5	8-5	7-3	6-10	5-7	5-5
	2-#5	40,000	10-8	12-0	9-3	9-7	8-7	7-5	6-11	5-6	5-4
		60,000	12-10	14-5	11-1	11-6	10-4	8-11	8-4	6-7	6-4
	2-#6	40,000	12-7	14-2	10-10	11-3	10-2	8-3	7-6	5-6	5-4
		60,000	DR	DR	DR	DR	DR	DR	DR	DR	DR
	Center distance A[k, l]		3-2	4-0	2-4	2-6	2-0	1-6	1-4	0-11	0-10
16	Span without stirrups[i, j]		6-5	7-9	5-7	5-10	5-2	4-5	4-2	3-7	3-6
	1-#4	40,000	6-2	7-1	5-6	5-8	5-1	4-5	4-2	3-5	3-4
		60,000	7-6	8-8	6-8	6-11	6-3	5-5	5-1	4-2	4-0
	1-#5	40,000	7-8	8-10	6-10	7-1	6-4	5-6	5-2	4-3	4-1
		60,000	9-4	10-9	8-4	8-7	7-9	6-8	6-3	5-2	5-0
	2-#4 1-#6	40,000	8-8	10-0	7-8	8-0	7-2	6-2	5-10	4-9	4-8
		60,000	12-0	13-11	10-9	11-2	10-0	8-8	8-1	6-8	6-6
	2-#5	40,000	12-3	14-2	11-0	11-4	10-3	8-10	8-3	6-9	6-7
		60,000	14-10	17-2	13-3	13-8	12-4	10-8	10-0	7-11	7-8
	2-#6	40,000	14-6	16-10	13-0	13-5	12-1	10-1	9-2	6-11	6-8
		60,000	17-5	20-2	15-7	16-1	14-6	11-10	10-8	7-11	7-8
	Center distance[k, l]		4-1	5-5	3-3	3-6	2-10	2-1	1-10	1-3	1-2

(continued)

TABLE R611.8(4)—continued
MAXIMUM ALLOWABLE CLEAR SPANS FOR 8-INCH-NOMINAL THICK FLAT LINTELS IN LOAD-BEARING WALLS[a, b, c, d, e, f, m]
ROOF CLEAR SPAN 40 FEET AND FLOOR CLEAR SPAN 32 FEET

LINTEL DEPTH, D[g] (inches)	NUMBER OF BARS AND BAR SIZE IN TOP AND BOTTOM OF LINTEL	STEEL YIELD STRENGTH[h], f_y (psi)	DESIGN LOADING CONDITION DETERMINED FROM TABLE R611.8(1)								
			1	2		3		4		5	
			Maximum ground snow load (psf)								
			—	30	70	30	70	30	70	30	70
			Maximum clear span of lintel (feet - inches)								
20	Span without stirrups[i, j]		7-10	9-10	7-1	7-5	6-7	5-8	5-4	4-7	4-6
	1-#5	40,000	8-4	9-11	7-8	8-0	7-2	6-3	5-10	4-9	4-8
		60,000	10-2	12-1	9-5	9-9	8-9	7-7	7-1	5-10	5-8
	2-#4 1-#6	40,000	9-5	11-3	8-8	9-0	8-1	7-0	6-7	5-5	5-3
		60,000	11-6	13-8	10-7	11-0	9-11	8-7	8-0	6-7	6-5
	2-#5	40,000	11-9	13-11	10-10	11-2	10-1	8-9	8-2	6-8	6-7
		60,000	16-4	19-5	15-0	15-7	14-0	12-2	11-4	9-3	9-0
	2-#6	40,000	16-0	19-0	14-9	15-3	13-9	11-10	10-10	8-3	8-0
		60,000	19-3	22-11	17-9	18-5	16-7	13-7	12-4	9-3	9-0
	Center distance A[k, l]		4-10	6-10	4-1	4-5	3-7	2-8	2-4	1-7	1-6
24	Span without stirrups[i, j]		9-2	11-9	8-7	8-11	8-0	6-11	6-6	5-7	5-6
	1-#5	40,000	8-11	10-10	8-6	8-9	7-11	6-10	6-5	5-3	5-2
		60,000	10-11	13-3	10-4	10-8	9-8	8-4	7-10	6-5	6-3
	2-#4 1-#6	40,000	10-1	12-3	9-7	9-11	8-11	7-9	7-3	6-0	5-10
		60,000	12-3	15-0	11-8	12-1	10-11	9-5	8-10	7-3	7-1
	2-#5	40,000	12-6	15-3	11-11	12-4	11-1	9-7	9-0	7-5	7-3
		60,000	17-6	21-3	16-7	17-2	15-6	13-5	12-7	10-4	10-1
	2-#6	40,000	17-2	20-11	16-3	16-10	15-3	13-2	12-4	9-7	9-4
		60,000	20-9	25-3	19-8	20-4	18-5	15-4	14-0	10-7	10-3
	Center distance A[k, l]		5-6	8-1	4-11	5-3	4-4	3-3	2-10	1-11	1-10

For SI: 1 inch = 25.4 mm, 1 foot = 304.8 mm, 1 pound per square foot = 0.0479 kPa, Grade 40 = 280 MPa; Grade 60 = 420 MPa.

Note: Top and bottom reinforcement for lintels without stirrups shown in shaded cells shall be equal to or greater than that required for lintel of the same depth and loading condition that has an allowable clear span that is equal to or greater than that of the lintel without stirrups.

a. See Table R611.3 for tolerances permitted from nominal thickness.

b. Table values are based on concrete with a minimum specified compressive strength of 2,500 psi. See Note j.

c. Table values are based on uniform loading. See Section R611.8.2 for lintels supporting concentrated loads.

d. Deflection criterion is $L/240$, where L is the clear span of the lintel in inches, or $^1/_2$-inch, whichever is less.

e. Linear interpolation is permitted between ground snow loads and between lintel depths.

f. DR indicates design required.

g. Lintel depth, D, is permitted to include the available height of wall located directly above the lintel, provided that the increased lintel depth spans the entire length of the lintel.

h. Stirrups shall be fabricated from reinforcing bars with the same yield strength as that used for the main longitudinal reinforcement.

i. Allowable clear span without stirrups applicable to all lintels of the same depth, D. Top and bottom reinforcement for lintels without stirrups shall not be less than the least amount of reinforcement required for a lintel of the same depth and loading condition with stirrups. All other spans require stirrups spaced at not more than $d/2$.

j. Where concrete with a minimum specified compressive strength of 3,000 psi is used, clear spans for lintels without stirrups shall be permitted to be multiplied by 1.05. If the increased span exceeds the allowable clear span for a lintel of the same depth and loading condition with stirrups, the top and bottom reinforcement shall be equal to or greater than that required for a lintel of the same depth and loading condition that has an allowable clear span that is equal to or greater than that of the lintel without stirrups that has been increased.

k. Center distance, A, is the center portion of the clear span where stirrups are not required. This is applicable to all longitudinal bar sizes and steel yield strengths.

l. Where concrete with a minimum specified compressive strength of 3,000 psi is used, center distance, A, shall be permitted to be multiplied by 1.10.

m. The maximum clear opening width between two solid wall segments shall be 18 feet. See Section R611.7.2.1. Lintel clear spans in the table greater than 18 feet are shown for interpolation and information only.

❖ See the commentary Section R611.8.2.1.

TABLE R611.8(5)
MAXIMUM ALLOWABLE CLEAR SPANS FOR 10-INCH-NOMINAL THICK FLAT LINTELS IN LOAD-BEARING WALLS[a, b, c, d, e, f, m]
ROOF CLEAR SPAN 40 FEET AND FLOOR CLEAR SPAN 32 FEET

LINTEL DEPTH, D[g] (inches)	NUMBER OF BARS AND BAR SIZE IN TOP AND BOTTOM OF LINTEL	STEEL YIELD STRENGTH[h], f_y (psi)	DESIGN LOADING CONDITION DETERMINED FROM TABLE R611.8(1)								
			1	2		3		4		5	
			Maximum ground snow load (psf)								
			—	30	70	30	70	30	70	30	70
			Maximum clear span of lintel (feet - inches)								
8	Span without stirrups[i, j]		6-0	7-2	4-7	4-10	4-1	3-1	2-11	2-3	2-2
	1-#4	40,000	4-3	4-9	3-7	3-9	3-4	2-9	2-7	2-1	2-1
		60,000	5-11	6-7	5-0	5-3	4-8	3-10	3-8	2-11	2-11
	1-#5	40,000	6-1	6-9	5-2	5-4	4-9	3-11	3-9	3-0	2-11
		60,000	7-4	8-1	6-3	6-5	5-9	4-9	4-6	3-7	3-7
	2-#4 1-#6	40,000	6-10	7-6	5-9	6-0	5-5	4-5	4-2	3-4	3-4
		60,000	8-2	9-1	6-11	7-2	6-6	5-4	5-0	4-1	4-0
	2-#5	40,000	8-4	9-3	7-1	7-4	6-7	5-5	5-1	4-1	4-0
		60,000	9-11	11-0	8-5	8-9	7-10	6-6	6-1	4-8	4-6
	2-#6	40,000	9-9	10-10	8-3	8-7	7-9	6-4	5-10	4-1	4-0
		60,000	DR	DR	DR	DR	DR	DR	DR	DR	DR
	Center distance A[k, l]		2-6	3-1	1-10	1-11	1-7	1-1	0-11	0-7	0-7
12	Span without stirrups[i, j]		5-5	6-7	4-7	4-10	4-3	3-5	3-3	2-8	2-8
	1-#4	40,000	5-3	6-0	4-8	4-10	4-4	3-7	3-4	2-9	2-8
		60,000	6-5	7-4	5-8	5-10	5-3	4-4	4-1	3-4	3-3
	1-#5	40,000	6-6	7-6	5-9	6-0	5-5	4-5	4-2	3-5	3-4
		60,000	7-11	9-1	7-0	7-3	6-7	5-5	5-1	4-2	4-0
	2-#4 1-#6	40,000	7-4	8-5	6-6	6-9	6-1	5-0	4-9	3-10	3-9
		60,000	10-3	11-9	9-1	9-5	8-6	7-0	6-7	5-4	5-3
	2-#5	40,000	10-5	12-0	9-3	9-7	8-8	7-2	6-9	5-5	5-4
		60,000	12-7	14-5	11-2	11-6	10-5	8-7	8-1	6-6	6-4
	2-#6	40,000	12-4	14-2	10-11	11-4	10-2	8-5	7-8	5-7	5-5
		60,000	14-9	17-0	13-1	13-6	12-2	10-0	9-1	6-6	6-4
	Center distance A[k, l]		3-9	4-11	2-11	3-2	2-7	1-9	1-7	1-0	1-0
16	Span without stirrups[i, j]		7-1	9-0	6-4	6-8	5-10	4-9	4-6	3-9	3-8
	1-#4	40,000	5-11	7-0	5-5	5-8	5-1	4-3	4-0	3-3	3-2
		60,000	7-3	8-7	6-8	6-11	6-3	5-2	4-10	3-11	3-10
	1-#5	40,000	7-4	8-9	6-9	7-0	6-4	5-3	4-11	4-0	3-11
		60,000	9-0	10-8	8-3	8-7	7-9	6-5	6-0	4-11	4-9
	2-#4 1-#6	40,000	8-4	9-11	7-8	7-11	7-2	5-11	5-7	4-6	4-5
		60,000	10-2	12-0	9-4	9-8	8-9	7-3	6-10	5-6	5-5
	2-#5	40,000	10-4	12-3	9-6	9-10	8-11	7-4	6-11	5-8	5-6
		60,000	14-4	17-1	13-3	13-8	12-4	10-3	9-8	7-10	7-8
	2-#6	40,000	14-1	16-9	13-0	13-5	12-2	10-1	9-6	7-0	6-10
		60,000	17-0	20-2	15-8	16-2	14-7	12-0	10-11	8-0	7-9
	Center distance[k, l]		4-9	6-8	4-0	4-4	3-6	2-5	2-2	1-5	1-4

(continued)

TABLE R611.8(5)—continued
MAXIMUM ALLOWABLE CLEAR SPANS FOR 10-INCH-NOMINAL THICK FLAT LINTELS IN LOAD-BEARING WALLS[a, b, c, d, e, f, m]
ROOF CLEAR SPAN 40 FEET AND FLOOR CLEAR SPAN 32 FEET

LINTEL DEPTH, D^g (inches)	NUMBER OF BARS AND BAR SIZE IN TOP AND BOTTOM OF LINTEL	STEEL YIELD STRENGTH[h], f_y (psi)	DESIGN LOADING CONDITION DETERMINED FROM TABLE R611.8(1)								
			1	2		3		4		5	
			Maximum ground snow load (psf)								
			—	30	70	30	70	30	70	30	70
			Maximum clear span of lintel (feet - inches)								
20	Span without stirrups[i, j]		8-7	11-4	8-1	8-5	7-5	6-1	5-9	4-10	4-9
	1-#4	40,000	6-5	7-10	6-2	6-4	5-9	4-9	4-6	3-8	3-7
		60,000	7-10	9-7	7-6	7-9	7-0	5-10	5-6	4-5	4-4
	1-#5	40,000	8-0	9-9	7-8	7-11	7-2	5-11	5-7	4-6	4-5
		60,000	9-9	11-11	9-4	9-8	8-9	7-3	6-10	5-6	5-5
	2-#4 1-#6	40,000	9-0	11-1	8-8	8-11	8-1	6-9	6-4	5-2	5-0
		60,000	11-0	13-6	10-6	10-11	9-10	8-2	7-9	6-3	6-2
	2-#5	40,000	11-3	13-9	10-9	11-1	10-0	8-4	7-10	6-5	6-3
		60,000	15-8	19-2	15-0	15-6	14-0	11-8	11-0	8-11	8-9
	2-#6	40,000	15-5	18-10	14-8	15-2	13-9	11-5	10-9	8-6	8-3
		60,000	18-7	22-9	17-9	18-5	16-7	13-10	12-9	9-5	9-2
	Center distance $A^{k, l}$		5-7	8-4	5-1	5-5	4-5	3-1	2-9	1-10	1-9
24	Span without stirrups[i, j]		9-11	13-7	9-9	10-2	9-0	7-5	7-0	5-10	5-9
	1-#5	40,000	8-6	10-8	8-5	8-8	7-10	6-6	6-2	5-0	4-11
		60,000	10-5	13-0	10-3	10-7	9-7	8-0	7-6	6-1	6-0
	2-#4 1-#6	40,000	9-7	12-1	9-6	9-9	8-10	7-5	7-0	5-8	5-6
		60,000	11-9	14-9	11-7	11-11	10-10	9-0	8-6	6-11	6-9
	2-#5	40,000	12-0	15-0	11-9	12-2	11-0	9-2	8-8	7-1	6-11
		60,000	14-7	18-3	14-4	14-10	13-5	11-2	10-7	8-7	8-5
	2-#6	40,000	14-3	17-11	14-1	14-7	13-2	11-0	10-4	8-5	8-3
		60,000	19-11	25-0	19-7	20-3	18-4	15-3	14-5	10-10	10-7
	Center distance $A^{k, l}$		6-3	9-11	6-1	6-6	5-4	3-9	3-4	2-2	2-1

For SI: 1 inch = 25.4 mm, 1 foot = 304.8 mm, 1 pound per square foot = 0.0479 kPa, Grade 40 = 280 MPa, Grade 60 = 420 MPa.

Note: Top and bottom reinforcement for lintels without stirrups shown in shaded cells shall be equal to or greater than that required for lintel of the same depth and loading condition that has an allowable clear span that is equal to or greater than that of the lintel without stirrups.

a. See Table R611.3 for tolerances permitted from nominal thickness.

b. Table values are based on concrete with a minimum specified compressive strength of 2,500 psi. See Note j.

c. Table values are based on uniform loading. See Section R611.8.2 for lintels supporting concentrated loads.

d. Deflection criterion is *L*/240, where *L* is the clear span of the lintel in inches, or $^1/_2$-inch, whichever is less.

e. Linear interpolation is permitted between ground snow loads and between lintel depths.

f. DR indicates design required.

g. Lintel depth, *D*, is permitted to include the available height of wall located directly above the lintel, provided that the increased lintel depth spans the entire length of the lintel.

h. Stirrups shall be fabricated from reinforcing bars with the same yield strength as that used for the main longitudinal reinforcement.

i. Allowable clear span without stirrups applicable to all lintels of the same depth, *D*. Top and bottom reinforcement for lintels without stirrups shall not be less than the least amount of reinforcement required for a lintel of the same depth and loading condition with stirrups. All other spans require stirrups spaced at not more than *d*/2.

j. Where concrete with a minimum specified compressive strength of 3,000 psi is used, clear spans for lintels without stirrups shall be permitted to be multiplied by 1.05. If the increased span exceeds the allowable clear span for a lintel of the same depth and loading condition with stirrups, the top and bottom reinforcement shall be equal to or greater than that required for a lintel of the same depth and loading condition that has an allowable clear span that is equal to or greater than that of the lintel without stirrups that has been increased.

k. Center distance, *A*, is the center portion of the clear span where stirrups are not required. This is applicable to all longitudinal bar sizes and steel yield strengths.

l. Where concrete with a minimum specified compressive strength of 3,000 psi is used, center distance, *A*, shall be permitted to be multiplied by 1.10.

m. The maximum clear opening width between two solid wall segments shall be 18 feet. See Section R611.7.2.1. Lintel clear spans in the table greater than 18 feet are shown for interpolation and information only.

❖ See the commentary for Section R611.8.2.1.

TABLE R611.8(6)
MAXIMUM ALLOWABLE CLEAR SPANS FOR 6-INCH-THICK WAFFLE-GRID LINTELS IN LOAD-BEARING WALLS[a, b, c, d, e, f, o]
MAXIMUM ROOF CLEAR SPAN 40 FEET AND MAXIMUM FLOOR SPAN 32 FEET

LINTEL DEPTH, D[g] (inches)	NUMBER OF BARS AND BAR SIZE IN TOP AND BOTTOM OF LINTEL	STEEL YIELD STRENGTH[h], f_y (psi)	DESIGN LOADING CONDITION DETERMINED FROM TABLE R611.8(1)								
			1	2		3		4		5	
			Maximum ground snow load (psf)								
			—	30	70	30	70	30	70	30	70
			Maximum clear span of lintel (feet - inches)								
8[i]	Span without stirrups[k, l]		2-7	2-9	2-0	2-1	2-0	2-0	2-0	2-0	2-0
	1-#4	40,000	5-2	5-5	4-0	4-3	3-7	3-3	2-11	2-4	2-3
		60,000	5-9	6-3	4-0	4-3	3-7	3-3	2-11	2-4	2-3
	1-#5	40,000	5-9	6-3	4-0	4-3	3-7	3-3	2-11	2-4	2-3
		60,000	5-9	6-3	4-0	4-3	3-7	3-3	2-11	2-4	2-3
	2-#4 1-#6	40,000	5-9	6-3	4-0	4-3	3-7	3-3	2-11	2-4	2-3
		60,000	DR	DR	DR	DR	DR	DR	DR	DR	DR
	Center distance A[m, n]		0-9	0-10	0-6	0-6	0-5	0-5	0-4	STL	STL
12[i]	Span without stirrups[k, l]		2-11	3-1	2-6	2-7	2-5	2-4	2-3	2-1	2-0
	1-#4	40,000	5-9	6-2	4-8	4-10	4-4	4-1	3-9	3-2	3-1
		60,000	8-0	8-7	6-6	6-9	6-0	5-5	4-11	3-11	3-10
	1-#5	40,000	8-1	8-9	6-8	6-11	6-0	5-5	4-11	3-11	3-10
		60,000	9-1	10-3	6-8	7-0	6-0	5-5	4-11	3-11	3-10
	2-#4 1-#6	40,000	9-1	9-9	6-8	7-0	6-0	5-5	4-11	3-11	3-10
	Center distance A[m, n]		1-3	1-5	0-10	0-11	0-9	0-8	0-6	STL	STL
16[i]	Span without stirrups[k, l]		4-0	4-4	3-6	3-7	3-4	3-3	3-1	2-10	2-10
	1-#4	40,000	6-7	7-3	5-6	5-9	5-2	4-10	4-6	3-9	3-8
		60,000	8-0	8-10	6-9	7-0	6-3	5-11	5-5	4-7	4-5
	1-#5	40,000	8-2	9-0	6-11	7-2	6-5	6-0	5-7	4-8	4-6
		60,000	11-5	12-6	9-3	9-9	8-4	7-7	6-10	5-6	5-4
	2-#4 1-#6	40,000	10-7	11-7	8-11	9-3	8-3	7-7	6-10	5-6	5-4
		60,000	12-2	14-0	9-3	9-9	8-4	7-7	6-10	5-6	5-4
	2-#5	40,000	12-2	14-2	9-3	9-9	8-4	7-7	6-10	5-6	5-4
		60,000	DR	DR	DR	DR	DR	DR	DR	DR	DR
	Center distance A[m, n]		1-8	2-0	1-2	1-3	1-0	0-11	0-9	STL	STL
20[i]	Span without stirrups[k, l]		5-0	5-6	4-6	4-7	4-3	4-1	4-0	3-8	3-8
	1-#4	40,000	7-2	8-2	6-3	6-6	5-10	5-6	5-1	4-3	4-2
		60,000	8-11	9-11	7-8	7-11	7-1	6-8	6-2	5-2	5-0
	1-#5	40,000	9-1	10-2	7-9	8-1	7-3	6-10	6-4	5-4	5-2
		60,000	12-8	14-2	10-11	11-3	10-2	9-6	8-9	7-1	6-10
	2-#4 1-#6	40,000	10-3	11-5	8-9	9-1	8-2	7-8	7-1	6-0	5-10
		60,000	14-3	15-11	11-9	12-5	10-8	9-9	8-9	7-1	6-10
	2-#5	40,000	14-6	16-3	11-6	12-1	10-4	9-6	8-6	6-11	6-8
		60,000	DR	DR	DR	DR	DR	DR	DR	DR	DR
	Center distance A[m, n]		2-0	2-6	1-6	1-7	1-3	1-1	1-0	STL	STL

(continued)

TABLE R611.8(6)—continued
MAXIMUM ALLOWABLE CLEAR SPANS FOR 6-INCH-THICK WAFFLE-GRID LINTELS IN LOAD-BEARING WALLS[a, b, c, d, e, f, o]
MAXIMUM ROOF CLEAR SPAN 40 FEET AND MAXIMUM FLOOR SPAN 32 FEET

LINTEL DEPTH, D[g] (inches)	NUMBER OF BARS AND BAR SIZE IN TOP AND BOTTOM OF LINTEL	STEEL YIELD STRENGTH[h], f_y (psi)	DESIGN LOADING CONDITION DETERMINED FROM TABLE R611.8(1)								
			1	2		3		4		5	
			Maximum ground snow load (psf)								
			—	30	70	30	70	30	70	30	70
			Maximum clear span of lintel (feet - inches)								
24w[j]	Span without stirrups[k, l]		6-0	6-8	5-5	5-7	5-3	5-0	4-10	4-6	4-5
	1-#4	40,000	7-11	9-0	6-11	7-2	6-5	6-0	5-7	4-8	4-7
		60,000	9-8	10-11	8-5	8-9	7-10	7-4	6-10	5-9	5-7
	1-#5	40,000	9-10	11-2	8-7	8-11	8-0	7-6	7-0	5-10	5-8
		60,000	12-0	13-7	10-6	10-10	9-9	9-2	8-6	7-2	6-11
	2-#4 1-#6	40,000	11-1	12-7	9-8	10-1	9-1	8-6	7-10	6-7	6-5
		60,000	15-6	17-7	13-6	14-0	12-8	11-10	10-8	8-7	8-4
	2-#5	40,000	15-6	17-11	12-8	13-4	11-6	10-7	9-7	7-10	7-7
		60,000	DR	DR	DR	DR	DR	DR	DR	DR	DR
	Center distance A[m, n]		2-4	3-0	1-9	1-11	1-6	1-4	1-2	STL	STL

For SI: 1 inch = 25.4 mm, 1 pound per square foot = 0.0479 kPa, 1 foot = 304.8 mm, Grade 40 = 280 MPa, Grade 60 = 420 MPa.

a. Where lintels are formed with waffle-grid forms, form material shall be removed, if necessary, to create top and bottom flanges of the lintel that are not less than 3 inches in depth (in the vertical direction), are not less than 5 inches in width for 6-inch-nominal waffle-grid forms and not less than 7 inches in width for 8-inch-nominal waffle-grid forms. See Figure R611.8(3). Flat form lintels shall be permitted in place of waffle-grid lintels. See Tables R611.8(2) through R611.8(5).

b. See Table R611.3 for tolerances permitted from nominal thicknesses and minimum dimensions and spacing of cores.

c. Table values are based on concrete with a minimum specified compressive strength of 2,500 psi. See Notes l and n. Table values are based on uniform loading. See Section R611.8.2 for lintels supporting concentrated loads.

d. Deflection criterion is $L/240$, where L is the clear span of the lintel in inches, or $^1/_2$-inch, whichever is less.

e. Linear interpolation is permitted between ground snow loads.

f. DR indicates design required. STL - stirrups required throughout lintel.

g. Lintel depth, D, is permitted to include the available height of wall located directly above the lintel, provided that the increased lintel depth spans the entire length of the lintel.

h. Stirrups shall be fabricated from reinforcing bars with the same yield strength as that used for the main longitudinal reinforcement.

i. Lintels less than 24 inches in depth with stirrups shall be formed from flat-walls forms [see Tables R611.8(2) through R611.8(5)], or, if necessary, form material shall be removed from waffle-grid forms so as to provide the required cover for stirrups. Allowable spans for lintels formed with flat-wall forms shall be determined from Tables R611.8(2) through R611.8(5).

j. Where stirrups are required for 24-inch deep lintels, the spacing shall not exceed 12 inches on center.

k. Allowable clear span without stirrups applicable to all lintels of the same depth, D. Top and bottom reinforcement for lintels without stirrups shall not be less than the least amount of reinforcement required for a lintel of the same depth and loading condition with stirrups. All other spans require stirrups spaced at not more than $d/2$.

l. Where concrete with a minimum specified compressive strength of 3,000 psi is used, clear spans for lintels without stirrups shall be permitted to be multiplied by 1.05. If the increased span exceeds the allowable clear span for a lintel of the same depth and loading condition with stirrups, the top and bottom reinforcement shall be equal to or greater than that required for a lintel of the same depth and loading condition that has an allowable clear span that is equal to or greater than that of the lintel without stirrups that has been increased.

m. Center distance, A, is the center portion of the span where stirrups are not required. This is applicable to all longitudinal bar sizes and steel yield strengths.

n. Where concrete with a minimum specified compressive strength of 3,000 psi is used, center distance, A, shall be permitted to be multiplied by 1.10.

o. The maximum clear opening width between two solid wall segments shall be 18 feet. See Section R611.7.2.1. Lintel spans in the table greater than 18 feet are shown for interpolation and information only.

❖ See the commentary for Section R611.8.2.1.

TABLE R611.8(7)
MAXIMUM ALLOWABLE CLEAR SPANS FOR 8-INCH-THICK WAFFLE-GRID LINTELS IN LOAD-BEARING WALLS[a, b, c, d, e, f, o]
MAXIMUM ROOF CLEAR SPAN 40 FEET AND MAXIMUM FLOOR CLEAR SPAN 32 FEET

LINTEL DEPTH, D[g] (inches)	NUMBER OF BARS AND BAR SIZE IN TOP AND BOTTOM OF LINTEL	STEEL YIELD STRENGTH[h], f_y (psi)	DESIGN LOADING CONDITION DETERMINED FROM TABLE R611.8(1)								
			1	2		3		4		5	
				Maximum ground snow load (psf)							
				30	70	30	70	30	70	30	70
			Maximum clear span of lintel (feet - inches)								
8[i]	Span with stirrups[k, l]		2-6	2-9	2-0	2-1	2-0	2-0	2-0	2-0	2-0
	1-#4	40,000	4-5	4-9	3-7	3-9	3-4	3-0	2-10	2-3	2-2
		60,000	5-6	6-2	4-0	4-3	3-7	3-1	2-10	2-3	2-2
	1-#5	40,000	5-6	6-2	4-0	4-3	3-7	3-1	2-10	2-3	2-2
	Center distance A[m, n]		0-9	0-10	0-6	0-6	0-5	0-4	0-4	STL	STL
12[i]	Span without stirrups[k, l]		2-10	3-1	2-6	2-7	2-5	2-3	2-2	2-0	2-0
	1-#4	40,000	5-7	6-1	4-8	4-10	4-4	3-11	3-8	3-0	2-11
		60,000	6-9	7-5	5-8	5-11	5-4	4-9	4-5	3-8	3-7
	1-#5	40,000	6-11	7-7	5-10	6-0	5-5	4-10	4-6	3-9	3-7
		60,000	8-8	10-1	6-7	7-0	5-11	5-2	4-8	3-9	3-7
	2-#4 1-#6	40,000	8-8	9-10	6-7	7-0	5-11	5-2	4-8	3-9	3-7
		60,000	8-8	10-1	6-7	7-0	5-11	5-2	4-8	3-9	3-7
	Center distance A[m, n]		1-2	1-5	0-10	0-11	0-9	0-7	0-6	STL	STL
16[i]	Span without stirrups[k, l]		3-10	4-3	3-6	3-7	3-4	3-2	3-0	2-10	2-9
	1-#4	40,000	6-5	7-2	5-6	5-9	5-2	4-8	4-4	3-7	3-6
		60,000	7-9	8-9	6-9	7-0	6-3	5-8	5-3	4-4	4-3
	1-#5	40,000	7-11	8-11	6-10	7-1	6-5	5-9	5-4	4-5	4-4
		60,000	9-8	10-11	8-4	8-8	7-10	7-0	6-6	5-2	5-1
	2-#4 1-#6	40,000	9-0	10-1	7-9	8-0	7-3	6-6	6-1	5-0	4-11
		60,000	11-5	13-10	9-2	9-8	8-3	7-2	6-6	5-2	5-1
	Center distance A[m, n]		1-6	1-11	1-2	1-3	1-0	0-10	0-8	STL	STL
20[i]	Span without stirrups[k, l]		4-10	5-5	4-5	4-7	4-3	4-0	3-11	3-7	3-7
	1-#4	40,000	7-0	8-1	6-3	6-5	5-10	5-3	4-11	4-1	3-11
		60,000	8-7	9-10	7-7	7-10	7-1	6-5	6-0	4-11	4-10
	1-#5	40,000	8-9	10-1	7-9	8-0	7-3	6-6	6-1	5-1	4-11
		60,000	10-8	12-3	9-6	9-10	8-10	8-0	7-5	6-2	6-0
	2-#4 1-#6	40,000	9-10	11-4	8-9	9-1	8-2	7-4	6-10	5-8	5-7
		60,000	12-0	13-10	10-8	11-0	9-11	9-0	8-4	6-8	6-6
	2-#5	40,000	12-3	14-1	10-10	11-3	10-2	8-11	8-1	6-6	6-4
		60,000	14-0	17-6	11-8	12-3	10-6	9-1	8-4	6-8	6-6
	Center distance A[m, n]		1-10	2-5	1-5	1-7	1-3	1-0	0-11	STL	STL

(continued)

TABLE R611.8(7)—continued
MAXIMUM ALLOWABLE CLEAR SPANS FOR 8-INCH-THICK WAFFLE-GRID LINTELS IN LOAD-BEARING WALLS[a, b, c, d, e, f, o]
MAXIMUM ROOF CLEAR SPAN 40 FEET AND MAXIMUM FLOOR CLEAR SPAN 32 FEET

LINTEL DEPTH, D[g] (inches)	NUMBER OF BARS AND BAR SIZE IN TOP AND BOTTOM OF LINTEL	STEEL YIELD STRENGTH[h], f_y (psi)	DESIGN LOADING CONDITION DETERMINED FROM TABLE R611.8(1)								
			1	2		3		4		5	
				Maximum ground snow load (psf)							
				30	70	30	70	30	70	30	70
			Maximum clear span of lintel (feet - inches)								
24[j]	Span without stirrups[k, l]		5-9	6-7	5-5	5-6	5-2	4-11	4-9	4-5	4-4
	1-#4	40,000	7-6	8-10	6-10	7-1	6-5	5-9	5-5	4-6	4-4
		60,000	9-2	10-9	8-4	8-8	7-10	7-1	6-7	5-6	5-4
	1-#5	40,000	9-5	11-0	8-6	8-10	8-0	7-2	6-8	5-7	5-5
		60,000	11-5	13-5	10-5	10-9	9-9	8-9	8-2	6-10	6-8
	2-#4 1-#6	40,000	10-7	12-5	9-8	10-0	9-0	8-1	7-7	6-3	6-2
		60,000	12-11	15-2	11-9	12-2	11-0	9-11	9-3	7-8	7-6
	2-#5	40,000	13-2	15-6	12-0	12-5	11-2	9-11	9-2	7-5	7-3
		60,000	16-3	21-0	14-1	14-10	12-9	11-1	10-1	8-1	7-11
	2-#6	40,000	14-4	18-5	12-6	13-2	11-5	9-11	9-2	7-5	7-3
	Center distance A[m, n]		2-1	2-11	1-9	1-10	1-6	1-3	1-1	STL	STL

For SI: 1 inch = 25.4 mm, 1 pound per square foot = 0.0479 kPa, 1 foot = 304.8 mm, Grade 40 = 280 MPa, Grade 60 = 420 MPa.

a. Where lintels are formed with waffle-grid forms, form material shall be removed, if necessary, to create top and bottom flanges of the lintel that are not less than 3 inches in depth (in the vertical direction), are not less than 5 inches in width for 6-inch-nominal waffle-grid forms and not less than 7 inches in width for 8-inch-nominal waffle-grid forms. See Figure R611.8(3). Flat form lintels shall be permitted in lieu of waffle-grid lintels. See Tables R611.8(2) through R611.8(5).

b. See Table R611.3 for tolerances permitted from nominal thicknesses and minimum dimensions and spacing of cores.

c. Table values are based on concrete with a minimum specified compressive strength of 2,500 psi. See Notes l and n. Table values are based on uniform loading. See Section R611.8.2 for lintels supporting concentrated loads.

d. Deflection criterion is $L/240$, where L is the clear span of the lintel in inches, or $^1/_2$-inch, whichever is less.

e. Linear interpolation is permitted between ground snow loads.

f. DR indicates design required. STL - stirrups required throughout lintel.

g. Lintel depth, D, is permitted to include the available height of wall located directly above the lintel, provided that the increased lintel depth spans the entire length of the lintel.

h. Stirrups shall be fabricated from reinforcing bars with the same yield strength as that used for the main longitudinal reinforcement.

i. Lintels less than 24 inches in depth with stirrups shall be formed from flat-walls forms [see Tables R611.8(2) through R611.8(5)], or, if necessary, form material shall be removed from waffle-grid forms so as to provide the required cover for stirrups. Allowable spans for lintels formed with flat-wall forms shall be determined from Tables R611.8(2) through R611.8(5).

j. Where stirrups are required for 24-inch deep lintels, the spacing shall not exceed 12 inches on center.

k. Allowable clear span without stirrups applicable to all lintels of the same depth, D. Top and bottom reinforcement for lintels without stirrups shall not be less than the least amount of reinforcement required for a lintel of the same depth and loading condition with stirrups. All other spans require stirrups spaced at not more than $d/2$.

l. Where concrete with a minimum specified compressive strength of 3,000 psi is used, clear spans for lintels without stirrups shall be permitted to be multiplied by 1.05. If the increased span exceeds the allowable clear span for a lintel of the same depth and loading condition with stirrups, the top and bottom reinforcement shall be equal to or greater than that required for a lintel of the same depth and loading condition that has an allowable clear span that is equal to or greater than that of the lintel without stirrups that has been increased.

m. Center distance, A, is the center portion of the span where stirrups are not required. This is applicable to all longitudinal bar sizes and steel yield strengths.

n. Where concrete with a minimum specified compressive strength of 3,000 psi is used, center distance, A, shall be permitted to be multiplied by 1.10.

o. The maximum clear opening width between two solid wall segments shall be 18 feet. See Section R611.7.2.1. Lintel spans in the table greater than 18 feet are shown for interpolation and information only.

❖ See the commentary for Section R611.8.2.1.

TABLE R611.8(8)
MAXIMUM ALLOWABLE CLEAR SPANS FOR 6-INCH-THICK SCREEN-GRID LINTELS IN LOAD-BEARING WALLS[a, b, c, d, e, f, p]
ROOF CLEAR SPAN 40 FEET AND FLOOR CLEAR SPAN 32 FEET

LINTEL DEPTH, D[g] (inches)	NUMBER OF BARS AND BAR SIZE IN TOP AND BOTTOM OF LINTEL	STEEL YIELD STRENGTH[h], f_y (psi)	DESIGN LOADING CONDITION DETERMINED FROM TABLE R611.8(1)								
			1	2		3		4		5	
				Maximum ground snow load (psf)							
				30	70	30	70	30	70	30	70
			Maximum clear span of lintel (feet - inches)								
12[i,j]	Span without stirrups		2-9	2-11	2-4	2-5	2-3	2-3	2-2	2-0	2-0
16[i,j]	Span without stirrups		3-9	4-0	3-4	3-5	3-2	3-1	3-0	2-9	2-9
20[i,j]	Span without stirrups		4-9	5-1	4-3	4-4	4-1	4-0	3-10	3-7	3-7
24[k]	Span without stirrups[l, m]		5-8	6-3	5-2	5-3	5-0	4-10	4-8	4-4	4-4
	1-#4	40,000	7-11	9-0	6-11	7-2	6-5	6-1	5-8	4-9	4-7
		60,000	9-9	11-0	8-5	8-9	7-10	7-5	6-10	5-9	5-7
	1-#5	40,000	9-11	11-2	8-7	8-11	8-0	7-7	7-0	5-11	5-9
		60,000	12-1	13-8	10-6	10-10	9-9	9-3	8-6	7-2	7-0
	2-#4 1-#6	40,000	11-2	12-8	9-9	10-1	9-1	8-7	7-11	6-8	6-6
		60,000	15-7	17-7	12-8	13-4	11-6	10-8	9-8	7-11	7-8
	2-#5	40,000	14-11	18-0	12-2	12-10	11-1	10-3	9-4	7-8	7-5
		60,000	DR	DR	DR	DR	DR	DR	DR	DR	DR
	Center distance A[n, o]		2-0	2-6	1-6	1-7	1-4	1-2	1-0	STL	STL

For SI: 1 inch = 25.4 mm, 1 pound per square foot = 0.0479 kPa, 1 foot = 304.8 mm, Grade 40 = 280 MPa, Grade 60 = 420 MPa.

a. Where lintels are formed with screen-grid forms, form material shall be removed if necessary to create top and bottom flanges of the lintel that are not less than 5 inches in width and not less than 2.5 inches in depth (in the vertical direction). See Figure R611.8(4). Flat form lintels shall be permitted in lieu of screen-grid lintels. See Tables R611.8(2) through R611.8(5).

b. See Table R611.3 for tolerances permitted from nominal thickness and minimum dimensions and spacings of cores.

c. Table values are based on concrete with a minimum specified compressive strength of 2,500 psi. See Notes m and o. Table values are based on uniform loading. See Section R611.7.2.1 for lintels supporting concentrated loads.

d. Deflection criterion is $L/240$, where L is the clear span of the lintel in inches, or $^1/_2$-inch, whichever is less.

e. Linear interpolation is permitted between ground snow loads.

f. DR indicates design required. STL indicates stirrups required throughout lintel.

g. Lintel depth, D, is permitted to include the available height of wall located directly above the lintel, provided that the increased lintel depth spans the entire length of the lintel.

h. Stirrups shall be fabricated from reinforcing bars with the same yield strength as that used for the main longitudinal reinforcement.

i. Stirrups are not required for lintels less than 24 inches in depth fabricated from screen-grid forms. Top and bottom reinforcement shall consist of a No. 4 bar having a yield strength of 40,000 psi or 60,000 psi.

j. Lintels between 12 and 24 inches in depth with stirrups shall be formed from flat-wall forms [see Tables R611.8(2) through R611.8(5)], or form material shall be removed from screen-grid forms to provide a concrete section comparable to that required for a flat wall. Allowable spans for flat lintels with stirrups shall be determined from Tables R611.8(2) through R6111.8(5).

k. Where stirrups are required for 24-inch deep lintels, the spacing shall not exceed 12 inches on center.

l. Allowable clear span without stirrups applicable to all lintels of the same depth, D. Top and bottom reinforcement for lintels without stirrups shall not be less than the least amount of reinforcement required for a lintel of the same depth and loading condition with stirrups. All other spans require stirrups spaced at not more than 12 inches.

m. Where concrete with a minimum specified compressive strength of 3,000 psi is used, clear spans for lintels without stirrups shall be permitted to be multiplied by 1.05. If the increased span exceeds the allowable clear span for a lintel of the same depth and loading condition with stirrups, the top and bottom reinforcement shall be equal to or greater than that required for a lintel of the same depth and loading condition that has an allowable clear span that is equal to or greater than that of the lintel without stirrups that has been increased.

n. Center distance, A, is the center portion of the span where stirrups are not required. This is applicable to all longitudinal bar sizes and steel yield strengths.

o. Where concrete with a minimum specified compressive strength of 3,000 psi is used, center distance, A, shall be permitted to be multiplied by 1.10.

p. The maximum clear opening width between two solid wall segments shall be 18 feet. See Section R611.7.2.1. Lintel spans in the table greater than 18 feet are shown for interpolation and information only.

❖ See the commentary for Section R611.8.2.1.

TABLE R611.8(9)
MAXIMUM ALLOWABLE CLEAR SPANS FOR FLAT LINTELS WITHOUT STIRRUPS IN NONLOAD-BEARING WALLS[a, b, c, d, e, g,]

LINTEL DEPTH, D^f (inches)	NUMBER OF BARS AND BAR SIZE	STEEL YIELD STRENGTH, f_y (psi)	NOMINAL WALL THICKNESS (inches)							
			4		6		8		10	
			Lintel Supporting							
			Concrete Wall	Light-framed Gable	Concrete Wall	Light-framed Gable	Concrete Wall	Light-framed Gable	Concrete Wall	Light-framed Gable
			Maximum Clear Span of Lintel (feet - inches)							
8	1-#4	40,000	10-11	11-5	9-7	11-2	7-10	9-5	7-3	9-2
		60,000	12-5	11-7	10-11	13-5	9-11	13-2	9-3	12-10
	1-#5	40,000	12-7	11-7	11-1	13-8	10-1	13-5	9-4	13-1
		60,000	DR	DR	12-7	16-4	11-6	14-7	10-9	14-6
	2-#4 1-#6	40,000	DR	DR	12-0	15-3	10-11	15-0	10-2	14-8
		60,000	DR	DR	DR	DR	12-2	15-3	11-7	15-3
	2-#5	40,000	DR	DR	DR	DR	12-7	16-7	11-9	16-7
		60,000	DR	DR	DR	DR	DR	DR	13-3	16-7
	2-#6	40,000	DR	DR	DR	DR	DR	DR	13-2	17-8
		60,000	DR	DR	DR	DR	DR	DR	DR	DR
12	1-#4	40,000	11-5	9-10	10-6	12-0	9-6	11-6	8-9	11-1
		60,000	11-5	9-10	11-8	13-3	10-11	14-0	10-1	13-6
	1-#5	40,000	11-5	9-10	11-8	13-3	11-1	14-4	10-3	13-9
		60,000	11-5	9-10	11-8	13-3	11-10	16-0	11-9	16-9
	2-#4 1-#6	40,000	DR	DR	11-8	13-3	11-10	16-0	11-2	15-6
		60,000	DR	DR	11-8	13-3	11-10	16-0	11-11	18-4
	2-#5	40,000	DR	DR	11-8	13-3	11-10	16-0	11-11	18-4
		60,000	DR	DR	11-8	13-3	11-10	16-0	11-11	18-4
16	1-#4	40,000	13-6	13-0	11-10	13-8	10-7	12-11	9-11	12-4
		60,000	13-6	13-0	13-8	16-7	12-4	15-9	11-5	15-0
	1-#5	40,000	13-6	13-0	13-10	17-0	12-6	16-1	11-7	15-4
		60,000	13-6	13-0	13-10	17-1	14-0	19-7	13-4	18-8
	2-#4 1-#6	40,000	13-6	13-0	13-10	17-1	13-8	18-2	12-8	17-4
		60,000	13-6	13-0	13-10	17-1	14-0	20-3	14-1	—
	2-#5	40,000	13-6	13-0	13-10	17-1	14-0	20-3	14-1	—
		60,000	DR	DR	13-10	17-1	14-0	20-3	14-1	—
20	1-#4	40,000	14-11	15-10	13-0	14-10	11-9	13-11	10-10	13-2
		60,000	15-3	15-10	14-11	18-1	13-6	17-0	12-6	16-2
	1-#5	40,000	15-3	15-10	15-2	18-6	13-9	17-5	12-8	16-6
		60,000	15-3	15-10	15-8	20-5	15-9	—	14-7	20-1
	2-#4 1-#6	40,000	15-3	15-10	15-8	20-5	14-11	—	13-10	—
		60,000	15-3	15-10	15-8	20-5	15-10	—	15-11	—
	2-#5	40,000	15-3	15-10	15-8	20-5	15-10	—	15-11	—
		60,000	15-3	15-10	15-8	20-5	15-10	—	15-11	—

(continued)

TABLE R611.8(9)—continued
MAXIMUM ALLOWABLE CLEAR SPANS FOR FLAT LINTELS WITHOUT STIRRUPS IN NONLOAD-BEARING WALLS[a, b, c, d, e, g, h]

LINTEL DEPTH, D[f] (inches)	NUMBER OF BARS AND BAR SIZE	STEEL YIELD STRENGTH, f_y (psi)	NOMINAL WALL THICKNESS (inches)							
			4		6		8		10	
			Lintel Supporting							
			Concrete Wall	Light-framed Gable	Concrete Wall	Light-framed Gable	Concrete Wall	Light-framed Gable	Concrete Wall	Light-framed Gable
			Maximum Clear Span of Lintel (feet - inches)							
24	1-#4	40,000	16-1	17-1	13-11	15-10	12-7	14-9	11-8	13-10
		60,000	16-11	18-5	16-1	19-3	14-6	18-0	13-5	17-0
	1-#5	40,000	16-11	18-5	16-3	19-8	14-9	18-5	13-8	17-4
		60,000	16-11	18-5	17-4	—	17-0	—	15-8	—
	2-#4 1-#6	40,000	16-11	18-5	17-4	—	16-1	—	14-10	—
		60,000	16-11	18-5	17-4	—	17-6	—	17-1	—
	2-#5	40,000	16-11	18-5	17-4	—	17-6	—	17-4	—
		60,000	16-11	18-5	17-4	—	17-6	—	17-8	—

For SI: 1 inch = 25.4 mm, 1 foot = 304.8 mm, Grade 40 = 280 MPa, Grade 60 = 420 MPa.

a. See Table R611.3 for tolerances permitted from nominal thickness.
b. Table values are based on concrete with a minimum specified compressive strength of 2,500 psi. See Note e.
c. Deflection criterion is $L/240$, where L is the clear span of the lintel in inches, or $^1/_2$ inch, whichever is less.
d. Linear interpolation between lintels depths, D, is permitted provided the two cells being used to interpolate are shaded.
e. Where concrete with a minimum specified compressive strength of 3,000 psi is used, spans in cells that are shaded shall be permitted to be multiplied by 1.05.
f. Lintel depth, D, is permitted to include the available height of wall located directly above the lintel, provided that the increased lintel depth spans the entire length of the lintel.
g. DR indicates design required.
h. The maximum clear opening width between two solid wall segments shall be 18 feet. See Section R611.7.2.1. Lintel spans in the table greater than 18 feet are shown for interpolation and information purposes only.

❖ See the commentary for Section R611.8.2.3.

TABLE R611.8(10)
MAXIMUM ALLOWABLE CLEAR SPANS FOR WAFFLE-GRID AND SCREEN-GRID LINTELS WITHOUT STIRRUPS IN NONLOAD-BEARING WALLS[c, d, e, f, g]

LINTEL DEPTH[h], D (inches)	FORM TYPE AND NOMINAL WALL THICKNESS (inches)					
	6-inch Waffle-grid[a]		8-inch Waffle-grid[a]		6-inch Screen-grid[b]	
	Lintel supporting					
	Concrete Wall	Light-framed Gable	Concrete Wall	Light-framed Gable	Concrete Wall	Light-framed Gable
	Maximum Clear Span of Lintel (feet - inches)					
8	10-3	8-8	8-8	8-3	—	—
12	9-2	7-6	7-10	7-1	8-8	6-9
16	10-11	10-0	9-4	9-3	—	—
20	12-5	12-2	10-7	11-2	—	—
24	13-9	14-2	11-10	12-11	13-0	12-9

For SI: 1 inch = 25.4 mm, 1 foot = 304.8 mm, Grade 40 = 280 MPa, Grade 60 = 420 MPa.

a. Where lintels are formed with waffle-grid forms, form material shall be removed, if necessary, to create top and bottom flanges of the lintel that are not less than 3 inches in depth (in the vertical direction), are not less than 5 inches in width for 6-inch waffle-grid forms and not less than 7 inches in width for 8-inch waffle-grid forms. See Figure R611.8(3). Flat form lintels shall be permitted in lieu of waffle-grid lintels. See Tables R611.8(2) through R611.8(5).
b. Where lintels are formed with screen-grid forms, form material shall be removed if necessary to create top and bottom flanges of the lintel that are not less than 5 inches in width and not less than 2.5 inches in depth (in the vertical direction). See Figure R611.8(4). Flat form lintels shall be permitted in lieu of screen-grid lintels. See Tables R611.8(2) through R611.8(5).
c. See Table R611.3 for tolerances permitted from nominal thickness and minimum dimensions and spacing of cores.
d. Table values are based on concrete with a minimum specified compressive strength of 2,500 psi. See Note g.
e. Deflection criterion is $L/240$, where L is the clear span of the lintel in inches, or $^1/_2$-inch, whichever is less.
f. Top and bottom reinforcement shall consist of a No. 4 bar having a minimum yield strength of 40,000 psi.
g. Where concrete with a minimum specified compressive strength of 3,000 psi is used, spans in shaded cells shall be permitted to be multiplied by 1.05.
h. Lintel depth, D, is permitted to include the available height of wall located directly above the lintel, provided that the increased lintel depth spans the entire length of the lintel.

❖ See the commentary for Section R611.8.2.3.

R611.8.2 Lintels. Lintels shall be provided over all openings equal to or greater than 2 feet (610 mm) in width. Lintels with uniform loading shall conform to Sections R611.8.2.1 and R611.8.2.2, or Section R611.8.2.3. Lintels supporting concentrated loads, such as from roof or floor beams or girders, shall be designed in accordance with ACI 318.

❖ To support the loads from above, any wall opening of 2 feet (610 mm) or more requires a lintel in accordance with Section R611.8.2. Figures R611.8(2) through R611.8(4) illustrate these requirements. Openings smaller than 2 feet (610 mm) can be installed without the use of a lintel to transfer loads from above as a beam. The referenced figures indicate the need for horizontal reinforcement and stirrups for transverse reinforcement depending on the type of wall being constructed and the size of the opening. Commentary Table R611.8.2 provides a summary of the referenced figures and tables in Sections R611.8.2.1 through R611.8.2.3.

The requirements of Section R611.8.2 are for lintels subjected to uniform loading. Lintels are either load bearing (LB) or nonload bearing (NLB). Lintels subjected to concentrated loads must be designed in accordance with ACI 318.

Table R611.8.2
APPLICABLE REQUIREMENTS FOR LINTELS

TYPE OF WALL	APPLICABLE FIGURES	APPLICABLE TABLES
Flat	R611.8(2)	Load bearing: R611.8(2) through R611.8(5) Nonload bearing: R611.8(5)
Waffle-grid	R611.8(3)	Load bearing: R611.8(6) and R611.8(7) Nonload bearing: R611.8(10)
Screen-grid	R611.8(4)	Load bearing: R611.8(8) Nonload bearing: R611.8(10)

R611.8.2.1 Lintels designed for gravity load-bearing conditions. Where a lintel will be subjected to gravity load condition 1 through 5 of Table R611.8(1), the clear span of the lintel shall not exceed that permitted by Tables R611.8(2) through R611.8(8). The maximum clear span of lintels with and without stirrups in flat walls shall be determined in accordance with Tables R611.8(2) through R611.8(5), and constructed in accordance with Figure R611.8(2). The maximum clear span of lintels with and without stirrups in waffle-grid walls shall be determined in accordance with Tables R611.8(6) and R611.8(7), and constructed in accordance with Figure R611.8(3). The maximum clear span of lintels with and without stirrups in screen-grid walls shall be determined in accordance with Table R611.8(8), and constructed in accordance with Figure R611.8(4).

Where required by the applicable table, No. 3 stirrups shall be installed in lintels at a maximum spacing of *d/2* where d equals the depth of the lintel, *D*, less the cover of the concrete as shown in Figures R611.8(2) through R611.8(4). The smaller value of *d* computed for the top and bottom bar shall be used to determine the maximum stirrup spacing. Where stirrups are required in a lintel with a single bar or two bundled bars in the top and bottom, they shall be fabricated like the letter "c" or "s" with 135-degree (2.36 rad) standard hooks at each end that comply with Section R611.5.4.5 and Figure R611.5.4(3) and installed as shown in Figures R611.8(2) through R611.8(4). Where two bars are required in the top and bottom of the lintel and the bars are not bundled, the bars shall be separated by a minimum of 1 inch (25 mm). The free end of the stirrups shall be fabricated with 90- or 135-degree (1.57 or 2.36 rad) standard hooks that comply with Section R611.5.4.5 and Figure R611.5.4(3) and installed as shown in Figures R611.8(2) and R611.8(3). For flat, waffle-grid and screen-grid lintels, stirrups are not required in the center distance, A, portion of spans in accordance with Figure R611.8(1) and Tables R611.8(2) through R611.8(8). See Section R611.8.2.2, Item 5, for requirement for stirrups through out lintels with bundled bars.

❖ In developing the loading conditions of Table R611.8(1), arching action above the opening was considered. Based on this concept, the lintel only supports the weight of the concrete above the lintel bounded by a triangle with its base corresponding to the clear span of the lintel, and sides which start at the top of the lintel at the edges of the opening and extend upward and inward toward the center of the opening at an angle of 45 degrees (0.79 rad). Given this geometry, the apex of the triangle is on the centerline of the opening and the height of the triangle is one-half the opening width. Therefore, in order to consider arching action, the height of the wall above the top of the lintel must be greater than one-half the opening width. Where arching action can be considered and there are no loads introduced into the triangular area, Table R611.8(1) indicates that the lintel is to be designed for a nonload-bearing condition (NLB). If loads are imposed on the wall within the triangular area beneath the arch, such as from a floor ledger board bolted to the side of the wall the lintel must be designed for load-bearing conditions (LB). Where a lintel is subjected to load-bearing conditions (LB) 1 through 5, as shown in Table R611.8(1), the lintel span and reinforcement must be in accordance with this section and Tables R611.8(2) through R611.8(8). The five (5) loading conditions can be summarized as follows:

1. Design for floor load only
2. Design for roof load only
3. Design for loads from roof, attic, light-framed second-story exterior wall and second floor
4. Design for loads from roof, attic, second-story exterior concrete wall and second floor, or in the case of a one-story house, design for loads from roof, attic, first-story exterior concrete wall and first floor
5. Design for loads from roof, attic, first- and second-story exterior concrete walls and first and second floor

Tables R611.8(2) through R611.8(8) provide designs for lintels with and without stirrups. Where stirrups are required, No. 3 stirrups must be installed in accordance with Section R611.8.2.1 and Figures R611.8(1) through R611.8(4). Where stirrups are required in a lintel with a single bar in the top and bottom, they are normally fabricated like the letter "c" or "s" with 135-degree (2.36 rad) standard hooks as shown in Figure R611.5.4(3) and installed as shown in Figures R611.8(2) through R611.8(4). Where two bars are required in the top and bottom of the lintel, stirrups must be fabricated with 90- or 135-degree (1.57 or 2.36 rad) standard hooks as shown in Figure R611.5.4(3) and installed as shown in Figures R611.8(2) and R611.8(3) unless the bars are bundled.

R611.8.2.2 Bundled bars in lintels. It is permitted to bundle two bars in contact with each other in lintels if all of the following are observed:

1. Bars no larger than No. 6 are bundled.
2. Where the wall thickness is not sufficient to provide not less than 3 inches (76 mm) of clear space beside bars (total on both sides) oriented horizontally in a bundle, the bundled bars shall be oriented in a vertical plane.
3. Where vertically oriented bundled bars terminate with standard hooks to develop the bars in tension beyond the support (see Section R611.5.4.4), the hook extensions shall be staggered to provide a minimum of 1 inch (25 mm) clear spacing between the extensions.
4. Bundled bars shall not be lap spliced within the lintel span and the length on each end of the lintel that is required to develop the bars in tension.
5. Bundled bars shall be enclosed within stirrups throughout the length of the lintel. Stirrups and the installation thereof shall comply with Section R611.8.2.1.

❖ Section 7.6.6 of ACI 318 permits bars to be bundled if provisions applicable to the bundled bars are followed. The number of bars being bundled is limited to two to increase the likelihood that the bundle will be properly encased by concrete, and to avoid having to increase development lengths required by Section 12.4.1 of ACI 318 for 3 or 4 bars in a bundle. The limitation on bar size in Item 1 is based on the maximum bar size shown in the lintel tables and is consistent with residential construction practices. The requirement in Item 2 that the wall be thick enough where horizontally oriented bundled bars are used to provide a minimum of 3 inches (76 mm) of clear space beside the bars is based on the use of two No. 4 bars bundled horizontally in a 4-inch (102 mm) wall. This requirement is to ensure that the bundle is properly encased in concrete and to facilitate getting concrete to portions of the lintel below the bundle. The requirement in Item 3 that hook extensions in vertically oriented bundles be separated by at least 1 inch (25 mm) is based on recommendations in Commentary Section R7.6.6 of ACI 318. The limitation in Item 4 on location of lap splices of bundled bars is to make sure that congestion due to the laps will not occur in the critical stress areas of the lintel. Item 5 is consistent with ACI 318 requirements.

R611.8.2.3 Lintels without stirrups designed for nonload-bearing conditions. The maximum clear span of lintels without stirrups designed for nonload-bearing conditions of Table R611.8(1).1 shall be determined in accordance with this section. The maximum clear span of lintels without stirrups in flat walls shall be determined in accordance with Table R611.8(9), and the maximum clear span of lintels without stirrups in walls of waffle-grid or screen-grid construction shall be determined in accordance with Table R611.8(10).

❖ Lintels in nonload-bearing walls are horizontal members used to transfer wall dead loads from above around openings. Lintels are divided into two categories as follows:

1. Lintels at the top of the wall in a one-story building or the second story of a two-story building where the gable portion of the endwall is light-framed construction (supporting light-framed gable); and
2. Lintels in concrete walls, including the gable (supporting concrete wall).

The following design assumptions were made in analyzing the lintels in nonload-bearing walls:

1. Lintels have fixed end restraints since the walls and lintels are cast monolithically. The negative moment at the support ($wl^2/_{12}$) is the governing moment.
2. In the case of waffle-grid and screen-grid walls, a vertical core occurs at each end of the lintel for proper bearing.
3. Lateral restraint (out-of-plane) is provided for the lintel by the floor or roof system above.
4. Allowable deflection criteria is the clear span of the lintel, in inches, divided by 240, or $^1/_2$-inch (12.7 mm), whichever is smaller.
5. Lintels support only dead loads from the wall above.

R611.9 Requirements for connections–general. Concrete walls shall be connected to footings, floors, ceilings and roofs in accordance with this section.

❖ This section prescribes the required connections between concrete walls and light-framed floor, ceiling and roof systems. The concrete wall connection to the footing must be in accordance with Sections R611.6 and R611.7.

R611.9.1 Connections between concrete walls and light-framed floor, ceiling and roof systems. Connections between concrete walls and light-framed floor, ceiling and roof systems using the prescriptive details of Figures R611.9(1) through R611.9(12) shall comply with this section and Sections R611.9.2 and R611.9.3.

❖ This section prescribes the anchor bolt requirements for face-mounted wall ledger board or cold-formed steel tracks and sill plates as shown in Figures R611.9(1) through R611.9(12). The anchor bolts must be headed bolts or threaded rod with hex or square nut. J- or S-hooks are not permitted.

Stay-in-place form material at each anchor bolt for face-mounted wood ledger board and cold-formed steel tracks must be removed such that each wood ledger board or steel track is in direct contact with the concrete at each bolt location. The exception permits a vapor retarder to be installed between the concrete and the wood or cold-formed steel.

R611.9.1.1 Anchor bolts. Anchor bolts used to connect light-framed floor, ceiling and roof systems to concrete walls in accordance with Figures R611.9(1) through R611.9(12) shall have heads, or shall be rods with threads on both ends with a hex or square nut on the end embedded in the concrete. Bolts and threaded rods shall comply with Section R611.5.2.2. Anchor bolts with J- or L-hooks shall not be used where the connection details in these figures are used.

❖ See the commentary for Section R611.9.1.

R611.9.1.2 Removal of stay-in-place form material at bolts. Holes in stay-in-place forms for installing bolts for attaching face-mounted wood ledger boards to the wall shall be a minimum of 4 inches (102 mm) in diameter for forms not greater than $1^1/_2$ inches (38 mm) in thickness, and increased 1 inch (25 mm) in diameter for each $^1/_2$-inch (13 mm) increase in form thickness. Holes in stay-in-place forms for installing bolts for attaching face-mounted cold-formed steel tracks to the wall shall be a minimum of 4 inches (102 mm) square. The wood ledger board or steel track shall be in direct contact with the concrete at each bolt location.

> **Exception:** A vapor retarder or other material less than or equal to $^1/_{16}$ inch (1.6 mm) in thickness is permitted to be installed between the wood ledger or cold-formed track and the concrete.

❖ See the commentary for Section R611.9.1.

R611.9.2 Connections between concrete walls and light-framed floor systems. Connections between concrete walls and light-framed floor systems shall be in accordance with one of the following:

1. For floor systems of wood frame construction, the provisions of Section R611.9.1 and the prescriptive details of Figures R611.9(1) through R611.9(4), where permitted by the tables accompanying those figures. Portions of connections of wood-framed floor systems not noted in the figures shall be in accordance with Section R502, or AF&PA/WFCM, if applicable.
2. For floor systems of cold-formed steel construction, the provisions of Section R611.9.1 and the prescriptive details of Figures R611.9(5) through R611.9(8), where permitted by the tables accompanying those figures. Portions of connections of cold-formed-steel framed floor systems not noted in the figures shall be in accordance with Section R505, or AISI S230, if applicable.
3. Proprietary connectors selected to resist loads and load combinations in accordance with Appendix A (ASD) or Appendix B (LRFD) of PCA 100.
4. An engineered design using loads and load combinations in accordance with Appendix A (ASD) or Appendix B (LRFD) of PCA 100.
5. An engineered design using loads and material design provisions in accordance with this code, or in accordance with ASCE 7, ACI 318, and AF&PA/NDS for wood frame construction or AISI S100 for cold-formed steel frame construction.

❖ Section R611.9.2 provides four approaches for designing anchorages between concrete walls and light-framed floor construction. First, prescriptive details are provided in Figures R611.9(1) through R611.9(4) for floors of wood framed construction and in Figures R611.9(5) through R611.9(8) for cold-formed steel framed floor systems. Each prescriptive detail may only be used where permitted by the table accompanying that figure. The prescriptive details are based on ASD loads and addresses loads in three directions occurring simultaneously (horizontal in the plane of the wall, horizontal out-of-plane and vertical). The second approach (Item 3) allows selection of proprietary connection hardware based on loads given in Appendix A or Appendix B of PCA 100. The third approach (Item 4) allows engineered anchorage design to meet the loads tabulated in Appendix A or Appendix B of PCA 100. The fourth approach (Item 5) involves an engineered design for the specific building configuration.

R611.9.3 Connections between concrete walls and light-framed ceiling and roof systems. Connections between concrete walls and light-framed ceiling and roof systems shall be in accordance with one of the following:

1. For ceiling and roof systems of wood frame construction, the provisions of Section R611.9.1 and the prescriptive details of Figures R611.9(9) and R611.9(10), where permitted by the tables accompanying those figures. Portions of connections of wood-framed ceiling and roof systems not noted in the figures shall be in accordance with Section R802, or AF&PA/WFCM, if applicable.
2. For ceiling and roof systems of cold-formed-steel construction, the provisions of Section R611.9.1 and the prescriptive details of Figures R611.9(11) and R611.9(12), where permitted by the tables accompanying those figures. Portions of connections of cold-formed-steel framed ceiling and roof systems not noted in the figures shall be in accordance with Section R804, or AISI S230, if applicable.
3. Proprietary connectors selected to resist loads and load combinations in accordance with Appendix A (ASD) or Appendix B (LRFD) of PCA 100.
4. An engineered design using loads and load combinations in accordance with Appendix A (ASD) or Appendix B (LRFD) of PCA 100.
5. An engineered design using loads and material design provisions in accordance with this code, or in accordance with ASCE 7, ACI 318, and AF&PA/NDS for

wood-frame construction or AISI S100 for cold-formed-steel frame construction.

❖ Section R611.9.3 provides four approaches for designing anchorages between concrete walls and light-framed ceiling and roof construction. First, prescriptive details are provided in Figures R611.9(9) and R611.9(10) for roofs of wood framed construction and in Figures R611.9(11) and R611.9(12) for cold-formed steel framed roof systems. Each prescriptive detail may only be used where permitted by the table accompanying that figure. The prescriptive details are based on the tabulated ASD loads in Appendix A of PCA 100 and address loads in three directions occurring simultaneously (horizontal in the plane of the wall, horizontal out-of-plane and vertical). The second approach (Item 3) allows selection of proprietary connection hardware based on loads given in Appendix A or B of PCA 100. The fourth approach (Item 5) involves an engineered design for the specific building configuration.

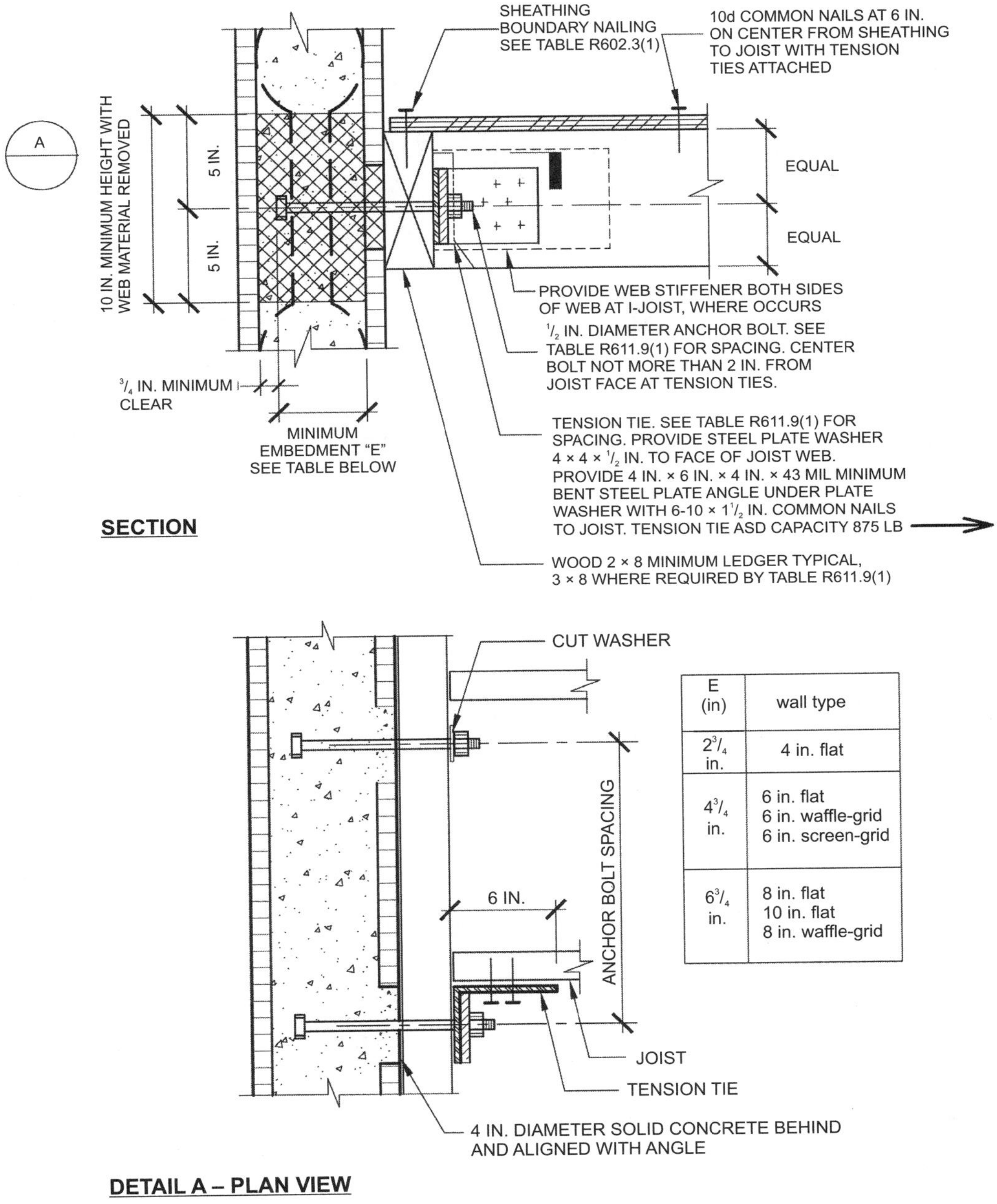

E (in)	wall type
2 3/4 in.	4 in. flat
4 3/4 in.	6 in. flat 6 in. waffle-grid 6 in. screen-grid
6 3/4 in.	8 in. flat 10 in. flat 8 in. waffle-grid

For SI: 1 mil = 0.0254 mm, 1 inch 25.4 mm, 1 pound-force = 4.448 N.

FIGURE R611.9(1)
WOOD-FRAMED FLOOR TO SIDE OF CONCRETE WALL, FRAMING PERPENDICULAR

❖ See the commentary for Sections R611.9.1 and R611.9.2.

R611.10 Floor, roof and ceiling diaphragms. Floors and roofs in all buildings with exterior walls of concrete shall be designed and constructed as *diaphragms*. Where gable-end walls occur, ceilings shall also be designed and constructed as *diaphragms*. The design and construction of floors, roofs and ceilings of wood framing or cold-formed-steel framing serving as *diaphragms* shall comply with the applicable requirements of this code, or AF&PA/WFCM or AISI S230, if applicable.

❖ Resistance for wind and seismic forces relies on wood structural panel diaphragms at the roof and floors. In some cases, blocked diaphragms are necessary in order to develop the shear strength required. Where gable endwalls occur, wood structural panel ceiling diaphragms are relied on to support the top of the concrete gable endwall, in addition to the support provided by the roof diaphragm. In addition to basic diaphragm construction, wood structural panel sheathing needs to be fastened at the noted edge-nail or edge-screw spacing to the framing members to which tension ties are attached.

TABLE R611.9(1)
WOOD-FRAMED FLOOR TO SIDE OF CONCRETE WALL, FRAMING PERPENDICULAR[a, b, c]

ANCHOR BOLT SPACING (inches)	TENSION TIE SPACING (inches)	BASIC WIND SPEED (mph)					
		85B	90B	100B	110B	120B	130B
				85C	90C	100C	110C
					85D	90D	100D
12	12						
12	24						
12	36						
12	48						
16	16					A	A
16	32						
16	48						
19.2	19.2	A	A	A	A	A	
19.2	38.4	A	A	A			

For SI: 1 inch = 25.4 mm, 1 mile per hour = 0.447 m/s.

a. This table is for use with the detail in Figure R611.9(1). Use of this detail is permitted where a cell is not shaded and prohibited where shaded.

b. Wall design per other provisions of Section R611 is required.

c. Letter "A" indicates that a minimum nominal 3 × 8 ledger is required.

❖ See the commentary for Sections R611.9.1 and R611.9.2.

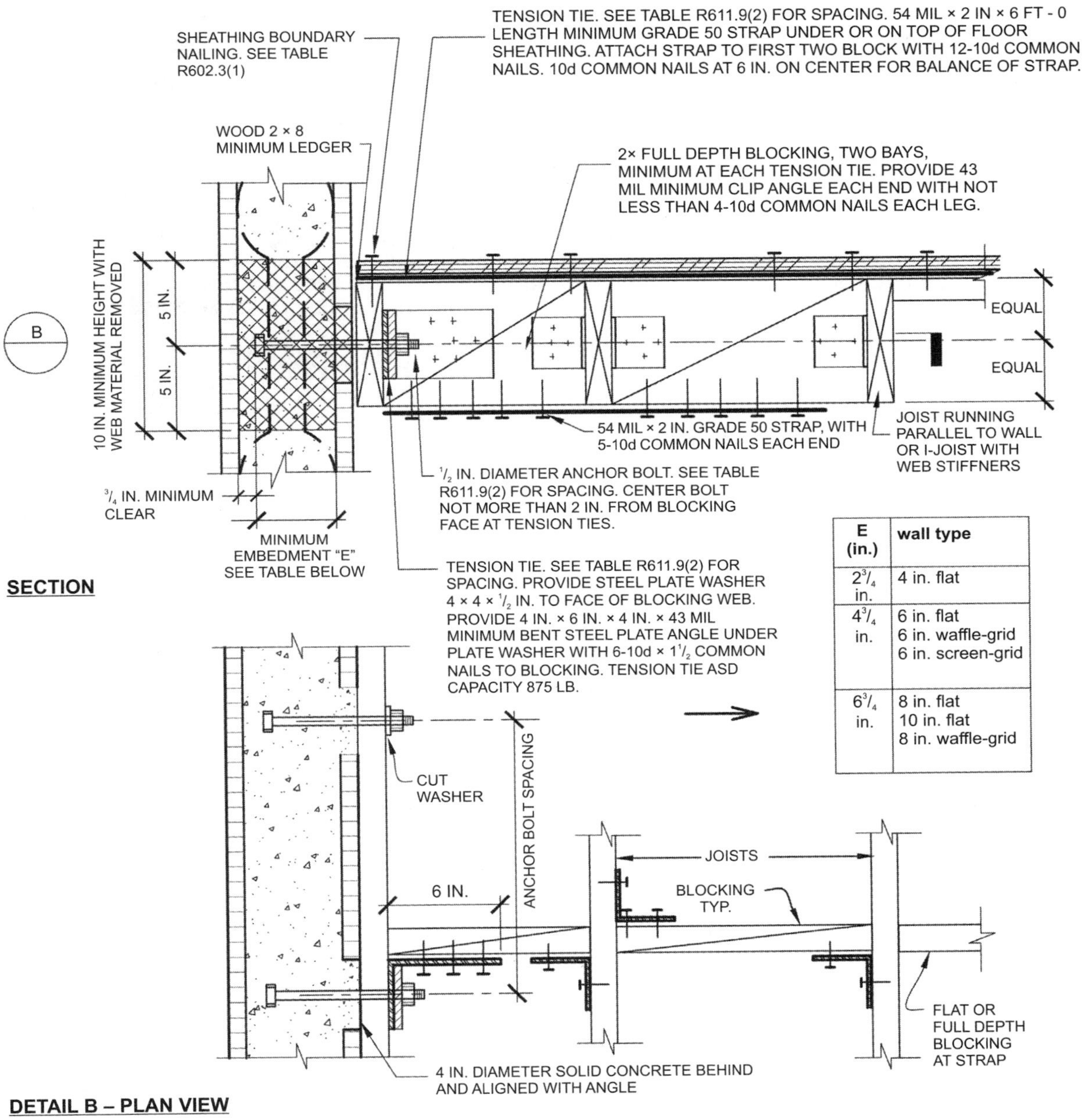

E (in.)	wall type
2 3/4 in.	4 in. flat
4 3/4 in.	6 in. flat 6 in. waffle-grid 6 in. screen-grid
6 3/4 in.	8 in. flat 10 in. flat 8 in. waffle-grid

For SI: 1 mil = 0.0254 mm, 1 inch = 25.4 mm, 1 foot = 304.8 mm, 1 pound-force = 4.448 N.

FIGURE R611.9(2)
WOOD-FRAMED FLOOR TO SIDE OF CONCRETE WALL FRAMING PARALLEL

❖ See the commentary for Sections R611.9.1 and R611.9.2.

TABLE R611.9(2)
WOOD-FRAMED FLOOR TO SIDE OF CONCRETE WALL, FRAMING PARALLEL[a, b]

ANCHOR BOLT SPACING (inches)	TENSION TIE SPACING (inches)	BASIC WIND SPEED (mph) AND WIND EXPOSURE CATEGORY					
		85b	90B	100B	110B	120B	130B
				85C	90C	100C	110C
					85D	90D	100D
12	12						
12	24						
12	36				■	■	■
12	48			■	■	■	■
16	16						
16	32					■	■
16	48			■	■	■	■
19.2	19.2						
19.2	38.4				■	■	■
24	24						
24	48			■	■	■	■

For SI: 1 inch = 25.4 mm, 1 mile per hour = 0.447 m/s.

a. This table is for use with the detail in Figure R611.9(2). Use of this detail is permitted where a cell is not shaded and prohibited where shaded.

b. Wall design per other provisions of Section R611 is required.

❖ See the commentary for Sections R611.9.1 and R611.9.2.

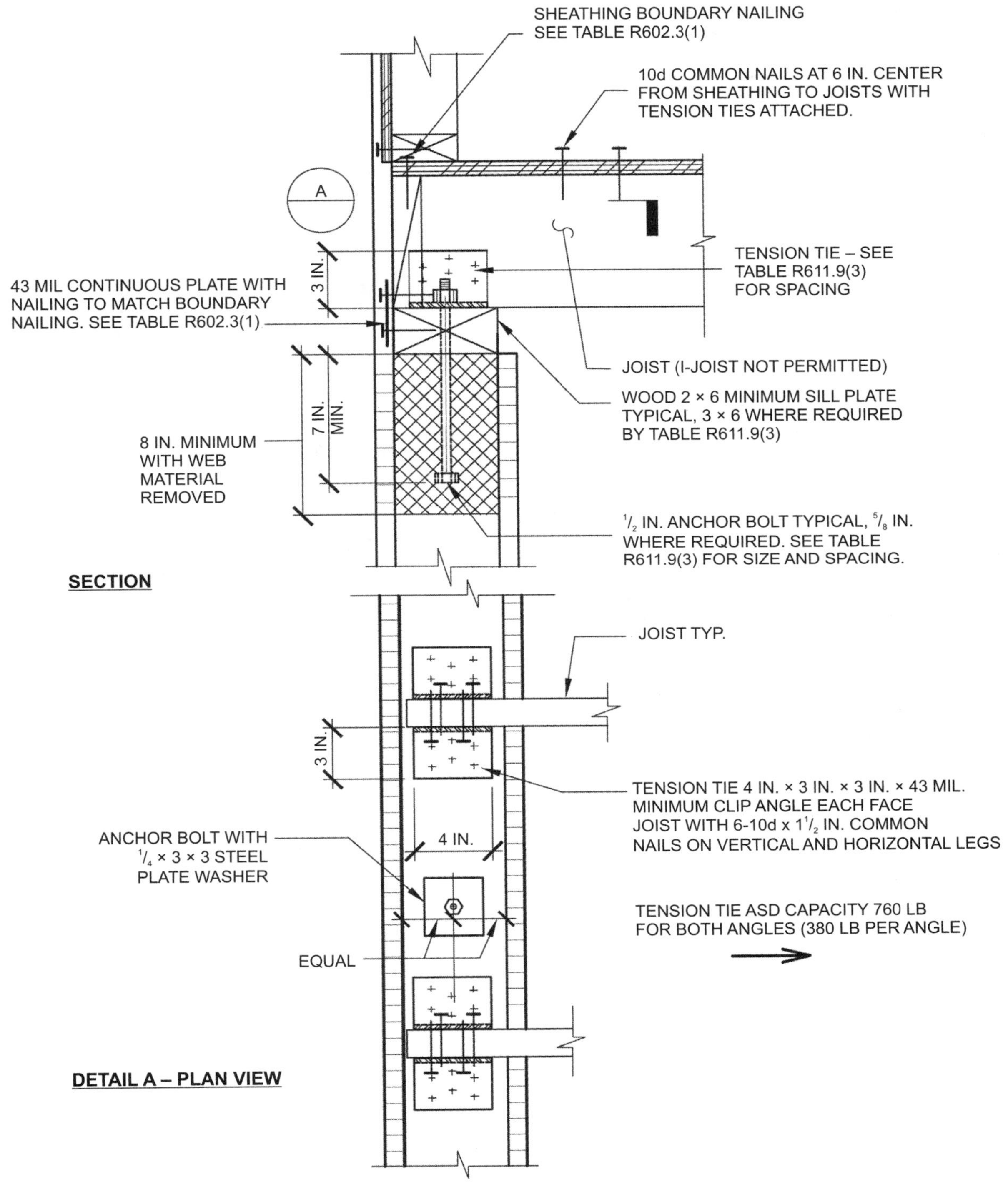

For SI: 1 mil = 0.0254 mm, 1 inch = 25.4 mm, 1 pound-force = 4.448 N.

FIGURE R611.9(3)
WOOD-FRAMED FLOOR TO TOP OF CONCRETE WALL FRAMING PERPENDICULAR

❖ See the commentary for Sections R611.9.1 and R611.9.2.

TABLE R611.9(3)
WOOD-FRAMED FLOOR TO TOP OF CONCRETE WALL, FRAMING PERPENDICULAR[a, b, c, d, e]

ANCHOR BOLT SPACING (inches)	TENSION TIE SPACING (inches)	BASIC WIND SPEED (mph) AND WIND EXPOSURE CATEGORY					
		85B	90B	100B	110B	120B	130B
				85C	90C	100C	110C
					85D	90D	100D
12	12						
12	24						
12	36						
12	48						
16	16					6 A	6 B
16	32					6 A	6 B
16	48						
19.2	19.2				6 A	6 A	6 B
19.2	38.4				6 A	6 A	
24	24			6 A	6 B	6 A	
24	48			6 A			

For SI: 1 inch = 25.4 mm, 1 mile per hour = 0.447 m/s.

a. This table is for use with the detail in Figure R611.9(3). Use of this detail is permitted where cell is not shaded, prohibited where shaded.

b. Wall design per other provisions in Section R611 is required.

c. For wind design, minimum 4-inch-nominal wall is permitted in unshaded cells with no number.

d. Number 6 indicates minimum permitted nominal wall thickness in inches necessary to develop required strength (capacity) of connection. As a minimum, this nominal thickness shall occur in the portion of the wall indicated by the cross-hatching in Figure R611.9(3). For the remainder of the wall, see Note b.

e. Letter "A" indicates that a minimum nominal 3 × 6 sill plate is required. Letter "B" indicates that a $^5/_8$-inch-diameter anchor bolt and a minimal nominal 3 × 6 sill plate are required.

❖ See the commentary for Sections R611.9.1 and R611.9.2.

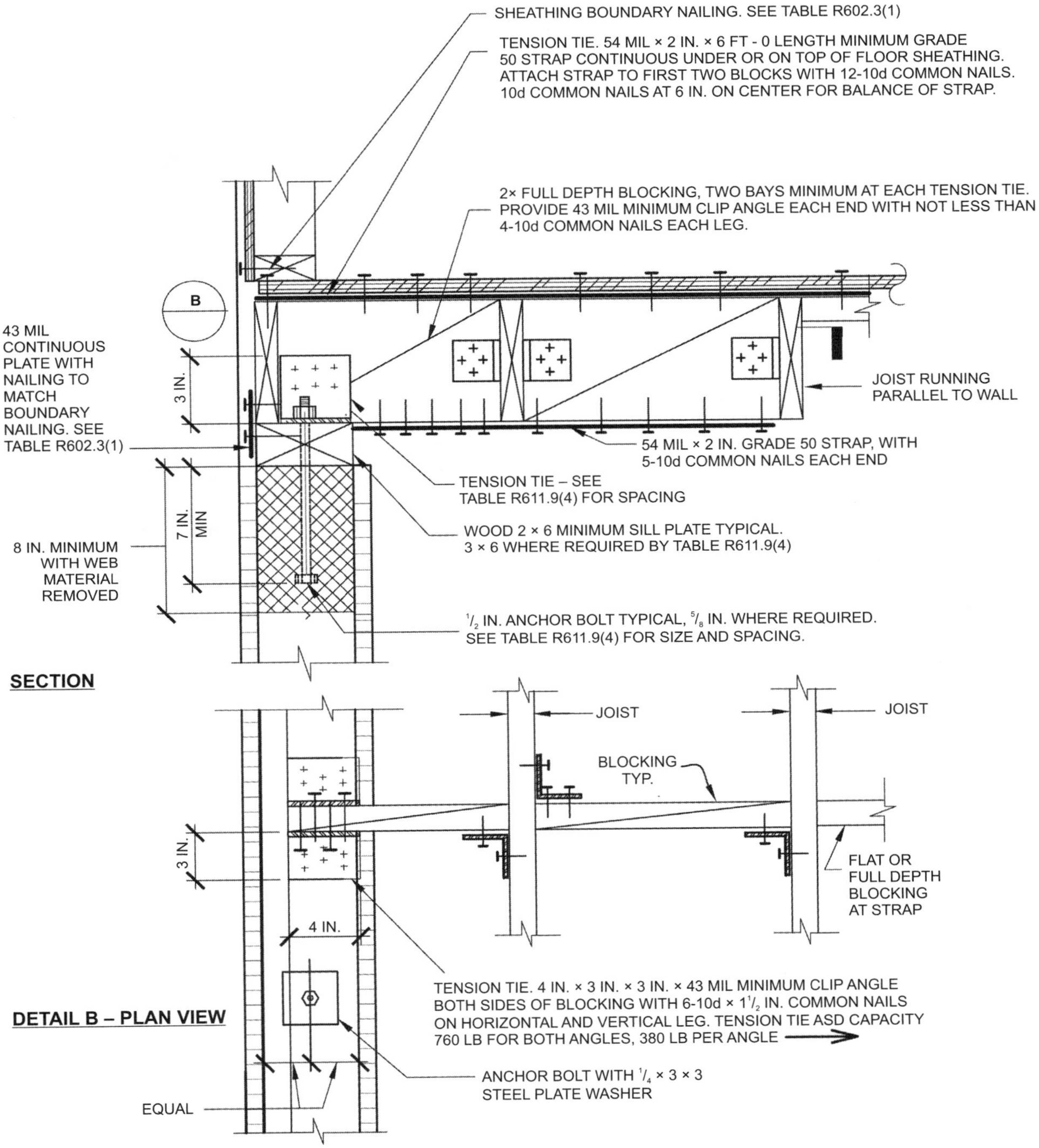

For SI: 1 mil = 0.0254 mm, 1 inch = 25.4 mm, 1 foot = 304.8 mm, 1 pound-force = 4.448 N.

FIGURE R611.9(4)
WOOD-FRAMED FLOOR TO TOP OF CONCRETE WALL FRAMING PARALLEL

❖ See the commentary for Section R611.9.1 and R611.9.2.

TABLE R611.9(4)
WOOD-FRAMED FLOOR TO TOP OF CONCRETE WALL, FRAMING PARALLEL[a, b, c, d, e]

ANCHOR BOLT SPACING (inches)	TENSION TIE SPACING (inches)	BASIC WIND SPEED (mph) AND WIND EXPOSURE CATEGORY					
		85B	90B	100B	110B	120B	130B
				85C	90C	100C	110C
					85D	90D	100D
	12						
12	24						
12	36						
12	48						
16	16					6 A	6 B
16	32					6 A	6 B
16	48						
19.2	19.2				6 A	6 A	6 B
19.2	38.4				6 A	6 A	
24	24			6 A	6 B	6 B	
24	48			6 A			

For SI: 1 inch = 25.4 mm, 1 mile per hour = 0.447 m/s.

a. This table is for use with the detail in Figure R611.9(4). Use of this detail is permitted where a cell is not shaded, prohibited where shaded.
b. Wall design per other provisions of Section R611 is required.
c. For wind design, minimum 4-inch-nominal wall is permitted in unshaded cells with no number.
d. Number 6 indicates minimum permitted nominal wall thickness in inches necessary to develop required strength (capacity) of connection. As a minimum, this nominal thickness shall occur in the portion of the wall indicated by the cross-hatching in Figure R611.9(4). For the remainder of the wall, see Note b.
e. Letter "A" indicates that a minimum nominal 3 × 6 sill plate is required. Letter "B" indicates that a $^5/_8$-inch-diameter anchor bolt and a minimal nominal 3 × 6 sill plate are required.

❖ See the commentary for Sections R611.9.1 and R611.9.2.

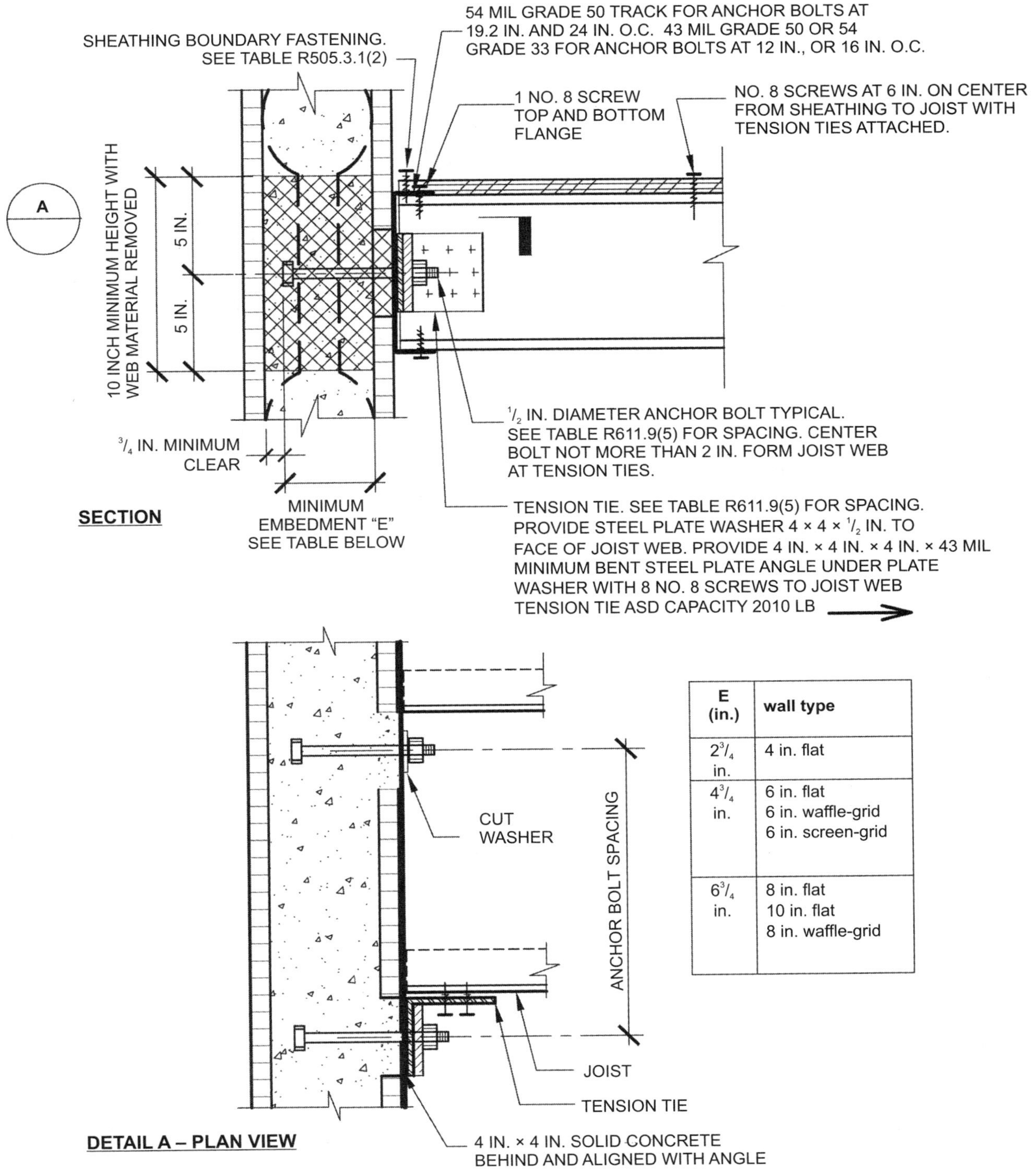

E (in.)	wall type
$2^3/_4$ in.	4 in. flat
$4^3/_4$ in.	6 in. flat 6 in. waffle-grid 6 in. screen-grid
$6^3/_4$ in.	8 in. flat 10 in. flat 8 in. waffle-grid

For SI: 1 mil = 0.0254 mm, 1 inch = 25.4 mm, 1 pound-force = 4.448 N.

FIGURE R611.9(5)
COLD-FORMED STEEL FLOOR TO SIDE OF CONCRETE WALL, FRAMING PERPENDICULAR

❖ See the commentary for Section R611.9.1 and R611.9.2.

TABLE R611.9(5)
COLD-FORMED STEEL-FRAMED FLOOR TO SIDE OF CONCRETE WALL, FRAMING PERPENDICULAR[a, b, c, d]

ANCHOR BOLT SPACING (inches)	TENSION TIE SPACING (inches)	BASIC WIND SPEED (mph) AND WIND EXPOSURE CATEGORY					
		85B	90B	100B	110B	120B	130B
				85C	90C	100C	110C
					85D	90D	100D
12	12						
12	24						
12	36						6
12	48					6	6
16	16						
16	32						
16	48					6	6
19.2	19.2						
19.2	38.4						6
24	24						
24	48					6	6

For SI: 1 inch = 25.4 mm, 1 mile per hour = 0.4470 m/s.

a. This table is for use with the detail in Figure R611.9(5). Use of this detail is permitted where a cell is not shaded.

b. Wall design per other provisions of Section R611 is required.

c. For wind design, minimum 4-inch-nominal wall is permitted in unshaded cells with no number.

d. Number 6 indicates minimum permitted nominal wall thickness in inches necessary to develop required strength (capacity) of connection. As a minimum, this nominal thickness shall occur in the portion of the wall indicated by the cross-hatching in Figure R611.9(5). For the remainder of the wall, see Note b.

❖ See the commentary for Sections R611.9.1 and R611.9.2.

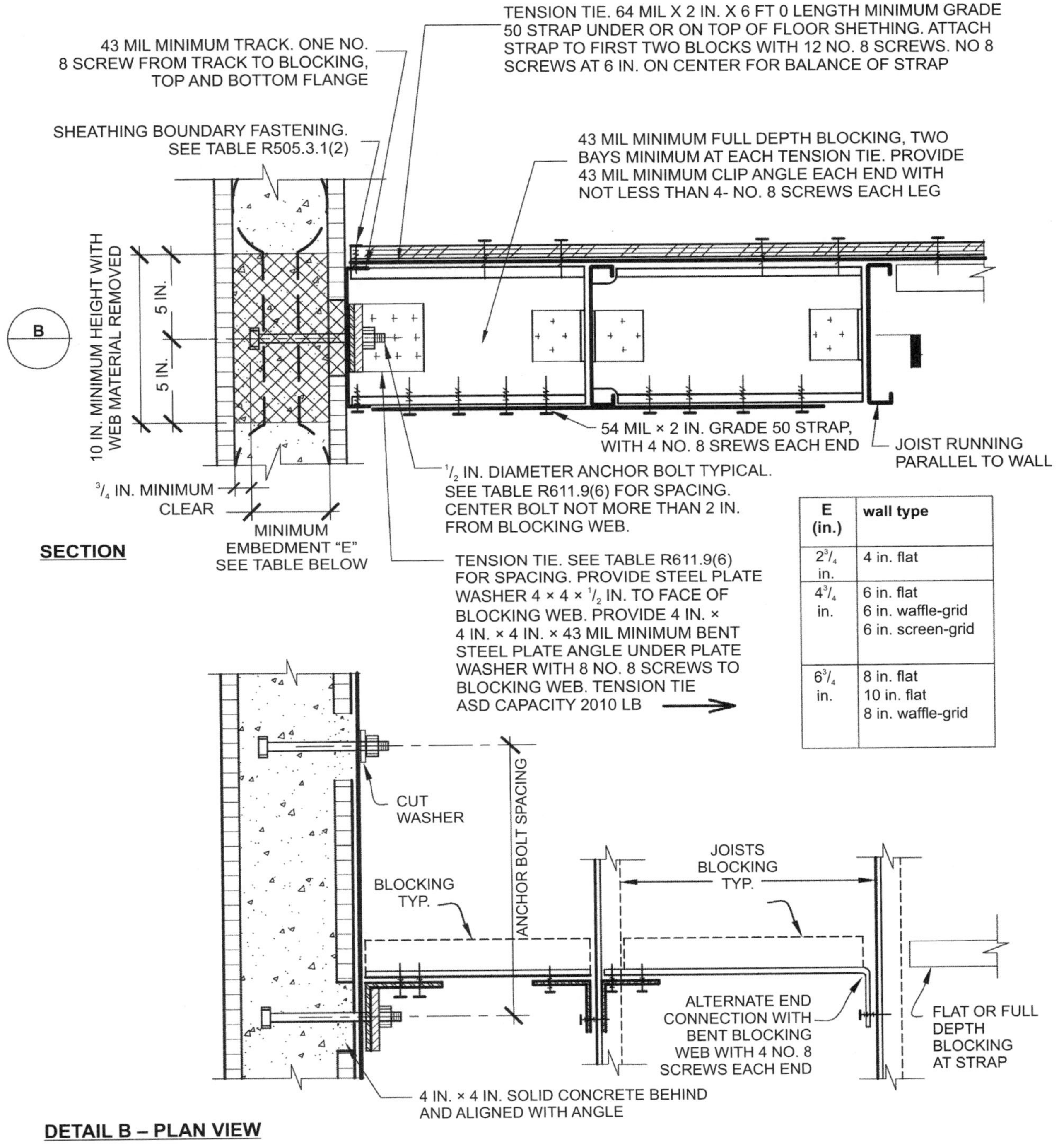

E (in.)	wall type
$2^{3}/_{4}$ in.	4 in. flat
$4^{3}/_{4}$ in.	6 in. flat 6 in. waffle-grid 6 in. screen-grid
$6^{3}/_{4}$ in.	8 in. flat 10 in. flat 8 in. waffle-grid

For SI: 1 mil = 0.0254 mm, 1 inch = 25.4 mm, 1 pound-force = 4.448 N.

FIGURE R611.9(6)
COLD-FORMED STEEL FLOOR TO SIDE OF CONCRETE WALL, FRAMING PARALLEL

❖ See the commentary for Sections R611.9.1 and R611.9.2.

TABLE R611.9(6)
COLD-FORMED STEEL-FRAMED FLOOR TO SIDE OF CONCRETE WALL, FRAMING PARALLEL[a, b, c, d]

ANCHOR BOLT SPACING (inches)	TENSION TIE SPACING (inches)	BASIC WIND SPEED (mph) AND WIND EXPOSURE CATEGORY					
		85B	90B	100B	110B	120B	130B
				85C	90C	100C	110C
					85D	90D	100D
12	12						
12	24						
12	36						6
12	48					6	6
16	16						
16	32						
16	48					6	6
19.2	19.2						
19.2	38.4						6
24	24						
24	48					6	6

For SI: 1 inch = 25.4 mm, 1 mile per hour = 0.447 m/s.

a. This table is for use with the detail in Figure R611.9(6). Use of this detail is permitted where a cell is not shaded.

b. Wall design per other provisions of Section R611 is required.

c. For wind design, minimum 4-inch-nominal wall is permitted in unshaded cells with no number.

d. Number 6 indicates minimum permitted nominal wall thick ness in inches necessary to develop required strength (capacity) of connection. As a minimum, this nominal thickness shall occur in the portion of the wall indicated by the cross-hatching in Figure R611.9(6). For the remainder of the wall, see Note b.

❖ See the commentary for Sections R611.9.1 and R611.9.2.

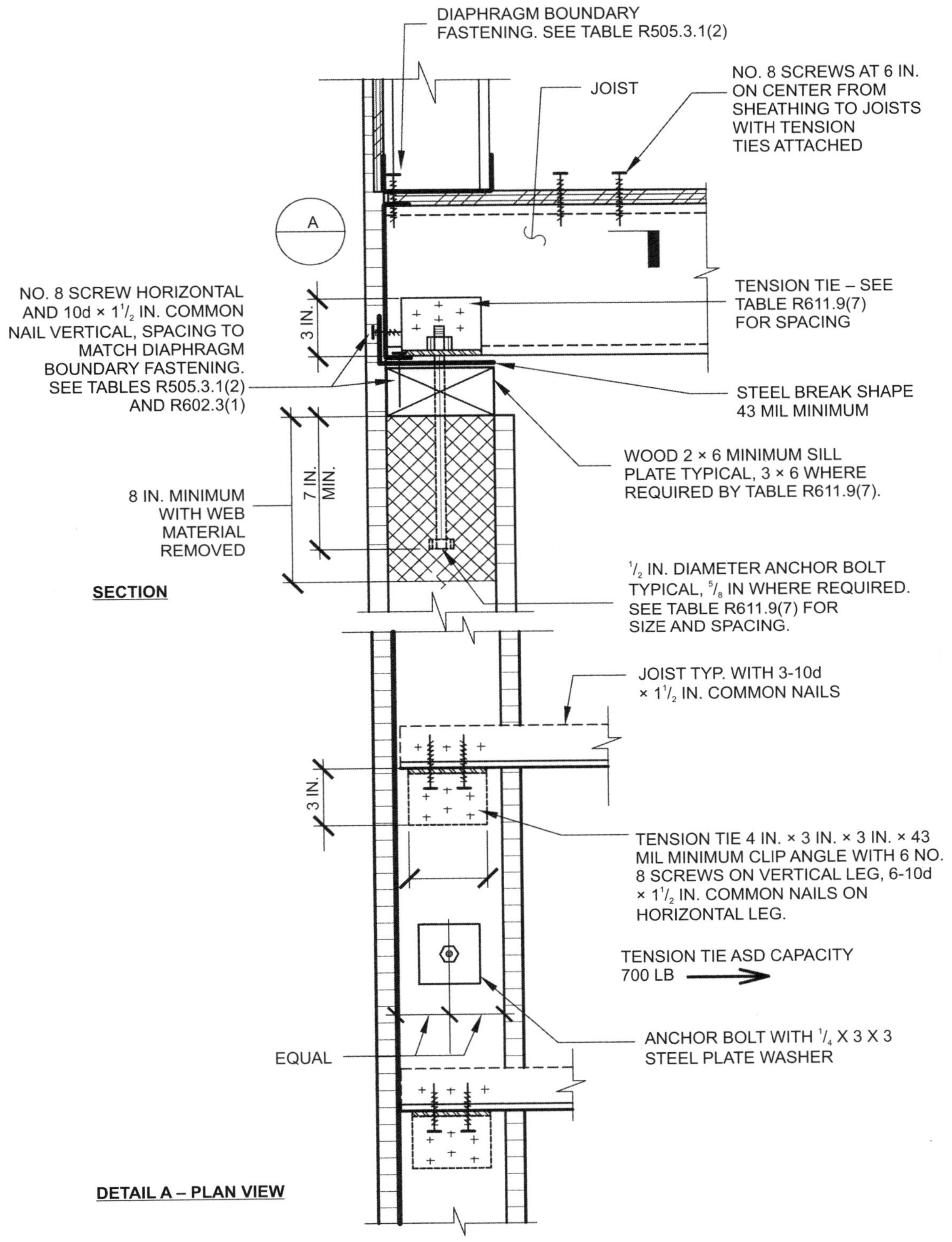

For SI: 1 mil = 0.0254 mm, 1 inch = 25.4 mm, 1 pound-force = 4.448 N.

FIGURE R611.9(7)
COLD-FORMED STEEL FLOOR TO TOP OF CONCRETE WALL FRAMING PERPENDICULAR

❖ See the commentary for Sections R611.9.1 and R611.9.2.

TABLE R611.9(7)
COLD-FORMED STEEL-FRAMED FLOOR TO TOP OF CONCRETE WALL, FRAMING PERPENDICULAR[a, b, c, d, e]

ANCHOR BOLT SPACING (inches)	TENSION TIE SPACING (inches)	BASIC WIND SPEED (mph) AND WIND EXPOSURE CATEGORY					
		85B	90B	100B	110B	120B	130B
				858C	90C	100C	110C
					85D	90D	100D
12	12						
12	24						
16	16					6 A	6 B
16	32					6 A	6 B
19.2	19.2				6 A	8 B	8 B
19.2	38.4				6 A	8 B	8 B
24	24			6 A	8 B	8 B	

For SI: 1 inch = 25.4 mm, 1 mile per hour = 0.447 m/s.

a. This table is for use with the detail in Figure R611.9(7). Use of this detail is permitted where a cell is not shaded, prohibited where shaded.

b. Wall design per other provisions of Section R611 is required.

c. For wind design, minimum 4-inch-nominal wall is permitted in unshaded cells with no number.

d. Numbers 6 and 8 indicate minimum permitted nominal wall thickness in inches necessary to develop required strength (capacity) of connection. As a minimum, this nominal thickness shall occur in the portion of the wall indicated by the cross-hatching in Figure R611.9(7). For the remainder of the wall, see Note b.

e. Letter "A" indicates that a minimum nominal 3 × 6 sill plate is required. Letter "B" indicates that a $^5/_8$-inch-diameter anchor bolt and a minimum nominal 3 × 6 sill plate are required.

❖ See the commentary for Sections R611.9.1 and R611.9.2.

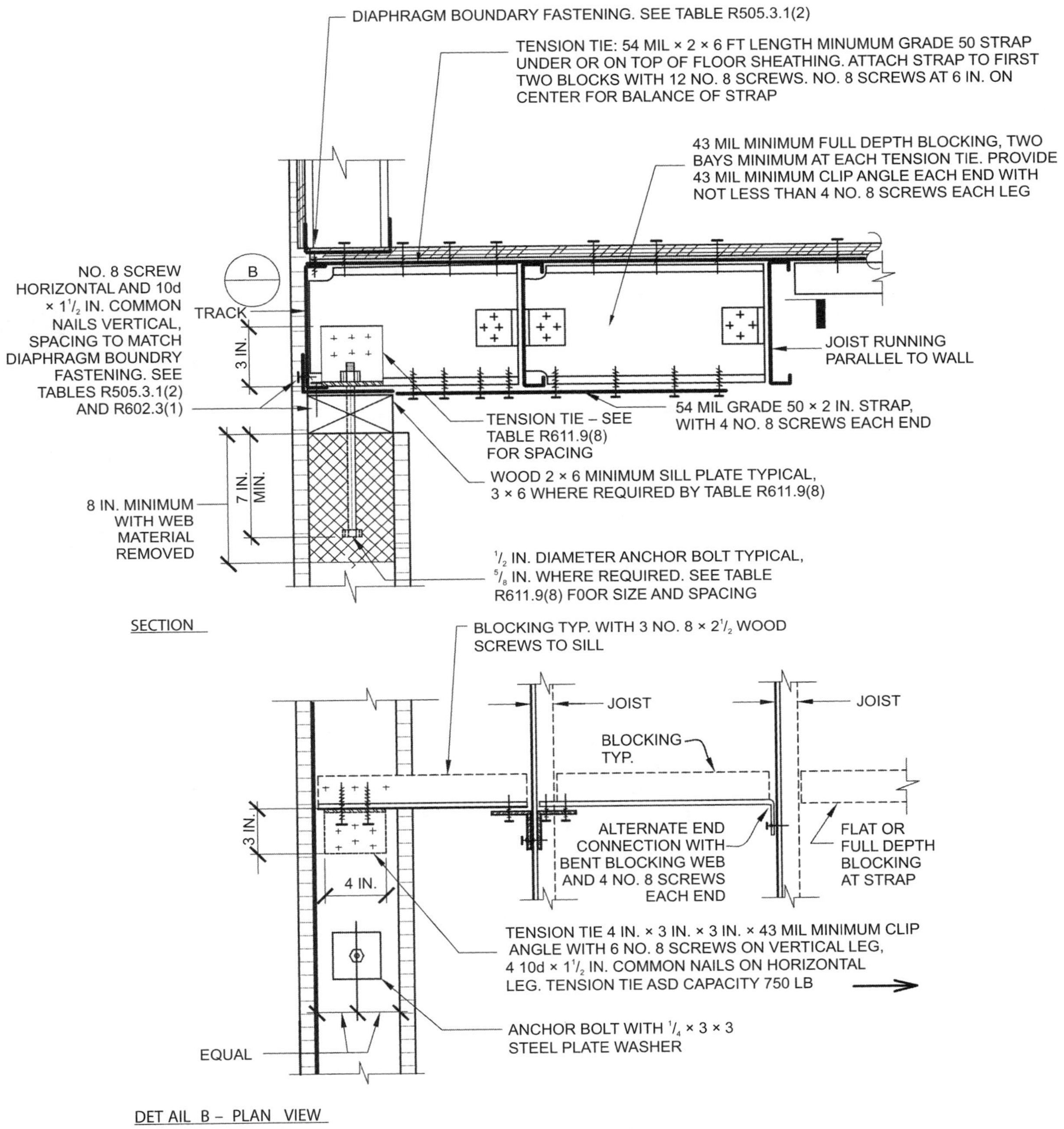

For SI: 1 mil = 0.0254 mm, 1 inch = 25.4 mm, 1 pound-force = 4.448 N.

FIGURE R611.9(8)
COLD-FORMED STEEL FLOOR TO TOP OF CONCRETE WALL, FRAMING PARALLEL

❖ See the commentary for Sections R611.9.1 and R611.9.2.

TABLE R611.9(8)
COLD-FORMED STEEL-FRAMED FLOOR TO TOP OF CONCRETE WALL, FRAMING PARALLEL[a, b, c, d, e]

ANCHOR BOLT SPACING (inches)	TENSION TIE SPACING (inches)	BASIC WIND SPEED (mph) AND WIND EXPOSURE CATEGORY					
		85B	90B	100B	110B	120B	130B
				85C	90C	100C	110C
					85D	90D	100D
12	12						
12	24						
16	16					6 A	6 B
16	32					6 A	6 B
19.2	19.2				6 A	8 B	8 B
19.2	38.4				6 A	8 B	8 B
24	24			6 A	8 B	8 B	

For SI: 1 inch = 25.4 mm, 1 mile per hour = 0.447 m/s.

a. This table is for use with the detail in Figure R611.9(8). Use of this detail is permitted where a cell is not shaded, prohibited where shaded.

b. Wall design per other provisions of Section R611 is required.

c. For wind design, minimum 4-inch-nominal wall is permitted in unshaded cells with no number.

d. Numbers 6 and 8 indicate minimum permitted nominal wall thickness in inches necessary to develop required strength (capacity) of connection. As a minimum, this nominal thickness shall occur in the portion of the wall indicated by the cross-hatching in Figure R611.9(8). For the remainder of the wall, see Note b.

e. Letter "A" indicates that a minimum nominal 3 × 6 sill plate is required. Letter "B" indicates that a $^5/_8$-inch-diameter anchor bolt and a minimum nominal 3 × 6 sill plate are required.

❖ See the commentary for Sections R611.9.1 and R611.9.2.

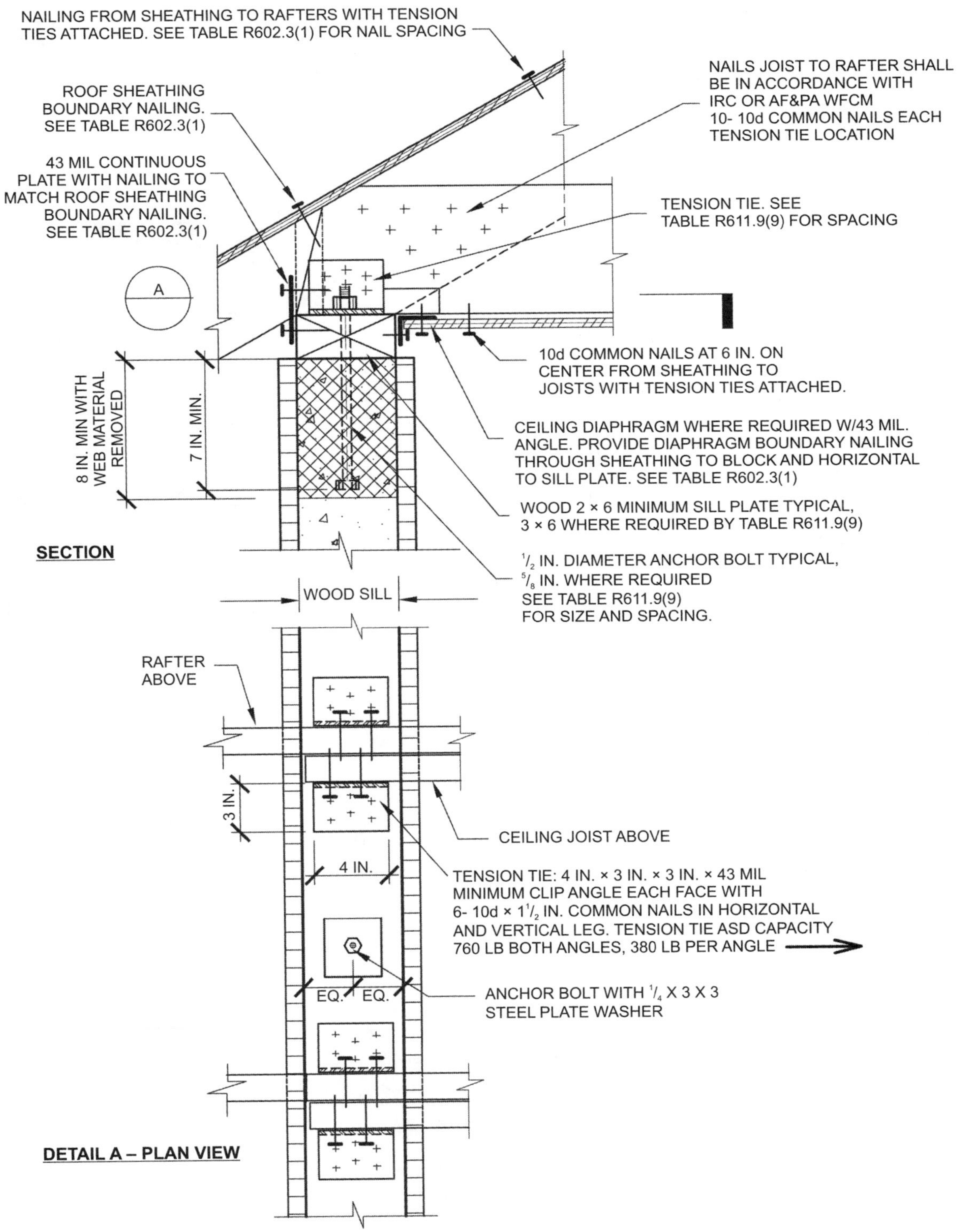

For SI: 1 mil = 0.0254 mm, 1 inch = 25.4 mm, 1 pound-force = 4.448 N.

FIGURE R611.9(9)
WOOD-FRAMED ROOF TO TOP OF CONCRETE WALL, FRAMING PERPENDICULAR

❖ See the commentary for Sections R611.9.1 and R611.9.3.

TABLE R611.9(9)
WOOD-FRAMED ROOF TO TOP OF CONCRETE WALL, FRAMING PERPENDICULAR[a, b, c, d, e]

ANCHOR BOLT SPACING (inches)	TENSION TIE SPACING (inches)	BASIC WIND SPEED (mph) AND WIND EXPOSURE CATEGORY					
		85B	90B	100B	110B	120B	130B
				85C	90C	100C	110C
					85D	90D	100D
12	12						
12	24						
12	36						
12	48						
16	16						6
16	32						6
16	48						
19.2	19.2					6	6 A
19.2	38.4					6	
24	24				6 A	6 A	6 B
24	48						

For SI: 1 inch = 25.4 mm, 1 mile per hour = 0.447 m/s.

a. This table is for use with the detail in Figure R611.9(9). Use of this detail is permitted where cell a is not shaded, prohibited where shaded.

b. Wall design per other provisions of Section R611 is required.

c. For wind design, minimum 4-inch-nominal wall is permitted in unshaded cells with no number.

d. Number 6 indicates minimum permitted nominal wall thickness in inches necessary to develop required strength (capacity) of connection. As a minimum, this nominal thickness shall occur in the portion of the wall indicated by the cross-hatching in Figure R611.9(9). For the remainder of the wall, see Note b.

e. Letter "A" indicates that a minimum nominal 3 × 6 sill plate is required. Letter "B" indicates that a $^{5}/_{8}$-inch-diameter anchor bolt and a minimum nominal 3 × 6 sill plate are required.

❖ See the commentary for Sections R611.9.1 and R611.9.3.

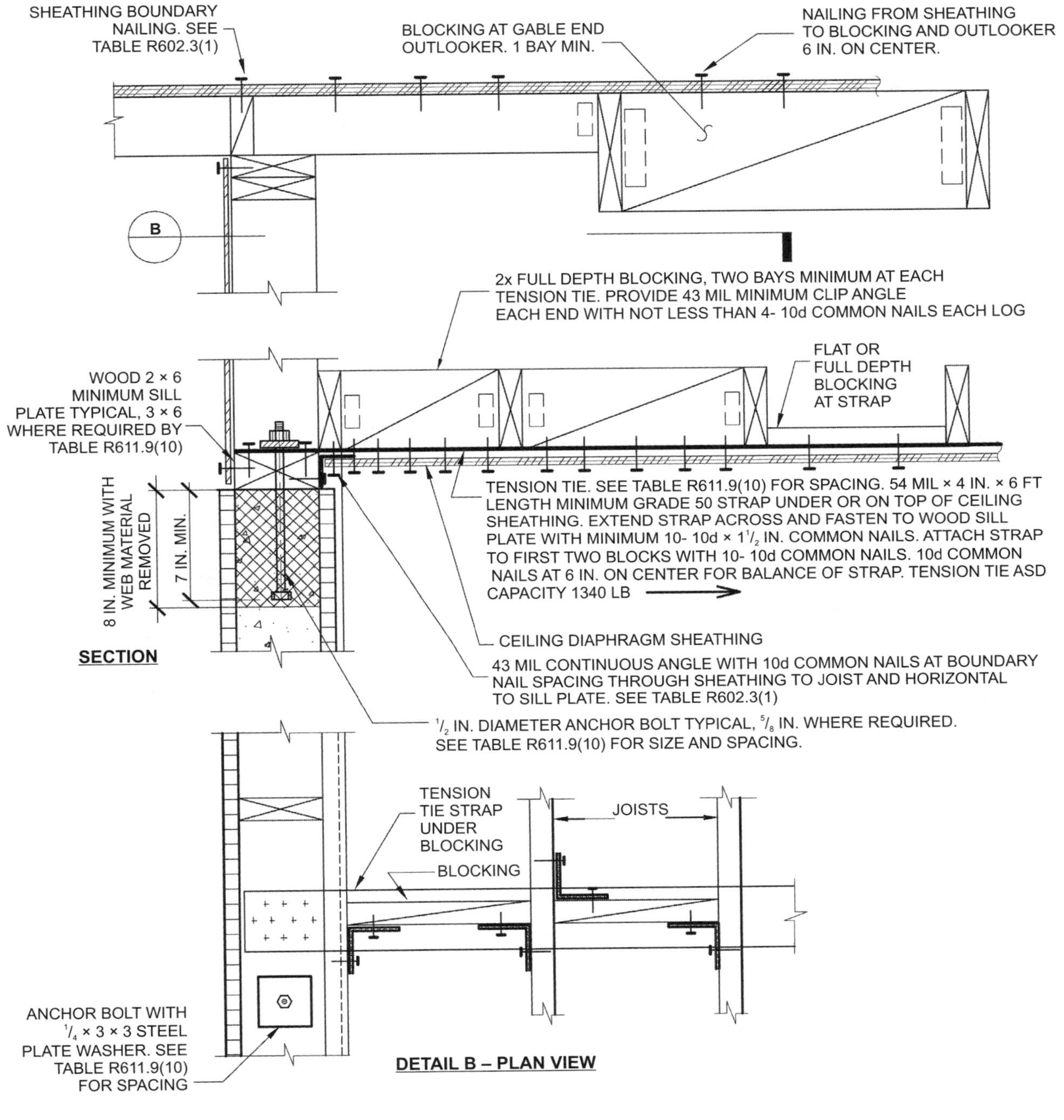

For SI: 1 mil = 0.0254 mm, 1 inch = 25.4 mm, 1 pound-force = 4.448 N.

FIGURE R611.9(10)
WOOD-FRAMED ROOF TO TOP OF CONCRETE WALL FRAMING PARALLEL

❖ See the commentary for Sections R611.9.1 and R611.9.3.

TABLE R611.9(10)
WOOD-FRAMED ROOF TO TOP OF CONCRETE WALL, FRAMING PARALLEL[a, b, c, d, e]

ANCHOR BOLT SPACING (inches)	TENSION TIE SPACING (inches)	BASIC WIND SPEED (mph) AND WIND EXPOSURE CATEGORY					
		85B	90B	100B	110B	120B	130B
				85C	90C	100C	110C
					85D	90D	100D
12	12						
12	24						
12	36						
12	48						
16	16					6	6
16	32					6	6
16	48					6	6
19.2	19.2				6	6	6 A
19.2	38.4				6	6	6 A
24	24			6	6 A	6 A	6 B
24	48			6	6 A	6 B	6 B

For SI: 1 inch = 25.4 mm, 1 mile per hour = 0.447 m/s.

a. This table is for use with the detail in Figure R611.9(10). Use of this detail is permitted where a cell is not shaded.

b. Wall design per other provisions of Section R611 is required.

c. For wind design, minimum 4-inch-nominal wall is permitted in cells with no number.

d. Number 6 indicates minimum permitted nominal wall thickness in inches necessary to develop required strength (capacity) of connection. As a minimum, this nominal thickness shall occur in the portion of the wall indicated by the cross-hatching in Figure R611.9(10). For the remainder of the wall, see Note b.

e. Letter “A” indicates that a minimum nominal 3 × 6 sill plate is required. Letter “B” indicates that a $^5/_8$-inch-diameter anchor bolt and a minimum nominal 3 × 6 sill plate are required.

❖ See the commentary for Sections R611.9.1 and R611.9.3.

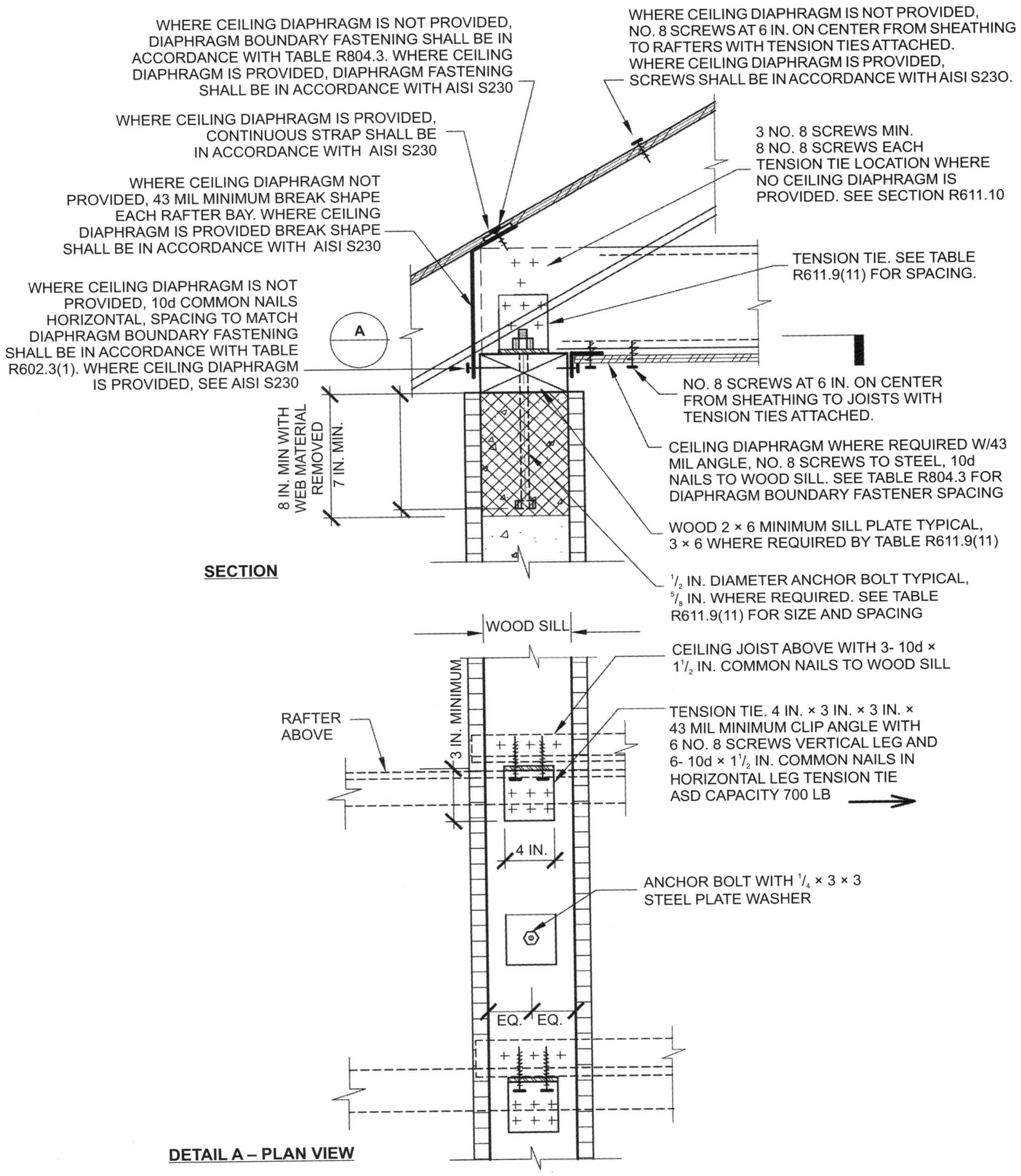
WHERE CEILING DIAPHRAGM IS NOT PROVIDED, DIAPHRAGM BOUNDARY FASTENING SHALL BE IN ACCORDANCE WITH TABLE R804.3. WHERE CEILING DIAPHRAGM IS PROVIDED, DIAPHRAGM FASTENING SHALL BE IN ACCORDANCE WITH AISI S230
WHERE CEILING DIAPHRAGM IS NOT PROVIDED, NO. 8 SCREWS AT 6 IN. ON CENTER FROM SHEATHING TO RAFTERS WITH TENSION TIES ATTACHED. WHERE CEILING DIAPHRAGM IS PROVIDED, SCREWS SHALL BE IN ACCORDANCE WITH AISI S23O.
WHERE CEILING DIAPHRAGM IS PROVIDED, CONTINUOUS STRAP SHALL BE IN ACCORDANCE WITH AISI S230
3 NO. 8 SCREWS MIN. 8 NO. 8 SCREWS EACH TENSION TIE LOCATION WHERE NO CEILING DIAPHRAGM IS PROVIDED. SEE SECTION R611.10
WHERE CEILING DIAPHRAGM NOT PROVIDED, 43 MIL MINIMUM BREAK SHAPE EACH RAFTER BAY. WHERE CEILING DIAPHRAGM IS PROVIDED BREAK SHAPE SHALL BE IN ACCORDANCE WITH AISI S230
TENSION TIE. SEE TABLE R611.9(11) FOR SPACING.
WHERE CEILING DIAPHRAGM IS NOT PROVIDED, 10d COMMON NAILS HORIZONTAL, SPACING TO MATCH DIAPHRAGM BOUNDARY FASTENING SHALL BE IN ACCORDANCE WITH TABLE R602.3(1). WHERE CEILING DIAPHRAGM IS PROVIDED, SEE AISI S230
A
NO. 8 SCREWS AT 6 IN. ON CENTER FROM SHEATHING TO JOISTS WITH TENSION TIES ATTACHED.
8 IN. MIN WITH WEB MATERIAL REMOVED
7 IN. MIN.
CEILING DIAPHRAGM WHERE REQUIRED W/43 MIL ANGLE, NO. 8 SCREWS TO STEEL, 10d NAILS TO WOOD SILL. SEE TABLE R804.3 FOR DIAPHRAGM BOUNDARY FASTENER SPACING
WOOD 2 × 6 MINIMUM SILL PLATE TYPICAL, 3 × 6 WHERE REQUIRED BY TABLE R611.9(11)
SECTION
1/2 IN. DIAMETER ANCHOR BOLT TYPICAL, 5/8 IN. WHERE REQUIRED. SEE TABLE R611.9(11) FOR SIZE AND SPACING
WOOD SILL
CEILING JOIST ABOVE WITH 3- 10d × 1 1/2 IN. COMMON NAILS TO WOOD SILL
3 IN. MINIMUM
RAFTER ABOVE
TENSION TIE. 4 IN. × 3 IN. × 3 IN. × 43 MIL MINIMUM CLIP ANGLE WITH 6 NO. 8 SCREWS VERTICAL LEG AND 6- 10d × 1 1/2 IN. COMMON NAILS IN HORIZONTAL LEG TENSION TIE ASD CAPACITY 700 LB
4 IN.
ANCHOR BOLT WITH 1/4 × 3 × 3 STEEL PLATE WASHER
EQ.
EQ.
DETAIL A – PLAN VIEW

For SI: 1 mil = 0.0254 mm, 1 inch = 25.4 mm, 1 foot = 304.8 mm, 1 pound-force = 4.448 N.

FIGURE R611.9(11)
COLD-FORMED STEEL ROOF TO TOP OF CONCRETE WALL, FRAMING PERPENDICULAR

❖ See the commentary for Sections R611.9.1 and R611.9.3.

TABLE R611.9(11)
WOOD-FRAMED ROOF TO TOP OF CONCRETE WALL, FRAMING PERPENDICULAR[a, b, c, d, e]

ANCHOR BOLT SPACING (inches)	TENSION TIE SPACING (inches)	BASIC WIND SPEED (mph) AND WIND EXPOSURE CATEGORY					
		85B	90B	100B	110B	120B	130B
				85C	90C	100C	110C
					85D	90D	100D
12	12						
12	24						
16	16					6	6
16	32					6	6
19.2	19.2				6	6	8 B
19.2	38.4				6	6	8 B
24	24			6	6	8 B	

For SI: 1 inch = 25.4 mm, 1 mile per hour = 0.447 m/s.

a. This table is for use with the detail in Figure R611.9(11). Use of this detail is permitted where a cell is not shaded, prohibited where shaded.

b. Wall design per other provisions of Section R611 is required.

c. For wind design, minimum 4-inch-nominal wall is permitted in unshaded cells with no number.

d. Numbers 6 and 8 indicate minimum permitted nominal wall thickness in inches necessary to develop required strength (capacity) of connection. As a minimum, this nominal thick ness shall occur in the portion of the wall indicated by the cross-hatching in Figure R611.9(11). For the remainder of the wall, see Note b.

e. Letter "B" indicates that a $^{5}/_{8}$-inch-diameter anchor bolt and a minimum nominal 3 × 6 sill plate are required.

❖ See the commentary for Sections R611.9.1 and R611.9.3.

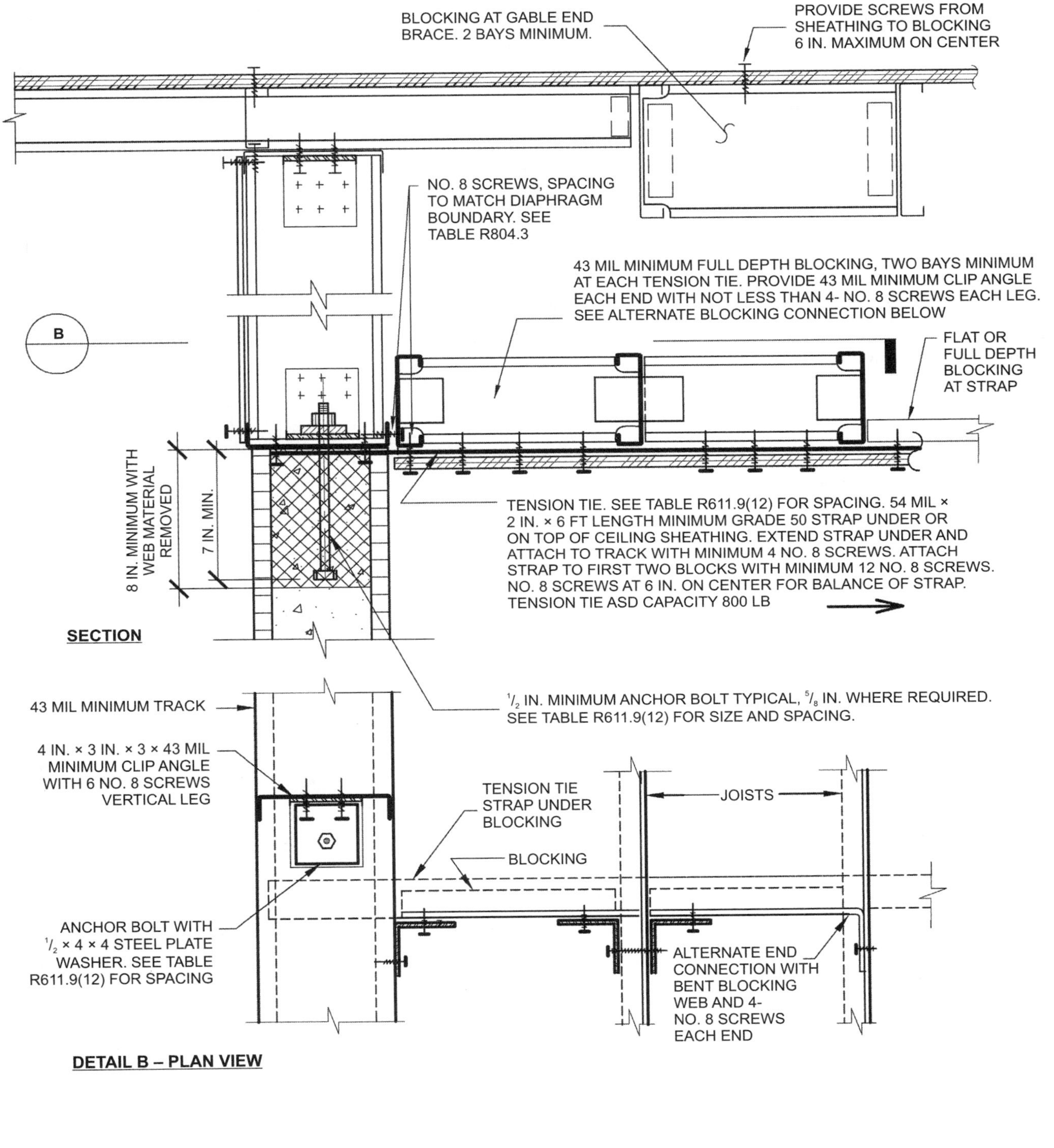

For SI: 1 mil = 0.0254 mm, 1 inch = 25.4 mm, 1 pound-force = 4.448 N.

FIGURE R611.9(12)
COLD-FORMED STEEL ROOF TO TOP OF CONCRETE WALL, FRAMING PARALLEL

❖ See the commentary for Sections R611.9.1 and R611.9.3.

TABLE R611.9(12)
COLD-FORMED STEEL ROOF TO TOP OF CONCRETE WALL, FRAMING PARALLEL[a, b, c, d, e]

ANCHOR BOLT SPACING (inches)	TENSION TIE SPACING (inches)	BASIC WIND SPEED (mph) AND WIND EXPOSURE CATEGORY					
		85B	90B	100B	110B	120B	130B
				85C	90C	100C	110C
					85D	90D	100D
12	12						
12	24						
16	16						
16	32						
19.2	19.2					6	6
19.2	38.4					6	6
24	24			6	6	8 B	8 B

For SI: 1 inch = 25.4 mm, 1 mile per hour = 0.447 m/s.

a. This table is for use with the detail in Figure R611.9(12). Use of this detail is permitted where a cell is not shaded.

b. Wall design per other provisions of Section R611 is required.

c. For wind design, minimum 4-inch-nominal wall is permitted in cells with no number.

d. Numbers 6 and 8 indicate minimum permitted nominal wall thickness in inches necessary to develop required strength (capacity) of connection. As a minimum, this nominal thickness shall occur in the portion of the wall indicated by the cross-hatching in Figure R611.9(12). For the remainder of the wall, see Note b.

e. Letter "B" indicates that a $^5/_8$-inch-diameter anchor bolt is required.

❖ See the commentary for Sections R611.9.1 and R611.9.3.

SECTION R612 EXTERIOR WINDOWS AND DOORS

R612.1 General. This section prescribes performance and construction requirements for exterior window and door installed in wall. Windows and doors shall be installed and flashed in accordance with the fenestration manufacturer's written installation instructions. Window and door openings shall be flashed in accordance with Section R703.8. Written installation instructions shall be provided by the fenestration manufacturer for each window or door.

❖ Doors and windows are components of the exterior wall. Accordingly, this section specifies performance criteria for exterior windows and doors as well as their supporting elements to protect against high wind pressure and water intrusion.

R612.2 Performance. Exterior windows and doors shall be designed to resist the design wind loads specified in Table R301.2(2) adjusted for height and exposure in accordance with Table R301.2(3).

❖ Exterior glazing is subject to wind loading as specified in Section R301.2.1. The code requires use of the component and cladding pressures of Table R301.2(2), illustrated in Figure R301.2(7).

R612.3 Testing and labeling. Exterior windows and sliding doors shall be tested by an *approved* independent laboratory, and bear a *label* identifying manufacturer, performance characteristics and *approved* inspection agency to indicate compliance with AAMA/WDMA/CSA 101/I.S.2/A440. Exterior side-hinged doors shall be tested and *labeled* as conforming to AAMA/WDMA/CSA 101/I.S.2/A440 or comply with Section R612.5.

Exception: Decorative glazed openings.

❖ AAMA/WDMA/CSA 101/I.S.2/A440 is the standard referenced by the code for evaluating exterior glazing components. Section R612.5 is an exception to the original scope of Section R612.3, giving requirements for all exterior window, skylight and door assemblies regardless of whether they are glass or nonglass. All of these doors, skylights and windows must be tested in accordance with ASTM E 330, as referenced in Section R612.5, because they are outside the scope of the referenced standard AAMA/WDMA/CSA 101/I.S.2/A440. ASTM E 330 will require testing to 1.5 times the design pressure.

R612.3.1 Comparative analysis. Structural wind load design pressures for window and door units smaller than the size tested in accordance with Section R612.3 shall be permitted to be higher than the design value of the tested unit provided such higher pressures are determined by accepted engineering analysis. All components of the small unit shall be the same as those of the tested unit. Where such calculated design pressures are used, they shall be validated by an additional test of the window or door unit having the highest allowable design pressure.

❖ Where a window, skylight or door is tested in accordance with the requirements of Section R612.6, smaller units of identical construction need not be tested. Instead, their properties may be determined by comparison to the tested unit and the application of accepted engineering analysis. However, where the analysis determines that the smaller units have a higher allowable design pressure, the unit with the highest allowable calculated design pressure must be tested to validate the findings, or the design pressure of the largest tested unit must be used.

R612.4 Garage doors. Garage doors shall be tested in accordance with either ASTM E 330 or ANSI/ DASMA 108, and shall meet the acceptance criteria of ANSI/DASMA 108.

❖ Overhead garage doors for vehicle access must be tested in accordance with either one of the standards indicated. In addition, they must also meet the acceptance criteria of ANSI/DASMA 108. See Section R612.5 for an exception to these requirements.

R612.5 Other exterior window and door assemblies. Exterior windows and door assemblies not included within the scope of Section R612.3 or Section R612.4 shall be tested in accordance with ASTM E 330. Glass in assemblies covered by this exception shall comply with Section R308.5.

❖ This section is, in essence, an exception to Sections R612.3 and R612.4 which allows ASTM E 330 as an alternate test standard. ASTM E 330 will require testing to 1.5 times the design pressure.

R612.6 Wind-borne debris protection. Protection of exterior windows and glass doors in buildings located in windborne debris regions shall be in accordance with Section R301.2.1.2.

❖ See the commentary for Sections R301.2.1.2 and R612.6.1.

R612.6.1 Fenestration testing and labeling. Fenestration shall be tested by an *approved* independent laboratory, listed by an *approved* entity, and bear a *label* identifying manufacturer, performance characteristics, and *approved* inspection agency to indicate compliance with the requirements of the following specification:

1. ASTM E 1886 and ASTM E 1996; or
2. AAMA 506.

❖ In buildings located in wind-borne debris regions as established in accordance with Section R301.2.1, exterior windows, skylights and doors must be protected in accordance with Section R301.2.1.2, which mirrors some of the requirements of Section R612.6.1. Fenestration regulated by Section R612.6.1 includes storm shutters not in accordance with the prescriptive requirements of the exception to Section R301.2.1.2, and exterior windows, skylights and doors not equipped with storm shutters. Such fenestrations must bear a label indicating that they have been tested in accordance with AAMA 506, or both ASTM E 1886 and ASTM E 1996. These tests determine the ability of the fenestration to resist cyclic pressures and impact by wind-borne debris (also see commentary, Section R301.2.1.2).

R612.7 Anchorage methods. The methods cited in this section apply only to anchorage of window and glass door assemblies to the main force-resisting system.

❖ These requirements provide a mechanism for transferring the wind forces from the door or window to a lateral-force-resisting element. Anchorage should be according to the manufacturer's recommendations. Variations should provide equal or greater anchorage performance.

R612.7.1 Anchoring requirements. Window and glass door assemblies shall be anchored in accordance with the published manufacturer's recommendations to achieve the design pressure specified. Substitute anchoring systems used for substrates not specified by the fenestration manufacturer shall provide equal or greater anchoring performance as demonstrated by accepted engineering practice.

❖ See the commentary for Section R612.7.

R612.7.2 Anchorage details. Products shall be anchored in accordance with the minimum requirements illustrated in Figures R612.7.2(1), R612.7.2(2), R612.7.2(3), R612.7.2(4), R612.7.2(5), R612.7.2(6), R612.7.2(7) and R612.7.2(8).

❖ The code incorporates prescriptive anchorage methods that provide for transfer of wind loads to the supporting construction (see the commentary, Section R612.7).

R612.7.2.1 Masonry, concrete or other structural substrate. Where the wood shim or buck thickness is less than $1^1/_2$ inches (38 mm), window and glass door assemblies shall be anchored through the jamb, or by jamb clip and anchors shall be embedded directly into the masonry, concrete or other substantial substrate material. Anchors shall adequately transfer load from the window or door frame into the rough opening substrate [see Figures R612.7.2(1) and R612.7.2(2)].

Where the wood shim or buck thickness is $1^1/_2$ inches (38 mm) or more, the buck is securely fastened to the masonry, concrete or other substantial substrate, and the buck extends beyond the interior face of the window or door frame, window and glass door assemblies shall be anchored through the jamb, or by jamb clip, or through the flange to the secured wood buck. Anchors shall be embedded into the secured wood buck to adequately transfer load from the window or door frame assembly [Figures R612.7.2(3), R612.7.2(4) and R612.7.2(5)].

❖ When a wood member at least 1.5 inches (38 mm) thick is used as a shim, it must be anchored to the substrate. Otherwise window and door assemblies must be anchored directly to the substrate [see Figures R612.7.2(1) through R612.7.2(5)].

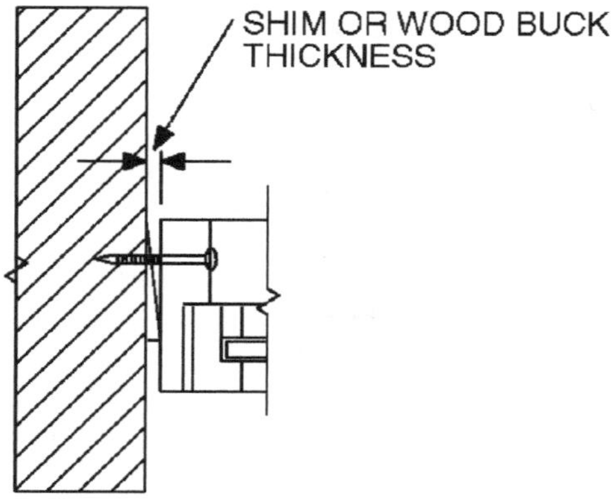

FIGURE R612.7.2(1)
THROUGH THE FRAME

❖ See the commentary for Sections R612.7 and R612.7.2.1.

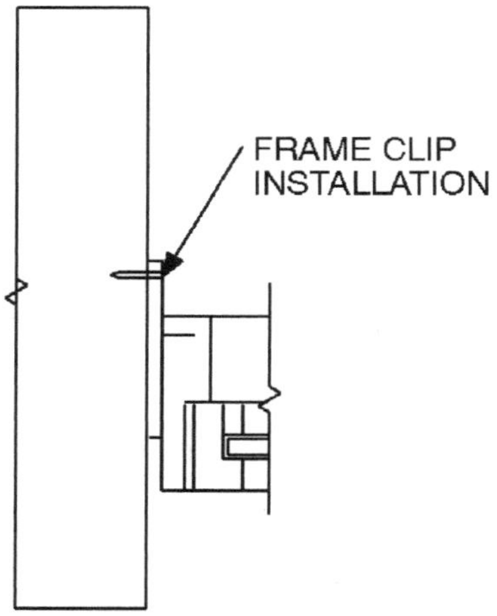

FIGURE R612.7.2(2)
FRAME CLIP

❖ See the commentary for Sections R612.7 and R612.7.2.1.

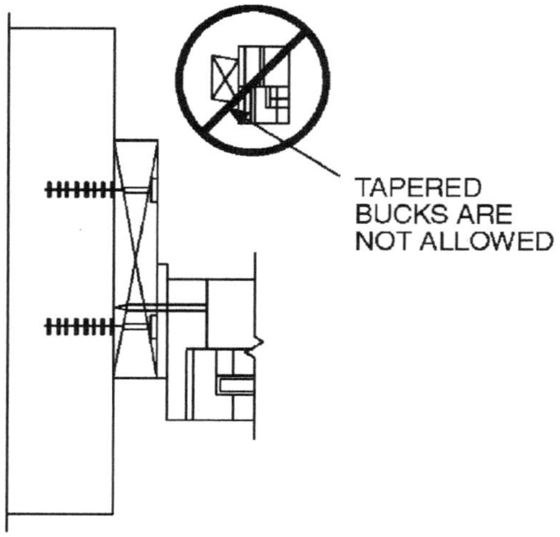

FIGURE R612.7.2(3)
THROUGH THE FRAME

❖ See the commentary for Sections R612.7 and R612.7.2.1.

R612.7.2.2 Wood or other approved framing material. Where the framing material is wood or other *approved* framing material, window and glass door assemblies shall be anchored through the frame, or by frame clip, or through the flange. Anchors shall be embedded into the frame construction to adequately transfer load [Figures R612.7.2(6), R612.7.2(7) and R612.7.2(8)].

❖ Door and window assemblies must be anchored directly to wood framing [see Figures R612.7.2(6) through R612.7.2(8)].

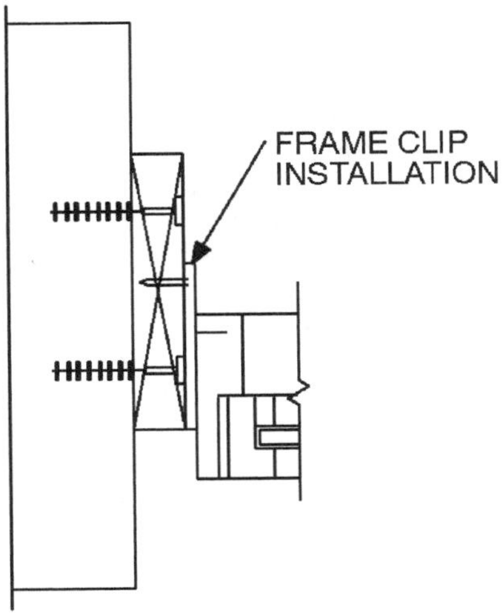

FIGURE R612.7.2(4)
FRAME CLIP

❖ See the commentary for Sections R612.7 and R612.7.2.1.

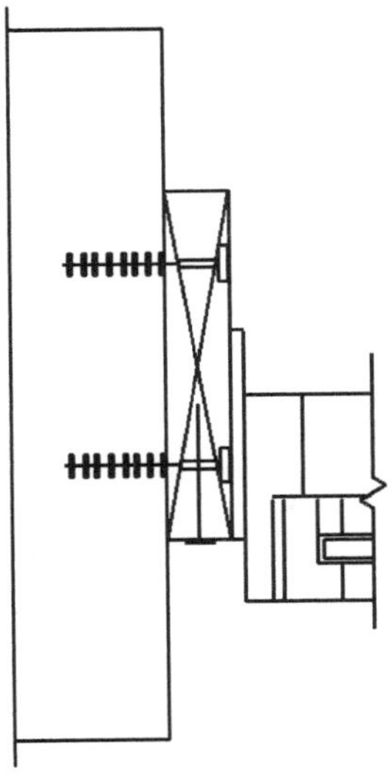

FIGURE R612.7.2(5)
THROUGH THE FLANGE

❖ See the commentary for Sections R612.7 and R612.7.2.1.

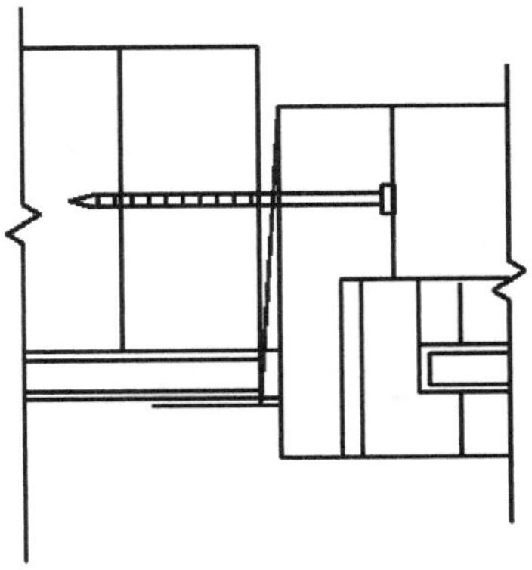

FIGURE R612.7.2(6)
THROUGH THE FLANGE

❖ See the commentary for Sections R612.7 and R612.7.2.2.

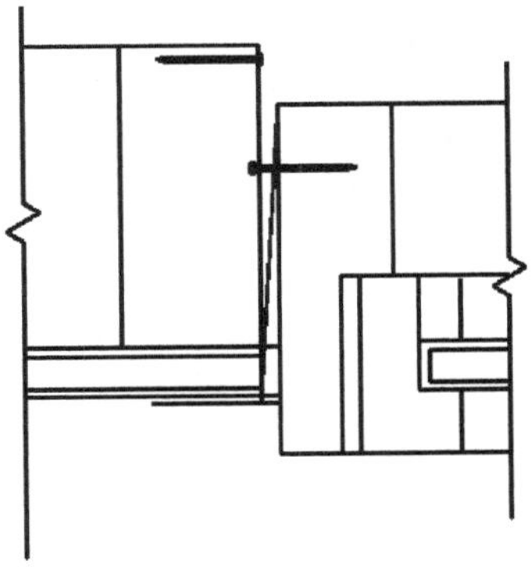

FIGURE R612.7.2(7)
FRAME CLIP

❖ See the commentary for Sections R612.7 and R612.7.2.2.

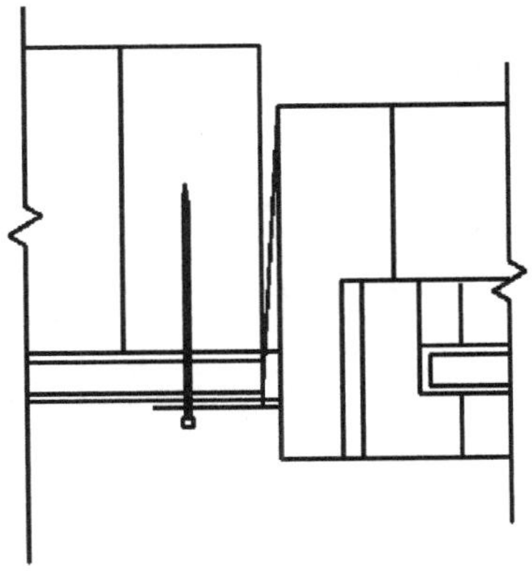

FIGURE R612.7.2(8)
THROUGH THE FLANGE

❖ See the commentary for Sections R612.7 and R612.7.2.2.

R612.8 Mullions. Mullions shall be tested by an *approved* testing laboratory in accordance with AAMA 450, or be engineered in accordance with accepted engineering practice. Mullions tested as stand-alone units or qualified by engineering shall use performance criteria cited in Sections R612.8.1, R612.8.2 and R612.8.3. Mullions qualified by an actual test of an entire assembly shall comply with Sections R612.8.1 and R612.8.3.

❖ Mullions provide support for adjacent window and/or door assemblies. Generally, they must be tested and perform as required for windows and doors. Deflection limits and a factor of safety maintain serviceability and provide minimum strength.

R612.8.1 Load transfer. Mullions shall be designed to transfer the design pressure loads applied by the window and door assemblies to the rough opening substrate.

❖ See the commentary for Section R612.8.

R612.8.2 Deflection. Mullions shall be capable of resisting the design pressure loads applied by the window and door assemblies to be supported without deflecting more than *L*/175, where *L* is the span of the mullion in inches.

❖ See the commentary for Section R612.8.

R612.8.3 Structural safety factor. Mullions shall be capable of resisting a load of 1.5 times the design pressure loads applied by the window and door assemblies to be supported without exceeding the appropriate material stress levels. If tested by an *approved* laboratory, the 1.5 times the design pressure load shall be sustained for 10 seconds, and the permanent deformation shall not exceed 0.4 percent of the mullion span after the 1.5 times design pressure load is removed.

❖ See the commentary for Section R612.8.

SECTION R613 STRUCTURAL INSULATED PANEL WALL CONSTRUCTION

R613.1 General. Structural insulated panel (SIP) walls shall be designed in accordance with the provisions of this section. When the provisions of this section are used to design structural insulated panel walls, project drawings, typical details and specifications are not required to bear the seal of the architect or engineer responsible for design, unless otherwise required by the state law of the *jurisdiction* having authority.

❖ The purpose of Section R613 is to provide a prescriptive method for the design and installation of structural insulated panel (SIP) walls, which otherwise would require an engineered design. It establishes minimum requirements for the component materials such as the facing, core and adhesive (see the typical panel illustrated in Figure R613.4). SIPs consist primarily of a foam plastic core laminated between wood structural panel facings. Structurally, SIPs can be thought of as analogous to a steel wide-flange beam, with the facing serving as the beam flanges and the core serving as the web with the composite assembly producing the desired strength and stiffness. In engineering design of such panels, axial forces are carried by the facings and bending is resisted by the internal force couple in the facings.

The benefits that can be provided by SIP construction include improved thermal efficiency and reduced air infiltration. Manufacturing SIPs in a factory-controlled environment provides improved tolerances when compared to conventional wood wall framing. SIPs also minimize the amount of wood framing required for wall construction. The less framing that is necessary, the lower the framing factor and the result is that less energy is lost due to thermal bridging.

TABLE R613.3.1
MINIMUM PROPERTIES FOR POLYURETHANE INSULATION USED AS SIPS CORE

PHYSICAL PROPERTY	POLYURETHANE
Density, core nominal (ASTM D 1622)	2.2 lb/ft^3
Compressive resistance at yield or 10% deformation, whichever occurs first (ASTM D 1621)	19 psi (perpendicular to rise)
Flexural strength, min. (ASTM C 203)	30 psi
Tensile strength, min. (ASTM D 1623)	35 psi
Shear strength, min. (ASTM C 273)	25 psi
Substrate adhesion, min. (ASTM D 1623)	22 psi
Water vapor permeance of 1.00-in. thickness, max. (ASTM E 96)	2.3 perm
Water absorption by total immersion, max. (ASTM C 272)	4.3% (volume)
Dimensional stability (change in dimensions), max. [ASTM D 2126 (7 days at 158°F/100% humidity and 7 days at -20°F)]	2%

For SI: 1 pound per cubic foot = 16.02 kg/m^3, 1 pound per square inch = 6.895 kPa, °C = [(°F) - 32]1.8.

❖ This table contains the specific material properties that must be met in order to qualify polyurethane as a core material for SIPs.

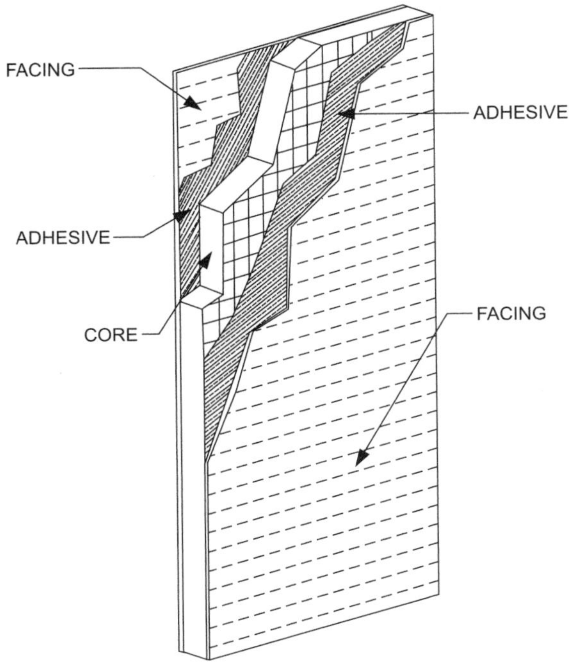

FIGURE R613.4
SIP WALL PANEL

❖ See the commentary for Section R613.4.

R613.2 Applicability limits. The provisions of this section shall control the construction of exterior structural insulated panel walls and interior load-bearing structural insulated panel walls for buildings not greater than 60 feet (18 288 mm) in length perpendicular to the joist or truss span, not greater than 40 feet (12 192 mm) in width parallel to the joist or truss span and not greater than two stories in height with each wall not greater than 10 feet (3048 mm) high. All exterior walls installed in accordance with the provisions of this section shall be considered as load-bearing walls. Structural insulated panel walls constructed in accordance with the provisions of this section shall be limited to sites subjected to a maximum design wind speed of 120 miles per hour (54 m/s), Exposure A or B or 110 miles per hour (49 m/s) Exposure C, and a maximum ground snow load of 70 pounds per foot (3.35 kPa), and Seismic Design Categories A, B and C.

❖ This section establishes the bounds for these prescriptive SIP walls. SIP construction that exceeds these limits should be substantiated as an alternative method of design and construction in accordance with Section R104.11. These limits for snow load and wind speed as well as the building plan dimensions, wall height and number of stories are all reflected in the required wall construction that is determined under Section R613.5. Also note that for earthquake concerns these provisions are restricted to SDC A, B or C.

R613.3 Materials. SIPs shall comply with the following criteria:

❖ This section establishes minimum requirements for the component materials such as the facing, core, adhesive, lumber and fasteners. For a manufacturer's product to be covered by this section of the code, all materials used in SIP production must meet or exceed these criteria. Using these minimum material properties, the capacity of SIP walls was established by testing which was then used to develop the wall construction requirements in Section R613.5.

R613.3.1 Core. The core material shall be composed of foam plastic insulation meeting one of the following requirements:

1. ASTM C 578 and have a minimum density of 0.90 pounds per cubic feet (14.4 kg/m^3); or
2. Polyurethane meeting the physical properties shown in Table R613.3.1, or;
3. An *approved* alternative.

All cores shall meet the requirements of Section R316.

❖ The middle section of the SIP is referred to as the core and it consists of light-weight foam plastic insulation. Two types of foam plastic insulation are recognized. For either type a minimum density requirement provides for a continuous, uniform layer of insulation. Item 1 refers to rigid, cellular polystyrene that complies with ASTM C 578 and Item 2 allows polyurethane insulation that complies with the material criteria listed in Table R613.3.1. The cross reference to Section R316 points to the provisions governing all foam plastic insulation which provide other necessary material criteria such as flame spread and smoke-developed indexes.

R613.3.2 Facing. Facing materials for SIPs shall be wood structural panels conforming to DOC PS 1 or DOC PS 2, each having a minimum nominal thickness of $^7/_{16}$ inch (11 mm) and shall meet the additional minimum properties specified in Table R613.3.2. Facing shall be identified by a grade mark or certificate of inspection issued by an *approved* agency.

❖ In addition to conforming to DOC PS 1 or DOC PS 2, additional wood structural panel properties are listed in Table R613.3.2. It should be noted that these properties exceed DOC PS 1 and DOC PS 2. Thus, the wood structural panel supplier should provide verification that their product meets these enhanced requirements for SIP facing in order to comply with this section.

R613.3.3 Adhesive. Adhesives used to structurally laminate the foam plastic insulation core material to the structural wood facers shall conform to ASTM D 2559 or *approved* alternative specifically intended for use as an adhesive used in the lamination of structural insulated panels. Each container of adhesive shall bear a *label* with the adhesive manufacturer's name, adhesive name and type and the name of the quality assurance agency.

❖ Structural adhesive bonds the foam plastic insulation to the wood structural panel facing in the lamination process and it is key to the SIP performance. This section recognizes ASTM D 2559 for this purpose which provides an adhesive rated for exterior use as is appropriate for exterior walls. This section also brings up "approved alternatives." Under Section R104.11, an alternative design may be based on ICC ES AC05 *Acceptance Criteria for Sandwich Panel Adhesives*, which would require a classification of Type II, Class 2, in order to provide high moisture resistance in a load-bearing application.

R613.3.4 Lumber. The minimum lumber framing material used for SIPs prescribed in this document is NLGA graded No. 2 Spruce-pine-fir. Substitution of other wood species/grades that meet or exceed the mechanical properties and specific gravity of No. 2 Spruce-pine-fir shall be permitted.

❖ This section establishes the minimum requirements for wood framing that is utilized in SIP wall construction.

R613.3.5 SIP screws. Screws used for the erection of SIPs as specified in Section R613.5 shall be fabricated from steel, shall be provided by the SIPs manufacturer and shall be sized to penetrate the wood member to which the assembly is being attached by a minimum of 1 inch (25 mm). The screws shall be corrosion resistant and have a minimum shank diameter of 0.188 inch (4.7 mm) and a minimum head diameter of 0.620 inch (15.5 mm).

❖ SIP screws must be steel and be corrosion resistant. The only required use of this fastener type is in the corner detail that is illustrated in Figure R613.9.

R613.3.6 Nails. Nails specified in Section R613 shall be common or galvanized box unless otherwise stated.

❖ Nailing of SIP wall assemblies is specified in Figures R613.5(3), R613.5(4), R613.5(5), R613.5.2, R613.8 and R613.9. This section clarifies the style of nail that must be used in these connections. For additional fastening requirements, see Section R613.5.

R613.4 SIP wall panels. SIPs shall comply with Figure R613.4 and shall have minimum panel thickness in accordance with Tables R613.5(1) and R613.5(2) for above-grade walls. All SIPs shall be identified by grade mark or certificate of inspection issued by an *approved* agency.

❖ Figure R613.4 illustrates the primary components of the SIP wall panel. Manufacturing these panels in a factory-controlled environment provides improved tolerances when compared to conventional wood wall framing. SIP panels minimize the amount of wood framing required in wall construction. Also, the core may be recessed as necessary around the panel edges to accept dimension lumber, such as at the top plate connection (see Section R613.5.1). In order to verify a SIP wall panel's suitability in a particular application it must be identified by either a grade mark or certificate of inspection.

R613.4.1 Labeling. All panels shall be identified by grade mark or certificate of inspection issued by an *approved* agency. Each (SIP) shall bear a stamp or *label* with the following minimum information:

1. Manufacturer name/logo.
2. Identification of the assembly.
3. Quality assurance agency.

❖ Third-party labeling is required for the entire SIP assembly once fabrication is completed. This requirement provides the ability to verify the panel's suitability for a particular application. The label must identify the quality assurance agency. This agency provides inspections to verify adequate quality control in the manufacturing process.

TABLE R613.3.2
MINIMUM PROPERTIES[a] FOR ORIENTED STRAND BOARD FACER MATERIAL IN SIP WALLS

Thickness (in.)	Product	Flatwise Stiffness[b] (lbf-in^2/ft)		Flatwise Strength [c] (lbf-in/ft)		Tension [c] (lbf/ft)		Density[d] (pcf)
		Along	Across	Along	Across	Along	Across	
7/16	Sheathing	55,600	16,500	1,040	460	7,450	5,800	34

For SI: 1 inch = 25.4 mm, 1 lbf-in^2/ft = 9.415 × 10^{-6} kPa/m, 1 lbf-in/ft = 3.707 × 10^{-4} kN/m, 1 lbf/ft = 0.0146 N/mm, 1 pound per cubic foot = 16.018 kg/m^3.

a. Values listed in Table R613.3.2 are qualification test values and are not to be used for design purposes.

b. Mean test value shall be in accordance with Section 7.6 of DOC PS 2.

c. Characteristic test value (5th percent with 75% confidence).

d. Density shall be based on oven-dry weight and oven-dry volume.

❖ The minimum properties for facing materials in Table R613.3.2 were established by the SIPs industry with a specific grade of wood structural panels. Panel properties for the wood structural panel facing materials in Table R613.3.2 of the 2009 IRC do not reflect the facing materials commonly available in the marketplace, which typically have higher properties in the along direction and lower properties in the across direction.

As a result, the Structural Insulated Panel Association (SIPA) worked with APA - The Engineered Wood Association to re-evaluate the performance of SIPs using readily available facing materials. Results of this re-evaluation are documented in APA Report T2009P-28, which shows no performance difference for SIP applications covered in IRC Section R613 when using the new facing materials with higher properties in the along direction and lower properties in the across direction, as compared to the 2009 IRC.

The facer material table has therefore been updated to reflect material properties of the wood structural panels currently used in the production of SIPs.

R613.5 Wall construction. Exterior walls of SIP construction shall be designed and constructed in accordance with the provisions of this section and Tables R613.5(1) and R613.5(2) and Figures R613.5(1) through R613.5(5). SIP walls shall be fastened to other wood building components in accordance with Tables R602.3(1) through R602.3(4).

Framing shall be attached in accordance with Table R602.3(1) unless otherwise provided for in Section R613.

❖ Based on the minimum material properties established in Section R613.3, the capacity of SIP walls was established by testing which was then used to develop the tabulated wall thickness requirements of this section. Other than the specific nailing requirements in Section R613, the fastening of wood components must be in accordance with the fastening schedule of Table R602.3(1). This section also makes it clear that the connection of the SIP wall assembly to other wood elements must be in accordance with the applicable requirements for fastening wood members. Figures R613.5(3), R613.5(4), R613.5(5), R613.8 and R613.9 indicate where SIP connections require sealant. With no particular sealant criteria given in this section, sealants should be according to the panel manufacturer's recommendations.

Because SIP walls do not contain stud framing, an adjusted fastener schedule for attaching an exterior wall covering may be necessary unless the manufac-

TABLE R613.5(1)
MINIMUM THICKNESS FOR SIP WALL SUPPORTING SIP OR LIGHT-FRAME ROOF ONLY (inches)[a]

Wind Speed (3-second gust)		Snow Load (psf)	Building Width (ft)														
			24			28			32			36			40		
			Wall Height (feet)			Wall Height (feet)			Wall Height (feet)			Wall Height (feet)			Wall Height (feet)		
Exp. A/B	Exp. C		8	9	10	8	9	10	8	9	10	8	9	10	8	9	10
85	—	20	4.5	4.5	4.5	4.5	4.5	4.5	4.5	4.5	4.5	4.5	4.5	4.5	4.5	4.5	4.5
		30	4.5	4.5	4.5	4.5	4.5	4.5	4.5	4.5	4.5	4.5	4.5	4.5	4.5	4.5	4.5
		50	4.5	4.5	4.5	4.5	4.5	4.5	4.5	4.5	4.5	4.5	4.5	4.5	4.5	4.5	4.5
		70	4.5	4.5	4.5	4.5	4.5	4.5	4.5	4.5	4.5	4.5	4.5	4.5	4.5	4.5	4.5
100	85	20	4.5	4.5	4.5	4.5	4.5	4.5	4.5	4.5	4.5	4.5	4.5	4.5	4.5	4.5	4.5
		30	4.5	4.5	4.5	4.5	4.5	4.5	4.5	4.5	4.5	4.5	4.5	4.5	4.5	4.5	4.5
		50	4.5	4.5	4.5	4.5	4.5	4.5	4.5	4.5	4.5	4.5	4.5	4.5	4.5	4.5	4.5
		70	4.5	4.5	4.5	4.5	4.5	4.5	4.5	4.5	4.5	4.5	4.5	6.5	4.5	4.5	N/A
110	100	20	4.5	4.5	4.5	4.5	4.5	4.5	4.5	4.5	4.5	4.5	4.5	4.5	4.5	4.5	4.5
		30	4.5	4.5	4.5	4.5	4.5	4.5	4.5	4.5	4.5	4.5	4.5	4.5	4.5	4.5	6.5
		50	4.5	4.5	4.5	4.5	4.5	6.5	4.5	4.5	6.5	4.5	4.5	N/A	4.5	4.5	N/A
		70	4.5	4.5	6.5	4.5	4.5	N/A	4.5	4.5	N/A	4.5	6.5	N/A	4.5	N/A	N/A
120	110	20	4.5	4.5	N/A	4.5	4.5	N/A	4.5	4.5	N/A	4.5	4.5	N/A	4.5	4.5	N/A
		30	4.5	4.5	N/A	4.5	4.5	N/A	4.5	4.5	N/A	4.5	4.5	N/A	4.5	6.5	N/A
		50	4.5	4.5	N/A	4.5	6.5	N/A	4.5	N/A	N/A	4.5	N/A	N/A	4.5	N/A	N/A
		70	4.5	N/A	N/A	4.5	N/A	N/A	4.5	N/A	N/A	N/A	N/A	N/A	N/A	N/A	N/A

For SI: 1 inch = 25.4 mm, 1 foot = 304.8 mm, 1 pound per square foot = 0.0479 kPa.

N/A = Not Applicable.

a. Design assumptions:

Deflection criteria: *L*/240.

Roof load: 7 psf.

Ceiling load: 5 psf.

Wind loads based on Table R301.2 (2).

Strength axis of facing materials applied vertically.

❖ The tabulated SIP wall thicknesses are given based on specific values of snow load and wind criteria as well as the building plan dimensions, wall height and number of stories all of which reflect the applicability limits of Section R613.2. Additional criteria such as deflection limit, dead loads, ceiling and floor live loads are noted under the table. Typical wall heights of up to 10 feet (305 mm) per story are provided for as this is the limit that is established in Section R613.2. Note that the story height allows an additional 16 inches (406 mm) for the depth of floor framing (see Section R301.3).

While there were no performance issues, the assumptions used in generating Tables R613.5(1) and R613.5(2) were also reviewed by the SIPA Technical Advisory Committee, which suggested more stringent criteria by not allowing any load duration increase, including wind load, for SIPs.

turer of the wall covering provides recommendations for attaching their product to SIP walls. A design may be required to determine the appropriate fastening requirements when a wall covering manufacturer does not provide attachment recommendations.

R613.5.1 Top plate connection. SIP walls shall be capped with a double top plate installed to provide overlapping at corner, intersections and splines in accordance with Figure R613.5.1. The double top plates shall be made up of a single 2 by top plate having a width equal to the width of the panel core, and shall be recessed into the SIP below. Over this top plate a cap plate shall be placed. The cap plate width shall match the SIP thickness and overlap the facers on both sides of the panel. End joints in top plates shall be offset at least 24 inches (610 mm).

❖ The double top plate consists of a cap plate which is sized to match the width of the SIP and a lower plate which is sized to match the width of the core allowing it to be recessed into the top edge of the panel. By matching the width of the panel, the cap plate delivers loads to the panel facings rather than the core of the panel. This double top plate configuration is depicted in Figures R613.5(3) and R613.5(4).

R613.5.2 Bottom (sole) plate connection. SIP walls shall have full bearing on a sole plate having a width equal to the nominal width of the foam core. When SIP walls are supported directly on continuous foundations, the wall wood sill plate shall be anchored to the foundation in accordance with Figure R613.5.2 and Section R403.1.

❖ Like the lower top plate, the bottom (sole) plate must match the width of the core so that it can be recessed within the panel as shown in Figures R613.5(4) and R613.5(5). Figure R613.5.2 illustrates a bottom plate connection to a foundation.

TABLE R613.5(2)
MINIMUM THICKNESS FOR SIP WALLS SUPPORTING SIP OR LIGHT-FRAME ONE STORY AND ROOF (inches)[a]

Wind Speed (3-second gust)		Snow Load (psf)	Building Width (ft)														
			24			28			32			36			40		
			Wall Height (feet)			Wall Height (feet)			Wall Height (feet)			Wall Height (feet)			Wall Height (feet)		
Exp. A/B	Exp. C		8	9	10	8	9	10	8	9	10	8	9	10	8	9	10
85	—	20	4.5	4.5	4.5	4.5	4.5	4.5	4.5	4.5	4.5	4.5	4.5	4.5	4.5	4.5	4.5
		30	4.5	4.5	4.5	4.5	4.5	4.5	4.5	4.5	4.5	4.5	4.5	4.5	4.5	4.5	4.5
		50	4.5	4.5	4.5	4.5	4.5	4.5	4.5	4.5	4.5	4.5	4.5	4.5	4.5	4.5	N/A
		70	4.5	4.5	4.5	4.5	4.5	4.5	4.5	4.5	6.5	4.5	4.5	N/A	4.5	N/A	N/A
100	85	20	4.5	4.5	4.5	4.5	4.5	4.5	4.5	4.5	6.5	4.5	4.5	N/A	4.5	4.5	N/A
		30	4.5	4.5	4.5	4.5	4.5	4.5	4.5	4.5	N/A	4.5	4.5	N/A	4.5	N/A	N/A
		50	4.5	4.5	6.5	4.5	4.5	N/A	4.5	4.5	N/A	4.5	N/A	N/A	N/A	N/A	N/A
		70	4.5	4.5	N/A	4.5	6.5	N/A	4.5	N/A	N/A	N/A	N/A	N/A	N/A	N/A	N/A
110	100	20	4.5	4.5	N/A	4.5	4.5	N/A	4.5	6.5	N/A	4.5	N/A	N/A	N/A	N/A	N/A
		30	4.5	4.5	N/A	4.5	4.5	N/A	4.5	N/A	N/A	4.5	N/A	N/A	N/A	N/A	N/A
		50	4.5	6.5	N/A	4.5	N/A	N/A	N/A	N/A	N/A	N/A	N/A	N/A	N/A	N/A	N/A
		70	4.5	N/A	N/A	N/A	N/A	N/A	N/A	N/A	N/A	N/A	N/A	N/A	N/A	N/A	N/A
120	110	20	4.5	N/A	N/A	4.5	N/A	N/A	N/A	N/A	N/A	N/A	N/A	N/A	N/A	N/A	N/A
		30	4.5	N/A	N/A	N/A	N/A	N/A	N/A	N/A	N/A	N/A	N/A	N/A	N/A	N/A	N/A
		50	N/A	N/A	N/A	N/A	N/A	N/A	N/A	N/A	N/A	N/A	N/A	N/A	N/A	N/A	N/A
		70	N/A	N/A	N/A	N/A	N/A	N/A	N/A	N/A	N/A	N/A	N/A	N/A	N/A	N/A	N/A

For SI: 1 inch = 25.4 mm, 1 foot = 304.8 mm, 1 pound per square foot = 0.0479 kPa.

N/A = Not Applicable.

a. Design assumptions:

Deflection criteria: *L*/240.

Roof load: 7 psf.

Ceiling load: 5 psf.

Second floor live load: 30 psf.

Second floor dead load: 10 psf.

Second floor dead load from walls: 10 psf.

Wind loads based on Table R301.2(2).

Strength axis of facing materials applied vertically.

❖ See the commentary to Table R613.5(1).

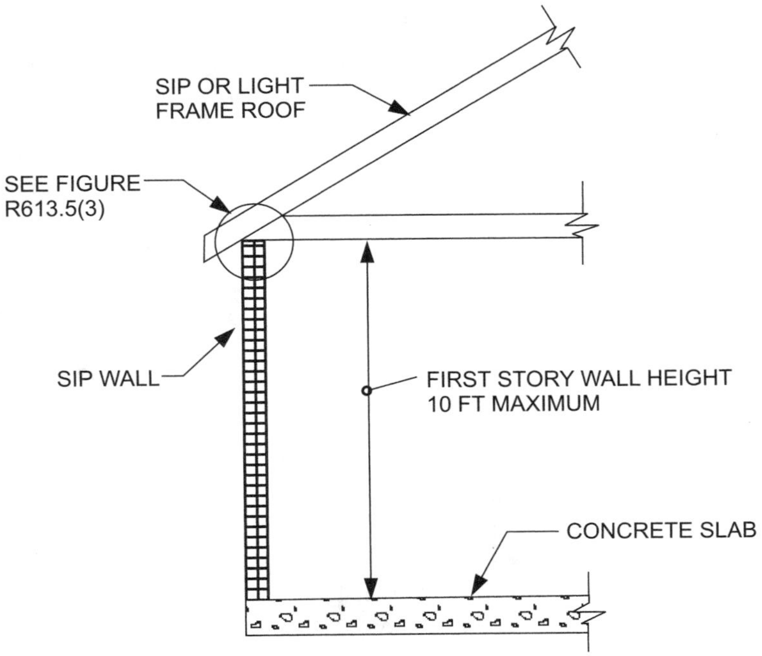

For SI: 1 foot = 304.8 mm.

FIGURE R613.5(1)
MAXIMUM ALLOWABLE HEIGHT OF SIP WALLS

❖ See the commentary to Table R613.5(1).

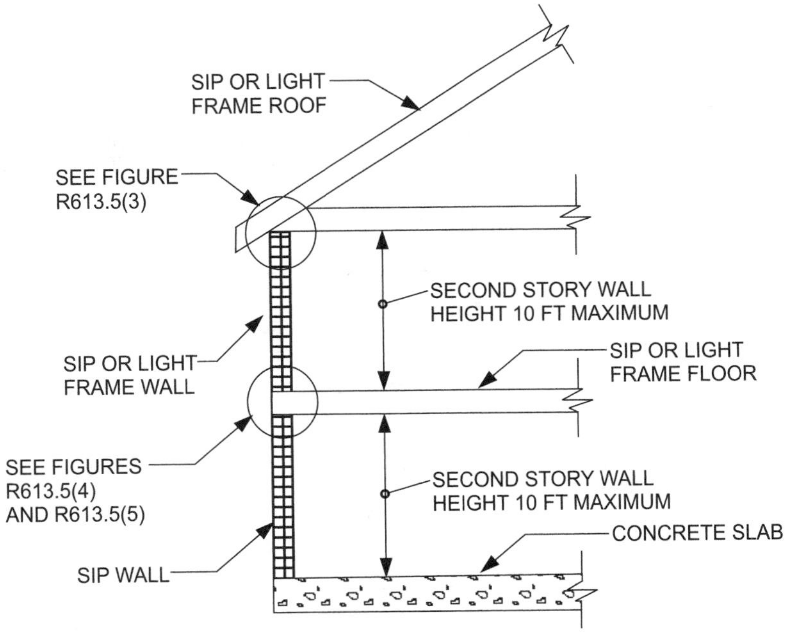

For SI: 1 foot = 304.8 mm.

FIGURE R613.5(2)
MAXIMUM ALLOWABLE HEIGHT OF SIP WALLS

❖ See the commentary to Table R613.5(1).

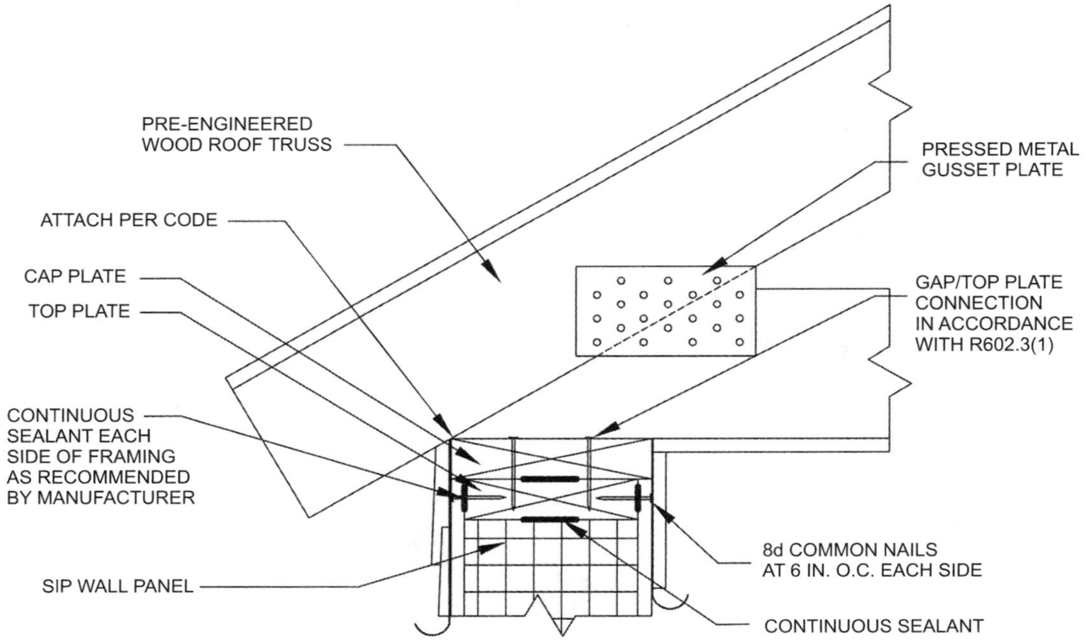

For SI: 1 inch = 25.4 mm.

FIGURE R613.5(3)
TRUSSED ROOF TO TOP PLATE CONNECTION

❖ See the commentary to Sections R613.3.6 and R613.5.

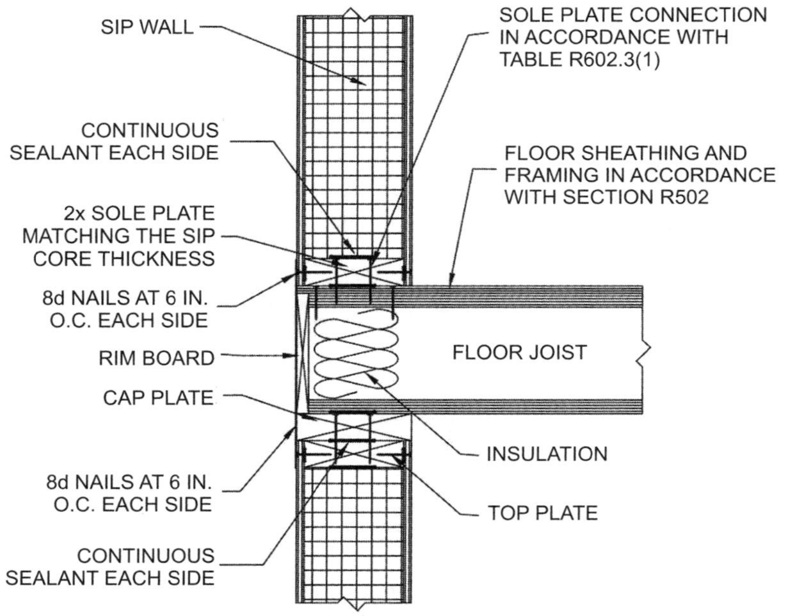

For SI: 1 inch = 25.4 mm.
Note: Figures illustrate SIP-specific attachment requirements. Other connections shall be made in accordance with Table R602.3(1) and (2) as appropriate.

FIGURE R613.5(4)
SIP WALL TO WALL PLATFORM FRAME CONNECTION

❖ As the note cautions, only the fastening for the SIP installation is illustrated, and other applicable fastening requirements for wood framing must be satisfied. Also see the commentary to Sections R613.3.6 and R613.5.

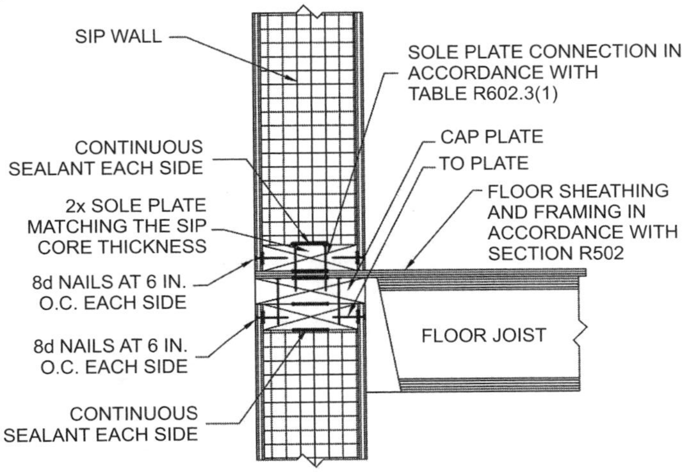

For SI: 1 inch = 25.4 mm.

Note: Figures illustrate SIP-specific attachment requirements. Other connections shall be made in accordance with Tables R602.3(1) and (2), as appropriate.

FIGURE R613.5(5)
SIP WALL TO WALL BALLOON FRAME CONNECTION
(I-Joist floor shown for Illustration only)

❖ As the note cautions, only the fastening for the SIP installation is illustrated, and other applicable fastening requirements for wood framing must be satisfied. Also see the commentary to Sections R613.3.6 and R613.5.

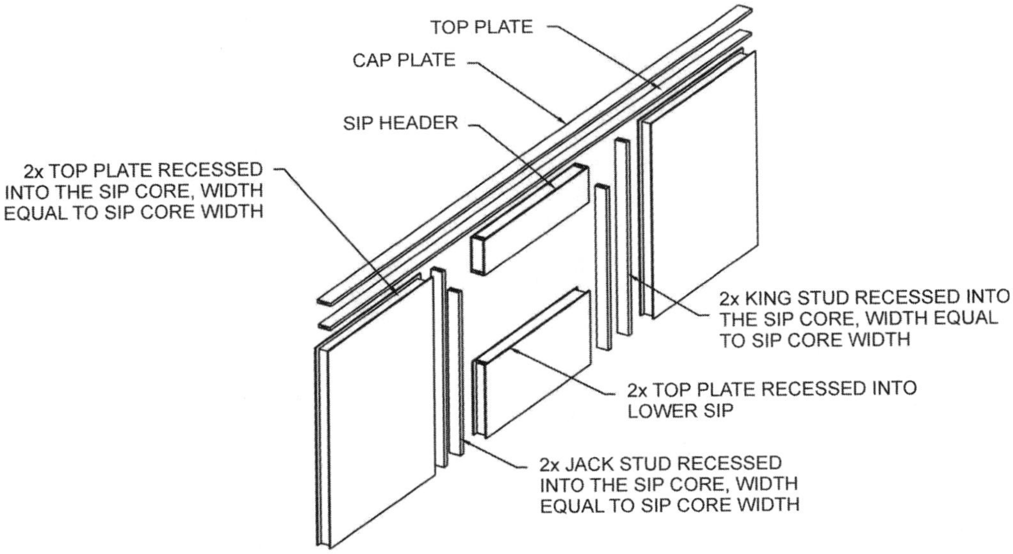

For SI: 1 inch = 25.4 mm.

Notes:

1. Top plates shall be continuous over header.
2. Lower 2x top plate shall have a width equal to the SIP core width and shall be recessed into the top edge of the panel. Cap plate shall be placed over the recessed top plate and shall have a width equal to the SIPs width.
3. SIP facing surfaces shall be nailed to framing and cripples with 8d common or galvanized box nails spaced 6 inches on center.
4. Galvanized nails shall be hot-dipped or tumbled. Framing shall be attached in accordance to Section R602.3(1) unless otherwise provide for in Section R613.

FIGURE R613.5.1
SIP WALL FRAMING CONFIGURATION

❖ Note 2 clarifies that the cap plate is placed over the recessed lower top plate. This figure also illustrates the requirements for headers (see Section R613.10).

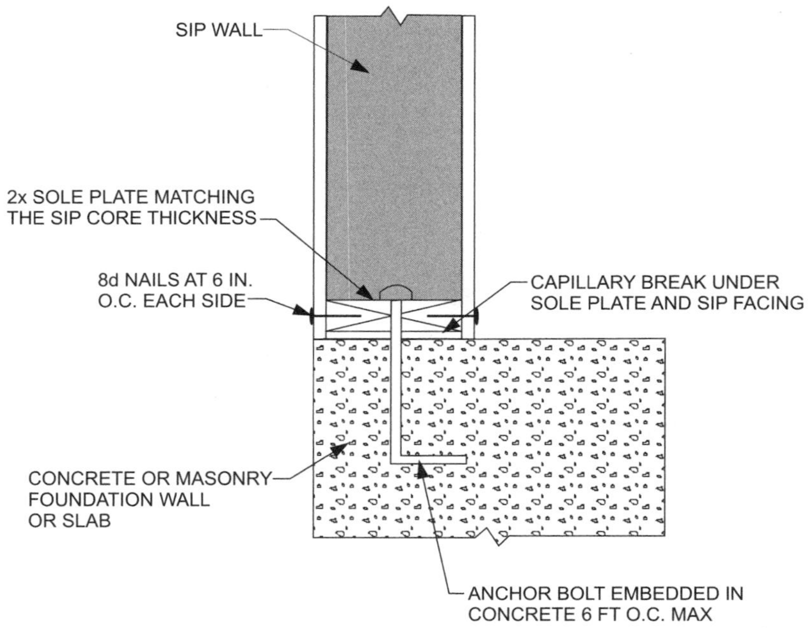

For SI: 1 inch = 25.4 mm, 1 foot = 304.8 mm.

FIGURE R613.5.2
SIP WALL TO CONCRETE SLAB FOR FOUNDATION WALL ATTACHMENT

❖ See the commentary to Section R613.5.2.

R613.5.3 Wall bracing. SIP walls shall be braced in accordance with Section R602.10. SIP walls shall be considered continuous wood structural panel sheathing for purposes of computing required bracing. SIP walls shall meet the requirements of Section R602.10.4.2 except that SIPs corners shall be fabricated as shown in Figure R613.9. When SIP walls are used for wall bracing, the SIP bottom plate shall be attached to wood framing below in accordance with Table R602.3(1).

❖ This section clarifies that the wood frame wall bracing requirements apply and where the SIP wall construction is used as wall bracing, it must be evaluated as a continuously sheathed braced wall line. When the SIP wall is required to serve as wall bracing, this section also makes it clear that the connection of the bottom (sole) plate must be in accordance with braced wall panel requirements in the general fastening schedule for wood construction.

R613.6 Interior load-bearing walls. Interior load-bearing walls shall be constructed as specified for exterior walls.

❖ This section directs the reader to exterior wall construction (see Section R613.5) for determining requirements for interior load-bearing walls.

R613.7 Drilling and notching. The maximum vertical chase penetration in SIPs shall have a maximum side dimension of 2 inches (51 mm) centered in the panel core. Vertical chases shall have a minimum spacing of 24-inches (610 mm) on center. Maximum of two horizontal chases shall be permitted in each wall panel, one at 14 inches (360 mm) from the bottom of the panel and one at mid-height of the wall panel. The maximum allowable penetration size in a wall panel shall be circular or rectangular with a maximum dimension of 12 inches (305 mm). Overcutting of holes in facing panels shall not be permitted.

❖ This section permits SIP manufacturers to form chases in the panel core both horizontally and vertically within the stated location and dimensional limits. Doing so during the fabrication process enables utilities to be installed in the finished walls without compressing insulation or having to field-drill through framing members.

R613.8 Connection. SIPs shall be connected at vertical in-plane joints in accordance with Figure R613.8 or by other *approved* methods.

❖ Prescriptive details for joining adjacent panels are provided in Figure R613.8. This connection is required for serviceability purposes and it is not intended to provide load transfer between the panels.

R613.9 Corner framing. Corner framing of SIP walls shall be constructed in accordance with Figure R613.9.

❖ The corner connection of SIP wall panels requires the use of SIP screws (see Section R613.3.5) as illustrated in Figure R613.9. Where SIP walls are used as wall bracing (see Section R613.5.3) this connection supersedes the corner details for continuously sheathed wall bracing in Table R602.3(1) and Figure R602.10.7.

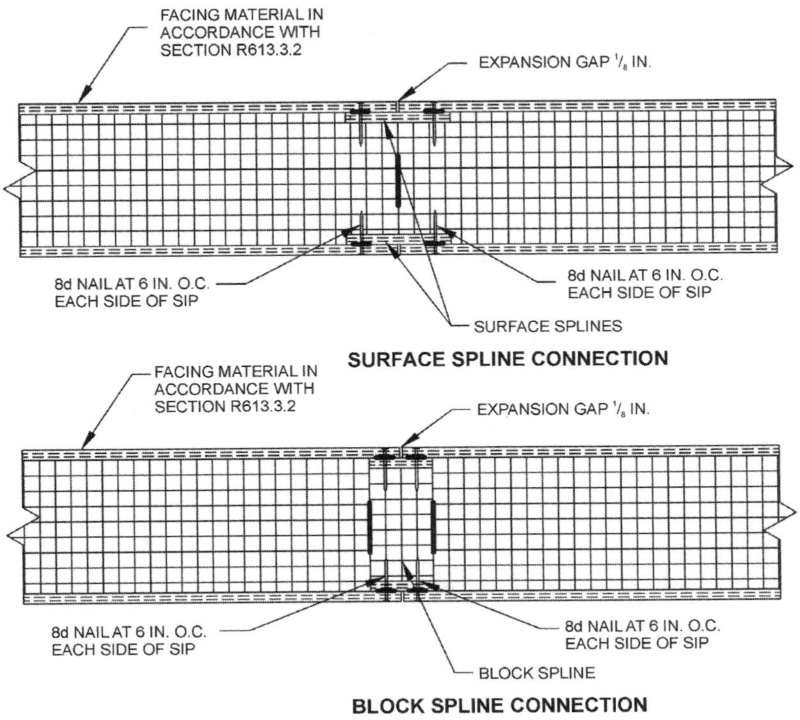

For SI: 1 inch = 25.4 mm.

FIGURE R613.8
TYPICAL SIP CONNECTION DETAILS FOR VERTICAL IN-PLANE JOINTS

❖ Two types of connection details for joining adjacent panels are furnished in this figure. These connections are intended primarily for serviceability purposes and are not designed to transfer stresses between the panels.

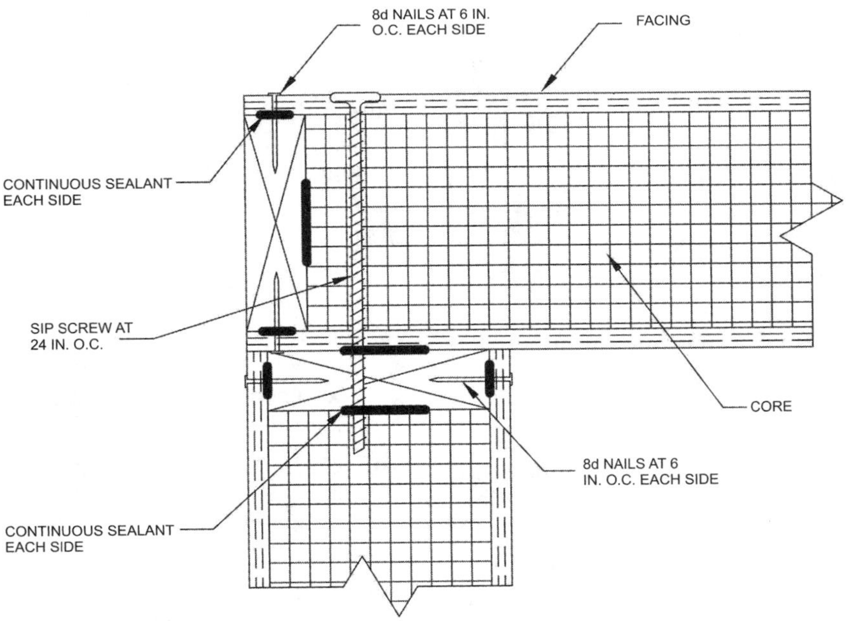

For SI: 1 inch = 25.4 mm.

FIGURE R613.9
SIP CORNER FRAMING DETAIL

❖ See the commentary to Section R613.9.

TABLE R613.10
MAXIMUM SPANS FOR $11^{7}/_{8}$-INCH-DEEP SIP HEADERS (feet)[a]

LOAD CONDITION	SNOW LOAD (psf)	BUILDING width (feet)				
		24	28	32	36	40
Supporting roof only	20	4	4	4	4	2
	30	4	4	4	2	2
	50	2	2	2	2	2
	70	2	2	2	N/A	N/A
Supporting roof and one-story	20	2	2	N/A	N/A	N/A
	30	2	2	N/A	N/A	N/A
	50	2	N/A	N/A	N/A	N/A
	70	N/A	N/A	N/A	N/A	N/A

For SI: 1 inch = 25.4 mm, 1 foot = 304.8 mm, 1 pound per square foot = 0.0479 kPa.
N/A = Not Applicable.
a. Design assumptions:
Maximum deflection criterion: *L*/360.
Maximum roof dead load: 10 psf.
Maximum ceiling load: 5 psf.
Maximum second floor live load: 30 psf.
Maximum second floor dead load: 10 psf.
Maximum second floor dead load from walls: 10 psf.

❖ See the commentary to Section R613.10.

R613.10 Headers. SIP headers shall be designed and constructed in accordance with Table R613.10 and Figure R613.5.1. SIPs headers shall be continuous sections without splines. Headers shall be at least $11^{7}/_{8}$ inches (302 mm) deep. Headers longer than 4 feet (1219 mm) shall be constructed in accordance with Section R602.7.

❖ The minimum depth of SIP headers is given and the permitted spans are tabulated. Where greater header spans are desired, the general requirements for wood headers are applicable, including those for wood structural panel box headers.

R613.10.1 Wood structural panel box headers. Wood structural panel box headers shall be allowed where SIP headers are not applicable. Wood structural panel box headers shall be constructed in accordance with Figure R602.7.2 and Table R602.7.2.

❖ See the commentary to Sections R613.10 and R602.7.1.

Bibliography

The following resource materials were used in the preparation of the commentary for this chapter of the code:

AAMA/WDMA/CSA 101/I.S.2/A440-08, *North American Fenestration Standard/Specification for Windows, Doors and Skylights.* Schaumburg, IL: American Architectural Manufacturers Association, 2008.

AAMA 506-06, *Voluntary Specifications for Hurricane Impact and Cycle Testing of Fenestration Products.* Schaumburg, IL: American Architectural Manufacturers Association, 2006.

ACI 318-08, *Building Code Requirements for Structural Concrete.* Farmington Hills, MI: American Concrete Institute, 2008.

ACI 530-08, *Building Code Requirements for Masonry Structures.* Farmington Hills, MI: American Concrete Institute, 2008.

AFPA NDS-05, *National Design Specification (NDS) for Wood Construction—with 2005 Supplement.* Washington, DC: American Forest and Paper Association, 2005.

AFPA WFCM-01, *Wood Frame Construction Manual for One- and Two-family Dwellings.* Washington, DC: American Forest and Paper Association, 2001.

AISI S100-07, *North American Specification for the Design of Cold-formed Steel Structural Members.* Washington, DC: American Iron and Steel Institute, 2007.

AISI S230-07, *Standard for Cold-formed Steel Framing-prescriptive Method for One- and Two-family Dwellings.* Washington, DC: American Iron and Steel Institute, 2007.

APA Report T2009P-28, *Testing of Structural Insulated Panels (SIPs) with New Facer Design Properties for the Structural Insulated Panel Association, Gig Harbor, Washington.* Tacoma, WA: APA – The Engineered Wood Association, 2009.

ASCE/SEI 5-08, *Building Code Requirements for Masonry Structures.* Reston, VA: American Society of Civil Engineers, 2008.

ASCE/SEI 7-05, *Minimum Design Loads for Buildings and Other Structures.* Reston, VA: American Society of Civil Engineers, 2005.

ASTM A 996/A 996M-06a, *Standard Specification for Rail-steel and Axle-steel Deformed Bars for Concrete Reinforcement.* West Conshohocken, PA: ASTM International, 2006.

ASTM C 94/C 94M-07, *Standard Specification for Ready-mixed Concrete.* West Conshohocken, PA: ASTM International, 2007.

ASTM C 270-07, *Standard Specification for Mortar for Unit Masonry.* West Conshohocken, PA: ASTM International, 2007.

ASTM C 476-02, *Specification for Grout for Masonry.* West Conshohocken, PA: ASTM International, 2002.

ASTM C 578-07, *Standard Specification for Rigid, Cellular Polystyrene Thermal Insulation.* West Conshohocken, PA: ASTM International, 2007.

ASTM C 685-01, *Standard Specification for Concrete Made by Volumetric Batching and Continuous Mixing.* West Conshohocken, PA: ASTM International, 2001.

ASTM D 2559-04, *Standard Specification for Adhesives for Structural Laminated Wood Products for Use Under Exterior (West Use) Exposure Conditions.* West Conshohocken, PA: ASTM International, 2004.

ASTM D 3957-06, *Standard Practices for Establishing Stress Grades for Structural Members Used in Log Buildings.* West Conshohocken, PA: ASTM International, 2006.

ASTM D 5456-09, *Standard Specification for Evaluation of Structural Composite Lumber Products.* West Conshohocken, PA: ASTM International, 2009.

ASTM E 330-02, *Standard Test Method for Structural Performance of Exterior Windows, Curtain Walls, Doors, Skylights by Uniform Static Air Pressure Difference.* West Conshohocken, PA: ASTM International, 2002.

ASTM E 1886-06, *Standard Test Method for Performance of Exterior Windows, Curtain Walls, Doors, and Impact Protective Systems Impacted by Missile(s) and Exposed to Cyclic Pressure Differentials.* West Conshohocken, PA: ASTM International, 2006.

ASTM E 1996-06, *Standard Specification for Performance of Exterior Windows, Curtain Walls, Doors and Impact Protective Systems Impacted by Windborne Debris in Hurricanes.* West Conshohocken, PA: ASTM International, 2006.

ASTM E 2634-08, *Standard Specification for Flat Wall Insulating Concrete Form (ICF) Systems.* West Conshohocken, PA: ASTM International, 2008.

DASMA 108-05, *Standard Method for Testing Sectional Garage Doors and Rolling Doors: Determination of Structural Performance Under Uniform Static Air Pressure Difference.* Cleveland, OH: Door and Access Systems Manufacturers Association International, 2005.

DOC PS 1-07, *Structural Plywood.* Washington, DC: United States Department of Commerce, 2007.

DOC PS 2-04, *Performance Standard for Wood-based Structural-use Panels.* Washington, DC: United States Department of Commerce, 2004.

ICC/APA-*The Engineered Wood Association, A Guide to the 2009 IRC Wood Wall Bracing Provisions.* Washington, DC: International Code Council, Inc., 2009.

ICC ES AC 05, *Acceptance Criteria for Sandwich Panel Adhesives.* Country Club Hills, IL. ICC Evaluation Service, Inc., 2010.

ICC IBC-12, *International Building Code.* Washington, DC. International Code Council, Inc., 2011.

PCA 100-07, *Prescriptive Design of Exterior Concrete Walls for One- and Two-family Dwellings (Pub. No. EB241).* Skokie, IL: Portland Cement Association, 2007.

TMS 402-05, *Building Code Requirements for Masonry Structures.* Boulder, CO: The Masonry Society, 2005.

Chapter 7: Wall Covering

General Comments

Interior wall coverings are used for a variety of purposes. Often designed simply as an aesthetic element to finish the interior of the building, these wall coverings may also protect the structural elements from impact or moisture damage. Exterior wall coverings, in great part, provide the weather-resistant exterior envelope that protects the building's interior from the elements. As evidenced by the common use of exterior insulation and finish systems (EIFS) in various regions of the country, new methods and materials are constantly introduced to provide for different appearances, improved insulating quality, sound transmission control and fire resistance. The code has developed prescriptive and performance regulations to control these aspects and the types and thickness of exterior wall coverings.

This chapter contains provisions for the design and construction of interior and exterior wall coverings. In addition to identifying the various types of wall covering regulated by the code, Chapter 7 references a number of material standards related to the specific types of wall covering materials. Of primary consideration are the application methods prescribed by the code for those wall covering materials conforming to their applicable referenced standard.

Section R701 establishes the scope of the chapter and states a general requirement dealing with weather protection for wall covering materials sensitive to adverse weather conditions. Section R702 establishes the various types of materials, materials standards and methods of application permitted for use as interior coverings, including interior plaster, gypsum board, ceramic tile, wood veneer paneling, hardboard paneling, wood shakes and wood shingles. The requirements for the use of vapor retarders for wall moisture control are included in this section. Section R703 addresses exterior wall coverings as well as the water-resistive barrier required beneath the exterior materials. Exterior wall coverings regulated by this section include aluminum, stone and masonry veneer, wood, hardboard, particleboard, wood structural panel siding, wood shakes and shingles, exterior plaster, steel, vinyl, fiber cement and exterior insulation finish.

Purpose

This chapter provides the minimum requirements applicable to wall covering materials used both in exterior and interior applications. It specifies the types of wall coverings addressed by the code, including such common materials as gypsum board and ceramic tile used on a building's interior, as well as horizontal siding and wood structural panel siding installed as the exterior membrane.

SECTION R701 GENERAL

R701.1 Application. The provisions of this chapter shall control the design and construction of the interior and exterior wall covering for all buildings.

❖ Chapter 7 deals with both interior and exterior wall coverings. A variety of interior coverings are regulated, including interior plaster, gypsum board, ceramic tile, wood veneer paneling, hardboard paneling, wood shakes and wood shingles. Wood siding, hardboard siding, wood structural panel siding, wood shakes, wood shingles, exterior plaster, stone veneer and masonry veneer are addressed where used as exterior wall coverings.

R701.2 Installation. Products sensitive to adverse weather shall not be installed until adequate weather protection for the installation is provided. Exterior sheathing shall be dry before applying exterior cover.

❖ Because many materials are subject to deterioration from the effects of moisture or other adverse weather conditions, the code limits the installation of such materials until proper weather protection is in place. Gypsum plaster systems, gypsum board and interior-grade plywood are particularly sensitive to moisture. It is also critical that exterior sheathing be dry prior to installation of the surface cover materials. Entrapment of moisture within the exterior wall assembly can result in a reduction in the effectiveness of the wall over time.

To assist in the application of the provisions of this chapter, it is important to understand the terminology that is used. Commentary Figures R701.2(1) and R701.2(2) provide pictorial descriptions of several of these terms.

SECTION R702 INTERIOR COVERING

R702.1 General. Interior coverings or wall finishes shall be installed in accordance with this chapter and Table R702.1(1), Table R702.1(2), Table R702.1(3) and Table R702.3.5. Interior masonry veneer shall comply with the requirements of Section R703.7.1 for support and Section R703.7.4 for anchorage, except an air space is not required.

Interior finishes and materials shall conform to the flame spread and smoke-development requirements of Section R302.9.

❖ This section contains the installation requirements for interior wall coverings. Interior wall coverings of plaster and gypsum board must be installed in accordance with the applicable tables. The installation requirements include material orientation, backing support, spacing and size as well as method of attachment to supports. These requirements can result in wall coverings that will perform as intended in regard to durability and appearance. Interior wall coverings of masonry veneer must be supported in accordance with Section R703.7.1 and anchored in accordance with Section R703.7.4. No airspace is required for interior masonry veneer since it is not required to be weather resistant. All interior finishes must meet the flame spread and smoke-development provisions of Section R302.9.

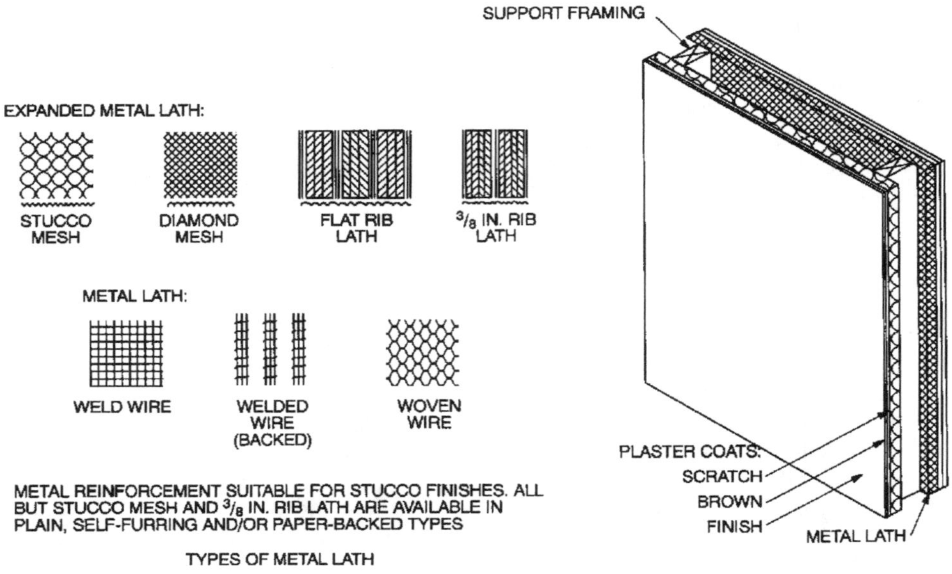

For SI: 1 inch = 25.4 mm.

Figure R701.2(1)
PLASTER WALL COVERING—APPLICABLE TERMS

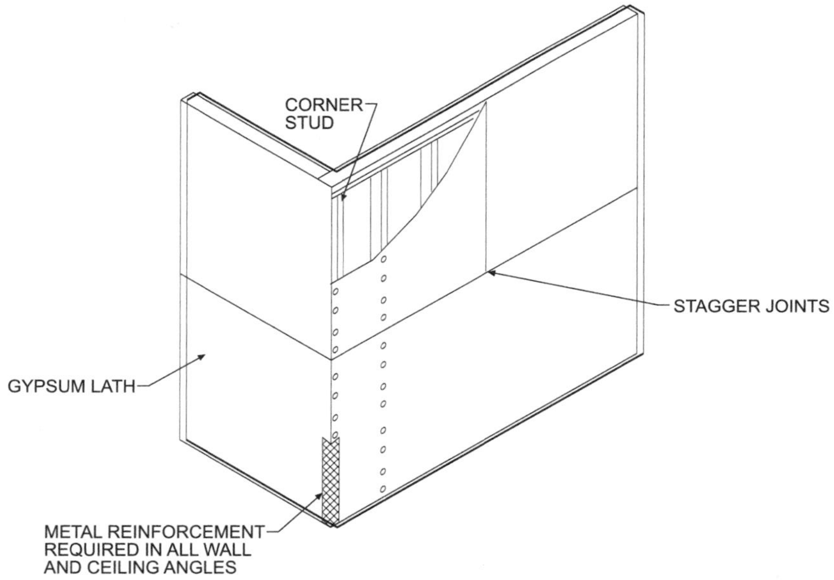

Figure R701.2(2)
GYPSUM APPLICATION—TERMS

TABLE R702.1(1)
THICKNESS OF PLASTER

PLASTER BASE	FINISHED THICKNESS OF PLASTER FROM FACE OF LATH, MASONRY, CONCRETE (inches)	
	Gypsum Plaster	Cement Plaster
Expanded metal lath	$^5/_8$, minimum[a]	$^5/_8$, minimum[a]
Wire lath	$^5/_8$, minimum[a]	$^3/_4$, minimum (interior)[b] $^7/_8$, minimum (exterior)[b]
Gypsum lath[g]	$^1/_2$, minimum	$^3/_4$, minimum (interior)[b]
Masonry walls[c]	$^1/_2$, minimum	$^1/_2$, minimum
Monolithic concrete walls[c, d]	$^5/_8$, maximum	$^7/_8$, maximum
Monolithic concrete ceilings[c, d]	$^3/_8$, maximum[e]	$^1/_2$, maximum
Gypsum veneer base[f, g]	$^1/_{16}$, minimum	$^3/_4$, minimum (interior)[b]
Gypsum sheathing[g]	—	$^3/_4$, minimum (interior)[b] $^7/_8$, minimum (exterior)[b]

For SI: 1 inch = 25.4 mm.

a. When measured from back plane of expanded metal lath, exclusive of ribs, or self-furring lath, plaster thickness shall be $^3/_4$ inch minimum.
b. When measured from face of support or backing.
c. Because masonry and concrete surfaces may vary in plane, thickness of plaster need not be uniform.
d. When applied over a liquid bonding agent, finish coat may be applied directly to concrete surface.
e. Approved acoustical plaster may be applied directly to concrete or over base coat plaster, beyond the maximum plaster thickness shown.
f. Attachment shall be in accordance with Table R702.3.5.
g. Where gypsum board is used as a base for cement plaster, a water-resistive barrier complying with Section R703.2 shall be provided.

❖ Based on several variables, this table sets forth the required plaster thickness measured from the face of the lath, masonry or concrete to the plaster surface. Both gypsum plaster and portland cement mortar are addressed, with the option of the plaster application occurring over a variety of plaster bases. In most cases, the minimum plaster thickness is mandated; however, the maximum thickness is regulated where plaster is applied over a monolithic concrete wall or ceiling.

TABLE R702.1(2)
GYPSUM PLASTER PROPORTIONS[a]

NUMBER	COAT	PLASTER BASE OR LATH	MAXIMUM VOLUME AGGREGATE PER 100 POUNDS NEAT PLASTER[b] (cubic feet)	
			Damp Loose Sand[a]	Perlite or Vermiculite[c]
Two-coat work	Base coat	Gypsum lath	2.5	2
	Base coat	Masonry	3	3
Three-coat work	First coat	Lath	2[d]	2
	Second coat	Lath	3[d]	2[e]
	First and second coats	Masonry	3	3

For SI: 1 inch = 25.4 mm, 1 cubic foot = 0.0283 m^3, 1 pound = 0.454 kg.

a. Wood-fibered gypsum plaster may be mixed in the proportions of 100 pounds of gypsum to not more than 1 cubic foot of sand where applied on masonry or concrete.
b. When determining the amount of aggregate in set plaster, a tolerance of 10 percent shall be allowed.
c. Combinations of sand and lightweight aggregate may be used, provided the volume and weight relationship of the combined aggregate to gypsum plaster is maintained.
d. If used for both first and second coats, the volume of aggregate may be 2.5 cubic feet.
e. Where plaster is 1 inch or more in total thickness, the proportions for the second coat may be increased to 3 cubic feet.

❖ This table covers the maximum permitted amount of aggregate for gypsum plaster, measured in cubic feet per 100 pounds (16 oz kg/m). Applicable for both damp, loose sand, perlite and vermiculite, the limits vary based on the type of plaster base or lath. The maximum proportions for each coat of both two-coat and three-coat plaster systems are identified.

TABLE R702.1(3)
CEMENT PLASTER PROPORTIONS, PARTS BY VOLUME

COAT	CEMENT PLASTER TYPE	CEMENTITIOUS MATERIALS: Portland Cement Type I, II or III or Blended Cement Type IP, I (PM), IS or I (SM)	CEMENTITIOUS MATERIALS: Plastic Cement	CEMENTITIOUS MATERIALS: Masonry Cement Type M, S or N	CEMENTITIOUS MATERIALS: Lime	VOLUME OF AGGREGATE PER SUM OF SEPARATE VOLUMES OF CEMENTITIOUS MATERIALS[b]
First	Portland or blended	1			$^3/_4$ - $1^1/_2$[a]	$2^1/_2$ - 4
	Masonry				1	$2^1/_2$ - 4
	Plastic		1			$2^1/_2$ - 4
Second	Portland or blended	1			$^3/_4$ - $1^1/_2$	3 - 5
	Masonry			1		3 - 5
	Plastic		1			3 - 5
Finish	Portland or blended	1			$^3/_4$ - 2	$1^1/_2$ - 3
	Masonry			1		$1^1/_2$ - 3
	Plastic		1			$1^1/_2$ - 3

For SI: 1 inch = 25.4 mm, 1 pound = 0.545 kg.

a. Lime by volume of 0 to $^3/_4$ shall be used when the plaster will be placed over low-absorption surfaces such as dense clay tile or brick.

b. The same or greater sand proportion shall be used in the second coat than used in the first coat.

❖ Where cement plaster is used rather than gypsum plaster, this table is applicable for determining the maximum permitted volume of portland, blended, plastic or masonry cement, lime, and aggregate. In addition, the minimum period for moist curing is established for interior cement plaster as is the minimum required interval between coats.

R702.2 Interior plaster.

❖ Multicoat plastering has been commonplace for the past century. The industry generally views multicoat work as necessary for control of plaster thickness and density, particularly when applied by hand. Because of the uniformity created where plaster is applied in thin, successive layers, the code requires three-coat plastering over metal lath and two-coat work over other plaster bases approved for use by the code. Reducing the requirement for plaster bases other than metal lath depends on the rigidity of the plaster base itself. More-rigid plaster bases are not as susceptible to variations in thickness and flatness of the surface. In fact, the first coat applied in three-coat work on a flexible base, such as metal lath, is used to stiffen that base to provide the rigidity necessary to attain uniform thickness and surface flatness.

Both a gypsum plaster and a cement plaster interior wall covering system are shown in Commentary Figure R702.2(1). The finished thicknesses for both plastering methods determined by the type of base materials are set forth in Table R702.1(1). Commentary Figure R702.2(2) illustrates the veneer plaster interior wall covering system. This section also cites the materials standards to which gypsum plaster and cement plaster must conform.

R702.2.1 Gypsum plaster. Gypsum plaster materials shall conform to ASTM C 5, C 22, C 28, C 35, C 59, C 61, C 587, C 631, C 847, C 933, C 1032 and C 1047, and shall be installed or applied in compliance with ASTM C 843 and C 844. Gypsum lath or gypsum base for veneer plaster shall conform to ASTM C 1396. Plaster shall not be less than three coats when applied over metal lath and not less than two coats when applied over other bases permitted by this section, except that veneer plaster may be applied in one coat not to exceed $^3/_{16}$ inch (4.76 mm) thickness, provided the total thickness is in accordance with Table R702.1(1).

❖ This section sets forth the standards applicable to gypsum plaster materials. These materials are to be installed according to ASTM C 843 and C 844. The minimum number of coats of gypsum plaster is established according to the type of base upon which they are installed.

R702.2.2 Cement plaster. Cement plaster materials shall conform to ASTM C 91 (Type M, S or N), C 150 (Type I, II and III), C 595 [Type IP, I (PM), IS and I (SM), C 847, C 897, C 926, C 933, C 1032, C 1047 and C 1328, and shall be installed or applied in compliance with ASTM C 1063. Gypsum lath shall conform to ASTM C 1396. Plaster shall not be less than three coats when applied over metal lath and not less than two coats when applied over other bases permitted by this section, except that veneer plaster may be applied in one coat not to exceed $^3/_{16}$ inch (4.76 mm) thickness, provided the total thickness is in accordance with Table R702.1(1).

❖ This section sets forth the standards applicable to cement plaster materials. These materials are to be installed according to ASTM C 1063. The minimum number of coats of cement plaster is established according to the type of base upon which they are installed.

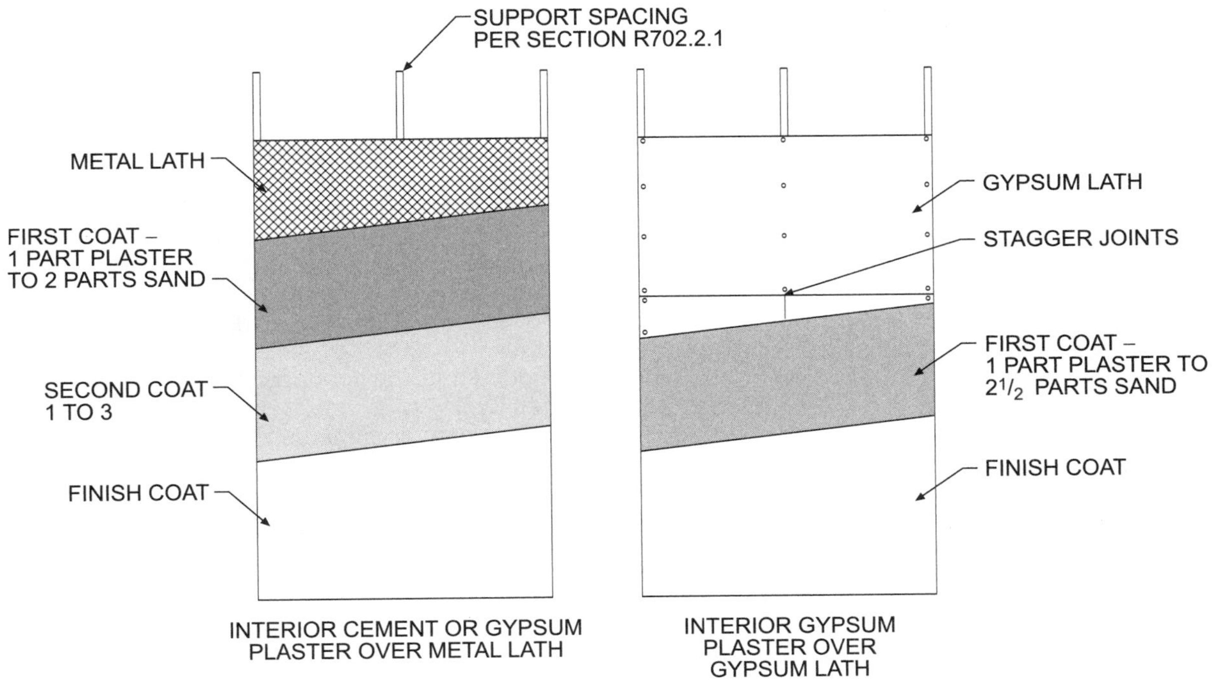

Figure R702.2(1)
INTERIOR PLASTERING SYSTEM

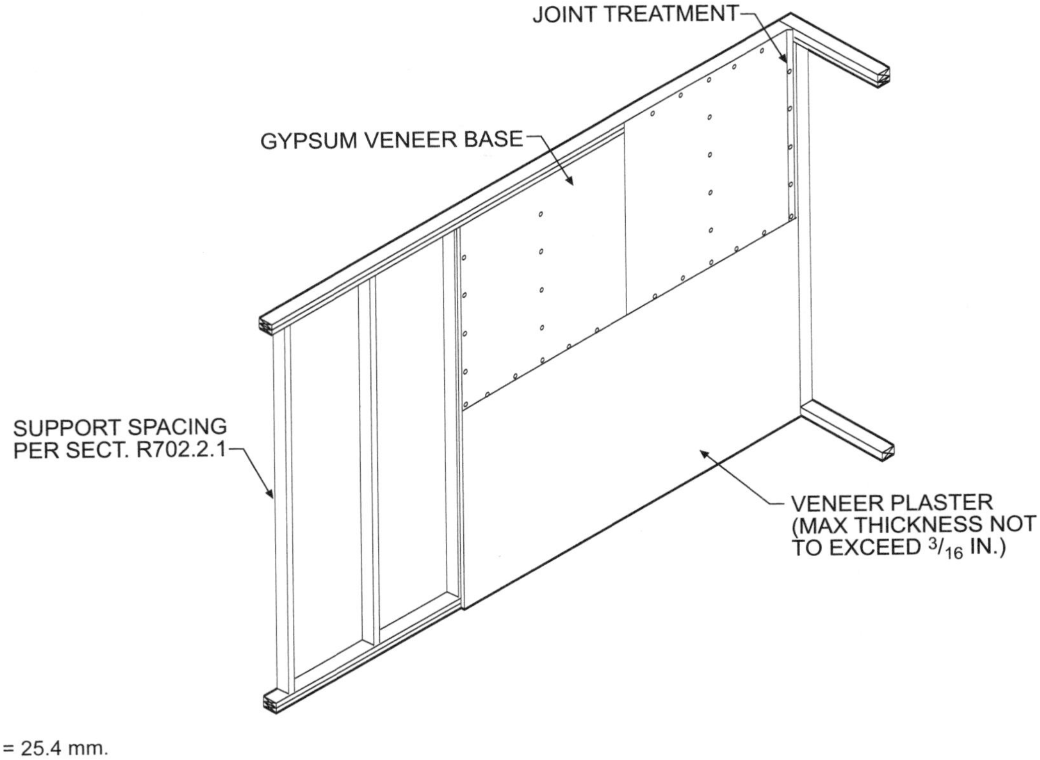

For SI: 1 inch = 25.4 mm.

Figure R702.2(2)
VENEER PLASTER SYSTEM

R702.2.2.1 Application. Each coat shall be kept in a moist condition for at least 24 hours prior to application of the next coat.

Exception: Applications installed in accordance with ASTM C 926.

❖ This section establishes the minimum amount of time the previous coat of plaster must remain moist before applying another coat.

R702.2.2.2 Curing. The finish coat for two-coat cement plaster shall not be applied sooner than 48 hours after application of the first coat. For three coat cement plaster the second coat shall not be applied sooner than 24 hours after application of the first coat. The finish coat for three-coat cement plaster shall not be applied sooner than 48 hours after application of the second coat.

❖ This section establishes the minimum amount of time the previous coat of plaster must cure before applying another coat.

R702.2.3 Support. Support spacing for gypsum or metal lath on walls or ceilings shall not exceed 16 inches (406 mm) for $^3/_8$-inch-thick (9.5 mm) or 24 inches (610 mm) for $^1/_2$-inch-thick (12.7 mm) plain gypsum lath. Gypsum lath shall be installed at right angles to support framing with end joints in adjacent courses staggered by at least one framing space.

❖ To provide a relatively firm base for the first layer of plaster, the code mandates a maximum span between supporting elements of the lath. Applicable to both gypsum and metal lath for either wall or ceiling installations, the maximum spacing of supports is based on the lath thickness. Gypsum lath must be installed perpendicular to the direction of the support framing, with the end joints staggered in adjacent courses.

R702.3 Gypsum board.

❖ Gypsum board is by far the most commonly used as an interior covering and is available in a variety of types and sizes. This section sets forth the criteria for the application of gypsum board in wall and ceiling applications.

R702.3.1 Materials. All gypsum board materials and accessories shall conform to ASTM C 22, C 475, C 514, C 1002, C 1047, C 1177, C 1178, C 1278, C 1396 or C 1658 and shall be installed in accordance with the provisions of this section. Adhesives for the installation of gypsum board shall conform to ASTM C 557.

❖ This section sets forth the standards applicable to gypsum board materials and accessories. Included in the list of standards is ASTM C 557, which deals with the use of adhesives to attach gypsum board to wood framing members.

R702.3.2 Wood framing. Wood framing supporting gypsum board shall not be less than 2 inches (51 mm) nominal thickness in the least dimension except that wood furring strips not less than 1-inch by 2-inch (25 mm by 51 mm) nominal dimension may be used over solid backing or framing spaced not more than 24 inches (610 mm) on center.

❖ For adequate structural backing, vertical support for gypsum board used as an interior wall covering is required to be no less than 2 inches (51 mm) nominal thickness. Dimensional lumber meeting this 2-inch (51 mm) nominal dimension would have an actual thickness of approximately $1^1/_2$ inches (38 mm).

Wood stripping used as furring over vertical supports on which the interior wall covering is to be applied is also required to meet the 2-inch (51 mm) nominal dimension in any direction unless the framing members are spaced at 24 inches (610 mm) on center or less. Where the spacing of the framing does not exceed 24 inches (610 mm) or where the furring strips are placed on solid backing, the minimum size requirements are reduced to 1 inch by 2 inches (25.4 mm by 51 mm).

Commentary Figures R702.3.2(1), R702.3.2(2) and R702.3.2(3) illustrate the dimensional constraints.

R702.3.3 Cold-formed steel framing. Cold-formed steel framing supporting gypsum board shall not be less than $1^1/_4$ inches (32 mm) wide in the least dimension. Nonload-bearing cold-formed steel framing shall comply with ASTM C 645. Load-bearing cold-formed steel framing and all cold-formed steel framing from 0.033 inch to 0.112 inch (1 mm to 3 mm) thick shall comply with ASTM C 955.

❖ In the same manner as wood framing, cold-formed steel framing must be of a minimum cross-sectional dimension where used to support gypsum board. For cold-formed steel members, the minimum dimension must be $1^1/_4$ inches (32 mm).

This section also references two standards covering cold-formed steel framing members. ASTM C 645 addresses nonstructural cold-formed steel framing members, while ASTM C 955 addresses load-bearing cold-formed steel studs, tracks and bracing for gypsum panels.

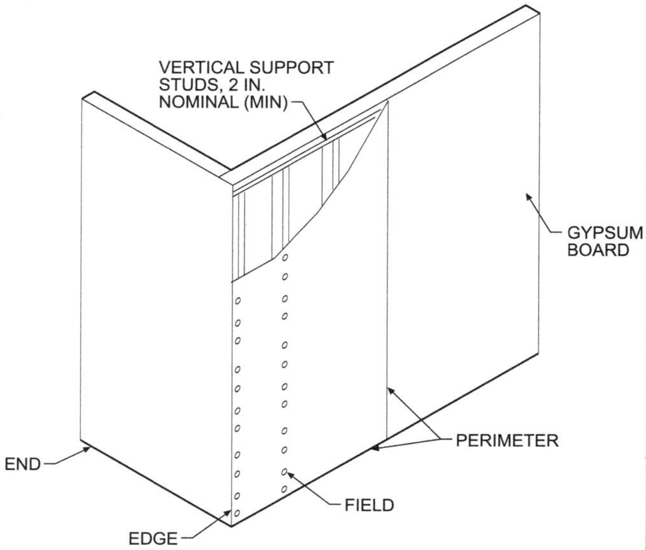

For SI: 1 inch = 25.4 mm.

Figure R702.3.2(1)
INTERIOR WALL COVERING—DIRECT APPLICATION

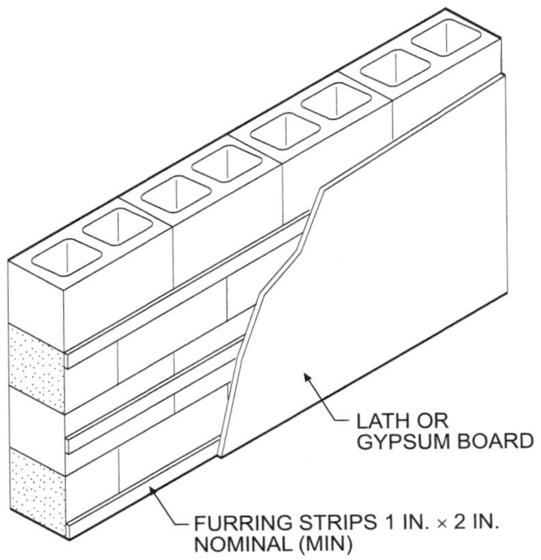

For SI: 1 inch = 1 inch = 25.4 mm.

Figure R702.3.2(2)
INTERIOR WALL COVERING-FURRED APPLICATION

R702.3.4 Insulating concrete form walls. Foam plastics for insulating concrete form walls constructed in accordance with Sections R404.1.2 and R611 on the interior of *habitable spaces* shall be protected in accordance with Section R316.4. Use of adhesives in conjunction with mechanical fasteners is permitted. Adhesives used for interior and exterior finishes shall be compatible with the insulating form materials.

❖ Reinforcing the provisions of Section R314.4 regarding the thermal barrier requirement for foam plastic insulation, this section references the requirements in Chapters 4 and 6 regulating foundation walls and insulating concrete form (ICF) wall systems. Foam plastic insulation used in these wall applications must be separated from the interior of the building by a minimum $^1/_2$-inch (12.7 mm) gypsum board or equivalent material.

Adhesives may be used to fasten the protective membrane (typically gypsum board) over the ICF wall. In all cases, however, mechanical fasteners must supplement the adhesive in accordance with Section R314.4. A total reliance on the adhesive to ensure the continued protection afforded by the gypsum board during a fire is misguided. Where used in conjunction with mechanical fasteners, adhesives must also be of a type that is compatible with the insulating form materials.

R702.3.5 Application. Maximum spacing of supports and the size and spacing of fasteners used to attach gypsum board shall comply with Table R702.3.5. Gypsum sheathing shall be attached to exterior walls in accordance with Table R602.3(1). Gypsum board shall be applied at right angles or parallel to framing members. All edges and ends of gypsum board shall occur on the framing members, except those edges and ends that are perpendicular to the framing members. Interior gypsum board shall not be installed where it is directly exposed to the weather or to water.

❖ To provide adequate backing for nailing the gypsum board to the support framing, all edges and ends, except those that are perpendicular to framing support members, must occur on framing members. Gypsum board may be installed in either a perpendicular or parallel manner. This is illustrated in Commentary Figure R702.3.5(1). Already addressed in a general sense in Section R701.2, the installation of interior gypsum board must not be done while the interior of the building is exposed to weather. Direct exposure of interior gypsum board to weather or water is not permitted.

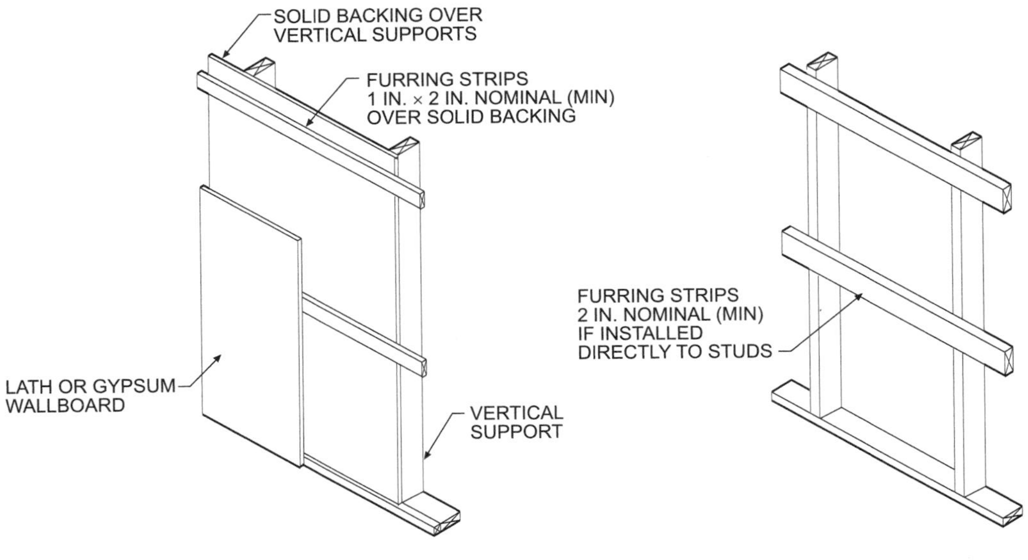

For SI: 1 inch = 25.4 mm.

Figure R702.3.2(3)
INTERIOR WALL COVERING-FURRED APPLICATION OVER SOLID BACKING OR OVER STUDS

TABLE R702.3.5
MINIMUM THICKNESS AND APPLICATION OF GYPSUM BOARD

THICKNESS OF GYPSUM BOARD (inches)	APPLICATION	ORIENTATION OF GYPSUM BOARD TO FRAMING	MAXIMUM SPACING OF FRAMING MEMBERS (inches o.c.)	MAXIMUM SPACING OF FASTENERS (inches)		SIZE OF NAILS FOR APPLICATION TO WOOD FRAMING[c]
				Nails[a]	Screws[b]	
Application without adhesive						
$^3/_8$	Ceiling[d]	Perpendicular	16	7	12	13 gage, $1^1/_4$″ long, $^{19}/_{64}$″ head; 0.098″ diameter, $1^1/_4$″ long, annular-ringed; or 4d cooler nail, 0.080″ diameter, $1^3/_8$″ long, $^7/_{32}$″ head.
	Wall	Either direction	16	8	16	
$^1/_2$	Ceiling	Either direction	16	7	12	13 gage, $1^3/_8$″ long, $^{19}/_{64}$″ head; 0.098″ diameter, $1^1/_4$″ long, annular-ringed; 5d cooler nail, 0.086″ diameter, $1^5/_8$″ long, $^{15}/_{64}$″ head; or gypsum board nail, 0.086? diameter, $1\ ^5/_8$″ long, $^9/_{32}$″ head.
	Ceiling[d]	Perpendicular	24	7	12	
	Wall	Either direction	24	8	12	
	Wall	Either direction	16	8	16	
$^5/_8$	Ceiling	Either direction	16	7	12	13 gage, $1^5/_8$″ long, $^{19}/_{64}$″ head; 0.098″ diameter, $1^3/_8$″ long, annular-ringed; 6d cooler nail, 0.092″ diameter, $1^7/_8$″ long, $^1/_4$″ head; or gypsum board nail, 0.0915″ diameter, $1^7/_8$″ long, $^{19}/_{64}$″ head.
	Ceiling[e]	Perpendicular	24	7	12	
	Wall	Either direction	24	8	12	
	Wall	Either direction	16	8	16	
Application with adhesive						
$^3/_8$	Ceiling[d]	Perpendicular	16	16	16	Same as above for $^3/_8$″ gypsum board
	Wall	Either direction	16	16	24	
$^1/_2$ or $^5/_8$	Ceiling	Either direction	16	16	16	Same as above for $^1/_2$″ and $^5/_8$″ gypsum board, respectively
	Ceiling[d]	Perpendicular	24	12	16	
	Wall	Either direction	24	16	24	
Two $^3/_8$ layers	Ceiling	Perpendicular	16	16	16	Base ply nailed as above for $^1/_2$″ gypsum board; face ply installed with adhesive
	Wall	Either direction	24	24	24	

For SI: 1 inch = 25.4 mm.

a. For application without adhesive, a pair of nails spaced not less than 2 inches apart or more than $2^1/_2$ inches apart may be used with the pair of nails spaced 12 inches on center.

b. Screws shall be in accordance with Section R702.3.6. Screws for attaching gypsum board to structural insulated panels shall penetrate the wood structural panel facing not less than $^7/_{16}$ inch.

c. Where cold-formed steel framing is used with a clinching design to receive nails by two edges of metal, the nails shall be not less than $^5/_8$ inch longer than the gypsum board thickness and shall have ringed shanks. Where the cold-formed steel framing has a nailing groove formed to receive the nails, the nails shall have barbed shanks or be 5d, $13^1/_2$ gage, $^{15}/_8$ inches long, $^{15}/_{64}$-inch head for $^1/_2$-inch gypsum board; and 6d, 13 gage, $1^7/_8$ inches long, $^{15}/_{64}$-inch head for $^5/_8$-inch gypsum board.

d. Three-eighths-inch-thick single-ply gypsum board shall not be used on a ceiling where a water-based textured finish is to be applied, or where it will be required to support insulation above a ceiling. On ceiling applications to receive a water-based texture material, either hand or spray applied, the gypsum board shall be applied perpendicular to framing. When applying a water-based texture material, the minimum gypsum board thickness shall be increased from $^3/_8$ inch to $^1/_2$ inch for 16-inch on center framing, and from $^1/_2$ inch to $^5/_8$ inch for 24-inch on center framing or $^1/_2$-inch sag-resistant gypsum ceiling board shall be used.

e. Type X gypsum board for garage ceilings beneath habitable rooms shall be installed perpendicular to the ceiling framing and shall be fastened at maximum 6 inches o.c. by minimum $1^7/_8$ inches 6d coated nails or equivalent drywall screws.

❖ Table R702.3.5 is a comprehensive table identifying the minimum thickness and fastening requirements for gypsum board. The gypsum board thickness [$^3/_8$-inch, $^1/_2$-inch or $^5/_8$-inch (10 mm, 12.7 mm or 16 mm)], location of the gypsum board (wall or ceiling), orientation of the gypsum board to the framing members (parallel or perpendicular), spacing of framing members [16 inches or 24 inches (406 mm or 610 mm)] on center and type of fasteners (nails or screws) are all set forth in the table. Where gypsum sheathing is to be attached to exterior walls, the provisions of Table R602.3(1) apply. Commentary Figure R702.3.5(2) shows a single-nailing method of attachment; however, Note a of Table R702.3.5 permits a double-nailing method as illustrated in Commentary Figure 702.3.5(3).

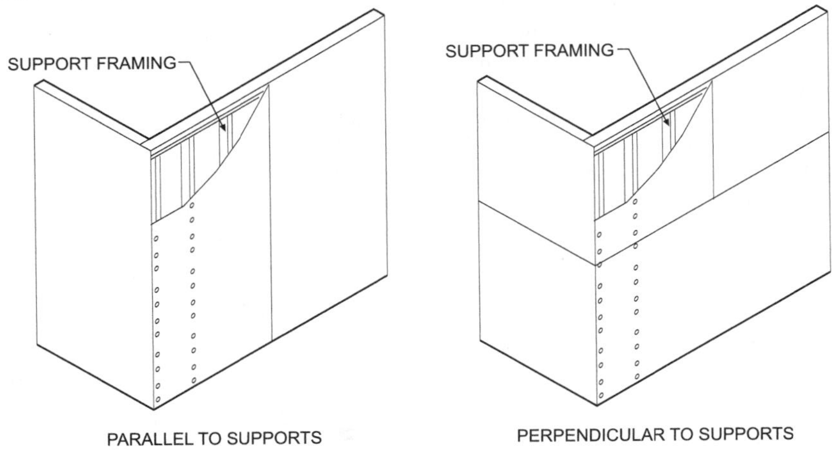

Figure R702.3.5(1)
GYPSUM BOARD APPLICATION

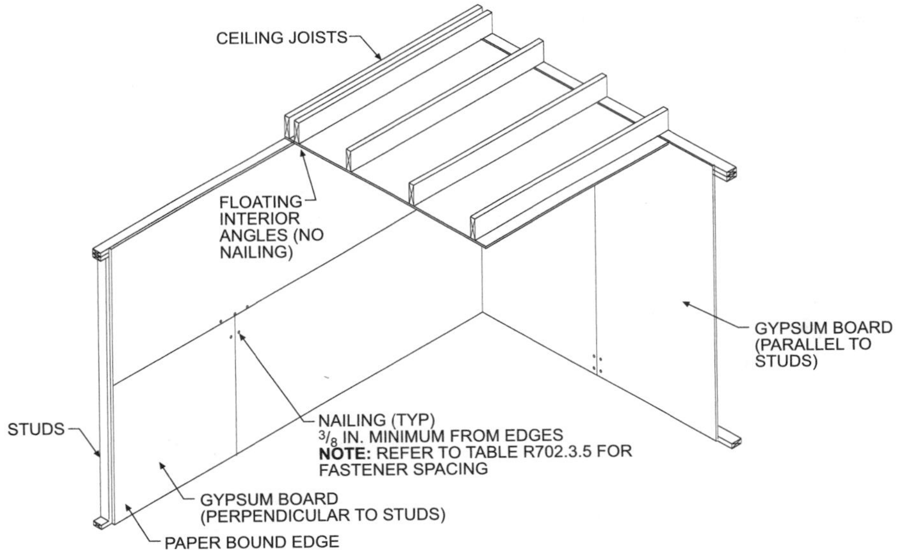

Figure R702.3.5(2)
GYPSUM BOARD—SINGLE-NAILING APPLICATION

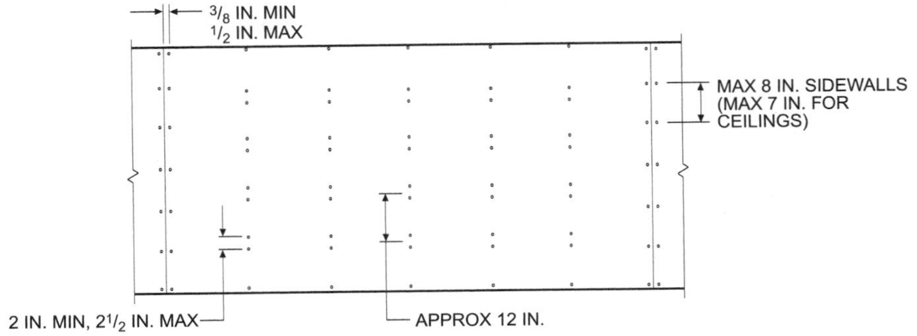

For SI: 1 inch = 25.4 mm.

Figure R702.3.5(3)
GYPSUM BOARD—DOUBLE-NAILING APPLICATION

R702.3.6 Fastening. Screws for attaching gypsum board to wood framing shall be Type W or Type S in accordance with ASTM C 1002 and shall penetrate the wood not less than $^5/_8$ inch (16 mm). Gypsum board shall be attached to cold-formed steel framing with minimum No. 6 screws. Screws for attaching gypsum board to cold-formed steel framing less than 0.033 inch (1 mm) thick shall be Type S in accordance with ASTM C 1002 or bugle head style in accordance with ASTM C 1513 and shall penetrate the steel not less than $^3/_8$ inch (9.5 mm). Screws for attaching gypsum board to cold-formed steel framing 0.033 inch to 0.112 inch (1 mm to 3 mm) thick shall be in accordance with ASTM C 954 or bugle head style in accordance with ASTM C 1513. Screws for attaching gypsum board to structural insulated panels shall penetrate the wood structural panel facing not less than $^7/_{16}$ inch (11 mm).

❖ Screws may be used in place of nails in the application of gypsum board. This section restates the provisions of Table R702.3.5 regarding the use of screws. Where gypsum board is attached to wood framing members, screws complying with ASTM C 1002 Type S or W are to be spaced in accordance with the table and must penetrate wood framing members a minimum of $^5/_8$ inch (15.9 mm). Where gypsum board is attached to cold-formed steel framing members, a minimum No. 6 screw is required. For cold-formed steel framing members less than 0.033 inch (1 mm) thick, screws must comply with either ASTM C 1002 Type S or ASTM C 1513 bugle-head style and must penetrate a minimum of $^3/_8$ inch (9.5 mm) into the framing. For cold-formed steel framing members between 0.033 inch to 0.112 inch (1 mm to 3 mm) thick, screws must comply with either ASTM C1002 Type S or ASTM C 1513 bugle-head style. Where gypsum board is attached to wood structural panels, the screws must penetrate a minimum of $^7/_{16}$ inch (11 mm) into the panel.

R702.3.7 Horizontal gypsum board diaphragm ceilings. Use of gypsum board shall be permitted on wood joists to create a horizontal *diaphragm* in accordance with Table R702.3.7. Gypsum board shall be installed perpendicular to ceiling framing members. End joints of adjacent courses of board shall not occur on the same joist. The maximum allowable *diaphragm* proportions shall be $1^1/_2$:1 between shear resisting elements. Rotation or cantilever conditions shall not be permitted. Gypsum board shall not be used in *diaphragm* ceilings to resist lateral forces imposed by masonry or concrete construction. All perimeter edges shall be blocked using wood members not less than 2-inch by 6-inch (51 mm by 152 mm) nominal dimension. Blocking material shall be installed flat over the top plate of the wall to provide a nailing surface not less than 2 inches (51 mm) in width for the attachment of the gypsum board.

❖ Generally a gypsum board ceiling does not serve as a load-carrying structural element for other than its own weight. This section provides installation requirements [refer to Commentary Figure R702.3.7(1)] that allow gypsum board to be used as a membrane in horizontal diaphragm ceilings. When installed according to this section, it provides an economical and aesthetically pleasing finished surface that can also resist horizontal shear and wind forces.

Table R702.3.7 provides the shear capacity, maximum fastener spacing and minimum edge distance that is to be used in designing a horizontal gypsum board ceiling diaphragm. Because the material thickness and fastener size do not vary, the capacity is strictly a function of the framing member spacing. Further qualifications on the tabulated shear capacities that are stated in the footnotes include: ceiling diaphragm capacities are not cumulative with other diaphragm capacities; the capacities given are only for "short-term" loads due to wind and earthquakes; in structures classified as Seismic Design Category D_0, D_1, D_2, or E the capacities must be reduced by 50 percent.

Limiting the diaphragm proportions and prohibiting cantilevers and rotation controls the deflection and distortion of the ceiling diaphragm [see Commentary Figure R702.3.7(1)].

The use of a gypsum board ceiling diaphragm to resist lateral loads from masonry or concrete is prohibited.

All perimeter edges must be blocked. See Commentary Figure R702.3.7(2).

TABLE R702.3.7
SHEAR CAPACITY FOR HORIZONTAL WOOD-FRAMED GYPSUM BOARD DIAPHRAGM CEILING ASSEMBLIES

MATERIAL	THICKNESS OF MATERIAL (min.) (inch)	SPACING OF FRAMING MEMBERS (max.) (inch)	SHEAR VALUE[a, b] (plf of ceiling)	MINIMUM FASTENER SIZE[c, d]
Gypsum board	$^1/_2$	16 o.c.	90	5d cooler or wallboard nail; $1^5/_8$-inch long; 0.086-inch shank; $^{15}/_{64}$-inch head
Gypsum board	$^1/_2$	24 o.c.	70	5d cooler or wallboard nail; $1^5/_8$-inch long; 0.086-inch shank; $^{15}/_{64}$-inch head

For SI: 1 inch = 25.4 mm, 1 pound per linear foot = 1.488 kg/m.

a. Values are not cumulative with other horizontal diaphragm values and are for short-term loading caused by wind or seismic loading. Values shall be reduced 25 percent for normal loading.

b. Values shall be reduced 50 percent in Seismic Design Categories D_0, D_1, D_2 and E.

c. $1^1/_4$-inch, #6 Type S or W screws may be substituted for the listed nails.

d. Fasteners shall be spaced not more than 7 inches on center at all supports, including perimeter blocking, and not less than $^3/_8$ inch from the edges and ends of the gypsum board.

❖ See the commentary to Section R702.3.7.

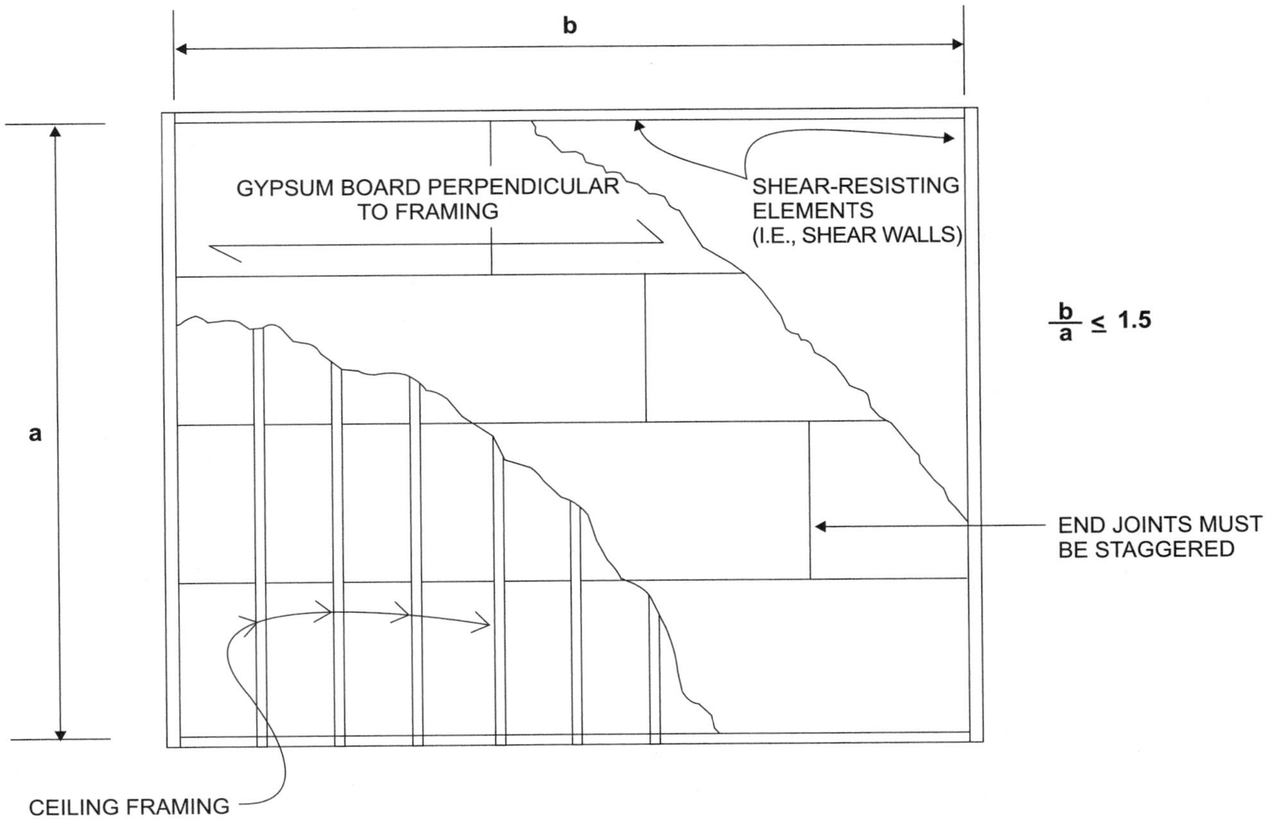

Figure R702.3.7(1)
HORIZONTAL GYPSUM BOARD CEILING DIAPHRAGM CONSTRUCTION

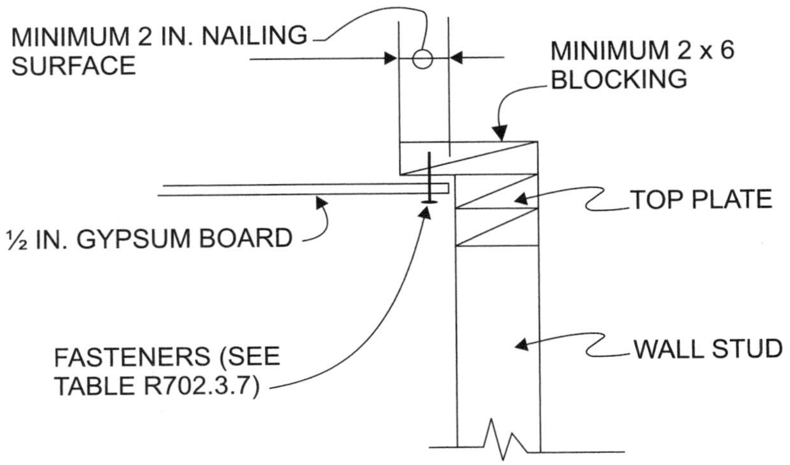

For SI: 1 inch = 25.4 mm.

Figure R702.3.7(2)
PERIMETER BLOCKING

R702.3.8 Water-resistant gypsum backing board. Gypsum board used as the base or backer for adhesive application of ceramic tile or other required nonabsorbent finish material shall conform to ASTM C 1396, C 1178 or C1278. Use of water-resistant gypsum backing board shall be permitted on ceilings where framing spacing does not exceed 12 inches (305 mm) on center for $^1/_2$-inch-thick (12.7 mm) or 16 inches (406 mm) for $^5/_8$-inch-thick (16 mm) gypsum board. Water-resistant gypsum board shall not be installed over a Class I or II vapor retarder in a shower or tub compartment. Cut or exposed edges, including those at wall intersections, shall be sealed as recommended by the manufacturer.

❖ Water-resistant gypsum backing board must be used as a base for ceramic tile or other required nonabsorbent finish materials where attached with an adhesive. It must be used in conjunction with the tile or wall panels because of its moisture-resistant qualities.

The use of water-resistant backing board is limited in a ceiling application, because $^1/_2$-inch-thick (12.7 mm) board is prohibited on ceilings where the framing members are spaced in excess of 12 inches (305 mm) on center. In addition, $^5/_8$-inch-thick water-resistant gypsum backing board is permitted only where ceiling members are spaced at 16 inches (406 mm) on center.

Gypsum board installed on the walls and ceilings at shower and bathtub areas must be finished to prevent moisture from penetrating the wall or ceiling finish and contacting the gypsum board. The finish applied to the exposed face of the gypsum board must create a water-resistant barrier that not only stops water from getting to the gypsum board but also prevents the release of moisture from within the wall or the gypsum board itself. For this reason, gypsum board must not be installed over the outboard side of any vapor barrier or retarder. This will create a waterproof membrane on both faces of the gypsum board, causing moisture to be trapped in the gypsum board that will ultimately cause it to decompose and fail.

R702.3.8.1 Limitations. Water resistant gypsum backing board shall not be used where there will be direct exposure to water, or in areas subject to continuous high humidity.

❖ Although many gypsum board sheet products are manufactured and approved for use in wet areas or areas exposed to moisture or humidity, there are still some extreme conditions where even water-resistant gypsum board will not provide the level of moisture protection necessary.

Water-resistant gypsum board is not to be used in areas that will be subject to a continuous exposure to moisture or humidity at locations such as saunas, steam rooms, gang showers or indoor pools. Gypsum board products, including the water-resistant type, are not intended for these extreme conditions and will not perform satisfactorily. Nongypsum wall and ceiling materials such as concrete masonry, ceramic tile on cement backer board, cement plaster (stucco) or other materials designed and recommended for high-moisture exposure must be used in these locations.

R702.4 Ceramic tile.

❖ Based on the code's requirement for nonabsorbent surfaces in bathtub and shower areas, ceramic tile is one of several materials used in such areas subject to water splash. Ceramic tile is regulated wherever installed, particularly when gypsum board is used as a base or backer board. This section regulates the installation of ceramic tile in residential construction.

R702.4.1 General. Ceramic tile surfaces shall be installed in accordance with ANSI A108.1, A108.4, A108.5, A108.6, A108.11, A118.1, A118.3, A136.1 and A137.1.

❖ This section references a number of ANSI standards that apply to the installation of ceramic tile. Commentary Figure R702.4.1 illustrates two possible methods of ceramic tile application.

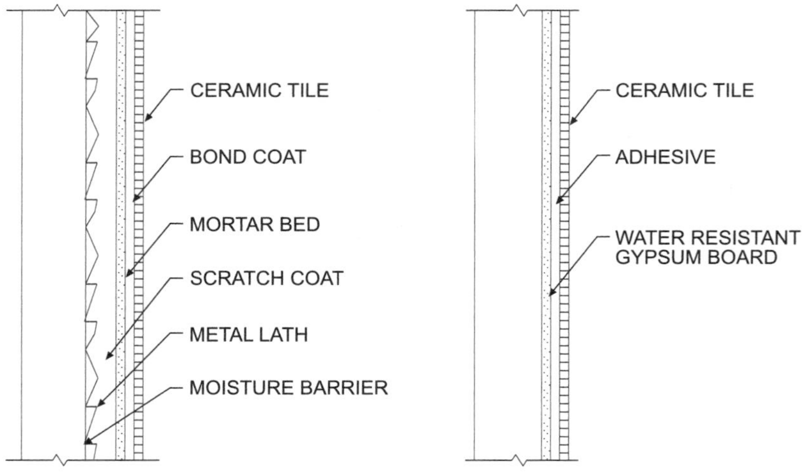

Figure R702.4.1
INTERIOR WALL TILE INSTALLATION

R702.4.2 Fiber-cement, fiber-mat reinforced cementitious backer units, glass mat gypsum backers and fiber-reinforced gypsum backers. Fiber-cement, fiber-mat reinforced cementitious backer units, glass mat gypsum backers or fiber-reinforced gypsum backers in compliance with ASTM C 1288, C 1325, C 1178 or C 1278, respectively, and installed in accordance with manufacturers' recommendations shall be used as backers for wall tile in tub and shower areas and wall panels in shower areas.

❖ Tile can be installed in a wet area with one of the following backing materials as illustrated in Commentary Figure 702.4.2: fiber-cement, fiber-mat reinforced cementitious backer units, glass mat gypsum backers, or fiber-reinforced gypsum backers. Paper-faced gypsum board (i.e. green board) must not be used in areas that are in direct contact with water. When this material gets wet through cracks in the grout joints, deteriorated caulking between the tile and tub assembly or improper flashing between the tub and tile interface, the water is absorbed into the material and can cause the paper facing to delaminate. The delamination can cause damage to the tile surface. In addition, water that has been trapped in paper-faced gypsum board has been known to create mold problems.

R702.5 Other finishes. Wood veneer paneling and hardboard paneling shall be placed on wood or cold-formed steel framing spaced not more than 16 inches (406 mm) on center. Wood veneer and hard board paneling less than $^1/_4$-inch (6 mm) nominal thickness shall not have less than a $^3/_8$-inch (10 mm) gypsum board backer. Wood veneer paneling not less than $^1/_4$-inch (6 mm) nominal thickness shall conform to ANSI/HPVA HP-1. Hardboard paneling shall conform to CPA/ANSI A135.5.

❖ This section contains requirements for wood veneer and hardboard paneling that are not addressed in previous sections. Modular home construction will generally use hardboard paneling applied directly to the studs. For this reason, there are specific provisions for wood veneer and hardboard siding. Wall finishes with adequate thickness will resist excessive bending and possible failure.

R702.6 Wood shakes and shingles. Wood shakes and shingles shall conform to CSSB *Grading Rules for Wood Shakes and Shingles* and shall be permitted to be installed directly to the studs with maximum 24 inches (610 mm) on-center spacing.

❖ Complying wood shakes and shingles may be attached directly to the studs in an interior application provided the studs are spaced at no more than 24 inches (610 mm) on center. The shakes and shingles must conform to the appropriate grading rules set forth by the Cedar Shake and Shingle Bureau.

R702.6.1 Attachment. Nails, staples or glue are permitted for attaching shakes or shingles to the wall, and attachment of the shakes or shingles directly to the surface shall be permitted provided the fasteners are appropriate for the type of wall surface material. When nails or staples are used, two fasteners shall be provided and shall be placed so that they are covered by the course above.

❖ Because there is no concern for weather protection, the shakes or shingles may be attached directly to

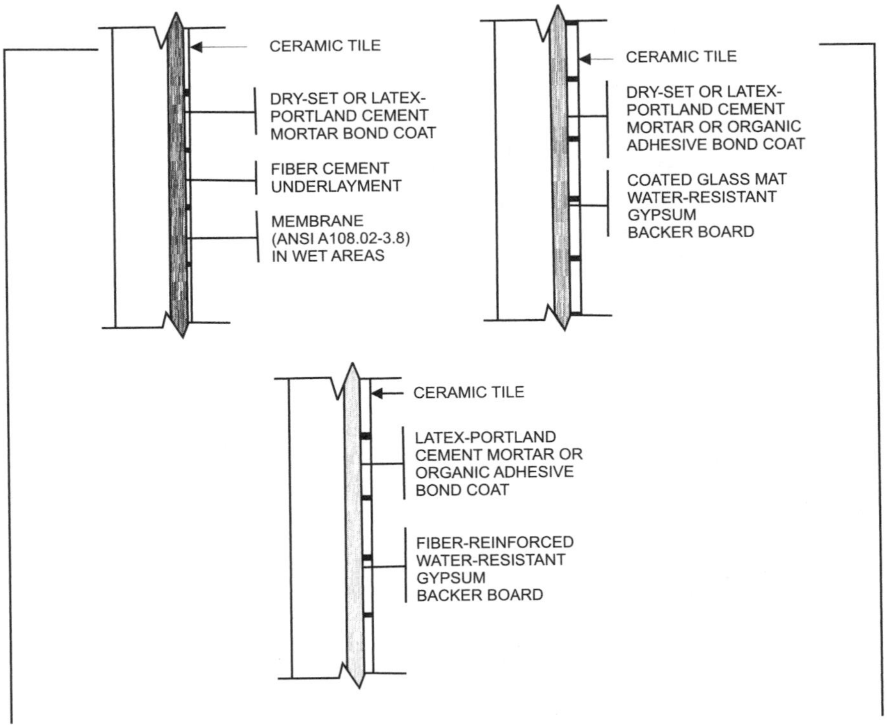

Figure R702.4.2
EXAMPLE OF WALL TILE INSTALLATION TUB AND SHOWER AREA

the surface with nails, staples or glue. At least two fasteners are required per shake or shingle where nails or staples are used, located so that the course above will cover the fasteners.

R702.6.2 Furring strips. Where furring strips are used, they shall be 1 inch by 2 inches or 1 inch by 3 inches (25 mm by 51 mm or 25 mm by 76 mm), spaced a distance on center equal to the desired exposure, and shall be attached to the wall by nailing through other wall material into the studs.

❖ In addition to direct attachment of wood shakes and shingles to studs or a wall surface, the use of furring strips is permitted under specific conditions. Furring strips should be either 1 inch by 2 inches (25 mm by 51 mm) or 1 inch by 3 inches (25 mm by 76 mm). The spacing of the strips is determined by the desired exposure of the shake or shingle, attached by nailing through other wall surface materials into the studs.

R702.7 Vapor retarders. Class I or II vapor retarders are required on the interior side of frame walls in Climate Zones 5, 6, 7, 8 and Marine 4.

Exceptions:

1. Basement walls.
2. Below grade portion of any wall.
3. Construction where moisture or its freezing will not damage the materials.

❖ The purpose of this section is to provide prescriptive methods for moisture control. The code contains three different vapor retarder classes, based upon the vapor permeability of the material. These are defined in Chapter 2 of the code. The basic requirement is for Class I or II vapor retarders to be installed on the interior side of framed walls in Climate Zones 5, 6, 7 and 8 and Marine 4. These climate zones are defined in Chapter 11 of the code. Figure N1101.10 and Table N1101.10 of Chapter 11 provide information regarding the climate zones for all geographic locations in the United States.

In cold climates, warm, moist air inside the building can migrate through the building envelope and condense on building surfaces or within building component cavities as the migrating air is cooled. Vapor retarders can help protect insulation and building materials from moisture damage, degradation and decay by preventing the water vapor from entering the external building envelope component cavities.

As can be seen by studying Figure N1101.10, Climate Zones 5, 6, 7 and 8 and Marine 4 are in the middle to northern portions of the continental United States and Alaska. These are areas where colder temperatures can be expected in winter months, which cause moisture from the interior of the building to condensate in the exterior walls of the building. A Class I or II vapor retarder is therefore called for on the interior side of the exterior wall. See the example as shown in Commentary Figure R702.7. A portion of the southern United States (along the Gulf of Mexico and the lower Atlantic coast) is a humid climate. In these locations, identified as Climate Zones 1 through 4 in Table N1101.10, no vapor retarder is required, as the exterior weather/water-resistant barrier provides the necessary protection.

R702.7.1 Class III vapor retarders. Class III vapor retarders shall be permitted where any one of the conditions in Table R702.7.1 is met.

❖ Wall assemblies can be designed and constructed to dry inwards, outwards and to both sides in all climate zones. This section allows more flexibility in the design and construction of moisture-forgiving wall systems. These requirements recognize that many common materials function to various degrees to slow the passage of moisture. In many situations, common materials such as the kraft facing on a fiberglass batt or latex paint may serve to retard moisture sufficiently. In particular, the "standard" sheet of polyethylene is usually not required as a vapor retarder in walls. This section therefore allows the use of Class III vapor retarders in lieu of the Class I or II otherwise called out in Section R702.7. Classes of vapor retarders are defined in Chapter 2.

The Class III vapor retarders allow more moisture vapor to pass through them. These are allowed to be used with exterior wall assemblies described in Table R702.7.1 that can be expected to dry. Section R702.7.2 describes materials that are deemed to be Class III vapor retarders, and Section R702.7.3 describes the attributes of cladding that make it "vented" as called out in Table R702.7.1.

TABLE R702.7.1
CLASS III VAPOR RETARDERS

CLIMATE-ZONE	CLASS III VAPOR RETARDERS PERMITTED FOR:[a]
Marine 4	Vented cladding over wood structural panels.
	Vented cladding over fiberboard.
	Vented cladding over gypsum.
	Insulated sheathing with R-value ≥ 2.5 over 2 × 4 wall.
	Insulated sheathing with R-value ≥ 3.75 over 2 × 6 wall.
5	Vented cladding over wood structural panels.
	Vented cladding over fiberboard.
	Vented cladding over gypsum.
	Insulated sheathing with R-value ≥ 5 over 2 × 4 wall.
	Insulated sheathing with R-value ≥ 7.5 over 2 × 6 wall.
6	Vented cladding over fiberboard.
	Vented cladding over gypsum.
	Insulated sheathing with R-value ≥ 7.5 over 2 × 4 wall.
	Insulated sheathing with R-value ≥ 11.25 over 2 × 6 wall.
7 and 8	Insulated sheathing with R-value ≥ 10 over 2 × 4 wall.
	Insulated sheathing with R-value ≥ 15 over 2 × 6 wall.

For SI: 1 pound per cubic foot = 16 kg/m^3.

a. Spray foam with a minimum density of 2 lb/ft^3 applied to the interior cavity side of wood structural panels, fiberboard, insulating sheathing or gypsum is deemed to meet the insulating sheathing requirement where the spray foam R-value meets or exceeds the specified insulating sheathing R-value.

❖ This table describes the situations where it is permissible to use Class III vapor retarders instead of Class

I or II. Class III vapor retarders allow more moisture vapor to pass through them. These are allowed to be used with exterior wall assemblies described in Table R702.7.1 that can be expected to dry. The table provides types of wall assemblies that can be used with Class III vapor retarders in different climate zones. The climate zones are determined from Chapter 11 of the code.

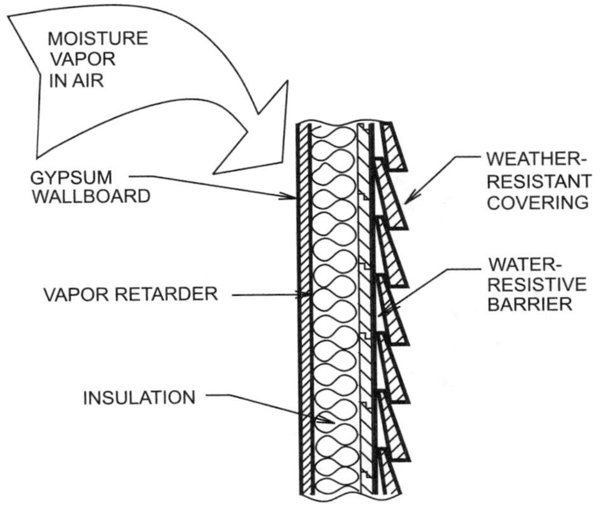
MOISTURE
VAPOR
IN AIR
GYPSUM
WALLBOARD
VAPOR RETARDER
INSULATION
WEATHER-
RESISTANT
COVERING
WATER-
RESISTIVE
BARRIER
IF UNCHECKED, MOISTURE VAPOR MOVES
FROM HUMID INTERIOR OF HOUSE
TOWARD COLDER EXTERIOR, CONDENSING WITHIN
SIDEWALL WHICH CAN CAUSE FINISH AND
INTERNAL DETERIORATION PROBLEMS.
NOTE: IN SOME CLIMATES THIS PROCESS OCCURS IN
REVERSE, WITH THE MOISTURE VAPOR MOVING FROM
THE EXTERIOR TOWARD THE INTERIOR.

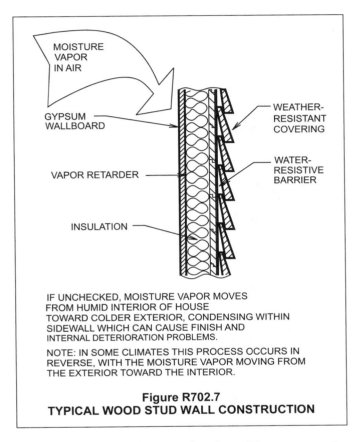

Figure R702.7
TYPICAL WOOD STUD WALL CONSTRUCTION

R702.7.2 Material vapor retarder class. The vapor retarder class shall be based on the manufacturer's certified testing or a tested assembly.

The following shall be deemed to meet the class specified:

Class I: Sheet polyethylene, unperforated aluminum foil.

Class II: Kraft-faced fiberglass batts.

Class III: Latex or enamel paint.

❖ The vapor retarder class is defined in Section 202 of the code in terms of vapor permeability. The test method called out for determination of the perm value is ASTM E 96. However, this section provides a list of materials that can be used for each class of vapor retarder and are deemed to comply with the test standard. No testing is required for these materials. All other materials are required to be tested.

R702.7.3 Minimum clear air spaces and vented openings for vented cladding. For the purposes of this section, vented cladding shall include the following minimum clear air spaces. Other openings with the equivalent vent area shall be permitted.

1. Vinyl lap or horizontal aluminum siding applied over a weather resistive barrier as specified in Table R703.4.
2. Brick veneer with a clear airspace as specified in Table R703.7.4.
3. Other approved vented claddings.

❖ The purpose of this section is to define what is intended by "vented cladding" as called out in Table R702.7.1. As can be seen, vented cladding is material commonly used in the code, including vinyl siding, aluminum siding, and brick with a clear airspace.

By "other approved vented claddings," the code intends materials that can be shown to allow drying of the components behind them in the same manner and at the same rate that the specific items in No. 1 or 2 would allow drying.

SECTION R703
EXTERIOR COVERING

R703.1 General. Exterior walls shall provide the building with a weather-resistant exterior wall envelope. The exterior wall envelope shall include flashing as described in Section R703.8.

❖ Exterior walls of buildings must be protected against damage caused by precipitation, wind and other weather conditions. This section requires that flashing be installed in the exterior wall at penetrations and terminations of the exterior wall covering. Section R703.8 is referenced for flashings requirements (see commentary, Section R703.8).

R703.1.1 Water resistance. The exterior wall envelope shall be designed and constructed in a manner that prevents the accumulation of water within the wall assembly by providing a water-resistant barrier behind the exterior veneer as required by Section R703.2 and a means of draining to the exterior water that enters the assembly. Protection against condensation in the exterior wall assembly shall be provided in accordance with Section R702.7 of this code.

Exceptions:

1. A weather-resistant exterior wall envelope shall not be required over concrete or masonry walls designed in accordance with Chapter 6 and flashed according to Section R703.7 or R703.8.
2. Compliance with the requirements for a means of drainage, and the requirements of Sections R703.2 and R703.8, shall not be required for an exterior wall envelope that has been demonstrated to resist wind-driven rain through testing of the exterior wall envelope, including joints, penetrations and intersections with dissimilar materials, in accordance with ASTM E 331 under the following conditions:
   2.1. Exterior wall envelope test assemblies shall include at least one opening, one control joint, one wall/eave interface and one wall sill. All

tested openings and penetrations shall be representative of the intended end-use configuration.

2.2. Exterior wall envelope test assemblies shall be at least 4 feet by 8 feet (1219 mm by 2438 mm) in size.

2.3. Exterior wall assemblies shall be tested at a minimum differential pressure of 6.24 pounds per square foot (299 Pa).

2.4. Exterior wall envelope assemblies shall be subjected to the minimum test exposure for a minimum of 2 hours.

The exterior wall envelope design shall be considered to resist wind-driven rain where the results of testing indicate that water did not penetrate control joints in the exterior wall envelope, joints at the perimeter of openings penetration or intersections of terminations with dissimilar materials.

❖ As part of providing a weather-resistant exterior envelope, exterior walls are required to provide water resistance. This section prescribes two basic components of providing water resistance for an exterior wall assembly: a water-resistive barrier installed between the covering and substrate of the exterior wall and a means of draining moisture that may penetrate behind the exterior wall assembly back to the exterior. Section R703.2 is referenced for the requirements of the water-resistive barrier (see commentary, Section R703.2). This section does not, however, contain a prescriptive requirement for the means of drainage. The method to provide the means of drainage is a performance criterion and must be evaluated based on the ability to allow moisture that may penetrate behind the exterior wall covering to effectively drain back to the exterior. This can be achieved in many ways including, but not limited to, providing a rain-screen pressure-equalized type of exterior assembly, or providing discontinuities or gaps between the surface of the substrate and the back side of the finish, such as through the use of noncorrodible furring or two layers of Grade D paper.

For common types of construction, such as vinyl siding or brick veneer, the typical practice of installing building paper, flashing and weeps will comply with the intent of this section. For stucco or adhered masonry veneer, installing two layers of Grade D paper and flashing will comply with the intent of this section. Discontinuities between the exterior wall covering and substrate must be such that they encourage the flow of moisture via gravity or capillary action to a location where the water may exit, such as at flashings and weeps. The absence of a means of drainage may result in the accumulation of moisture that becomes trapped between the wall covering and the substrate. Over time, extended exposure to moisture may contribute to the degradation of the wall covering, building substrate or even the structural elements of the exterior wall.

Exception 1 states that where the exterior wall envelope is designed and constructed of concrete or masonry materials in accordance with the requirements of Chapter 6 and flashed in accordance with Section R703.7 or R703.8, the water-resistive barrier and a means of drainage may be omitted. This is because the penetration of moisture behind the exterior wall covering is not detrimental to concrete and masonry substrates.

Exception 2 permits the use of exterior wall coverings that do not have a means of draining water or meet the prescriptive requirements of Sections R703.2 and R703.8, provided that the exterior wall envelope, with penetration details, demonstrates wind-driven rain resistance when tested. The test specimen(s) must incorporate the penetration and termination details intended for use. Only details that have not allowed water to penetrate will be permitted to be used in the exterior wall envelope. The minimum panel size specified represents that which is commonly used in testing to ASTM E 331; however, this does not preclude the testing of larger panels if desired. The modifications to the test pressure differential and test duration are intended to represent more closely conditions that will be encountered in service. The pass/fail criterion is based on the visual observation of moisture on the interior side of the wall assembly. For frame-type wall assemblies, such as stud walls, this requires the observation of locations such as the interior face of the exterior wall sheathing and wall framing members for the presence of moisture during the test. The test method is intended to assess the performance of the method(s) and material(s) used to seal the interface between the termination of the exterior wall covering and the penetrating item(s) or abutting construction. The test is not necessarily intended to test the performance of the penetrating item.

Walls designed and constructed in accordance with this chapter must also comply with the requirements of Section R702.7 of the code. This requires that frame-type wall assemblies be protected from moisture infiltration from the building interior through the use of a vapor retarder (see Commentary Figure R702.7) or by the ventilation of the wall cavity within a frame-type wall assembly.

R703.1.2 Wind resistance. Wall coverings, backing materials and their attachments shall be capable of resisting wind loads in accordance with Tables R301.2(2) and R301.2(3). Wind-pressure resistance of the siding and backing materials shall be determined by ASTM E 330 or other applicable standard test methods. Where wind-pressure resistance is determined by design analysis, data from approved design standards and analysis conforming to generally accepted engineering practice shall be used to evaluate the siding and backing material and its fastening. All applicable failure modes including bending rupture of siding, fastener withdrawal and fastener head pull-through shall be considered in the testing or design analysis. Where the wall covering and the backing material resist wind load as an assembly, use of the design capacity of the assembly shall be permitted.

❖ All the components in an exterior wall assembly must be able to resist wind loads. This includes the exterior wall covering as well as the substrate backing portion of the wall assembly. This section references component and cladding wind loads and height and exposure adjustment factors for those loads given in Tables R301.2(2) and R301.2(3), respectively. The resistance of the exterior wall assembly can be determined by testing or design analysis. Testing should conform with ASTM E 330 or other standard test methods which are applicable. Design analysis should be conducted in accordance with design standards and generally accepted engineering principles. Testing or design analysis should consider all failure modes including flexural failure of wall covering and fastener withdrawal and head pull-through. When the entire exterior wall assembly, consisting of the wall covering and the substrate, contributes to its ability to resist wind loads, the structural design capacity of the assembly is permitted to be taken into account.

R703.2 Water-resistive barrier. One layer of No. 15 asphalt felt, free from holes and breaks, complying with ASTM D 226 for Type 1 felt or other approved water-resistive barrier shall be applied over studs or sheathing of all exterior walls. Such felt or material shall be applied horizontally, with the upper layer lapped over the lower layer not less than 2 inches (51 mm). Where joints occur, felt shall be lapped not less than 6 inches (152 mm). The felt or other approved material shall be continuous to the top of walls and terminated at penetrations and building appendages in a manner to meet the requirements of the exterior wall envelope as described in Section R703.1.

Exception: Omission of the water-resistive barrier is permitted in the following situations:

1. In detached accessory buildings.
2. Under exterior wall finish materials as permitted in Table R703.4.
3. Under paperbacked stucco lath when the paper backing is an approved water-resistive barrier.

❖ Asphalt-saturated felt or any other approved water-resistive material is required behind all types of materials used as exterior wall coverings because of the possibility of moisture penetrating the substrate behind it. This felt or other material protects the wall construction from potential rotting.

The water-resistive membrane may be omitted where there is a low possibility of moisture penetration or the potential for moisture penetration is not a great concern. The water-resistive membrane may be eliminated when approved paperbacked stucco lath is used; the paper backing functions as the membrane if the backing paper is not punctured. The paper backing must be an approved water-resistive sheathing paper.

R703.3 Wood, hardboard and wood structural panel siding.

❖ This section addresses the use of wood siding products as exterior wall coverings. Both horizontal siding and panel siding are regulated.

R703.3.1 Panel siding. Joints in wood, hardboard or wood structural panel siding shall be made as follows unless otherwise approved. Vertical joints in panel siding shall occur over framing members, unless wood or wood structural panel sheathing is used, and shall be shiplapped or covered with a batten. Horizontal joints in panel siding shall be lapped a minimum of 1 inch (25 mm) or shall be shiplapped or shall be flashed with Z-flashing and occur over solid blocking, wood or wood structural panel sheathing.

❖ Plywood is acceptable as an exterior wall covering if the plywood is approved for use in an exterior location and the joints are made waterproof. Typical joint treatments include: (1) lapping joints horizontally, and (2) having the joints occur over framing and protecting those joints with a continuous wood batten using approved caulking, flashing or vertical or horizontal shiplaps. See Commentary Figures R703.3.1(1) and R703.3.1(2) for illustrations of joint treatment.

R703.3.2 Horizontal siding. Horizontal lap siding shall be installed in accordance with the manufacturer's recommendations. Where there are no recommendations the siding shall be lapped a minimum of 1 inch (25 mm), or $^1/_2$ inch (13 mm) if rabbeted, and shall have the ends caulked, covered with a batten or sealed and installed over a strip of flashing.

❖ To maintain the necessary weather protection, horizontal siding is required to be installed using the manufacturer's recommendations. Where a manufacturer does not provide installation recommendations, horizontal siding must have a minimum lap of at least 1 inch (25 mm), unless the siding is rabbeted, in which case the joint need be only $^1/_2$ inch (12.7 mm). Various methods can be used to protect the ends of horizontal siding, including the use of flashing, caulking or battens.

R703.4 Attachments. Unless specified otherwise, all wall coverings shall be securely fastened in accordance with Table R703.4 or with other *approved* aluminum, stainless steel, zinc-coated or other *approved* corrosion-resistive fasteners. Where the basic wind speed in accordance with Figure R301.2(4)A is 110 miles per hour (49 m/s) or higher, the attachment of wall coverings shall be designed to resist the component and cladding loads specified in Table R301.2(2), adjusted for height and exposure in accordance with Table R301.2(3).

❖ The presence of frequent moisture in the atmosphere necessitates that wall covering be fastened with approved corrosion-resistant fasteners such as aluminum, stainless steel, zinc or zinc-coated fasteners. Table R703.4 establishes the size of fasteners and other attachment requirements. The attachment provisions in Table R703.4 are not intended for construction where wind speeds meet or exceed 110 mph (49 m/s) unless verified through design.

For wind speeds 110 mph (49 m/s) or greater, Tables R301.2(2) and R301.2(3) must be used for wind loads for the attachment of wall coverings.

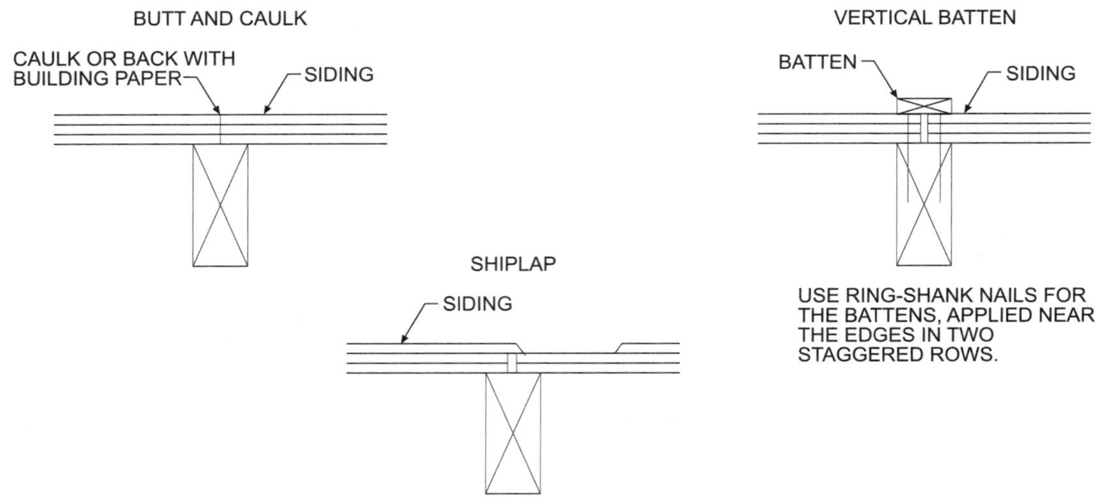

Figure R703.3.1(1)
VERTICAL JOINT TREATMENT OF PANEL SIDING

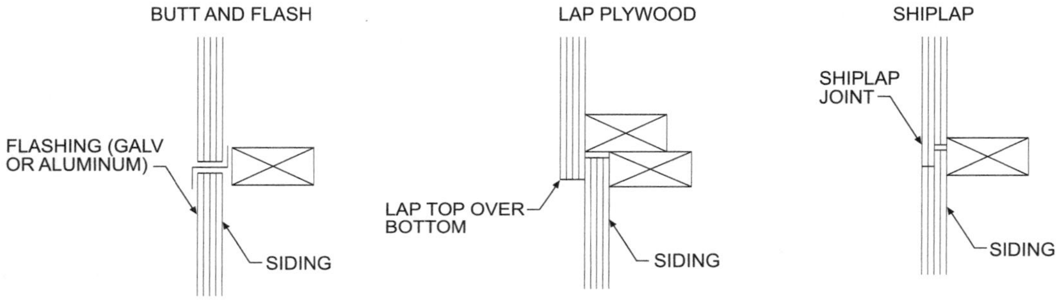

Figure R703.3.1(2)
HORIZONTAL JOINT TREATMENT OF PANEL SIDING

R703.5 Wood shakes and shingles. Wood shakes and shingles shall conform to CSSB *Grading Rules for Wood Shakes and Shingles.*

❖ Grading rules established by the Cedar Shake and Shingle Bureau are applicable to wood shakes and shingles used as exterior wall covering materials. The provisions of this section deal with the specific application and attachment of wood shakes and shingles to the exterior of a building.

R703.5.1 Application. Wood shakes or shingles shall be applied either single-course or double-course over nominal $^1/_2$-inch (13 mm) wood-based sheathing or to furring strips over $^1/_2$-inch (13 mm) nominal nonwood sheathing. A permeable water-resistive barrier shall be provided over all sheathing, with horizontal overlaps in the membrane of not less than 2 inches (51 mm) and vertical overlaps of not less than 6 inches (152 mm). Where furring strips are used, they shall be 1 inch by 3 inches or 1 inch by 4 inches (25 mm by 76 mm or 25 mm by 102 mm) and shall be fastened horizontally to the studs with 7d or 8d box nails and shall be spaced a distance on center equal to the actual weather exposure of the shakes or shingles, not to exceed the maximum exposure specified in Table R703.5.2. The spacing between adjacent shingles to allow for expansion shall not exceed $^1/_4$ inch (6 mm), and between adjacent shakes, it shall not exceed $^1/_2$ inch (13 mm). The offset spacing between joints in adjacent courses shall be a minimum of $1^1/_2$ inches (38 mm).

❖ Where wood shakes and shingles are applied as an exterior wall covering, the sheathing must be covered with a water-resistive permeable membrane prior to shingle or shake installation. The application of wood shakes and shingles may be over either solid sheathing or furring strips applied directly to the studs. The minimum sheathing thickness and furring strip size and fastening are prescribed by the code. To allow for expansion, a small space should be established between each shake or shingle. The spacing is limited to $^1/_4$ inch (6.4 mm) between adjacent shingles and $^1/_2$ inch (12.7 mm) between adjacent shakes. To avoid channels between the shakes or shingles where water penetration might occur, joints must be offset at least $1^1/_2$ inches (38 mm) in adjacent courses.

TABLE R703.4
WEATHER-RESISTANT SIDING ATTACHMENT AND MINIMUM THICKNESS

SIDING MATERIAL		NOMINAL THICKNESS[a] (inches)	JOINT TREATMENT	WATER-RESISTIVE BARRIER REQUIRED	TYPE OF SUPPORTS FOR THE SIDING MATERIAL AND FASTENERS[b, c, d]					
					Wood or wood structural panel sheathing into stud	Fiberboard sheathing into stud	Gypsum sheathing into stud	Foam plastic sheathing into stud	Direct to studs	Number or spacing of fasteners
Horizontal aluminum[e]	Without insulation	0.019[f] 0.024	Lap	Yes	0.120 nail $1^{1}/_{2}''$ long	0.120 nail 2″ long	0.120 nail 2″ long	0.120 nail[y]	Not allowed	Same as stud spacing
			Lap	Yes	0.120 nail $1^{1}/_{2}''$ long	0.120 nail 2″ long	0.120 nail 2″ long	0.120 nail[y]	Not allowed	
	With insulation	0.019	Lap	Yes	0.120 nail $1^{1}/_{2}''$ long	0.120 nail $2^{1}/_{2}''$ long	0.120 nail $2^{1}/_{2}''$ long	0.120 nail[y]	0.120 nail $1^{1}/_{2}''$ long	
Anchored veneer: brick, concrete, masonry or stone		2	Section R703	Yes	See Section R703 and Figure R703.7[g]					
Adhered veneer: concrete, stone or masonry[w]		—	Section R703	Yes Note w	See Section R703.6.1[g] or in accordance with the manufacturer's instructions.					
Hardboard[k] Panel siding-vertical		$^{7}/_{16}$	—	Yes	Note m	Note m	Note m	Note m	Note m	6″ panel edges 12″ inter. sup.[n]
Hardboard[k] Lap-siding-horizontal		$^{7}/_{16}$	Note p	Yes	Note o	Note o	Note o	Note o	Note o	Same as stud spacing 2 per bearing
Steel[h]		29 ga.	Lap	Yes	0.113 nail $1^{3}/_{4}''$ Staple-$1^{3}/_{4}''$	0.113 nail $2^{3}/_{4}''$ Staple-$2^{1}/_{2}''$	0.113 nail $2^{1}/_{2}''$ Staple-$2^{1}/_{4}''$	0.113 nail[v] Staple[v]	Not allowed	Same as stud spacing
Particleboard panels		$^{3}/_{8}$ - $^{1}/_{2}$	—	Yes	6d box nail (2″ × 0.099″)	6d box nail (2″ × 0.099″)	6d box nail (2″ × 0.099″)	box nail[v]	6d box nail (2″ × 0.099″), $^{3}/_{8}$ not allowed	6″ panel edge, 12" inter. sup.
		$^{5}/_{8}$	—	Yes	6d box nail (2″ × 0.099″)	8d box nail ($2^{1}/_{2}''$ × 0.113″)	8d box nail ($2^{1}/_{2}''$ × 0.113″)	box nail[v]	6d box nail (2″ × 0.099″)	
Wood structural panel[i] ANSI/APA-PRP 210 siding[i] (exterior grade)		$^{3}/_{8}$ - $^{1}/_{2}$	Note p	Yes	0.099 nail-2″	0.113 nail-$2^{1}/_{2}''$	0.113 nail-2 $^{1}/_{2}''$	0.113 nail[v]	0.099 nail-2″	6″ panel edges, 12″ inter. sup.
Wood structural panel lapsiding		$^{3}/_{8}$ - $^{1}/_{2}$	Note p Note x	Yes	0.099 nail-2″	0.113 nail-$2^{1}/_{2}''$	0.113 nail-2 $^{1}/_{2}''$	0.113 nail[x]	0.099 nail-2″	8″ along bottom edge
Vinyl siding[l]		0.035	Lap	Yes	0.120 nail (shank) with a 0.313 head or 16-gage staple with $^{3}/_{8}$ to $^{1}/_{2}$-inch crown[y, z]	0.120 nail (shank) with a 0.313 head or 16-gage staple with $^{3}/_{8}$ to $^{1}/_{2}$-inch crown[y]	0.120 nail (shank) with a 0.313 head or 16-gage staple with $^{3}/_{8}$ to $^{1}/_{2}$-inch crown[y]	0.120 nail (shank) with a 0.313 head per Section R703.11.2	Not allowed	16 inches on center or specified by the manufacturer instructions or test report
Wood[j] rustic, drop		$^{3}/_{8}$ Min	Lap	Yes	Fastener penetration into stud-1″				0.113 nail-$2^{1}/_{2}''$ Staple-2″	Face nailing up to 6″ widths, 1 nail per bearing; 8″ widths and over, 2 nails per bearing

(continued)

TABLE R703.4—continued
WEATHER-RESISTANT SIDING ATTACHMENT AND MINIMUM THICKNESS

SIDING MATERIAL	NOMINAL THICKNESS[a] (inches)	JOINT TREATMENT	WATER-RESISTIVE BARRIER REQUIRED	TYPE OF SUPPORTS FOR THE SIDING MATERIAL AND FASTENERS[b, c, d]					
				Wood or wood structural panel sheathing into stud	Fiberboard sheathing into stud	Gypsum sheathing into stud	Foam plastic sheathing into stud	Direct to studs	Number or spacing of fasteners
Shiplap	$^{19}/_{32}$ Average	Lap	Yes	Fastener penetration into stud-1″				0.113 nail-2$^{1}/_{2}$″ Staple-2″	Face nailing up to 6″ widths, 1 nail per bearing; 8″ widths and over, 2 nails per bearing
Bevel	$^{7}/_{16}$								
Butt tip	$^{3}/_{16}$	Lap	Yes						
Fiber cement panel siding[q]	$^{5}/_{16}$	Note q	Yes Note u	6d common corrosion-resistant nail[r]	6d common corrosion-resistant nail[r]	6d common corrosion-resistant nail[r]	6d common corrosion-resistant nail[r, v]	4d common corrosion-resistant nail[r]	6″ o.c. on edges, 12″ o.c. on intermed. studs
Fiber cement lap siding[s]	$^{5}/_{16}$	Note s	Yes Note u	6d common corrosion-resistant nail[r]	6d common corrosion-resistant nail[r]	6d common corrosion-resistant nail[r]	6d common corrosion-resistant nail[r, v]	6d common corrosion-resistant nail or 11-gage roofing nail[r]	Note t

For SI: 1 inch = 25.4 mm.

a. Based on stud spacing of 16 inches on center where studs are spaced 24 inches, siding shall be applied to sheathing approved for that spacing.
b. Nail is a general description and shall be T-head, modified round head, or round head with smooth or deformed shanks.
c. Staples shall have a minimum crown width of $^{7}/_{16}$-inch outside diameter and be manufactured of minimum 16-gage wire.
d. Nails or staples shall be aluminum, galvanized, or rust-preventative coated and shall be driven into the studs where fiberboard, gypsum, or foam plastic sheathing backing is used. Where wood or wood structural panel sheathing is used, fasteners shall be driven into studs unless otherwise permitted to be driven into sheathing in accordance with the siding manufacturer's installation instructions.
e. Aluminum nails shall be used to attach aluminum siding.
f. Aluminum (0.019 inch) shall be unbacked only when the maximum panel width is 10 inches and the maximum flat area is 8 inches. The tolerance for aluminum siding shall be +0.002 inch of the nominal dimension.
g. All attachments shall be coated with a corrosion-resistant coating.
h. Shall be of approved type.
i. Three-eighths-inch plywood shall not be applied directly to studs spaced more than 16 inches on center when long dimension is parallel to studs. Plywood $^{1}/_{2}$-inch or thinner shall not be applied directly to studs spaced more than 24 inches on center. The stud spacing shall not exceed the panel span rating provided by the manufacturer unless the panels are installed with the face grain perpendicular to the studs or over sheathing approved for that stud spacing.
j. Wood board sidings applied vertically shall be nailed to horizontal nailing strips or blocking set 24 inches on center. Nails shall penetrate 1$^{1}/_{2}$ inches into studs, studs and wood sheathing combined or blocking.
k. Hardboard siding shall comply with CPA/ANSI A135.6.
l. Vinyl siding shall comply with ASTM D 3679.
m. Minimum shank diameter of 0.092 inch, minimum head diameter of 0.225 inch, and nail length must accommodate sheathing and penetrate framing 1$^{1}/_{2}$ inches.
n. When used to resist shear forces, the spacing must be 4 inches at panel edges and 8 inches on interior supports.
o. Minimum shank diameter of 0.099 inch, minimum head diameter of 0.240 inch, and nail length must accommodate sheathing and penetrate framing 1$^{1}/_{2}$ inches.
p. Vertical end joints shall occur at studs and shall be covered with a joint cover or shall be caulked.
q. See Section R703.10.1.
r. Fasteners shall comply with the nominal dimensions in ASTM F 1667.
s. See Section R703.10.2.
t. Face nailing: one 6d common nail through the over lap ping planks at each stud. Concealed nailing: one 11 gage 1$^{1}/_{2}$ inch long galv. roofing nail through the top edge of each plank at each stud.
u. See Section R703.2 exceptions.
v. Minimum nail length must accommodate sheathing and penetrate framing 1$^{1}/_{2}$ inches.
w. Adhered masonry veneer shall comply with the requirements of Section R703.6.3 and shall comply with the requirements in Sections 6.1 and 6.3 of TMS-402 ACI 530/ASCE 5.
x. Vertical joints, if staggered shall be permitted to be away from studs if applied over wood structural panel sheathing.
y. Minimum fastener length must accommodate sheathing and penetrate framing 0.75 inches or in accordance with the manufacturer's installation instructions.
z. Where approved by the manufacturer's instructions or test report siding shall be permitted to be installed with fasteners penetrating not less than 0.75 inches through wood or wood structural sheathing with or without penetration into the framing.

❖ This table should be used in addition to all other applicable requirements of the code for the specific material under consideration. Testing and experience have determined that the minimum thicknesses tabulated will be durable and protect the building against the elements for relatively long periods when the siding is attached and maintained as indicated.

R703.5.2 Weather exposure. The maximum weather exposure for shakes and shingles shall not exceed that specified in Table R703.5.2.

❖ Table R703.5.2 specifies the maximum weather exposure permitted for wood shakes and shingles installed on exterior walls. Where a single course of wood shingles is used, the maximum exposure varies from $7^1/_2$ inches to $11^1/_2$ inches (190.5 mm to 292 mm), depending on the length of the shingles. Maximum wood shake exposures are either $8^1/_2$ inches or $11^1/_2$ inches (216 mm to 292 mm) for a single course application. In double course applications, the maximum exposures are increased for both shingles and shakes.

TABLE R703.5.2
MAXIMUM WEATHER EXPOSURE FOR WOOD SHAKES AND SHINGLES ON EXTERIOR WALLS[a, b, c]
(Dimensions are in inches)

LENGTH	EXPOSURE FOR SINGLE COURSE	EXPOSURE FOR DOUBLE COURSE
Shingles[a]		
16	$7^1/_2$	12[b]
18	$8^1/_2$	14[c]
24	$11^1/_2$	16
Shakes[a]		
18	$8^1/_2$	14
24	$11^1/_2$	18

For SI: 1 inch = 25.4 mm.
a. Dimensions given are for No. 1 grade.
b. A maximum 10-inch exposure is permitted for No. 2 grade.
c. A maximum 11-inch exposure is permitted for No. 2 grade.

❖ Depending on the length of the shingle and the number of courses, the table specifies the maximum length of exposure for weathering purposes. Shingles are available in three different lengths, while shakes are regulated for two lengths. Both a single-course application and double-course application are addressed.

R703.5.3 Attachment. Each shake or shingle shall be held in place by two hot-dipped zinc-coated, stainless steel, or aluminum nails or staples. The fasteners shall be long enough to penetrate the sheathing or furring strips by a minimum of $^1/_2$ inch (13 mm) and shall not be overdriven.

❖ Hot-dipped zinc-coated, stainless steel, or aluminum fasteners are to be used to attach wood shakes and shingles. Either nails or staples are permitted, with two fasteners used for each shake or shingle. The length of the nails or staples must be adequate to penetrate sheathing or furring strips a minimum of $^1/_2$ inch (12.7 mm). As the nails or staples are driven, it is important that they do not excessively penetrate the shake or shingle, which could result in reduced holding force.

R703.5.3.1 Staple attachment. Staples shall not be less than 16 gage and shall have a crown width of not less than $^7/_{16}$ inch (11 mm), and the crown of the staples shall be parallel with the butt of the shake or shingle. In single-course application, the fasteners shall be concealed by the course above and shall be driven approximately 1 inch (25 mm) above the butt line of the succeeding course and $^3/_4$ inch (19 mm) from the edge. In double-course applications, the exposed shake or shingle shall be face-nailed with two casing nails, driven approximately 2 inches (51 mm) above the butt line and $^3/_4$ inch (19 mm) from each edge. In all applications, staples shall be concealed by the course above. With shingles wider than 8 inches (203 mm) two additional nails shall be required and shall be nailed approximately 1 inch (25 mm) apart near the center of the shingle.

❖ This section describes in detail the proper attachment of either staples or nails where used as fasteners for wood shakes and shingles applied to exterior walls. Both single-course and double-course application methods are described. The general requirement of Section R703.5.3 for two fasteners per shingle is modified by this section, as two additional fasteners are needed for shingles having a width of more than 8 inches (203 mm).

R703.5.4 Bottom courses. The bottom courses shall be doubled.

❖ In all cases, the bottom course of shakes and shingles in an exterior wall application is to be doubled.

R703.6 Exterior plaster. Installation of these materials shall be in compliance with ASTM C 926 and ASTM C 1063 and the provisions of this code.

❖ Portland cement plaster is the only material approved by the code for exterior plaster. Gypsum plaster deteriorates under conditions of weather and moisture, which are prevalent on the exterior surfaces of buildings. Commentary Figures R703.6(1) and R703.6(2) illustrate exterior plastering systems. ASTM C 926 is the standard specification for the installation of portland-cement-based plaster. ASTM C 926 provides the minimum requirements for the application of portland cement-based plaster for exterior stucco) along with the tables necessary for proportioning various plaster mixes and thicknesses. ASTM C 1063 is the standard specification for exterior lathing and furring for portland-cement-based plastering, as specified in ASTM C 926. Installation of exterior plaster must also comply with the provisions of this code.

R703.6.1 Lath. All lath and lath attachments shall be of corrosion-resistant materials. Expanded metal or woven wire lath shall be attached with $1^1/_2$-inch-long (38 mm), 11 gage nails having a $^7/_{16}$-inch (11.1 mm) head, or $^7/_8$-inch-long (22.2 mm), 16 gage staples, spaced at no more than 6 inches (152 mm), or as otherwise *approved.*

❖ Lath and its attachments must be of corrosion-resistant materials. Unless another fastening method is approved, lath must be attached to framing members with the nails or staples specified in this section and spaced at a maximum of 6 inches (152 mm) on center along the framing member.

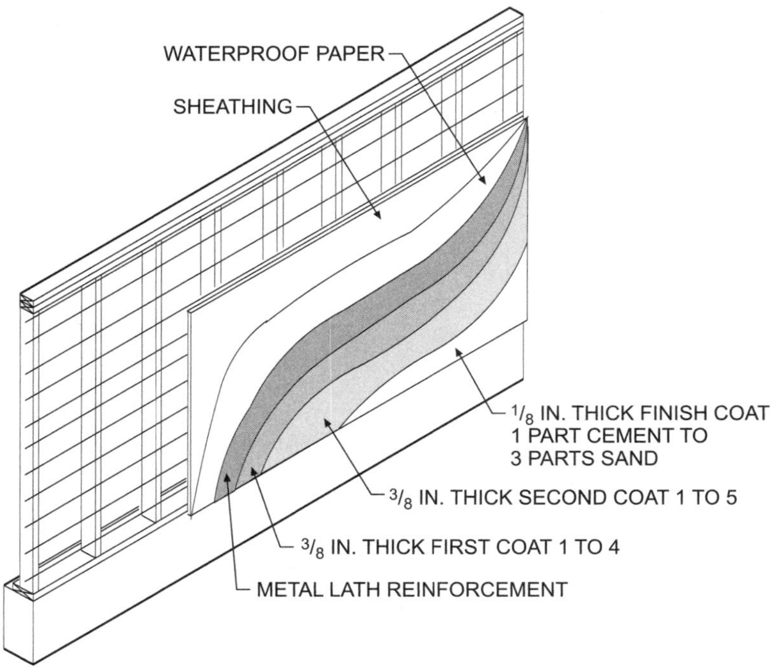

For SI: 1 inch = 25.4 mm.

Figure R703.6(1)
EXTERIOR PLASTERING SYSTEM

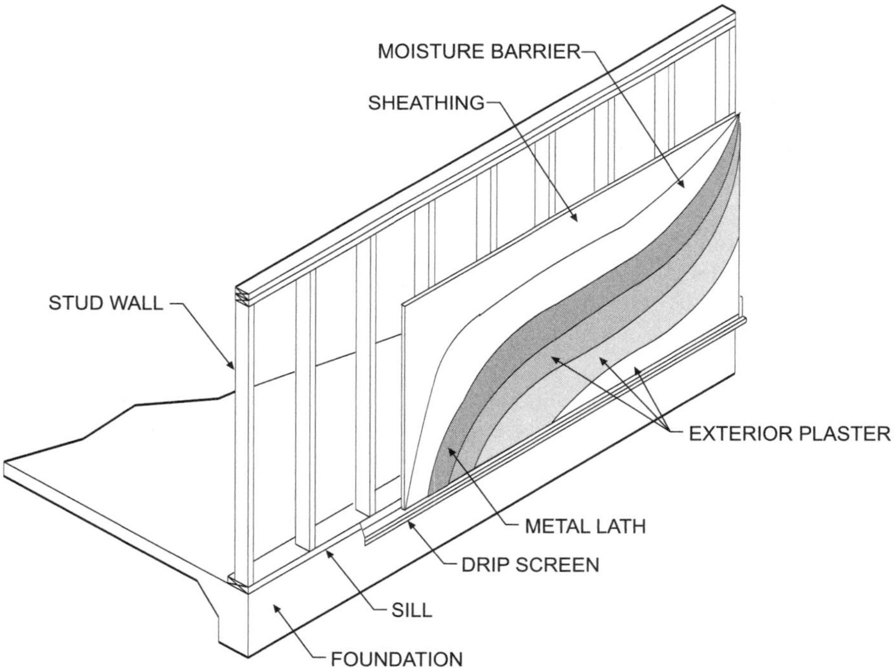

Figure R703.6(2)
EXTERIOR PLASTER-WOOD FRAME ON CONCRETE FLOOR SLAB

R703.6.2 Plaster. Plastering with portland cement plaster shall be not less than three coats when applied over metal lath or wire lath and shall be not less than two coats when applied over masonry, concrete, pressure-preservative treated wood or decay-resistant wood as specified in Section R317.1 or gypsum backing. If the plaster surface is completely covered by veneer or other facing material or is completely concealed, plaster application need be only two coats, provided the total thickness is as set forth in Table R702.1(1).

On wood-frame construction with an on-grade floor slab system, exterior plaster shall be applied to cover, but not extend below, lath, paper and screed.

The proportion of aggregate to cementitious materials shall be as set forth in Table R702.1(3).

❖ The code requires that exterior portland cement plaster be applied in not less than three coats when applied over metal or wire-fabric lath, for the same reasons as discussed for interior plaster. Where the portland cement plaster is applied over other approved plaster bases, the code requires only two-coat work. The code permits plaster work that is completely concealed to be of only two coats provided the total thickness is that required by Table R702.1(1), because the finish coat of plaster is to provide a surface for exterior finishes and to provide an aesthetic appearance. Thus, where the plaster surface is to be completely concealed, it is not necessary to apply a finish coat.

The code requires that the exterior plaster be installed to completely cover, but not extend below, the lath and paper-on-wood exterior wall construction supported by an on-grade concrete slab. This requirement, combined with the presence of a weep screed, prevents the entrapment of free moisture and the subsequent channeling of the moisture to the interior of the building.

R703.6.2.1 Weep screeds. A minimum 0.019-inch (0.5 mm) (No. 26 galvanized sheet gage), corrosion-resistant weep screed or plastic weep screed, with a minimum vertical attachment flange of $3^1/_2$ inches (89 mm) shall be provided at or below the foundation plate line on exterior stud walls in accordance with ASTM C 926. The weep screed shall be placed a minimum of 4 inches (102 mm) above the earth or 2 inches (51 mm) above paved areas and shall be of a type that will allow trapped water to drain to the exterior of the building. The weather-resistant barrier shall lap the attachment flange. The exterior lath shall cover and terminate on the attachment flange of the weep screed.

❖ Water and moisture can penetrate an exterior plaster wall for a variety of reasons and in a number of ways. Some moisture will penetrate the plaster in an exterior wall; therefore, the design of the wall should include a weep screed, which will provide a way to release the moisture [see Commentary Figure R703.6(3)]. Once water or moisture penetrates the plaster, it will migrate down the exterior face of the weather-resistive barrier until it reaches the sill plate or mud sill. At this point, the water will seek a way out of the wall. If the exterior plaster system is not detailed and constructed with provisions to allow the moisture to escape to the exterior, it will find its own way out. This exit will almost certainly be through the interior of the wall and cause leaking, and therefore damage, to the interior of the building. For this reason, the code requires a continuous weep screed at the bottom of exterior walls to permit the moisture to escape to the exterior of the building.

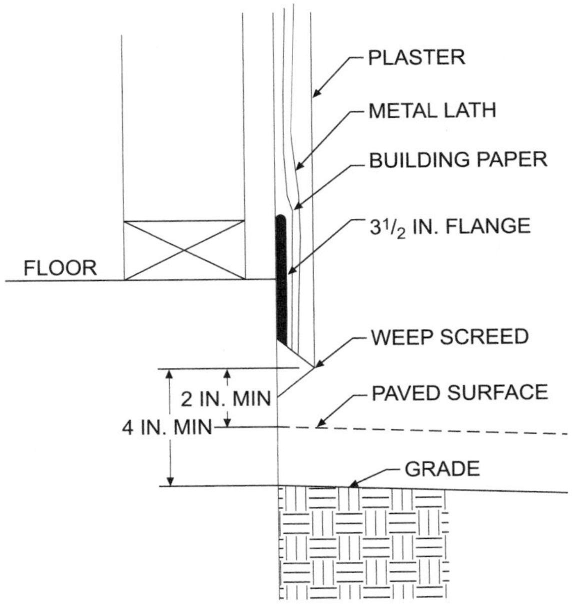

Figure R703.6(3)
WEEP SCREED FOR EXTERIOR PLASTER ON STUD

R703.6.3 Water-resistive barriers. Water-resistive barriers shall be installed as required in Section R703.2 and, where applied over wood-based sheathing, shall include a water-resistive vapor-permeable barrier with a performance at least equivalent to two layers of Grade D paper. The individual layers shall be installed independently such that each layer provides a separate continuous plane and any flashing (installed in accordance with Section R703.8) intended to drain to the water-resistive barrier is directed between the layers.

Exception: Where the water-resistive barrier that is applied over wood-based sheathing has a water resistance equal to or greater than that of 60-minute Grade D paper and is separated from the stucco by an intervening, substantially non-water-absorbing layer or designed drainage space.

❖ The code requires that a water-resistive barrier be installed behind exterior plaster for the reasons provided in Section R703.2. The code also requires that when the barrier is applied over wood-based sheathing such as plywood, the barrier is to be two layers of Grade D building paper.

This requirement is based on the observed problems that occur when one layer is applied over wood sheathing. The wood sheathing eventually exhibits dry rot due to the penetration of moisture. Cracking is then created in the plaster due to movement of the sheathing caused by alternate expansion and contraction. Field experience has shown that where two layers of building paper are used, the penetration of moisture is considerably decreased, as is the cracking of the plaster due to movement of the sheathing caused by the wet and dry cycles.

The code also requires each layer of the water-resistive barrier to be independently installed in a manner that provides a continuous drainage plane. The primary function of the inboard layer is to keep water away from the sheathing and also from penetrating into the stud cavity. This layer should be integrated with window and door flashings, the weep screed at the bottom of the wall as well as any through-wall flashing or expansion joints. The primary function of the outboard layer that comes in contact with the stucco is to separate the stucco from the inboard layer of the water-resistive barrier. This layer is at times referred to as a "sacrificial layer," an "intervening layer" or a "bond break layer."

This provision includes a reference to "Grade D paper" without defining the term or providing a reference to a standard. The reference to Grade D paper actually refers to Federal Specification UU-B-790a, *Federal Specifications for Building Paper, Vegetable Fiber—Kraft, Waterproofed, Water Repellent, and Fire Resistant*. This reference can be traced back to the *Uniform Building Code*® (UBC®) and UBC Standard 14-1, which was based on Federal Specification UU-B-790a. In Section 3.6.4 of that specification, the physical properties for Grade D paper are established. The reference to Kraft waterproof building paper is used because it has the appropriate water vapor permeability to prevent entrapment of moisture between the paper and the sheathing.

The exception permits the application of alternative standard practices by recognizing stucco systems in which one of the two layers of water-resistive barrier is replaced by a layer that although not a "Grade D paper" provides separation from the wet stucco application and provides a barrier to moisture from the stucco to the water-resistive barrier.

R703.6.4 Application. Each coat shall be kept in a moist condition for at least 48 hours prior to application of the next coat.

Exception: Applications installed in accordance with ASTM C 926.

❖ Plaster should be kept moist prior to the application of the next coat. This is usually accomplished by fogging the surface with water at the start and again at the end of the workday. If it is excessively hot, dry and windy, more frequent moistening and covering with polyethylene plastic may be required to keep the plaster in a moist condition.

R703.6.5 Curing. The finish coat for two-coat cement plaster shall not be applied sooner than seven days after application of the first coat. For three-coat cement plaster, the second coat shall not be applied sooner than 48 hours after application of the first coat. The finish coat for three-coat cement plaster shall not be applied sooner than seven days after application of the second coat.

❖ Plaster must be allowed to cure sufficiently between coats. This promotes more intimate contact between the coats and allows a more complete curing of the coats already applied beneath. The finish plaster coat should only be applied after allowing the previous coat to cure seven days.

R703.7 Stone and masonry veneer, general. Stone and masonry veneer shall be installed in accordance with this chapter, Table R703.4 and Figure R703.7. These veneers installed over a backing of wood or cold-formed steel shall be limited to the first *story* above-grade plane and shall not exceed 5 inches (127 mm) in thickness. See Section R602.10 for wall bracing requirements for masonry veneer for wood-framed construction and Section R603.9.5 for wall bracing requirements for masonry veneer for cold-formed steel construction.

Exceptions:

1. For all buildings in Seismic Design Categories A, B and C, exterior stone or masonry veneer, as specified in Table R703.7(1), with a backing of wood or steel framing shall be permitted to the height specified in Table R703.7(1) above a noncombustible foundation.
2. For detached one- or two-family *dwellings* in Seismic Design Categories D_0, D_1 and D_2, exterior stone or masonry veneer, as specified in Table R703.7(2), with a backing of wood framing shall be permitted to the height specified in Table R703.7(2) above a noncombustible foundation.

❖ For the purposes of this section, stone and masonry veneer are nonstructural facing materials of natural or manufactured stone or clay or concrete brick installed as an anchored veneer which provide ornamentation, protection or insulation. To be considered a veneer, the material cannot act structurally with the backing insofar as the structural strength of the assembly is concerned. Stone and masonry veneer can be used as either an interior or exterior wall finish.

In addition to the provisions of this section, Table R703.4 and Figure R703.7 provide details for the installation of stone or masonry veneer. Because of their weight, stone and masonry veneer can impose lateral loads upon the structure from seismic events that are more than the prescriptive loading permitted by Section R301.2.2.2.1 for wood or cold-formed steel construction. For this reason, stone or masonry veneer installed with a backing of wood or cold-formed steel is limited to the first story above grade and a maximum thickness of 5 inches (127 mm) unless it meets Exception 1 or 2.

Exception 1 allows stone or masonry veneer to be constructed to either two or three stories above grade on structures located in Seismic Design Category A, B, or C when installed with wood or cold-formed steel backing and with increased wall bracing (see Table R703.7.1 and Sections R602.12 and R603.9.5).

Exception 1 applies to townhouses or detached one- or two-family dwellings with a backing of wood or cold-formed steel frame which include a stone or masonry veneer.

Exception 2 applies only to detached one- and two-family dwellings with a backing of wood framing which include a stone or masonry veneer. Exception 2 allows stone or masonry veneer with wood backing, up to three stories above grade on structures located in Seismic Design Category D_0 or D_1 and two stories above grade for structures located in Seismic Design Category D_2 under certain conditions that include increasing the amount of wall bracing and providing hold-down connectors [see Table R703.7(2) and Sections R602.12 and R603.9.5].

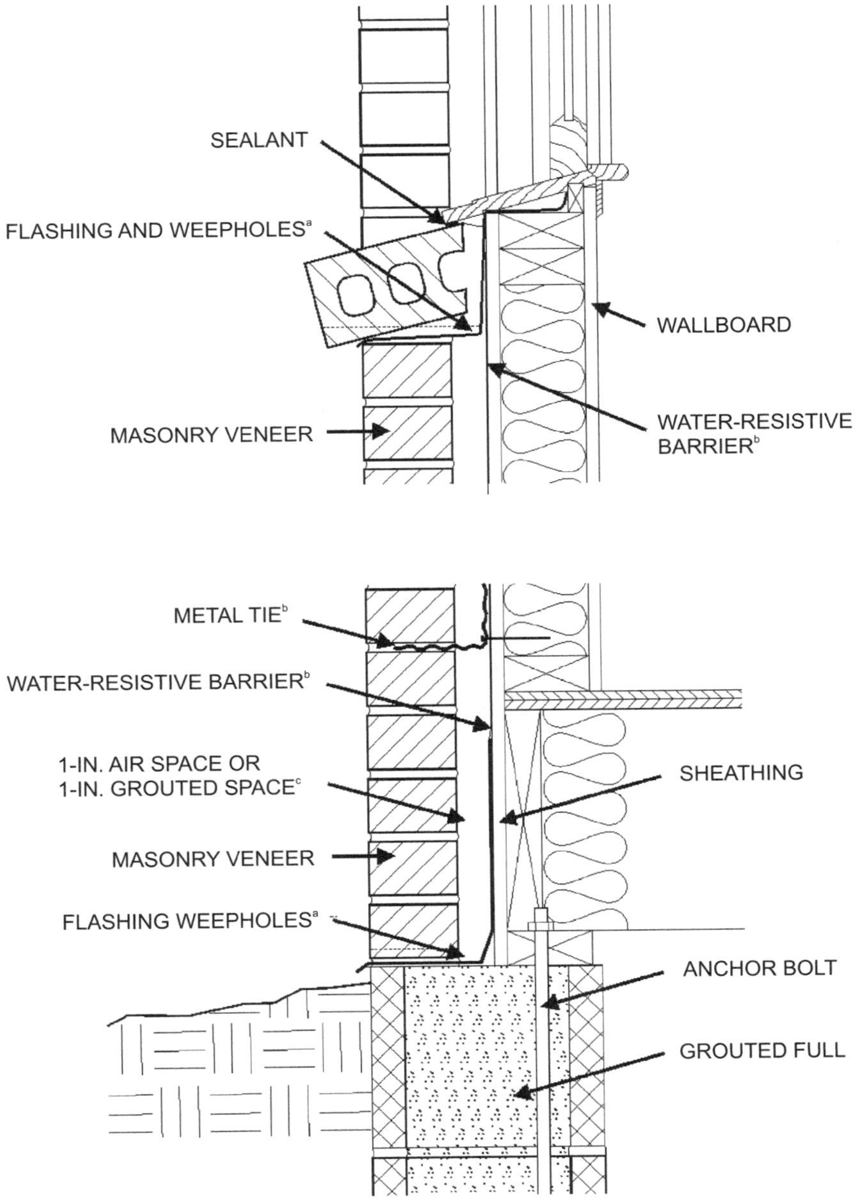

For SI: 1 inch = 24.5 mm.

FIGURE R703.7
MASONRY VENEER WALL DETAILS

(continued)

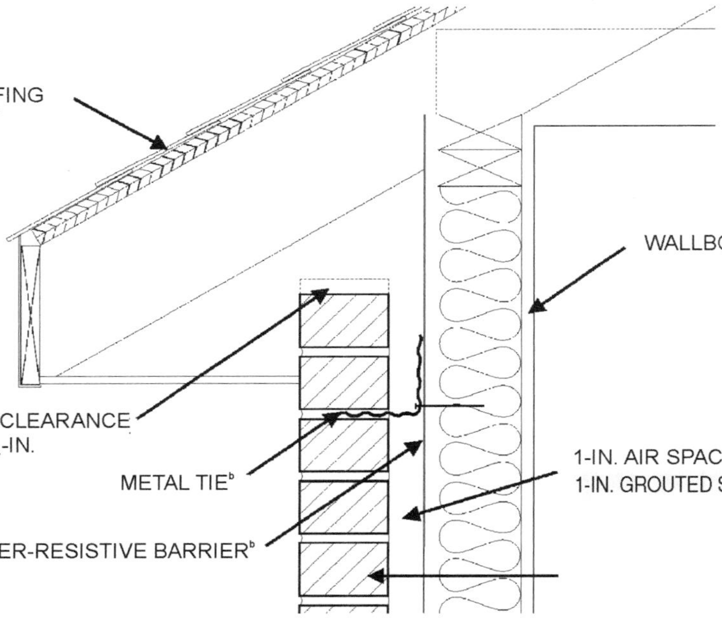

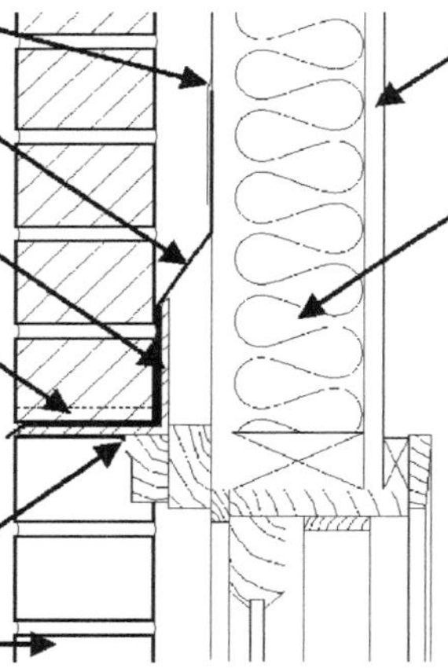

For SI: 1 inch = 25.4 mm.

a. See Sections R703.7.5, R703.7.6 and R703.8.

b. See Sections R703.2 and R703.7.4.

c. See Section R703.7.4.2 and Table R703.7.4.

d. See Section R703.7.3.

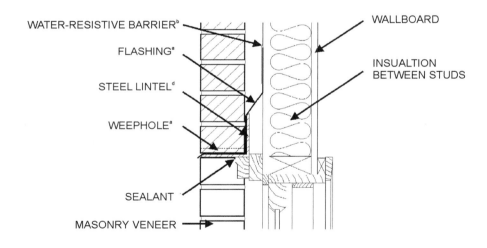

FIGURE R703.7—continued
MASONRY VENEER WALL DETAILS

❖ This figure outlines a number of methods for attaching anchored masonry veneer to framed walls. The fundamental application and attachment methods are listed, including the use of a water-resistive barrier or water-repellant sheathing, the provision for weep holes and the need for an air space unless mortared. References to the appropriate code requirements are listed in the notes.

TABLE R703.7(1)
STONE OR MASONRY VENEER LIMITATIONS AND REQUIREMENTS, WOOD OR STEEL FRAMING, SEISMIC DESIGN CATEGORIES A, B AND C

SEISMIC DESIGN CATEGORY	NUMBER OF WOOD OR STEEL-FRAMED STORIES	MAXIMUM HEIGHT OF VENEER ABOVE NONCOMBUSTIBLE FOUNDATION[a] (feet)	MAXIMUM NOMINAL THICKNESS OF VENEER (inches)	MAXIMUM WEIGHT OF VENEER (psf)[b]	WOOD OR STEEL-FRAMED STORY
A or B	Steel: 1 or 2 Wood: 1, 2 or 3	30	5	50	all
C	1	30	5	50	1 only
	2	30	5	50	top
					bottom
	Wood only: 3	30	5	50	top
					middle
					bottom

For SI: 1 inch = 25.4 mm, 1 foot = 304.8 mm, 1 pound per square foot = 0.479 kPa.

a. An additional 8 feet is permitted for gable end walls. See also story height limitations of Section R301.3.

b. Maximum weight is installed weight and includes weight of mortar, grout, lath and other materials used for installation. Where veneer is placed on both faces of a wall, the combined weight shall not exceed that specified in this table.

❖ See the commentary to Section R703.7.

TABLE R703.7(2)
STONE OR MASONRY VENEER LIMITATIONS AND REQUIREMENTS, ONE- AND TWO-FAMILY DETACHED DWELLINGS, WOOD FRAMING, SEISMIC DESIGN CATEGORIES D_0, D_1 AND D_2

SEISMIC DESIGN CATEGORY	NUMBER OF WOOD FRAMED STORIES[a]	MAXIMUM HEIGHT OF VENEER ABOVE NONCOMBUSTIBLE FOUNDATION OR FOUNDATION WALL (feet)	MAXIMUM NOMINAL THICKNESS OF VENEER (inches)	MAXIMUM WEIGHT OF VENEER (psf)[b]
D_0	1	20[c]	4	40
	2	20[c]	4	40
	3	30[d]	4	40
D_1	1	20[c]	4	40
	2	20[c]	4	40
	3	20[c]	4	40
D_2	1	20[c]	3	30
	2	20[c]	3	30

For SI: 1 inch = 25.4 mm, 1 foot = 304.8 mm, 1 pound per square foot = 0.479 kPa, 1 pound-force = 4.448 N.

a. Cripple walls are not permitted in Seismic Design Categories D_0, D_1 and D_2.

b. Maximum weight is installed weight and includes weight of mortar, grout and lath, and other materials used for installation.

c. The veneer shall not exceed 20 feet in height above a noncombustible foundation, with an additional 8 feet permitted for gable end walls, or 30 feet in height with an additional 8 feet for gable end walls where the lower 10 feet has a backing of concrete or masonry wall. See also story height limitations of Section R301.3.

d. The veneer shall not exceed 30 feet in height above a noncombustible foundation, with an additional 8 feet permitted for gable end walls. See also story height limitations of Section R301.3.

❖ See the commentary to Section R703.7.

R703.7.1 Interior veneer support. Veneers used as interior wall finishes shall be permitted to be supported on wood or cold-formed steel floors that are designed to support the loads imposed.

❖ Where designed to support the weight of the veneer, a wood-framed or cold-formed steel floor system may be used to support stone or masonry veneer used in an interior application.

R703.7.2 Exterior veneer support. Except in Seismic Design Categories D_0, D_1 and D_2, exterior masonry veneers having an installed weight of 40 pounds per square foot (195 kg/m^2) or less shall be permitted to be supported on wood or cold-formed steel construction. When masonry veneer supported by wood or cold-formed steel construction adjoins masonry veneer supported by the foundation, there shall be a movement joint between the veneer supported by the wood or cold-formed steel construction and the veneer supported by the foundation. The wood or cold-formed steel construction supporting the masonry veneer shall be designed to limit the deflection to $^1/_{600}$ of the span for the supporting members. The design of the wood or cold-formed steel construction shall consider the weight of the veneer and any other loads.

❖ Two criteria must be met to support exterior masonry veneer directly on wood or cold-formed steel construction rather than on a concrete or masonry foundation wall. First, the construction must be located in Seismic Design Category A, B or C. Second, the total installed weight of the veneer must not exceed 40 pounds per square foot (psf) (195 kg/m^2). A movement joint must be provided between the exterior veneer supported by wood or steel and the adjacent exterior veneer supported by the foundation. The wood or cold-formed steel members supporting the masonry veneer are to be designed to have a maximum deflection of $^{1}/_{600}$ of the span of the supporting members.

R703.7.2.1 Support by steel angle. A minimum 6 inches by 4 inches by $^{5}/_{16}$ inch (152 mm by 102 mm by 8 mm) steel angle, with the long leg placed vertically, shall be anchored to double 2 inches by 4 inches (51 mm by 102 mm) wood studs at a maximum on-center spacing of 16 inches (406 mm). Anchorage of the steel angle at every double stud spacing shall be a minimum of two $^{7}/_{16}$ inch (11 mm) diameter by 4 inch (102 mm) lag screws. The steel angle shall have a minimum clearance to underlying construction of $^{1}/_{16}$ inch (2 mm). A minimum of two-thirds the width of the masonry veneer thickness shall bear on the steel angle. Flashing and weep holes shall be located in the masonry veneer wythe in accordance with Figure R703.7.2.1. The maximum height of masonry veneer above the steel angle support shall be 12 feet, 8 inches (3861 mm). The air space separating the masonry veneer from the wood backing shall be in accordance with Sections R703.7.4 and R703.7.4.2. The method of support for the masonry veneer on wood construction shall be constructed in accordance with Figure R703.7.2.1.

The maximum slope of the roof construction without stops shall be 7:12. Roof construction with slopes greater than 7:12 but not more than 12:12 shall have stops of a minimum 3 inch by 3 inch by $^{1}/_{4}$ inch (76 mm by 76 mm by 6 mm) steel plate welded to the angle at 24 inches (610 mm) on center along the angle or as *approved* by the *building official*.

❖ As illustrated in Figure R703.7.2.1, a steel angle may be used to support exterior masonry veneer in wood-frame construction. The minimum size of the angle is specified, as are the details of anchorage and the minimum bearing length. Two limits are placed on this method of veneer support: the maximum veneer height bearing on the steel angle must be 12 feet, 8 inches (3861 mm), and the roof slope is limited to 7:12. A roof slope greater than 7:12 is permitted if stops are welded to the steel angle to prevent the veneer from sliding down the angle.

R703.7.2.2 Support by roof construction. A steel angle shall be placed directly on top of the roof construction. The roof supporting construction for the steel angle shall consist of a minimum of three 2 inch by 6 inch (51 mm by 152 mm) wood members. The wood member abutting the vertical wall stud construction shall be anchored with a minimum of three $^{5}/_{8}$-inch (16 mm) diameter by 5-inch (127 mm) lag screws to every wood stud spacing. Each additional roof member shall be anchored by the use of two 10d nails at every wood stud spacing. A minimum of two-thirds the width of the masonry veneer thickness shall bear on the steel angle. Flashing and weep holes shall be located in the masonry veneer wythe in accordance with Figure R703.7.2.2. The maximum height of

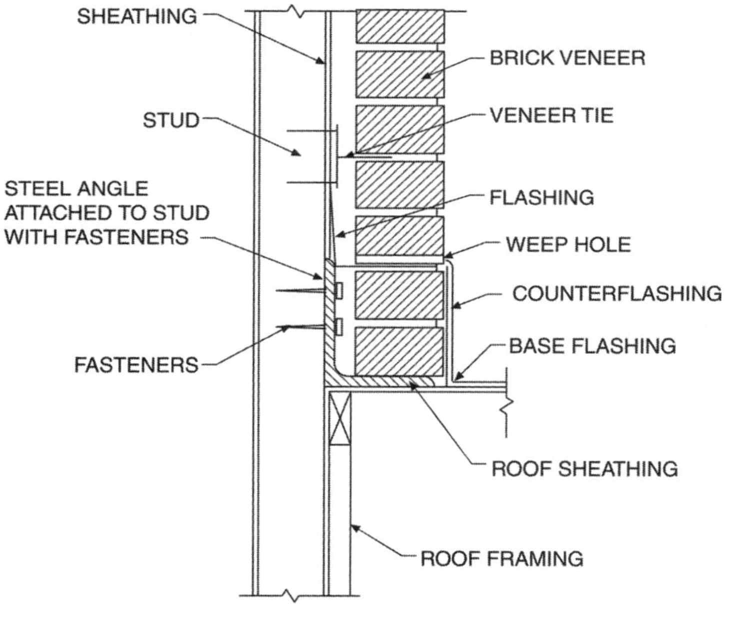

SUPPORT BY STEEL ANGLE

FIGURE R703.7.2.1
EXTERIOR MASONRY VENEER SUPPORT BY STEEL ANGLES

❖ Under most conditions, exterior masonry veneer will be supported by the foundation of the building. Where support of the veneer is by wood construction, the code describes special methods of veneer support. Figures R703.7.2.1 and R703.7.2.2 illustrate support by either steel angle or roof members. The figures show the fundamental elements of attachment, while Sections R703.7.2.1 and R703.7.2.2 contain the details of the installation materials and methods.

the masonry veneer above the steel angle support shall be 12 feet, 8 inches (3861 mm). The air space separating the masonry veneer from the wood backing shall be in accordance with Sections R703.7.4 and R703.7.4.2. The support for the masonry veneer on wood construction shall be constructed in accordance with Figure R703.7.2.2.

The maximum slope of the roof construction without stops shall be 7:12. Roof construction with slopes greater than 7:12 but not more than 12:12 shall have stops of a minimum 3 inch by 3 inch by $^1/_4$ inch (76 mm by 76 mm by 6 mm) steel plate welded to the angle at 24 inches (610 mm) on center along the angle or as *approved* by the *building official*.

❖ In addition to the option of having a steel angle support exterior masonry veneer, the code also permits the wood roof construction itself to support the exterior veneer. As detailed in Figure R703.7.2.2, the masonry veneer must be supported by a steel angle which in turn distributes the veneer load to at least three 2-inch by 6-inch (51 mm by 152 mm) wood members. This section also addresses the attachment details and bearing conditions. As described for masonry veneer supported by steel angles, the maximum veneer height and maximum roof slope are also regulated. The roof slope is limited to 7:12; however, a roof slope greater than 7:12 is permitted if stops are welded to the steel angle to prevent the veneer from sliding down the angle.

R703.7.3 Lintels. Masonry veneer shall not support any vertical load other than the dead load of the veneer above. Veneer above openings shall be supported on lintels of noncombustible materials. The lintels shall have a length of bearing not less than 4 inches (102 mm). Steel lintels shall be shop coated with a rust-inhibitive paint, except for lintels made of corrosion-resistant steel or steel treated with coatings to provide corrosion resistance. Construction of openings shall comply with either Section R703.7.3.1 or 703.7.3.2.

❖ Masonry veneer, like all veneers, is a nonload-bearing wall covering. It is not intended to support any loads, except for the dead load of the masonry veneer above. A noncombustible lintel is required above an opening through the masonry veneer. The lintel must be protected from rust. Lintels must be sized and installed in accordance with Section R703.7.3.1 or R703.7.3.2.

R703.7.3.1 Allowable span. The allowable span shall not exceed the values set forth in Table R703.7.3.1.

❖ A lintel meeting this section must meet Table R703.7.3.1. This table lists the maximum size opening a certain lintel size is allowed to span based on the number of stories above the lintel. Note c stipulates that other steel members can be used as long as they meet the structural design requirements using standard engineering practice.

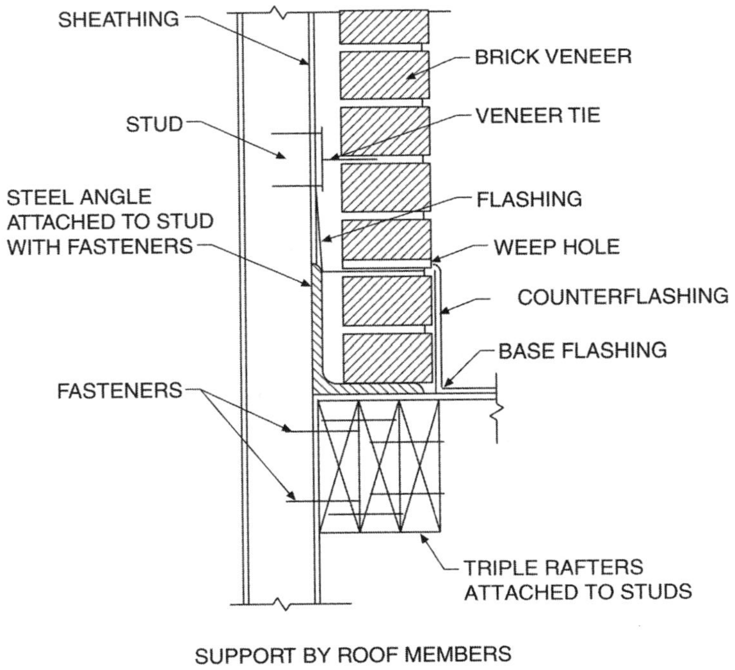

FIGURE R703.7.2.2
EXTERIOR MASONRY VENEER SUPPORT BY ROOF MEMBERS

❖ See the commentary to Figure R703.7.2.1.

TABLE R703.7.3.1
ALLOWABLE SPANS FOR LINTELS SUPPORTING MASONRY VENEER[a, b, c, d]

SIZE OF STEEL ANGLE[a, c, d] (inches)	NO STORY ABOVE	ONE STORY ABOVE	TWO STORIES ABOVE	NO. OF $^1/_2$-INCH OR EQUIVALENT REINFORCING BARS IN REINFORCED LINTEL[b, d]
$3 \times 3 \times ^1/_4$	6′-0″	4′-6″	3′-0″	1
$4 \times 3 \times ^1/_4$	8′-0″	6′-0″	4′-6″	1
$5 \times 3^1/_2 \times ^5/_{16}$	10′-0″	8′-0″	6′-0″	2
$6 \times 3^1/_2 \times ^5/_{16}$	14′-0″	9′-6″	7′-0″	2
$2\text{-}6 \times 3^1/_2 \times ^5/_{16}$	20′-0″	12′-0″	9′-6″	4

For SI: 1 inch = 25.4 mm, 1 foot = 304.8 mm.

a. Long leg of the angle shall be placed in a vertical position.

b. Depth of reinforced lintels shall not be less than 8 inches and all cells of hollow masonry lintels shall be grouted solid. Reinforcing bars shall extend not less than 8 inches into the support.

c. Steel members indicated are adequate typical examples; other steel members meeting structural design requirements may be used.

d. Either steel angle or reinforced lintel shall span opening.

❖ See the commentary to Section R703.7.3.1.

R703.7.3.2 Maximum span. The allowable span shall not exceed 18 feet 3 inches (5562 mm) and shall be constructed to comply with Figure R703.7.3.2 and the following:

1. Provide a minimum length of 18 inches (457 mm) of masonry veneer on each side of opening as shown in Figure R703.7.3.2.
2. Provide a minimum 5-inch by $3^1/_2$-inch by $^5/_{16}$-inch (127 mm by 89 mm by 7.9 mm) steel angle above the opening and shore for a minimum of 7 days after installation.
3. Provide double-wire joint reinforcement extending 12 inches (305 mm) beyond each side of the opening. Lap splices of joint reinforcement a minimum of 12 inches (305 mm). Comply with one of the following:
   3.1. Double-wire joint reinforcement shall be $^3/_{16}$-inch (4.8 mm) diameter and shall be placed in the first two bed joints above the opening.
   3.2. Double-wire joint reinforcement shall be 9 gauge (0.144 inch or 3.66 mm diameter) and shall be placed in the first three bed joints above the opening.
4. Provide the height of masonry veneer above opening, in accordance with Table R703.7.3.2.

❖ A lintel meeting this section can span an opening no longer than 18 feet 3 inches (5563 mm). Such openings are generally installed to accommodate a two-car garage door. A minimum-size steel angle must be installed with the long leg in the vertical position below the masonry veneer and a minimum size and amount of joint reinforcement must be installed in the courses above the veneer. In addition, the masonry veneer opening must comply with the requirements in Figure R703.7.3.2. There are no explicit limitations given on the height of masonry for the prescriptive lintel in Figure R703.7.3.2. The limits for the maximum and minimum height of masonry veneer as shown in Figure R703.7.3.2 must be in accordance with Table R703.7.3.2.

TABLE R703.7.3.2
HEIGHT OF MASONRY VENEER ABOVE OPENING

MINIMUM HEIGHT OF MASONRY VENEER ABOVE OPENING (INCH)	MAXIMUM HEIGHT OF MASONRY VENEER ABOVE OPENING (FEET)
13	< 5
24	5 to < 12
60	12 to height above support allowed by Section R703.7

For SI: 1 inch = 25.4 mm, 1 foot = 304.8 mm.

❖ See the commentary to Section R703.7.3.2.

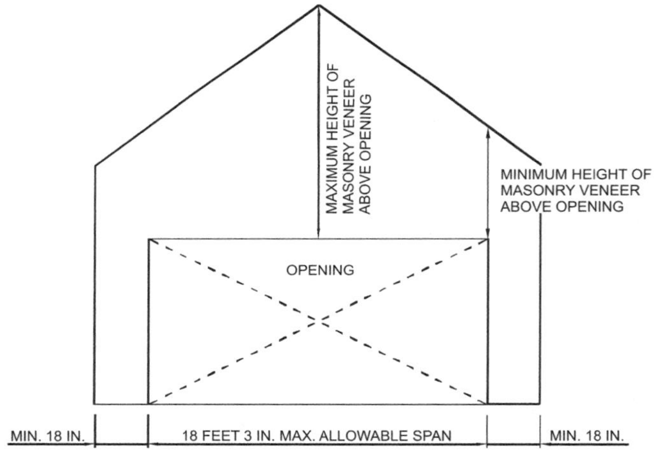

For SI: 1 inch = 25.4 mm, 1 foot = 304.8 mm.

FIGURE R703.7.3.2
MASONRY VENEER OPENING

❖ See the commentary to Section R703.7.3.2.

R703.7.4 Anchorage. Masonry veneer shall be anchored to the supporting wall studs with corrosion-resistant metal ties embedded in mortar or grout and extending into the veneer a minimum of $1^1/_2$ inches (38 mm), with not less than $^5/_8$-inch (15.9 mm) mortar or grout cover to outside face. Masonry veneer shall conform to Table R703.7.4.

❖ To maintain the attachment of the masonry veneer to the wall studs, proper anchorage must be provided. In all cases, corrosion-resistant metal ties must be used and be embedded into the back of the veneer a minimum distance. A minimum extent of mortar or grout is required to cover the tie from the outside face of the wall. As specified in Table R703.7.4, two types of ties are addressed for attachment to wood construction: corrugated sheet metal ties and metal strand wire ties. The distance between the veneer and the sheathing is limited and varies with the type of tie used. For corrugated sheet metal ties, this dimension is given as nominal since the back of the masonry veneer may not be true to a line due to the variation in masonry unit width allowed within most masonry unit standards and the alignment of the wall elements within the backing wall. Only adjustable metal strand wire ties may be used where the backing is of cold-formed steel studs. The distance between the steel backing and the masonry veneer is also limited.

Table R703.7.4 also specifies the fastening requirements of metal ties to the studs. Nails are not permitted for fastening of the ties in Seismic Design Categories D_0, D_1 or D_2. This is based on recent full-scale building shaking-table testing conducted at the University of California, San Diego. The report by Richard E. Klinger, published in the *TMS Journal*, found that "fasteners on one side of the specimen failed by extraction of nails under dynamic tensile loads, at levels of shaking less than the Design Basis Earthquake (DBE)."

R703.7.4.1 Size and spacing. Veneer ties, if strand wire, shall not be less in thickness than No. 9 U.S. gage [(0.148 inch) (4 mm)] wire and shall have a hook embedded in the mortar joint, or if sheet metal, shall be not less than No. 22 U.S. gage by [(0.0299 inch) (0.76 mm)] $^7/_8$ inch (22 mm) corrugated. Each tie shall support not more than 2.67 square feet (0.25 m^2) of wall area and shall be spaced not more than 32 inches (813 mm) on center horizontally and 24 inches (635 mm) on center vertically.

Exception: In Seismic Design Category D_0, D_1 or D_2 or townhouses in Seismic Design Category C or in wind areas of more than 30 pounds per square foot pressure (1.44 kPa), each tie shall support not more than 2 square feet (0.2 m^2) of wall area.

❖ This section sets forth the minimum size of metal veneer ties for use with masonry veneer applications. The maximum amount of wall area that can be supported by a veneer tie is 2.67 square feet (0.248 m^2). The spacing between the ties is also limited to a maximum horizontal distance of 32 inches (813 mm) and vertical distance of 24 inches (610 mm). An increased number of ties are required for buildings constructed in a high seismic design category or where a high wind load exists.

R703.7.4.1.1 Veneer ties around wall openings. Additional metal ties shall be provided around all wall openings greater than 16 inches (406 mm) in either dimension. Metal ties around the perimeter of openings shall be spaced not more than 3 feet (9144 mm) on center and placed within 12 inches (305 mm) of the wall opening.

❖ When an opening with at least one dimension larger than 16 inches (406 mm) is located through a masonry veneer, an increase in the number of metal ties around the opening is required. Openings of this size are often for windows and doors. This section contains requirements for the maximum spacing between ties and the maximum distance from the opening to the ties adjacent to it. Openings less than 16 inches (406 mm) in either dimension are exempt from these provisions.

R703.7.4.2 Grout fill. As an alternative to the air space required by Table R703.7.4, grout shall be permitted to fill the air space. When the air space is filled with grout, a water-resistive barrier is required over studs or sheathing. When filling the air space, replacing the sheathing and water-resistive barrier with a wire mesh and *approved* water-resistive barrier or an *approved* water-resistive barrier-backed reinforcement attached directly to the studs is permitted.

❖ The air space required by Section R703.7.4 is permitted to be filled with grout when a water-resistive barrier is applied over studs and sheathing and the air space is completely filled with the grout. The sheathing and water-resistive barrier are permitted to be replaced with a wire mesh and approved water-resistive barrier or an approved water-resistive barrier-backed reinforcement which is attached directly to the studs.

TABLE R703.7.4
TIE ATTACHMENT AND AIR SPACE REQUIREMENTS

BACKING AND TIE	MINIMUM TIE	MINIMUM TIE FASTENER[a]	AIR SPACE	
Wood stud backing with corrugated sheet metal	22 U.S. gage (0.0299 in.) × $^7/_8$ in. wide	8d common nail [b] ($2^1/_2$ in. × 0.131 in.)	Nominal 1 in. between sheathing and veneer	
Wood stud backing with metal strand wire	W1.7 (No. 9 U.S. gage; 0.148 in.) with hook embedded in mortar joint	8d common nail [b] ($2^1/_2$ in. × 0.131 in.)	Minimum nominal 1 in. between sheathing and veneer	Maximum $4^1/_2$ in. between backing and veneer
Cold-formed steel stud backing with adjustable metal strand wire	W1.7 (No. 9 U.S. gage; 0.148 in.) with hook embedded in mortar joint	No. 10 screw extending through the steel framing a minimum of three exposed threads	Minimum nominal 1 in. between sheathing and veneer	Maximum $4^1/_2$ in. between backing and veneer

For SI: 1 inch = 25.4 mm.

a. In Seismic Design Category D_0, D_1 or D_2, the minimum tie fastener shall be an 8d ring-shank nail ($2^1/_2$ in. × 0.131 in.) or a No. 10 screw extending through the steel framing a minimum of three exposed threads.

b. All fasteners shall have rust-inhibitive coating suitable for the installation in which they are being used, or be manufactured from material not susceptible to corrosion.

❖ See the commentary to Section R703.7.4.

R703.7.5 Flashing. Flashing shall be located beneath the first course of masonry above finished ground level above the foundation wall or slab and at other points of support, including structural floors, shelf angles and lintels when masonry veneers are designed in accordance with Section R703.7. See Section R703.8 for additional requirements.

❖ Flashing is necessary to direct moisture in the air space behind the masonry veneer toward the exterior. Flashing is required to be located at the first course of masonry above the finished ground level, as well as at other points of support such as at shelf angles and lintels. Flashing beneath the first course of masonry above the finished ground level is also necessary to prevent water entry into the wall below. Flashing must be made from an approved corrosion-resistant material and comply with Section R703.8.

R703.7.6 Weepholes. Weepholes shall be provided in the outside wythe of masonry walls at a maximum spacing of 33 inches (838 mm) on center. Weepholes shall not be less than $^3/_{16}$ inch (5 mm) in diameter. Weepholes shall be located immediately above the flashing.

❖ Weepholes must be provided directly above the flashing mandated by Section R703.7.5 to allow for the escape of any moisture that may have penetrated the masonry veneer. Because moisture will adversely affect the integrity of the wall if not removed from the wall assembly, weep holes must be installed within the maximum spacing specified in this section. The minimum diameter of the weep holes is also stipulated.

R703.8 Flashing. *Approved* corrosion-resistant flashing shall be applied shingle-fashion in a manner to prevent entry of water into the wall cavity or penetration of water to the building structural framing components. Self-adhered membranes used as flashing shall comply with AAMA 711. The flashing shall extend to the surface of the exterior wall finish. *Approved* corrosion-resistant flashings shall be installed at all of the following locations:

1. Exterior window and door openings. Flashing at exterior window and door openings shall extend to the surface of the exterior wall finish or to the water-resistive barrier for subsequent drainage. Flashing at exterior window and door openings shall be installed in accordance with one or more of the following:
   - 1.1. The fenestration manufacturer's installation and flashing instructions, or for applications not addressed in the fenestration manufacturer's instructions, in accordance with the flashing manufacturer's instructions. Where flashing instructions or details are not provided, pan flashing shall be installed at the sill of exterior window and door openings. Pan flashing shall be sealed or sloped in such a manner as to direct water to the surface of the exterior wall finish or to the water-resistive barrier for subsequent drainage. Openings using pan flashing shall also incorporate flashing or protection at the head and sides.
   - 1.2. In accordance with the flashing design or method of a registered design professional.
   - 1.3. In accordance with other approved methods.
2. At the intersection of chimneys or other masonry construction with frame or stucco walls, with projecting lips on both sides under stucco copings.
3. Under and at the ends of masonry, wood or metal copings and sills.
4. Continuously above all projecting wood trim.
5. Where exterior porches, decks or stairs attach to a wall or floor assembly of wood-frame construction.
6. At wall and roof intersections.
7. At built-in gutters.

❖ The code requires that all points subject to the entry of moisture be appropriately flashed. Roof and wall intersections and parapets create significant challenges, as do exterior wall openings exposed to the weather. Where wind-driven rain is expected, the concerns are even greater. Self-adhered flashing is required to comply with AAMA 711. Window and door manufacturers are required by Section R612.1 of the IRC to provide installation instructions for each window and door. Most window and door manufacturers require installation per their instructions and many window and door manufacturers are incorporating a pan flashing in their window and door installation instructions. Window and door manufacturers create installation and flashing instructions for a wide variety of wall conditions, but are unable to create installation instructions for every conceivable wall condition. Items 1.2 and 1.3 of the flashing methods allow necessary flexibility while retaining the performance requirements of Section R703.8. Although the code identifies a number of locations where flashing is specifically required, the entire exterior envelope must be weather tight to protect the interior from weather. Therefore, any location on the exterior envelope that provides a route for the admission of water or moisture into the building must be properly protected. Commentary Figure R703.8 illustrates examples of flashing.

R703.9 Exterior insulation and finish system (EIFS)/EIFS with drainage. Exterior Insulation and Finish System (EIFS) shall comply with this chapter and Sections R703.9.1 and R703.9.3. EIFS with drainage shall comply with this chapter and Sections R703.9.2, R703.9.3 and R703.9.4.

❖ This section provides the requirements for Exterior Insulation and Finish Systems (EIFS) and Exterior Insulation and Finish Systems with Drainage (EIFS with Drainage). See Section R202 for the definition for both.

- EIFS must comply with ASTM E 2568. EIFS with Drainage must comply with ASTM E 2568 and ASTM E 2273 in accordance with Section R703.9.2.

- EIFS and EIFS with Drainage must be installed in accordance with the manufacturers' instruction.
- EIFS with Drainage requires a water-resistive barrier in accordance with Sections R703.9.2.1 and R703.9.2.2.
- EIFS and EIFS with Drainage must not be located within 6 inches (152 mm) of the finished ground level. Additionally, decorative trim must not be face nailed through the EIFS or EIFS with Drainage.

R703.9.1 Exterior insulation and finish system (EIFS). EIFS shall comply with ASTM E 2568.

❖ See the commentary to Section R703.9.

R703.9.2 Exterior insulation and finish system (EIFS) with drainage. EIFS with drainage shall comply with ASTM E 2568 and shall have an average minimum drainage efficiency of 90 percent when tested in accordance with ASTM E 2273.

❖ See the commentary to Section R703.9.

R703.9.2.1 Water-resistive barrier. The water-resistive barrier shall comply with Section R703.2 or ASTM E 2570.

❖ See the commentary to Section R703.9.

R703.9.2.2 Installation. The water-resistive barrier shall be applied between the EIFS and the wall sheathing.

❖ See the commentary to Section R703.9.

R703.9.3 Flashing, general. Flashing of EIFS shall be provided in accordance with the requirements of Section R703.8.

❖ See the commentary to Section R703.9.

R703.9.4 EIFS/EIFS with drainage installation. All EIFS shall be installed in accordance with the manufacturer's installation instructions and the requirements of this section.

❖ See the commentary to Section R703.9.

R703.9.4.1 Terminations. The EIFS shall terminate not less than 6 inches (152 mm) above the finished ground level.

❖ See the commentary to Section R703.9.

R703.9.4.2 Decorative trim. Decorative trim shall not be face nailed though the EIFS.

❖ See the commentary to Section R703.9.

R703.10 Fiber cement siding.

❖ Fiber cement siding is manufactured from fiber-reinforced cement and is permitted as a weather-resistant siding in either panel or horizontal lap form.

R703.10.1 Panel siding. Fiber-cement panels shall comply with the requirements of ASTM C 1186, Type A, minimum Grade II. Panels shall be installed with the long dimension either parallel or perpendicular to framing. Vertical and horizontal joints shall occur over framing members and shall be sealed with caulking, covered with battens or shall be designed to comply with Section R703.1. Panel siding shall

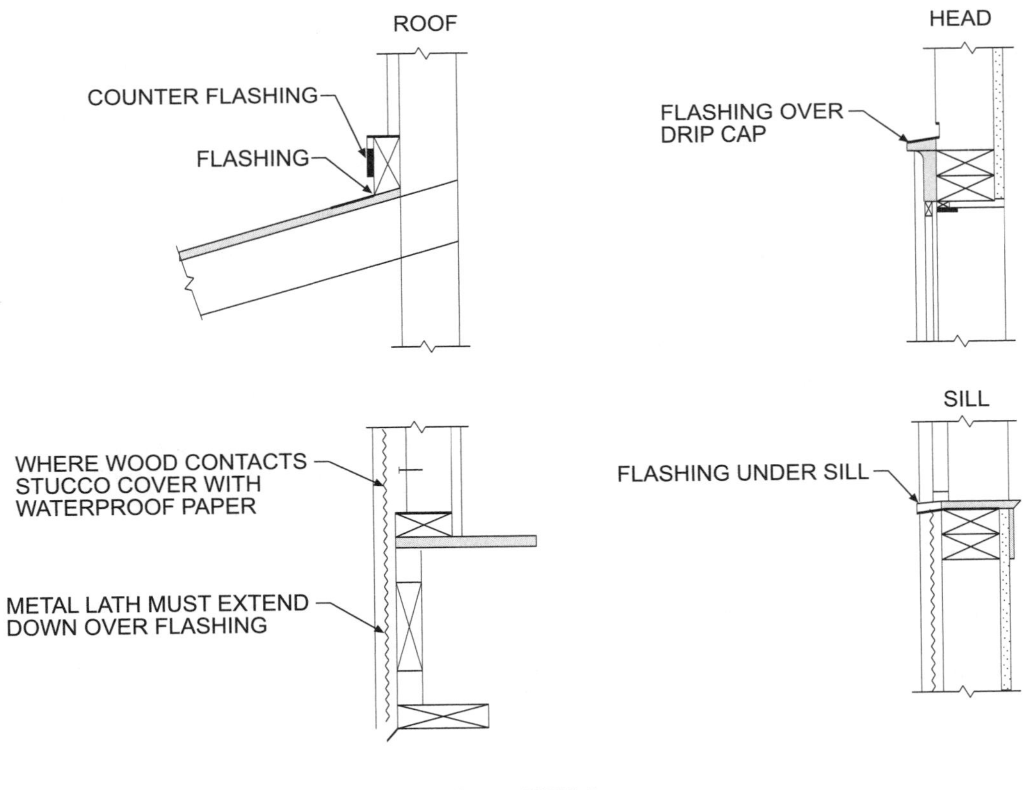

Figure R703.8
FLASHING EXAMPLES

be installed with fasteners according to Table R703.4 or *approved* manufacturer's installation instructions.

❖ Panel siding of fiber cement must meet the requirements of ASTM C 1186, Type A, minimum Grade II. Panels can be installed vertically or horizontally. All joints are required to be located over framing members. Joints must be sealed with caulking and covered with continuous battens or otherwise installed to comply with Section R703.1 to result in a weather-resistant exterior wall envelope. Panel fasteners must meet the requirements of Table R703.4 or manufacturer's installation instructions.

R703.10.2 Lap siding. Fiber-cement lap siding having a maximum width of 12 inches shall comply with the requirements of ASTM C 1186, Type A, minimum Grade II. Lap siding shall be lapped a minimum of $1^1/_4$ inches (32 mm) and lap siding not having tongue-and-groove end joints shall have the ends sealed with caulking, installed with an H-section joint cover, located over a strip of flashing or shall be designed to comply with Section R703.1. Lap siding courses may be installed with the fastener heads exposed or concealed, according to Table R703.4 or *approved* manufacturers' installation instructions.

❖ To maintain the necessary weather protection, horizontal siding must have a minimum lap of at least $1^1/_4$ inches (32 mm). Lap siding without tongue-and-groove end joints must be treated with one of the following methods: (1) sealed with caulk, (2) covered with an H-section joint cover or (3) otherwise installed to comply with Section R703.1 to result in a weather-resistant exterior wall envelope. Fastener heads may be exposed or concealed in accordance with approved manufacturers' installation instructions.

R703.11 Vinyl siding. Vinyl siding shall be certified and *labeled* as conforming to the requirements of ASTM D 3679 by an *approved* quality control agency.

❖ Polyvinyl chloride (PVC) siding is specifically addressed here as an exterior wall covering. PVC siding must conform to the provisions of ASTM D 3679. The vinyl siding must bear a label, which means that the manufacturer must have regular inspections by a third-party quality control agency.

R703.11.1 Installation. Vinyl siding, soffit and accessories shall be installed in accordance with the manufacturer's installation instructions.

❖ Vinyl siding must be applied to conform to the water-resistive barrier requirements of Sections R703.1 and R703.2 and the attachment requirements of Section R703.4 (see commentary, Sections R703.1, R703.2 and R703.4).

The installation must also comply with the manufacturer's instructions. ASTM D 3679 requires installation of the siding in accordance with Practice D 4756 (ASTM D 4756) and the manufacturer's instructions. These instructions are necessary for compliance with the performance requirements of this chapter.

R703.11.1.1 Vinyl soffit panels. Soffit panels shall be individually fastened to a supporting component such as a nailing strip, fascia or subfascia component or as specified by the manufacturer's instructions.

❖ Polyvinyl chloride (PVC) may be used to cover the soffit. The material must be installed as panels and fastened to a supporting component or in accordance with the manufacturer's installation instructions.

R703.11.2 Foam plastic sheathing. Vinyl siding used with foam plastic sheathing shall be installed in accordance with Section R703.11.2.1, R703.11.2.2, or R703.11.2.3.

Exception: Where the foam plastic sheathing is applied directly over wood structural panels, fiberboard, gypsum sheathing or other *approved* backing capable of independently resisting the design wind pressure, the vinyl siding shall be installed in accordance with Section R703.11.1.

❖ This section establishes a proper basis for vinyl siding applications with foam sheathing by applying appropriate adjustment factors to vinyl siding wind pressure ratings to address this specific assembly condition. Because the vinyl and foam sheathing assembly serve as the primary weather barrier or envelope for the building (when no additional structural sheathing is applied), the vinyl siding pressure rating values have been factored to provide a net safety factor of 2.0 instead of 1.5 as required by ASTM D 3679 for applications of vinyl siding over "solid walls." The adjustment factors specified in Section R703.11.2.2 also account for difference in pressure equalization effects addressed in ASTM D 3679, Annex A for the specific wall assembly conditions where vinyl siding is used with a foam sheathing backing material.

The requirements of Section R703.11.2 are intended to improve the wind-resistant performance of combinations of foam sheathing and vinyl siding commonly used to meet or exceed energy code requirements and newer green building guidelines or standards. The prescriptive requirements and adjustment factors are based on certified testing of various combinations of foam sheathing and vinyl siding products conducted at the NAHB Research Center, Inc. and also testing serving as the basis for the wind pressure rating method for vinyl siding as explained in ASTM D 3679, Annex A.

An exception exempts foam plastic sheathing from having to contribute to resisting wind if the backing behind it can resist the wind load by itself. If it cannot, then the foam plastic sheathing and vinyl siding together must comply with one of the following three provisions:

1. If the structure is located where the Basic Wind Speed does not exceed 90 mph (40 m/s) and is in an Exposure Category B or less, the installation must meet Section R703.11.2.1.

2. If the structure is located where the Basic Wind Speed exceeds 90 mph (40 m/s) or is in an Exposure Category C or D or otherwise cannot fully comply with all the requirements of Section

R703.11.2.1, the installation must meet Section R703.11.2.2.

3. If the vinyl siding manufacturer's product specifications indicate that an approved design wind pressure rating when installed over foam plastic sheathing is provided, the vinyl siding and foam plastic sheathing are permitted to be installed in accordance with the manufacturer's installation instructions as stipulated by Section R703.11.2.3.

R703.11.2.1 Basic wind speed not exceeding 90 miles per hour and Exposure Category B. Where the basic wind speed does not exceed 90 miles per hour (40 m/s), the Exposure Category is B and gypsum wall board or equivalent is installed on the side of the wall opposite the foam plastic sheathing, the minimum siding fastener penetration into wood framing shall be $1^1/_4$ inches (32 mm) using minimum 0.120-inch diameter nail (shank) with a minimum 0.313-inch diameter head, 16 inches on center. The foam plastic sheathing shall be minimum $^1/_2$-inch-thick (12.7 mm) (nominal) extruded polystyrene per ASTM C 578, $^1/_2$-inch-thick (12.7 mm) (nominal) polyisocyanurate per ASTM C 1289, or 1-inch-thick (25 mm) (nominal) expanded polystyrene per ASTM C 578.

❖ See the commentary to Section R703.11.2.

R703.11.2.2 Basic wind speed exceeding 90 miles per hour or Exposure Categories C and D. Where the basic wind speed exceeds 90 miles per hour (40 m/s) or the Exposure Category is C or D, or all conditions of Section R703.11.2.1 are not met, the adjusted design pressure rating for the assembly shall meet or exceed the loads listed in Tables R301.2(2) adjusted for height and exposure using Table R301.2(3). The design wind pressure rating of the vinyl siding for installation over solid sheathing as provided in the vinyl siding manufacturer's product specifications shall be adjusted for the following wall assembly conditions:

1. For wall assemblies with foam plastic sheathing on the exterior side and gypsum wall board or equivalent on the interior side of the wall, the vinyl siding's design wind pressure rating shall be multiplied by 0.39.
2. For wall assemblies with foam plastic sheathing on the exterior side and no gypsum wall board or equivalent on the interior side of wall, the vinyl siding's design wind pressure rating shall be multiplied by 0.27.

❖ See the commentary to Section R703.11.2.

R703.11.2.3 Manufacturer specification. Where the vinyl siding manufacturer's product specifications provide an *approved* design wind pressure rating for installation over foam plastic sheathing, use of this design wind pressure rating shall be permitted and the siding shall be installed in accordance with the manufacturer's installation instructions.

❖ See the commentary to Section R703.11.2.

R703.12 Adhered masonry veneer installation. Adhered masonry veneer shall be installed in accordance with the manufacturer's instructions.

❖ Adhered masonry veneer must be installed according to the instructions of the manufacturer. Adhered masonry veneer must also meet the requirements listed in Table R703.4 which require that it be installed with a water-resistive barrier complying with Section R703.6.3 and that it meet the requirements of Sections 6.1 and 6.3 of TMS 402/ACI 530/ASCE 5. Where no instructions from the manufacturer are provided, Table R703.4 further requires that adhered masonry veneer be supported and fastened in accordance with the lath requirements in Section R703.6.1.

R703.12.1 Clearances. On exterior stud walls, adhered masonry veneer shall be installed:

1. Minimum of 4 inches (102 mm) above the earth;
2. Minimum of 2 inches (51 mm) above paved areas; or
3. Minimum of $^1/_2$ inch (12 mm) above exterior walking surfaces which are supported by the same foundation that supports the exterior wall.

❖ The clearance requirements for adhered masonry veneer installed on exterior stud walls are consistent with stucco applications, but goes one step further by specifying a minimum of $^1/_2$-inch (12.7 mm) clearance to exterior walking surfaces that are supported by the same foundation that supports the wall to which the exterior veneer is adhered. The requirement that both the wall and the walking surface be supported by the same foundation, along with the flashing performance requirements of Section R703.8 for exterior wall intersections with porches, decks or stairs, limits this $^1/_2$-inch (12.7 mm) clearance to building elements stable to each other and required to be flashed to manage water. This $^1/_2$-inch (12.7 mm) clearance requirement allows for architectural and aesthetic improvements in the installation of adhered masonry veneer.

R703.12.2 Flashing at foundation. A corrosion-resistant screed or flashing of a minimum 0.019-inch (0.48 mm) or 26-gage galvanized or plastic with a minimum vertical attachment flange of $3^1/_2$ inches (89 mm) shall be installed to extend a minimum of 1 inch (25 mm) below the foundation plate line on exterior stud walls in accordance with Section R703.8. The water-resistive barrier, as required by Table R703.4, Footnote w, shall lap over the exterior of the attachment flange of the screed or flashing.

❖ The requirements for flashing at the foundation for adhered masonry veneer are similar to the weep screed requirements for stucco and complement the flashing performance requirements of Section R703.8, while at the same time allowing for effective alternatives to the stucco-specific weep screed.

Bibliography

The following resource materials were used in the preparation of the commentary for this chapter of the code:

AAMA 711-07, *Voluntary Specification for Self Adhering Flashing Used for Installation of Exterior Wall Fenestration Products*. Schaumburg, IL: American Architectural Manufacturers Association, 2007.

ASTM C 557-03e01, *Specification for Adhesives for Fastening Gypsum Wallboard to Wood Framing*. West Conshohocken, PA: ASTM International, 2003.

ASTM C 645-08a, *Specification for Nonstructural Steel Framing Members*. West Conshohocken, PA: ASTM International, 2008.

ASTM C 843-99(2006), *Specification for Application of Gypsum Veneer Plaster*. West Conshohocken, PA: ASTM International, 2006.

ASTM C 844-04, *Specification for Application of Gypsum Base to Receive Gypsum Veneer Plaster*. West Conshohocken, PA: ASTM International, 2004.

ASTM C 926-06, *Specification for Application of Portland Cement Based-Plaster*. West Conshohocken, PA: ASTM International, 2006.

ASTM C 955-09, *Specification for Load-bearing (Transverse and Axial) Steel Studs, Runners (Tracks), and Bracing or Bridging for Screw Application of Gypsum Panel Products and Metal Plaster Bases*. West Conshohocken, PA: ASTM International, 2009.

ASTM C 1002-07, *Specification for Steel Drill Screws for the Application of Gypsum Panel Products or Metal Plaster Bases*. West Conshohocken, PA: ASTM International, 2007.

ASTM C 1063-08, *Specification for Installation of Lathing and Furring to Receive Interior and Exterior Portland Cement-Based Plaster*. West Conshohocken, PA: ASTM International, 2008.

ASTM C 1186-08, *Specification for Flat Nonasbestos Fiber Cement Sheets*. West Conshohocken, PA: ASTM International, 2008.

ASTM C 1513-04, *Standard Specification for Steel Tapping Screws for Cold-formed Steel Framing Connections*. West Conshohocken, PA: ASTM International, 2004.

ASTM D 3679-09, *Specification for Rigid Poly (Vinyl Chloride) (PVC) Siding*. West Conshohocken, PA: ASTM International, 2009.

ASTM D 4756-06, *Practice for the Installation of Rigid Poly (Vinyl-Chloride) (PVC) Siding and Soffits*. West Conshohocken, PA: ASTM International, 2006.

ASTM E 96/E 96M-05, *Standard Test Methods for Water Vapor Transmission of Materials*. West Conshohocken, PA: ASTM International, 2005.

ASTM E 330-02, *Test Method for Structural Performance of Exterior Windows, Curtain Walls and Doors by Uniform Static Air Pressure Difference*. West Conshohocken, PA: ASTM International, 2002.

ASTM E 331-09, *Test Method for Water Penetration of Exterior Windows, Skylights, Doors and Curtain Walls by Uniform Static Air Pressure Difference*. West Conshohocken, PA: ASTM International, 2009.

ASTM E 2273-03, *Standard Test Method for Determining the Drainage Efficiency of Exterior Insulation and Finish Systems (EIFS) Clad Wall Assemblies*. West Conshohocken, PA: ASTM International, 2003.

ASTM E 2568-09e1, *Standard Specification for PB Exterior Insulation and Finish Systems (EIFS)*. West Conshohocken, PA: ASTM International, 2009.

CSSB-97, *Grading and Packing Rules for Western Red Cedar Shakes and Western Red Shingles of the Cedar Shake and Shingle Bureau*. Sumas, WA: Cedar Shake & Shingle Bureau, 1997.

Federal Specification UU-B-790a, *Specifications for Building Paper, Vegetable Fiber (Kraft, Waterproofed, Water Repellent, and Fire Resistant*. Washington, DC: United States General Services Administration, 1992.

Klinger, Richard E. "Behavior of Anchored Masonry Veneer with Light Wood Stud-framing or Masonry Backing in Full-Scale Whole-building Shaking Table Test." *TMS Journal.* Boulder, CO: The Masonry Society, June 2009.

TMS 402/ACI 530/ASCE 5-05, *Building Code Requirements for Masonry Structures*. Boulder, CO: The Masonry Society, 2005.

Chapter 8: Roof-ceiling Construction

General Comments

Section R801 establishes the scope of the chapter as well as performance requirements for roof and ceiling construction. There are two types of roof-ceiling framing systems: wood framing and steel framing. Section R802 addresses wood roof and ceiling framing, while Section R804 deals with steel roof and ceiling framing. Section R803 specifies requirements for roof sheathing to be used with either of these framing systems. Other topics covered in this chapter are the application of ceiling finishes in Section R805, the proper ventilation of concealed spaces in roofs (e.g., enclosed attics and rafter spaces) and unvented attic assemblies in Section R806, and the "access into attic" criteria in Section R807.

Purpose

Chapter 8 regulates the design and construction of roof-ceiling systems. Proper roof-member design provides for the support of the required design live and snow loads as well as providing for the transfer of these loads to supporting walls. Attics must be identified as "no storage" or "limited storage" zones so that ceiling joists can be appropriately sized. Allowable span tables are provided to simplify the selection of rafter and ceiling joist size for wood roof framing and cold-formed steel framing.

Roof systems must resist wind uplift, and tiedowns must be installed when necessary. Also, the roof system comprised of sheathing fastened to either wood or steel framing serves as a diaphragm, which is a key lateral load path element in resisting the forces of wind and earthquakes.

Chapter 8 also provides requirements for the application of ceiling finishes, the proper ventilation of concealed spaces in roofs (e.g., enclosed attics and rafter spaces), unvented attic assemblies and attic access.

SECTION R801 GENERAL

R801.1 Application. The provisions of this chapter shall control the design and construction of the roof-ceiling system for all buildings.

❖ The provisions of Chapter 8 address conventionally framed roof-ceiling construction.

R801.2 Requirements. Roof and ceiling construction shall be capable of accommodating all loads imposed according to Section R301 and of transmitting the resulting loads to the supporting structural elements.

❖ This section states the performance expectations for roof and ceiling construction.

R801.3 Roof drainage. In areas where expansive or collapsible soils are known to exist, all *dwellings* shall have a controlled method of water disposal from roofs that will collect and discharge roof drainage to the ground surface at least 5 feet (1524 mm) from foundation walls or to an *approved* drainage system.

❖ Saturated expansive or collapsible soils can lead to foundation failures because their additional loads are imposed on the foundation wall. To minimize the potential for the soil adjacent to the foundation wall to become saturated by roof drainage, the code requires that the roofs drain 5 feet (1524 mm) from the foundation. This requirement applies to foundation walls only and does not apply to slab-on-grade foundations.

Although this section is specific to discharging the roof drainage away from foundation walls, discharging the roof drainage away from other types of foundations, such as slab-on-grade on expansive or collapsible soil, may need consideration. Industry standards and/or the engineered design for other types of foundations may also require roof drainage.

SECTION R802 WOOD ROOF FRAMING

R802.1 Identification. Load-bearing dimension lumber for rafters, trusses and ceiling joists shall be identified by a grade mark of a lumber grading or inspection agency that has been approved by an accreditation body that complies with DOC PS 20. In lieu of a grade mark, a certificate of inspection issued by a lumber grading or inspection agency meeting the requirements of this section shall be accepted.

❖ See the commentary to Section R502.1 and Figure R502.1.

R802.1.1 Blocking. Blocking shall be a minimum of utility grade lumber.

❖ Because blocking is a less demanding task than load carrying, the code permits the use of utility-grade lumber.

R802.1.2 End-jointed lumber. *Approved* end-jointed lumber identified by a grade mark conforming to Section R802.1 may be used interchangeably with solid-sawn members of the same species and grade. End-jointed lumber used in an assembly required elsewhere in this code to have a fire-resistance rating shall have the designation "Heat-Resistant Adhesive" or "HRA" included in its grade mark.

❖ See the commentary to Section R502.1.3.

R802.1.3 Fire-retardant-treated wood. Fire-retardant-treated wood (FRTW) is any wood product which, when impregnated with chemicals by a pressure process or other means during manufacture, shall have, when tested in accordance with ASTM E 84 or UL 723, a listed flame spread index of 25 or less and shows no evidence of significant progressive combustion when the test is continued for an additional 20-minute period. In addition, the flame front shall not progress more than 10.5 feet (3200 mm) beyond the center line of the burners at any time during the test.

❖ Fire-retardant-treated wood (FRTW) is plywood and lumber that has been pressure impregnated with chemicals to improve its flame spread characteristics beyond those of untreated wood. The effectiveness of the pressure-impregnated fire-retardant treatment is determined by subjecting the material to tests conducted in accordance with ASTM E 84 or UL 723, with the modification that the test is extended to 30 minutes rather than 15 minutes. Using this procedure, a flame spread index is established during the standard 10-minute test period. The test is continued for an additional 20 minutes. During this added time period, there must not be any significant flame spread. At no time must the flame spread more than $10^1/_2$ feet (3200 mm) past the centerline of the burners.

The result of impregnating wood with fire-retardant chemicals is a chemical reaction at certain temperature ranges. This reaction reduces the release of certain intermediate products that contribute to the flaming of wood, and the reaction also results in the formation of a greater percentage of charcoal and water. Some chemicals are effective in reducing the oxidation rate for charcoal residue. Fire-retardant chemicals also reduce the heat release rate of FRTW when it is burning over a wide range of temperatures. This section gives provisions for the treatment and use of FRTW.

R802.1.3.1 Pressure process. For wood products impregnated with chemicals by a pressure process, the process shall be performed in closed vessels under pressures not less than 50 pounds per square inch gauge (psig) (344.7 kPa).

❖ This section elaborates on the requirements of treatment using a pressure process.

R802.1.3.2 Other means during manufacture. For wood products produced by other means during manufacture the treatment shall be an integral part of the manufacturing process of the wood product. The treatment shall provide permanent protection to all surfaces of the wood product.

❖ This section elaborates on the requirements of treatment using other means during manufacture.

R802.1.3.3 Testing. For wood products produced by other means during manufacture, other than a pressure process, all sides of the wood product shall be tested in accordance with and produce the results required in Section R802.1.3. Testing of only the front and back faces of wood structural panels shall be permitted.

❖ This section provides for added testing of wood products with treatments that are not impregnated by a pressure process. Requiring equivalent performance from all sides of the wood product eliminates any concern over the orientation when it is installed.

R802.1.3.4 Labeling. Fire-retardant-treated lumber and wood structural panels shall be *labeled.* The *label* shall contain:

1. The identification *mark* of an *approved agency* in accordance with Section 1703.5 of the *International Building Code.*
2. Identification of the treating manufacturer.
3. The name of the fire-retardant treatment.
4. The species of wood treated.
5. Flame spread index and smoke-developed index.
6. Method of drying after treatment.
7. Conformance to applicable standards in accordance with Sections R802.1.3.5 through R802.1.3.8.
8. For FRTW exposed to weather, or a damp or wet location, the words "No increase in the listed classification when subjected to the Standard Rain Test" (ASTM D 2898).

❖ For continued quality, each piece of fire-retardant-treated wood must be identified by an approved agency having a reinspection service. The identification must show the performance rating of the material, including the 30-minute ASTM E 84 or UL 723 test results determined in Section R802.1.3 and the design adjustment values determined in Section R802.1.3.2. The third-party agency that provides the FRTW label is also required to state on the label that the FRTW complies with the requirements of Section R802.1.3 and that design adjustment values have been determined for the FRTW in compliance with the provisions of Section R802.1.3.2.

The FRTW label must be distinct from the grading label to avoid confusion between the two. The grading label provides information about the properties of wood before it is fire-retardant treated; the FRTW label shows properties of the wood after FRTW treatment. It is imperative that the FRTW label be presented in a manner that complements the grading label and does not create confusion over which label takes precedence.

R802.1.3.5 Strength adjustments. Design values for untreated lumber and wood structural panels as specified in Section R802.1 shall be adjusted for fire-retardant-treated wood. Adjustments to design values shall be based upon an *approved* method of investigation which takes into consideration the effects of the anticipated temperature and humidity to which the fire-retardant-treated wood will be subjected, the type of treatment and redrying procedures.

❖ Experience has shown that certain factors can affect the physical properties of FRTW. Among these factors are the pressure treatment and redrying processes used and the extremes of temperature and humidity that the FRTW will be subjected to once installed. The design values for all FRTW must be adjusted for the effects of the treatment and environmental conditions, such as high temperature and humidity in attic installations. This section requires the determination of these design adjustment values, based on an investigation procedure that includes subjecting the FRTW to similar temperatures and humidities and that has been approved by the code official. The FRTW tested must be identical to that which is produced. Items to be considered by the code official reviewing the test procedure include species and grade of the untreated wood and conditioning of the wood, such as drying before the fire-retardant treatment process. A fire-retardant wood treater may choose to have its treatment process evaluated by model code evaluation services.

The FRTW is required by Section R802.1.3.1 to be labeled with the design adjustment values. These can take the form of factors that are multiplied by the original design values of the untreated wood to determine its allowable stresses or new allowable stresses that have already been factored down in consideration of the FRTW treatment.

R802.1.3.5.1 Wood structural panels. The effect of treatment and the method of redrying after treatment, and exposure to high temperatures and high humidities on the flexure properties of fire-retardant-treated softwood plywood shall be determined in accordance with ASTM D 5516. The test data developed by ASTM D 5516 shall be used to develop adjustment factors, maximum loads and spans, or both for untreated plywood design values in accordance with ASTM D 6305. Each manufacturer shall publish the allowable maximum loads and spans for service as floor and roof sheathing for their treatment.

❖ This section references the test standard developed to evaluate the flexural properties of fire-retardant-treated plywood that is exposed to high temperatures. Note that while the section title refers to wood structural panels, the standard is specifically for softwood plywood. Therefore, judgment is required in determining the effects of elevated temperature and humidity on other types of wood structural panels.

R802.1.3.5.2 Lumber. For each species of wood treated, the effect of the treatment and the method of redrying after treatment and exposure to high temperatures and high humidities on the allowable design properties of fire-retardant-treated lumber shall be determined in accordance with ASTM D 5664. The test data developed by ASTM D 5664 shall be used to develop modification factors for use at or near room temperature and at elevated temperatures and humidity in accordance with ASTM D 6841. Each manufacturer shall publish the modification factors for service at temperatures of not less than 80°F (27°C) and for roof framing. The roof framing modification factors shall take into consideration the climatological location.

❖ This section references the test standards developed to determine the necessary adjustments to design values for lumber that has been fire-retardant treated and includes the effects of elevated temperatures and humidity.

R802.1.3.6 Exposure to weather. Where fire-retardant-treated wood is exposed to weather or damp or wet locations, it shall be identified as "Exterior" to indicate there is no increase in the listed flame spread index as defined in Section R802.1.3 when subjected to ASTM D 2898.

❖ Some fire-retardant treatments are soluble when exposed to the weather or used under high-humidity conditions. The humidity threshold established for interior applications in Section R802.1.3.4 is 92 percent. Therefore FRTW used in an interior location that will exceed this threshold must comply with this section. When a FRTW product is to be exposed to any of the conditions noted in this section, it must be further tested in accordance with ASTM D 2898. Testing requires the material to meet the performance criteria listed in Section R802.1.3. The material is then subjected to the ASTM weathering test and retested after drying. There must not be any significant differences in the performance recorded before and after the weathering test.

R802.1.3.7 Interior applications. Interior fire-retardant-treated wood shall have a moisture content of not over 28 percent when tested in accordance with ASTM D 3201 procedures at 92 percent relative humidity. Interior fire-retardant-treated wood shall be tested in accordance with Section R802.1.3.5.1 or R802.1.3.5.2. Interior fire-retardant-treated wood designated as Type A shall be tested in accordance with the provisions of this section.

❖ The environment in which the FRTW is used can affect its performance. To make sure that performance will be adequate in a humid interior condition, testing in accordance with ASTM D 3201 as well as testing specified in Section R802.1.3.2.1 or R802.1.3.2.2 is required for all interior wood. Requiring all interior wood to be tested for the effects of high temperature and humidity reduces the chance of a premature failure resulting from improper use of FRTW.

R802.1.3.8 Moisture content. Fire-retardant-treated wood shall be dried to a moisture content of 19 percent or less for lumber and 15 percent or less for wood structural panels before use. For wood kiln dried after treatment (KDAT) the kiln temperatures shall not exceed those used in kiln drying the lumber and plywood submitted for the tests described in Section R802.1.3.5.1 for plywood and R802.1.3.5.2 for lumber.

❖ These moisture content thresholds are necessary to prevent leaching of the fire retardant from the wood. Section R802.1.3.2 requires that the strength adjustments consider the redrying procedure, and this section clarifies that the drying temperatures for wood kiln dried after treatment must be consistent with those on which the adjustment factors were based.

R802.1.4 Structural glued laminated timbers. Glued laminated timbers shall be manufactured and identified as required in ANSI/AITC A190.1 and ASTM D 3737.

❖ This section states that glulam timbers must be manufactured following ANSI/AITC A190.1 and ASTM D 3737. Knowing the standard these products must meet makes it easier to determine that the product found in the field will meet the design requirements.

R802.1.5 Structural log members. Stress grading of structural log members of nonrectangular shape, as typically used in log buildings, shall be in accordance with ASTM D 3957. Such structural log members shall be identified by the grade mark of an *approved* lumber grading or inspection agency. In lieu of a grade mark on the material, a certificate of inspection as to species and grade issued by a lumber-grading or inspection agency meeting the requirements of this section shall be permitted to be accepted.

❖ This section addresses grading requirements for logs used as structural members. This subsection specifies the reference for acceptable methods for establishing structural capacities of logs and specifies the requirement for a grading stamp or alternate certification on structural logs. Structural log members must be graded in accordance with ASTM D 3957.

R802.1.6 Structural composite lumber. Structural capacities for structural composite lumber shall be established and monitored in accordance with ASTM D 5456.

❖ See the commentary to Section R502.1.7.

R802.2 Design and construction. The framing details required in Section R802 apply to roofs having a minimum slope of three units vertical in 12 units horizontal (25-percent slope) or greater. Roof-ceilings shall be designed and constructed in accordance with the provisions of this chapter and Figures R606.11(1), R606.11(2) and R606.11(3) or in accordance with AFPA/NDS. Components of roof-ceilings shall be fastened in accordance with Table R602.3(1).

❖ The prescriptive framing requirements of Section R802 apply only to roofs having a minimum slope of 3 units vertical in 12 units horizontal (25-percent slope). Wood roof-ceiling construction must comply with Section R802 or AF&PA/NDS, *National Design Specification for Wood Construction*. The references to Chapter 6 figures are meant for buildings using masonry wall construction.

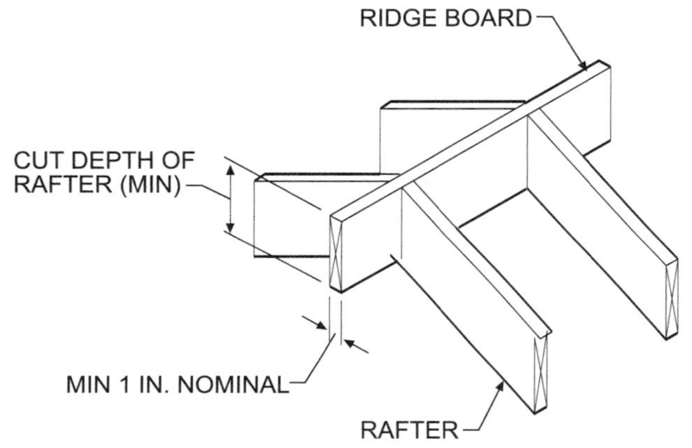

For SI: 1 inch = 25.4 mm.

Figure R802.3(1)
RIDGE BOARD

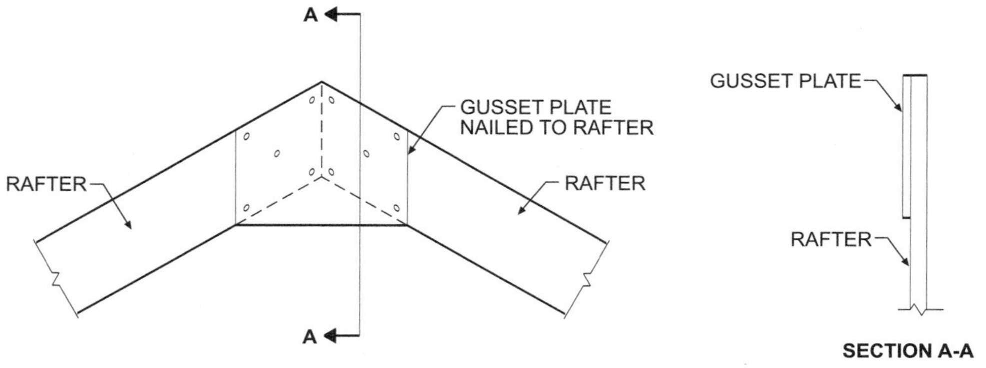

Figure R802.3(2)
RIDGE ALTERNATIVE—GUSSET PLATE TIE

R802.3 Framing details. Rafters shall be framed to ridge board or to each other with a gusset plate as a tie. Ridge board shall be at least 1-inch (25 mm) nominal thickness and not less in depth than the cut end of the rafter. At all valleys and hips there shall be a valley or hip rafter not less than 2-inch (51 mm) nominal thickness and not less in depth than the cut end of the rafter. Hip and valley rafters shall be supported at the ridge by a brace to a bearing partition or be designed to carry and distribute the specific load at that point. Where the roof pitch is less than three units vertical in 12 units horizontal (25-percent slope), structural members that support rafters and ceiling joists, such as ridge beams, hips and valleys, shall be designed as beams.

❖ Traditional practice is to provide a ridgeboard between opposite rafters as a nailing base and to provide a full bearing for the rafter. Rafters must be placed directly opposite each other, and the ridgeboard must have a depth equal to the end of the rafter as illustrated in Commentary Figure R802.3(1). Commentary Figure R802.3(2) shows an option for framing opposing rafters using a gusset plate.

R802.3.1 Ceiling joist and rafter connections. Ceiling joists and rafters shall be nailed to each other in accordance with Table R802.5.1(9), and the rafter shall be nailed to the top wall plate in accordance with Table R602.3(1). Ceiling joists shall be continuous or securely joined in accordance with Table R802.5.1(9) where they meet over interior partitions and are nailed to adjacent rafters to provide a continuous tie across the building when such joists are parallel to the rafters.

Where ceiling joists are not connected to the rafters at the top wall plate, joists connected higher in the *attic* shall be installed as rafter ties, or rafter ties shall be installed to provide a continuous tie. Where ceiling joists are not parallel to rafters, rafter ties shall be installed. Rafter ties shall be a minimum of 2 inches by 4 inches (51 mm by 102 mm) (nominal), installed in accordance with the connection requirements in Table R802.5.1(9), or connections of equivalent capacities shall be provided. Where ceiling joists or rafter ties are not provided, the ridge formed by these rafters shall be supported by a wall or girder designed in accordance with accepted engineering practice.

Collar ties or ridge straps to resist wind uplift shall be connected in the upper third of the *attic* space in accordance with Table R602.3(1).

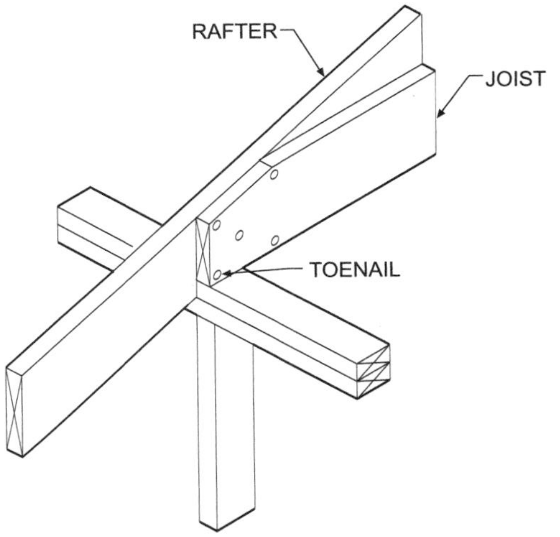

Figure R802.3.1(1)
CEILING JOIST AT SUPPORTS

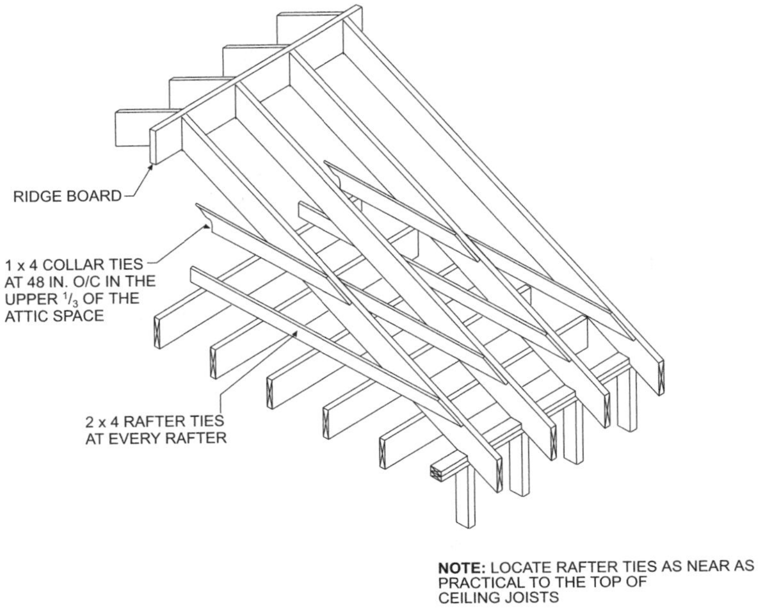

For SI: 1 inch = 25.4 mm.

Figure R802.3.1(2)
ROOF FRAMING WITH CEILING JOISTS NOT PARALLEL TO RAFTERS

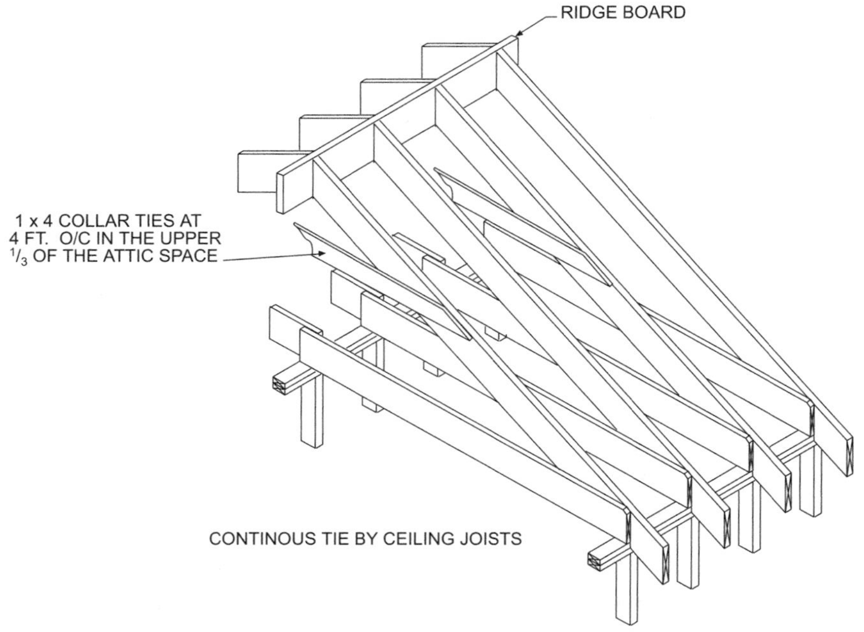

Figure R802.3.1(3)
ROOF FRAMING WITH CEILING JOISTS PARALLEL TO RAFTERS

Collar ties shall be a minimum of 1 inch by 4 inches (25 mm by 102 mm) (nominal), spaced not more than 4 feet (1219 mm) on center.

❖ The requirements of this section describe connections to resist horizontal thrust from gravity loads and ridge uplift from wind loads. Ceiling joists or rafter ties are the framing members used to resist the horizontal thrust from gravity loads. A ceiling joist or rafter tie is required at every rafter. Collar ties or ridge straps are used to resist the ridge uplift caused by the wind load. Collar ties or ridge straps must be spaced not further than every 4 feet (1219 mm) on center.

Ceiling joists located at the top plate and parallel to the rafters must be connected as shown in Commentary Figure R802.3.1(1). The connection must be in accordance with Table R802.5.1(9).

Ceiling joists located at the top plate and parallel to the rafters must be continuous or lap joined as shown in Commentary Figures R802.3.1(3) and R802.3.2(1). The lap splice connection must be in accordance with Table R802.5.1(9), Note e.

Where ceiling joists or rafter ties are located above the top plate or the ceiling joists are not parallel to the rafters, as shown in Commentary Figure R802.3.1(2), the connections must be in accordance with Table R802.5.1(9).

Cathedral ceilings (where ceiling joists or rafter ties are not installed) must be designed so the wall or ridge beam carries the full load of the roof. This would require the walls to be supported by a continuous foundation and/or beams and girders. The ridge beam must also be capable of supporting the full load exerted by the tributary area of the rafters. The ridge beam must be stiff enough to minimize the deflection from the rafter loads so that rafter thrust does not displace the walls.

Collar ties or ridge straps must be located in the upper one-third of the attic space, be a minimum of 1 inch by 4 inches (25 mm by 102 mm), be spaced not more than 4 feet (1219 mm) on center and connected in accordance with Table R602.3(1). The minimum prescriptive connections in Table R602.3(1) for collar ties and ridge straps are based on minimum connection requirements using the 2001 *Wood Frame Construction Manual for Slopes Greater than 3:12, at a Wind Speed of 100 mph (45 m/s) or Less, for a Roof Span of 36 Feet (10 973 mm) or Less* [see Commentary Figures R802.3.1(2) and R802.3.1(3)].

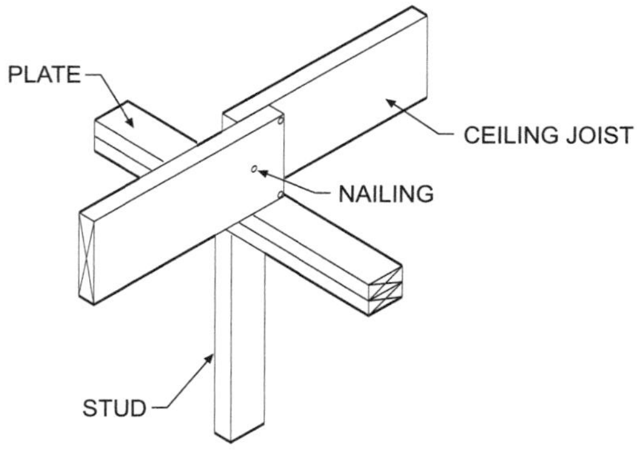

Figure R802.3.2(1)
LAPPED CEILING JOIST CONNECTION PROVIDING RESISTANCE TO RAFTER THRUST

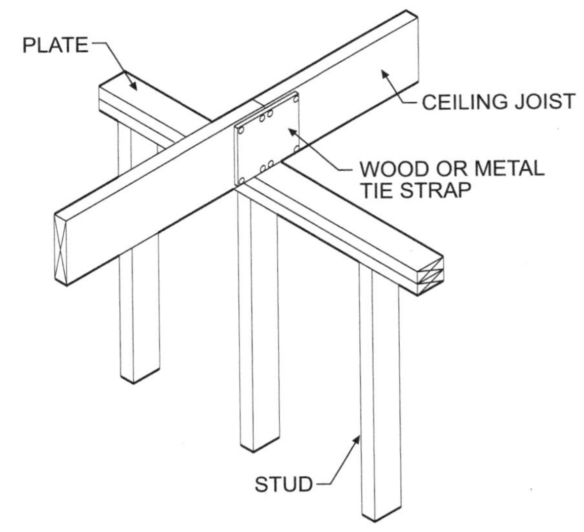

Figure R802.3.2(2)
BUTTED CEILING JOIST CONNECTION PROVIDING RESISTANCE TO RAFTER THRUST

R802.3.2 Ceiling joists lapped. Ends of ceiling joists shall be lapped a minimum of 3 inches (76 mm) or butted over bearing partitions or beams and toenailed to the bearing member. Where ceiling joists are used to provide resistance to rafter thrust, lapped joists shall be nailed together in accordance with Table R802.5.1(9) and butted joists shall be tied together in a manner to resist such thrust. Joists that do not resist thrust shall be permitted to be nailed in accordance with Table R602.3(1).

❖ Where ceiling joists function as a structural element, providing resistance to rafter thrust, proper connection through a lapped joint or butted joint connection to form a continuous tie is required [see Commentary Figures R802.3.2(1) and R802.3.2(2)].

The lap splice nailing of Table R602.3(1) applies only if the ceiling joist is not attached to parallel rafters. For a ceiling joist attached parallel to the rafter, the connection must be in accordance with Note e of Table R802.5.1(9) (see commentary, Section R802.3.1).

R802.4 Allowable ceiling joist spans. Spans for ceiling joists shall be in accordance with Tables R802.4(1) and R802.4(2). For other grades and species and for other loading conditions, refer to the AF&PA *Span Tables for Joists and Rafters.*

❖ Tables R802.4(1) and R802.4(2) list allowable ceiling joist spans for common lumber sizes, species and grades based on spacing and design loads. These tables are similar to the rafter tables explained in the commentary for Section R802.5. The tables provide ceiling joist spans for live loads of 10 and 20 pounds per square foot (psf) (479 Pa and 958 Pa) and dead loads of 5 and 10 psf (239 and 479 Pa). The weight of the ceiling joist is included in the 5 or 10 psf (239 or 479 Pa) dead load.

The dead load of the combined roof and ceiling assemblies must not exceed 15 psf (718.2 Pa) for townhouses in Seismic Design Category C and all structures in Seismic Design Categories D_0, D_1 and D_2. Section R301.2.2.2.1 provides an exception for exceeding the 15 psf restriction (see commentary, Section R301.2.2.2.1).

R802.5 Allowable rafter spans. Spans for rafters shall be in accordance with Tables R802.5.1(1) through R802.5.1(8). For other grades and species and for other loading conditions, refer to the AF&PA *Span Tables for Joists and Rafters*. The span of each rafter shall be measured along the horizontal projection of the rafter.

❖ Tables R802.5.1(1) through R802.5.1(8) list allowable rafter spans for common lumber sizes, species and grades based on spacing and design loads. Tables R802.5.1(2) through R802.5.1(8) are based on ground snow loads of 30 psf (1.44 kPa), 50 psf (2.39 kPa) and 70 psf (3.35 kPa). The snow loads are based on ASCE 7-95 criteria as follows:

for balanced condition: $p_s = C_s \times p_f = 0.7\ C_e \times C_t \times C_s \times p_g = 0.77 \times p_g$ where $C_e = 1.9$, $C_t = 1.1$, and $C_s = 1.0$.

for unbalanced condition: ps = $1.3\ p_f/C_e = 0.7 \times 1.3 \times C_t \times C_s \times p_g = 1.00 \times p_g$ where $C_t = 1.1$ and $C_s = 1.0$.

The unbalanced condition controls. By inspection, the unbalanced condition is equal to the ground snow load.

Tables R802.5.1(1) through R802.5.1(8) provide rafter spans for dead loads of 10 and 20 psf (479 Pa and 958 Pa). The weight of the rafter is included in the 10 or 20 psf (479 Pa or 958 Pa) dead load.

The dead load of the combined roof and ceiling assemblies must not exceed 15 psf (718 Pa) for townhouses in Seismic Design Category C and all structures in Seismic Design Categories D_0, D_1 and D_2. Section R301.2.2.2.1 provides an exception for exceeding the 15 psf (718 Pa) restriction (see commentary, Section R301.2.2.2.1).

Example 1 in the commentary for Table R802.5.1(1) explains the allowable spans for clear-span roof rafters using the code tables. As illustrated in Figure R802.5.1, the horizontal projection of the rafter must be used for the span length rather than using the true length of the rafter measured along the slope of the roof.

R802.5.1 Purlins. Installation of purlins to reduce the span of rafters is permitted as shown in Figure R802.5.1. Purlins shall be sized no less than the required size of the rafters that they support. Purlins shall be continuous and shall be supported by 2-inch by 4-inch (51 mm by 102 mm) braces installed to bearing walls at a slope not less than 45 degrees (0.785 rad) from the horizontal. The braces shall be spaced not more than 4 feet (1219 mm) on center and the unbraced length of braces shall not exceed 8 feet (2438 mm).

❖ This section contains specific instructions for the installation of purlins that will provide intermediate support for the rafters and thus reduce the rafter span. The purlins must be at least the same size as the rafters which they support. The rafters must bear on the purlin to provide proper support for the rafters. See

Commentary Figure R802.5.1(1) for an example of typical installation of purlins. Example 2 in the commentary for Table R802.5.1(1) shows the effect on rafter design and roof purlin requirements when braces to interior bearing walls support the roof purlins.

TABLE R802.5.1(1). See page 8-14.

❖ Adjustment factors, in Note a, for rafter span Tables R802.5(1-8) are limited to cases where the ceiling joists or rafter ties are in the lower third of the attic space. When the ceiling joist or rafter ties are located higher in the attic space, lateral deflection of the rafter below the rafter ties can become excessive and require additional engineering analysis. The following examples assume that the ceiling joists are located at the top plate.

Example 1:

You wish to design a rafter system using Douglas-fir-larch #2 lumber for a 20 psf (958 Pa) live load, with rafters spaced 24 inches (610 mm) on center and having a horizontal projection of the rafter span of 14 feet, 6 inches (4420 mm). The rafters do not support a ceiling, and the roof slope is $2^1/_2$ units vertical in 12 units horizontal. Assume the roof dead load is 10 psf (479 Pa).

Solution:

The object is to locate a tabulated span value of 14 feet, 6 inches (4420 mm) or more. Going to Table R802.5.1 (1) of the code under the portion headed "DEAD LOAD = 10 psf," note that 2-inch by 8-inch (51 mm by 208 mm) rafters at 24 inches (610 mm) on center will span 14 feet, 10 inches (4521 mm).

Example 2:

To find a substitute for the 2-inch by 8-inch (51 mm by 208 mm) rafters of Example 1, determine the required rafter size for a condition that uses a purlin and 2-inch by 4-inch (51 mm by 102 mm)

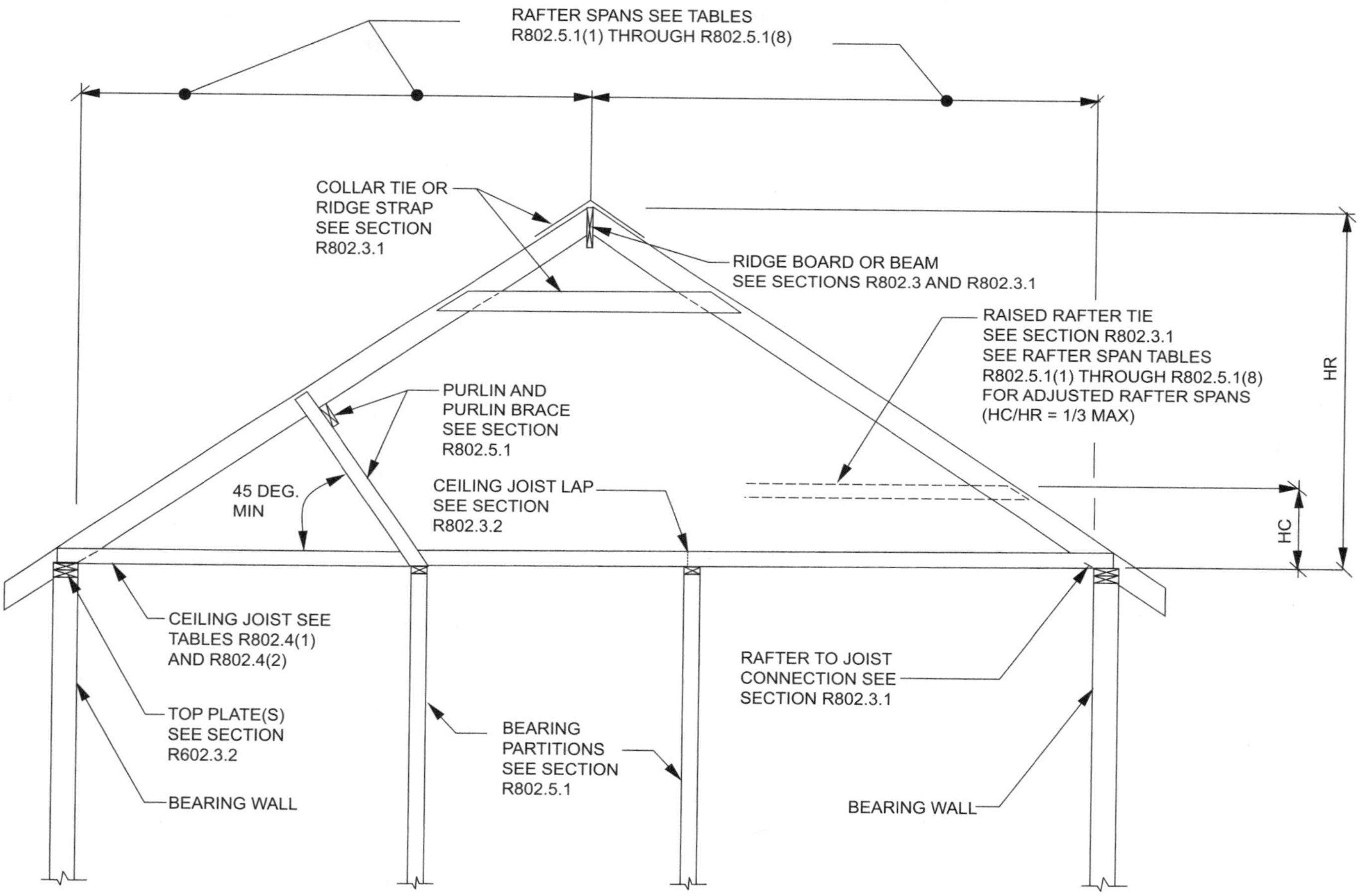

For SI: 1 inch = 25.4 mm, 1 foot = 305.8 mm, 1 degree = 0.018 rad.

Note: Where ceiling joists run perpendicular to the rafter, rafter ties shall be installed in accordance with Section R802.3.1.

H_c = Height of ceiling joists or rafter ties measured vertically above the top of rafter support walls.

H_R = Height of roof ridge measured vertically above the top of the rafter support walls.

FIGURE R802.5.1
BRACED RAFTER CONSTRUCTION

❖ This figure illustrates the use of continuous purlins to reduce rafter spans as described in Section R802.5.1.

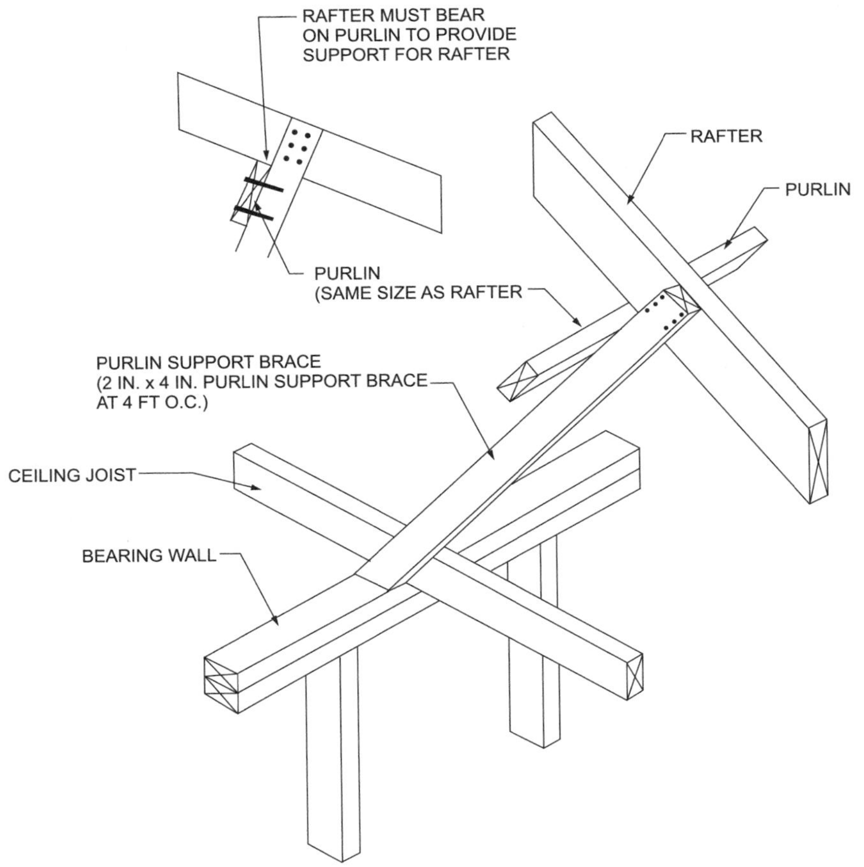

Figure R802.5.1(1)
EXAMPLE OF PURLIN INSTALLATION

braces (located at mid span of the rafters) at 4 feet (1219 mm) on center as depicted in Figure R802.5.1. Assume the same loads, rafter spacing, and 14-foot, 6-inch (4420 mm) horizontal rafter projection. Between the exterior wall and ridge, use Hem-fir #3 lumber.

Solution:

The rafter span, identified in Figure R802.5.1, is reduced to 7 feet, 3 inches (2210 mm) by adding the purlin at midspan. In Table R802.5. 1(1) of the code, 2-inch by 6-inch (51 mm by 152 mm) rafters at 24 inches (610 mm) on center will span 8 feet, 10 inches (2692 mm).

R802.6 Bearing. The ends of each rafter or ceiling joist shall have not less than $1^1/_2$ inches (38 mm) of bearing on wood or metal and not less than 3 inches (76 mm) on masonry or concrete. The bearing on masonry or concrete shall be direct, or a sill plate of 2-inch (51 mm) minimum nominal thickness shall be provided under the rafter or ceiling joist. The sill plate shall provide a minimum nominal bearing area of 48 square inches (30 865 mm^2).

❖ The requirements for roof rafters and ceiling joists bearing on wood, metal and masonry are similar to the requirements for floor joists. See the commentary for Section R502.6 relative to the bearing requirements for wood, metal and masonry.

R802.6.1 Finished ceiling material. If the finished ceiling material is installed on the ceiling prior to the attachment of the ceiling to the walls, such as in construction at a factory, a compression strip of the same thickness as the finish ceiling material shall be installed directly above the top plate of bearing walls if the compressive strength of the finish ceiling material is less than the loads it will be required to withstand. The compression strip shall cover the entire length of such top plate and shall be at least one-half the width of the top plate. It shall be of material capable of transmitting the loads transferred through it.

❖ In prefabricated (panelized) construction where the finished ceiling material is attached to the roof assembly before it is attached to the walls, care must be taken to provide a compression strip to line up with the top plate of the supporting walls in order to have an adequate bearing surface.

TABLE R802.4(1)
CEILING JOIST SPANS FOR COMMON LUMBER SPECIES
(Uninhabitable attics without storage, live load = 10 psf, L/Δ = 240)

CEILING JOIST SPACING (inches)	SPECIES AND GRADE		DEAD LOAD = 5 psf			
			2 × 4	2 × 6	2 × 8	2 × 10
			Maximum ceiling joist spans			
			(feet - inches)	(feet - inches)	(feet - inches)	(feet - inches)
12	Douglas fir-larch	SS	13-2	20-8	Note a	Note a
	Douglas fir-larch	#1	12-8	19-11	Note a	Note a
	Douglas fir-larch	#2	12-5	19-6	25-8	Note a
	Douglas fir-larch	#3	10-10	15-10	20-1	24-6
	Hem-fir	SS	12-5	19-6	25-8	Note a
	Hem-fir	#1	12-2	19-1	25-2	Note a
	Hem-fir	#2	11-7	18-2	24-0	Note a
	Hem-fir	#3	10-10	15-10	20-1	24-6
	Southern pine	SS	12-11	20-3	Note a	Note a
	Southern pine	#1	12-8	19-11	Note a	Note a
	Southern pine	#2	12-5	19-6	25-8	Note a
	Southern pine	#3	11-6	17-0	21-8	25-7
	Spruce-pine-fir	SS	12-2	19-1	25-2	Note a
	Spruce-pine-fir	#1	11-10	18-8	24-7	Note a
	Spruce-pine-fir	#2	11-10	18-8	24-7	Note a
	Spruce-pine-fir	#3	10-10	15-10	20-1	24-6
16	Douglas fir-larch	SS	11-11	18-9	24-8	Note a
	Douglas fir-larch	#1	11-6	18-1	23-10	Note a
	Douglas fir-larch	#2	11-3	17-8	23-0	Note a
	Douglas fir-larch	#3	9-5	13-9	17-5	21-3
	Hem-fir	SS	11-3	17-8	23-4	Note a
	Hem-fir	#1	11-0	17-4	22-10	Note a
	Hem-fir	#2	10-6	16-6	21-9	Note a
	Hem-fir	#3	9-5	13-9	17-5	21-3
	Southern pine	SS	11-9	18-5	24-3	Note a
	Southern pine	#1	11-6	18-1	23-1	Note a
	Southern pine	#2	11-3	17-8	23-4	Note a
	Southern pine	#3	10-0	14-9	18-9	22-2
	Spruce-pine-fir	SS	11-0	17-4	22-10	Note a
	Spruce-pine-fir	#1	10-9	16-11	22-4	Note a
	Spruce-pine-fir	#2	10-9	16-11	22-4	Note a
	Spruce-pine-fir	#3	9-5	13-9	17-5	21-3

(continued)

TABLE R802.4(1)—continued
CEILING JOIST SPANS FOR COMMON LUMBER SPECIES
(Uninhabitable attics without storage, live load = 10 psf, L/Δ = 240)

CEILING JOIST SPACING (inches)	SPECIES AND GRADE		DEAD LOAD = 5 psf			
			2 × 4	2 × 6	2 × 8	2 × 10
			Maximum ceiling joist spans			
			(feet - inches)	(feet - inches)	(feet - inches)	(feet - inches)
19.2	Douglas fir-larch	SS	11-3	17-8	23-3	Note a
	Douglas fir-larch	#1	10-10	17-0	22-5	Note a
	Douglas fir-larch	#2	10-7	16-7	21-0	25-8
	Douglas fir-larch	#3	8-7	12-6	15-10	19-5
	Hem-fir	SS	10-7	16-8	21-11	Note a
	Hem-fir	#1	10-4	16-4	21-6	Note a
	Hem-fir	#2	9-11	15-7	20-6	25-3
	Hem-fir	#3	8-7	12-6	15-10	19-5
	Southern -pine	SS	11-0	17-4	22-10	Note a
	Southern pine	#1	10-10	17-0	22-5	Note a
	Southern pine	#2	10-7	16-8	21-11	Note a
	Southern pine	#3	9-1	13-6	17-2	20-3
	Spruce-pine-fir	SS	10-4	16-4	21-6	Note a
	Spruce-pine-fir	#1	10-2	15-11	21-0	25-8
	Spruce-pine-fir	#2	10-2	15-11	21-0	25-8
	Spruce-pine-fir	#3	8-7	12-6	15-10	19-5
24	Douglas fir-larch	SS	10-5	16-4	21-7	Note a
	Douglas fir-larch	#1	10-0	15-9	20-1	24-6
	Douglas fir-larch	#2	9-10	14-10	18-9	22-11
	Douglas fir-larch	#3	7-8	11-2	14-2	17-4
	Hem-fir	SS	9-10	15-6	20-5	Note a
	Hem-fir	#1	9-8	15-2	19-7	23-11
	Hem-fir	#2	9-2	14-5	18-6	22-7
	Hem-fir	#3	7-8	11-2	14-2	17-4
	Southern pine	SS	10-3	16-1	21-2	Note a
	Southern pine	#1	10-0	15-9	20-10	Note a
	Southern pine	#2	9-10	15-6	20-1	23-11
	Southern pine	#3	8-2	12-0	15-4	18-1
	Spruce-pine-fir	SS	9-8	15-2	19-11	25-5
	Spruce-pine-fir	#1	9-5	14-9	18-9	22-11
	Spruce-pine-fir	#2	9-5	14-9	18-9	22-11
	Spruce-pine-fir	#3	7-8	11-2	14-2	17-4

Check sources for availability of lumber in lengths greater than 20 feet.

For SI: 1 inch = 25.4 mm, 1 foot = 304.8 mm, 1 pound per square foot = 0.0479kPa.

a. Span exceeds 26 feet in length.

❖ See the commentary to Section R802.4.

TABLE R802.4(2)
CEILING JOIST SPANS FOR COMMON LUMBER SPECIES
(Uninhabitable attics with limited storage, live load = 20 psf, L/Δ = 240)

CEILING JOIST SPACING (inches)	SPECIES AND GRADE		DEAD LOAD = 10 psf			
			2 × 4	2 × 6	2 × 8	2 × 10
			Maximum ceiling joist spans			
			(feet - inches)	(feet - inches)	(feet - inches)	(feet - inches)
12	Douglas fir-larch	SS	10-5	16-4	21-7	Note a
	Douglas fir-larch	#1	10-0	15-9	20-1	24-6
	Douglas fir-larch	#2	9-10	14-10	18-9	22-11
	Douglas fir-larch	#3	7-8	11-2	14-2	17-4
	Hem-fir	SS	9-10	15-6	20-5	Note a
	Hem-fir	#1	9-8	15-2	19-7	23-11
	Hem-fir	#2	9-2	14-5	18-6	22-7
	Hem-fir	#3	7-8	11-2	14-2	17-4
	Southern pine	SS	10-3	16-1	21-2	Note a
	Southern pine	#1	10-0	15-9	20-10	Note a
	Southern pine	#2	9-10	15-6	20-1	23-11
	Southern pine	#3	8-2	12-0	15-4	18-1
	Spruce-pine-fir	SS	9-8	15-2	19-11	25-5
	Spruce-pine-fir	#1	9-5	14-9	18-9	22-11
	Spruce-pine-fir	#2	9-5	14-9	18-9	22-11
	Spruce-pine-fir	#3	7-8	11-2	14-2	17-4
16	Douglas fir-larch	SS	9-6	14-11	19-7	25-0
	Douglas fir-larch	#1	9-1	13-9	17-5	21-3
	Douglas fir-larch	#2	8-9	12-10	16-3	19-10
	Douglas fir-larch	#3	6-8	9-8	12-4	15-0
	Hem-fir	SS	8-11	14-1	18-6	23-8
	Hem-fir	#1	8-9	13-5	16-10	20-8
	Hem-fir	#2	8-4	12-8	16-0	19-7
	Hem-fir	#3	6-8	9-8	12-4	15-0
	Southern pine	SS	9-4	14-7	19-3	24-7
	Southern pine	#1	9-1	14-4	18-11	23-1
	Southern pine	#2	8-11	13-6	17-5	20-9
	Southern pine	#3	7-1	10-5	13-3	15-8
	Spruce-pine-fir	SS	8-9	13-9	18-1	23-1
	Spruce-pine-fir	#1	8-7	12-10	16-3	19-10
	Spruce-pine-fir	#2	8-7	12-10	16-3	19-10
	Spruce-pine-fir	#3	6-8	9-8	12-4	15-0

(continued)

TABLE R802.4(2)—continued
CEILING JOIST SPANS FOR COMMON LUMBER SPECIES
(Uninhabitable attics with limited storage, live load = 20 psf, L/Δ = 240)

CEILING JOIST SPACING (inches)	SPECIES AND GRADE		DEAD LOAD = 10 psf			
			2 × 4	2 × 6	2 × 8	2 × 10
			Maximum ceiling joist spans			
			(feet - inches)	(feet - inches)	(feet - inches)	(feet - inches)
19.2	Douglas fir-larch	SS	8-11	14-0	18-5	23-4
	Douglas fir-larch	#1	8-7	12-6	15-10	19-5
	Douglas fir-larch	#2	8-0	11-9	14-10	18-2
	Douglas fir-larch	#3	6-1	8-10	11-3	13-8
	Hem-fir	SS	8-5	13-3	17-5	22-3
	Hem-fir	#1	8-3	12-3	15-6	18-11
	Hem-fir	#2	7-10	11-7	14-8	17-10
	Hem-fir	#3	6-1	8-10	11-3	13-8
	Southern pine	SS	8-9	13-9	18-1	23-1
	Southern pine	#1	8-7	13-6	17-9	21-1
	Southern pine	#2	8-5	12-3	15-10	18-11
	Southern pine	#3	6-5	9-6	12-1	14-4
	Spruce-pine-fir	SS	8-3	12-11	17-1	21-8
	Spruce-pine-fir	#1	8-0	11-9	14-10	18-2
	Spruce-pine-fir	#2	8-0	11-9	14-10	18-2
	Spruce-pine-fir	#3	6-1	8-10	11-3	13-8
24	Douglas fir-larch	SS	8-3	13-0	17-1	20-11
	Douglas fir-larch	#1	7-8	11-2	14-2	17-4
	Douglas fir-larch	#2	7-2	10-6	13-3	16-3
	Douglas fir-larch	#3	5-5	7-11	10-0	12-3
	Hem-fir	SS	7-10	12-3	16-2	20-6
	Hem-fir	#1	7-6	10-11	13-10	16-11
	Hem-fir	#2	7-1	10-4	13-1	16-0
	Hem-fir	#3	5-5	7-11	10-0	12-3
	Southern pine	SS	8-1	12-9	16-10	21-6
	Southern pine	#1	8-0	12-6	15-10	18-10
	Southern pine	#2	7-8	11-0	14-2	16-11
	Southern pine	#3	5-9	8-6	10-10	12-10
	Spruce-pine-fir	SS	7-8	12-0	15-10	19-5
	Spruce-pine-fir	#1	7-2	10-6	13-3	16-3
	Spruce-pine-fir	#2	7-2	10-6	13-3	16-3
	Spruce-pine-fir	#3	5-5	7-11	10-0	12-3

Check sources for availability of lumber in lengths greater than 20 feet.

For SI: 1 inch = 25.4 mm, 1 foot = 304.8 mm, 1 pound per square foot = 0.0479kPa.

a. Span exceeds 26 feet in length.

❖ See the commentary to Section R802.4.

TABLE R802.5.1(1)
RAFTER SPANS FOR COMMON LUMBER SPECIES
(Roof live load=20 psf, ceiling not attached to rafters, L/Δ = 180)

RAFTER SPACING (inches)	SPECIES AND GRADE		DEAD LOAD = 10 psf					DEAD LOAD = 20 psf				
			2 × 4	2 × 6	2 × 8	2 × 10	2 × 12	2 × 4	2 × 6	2 × 8	2 × 10	2 × 12
			Maximum rafter spans[a]									
			(feet - inches)	(feet - inches)	(feet - inches)	(feet - inches)	(feet - inches)	(feet - inches)	(feet - inches)	(feet - inches)	(feet - inches)	(feet - inches)
12	Douglas fir-larch	SS	11-6	18-0	23-9	Note b	Note b	11-6	18-0	23-5	Note b	Note b
	Douglas fir-larch	#1	11-1	17-4	22-5	Note b	Note b	10-6	15-4	19-5	23-9	Note b
	Douglas fir-larch	#2	10-10	16-7	21-0	25-8	Note b	9-10	14-4	18-2	22-3	25-9
	Douglas fir-larch	#3	8-7	12-6	15-10	19-5	22-6	7-5	10-10	13-9	16-9	19-6
	Hem-fir	SS	10-10	17-0	22-5	Note b	Note b	10-10	17-0	22-5	Note b	Note b
	Hem-fir	#1	10 -7	16-8	21-10	Note b	Note b	10-3	14-11	18-11	23-2	Note b
	Hem-fir	#2	10-1	15-11	20-8	25-3	Note b	9-8	14-2	17-11	21-11	25-5
	Hem-fir	#3	8-7	12-6	15-10	19-5	22-6	7-5	10-10	13-9	16-9	19-6
	Southern pine	SS	11-3	17-8	23-4	Note b	Note b	11-3	17-8	23-4	Note b	Note b
	Southern pine	#1	11-1	17-4	22-11	Note b	Note b	11-1	17-3	21-9	25-10	Note b
	Southern pine	#2	10-10	17-0	22-5	Note b	Note b	10-6	15-1	19-5	23-2	Note b
	Southern pine	#3	9-1	13-6	17-2	20-3	24-1	7-11	11-8	14-10	17-6	20-11
	Spruce-pine-fir	SS	10-7	16-8	21-11	Note b	Note b	10-7	16-8	21-9	Note b	Note b
	Spruce-pine-fir	#1	10-4	16-3	21-0	25-8	Note b	9-10	14-4	18-2	22-3	25-9
	Spruce-pine-fir	#2	10-4	16-3	21-0	25-8	Note b	9-10	14-4	18-2	22-3	25-9
	Spruce-pine-fir	#3	8-7	12-6	15-10	19-5	22-6	7-5	10-10	13-9	16-9	19-6
16	Douglas fir-larch	SS	10-5	16-4	21-7	Note b	Note b	10-5	16-0	20-3	24-9	Note b
	Douglas fir-larch	#1	10-0	15-4	19-5	23-9	Note b	9-1	13-3	16-10	20-7	23-10
	Douglas fir-larch	#2	9-10	14-4	18-2	22-3	25-9	8-6	12-5	15-9	19-3	22-4
	Douglas fir-larch	#3	7-5	10-10	13-9	16-9	19-6	6-5	9-5	11-11	14-6	16-10
	Hem-fir	SS	9-10	15-6	20-5	Note b	Note b	9-10	15-6	19-11	24-4	Note b
	Hem-fir	#1	9-8	14-11	18-11	23-2	Note b	8-10	12-11	16-5	20-0	23-3
	Hem-fir	#2	9-2	14-2	17-11	21-11	25-5	8-5	12-3	15-6	18-11	22-0
	Hem-fir	#3	7-5	10-10	13-9	16-9	19-6	6-5	9-5	11-11	14-6	16-10
	Southern pine	SS	10-3	16-1	21-2	Note b	Note b	10-3	16-1	21-2	Note b	Note b
	Southern pine	#1	10-0	15-9	20-10	25-10	Note b	10-0	15-0	18-10	22-4	Note b
	Southern pine	#2	9-10	15-1	19-5	23-2	Note b	9-1	13-0	16-10	20-1	23-7
	Southern pine	#3	7-11	11-8	14-10	17-6	20-11	6-10	10-1	12-10	15-2	18-1
	Spruce-pine-fir	SS	9-8	15-2	19-11	25-5	Note b	9-8	14-10	18-10	23-0	Note b
	Spruce-pine-fir	#1	9-5	14-4	18-2	22-3	25-9	8-6	12-5	15-9	19-3	22-4
	Spruce-pine-fir	#2	9-5	14-4	18-2	22-3	25-9	8-6	12-5	15-9	19-3	22-4
	Spruce-pine-fir	#3	7-5	10-10	13-9	16-9	19-6	6-5	9-5	11-11	14-6	16-10
19.2	Douglas fir-larch	SS	9-10	15-5	20-4	25-11	Note b	9-10	14-7	18-6	22-7	Note b
	Douglas fir-larch	#1	9-5	14-0	17-9	21-8	25-2	8-4	12-2	15-4	18-9	21-9
	Douglas fir-larch	#2	8-11	13-1	16-7	20-3	23-6	7-9	11-4	14-4	17-7	20-4
	Douglas fir-larch	#3	6-9	9-11	12-7	15-4	17-9	5-10	8-7	10-10	13-3	15-5
	Hem-fir	SS	9-3	14-7	19-2	24-6	Note b	9-3	14-4	18-2	22-3	25-9
	Hem-fir	#1	9-1	13-8	17-4	21-1	24-6	8-1	11-10	15-0	18-4	21-3
	Hem-fir	#2	8-8	12-11	16-4	20-0	23-2	7-8	11-2	14-2	17-4	20-1
	Hem-fir	#3	6-9	9-11	12-7	15-4	17-9	5-10	8-7	10-10	13-3	15-5
	Southern pine	SS	9-8	15-2	19-11	25-5	Note b	9-8	15-2	19-11	25-5	Note b
	Southern pine	#1	9-5	14-10	19-7	23-7	Note b	9-3	13-8	17-2	20-5	24-4
	Southern pine	#2	9-3	13-9	17-9	21-2	24-10	8-4	11-11	15-4	18-4	21-6
	Southern pine	#3	7-3	10-8	13-7	16-0	19-1	6-3	9-3	11-9	13-10	16-6
	Spruce-pine-fir	SS	9-1	14-3	18-9	23-11	Note b	9-1	13-7	17-2	21-0	24-4
	Spruce-pine-fir	#1	8-10	13-1	16-7	20-3	23-6	7-9	11-4	14-4	17-7	20-4
	Spruce-pine-fir	#2	8-10	13-1	16-7	20-3	23-6	7-9	11-4	14-4	17-7	20-4
	Spruce-pine-fir	#3	6-9	9-11	12-7	15-4	17-9	5-10	8-7	10-10	13-3	15-5

(continued)

TABLE R802.5.1(1)—continued
RAFTER SPANS FOR COMMON LUMBER SPECIES
(Roof live load=20 psf, ceiling not attached to rafters, L/Δ = 180)

RAFTER SPACING (inches)	SPECIES AND GRADE		DEAD LOAD = 10 psf					DEAD LOAD = 20 psf				
			2 × 4	2 × 6	2 × 8	2 × 10	2 × 12	2 × 4	2 × 6	2 × 8	2 × 10	2 × 12
			Maximum rafter spans[a]									
			(feet - inches)	(feet - inches)	(feet - inches)	(feet - inches)	(feet - inches)	(feet - inches)	(feet - inches)	(feet - inches)	(feet - inches)	(feet - inches)
24	Douglas fir-larch	SS	9-1	14-4	18-10	23-4	Note b	8-11	13-1	16-7	20-3	23-5
	Douglas fir-larch	#1	8-7	12-6	15-10	19-5	22-6	7-5	10-10	13-9	16-9	19-6
	Douglas fir-larch	#2	8-0	11-9	14-10	18-2	21-0	6-11	10-2	12-10	15-8	18-3
	Douglas fir-larch	#3	6-1	8-10	11-3	13-8	15-11	5-3	7-8	9-9	11-10	13-9
	Hem-fir	SS	8-7	13-6	17-10	22-9	Note b	8-7	12-10	16-3	19-10	23-0
	Hem-fir	#1	8-4	12-3	15-6	18-11	21-11	7-3	10-7	13-5	16-4	19-0
	Hem-fir	#2	7-11	11-7	14-8	17-10	20-9	6-10	10-0	12-8	15-6	17-11
	Hem-fir	#3	6-1	8-10	11-3	13-8	15-11	5-3	7-8	9-9	11-10	13-9
	Southern pine	SS	8-11	14-1	18-6	23-8	Note b	8-11	14-1	18-6	22-11	Note b
	Southern pine	#1	8-9	13-9	17-9	21-1	25-2	8-3	12-3	15-4	18-3	21-9
	Southern pine	#2	8-7	12-3	15-10	18-11	22-2	7-5	10-8	13-9	16-5	19-3
	Southern pine	#3	6-5	9-6	12-1	14-4	17-1	5-7	8-3	10-6	12-5	14-9
	Spruce-pine-fir	SS	8-5	13-3	17-5	21-8	25-2	8-4	12-2	15-4	18-9	21-9
	Spruce-pine-fir	#1	8-0	11-9	14-10	18-2	21-0	6-11	10-2	12-10	15-8	18-3
	Spruce-pine-fir	#2	8-0	11-9	14-10	18-2	21-0	6-11	10-2	12-10	15-8	18-3
	Spruce-pine-fir	#3	6-1	8-10	11-3	13-8	15-11	5-3	7-8	9-9	11-10	13-9

Check sources for availability of lumber in lengths greater than 20 feet.

For SI: 1 inch = 25.4 mm, 1 foot = 304.8 mm, 1 pound per square foot = 0.0479 kPa.

a. The tabulated rafter spans assume that ceiling joists are located at the bottom of the attic space or that some other method of resisting the outward push of the rafters on the bearing walls, such as rafter ties, is provided at that location. When ceiling joists or rafter ties are located higher in the attic space, the rafter spans shall be multiplied by the factors given below:

H_C/H_R	Rafter Span Adjustment Factor
1/3	0.67
1/4	0.76
1/5	0.83
1/6	0.90
1/7.5 or less	1.00

where:

H_C = Height of ceiling joists or rafter ties measured vertically above the top of the rafter support walls.

H_R = Height of roof ridge measured vertically above the top of the rafter support walls.

b. Span exceeds 26 feet in length.

TABLE R802.5.1(2)
RAFTER SPANS FOR COMMON LUMBER SPECIES
(Roof live load=20 psf, ceiling attached to rafters, L/Δ = 240)

RAFTER SPACING (inches)	SPECIES AND GRADE		DEAD LOAD = 10 psf					DEAD LOAD = 20 psf				
			2 × 4	2 × 6	2 × 8	2 × 10	2 × 12	2 × 4	2 × 6	2 × 8	2 × 10	2 × 12
			Maximum rafter spans[a]									
			(feet - inches)	(feet - inches)	(feet - inches)	(feet - inches)	(feet - inches)	(feet - inches)	(feet - inches)	(feet - inches)	(feet - inches)	(feet - inches)
12	Douglas fir-larch	SS	10-5	16-4	21-7	Note b	Note b	10-5	16-4	21-7	Note b	Note b
	Douglas fir-larch	#1	10-0	15-9	20-10	Note b	Note b	10-0	15-4	19-5	23-9	Note b
	Douglas fir-larch	#2	9-10	15-6	20-5	25-8	Note b	9-10	14-4	18-2	22-3	25-9
	Douglas fir-larch	#3	8-7	12-6	15-10	19-5	22-6	7-5	10-10	13-9	16-9	19-6
	Hem-fir	SS	9-10	15-6	20-5	Note b	Note b	9-10	15-6	20-5	Note b	Note b
	Hem-fir	#1	9-8	15-2	19-11	25-5	Note b	9-8	14-11	18-11	23-2	Note b
	Hem-fir	#2	9-2	14-5	19-0	24-3	Note b	9-2	14-2	17-11	21-11	25-5
	Hem-fir	#3	8-7	12-6	15-10	19-5	22-6	7-5	10-10	13-9	16-9	19-6
	Southern pine	SS	10-3	16-1	21-2	Note b	Note b	10-3	16-1	21-2	Note b	Note b
	Southern pine	#1	10-0	15-9	20-10	Note b	Note b	10-0	15-9	20-10	25-10	Note b
	Southern pine	#2	9-10	15-6	20-5	Note b	Note b	9-10	15-1	19-5	23-2	Note b
	Southern pine	#3	9-1	13-6	17-2	20-3	24-1	7-11	11-8	14-10	17-6	20-11
	Spruce-pine-fir	SS	9-8	15-2	19-11	25-5	Note b	9-8	15-2	19-11	25-5	Note b
	Spruce-pine-fir	#1	9-5	14-9	19-6	24-10	Note b	9-5	14-4	18-2	22-3	25-9
	Spruce-pine-fir	#2	9-5	14-9	19-6	24-10	Note b	9-5	14-4	18-2	22-3	25-9
	Spruce-pine-fir	#3	8-7	12-6	15-10	19-5	22-6	7-5	10-10	13-9	16-9	19-6
16	Douglas fir-larch	SS	9-6	14-11	19-7	25-0	Note b	9-6	14-11	19-7	24-9	Note b
	Douglas fir-larch	#1	9-1	14-4	18-11	23-9	Note b	9-1	13-3	16-10	20-7	23-10
	Douglas fir-larch	#2	8-11	14-1	18-2	22-3	25-9	8-6	12-5	15-9	19-3	22-4
	Douglas fir-larch	#3	7-5	10-10	13-9	16-9	19-6	6-5	9-5	11-11	14-6	16-10
	Hem-fir	SS	8-11	14-1	18-6	23-8	Note b	8-11	14-1	18-6	23-8	Note b
	Hem-fir	#1	8-9	13-9	18-1	23-1	Note b	8-9	12-11	16-5	20-0	23-3
	Hem-fir	#2	8-4	13-1	17-3	21-11	25-5	8-4	12-3	15-6	18-11	22-0
	Hem-fir	#3	7-5	10-10	13-9	16-9	19-6	6-5	9-5	11-11	14-6	16-10
	Southern pine	SS	9-4	14-7	19-3	24-7	Note b	9-4	14-7	19-3	24-7	Note b
	Southern pine	#1	9-1	14-4	18-11	24-1	Note b	9-1	14-4	18-10	22-4	Note b
	Southern pine	#2	8-11	14-1	18-6	23-2	Note b	8-11	13-0	16-10	20-1	23-7
	Southern pine	#3	7-11	11-8	14-10	17-6	20-11	6-10	10-1	12-10	15-2	18-1
	Spruce-pine-fir	SS	8-9	13-9	18-1	23-1	Note b	8-9	13-9	18-1	23-0	Note b
	Spruce-pine-fir	#1	8-7	13-5	17-9	22-3	25-9	8-6	12-5	15-9	19-3	22-4
	Spruce-pine-fir	#2	8-7	13-5	17-9	22-3	25-9	8-6	12-5	15-9	19-3	22-4
	Spruce-pine-fir	#3	7-5	10-10	13-9	16-9	19-6	6-5	9-5	11-11	14-6	16-10
19.2	Douglas fir-larch	SS	8-11	14-0	18-5	23-7	Note b	8-11	14-0	18-5	22-7	Note b
	Douglas fir-larch	#1	8-7	13-6	17-9	21-8	25-2	8-4	12-2	15-4	18-9	21-9
	Douglas fir-larch	#2	8-5	13-1	16-7	20-3	23-6	7-9	11-4	14-4	17-7	20-4
	Douglas fir-larch	#3	6-9	9-11	12-7	15-4	17-9	5-10	8-7	10-10	13-3	15-5
	Hem-fir	SS	8-5	13-3	17-5	22-3	Note b	8-5	13-3	17-5	22-3	25-9
	Hem-fir	#1	8-3	12-11	17-1	21-1	24-6	8-1	11-10	15-0	18-4	21-3
	Hem-fir	#2	7-10	12-4	16-3	20-0	23-2	7-8	11-2	14-2	17-4	20-1
	Hem-fir	#3	6-9	9-11	12-7	15-4	17-9	5-10	8-7	10-10	13-3	15-5

(continued)

TABLE R802.5.1(2)—continued
RAFTER SPANS FOR COMMON LUMBER SPECIES
(Roof live load=20 psf, ceiling attached to rafters, L/Δ = 240)

RAFTER SPACING (inches)	SPECIES AND GRADE		DEAD LOAD = 10 psf					DEAD LOAD = 20 psf				
			2 × 4	2 × 6	2 × 8	2 × 10	2 × 12	2 × 4	2 × 6	2 × 8	2 × 10	2 × 12
			Maximum rafter spans[a]									
			(feet - inches)	(feet - inches)	(feet - inches)	(feet - inches)	(feet - inches)	(feet - inches)	(feet - inches)	(feet - inches)	(feet - inches)	(feet - inches)
19.2	Southern pine	SS	8-9	13-9	18-1	23-1	Note b	8-9	13-9	18-1	23-1	Note b
	Southern pine	#1	8-7	13-6	17-9	22-8	Note b	8-7	13-6	17-2	20-5	24-4
	Southern pine	#2	8-5	13-3	17-5	21-2	24-10	8-4	11-11	15-4	18-4	21-6
	Southern pine	#3	7-3	10-8	13-7	16-0	19-1	6-3	9-3	11-9	13-10	16-6
	Spruce-pine-fir	SS	8-3	12-11	17-1	21-9	Note b	8-3	12-11	17-1	21-0	24-4
	Spruce-pine-fir	#1	8-1	12-8	16-7	20-3	23-6	7-9	11-4	14-4	17-7	20-4
	Spruce-pine-fir	#2	8-1	12-8	16-7	20-3	23-6	7-9	11-4	14-4	17-7	20-4
	Spruce-pine-fir	#3	6-9	9-11	12-7	15-4	17-9	5-10	8-7	10-10	13-3	15-5
24	Douglas fir-larch	SS	8-3	13-0	17-2	21-10	Note b	8-3	13-0	16-7	20-3	23-5
	Douglas fir-larch	#1	8-0	12-6	15-10	19-5	22-6	7-5	10-10	13-9	16-9	19-6
	Douglas fir-larch	#2	7-10	11-9	14-10	18-2	21-0	6-11	10-2	12-10	15-8	18-3
	Douglas fir-larch	#3	6-1	8-10	11-3	13-8	15-11	5-3	7-8	9-9	11-10	13-9
	Hem-fir	SS	7-10	12-3	16-2	20-8	25-1	7-10	12-3	16-2	19-10	23-0
	Hem-fir	#1	7-8	12-0	15-6	18-11	21-11	7-3	10-7	13-5	16-4	19-0
	Hem-fir	#2	7-3	11-5	14-8	17-10	20-9	6-10	10-0	12-8	15-6	17-11
	Hem-fir	#3	6-1	8-10	11-3	13-8	15-11	5-3	7-8	9-9	11-10	13-9
	Southern pine	SS	8-1	12-9	16-10	21-6	Note b	8-1	12-9	16-10	21-6	Note b
	Southern pine	#1	8-0	12-6	16-6	21-1	25-2	8-0	12-3	15-4	18-3	21-9
	Southern pine	#2	7-10	12-3	15-10	18-11	22-2	7-5	10-8	13-9	16-5	19-3
	Southern pine	#3	6-5	9-6	12-1	14-4	17-1	5-7	8-3	10-6	12-5	14-9
	Spruce-pine-fir	SS	7-8	12-0	15-10	20-2	24-7	7-8	12-0	15-4	18-9	21-9
	Spruce-pine-fir	#1	7-6	11-9	14-10	18-2	21-0	6-11	10-2	12-10	15-8	18-3
	Spruce-pine-fir	#2	7-6	11-9	14-10	18-2	21-0	6-11	10-2	12-10	15-8	18-3
	Spruce-pine-fir	#3	6-1	8-10	11-3	13-8	15-11	5-3	7-8	9-9	11-10	13-9

Check sources for availability of lumber in lengths greater than 20 feet.

For SI: 1 inch = 25.4 mm, 1 foot = 304.8 mm, 1 pound per square foot = 0.0479 kPa.

a. The tabulated rafter spans assume that ceiling joists are located at the bottom of the attic space or that some other method of resisting the outward push of the rafters on the bearing walls, such as rafter ties, is provided at that location. When ceiling joists or rafter ties are located higher in the attic space, the rafter spans shall be multiplied by the factors given below:

H_C/H_R	Rafter Span Adjustment Factor
1/3	0.67
1/4	0.76
1/5	0.83
1/6	0.90
1/7.5 or less	1.00

where:

H_C = Height of ceiling joists or rafter ties measured vertically above the top of the rafter support walls.

H_R = Height of roof ridge measured vertically above the top of the rafter support walls.

b. Span exceeds 26 feet in length.

❖ See the commentary to Sections R802.5 and R802.5.1 and Table R802.5.1(1).

TABLE R802.5.1(3)
RAFTER SPANS FOR COMMON LUMBER SPECIES
(Ground snow load=30 psf, ceiling not attached to rafters, L/Δ = 180)

RAFTER SPACING (inches)	SPECIES AND GRADE		DEAD LOAD = 10 psf					DEAD LOAD = 20 psf				
			2 × 4	2 × 6	2 × 8	2 × 10	2 × 12	2 × 4	2 × 6	2 × 8	2 × 10	2 × 12
			Maximum rafter spans[a]									
			(feet - inches)	(feet - inches)	(feet - inches)	(feet - inches)	(feet - inches)	(feet - inches)	(feet - inches)	(feet - inches)	(feet - inches)	(feet - inches)
12	Douglas fir-larch	SS	10-0	15-9	20-9	Note b	Note b	10-0	15-9	20-1	24-6	Note b
	Douglas fir-larch	#1	9-8	14-9	18-8	22-9	Note b	9-0	13-2	16-8	20-4	23-7
	Douglas fir-larch	#2	9-5	13-9	17-5	21-4	24-8	8-5	12-4	15-7	19-1	22-1
	Douglas fir-larch	#3	7-1	10-5	13-2	16-1	18-8	6-4	9-4	11-9	14-5	16-8
	Hem-fir	SS	9-6	14-10	19-7	25-0	Note b	9-6	14-10	19-7	24-1	Note b
	Hem-fir	#1	9-3	14-4	18-2	22-2	25-9	8-9	12-10	16-3	19-10	23-0
	Hem-fir	#2	8-10	13-7	17-2	21-0	24-4	8-4	12-2	15-4	18-9	21-9
	Hem-fir	#3	7-1	10-5	13-2	16-1	18-8	6-4	9-4	11-9	14-5	16-8
	Southern pine	SS	9-10	15-6	20-5	Note b	Note b	9-10	15-6	20-5	Note b	Note b
	Southern pine	#1	9-8	15-2	20-0	24-9	Note b	9-8	14-10	18-8	22-2	Note b
	Southern pine	#2	9-6	14-5	18-8	22-3	Note b	9-0	12-11	16-8	19-11	23-4
	Southern pine	#3	7-7	11-2	14-3	16-10	20-0	6-9	10-0	12-9	15-1	17-11
	Spruce-pine-fir	SS	9-3	14-7	19-2	24-6	Note b	9-3	14-7	18-8	22-9	Note b
	Spruce-pine-fir	#1	9-1	13-9	17-5	21-4	24-8	8-5	12-4	15-7	19-1	22-1
	Spruce-pine-fir	#2	9-1	13-9	17-5	21-4	24-8	8-5	12-4	15-7	19-1	22-1
	Spruce-pine-fir	#3	7-1	10-5	13-2	16-1	18-8	6-4	9-4	11-9	14-5	16-8
16	Douglas fir-larch	SS	9-1	14-4	18-10	23-9	Note b	9-1	13-9	17-5	21-3	24-8
	Douglas fir-larch	#1	8-9	12-9	16-2	19-9	22-10	7-10	11-5	14-5	17-8	20-5
	Douglas fir-larch	#2	8-2	11-11	15-1	18-5	21-5	7-3	10-8	13-6	16-6	19-2
	Douglas fir-larch	#3	6-2	9-0	11-5	13-11	16-2	5-6	8-1	10-3	12-6	14-6
	Hem-fir	SS	8-7	13-6	17-10	22-9	Note b	8-7	13-6	17-1	20-10	24-2
	Hem-fir	#1	8-5	12-5	15-9	19-3	22-3	7-7	11-1	14-1	17-2	19-11
	Hem-fir	#2	8-0	11-9	14-11	18-2	21-1	7-2	10-6	13-4	16-3	18-10
	Hem-fir	#3	6-2	9-0	11-5	13-11	16-2	5-6	8-1	10-3	12-6	14-6
	Southern pine	SS	8-11	14-1	18-6	23-8	Note b	8-11	14-1	18-6	23-8	Note b
	Southern pine	#1	8-9	13-9	18-1	21-5	25-7	8-8	12-10	16-2	19-2	22-10
	Southern pine	#2	8-7	12-6	16-2	19-3	22-7	7-10	11-2	14-5	17-3	20-2
	Southern pine	#3	6-7	9-8	12-4	14-7	17-4	5-10	8-8	11-0	13-0	15-6
	Spruce-pine-fir	SS	8-5	13-3	17-5	22-1	25-7	8-5	12-9	16-2	19-9	22-10
	Spruce-pine-fir	#1	8-2	11-11	15-1	18-5	21-5	7-3	10-8	13-6	16-6	19-2
	Spruce-pine-fir	#2	8-2	11-11	15-1	18-5	21-5	7-3	10-8	13-6	16-6	19-2
	Spruce-pine-fir	#3	6-2	9-0	11-5	13-11	16-2	5-6	8-1	10-3	12-6	14-6
19.2	Douglas fir-larch	SS	8-7	13-6	17-9	21-8	25-2	8-7	12-6	15-10	19-5	22-6
	Douglas fir-larch	#1	7-11	11-8	14-9	18-0	20-11	7-1	10-5	13-2	16-1	18-8
	Douglas fir-larch	#2	7-5	10-11	13-9	16-10	19-6	6-8	9-9	12-4	15-1	17-6
	Douglas fir-larch	#3	5-7	8-3	10-5	12-9	14-9	5-0	7-4	9-4	11-5	13-2
	Hem-fir	SS	8-1	12-9	16-9	21-4	24-8	8-1	12-4	15-7	19-1	22-1
	Hem-fir	#1	7-9	11-4	14-4	17-7	20-4	6-11	10-2	12-10	15-8	18-2
	Hem-fir	#2	7-4	10-9	13-7	16-7	19-3	6-7	9-7	12-2	14-10	17-3
	Hem-fir	#3	5-7	8-3	10-5	12-9	14-9	5-0	7-4	9-4	11-5	13-2

(continued)

TABLE R802.5.1(3)—continued
RAFTER SPANS FOR COMMON LUMBER SPECIES
(Ground snow load=30 psf, ceiling not attached to rafters, L/Δ = 180)

RAFTER SPACING (inches)	SPECIES AND GRADE		DEAD LOAD = 10 psf					DEAD LOAD = 20 psf				
			2 × 4	2 × 6	2 × 8	2 × 10	2 × 12	2 × 4	2 × 6	2 × 8	2 × 10	2 × 12
			Maximum rafter spans[a]									
			(feet - inches)	(feet - inches)	(feet - inches)	(feet - inches)	(feet - inches)	(feet - inches)	(feet - inches)	(feet - inches)	(feet - inches)	(feet - inches)
19.2	Southern pine	SS	8-5	13-3	17-5	22-3	Note b	8-5	13-3	17-5	22-0	25-9
	Southern pine	#1	8-3	13-0	16-6	19-7	23-4	7-11	11-9	14-9	17-6	20-11
	Southern pine	#2	7-11	11-5	14-9	17-7	20-7	7-1	10-2	13-2	15-9	18-5
	Southern pine	#3	6-0	8-10	11-3	13-4	15-10	5-4	7-11	10-1	11-11	14-2
	Spruce-pine-fir	SS	7-11	12-5	16-5	20-2	23-4	7-11	11-8	14-9	18-0	20-11
	Spruce-pine-fir	#1	7-5	10-11	13-9	16-10	19-6	6-8	9-9	12-4	15-1	17-6
	Spruce-pine-fir	#2	7-5	10-11	13-9	16-10	19-6	6-8	9-9	12-4	15-1	17-6
	Spruce-pine-fir	#3	5-7	8-3	10-5	12-9	14-9	5-0	7-4	9-4	11-5	13-2
24	Douglas fir-larch	SS	7-11	12-6	15-10	19-5	22-6	7-8	11-3	14-2	17-4	20-1
	Douglas fir-larch	#1	7-1	10-5	13-2	16-1	18-8	6-4	9-4	11-9	14-5	16-8
	Douglas fir-larch	#2	6-8	9-9	12-4	15-1	17-6	5-11	8-8	11-0	13-6	15-7
	Douglas fir-larch	#3	5-0	7-4	9-4	11-5	13-2	4-6	6-7	8-4	10-2	11-10
	Hem-fir	SS	7-6	11-10	15-7	19-1	22-1	7-6	11-0	13-11	17-0	19-9
	Hem-fir	#1	6-11	10-2	12-10	15-8	18-2	6-2	9-1	11-6	14-0	16-3
	Hem-fir	#2	6-7	9-7	12-2	14-10	17-3	5-10	8-7	10-10	13-3	15-5
	Hem-fir	#3	5-0	7-4	9-4	11-5	13-2	4-6	6-7	8-4	10-2	11-10
	Southern pine	SS	7-10	12-3	16-2	20-8	25-1	7-10	12-3	16-2	19-8	23-0
	Southern pine	#1	7-8	11-9	14-9	17-6	20-11	7-1	10-6	13-2	15-8	18-8
	Southern pine	#2	7-1	10-2	13-2	15-9	18-5	6-4	9-2	11-9	14-1	16-6
	Southern pine	#3	5-4	7-11	10-1	11-11	14-2	4-9	7-1	9-0	10-8	12-8
	Spruce-pine-fir	SS	7-4	11-7	14-9	18-0	20-11	7-1	10-5	13-2	16-1	18-8
	Spruce-pine-fir	#1	6-8	9-9	12-4	15-1	17-6	5-11	8-8	11-0	13-6	15-7
	Spruce-pine-fir	#2	6-8	9-9	12-4	15-1	17-6	5-11	8-8	11-0	13-6	15-7
	Spruce-pine-fir	#3	5-0	7-4	9-4	11-5	13-2	4-6	6-7	8-4	10-2	11-10

Check sources for availability of lumber in lengths greater than 20 feet.

For SI: 1 inch = 25.4 mm, 1 foot = 304.8 mm, 1 pound per square foot = 0.0479 kPa.

a. The tabulated rafter spans assume that ceiling joists are located at the bottom of the attic space or that some other method of resisting the outward push of the rafters on the bearing walls, such as rafter ties, is provided at that location. When ceiling joists or rafter ties are located higher in the attic space, the rafter spans shall be multiplied by the factors given below:

H_C/H_R	Rafter Span Adjustment Factor
1/3	0.67
1/4	0.76
1/5	0.83
1/6	0.90
1/7.5 or less	1.00

where:

H_C = Height of ceiling joists or rafter ties measured vertically above the top of the rafter support walls.

H_R = Height of roof ridge measured vertically above the top of the rafter support walls.

b. Span exceeds 26 feet in length.

❖ See the commentary to Sections R802.5 and R802.5.1 and Table 802.5.1(1).

TABLE R802.5.1(4)
RAFTER SPANS FOR COMMON LUMBER SPECIES
(Ground snow load=50 psf, ceiling not attached to rafters, L/Δ = 180)

RAFTER SPACING (inches)	SPECIES AND GRADE		DEAD LOAD = 10 psf					DEAD LOAD = 20 psf				
			2 × 4	2 × 6	2 × 8	2 × 10	2 × 12	2 × 4	2 × 6	2 × 8	2 × 10	2 × 12
			Maximum rafter spans[a]									
			(feet - inches)	(feet - inches)	(feet - inches)	(feet - inches)	(feet - inches)	(feet - inches)	(feet - inches)	(feet - inches)	(feet - inches)	(feet - inches)
12	Douglas fir-larch	SS	8-5	13-3	17-6	22-4	26-0	8-5	13-3	17-0	20-9	24-0
	Douglas fir-larch	#1	8-2	12-0	15-3	18-7	21-7	7-7	11-2	14-1	17-3	20-0
	Douglas fir-larch	#2	7-8	11-3	14-3	17-5	20-2	7-1	10-5	13-2	16-1	18-8
	Douglas fir-larch	#3	5-10	8-6	10-9	13-2	15-3	5-5	7-10	10-0	12-2	14-1
	Hem-fir	SS	8-0	12-6	16-6	21-1	25-6	8-0	12-6	16-6	20-4	23-7
	Hem-fir	#1	7-10	11-9	14-10	18-1	21-0	7-5	10-10	13-9	16-9	19-5
	Hem-fir	#2	7-5	11-1	14-0	17-2	19-11	7-0	10-3	13-0	15-10	18-5
	Hem-fir	#3	5-10	8-6	10-9	13-2	15-3	5-5	7-10	10-0	12-2	14-1
	Southern pine	SS	8-4	13-0	17-2	21-11	Note b	8-4	13-0	17-2	21-11	Note b
	Southern pine	#1	8-2	12-10	16-10	20-3	24-1	8-2	12-6	15-9	18-9	22-4
	Southern pine	#2	8-0	11-9	15-3	18-2	21-3	7-7	10-11	14-1	16-10	19-9
	Southern pine	#3	6-2	9-2	11-8	13-9	16-4	5-9	8-5	10-9	12-9	15-2
	Spruce-pine-fir	SS	7-10	12-3	16-2	20-8	24-1	7-10	12-3	15-9	19-3	22-4
	Spruce-pine-fir	#1	7-8	11-3	14-3	17-5	20-2	7-1	10-5	13-2	16-1	18-8
	Spruce-pine-fir	#2	7-8	11-3	14-3	17-5	20-2	7-1	10-5	13-2	16-1	18-8
	Spruce-pine-fir	#3	5-10	8-6	10-9	13-2	15-3	5-5	7-10	10-0	12-2	14-1
16	Douglas fir-larch	SS	7-8	12-1	15-10	19-5	22-6	7-8	11-7	14-8	17-11	20-10
	Douglas fir-larch	#1	7-1	10-5	13-2	16-1	18-8	6-7	9-8	12-2	14-11	17-3
	Douglas fir-larch	#2	6-8	9-9	12-4	15-1	17-6	6-2	9-0	11-5	13-11	16-2
	Douglas fir-larch	#3	5-0	7-4	9-4	11-5	13-2	4-8	6-10	8-8	10-6	12-3
	Hem-fir	SS	7-3	11-5	15-0	19-1	22-1	7-3	11-5	14-5	17-8	20-5
	Hem-fir	#1	6-11	10-2	12-10	15-8	18-2	6-5	9-5	11-11	14-6	16-10
	Hem-fir	#2	6-7	9-7	12-2	14-10	17-3	6-1	8-11	11-3	13-9	15-11
	Hem-fir	#3	5-0	7-4	9-4	11-5	13-2	4-8	6-10	8-8	10-6	12-3
	Southern pine	SS	7-6	11-10	15-7	19-11	24-3	7-6	11-10	15-7	19-11	23-10
	Southern pine	#1	7-5	11-7	14-9	17-6	20-11	7-4	10-10	13-8	16-2	19-4
	Southern pine	#2	7-1	10-2	13-2	15-9	18-5	6-7	9-5	12-2	14-7	17-1
	Southern pine	#3	5-4	7-11	10-1	11-11	14-2	4-11	7-4	9-4	11-0	13-1
	Spruce-pine-fir	SS	7-1	11-2	14-8	18-0	20-11	7-1	10-9	13-8	15-11	19-4
	Spruce-pine-fir	#1	6-8	9-9	12-4	15-1	17-6	6-2	9-0	11-5	13-11	16-2
	Spruce-pine-fir	#2	6-8	9-9	12-4	15-1	17-6	6-2	9-0	11-5	13-11	16-2
	Spruce-pine-fir	#3	5-0	7-4	9-4	11-5	13-2	4-8	6-10	8-8	10-6	12-3
19.2	Douglas fir-larch	SS	7-3	11-4	14-6	17-8	20-6	7-3	10-7	13-5	16-5	19-0
	Douglas fir-larch	#1	6-6	9-6	12-0	14-8	17-1	6-0	8-10	11-2	13-7	15-9
	Douglas fir-larch	#2	6-1	8-11	11-3	13-9	15-11	5-7	8-3	10-5	12-9	14-9
	Douglas fir-larch	#3	4-7	6-9	8-6	10-5	12-1	4-3	6-3	7-11	9-7	11-2
	Hem-fir	SS	6-10	10-9	14-2	17-5	20-2	6-10	10-5	13-2	16-1	18-8
	Hem-fir	#1	6-4	9-3	11-9	14-4	16-7	5-10	8-7	10-10	13-3	15-5
	Hem-fir	#2	6-0	8-9	11-1	13-7	15-9	5-7	8-1	10-3	12-7	14-7
	Hem-fir	#3	4-7	6-9	8-6	10-5	12-1	4-3	6-3	7-11	9-7	11-2

(continued)

TABLE R802.5.1(4)—continued
RAFTER SPANS FOR COMMON LUMBER SPECIES
(Ground snow load=50 psf, ceiling not attached to rafters, L/Δ = 180)

RAFTER SPACING (inches)	SPECIES AND GRADE		DEAD LOAD = 10 psf					DEAD LOAD = 20 psf				
			2 × 4	2 × 6	2 × 8	2 × 10	2 × 12	2 × 4	2 × 6	2 × 8	2 × 10	2 × 12
			Maximum rafter spans[a]									
			(feet - inches)	(feet - inches)	(feet - inches)	(feet - inches)	(feet - inches)	(feet - inches)	(feet - inches)	(feet - inches)	(feet - inches)	(feet - inches)
19.2	Southern pine	SS	7-1	11-2	14-8	18-9	22-10	7-1	11-2	14-8	18 7	21-9
	Southern pine	#1	7-0	10-8	13-5	16-0	19-1	6-8	9-11	12-5	14-10	17-8
	Southern pine	#2	6-6	9-4	12-0	14-4	16-10	6-0	8-8	11-2	13-4	15-7
	Southern pine	#3	4-11	7-3	9-2	10-10	12-11	4-6	6-8	8-6	10-1	12-0
	Spruce-pine-fir	SS	6-8	10-6	13-5	16-5	19-1	6-8	9-10	12-5	15-3	17-8
	Spruce-pine-fir	#1	6-1	8-11	11-3	13-9	15-11	5-7	8-3	10-5	12-9	14-9
	Spruce-pine-fir	#2	6-1	8-11	11-3	13-9	15-11	5-7	8-3	10-5	12-9	14-9
	Spruce-pine-fir	#3	4-7	6-9	8-6	10-5	12-1	4-3	6-3	7-11	9-7	11-2
24	Douglas fir-larch	SS	6-8	10-	13-0	15-10	18-4	6-6	9-6	12-0	14-8	17-0
	Douglas fir-larch	#1	5-10	8-6	10-9	13-2	15-3	5-5	7-10	10-0	12-2	14-1
	Douglas fir-larch	#2	5-5	7-11	10-1	12-4	14-3	5-0	7-4	9-4	11-5	13-2
	Douglas fir-larch	#3	4-1	6-0	7-7	9-4	10-9	3-10	5-7	7-1	8-7	10-0
	Hem-fir	SS	6-4	9-11	12-9	15-7	18-0	6-4	9-4	11-9	14-5	16-8
	Hem-fir	#1	5-8	8-3	10-6	12-10	14-10	5-3	7-8	9-9	11-10	13-9
	Hem-fir	#2	5-4	7-10	9-11	12-1	14-1	4-11	7-3	9-2	11-3	13-0
	Hem-fir	#3	4-1	6-0	7-7	9-4	10-9	3-10	5-7	7-1	8-7	10-0
	Southern pine	SS	6-7	10-4	13-8	17-5	21-0	6-7	10-4	13-8	16-7	19-5
	Southern pine	#1	6-5	9-7	12-0	14-4	17-1	6-0	8-10	11-2	13-3	15-9
	Southern pine	#2	5-10	8-4	10-9	12-10	15-1	5-5	7-9	10-0	11-11	13-11
	Southern pine	#3	4-4	6-5	8-3	9-9	11-7	4-1	6-0	7-7	9-0	10-8
	Spruce-pine-fir	SS	6-2	9-6	12-0	14-8	17-1	6-0	8-10	11-2	13-7	15-9
	Spruce-pine-fir	#1	5-5	7-11	10-1	12-4	14-3	5-0	7-4	9-4	11-5	13-2
	Spruce-pine-fir	#2	5-5	7-11	10-1	12-4	14-3	5-0	7-4	9-4	11-5	13-2
	Spruce-pine-fir	#3	4-1	6-0	7-7	9-4	10-9	3-10	5-7	7-1	8-7	10-0

Check sources for availability of lumber in lengths greater than 20 feet.

For SI: 1 inch = 25.4 mm, 1 foot = 304.8 mm, 1 pound per square foot = 0.0479 kPa

a. The tabulated rafter spans assume that ceiling joists are located at the bottom of the attic space or that some other method of resisting the outward push of the rafters on the bearing walls, such as rafter ties, is provided at that location. When ceiling joists or rafter ties are located higher in the attic space, the rafter spans shall be multiplied by the factors given below:

H_C/H_R	Rafter Span Adjustment Factor
1/3	0.67
1/4	0.76
1/5	0.83
1/6	0.90
1/7.5 or less	1.00

where:

H_C = Height of ceiling joists or rafter ties measured vertically above the top of the rafter support walls.

H_R = Height of roof ridge measured vertically above the top of the rafter support walls.

b. Span exceeds 26 feet in length.

❖ See the commentary to Sections R802.5 and R802.5.1 and Table 802.5.1(1).

TABLE R802.5.1(5)
RAFTER SPANS FOR COMMON LUMBER SPECIES
(Ground snow load=30 psf, ceiling attached to rafters, L/Δ = 240)

RAFTER SPACING (inches)	SPECIES AND GRADE		DEAD LOAD = 10 psf					DEAD LOAD = 20 psf				
			2 × 4	2 × 6	2 × 8	2 × 10	2 × 12	2 × 4	2 × 6	2 × 8	2 × 10	2 × 12
			Maximum rafter spans[a]									
			(feet - inches)	(feet - inches)	(feet - inches)	(feet - inches)	(feet - inches)	(feet - inches)	(feet - inches)	(feet - inches)	(feet - inches)	(feet - inches)
12	Douglas fir-larch	SS	9-1	14-4	18-10	24-1	Note b	9-1	14-4	18-10	24-1	Note b
	Douglas fir-larch	#1	8-9	13-9	18-2	22-9	Note b	8-9	13-2	16-8	20-4	23-7
	Douglas fir-larch	#2	8-7	13-6	17-5	21-4	24-8	8-5	12-4	15-7	19-1	22-1
	Douglas fir-larch	#3	7-1	10-5	13-2	16-1	18-8	6-4	9-4	11-9	14-5	16-8
	Hem-fir	SS	8-7	13-6	17-10	22-9	Note b	8-7	13-6	17-10	22-9	Note b
	Hem-fir	#1	8-5	13-3	17-5	22-2	25-9	8-5	12-10	16-3	19-10	23-0
	Hem-fir	#2	8-0	12-7	16-7	21-0	24-4	8-0	12-2	15-4	18-9	21-9
	Hem-fir	#3	7-1	10-5	13-2	16-1	18-8	6-4	9-4	11-9	14-5	16-8
	Southern pine	SS	8-11	14-1	18-6	23-8	Note b	8-11	14-1	18-6	23-8	Note b
	Southern pine	#1	8-9	13-9	18-2	23-2	Note b	8-9	13-9	18-2	22-2	Note b
	Southern pine	#2	8-7	13-6	17-10	22-3	Note b	8-7	12-11	16-8	19-11	23-4
	Southern pine	#3	7-7	11-2	14-3	16-10	20-0	6-9	10-0	12-9	15-1	17-11
	Spruce-pine-fir	SS	8-5	13-3	17-5	22-3	Note b	8-5	13-3	17-5	22-3	Note b
	Spruce-pine-fir	#1	8-3	12-11	17-0	21-4	24-8	8-3	12-4	15-7	19-1	22-1
	Spruce-pine-fir	#2	8-3	12-11	17-0	21-4	24-8	8-3	12-4	15-7	19-1	22-1
	Spruce-pine-fir	#3	7-1	10-5	13-2	16-1	18-8	6-4	9-4	11-9	14-5	16-8
16	Douglas fir-larch	SS	8-3	13-0	17-2	21-10	Note b	8-3	13-0	17-2	21-3	24-8
	Douglas fir-larch	#1	8-0	12-6	16-2	19-9	22-10	7-10	11-5	14-5	17-8	20-5
	Douglas fir-larch	#2	7-10	11-11	15-1	18-5	21-5	7-3	10-8	13-6	16-6	19-2
	Douglas fir-larch	#3	6-2	9-0	11-5	13-11	16-2	5-6	8-1	10-3	12-6	14-6
	Hem-fir	SS	7-10	12-3	16-2	20-8	25-1	7-10	12-3	16-2	20-8	24-2
	Hem-fir	#1	7-8	12-0	15-9	19-3	22-3	7-7	11-1	14-1	17-2	19-11
	Hem-fir	#2	7-3	11-5	14-11	18-2	21-1	7-2	10-6	13-4	16-3	18-10
	Hem-fir	#3	6-2	9-0	11-5	13-11	16-2	5-6	8-1	10-3	12-6	14-6
	Southern pine	SS	8-1	12-9	16-10	21-6	Note b	8-1	12-9	16-10	21-6	Note b
	Southern pine	#1	8-0	12-6	16-6	21-1	25-7	8-0	12-6	16-2	19-2	22-10
	Southern pine	#2	7-10	12-3	16-2	19-3	22-7	7-10	11-2	14-5	17-3	20-2
	Southern pine	#3	6-7	9-8	12-4	14-7	17-4	5-10	8-8	11-0	13-0	15-6
	Spruce-pine-fir	SS	7-8	12-0	15-10	20-2	24-7	7-8	12-0	15-10	19-9	22-10
	Spruce-pine-fir	#1	7-6	11-9	15-1	18-5	21-5	7-3	10-8	13-6	16-6	19-2
	Spruce-pine-fir	#2	7-6	11-9	15-1	18-5	21-5	7-3	10-8	13-6	16-6	19-2
	Spruce-pine-fir	#3	6-2	9-0	11-5	13-11	16-2	5-6	8-1	10-3	12-6	14-6
19.2	Douglas fir-larch	SS	7-9	12-3	16-1	20-7	25-0	7-9	12-3	15-10	19-5	22-6
	Douglas fir-larch	#1	7-6	11-8	14-9	18-0	20-11	7-1	10-5	13-2	16-1	18-8
	Douglas fir-larch	#2	7-4	10-11	13-9	16-10	19-6	6-8	9-9	12-4	15-1	17-6
	Douglas fir-larch	#3	5-7	8-3	10-5	12-9	14-9	5-0	7-4	9-4	11-5	13-2
	Hem-fir	SS	7-4	11-7	15-3	19-5	23-7	7-4	11-7	15-3	19-1	22-1
	Hem-fir	#1	7-2	11-4	14-4	17-7	20-4	6-11	10-2	12-10	15-8	18-2
	Hem-fir	#2	6-10	10-9	13-7	16-7	19-3	6-7	9-7	12-2	14-10	17-3
	Hem-fir	#3	5-7	8-3	10-5	12-9	14-9	5-0	7-4	9-4	11-5	13-2

(continued)

TABLE R802.5.1(5)—continued
RAFTER SPANS FOR COMMON LUMBER SPECIES
(Ground snow load=30 psf, ceiling attached to rafters, L/Δ = 240)

RAFTER SPACING (inches)	SPECIES AND GRADE		DEAD LOAD = 10 psf					DEAD LOAD = 20 psf				
			2 × 4	2 × 6	2 × 8	2 × 10	2 × 12	2 × 4	2 × 6	2 × 8	2 × 10	2 × 12
			Maximum rafter spans[a]									
			(feet - inches)	(feet - inches)	(feet - inches)	(feet - inches)	(feet - inches)	(feet - inches)	(feet - inches)	(feet - inches)	(feet - inches)	(feet - inches)
19.2	Southern pine	SS	7-8	12-0	15-10	20-2	24-7	7-8	12-0	15-10	20-2	24-7
	Southern pine	#1	7-6	11-9	15-6	19-7	23-4	7-6	11-9	14-9	17-6	20-11
	Southern pine	#2	7-4	11-5	14-9	17-7	20-7	7-1	10-2	13-2	15-9	18-5
	Southern pine	#3	6-0	8-10	11-3	13-4	15-10	5-4	7-11	10-1	11-11	14-2
	Spruce-pine-fir	SS	7-2	11-4	14-11	19-0	23-1	7-2	11-4	14-9	18-0	20-11
	Spruce-pine-fir	#1	7-0	10-11	13-9	16-10	19-6	6-8	9-9	12-4	15-1	17-6
	Spruce-pine-fir	#2	7-0	10-11	13-9	16-10	19-6	6-8	9-9	12-4	15-1	17-6
	Spruce-pine-fir	#3	5-7	8-3	10-5	12-9	14-9	5-0	7-4	9-4	11-5	13-2
24	Douglas fir-larch	SS	7-3	11-4	15-0	19-1	22-6	7-3	11-3	14-2	17-4	20-1
	Douglas fir-larch	#1	7-0	10-5	13-2	16-1	18-8	6-4	9-4	11-9	14-5	16-8
	Douglas fir-larch	#2	6-8	9-9	12-4	15-1	17-6	5-11	8-8	11-0	13-6	15-7
	Douglas fir-larch	#3	5-0	7-4	9-4	11-5	13-2	4-6	6-7	8-4	10-2	11-10
	Hem-fir	SS	6-10	10-9	14-2	18-0	21-11	6-10	10-9	13-11	17-0	19-9
	Hem-fir	#1	6-8	10-2	12-10	15-8	18-2	6-2	9-1	11-6	14-0	16-3
	Hem-fir	#2	6-4	9-7	12-2	14-10	17-3	5-10	8-7	10-10	13-3	15-5
	Hem-fir	#3	5-0	7-4	9-4	11-5	13-2	4-6	6-7	8-4	10-2	11-10
	Southern pine	SS	7-1	11-2	14-8	18-9	22-10	7-1	11-2	14-8	18-9	22-10
	Southern pine	#1	7-0	10-11	14-5	17-6	20-11	7-0	10-6	13-2	15-8	18-8
	Southern pine	#2	6-10	10-2	13-2	15-9	18-5	6-4	9-2	11-9	14-1	16-6
	Southern pine	#3	5-4	7-11	10-1	11-11	14-2	4-9	7-1	9-0	10-8	12-8
	Spruce-pine-fir	SS	6-8	10-6	13-10	17-8	20-11	6-8	10-5	13-2	16-1	18-8
	Spruce-pine-fir	#1	6-6	9-9	12-4	15-1	17-6	5-11	8-8	11-0	13-6	15-7
	Spruce-pine-fir	#2	6-6	9-9	12-4	15-1	17-6	5-11	8-8	11-0	13-6	15-7
	Spruce-pine-fir	#3	5-0	7-4	9-4	11-5	13-2	4-6	6-7	8-4	10-2	11-10

Check sources for availability of lumber in lengths greater than 20 feet.

For SI: 1 inch = 25.4 mm, 1 foot = 304.8 mm, 1 pound per square foot = 0.0479 kPa.

a. The tabulated rafter spans assume that ceiling joists are located at the bottom of the attic space or that some other method of resisting the outward push of the rafters on the bearing walls, such as rafter ties, is provided at that location. When ceiling joists or rafter ties are located higher in the attic space, the rafter spans shall be multiplied by the factors given below:

H_C/H_R	Rafter Span Adjustment Factor
1/3	0.67
1/4	0.76
1/5	0.83
1/6	0.90
1/7.5 or less	1.00

where:

H_C = Height of ceiling joists or rafter ties measured vertically above the top of the rafter support walls.

H_R = Height of roof ridge measured vertically above the top of the rafter support walls.

b. Span exceeds 26 feet in length.

❖ See the commentary to Sections R802.5 and R802.5.1 and Table R802.5.1(1).

TABLE R802.5.1(6)
RAFTER SPANS FOR COMMON LUMBER SPECIES
(Ground snow load=50 psf, ceiling attached to rafters, L/Δ = 240)

RAFTER SPACING (inches)	SPECIES AND GRADE		DEAD LOAD = 10 psf					DEAD LOAD = 20 psf				
			2 × 4	2 × 6	2 × 8	2 × 10	2 × 12	2 × 4	2 × 6	2 × 8	2 × 10	2 × 12
			Maximum rafter spans[a]									
			(feet-inches)	(feet-inches)	(feet-inches)	(feet-inches)	(feet-inches)	(feet-inches)	(feet-inches)	(feet-inches)	(feet-inches)	(feet-inches)
12	Douglas fir-larch	SS	7-8	12-1	15-11	20-3	24-8	7-8	12-1	15-11	20-3	24-0
	Douglas fir-larch	#1	7-5	11-7	15-3	18-7	21-7	7-5	11-2	14-1	17-3	20-0
	Douglas fir-larch	#2	7-3	11-3	14-3	17-5	20-2	7-1	10-5	13-2	16-1	18-8
	Douglas fir-larch	#3	5-10	8-6	10-9	13-2	15-3	5-5	7-10	10-0	12-2	14-1
	Hem-fir	SS	7-3	11-5	15-0	19-2	23-4	7-3	11-5	15-0	19-2	23-4
	Hem-fir	#1	7-1	11-2	14-8	18-1	21-0	7-1	10-10	13-9	16-9	19-5
	Hem-fir	#2	6-9	10-8	14-0	17-2	19-11	6-9	10-3	13-0	15-10	18-5
	Hem-fir	#3	5-10	8-6	10-9	13-2	15-3	5-5	7-10	10-0	12-2	14-1
	Southern pine	SS	7-6	11-10	15-7	19-11	24-3	7-6	11-10	15-7	19-11	24-3
	Southern pine	#1	7-5	11-7	15-4	19-7	23-9	7-5	11-7	15-4	18-9	22-4
	Southern pine	#2	7-3	11-5	15-0	18-2	21-3	7-3	10-11	14-1	16-10	19-9
	Southern pine	#3	6-2	9-2	11-8	13-9	16-4	5-9	8-5	10-9	12-9	15-2
	Spruce-pine-fir	SS	7-1	11-2	14-8	18-9	22-10	7-1	11-2	14-8	18-9	22-4
	Spruce-pine-fir	#1	6-11	10-11	14-3	17-5	20-2	6-11	10-5	13-2	16-1	18-8
	Spruce-pine-fir	#2	6-11	10-11	14-3	17-5	20-2	6-11	10-5	13-2	16-1	18-8
	Spruce-pine-fir	#3	5-10	8-6	10-9	13-2	15-3	5-5	7-10	10-0	12-2	14-1
16	Douglas fir-larch	SS	7-0	11-0	14-5	18-5	22-5	7-0	11-0	14-5	17-11	20-10
	Douglas fir-larch	#1	6-9	10-5	13-2	16-1	18-8	6-7	9-8	12-2	14-11	17-3
	Douglas fir-larch	#2	6-7	9-9	12-4	15-1	17-6	6-2	9-0	11-5	13-11	16-2
	Douglas fir-larch	#3	5-0	7-4	9-4	11-5	13-2	4-8	6-10	8-8	10-6	12-3
	Hem-fir	SS	6-7	10-4	13-8	17-5	21-2	6-7	10-4	13-8	17-5	20-5
	Hem-fir	#1	6-5	10-2	12-10	15-8	18-2	6-5	9-5	11-11	14-6	16-10
	Hem-fir	#2	6-2	9-7	12-2	14-10	17-3	6-1	8-11	11-3	13-9	15-11
	Hem-fir	#3	5-0	7-4	9-4	11-5	13-2	4-8	6-10	8-8	10-6	12-3
	Southern pine	SS	6-10	10-9	14-2	18-1	22-0	6-10	10-9	14-2	18-1	22-0
	Southern pine	#1	6-9	10-7	13-11	17-6	20-11	6-9	10-7	13-8	16-2	19-4
	Southern pine	#2	6-7	10-2	13-2	15-9	18-5	6-7	9-5	12-2	14-7	17-1
	Southern pine	#3	5-4	7-11	10-1	11-11	14-2	4-11	7-4	9-4	11-0	13-1
	Spruce-pine-fir	SS	6-5	10-2	13-4	17-0	20-9	6-5	10-2	13-4	16-8	19-4
	Spruce-pine-fir	#1	6-4	9-9	12-4	15-1	17-6	6-2	9-0	11-5	13-11	16-2
	Spruce-pine-fir	#2	6-4	9-9	12-4	15-1	17-6	6-2	9-0	11-5	13-11	16-2
	Spruce-pine-fir	#3	5-0	7-4	9-4	11-5	13-2	4-8	6-10	8-8	10-6	12-3
19.2	Douglas fir-larch	SS	6-7	10-4	13-7	17-4	20-6	6-7	10-4	13-5	16-5	19-0
	Douglas fir-larch	#1	6-4	9-6	12-0	14-8	17-1	6-0	8-10	11-2	13-7	15-9
	Douglas fir-larch	#2	6-1	8-11	11-3	13-9	15-11	5-7	8-3	10-5	12-9	14-9
	Douglas fir-larch	#3	4-7	6-9	8-6	10-5	12-1	4-3	6-3	7-11	9-7	11-2
	Hem-fir	SS	6-2	9-9	12-10	16-5	19-11	6-2	9-9	12-10	16-1	18-8
	Hem-fir	#1	6-1	9-3	11-9	14-4	16-7	5-10	8-7	10-10	13-3	15-5
	Hem-fir	#2	5-9	8-9	11-1	13-7	15-9	5-7	8-1	10-3	12-7	14-7
	Hem-fir	#3	4-7	6-9	8-6	10-5	12-1	4-3	6-3	7-11	9-7	11-2

(continiued)

TABLE R802.5.1(6)—continued
RAFTER SPANS FOR COMMON LUMBER SPECIES
(Ground snow load=50 psf, ceiling attached to rafters, L/Δ = 240)

RAFTER SPACING (inches)	SPECIES AND GRADE		DEAD LOAD = 10 psf					DEAD LOAD = 20 psf				
			2 × 4	2 × 6	2 × 8	2 × 10	2 × 12	2 × 4	2 × 6	2 × 8	2 × 10	2 × 12
			Maximum rafter spans[a]									
			(feet-inches)	(feet-inches)	(feet-inches)	(feet-inches)	(feet-inches)	(feet-inches)	(feet-inches)	(feet-inches)	(feet-inches)	(feet-inches)
19.2	Southern pine	SS	6-5	10-2	13-4	17-0	20-9	6-5	10-2	13-4	17-0	20-9
	Southern pine	#1	6-4	9-11	13-1	16-0	19-1	6-4	9-11	12-5	14-10	17-8
	Southern pine	#2	6-2	9-4	12-0	14-4	16-10	6-0	8-8	11-2	13-4	15-7
	Southern pine	#3	4-11	7-3	9-2	10-10	12-11	4-6	6-8	8-6	10-1	12-0
	Spruce-pine-fir	SS	6-1	9-6	12-7	16-0	19-1	6-1	9-6	12-5	15-3	17-8
	Spruce-pine-fir	#1	5-11	8-11	11-3	13-9	15-11	5-7	8-3	10-5	12-9	14-9
	Spruce-pine-fir	#2	5-11	8-11	11-3	13-9	15-11	5-7	8-3	10-5	12-9	14-9
	Spruce-pine-fir	#3	4-7	6-9	8-6	10-5	12-1	4-3	6-3	7-11	9-7	11-2
24	Douglas fir-larch	SS	6-1	9-7	12-7	15-10	18-4	6-1	9-6	12-0	14-8	17-0
	Douglas fir-larch	#1	5-10	8-6	10-9	13-2	15-3	5-5	7-10	10-0	12-2	14-1
	Douglas fir-larch	#2	5-5	7-11	10-1	12-4	14-3	5-0	7-4	9-4	11-5	13-2
	Douglas fir-larch	#3	4-1	6-0	7-7	9-4	10-9	3-10	5-7	7-1	8-7	10-0
	Hem-fir	SS	5-9	9-1	11-11	15-2	18-0	5-9	9-1	11-9	14-5	15-11
	Hem-fir	#1	5-8	8-3	10-6	12-10	14-10	5-3	7-8	9-9	11-10	13-9
	Hem-fir	#2	5-4	7-10	9-11	12-1	14-1	4-11	7-3	9-2	11-3	13-0
	Hem-fir	#3	4-1	6-0	7-7	9-4	10-9	3-10	5-7	7-1	8-7	10-0
	Southern pine	SS	6-0	9-5	12-5	15-10	19-3	6-0	9-5	12-5	15-10	19-3
	Southern pine	#1	5-10	9-3	12-0	14-4	17-1	5-10	8-10	11-2	13-3	15-9
	Southern pine	#2	5-9	8-4	10-9	12-10	15-1	5-5	7-9	10-0	11-11	13-11
	Southern pine	#3	4-4	6-5	8-3	9-9	11-7	4-1	6-0	7-7	9-0	10-8
	Spruce-pine-fir	SS	5-8	8-10	11-8	14-8	17-1	5-8	8-10	11-2	13-7	15-9
	Spruce-pine-fir	#1	5-5	7-11	10-1	12-4	14-3	5-0	7-4	9-4	11-5	13-2
	Spruce-pine-fir	#2	5-5	7-11	10-1	12-4	14-3	5-0	7-4	9-4	11-5	13-2
	Spruce-pine-fir	#3	4-1	6-0	7-7	9-4	10-9	3-10	5-7	7-1	8-7	10-0

Check sources for availability of lumber in lengths greater than 20 feet.

For SI: 1 inch = 25.4 mm, 1 foot = 304.8 mm, 1 pound per square foot = 0.0479 kPa.

a. The tabulated rafter spans assume that ceiling joists are located at the bottom of the attic space or that some other method of resisting the outward push of the rafters on the bearing walls, such as rafter ties, is provided at that location. When ceiling joists or rafter ties are located higher in the attic space, the rafter spans shall be multiplied by the factors given below:

H_C/H_R	Rafter Span Adjustment Factor
1/3	0.67
1/4	0.76
1/5	0.83
1/6	0.90
1/7.5 or less	1.00

where:

H_C = Height of ceiling joists or rafter ties measured vertically above the top of the rafter support walls.

H_R = Height of roof ridge measured vertically above the top of the rafter support walls.

❖ See the commentary to Sections R802.5 and R802.5.1 and Table R802.5.1(1).

TABLE R802.5.1(7)
RAFTER SPANS FOR 70 PSF GROUND SNOW LOAD
(Ceiling not attached to rafters, L/Δ = 180)

RAFTER SPACING (inches)	SPECIES AND GRADE		DEAD LOAD = 10 psf					DEAD LOAD = 20 psf				
			2 × 4	2 × 6	2 × 8	2 × 10	2 × 12	2 × 4	2 × 6	2 × 8	2 × 10	2 × 12
			Maximum Rafter Spans[a]									
			(feet-inches)	(feet-inches)	(feet-inches)	(feet-inches)	(feet-inches)	(feet-inches)	(feet-inches)	(feet-inches)	(feet-inches)	(feet-inches)
12	Douglas fir-larch	SS	7-7	11-10	15-8	19-5	22-6	7-7	11-10	15-0	18-3	21-2
	Douglas fir-larch	#1	7-1	10-5	13-2	16-1	18-8	6-8	9-10	12-5	15-2	17-7
	Douglas fir-larch	#2	6-8	9-9	12-4	15-1	17-6	6-3	9-2	11-8	14-2	16-6
	Douglas fir-larch	#3	5-0	7-4	9-4	11-5	13-2	4-9	6-11	8-9	10-9	12-5
	Hem-fir	SS	7-2	11-3	14-9	18-10	22-1	7-2	11-3	14-8	18-0	20-10
	Hem-fir	#1	6-11	10-2	12-10	15-8	18-2	6-6	9-7	12-1	14-10	17-2
	Hem-fir	#2	6-7	9-7	12-2	14-10	17-3	6-2	9-1	11-5	14-0	16-3
	Hem-fir	#3	5-0	7-4	9-4	11-5	13-2	4-9	6-11	8-9	10-9	12-5
	Southern pine	SS	7-5	11-8	15-4	19-7	23-10	7-5	11-8	15-4	19-7	23-10
	Southern pine	#1	7-3	11-5	14-9	17-6	20-11	7-3	11-1	13-11	16-6	19-8
	Southern pine	#2	7-1	10-2	13-2	15-9	18-5	6-8	9-7	12-5	14-10	17-5
	Southern pine	#3	5-4	7-11	10-1	11-11	14-2	5-1	7-5	9-6	11-3	13-4
	Spruce-pine-fir	SS	7-0	11-0	14-6	18-0	20-11	7-0	11-0	13-11	17-0	19-8
	Spruce-pine-fir	#1	6-8	9-9	12-4	15-1	17-6	6-3	9-2	11-8	14-2	16-6
	Spruce-pine-fir	#2	6-8	9-9	12-4	15-1	17-6	6-3	9-2	11-8	14-2	16-6
	Spruce-pine-fir	#3	5-0	7-4	9-4	11-5	13-2	4-9	6-11	8-9	10-9	12-5
16	Douglas fir-larch	SS	6-10	10-9	13-9	16-10	19-6	6-10	10-3	13-0	15-10	18-4
	Douglas fir-larch	#1	6-2	9-0	11-5	13-11	16-2	5-10	8-6	10-9	13-2	15-3
	Douglas fir-larch	#2	5-9	8-5	10-8	13-1	15-2	5-5	7-11	10-1	12-4	14-3
	Douglas fir-larch	#3	4-4	6-4	8-1	9-10	11-5	4-1	6-0	7-7	9-4	10-9
	Hem-fir	SS	6-6	10-2	13-5	16-6	19-2	6-6	10-1	12-9	15-7	18-0
	Hem-fir	#1	6-0	8-9	11-2	13-7	15-9	5-8	8-3	10-6	12-10	14-10
	Hem-fir	#2	5-8	8-4	10-6	12-10	14-11	5-4	7-10	9-11	12-1	14-1
	Hem-fir	#3	4-4	6-4	8-1	9-10	11-5	4-1	6-0	7-7	9-4	10-9
	Southern pine	SS	6-9	10-7	14-0	17-10	21-8	6-9	10-7	14-0	17-10	21-0
	Southern pine	#1	6-7	10-2	12-9	15-2	18-1	6-5	9-7	12-0	14-4	17-1
	Southern pine	#2	6-2	8-10	11-5	13-7	16-0	5-10	8-4	10-9	12-10	15-1
	Southern pine	#3	4-8	6-10	8-9	10-4	12-3	4-4	6-5	8-3	9-9	11-7
	Spruce-pine-fir	SS	6-4	10-0	12-9	15-7	18-1	6-4	9-6	12-0	14-8	17-1
	Spruce-pine-fir	#1	5-9	8-5	10-8	13-1	15-2	5-5	7-11	10-1	12-4	14-3
	Spruce-pine-fir	#2	5-9	8-5	10-8	13-1	15-2	5-5	7-11	10-1	12-4	14-3
	Spruce-pine-fir	#3	4-4	6-4	8-1	9-10	11-5	4-1	6-0	7-7	9-4	10-9
19.2	Douglas fir-larch	SS	6-5	9-11	12-7	15-4	17-9	6-5	9-4	11-10	14-5	16-9
	Douglas fir-larch	#1	5-7	8-3	10-5	12-9	14-9	5-4	7-9	9-10	12-0	13-11
	Douglas fir-larch	#2	5-3	7-8	9-9	11-11	13-10	5-0	7-3	9-2	11-3	13-0
	Douglas fir-larch	#3	4-0	5-10	7-4	9-0	10-5	3-9	5-6	6-11	8-6	9-10
	Hem-fir	SS	6-1	9-7	12-4	15-1	17-4	6-1	9-2	11-8	14-2	15-5
	Hem-fir	#1	5-6	8-0	10-2	12-5	14-5	5-2	7-7	9-7	11-8	13-7
	Hem-fir	#2	5-2	7-7	9-7	11-9	13-7	4-11	7-2	9-1	11-1	12-10
	Hem-fir	#3	4-0	5-10	7-4	9-0	10-5	3-9	5-6	6-11	8-6	9-10

(continued)

TABLE R802.5.1(7)—continued
RAFTER SPANS FOR 70 PSF GROUND SNOW LOAD
(Ceiling not attached to rafters, L/Δ = 180)

RAFTER SPACING (inches)	SPECIES AND GRADE		DEAD LOAD = 10 psf					DEAD LOAD = 20 psf				
			2 × 4	2 × 6	2 × 8	2 × 10	2 × 12	2 × 4	2 × 6	2 × 8	2 × 10	2 × 12
			Maximum Rafter Spans[a]									
			(feet-inches)	(feet-inches)	(feet-inches)	(feet-inches)	(feet-inches)	(feet-inches)	(feet-inches)	(feet-inches)	(feet-inches)	(feet-inches)
19.2	Southern pine	SS	6-4	10-0	13-2	16-9	20-4	6-4	10-0	13-2	16-5	19-2
	Southern pine	#1	6-3	9-3	11-8	13-10	16-6	5-11	8-9	11-0	13-1	15-7
	Southern pine	#2	5-7	8-1	10-5	12-5	14-7	5-4	7-7	9-10	11-9	13-9
	Southern pine	#3	4-3	6-3	8-0	9-5	11-2	4-0	5-11	7-6	8-10	10-7
	Spruce-pine-fir	SS	6-0	9-2	11-8	14-3	16-6	5-11	8-8	11-0	13-5	15-7
	Spruce-pine-fir	#1	5-3	7-8	9-9	11-11	13-10	5-0	7-3	9-2	11-3	13-0
	Spruce-pine-fir	#2	5-3	7-8	9-9	11-11	13-10	5-0	7-3	9-2	11-3	13-0
	Spruce-pine-fir	#3	4-0	5-10	7-4	9-0	10-5	3-9	5-6	6-11	8-6	9-10
24	Douglas fir-larch	SS	6-0	8-10	11-3	13-9	15-11	5-9	8-4	10-7	12-11	15-0
	Douglas fir-larch	#1	5-0	7-4	9-4	11-5	13-2	4-9	6-11	8-9	10-9	12-5
	Douglas fir-larch	#2	4-8	6-11	8-9	10-8	12-4	4-5	6-6	8-3	10-0	11-8
	Douglas fir-larch	#3	3-7	5-2	6-7	8-1	9-4	3-4	4-11	6-3	7-7	8-10
	Hem-fir	SS	5-8	8-8	11-0	13-6	13-11	5-7	8-3	10-5	12-4	12-4
	Hem-fir	#1	4-11	7-2	9-1	11-1	12-10	4-7	6-9	8-7	10-6	12-2
	Hem-fir	#2	4-8	6-9	8-7	10-6	12-2	4-4	6-5	8-1	9-11	11-6
	Hem-fir	#3	3-7	5-2	6-7	8-1	9-4	3-4	4-11	6-3	7-7	8-10
	Southern pine	SS	5-11	9-3	12-2	15-7	18-2	5-11	9-3	12-2	14-8	17-2
	Southern pine	#1	5-7	8-3	10-5	12-5	14-9	5-3	7-10	9-10	11-8	13-11
	Southern pine	#2	5-0	7-3	9-4	11-1	13-0	4-9	6-10	8-9	10-6	12-4
	Southern pine	#3	3-9	5-7	7-1	8-5	10-0	3-7	5-3	6-9	7-11	9-5
	Spruce-pine-fir	SS	5-6	8-3	10-5	12-9	14-9	5-4	7-9	9-10	12-0	12-11
	Spruce-pine-fir	#1	4-8	6-11	8-9	10-8	12-4	4-5	6-6	8-3	10-0	11-8
	Spruce-pine-fir	#2	4-8	6-11	8-9	10-8	12-4	4-5	6-6	8-3	10-0	11-8
	Spruce-pine-fir	#3	3-7	5-2	6-7	8-1	9-4	3-4	4-11	6-3	7-7	8-10

Check sources for availability of lumber in lengths greater than 20 feet.

For SI: 1 inch = 25.4 mm, 1 foot = 304.8 mm, 1 pound per square foot = 0.0479 kPa.

a. The tabulated rafter spans assume that ceiling joists are located at the bottom of the attic space or that some other method of resisting the outward push of the rafters on the bearing walls, such as rafter ties, is provided at that location. When ceiling joists or rafter ties are located higher in the attic space, the rafter spans shall be multiplied by the factors given below:

H_C/H_R	Rafter Span Adjustment Factor
1/3	0.67
1/4	0.76
1/5	0.83
1/6	0.90
1/7.5 or less	1.00

where:

H_C = Height of ceiling joists or rafter ties measured vertically above the top of the rafter support walls.

H_R = Height of roof ridge measured vertically above the top of the rafter support walls.

❖ See the commentary to Sections R802.5 and R802.5.1 and Table R802.5.1(1).

TABLE R802.5.1(8)
RAFTER SPANS FOR 70 PSF GROUND SNOW LOAD
(Ceiling attached to rafters, L/Δ = 240)

RAFTER SPACING (inches)	SPECIES AND GRADE		DEAD LOAD = 10 psf					DEAD LOAD = 20 psf				
			2 × 4	2 × 6	2 × 8	2 × 10	2 × 12	2 × 4	2 × 6	2 × 8	2 × 10	2 × 12
			Maximum rafter spans[a]									
			(feet - inches)	(feet - inches)	(feet - inches)	(feet - inches)	(feet - inches)	(feet - inches)	(feet - inches)	(feet - inches)	(feet - inches)	(feet - inches)
12	Douglas fir-larch	SS	6-10	10-9	14-3	18-2	22-1	6-10	10-9	14-3	18-2	21-2
	Douglas fir-larch	#1	6-7	10-5	13-2	16-1	18-8	6-7	9-10	12-5	15-2	17-7
	Douglas fir-larch	#2	6-6	9-9	12-4	15-1	17-6	6-3	9-2	11-8	14-2	16-6
	Douglas fir-larch	#3	5-0	7-4	9-4	11-5	13-2	4-9	6-11	8-9	10-9	12-5
	Hem-fir	SS	6-6	10-2	13-5	17-2	20-10	6-6	10-2	13-5	17-2	20-10
	Hem-fir	#1	6-4	10-0	12-10	15-8	18-2	6-4	9-7	12-1	14-10	17-2
	Hem-fir	#2	6-1	9-6	12-2	14-10	17-3	6-1	9-1	11-5	14-0	16-3
	Hem-fir	#3	5-0	7-4	9-4	11-5	13-2	4-9	6-11	8-9	10-9	12-5
	Southern pine	SS	6-9	10-7	14-0	17-10	21-8	6-9	10-7	14-0	17-10	21-8
	Southern pine	#1	6-7	10-5	13-8	17-6	20-11	6-7	10-5	13-8	16-6	19-8
	Southern pine	#2	6-6	10-2	13-2	15-9	18-5	6-6	9-7	12-5	14-10	17-5
	Southern pine	#3	5-4	7-11	10-1	11-11	14-2	5-1	7-5	9-6	11-3	13-4
	Spruce-pine-fir	SS	6-4	10-0	13-2	16-9	20-5	6-4	10-0	13-2	16-9	19-8
	Spruce-pine-fir	#1	6-2	9-9	12-4	15-1	17-6	6-2	9-2	11-8	14-2	16-6
	Spruce-pine-fir	#2	6-2	9-9	12-4	15-1	17-6	6-2	9-2	11-8	14-2	16-6
	Spruce-pine-fir	#3	5-0	7-4	9-4	11-5	13-2	4-9	6-11	8-9	10-9	12-5
16	Douglas fir-larch	SS	6-3	9-10	12-11	16-6	19-6	6-3	9-10	12-11	15-10	18-4
	Douglas fir-larch	#1	6-0	9-0	11-5	13-11	16-2	5-10	8-6	10-9	13-2	15-3
	Douglas fir-larch	#2	5-9	8-5	10-8	13-1	15-2	5-5	7-11	10-1	12-4	14-3
	Douglas fir-larch	#3	4-4	6-4	8-1	9-10	11-5	4-1	6-0	7-7	9-4	10-9
	Hem-fir	SS	5-11	9-3	12-2	15-7	18-11	5-11	9-3	12-2	15-7	18-0
	Hem-fir	#1	5-9	8-9	11-2	13-7	15-9	5-8	8-3	10-6	12-10	14-10
	Hem-fir	#2	5-6	8-4	10-6	12-10	14-11	5-4	7-10	9-11	12-1	14-1
	Hem-fir	#3	4-4	6-4	8-1	9-10	11-5	4-1	6-0	7-7	9-4	10-9
	Southern pine	SS	6-1	9-7	12-8	16-2	19-8	6-1	9-7	12-8	16-2	19-8
	Southern pine	#1	6-0	9-5	12-5	15-2	18-1	6-0	9-5	12-0	14-4	17-1
	Southern pine	#2	5-11	8-10	11-5	13-7	16-0	5-10	8-4	10-9	12-10	15-1
	Southern pine	#3	4-8	6-10	8-9	10-4	12-3	4-4	6-5	8-3	9-9	11-7
	Spruce-pine-fir	SS	5-9	9-1	11-11	15-3	18-1	5-9	9-1	11-11	14-8	17-1
	Spruce-pine-fir	#1	5-8	8-5	10-8	13-1	15-2	5-5	7-11	10-1	12-4	14-3
	Spruce-pine-fir	#2	5-8	8-5	10-8	13-1	15-2	5-5	7-11	10-1	12-4	14-3
	Spruce-pine-fir	#3	4-4	6-4	8-1	9-10	11-5	4-1	6-0	7-7	9-4	10-9
19.2	Douglas fir-larch	SS	5-10	9-3	12-2	15-4	17-9	5-10	9-3	11-10	14-5	16-9
	Douglas fir-larch	#1	5-7	8-3	10-5	12-9	14-9	5-4	7-9	9-10	12-0	13-11
	Douglas fir-larch	#2	5-3	7-8	9-9	11-11	13-10	5-0	7-3	9-2	11-3	13-0
	Douglas fir-larch	#3	4-0	5-10	7-4	9-0	10-5	3-9	5-6	6-11	8-6	9-10
	Hem-fir	SS	5-6	8-8	11-6	14-8	17-4	5-6	8-8	11-6	14-2	15-5
	Hem-fir	#1	5-5	8-0	10-2	12-5	14-5	5-2	7-7	9-7	11-8	13-7
	Hem-fir	#2	5-2	7-7	9-7	11-9	13-7	4-11	7-2	9-1	11-1	12-10
	Hem-fir	#3	4-0	5-10	7-4	9-0	10-5	3-9	5-6	6-11	8-6	9-10

(continued)

TABLE R802.5.1(8)—continued
RAFTER SPANS FOR 70 PSF GROUND SNOW LOAD
(Ceiling attached to rafters, L/Δ = 240)

RAFTER SPACING (inches)	SPECIES AND GRADE		DEAD LOAD = 10 psf					DEAD LOAD = 20 psf				
			2 × 4	2 × 6	2 × 8	2 × 10	2 × 12	2 × 4	2 × 6	2 × 8	2 × 10	2 × 12
			Maximum rafter spans[a]									
			(feet - inches)	(feet - inches)	(feet - inches)	(feet - inches)	(feet - inches)	(feet - inches)	(feet - inches)	(feet - inches)	(feet - inches)	(feet - inches)
19.2	Southern pine	SS	5-9	9-1	11-11	15-3	18-6	5-9	9-1	11-11	15-3	18-6
	Southern pine	#1	5-8	8-11	11-8	13-10	16-6	5-8	8-9	11-0	13-1	15-7
	Southern pine	#2	5-6	8-1	10-5	12-5	14-7	5-4	7-7	9-10	11-9	13-9
	Southern pine	#3	4-3	6-3	8-0	9-5	11-2	4-0	5-11	7-6	8-10	10-7
	Spruce-pine-fir	SS	5-5	8-6	11-3	14-3	16-6	5-5	8-6	11-0	13-5	15-7
	Spruce-pine-fir	#1	5-3	7-8	9-9	11-11	13-10	5-0	7-3	9-2	11-3	13-0
	Spruce-pine-fir	#2	5-3	7-8	9-9	11-11	13-10	5-0	7-3	9-2	11-3	13-0
	Spruce-pine-fir	#3	4-0	5-10	7-4	9-0	10-5	3-9	5-6	6-11	8-6	9-10
24	Douglas fir-larch	SS	5-5	8-7	11-3	13-9	15-11	5-5	8-4	10-7	12-11	15-0
	Douglas fir-larch	#1	5-0	7-4	9-4	11-5	13-2	4-9	6-11	8-9	10-9	12-5
	Douglas fir-larch	#2	4-8	6-11	8-9	10-8	12-4	4-5	6-6	8-3	10-0	11-8
	Douglas fir-larch	#3	3-7	5-2	6-7	8-1	9-4	3-4	4-11	6-3	7-7	8-10
	Hem-fir	SS	5-2	8-1	10-8	13-6	13-11	5-2	8-1	10-5	12-4	12-4
	Hem-fir	#1	4-11	7-2	9-1	11-1	12-10	4-7	6-9	8-7	10-6	12-2
	Hem-fir	#2	4-8	6-9	8-7	10-6	12-2	4-4	6-5	8-1	9-11	11-6
	Hem-fir	#3	3-7	5-2	6-7	8-1	9-4	3-4	4-11	6-3	7-7	8-10
	Southern pine	SS	5-4	8-5	11-1	14-2	17-2	5-4	8-5	11-1	14-2	17-2
	Southern pine	#1	5-3	8-3	10-5	12-5	14-9	5-3	7-10	9-10	11-8	13-11
	Southern pine	#2	5-0	7-3	9-4	11-1	13-0	4-9	6-10	8-9	10-6	12-4
	Southern pine	#3	3-9	5-7	7-1	8-5	10-0	3-7	5-3	6-9	7-11	9-5
	Spruce-pine-fir	SS	5-0	7-11	10-5	12-9	14-9	5-0	7-9	9-10	12-0	12-11
	Spruce-pine-fir	#1	4-8	6-11	8-9	10-8	12-4	4-5	6-6	8-3	10-0	11-8
	Spruce-pine-fir	#2	4-8	6-11	8-9	10-8	12-4	4-5	6-6	8-3	10-0	11-8
	Spruce-pine-fir	#3	3-7	5-2	6-7	8-1	9-4	3-4	4-11	6-3	7-7	8-10

Check sources for availability of lumber in lengths greater than 20 feet.

For SI: 1 inch = 25.4 mm, 1 foot = 304.8 mm, 1 pound per square foot = 0.0479 kPa.

a. The tabulated rafter spans assume that ceiling joists are located at the bottom of the attic space or that some other method of resisting the outward push of the rafters on the bearing walls, such as rafter ties, is provided at that location. When ceiling joists or rafter ties are located higher in the attic space, the rafter spans shall be multiplied by the factors given below:

H_C/H_R	Rafter Span Adjustment Factor
1/3	0.67
1/4	0.76
1/5	0.83
1/6	0.90
1/7.5 or less	1.00

where:

H_C = Height of ceiling joists or rafter ties measured vertically above the top of the rafter support walls.

H_R = Height of roof ridge measured vertically above the top of the rafter support walls.

❖ See the commentary to Sections R802.5 and R802.5.1 and Table R802.5.1(1).

TABLE R802.5.1(9)
RAFTER/CEILING JOIST HEEL JOINT CONNECTIONS[a, b, c, d, e, f, h]

RAFTER SLOPE	RAFTER SPACING (inches)	GROUND SNOW LOAD (psf)															
		20[g]				30				50				70			
		Roof span (feet)															
		12	20	28	36	12	20	28	36	12	20	28	36	12	20	28	36
		Required number of 16d common nails[a, b] per heel joint splices[c, d, e, f]															
3:12	12	4	6	8	10	4	6	8	11	5	8	12	15	6	11	15	20
	16	5	8	10	13	5	8	11	14	6	11	15	20	8	14	20	26
	24	7	11	15	19	7	11	16	21	9	16	23	30	12	21	30	39
4:12	12	3	5	6	8	3	5	6	8	4	6	9	11	5	8	12	15
	16	4	6	8	10	4	6	8	11	5	8	12	15	6	11	15	20
	24	5	8	12	15	5	9	12	16	7	12	17	22	9	16	23	29
5:12	12	3	4	5	6	3	4	5	7	3	5	7	9	4	7	9	12
	16	3	5	6	8	3	5	7	9	4	7	9	12	5	9	12	16
	24	4	7	9	12	4	7	10	13	6	10	14	18	7	13	18	23
7:12	12	3	4	4	5	3	3	4	5	3	4	5	7	3	5	7	9
	16	3	4	5	6	3	4	5	6	3	5	7	9	4	6	9	11
	24	3	5	7	9	3	5	7	9	4	7	10	13	5	9	13	17
9:12	12	3	3	4	4	3	3	3	4	3	3	4	5	3	4	5	7
	16	3	4	4	5	3	3	4	5	3	4	5	7	3	5	7	9
	24	3	4	6	7	3	4	6	7	3	6	8	10	4	7	10	13
12:12	12	3	3	3	3	3	3	3	3	3	3	3	4	3	3	4	5
	16	3	3	4	4	3	3	3	4	3	3	4	5	3	4	5	7
	24	3	4	4	5	3	3	4	6	3	4	6	8	3	6	8	10

For SI: 1 inch = 25.4 mm, 1 foot = 304.8 mm, 1 pound per square foot = 0.0479 kPa.

a. 40d box nails shall be permitted to be substituted for 16d common nails.

b. Nailing requirements shall be permitted to be reduced 25 percent if nails are clinched.

c. Heel joint connections are not required when the ridge is supported by a load-bearing wall, header or ridge beam.

d. When intermediate support of the rafter is provided by vertical struts or purlins to a load-bearing wall, the tabulated heel joint connection requirements shall be permitted to be reduced proportionally to the reduction in span.

e. Equivalent nailing patterns are required for ceiling joist to ceiling joist lap splices.

f. When rafter ties are substituted for ceiling joists, the heel joint connection requirement shall be taken as the tabulated heel joint connection requirement for two-thirds of the actual rafter slope.

g. Applies to roof live load of 20 psf or less.

h. Tabulated heel joint connection requirements assume that ceiling joists or rafter ties are located at the bottom of the attic space. When ceiling joists or rafter ties are located higher in the attic, heel joint connection requirements shall be increased by the following factors:

H_C/H_R	Heel Joint Connection Adjustment Factor
1/3	1.5
1/4	1.33
1/5	1.25
1/6	1.2
1/10 or less	1.11

where:

H_C = Height of ceiling joists or rafter ties measured vertically above the top of the rafter support walls.

H_R = Height of roof ridge measured vertically above the top of the rafter support walls.

❖ This table contains requirements for the connection between roof rafters and ceiling joists to be used in areas with ground snow loads of 20 psf (1.44 kPa) up to 70 psf (3.35 kPa). The connection is essential to resist the thrust in the roof rafter where the ridge board does not provide vertical support of the roof rafter. Section R802.3 requires a designed ridge beam if the roof slope is less than 3:12. The 20 psf (1.44 kPa) ground snow load connection must also be used where the minimum roof live load is 20 psf (1.44 kPa) or less. See Note e of this table.

Adjustment factors for the heel joint connection in Note h are limited to cases where the ceiling joists or rafter ties are in the lower third of the attic space. When the ceiling joists or rafter ties are located higher in the attic space, lateral deflection of the rafter below the rafter ties can become excessive and require additional engineering analysis (see commentary, Section R802.3.1).

R802.7 Cutting, drilling and notching. Structural roof members shall not be cut, bored or notched in excess of the limitations specified in this section.

❖ The limitations on cutting and notching roof rafters and ceiling joists are similar to those for cutting and notching of floor joists (see commentary, Section R502.8).

R802.7.1 Sawn lumber. Cuts, notches, and holes in solid lumber joists, rafters, blocking and beams shall comply with the provisions of R502.8.1 except that cantilevered portions of rafters shall be permitted in accordance with Section R802.7.1.1.

❖ Figure R502.8 illustrates cutting and notching limitations. The tension side of a member 4 inches (102 mm) or more in thickness can be notched only at the ends (also see commentary, Section R502.8).

R802.7.1.1 Cantilevered portions of rafters. Notches on cantilevered portions of rafters are permitted provided the dimension of the remaining portion of the rafter is not less than $3^1/_2$ inches (89 mm) and the length of the cantilever does not exceed 24 inches (610 mm) in accordance with Figure R802.7.1.1.

❖ Where a cantilevered rafter is notched at the top plate, the rafter must have a minimum depth at the notch of $3^1/_2$ inches (89 mm), as shown in Figure R802.7.1.1.

R802.7.1.2 Ceiling joist taper cut. Taper cuts at the ends of the ceiling joist shall not exceed one-fourth the depth of the member in accordance with Figure R802.7.1.2.

❖ The depth of the taper at the ends of ceiling joists is measured at the inside face of the support, as shown in Figure R802.7.1.2.

R802.7.2 Engineered wood products. Cuts, notches and holes bored in trusses, structural composite lumber, structural glue-laminated members or I-joists are prohibited except where permitted by the manufacturer's recommendations or where the effects of such *alterations* are specifically considered in the design of the member by a registered *design professional.*

❖ The cutting and notching limitations for sawn lumber do not apply to engineered wood products. Structural composite lumber is a generic term which encompasses a variety of engineered composite wood products, including laminated veneer lumber (LVL). Also, included in the term are laminated strand lumber (LSL), parallel strand lumber (PSL) and oriented strand lumber (OSL). The prohibitions in this section apply to all of these products.

Engineered wood products must not be cut, notched or bored unless those alterations are considered in the design of the member. That consideration is to be made either by the manufacturer and reflected in use recommendations (which is common in I-joists, permitting some limited alterations to webs) or by a registered design professional.

R802.8 Lateral support. Roof framing members and ceiling joists having a depth-to-thickness ratio exceeding 5 to 1

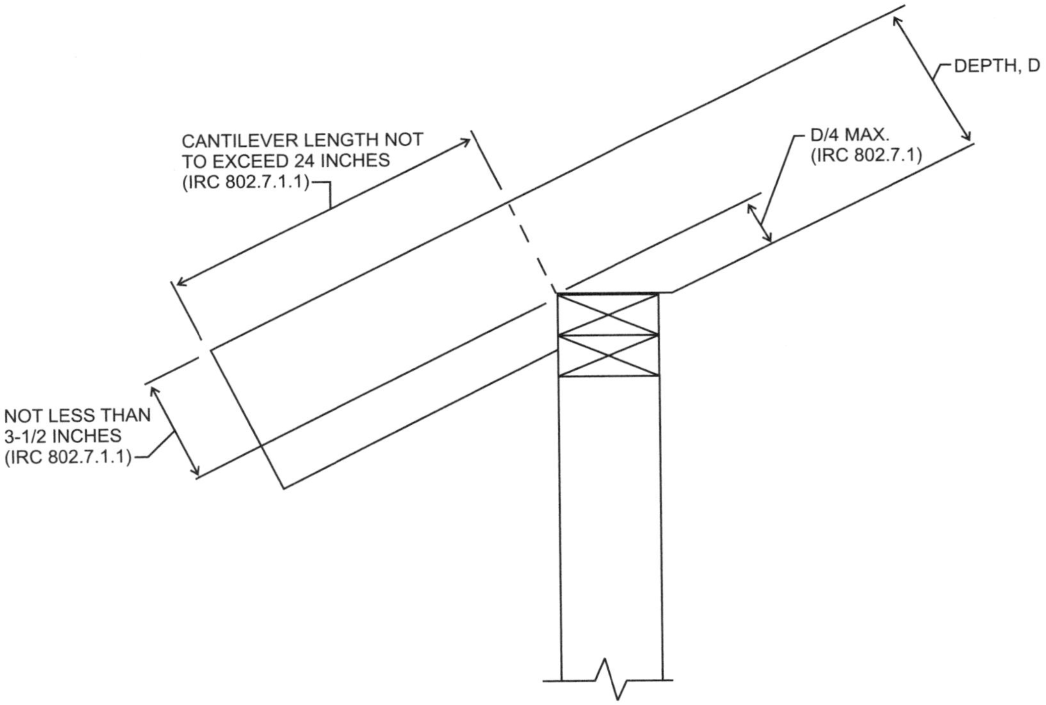

For SI: 1 inch = 25.4 mm.

FIGURE R802.7.1.1
RAFTER NOTCH

❖ See the commentary to Section R802.7.1.1.

based on nominal dimensions shall be provided with lateral support at points of bearing to prevent rotation. For roof rafters with ceiling joists attached per Table R602.3(1), the depth-to-thickness ratio for the total assembly shall be determined using the combined thickness of the rafter plus the attached ceiling joist.

Exception: Roof trusses shall be braced in accordance with Section R802.10.3.

❖ Rafters and ceiling joists must be laterally supported to prevent twisting at the supports. See the commentary to Section R802.10.3 for bracing for roof trusses.

R802.8.1 Bridging. Rafters and ceiling joists having a depth-to-thickness ratio exceeding 6 to 1 based on nominal dimensions shall be supported laterally by solid blocking, diagonal bridging (wood or metal) or a continuous 1-inch by 3-inch (25 mm by 76 mm) wood strip nailed across the rafters or ceiling joists at intervals not exceeding 8 feet (2438 mm).

❖ Bridging must be installed for rafters and ceiling joists. This is similar to the requirements for floor joists in Section R502.7.1.

R802.9 Framing of openings. Openings in roof and ceiling framing shall be framed with header and trimmer joists. When the header joist span does not exceed 4 feet (1219 mm), the header joist may be a single member the same size as the ceiling joist or rafter. Single trimmer joists may be used to carry a single header joist that is located within 3 feet (914 mm) of the trimmer joist bearing. When the header joist span exceeds 4 feet (1219 mm), the trimmer joists and the header joist shall be doubled and of sufficient cross section to support the ceiling joists or rafter framing into the header. *Approved* hangers shall be used for the header joist to trimmer joist connections when the header joist span exceeds 6 feet (1829 mm). Tail joists over 12 feet (3658 mm) long shall be supported at the header by framing anchors or on ledger strips not less than 2 inches by 2 inches (51 mm by 51 mm).

❖ Requirements for framing at roof openings are similar to those of Section R502.10 for floor openings.

R802.10 Wood trusses.

R802.10.1 Truss design drawings. Truss design drawings, prepared in conformance to Section R802.10.1, shall be provided to the *building official* and *approved* prior to installation. Truss design drawings shall include, at a minimum, the information specified below. Truss design drawings shall be provided with the shipment of trusses delivered to the jobsite.

1. Slope or depth, span and spacing.
2. Location of all joints.
3. Required bearing widths.
4. Design loads as applicable.
   - 4.1. Top chord live load (as determined from Section R301.6).
   - 4.2. Top chord dead load.
   - 4.3. Bottom chord live load.
   - 4.4. Bottom chord dead load.

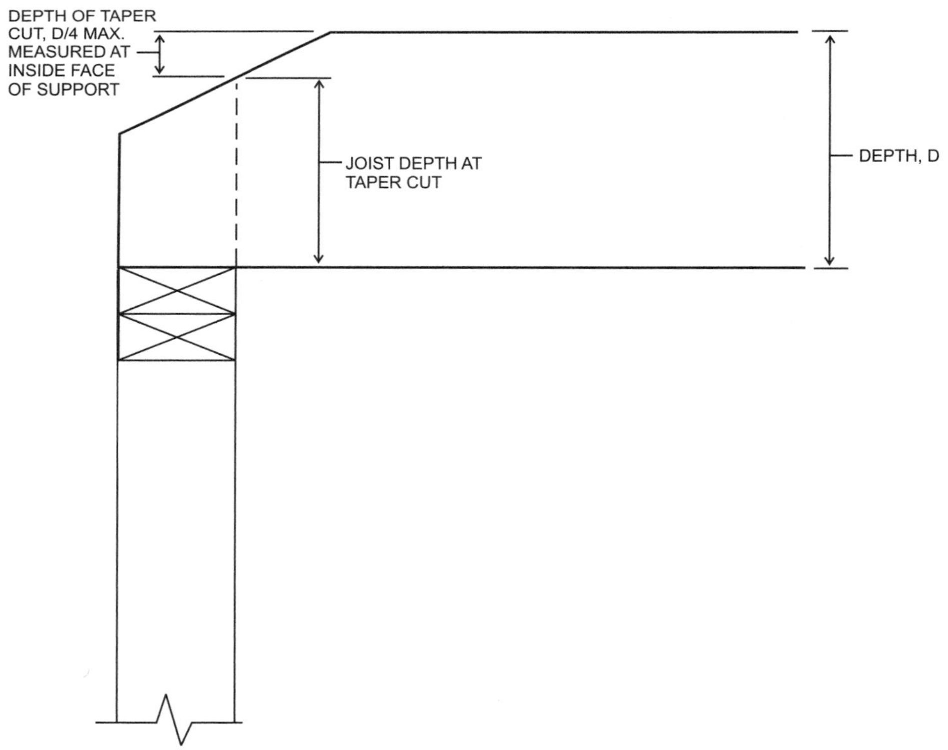

FIGURE R802.7.1.2
CEILING JOIST TAPER CUT

❖ See the commentary to Section R802.7.1.2.

4.5. Concentrated loads and their points of application.

4.6. Controlling wind and earthquake loads.

5. Adjustments to lumber and joint connector design values for conditions of use.
6. Each reaction force and direction.
7. Joint connector type and description (e.g., size, thickness or gage) and the dimensioned location of each joint connector except where symmetrically located relative to the joint interface.
8. Lumber size, species and *grade for each member.*
9. Connection requirements for:

9.1. Truss to girder-truss.

9.2. Truss ply to ply.

9.3. Field splices.

10. Calculated deflection ratio and/or maximum description for live and total load.
11. Maximum axial compression forces in the truss members to enable the building designer to design the size, connections and anchorage of the permanent continuous lateral bracing. Forces shall be shown on the truss design drawing or on supplemental documents.
12. Required permanent truss member bracing location.

❖ See the definition in Chapter 2 of "Truss design drawing." Also see the commentary to Section R802.10.2.

R802.10.2 Design. Wood trusses shall be designed in accordance with accepted engineering practice. The design and manufacture of metal-plate-connected wood trusses shall comply with ANSI/TPI 1. The truss design drawings shall be prepared by a registered professional where required by the statutes of the *jurisdiction* in which the project is to be constructed in accordance with Section R106.1.

❖ The code contains no prescriptive provisions for the design and installation of wood trusses. A design is required in accordance with accepted engineering practice. For snow load, the truss design must be checked for both the balanced and unbalanced condition. Although the unbalanced condition controls for rafter design, that is not necessarily the case for truss design. The balanced condition could control. See the commentary for Section R802.5 for the snow load criteria. In recognition of the extensive use of trusses in residential construction, the code references ANSI/TPI 1, *National Design Standard for Metal-plate-connected Wood Truss Construction*. This standard provides regulations for the design and installation of metal-plate-connected wood trusses, including the procedures for full-scale tests and testing methods for evaluating metal-plate connectors. In addition to adequate design, the trusses must be handled and erected properly so the performance capabilities of the trusses are not compromised. A truss member should never be cut without approval from the design engineer.

Usually trusses are delivered in bundles, which reduces the potential for damage. Sufficient slings and spreader bars should be used to reduce stresses caused by sway and bending.

Shop drawings showing the lumber schedule, design loads and panel point details (size, location and attachment of plates) should be filed with the building permit application and should be available at the time of inspection.

R802.10.2.1 Applicability limits. The provisions of this section shall control the design of truss roof framing when snow controls for buildings not greater than 60 feet (18 288 mm) in length perpendicular to the joist, rafter or truss span, not greater than 36 feet (10 973 mm) in width parallel to the joist, rafter or truss span, not more than three stories above *grade plane* in height, and roof slopes not smaller than 3:12 (25 percent slope) or greater than 12:12 (100 percent slope). Truss roof framing constructed in accordance with the provisions of this section shall be limited to sites subjected to a maximum design wind speed of 110 miles per hour (49 m/s), Exposure A, B or C, and a maximum ground snow load of 70 psf (3352 Pa). For consistent loading of all truss types, roof snow load is to be computed as: 0.7 p_g.

❖ This section defines the applicability limits of the snow load for wood trusses to be consistent with the snow load applicable to wood or steel rafters.

The roof snow load is based on a maximum ground snow load of 70 pounds per square foot (3.35 kPa). The roof snow load is taken as 0.7 p_g, where p_g is equal to the ground snow load.

Applied roof snow loads were calculated by multiplying the ground snow load by an 0.7 conversion factor in accordance with ASCE 7 (ASCE, 1998). No further reductions were made for special cases.

The sloped roof snow load, $p_s = C_s \times p_f$, where p_f is the flat roof snow load. $p_f = 0.7 \times C_e \times C_t \times I \times p_g$.

Unbalanced snow loads, sliding snow loads, and snow drifts on lower roofs were not considered due to the lack of evidence for damage from unbalanced loads on homes and the lack of data to typify the statistical uncertainties associated with this load pattern on residential structures. Rain-on-snow surcharge load was also not considered in the calculations. Roof slopes in this document exceed the $^1/_2$ inch (12.7 mm) per foot requirement by ASCE 7 for the added load to be considered. Therefore, roof snow load was computed as: $1.0 \times 0.7 \times 1.0 \times 1.0 \times p_g = 0.7\ p_g$.

R802.10.3 Bracing. Trusses shall be braced to prevent rotation and provide lateral stability in accordance with the requirements specified in the *construction documents* for the building and on the individual truss design drawings. In the absence of specific bracing requirements, trusses shall be braced in accordance with accepted industry practice such as the SBCA *Building Component Safety Information (BCSI) Guide to Good Practice for Handling, Installing & Bracing of Metal Plate Connected Wood Trusses.*

❖ To prevent their collapse during construction, and until permanent bracing is installed, trusses should be adequately braced temporarily. When braced for use, trusses should be positioned as close to vertical as possible; tilted trusses will not perform as required.

The construction documents and the individual truss design drawings should specify the bracing.

R802.10.4 Alterations to trusses. Truss members shall not be cut, notched, drilled, spliced or otherwise altered in any way without the approval of a registered *design professional.* Alterations resulting in the addition of load (e.g., HVAC equipment, water heater) that exceeds the design load for the truss shall not be permitted without verification that the truss is capable of supporting such additional loading.

❖ The addition of loads in excess of the design load is allowed only if the truss is shown to have the additional capacity. See the commentary for Section R502.8.2.

R802.11 Roof tie-down.

❖ Because roof uplift caused by wind can be a significant factor in roof-system damage, the code requires roof-to-wall connections capable of resisting this force.

R802.11.1 Uplift resistance. Roof assemblies shall have uplift resistance in accordance with Sections R802.11.1.2 and R802.11.1.3.

Where the uplift force does not exceed 200 pounds, rafters and trusses spaced not more than 24 inches (610 mm) on center shall be permitted to be attached to their supporting wall assemblies in accordance with Table R602.3(1).

Where the basic wind speed does not exceed 90 mph, the wind exposure category is B, the roof pitch is 5:12 or greater, and the roof span is 32 feet (9754 mm) or less, rafters and trusses spaced not more than 24 inches (610 mm) on center shall be permitted to be attached to their supporting wall assemblies in accordance with Table R602.3(1).

❖ In Table R602.3(1), the nailing of the rafter or roof truss to the top plate provides a 200-pound maximum uplift capacity where the rafter or truss is spaced not more than 24 inches (610 mm) on center. The basis of the capacity of this nailing is calculated from AF&PA NDS. The 200-pound uplift capacity is adequate for a basic wind speed of 90 miles per hour, Exposure B with a 32-foot (9754 mm) roof span and a 5:12 or greater pitch.

For trusses, there are three options provided for determining the roof uplift: the truss design drawings, Table R802.11 or an engineered design. In many jurisdictions (particularly rural ones), an engineered truss design is not required and the local truss fabricator will run the software from the plate company. These jurisdictions may also have limited or no plan review. Thus, there is less opportunity to ensure that the proper wind speed, building dimensions, mean roof height, etc., are used, and a possibility that overly conservative roof uplift loads will be generated on the truss design drawing. Hence, the ability to determine an uplift load from Table R802.11, even when there are truss drawings, is provided; however, use of Table R802.11 is limited to roof rafters and single-ply trusses within the applicability limits of Section R802.10.1.2. Girder trusses and roof beams require engineered connections and/or use of the truss design drawing values.

For rafters, there are two options provided for determining the roof uplift: Table R802.11 or an engineered design.

Table R802.11 is based on Table 2.2A of the WFCM, which is based on the latest ASCE 7 wind-load provisions. The new table expands upon the WFCM table by incorporating values for high-slope roofs. These factors were derived using the ASCE 7 wind provisions and the calculation method used to develop Table 2.2A of the WFCM. A factor for hip roofs is also added, as hip roofs have seen similar improved performance in high-wind events.

Determination of the uplift connection forces is as follows: Basic wind speed is as set forth in Table R301.2(1) and the wind exposure category is determined in accordance with Section R301.2.1.4. Using this wind speed and exposure, the designer selects the uplift forces from Table R802.11 based on the rafter or truss spacing, roof span and pitch. The uplift connection of the rafter or truss to the wall must have the capacity to resist the uplift force specified in Table R802.11.

R802.11.1.2 Truss uplift resistance. Trusses shall be attached to supporting wall assemblies by connections capable of resisting uplift forces as specified on the truss design drawings. Uplift forces shall be permitted to be determined as specified by Table R802.11, if applicable, or as determined by accepted engineering practice.

❖ See the commentary to Section R802.11.1.

R802.11.1.3 Rafter uplift resistance. Individual rafters shall be attached to supporting wall assemblies by connections capable of resisting uplift forces as determined by Table R802.11 or as determined by accepted engineering practice. Connections for beams used in a roof system shall be designed in accordance with accepted engineering practice.

❖ See the commentary to Section R802.11.1.

SECTION R803
ROOF SHEATHING

R803.1 Lumber sheathing. Allowable spans for lumber used as roof sheathing shall conform to Table R803.1. Spaced lumber sheathing for wood shingle and shake roofing shall conform to the requirements of Sections R905.7 and R905.8. Spaced lumber sheathing is not allowed in Seismic Design Category D_2.

❖ Table R803.1 specifies the minimum thickness of lumber roof sheathing based on rafter spacing. Spaced lumber sheathing used in conjunction with wood shingle or shake roofs is permitted when installed in accordance with Section R905. Because spaced lumber sheathing does not provide a roof diaphragm capable of transferring high in-plane loads to the supporting shear walls or braced wall lines, it is prohibited in structures classified as Seismic Design Category D_2.

TABLE R802.11
RAFTER OR TRUSS UPLIFT CONNECTION FORCES FROM WIND (POUNDS PER CONNECTION)[a, b, c, d, e, f, g, h]

RAFTER OR TRUSS SPACING	ROOF SPAN (feet)	EXPOSURE B							
		Basic Wind Speed (mph)							
		85		90		100		110	
		Roof Pitch		Roof Pitch		Roof Pitch		Roof Pitch	
		< 5:12	≥ 5:12	< 5:12	≥ 5:12	< 5:12	≥ 5:12	< 5:12	≥ 5:12
12″ o.c.	12	47	41	62	54	93	81	127	110
	18	59	51	78	68	119	104	165	144
	24	70	61	93	81	145	126	202	176
	28	77	67	104	90	163	142	227	197
	32	85	74	115	100	180	157	252	219
	36	93	81	126	110	198	172	277	241
	42	105	91	143	124	225	196	315	274
	48	116	101	159	138	251	218	353	307
16″ o.c.	12	63	55	83	72	124	108	169	147
	18	78	68	103	90	159	138	219	191
	24	93	81	124	108	193	168	269	234
	28	102	89	138	120	217	189	302	263
	32	113	98	153	133	239	208	335	291
	36	124	108	168	146	264	230	369	321
	42	139	121	190	165	299	260	420	365
	48	155	135	212	184	335	291	471	410
24″ o.c.	12	94	82	124	108	186	162	254	221
	18	117	102	155	135	238	207	329	286
	24	140	122	186	162	290	252	404	351
	28	154	134	208	181	326	284	454	395
	32	170	148	230	200	360	313	504	438
	36	186	162	252	219	396	345	554	482
	42	209	182	285	248	449	391	630	548
	48	232	202	318	277	502	437	706	614

RAFTER OR TRUSS SPACING	ROOF SPAN (feet)	EXPOSURE C							
		Basic Wind Speed (mph)							
		85		90		100		110	
		Roof Pitch		Roof Pitch		Roof Pitch		Roof Pitch	
		< 5:12	≥ 5:12	< 5:12	≥ 5:12	< 5:12	≥ 5:12	< 5:12	≥ 5:12
12″ o.c.	12	94	82	114	99	157	137	206	179
	18	120	104	146	127	204	177	268	233
	24	146	127	179	156	251	218	330	287
	28	164	143	201	175	283	246	372	324
	32	182	158	224	195	314	273	414	360
	36	200	174	246	214	346	301	456	397
	42	227	197	279	243	394	343	520	452
	48	254	221	313	272	441	384	583	507

(continued)

TABLE R802.11—continued
RAFTER OR TRUSS UPLIFT CONNECTION FORCES FROM WIND (POUNDS PER CONNECTION)[a, b, c, d, e, f, g, h]

RAFTER OR TRUSS SPACING	ROOF SPAN (feet)	EXPOSURE C							
		Basic Wind Speed (mph)							
		85		90		100		110	
		Roof Pitch		Roof Pitch		Roof Pitch		Roof Pitch	
		< 5:12	≥ 5:12	< 5:12	≥ 5:12	< 5:12	≥ 5:12	< 5:12	≥ 5:12
16″ o.c.	12	125	109	152	132	209	182	274	238
	18	160	139	194	169	271	236	356	310
	24	194	169	238	207	334	291	439	382
	28	218	190	267	232	376	327	495	431
	32	242	211	298	259	418	364	551	479
	36	266	231	327	284	460	400	606	527
	42	302	263	372	324	524	456	691	601
	48	338	294	416	362	587	511	775	674
24″ o.c.	12	188	164	228	198	314	273	412	358
	18	240	209	292	254	408	355	536	466
	24	292	254	358	311	502	437	660	574
	28	328	285	402	350	566	492	744	647
	32	364	317	448	390	628	546	828	720
	36	400	348	492	428	692	602	912	793
	42	454	395	558	485	786	684	1040	905
	48	508	442	626	545	882	767	1166	1014

For SI: 1 inch = 25.4 mm, 1 foot = 304.8 mm, 1 mile per hour = 0.447 m/s, 1 pound = 0.454 kg, 1 pound per linear foot = 14.5 N/m.

a. The uplift connection forces are based on a maximum 33-foot mean roof height and Wind Exposure Category B or C. For Exposure D, the uplift connection force shall be selected from the Exposure C portion of the table using the next highest tabulated basic wind speed. The Adjustment Coefficients in Table R301.2(3) shall not be used to multiply the above forces for Exposures C and D or for other mean roof heights.

b. The uplift connection forces include an allowance for roof and ceiling assembly dead load of 15 psf.

c. The tabulated uplift connection forces are limited to a maximum roof overhang of 24 inches.

d. The tabulated uplift connection forces shall be permitted to be multiplied by 0.75 for connections not located within 8 feet of building corners.

e. For buildings with hip roofs with 5:12 and greater pitch, the tabulated uplift connection forces shall be permitted to be multiplied by 0.70. This reduction shall not be combined with any other reduction in tabulated forces.

f. For wall-to-wall and wall-to-foundation connections, the uplift connection force shall be permitted to be reduced by 60 plf for each full wall above.

g. Linear interpolation between tabulated roof spans and wind speeds shall be permitted.

h. The tabulated forces for a 12-inch on-center spacing shall be permitted to be used to determine the uplift load in pounds per linear foot.

❖ See the commentary to Section R802.11.1.

TABLE R803.1
MINIMUM THICKNESS OF LUMBER ROOF SHEATHING

RAFTER OR BEAM SPACING (inches)	MINIMUM NET THICKNESS (inches)
24	$^5/_8$
48[a]	$1^1/_2$ T & G
60[b]	
72[c]	

For SI: 1 inch = 25.4 mm.

a. Minimum 270 F_b, 340,000 E.

b. Minimum 420 F_b, 660,000 E.

c. Minimum 600F_b, 1,150,000 E.

❖ See the commentary to Section R803.1.

R803.2 Wood structural panel sheathing.

R803.2.1 Identification and grade. Wood structural panels shall conform to DOC PS 1, DOC PS 2 or, when manufactured in Canada, CSA O437 or CSA O325, and shall be identified for grade, bond classification, and Performance Category by a grade mark or certificate of inspection issued by an *approved* agency. Wood structural panels shall comply with the grades specified in Table R503.2.1.1(1).

❖ Span rating and installation requirements for wood structural panels used as roof sheathing are similar to those of Section R503.2 for floor sheathing.

R803.2.1.1 Exposure durability. All wood structural panels, when designed to be permanently exposed in outdoor applications, shall be of an exterior exposure durability. Wood structural panel roof sheathing exposed to the underside may be of interior type bonded with exterior glue, identified as Exposure 1.

❖ Exterior-type wood structural panels are required when the wood structural panels are exposed to the environment. Exterior wood structural panels use a moisture-resistant glue that may penetrate the outer veneer lamination. Additionally, exterior wood structural panels typically use a higher grade of veneer for the inner plies, which are compatible with the exterior glue-line performance. Wood structural panel roof sheathing exposed only on the underside, such as eave overhangs, is permitted to be of the interior type with exterior glue because of the decrease in exposure hazard relative to moisture.

R803.2.1.2 Fire-retardant-treated plywood. The allowable unit stresses for fire-retardant-treated plywood, including fastener values, shall be developed from an *approved* method of investigation that considers the effects of anticipated temperature and humidity to which the fire-retardant-treated plywood will be subjected, the type of treatment and redrying process. The fire-retardant-treated plywood shall be graded by an *approved agency.*

❖ See the commentary to Section R802.1.3.

R803.2.2 Allowable spans. The maximum allowable spans for wood structural panel roof sheathing shall not exceed the values set forth in Table R503.2.1.1(1), or APA E30.

❖ Table 503.2.1.1(1) and APA E30 list allowable spans and loads for wood structural panel roof sheathing. Also see the commentary for Section R503.2.2.

R803.2.3 Installation. Wood structural panel used as roof sheathing shall be installed with joints staggered or not staggered in accordance with Table R602.3(1), or APA E30 for wood roof framing or with Table R804.3 for steel roof framing.

❖ This section refers to the fastening schedule in Chapter 6 for attachment to wood roof framing. Notes f and g to Table R602.3(1) establish zones requiring increased nailing based on criteria such as basic wind speed and/or mean roof height. APA E30, *APA Design and Construction Guide* is referenced as an alternative to the Chapter 6 fastening schedule.

This section also gives a cross reference to the Roof Framing Fastening Schedule, Table R804.3, for attachments to cold-formed steel roof framing.

SECTION R804 STEEL ROOF FRAMING

R804.1 General. Elements shall be straight and free of any defects that would significantly affect their structural performance. Cold-formed steel roof framing members shall comply with the requirements of this section.

❖ The provisions of this section govern cold-formed steel roof and ceiling framing.

R804.1.1 Applicability limits. The provisions of this section shall control the construction of cold-formed steel roof framing for buildings not greater than 60 feet (18 288 mm) perpendicular to the joist, rafter or truss span, not greater than 40 feet (12 192 mm) in width parallel to the joist span or truss, less than or equal to three stories above *grade* plane and with roof slopes not less than 3:12 (25-percent slope) or greater than 12:12 (100-percent slope). Cold-formed steel roof framing constructed in accordance with the provisions of this section shall be limited to sites subjected to a maximum design wind speed of 110 miles per hour (49 m/s), Exposure B or C, and a maximum ground snow load of 70 pounds per square foot (3350 Pa).

❖ Other than providing roof slope limits, this section is identical to Section R505.1.1 and the limitations illustrated in Commentary Figure R505.1.1.

R804.1.2 In-line framing. Cold-formed steel roof framing constructed in accordance with Section R804 shall be located in line with load-bearing studs in accordance with Figure R804.1.2 and the tolerances specified as follows:

1. The maximum tolerance shall be $^3/_4$ inch (19.1 mm) between the centerline of the horizontal framing member and the centerline of the vertical framing member.
2. Where the centerline of the horizontal framing member and bearing stiffener are located to one side of the center line of the vertical framing member, the maximum tolerance shall be $^1/_8$ inch (3 mm) between the web of the horizontal framing member and the edge of the vertical framing member.

❖ The code does not anticipate loads on the top plate of a wall that would cause bending of the top plate. Therefore, roof-framing members must be located directly in line with load-bearing studs of the supporting steel-framed walls as indicated in Figure R804.1.2. The permitted offset provides some tolerance for rafter location while not adversely affecting the top plate. See the commentary for Section R505.1.2.

R804.2 Structural framing. Load-bearing, cold-formed steel roof framing members shall comply with Figure R804.2(1) and with the dimensional and minimum thickness requirements specified in Tables R804.2(1) and R804.2(2). Tracks shall comply with Figure R804.2(2) and shall have a minimum flange width of $1^1/_4$ inches (32 mm).

❖ Figure R804.2(1) shows a typical load-bearing steel roof or ceiling member, which must comply with the dimensional and minimum thickness requirements specified in Tables R804.2(1) and R804.2(2). These tables are similar to those in Section R603 for walls and those in Section R505 for floors. Note a to Table R804.2(1) explains the member designation used for all steel framing members in the code. Steel thickness is expressed in mils, and Table R804.2(2) gives the equivalent thickness in inches. The referenced thickness pertains to the base metal thickness measured prior to painting or the application of galvanized coatings. The base metal thickness is typically stamped or embossed on the member by the manufacturer.

R804.2.1 Material. Load-bearing, cold-formed steel framing members shall be cold-formed to shape from structural quality sheet steel complying with the requirements of one of the following:

1. ASTM A 653: *Grades* 33 and 50 (Class 1 and 3).
2. ASTM A 792: *Grades* 33 and 50A.

3. ASTM A 1003: Structural *Grades* 33 Type H and 50 Type H.

❖ Load-bearing steel framing members must be cold formed to shape from structural quality sheet steel complying with the applicable material standards. The steel grades specified have the ductility and strength to meet the intent of these provisions.

R804.2.2 Identification. Load-bearing, cold-formed steel framing members shall have a legible *label,* stencil, stamp or embossment with the following information as a minimum:

1. Manufacturer's identification.
2. Minimum base steel thickness in inches (mm).
3. Minimum coating designation.
4. Minimum yield strength, in kips per square inch (ksi) (MPa).

❖ Load-bearing steel framing members must have a legible label, stencil, stamp or embossment. This identification allows for verification that materials that are installed are consistent with the design and meet the intent of the code.

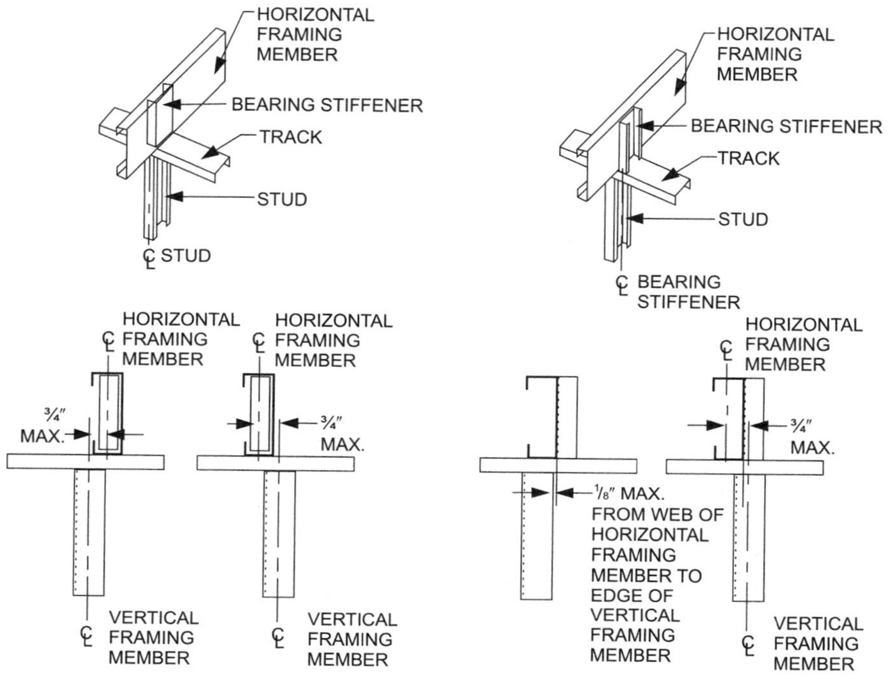

For SI: 1 inch = 25.4 mm.

FIGURE R804.1.2
IN-LINE FRAMING

❖ See commentary to Section R804.1.2.

TABLE R804.2(1)
LOAD-BEARING COLD-FORMED STEEL MEMBER SIZES

NOMINAL MEMBER SIZE MEMBER DESIGNATION[a]	WEB DEPTH (inches)	MINIMUM FLANGE WIDTH (inches)	MAXIMUM FLANGE WIDTH (inches)	MINIMUM LIP SIZE (inches)
350S162-t	3.5	1.625	2	0.5
550S162-t	5.5	1.625	2	0.5
800S162-t	8	1.625	2	0.5
1000S162-t	10	1.625	2	0.5
1200S162-t	12	1.625	2	0.5

For SI: 1 inch = 25.4 mm.

a. The member designation is defined by the first number representing the member depth in hundredths of an inch, the letter "S" representing a stud or joist member, the second number representing the flange width in hundredths of an inch, and the letter "t" shall be a number representing the minimum base metal thickness in mils [see Table R804.2(2)].

❖ See the commentary to Section R804.2.

TABLE R804.2(2)
MINIMUM THICKNESS OF COLD-FORMED STEEL MEMBERS

DESIGNATION THICKNESS (mils)	MINIMUM BASE STEEL THICKNESS (inch)
33	0.0329
43	0.0428
54	0.0538
68	0.0677
97	0.0966

For SI: 1 inch = 25.4 mm, 1 mil = 0.0254 mm.

❖ See the commentary to Section R804.2.

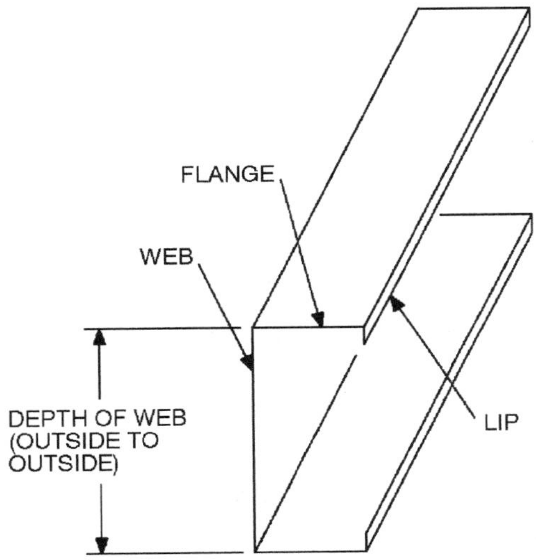

FIGURE R804.2(1)
C-SHAPED SECTION

❖ See the commentary to Section R804.2.

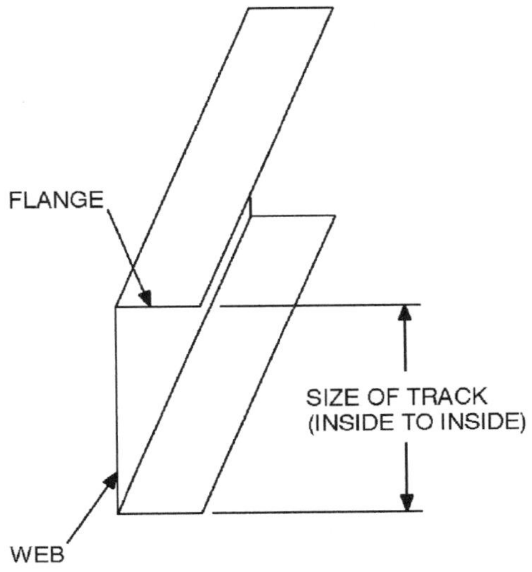

FIGURE R804.2(2)
TRACKSECTION

❖ See the commentary to Section R804.2.

R804.2.3 Corrosion protection. Load-bearing, cold-formed steel framing shall have a metallic coating complying with ASTM A 1003 and one of the following:

1. A minimum of G 60 in accordance with ASTM A 653.
2. A minimum of AZ 50 in accordance with ASTM A 792.

❖ The metallic coatings specified correspond to requirements in the referenced material standards in Section R804.2.1.

R804.2.4 Fastening requirements. Screws for steel-to-steel connections shall be installed with a minimum edge distance and center-to-center spacing of $^1/_2$ inch (13 mm), shall be self-drilling tapping, and shall conform to ASTM C 1513. Structural sheathing shall be attached to cold-formed steel roof rafters with minimum No. 8 self-drilling tapping screws that conform to ASTM C 1513. Screws for attaching structural sheathing to cold-formed steel roof framing shall have a minimum head diameter of 0.292 inch (7.4 mm) with countersunk heads and shall be installed with a minimum edge distance of $^3/_8$ inch (10 mm). Gypsum board ceilings shall be attached to cold-formed steel joists with minimum No. 6 screws conforming to ASTM C 954 or ASTM C 1513 with a bugle-head style and shall be installed in accordance with Section R805. For all connections, screws shall extend through the steel a minimum of three exposed threads. All fasteners shall have rust-inhibitive coating suitable for the installation in which they are being used, or be manufactured from material not susceptible to corrosion.

Where No. 8 screws are specified in a steel-to-steel connection, reduction of the required number of screws in the connection is permitted in accordance with the reduction factors in Table R804.2.4 when larger screws are used or when one of the sheets of steel being connected is thicker than 33 mils (0.84 mm). When applying the reduction factor, the resulting number of screws shall be rounded up.

❖ Fasteners meeting the referenced standards and installed in accordance with this section will provide a load capacity consistent with these prescriptive provisions (see commentary, Section R505.2.4).

TABLE R804.2.4
SCREW SUBSTITUTION FACTOR

SCREW SIZE	THINNEST CONNECTED STEEL SHEET (mils)	
	33	43
#8	1.0	0.67
#10	0.93	0.62
#12	0.86	0.56

For SI: 1 mil = 0.0254 mm.

❖ See the commentary to Table R505.2.4.

R804.2.5 Web holes, web hole reinforcing and web hole patching. Web holes, web hole reinforcing, and web hole patching shall be in accordance with this section.

❖ Sections R804.2.5.1 through R804.2.5.1.3 specify the criteria for web holes, web hole reinforcing and web hole patching in roof framing members.

R804.2.5.1 Web holes. Web holes in roof framing members shall comply with all of the following conditions:

1. Holes shall conform to Figure R804.2.5.1;
2. Holes shall be permitted only along the centerline of the web of the framing member;
3. Center-to-center spacing of holes shall not be less than 24 inches (610 mm);
4. The web hole width shall not be greater than one-half the member depth, or $2^1/_2$ inches (64.5 mm);
5. Holes shall have a web hole length not exceeding $4^1/_2$ inches (114 mm); and
6. The minimum distance between the edge of the bearing surface and the edge of the web hole shall not be less than 10 inches (254 mm).

Framing members with web holes not conforming to the above requirements shall be reinforced in accordance with Section R804.2.5.2, patched in accordance with Section R804.2.5.3 or designed in accordance with accepted engineering practices.

❖ To allow for routing utilities through steel-framed roof framing members, holes (also referred to as penetrations, utility holes and punchouts) are permitted in the webs of roof framing members. To avoid adversely affecting the strength of floor joists, the web holes are limited to the locations along the centerline of the web of the framing member as illustrated in Figure R804.2.5.1.

To avoid overstressing the joist web through high shear, these web holes must occur more than 10 inches (254 mm) from the edge of a load-bearing surface. Patching of web holes must extend at least 1 inch (25 mm) beyond all edges of the hole, and it must be fastened to the web with No. 8 screws having a minimum edge distance of $^1/_2$ inch (12.7 mm). Framing members with web holes exceeding these limits must be reinforced, patched or designed in accordance with accepted engineering practice (see commentary, Sections R804.2.5.2 and R804.2.5.3). Holes are permitted only along the centerline of the web of roof-framing members. Their location and size are limited as shown in Figure R804.2.5.1. Framing members with web holes exceeding these limits must be patched or designed in accordance with accepted engineering practices (see commentary, Sections R804.2.5.2 and R804.2.5.3).

R804.2.5.2 Web hole reinforcing. Reinforcement of web holes in ceiling joists not conforming to the requirements of Section R804.2.5.1 shall be permitted if the hole is located fully within the center 40 percent of the span and the depth and length of the hole does not exceed 65 percent of the flat width of the web. The reinforcing shall be a steel plate or C-shape section with a hole that does not exceed the web hole size limitations of Section R804.2.5.1 for the member being reinforced. The steel reinforcing shall be the same thickness as the receiving member and shall extend at least 1 inch (25.4 mm) beyond all edges of the hole. The steel reinforcing shall be fastened to the web of the receiving member with No. 8

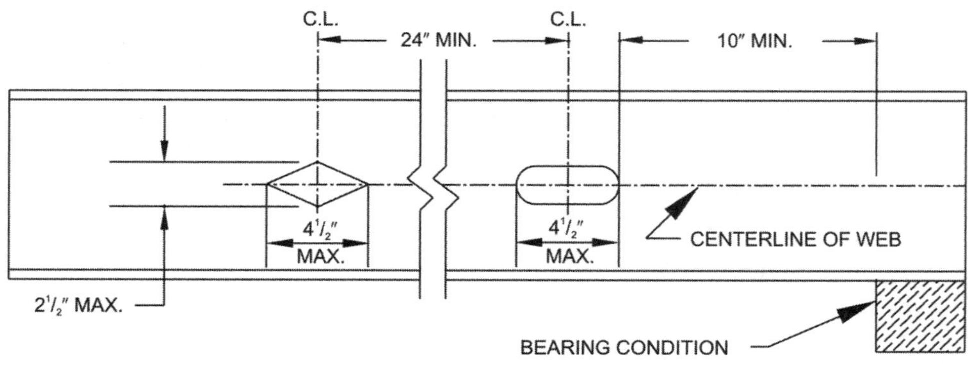

For SI: 1 inch = 25.4 mm.

FIGURE R804.2.5.1
WEB HOLES

❖ See the commentary to Section R804.2.5.1.

screws spaced no greater than 1 inch (25.4 mm) center-to-center along the edges of the patch with minimum edge distance of $^1/_2$ inch (13 mm).

❖ Web holes that are not within the criteria of Section R804.2.5.1 are permitted if they are reinforced and comply with the criteria of this section.

The reinforcing provisions of this section were developed based on engineering judgement and accepted engineering practice and confirmed by testing.

R804.2.5.3 Hole patching. Patching of web holes in roof framing members not conforming to the requirements in Section R804.2.5.1 shall be permitted in accordance with either of the following methods:

1. Framing members shall be replaced or designed in accordance with accepted engineering practices where web holes exceed the following size limits:
   1.1. The depth of the hole, measured across the web, exceeds 70 percent of the flat width of the web; or
   1.2. The length of the hole measured along the web, exceeds 10 inches (254 mm) or the depth of the web, whichever is greater.
2. Web holes not exceeding the dimensional requirements in Section R804.2.5.3, Item 1, shall be patched with a solid steel plate, stud section or track section in accordance with Figure R804.2.5.3. The steel patch shall, as a minimum, be the same thickness as the receiving member and shall extend at least 1 inch (25 mm) beyond all edges of the hole. The steel patch shall be fastened to the web of the receiving member with No.8 screws spaced no greater than 1 inch (25 mm) center-to-center along the edges of the patch with minimum edge distance of $^1/_2$ inch (13 mm).

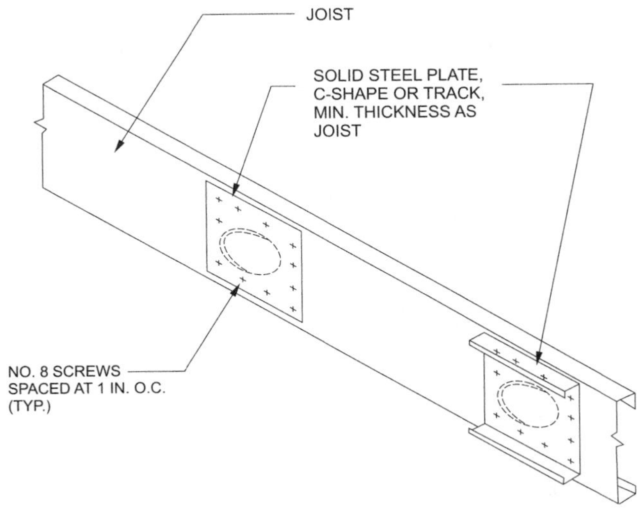

For SI: 1 inch = 25.4 mm.

FIGURE R804.2.5.3
WEB HOLE PATCH

❖ See the commentary to Section R804.2.5.3.

❖ Framing members with web holes exceeding the limits, outlined in Item 1 of this section, must be replaced or designed in accordance with accepted engineering practice. Framing members with web holes within the limits of Item 1, but exceeding the limits of Section R804.2.5.1, must be patched in accordance with Item 2 of this section (see commentary, Section R804.2.5.1).

R804.3 Roof construction. Cold-formed steel roof systems constructed in accordance with the provisions of this section shall consist of both ceiling joists and rafters in accordance with Figure R804.3 and fastened in accordance with Table R804.3, and hip framing in accordance with Section R804.3.3.

❖ Figure R804.3 illustrates steel roof system components. The roof-framing fastening schedule in Table R804.3 specifies the minimum connection requirements for components of the roof-framing system. This section contains provisions for both gable and hip roofs. Hip roof framing is in accordance with Section R804.3.3.

R804.3.1 Ceiling joists. Cold-formed steel ceiling joists shall be in accordance with this section.

❖ Sections R804.3.1.1 through R804.3.1.5 provide the requirements for ceiling joist size, bearing stiffeners, bottom flange bracing, top flange bracing, and splicing.

R804.3.1.1 Minimum ceiling joist size. Ceiling joist size and thickness shall be determined in accordance with the limits set forth in Tables R804.3.1.1(1) through R804.3.1.1(8). When determining the size of ceiling joists, the lateral support of the top flange shall be classified as unbraced, braced at mid-span or braced at third points in accordance with Section R804.3.1.4. Where sheathing material is attached to the top flange of ceiling joists or where the bracing is spaced closer than third point of the joists, the "third point" values from Tables R804.3.1.1(1) through R804.3.1.1(8) shall be used.

Ceiling joists shall have a bearing support length of not less than $1^1/_2$ inches (38 mm) and shall be connected to roof rafters (heel joint) with No. 10 screws in accordance with Figures R804.3.1.1(1) and R804.3.1.1(2) and Table 804.3.1.1(9).

When continuous joists are framed across interior bearing supports, the interior bearing supports shall be located within 24 inches (610 mm) of midspan of the ceiling joist, and the individual spans shall not exceed the applicable spans in Tables R804.3.1.1(2), R804.3.1.1(4), R804.3.1.1(6) and R804.3.1.1(8).

When the *attic* is to be used as an *occupied space,* the ceiling joists shall be designed in accordance with Section R505.

❖ Allowable spans for ceiling joists are given in Tables R804.3.1.1(1), R804.3.1.1(2), R804.3.1.1(5) and R804.3.1.1(6) for live loads of 10 psf (479 Pa) and Tables R804.3.1.1(3), R804.3.1.1(4), R804.3.1.1(7) and R804.3.1.1(8) for live loads of 20 psf (958 Pa). An attic with occupied space requires the use of higher design live loads, and the ceiling joists must be designed in accordance with Section R505.3.2.

Ceiling joists must be connected to rafters in accordance with Figures R804.3.1.1(1) and R804.3.1.1(2) and Table R804.3.1.1(9). The use of the allowable span tables is illustrated below.

Example:

Steel ceiling joists at 16 inches (406 mm) on center support a ceiling with limited attic storage and are braced at mid span. Their span is 17 feet, 6 inches (5334 mm) in a 35-foot-wide (10 668 mm) building with a roof slope of 8:12 and 30 psf (1436 Pa) snow load. Determine the required joist size and the required connection to rafters.

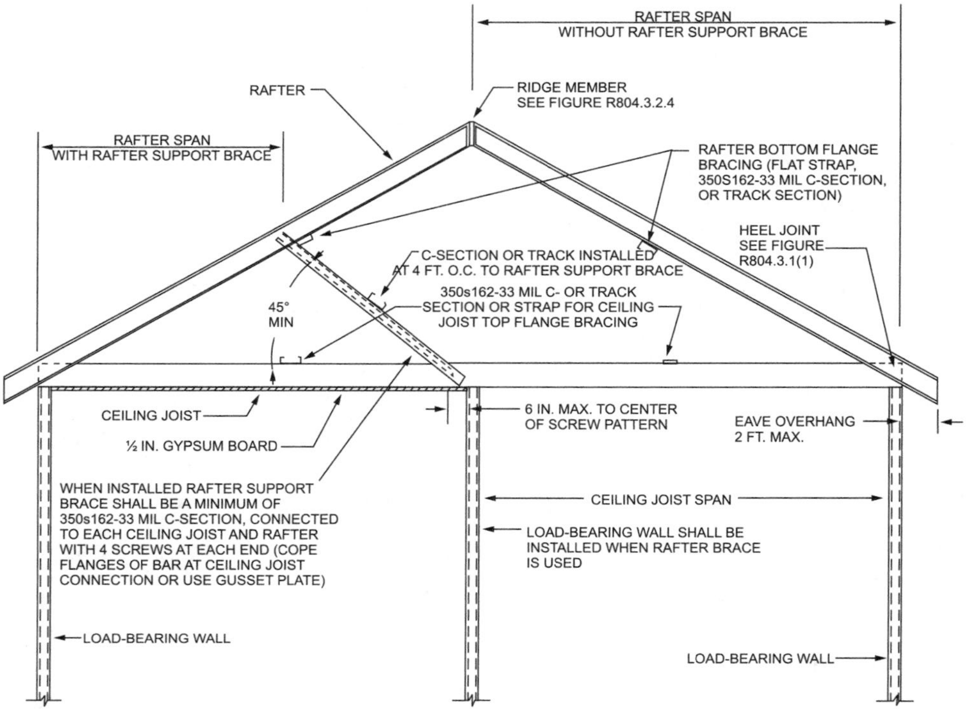

For SI: 1 inch = 25.4 mm, 1 foot = 304.8 mm, 1 mil = 0.0254 mm.

FIGURE R804.3
STEEL ROOF CONSTRUCTION

❖ This figure is a schematic of typical steel-framed room construction, providing section references to the code requirements for many of the roof components.

TABLE R804.3
ROOF FRAMING FASTENING SCHEDULE[a, b]

DESCRIPTION OF BUILDING ELEMENTS	NUMBER AND SIZE OF FASTENERS	SPACING OF FASTENERS
Ceiling joist to top track of load-bearing wall	2 No. 10 screws	Each joist
Roof sheathing (oriented strand board or plywood) to rafter	No. 8 screws	6″ o.c. on edges and 12″ o.c. at interior supports. 6″ o.c. at gable end truss
Truss to bearing wall[a]	2 No. 10 screws	Each truss
Gable end truss to end wall top track	No. 10 screws	12″ o.c.
Rafter to ceiling joist	Minimum No. 10 screws, per Table R804.3.1.1(9)	Evenly spaced, not less than $^1/_2$″ from all edges

For SI: 1 inch = 25.4 mm, 1 foot = 304.8 mm, 1 pound per square foot = 0.0479 kPa, 1 mil = 0.0254 mm.

a. Screws shall be applied through the flanges of the truss or ceiling joist or a 54-mil clip angle shall be used with two No. 10 screws in each leg. See Section R804.3.9 for additional requirements to resist uplift forces.

b. Spacing of fasteners on roof sheathing panel edges applies to panel edges supported by framing members and at all roof plane perimeters. Blocking of roof sheathing panel edges perpendicular to the framing members shall not be required except at the intersection of adjacent roof planes. Roof perimeter shall be supported by framing members or cold-formed blocking of the same depth and gage as the floor members.

❖ See the commentary to Section R804.3.

Solution:

The live load is 20 psf (958 Pa) for limited attic storage. In Table R804.3.1.1(3), under the column headed "Mid-span Bracing," look under 16-inch joist spacing. Note that 550S162-97 has an allowable span of 18 feet, 5 inches (5613 mm), which is greater than the required span.

The joist-rafter connection in Table R804.3.1 requires four No. 10 screws fastened in accordance with Figure R804.3.1(1).

When continuous joists are framed across interior bearing supports, such supports must be located within 24 inches (610 mm) of mid span of the ceiling joist, and the individual spans must not exceed the applicable spans in Table R804.3.1.1(2), R804.3.1.1(4), R804.3.1.1(6) or R804.3.1.1(8).

Bearing stiffeners are required at each bearing point as well as at concentrated loads. See Note c to Tables R804.3.1.1(1) through R804.3.1.1(8). Bearing stiffeners must be installed in accordance with Section R804.3.1.2.

The tables provide ceiling joist spans for a ceiling dead load of 5 psf (239 Pa).

The dead load of the combined roof and ceiling assemblies must not exceed 15 psf for townhouses in Seismic Design Category C and all structures in Seismic Design Categories D_0, D_1 and D_2. Section R301.2.2.2.1 provides an exception for exceeding the 15 psf (718 Pa) restriction (see commentary, Section R301.2.2.2.1).

TABLE R804.3.1.1(1)
CEILING JOIST SPANS
SINGLE SPANS WITH BEARING STIFFENERS
10 PSF LIVE LOAD (NO ATTIC STORAGE)[a, b, c] 33 KSI STEEL

MEMBER DESIGNATION	ALLOWABLE SPAN (feet-inches)					
	Lateral Support of Top (Compression) Flange					
	Unbraced		Mid-span Bracing		Third-point Bracing	
	Ceiling Joist Spacing (inches)					
	16	24	16	24	16	24
350S162-33	9′-5″	8′-6″	12′-2″	10′-4″	12′-2″	10′-7″
350S162-43	10′-3″	9′-2″	12′-10″	11′-2″	12′-10″	11′-2″
350S162-54	11′-1″	9′-11″	13′-9″	12′-0″	13′-9″	12′-0″
350S162-68	12′-1″	10′-9″	14′-8″	12′-10″	14′-8″	12′-10″
350S162-97	14′-4″	12′-7″	16′-4″	14′-3″	16′-4″	14′-3″
550S162-33	10′-7″	9′-6″	14′-10″	12′-10″	15′-11″	13′-4″
550S162-43	11′-8″	10′-6″	16′-4″	14′-3″	17′-10″	15′-3″
550S162-54	12′-6″	11′-2″	17′-7″	15′-7″	19′-5″	16′-10″
550S162-68	13′-6″	12′-1″	19′-2″	17′-1″	21′-0″	18′-4″
550S162-97	15′-9″	13′-11″	21′-8″	19′-3″	23′-5″	20′-5″
800S162-33	12′-2″	10′-11″	17′-8″	15′-10″	19′-10″	17′-1″
800S162-43	13′-0″	11′-9″	18′-10″	17′-0″	21′-6″	19′-1″
800S162-54	13′-10″	12′-5″	20′-0″	18′-0″	22′-9″	20′-4″
800S162-68	14′-11″	13′-4″	21′-3″	19′-1″	24′-1″	21′-8″
800S162-97	17′-1″	15′-2″	23′-10″	21′-3″	26′-7″	23′-10″
1000S162-43	13′-11″	12′-6″	20′-2″	18′-3″	23′-1″	20′-9″
1000S162-54	14′-9″	13′-3″	21′-4″	19′-3″	24′-4″	22′-0″
1000S162-68	15′-10″	14′-2″	22′-8″	20′-5″	25′-9″	23′-2″
1000S162-97	18′-0″	16′-0″	25′-3″	22′-7″	28′-3″	25′-4″
1200S162-43	14′-8″	13′-3″	21′-4″	19′-3″	24′-5″	21′-8″
1200S162-54	15′-7″	14′-0″	22′-6″	20′-4″	25′-9″	23′-2″
1200S162-68	16′-8″	14′-11″	23′-11″	21′-6″	27′-2″	24′-6″
1200S162-97	18′-9″	16′-9″	26′-6″	23′-8″	29′-9″	26′-9″

For SI: 1 inch = 25.4 mm, 1 foot = 304.8 mm, 1 pound per square foot = 0.0479 kPa.

a. Deflection criterion: *L*/240 for total loads.

b. Ceiling dead load = 5 psf.

c. Bearing stiffeners are required at all bearing points and concentrated load locations.

❖ See the commentary to Section R804.3.1.1.

TABLE R804.3.1.1(2)
CEILING JOIST SPANS
TWO EQUAL SPANS WITH BEARING STIFFENERS
10 PSF LIVE LOAD (NO ATTIC STORAGE)[a, b, c] 33 KSI STEEL

MEMBER DESIGNATION	ALLOWABLE SPAN (feet-inches)					
	Lateral Support of Top (Compression) Flange					
	Unbraced		Mid-span Bracing		Third-point Bracing	
	Ceiling Joist Spacing (inches)					
	16	24	16	24	16	24
350S162-33	12′-11″	10′-11″	13′-5″	10′-11″	13′-5″	10′-11″
350S162-43	14′-2″	12′-8″	15′-10″	12′-11″	15′-10″	12′-11″
350S162-54	15′-6″	13′-10″	17′-1″	14′-6″	17′-9″	14′-6″
350S162-68	17′-3″	15′-3″	18′-6″	16′-1″	19′-8″	16′-1″
350S162-97	20′-10″	18′-4″	21′-5″	18′-10″	21′-11″	18′-10″
550S162-33	14′-4″	12′-11″	16′-7″	14′-1″	17′-3″	14′-1″
550S162-43	16′-0″	14′-1″	17′-11″	16′-1″	20′-7″	16′-10″
550S162-54	17′-4″	15′-6″	19′-5″	17′-6″	23′-2″	19′-0″
550S162-68	19′-1″	16′-11″	20′-10″	18′-8″	25′-2″	21′-5″
550S162-97	22′-8″	19′-9″	23′-6″	20′-11″	27′-11″	25′-1″
800S162-33	16′-5″	14′-10″	19′-2″	17′-3″	23′-1″	18′-3″
800S162-43	17′-9″	15′-11″	20′-6″	18′-5″	25′-0″	22′-6″
800S162-54	19′-1″	17′-1″	21′-8″	19′-6″	26′-4″	23′-9″
800S162-68	20′-9″	18′-6″	23′-1″	20′-9″	28′-0″	25′-2″
800S162-97	24′-5″	21′-6″	26′-0″	23′-2″	31′-1″	27′-9″
1000S162-43	18′-11″	17′-0″	21′-11″	19′-9″	26′-8″	24′-1″
1000S162-54	20′-3″	18′-2″	23′-2″	20′-10″	28′-2″	25′-5″
1000S162-68	21′-11″	19′-7″	24′-7″	22′-2″	29′-10″	26′-11″
1000S162-97	25′-7″	22′-7″	27′-6″	24′-6″	33′-0″	29′-7″
1200S162-43	19′-11″	17′-11″	23′-1″	20′-10″	28′-3″	25′-6″
1200S162-54	21′-3″	19′-1″	24′-5″	22′-0″	29′-9″	26′-10″
1200S162-68	23′-0″	20′-7″	25′-11″	23′-4″	31′-6″	28′-4″
1200S162-97	26′-7″	23′-6″	28′-9″	25′-10″	34′-8″	31′-1″

For SI: 1 inch = 25.4 mm, 1 foot = 304.8 mm, 1 pound per square foot = 0.0479 kPa.

a. Deflection criterion: *L*/240 for total loads.

b. Ceiling dead load = 5 psf.

c. Bearing stiffeners are required at all bearing points and concentrated load locations.

❖ See the commentary to Section R804.3.1.1.

TABLE R804.3.1.1(3)
CEILING JOIST SPANS
SINGLE SPANS WITH BEARING STIFFENERS
20 PSF LIVE LOAD (LIMITED ATTIC STORAGE)[a, b, c] 33 KSI STEEL

MEMBER DESIGNATION	ALLOWABLE SPAN (feet-inches)					
	Lateral Support of Top (Compression) Flange					
	Unbraced		Mid-span Bracing		Third-point Bracing	
	Ceiling Joist Spacing (inches)					
	16	24	16	24	16	24
350S162-33	8′-2″	7′-2″	9′-9″	8′-1″	9′-11″	8′-1″
350S162-43	8′-10″	7′-10″	11′-0″	9′-5″	11′-0″	9′-7″
350S162-54	9′-6″	8′-6″	11′-9″	10′-3″	11′-9″	10′-3″
350S162-68	10′-4″	9′-2″	12′-7″	11′-0″	12′-7″	11′-0″
350S162-97	12′-1″	10′-8″	14′-0″	12′-0″	14′-0″	12′-0″
550S162-33	9′-2″	8′-3″	12′-2″	10′-2″	12′-6″	10′-5″
550S162-43	10′-1″	9′-1″	13′-7″	11′-7″	14′-5″	12′-2″
550S162-54	10′-9″	9′-8″	14′-10″	12′-10″	15′-11″	13′-6″
550S162-68	11′-7″	10′-4″	16′-4″	14′-0″	17′-5″	14′-11″
550S162-97	13′-4″	11′-10″	18′-5″	16′-2″	20′-1″	17′-1″
800S162-33	10′-7″	9′-6″	15′-1″	13′-0″	16′-2″	13′-7″
800S162-43	11′-4″	10′-2″	16′-5″	14′-6″	18′-2″	15′-9″
800S162-54	12′-0″	10′-9″	17′-4″	15′-6″	19′-6″	17′-0″
800S162-68	12′-10″	11′-6″	18′-5″	16′-6″	20′-10″	18′-3″
800S162-97	14′-7″	12′-11″	20′-5″	18′-3″	22′-11″	20′-5″
1000S162-43	12′-1″	10′-11″	17′-7″	15′-10″	19′-11″	17′-3″
1000S162-54	12′-10″	11′-6″	18′-7″	16′-9″	21′-2″	18′-10″
1000S162-68	13′-8″	12′-3″	19′-8″	17′-8″	22′-4″	20′-1″
1000S162-97	15′-4″	13′-8″	21′-8″	19′-5″	24′-5″	21′-11″
1200S162-43	12′-9″	11′-6″	18′-7″	16′-6″	20′-9″	18′-2″
1200S162-54	13′-6″	12′-2″	19′-7″	17′-8″	22′-5″	20′-2″
1200S162-68	14′-4″	12′-11″	20′-9″	18′-8″	23′-7″	21′-3″
1200S162-97	16′-1″	14′-4″	22′-10″	20′-6″	25′-9″	23′-2″

For SI: 1 inch = 25.4 mm, 1 foot = 304.8 mm, 1 pound per square foot = 0.0479 kPa.

a. Deflection criterion: *L*/240 for total loads.

b. Ceiling dead load = 5 psf.

c. Bearing stiffeners are required at all bearing points and concentrated load locations.

❖ See the commentary to Section R804.3.1.1.

TABLE R804.3.1.1(4)
CEILING JOIST SPANS
TWO EQUAL SPANS WITH BEARING STIFFENERS
20 PSF LIVE LOAD (LIMITED ATTIC STORAGE)[a, b, c] 33 KSI STEEL

MEMBER DESIGNATION	ALLOWABLE SPAN (feet-inches)					
	Lateral Support of Top (Compression) Flange					
	Unbraced		Mid-span Bracing		Third-point Bracing	
	Ceiling Joist Spacing (inches)					
	16	24	16	24	16	24
350S162-33	10′-2″	8′-4″	10′-2″	8′-4″	10′-2″	8′-4″
350S162-43	12′-1″	9′-10″	12′-1″	9′-10″	12′-1″	9′-10″
350S162-54	13′-3″	11′-0″	13′-6″	11′-0″	13′-6″	11′-0″
350S162-68	14′-7″	12′-3″	15′-0″	12′-3″	15′-0″	12′-3″
350S162-97	17′-6″	14′-3″	17′-6″	14′-3″	17′-6″	14′-3″
550S162-33	12′-5″	10′-9″	13′-2″	10′-9″	13′-2″	10′-9″
550S162-43	13′-7″	12′-1″	15′-6″	12′-9″	15′-8″	12′-9″
550S162-54	14′-11″	13′-4″	16′-10″	14′-5″	17′-9″	14′-5″
550S162-68	16′-3″	14′-5″	18′-0″	16′-1″	20′-0″	16′-4″
550S162-97	19′-1″	16′-10″	20′-3″	18′-0″	23′-10″	19′-5″
800S162-33	14′-3″	12′-4″	16′-7″	12′-4″	16′-7″	12′-4″
800S162-43	15′-4″	13′-10″	17′-9″	16′-0″	21′-8″	17′-9″
800S162-54	16′-5″	14′-9″	18′-10″	16′-11″	22′-11″	20′-6″
800S162-68	17′-9″	15′-11″	20′-0″	18′-0″	24′-3″	21′-10″
800S162-97	20′-8″	18′-3″	22′-3″	19′-11″	26′-9″	24′-0″
1000S162-43	16′-5″	14′-9″	19′-0″	17′-2″	23′-3″	18′-11″
1000S162-54	17′-6″	15′-8″	20′-1″	18′-1″	24′-6″	22′-1″
1000S162-68	18′-10″	16′-10″	21′-4″	19′-2″	25′-11″	23′-4″
1000S162-97	21′-8″	19′-3″	23′-7″	21′-2″	28′-5″	25′-6″
1200S162-43	17′-3″	15′-7″	20′-1″	18′-2″	24′-6″	18′-3″
1200S162-54	18′-5″	16′-6″	21′-3″	19′-2″	25′-11″	23′-5″
1200S162-68	19′-9″	17′-8″	22′-6″	20′-3″	27′-4″	24′-8″
1200S162-97	22′-7″	20′-1″	24′-10″	22′-3″	29′-11″	26′-11″

For SI: 1 inch = 25.4 mm, 1 foot = 304.8 mm, 1 pound per square foot = 0.0479 kPa.
a. Deflection criterion: *L*/240 for total loads.
b. Ceiling dead load = 5 psf.
c. Bearing stiffeners are required at all bearing points and concentrated load locations.

❖ See the commentary to Section R804.3.1.1.

TABLE R804.3.1.1(5)
CEILING JOIST SPANS
SINGLE SPANS WITHOUT BEARING STIFFENERS
10 PSF LIVE LOAD (NO ATTIC STORAGE)[a, b] 33 KSI STEEL

MEMBER DESIGNATION	ALLOWABLE SPAN (feet-inches)					
	Lateral Support of Top (Compression) Flange					
	Unbraced		Mid-span Bracing		Third-point Bracing	
	Ceiling Joist Spacing (inches)					
	16	24	16	24	16	24
350S162-33	9′-5″	8′-6″	12′-2″	10′-4″	12′-2″	10′-7″
350S162-43	10′-3″	9′-12″	13′-2″	11′-6″	13′-2″	11′-6″
350S162-54	11′-1″	9′-11″	13′-9″	12′-0″	13′-9″	12′-0″
350S162-68	12′-1″	10′-9″	14′-8″	12′-10″	14′-8″	12′-10″
350S162-97	14′-4″	12′-7″	16′-10″	14′-3″	16′-4″	14′-3″
550S162-33	10′-7″	9′-6″	14′-10″	12′-10″	15′-11″	13′-4″
550S162-43	11′-8″	10′-6″	16′-4″	14′-3″	17′-10″	15′-3″
550S162-54	12′-6″	11′-2″	17′-7″	15′-7″	19′-5″	16′-10″
550S162-68	13′-6″	12′-1″	19′-2″	17′-0″	21′-0″	18′-4″
550S162-97	15′-9″	13′-11″	21′-8″	19′-3″	23′-5″	20′-5″
800S162-33	—	—	—	—	—	—
800S162-43	13′-0″	11′-9″	18′-10″	17′-0″	21′-6″	19′-0″
800S162-54	13′-10″	12′-5″	20′-0″	18′-0″	22′-9″	20′-4″
800S162-68	14′-11″	13′-4″	21′-3″	19′-1″	24′-1″	21′-8″
800S162-97	17′-1″	15′-2″	23′-10″	21′-3″	26′-7″	23′-10″
1000S162-43	—	—	—	—	—	—
1000S162-54	14′-9″	13′-3″	21′-4″	19′-3″	24′-4″	22′-0″
1000S162-68	15′-10″	14′-2″	22′-8″	20′-5″	25′-9″	23′-2″
1000S162-97	18′-0″	16′-0″	25′-3″	22′-7″	28′-3″	25′-4″
1200S162-43	—	—	—	—	—	—
1200S162-54	—	—	—	—	—	—
1200S162-68	16′-8″	14′-11″	23′-11″	21′-6″	27′-2″	24′-6″
1200S162-97	18′-9″	16′-9″	26′-6″	23′-8″	29′-9″	26′-9″

For SI: 1 inch = 25.4 mm, 1 foot = 304.8 mm, 1 pound per square foot = 0.0479 kPa.

a. Deflection criterion: *L*/240 for total loads.

b. Ceiling dead load = 5 psf.

❖ See the commentary to Section R804.3.1.1.

TABLE R804.3.1.1(6)
CEILING JOIST SPANS
TWO EQUAL SPANS WITHOUT BEARING STIFFENERS
10 PSF LIVE LOAD (NO ATTIC STORAGE)[a,b] 33 KSI STEEL

MEMBER DESIGNATION	ALLOWABLE SPAN (feet-inches)					
	Lateral Support of Top (Compression) Flange					
	Unbraced		Mid-span Bracing		Third-point Bracing	
	Ceiling Joist Spacing (inches)					
	16	24	16	24	16	24
350S162-33	11′-9″	8′-11″	11′-9″	8′-11″	11′-9″	8′-11″
350S162-43	14′-2″	11′-7″	14′-11″	11′-7″	14′-11″	11′-7″
350S162-54	15′-6″	13′-10″	17′-1″	13′-10″	17′-7″	13′-10″
350S162-68	17′-3″	15′-3″	18′-6″	16′-1″	19′-8″	16′-1″
350S162-97	20′-10″	18′-4″	21′-5″	18′-9″	21′-11″	18′-9″
550S162-33	13′-4″	9′-11″	13′-4″	9′-11″	13′-4″	9′-11″
550S162-43	16′-0″	13′-6″	17′-9″	13′-6″	17′-9″	13′-6″
550S162-54	17′-4″	15′-6″	19′-5″	16′-10″	21′-9″	16′-10″
550S162-68	19′-1″	16′-11″	20′-10″	18′-8″	24′-11″	20′-6″
550S162-97	22′-8″	20′-0″	23′-9″	21′-1″	28′-2″	25′-1″
800S162-33	—	—	—	—	—	—
800S162-43	17′-9″	15′-7″	20′-6″	15′-7″	21′-0″	15′-7″
800S162-54	19′-1″	17′-1″	21′-8″	19′-6″	26′-4″	23′-10″
800S162-68	20′-9″	18′-6″	23′-1″	20′-9″	28′-0″	25′-2″
800S162-97	24′-5″	21′-6″	26′-0″	23′-2″	31′-1″	27′-9″
1000S162-43	—	—	—	—	—	—
1000S162-54	20′-3″	18′-2″	23′-2″	20′-10″	28′-2″	21′-2″
1000S162-68	21′-11″	19′-7″	24′-7″	22′-2″	29′-10″	26′-11″
1000S162-97	25′-7″	22′-7″	27′-6″	24′-6″	33′-0″	29′-7″
1200S162-43	—	—	—	—	—	—
1200S162-54	—	—	—	—	—	—
1200S162-68	23′-0″	20′-7″	25′-11″	23′-4″	31′-6″	28′-4″
1200S162-97	26′-7″	23′-6″	28′-9″	25′-10″	34′-8″	31′-1″

For SI: 1 inch = 25.4 mm, 1 foot = 304.8 mm, 1 pound per square foot = 0.0479 kPa.

a. Deflection criterion: *L*/240 for total loads.

b. Ceiling dead load = 5 psf.

❖ See the commentary to Section R804.3.1.1.

TABLE R804.3.1.1(7)
CEILING JOIST SPANS
SINGLE SPANS WITHOUT BEARING STIFFENERS
20 PSF LIVE LOAD (LIMITED ATTIC STORAGE)[a, b] 33 KSI STEEL

MEMBER DESIGNATION	ALLOWABLE SPAN (feet-inches)					
	Lateral Support of Top (Compression) Flange					
	Unbraced		Mid-span Bracing		Third-point Bracing	
	Ceiling Joist Spacing (inches)					
	16	24	16	24	16	24
350S162-33	8′-2″	6′-10″	9′-9″	6′-10″	9′-11″	6′-10″
350S162-43	8′-10″	7′-10″	11′-0″	9′-5″	11′-0″	9′-7″
350S162-54	9′-6″	8′-6″	11′-9″	10′-3″	11′-9″	10′-3″
350S162-68	10′-4″	9′-2″	12′-7″	11′-0″	12′-7″	11′-0″
350S162-97	12′-10″	10′-8″	13′-9″	12′-0″	13′-9″	12′-0″
550S162-33	9′-2″	8′-3″	12′-2″	8′-5″	12′-6″	8′-5″
550S162-43	10′-1″	9′-1″	13′-7″	11′-8″	14′-5″	12′-2″
550S162-54	10′-9″	9′-8″	14′-10″	12′-10″	15′-11″	13′-6″
550S162-68	11′-7″	10′-4″	16′-4″	14′-0″	17′-5″	14′-11″
550S162-97	13′-4″	11′-10″	18′-5″	16′-2″	20′-1″	17′-4″
800S162-33	—	—	—	—	—	—
800S162-43	11′-4″	10′-1″	16′-5″	13′-6″	18′-1″	13′-6″
800S162-54	20′-0″	10′-9″	17′-4″	15′-6″	19′-6″	27′-0″
800S162-68	12′-10″	11′-6″	18′-5″	16′-6″	20′-10″	18′-3″
800S162-97	14′-7″	12′-11″	20′-5″	18′-3″	22′-11″	20′-5″
1000S162-43	—	—	—	—	—	—
1000S162-54	12′-10″	11′-6″	18′-7″	16′-9″	21′-2″	15′-5″
1000S162-68	13′-8″	12′-3″	19′-8″	17′-8″	22′-4″	20′-1″
1000S162-97	15′-4″	13′-8″	21′-8″	19′-5″	24′-5″	21′-11″
1200S162-43	—	—	—	—	—	—
1200S162-54	—	—	—	—	—	—
1200S162-68	14′-4″	12′-11″	20′-9″	18′-8″	23′-7″	21′-3″
1200S162-97	16′-1″	14′-4″	22′-10″	20′-6″	25′-9″	23′-2″

For SI: 1 inch = 25.4 mm, 1 foot = 304.8 mm, 1 pound per square foot = 0.0479 kPa.

a. Deflection criterion: $L/240$ for total loads.

b. Ceiling dead load = 5 psf.

❖ See the commentary to Section R804.3.1.1.

**TABLE R804.3.1.1(8)
CEILING JOIST SPANS
TWO EQUAL SPANS WITHOUT BEARING STIFFENERS
20 PSF LIVE LOAD (LIMITED ATTIC STORAGE)[a, b] 33 KSI STEEL**

MEMBER DESIGNATION	ALLOWABLE SPAN (feet-inches)					
	Lateral Support of Top (Compression) Flange					
	Unbraced		Mid-span Bracing		Third-point Bracing	
	Ceiling Joist Spacing (inches)					
	16	24	16	24	16	24
350S162-33	8′-1″	6′-1″	8′-1″	6′-1″	8′-1″	6′-1″
350S162-43	10′-7″	8′-1″	10′-7″	8′-1″	10′-7″	8′-1″
350S162-54	12′-8″	9′-10″	12′-8″	9′-10″	12′-8″	9′-10″
350S162-68	14′-7″	11′-10″	14′-11″	11′-10″	14′-11″	11′-10″
350S162-97	17′-6″	14′-3″	17′-6″	14′-3″	17′-6″	14′-3″
550S162-33	8′-11″	6′-8″	8′-11″	6′-8″	8′-11″	6′-8″
550S162-43	12′-3″	9′-2″	12′-3″	9′-2″	12′-3″	9′-2″
550S162-54	14′-11″	11′-8″	15′-4″	11′-8″	15′-4″	11′-8″
550S162-68	16′-3″	14′-5″	18′-0″	15′-8″	18′-10″	14′-7″
550S162-97	19′-1″	16′-10″	20′-3″	18′-0″	23′-9″	19′-5″
800S162-33	—	—	—	—	—	—
800S162-43	13′-11″	9′-10″	13′-11″	9′-10″	13′-11″	9′-10″
800S162-54	16′-5″	13′-9″	18′-8″	13′-9″	18′-8″	13′-9″
800S162-68	17′-9″	15′-11″	20′-0″	18′-0″	24′-1″	18′-3″
800S162-97	20′-8″	18′-3″	22′-3″	19′-11″	26′-9″	24′-0″
1000S162-43	—	—	—	—	—	—
1000S162-54	17′-6″	13′-11″	19′-1″	13′-11″	19′-1″	13′-11″
1000S162-68	18′-10″	16′-10″	21′-4″	19′-2″	25′-11″	19′-7″
1000S162-97	21′-8″	19′-3″	23′-7″	21′-2″	28′-5″	25′-6″
1200S162-43	—	—	—	—	—	—
1200S162-54	—	—	—	—	—	—
1200S162-68	19′-9″	17′-8″	22′-6″	19′-8″	26′-8″	19′-8″
1200S162-97	22′-7″	20′-1″	24′-10″	22′-3″	29′-11″	26′-11″

For SI: 1 inch = 25.4 mm, 1 foot = 304.8 mm, 1 pound per square foot = 0.0479 kPa.

a. Deflection criterion: *L*/240 for total loads.

b. Ceiling dead load = 5 psf.

❖ See the commentary to Section R804.3.1.1.

TABLE R804.3.1.1(9)
NUMBER OF SCREWS REQUIRED FOR CEILING JOIST TO ROOF RAFTER CONNECTION[a]

ROOF SLOPE	NUMBER OF SCREWS																			
	Building width (feet)																			
	24				28				32				36				40			
	Ground snow load (psf)																			
	20	30	50	70	20	30	50	70	20	30	50	70	20	30	50	70	20	30	50	70
3/12	5	6	9	11	5	7	10	13	6	8	11	15	7	8	13	17	8	9	14	19
4/12	4	5	7	9	4	5	8	10	5	6	9	12	5	7	10	13	6	7	11	14
5/12	3	4	6	7	4	4	6	8	4	5	7	10	5	5	8	11	5	6	9	12
6/12	3	3	5	6	3	4	6	7	4	4	6	8	4	5	7	9	4	5	8	10
7/12	3	3	4	6	3	3	5	7	3	4	6	7	4	4	6	8	4	5	7	9
8/12	2	3	4	5	3	3	5	6	3	4	5	7	3	4	6	8	4	4	6	8
9/12	2	3	4	5	3	3	4	6	3	3	5	6	3	4	5	7	3	4	6	8
10/12	2	2	4	5	2	3	4	5	3	3	5	6	3	3	5	7	3	4	6	7
11/12	2	2	3	4	2	3	4	5	3	3	4	6	3	3	5	6	3	4	5	7
12/12	2	2	3	4	2	3	4	5	2	3	4	5	3	3	5	6	3	4	5	7

For SI: 1 inch = 25.4 mm, 1 foot = 304.8 mm, 1 pound per square foot = 0.0479kPa.

a. Screws shall be No. 10.

❖ See the commentary to Section R804.3.2.4.

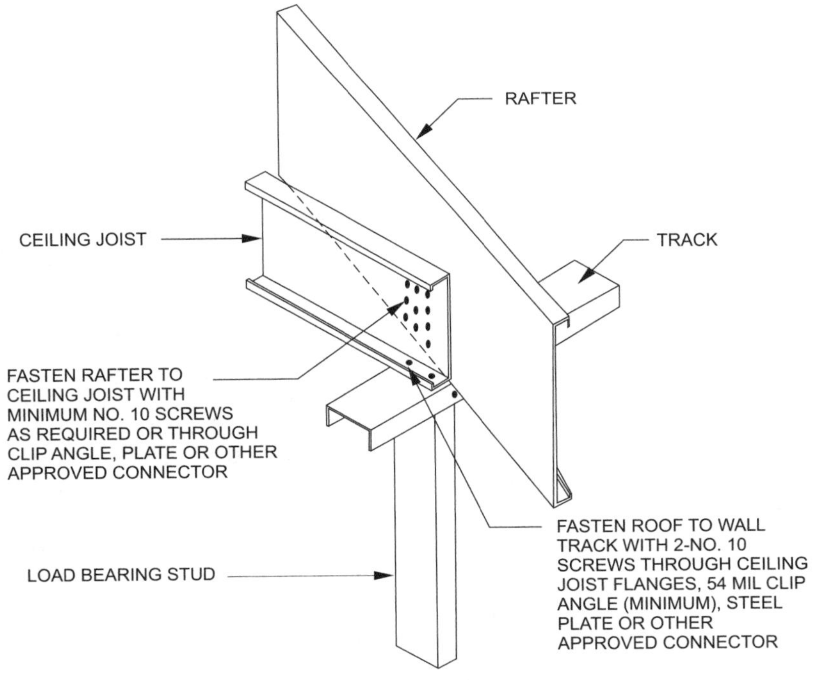

For SI: 1 mil = 0.0254 mm.

FIGURE R804.3.1.1(1)
JOIST TO RAFTER CONNECTION

❖ The fastening to the wall track is illustrated. See the commentary to Section R804.3.2.4 for a discussion of the joist-to-rafter connection.

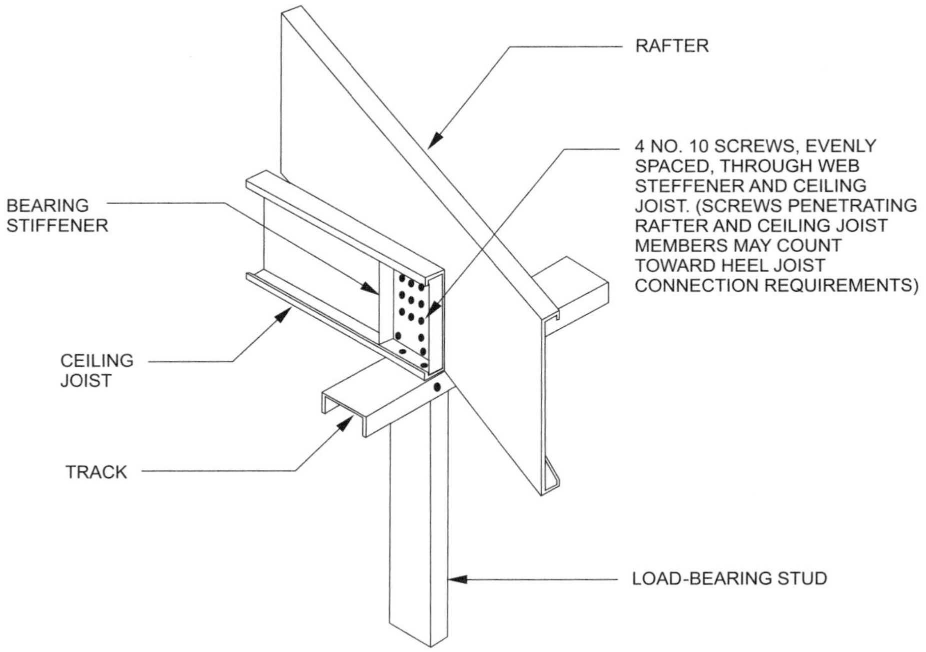

FIGURE R804.3.1.1(2)
BEARING STIFFENER

❖ See the commentary to Section R804.3.2.4.

R804.3.1.2 Ceiling joist bearing stiffeners. Where required in Tables R804.3.1.1(1) through R804.3.1.1(8), bearing stiffeners shall be installed at each bearing support in accordance with Figure R804.3.1.1(2). Bearing stiffeners shall be fabricated from a C-shaped or track member in accordance with the one of following:

1. C-shaped bearing stiffeners shall be a minimum 33 mils (0.84 mm) thick.
2. Track bearing stiffener shall be a minimum 43 mils (1.09 mm) thick.

The minimum length of a bearing stiffener shall be the depth of member being stiffened minus $^3/_8$ inch (9.5 mm). Each stiffener shall be fastened to the web of the ceiling joist with a minimum of four No. 8 screws equally spaced as shown in Figure R804.3.1.1(2). Installation of stiffeners shall be permitted on either side of the web.

❖ Stiffeners (also referred to as transverse stiffeners or web stiffeners) at points of bearing provide resistance to web crippling in the joist. This section provides the option to use a 33-mils-thick (0.84 mm) C-shaped section or 43-mils-thick (1.09 mm) track section as stiffening. The stiffener is illustrated in Figure R804.3.1.1(2). All bearing stiffeners are to be fastened in accordance with Figure R804.3.1.1(2).

R804.3.1.3 Ceiling joist bottom flange bracing. The bottom flanges of ceiling joists shall be laterally braced by the application of gypsum board or continuous steel straps installed perpendicular to the joist run in accordance with one of the following:

1. Gypsum board shall be fastened with No. 6 screws in accordance with Section R702.
2. Steel straps with a minimum size of $1^1/_2$ inches by 33 mils (38 mm by 0.84 mm) shall be installed at a maximum spacing of 4 feet (1219 mm). Straps shall be fastened to the bottom flange at each joist with one No. 8 screw and shall be fastened to blocking with two No. 8 screws. Blocking shall be installed between joists at a maximum spacing of 12 feet (3658 mm) measured along a line of continuous strapping (perpendicular to the joist run). Blocking shall also be located at the termination of all straps.

❖ Steel ceiling joists must have the bottom flanges laterally braced either with gypsum board or continuous steel strapping. Strap bracing requires the installation of solid blocking in line with the straps at the termination of all straps and at a maximum spacing of 12 feet (3658 mm) measured perpendicular to the joist run.

R804.3.1.4 Ceiling joist top flange bracing. The top flanges of ceiling joists shall be laterally braced as required by Tables R804.3.1.1(1) through R804.3.1.1(8), in accordance with one of the following:

1. Minimum 33-mil (0.84 mm) C-shaped member in accordance with Figure R804.3.1.4(1).
2. Minimum 33-mil (0.84 mm) track section in accordance with Figure R804.3.1.4(1).
3. Minimum 33-mil (0.84 mm) hat section in accordance with Figure R804.3.1.4(1).

4. Minimum 54-mil (1.37 mm) $1^1/_2$-inch cold-rolled channel section in accordance with Figure R804.3.1.4(1).

5. Minimum $1^1/_2$-inch by 33-mil (38 mm by 0.84 mm) continuous steel strap in accordance with Figure R804.3.1.4(2).

Lateral bracing shall be installed perpendicular to the ceiling joists and shall be fastened to the top flange of each joist with one No. 8 screw. Blocking shall be installed between joists in line with bracing at a maximum spacing of 12 feet (3658 mm) measured perpendicular to the joists. Ends of lateral bracing shall be attached to blocking or anchored to a stable building component with two No. 8 screws.

❖ The application of lateral support, to the top flanges of the ceiling joists, at mid span or at third-point of the span will permit longer spans in accordance with Tables R804.3.1.1(1) through R804.3.1.1(8). This section permits the top flange bracing to be C-section, track section, hat section, cold-rolled channel section or steel strap [see Figures R804.3.1.4(1) and R804.3.1.4(2) for details]. In addition to the top flange bracing blocking must be installed in line with the bracing at a maximum spacing of 12 feet (3658 mm) measured perpendicular to the joist run. Splice for the steel strap bracing must occur at blocking [see Figure R804.3.1.4(2)].

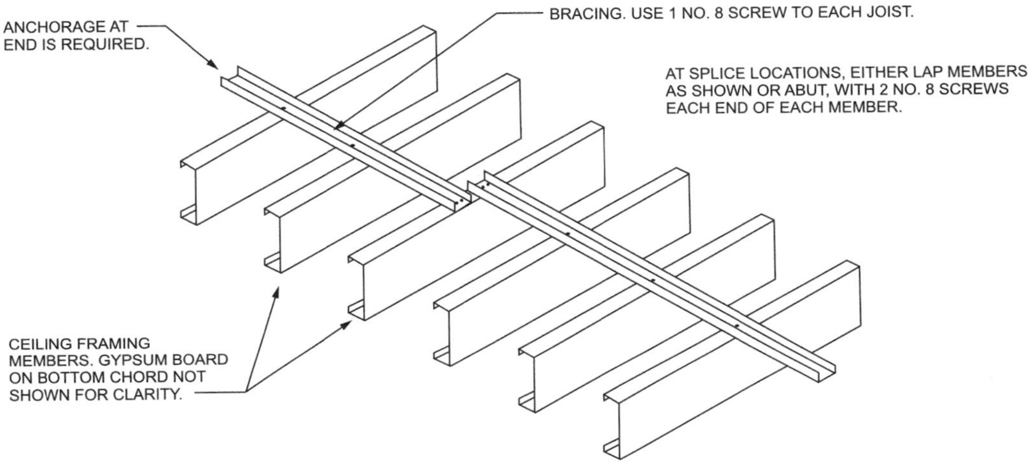

FIGURE R804.3.1.4(1)
CEILING JOIST TOP FLANGE BRACING WITH C-SHAPE, TRACK OR COLD-ROLLED CHANNEL

❖ See the commentary to Section R804.3.1.4. Blocking must be installed in line with the bracing at a maximum spacing of 12 feet (305 mm).

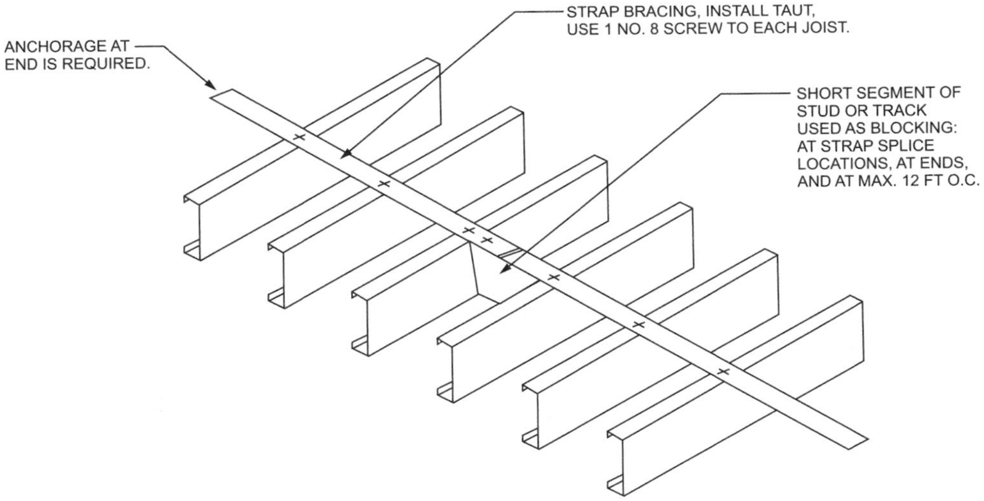

For SI: 1foot = 304.8 mm.

FIGURE R804.3.1.4(2)
CEILING JOIST TOP FLANGE BRACING WITH CONTINUOUS STEEL STRAP AND BLOCKING

❖ The splice for the flat strap must occur at blocking. See the commentary to Section R804.3.1.4.

R804.3.1.5 Ceiling joist splicing. Splices in ceiling joists shall be permitted, if ceiling joist splices are supported at interior bearing points and are constructed in accordance with Figure R804.3.1.5. The number of screws on each side of the splice shall be the same as required for the heel joint connection in Table R804.3.1.1(9).

❖ Steel ceiling joists are permitted to be spliced only at supports at interior bearing point (see Figure R804.3.1.5 for details).

R804.3.2 Roof rafters. Cold-formed steel roof rafters shall be in accordance with this section.

❖ Sections R804.3.2.1 through R804.3.2.5 provide the requirements for roof rafter size, support brace, splice, connection to ceiling joist and ridge and bottom flange bracing.

R804.3.2.1 Minimum roof rafter sizes. Roof rafter size and thickness shall be determined in accordance with the limits set forth in Tables R804.3.2.1(1) and R804.3.2.1(2) based on the horizontal projection of the roof rafter span. For determination of roof rafter sizes, reduction of roof spans shall be permitted when a roof rafter support brace is installed in accordance with Section R804.3.2.2. The reduced roof rafter span shall be taken as the larger of the distance from the roof rafter support brace to the ridge or to the heel measured horizontally.

For the purpose of determining roof rafter sizes in Tables R804.3.2.1(1) and R804.3.2.1(2), wind speeds shall be converted to equivalent ground snow loads in accordance with Table R804.3.2.1(3). Roof rafter sizes shall be based on the higher of the ground snow load or the equivalent snow load converted from the wind speed.

❖ Tables R804.3.2.1(1) and R804.3.2.1(2) list allowable rafter spans based on ground snow loads. To account for wind loading on rafters, Table R804.3.2.1(3) must be used to convert wind speed to an equivalent ground snow load. Then the allowable rafter span is determined from Tables R804.3.2.1(1) and R804.3.2.1(2) based on the larger of either the ground snow load or the equivalent snow load converted from the wind speed. The rafter span is permitted to be reduced when a support brace is installed (see Section R804.3.2.2). The reduced rafter span is the larger of the distance from the ridge to the support brace or from the support brace to the heel.

Tables R804.3.2.1(1) and R804.3.2.1(2) provide rafter spans for roof dead load of 12 psf (575 Pa).

The dead load of the combined roof and ceiling assemblies must not exceed 15 psf (718 Pa) for townhouses in Seismic Design Category C and all structures in Seismic Design Categories D_0, D_1 and D_2. Section R301.2.2.2.1 provides an exception for exceeding the 15 psf (718 Pa) restriction (see commentary, Section R301.2.2.2.1).

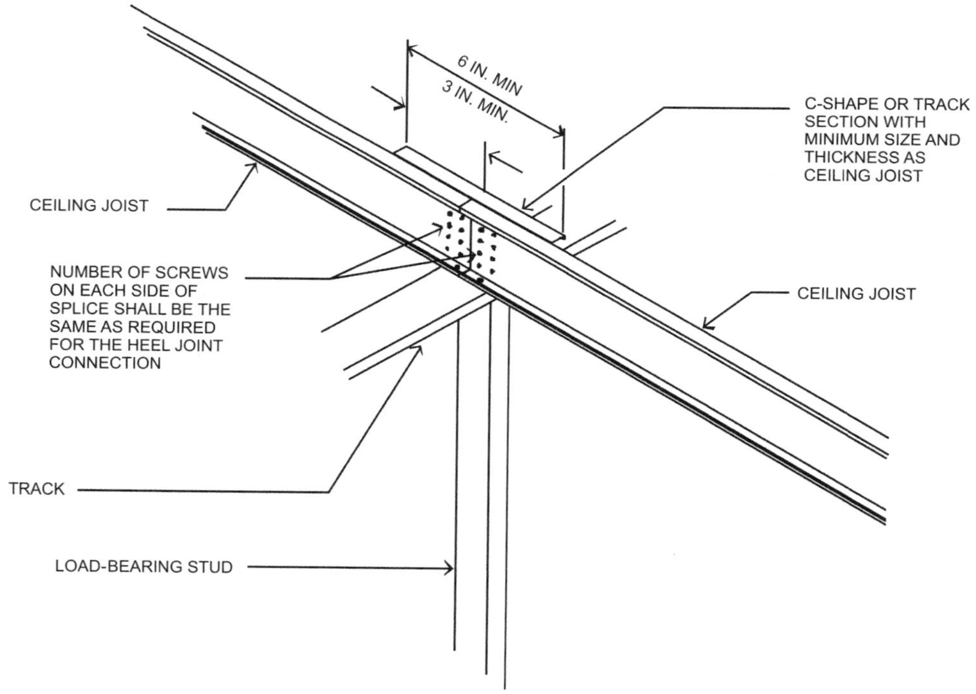

For SI: 1 inch = 25.4 mm.

FIGURE R804.3.1.5
SPLICED CEILING JOISTS

❖ See the commentary to Section R804.3.1.5.

Use of the allowable span tables is illustrated below.

Example 1:

You wish to design a rafter system using 33 ksi (228 MPa) steel for a 20 psf (958 Pa) ground snow load, and wind speed of 100 mph, Exposure B. The rafter spacing is 24 inches (610 mm) on center with a horizontal projection of the rafter span of 16 feet, 4 inches. The roof slope is 12 units vertical in 12 units horizontal (100-percent slope).

Solution 1:

Convert the wind speed to an equivalent ground snow load from Table R804.3.2.1(3). For a basic wind speed of 100 mph (44 m/s), Exposure B and a roof slope of 12:12, the equivalent ground snow load is 50 psf (2.4 kPa). The 50 psf (2.4 kPa) snow load is greater than the 20 psf (0.96 kPa); therefore you must use the 50 psf ground snow load.

The object is to locate a tabulated span value of 16 feet, 4 inches or more. Going to Table R804.3.2.1(1) of the code under the portion headed "Ground snow load 50" and "Rafter spacing 24" note that rafter size 1000S162-68 will span 16 feet, 4 inches (4978 mm).

Example 2:

To find a substitute for the 1000S162-68 rafters of Example 1, determine the required rafter size for a condition that uses a rafter brace located at mid span of each rafter as depicted in Figure R804.3. Assume the same loads, rafter spacing, and 16-foot, 4-inch (4978 mm) horizontal rafter span between the exterior wall and ridge. Use 33 ksi (228 MPa) steel.

Solution 2:

The rafter span, identified in Figure R804.3, is reduced to 8 feet, 2 inches (2489 mm) by adding the rafter brace at mid span.

The object is to locate a tabulated span value of 8 feet, 2 inches (51 mm) or more. Going to Table R804.3.2.1(1) of the code under the portion headed "Ground snow load 50" and "Rafter spacing 24," note that rafter size 550S162-43 will span 9 feet, 5 inches.

R804.3.2.1.1 Eave overhang. Eave overhangs shall not exceed 24 inches (610 mm) measured horizontally.

❖ Figure R804.3 shows the permitted eave overhang. In addition to limiting the gravity load, this provision places a cap on the amount of wind uplift on the eave overhang that must be considered when determining the roof tie-down.

R804.3.2.1.2 Rake overhangs. Rake overhangs shall not exceed 12 inches (305 mm) measured horizontally. Outlookers at gable endwalls shall be installed in accordance with Figure R804.3.2.1.2.

❖ Rake overhang at cable endwalls is limited to 12 inches (305 mm) and must be constructed as outlook rafters as shown in Option 1 or Option 2 in Figure R804.3.2.1.2.

R804.3.2.2 Roof rafter support brace. When used to reduce roof rafter spans in determining roof rafter sizes, a roof rafter support brace shall meet all of the following conditions:

1. Minimum 350S162-33 C-shaped brace member with maximum length of 8 feet (2438 mm).
2. Minimum brace member slope of 45 degrees (0.785 rad) to the horizontal.
3. Minimum connection of brace to a roof rafter and ceiling joist with four No.10 screws at each end.
4. Maximum 6 inches (152 mm) between brace/ceiling joist connection and load-bearing wall below.
5. Each roof rafter support brace greater than 4 feet (1219 mm) in length, shall be braced with a supplemental brace having a minimum size of 350S162-33 or 350T162-33 such that the maximum unsupported length of the roof rafter support brace is 4 feet (1219 mm). The supplemental brace shall be continuous and shall be connected to each roof rafter support brace using two No.8 screws.

❖ This section provides the requirements and installation details of a rafter support brace. The rafter support brace is to be installed at each rafter and must be no greater than 8 feet (2438 mm) in length. Each rafter support brace that is greater than 4 feet (1219 mm) in length must have a continuous horizontal brace (purlin) perpendicular to the rafter support brace span such that the maximum unsupported length of the rafter support brace is 4 feet (1219 mm) (see Figure R804.3).

R804.3.2.3 Roof rafter splice. Roof rafters shall not be spliced.

❖ Splicing of roof rafters is not permitted.

R804.3.2.4 Roof rafter to ceiling joist and ridge member connection. Roof rafters shall be connected to a parallel ceiling joist to form a continuous tie between exterior walls in accordance with Figure R804.3.1.1(1) or R804.3.1.1(2) and Table R804.3.1.1(9). Ceiling joists shall be connected to the top track of the load-bearing wall in accordance with Table R804.3, either with two No. 10 screws applied through the flange of the ceiling joist or by using a 54-mil (1.37 mm) clip angle with two No.10 screws in each leg. Roof rafters shall be connected to a ridge member with a minimum 2-inch by 2-inch (51 mm by 51 mm) clip angle fastened with No. 10 screws to the ridge member in accordance with Figure R804.3.2.4 and Table R804.3.2.4. The clip angle shall have a steel thickness equivalent to or greater than the roof rafter thickness and shall extend the depth of the roof rafter member to the extent possible. The ridge member shall be fabricated from a C-shaped member and a track section, which shall have a minimum size and steel thickness equivalent to or greater than that of adjacent roof rafters and shall be installed in accordance with Figure R804.3.2.4. The ridge member shall extend the full depth of the sloped roof rafter cut.

❖ The requirements of this section describe connections to resist horizontal thrust from gravity loads and ridge uplift from wind loads. Ceiling joists or rafter ties are the framing members used to resist the horizontal thrust from gravity loads. The ridge member connections are used to resist the ridge uplift caused by the wind load.

Ceiling joists must be located at the top track of the exterior walls and parallel to the rafters.

A ceiling joist or rafter tie is required to be connected to the top track and to each rafter [see Figures R804.3.1.1(1) and R804.3.1.1(2)]. Each rafter must be connected to a ridge member. The ridge member depth must equal or exceed the depth of the sloped rafter cut (see Figure R804.3.2.4).

TABLE R804.3.2.1(1)
ROOF RAFTER SPANS[a, b, c]
33 KSI STEEL

MEMBER DESIGNATION	ALLOWABLE SPAN MEASURED HORIZONTALLY (feet-inches)							
	Ground snow load (psf)							
	20		30		50		70	
	Rafter spacing (inches)							
	16	24	16	24	16	24	16	24
550S162-33	14′-0″	11′-6″	11′-11″	9′-7″	9′-6″	7′-9″	8′-2″	6′-8″
550S162-43	16′-8″	13′-11″	14′-5″	11′-9″	11′-6″	9′-5″	9′-10″	8′-0″
550S162-54	17′-11″	15′-7″	15′-7″	13′-3″	12′-11″	10′-7″	11′-1″	9′-1″
550S162-68	19′-2″	16′-9″	16′-9″	14′-7″	14′-1″	11′-10″	12′-6″	10′-2″
550S162-97	21′-3″	18′-6″	18′-6″	16′-2″	15′-8″	13′-8″	14′-0″	12′-2″
800S162-33	16′-5″	13′-5″	13′-11″	11′-4″	11′-1″	8′-2″	9′-0″	6′-0″
800S162-43	19′-9″	16′-1″	16′-8″	13′-7″	13′-4″	10′-10″	11′-5″	9′-4″
800S162-54	22′-8″	18′-6″	19′-2″	15′-8″	15′-4″	12′-6″	13′-1″	10′-8″
800S162-68	25′-10″	21′-2″	21′-11″	17′-10″	17′-6″	14′-4″	15′-0″	12′-3″
800S162-97	21′-3″	18′-6″	18′-6″	16′-2″	15′-8″	13′-8″	14′-0″	12′-2″
1000S162-43	22′-3″	18′-2″	18′-9″	15′-8″	15′-0″	12′-3″	12′-10″	10′-6″
1000S162-54	25′-8″	20′-11″	21′-8″	17′-9″	17′-4″	14′-2″	14′-10″	12′-1″
1000S162-68	29′-7″	24′-2″	25′-0″	20′-5″	20′-0″	16′-4″	17′-2″	14′-0″
1000S162-97	34′-8″	30′-4″	30′-4″	25′-10″	25′-3″	20′-8″	21′-8″	17′-8″
1200S162-54	28′-3″	23′-1″	23′-11″	19′-7″	19′-2″	15′-7″	16′-5″	13′-5″
1200S162-68	32′-10″	26′-10″	27′-9″	22′-8″	22′-2″	18′-1″	19′-0″	15′-6″
1200S162-97	40′-6″	33′-5″	34′-6″	28′-3″	27′-7″	22′-7″	23′-8″	19′-4″

For SI: 1 inch = 25.4 mm, 1 foot = 304.8 mm, 1 pound per square foot = 0.0479 kPa.

a. Table provides maximum horizontal rafter spans in feet and inches for slopes between 3:12 and 12:12.

b. Deflection criterion: *L*/240 for live loads and *L*/180 for total loads.

c. Roof dead load = 12 psf.

❖ See the commentary to Section R804.3.2.1.

TABLE R804.3.2.1(2)
ROOF RAFTER SPANS[a, b, c]
50 KSI STEEL

MEMBER DESIGNATION	ALLOWABLE SPAN MEASURED HORIZONTALLY (feet-inches)							
	Equivalent ground snow load (psf)							
	20		30		50		70	
	Rafter spacing (inches)							
	16	24	16	24	16	24	16	24
550S162-33	15′-4″	12′-11″	13′-4″	10′-11″	10′-9″	8′-9″	9′-2″	7′-6″
550S162-43	16′-8″	14′-7″	14′-7″	12′-9″	12′-3″	10′-6″	11′-0″	9′-0″
550S162-54	17′-11″	15′-7″	15′-7″	13′-8″	13′-2″	11′-6″	11′-9″	10′-3″
550S162-68	19′-2″	16′-9″	16′-9″	14′-7″	14′-1″	12′-4″	12′-7″	11′-0″
550S162-97	21′-3″	18′-6″	18′-6″	16′-2″	15′-8″	13′-8″	14′-0″	12′-3″
800S162-33	18′-10″	15′-5″	15′-11″	12′-9″	12′-3″	8′-2″	9′-0″	6′-0″
800S162-43	22′-3″	18′-2″	18′-10″	15′-5″	15′-1″	12′-3″	12′-11″	10′-6″
800S162-54	24′-2″	21′-2″	21′-1″	18′-5″	17′-10″	14′-8″	15′-5″	12′-7″
800S162-68	25′-11″	22′-8″	22′-8″	19′-9″	19′-1″	16′-8″	17′-1″	14′-9″
800S162-97	28′-10″	25′-2″	25′-2″	22′-0″	21′-2″	18′-6″	19′-0″	16′-7″
1000S162-43	25′-2″	20′-7″	21′-4″	17′-5″	17′-0″	13′-11″	14′-7″	10′-7″
1000S162-54	29′-0″	24′-6″	25′-4″	20′-9″	20′-3″	16′-7″	17′-5″	14′-2″
1000S162-68	31′-2″	27′-3″	27′-3″	23′-9″	20′-0″	19′-6″	20′-6″	16′-8″
1000S162-97	34′-8″	30′-4″	30′-4″	26′-5″	25′-7″	22′-4″	22′-10″	20′-0″
1200S162-54	33′-2″	27′-1″	28′-1″	22′-11″	22′-5″	18′-4″	19′-3″	15′-8″
1200S162-68	36′-4″	31′-9″	31′-9″	27′-0″	26′-5″	21′-6″	22′-6″	18′-6″
1200S162-97	40′-6″	35′-4″	35′-4″	30′-11″	29′-10″	26′-1″	26′-8″	23′-1″

For SI: 1 inch = 25.4 mm, 1 foot = 304.8 mm, 1 pound per square foot = 0.0479 kPa.

a. Table provides maximum horizontal rafter spans in feet and inches for slopes between 3:12 and 12:12.

b. Deflection criterion: *L*/240 for live loads and *L*/180 for total loads.

c. Roof dead load = 12 psf.

❖ See the commentary to Section R804.3.2.1.

TABLE R804.3.2.1(3)
BASIC WIND SPEED TO EQUIVALENT SNOW LOAD CONVERSION

BASIC WIND SPEED AND EXPOSURE		EQUIVALENT GROUND SNOW LOAD (psf)									
		Roof slope									
Exp. B	Exp. C	3:12	4:12	5:12	6:12	7:12	8:12	9:12	10:12	11:12	12:12
85 mph	—	20	20	20	20	20	20	30	30	30	30
100 mph	85 mph	20	20	20	20	30	30	30	30	50	50
110 mph	100 mph	20	20	20	20	30	50	50	50	50	50
—	110 mph	30	30	30	50	50	50	70	70	70	—

For SI: 1 mile per hour = 0.447 m/s, 1 pound per square foot = 0.0479 kPa.

❖ See the commentary to Section R804.3.2.1.

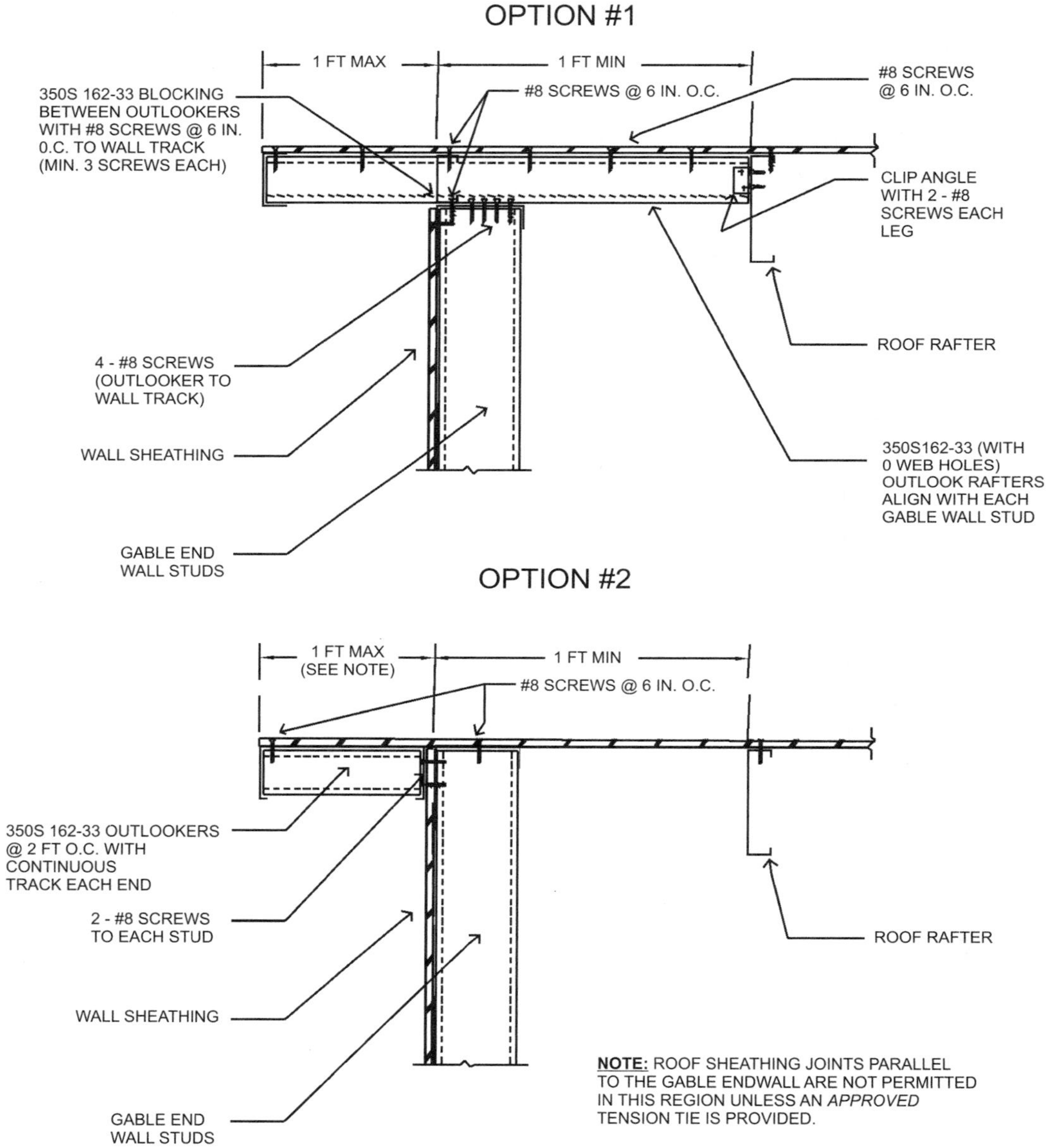

For SI: 1 inch = 25.4 mm, 1 foot = 304.8 mm.

FIGURE R804.3.2.1.2.
GABLE ENDWALL OVERHANG DETAILS

❖ See the commentary to Section R804.3.2.1.2.

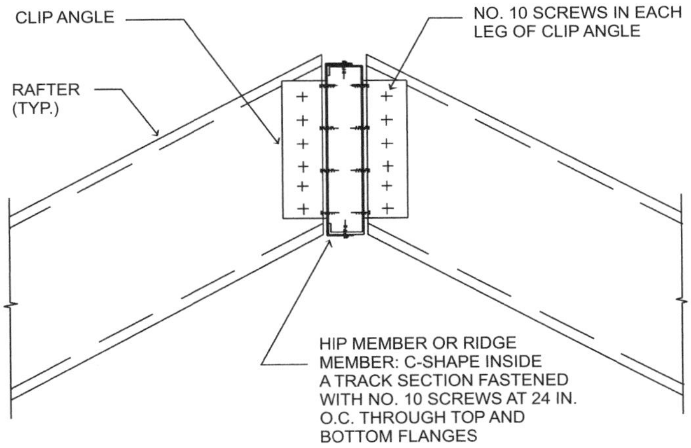

For SI: 1 inch = 25.4 mm.

FIGURE R804.3.2.4
HIP MEMBER OR RIDGE MEMBER CONNECTION

❖ See the commentary to Sections R804.3.2.4 and R804.3.3.2.

TABLE R804.3.2.4
SCREWS REQUIRED AT EACH LEG OF CLIP ANGLE FOR HIP RAFTER TO HIP MEMBER OR ROOF RAFTER TO RIDGE MEMBER CONNECTION[a]

BUILDING WIDTH (feet)	NUMBER OF SCREWS			
	Ground snow load (psf)			
	0 to 20	21 to 30	31 to 50	51 to 70
24	2	2	3	4
28	2	3	4	5
32	2	3	4	5
36	3	3	5	6
40	3	4	5	7

For SI: 1 inch = 25.4 mm, 1 foot = 304.8 mm, 1 pound per square foot = 0.0479 kPa.

a. Screws shall be No. 10 minimum.

❖ See the commentary to Section R804.3.2.4.

R804.3.2.5 Roof rafter bottom flange bracing. The bottom flanges of roof rafters shall be continuously braced, at a maximum spacing of 8 feet (2440 mm) as measured parallel to the roof rafters, with one of the following members:

1. Minimum 33-mil (0.84 mm) C-shaped member.
2. Minimum 33-mil (0.84 mm) track section.
3. Minimum $1^1/_2$-inch by 33-mil (38 mm by 0.84 mm) steel strap.

The bracing element shall be fastened to the bottom flange of each roof rafter with one No. 8 screw and shall be fastened to blocking with two No. 8 screws. Blocking shall be installed between roof rafters in-line with the continuous bracing at a maximum spacing of 12 feet (3658 mm) measured perpendicular to the roof rafters. The ends of continuous bracing shall be fastened to blocking or anchored to a stable building component with two No. 8 screws.

❖ Installation of bracing to the rafter bottom flange is necessary to provide resistance to twisting and/or lateral displacement of the rafters. The rafter bottom flange bracing must be spaced no greater than 8 feet (2440 mm) as measured parallel to the rafter. The rafter bottom flange bracing is permitted to be either a C-section, track section or steel strap. In addition to the bottom flange bracing, blocking must be installed in line with the bracing, at the termination of all bracing and at a maximum spacing of 12 feet (3658 mm), measured perpendicular to the joist run.

R804.3.3 Hip framing. Hip framing shall consist of jack-rafters, hip members, hip support columns and connections in accordance with this section, or shall be in accordance with an *approved* design. The provisions of this section for hip members and hip support columns shall apply only where the jack rafter slope is greater than or equal to the roof slope. For the purposes of determining member sizes in this section, wind speeds shall be converted to equivalent ground snow load in accordance with Table R804.3.2.1(3).

❖ This section provides the prescriptive requirements for hip roof framing.

Sections R804.3.3.1 through R804.3.3.5 provide the requirements for jack rafters, hip members, hip support columns and framing connections. Wind speed must be converted to equivalent snow load to determine the member sizes (see commentary, Section R804.3.2.1).

R804.3.3.1 Jack rafters. Jack rafters shall meet the requirements for roof rafters in accordance with Section R804.3.2, except that the requirements in Section R804.3.2.4 shall not apply.

❖ See the commentary to Section R804.3.2. All sections of R804.3.2 apply except Section R804.3.2.4.

R804.3.3.2 Hip members. Hip members shall be fabricated from C-shape members and track section, which shall have minimum sizes determined in accordance with Table R804.3.3.2. The C-shape member and track section shall be connected at a maximum spacing of 24 inches (610 mm) using No. 10 screws through top and bottom flanges in accordance with Figure R804.3.2.4. The depth of the hip member shall match that of the roof rafters and jack rafters, or shall be based on an *approved* design for a beam pocket at the corner of the supporting wall.

❖ Hip members consist of a C-section attached to the inside of a track section in accordance with Figure R804.3.2.4. The C-section and track section size in accordance with Table R804.3.3.2.

The hip member depth must equal or exceed the depth of the sloped rafter or jack rafter cut as shown in Figure R804.3.2.4.

R804.3.3.3 Hip support columns. Hip support columns shall be used to support hip members at the ridge. A hip support column shall consist of a pair of C-shape members, with a minimum size determined in accordance with Table R804.3.3.3. The C-shape members shall be connected at a maximum spacing of 24 inches (610 mm) on center to form a box using minimum 3-inch by 33-mil (76 mm by 0.84 mm) strap connected to each of the flanges of the C-shape members with three-No. 10 screws. Hip support columns shall have a continuous load path to the foundation and shall be

TABLE R804.3.3.2
HIP MEMBER SIZES, 33 ksi STEEL

BUILDING WIDTH (feet)	HIP MEMBER DESIGNATION[a]			
	Equivalent ground snow load (psf)			
	0 to 20	21 to 30	31 to 50	51 to 70
24	800S162-68	800S162-68	800S162-97	1000S162-97
	800T150-68	800T150-68	800T150-97	1000T150-97
28	1000S162-68	1000S162-68	1000S162-97	1200S162-97
	1000T150-68	1000T150-68	1000T150-97	1200T150-97
32	1000S162-97	1000S162-97	1200S162-97	—
	1000T150-97	1000T150-97	1200T150-97	
36	1200S162-97 1200T150-97	—	—	—
40	—	—	—	—

For SI: 1 foot = 304.8 mm, 1 pound per square foot = 0.0479 kPa.

a. The web depth of the roof rafters and jack rafters is to match at the hip or they shall be installed in accordance with an approved design.

❖ See the commentary to Section R804.3.3.2.

TABLE R804.3.3.3
HIP SUPPORT COLUMN SIZES

BUILDING WIDTH (feet)	HIP SUPPORT COLUMN DESIGNATION[a, b]			
	Equivalent ground snow load (psf)			
	0 to 20	21 to 30	31 to 50	51 to 70
24	2-350S162-33	2-350S162-33	2-350S162-43	2-350S162-54
28	2-350S162-54	2-550S162-54	2-550S162-68	2-550S162-68
32	2-550S162-68	2-550S162-68	2-550S162-97	—
36	2-550S162-97	—	—	—
40	—	—	—	—

For SI: 1 foot = 304.8 mm, 1 pound per square foot = 0.0479 kPa.

a. Box shape column only in accordance with Figure R804.3.3.4(2).

b. 33-ksi steel for 33- and 43-mil material; 50-ksi steel for thicker material.

❖ See the commentary to Section R804.3.3.3.

supported at the ceiling line by an interior wall or by an *approved* design for a supporting element.

❖ The hip support column is a box section comprised of two C-sections assembled flange to flange as shown in Figure R804.3.3.4(2). The minimum C-section sizes used to form the box section are determined from Table R804.3.3.3 based on equivalent ground snow load and building width. The hip support column supports the hip members at the ridge as shown in Figure R804.3.3.4(3). The column must have a continuous load path to the foundation.

R804.3.3.4 Hip framing connections. Hip rafter framing connections shall be installed in accordance with the following:

1. Jack rafters shall be connected at the eave to a parallel C-shape blocking member in accordance with Figure R804.3.3.4(1). The C-shape blocking member shall be attached to the supporting wall track with minimum two No. 10 screws.
2. Jack rafters shall be connected to a hip member with a minimum 2-inch by 2-inch (51 mm by 51 mm) clip angle fastened with No.10 screws to the hip member in accordance with Figure R804.3.2.4 and Table R804.3.2.4. The clip angle shall have a steel thickness equivalent to or greater than the jack rafter thickness and shall extend the depth of the jack rafter member to the extent possible.
3. The connection of the hip support columns at the ceiling line shall be in accordance with Figure R804.3.3.4(2), with an uplift strap sized in accordance with Table R804.3.3.4(1).
4. The connection of hip support members, ridge members and hip support columns at the ridge shall be in

TABLE R804.3.3.4(1)
UPLIFT STRAP CONNECTION REQUIREMENTS HIP SUPPORT COLUMN AT CEILING LINE

BUILDING WIDTH (feet)	BASIC WIND SPEED (mph) EXPOSURE B				
	85	100	110	—	—
	BASIC WIND SPEED (mph) EXPOSURE C				
	—	85	—	100	110
	Number of No. 10 screws in each end of each 3-inch by 54-mil steel strap[a, b, c]				
24	3	4	4	6	7
28	4	6	6	8	10
32	5	8	8	11	13
36	7	10	11	14	17
40	—	—	—	—	—

For SI: 1 foot = 304.8 mm, 1 pound per square foot = 0.0479 kPa, 1 mil = 0.0254 mm.
a. Two straps are required, one each side of the column.
b. Space screws at $^3/_4$ inch on center and provide $^3/_4$-inch end distance.
c. 50-ksi steel strap.

❖ See the commentary to Section R804.3.3.4.

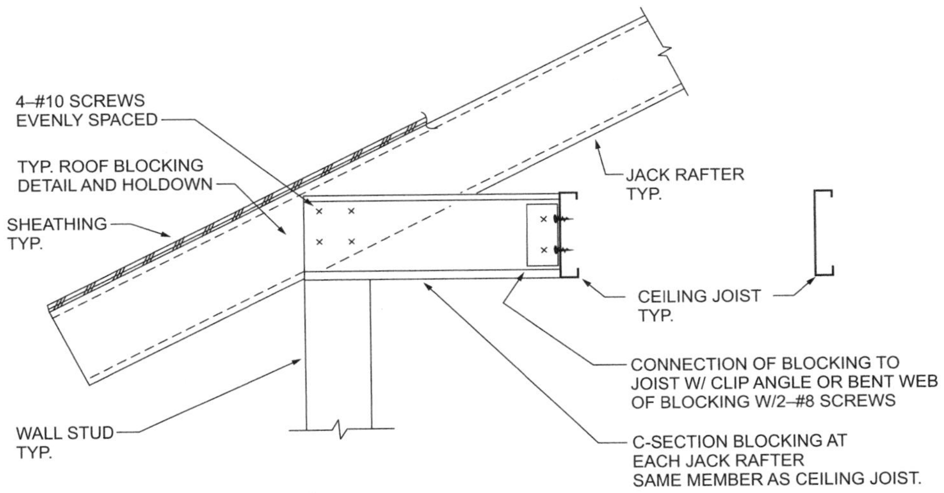

FIGURE R804.3.3.4(1)
JACK RAFTER CONNECTION AT EAVE

❖ See the commentary to Section R804.3.3.4.

accordance with Figures R804.3.3.4(3) and R804.3.3.4(4) and Table R804.3.3.4(2).

5. The connection of hip members to the wall corner shall be in accordance with Figure R804.3.3.4(5) and Table R804.3.3.4(3).

❖ This section prescribes the connection details for jack rafter at eave, jack rafter to hip member, hip support column at ceiling, hip support column at ridge and hip member to top of exterior wall.

The details are as depicted in Figure R804.3.2.4 and Figures R804.3.3.4(1) through R804.3.3.4(5).

R804.3.4 Cutting and notching. Flanges and lips of load-bearing, cold-formed steel roof framing members shall not be cut or notched.

❖ Flanges and lips of load-bearing steel roof framing members must not be cut or notched because this would affect their structural integrity.

R804.3.5 Headers. Roof-ceiling framing above wall openings shall be supported on headers. The allowable spans for headers in load-bearing walls shall not exceed the values set forth in Section R603.6 and Tables R603.6(1) through R603.6(24).

❖ Headers must be located immediately below the ceiling or roof framing (for example, at the top of the track), above wall openings, in all exterior walls and in interior load-bearing walls in accordance with Section R603.6. Tables R603.6(1) through R603.6.1(24) gives the allowable spans for headers in bearing walls. See the commentary to Section R603.6.

TABLE R804.3.3.4(2)
CONNECTION REQUIREMENTS HIP MEMBER TO HIP SUPPORT COLUMN

BUILDING WIDTH (feet)	NUMBER OF NO. 10 SCREWS IN EACH FRAMING ANGLE[a, b, c]			
	Equivalent ground snow load (psf)			
	0 to 20	21 to 30	31 to 50	51 to 70
24	10	10	10	12
28	10	10	14	18
32	10	12	—	—
36	14	—	—	—
40	—	—	—	—

For SI: 1 foot = 304.8 mm, 1 pound per square foot = 0.0479 kPa.

a. Screws to be divided equally between the connection to the hip member and the column. Refer to Figures R804.3.3.4(3) and R804.3.3.4(4).

b. The number of screws required in each framing angle is not to be less than shown in Table R804.3.3.4(1).

c. 50-ksi steel from the framing angle.

❖ See the commentary to Section R804.3.3.4.

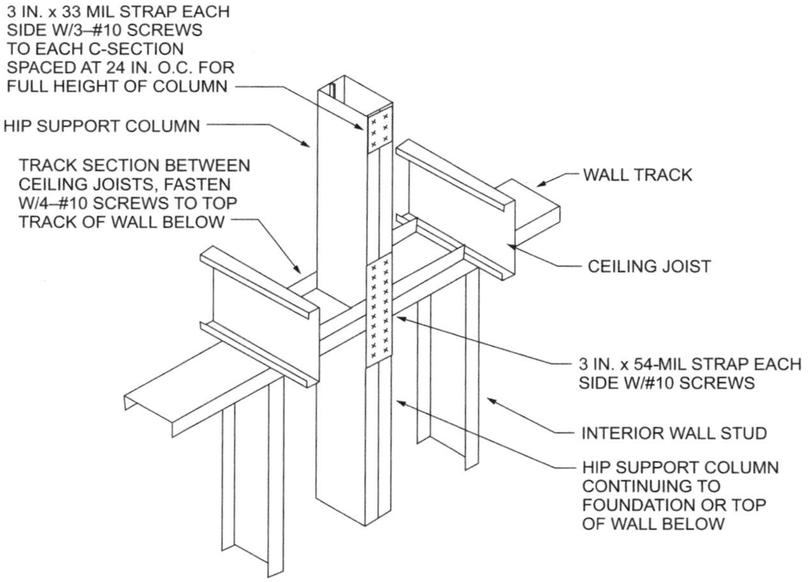

For SI: 1inch = 25.4 mm, 1 mil = 0.0254 mm.

FIGURE R804.3.3.4(2)
HIP SUPPORT COLUMN

❖ See the commentary to Section R804.3.3.4.

TABLE R804.3.3.4(3)
UPLIFT STRAP CONNECTION REQUIREMENTS HIP MEMBER TO WALL

BUILDING WIDTH (feet)	BASIC WIND SPEED (mph) EXPOSURE B				
	85	100	110	—	—
	BASIC WIND SPEED (mph) EXPOSURE C				
	—	85	—	100	110
	Number of No. 10 screws in each end of each 3-inch by 54-mil steel strap[a, b, c]				
24	2	2	3	3	4
28	2	3	3	4	5
32	3	4	4	6	7
36	3	5	5	7	8
40	—	—	—	—	—

For SI: 1 foot = 304.8 mm,1 pound per square foot = 0.0479 kPa.

a. Two straps are required, one each side of the column.

b. Space screws at $^3/_4$ inches on center and provide $^3/_4$-inch end distance.

c. 50-ksi steel strap.

❖ See the commentary to Section R804.3.3.4.

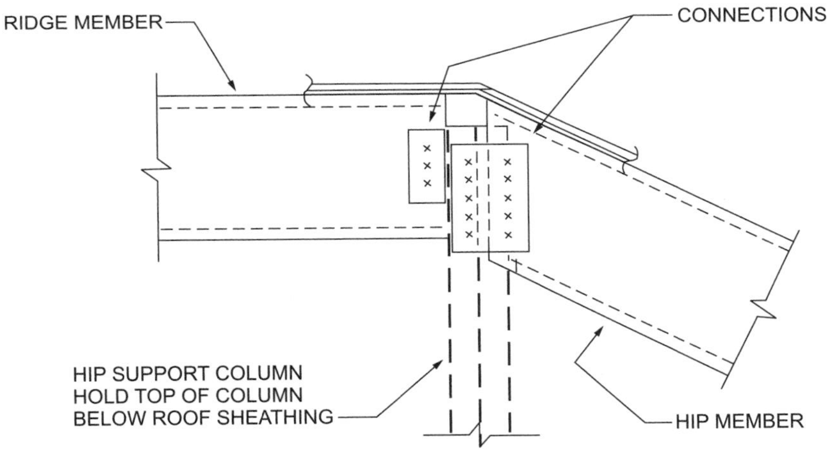

FIGURE R804.3.3.4(3)
HIP CONNECTIONS AT RIDGE

❖ See the commentary to Section R804.3.3.4.

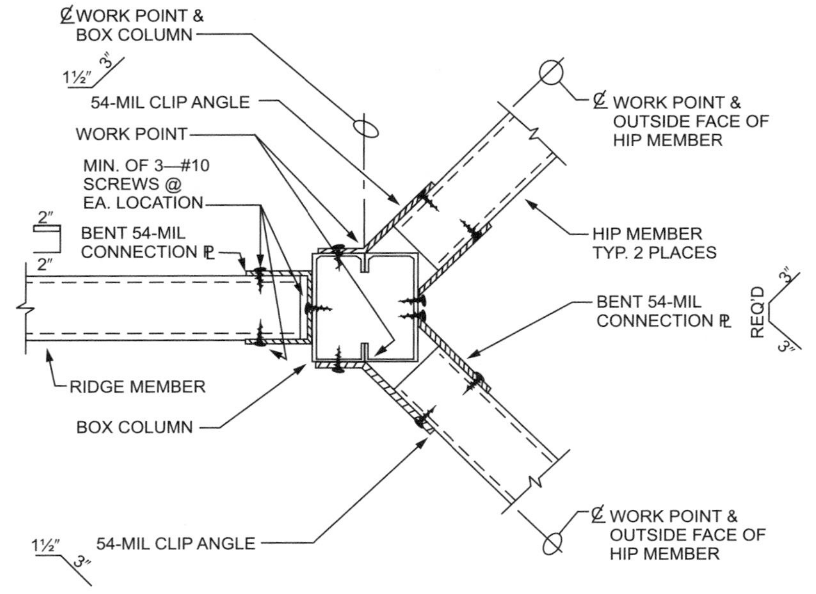

NOTE: RAFTERS NOT SHOWN FOR CLARITY

CONNECTION @ 3½″ BOX COLUMN

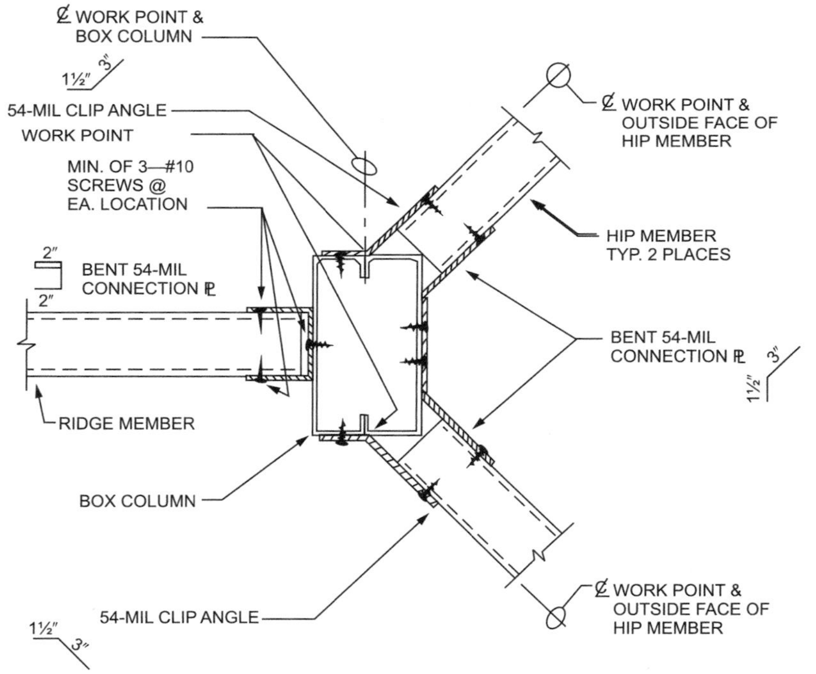

NOTE: RAFTERS NOT SHOWN FOR CLARITY

CONNECTION @ 5½″ BOX COLUMN

For SI: 1 inch = 25.4 mm, 1 mil = 0.0254 mm.

FIGURE R804.3.3.4(4)
HIP CONNECTIONS AT RIDGE AND BOX COLUMN

❖ See the commentary to Section R804.3.3.4.

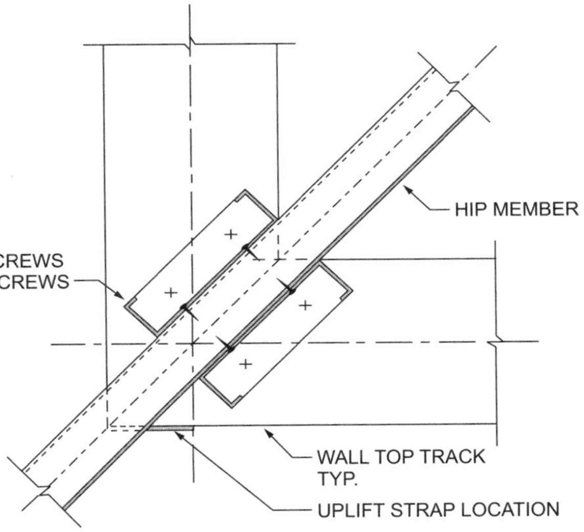

FIGURE R804.3.3.4(5)
HIP MEMBER CONNECTION AT WALL CORNER

❖ See the commentary to Section R804.3.3.4.

R804.3.6 Framing of openings in roofs and ceilings. Openings in roofs and ceilings shall be framed with header and trimmer joists. Header joist spans shall not exceed 4 feet (1219 mm) in length. Header and trimmer joists shall be fabricated from joist and track members having a minimum size and thickness at least equivalent to the adjacent ceiling joists or roof rafters and shall be installed in accordance with Figures R804.3.6(1) and R804.3.6(2). Each header joist shall be connected to trimmer joists with a minimum of four 2-inch by 2-inch (51 by 51 mm) clip angles. Each clip angle shall be fastened to both the header and trimmer joists with four No. 8 screws, evenly spaced, through each leg of the clip angle. The steel thickness of the clip angles shall be not less than that of the ceiling joist or roof rafter. Each track section for a built-up header or trimmer joist shall extend the full length of the joist (continuous).

❖ Roof openings must be framed as illustrated in Figures R804.3.6(1) and R804.3.6(2). Each header joist must be connected to the trimmer joists with a minimum of four 2-inch by 2-inch (51 mm by 51 mm) clip angles, which must be fastened to both the header and trimmer joists with four No. 8 screws, evenly spaced, through each leg. The clip angle's thickness must not be less than that of the ceiling joist or roof rafter.

R804.3.7 Roof trusses. Cold-formed steel trusses shall be designed and installed in accordance with AISI S100, Section D4. In the absence of specific bracing requirements, trusses shall be braced in accordance with accepted industry practices, such as the SBCA *Cold-Formed Steel Building Component Safety Information (CFSBCSI) Guide to Good Practice for Handling, Installing & Bracing of Cold-Formed Steel Trusses*. Trusses shall be connected to the top track of the load-bearing wall in accordance with Table R804.3, either with two No. 10 screws applied through the flange of the truss or by using a 54-mil (1.37 mm) clip angle with two No. 10 screws in each leg.

❖ The code contains no prescriptive provisions for the design and installation of cold-formed steel trusses. A design is required that complies with accepted engineering practice. Recognizing the extensive use of trusses in residential construction, the code references AISI S100, Section D4. Section D4 of AISI S100 specifies the criteria for the design of cold-formed steel light-frame construction and requires trusses to be designed in accordance with AISI S214. The standard provides regulations for the design and installation of cold-formed steel trusses, including quality assurance and the procedures for full-scale tests. In addition to adequate design, it is important that the handling and erection of the trusses be performed properly so that the performance capability of the trusses is not compromised. A truss member must never be cut, notched or altered without approval from the design professional.

R804.3.8 Ceiling and roof diaphragms. Ceiling and roof diaphragms shall be in accordance with this section.

❖ This section provides the prescriptive requirements for ceiling and roof diaphragms. Sections R804.3.8.1 and R804.3.8.2 provide the details for ceiling diaphragms of gypsum board or wood structural panel and roof diaphragms of wood structural panel.

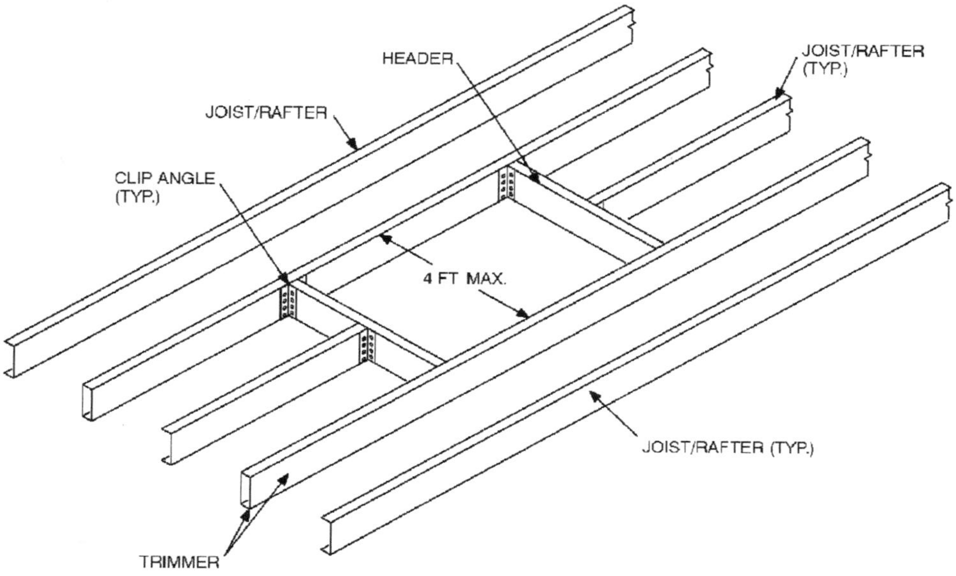

For SI: 1 foot = 304.8 mm.

FIGURE R804.3.6(1)
ROOF OR CEILING OPENING

❖ See the commentary to Section R804.3.6.

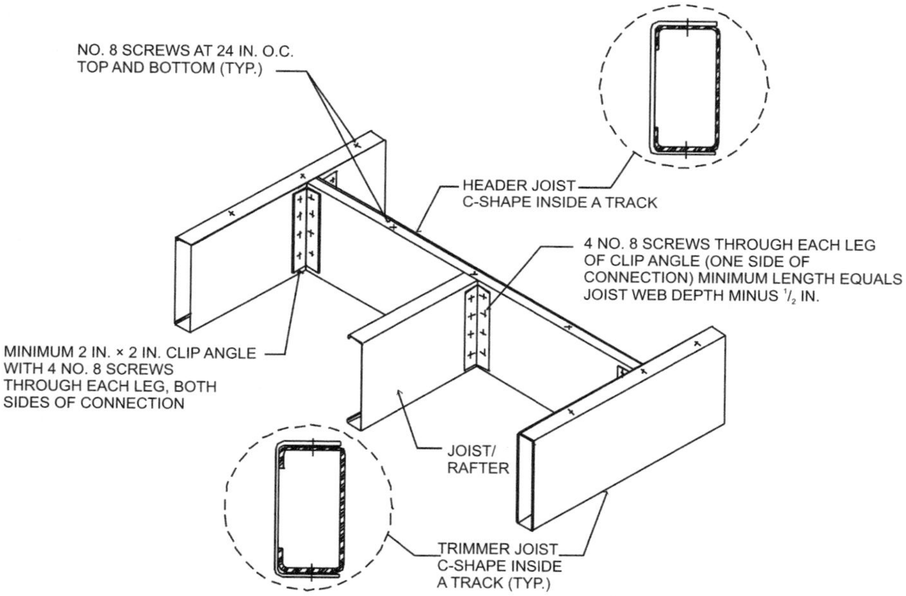

For SI: 1 inch = 25.4 mm.

FIGURE R804.3.6(2)
HEADER TO TRIMMER CONNECTION

❖ See the commentary to Section R804.3.6.

R804.3.8.1 Ceiling diaphragms. At gable endwalls a ceiling *diaphragm* shall be provided by attaching a minimum $^1/_2$-inch (12.7 mm) gypsum board in accordance with Tables R804.3.8(1) and R804.3.8(2) or a minimum $^3/_8$-inch (9.5 mm) wood structural panel sheathing, which complies with Section R803, in accordance with Table R804.3.8(3) to the bottom of ceiling joists or roof trusses and connected to wall framing in accordance with Figures R804.3.8(1) and R804.3.8(2), unless studs are designed as full height without bracing at the ceiling. Flat blocking shall consist of C-shape or track section with a minimum thickness of 33 mils (0.84 mm).

The ceiling *diaphragm* shall be secured with screws spaced at a maximum 6 inches (152 mm) o.c. at panel edges and a maximum 12 inches (305 mm) o.c. in the field. Multiplying the required lengths in Tables R804.3.8(1) and R804.3.8(2) for gypsum board sheathed ceiling diaphragms shall be permitted to be multiplied by 0.35 shall be permitted if all panel edges are blocked. Multiplying the required lengths in Tables R804.3.8(1) and R804.3.8(2) for gypsum board sheathed ceiling diaphragms by 0.9 shall be permitted if all panel edges are secured with screws spaced at 4 inches (102 mm) o.c.

❖ A ceiling diaphragm is required unless the gable endwall studs are designed to extend full height to the roof without bracing at the ceiling line. The ceiling diaphragm is constructed of either gypsum board or wood panel sheathing in accordance with Tables R804.3.8(1) through R804.3.8(3). See Figures R804.3.8(1) and R804.3.8(2) for connection to the gable endwall and sidewall.

The required length of gypsum board ceiling diaphragm is permitted to be reduced, to 35 percent of that specified in the tables, if all panel edges are blocked.

The required length of gypsum board ceiling diaphragm is permitted to be reduced to 90 percent of that specified in the tables, if panel edge screws are spaced at 4 inches (102 mm).

R804.3.8.2 Roof diaphragm. A roof *diaphragm* shall be provided by attaching a minimum of $^3/_8$-inch (9.5 mm) wood structural panel which complies with Section R803 to roof rafters or truss top chords in accordance with Table R804.3. Buildings with 3:1 or larger plan *aspect ratio* and with roof rafter slope (pitch) of 9:12 or larger shall have the roof rafters and ceiling joists blocked in accordance with Figure R804.3.8(3).

❖ A roof diaphragm consisting of minimum $^3/_8$-inch-thick (9.5 mm) wood structural panel sheathing is required in accordance with this section. Roof blocking is required where the building plan aspect ratio is equal to or greater than 3:1 and the roof pitch is equal to or greater than 9:12 [see Figure R804.3.8(3) for roof blocking details].

TABLE R804.3.8(1)
REQUIRED LENGTHS FOR CEILING DIAPHRAGMS AT GABLE ENDWALLS
GYPSUM BOARD SHEATHED, CEILING HEIGHT = 8 FEET[a, b, c, d, e, f]

Exposure B		BASIC WIND SPEED (mph)				
		85	100	110	—	—
Exposure C		—	85	—	100	110
Roof pitch	Building endwall width (feet)	Minimum diaphragm length (feet)				
3:12 to 6:12	24 - 28	14	20	22	28	32
	28 - 32	16	22	28	32	38
	32 - 36	20	26	32	38	44
	36 - 40	22	30	36	44	50
6:12 to 9:12	24 - 28	16	22	26	32	36
	28 - 32	20	26	32	38	44
	32 - 36	22	32	38	44	52
	36 - 40	26	36	44	52	60
9:12 to 12:12	24 - 28	18	26	30	36	42
	28 - 32	22	30	36	42	50
	32 - 36	26	36	42	50	60
	36 - 40	30	42	50	60	70

For SI: 1 inch = 25.4 mm, 1 pound per square foot = 0.0479 kPa, 1 mile per hour = 0.447 m/s, 1 foot = 304.8 mm, 1 mil = 0.0254 mm.

a. Ceiling diaphragm is composed of $^1/_2$-inch gypsum board (min. thickness) secured with screws spaced at 6 inches o.c. at panel edges and 12 inches o.c. infield. Use No. 8 screws (min.) when framing members have a designation thickness of 54 mils or less and No. 10 screws (min.) when framing members have a designation thickness greater than 54 mils.
b. Maximum aspect ratio (length/width) of diaphragms is 2:1.
c. Building width is in the direction of horizontal framing members supported by the wall studs.
d. Required diaphragm lengths are to be provided at each end of the structure.
e. Multiplying required diaphragm lengths by 0.35 is permitted if all panel edges are blocked.
f. Multiplying required diaphragm lengths by 0.9 is permitted if all panel edges are secured with screws spaced at 4 inches o.c.

❖ See the commentary to Section R804.3.8.1.

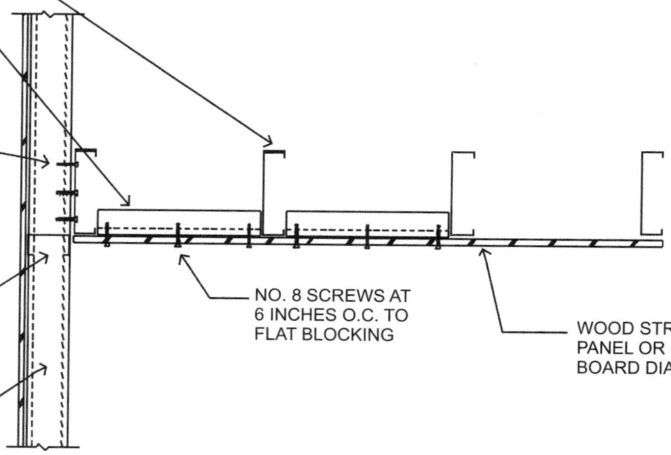

For SI: 1 inch = 25.4 mm.

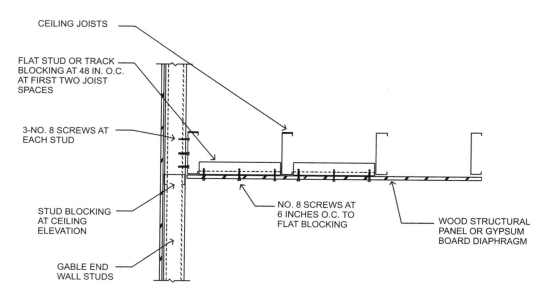

FIGURE R804.3.8(1)
CEILING DIAPHRAGM TO GABLE ENDWALL DETAIL

❖ See the commentary to Section R804.3.8. 1.

TABLE R804.3.8(2)
REQUIRED LENGTHS FOR CEILING DIAPHRAGMS AT GABLE ENDWALLS
GYPSUM BOARD SHEATHED CEILING HEIGHT = 9 OR 10 FEET[a, b, c, d, e, f]

Exposure B		BASIC WIND SPEED (mph)				
		85	100	110	—	—
Exposure C		—	85	—	100	110
Roof pitch	Building endwall width (feet)	Minimum diaphragm length (feet)				
3:12 to 6:12	24 - 28	16	22	26	32	38
	28 - 32	20	26	32	38	44
	32 - 36	22	30	36	44	50
	36 - 40	26	36	42	50	58
6:12 to 9:12	24 - 28	18	26	30	36	42
	28 - 32	22	30	36	42	50
	32 - 36	26	36	42	50	58
	36 - 40	30	42	48	58	68
9:12 to 12:12	24 - 28	20	28	34	40	46
	28 - 32	24	34	40	48	56
	32 - 36	28	40	48	56	66
	36 - 40	34	46	56	66	78

For SI: 1 inch = 25.4 mm, 1 pound per square foot = 0.0479 kPa, 1 mile per hour = 0.447 m/s, 1 foot = 304.8 mm, 1 mil = 0.0254 mm.

a. Ceiling diaphragm is composed of $^1/_2$-inch gypsum board (min. thickness) secured with screws spaced at 6 inches o.c. at panel edges and 12 inches o.c. infield. Use No. 8 screws (min.) when framing members have a designation thickness of 54 mils or less and No. 10 screws (min.) when framing members have a designation thickness greater than 54 mils.

b. Maximum aspect ratio (length/width) of diaphragms is 2:1.

c. Building width is in the direction of horizontal framing members supported by the wall studs.

d. Required diaphragm lengths are to be provided at each end of the structure.

e. Required diaphragm lengths are permitted to be multiplied by 0.35 if all panel edges are blocked.

f. Required diaphragm lengths are permitted to be multiplied by 0.9 if all panel edges are secured with screws spaced at 4 inches o.c.

❖ See the commentary to Section R804.3.8.1.

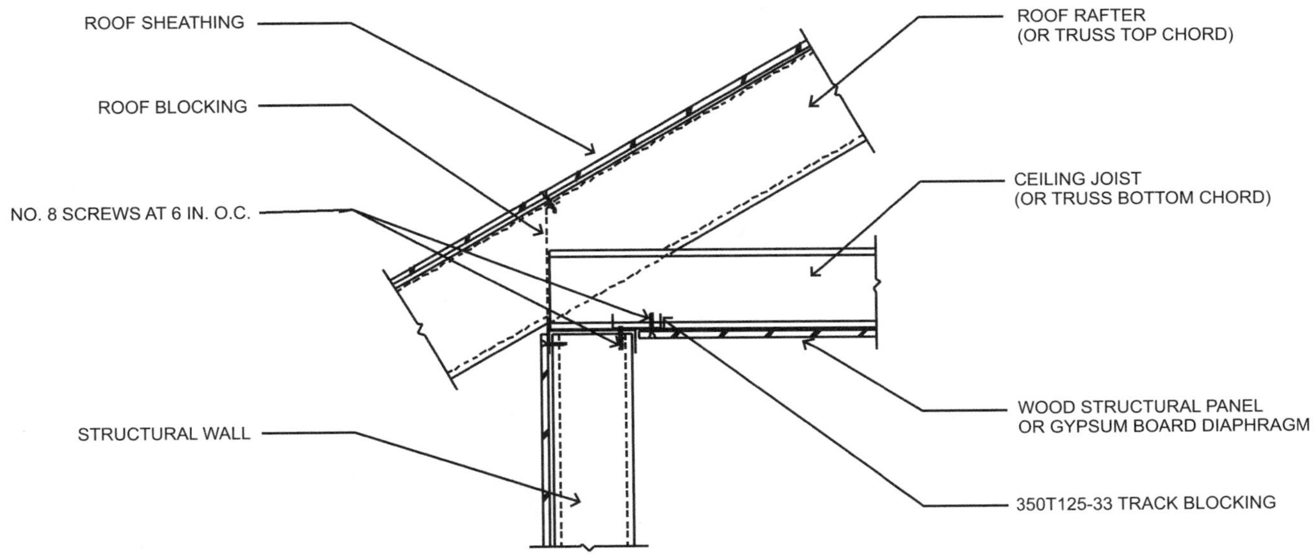

For SI: 1 inch = 25.4 mm.

FIGURE R804.3.8(2)
CEILING DIAPHRAGM TO SIDEWALL DETAIL

❖ See the commentary to Section R804.3.8.1.

TABLE R804.3.8(3)
REQUIRED LENGTHS FOR CEILING DIAPHRAGMS AT GABLE ENDWALLS
WOOD STRUCTURAL PANEL SHEATHED CEILING HEIGHT = 8, 9 OR 10 FEET[a, b, c, d]

Exposure B		BASIC WIND SPEED (mph)				
		85	100	110	—	—
Exposure C		—	85	—	100	110
Roof pitch	Building endwall width (feet)	Minimum diaphragm length (feet)				
3:12 to 6:12	24 - 28	10	10	10	10	10
	28 - 32	12	12	12	12	12
	32 - 36	12	12	12	12	12
	36 - 40	14	14	14	14	14
6:12 to 9:12	24 - 28	10	10	10	10	10
	28 - 32	12	12	12	12	12
	32 - 36	12	12	12	12	12
	36 - 40	14	14	14	14	14
9:12 to 12:12	24 - 28	10	10	10	10	10
	28 - 32	12	12	12	12	12
	32 - 36	12	12	12	12	12
	36 - 40	14	14	14	14	14

For SI: 1 inch = 25.4 mm, 1 pound per square foot = 0.0479 kPa, 1 mile per hour = 0.447 m/s, 1 foot = 304.8 mm, 1 mil = 0.0254 mm.

a. Ceiling diaphragm is composed of $^3/_8$-inch wood structural panel sheathing (min. thickness) secured with screws spaced at 6 inches o.c. at panel edges and in field.
Use No. 8 screws (min.) when framing members have a designation thickness of 54 mils or less and No. 10 screws (min.) when framing members have a designation thickness greater than 54 mils.

b. Maximum aspect ratio (length/width) of diaphragms is 3:1.

c. Building width is in the direction of horizontal framing members supported by the wall studs.

d. Required diaphragm lengths are to be provided at each end of the structure.

❖ See the commentary to Section R804.3.8.1.

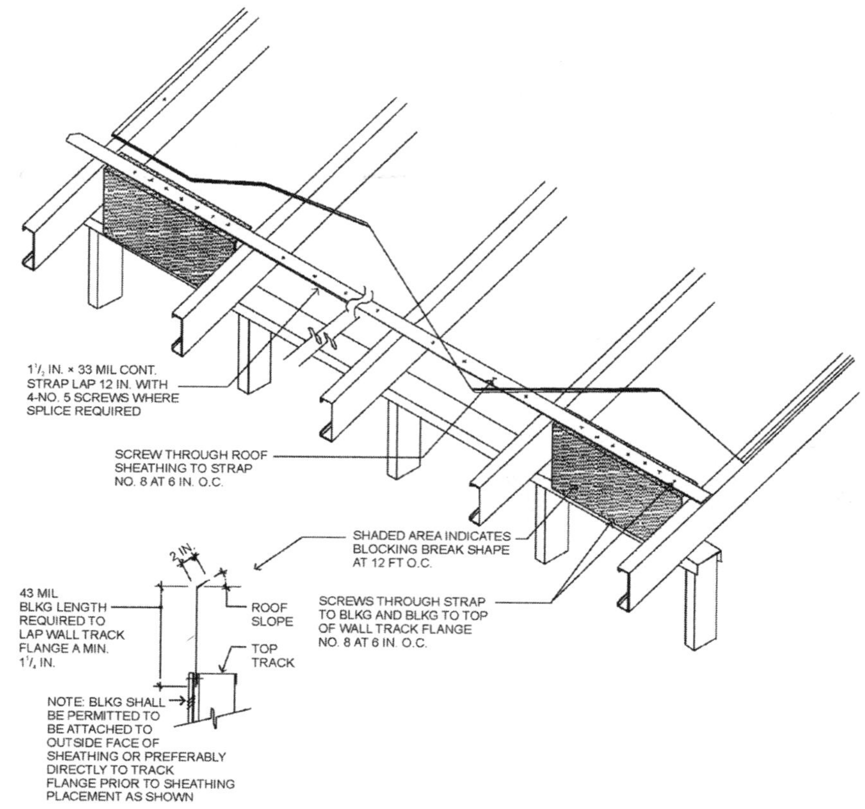

For SI: 1 mil = 0.0254 mm, 1 inch = 25.4 mm.

FIGURE R804.3.8(3)
ROOF BLOCKING DETAIL

❖ See the commentary to Section R804.3.8.2.

R804.3.9 Roof tie-down. Roof assemblies subject to wind uplift pressures of 20 pounds per square foot (0.96 kPa) or greater, as established in Table R301.2(2), shall have rafter-to-bearing wall ties provided in accordance with Table R802.11.

❖ Steel-framed roof assemblies require rafter-to-bearing wall ties with a capacity determined in the manner described in Section R802.11.

SECTION R805
CEILING FINISHES

R805.1 Ceiling installation. Ceilings shall be installed in accordance with the requirements for interior wall finishes as provided in Section R702.

❖ This section provides a reference to the ceiling installation requirements found in Section R702.

SECTION R806
ROOF VENTILATION

R806.1 Ventilation required. Enclosed *attics* and enclosed rafter spaces formed where ceilings are applied directly to the underside of roof rafters shall have cross ventilation for each separate space by ventilating openings protected against the entrance of rain or snow. Ventilation openings shall have a least dimension of $^1/_{16}$ inch (1.6 mm) minimum and $^1/_4$ inch (6.4 mm) maximum. Ventilation openings having a least dimension larger than $^1/_4$ inch (6.4 mm) shall be provided with corrosion-resistant wire cloth screening, hardware cloth, or similar material with openings having a least dimension of $^1/_{16}$ inch (1.6 mm) minimum and $^1/_4$ inch (6.4 mm) maximum. Openings in roof framing members shall conform to the requirements of Section R802.7. Required ventilation openings shall open directly to the outside air.

Exception: Attic ventilation shall not be required when determined not necessary by the code official due to atmospheric or climatic conditions.

❖ Large amounts of water vapor migrate by air movement or diffusion through the building envelope materials because of a vapor pressure difference. The sources of water vapor include cooking, laundering, bathing and human breathing and perspiration. These can account for an average daily production of 25 pounds (11.3 kg) of water vapor in a typical family-of-four dwelling. The average can be much higher where appliances such as humidifiers, washers and dryers are used.

As the vapor moves into the attic, it may reach its dew point, condensing on wood roof components. This wetting and drying action will cause rotting and decay. To avoid this, the attic must be ventilated to prevent the accumulation of water on building components. The installation of a vapor retarder acts to prevent the passage of moisture to the attic, and an effective vapor retarder allows a decrease in ventilation. Vapor retarders are ineffective when openings in the barrier allow moisture to be carried by air into the attic. This is also the reason exhaust fans must terminate outdoors and not in the attic. Care should be exercised to assure that attic vent openings remain unobstructed.

To minimize condensation problems within attic and enclosed rafter spaces, free-flow ventilation of such spaces is required. Ventilation openings must be screened to prevent the entry of animals. Ventilation openings requiring cutting or notching of the roof framing member must comply with Section R802.7.

The exception provides for the building official to waive the attic ventilation requirements based on local atmospheric or climatic conditions.

R806.2 Minimum vent area. The minimum net free ventilating area shall be $^{1}/_{150}$ of the area of the vented space.

Exception: The minimum net free ventilation area shall be $^{1}/_{300}$ of the vented space provided one or more of the following conditions are met:

1. In Climate Zones 6, 7 and 8, a Class I or II vapor retarder is installed on the warm-in-winter side of the ceiling.
2. At least 40 percent and not more than 50 percent of the required ventilating area is provided by ventilators located in the upper portion of the attic or rafter space. Upper ventilators shall be located no more than 3 feet (914 mm) below the ridge or highest point of the space, measured vertically, with the balance of the required ventilation provided by eave or cornice vents. Where the location of wall or roof framing members conflicts with the installation of upper ventilators, installation more than 3 feet (914 mm) below the ridge or highest point of the space shall be permitted.

❖ The attic vent size required by the code should not be overlooked. The net-free area can be as much as 50 percent less than the gross opening area. For example, one manufacturer's 24-inch square (610 mm) gable vent [gross area equals 576 square inches (0.37 m^2)] is listed in their catalog as having a net free area of 308 square inches (0.20 m^2), which is about 53 percent of the gross area. The manufacturer's literature should be consulted to obtain free-area information.

The exception permits a reduction in the required venting area if one of the two conditions is met. Condition 1 recognizes that in cold climates, a vapor retarder is required and permits a reduction based on the vapor retarder being installed on the warm-in-winter side of the ceiling. Condition 2 allows a reduction, provided cross ventilation is supplied as specified.

R806.3 Vent and insulation clearance. Where eave or cornice vents are installed, insulation shall not block the free flow of air. A minimum of a 1-inch (25 mm) space shall be provided between the insulation and the roof sheathing and at the location of the vent.

❖ Vent openings must be maintained clear, and they must not block the free flow of air; therefore, the code requires that insulation be held back from the vent opening a minimum of 1 inch (25 mm). The 1-inch (25 mm) clearance must be maintained not only at the vent but throughout the attic and rafter spaces.

R806.4 Installation and weather protection. Ventilators shall be installed in accordance with manufacturer's installation instructions. Installation of ventilators in roof systems shall be in accordance with the requirements of Section R903. Installation of ventilators in wall systems shall be in accordance with the requirements of Section R703.1.

❖ The section requires that ventilators be installed in accordance with the manufacturer's installation instructions. This is essential so that ventilators provide the proper cross ventilation and perform as intended. Additionally, this section requires roof or wall ventilator installations to comply with the proper code-required weather protection, including flashing.

R806.5 Unvented attic and unvented enclosed rafter assemblies. Unvented *attic* assemblies (spaces between the ceiling joists of the top *story* and the roof rafters) and unvented enclosed rafter assemblies (spaces between ceilings that are applied directly to the underside of roof framing members/rafters and the structural roof sheathing at the top of the roof framing members/rafters) shall be permitted if all the following conditions are met:

1. The unvented *attic* space is completely contained within the *building thermal envelope.*
2. No interior Class I vapor retarders are installed on the ceiling side (*attic* floor) of the unvented *attic* assembly or on the ceiling side of the unvented enclosed rafter assembly.
3. Where wood shingles or shakes are used, a minimum $^{1}/_{4}$-inch (6 mm) vented air space separates the shingles or shakes and the roofing underlayment above the structural sheathing.
4. In Climate Zones 5, 6, 7 and 8, any *air-impermeable insulation* shall be a Class II vapor retarder, or shall have a Class III vapor retarder coating or covering in direct contact with the underside of the insulation.
5. Either Items 5.1, 5.2 or 5.3 shall be met, depending on the air permeability of the insulation directly under the structural roof sheathing.

   5.1. *Air-impermeable insulation* only. Insulation shall be applied in direct contact with the underside of the structural roof sheathing.

   5.2. Air-permeable insulation only. In addition to the air-permeable insulation installed directly below the structural sheathing, rigid board or sheet insulation shall be installed directly above the structural roof sheathing as specified in Table R806.5 for condensation control.

5.3. Air-impermeable and air-permeable insulation. The *air-impermeable insulation* shall be applied in direct contact with the underside of the structural roof sheathing as specified in Table R806.5 for condensation control. The air-permeable insulation shall be installed directly under the *air-impermeable insulation.*

5.4. Where preformed insulation board is used as the air-impermeable insulation layer, it shall be sealed at the perimeter of each individual sheet interior surface to form a continuous layer.

❖ Unvented attics are attics where the insulation and air barrier boundary are moved to be directly above the attic space, instead of on top of the ceiling. Unvented attics eliminate the extreme temperatures of the attic, thereby placing the HVAC, ducts, pipes, and anything in the attic space into a more favorable environment. Unvented attics increase energy efficiency and decrease wear and tear on equipment in the attic. This section describes attics where the insulation and air barrier are above instead of below the attic space. Moving the insulation and placing an air-impermeable barrier above the attic moderates attic conditions so they are similar to the conditions of the residential space below. The primary benefit of having the insulation and air barrier above the attic is that ducts and/or HVAC equipment in the attic are not delivering cooled air through a hot summer attic and heated air through a cold winter attic. Another benefit is to eliminate the attic vents that sometimes allow moisture to condense inside the attic, admit rain during extreme weather and possibly admit sparks in fires.

Because this space is inside the building's thermal envelope, the traditional attic ventilation required by Sections R806.1 and R806.2 is not required. Unvented attics require water/moisture control. Water moves in (or out) of buildings three main ways. The greatest amount of moisture is moved as bulk water (rain or any kind of water flow). Less moisture is moved by moving moist air, such as with infiltration. The least amount of moisture is moved by moisture migration through materials. As with any attic, the roof itself is the main barrier for keeping water from entering the attic.

Unvented cathedral ceilings (enclosed rafter assemblies) are permitted under this section. Unvented attics and unvented cathedral ceilings are similar. The governing physics are identical so these requirements work for both.

The provisions of this section can be applied to any attic area which is in compliance with this section. The attic is a traditional attic space, with the exception that it need not be ventilated and it will not get as hot or as cold as an attic that is open to the exterior.

It is very important that all of the five listed conditions be reviewed and considered for each building that uses the provisions of this section.

Item 1 requires that the attic space be completely contained within the building thermal envelope.

Item 2, which applies to all climate zones, prohibits the installation of a vapor retarder where it is typically installed at the ceiling level (attic floor) of a traditional ventilated attic. This assures that no barrier is installed which would separate the conditioned attic area from the remaining portion of the home. This requirement gives the attic space a limited potential to dry into the space beneath the attic so that small amounts of excess moisture can be removed from the attic. A sheet of polyurethane film or any material with a foil film facing are examples of vapor barriers that are not permitted on the attic floor.

Item 3 applies to all climate zones and contains special requirements which apply to wood shingles and shakes roof coverings. Wood shakes and shingles require a vented space under them to allow the wood to dry after it gets wet from rain.

Item 4 applies only to Climate Zones 5 or higher. The air-impermeable insulation must qualify as a vapor retarder or have a vapor retarder in direct contact with the insulation on the underside (interior) face of the insulation.

Item 5 requires sufficient insulation to keep moisture from condensing on the "condensing surface" inside the attic in "average conditions." The insulation works to prevent condensation by keeping the condensing surface above the temperature where condensation will occur. Small amounts of condensation may occasionally occur at more extreme conditions; however, this is not a concern. The condensing surface is the interior side of the roof deck for air-permeable insulation and the interior of the insulation for air-impermeable insulation. The condensing surfaces differ because attic air can circulate through air-permeable insulation to contact the roof deck but can get only to the interior of air-impermeable insulation. The requirement for "air-impermeable" insulation will assure that air and the moisture that it can contain will not pass through the insulation to reach a point where it could condense because of the temperature. Item 5 specifies that air-impermeable insulation be in direct contact with the interior side of the roof deck. Air-impermeable insulation prevents the movement of moist air that comes in as infiltration through the roof into the interior of the attic. Air-impermeable insulation is defined as having an air permanence of 0.02 L/s-m^2 (at 75 Pa pressure) or less. Expanding spray foams and insulated sheathing (hard-foam sheathing board) are common types of air-impermeable insulation. When using insulated sheathing, attention to the details of completing the air sealing is required as the sheathing is installed in the roof. Fiberglass and cellulose are common types of air-permeable insulation.

Note that the insulation required by Item 5 may be more or less than the insulation required for energy efficiency in Chapter 11. If the amount of insulation required by Item 5 varies from the amount specified in Chapter 11, the provisions of Section R1102.1 would apply, and the higher insulation value would be required. The insulation provided to comply with Item 5 may be considered as contributing to the insulation required in Chapter 11 (see Section N1102.1.1).

It is important to realize that this section cannot be viewed as modifying or eliminating requirements found elsewhere in the code. Examples of sections which still affect these attic areas include Sections R302.10, R302.13, R316, R807, N1102.1 and others. Because the insulation typically used with this provision is some type of foam plastic, the requirements of Section R316 must be applied. The provisions of Section R806.4 do not in any way modify or eliminate the requirements for a thermal barrier (Section R316.4) or protection from ignition (Section R316.5.3). See the commentary for Sections R302.10 and R316 for a complete discussion regarding these requirements and the options available. Ducts in this unvented attic construction would be considered as being inside the building thermal envelope and would not require insulation (see Section N1103.2.1).

The provisions of this section consider the attic assembly as a "conditioned" space; there is no requirement for the space to be provided with conditioned air supply. The attic space is considered indirectly conditioned because of omission of the air barrier, insulation at the ceiling and leakage around the attic access opening. An attic assembly complying with Section R806.4 will generally fall within the temperature ranges specified in the definition of "Conditioned space."

The key concept of this section is to move the thermal envelope (insulation) above the attic, resulting in the attic being in a conditioned (or sometimes semi-conditioned) space. Direct air supply to the attic is not required if the attic floor is not insulated; the attic temperature would be similar to interior conditioned spaces. Ducts and/or HVAC equipment in the attic also help moderate the attic conditions.

TABLE R806.5
INSULATION FOR CONDENSATION CONTROL

CLIMATE ZONE	MINIMUM RIGID BOARD ON AIR-IMPERMEABLE INSULATION *R*-VALUE[a]
2B and 3B tile roof only	0 (none required)
1, 2A, 2B, 3A, 3B, 3C	R-5
4C	R-10
4A, 4B	R-15
5	R-20
6	R-25
7	R-30
8	R-35

a. Contributes to but does not supersede the requirements in Section N1103.2.1.

❖ See the commentary to Section R806.5.

SECTION R807
ATTIC ACCESS

R807.1 Attic access. Buildings with combustible ceiling or roof construction shall have an *attic* access opening to *attic* areas that exceed 30 square feet (2.8 m^2) and have a vertical height of 30 inches (762 mm) or greater. The vertical height shall be measured from the top of the ceiling framing members to the underside of the roof framing members.

The rough-framed opening shall not be less than 22 inches by 30 inches (559 mm by 762 mm) and shall be located in a hallway or other readily accessible location. When located in a wall, the opening shall be a minimum of 22 inches wide by 30 inches high (559 mm wide by 762 mm high). When the access is located in a ceiling, minimum unobstructed headroom in the *attic* space shall be 30 inches (762 mm) at some point above the access measured vertically from the bottom of ceiling framing members. See Section M1305.1.3 for access requirements where mechanical *equipment* is located in *attics*.

❖ The requirement for an attic access is predicated on the likelihood that during the life of the structure, access to an attic space for repair of piping, electrical and mechanical systems will be required.

Bibliography

The following resource materials were used in the preparation of the commentary for this chapter of the code:

AFPA 93, *Span Tables for Joists and Rafters*. Washington, DC: American Forest and Paper Association, 1993.

AFPA/NDS-05, *National Design Specification (NDS) for Wood Construction–with 2005 Supplement*. Washington, DC: American Forest and Paper Association, 2005.

AFPA/WFCM-01, AF&PA *Wood Frame Construction Manual (WFCM) for One- and Two-family Dwellings*. Leesburg, VA: American Wood Council, 2001.

AISI S100-07/S1-10, *North American Specification for Design of Cold-formed Steel Structural Members with Supplement 1, dated 2010*. Washington, DC: American Iron and Steel Institute, 2010.

AISI S214-07, *North American Standard for Cold-formed Steel Framing-truss*. Washington, DC: American Iron and Steel Institute, 2007.

ANSI/AITC A 190.1-07, *Structural Glued Laminated Timber*. Centennial, CO: American Institute of Timber Construction, 2007.

ANSI/TPI 1-2007, *National Design Standard for Metal-plate-connected Wood Truss Construction*. Madison, WI: Truss Plate Institute, 2007.

APA E30-03, *Engineered Wood Construction Guide*. Tacoma, WA: APA-The Engineered Wood Association, 2003.

ASCE/SEI 7-10, *Minimum Design Loads for Buildings and Other Structures*. Reston, VA: American Society of Civil Engineers/Structural Engineering Institute, 2010.

ASTM D 2898-08e01, *Test Methods for Accelerated Weathering of Fire-retardant-treated Wood for Fire Testing*. West Conshohocken, PA: ASTM International, 2008.

ASTM D 3201-08a, *Test Method for Hygroscopic Properties of Fire-retardant Wood and Wood-base Products*. West Conshohocken, PA: ASTM International, 2008.

ASTM D 3737-08, *Practice for Establishing Allowable Properties for Structural Glued Laminated Timber (Glulam)*. West Conshohocken, PA: ASTM International, 2008.

ASTM D 3957-06, *Standard Practices for Establishing Stress Grades for Structural Members Used in Log Buildings*. West Conshohocken, PA: ASTM International, 2006.

ASTM E 84-09, *Test Method for Surface Burning Characteristics of Building Materials*. West Conshohocken, PA: ASTM International, 2007. West Conshohocken, PA: ASTM International, 2009.

UL 723-08, *Standard for Test for Surface Burning Characteristics of Building Materials*. Northbrook, IL: Underwriters Laboratories Inc., 2008.

Chapter 9: Roof Assemblies

General Comments

Although a small portion of this chapter regulates the fire classification of roof covering materials, the major emphasis is on the materials and installation methods that will result in a weather-tight exterior envelope. The code limits the requirements for fire-retardant roof coverings to two scenarios: (1) where the local or state ordinances in effect mandate a classified roof covering and (2) where the edge of the roof is in close proximity to a property line. Otherwise, the provisions of the chapter focus on the weather-protection features of a roof assembly.

The chapter contains a number of specific code requirements for various types of roof covering materials. These requirements may place limitations on roof slope, identify the proper attachment methods, mandate the use of underlayment and address flashing at appropriate points. In addition, the code often references a specification or installation standard for a particular type of roof covering. In all cases, it is important that the manufacturer's installation instructions be followed.

This chapter contains provisions for the design and construction of roof assemblies, primarily focusing on the materials and installation of roof coverings. The general requirements for roof assemblies found in the code deal with the role of the roof as a portion of the exterior envelope that provides weather protection for the building. Fire classifications, roof drainage, roof insulation and reroofing operations are also addressed.

Section R901 states the scope of the chapter, including both the materials and the quality of roof construction. Section R902 deals with fire-classification requirements for roof covering materials where the roof extends very close to an adjoining property line. Section R903 establishes the general requirements for weather protection, including provisions for flashing and roof drainage.

Section R904 establishes general requirements for roofing materials. Section R905 discusses the many different types of roof covering materials regulated by the code, including asphalt shingles, clay and concrete tile, metal roof shingles, mineral-surfaced roll roofing, wood shakes and shingles, slate and slate-type shingles, built-up roofs, metal roof panels, modified bitumen roofing, thermoset and thermoplastic single-ply roofing, sprayed polyurethane foam roofing and liquid-applied coatings and photovoltaic modules/shingles. Section R906 allows the installation of above-deck thermal installation where appropriately listed, and Section R907 establishes the methods and materials acceptable for recovering or replacing an existing roof.

Purpose

Chapter 9 regulates the design and construction of roof assemblies. A roof assembly includes the roof deck, vapor retarder, substrate or thermal barrier, insulation, vapor retarder and roof covering. This chapter provides the requirement for wind resistance of roof coverings.

Roof assemblies fulfill a variety of functions; however, this chapter deals primarily with roof coverings for weather protection. The requirements address the design and construction of roof assemblies, with a focus on the materials and quality of installation. Although a number of requirements are specified directly from the code, the majority of the regulations applicable to roof coverings come from the material standards. In addition, the manufacturer's installation instructions should always be followed for the specific roof covering material under consideration. Flashing and roof drainage criteria are also important aspects of the weather-protection features. Many of the common roof coverings are addressed by detailed provisions regulating the roof deck, roof slope, underlayment, application and attachment. Where a new roof is required for an existing building, the code sets specific guidelines for either recovering or replacing the existing roof covering.

SECTION R901 GENERAL

R901.1 Scope. The provisions of this chapter shall govern the design, materials, construction and quality of roof assemblies.

❖ The focus of this chapter is on roof coverings. A variety of roof covering materials are included, such as asphalt shingles, clay and concrete tile, metal roof shingles, wood shingles and shakes and metal roof panels. Roof classification, drainage, insulation and flashing are also regulated, as is the reroofing of an existing roof.

SECTION R902 ROOF CLASSIFICATION

R902.1 Roofing covering materials. Roofs shall be covered with materials as set forth in Sections R904 and R905. Class A, B or C roofing shall be installed in areas designated by law as requiring their use or when the edge of the roof is less than

3 feet (914 mm) from a lot line. Classes A, B and C roofing required by this section to be listed shall be tested in accordance with UL 790 or ASTM E 108.

Exceptions:

1. Class A roof assemblies include those with coverings of brick, masonry and exposed concrete roof deck.
2. Class A roof assemblies also include ferrous or copper shingles or sheets, metal sheets and shingles, clay or concrete roof tile, or slate installed on noncombustible decks.
3. Class A roof assemblies include minimum 16 oz/ft^2 copper sheets installed over combustible decks.

❖ The code identifies two conditions under which roof assembly classifications A, B, or C are required. First, where local, state or other governing ordinance requires the use of Class A, B or C roofing, it must be installed as designated by the law. Second, on roofs that project within 3 feet (914 mm) of the lot line, a Class A, B or C roof must be installed. Testing for such roofing must be in accordance with either ASTM E 108 or UL 790.

Roof coverings classified as Class A, B or C have been shown through testing to provide protection of the roof against severe, moderate and light fire exposures, respectively. These exposures are external and are generally created by fires in adjoining structures, wild fires and fires from the subject building that extend up the exterior and onto the top surface of the roof. Wild fires and some structure fires create flying and flaming brands that can ignite nonclassified roof coverings.

By their nature, roof coverings composed of brick, masonry, exposed concrete roof deck, slate, clay tile, concrete tile, ferrous or copper shingles or sheets and metal sheets or shingles are considered excellent materials for resisting fire spread. The exceptions permit these materials to be considered Class A and exempt from testing to UL 790 or ASTM E 108. The materials in Exception 2 must be installed on noncombustible decks to be considered Class A and exempt from testing.

Where installed on a combustible deck, under brush fire conditions in the field, roof coverings can be exposed to burning brands that may break slate or clay or concrete roof tile, since they are brittle materials, and expose the roof deck to the fire; or the high winds caused by the brush fire can lift the butt ends of slate or concrete or clay roof tiles, allowing the entry of embers under the roof covering and igniting the combustible deck.

These materials in Exception 2 must be tested if installed on a combustible deck.

Exception 3 permits 16 ounce/square foot copper sheets to be considered Class A and exempt from testing where installed over combustible decks.

R902.2 Fire-retardant-treated shingles and shakes. Fire-retardant-treated wood shakes and shingles shall be treated by impregnation with chemicals by the full-cell vacuum-pressure process, in accordance with AWPA C1. Each bundle shall be marked to identify the manufactured unit and the manufacturer, and shall also be *labeled* to identify the classification of the material in accordance with the testing required in Section R902.1, the treating company and the quality control agency.

❖ This section recognizes the process of impregnating wood shakes and shingles with chemicals in order to make them fire retardant. AWPA C1, regulating the preservative treatment of timber products by pressure processes, is the referenced standard. During the impregnation process the cells of wood shakes and shingles are exposed to a vacuum that draws all of the air and moisture from the material. Fire-retardant chemicals are then impregnated into the shakes and shingles by pressurization. After proper curing, the shakes and shingles are then classified. The fire classification must be marked on each bundle of materials, along with other manufacturing information. Identification will also include the name of the company performing the pressure-treatment process, as well as the quality control agency reviewing for compliance with the test standard.

SECTION R903
WEATHER PROTECTION

R903.1 General. Roof decks shall be covered with *approved* roof coverings secured to the building or structure in accordance with the provisions of this chapter. Roof assemblies shall be designed and installed in accordance with this code and the *approved* manufacturer's installation instructions such that the roof assembly shall serve to protect the building or structure.

❖ In all cases, a roof must be designed to provide protection from the elements. For the roof to adequately perform this function, it must be designed in accordance with this chapter. This section requires flashing where the roof intersects vertical elements such as walls, chimneys, dormers, plumbing stacks, plumbing vents and other penetrations of the weather-protective barrier. Roof drainage to remove water from the roof to an approved location is also regulated.

Materials and installation methods for roof decks and supporting elements are regulated by Chapter 8 of the code. A complying roof deck must then be covered in an approved manner to provide the necessary weather protection mandated by the code. This chapter includes the necessary provisions for such a weatherproof barrier. In addition, and just as important, the approved manufacturer's installation instructions must be adhered to.

R903.2 Flashing. Flashings shall be installed in a manner that prevents moisture from entering the wall and roof through joints in copings, through moisture permeable materials and at intersections with parapet walls and other penetrations through the roof plane.

❖ As with flashing required for other locations at the building's exterior, it is critical that those locations where moisture may penetrate the exterior membrane at the roof be adequately protected. General locations identified include joints at the coping, intersections with parapet walls and other penetrations through the roof plane.

R903.2.1 Locations. Flashings shall be installed at wall and roof intersections, wherever there is a change in roof slope or direction and around roof openings. A flashing shall be installed to divert the water away from where the eave of a sloped roof intersects a vertical sidewall. Where flashing is of metal, the metal shall be corrosion resistant with a thickness of not less than 0.019 inch (0.5 mm) (No. 26 galvanized sheet).

❖ This section identifies specific roof locations for the installation of flashings. In those cases where metal is used as the flashing material, it must be corrosion resistant and a minimum No. 26 galvanized sheet.

R903.2.2 Crickets and saddles. A cricket or saddle shall be installed on the ridge side of any chimney or penetration more than 30 inches (762 mm) wide as measured perpendicular to the slope. Cricket or saddle coverings shall be sheet metal or of the same material as the roof covering.

Exception: Unit skylights installed in accordance with Section R308.6 and flashed in accordance with the manufacturer's instructions shall be permitted to be installed without a cricket or saddle.

❖ Where a chimney is more than 30 inches (762 mm) wide, as measured perpendicular to the slope, a cricket or saddle must be installed on the ridge side of the chimney to divert water away from the chimney penetration of the roof. A chimney of such width creates the potential for standing water that, despite appropriate flashing, can be detrimental to the roof's performance.

The exception permits engineered skylights that are designed to prevent water infiltration without the use of a cricket.

R903.3 Coping. Parapet walls shall be properly coped with noncombustible, weatherproof materials of a width no less than the thickness of the parapet wall.

❖ The top of an exterior wall, typically a masonry parapet wall, must have a coping. It is typically sloped to prevent standing water, and it must be of noncombustible, weatherproof materials. To maintain protection at the top of the wall, the coping must extend the full thickness of the parapet wall.

R903.4 Roof drainage. Unless roofs are sloped to drain over roof edges, roof drains shall be installed at each low point of the roof.

❖ In most residential construction, drainage water from the roof simply flows down the roof's slope to and over the roof edge, often to a gutter and downspout system that carries the water away from the building. In situations where the roof design does not allow for water flow over the roof edges, roof drains must be installed in the roof surface. The drains will be at the low points of the roof, sized and located to remove all roof water. In a case where the low points occur at parapets of the exterior walls, scuppers through the parapets must be provided, constructed with the flow points level with the roof surface at the low points and located based on the roof slope and contributing roof area.

R903.4.1 Secondary (emergency overflow) drains or scuppers. Where roof drains are required, secondary emergency overflow roof drains or scuppers shall be provided where the roof perimeter construction extends above the roof in such a manner that water will be entrapped if the primary drains allow buildup for any reason. Overflow drains having the same size as the roof drains shall be installed with the inlet flow line located 2 inches (51 mm) above the low point of the roof, or overflow scuppers having three times the size of the roof drains and having a minimum opening height of 4 inches (102 mm) shall be installed in the adjacent parapet walls with the inlet flow located 2 inches (51 mm) above the low point of the roof served. The installation and sizing of overflow drains, leaders and conductors shall comply with Sections 1106 and 1108 as applicable of the *International Plumbing Code*.

Overflow drains shall discharge to an *approved* location and shall not be connected to roof drain lines.

❖ It is quite possible that roof drains may become clogged or blocked by debris, resulting in insufficient drainage. This could cause structural failure of portions of the roof system resulting from the increased weight on the roof caused by the impounded water. This section addresses such concerns by requiring overflow drains or overflow scuppers for those roofs that do not drain over the roof edges.

Although the required roof drains must be located at the low points of the roof, the overflow drain inlets need to be located 2 inches (51 mm) vertically above the low points. The same criterion applies where overflow scuppers are provided. In addition, overflow scuppers must be sized with a minimum opening height of 4 inches (102 mm) and have an opening area at least three times that of the roof drains. The *International Plumbing Code*® (IPC®) sets forth the specific details for the installation and sizing of overflow drains, leaders and conductors.

The overflow drains must not be connected to the roof drains. In the event a blockage occurs in the drainage line beyond the point of connection, both the roof drain and the overflow drain would be affected. The overflow drain should discharge to an approved location, typically a highly visible location that would indicate the overflow system is in use. This will provide an alert that there is some sort of blockage in the primary roof drain system.

SECTION R904
MATERIALS

R904.1 Scope. The requirements set forth in this section shall apply to the application of roof covering materials specified herein. Roof assemblies shall be applied in accordance with this chapter and the manufacturer's installation instructions. Installation of roof assemblies shall comply with the applicable provisions of Section R905.

❖ Roof assemblies and roof covering materials must conform to the provisions of this chapter, the applicable referenced standards listed in Chapter 44 and the manufacturer's installation instructions. This section provides general conditions for the application of roof covering materials referenced elsewhere in Chapter 9. A cross reference to Section R905 is provided for requirements dealing with the installation of specific roof coverings.

R904.2 Compatibility of materials. Roof assemblies shall be of materials that are compatible with each other and with the building or structure to which the materials are applied.

❖ The materials that comprise a roof assembly must be compatible in order to avoid a decrease or failure in the performance of the roof. Not only must the roofing materials be compatible with each other, they must also be suitable for the building as a whole.

R904.3 Material specifications and physical characteristics. Roof covering materials shall conform to the applicable standards listed in this chapter. In the absence of applicable standards or where materials are of questionable suitability, testing by an *approved* testing agency shall be required by the *building official* to determine the character, quality and limitations of application of the materials.

❖ In most cases, materials used as roof coverings are regulated by an applicable standard. Identified in this chapter, they are also listed in Chapter 44. If a roof covering material is not regulated by a specific standard, the building official must have substantiating data to show that the materials are consistent with the intent of the code regarding quality and application limitations. Typically, the roof covering material under consideration must be tested to provide the information necessary to show code compliance.

R904.4 Product identification. Roof covering materials shall be delivered in packages bearing the manufacturer's identifying marks and *approved* testing agency *labels* when required. Bulk shipments of materials shall be accompanied by the same information issued in the form of a certificate or on a bill of lading by the manufacturer.

❖ Identification of roof covering materials provides much of the information necessary for acceptance of the materials by the building official. Because it is often difficult, if not impossible, to identify complying roof covering materials by visual observation, other proper identification of the materials is mandatory. Packaging of roof covering materials must include the manufacturer's identification marks. If the code or applicable standard requires review by an approved testing agency, the identifying marks of the testing agency must also be shown. The code also addresses the identification requirements for bulk shipments of materials.

SECTION R905
REQUIREMENTS FOR ROOF COVERINGS

R905.1 Roof covering application. Roof coverings shall be applied in accordance with the applicable provisions of this section and the manufacturer's installation instructions. Unless otherwise specified in this section, roof coverings shall be installed to resist the component and cladding loads specified in Table R301.2(2), adjusted for height and exposure in accordance with Table R301.2(3).

❖ In addition to the minimum requirements specified for roof coverings in the code, manuals published by various associations provide detailed discussions of the proper methods of installing roof coverings. These methods have been established based on many years of experience with the materials and their performance. Although the provisions in these documents are not specific code requirements, this section mandates the use of the manufacturer's installation instructions. It is important that roof coverings remain intact and in place when subjected to wind. Without an intact roof covering, the building would be subjected to either water damage, which could reduce its structural stability, or to higher wind pressures than the building is designed for. Unless Section R905 specifies otherwise, roof coverings must be installed to resist the wind pressures determined from Table R301.2(2) [see commentary, Tables R301.2(2) and R301.2(3)].

R905.2 Asphalt shingles. The installation of asphalt shingles shall comply with the provisions of this section.

❖ This section regulates asphalt shingles composed of organic felt or glass felt and coated with mineral granules. Provisions address requirements for sheathing, roof slope, underlayment, fasteners and attachment.

R905.2.1 Sheathing requirements. Asphalt shingles shall be fastened to solidly sheathed decks.

❖ The code requires a solid roof surface for the installation of asphalt shingles. Section R803 regulates solid sheathing.

R905.2.2 Slope. Asphalt shingles shall be used only on roof slopes of two units vertical in 12 units horizontal (2:12) or greater. For roof slopes from two units vertical in 12 units horizontal (2:12) up to four units vertical in 12 units horizontal (4:12), double underlayment application is required in accordance with Section R905.2.7.

❖ The performance of all roof coverings is based in part on the slope of the roof surface. As the slope of the roof decreases, water drainage is slowed, and the potential for water intrusion increases because of the greater potential of water back-up under the roofing. Asphalt shingles, because of their configuration and installation methods, are restricted to use on roofs having a minimum slope of 2:12. Where the slope is no steeper than 4:12, the underlayment must be doubled to provide a greater barrier to leakage. Section R905.2.7 specifies the method of such double underlayment. Commentary Figure R905.2.2(1) shows an example of asphalt roof shingles installed on a high-slope roof (4:12 minimum), while Commentary Figure R905.2.2(2) depicts a low-slope roof installation (between 2:12 and 4:12).

R905.2.3 Underlayment. Unless otherwise noted, required underlayment shall conform to ASTM D 226 Type I, ASTM D 4869 Type I, or ASTM D 6757.

Self-adhering polymer modified bitumen sheet shall comply with ASTM D 1970.

❖ Four types of underlayment are recognized for use with asphalt shingle roof coverings: asphalt-saturated organic felts as regulated by ASTM D 226, Type I; asphalt-saturated organic felt shingles, per ASTM D 4869, Type I; inorganic underlayment for use with steep slope roof, per ASTM D 6757; and self-adhering polymer modified bitumen sheet materials are addressed in ASTM D 1970. Section R905.2.7 contains the methods prescribed for the installation of underlayment for asphalt shingles.

R905.2.4 Asphalt shingles. Asphalt shingles shall comply with ASTM D 225 or D 3462.

❖ Two test standards regulate asphalt shingles. ASTM D 225 addresses asphalt shingles made from organic

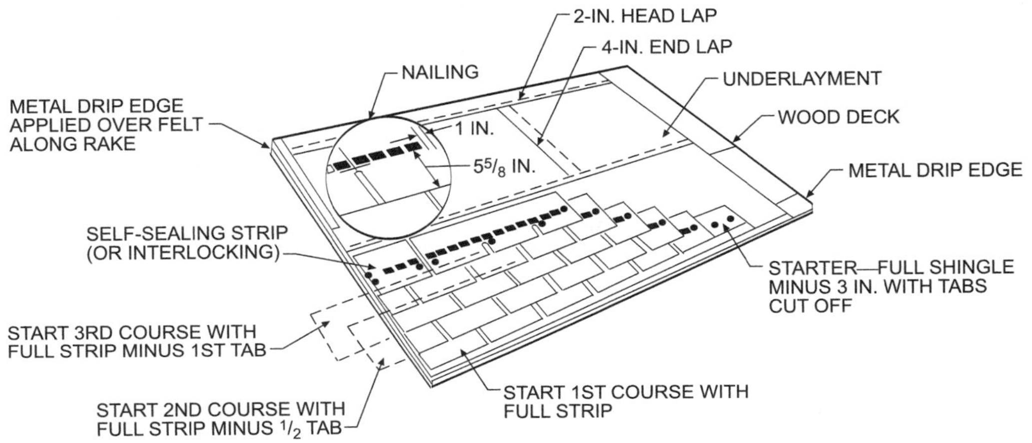

For SI: 1 inch = 25.4 mm.

Figure R905.2.2(1)
ASPHALT ROOFING SINGLES APPLICATION HIGH SLOPE (4:12 MINIMUM)
(Source NRCA)

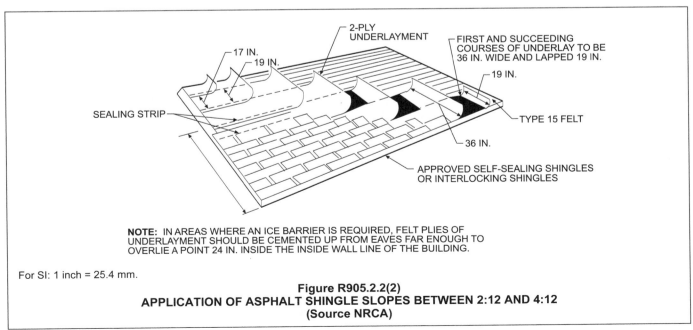

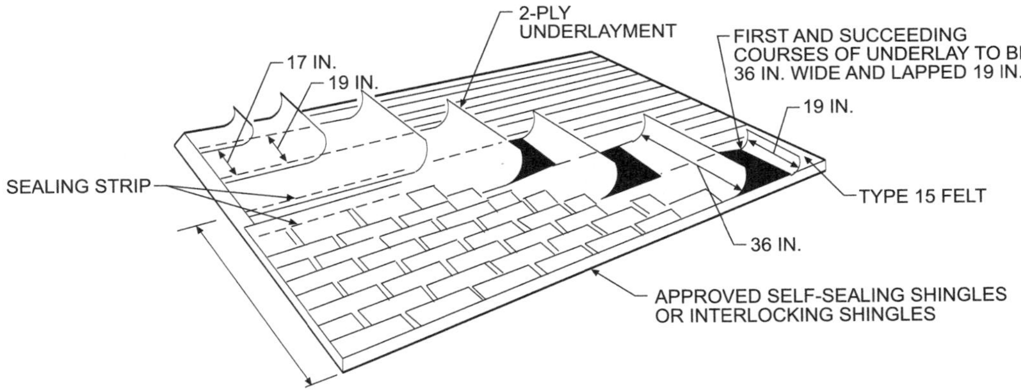

NOTE: IN AREAS WHERE AN ICE BARRIER IS REQUIRED, FELT PLIES OF UNDERLAYMENT SHOULD BE CEMENTED UP FROM EAVES FAR ENOUGH TO OVERLIE A POINT 24 IN. INSIDE THE INSIDE WALL LINE OF THE BUILDING.

For SI: 1 inch = 25.4 mm.

Figure R905.2.2(2)
APPLICATION OF ASPHALT SHINGLE SLOPES BETWEEN 2:12 AND 4:12
(Source NRCA)

felt, while ASTM D 3462 deals with shingles made from glass felt. Both shingle types are surfaced with mineral granules.

R905.2.4.1 Wind resistance of asphalt shingles. Asphalt shingles shall be tested in accordance with ASTM D 7158. Asphalt shingles shall meet the classification requirements of Table R905.2.4.1(1) for the appropriate maximum basic wind speed. Asphalt shingle packaging shall bear a *label* to indicate compliance with ASTM D 7158 and the required classification in Table R905.2.4.1(1).

Exception: Asphalt shingles not included in the scope of ASTM D 7158 shall be tested and *labeled* to indicate compliance with ASTM D 3161 and the required classification in Table R905.2.4.1(2).

❖ This section provides requirements for the testing, classification and labeling of asphalt shingles to demonstrate resistance to wind forces. The asphalt shingles must meet the classification requirement for the appropriate maximum basic wind speed and comply with the appropriate product standard. The required standard is ASTM D 7158 with an exception for products outside the scope of ASTM D 7158. ASTM D 7158 provides a method of testing that is appropriate for sealed asphalt shingles. The exception references ASTM D 3161 and is necessary for unsealed shingles. Tables R905.2.4.1(1) and R905.2.4.1(2) provide the proper application of the two standards. The tables will assist in the proper selection of asphalt shingles based upon the appropriate basic wind speed and the applicable standard.

R905.2.5 Fasteners. Fasteners for asphalt shingles shall be galvanized steel, stainless steel, aluminum or copper roofing nails, minimum 12 gage [0.105 inch (3 mm)] shank with a minimum $^3/_8$-inch-diameter (10 mm) head, ASTM F 1667, of a length to penetrate through the roofing materials and a minimum of $^3/_4$ inch (19 mm) into the roof sheathing. Where the roof sheathing is less than $^3/_4$ inch (19 mm) thick, the fasteners shall penetrate through the sheathing. Fasteners shall comply with ASTM F 1667.

❖ Roofing nails must be of a corrosion-resistant material, specified in this section as either galvanized steel, stainless steel, aluminum or copper. A roofing nail must have a minimum 12 gage shank with a head at least $^3/_8$ inch (9.5 mm) in diameter. To provide the necessary holding power, roofing nails must penetrate the roof sheathing a minimum of $^3/_4$ inch (19.1 mm). For roof sheathing having a thickness less than $^3/_4$ inch (19.1 mm), the nails must penetrate completely through the sheathing. The material specification standard for roofing nails is ASTM F 1667.

TABLE R905.2.4.1(1)
CLASSIFICATION OF ASPHALT ROOF SHINGLES PER ASTM D 7158

MAXIMUM BASIC WIND SPEED FROM FIGURE 301.2(4)A (mph)	CLASSIFICATION REQUIREMENT
85	D, G or H
90	D, G or H
100	G or H
110	G or H
120	G or H
130	H
140	H
150	H

For SI:1 mile per hour = 0.447 m/s.

❖ See the commentary to Section 905.2.4.1.

TABLE R905.2.4.1(2)
CLASSIFICATION OF ASPHALT SHINGLES PER ASTM D 3161

MAXIMUM BASIC WIND SPEED FROM FIGURE 301.2(4)A (mph)	CLASSIFICATION REQUIREMENT
85	A, D or F
90	A, D or F
100	A, D or F
110	F
120	F
130	F
140	F
150	F

For SI: 1 mile per hour = 0.447 m/s.

❖ See the commentary to Section 905.2.4.1.

R905.2.6 Attachment. Asphalt shingles shall have the minimum number of fasteners required by the manufacturer, but not less than four fasteners per strip shingle or two fasteners per individual shingle. Where the roof slope exceeds 21 units vertical in 12 units horizontal (21:12, 175-percent slope), shingles shall be installed as required by the manufacturer.

❖ Section R905.2.4.1 requires the wind resistance of asphalt shingles to be determined by ASTM D 7158 or ASTM D 3161.The ASTM D 7158 test method covers the procedure for calculating the wind resistance of asphalt shingles when applied in accordance with the manufacturer's instruction and sealed under defined conditions. The ASTM D 3161 test method covers the procedure for testing asphalt shingles that are resistant to wind blow-up or blow-off when applied on low slopes in accordance with the manufacturer's instructions. This test method may be used to test self-sealing or interlocked shingles.

Neither of these tests is a test of the fasteners. However, fastening in accordance with the manufacturer's instruction is required for shingles that pass these tests.

The minimum number of fasteners for asphalt shingles must be the number required by the manufacturer's instructions. Except for steep roof slopes, the minimum quantity of fasteners per shingle is four for strip shingles and two for individual shingles.

For steep roof slopes (slopes greater than 21:12), the manufacturer's installation instructions and minimum quantity of fasteners per shingles must be followed.

R905.2.7 Underlayment application. For roof slopes from two units vertical in 12 units horizontal (17-percent slope), up to four units vertical in 12 units horizontal (33-percent slope), underlayment shall be two layers applied in the following manner. Apply a 19-inch (483 mm) strip of underlayment felt parallel to and starting at the eaves, fastened sufficiently to hold in place. Starting at the eave, apply 36-inch-wide (914 mm) sheets of underlayment, overlapping successive sheets 19 inches (483 mm), and fastened sufficiently to hold in place. Distortions in the underlayment shall not interfere with the ability of the shingles to seal. For roof slopes of four units vertical in 12 units horizontal (33-percent slope) or greater, underlayment shall be one layer applied in the following manner. Underlayment shall be applied shingle fashion, parallel to and starting from the eave and lapped 2 inches (51 mm), fastened sufficiently to hold in place. Distortions in the underlayment shall not interfere with the ability of the shingles to seal. End laps shall be offset by 6 feet (1829 mm).

❖ This section specifies the underlayment requirements for asphalt shingles installed on both low-slope and high-slope roofs. Low-slope roofs present a potential problem because water drains slowly, and this creates the opportunity for water back-up. Wind-driven rain can also pose a problem. Therefore, a special underlayment application method is used so that the roof remains weathertight. All portions of the roof will be protected by a minimum of two layers of underlayment if the installer follows the instructions found in this section. Commentary Figure R905.2.7 illustrates the required application of underlayment for a low-slope application. Only one layer of underlayment, applied in shingle fashion in accordance with this sec-

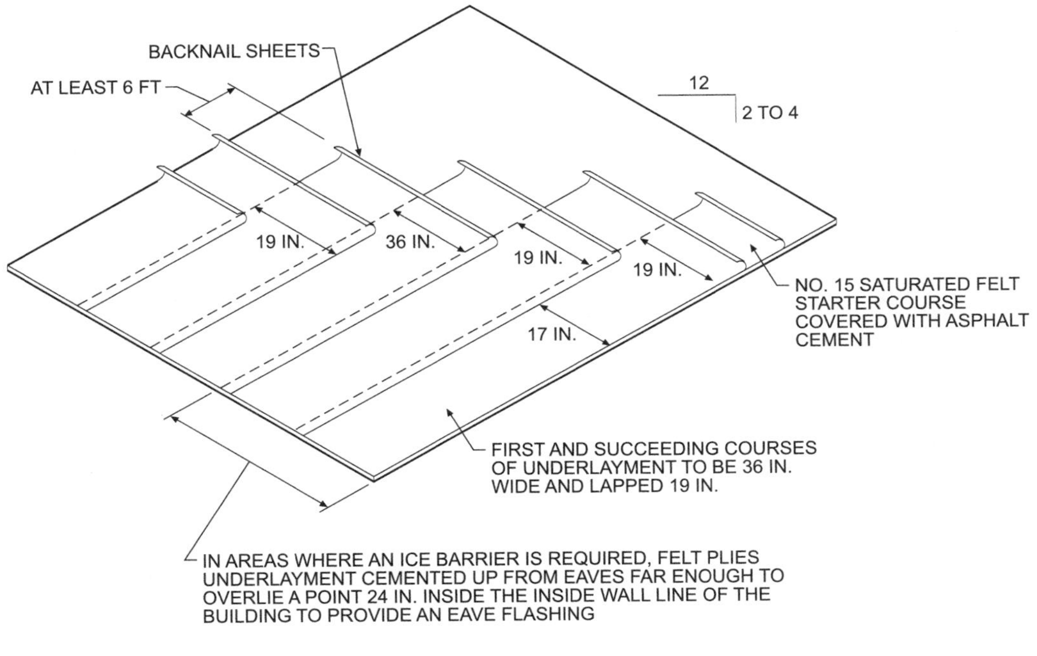

For SI: 1inch = 25.4 mm, 1 foot = 304.8 mm.

Figure R905.2.7
LOW-SLOPE DOUBLE-PLY UNDERLAYMENT APPICATION

tion, is mandated for high-slope roofs of asphalt shingles. For both low-slope and high-slope roofs the underlayment must not have distortions that will interfere with the ability of the shingles to lie flat and seal. Installation of shingles over a distorted surface can result in reduced wind resistance and unacceptable aesthetics.

R905.2.7.1 Ice barrier. In areas where there has been a history of ice forming along the eaves causing a backup of water as designated in Table R301.2(1), an ice barrier that consists of a least two layers of underlayment cemented together or of a self-adhering polymer modified bitumen sheet, shall be used in lieu of normal underlayment and extend from the lowest edges of all roof surfaces to a point at least 24 inches (610 mm) inside the exterior wall line of the building.

Exception: Detached *accessory structures* that contain no *conditioned floor area*.

❖ Where ice dams may be formed along the eave because snow continually freezes and thaws or frozen slush backs up in gutters, the underlayment application in the area of the eaves must be modified to prevent ice dams from forcing water under the roofing, which could damage ceilings, walls and insulation [see Commentary Figure R905.2.7.1(1)]. Two layers of underlayment should be cemented together with asphalt cement from the lowest edge of the roof and continue up the roof to a point that is at least 24 inches (610 mm) inside the interior wall line of the building as shown in Commentary Figure R905.2.7.1(2). The environment within the envelope of the building provides adequate warmth to prevent ice dams from forming above the heated space; therefore, the two layers of cemented underlayment are permitted to terminate 24 inches (610 mm) inside the interior wall line of the building. The local jurisdiction is responsible for determining whether the ice barrier is required based on weather records, and it must so indicate in Table R301.2(1).

An exception to this section exempts accessory buildings from such restrictions because they are unheated structures where the need for protection against ice dams is unnecessary. The same exception is found in Sections R905.4.3.1, R905.5.3.1, R905.6.3.1, R905.7.3.1 and R905.8.3.1.

R905.2.7.2 Underlayment and high winds. Underlayment applied in areas subject to high winds [above 110 mph (49 m/s) in accordance with Figure R301.2(4)A] shall be applied with corrosion-resistant fasteners in accordance with manufacturer's installation instructions. Fasteners are to be applied along the overlap not farther apart than 36 inches (914 mm) on center.

Underlayment installed where the basic wind speed equals or exceeds 120 mph (54 m/s) shall comply with ASTM D 226 Type II, ASTM D 4869 Type IV, or ASTM D 6757. The underlayment shall be attached in a grid pattern of 12 inches (305 mm) between side laps with a 6-inch (152 mm) spacing at the side laps. Underlayment shall be applied in accordance with Section R905.2.7 except all laps shall be a minimum of 4 inches (102 mm). Underlayment shall be attached using metal or plastic cap nails with a head diameter of not less than 1 inch (25.4 mm) with a thickness of at least 32-gauge sheet metal. The cap-nail shank shall be a minimum of 12 gauge (0.105 inches) with a length to penetrate through the roof sheathing or a minimum of $^3/_4$ inch (19 mm) into the roof sheathing.

Exception: As an alternative, adhered underlayment complying with ASTM D 1970 shall be permitted.

❖ In high-wind areas, corrosion-resistant fasteners must be located in accordance with the manufacturer's installation instructions, but in no case can they be more than 36 inches (914 mm) on center. The general requirement for sufficient fastening to hold the underlayment in place is not adequate where increased wind loads are anticipated.

In areas subject to high wind speeds, ASTM D 226, Type II (No.30) must be used. Tests and studies have shown that ASTM D 226, Type I (No. 15) underlayment performs very poorly in areas subject to wind speeds of 120 miles per hour (mph) (54 m/s) or greater. Observations of roof underlayment performance following Hurricane Ike in Texas and in two sets of tests conducted at the University of Florida and Florida International University demonstrated that ASTM D 226, Type I underlayment performed poorly when subjected to wind over about 110 mph. In the laboratory tests, specimens covered with ASTM D 226, Type I and Type II underlayment performed dramatically differently. ASTM D 2267, Type I felt (No.15) material completely blew off some portions of the specimen as winds exceeded 110 mph and pulled over the plastic caps on other parts of the specimen. In contrast, the ASTM 226, Type II (No. 30) material remained in place and showed very few signs of distress. Plastic caps deformed more than the metal caps in several installations.

R905.2.8 Flashing. Flashing for asphalt shingles shall comply with this section.

❖ This section sets forth the special conditions for flashing installed as a part of an asphalt shingle roof system. Specific flashing locations addressed include base and cap flashing, valleys, crickets and saddles and sidewall flashing.

R905.2.8.1 Base and cap flashing. Base and cap flashing shall be installed in accordance with manufacturer's installation instructions. Base flashing shall be of either corrosion-resistant metal of minimum nominal 0.019-inch (0.5 mm) thickness or mineral surface roll roofing weighing a minimum of 77 pounds per 100 square feet (4 kg/m^2). Cap flashing shall be corrosion-resistant metal of minimum nominal 0.019-inch (0.5 mm) thickness.

❖ If metal is used as cap or base flashing, it must be corrosion resistant and have a minimum nominal thickness of 0.019-inch (0.483 mm). Mineral-surfaced roll roofing may also be used as base flashing, if it has a minimum weight of 77 pounds per 100 square feet (4 kg/m^2).

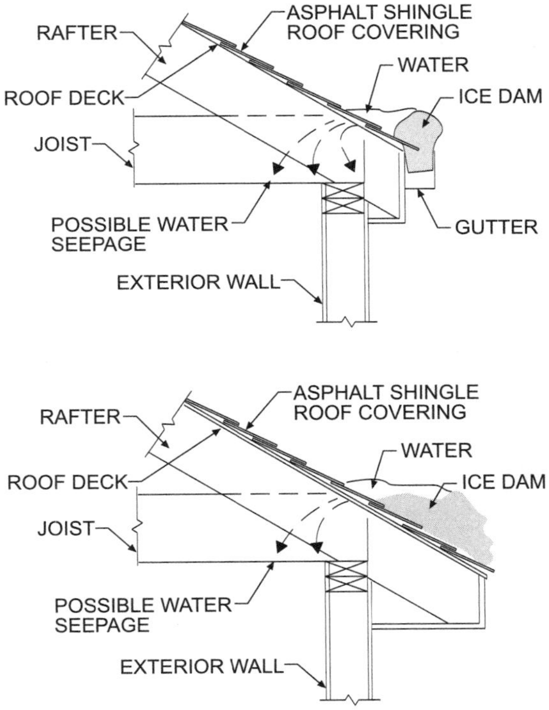

Figure R905.2.7.1(1)
POSSIBLE WATER DAMAGE FROM AN ICE DAM

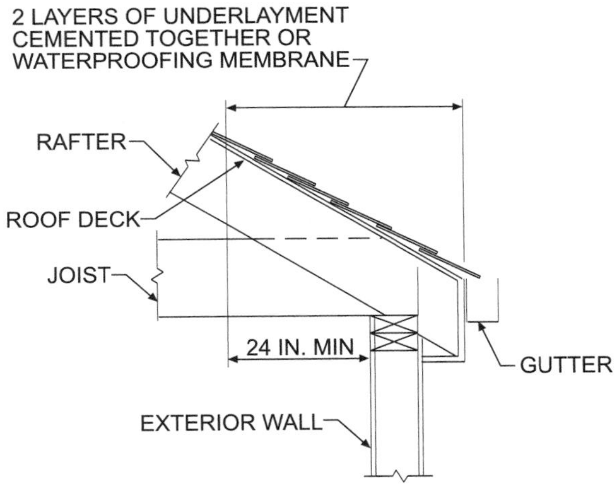

For SI: 1 inch = 25.4 mm.

Figure R905.2.7.1(2)
PROTECTIVE ICE BARRIER

R905.2.8.2 Valleys. Valley linings shall be installed in accordance with the manufacturer's installation instructions before applying shingles. Valley linings of the following types shall be permitted:

1. For open valleys (valley lining exposed) lined with metal, the valley lining shall be at least 24 inches (610 mm) wide and of any of the corrosion-resistant metals in Table R905.2.8.2.
2. For open valleys, valley lining of two plies of mineral surfaced roll roofing, complying with ASTM D 3909 or ASTM D 6380 Class M, shall be permitted. The bottom layer shall be 18 inches (457 mm) and the top layer a minimum of 36 inches (914 mm) wide.

3. For closed valleys (valley covered with shingles), valley lining of one ply of smooth roll roofing complying with ASTM D 6380 and at least 36 inches wide (914 mm) or valley lining as described in Item 1 or 2 above shall be permitted. Self-adhering polymer modified bitumen underlayment complying with ASTM D 1970 shall be permitted in lieu of the lining material.

❖ Open valley linings may be of either metal or mineral-surfaced roll roofing as set forth in this section. Closed valleys are also permitted with a number of lining alternatives available. Commentary Figures R905.2.8.2(1) and R905.2.8.2(2) illustrate typical valley flashings for open and closed (woven) valleys, respectively.

TABLE R905.2.8.2
VALLEY LINING MATERIAL

MATERIAL	MINIMUM THICKNESS (inches)	GAGE	WEIGHT (pounds)
Cold-rolled copper	0.0216 nominal	—	ASTM B 370, 16 oz. per square foot
Lead-coated copper	0.0216 nominal	—	ASTM B 101, 16 oz. per square foot
High-yield copper	0.0162 nominal	—	ASTM B 370, 12 oz. per square foot
Lead-coated high-yield copper	0.0162 nominal	—	ASTM B 101, 12 oz. per square foot
Aluminum	0.024	—	—
Stainless steel	—	28	—
Galvanized steel	0.0179	26 (zinc coated G90)	—
Zinc alloy	0.027	—	—
Lead	—	—	$2^1/_2$
Painted terne	—	—	20

For SI: 1 inch = 25.4 mm, 1 pound = 0.454 kg.

❖ If exposed corrosion-resistant metal materials are used as valley linings, they are regulated by this table. The table identifies a variety of acceptable metal linings, including copper, aluminum, stainless steel, galvanized steel, zinc alloy, lead and painted terne. The materials are regulated by their thickness, gage or weight.

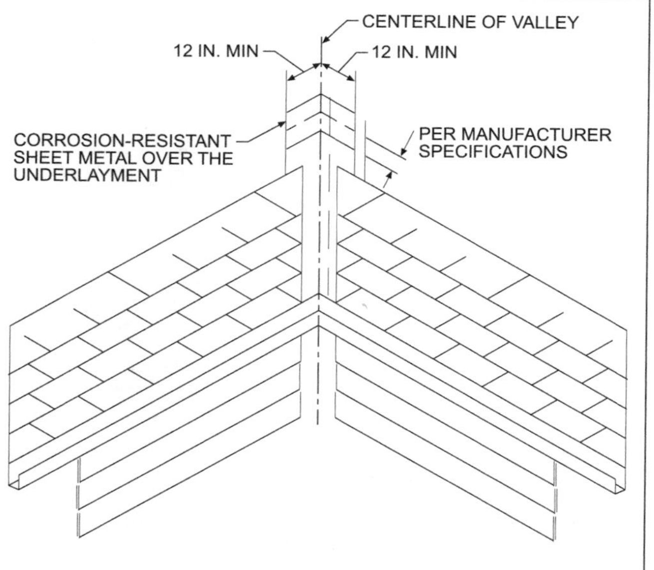

For SI: 1 inch = 25.4 mm.

Figure R905.2.8.2(1)
FLASHING FOR OPEN ROOF VALLEY

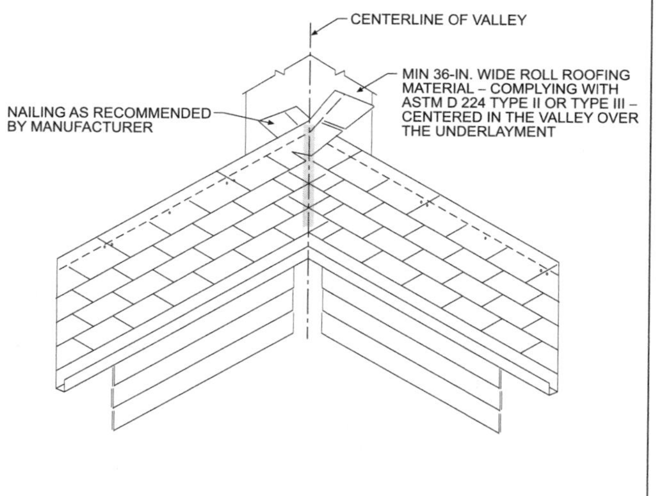

For SI: 1 inch = 25.4 mm.

Figure R905.2.8.2(2)
FLASHING FOR WOVEN VALLEY

R905.2.8.3 Sidewall flashing. Base flashing against a vertical sidewall shall be continuous or step flashing and shall be a minimum of 4 inches (102 mm) in height and 4 inches (102 mm) in width and shall direct water away from the vertical sidewall onto the roof and/or into the gutter. Where siding is provided on the vertical sidewall, the vertical leg of the flashing shall be continuous under the siding. Where anchored masonry veneer is provided on the vertical sidewall, the base flashing shall be provided in accordance with this section and counterflashing shall be provided in accordance with Section R703.7.2.2. Where exterior plaster or adhered masonry veneer is provided on the vertical sidewall, the base flashing shall be provided in accordance with this section and Section R703.6.3.

❖ Continuous or step flashing must be used to flash along the intersection of a roof and a vertical sidewall. Commentary Figure R905.2.8.3 shows an example of flashing at a sidewall.

R905.2.8.4 Other flashing. Flashing against a vertical front wall, as well as soil stack, vent pipe and chimney flashing, shall be applied according to the asphalt shingle manufacturer's printed instructions.

❖ The installation of flashing at vertical roof penetrations such as chimneys, soil stacks and vent pipes must be in conformance to the instructions provided by the asphalt shingle manufacturer. Commentary Figure R905.2.8.4 shows an example of chimney flashing.

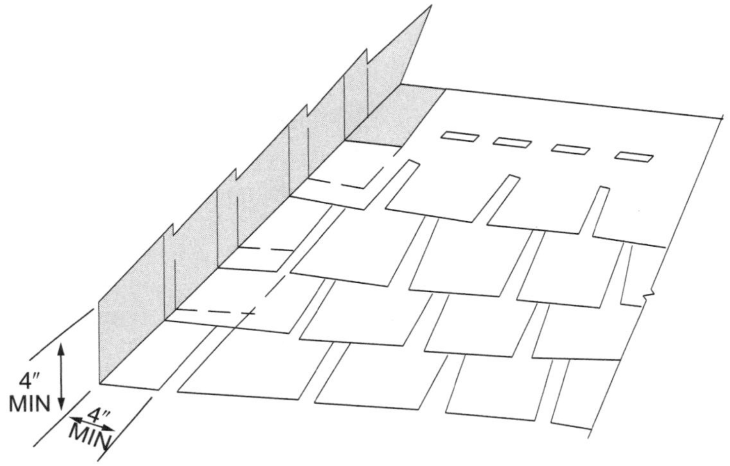

For SI: 1 inch = 25.4 mm.

Figure R905.2.8.3
FLASHING AT SIDE WALL

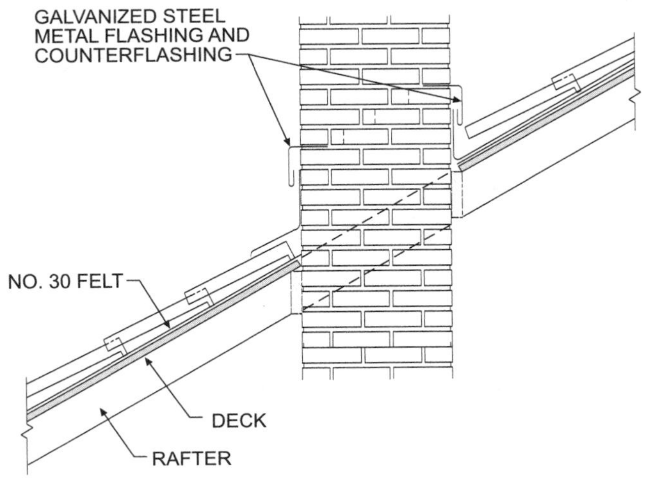

Figure R905.2.8.4
CHIMNEY FLASHING DETAIL

R905.2.8.5 Drip edge. A drip edge shall be provided at eaves and gables of shingle roofs. Adjacent pieces of drip edge shall be overlapped a minimum of 2 inches (51 mm). Drip edges shall extend a minimum of 0.25 inch (6.4 mm) below the roof sheathing and extend up the roof deck a minimum of 2 inches (51 mm). Drip edges shall be mechanically fastened to the roof deck at a maximum of 12 inches (305 mm) o.c. with fasteners as specified in Section R905.2.5. Underlayment shall be installed over the drip edge along eaves and under the underlayment on gables. Unless specified differently by the shingle manufacturer, shingles are permitted to be flush with the drip edge.

❖ Drip edge is a metal flashing or other overhanging component applied at the roof edge that is intended to control the direction of dripping water and help protect the underlying building components. Drip edge has an outward projecting lower edge to direct the water away from the building. It is also used to break the continuity of contact between the roof perimeter and the wall components to help prevent capillary action.

Metal drip edge is a formed metal flashing that extends back from and bends down over the roof edge. Along the roof gables, rakes and the eave, the drip edge is applied over the underlayment.

R905.3 Clay and concrete tile. The installation of clay and concrete tile shall comply with the provisions of this section.

❖ This section addresses both concrete tile and clay tile. Commentary Figure R905.3 depicts several examples of clay and concrete roof tile.

R905.3.1 Deck requirements. Concrete and clay tile shall be installed only over solid sheathing or spaced structural sheathing boards.

❖ Both solid sheathing and spaced structural sheathing boards are permitted as a base for the installation of clay or concrete roof tile.

R905.3.2 Deck slope. Clay and concrete roof tile shall be installed on roof slopes of two and one-half units vertical in 12 units horizontal ($2^1/_2$:12) or greater. For roof slopes from two and one-half units vertical in 12 units horizontal ($2^1/_2$:12) to four units vertical in 12 units horizontal (4:12), double underlayment application is required in accordance with Section R905.3.3.

❖ To keep the roof covering weather tight and promote proper drainage, clay and concrete tile cannot be installed on a roof having a slope of less than $2^1/_2$:12. The application of two layers of underlayment is mandated on roofs of 4:12 and less. The provisions of Section R905.3.3 address the full requirements for underlayment on concrete or clay tile roofs.

R905.3.3 Underlayment. Unless otherwise noted, required underlayment shall conform to ASTM D 226 Type II; ASTM D 2626 Type I; or ASTM D 6380 Class M mineral surfaced roll roofing.

❖ The code recognizes three types of underlayment for use with clay or concrete tile roof covering: asphalt-saturated organic felts as regulated by ASTM D 226, Type II; asphalt-saturated and coated organic felt base sheets per ASTM D 2626, Type I, and asphalt

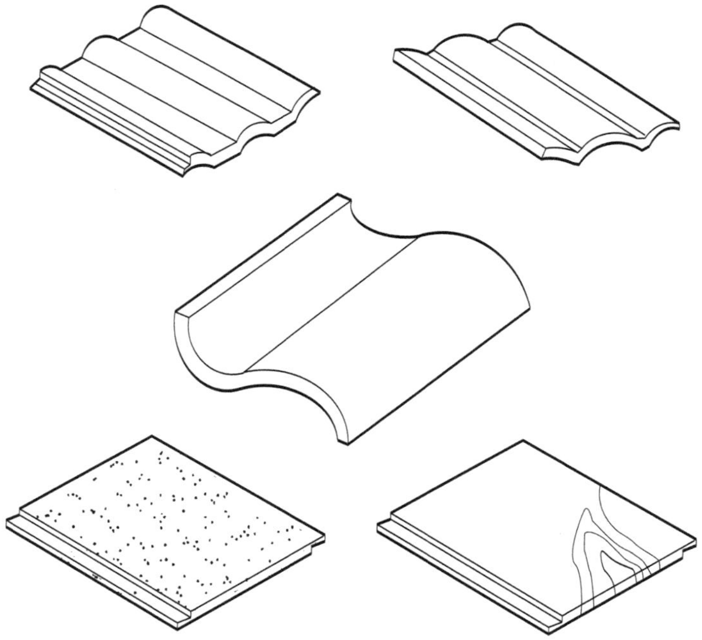

Figure R905.3
ROLL AND FLAT TILE EXAMPLES

organic felt roll roofing surfaced with mineral granules addressed in ASTM D 6380, Class M. This section also contains the methods prescribed for the installation of underlayment for clay and concrete tile.

R905.3.3.1 Low slope roofs. For roof slopes from two and one-half units vertical in 12 units horizontal ($2^1/_2$:12), up to four units vertical in 12 units horizontal (4:12), underlayment shall be a minimum of two layers underlayment applied as follows:

1. Starting at the eave, a 19-inch (483 mm) strip of underlayment shall be applied parallel with the eave and fastened sufficiently in place.
2. Starting at the eave, 36-inch-wide (914 mm) strips of underlayment felt shall be applied, overlapping successive sheets 19 inches (483 mm), and fastened sufficiently in place.

❖ Where the roof covering consists of clay tile or concrete tile, the roof is considered low-slope if the slope is at least $2^1/_2$:12 but no steeper than 4:12. Low-slope roofs present a potential problem because water drains slowly; this creates an opportunity for water back-up. Wind-driven rain can also pose a problem. Therefore, a special underlayment application method is used so that the roof remains weather tight. All portions of the roof will be protected by a minimum of two layers of underlayment if the installer follows the instructions found in this section.

R905.3.3.2 High slope roofs. For roof slopes of four units vertical in 12 units horizontal (4:12) or greater, underlayment shall be a minimum of one layer of underlayment felt applied shingle fashion, parallel to and starting from the eaves and lapped 2 inches (51 mm), fastened sufficiently in place.

❖ Where the roof covering consists of clay tile or concrete tile, the roof is considered high-slope if the slope is at least 4:12. Only one layer of underlayment felt, applied in shingle fashion in accordance with this section, is mandated for high-slope roofs of concrete or clay tile.

R905.3.3.3 Underlayment and high winds. Underlayment applied in areas subject to high wind [above 110 miles per hour (49 m/s) in accordance with Figure R301.2(4)A] shall be applied with corrosion-resistant fasteners in accordance with manufacturer's installation instructions. Fasteners are to be applied along the overlap not farther apart than 36 inches (914 mm) on center.

Underlayment installed where the basic wind speed equals or exceeds 120 mph (54 m/s) shall be attached in a grid pattern of 12 inches (305 mm) between side laps with a 6-inch (152 mm) spacing at the side laps. Underlayment shall be applied in accordance with Section R905.2.7 except all laps shall be a minimum of 4 inches (102 mm). Underlayment shall be attached using metal or plastic cap nails with a head diameter of not less than 1 inch (25.4 mm) with a thickness of at least 32-gauge sheet metal. The cap-nail shank shall be a minimum of 12 gauge (0.105 inches) with a length to penetrate through the roof sheathing or a minimum of $^3/_4$-inch (19 mm) into the roof sheathing.

Exception: As an alternative, adhered underlayment complying with ASTM D 1970 shall be permitted.

❖ In high-wind areas, corrosion-resistant fasteners must be located in accordance with the manufacturer's installation instructions, but in no case can they be more than 36 inches (914 mm) on center. The general requirement for sufficient fastening to hold the underlayment in place is not adequate where increased wind loads are anticipated (see commentary, Section R905.2.7.2).

R905.3.4 Clay tile. Clay roof tile shall comply with ASTM C 1167.

❖ ASTM C 1167 regulates clay roof tiles manufactured from clay, shale or similar natural earthy substances. The standard addresses a variety of performance requirements, including durability, strength and permeability.

R905.3.5 Concrete tile. Concrete roof tile shall comply with ASTM C 1492.

❖ ASTM C 1492 regulates concrete roof tiles manufactured from portland cement, water and mineral aggregates with or without the inclusion of other materials. Lightweight or normal weight aggregates or both are used in the manufacture of concrete tiles. The tiles are shaped during manufacturing by molding, pressing or extrusion. The tiles are usually planar or undulating rectangular shapes available in a variety of cross-sectional areas, profiles, shapes, sizes, surface textures and colors. The standard addresses a variety of performance tests including dimensional tolerances, freeze/thaw, transverse strength, permeability and water absorption.

R905.3.6 Fasteners. Nails shall be corrosion resistant and not less than 11 gage, $^5/_{16}$-inch (11 mm) head, and of sufficient length to penetrate the deck a minimum of $^3/_4$ inch (19 mm) or through the thickness of the deck, whichever is less. Attaching wire for clay or concrete tile shall not be smaller than 0.083 inch (2 mm). Perimeter fastening areas include three tile courses but not less than 36 inches (914 mm) from either side of hips or ridges and edges of eaves and gable rakes.

❖ Roofing nails for fastening clay or concrete tiles must be corrosion resistant. A roofing nail must have a minimum 11 gage shank with a head at least $^5/_{16}$ inch (10.6 mm) in diameter. To provide the necessary holding power, roofing nails must penetrate the roof sheathing a minimum of $^3/_4$ inch (19.1 mm). For roof sheathing having a thickness less than $^3/_4$ inch (19.1 mm), the nails must penetrate completely through the sheathing. The portions of the roof considered as perimeter fastening areas include a minimum of three tile courses, but in no case are they less than 36 inches (914 mm) from the edges of eaves and gable rakes or from either side of hips and ridges.

R905.3.7 Application. Tile shall be applied in accordance with this chapter and the manufacturer's installation instructions, based on the following:

1. Climatic conditions.
2. Roof slope.
3. Underlayment system.
4. Type of tile being installed.

Clay and concrete roof tiles shall be fastened in accordance with this section and the manufacturer's installation instructions. Perimeter tiles shall be fastened with a minimum of one fastener per tile. Tiles with installed weight less than 9 pounds per square foot (0.4 kg/m^2) require a minimum of one fastener per tile regardless of roof slope. Clay and concrete roof tile attachment shall be in accordance with the manufacturer's installation instructions where applied in areas where the wind speed exceeds 100 miles per hour (45 m/s) and on buildings where the roof is located more than 40 feet (12 192 mm) above *grade*. In areas subject to snow, a minimum of two fasteners per tile is required. In all other areas, clay and concrete roof tiles shall be attached in accordance with Table R905.3.7.

❖ This section identifies four key issues that must be taken into consideration when using the code and the manufacturer's installation instructions in the proper application of clay and concrete roof tile. At least one fastener is required for each roof tile for those tiles at the perimeter of the roof, as well as any tiles weighing less than 9 pounds per square foot (0.4 kg/m^2). In areas where the roof is exposed to a roof snow load, at least two fasteners are required for each tile. Because of the special concerns in high-wind areas, or where the roof height is well above grade, the manufacturer's installation instructions must be referred to in the determination of the minimum number of fasteners.

If none of the above-mentioned conditions occur, Table R905.3.7 must be used to determine proper fastening. The type of sheathing (spaced or solid), the presence of battens and the slope of the roof all impact the minimum number of fasteners required. Commentary Figures R905.3.7(1) through (4) illustrate some of the common shapes of tile and installation techniques.

TABLE R905.3.7
CLAY AND CONCRETE TILE ATTACHMENT

SHEATHING	ROOF SLOPE	NUMBER OF FASTENERS
Solid without battens	All	One per tile
Spaced or solid with battens and slope < 5:12	Fasteners not required	—
Spaced sheathing without battens	5:12 ≤ slope < 12:12	One per tile/every other row
	12:12 ≤ slope < 24:12	One per tile

❖ This table sets forth the number of fasteners required to attach clay or concrete tile to various types of sheathing. The table is not applicable where the wind speed exceeds 100 mph (45 m/s), where the building's roof is located more than 40 feet (12 192 mm) above grade or in locales subject to snow. Under all other conditions, the table indicates the number of fasteners required based on the type of sheathing and, in some cases, the slope of the roof.

R905.3.8 Flashing. At the juncture of roof vertical surfaces, flashing and counterflashing shall be provided in accordance with this chapter and the manufacturer's installation instructions and, where of metal, shall not be less than 0.019 inch (0.5 mm) (No. 26 galvanized sheet gage) corrosion-resistant metal. The valley flashing shall extend at least 11 inches (279 mm) from the centerline each way and have a splash diverter rib not less than 1 inch (25 mm) high at the flow line formed as part of the flashing. Sections of flashing shall have an end lap of not less than 4 inches (102 mm). For roof slopes of three units vertical in 12 units horizontal (25-percent slope) and greater, valley flashing shall have a 36-inch-wide (914 mm) underlayment of one layer of Type I underlayment running the full length of the valley, in addition to other required underlayment. In areas where the average daily temperature in January is 25°F (-4°C) or less, metal valley flashing underlayment shall be solid-cemented to the roofing underlayment for slopes less than seven units vertical in 12 units horizontal (58-percent slope) or be of self-adhering polymer modified bitumen sheet.

❖ Flashing for clay and concrete roof tile is important because the shape of the tile may terminate 2 or 3 inches (51 or 76 mm) above the roof sheathing. Flashing details, therefore, must be constructed in accordance with the code and the manufacturer's installation instructions for the tile shape being used. Some common methods of providing flashing at different locations are illustrated in Commentary Figures R905.3.8(1) through (4).

R905.4 Metal roof shingles. The installation of metal roof shingles shall comply with the provisions of this section.

❖ Metal roof shingles differ from the metal roof panels regulated by Section R905.10. By definition, a metal roof shingle must have an installed weather exposure of less than 3 square feet (0.28 m^2) per shingle or sheet. While decking and slope limitations differ to some degree, the provisions for underlayment and flashing are very similar to those for other types of shingle applications.

R905.4.1 Deck requirements. Metal roof shingles shall be applied to a solid or closely fitted deck, except where the roof covering is specifically designed to be applied to spaced sheathing.

❖ Unless manufacturers of metal shingles indicate that their particular shingle may be placed over a roof deck having spaced sheathing, metal roof shingles must be installed on a solid or closely fitted deck.

R905.4.2 Deck slope. Metal roof shingles shall not be installed on roof slopes below three units vertical in 12 units horizontal (25-percent slope).

❖ Based on the characteristics of shape, size and method of attachment, metal roof shingles must not be applied to roofs having a slope of less than 3:12. A lesser slope increases the potential for leakage to an unacceptable level.

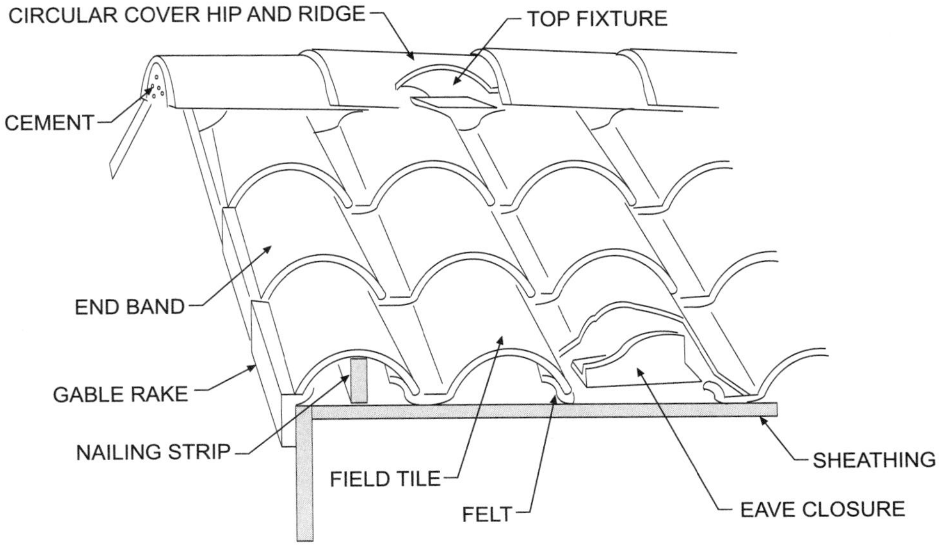

Figure R905.3.7(1)
APPLICATION OF ROLL TILE

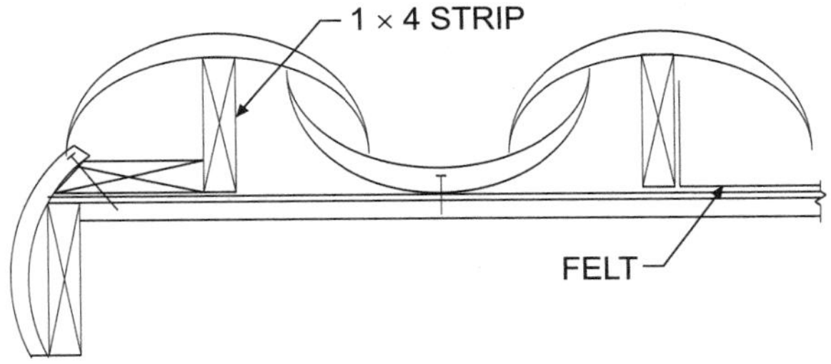

For SI: 1 inch = 25.4 mm.

Figure R905.3.7(2)
TILE ROOF GABLE RAKE SECTION

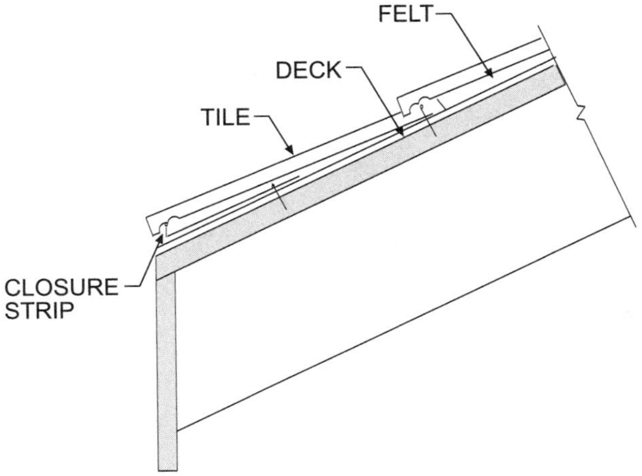

Figure R905.3.7(3)
TILE ROOF EAVE SECTION

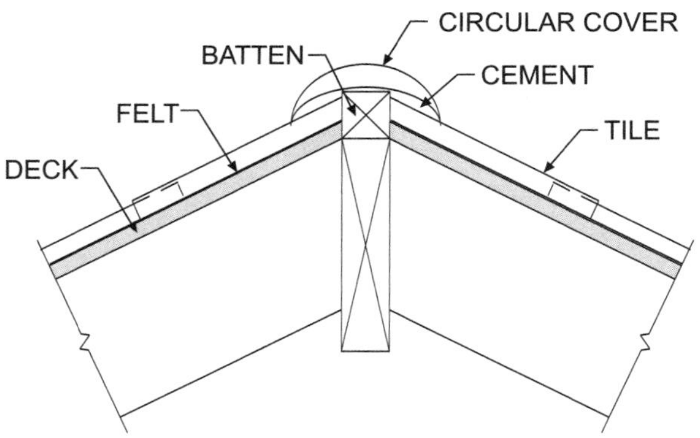

Figure R905.3.7(4)
TILE ROOF RIDGE SECTION

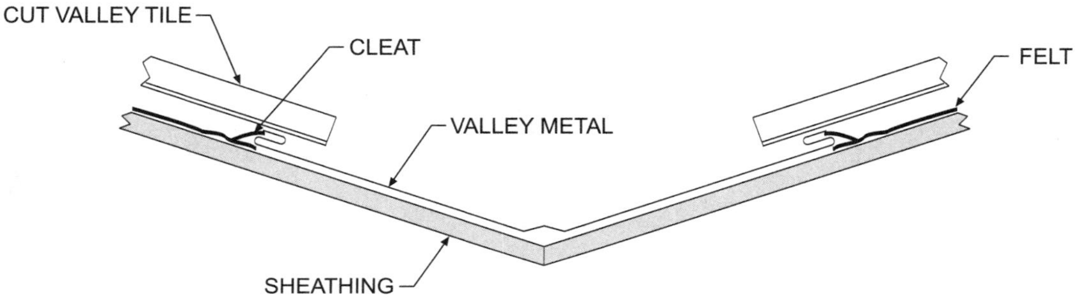

Figure R905.3.8(1)
TILE ROOF FLASHING AT VALLEY

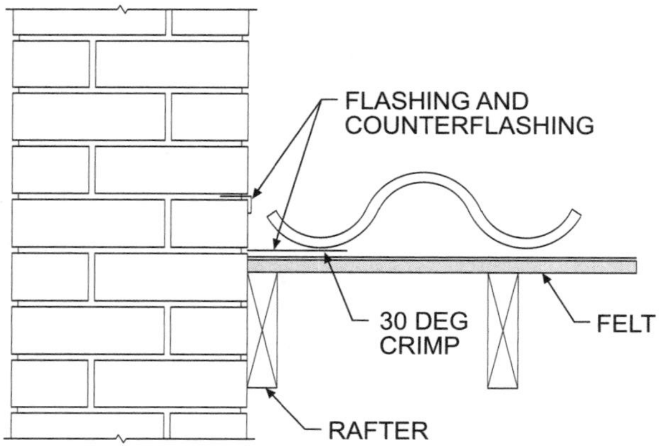

For SI: 1 degree = 0.01745 rad.

Figure R905.3.8(2)
TILE ROOF FLASHING AT ROOF ABUTTING CHIMNEY/WALL

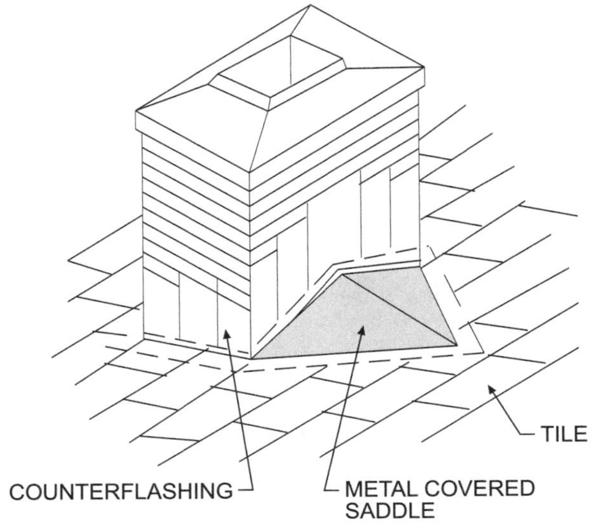

Figure R905.3.8(3)
TILE ROOF CHIMNEY FLASHING DETAIL

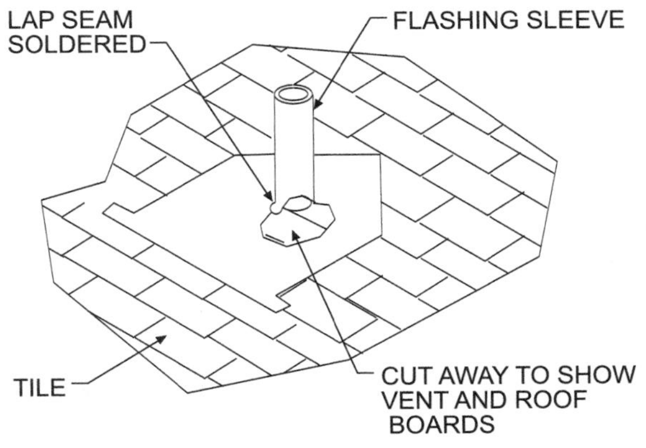

Figure R905.3.8(4)
TILE ROOF FLASHING FOR A VENT PIPE

R905.4.3 Underlayment. Underlayment shall comply with ASTM D 226, Type I or Type II, ASTM D 4869, Type I or Type II, or ASTM D 1970. Underlayment shall be installed in accordance with the manufacturer's installation instructions.

❖ Three types of underlayment are recognized for use with metal shingle roof coverings: asphalt-saturated organic felt as regulated by ASTM D 226, Type I or II, asphalt-saturated organic felt shingles per ASTM D 4869, Type I or II and self-adhering polymer modified bitumen shingles per ASTM D 1970.

R905.4.3.1 Ice barrier. In areas where there has been a history of ice forming along the eaves causing a backup of water as designated in Table R301.2(1), an ice barrier that consists of at least two layers of underlayment cemented together or a self-adhering polymer modified bitumen sheet shall be used in place of normal underlayment and extend from the lowest edges of all roof surfaces to a point at least 24 inches (610 mm) inside the exterior wall line of the building.

Exception: Detached *accessory structures* that contain no *conditioned floor area*.

❖ Where ice dams may be formed along the eave because snow continually freezes and thaws or frozen slush backs up in gutters, the underlayment application in the area of the eaves must be modified to prevent ice dams from forcing water under the roofing, which could damage ceilings, walls and insulation. Two layers of underlayment should be cemented together with asphalt cement from the lowest edge of the roof and continue up the roof to a point that is at least 24 inches (610 mm) inside the interior wall line of the building as shown in Commentary Figure R905.2.7.1(2). The environment within the envelope of the building provides adequate warmth to prevent ice dams from forming above the heated space; therefore, the two layers of cemented underlayment are permitted to terminate 24 inches (610 mm) inside the interior wall line of the building. The local jurisdiction is responsible for determining whether the ice barrier is required based on weather records, and it must so indicate in Table R301.2(1).

The exception is used throughout Chapter 9 for unconditioned accessory buildings (see commentary, Section R905.2.7.1).

R905.4.3.2 Underlayment and high winds. Underlayment applied in areas subject to high winds [above 110 mph (49 m/s) in accordance with Figure R301.2(4)A] shall be applied with corrosion-resistant fasteners in accordance with manufacturer's installation instructions. Fasteners are to be applied along the overlap not farther apart than 36 inches (914 mm) on center.

Underlayment installed where the basic wind speed equals or exceeds 120 mph (54 m/s) shall comply with ASTM D 226 Type II, ASTM D 4869 Type IV, or ASTM D 1970. The underlayment shall be attached in a grid pattern of 12 inches (305 mm) between side laps with a 6-inch (152 mm) spacing at the side laps. Underlayment shall be applied in accordance with Section R905.2.7 except all laps shall be a minimum of 4 inches (102 mm). Underlayment shall be attached using metal or plastic cap nails with a head diameter of not less than 1 inch (25.4 mm) with a thickness of at least 32 gauge sheet metal. The cap-nail shank shall be a minimum of 12 gauge (0.105 inches) with a length to penetrate through the roof sheathing or a minimum of $^3/_4$ inch (19 mm) into the roof sheathing.

Exception: As an alternative, adhered underlayment complying with ASTM D 1970 shall be permitted.

❖ See the commentary to Section R905.2.7.2.

R905.4.4 Material standards. Metal roof shingle roof coverings shall comply with Table R905.10.3(1). The materials used for metal roof shingle roof coverings shall be naturally corrosion resistant or be made corrosion resistant in accordance with the standards and minimum thicknesses listed in Table R905.10.3(2).

❖ Table R905.10.3(1) lists the metal roof covering types and their material standards. Many of the materials listed are inherently corrosion resistant, such as aluminum, copper and lead. For these materials, listing a material standard for the base material is sufficient. However, for steel roofing products, coatings are added to the base material to provide the necessary corrosion resistance.

Table R905.10.3(2) lists the minimum coating thickness required to provide minimum corrosion resistance for the base steel. Note a to Table R905.10.3(2) specifies that the paint systems are intended to be applied to steel having one of the corrosion-resistant coatings listed.

R905.4.5 Application. Metal roof shingles shall be secured to the roof in accordance with this chapter and the *approved* manufacturer's installation instructions.

❖ The proper application of metal roof shingles is typically based on the installation instructions of the shingle manufacturer. Where the code sets forth specific requirements for the attachment of metal shingles, those requirements must be met.

R905.4.6 Flashing. Roof valley flashing shall be of corrosion-resistant metal of the same material as the roof covering or shall comply with the standards in Table R905.10.3(1). The valley flashing shall extend at least 8 inches (203 mm) from the centerline each way and shall have a splash diverter rib not less than $^3/_4$ inch (19 mm) high at the flow line formed as part of the flashing. Sections of flashing shall have an end lap of not less than 4 inches (102 mm). The metal valley flashing shall have a 36-inch-wide (914 mm) underlayment directly under it consisting of one layer of underlayment running the full length of the valley, in addition to underlayment required for metal roof shingles. In areas where the average daily temperature in January is 25°F (-4°C) or less, the metal valley flashing underlayment shall be solid cemented to the roofing underlayment for roof slopes under seven units vertical in 12 units horizontal (58-percent slope) or self-adhering polymer modified bitumen sheet.

❖ This section specifies two approaches to roof valley flashing for metal-shingle roofs. The first is the use of material consistent with that of the shingles for the flashing. As an option, the flashing must be accomplished in compliance with the applicable material standard listed in Table R905.10.3. This table identifies the various types of metal roof covering and the corresponding application standard. This section specifies installation details for the valley flashing as well as the requirement for underlayment running the full length of the valley. In cold-weather areas, the special underlayment techniques to eliminate ice dams apply to valley flashing underlayment.

R905.5 Mineral-surfaced roll roofing. The installation of mineral-surfaced roll roofing shall comply with this section.

❖ This section addresses the decking requirements and slope, underlayment, materials standards and application methods for mineral-surfaced roll roofing.

R905.5.1 Deck requirements. Mineral-surfaced roll roofing shall be fastened to solidly sheathed roofs.

❖ The deck must be solidly sheathed for mineral-surfaced roll roofing to be installed.

R905.5.2 Deck slope. Mineral-surfaced roll roofing shall not be applied on roof slopes below one unit vertical in 12 units horizontal (8-percent slope).

❖ Although mineral-surfaced roll roofing provides a virtually weather-tight roof surface, it is still necessary that water flow be positive toward the drainage points. Therefore, a minimum slope of 1:12 is mandated for roof decks that are covered with mineral-surfaced roll roofing.

R905.5.3 Underlayment. Underlayment shall comply with ASTM D 226, Type I or ASTM D 4869, Type I or II.

❖ Two types of underlayment are recognized for use with mineral-surfaced roll roof coverings; asphalt-saturated organic felt as regulated by ASTM D 226, Type I, or asphalt-saturated organic felt shingles per ASTM D 4869, Type I or II.

R905.5.3.1 Ice barrier. In areas where there has been a history of ice forming along the eaves causing a backup of water as designated in Table R301.2(1), an ice barrier that consists of at least two layers of underlayment cemented together or a self-adhering polymer modified bitumen sheet shall be used in place of normal underlayment and extend from the lowest edges of all roof surfaces to a point at least 24 inches (610 mm) inside the exterior wall line of the building.

Exception: Detached *accessory structures* that contain no *conditioned floor area*.

❖ Where ice dams may be formed along the eave because snow continually freezes and thaws or frozen slush backs up in gutters, the underlayment application in the area of the eaves is modified to prevent ice dams from forcing water under the roofing, which could damage ceilings, walls and insulation. Two layers of underlayment should be cemented together with asphalt cement from the lowest edge of the roof and continue up the roof to a point that is at least 24 inches (610 mm) inside the interior wall line of the building as shown in Commentary Figure R905.2.7.1(2). The environment within the envelope of the building provides adequate warmth to prevent ice dams from forming above the heated space; therefore, the two layers of cemented underlayment are permitted to terminate 24 inches (610 mm) inside the interior wall line of the building. The local jurisdiction is responsible for determining whether the ice barrier is required based on weather records, and it must so indicate in Table R301.2(1).

The exception is used throughout Chapter 9 for unconditioned accessory buildings (see commentary, Section R905.2.7.1).

R905.5.3.2 Underlayment and high winds. Underlayment applied in areas subject to high winds [above 110 mph (49 m/s) in accordance with Figure R301.2(4)A] shall be applied with corrosion-resistant fasteners in accordance with manufacturer's installation instructions. Fasteners are to be applied along the overlap not farther apart than 36 inches (914 mm) on center.

Underlayment installed where the basic wind speed equals or exceeds 120 mph (54 m/s) shall comply with ASTM D 226 Type II or ASTM D 4869 Type IV. The underlayment shall be attached in a grid pattern of 12 inches (305 mm) between side laps with a 6-inch (152 mm) spacing at the side laps. Underlayment shall be applied in accordance with Section R905.2.7 except all laps shall be a minimum of 4 inches (102 mm). Underlayment shall be attached using metal or plastic cap nails with a head diameter of not less than 1 inch (25.4 mm) with a thickness of at least 32-gauge sheet metal. The cap-nail shank shall be a minimum of 12 gauge (0.105 inches) with a length to penetrate through the roof sheathing or a minimum of $^3/_4$ inch (19 mm) into the roof sheathing.

Exception: As an alternative, adhered underlayment complying with ASTM D 1970 shall be permitted.

❖ See the commentary to Section R905.2.7.2.

R905.5.4 Material standards. Mineral-surfaced roll roofing shall conform to ASTM D 3909 or ASTM D 6380, Class M.

❖ The code recognizes two materials standards for the regulation of mineral-surfaced roll roofing, ASTM D 3909 for asphalt roll roofing of glass felt surfaced with mineral granules and ASTM D 6380, Class M for asphalt roll roofing of organic felt surfaced with mineral granules.

R905.5.5 Application. Mineral-surfaced roll roofing shall be installed in accordance with this chapter and the manufacturer's installation instructions.

❖ The general provisions of this chapter apply to the installation of mineral-surfaced roll roofing, as do the installation instructions of the manufacturer of the specific roofing system.

R905.6 Slate and slate-type shingles. The installation of slate and slate-type shingles shall comply with the provisions of this section.

❖ This section establishes the materials standards and installation criteria for shingles of slate and slate-type materials. It includes specific provisions related to deck slope, underlayment, shingle application and flashing.

R905.6.1 Deck requirements. Slate shingles shall be fastened to solidly sheathed roofs.

❖ Only solidly-sheathed roofs can be used as the deck for the installation of slate and slate-type shingles.

R905.6.2 Deck slope. Slate shingles shall be used only on slopes of four units vertical in 12 units horizontal (33-percent slope) or greater.

❖ Shingles of slate or slate-type materials pose special problems because of the inconsistency of the shingles' shape and size. To keep the roof covering weather- tight and promote proper drainage, slate and slate-type shingles cannot be installed on a roof having a slope of less than 4:12. Lower-sloped roofs present a potential problem because water drains slowly and, thus, creates the opportunity for water back-up. Wind-driven rain can also pose a problem.

R905.6.3 Underlayment. Underlayment shall comply with ASTM D 226, Type I, or ASTM D 4869, Type I or II. Underlayment shall be installed in accordance with the manufacturer's installation instructions.

❖ Two types of underlayment are recognized for use with slate and slate-type shingles roof coverings; asphalt saturated organic felt as regulated by ASTM D 226, Type I, or asphalt-saturated organic felt shingles per ASTM D 4869, Type I or II.

R905.6.3.1 Ice barrier. In areas where there has been a history of ice forming along the eaves causing a backup of water as designated in Table R301.2(1), an ice barrier that consists of at least two layers of underlayment cemented together or a self-adhering polymer modified bitumen sheet shall be used in lieu of normal underlayment and extend from the lowest edges of all roof surfaces to a point at least 24 inches (610 mm) inside the exterior wall line of the building.

Exception: Detached *accessory structures* that contain no *conditioned floor area.*

❖ Where ice dams may be formed along the eave because snow continually freezes and thaws or frozen slush backs up in gutters, the underlayment application in the area of the eaves is modified to prevent ice dams from forcing water under the roofing, which could damage ceilings, walls and insulation. Two layers of underlayment should be cemented together with asphalt cement from the lowest edge of the roof and continue up the roof to a point that is at least 24 inches (610 mm) inside the interior wall line of the building as shown in Commentary Figure R905.2.7.1(2). The required underlayment must comply with ASTM D 226, Type II. This is the rare condition under which Type II underlayment is mandated. The referenced underlayment for most other types of roof covering materials is Type I. The environment within the envelope of the building provides adequate warmth to prevent ice dams from forming above the heated space; therefore, the two layers of cemented underlayment are permitted to terminate 24 inches (610 mm) inside the interior wall line of the building. The local jurisdiction is responsible for determining whether the ice barrier is required based on weather records, and it must so indicate in Table R301.2(1).

The exception is used throughout Chapter 9 for unconditioned accessory buildings (see commentary, Section R905.2.7.1).

R905.6.3.2 Underlayment and high winds. Underlayment applied in areas subject to high winds [above 110 mph (49 m/s) in accordance with Figure R301.2(4)A] shall be applied with corrosion-resistant fasteners in accordance with manufacturer's installation instructions. Fasteners are to be applied along the overlap not farther apart than 36 inches (914 mm) on center.

Underlayment installed where the basic wind speed equals or exceeds 120 mph (54 m/s) shall comply with ASTM D 226 Type II or ASTM D 4869 Type IV. The underlayment shall be attached in a grid pattern of 12 inches (305 mm) between side laps with a 6-inch (152 mm) spacing at the side laps. Underlayment shall be applied in accordance with Section R905.2.7 except all laps shall be a minimum of 4 inches (102 mm). Underlayment shall be attached using metal or plastic cap nails with a head diameter of not less than 1 inch (25.4 mm) with a thickness of at least 32-gauge sheet metal. The cap-nail shank shall be a minimum of 12 gauge (0.105 inches) with a length to penetrate through the roof sheathing or a minimum of $^3/_4$ inch (19 mm) into the roof sheathing.

Exception: As an alternative, adhered underlayment complying with ASTM D 1970 shall be permitted.

❖ See the commentary to Section R905.2.7.2.

R905.6.4 Material standards. Slate shingles shall comply with ASTM C 406.

❖ The standard applicable to roofing slate is ASTM C 406.

R905.6.5 Application. Minimum headlap for slate shingles shall be in accordance with Table R905.6.5. Slate shingles shall be secured to the roof with two fasteners per slate. Slate shingles shall be installed in accordance with this chapter and the manufacturer's installation instructions.

❖ The general provisions of this chapter apply to the installation of slate and slate-type shingles, as do the installation instructions of the manufacturer of the specific roofing materials. In addition, this section sets forth two requirements specific to slate roofing. First, two fasteners must be used to secure each slate shingle in place. Second, Table R905.6.5 mandates the minimum headlap at each shingle course. Taking into account the effects of both water traveling down the roof's surface and wind-driven rain, the minimum headlap is reduced as the slope of the roof increases.

TABLE R905.6.5
SLATE SHINGLE HEADLAP

SLOPE	HEADLAP (inches)
4:12 ≤ slope < 8:12	4
8:12 ≤ slope < 20:12	3
Slope ≤ 20:12	2

For SI: 1 inch = 25.4 mm.

❖ If slate shingles are installed, they must have a minimum lap at each course as required by this table. Because of the potential for water infiltration, as the slope of the roof decreases, the minimum headlap increases. The extended lap aids in resisting the effects of slow-traveling water and wind-driven rain.

R905.6.6 Flashing. Flashing and counterflashing shall be made with sheet metal. Valley flashing shall be a minimum of 15 inches (381 mm) wide. Valley and flashing metal shall be a minimum uncoated thickness of 0.0179-inch (0.5 mm) zinc coated G90. Chimneys, stucco or brick walls shall have a minimum of two plies of felt for a cap flashing consisting of a 4-inch-wide (102 mm) strip of felt set in plastic cement and extending 1 inch (25 mm) above the first felt and a top coating of plastic cement. The felt shall extend over the base flashing 2 inches (51 mm).

❖ Special requirements are necessary for the sheet metal flashing and counterflashing of roofs covered with slate or slate-type shingles. The provisions address flashing for roof valleys and for the roof intersections with chimneys and stucco or brick walls.

R905.7 Wood shingles. The installation of wood shingles shall comply with the provisions of this section.

❖ Wood shingles differ from wood shakes in that wood shingles are sawed materials and have a uniform butt thickness across the individual shingle length. Wood shakes are split from logs, shaped by the manufacturer, and have varying butt thicknesses. This section establishes the material standards and installation criteria for wood shingles. Also included are specific provisions related to deck requirements and slope, underlayment, shingle application and valley flashing.

R905.7.1 Deck requirements. Wood shingles shall be installed on solid or spaced sheathing. Where spaced sheathing is used, sheathing boards shall not be less than 1-inch by 4-inch (25.4 mm by 102 mm) nominal dimensions and shall be spaced on centers equal to the weather exposure to coincide with the placement of fasteners.

❖ Either solid or spaced sheathing is permitted as a deck for the installation of wood shingles. If spaced sheathing is used, the minimum 1-inch by 4-inch (25.4 mm by 102 mm) sheathing boards are required. The placement of the spaced sheathing is important, with the center-to-center spacing of the sheathing boards equivalent to the weather exposure of the wood shingles. This method allows a consistent fastening pattern for applying the wood-shingle roof covering materials. Table R905.7.5 contains the maximum permitted weather exposure for wood shingles.

R905.7.1.1 Solid sheathing required. In areas where the average daily temperature in January is 25°F (-4°C) or less, solid sheathing is required on that portion of the roof requiring the application of an ice barrier.

❖ In cold-weather climates where an ice shield is required in accordance with Section R905.7.3, solid sheathing must be installed for decking. The solid sheathing need be located only where the ice shield is applied, typically a point from the eave's edge to at least 24 inches (610 mm) inside the exterior wall line of the building.

R905.7.2 Deck slope. Wood shingles shall be installed on slopes of three units vertical in 12 units horizontal (25-percent slope) or greater.

❖ Wood shingle roofs pose special problems for weather protection because of the texture of the materials. To keep the roof covering weather tight and to promote proper drainage, wood shingles cannot be installed on a roof having a slope of less than 3:12. Lower-sloped roofs present a potential problem because water drains slowly, creating an opportunity for water back-up. Wind-driven rain can also pose a problem.

R905.7.3 Underlayment. Underlayment shall comply with ASTM D 226, Type I or ASTM D 4869, Type I or II.

❖ Two types of underlayment are recognized for use with wood shingle roof coverings: asphalt-saturated organic felt as regulated by ASTM D 226, Type I, or asphalt-saturated organic felt shingles per ASTM D 4869, Type I or II.

R905.7.3.1 Ice barrier. In areas where there has been a history of ice forming along the eaves causing a backup of water as designated in Table R301.2(1), an ice barrier that consists of at least two layers of underlayment cemented together or a self-adhering polymer modified bitumen sheet shall be used in lieu of normal underlayment and extend from the lowest edges of all roof surfaces to a point at least 24 inches (610 mm) inside the exterior wall line of the building.

Exception: Detached *accessory structures* that contain no *conditioned floor area*.

❖ Where ice dams may be formed along the eave because snow continually freezes and thaws or frozen slush backs up in gutters, the underlayment application in the area of the eaves is modified to prevent ice dams from forcing water under the roofing, which could damage ceilings, walls and insulation. Two layers of underlayment should be cemented together with asphalt cement from the lowest edge of the roof and continue up the roof to a point that is at least 24 inches (610 mm) inside the interior wall line of the building as shown in Commentary Figure R905.2.7.1(2). The environment within the envelope of the building provides adequate warmth to prevent ice dams from forming above the heated space; therefore, the two layers of cemented underlayment are permitted to terminate 24 inches (610 mm) inside the interior wall line of the building. The local jurisdiction is responsible for determining whether the ice barrier is required based on weather records, and it must so indicate in Table R301.2(1).

The exception is used throughout Chapter 9 for unconditioned accessory buildings (see commentary, Section R905.2.7.1).

R905.7.3.2 Underlayment and high winds. Underlayment applied in areas subject to high winds [above 110 mph (49 m/s) in accordance with Figure R301.2(4)A] shall be applied with corrosion-resistant fasteners in accordance with manufacturer's installation instructions. Fasteners are to be applied along the overlap not farther apart than 36 inches (914 mm) on center.

Underlayment installed where the basic wind speed equals or exceeds 120 mph (54 m/s) shall comply with ASTM D 226 Type II or ASTM D 4869 Type IV. The underlayment shall be attached in a grid pattern of 12 inches (305 mm) between side laps with a 6-inch (152 mm) spacing at the side laps. Underlayment shall be applied in accordance with Section R905.2.7 except all Head laps shall be a minimum of 4 inches (102 mm). Underlayment shall be attached using metal or plastic cap nails with a head diameter of not less than 1 inch (25.4 mm) with a thickness of at least 32-gauge sheet metal. The cap-nail shank shall be a minimum of 12 gauge (0.105 inches) with a length to penetrate through the roof sheathing or a minimum of $^{3}/_{4}$ inch (19 mm) into the roof sheathing.

Exception: As an alternative, adhered underlayment complying with ASTM D 1970 shall be permitted.

❖ See the commentary to Section R905.2.7.2.

R905.7.4 Material standards. Wood shingles shall be of naturally durable wood and comply with the requirements of Table R905.7.4.

❖ Wood shingles are manufactured in 24-, 18- and 16-inch (610, 457 and 406 mm) lengths and are graded into three categories: 1, 2 and 3. The Cedar Shake and Shingle Bureau establishes the grading rules for wood shingles. Table R905.7.4 describes each grade, and Commentary Figure R905.7.4 shows an example label.

TABLE R905.7.4
WOOD SHINGLE MATERIAL REQUIREMENTS

MATERIAL	MINIMUM GRADES	APPLICABLE GRADING RULES
Wood shingles of naturally durable wood	1, 2 or 3	Cedar Shake and Shingle Bureau

❖ This table identifies the basic material requirements for wood roof shingles. It states that the shingles must be of naturally durable wood; graded as No. 1, No. 2 or No. 3 and subject to the applicable grading rules of the Cedar Shake and Shingle Bureau. Each grade is mentioned in Table R905.7.4. Commentary Figure R905.7.4 is an example of the required identification label of an approved grading or inspection bureau or agency.

Figure R905.7.4
WOOD SHINGLE GRADE DESCRIPTION AND EXAMPLE LABEL

R905.7.5 Application. Wood shingles shall be installed according to this chapter and the manufacturer's installation instructions. Wood shingles shall be laid with a side lap not less than $1^1/_2$ inches (38 mm) between joints in courses, and no two joints in any three adjacent courses shall be in direct alignment. Spacing between shingles shall not be less than $^1/_4$ inch to $^3/_8$ inch (6 mm to 10 mm). Weather exposure for wood shingles shall not exceed those set in Table R905.7.5. Fasteners for wood shingles shall be corrosion resistant with a minimum penetration of $^1/_2$ inch (13 mm) into the sheathing. For sheathing less than $^1/_2$ inch (13 mm) in thickness, the fasteners shall extend through the sheathing. Wood shingles shall be attached to the roof with two fasteners per shingle, positioned no more than $^3/_4$ inch (19 mm) from each edge and no more than 1 inch (25 mm) above the exposure line.

❖ The general provisions of this chapter apply to the installation of wood shingles, as do the installation instructions of the manufacturer of the specific roofing materials. In addition, this section sets forth specific requirements regulating the installation method for wood shingles. Many of these requirements are shown in Commentary Figure R905.7.5(1).

Fasteners for applying wood shingles must be corrosion resistant. Only two nails or staples are to be used for each shingle, and they must be placed approximately $^3/_4$ inch (19 mm) from each side edge and 1 inch (25 mm) above the exposure line. The depth of fastener penetration must be a minimum of $^1/_2$ inch (12.7 mm) into the sheathing or completely through sheathing less than $^1/_2$ inch (12.7 mm) thick. Care in spacing and driving the fasteners is necessary to provide maximum service from the roof covering [see Commentary Figure R905.7.5(2)].

TABLE R905.7.5
WOOD SHINGLE WEATHER EXPOSURE AND ROOF SLOPE

ROOFING MATERIAL	LENGTH (inches)	GRADE	EXPOSURE (inches) 3:12 pitch to < 4:12	EXPOSURE (inches) 4:12 pitch or steeper
Shingles of naturally durable wood	16	No. 1	$3^3/_4$	5
		No. 2	$3^1/_2$	4
		No. 3	3	$3^1/_2$
	18	No. 1	$4^1/_4$	$5^1/_2$
		No. 2	4	$4^1/_2$
		No. 3	$3^1/_2$	4
	24	No. 1	$5^3/_4$	$7^1/_2$
		No. 2	$5^1/_2$	$6^1/_2$
		No. 3	5	$5^1/_2$

For SI: 1 inch = 25.4 mm.

❖ Table R905.7.5 specifies wood shingle exposure. Depending on the grade of the material, the total shingle length and the slope of the roof deck, the table specifies the maximum length of exposure for weathering purposes. The grade of the shingle is listed in the label required for wood shingles. Each of the three different grades is available in three different lengths. For lower-sloped roofs (3:12 to 4:12) where wind uplift is more significant, the maximum weather exposure is always less than for the steeper sloped roofs (4:12 and greater).

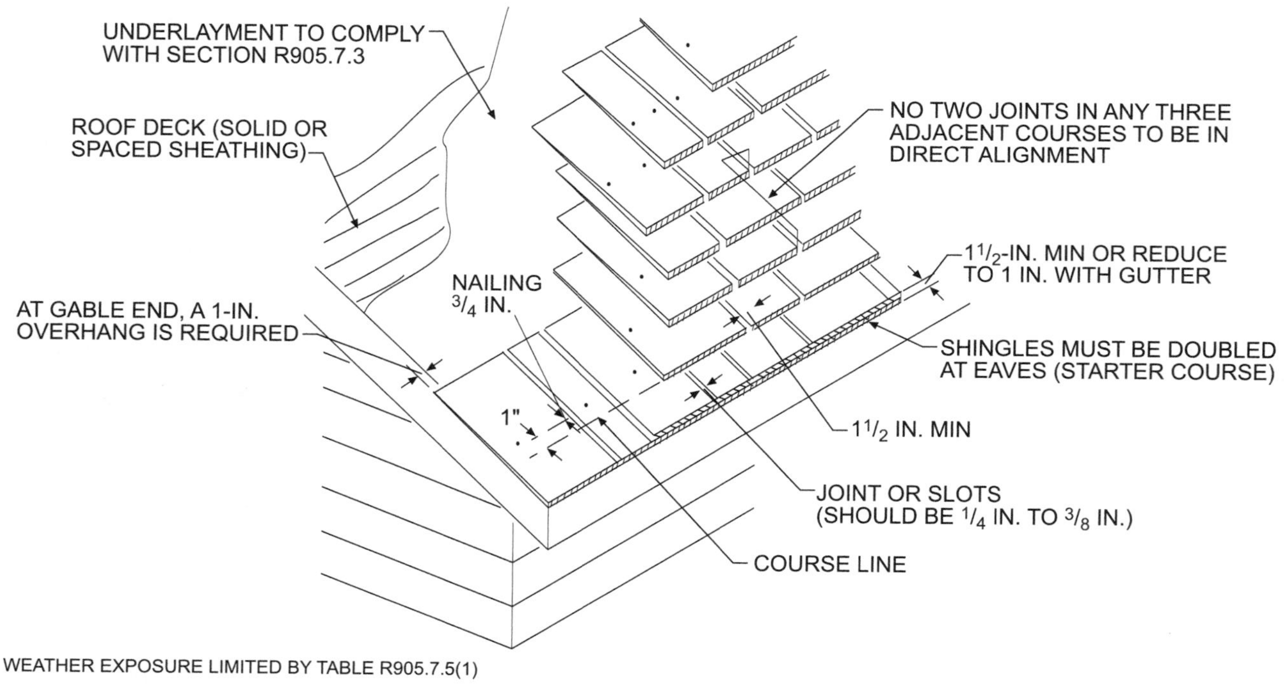

For SI: 1 inch = 25.4 mm.

Figure R905.7.5(1)
WOOD SHINGLE APPLICATION

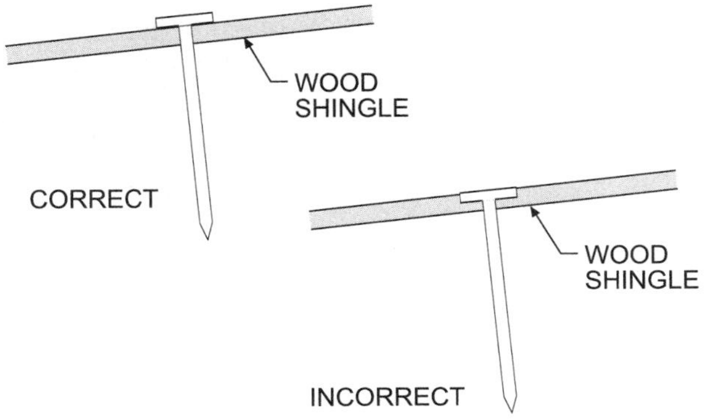

Figure R905.7.5(2)
FASTENER PENETRATION

R905.7.6 Valley flashing. Roof flashing shall be not less than No. 26 gage [0.019 inches (0.5 mm)] corrosion-resistant sheet metal and shall extend 10 inches (254 mm) from the centerline each way for roofs having slopes less than 12 units vertical in 12 units horizontal (100-percent slope), and 7 inches (178 mm) from the centerline each way for slopes of 12 units vertical in 12 units horizontal and greater. Sections of flashing shall have an end lap of not less than 4 inches (102 mm).

❖ Valleys formed by the intersection of two sloping roofs are vulnerable to leakage because of the high concentration of water; therefore, valley flashing must be maintained to prevent blockage. This section requires the use of minimum No. 26 gage (0.019 inch/0.48 mm) corrosion-resistant sheet metal for this purpose. Commentary Figure R905.7.6 illustrates the code provisions for valley flashing.

R905.7.7 Label required. Each bundle of shingles shall be identified by a *label* of an *approved* grading or inspection bureau or agency.

❖ To verify compliance with the material standards set forth in Section R905.7.4, every bundle of wood shingles must be labeled. The label, a sample of which is shown in Commentary Figure R905.7.4, must indicate the name of the approved grading or inspection bureau or agency.

R905.8 Wood shakes. The installation of wood shakes shall comply with the provisions of this section.

❖ Wood shakes differ from wood shingles in that wood shakes are split from logs, shaped by the manufacturer and have varying butt thicknesses. Wood shingles are sawed materials and have a uniform butt thickness across the individual shingle length. This section establishes the materials standards and installation criteria for wood shakes. Also included are specific provisions related to deck requirements and slope, underlayment, interlayment, shingle application and valley flashing.

R905.8.1 Deck requirements. Wood shakes shall be used only on solid or spaced sheathing. Where spaced sheathing is used, sheathing boards shall not be less than 1-inch by 4-inch (25 mm by 102 mm) nominal dimensions and shall be spaced on centers equal to the weather exposure to coincide with the placement of fasteners. Where 1-inch by 4-inch (25 mm by 102 mm) spaced sheathing is installed at 10 inches (254 mm) on center, additional 1-inch by 4-inch (25 mm by 102 mm) boards shall be installed between the sheathing boards.

❖ Either solid or spaced sheathing is permitted as a deck for the installation of wood shakes. If spaced sheathing is used, the minimum 1-inch by 4-inch (25 mm by 102 mm) sheathing boards are required. The placement of the spaced sheathing is important, with the center-to-center spacing of the sheathing boards equivalent to the weather exposure of the wood shakes. This method allows for a consistent fastening pattern for applying wood shake roof covering materials. Table R905.8.6 contains the maximum permitted weather exposure for wood shakes. If the greatest allowable exposure of 10 inches (254 mm) is used per Table R905.8.6, an additional 1-inch by 4-inch (25 mm by 102 mm) board must be installed between the sheathing boards.

R905.8.1.1 Solid sheathing required. In areas where the average daily temperature in January is 25°F (-4°C) or less, solid sheathing is required on that portion of the roof requiring an ice barrier.

❖ In cold-weather climates where an ice shield is required in accordance with Section R905.7.3, solid sheathing must be installed for decking. The solid sheathing need be located only where the ice shield is applied, typically a point from the eave's edge to at least 24 inches (610 mm) inside the exterior wall line of the building.

R905.8.2 Deck slope. Wood shakes shall only be used on slopes of three units vertical in 12 units horizontal (25-percent slope) or greater.

❖ Wood shake roofs pose special problems for weather protection because of the texture of the materials. To keep the roof covering weather tight and to promote proper drainage, wood shakes cannot be installed on a roof having a slope of less than 3:12. Lower-sloped roofs present a potential problem because water drains slowly, creating an opportunity for water back-up. Wind-driven rain can also pose a problem. The use of an interlayment as installed per Section R905.8.7 also plays an important part in maintaining the weathertightness of the roof.

R905.8.3 Underlayment. Underlayment shall comply with ASTM D 226, Type I or ASTM D 4869, Type I or II.

❖ Two types of underlayment are recognized for use with wood-shingle roof coverings: asphalt-saturated organic felt as regulated by ASTM D 226, Type I, or asphalt-saturated organic felt shingles per ASTM D 4869, Type I or II.

R905.8.3.1 Ice barrier. In areas where there has been a history of ice forming along the eaves causing a backup of water as designated in Table R301.2(1), an ice barrier that consists of at least two layers of underlayment cemented together or a self-adhering polymer modified bitumen sheet shall be used in place of normal underlayment and extend from the lowest edges of all roof surfaces to a point at least 24 inches (610 mm) inside the exterior wall line of the building.

Exception: Detached *accessory structures* that contain no *conditioned floor area*.

❖ Where ice dams may be formed along the eave because snow continually freezes and thaws or frozen slush backs up in gutters, the underlayment application in the area of the eaves is modified to prevent ice dams from forcing water under the roofing, which could damage ceilings, walls and insulation. Two layers of underlayment should be cemented together with asphalt cement from the lowest edge of the roof and continue up the roof to a point that is at least 24 inches (610 mm) inside the interior wall line of the building as shown in Commentary Figure R905.2.7.1(2). The environment within the envelope of the building provides adequate warmth to prevent ice dams from forming above the heated space; therefore, the two layers of cemented underlayment are permitted to terminate 24 inches (610 mm) inside the interior wall line of the building. The local jurisdiction is responsible for determining whether the ice barrier is required based on weather records, and it must so indicate in Table R301.2(1).

The exception is used throughout Chapter 9 for unconditioned accessory buildings (see commentary, Section R905.2.7.1).

R905.8.3.2 Underlayment and high winds. Underlayment applied in areas subject to high winds [above 110 mph (49 m/s) in accordance with Figure R301.2(4)A] shall be applied with corrosion-resistant fasteners in accordance with manufacturer's installation instructions. Fasteners are to be applied along the overlap not farther apart than 36 inches (914 mm) on center.

Underlayment installed where the basic wind speed equals or exceeds 120 mph (54 m/s) shall comply with ASTM D 226 Type II or ASTM D 4869 Type IV. The underlayment shall be attached in a grid pattern of 12 inches (305 mm) between side laps with a 6-inch (152 mm) spacing at the side laps. Underlayment shall be applied in accordance with Section R905.2.7 except all laps shall be a minimum of 4 inches (102 mm). Underlayment shall be attached using metal or plastic

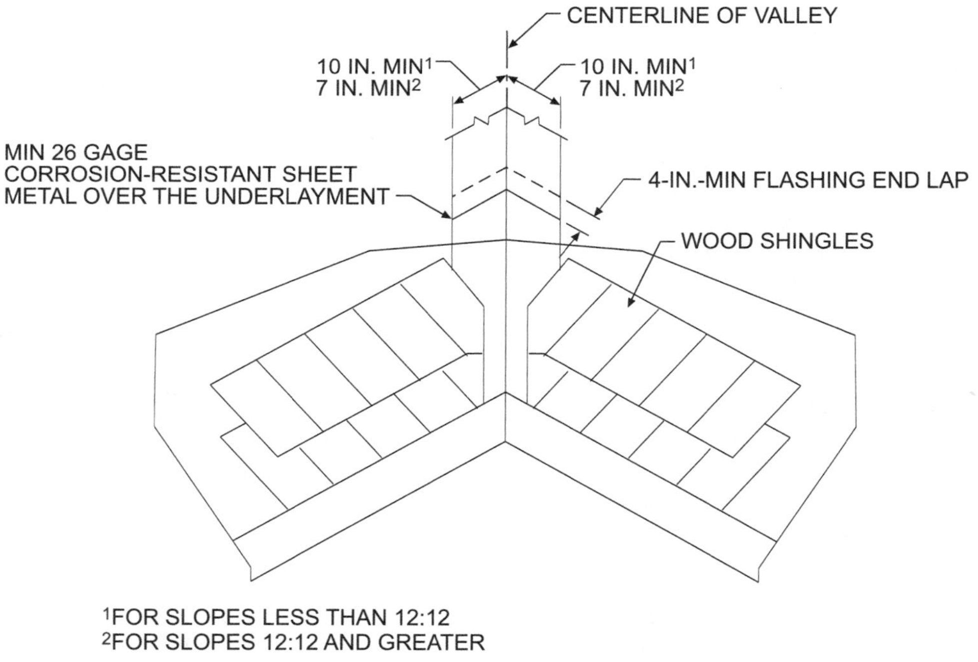

For SI: 1 inch = 25.4 mm.

Figure R905.7.6
WOOD SHINGLE VALLEY FLASHING

cap nails with a head diameter of not less than 1 inch (25.4 mm) with a thickness of at least 32-gauge sheet metal. The cap-nail shank shall be a minimum of 12 gauge (0.105 inches) with a length to penetrate through the roof sheathing or a minimum of $^3/_4$ inch (19 mm) into the roof sheathing.

Exception: As an alternative, adhered underlayment complying with ASTM D 1970 shall be permitted.

❖ See the commentary to Section R905.2.7.2.

R905.8.4 Interlayment. Interlayment shall comply with ASTM D 226, Type I.

❖ Wood shakes must be installed with an interlayment of minimum No. 30 felt shingled between each course in the manner described by Section R905.8.7. The interlayment must be in compliance with ASTM D 226, Type I, for asphalt-saturated organic felt.

R905.8.5 Material standards. Wood shakes shall comply with the requirements of Table R905.8.5.

❖ Wood shakes are divided into five material classifications in the code, identified in Table R905.8.5. There are two grades of wood shakes: No. 1 and No. 2. The shakes are cut into 18-inch (457 mm) and 24-inch (610 mm) lengths.

TABLE R905.8.5
WOOD SHAKE MATERIAL REQUIREMENTS

MATERIAL	MINIMUM GRADES	APPLICABLE GRADING RULES
Wood shakes of naturally durable wood	1	Cedar Shake and Shingle Bureau
Taper sawn shakes of naturally durable wood	1 or 2	Cedar Shake and Shingle Bureau
Preservative-treated shakes and shingles of naturally durable wood	1	Cedar Shake and Shingle Bureau
Fire-retardant-treated shakes and shingles of naturally durable wood	1	Cedar Shake and Shingle Bureau
Preservative-treated taper sawn shakes of Southern pine treated in accordance with AWPA Standard U1 (Commodity Specification A, Use Category 3B and Section 5.6)	1 or 2	Forest Products Laboratory of the Texas Forest Services

❖ Wood shakes are regulated based on the type of shake material, the treatment applied to the shake and the cut of the shake. Most of the grading rules applicable to wood shakes come from the Cedar Shake and Shingle Bureau. If preservative-treated tapersawn southern yellow pine wood shakes are to be installed, they are governed under the grading rules of the Forest Products Laboratory of the Texas Forest Services. The minimum grade required by the table must be shown on the approved label required on each bundle of wood shakes.

R905.8.6 Application. Wood shakes shall be installed according to this chapter and the manufacturer's installation instructions. Wood shakes shall be laid with a side lap not less than $1^1/_2$ inches (38 mm) between joints in adjacent courses. Spacing between shakes in the same course shall be $^3/_8$ inch to $^5/_8$ inch (9.5 mm to 15.9 mm) for shakes and tapersawn shakes of naturally durable wood and shall be $^3/_8$ inch to $^5/_8$ inch (9.5 mm to 15.9 mm) for preservative-treated taper sawn shakes. Weather exposure for wood shakes shall not exceed those set forth in Table R905.8.6. Fasteners for wood shakes shall be corrosion-resistant, with a minimum penetration of $^1/_2$ inch (12.7 mm) into the sheathing. For sheathing less than $^1/_2$ inch (12.7 mm) thick, the fasteners shall extend through the sheathing. Wood shakes shall be attached to the roof with two fasteners per shake, positioned no more than 1 inch (25 mm) from each edge and no more than 2 inches (51 mm) above the exposure line.

❖ The general provisions of this chapter apply to the installation of wood shakes, as do the installation instructions of the manufacturer of the specific roofing materials. In addition, this section sets forth specific requirements regulating the installation method for wood shakes. The minimum spacing of $^3/_8$ inch between shakes is necessary to prevent the accumulation of leaves and/or needles from evergreen trees in the spaces. A smaller space would cause accumulation and premature aging of the shakes. Commentary Figure R905.8.6 shows many of these requirements.

Fasteners for applying wood shakes must be corrosion resistant. Only two nails or staples are to be used for each shake, and they must be placed approximately 1 inch (25 mm) from each side edge and no more than 2 inches (51 mm) above the exposure line. The depth of fastener penetration must be a minimum of $^1/_2$ inch (12.7 mm) into the sheathing, or completely through sheathing less than $^1/_2$ inch (12.7 mm) thick. Care in spacing and driving is necessary to obtain maximum service from the roof covering; see Commentary Figure R905.7.5(2).

TABLE R905.8.6
WOOD SHAKE WEATHER EXPOSURE AND ROOF SLOPE

ROOFING MATERIAL	LENGTH (inches)	GRADE	EXPOSURE (inches)
			4:12 pitch or steeper
Shakes of naturally durable wood	18	No. 1	$7^1/_2$
	24	No. 1	10[a]
Preservative-treated taper sawn shakes of Southern Yellow Pine	18	No. 1	$7^1/_2$
	24	No. 1	10
	18	No. 2	$5^1/_2$
	24	No. 2	$7^1/_2$
Taper-sawn shakes of naturally durable wood	18	No. 1	$7^1/_2$
	24	No. 1	10
	18	No. 2	$5^1/_2$
	24	No. 2	$7^1/_2$

For SI: 1 inch = 25.4 mm.

a. For 24-inch by $^3/_8$-inch handsplit shakes, the maximum exposure is $7^1/_2$ inches.

❖ Table R905.8.6 specifies wood shake exposure. Depending on the grade of the material and the total shingle length, the table specifies the maximum length of exposure for weathering purposes. The grade of the shingle will be listed in the label required for wood shingles.

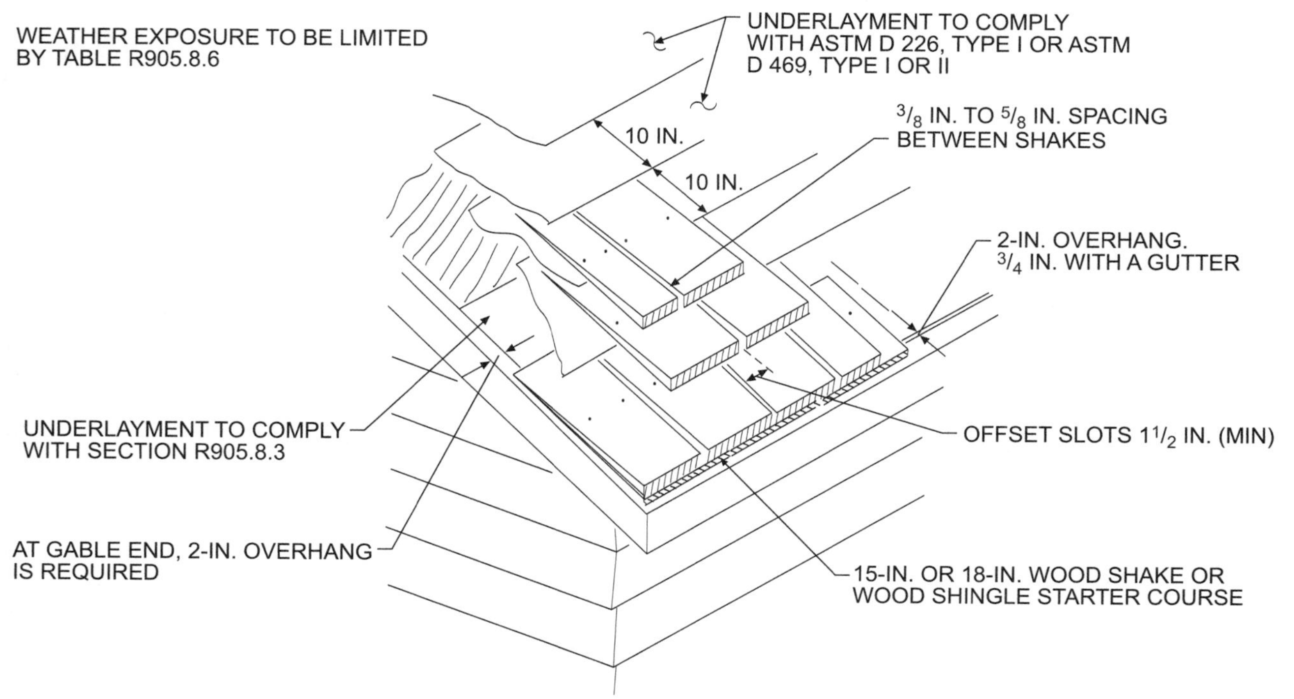

For SI: 1 inch = 25.4 mm.

Figure R905.8.6
WOOD SHAKE APPLICATION

R905.8.7 Shake placement. The starter course at the eaves shall be doubled and the bottom layer shall be either 15-inch (381 mm), 18-inch (457 mm) or 24-inch (610 mm) wood shakes or wood shingles. Fifteen-inch (381 mm) or 18-inch (457 mm) wood shakes may be used for the final course at the ridge. Shakes shall be interlaid with 18-inch-wide (457 mm) strips of not less than No. 30 felt shingled between each course in such a manner that no felt is exposed to the weather by positioning the lower edge of each felt strip above the butt end of the shake it covers a distance equal to twice the weather exposure.

❖ To provide as weather tight a roof as possible where using wood shakes as the roof covering material, a special installation method is described in this section. At the eaves, the starter course must be two layers of shakes. Going up the roof toward the ridge, the shakes must be supplemented with 18-inch (457 mm) interlayment strips of minimum No. 30 felt. The manner of interlayment must prevent any felt being exposed to the weather. This is illustrated in Commentary Figure R905.8.7.

R905.8.8 Valley flashing. Roof valley flashing shall not be less than No. 26 gage [0.019 inch (0.5 mm)] corrosion-resistant sheet metal and shall extend at least 11 inches (279 mm) from the centerline each way. Sections of flashing shall have an end lap of not less than 4 inches (102 mm).

❖ Valleys formed by the intersection of two sloping roofs are vulnerable to leakage because of the high concentration of water; therefore, valley flashing must be maintained to prevent blockage. This section requires the use of minimum No. 26 gage (0.019 inch/0.48 mm) corrosion-resistant sheet metal for this purpose.

R905.8.9 Label required. Each bundle of shakes shall be identified by a *label* of an *approved* grading or inspection bureau or agency.

❖ To verify compliance with the material standards set forth in Section R905.7.4, every bundle of wood shakes must be labeled. The label, a sample of which is shown in Commentary Figure R905.8.9, must show the name of the approved grading or inspection bureau or agency.

R905.9 Built-up roofs. The installation of built-up roofs shall comply with the provisions of this section.

❖ A built-up roof is a roof covering system built at the job site. The provisions of this section deal with the fundamental requirements for such roofs. With all of the different products available, the possible combination of built-up roof systems is almost endless.

Asphalt and coal tar are the bitumens used for a built-up roof. They become fluid when heated and bond the felt layers together. To fuse the roof felts properly, the bitumens must be applied at the proper temperature. Too low a temperature will cause inadequate fusion, and overheating may change the properties of the bitumens or cause a fire hazard.

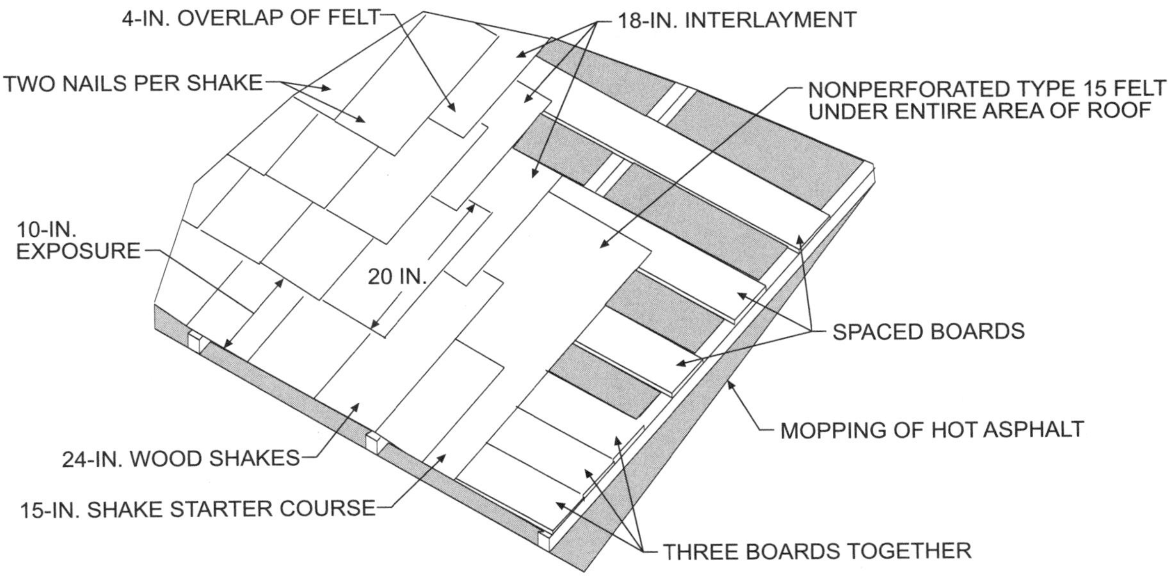

For SI: 1 inch = 25.4 mm.

Figure R905.8.7
WOOD SHAKE APPLICATION ON ROOFS WITH SLOPES NOT LESS THAN 3 IN 12

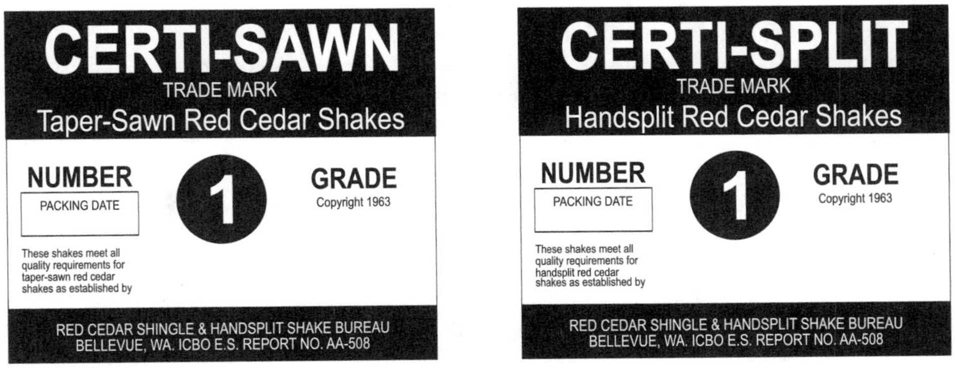

Figure R905.8.9
WOOD SHAKE DESCRIPTION AND EXAMPLE LABELS

R905.9.1 Slope. Built-up roofs shall have a design slope of a minimum of one-fourth unit vertical in 12 units horizontal (2-percent slope) for drainage, except for coal-tar built-up roofs, which shall have a design slope of a minimum one-eighth unit vertical in 12 units horizontal (1-percent slope).

❖ Built-up roofs are often chosen for very low-slope applications because they can be installed on roof decks having a minimum slope of $^1/_4$ inch per foot ($^1/_4$:12), just enough to allow positive drainage. Only where coal-tar built-up roofs are installed must the roof deck slope to a greater degree. A minimum roof slope of 1:12 is required for coal-tar built-up roofs.

R905.9.2 Material standards. Built-up roof covering materials shall comply with the standards in Table R905.9.2 or UL 55A.

❖ Table R905.9.2 of the code sets forth the appropriate material standards for the various types of built-up roofs. Eighteen different types of roofing systems are identified, and their corresponding specification standard is noted.

Also, as an alternative to the material standards in Table R905.9.2, the code permits compliance with UL 55A. UL 55A has been in use since 1919 and is still used to evaluate the following materials used in the construction of built-up roof coverings: hot-mopping asphalt; asphalt-saturated and organic felt; coal-tar pitch; coal-tar saturated organic felt; and asphalt-coated glass-fire mat (felt). Several listed roofing systems, evaluated to UL 790 or ASTM E 108, are based on compliance with UL 55A.

TABLE R905.9.2. See below.

❖ The types of materials used in built-up roofing must comply with the following standards:

- ASTM D 6083, which covers a liquid-applied water-dispersed 100-percent acrylic elastomeric latex coating used as a protective coating for roofs.
- ASTM D 1863, which covers the quality and grading of crushed stone, crushed slag and water-worn gravel suitable for use as aggregate surfacing on built-up roofs.
- ASTM D 3747, which covers emulsified asphalt adhesive for use in adhering preformed roof insulation to steel roof decks with inclines up to 33 percent. When applied as a continuous film over an acceptable deck surface, the emulsion functions as both an adhesive and a vapor retarder.
- ASTM D 3019, D 2822 and D 4586, which cover asphalt cements used in roofing.
  - ASTM D 3019 covers lap cement consisting of asphalt dissolved in a volatile petroleum solvent with or without mineral or other stabilizers, or both, for use with roll roofing.
  - ASTM D 2822 covers asphalt roof cement used for trowel application to roofings and flashings.
  - ASTM D 4586 covers asbestos-free asphalt roof cement suitable for trowel application to roofings and flashings.
- ASTM D 4601, which covers asphalt-impregnated and coated glass fiber base sheet, with or without perforations, for use as the first ply of the built-up roofing. When not perforated, this sheet may be used as a vapor retarder under or between roof insulation with a solid top coating of asphaltic material.
- ASTM D 1227, D 2823, D 2824 and D 4479, which cover asphalt coatings used in roofing.
  - ASTM D1227 covers emulsified asphalt suitable for use as a protective coating for built-up roofs and other exposed surfaces with inclines of not less than 4 percent.
  - ASTM D 2823 covers asphalt roof coatings of brushing or spraying consistency.
  - ASTM D 2824 covers aluminum-pigmented asphalt roof coating, including nonfibered, asbestos-fibered and fibered without asbestos.
  - ASTM D 4479 covers asbestos-free asphalt roof coating of brushing or spraying consistency.
- ASTM D 2178, which covers glass felt impregnated to varying degrees with asphalt, which

TABLE R905.9.2
BUILT-UP ROOFING MATERIAL STANDARDS

MATERIAL STANDARD	STANDARD
Acrylic coatings used in roofing	ASTM D 6083
Aggregate surfacing	ASTM D 1863
Asphalt adhesive used in roofing	ASTM D 3747
Asphalt cements used in roofing	ASTM D 2822; D 3019; D 4586
Asphalt-coated glass fiber base sheet	ASTM D 4601
Asphalt coatings used in roofing	ASTM D 1227; D 2823; D 2824; D 4479
Asphalt glass felt	ASTM D 2178
Asphalt primer used in roofing	ASTM D 41
Asphalt-saturated and asphalt-coated organic felt base sheet	ASTM D 2626
Asphalt-saturated organic felt (perforated)	ASTM D 226
Asphalt used in roofing	ASTM D 312
Coal-tar cements used in roofing	ASTM D 4022; D 5643
Coal-tar primer used in roofing, dampproofing and waterproofing	ASTM D 43
Coal-tar saturated organic felt	ASTM D 227
Coal-tar used in roofing	ASTM D 450, Type I or II
Glass mat, coal tar	ASTM D 4990
Glass mat, venting type	ASTM D 4897
Mineral-surfaced inorganic cap sheet	ASTM D 3909
Thermoplastic fabrics used in roofing	ASTM D 5665; D 5726

may be used with asphalts conforming to the requirements of ASTM D 312 in the construction of built-up roofs, and with asphalts conforming to the requirements of ASTM D 449 in the membrane system of waterproofing.

- ASTM D 41, which covers asphaltic primer suitable for use with asphalt in roofing, dampproofing and waterproofing below or above ground level, for application to concrete, masonry, metal and asphalt surfaces.
- ASTM D 2626, which covers base sheet with fine mineral surfacing on the top side, with or without perforations, for use as the first ply of a built-up roof. When not perforated, this sheet may be used as a vapor retarder under roof insulation.
- ASTM D 226, which covers asphalt-saturated organic felts, with perforations, that may be used with asphalts conforming to the requirements of ASTM D 312 in the construction of built-up roofs, and with asphalts conforming to the requirements of ASTM D 449 in the membrane system of waterproofing.
- ASTM D 312, which covers four types of asphalt intended for use in built-up roof construction. The specification is for general classification purposes only and does not imply restrictions on the slope at which an asphalt must be used. There are four classifications:
  - Type I includes asphalts that are relatively susceptible to flow at roof temperatures with good adhesive and self-sealing properties.
  - Type II includes asphalts that are moderately susceptible to flow at roof temperatures.
  - Type III includes asphalts that are relatively not susceptible to flow at roof temperatures for use in built-up roof construction on slope inclines from 8.3 percent to 25 percent.
  - Type IV includes asphalts that are generally not susceptible to flow at roof temperatures for use in built-up roof construction on slope inclines from approximately 16.7 percent to 50 percent.
- ASTM D 4022 and D 5643, which cover coal-tar cements used in roofing.
  - ASTM D 4022 covers coal-tar roof cement suitable for trowel application in coal-tar roofing and flashing systems.
  - ASTM D 5643 covers asbestos-free coal-tar roof cement suitable for trowel application in coal-tar roofing and flashing systems.
- ASTM D 227, which covers coal-tar saturated organic felt that may be used with coal-tar pitches conforming to the requirements of ASTM D 450 in the construction of built-up roofs and in the membrane system of waterproofing.
- ASTM D 43 covers coal-tar primer suitable for use with coal-tar pitch in roofing, dampproofing and waterproofing below or above ground level for application to concrete, masonry and coal tar surfaces.
- ASTM D 450, Types I and II, which covers coal-tar pitch suitable for use in the construction of built-up roofing, dampproofing and membrane waterproofing systems.
  - Type I is suitable for use in built-up roofing systems with felts conforming to the requirements of ASTM D 227 or as specified by the manufacturer.
  - Type II is suitable for use in dampproofing and in membrane waterproofing systems.
- ASTM D 4990, which covers glass felt impregnated with coal tar intended to be used with coal-tar pitch conforming to the requirements of ASTM D 450 in construction of built-up roofs and waterproofing systems.
- ASTM D 4897, which covers asphalt-impregnated and coated glass fiber base sheet with mineral surfacing on the top side and coarse mineral granules on the bottom side for use as the first ply of a roofing membrane. These base sheets provide for the lateral release of pressure in roofing systems because they are not solidly attached, and the coarse granular surface provides an open, porous channel in the horizontal plane beneath the membrane. The base sheets can be with or without perforations or embossings.
- ASTM D 3909, which covers asphalt-impregnated and coated glass felt roll roofing surfaced on the weather side with mineral granules, for use as a cap sheet in the construction of built-up roofs.
- ASTM D 5665 and D 5726, which cover thermoplastic fabrics used in roofing.
  - ASTM D 5665 covers thermoplastic fabrics such as polyester, polyester/polyamide bicomponent or composites with fiberglass or polyester scrims that can be used during the construction of cold-applied roofing and waterproofing.
  - ASTM D 5726 covers thermoplastic fabrics such as polyester, polyester/polyamide bicomponent or composites with fiberglass or polyester scrims that can be used during the construction of hot-applied roofing and waterproofing.

R905.9.3 Application. Built-up roofs shall be installed according to this chapter and the manufacturer's installation instructions.

❖ The general provisions of this chapter apply to the installation of built-up roofs, as do the installation instructions of the manufacturer of the specific roofing system. Commentary Figures R905.9.3(1) through (3) illustrate examples of the application of the base ply to nailable and nonnailable decks.

R905.10 Metal roof panels. The installation of metal roof panels shall comply with the provisions of this section.

❖ Metal roof panels differ from metal roof shingles regulated by Section R905.4. By definition, a metal roof panel must have an installed weather exposure of at least 3 square feet (0.28 m^2) per shingle or sheet. This section addresses the deck requirements, deck slope and attachment requirements for metal roof panels.

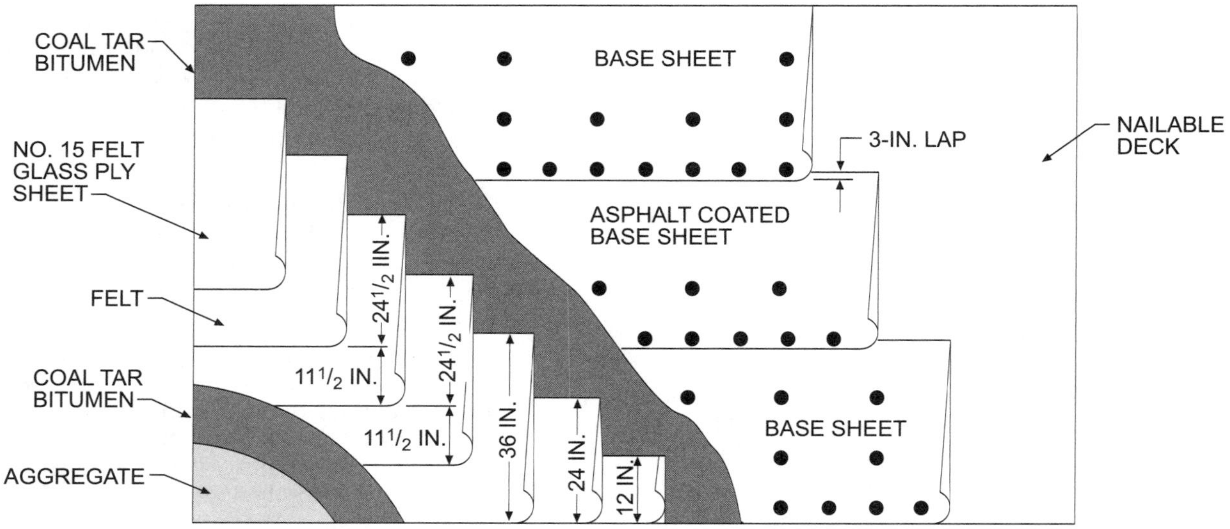

For SI: 1 inch = 25.4 mm.

Figure R905.9.3(1)
BUILT-UP ROOF—NAILABLE DECK APPLICATION

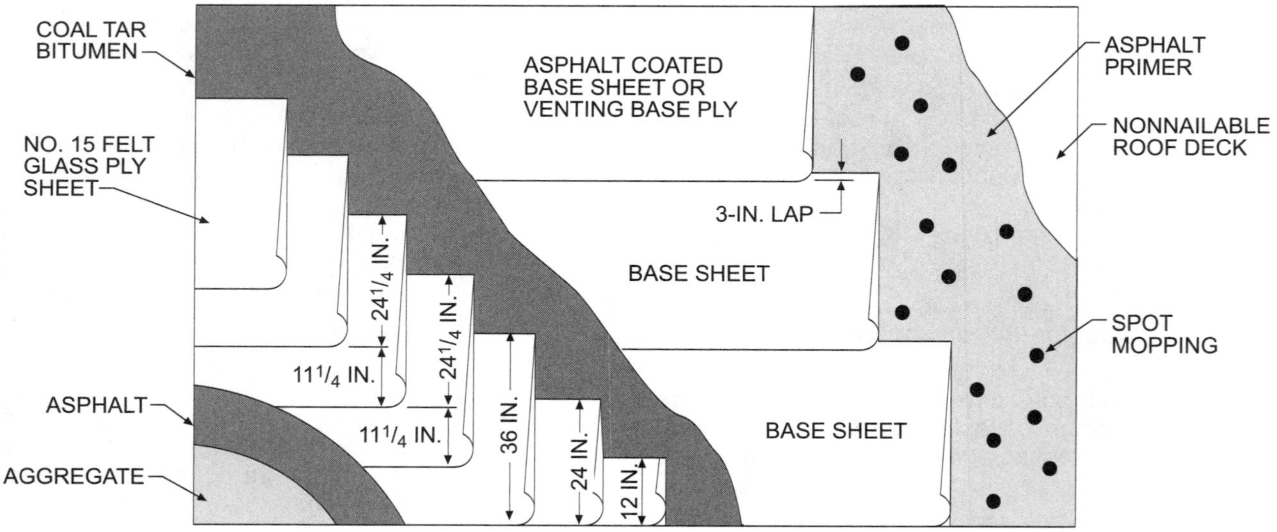

For SI: 1 inch = 25.4 mm.

Figure R905.9.3(2)
BUILT-UP ROOF—APPLICATION TO NONNAILABLE DECK BY SPOT MOPPING

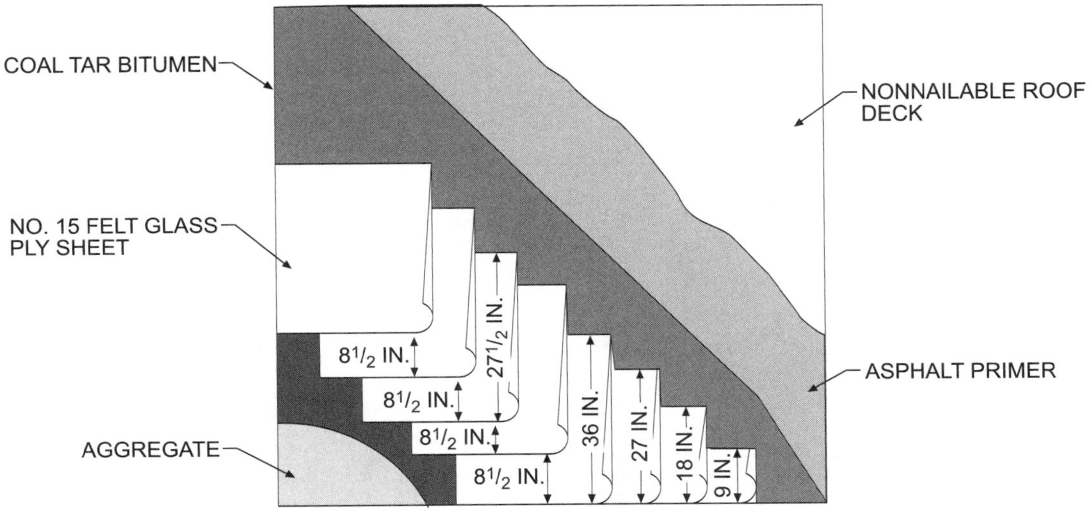

For SI: 1 inch = 25.4 mm.

Figure R905.9.3(3)
BUILT-UP-APPLICATION TO NONNAILABLE DECK SOLID MOPPING

R905.10.1 Deck requirements. Metal roof panel roof coverings shall be applied to solid or spaced sheathing, except where the roof covering is specifically designed to be applied to spaced supports.

❖ Unless specifically limited, metal roof panels may be applied to either solidly sheathed decks or spaced sheathing. In those cases where the specific metal panels are designed to be installed only on spaced supports, the manufacturer's installation instructions govern.

R905.10.2 Slope. Minimum slopes for metal roof panels shall comply with the following:

1. The minimum slope for lapped, nonsoldered-seam metal roofs without applied lap sealant shall be three units vertical in 12 units horizontal (25-percent slope).
2. The minimum slope for lapped, nonsoldered-seam metal roofs with applied lap sealant shall be one-half vertical unit in 12 units horizontal (4-percent slope). Lap sealants shall be applied in accordance with the *approved* manufacturer's installation instructions.
3. The minimum slope for standing-seam roof systems shall be one-quarter unit vertical in 12 units horizontal (2-percent slope).

❖ The minimum slope permitted for a roof deck supporting metal roof panels varies based on the type of panels being installed. Three different types of metal roof panel systems are regulated: lapped, nonsoldered seam metal roofs without applied lap sealant; lapped, nonsoldered seam metal roofs with applied lap sealant and standing seam metal roof systems. The slope limitations vary based on the ability of the metal panel roof system to shed water without water intrusion. This section allows lower minimum roof slopes for lapped nonsoldered seam metal roofs when lap sealants are used. To help ensure that lap sealants perform as part of the overall roofing system the lap sealant must be installed in accordance with the approved manufacturer's installation instructions.

R905.10.3 Material standards. Metal-sheet roof covering systems that incorporate supporting structural members shall be designed in accordance with the *International Building Code*. Metal-sheet roof coverings installed over structural decking shall comply with Table R905.10.3(1). The materials used for metal-sheet roof coverings shall be naturally corrosion resistant or provided with corrosion resistance in accordance with the standards and minimum thicknesses shown in Table R905.10.3(2).

❖ If the metal roof panel system uses the structural members of the roof construction for support, it is necessary to design the roof system is accordance with the *International Building Code®* (IBC®). Only where the metal roof coverings are installed and supported by structural decking may the provisions of the code be used. Under such conditions, Tables R905.10.3(1) and R905.10.3(2) must be consulted.

Table R905.10.3(1) lists the metal roof covering types and their material standards. Many of the materials listed are inherently corrosion resistant, such as aluminum, copper and lead. For these materials, listing a material standard for the base material is sufficient. However, for steel roofing products, coatings are added to the base material to provide the necessary corrosion resistance.

Table R905.10.3(2) lists the minimum coating thickness required to provide minimum corrosion resistance for the base steel. Note a to Table R905.10.3(2) specifies that the paint systems are intended to be applied to steel having one of the corrosion-resistant coatings listed.

TABLE R905.10.3(1)
METAL ROOF COVERING STANDARDS

ROOF COVERING TYPE	STANDARD APPLICATION RATE/THICKNESS
Galvanized steel	ASTM A 653 G90 Zinc coated
Stainless steel	ASTM A 240, 300 Series alloys
Steel	ASTM A 924
Lead-coated copper	ASTM B 101
Cold-rolled copper	ASTM B 370 minimum 16 oz/sq ft and 12 oz/sq ft high-yield copper for metal-sheet roof-covering systems; 12 oz/sq ft for preformed metal shingle systems.
Hard lead	2 lb/sq ft
Soft lead	3 lb/sq ft
Aluminum	ASTM B 209, 0.024 minimum thickness for roll-formed panels and 0.019-inch minimum thickness for pressformed shingles.
Terne (tin) and terne-coated stainless	Terne coating of 40 lb per double base box, field painted where applicable in accordance with manufacturer's installation instructions.
Zinc	0.027 inch minimum thickness: 99.995% electrolytic high-grade zinc with alloy additives of copper (0.08 - 0.20%), titanium (0.07% - 0.12%) and aluminum (0.015%).

For SI: 1 ounce per square foot = 0.305 kg/m^2, 1 pound per square foot = 4.214 kg/m^2, 1 inch = 25.4 mm, 1 pound = 0.454 kg.

❖ See the commentary to Sections R905.10.3 and R905.4.4. CDA 4115-1929, *Copper in Architecture—Design Handbook*, is no longer listed as a reference for copper roof covering; however, the provisions of CDA 4115-1929 may be useful as a guideline.

TABLE R905.10.3(2)
MINIMUM CORROSION RESISTANCE

55% aluminum-zinc alloy coated steel	ASTM A 792 AZ 50
5% aluminum alloy-coated steel	ASTM A 875 GF60
Aluminum-coated steel	ASTM A 463 T2 65
Galvanized steel	ASTM A 653 G-90
Prepainted steel	ASTM A 755[a]

a. Paint systems in accordance with ASTM A 755 shall be applied over steel products with corrosion-resistant coatings complying with ASTM A 792, ASTM A 875, ASTM A 463, or ASTM A 653.

❖ See the commentary to Sections R905.10.3 and R905.4.4.

R905.10.4 Attachment. Metal roof panels shall be secured to the supports in accordance with this chapter and the manufacturer's installation instructions. In the absence of manufacturer's installation instructions, the following fasteners shall be used:

1. Galvanized fasteners shall be used for steel roofs.
2. Copper, brass, bronze, copper alloy and 300-series stainless steel fasteners shall be used for copper roofs.
3. Stainless steel fasteners are acceptable for metal roofs.

❖ The general provisions of this chapter apply to the installation of metal panel roofing, as do the installation instructions of the manufacturer of the specific roofing system. The code requires various types of fasteners based on specific types of metal roofs unless the manufacturer's installation instructions specify otherwise.

R905.10.5 Underlayment. Underlayment shall be installed in accordance with the manufacturer's installation instructions.

❖ The code requires the underlayment for metal roof panels to be installed in accordance with the manufacturer's installation instructions.

R905.10.5.1 Underlayment and high winds. Underlayment applied in areas subject to high winds [above 110 mph (49 m/s) in accordance with Figure R301.2(4)A] shall be applied with corrosion-resistant fasteners in accordance with manufacturer's installation instructions. Fasteners are to be applied along the overlap not farther apart than 36 inches (914 mm) on center.

Underlayment installed where the basic wind speed equals or exceeds 120 mph (54 m/s) shall comply with ASTM D 226 Type II. The underlayment shall be attached in a grid pattern of 12 inches (305 mm) between side laps with a 6-inch (152 mm) spacing at the side laps. Underlayment shall be applied in accordance with Section R905.2.7 except all laps shall be a minimum of 4 inches (102 mm).Underlayment shall be attached using metal or plastic cap nails with a head diameter of not less than 1 inch (25.4 mm) with a thickness of at least 32-gauge sheet metal. The cap-nail shank shall be a minimum of 12 gauge (0.105 inches) with a length to penetrate through the roof sheathing or a minimum of $^3/_4$ inch (19 mm) into the roof sheathing.

Exception: As an alternative, adhered underlayment complying with ASTM D 1970 shall be permitted.

❖ See the commentary to Section R905.2.7.2.

R905.11 Modified bitumen roofing. The installation of modified bitumen roofing shall comply with the provisions of this section.

❖ This section establishes the materials standards and installation criteria for modified bitumen roofing systems.

R905.11.1 Slope. Modified bitumen membrane roofs shall have a design slope of a minimum of one-fourth unit vertical in 12 units horizontal (2-percent slope) for drainage.

❖ Where a modified bitumen roofing system is to be installed, the roof design slope must be a minimum of 1:48, more often recognized as $^1/_4$ inch per foot.

R905.11.2 Material standards. Modified bitumen roof coverings shall comply with the standards in Table R905.11.2.

❖ Table R905.11.2 must be referenced for the installer to determine the appropriate specification standard for various materials that may be used in the application of modified bitumen roofing. The table identifies referenced standards for the membrane, primer, cement, adhesive and coatings that may be a part of the roof installation.

TABLE R905.11.2
MODIFIED BITUMEN ROOFING MATERIAL STANDARDS

MATERIAL	STANDARD
Acrylic coating	ASTM D 6083
Asphalt adhesive	ASTM D 3747
Asphalt cement	ASTM D 3019
Asphalt coating	ASTM D 1227; D 2824
Asphalt primer	ASTM D 41
Modified bitumen roof membrane	ASTM D 6162; D 6163; D 6164; D 6222; D 6223; D 6298; CGSB 37-GP-56M

❖ The materials used in modified bitumen roof coverings must comply with the following standards: ASTM D 6162, D 6163, D 6164, D 6222, D 6223, D 6298 and CGSB 37-56M, which cover modified bitumen roof membranes.

- ASTM D 6162 covers prefabricated modified bituminous sheet materials reinforced with a combination of polyester fabric and glass fiber, with or without granules, which use styrene-butadiene-styrene (SBS) thermoplastic elastomer as the primary modifier and are intended for use in the fabrication of multiple-ply roofing and waterproofing membranes.
  - ASTM D 6163 covers prefabricated modified bituminous sheet materials with glass fiber reinforcement, with or without granules, which use SBS thermoplastic elastomer as the primary modifier and are intended for use in the fabrication of multiple-ply roofing and waterproofing membranes.
  - ASTM D 6164 covers prefabricated modified bituminous sheet materials reinforced with polyester fabric, with or without granules, which use SBS thermoplastic elastomer as the primary modifier and are intended for use in the fabrication of multiple-ply roofing and waterproofing membranes.
  - ASTM D 6222 covers atactic polypropylene (APP) modified bituminous sheet materials using polyester reinforcement.
  - ASTM D 6223 covers APP modified bituminous sheet materials using a combination of polyester and glass fiber reinforcement.
  - ASTM D 6298 covers fiberglass reinforced modified bituminous sheet materials that use SBS thermoplastic elastomer as the primary modifier and are surfaced with a factory-applied continuous metal foil.
  - CGSB 37-56M covers prefabricated modified bituminous membrane reinforced for roofing applications.
- ASTM D 41 covers asphaltic primer suitable for use with asphalt in roofing, dampproofing and waterproofing below or above ground level, for application to concrete, masonry, metal and asphalt surfaces.
- ASTM D 3019 covers lap cement consisting of asphalt dissolved in a volatile petroleum solvent with or without mineral or other stabilizers, or both, for use with roll roofing.
- ASTM D 3747 covers emulsified asphalt adhesive for use in adhering preformed roof insulation to steel roof decks with inclines up to 33 percent. When applied as a continuous film over an acceptable deck surface, the emulsion functions as both an adhesive and a vapor retarder.
- ASTM D 1227 and D 2824 cover asphalt roof coatings.
  - ASTM D 1227 covers emulsified asphalt suitable for use as a protective coating for built-up roofs and other exposed surfaces with inclines of not less than 4 percent.
  - ASTM D 2824 covers aluminum-pigmented asphalt roof coatings, including nonfibered, asbestos-fibered, and asbestos-free fibered materials.
- ASTM D 6083 covers a liquid-applied water-dispersed 100-percent acrylic elastomeric latex coating used as a protective coating for roofs.

R905.11.3 Application. Modified bitumen roofs shall be installed according to this chapter and the manufacturer's installation instructions.

❖ The general provisions of this chapter apply to the installation of modified bitumen roofing, as do the installation instructions of the manufacturer of the specific roofing materials or system.

R905.12 Thermoset single-ply roofing. The installation of thermoset single-ply roofing shall comply with the provisions of this section.

❖ This section establishes the materials standards and installation criteria for thermoset single-ply roofing systems.

R905.12.1 Slope. Thermoset single-ply membrane roofs shall have a design slope of a minimum of one-fourth unit vertical in 12 units horizontal (2-percent slope) for drainage.

❖ If a thermoset single-ply roof system is to be installed, the roof design slope must be a minimum of 1:48, more often recognized as $^1/_4$-inch per foot.

R905.12.2 Material standards. Thermoset single-ply roof coverings shall comply with ASTM D 4637, ASTM D 5019 or CGSB 37-GP-52M.

❖ The code recognizes three materials standards for the regulation of thermoset single-ply membrane roofs. The standards include ASTM D 4637 for ethylene-propylene-diene-terpolymer (EPDM) sheets, ASTM D 5019 for reinforced nonvulcanized polymeric sheet made from chlorosulfonated polyethylene (CSPE) and polyisobutylene (PIB) and CGSB 37-GP-52M for sheet-applied elastomeric roofing and waterproofing membrane.

R905.12.3 Application. Thermoset single-ply roofs shall be installed according to this chapter and the manufacturer's installation instructions.

❖ The general provisions of this chapter apply to the installation of thermoset single-ply roofing, as do the installation instructions of the manufacturer of the specific roofing system.

R905.13 Thermoplastic single-ply roofing. The installation of thermoplastic single-ply roofing shall comply with the provisions of this section.

❖ This section sets forth the materials standards and installation criteria for single-ply roof systems consisting of polyvinyl chloride material.

R905.13.1 Slope. Thermoplastic single-ply membrane roofs shall have a design slope of a minimum of one-fourth unit vertical in 12 units horizontal (2-percent slope).

❖ If a thermoplastic single-ply roof system is to be installed, the roof design slope must be a minimum of 1:48, more often recognized as $^1/_4$-inch per foot (21 mm/m).

R905.13.2 Material standards. Thermoplastic single-ply roof coverings shall comply with ASTM D 4434, ASTM D 6754, ASTM D 6878 or CGSB CAN/CGSB 37.54.

❖ Four materials standards, ASTM D 4434, ASTM D 6754, ASTM D 6878 and CGSB CAN/CGSB 37.54, are recognized for the regulation of thermoplastic single-ply membrane roofs and are acceptable standards in determining the appropriateness of the roofing system.

R905.13.3 Application. Thermoplastic single-ply roofs shall be installed according to this chapter and the manufacturer's installation instructions.

❖ The general provisions of this chapter apply to the installation of thermoplastic single-ply roofing, as do the installation instructions of the manufacturer of the roofing system.

R905.14 Sprayed polyurethane foam roofing. The installation of sprayed polyurethane foam roofing shall comply with the provisions of this section.

❖ This section establishes the materials standards and installation criteria for sprayed polyurethane foam roofing systems.

R905.14.1 Slope. Sprayed polyurethane foam roofs shall have a design slope of a minimum of one-fourth unit vertical in 12 units horizontal (2-percent slope) for drainage.

❖ Where a sprayed polyurethane foam roof system is to be installed, the roof design slope must be a minimum of 1:48, more often recognized as $^1/_4$-inch per foot (21 mm/m).

R905.14.2 Material standards. Spray-applied polyurethane foam insulation shall comply with ASTM C 1029, Type III or IV.

❖ ASTM C 1029, Type III or IV, is the standard for the regulation of spray-applied rigid cellular polyurethane thermal insulation.

R905.14.3 Application. Foamed-in-place roof insulation shall be installed in accordance with this chapter and the manufacturer's installation instructions. A liquid-applied protective coating that complies with Table R905.14.3 shall be applied no less than 2 hours nor more than 72 hours following the application of the foam.

❖ The general provisions of this chapter address the installation of spray-applied rigid cellular polyurethane thermal membrane roofing, as do the installation instructions of the manufacturer of the roofing system. In addition, a liquid-applied protective coating must be provided to protect the exterior surface of the foam. The protective coating specified in Table R905.14.3 must be applied between 2 and 72 hours following the installation of the foam insulation.

TABLE R905.14.3
PROTECTIVE COATING MATERIAL STANDARDS

MATERIAL	STANDARD
Acrylic coating	ASTM D 6083
Silicone coating	ASTM D 6694
Moisture-cured polyurethane coating	ASTM D 6947

❖ See the commentary to Section 905.14.3.

R905.14.4 Foam plastics. Foam plastic materials and installation shall comply with Section R316.

❖ The foam-plastics provisions of Section R316, particularly those of Section R316.5.2 specific to roof covering assemblies, are applicable to roof assemblies containing sprayed polyurethane foam. The use of wood structural-panel roof sheathing is permitted in lieu of a thermal barrier for separating the interior of the building from the sprayed polyurethane foam-plastic roofing material. Sections R316.5.2 and R803 contain specific requirements for wood structural-panel sheathing.

R905.15 Liquid-applied roofing. The installation of liquid-applied roofing shall comply with the provisions of this section.

❖ This section establishes the material standards and installation criteria for liquid-applied roofing.

R905.15.1 Slope. Liquid-applied roofing shall have a design slope of a minimum of one-fourth unit vertical in 12 units horizontal (2-percent slope).

❖ If a liquid-applied roofing is to be installed, the roof design slope must be a minimum of 1:48, more often recognized as $^1/_4$-inch per foot (21 mm/m).

R905.15.2 Material standards. Liquid-applied roofing shall comply with ASTM C 836, C 957, D 1227, D 3468, D 6083, D 6694 or D 6947.

❖ The code recognizes several materials standards for the regulation of liquid-applied roofing. They include ASTM C 836 for cold-liquid applied elastomeric waterproofing membranes that require the installation of separate wearing surfaces, ASTM C 957 for cold-liquid applied elastomeric waterproofing membranes that provide integral wearing surfaces, ASTM D 1227 for emulsified asphalt used as a protective coating for built-up roofing systems, ASTM D 3468 for liquid-applied neoprene and chlorosulfanated polyethylene materials used in roofing and waterproofing, ASTM D 6083 for liquid-applied acrylic roof coatings, ASTM D 6694 for silicone-based elastomeric coating for spray polyethylene foam insulation and ASTM D 6947 for liquid-applied moisture-cured polyurethane.

R905.15.3 Application. Liquid-applied roofing shall be installed according to this chapter and the manufacturer's installation instructions.

❖ The general provisions of this chapter apply to the installation of liquid-applied roofing, as do the installation instructions of the manufacturer of the roofing system.

R905.16 Photovoltaic modules/shingles. The installation of photovoltaic modules/shingles shall comply with the provisions of this section.

❖ The ever increasing number of installations of photovoltaic panels and modules raises concerns about the safety of these installations. This section requires these products to be listed and installed in accordance with the manufacturer's instructions. UL 1703 is the standard used to investigate photovoltaic modules and panels, and includes construction and performance requirements to address potential safety hazards. Over 60 companies currently have UL 1703 listings for photovoltaic modules and panels.

The section provides guidance for installers and code officials regarding the installation of photovoltaic modules/shingles. These shingles are integrated with the building and provide both a roof covering and source of electrical power. The appropriate design slope and fastening of the shingles are different for each manufacturer's product. For wind resistance, the procedures used in ASTM D 3161 for asphalt shingles are appropriate to use when adapted for these types of shingles.

R905.16.1 Material standards. Photovoltaic modules/shingles shall be listed and labeled in accordance with UL 1703.

❖ See the commentary to Section R905.16.

R905.16.2 Attachment. Photovoltaic modules/shingles shall be attached in accordance with the manufacturer's installation instructions.

❖ See the commentary to Section R905.16.

R905.16.3 Wind resistance. Photovoltaic modules/shingles shall be tested in accordance with procedures and acceptance criteria in ASTM D 3161. Photovoltaic modules/shingles shall comply with the classification requirements of Table R905.2.4.1(2) for the appropriate maximum basic wind speed. Photovoltaic modules/shingle packaging shall bear a label to indicate compliance with the procedures in ASTM D 3161 and the required classification from Table R905.2.4.1(2).

❖ See the commentary to Section R905.16.

SECTION R906
ROOF INSULATION

R906.1 General. The use of above-deck thermal insulation shall be permitted provided such insulation is covered with an *approved* roof covering and complies with FM 4450 or UL 1256.

❖ During these days of energy consciousness, roof insulation has become more and more prevalent. It has distinct benefits not only in energy conservation but also in building-occupant comfort. Insulation also provides a smooth, uniform substrate for application of the roofing materials. The code requires that above-deck thermal insulation be covered by an approved roof covering and be in compliance with FM 4450 or UL 1256. Although UL 1256 specifies the testing methods for determining the fire resistance of roof covering materials, FM 4450 deals with Class I insulated steel deck roofs only.

R906.2 Material standards. Above-deck thermal insulation board shall comply with the standards in Table R906.2.

❖ The referenced material standards provide additional guidance on the physical properties of roof insulation when used as above-deck components of roof assemblies.

TABLE R906.2
MATERIAL STANDARDS FOR ROOF INSULATION

Cellular glass board	ASTM C 552
Composite boards	ASTM C 1289, Type III, IV, V or VI
Expanded polystyrene	ASTM C 578
Extruded polystyrene board	ASTM C 578
Perlite board	ASTM C 728
Polyisocyanurate board	ASTM C 1289, Type I or II
Wood fiberboard	ASTM C 208

❖ This table incorporates industry-recognized material standards into the code for materials commonly used in above roof deck insulation practices.

SECTION R907
REROOFING

R907.1 General. Materials and methods of application used for recovering or replacing an existing roof covering shall comply with the requirements of Chapter 9.

Exception: Reroofing shall not be required to meet the minimum design slope requirement of one-quarter unit vertical in 12 units horizontal (2-percent slope) in Section R905 for roofs that provide positive roof drainage.

❖ This section addresses the concerns associated with unregulated reroofing operations. The provisions require that when an existing building is reroofed, the existing roof be structurally sound and in a proper condition to receive the new roofing. It is often necessary to remove the existing roof covering prior to installing the new roofing materials. This section identifies those situations where the reroofing cannot occur over an existing roof covering.

If new roof covering materials are to be installed as a replacement for an existing roof covering, or if the new roof covering is applied directly over an existing roof, the roof covering materials and the application process must conform to the provisions for a new roof covering installation.

There may be occasions where only a small portion of the existing roof covering is in need of replacement or a limited amount of the roof is in need of repair. In those situations, it is acceptable to use the provisions for existing buildings in Chapter 34 of the IBC, particularly Section 3401.4. This section of the IBC permits the alteration or repair of nonstructural elements using the same materials that are already in place. In other words, the repair or replacement of existing roof covering may be made using the same materials and methods found on the existing roof. It is assumed that the repairs will not cause the building to be reduced in code compliance from the previous level.

R907.2 Structural and construction loads. The structural roof components shall be capable of supporting the roof covering system and the material and equipment loads that will be encountered during installation of the roof covering system.

❖ A fundamental requirement throughout the code is the recognition of the impact of any new repair or replacement work on the structural system. If a new roof covering system is installed, the structural members of the roof and any additional structural members that carry roof loads must be reviewed for their ability to support the loads that will be imposed during the installation process. This includes the weight of the new roof covering materials as well as any installation equipment that is placed on the roof.

R907.3 Recovering versus replacement. New roof coverings shall not be installed without first removing all existing layers of roof coverings where any of the following conditions exist:

1. Where the existing roof or roof covering is water-soaked or has deteriorated to the point that the existing roof or roof covering is not adequate as a base for additional roofing.
2. Where the existing roof covering is wood shake, slate, clay, cement or asbestos-cement tile.
3. Where the existing roof has two or more applications of any type of roof covering.

Exceptions:

1. Complete and separate roofing systems, such as standing-seam metal roof systems, that are designed to transmit the roof loads directly to the building's structural system and that do not rely on existing roofs and roof coverings for support, shall not require the removal of existing roof coverings.
2. Installation of metal panel, metal shingle and concrete and clay tile roof coverings over existing wood shake roofs shall be permitted when the application is in accordance with Section R907.4.
3. The application of new protective coating over existing spray polyurethane foam roofing systems shall be permitted without tear-off of existing roof coverings.
4. Where the existing roof assembly includes an ice barrier membrane that is adhered to the roof deck, the existing ice barrier membrane shall be permitted to remain in place and covered with an additional layer of ice barrier membrane in accordance with Section R905.

❖ The base for application of new roofing materials must provide a sound and consistent surface on which to install the new materials. The code will not permit the installation of new roof coverings over existing roof coverings where there is a potential for future problems with the roof's effectiveness. Therefore, the code lists four specific conditions where all existing layers of the roof covering materials must be removed prior to the installation of new roofing.

If the existing roof or roof covering is water soaked, the concealment of the existing roof in such a condition will lead to problems of deterioration and failure in the future. Any situation where the existing roof construction does not provide for an acceptable base is reason to remove all existing roof covering materials prior to reroofing. In addition, roofing materials such as wood shakes, clay tiles and similar materials are not generally acceptable as a base for a new roof covering because of the variation in their surfaces. Exception 2 permits the installation of a new roof covering over wood shakes or shingle roofs only if the surface below the new roofing is properly protected in accordance with Section R907.4.

A maximum of two roof coverings is permitted by the code, based primarily on the dead load weight that is accumulated on the roof for every layer of roofing materials. Where two or more applications of any roof covering are present, the existing materials must be removed prior to installation of the new roof covering. Asphalt shingles present a special concern in those areas likely to have hail. Where moderate or severe hail damage can be expected, based on the map shown in Figure R903.5, it is always necessary to remove an existing asphalt shingle roof covering prior to the application of any new roof covering (see commentary, Section R903.5).

Exception 1 states that new roofing systems that are designed to transmit all roof loads directly to the structural supports of the building do not necessitate that the existing roofing system be removed.

Exception 3 permits recoating of an existing spray polyurethane foam roofing system without removal of the spray polyurethane foam roof covering. Recoating does not add significant weight to the roof assembly or compromise the long-term performance of the roofing assembly. Industry practices for the recoat of an existing spray polyurethane foam roofing system are detailed in ASTM D 6705, *Standard Guide for the Repair and Recoat of Spray Polyurethane Foam Roofing Systems*.

Exception 4 permits an adhered ice barrier membrane to remain in place. In roof removal situations where an existing ice barrier membrane is adhered to the existing roof deck, it is oftentimes difficult, if not impossible, to remove the existing layer of the adhered ice barrier membrane without damaging or replacing the roof deck. This exception is intended to account for this situation by allowing the existing adhered ice barrier membrane to remain in place and be covered with a new ice barrier membrane as required in Section R905, followed by the installation of the new primary roof covering material.

R907.4 Roof recovering. Where the application of a new roof covering over wood shingle or shake roofs creates a combustible concealed space, the entire existing surface shall be covered with gypsum board, mineral fiber, glass fiber or other *approved* materials securely fastened in place.

❖ A new roof covering can be applied over the top of an existing roof of wood shakes or shingles only if the existing roof surface of shakes or shingles is protected to address the concern of concealed combustible spaces. The application of a new roof over wood shakes or shingles creates an extensive amount of concealed area, all with a high degree of combustible materials. By using gypsum board, mineral fiber, glass fiber or other similar materials in the reroofing installation, the combustible shakes or shingles will be protected by materials suitable for fireblocking.

R907.5 Reinstallation of materials. Existing slate, clay or cement tile shall be permitted for reinstallation, except that damaged, cracked or broken slate or tile shall not be reinstalled. Any existing flashings, edgings, outlets, vents or similar devices that are a part of the assembly shall be replaced when rusted, damaged or deteriorated. Aggregate surfacing materials shall not be reinstalled.

❖ Because all roofing materials should make a building weathertight, their reuse is strictly limited. Unless damaged, existing tile of cement, slate or clay may be reused. The reuse of all other roofing materials is prohibited. Roof accessories, such as metal edgings, flashing, drain outlets and collars, may be reinstalled only if they are in a suitable condition.

R907.6 Flashings. Flashings shall be reconstructed in accordance with *approved* manufacturer's installation instructions. Metal flashing to which bituminous materials are to be adhered shall be primed prior to installation.

❖ During reroofing operations, all flashings that are to remain must be reconstructed in a manner consistent with the manufacturer's instructions. In those cases where bituminous materials are applied to any existing metal flashing, a primer must be applied to the flashing to increase adhesion to the surface.

Bibliography

The following resource materials were used in the preparation of the commentary for this chapter of the code:

ASTM C 836-06, *Specification for High Solids Content, Cold Liquid-applied Elastomeric Waterproofing Membrane for Use with Separate Wearing Course*. West Conshohocken, PA: ASTM International, 2006.

ASTM C 957-06, *Specification for High Solids Content, Cold Liquid-applied Elastomeric Waterproofing Membrane for Use with Integral Wearing Surface*. West Conshohocken, PA: ASTM International, 2006.

ASTM C 1167-03, *Specification for Clay Roof Tiles*. West Conshohocken, PA: ASTM International, 2003.

ASTM C 1492-03, *Specification for Concrete Roof Tile*. West Conshohocken, PA: ASTM International, 2003.

ASTM D 41-05, *Specification for Asphalt Primer Used in Roofing, Dampproofing, and Waterproofing*. West Conshohocken, PA: ASTM International, 2005.

ASTM D 43-00 (2006), *Specification for Coal Tar Primer Used in Roofing, Dampproofing and Waterproofing*. West Conshohocken, PA: ASTM International, 2006.

ASTM D 225-07, *Specification for Asphalt Shingles (Organic Felt) Surfaced with Mineral Granules*. West Conshohocken, PA: ASTM International, 2007.

ASTM D 226-06, *Specification for Asphalt-saturated (Organic Felt) Used in Roofing and Waterproofing*. West Conshohocken, PA: ASTM International, 2006.

ASTM D 227-03, *Specification for Coal-tar Saturated (Organic Felt) Used in Roofing and Waterproofing*. West Conshohocken, PA: ASTM International, 2003.

ASTM D 312-00(2006), *Specification for Asphalt Used in Roofing*. West Conshohocken, PA: ASTM International, 2006.

ASTM D 449-08, *Specification for Asphalt Used in Dampproofing and Waterproofing*. West Conshohocken, PA: ASTM International, 2008.

ASTM D 450-07, *Specification for Coal-tar Pitch Used in Roofing, Dampproofing and Waterproofing*. West Conshohocken, PA: ASTM International, 2007.

ASTM D 1227-95(2007), *Specification for Emulsified Asphalt Used as a Protective Coating for Roofing*. West Conshohocken, PA: ASTM International, 2007.

ASTM D 1863-05, *Specification for Mineral Aggregate Used in Built-up Roofs*. West Conshohocken, PA: ASTM International, 2005.

ASTM D 1970-09, *Specification for Self-adhering Polymer Modified Bitumen Sheet Materials Used as Steep Roofing Underlayment for Ice-dam Protection*. West Conshohocken, PA: ASTM International, 2009.

ASTM D 2178-2004, *Specification for Asphalt Glass Felt Used in Roofing and Waterproofing*. West Conshohocken, PA: ASTM International, 2004.

ASTM D 2626-04, *Specification for Asphalt-saturated and Coated Organic Felt Base Sheet Used in Roofing*. West Conshohocken, PA: ASTM International, 2004.

ASTM D 2822-05, *Specification for Asphalt Roof Cement, Asbestos Containing*. West Conshohocken, PA: ASTM International, 2005.

ASTM D 2823-05, *Specification for Asphalt Roof Coatings, Asbestos Containing*. West Conshohocken, PA: ASTM International, 2005.

ASTM D 2824-06, *Specification for Aluminum-pigmented Asphalt Roof Coatings, Nonfibered Asbestos Fibered, and Fibered without Asbestos*. West Conshohocken, PA: ASTM International, 2006.

ASTM D 3019-08, *Specification for Lap Cement Used with Asphalt Roll Roofing, Nonfibered, Asbestos Fibered, and Nonasbestos Fibered*. West Conshohocken, PA: ASTM International, 2008.

ASTM D 3161-09, *Test Method for Wind Resistance of Asphalt Shingles (Fan Induced Method)*. West Conshohocken, PA: ASTM International, 2009.

ASTM D 3462-09, *Specification for Asphalt Shingles Made From Glass Felt and Surfaced with Mineral Granules*. West Conshohocken, PA: ASTM International, 2009.

ASTM D 3468-99(2006)e01, *Specification for Liquid-applied Neoprene and Chlorosulfanated Polyethylene Used in Roofing and Waterproofings*. West Conshohocken, PA: ASTM International, 2006.

ASTM D 3747-08, *Specification for Emulsified Asphalt Adhesive for Adhering Roof Insulation*. West Conshohocken, PA: ASTM International, 2008.

ASTM D 3909-97b(2004)e01, *Specification for Asphalt Roll Roofing (Glass Felt) Surfaced with Mineral Granules*. West Conshohocken, PA: ASTM International, 2004.

ASTM D 4022-07, *Specification for Coal Tar Roof Cement, Asbestos Containing*. West Conshohocken, PA: ASTM International, 2007.

ASTM D 4434/4434M-09, *Specification for Poly (Vinyl Chloride) Sheet Roofing*. West Conshohocken, PA: ASTM International, 2009.

ASTM D 4479-07, *Specification for Asphalt Roof Coatings, Asbestos-free*. West Conshohocken, PA: ASTM International, 2007.

ASTM D 4586-07, *Specification for Asphalt Roof Cement, Asbestos-free*. West Conshohocken, PA: ASTM International, 2007.

ASTM D 4601-04, *Specification for Asphalt-Coated Glass Fiber Base Sheet Used in Roofing*. West Conshohocken, PA: ASTM International, 2004.

ASTM D 4637-08, *Specification for EPDM Sheet Used in Single-ply Roof Membrane*. West Conshohocken, PA: ASTM International, 2008.

ASTM D 4869-05e01, *Specification for Asphalt-Saturated (Organic Felt) Underlayment Used in Steep Slope Roofing*. West Conshohocken, PA: ASTM International, 2005.

ASTM D 4897-01, *Specification for Asphalt Coated Glass-fiber Venting Base Sheet Used in Roofing*. West Conshohocken, PA: ASTM International, 2001.

ASTM D 4990-97a(2005)e01, *Specification for Coal Tar Glass Felt Used in Roofing and Waterproofing*. West Conshohocken, PA: ASTM International, 2005.

ASTM D 5019-07a, *Specification for Reinforced Nonvulcanized Polymeric Sheet Used in Roofing Membrane*. West Conshohocken, PA: ASTM International, 2007.

ASTM D 5643-06, *Specification for Coal Tar Roof Cement Asbestos-free*. West Conshohocken, PA: ASTM International, 2006.

ASTM D 5665-99a(2006), *Specification for Thermoplastic Fabrics Used in Cold-applied Roofing and Waterproofing*. West Conshohocken, PA: ASTM International, 2006.

ASTM D 5726-98(2005), *Specification for Thermoplastic Fabrics Used in Hot-applied Roofing and Waterproofing*. West Conshohocken, PA: ASTM International, 2005.

ASTM D 6083-05e01, *Specification for Liquid Applied Acrylic Coating Used in Roofing*. West Conshohocken, PA: ASTM International, 2005.

ASTM D 6162-00a(2008), *Specification for Styrene Butadiene Styrene (SBS) Modified Bituminous Sheet Materials Using a Combination of Polyester and Glass Fiber Reinforcements*. West Conshohocken, PA: ASTM International, 2008.

ASTM D 6163-00(2008), *Specification for Styrene Butadiene Styrene (SBS) Modified Bituminous Sheet Materials Using Glass Fiber Reinforcements*. West Conshohocken, PA: ASTM International, 2008.

ASTM D 6164-05e1, *Specification for Styrene Butadiene Styrene (SBS) Modified Bituminous Sheet Materials Using Polyester Reinforcements*. West Conshohocken, PA: ASTM International, 2005.

ASTM D 6222-08, *Specification for Atactic Polypropelene (APP) Modified Bituminous Sheet Materials Using Polyester Reinforcement*. West Conshohocken, PA: ASTM International, 2008.

ASTM D 6223-02e02, *Specification for Atactic Polypropelene (APP) Modified Bituminous Sheet Materials Using a Combination of Polyester and Glass Fiber Reinforcement*. West Conshohocken, PA: ASTM International, 2002.

ASTM D 6298-05e1, *Specification for Fiberglass Reinforced Styrene-butadiene-styrene (SBS) Modified Bituminous Sheet with a Factory Applied Metal Surface*. West Conshohocken, PA: ASTM International, 2005.

ASTM D 6380-03(2009), *Standard Specification for Asphalt Roll Roofing (Organic Felt)*. West Conshohocken, PA: ASTM International, 2009.

ASTM D 6694-08, *Standard Specification Liquid-applied Silicone Coating Used in Spray Polurethane Foam Roofing*. West Conshohocken, PA: ASTM International, 2008.

ASTM D 6757-07, *Standard Specification for Inorganic Underlayment for Use with Steep Slope Roofing Products*. West Conshohocken, PA: ASTM International, 2007.

ASTM D 6878-08e1, *Standard Specification for Thermoplastic Polyolefin Based Sheet Roofing*. West Conshohocken, PA: ASTM International, 2008.

ASTM D 6947-07, *Standard Specification for Liquid Applied Moisture Cured Polyurethane Coating Used in Spray Polyurethane Foam Roofing System*. West Conshohocken, PA: ASTM International, 2007.

ASTM D 7158-08d, *Standard Test Method for Wind Resistance of Sealed Asphalt Shingles (Uplift Force/Uplift Resistance Method)*. West Conshohocken, PA: ASTM International, 2008.

ASTM E 108-07a, *Test Method for Fire Tests of Roof Coverings*. West Conshohocken, PA: ASTM International, 2007.

ASTM F 1667-05, *Specification for Driven Fasteners, Nails, Spikes, and Staples*. West Conshohocken, PA: ASTM International, 2005.

AWPA C1-03, *All Timber Products—Preservative Treatment by Pressure Processes*. Birmingham, AL: American Wood–Preservers' Association, 2003.

CDA 4115-1929, *Copper in Architecture—Design Handbook*. United Kingdom: The Copper Development Association, 1929.

CGSB 37-GP-56M-80, *Membrane, Modified Bituminous, Prefabricated and Reinforced for Roofing—with December 1985 Amendment*. Gatineau, Quebec: Canadian General Standards Board, 1980.

CGSB CAN/CGSB 37.54-95, *Polyvinyl Chloride Roofing and Waterproofing Membrane*. Gatineau, Quebec: Canadian General Standards Board, 1995.

CSSB-97, *Grading and Packing Rules for Western Red Cedar Shakes and Western Red Shingles of the Cedar Shake and Shingle Bureau*. Sumas, WA: Cedar Shake & Shingle Bureau, 1997.

FM 4450-(1989), *Approved Standard for Class I Insulated Steel Deck Roofs—Supplements through July 1992*. Johnson, RI: Factory Mutual, 1989.

IBC-12, *International Building Code*. Washington, DC: International Code Council, 2011.

IPC-12, *International Plumbing Code*. Washington, DC: International Code Council, 2011.

UL 55A-04, *Materials for Built-Up Roof Coverings*. Northbrook, IL: Underwriters Laboratories Inc., 2004.

UL 790-04, *Standard Test Methods for Fire Tests of Roof Coverings with Revisions through October 2008*. Northbrook, IL: Underwriters Laboratories Inc., 2004.

UL 1256-02, *Fire Test of Roof Deck Construction*. Northbrook, IL: Underwriters Laboratories Inc., 2002.

UL 1703-02, *Flat-Plate Photovoltaic Modules and Panels with Revisions through April 2008*. Northbrook, IL: Underwriters Laboratories Inc., 2002.

Chapter 10: Chimneys and Fireplaces

General Comments

Chapter 10 regulates two basic types of chimneys and fireplaces: factory-built and those constructed on site of masonry and other approved materials. Chimneys and fireplaces constructed of masonry rely on prescriptive requirements for the details of their construction; the factory-built type rely on the listing and labeling method of approval. This chapter also contains provisions for unvented gas log heaters. Seismic issues related to fireplaces and chimneys, including reinforcing requirements, are also addressed in Chapter 10.

Section R1001 regulates the construction of masonry fireplaces. Section R1002 establishes the standards for the use and installation of masonry heaters. Section R1 003 regulates the construction of masonry chimneys.

Section R1004 establishes the standards for the use and installation of factory-built fireplaces including the use of unvented gas log heaters. Section R1005 establishes the standards for the use and installation of factory-built chimneys. Section R1006 requires the installation of an exterior air supply for use with both factory-built and masonry fireplaces.

Purpose

Chapter 10 contains requirements for the safe construction of masonry chimneys and fireplaces and establishes the standards for the use and installation of factory-built chimneys, fireplaces and masonry heaters. Chimneys and fireplaces constructed of masonry rely on prescriptive requirements for the details of their construction; the factory-built type relies on the listing and labeling method of approval. This chapter also provides the requirements for seismic reinforcing and anchorage of masonry fireplaces and chimneys.

SECTION R1001 MASONRY FIREPLACES

R1001.1 General. Masonry fireplaces shall be constructed in accordance with this section and the applicable provisions of Chapters 3 and 4.

❖ This section covers details for masonry fireplaces. Table R1001.1 and Figure R1001.1 show details for masonry fireplaces and chimneys.

R1001.2 Footings and foundations. Footings for masonry fireplaces and their chimneys shall be constructed of concrete or *solid masonry* at least 12 inches (305 mm) thick and shall extend at least 6 inches (152 mm) beyond the face of the fireplace or foundation wall on all sides. Footings shall be founded on natural, undisturbed earth or engineered fill below frost depth. In areas not subjected to freezing, footings shall be at least 12 inches (305 mm) below finished *grade*.

❖ The dead-load bearing pressure for a building structure, particularly if it is a light wood-frame building, is usually quite low, even though the bearing pressure on the foundation for a masonry or concrete fireplace and chimney can be several times higher. It is good practice to proportion the foundation of a masonry fireplace and chimney to have approximately the same bearing pressure as is present under the building structure itself, although the codes do not require such a design. Where the soil is compressible, differential settlements between the fireplace and chimney and its surrounding structure can cause cracking of the finish materials in the vicinity of the fireplace and chimney. Also, firestopping may be displaced so that a draft opening is created.

R1001.2.1 Ash dump cleanout. Cleanout openings located within foundation walls below fireboxes, when provided, shall be equipped with ferrous metal or masonry doors and frames constructed to remain tightly closed except when in use. Cleanouts shall be accessible and located so that ash removal will not create a hazard to combustible materials.

❖ Noncombustible, tightly sealed cleanout doors must reduce the danger of fire spread through the cleanout openings. Cleanout openings must be easily accessible to allow ash removal. The ashes and other materials left in the fireplace may remain hot for some time. Cleanout doors must be tight closing to prevent air from entering the firebox and causing the unburned materials to reignite. The cleanout door must be located so that ashes and other unburned materials that may still be hot can be removed without creating a hazard to combustible materials.

R1001.3 Seismic reinforcing. Masonry or concrete chimneys in Seismic Design Category D_0, D_1 or D_2 shall be reinforced. Reinforcing shall conform to the requirements set forth in Table R1001.1 and Section R609, Grouted Masonry.

❖ Masonry and concrete have inherently brittle natures, and they are therefore weak when subjected to tensile forces. Using steel, an inherently ductile material, provides ductility in masonry and concrete construction. Chimneys in general and unreinforced masonry and concrete chimneys in particular do not perform well in earthquakes because of their tall and slender geometry. Observation of buildings after earthquakes has shown that chimneys suffered severe damage and sometimes completely collapsed. To improve the performance of masonry and concrete chimneys under severe earthquake loading, steel-reinforcing bars are required as shown in Table R1001.1. This reinforcement is required in Seismic Design Categories D_0, D_1 and D_2 only. Reinforcement is not required in Seismic Design Categories A, B, and C, and buildings in seismic design categories higher than D must be designed in accordance with the *International Building Code*® (IBC®) (see Section R301.2.2).

R1001.3.1 Vertical reinforcing. For chimneys up to 40 inches (1016 mm) wide, four No. 4 continuous vertical bars shall be placed between wythes of *solid masonry* or within the cells of hollow unit masonry and grouted in accordance with Section R609. Grout shall be prevented from bonding with the flue liner so that the flue liner is free to move with thermal expansion. For chimneys more than 40 inches (1016 mm) wide, two additional No. 4 vertical bars shall be provided for each additional flue incorporated into the chimney or for each additional 40 inches (1016 mm) in width or fraction thereof.

❖ To keep the chimney structure together and prevent severe damage or fracture, four No. 4 bars, continuous from bottom to top, are required. The bars are typically placed at the four corners, similar to the placement of vertical bars in concrete columns. The hollow cells in hollow unit masonry are used for placement of vertical bars, and if solid masonry is used, the reinforcing steel must be placed between the wythes. For wider chimneys or chimneys with more than one flue, four No. 4 bars are not adequate, and reinforcing bars in increments of two are added for each additional flue or each additional 40 inches (1016 mm) of chimney width.

R1001.3.2 Horizontal reinforcing. Vertical reinforcement shall be placed within $^1/_4$-inch (6 mm) ties, or other reinforcing of equivalent net cross-sectional area, placed in the bed joints according to Section R607 at a minimum of every 18 inches (457 mm) of vertical height. Two such ties shall be installed at each bend in the vertical bars.

❖ Vertical reinforcement under high seismic loading tends to buckle, then burst out of the concrete or masonry. To prevent this action, the vertical reinforcing required under Section R1001.3.1 must be placed in $^1/_4$-inch (6.4 mm) ties that are placed in the bed joints in accordance with Section R607. The ties are required at no more than 18-inch (457 mm) intervals vertically to form an effective cage for the vertical reinforcement. Where the vertical reinforcing bars have bends and abrupt changes in direction as a result of changes in sections of the chimney, two horizontal ties are required at each bend to handle the additional effects of stress concentration.

R1001.4 Seismic anchorage. Masonry or concrete chimneys in Seismic Design Category D_0, D_1 or D_2 shall be anchored at each floor, ceiling or roof line more than 6 feet (1829 mm) above *grade*, except where constructed completely within the exterior walls. Anchorage shall conform to the requirements of Section R1001.4.1.

❖ The sway of a chimney under earthquake lateral movement may cause it to impact the roof and floor framing and ultimately pull away from the framing. To control or prevent this action in Seismic Design Categories D_0, D_1 and D_2, masonry and concrete chimneys must be anchored at each floor, ceiling or roof line to provide lateral bracing. Such anchorage is not required in three situations: (1) in Seismic Design Categories A, B, and C; (2) at floor, ceiling or roof lines 6 feet (1829 mm) or less above grade and (3) where the chimney is constructed completely within the exterior walls and is therefore braced by the building elements surrounding it.

In Seismic Design Categories E and F, the reinforcing and anchorage must be designed in accordance with the IBC.

R1001.4.1 Anchorage. Two $^3/_{16}$-inch by 1-inch (5 mm by 25 mm) straps shall be embedded a minimum of 12 inches (305 mm) into the chimney. Straps shall be hooked around the outer bars and extend 6 inches (152 mm) beyond the bend. Each strap shall be fastened to a minimum of four floor ceiling or floor joists or rafters with two $^1/_2$-inch (13 mm) bolts.

❖ The anchors required under Section R1001.4 consist of a minimum of two straps, each $^3/_{16}$ inch by 1 inch (4.8 mm by 25 mm). The straps must be firmly connected to the building framing and chimney to control or prevent the chimney from pulling away or buckling. To accomplish this firm connection, the straps must be bolted to at least four joists or rafters using $^1/_2$-inch (12.7 mm) bolts. At the other end they must be embedded at least 12 inches (305 mm) into the chimney and hooked around the outer vertical reinforcing bars. Short hooks might pull out in earthquakes, and for this reason the minimum length of the hook around the vertical bar is 6 inches (152 mm).

R1001.5 Firebox walls. Masonry fireboxes shall be constructed of *solid masonry* units, hollow masonry units grouted solid, stone or concrete. When a lining of firebrick at least 2 inches (51 mm) thick or other *approved* lining is provided, the minimum thickness of back and side walls shall each be 8 inches (203 mm) of *solid masonry*, including the lining. The width of joints between firebricks shall not be greater than $^1/_4$ inch (6 mm). When no lining is provided, the total minimum thickness of back and side walls shall be 10 inches (254 mm) of *solid masonry*. Firebrick shall conform to ASTM C 27 or C 1261 and shall be laid with medium duty refractory mortar conforming to ASTM C 199.

❖ This section specifies the minimum thicknesses of refractory brick or solid masonry necessary to contain the generated heat.

Solid masonry walls forming the firebox must have a minimum total thickness of 8 inches (204 mm), including the refractory lining.

The refractory lining is to consist of a low-duty, fireclay refractory brick with a minimum thickness of 2 inches (51 mm), laid with medium-duty refractory mortar. Mortar joints are generally $^1/_{16}$ to $^3/_{16}$ inch (1.6 to 4.8 mm) thick, but not thicker than $^1/_4$ inch (6.4 mm), to reduce thermal movements and prevent joint deterioration.

Where a firebrick lining is not used in firebox construction, the wall thickness is not to be less than 10 inches (254 mm) of solid masonry. Firebrick must conform to ASTM C 27 or ASTM C 1261 and must be laid with medium-duty refractory mortar conforming to ASTM C 199.

R1001.5.1 Steel fireplace units. Installation of steel fireplace units with *solid masonry* to form a masonry fireplace is permitted when installed either according to the requirements of their listing or according to the requirements of this section. Steel fireplace units incorporating a steel firebox lining, shall be constructed with steel not less than $^{1}/_{4}$ inch (6 mm) thick, and an air-circulating chamber which is ducted to the interior of the building. The firebox lining shall be encased with *solid masonry* to provide a total thickness at the back and sides of not less than 8 inches (203 mm), of which not less than 4 inches (102 mm) shall be of *solid masonry* or concrete. Circulating air ducts used with steel fireplace units shall be constructed of metal or masonry.

❖ Steel fireplace units and an air chamber may be used as part of the required thickness of masonry fireplaces so that the combined thickness of the masonry and steel is 8 inches (203 mm). The steel may not be less than $^{1}/_{4}$ inch (6.4 mm) thick, and the masonry must be at least 4 inches thick. Where ducts are used as part of a steel fireplace unit to circulate warm air, they must be constructed of metal or masonry.

TABLE R1001.1
SUMMARY OF REQUIREMENTS FOR MASONRY FIREPLACES AND CHIMNEYS

ITEM	LETTER[a]	REQUIREMENTS
Hearth slab thickness	A	4″
Hearth extension (each side of opening)	B	8″ fireplace opening < 6 square foot. 12″ fireplace opening ≥ 6 square foot.
Hearth extension (front of opening)	C	16″ fireplace opening < 6 square foot. 20″ fireplace opening ≥ 6 square foot.
Hearth slab reinforcing	D	Reinforced to carry its own weight and all imposed loads.
Thickness of wall of firebox	E	10″ solid brick or 8″ where a firebrick lining is used. Joints in firebrick $^{1}/_{4}$″ maximum.
Distance from top of opening to throat	F	8″
Smoke chamber wall thickness Unlined walls	G	6″ 8″
Chimney Vertical reinforcing[b]	H	Four No. 4 full-length bars for chimney up to 40″ wide. Add two No. 4 bars for each additional 40″ or fraction of width or each additional flue.
Horizontal reinforcing	J	$^{1}/_{4}$″ ties at 18″ and two ties at each bend in vertical steel.
Bond beams	K	No specified requirements.
Fireplace lintel	L	Noncombustible material.
Chimney walls with flue lining	M	Solid masonry units or hollow masonry units grouted solid with at least 4-inch nominal thickness.
Distances between adjacent flues	—	See Section R1003.13.
Effective flue area (based on area of fireplace opening)	P	See Section R1003.15.
Clearances Combustible material Mantel and trim Above roof	R	See Sections R1001.11 and R1003.18. See Section R1001.11, Exception 4. 3′ at roofline and 2′ at 10′.
Anchorage[b] Strap Number Embedment into chimney Fasten to Bolts	S	$^{3}/_{16}$″ × 1″ Two 12″ hooked around outer bar with 6″ extension. 4 joists Two $^{1}/_{2}$″ diameter.
Footing Thickness Width	T	12″ min. 6″ each side of fireplace wall.

For SI: 1 inch = 25.4 mm, 1 foot = 304.8 mm, 1 square foot = 0.0929 m^2.

Note: This table provides a summary of major requirements for the construction of masonry chimneys and fireplaces. Letter references are to Figure R1001.1, which shows examples of typical construction. This table does not cover all requirements, nor does it cover all aspects of the indicated requirements. For the actual mandatory requirements of the code, see the indicated section of text.

a. The letters refer to Figure R1001.1.

b. Not required in Seismic Design Category A, B or C.

❖ This table provides a summary of the requirements for masonry fireplaces and chimneys directly correlated to Figure R1001.1. The first column lists various parts of the fireplace. The second column shows a letter designation corresponding to the item in the figure with the same letter. The third column contains the requirements for that particular fireplace or chimney.

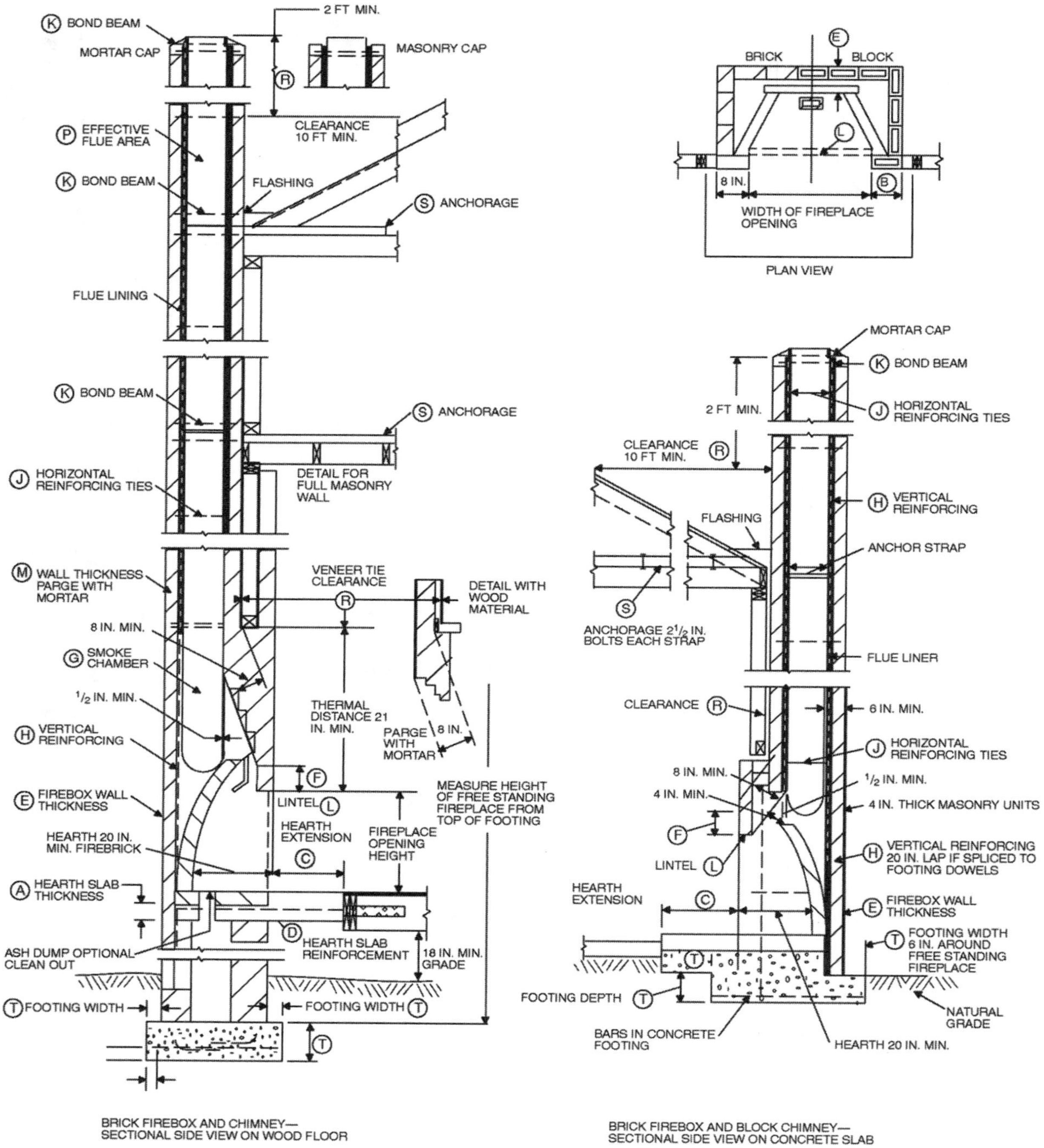

For SI: 1 inch = 25.4 mm, 1 foot = 304.8 mm.

FIGURE R1001.1
FIREPLACE AND CHIMNEY DETAILS

❖ See the commentary to Table R1001.1.

R1001.6 Firebox dimensions. The firebox of a concrete or masonry fireplace shall have a minimum depth of 20 inches (508 mm). The throat shall not be less than 8 inches (203 mm) above the fireplace opening. The throat opening shall not be less than 4 inches (102 mm) deep. The cross-sectional area of the passageway above the firebox, including the throat, damper and smoke chamber, shall not be less than the cross-sectional area of the flue.

> **Exception:** Rumford fireplaces shall be permitted provided that the depth of the fireplace is at least 12 inches (305 mm) and at least one-third of the width of the fireplace opening, that the throat is at least 12 inches (305 mm) above the lintel and is at least $^{1}/_{20}$ the cross-sectional area of the fireplace opening.

❖ The proper functioning of the fireplace depends on the size of the face opening and the chimney dimensions, which in turn are related to the room size [see Commentary Figure R1001.6(1)]. This section specifies a minimum depth of 20 inches (508 mm) for the combustion chamber because that depth influences the draft requirement. The dimensions of the firebox (depth, opening size and shape) are usually based on two considerations: aesthetics and the need to prevent the room from overheating. Suggested dimensions for single-opening fireboxes are given in technical publications of the Brick Institute of America (BIA) and the National Concrete Masonry Association (NCMA).

This section also contains additional criteria for the throat's location and minimum cross-sectional area. Those criteria are based on many years of construction of successfully functioning fireplaces. The exception permits the use of Rumford fireplaces, which are tall, shallow fireplaces that can radiate a large amount of heat into a room.

The code reference to "depth of fireplace" is interpreted as the depth of the firebox [see Commentary Figure R1001.6(2)]. The throat must be made at least 12 inches (305 mm) above the lintel and at least 5 percent ($^{1}/_{20}$) of the cross-sectional area of the fireplace opening. Smoke chambers and flues for Rumford fireplaces should be sized and built like those of other masonry fireplaces. Even though those who build Rumford fireplaces do not totally agree about how they work, many books and guides address their construction.

R1001.7 Lintel and throat. Masonry over a fireplace opening shall be supported by a lintel of noncombustible material. The minimum required bearing length on each end of the fireplace opening shall be 4 inches (102 mm). The fireplace throat or damper shall be located a minimum of 8 inches (203 mm) above the lintel.

❖ A noncombustible lintel with a minimum bearing length of 4 inches (102 mm) on each end must be provided to support masonry above the fireplace opening. Provisions should be made to allow for expansion of the steel lintels when they are heated.

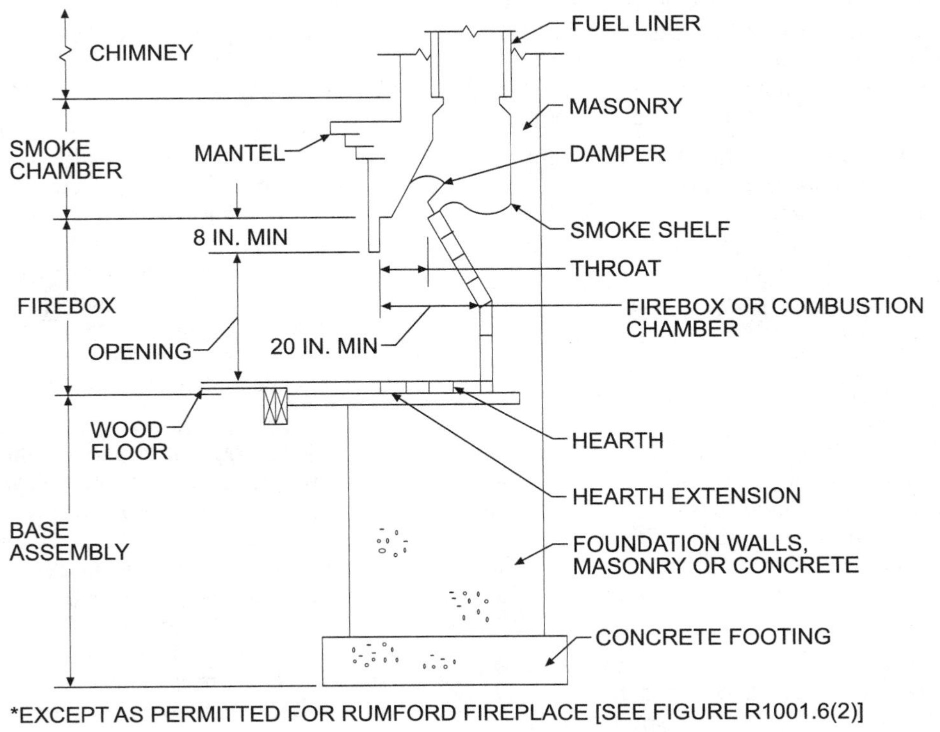

For SI: 1 inch = 25.4 mm.

Figure R1001.6(1)
SECTION THROUGH FIREPLACE

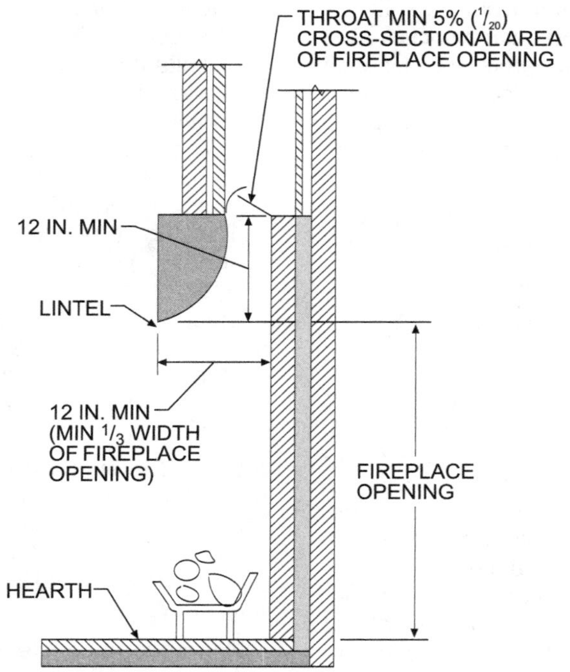

For SI: 1 inch = 25.4 mm.

Figure R1001.6(2)
RUMFORD FIREPLACE

R1001.7.1 Damper. Masonry fireplaces shall be equipped with a ferrous metal damper located at least 8 inches (203 mm) above the top of the fireplace opening. Dampers shall be installed in the fireplace or the chimney venting the fireplace, and shall be operable from the room containing the fireplace.

❖ A damper is used to close the chimney flue when the fireplace is not in use. This section provides guidance on its location and construction.

R1001.8 Smoke chamber. Smoke chamber walls shall be constructed of *solid masonry* units, hollow masonry units grouted solid, stone or concrete. The total minimum thickness of front, back and side walls shall be 8 inches (203 mm) of *solid masonry*. The inside surface shall be parged smooth with refractory mortar conforming to ASTM C 199. When a lining of firebrick at least 2 inches (51 mm) thick, or a lining of vitrified clay at least $^5/_8$ inch (16 mm) thick, is provided, the total minimum thickness of front, back and side walls shall be 6 inches (152 mm) of *solid masonry*, including the lining. Firebrick shall conform to ASTM C 1261 and shall be laid with medium duty refractory mortar conforming to ASTM C 199. Vitrified clay linings shall conform to ASTM C 315.

❖ The smoke chamber is located directly above the firebox and must be constructed of solid masonry. A smoke shelf located behind the damper deflects downdrafts. As the downdraft hits the smoke shelf, it is turned upward by the damper assembly. Curved smoke shelves perform better; however, flat smoke shelves are acceptable.

Parging the inside surface smooth with refractory mortar meeting ASTM C 199 will provide a liner with the ability to withstand 1800°F (983°C) as required for other lining materials.

R1001.8.1 Smoke chamber dimensions. The inside height of the smoke chamber from the fireplace throat to the beginning of the flue shall not be greater than the inside width of the fireplace opening. The inside surface of the smoke chamber shall not be inclined more than 45 degrees (0.79 rad) from vertical when prefabricated smoke chamber linings are used or when the smoke chamber walls are rolled or sloped rather than corbeled. When the inside surface of the smoke chamber is formed by corbeled masonry, the walls shall not be corbeled more than 30 degrees (0.52 rad) from vertical.

❖ This section provides specific dimensions for the smoke chamber of masonry fireplaces, which are necessary to provide proper flow of the smoke and products of combustion out of the fireplace and through the chimney. The height of the smoke chamber cannot exceed the inside width of the fireplace opening. The inside surface of the smoke chamber may be sloped 45 degrees (0.79 rad) from the vertical when constructed of a prefabricated smoke chamber lining or the smoke chamber walls are rolled or sloped. If the smoke chamber is sloped using corbeled masonry, the slope is limited to be no more than 30 degrees (0.52 rad) from the vertical. The slope is restricted more when using corbeled masonry because the surface is not smooth and will tend to create turbulence within the smoke chambers, thus affecting the flow of smoke to the flue.

R1001.9 Hearth and hearth extension. Masonry fireplace hearths and hearth extensions shall be constructed of concrete or masonry, supported by noncombustible materials, and reinforced to carry their own weight and all imposed loads. No combustible material shall remain against the underside of hearths and hearth extensions after construction.

❖ The hearth includes both the floor of the firebox and the projection in front of it. The hearth extension protects wood, carpet and combustible materials from being ignited by sparks, hot embers or ashes that may fall from the fire box.

R1001.9.1 Hearth thickness. The minimum thickness of fireplace hearths shall be 4 inches (102 mm).

❖ The minimum thickness of fireplace hearths is 4 inches (102 mm).

R1001.9.2 Hearth extension thickness. The minimum thickness of hearth extensions shall be 2 inches (51 mm).

Exception: When the bottom of the firebox opening is raised at least 8 inches (203 mm) above the top of the hearth extension, a hearth extension of not less than $^{3}/_{8}$-inch-thick (10 mm) brick, concrete, stone, tile or other *approved* noncombustible material is permitted.

❖ The hearth extension must be 2 inches (51 mm) thick unless the bottom of the fireplace opening is raised at least 8 inches (203 mm) above the top of the hearth extension. Where the bottom of the fireplace opening is raised at least 8 inches (203 mm) above the hearth extension, the hearth extension can be reduced to a thickness of $^{3}/_{8}$-inch-thick (9.5 mm) brick, concrete, stone, tile or other approved noncombustible material.

R1001.10 Hearth extension dimensions. Hearth extensions shall extend at least 16 inches (406 mm) in front of and at least 8 inches (203 mm) beyond each side of the fireplace opening. Where the fireplace opening is 6 square feet (0.6 m^2) or larger, the hearth extension shall extend at least 20 inches (508 mm) in front of and at least 12 inches (305 mm) beyond each side of the fireplace opening.

❖ The hearth must extend in front of and to the side of the firebox opening a sufficient dimension to prevent burning embers from landing on combustible surfaces.

R1001.11 Fireplace clearance. All wood beams, joists, studs and other combustible material shall have a clearance of not less than 2 inches (51 mm) from the front faces and sides of masonry fireplaces and not less than 4 inches (102 mm) from the back faces of masonry fireplaces. The air space shall not be filled, except to provide fire blocking in accordance with Section R1001.12.

Exceptions:

1. Masonry fireplaces *listed* and *labeled* for use in contact with combustibles in accordance with UL 127 and installed in accordance with the manufacturer's installation instructions are permitted to have combustible material in contact with their exterior surfaces.
2. When masonry fireplaces are part of masonry or concrete walls, combustible materials shall not be in contact with the masonry or concrete walls less than 12 inches (306 mm) from the inside surface of the nearest firebox lining.
3. Exposed combustible trim and the edges of sheathing materials such as wood siding, flooring and drywall shall be permitted to abut the masonry fireplace side walls and hearth extension in accordance with Figure R1001.11, provided such combustible trim or sheathing is a minimum of 12 inches (305 mm) from the inside surface of the nearest firebox lining.
4. Exposed combustible mantels or trim may be placed directly on the masonry fireplace front surrounding

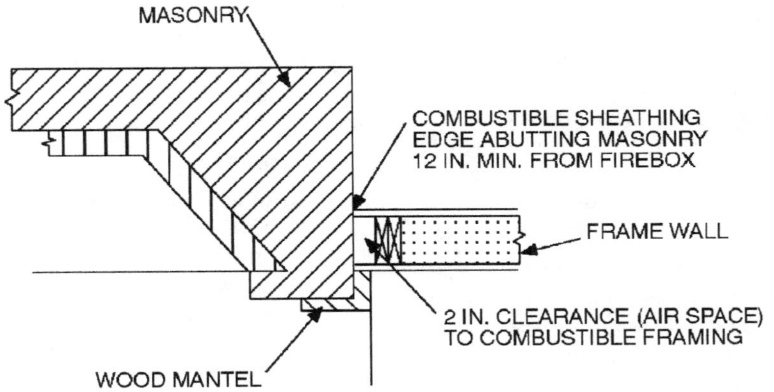

For SI: 1 inch = 25.4 mm.

FIGURE R1001.11
CLEARANCE FROM COMBUSTIBLES

❖ See the commentary to Section R1001.11.

the fireplace opening providing such combustible materials are not placed within 6 inches (152 mm) of a fireplace opening. Combustible material within 12 inches (306 mm) of the fireplace opening shall not project more than $^1/_8$ inch (3 mm) for each 1-inch (25 mm) distance from such an opening.

❖ A 2-inch (51 mm) clearance is required from the front and side surfaces of the fireplace. A 4-inch (102 mm) clearance is required from the back of the fireplace. This airspace created around the fireplace cannot be filled, except for approved fireblocking (see Figure R1001.11).

There are four exceptions to this section that allow combustible materials to come in contact with the fireplace. The first allows factory-built fireplaces constructed in accordance with UL 127 to be in contact with combustibles. The installation must carefully follow the manufacturer's instructions.

The second exception applies to masonry fireplaces constructed in concrete or masonry walls. In this case combustible materials must keep at least 12 inches (305 mm) from the inside surface of the firebox lining because of the conductive nature of the concrete or masonry walls.

Exception 3 permits combustible trim and edges of certain sheathing materials to abut the fireplace in accordance with Figure R1001.11. This exception permits the combustible trim to cover openings that would otherwise occur between the fireplace and other construction.

Exception 4 provides specific dimensions that protect combustible trim from radiant heat.

R1001.12 Fireplace fireblocking. Fireplace fireblocking shall comply with the provisions of Section R602.8.

❖ Refer to the commentary to Section R602.8.

SECTION R1002 MASONRY HEATERS

R1002.1 Definition. A masonry heater is a heating *appliance* constructed of concrete or *solid masonry*, hereinafter referred to as masonry, which is designed to absorb and store heat from a solid-fuel fire built in the firebox by routing the exhaust gases through internal heat exchange channels in which the flow path downstream of the firebox may include flow in a horizontal or downward direction before entering the chimney and which delivers heat by radiation from the masonry surface of the heater.

❖ Masonry heaters are appliances designed to absorb and store heat from a relatively small fire and to radiate that heat into a building interior. They are thermally more efficient than traditional fireplaces because of their design. Interior passageways through the heater allow hot exhaust gases from the fire to transfer heat into the masonry, which then radiates into the building.

R1002.2 Installation. Masonry heaters shall be installed in accordance with this section and comply with one of the following:

1. Masonry heaters shall comply with the requirements of ASTM E 1602; or
2. Masonry heaters shall be *listed* and *labeled* in accordance with UL 1482 and installed in accordance with the manufacturer's installation instructions.

❖ ASTM E 1602 and UL 1482 contain guidelines for the installation of masonry heaters.

R1002.3 Footings and foundation. The firebox floor of a masonry heater shall be a minimum thickness of 4 inches (102 mm) of noncombustible material and be supported on a noncombustible footing and foundation in accordance with Section R1003.2.

❖ This section prescribes the minimum foundation requirements that will adequately support a masonry heating appliance. Section R1003.2 goes into specific detail on footings and foundations for masonry chimneys.

R1002.4 Seismic reinforcing. In Seismic Design Categories D_0, D_1 and D_2, masonry heaters shall be anchored to the masonry foundation in accordance with Section R1003.3. Seismic reinforcing shall not be required within the body of a masonry heater whose height is equal to or less than 3.5 times its body width and where the masonry chimney serving the heater is not supported by the body of the heater. Where the masonry chimney shares a common wall with the facing of the masonry heater, the chimney portion of the structure shall be reinforced in accordance with Section R1003.

❖ Because of the large bulk and squat geometry of these heaters, seismic reinforcement is not typically required. Flexural tensile stresses, which typically cause damage to unreinforced masonry, rarely occur. Where chimneys extend above these heaters, however, seismic reinforcement is required by Section R1003.3. See the commentary to Sections R1003.3 and R1003.4 for seismic reinforcements and anchorage for chimneys.

R1002.5 Masonry heater clearance. Combustible materials shall not be placed within 36 inches (914 mm) of the outside surface of a masonry heater in accordance with NFPA 211 Section 8-7 (clearances for solid-fuel-burning *appliances*), and the required space between the heater and combustible material shall be fully vented to permit the free flow of air around all heater surfaces.

Exceptions:

1. When the masonry heater wall is at least 8 inches (203 mm) thick of *solid masonry* and the wall of the heat exchange channels is at least 5 inches (127 mm) thick of *solid masonry*, combustible materials shall not be placed within 4 inches (102 mm) of the outside surface of a masonry heater. A clearance of at least 8 inches (203 mm) shall be provided between the gas-tight capping slab of the heater and a combustible ceiling.
2. Masonry heaters listed and labeled in accordance with UL 1482 may be installed in accordance with the listing specifications and the manufacturer's written instructions.

❖ Heat conducted through masonry heater walls can ignite combustible structural materials in contact with these walls. For this reason, a minimum required clearance to combustibles from masonry heaters has been established. Because masonry heaters typically generate more heat for a longer period of time than traditional fireplaces, greater clearances to combustible materials are needed to reduce the risk of fire.

SECTION R1003
MASONRY CHIMNEYS

R1003.1 Definition. A masonry chimney is a chimney constructed of *solid masonry* units, hollow masonry units grouted solid, stone or concrete, hereinafter referred to as masonry. Masonry chimneys shall be constructed, anchored, supported and reinforced as required in this chapter.

❖ A masonry chimney is a field-constructed assembly that can consist of solid masonry units, hollow masonry units grouted solid, concrete or rubble stone. A masonry chimney is permitted to serve residential (low-heat), medium-heat and high-heat appliances. This section outlines the general code requirements regarding construction details for all masonry chimneys, including those serving masonry fireplaces regulated by Section R1001.

R1003.2 Footings and foundations. Footings for masonry chimneys shall be constructed of concrete or *solid masonry* at least 12 inches (305 mm) thick and shall extend at least 6 inches (152 mm) beyond the face of the foundation or support wall on all sides. Footings shall be founded on natural undisturbed earth or engineered fill below frost depth. In areas not subjected to freezing, footings shall be at least 12 inches (305 mm) below finished *grade*.

❖ Masonry fireplaces and chimneys must be supported on adequate foundations because of their weight and the forces imposed on them by wind, earthquakes and other effects. This section prescribes minimum foundation requirements that are typically adequate to support a standard chimney (see Commentary Figure R1003.2).

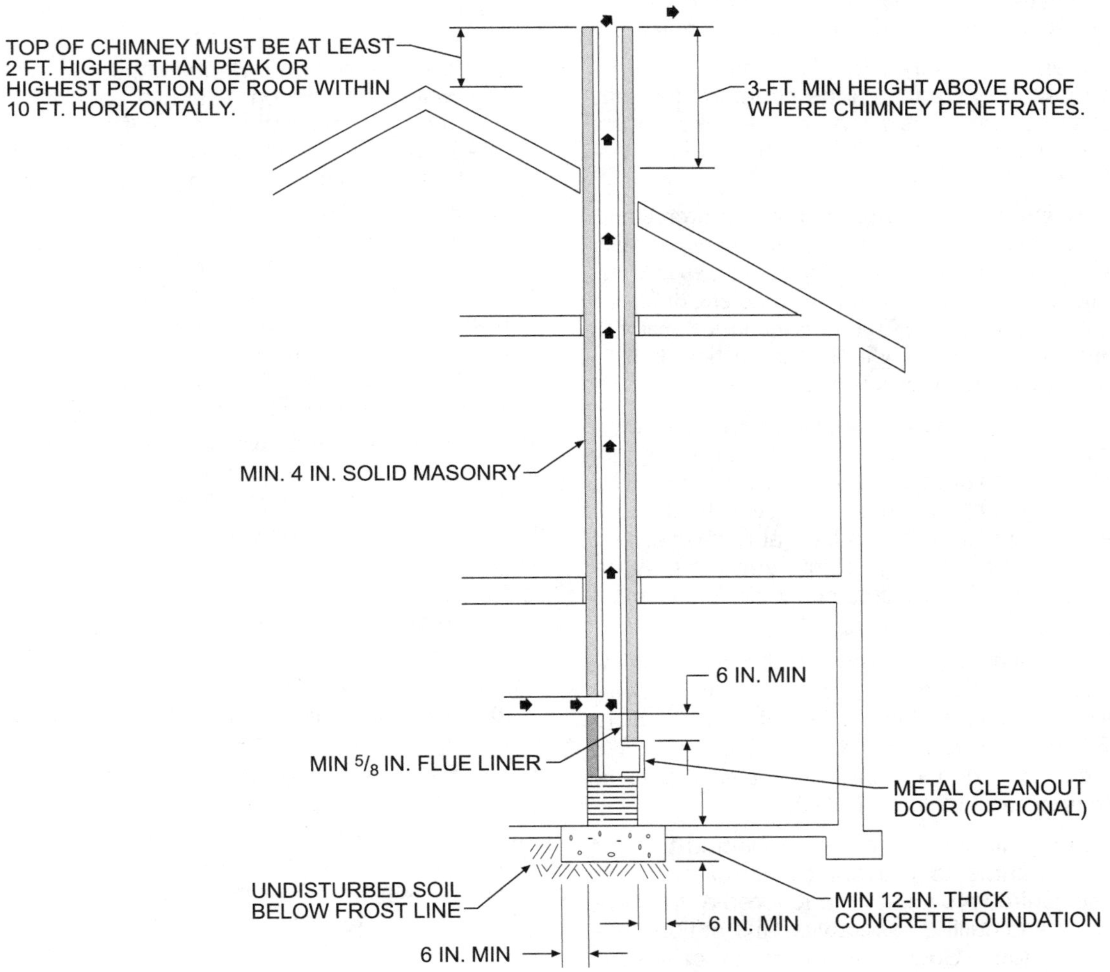

For SI: 1 inch = 25.4 mm, 1 foot = 304.8 mm.

Figure R1003.2
TYPICAL MASONRY CHIMNEY

R1003.3 Seismic reinforcing. Masonry or concrete chimneys shall be constructed, anchored, supported and reinforced as required in this chapter. In Seismic Design Category D_0, D_1 or D_2 masonry and concrete chimneys shall be reinforced and anchored as detailed in Section R1003.3.1, R1003.3.2 and R1003.4. In Seismic Design Category A, B or C, reinforcement and seismic anchorage is not required.

❖ Unreinforced fireplaces and chimneys subjected to strong ground motion have been severely damaged in past earthquakes. The requirements in this section provide minimum reinforcement in an effort to keep these structures together during such events. More substantial reinforcement, however, may be required in areas of high seismicity or for atypical chimneys.

R1003.3.1 Vertical reinforcing. For chimneys up to 40 inches (1016 mm) wide, four No. 4 continuous vertical bars, anchored in the foundation, shall be placed in the concrete, or between wythes of *solid masonry*, or within the cells of hollow unit masonry, and grouted in accordance with Section R609.1.1. Grout shall be prevented from bonding with the flue liner so that the flue liner is free to move with thermal expansion. For chimneys more than 40 inches (1016 mm) wide, two additional No. 4 vertical bars shall be installed for each additional 40 inches (1016 mm) in width or fraction thereof.

❖ These requirements are traditional minimum prescriptive provisions to help maintain the structural integrity of fireplaces and chimneys during earthquakes. More reinforcement may be required in areas of high seismicity or for atypical chimneys.

R1003.3.2 Horizontal reinforcing. Vertical reinforcement shall be placed enclosed within $^1/_4$-inch (6 mm) ties, or other reinforcing of equivalent net cross-sectional area, spaced not to exceed 18 inches (457 mm) on center in concrete, or placed in the bed joints of unit masonry, at a minimum of every 18 inches (457 mm) of vertical height. Two such ties shall be installed at each bend in the vertical bars.

❖ These requirements are traditional minimum prescriptive provisions to help maintain the structural integrity of fireplaces and chimneys during earthquakes. The vertical reinforcement required by Section R1003.3.1 must be enclosed within the horizontal reinforcement required by this section. More reinforcement may be required in areas of high seismicity or for atypical chimneys.

R1003.4 Seismic anchorage. Masonry and concrete chimneys and foundations in Seismic Design Category D_0, D_1 or D_2 shall be anchored at each floor, ceiling or roof line more than 6 feet (1829 mm) above *grade*, except where constructed completely within the exterior walls. Anchorage shall conform to the requirements in Section R1003.4.1.

❖ Fireplaces and chimneys must be connected to floor and roof diaphragms to prevent overturning during earthquakes. Chimneys must be anchored at the ceiling line of roof and ceiling assemblies and at floor levels below the roof. Such anchorage is of lesser importance where the floor assembly is 6 feet (1829 mm) or less above grade.

R1003.4.1 Anchorage. Two $^3/_{16}$-inch by 1-inch (5 mm by 25 mm) straps shall be embedded a minimum of 12 inches (305 mm) into the chimney. Straps shall be hooked around the outer bars and extend 6 inches (152 mm) beyond the bend. Each strap shall be fastened to a minimum of four floor joists with two $^1/_2$-inch (13 mm) bolts.

❖ The prescriptive requirements in this section are traditional for typical fireplaces and chimneys. More substantial anchorage may be required in areas of high seismicity, for large fireplaces or where the distance between floor and roof diaphragms is large.

R1003.5 Corbeling. Masonry chimneys shall not be corbeled more than one-half of the chimney's wall thickness from a wall or foundation, nor shall a chimney be corbeled from a wall or foundation that is less than 12 inches (305 mm) thick unless it projects equally on each side of the wall, except that on the second *story* of a two-story *dwelling*, corbeling of chimneys on the exterior of the enclosing walls may equal the wall thickness. The projection of a single course shall not exceed one-half the unit height or one-third of the unit bed depth, whichever is less.

❖ Corbeling is the projection of masonry from the surface of the wall or fireplace in small increments for each course of masonry. The chimney should not be corbeled more than one-half of its wall thickness from the wall or foundation. A single course is not permitted to project more than one-half of the individual unit height or more than one-third of the individual unit depth. Exceeding these amounts could cause the unit to fail because it might not be able to carry the intended load [see Commentary Figure R1003.5(1)].

Additionally, masonry walls are not to be corbeled more than one-half of the chimney's wall thickness from the wall or foundation unless it projects equally on each side of the wall. If the corbeling is not equal on each side, the chimney might overturn or crack [see Commentary Figure R1003.5(2)].

R1003.6 Changes in dimension. The chimney wall or chimney flue lining shall not change in size or shape within 6 inches (152 mm) above or below where the chimney passes through floor components, ceiling components or roof components.

❖ Changes in the size or shape of a chimney change the stiffness of the chimney, and these areas have more potential for leaks or cracks in the chimney. The code prohibits these changes within 6 inches (152 mm) of where the chimney passes through the floor, ceiling or roof components.

R1003.7 Offsets. Where a masonry chimney is constructed with a fireclay flue liner surrounded by one wythe of masonry, the maximum offset shall be such that the centerline of the flue above the offset does not extend beyond the center of the chimney wall below the offset. Where the chimney offset is supported by masonry below the offset in an *approved* manner, the maximum offset limitations shall not apply. Each individual corbeled masonry course of the offset shall not exceed the projection limitations specified in Section R1003.5.

❖ Limitations on chimney offsets provide a gradual transition to prevent critical stress concentrations at the bottom of the offset. The maximum offset of the centerline of the flue to the centerline of the wall of the chimney below maintains structural stability for vertical loads. The slope of the offset is limited to that permitted for corbels in Section R1003.5. The offset limitation does not apply in cases where the offset is supported by masonry below the offset and in a manner approved by the building official.

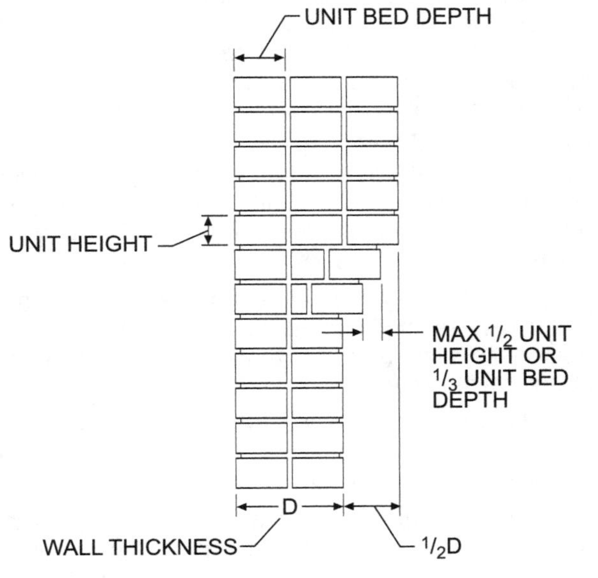

Figure R1003.5(1)
INDIVIDUAL CORBELS

R1003.8 Additional load. Chimneys shall not support loads other than their own weight unless they are designed and constructed to support the additional load. Construction of masonry chimneys as part of the masonry walls or reinforced concrete walls of the building shall be permitted.

❖ Chimneys are subject to considerable stresses resulting from thermal effects and therefore should not support any structural load other than their own weight, unless specifically designed as a supporting member for the additional load. Also, because of its heavy mass, a chimney will tend to settle more than the building structures. As a consequence, the chimney and any part of the building that it supports will settle at a greater rate and to a greater degree than the rest of the building, resulting in damage. If a chimney is subject to structural loads for which it is not designed, the additional stresses created could lead to cracks in the chimney. Thus, the chimney would become a hazard because of the potential of flames from the firebox penetrating the cracks and igniting combustible construction. Moreover, buildups of the products of combustion that leak through the cracks create hazardous conditions inside the building.

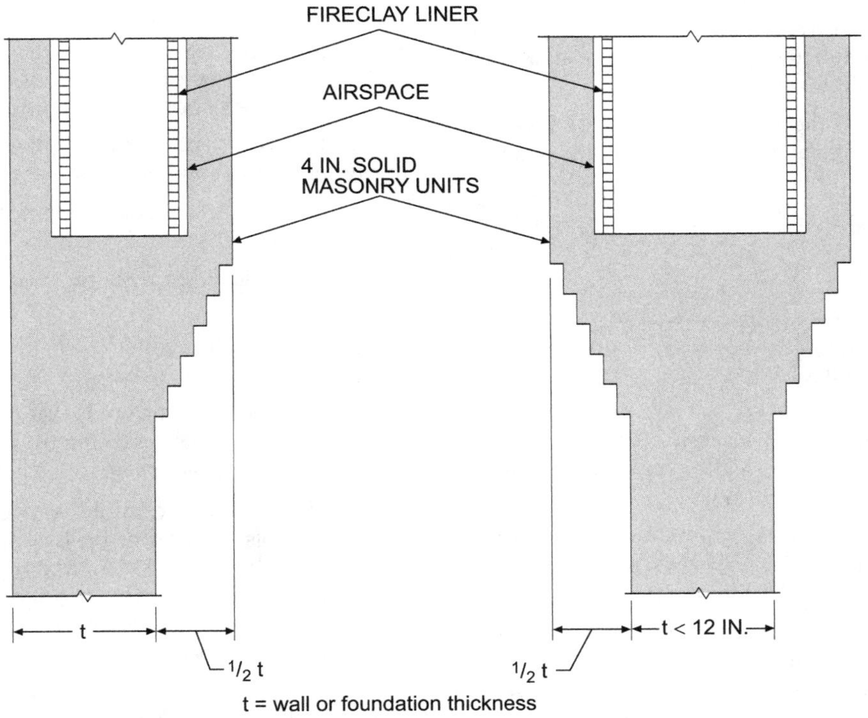

For SI: 1 inch = 25.4 mm.

Figure R1003.5(2)
CORBELING FROM FOUNDATION WALL

R1003.9 Termination. Chimneys shall extend at least 2 feet (610 mm) higher than any portion of a building within 10 feet (3048 mm), but shall not be less than 3 feet (914 mm) above the highest point where the chimney passes through the roof.

❖ The provisions for the termination height of chimneys above the roof provide for the necessary upward draft in the chimney. Experience has indicated that the required heights produce satisfactory operation of the chimney (see Commentary Figure R1003.2).

R1003.9.1 Chimney caps. Masonry chimneys shall have a concrete, metal or stone cap, sloped to shed water, a drip edge and a caulked bond break around any flue liners in accordance with ASTM C 1283.

❖ The section requires masonry chimneys to have caps and references ASTM C 1283, which covers the minimum requirements for installing a clay flue lining for residential concrete or masonry chimneys. A chimney cap protects the top of the masonry surrounding the flue. Although masonry chimneys typically have some type of weatherproof cap, this code section specifically requires a chimney cap and prescribes the minimum criteria for its construction. The cap must be sloped to the outside, overhang the face of the masonry chimney and provide a drip edge. The prescribed caulking of the joint between the masonry and the flue serves as both a sealant and a bond break for any differential movement. The requirements for chimney caps are consistent with the referenced standards ASTM C 1283, *Standard Practice for Installing Clay Flue Lining,* and ASTM C 315, *Standard Specification for Clay Flue Liners and Chimney Pots*.

R1003.9.2 Spark arrestors. Where a spark arrestor is installed on a masonry chimney, the spark arrestor shall meet all of the following requirements:

1. The net free area of the arrestor shall not be less than four times the net free area of the outlet of the chimney flue it serves.
2. The arrestor screen shall have heat and corrosion resistance equivalent to 19-gage galvanized steel or 24-gage stainless steel.
3. Openings shall not permit the passage of spheres having a diameter greater than $^1/_2$ inch (13 mm) nor block the passage of spheres having a diameter less than $^3/_8$ inch (10 mm).
4. The spark arrestor shall be accessible for cleaning and the screen or chimney cap shall be removable to allow for cleaning of the chimney flue.

❖ This section contains specifications for spark arrestors, if they are provided. Their use is not mandated by the code, but owners and builders often install them.

R1003.9.3 Rain caps. Where a masonry or metal rain cap is installed on a masonry chimney, the net free area under the cap shall not be less than four times the net free area of the outlet of the chimney flue it serves.

❖ Rain caps are installed a prescribed distance above a chimney flue termination to limit the amount of rain entering the flue. Because an improperly designed chimney rain cap can reduce the draft of a chimney, this section requires the net free area under the rain cap to be at least four times the area of the outlet of the flue opening. Rain caps are also used to keep animals such as rodents and birds from nesting in the chimney or fireplace. As the language indicates, rain caps are not required by the code, but when they are installed they must have the minimum clearance above the flue termination to provide the minimum net free area required by this section.

R1003.10 Wall thickness. Masonry chimney walls shall be constructed of *solid masonry* units or hollow masonry units grouted solid with not less than a 4-inch (102 mm) nominal thickness.

❖ Masonry chimney walls must have a minimum nominal thickness of 4 inches (102 mm) of either solid masonry units or hollow masonry grouted solid for structural stability.

R1003.10.1 Masonry veneer chimneys. Where masonry is used to veneer a frame chimney, through-flashing and weep holes shall be installed as required by Section R703.

❖ Masonry veneer is a nonstructural facing material which provides ornamentation, protection or insulation. To be considered a veneer, the material cannot act structurally with the backing insofar as the structural strength of the assembly is concerned.

R1003.11 Flue lining (material). Masonry chimneys shall be lined. The lining material shall be appropriate for the type of *appliance* connected, according to the terms of the *appliance* listing and manufacturer's instructions.

❖ Masonry chimneys must be properly lined to allow for the smooth flow of products of combustion. This section regulates specific flue linings.

R1003.11.1 Residential-type appliances (general). Flue lining systems shall comply with one of the following:

1. Clay flue lining complying with the requirements of ASTM C 315.
2. Listed and labeled chimney lining systems complying with UL 1777.
3. Factory-built chimneys or chimney units listed for installation within masonry chimneys.
4. Other *approved* materials that will resist corrosion, erosion, softening or cracking from flue gases and condensate at temperatures up to 1,800°F (982°C).

❖ Flue linings must comply with one of the standards listed in this section or be listed for installation within masonry chimneys. Other materials may be used, provided they will resist flue gases and condensate at temperatures up to 1,800°F (982°C) and will not crack, soften or corrode.

R1003.11.2 Flue linings for specific appliances. Flue linings other than these covered in Section R1003.11.1, intended for use with specific types of *appliances*, shall comply with Sections R1003.11.3 through R1003.11.6.

❖ Except as provided in Section R1003.11.1, flue linings for use with specific types of appliances must comply with Sections R1003.11.3 through R1003.11.6.

R1003.11.3 Gas appliances. Flue lining systems for gas *appliances* shall be in accordance with Chapter 24.

❖ Chapter 24 of the code covers flue linings for gas appliances.

R1003.11.4 Pellet fuel-burning appliances. Flue lining and vent systems for use in masonry chimneys with pellet fuel-burning *appliances* shall be limited to the following:

1. Flue lining systems complying with Section R1003.11.1.
2. Pellet vents listed for installation within masonry chimneys. (See Section R1003.11.6 for marking.)

❖ This section's provisions limit flue lining and vent systems.

R1003.11.5 Oil-fired appliances approved for use with Type L vent. Flue lining and vent systems for use in masonry chimneys with oil-fired *appliances approved* for use with Type L vent shall be limited to the following:

1. Flue lining systems complying with Section R1003.11.1.
2. Listed chimney liners complying with UL 641. (See Section R1003.11.6 for marking.)

❖ Flue lining and vent systems for use in masonry chimneys with oil-fired appliances approved for use with a Type L vent are limited to either flue lining systems complying with Section R1003.11.1 or listed chimney liners complying with UL 641.

R1003.11.6 Notice of usage. When a flue is relined with a material not complying with Section R1003.11.1, the chimney shall be plainly and permanently identified by a *label* attached to a wall, ceiling or other conspicuous location adjacent to where the connector enters the chimney. The *label* shall include the following message or equivalent language:

THIS CHIMNEY FLUE IS FOR USE ONLY WITH [TYPE OR CATEGORY OF *APPLIANCE*] *APPLIANCES* THAT BURN [TYPE OF FUEL]. DO NOT CONNECT OTHER TYPES OF *APPLIANCES*.

❖ Flues relined with materials not conforming to Section R1003.11.1 must be permanently identified in a conspicuous location adjacent to the connector's entrance into the building. Thus, current and future occupants of the building will know the type of appliance that may be attached to the flue.

R1003.12 Clay flue lining (installation). Clay flue liners shall be installed in accordance with ASTM C 1283 and extend from a point not less than 8 inches (203 mm) below the lowest inlet or, in the case of fireplaces, from the top of the smoke chamber to a point above the enclosing walls. The lining shall be carried up vertically, with a maximum slope no greater than 30 degrees (0.52 rad) from the vertical.

Clay flue liners shall be laid in medium-duty water insoluble refractory mortar conforming to ASTM C 199 with tight mortar joints left smooth on the inside and installed to maintain an air space or insulation not to exceed the thickness of the flue liner separating the flue liners from the interior face of the chimney masonry walls. Flue liners shall be supported on all sides. Only enough mortar shall be placed to make the joint and hold the liners in position.

❖ Clay flue linings must be installed so there is no leakage through the liner to adjacent spaces, which would result in contaminated air. Flue linings must be installed in accordance with ASTM C 1283, beginning at least 8 inches (203 mm) below the lowest inlet and extending to a point above the top of the chimney walls. Masonry chimneys atop masonry fireplaces should have the lining start at the top of the smoke chamber of the fireplace. If the chimney is corbeled, the lining may not slope more than 30 degrees (0.52 rad) from the vertical. A vent space is required around the lining and is not to be obstructed by the mortar used to set the liner. Only the amount of mortar needed to set the liners and hold them in place should be used. The mortar used must be water insoluble since it will be exposed to weather. This airspace will help limit the transfer of the heat build-up in the chimney to the masonry wall. This air space cannot be used for venting other appliances (see Commentary Figure R1003.12).

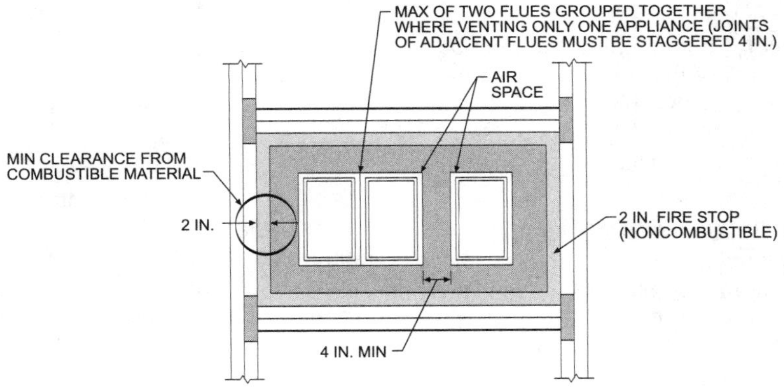

For SI: 1 inch = 25.4 mm.

Figure R1003.12
MULTIPLE FLUES

R1003.12.1 Listed materials. *Listed* materials used as flue linings shall be installed in accordance with the terms of their listings and manufacturer's instructions.

❖ Materials used as flue liners must be installed per the manufacturer's listing and instructions so the flue gases will be properly removed from the structure and not leak back in.

R1003.12.2 Space around lining. The space surrounding a chimney lining system or vent installed within a masonry chimney shall not be used to vent any other *appliance*.

Exception: This shall not prevent the installation of a separate flue lining in accordance with the manufacturer's installation instructions.

❖ The open space around a chimney lining system or vent that is installed within a masonry chimney is not to be used to vent any other appliance. This unused space is otherwise not designed or constructed to vent flue gases.

R1003.13 Multiple flues. When two or more flues are located in the same chimney, masonry wythes shall be built between adjacent flue linings. The masonry wythes shall be at least 4 inches (102 mm) thick and bonded into the walls of the chimney.

Exception: When venting only one *appliance*, two flues may adjoin each other in the same chimney with only the flue lining separation between them. The joints of the adjacent flue linings shall be staggered at least 4 inches (102 mm).

❖ Each separate fireplace should have its own flue. Even outdoor barbecues connected to a chimney should have their own separate flue. Flues in the same chimney must be separated by a 4-inch (102 mm) masonry wythe bonded into the walls of the chimney. Failure to properly separate flues within a common chimney could result in transfer of smoke or other products of combustion from one flue to the other as a result of the downdraft created by interior suction of the inactive flue. The draft could also be created by external wind effects forcing the smoke down the inactive flues as it exhausts the active flue. Two flues in the same chimney venting only one appliance need not be separated from each other; however, the flue joints must be staggered at least 4 inches (102 mm) (see Commentary Figure R1003.12).

R1003.14 Flue area (appliance). Chimney flues shall not be smaller in area than that of the area of the connector from the *appliance* [see Tables R1003.14(1) and R1003.14(2)]. The sizing of a chimney flue to which multiple *appliance* venting systems are connected shall be in accordance with Section M1805.3.

❖ Tables R1001.14(1) and R1001.14(2) determine the sizes of chimney flues for appliances. In no case may the flue of the chimney have a smaller area than the area of the connector from the appliance. If the appliance is connected to a flue smaller than the appliance's flue, a backdraft can occur, which will result in the appliance not operating properly and the products of combustion not being vented to the outside.

TABLE R1003.14(1)
NET CROSS-SECTIONAL AREA OF ROUND FLUE SIZES[a]

FLUE SIZE, INSIDE DIAMETER (inches)	CROSS-SECTIONAL AREA (square inches)
6	28
7	38
8	50
10	78
$10^3/_4$	90
12	113
15	176
18	254

For SI: 1 inch = 25.4 mm, 1 square inch = 645.16 mm^2.
a. Flue sizes are based on ASTM C 315.

❖ Table R1003.14(1) lists the net cross-sectional area of round flues. The first column provides the interior flue diameter; the second column provides the cross-sectional area. For example, an 8-inch (203 mm) round flue requires a cross-sectional area of at least 50 square inches (32 258 mm^2). Flue sizes are based on ASTM C 315.

TABLE R1003.14(2)
NET CROSS-SECTIONAL AREA OF SQUARE AND RECTANGULAR FLUE SIZES

FLUE SIZE, OUTSIDE NOMINAL DIMENSIONS (inches)	CROSS-SECTIONAL AREA (square inches)
4.5 × 8.5	23
4.5 × 13	34
8 × 8	42
8.5 × 8.5	49
8 × 12	67
8.5 × 13	76
12 × 12	102
8.5 × 18	101
13 × 13	127
12 × 16	131
13 × 18	173
16 × 16	181
16 × 20	222
18 × 18	233
20 × 20	298
20 × 24	335
24 × 24	431

For SI: 1 inch = 25.4 mm, 1 square inch = 645.16 mm^2.

❖ Table R1003.14(2) lists the net cross-sectional area of rectangular flues. The first column provides the outside dimensions of the flue. The second column provides the cross-sectional area for the flue. For example, a rectangular flue with an exterior dimension of $11^1/_2$ inches by $15^1/_2$ inches (292 mm by 394 mm) requires a net cross-sectional area of at least 124 square inches (80 000 mm^2). Flue sizes are based on ASTM C 315.

R1003.15 Flue area (masonry fireplace). Flue sizing for chimneys serving fireplaces shall be in accordance with Section R1003.15.1 or Section R1003.15.2.

❖ There are two methods for determining the flue size for masonry fireplaces. Option 1 uses a fraction of the area of the fireplace opening. Option 2 allows the use of a chart in determining the flue size.

R1003.15.1 Option 1. Round chimney flues shall have a minimum net cross-sectional area of at least $^1/_{12}$ of the fireplace opening. Square chimney flues shall have a minimum net cross-sectional area of $^1/_{10}$ of the fireplace opening. Rectangular chimney flues with an *aspect ratio* less than 2 to 1 shall have a minimum net cross-sectional area of $^1/_{10}$ of the fireplace opening. Rectangular chimney flues with an *aspect ratio* of 2 to 1 or more shall have a minimum net cross-sectional area of $^1/_8$ of the fireplace opening. Cross-sectional areas of clay flue linings are shown in Tables R1003.14(1) and R1003.14(2) or as provided by the manufacturer or as measured in the field.

❖ Option 1 indicates the use of different fractional amounts of the fireplace opening depending on the shape of the flue. Round chimney flues must have a net cross-sectional area of $^1/_{12}$ of the fireplace opening, and square flues must have a net cross-sectional area of $^1/_{12}$ of the fireplace opening. Rectangular flues need either $^1/_8$ or $^1/_{10}$ of the net cross-sectional area of the fireplace opening depending on the aspect ratio of the flue. These fractional amounts vary because the effective flue area (EFA) of a square, smooth flue is equal to that of a smooth, round flue where the diameter of the round flue is the same as one side of the square flue (see Commentary Figure R1003.15.1). The cross-hatched areas within the flue of the square or rectangular vent provide no draft.

R1003.15.2 Option 2. The minimum net cross-sectional area of the chimney flue shall be determined in accordance with Figure R1003.15.2. A flue size providing at least the equivalent net cross-sectional area shall be used. Cross-sectional areas of clay flue linings are shown in Tables R1003.14(1) and R1003.14(2) or as provided by the manufacturer or as measured in the field. The height of the chimney shall be measured from the firebox floor to the top of the chimney flue.

❖ Option 2 entails using the chart in Figure R1003.15.2 to determine the required flue area. The flue must have the equivalent net cross-sectional area as determined by the chart. The cross-sectional area of clay flue linings is given in Tables R1003.14(1) and R1003.14(2) or as provided by the manufacturer. Alternately, the cross-sectional area of the flue may be field measured.

R1003.16 Inlet. Inlets to masonry chimneys shall enter from the side. Inlets shall have a thimble of fireclay, rigid refractory material or metal that will prevent the connector from pulling out of the inlet or from extending beyond the wall of the liner.

❖ Chimney inlet thimbles must not project into the flue. Inlets must have a thimble of fire clay, rigid refractory material or other metal. Thimbles must be installed so they will not dislodge from the inlet opening (see Commentary Figure R1003.16).

R1003.17 Masonry chimney cleanout openings. Cleanout openings shall be provided within 6 inches (152 mm) of the base of each flue within every masonry chimney. The upper edge of the cleanout shall be located at least 6 inches (152 mm) below the lowest chimney inlet opening. The height of the opening shall be at least 6 inches (152 mm). The cleanout shall be provided with a noncombustible cover.

> **Exception:** Chimney flues serving masonry fireplaces where cleaning is possible through the fireplace opening.

❖ The code requires that cleanout openings be provided within 6 inches (152 mm) of the base of the flue. Additionally, the upper edge of the cleanout opening must be at least 6 inches (152 mm) below the lowest chimney outlet. The cleanout openings should be at least 6 inches (152 mm) high and have a noncombustible cover. The interior of the chimney

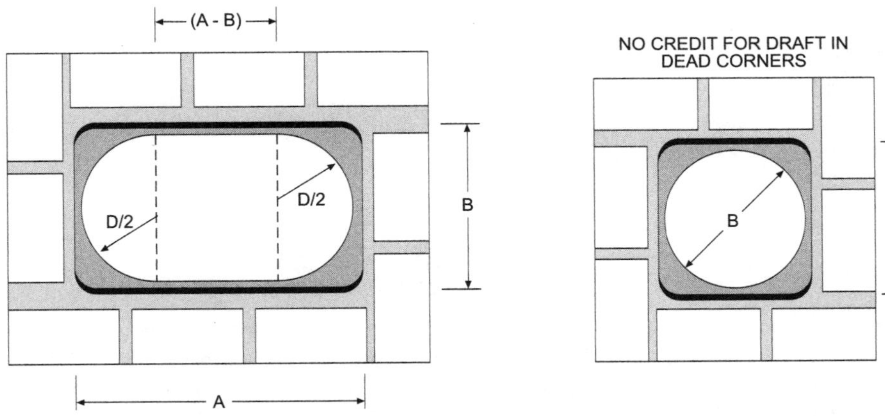

Figure R1003.15.1
MULTIPLE FLUES

may become coated with carbonaceous deposits and acids, and as a result the brick in the mortar will be chemically attacked, causing a gradual deterioration. The deposits will fall to the bottom of the chimney, and a cleanout will be necessary to remove them. Additionally, chimney-cleaning operations, which should be done periodically, will result in the accumulation of deposits at the base of the chimney, necessitating a cleanout opening. In the case of a fireplace chimney, the fireplace opening itself provides the cleanout, and no further opening is required.

R1003.18 Chimney clearances. Any portion of a masonry chimney located in the interior of the building or within the exterior wall of the building shall have a minimum air space clearance to combustibles of 2 inches (51 mm). Chimneys located entirely outside the exterior walls of the building, including chimneys that pass through the soffit or cornice, shall have a minimum air space clearance of 1 inch (25 mm). The air space shall not be filled, except to provide fire blocking in accordance with Section R1003.19.

Exceptions:

1. Masonry chimneys equipped with a chimney lining system listed and *labeled* for use in chimneys in contact with combustibles in accordance with UL 1777 and installed in accordance with the manufacturer's installation instructions are permitted to have

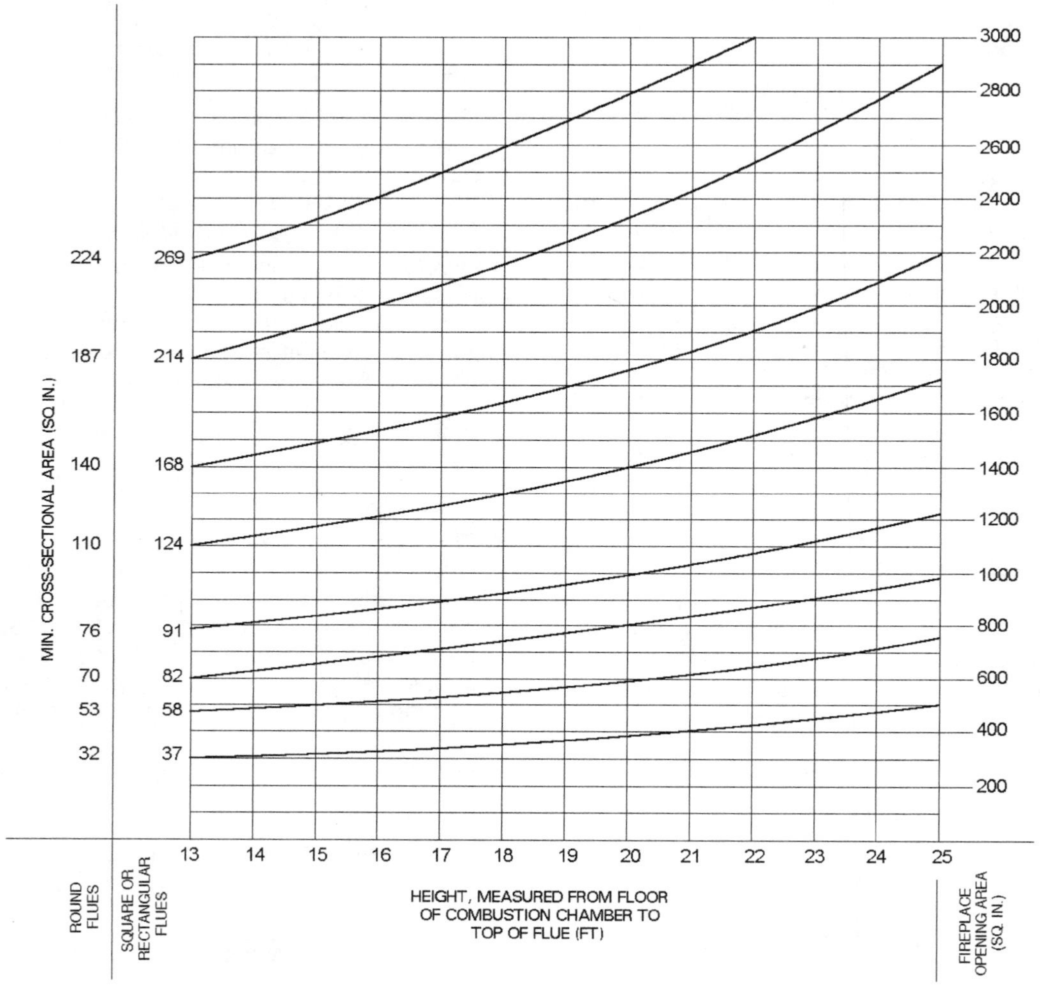

For SI: 1 foot = 304.8 mm, 1 square inch = 645.16 mm^2.

FIGURE R1003.15.2
FLUE SIZES FOR MASONRY CHIMNEYS

❖ Figure R1003.15.2 is used to determine the minimum flue size required for masonry chimneys. Two factors contribute to the determination of the flue size: (1) the height of the flue measured from the floor of the combustion chamber to the top of the flue and (2) the fireplace opening dimension. In using the chart, one would go to the intersection of the x-y axis of the flue height and the fireplace opening size and then to the curved line above. Follow the curved line to the left to determine the minimum flue size. For example, for a flue with a height of 20 feet (6096 mm) and a fireplace opening of 1200 square inches (0.77 m^2), the required net cross-sectional area would be 110 square inches (70 968 mm^2) for a round flue or 124 square inches (80 000 mm^2) for a rectangular flue.

combustible material in contact with their exterior surfaces.

2. When masonry chimneys are constructed as part of masonry or concrete walls, combustible materials shall not be in contact with the masonry or concrete wall less than 12 inches (305 mm) from the inside surface of the nearest flue lining.

3. Exposed combustible trim and the edges of sheathing materials, such as wood siding and flooring, shall be permitted to abut the masonry chimney side walls, in accordance with Figure R1003.18, provided such combustible trim or sheathing is a minimum of 12 inches (305 mm) from the inside surface of the nearest flue lining. Combustible material and trim shall not overlap the corners of the chimney by more than 1 inch (25 mm).

❖ Because of the passage of hot gases and products of combustion, the chimney becomes heated. The code requires that combustibles not be placed directly against the chimney to avoid potential heat transfer and eventual combustion of the combustibles. Masonry chimneys located on the interior of a structure or within an exterior wall require a 2-inch (51 mm) airspace clearance to combustibles. Masonry chimneys located entirely outside of the exterior walls are permitted to have a 1-inch (25 mm) clearance to combustibles (see Figure R1003.18). Three exceptions permit reduced clearances.

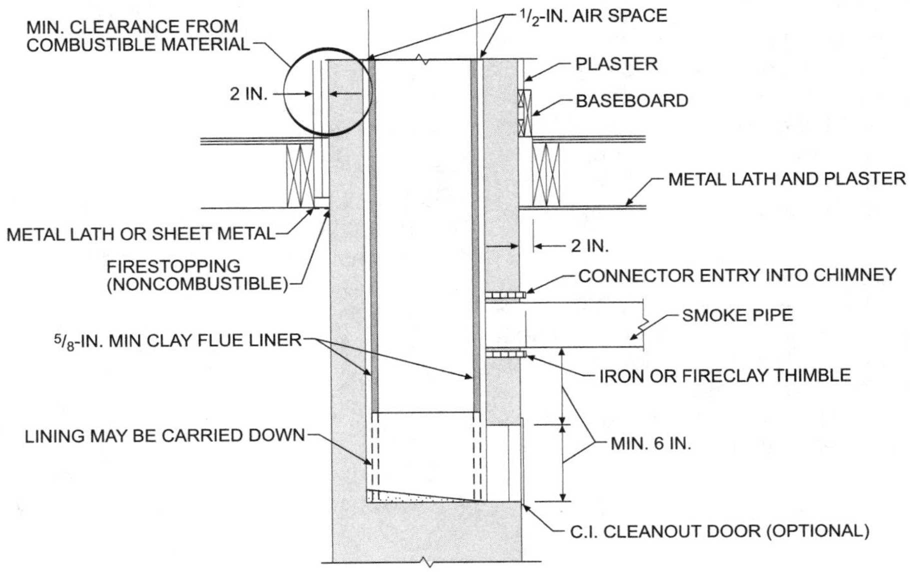

For SI: 1 inch = 25.4 mm.

Figure R1003.16
CHIMNEY WITH INLET—MINIMUM CLEARANCES TO COMBUSTIBLE FIRESTOPPING

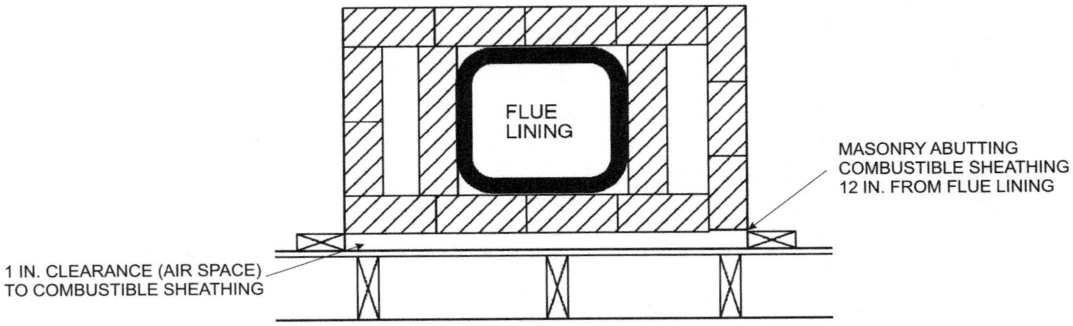

For SI: 1 inch = 25.4 mm.

FIGURE R1003.18
CLEARANCE FROM COMBUSTIBLES

❖ See the commentary to Section 1003.18.

R1003.19 Chimney fireblocking. All spaces between chimneys and floors and ceilings through which chimneys pass shall be fireblocked with noncombustible material securely fastened in place. The fireblocking of spaces between chimneys and wood joists, beams or headers shall be self-supporting or be placed on strips of metal or metal lath laid across the spaces between combustible material and the chimney.

❖ As in concealed spaces in wood frame construction, spaces such as floors and ceilings that chimneys pass through must be fireblocked. Section R1003.18 prohibits chimneys from abutting combustible material. Therefore, the 1-inch (25 mm) to 2-inch (51 mm) airspace required to separate the chimney from combustible material must be fireblocked. The fireblocking must be of noncombustible material to a depth of 1 inch (25 mm) and be supported by noncombustible material such as metal lath.

R1003.20 Chimney crickets. Chimneys shall be provided with crickets when the dimension parallel to the ridgeline is greater than 30 inches (762 mm) and does not intersect the ridgeline. The intersection of the cricket and the chimney shall be flashed and counterflashed in the same manner as normal roof-chimney intersections. Crickets shall be constructed in compliance with Figure R1003.20 and Table R1003.20.

❖ Chimney crickets are required when the chimney width parallel to the ridgeline is more than 30 inches (762 mm), provided the chimney does not intersect the ridgeline. The cricket diverts the flow of water from the roof around the chimney. Flashing is required between the cricket and the chimney, as it is in other roof-chimney intersections.

TABLE R1003.20
CRICKET DIMENSIONS

ROOF SLOPE	H
12 - 12	$^1/_2$ of W
8 - 12	$^1/_3$ of W
6 - 12	$^1/_4$ of W
4 - 12	$^1/_6$ of W
3 - 12	$^1/_8$ of W

❖ Table R1003.20 establishes the height of chimney crickets as depicted in Figure R1003.20. For example, a cricket installed at the face of a chimney that is 4 feet wide (1219 mm) on a 6:12 roof slope, would need to be 1 foot high (305 mm) ($^1/_4$ of 4 feet).

SECTION R1004
FACTORY-BUILT FIREPLACES

R1004.1 General. Factory-built fireplaces shall be *listed* and *labeled* and shall be installed in accordance with the conditions of the *listing*. Factory-built fireplaces shall be tested in accordance with UL 127.

❖ The code requires that factory-built fireplaces be listed and labeled. In addition, factory-built fireplaces must be tested by an approved testing lab in accordance with UL 127 (see commentary, Section R1005).

R1004.2 Hearth extensions. Hearth extensions of *approved* factory-built fireplaces shall be installed in accordance with the *listing* of the fireplace. The hearth extension shall be readily distinguishable from the surrounding floor area. Listed and labeled hearth extensions shall comply with UL 1618.

❖ The primary requirements here are that the hearth extension must be installed in accordance with the

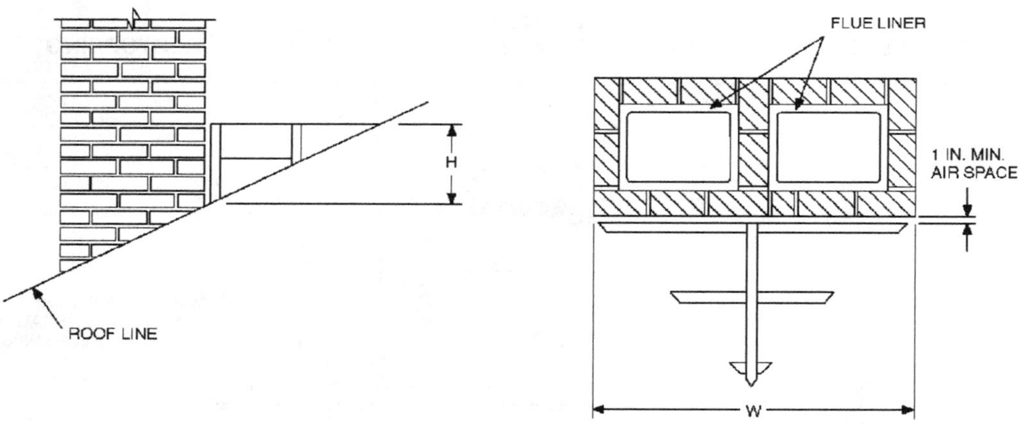

For SI: 1 inch = 25.4 mm.

FIGURE R1003.20
CHIMNEY CRICKET

❖ Figure R1003.20 depicts the height of chimney crickets and the clearance of combustibles to the chimney.

listing of the fireplace, and it must be easily distinguished from the surrounding floor areas. In lieu of a site-built hearth extension, a listed and labeled hearth extension complying with UL 1618 may be used. UL 1618 includes a comprehensive set of construction and performance requirements for the hearth extension.

R1004.3 Decorative shrouds. Decorative shrouds shall not be installed at the termination of chimneys for factory-built fireplaces except where the shrouds are listed and *labeled* for use with the specific factory-built fireplace system and installed in accordance with the manufacturer's installation instructions.

❖ These provisions are the same as found in Section R1005.2 (see commentary, Section R1005.2).

R1004.4 Unvented gas log heaters. An unvented gas log heater shall not be installed in a factory-built fireplace unless the fireplace system has been specifically tested, *listed* and *labeled* for such use in accordance with UL 127.

❖ Unvented gas log heaters must not be installed in factory-built fireplaces if the fireplace has not been specifically tested, listed, and labeled for such use. Without such testing, there is no way to know how the unvented gas log heater will function in the fireplace, and it may cause harm to both the appliance and the occupants of the building.

SECTION R1005
FACTORY-BUILT CHIMNEYS

R1005.1 Listing. Factory-built chimneys shall be *listed* and *labeled* and shall be installed and terminated in accordance with the manufacturer's installation instructions.

❖ Factory-built fireplaces can be used if they are listed and labeled and are installed following the manufacturer's installation instructions. "Listed" and "Labeled" are defined in Chapter 2.

R1005.2 Decorative shrouds. Decorative shrouds shall not be installed at the termination of factory-built chimneys except where the shrouds are *listed* and *labeled* for use with the specific factory-built chimney system and installed in accordance with the manufacturer's installation instructions.

❖ In general, decorative shrouds are not permitted on top of factory-built chimneys. Installing a shroud on top of a chimney that is not designed for one may result in improper venting of the appliance attached to the chimney. Decorative shrouds are permitted if they are listed and labeled for use with the specific chimney and are installed in accordance with the manufacturers' installation instructions.

R1005.3 Solid-fuel appliances. Factory-built chimneys installed in *dwelling units* with solid-fuel-burning *appliances* shall comply with the Type HT requirements of UL 103 and shall be marked "Type HT and "Residential Type and Building Heating *Appliance* Chimney."

Exception: Chimneys for use with open combustion chamber fireplaces shall comply with the requirements of UL 103 and shall be marked "Residential Type and Building Heating *Appliance* Chimney."

Chimneys for use with open combustion chamber *appliances* installed in buildings other than *dwelling units* shall comply with the requirements of UL 103 and shall be marked "Building Heating *Appliance* Chimney" or "Residential Type and Building Heating *Appliance* Chimney."

❖ Factory-built chimneys must withstand the high temperatures associated with solid fuel burning. For that reason the chimney must be clearly marked as Type HT (High Temperature) with the language "Residential Type and Building Heating Appliance Chimney." Factory-built chimneys for use with solid-fuel-burning appliances must comply with Type HT requirements of UL 103. The exception allows these chimneys to be used with open combustion chamber fireplace stoves that comply with UL 103.

R1005.4 Factory-built fireplaces. Chimneys for use with factory-built fireplaces shall comply with the requirements of UL 127.

❖ Chimneys installed and used for factory-built fireplaces must comply with UL 127.

R1005.5 Support. Where factory-built chimneys are supported by structural members, such as joists and rafters, those members shall be designed to support the additional load.

❖ Joists and rafters supporting factory-built chimneys must be designed to carry the load of the chimney in addition to other loads that the members carry.

R1005.6 Medium-heat appliances. Factory-built chimneys for medium-heat *appliances* producing flue gases having a temperature above 1,000°F (538°C), measured at the entrance to the chimney shall comply with UL 959.

❖ Factory-built chimneys for medium-heat appliances that produce flue gases with a temperature above 1,000°F (538°C) at the entry to the chimney must comply with UL 959.

Factory-built fireplaces are permitted when they are listed and labeled and are installed in accordance with the manufacturer's installation instructions. The word "approved" is defined in Chapter 2. Both factory-built fireplaces and chimneys for solid-fuel-burning appliances must meet the applicable UL standard as listed in the code. Agencies such as UL provide listing services for factory-built fireplaces and chimneys. The listing includes a description of the manufactured unit and requirements for clearance from combustible materials, hearth details if necessary, and other limitations placed on the unit to result in a safe installation. Factory-built fireplaces are installed in conjunction with factory-built chimneys; thus, the complete installation of fireplace and chimney is factory built (see Commentary Figure R1005.6).

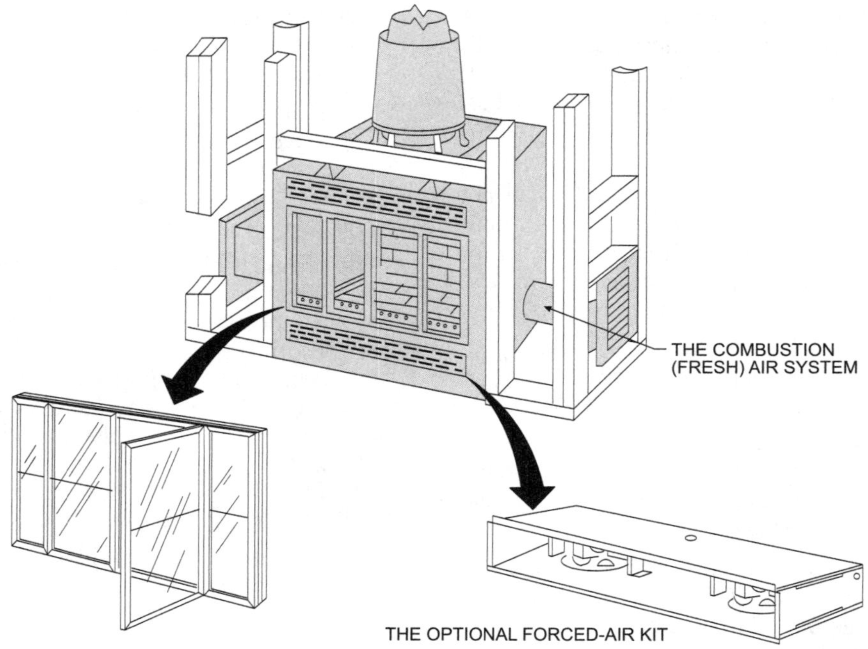

Figure R1005.6
EXAMPLE OF FACTORY-BUILT FIREPLACE AND CHIMENY

R1005.7 Factory-built chimney offsets. Where a factory-built chimney assembly incorporates offsets, no part of the chimney shall be at an angle of more than 30 degrees from vertical at any point in the assembly and the chimney assembly shall not include more than four elbows.

❖ Section R1005.3 requires factory-built chimneys for solid-fuel-burning appliances to be listed and labeled in accordance with UL 103, *Factory-Built Chimneys for Residential Type and Building Heating Appliances,* and installed in accordance with the manufacturer's installation instructions. Ideally, factory-built chimneys should be vertical for the entire length from the fireplace to the roof termination; however, offsets often are necessary to avoid structural components or to place the termination in the preferred location. Typically, the manufacturer's installation instructions include limitations on the maximum offset from vertical and the maximum number of elbows in the entire length of the chimney in accordance with UL 103. Section R1005.7 comes directly from UL 103 and intends to make this information readily available to the inspector, who may not have access to the standard or the manufacturer's installation instructions. This section also serves to emphasize that specific limitations apply to factory-built chimneys that are different from common venting systems such as Type B gas vents. To safely vent combustion products, factory-built chimneys must be installed so that no portion of the chimney exceeds an angle of 30 degrees from vertical. The code also limits the number of offsets by setting the maximum number of elbow fittings to four.

SECTION R1006
EXTERIOR AIR SUPPLY

R1006.1 Exterior air. Factory-built or masonry fireplaces covered in this chapter shall be equipped with an exterior air supply to assure proper fuel combustion unless the room is mechanically ventilated and controlled so that the indoor pressure is neutral or positive.

❖ An adequate supply of combustion air must be provided for the fireplace. The code requires that both masonry and factory-built fireplaces have an exterior air supply. The air ducts used in either type of fireplace must be listed.

R1006.1.1 Factory-built fireplaces. Exterior *combustion air* ducts for factory-built fireplaces shall be a *listed* component of the fireplace and shall be installed according to the fireplace manufacturer's instructions.

❖ An exterior combustion air supply for factory-built fireplaces must be listed components, and the ducts must be installed in accordance with the manufacturer's instructions.

R1006.1.2 Masonry fireplaces. *Listed combustion air* ducts for masonry fireplaces shall be installed according to the terms of their *listing* and the manufacturer's instructions.

❖ If a masonry fireplace uses a listed combustion air duct, the air duct must be installed per its listing.

R1006.2 Exterior air intake. The exterior air intake shall be capable of supplying all *combustion air* from the exterior of the *dwelling* or from spaces within the *dwelling* ventilated with outside air such as nonmechanically ventilated crawl or *attic* spaces. The exterior air intake shall not be located within the garage or *basement* of the *dwelling* nor shall the air intake be located at an elevation higher than the firebox. The exterior air intake shall be covered with a corrosion-resistant screen of $^1/_4$-inch (6 mm) mesh.

❖ The exterior air supply must be capable of providing all of the necessary combustion air. This air must be taken from the exterior of the building or from spaces such as crawl or attic spaces that are adequately ventilated. The crawl or attic spaces cannot be mechanically ventilated. Attic and/or crawl space mechanical ventilating systems are primarily used to remove air from those areas.

They do this by exhausting the unwanted air or creating a negative pressure in those areas. If an air intake for a fireplace terminates in a crawl or attic space that has a mechanical ventilation system there is a potential for the air intake to perform exactly opposite of its designed intent. The exterior air intake cannot be in the garage or basement of the dwelling unit. The air intake must be lower than the fire box so that the firebox will properly draw in the combustion air. Where combustion air openings are located inside the firebox, the air intake opening on the outside of the dwelling cannot be located higher than the firebox. Such an installation could create a chimney effect, drawing the products of combustion up through the combustion air ducts. These ducts are not generally constructed of materials which can withstand the heat and sparks that could be drawn through them. Interior combustion air openings located outside the firebox are not subject to this requirement. The exterior air intake must be covered with a corrosion-resistant screen of $^1/_4$-inch (6.4 mm) mesh.

R1006.3 Clearance. Unlisted *combustion air* ducts shall be installed with a minimum 1-inch (25 mm) clearance to combustibles for all parts of the duct within 5 feet (1524 mm) of the duct outlet.

❖ Where masonry fireplaces use unlisted combustion air ducts, the ducts must be installed with a minimum clearance of 1 inch (25 mm) to combustibles for all parts of the ducts within 5 feet (1524 mm) of the outlets.

R1006.4 Passageway. The *combustion air* passageway shall be a minimum of 6 square inches (3870 mm^2) and not more than 55 square inches (0.035 m^2), except that *combustion air* systems for listed fireplaces shall be constructed according to the fireplace manufacturer's instructions.

❖ The combustion air passageway must be a minimum of 6 square inches (3870 mm^2) and no more than 55 square inches (0.035 m^2). Combustion air systems for listed fireplaces must have combustion air passageways in a size that is established by the manufacturer, which may be different from the dimensions indicated in this section. Air passageways supply combustion air to the firebox without relying on the air within the structure. Without this combustion air, the oxygen in the structure could be depleted to a level unacceptable for the occupants of the building.

R1006.5 Outlet. Locating the exterior air outlet in the back or sides of the firebox chamber or within 24 inches (610 mm) of the firebox opening on or near the floor is permitted. The outlet shall be closable and designed to prevent burning material from dropping into concealed combustible spaces.

❖ The combustion air opening may be located in the back or sides of the firebox chamber, or it may be located within 24 inches (610 mm) of the firebox opening on or near the floor. If the opening is on or near the floor, the outlet must be closeable and designed to prevent burning material from dropping into concealed combustible spaces. A noncombustible screen could be used for this purpose, if the mesh is fine enough to prevent burning embers from penetrating the screen.

Bibliography

The following resource materials were used in the preparation of the commentary for this chapter of the code:.

ASTM C 27-98(2008), *Specification for Standard Classification of Fireclay and High-Alumina Refractory Brick*. West Conshohocken, PA: ASTM International, 2008.

ASTM C 199-84(2005), *Test Method for Pier Test for Refractory Mortar*. West Conshohocken, PA: ASTM International, 2005.

ASTM C 315-07, *Specification for Clay Flue Linings and Chimney Pots*. West Conshohocken, PA: ASTM International, 2007.

ASTM C 1261-07, *Specification for Firebox Brick for Residential Fireplaces*. West Conshohocken, PA: ASTM International, 2007.

ASTM C 1283-07a, *Practice for Installing Clay Flue Lining*. West Conshohocken, PA: ASTM International, 2007.

ASTM E 1602-03, *Guide for Construction of Solid Fuel Burning Masonry Heaters*. West Conshohocken, PA: ASTM International, 2003.

IBC-12, *International Building Code*. Washington, DC: International Code Council, 2011.

UL 103-01, *Factory-built Chimneys for Residential Type and Building Heating Appliances—with Revisions through June 2006*. Northbrook, IL: Underwriters Laboratories Inc., 2006.

UL 127-08, *Factory-built Fireplaces*. Northbrook, IL: Underwriters Laboratories Inc., 2008.

UL 641-95, *Type L, Low-temperature Venting Systems-with Revisions through August 2005, Type L*. Northbrook, IL: Underwriters Laboratories Inc., 2005.

UL 959-01, *Medium-heat Appliance Factory-Built Chimneys—with Revisions through September 2006*. Northbrook, IL: Underwriters Laboratories Inc., 2006.

UL 1482-98, *Solid-fuel Type Room Heaters with Revisions through January 2000*. Northbrook, IL: Underwriters Laboratories Inc., 2000.

UL 1618-09, *Wall Protectors, Floor Protectors, and Hearth Extension*. Northbrook, IL. Underwriters Laboratories Inc., 2009.

Part IV—Energy Conservation

Chapter 11 [RE]: Energy Efficiency

General Comments

Chapter 11 contains the energy efficiency-related requirements for the design and construction of buildings regulated under the code. The applicable portions of the building must comply with the provisions within this chapter for energy efficiency.

Section N1101 contains the scope and application of the chapter and also regulates material identification and labeling. Section N1102 contains the insulation *R*-value requirements and window *U*-factor requirements for the building envelope, which includes the roof/ceiling, wall and floor assembly. Section N1103 contains the requirements for heating and cooling systems, and includes requirements for equipment efficiency, duct installation, piping insulation and the requirements for service water heating performance. Section N1104 deals with lighting equipment.

Purpose

This chapter defines requirements for the portions of the building and building systems that impact energy use in new construction and promotes the effective use of energy. The provisions within the chapter promote energy efficiency in the building envelope, the heating and cooling system, and the service water heating system of the building. Compliance with this chapter will provide a minimum level of energy efficiency for new construction. Greater levels of efficiency can be installed to decrease the energy use of new construction.

SECTION N1101 GENERAL

N1101.1 Scope. This chapter regulates the energy efficiency for the design and construction of buildings regulated by this code.

Note: *The text of the following Sections N1101.2 through N1105 is extracted from the 2012 edition of the* International Energy Conservation Code—Residential Provisions *and has been editorially revised to conform to the scope and application of this code. The section numbers appearing in parenthesis after each section number are the section numbers of the corresponding text in the* International Energy Conservation Code—Residential Provisions.

❖ Chapter 11 applies to portions of the building thermal envelope that enclose conditioned space, as shown in Commentary Figure N1101.1(1). Conditioned space is the area provided with heating or cooling either directly, through a positive heating/cooling supply system (such as registers located in the space), or indirectly through an opening that allows heated or cooled air to communicate directly with the space. For example, a walk-in closet connected to a master bedroom suite may not contain a positive heating supply through a register, but it would be conditioned indirectly by the free passage of heated or cooled air into the space from the bedroom.

A good example of the exception would be an unconditioned garage or attic space. In the case of a garage, if the unconditioned garage area is separated from the conditioned portions of the residence by an assembly that meets the "building envelope" criteria (meaning that the wall between them is insulated), the exterior walls of the garage would not need to be insulated to separate the garage from the exterior climate.

The building thermal envelope consists of the wall, roof/ceiling and floor assemblies that surround the conditioned space. Raised floors over a crawl space or garage, or directly exposed to the outside air are considered to be part of the floor assembly. Walls surrounding a conditioned basement (in addition to surrounding conditioned spaces above grade) are part of the building envelope. The code defines above-grade walls surrounding conditioned spaces as exterior walls. This definition includes walls between the conditioned space and unconditioned garage; roof and basement knee, dormer, gable-end walls; and walls enclosing a mansard roof and basement walls with an average below-grade area that is less than 50 percent of the total basement gross wall area. This definition would not include walls separating an unconditioned garage from the outdoors. The roof/ceiling assembly is the surface where insulation will be installed, typically on top of the gypsum board [see Commentary Figure N1101.1(2)].

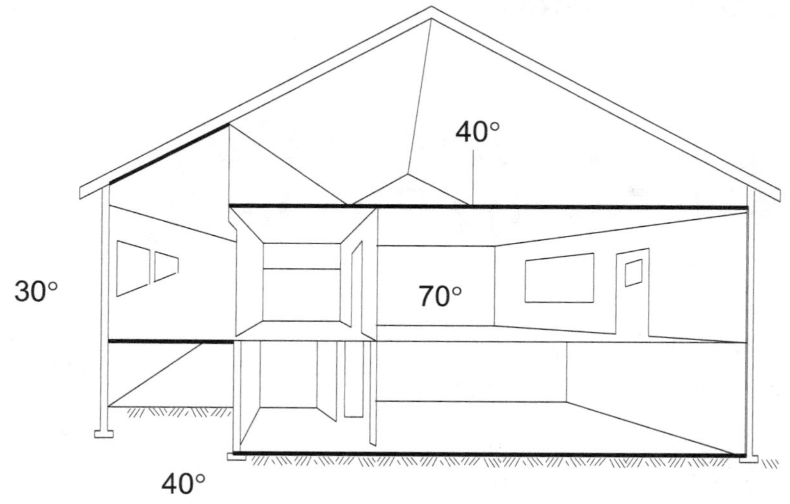

Figure N1101.1(1)
CONDITIONED ENVELOPE

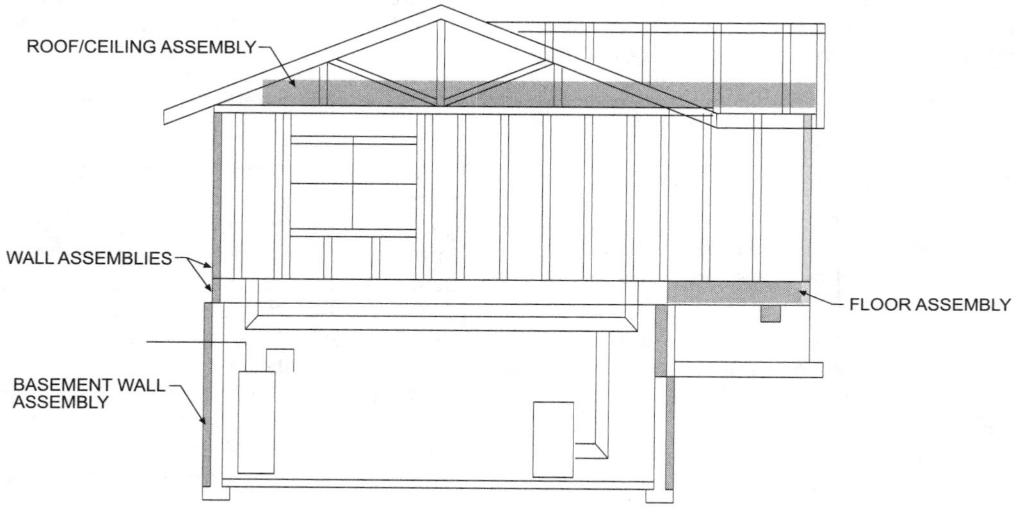

Figure N1101.1(2)
BUILDING ENVELOPE

N1101.2 (R101.3) Intent. This code shall regulate the design and construction of buildings for the effective use and conservation of energy over the useful life of each building. This code is intended to provide flexibility to permit the use of innovative approaches and techniques to achieve this objective. This code is not intended to abridge safety, health or environmental requirements contained in other applicable codes or ordinances.

❖ This chapter is broad in its application, yet specific to regulating the use of energy in buildings where that energy is used primarily for human comfort or heating and cooling to protect the contents. In general, the requirements of the code address the design of all building systems that affect the visual and thermal comfort of the occupants, including:

- Lighting systems and controls.
- Wall, roof and floor insulation.
- Windows and skylights.
- Cooling equipment (air conditioners, chillers and cooling towers).
- Heating equipment (boilers, furnaces and heat pumps).
- Pumps, piping and liquid circulation systems.

- Supply and return fans.
- Service hot water systems (kitchens and lavatories).

The intent of the chapter is to define requirements for the portions of a building and building systems that affect energy use in new construction and to promote the effective use of energy. Where a code application for a specific situation is in question, the authority having jurisdiction for the building should favor the action that will promote the effective use of energy. The code official may also consider the cost of the required action compared to the energy that will be saved over the life of that action.

This section of the code supports flexibility in application of the code requirements. Although many of the requirements are given in a prescriptive format for ease of use, it is not the intent of the code to stifle innovation—especially innovative techniques that conserve energy. Innovative approaches that lead to energy efficiency should be encouraged, even if the approach is not specifically listed in the code or does not meet the strict letter of the code. This principle should be applied to methods for determining compliance with the code and the building construction techniques used to meet the code.

Any design should first be evaluated to see whether it meets the code requirements directly. If an innovative approach is preferred, the applicant is responsible for demonstrating that the innovative concept promotes energy efficiency. Where the literal code requirements have not been satisfied but the applicant claims to meet the intent, the code official will likely have to exercise professional judgment to determine whether the proposed design meets the intent of the code in the interest of energy efficiency (see commentary, Section R103).

N1101.3 (R101.4.3) Additions, alterations, renovations or repairs. Additions, alterations, renovations or repairs to an existing building, building system or portion thereof shall conform to the provisions of this code as they relate to new construction without requiring the unaltered portion(s) of the existing building or building system to comply with this code. Additions, alterations, renovations or repairs shall not create an unsafe or hazardous condition or overload existing building systems. An addition shall be deemed to comply with this code if the addition alone complies or if the existing building and addition comply with this code as a single building.

Exception: The following need not comply provided the energy use of the building is not increased:

1. Storm windows installed over existing fenestration.
2. Glass only replacements in an existing sash and frame.
3. Existing ceiling, wall or floor cavities exposed during construction provided that these cavities are filled with insulation.
4. Construction where the existing roof, wall or floor cavity is not exposed.
5. Reroofing for roofs where neither the sheathing nor the insulation is exposed. Roofs without insulation in the cavity and where the sheathing or insulation is exposed during reroofing shall be insulated either above or below the sheathing.
6. Replacement of existing doors that separate *conditioned space* from the exterior shall not require the installation of a vestibule or revolving door, provided, however, that an existing vestibule that separates a *conditioned space* from the exterior shall not be removed.
7. Alterations that replace less than 50 percent of the luminaires in a space, provided that such alterations do not increase the installed interior lighting power.
8. Alterations that replace only the bulb and ballast within the existing luminaires in a space provided that the *alteration* does not increase the installed interior lighting power.

❖ Simply stated, new work must comply with the code requirements for new construction. Any alteration or addition to an existing system involving new work is subject to the requirements of the code. Additions or alterations can place additional loads or different demands on an existing system and those loads or demands could necessitate changing all or part of the existing system. Additions and alterations must not cause an existing system to be any less in compliance with the code than it was before the changes.

Additions to existing buildings must comply with the code when the addition is within the scope of the code and would not otherwise be exempted (see code text and commentary, Section N1101.6). Additions include new construction, such as a conditioned bedroom, sun space or enclosed porch added to an existing building. Additions also include existing spaces converted from unconditioned or exempt spaces to conditioned spaces. For example, a finished basement, an attic converted to a bedroom or a carport converted to a den are additions. The addition of an unconditioned garage would not be considered within the scope of the code because the code applies to heated or cooled (conditioned) spaces.

Although not specifically defined in the code, building codes typically define an "addition" as any increase in a building's habitable floor area (which can be interpreted as any increase in the conditioned floor area). For example, an unconditioned garage converted to a bedroom is an addition. If a conditioned floor area is expanded, such as a room made larger by moving out a wall, only the newly conditioned space must meet the code. A flat window added to a room does not increase the conditioned space and thus is not an addition by this definition. If several changes are made to a building at the same time, only the changes that expand the conditioned floor area are required to meet the code. The addition (the newly conditioned floor space) complies with the code if it complies with all of the applicable requirements in this chapter. For example, requirements applicable to the addition of a new room would most likely include insulating the exterior walls, ceiling and floor to the levels specified in the code; sealing all joints and penetrations; installing a vapor retarder in unventilated frame walls, floors and ceilings; identifying installed insulation *R*-values and window *U*-factors and insulating and sealing any ducts passing through unconditioned portions or within exterior envelope components (walls, ceilings or floors) of the new space. Compliance approaches for additions include:

1. The entire building (the existing building plus the addition) complies with the code. If the building inclusive of the addition complies with the code, the addition will also comply, regardless of whether the addition complies alone. For example, a sunroom that does not comply with the code is added to a house. If the entire house (with the sunroom) complies, the addition also complies.
2. Where approved by the code official, the addition, including possible concurrent renovation, does not result in any increase in the building's overall area-weighted thermal transmittance (UA), or otherwise any increase in annual demand for either fossil fuel or electrical energy supply. The change in UA or energy use can be quantified using any of the commonly used hourly, full-year simulation tools. For example, additions that add rooms while simultaneously upgrading existing heating, ventilating and air-conditioning (HVAC) systems, windows and insulation often reduce the annual energy use or UA of the existing part of the home more than offsetting the energy use attributed to the added space in the home.
3. The addition itself can comply with the prescriptive methods found within this chapter. The components of the building addition must meet the insulation *R*-values, fenestration *U*-factors and SHGC requirements shown in Table N1102.1.1.

An existing energy-using system (envelope, mechanical, service water heating, electrical distribution or lighting) is generally considered to be "grandfathered" in with the code adoption if the criteria for this level are the regulations (or code) under which the existing building was originally constructed. It should be noted that a specific level of safety is dictated by provisions dealing with hazard abatement in existing buildings and maintenance provisions, as contained in the code, the *International Property Maintenance Code*® (IPMC®) and the *International Fire Code*® (IFC®).

The exceptions address situations where the alteration or repair of a structure or element is not required to comply with the provisions of the code. Some of these situations are typically either a normal part of ongoing maintenance of the building, would improve the performance or would not present an opportunity for improved energy savings. All of these exceptions are tied to the fact that they are permitted, provided "the energy use of the building is not increased."

Exception 1 is a fairly self-evident provision. First of all, due to the limited nature of the work, there is little opportunity to make additional changes. This helps to reinforce the statement from the main paragraph that the intent is not to make "the unaltered portions(s) of the existing building or building system" comply with the code. In this case, the addition of a storm window over an existing window will only improve performance of the existing fenestration.

Exception 2 addresses situations such as when a child hits a baseball through another neighbor's window. When the glass is being replaced, the code does not require the owner to change out existing window assemblies. While the existing window may not meet the correct *U*-factor for the climate zone, a replacement in kind should be permitted. Although this provision does not place any requirement on the replacement glass, the glass must meet requirements for new glazing in accordance with *International Building Code*® (IBC®), Chapter 24.

Exception 3 is important for a couple of the limitations that it contains. First of all, the provision only applies when the ceiling, wall or floor cavity is "exposed during construction." If the cavity is not opened up, then there is no requirement to do anything. If the cavity is exposed, the requirement will only be to "fill" it with insulation. Therefore, the level of insulation is not required to comply with the building thermal envelope requirements, but is instead only required to be "filled" with any type of insulation and not to any specific *R*-value.

Exception 4 will exempt the need to make changes to the building thermal envelope because the building cavities are not exposed.

Exception 5 applies to roofs that are part of the building envelope and typically would have below-deck or above-deck insulation. The second sentence of Exception 5 permits the code-required insulation to be above or below the deck. For a typical single-family home (with a nonconditioned space), the ceiling is the building thermal envelope and the roof is not and, therefore, Exception 5 would not apply. However, if during reroofing the existing ceiling cavities are exposed, then Exception 3 would apply.

Exception 6 is a recognition that the use of vestibules for entry doors is relatively new for energy conservation.

Exceptions 7 and 8 are intended to allow the changing of light fixtures to an extent that does not require complying with this code.

N1101.4 (R101.4.5) Change in space conditioning. Any nonconditioned space that is altered to become *conditioned space* shall be required to be brought into full compliance with this chapter.

❖ When nonconditioned spaces are converted to conditioned spaces, the impacts on the community energy resources are the same as new construction. As such, they should be required to meet the minimum standards set by the code for new construction.

N1101.5 (R101.5.1) Compliance materials. The *building official* shall be permitted to approve specific computer software, worksheets, compliance manuals and other similar materials that meet the intent of this code.

❖ The code intends to permit the use of innovative approaches and techniques, provided that they result in the effective use of energy. This section simply recognizes that there are many federal, state and local programs as well as computer software that deal with energy efficiency. Therefore, the code simply states that the code official does have the authority to accept those methods of compliance, provided that they meet the intent of the code. Some of the easiest examples to illustrate this provision are the REScheck™ and COMcheck™ software that are put out by the U.S. Department of Energy (DOE). An example of another program that may be deemed acceptable by the code official is the ENERGY STAR program.

N1101.6 (R101.5.2) Low-energy buildings. The following buildings, or portions thereof, separated from the remainder of the building by *building thermal envelope* assemblies complying with this code shall be exempt from the *building thermal envelope* provisions of this code:

1. Those with a peak design rate of energy usage less than 3.4 Btu/h·ft^2 (10.7 W/m^2) or 1.0 watt/ft^2 (10.7 W/m^2) of floor area for space conditioning purposes.
2. Those that do not contain *conditioned space*.

❖ This section stipulates the conditions that permit a building or structure or portion of a building or structure to be exempt from the code based on the marginal energy-savings potential of such low-energy-use structures. This section also shows that there is no justification for exempting building, mechanical, service water heating or lighting systems from the applicable criteria of the code simply because the building is not heated or cooled or is partially conditioned. Building, mechanical, service water heating and lighting systems and their subsystems are no less an energy conservation opportunity just because the building or space is unconditioned. Thus, for the buildings that are discussed in Items 1 and 2, the exemption is intended to apply only to the building thermal envelope requirements.

Item 1 exempts buildings and portions of buildings with low summer and winter peak rates of energy use [below 3.4 Btu/h • ft^2 or 1.0 W/ft^2 (10.7 W/m^2)]. The phraseology "a peak design rate of energy usage for space conditioning purposes" refers to the total peak primary energy used for space conditioning, service water heating and lighting for all fuels (electrical, gas, oil, propane, hydrogen, etc.). For example, consider a 100-square-foot (9.3 m^2) building with no space conditioning system and having a 100-watt lightbulb installed for interior lighting. This building is right at the threshold of 1.0 W/ft^2 (100 watts/100 ft^2 = 1.0 W/ft^2).

Thus, the addition of any space-conditioning equipment that uses more energy than the 100-watt light bulb would require code compliance. The peak rating of an appliance or piece of equipment can be determined by its nameplate rating or the manufacturer's literature.

Energy from on-site solar, hydroelectric, wind or other nondepletable, renewable source producing energy at the end-user's facility (or site) is excluded from the peak rate of energy use. (Renewable energy is considered energy that is not purchased. Nonrenewable, or conventional energy, is energy that is purchased, often from a utility service provider, co-op or municipal power authority. See the definition of "Energy cost" in Chapter 2[RE]). Conventional energy associated with the collection of renewable energy, such as energy used by the pumps and fans serving a solar collector, is included in the peak rate of energy use. When a home has both renewable and conventional systems, the peak rate includes the conventional systems even if the occupants primarily intend to use the renewable systems.

Few buildings designed for human occupancy will qualify for this exemption. The exemption generally applies only to buildings without heating or cooling systems or portions of buildings that are not heated or cooled, such as unconditioned garages and storage facilities (see commentary, Item 2 below). If an exemption is claimed for a building, the permittee should provide enough supporting documentation to validate the claim. A list detailing all mechanical equipment, appliances and lighting must be submitted to justify exemption under this section. The list should specifically note the energy sources for heating, cooling, lighting and water heating, including the nameplate input capacities for HVAC and water-heating equipment.

Portions of buildings can also qualify for this exemption. Where a portion of a building meets the criteria for this exemption, that portion is not required to comply with the requirements of the code to the extent that Section N1101.6 permits. Other portions of the building, including the construction assemblies separating conditioned and unconditioned portions, define the limits of the building that must meet the code requirements.

Item 2 indicates that the thermal envelope requirements of the code do not apply to buildings or portions of buildings that are neither heated nor cooled to create a "conditioned space." Though not stated directly in the code, buildings with space-conditioning systems that use energy entirely from nondepletable, renewable sources are also exempt.

For a room or portion of a building to be considered neither heated nor cooled, the space must not contain:

1. A space-conditioning system designed to serve that space;
2. A space-conditioning register/diffuser or hydronic terminal unit serving the space; or
3. An uninsulated duct or pipe where one would normally be required to be insulated.

In the past, the code only considered a space as being "conditioned" when it was being heated or cooled to keep the temperature within the human comfort range. However, based on the definition for "Conditioned space," even a space that is heated only to a level to prevent the freezing of the contents would still be considered as a conditioned space and, therefore, unable to use the exemption this section provides. The space must also be physically separated from conditioned spaces by the building's thermal envelope. For example, a sunroom separated from the main house by an insulated door and wall is physically separated from the conditioned space. In this case, the door and wall separating the conditioned space from the sunroom are part of the building thermal envelope and must meet the code requirements. In the case of a sun room, Florida room, three-season room, etc., even a statement by the permittee that the space-conditioning system will not be used is not sufficient to demonstrate that a space qualifies for the "unconditioned" exception. Any type of added-space conditioning system, such as a small portable heater, would affect compliance. See Chapter 2[RE] and the definition for "Conditioned space."

N1101.7 (R102.1.1) Above code programs. The *building official* or other authority having jurisdiction shall be permitted to deem a national, state or local energy-efficiency program to exceed the energy efficiency required by this code. Buildings *approved* in writing by such an energy-efficiency program shall be considered in compliance with this code. The requirements identified as "mandatory" in Chapters 4 and 5 of this code, as applicable, shall be met.

❖ The purpose of this section is to specifically state that the code official does have the authority to review and accept compliance with another energy program that may exceed that required by the code, as long as the minimum "mandatory" requirements of this code are met. This provision is really a continuation of the provision stated in Section N1101.2 and the fact that the code does intend to accept alternatives as long as the end result is an energy-efficient building that is comparable to or better than that required by the code.

This is also a good section to help reinforce the fact that this code as a model code is a "minimum" code. Therefore, it establishes the minimum requirement that must be met and that anything that exceeds the level is permitted.

While "above code programs" are acceptable because they do exceed the "minimum" requirements of the code, it would not be proper to require compliance with such "above code programs." Besides the code being the minimum level of acceptable energy efficiency, it is also the maximum efficiency that the code official can require. A building built to the absolute minimum requirement is also the maximum that the code official can demand. It is perfectly acceptable for a designer or builder to exceed the code requirements, but it is not proper for the code official to demand such higher performance.

This section also contains language to ensure that the "mandatory" requirements of the code, such as sealing the building envelope (Section N1102.4) and sealing ducts (Section N1103.2.2), are complied with for all buildings. Since the code has deemed that the mandatory requirements should apply to all buildings, it is reasonable that "above code programs" not be allowed to bypass these requirements.

N1101.8 (R103.2) Information on construction documents. Construction documents shall be drawn to scale upon suitable material. Electronic media documents are permitted to be submitted when *approved* by the *building official*. Construction documents shall be of sufficient clarity to indicate the location, nature and extent of the work proposed, and show in sufficient detail pertinent data and features of the building, systems and equipment as herein governed. Details shall include, but are not limited to, as applicable, insulation materials and their *R*-values; fenestration *U*-factors and SHGCs; area-weighted *U*-factor and SHGC calculations; mechanical system design criteria; mechanical and service water heating system and equipment types, sizes and efficiencies; economizer description; equipment and systems controls; fan motor horsepower (hp) and controls; duct sealing, duct and pipe insulation and location; lighting fixture schedule with wattage and control narrative; and air sealing details.

❖ For a comprehensive plan review, all code requirements should be incorporated in the design and construction documents. All of the project information, including specifications, project scope, calculations, and detailed drawings, should be submitted to the code official so that code compliance can be verified. All parties should clearly understand what the project entails. A good plan review is essential to ensure code compliance and a successful project. A statement on the construction documents, such as, "All insulation levels shall comply with the 2012 edition of the IECC," is not an acceptable substitute for showing the required information. Note also that the code official is authorized to require additional project and code-related information as necessary.

For example, insulation *R*-values and glazing and door *U*-factors must be clearly marked on the building plans, specifications or forms used to show compliance. Where two or more different insulation levels exist for the same component (two insulation levels are used in ceilings), the permit applicant must record each level separately on the plans or specifications and clarify where in the building each level of insulation will be installed.

The following discussion is presented for the benefit of both the applicant and the plans examiner. This is not an all-inclusive list, but rather is intended to reflect the minimum scope of information needed to determine energy code compliance:

Permit Applicant's Responsibilities. At permit application, the goal of the applicant is to provide all necessary information to show compliance with the code. If the plans examiner is able to verify compliance in a single review, the permit can be issued and construction may be started without delay.

Depending on whether the prescriptive or performance methods of compliance are used, the amount and detail of the required information may vary. For example, if using the prescriptive method of compliance, the *U*-factor and SHGC may be the only information that is needed to verify fenestration compliance. If the "total UA alternative" (Section N1102.1.4) or the performance option (Section N1105) is used, then additional information, such as the fenestration sizes and orientation, may be needed to demonstrate compliance. The envelope information that needs to be on the plans can be presented in a number of ways:

- *On the drawings.* Include elevations that indicate window, door and skylight areas and sections that show insulation position and thickness.
- *On sections and in schedules.* For instance, list *R*-values of insulation on sections and include *U*-factors, shading coefficient, visible light transmittance and air infiltration on fenestration and opaque door schedules.
- *Through notes and callouts.* Note that all exterior joints are to be caulked, gasketed, weatherstripped or otherwise sealed.
- *Through supplementary worksheets or calculations.* Provide area-weighted calculations where required, such as for projection factors and heat capacity. The permit applicant may include these calculations on the drawings, incorporate as additional columns in the schedule or submit completed code compliance worksheets provided by the jurisdiction.

Incorrect information may be caused by a lack of understanding of the code. More likely, it indicates that the code has changed since the last project. The applicant can use a correction list as a reminder to update the office specifications to avoid receiving this same correction again in the future.

Plans Examiner's Responsibilities. The plans examiner must review each permit application for code compliance before a permit is issued. By letting the designer and contractor know what's expected of them early in the process, the building department can increase the likelihood that the approved drawings will comply with the code. This helps the inspector avoid the headache of correcting a contractor who is following drawings that do not meet the code requirements.

The biggest challenge for the plans examiner is often determining where the necessary information is and whether the drawings are complete. The plans examiner should make sure the applicant includes a summary or checklist as part of the submittal package. When building envelope information is provided on the construction documents, it makes the job of the plans examiner easier, generally making for a more thorough review and reducing turnaround time.

A complete building envelope plan review covers all the requirements specific to the architectural building shell, but the electrical drawings may also need to be included and reviewed if the applicant seeks credit for automatic daylighting control for skylights or fenestration.

- Check that duct insulation thickness and conductivity (*k*-value) are on the drawings and comply with the code.
- Check that the duct insulation *R*-value is on the drawings and complies with the code.
- Check that there is a note indicating that ducts are to be constructed and sealed in accordance with the IMC.
- Check that there is a note indicating that operating and equipment maintenance manuals will be supplied to the owner, that air and hydronic systems will be balanced and that the control system will be tested and calibrated.

N1101.9 (R202) Defined terms. The following words and terms shall, for the purposes of this chapter, have the meanings shown herein.

❖ For the purposes of this code, certain abbreviations, terms, phrases, words and their derivatives have the meanings given in Chapter 2 or in this section. The code, with its broad scope of applicability, includes terms used in a variety of construction and energy-related disciplines. These terms can often have multiple meanings, depending on their context or discipline. Therefore, Chapter 2 and this section establish specific meanings for these terms.

ABOVE-GRADE WALL. A wall more than 50 percent above grade and enclosing *conditioned space*. This includes between-floor spandrels, peripheral edges of floors, roof and basement knee walls, dormer walls, gable end walls, walls enclosing a mansard roof and skylight shafts.

❖ This definition provides the details of which walls must be treated as above-grade walls. This will help to make the distinction between these walls and basement walls (see definition and commentary, "Basement wall"). These two wall types face a different amount of energy transfer and, therefore, have different insulation requirements. In order to determine the proper insulation requirements for the various walls, both of the definitions should be reviewed. The definition includes any wall that is a part of the building thermal envelope ("enclosing conditioned space") and meets the area requirements. For example, the wall between a dwelling and an unconditioned garage would be included within this definition (see commentary, "Building thermal envelope").

ACCESSIBLE. Admitting close approach as a result of not being guarded by locked doors, elevation or other effective means (see "Readily *accessible*").

❖ Providing access to mechanical equipment and appliances is necessary to facilitate inspection, observation, maintenance, adjustment, repair or replacement. Access to equipment means the equipment can be physically reached without having to remove a permanent portion of the structure. It is acceptable, for example, to install equipment in an interstitial space that would require removal of lay-in suspended ceiling panels to gain access. Equipment would not be considered accessible if it were necessary to remove or open any portion of a structure other than panels, doors, covers or similar obstructions intended to be removed or opened (see the definition of "Readily accessible"). Access can be described as the capability of being reached or approached for the purpose of inspection, observation, maintenance, adjustment, repair or replacement. Achieving access may first require the removal or opening of a panel, door or similar obstruction, and may require the overcoming of an obstacle such as elevation.

ADDITION. An extension or increase in the *conditioned space* floor area or height of a building or structure.

❖ The code uses this term to reflect new construction that is being added to an existing building. This definition is important when determining the applicability of the code provisions (see Section R101.4.3).

AIR BARRIER. Material(s) assembled and joined together to provide a barrier to air leakage through the building envelope. An air barrier may be a single material or a combination of materials.

❖ Building tightness against air infiltration is an important aspect of energy conservation. The term "air barrier" is defined to support the provisions of Section N1102.4 regarding air leakage and building tightness. Note that an air barrier is not a single membrane, but rather the system of sealants, seals, insulation and wall sheathing that prevent air infiltration. Although the evaluation and report are prepared by ICC, the code official is not obligated to accept or approve the product based on the evaluation report. The term "approved" is always tied to the code official's approval of the product or project. ICC's publishing of an evaluation report does not take away the fact that the approval of the code official is still needed for the material or method described in the report.

AUTOMATIC. Self-acting, operating by its own mechanism when actuated by some impersonal influence, as, for example, a change in current strength, pressure, temperature or mechanical configuration (see "Manual").

❖ Operation or control devices or systems operating automatically, as opposed to manually, are designed to operate safely with only periodic human intervention or supervision. A thermostat would be an example of something that is automatic. While a person would set the thermostat to the desired temperature, the thermostat would cycle the heating or cooling system on or off on its own once the temperature hits the established set points.

BASEMENT WALL. A wall 50 percent or more below grade and enclosing *conditioned space*.

❖ Because basement walls are in contact with the ground and ground temperatures differ from air temperatures, the amount of energy transferred through a basement wall is different than the energy transfer through a wall predominantly above grade. Therefore, the code provides different thermal requirements for basement walls and above-grade walls. An individual wall enclosing conditioned space is classified as a "basement wall" where the gross wall area is 50 percent or more below grade and is bounded by soil; otherwise, the wall is classified as an "above-grade wall." This definition includes a below-grade interior wall separating a basement from a crawl space that meets the percentage requirement. This basement wall classification applies to the whole wall area, even if a portion of the individual wall is not below grade. Therefore, the above-grade portion of the wall is considered as part of the basement wall where the total wall is 50 percent or more below grade. Both sections of the wall (above grade and below grade) are then insulated as a "basement wall." Likewise, where an exterior wall is less than 50-percent below grade, the whole wall area is classified as an above-grade wall, including the portion underground [see Commentary Figure N1101.9]. For example, the wall of a walk-out basement that is entirely above grade is not considered a basement wall. The wall area of the walk-out wall must be considered as an above-grade wall and compared to the code requirements for above-grade walls. Thus, where the average below-grade depth of the side walls is 50 percent or greater, they are basement walls. If not, they are above-grade walls. The intent of this definition is to apply the provision to each wall enclosing the space. It is not intended to be applied to the aggregate of all of the walls of the basement. Therefore, this classification is done for each individual wall segment and not on an aggregate basis. The basement wall requirements apply only to the opaque basement wall area, excluding windows and doors. For purposes of meeting the code requirements, windows and doors in a basement wall are regulated as any other fenestration opening.

BUILDING. Any structure used or intended for supporting or sheltering any use or occupancy, including any mechanical systems, service water heating systems and electric power and lighting systems located on the building site and supporting the building.

❖ This definition indicates that where this term is used in the code, it means a structure that is intended to provide shelter or support for some activity or occupancy. Though not addressed in the code, it is important to note that the IBC does permit that a fire wall

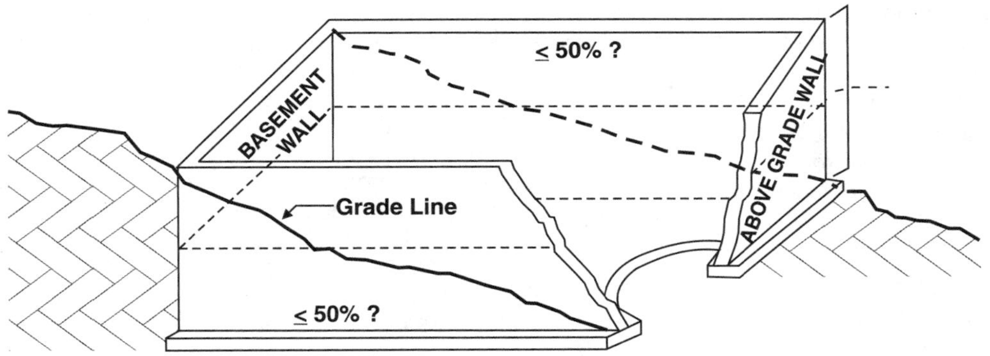

Figure N1101.9
BASEMENT WALLS

forms a demarcation in between two separate, structurally independent buildings. Therefore, the code provisions could be applied to each building separately or to the structure as a whole. This would be a designer's decision to make.

BUILDING SITE. A continguous area of land that is under the ownership or control of one entity.

❖ "Building site" is a key term used throughout this chapter. It is the area of land that is under the ownership of one entity.

BUILDING THERMAL ENVELOPE. The basement walls, exterior walls, floor, roof, and any other building elements that enclose *conditioned space* or provides a boundary between *conditioned space* and exempt or unconditioned space.

❖ The "building thermal envelope" is a key term and resounding theme used throughout the code. It defines what portions of the building form a structurally bound conditioned space and are thereby covered by the insulation and infiltration (air leakage) requirements of the code. The building thermal envelope includes all building components separating conditioned spaces (see commentary, "Conditioned space") from unconditioned spaces or outside ambient conditions and through which heat is transferred. For example, the walls and doors separating an unheated garage (unconditioned space) from a living area (conditioned space) are part of the building thermal envelope. The walls and doors separating an unheated garage from the outdoors are not part of the building thermal envelope. Walls, floors and other building components separating two conditioned spaces are not part of the building thermal envelope. For example, interior partition walls, the common or party walls separating dwelling units in multiple-family buildings and the wall between a new conditioned addition and the existing conditioned space are not considered part of the building thermal envelope. Unconditioned spaces (areas having no heating or cooling sources) are placed outside the building envelope. A space is conditioned if it is heated or cooled directly or where a space is indirectly supplied with heating or cooling through uninsulated walls, floors or uninsulated ducts or HVAC piping. Boundaries that define the building envelope include the following:

- Building assemblies separating a conditioned space from outdoor ambient weather conditions.
- Building assemblies separating a conditioned space from the ground under or around that space, such as the ground around the perimeter of a slab or the soil at the exterior of a conditioned basement wall. Note that the code does not specify requirements for insulating basement floors or underneath slab floors (except at the perimeter edges).
- Building assemblies separating a conditioned space from an unconditioned garage, unconditioned sunroom or similar unheated/cooled area.

The code specifies requirements for ceiling, wall, floor, basement wall, slab-edge and crawl space wall components of the building envelope. In some cases, it may be unclear how to classify a particular part of a building. For example, skylight shafts have properties of a wall assembly but are located in the ceiling assembly. Table 202(1) shows some examples or suggestions of how building envelope components are generally classified and can be used to help determine which code requirement applies to a given building envelope component. Because many of these items are not addressed specifically within the code, the code official should make the determination as to the appropriate classification and construction. When no distinction exists between roof and wall, such as in an A-frame structure, the code official should determine the appropriate classification. Historically, some codes have designated a wall as having a slope of 60 degrees or greater from the horizontal plane. In such situations, if the wall slope is less than 60 degrees, then classification as a "roof" is appropriate. Because the code is silent on this issue, other options such as stating that the roof could be considered to begin at a point 8 feet (2439 mm) above the floor surface of the uppermost story could be used. The reference to "exempted spaces" in the definition of "Building thermal envelope" refers to spaces identified as exempt from this scope of the code.

***C*-FACTOR (THERMAL CONDUCTANCE).** The coefficient of heat transmission (surface to surface) through a building component or assembly, equal to the time rate of heat flow per unit area and the unit temperature difference between the warm side and cold side surfaces (Btu/h · ft^2 · °F) [W/(m^2 · K)].

❖ This definition is included to define a term needed in the provisions of the code for the *U*-factor alternative given in Section N1102.1.2. The term appears in Table N1102.1.2.

CONDITIONED FLOOR AREA. The horizontal projection of the floors associated with the *conditioned space*.

❖ The conditioned floor area is the total area of all floors in the conditioned space of the building.

CONDITIONED SPACE. An area or room within a building being heated or cooled, containing uninsulated ducts, or with a fixed opening directly into an adjacent *conditioned space*.

❖ A conditioned space is typically any space that does not communicate directly to the outside; that is, a space not directly ventilated to the outdoors and meets one of the following criteria:

1. The space has a heating or cooling supply register.
2. The space has heating or cooling equipment designed to heat or cool the space, or both, such as a radiant heater built into the ceiling, a baseboard heater or a wall-mounted gas heater.
3. The space contains uninsulated ducts or uninsulated hydronic heating surfaces.
4. The space is inside the building thermal envelope. For example:
   - A basement with insulated walls but without insulation on the basement ceiling;
   - A closet on a home's exterior wall that is insulated on the exterior surface of the closet wall;
   - A space adjacent to and not physically separated from a conditioned space (such as a room adjacent to another room with a heating duct but without a door that can be closed between the rooms); or
   - A room completely surrounded by conditioned spaces.

The builder/designer has some flexibility in defining the bounds of the conditioned space as long as the building envelope requirements are met. Spaces that are not conditioned directly but have uninsulated surfaces separating them from conditioned spaces are included within the insulated envelope of the building. For example, an unventilated crawl space below an uninsulated floor is considered part of the conditioned space, even where no heat is directly supplied to the crawl space area. Where the crawl space is included as a conditioned space, the builder must insulate the exterior crawl space walls instead of the floor above. The task of defining the building envelope is left to the permit applicant.

Examples of unconditioned spaces include garages and basements that are neither heated nor cooled if all duct surfaces running through these spaces are insulated; attached sunrooms that are neither heated nor cooled and have insulated/weatherstripped doors to separate the sun space from the conditioned space; attics and ventilated crawl spaces. Note that the boundary between the conditioned and unconditioned space is subject to the infiltration control requirements of the code (see commentary, Section N1102.4).

Historically, the code tied this definition to "conditioning for human comfort" or by specifying that the conditioning fall within a specific range. Starting with the 2006 edition, the code considers any type of conditioning as creating a "conditioned space." Therefore, providing heating within a storage building to keep the stock from freezing would still be considered as creating a conditioned space (see Section R101.5.2, Item 2).

CONTINUOUS AIR BARRIER. A combination of materials and assemblies that restrict or prevent the passage of air through the building thermal envelope.

❖ This term is used throughout the code and the language was modified to clarify the options for meeting the continuous air-barrier requirements.

CRAWL SPACE WALL. The opaque portion of a wall that encloses a crawl space and is partially or totally below grade.

❖ Because exterior crawl space walls may be in contact with the ground, crawl spaces exhibit different heat transfer properties than those of exterior above-grade walls or even basement walls. Thus, the code includes different thermal requirements for exterior crawl space walls.

This definition and distinction is also to help coordinate with Section N1102.2.8. While a basement is defined as a "conditioned space," crawl spaces may be designed to be either conditioned or unconditioned spaces.

DEMAND RECIRCULATION WATER SYSTEM. A water distribution system where pump(s) prime the service hot water piping with heated water upon demand for hot water.

❖ This is essential in the code to add a water distribution system that is insulated to retain the heated water upon demand for hot water and conserve energy usage.

DUCT. A tube or conduit utilized for conveying air. The air passages of self-contained systems are not to be construed as air ducts.

❖ Ducts can be factory manufactured or field constructed of sheet metal, gypsum board, fibrous glass board or other approved materials. Ducts are used in air distribution systems, exhaust systems, smoke control systems and combustion air-supply systems. Air passageways that are integral parts of an air handler, packaged air-conditioning unit or similar piece of self-contained, factory-built equipment are not considered ducts in the context of the code.

DUCT SYSTEM. A continuous passageway for the transmission of air that, in addition to ducts, includes duct fittings, dampers, plenums, fans and accessory air-handling equipment and appliances.

❖ Duct systems are part of an air distribution system and include supply, return, transfer and relief/exhaust air systems.

ENCLOSED SPACE. A volume surrounded by solid surfaces such as walls, floors, roofs, and openable devices such as doors and operable windows.

❖ This term is used throughout the code and the language was added to note openable devices such as doors and operable windows are included.

ENERGY ANALYSIS. A method for estimating the annual energy use of the *proposed design* and *standard reference design* based on estimates of energy use.

❖ Designs founded on total building performance (see Section N1105) use energy analysis when the proposed building cannot satisfy the prescriptive criteria in Section N1102, N1103, N1104 or N1105 or when the design professional requires more flexibility for a sophisticated or innovative design. Using the total-building-performance design methodology, the proposed design is evaluated based on the cost of various types of energy used rather than the units of energy used (Btu, kWh). That cost must be established using an hour-by-hour, full-year (8760 hours) simulation tool capable of simulating the performance of both the proposed and standard designs (see commentary, "Energy simulation tool"). The simulation must be capable of converting calculated energy demand and consumption into utility costs using the actual utility rate schedules rather than the average cost of electricity or gas.

ENERGY COST. The total estimated annual cost for purchased energy for the building functions regulated by this code, including applicable demand charges.

❖ The total annual cost for purchased energy includes demand, power and fuel adjustment charges and the impact of special rate programs for large-volume customers. A thorough evaluation of existing tariffs and fee schedules may uncover substantial savings opportunities. In some states, for example, manufacturing customers are exempt from sales taxes for energy. In other states, utilities may have multiple tariff options.

ENERGY SIMULATION TOOL. An *approved* software program or calculation-based methodology that projects the annual energy use of a building.

❖ An energy simulation tool is typically a software package incorporating, among other features, an hour-by-hour, full-year (8760 hours), multiple-zone program to simulate the performance of both proposed and standard design buildings. It is possible to use other types of simulation tools that approximate the dynamics of the hourly energy programs, which can be shown to produce equivalent results for the type of building and HVAC systems under consideration. However, the simulation must be capable of converting calculated energy demand and consumption into utility costs using the actual utility rate schedules (rather than average cost of electricity or gas). Some examples of when an hour-by-hour, full-year type of program is required are:

- When the features intended to reduce energy consumption require time-of-day interactions between weather, loads and operating criteria. Examples include: night ventilation or building thermal storage; chilled water or ice storage; heat recovery; daylighting and water economizer cooling.
- When utility rates are time-of-day sensitive and the proposed design uses time-of-day load shifting between different types of mechanical plant components.

Another distinguishing feature among simulation tools is their sophistication in modeling HVAC systems and plant equipment. Basically, three levels of complexity are used: constant efficiency models; models with simple part-load efficiency adjustment and models with complex part-load efficiency adjustment. Simulation tools in the first category simply calculate hourly equipment input power requirements at part load by applying the full-load efficiency at any given hour. These programs should be avoided for all but constant load applications. Simulated tools with simple part-load efficiency adjustments use a profile of percent-rated input power versus percent-rated load. At each hour, these programs calculate input power to each piece of equipment. These tools are far more accurate than the constant-efficiency models, but still lack accurate compensation for environmental variables.

Building energy simulation tools, available for analyzing daylighting, passive solar design and solar systems, are described in more detail at www.eere.energy.gov/buildings/tools_directory/. This is a portion of the U.S. DOE's Energy Efficiency and Renewable Energy website.

The information includes program uses, computer hardware required, price and contact information. The most sophisticated simulation tools incorporate a number of profiles for each and every piece of equipment. For variable flow fans, this might be as simple as a single profile of percent-rated input power versus percent-rated airflow. For more complex equipment, such as a cooling tower, the program considers such variables as the wet-bulb temperature, the approach (difference between the condenser water supply temperature and the wet-bulb temperature) and the range (difference between the condenser water entering and leaving temperatures). Each of these variables is used to adjust both the hourly capacity of the tower and the hourly operation (one fan, two fans or no fans). For all but the simplest systems, programs of this category must be used to obtain accurate results. Questions regarding a particular tool's ability to model the building on an hour-by-hour, full-year basis should be addressed to the proprietor or distributor.

EXTERIOR WALL. Walls including both above-grade walls and basement walls.

❖ Wall insulation requirements defined within the code include almost all opaque exterior construction-bounding conditioned space. Depending on the compliance method, the wall type, glazing percentage and whether the wall is on the exterior or just separating conditioned from unconditioned space can affect the wall insulation requirement. Note also that doors (both glazed and opaque) are considered fenestration (see commentary, "Fenestration").

In earlier editions, the code included the limitation that an exterior wall is vertical or sloped at an angle of 60 degrees (1.1 rad) or greater from the horizontal. This limitation may still be helpful to consider if dealing with unusual situations such as an A-frame building. Where a determination is needed to decide whether the roof or wall provisions are appropriate, this limitation could be helpful.

FENESTRATION. Skylights, roof windows, vertical windows (fixed or moveable), opaque doors, glazed doors, glazed block and combination opaque/glazed doors. Fenestration includes products with glass and nonglass glazing materials.

❖ The term "fenestration" refers both to opaque and glazed doors and the light-transmitting areas of a wall or roof, but primarily windows and skylights. The code sets performance requirements for fenestration by establishing separate requirements that differ from the wall and roof requirements based on the type of fenestration and, in the case of the prescriptive commercial requirements, by limiting the fenestration area. In some of the compliance options, the fenestration type and area allowed depends on the shading coefficient, the size of overhangs, the thermal performance (U-factor) and whether daylighting controls are installed.

Site-built windows, doors, and skylights are exempt from the air leakage testing requirements given in Section N1102.4.3. This definition provides information on what exactly site-built means, to define the scope of the exception given in Section N1102.4.3.

HEATED SLAB. Slab-on-grade construction in which the heating elements, hydronic tubing, or hot air distribution system is in contact with, or placed within or under, the slab.

❖ The space above a heated slab is always conditioned (heated). The definition clarifies that certain slabs are the heating source and, as covered in the code, may require more insulation than unheated slabs. The installation of a radiant heat source in a space does not, in and of itself, qualify a slab as heated.

HIGH-EFFICACY LAMPS. Compact fluorescent lamps, T-8 or smaller diameter linear fluorescent lamps, or lamps with a minimum efficacy of:

1. 60 lumens per watt for lamps over 40 watts;
2. 50 lumens per watt for lamps over 15 watts to 40 watts; and
3. 40 lumens per watt for lamps 15 watts or less.

❖ The code contains requirements that a percentage of permanent lighting fixtures have high-efficacy lamps. Therefore, this definition is necessary to indicate what is required to qualify as a high-efficacy lamp.

INFILTRATION. The uncontrolled inward air leakage into a building caused by the pressure effects of wind or the effect of differences in the indoor and outdoor air density or both.

❖ Air leakage is random movement of air into and out of a building through cracks and holes in the building envelope. In technical terms, air leakage is called "infiltration" (air moving into a building) or "exfiltration" (air moving out of a building). In nontechnical terms, air leaks are often referred to as "drafts." Infiltration may be reduced by either reducing the sources of air leakage (joints, penetrations and holes in the building envelope) or by reducing the pressures driving the airflow.

INSULATING SHEATHING. An insulating board with a core material having a minimum *R*-value of R-2.

❖ Some exterior hardboard and vinyl siding products are not recommended for use in direct contact with aluminum-foil-faced sheathing products (check with the product manufacturer). Other sheathing products are uniquely designed for use in direct contact with wood, brick, vinyl, aluminum and hardboard-based sidings. To be considered insulating sheathing, the product must have an *R*-value no less than R-2.

LOW-VOLTAGE LIGHTING. Lighting equipment powered through a transformer such as a cable conductor, a rail conductor and track lighting.

❖ Track lighting is a form of low-voltage lighting featuring a continuous-powered track that accepts fixtures (called heads) anywhere along its length. The heads and the wide variety of lamps (bulbs) they accept suit virtually every lighting situation and most heads are also adjustable. One-, two-, three- and four-circuit tracks let the lighting design professional mix standard and low-voltage heads or have multiple switching options. Track systems are easily modified or added to as lighting needs change. Although tracks can be recessed flush with a ceiling surface or suspended from posts, they are typically fastened to the surface either directly or with mounting clips. Controls ranging from a simple on/off switch to elaborate programmable dimming devices are widely available, making the system extremely flexible.

MANUAL. Capable of being operated by personal intervention (see "Automatic").

❖ Devices, systems or equipment having manual controls or overrides are designed to operate safely with only human intervention instead of having an automatic operation or control system (see commentary, "Automatic").

A typical light switch would be an example of a manual device. The lights that the switch controls would stay either off or on until a person took the action of flipping the switch to the appropriate position.

PROPOSED DESIGN. A description of the proposed building used to estimate annual energy use for determining compliance based on total building performance.

❖ The proposed design is simply a description of the proposed building that is being used to estimate annual energy costs for determining compliance based on total building performance. The proposed design is effectively the subject building intended to be built. The performance of the proposed design (the building exactly as it is anticipated to be constructed) is then compared to the standard reference design (a similar building that is assumed to be built to a prescriptive set of minimum code requirements). Although the simulated performance alternative method permits tradeoffs in energy use between different systems, the proposed design must still comply with the mandatory requirements. General ("mandatory") requirements are separately specified in this chapter. For instance, the air-leakage requirements are general requirements that apply, even if the project uses any one of the two performance methods referenced above.

READILY ACCESSIBLE. Capable of being reached quickly for operation, renewal or inspection without requiring those to whom ready access is requisite to climb over or remove obstacles or to resort to portable ladders or access equipment (see "*Accessible*").

❖ Readily accessible can be described as the capability of being quickly reached or approached for operation, inspection, observation or emergency action. Ready access does not require the removal or movement of any door panel or similar obstruction or overcoming physical obstructions or obstacles, including differential elevation.

REPAIR. The reconstruction or renewal of any part of an existing building.

❖ The repair of an item, appliance, energy-using subsystem or other piece of equipment typically does not require a permit. This definition makes it clear that a repair is limited to work on the item and does not include its replacement or other new work.

This definition is important when applying the provisions of Section C101.4.3 and determining how the code will apply to existing buildings and installations.

RESIDENTIAL BUILDING. For this code, includes detached one- and two-family dwellings and multiple single-family dwellings (townhouses) as well as Group R-2, R-3 and R-4 buildings three stories or less in height above grade plane.

❖ The definition of a residential building is important not only for what it does include, but also for what it does not. One of the primary limitations of this definition is the fact that this term will only include the Group R-2, R-3 and R-4 occupancies when they are three stories or less in height. Therefore, if a Group R-2, R3 or R-4 occupancy is over three stories in height, it would be defined as a "Commercial building" (see definition, "Commercial building") and be required to comply with the requirements of Chapter 4[CE] instead of the residential provisions of Chapter 4[RE]. Buildings that are classified as a Group R-3 are not affected by the three-story limitation. Because of this, a one- or two-family dwelling would always be considered as a residential building regardless of the number of stories.

It should be noted that a Group R-1 building is not included within this definition. Therefore, any hotel, motel or similar use that is classified as a Group R-1 occupancy would need to comply with the commercial building requirements. This would apply even if the hotel or motel was three stories or less in height.

Though not specifically addressed within this definition, any building that is built under the provisions of the *International Residential Code*® (IRC®) would also be considered as being a "residential building."

***R*-VALUE (THERMAL RESISTANCE).** The inverse of the time rate of heat flow through a body from one of its bounding surfaces to the other surface for a unit temperature difference between the two surfaces, under steady state conditions, per unit area ($h \cdot ft^2 \cdot °F/Btu$) [$(m^2 \cdot K)/W$].

❖ Thermal resistance measures how well a material or series of materials retards heat flow. Insulation thermal resistance is rated using *R*-values. As the *R*-value of an element or assembly increases, the heat loss or gain through that element or assembly decreases. Thus, a higher *R*-value is considered better than a lower *R*-value.

SERVICE WATER HEATING. Supply of hot water for purposes other than comfort heating.

❖ Although the definition makes it clear that the code requirement applies to equipment used to produce and distribute hot water for purposes other than comfort heating, the definition also applies to energy-efficient process water heating systems and equipment. Equipment providing or distributing hot water for uses such as restrooms; showers; laundries; kitchens; pools and spas; defrosting of sidewalks and driveways; carwashes; beauty salons and other commercial enterprises are included. Space-conditioning boilers and distribution systems are not considered service water heating components.

SKYLIGHT. Glass or other transparent or translucent glazing material installed at a slope of less than 60 degrees (1.05 rad) from horizontal. Glazing material in skylights, including unit skylights, solariums, sunrooms, roofs and sloped walls is included in this definition.

❖ A skylight is a glazed opening in a roof that admits daylight. Skylights are often the only method of bringing natural light into an interior, enclosed area. Unfortunately, these fixtures often do their job too well. Installing too many skylights or ones too large for the room can lead to overheating during warm-weather months. Also, choosing the most energy-efficient models can compromise light transmission—the reason people buy skylights in the first place.

Skylights are available in a variety of sizes and shapes, though rectangular units are the most common. Although most skylights are fixed or inoperable, others can be opened and shut like a window or have hidden ventilating systems. Large operable skylights designed for the sloping ceilings of attic rooms are even marketed as "roof windows." These approaches help cool the room in warm weather by venting hot air. The IBC requires either tempered or laminated glass in skylights. Both types are designed to stand up to snow loads and provide protection against falling objects. Tempered glass breaks into small pieces, rather than large shards, if damaged. Laminated glass, which is fused with a thin layer of plastic, stays in place for added safety if broken. Laminated glass is also better at keeping out sound and is slightly more energy efficient, though also slightly more expensive. Skylights are not energy efficient. They collect little heat during the winter, which is when it is needed most—when the sun is low in the sky. Worse yet, because they're located where the pressure difference between the inside and outside of the house is greatest, skylights are an easy escape route for heated air. Also, in the summer they can heat up the home quickly—when it is not needed.

Like windows, skylights offer a variety of energy-efficient glazing options, including low-E and tinted glass. Green tints are better than bronze tints for reducing solar heat gain while letting in plenty of visible light. And because skylights usually are not visible from the street, tinted glazings are less likely to affect aesthetics.

It is important to note that the term "skylight" can also include glazed roofs and sloped walls. This distinction can affect the proper application of the code requirements. To help determine the appropriate *U*-factor, this definition includes the slope limitation. If the slope of the roof or sloped wall is 60 degrees or less from the horizontal, then the skylight *U*-factor is appropriate. If the slope is greater than 60 degrees, then the glazing would be considered as a fenestration in a wall. As an example using the requirements of Table N1102.1.1 for Climate Zone 1, a skylight would require a *U*-factor of 0.75, while the fenestration in a wall would require a *U*-factor of 1.20. Both the skylight and the more vertical wall fenestration would require an SHGC of 0.30 (see Table N1102.1.1, Note b).

SOLAR HEAT GAIN COEFFICIENT (SHGC). The ratio of the solar heat gain entering the space through the fenestration assembly to the incident solar radiation. Solar heat gain includes directly transmitted solar heat and absorbed solar radiation which is then reradiated, conducted or convected into the space.

❖ The SHGC is the fraction of incident solar radiation admitted through a window or skylight. This includes the solar radiation that is both directly transmitted and that which is absorbed and subsequently released inward. Therefore, the SHGC measures how well a window blocks the heat from sunlight. The SHGC is the fraction of the heat from the sun that enters through a window. SHGC is expressed as a number between 0 and 1. The lower a window's SHGC, the less solar heat it transmits.

In the warmer climate zones, where cooling is the dominant requirement, the code will generally impose a limitation on the amount of solar heat gain permitted. In colder climate zones, the code will generally not place any SHGC requirement on the fenestration. Thus, it will depend upon the climate zone as to whether a higher or lower SHGC is best. While windows with lower SHGC values reduce summer cooling and overheating, they also reduce free winter solar heat gain.

STANDARD REFERENCE DESIGN. A version of the *proposed design* that meets the minimum requirements of this code and is used to determine the maximum annual energy use requirement for compliance based on total building performance.

❖ The standard reference design is simply the same building design as that intended to be built (see commentary, "Proposed design"), except the energy conservation features required by the code (insulation, windows, infiltration, mechanical, lighting and service water heating systems) are modified to meet the minimum prescriptive requirements, as applicable. Note that the standard reference design is not truly a separate building and it is never actually built. It is the baseline against which the proposed design is measured.

The performance of the proposed design (the building exactly as it is anticipated to be constructed) is then compared to the standard reference design (a similar building that is assumed to be built to a prescriptive set of minimum code requirements). Under the performance paths, if the energy efficiency of the proposed design is equal to or better than that of the standard reference design, then the proposed design is acceptable and in compliance with the code requirements.

SUNROOM. A one-story structure attached to a dwelling with a glazing area in excess of 40 percent of the gross area of the structure's exterior walls and roof.

❖ Sunrooms are unique rooms that create a space that differs in character from that provided by conventional portions of a dwelling. Sunrooms and other highly glazed structures are sometimes called conservatories or solariums. These sunrooms are often added onto an existing building but there is nothing in the code that would prohibit them from being constructed as a part of a new dwelling.

This definition distinguishes sunrooms from other conventional spaces because they are limited to one story in height and the total glazing area needs to be at least 40 percent of the exterior wall and roof area of the sunroom. This definition and its use in Sections

N1102.2.2 and N1102.3.5 are important since most dwellings do not have such large amounts of glazing.

Many of these sunrooms are constructed to be unconditioned spaces and will not have any type of heating or cooling. Other sunrooms may be conditioned either indirectly by openings from the adjacent conditioned space of the dwelling, be provided with a separate space conditioning system or they can be served as a separate zone from the dwelling's space-conditioning system. Due to the large amount of glazing and, therefore, the probability that the sunroom will either be hot in the summer or cold in the winter and may be unused during such times, the code provisions address the thermal isolation of these spaces (see definition, "Thermal isolation").

When addressing the thermal isolation of the sunrooms, the provisions of Section N1102.2.2 need to be reviewed and code users need to realize how they coordinate with Section N1101.3. Section N1102.2.10 will require that "new" walls that separate a sunroom from a conditioned space must comply with the building thermal envelope requirements. Therefore, if a sunroom is built as an addition to an existing dwelling, the existing wall between the sunroom and any conditioned space is not regulated. The same concept applies to the provisions of Section N1102.3.6. Any "new windows and doors" between the sunroom and the dwelling will be regulated, but existing fenestration could remain unchanged.

THERMAL ISOLATION. Physical and space conditioning separation from *conditioned space(s)*. The *conditioned space*(s) shall be controlled as separate zones for heating and cooling or conditioned by separate equipment.

❖ This term is somewhat conceptually similar to the separation provided by the building envelope, but instead of being between the conditioned space and the exterior or conditioned space and unconditioned space, this separation occurs between two conditioned spaces. In this situation it is the separation between the dwelling and the sunroom. Therefore, where the sunroom is an addition, the existing exterior wall of the dwelling will generally provide the thermal isolation between the dwelling and the sunroom at the point where the sunroom addition is attached. If a new door or window opening is added to permit passage between or to provide a connection with the dwelling and the sunroom, that new door or window is required to comply with the fenestration *U*-factor specified in Table N1102.1.1. By requiring the maximum *U*-factor, it ensures a reasonable level of energy conservation between the dwelling and the sunroom so they would not need to be brought into compliance with the *U*-factor specified in the table.

This exemption for the existing doors and windows is based not only on the language in this definition, but also on the scoping requirements found in Sections N1102.3.6. Since these existing doors and windows would be serving as a part of the building envelope for the dwelling prior to the addition of the sunroom, they may be accepted as a part of the thermal isolation between the dwelling and any new sunroom addition (see commentary, "Sunroom" and Section N1102.3.5).

This definition is also important due to the restriction that it places on the heating and cooling system for the various spaces. In order to be thermally isolated, the two spaces are either served by separate systems or as separate zones on a common system.

THERMOSTAT. An automatic control device used to maintain temperature at a fixed or adjustable set point.

❖ Thermostats combine control and sensing functions in a single device. Thermostats may signal other control devices to trigger an action when certain temperatures are reached or surpassed. Because thermostats are so prevalent, the various types and their operating characteristics are described here.

The occupied/unoccupied or dual-temperature room thermostat reduces temperature at night. It may be indexed (changed from occupied to unoccupied operation or vice versa) individually from a remote point or in a group by a manual or time switch. Some types have an individual clock and switch built in.

The pneumatic day-night thermostat uses a two-pressure air-supply system, where changing the pressure at a central point from one value to the other actuates switching devices in the thermostat and indexes it.

The heating/cooling or summer/winter thermostat can have its action reversed and its set point changed in response to outdoor and comfort conditions. It is used to actuate controlled devices, such as valves or dampers, that regulate a heating source at one time and a cooling source at another.

The pneumatic heating-cooling thermostat uses a two-pressure air supply similar to that described for occupied/unoccupied thermostats.

Multistage thermostats are arranged to operate two or more successive steps in sequence.

A submaster thermostat has its set point raised or lowered over a predetermined range, in accordance with variations in output from a master controller. The master controller can be a thermostat, manual switch, pressure controller or similar device. For example, a master thermostat measuring outdoor air temperature can be used to readjust the set point of a submaster thermostat that controls the water temperature in a heating system.

A wet-bulb thermostat is often used (in combination with a dry-bulb thermostat) for humidity control. Using a wick or another means of keeping the bulb wet with pure (distilled) water and rapid air motion to ensure a true wet-bulb measurement is essential. Wet-bulb thermostats are seldom used.

A dew-point thermostat is designed to control dew-point temperatures.

***U*-FACTOR (THERMAL TRANSMITTANCE).** The coefficient of heat transmission (air to air) through a building component or assembly, equal to the time rate of heat flow per unit area and unit temperature difference between the warm side and cold side air films (Btu/h · ft^2 · °F) [W/(m^2 · K)].

❖ Thermal transmittance (U-factor) is a measure of how well a material or series of materials conducts heat. U-factors for window and door assemblies are the reciprocal of the assembly R-value:

$$U\text{-factor} = \frac{1}{R\text{-value}}$$

For other building assemblies, such as a wall or roof/ceiling, the R-value used in the above equation is the R-value of the entire assembly, not just the insulation. This distinction is important and reflects the provisions of Sections N1102.1.2 and N1102.1.3. It also explains why there are differences between the comparable values of Tables N1102.1.1 and N1102.1.3.

The U-factors will be used in a number of locations of the code. When using the performance options or the "total UA alternative" (see Section N1102.1.4), the individual thermal transmittance (U-factor) of each element is multiplied by the area of each envelope component (walls, floors and ceilings) and the area of each fenestration element (doors, windows and skylights). Therefore, the UA is simply the U-factor times the area. For example, a 400-square-foot (37 m^2) wall with a U-factor of 0.082 would result in a UA of 32.8 ($400 \times 0.082 = 32.8$). The total UA would simply be the sum of all of the individual UAs for each building element (walls, floors, ceilings, doors, windows and skylights).

VENTILATION AIR. That portion of supply air that comes from outside (outdoors) plus any recirculated air that has been treated to maintain the desired quality of air within a designated space.

❖ Ventilation air can be used for comfort cooling, control of air contaminants, equipment cooling and replenishing oxygen levels (see commentary, "Ventilation").

VISIBLE TRANSMITTANCE [VT]. The ratio of visible light entering the space through the fenestration product assembly to the incident visible light, visible transmittance, includes the effects of glazing material and frame and is expressed as a number between 0 and 1.

❖ The required SHGC rating for glazed fenestration has been continuously lowered such that the amount of light admitted by the glazing is lower. With the advent of requirements for daylighting, the need is for fenestration to admit light; thus, there are opposing needs in the code. The visible transmittance of glazing is now an issue and is therefore defined in the code.

WHOLE HOUSE MECHANICAL VENTILATION SYSTEM. An exhaust system, supply system, or combination thereof that is designed to mechanically exchange indoor air with outdoor air when operating continuously or through a programmed intermittent schedule to satisfy the whole house ventilation rates.

❖ Because the construction of buildings is increasingly more air tight, the IMC now requires whole house mechanical ventilation when the measured air leakage rate is lower than a threshold amount. The IECC contains requirements for fan efficacy.

ZONE. A space or group of spaces within a building with heating or cooling requirements that are sufficiently similar so that desired conditions can be maintained throughout using a single controlling device.

❖ The simplest all-air system is a supply unit serving a single-temperature control zone. Ideally, this system responds completely to the space needs and well-designed control systems maintain temperature and humidity closely and efficiently. Single-zone systems can be shut down when not required without affecting the operation of adjacent areas. Thus, the concept of "zoning" for temperature control requires discussion.

Exterior Zoning. Exterior zones are affected by varying weather conditions—wind, temperature and sun—and, depending on the geographic area and season, require both heating and cooling. This variation gives the designer considerable flexibility in choosing a system and results in the greatest advantages. The need for separate perimeter zone heating is determined by:

- Severity of the heating load (i.e., geographic location).
- Nature and orientation of the building envelope.
- Effects of downdraft at windows and the radiant effect of the cold glass surface (type of glass, area, height and U-factor).
- Type of occupancy (sedentary versus transient).
- Operating costs (in buildings such as offices and schools that are occupied for considerable periods). Fan-operating costs can be reduced by heating with perimeter radiation during unoccupied periods, rather than operating the main supply fans or local unit fans.

Interior Zoning. Conditions in interior spaces are relatively constant because they are isolated from external influences. Usually, interior spaces require cooling throughout the year. Interior spaces with a roof exposure, however, may require similar treatment to perimeter spaces requiring heat. To summarize, zone control is required when the conditions at the thermostat are not representative of all the rooms or the entire exposure. This situation will almost certainly occur if any of the following conditions exist:

- The building has more than one level.
- One or more spaces are used for entertaining large groups.
- One or more spaces have large glass areas.
- The building has an indoor swimming pool or hot tub.
- The building has a solarium or atrium. In addition, zoning may be required when several rooms or spaces are isolated from each other and from the thermostat.

- The building spreads out in many directions (wings). Some spaces are distinctly isolated from the rest of the building.
- The envelope only has one or two exposures.
- The building has a room or rooms in a basement.
- The building has a room or rooms in an attic space.
- The building has one or more rooms with a slab or exposed floor.
- There are discrete heating/cooling duct systems for each zone requiring control.
- There are automatic zone damper systems in a single heating/cooling duct system.

N1101.10 (R301.1) Climate zones. Climate zones from Figure N1101.10 or Table N1101.10 shall be used in determining the applicable requirements in Sections N1101 through N1105. Locations not in Table N1101.10 (outside the United States) shall be assigned a climate zone based on Section N101.10.2.

❖ Climate involves temperature, moisture, wind and sun and also includes both daily and seasonal patterns of variation of the parameters. To account for these variations, the code establishes climate zones that serve as the basis for the code provisions.

This section serves as the starting point for determining virtually all of the code requirements, especially under the prescriptive compliance paths. Because of their easy-to-understand graphic nature, maps have proven useful over the years as an effective way to enable code users to determine climate-dependent requirements. Therefore, for the United States, the climate zones are shown in the map in Figure N1101.10. Because of the limited size of the map, the code also includes a listing of the climate zones by states and counties in Table N1101.10. Table N1101.10 will allow users to positively identify climate-zone assignments in those few locations for which the map interpretation may be difficult. Whether the map or the county list is used, the climate classification for each area will be the same.

When dealing with the prescriptive compliance paths, the code user would simply look at the map or listing and select the proper climate zone based upon the location of the building. When using a performance approach, additional climatic data may be needed.

Virtually every building energy code that has been developed for use in the United States has included a performance-based compliance path, which allows users to perform an energy analysis and demonstrate compliance based on equivalence with the prescriptive requirements. To perform these analyses, users must select appropriate weather data for their given project's location. The selection of appropriate weather data is straightforward for any project located in or around one of the various weather stations within the United States. For other locations, selecting the most appropriate weather site can be problematic. The codes themselves provide little guidance to help with this selection process. During the development of the new climate zones, the developers mapped every county in the United States to the most appropriate SAMSON station (National Climatic Data Center "Solar and Meteorological Surface Observation Network" station) for each county as a whole. This mapping is not included within the code but may be used in some compliance software. Designating an appropriate SAMSON station should not be considered to be the only climate data permitted for a given county. It could, however, be used in the absence of better information. Where local data better reflects regional or microclimatic conditions of an area, they would be appropriate to use. For example, elevation has a large impact on climate and can vary dramatically within individual counties, especially in the western United States. Where elevation differences are significant, code officials may require use of sites that differ from the sites designated as being the most appropriate for the county. For additional information on this topic, review the paper "Climate Classification for Building Energy Codes and Standards."

N1101.10.1 (R301.2) Warm humid counties. Warm humid counties are identified in Table N1101.10 by an asterisk.

❖ Table N1101.10 provides a listing of the counties within the southeastern United States that fall below the white-dashed line that appears in the map in Figure N1101.10. The warm-humid climate designation includes parts of eight states and also covers all of Florida, Hawaii and the U.S. territories. Table N1101.10.2(1) provides the details that were used to determine the classification of the warm-humid designation for the counties.

There currently are very few requirements within the code that are specifically tied to the warm-humid climate criteria. Although not tied directly to the warm-humid designation, many other code sections, such as those addressing moisture control and energy recovery ventilation systems, do take these climatic features into account.

N1101.10.2 (R301.3) International climate zones. The climate zone for any location outside the United States shall be determined by applying Table N1101.10.2(1) and then Table N1101.10.2(2).

❖ Although the code and the new climate zone classifications it includes are predominately used within the United States, they can be used in any location. Because the mapping and decisions that were made during the development of the new climate zones focused primarily upon the United States, this section provides the details of how to properly classify the climate zones based upon the thermal criteria [Table N1101.10.2(2)], the major climate types [Table N1101.10.2(1)] and the warm-humid criteria (Section N1101.10.2) for locations outside of the United States.

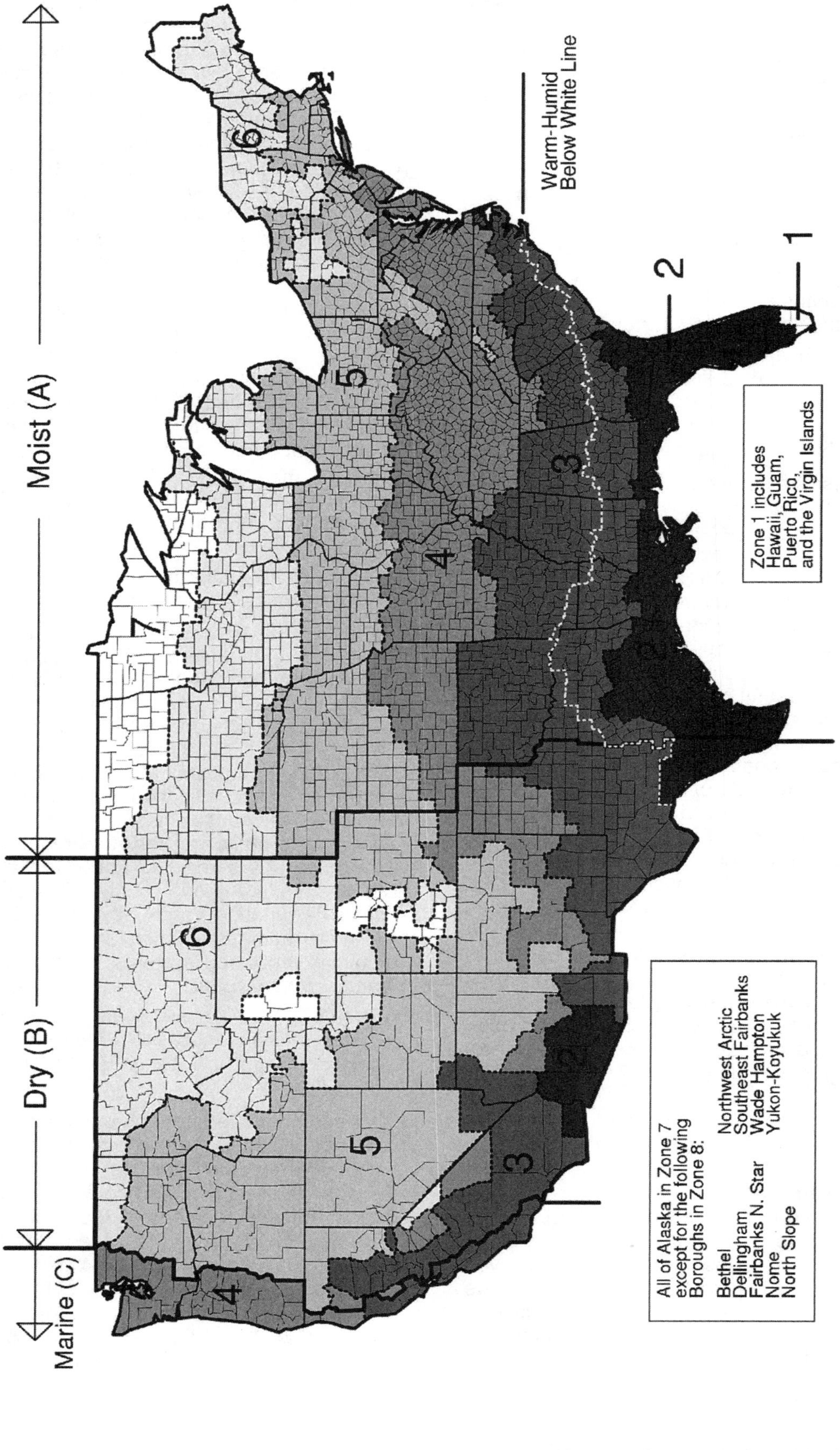

FIGURE N1101.10 (R301.1)
CLIMATE ZONES

**TABLE N1101.10 (R301.1)
CLIMATE ZONES, MOISTURE REGIMES, AND WARM-HUMID
DESIGNATIONS BY STATE, COUNTY AND TERRITORY**

Key: A – Moist, B – Dry, C – Marine. Absence of moisture designation indicates moisture regime is irrelevant. Asterisk (*) indicates a warm-humid location.

US STATES

ALABAMA

3A Autauga*
2A Baldwin*
3A Barbour*
3A Bibb
3A Blount
3A Bullock*
3A Butler*
3A Calhoun
3A Chambers
3A Cherokee
3A Chilton
3A Choctaw*
3A Clarke*
3A Clay
3A Cleburne
3A Coffee*
3A Colbert
3A Conecuh*
3A Coosa
3A Covington*
3A Crenshaw*
3A Cullman
3A Dale*
3A Dallas*
3A DeKalb
3A Elmore*
3A Escambia*
3A Etowah
3A Fayette
3A Franklin
3A Geneva*
3A Greene
3A Hale
3A Henry*
3A Houston*
3A Jackson
3A Jefferson
3A Lamar
3A Lauderdale
3A Lawrence
3A Lee
3A Limestone
3A Lowndes*
3A Macon*
3A Madison
3A Marengo*
3A Marion
3A Marshall
2A Mobile*
3A Monroe*
3A Montgomery*
3A Morgan
3A Perry*
3A Pickens
3A Pike*
3A Randolph
3A Russell*
3A Shelby
3A St. Clair
3A Sumter
3A Talladega
3A Tallapoosa
3A Tuscaloosa
3A Walker
3A Washington*
3A Wilcox*
3A Winston

ALASKA

7 Aleutians East
7 Aleutians West
7 Anchorage
8 Bethel
7 Bristol Bay
7 Denali
8 Dillingham
8 Fairbanks North Star
7 Haines
7 Juneau
7 Kenai Peninsula
7 Ketchikan Gateway
7 Kodiak Island
7 Lake and Peninsula
7 Matanuska-Susitna
8 Nome
8 North Slope
8 Northwest Arctic
7 Prince of Wales Outer Ketchikan
7 Sitka
7 Skagway-Hoonah-Angoon
8 Southeast Fairbanks
7 Valdez-Cordova
8 Wade Hampton
7 Wrangell-Petersburg
7 Yakutat
8 Yukon-Koyukuk

ARIZONA

5B Apache
3B Cochise
5B Coconino
4B Gila
3B Graham
3B Greenlee
2B La Paz
2B Maricopa
3B Mohave
5B Navajo
2B Pima
2B Pinal
3B Santa Cruz
4B Yavapai
2B Yuma

ARKANSAS

3A Arkansas
3A Ashley
4A Baxter
4A Benton
4A Boone
3A Bradley
3A Calhoun
4A Carroll
3A Chicot
3A Clark
3A Clay
3A Cleburne
3A Cleveland
3A Columbia*
3A Conway
3A Craighead
3A Crawford
3A Crittenden
3A Cross
3A Dallas
3A Desha
3A Drew
3A Faulkner
3A Franklin
4A Fulton
3A Garland
3A Grant
3A Greene
3A Hempstead*
3A Hot Spring
3A Howard
3A Independence
4A Izard
3A Jackson
3A Jefferson
3A Johnson
3A Lafayette*
3A Lawrence
3A Lee
3A Lincoln
3A Little River*
3A Logan
3A Lonoke
4A Madison
4A Marion
3A Miller*
3A Mississippi
3A Monroe
3A Montgomery
3A Nevada
4A Newton
3A Ouachita
3A Perry
3A Phillips
3A Pike
3A Poinsett
3A Polk
3A Pope
3A Prairie
3A Pulaski
3A Randolph
3A Saline
3A Scott
4A Searcy
3A Sebastian
3A Sevier*
3A Sharp
3A St. Francis
4A Stone
3A Union*
3A Van Buren
4A Washington
3A White
3A Woodruff
3A Yell

CALIFORNIA

3C Alameda
6B Alpine
4B Amador
3B Butte
4B Calaveras
3B Colusa
3B Contra Costa
4C Del Norte
4B El Dorado
3B Fresno
3B Glenn

(continued)

**TABLE N1101.10 (R301.1)—continued
CLIMATE ZONES, MOISTURE REGIMES, AND WARM-HUMID
DESIGNATIONS BY STATE, COUNTY AND TERRITORY**

4C Humboldt
2B Imperial
4B Inyo
3B Kern
3B Kings
4B Lake
5B Lassen
3B Los Angeles
3B Madera
3C Marin
4B Mariposa
3C Mendocino
3B Merced
5B Modoc
6B Mono
3C Monterey
3C Napa
5B Nevada
3B Orange
3B Placer
5B Plumas
3B Riverside
3B Sacramento
3C San Benito
3B San Bernardino
3B San Diego
3C San Francisco
3B San Joaquin
3C San Luis Obispo
3C San Mateo
3C Santa Barbara
3C Santa Clara
3C Santa Cruz
3B Shasta
5B Sierra
5B Siskiyou
3B Solano
3C Sonoma
3B Stanislaus
3B Sutter
3B Tehama
4B Trinity
3B Tulare
4B Tuolumne
3C Ventura
3B Yolo
3B Yuba

COLORADO

5B Adams
6B Alamosa
5B Arapahoe
6B Archuleta
4B Baca
5B Bent
5B Boulder
6B Chaffee
5B Cheyenne
7 Clear Creek
6B Conejos
6B Costilla
5B Crowley
6B Custer
5B Delta
5B Denver
6B Dolores
5B Douglas
6B Eagle
5B Elbert
5B El Paso
5B Fremont
5B Garfield
5B Gilpin
7 Grand
7 Gunnison
7 Hinsdale
5B Huerfano
7 Jackson
5B Jefferson
5B Kiowa
5B Kit Carson
7 Lake
5B La Plata
5B Larimer
4B Las Animas
5B Lincoln
5B Logan
5B Mesa
7 Mineral
6B Moffat
5B Montezuma
5B Montrose
5B Morgan
4B Otero
6B Ouray
7 Park
5B Phillips
7 Pitkin
5B Prowers
5B Pueblo
6B Rio Blanco
7 Rio Grande
7 Routt
6B Saguache
7 San Juan
6B San Miguel
5B Sedgwick
7 Summit
5B Teller
5B Washington
5B Weld
5B Yuma

CONNECTICUT

5A (all)

DELAWARE

4A (all)

DISTRICT OF COLUMBIA

4A (all)

FLORIDA

2A Alachua*
2A Baker*
2A Bay*
2A Bradford*
2A Brevard*
1A Broward*
2A Calhoun*
2A Charlotte*
2A Citrus*
2A Clay*
2A Collier*
2A Columbia*
2A DeSoto*
2A Dixie*
2A Duval*
2A Escambia*
2A Flagler*
2A Franklin*
2A Gadsden*
2A Gilchrist*
2A Glades*
2A Gulf*
2A Hamilton*
2A Hardee*
2A Hendry*
2A Hernando*
2A Highlands*
2A Hillsborough*
2A Holmes*
2A Indian River*
2A Jackson*
2A Jefferson*
2A Lafayette*
2A Lake*
2A Lee*
2A Leon*
2A Levy*
2A Liberty*
2A Madison*
2A Manatee*
2A Marion*
2A Martin*
1A Miami-Dade*
1A Monroe*
2A Nassau*
2A Okaloosa*
2A Okeechobee*
2A Orange*
2A Osceola*
2A Palm Beach*
2A Pasco*
2A Pinellas*
2A Polk*
2A Putnam*
2A Santa Rosa*
2A Sarasota*
2A Seminole*
2A St. Johns*
2A St. Lucie*
2A Sumter*
2A Suwannee*
2A Taylor*
2A Union*
2A Volusia*
2A Wakulla*
2A Walton*
2A Washington*

GEORGIA

2A Appling*
2A Atkinson*
2A Bacon*
2A Baker*
3A Baldwin
4A Banks
3A Barrow
3A Bartow
3A Ben Hill*
2A Berrien*
3A Bibb
3A Bleckley*
2A Brantley*
2A Brooks*
2A Bryan*
3A Bulloch*
3A Burke
3A Butts
3A Calhoun*
2A Camden*
3A Candler*
3A Carroll
4A Catoosa
2A Charlton*
2A Chatham*
3A Chattahoochee*
4A Chattooga
3A Cherokee
3A Clarke
3A Clay*
3A Clayton
2A Clinch*
3A Cobb
3A Coffee*
2A Colquitt*
3A Columbia
2A Cook*
3A Coweta

(continued)

TABLE N1101.10 (R301.1)—continued
CLIMATE ZONES, MOISTURE REGIMES, AND WARM-HUMID DESIGNATIONS BY STATE, COUNTY AND TERRITORY

3A Crawford
3A Crisp*
4A Dade
4A Dawson
2A Decatur*
3A DeKalb
3A Dodge*
3A Dooly*
3A Dougherty*
3A Douglas
3A Early*
2A Echols*
2A Effingham*
3A Elbert
3A Emanuel*
2A Evans*
4A Fannin
3A Fayette
4A Floyd
3A Forsyth
4A Franklin
3A Fulton
4A Gilmer
3A Glascock
2A Glynn*
4A Gordon
2A Grady*
3A Greene
3A Gwinnett
4A Habersham
4A Hall
3A Hancock
3A Haralson
3A Harris
3A Hart
3A Heard
3A Henry
3A Houston*
3A Irwin*
3A Jackson
3A Jasper
2A Jeff Davis*
3A Jefferson
3A Jenkins*
3A Johnson*
3A Jones
3A Lamar
2A Lanier*
3A Laurens*
3A Lee*
2A Liberty*
3A Lincoln
2A Long*
2A Lowndes*
4A Lumpkin
3A Macon*
3A Madison
3A Marion*
3A McDuffie
2A McIntosh*
3A Meriwether
2A Miller*
2A Mitchell*
3A Monroe
3A Montgomery*
3A Morgan
4A Murray
3A Muscogee
3A Newton
3A Oconee
3A Oglethorpe
3A Paulding
3A Peach*
4A Pickens
2A Pierce*
3A Pike
3A Polk
3A Pulaski*
3A Putnam
3A Quitman*
4A Rabun
3A Randolph*
3A Richmond
3A Rockdale
3A Schley*
3A Screven*
2A Seminole*
3A Spalding
4A Stephens
3A Stewart*
3A Sumter*
3A Talbot
3A Taliaferro
2A Tattnall*
3A Taylor*
3A Telfair*
3A Terrell*
2A Thomas*
3A Tift*
2A Toombs*
4A Towns
3A Treutlen*
3A Troup
3A Turner*
3A Twiggs*
4A Union
3A Upson
4A Walker
3A Walton
2A Ware*
3A Warren
3A Washington
2A Wayne*
3A Webster*
3A Wheeler*
4A White
4A Whitfield
3A Wilcox*
3A Wilkes
3A Wilkinson
3A Worth*

HAWAII

1A (all)*

IDAHO

5B Ada
6B Adams
6B Bannock
6B Bear Lake
5B Benewah
6B Bingham
6B Blaine
6B Boise
6B Bonner
6B Bonneville
6B Boundary
6B Butte
6B Camas
5B Canyon
6B Caribou
5B Cassia
6B Clark
5B Clearwater
6B Custer
5B Elmore
6B Franklin
6B Fremont
5B Gem
5B Gooding
5B Idaho
6B Jefferson
5B Jerome
5B Kootenai
5B Latah
6B Lemhi
5B Lewis
5B Lincoln
6B Madison
5B Minidoka
5B Nez Perce
6B Oneida
5B Owyhee
5B Payette
5B Power
5B Shoshone
6B Teton
5B Twin Falls
6B Valley
5B Washington

ILLINOIS

5A Adams
4A Alexander
4A Bond
5A Boone
5A Brown
5A Bureau
5A Calhoun
5A Carroll
5A Cass
5A Champaign
4A Christian
5A Clark
4A Clay
4A Clinton
5A Coles
5A Cook
4A Crawford
5A Cumberland
5A DeKalb
5A De Witt
5A Douglas
5A DuPage
5A Edgar
4A Edwards
4A Effingham
4A Fayette
5A Ford
4A Franklin
5A Fulton
4A Gallatin
5A Greene
5A Grundy
4A Hamilton
5A Hancock
4A Hardin
5A Henderson
5A Henry
5A Iroquois
4A Jackson
4A Jasper
4A Jefferson
5A Jersey
5A Jo Daviess
4A Johnson
5A Kane
5A Kankakee
5A Kendall
5A Knox
5A Lake
5A La Salle
4A Lawrence
5A Lee
5A Livingston
5A Logan
5A Macon
4A Macoupin
4A Madison
4A Marion
5A Marshall
5A Mason
4A Massac
5A McDonough
5A McHenry

(continued)

TABLE N1101.10 (R301.1)—continued
CLIMATE ZONES, MOISTURE REGIMES, AND WARM-HUMID DESIGNATIONS BY STATE, COUNTY AND TERRITORY

5A McLean
5A Menard
5A Mercer
4A Monroe
4A Montgomery
5A Morgan
5A Moultrie
5A Ogle
5A Peoria
4A Perry
5A Piatt
5A Pike
4A Pope
4A Pulaski
5A Putnam
4A Randolph
4A Richland
5A Rock Island
4A Saline
5A Sangamon
5A Schuyler
5A Scott
4A Shelby
5A Stark
4A St. Clair
5A Stephenson
5A Tazewell
4A Union
5A Vermilion
4A Wabash
5A Warren
4A Washington
4A Wayne
4A White
5A Whiteside
5A Will
4A Williamson
5A Winnebago
5A Woodford

INDIANA

5A Adams
5A Allen
5A Bartholomew
5A Benton
5A Blackford
5A Boone
4A Brown
5A Carroll
5A Cass
4A Clark
5A Clay
5A Clinton
4A Crawford
4A Daviess
4A Dearborn
5A Decatur
5A De Kalb
5A Delaware
4A Dubois
5A Elkhart
5A Fayette
4A Floyd
5A Fountain
5A Franklin
5A Fulton
4A Gibson
5A Grant
4A Greene
5A Hamilton
5A Hancock
4A Harrison
5A Hendricks
5A Henry
5A Howard
5A Huntington
4A Jackson
5A Jasper
5A Jay
4A Jefferson
4A Jennings
5A Johnson
4A Knox
5A Kosciusko
5A Lagrange
5A Lake
5A La Porte
4A Lawrence
5A Madison
5A Marion
5A Marshall
4A Martin
5A Miami
4A Monroe
5A Montgomery
5A Morgan
5A Newton
5A Noble
4A Ohio
4A Orange
5A Owen
5A Parke
4A Perry
4A Pike
5A Porter
4A Posey
5A Pulaski
5A Putnam
5A Randolph
4A Ripley
5A Rush
4A Scott
5A Shelby
4A Spencer
5A Starke
5A Steuben
5A St. Joseph
4A Sullivan
4A Switzerland
5A Tippecanoe
5A Tipton
5A Union
4A Vanderburgh
5A Vermillion
5A Vigo
5A Wabash
5A Warren
4A Warrick
4A Washington
5A Wayne
5A Wells
5A White
5A Whitley

IOWA

5A Adair
5A Adams
6A Allamakee
5A Appanoose
5A Audubon
5A Benton
6A Black Hawk
5A Boone
6A Bremer
6A Buchanan
6A Buena Vista
6A Butler
6A Calhoun
5A Carroll
5A Cass
5A Cedar
6A Cerro Gordo
6A Cherokee
6A Chickasaw
5A Clarke
6A Clay
6A Clayton
5A Clinton
5A Crawford
5A Dallas
5A Davis
5A Decatur
6A Delaware
5A Des Moines
6A Dickinson
5A Dubuque
6A Emmet
6A Fayette
6A Floyd
6A Franklin
5A Fremont
5A Greene
6A Grundy
5A Guthrie
6A Hamilton
6A Hancock
6A Hardin
5A Harrison
5A Henry
6A Howard
6A Humboldt
6A Ida
5A Iowa
5A Jackson
5A Jasper
5A Jefferson
5A Johnson
5A Jones
5A Keokuk
6A Kossuth
5A Lee
5A Linn
5A Louisa
5A Lucas
6A Lyon
5A Madison
5A Mahaska
5A Marion
5A Marshall
5A Mills
6A Mitchell
5A Monona
5A Monroe
5A Montgomery
5A Muscatine
6A O'Brien
6A Osceola
5A Page
6A Palo Alto
6A Plymouth
6A Pocahontas
5A Polk
5A Pottawattamie
5A Poweshiek
5A Ringgold
6A Sac
5A Scott
5A Shelby
6A Sioux
5A Story
5A Tama
5A Taylor
5A Union
5A Van Buren
5A Wapello
5A Warren
5A Washington
5A Wayne
6A Webster
6A Winnebago

(continued)

**TABLE N1101.10 (R301.1)—continued
CLIMATE ZONES, MOISTURE REGIMES, AND WARM-HUMID
DESIGNATIONS BY STATE, COUNTY AND TERRITORY**

6A Winneshiek
5A Woodbury
6A Worth
6A Wright

KANSAS

4A Allen
4A Anderson
4A Atchison
4A Barber
4A Barton
4A Bourbon
4A Brown
4A Butler
4A Chase
4A Chautauqua
4A Cherokee
5A Cheyenne
4A Clark
4A Clay
5A Cloud
4A Coffey
4A Comanche
4A Cowley
4A Crawford
5A Decatur
4A Dickinson
4A Doniphan
4A Douglas
4A Edwards
4A Elk
5A Ellis
4A Ellsworth
4A Finney
4A Ford
4A Franklin
4A Geary
5A Gove
5A Graham
4A Grant
4A Gray
5A Greeley
4A Greenwood
5A Hamilton
4A Harper
4A Harvey
4A Haskell
4A Hodgeman
4A Jackson
4A Jefferson
5A Jewell
4A Johnson
4A Kearny
4A Kingman
4A Kiowa
4A Labette
5A Lane
4A Leavenworth
4A Lincoln
4A Linn
5A Logan
4A Lyon
4A Marion
4A Marshall
4A McPherson
4A Meade
4A Miami
5A Mitchell
4A Montgomery
4A Morris
4A Morton
4A Nemaha
4A Neosho
5A Ness
5A Norton
4A Osage
5A Osborne
4A Ottawa
4A Pawnee
5A Phillips
4A Pottawatomie
4A Pratt
5A Rawlins
4A Reno
5A Republic
4A Rice
4A Riley
5A Rooks
4A Rush
4A Russell
4A Saline
5A Scott
4A Sedgwick
4A Seward
4A Shawnee
5A Sheridan
5A Sherman
5A Smith
4A Stafford
4A Stanton
4A Stevens
4A Sumner
5A Thomas
5A Trego
4A Wabaunsee
5A Wallace
4A Washington
5A Wichita
4A Wilson
4A Woodson
4A Wyandotte

KENTUCKY

4A (all)

LOUISIANA

2A Acadia*
2A Allen*
2A Ascension*
2A Assumption*
2A Avoyelles*
2A Beauregard*
3A Bienville*
3A Bossier*
3A Caddo*
2A Calcasieu*
3A Caldwell*
2A Cameron*
3A Catahoula*
3A Claiborne*
3A Concordia*
3A De Soto*
2A East Baton Rouge*
3A East Carroll
2A East Feliciana*
2A Evangeline*
3A Franklin*
3A Grant*
2A Iberia*
2A Iberville*
3A Jackson*
2A Jefferson*
2A Jefferson Davis*
2A Lafayette*
2A Lafourche*
3A La Salle*
3A Lincoln*
2A Livingston*
3A Madison*
3A Morehouse
3A Natchitoches*
2A Orleans*
3A Ouachita*
2A Plaquemines*
2A Pointe Coupee*
2A Rapides*
3A Red River*
3A Richland*
3A Sabine*
2A St. Bernard*
2A St. Charles*
2A St. Helena*
2A St. James*
2A St. John the Baptist*
2A St. Landry*
2A St. Martin*
2A St. Mary*
2A St. Tammany*
2A Tangipahoa*
3A Tensas*
2A Terrebonne*
3A Union*
2A Vermilion*
3A Vernon*
2A Washington*
3A Webster*
2A West Baton Rouge*
3A West Carroll
2A West Feliciana*
3A Winn*

MAINE

6A Androscoggin
7 Aroostook
6A Cumberland
6A Franklin
6A Hancock
6A Kennebec
6A Knox
6A Lincoln
6A Oxford
6A Penobscot
6A Piscataquis
6A Sagadahoc
6A Somerset
6A Waldo
6A Washington
6A York

MARYLAND

4A Allegany
4A Anne Arundel
4A Baltimore
4A Baltimore (city)
4A Calvert
4A Caroline
4A Carroll
4A Cecil
4A Charles
4A Dorchester
4A Frederick
5A Garrett
4A Harford
4A Howard
4A Kent
4A Montgomery
4A Prince George's
4A Queen Anne's
4A Somerset
4A St. Mary's
4A Talbot
4A Washington
4A Wicomico
4A Worcester

MASSACHSETTS

5A (all)

MICHIGAN

6A Alcona
6A Alger

(continued)

TABLE N1101.10 (R301.1)—continued CLIMATE ZONES, MOISTURE REGIMES, AND WARM-HUMID DESIGNATIONS BY STATE, COUNTY AND TERRITORY

5A Allegan
6A Alpena
6A Antrim
6A Arenac
7 Baraga
5A Barry
5A Bay
6A Benzie
5A Berrien
5A Branch
5A Calhoun
5A Cass
6A Charlevoix
6A Cheboygan
7 Chippewa
6A Clare
5A Clinton
6A Crawford
6A Delta
6A Dickinson
5A Eaton
6A Emmet
5A Genesee
6A Gladwin
7 Gogebic
6A Grand Traverse
5A Gratiot
5A Hillsdale
7 Houghton
6A Huron
5A Ingham
5A Ionia
6A Iosco
7 Iron
6A Isabella
5A Jackson
5A Kalamazoo
6A Kalkaska
5A Kent
7 Keweenaw
6A Lake
5A Lapeer
6A Leelanau
5A Lenawee
5A Livingston
7 Luce
7 Mackinac
5A Macomb
6A Manistee
6A Marquette
6A Mason
6A Mecosta
6A Menominee
5A Midland
6A Missaukee
5A Monroe
5A Montcalm
6A Montmorency
5A Muskegon
6A Newaygo
5A Oakland
6A Oceana
6A Ogemaw
7 Ontonagon
6A Osceola
6A Oscoda
6A Otsego
5A Ottawa
6A Presque Isle
6A Roscommon
5A Saginaw
6A Sanilac
7 Schoolcraft
5A Shiawassee
5A St. Clair
5A St. Joseph
5A Tuscola
5A Van Buren
5A Washtenaw
5A Wayne
6A Wexford

MINNESOTA

7 Aitkin
6A Anoka
7 Becker
7 Beltrami
6A Benton
6A Big Stone
6A Blue Earth
6A Brown
7 Carlton
6A Carver
7 Cass
6A Chippewa
6A Chisago
7 Clay
7 Clearwater
7 Cook
6A Cottonwood
7 Crow Wing
6A Dakota
6A Dodge
6A Douglas
6A Faribault
6A Fillmore
6A Freeborn
6A Goodhue
7 Grant
6A Hennepin
6A Houston
7 Hubbard
6A Isanti
7 Itasca
6A Jackson
7 Kanabec
6A Kandiyohi
7 Kittson
7 Koochiching
6A Lac qui Parle
7 Lake
7 Lake of the Woods
6A Le Sueur
6A Lincoln
6A Lyon
7 Mahnomen
7 Marshall
6A Martin
6A McLeod
6A Meeker
7 Mille Lacs
6A Morrison
6A Mower
6A Murray
6A Nicollet
6A Nobles
7 Norman
6A Olmsted
7 Otter Tail
7 Pennington
7 Pine
6A Pipestone
7 Polk
6A Pope
6A Ramsey
7 Red Lake
6A Redwood
6A Renville
6A Rice
6A Rock
7 Roseau
6A Scott
6A Sherburne
6A Sibley
6A Stearns
6A Steele
6A Stevens
7 St. Louis
6A Swift
6A Todd
6A Traverse
6A Wabasha
7 Wadena
6A Waseca
6A Washington
6A Watonwan
7 Wilkin
6A Winona
6A Wright
6A Yellow Medicine

MISSISSIPPI

3A Adams*
3A Alcorn
3A Amite*
3A Attala
3A Benton
3A Bolivar
3A Calhoun
3A Carroll
3A Chickasaw
3A Choctaw
3A Claiborne*
3A Clarke
3A Clay
3A Coahoma
3A Copiah*
3A Covington*
3A DeSoto
3A Forrest*
3A Franklin*
3A George*
3A Greene*
3A Grenada
2A Hancock*
2A Harrison*
3A Hinds*
3A Holmes
3A Humphreys
3A Issaquena
3A Itawamba
2A Jackson*
3A Jasper
3A Jefferson*
3A Jefferson Davis*
3A Jones*
3A Kemper
3A Lafayette
3A Lamar*
3A Lauderdale
3A Lawrence*
3A Leake
3A Lee
3A Leflore
3A Lincoln*
3A Lowndes
3A Madison
3A Marion*
3A Marshall
3A Monroe
3A Montgomery
3A Neshoba
3A Newton
3A Noxubee
3A Oktibbeha
3A Panola
2A Pearl River*
3A Perry*
3A Pike*

(continued)

TABLE N1101.10 (R301.1)—continued
CLIMATE ZONES, MOISTURE REGIMES, AND WARM-HUMID DESIGNATIONS BY STATE, COUNTY AND TERRITORY

3A Pontotoc
3A Prentiss
3A Quitman
3A Rankin*
3A Scott
3A Sharkey
3A Simpson*
3A Smith*
2A Stone*
3A Sunflower
3A Tallahatchie
3A Tate
3A Tippah
3A Tishomingo
3A Tunica
3A Union
3A Walthall*
3A Warren*
3A Washington
3A Wayne*
3A Webster
3A Wilkinson*
3A Winston
3A Yalobusha
3A Yazoo

MISSOURI

5A Adair
5A Andrew
5A Atchison
4A Audrain
4A Barry
4A Barton
4A Bates
4A Benton
4A Bollinger
4A Boone
5A Buchanan
4A Butler
5A Caldwell
4A Callaway
4A Camden
4A Cape Girardeau
4A Carroll
4A Carter
4A Cass
4A Cedar
5A Chariton
4A Christian
5A Clark
4A Clay
5A Clinton
4A Cole
4A Cooper
4A Crawford
4A Dade
4A Dallas
5A Daviess
5A DeKalb
4A Dent
4A Douglas
4A Dunklin
4A Franklin
4A Gasconade
5A Gentry
4A Greene
5A Grundy
5A Harrison
4A Henry
4A Hickory
5A Holt
4A Howard
4A Howell
4A Iron
4A Jackson
4A Jasper
4A Jefferson
4A Johnson
5A Knox
4A Laclede
4A Lafayette
4A Lawrence
5A Lewis
4A Lincoln
5A Linn
5A Livingston
5A Macon
4A Madison
4A Maries
5A Marion
4A McDonald
5A Mercer
4A Miller
4A Mississippi
4A Moniteau
4A Monroe
4A Montgomery
4A Morgan
4A New Madrid
4A Newton
5A Nodaway
4A Oregon
4A Osage
4A Ozark
4A Pemiscot
4A Perry
4A Pettis
4A Phelps
5A Pike
4A Platte
4A Polk
4A Pulaski
5A Putnam
5A Ralls
4A Randolph
4A Ray
4A Reynolds
4A Ripley
4A Saline
5A Schuyler
5A Scotland
4A Scott
4A Shannon
5A Shelby
4A St. Charles
4A St. Clair
4A Ste. Genevieve
4A St. Francois
4A St. Louis
4A St. Louis (city)
4A Stoddard
4A Stone
5A Sullivan
4A Taney
4A Texas
4A Vernon
4A Warren
4A Washington
4A Wayne
4A Webster
5A Worth
4A Wright

MONTANA

6B (all)

NEBRASKA

5A (all)

NEVADA

5B Carson City (city)
5B Churchill
3B Clark
5B Douglas
5B Elko
5B Esmeralda
5B Eureka
5B Humboldt
5B Lander
5B Lincoln
5B Lyon
5B Mineral
5B Nye
5B Pershing
5B Storey
5B Washoe
5B White Pine

NEW HAMPSHIRE

6A Belknap
6A Carroll
5A Cheshire
6A Coos
6A Grafton
5A Hillsborough
6A Merrimack
5A Rockingham
5A Strafford
6A Sullivan

NEW JERSEY

4A Atlantic
5A Bergen
4A Burlington
4A Camden
4A Cape May
4A Cumberland
4A Essex
4A Gloucester
4A Hudson
5A Hunterdon
5A Mercer
4A Middlesex
4A Monmouth
5A Morris
4A Ocean
5A Passaic
4A Salem
5A Somerset
5A Sussex
4A Union
5A Warren

NEW MEXICO

4B Bernalillo
5B Catron
3B Chaves
4B Cibola
5B Colfax
4B Curry
4B DeBaca
3B Dona Ana
3B Eddy
4B Grant
4B Guadalupe
5B Harding
3B Hidalgo
3B Lea
4B Lincoln
5B Los Alamos
3B Luna
5B McKinley
5B Mora
3B Otero
4B Quay
5B Rio Arriba
4B Roosevelt
5B Sandoval
5B San Juan
5B San Miguel
5B Santa Fe
4B Sierra
4B Socorro

(continued)

TABLE N1101.10 (R301.1)—continued
CLIMATE ZONES, MOISTURE REGIMES, AND WARM-HUMID DESIGNATIONS BY STATE, COUNTY AND TERRITORY

5B Taos
5B Torrance
4B Union
4B Valencia

NEW YORK

5A Albany
6A Allegany
4A Bronx
6A Broome
6A Cattaraugus
5A Cayuga
5A Chautauqua
5A Chemung
6A Chenango
6A Clinton
5A Columbia
5A Cortland
6A Delaware
5A Dutchess
5A Erie
6A Essex
6A Franklin
6A Fulton
5A Genesee
5A Greene
6A Hamilton
6A Herkimer
6A Jefferson
4A Kings
6A Lewis
5A Livingston
6A Madison
5A Monroe
6A Montgomery
4A Nassau
4A New York
5A Niagara
6A Oneida
5A Onondaga
5A Ontario
5A Orange
5A Orleans
5A Oswego
6A Otsego
5A Putnam
4A Queens
5A Rensselaer
4A Richmond
5A Rockland
5A Saratoga
5A Schenectady
6A Schoharie
6A Schuyler
5A Seneca
6A Steuben
6A St. Lawrence
4A Suffolk
6A Sullivan
5A Tioga
6A Tompkins
6A Ulster
6A Warren
5A Washington
5A Wayne
4A Westchester
6A Wyoming
5A Yates

NORTH CAROLINA

4A Alamance
4A Alexander
5A Alleghany
3A Anson
5A Ashe
5A Avery
3A Beaufort
4A Bertie
3A Bladen
3A Brunswick*
4A Buncombe
4A Burke
3A Cabarrus
4A Caldwell
3A Camden
3A Carteret*
4A Caswell
4A Catawba
4A Chatham
4A Cherokee
3A Chowan
4A Clay
4A Cleveland
3A Columbus*
3A Craven
3A Cumberland
3A Currituck
3A Dare
3A Davidson
4A Davie
3A Duplin
4A Durham
3A Edgecombe
4A Forsyth
4A Franklin
3A Gaston
4A Gates
4A Graham
4A Granville
3A Greene
4A Guilford
4A Halifax
4A Harnett
4A Haywood
4A Henderson
4A Hertford
3A Hoke
3A Hyde
4A Iredell
4A Jackson
3A Johnston
3A Jones
4A Lee
3A Lenoir
4A Lincoln
4A Macon
4A Madison
3A Martin
4A McDowell
3A Mecklenburg
5A Mitchell
3A Montgomery
3A Moore
4A Nash
3A New Hanover*
4A Northampton
3A Onslow*
4A Orange
3A Pamlico
3A Pasquotank
3A Pender*
3A Perquimans
4A Person
3A Pitt
4A Polk
3A Randolph
3A Richmond
3A Robeson
4A Rockingham
3A Rowan
4A Rutherford
3A Sampson
3A Scotland
3A Stanly
4A Stokes
4A Surry
4A Swain
4A Transylvania
3A Tyrrell
3A Union
4A Vance
4A Wake
4A Warren
3A Washington
5A Watauga
3A Wayne
4A Wilkes
3A Wilson
4A Yadkin
5A Yancey

NORTH DAKOTA

6A Adams
7 Barnes
7 Benson
6A Billings
7 Bottineau
6A Bowman
7 Burke
6A Burleigh
7 Cass
7 Cavalier
6A Dickey
7 Divide
6A Dunn
7 Eddy
6A Emmons
7 Foster
6A Golden Valley
7 Grand Forks
6A Grant
7 Griggs
6A Hettinger
7 Kidder
6A LaMoure
6A Logan
7 McHenry
6A McIntosh
6A McKenzie
7 McLean
6A Mercer
6A Morton
7 Mountrail
7 Nelson
6A Oliver
7 Pembina
7 Pierce
7 Ramsey
6A Ransom
7 Renville
6A Richland
7 Rolette
6A Sargent
7 Sheridan
6A Sioux
6A Slope
6A Stark
7 Steele
7 Stutsman
7 Towner
7 Traill
7 Walsh
7 Ward
7 Wells
7 Williams

OHIO

4A Adams
5A Allen

(continued)

**TABLE N1101.10 (R301.1)—continued
CLIMATE ZONES, MOISTURE REGIMES, AND WARM-HUMID
DESIGNATIONS BY STATE, COUNTY AND TERRITORY**

5A Ashland
5A Ashtabula
5A Athens
5A Auglaize
5A Belmont
4A Brown
5A Butler
5A Carroll
5A Champaign
5A Clark
4A Clermont
5A Clinton
5A Columbiana
5A Coshocton
5A Crawford
5A Cuyahoga
5A Darke
5A Defiance
5A Delaware
5A Erie
5A Fairfield
5A Fayette
5A Franklin
5A Fulton
4A Gallia
5A Geauga
5A Greene
5A Guernsey
4A Hamilton
5A Hancock
5A Hardin
5A Harrison
5A Henry
5A Highland
5A Hocking
5A Holmes
5A Huron
5A Jackson
5A Jefferson
5A Knox
5A Lake
4A Lawrence
5A Licking
5A Logan
5A Lorain
5A Lucas
5A Madison
5A Mahoning
5A Marion
5A Medina
5A Meigs
5A Mercer
5A Miami
5A Monroe
5A Montgomery
5A Morgan
5A Morrow
5A Muskingum
5A Noble
5A Ottawa
5A Paulding
5A Perry
5A Pickaway
4A Pike
5A Portage
5A Preble
5A Putnam
5A Richland
5A Ross
5A Sandusky
4A Scioto
5A Seneca
5A Shelby
5A Stark
5A Summit
5A Trumbull
5A Tuscarawas
5A Union
5A Van Wert
5A Vinton
5A Warren
4A Washington
5A Wayne
5A Williams
5A Wood
5A Wyandot

OKLAHOMA

3A Adair
3A Alfalfa
3A Atoka
4B Beaver
3A Beckham
3A Blaine
3A Bryan
3A Caddo
3A Canadian
3A Carter
3A Cherokee
3A Choctaw
4B Cimarron
3A Cleveland
3A Coal
3A Comanche
3A Cotton
3A Craig
3A Creek
3A Custer
3A Delaware
3A Dewey
3A Ellis
3A Garfield
3A Garvin
3A Grady
3A Grant
3A Greer
3A Harmon
3A Harper
3A Haskell
3A Hughes
3A Jackson
3A Jefferson
3A Johnston
3A Kay
3A Kingfisher
3A Kiowa
3A Latimer
3A Le Flore
3A Lincoln
3A Logan
3A Love
3A Major
3A Marshall
3A Mayes
3A McClain
3A McCurtain
3A McIntosh
3A Murray
3A Muskogee
3A Noble
3A Nowata
3A Okfuskee
3A Oklahoma
3A Okmulgee
3A Osage
3A Ottawa
3A Pawnee
3A Payne
3A Pittsburg
3A Pontotoc
3A Pottawatomie
3A Pushmataha
3A Roger Mills
3A Rogers
3A Seminole
3A Sequoyah
3A Stephens
4B Texas
3A Tillman
3A Tulsa
3A Wagoner
3A Washington
3A Washita
3A Woods
3A Woodward

OREGON

5B Baker
4C Benton
4C Clackamas
4C Clatsop
4C Columbia
4C Coos
5B Crook
4C Curry
5B Deschutes
4C Douglas
5B Gilliam
5B Grant
5B Harney
5B Hood River
4C Jackson
5B Jefferson
4C Josephine
5B Klamath
5B Lake
4C Lane
4C Lincoln
4C Linn
5B Malheur
4C Marion
5B Morrow
4C Multnomah
4C Polk
5B Sherman
4C Tillamook
5B Umatilla
5B Union
5B Wallowa
5B Wasco
4C Washington
5B Wheeler
4C Yamhill

PENNSYLVANIA

5A Adams
5A Allegheny
5A Armstrong
5A Beaver
5A Bedford
5A Berks
5A Blair
5A Bradford
4A Bucks
5A Butler
5A Cambria
6A Cameron
5A Carbon
5A Centre
4A Chester
5A Clarion
6A Clearfield
5A Clinton
5A Columbia
5A Crawford
5A Cumberland
5A Dauphin
4A Delaware
6A Elk
5A Erie
5A Fayette
5A Forest
5A Franklin
5A Fulton
5A Greene

(continued)

TABLE N1101.10 (R301.1)—continued
CLIMATE ZONES, MOISTURE REGIMES, AND WARM-HUMID DESIGNATIONS BY STATE, COUNTY AND TERRITORY

5A Huntingdon
5A Indiana
5A Jefferson
5A Juniata
5A Lackawanna
5A Lancaster
5A Lawrence
5A Lebanon
5A Lehigh
5A Luzerne
5A Lycoming
6A McKean
5A Mercer
5A Mifflin
5A Monroe
4A Montgomery
5A Montour
5A Northampton
5A Northumberland
5A Perry
4A Philadelphia
5A Pike
6A Potter
5A Schuylkill
5A Snyder
5A Somerset
5A Sullivan
6A Susquehanna
6A Tioga
5A Union
5A Venango
5A Warren
5A Washington
6A Wayne
5A Westmoreland
5A Wyoming
4A York

RHODE ISLAND

5A (all)

SOUTH CAROLINA

3A Abbeville
3A Aiken
3A Allendale*
3A Anderson
3A Bamberg*
3A Barnwell*
3A Beaufort*
3A Berkeley*
3A Calhoun
3A Charleston*
3A Cherokee
3A Chester
3A Chesterfield
3A Clarendon
3A Colleton*
3A Darlington
3A Dillon
3A Dorchester*
3A Edgefield
3A Fairfield
3A Florence
3A Georgetown*
3A Greenville
3A Greenwood
3A Hampton*
3A Horry*
3A Jasper*
3A Kershaw
3A Lancaster
3A Laurens
3A Lee
3A Lexington
3A Marion
3A Marlboro
3A McCormick
3A Newberry
3A Oconee
3A Orangeburg
3A Pickens
3A Richland
3A Saluda
3A Spartanburg
3A Sumter
3A Union
3A Williamsburg
3A York

SOUTH DAKOTA

6A Aurora
6A Beadle
5A Bennett
5A Bon Homme
6A Brookings
6A Brown
6A Brule
6A Buffalo
6A Butte
6A Campbell
5A Charles Mix
6A Clark
5A Clay
6A Codington
6A Corson
6A Custer
6A Davison
6A Day
6A Deuel
6A Dewey
5A Douglas
6A Edmunds
6A Fall River
6A Faulk
6A Grant
5A Gregory
6A Haakon
6A Hamlin
6A Hand
6A Hanson
6A Harding
6A Hughes
5A Hutchinson
6A Hyde
5A Jackson
6A Jerauld
6A Jones
6A Kingsbury
6A Lake
6A Lawrence
6A Lincoln
6A Lyman
6A Marshall
6A McCook
6A McPherson
6A Meade
5A Mellette
6A Miner
6A Minnehaha
6A Moody
6A Pennington
6A Perkins
6A Potter
6A Roberts
6A Sanborn
6A Shannon
6A Spink
6A Stanley
6A Sully
5A Todd
5A Tripp
6A Turner
5A Union
6A Walworth
5A Yankton
6A Ziebach

TENNESSEE

4A Anderson
4A Bedford
4A Benton
4A Bledsoe
4A Blount
4A Bradley
4A Campbell
4A Cannon
4A Carroll
4A Carter
4A Cheatham
3A Chester
4A Claiborne
4A Clay
4A Cocke
4A Coffee
3A Crockett
4A Cumberland
4A Davidson
4A Decatur
4A DeKalb
4A Dickson
3A Dyer
3A Fayette
4A Fentress
4A Franklin
4A Gibson
4A Giles
4A Grainger
4A Greene
4A Grundy
4A Hamblen
4A Hamilton
4A Hancock
3A Hardeman
3A Hardin
4A Hawkins
3A Haywood
3A Henderson
4A Henry
4A Hickman
4A Houston
4A Humphreys
4A Jackson
4A Jefferson
4A Johnson
4A Knox
3A Lake
3A Lauderdale
4A Lawrence
4A Lewis
4A Lincoln
4A Loudon
4A Macon
3A Madison
4A Marion
4A Marshall
4A Maury
4A McMinn
3A McNairy
4A Meigs
4A Monroe
4A Montgomery
4A Moore
4A Morgan
4A Obion
4A Overton
4A Perry
4A Pickett
4A Polk
4A Putnam
4A Rhea

(continued)

**TABLE N1101.10 (R301.1)—continued
CLIMATE ZONES, MOISTURE REGIMES, AND WARM-HUMID
DESIGNATIONS BY STATE, COUNTY AND TERRITORY**

4A Roane
4A Robertson
4A Rutherford
4A Scott
4A Sequatchie
4A Sevier
3A Shelby
4A Smith
4A Stewart
4A Sullivan
4A Sumner
3A Tipton
4A Trousdale
4A Unicoi
4A Union
4A Van Buren
4A Warren
4A Washington
4A Wayne
4A Weakley
4A White
4A Williamson
4A Wilson

TEXAS

2A Anderson*
3B Andrews
2A Angelina*
2A Aransas*
3A Archer
4B Armstrong
2A Atascosa*
2A Austin*
4B Bailey
2B Bandera*
2A Bastrop*
3B Baylor
2A Bee*
2A Bell*
2A Bexar*
3A Blanco*
3B Borden
2A Bosque*
3A Bowie*
2A Brazoria*
2A Brazos*
3B Brewster
4B Briscoe
2A Brooks*
3A Brown*
2A Burleson*
3A Burnet*
2A Caldwell*
2A Calhoun*
3B Callahan
2A Cameron*
3A Camp*
4B Carson
3A Cass*
4B Castro
2A Chambers*
2A Cherokee*
3B Childress
3A Clay
4B Cochran
3B Coke
3B Coleman
3A Collin*
3B Collingsworth
2A Colorado*
2A Comal*
3A Comanche*
3B Concho
3A Cooke
2A Coryell*
3B Cottle
3B Crane
3B Crockett
3B Crosby
3B Culberson
4B Dallam
3A Dallas*
3B Dawson
4B Deaf Smith
3A Delta
3A Denton*
2A DeWitt*
3B Dickens
2B Dimmit*
4B Donley
2A Duval*
3A Eastland
3B Ector
2B Edwards*
3A Ellis*
3B El Paso
3A Erath*
2A Falls*
3A Fannin
2A Fayette*
3B Fisher
4B Floyd
3B Foard
2A Fort Bend*
3A Franklin*
2A Freestone*
2B Frio*
3B Gaines
2A Galveston*
3B Garza
3A Gillespie*
3B Glasscock
2A Goliad*
2A Gonzales*
4B Gray
3A Grayson
3A Gregg*
2A Grimes*
2A Guadalupe*
4B Hale
3B Hall
3A Hamilton*
4B Hansford
3B Hardeman
2A Hardin*
2A Harris*
3A Harrison*
4B Hartley
3B Haskell
2A Hays*
3B Hemphill
3A Henderson*
2A Hidalgo*
2A Hill*
4B Hockley
3A Hood*
3A Hopkins*
2A Houston*
3B Howard
3B Hudspeth
3A Hunt*
4B Hutchinson
3B Irion
3A Jack
2A Jackson*
2A Jasper*
3B Jeff Davis
2A Jefferson*
2A Jim Hogg*
2A Jim Wells*
3A Johnson*
3B Jones
2A Karnes*
3A Kaufman*
3A Kendall*
2A Kenedy*
3B Kent
3B Kerr
3B Kimble
3B King
2B Kinney*
2A Kleberg*
3B Knox
3A Lamar*
4B Lamb
3A Lampasas*
2B La Salle*
2A Lavaca*
2A Lee*
2A Leon*
2A Liberty*
2A Limestone*
4B Lipscomb
2A Live Oak*
3A Llano*
3B Loving
3B Lubbock
3B Lynn
2A Madison*
3A Marion*
3B Martin
3B Mason
2A Matagorda*
2B Maverick*
3B McCulloch
2A McLennan*
2A McMullen*
2B Medina*
3B Menard
3B Midland
2A Milam*
3A Mills*
3B Mitchell
3A Montague
2A Montgomery*
4B Moore
3A Morris*
3B Motley
3A Nacogdoches*
3A Navarro*
2A Newton*
3B Nolan
2A Nueces*
4B Ochiltree
4B Oldham
2A Orange*
3A Palo Pinto*
3A Panola*
3A Parker*
4B Parmer
3B Pecos
2A Polk*
4B Potter
3B Presidio
3A Rains*
4B Randall
3B Reagan
2B Real*
3A Red River*
3B Reeves
2A Refugio*
4B Roberts
2A Robertson*
3A Rockwall*
3B Runnels
3A Rusk*
3A Sabine*
3A San Augustine*
2A San Jacinto*
2A San Patricio*

(continued)

TABLE N1101.10 (R301.1)—continued
CLIMATE ZONES, MOISTURE REGIMES, AND WARM-HUMID DESIGNATIONS BY STATE, COUNTY AND TERRITORY

3A San Saba*
3B Schleicher
3B Scurry
3B Shackelford
3A Shelby*
4B Sherman
3A Smith*
3A Somervell*
2A Starr*
3A Stephens
3B Sterling
3B Stonewall
3B Sutton
4B Swisher
3A Tarrant*
3B Taylor
3B Terrell
3B Terry
3B Throckmorton
3A Titus*
3B Tom Green
2A Travis*
2A Trinity*
2A Tyler*
3A Upshur*
3B Upton
2B Uvalde*
2B Val Verde*
3A Van Zandt*
2A Victoria*
2A Walker*
2A Waller*
3B Ward
2A Washington*
2B Webb*
2A Wharton*
3B Wheeler
3A Wichita
3B Wilbarger
2A Willacy*
2A Williamson*
2A Wilson*
3B Winkler
3A Wise
3A Wood*
4B Yoakum
3A Young
2B Zapata*
2B Zavala*

UTAH

5B Beaver
6B Box Elder
6B Cache
6B Carbon
6B Daggett
5B Davis
6B Duchesne
5B Emery
5B Garfield
5B Grand
5B Iron
5B Juab
5B Kane
5B Millard
6B Morgan
5B Piute
6B Rich
5B Salt Lake
5B San Juan
5B Sanpete
5B Sevier
6B Summit
5B Tooele
6B Uintah
5B Utah
6B Wasatch
3B Washington
5B Wayne
5B Weber

VERMONT

6A (all)

VIRGINIA

4A (all)

WASHINGTON

5B Adams
5B Asotin
5B Benton
5B Chelan
4C Clallam
4C Clark
5B Columbia
4C Cowlitz
5B Douglas
6B Ferry
5B Franklin
5B Garfield
5B Grant
4C Grays Harbor
4C Island
4C Jefferson
4C King
4C Kitsap
5B Kittitas
5B Klickitat
4C Lewis
5B Lincoln
4C Mason
6B Okanogan
4C Pacific
6B Pend Oreille
4C Pierce
4C San Juan
4C Skagit
5B Skamania
4C Snohomish
5B Spokane
6B Stevens
4C Thurston
4C Wahkiakum
5B Walla Walla
4C Whatcom
5B Whitman
5B Yakima

WEST VIRGINIA

5A Barbour
4A Berkeley
4A Boone
4A Braxton
5A Brooke
4A Cabell
4A Calhoun
4A Clay
5A Doddridge
5A Fayette
4A Gilmer
5A Grant
5A Greenbrier
5A Hampshire
5A Hancock
5A Hardy
5A Harrison
4A Jackson
4A Jefferson
4A Kanawha
5A Lewis
4A Lincoln
4A Logan
5A Marion
5A Marshall
4A Mason
4A McDowell
4A Mercer
5A Mineral
4A Mingo
5A Monongalia
4A Monroe
4A Morgan
5A Nicholas
5A Ohio
5A Pendleton
4A Pleasants
5A Pocahontas
5A Preston
4A Putnam
5A Raleigh
5A Randolph
4A Ritchie
4A Roane
5A Summers
5A Taylor
5A Tucker
4A Tyler
5A Upshur
4A Wayne
5A Webster
5A Wetzel
4A Wirt
4A Wood
4A Wyoming

WISCONSIN

6A Adams
7 Ashland
6A Barron
7 Bayfield
6A Brown
6A Buffalo
7 Burnett
6A Calumet
6A Chippewa
6A Clark
6A Columbia
6A Crawford
6A Dane
6A Dodge
6A Door
7 Douglas
6A Dunn
6A Eau Claire
7 Florence
6A Fond du Lac
7 Forest
6A Grant
6A Green
6A Green Lake
6A Iowa
7 Iron
6A Jackson
6A Jefferson
6A Juneau
6A Kenosha
6A Kewaunee
6A La Crosse
6A Lafayette
7 Langlade
7 Lincoln
6A Manitowoc
6A Marathon
6A Marinette
6A Marquette
6A Menominee
6A Milwaukee
6A Monroe
6A Oconto
7 Oneida
6A Outagamie

(continued)

TABLE N1101.10 (R301.1)—continued
CLIMATE ZONES, MOISTURE REGIMES, AND WARM-HUMID DESIGNATIONS BY STATE, COUNTY AND TERRITORY

6A Ozaukee
6A Pepin
6A Pierce
6A Polk
6A Portage
7 Price
6A Racine
6A Richland
6A Rock
6A Rusk
6A Sauk
7 Sawyer
6A Shawano
6A Sheboygan
6A St. Croix
7 Taylor
6A Trempealeau
6A Vernon
7 Vilas
6A Walworth
7 Washburn
6A Washington
6A Waukesha
6A Waupaca
6A Waushara
6A Winnebago
6A Wood

WYOMING

6B Albany
6B Big Horn
6B Campbell
6B Carbon
6B Converse
6B Crook
6B Fremont
5B Goshen
6B Hot Springs
6B Johnson
6B Laramie
7 Lincoln
6B Natrona
6B Niobrara
6B Park
5B Platte
6B Sheridan
7 Sublette
6B Sweetwater
7 Teton
6B Uinta
6B Washakie
6B Weston

US TERRITORIES

AMERICAN SAMOA

1A (all)*

GUAM

1A (all)*

NORTHERN MARIANA ISLANDS

1A (all)*

PUERTO RICO

1A (all)*

VIRGIN ISLANDS

1A (all)*

TABLE N1101.10.2(1) [R302.3(1)]
INTERNATIONAL CLIMATE ZONE DEFINITIONS

MAJOR CLIMATE TYPE DEFINITIONS
Marine (C) Definition—Locations meeting all four criteria: 1. Mean temperature of coldest month between -3°C (27°F) and 18°C (65°F). 2. Warmest month mean < 22°C (72°F). 3. At least four months with mean temperatures over 10°C (50°F). 4. Dry season in summer. The month with the heaviest precipitation in the cold season has at least three times as much precipitation as the month with the least precipitation in the rest of the year. The cold season is October through March in the Northern Hemisphere and April through September in the Southern Hemisphere.
Dry (B) Definition—Locations meeting the following criteria: Not marine and $P_{in} < 0.44 \times (TF - 19.5)$ [$P_{cm} < 2.0 \times (TC + 7)$ in SI units] where: P_{in} = Annual precipitation in inches (cm) T = Annual mean temperature in °F (°C)
Moist (A) Definition—Locations that are not marine and not dry.
Warm-humid Definition—Moist (A) locations where either of the following wet-bulb temperature conditions shall occur during the warmest six consecutive months of the year: 1. 67°F (19.4°C) or higher for 3,000 or more hours; or 2. 73°F (22.8°C) or higher for 1,500 or more hours.

For SI: °C = [(°F)-32]/1.8, 1 inch = 2.54 cm.

TABLE N1101.10.2(2) [R301.3(2)]
INTERNATIONAL CLIMATE ZONE DEFINITIONS

ZONE NUMBER	THERMAL CRITERIA	
	IP Units	SI Units
1	9000 < CDD50°F	5000 < CDD10°C
2	6300 < CDD50°F ≤ 9000	3500 < CDD10°C ≤ 5000
3A and 3B	4500 < CDD50°F ≤ 6300 AND HDD65°F ≤ 5400	2500 < CDD10°C ≤ 3500 AND HDD18°C ≤ 3000
4A and 4B	CDD50°F ≤ 4500 AND HDD65°F ≤ 5400	CDD10°C ≤ 2500 AND HDD18°C ≤ 3000
3C	HDD65°F ≤ 3600	HDD18°C ≤ 2000
4C	3600 < HDD65°F ≤ 5400	2000 < HDD18°C ≤ 3000
5	5400 < HDD65°F ≤ 7200	3000 < HDD18°C ≤ 4000
6	7200 < HDD65°F ≤ 9000	4000 < HDD18°C ≤ 5000
7	9000 < HDD65°F ≤ 12600	5000 < HDD18°C ≤ 7000
8	12600 < HDD65°F	7000 < HDD18°C

For SI: °C = [(°F)-32]/1.8.

N1101.11 (R302.1) Interior design conditions. The interior design temperatures used for heating and cooling load calculations shall be a maximum of 72°F (22°C) for heating and minimum of 75°F (24°C) for cooling.

❖ While the previous sections of Chapter 3 address outdoor design conditions, this section provides the interior conditions that will be used for properly sizing the mechanical equipment. The proper sizing of mechanical equipment (see Section N1103.6) can vary depending upon the selected design conditions. While the code does address oversizing equipment, it is not enforceable without establishing the exact design parameters. This section is included in the code only for system sizing, and it does not affect the interior design temperatures required by other codes such as Section 1204 of the IBC or Section 602.2 of the IPMC.

The 75°F (24°C) design temperature for heating was used in the code so that it coordinated with both the ASHRAE *Handbook of Fundamentals* and the Air Conditioning Contractors of America (ACCA) Manuals S and J, which are established standards that deal with equipment sizing.

N1101.12 (R303.1) Identification. Materials, systems and equipment shall be identified in a manner that will allow a determination of compliance with the applicable provisions of this code.

❖ The intent of this section is to make certain that sufficient information exists to determine compliance with the code during the plan review and field inspection phases. The permittee can submit the required equipment and materials information on the building plans, specification sheets or schedules or in any other way that allows the code official to clearly identify which specifications apply to which portions of the building; that is, which parts of the building are insulated to the levels listed. Materials information includes envelope insulation levels, glazing assembly *U*-factors and duct and piping insulation levels. Equipment information includes heating and cooling equipment and appliance efficiencies where high-efficiency equipment is claimed to meet code requirements.

This section contains specific material, equipment and system identification requirements for the approval and installation of the items required by the code. Although the means for permanent marking (tag, stencil, label, stamp, sticker, bar code, etc.) is often determined and applied by the manufacturer, if there is any uncertainty about the product, the mark is subject to the approval of the code official.

N1101.12.1 (R303.1.1) Building thermal envelope insulation. An *R*-value identification mark shall be applied by the manufacturer to each piece of *building thermal envelope* insulation 12 inches (305 mm) or greater in width. Alternately, the insulation installers shall provide a certification listing the type, manufacturer and *R*-value of insulation installed in each element of the *building thermal envelope*. For blown or sprayed insulation (fiberglass and cellulose), the initial installed thickness, settled thickness, settled *R*-value, installed density, coverage area and number of bags installed shall be *listed* on the certification. For sprayed polyurethane foam (SPF) insulation, the installed thickness of the areas covered and *R*-value of installed thickness shall be *listed* on the certification. The insulation installer shall sign, date and post the certification in a conspicuous location on the job site.

❖ The thermal performance of insulation is rated in terms of *R*-value. For products lacking an *R*-value identification, the installer (or builder) must provide the insulation performance data. For example, some insulation materials, such as foamed-in-place urethane, can be installed in wall, floor and cathedral ceiling cavities. These products are not labeled, as is batt insulation, nor is it appropriate for them to be evaluated as required in the code for blown or sprayed insulation; however, the installer must certify the type, thickness and *R*-value of these materials.

The *R*-value of loose-fill insulation (blown or sprayed) is dependent on both the installed thickness and density (number of bags used). Therefore, loose-fill insulation cannot be directly labeled by the manufacturer. Many blown insulation products carry a manufacturer's *R*-value guarantee when installed to a designated thickness, "inches = *R*-value." Blown insulation products lacking this manufacturer's guarantee can be subjected to special inspection and testing; what is referred to as "cookie cutting." Cookie cutting involves extracting a column of insulation with a cylinder to determine its density. The insulation depth and density must yield the specified *R*-value according to the manufacturer's bag label specification.

The code and Federal Trade Commission Rule 460 require that installers of insulation in homes, apartments and manufactured housing units report this information to the authority having jurisdiction in the form of a certification posted in a conspicuous location (see Commentary Figure N1101.12.1).

N1101.12.1.1 (R303.1.1.1) Blown or sprayed roof/ceiling insulation. The thickness of blown-in or sprayed roof/ceiling insulation (fiberglass or cellulose) shall be written in inches (mm) on markers that are installed at least one for every 300 square feet (28 m^2) throughout the attic space. The markers shall be affixed to the trusses or joists and marked with the minimum initial installed thickness with numbers a minimum of 1 inch (25 mm) in height. Each marker shall face the attic access opening. Spray polyurethane foam thickness and installed *R*-value shall be *listed* on certification provided by the insulation installer.

❖ To help verify the installed *R*-value of blown-in or spray-applied insulation, the installer must certify the following information in a signed statement posted in a conspicuous place (see Section N1101.12.1):

- The type of insulation used and manufacturer.
- The insulation's coverage per bag (the number of bags required to result in a given *R*-value for a given area), as well as the settled *R*-value.
- The initial and settled thickness.
- The number of bags installed.

Under circumstances where the insulation *R*-value is guaranteed, only the initial thickness is required on the certification.

This Attic Has Been Insulated To

R- ______

By A Professional Insulation Contractor

The insulation in this attic was installed by a qualified professional contractor to the *R*-value stated above

Certificate of Insulation

BUILDING ADDRESS: ____________

CONTRACTOR: ____________

Installation Date ____________

License# ____________

Area Insulated	R-Value	Installed Thickness	Settled Thickness	Installed Density	No. Bags	Sq. Ft.
Attic						
Walls						
Floors						

I, ______________________, (print name) certify that this residence/building has been insulated to the stated R-value and that the installation is in conformance with all applicable codes, standards, regulations and specifications.

Authorized Signature ______________________ Date ________

Figure N1101.12.1
SAMPLE CERTIFICATE OF INSULATION

(Logos courtesy of Cellulose Insulation Manufacturer's Association, http://cellulose.org, Insulation Contractors Association of America, www.insulate.org, and North American Insulation Manufacturers Association, www.NAIMA.org)

This section helps demonstrate compliance and enforcement of the provisions found in Section N1101.12.1. To assist with application and enforcement, loose-fill ceiling insulation also requires thickness markers that are attached to the framing and face the attic access.

In a large space, markers placed evenly about every 17 feet (5182 mm) (with some markers at the edge of the space) will meet this requirement. For sprayed polyurethane, such markers are not effective. When using this product, the code requires that the measured thickness and *R*-value be recorded on the certificate.

N1101.12.2 (R303.1.2) Insulation mark installation. Insulating materials shall be installed such that the manufacturer's *R*-value mark is readily observable upon inspection.

❖ For batt insulation, manufacturers' *R*-value designations and stripe codes are often printed directly on the insulation. Where possible, the insulation must be installed so these designations are readable. Backed floor batts can be installed with the designation against the underfloor, which means it would not be visible. In those cases, the *R*-value must be certified by the installer or be validated by some other means (see commentary, Section R303.1.1 of the IECC).

N1101.12.3 (R303.1.3) Fenestration product rating. *U*-factors of fenestration products (windows, doors and skylights) shall be determined in accordance with NFRC 100 by an accredited, independent laboratory, and labeled and certified by the manufacturer. Products lacking such a labeled *U*-factor shall be assigned a default *U*-factor from Table N1101.12.3(1) or N1101.12.3(2). The solar heat gain coefficient (SHGC) and *visible transmittance* (VT) of glazed fenestration products (windows, glazed doors and skylights) shall be determined in accordance with NFRC 200 by an accredited, independent laboratory, and labeled and certified by the manufacturer. Products lacking such a labeled SHGC or VT shall be assigned a default SHGC or VT from Table N1101.12.3(3).

❖ Until recently, the buyers of fenestration products received energy performance information in a variety of ways. Some manufacturers described performance by showing *R*-values of the glass. While the glass might have been a good performer, the rating did not include the effects of the frame. Other manufacturers touted the insulating value of different window components, but these, too, did not reflect the total window system performance. When manufacturers rated the entire product, some used test laboratory measurements and others used computer calculations. Even among those using test laboratory reports, the test laboratories often tested the products under different procedures, making an "apples-to-apples" comparison difficult. The different rating methods confused builders and consumers. They also created headaches for manufacturers trying to differentiate the performance of their products from the performance of their competitors' products.

The NFRC has developed a fenestration energy rating system based on whole-product performance. This accurately accounts for the energy-related effects of all the product's component parts and prevents information about a single component from being compared in a misleading way to other whole-product properties. With energy ratings based on whole-product performance, NFRC helps builders, designers and consumers directly compare products with different construction details and attributes.

Products that have been rated by NFRC-approved testing laboratories and certified by NFRC-accredited independent certification and inspection agencies carry a temporary and permanent label featuring the "NFRC-certified" mark. With this mark, the manufacturer stipulates that the energy performance of the product was determined according to NFRC rules and procedures.

By certifying and labeling their products, manufacturers are demonstrating their commitment to providing accurate energy and energy-related performance information. The code purposely sets the default values fairly high. This helps to encourage the use of products that have been tested and also ensures that products that have little energy-saving values are not used inappropriately in the various climate zones. By setting the default value so high, it will also prevent someone from removing the label from a tested window and then using the default values. Therefore, the default values are most representative of the lower end of the energy-efficient products.

Products that are not NFRC certified and do not exactly match the specifications in Tables N1101.12.3(1) and N1101.12.3(2) must use the tabular specification for the product most closely resembling it. In the absence of tested *U*-factors, the default *U*-factor for doors containing glazing can be a combination of the glazing and door *U*-factor as described in the definition for "*U*-factor" (see commentary, Section 202, "*U*-factor"). The NFRC procedures determine *U*-factor and SHGC ratings based on the whole fenestration assembly [untested fenestration products have default *U*-factors and SHGCs assigned as described in the commentary to Tables N1101.12.3(1) through N1101.12.3(3)]. During construction inspection, the label on each glazing assembly should be checked for conformance to the *U*-factor specified on the approved plans. These labels must be left on the glazing until after the building has been inspected for compliance. A sample NFRC label is shown in Commentary Figure N1101.12.3(1).

Products certified according to NFRC procedures are listed in the *Certified Products Directory*. The directory is published annually and contains energy performance information for over 1.4 million fenestration product options listed by over 450 manufacturers. When using the directory or shopping for NFRC-certified products, it is important to note:

1. A product is considered to be NFRC certified only if it carries the NFRC label. Simply being listed in this directory is not enough.

2. The NFRC-certified mark does not signify that the product meets any energy-efficiency standards or criteria.

3. NFRC neither sets minimum performance standards nor mandates specific performance levels. Rather, NFRC ratings can be used to determine whether a product meets a state or local code or other performance requirement and to compare the energy performance of different products during plan review. For questions about NFRC and its rating and labeling system, more information is available on the organization's website at www.nfrc.org. NFRC adopted a new energy performance label in 2005. It lists the manufacturer, describes the product, provides a source for additional information and includes ratings for one or more energy performance characteristics.

The IECC offers an alternative to NFRC-certified glazed fenestration product *U*-factor ratings. In the absence of *U*-factors based on NFRC test procedures, the default *U*-factors in Table N1101.12.3(1) must be used. When a composite of materials from two different product types is used, the code official should be consulted regarding how the product will be rated. Generally, the product must be assigned the higher *U*-factor, although an average based on the *U*-factors and areas may be acceptable in some cases.

The product cannot receive credit for a feature that cannot be seen. Because performance features such as argon fill and low-emissivity coatings for glass are not visually verifiable, they do not receive credit in the default tables. Tested *U*-factors for these windows are often lower, so using tested *U*-factors is to the applicant's advantage. Commentary Figure N1101.12.3(2) illustrates visually verifiable window characteristics, among other various window performance, function and cost considerations.

A single-glazed window with an installed storm window may be considered a double-glazed assembly and use the corresponding *U*-factor from the default table. For example, the *U*-factor 0.80 in Table N1101.12.3(1) applies to a single-glazed, metal window without a thermal break (but with an installed storm window). If the storm window was not installed, the *U*-factor would be 1.20.

TABLE N1101.12.3(1) [R303.1.3(1)]
DEFAULT GLAZED FENESTRATION *U*-FACTOR

FRAME TYPE	SINGLE PANE	DOUBLE PANE	SKYLIGHT	
			Single	Double
Metal	1.20	0.80	2.00	1.30
Metal with Thermal Break	1.10	0.65	1.90	1.10
Nonmetal or Metal Clad	0.95	0.55	1.75	1.05
Glazed Block	0.60			

TABLE N1101.12.3(2) [R303.1.3(2)]
DEFAULT DOOR *U*-FACTORS

DOOR TYPE	*U*-FACTOR
Uninsulated Metal	1.20
Insulated Metal	0.60
Wood	0.50
Insulated, nonmetal edge, max 45% glazing, any glazing double pane	0.35

❖ Door *U*-factors in Table N1101.12.3(2) should be used wherever NFRC-certified ratings are not available.

There are a few other aspects to note about doors. Opaque door *U*-factors must include the effects of the door edge and the frame. Calculating *U*-factors based on a cross section through the insulated portion is not acceptable. To take credit for a thermal break, the door must have one in both the door slab and in the frame. The values in the table are founded on principles established in the 1997 *ASHRAE Handbook of Fundamentals*.

TABLE N1101.12.3(3) [R303.1.3(3)]
DEFAULT GLAZED FENESTRATION SHGC AND VT

	SINGLE GLAZED		DOUBLE GLAZED		GLAZED BLOCK
	Clear	Tinted	Clear	Tinted	
SHGC	0.8	0.7	0.7	0.6	0.6
VT	0.6	0.3	0.6	0.3	0.6

❖ This table offers an alternative to NFRC-certified SHGC and visible transmittance (VT) values based on visually verifiable characteristics of the fenestration product. The SHGC is the fraction of incident solar radiation absorbed and directly transmitted by the window area, then subsequently reradiated, conducted or convected inward. SHGC is a ratio, expressed as a number between 0 and 1. The lower a window's SHGC, the less solar heat it transmits. The VT is the ratio of visible light entering the space through the fenestration product assembly to the incident visible light. Visible transmittance includes the effects of glazing material and frame and is expressed as a number between 0 and 1.

An SHGC of 0.40 or less is recommended in cooling-dominated climates (Climate Zones 1-3). In heating-dominated climates, a high SHGC increases passive solar gain for the heating but reduces cooling-season performance. A low SHGC improves cooling-season performance but reduces passive solar gains for heating.

N1101.12.4 (R303.1.4) Insulation product rating. The thermal resistance (*R*-value) of insulation shall be determined in accordance with the U.S. Federal Trade Commission *R*-value rule (CFR Title 16, Part 460) in units of h × ft^2 × °F/Btu at a mean temperature of 75°F (24°C).

❖ This section brings two important requirements to the code.

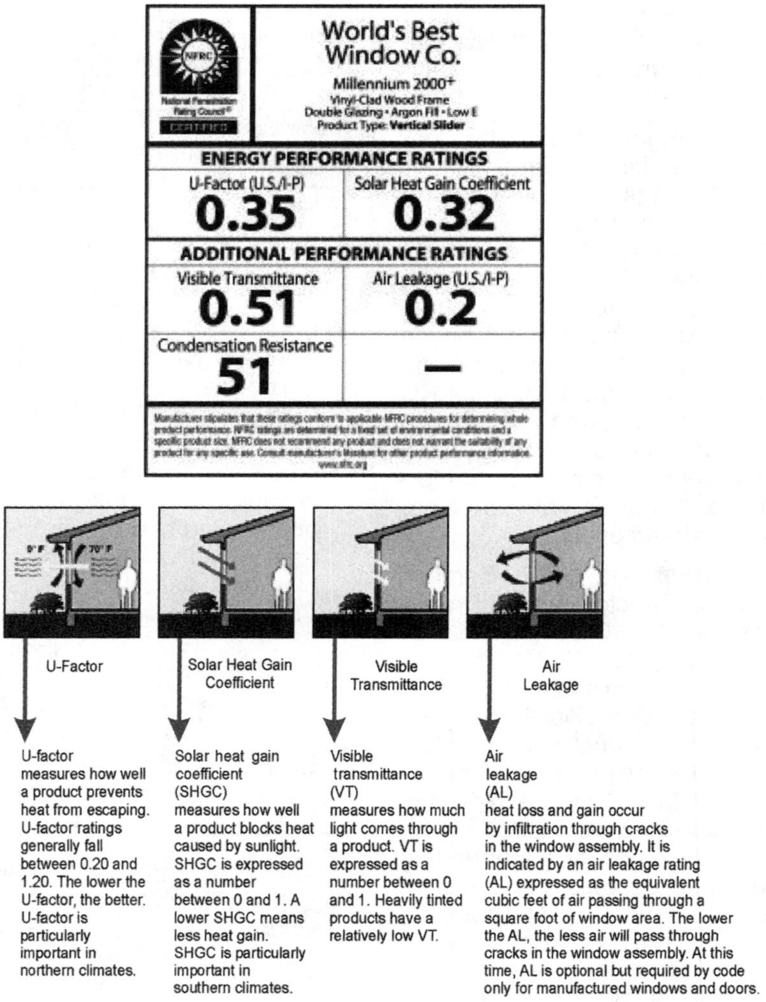

Figure N1101.12.3(1)
SAMPLE NATIONAL FENESTRATION RATING COUNCIL (NFRC) LABEL

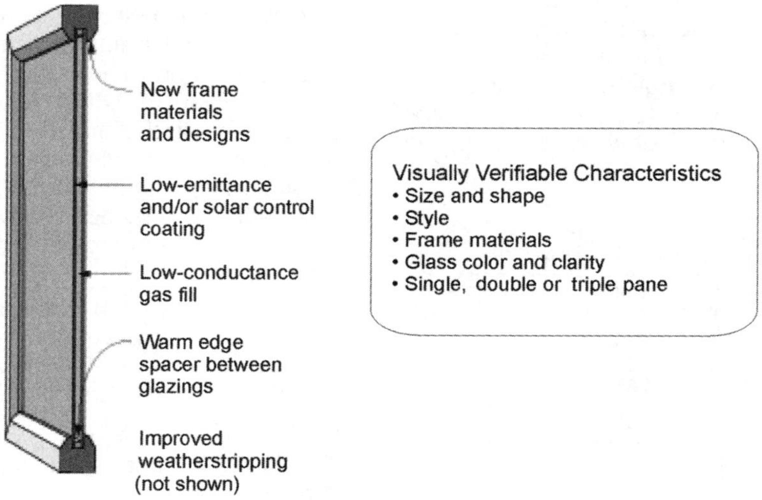

Figure N1101.12.3(2)
VISUALLY VERIFIABLE WINDOW CHARACTERISTICS

First, the Federal Trade Commission *R*-value rule details specific test standards for insulation. The test standards are specific to the type of insulation and intended use. This clarifies any questions on the rating conditions to be used for insulation materials.

Second, the text above specifies the rating temperature to be used when evaluating the *R*-value of the product, providing consistency not currently in the code. Insulation products sometimes list several *R*-values based on different test temperatures. This eliminates any question as to which *R*-value to use. The temperature selected is a standard rating condition.

N1101.13 (R303.2) Installation. All materials, systems and equipment shall be installed in accordance with the manufacturer's installation instructions and this code.

❖ Manufacturers' installation instructions are thoroughly evaluated by the listing agency, verifying that a safe installation is prescribed. When an appliance is tested to obtain a listing and label, the approval agency installs the appliance in accordance with the manufacturer's instructions. The appliance is tested under these conditions; thus, the installation instructions become an integral part of the labeling process. The listing agency can require that the manufacturer alter, delete or add information to the instructions as necessary to achieve compliance with applicable standards and code requirements.

Manufacturers' installation instructions are an enforceable extension of the code and must be in the hands of the code official when an inspection takes place. Inspectors must carefully and completely read and comprehend the manufacturer's instructions in order to properly perform an installation inspection. In some cases, the code will specifically address an installation requirement that is also addressed in the manufacturer's installation instructions. The code requirement may be the same or may exceed the requirement in the manufacturer's installation instructions. The manufacturer's installation instructions could contain requirements that exceed those in the code. In such cases, the more restrictive requirements would apply (see commentary, Section 106).

Even if an installation appears to be in compliance with the manufacturer's instructions, the installation cannot be completed or approved until all associated components, connections and systems that serve the appliance or equipment are also in compliance with the requirements of the applicable *International Codes*® (I-Codes®) of reference. For example, a gas-fired boiler installation must not be approved if the boiler is connected to a deteriorated, undersized or otherwise unsafe chimney or vent. Likewise, the same installation must not be approved if the existing gas piping has insufficient capacity to supply the boiler load or if the electrical supply circuit is inadequate or unsafe.

Manufacturers' installation instructions are often updated and changed for various reasons, such as changes in the appliance, equipment or material design; revisions to the product standards and as a result of field experiences related to existing installations. The code official should stay abreast of any changes by reviewing the manufacturer's instructions for every installation.

N1101.13.1 (R303.2.1) Protection of exposed foundation insulation. Insulation applied to the exterior of basement walls, crawlspace walls and the perimeter of slab-on-grade floors shall have a rigid, opaque and weather-resistant protective covering to prevent the degradation of the insulation's thermal performance. The protective covering shall cover the exposed exterior insulation and extend a minimum of 6 inches (153 mm) below grade.

❖ The ultimate performance of insulation material is directly proportional to the workmanship involved in the materials' initial installation, as well as the materials' integrity over the life of the structure. Accordingly, foundation wall and slab-edge insulation materials installed in the vicinity of the exterior grade line require protection from damage that could occur from contact by lawn-mowing and maintenance equipment, garden hoses, garden tools, perimeter landscape materials, etc. In addition, the long-term thermal performance of foam-plastic insulation materials is adversely affected by direct exposure to the sun. To protect the insulation from sunlight and physical damage, it must have a protective covering that is inflexible, puncture resistant, opaque and weather resistant.

N1101.14 (R303.3) Maintenance information. Maintenance instructions shall be furnished for equipment and systems that require preventive maintenance. Required regular maintenance actions shall be clearly stated and incorporated on a readily accessible label. The label shall include the title or publication number for the operation and maintenance manual for that particular model and type of product.

❖ This section establishes an owner's responsibility for maintaining the building in accordance with the requirements of the code and other referenced standards. This section requires, among others, that mechanical and service water heating equipment and appliance maintenance information be made available to the owner/operator. This section does not require that labels be added to existing equipment; having the manufacturer's maintenance literature is usually sufficient to meet this requirement. During final occupancy inspection, the mechanical equipment and water heater should be inspected to verify that the information is taped to each unit or referenced on a label mounted in a conspicuous location on the units.

The code official has the authority to rule on the performance of maintenance work when equipment functions would be affected by such work. He or she also has the authority to require a building and its energy-using systems to be maintained in compliance with the public health and safety provisions required by other I-Codes.

N1101.15 (R401.2) Compliance. Projects shall comply with Sections identified as "mandatory" and with either sections identified as "prescriptive" or the performance approach in Section N1105.

❖ This section allows residential buildings to comply with either the prescriptive building thermal envelope requirements of Sections N1102.1 through N1102.3, N1103.2.1, N1103.4.2 and N1104.1 or the performance options that are provided in Section N1105. Under either option, the building must comply with the mandatory requirements that are found in Sections N1101.3, N1102.4, N1102.5, N1103.1, N1103.1.2, N1103.2.2, N1103.2.3, N1103.3, N1103.4, N1103.5, N1103.7, N1103.8, N1103.9 and N1104. A code user may evaluate both options and use the one that fits the project best since these two differing methods can result in different requirements. Most requirements are given prescriptively. Alternative tradeoffs are specified for many requirements, such as for the building thermal envelope requirements. For requirements specified by *U*-factors, an overall UA (*U*-factor times the area) can be used to show equivalence. A performance-based annual energy calculation can also be met by showing overall energy equivalence.

The majority of the requirements of this chapter are based upon the climate zone where the project is being built. The appropriate climate zone can be found in Chapter 3[RE] of the IECC. Zones 1 through 7 apply to various parts of the continental United States and are defined by county lines. Zones 7 and 8 apply to various parts of Alaska, and Hawaii is classified as Zone 1. The climate zones have been divided into "marine," "dry" and "moist" to deal with levels of humidity. For more details and background on the development of the new climate zones, see the commentary in Chapter 3[RE] of the IECC.

N1101.16 (R401.3) Certificate (Mandatory). A permanent certificate shall be completed and posted on or in the electrical distribution panel by the builder or registered design professional. The certificate shall not cover or obstruct the visibility of the circuit directory label, service disconnect label or other required labels. The certificate shall list the predominant *R*-values of insulation installed in or on ceiling/roof, walls, foundation (slab, *basement wall,* crawl space wall and/or floor) and ducts outside conditioned spaces; *U*-factors for fenestration and the solar heat gain coefficient (SHGC) of fenestration, and the results from any required duct system and building envelope air leakage testing done on the building. Where there is more than one value for each component, the certificate shall list the value covering the largest area. The certificate shall list the types and efficiencies of heating, cooling and service water heating equipment. Where a gas-fired unvented room heater, electric furnace, or baseboard electric heater is installed in the residence, the certificate shall list "gas-fired unvented room heater," "electric furnace" or "baseboard electric heater," as appropriate. An efficiency shall not be *listed* for gas-fired unvented room heaters, electric furnaces or electric baseboard heaters.

❖ This section is intended to increase the consumer's awareness of the energy-efficiency ratings for the various building elements in his or her home. The builder or registered design professional has to complete the certificate and place it on or inside the electrical panel (see Commentary Figure N1101.16). The permanent certificate shall not cover or obstruct the visibility of the circuit directory label, service disconnect label or other required labels.

The certificate must disclose the building's *R*-values, fenestration *U*-factors and fenestration SHGC, HVAC equipment types and efficiencies. The energy

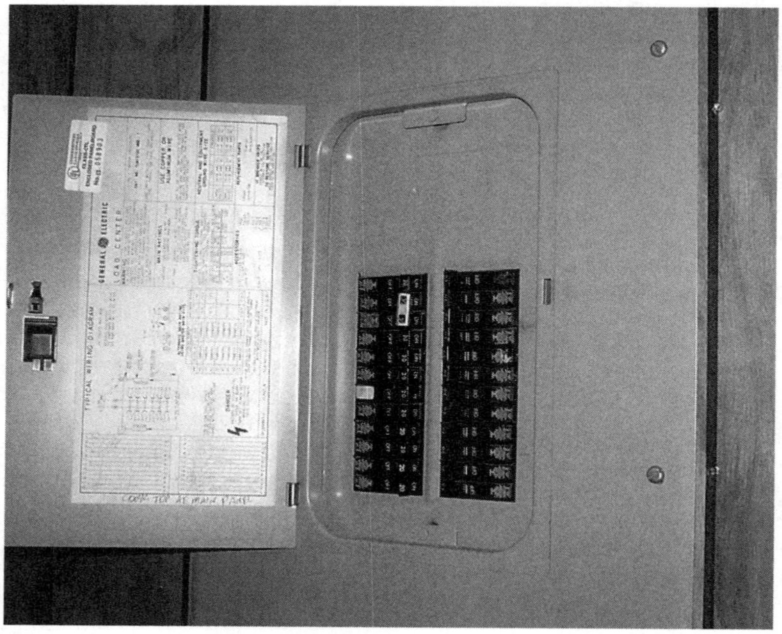

Figure N1101.16
CERTIFICATE

efficiency of a building as a system is a function of many elements considered as separate parts of the whole. It is difficult to have a proper identification and analysis of a building's energy efficiency once the building is completed because many of the elements may not be readily accessible.

This information is also valuable for existing structures undergoing alterations and additions to help determine the appropriate sizing for the mechanical systems. This is meant to be a simple certificate that is easy to read. The certificate does not contain all the information required for compliance and cannot be substituted for information on the required construction documents. Instead, the certificate is meant to provide the housing owner, occupant or buyer with a simple-to-understand overview of the home's energy efficiency. Where there is a mixture of insulation and fenestration values, the value applying to the largest area is specified. For example, if most of the wall insulation was R-19, but a limited area bordering the garage was R-13, the certificate would specify R-19 for the walls. (In contrast, plans and overall compliance would need to account for both *R*-values.)

The code specifies the minimum information on the certificate, but does not prohibit additional information being added so long as the required information is clearly visible. For example, a builder might choose to list energy-efficiency features beyond those required by the code.

SECTION N1102 BUILDING THERMAL ENVELOPE

N1102.1 (R402.1) General (Prescriptive). The *building thermal envelope* shall meet the requirements of Sections N1102.1.1 through N1102.1.4.

❖ The provisions of Section N1102 are the detailed requirements of the levels of insulation, the performance of openings (fenestrations) and air-leakage and moisture-control provisions that serve to establish the building's energy efficiency. When combined with the "systems" requirements (Section N1103 and the electrical power requirements in Section N1104), these three sections will provide the total package of energy conservation that the code requires.

The term "building thermal envelope" is defined in Section N1101.9 as being "the basement walls, exterior walls, floor, roof and any other building elements that enclose conditioned spaces. This boundary also includes the boundary between conditioned space or provides a boundary between conditioned space and exempt or unconditioned space." Therefore, when combined with the definition of "Conditioned space," the code has defined the boundaries of the building that will be regulated by this section. The building thermal envelope is a key term and resounding theme used throughout the energy requirements. It defines what portions of the building structure bound conditioned space and are thereby covered by the insulation and infiltration (air leakage) requirements of the code. The building thermal envelope includes all building components separating conditioned spaces (see commentary, "Conditioned space") from unconditioned spaces or outside ambient conditions and through which heat is transferred. For example, the walls and doors separating an unheated garage (unconditioned space) from a living area (conditioned space) are part of the building envelope. The walls and doors separating an unheated garage from the outdoors are not part of the building thermal envelope. Walls, floors and other building components separating two conditioned spaces are not part of the building envelope. For example, interior partition walls, the common or party walls separating dwelling units in multiple-family buildings and the wall between a new conditioned addition and the existing conditioned space are not considered part of the building envelope.

Unconditioned spaces (areas having no heating or cooling sources) are considered outside the building thermal envelope and are exempt from these requirements (see Section N1101.6). A space is conditioned if it is heated or cooled directly; communicates directly with a conditioned space or where a space is indirectly supplied with heating, cooling or both through uninsulated walls, floors or uninsulated ducts or HVAC piping. Boundaries that define the building envelope include the following:

- Building assemblies separating a conditioned space from outdoor ambient weather conditions.
- Building assemblies separating a conditioned space from the ground under or around that space, such as the ground around the perimeter of a slab or the soil at the exterior of a conditioned basement wall. Note that the code does not specify requirements for insulating basement floors or underneath slab floors (except at the perimeter edges).
- Building assemblies separating a conditioned space from an unconditioned garage, unconditioned sunroom or similar unheated/cooled area.

The code specifies requirements for ceiling, wall, floor, basement wall, slab-edge and crawl space wall components of the building envelope. In some cases, it may be unclear how to classify a particular part of a building. For example, skylight shafts have properties of a wall assembly but are located in the ceiling assembly. In these situations, a determination needs to be made and approved by the code official prior to construction so that the proper level of insulation can be installed to complete the building thermal envelope. Generally, skylight shafts and other items that are vertical or at an angle of greater than 60 degrees (1.1 rad) from the horizontal would typically use the wall insulation value.

N1102.1.1 (R402.1.1) Insulation and fenestration criteria. The *building thermal envelope* shall meet the requirements of Table N1102.1.1 based on the climate zone specified in Section N1101.10.

❖ This section serves as the basis for the code's general insulation and fenestration requirements. Therefore, this is the first place to determine what the requirements for the building thermal envelope will be. There are specific requirements for certain assemblies and locations that are addressed in Sections N1102.2 and N1102.3. This section begins by establishing the requirements for the building thermal envelope by requiring compliance with the proper component insulation and fenestration requirements of Table N1102.1.1. However, once that general requirement is established, Sections N1102.1.2, N1102.1.3 and N1102.1.4 will provide three possible means of showing that the building thermal envelope will comply. Any of the three methods may be used at the discretion of the designer. The three options and their advantages are discussed in the commentary with the subsections. In general, the later subsections will provide the designer with more options and flexibility, but they are also more complex than using Table N1102.1.1 on an individual component basis.

Table N1102.1.1 lists the minimum *R*-value and maximum *U*-factor and SHGC requirements for different portions of the building thermal envelope, including basement and exterior walls, floor, ceiling and any other building elements that enclose conditioned space. Using the table begins with determining the climate zone for the proposed location from Table R301.1 or Figure R301.1. Once the climate zone has been determined, each of the *R*-value, *U*-factor or SHGC requirements must be met for the applicable component (e.g., ceilings, walls, floors, etc.).

Maximum fenestration *U*-factor is the first column in Table N1102.1.1 that must be complied with (see definition for "Fenestration" in Chapter 2). Except as modified or exempted by Section N1102.3, each fenestration product in the proposed building must not exceed the maximum *U*-factor requirement presented in the table for a particular climate zone. For example, a single-family residence located in Climate Zone 5 would require installation of glazed fenestration products with a maximum *U*-factor of 0.32. This would include all glazing in the walls of the building thermal envelope (e.g., vertical windows) and skylights in the roof would be limited to a maximum *U*-factor of 0.55. The proposed glazing *U*-factor should be called out in the building plans either on the floor plan or within a window schedule. This will provide the necessary information to the field inspector, who will then need to verify that what is on the plans is installed in the field.

Fenestration products that do not have labels on them must use the default *U*-factors contained in Table

TABLE N1102.1.1 (R402.1.1)
INSULATION AND FENESTRATION REQUIREMENTS BY COMPONENT[a]

CLIMATE ZONE	FENESTRATION *U*-FACTOR[b]	SKYLIGHT[b] *U*-FACTOR	GLAZED FENESTRATION SHGC[b, e]	CEILING *R*-VALUE	WOOD FRAME WALL *R*-VALUE	MASS WALL *R*-VALUE[i]	FLOOR *R*-VALUE	BASEMENT[c] WALL *R*-VALUE	SLAB[d] *R*-VALUE & DEPTH	CRAWL SPACE[c] WALL *R*-VALUE
1	NR	0.75	0.25	30	13	3/4	13	0	0	0
2	0.40	0.65	0.25	38	13	4/6	13	0	0	0
3	0.35	0.55	0.25	38	20 or 13 + 5[h]	8/13	19	5/13[f]	0	5/13
4 except Marine	0.35	0.55	0.40	49	20 or 13 + 5[h]	8/13	19	10 /13	10, 2 ft	10/13
5 and Marine 4	0.32	0.55	NR	49	20 or 13 + 5[h]	13/17	30[g]	15/19	10, 2 ft	15/19
6	0.32	0.55	NR	49	20 + 5 or 13 + 10[h]	15/20	30[g]	15/19	10, 4 ft	15/19
7 and 8	0.32	0.55	NR	49	20 + 5 or 13 + 10[h]	19/21	38[g]	15/19	10, 4 ft	15/19

For SI: 1 foot = 304.8 mm.

a. *R*-values are minimums. *U*-factors and SHGC are maximums. When insulation is installed in a cavity which is less than the label or design thickness of the insulation, the installed *R*-value of the insulation shall not be less than the *R*-value specified in the table.

b. The fenestration *U*-factor column excludes skylights. The SHGC column applies to all glazed fenestration.
Exception: Skylights may be excluded from glazed fenestration SHGC requirements in Climate Zones 1 through 3 where the SHGC for such skylights does not exceed 0.30.

c. "15/19" means R-15 continuous insulation on the interior or exterior of the home or R-19 cavity insulation at the interior of the basement wall. "15/19" shall be permitted to be met with R-13 cavity insulation on the interior of the basement wall plus R-5 continuous insulation on the interior or exterior of the home. "10/13" means R-10 continuous insulation on the interior or exterior of the home or R-13 cavity insulation at the interior of the basement wall.

d. R-5 shall be added to the required slab edge *R*-values for heated slabs. Insulation depth shall be the depth of the footing or 2 feet, whichever is less in Zones 1 through 3 for heated slabs.

e. There are no SHGC requirements in the Marine Zone.

f. Basement wall insulation is not required in warm-humid locations as defined by Figure N1101.10 and Table N1101.10.

g. Or insulation sufficient to fill the framing cavity, R-19 minimum.

h. First value is cavity insulation, second is continuous insulation or insulated siding, so "13 + 5" means R-13 cavity insulation plus R-5 continuous insulation or insulated siding. If structural sheathing covers 40 percent or less of the exterior, continuous insulation *R*-value shall be permitted to be reduced by no more than R-3 in the locations where structural sheathing is used – to maintain a consistent total sheathing thickness.

i. The second *R*-value applies when more than half the insulation is on the interior of the mass wall.

N1101.12.3(1) or N1101.12.3(3) (see commentary, Section N1101.12.3). Note that the lowest default *U*-factor included in the table for glazed fenestration is listed at 0.55 for a "nonmetal or metal-clad double-pane window." This *U*-factor will not meet the requirements of the code in Climate Zones 2 and higher.

TABLE N1102.1.1. See page 11-40.

❖ Table N1102.1.1 serves as the basis for establishing the building thermal envelope requirements based on the text of Section N1102.1.1 and sets the performance level for each of the individual components listed. See the commentary for Sections N1102.1.1, N1102.2 and N1102.3 for additional discussion related to the components in the table. The simplest compliance approach is to meet these requirements directly. Note that the requirements do not change based on the area of the components of the residence. These same requirements apply to changes in existing buildings; for example, additions.

A few specifics of Table N1102.1.1 may benefit from clarification.

When applying the fenestration requirements of this table, it is important to remember the definition of "Fenestration" and that it does include items such as doors, glass blocks and other items as well as windows. Therefore, any door located in the building thermal envelope would still be subject to these limitations. Although vertical fenestration (vertical windows and doors) and skylights have a separate column for *U*-factor, the SHGC applies to both. This is reinforced by the provisions of Note b.

The ceiling *R*-value requirements are precalculated for insulation only and already assume a credible *R*-value for other building materials such as air films, interior sheathing and exterior sheathing. The only *R*-value for ceiling insulation that may be used to meet the requirements is that installed between the conditioned space and the vented airspace in the roof/ceiling assembly. This is typically not an issue because most insulation is installed directly on top of the gypsum board ceiling and the ceiling location represents the building thermal envelope. Insulation installed in the ceiling must meet or exceed the required insulation level. The "conditioned attic" requirements of the code may be viewed as an acceptable alternative if approved by the code official. These minimum ceiling *R*-values would still be applicable where the provisions of Section R806.4 of the code were used to create a conditioned attic assembly. In those cases, the location of the insulation and air barrier (building thermal envelope) are simply located at the roof instead of at the ceiling line. See the commentary to Sections N1102.2.1 and N1102.2.2 for additional information regarding ceiling insulation requirements.

The *R*-values presented under the "Walls" column represent the sum of the insulation materials installed between the framing cavity and, if used, the insulating sheathing. See Section 402.1.2 regarding how to compute the *R*-value. Insulating sheathing must have an *R*-value of at least R-2 to be considered. The R-2 limitation comes from the definition of "Insulating sheathing." The *R*-value of noninsulative interior finishes, such as sheet rock, or exterior coverings like wood structural panel siding is not considered when determining whether the proposed wall assembly meets the requirements. For example, in Commentary Figure N1102.1.2, the *R*-value of the cavity insulation installed between framing (R-13) is added to the insulating sheathing installed on the outside of the studs (R-6) resulting in an R-19 wall. The R-19 total insulation value can then be compared to the *R*-value requirement for the specific climate zone in Table N1102.1.1 to determine compliance.

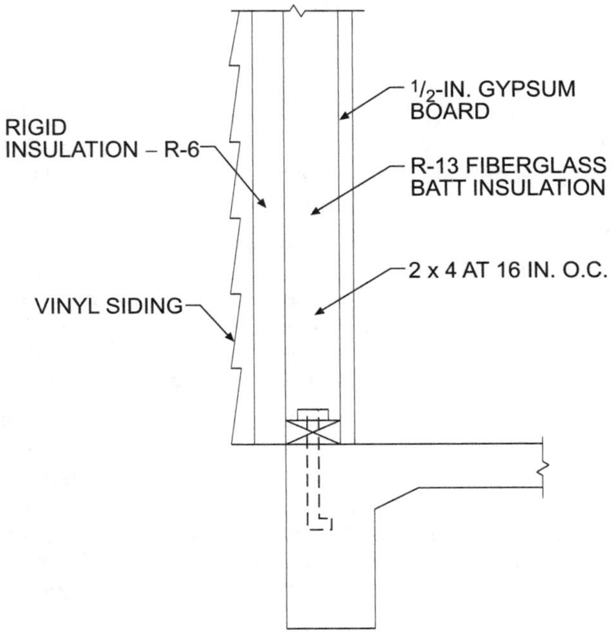

For SI: 1 inch = 25.4 mm.

Figure N1102.1.2
EXTERIOR *R*-VALUES

The insulation *R*-value requirement for exterior walls assumes wood framing. Walls framed using steel studs or constructed of materials such as a concrete masonry unit (CMU) are addressed in Sections N1102.2.5 and N1102.2.6. Whenever a residence has more than one type of wall (frame or mass) or more than one type of below-grade wall (conditioned basement or crawl space), the requirement for each component is taken from the appropriate column in Table N1102.1.1.

Mass walls are defined and have additional requirements within Section 402.2.5. Mass walls are intended to be above-grade walls and do not include basement walls, which have a separate entry within the table.

Note a reminds the code user which level of performance is required. Therefore, when dealing with *R*-values, a higher number would be better. When dealing with *U*-factors, the lower the number, the better the performance.

In accordance with Note c, for basement walls and crawl space walls, the two numbers separated by a "/" represent the values for continuous and cavity insulation; either will meet the code's requirements. For example, in Climate Zone 6, the wall can either be covered with continuous insulation to a minimum level of R-15 or, if some type of framing is used (such as a wood-frame wall used to finish out a basement), R-19 insulation must be installed within the cavity. This higher level of cavity insulation adjusts for the bridging or reduction in energy efficiency that the framing elements would create.

In accordance with Note d, heated slabs require R-5 insulation in Climate Zones 1, 2 and 3 and R-15 slab edge insulation in Climate Zones 4 and above. This R-15 insulation is the result of R-5 being "added" to the R-10 insulation level specified in the table for Climate Zones 4 through 8.

In accordance with Note g, where R-30 under-floor insulation is required, less insulation may be used if the framing cavity is filled, down to a minimum of R-19. This recognizes that extending the framing solely to hold more insulation can cost more than it is worth.

N1102.1.2 (R402.1.2) *R*-value computation. Insulation material used in layers, such as framing cavity insulation and insulating sheathing, shall be summed to compute the component *R*-value. The manufacturer's settled *R*-value shall be used for blown insulation. Computed *R*-values shall not include an *R*-value for other building materials or air films.

❖ This section indicates how the *R*-value in Table N1102.1.1 is to be determined. Table N1102.1.1 specifies the required *R*-values for the insulation products, the nominal *R*-value. This is the *R*-value of the insulation products only. Although other products and features such as finish materials, air films and airspaces may contribute to overall energy efficiency, when determining the *R*-value in the code, these additional items are not considered and do not contribute to the nominal *R*-value. For example, if a wall had R-13 cavity insulation, gypsum board with an *R*-value of almost R-1 and exterior siding that has an *R*-value of R-1, the overall wall *R*-value is simply R-13 because the gypsum board and the exterior siding do not contribute to the *R*-value for purposes of determining code compliance. Where there is more than one layer of insulation, the *R*-values for the layers are summed. For example, a wall with R-13 batts within the framing cavity and R-4 insulated sheathing would be treated as an R-17 wall (13 + 4 = 17). It is only insulation materials that may be summed to determine the component's *R*-value.

N1102.1.3 (R402.1.3) *U*-factor alternative. An assembly with a *U*-factor equal to or less than that specified in Table N1102.1.3 shall be permitted as an alternative to the *R*-value in Table N1102.1.1.

❖ For residences built with common insulation products, the most direct method of compliance is often the *R*-values in Table N1102.1.1. As an alternative, compliance can be demonstrated by calculating the *U*-factor for a component. Table N1102.1.3 gives *U*-factors that are deemed to be equivalent to the *R*-values in the prescriptive tables. Unlike the *R*-values in Table N1102.1.1, which consider only the insulation, *U*-factors consider all the parts of the construction. *U*-factors for a wall might include exterior siding, gypsum board and air films, all of which would be excluded from the *R*-value computation by Section N1102.1.2; for example, whether wall framing is 16 or 24 inches (406 or 610 mm) on center matters in computing the *U*-factor. Whether framing is metal or wood can also have a significant impact on *U*-factor.

U-factors are well suited to several applications. Construction types that limit the amount of framing or include thermal breaks as part of their design may benefit from *U*-factor calculations. Components with complex or nonuniform geometries can use testing to establish *U*-factors. Compliance with the "total UA alternative" or tradeoff approach in Section N1102.1.4 requires the use of the *U*-factor tables.

TABLE N1102.1.3 (R402.1.3)
EQUIVALENT *U*-FACTORS[a]

CLIMATE ZONE	FENESTRATION *U*-FACTOR	SKYLIGHT *U*-FACTOR	CEILING *U*-FACTOR	FRAME WALL *U*-FACTOR	MASS WALL *U*-FACTOR[b]	FLOOR *U*-FACTOR	BASEMENT WALL *U*-FACTOR	CRAWL SPACE WALL *U*-FACTOR
1	0.50	0.75	0.035	0.082	0.197	0.064	0.360	0.477
2	0.40	0.65	0.030	0.082	0.165	0.064	0.360	0.477
3	0.35	0.55	0.030	0.057	0.098	0.047	0.091[c]	0.136
4 except Marine	0.35	0.55	0.026	0.057	0.098	0.047	0.059	0.065
5 and Marine 4	0.32	0.55	0.026	0.057	0.082	0.033	0.050	0.055
6	0.32	0.55	0.026	0.048	0.060	0.033	0.050	0.055
7 and 8	0.32	0.55	0.026	0.048	0.057	0.028	0.050	0.055

a. Nonfenestration *U*-factors shall be obtained from measurement, calculation or an approved source.

b. When more than half the insulation is on the interior, the mass wall *U*-factors shall be a maximum of 0.17 in Zone 1, 0.14 in Zone 2, 0.12 in Zone 3, 0.087 in Zone 4 except Marine, 0.065 in Zone 5 and Marine 4, and 0.057 in Zones 6 through 8.

c. Basement wall *U*-factor of 0.360 in warm-humid locations as defined by Figure 301.1 and Table 301.1.

❖ This table provides the equivalent *U*-factors that may be used under Sections N1102.1.3 and N1102.1.4. See the commentary for Section N1102.1.3 for discussion related to this table.

N1102.1.4 (R402.1.4) Total UA alternative. If the total *building thermal envelope* UA (sum of *U*-factor times assembly area) is less than or equal to the total UA resulting from using the *U*-factors in Table N1102.1.3 (multiplied by the same assembly area as in the proposed building), the building shall be considered in compliance with Table N1102.1.1. The UA calculation shall be done using a method consistent with the ASHRAE *Handbook of Fundamentals* and shall include the thermal bridging effects of framing materials. The SHGC requirements shall be met in addition to UA compliance.

❖ This alternative allows one portion of the building to make up for another. It recognizes that there may be reasons for less insulation in some parts of the building, which can be compensated for by more insulation in other parts of the residence. The key concept is that the overall building thermal flow (UA) meets the code. This concept could allow a ceiling to make up for a wall or vice versa. As a practical matter, whether a building will comply by this method can sometimes be estimated quickly. A large area that is significantly over the required *R*-value will make up for a small area only mildly under the required *R*-value. Likewise, it will sometimes be obvious that a small area that mildly exceeds the requirement will not make up for a large area well below the requirement.

This section will allow for such tradeoffs but only if the total UA for the proposed building is below the aggregate UA calculation using the required values in Table N1102.1.3 and the same assembly areas as the actual building. In other words, under this alternative, components with varying insulating values can be "traded off" with one another as the builder sees fit just as long as the total UA calculation for the entire building complies with a calculation for that same house that uses the same assembly areas and the maximum UA values from Table N1102.1.3.

The UA is the sum of the component *U*-factors times each assembly area. The maximum allowable UA is the UA for a proposed design as if it was insulated to meet exactly the individual component *U*-factor requirements. This tradeoff provision allows the type of insulation and installed fenestration to vary, which permits significant design flexibility. The desire for tradeoffs in construction is common because of unexpected problems or design conflicts, and a UA tradeoff analysis is usually calculated with the assistance of electronic compliance tools, depending on the jurisdiction. (For example, the DOE has online compliance software for the code called REScheck, which can be downloaded from the DOE website at www.energycodes.gov. REScheck, if approved by the jurisdiction as compliant with the code, can be used to perform a UA tradeoff analysis.)

This section explicitly prohibits the tradeoff of SHGC requirements, requiring that the "SHGC requirements be met in addition to UA compliance." As a result, glazed fenestration must comply with the SHGC values shown in Table N1102.1.1 even if the *U*-factor is modified by trading off against some other component.

The requirements of this section establish specific additional requirements for any tradeoff. First, the baseline house must have the same assembly areas as the proposed house (e.g., the same area of each assembly—fenestration, skylights, ceiling, wall and floor). Second, the calculation should be done consistent with the *ASHRAE Handbook of Fundamentals*. Third, the calculation must include the thermal bridging effects of framing materials. To meet these requirements, the calculation method must either specifically combine the actual framing and insulation paths (with their specific areas and *U*-values) or use framing factors such as those found in the *ASHRAE Handbook of Fundamentals* for all framed building components (Note: this is not necessary for fenestration, which is a whole product value). To illustrate this approach, assume 1,000 square feet (93 m^2) of wall, of which 250 square feet (23 m^2) is framing (assuming 0.81 *U*-factor) and 750 square feet (70 m^2) is cavity (R-13 insulation). The baseline (general code requirement) and proposed opaque wall UAs are computed as follows:

- Baseline Opaque Wall for Climate Zone 2 = (0.082 x 1000) = 82. (The 0.082 value used in the calculation is taken from Table 402.1.3.)
- Proposed Opaque Wall UA = (0.81 x 250) + (0.077 x 750) = 78. (The 0.81 value used in the calculation was given above as an assumption. The 0.077 value is determined based on the *R*-value of 13; that is, 1 ÷ 13 = 0.077.)

Similar computations would be done for each assembly (such as fenestration or ceilings) and the baseline and proposed values then totaled and compared. If the baseline is greater than or equal to the proposed values, the house satisfies the UA alternative. The home will still need to meet all other prescriptive requirements, including the fenestration SHGC in Table N1102.1.1, the air leakage requirements of Section N1102.4 and the moisture-control requirements of Section N1102.5.

It also should be noted that this alternative is limited to UA tradeoffs for the building's thermal envelope. The 2012 code does not authorize or establish any basis for HVAC tradeoffs associated with its UA tradeoff option. HVAC performance is simply not addressed under the code prescriptive or UA tradeoff paths. It is addressed only in Section 405 under the "Simulated Performance Alternative." As a result, the code elected to limit these simplified tradeoffs to UA envelope tradeoffs and defer any more complex tradeoffs exclusively to the "Simulated Performance Alternative" in Section N1105. If the builder wishes to factor in HVAC performance for tradeoffs, Section N1105 can be used and is permitted based on Section N1101.2, accepting Section N1105 as a compliance option.

Documentation acceptable to the local building department generally must be submitted to the appropriate authority to certify acceptable component UA tradeoffs.

UA alternative example (for calculating standard and proposed designs):

Building areas and *U*-factors (for single-family, Zone 5 from Table N1102.1.3)

Standard (Code) Design	Proposed Design
Exterior Wall 1050 ft^2 U_w = -0.057	Exterior Wall 1050 ft^2 U = 0.0451
Glazed doors and windows 192 ft^2 U_g = 0.32	Glazed doors and windows 192 ft^2 U = 0.40
Opaque exterior door 38 ft^2 U_g = 0.32	Opaque exterior door 38 ft^2 U = 0.35
Roof 1500 ft^2 U_r = 0.026	Roof 1500 ft^2 U_r = 0.026
Floor (slab) 1500 ft^2 U = 0.033	Floor (slab) 1500 ft^2 U_{fe} = 0.033

UA Standard = $(0.057 \times 1050) + (0.32 \times 192) + (0.32 \times 38) + (0.026 \times 1500) + (0.033 \times 1500) = 222$

UA Proposed = $(0.0451 \times 1050) + (0.40 \times 192) + (0.35 \times 38) + (0.026 \times 1500) + (0.033 \times 1500) = 225$

UA Proposed > UA Standard; therefore design fails to meet code requirements.

N1102.2 (R402.2) Specific insulation requirements (Prescriptive). In addition to the requirements of Section N1102.1, insulation shall meet the specific requirements of Sections N1102.2.1 through N1102.2.12.

❖ This section contains specific requirements to be followed for the individual items listed within the subsections. Although Section N1102.1 and Tables N1102.1.1 and N1102.1.3 provide the general basis for complying with the energy requirements of this chapter, Section N1102.2 provides additional details regarding the actual construction of the assemblies or modifications that may affect the general requirements. Relying on the principle that specific requirements apply over general requirements will help ensure that these specific provisions are properly followed.

N1102.2.1 (R402.2.1) Ceilings with attic spaces. When Section N1102.1.1 would require R-38 in the ceiling, R-30 shall be deemed to satisfy the requirement for R-38 wherever the full height of uncompressed R-30 insulation extends over the wall top plate at the eaves. Similarly, R-38 shall be deemed to satisfy the requirement for R-49 wherever the full height of uncompressed R-38 insulation extends over the wall top plate at the eaves. This reduction shall not apply to the *U*-factor alternative approach in Section N1102.1.3 and the total UA alternative in Section N1102.1.4.

❖ The required ceiling *R*-value found in the code is based on the assumption that standard truss or rafter construction was being used. When raised-heel trusses or other methods of framing that would not permit the ceiling insulation to be installed to its full depth over the entire area are used, then the code would permit the installation of a lower *R*-value insulation. The general assumption is that ceiling insulation will be compressed at the edges and, if the special construction techniques are used, that the level of insulation required can be reduced.

Insulation installed in a typical roof assembly will be full height throughout the center portions of the assembly and will taper (be compressed) at the edges as the roof nears the top plate of the exterior wall system [see Commentary Figure N1102.2.1(1)]. The slope of the roof causes this tapering, which is further amplified by any baffling installed to direct ventilation air from the eave vents up and over the insulation. Because of this tapering, the installed *R*-value near the plate lines will be less than the rated *R*-value for the insulation. This is caused by compression (compressed insulation has a lower *R*-value than insulation installed to its full thickness) and the limited space between the floor of the attic and the roof sheathing near the exterior plate line. Thus, a typical installation, on average, will have a lower *R*-value than that of the rated insulation. Because of this, the code will allow installation of a lower insulation value if it can be installed full thickness, to its rated *R*-value, over the plate line of the exterior wall. This allowance recognizes that a partial thermal "bypass" has been made more efficient by using insulation with the full *R*-value at the eaves. The full insulation *R*-value is sometimes achieved by what is termed an "energy truss" or "advanced framing." This can be achieved by using an oversized truss or raised-heel truss as shown in Commentary Figure N1102.2.1(2). Another way to achieve the full *R*-value would be by use of insulation with a higher *R*-value per inch at the eaves. The use of the options permitted by this section allows substituting R-30 for R-38 insulation, and R-38 may be substituted for R-49 insulation to meet the requirements of the code. When using the conditioned attic requirements of the code, the same option of using a reduced *R*-value would apply if the insulation was installed directly under the roof deck, rather than on the attic floor (see commentary, Section R806.4). The insulation would be permitted to meet the lesser *R*-value, presuming the full *R*-value was met over the eaves. Of course, this situation would also presume that the attic space beneath the insulation was not vented.

Note that this text applies only to the *R*-value portion of the code; there is no reduction in requirements if the *U*-factor alternative or the total UA alternative is used. In those cases, the reduced thickness must be accounted for in the calculations. In addition, if the residence had more than one separate attic space, it is possible this section could apply to one attic space, but not another.

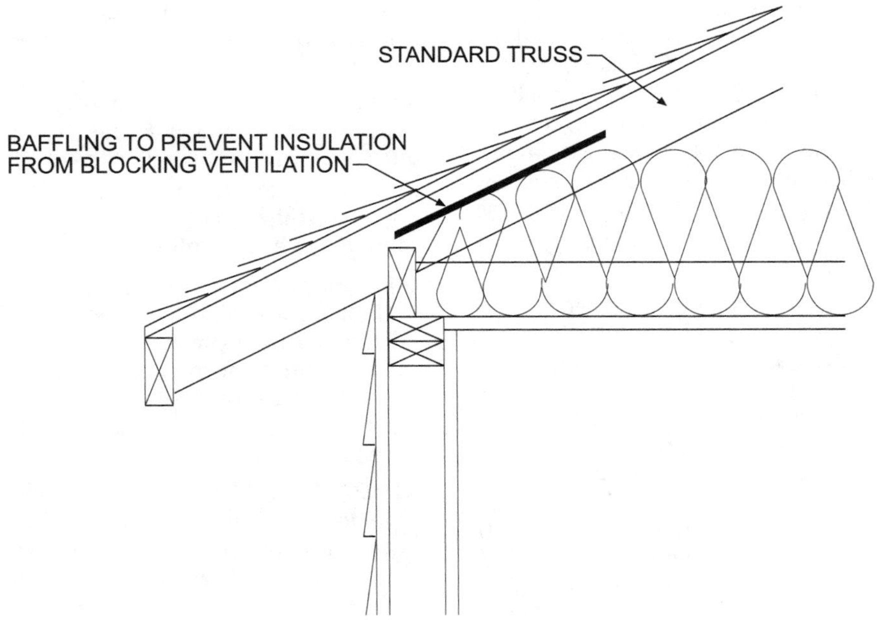

Figure N1102.2.1(1)
TYPICAL ROOF ASSEMBLY

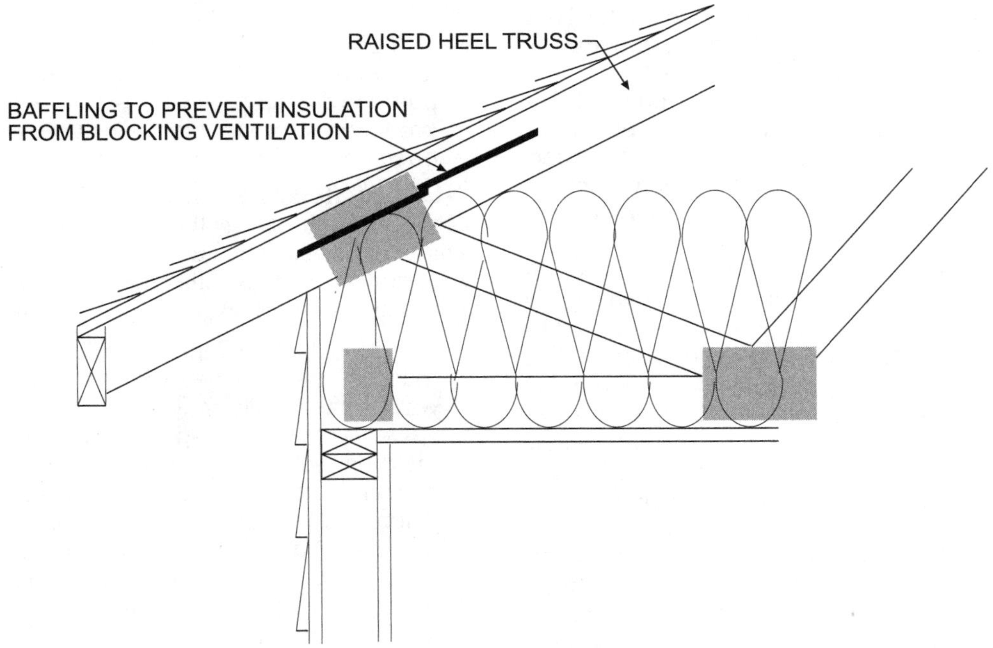

Figure N1102.2.1(2)
RAISED-HEEL TRUSS

N1102.2.2 (R402.2.2) Ceilings without attic spaces. Where Section N1102.1.1 would require insulation levels above R-30 and the design of the roof/ceiling assembly does not allow sufficient space for the required insulation, the minimum required insulation for such roof/ceiling assemblies shall be R-30. This reduction of insulation from the requirements of Section N1102.1.1 shall be limited to 500 square feet (46 m^2) or 20 percent of the total insulated ceiling area, whichever is less. This reduction shall not apply to the *U*-factor alternative approach in Section N1102.1.3 and the total UA alternative in Section N1102.1.4.

❖ In situations where the ceiling is installed directly onto the roof rafters and no attic space is created, this section will allow a reduced level of ceiling insulation for a limited area. This section addresses the construction of what is typically called "cathedral" or "vaulted" ceilings, which therefore result in that portion of the home not having an attic area above the ceiling. See the definition of "Attic" in Chapter 2 of the code. Based on the use of solid sawn lumber (2 × 8, 2 × 10 or 2 × 12) in conventional construction, as the ceiling *R*-value requirement increases, it may be impossible to install the required ceiling insulation from Table N1102.1.1 within the available cavity depth. In addition, the ventilation requirements of Section R806.3 will further reduce the available space by requiring a minimum space of 1 inch (25 mm) between the insulation and the roof sheathing. Therefore, when the depth of the cavity will not permit the required insulation level, this section permits a reduction to R-30 ceiling insulation instead of the normally required R-38 or R-49 requirement from Table N1102.1.1. This will generally result in reducing the required insulation level instead of having to increase the depth of the framing members. This section in the code recognizes that increases in framing size done only to accommodate higher *R*-values are an expensive way to achieve a limited increase in *R*-value.

It is important to note that this section applies only to areas that have a required insulation level above R-30 (Climate Zones 4 through 8). Further, the reduction is limited to portions of ceiling assemblies that do not exceed 500 square feet (46 m^2) or 20 percent of the ceiling area. The intent is that the 500-square-foot (46 m^2) or 20-percent limitation be the total aggregate exempted amount of the building's thermal envelope (ceiling) that can use this reduction. It is not the intent that a home could have multiple areas that were each under the 500-square-foot (46 m^2) or 20-percent limit but would aggregate to more than that amount. In situations that cannot meet these limitations [homes in Climate Zones 1 through 3 or homes needing more than 500 square feet (46 m^2) of reduced ceiling insulation], the depth of the rafters would have to be increased to meet the table's required insulation level or other design changes or means of compliance would be necessary. This provision does not apply if the *U*-factor alternative or the total UA alternative is being used. In those cases, the reduced insulation level might be accomplished, but it would be accounted for with additional insulating components in the case of the *U*-factor alternative or with added insulation elsewhere in the case of the total UA alternative.

N1102.2.3 (R402.2.3) Eave baffle. For air permeable insulations in vented attics, a baffle shall be installed adjacent to soffit and eave vents. Baffles shall maintain an opening equal or greater than the size of the vent. The baffle shall extend over the top of the attic insulation. The baffle shall be permitted to be any solid material.

❖ For air permeable insulations in vented attics, a baffle shall be installed adjacent to soffit and eave vents. This will help prevent wind from degrading the attic insulation performance. Baffles serve to keep vents open, insulation in place and keep the wind from blowing through the insulation and reducing its effectiveness.

N1102.2.4 (R402.2.4) Access hatches and doors. Access doors from conditioned spaces to unconditioned spaces (e.g., attics and crawl spaces) shall be weatherstripped and insulated to a level equivalent to the insulation on the surrounding surfaces. Access shall be provided to all equipment that prevents damaging or compressing the insulation. A wood framed or equivalent baffle or retainer is required to be provided when loose fill insulation is installed, the purpose of which is to prevent the loose fill insulation from spilling into the living space when the attic access is opened, and to provide a permanent means of maintaining the installed *R*-value of the loose fill insulation.

❖ The portion of a ceiling used for an attic access door is a weak part of the building thermal envelope. The purpose of this provision is to ensure that measures are taken to prevent large loss of energy through this opening.

N1102.2.5 (R402.2.5) Mass walls. Mass walls for the purposes of this chapter shall be considered above-grade walls of concrete block, concrete, insulated concrete form (ICF), masonry cavity, brick (other than brick veneer), earth (adobe, compressed earth block, rammed earth) and solid timber/logs.

❖ The code uses a simple definition for mass walls. Walls made of the specified materials and mass are mass walls. Mass walls may meet the lower mass wall *R*-value (as compared to frame wall values) specified in their respective climate zones because of the energy-conserving characteristics of mass walls. Note that the difference between the wood-frame *R*-value is greatest in southern climates. This recognizes that the thermal "averaging" provided by mass walls is most effective in warmer climates. In the very northern climates where there is almost continual heating during parts of the year, the thermal mass is of limited value. The code simply lists the types of walls that are considered as being "mass" walls. There is no additional limitation or characteristics specified for the walls that would be applicable when using the code.

In general terms, the heat capacity is a measure of how well a material stores heat. The higher the heat capacity, the greater the amount of heat stored in the material. For example, a 6-inch (152 mm) heavyweight concrete wall has a heat capacity of 14 Btu/ft^2 × °F (620 J/m^2 × K) compared to a conventional 2 by 4 wood-framed wall with a heat capacity of approximately 3 Btu/ft^2 × °F (138 J/m^2 × K). Tables N1102.1.1 and N1102.1.3 are used to determine the insulation or equivalent *U*-factor requirements for mass walls. To use the tables, first consult either Table R301.1 or Figure R301.1 to determine the climate zone of the proposed project. The *R*-value for the assembly is then given for the mass wall in Table N1102.1.1 or the *U*-factor is given in Table N1102.1.3.

When dealing with the mass walls, the insulation location is important. Note i of Table N1102.1.1 indicates that the second *R*-value, which is the higher *R*-value for mass walls in each climate zone, is applicable when more than half of the insulation is on the interior of the mass wall. The first values listed in Table N1102.1.1 are, therefore, based on the installation of the majority of the insulation ("at least 50 percent") being located on the exterior side of the wall or being integral to the wall. Likewise, Note b of Table N1102.1.3 provides lower maximum *U*-factors for mass walls with more than 50 percent of the insulation on the interior of the wall.

For an example of a wall assembly that has insulation on the exterior of the wall (between the mass wall and the exterior), see Commentary Figure N1102.2.5(1). Concrete masonry units with insulated cores or masonry cavity walls are examples of integral insulation. For mass walls with the insulation installed on the interior (an insulated furred wall located between the conditioned space and the mass wall), see Commentary Figure N1102.2.5(2). As shown, these two figures would be examples of mass walls that meet or exceed the general requirements of this section and Table N1102.1.1 for Climate Zones 1 through 4, except for those in Marine Zone 4.

N1102.2.6 (R402.2.6) Steel-frame ceilings, walls, and floors. Steel-frame ceilings, walls, and floors shall meet the insulation requirements of Table N1102.2.6 or shall meet the *U*-factor requirements of Table N1102.1.3. The calculation of the *U*-factor for a steel-frame envelope assembly shall use a series-parallel path calculation method.

❖ The insulation requirements of Table N1102.1.1 are based on conventional wood-frame construction methods. Because the code also includes provisions applicable to steel-framing methods, this chapter has been written to include this material. Table N1102.2.6 specifies combinations of cavity and continuous insulation for steel framing that are equivalent to the specified wood-frame component *R*-values in Table N1102.1.1.

Table N1102.1.1 cannot be used directly for steel-frame components. When using Table N1102.2.6, all listed options comply; the code user can choose any option on the list that corresponds to the correct *R*-value in the left-hand column. Steel has a much higher thermal conductivity (ability to transfer heat) than wood. Therefore, steel-frame cavities require a higher *R*-value or include a requirement for insulated sheathing that acts as a thermal break or both.

Instead of using this table, the code user could choose to calculate or measure a *U*-factor for a building component and show compliance based on meeting the *U*-factor requirement in Table N1102.1.3. The code user could use the total UA tradeoff in Section N1102.1.4 or even the performance-based approach in Section N1105 to show compliance based on the overall building, even if the steel-frame wall did not

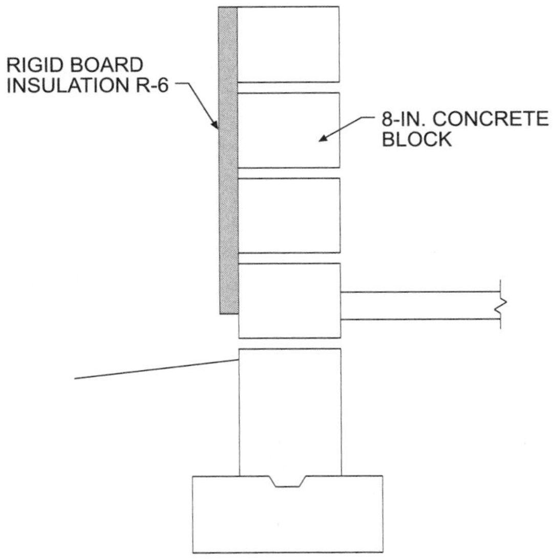

For SI: 1 inch = 25.4 mm.

Figure 1102.2.5(1)
EXTERIOR INSULATION

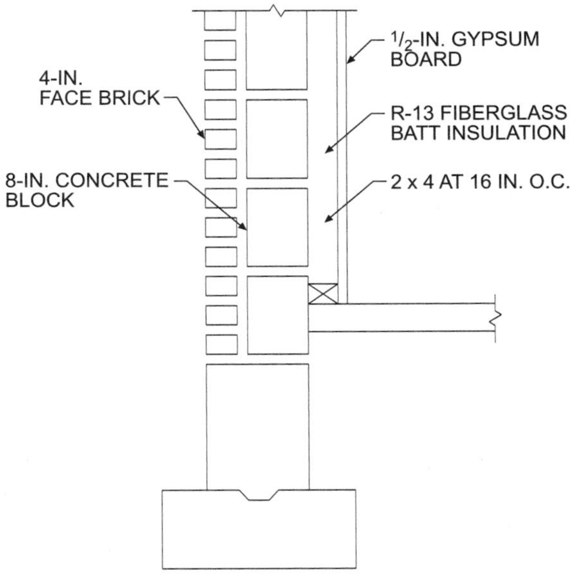

For SI: 1 inch = 25.4 mm.

Figure 1102.2.5(2)
INTERIOR INSULATION

meet the code requirements directly. Other combinations of cavity insulation and continuous insulation not shown in Table N1102.2.6 would also be allowed; however, if a combination from the table is used, no additional calculation is necessary.

TABLE N1102.2.6 (R402.2.6)
STEEL-FRAME CEILING, WALL AND FLOOR INSULATION (*R*-VALUE)

WOOD FRAME *R*-VALUE REQUIREMENT	COLD-FORMED STEEL EQUIVALENT *R*-VALUE[a]
Steel Truss Ceilings[b]	
R-30	R-38 or R-30 + 3 or R-26 + 5
R-38	R-49 or R-38 + 3
R-49	R-38 + 5
Steel Joist Ceilings[b]	
R-30	R-38 in 2 × 4 or 2 × 6 or 2 × 8 R-49 in any framing
R-38	R-49 in 2 × 4 or 2 × 6 or 2 × 8 or 2 × 10
Steel-Framed Wall, 16″ o.c.	
R-13	R-13 + 4.2 or R-19 + 2.1 or R-21 + 2.8 or R-0 + 9.3 or R-15 + 3.8 or R-21 + 3.1
R-13 + 3	R-0 + 11.2 or R-13 + 6.1 or R-15 + 5.7 or R-19 + 5.0 or R-21 + 4.7
R-20	R-0 + 14.0 or R-13 + 8.9 or R-15 + 8.5 or R-19 + 7.8 or R-19 + 6.2 or R-21 + 7.5
R-20 + 5	R-13 + 12.7 or R-15 + 12.3 or R-19 + 11.6 or R-21 + 11.3 or R-25 + 10.9
R-21	R-0 + 14.6 or R-13 + 9.5 or R-15 + 9.1 or R-19 + 8.4 or R-21 + 8.1 or R-25 + 7.7
Steel-Framed Wall, 24″ o.c.	
R-13	R-0 + 9.3 or R-13 + 3.0 or R-15 + 2.4
R-13 + 3	R-0 + 11.2 or R-13 + 4.9 or R-15 + 4.3 or R-19 + 3.5 or R-21 + 3.1
R-20	R-0 + 14.0 or R-13 + 7.7 or R-15 + 7.1 or R-19 + 6.3 or R-21 + 5.9
R-20 + 5	R-13 + 11.5 or R-15 + 10.9 or R-19 + 10.1 or R-21 + 9.7 or R-25 + 9.1
R-21	R-0 + 14.6 or R-13 + 8.3 or R-15 + 7.7 or R-19 + 6.9 or R-21 + 6.5 or R-25 + 5.9
Steel Joist Floor	
R-13	R-19 in 2 × 6, or R-19 + 6 in 2 × 8 or 2 × 10
R-19	R-19 + 6 in 2 × 6, or R-19 + 12 in 2 × 8 or 2 × 10

a. Cavity insulation *R*-value is listed first, followed by continuous insulation *R*-value.
b. Insulation exceeding the height of the framing shall cover the framing.

❖ See the commentary to Section N1102.2.6.

N1102.2.7 (R402.2.7) Floors. Floor insulation shall be installed to maintain permanent contact with the underside of the subfloor decking.

❖ Floors that are a part of the building thermal envelope, such as those over a crawl space or an unconditioned garage, are required to meet or exceed the floor *R*-value requirements listed in Table N1102.1.1. The insulation *R*-value requirements range from R-13 in warm climates to R-38 in extremely cold climates. Insulation must be installed between the floor joists and must be well supported with netting, wire, wood strips or another method of support so that the insulation does not droop or fall out of the joist cavities over time. Some floor insulation has a tendency to sag or drop with time. This sag or drop exposes the subfloor directly to the temperature beneath the floor. Sagging also has a tendency to open airflow paths to parts of the floor, producing cold spots and negating the value of the floor insulation in the effected section of the floor. Even small areas that lack insulation or allow air circulation between the floor insulation and the subfloor can have a marked effect on the energy efficiency of the floor. This section specifies that floor insulation must be installed so it will maintain "permanent contact" with the subfloor (meaning over the useful life of the residence).

Earlier editions of the code and other energy codes included a provision requiring that floor assemblies over outside air (not including a vented crawl space or an unconditioned garage) be insulated to the higher *R*-value requirements for ceilings. This typically applied only if over 25 percent of the floor assembly area was directly exposed to the outside air. With the changes made in the 2006 code, this requirement no longer exists. Therefore, if a floor of a building extends over outside air, such as over an open carport, the floor is still insulated to the "floor" requirement of Table N1102.1.1 and the direct contact requirements of this section become even more important.

N1102.2.8 (R402.2.8) Basement walls. Walls associated with conditioned basements shall be insulated from the top of the *basement wall* down to 10 feet (3048 mm) below grade or to the basement floor, whichever is less. Walls associated with unconditioned basements shall meet this requirement unless the floor overhead is insulated in accordance with Sections N1102.1.1 and N1102.2.7.

❖ The walls of conditioned basements must be insulated to meet the requirements of Table N1102.1.1. Each wall of a basement must be considered separately to determine whether it is a basement wall or an exterior wall. It is a basement wall if it has an average below-grade wall area of 50 percent or more. A wall that is less than 50 percent below grade is an exterior wall and must meet the insulation requirements for walls. Most walls associated with basements will be at least 50-percent below grade and, therefore, must meet basement wall requirements.

Walkout basements offer a challenge in determining compliance with the code. A walkout basement [see Commentary Figures N1102.2.8(1) and (2)] may have a back wall that is entirely below grade, a front wall or the walkout portion that is entirely above grade and two sidewalls with a grade line running diagonally. In this case, the back wall must meet the requirements for basement walls, the front wall would need to meet the requirements for walls (either framed wall or mass wall, as applicable) and the side walls would need to be evaluated to determine whether they were 50 percent or more below grade and, therefore, basement walls.

Basement insulation must extend up to 10 feet (3048 mm) under the ground, or at least as far as the basement wall extends under the ground. Heat flow into the ground occurs all along the buried portion of the wall, as well as along the above-ground portion of the wall. Heat flow below 10 feet (3048 mm) is greatly diminished so the code requires basement insulation only down to 10 feet (3048 mm), or the depth of the basement wall.

The code does not specify whether the insulation is to be placed on the inside or outside of a basement wall. In some localities moisture considerations may suggest the type and location for insulation.

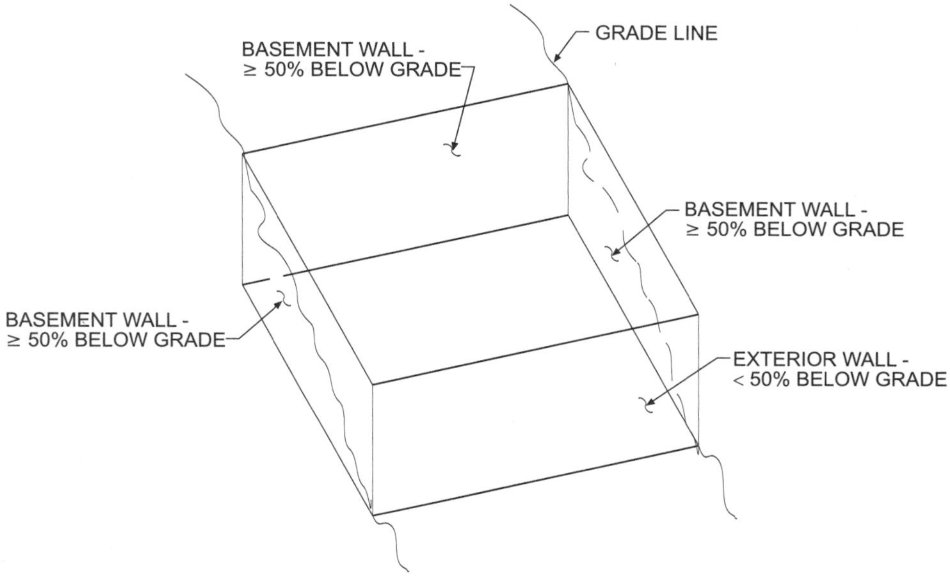

Figure N1102.2.8(1)
WALK-OUT BASEMENT

Figure N1102.2.8(2)
WALK-OUT BASEMENT

The last part of this section allows insulating unconditioned basement walls as an alternative to insulating the floor above the unconditioned basement. Therefore, it essentially shifts the location of the building thermal envelope from the floor to the basement walls. Although not required, insulating the unconditioned basement walls makes a good deal of sense if a basement is likely to be conditioned at some time after construction.

Because the rim joist between floors is a part of the building envelope, this must be insulated also if the basement is conditioned.

When applying the provisions of this section, it is important to review the definitions for both "Above grade wall" and "Basement wall" in Section N1101.9. For code users who also deal with commercial buildings, remember that the rule for residential buildings is 50 percent above or below grade and not the 85-percent minimum below grade with 15-percent maximum above-grade area that is applicable under Section C402.2.2 of the IECC for commercial construction.

N1102.2.9 (R402.2.9) Slab-on-grade floors. Slab-on-grade floors with a floor surface less than 12 inches (305 mm) below grade shall be insulated in accordance with Table N1102.1.1. The insulation shall extend downward from the top of the slab on the outside or inside of the foundation wall. Insulation located below grade shall be extended the distance provided in Table N1102.1.1 by any combination of vertical insulation, insulation extending under the slab or insulation extending out from the building. Insulation extending away from the building shall be protected by pavement or by a minimum of 10 inches (254 mm) of soil. The top edge of the insulation installed between the *exterior wall* and the edge of the interior slab shall be permitted to be cut at a 45-degree (0.79 rad) angle away from the *exterior wall.* Slab-edge insulation is not required in jurisdictions designated by the *building official* as having a very heavy termite infestation.

❖ The perimeter edges of slab-on-grade floors must be insulated to the *R*-values listed in Table N1102.1.1. These requirements apply only to slabs 12 inches (305 mm) or less below grade. The listed *R*-value requirements in the table are for unheated slabs. A heated slab must add another R-5 to the required insulation levels based on Note d of Table N1102.1.1.

The insulation must extend downward from the top of the slab or downward to the bottom of the slab and then horizontally in either direction until the distance listed in Table N1102.1.1 is reached. See Commentary Figure N1102.2.9 for examples on how the distance is measured. Most of the heat loss from a slab will occur in the edge that is exposed directly to the outside air. The insulation must be installed to the top of the slab edge to prevent this heat loss. Slab insulation may be installed on the exterior of the slab edge or between the interior wall and the edge of the interior slab as in a nonmonolithic slab. In this type of installation, the exposed insulation could cause problems with tack strips for carpeting. Therefore, the insulation is allowed to be cut at a 45-degree angle away from the exterior of the wall. If a monolithic slab and foundation is being used, the required insulation would obviously need to be installed on the exterior and then either extended to the required depth or turned out to the exterior and be protected by either some type of pavement or a minimum of 10 inches

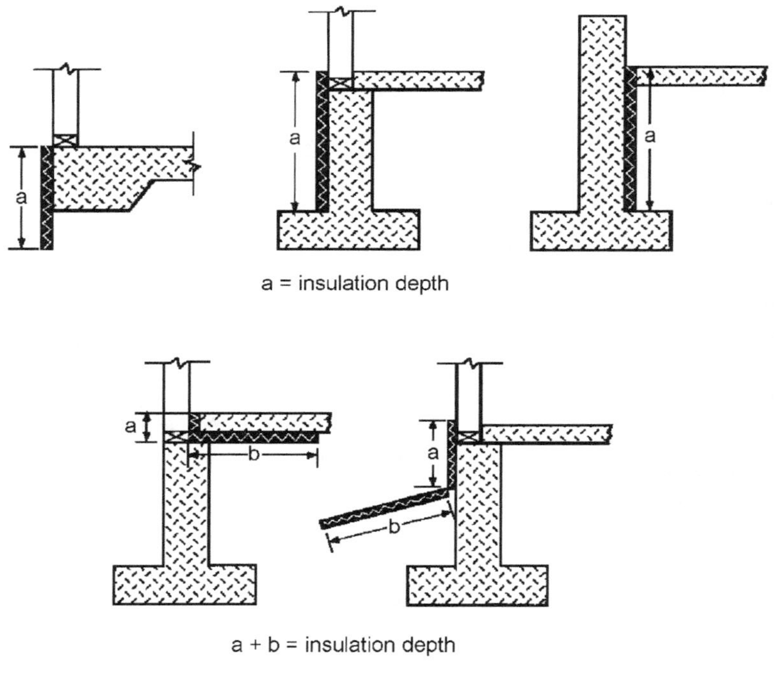

Figure N1102.2.9
SLAB INSULATION METHODS

(254 mm) of soil. Insulation that is exposed on or near the surface is easily damaged. This protection method ensures that the insulation remains in place and provides the intended energy savings.

In areas with very heavy termite infestation, slab perimeter insulation need not be installed in accordance with Table N1102.1.1. These areas are identified in Figure R301.2(6) or the jurisdiction may base its determination on the local history and situation. It is important to understand that the revisions of the 2006 code energy requirements provide this exemption from the slab insulation provisions for any area with heavy termite infestations. The fact that this is an exemption and does not contain any requirement for a compensating increase of insulation at other locations is important.

The requirements of Section N1102.1.4 could still be used in areas that do not have a heavy termite infestation to eliminate the slab edge insulation if desired. Typically, slab perimeter insulation can be traded off entirely in these climates by increasing the ceiling or wall insulation *R*-values or by using glazing with a lower *U*-factor.

N1102.2.10 (R402.2.10) Crawl space walls. As an alternative to insulating floors over crawl spaces, crawl space walls shall be permitted to be insulated when the crawl space is not vented to the outside. Crawl space wall insulation shall be permanently fastened to the wall and extend downward from the floor to the finished grade level and then vertically and/or horizontally for at least an additional 24 inches (610 mm). Exposed earth in unvented crawl space foundations shall be covered with a continuous Class I vapor retarder in accordance with this code. All joints of the vapor retarder shall overlap by 6 inches (153 mm) and be sealed or taped. The edges of the vapor retarder shall extend at least 6 inches (153 mm) up the stem wall and shall be attached to the stem wall.

❖ The code allows for the insulation of crawl space walls in crawl spaces instead of insulating the floor between the crawl space and the conditioned space. In essence, the code user is defining the thermal boundary as either the floor or the crawl space wall. Because the ground under the crawl space is tempered by the thermal mass of the dirt, the temperature of the crawl space is usually more favorable than the outside temperature. This is a popular practice for freeze protection of plumbing pipes in colder climates because it is common to install plumbing in the crawl space. The heat transferred through the uninsulated floor to the crawl space helps keep the crawl space temperature above freezing when the outside air temperature drops below freezing. To comply with this provision, the crawl space must be mechanically vented or supplied with conditioned air from the living space. Section 1203.3 of the IBC and its subsections address this crawl space ventilation requirement.

The code also requires installation of insulation from the sill plate downward to the exterior finished grade level and then an additional 24 inches (610 mm) either vertically or horizontally (see Commentary Figure N1102.2.10). Under this insulation scenario, the rim joist is considered part of the conditioned envelope and must be insulated to the same level as the exterior wall. The insulation must be attached securely to the crawl space wall so that it does not fall off. The code also requires installing a continuous Class I vapor retarder on the floor of the crawl space to prevent ground-water vapor from entering the crawl space. A Class I vapor retarder is defined in the IBC as a material with a permeance of 0.1 perm or less (see the definition of "Vapor retarder class" in the IBC). The vapor retarder is to be installed with all joints overlapped and sealed or taped to provide con-

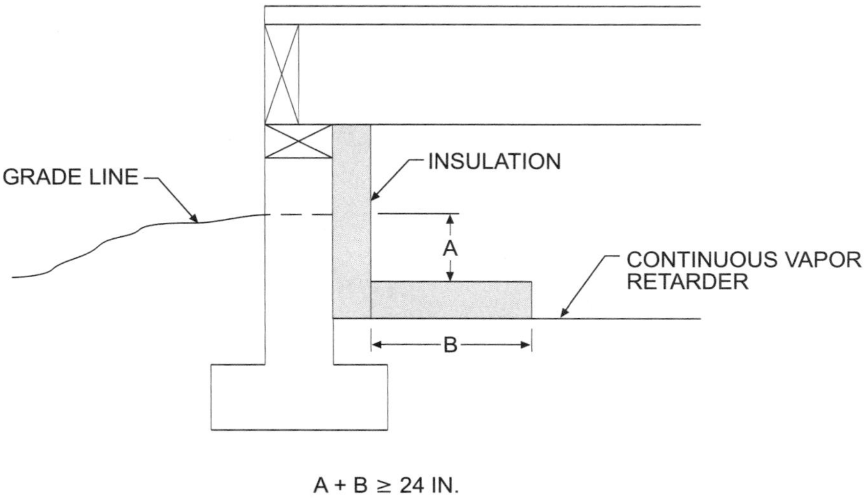

For SI: 1 inch = 25.4 mm.

Figure N1102.2.10
CRAWL SPACE WALL INSULATION

tinuity. Also, the vapor retarder must extend up the crawl space wall and be secured to the wall with an appropriate attachment, such as either an approved mastic or a treated wood nailer.

N1102.2.11 (R402.2.11) Masonry veneer. Insulation shall not be required on the horizontal portion of the foundation that supports a masonry veneer.

❖ For exterior foundation insulation, the horizontal portion of the foundation that supports a masonry veneer need not be insulated. For slab-edge insulation installed on the exterior of the slab, the code allows the insulation to start at the bottom of the masonry veneer and extend downward. This is essentially a matter of practicality and accommodates the construction of a "brick ledge" without the need for insulating the foundation at the point where the masonry would bear upon it.

N1102.2.12 (R402.2.12) Sunroom insulation. All *sunrooms* enclosing conditioned spaces shall meet the insulation requirements of this code.

Exception: For *sunrooms* with *thermal isolation*, and enclosing conditioned spaces, the following exceptions to the insulation *requirements* of this code shall apply:

1. The minimum ceiling insulation *R*-values shall be R-19 in Zones 1 through 4 and R-24 in Zones 5 through 8; and
2. The minimum wall *R*-value shall be R-13 in all zones. Wall(s) separating a *sunroom* with a *thermal isolation* from *conditioned space* shall meet the *building thermal envelope* requirements of this code.

❖ With the amount of glass in sunrooms, it is imperative the insulation requirements of the code be adhered to ensure minimal current requirements. The exceptions also tighten the requirements per zone to minimize energy usage.

N1102.3 (R402.3) Fenestration (Prescriptive). In addition to the requirements of Section N1102, fenestration shall comply with Sections N1102.3.1 through N1102.3.6.

❖ This section contains specific requirements that affect the requirements for the individual items listed within the subsections. Although Section N1102.1 and Tables N1102.1.1 and N1102.1.3 provide the general basis for complying with the energy requirements of this chapter, Section N1102.3 provides additional details regarding the application of the provisions or modifications that may affect the general fenestration requirements. Relying on the general code policy that specific requirements apply over general requirements will help ensure that these specific provisions are properly applied.

The term "fenestration" in this section refers to opaque doors and the light-transmitting areas of a residential building's wall, floor or roof, generally window, skylight and nonopaque door products (see the definition of "Fenestration" in Chapter 2). Prior to the 2006 edition, the residential energy provisions of the code applied only to buildings with 15 percent or less of glazing areas (fenestration). The code's prescriptive requirements varied depending on the type of residential occupancy, but did include limitations on the amount of fenestration (maximum of 25 percent for one- and two-family dwellings and 30 percent for R-2, R-4 or townhouses). These earlier versions of the code established whole building performance requirements, with fenestration performance requirements as a derivative value dependent on window area, overall envelope area and the performance of other assemblies (e.g., walls, ceilings, floors). The code establishes specific simplified prescriptive requirements (without area considerations) for fenestration products in Table N1102.1.1—specifically, fenestration *U*-factors, skylight *U*-factors and glazed fenestration SHGCs. The elimination of these fenestration area limitations helped to greatly simplify the application of the code's envelope requirements.

The fenestration requirements of the code are critical to the overall energy efficiency of the residence. First, unlike opaque assemblies, glazed fenestration can transmit a substantial amount of heat through the glazing into the living space in both the summer and winter, resulting in a unique concern about solar heat gain (and, as a result, necessitating SHGC requirements). Second, the insulating value (*U*-factor) of typical fenestration is much higher than that of a typical wall. For example, a good low emissivity (low-E), insulated glass wood or vinyl fenestration product will have less than one-fourth the insulating value of an equivalent area of R-13 insulated opaque wall. These issues have an effect not only on energy use, but overall occupant comfort, condensation and other issues.

In accordance with Section N1101.12.3, the *U*-factor and SHGC for each fenestration product must be obtained from a label attached to the product certifying that the values were determined in accordance with NFRC procedures by an accredited, independent lab or from a limited default table.

N1102.3.1 (R402.3.1) *U*-factor. An area-weighted average of fenestration products shall be permitted to satisfy the *U*-factor requirements.

❖ Section N1102.3.1 permits using the calculated "area-weighted average" *U*-factor of all fenestration products in the building to satisfy the fenestration *U*-factor requirements set by Table N1102.1.1 or N1102.1.3. As a result, if all fenestration products (window, door or skylight) do not meet the specific value, the user can still achieve compliance if the weighted average of all products is equal to or less than the specified value.

This option permits the use of some windows that have values lower than the prescriptive general requirement, as long as these poorly performing windows are offset by windows with values better than the requirement.

When applying this "area-weighted" option, it is important to remember that the term "fenestration" includes windows, skylights, doors with glazing and opaque doors, all of which would be included in the average calculation.

Using the *U*-factor requirement of 0.35 for Climate Zone 6 as an illustration, this section provides two options for compliance. The simplest option is to ensure that all windows and doors have labeled NFRC values of 0.35 or less. This approach is also more likely to ensure adequate performance and comfort throughout the home. Alternatively, a weighted average may be taken of the values from all windows and doors to see if the "weighted average" is less than or equal to 0.35.

As a simple example, assume 100 square feet (9.3 m^2) of 0.32 windows, 100 square feet (9.3 m^2) of 0.36 windows and one 20-square-foot (1.8 m^2) 0.40 *U*-factor door [(100 × 0.32) + (100 × 0.36) + (20 × 0.40)]/(100 + 100 + 20) = 0.345 (weighted average *U*-factor). Therefore, because the weighted average *U*-factor is less than the required 0.35, the fenestration in this example would be in compliance with the code.

In accordance with Section 303.1.2, the *U*-factor for each fenestration product must be obtained from a label attached to the product certifying that the *U*-factor was determined in accordance with NFRC procedures by an accredited, independent lab. In the absence of an NFRC-labeled *U*-factor, a value from the limited default tables [Tables 303.1.3(1) and 303.1.3(2)] must be used. The provisions of Section 402.5 should be reviewed when using the area-weighted average in the tradeoff options (see commentary, Section 402.5).

N1102.3.2 (R402.3.2) Glazed fenestration SHGC. An area-weighted average of fenestration products more than 50-percent glazed shall be permitted to satisfy the SHGC requirements.

❖ Under Table N1102.1.1, all glazed fenestration products in Climate Zones 1 through 3 must have an SHGC equal to or less than 0.30 (there is no requirement in Climate Zones 4 through 8). This requirement is intended to control unwanted solar gain in cooling-dominated climates to increase comfort, reduce air-conditioning energy and peaks, reduce HVAC sizing and reduce energy costs.

Similar to Section N1102.3.1, Section N1102.3.2 allows some latitude for individual product variability by permitting this performance requirement to be met using an area-weighted average of all of the fenestration products that are more than 50-percent glazed. The 50-percent glazing threshold is established to exclude doors or other fenestration products that are either completely or largely opaque from the equation. The reason for this exclusion is that opaque elements do not allow solar heat gain as glazing does. The area-weighted calculation approach is explained above in an example with Section N1102.3.1. An additional example for SHGC is as follows:

Window 1	SHGC - 0.24	200 ft^2
Window 2	SHGC - 0.32	100 ft^2
Window 3	SHGC - 0.30	100 ft^2
Sliding glass door	SHGC - 0.40	40 ft^2

[(200 ft^2 × 0.24) + (100 ft^2 × 0.32) + (100 ft^2 × 0.30) + (40 ft^2 × 0.40)]/440 ft^2

440 ft^2 = SHGC - 0.29 Average

Using the figures in the example, even though Window 2 and the sliding glass door both have SHGC values that exceed the 0.25 limitation of Table N1102.1.1 for Climate Zones 1 through 3, this design will comply because the weighted average is 0.34.

In accordance with Section R303.1.3, the SHGC for each glazed fenestration product must be obtained from a label attached to the product certifying that the SHGC was determined in accordance with NFRC procedures by an accredited, independent lab. In the absence of an NFRC-labeled SHGC, a value from the limited default table [Table R303.1.3(1)] must be used.

It is important to note that the SHGC requirement must be met by the fenestration product on a stand-alone basis. The code does not permit the alternative of a "permanent solar shading device" such as eave overhangs or awnings, as was permitted by other energy codes and previous versions of the code, to assist in code compliance.

N1102.3.3 (R402.3.3) Glazed fenestration exemption. Up to 15 square feet (1.4 m^2) of glazed fenestration per dwelling unit shall be permitted to be exempt from *U*-factor and SHGC requirements in Section N1102.1.1. This exemption shall not apply to the *U*-factor alternative approach in Section N1102.1.3 and the Total UA alternative in Section N1102.1.4.

❖ In addition to using the area-weighted average approach (Sections N1102.3.1 and N1102.3.2) to allow maximum compliance flexibility for builders, the code allows up to 15 square feet (1.4 m^2) of the building's total glazed fenestration area to be exempt from the *U*-factor and SHGC requirements listed in Table N1102.1.1. All other glazing must meet or exceed the designated *U*-factor and SHGC requirements. The exempted glazing area should be designated on the building plan, either on the floor plan or within a window schedule. This will give the necessary information to the field inspector who will then need to verify that what is on the plans is installed in the field. This exemption allows the use of ornate or unique window, skylight or glazed door assemblies in a building without going to another compliance approach. The area, the *U*-factor and SHGC of the exempt product(s) should be excluded from the area-weighted calculations that may be performed under Sections N1102.3.1 and N1102.3.2. In addition, the exception provided by this section would also allow this 15 square feet (1.4 m^2) of glazing to be excluded from the limits of Section N1102.5. Note that this exemption does not apply where the basis for the thermal envelope was the *U*-factor alternative in Section N1102.1.3 or the total UA alternative in Section N1102.1.4. In these cases, the amount of insulation does not need to be exempted.

N1102.3.4 (R402.3.4) Opaque door exemption. One side-hinged opaque door assembly up to 24 square feet (2.22 m^2) in area is exempted from the *U*-factor requirement in Section N1102.1.1. This exemption shall not apply to the *U*-factor alternative approach in Section N1102.1.3 and the total UA alternative in Section N1102.1.4.

❖ Similar to the exemption provided in Section N1102.3.3 to enhance design flexibility, the code allows one side-hinged opaque door to be exempt from fenestration *U*-factor requirements as contained in Table N1102.1.1, as well as the limitations of Section N1102.5. Although the code does not define it, an opaque door is generally considered to be a fenestration product with an overall glazing area of less than 50 percent. The opaque door exemption allows builders to use an ornate or otherwise *U*-factor noncompliant entrance door assembly in a building without going to another compliance approach. The area and the *U*-factor of the exempt product should be excluded from the area-weighted calculations under Section N1102.3.1.

N1102.3.5 (R402.3.5) Sunroom *U*-factor. All *sunrooms* enclosing conditioned spaces shall meet the fenestration requirements of this code.

Exception: For *sunrooms* with *thermal isolation* and enclosing conditioned spaces, in Zones 4 through 8, the following exceptions to the fenestration requirements of this code shall apply:

1. The maximum fenestration *U*-factor shall be 0.45; and
2. The maximum skylight *U*-factor shall be 0.70. New fenestration separating the *sunroom* with *thermal isolation* from *conditioned space* shall meet the *building thermal envelope* requirements of this code.

❖ This section simply reminds the user that sunrooms enclosing conditioned spaces are subject to code requirements. The exception tightens the maximum fenestration *U*-factor for Climate Zones 4 through 8. The maximum skylight *U*-factor shall be 0.70. Given the amount of glass in a sunroom, this section of the code tightens up the fenestration requirements and minimizes energy usage.

N1102.3.6 (R402.3.6) Replacement fenestration. Where some or all of an existing fenestration unit is replaced with a new fenestration product, including sash and glazing, the replacement fenestration unit shall meet the applicable requirements for *U*-factor and SHGC in Table N1102.1.1.

❖ Replacing only a glass pane in an existing sash and frame would not fall under this provision if the *U*-factor and SHGC will be equal to or lower than the values prior to the replacement. In situations where the existing values are not known or where the replacement is more than just replacing the glass pane, the provisions may be applicable. It is often common practice when fenestration is replaced to remove only the sash and glazing of an existing window and replace them with an entirely new fenestration product. Sometimes during the process, the existing frame is also removed, but many times the new fenestration product is custom made to fit in the existing space left after the sash and glazed portions are removed. In essence, the new fenestration is installed in or over the existing frame. Whether the existing frame is removed or not, these types of replacements are regulated by this section.

Section N1102.3.6 requires that each fenestration unit replaced in a residence not exceed the maximum fenestration *U*-factor and SHGC for the applicable climate zone. This requirement applies to all replacement windows, even if the existing frame is not removed (e.g., the new window is placed inside the old frame), so long as the sash and glazing is replaced. In addition, remember that the definition of "Fenestration" includes doors, which must meet the same *U*-factor requirements as windows. Therefore, the replacement of a door would also have to meet these requirements.

When dealing with replacement fenestration, the code official should be consulted to explain how this requirement will be applied. For simple ease of application for both the code official and the installer, the *U*-factor and SHGC requirements could simply be applied to each fenestration unit. Therefore, the *U*-factor and SHGC required for each unit would be the values listed directly in Table N1102.1.1. If acceptable, however, to the code official and additional information is available regarding the performance of the remaining existing windows in the home, it may be reasonable to permit the use of the area-weighted values of Sections N1102.3.1 and N1102.3.2 or even the exemptions of Sections N1102.3.3 and N1102.3.4 to the replaced fenestration unit.

In accordance with Section R102.1, the *U*-factor and SHGC for each replacement fenestration product must be obtained from a label attached to the product certifying that the values were determined in accordance with NFRC procedures by an accredited, independent lab. In the absence of an NFRC-labeled *U*-factor or SHGC, a value from the limited default tables [Tables R303.1.3(1) through R303.1.3(3)] must be used. The NFRC procedures do include applicable methods to test various replacement products.

N1102.4 (R402.4) Air leakage (Mandatory). The building thermal envelope shall be constructed to limit air leakage in accordance with the requirements of Sections N1102.4.1 through N1102.4.4.

❖ Sealing the building envelope is critical to good thermal performance of the building. The seal will prevent warm, conditioned air from leaking out around doors, windows and other cracks during the heating season, thereby reducing the cost of heating the residence. During hot summer months, a proper seal will stop hot air from entering the residence, helping to reduce the air-conditioning load on the building. Any penetration in the building envelope must be thoroughly sealed during the construction process, including holes made for the installation of plumbing, electrical and heating and cooling systems (see Commentary Figure N1102.4). The code lists several areas that must be caulked, gasketed, weatherstripped, wrapped or otherwise sealed to limit uncontrolled air movement. Most of the air sealing will be done prior to the installation of an interior wall covering because any penetration will be noticeable and accessible at this time. The code allows the use of airflow retarders (house wraps) or other solid materials as an acceptable method to meet this requirement. To be effective, the building thermal envelope seal must be:

- Impermeable to airflow.
- Continuous over the entire building envelope.
- Able to withstand the forces that may act on it during and after construction.
- Durable over the expected lifetime of the building.

It is unlikely that the same type of barrier will be used on all portions of the building's thermal envelope. Therefore, joints between the various elements, as well as joints or splices within products (such as the overlap in separate pieces of house wrap), must be effectively addressed to provide the continuity needed to perform as desired.

N1102.4.1 (R402.4.1) Building thermal envelope. The *building thermal envelope* shall comply with Sections N1102.4.1.1 and N1102.4.1.2. The sealing methods between dissimilar materials shall allow for differential expansion and contraction.

❖ Air infiltration is a major source of energy use because the incoming air usually requires conditioning. The uncontrolled introduction of outside air (infiltration) creates a load that varies with time. Ventilation, the controlled introduction of fresh air, is more manageable and provides a more controlled air quality. Uncontrolled infiltration also has a tendency to create or aggravate moisture problems, providing an additional reason to limit infiltration.

N1102.4.1.1 (R402.4.1.1) Installation. The components of the *building thermal envelope* as listed in Table N1102.4.1.1 shall be installed in accordance with the manufacturer's instructions and the criteria listed in Table N1102.4.1.1, as applicable to the method of construction. Where required by the *building official*, an *approved* third party shall inspect all components and verify compliance.

❖ The provisions of this section are intended to reduce the energy loss to infiltration and to improve insulation installation.

The code allows all materials that are commonly used as sheathing to be part of the thermal envelope, including interior drywall. By the definition of an air barrier, gypsum board should be considered as should exterior sheathing. The solid sheathing is not enough —for the air barrier to be effective the joints and openings must be sealed. The code does not require the air barrier on the inside of the air permeable insulation.

In some cases, inspection of the air barrier is better performed by individuals with more expertise than the staff of the authority having jurisdiction might have. Therefore, the code gives the code official authority to require third-party inspectors.

Note that the line item in the table regarding fireplaces requires gaskets on fireplace doors. All wood-burning fireplaces are a source of air leakage. Loss of

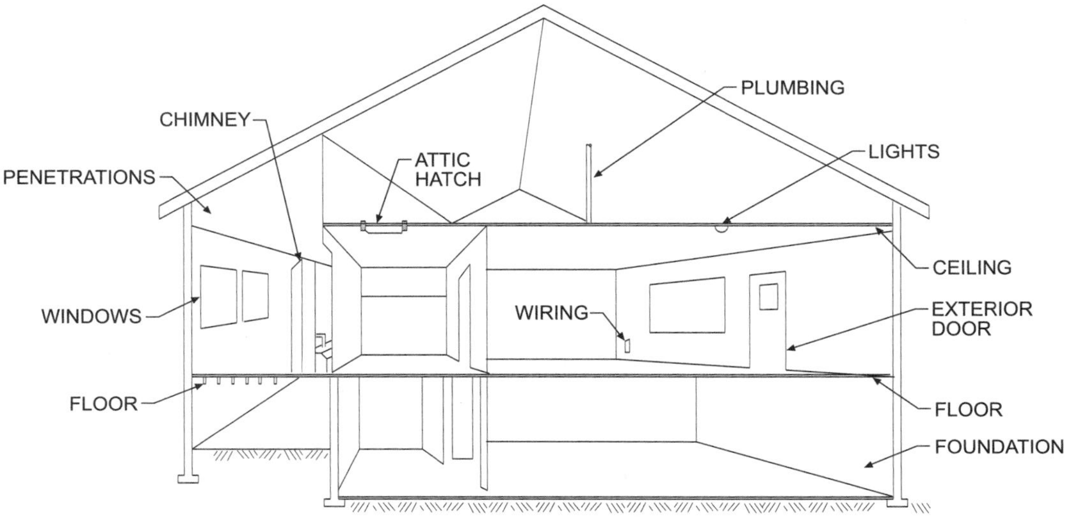

Figure N1102.4
TYPICAL SOURCES OF AIR LEAKAGE

energy through these units can be reduced with gasketed doors and a requirement that combustion air be brought directly from the outdoors to the firebox. Gaskets on the fireplace doors will help to minimize air leakage into the firebox. Air that leaks past a poorly gasketed fireplace door, or a door that is simply left open, will flow up the chimney aided by the chimney draft. During the majority of the year the fireplace will not be operating. The combination of well-gasketed doors and a well-sealed flue damper will prevent air leakage through what is effectively an enormous hole in the thermal envelope of the building.

However, it should be noted that the difficulty with this requirement is that most factory-built fireplaces tested in accordance with the code required test standard for factory-built fireplaces, UL127, are not tested

TABLE N1102.4.1.1 (R402.4.1.1)
AIR BARRIER AND INSULATION INSTALLATION

COMPONENT	CRITERIA[a]
Air barrier and thermal barrier	A continuous air barrier shall be installed in the building envelope. Exterior thermal envelope contains a continuous air barrier. Breaks or joints in the air barrier shall be sealed. Air-permeable insulation shall not be used as a sealing material.
Ceiling/attic	The air barrier in any dropped ceiling/soffit shall be aligned with the insulation and any gaps in the air barrier sealed. Access openings, drop down stair or knee wall doors to unconditioned attic spaces shall be sealed.
Walls	Corners and headers shall be insulated and the junction of the foundation and sill plate shall be sealed. The junction of the top plate and top of exterior walls shall be sealed. Exterior thermal envelope insulation for framed walls shall be installed in substantial contact and continuous alignment with the air barrier. Knee walls shall be sealed.
Windows, skylights and doors	The space between window/door jambs and framing and skylights and framing shall be sealed.
Rim joists	Rim joists shall be insulated and include the air barrier.
Floors (including above-garage and cantilevered floors)	Insulation shall be installed to maintain permanent contact with underside of subfloor decking. The air barrier shall be installed at any exposed edge of insulation.
Crawl space walls	Where provided in lieu of floor insulation, insulation shall be permanently attached to the crawlspace walls. Exposed earth in unvented crawl spaces shall be covered with a Class I vapor retarder with overlapping joints taped.
Shafts, penetrations	Duct shafts, utility penetrations, and flue shafts opening to exterior or unconditioned space shall be sealed.
Narrow cavities	Batts in narrow cavities shall be cut to fit, or narrow cavities shall be filled by insulation that on installation readily conforms to the available cavity space.
Garage separation	Air sealing shall be provided between the garage and conditioned spaces.
Recessed lighting	Recessed light fixtures installed in the building thermal envelope shall be air tight, IC rated, and sealed to the drywall.
Plumbing and wiring	Batt insulation shall be cut neatly to fit around wiring and plumbing in exterior walls, or insulation that on installation readily conforms to available space shall extend behind piping and wiring.
Shower/tub on exterior wall	Exterior walls adjacent to showers and tubs shall be insulated and the air barrier installed separating them from the showers and tubs.
Electrical/phone box on exterior walls	The air barrier shall be installed behind electrical or communication boxes or air-sealed boxes shall be installed.
HVAC register boots	HVAC register boots that penetrate building thermal envelope shall be sealed to the sub-floor or drywall.
Fireplace	An air barrier shall be installed on fireplace walls. Fireplaces shall have gasketed doors.

a. In addition, inspection of log walls shall be in accordance with the provisions of ICC-400.

❖ This table contains the list of items that are required to be visually inspected if the visual inspection option for demonstrating building air tightness given in Section N1102.4.1.1 is chosen.

and listed with doors. Therefore, there is a significant safety hazard in adding gasketed doors to a factory-built fireplace tested and manufactured in accordance with UL127, without doors. That hazard would be overheating of the unit and high likelihood of causing fires. This provision in Table N1102.4.1.1 for gasketed fireplace doors is a general provision to prevent air leakage and was not intended to prohibit the use of factory-built fireplaces listed for use without doors, in violation of that listing. The specific requirements of the test standard UL127, the product listing and the manufacturer's instructions would prevail for factory-built fireplaces. Keep in mind that Section N1102.4.3 requires tight fitting dampers for all fireplaces as well, which will prevent air leakage.

N1102.4.1.2 (R402.4.1.2) Testing. The building or dwelling unit shall be tested and verified as having an air leakage rate of not exceeding 5 air changes per hour in Zones 1 and 2, and 3 air changes per hour in Zones 3 through 8. Testing shall be conducted with a blower door at a pressure of 0.2 inches w.g. (50 Pascals). Where required by the *building official*, testing shall be conducted by an *approved* third party. A written report of the results of the test shall be signed by the party conducting the test and provided to the *building official*. Testing shall be performed at any time after creation of all penetrations of the *building thermal envelope*.

During testing:

1. Exterior windows and doors, fireplace and stove doors shall be closed, but not sealed, beyond the intended weatherstripping or other infiltration control measures;
2. Dampers including exhaust, intake, makeup air, back-draft and flue dampers shall be closed, but not sealed beyond intended infiltration control measures;
3. Interior doors, if installed at the time of the test, shall be open;
4. Exterior doors for continuous ventilation systems and heat recovery ventilators shall be closed and sealed;
5. Heating and cooling systems, if installed at the time of the test, shall be turned off; and
6. Supply and return registers, if installed at the time of the test, shall be fully open.

❖ The purpose of this code section is to test the building or dwelling unit to demonstrate the building's air tightness. A blower door test, which is a house pressurization test, should be done with a blower door at a pressure of 0.2 inches water gauge [50 Pascals (1psf)]. The building or dwelling unit shall be tested and verified as having an air leakage rate of not exceeding 5 ACH50, or five air changes per hour at 50 Pascals (1 psf) in Climate Zones 1 and 2 and 3 ACH50, or three air changes per hour at 50 Pascals (1 psf), in Climate Zones 3 through 8. The ACH50 is a common measurement made when doing air infiltration tests and, therefore, a reasonable metric for use in the code. Testing can be conducted by an approved third party, if allowed by the code official. Section N1102.4.2.1 requires that HVAC ducts not be sealed during the test. In this context, "sealed" is intended to mean sealed off from the interior of the house. The maximum is 5ACH50, or five air changes per hour at 50 Pascals (1 psf).

N1102.4.2 (R402.4.2) Fireplaces. New wood-burning fireplaces shall have tight-fitting flue dampers and outdoor combustion air.

❖ All wood-burning fireplaces are a source of air leakage. Loss of energy through these units can be reduced with tight-fitting flue dampers and a requirement that combustion air be brought directly from the outdoors to the firebox. A well-sealed flue damper will prevent air leakage through what is effectively an enormous hole in the thermal envelope of the building.

Table N1102.4.1.1 also requires gaskets on fireplace doors. This is a general requirement that can be problematic for factory-built fireplaces (see the commentary to Table N1102.4.1.1).

N1102.4.3 (R402.4.3) Fenestration air leakage. Windows, skylights and sliding glass doors shall have an air infiltration rate of no more than 0.3 cfm per square foot (1.5 L/s/m^2), and swinging doors no more than 0.5 cfm per square foot (2.6 L/s/m^2), when tested according to NFRC 400 or AAMA/WDMA/CSA 101/I.S.2/A440 by an accredited, independent laboratory and *listed* and *labeled* by the manufacturer.

Exception: Site-built windows, skylights and doors.

❖ Windows, skylights and doors should be tested and labeled by the manufacturer as meeting the air infiltration requirements. The intent of this section is to effectively complete the sealing of the building's thermal envelope by providing specific testing and performance criteria for windows, skylights and doors. This testing and labeling requirement provides an easy method for both the builder and the inspector to demonstrate compliance with the code. While "site built" fenestration is exempted from these requirements, units would have to be "durably sealed" to limit infiltration according to the requirements in Section N1102.4.1.

N1102.4.4 (R402.4.4) Recessed lighting. Recessed luminaires installed in the *building thermal envelope* shall be sealed to limit air leakage between conditioned and unconditioned spaces. All recessed luminaires shall be IC-rated and *labeled* as having an air leakage rate not more than 2.0 cfm (0.944 L/s) when tested in accordance with ASTM E 283 at a 1.57 psf (75 Pa) pressure differential. All recessed luminaires shall be sealed with a gasket or caulk between the housing and the interior wall or ceiling covering.

❖ To correctly apply this provision, it is important to realize that it deals only with recessed lights that are installed in the building thermal envelope. Therefore, lights that are located so that all sides of the luminaire are surrounded by conditioned space would not fall under this section's requirements. For example, a light located in a soffit would not be regulated if the soffit was below a ceiling that served as the building's thermal envelope. Additionally, a light installed in the floor/

ceiling assembly between a first-floor living room and a bedroom above it would be exempted.

Because of their typical location of installation, recessed lighting fixtures pose a potential fire hazard if incorrectly covered with insulation. In addition, the ceiling or a barrier directly above it often serves as the air leakage or moisture barrier for the home.

Therefore, holes cut through the ceiling to install these fixtures also act as chimneys that transfer heat loss and moisture through the building envelope into attic spaces. The heat loss resulting from improperly insulated recessed lighting fixtures can be significant.

Recessed lighting fixtures must be insulation contact (IC) rated lights, which are typically double-can fixtures, with one can inside another (see Commentary Figure N1102.4.3). The outer can (in contact with the insulation) is tested to make sure it remains cool enough to avoid a fire hazard. An IC-rated fixture should have the IC rating stamped on the fixture or printed on an attached label.

Recessed lights must also be tightly sealed or gasketed to prevent air leakage through the fixture into the ceiling cavity.

N1102.5 (R402.5) Maximum fenestration *U*-factor and SHGC (Mandatory). The area-weighted average maximum fenestration *U*-factor permitted using tradeoffs from Section N1102.1.4 or N1105 shall be 0.48 in Zones 4 and 5 and 0.40 in Zones 6 through 8 for vertical fenestration, and 0.75 in Zones 4 through 8 for skylights. The area-weighted average maximum fenestration SHGC permitted using tradeoffs from Section N1105 in Zones 1 through 3 shall be 0.50.

❖ This section is intended to clarify the application of the fenestration performance maximums and to set reasonable performance levels for these products when using the tradeoffs of Section N1102.1.4 or N1105.

This section does not define the minimum code requirements for fenestration *U*-factors that are set in Table N1102.1.1. Rather, this section sets limits on the tradeoffs allowed based on the total UA calculations (see Section N1102.1.4) or the simulated performance option found in Section N1105. This section is in contrast to the basic principle used in Section N1102.1.4. Although that section will allow a *U*-factor that does not meet the general code requirements offset by the increased efficiency of another part of the residence, this section establishes a limitation on the level of efficiency that can be compensated for.

Note that this section does not set a limit on individual products because Sections N1102.3.1 and N1102.3.2 are not affected. Rather, this section sets an overall limit on the weighted average of those products. These limits are often called "hard limits" or "maximum tradeoff" limits.

In situations where the code is applied to only a portion of the residence, such as only to the addition but not the existing residence, the weighted average could be calculated for just the addition (see commentary, Sections N1102.3.1 and N1102.3.2).

The thermal properties (*U*-factor) of skylights are uniquely different from window products. After installation they perform differently than windows and are, therefore, rated differently. As a result, even the highest rated skylights cannot achieve the same level of *U*-factor performance as windows and glass doors. Therefore, the code provides different values for these "maximum" limits in the tradeoff.

The SHGC provisions are limited in the cooling-dominated climate zones. Therefore, where solar heat gain through the fenestration can affect the cooling load, comfort and efficiency, a "hard limit" of 0.5 would be applied when using Section N1105.

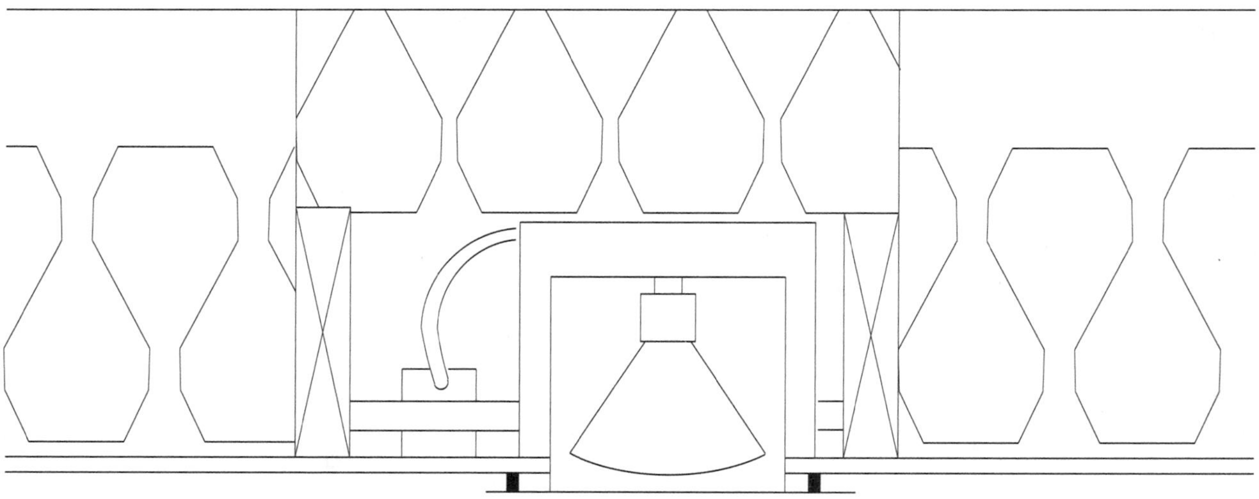

Figure N1102.4.3
IC-RATED RECESSED LIGHT

SECTION N1103
SYSTEMS

N1103.1 (R403.1) Controls (Mandatory). At least one thermostat shall be provided for each separate heating and cooling system.

❖ This provision ensures that a separate thermostat is installed for each system. As an example, if separate systems are installed so that one serves the downstairs and one serves the upstairs of a two-story residence, two separate thermostats would be required, one regulating each level. This allows for greater flexibility, control and energy savings than would be possible if both systems were controlled by a single thermostat.

N1103.1.1 (R403.1.1) Programmable thermostat. Where the primary heating system is a forced-air furnace, at least one thermostat per dwelling unit shall be capable of controlling the heating and cooling system on a daily schedule to maintain different temperature set points at different times of the day. This thermostat shall include the capability to set back or temporarily operate the system to maintain zone temperatures down to 55°F (13°C) or up to 85°F (29°C). The thermostat shall initially be programmed with a heating temperature set point no higher than 70°F (21°C) and a cooling temperature set point no lower than 78°F (26°C).

❖ This code section provides a requirement that gives each household an opportunity for energy savings by requiring a programmable thermostat that allows changing the temperature setpoints automatically throughout the day.

N1103.1.2 (R403.1.2) Heat pump supplementary heat (Mandatory). Heat pumps having supplementary electric-resistance heat shall have controls that, except during defrost, prevent supplemental heat operation when the heat pump compressor can meet the heating load.

❖ Heat pump systems must have controls that prevent supplementary electric resistance heater operation when the heating load can be met by the heat pump alone. Typically, these controls will be thermostats that will have a "heat pump" designation on them. The make and model of the thermostat should be called out on the building plans so the inspector can verify that what is installed in the field matches the plans. Because change-outs in the field are common, the instructions that come with the thermostat can also be checked to verify that the thermostat is designed for use with a particular heat pump. To limit the hours of use of the electric-resistance heating unit and provide the most cost-efficient operation of the equipment, the specific control language is included in this section.

N1103.2 (R403.2) Ducts. Ducts and air handlers shall be in accordance with Sections N1103.2.1 through N1103.2.3.

❖ The four subsections of Section N1103.2 provide the requirements of this chapter, which apply to the ducts and air handlers used in residential buildings.

N1103.2.1 (R403.2.1) Insulation (Prescriptive). Supply ducts in attics shall be insulated to a minimum of R-8. All other ducts shall be insulated to a minimum of R-6.

Exception: Ducts or portions thereof located completely inside the *building thermal envelope.*

❖ HVAC ductwork located outside of the conditioned space must be insulated minimum *R*-values. This includes both supply and return ducts. The exception addresses ductwork that is located in part or completely within a conditioned space. This ductwork need not be insulated because, by definition, it is already within and protected by the building's thermal insulation envelope.

N1103.2.2 (R403.2.2) Sealing (Mandatory). Ducts, air handlers, and filter boxes shall be sealed. Joints and seams shall comply with Section M1601.4.1 of this code.

Exceptions:

1. Air-impermeable spray foam products shall be permitted to be applied without additional joint seals.
2. Where a duct connection is made that is partially inaccessible, three screws or rivets shall be equally spaced on the exposed portion of the joint so as to prevent a hinge effect.
3. Continuously welded and locking-type longitudinal joints and seams in ducts operating at static pressures less than 2 inches of water column (500 Pa) pressure classification shall not require additional closure systems.

Duct tightness shall be verified by either of the following:

1. Postconstruction test: Total leakage shall be less than or equal to 4 cfm (113.3 L/min) per 100 square feet (9.29 m^2) of conditioned floor area when tested at a pressure differential of 0.1 inches w.g. (25 Pa) across the entire system, including the manufacturer's air handler enclosure. All register boots shall be taped or otherwise sealed during the test.
2. Rough-in test: Total leakage shall be less than or equal to 4 cfm (113.3 L/min) per 100 ft^2 (9.29 m^2) of conditioned floor area when tested at a pressure differential of 0.1 inches w.g. (25 Pa) across the system, including the manufacturer's air handler enclosure. All registers shall be taped or otherwise sealed during the test. If the air handler is not installed at the time of the test, total leakage shall be less than or equal to 3 cfm (85 L/min) per 100 square feet (9.29 m^2) of conditioned floor area.

   Exception: The total leakage test is not required for ducts and air handlers located entirely within the building thermal envelope.

❖ Ducts must be sealed in accordance with the duct sealing requirements of either the IMC or the code. Joints and seams that fail in a duct system can result in increased energy use because the conditioned air will be delivered to an unconditioned space or into a building space where it is not needed, such as a wall or floor assembly, crawl space or attic, instead of to a room or area inside the conditioned envelope.

Return-air ductwork installed in basements or concealed building spaces may conduct chemicals or other products that produce potentially harmful fumes. Because the return air operates under negative pressure, any leaks could draw fumes, moisture, soil gases or odors from the surrounding area and direct them into the house. Therefore, sealing of return-air ductwork is also a requirement.

Duct tightness must be checked by leakage testing of the ducts at the end of construction or at the time of rough in. The pass/fail criteria for leakage rate is more stringent if the rough-in test time is chosen. This is simply because it is probably that some of the ductwork will be disturbed during subsequent construction and, therefore, have a higher leakage rate than what was tested.

N1103.2.2.1 (R403.2.2.1) Sealed air handler. Air handlers shall have a manufacturer's designation for an air leakage of no more than 2 percent of the design air flow rate when tested in accordance with ASHRAE 193.

❖ This section of the code requires that air handlers have a manufacturer's designation for an air leakage of no more than 2 percent of the design flow rate when tested in accordance with ASHRAE 193. Energy conservation measures in the air-conditioning industry have driven the manufacturers of systems and components to establish compliance with leakage limits in ducts and air-handling units. The standards set by the American Society of Heating, Refrigerating and Air-conditioning Engineers (ASHRAE) form the basis for testing. Establishing an air-handler leakage rate, given the availability of a uniform test procedure, is prudent since any leakage in the air-handling unit contributes to waste of energy. The magnitude of leakage has a direct bearing on energy use and indoor air quality.

N1103.2.3 (R403.2.3) Building cavities (Mandatory). Building framing cavities shall not be used as ducts or plenums.

❖ Building framing cavities shall not be used as ducts or plenums. Stud bays and other building cavities that are exposed to the differing outside temperatures cannot be used as supply air ducts. In addition, these spaces are limited to use for return air only because the negative pressures within the return air plenum with respect to surrounding spaces will decrease the likelihood of spreading smoke to other spaces via the plenum. In addition, for many of the same reasons that sealing (Section N1103.2.2) is required, building cavities are a very inefficient means of distributing air.

N1103.3 (R403.3) Mechanical system piping insulation (Mandatory). Mechanical system piping capable of carrying fluids above 105°F (41°C) or below 55°F (13°C) shall be insulated to a minimum of R-3.

❖ Heat losses during mechanical fluid distribution impact building energy use both in the energy required to make up for the lost heat and in the additional load that can be placed on the space-cooling system if the heat is released to air-conditioned space. These losses can be effectively limited by insulating the mechanical system piping that conveys fluids at extreme temperatures.

N1103.3.1 (R403.3.1) Protection of piping insulation. Piping insulation exposed to weather shall be protected from damage, including that caused by sunlight, moisture, equipment maintenance, and wind, and shall provide shielding from solar radiation that can cause degradation of the material. Adhesive tape shall not be permitted.

❖ Proper protection of exposed piping insulation to the weather elements is necessary to protect it from damage, including that due to sunlight, moisture, equipment maintenance and wind.

N1103.4 (R403.4) Service hot water systems. Energy conservation measures for service hot water systems shall be in accordance with Sections N1103.4.1 and N1103.4.2.

❖ A review of the definition of "Service water heating" in Chapter 2 is appropriate before applying these requirements. Any time the system is set up to circulate the water through it, the piping is required to be insulated as listed in Section N1103.4.2.

N1103.4.1 (R403.4.1) Circulating hot water systems (Mandatory). Circulating hot water systems shall be provided with an automatic or readily *accessible* manual switch that can turn off the hot-water circulating pump when the system is not in use.

❖ When the distribution piping is heated to maintain usage temperatures, such as in circulating hot water systems or systems using pipe heating cable, the system pump or heat trace cable must have conveniently located manual or automatic switches or other controls that can be set to optimize system operation or turn off the system during periods of reduced demand. The simplest of these devices is an automatic time clock. Notice, however, that the code will accept either a readily accessible manual switch or an automatic means to turn off the circulating pump.

N1103.4.2 (R403.4.2) Hot water pipe insulation (Prescriptive). Insulation for hot water pipe with a minimum thermal resistance (*R*-value) of R-3 shall be applied to the following:

1. Piping larger than $^3/_4$-inch nominal diameter.
2. Piping serving more than one dwelling unit.
3. Piping from the water heater to kitchen outlets.
4. Piping located outside the conditioned space.
5. Piping from the water heater to a distribution manifold.
6. Piping located under a floor slab.
7. Buried piping.
8. Supply and return piping in recirculation systems other than demand recirculation systems.
9. Piping with run lengths greater than the maximum run lengths for the nominal pipe diameter given in Table N1103.4.2.

All remaining piping shall be insulated to at least R-3 or meet the run length requirements of Table N1103.4.2.

❖ Hot water piping shall be insulated with a minimum thermal resistance *R*-value of R-3 or meet the run length requirements of Table N1103.4.2. Applicable piping that needs to be insulated is listed in the code.

TABLE N1103.4.2 (R403.4.2)
MAXIMUM RUN LENGTH (feet)[a]

Nominal pipe diameter of largest diameter pipe in the run (inch)	$^3/_8$	$^1/_2$	$^3/_4$	$> ^3/_4$
Maximum run length	30	20	10	5

For SI: 1 inch = 25.4 mm, 1 foot = 304.8 mm.

a. Total length of all piping from the distribution manifold or the recirculation loop to a point of use.

❖ See the commentary to Section N1103.4.2.

N1103.5 (R403.5) Mechanical ventilation (Mandatory). The building shall be provided with ventilation that meets the requirements of Section M1507 of this code or with other approved means of ventilation. Outdoor air intakes and exhausts shall have automatic or gravity dampers that close when the ventilation system is not operating.

❖ Mechanical ventilation is the alternative to having natural ventilation. Both natural and mechanical ventilation can be provided to a space. Unlike natural ventilation, mechanical ventilation does not depend on unpredictable air pressure differentials between the indoors and outdoors to create airflow. The volume of air supplied to a space must be approximately equal to the volume of the air removed from the space. Otherwise, the space will be either positively or negatively pressurized and the actual ventilation flow rate will be equivalent to the lower rate of either the air supply or air exhaust.

The requirements of this section are intended to reduce infiltration into the building when ventilation systems are off. Infiltration speeds up natural cooling or warming of the space during off-hours and can increase the energy use required to maintain normal temperatures.

Any outdoor air inlets or outlets serving fans, boilers and other HVAC systems equipped with an on/off switch that either introduces outside air into a building or exhaust air outside of a building must have dampers that automatically close when the fan is shut off. These dampers may be either gravity type or motorized, regardless of whether the fan is supplying or exhausting air. One of the most common such damper that is seen on homes is the gravity-type backdraft damper on a clothes dryer exhaust duct.

N1103.5.1 (R403.5.1) Whole-house mechanical ventilation system fan efficacy. Mechanical ventilation system fans shall meet the efficacy requirements of Table N1103.5.1.

Exception: Where mechanical ventilation fans are integral to tested and listed HVAC equipment, they shall be powered by an electronically commutated motor.

❖ This section of the code is applicable to whole-house mechanical ventilation systems that meet the efficacy requirements of Table N1103.5.1. Findings from a recent study commissioned by the U.S. DOE and the California Energy Commission identified that energy consumption of whole-house mechanical ventilation systems is significant. The study revealed that large disparities exist in the energy consumption and associated operating costs of whole-house mechanical ventilation systems no matter what the climate is. To reduce the amount of energy consumed by residential mechanical ventilation systems is to address the power consumption of the fans that are powering the system. This is important because these fans will operate many hours per day. The table offers energy efficacy levels for exhaust fans that are the same levels as the current ENERGY STAR ventilation fan specifications.

N1103.6 (R403.6) Equipment sizing (Mandatory). Heating and cooling equipment shall be sized in accordance with ACCA Manual S based on building loads calculated in accordance with ACCA Manual J or other *approved* heating and cooling calculation methodologies.

❖ Once the building's thermal envelope is properly insulated and sealed, it will oftentimes allow for the reduction of equipment size from what has typically been installed using a "rule-of-thumb" method or other means of estimating. A properly sized system will operate more efficiently and help to improve the occupant comfort and extend the equipment's service life. Section M1401.3 stipulates that the heating and cooling equipment must be sized based on the building loads calculated in accordance with the ACCA Manual J. This manual contains a simplified method of calculating heating and cooling loads. It includes a room-by-room calculation method that allows the designer to determine the required capacity of the heating and cooling equipment. In addition, it provides a means to estimate the airflow requirements

TABLE N1103.5.1 (R403.5.1)
MECHANICAL VENTILATION SYSTEM FAN EFFICACY

FAN LOCATION	AIRFLOW RATE MINIMUM (CFM)	MINIMUM EFFICACY (CFM/WATT)	AIRFLOW RATE MAXIMUM (CFM)
Range hoods	Any	2.8 cfm/watt	Any
In-line fan	Any	2.8 cfm/watt	Any
Bathroom, utility room	10	1.4 cfm/watt	< 90
Bathroom, utility room	90	2.8 cfm/watt	Any

For SI: 1 cubic foot per minute = 28.3 L/min.

❖ See the commentary to Section N1103.5.1.

for each of the areas in the house. This estimate can be used in sizing the duct system for the types of heating and cooling units that use air as the medium for heat transfer. Other approved methods may be used with the code official's approval.

Though not required by the code, the calculated airflows would provide a means to evaluate the installation of the mechanical system. By providing the proper airflow to the various portions of the building, the system can operate more efficiently and can help prevent spaces being too hot or too cold.

N1103.7 (R403.7) Systems serving multiple dwelling units (Mandatory). Systems serving multiple dwelling units shall comply with Sections C403 and C404 of the IECC—Commercial Provisions in lieu of Section N1103.

❖ The criteria in Section N1103 primarily address standalone mechanical systems in single-family houses. However, Chapter 11 also includes townhouses up to three stories. Some of these building projects will have more complicated mechanical systems that may consist of a single system serving multiple dwelling units.

N1103.8 (R403.8) Snow melt system controls (Mandatory). Snow- and ice-melting systems, supplied through energy service to the building, shall include automatic controls capable of shutting off the system when the pavement temperature is above 50°F (10°C), and no precipitation is falling and an automatic or manual control that will allow shutoff when the outdoor temperature is above 40°F (4.8°C).

❖ Snow melt equipment is being installed at a greater frequency in residential projects in communities with a high snow accumulation. Previously, the code only required that the building be built to a certain level of efficiency; however, there was no limit placed on the energy use for snow melt, which can be twice the energy use per square foot of the building.

This section does not restrict the use or sizing of snow melt, but it does require that controls be installed on the equipment so that the system will operate more efficiently. The automatic controls provide efficient operation by keeping the system in an idle mode until light snow begins to fall and allowing adequate warm-up before a heavy snowfall. Systems that only use manual controls require the building owner to manually turn on the system when it starts to snow or to leave the system running in the snow-melting mode, using significantly more energy. Chapter 50, Snow Melting and Freeze Protection, 2003 *ASHRAE Applications Handbook* states that using a manual switch to operate snow melt equipment may not melt snow effectively; thus, snow will accumulate. This requirement is also referenced in ANSI/ASHRAE/IESNA Standard 90.1, Section 6.4.3.8, Freeze protection and Snow/Ice Melting Systems.

N1103.9 (R403.9) Pools and inground permanently installed spas (Mandatory). Pools and inground permanently installed spas shall comply with Sections N1103.9.1 through N1103.9.3.

❖ Because of the heating and filtering operations involved with pools and inground permanently installed spas, they provide a good opportunity to save energy by limiting heat loss or pump operation. This section provides the scoping requirements for the pool and spa heaters, time switches and pool covers that can make the operation more energy efficient. These features would be able to provide energy savings for residential pools if their owners wish to use them. The requirements of this section were added into the code in order to help coordinate with requirements found within ASHRAE 90.1 and to help reduce the energy used by these pools and inground permanently installed spa systems.

N1103.9.1 (R403.9.1) Heaters. All heaters shall be equipped with a readily *accessible* on-off switch that is mounted outside of the heater to allow shutting off the heater without adjusting the thermostat setting. Gas-fired heaters shall not be equipped with constant burning pilot lights.

❖ The accessible on-off switch allows the heaters to be completely turned off where there are periods that the heat is not needed or when the pool or spa may not be used for a period of time.

N1103.9.2 (R403.9.2) Time switches. Time switches or other control method that can automatically turn off and on heaters and pumps according to a preset schedule shall be installed on all heaters and pumps. Heaters, pumps and motors that have built in timers shall be deemed in compliance with this requirement.

Exceptions:

1. Where public health standards require 24-hour pump operation.
2. Where pumps are required to operate solar-and waste-heat-recovery pool heating systems.

❖ The use of a time switch or other control method to control the heater and pumps provides an easy system for pool and spa operations and energy savings. The application of Exception 1 is dependent on the requirements of the local health department in the jurisdiction. Because these are often public pools and spas, the health department may require continuous filtering or circulation. Exception 2 grants a credit for using other systems that help the pool and spa operate more efficiently. Therefore, when solar- and waste-heat recovery systems are used to heat the pool, the exception eliminates the time-switch requirement.

N1103.9.3 (R403.9.3) Covers. Heated pools and inground permanently installed spas shall be provided with a vapor-retardant cover.

Exception: Pools deriving over 70 percent of the energy for heating from site-recovered energy, such as a heat pump or solar energy source computed over an operating season.

❖ When energy is used to heat a pool or a spa, a cover is required to help hold in the heat and keep it from being lost to the surrounding air. The level of protection or insulation that the cover must provide depends upon the temperature that the pool is heated to. Any time a pool or spa is heated, the code will require a vapor-retardant pool cover. This type of cover is not required to provide any minimum level of insulation value. It simply will help hold some of the heat in, much like placing a lid on a pot. In situations where the pool is heated above 90°F (32°C), the cover must be insulated to the specified R-12 level. The exception is similar to that found in Section N1103.9.2.

SECTION N1104 ELECTRICAL POWER AND LIGHTING SYSTEMS (MANDATORY)

N1104.1 (R404.1) Lighting equipment (Mandatory). A minimum of 75 percent of the lamps in permanently installed lighting fixtures shall be high-efficacy lamps or a minimum of 75 percent of the permanently installed lighting fixtures shall contain only high-efficacy lamps.

Exception: Low-voltage lighting shall not be required to utilize high-efficiency lamps.

❖ Lighting comprises about 12 percent of primary residential energy, making this requirement a substantial energy saver. The overwhelming majority of residential lighting is incandescent—the least energy efficient of all light types. More efficient lighting options are available.

One more efficient lighting option is the compact fluorescent light (CFL). CFLs use about 80 percent less energy than standard incandescent lighting. Limiting this requirement to 75 percent of the lamps in a residence ensures there will be plenty of exceptions for situations where a CFL might not work as well, such as dimmable fixtures.

N1104.1.1 (R404.1.1) Lighting equipment (Mandatory). Fuel gas lighting systems shall not have continuously burning pilot lights.

❖ Continuously burning pilots waste energy; therefore, the code does not use them.

SECTION N1105 SIMULATED PERFORMANCE ALTERNATIVE (PERFORMANCE)

❖ Section N1105 describes an alternative way to meet the code's goal of effective use of energy based on showing that the predicted annual energy use of a proposed design is less than or equal to that of the same home if it had been built to meet the prescriptive criteria in Sections N1102 and N1103. Section N1106 does not prescribe a single set of requirements; rather, it provides a process to reach the energy-efficiency goal based on establishing equivalence with the intent of the code. Because of the level of detail required in the analysis, this method of design is not often used for residential buildings; however, with the changes that have been made in Section N1102 and in newer editions of the DOE's REScheck software, Section N1105 may become a more popular means of demonstrating code compliance. This section may allow designers to show that many of their current plans meet the overall code requirements even though individual components of the home fall below the required code compliance levels.

When using this section and other performance options, the term "standard reference design" and "proposed design" are used extensively. These terms can create confusion for some users, but really are very easy to understand. The "proposed design" is exactly that. It is the building as it is proposed to be built, including building size and room configurations, window and wall assemblies, mechanical equipment, orientation, etc. It is essentially the building as it is shown on the construction documents (plans and specifications). The "standard reference design" is the same building configuration and orientation as the proposed design, but instead of matching the plans for the building as it will actually be built, the standard design is shown to only meet the minimum requirements of the code. For example, the standard reference design building in Climate Zone 4 would require that the walls and roof be insulated to a level of R-13 and R-38, respectively, based on the requirements in Table N1102.1.1 (*U*-factors of 0.082 and 0.030 in accordance with Table N1102.1.3). On the other hand, the proposed design may show that the intent is to install R-19 insulation in the walls and R-49 in the roof. Therefore, if everything else in the two buildings was identical, it would be easy to see in this example that the proposed design does exceed what is required by the code in the standard reference design.

There are two fundamental requirements for using this code. First, Section N1105 compliance is based on total estimated annual energy usage across the major energy-using systems in a residential building: envelope, mechanical and service water heating. Note that Section N1105 does not include lighting within the analysis process. Second, Section N1105 compares the energy use of the proposed design to that of a standard reference design. As mentioned above, the standard reference design is the same building design as that proposed, except that the energy features required by the code (insulation, windows, HVAC, infiltration) are modified to meet the minimum prescriptive requirements in Sections N1102 and N1103. The standard reference design is used only for comparison and is never actually built.

Section N1105 sets both general principles and specific guidelines for use in computing the estimated annual energy cost of the proposed and standard reference designs.

These guidelines constitute a large portion of Section N1105 and are easily seen in Table N1105.5.2(1), but are necessary to maintain fairness and consistency between the proposed and standard designs. Although the simulated performance alternative method is the most complex method, it gives the design professional the flexibility to introduce exterior walls, roof/ceiling components, etc., that do not meet the requirements of the prescriptive performance approach in Sections N1102 and N1103, but are considered acceptable where the annual energy use of the proposed building is equal to or less than that of the standard reference design building. Envelope features that lower energy consumption (window orientation, passive-solar features or the use of "cool" reflective roofing products in cooling-dominated climates) and mechanical and service water heating systems that are more efficient than those required by the minimum prescriptive requirements in Sections N1102 and N1103 are used to offset the potentially high thermal transmittance of an innovative exterior envelope design in this instance.

The simulated performance alternative also allows energy supplied by renewable energy sources on the building site to be discounted from the total energy consumption of the proposed design building. Because renewable energy obtained on the site comes from nondepletable sources such as solar radiation, wind, plant byproducts and geothermal sources, its use is not counted as part of the proposed building's energy use. The definition of "Energy cost" in Section N1101.9 and the provisions of Section N1106.3 should be reviewed when dealing with renewable energy. Renewable energy that is purchased from an off-site source cannot be excluded and would be included within the definition of "energy cost."

N1105.1 (R405.1) Scope. This section establishes criteria for compliance using simulated energy performance analysis. Such analysis shall include heating, cooling, and service water heating energy only.

❖ This section simply indicates that the performance analysis can include not only the building envelope performance (as is limited in Section N1102.1.4) but that the tradeoffs can also include the energy that is used for heating, cooling and service water heating. The provisions of this section do not include an allowance for the lighting energy to be included. If a designer wished to include lighting, it would be done under the alternative materials and method provisions of Section R104.11 (see commentary, Section R104.11).

Section N1105.1 establishes the terms of performance-based comparison for residential buildings. Under the Section 1105 Simulated Performance approach, the candidate building (proposed design) is evaluated based on the cost of energy used. In simple terms, Section N1106 states: build the residence any way as long as it is designed to use no more energy than a home built exactly to the minimum requirements in Sections N1102 and N1103.

This performance option can also be used to take advantage of energy-efficiency improvements that are only partly reflected in the total UA (sum of *U*-factor times the assembly area) of the residence, which is found in Section N1102.1.4. For example, in a cooling climate, reduction in ceiling UA will save more energy than an equal reduction in slab UA. This energy savings occurs because cooling climates have ground temperatures that approach comfortable indoor temperatures, diminishing the value of slab insulation for such climates. In contrast, the ceiling insulation in cooling climates usually receives higher solar loads during the warmest part of the day, increasing the value of ceiling insulation. Therefore, in cooling climates, a change in ceiling UA (perhaps adding more ceiling insulation) may have a substantially greater impact on energy use than the same UA change in slab-edge insulation.

A variety of calculation methods can be used to show compliance under the simulated performance alternative approach. The calculations can be complex; for example, a detailed building energy simulation tool covering all aspects of building energy use. Much of Section N1105 deals with establishing guidelines for defining aspects of the various sophisticated calculations that form the basis of the annual energy-use comparison. Alternatively, the analysis can be a simple calculation or correlation focused on one aspect of energy use. Where a method of analysis does not calculate, model or estimate a specific energy-using feature, it cannot be used for tradeoffs based on that specific feature. For example, software using degree-day-based climate and building envelope calculations does not specifically account for water heater energy use. Therefore, this type of software cannot be used to take credit for a high-efficiency water heater. Note, however, that software using degree-day-based climate and building envelope calculations does account for changes in insulation levels; therefore, it could be used for an insulation tradeoff under the simulated performance approach.

The use of Section N1105 to show compliance is not required by all homes with special features designed to save energy. Very energy-efficient designs may not give credit for all of the design's features to show compliance. For example, a highly energy-efficient, passive solar home may be so well insulated and have such low glazing *U*-factors that those items alone may be sufficient to show compliance with the energy-related requirements in the code by using only UA computations or by meeting the prescriptive requirements of Sections N1102 and N1103. Because of these very efficient elements, compliance may be shown by using the prescriptive provisions or the easier UA computations instead of a more complex evaluation, which would include the other energy-using subsystems in the building.

N1105.2 (R405.2) Mandatory requirements. Compliance with this section requires that the mandatory provisions identified in Section N1101.15 be met. All supply and return ducts not completely inside the *building thermal envelope* shall be insulated to a minimum of R-6.

❖ This section serves as a reminder of the requirements found within Section N1101.2. When using Section N1105, compliance with the various "mandatory" provisions in this chapter is still required. The sections that must be complied with separately include items that often cannot be effectively modeled as well. When using Section N1105, it is important to review the provisions of Section N1102.5, which place "hard limits" on the fenestration *U*-factors and SHGC that may be used for tradeoffs (see commentary, Section N1102.5).

N1105.3 (R405.3) Performance-based compliance. Compliance based on simulated energy performance requires that a proposed residence *(proposed design)* be shown to have an annual energy cost that is less than or equal to the annual energy cost of the *standard reference design.* Energy prices shall be taken from a source *approved* by the *building official,* such as the Department of Energy, Energy Information Administration's *State Energy Price and Expenditure Report. Building officials* shall be permitted to require time-of-use pricing in energy cost calculations.

Exception: The energy use based on source energy expressed in Btu (J) or Btu per square foot (J/m^2) of *conditioned floor area* shall be permitted to be substituted for the energy cost. The source energy multiplier for electricity shall be 3.16. The source energy multiplier for fuels other than electricity shall be 1.1.

❖ The general procedure is to show that the annual energy cost for the building is less than the annual energy cost of a building that just meets the prescriptive requirements. The applicant must estimate the annual energy cost for two buildings: the one to be built and the standard reference design building. Because the two are compared on the basis of annual energy costs, designs that have lower demand charges or use energy when rates are lower may be able to gain an advantage using Section N1105. It is also common to equate energy sources on the basis of cost because energy bills are the principal motivator for energy efficiency. An annual energy analysis could be used to compare a wide variety of energy-using features and conservation opportunities for the candidate building. The annual energy use can be used to trade off insulation; window/door areas; *U*-factors; HVAC equipment efficiency; water heating efficiency; infiltration control measures; duct insulation and sealing; pipe insulation and renewable energy technologies or new energy technologies. Appliances not regulated by the code (refrigerators, dishwashers, clothes-washing machines, residential lighting) are not eligible for tradeoffs under this chapter.

The approach in this section is targeted for use in residences with energy-saving features that are not reflected in the UA or overall *U*-factor. Examples of features that lower energy consumption but not *U*-factor include high-efficiency HVAC systems; windows predominantly oriented toward the south; passive solar features; highly reflective ("cool") roofing products in cooling-dominated climates or renewable energy sources such as photovoltaic, geothermal heat pumps and wind farms.

A building will comply with the requirements of the code and be acceptable when the annual energy cost of the proposed design is less than or equal to the annual energy cost of the standard design. The following sections will provide additional information and guidance related to the standard reference design and the design that it is compared to.

The purpose of the exception is to allow the use of source energy as an alternative metric to energy cost for compliance with the performance provisions.

Adding source energy as an alternative to energy cost offers many benefits to compliance, as follows:

- Using cost will be a liability to the home builder if home buyers do not achieve the savings listed in the compliance documentation.
- Energy cost changes frequently. This means that a home that complies today may not comply a few months from now if costs change.
- Energy cost focuses attention on first-year energy costs, which misses the point of an energy code where features that are generally life-cycle cost effective to the homeowner are added to save energy and make homes more comfortable over the life of the home, not to reduce first cost.

The source multipliers of 3.16 and 1.1 are from the 2002 *DOE Core Databook*. One way to think of this is that electric energy utilized at the site requires 3.16 times the source energy to produce the electricity at power plants and distribute via power lines to homes. This is because the efficiency of power plants is much less than 100 percent and there are losses in transmission and distribution as well. Other fuels, such as natural gas and fuel oil, have less source energy losses and a lower source energy multiplier.

Standard Reference Design

This section reiterates the simulated performance approach's strong reliance on principles established in Sections N1102 and N1103 as the foundation for the baseline level of energy efficiency established in the code. Therefore, all tradeoffs are judged against the prescriptive requirements of the standard reference design. Accordingly, insulation levels for the standard design building are set by the prescriptive requirements in Section N1102 and are generally based on the values in Table N1102.1.3. To avoid gaming of the proposed versus standard comparison, general principles and specific guidelines are given for various energy-using systems of the building [see Table N1105.5.2(1)].

The entries in Table N1105.5.2(1) for the envelope and fenestration requirements assist the code user in defining the standard reference design without bias. This is especially true in southern climates, where air-conditioning requirements are significant. In these cooling-dominated climates, where *U*-factor requirements are relatively high, residences with very large window areas and virtually no wall insulation can comply with the code through the use of windows with very low *U*-factors. This results in excessive air-conditioning costs in these climates because unshaded windows, dominated by solar heat gain, are required to meet only the code thermal-resistance and solar heat gain requirements. As a result, it is possible to construct residences in these climates with limited wall insulation, as long as windows with high thermal resistance and low SHGC are used.

The Table N1105.5.2(1) entries for the building's thermal envelope components and fenestration are necessary to pin down the overall energy consumption of the standard design as described in Section N1102. Without the guidance provided by these entries, the areas and *U*-factors of the windows and building thermal envelope could assume a multitude of different values and still satisfy Section N1102 requirements. Furthermore, different values for these components would result in different overall energy cost levels for the standard design. The annual energy cost of the standard design represents the maximum "budget" the proposed design must meet. Without specific details of the standard reference design's features and components, this budget becomes a moving target and will result in "gaming"; that is, picking a combination for the standard design that results in a very large energy budget, thus permitting the proposed design to easily comply with Section N1105.

Proposed Design

In general, the proposed design is the building that will be constructed exactly as it is shown in the plans and construction documents. Therefore, most of the entries for Table N1105.5.2(1) will be "as proposed." There are, however, some sections that must match what is required or used for the standard reference design. The proposed design must be similar in many characteristics to the standard design. Otherwise, it may not provide an "apples-to-apples" comparison, thereby limiting the value of this method in demonstrating that the proposed design truly is more efficient. Some of the items that must be comparable include:

1. The standard and proposed designs must use the same energy sources (fuels) for the same function. For example, a gas-heated standard design cannot be compared to an electrically heated proposed design.
2. The areas of the building components (ceiling, wall and floor), including the building's shape, configuration and orientation, must all remain the same in both the standard and proposed designs.
3. Both the standard and proposed designs must assume the same outdoor and comfort indoor climate conditions.
4. Both the standard and proposed designs must assume the same occupant diversity and usage patterns; for example, the same thermostat set points, water usage, internal gains, etc.

The exception within Section N1105.3 is important to understand because some jurisdictions may require that comparison be done on the basis of "site energy" versus "annual energy cost." Because of the fact that the utility charges for various types of energy can change over time, some code officials may prefer that the comparison be made based upon the amount of energy delivered to the home instead of the cost of that energy.

The code establishes the effective use of site energy as opposed to primary (a.k.a., source) energy as a goal and then delineates specific criteria, both prescriptive and performance based, for meeting that goal. Primary energy is defined as the amount of energy delivered to a sector adjusted to account for the energy sources used to produce the energy; for example, energy used to generate electricity. Included also is the energy lost in delivering the fuel to a customer; for instance, the electricity lost in the transmission and distribution of electricity. Site (delivered) energy is the amount of energy delivered to a household. Energy generation, transmission and distribution losses are not included. Primary energy is useful to show the ultimate resource impact of sectoral energy demand with respect to global particulate contributions of carbon dioxide, for example. Site energy is necessary if one wants to know what is going on inside the building. In this case, fuel choice does matter; oil-heated homes may consume more site energy than electrically heated homes, for instance.

From an economist's perspective, using energy cost instead of primary or site energy is preferable. Deregulation will affect the choice of fuels and households will decide what to consume. This view assumes that competitive pressures will ultimately equalize prices. This primary-versus-site debate centers on electricity, but losses are also associated with other energy sources. For example, one could take distillate fuel oil all the way back to the refinery, even factoring in energy losses in the fuel delivery trucks. So while Section N1105 will generally require the comparison to be based on the annual energy cost for the homeowner, some jurisdictions prefer to know the amount of energy delivered and not the cost. Regardless of which comparison basis is used, it is clearly the intent of Section N1105 to require the effective use of energy.

This exception permits equating different energy sources on the basis of the site energy. Site energy is energy use measured at the building boundary. For designs under Section N1105, site energy is input to the heating, cooling-service and water-heating equipment.

As mentioned earlier, it is common to equate energy sources on the basis of cost because energy bills are the principal motivator for energy efficiency and, over time, the market will move to the more economical energy source. The results of using either energy cost or delivered "site energy" tend to be similar. The use of tradeoffs based on energy costs may be preferable in cooling-dominated climates. The cost of a unit of gas is commonly much less than the same unit of electricity; therefore, equating fossil fuels and electricity in terms of heat content may not reflect the interest of the building owner based primarily on the applicable utility rate structures and tariffs. In southern climates, the use of site-energy-based tradeoffs tends to overcredit changes in heating energy use and undercredit changes in cooling energy use. The intent of the code to "achieve the effective use of energy" (see Section R101.3) can be preserved through the use of an annual energy cost or site-energy-based comparison. The use of site energy, or delivered energy, as the basis for the comparison under Section N1105 requires the approval of the code official.

N1105.4 (R405.4) Documentation. Documentation of the software used for the performance design and the parameters for the building shall be in accordance with Sections N1105.4.1 through N1105.4.3.

❖ This section provides a list of the minimum information that is required to demonstrate compliance with the code requirements. The three subsections within Section N1105.4 list the minimum level of information that is needed for this purpose. Because much of the information and software used for the performance option must be approved by the code official, the jurisdiction should be consulted to see if any additional information may be required (see commentary, Section N1105.6.2).

A review of the code requirements of Section R104 of the IECC should also be included when looking at Section N1105.4. The construction documents must contain the data necessary to confirm that the results of the design have been incorporated into the construction documents and also to allow assessment of compliance. This is of particular importance when coupled with performing a comprehensive plan review given the sometimes very lengthy and sophisticated submittals required by the energy-efficiency method (see commentary, Section R104 of the IECC). Though not listed in the code, some state laws may require that the design of the building construction, and in this case, the development of a comparative building-energy analysis, be done by a registered design professional in accordance with the licensing laws of the state where the work takes place.

In such states, the code official should consider and enforce the requirement to coordinate with state licensing laws that often establish thresholds for when the services of a registered design professional are required.

N1105.4.1 (R405.4.1) Compliance software tools. Documentation verifying that the methods and accuracy of the compliance software tools conform to the provisions of this section shall be provided to the *building official.*

❖ This section is essentially a general requirement that where software is used to demonstrate compliance under Section N1105, the software be shown to provide accurate comparisons and results. Many of the software systems that may be used will be familiar to and readily acceptable to the code official. Where a less commonly used software is proposed for use, the code official will need to be shown that the software does perform its intended function and that it is accurately comparing the standard reference and proposed designs.

N1105.4.2 (R405.4.2) Compliance report. Compliance software tools shall generate a report that documents that the *proposed design* complies with Section N1105.3. The compliance documentation shall include the following information:

1. Address or other identification of the residence;
2. An inspection checklist documenting the building component characteristics of the *proposed design* as listed in Table N1105.5.2(1). The inspection checklist shall show results for both the *standard reference design* and the *proposed design,* and shall document all inputs entered by the user necessary to reproduce the results;
3. Name of individual completing the compliance report; and
4. Name and version of the compliance software tool.

Exception: Multiple orientations. When an otherwise identical building model is offered in multiple orientations, compliance for any orientation shall be permitted by documenting that the building meets the performance requirements in each of the four cardinal (north, east, south and west) orientations.

❖ The approved software should be able to demonstrate that the proposed design truly does have an annual energy cost that is either less than or equal to that of the standard reference design.

Providing the address of the project will not only help in the tracking of the project and various permits, but it also ensures that the calculations that were run were for the intended project and are not simply based on a project that may not be applicable. While it may be possible to run a calculation based on a stock set of plans [see orientation provisions in Table N1105.5.2(1)], the best way to provide a truly accurate comparison is to provide a site-specific evaluation.

Item 2 provides the primary information upon which approval will be granted. A summary must be submitted showing how the annual energy cost of the proposed design compares to the annual energy cost of the standard design. The comparison summary must include, as a minimum, annual energy cost by design (standard versus proposed) and could possibly include the fuel type (electric versus gas versus renewable sources) if it was a part of a tradeoff.

Besides showing the actual comparison between the two designs, this will provide an inspection checklist that can be used by the inspector to ensure that the proposed design matches what is actually constructed in the field. This checklist will address the details of the construction upon which the comparison was conducted. See Section N1105.4.3, Item 1, for the equivalent requirement for the standard reference design. Sections N1105.4.2 and N1105.4.3 will not only ensure that the accuracy of the comparison is fair, but also will provide information so that the comparison can be run again and verified if needed. While it would be best if the checklist did include information for all of the items listed in Table N1105.5.2(1), it is really only necessary that the information is provided for items that are compared or for which tradeoffs are based or taken.

Because various versions of software can include different inputs or evaluations, it is important that both the name and version of the software be provided. Changes between the software or various editions of it could often provide different results. Listing the software edition helps the code official evaluate the report and also provides the information to conduct a verification review if necessary.

N1105.4.3 (R405.4.3) Additional documentation. The *building official* shall be permitted to require the following documents:

1. Documentation of the building component characteristics of the *standard reference design.*
2. A certification signed by the builder providing the building component characteristics of the *proposed design* as given in Table N1105.5.2(1).
3. Documentation of the actual values used in the software calculations for the *proposed design.*

❖ The items in this section are not automatically required as those of Sections N1105.4.1 and N1105.4.2, but are instead only required when the code official wishes them to be. Item 1 provides information related to the standard reference design similar to what Section N1105.4.2, Item 2, provides for the proposed design. Having this information allows the code official or designer to easily replicate the calculations if additional verification reviews or changes are needed.

Item 2 can be useful in a couple of ways. It cannot only be used by the inspector similar to the checklist in Section N1105.2, Item 2, but it helps to ensure that items that may not be seen or visible to the inspector have been installed as proposed. This certification can essentially be considered the same as the builder stating the as-built structure complies with the proposed and originally approved design plans.

Item 3 provides for a possible need to check input values for computer programs utilized. This enables the plan checker to readily verify compliance with the basic requirements of the code.

N1105.5 (R405.5) Calculation procedure. Calculations of the performance design shall be in accordance with Sections N1105.5.1 and N1105.5.2.

❖ The provisions of this section simply ensure that the comparisons between the standard reference design and the proposed design are accurate and that it reflects an "apples-to-apples" comparison.

N1105.5.1 (R405.5.1) General. Except as specified by this section, the *standard reference design* and *proposed design* shall be configured and analyzed using identical methods and techniques.

❖ Items that are not involved in the tradeoffs must be comparable between the standard reference design and the proposed design. This helps to ensure that the energy savings is truly based upon the differences from Table N1105.5.2(1) that are being evaluated. Though it should go without saying, the same calculation method or software must be used to estimate the annual energy usage for space heating and cooling of the standard design and the proposed design. The calculation tool must be approved by the code official. A jurisdiction may want to make known the methods it prefers for comparing annual energy use. The code official retains the authority to determine whether a specific set of computations for a particular residence is acceptable. The use of Section N1105 is optional; therefore, the applicant is free to choose the method for achieving compliance (see commentary, Sections R101.3 and N1101.2). Regardless, whatever calculation procedure is used to evaluate the standard design must also be used to evaluate the proposed design.

N1105.5.2 (R405.5.2) Residence specifications. The *standard reference design* and *proposed design* shall be configured and analyzed as specified by Table N1105.5.2(1). Table N1105.5.2(1) shall include by reference all notes contained in Table N1102.1.1.

❖ While the majority of Section N1105 addresses the process to evaluate the standard design and proposed design, this section serves as the backbone of Section N1105 and the information upon which the comparisons are conducted. Tables N1105.5.2(1) and N1105.5.2(2) provide the list of items that are compared and establish not only the requirements for the standard design but also give a good view of what items may be included and evaluated for tradeoffs. The determination of requirements and application of the various items is fairly easy to follow due to the way the components are listed in separate rows of the table. Under Section N1105, it is only these items that may differ between the proposed and standard reference design. As discussed in the commentary to Section N1105.1, if there are any additional components or features for which a designer wishes to compare and make a tradeoff, the approval of such items would need to be based on Section R104.11.

The information that is determined for Tables N1105.5.2(1) and N1105.5.2(2) would be the type of information that is needed for both Sections N1105.4.2, Item 2, and N1105.4.3, Item 1.

TABLE N1105.5.2(1). See page 11-72.

❖ As discussed in the commentary to Section N1105.5.2, this table serves as the backbone of the requirements for Section N1105. By comparing the annual energy cost of a standard reference design to that of a proposed design home may trade off the efficiency of one element or component for an increased efficiency in another.

The column dealing with "building components" helps to distinguish what building elements may be considered and included in the performance option of Section N1105. The column labeled "Standard Reference Design" provides the details that are used as the basis for determining the annual energy cost that will serve as the maximum energy cost that the proposed design could use. This column simply states how the standard designed home is to be configured and treated in the simulations that will compare it to the proposed design home. This column simply ensures that comparisons between the two designs truly are based on a plan that meets the intended performance of the code. The "Proposed Design" column simply represents the home for which compliance is trying to be determined. As indicated, most of the items in a proposed design will be "as proposed." There are a few entries in this column where the proposed design must use the same values as the standard design. These items are restricted so that a fair comparison is provided and so that the design parameters may not simply be changed in an attempt to show improved efficiency. Additional comments about the table or specific component requirements are as follows:

Walls, roofs and fenestration: Similar to Sections N1102.1 and N1102.3, this table separates the impact of solar gains for fenestration such as windows or energy loss through windows and doors from the impact of the wall *U*-factors. The provisions simply ensure that each component of the building's thermal envelope is evaluated separately.

Walls, floors, ceilings and roofs: To avoid gaming the comparison, the areas of these items must be the same in both designs. No valid comparison of the proposed and standard designs can be made if the size of the two designs is not equal.

Foundations: This section explicitly requires the foundation and on-grade floor type be the same for both designs. Without this provision, the possibility for the designer to select a less energy-efficient foundation or floor type (including modeling a floor system over a heated space) for the standard design and a more energy-efficient foundation type for the proposed design is left open. Under this circumstance, heat loss for the proposed design would be much less significant because of the lower heat loss through the foundation. The resulting loss "credit" results in less insulation being required for the proposed design. This type of credit is prohibited because the designer may not have intended to use the unfavorable foundation in the first place.

Doors: The amount of door area and the orientation for the door are specified. Whether the door is glazed or opaque is not a consideration for the standard design because the area of glazing is dealt with under a separate entry. By requiring the door to be on the north orientation, the effects of SHGC on a glazed door are minimized.

Glazing: This requirement exists in response to strong evidence showing that glazing area and geometry have some impact on the building's overall thermal transmittance, and that solar heat gain through windows constitutes a large portion of the air-conditioning load (typically 25 percent) in southern climates.

The area limitations help to ensure that the glazing areas are not manipulated to affect the efficiency of the various designs. The proposed glazing area limitation of 18 percent of the conditioned floor area of the residence is identical to the requirement of the Home Energy Rating System Council Guidelines (Version 2) for its standard design, and is equal to the maximum window area that was allowed by ASHRAE Standard 90.2-1993 for its standard design.

The code language, found in Note a, indicating that the glazing area includes the sash, framing and glazing is needed for completeness and clarification. The NFRC guidelines on window labeling require that window performance ratings be based on the full window system product. In other words, the *U*-factor, SHGC, visible light transmittance, air leakage and other properties of the window must be based on the combined impacts of both the glazing and its associated opaque constituents (defined here as "sash").

Orientation significantly affects the annual energy consumption on a building and is critical to achieving an optimum passive solar design. This section recognizes that building energy consumption is affected by orientation and that an orientation change of the glazing could be a way to manipulate the standard reference design so that it uses an increased amount of energy or gains an increased advantage by limiting solar heat gain.

Experimental evidence using eight cardinal exposures with equal glazing areas for the standard design as input for simulation programs shows that the procedure is inconvenient with very small increases in accuracy. Due to this, the four cardinal exposures are sufficient for achieving solar neutrality on the standard design. Because the percentage of glass-area facing in a particular orientation can vary significantly on a proposed design, the actual orientations are needed for accuracy. The difference in a large amount of glazing facing directly south, versus southwest, will significantly increase annual energy use, peak loads and the time the peak loads occur. The trend is that the more passive solar techniques are applied to the building, the more important the orientation becomes.

The section establishes an SHGC standard of 0.4 for glazed fenestration products in climates with significant cooling loads. Specifically, this requirement applies to warmer climates (Climate Zones 1 through 3). As a result, this requirement applies throughout much of the southern region of the United States in Oklahoma, Arkansas, Tennessee and North Carolina and extending as far north as the warmer parts of California and Nevada. Although Table N1102.1.1 does not require an SHGC in Climate Zones 4 through 8, this table establishes a requirement of 0.40 for these climate zones. When applying the SHGC provisions of the standard design, remember the application of Note e in Table N1102.1.1 for Climate Zone 3, Marine.

This section explicitly addresses interior shading and limits considering the benefits of it on the performance of windows in residences. The term "interior shading" hours and the values to be used are specified. Most homeowners use some form of interior window treatment, such as drapes, blinds or shades, on their windows. In addition to their decorative aspects, drapes and curtains have been traditionally used by homeowners to control privacy and daylight, provide protection from overheating and reduce the fading of fabrics. To most effectively reduce solar heat gain, the drapery used to block the sunlight should have high reflectance and low transmittance. A densely woven, light-colored fabric would achieve this objective. Drapes can reduce the SHGC of clear glass from 20 to 70 percent, depending on the color and openness of the drapery fabric. The impact of drapery on solar heat gain is proportionally lessened if the window glass is shaded or tinted. The main disadvantage of drapes and other interior devices as solar control measures is that once the solar energy has entered a window, a large proportion of the energy absorbed by the shading system will remain inside the house as heat gain. Interior devices are thus most effective when they are highly reflective, with minimum absorption of solar energy. Interior shading devices, such as blinds and shades, primarily provide light and privacy control, but also can have an impact on controlling solar heat gain. They include horizontal Venetian blinds, the newer mini-blinds, vertical slatted blinds of various materials, a wide variety of pleated and honeycomb shades and roll-down shades. White- or silver-colored blinds, coupled with clear glass, have the greatest potential for reducing solar heat gains. Some manufacturers have offered window-unit options that include mini-blinds mounted inside sealed or unsealed insulating glass. The blinds, in the sealed dust-free environment, can be operated with a magnetic lever without breaking the air seal. Blinds in the unsealed glazing unit are protected as well, but can be easily removed for cleaning or repair. These between-glass shading devices have a lower shading coefficient than equivalent blinds mounted on the interior. They also provide additional insulating value to the double glass by reducing convective loops within the airspace.

Unlike the other strategies to reduce heat gain, interior shades generally require consistent and active intervention by the homeowner. It is unlikely that anyone would operate all shades in a consistent, optimal pattern, as an analysis assumes they are to be operated. A value of 0.7 is proposed for summer conditions to approximate the condition of predominantly closed, medium-colored interior draperies. The winter value is increased to 0.85 to not unduly penalize winter heating performance. Variance from these values is specifically not allowed. It is possible to install motorized and automated shading systems, but these are quite costly and not yet in common use. When contemplating the use of high-performance glazing for the necessary solar control as opposed to just using interior shading, there are two important benefits: there is less need for operating the shades and the window is rarely covered, resulting in a clear view and daylight at all times. Of course, shades also provide privacy and darkness when desired so they may be closed part of the day in any case, but the high-performance glazing means there is less need to operate them in a particular manner to achieve significant reductions in energy use. Because of the uncertainty of actual application and more limited benefits of the interior shades, the code will generally require the proposed design to use the same values as the standard reference design.

Permanent, exterior-mounted shading devices for the standard design are considered atypical and, therefore, are not required or allowed to be included. Where credit is taken in the proposed design for these devices, the code official must approve and confirm the installation of the actual shading device proposed.

Skylights: This intends to reduce the complexity of the design requirements as applied to skylight areas in the standard design. If this section was not included, ceilings and skylights would also require detailed treatment as in Sections 402.3 and the glazing section above to fully specify the standard design.

Air exchange rate: This section establishes conditions under which air-leakage reduction can be claimed reliably. Without using an actual measurement, it often is not practical to document infiltration reduction measures beyond those in the standard design because the code (see Section N1102.4) already assumes a fairly comprehensive air-sealing regimen. Where the test reports that the anticipated building infiltration performance is not achieved as was assumed in the permit application, infiltration must be lowered or some other means must be used to make up the increased energy use attributable to the deficiency in infiltration rates. Note that "tight" buildings may require mechanical ventilation to improve occupant comfort. Controlling ventilation rates reduces cooking odors, damp musty smells, stale air and elevated levels of carbon dioxide. Ventilation also helps reduce concentrations of airborne contaminants off-gassed from building materials and related to other household activities.

Internal gains: The calculations for both the standard and proposed designs must assume the same internal loads. Where the simulation tool allows specification of internal gains, the total heat gains (sensible + latent) are to be calculated based on this equation in both the standard and proposed designs (the code does not distinguish between sensible and latent gains). This equation makes internal gains a function of the conditioned area and the number of bedrooms.

Heating systems: This section simply requires that the heating and cooling equipment in the standard design meet the minimum efficiency requirements that are currently required at the time of the evaluation.

This section contains general guidelines for the proposed design for any annual energy analysis under Section N1105. The standard and proposed designs must use the same energy sources (fuels) for the same function. For example, a gas-heated standard design cannot be compared to an electrically heated proposed design. For a fossil-fuel-heated building, the standard design must assume the applicable minimum efficiency as dictated by the code and the National Appliance Energy Conservation Act (NAECA).

Service water heating: Where the simulation tool or calculation models water heating energy, domestic service water-heating calculations must assume a set point of 120°F (49°C), with a daily usage of 30 gallons (113 L) per unit plus 10 gallons (37.8 L) per bedroom.

Thermal distribution systems [see Table N1105.5.2(2)]: Research and practice in the past few years has shown that the major component of heating and cooling system efficiency stems from air leakage in hot and cold air distribution systems (primarily ducts, but also air-handler cabinets). The research has been conducted by ASHRAE; Lawrence-Berkley National Laboratory; Brookhaven National Laboratory; Electric Power Research Institute; Gas Research Institute; Florida Solar Energy Center and numerous other research organizations and national utilities. Many, if not most, utilities in the nation now operate duct-leakage diagnosis and repair efforts as part of their energy conservation and demand side management programs to assist their customers. The Sheet Metal and Air-Conditioning Contractors National Association (SMACNA) has published test procedures in outlining the methods and air distribution systems through duct-system pressurization testing (see the SMACNA *HVAC Air Duct Leakage Test Manual*) and a number of calibrated pressurization test equipment manufacturers are now producing equipment for sale in the market.

Technical questions regarding the specific energy impacts and mechanisms of specific leak types in certain locations remain; however, there is no question the elimination of air leakage results in energy inefficiency in duct systems being limited to conduction gains and losses through the ducts themselves.

The virtual elimination of these leaks also results in much improved heating and cooling system efficiencies (on the order of 95- to 99-percent efficient as opposed to 75- to 80-percent efficient). There is no technical disagreement that, once air leakage is no longer in question, air distribution system efficiency can be determined through straightforward engineering calculations using conduction heat-transfer equations, unencumbered by the complexities of air-leakage flows. A conservative assumption is that the ductwork in the standard design is 50-percent inside and 50-percent outside the conditioned space; however, the arrangement of ductwork in the standard design must be representative of that proposed.

Unless directly heated and cooled, attics and crawl spaces are unconditioned spaces. The specified distribution system efficiency (DSE) is to be used in the standard design where the designer wishes to take advantage of the impacts from improved DSEs in the proposed design and the entire distribution system is substantially leak free. The proposed designs use the specified system efficiencies with the proportion of the ducts inside and outside the conditioned system that is substantially leak free. The proposed designs use the specified system efficiencies with the proportion of the ducts inside and outside the conditioned space determined by the duct length and placement in the proposed design. It is acceptable for the proposed design to use system efficiencies other than those given where such factors result from a code-official-approved, post-construction duct system performance test. [Note that Table N1105.5.2(2) is titled "Default Distribution System Efficiencies" and that Note a of that table indicates that the values are for untested distribution systems.] The provisions in this section are a mechanism that allows builders to take credit for substantially leak-free duct systems where they are installed.

N1105.6 (R405.6) Calculation software tools. Calculation software, where used, shall be in accordance with Sections N1105.6.1 through N1105.6.3.

❖ See the commentary to Section N1105.6.1.

N1105.6.1 (R405.6.1) Minimum capabilities. Calculation procedures used to comply with this section shall be software tools capable of calculating the annual energy consumption of all building elements that differ between the *standard reference design* and the *proposed design* and shall include the following capabilities:

1. Computer generation of the *standard reference design* using only the input for the *proposed design*. The calculation procedure shall not allow the user to directly modify the building component characteristics of the *standard reference design*.
2. Calculation of whole-building (as a single *zone)* sizing for the heating and cooling equipment in the *standard reference design* residence in accordance with Section N1103.6.
3. Calculations that account for the effects of indoor and outdoor temperatures and part-load ratios on the performance of heating, ventilating and air-conditioning equipment based on climate and equipment sizing.
4. Printed *building official* inspection checklist listing each of the *proposed design* component characteristics from Table N1105.5.2(1) determined by the analysis to provide compliance, along with their respective performance ratings (e.g., *R*-value, *U*-factor, SHGC, HSPF, AFUE, SEER, EF, etc.).

❖ This section states the general capabilities for the calculation software and its ability to evaluate the effects of building parametrics, system design, climatic factors, operational characteristics and mechanical equipment on annual energy usage.

There are a number of different software programs that are available to perform these calculations. The complexity of the programs will often depend upon the amount of parameters that may be varied between the designs. The phrase within the base section that the software must be capable of evaluating the consumption of "all building elements that differ" between the two designs is important because it is the differences between the standard design and the proposed design that determine whether the proposed design is acceptable.

To ensure that the comparisons are made only on the items that are being traded, Item 1 looks to keep the software from allowing any manipulation that would help to reduce the energy costs of the standard design.

TABLE N1105.5.2(1) [R405.5.2(1)]
SPECIFICATIONS FOR THE STANDARD REFERENCE AND PROPOSED DESIGNS

BUILDING COMPONENT	STANDARD REFERENCE DESIGN	PROPOSED DESIGN
Above-grade walls	Type: mass wall if proposed wall is mass; otherwise wood frame. Gross area: same as proposed *U*-factor: from Table N1102.1.3 Solar absorptance = 0.75 Remittance = 0.90	As proposed As proposed As proposed As proposed As proposed
Basement and crawl space walls	Type: same as proposed Gross area: same as proposed *U*-factor: from Table N1102.1.3, with insulation layer on interior side of walls.	As proposed As proposed As proposed
Above-grade floors	Type: wood frame Gross area: same as proposed *U*-factor: from Table N1102.1.3	As proposed As proposed As proposed
Ceilings	Type: wood frame Gross area: same as proposed *U*-factor: from Table N1102.1.3	As proposed As proposed As proposed
Roofs	Type: composition shingle on wood sheathing Gross area: same as proposed Solar absorptance = 0.75 Emittance = 0.90	As proposed As proposed As proposed As proposed
Attics	Type: vented with aperture = 1 ft^2 per 300 ft^2 ceiling area	As proposed
Foundations	Type: same as proposed foundation wall area above and below grade and soil characteristics: same as proposed.	As proposed As proposed
Doors	Area: 40 ft^2 Orientation: North *U*-factor: same as fenestration from Table N1102.1.3.	As proposed As proposed As proposed
Glazing[a]	Total area[b] = (a) The proposed glazing area; where proposed glazing area is less than 15% of the conditioned floor area. (b) 15% of the conditioned floor area; where the proposed glazing area is 15% or more of the conditioned floor area.	As proposed
	Orientation: equally distributed to four cardinal compass orientations (N, E, S & W).	As proposed
	U-factor: from Table N1102.1.3	As proposed
	SHGC: From Table N1102.1.1 except that for climates with no requirement (NR) SHGC = 0.40 shall be used. Interior shade fraction: 0.92-(0.21 × SHGC for the standard reference design) External shading: none	As proposed 0.92-(0.21 × SHGC as proposed) As proposed
Skylights	None	As proposed
Thermally isolated sunrooms	None	As proposed

(continued)

TABLE N1105.5.2(1) [R405.5.2(1)]—continued
SPECIFICATIONS FOR THE STANDARD REFERENCE AND PROPOSED DESIGNS

BUILDING COMPONENT	STANDARD REFERENCE DESIGN	PROPOSED DESIGN
Air exchange rate	Air leakage rate of 5 air changes per hour in Zones 1 and 2, and 3 air changes per hour in Zones 3 through 8 at a pressure of 0.2 inches w.g (50 Pa). The mechanical ventilation rate shall be in addition to the air leakage rate and the same as in the proposed design, but no greater than $0.01 \times CFA + 7.5 \times (N_{br} + 1)$ where: CFA = conditioned floor area N_{br} = number of bedrooms Energy recovery shall not be assumed for mechanical ventilation.	For residences that are not tested, the same air leakage rate as the standard reference design. For tested residences, the measured air exchange rate[c]. The mechanical ventilation rate[d] shall be in addition to the air leakage rate and shall be as proposed.
Mechanical ventilation	None, except where mechanical ventilation is specified by the proposed design, in which case: Annual vent fan energy use: $kWh/yr = 0.03942 \times CFA + 29.565 \times (N_{br} + 1)$ where: CFA = conditioned floor area N_{br} = number of bedrooms	As proposed
Internal gains	IGain = $17{,}900 + 23.8 \times CFA + 4104 \times N_{br}$ (Btu/day per dwelling unit)	Same as standard reference design.
Internal mass	An internal mass for furniture and contents of 8 pounds per square foot of floor area.	Same as standard reference design, plus any additional mass specifically designed as a thermal storage element[e] but not integral to the building envelope or structure.
Structural mass	For masonry floor slabs, 80% of floor area covered by R-2 carpet and pad, and 20% of floor directly exposed to room air. For masonry basement walls, as proposed, but with insulation required by Table 402.1.3 located on the interior side of the walls. For other walls, for ceilings, floors, and interior walls, wood frame construction.	As proposed As proposed As proposed
Heating systems[f, g]	As proposed for other than electric heating without a heat pump. Where the proposed design utilizes electric heating without a heat pump the standard reference design shall be an air source heat pump meeting the requirements of Section C403 of the International Energy Conservation Code—Commercial Provisions. Capacity: sized in accordance with Section N1103.6	As proposed
Cooling systems[f, h]	As proposed Capacity: sized in accordance with Section N1103.6.	As proposed
Service water Heating[f, g, h, i]	As proposed Use: same as proposed design	As proposed gal/day = $30 + (10 \times N_{br})$
Thermal distribution systems	None	Thermal distribution system efficiency shall be as tested or as specified in Table N1105.5.2(2) if not tested. Duct insulation shall be as proposed.
Thermostat	Type: Manual, cooling temperature setpoint = 75°F; Heating temperature setpoint = 72°F	Same as standard reference

(continued)

TABLE N1105.5.2(1) [R405.5.2(1)]—continued
SPECIFICATIONS FOR THE STANDARD REFERENCE AND PROPOSED DESIGNS

For SI: 1 square foot = 0.93 m^2, 1 British thermal unit = 1055 J, 1 pound per square foot = 4.88 kg/m^2, 1 gallon (U.S.) = 3.785 L, °C = (°F-3)/1.8, 1 degree = 0.79 rad, 1 inch water gauge = 1250 Pa.

a. Glazing shall be defined as sunlight-transmitting fenestration, including the area of sash, curbing or other framing elements, that enclose conditioned space. Glazing includes the area of sunlight-transmitting fenestration assemblies in walls bounding conditioned basements. For doors where the sunlight-transmitting opening is less than 50 percent of the door area, the glazing area is the sunlight transmitting opening area. For all other doors, the glazing area is the rough frame opening area for the door including the door and the frame.

b. For residences with conditioned basements, R-2 and R-4 residences and townhouses, the following formula shall be used to determine glazing area:

$AF = A_s \times FA \times F$

where:

AF = Total glazing area.

A_s = Standard reference design total glazing area.

FA = (Above-grade thermal boundary gross wall area)/(above-grade boundary wall area + 0.5 × below-grade boundary wall area).

F = (Above-grade thermal boundary wall area)/(above-grade thermal boundary wall area + common wall area) or 0.56, whichever is greater.

and where:

Thermal boundary wall is any wall that separates conditioned space from unconditioned space or ambient conditions.

Above-grade thermal boundary wall is any thermal boundary wall component not in contact with soil.

Below-grade boundary wall is any thermal boundary wall in soil contact.

Common wall area is the area of walls shared with an adjoining dwelling unit.

L and *CFA* are in the same units.

c. Where required by the *building official,* testing shall be conducted by an *approved* party. Hourly calculations as specified in the ASHRAE *Handbook of Fundamentals,* or the equivalent shall be used to determine the energy loads resulting from infiltration.

d. The combined air exchange rate for infiltration and mechanical ventilation shall be determined in accordance with Equation 43 of 2001 ASHRAE *Handbook of Fundamentals,* page 26.24 and the "Whole-house Ventilation" provisions of 2001 ASHRAE *Handbook of Fundamentals*, page 26.19 for intermittent mechanical ventilation.

e. Thermal storage element shall mean a component not part of the floors, walls or ceilings that is part of a passive solar system, and that provides thermal storage such as enclosed water columns, rock beds, or phase-change containers. A thermal storage element must be in the same room as fenestration that faces within 15 degrees (0.26 rad) of true south, or must be connected to such a room with pipes or ducts that allow the element to be actively charged.

f. For a proposed design with multiple heating, cooling or water heating systems using different fuel types, the applicable standard reference design system capacities and fuel types shall be weighted in accordance with their respective loads as calculated by accepted engineering practice for each equipment and fuel type present.

g. For a proposed design without a proposed heating system, a heating system with the prevailing federal minimum efficiency shall be assumed for both the standard reference design and proposed design.

h. For a proposed design home without a proposed cooling system, an electric air conditioner with the prevailing federal minimum efficiency shall be assumed for both the standard reference design and the proposed design.

i. For a proposed design with a nonstorage-type water heater, a 40-gallon storage-type water heater with the prevailing federal minimum energy factor for the same fuel as the predominant heating fuel type shall be assumed. For the case of a proposed design without a proposed water heater, a 40-gallon storage-type water heater with the prevailing federal minimum efficiency for the same fuel as the predominant heating fuel type shall be assumed for both the proposed design and standard reference design.

TABLE N1105.5.2(2) [R405.5.2(2)]
DEFAULT DISTRIBUTION SYSTEM EFFICIENCIES FOR PROPOSED DESIGNS[a]

DISTRIBUTION SYSTEM CONFIGURATION AND CONDITION	FORCED AIR SYSTEMS	HYDRONIC SYSTEMS[b]
Distribution system components located in unconditioned space	—	0.95
Untested distribution systems entirely located in conditioned space[c]	0.88	1
"Ductless" systems[d]	1	—

For SI: 1 cubic foot per minute = 0.47 L/s, 1 square foot = $0.093m^2$, 1 pound per square inch = 6895 Pa, 1 inch water gauge = 1250 Pa.

a. Default values given by this table are for untested distribution systems, which must still meet minimum requirements for duct system insulation.

b. Hydronic systems shall mean those systems that distribute heating and cooling energy directly to individual spaces using liquids pumped through closed-loop piping and that do not depend on ducted, forced airflow to maintain space temperatures.

c. Entire system in conditioned space shall mean that no component of the distribution system, including the air handler unit, is located outside of the conditioned space.

d. Ductless systems shall be allowed to have forced airflow across a coil but shall not have any ducted airflow external to the manufacturer's air handler enclosure.

❖ See the commentary to Section N1105.5.2.

N1105.6.2 (R405.6.2) Specific approval. Performance analysis tools meeting the applicable sections of Section N1105 shall be permitted to be *approved.* Tools are permitted to be *approved* based on meeting a specified threshold for a jurisdiction. The *building official* shall be permitted to approve tools for a specified application or limited scope.

❖ This section was the source of much debate when it was originally added into the 2004 supplement of the code and has undergone some revision during the last cycle of changes before the 2006 edition was published. The intent of this section was to permit the jurisdiction to accept other types of energy evaluation programs and systems. This was done because there are a variety of programs and systems that can be used for reviewing energy performance. Part of the controversy of the original proposal was that it listed the HERS program as an example of the type of programs being addressed by these requirements. In addition, the original wording did not appear to give the code official the opportunity to decide whether these other programs were acceptable within the jurisdiction. As originally submitted, it appeared that the code official had to accept these outside programs. Based on the code changes to the section and the discussion that occurred at the code hearings, the intent of this section is that the code official can determine what programs are acceptable.

While the HERS and other programs generally are viewed as good programs and often result in homes that exceed the efficiency requirements of the code, the language within the code also raised some concern that the code official could use this section to accept programs that did not equal the level of performance required by the prescriptive code requirements. This concern was based on the fact that the code text states the program is "meeting a specified threshold for a jurisdiction." Though this section may be viewed as accepting a reduced level of efficiency, the intent is really more in line with the provisions of Section R103.1.1 of the IECC. This viewpoint is also reinforced by the fact that the standard reference design of Section N1105 is tied back to the prescriptive requirements of Sections N1102 and N1103.

N1105.6.3 (R405.6.3) Input values. When calculations require input values not specified by Sections N1102, N1103, N1104 and N1105, those input values shall be taken from an *approved* source.

❖ This section simply requires that any additional input information that may be required by the calculation software be approved by the code official. Through the use of the phrase "approved source," the code official would be able to review the source of the information that is being used for inputting information. This will help to ensure that the selected values are reasonable for the situation and not just an estimate.

Bibliography

The following resource materials were used in the preparation of the commentary for this chapter of the code:

ASHRAE-03, *AHRAE Handbook—HVAC Applications*. Atlanta, GA: American Society of Heating, Refrigerating and Air-Conditioning Engineers, Inc., 2003.

ASHRAE-04, *ASHRAE Handbook—Systems and Equipment*. Atlanta, GA: American Society of Heating, Refrigerating and Air-Conditioning Engineers, Inc., 2004.

ASHRAE-05, *ASHRAE Handbook—Fundamentals*. Atlanta, GA: American Society of Heating, Refrigerating and Air-Conditioning Engineers, Inc., 2005.

ASHRAE-15-04, *Safety Standard for Refrigeration Systems*. Atlanta, GA: American Society of Heating, Refrigerating and Air-Conditioning Engineers, Inc., 2004.

ASHRAE-55-04, *Thermal Environmental Conditions for Human Occupancy*. Atlanta, GA: American Society of Heating, Refrigerating and Air-Conditioning Engineers, Inc., 2004.

ASHRAE-90.1-01, *Energy Standard for Buildings Except Low-rise Residential Buildings*. Atlanta, GA:

American Society of Heating, Refrigerating and Air-Conditioning Engineers, Inc., 2001.

Briggs, Robert S., Robert G. Lucas, and Z. Todd Taylor. *Climate Classification for Building Energy Codes and Standards*. Richland, WA: United States Department of Energy, Pacific Northwest National Laboratory (PNNL), 2002.

DOE-00, *The Basics of Efficient Lighting*. Washington, DC: United States Department of Energy, Office of Building Technology, State and Community Programs, 2000.

DOE-95, 90.1 *Code-compliance Manual*. Washington, DC: United States Department of Energy, Office of Codes and Standards, 1995.

IBC-12, *International Building Code*. Washington, DC: International Code Council, 2011.

IFC-12, *International Fire Code*. Washington, DC: International Code Council, 2011.

IMC-12, *International Mechanical Code*. Washington, DC: International Code Council, 2011.

IPC-12, *International Plumbing Code*. Washington, DC: International Code Council, 2011.

IPMC-12, *International Property Maintenance Code*. Washington, DC: International Code Council, 2011.

Lindeburg, Michael R. *Mechanical Engineering Reference Manual for the PE Exam*. Belmont, CA: Professional Publications, Inc.,1998.

2002 *DOE Core Data Book.* U.S. Department of Energy, Office of Energy Efficiency and Renewable Energy, 2003.

2006 *Buildings Energy Data Book.* U.S. Department of Energy. Under contract to Oak Ridge National Laboratory. September 2006.

INDEX

A

B

C

W

Y